COLLEGE MATHEMATICS

FOR BUSINESS, ECONOMICS, LIFE SCIENCES, AND SOCIAL SCIENCES

Thirteenth Edition

RAYMOND A. BARNETT Merritt College

MICHAEL R. ZIEGLER Marquette University

KARL E. BYLEEN Marquette University

Boston   Columbus   Indianapolis   New York   San Francisco   Upper Saddle River
Amsterdam   Cape Town   Dubai   London   Madrid   Milan   Munich   Paris   Montréal   Toronto
Delhi   Mexico City   São Paulo   Sydney   Hong Kong   Seoul   Singapore   Taipei   Tokyo

Editor in Chief: Deirdre Lynch
Executive Editor: Jennifer Crum
Project Manager: Kerri Consalvo
Editorial Assistant: Joanne Wendelken
Senior Managing Editor: Karen Wernholm
Senior Production Supervisor: Ron Hampton
Interior and Cover Design: Beth Paquin
Executive Manager, Course Production: Peter Silvia
Associate Media Producer: Christina Maestri
Digital Assets Manager: Marianne Groth
Executive Marketing Manager: Jeff Weidenaar
Marketing Assistant: Brooke Smith
Rights and Permissions Advisor: Joseph Croscup
Senior Manufacturing Buyer: Carol Melville
Production Coordination and Composition: Integra

Cover photo: Leigh Prather/Shutterstock; Dmitriy Raykin/Shutterstock;
         Peter Waters/Shutterstock; Anna Subbotina/Shutterstock
Photo credits: page 2, iStockphoto/Thinkstock; page 42, Purestock/Thinstock; page 126,
         Fuse/Thinkstock; page 173, iStockphoto/Thinkstock; page 255, Glen Gaffney/
         Shutterstock; page 285, Deusexlupus/Fotolia; page 345, Phil Date/Shutterstock;
         page 385, Mark Thomas/Alamy; page 447, Sritangphoto/Shutterstock; page 488,
         Purestock/Thinkstock; page 574, Vario Images/Alamy; page 631, P. Amedzro/
         Alamy; page 713, Anonymous Donor/Alamy; page 775, Shime/Fotolia; page 818,
         Aurora Photos/Alamy

Many of the designations used by manufacturers and sellers to distinguish their products are claimed as trademarks. Where those designations appear in this book, and Pearson was aware of a trademark claim, the designations have been printed in initial caps or all caps.

Library of Congress Cataloging-in-Publication Data

College mathematics for business, economics, life sciences, and social
sciences / Raymond A. Barnett … [et al.].—13th ed.
    p. cm.
Includes index.
ISBN-13: 978-0-321-94551-8
ISBN-10: 0-321-94551-4
1. Mathematics.   2. Social sciences—Mathematics.   3. Biomathematics.   I. Barnett, Raymond A.
QA37.3.B37   2014
510—dc23   2013023209

3 4 5 6 7 8 9 10—V011—17 16 15 14

www.pearsonhighered.com

ISBN-10: 0-321-94551-4
ISBN-13: 978-0-321-94551-8

CONTENTS

Preface viii

Diagnostic Prerequisite Test. xix

PART 1   A LIBRARY OF ELEMENTARY FUNCTIONS

Chapter 1   Linear Equations and Graphs. 2
1.1 Linear Equations and Inequalities 3
1.2 Graphs and Lines. 12
1.3 Linear Regression 26
Chapter 1 Summary and Review 38
Review Exercises 39

Chapter 2   Functions and Graphs 42
2.1 Functions. 43
2.2 Elementary Functions: Graphs and Transformations 57
2.3 Quadratic Functions 69
2.4 Polynomial and Rational Functions 84
2.5 Exponential Functions 95
2.6 Logarithmic Functions 106
Chapter 2 Summary and Review 117
Review Exercises 120

PART 2   FINITE MATHEMATICS

Chapter 3   Mathematics of Finance 126
3.1 Simple Interest 127
3.2 Compound and Continuous Compound Interest 134
3.3 Future Value of an Annuity; Sinking Funds 147
3.4 Present Value of an Annuity; Amortization 155
Chapter 3 Summary and Review 167
Review Exercises 169

Chapter 4   Systems of Linear Equations; Matrices. 173
4.1 Review: Systems of Linear Equations in Two Variables 174
4.2 Systems of Linear Equations and Augmented Matrices 187
4.3 Gauss–Jordan Elimination 196
4.4 Matrices: Basic Operations 210
4.5 Inverse of a Square Matrix 222
4.6 Matrix Equations and Systems of Linear Equations. 234
4.7 Leontief Input–Output Analysis 242
Chapter 4 Summary and Review 250
Review Exercises 251

Chapter 5   Linear Inequalities and Linear Programming **255**

5.1   Linear Inequalities in Two Variables 256
5.2   Systems of Linear Inequalities in Two Variables 263
5.3   Linear Programming in Two Dimensions: A Geometric Approach . 270
Chapter 5 Summary and Review 282
Review Exercises 283

Chapter 6   Linear Programming: The Simplex Method **285**

6.1   The Table Method: An Introduction to the Simplex Method 286
6.2   The Simplex Method:
      Maximization with Problem Constraints of the Form $\leq$ 297
6.3   The Dual Problem:
      Minimization with Problem Constraints of the Form $\geq$ 313
6.4   Maximization and Minimization with
      Mixed Problem Constraints 326
Chapter 6 Summary and Review 341
Review Exercises 342

Chapter 7   Logic, Sets, and Counting **345**

7.1   Logic 346
7.2   Sets 354
7.3   Basic Counting Principles 361
7.4   Permutations and Combinations 369
Chapter 7 Summary and Review 380
Review Exercises 382

Chapter 8   Probability **385**

8.1   Sample Spaces, Events, and Probability 386
8.2   Union, Intersection, and Complement of Events; Odds 399
8.3   Conditional Probability, Intersection, and Independence 411
8.4   Bayes' Formula 425
8.5   Random Variable, Probability Distribution, and Expected Value . . 432
Chapter 8 Summary and Review 441
Review Exercises 443

Chapter 9   Markov Chains **447**

9.1   Properties of Markov Chains 448
9.2   Regular Markov Chains 459
9.3   Absorbing Markov Chains 469
Chapter 9 Summary and Review 483
Review Exercises 484

PART 3   CALCULUS

Chapter 10   Limits and the Derivative 488
10.1   Introduction to Limits 489
10.2   Infinite Limits and Limits at Infinity 503
10.3   Continuity. 515
10.4   The Derivative. 526
10.5   Basic Differentiation Properties 541
10.6   Differentials 550
10.7   Marginal Analysis in Business and Economics 557
Chapter 10 Summary and Review 568
Review Exercises 569

Chapter 11   Additional Derivative Topics 574
11.1   The Constant e and Continuous Compound Interest 575
11.2   Derivatives of Exponential and Logarithmic Functions 581
11.3   Derivatives of Products and Quotients 590
11.4   The Chain Rule 598
11.5   Implicit Differentiation 608
11.6   Related Rates 614
11.7   Elasticity of Demand 620
Chapter 11 Summary and Review 627
Review Exercises 629

Chapter 12   Graphing and Optimization 631
12.1   First Derivative and Graphs 632
12.2   Second Derivative and Graphs 648
12.3   L'Hôpital's Rule 665
12.4   Curve-Sketching Techniques 674
12.5   Absolute Maxima and Minima 687
12.6   Optimization 695
Chapter 12 Summary and Review 708
Review Exercises 709

Chapter 13   Integration 713
13.1   Antiderivatives and Indefinite Integrals 714
13.2   Integration by Substitution 725
13.3   Differential Equations; Growth and Decay 736
13.4   The Definite Integral 747
13.5   The Fundamental Theorem of Calculus 757
Chapter 13 Summary and Review 769
Review Exercises 771

Chapter 14   Additional Integration Topics 775
            14.1 Area Between Curves 776
            14.2 Applications in Business and Economics 785
            14.3 Integration by Parts 797
            14.4 Other Integration Methods 803
            Chapter 14 Summary and Review 814
            Review Exercises 815

Chapter 15   Multivariable Calculus 818
            15.1 Functions of Several Variables 819
            15.2 Partial Derivatives 828
            15.3 Maxima and Minima 837
            15.4 Maxima and Minima Using Lagrange Multipliers 845
            15.5 Method of Least Squares 854
            15.6 Double Integrals over Rectangular Regions 864
            15.7 Double Integrals over More General Regions 874
            Chapter 15 Summary and Review 882
            Review Exercises 885

Appendix A   Basic Algebra Review 888
            A.1 Real Numbers 888
            A.2 Operations on Polynomials 894
            A.3 Factoring Polynomials 900
            A.4 Operations on Rational Expressions 906
            A.5 Integer Exponents and Scientific Notation 912
            A.6 Rational Exponents and Radicals 916
            A.7 Quadratic Equations 922

Appendix B   Special Topics 931
            B.1 Sequences, Series, and Summation Notation 931
            B.2 Arithmetic and Geometric Sequences 937
            B.3 Binomial Theorem 943

Appendix C   Tables 947

            Answers A-1

            Index I-1

            Index of Applications I-12

Available separately: Calculus Topics to Accompany Calculus, 13e,
                    and College Mathematics, 13e

Chapter 1   Differential Equations
            1.1 Basic Concepts
            1.2 Separation of Variables
            1.3 First-Order Linear Differential Equations
            Chapter 1 Review
            Review Exercises

Chapter 2   Taylor Polynomials and Infinite Series
          2.1 Taylor Polynomials
          2.2 Taylor Series
          2.3 Operations on Taylor Series
          2.4 Approximations Using Taylor Series
          Chapter 2 Review
          Review Exercises

Chapter 3   Probability and Calculus
          3.1 Improper Integrals
          3.2 Continuous Random Variables
          3.3 Expected Value, Standard Deviation, and Median
          3.4 Special Probability Distributions
          Chapter 3 Review
          Review Exercises

Appendixes A and B   (Refer to back of *College Mathematics for Business, Economics, Life Sciences, and Social Sciences,* 13e)

Appendix C   Tables
          Table III Area Under the Standard Normal Curve

Appendix D   Special Calculus Topic
          D.1 Interpolating Polynomials and Divided Differences

          Answers
          Solutions to Odd-Numbered Exercises
          Index
          Applications Index

PREFACE

The thirteenth edition of *College Mathematics for Business, Economics, Life Sciences, and Social Sciences* is designed for a two-term (or condensed one-term) course in finite mathematics and calculus for students who have had one to two years of high school algebra or the equivalent. The book's overall approach, refined by the authors' experience with large sections of college freshmen, addresses the challenges of teaching and learning when prerequisite knowledge varies greatly from student to student.

The authors had three main goals when writing this text:

▶ To write a text that students can easily comprehend

▶ To make connections between what students are learning and how they may apply that knowledge

▶ To give flexibility to instructors to tailor a course to the needs of their students.

Many elements play a role in determining a book's effectiveness for students. Not only is it critical that the text be accurate and readable, but also, in order for a book to be effective, aspects such as the page design, the interactive nature of the presentation, and the ability to support and challenge all students have an incredible impact on how easily students comprehend the material. Here are some of the ways this text addresses the needs of students at all levels:

▶ Page layout is clean and free of potentially distracting elements.

▶ *Matched Problems* that accompany each of the completely worked examples help students gain solid knowledge of the basic topics and assess their own level of understanding before moving on.

▶ Review material (Appendix A and Chapters 1 and 2) can be used judiciously to help remedy gaps in prerequisite knowledge.

▶ A *Diagnostic Prerequisite Test* prior to Chapter 1 helps students assess their skills, while the *Basic Algebra Review* in Appendix A provides students with the content they need to remediate those skills.

▶ *Explore and Discuss* problems lead the discussion into new concepts or build upon a current topic. They help students of all levels gain better insight into the mathematical concepts through thought-provoking questions that are effective in both small and large classroom settings.

▶ Instructors are able to easily craft homework assignments that best meet the needs of their students by taking advantage of the variety of types and difficulty levels of the exercises. Exercise sets at the end of each section consist of a *Skills Warm-up* (four to eight problems that review prerequisite knowledge specific to that section) followed by problems divided into categories A, B, and C by level of difficulty, with level-C exercises being the most challenging.

▶ The MyMathLab course for this text is designed to help students help themselves and provide instructors with actionable information about their progress. The immediate feedback students receive when doing homework and practice in MyMathLab is invaluable, and the easily accessible e-book enhances student learning in a way that the printed page sometimes cannot.

Most important, all students get substantial experience in modeling and solving real-world problems through application examples and exercises chosen from business and economics, life sciences, and social sciences. Great care has been taken to write a book that is mathematically correct, with its emphasis on computational skills, ideas, and problem solving rather than mathematical theory.

Finally, the choice and independence of topics make the text readily adaptable to a variety of courses (see the chapter dependencies chart on page xiii). This text is one of three books in the authors' college mathematics series. The others are *Finite Mathematics for Business, Economics, Life Sciences, and Social Sciences*, and *Calculus for Business, Economics, Life Sciences, and Social Sciences*. *Additional Calculus Topics*, a supplement written to accompany the Barnett/Ziegler/Byleen series, can be used in conjunction with any of these books.

New to This Edition

Fundamental to a book's effectiveness is classroom use and feedback. Now in its thirteenth edition, *College Mathematics for Business, Economics, Life Sciences, and Social Sciences* has had the benefit of a substantial amount of both. Improvements in this edition evolved out of the generous response from a large number of users of the last and previous editions as well as survey results from instructors, mathematics departments, course outlines, and college catalogs. In this edition,

▶ The Diagnostic Prerequisite Test has been revised to identify the specific deficiencies in prerequisite knowledge that cause students the most difficulty with finite mathematics and calculus.

▶ Most exercise sets now begin with a *Skills Warm-up*—four to eight problems that review prerequisite knowledge specific to that section in a just-in-time approach. References to review material are given for the benefit of students who struggle with the warm-up problems and need a refresher.

▶ Section 6.1 has been rewritten to better motivate and introduce the simplex method and associated terminology.

▶ Section 14.4 has been rewritten to cover the trapezoidal rule and Simpson's rule.

▶ Examples and exercises have been given up-to-date contexts and data.

▶ Exposition has been simplified and clarified throughout the book.

▶ An Annotated Instructor's Edition is now available, providing answers to exercises directly on the page (whenever possible). *Teaching Tips* provide less-experienced instructors with insight on common student pitfalls, suggestions for how to approach a topic, or reminders of which prerequisite skills students will need. Lastly, the difficulty level of exercises is indicated only in the AIE so as not to discourage students from attempting the most challenging "C" level exercises.

▶ *MyMathLab* for this text has been enhanced greatly in this revision. Most notably, a "Getting Ready for Chapter X" has been added to each chapter as an optional resource for instructors and students as a way to address the prerequisite skills that students need, and are often missing, for each chapter. Many more improvements have been made. See the detailed description on pages xvii and xviii for more information.

Trusted Features

Emphasis and Style

As was stated earlier, this text is written for student comprehension. To that end, the focus has been on making the book both mathematically correct and accessible to students. Most derivations and proofs are omitted, except where their inclusion adds significant insight into a particular concept as the emphasis is on computational skills, ideas, and problem solving rather than mathematical theory. General concepts and results are typically presented only after particular cases have been discussed.

Design

One of the hallmark features of this text is the **clean, straightforward design** of its pages. Navigation is made simple with an obvious hierarchy of key topics and a judicious use of call-outs and pedagogical features. We made the decision to maintain a two-color design to help students stay focused on the mathematics and applications. Whether students start in the chapter opener or in the exercise sets, they can easily reference the content, examples, and *Conceptual Insights* they need to understand the topic at hand. Finally, a functional use of color improves the clarity of many illustrations, graphs, and explanations, and guides students through critical steps (see pages 61, 108, and 402).

Examples and Matched Problems

More than 490 completely worked examples are used to introduce concepts and to demonstrate problem-solving techniques. Many examples have multiple parts, significantly increasing the total number of worked examples. The examples are annotated using blue text to the right of each step, and the problem-solving steps are clearly identified. **To give students extra help** in working through examples, dashed boxes are used to enclose steps that are usually performed mentally and rarely mentioned in other books (see Example 2 on page 4). Though some students may not need these additional steps, many will appreciate the fact that the authors do not assume too much in the way of prior knowledge.

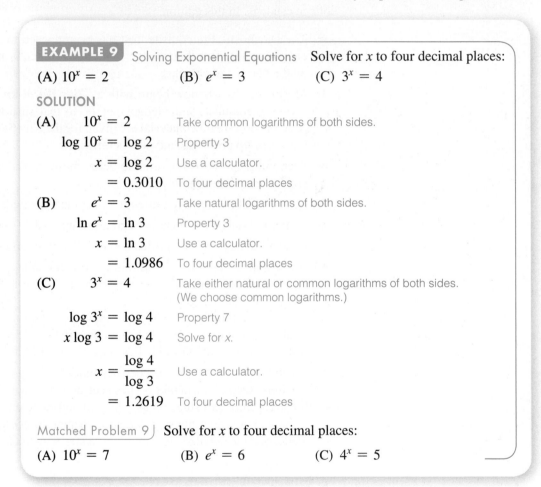

EXAMPLE 9   Solving Exponential Equations   Solve for x to four decimal places:

(A) $10^x = 2$ (B) $e^x = 3$ (C) $3^x = 4$

SOLUTION

(A) $10^x = 2$ Take common logarithms of both sides.

$\log 10^x = \log 2$ Property 3

$x = \log 2$ Use a calculator.

$= 0.3010$ To four decimal places

(B) $e^x = 3$ Take natural logarithms of both sides.

$\ln e^x = \ln 3$ Property 3

$x = \ln 3$ Use a calculator.

$= 1.0986$ To four decimal places

(C) $3^x = 4$ Take either natural or common logarithms of both sides. (We choose common logarithms.)

$\log 3^x = \log 4$ Property 7

$x \log 3 = \log 4$ Solve for x.

$x = \dfrac{\log 4}{\log 3}$ Use a calculator.

$= 1.2619$ To four decimal places

Matched Problem 9   Solve for x to four decimal places:

(A) $10^x = 7$ (B) $e^x = 6$ (C) $4^x = 5$

     Each example is followed by a similar *Matched Problem* for the student to work while reading the material. This actively involves the student in the learning process. The answers to these matched problems are included at the end of each section for easy reference.

Explore and Discuss

Most every section contains *Explore and Discuss* problems at appropriate places to encourage students to think about a relationship or process before a result is stated or to investigate additional consequences of a development in the text. This serves to foster critical thinking and communication skills. The Explore and Discuss material can be used for in-class discussions or out-of-class group activities and is effective in both small and large class settings.

Explore and Discuss 2   How many x intercepts can the graph of a quadratic function have? How many y intercepts? Explain your reasoning.

New to this edition, annotations in the instructor's edition provide tips for less-experienced instructors on how to engage students in these Explore and Discuss activities, expand on the topic, or simply guide student responses.

Exercise Sets

The book contains over 6,500 carefully selected and graded exercises. Many problems have multiple parts, significantly increasing the total number of exercises. Exercises are paired so that consecutive odd- and even-numbered exercises are of the same type and difficulty level. Each exercise set is designed to allow instructors to craft just the right assignment for students. Exercise sets are categorized as Skills Warm-up (review of prerequisite knowledge), and within the Annotated Instructor's Edition only, as A (routine easy mechanics), B (more difficult mechanics), and C (difficult mechanics and some theory) to make it easy for instructors to create assignments that are appropriate for their classes. The *writing exercises*, indicated by the icon ✎, provide students with an opportunity to express their understanding of the topic in writing. Answers to all odd-numbered problems are in the back of the book. Answers to application problems in linear programming include both the mathematical model and the numeric answer.

Applications

A major objective of this book is to give the student substantial experience in modeling and solving real-world problems. Enough applications are included to convince even the most skeptical student that mathematics is really useful (see the Index of Applications at the back of the book). Almost every exercise set contains application problems, including applications from business and economics, life sciences, and social sciences. An instructor with students from all three disciplines can let them choose applications from their own field of interest; if most students are from one of the three areas, then special emphasis can be placed there. Most of the applications are simplified versions of actual real-world problems inspired by professional journals and books. No specialized experience is required to solve any of the application problems.

Additional Pedagogical Features

The following features, while helpful to any student, are particularly helpful to students enrolled in a large classroom setting where access to the instructor is more challenging or just less frequent. These features provide much-needed guidance for students as they tackle difficult concepts.

▶ **Call-out boxes** highlight important definitions, results, and step-by-step processes (see pages 90, 96–97).

▶ **Caution statements** appear throughout the text where student errors often occur (see pages 138, 143, and 176).

⚠ **CAUTION**   Note that in Example 11 we let $x = 0$ represent 1900. If we let $x = 0$ represent 1940, for example, we would obtain a different logarithmic regression equation, but the prediction for 2015 would be the same. We would *not* let $x = 0$ represent 1950 (the first year in Table 1) or any later year, because logarithmic functions are undefined at 0.   ▲

▶ **Conceptual Insights**, appearing in nearly every section, often make explicit connections to previous knowledge, but sometimes encourage students to think beyond the particular skill they are working on and see a more enlightened view of the concepts at hand (see pages 59, 140, 216).

> **CONCEPTUAL INSIGHT**
>
> The notation (2.7) has two common mathematical interpretations: the ordered pair with first coordinate 2 and second coordinate 7, and the open interval consisting of all real numbers between 2 and 7. The choice of interpretation is usually determined by the context in which the notation is used. The notation $(2, -7)$ could be interpreted as an ordered pair but not as an interval. In interval notation, the left endpoint is always written first. So, $(-7, 2)$ is correct interval notation, but $(2, -7)$ is not.

▶ The newly revised **Diagnostic Prerequisite Test**, located at the front of the book, provides students with a tool to assess their prerequisite skills prior to taking the course. The **Basic Algebra Review**, in Appendix A, provides students with seven sections of content to help them remediate in specific areas of need. Answers to the Diagnostic Prerequisite Test are at the back of the book and reference specific sections in the Basic Algebra Review or Chapter 1 for students to use for remediation.

Graphing Calculator and Spreadsheet Technology

Although access to a graphing calculator or spreadsheets is not assumed, it is likely that many students will want to make use of this technology. To assist these students, optional graphing calculator and spreadsheet activities are included in appropriate places. These include brief discussions in the text, examples or portions of examples solved on a graphing calculator or spreadsheet, and exercises for the student to solve. For example, linear regression is introduced in Section 1.3, and regression techniques on a graphing calculator are used at appropriate points to illustrate mathematical modeling with real data. All the optional graphing calculator material is clearly identified with the icon ⊞ and can be omitted without loss of continuity, if desired. Optional spreadsheet material is identified with the icon ▦. Graphing calculator screens displayed in the text are actual output from the TI-84 Plus graphing calculator.

Chapter Reviews

Often it is during the preparation for a chapter exam that concepts gel for students, making the chapter review material particularly important. The chapter review sections in this text include a comprehensive summary of important terms, symbols, and concepts, keyed to completely worked examples, followed by a comprehensive set of Review Exercises. Answers to Review Exercises are included at the back of the book; *each answer contains a reference to the section in which that type of problem is discussed* so students can remediate any deficiencies in their skills on their own.

Content

The text begins with the development of a library of elementary functions in **Chapters 1 and 2**, including their properties and applications. Many students will be familiar with most, if not all, of the material in these introductory chapters. Depending on students'

Chapter Dependencies

Diagnostic
Prerequisite Test

PART ONE: A LIBRARY OF ELEMENTARY FUNCTIONS*

1 Linear Equations and Graphs → 2 Functions and Graphs

PART TWO: FINITE MATHEMATICS

3 Mathematics of Finance

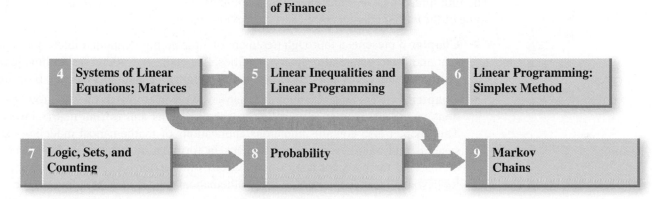

4 Systems of Linear Equations; Matrices → 5 Linear Inequalities and Linear Programming → 6 Linear Programming: Simplex Method

7 Logic, Sets, and Counting → 8 Probability → 9 Markov Chains

PART THREE: CALCULUS

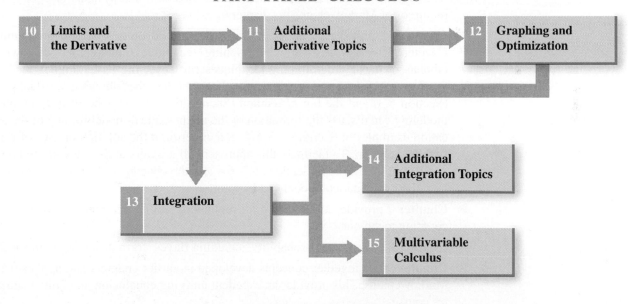

10 Limits and the Derivative → 11 Additional Derivative Topics → 12 Graphing and Optimization

13 Integration → 14 Additional Integration Topics

13 Integration → 15 Multivariable Calculus

APPENDIXES

A Basic Algebra Review

B Special Topics

*Selected topics from Part One may be referred to as needed in Parts Two or Three or reviewed systematically before starting Part Two.

preparation and the course syllabus, an instructor has several options for using the first two chapters, including the following:

(i) Skip Chapters 1 and 2 and refer to them only as necessary later in the course;

(ii) Cover Chapter 1 quickly in the first week of the course, emphasizing price–demand equations, price–supply equations, and linear regression, but skip Chapter 2;

(iii) Cover Chapters 1 and 2 systematically before moving on to other chapters.

The material in Part Two (Finite Mathematics) can be thought of as four units:

1. Mathematics of finance (Chapter 3)
2. Linear algebra, including matrices, linear systems, and linear programming (Chapters 4, 5, and 6)
3. Probability and statistics (Chapters 7 and 8)
4. Applications of linear algebra and probability to Markov chains (Chapter 9)

The first three units are independent of each other, while the fourth unit is dependent on some of the earlier chapters (see chart on previous page).

▶ **Chapter 3** presents a thorough treatment of simple and compound interest and present and future value of ordinary annuities. Appendix B.1 addresses arithmetic and geometric sequences and can be covered in conjunction with this chapter, if desired.

▶ **Chapter 4** covers linear systems and matrices with an emphasis on using row operations and Gauss–Jordan elimination to solve systems and to find matrix inverses. This chapter also contains numerous applications of mathematical modeling using systems and matrices. To assist students in formulating solutions, all answers at the back of the book for application exercises in Sections 4.3, 4.5, and the chapter Review Exercises contain both the mathematical model and its solution. The row operations discussed in Sections 4.2 and 4.3 are required for the simplex method in Chapter 6. Matrix multiplication, matrix inverses, and systems of equations are required for Markov chains in Chapter 9.

▶ **Chapters 5 and 6** provide a broad and flexible coverage of linear programming. Chapter 5 covers two-variable graphing techniques. Instructors who wish to emphasize linear programming techniques can cover the basic simplex method in Sections 6.1 and 6.2 and then discuss either or both of the following: the dual method (Section 6.3) and the big M method (Section 6.4). Those who want to emphasize modeling can discuss the formation of the mathematical model for any of the application examples in Sections 6.2–6.4, and either omit the solution or use software to find the solution. To facilitate this approach, all answers at the back of the book for application exercises in Sections 6.2–6.4 and the chapter Review Exercises contain both the mathematical model and its solution.

▶ **Chapter 7** provides a foundation for probability with a treatment of logic, sets, and counting techniques.

▶ **Chapter 8** covers basic probability, including Bayes' formula and random variables.

▶ **Chapter 9** ties together concepts developed in earlier chapters and applies them to Markov chains. This provides an excellent unifying conclusion to a finite mathematics course.

The material in Part Three (Calculus) consists of differential calculus (Chapters 10–12), integral calculus (Chapters 13 and 14), multivariable calculus (Chapter 15). In general, Chapters 10–12 must be covered in sequence; however, certain sections can be omitted or given brief treatments, as pointed out in the discussion that follows (see the Chapter Dependencies chart on page xiii).

▶ **Chapter 10** introduces the derivative. The first three sections cover limits (including infinite limits and limits at infinity), continuity, and the limit properties that are essential to understanding the definition of the derivative in Section 10.4. The remaining sections of the chapter cover basic rules of differentiation, differentials, and applications of derivatives in business and economics. The interplay between graphical, numerical, and algebraic concepts is emphasized here and throughout the text.

▶ In **Chapter 11** the derivatives of exponential and logarithmic functions are obtained before the product rule, quotient rule, and chain rule are introduced. Implicit differentiation is introduced in Section 11.5 and applied to related rates problems in Section 11.6. Elasticity of demand is introduced in Section 11.7. The topics in these last three sections of Chapter 11 are not referred to elsewhere in the text and can be omitted.

▶ **Chapter 12** focuses on graphing and optimization. The first two sections cover first-derivative and section-derivative graph properties. L'Hôpital's rule is discussed in Section 12.3. A graphing strategy is presented and illustrated in Section 12.4. Optimization is covered in Sections 12.5 and 12.6, including examples and problems involving end-point solutions.

▶ **Chapter 13** introduces integration. The first two sections cover antidifferentiation techniques essential to the remainder of the text. Section 13.3 discusses some applications involving differential equations that can be omitted. The definite integral is defined in terms of Riemann sums in Section 13.4 and the fundamental theorem of calculus is discussed in Section 13.5. As before, the interplay between graphical, numerical, and algebraic properties is emphasized. These two sections are also required for the remaining chapters in the text.

▶ **Chapter 14** covers additional integration topics and is organized to provide maximum flexibility for the instructor. The first section extends the area concepts introduced in Chapter 14 to the area between two curves and related applications. Section 14.2 covers three more applications of integration, and Sections 14.3 and 14.4 deal with additional methods of integration, including integration by parts, the trapezoidal rule, and Simpson's rule. Any or all of the topics in Chapter 14 can be omitted.

▶ **Chapter 15** deals with multivariable calculus. The first five sections can be covered any time after Section 12.6 has been completed. Sections 15.6 and 15.7 require the integration concepts discussed in Chapter 13.

▶ **Appendix A** contains a concise review of basic algebra that may be covered as part of the course or referenced as needed. As mentioned previously, **Appendix B** contains additional topics that can be covered in conjunction with certain sections in the text, if desired.

Accuracy Check

Because of the careful checking and proofing by a number of mathematics instructors (acting independently), the authors and publisher believe this book to be substantially error free. If an error should be found, the authors would be grateful if notification were sent to Karl E. Byleen, 9322 W. Garden Court, Hales Corners, WI 53130; or by e-mail to kbyleen@wi.rr.com.

Student Supplements

Student's Solutions Manual

▶ By Garret J. Etgen, University of Houston

▶ This manual contains detailed, carefully worked-out solutions to all odd-numbered section exercises and all Chapter Review exercises. Each section begins with Things to Remember, a list of key material for review.

▶ ISBN-13: 978-0-321-94677-5

Additional Calculus Topics to Accompany Calculus, 13e, and College Mathematics, 13e

▶ This separate book contains three unique chapters: Differential Equations, Taylor Polynomials and Infinite Series, and Probability and Calculus.

▶ ISBN 13: 978-0-321-93169-6; ISBN 10: 0-321-931696

Graphing Calculator Manual for Applied Math

▶ By Victoria Baker, Nicholls State University

▶ This manual contains detailed instructions for using the TI-83/TI-83 Plus/TI-84 Plus C calculators with this textbook. Instructions are organized by mathematical topics.

▶ Available in MyMathLab.

Excel Spreadsheet Manual for Applied Math

▶ By Stela Pudar-Hozo, Indiana University–Northwest

▶ This manual includes detailed instructions for using Excel spreadsheets with this textbook. Instructions are organized by mathematical topics.

▶ Available in MyMathLab.

Guided Lecture Notes

▶ By Salvatore Sciandra,
Niagara County Community College

▶ These worksheets for students contain unique examples to enforce what is taught in the lecture and/or material covered in the text. Instructor worksheets are also available and include answers.

▶ Available in MyMathLab or through Pearson Custom Publishing.

Videos with Optional Captioning

▶ The video lectures with optional captioning for this text make it easy and convenient for students to watch videos from a computer at home or on campus. The complete set is ideal for distance learning or supplemental instruction.

▶ Every example in the text is represented by a video.

▶ Available in MyMathLab.

Instructor Supplements

New! Annotated Instructor's Edition

▶ This book contains answers to all exercises in the text on the same page as the exercises whenever possible. In addition, Teaching Tips are provided for less-experienced instructors. Exercises are coded by level of difficulty only in the AIE so students are not dissuaded from trying more challenging exercises.

▶ ISBN-13: 978-0-321-94616-4

Online Instructor's Solutions Manual (downloadable)

▶ By Garret J. Etgen, University of Houston

▶ This manual contains detailed solutions to all even-numbered section problems.

▶ Available in MyMathLab or through http://www.pearsonhighered.com/educator.

Mini Lectures (downloadable)

▶ By Salvatore Sciandra,
Niagara County Community College

▶ Mini Lectures are provided for the teaching assistant, adjunct, part-time or even full-time instructor for lecture preparation by providing learning objectives, examples (and answers) not found in the text, and teaching notes.

▶ Available in MyMathLab or through http://www.pearsonhighered.com/educator.

PowerPoint® Lecture Slides

▶ These slides present key concepts and definitions from the text. They are available in MyMathLab or at http://www.pearsonhighered.com/educator.

Technology Resources

MyMathLab® Online Course
(access code required)

MyMathLab delivers **proven results** in helping individual students succeed.

▶ MyMathLab has a consistently positive impact on the quality of learning in higher education math instruction. MyMathLab can be successfully implemented in any environment—lab based, hybrid, fully online, traditional—and demonstrates the quantifiable difference that integrated usage has on student retention, subsequent success, and overall achievement.

▶ MyMathLab's comprehensive online gradebook automatically tracks your students' results on tests, quizzes, homework, and in the study plan. You can use the gradebook to quickly intervene if your students have trouble or to provide positive feedback on a job well done. The data within MyMathLab is easily exported to a variety of spreadsheet programs, such as Microsoft Excel. You can determine which points of data you want to export and then analyze the results to determine success.

MyMathLab provides **engaging experiences** that personalize, stimulate, and measure learning for each student.

▶ **Personalized Learning:** MyMathLab offers two important features that support adaptive learning—personalized homework and the adaptive study plan. These features allow your students to work on what they need to learn when it makes the most sense, maximizing their potential for understanding and success.

▶ **Exercises:** The homework and practice exercises in MyMathLab are correlated to the exercises in the textbook, and they regenerate algorithmically to give students unlimited opportunity for practice and mastery. The software offers immediate, helpful feedback when students enter incorrect answers.

▶ **Chapter-Level, Just-in-Time Remediation:** The MyMathLab course for these texts includes a short diagnostic, called Getting Ready, prior to each chapter to assess students' prerequisite knowledge. This diagnostic can then be tied to personalized homework so that each student receives a homework assignment specific to his or her prerequisite skill needs.

▶ **Multimedia Learning Aids:** Exercises include guided solutions, sample problems, animations, videos, and eText access for extra help at the point of use.

And, MyMathLab comes from an **experienced partner** with educational expertise and an eye on the future.

▶ Knowing that you are using a Pearson product means that you are using quality content. That means that our eTexts are accurate and our assessment tools work. It means we are committed to making MyMathLab as accessible as possible. MyMathLab is compatible with the JAWS 12/13 screen reader, and enables multiple-choice and free-response problem types to be read and interacted with via keyboard controls and math notation input. More information on this functionality is available at http://mymathlab. com/accessibility.

▶ Whether you are just getting started with MyMathLab or you have a question along the way, we're here to help you learn about our technologies and how to incorporate them into your course.

▶ To learn more about how MyMathLab combines proven learning applications with powerful assessment and continuously adaptive capabilities, visit www. mymathlab.com or contact your Pearson representative.

MyMathLab® Ready-to-Go Course
(access code required)

These new Ready-to-Go courses provide students with all the same great MyMathLab features but make it easier for instructors to get started. Each course includes preassigned homework and quizzes to make creating a course even simpler. In addition, these prebuilt courses include a course-level Getting Ready diagnostic that helps pinpoint student weaknesses in prerequisite skills. Ask your Pearson representative about the details for this particular course or to see a copy of this course.

MyLabsPlus®

MyLabsPlus combines proven results and engaging experiences from MyMathLab® and MyStatLab™ with convenient management tools and a dedicated services team. Designed to support growing math and statistics programs, it includes additional features such as

▶ **Batch Enrollment:** Your school can create the login name and password for every student and instructor, so everyone can be ready to start class on the first day. Automation of this process is also possible through integration with your school's Student Information System.

▶ **Login from your campus portal:** You and your students can link directly from your campus portal into your MyLabsPlus courses. A Pearson service team works with your institution to create a single sign-on experience for instructors and students.

▶ **Advanced Reporting:** MyLabsPlus advanced reporting allows instructors to review and analyze students' strengths and weaknesses by tracking their performance on tests, assignments, and tutorials. Administrators can review grades and assignments across all courses on your MyLabsPlus campus for a broad overview of program performance.

▶ **24/7 Support:** Students and instructors receive 24/7 support, 365 days a year, by email or online chat.

MyLabsPlus is available to qualified adopters. For more information, visit our website at www.mylabsplus.com or contact your Pearson representative.

MathXL® Online Course
(access code required)

MathXL is the homework and assessment engine that runs MyMathLab. (MyMathLab is MathXL plus a learning-management system.)

With MathXL, instructors can

▶ Create, edit, and assign online homework and tests using algorithmically generated exercises correlated at the objective level to the textbook.

▶ Create and assign their own online exercises and import TestGen tests for added flexibility.

▶ Maintain records of all student work tracked in MathXL's online gradebook.

With MathXL, students can

▶ Take chapter tests in MathXL and receive personalized study plans and/or personalized homework assignments based on their test results.

▶ Use the study plan and/or the homework to link directly to tutorial exercises for the objectives they need to study.

▶ Access supplemental animations and video clips directly from selected exercises.

MathXL is available to qualified adopters. For more information, visit our website at www.mathxl.com or contact your Pearson representative.

TestGen®

TestGen (www.pearsoned.com/testgen) enables instructors to build, edit, print, and administer tests using a computerized bank of questions developed to cover all the objectives of the text. TestGen is algorithmically based, allowing instructors to create multiple, but equivalent, versions of the same question or test with the click of a button. Instructors can also modify test bank questions or add new questions. The software and test bank are available for download from Pearson Education's online catalog.

Acknowledgments

In addition to the authors many others are involved in the successful publication of a book. We wish to thank the following reviewers:

Mark Barsamian, *Ohio University*
Britt Cain, *Austin Community College*
Florence Chambers, *Southern Maine Community College*
Kathleen Coskey, *Boise State University*
Tim Doyle, *DePaul University*
J. Robson Eby, *Blinn College–Bryan Campus*
Irina Franke, *Bowling Green State University*
Jerome Goddard II, *Auburn University–Montgomery*
Andrew J. Hetzel, *Tennessee Tech University*
Fred Katiraie, *Montgomery College*
Timothy Kohl, *Boston University*

Dan Krulewich, *University of Missouri, Kansas City*
Rebecca Leefers, *Michigan State University*
Scott Lewis, *Utah Valley University*
Bishnu Naraine, *St. Cloud State University*
Kevin Palmowski, *Iowa State University*
Saliha Shah, *Ventura College*
Alexander Stanoyevitch,
    California State University–Dominguez Hills
Mary Ann Teel, *University of North Texas*
Jerimi Ann Walker, *Moraine Valley Community College*
Hong Zhang, *University of Wisconsin, Oshkosh*

We also express our thanks to

Damon Demas, Mark Barsamian, Theresa Schille, J. Robson Eby, John Samons, and Gary Williams for providing a careful and thorough accuracy check of the text, problems, and answers.

Garret Etgen, Salvatore Sciandra, Victoria Baker, and Stela Pudar-Hozo for developing the supplemental materials so important to the success of a text.

All the people at Pearson Education who contributed their efforts to the production of this book.

Diagnostic Prerequisite Test

Work all of the problems in this self-test without using a calculator. Then check your work by consulting the answers in the back of the book. Where weaknesses show up, use the reference that follows each answer to find the section in the text that provides the necessary review.

1. Replace each question mark with an appropriate expression that will illustrate the use of the indicated real number property:

   (A) Commutative $(\cdot): x(y + z) = ?$

   (B) Associative $(+): 2 + (x + y) = ?$

   (C) Distributive: $(2 + 3)x = ?$

Problems 2–6 refer to the following polynomials:

   (A) $3x - 4$ (B) $x + 2$

   (C) $2 - 3x^2$ (D) $x^3 + 8$

2. Add all four.

3. Subtract the sum of (A) and (C) from the sum of (B) and (D).

4. Multiply (C) and (D).

5. What is the degree of each polynomial?

6. What is the leading coefficient of each polynomial?

In Problems 7 and 8, perform the indicated operations and simplify.

7. $5x^2 - 3x[4 - 3(x - 2)]$

8. $(2x + y)(3x - 4y)$

In Problems 9 and 10, factor completely.

9. $x^2 + 7x + 10$

10. $x^3 - 2x^2 - 15x$

11. Write 0.35 as a fraction reduced to lowest terms.

12. Write $\dfrac{7}{8}$ in decimal form.

13. Write in scientific notation:

   (A) 4,065,000,000,000   (B) 0.0073

14. Write in standard decimal form:

   (A) 2.55×10^8 (B) 4.06×10^{-4}

15. Indicate true (T) or false (F):

   (A) A natural number is a rational number.

   (B) A number with a repeating decimal expansion is an irrational number.

16. Give an example of an integer that is not a natural number.

In Problems 17–24, simplify and write answers using positive exponents only. All variables represent positive real numbers.

17. $6(xy^3)^5$

18. $\dfrac{9u^8v^6}{3u^4v^8}$

19. $(2 \times 10^5)(3 \times 10^{-3})$

20. $(x^{-3}y^2)^{-2}$

21. $u^{5/3}u^{2/3}$

22. $(9a^4b^{-2})^{1/2}$

23. $\dfrac{5^0}{3^2} + \dfrac{3^{-2}}{2^{-2}}$

24. $(x^{1/2} + y^{1/2})^2$

In Problems 25–30, perform the indicated operation and write the answer as a simple fraction reduced to lowest terms. All variables represent positive real numbers.

25. $\dfrac{a}{b} + \dfrac{b}{a}$

26. $\dfrac{a}{bc} - \dfrac{c}{ab}$

27. $\dfrac{x^2}{y} \cdot \dfrac{y^6}{x^3}$

28. $\dfrac{x}{y^3} \div \dfrac{x^2}{y}$

29. $\dfrac{\dfrac{1}{7 + h} - \dfrac{1}{7}}{h}$

30. $\dfrac{x^{-1} + y^{-1}}{x^{-2} - y^{-2}}$

31. Each statement illustrates the use of one of the following real number properties or definitions. Indicate which one.

Commutative $(+, \cdot)$	Associative $(+, \cdot)$	Distributive
Identity $(+, \cdot)$	Inverse $(+, \cdot)$	Subtraction
Division	Negatives	Zero

   (A) $(-7) - (-5) = (-7) + [-(-5)]$

   (B) $5u + (3v + 2) = (3v + 2) + 5u$

   (C) $(5m - 2)(2m + 3) = (5m - 2)2m + (5m - 2)3$

   (D) $9 \cdot (4y) = (9 \cdot 4)y$

   (E) $\dfrac{u}{-(v - w)} = \dfrac{u}{w - v}$

   (F) $(x - y) + 0 = (x - y)$

32. Round to the nearest integer:

   (A) $\dfrac{17}{3}$ (B) $-\dfrac{5}{19}$

33. Multiplying a number x by 4 gives the same result as subtracting 4 from x. Express as an equation, and solve for x.

34. Find the slope of the line that contains the points $(3, -5)$ and $(-4, 10)$.

35. Find the x and y coordinates of the point at which the graph of $y = 7x - 4$ intersects the x axis.

36. Find the x and y coordinates of the point at which the graph of $y = 7x - 4$ intersects the y axis.

In Problems 37 and 38, factor completely.

37. $x^2 - 3xy - 10y^2$

38. $6x^2 - 17xy + 5y^2$

In Problems 39–42, write in the form $ax^p + by^q$ where a, b, p, and q are rational numbers.

39. $\dfrac{3}{x} + 4\sqrt{y}$

40. $\dfrac{8}{x^2} - \dfrac{5}{y^4}$

41. $\dfrac{2}{5x^{3/4}} - \dfrac{7}{6y^{2/3}}$

42. $\dfrac{1}{3\sqrt{x}} + \dfrac{9}{\sqrt[3]{y}}$

In Problems 43 and 44, write in the form $a + b\sqrt{c}$ where a, b, and c are rational numbers.

43. $\dfrac{1}{4 - \sqrt{2}}$

44. $\dfrac{5 - \sqrt{3}}{5 + \sqrt{3}}$

In Problems 45–50, solve for x.

45. $x^2 = 5x$

46. $3x^2 - 21 = 0$

47. $x^2 - x - 20 = 0$

48. $-6x^2 + 7x - 1 = 0$

49. $x^2 + 2x - 1 = 0$

50. $x^4 - 6x^2 + 5 = 0$

PART 1

A LIBRARY OF ELEMENTARY FUNCTIONS

1

Linear Equations and Graphs

1.1 Linear Equations and Inequalities

1.2 Graphs and Lines

1.3 Linear Regression

Chapter 1 Summary and Review

Review Exercises

Introduction

We begin by discussing some algebraic methods for solving equations and inequalities. Next, we introduce coordinate systems that allow us to explore the relationship between algebra and geometry. Finally, we use this algebraic–geometric relationship to find equations that can be used to describe real-world data sets. For example, in Section 1.3 you will learn how to find the equation of a line that fits data on winning times in an Olympic swimming event (see Problems 27 and 28 on page 37). We also consider many applied problems that can be solved using the concepts discussed in this chapter.

1.1 Linear Equations and Inequalities

- Linear Equations
- Linear Inequalities
- Applications

The equation

$$3 - 2(x + 3) = \frac{x}{3} - 5$$

and the inequality

$$\frac{x}{2} + 2(3x - 1) \geq 5$$

are both first degree in one variable. In general, a **first-degree**, or **linear**, **equation** in one variable is any equation that can be written in the form

$$\text{Standard form:} \quad ax + b = 0 \quad a \neq 0 \tag{1}$$

If the equality symbol, $=$, in (1) is replaced by $<$, $>$, $\leq$, or $\geq$, the resulting expression is called a **first-degree**, or **linear**, **inequality**.

A **solution** of an equation (or inequality) involving a single variable is a number that when substituted for the variable makes the equation (or inequality) true. The set of all solutions is called the **solution set**. When we say that we **solve an equation** (or inequality), we mean that we find its solution set.

Knowing what is meant by the solution set is one thing; finding it is another. We start by recalling the idea of equivalent equations and equivalent inequalities. If we perform an operation on an equation (or inequality) that produces another equation (or inequality) with the same solution set, then the two equations (or inequalities) are said to be **equivalent**. The basic idea in solving equations or inequalities is to perform operations that produce simpler equivalent equations or inequalities and to continue the process until we obtain an equation or inequality with an obvious solution.

Linear Equations

Linear equations are generally solved using the following equality properties.

THEOREM 1 Equality Properties

An equivalent equation will result if

1. The same quantity is added to or subtracted from each side of a given equation.
2. Each side of a given equation is multiplied by or divided by the same nonzero quantity.

EXAMPLE 1 Solving a Linear Equation Solve and check:

$$8x - 3(x - 4) = 3(x - 4) + 6$$

SOLUTION

$8x - 3(x - 4) = 3(x - 4) + 6$	Use the distributive property.
$8x - 3x + 12 = 3x - 12 + 6$	Combine like terms.
$5x + 12 = 3x - 6$	Subtract $3x$ from both sides.
$2x + 12 = -6$	Subtract 12 from both sides.
$2x = -18$	Divide both sides by 2.
$x = -9$	

CHECK

$$8x - 3(x - 4) = 3(x - 4) + 6$$
$$8(-9) - 3[(-9) - 4] \stackrel{?}{=} 3[(-9) - 4] + 6$$
$$-72 - 3(-13) \stackrel{?}{=} 3(-13) + 6$$
$$-33 \stackrel{\checkmark}{=} -33$$

Matched Problem 1 Solve and check: $3x - 2(2x - 5) = 2(x + 3) - 8$

Explore and Discuss 1   According to equality property 2, multiplying both sides of an equation by a nonzero number always produces an equivalent equation. What is the smallest positive number that you could use to multiply both sides of the following equation to produce an equivalent equation without fractions?

$$\frac{x+1}{3} - \frac{x}{4} = \frac{1}{2}$$

EXAMPLE 2   Solving a Linear Equation   Solve and check: $\dfrac{x+2}{2} - \dfrac{x}{3} = 5$

SOLUTION   What operations can we perform on

$$\frac{x+2}{2} - \frac{x}{3} = 5$$

to eliminate the denominators? If we can find a number that is exactly divisible by each denominator, we can use the multiplication property of equality to clear the denominators. The LCD (least common denominator) of the fractions, 6, is exactly what we are looking for! Actually, any common denominator will do, but the LCD results in a simpler equivalent equation. So, we multiply both sides of the equation by 6:

$$6\left(\frac{x+2}{2} - \frac{x}{3}\right) = 6 \cdot 5 \quad *$$

$$\overset{3}{6} \cdot \frac{(x+2)}{2} - \overset{2}{6} \cdot \frac{x}{3} = 30$$

$$3(x+2) - 2x = 30 \qquad \text{Use the distributive property.}$$
$$3x + 6 - 2x = 30 \qquad \text{Combine like terms.}$$
$$x + 6 = 30 \qquad \text{Subtract 6 from both sides.}$$
$$x = 24$$

CHECK
$$\frac{x+2}{2} - \frac{x}{3} = 5$$

$$\frac{24+2}{2} - \frac{24}{3} \overset{?}{=} 5$$

$$13 - 8 \overset{?}{=} 5$$

$$5 \overset{\checkmark}{=} 5$$

Matched Problem 2   Solve and check: $\dfrac{x+1}{3} - \dfrac{x}{4} = \dfrac{1}{2}$

In many applications of algebra, formulas or equations must be changed to alternative equivalent forms. The following example is typical.

EXAMPLE 3   Solving a Formula for a Particular Variable   If you deposit a principal P in an account that earns simple interest at an annual rate r, then the amount A in the account after t years is given by $A = P + Prt$. Solve for

(A) r in terms of A, P, and t

(B) P in terms of A, r, and t

---

*Dashed boxes are used throughout the book to denote steps that are usually performed mentally.

SOLUTION   (A)

$$A = P + Prt \qquad \text{Reverse equation.}$$
$$P + Prt = A \qquad \text{Subtract } P \text{ from both sides.}$$
$$Prt = A - P \qquad \text{Divide both members by } Pt.$$
$$r = \frac{A - P}{Pt}$$

(B)

$$A = P + Prt \qquad \text{Reverse equation.}$$
$$P + Prt = A \qquad \text{Factor out } P \text{ (note the use of the distributive property).}$$
$$P(1 + rt) = A \qquad \text{Divide by } (1 + rt).$$
$$P = \frac{A}{1 + rt}$$

Matched Problem 3 ⟩ If a cardboard box has length L, width W, and height H, then its surface area is given by the formula $S = 2LW + 2LH + 2WH$. Solve the formula for

(A) L in terms of S, W, and H (B) H in terms of S, L, and W

Linear Inequalities

Before we start solving linear inequalities, let us recall what we mean by $<$ (less than) and $>$ (greater than). If a and b are real numbers, we write

$$a < b \qquad a \text{ is less than } b$$

if there exists a positive number p such that $a + p = b$. Certainly, we would expect that if a positive number was added to any real number, the sum would be larger than the original. That is essentially what the definition states. If $a < b$, we may also write

$$b > a \qquad b \text{ is greater than } a.$$

EXAMPLE 4   Inequalities

(A) $3 < 5$ Since $3 + 2 = 5$

(B) $-6 < -2$ Since $-6 + 4 = -2$

(C) $0 > -10$ Since $-10 < 0$ (because $-10 + 10 = 0$)

Matched Problem 4 ⟩ Replace each question mark with either $<$ or $>$.

(A) $2 \; ? \; 8$ (B) $-20 \; ? \; 0$ (C) $-3 \; ? \; -30$

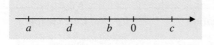

Figure 1 $a < b, c > d$

The inequality symbols have a very clear geometric interpretation on the real number line. If $a < b$, then a is to the left of b on the number line; if $c > d$, then c is to the right of d on the number line (Fig. 1). Check this geometric property with the inequalities in Example 4.

Explore and Discuss 2   Replace ? with $<$ or $>$ in each of the following:

(A) $-1 \; ? \; 3$ and $2(-1) \; ? \; 2(3)$

(B) $-1 \; ? \; 3$ and $-2(-1) \; ? \; -2(3)$

(C) $12 \; ? \; -8$ and $\dfrac{12}{4} \; ? \; \dfrac{-8}{4}$

(D) $12 \; ? \; -8$ and $\dfrac{12}{-4} \; ? \; \dfrac{-8}{-4}$

Based on these examples, describe the effect of multiplying both sides of an inequality by a number.

The procedures used to solve linear inequalities in one variable are almost the same as those used to solve linear equations in one variable, but with one important exception, as noted in item 3 of Theorem 2.

THEOREM 2 Inequality Properties

An equivalent inequality will result, and the **sense or direction will remain the same** if each side of the original inequality

1. has the same real number added to or subtracted from it.
2. is multiplied or divided by the same *positive* number.

An equivalent inequality will result, and the **sense or direction will reverse** if each side of the original inequality

3. is multiplied or divided by the same *negative* number.

Note: Multiplication by 0 and division by 0 are not permitted.

Therefore, we can perform essentially the same operations on inequalities that we perform on equations, with the exception that **the sense of the inequality reverses if we multiply or divide both sides by a negative number.** Otherwise, the sense of the inequality does not change. For example, if we start with the true statement

$$-3 > -7$$

and multiply both sides by 2, we obtain

$$-6 > -14$$

and the sense of the inequality stays the same. But if we multiply both sides of $-3 > -7$ by -2, the left side becomes 6 and the right side becomes 14, so we must write

$$6 < 14$$

to have a true statement. The sense of the inequality reverses.

If $a < b$, the **double inequality** $a < x < b$ means that $a < x$ and $x < b$; that is, x is between a and b. **Interval notation** is also used to describe sets defined by inequalities, as shown in Table 1.

The numbers a and b in Table 1 are called the **endpoints** of the interval. An interval is **closed** if it contains all its endpoints and **open** if it does not contain any of its endpoints. The intervals $[a, b]$, $(-\infty, a]$, and $[b, \infty)$ are closed, and the intervals (a, b), $(-\infty, a)$,

Table 1 Interval Notation

Interval Notation	Inequality Notation	Line Graph
$[a, b]$	$a \leq x \leq b$	$\overset{\quad a \qquad b \quad}{\longmapsto\!\dashv} x$
$[a, b)$	$a \leq x < b$	$\overset{\quad a \qquad b \quad}{\longmapsto\!) } x$
$(a, b]$	$a < x \leq b$	$\overset{\quad a \qquad b \quad}{(\!\longmapsto\!\dashv} x$
(a, b)	$a < x < b$	$\overset{\quad a \qquad b \quad}{(\!\longmapsto\!)} x$
$(-\infty, a]$	$x \leq a$	$\overset{\qquad a \quad}{\longleftarrow\!\dashv} x$
$(-\infty, a)$	$x < a$	$\overset{\qquad a \quad}{\longleftarrow\!)} x$
$[b, \infty)$	$x \geq b$	$\overset{\quad b \qquad}{\vdash\!\longrightarrow} x$
(b, ∞)	$x > b$	$\overset{\quad b \qquad}{(\!\longrightarrow} x$

and (b, ∞) are open. Note that the symbol ∞ (read infinity) is not a number. When we write $[b, \infty)$, we are simply referring to the interval that starts at b and continues indefinitely to the right. We never refer to ∞ as an endpoint, and we never write $[b, \infty]$. The interval $(-\infty, \infty)$ is the entire real number line.

Note that an endpoint of a line graph in Table 1 has a square bracket through it if the endpoint is included in the interval; a parenthesis through an endpoint indicates that it is not included.

CONCEPTUAL INSIGHT

The notation $(2, 7)$ has two common mathematical interpretations: the ordered pair with first coordinate 2 and second coordinate 7, and the open interval consisting of all real numbers between 2 and 7. The choice of interpretation is usually determined by the context in which the notation is used. The notation $(2, -7)$ could be interpreted as an ordered pair but not as an interval. In interval notation, the left endpoint is always written first. So, $(-7, 2)$ is correct interval notation, but $(2, -7)$ is not.

EXAMPLE 5   Interval and Inequality Notation, and Line Graphs

(A) Write $[-2, 3)$ as a double inequality and graph.

(B) Write $x \geq -5$ in interval notation and graph.

SOLUTION (A) $[-2, 3)$ is equivalent to $-2 \leq x < 3$.

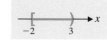

(B) $x \geq -5$ is equivalent to $[-5, \infty)$.

Matched Problem 5

(A) Write $(-7, 4]$ as a double inequality and graph.

(B) Write $x < 3$ in interval notation and graph.

Explore and Discuss 3   The solution to Example 5B shows the graph of the inequality $x \geq -5$. What is the graph of $x < -5$? What is the corresponding interval? Describe the relationship between these sets.

EXAMPLE 6   Solving a Linear Inequality   Solve and graph:

$$2(2x + 3) < 6(x - 2) + 10$$

SOLUTION

$2(2x + 3) < 6(x - 2) + 10$	Remove parentheses.
$4x + 6 < 6x - 12 + 10$	Combine like terms.
$4x + 6 < 6x - 2$	Subtract $6x$ from both sides.
$-2x + 6 < -2$	Subtract 6 from both sides.
$-2x < -8$	Divide both sides by -2 and reverse the sense of the inequality.

$$x > 4 \quad \text{or} \quad (4, \infty)$$

Notice that in the graph of $x > 4$, we use a parenthesis through 4, since the point 4 is not included in the graph.

Matched Problem 6   Solve and graph: $3(x - 1) \leq 5(x + 2) - 5$

EXAMPLE 7 Solving a Double Inequality Solve and graph: $-3 < 2x + 3 \le 9$

SOLUTION We are looking for all numbers x such that $2x + 3$ is between -3 and 9, including 9 but not -3. We proceed as before except that we try to isolate x in the middle:

$$-3 < 2x + 3 \le 9$$
$$-3 - 3 < 2x + 3 - 3 \le 9 - 3$$
$$-6 < 2x \le 6$$
$$\frac{-6}{2} < \frac{2x}{2} \le \frac{6}{2}$$
$$-3 < x \le 3 \quad \text{or} \quad (-3, 3]$$

Matched Problem 7 Solve and graph: $-8 \le 3x - 5 < 7$

Note that a linear equation usually has exactly one solution, while a linear inequality usually has infinitely many solutions.

Applications

To realize the full potential of algebra, we must be able to translate real-world problems into mathematics. In short, we must be able to do word problems.

Here are some suggestions that will help you get started:

PROCEDURE For Solving Word Problems

1. Read the problem carefully and introduce a variable to represent an unknown quantity in the problem. Often the question asked in a problem will indicate the unknown quantity that should be represented by a variable.
2. Identify other quantities in the problem (known or unknown), and whenever possible, express unknown quantities in terms of the variable you introduced in Step 1.
3. Write a verbal statement using the conditions stated in the problem and then write an equivalent mathematical statement (equation or inequality).
4. Solve the equation or inequality and answer the questions posed in the problem.
5. Check the solution(s) in the original problem.

EXAMPLE 8 Purchase Price Alex purchases a plasma TV, pays 7% state sales tax, and is charged $65 for delivery. If Alex's total cost is $1,668.93, what was the purchase price of the TV?

SOLUTION

Step 1 **Introduce a variable for the unknown quantity.** After reading the problem, we decide to let x represent the purchase price of the TV.

Step 2 **Identify quantities in the problem.**

Delivery charge: $65

Sales tax: $0.07x$

Total cost: $1,668.93

Step 3 **Write a verbal statement and an equation.**

Price + Delivery Charge + Sales Tax = Total Cost

$$x \quad + \quad\quad 65 \quad + \quad 0.07x \ = \ 1{,}668.93$$

Step 4 **Solve the equation and answer the question.**

$$x + 65 + 0.07x = 1{,}668.93 \quad \text{Combine like terms.}$$
$$1.07x + 65 = 1{,}668.93 \quad \text{Subtract 65 from both sides.}$$
$$1.07x = 1{,}603.93 \quad \text{Divide both sides by 1.07.}$$
$$x = 1{,}499$$

The price of the TV is $1,499.

Step 5 **Check the answer in the original problem.**

$$\text{Price} = \$1{,}499.00$$
$$\text{Delivery charge} = \$\ \ \ 65.00$$
$$\underline{\text{Tax} = 0.07 \cdot 1{,}499 = \$\ \ 104.93}$$
$$\text{Total} = \$1{,}668.93$$

Matched Problem 8) Mary paid 8.5% sales tax and a $190 title and license fee when she bought a new car for a total of $28,400. What is the purchase price of the car?

The next example involves the important concept of **break-even analysis**, which is encountered in several places in this text. Any manufacturing company has **costs**, C, and **revenues**, R. The company will have a **loss** if $R < C$, will **break even** if $R = C$, and will have a **profit** if $R > C$. Costs involve **fixed costs**, such as plant overhead, product design, setup, and promotion, and **variable costs**, which are dependent on the number of items produced at a certain cost per item.

EXAMPLE 9 Break-Even Analysis A multimedia company produces DVDs. Onetime fixed costs for a particular DVD are $48,000, which include costs such as filming, editing, and promotion. Variable costs amount to $12.40 per DVD and include manufacturing, packaging, and distribution costs for each DVD actually sold to a retailer. The DVD is sold to retail outlets at $17.40 each. How many DVDs must be manufactured and sold in order for the company to break even?

SOLUTION

Step 1 Let $x =$ number of DVDs manufactured and sold.

Step 2
$$C = \text{cost of producing } x \text{ DVDs}$$
$$R = \text{revenue (return) on sales of } x \text{ DVDs}$$
$$\text{Fixed costs} = \$48{,}000$$
$$\text{Variable costs} = \$12.40x$$
$$C = \text{Fixed costs} + \text{variable costs}$$
$$= \$48{,}000 + \$12.40x$$
$$R = \$17.40x$$

Step 3 The company breaks even if $R = C$; that is, if

$$\$17.40x = \$48{,}000 + \$12.40x$$

Step 4
$$17.4x = 48{,}000 + 12.4x \quad \text{Subtract 12.4x from both sides.}$$
$$5x = 48{,}000 \quad \text{Divide both sides by 5.}$$
$$x = 9{,}600$$

The company must make and sell 9,600 DVDs to break even.

Step 5 Check:

Costs	Revenue
$48{,}000 + 12.4(9{,}600)$	$17.4(9{,}600)$
$= \$167{,}040$	$= \$167{,}040$

Matched Problem 9 ⟩ How many DVDs would a multimedia company have to make and sell to break even if the fixed costs are $36,000, variable costs are $10.40 per DVD, and the DVDs are sold to retailers for $15.20 each?

Table 2 **CPI (1982–1984 = 100)**

Year	Index
1960	29.6
1973	44.4
1986	109.6
1999	156.9
2012	229.6

EXAMPLE 10 Consumer Price Index   The Consumer Price Index (CPI) is a measure of the average change in prices over time from a designated reference period, which equals 100. The index is based on prices of basic consumer goods and services. Table 2 lists the CPI for several years from 1960 to 2012. What net annual salary in 2012 would have the same purchasing power as a net annual salary of $13,000 in 1960? Compute the answer to the nearest dollar. (*Source*: U.S. Bureau of Labor Statistics)

SOLUTION

Step 1 Let x = the purchasing power of an annual salary in 2012.

Step 2 Annual salary in 1960 = $13,000

$$\text{CPI in 1960} = 29.6$$
$$\text{CPI in 2012} = 229.6$$

Step 3 The ratio of a salary in 2012 to a salary in 1960 is the same as the ratio of the CPI in 2012 to the CPI in 1960.

$$\frac{x}{13,000} = \frac{229.6}{29.6} \qquad \text{Multiply both sides by 13,000.}$$

Step 4
$$x = 13,000 \cdot \frac{229.6}{29.6}$$
$$= \$100,838 \text{ per year}$$

Step 5 To check the answer, we confirm that the salary ratio agrees with the CPI ratio:

Salary Ratio	**CPI Ratio**
$\dfrac{100,838}{13,000} = 7.757$	$\dfrac{229.6}{29.6} = 7.757$

Matched Problem 10 ⟩ What net annual salary in 1973 would have had the same purchasing power as a net annual salary of $100,000 in 2012? Compute the answer to the nearest dollar.

Exercises 1.1

Solve Problems 1–6.

1. $2m + 9 = 5m - 6$ **2.** $3y - 4 = 6y - 19$

3. $2x + 3 < -4$ **4.** $5x + 2 > 1$

5. $-3x \geq -12$ **6.** $-4x \leq 8$

Solve Problems 7–10 and graph.

7. $-4x - 7 > 5$ **8.** $-2x + 8 < 4$

9. $2 \leq x + 3 \leq 5$ **10.** $-4 < 2y - 3 < 9$

Solve Problems 11–24.

11. $\dfrac{x}{4} + \dfrac{1}{2} = \dfrac{1}{8}$ **12.** $\dfrac{m}{3} - 4 = \dfrac{2}{3}$

13. $\dfrac{y}{-5} > \dfrac{3}{2}$ **14.** $\dfrac{x}{-4} < \dfrac{5}{6}$

15. $2u + 4 = 5u + 1 - 7u$ **16.** $-3y + 9 + y = 13 - 8y$

17. $10x + 25(x - 3) = 275$ **18.** $-3(4 - x) = 5 - (x + 1)$

19. $3 - y \leq 4(y - 3)$ **20.** $x - 2 \geq 2(x - 5)$

21. $\dfrac{x}{5} - \dfrac{x}{6} = \dfrac{6}{5}$ **22.** $\dfrac{y}{4} - \dfrac{y}{3} = \dfrac{1}{2}$

23. $\dfrac{m}{5} - 3 < \dfrac{3}{5} - \dfrac{m}{2}$ **24.** $\dfrac{u}{2} - \dfrac{2}{3} < \dfrac{u}{3} + 2$

Solve Problems 25–28 and graph.

25. $2 \leq 3x - 7 < 14$ **26.** $-4 \leq 5x + 6 < 21$

27. $-4 \leq \frac{9}{5}C + 32 \leq 68$ **28.** $-1 \leq \frac{2}{3}t + 5 \leq 11$

Solve Problems 29–34 for the indicated variable.

29. $3x - 4y = 12$; for y **30.** $y = -\frac{2}{3}x + 8$; for x

31. $Ax + By = C$; for y $(B \neq 0)$

32. $y = mx + b$; for m **33.** $F = \frac{9}{5}C + 32$; for C

34. $C = \frac{5}{9}(F - 32)$; for F

Solve Problems 35 and 36 and graph.

35. $-3 \leq 4 - 7x < 18$ **36.** $-10 \leq 8 - 3u \leq -6$

37. What can be said about the signs of the numbers a and b in each case?

(A) $ab > 0$ (B) $ab < 0$

(C) $\dfrac{a}{b} > 0$ (D) $\dfrac{a}{b} < 0$

38. What can be said about the signs of the numbers a, b, and c in each case?

(A) $abc > 0$ (B) $\dfrac{ab}{c} < 0$

(C) $\dfrac{a}{bc} > 0$ (D) $\dfrac{a^2}{bc} < 0$

39. If both a and b are positive numbers and b/a is greater than 1, then is $a - b$ positive or negative?

40. If both a and b are negative numbers and b/a is greater than 1, then is $a - b$ positive or negative?

In Problems 41–46, discuss the validity of each statement. If the statement is true, explain why. If not, give a counterexample.

41. If the intersection of two open intervals is nonempty, then their intersection is an open interval.

42. If the intersection of two closed intervals is nonempty, then their intersection is a closed interval.

43. The union of any two open intervals is an open interval.

44. The union of any two closed intervals is a closed interval.

45. If the intersection of two open intervals is nonempty, then their union is an open interval.

46. If the intersection of two closed intervals is nonempty, then their union is a closed interval.

Applications

47. Ticket sales. A rock concert brought in $432,500 on the sale of 9,500 tickets. If the tickets sold for $35 and $55 each, how many of each type of ticket were sold?

48. Parking meter coins. An all-day parking meter takes only dimes and quarters. If it contains 100 coins with a total value of $14.50, how many of each type of coin are in the meter?

49. IRA. You have $500,000 in an IRA (Individual Retirement Account) at the time you retire. You have the option of investing this money in two funds: Fund A pays 5.2% annually and Fund B pays 7.7% annually. How should you divide your money between Fund A and Fund B to produce an annual interest income of $34,000?

50. IRA. Refer to Problem 49. How should you divide your money between Fund A and Fund B to produce an annual interest income of $30,000?

51. Car prices. If the price change of cars parallels the change in the CPI (see Table 2 in Example 10), what would a car sell for (to the nearest dollar) in 2012 if a comparable model sold for $10,000 in 1999?

52. Home values. If the price change in houses parallels the CPI (see Table 2 in Example 10), what would a house valued at $200,000 in 2012 be valued at (to the nearest dollar) in 1960?

53. Retail and wholesale prices. Retail prices in a department store are obtained by marking up the wholesale price by 40%. That is, the retail price is obtained by adding 40% of the wholesale price to the wholesale price.

(A) What is the retail price of a suit if the wholesale price is $300?

(B) What is the wholesale price of a pair of jeans if the retail price is $77?

54. Retail and sale prices. Sale prices in a department store are obtained by marking down the retail price by 15%. That is, the sale price is obtained by subtracting 15% of the retail price from the retail price.

(A) What is the sale price of a hat that has a retail price of $60?

(B) What is the retail price of a dress that has a sale price of $136?

55. Equipment rental. A golf course charges $52 for a round of golf using a set of their clubs, and $44 if you have your own clubs. If you buy a set of clubs for $270, how many rounds must you play to recover the cost of the clubs?

56. Equipment rental. The local supermarket rents carpet cleaners for $20 a day. These cleaners use shampoo in a special cartridge that sells for $16 and is available only from the supermarket. A home carpet cleaner can be purchased for $300. Shampoo for the home cleaner is readily available for $9 a bottle. Past experience has shown that it takes two shampoo cartridges to clean the 10-foot-by-12-foot carpet in your living room with the rented cleaner. Cleaning the same area with the home cleaner will consume three bottles of shampoo. If you buy the home cleaner, how many times must you clean the living-room carpet to make buying cheaper than renting?

57. Sales commissions. One employee of a computer store is paid a base salary of $2,000 a month plus an 8% commission on all sales over $7,000 during the month. How much must the employee sell in one month to earn a total of $4,000 for the month?

58. Sales commissions. A second employee of the computer store in Problem 57 is paid a base salary of $3,000 a month plus a 5% commission on all sales during the month.

(A) How much must this employee sell in one month to earn a total of $4,000 for the month?

(B) Determine the sales level at which both employees receive the same monthly income.

(C) If employees can select either of these payment methods, how would you advise an employee to make this selection?

59. Break-even analysis. A publisher for a promising new novel figures fixed costs (overhead, advances, promotion, copy editing, typesetting) at $55,000, and variable costs (printing, paper, binding, shipping) at $1.60 for each book produced. If the book is sold to distributors for $11 each, how many must be produced and sold for the publisher to break even?

60. Break-even analysis. The publisher of a new book figures fixed costs at $92,000 and variable costs at $2.10 for each book produced. If the book is sold to distributors for $15 each, how many must be sold for the publisher to break even?

61. Break-even analysis. The publisher in Problem 59 finds that rising prices for paper increase the variable costs to $2.10 per book.

✎ (A) Discuss possible strategies the company might use to deal with this increase in costs.

(B) If the company continues to sell the books for $11, how many books must they sell now to make a profit?

(C) If the company wants to start making a profit at the same production level as before the cost increase, how much should they sell the book for now?

62. Break-even analysis. The publisher in Problem 60 finds that rising prices for paper increase the variable costs to $2.70 per book.

✎ (A) Discuss possible strategies the company might use to deal with this increase in costs.

(B) If the company continues to sell the books for $15, how many books must they sell now to make a profit?

(C) If the company wants to start making a profit at the same production level as before the cost increase, how much should they sell the book for now?

63. Wildlife management. A naturalist estimated the total number of rainbow trout in a certain lake using the capture–mark–recapture technique. He netted, marked, and released 200 rainbow trout. A week later, allowing for thorough mixing, he again netted 200 trout, and found 8 marked ones among them. Assuming that the proportion of marked fish in the second sample was the same as the proportion of all marked fish in the total population, estimate the number of rainbow trout in the lake.

64. Temperature conversion. If the temperature for a 24-hour period at an Antarctic station ranged between $-49°F$ and $14°F$ (that is, $-49 \leq F \leq 14$), what was the range in degrees Celsius? [*Note:* $F = \frac{9}{5}C + 32$.]

65. Psychology. The IQ (intelligence quotient) is found by dividing the mental age (MA), as indicated on standard tests, by the chronological age (CA) and multiplying by 100. For example, if a child has a mental age of 12 and a chronological age of 8, the calculated IQ is 150. If a 9-year-old girl has an IQ of 140, compute her mental age.

66. Psychology. Refer to Problem 65. If the IQ of a group of 12-year-old children varies between 80 and 140, what is the range of their mental ages?

Answers to Matched Problems

1. $x = 4$ **2.** $x = 2$

3. (A) $L = \dfrac{S - 2WH}{2W + 2H}$ (B) $H = \dfrac{S - 2LW}{2L + 2W}$

4. (A) $<$ (B) $<$ (C) $>$

5. (A) $-7 < x \leq 4$;   (B) $(-\infty, 3)$

         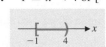

6. $x \geq -4$ or $[-4, \infty)$ **7.** $-1 \leq x < 4$ or $[-1, 4)$

8. $26,000   **9.** 7,500 DVDs   **10.** $19,338

1.2 Graphs and Lines

- Cartesian Coordinate System
- Graphs of $Ax + By = C$
- Slope of a Line
- Equations of Lines: Special Forms
- Applications

In this section, we will consider one of the most basic geometric figures—a line. When we use the term *line* in this book, we mean *straight line*. We will learn how to recognize and graph a line, and how to use information concerning a line to find its equation. Examining the graph of any equation often results in additional insight into the nature of the equation's solutions.

Cartesian Coordinate System

Recall that to form a **Cartesian** or **rectangular coordinate system**, we select two real number lines—one horizontal and one vertical—and let them cross through their origins as indicated in Figure 1. Up and to the right are the usual choices for the positive directions. These two number lines are called the **horizontal axis** and the **vertical**

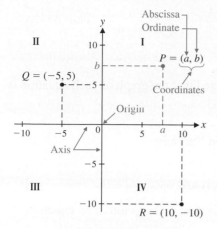

Figure 1 **The Cartesian (rectangular) coordinate system**

axis, or, together, the **coordinate axes**. The horizontal axis is usually referred to as the *x* **axis** and the vertical axis as the *y* **axis**, and each is labeled accordingly. The coordinate axes divide the plane into four parts called **quadrants**, which are numbered counterclockwise from I to IV (see Fig. 1).

Now we want to assign *coordinates* to each point in the plane. Given an arbitrary point *P* in the plane, pass horizontal and vertical lines through the point (Fig. 1). The vertical line will intersect the horizontal axis at a point with coordinate *a*, and the horizontal line will intersect the vertical axis at a point with coordinate *b*. These two numbers, written as the **ordered pair** (a, b),* form the **coordinates** of the point *P*. The first coordinate, *a*, is called the **abscissa** of *P*; the second coordinate, *b*, is called the **ordinate** of *P*. The abscissa of *Q* in Figure 1 is −5, and the ordinate of *Q* is 5. The coordinates of a point can also be referenced in terms of the axis labels. The *x* **coordinate** of *R* in Figure 1 is 10, and the *y* **coordinate** of *R* is −10. The point with coordinates $(0, 0)$ is called the **origin**.

The procedure we have just described assigns to each point *P* in the plane a unique pair of real numbers (a, b). Conversely, if we are given an ordered pair of real numbers (a, b), then, reversing this procedure, we can determine a unique point *P* in the plane. Thus,

> There is a one-to-one correspondence between the points in a plane and the elements in the set of all ordered pairs of real numbers.

This is often referred to as the **fundamental theorem of analytic geometry**.

Graphs of Ax + By = C

In Section 1.1, we called an equation of the form $ax + b = 0$ $(a \neq 0)$ a linear equation in one variable. Now we want to consider linear equations in two variables:

> **DEFINITION** Linear Equations in Two Variables
> A **linear equation in two variables** is an equation that can be written in the **standard form**
> $$Ax + By = C$$
> where *A*, *B*, and *C* are constants (*A* and *B* not both 0), and *x* and *y* are variables.

A **solution** of an equation in two variables is an ordered pair of real numbers that satisfies the equation. For example, $(4, 3)$ is a solution of $3x − 2y = 6$. The **solution set** of an equation in two variables is the set of all solutions of the equation. The **graph** of an equation is the graph of its solution set.

Explore and Discuss 1 (A) As noted earlier, $(4, 3)$ is a solution of the equation

$$3x − 2y = 6$$

Find three more solutions of this equation. Plot these solutions in a Cartesian coordinate system. What familiar geometric shape could be used to describe the solution set of this equation?

---

*Here we use (a, b) as the coordinates of a point in a plane. In Section 1.1, we used (a, b) to represent an interval on a real number line. These concepts are not the same. You must always interpret the symbol (a, b) in terms of the context in which it is used.

(B) Repeat part (A) for the equation $x = 2$.

(C) Repeat part (A) for the equation $y = -3$.

In Explore and Discuss 1, you may have recognized that the graph of each equation is a (straight) line. Theorem 1 confirms this fact.

THEOREM 1   Graph of a Linear Equation in Two Variables

The graph of any equation of the form

$$Ax + By = C \quad (A \text{ and } B \text{ not both } 0) \tag{1}$$

is a line, and any line in a Cartesian coordinate system is the graph of an equation of this form.

If $A \neq 0$ and $B \neq 0$, then equation (1) can be written as

$$y = -\frac{A}{B}x + \frac{C}{B} = mx + b, m \neq 0$$

If $A = 0$ and $B \neq 0$, then equation (1) can be written as

$$y = \frac{C}{B}$$

and its graph is a **horizontal line**. If $A \neq 0$ and $B = 0$, then equation (1) can be written as

$$x = \frac{C}{A}$$

and its graph is a **vertical line**. To graph equation (1), or any of its special cases, plot any two points in the solution set and use a straightedge to draw the line through these two points. The points where the line crosses the axes are often the easiest to find. The y intercept* is the y coordinate of the point where the graph crosses the y axis, and the x intercept is the x coordinate of the point where the graph crosses the x axis. To find the y intercept, let $x = 0$ and solve for y. To find the x intercept, let $y = 0$ and solve for x. It is a good idea to find a third point as a check point.

EXAMPLE 1 Using Intercepts to Graph a Line   Graph: $3x - 4y = 12$

SOLUTION

x	y	
0	-3	y intercept
4	0	x intercept
8	3	Check point

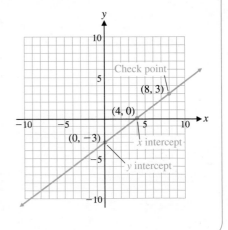

Matched Problem 1 Graph: $4x - 3y = 12$

---

*If the x intercept is a and the y intercept is b, then the graph of the line passes through the points $(a, 0)$ and $(0, b)$. It is common practice to refer to both the numbers a and b and the points $(a, 0)$ and $(0, b)$ as the x and y intercepts of the line.

The icon in the margin is used throughout this book to identify optional graphing calculator activities that are intended to give you additional insight into the concepts under discussion. You may have to consult the manual for your calculator* for the details necessary to carry out these activities.

EXAMPLE 2 Using a Graphing Calculator Graph $3x - 4y = 12$ on a graphing calculator and find the intercepts.

SOLUTION First, we solve $3x - 4y = 12$ for y.

$$3x - 4y = 12 \qquad \text{Add } -3x \text{ to both sides.}$$
$$-4y = -3x + 12 \qquad \text{Divide both sides by } -4.$$
$$y = \frac{-3x + 12}{-4} \qquad \text{Simplify.}$$
$$y = \frac{3}{4}x - 3 \qquad\qquad\qquad\qquad (2)$$

Now we enter the right side of equation (2) in a calculator (Fig. 2A), enter values for the window variables (Fig. 2B), and graph the line (Fig. 2C). (The numerals to the left and right of the screen in Figure 2C are Xmin and Xmax, respectively. Similarly, the numerals below and above the screen are Ymin and Ymax.)

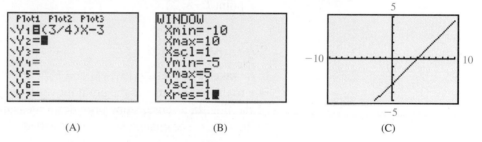

(A)   (B)   (C)

Figure 2 **Graphing a line on a graphing calculator**

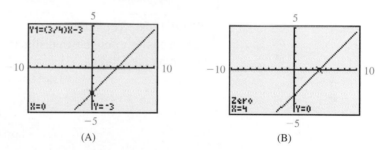

(A)   (B)

Figure 3 **Using TRACE and zero on a graphing calculator**

Next we use two calculator commands to find the intercepts: TRACE (Fig. 3A) and zero (Fig. 3B). The y intercept is -3 (Fig. 3A) and the x intercept is 4 (Fig. 3B).

Matched Problem 2 Graph $4x - 3y = 12$ on a graphing calculator and find the intercepts.

EXAMPLE 3 Horizontal and Vertical Lines
(A) Graph $x = -4$ and $y = 6$ simultaneously in the same rectangular coordinate system.
(B) Write the equations of the vertical and horizontal lines that pass through the point $(7, -5)$.

---

*We used a Texas Instruments graphing calculator from the TI-83/84 family to produce the graphing calculator screens in the book. Manuals for most graphing calculators are readily available on the Internet.

SOLUTION

(A)

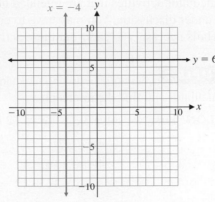

(B) Horizontal line through $(7, -5)$: $y = -5$
Vertical line through $(7, -5)$: $x = 7$

Matched Problem 3

(A) Graph $x = 5$ and $y = -3$ simultaneously in the same rectangular coordinate system.

(B) Write the equations of the vertical and horizontal lines that pass through the point $(-8, 2)$.

Slope of a Line

If we take two points, $P_1(x_1, y_1)$ and $P_2(x_2, y_2)$, on a line, then the ratio of the change in y to the change in x as the point moves from point P_1 to point P_2 is called the **slope** of the line. In a sense, slope provides a measure of the "steepness" of a line relative to the x axis. The change in x is often called the **run**, and the change in y is the **rise**.

DEFINITION Slope of a Line

If a line passes through two distinct points, $P_1(x_1, y_1)$ and $P_2(x_2, y_2)$, then its slope is given by the formula

$$m = \frac{y_2 - y_1}{x_2 - x_1} \qquad x_1 \neq x_2$$

$$= \frac{\text{vertical change (rise)}}{\text{horizontal change (run)}}$$

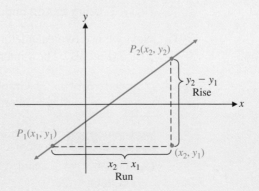

For a horizontal line, y does not change; its slope is 0. For a vertical line, x does not change; $x_1 = x_2$ so its slope is not defined. In general, the slope of a line may be positive, negative, 0, or not defined. Each case is illustrated geometrically in Table 1.

Table 1 **Geometric Interpretation of Slope**

Line	Rising as x moves from left to right	Falling as x moves from left to right	Horizontal	Vertical
Slope	Positive	Negative	0	Not defined
Example				

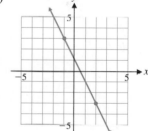

$$m = \frac{b}{a} = \frac{b'}{a'}$$

Figure 4

CONCEPTUAL INSIGHT

One property of real numbers discussed in Appendix A, Section A.1, is

$$\frac{-a}{-b} = -\frac{-a}{b} = -\frac{a}{-b} = \frac{a}{b}, \quad b \neq 0$$

This property implies that it does not matter which point we label as P_1 and which we label as P_2 in the slope formula. For example, if $A = (4, 3)$ and $B = (1, 2)$, then

$$B = P_2 = (1, 2) \qquad A = P_2 = (4, 3)$$
$$A = P_1 = (4, 3) \qquad B = P_1 = (1, 2)$$
$$m = \frac{2 - 3}{1 - 4} = \frac{-1}{-3} = \frac{1}{3} = \frac{3 - 2}{4 - 1}$$

A property of similar triangles (see Table I in Appendix C) ensures that the slope of a line is the same for any pair of distinct points on the line (Fig. 4).

EXAMPLE 4 Finding Slopes Sketch a line through each pair of points, and find the slope of each line.

(A) $(-3, -2), (3, 4)$ (B) $(-1, 3), (2, -3)$

(C) $(-2, -3), (3, -3)$ (D) $(-2, 4), (-2, -2)$

SOLUTION

(A)

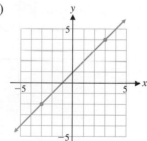

$$m = \frac{4 - (-2)}{3 - (-3)} = \frac{6}{6} = 1$$

(B)

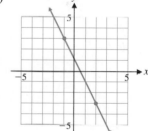

$$m = \frac{-3 - 3}{2 - (-1)} = \frac{-6}{3} = -2$$

(C)

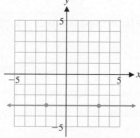

$$m = \frac{-3 - (-3)}{3 - (-2)} = \frac{0}{5} = 0$$

(D)

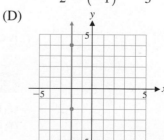

$$m = \frac{-2 - 4}{-2 - (-2)} = \frac{-6}{0}$$

Slope is not defined.

Matched Problem 4 Find the slope of the line through each pair of points.

(A) $(-2, 4), (3, 4)$ (B) $(-2, 4), (0, -4)$

(C) $(-1, 5), (-1, -2)$ (D) $(-1, -2), (2, 1)$

Equations of Lines: Special Forms

Let us start by investigating why $y = mx + b$ is called the *slope-intercept form* for a line.

Explore and Discuss 2   (A) Graph $y = x + b$ for $b = -5, -3, 0, 3$, and 5 simultaneously in the same coordinate system. Verbally describe the geometric significance of b.

(B) Graph $y = mx - 1$ for $m = -2, -1, 0, 1$, and 2 simultaneously in the same coordinate system. Verbally describe the geometric significance of m.

 (C) Using a graphing calculator, explore the graph of $y = mx + b$ for different values of m and b.

As you may have deduced from Explore and Discuss 2, constants m and b in $y = mx + b$ have the following geometric interpretations.

If we let $x = 0$, then $y = b$. So the graph of $y = mx + b$ crosses the y axis at $(0, b)$. The constant b is the y *intercept*. For example, the y intercept of the graph of $y = -4x - 1$ is -1.

To determine the geometric significance of m, we proceed as follows: If $y = mx + b$, then by setting $x = 0$ and $x = 1$, we conclude that $(0, b)$ and $(1, m + b)$ lie on its graph (Fig. 5). The slope of this line is given by:

$$\text{Slope} = \frac{y_2 - y_1}{x_2 - x_1} = \frac{(m + b) - b}{1 - 0} = m$$

So m is the slope of the line given by $y = mx + b$.

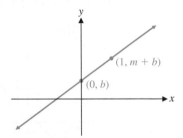

Figure 5

DEFINITION   Slope-Intercept Form

The equation

$$y = mx + b \quad\quad m = \text{slope}, b = y\text{ intercept} \tag{3}$$

is called the **slope-intercept form** of an equation of a line.

EXAMPLE 5   Using the Slope-Intercept Form

(A) Find the slope and y intercept, and graph $y = -\frac{2}{3}x - 3$.

(B) Write the equation of the line with slope $\frac{2}{3}$ and y intercept -2.

SOLUTION

(A) Slope $= m = -\frac{2}{3}$ (B) $m = \frac{2}{3}$ and $b = -2$;

     y intercept $= b = -3$ so, $y = \frac{2}{3}x - 2$

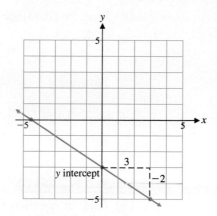

Matched Problem 5 Write the equation of the line with slope $\frac{1}{2}$ and y intercept -1. Graph.

Suppose that a line has slope m and passes through a fixed point (x_1, y_1). If the point (x, y) is any other point on the line (Fig. 6), then

$$\frac{y - y_1}{x - x_1} = m$$

That is,

$$y - y_1 = m(x - x_1) \qquad (4)$$

We now observe that (x_1, y_1) also satisfies equation (4) and conclude that equation (4) is an equation of a line with slope m that passes through (x_1, y_1).

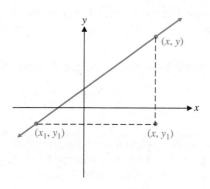

Figure 6

> **DEFINITION Point-Slope Form**
>
> An equation of a line with slope m that passes through (x_1, y_1) is
>
> $$y - y_1 = m(x - x_1) \qquad (4)$$
>
> which is called the **point-slope form** of an equation of a line.

The point-slope form is extremely useful, since it enables us to find an equation for a line if we know its slope and the coordinates of a point on the line or if we know the coordinates of two points on the line.

EXAMPLE 6 Using the Point-Slope Form

(A) Find an equation for the line that has slope $\frac{1}{2}$ and passes through $(-4, 3)$. Write the final answer in the form $Ax + By = C$.

(B) Find an equation for the line that passes through the points $(-3, 2)$ and $(-4, 5)$. Write the resulting equation in the form $y = mx + b$.

SOLUTION

(A) Use $y - y_1 = m(x - x_1)$. Let $m = \frac{1}{2}$ and $(x_1, y_1) = (-4, 3)$. Then

$$y - 3 = \tfrac{1}{2}[x - (-4)]$$
$$y - 3 = \tfrac{1}{2}(x + 4) \qquad \text{Multiply both sides by 2.}$$
$$2y - 6 = x + 4$$
$$-x + 2y = 10 \quad \text{or} \quad x - 2y = -10$$

(B) First, find the slope of the line by using the slope formula:

$$m = \frac{y_2 - y_1}{x_2 - x_1} = \frac{5 - 2}{-4 - (-3)} = \frac{3}{-1} = -3$$

Now use $y - y_1 = m(x - x_1)$ with $m = -3$ and $(x_1, y_1) = (-3, 2)$:

$$y - 2 = -3[x - (-3)]$$
$$y - 2 = -3(x + 3)$$
$$y - 2 = -3x - 9$$
$$y = -3x - 7$$

Matched Problem 6

(A) Find an equation for the line that has slope $\frac{2}{3}$ and passes through $(6, -2)$. Write the resulting equation in the form $Ax + By = C, A > 0$.

(B) Find an equation for the line that passes through $(2, -3)$ and $(4, 3)$. Write the resulting equation in the form $y = mx + b$.

The various forms of the equation of a line that we have discussed are summarized in Table 2 for quick reference.

Table 2 **Equations of a Line**

Standard form	$Ax + By = C$	A and B not both 0
Slope-intercept form	$y = mx + b$	Slope: m; y intercept: b
Point-slope form	$y - y_1 = m(x - x_1)$	Slope: m; point: (x_1, y_1)
Horizontal line	$y = b$	Slope: 0
Vertical line	$x = a$	Slope: undefined

Applications

We will now see how equations of lines occur in certain applications.

EXAMPLE 7 Cost Equation The management of a company that manufactures skateboards has fixed costs (costs at 0 output) of $300 per day and total costs of $4,300 per day at an output of 100 skateboards per day. Assume that cost C is linearly related to output x.

(A) Find the slope of the line joining the points associated with outputs of 0 and 100; that is, the line passing through $(0, 300)$ and $(100, 4,300)$.

(B) Find an equation of the line relating output to cost. Write the final answer in the form $C = mx + b$.

(C) Graph the cost equation from part (B) for $0 \le x \le 200$.

SOLUTION

(A) $m = \dfrac{y_2 - y_1}{x_2 - x_1}$

$= \dfrac{4,300 - 300}{100 - 0}$

$= \dfrac{4,000}{100} = 40$

(B) We must find an equation of the line that passes through $(0, 300)$ with slope 40. We use the slope-intercept form:

$$C = mx + b$$
$$C = 40x + 300$$

(C)

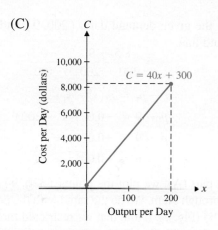

In Example 7, the *fixed cost* of $300 per day covers plant cost, insurance, and so on. This cost is incurred whether or not there is any production. The *variable cost* is $40x$, which depends on the day's output. Since increasing production from x to $x + 1$ will increase the cost by $40 (from $40x + 300$ to $40x + 340$), the slope 40 can be interpreted as the **rate of change** of the cost function with respect to production x.

Matched Problem 7) Answer parts (A) and (B) in Example 7 for fixed costs of $250 per day and total costs of $3,450 per day at an output of 80 skateboards per day.

In a free competitive market, the price of a product is determined by the relationship between supply and demand. If there is a surplus—that is, the supply is greater than the demand—the price tends to come down. If there is a shortage—that is, the demand is greater than the supply—the price tends to go up. The price tends to move toward an equilibrium price at which the supply and demand are equal. Example 8 introduces the basic concepts.

EXAMPLE 8   Supply and Demand   At a price of $9.00 per box of oranges, the supply is 320,000 boxes and the demand is 200,000 boxes. At a price of $8.50 per box, the supply is 270,000 boxes and the demand is 300,000 boxes.

(A) Find a price–supply equation of the form $p = mx + b$, where p is the price in dollars and x is the corresponding supply in thousands of boxes.

(B) Find a price–demand equation of the form $p = mx + b$, where p is the price in dollars and x is the corresponding demand in thousands of boxes.

(C) Graph the price–supply and price–demand equations in the same coordinate system and find their point of intersection.

SOLUTION

(A) To find a price–supply equation of the form $p = mx + b$, we must find two points of the form (x, p) that are on the supply line. From the given supply data, $(320, 9)$ and $(270, 8.5)$ are two such points. First, find the slope of the line:

$$m = \frac{9 - 8.5}{320 - 270} = \frac{0.5}{50} = 0.01$$

Now use the point-slope form to find the equation of the line:

$$p - p_1 = m(x - x_1) \qquad (x_1, p_1) = (320, 9)$$
$$p - 9 = 0.01(x - 320)$$
$$p - 9 = 0.01x - 3.2$$
$$p = 0.01x + 5.8 \qquad \text{Price–supply equation}$$

(B) From the given demand data, $(200, 9)$ and $(300, 8.5)$ are two points on the demand line.

$$m = \frac{8.5 - 9}{300 - 200} = \frac{-0.5}{100} = -0.005$$

$$p - p_1 = m(x - x_1) \qquad (x_1, p_1) = (200, 9)$$
$$p - 9 = -0.005(x - 200)$$
$$p - 9 = -0.005x + 1$$
$$p = -0.005x + 10 \qquad \text{Price–demand equation}$$

(C) From part (A), we plot the points $(320, 9)$ and $(270, 8.5)$ and then draw the line through them. We do the same with the points $(200, 9)$ and $(300, 8.5)$ from part (B) (Fig. 7). (Note that we restricted the axes to intervals that contain these data points.) To find the intersection point of the two lines, we equate the right-hand sides of the price–supply and price–demand equations and solve for x:

$$\text{Price–supply} \quad \text{Price–demand}$$
$$0.01x + 5.8 = -0.005x + 10$$
$$0.015x = 4.2$$
$$x = 280$$

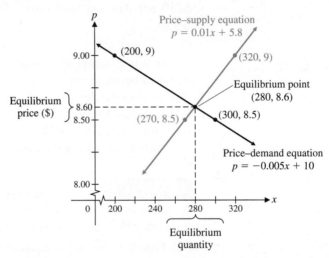

Figure 7 **Graphs of price–supply and price–demand equations**

Now use the price–supply equation to find p when $x = 280$:

$$p = 0.01x + 5.8$$
$$p = 0.01(280) + 5.8 = 8.6$$

As a check, we use the price–demand equation to find p when $x = 280$:

$$p = -0.005x + 10$$
$$p = -0.005(280) + 10 = 8.6$$

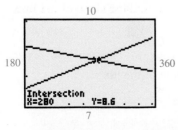

Figure 8 **Finding an intersection point**

The lines intersect at $(280, 8.6)$. The intersection point of the price–supply and price–demand equations is called the **equilibrium point**, and its coordinates are the **equilibrium quantity** (280) and the **equilibrium price** ($8.60). These terms are illustrated in Figure 7. The intersection point can also be found by using the INTERSECT command on a graphing calculator (Fig. 8). To summarize, the price of a box of oranges tends toward the equilibrium price of $8.60, at which the supply and demand are both equal to 280,000 boxes.

Matched Problem 8) At a price of $12.59 per box of grapefruit, the supply is 595,000 boxes and the demand is 650,000 boxes. At a price of $13.19 per box, the supply is 695,000 boxes and the demand is 590,000 boxes. Assume that the relationship between price and supply is linear and that the relationship between price and demand is linear.

(A) Find a price–supply equation of the form $p = mx + b$.

(B) Find a price–demand equation of the form $p = mx + b$.

(C) Find the equilibrium point.

Exercises 1.2

Problems 1–4 refer to graphs (A)–(D).

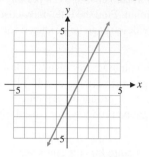

(A)

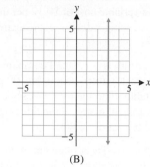

(B)

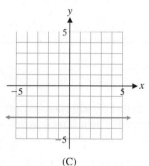

(C)

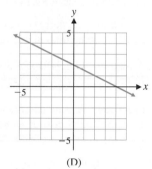

(D)

1. Identify the graph(s) of lines with a negative slope.

2. Identify the graph(s) of lines with a positive slope.

3. Identify the graph(s) of any lines with slope zero.

4. Identify the graph(s) of any lines with undefined slope.

In Problems 5–8, sketch a graph of each equation in a rectangular coordinate system.

5. $y = 2x - 3$

6. $y = \dfrac{x}{2} + 1$

7. $2x + 3y = 12$

8. $8x - 3y = 24$

In Problems 9–14, find the slope and y intercept of the graph of each equation.

9. $y = 5x - 7$

10. $y = 3x + 2$

11. $y = -\dfrac{5}{2}x - 9$

12. $y = -\dfrac{10}{3}x + 4$

13. $y = \dfrac{x}{4} + \dfrac{2}{3}$

14. $y = \dfrac{x}{5} - \dfrac{1}{2}$

In Problems 15–18, write an equation of the line with the indicated slope and y intercept.

15. Slope $= 2$
y intercept $= 1$

16. Slope $= 1$
y intercept $= 5$

17. Slope $= -\dfrac{1}{3}$
y intercept $= 6$

18. Slope $= \dfrac{6}{7}$
y intercept $= -\dfrac{9}{2}$

In Problems 19–22, use the graph of each line to find the x intercept, y intercept, and slope. Write the slope-intercept form of the equation of the line.

19.

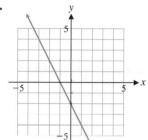

20.

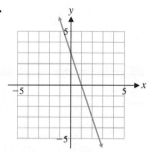

21.

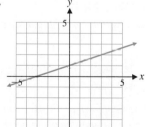

22.

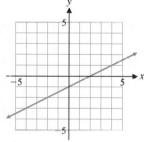

Sketch a graph of each equation or pair of equations in Problems 23–28 in a rectangular coordinate system.

23. $y = -\dfrac{2}{3}x - 2$

24. $y = -\dfrac{3}{2}x + 1$

25. $3x - 2y = 10$

26. $5x - 6y = 15$

27. $x = 3; y = -2$

28. $x = -3; y = 2$

In Problems 29–34, find the slope of the graph of each equation.

29. $4x + y = 3$

30. $5x - y = -2$

31. $3x + 5y = 15$

32. $2x - 3y = 18$

33. $-4x + 2y = 9$

34. $-x + 8y = 4$

35. Given $Ax + By = 12$, graph each of the following three cases in the same coordinate system.

(A) $A = 2$ and $B = 0$

(B) $A = 0$ and $B = 3$

(C) $A = 3$ and $B = 4$

36. Given $Ax + By = 24$, graph each of the following three cases in the same coordinate system.

(A) $A = 6$ and $B = 0$

(B) $A = 0$ and $B = 8$

(C) $A = 2$ and $B = 3$

37. Graph $y = 25x + 200$, $x \geq 0$.

38. Graph $y = 40x + 160$, $x \geq 0$.

39. (A) Graph $y = 1.2x - 4.2$ in a rectangular coordinate system.

(B) Find the x and y intercepts algebraically to one decimal place.

(C) Graph $y = 1.2x - 4.2$ in a graphing calculator.

(D) Find the x and y intercepts to one decimal place using TRACE and the `zero` command.

40. (A) Graph $y = -0.8x + 5.2$ in a rectangular coordinate system.

(B) Find the x and y intercepts algebraically to one decimal place.

(C) Graph $y = -0.8x + 5.2$ in a graphing calculator.

(D) Find the x and y intercepts to one decimal place using TRACE and the `zero` command.

(E) Using the results of parts (A) and (B), or (C) and (D), find the solution set for the linear inequality

$$-0.8x + 5.2 < 0$$

In Problems 41–44, write the equations of the vertical and horizontal lines through each point.

41. $(4, -3)$

42. $(-5, 6)$

43. $(-1.5, -3.5)$

44. $(2.6, 3.8)$

In Problems 45–52, write the slope-intercept form of the equation of the line with the indicated slope that goes through the given point.

45. $m = 5$; $(3, 0)$

46. $m = 4$; $(0, 6)$

47. $m = -2$; $(-1, 9)$

48. $m = -10$; $(2, -5)$

49. $m = \dfrac{1}{3}$; $(-4, -8)$

50. $m = \dfrac{2}{7}$; $(7, 1)$

51. $m = -3.2$; $(5.8, 12.3)$

52. $m = 0.9$; $(2.3, 6.7)$

In Problems 53–60,

(A) *Find the slope of the line that passes through the given points.*

(B) *Find the standard form of the equation of the line.*

(C) *Find the slope-intercept form of the equation of the line.*

53. $(2, 5)$ and $(5, 7)$

54. $(1, 2)$ and $(3, 5)$

55. $(-2, -1)$ and $(2, -6)$

56. $(2, 3)$ and $(-3, 7)$

57. $(5, 3)$ and $(5, -3)$

58. $(1, 4)$ and $(0, 4)$

59. $(-2, 5)$ and $(3, 5)$

60. $(2, 0)$ and $(2, -3)$

61. Discuss the relationship among the graphs of the lines with equation $y = mx + 2$, where m is any real number.

62. Discuss the relationship among the graphs of the lines with equation $y = -0.5x + b$, where b is any real number.

Applications

63. Cost analysis. A donut shop has a fixed cost of $124 per day and a variable cost of $0.12 per donut. Find the total daily cost of producing x donuts. How many donuts can be produced for a total daily cost of $250?

64. Cost analysis. A small company manufactures picnic tables. The weekly fixed cost is $1,200 and the variable cost is $45 per table. Find the total weekly cost of producing x picnic tables. How many picnic tables can be produced for a total weekly cost of $4,800?

65. Cost analysis. A plant can manufacture 80 golf clubs per day for a total daily cost of $7,647 and 100 golf clubs per day for a total daily cost of $9,147.

(A) Assuming that daily cost and production are linearly related, find the total daily cost of producing x golf clubs.

(B) Graph the total daily cost for $0 \leq x \leq 200$.

(C) Interpret the slope and y intercept of this cost equation.

66. Cost analysis. A plant can manufacture 50 tennis rackets per day for a total daily cost of $3,855 and 60 tennis rackets per day for a total daily cost of $4,245.

(A) Assuming that daily cost and production are linearly related, find the total daily cost of producing x tennis rackets.

(B) Graph the total daily cost for $0 \leq x \leq 100$.

(C) Interpret the slope and y intercept of this cost equation.

67. Business—Markup policy. A drugstore sells a drug costing $85 for $112 and a drug costing $175 for $238.

(A) If the markup policy of the drugstore is assumed to be linear, write an equation that expresses retail price R in terms of cost C (wholesale price).

(B) What does a store pay (to the nearest dollar) for a drug that retails for $185?

68. Business—Markup policy. A clothing store sells a shirt costing $20 for $33 and a jacket costing $60 for $93.

(A) If the markup policy of the store is assumed to be linear, write an equation that expresses retail price R in terms of cost C (wholesale price).

(B) What does a store pay for a suit that retails for $240?

69. Business—Depreciation. A farmer buys a new tractor for $157,000 and assumes that it will have a trade-in value of $82,000 after 10 years. The farmer uses a constant rate of depreciation (commonly called **straight-line**

depreciation—one of several methods permitted by the IRS) to determine the annual value of the tractor.

(A) Find a linear model for the depreciated value *V* of the tractor *t* years after it was purchased.

(B) What is the depreciated value of the tractor after 6 years?

(C) When will the depreciated value fall below $70,000?

(D) Graph *V* for $0 \le t \le 20$ and illustrate the answers from parts (B) and (C) on the graph.

70. Business—Depreciation. A charter fishing company buys a new boat for $224,000 and assumes that it will have a trade-in value of $115,200 after 16 years.

(A) Find a linear model for the depreciated value *V* of the boat *t* years after it was purchased.

(B) What is the depreciated value of the boat after 10 years?

(C) When will the depreciated value fall below $100,000?

(D) Graph *V* for $0 \le t \le 30$ and illustrate the answers from (B) and (C) on the graph.

71. Boiling point. The temperature at which water starts to boil is called its **boiling point** and is linearly related to the altitude. Water boils at 212°F at sea level and at 193.6°F at an altitude of 10,000 feet. (*Source*: biggreenegg.com)

(A) Find a relationship of the form $T = mx + b$ where *T* is degrees Fahrenheit and *x* is altitude in thousands of feet.

(B) Find the boiling point at an altitude of 3,500 feet.

(C) Find the altitude if the boiling point is 200°F.

(D) Graph *T* and illustrate the answers to (B) and (C) on the graph.

72. Boiling point. The temperature at which water starts to boil is also linearly related to barometric pressure. Water boils at 212°F at a pressure of 29.9 inHg (inches of mercury) and at 191°F at a pressure of 28.4 inHg. (*Source*: biggreenegg.com)

(A) Find a relationship of the form $T = mx + b$, where *T* is degrees Fahrenheit and *x* is pressure in inches of mercury.

(B) Find the boiling point at a pressure of 31 inHg.

(C) Find the pressure if the boiling point is 199°F.

(D) Graph *T* and illustrate the answers to (B) and (C) on the graph.

73. Flight conditions. In stable air, the air temperature drops about 3.6°F for each 1,000-foot rise in altitude. (*Source*: Federal Aviation Administration)

(A) If the temperature at sea level is 70°F, write a linear equation that expresses temperature *T* in terms of altitude *A* in thousands of feet.

(B) At what altitude is the temperature 34°F?

74. Flight navigation. The airspeed indicator on some aircraft is affected by the changes in atmospheric pressure at different altitudes. A pilot can estimate the true airspeed by observing the indicated airspeed and adding to it about 1.6% for every 1,000 feet of altitude. (*Source:* Megginson Technologies Ltd.)

(A) A pilot maintains a constant reading of 200 miles per hour on the airspeed indicator as the aircraft climbs from sea level to an altitude of 10,000 feet. Write a linear equation that expresses true airspeed *T* (in miles per hour) in terms of altitude *A* (in thousands of feet).

(B) What would be the true airspeed of the aircraft at 6,500 feet?

75. Demographics. The average number of persons per household in the United States has been shrinking steadily for as long as statistics have been kept and is approximately linear with respect to time. In 1980 there were about 2.76 persons per household, and in 2012 about 2.55. (*Source:* U.S. Census Bureau)

(A) If *N* represents the average number of persons per household and *t* represents the number of years since 1980, write a linear equation that expresses *N* in terms of *t*.

(B) Use this equation to estimate household size in the year 2030.

76. Demographics. The **median** household income divides the households into two groups: the half whose income is less than or equal to the median, and the half whose income is greater than the median. The median household income in the United States grew from about $30,000 in 1990 to about $53,000 in 2010. (*Source*: U.S. Census Bureau)

(A) If *I* represents the median household income and *t* represents the number of years since 1990, write a linear equation that expresses *I* in terms of *t*.

(B) Use this equation to estimate median household income in the year 2030.

77. Cigarette smoking. The percentage of female cigarette smokers in the United States declined from 21.0% in 2000 to 17.3% in 2010. (*Source*: Centers for Disease Control)

(A) Find a linear equation relating percentage of female smokers (*f*) to years since 2000 (*t*).

(B) Find the year in which the percentage of female smokers falls below 12%.

78. Cigarette smoking. The percentage of male cigarette smokers in the United States declined from 25.7% in 2000 to 21.5% in 2010. (*Source:* Centers for Disease Control)

(A) Find a linear equation relating percentage of male smokers (*m*) to years since 2000 (*t*).

(B) Find the year in which the percentage of male smokers falls below 12%.

79. Supply and demand. At a price of $2.28 per bushel, the supply of barley is 7,500 million bushels and the demand is 7,900 million bushels. At a price of $2.37 per bushel, the supply is 7,900 million bushels and the demand is 7,800 million bushels.

(A) Find a price–supply equation of the form $p = mx + b$.

(B) Find a price–demand equation of the form $p = mx + b$.

(C) Find the equilibrium point.

(D) Graph the price–supply equation, price–demand equation, and equilibrium point in the same coordinate system.

80. Supply and demand. At a price of $1.94 per bushel, the supply of corn is 9,800 million bushels and the demand is 9,300 million bushels. At a price of $1.82 per bushel, the supply is 9,400 million bushels and the demand is 9,500 million bushels.

(A) Find a price–supply equation of the form $p = mx + b$.

(B) Find a price–demand equation of the form $p = mx + b$.

(C) Find the equilibrium point.

(D) Graph the price–supply equation, price–demand equation, and equilibrium point in the same coordinate system.

81. Physics. Hooke's law states that the relationship between the stretch s of a spring and the weight w causing the stretch is linear. For a particular spring, a 5-pound weight causes a stretch of 2 inches, while with no weight, the stretch of the spring is 0.

(A) Find a linear equation that expresses s in terms of w.

(B) What is the stretch for a weight of 20 pounds?

(C) What weight will cause a stretch of 3.6 inches?

82. Physics. The distance d between a fixed spring and the floor is a linear function of the weight w attached to the bottom of the spring. The bottom of the spring is 18 inches from the floor when the weight is 3 pounds, and 10 inches from the floor when the weight is 5 pounds.

(A) Find a linear equation that expresses d in terms of w.

(B) Find the distance from the bottom of the spring to the floor if no weight is attached.

(C) Find the smallest weight that will make the bottom of the spring touch the floor. (Ignore the height of the weight.)

Answers to Matched Problems

1.

2. y intercept $= -4$, x intercept $= 3$
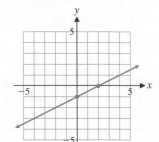

3. (A)
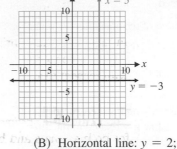

(B) Horizontal line: $y = 2$; vertical line: $x = -8$

4. (A) 0   (B) -4
   (C) Not defined   (D) 1

5. $y = \frac{1}{2}x - 1$

6. (A) $2x - 3y = 18$ **7.** (A) $m = 40$
   (B) $y = 3x - 9$ (B) $C = 40x + 250$

8. (A) $p = 0.006x + 9.02$
   (B) $p = -0.01x + 19.09$
   (C) $(629, 12.80)$

1.3   Linear Regression

- Slope as a Rate of Change
- Linear Regression

Mathematical modeling is the process of using mathematics to solve real-world problems. This process can be broken down into three steps (Fig. 1):

Step 1   *Construct* the **mathematical model** (that is, a mathematics problem that, when solved, will provide information about the real-world problem).

Step 2   *Solve* the mathematical model.

Step 3   *Interpret* the solution to the mathematical model in terms of the original real-world problem.

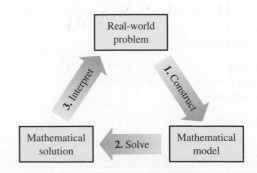

Figure 1

In more complex problems, this cycle may have to be repeated several times to obtain the required information about the real-world problem. In this section, we will discuss one of the simplest mathematical models, a linear equation. With the aid of a graphing calculator or computer, we also will learn how to analyze a linear model based on real-world data.

Slope as a Rate of Change

If x and y are related by the equation $y = mx + b$, where m and b are constants with $m \neq 0$, then x and y are **linearly related**. If (x_1, y_1) and (x_2, y_2) are two distinct points on this line, then the slope of the line is

$$m = \frac{y_2 - y_1}{x_2 - x_1} = \frac{\text{Change in } y}{\text{Change in } x} \tag{1}$$

In applications, ratio (1) is called the **rate of change** of y with respect to x. Since the slope of a line is unique, **the rate of change of two linearly related variables is constant**. Here are some examples of familiar rates of change: miles per hour, revolutions per minute, price per pound, passengers per plane, and so on. If the relationship between x and y is not linear, ratio (1) is called the **average rate of change** of y with respect to x.

EXAMPLE 1 Estimating Body Surface Area Appropriate doses of medicine for both animals and humans are often based on body surface area (BSA). Since weight is much easier to determine than BSA, veterinarians use the weight of an animal to estimate BSA. The following linear equation expresses BSA for canines in terms of weight:

$$a = 16.12w + 375.6$$

where a is BSA in square inches and w is weight in pounds. (*Source:* Veterinary Oncology Consultants, PTY LTD)

(A) Interpret the slope of the BSA equation.

(B) What is the effect of a one-pound increase in weight?

SOLUTION

(A) The rate-of-change BSA with respect to weight is 16.12 square inches per pound.

(B) Since slope is the ratio of rise to run, increasing w by 1 pound (run) increases a by 16.12 square inches (rise).

Matched Problem 1 The equation $a = 28.55w + 118.7$ expresses BSA for felines in terms of weight, where a is BSA in square inches and w is weight in pounds.

(A) Interpret the slope of the BSA equation.

(B) What is the effect of a one-pound increase in weight?

Explore and Discuss 1 As illustrated in Example 1A, the slope m of a line with equation $y = mx + b$ has two interpretations:

1. m is the rate of change of y with respect to x.
2. Increasing x by one unit will change y by m units.

How are these two interpretations related?

Parachutes are used to deliver cargo to areas that cannot be reached by other means. The **rate of descent** of the cargo is the rate of change of altitude with respect to time. The absolute value of the rate of descent is called the **speed** of the cargo. At low altitudes, the altitude of the cargo and the time in the air are linearly related. The appropriate rate of descent varies widely with the item. Bulk food (rice, flour, beans, etc.) and clothing can tolerate nearly any rate of descent under 40 ft/sec. Machinery and electronics (pumps, generators, radios, etc.) should generally be dropped at 15 ft/sec or less. Butler Tactical Parachute Systems in Roanoke, Virginia, manufactures a variety of canopies for dropping cargo. The following example uses information taken from the company's brochures.

EXAMPLE 2 Finding the Rate of Descent A 100-pound cargo of delicate electronic equipment is dropped from an altitude of 2,880 feet and lands 200 seconds later. (*Source*: Butler Tactical Parachute Systems)

(A) Find a linear model relating altitude a (in feet) and time in the air t (in seconds).

(B) How fast is the cargo moving when it lands?

SOLUTION

(A) If $a = mt + b$ is the linear equation relating altitude a and time in air t, then the graph of this equation must pass through the following points:

$$(t_1, a_1) = (0, 2,880) \quad \text{Cargo is dropped from plane.}$$
$$(t_2, a_2) = (200, 0) \quad \text{Cargo lands.}$$

The slope of this line is

$$m = \frac{a_2 - a_1}{t_2 - t_1} = \frac{0 - 2,880}{200 - 0} = -14.4$$

and the equation of this line is

$$a - 0 = -14.4(t - 200)$$
$$a = -14.4t + 2,880$$

(B) The rate of descent is the slope $m = -14.4$, so the speed of the cargo at landing is $|-14.4| = 14.4$ ft/sec.

Matched Problem 2 A 400-pound load of grain is dropped from an altitude of 2,880 feet and lands 80 seconds later.

(A) Find a linear model relating altitude a (in feet) and time in the air t (in seconds).

(B) How fast is the cargo moving when it lands?

Linear Regression

In real-world applications, we often encounter numerical data in the form of a table. **Regression analysis** is a process for finding a function that provides a useful model for a set of data points. Graphs of equations are often called **curves**, and regression analysis is also referred to as **curve fitting**. In the next example, we use a linear model obtained by using **linear regression** on a graphing calculator.

Table 1 **Round-Shaped Diamond Prices**

Weight (carats)	Price
0.5	$2,790
0.6	$3,191
0.7	$3,694
0.8	$4,154
0.9	$5,018
1.0	$5,898

Source: www.tradeshop.com

EXAMPLE 3 Diamond Prices Prices for round-shaped diamonds taken from an online trader are given in Table 1.

(A) A linear model for the data in Table 1 is given by

$$p = 6,140c - 480 \qquad (2)$$

where p is the price of a diamond weighing c carats. (We will discuss the source of models like this later in this section.) Plot the points in Table 1 on a Cartesian coordinate system, producing a *scatter plot*, and graph the model on the same axes.

(B) Interpret the slope of the model in (2).

(C) Use the model to estimate the cost of a 0.85-carat diamond and the cost of a 1.2-carat diamond. Round answers to the nearest dollar.

(D) Use the model to estimate the weight of a diamond (to two decimal places) that sells for $4,000.

SOLUTION

(A) A **scatter plot** is simply a graph of the points in Table 1 (Fig. 2A). To add the graph of the model to the scatter plot, we find any two points that satisfy equation (2) [we choose $(0.4, 1,976)$ and $(1.1, 6,274)$]. Plotting these points and drawing a line through them gives us Figure 2B.

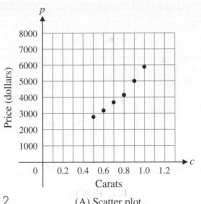

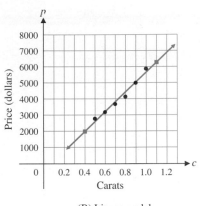

Figure 2   (A) Scatter plot   (B) Linear model

(B) The rate of change of the price of a diamond with respect to its weight is 6,140. Increasing the weight by one carat will increase the price by about $6,140.

(C) The graph of the model (Fig. 2B) does not pass through any of the points in the scatter plot, but it comes close to all of them. [Verify this by evaluating equation (2) at $c = 0.5, 0.6, \ldots, 1$.] So we can use equation (2) to approximate points not in Table 1.

$$c = 0.85 \qquad\qquad c = 1.2$$
$$p \approx 6,140(0.85) - 480 \qquad p \approx 6,140(1.2) - 480$$
$$= \$4,739 \qquad\qquad = \$6,888$$

A 0.85-carat diamond will cost about $4,739, and a 1.2-carat diamond will cost about $6,888.

(D) To find the weight of a $4,000 diamond, we solve the following equation for c:

$$6,140c - 480 = 4,000 \qquad \text{Add 480 to both sides.}$$
$$6,140c = 4,480 \qquad \text{Divide both sides by 6,140.}$$
$$c = \frac{4,480}{6,140} \approx 0.73 \qquad \text{Rounded to two decimal places.}$$

A $4,000 diamond will weigh about 0.73 carat.

Matched Problem 3 Prices for emerald-shaped diamonds from an online trader are given in Table 2. Repeat Example 3 for this data with the linear model

$$p = 5,600c - 1,100$$

where p is the price of an emerald-shaped diamond weighing c carats.

The model we used in Example 3 was obtained using a technique called **linear regression**, and the model is called the **regression line**. This technique produces a line that is the **best fit*** for a given data set. Although you can find a linear regression line

Table 2 **Emerald-Shaped Diamond Prices**

Weight (carats)	Price
0.5	$1,677
0.6	$2,353
0.7	$2,718
0.8	$3,218
0.9	$3,982
1.0	$4,510

Source: www.tradeshop.com

*The line of best fit is the line that minimizes the sum of the squares of the vertical distances from the data points to the line.

by hand, we prefer to leave the calculations to a graphing calculator or a computer. Don't be concerned if you don't have either of these electronic devices. We will supply the regression model in most of the applications we discuss, as we did in Example 3.

Explore and Discuss 2   As stated previously, we used linear regression to produce the model in Example 3. If you have a graphing calculator that supports linear regression, then you can find this model. The linear regression process varies greatly from one calculator to another. Consult the user's manual for the details of linear regression. The screens in Figure 3 are related to the construction of the model in Example 3 on a Texas Instruments TI-84 Plus.

(A) Produce similar screens on your graphing calculator.

(B) Do the same for Matched Problem 3.

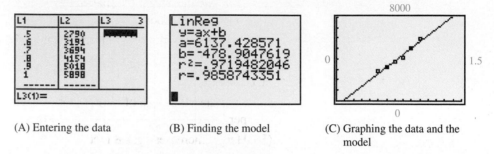

(A) Entering the data   (B) Finding the model   (C) Graphing the data and the model

Figure 3   **Linear regression on a graphing calculator**

In Example 3, we used the regression model to approximate points that were not given in Table 1 but would fit between points in the table. This process is called **interpolation**. In the next example, we use a regression model to approximate points outside the given data set. This process is called **extrapolation**, and the approximations are often referred to as **predictions**.

Table 3 **Atmospheric Concentration of CO_2 (parts per million)**

2000	2005	2007	2009	2011
369	380	384	387	392

Source: U.S. Department of Energy

EXAMPLE 4   Atmospheric Concentration of Carbon Dioxide   Table 3 contains information about the concentration of carbon dioxide (CO_2) in the atmosphere. The linear regression model for the data is

$$C = 369 + 2.05t$$

where C is the concentration (in parts per million) of carbon dioxide and t is the time in years with $t = 0$ corresponding to the year 2000.

(A) Interpret the slope of the regression line as a rate of change.

(B) Use the regression model to predict the concentration of CO_2 in the atmosphere in 2025.

SOLUTION

(A) The slope $m = 2.05$ is the rate of change of concentration of CO_2 with respect to time. Since the slope is positive, the concentration of CO_2 is increasing at a rate of 2.05 parts per million per year.

(B) If $t = 25$, then

$$C = 369 + 2.05(25) \approx 420$$

So the model predicts that the atmospheric concentration of CO_2 will be approximately 420 parts per million in 2025.

Matched Problem 4   Using the model of Example 4, estimate the concentration of carbon dioxide in the atmosphere in the year 1990.

Forest managers estimate growth, volume, yield, and forest potential. One common measure is the diameter of a tree at breast height (Dbh), which is defined as the diameter of the tree at a point 4.5 feet above the ground on the uphill side of the tree. Example 5 uses Dbh to estimate the height of balsam fir trees.

EXAMPLE 5   *Forestry*   A linear regression model for the height of balsam fir trees is
$$h = 3.8d + 18.73$$

where d is Dbh in inches and h is the height in feet.

(A) Interpret the slope of this model.

(B) What is the effect of a 1-inch increase in Dbh?

(C) Estimate the height of a balsam fir with a Dbh of 8 inches. Round your answer to the nearest foot.

(D) Estimate the Dbh of a balsam fir that is 30 feet tall. Round your answer to the nearest inch.

SOLUTION

(A) The rate of change of height with respect to breast height diameter is 3.8 feet per inch.

(B) Height increases by 3.8 feet.

(C) We must find h when $d = 8$:

$$h = 3.8d + 18.73 \qquad \text{Substitute } d = 8.$$
$$h = 3.8(8) + 18.73 \qquad \text{Evaluate.}$$
$$h = 49.13 \approx 49 \text{ ft}$$

(D) We must find d when $h = 30$:

$$h = 3.8d + 18.73 \qquad \text{Substitute } h = 30.$$
$$30 = 3.8d + 18.73 \qquad \text{Subtract 18.73 from both sides.}$$
$$11.27 = 3.8d \qquad \text{Divide both sides by 3.8.}$$
$$d = \frac{11.27}{3.8} \approx 3 \text{ in.}$$

The data used to produce the regression model in Example 5 are from the Jack Haggerty Forest at Lakehead University in Canada (Table 4). We used the popular

Table 4 **Height and Diameter of the Balsam Fir**

Dbh (in.)	Height (ft)	Dbh (in.)	Height (ft)	Dbh (in.)	Height (ft)	Dbh (in.)	Height (ft)
6.5	51.8	6.4	44.0	3.1	19.7	4.6	26.6
8.6	50.9	4.4	46.9	7.1	55.8	4.8	33.1
5.7	49.2	6.5	52.2	6.3	32.8	3.1	28.5
4.9	46.3	4.1	46.9	2.4	26.2	3.2	29.2
6.4	44.3	8.8	51.2	2.5	29.5	5.0	34.1
4.1	46.9	5.0	36.7	6.9	45.9	3.0	28.2
1.7	13.1	4.9	34.1	2.4	32.8	4.8	33.8
1.8	19.0	3.8	32.2	4.3	39.4	4.4	35.4
3.2	20.0	5.5	49.2	7.3	36.7	11.3	55.4
5.1	46.6	6.3	39.4	10.9	51.5	3.7	32.2

(*Source:* Jack Haggerty Forest, Lakehead University, Canada)

spreadsheet Excel to produce a scatter plot of the data in Table 4 and to find the regression model (Fig. 4).

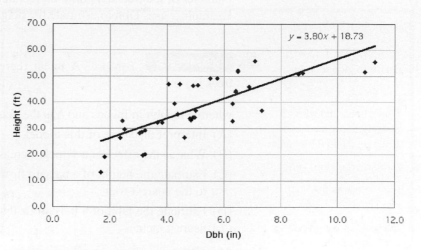

Figure 4 **Linear regression with a spreadsheet**

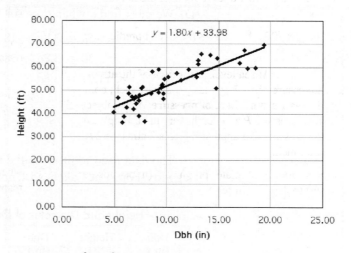

Matched Problem 5) Figure 5 shows the scatter plot for white spruce trees in the Jack Haggerty Forest at Lakehead University in Canada. A regression model produced by a spreadsheet (Fig. 5), after rounding, is

$$h = 1.8d + 34$$

where d is Dbh in inches and h is the height in feet.

Figure 5 **Linear regression for white spruce trees**

(A) Interpret the slope of this model.

(B) What is the effect of a 1-inch increase in Dbh?

(C) Estimate the height of a white spruce with a Dbh of 10 inches. Round your answer to the nearest foot.

(D) Estimate the Dbh of a white spruce that is 65 feet tall. Round your answer to the nearest inch.

Exercises 1.3

Applications

1. **Ideal weight.** Dr. J. D. Robinson published the following estimate of the ideal body weight of a woman:

   49 kg + 1.7 kg for each inch over 5 ft

   (A) Find a linear model for Robinson's estimate of the ideal weight of a woman using w for ideal body weight (in kilograms) and h for height over 5 ft (in inches).

   (B) Interpret the slope of the model.

   (C) If a woman is 5′4″ tall, what does the model predict her weight to be?

   (D) If a woman weighs 60 kg, what does the model predict her height to be?

2. **Ideal weight.** Dr. J. D. Robinson also published the following estimate of the ideal body weight of a man:

   52 kg + 1.9 kg for each inch over 5 ft

   (A) Find a linear model for Robinson's estimate of the ideal weight of a man using w for ideal body weight (in kilograms) and h for height over 5 ft (in inches).

   (B) Interpret the slope of the model.

   (C) If a man is 5′8″ tall, what does the model predict his weight to be?

   (D) If a man weighs 70 kg, what does the model predict his height to be?

3. **Underwater pressure.** At sea level, the weight of the atmosphere exerts a pressure of 14.7 pounds per square inch, commonly referred to as 1 **atmosphere of pressure**. As an object descends in water, pressure P and depth d are linearly related. In salt water, the pressure at a depth of 33 ft is 2 atms, or 29.4 pounds per square inch.

   (A) Find a linear model that relates pressure P (in pounds per square inch) to depth d (in feet).

   (B) Interpret the slope of the model.

   (C) Find the pressure at a depth of 50 ft.

   (D) Find the depth at which the pressure is 4 atms.

4. **Underwater pressure.** Refer to Problem 3. In fresh water, the pressure at a depth of 34 ft is 2 atms, or 29.4 pounds per square inch.

   (A) Find a linear model that relates pressure P (in pounds per square inch) to depth d (in feet).

   (B) Interpret the slope of the model.

   (C) Find the pressure at a depth of 50 ft.

   (D) Find the depth at which the pressure is 4 atms.

5. **Rate of descent—Parachutes.** At low altitudes, the altitude of a parachutist and time in the air are linearly related. A jump at 2,880 ft using the U.S. Army's T-10 parachute system lasts 120 secs.

   (A) Find a linear model relating altitude a (in feet) and time in the air t (in seconds).

   (B) Find the rate of descent for a T-10 system.

   (C) Find the speed of the parachutist at landing.

6. **Rate of descent—Parachutes.** The U.S Army is considering a new parachute, the Advanced Tactical Parachute System (ATPS). A jump at 2,880 ft using the ATPS system lasts 180 secs.

   (A) Find a linear model relating altitude a (in feet) and time in the air t (in seconds).

   (B) Find the rate of descent for an ATPS system parachute.

   (C) Find the speed of the parachutist at landing.

7. **Speed of sound.** The speed of sound through air is linearly related to the temperature of the air. If sound travels at 331 m/sec at 0°C and at 343 m/sec at 20°C, construct a linear model relating the speed of sound (s) and the air temperature (t). Interpret the slope of this model. (*Source:* Engineering Toolbox)

8. **Speed of sound.** The speed of sound through sea water is linearly related to the temperature of the water. If sound travels at 1,403 m/sec at 0°C and at 1,481 m/sec at 20°C, construct a linear model relating the speed of sound (s) and the air temperature (t). Interpret the slope of this model. (*Source:* Engineering Toolbox)

9. **Energy production.** Table 5 lists U.S. fossil fuel production as a percentage of total energy production for selected years. A linear regression model for this data is

   $$y = -0.3x + 84.4$$

   where x represents years since 1985 and y represents the corresponding percentage of total energy production.

   Table 5 **U.S. Fossil Fuel Production**

Year	Production (%)
1985	85
1990	83
1995	81
2000	80
2005	79
2010	78

   Source: Energy Information Administration

   (A) Draw a scatter plot of the data and a graph of the model on the same axes.

   (B) Interpret the slope of the model.

(C) Use the model to predict fossil fuel production in 2025.

(D) Use the model to estimate the first year for which fossil fuel production is less than 70% of total energy production.

10. Energy consumption. Table 6 lists U.S. fossil fuel consumption as a percentage of total energy consumption for selected years. A linear regression model for this data is

$$y = -0.09x + 85.8$$

where x represents years since 1985 and y represents the corresponding percentage of fossil fuel consumption.

Table 6 **U.S. Fossil Fuel Consumption**

Year	Consumption (%)
1985	86
1990	85
1995	85
2000	84
2005	85
2010	83

Source: Energy Information Administration

(A) Draw a scatter plot of the data and a graph of the model on the same axes.

(B) Interpret the slope of the model.

(C) Use the model to predict fossil fuel consumption in 2025.

(D) Use the model to estimate the first year for which fossil fuel consumption is less than 80% of total energy consumption.

11. Cigarette smoking. The data in Table 7 shows that the percentage of female cigarette smokers in the U.S. declined from 22.1% in 1997 to 17.3% in 2010.

Table 7 **Percentage of Smoking Prevalence among U.S. Adults**

Year	Males (%)	Females (%)
1997	27.6	22.1
2000	25.7	21.0
2003	24.1	19.2
2006	23.9	18.0
2010	21.5	17.3

Source: Centers for Disease Control

(A) Applying linear regression to the data for females in Table 7 produces the model

$$f = -0.39t + 21.93$$

where f is percentage of female smokers and t is time in years since 1997. Draw a scatter plot of the female smoker data and a graph of the regression model on the same axes for $0 \le t \le 15$.

(B) Estimate the first year in which the percentage of female smokers is less than 10%.

12. Cigarette smoking. The data in Table 7 shows that the percentage of male cigarette smokers in the U.S. declined from 27.6% in 1997 to 21.5% in 2010.

(A) Applying linear regression to the data for males in Table 7 produces the model

$$m = -0.44t + 27.28$$

where m is percentage of male smokers and t is time in years since 1997. Draw a scatter plot of the male smoker data and a graph of the regression model for $0 \le t \le 15$.

(B) Estimate the first year in which the percentage of male smokers is less than 15%.

13. Undergraduate enrollment. Table 8 lists enrollment in U.S. degree-granting institutions for both undergraduate and graduate students. A linear regression model for undergraduate enrollment is

$$y = 0.23x + 9.56$$

where x represents years since 1980 and y is undergraduate enrollment in millions of students.

Table 8 **Fall Undergraduate and Graduate Enrollment (millions of students)**

Year	Undergraduate	Graduate
1980	10.48	1.34
1985	10.60	1.38
1990	11.96	1.59
1995	12.23	1.73
2000	13.16	1.85
2005	14.96	2.19
2010	18.08	2.94

Source: National Center for Education Statistics

(A) Draw a scatter plot of the undergraduate enrollment data and a graph of the model on the same axes.

(B) Predict the undergraduate student enrollment in 2025 (to the nearest 100,000).

(C) Interpret the slope of the model.

14. Graduate student enrollment. A linear regression model for the graduate student enrollment in Table 8 is

$$y = 0.048x + 1.14$$

where x represents years since 1980 and y is graduate enrollment in millions of students.

(A) Draw a scatter plot of the graduate enrollment data and a graph of the model on the same axes.

(B) Predict the graduate student enrollment in 2025 (to the nearest 100,000).

(C) Interpret the slope of the model.

15. Licensed drivers. Table 9 contains the state population and the number of licensed drivers in the state (both in millions) for the states with population under 1 million in 2010. The regression model for this data is

$$y = 0.75x$$

where x is the state population and y is the number of licensed drivers in the state.

Table 9 Licensed Drivers in 2010

State	Population	Licensed Drivers
Alaska	0.71	0.52
Delaware	0.90	0.70
Montana	0.99	0.74
North Dakota	0.67	0.48
South Dakota	0.81	0.60
Vermont	0.63	0.51
Wyoming	0.56	0.42

Source: Bureau of Transportation Statistics

(A) Draw a scatter plot of the data and a graph of the model on the same axes.

(B) If the population of Idaho in 2010 was about 1.6 million, use the model to estimate the number of licensed drivers in Idaho in 2010 to the nearest thousand.

(C) If the number of licensed drivers in Rhode Island in 2010 was about 0.75 million, use the model to estimate the population of Rhode Island in 2010 to the nearest thousand.

16. **Licensed drivers.** Table 10 contains the state population and the number of licensed drivers in the state (both in millions) for the states with population over 10 million in 2010. The regression model for this data is

$$y = 0.63x + 0.31$$

where x is the state population and y is the number of licensed drivers in the state.

Table 10 Licensed Drivers in 2010

State	Population	Licensed Drivers
California	37	24
Florida	19	14
Illinois	13	8
New York	19	11
Ohio	12	8
Pennsylvania	13	9
Texas	25	15

Source: Bureau of Transportation Statistics

(A) Draw a scatter plot of the data and a graph of the model on the same axes.

(B) If the population of Minnesota in 2010 was about 5.3 million, use the model to estimate the number of licensed drivers in Minnesota in 2010 to the nearest thousand.

(C) If the number of licensed drivers in Wisconsin in 2010 was about 4.1 million, use the model to estimate the population of Wisconsin in 2010 to the nearest thousand.

17. **Net sales.** A linear regression model for the net sales data in Table 11 is

$$S = 15.8t + 251$$

where S is net sales and t is time since 2000 in years.

Table 11 Walmart Stores, Inc.

Billions of U.S. Dollars	2008	2009	2010	2011	2012
Net sales	374	401	405	419	444
Operating income	21.9	22.8	24.0	25.5	26.6

Source: Walmart Stores, Inc.

(A) Draw a scatter plot of the data and a graph of the model on the same axes.

(B) Predict Walmart's net sales for 2022.

18. **Operating income.** A linear regression model for the operating income data in Table 11 is

$$I = 1.21t + 12.06$$

where I is operating income and t is time since 2000 in years.

(A) Draw a scatter plot of the data and a graph of the model on the same axes.

(B) Predict Walmart's annual operating income for 2024.

19. **Freezing temperature.** Ethylene glycol and propylene glycol are liquids used in antifreeze and deicing solutions. Ethylene glycol is listed as a hazardous chemical by the Environmental Protection Agency, while propylene glycol is generally regarded as safe. Table 12 lists the freezing temperature for various concentrations (as a percentage of total weight) of each chemical in a solution used to deice airplanes. A linear regression model for the ethylene glycol data in Table 12 is

$$E = -0.55T + 31$$

where E is the percentage of ethylene glycol in the deicing solution and T is the temperature at which the solution freezes.

Table 12 Freezing Temperatures

Freezing Temperature (°F)	Ethylene Glycol (% Wt.)	Propylene Glycol (% Wt.)
−50	56	58
−40	53	55
−30	49	52
−20	45	48
−10	40	43
0	33	36
10	25	29
20	16	19

Source: T. Labuza, University of Minnesota

(A) Draw a scatter plot of the data and a graph of the model on the same axes.

(B) Use the model to estimate the freezing temperature to the nearest degree of a solution that is 30% ethylene glycol.

(C) Use the model to estimate the percentage of ethylene glycol in a solution that freezes at 15°F.

20. Freezing temperature. A linear regression model for the propylene glycol data in Table 12 is

$$P = -0.54T + 34$$

where P is the percentage of propylene glycol in the deicing solution and T is the temperature at which the solution freezes.

(A) Draw a scatter plot of the data and a graph of the model on the same axes.

(B) Use the model to estimate the freezing temperature to the nearest degree of a solution that is 30% propylene glycol.

(C) Use the model to estimate the percentage of propylene glycol in a solution that freezes at 15°F.

21. Forestry. The figure contains a scatter plot of 100 data points for black spruce trees and the linear regression model for this data.

(A) Interpret the slope of the model.

(B) What is the effect of a 1-in. increase in Dbh?

(C) Estimate the height of a black spruce with a Dbh of 15 in. Round your answer to the nearest foot.

(D) Estimate the Dbh of a black spruce that is 25 ft tall. Round your answer to the nearest inch.

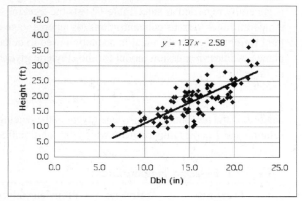

black spruce
Source: Lakehead University

22. Forestry. The figure contains a scatter plot of 100 data points for black walnut trees and the linear regression model for this data.

(A) Interpret the slope of the model.

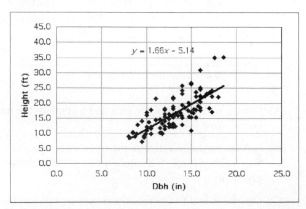

black walnut
Source: Kagen Research

(B) What is the effect of a 1-in. increase in Dbh?

(C) Estimate the height of a black walnut with a Dbh of 12 in. Round your answer to the nearest foot.

(D) Estimate the Dbh of a black walnut that is 25 ft tall. Round your answer to the nearest inch.

23. Cable television. Table 13 shows the increase in both price and revenue for cable television in the United States. The figure shows a scatter plot and a linear regression model for the average monthly price data in Table 13.

Table 13 **Cable Television Price and Revenue**

Year	Average Monthly Price (dollars)	Annual Total Revenue (billions of dollars)
2000	30.37	36.43
2002	34.71	47.99
2004	38.14	58.59
2006	41.17	71.89
2008	44.28	85.23
2010	47.89	93.37

Source: SNL Kagan

(A) Interpret the slope of the model.

(B) Use the model to predict the average monthly price (to the nearest dollar) in 2024.

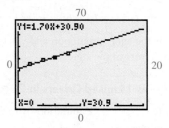

24. Cable television. The figure shows a scatter plot and a linear regression model for the annual revenue data in Table 13.

(A) Interpret the slope of the model.

(B) Use the model to predict the annual revenue (to the nearest billion dollars) in 2024.

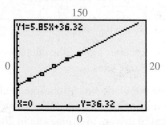

25. College enrollment. Table 14 lists the fall enrollment in degree-granting institutions by gender, and the figure contains a scatter plot and a regression line for each data set.

(A) Interpret the slope of each model.

(B) Predict both the male enrollment and the female enrollment in 2025.

Table 14 **Fall Enrollment (millions of students)**

Year	Male	Female
1970	5.04	3.54
1975	6.15	5.04
1980	5.87	6.22
1985	5.82	6.43
1990	6.28	7.53
1995	6.34	7.92
2000	6.72	8.59
2005	7.46	10.03
2010	9.04	11.97

Source: National Center for Education Statistics

(C) Estimate the first year for which female enrollment will exceed male enrollment by at least 5 million.

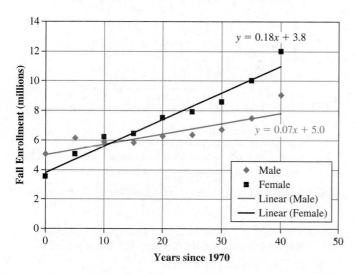

26. **Telephone expenditures.** Table 15 lists average annual telephone expenditures (in dollars) per consumer unit on residential phone service and cellular phone service, and the figure contains a scatter plot and regression line for each data set.

Table 15 **Telephone Expenditures**

Year	Residential Service ($)	Cellular Service ($)
2001	686	210
2003	620	316
2005	570	455
2007	482	608
2009	434	712

Source: Bureau of Labor Statistics

(A) Interpret the slope of each model.

(B) Predict (to the nearest dollar) the average annual residential and cellular expenditures in 2020.

(C) Would the linear regression models give reasonable predictions for the year 2025? Explain.

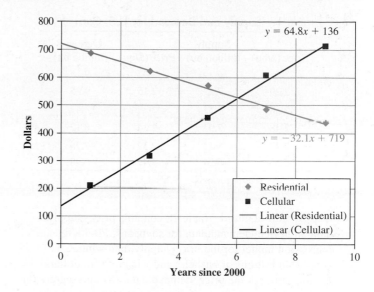

Problems 27–30 require a graphing calculator or a computer that can calculate the linear regression line for a given data set.

27. **Olympic Games.** Find a linear regression model for the men's 100-meter freestyle data given in Table 16, where x is years since 1990 and y is winning time (in seconds). Do the same for the women's 100-meter freestyle data. (Round regression coefficients to three decimal places.) Do these models indicate that the women will eventually catch up with the men?

Table 16 **Winning Times in Olympic Swimming Events**

	100-Meter Freestyle		200-Meter Backstroke	
	Men	Women	Men	Women
1992	49.02	54.65	1:58.47	2:07.06
1996	48.74	54.50	1.58.54	2:07.83
2000	48.30	53.83	1:56.76	2:08.16
2004	48.17	53.84	1:54.76	2:09.16
2008	47.21	53.12	1:53.94	2:05.24
2012	47.52	53.00	1:53.41	2:04.06

Source: www.infoplease.com

28. **Olympic Games.** Find a linear regression model for the men's 200-meter backstroke data given in Table 16, where x is years since 1990 and y is winning time (in seconds). Do the same for the women's 200-meter backstroke data. (Round regression coefficients to three decimal places.) Do these models indicate that the women will eventually catch up with the men?

29. **Supply and demand.** Table 17 contains price–supply data and price–demand data for corn. Find a linear regression model for the price–supply data where x is supply (in billions of bushels) and y is price (in dollars). Do the same for the price–demand data. (Round regression coefficients to two decimal places.) Find the equilibrium price for corn.

Table 17 **Supply and Demand for U.S. Corn**

Price ($/bu)	Supply (billion bu)	Price ($/bu)	Demand (billion bu)
2.15	6.29	2.07	9.78
2.29	7.27	2.15	9.35
2.36	7.53	2.22	8.47
2.48	7.93	2.34	8.12
2.47	8.12	2.39	7.76
2.55	8.24	2.47	6.98

Source: www.usda.gov/nass/pubs/histdata.htm

30. Supply and demand. Table 18 contains price–supply data and price–demand data for soybeans. Find a linear regression model for the price–supply data where x is supply (in billions of bushels) and y is price (in dollars). Do the same for the price–demand data. (Round regression coefficients to two decimal places.) Find the equilibrium price for soybeans.

Table 18 **Supply and Demand for U.S. Soybeans**

Price ($/bu)	Supply (billion bu)	Price ($/bu)	Demand (billion bu)
5.15	1.55	4.93	2.60
5.79	1.86	5.48	2.40
5.88	1.94	5.71	2.18
6.07	2.08	6.07	2.05
6.15	2.15	6.40	1.95
6.25	2.27	6.66	1.85

Source: www.usda.gov/nass/pubs/histdata.htm

Answers to Matched Problems

1. (A) The rate of change BSA with respect to weight is 28.55 square inches per pound.

   (B) Increasing w by 1 pound increases a by 28.55 square inches.

2. (A) $a = -36t + 2,880$ (B) 36 ft/sec

3. (A)

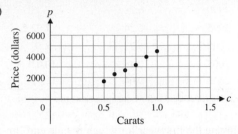

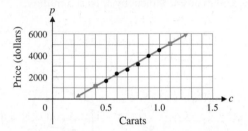

   (B) The rate of change of the price of a diamond with respect to its weight is $5,600. Increasing the weight by one carat will increase the price by about $5,600.

   (C) $3,660; $5,620   (D) 0.91 carat

4. Approximately 349 parts per million.

5. (A) The slope is 1.8, so the rate of change of height with respect to breast height diameter is 1.8 feet per inch.

   (B) Height increases by 1.8 feet.

   (C) 52 ft   (D) 17 in.

Chapter 1   Summary and Review

Important Terms, Symbols, and Concepts

1.1 Linear Equations and Inequalities

EXAMPLES

- A **first-degree**, or **linear**, **equation** in one variable is any equation that can be written in the form

  Standard form: $ax + b = 0$ $a \neq 0$

  If the equality sign in the standard form is replaced by $<, >, \leq,$ or $\geq$, the resulting expression is called a **first-degree**, or **linear**, **inequality**.

- A **solution** of an equation (or inequality) involving a single variable is a number that when substituted for the variable makes the equation (inequality) true. The set of all solutions is called the **solution set**.

- If we perform an operation on an equation (or inequality) that produces another equation (or inequality) with the same solution set, then the two equations (or inequalities) are **equivalent**. Equations are solved by adding or subtracting the same quantity to both sides, or by multiplying both sides by the same *nonzero* quantity until an equation with an obvious solution is obtained.

  Ex. 1, p. 3
  Ex. 2, p. 4

- The **interval notation** $[a, b)$, for example, represents the solution of the **double inequality** $a \leq x < b$.

  Ex. 5, p. 7

1.1 Linear Equations and Inequalities (*Continued*)

- Inequalities are solved in the same manner as equations with one important exception. If both sides of an inequality are multiplied by the same *negative* number or divided by the same *negative* number, then the direction or sense of the inequality will reverse ($<$ becomes $>$, $\geq$ becomes $\leq$, and so on). Ex. 6, p. 7
Ex. 7, p. 8

- A suggested strategy (p. 8) can be used to solve many word problems. Ex. 8, p. 8

- A company breaks even if revenues $R =$ costs C, makes a profit if $R > C$, and incurs a loss if $R < C$. Ex. 9, p. 9

1.2 Graphs and Lines

- A **Cartesian or rectangular coordinate system** is formed by the intersection of a horizontal real number line, usually called the **x axis**, and a vertical real number line, usually called the **y axis**, at their origins. The axes determine a plane and divide this plane into four **quadrants**. Each point in the plane corresponds to its **coordinates**—an ordered pair (a, b) determined by passing horizontal and vertical lines through the point. The **abscissa** or **x coordinate** a is the coordinate of the intersection of the vertical line and the x axis, and the **ordinate** or **y coordinate** b is the coordinate of the intersection of the horizontal line and the y axis. The point with coordinates $(0, 0)$ is called the **origin**. Fig 1, p. 13

- The **standard form** for a linear equation in two variables is $Ax + By = C$, with A and B not both zero. The graph of this equation is a line, and every line in a Cartesian coordinate system is the graph of a linear equation. Ex. 1, p. 14
Ex. 2, p. 15

- The graph of the equation $x = a$ is a **vertical line** and the graph of $y = b$ is a **horizontal line**. Ex. 3, p. 15

- If (x_1, y_1) and (x_2, y_2) are two distinct points on a line, then $m = (y_2 - y_1)/(x_2 - x_1)$ is the **slope** of the line. Ex. 4, p. 17

- The equation $y = mx + b$ is the **slope-intercept form** of the equation of the line with slope m and y intercept b. Ex. 5, p. 18

- The **point-slope form** of the equation of the line with slope m that passes through (x_1, y_1) is $y - y_1 = m(x - x_1)$. Ex. 6, p. 19

- In a competitive market, the intersection of the supply equation and the demand equation is called the **equilibrium point**, the corresponding price is called the **equilibrium price**, and the common value of supply and demand is called the **equilibrium quantity**. Ex. 8, p. 21

1.3 Linear Regression

- A **mathematical model** is a mathematics problem that, when solved, will provide information about a real-world problem.

- If the variables x and y are related by the equation $y = mx + b$, then x and y are **linearly related** and the slope m is the **rate of change** of y with respect to x. Ex. 1, p. 27
Ex. 2, p. 28

- A graph of the points in a data set is called a **scatter plot**. **Linear regression** is used to find the line that is the **best fit** for a data set. A regression model can be used to **interpolate** between points in a data set or to **extrapolate** or predict points outside the data set. Ex. 3, p. 29
Ex. 4, p. 30
Ex. 5, p. 31

Review Exercises

Work through all the problems in this chapter review and check answers in the back of the book. Following each answer you will find a number in italics indicating the section where that type of problem is discussed. Where weaknesses show up, review appropriate sections in the text.

1. Solve $2x + 3 = 7x - 11$.

2. Solve $\dfrac{x}{12} - \dfrac{x - 3}{3} = \dfrac{1}{2}$.

3. Solve $2x + 5y = 9$ for y.

4. Solve $3x - 4y = 7$ for x.

Solve Problems 5–7 and graph on a real number line.

5. $4y - 3 < 10$

6. $-1 < -2x + 5 \leq 3$

7. $1 - \dfrac{x - 3}{3} \leq \dfrac{1}{2}$

8. Sketch a graph of $3x + 2y = 9$.

9. Write an equation of a line with x intercept 6 and y intercept 4. Write the final answer in the form $Ax + By = C$.

10. Sketch a graph of $2x - 3y = 18$. What are the intercepts and slope of the line?

11. Write an equation in the form $y = mx + b$ for a line with slope $-\dfrac{2}{3}$ and y intercept 6.

12. Write the equations of the vertical line and the horizontal line that pass through $(-6, 5)$.

13. Write the equation of a line through each indicated point with the indicated slope. Write the final answer in the form $y = mx + b$.

   (A) $m = -\dfrac{2}{3}; (-3, 2)$ (B) $m = 0; (3, 3)$

14. Write the equation of the line through the two indicated points. Write the final answer in the form $Ax + By = C$.

   (A) $(-3, 5), (1, -1)$ (B) $(-1, 5), (4, 5)$

   (C) $(-2, 7), (-2, -2)$

Solve Problems 15–19.

15. $3x + 25 = 5x$ 16. $\dfrac{u}{5} = \dfrac{u}{6} + \dfrac{6}{5}$

17. $\dfrac{5x}{3} - \dfrac{4 + x}{2} = \dfrac{x - 2}{4} + 1$

18. $0.05x + 0.25(30 - x) = 3.3$

19. $0.2(x - 3) + 0.05x = 0.4$

Solve Problems 20–24 and graph on a real number line.

20. $2(x + 4) > 5x - 4$ 21. $3(2 - x) - 2 \le 2x - 1$

22. $\dfrac{x + 3}{8} - \dfrac{4 + x}{2} > 5 - \dfrac{2 - x}{3}$

23. $-5 \le 3 - 2x < 1$ 24. $-1.5 \le 2 - 4x \le 0.5$

25. Given $Ax + By = 30$, graph each of the following cases on the same coordinate axes.

   (A) $A = 5$ and $B = 0$ (B) $A = 0$ and $B = 6$

   (C) $A = 6$ and $B = 5$

26. Describe the graphs of $x = -3$ and $y = 2$. Graph both simultaneously in the same coordinate system.

27. Describe the lines defined by the following equations:

   (A) $3x + 4y = 0$ (B) $3x + 4 = 0$

   (C) $4y = 0$ (D) $3x + 4y - 36 = 0$

Solve Problems 28 and 29 for the indicated variable.

28. $A = \dfrac{1}{2}(a + b)h$; for $a\,(h \ne 0)$

29. $S = \dfrac{P}{1 - dt}$; for $d\,(dt \ne 1)$

30. For what values of a and b is the inequality $a + b < b - a$ true?

31. If a and b are negative numbers and $a > b$, then is a/b greater than 1 or less than 1?

32. Graph $y = mx + b$ and $y = -\dfrac{1}{m}x + b$ simultaneously in the same coordinate system for b fixed and several different values of $m, m \ne 0$. Describe the apparent relationship between the graphs of the two equations.

Applications

33. *Investing.* An investor has $300,000 to invest. If part is invested at 5% and the rest at 9%, how much should be invested at 5% to yield 8% on the total amount?

34. *Break-even analysis.* A producer of educational DVDs is producing an instructional DVD. She estimates that it will cost $90,000 to record the DVD and $5.10 per unit to copy and distribute the DVD. If the wholesale price of the DVD is $14.70, how many DVDs must be sold for the producer to break even?

35. *Sports medicine.* A simple rule of thumb for determining your maximum safe heart rate (in beats per minute) is to subtract your age from 220. While exercising, you should maintain a heart rate between 60% and 85% of your maximum safe rate.

   (A) Find a linear model for the minimum heart rate m that a person of age x years should maintain while exercising.

   (B) Find a linear model for the maximum heart rate M that a person of age x years should maintain while exercising.

   (C) What range of heartbeats should you maintain while exercising if you are 20 years old?

   (D) What range of heartbeats should you maintain while exercising if you are 50 years old?

36. *Linear depreciation.* A bulldozer was purchased by a construction company for $224,000 and has a depreciated value of $100,000 after 8 years. If the value is depreciated linearly from $224,000 to $100,000,

   (A) Find the linear equation that relates value V (in dollars) to time t (in years).

   (B) What would be the depreciated value after 12 years?

37. *Business—Pricing.* A sporting goods store sells tennis rackets that cost $130 for $208 and court shoes that cost $50 for $80.

   (A) If the markup policy of the store for items that cost over $10 is linear and is reflected in the pricing of these two items, write an equation that expresses retail price R in terms of cost C.

(B) What would be the retail price of a pair of in-line skates that cost $120?

(C) What would be the cost of a pair of cross-country skis that had a retail price of $176?

(D) What is the slope of the graph of the equation found in part (A)? Interpret the slope relative to the problem.

38. Income. A salesperson receives a base salary of $400 per week and a commission of 10% on all sales over $6,000 during the week. Find the weekly earnings for weekly sales of $4,000 and for weekly sales of $10,000.

39. Price–demand. The weekly demand for mouthwash in a chain of drug stores is 1,160 bottles at a price of $3.79 each. If the price is lowered to $3.59, the weekly demand increases to 1,320 bottles. Assuming that the relationship between the weekly demand x and price per bottle p is linear, express p in terms of x. How many bottles would the stores sell each week if the price were lowered to $3.29?

40. Freezing temperature. Methanol, also known as wood alcohol, can be used as a fuel for suitably equipped vehicles. Table 1 lists the freezing temperature for various concentrations (as a percentage of total weight) of methanol in water. A linear regression model for the data in Table 1 is

$$T = 40 - 2M$$

where M is the percentage of methanol in the solution and T is the temperature at which the solution freezes.

Table 1

Methanol (%Wt)	Freezing temperature (°F)
0	32
10	20
20	0
30	−15
40	−40
50	−65
60	−95

Source: Ashland Inc.

(A) Draw a scatter plot of the data and a graph of the model on the same axes.

(B) Use the model to estimate the freezing temperature to the nearest degree of a solution that is 35% methanol.

(C) Use the model to estimate the percentage of methanol in a solution that freezes at −50°F.

41. High school dropout rates. Table 2 gives U.S. high school dropout rates as percentages for selected years since 1980. A linear regression model for the data is

$$r = -0.198t + 14.2$$

where t represents years since 1980 and r is the dropout rate.

Table 2 **High School Dropout Rates (%)**

1980	1985	1990	1995	2000	2005	2010
14.1	12.6	12.1	12.0	10.9	9.4	7.4

(A) Interpret the slope of the model.

(B) Draw a scatter plot of the data and the model in the same coordinate system.

(C) Use the model to predict the first year for which the dropout rate is less than 5%.

42. Consumer Price Index. The U.S. Consumer Price Index (CPI) in recent years is given in Table 3. A scatter plot of the data and linear regression line are shown in the figure, where x represents years since 2000.

Table 3 **Consumer Price Index (1982–1984 = 100)**

Year	CPI
2000	172.2
2002	179.9
2004	188.9
2006	198.3
2008	211.1
2010	218.1

Source: U.S. Bureau of Labor Statistics

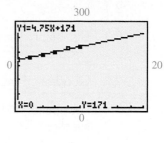

(A) Interpret the slope of the model.

(B) Predict the CPI in 2024.

43. Forestry. The figure contains a scatter plot of 20 data points for white pine trees and the linear regression model for this data.

(A) Interpret the slope of the model.

(B) What is the effect of a 1-in. increase in Dbh?

(C) Estimate the height of a white pine tree with a Dbh of 25 in. Round your answer to the nearest foot.

(D) Estimate the Dbh of a white pine tree that is 15 ft tall. Round your answer to the nearest inch.

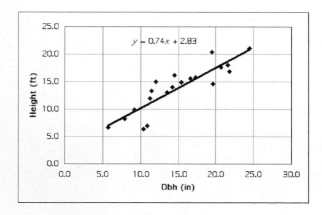

2

Functions and Graphs

2.1 Functions

2.2 Elementary Functions: Graphs and Transformations

2.3 Quadratic Functions

2.4 Polynomial and Rational Functions

2.5 Exponential Functions

2.6 Logarithmic Functions

Chapter 2 Summary and Review

Review Exercises

Introduction

The function concept is one of the most important ideas in mathematics. The study of mathematics beyond the elementary level requires a firm understanding of a basic list of elementary functions, their properties, and their graphs. See the inside back cover of this book for a list of the functions that form our library of elementary functions. Most functions in the list will be introduced to you by the end of Chapter 2. For example, in Section 2.3 you will learn how to apply quadratic functions to model the effect of tire pressure on mileage (see Problems 61 and 63 on pages 81 and 82).

2.1 Functions

- Equations in Two Variables
- Definition of a Function
- Functions Specified by Equations
- Function Notation
- Applications

We introduce the general notion of a *function* as a correspondence between two sets. Then we restrict attention to functions for which the two sets are both sets of real numbers. The most useful are those functions that are specified by equations in two variables. We discuss the terminology and notation associated with functions, graphs of functions, and applications.

Equations in Two Variables

In Chapter 1, we found that the graph of an equation of the form $Ax + By = C$, where A and B are not both zero, is a line. Because a line is determined by any two of its points, such an equation is easy to graph: Just plot *any* two points in its solution set and sketch the unique line through them.

More complicated equations in two variables, such as $y = 9 - x^2$ or $x^2 = y^4$, are more difficult to graph. To **sketch the graph** of an equation, we plot enough points from its solution set in a rectangular coordinate system so that the total graph is apparent, and then we connect these points with a smooth curve. This process is called **point-by-point plotting**.

EXAMPLE 1 Point-by-Point Plotting Sketch the graph of each equation.

(A) $y = 9 - x^2$ (B) $x^2 = y^4$

SOLUTION

(A) Make up a table of solutions—that is, ordered pairs of real numbers that satisfy the given equation. For easy mental calculation, choose integer values for x.

x	-4	-3	-2	-1	0	1	2	3	4
y	-7	0	5	8	9	8	5	0	-7

After plotting these solutions, if there are any portions of the graph that are unclear, plot additional points until the shape of the graph is apparent. Then join all the plotted points with a smooth curve (Fig. 1). Arrowheads are used to indicate that the graph continues beyond the portion shown here with no significant changes in shape.

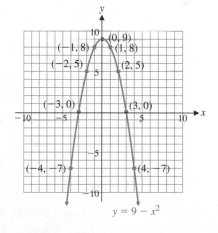

Figure 1 $y = 9 - x^2$

(B) Again we make a table of solutions—here it may be easier to choose integer values for y and calculate values for x. Note, for example, that if $y = 2$, then $x = \pm 4$; that is, the ordered pairs $(4, 2)$ and $(-4, 2)$ are both in the solution set.

x	± 9	± 4	± 1	0	± 1	± 4	± 9
y	-3	-2	-1	0	1	2	3

We plot these points and join them with a smooth curve (Fig. 2).

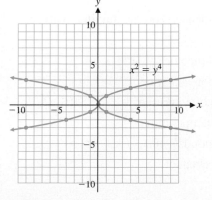

Figure 2 $x^2 = y^4$

Matched Problem 1 ⎤ Sketch the graph of each equation.

(A) $y = x^2 - 4$ (B) $y^2 = \dfrac{100}{x^2 + 1}$

Explore and Discuss 1   To graph the equation $y = -x^3 + 3x$, we use point-by-point plotting to obtain

x	y
-1	-2
0	0
1	2

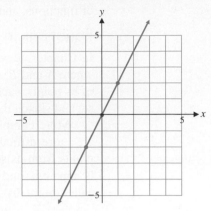

(A) Do you think this is the correct graph of the equation? Why or why not?

(B) Add points on the graph for $x = -2, -1.5, -0.5, 0.5, 1.5,$ and 2.

(C) Now, what do you think the graph looks like? Sketch your version of the graph, adding more points as necessary.

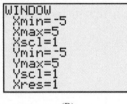 (D) Graph this equation on a graphing calculator and compare it with your graph from part (C).

(A)

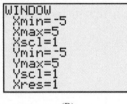

(B)

Figure 3

The icon in the margin is used throughout this book to identify optional graphing calculator activities that are intended to give you additional insight into the concepts under discussion. You may have to consult the manual for your graphing calculator for the details necessary to carry out these activities. For example, to graph the equation in Explore and Discuss 1 on most graphing calculators, you must enter the equation (Fig. 3A) and the window variables (Fig. 3B).

As Explore and Discuss 1 illustrates, the shape of a graph may not be apparent from your first choice of points. Using point-by-point plotting, it may be difficult to find points in the solution set of the equation, and it may be difficult to determine when you have found enough points to understand the shape of the graph. We will supplement the technique of point-by-point plotting with a detailed analysis of several basic equations, giving you the ability to sketch graphs with accuracy and confidence.

Definition of a Function

Central to the concept of function is correspondence. You are familiar with correspondences in daily life. For example,

To each person, there corresponds an annual income.

To each item in a supermarket, there corresponds a price.

To each student, there corresponds a grade-point average.

To each day, there corresponds a maximum temperature.

For the manufacture of x items, there corresponds a cost.

For the sale of x items, there corresponds a revenue.

To each square, there corresponds an area.

To each number, there corresponds its cube.

One of the most important aspects of any science is the establishment of correspondences among various types of phenomena. Once a correspondence is known, predictions can be made. A cost analyst would like to predict costs for various levels of output in a manufacturing process; a medical researcher would like to know the correspondence between heart disease and obesity; a psychologist would like to predict the level of performance after a subject has repeated a task a given number of times; and so on.

What do all of these examples have in common? Each describes the matching of elements from one set with the elements in a second set.

Consider Tables 1–3. Tables 1 and 2 specify functions, but Table 3 does not. Why not? The definition of the term *function* will explain.

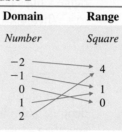

Table 1

Domain	Range
Number	*Cube*
−2	−8
−1	−1
0	0
1	1
2	8

Table 2

Domain	Range
Number	*Square*
−2	4
−1	1
0	0
1	
2	

Table 3

Domain	Range
Number	*Square root*
0	0
1	1
1	−1
4	2
4	−2
9	3
9	−3

> **DEFINITION** Function
> A **function** is a correspondence between two sets of elements such that to each element in the first set, there corresponds one and only one element in the second set.
> The first set is called the **domain**, and the set of corresponding elements in the second set is called the **range**.

Tables 1 and 2 specify functions since to each domain value, there corresponds exactly one range value (for example, the cube of −2 is −8 and no other number). On the other hand, Table 3 does not specify a function since to at least one domain value, there corresponds more than one range value (for example, to the domain value 9, there corresponds −3 and 3, both square roots of 9).

Explore and Discuss 2 Consider the set of students enrolled in a college and the set of faculty members at that college. Suppose we define a correspondence between the two sets by saying that a student corresponds to a faculty member if the student is currently enrolled in a course taught by that faculty member. Is this correspondence a function? Discuss.

Functions Specified by Equations

Most of the functions in this book will have domains and ranges that are (infinite) sets of real numbers. The **graph** of such a function is the set of all points (x, y) in the Cartesian plane such that x is an element of the domain and y is the corresponding element in the range. The correspondence between domain and range elements is often specified by an equation in two variables. Consider, for example, the equation for the area of a rectangle with width 1 inch less than its length (Fig. 4). If x is the length, then the area y is given by

$$y = x(x - 1) \qquad x \geq 1$$

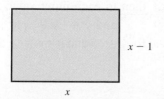

Figure 4

For each **input** x (length), we obtain an **output** y (area). For example,

If $x = 5$,   then $y = 5(5 - 1) = 5 \cdot 4 = 20$.

If $x = 1$,   then $y = 1(1 - 1) = 1 \cdot 0 = 0$.

If $x = \sqrt{5}$,   then $y = \sqrt{5}(\sqrt{5} - 1) = 5 - \sqrt{5}$

$$\approx 2.7639.$$

The input values are domain values, and the output values are range values. The equation assigns each domain value x a range value y. The variable x is called an *independent variable* (since values can be "independently" assigned to x from the domain), and y is called a *dependent variable* (since the value of y "depends" on the value assigned to x). In general, any variable used as a placeholder for domain values is called an **independent variable**; any variable that is used as a placeholder for range values is called a **dependent variable**.

When does an equation specify a function?

DEFINITION Functions Specified by Equations

If in an equation in two variables, we get exactly one output (value for the dependent variable) for each input (value for the independent variable), then the equation specifies a function. The graph of such a function is just the graph of the specifying equation.

If we get more than one output for a given input, the equation does not specify a function.

EXAMPLE 2   Functions and Equations   Determine which of the following equations specify functions with independent variable x.

(A) $4y - 3x = 8$, x a real number   (B) $y^2 - x^2 = 9$, x a real number

SOLUTION

(A) Solving for the dependent variable y, we have

$$4y - 3x = 8$$

$$4y = 8 + 3x \tag{1}$$

$$y = 2 + \frac{3}{4}x$$

Since each input value x corresponds to exactly one output value $\left(y = 2 + \frac{3}{4}x\right)$, we see that equation (1) specifies a function.

(B) Solving for the dependent variable y, we have

$$y^2 - x^2 = 9$$

$$y^2 = 9 + x^2 \tag{2}$$

$$y = \pm\sqrt{9 + x^2}$$

Since $9 + x^2$ is always a positive real number for any real number x, and since each positive real number has two square roots,* then to each input value x there corresponds two output values $\left(y = -\sqrt{9 + x^2} \text{ and } y = \sqrt{9 + x^2}\right)$. For example, if $x = 4$, then equation (2) is satisfied for $y = 5$ and for $y = -5$. So equation (2) does not specify a function.

---

*Recall that each positive real number N has two square roots: $\sqrt{N}$, the principal square root; and $-\sqrt{N}$, the negative of the principal square root (see Appendix A, Section A.6).

Matched Problem 2 Determine which of the following equations specify functions with independent variable x.

(A) $y^2 - x^4 = 9$, x a real number   (B) $3y - 2x = 3$, x a real number

Since the graph of an equation is the graph of all the ordered pairs that satisfy the equation, it is very easy to determine whether an equation specifies a function by examining its graph. The graphs of the two equations we considered in Example 2 are shown in Figure 5.

In Figure 5A, notice that any vertical line will intersect the graph of the equation $4y - 3x = 8$ in exactly one point. This shows that to each x value, there corresponds exactly one y value, confirming our conclusion that this equation specifies a function. On the other hand, Figure 5B shows that there exist vertical lines that intersect the graph of $y^2 - x^2 = 9$ in two points. This indicates that there exist x values to which there correspond two different y values and verifies our conclusion that this equation does not specify a function. These observations are generalized in Theorem 1.

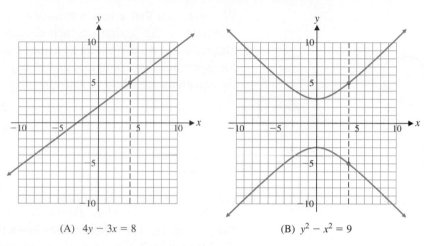

(A) $4y - 3x = 8$ (B) $y^2 - x^2 = 9$

Figure 5

THEOREM 1 Vertical-Line Test for a Function

An equation specifies a function if each vertical line in the coordinate system passes through, at most, one point on the graph of the equation.

If any vertical line passes through two or more points on the graph of an equation, then the equation does not specify a function.

The function graphed in Figure 5A is an example of a *linear function*. The vertical-line test implies that equations of the form $y = mx + b$, where $m \neq 0$, specify functions; they are called **linear functions**. Similarly, equations of the form $y = b$ specify functions; they are called **constant functions**, and their graphs are horizontal lines. The vertical-line test implies that equations of the form $x = a$ do not specify functions; note that the graph of $x = a$ is a vertical line.

In Example 2, the domains were explicitly stated along with the given equations. In many cases, this will not be done. Unless stated to the contrary, we shall adhere to the following convention regarding domains and ranges for functions specified by equations:

> If a function is specified by an equation and the domain is not indicated, then we assume that the domain is the set of all real-number replacements of the independent variable (inputs) that produce real values for the dependent variable (outputs). The range is the set of all outputs corresponding to input values.

EXAMPLE 3 Finding a Domain Find the domain of the function specified by the equation $y = \sqrt{4 - x}$, assuming that x is the independent variable.

SOLUTION For y to be real, $4 - x$ must be greater than or equal to 0; that is,

$$4 - x \geq 0$$
$$-x \geq -4$$
$$x \leq 4 \qquad \text{Sense of inequality reverses when both sides are divided by } -1.$$

Domain: $x \leq 4$ (inequality notation) or $(-\infty, 4]$ (interval notation)

Matched Problem 3 Find the domain of the function specified by the equation $y = \sqrt{x - 2}$, assuming x is the independent variable.

Function Notation

We have seen that a function involves two sets, a domain and a range, and a correspondence that assigns to each element in the domain exactly one element in the range. Just as we use letters as names for numbers, now we will use letters as names for functions. For example, f and g may be used to name the functions specified by the equations $y = 2x + 1$ and $y = x^2 + 2x - 3$:

$$f: \quad y = 2x + 1$$
$$g: \quad y = x^2 + 2x - 3 \qquad (3)$$

If x represents an element in the domain of a function f, then we frequently use the symbol

$$f(x)$$

in place of y to designate the number in the range of the function f to which x is paired (Fig. 6). This symbol does *not* represent the product of f and x. The symbol $f(x)$ is read as "f of x," "f at x," or "the value of f at x." Whenever we write $y = f(x)$, we assume that the variable x is an independent variable and that both y and $f(x)$ are dependent variables.

Using function notation, we can now write functions f and g in equation (3) as

$$f(x) = 2x + 1 \qquad \text{and} \qquad g(x) = x^2 + 2x - 3$$

Let us find $f(3)$ and $g(-5)$. To find $f(3)$, we replace x with 3 wherever x occurs in $f(x) = 2x + 1$ and evaluate the right side:

$$f(x) = 2x + 1$$
$$f(3) = 2 \cdot 3 + 1$$
$$= 6 + 1 = 7 \qquad \text{For input 3, the output is 7.}$$

Therefore,

$$f(3) = 7 \qquad \text{The function } f \text{ assigns the range value 7 to the domain value 3.}$$

To find $g(-5)$, we replace each x by -5 in $g(x) = x^2 + 2x - 3$ and evaluate the right side:

$$g(x) = x^2 + 2x - 3$$
$$g(-5) = (-5)^2 + 2(-5) - 3$$
$$= 25 - 10 - 3 = 12 \qquad \text{For input } -5, \text{ the output is 12.}$$

Therefore,

$$g(-5) = 12 \qquad \text{The function } g \text{ assigns the range value 12 to the domain value } -5.$$

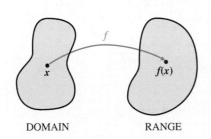

DOMAIN   RANGE

Figure 6

It is very important to understand and remember the definition of $f(x)$:

For any element x in the domain of the function f, the symbol $f(x)$ represents the element in the range of f corresponding to x in the domain of f. If x is an input value, then $f(x)$ is the corresponding output value. If x is an element that is not in the domain of f, then f is not defined at x and $f(x)$ does not exist.

EXAMPLE 4 Function Evaluation For $f(x) = 12/(x - 2)$, $g(x) = 1 - x^2$, and $h(x) = \sqrt{x - 1}$, evaluate:

(A) $f(6)$ (B) $g(-2)$ (C) $h(-2)$ (D) $f(0) + g(1) - h(10)$

SOLUTION (A) $f(6) = \dfrac{12}{6 - 2} \ {}^* = \dfrac{12}{4} = 3$

(B) $g(-2) = 1 - (-2)^2 = 1 - 4 = -3$

(C) $h(-2) = \sqrt{-2 - 1} = \sqrt{-3}$

But $\sqrt{-3}$ is not a real number. Since we have agreed to restrict the domain of a function to values of x that produce real values for the function, -2 is not in the domain of h, and $h(-2)$ does not exist.

(D) $f(0) + g(1) - h(10) = \dfrac{12}{0 - 2} + (1 - 1^2) - \sqrt{10 - 1}$

$= \dfrac{12}{-2} + 0 - \sqrt{9}$

$= -6 - 3 = -9$

Matched Problem 4 Use the functions in Example 4 to find

(A) $f(-2)$ (B) $g(-1)$ (C) $h(-8)$ (D) $\dfrac{f(3)}{h(5)}$

EXAMPLE 5 Finding Domains Find the domains of functions f, g, and h:

$$f(x) = \dfrac{12}{x - 2} \qquad g(x) = 1 - x^2 \qquad h(x) = \sqrt{x - 1}$$

SOLUTION *Domain of f:* $12/(x - 2)$ represents a real number for all replacements of x by real numbers except for $x = 2$ (division by 0 is not defined). Thus, $f(2)$ does not exist, and the domain of f is the set of all real numbers except 2. We often indicate this by writing

$$f(x) = \dfrac{12}{x - 2} \qquad x \neq 2$$

Domain of g: The domain is R, the set of all real numbers, since $1 - x^2$ represents a real number for all replacements of x by real numbers.

Domain of h: The domain is the set of all real numbers x such that $\sqrt{x - 1}$ is a real number, so

$$x - 1 \geq 0$$
$$x \geq 1 \qquad \text{or, in interval notation,} \qquad [1, \infty)$$

Matched Problem 5 Find the domains of functions F, G, and H:

$$F(x) = x^2 - 3x + 1 \qquad G(x) = \dfrac{5}{x + 3} \qquad H(x) = \sqrt{2 - x}$$

---

*Dashed boxes are used throughout the book to represent steps that are usually performed mentally.

In addition to evaluating functions at specific numbers, it is important to be able to evaluate functions at expressions that involve one or more variables. For example, the **difference quotient**

$$\frac{f(x + h) - f(x)}{h} \qquad x \text{ and } x + h \text{ in the domain of } f, h \neq 0$$

is studied extensively in calculus.

CONCEPTUAL INSIGHT

In algebra, you learned to use parentheses for grouping variables. For example,

$$2(x + h) = 2x + 2h$$

Now we are using parentheses in the function symbol $f(x)$. For example, if $f(x) = x^2$, then

$$f(x + h) = (x + h)^2 = x^2 + 2xh + h^2$$

Note that $f(x) + f(h) = x^2 + h^2 \neq f(x + h)$. That is, the function name f does not distribute across the grouped variables $(x + h)$, as the "2" does in $2(x + h)$ (see Appendix A, Section A.2).

EXAMPLE 6   Using Function Notation   For $f(x) = x^2 - 2x + 7$, find

(A) $f(a)$ (B) $f(a + h)$ (C) $f(a + h) - f(a)$ (D) $\dfrac{f(a + h) - f(a)}{h}$, $h \neq 0$

SOLUTION

(A) $f(a) = a^2 - 2a + 7$

(B) $f(a + h) = (a + h)^2 - 2(a + h) + 7 = a^2 + 2ah + h^2 - 2a - 2h + 7$

(C) $f(a + h) - f(a) = (a^2 + 2ah + h^2 - 2a - 2h + 7) - (a^2 - 2a + 7)$
$$= 2ah + h^2 - 2h$$

(D) $\dfrac{f(a + h) - f(a)}{h} = \dfrac{2ah + h^2 - 2h}{h} = \dfrac{h(2a + h - 2)}{h}$ Because $h \neq 0, \dfrac{h}{h} = 1.$
$$= 2a + h - 2$$

Matched Problem 6 ⌋   Repeat Example 6 for $f(x) = x^2 - 4x + 9$.

Applications

We now turn to the important concepts of **break-even** and **profit–loss** analysis, which we will return to a number of times in this book. Any manufacturing company has **costs**, C, and **revenues**, R. The company will have a **loss** if $R < C$, will **break even** if $R = C$, and will have a **profit** if $R > C$. Costs include **fixed costs** such as plant overhead, product design, setup, and promotion; and **variable costs**, which are dependent on the number of items produced at a certain cost per item. In addition, **price–demand** functions, usually established by financial departments using historical data or sampling techniques, play an important part in profit–loss analysis. We will let x, the number of units manufactured and sold, represent the independent variable. Cost functions, revenue functions, profit functions, and price–demand functions are often stated in the following forms, where a, b, m, and n are constants determined from the context of a particular problem:

Cost Function

$$C = (\text{fixed costs}) + (\text{variable costs})$$
$$= a + bx$$

Price–Demand Function

$$p = m - nx \quad \text{x is the number of items that can be sold at \$p per item.}$$

Revenue Function

$$R = (\text{number of items sold}) \times (\text{price per item})$$
$$= xp = x(m - nx)$$

Profit Function

$$P = R - C$$
$$= x(m - nx) - (a + bx)$$

Example 7 and Matched Problem 7 explore the relationships among the algebraic definition of a function, the numerical values of the function, and the graphical representation of the function. The interplay among algebraic, numeric, and graphic viewpoints is an important aspect of our treatment of functions and their use. In Example 7, we will see how a function can be used to describe data from the real world, a process that is often referred to as *mathematical modeling*. Note that the domain of such a function is determined by practical considerations within the problem.

EXAMPLE 7 Price–Demand and Revenue Modeling A manufacturer of a popular digital camera wholesales the camera to retail outlets throughout the United States. Using statistical methods, the financial department in the company produced the price–demand data in Table 4, where p is the wholesale price per camera at which x million cameras are sold. Notice that as the price goes down, the number sold goes up.

Table 4 **Price–Demand**

x (Millions)	p ($)
2	87
5	68
8	53
12	37

Using special analytical techniques (regression analysis), an analyst obtained the following price–demand function to model the Table 4 data:

$$p(x) = 94.8 - 5x \quad 1 \le x \le 15 \tag{4}$$

(A) Plot the data in Table 4. Then sketch a graph of the price–demand function in the same coordinate system.

(B) What is the company's revenue function for this camera, and what is its domain?

(C) Complete Table 5, computing revenues to the nearest million dollars.

(D) Plot the data in Table 5. Then sketch a graph of the revenue function using these points.

(E) Plot the revenue function on a graphing calculator.

Table 5 **Revenue**

x (Millions)	R(x) (Million $)
1	90
3	
6	
9	
12	
15	

SOLUTION   (A)

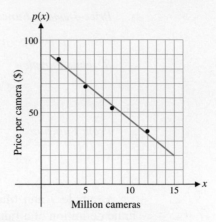

Figure 7 **Price–demand**

In Figure 7, notice that the model approximates the actual data in Table 4, and it is assumed that it gives realistic and useful results for all other values of x between 1 million and 15 million.

(B) $R(x) = xp(x) = x(94.8 - 5x)$ million dollars

Domain: $1 \leq x \leq 15$

[Same domain as the price-demand function, equation (4).]

(C)   Table 5 **Revenue**

x (Millions)	$R(x)$ (Million $)
1	90
3	239
6	389
9	448
12	418
15	297

(D)

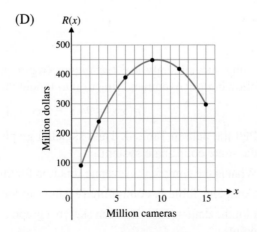

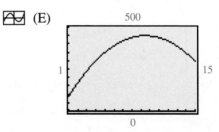 (E)

Matched Problem 7 The financial department in Example 7, using statistical techniques, produced the data in Table 6, where $C(x)$ is the cost in millions of dollars for manufacturing and selling x million cameras.

Table 6 **Cost Data**

x (Millions)	$C(x)$ (Million $)
1	175
5	260
8	305
12	395

Using special analytical techniques (regression analysis), an analyst produced the following cost function to model the Table 6 data:

$$C(x) = 156 + 19.7x \qquad 1 \le x \le 15 \qquad (5)$$

(A) Plot the data in Table 6. Then sketch a graph of equation (5) in the same coordinate system.

(B) What is the company's profit function for this camera, and what is its domain?

(C) Complete Table 7, computing profits to the nearest million dollars.

Table 7 **Profit**

x (Millions)	P(x) (Million $)
1	−86
3	
6	
9	
12	
15	

(D) Plot the data in Table 7. Then sketch a graph of the profit function using these points.

(E) Plot the profit function on a graphing calculator.

Exercises 2.1

In Problems 1–8, use point-by-point plotting to sketch the graph of each equation.

1. $y = x + 1$ **2.** $x = y + 1$

3. $x = y^2$ **4.** $y = x^2$

5. $y = x^3$ **6.** $x = y^3$

7. $xy = -6$ **8.** $xy = 12$

Indicate whether each table in Problems 9–14 specifies a function.

9.

Domain	Range
3 → 0	
5 → 1	
7 → 2	

10.

Domain	Range
−1 → 5	
−2 → 7	
−3 → 9	

11.

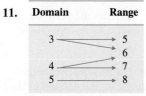

Domain	Range
3 → 5	
→ 6	
4 → 7	
5 → 8	

12.

Domain	Range
8 → 0	
9 → 1	
→ 2	
10 → 3	

13.

Domain	Range
3 →	
6 → 5	
9 → 6	
12 →	

14.

Domain	Range
−2 →	
−1 → 6	
0 →	
1 →	

Indicate whether each graph in Problems 15–20 specifies a function.

15.

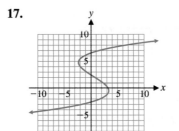

16.

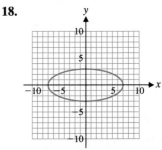

17.

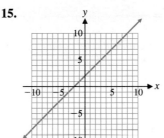

18.

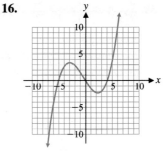

19.

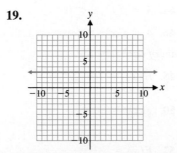

20.

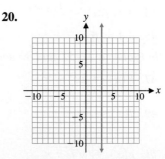

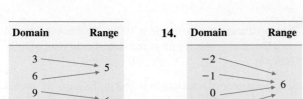

In Problems 21–28, each equation specifies a function with independent variable x. Determine whether the function is linear, constant, or neither.

21. $y - 2x = 7$

22. $y = 10 - 3x$

23. $xy - 4 = 0$

24. $x^2 - y = 8$

25. $y = 5x + \dfrac{1}{2}(7 - 10x)$

26. $y = \dfrac{2 + x}{3} + \dfrac{2 - x}{3}$

27. $3x + 4y = 5$

28. $9x - 2y + 6 = 0$

In Problems 29–36, use point-by-point plotting to sketch the graph of each function.

29. $f(x) = 1 - x$

30. $f(x) = \dfrac{x}{2} - 3$

31. $f(x) = x^2 - 1$

32. $f(x) = 3 - x^2$

33. $f(x) = 4 - x^3$

34. $f(x) = x^3 - 2$

35. $f(x) = \dfrac{8}{x}$

36. $f(x) = \dfrac{-6}{x}$

In Problems 37 and 38, the three points in the table are on the graph of the indicated function f. Do these three points provide sufficient information for you to sketch the graph of $y = f(x)$? Add more points to the table until you are satisfied that your sketch is a good representation of the graph of $y = f(x)$ on the interval $[-5, 5]$.

37.

x	−1	0	1
$f(x)$	−1	0	1

$f(x) = \dfrac{2x}{x^2 + 1}$

38.

x	0	1	2
$f(x)$	0	1	2

$f(x) = \dfrac{3x^2}{x^2 + 2}$

In Problems 39–46, use the following graph of a function f to determine x or y to the nearest integer, as indicated. Some problems may have more than one answer.

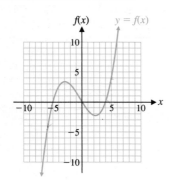

39. $y = f(-5)$

40. $y = f(4)$

41. $y = f(5)$

42. $y = f(-2)$

43. $0 = f(x)$

44. $3 = f(x), x < 0$

45. $-4 = f(x)$

46. $4 = f(x)$

In Problems 47–52, find the domain of each function.

47. $F(x) = 2x^3 - x^2 + 3$

48. $H(x) = 7 - 2x^2 - x^4$

49. $f(x) = \dfrac{x - 2}{x + 4}$

50. $g(x) = \dfrac{x + 1}{x - 2}$

51. $g(x) = \sqrt{7 - x}$

52. $F(x) = \dfrac{1}{\sqrt{5 + x}}$

In Problems 53–60, does the equation specify a function with independent variable x? If so, find the domain of the function. If not, find a value of x to which there corresponds more than one value of y.

53. $2x + 5y = 10$

54. $6x - 7y = 21$

55. $y(x + y) = 4$

56. $x(x + y) = 4$

57. $x^{-3} + y^3 = 27$

58. $x^2 + y^2 = 9$

59. $x^3 - y^2 = 0$

60. $\sqrt{x} - y^3 = 0$

In Problems 61–72, find and simplify the expression if $f(x) = x^2 - 4$.

61. $f(4)$

62. $f(-5)$

63. $f(x + 1)$

64. $f(x - 2)$

65. $f(-6x)$

66. $f(10x)$

67. $f(x^3)$

68. $f(\sqrt{x})$

69. $f(2) + f(h)$

70. $f(-3) + f(h)$

71. $f(2 + h)$

72. $f(-3 + h)$

73. $f(2 + h) - f(2)$

74. $f(-3 + h) - f(-3)$

In Problems 75–80, find and simplify each of the following, assuming $h \neq 0$ in (C).

(A) $f(x + h)$

(B) $f(x + h) - f(x)$

(C) $\dfrac{f(x + h) - f(x)}{h}$

75. $f(x) = 4x - 3$

76. $f(x) = -3x + 9$

77. $f(x) = 4x^2 - 7x + 6$

78. $f(x) = 3x^2 + 5x - 8$

79. $f(x) = x(20 - x)$

80. $f(x) = x(x + 40)$

Problems 81–84 refer to the area A and perimeter P of a rectangle with length l and width w (see the figure).

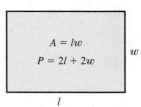

81. The area of a rectangle is 25 sq. in. Express the perimeter $P(w)$ as a function of the width w, and state the domain of this function.

82. The area of a rectangle is 81 sq. in. Express the perimeter $P(l)$ as a function of the length l, and state the domain of this function.

83. The perimeter of a rectangle is 100 m. Express the area $A(l)$ as a function of the length l, and state the domain of this function.

84. The perimeter of a rectangle is 160 m. Express the area $A(w)$ as a function of the width w, and state the domain of this function.

Applications

85. Price–demand. A company manufactures memory chips for microcomputers. Its marketing research department, using statistical techniques, collected the data shown in Table 8, where p is the wholesale price per chip at which x million chips can be sold. Using special analytical techniques (regression analysis), an analyst produced the following price–demand function to model the data:

$$p(x) = 75 - 3x \quad 1 \le x \le 20$$

Table 8 **Price–Demand**

x (millions)	p ($)
1	72
4	63
9	48
14	33
20	15

(A) Plot the data points in Table 8, and sketch a graph of the price–demand function in the same coordinate system.

(B) What would be the estimated price per chip for a demand of 7 million chips? For a demand of 11 million chips?

86. Price–demand. A company manufactures notebook computers. Its marketing research department, using statistical techniques, collected the data shown in Table 9, where p is the wholesale price per computer at which x thousand computers can be sold. Using special analytical techniques (regression analysis), an analyst produced the following price–demand function to model the data:

$$p(x) = 2,000 - 60x \quad 1 \le x \le 25$$

Table 9 **Price–Demand**

x (thousands)	p($)
1	1,940
8	1,520
16	1,040
21	740
25	500

(A) Plot the data points in Table 9, and sketch a graph of the price–demand function in the same coordinate system.

(B) What would be the estimated price per computer for a demand of 11,000 computers? For a demand of 18,000 computers?

87. Revenue.
(A) Using the price–demand function

$$p(x) = 75 - 3x \quad 1 \le x \le 20$$

from Problem 85, write the company's revenue function and indicate its domain.

(B) Complete Table 10, computing revenues to the nearest million dollars.

Table 10 **Revenue**

x (millions)	R(x) (million $)
1	72
4	
8	
12	
16	
20	

(C) Plot the points from part (B) and sketch a graph of the revenue function using these points. Choose millions for the units on the horizontal and vertical axes.

88. Revenue.
(A) Using the price–demand function

$$p(x) = 2,000 - 60x \quad 1 \le x \le 25$$

from Problem 86, write the company's revenue function and indicate its domain.

(B) Complete Table 11, computing revenues to the nearest thousand dollars.

Table 11 **Revenue**

x (thousands)	R(x) (thousand $)
1	1,940
5	
10	
15	
20	
25	

(C) Plot the points from part (B) and sketch a graph of the revenue function using these points. Choose thousands for the units on the horizontal and vertical axes.

89. Profit. The financial department for the company in Problems 85 and 87 established the following cost function for producing and selling x million memory chips:

$$C(x) = 125 + 16x \text{ million dollars}$$

(A) Write a profit function for producing and selling x million memory chips and indicate its domain.

(B) Complete Table 12, computing profits to the nearest million dollars.

Table 12 **Profit**

x (millions)	P(x) (million $)
1	−69
4	
8	
12	
16	
20	

(C) Plot the points in part (B) and sketch a graph of the profit function using these points.

90. Profit. The financial department for the company in Problems 86 and 88 established the following cost function for producing and selling x thousand notebook computers:

$$C(x) = 4,000 + 500x \text{ thousand dollars}$$

(A) Write a profit function for producing and selling x thousand notebook computers and indicate its domain.

(B) Complete Table 13, computing profits to the nearest thousand dollars.

Table 13   **Profit**

x (thousands)	$P(x)$ (thousand $)
1	−2,560
5	
10	
15	
20	
25	

(C) Plot the points in part (B) and sketch a graph of the profit function using these points.

91. Packaging. A candy box will be made out of a piece of cardboard that measures 8 by 12 in. Equal-sized squares x inches on a side will be cut out of each corner, and then the ends and sides will be folded up to form a rectangular box.

(A) Express the volume of the box $V(x)$ in terms of x.

(B) What is the domain of the function V (determined by the physical restrictions)?

(C) Complete Table 14.

Table 14   **Volume**

x	$V(x)$
1	
2	
3	

(D) Plot the points in part (C) and sketch a graph of the volume function using these points.

92. Packaging. Refer to Problem 91.

(A) Table 15 shows the volume of the box for some values of x between 1 and 2. Use these values to estimate to one decimal place the value of x between 1 and 2 that would produce a box with a volume of 65 cu. in.

Table 15   **Volume**

x	$V(x)$
1.1	62.524
1.2	64.512
1.3	65.988
1.4	66.976
1.5	67.5
1.6	67.584
1.7	67.252

(B) Describe how you could refine this table to estimate x to two decimal places.

(C) Carry out the refinement you described in part (B) and approximate x to two decimal places.

93. Packaging. Refer to Problems 91 and 92.

(A) Examine the graph of $V(x)$ from Problem 91D and discuss the possible locations of other values of x that would produce a box with a volume of 65 cu. in.

(B) Construct a table like Table 15 to estimate any such value to one decimal place.

(C) Refine the table you constructed in part (B) to provide an approximation to two decimal places.

94. Packaging. A parcel delivery service will only deliver packages with length plus girth (distance around) not exceeding 108 in. A rectangular shipping box with square ends x inches on a side is to be used.

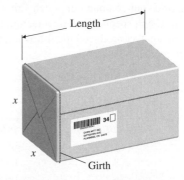

(A) If the full 108 in. is to be used, express the volume of the box $V(x)$ in terms of x.

(B) What is the domain of the function V (determined by the physical restrictions)?

(C) Complete Table 16.

Table 16   **Volume**

x	$V(x)$
5	
10	
15	
20	
25	

(D) Plot the points in part (C) and sketch a graph of the volume function using these points.

95. Muscle contraction. In a study of the speed of muscle contraction in frogs under various loads, British biophysicist A.W. Hill determined that the weight w (in grams) placed on the muscle and the speed of contraction v (in centimeters per second) are approximately related by an equation of the form

$$(w + a)(v + b) = c$$

where a, b, and c are constants. Suppose that for a certain muscle, $a = 15, b = 1$, and $c = 90$. Express v as a function of w. Find the speed of contraction if a weight of 16 g is placed on the muscle.

96. Politics. The percentage s of seats in the House of Representatives won by Democrats and the percentage v of votes cast for Democrats (when expressed as decimal fractions) are related by the equation

$$5v - 2s = 1.4 \quad 0 < s < 1, \quad 0.28 < v < 0.68$$

(A) Express v as a function of s and find the percentage of votes required for the Democrats to win 51% of the seats.

(B) Express s as a function of v and find the percentage of seats won if Democrats receive 51% of the votes.

Answers to Matched Problems

1. (A)

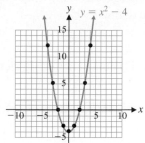

$y = x^2 - 4$

(B)

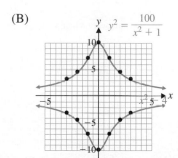

$y^2 = \dfrac{100}{x^2 + 1}$

2. (A) Does not specify a function   **(B)** Specifies a function

3. $x \geq 2$ (inequality notation) or $[2, \infty)$ (interval notation)

4. (A) -3 **(B)** 0   **(C)** Does not exist   **(D)** 6

5. Domain of F: R; domain of G: all real numbers except -3; domain of H: $x \leq 2$ (inequality notation) or $(-\infty, 2]$ (interval notation)

6. (A) $a^2 - 4a + 9$ **(B)** $a^2 + 2ah + h^2 - 4a - 4h + 9$

(C) $2ah + h^2 - 4h$ (D) $2a + h - 4$

7. (A)

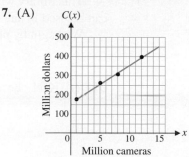

$C(x)$

Million dollars

Million cameras

(B) $P(x) = R(x) - C(x) = x(94.8 - 5x) - (156 + 19.7x)$; domain: $1 \leq x \leq 15$

(C) Table 7 **Profit**

x (millions)	$P(x)$ (million \$)
1	-86
3	24
6	115
9	115
12	25
15	-155

(D)

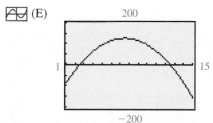

$P(x)$

Million dollars

Million cameras

(E)

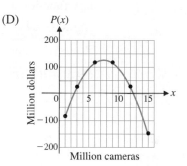

2.2 Elementary Functions: Graphs and Transformations

- A Beginning Library of Elementary Functions
- Vertical and Horizontal Shifts
- Reflections, Stretches, and Shrinks
- Piecewise-Defined Functions

Each of the functions

$$g(x) = x^2 - 4 \quad h(x) = (x - 4)^2 \quad k(x) = -4x^2$$

can be expressed in terms of the function $f(x) = x^2$:

$$g(x) = f(x) - 4 \quad h(x) = f(x - 4) \quad k(x) = -4f(x)^2$$

In this section, we will see that the graphs of functions g, h, and k are closely related to the graph of function f. Insight gained by understanding these relationships will help us analyze and interpret the graphs of many different functions.

A Beginning Library of Elementary Functions

As you progress through this book, you will repeatedly encounter a small number of elementary functions. We will identify these functions, study their basic properties, and include them in a library of elementary functions (see the inside back cover).

This library will become an important addition to your mathematical toolbox and can be used in any course or activity where mathematics is applied.

We begin by placing six basic functions in our library.

DEFINITION   Basic Elementary Functions

$$f(x) = x \qquad\qquad \text{Identity function}$$
$$h(x) = x^2 \qquad\qquad \text{Square function}$$
$$m(x) = x^3 \qquad\qquad \text{Cube function}$$
$$n(x) = \sqrt{x} \qquad\qquad \text{Square root function}$$
$$p(x) = \sqrt[3]{x} \qquad\qquad \text{Cube root function}$$
$$g(x) = |x| \qquad\qquad \text{Absolute value function}$$

These elementary functions can be evaluated by hand for certain values of x and with a calculator for all values of x for which they are defined.

EXAMPLE 1   Evaluating Basic Elementary Functions   Evaluate each basic elementary function at

(A) $x = 64$ 　　　　　　　　　　　　　　 (B) $x = -12.75$

Round any approximate values to four decimal places.

SOLUTION　(A)　$f(64) = 64$

$\qquad\qquad h(64) = 64^2 = 4{,}096$ 　　　　　　　Use a calculator.

$\qquad\qquad m(64) = 64^3 = 262{,}144$ 　　　　　Use a calculator.

$\qquad\qquad n(64) = \sqrt{64} = 8$

$\qquad\qquad p(64) = \sqrt[3]{64} = 4$

$\qquad\qquad g(64) = |64| = 64$

　　(B)　$f(-12.75) = -12.75$

$\qquad\qquad h(-12.75) = (-12.75)^2 = 162.5625$ 　　Use a calculator.

$\qquad\qquad m(-12.75) = (-12.75)^3 \approx -2{,}072.6719$　Use a calculator.

$\qquad\qquad n(-12.75) = \sqrt{-12.75}$ 　　　　　　　Not a real number.

$\qquad\qquad p(-12.75) = \sqrt[3]{-12.75} \approx -2.3362$ 　　Use a calculator.

$\qquad\qquad g(-12.75) = |-12.75| = 12.75$

Matched Problem 1   Evaluate each basic elementary function at

(A) $x = 729$ 　　　　　　　　　　　　　　 (B) $x = -5.25$

Round any approximate values to four decimal places.

Remark—Most computers and graphing calculators use ABS(x) to represent the absolute value function. The following representation can also be useful:

$$|x| = \sqrt{x^2}$$

Figure 1 shows the graph, range, and domain of each of the basic elementary functions.

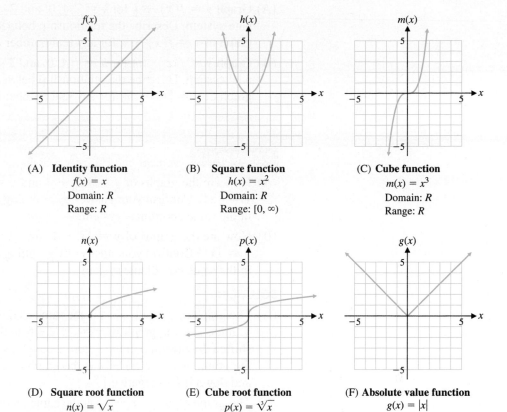

(A) **Identity function**
$f(x) = x$
Domain: R
Range: R

(B) **Square function**
$h(x) = x^2$
Domain: R
Range: $[0, \infty)$

(C) **Cube function**
$m(x) = x^3$
Domain: R
Range: R

(D) **Square root function**
$n(x) = \sqrt{x}$
Domain: $[0, \infty)$
Range: $[0, \infty)$

(E) **Cube root function**
$p(x) = \sqrt[3]{x}$
Domain: R
Range: R

(F) **Absolute value function**
$g(x) = |x|$
Domain: R
Range: $[0, \infty)$

Figure 1 **Some basic functions and their graphs***

CONCEPTUAL INSIGHT

Absolute Value In beginning algebra, absolute value is often interpreted as distance from the origin on a real number line (see Appendix A, Section A.1).

distance $= 6 = -(-6)$ distance $= 5$

$-10 \quad -5 \quad 0 \quad 5 \quad 10$

If $x < 0$, then $-x$ is the *positive* distance from the origin to x, and if $x > 0$, then x is the positive distance from the origin to x. Thus,

$$|x| = \begin{cases} -x & \text{if } x < 0 \\ x & \text{if } x \geq 0 \end{cases}$$

Vertical and Horizontal Shifts

If a new function is formed by performing an operation on a given function, then the graph of the new function is called a **transformation** of the graph of the original function. For example, graphs of $y = f(x) + k$ and $y = f(x + h)$ are transformations of the graph of $y = f(x)$.

Note: Letters used to designate these functions may vary from context to context; R is the set of all real numbers.

Explore and Discuss 1   Let $f(x) = x^2$.

   (A) Graph $y = f(x) + k$ for $k = -4, 0$, and 2 simultaneously in the same coordinate system. Describe the relationship between the graph of $y = f(x)$ and the graph of $y = f(x) + k$ for any real number k.

   (B) Graph $y = f(x + h)$ for $h = -4, 0$, and 2 simultaneously in the same coordinate system. Describe the relationship between the graph of $y = f(x)$ and the graph of $y = f(x + h)$ for any real number h.

EXAMPLE 2   Vertical and Horizontal Shifts

   (A) How are the graphs of $y = |x| + 4$ and $y = |x| - 5$ related to the graph of $y = |x|$? Confirm your answer by graphing all three functions simultaneously in the same coordinate system.

   (B) How are the graphs of $y = |x + 4|$ and $y = |x - 5|$ related to the graph of $y = |x|$? Confirm your answer by graphing all three functions simultaneously in the same coordinate system.

SOLUTION

   (A) The graph of $y = |x| + 4$ is the same as the graph of $y = |x|$ shifted upward 4 units, and the graph of $y = |x| - 5$ is the same as the graph of $y = |x|$ shifted downward 5 units. Figure 2 confirms these conclusions. [It appears that the graph of $y = f(x) + k$ is the graph of $y = f(x)$ shifted up if k is positive and down if k is negative.]

   (B) The graph of $y = |x + 4|$ is the same as the graph of $y = |x|$ shifted to the left 4 units, and the graph of $y = |x - 5|$ is the same as the graph of $y = |x|$ shifted to the right 5 units. Figure 3 confirms these conclusions. [It appears that the graph of $y = f(x + h)$ is the graph $y = f(x)$ shifted right if h is negative and left if h is positive–the opposite of what you might expect.]

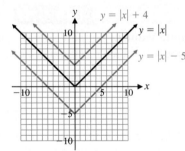

Figure 2 **Vertical shifts**

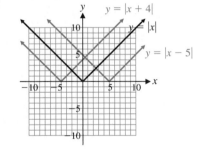

Figure 3 **Horizontal shifts**

Matched Problem 2

   (A) How are the graphs of $y = \sqrt{x} + 5$ and $y = \sqrt{x} - 4$ related to the graph of $y = \sqrt{x}$? Confirm your answer by graphing all three functions simultaneously in the same coordinate system.

   (B) How are the graphs of $y = \sqrt{x + 5}$ and $y = \sqrt{x - 4}$ related to the graph of $y = \sqrt{x}$? Confirm your answer by graphing all three functions simultaneously in the same coordinate system.

    Comparing the graphs of $y = f(x) + k$ with the graph of $y = f(x)$, we see that the graph of $y = f(x) + k$ can be obtained from the graph of $y = f(x)$ by **vertically translating** (shifting) the graph of the latter upward k units if k is positive and downward $|k|$ units if k is negative. Comparing the graphs of $y = f(x + h)$ with the graph of $y = f(x)$, we see that the graph of $y = f(x + h)$ can be obtained from the graph of $y = f(x)$ by **horizontally translating** (shifting) the graph of the latter h units to the left if h is positive and $|h|$ units to the right if h is negative.

EXAMPLE 3   Vertical and Horizontal Translations (Shifts)   The graphs in Figure 4 are either horizontal or vertical shifts of the graph of $f(x) = x^2$. Write appropriate equations for functions H, G, M, and N in terms of f.

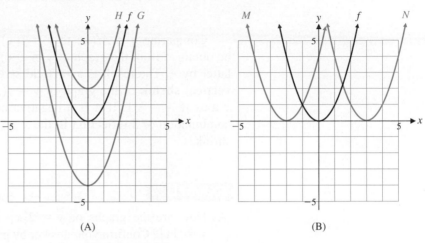

Figure 4 **Vertical and horizontal shifts**

SOLUTION   Functions H and G are vertical shifts given by

$$H(x) = x^2 + 2 \qquad G(x) = x^2 - 4$$

Functions M and N are horizontal shifts given by

$$M(x) = (x + 2)^2 \qquad N(x) = (x - 3)^2$$

Matched Problem 3   The graphs in Figure 5 are either horizontal or vertical shifts of the graph of $f(x) = \sqrt[3]{x}$. Write appropriate equations for functions H, G, M, and N in terms of f.

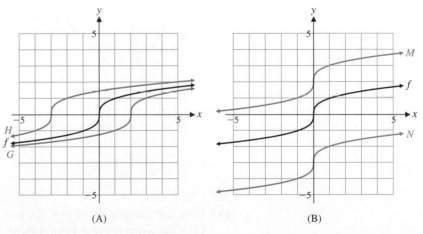

Figure 5 **Vertical and horizontal shifts**

Reflections, Stretches, and Shrinks

We now investigate how the graph of $y = Af(x)$ is related to the graph of $y = f(x)$ for different real numbers A.

Explore and Discuss 2

(A) Graph $y = Ax^2$ for $A = 1, 4$, and $\frac{1}{4}$ simultaneously in the same coordinate system.

(B) Graph $y = Ax^2$ for $A = -1, -4$, and $-\frac{1}{4}$ simultaneously in the same coordinate system.

(C) Describe the relationship between the graph of $h(x) = x^2$ and the graph of $G(x) = Ax^2$ for any real number A.

Comparing $y = Af(x)$ to $y = f(x)$, we see that the graph of $y = Af(x)$ can be obtained from the graph of $y = f(x)$ by multiplying each ordinate value of the latter by A. The result is a **vertical stretch** of the graph of $y = f(x)$ if $A > 1$, a **vertical shrink** of the graph of $y = f(x)$ if $0 < A < 1$, and a **reflection in the x axis** if $A = -1$. If A is a negative number other than -1, then the result is a combination of a reflection in the x axis and either a vertical stretch or a vertical shrink.

EXAMPLE 4   Reflections, Stretches, and Shrinks

(A) How are the graphs of $y = 2|x|$ and $y = 0.5|x|$ related to the graph of $y = |x|$? Confirm your answer by graphing all three functions simultaneously in the same coordinate system.

(B) How is the graph of $y = -2|x|$ related to the graph of $y = |x|$? Confirm your answer by graphing both functions simultaneously in the same coordinate system.

SOLUTION

(A) The graph of $y = 2|x|$ is a vertical stretch of the graph of $y = |x|$ by a factor of 2, and the graph of $y = 0.5|x|$ is a vertical shrink of the graph of $y = |x|$ by a factor of 0.5. Figure 6 confirms this conclusion.

(B) The graph of $y = -2|x|$ is a reflection in the x axis and a vertical stretch of the graph of $y = |x|$. Figure 7 confirms this conclusion.

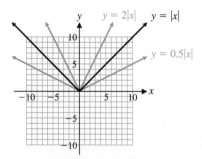

Figure 6 **Vertical stretch and shrink**

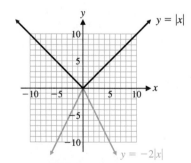

Figure 7 **Reflection and vertical stretch**

Matched Problem 4

(A) How are the graphs of $y = 2x$ and $y = 0.5x$ related to the graph of $y = x$? Confirm your answer by graphing all three functions simultaneously in the same coordinate system.

(B) How is the graph of $y = -0.5x$ related to the graph of $y = x$? Confirm your answer by graphing both functions in the same coordinate system.

The various transformations considered above are summarized in the following box for easy reference:

SUMMARY Graph Transformations

Vertical Translation:

$$y = f(x) + k \quad \begin{cases} k > 0 & \text{Shift graph of } y = f(x) \text{ up } k \text{ units.} \\ k < 0 & \text{Shift graph of } y = f(x) \text{ down } |k| \text{ units.} \end{cases}$$

Horizontal Translation:

$$y = f(x + h) \quad \begin{cases} h > 0 & \text{Shift graph of } y = f(x) \text{ left } h \text{ units.} \\ h < 0 & \text{Shift graph of } y = f(x) \text{ right } |h| \text{ units.} \end{cases}$$

Reflection:

$$y = -f(x) \quad \text{Reflect the graph of } y = f(x) \text{ in the } x \text{ axis.}$$

Vertical Stretch and Shrink:

$$y = Af(x) \quad \begin{cases} A > 1 & \text{Stretch graph of } y = f(x) \text{ vertically} \\ & \text{by multiplying each ordinate value by } A. \\ 0 < A < 1 & \text{Shrink graph of } y = f(x) \text{ vertically} \\ & \text{by multiplying each ordinate value by } A. \end{cases}$$

Explore and Discuss 3   Explain why applying any of the graph transformations in the summary box to a linear function produces another linear function.

EXAMPLE 5   Combining Graph Transformations   Discuss the relationship between the graphs of $y = -|x - 3| + 1$ and $y = |x|$. Confirm your answer by graphing both functions simultaneously in the same coordinate system.

SOLUTION   The graph of $y = -|x - 3| + 1$ is a reflection of the graph of $y = |x|$ in the x axis, followed by a horizontal translation of 3 units to the right, and a vertical translation of 1 unit upward. Figure 8 confirms this description.

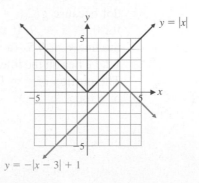

Figure 8 **Combined transformations**

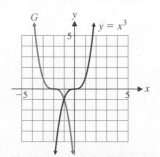

Figure 9 **Combined transformations**

Matched Problem 5   The graph of $y = G(x)$ in Figure 9 involves a reflection and a translation of the graph of $y = x^3$. Describe how the graph of function G is related to the graph of $y = x^3$ and find an equation of the function G.

Piecewise-Defined Functions

Earlier we noted that the absolute value of a real number x can be defined as

$$|x| = \begin{cases} -x & \text{if } x < 0 \\ x & \text{if } x \geq 0 \end{cases}$$

Notice that this function is defined by different rules for different parts of its domain. Functions whose definitions involve more than one rule are called **piecewise-defined functions**. Graphing one of these functions involves graphing each rule over the appropriate portion of the domain (Fig. 10). In Figure 10C, notice that an open dot is used to show that the point $(0, -2)$ is not part of the graph and a solid dot is used to show that $(0, 2)$ is part of the graph.

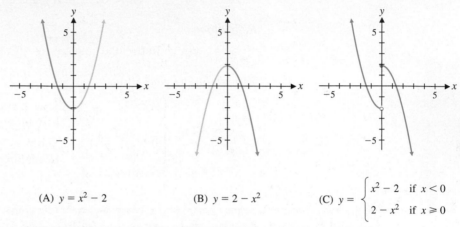

(A) $y = x^2 - 2$ (B) $y = 2 - x^2$ (C) $y = \begin{cases} x^2 - 2 & \text{if } x < 0 \\ 2 - x^2 & \text{if } x \geq 0 \end{cases}$

Figure 10   **Graphing a piecewise-defined function**

EXAMPLE 6   Graphing Piecewise-Defined Functions   Graph the piecewise-defined function

$$g(x) = \begin{cases} x + 1 & \text{if } 0 \leq x < 2 \\ 0.5x & \text{if } x \geq 2 \end{cases}$$

SOLUTION   If $0 \leq x < 2$, then the first rule applies and the graph of g lies on the line $y = x + 1$ (a vertical shift of the identity function $y = x$). If $x = 0$, then $(0, 1)$ lies on $y = x + 1$; we plot $(0, 1)$ with a solid dot (Fig. 11) because $g(0) = 1$. If $x = 2$, then $(2, 3)$ lies on $y = x + 1$; we plot $(2, 3)$ with an open dot because $g(2) \neq 3$. The line segment from $(0, 1)$ to $(2, 3)$ is the graph of g for $0 \leq x < 2$. If $x \geq 2$, then the second rule applies and the graph of g lies on the line $y = 0.5x$ (a vertical shrink of the identity function $y = x$). If $x = 2$, then $(2, 1)$ lies on the line $y = 0.5x$; we plot $(2, 1)$ with a solid dot because $g(2) = 1$. The portion of $y = 0.5x$ that starts at $(2, 1)$ and extends to the right is the graph of g for $x \geq 2$.

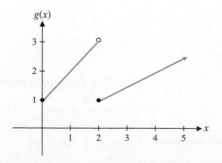

Figure 11

Matched Problem 6 Graph the piecewise-defined function

$$h(x) = \begin{cases} -2x + 4 & \text{if } 0 \le x \le 2 \\ x - 1 & \text{if } x > 2 \end{cases}$$

As the next example illustrates, piecewise-defined functions occur naturally in many applications.

EXAMPLE 7 Natural Gas Rates Easton Utilities uses the rates shown in Table 1 to compute the monthly cost of natural gas for each customer. Write a piecewise definition for the cost of consuming x CCF (cubic hundred feet) of natural gas and graph the function.

Table 1 **Charges per Month**

\$0.7866 per CCF for the first 5 CCF
\$0.4601 per CCF for the next 35 CCF
\$0.2508 per CCF for all over 40 CCF

SOLUTION If $C(x)$ is the cost, in dollars, of using x CCF of natural gas in one month, then the first line of Table 1 implies that

$$C(x) = 0.7866x \quad \text{if } 0 \le x \le 5$$

Note that $C(5) = 3.933$ is the cost of 5 CCF. If $5 < x \le 40$, then $x - 5$ represents the amount of gas that cost \$0.4601 per CCF, $0.4601(x - 5)$ represents the cost of this gas, and the total cost is

$$C(x) = 3.933 + 0.4601(x - 5)$$

If $x > 40$, then

$$C(x) = 20.0365 + 0.2508(x - 40)$$

where $20.0365 = C(40)$, the cost of the first 40 CCF. Combining all these equations, we have the following piecewise definition for $C(x)$:

$$C(x) = \begin{cases} 0.7866x & \text{if } 0 \le x \le 5 \\ 3.933 + 0.4601(x - 5) & \text{if } 5 < x \le 40 \\ 20.0365 + 0.2508(x - 40) & \text{if } x > 40 \end{cases}$$

To graph C, first note that each rule in the definition of C represents a transformation of the identity function $f(x) = x$. Graphing each transformation over the indicated interval produces the graph of C shown in Figure 12.

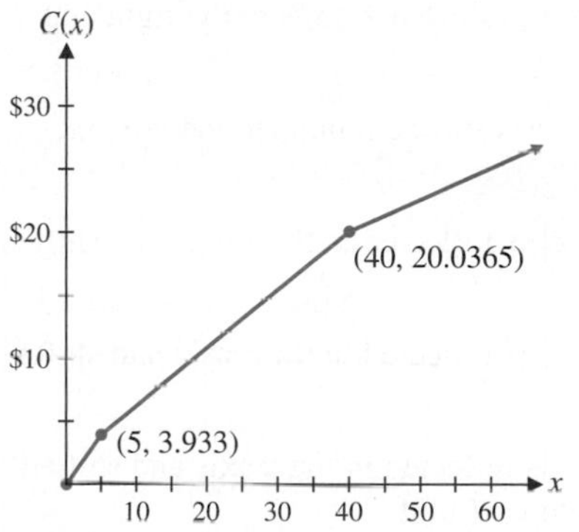

Figure 12 **Cost of purchasing x CCF of natural gas**

Matched Problem 7 Trussville Utilities uses the rates shown in Table 2 to compute the monthly cost of natural gas for residential customers. Write a piecewise definition for the cost of consuming x CCF of natural gas and graph the function.

Table 2 **Charges per Month**

\$0.7675 per CCF for the first 50 CCF
\$0.6400 per CCF for the next 150 CCF
\$0.6130 per CCF for all over 200 CCF

Exercise 2.2

In Problems 1–8, find the domain and range of the function.

1. $f(x) = 5x - 10$

2. $f(x) = -4x + 12$

3. $f(x) = 15 - \sqrt{x}$

4. $f(x) = 3 + \sqrt{x}$

5. $f(x) = 2|x| + 7$

6. $f(x) = -5|x| + 2$

7. $f(x) = \sqrt[3]{x} + 100$

8. $f(x) = 20 - 10\sqrt[3]{x}$

In Problems 9–24, graph each of the functions using the graphs of functions f and g below.

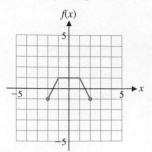

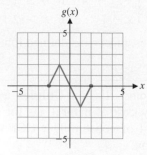

9. $y = f(x) + 2$

10. $y = g(x) - 1$

11. $y = f(x + 2)$

12. $y = g(x - 1)$

13. $y = g(x - 3)$

14. $y = f(x + 3)$

15. $y = g(x) - 3$

16. $y = f(x) + 3$

17. $y = -f(x)$

18. $y = -g(x)$

19. $y = 0.5g(x)$

20. $y = 2f(x)$

21. $y = 2f(x) + 1$

22. $y = -0.5g(x) + 3$

23. $y = 2(f(x) + 1)$

24. $y = -(0.5g(x) + 3)$

In Problems 25–32, indicate verbally how the graph of each function is related to the graph of one of the six basic functions in Figure 1 on page 59. Sketch a graph of each function.

25. $g(x) = -|x + 3|$

26. $h(x) = -|x - 5|$

27. $f(x) = (x - 4)^2 - 3$

28. $m(x) = (x + 3)^2 + 4$

29. $f(x) = 7 - \sqrt{x}$

30. $g(x) = -6 + \sqrt[3]{x}$

31. $h(x) = -3|x|$

32. $m(x) = -0.4x^2$

Each graph in Problems 33–40 is the result of applying a sequence of transformations to the graph of one of the six basic functions in Figure 1 on page 59. Identify the basic function and describe the transformation verbally. Write an equation for the given graph.

33.

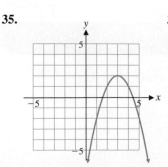

34.

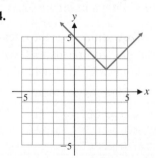

35.

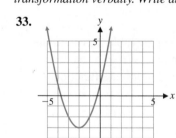

36.

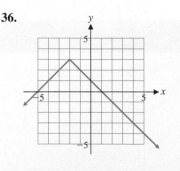

37.

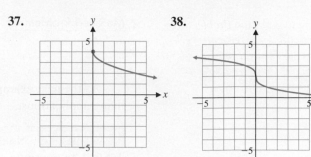

38.

39.

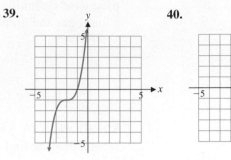

40.

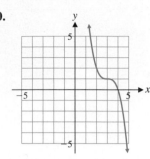

In Problems 41–46, the graph of the function g is formed by applying the indicated sequence of transformations to the given function f. Find an equation for the function g and graph g using $-5 \leq x \leq 5$ and $-5 \leq y \leq 5$.

41. The graph of $f(x) = \sqrt{x}$ is shifted 2 units to the right and 3 units down.

42. The graph of $f(x) = \sqrt[3]{x}$ is shifted 3 units to the left and 2 units up.

43. The graph of $f(x) = |x|$ is reflected in the x axis and shifted to the left 3 units.

44. The graph of $f(x) = |x|$ is reflected in the x axis and shifted to the right 1 unit.

45. The graph of $f(x) = x^3$ is reflected in the x axis and shifted 2 units to the right and down 1 unit.

46. The graph of $f(x) = x^2$ is reflected in the x axis and shifted to the left 2 units and up 4 units.

Graph each function in Problems 47–52.

47. $f(x) = \begin{cases} 2 - 2x & \text{if } x < 2 \\ x - 2 & \text{if } x \geq 2 \end{cases}$

48. $g(x) = \begin{cases} x + 1 & \text{if } x < -1 \\ 2 + 2x & \text{if } x \geq -1 \end{cases}$

49. $h(x) = \begin{cases} 5 + 0.5x & \text{if } 0 \leq x \leq 10 \\ -10 + 2x & \text{if } x > 10 \end{cases}$

50. $h(x) = \begin{cases} 10 + 2x & \text{if } 0 \leq x \leq 20 \\ 40 + 0.5x & \text{if } x > 20 \end{cases}$

51. $h(x) = \begin{cases} 2x & \text{if } 0 \leq x \leq 20 \\ x + 20 & \text{if } 20 < x \leq 40 \\ 0.5x + 40 & \text{if } x > 40 \end{cases}$

52. $h(x) = \begin{cases} 4x + 20 & \text{if } 0 \leq x \leq 20 \\ 2x + 60 & \text{if } 20 < x \leq 100 \\ -x + 360 & \text{if } x > 100 \end{cases}$

Each of the graphs in Problems 53–58 involves a reflection in the x axis and/or a vertical stretch or shrink of one of the basic functions in Figure 1 on page 59. Identify the basic function, and describe the transformation verbally. Write an equation for the given graph.

53.
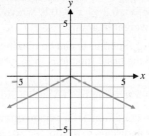

54.
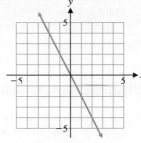

55.
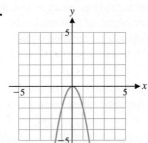

56.
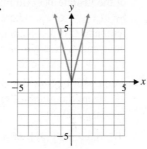

57.
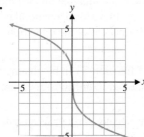

58.
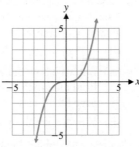

Changing the order in a sequence of transformations may change the final result. Investigate each pair of transformations in Problems 59–64 to determine if reversing their order can produce a different result. Support your conclusions with specific examples and/or mathematical arguments.

59. Vertical shift; horizontal shift

60. Vertical shift; reflection in *y* axis

61. Vertical shift; reflection in *x* axis

62. Vertical shift; vertical stretch

63. Horizontal shift; reflection in *y* axis

64. Horizontal shift; vertical shrink

Applications

65. Price-demand. A retail chain sells DVD players. The retail price $p(x)$ (in dollars) and the weekly demand x for a particular model are related by

$$p(x) = 115 - 4\sqrt{x} \qquad 9 \le x \le 289$$

(A) Describe how the graph of function p can be obtained from the graph of one of the basic functions in Figure 1 on page 59.

(B) Sketch a graph of function p using part (A) as an aid.

66. Price-supply. The manufacturers of the DVD players in Problem 65 are willing to supply x players at a price of $p(x)$ as given by the equation

$$p(x) = 4\sqrt{x} \qquad 9 \le x \le 289$$

(A) Describe how the graph of function p can be obtained from the graph of one of the basic functions in Figure 1 on page 59.

(B) Sketch a graph of function p using part (A) as an aid.

67. Hospital costs. Using statistical methods, the financial department of a hospital arrived at the cost equation

$$C(x) = 0.00048(x - 500)^3 + 60,000 \qquad 100 \le x \le 1,000$$

where $C(x)$ is the cost in dollars for handling x cases per month.

(A) Describe how the graph of function C can be obtained from the graph of one of the basic functions in Figure 1 on page 59.

(B) Sketch a graph of function C using part (A) and a graphing calculator as aids.

68. Price-demand. A company manufactures and sells in-line skates. Its financial department has established the price-demand function

$$p(x) = 190 - 0.013(x - 10)^2 \qquad 10 \le x \le 100$$

where $p(x)$ is the price at which x thousand pairs of in-line skates can be sold.

(A) Describe how the graph of function p can be obtained from the graph of one of the basic functions in Figure 1 on page 59.

(B) Sketch a graph of function p using part (A) and a graphing calculator as aids.

69. Electricity rates. Table 3 shows the electricity rates charged by Monroe Utilities in the summer months. The base is a fixed monthly charge, independent of the kWh (kilowatt-hours) used during the month.

(A) Write a piecewise definition of the monthly charge $S(x)$ for a customer who uses x kWh in a summer month.

(B) Graph $S(x)$.

Table 3 **Summer (July–October)**

Base charge, $8.50
First 700 kWh or less at 0.0650/kWh
Over 700 kWh at 0.0900/kWh

70. Electricity rates. Table 4 shows the electricity rates charged by Monroe Utilities in the winter months.

(A) Write a piecewise definition of the monthly charge $W(x)$ for a customer who uses x kWh in a winter month.

Table 4 **Winter (November–June)**

Base charge, $8.50
First 700 kWh or less at 0.0650/kWh
Over 700 kWh at 0.0530/kWh

(B) Graph $W(x)$.

71. State income tax. Table 5 shows a recent state income tax schedule for married couples filing a joint return in Kansas.

(A) Write a piecewise definition for the tax due $T(x)$ on an income of x dollars.

(B) Graph $T(x)$.

(C) Find the tax due on a taxable income of $40,000. Of $70,000.

Table 5 **Kansas State Income Tax**

SCHEDULE I—MARRIED FILING JOINT		
If taxable income is		
Over	But Not Over	Tax Due Is
$0	$30,000	3.50% of taxable income
$30,000	$60,000	$1,050 plus 6.25% of excess over $30,000
$60,000		$2,925 plus 6.45% of excess over $60,000

72. State income tax. Table 6 shows a recent state income tax schedule for individuals filing a return in Kansas.

Table 6 **Kansas State Income Tax**

SCHEDULE II—SINGLE, HEAD OF HOUSEHOLD, OR MARRIED FILING SEPARATE		
If taxable income is		
Over	But Not Over	Tax Due Is
$0	$15,000	3.50% of taxable income
$15,000	$30,000	$525 plus 6.25% of excess over $15,000
$30,000		$1,462.50 plus 6.45% of excess over $30,000

(A) Write a piecewise definition for the tax due $T(x)$ on an income of x dollars.

(B) Graph $T(x)$.

(C) Find the tax due on a taxable income of $20,000. Of $35,000.

(D) Would it be better for a married couple in Kansas with two equal incomes to file jointly or separately? Discuss.

73. Human weight. A good approximation of the normal weight of a person 60 inches or taller but not taller than 80 inches is given by $w(x) = 5.5x - 220$, where x is height in inches and $w(x)$ is weight in pounds.

(A) Describe how the graph of function w can be obtained from the graph of one of the basic functions in Figure 1, page 59.

(B) Sketch a graph of function w using part (A) as an aid.

74. Herpetology. The average weight of a particular species of snake is given by $w(x) = 463x^3$, $0.2 \leq x \leq 0.8$, where x is length in meters and $w(x)$ is weight in grams.

(A) Describe how the graph of function w can be obtained from the graph of one of the basic functions in Figure 1, page 59.

(B) Sketch a graph of function w using part (A) as an aid.

75. Safety research. Under ideal conditions, if a person driving a vehicle slams on the brakes and skids to a stop, the speed of the vehicle $v(x)$ (in miles per hour) is given approximately by $v(x) = C\sqrt{x}$, where x is the length of skid marks (in feet) and C is a constant that depends on the road conditions and the weight of the vehicle. For a particular vehicle, $v(x) = 7.08\sqrt{x}$ and $4 \leq x \leq 144$.

(A) Describe how the graph of function v can be obtained from the graph of one of the basic functions in Figure 1, page 59.

(B) Sketch a graph of function v using part (A) as an aid.

76. Learning. A production analyst has found that on average it takes a new person $T(x)$ minutes to perform a particular assembly operation after x performances of the operation, where $T(x) = 10 - \sqrt[3]{x}$, $0 \leq x \leq 125$.

(A) Describe how the graph of function T can be obtained from the graph of one of the basic functions in Figure 1, page 59.

(B) Sketch a graph of function T using part (A) as an aid.

Answers to Matched Problems

1. (A) $f(729) = 729$, $h(729) = 531,441$,
$m(729) = 387,420,489$, $n(729) = 27$, $p(729) = 9$,
$g(729) = 729$

(B) $f(-5.25) = -5.25$, $h(-5.25) = 27.5625$,
$m(-5.25) = -144.7031$, $n(-5.25)$ is not a real number,
$p(-5.25) = -1.7380$, $g(-5.25) = 5.25$

2. (A) The graph of $y = \sqrt{x} + 5$ is the same as the graph of $y = \sqrt{x}$ shifted upward 5 units, and the graph of $y = \sqrt{x} - 4$ is the same as the graph of $y = \sqrt{x}$ shifted downward 4 units. The figure confirms these conclusions.

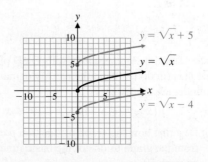

(B) The graph of $y = \sqrt{x + 5}$ is the same as the graph of $y = \sqrt{x}$ shifted to the left 5 units, and the graph of $y = \sqrt{x - 4}$ is the same as the graph of $y = \sqrt{x}$ shifted to the right 4 units. The figure confirms these conclusions.

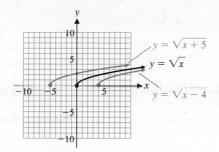

3. $H(x) = \sqrt[3]{x} + 3, G(x) = \sqrt[3]{x} - 2, M(x) = \sqrt[3]{x} + 2,$
$N(x) = \sqrt[3]{x} - 3$

4. (A) The graph of $y = 2x$ is a vertical stretch of the graph of $y = x$, and the graph of $y = 0.5x$ is a vertical shrink of the graph of $y = x$. The figure confirms these conclusions.

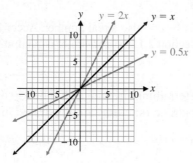

(B) The graph of $y = -0.5x$ is a vertical shrink and a reflection in the x axis of the graph of $y = x$. The figure confirms this conclusion.

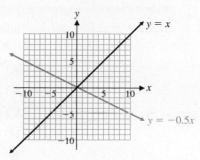

5. The graph of function G is a reflection in the x axis and a horizontal translation of 2 units to the left of the graph of $y = x^3$. An equation for G is $G(x) = -(x + 2)^3$.

6.

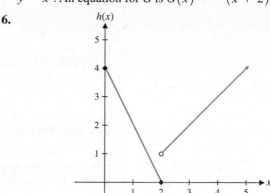

7. $C(x) = \begin{cases} 0.7675x & \text{if } 0 \le x \le 50 \\ 38.375 + 0.64(x - 50) & \text{if } 50 < x \le 200 \\ 134.375 + 0.613(x - 200) & \text{if } 200 < x \end{cases}$

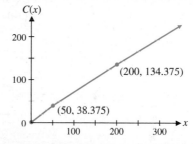

2.3 Quadratic Functions

- Quadratic Functions, Equations, and Inequalities
- Properties of Quadratic Functions and Their Graphs
- Applications

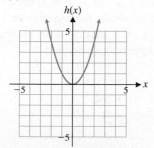

Figure 1 **Square function $h(x) = x^2$**

If the degree of a linear function is increased by one, we obtain a *second-degree function,* usually called a *quadratic function,* another basic function that we will need in our library of elementary functions. We will investigate relationships between quadratic functions and the solutions to quadratic equations and inequalities. Other important properties of quadratic functions will also be investigated, including maxima and minima. We will then be in a position to solve important practical problems such as finding production levels that will generate maximum revenue or maximum profit.

Quadratic Functions, Equations, and Inequalities

The graph of the square function $h(x) = x^2$ is shown in Figure 1. Notice that the graph is symmetric with respect to the y axis and that $(0, 0)$ is the lowest point on the graph. Let's explore the effect of applying a sequence of basic transformations to the graph of h.

Explore and Discuss 1

Indicate how the graph of each function is related to the graph of the function $h(x) = x^2$. Find the highest or lowest point, whichever exists, on each graph.

(A) $f(x) = (x - 3)^2 - 7 = x^2 - 6x + 2$
(B) $g(x) = 0.5(x + 2)^2 + 3 = 0.5x^2 + 2x + 5$
(C) $m(x) = -(x - 4)^2 + 8 = -x^2 + 8x - 8$
(D) $n(x) = -3(x + 1)^2 - 1 = -3x^2 - 6x - 4$

Graphing the functions in Explore and Discuss 1 produces figures similar in shape to the graph of the square function in Figure 1. These figures are called *parabolas*.* The functions that produce these parabolas are examples of the important class of *quadratic functions*.

DEFINITION   Quadratic Functions

If a, b, and c are real numbers with $a \neq 0$, then the function

$$f(x) = ax^2 + bx + c \quad \text{Standard form}$$

is a **quadratic function** and its graph is a **parabola**.

CONCEPTUAL INSIGHT

If x is any real number, then $ax^2 + bx + c$ is also a real number. According to the agreement on domain and range in Section 2.1, the domain of a quadratic function is R, the set of real numbers.

We will discuss methods for determining the range of a quadratic function later in this section. Typical graphs of quadratic functions are illustrated in Figure 2.

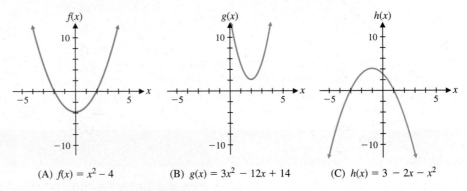

(A) $f(x) = x^2 - 4$ (B) $g(x) = 3x^2 - 12x + 14$ (C) $h(x) = 3 - 2x - x^2$

Figure 2 **Graphs of quadratic functions**

CONCEPTUAL INSIGHT

An x intercept of a function is also called a **zero** of the function. The x intercept of a linear function can be found by solving the linear equation $y = mx + b = 0$ for x, $m \neq 0$ (see Section 1.2). Similarly, the x intercepts of a quadratic function can be found by solving the quadratic equation $y = ax^2 + bx + c = 0$ for x, $a \neq 0$. Several methods for solving quadratic equations are discussed in Appendix A, Section A.7. The most popular of these is the **quadratic formula**.

If $ax^2 + bx + c = 0$, $a \neq 0$, then

$$x = \frac{-b \pm \sqrt{b^2 - 4ac}}{2a}, \text{ provided } b^2 - 4ac \geq 0$$

---

*The arc of a basketball shot is a parabola. Reflecting telescopes, solar furnaces, and automobile headlights are some of the many applications of parabolas.

EXAMPLE 1   Intercepts, Equations, and Inequalities

(A) Sketch a graph of $f(x) = -x^2 + 5x + 3$ in a rectangular coordinate system.

(B) Find x and y intercepts algebraically to four decimal places.

(C) Graph $f(x) = -x^2 + 5x + 3$ in a standard viewing window.

(D) Find the x and y intercepts to four decimal places using TRACE and ZERO on your graphing calculator.

(E) Solve the quadratic inequality $-x^2 + 5x + 3 \geq 0$ graphically to four decimal places using the results of parts (A) and (B) or (C) and (D).

(F) Solve the equation $-x^2 + 5x + 3 = 4$ graphically to four decimal places using INTERSECT on your graphing calculator.

SOLUTION

(A) Hand-sketching a graph of f:

x	y
-1	-3
0	3
1	7
2	9
3	9
4	7
5	3
6	-3

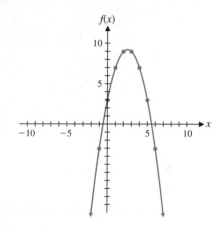

(B) Finding intercepts algebraically:

$$y \text{ intercept: } f(0) = -(0)^2 + 5(0) + 3 = 3$$

$$x \text{ intercepts: } f(x) = 0$$

$$-x^2 + 5x + 3 = 0 \quad \text{Quadratic equation}$$

$$x = \frac{-b \pm \sqrt{b^2 - 4ac}}{2a} \quad \text{Quadratic formula (see Appendix A.7)}$$

$$x = \frac{-(5) \pm \sqrt{5^2 - 4(-1)(3)}}{2(-1)}$$

$$= \frac{-5 \pm \sqrt{37}}{-2} = -0.5414 \quad \text{or} \quad 5.5414$$

(C) Graph in a graphing calculator:

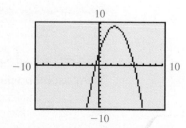

(D) Finding intercepts graphically using a graphing calculator:

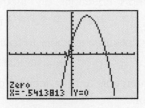 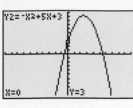

| x intercept: −0.5414 | x intercept: 5.5414 | y intercept: 3 |

(E) Solving $-x^2 + 5x + 3 \geq 0$ graphically: The quadratic inequality

$$-x^2 + 5x + 3 \geq 0$$

holds for those values of x for which the graph of $f(x) = -x^2 + 5x + 3$ in the figures in parts (A) and (C) is at or above the x axis. This happens for x between the two x intercepts [found in part (B) or (D)], including the two x intercepts. The solution set for the quadratic inequality is $-0.5414 \leq x \leq 5.5414$ or $[-0.5414, 5.5414]$.

(F) Solving the equation $-x^2 + 5x + 3 = 4$ using a graphing calculator:

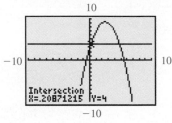 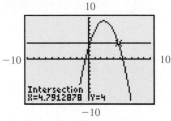

| $-x^2 + 5x + 3 = 4$ at $x = 0.2087$ | $-x^2 + 5x + 3 = 4$ at $x = 4.7913$ |

Matched Problem 1

(A) Sketch a graph of $g(x) = 2x^2 - 5x - 5$ in a rectangular coordinate system.

(B) Find x and y intercepts algebraically to four decimal places.

(C) Graph $g(x) = 2x^2 - 5x - 5$ in a standard viewing window.

(D) Find the x and y intercepts to four decimal places using TRACE and the ZERO command on your graphing calculator.

(E) Solve $2x^2 - 5x - 5 \geq 0$ graphically to four decimal places using the results of parts (A) and (B) or (C) and (D).

(F) Solve the equation $2x^2 - 5x - 5 = -3$ graphically to four decimal places using INTERSECT on your graphing calculator.

Explore and Discuss 2   How many x intercepts can the graph of a quadratic function have? How many y intercepts? Explain your reasoning.

Properties of Quadratic Functions and Their Graphs

Many useful properties of the quadratic function can be uncovered by transforming

$$f(x) = ax^2 + bx + c \quad a \neq 0$$

into the **vertex form**

$$f(x) = a(x - h)^2 + k$$

The process of *completing the square* (see Appendix A.7) is central to the transformation. We illustrate the process through a specific example and then generalize the results.

Consider the quadratic function given by

$$f(x) = -2x^2 + 16x - 24 \qquad (1)$$

We use completing the square to transform this function into vertex form:

$$f(x) = -2x^2 + 16x - 24$$

Factor the coefficient of x^2 out of the first two terms.

$$= -2(x^2 - 8x) - 24$$

$$= -2(x^2 - 8x + \, ?\,) - 24$$

$$= -2(x^2 - 8x + \mathbf{16}) - 24 + \mathbf{32}$$

Add 16 to complete the square inside the parentheses. Because of the -2 outside the parentheses, we have actually added -32, so we must add 32 to the outside.

$$= -2(x - 4)^2 + 8$$

The transformation is complete and can be checked by multiplying out.

Therefore,

$$f(x) = -2(x - 4)^2 + 8 \qquad (2)$$

If $x = 4$, then $-2(x - 4)^2 = 0$ and $f(4) = 8$. For any other value of x, the negative number $-2(x - 4)^2$ is added to 8, making it smaller. Therefore,

$$f(4) = 8$$

is the *maximum value* of $f(x)$ for all x. Furthermore, if we choose any two x values that are the same distance from 4, we will obtain the same function value. For example, $x = 3$ and $x = 5$ are each one unit from $x = 4$ and their function values are

$$f(3) = -2(3 - 4)^2 + 8 = 6$$
$$f(5) = -2(5 - 4)^2 + 8 = 6$$

Therefore, the vertical line $x = 4$ is a line of symmetry. That is, if the graph of equation (1) is drawn on a piece of paper and the paper is folded along the line $x = 4$, then the two sides of the parabola will match exactly. All these results are illustrated by graphing equations (1) and (2) and the line $x = 4$ simultaneously in the same coordinate system (Fig. 3).

From the preceding discussion, we see that as x moves from left to right, $f(x)$ is increasing on $(-\infty, 4]$, and decreasing on $[4, \infty)$, and that $f(x)$ can assume no value greater than 8. Thus,

$$\text{Range of } f: \quad y \le 8 \quad \text{or} \quad (-\infty, 8]$$

In general, the graph of a quadratic function is a parabola with line of symmetry parallel to the vertical axis. The lowest or highest point on the parabola, whichever exists, is called the **vertex**. The maximum or minimum value of a quadratic function always occurs at the vertex of the parabola. The line of symmetry through the vertex is called the **axis** of the parabola. In the example above, $x = 4$ is the axis of the parabola and $(4, 8)$ is its vertex.

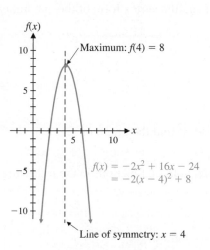

Figure 3 **Graph of a quadratic function**

In the figure: $f(x)$; Maximum: $f(4) = 8$; $f(x) = -2x^2 + 16x - 24 = -2(x - 4)^2 + 8$; Line of symmetry: $x = 4$

CONCEPTUAL INSIGHT

Applying the graph transformation properties discussed in Section 2.2 to the transformed equation,

$$f(x) = -2x^2 + 16x - 24$$
$$= -2(x - 4)^2 + 8$$

we see that the graph of $f(x) = -2x^2 + 16x - 24$ is the graph of $g(x) = x^2$ vertically stretched by a factor of 2, reflected in the x axis, and shifted to the right 4 units and up 8 units, as shown in Figure 4.

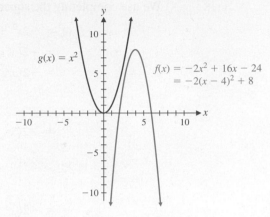

Figure 4   **Graph of *f* is the graph of *g* transformed**

Note the important results we have obtained from the vertex form of the quadratic function f:

- The vertex of the parabola
- The axis of the parabola
- The maximum value of $f(x)$
- The range of the function f
- The relationship between the graph of $g(x) = x^2$ and the graph of $f(x) = -2x^2 + 16x - 24$

The preceding discussion is generalized to all quadratic functions in the following summary:

SUMMARY   Properties of a Quadratic Function and Its Graph
Given a quadratic function and the vertex form obtained by completing the square

$$f(x) = ax^2 + bx + c \qquad a \neq 0 \qquad \text{Standard form}$$
$$= a(x - h)^2 + k \qquad\qquad \text{Vertex form}$$

we summarize general properties as follows:

1. The graph of f is a parabola:

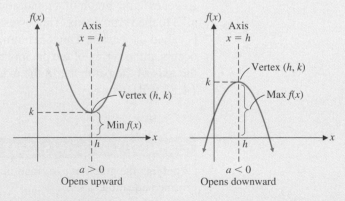

2. Vertex: (h, k) (parabola increases on one side of the vertex and decreases on the other)

3. Axis (of symmetry): $x = h$ (parallel to y axis)
4. $f(h) = k$ is the minimum if $a > 0$ and the maximum if $a < 0$
5. Domain: All real numbers.   Range: $(-\infty, k]$ if $a < 0$ or $[k, \infty)$ if $a > 0$
6. The graph of f is the graph of $g(x) = ax^2$ translated horizontally h units and vertically k units.

EXAMPLE 2   Analyzing a Quadratic Function   Given the quadratic function

$$f(x) = 0.5\,x^2 - 6x + 21$$

(A) Find the vertex form for f.

(B) Find the vertex and the maximum or minimum. State the range of f.

(C) Describe how the graph of function f can be obtained from the graph of $g(x) = x^2$ using transformations.

(D) Sketch a graph of function f in a rectangular coordinate system.

(E) Graph function f using a suitable viewing window.

(F) Find the vertex and the maximum or minimum using the appropriate graphing calculator command.

SOLUTION

(A) Complete the square to find the vertex form:

$$\begin{aligned}
f(x) &= 0.5\,x^2 - 6x + 21 \\
&= 0.5(x^2 - 12x + \,?\,) + 21 \\
&= 0.5(x^2 - 12x + 36) + 21 - 18 \\
&= 0.5(x - 6)^2 + 3
\end{aligned}$$

(B) From the vertex form, we see that $h = 6$ and $k = 3$. Thus, vertex: $(6,3)$; minimum: $f(6) = 3$; range: $y \geq 3$ or $[3, \infty)$.

(C) The graph of $f(x) = 0.5(x - 6)^2 + 3$ is the same as the graph of $g(x) = x^2$ vertically shrunk by a factor of 0.5, and shifted to the right 6 units and up 3 units.

(D) Graph in a rectangular coordinate system:

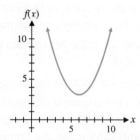

(E) Graph in a graphing calculator:

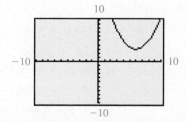

(F) Find the vertex and minimum using the minimum command:

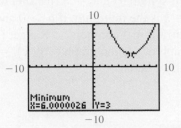

Vertex: $(6,3)$; minimum: $f(6) = 3$

Matched Problem 2 Given the quadratic function $f(x) = -0.25x^2 - 2x + 2$

(A) Find the vertex form for f.

(B) Find the vertex and the maximum or minimum. State the range of f.

(C) Describe how the graph of function f can be obtained from the graph of $g(x) = x^2$ using transformations.

(D) Sketch a graph of function f in a rectangular coordinate system.

(E) Graph function f using a suitable viewing window.

(F) Find the vertex and the maximum or minimum using the appropriate graphing calculator command.

Applications

EXAMPLE 3 Maximum Revenue This is a continuation of Example 7 in Section 2.1. Recall that the financial department in the company that produces a digital camera arrived at the following price–demand function and the corresponding revenue function:

$$p(x) = 94.8 - 5x \qquad \text{Price–demand function}$$

$$R(x) = xp(x) = x(94.8 - 5x) \qquad \text{Revenue function}$$

where $p(x)$ is the wholesale price per camera at which x million cameras can be sold and $R(x)$ is the corresponding revenue (in millions of dollars). Both functions have domain $1 \le x \le 15$.

(A) Find the value of x to the nearest thousand cameras that will generate the maximum revenue. What is the maximum revenue to the nearest thousand dollars? Solve the problem algebraically by completing the square.

(B) What is the wholesale price per camera (to the nearest dollar) that generates the maximum revenue?

(C) Graph the revenue function using an appropriate viewing window.

(D) Find the value of x to the nearest thousand cameras that will generate the maximum revenue. What is the maximum revenue to the nearest thousand dollars? Solve the problem graphically using the maximum command.

SOLUTION

(A) Algebraic solution:

$$\begin{aligned}
R(x) &= x(94.8 - 5x) \\
&= -5x^2 + 94.8x \\
&= -5(x^2 - 18.96x + ?) \\
&= -5(x^2 - 18.96x + 89.8704) + 449.352 \\
&= -5(x - 9.48)^2 + 449.352
\end{aligned}$$

The maximum revenue of 449.352 million dollars ($449,352,000) occurs when $x = 9.480$ million cameras (9,480,000 cameras).

(B) Finding the wholesale price per camera: Use the price-demand function for an output of 9.480 million cameras:

$$p(x) = 94.8 - 5x$$
$$p(9.480) = 94.8 - 5(9.480)$$
$$= \$47 \text{ per camera}$$

(C) Graph on a graphing calculator:

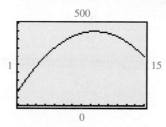

(D) Graphical solution using a graphing calculator:

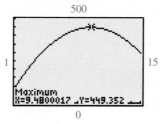

The manufacture and sale of 9.480 million cameras (9,480,000 cameras) will generate a maximum revenue of 449.352 million dollars ($449, 352, 000).

Matched Problem 3) The financial department in Example 3, using statistical and analytical techniques (see Matched Problem 7 in Section 2.1), arrived at the cost function

$$C(x) = 156 + 19.7x \quad \text{Cost function}$$

where $C(x)$ is the cost (in millions of dollars) for manufacturing and selling x million cameras.

(A) Using the revenue function from Example 3 and the preceding cost function, write an equation for the profit function.

(B) Find the value of x to the nearest thousand cameras that will generate the maximum profit. What is the maximum profit to the nearest thousand dollars? Solve the problem algebraically by completing the square.

(C) What is the wholesale price per camera (to the nearest dollar) that generates the maximum profit?

(D) Graph the profit function using an appropriate viewing window.

(E) Find the output to the nearest thousand cameras that will generate the maximum profit. What is the maximum profit to the nearest thousand dollars? Solve the problem graphically using the maximum command.

EXAMPLE 4 Break-Even Analysis Use the revenue function from Example 3 and the cost function from Matched Problem 3:

$$R(x) = x(94.8 - 5x) \quad \text{Revenue function}$$
$$C(x) = 156 + 19.7x \quad \text{Cost function}$$

Both have domain $1 \le x \le 15$.

(A) Sketch the graphs of both functions in the same coordinate system.

(B) **Break-even points** are the production levels at which $R(x) = C(x)$. Find the break-even points algebraically to the nearest thousand cameras.

(C) Plot both functions simultaneously in the same viewing window.

(D) Use INTERSECT to find the break-even points graphically to the nearest thousand cameras.

(E) Recall that a loss occurs if $R(x) < C(x)$ and a profit occurs if $R(x) > C(x)$. For what values of x (to the nearest thousand cameras) will a loss occur? A profit?

SOLUTION

(A) Sketch of functions:

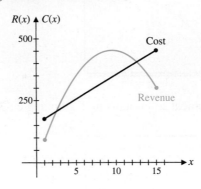

(B) Algebraic solution:

Find x such that $R(x) = C(x)$:

$$x(94.8 - 5x) = 156 + 19.7x$$
$$-5x^2 + 75.1x - 156 = 0$$
$$x = \frac{-75.1 \pm \sqrt{75.1^2 - 4(-5)(-156)}}{2(-5)}$$
$$= \frac{-75.1 \pm \sqrt{2{,}520.01}}{-10}$$
$$x = 2.490 \quad \text{and} \quad 12.530$$

The company breaks even at $x = 2.490$ million cameras (2,490,000 cameras) and at $x = 12.530$ million cameras (12,530,000 cameras).

(C) Graph on a graphing calculator:

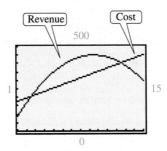

(D) Graphical solution:

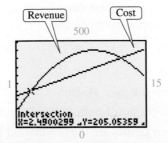 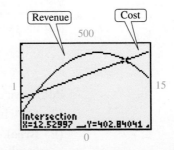

The company breaks even at $x = 2.490$ million cameras (2,490,000 cameras) and at $x = 12.530$ million cameras (12,530,000 cameras).

(E) Use the results from parts (A) and (B) or (C) and (D):

$$\text{Loss:} \quad 1 \le x < 2.490 \quad \text{or} \quad 12.530 < x \le 15$$
$$\text{Profit:} \quad 2.490 < x < 12.530$$

Matched Problem 4) Use the profit equation from Matched Problem 3:

$$P(x) = R(x) - C(x)$$
$$= -5x^2 + 75.1x - 156 \quad \text{Profit function}$$
$$\text{Domain:} \quad 1 \le x \le 15$$

(A) Sketch a graph of the profit function in a rectangular coordinate system.

(B) Break-even points occur when $P(x) = 0$. Find the break-even points algebraically to the nearest thousand cameras.

(C) Plot the profit function in an appropriate viewing window.

(D) Find the break-even points graphically to the nearest thousand cameras.

(E) A loss occurs if $P(x) < 0$, and a profit occurs if $P(x) > 0$. For what values of x (to the nearest thousand cameras) will a loss occur? A profit?

A visual inspection of the plot of a data set might indicate that a parabola would be a better model of the data than a straight line. In that case, rather than using linear regression to fit a linear model to the data, we would use **quadratic regression** on a graphing calculator to find the function of the form $y = ax^2 + bx + c$ that best fits the data.

EXAMPLE 5 Outboard Motors Table 1 gives performance data for a boat powered by an Evinrude outboard motor. Use quadratic regression to find the best model of the form $y = ax^2 + bx + c$ for fuel consumption y (in miles per gallon) as a function of speed x (in miles per hour). Estimate the fuel consumption (to one decimal place) at a speed of 12 miles per hour.

Table 1

rpm	mph	mpg
2,500	10.3	4.1
3,000	18.3	5.6
3,500	24.6	6.6
4,000	29.1	6.4
4,500	33.0	6.1
5,000	36.0	5.4
5,400	38.9	4.9

SOLUTION Enter the data in a graphing calculator (Fig. 5A) and find the quadratic regression equation (Fig. 5B). The data set and the regression equation are graphed in Figure 5C. Using TRACE, we see that the estimated fuel consumption at a speed of 12 mph is 4.5 mpg.

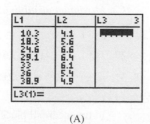

(A)

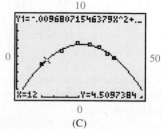

(C)

Figure 5

Matched Problem 5) Refer to Table 1. Use quadratic regression to find the best model of the form $y = ax^2 + bx + c$ for boat speed y (in miles per hour) as a function of engine speed x (in revolutions per minute). Estimate the boat speed (in miles per hour, to one decimal place) at an engine speed of 3,400 rpm.

Exercises 2.3

In Problems 1–8, find the vertex form of each quadratic function by completing the square.

1. $f(x) = x^2 - 10x$ **2.** $f(x) = x^2 + 16x$

3. $f(x) = x^2 + 20x + 50$ **4.** $f(x) = x^2 - 12x - 8$

5. $f(x) = -2x^2 + 4x - 5$ **6.** $f(x) = 3x^2 + 18x + 21$

7. $f(x) = 2x^2 + 2x + 1$ **8.** $f(x) = -5x^2 + 15x - 11$

In Problems 9–12, write a brief verbal description of the relationship between the graph of the indicated function and the graph of $y = x^2$.

9. $f(x) = x^2 - 4x + 3$ **10.** $g(x) = x^2 - 2x - 5$

11. $m(x) = -x^2 + 6x - 4$ **12.** $n(x) = -x^2 + 8x - 9$

13. Match each equation with a graph of one of the functions f, g, m, or n in the figure.

(A) $y = -(x + 2)^2 + 1$ (B) $y = (x - 2)^2 - 1$

(C) $y = (x + 2)^2 - 1$ (D) $y = -(x - 2)^2 + 1$

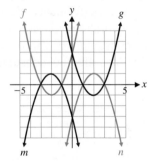

14. Match each equation with a graph of one of the functions f, g, m, or n in the figure.

(A) $y = (x - 3)^2 - 4$ (B) $y = -(x + 3)^2 + 4$

(C) $y = -(x - 3)^2 + 4$ (D) $y = (x + 3)^2 - 4$

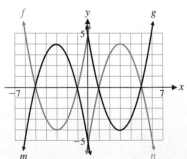

For the functions indicated in Problems 15–18, find each of the following to the nearest integer by referring to the graphs for Problems 13 and 14.

(A) *Intercepts*   (B) *Vertex*

(C) *Maximum or minimum*   (D) *Range*

15. Function n in the figure for Problem 13

16. Function m in the figure for Problem 14

17. Function f in the figure for Problem 13

18. Function g in the figure for Problem 14

In Problems 19–22, find each of the following:

(A) *Intercepts*   (B) *Vertex*

(C) *Maximum or minimum*   (D) *Range*

19. $f(x) = -(x - 3)^2 + 2$ **20.** $g(x) = -(x + 2)^2 + 3$

21. $m(x) = (x + 1)^2 - 2$ **22.** $n(x) = (x - 4)^2 - 3$

In Problems 23–26, write an equation for each graph in the form $y = a(x - h)^2 + k$, where a is either 1 or −1 and h and k are integers.

23.
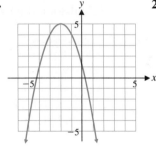

24.
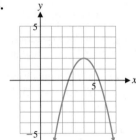

25.
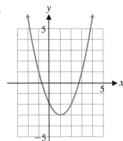

26.
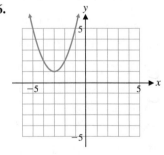

In Problems 27–32, find the vertex form for each quadratic function. Then find each of the following:

(A) *Intercepts*   (B) *Vertex*

(C) *Maximum or minimum*   (D) *Range*

27. $f(x) = x^2 - 8x + 12$ **28.** $g(x) = x^2 - 6x + 5$

29. $r(x) = -4x^2 + 16x - 15$ **30.** $s(x) = -4x^2 - 8x - 3$

31. $u(x) = 0.5x^2 - 2x + 5$ **32.** $v(x) = 0.5x^2 + 4x + 10$

33. Let $f(x) = 0.3x^2 - x - 8$. Solve each equation graphically to two decimal places.

(A) $f(x) = 4$ (B) $f(x) = -1$ (C) $f(x) = -9$

34. Let $g(x) = -0.6x^2 + 3x + 4$. Solve each equation graphically to two decimal places.

(A) $g(x) = -2$ (B) $g(x) = 5$ (C) $g(x) = 8$

35. Let $f(x) = 125x - 6x^2$. Find the maximum value of f to four decimal places graphically.

36. Let $f(x) = 100x - 7x^2 - 10$. Find the maximum value of f to four decimal places graphically.

In Problems 37–40, first write each function in vertex form; then find each of the following (to two decimal places):

(A) *Intercepts*   (B) *Vertex*

(C) *Maximum or minimum*   (D) *Range*

37. $g(x) = 0.25x^2 - 1.5x - 7$

38. $m(x) = 0.20x^2 - 1.6x - 1$

39. $f(x) = -0.12x^2 + 0.96x + 1.2$

40. $n(x) = -0.15x^2 - 0.90x + 3.3$

Solve Problems 41–46 graphically to two decimal places using a graphing calculator.

41. $2 - 5x - x^2 = 0$ **42.** $7 + 3x - 2x^2 = 0$

43. $1.9x^2 - 1.5x - 5.6 < 0$ **44.** $3.4 + 2.9x - 1.1x^2 \geq 0$

45. $2.8 + 3.1x - 0.9x^2 \leq 0$ **46.** $1.8x^2 - 3.1x - 4.9 > 0$

47. Given that f is a quadratic function with minimum $f(x) = f(2) = 4$, find the axis, vertex, range, and x intercepts.

48. Given that f is a quadratic function with maximum $f(x) = f(-3) = -5$, find the axis, vertex, range, and x intercepts.

In Problems 49–52,

(A) *Graph f and g in the same coordinate system.*

(B) *Solve $f(x) = g(x)$ algebraically to two decimal places.*

(C) *Solve $f(x) > g(x)$ using parts (A) and (B).*

(D) *Solve $f(x) < g(x)$ using parts (A) and (B).*

49. $f(x) = -0.4x(x - 10)$
$g(x) = 0.3x + 5$
$0 \leq x \leq 10$

50. $f(x) = -0.7x(x - 7)$
$g(x) = 0.5x + 3.5$
$0 \leq x \leq 7$

51. $f(x) = -0.9x^2 + 7.2x$
$g(x) = 1.2x + 5.5$
$0 \leq x \leq 8$

52. $f(x) = -0.7x^2 + 6.3x$
$g(x) = 1.1x + 4.8$
$0 \leq x \leq 9$

53. How can you tell from the graph of a quadratic function whether it has exactly one real zero?

54. How can you tell from the graph of a quadratic function whether it has no real zeros?

55. How can you tell from the standard form $y = ax^2 + bx + c$ whether a quadratic function has two real zeros?

56. How can you tell from the standard form $y = ax^2 + bx + c$ whether a quadratic function has exactly one real zero?

57. How can you tell from the vertex form $y = a(x - h)^2 + k$ whether a quadratic function has no real zeros?

58. How can you tell from the vertex form $y = a(x - h)^2 + k$ whether a quadratic function has two real zeros?

In Problems 59 and 60, assume that a, b, c, h, and k are constants with $a \neq 0$ such that

$$ax^2 + bx + c = a(x - h)^2 + k$$

for all real numbers x.

59. Show that $h = -\dfrac{b}{2a}$.   **60.** Show that $k = \dfrac{4ac - b^2}{4a}$.

Applications

61. **Tire mileage.** An automobile tire manufacturer collected the data in the table relating tire pressure x (in pounds per square inch) and mileage (in thousands of miles):

x	Mileage
28	45
30	52
32	55
34	51
36	47

A mathematical model for the data is given by

$$f(x) = -0.518x^2 + 33.3x - 481$$

(A) Complete the following table. Round values of $f(x)$ to one decimal place.

x	Mileage	$f(x)$
28	45	
30	52	
32	55	
34	51	
36	47	

(B) Sketch the graph of f and the mileage data in the same coordinate system.

(C) Use values of the modeling function rounded to two decimal places to estimate the mileage for a tire pressure of 31 lbs/sq in. and for 35 lbs/sq in.

(D) Write a brief description of the relationship between tire pressure and mileage.

62. **Automobile production.** The table shows the retail market share of passenger cars from Ford Motor Company as a percentage of the U.S. market.

Year	Market Share
1980	17.2%
1985	18.8%
1990	20.0%
1995	20.7%
2000	20.2%
2005	17.4%
2010	16.4%

A mathematical model for this data is given by

$$f(x) = -0.0169x^2 + 0.47x + 17.1$$

where $x = 0$ corresponds to 1980.

(A) Complete the following table. Round values of $f(x)$ to one decimal place.

x	Market Share	$f(x)$
0	17.2	
5	18.8	
10	20.0	
15	20.7	
20	20.2	
25	17.4	
30	16.4	

(B) Sketch the graph of f and the market share data in the same coordinate system.

(C) Use values of the modeling function f to estimate Ford's market share in 2020 and in 2025.

✎ (D) Write a brief verbal description of Ford's market share from 1980 to 2010.

63. **Tire mileage.** Using quadratic regression on a graphing calculator, show that the quadratic function that best fits the data on tire mileage in Problem 61 is

$$f(x) = -0.518x^2 + 33.3x - 481$$

64. **Automobile production.** Using quadratic regression on a graphing calculator, show that the quadratic function that best fits the data on market share in Problem 62 is

$$f(x) = -0.0169x^2 + 0.47x + 17.1$$

65. **Revenue.** The marketing research department for a company that manufactures and sells memory chips for microcomputers established the following price–demand and revenue functions:

$$p(x) = 75 - 3x \qquad \text{Price–demand function}$$
$$R(x) = xp(x) = x(75 - 3x) \qquad \text{Revenue function}$$

where $p(x)$ is the wholesale price in dollars at which x million chips can be sold, and $R(x)$ is in millions of dollars. Both functions have domain $1 \le x \le 20$.

(A) Sketch a graph of the revenue function in a rectangular coordinate system.

(B) Find the value of x that will produce the maximum revenue. What is the maximum revenue?

(C) What is the wholesale price per chip that produces the maximum revenue?

66. **Revenue.** The marketing research department for a company that manufactures and sells notebook computers established the following price–demand and revenue functions:

$$p(x) = 2,000 - 60x \qquad \text{Price–demand function}$$
$$R(x) = xp(x) \qquad \text{Revenue function}$$
$$\quad = x(2,000 - 60x)$$

where $p(x)$ is the wholesale price in dollars at which x thousand computers can be sold, and $R(x)$ is in thousands of dollars. Both functions have domain $1 \le x \le 25$.

(A) Sketch a graph of the revenue function in a rectangular coordinate system.

(B) Find the value of x that will produce the maximum revenue. What is the maximum revenue to the nearest thousand dollars?

(C) What is the wholesale price per computer (to the nearest dollar) that produces the maximum revenue?

67. **Break-even analysis.** Use the revenue function from Problem 65 and the given cost function:

$$R(x) = x(75 - 3x) \qquad \text{Revenue function}$$
$$C(x) = 125 + 16x \qquad \text{Cost function}$$

where x is in millions of chips, and $R(x)$ and $C(x)$ are in millions of dollars. Both functions have domain $1 \le x \le 20$.

(A) Sketch a graph of both functions in the same rectangular coordinate system.

(B) Find the break-even points to the nearest thousand chips.

(C) For what values of x will a loss occur? A profit?

68. **Break-even analysis.** Use the revenue function from Problem 66, and the given cost function:

$$R(x) = x(2,000 - 60x) \qquad \text{Revenue function}$$
$$C(x) = 4,000 + 500x \qquad \text{Cost function}$$

where x is thousands of computers, and $C(x)$ and $R(x)$ are in thousands of dollars. Both functions have domain $1 \le x \le 25$.

(A) Sketch a graph of both functions in the same rectangular coordinate system.

(B) Find the break-even points.

(C) For what values of x will a loss occur? A profit?

69. Profit-loss analysis. Use the revenue and cost functions from Problem 67:

$$R(x) = x(75 - 3x) \quad \text{Revenue function}$$
$$C(x) = 125 + 16x \quad \text{Cost function}$$

where x is in millions of chips, and $R(x)$ and $C(x)$ are in millions of dollars. Both functions have domain $1 \leq x \leq 20$.

(A) Form a profit function P, and graph R, C, and P in the same rectangular coordinate system.

(B) Discuss the relationship between the intersection points of the graphs of R and C and the x intercepts of P.

(C) Find the x intercepts of P and the break-even points to the nearest thousand chips.

(D) Refer to the graph drawn in part (A). Does the maximum profit appear to occur at the same value of x as the maximum revenue? Are the maximum profit and the maximum revenue equal? Explain.

(E) Verify your conclusion in part (D) by finding the value of x (to the nearest thousand chips) that produces the maximum profit. Find the maximum profit (to the nearest thousand dollars), and compare with Problem 65B.

70. Profit-loss analysis. Use the revenue and cost functions from Problem 66:

$$R(x) = x(2,000 - 60x) \quad \text{Revenue function}$$
$$C(x) = 4,000 + 500x \quad \text{Cost function}$$

where x is thousands of computers, and $R(x)$ and $C(x)$ are in thousands of dollars. Both functions have domain $1 \leq x \leq 25$.

(A) Form a profit function P, and graph R, C, and P in the same rectangular coordinate system.

(B) Discuss the relationship between the intersection points of the graphs of R and C and the x intercepts of P.

(C) Find the x intercepts of P and the break-even points.

(D) Refer to the graph drawn in part (A). Does the maximum profit appear to occur at the same value of x as the maximum revenue? Are the maximum profit and the maximum revenue equal? Explain.

(E) Verify your conclusion in part (D) by finding the value of x that produces the maximum profit. Find the maximum profit and compare with Problem 66B.

71. Medicine. The French physician Poiseuille was the first to discover that blood flows faster near the center of an artery than near the edge. Experimental evidence has shown that the rate of flow v (in centimeters per second) at a point x centimeters from the center of an artery (see the figure) is given by

$$v = f(x) = 1,000(0.04 - x^2) \quad 0 \leq x \leq 0.2$$

Find the distance from the center that the rate of flow is 20 centimeters per second. Round answer to two decimal places.

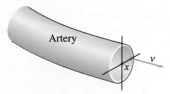

Figure for 71 and 72

72. Medicine. Refer to Problem 71. Find the distance from the center that the rate of flow is 30 centimeters per second. Round answer to two decimal places.

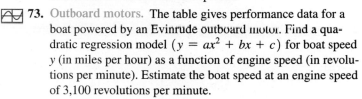 **73. Outboard motors.** The table gives performance data for a boat powered by an Evinrude outboard motor. Find a quadratic regression model ($y = ax^2 + bx + c$) for boat speed y (in miles per hour) as a function of engine speed (in revolutions per minute). Estimate the boat speed at an engine speed of 3,100 revolutions per minute.

Table for 73 and 74

rpm	mph	mpg
1,500	4.5	8.2
2,000	5.7	6.9
2,500	7.8	4.8
3,000	9.6	4.1
3,500	13.4	3.7

74. Outboard motors. The table gives performance data for a boat powered by an Evinrude outboard motor. Find a quadratic regression model ($y = ax^2 + bx + c$) for fuel consumption y (in miles per gallon) as a function of engine speed (in revolutions per minute). Estimate the fuel consumption at an engine speed of 2,300 revolutions per minute.

Answers to Matched Problems

1. (A)

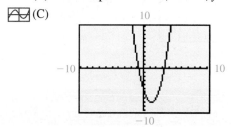

(B) x intercepts: $-0.7656, 3.2656$; y intercept: -5

(C)

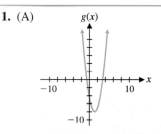

(D) x intercepts: $-0.7656, 3.2656$; y intercept: -5

(E) $x \leq -0.7656$ or $x \geq 3.2656$; or $(-\infty, -0.7656]$ or $[3.2656, \infty)$

(F) $x = -0.3508, 2.8508$

2. (A) $f(x) = -0.25(x + 4)^2 + 6$.

(B) Vertex: $(-4, 6)$; maximum: $f(-4) = 6$; range: $y \leq 6$ or $(-\infty, 6]$

(C) The graph of $f(x) = -0.25(x + 4)^2 + 6$ is the same as the graph of $g(x) = x^2$ vertically shrunk by a factor of 0.25, reflected in the x axis, and shifted 4 units to the left and 6 units up.

(D)

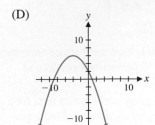

(E)

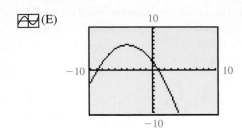

(F) Vertex: $(-4, 6)$; maximum: $f(-4) = 6$

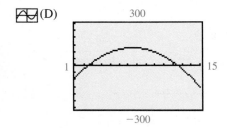

3. (A) $P(x) = R(x) - C(x) = -5x^2 + 75.1x - 156$

(B) $P(x) = R(x) - C(x) = -5(x - 7.51)^2 + 126.0005$; the manufacture and sale of 7,510,000 cameras will produce a maximum profit of $126,001,000.

(C) $p(7.510) = \$57$

(D)

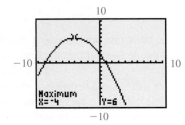

(E)

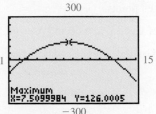

The manufacture and sale of 7,510,000 cameras will produce a maximum profit of $126,001,000. (Notice that maximum profit does not occur at the same value of x where maximum revenue occurs.)

4. (A)

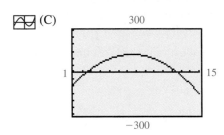

(B) $x = 2.490$ million cameras (2,490,000 cameras) and $x = 12.530$ million cameras. (12,530,000 cameras)

(C)

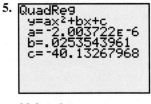

(D) $x = 2.490$ million cameras (2,490,000 cameras) and $x = 12.530$ million cameras (12,530,000 cameras)

(E) Loss: $1 \le x < 2.490$ or $12.530 < x \le 15$; profit: $2.490 < x < 12.530$

5.
```
QuadReg
y=ax²+bx+c
a=-2.003722ε-6
b=.0253543961
c=-40.13267968
```

22.9 mph

2.4 Polynomial and Rational Functions

- Polynomial Functions
- Regression Polynomials
- Rational Functions
- Applications

Linear and quadratic functions are special cases of the more general class of *polynomial functions*. Polynomial functions are a special case of an even larger class of functions, the *rational functions*. We will describe the basic features of the graphs of polynomial and rational functions. We will use these functions to solve real-world problems where linear or quadratic models are inadequate; for example, to determine the relationship between length and weight of a species of fish, or to model the training of new employees.

Polynomial Functions

A linear function has the form $f(x) = mx + b$ (where $m \neq 0$) and is a polynomial function of degree 1. A quadratic function has the form $f(x) = ax^2 + bx + c$ (where $a \neq 0$) and is a polynomial function of degree 2. Here is the general definition of a polynomial function.

DEFINITION Polynomial Function

A **polynomial function** is a function that can be written in the form

$$f(x) = a_n x^n + a_{n-1} x^{n-1} + \cdots + a_1 x + a_0$$

for n a nonnegative integer, called the **degree** of the polynomial. The coefficients $a_0, a_1, \ldots, a_n$ are real numbers with $a_n \neq 0$. The **domain** of a polynomial function is the set of all real numbers.

Figure 1 shows graphs of representative polynomial functions of degrees 1 through 6. The figure, which also appears on the inside back cover, suggests some general properties of graphs of polynomial functions.

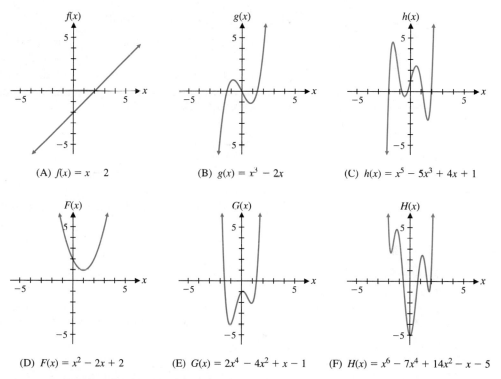

(A) $f(x) = x - 2$

(B) $g(x) = x^3 - 2x$

(C) $h(x) = x^5 - 5x^3 + 4x + 1$

(D) $F(x) = x^2 - 2x + 2$

(E) $G(x) = 2x^4 - 4x^2 + x - 1$

(F) $H(x) = x^6 - 7x^4 + 14x^2 - x - 5$

Figure 1 **Graphs of polynomial functions**

Notice that the odd-degree polynomial graphs start negative, end positive, and cross the x axis at least once. The even-degree polynomial graphs start positive, end positive, and may not cross the x axis at all. In all cases in Figure 1, the **leading coefficient**—that is, the coefficient of the highest-degree term—was chosen positive. If any leading coefficient had been chosen negative, then we would have a similar graph but reflected in the x axis.

A polynomial of degree n can have, at most, n linear factors. Therefore, the graph of a polynomial function of positive degree n can intersect the x axis at most n times. Note from Figure 1 that a polynomial of degree n may intersect the x axis fewer than n times. An x intercept of a function is also called a **zero**[*] or **root** of the function.

The graph of a polynomial function is **continuous,** with no holes or breaks. That is, the graph can be drawn without removing a pen from the paper. Also, the graph of a polynomial has no sharp corners. Figure 2 shows the graphs of two functions— one that is not continuous, and the other that is continuous but with a sharp corner. Neither function is a polynomial.

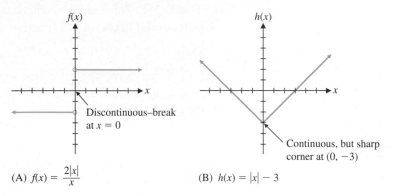

(A) $f(x) = \dfrac{2|x|}{x}$ (B) $h(x) = |x| - 3$

Figure 2 **Discontinuous and sharp-corner functions**

 Regression Polynomials

In Chapter 1, we saw that regression techniques can be used to fit a straight line to a set of data. Linear functions are not the only ones that can be applied in this manner. Most graphing calculators have the ability to fit a variety of curves to a given set of data. We will discuss polynomial regression models in this section and other types of regression models in later sections.

EXAMPLE 1 Estimating the Weight of a Fish Using the length of a fish to estimate its weight is of interest to both scientists and sport anglers. The data in Table 1 give the average weights of lake trout for certain lengths. Use the data and regression techniques to find a polynomial model that can be used to estimate the weight of a lake trout for any length. Estimate (to the near- est ounce) the weights of lake trout of lengths 39, 40, 41, 42, and 43 inches, respectively.

Table 1 **Lake Trout**

Length (in.) x	Weight (oz) y	Length (in.) x	Weight (oz) y
10	5	30	152
14	12	34	226
18	26	38	326
22	56	44	536
26	96		

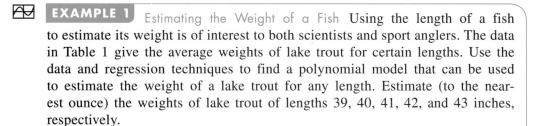

SOLUTION The graph of the data in Table 1 (Fig. 3A) indicates that a linear regression model would not be appropriate in this case. And, in fact, we would not expect a linear relationship between length and weight. Instead, it is more

---

[*]Only real numbers can be x intercepts. Functions may have complex zeros that are not real numbers, but such zeros, which are not x intercepts, will not be discussed in this book.

likely that the weight would be related to the cube of the length. We use a cubic regression polynomial to model the data (Fig. 3B). (Consult your manual for the details of calculating regression polynomials on your graphing calculator.) Figure 3C adds the graph of the polynomial model to the graph of the data. The graph in Figure 3C shows that this cubic polynomial does provide a good fit for the data. (We will have more to say about the choice of functions and the accuracy of the fit provided by regression analysis later in the book.) Figure 3D shows the estimated weights for the lengths requested.

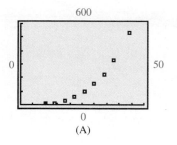
(A)

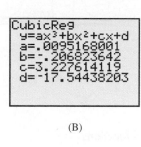
(B)

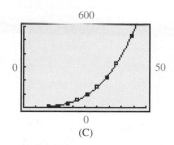
(C)

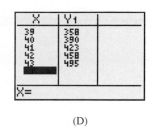
(D)

Figure 3

Matched Problem 1 ⏐ The data in Table 2 give the average weights of pike for certain lengths. Use a cubic regression polynomial to model the data. Estimate (to the nearest ounce) the weights of pike of lengths 39, 40, 41, 42, and 43 inches, respectively.

Table 2 **Pike**

Length (in.)	Weight (oz)	Length (in.)	Weight (oz)
x	y	x	y
10	5	30	108
14	12	34	154
18	26	38	210
22	44	44	326
26	72	52	522

Rational Functions

Just as rational numbers are defined in terms of quotients of integers, *rational functions* are defined in terms of quotients of polynomials. The following equations specify rational functions:

$$f(x) = \frac{1}{x} \quad g(x) = \frac{x-2}{x^2-x-6} \quad h(x) = \frac{x^3-8}{x}$$

$$p(x) = 3x^2 - 5x \quad q(x) = 7 \quad r(x) = 0$$

DEFINITION Rational Function

A **rational function** is any function that can be written in the form

$$f(x) = \frac{n(x)}{d(x)} \quad d(x) \neq 0$$

where $n(x)$ and $d(x)$ are polynomials. The **domain** is the set of all real numbers such that $d(x) \neq 0$.

Figure 4 shows the graphs of representative rational functions. Note, for example, that in Figure 4A the line $x = 2$ is a *vertical asymptote* for the function. The graph of f gets closer to this line as x gets closer to 2. The line $y = 1$ in Figure 4A is a *horizontal asymptote* for the function. The graph of f gets closer to this line as x increases or decreases without bound.

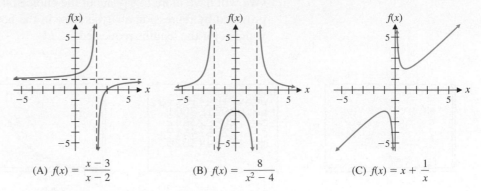

(A) $f(x) = \dfrac{x - 3}{x - 2}$ (B) $f(x) = \dfrac{8}{x^2 - 4}$ (C) $f(x) = x + \dfrac{1}{x}$

Figure 4 **Graphs of rational functions**

The number of vertical asymptotes of a rational function $f(x) = n(x)/d(x)$ is at most equal to the degree of $d(x)$. A rational function has at most one horizontal asymptote (note that the graph in Fig. 4C does not have a horizontal asymptote). Moreover, the graph of a rational function approaches the horizontal asymptote (when one exists) both as x increases and decreases without bound.

EXAMPLE 2   Graphing Rational Functions   Given the rational function:

$$f(x) = \frac{3x}{x^2 - 4}$$

(A) Find the domain.

(B) Find the x and y intercepts.

(C) Find the equations of all vertical asymptotes.

(D) If there is a horizontal asymptote, find its equation.

(E) Using the information from (A)–(D) and additional points as necessary, sketch a graph of f for $-7 \le x \le 7$.

SOLUTION

(A) $x^2 - 4 = (x - 2)(x + 2)$, so the denominator is 0 if $x = -2$ or $x = 2$. Therefore the domain is the set of all real numbers except -2 and 2.

(B) *x intercepts:* $f(x) = 0$ only if $3x = 0$, or $x = 0$. So the only x intercept is 0.

 y intercept:

$$f(0) = \frac{3 \cdot 0}{0^2 - 4} = \frac{0}{-4} = 0$$

So the y intercept is 0.

(C) Consider individually the values of x for which the denominator is 0, namely, 2 and -2, found in part (A).

 (i) If $x = 2$, the numerator is 6, and the denominator is 0, so $f(2)$ is undefined. But for numbers just to the right of 2 (like 2.1, 2.01, 2.001), the numerator is close to 6, and the denominator is a positive number close to 0, so the fraction $f(x)$ is large and positive. For numbers just to the left of 2 (like 1.9, 1.99, 1.999), the numerator is close to 6, and the denominator is a negative number close to 0, so the fraction $f(x)$ is large (in absolute value) and negative. Therefore, the line $x = 2$ is a vertical asymptote, and $f(x)$ is positive to the right of the asymptote, and negative to the left.

(ii) If $x = -2$, the numerator is -6, and the denominator is 0, so $f(2)$ is unde-fined. But for numbers just to the right of -2 (like $-1.9, -1.99, -1.999$), the numerator is close to -6, and the denominator is a negative number close to 0, so the fraction $f(x)$ is large and positive. For numbers just to the left of 2 (like $-2.1, -2.01, -2.001$), the numerator is close to -6, and the denomi-nator is a positive number close to 0, so the fraction $f(x)$ is large (in absolute value) and negative. Therefore, the line $x = -2$ is a vertical asymptote, and $f(x)$ is positive to the right of the asymptote and negative to the left.

(D) Rewrite $f(x)$ by dividing each term in the numerator and denominator by the highest power of x in $f(x)$.

$$f(x) = \frac{3x}{x^2 - 4} = \frac{\dfrac{3x}{x^2}}{\dfrac{x^2}{x^2} - \dfrac{4}{x^2}} = \frac{\dfrac{3}{x}}{1 - \dfrac{4}{x^2}}$$

As x increases or decreases without bound, the numerator tends to 0 and the denominator tends to 1; so, $f(x)$ tends to 0. The line $y = 0$ is a horizontal asymptote.

(E) Use the information from parts (A)–(D) and plot additional points as necessary to complete the graph, as shown in Figure 5.

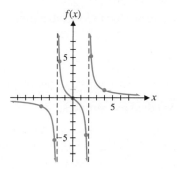

x	$f(x)$
-4	-1
-2.3	-5.3
-1.7	4.6
0	0
1.7	-4.6
2.3	5.3
4	1

Figure 5

Matched Problem 2 Given the rational function $g(x) = \dfrac{3x + 3}{x^2 - 9}$,

(A) Find the domain.

(B) Find the x and y intercepts.

(C) Find the equations of all vertical asymptotes.

(D) If there is a horizontal asymptote, find its equation.

(E) Using the information from parts (A)–(D) and additional points as necessary, sketch a graph of g for $-10 \le x \le 10$.

CONCEPTUAL INSIGHT

Consider the rational function

$$g(x) = \frac{3x^2 - 12x}{x^3 - 4x^2 - 4x + 16} = \frac{3x(x - 4)}{(x^2 - 4)(x - 4)}$$

The numerator and denominator of g have a common zero, $x = 4$. If $x \ne 4$, then we can cancel the factor $x - 4$ from the numerator and denominator, leaving the function $f(x)$ of Example 2. So the graph of g (Fig. 6) is identical to the graph of f (Fig. 5), except that the graph of g has an open dot at $(4, 1)$, indicating that 4 is not in the domain of g.

In particular, f and g have the same asymptotes. Note that the line $x = 4$ is *not* a vertical asymptote of g, even though 4 is a zero of its denominator.

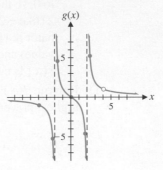

Figure 6

Graphing rational functions is aided by locating vertical and horizontal asymptotes first, if they exist. The following general procedure is suggested by Example 2 and the Conceptual Insight above.

PROCEDURE   Vertical and Horizontal Asymptotes of Rational Functions

Consider the rational function

$$f(x) = \frac{n(x)}{d(x)}$$

where $n(x)$ and $d(x)$ are polynomials.

Vertical asymptotes:

Case 1. Suppose $n(x)$ and $d(x)$ have no real zero in common. If c is a real number such that $d(c) = 0$, then the line $x = c$ is a vertical asymptote of the graph of f.

Case 2. If $n(x)$ and $d(x)$ have one or more real zeros in common, cancel common linear factors, and apply Case 1 to the reduced function. (The reduced function has the same asymptotes as f.)

Horizontal asymptote:

Case 1. If degree $n(x) <$ degree $d(x)$, then $y = 0$ is the horizontal asymptote.

Case 2. If degree $n(x) =$ degree $d(x)$, then $y = a/b$ is the horizontal asymptote, where a is the leading coefficient of $n(x)$, and b is the leading coefficient of $d(x)$.

Case 3. If degree $n(x) >$ degree $d(x)$, there is no horizontal asymptote.

Example 2 illustrates Case 1 of the procedure for horizontal asymptotes. Cases 2 and 3 are illustrated in Example 3 and Matched Problem 3.

EXAMPLE 3   Finding Asymptotes Find the vertical and horizontal asymptotes of the rational function

$$f(x) = \frac{3x^2 + 3x - 6}{2x^2 - 2}$$

SOLUTION Vertical asymptotes We factor the numerator $n(x)$ and the denominator $d(x)$:

$$n(x) = 3(x^2 + x - 2) = 3(x - 1)(x + 2)$$
$$d(x) = 2(x^2 - 1) = 2(x - 1)(x + 1)$$

The reduced function is

$$\frac{3(x + 2)}{2(x + 1)}$$

which, by the procedure, has the vertical asymptote $x = -1$. Therefore, $x = -1$ is the only vertical asymptote of f.

Horizontal asymptote   Both $n(x)$ and $d(x)$ have degree 2 (Case 2 of the procedure for horizontal asymptotes). The leading coefficient of the numerator $n(x)$ is 3, and the leading coefficient of the denominator $d(x)$ is 2. So $y = 3/2$ is the horizontal asymptote.

Matched Problem 3   Find the vertical and horizontal asymptotes of the rational function

$$f(x) = \frac{x^3 - 4x}{x^2 + 5x}$$

Explore and Discuss 1   A function f is **bounded** if the entire graph of f lies between two horizontal lines. The only polynomials that are bounded are the constant functions, but there are many rational functions that are bounded. Give an example of a bounded rational function, with domain the set of all real numbers, that is not a constant function.

Applications

Rational functions occur naturally in many types of applications.

EXAMPLE 4   Employee Training   A company that manufactures computers has established that, on the average, a new employee can assemble $N(t)$ components per day after t days of on-the-job training, as given by

$$N(t) = \frac{50t}{t + 4} \quad t \geq 0$$

Sketch a graph of N, $0 \leq t \leq 100$, including any vertical or horizontal asymptotes. What does $N(t)$ approach as t increases without bound?

SOLUTION   Vertical asymptotes   None for $t \geq 0$

Horizontal asymptote

$$N(t) = \frac{50t}{t + 4} = \frac{50}{1 + \dfrac{4}{t}}$$

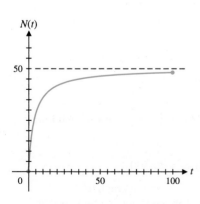

$N(t)$ approaches 50 (the leading coefficient of $50t$ divided by the leading coefficient of $t + 4$) as t increases without bound. So $y = 50$ is a horizontal asymptote.

Sketch of graph   The graph is shown in the margin.

$N(t)$ approaches 50 as t increases without bound. It appears that 50 components per day would be the upper limit that an employee would be expected to assemble.

Matched Problem 4   Repeat Example 4 for $N(t) = \dfrac{25t + 5}{t + 5} \quad t \geq 0$.

Exercises 2.4

In Problems 1–10, for each polynomial function find the following:

(A) Degree of the polynomial

(B) All x intercepts

(C) The y intercept

1. $f(x) = 50 - 5x$

2. $f(x) = 72 + 12x$

3. $f(x) = x^4(x - 1)$

4. $f(x) = x^3(x + 5)$

5. $f(x) = x^2 + 3x + 2$

6. $f(x) = x^2 - 4x - 5$

7. $f(x) = (x^2 - 1)(x^2 - 9)$

8. $f(x) = (x^2 - 4)(x^3 + 27)$

9. $f(x) = (2x + 3)^4(x - 5)^5$

10. $f(x) = (x + 3)^2(8x - 4)^6$

Each graph in Problems 11–18 is the graph of a polynomial function. Answer the following questions for each graph:

(A) What is the minimum degree of a polynomial function that could have the graph?

(B) Is the leading coefficient of the polynomial negative or positive?

11.
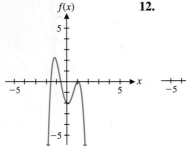

12.
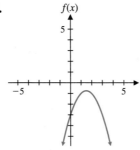

13.
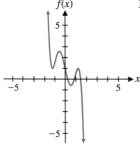

14.
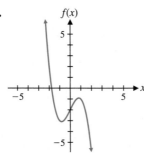

15.
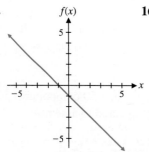

16.
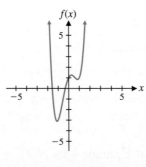

17.
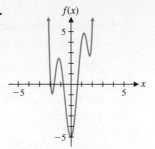

18.
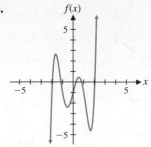

19. What is the maximum number of x intercepts that a polynomial of degree 10 can have?

20. What is the maximum number of x intercepts that a polynomial of degree 7 can have?

✎ 21. What is the minimum number of x intercepts that a polynomial of degree 9 can have? Explain.

✎ 22. What is the minimum number of x intercepts that a polynomial of degree 6 can have? Explain.

For each rational function in Problems 23–28,

(A) Find the intercepts for the graph.

(B) Determine the domain.

(C) Find any vertical or horizontal asymptotes for the graph.

(D) Sketch any asymptotes as dashed lines. Then sketch a graph of $y = f(x)$ for $-10 \le x \le 10$ and $-10 \le x \le 10$.

(E) Graph $y = f(x)$ in a standard viewing window using a graphing calculator.

23. $f(x) = \dfrac{x + 2}{x - 2}$

24. $f(x) = \dfrac{x - 3}{x + 3}$

25. $f(x) = \dfrac{3x}{x + 2}$

26. $f(x) = \dfrac{2x}{x - 3}$

27. $f(x) = \dfrac{4 - 2x}{x - 4}$

28. $f(x) = \dfrac{3 - 3x}{x - 2}$

29. Compare the graph of $y = 2x^4$ to the graph of $y = 2x^4 - 5x^2 + x + 2$ in the following two viewing windows:

(A) $-5 \le x \le 5, -5 \le y \le 5$

(B) $-5 \le x \le 5, -500 \le y \le 500$

30. Compare the graph of $y = x^3$ to the graph of $y = x^3 - 2x + 2$ in the following two viewing windows:

(A) $-5 \le x \le 5, -5 \le y \le 5$

(B) $-5 \le x \le 5, -500 \le y \le 500$

31. Compare the graph of $y = -x^5$ to the graph of $y = -x^5 + 4x^3 - 4x + 1$ in the following two viewing windows:

(A) $-5 \le x \le 5, -5 \le y \le 5$

(B) $-5 \le x \le 5, -500 \le y \le 500$

32. Compare the graph of $y = -x^5$ to the graph of $y = -x^5 + 5x^3 - 5x + 2$ in the following two viewing windows:

(A) $-5 \le x \le 5, -5 \le y \le 5$

(B) $-5 \le x \le 5, -500 \le y \le 500$

In Problems 33–40, find the equation of any horizontal asymptote.

33. $f(x) = \dfrac{5x^3 + 2x - 3}{6x^3 - 7x + 1}$

34. $f(x) = \dfrac{6x^4 - x^3 + 2}{4x^4 + 10x + 5}$

35. $f(x) = \dfrac{1 - 5x + x^2}{2 + 3x + 4x^2}$

36. $f(x) = \dfrac{8 - x^3}{1 + 2x^3}$

37. $f(x) = \dfrac{x^4 + 2x^2 + 1}{1 - x^5}$

38. $f(x) = \dfrac{3 + 5x}{x^2 + x + 3}$

39. $f(x) = \dfrac{x^2 + 6x + 1}{x - 5}$

40. $f(x) = \dfrac{x^2 + x^4 + 1}{x^3 + 2x - 4}$

In Problems 41–46, find the equations of any vertical asymptotes.

41. $f(x) = \dfrac{x^2 + 1}{(x^2 - 1)(x^2 - 9)}$

42. $f(x) = \dfrac{2x + 5}{(x^2 - 4)(x^2 - 16)}$

43. $f(x) = \dfrac{x^2 - x - 6}{x^2 - 3x - 10}$

44. $f(x) = \dfrac{x^2 - 8x + 7}{x^2 + 7x - 8}$

45. $f(x) = \dfrac{x^2 + 3x}{x^3 - 36x}$

46. $f(x) = \dfrac{x^2 + x - 2}{x^3 - 3x^2 + 2x}$

For each rational function in Problems 47–52,

(A) *Find any intercepts for the graph.*

(B) *Find any vertical and horizontal asymptotes for the graph.*

(C) *Sketch any asymptotes as dashed lines. Then sketch a graph of f for $-10 \le x \le 10$ and $-10 \le y \le 10$.*

(D) *Graph the function in a standard viewing window using a graphing calculator.*

47. $f(x) = \dfrac{2x^2}{x^2 - x - 6}$

48. $f(x) = \dfrac{3x^2}{x^2 + x - 6}$

49. $f(x) = \dfrac{6 - 2x^2}{x^2 - 9}$

50. $f(x) = \dfrac{3 - 3x^2}{x^2 - 4}$

51. $f(x) = \dfrac{-4x + 24}{x^2 + x - 6}$

52. $f(x) = \dfrac{5x - 10}{x^2 + x - 12}$

53. Write an equation for the lowest-degree polynomial function with the graph and intercepts shown in the figure.

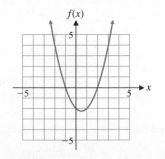

54. Write an equation for the lowest-degree polynomial function with the graph and intercepts shown in the figure.

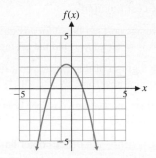

55. Write an equation for the lowest-degree polynomial function with the graph and intercepts shown in the figure.

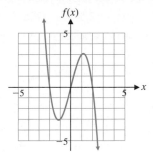

56. Write an equation for the lowest-degree polynomial function with the graph and intercepts shown in the figure.

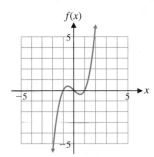

Applications

57. Average cost. A company manufacturing snowboards has fixed costs of $200 per day and total costs of $3,800 per day at a daily output of 20 boards.

(A) Assuming that the total cost per day, $C(x)$, is linearly related to the total output per day, x, write an equation for the cost function.

(B) The average cost per board for an output of x boards is given by $\overline{C}(x) = C(x)/x$. Find the average cost function.

(C) Sketch a graph of the average cost function, including any asymptotes, for $1 \le x \le 30$.

(D) What does the average cost per board tend to as production increases?

58. Average cost. A company manufacturing surfboards has fixed costs of $300 per day and total costs of $5,100 per day at a daily output of 20 boards.

(A) Assuming that the total cost per day, $C(x)$, is linearly related to the total output per day, x, write an equation for the cost function.

(B) The average cost per board for an output of x boards is given by $\overline{C}(x) = C(x)/x$. Find the average cost function.

(C) Sketch a graph of the average cost function, including any asymptotes, for $1 \le x \le 30$.

(D) What does the average cost per board tend to as production increases?

59. Replacement time. An office copier has an initial price of $2,500. A service contract costs $200 for the first year and increases $50 per year thereafter. It can be shown that the total cost of the copier after n years is given by

$$C(n) = 2,500 + 175n + 25n^2$$

The average cost per year for n years is given by $\overline{C}(n) = C(n)/n$.

(A) Find the rational function $\overline{C}$.

(B) Sketch a graph of $\overline{C}$ for $2 \le n \le 20$.

(C) When is the average cost per year at a minimum, and what is the minimum average annual cost? *[Hint: Refer to the sketch in part (B) and evaluate $\overline{C}(n)$ at appropriate integer values until a minimum value is found.]* The time when the average cost is minimum is frequently referred to as the **replacement time** for the piece of equipment.

(D) Graph the average cost function $\overline{C}$ on a graphing calculator and use an appropriate command to find when the average annual cost is at a minimum.

60. Minimum average cost. Financial analysts in a company that manufactures DVD players arrived at the following daily cost equation for manufacturing x DVD players per day:

$$C(x) = x^2 + 2x + 2,000$$

The average cost per unit at a production level of x players per day is $\overline{C}(x) = C(x)/x$.

(A) Find the rational function $\overline{C}$.

(B) Sketch a graph of $\overline{C}$ for $5 \le x \le 150$.

(C) For what daily production level (to the nearest integer) is the average cost per unit at a minimum, and what is the minimum average cost per player (to the nearest cent)? *[Hint: Refer to the sketch in part (B) and evaluate $\overline{C}(x)$ at appropriate integer values until a minimum value is found.]*

(D) Graph the average cost function $\overline{C}$ on a graphing calculator and use an appropriate command to find the daily production level (to the nearest integer) at which the average cost per player is at a minimum. What is the minimum average cost to the nearest cent?

61. Minimum average cost. A consulting firm, using statistical methods, provided a veterinary clinic with the cost equation

$$C(x) = 0.00048(x - 500)^3 + 60,000$$

$$100 \le x \le 1,000$$

where $C(x)$ is the cost in dollars for handling x cases per month. The average cost per case is given by $\overline{C}(x) = C(x)/x$.

(A) Write the equation for the average cost function $\overline{C}$.

(B) Graph $\overline{C}$ on a graphing calculator.

(C) Use an appropriate command to find the monthly caseload for the minimum average cost per case. What is the minimum average cost per case?

62. Minimum average cost. The financial department of a hospital, using statistical methods, arrived at the cost equation

$$C(x) = 20x^3 - 360x^2 + 2,300x - 1,000$$

$$1 \le x \le 12$$

where $C(x)$ is the cost in thousands of dollars for handling x thousand cases per month. The average cost per case is given by $\overline{C}(x) = C(x)/x$.

(A) Write the equation for the average cost function $\overline{C}$.

(B) Graph $\overline{C}$ on a graphing calculator.

(C) Use an appropriate command to find the monthly caseload for the minimum average cost per case. What is the minimum average cost per case to the nearest dollar?

63. Diet. Table 3 shows the per capita consumption of ice cream and eggs in the United States for selected years since 1980.

(A) Let x represent the number of years since 1980 and find a cubic regression polynomial for the per capita consumption of ice cream.

(B) Use the polynomial model from part (A) to estimate (to the nearest tenth of a pound) the per capita consumption of ice cream in 2025.

Table 3 **Per Capita Consumption of Ice Cream and Eggs**

Year	Ice Cream (pounds)	Eggs (number)
1980	17.5	266
1985	18.1	251
1990	15.8	231
1995	15.5	229
2000	16.5	247
2005	14.4	252
2010	13.3	242

Source: U.S Department of Agriculture

64. Diet. Refer to Table 3.

(A) Let x represent the number of years since 1980 and find a cubic regression polynomial for the per capita consumption of eggs.

(B) Use the polynomial model from part (A) to estimate (to the nearest integer) the per capita consumption of eggs in 2022.

65. Physiology. In a study on the speed of muscle contraction in frogs under various loads, researchers W. O. Fems and J. Marsh found that the speed of contraction decreases with increasing loads. In particular, they found that the relationship between speed of contraction v (in centimeters per second) and load x (in grams) is given approximately by

$$v(x) = \frac{26 + 0.06x}{x} \qquad x \ge 5$$

(A) What does $v(x)$ approach as x increases?

(B) Sketch a graph of function v.

66. Learning theory. In 1917, L. L. Thurstone, a pioneer in quantitative learning theory, proposed the rational function

$$f(x) = \frac{a(x + c)}{(x + c) + b}$$

to model the number of successful acts per unit time that a person could accomplish after x practice sessions. Suppose that for a particular person enrolled in a typing class,

$$f(x) = \frac{55(x + 1)}{(x + 8)} \quad x \geq 0$$

where $f(x)$ is the number of words per minute the person is able to type after x weeks of lessons.

(A) What does $f(x)$ approach as x increases?

(B) Sketch a graph of function f, including any vertical or horizontal asymptotes.

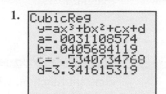 **67.** Marriage. Table 4 shows the marriage and divorce rates per 1,000 population for selected years since 1960.

(A) Let x represent the number of years since 1960 and find a cubic regression polynomial for the marriage rate.

(B) Use the polynomial model from part (A) to estimate the marriage rate (to one decimal place) for 2025.

Table 4 **Marriages and Divorces (per 1,000 Population)**

Date	Marriages	Divorces
1960	8.5	2.2
1970	10.6	3.5
1980	10.6	5.2
1990	9.8	4.7
2000	8.5	4.1
2010	6.8	3.6

Source: National Center for Health Statistics

68. Divorce. Refer to Table 4.

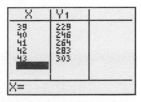 (A) Let x represent the number of years since 1960 and find a cubic regression polynomial for the divorce rate.

(B) Use the polynomial model from part (A) to estimate the divorce rate (to one decimal place) for 2025.

Answers to Matched Problems

1.

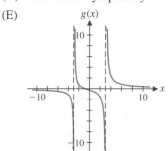

2. (A) Domain: all real numbers except -3 and 3

(B) x intercept: -1; y intercept: $-\dfrac{1}{3}$

(C) Vertical asymptotes: $x = -3$ and $x = 3$;

(D) Horizontal asymptote: $y = 0$

(E)

3. Vertical asymptote: $x = -5$

Horizontal asymptote: none

4. No vertical asymptotes for $t \geq 0$; $y = 25$ is a horizontal asymptote. $N(t)$ approaches 25 as t increases without bound. It appears that 25 components per day would be the upper limit that an employee would be expected to assemble.

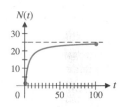

2.5 Exponential Functions

- Exponential Functions
- Base e Exponential Functions
- Growth and Decay Applications
- Compound Interest

This section introduces an important class of functions called *exponential functions*. These functions are used extensively in modeling and solving a wide variety of real-world problems, including growth of money at compound interest, growth of populations, radioactive decay, and learning associated with the mastery of such devices as a new computer or an assembly process in a manufacturing plant.

Exponential Functions

We start by noting that

$$f(x) = 2^x \quad \text{and} \quad g(x) = x^2$$

are not the same function. Whether a variable appears as an exponent with a constant base or as a base with a constant exponent makes a big difference. The function g is a quadratic function, which we have already discussed. The function f is a new type of function called an *exponential function*. In general,

> **DEFINITION   Exponential Function**
> The equation
>
> $$f(x) = b^x \quad b > 0, b \neq 1$$
>
> defines an **exponential function** for each different constant b, called the **base**. The **domain** of f is the set of all real numbers, and the **range** of f is the set of all positive real numbers.

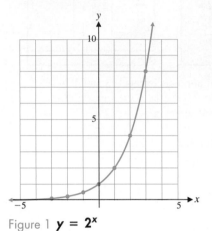

Figure 1 **$y = 2^x$**

We require the base b to be positive to avoid imaginary numbers such as $(-2)^{1/2} = \sqrt{-2} = i\sqrt{2}$. We exclude $b = 1$ as a base, since $f(x) = 1^x = 1$ is a constant function, which we have already considered.

When asked to hand-sketch graphs of equations such as $y = 2^x$ or $y = 2^{-x}$, many students do not hesitate. [*Note:* $2^{-x} = 1/2^x = (1/2)^x$.] They make tables by assigning integers to x, plot the resulting points, and then join these points with a smooth curve as in Figure 1. The only catch is that we have not defined 2^x for all real numbers. From Appendix A, Section A.6, we know what $2^5, 2^{-3}, 2^{2/3}, 2^{-3/5}, 2^{1.4}$, and $2^{-3.14}$ mean (that is, 2^p, where p is a rational number), but what does

$$2^{\sqrt{2}}$$

mean? The question is not easy to answer at this time. In fact, a precise definition of $2^{\sqrt{2}}$ must wait for more advanced courses, where it is shown that

$$2^x$$

names a positive real number for x any real number, and that the graph of $y = 2^x$ is as indicated in Figure 1.

It is useful to compare the graphs of $y = 2^x$ and $y = 2^{-x}$ by plotting both on the same set of coordinate axes, as shown in Figure 2A. The graph of

$$f(x) = b^x \quad b > 1 \text{ (Fig. 2B)}$$

looks very much like the graph of $y = 2^x$, and the graph of

$$f(x) = b^x \quad 0 < b < 1 \text{ (Fig. 2B)}$$

looks very much like the graph of $y = 2^{-x}$. Note that in both cases the x axis is a horizontal asymptote for the graphs.

The graphs in Figure 2 suggest the following general properties of exponential functions, which we state without proof:

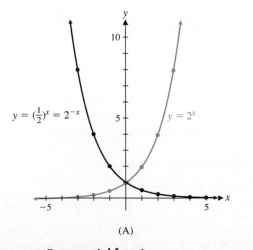

(A)

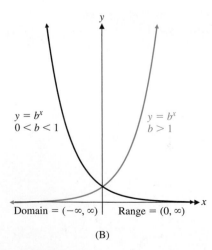

(B)

Figure 2 **Exponential functions**

> **THEOREM 1** Basic Properties of the Graph of $f(x) = b^x$, $b > 0$, $b \neq 1$
>
> 1. All graphs will pass through the point $(0, 1)$. $b^0 = 1$ for any permissible base b.
> 2. All graphs are continuous curves, with no holes or jumps.
> 3. The x axis is a horizontal asymptote.
> 4. If $b > 1$, then b^x increases as x increases.
> 5. If $0 < b < 1$, then b^x decreases as x increases.

CONCEPTUAL INSIGHT

Recall that the graph of a rational function has at most one horizontal asymptote and that it approaches the horizontal asymptote (if one exists) both as $x \to \infty$ *and* as $x \to -\infty$ (see Section 2.4). The graph of an exponential function, on the other hand, approaches its horizontal asymptote as $x \to \infty$ *or* as $x \to -\infty$, but not both. In particular, there is no rational function that has the same graph as an exponential function.

The use of a calculator with the key $\boxed{y^x}$, or its equivalent, makes the graphing of exponential functions almost routine. Example 1 illustrates the process.

EXAMPLE 1 Graphing Exponential Functions Sketch a graph of $y = \left(\frac{1}{2}\right)4^x$, $-2 \leq x \leq 2$.

SOLUTION Use a calculator to create the table of values shown. Plot these points, and then join them with a smooth curve as in Figure 3.

x	y
-2	0.031
-1	0.125
0	0.50
1	2.00
2	8.00

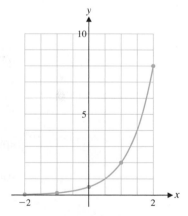

Figure 3 **Graph of $y = \left(\frac{1}{2}\right)4^x$**

Matched Problem 1 Sketch a graph of $y = \left(\frac{1}{2}\right)4^{-x}$, $-2 \leq x \leq 2$.

Exponential functions, whose domains include irrational numbers, obey the familiar laws of exponents discussed in Appendix A, Section A.6 for rational exponents. We summarize these exponent laws here and add two other important and useful properties.

THEOREM 2   Properties of Exponential Functions

For a and b positive, $a \neq 1$, $b \neq 1$, and x and y real,

1. Exponent laws:

$$a^x a^y = a^{x+y} \quad \frac{a^x}{a^y} = a^{x-y} \qquad\qquad \frac{4^{2y}}{4^{5y}} = 4^{2y-5y} = 4^{-3y}$$

$$(a^x)^y = a^{xy} \quad (ab)^x = a^x b^x \quad \left(\frac{a}{b}\right)^x = \frac{a^x}{b^x}$$

2. $a^x = a^y$ if and only if $x = y$

If $7^{5t+1} = 7^{3t-3}$, then
$5t + 1 = 3t - 3$, and $t = -2$.

3. For $x \neq 0$,

$\quad a^x = b^x$ if and only if $a = b$

If $a^5 = 2^5$, then $a = 2$.

Base e Exponential Functions

Of all the possible bases b we can use for the exponential function $y = b^x$, which ones are the most useful? If you look at the keys on a calculator, you will probably see $\boxed{10^x}$ and $\boxed{e^x}$. It is clear why base 10 would be important, because our number system is a base 10 system. But what is e, and why is it included as a base? It turns out that base e is used more frequently than all other bases combined. The reason for this is that certain formulas and the results of certain processes found in calculus and more advanced mathematics take on their simplest form if this base is used. This is why you will see e used extensively in expressions and formulas that model real-world phenomena. In fact, its use is so prevalent that you will often hear people refer to $y = e^x$ as *the* exponential function.

The base e is an irrational number, and like π, it cannot be represented exactly by any finite decimal or fraction. However, e can be approximated as closely as we like by evaluating the expression

$$\left(1 + \frac{1}{x}\right)^x \tag{1}$$

for sufficiently large values of x. What happens to the value of expression (1) as x increases without bound? Think about this for a moment before proceeding. Maybe you guessed that the value approaches 1, because

$$1 + \frac{1}{x}$$

approaches 1, and 1 raised to any power is 1. Let us see if this reasoning is correct by actually calculating the value of the expression for larger and larger values of x. Table 1 summarizes the results.

Table 1

x	$\left(1 + \dfrac{1}{x}\right)^x$
1	2
10	$2.593\,74\ldots$
100	$2.704\,81\ldots$
1,000	$2.716\,92\ldots$
10,000	$2.718\,14\ldots$
100,000	$2.718\,27\ldots$
1,000,000	$2.718\,28\ldots$

Interestingly, the value of expression (1) is never close to 1 but seems to be approaching a number close to 2.7183. In fact, as x increases without bound, the value of expression (1) approaches an irrational number that we call e. The irrational number e to 12 decimal places is

$$e = \textbf{2.718 281 828 459}$$

Compare this value of e with the value of e^1 from a calculator.

> **DEFINITION** Exponential Function with Base e
> Exponential function with base e and base $1/e$, respectively, are defined by
>
> $$y = e^x \quad \text{and} \quad y = e^{-x}$$
> $$\text{Domain: } (-\infty, \infty)$$
> $$\text{Range: } (0, \infty)$$
>
>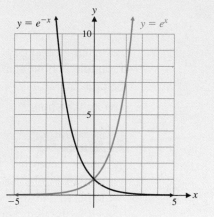

Explore and Discuss 1   Graph the functions $f(x) = e^x$, $g(x) = 2^x$, and $h(x) = 3^x$ on the same set of coordinate axes. At which values of x do the graphs intersect? For positive values of x, which of the three graphs lies above the other two? Below the other two? How does your answer change for negative values of x?

Growth and Decay Applications

Functions of the form $y = ce^{kt}$, where c and k are constants and the independent variable t represents time, are often used to model population growth and radioactive decay. Note that if $t = 0$, then $y = c$. So the constant c represents the initial population (or initial amount). The constant k is called the **relative growth rate** and has the following interpretation: Suppose that $y = ce^{kt}$ models the population growth of a country, where y is the number of persons and t is time in years. If the relative growth rate is $k = 0.02$, then at any time t, the population is growing at a rate of $0.02\,y$ persons (that is, 2% of the population) per year.

We say that **population is growing continuously at relative growth rate k** to mean that the population y is given by the model $y = ce^{kt}$.

EXAMPLE 2   Exponential Growth   Cholera, an intestinal disease, is caused by a cholera bacterium that multiplies exponentially. The number of bacteria grows continuously at relative growth rate 1.386, that is,

$$N = N_0 e^{1.386t}$$

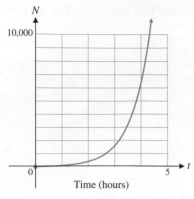

Figure 4

where N is the number of bacteria present after t hours and N_0 is the number of bacteria present at the start $(t = 0)$. If we start with 25 bacteria, how many bacteria (to the nearest unit) will be present:

(A) In 0.6 hour?   (B) In 3.5 hours?

SOLUTION Substituting $N_0 = 25$ into the preceding equation, we obtain

$$N = 25e^{1.386t} \quad \text{The graph is shown in Figure 4.}$$

(A) Solve for N when $t = 0.6$:

$$N = 25e^{1.386(0.6)} \quad \text{Use a calculator.}$$
$$= 57 \text{ bacteria}$$

(B) Solve for N when $t = 3.5$:

$$N = 25e^{1.386(3.5)} \quad \text{Use a calculator.}$$
$$= 3,197 \text{ bacteria}$$

Matched Problem 2 Refer to the exponential growth model for cholera in Example 2. If we start with 55 bacteria, how many bacteria (to the nearest unit) will be present

(A) In 0.85 hour?   (B) In 7.25 hours?

EXAMPLE 3 Exponential Decay Cosmic-ray bombardment of the atmosphere produces neutrons, which in turn react with nitrogen to produce radioactive carbon-14 (^{14}C). Radioactive ^{14}C enters all living tissues through carbon dioxide, which is first absorbed by plants. As long as a plant or animal is alive, ^{14}C is maintained in the living organism at a constant level. Once the organism dies, however, ^{14}C decays according to the equation

$$A = A_0e^{-0.000124t}$$

where A is the amount present after t years and A_0 is the amount present at time $t = 0$.

(A) If 500 milligrams of ^{14}C is present in a sample from a skull at the time of death, how many milligrams will be present in the sample in 15,000 years? Compute the answer to two decimal places.

(B) The **half-life** of ^{14}C is the time t at which the amount present is one-half the amount at time $t = 0$. Use Figure 5 to estimate the half-life of ^{14}C.

SOLUTION Substituting $A_0 = 500$ in the decay equation, we have

$$A = 500e^{-0.000124t} \quad \text{See the graph in Figure 5.}$$

(A) Solve for A when $t = 15,000$:

$$A = 500e^{-0.000124(15,000)} \quad \text{Use a calculator.}$$
$$= 77.84 \text{ milligrams}$$

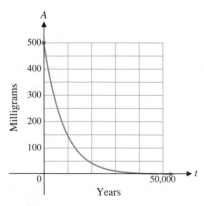

Figure 5

(B) Refer to Figure 5, and estimate the time t at which the amount A has fallen to 250 milligrams: $t \approx 6,000$ years. (Finding the intersection of $y_1 = 500e^{-0.000124x}$ and $y_2 = 250$ on a graphing calculator gives a better estimate: $t \approx 5,590$ years.)

Matched Problem 3 Refer to the exponential decay model in Example 3. How many milligrams of ^{14}C would have to be present at the beginning in order to have 25 milligrams present after 18,000 years? Compute the answer to the nearest milligram.

If you buy a new car, it is likely to depreciate in value by several thousand dollars during the first year you own it. You would expect the value of the car to decrease in each subsequent year, but not by as much as in the previous year. If you drive the car long enough, its resale value will get close to zero. An exponential decay function will often be a good model of depreciation; a linear or quadratic function would not be suitable (why?). We can use **exponential regression** on a graphing calculator to find the function of the form $y = ab^x$ that best fits a data set.

EXAMPLE 4   Depreciation   Table 2 gives the market value of a hybrid sedan (in dollars) x years after its purchase. Find an exponential regression model of the form $y = ab^x$ for this data set. Estimate the purchase price of the hybrid. Estimate the value of the hybrid 10 years after its purchase. Round answers to the nearest dollar.

Table 2

x	Value ($)
1	12,575
2	9,455
3	8,115
4	6,845
5	5,225
6	4,485

SOLUTION   Enter the data into a graphing calculator (Fig. 6A) and find the exponential regression equation (Fig. 6B). The estimated purchase price is $y_1(0) =$ $14,910. The data set and the regression equation are graphed in Figure 6C. Using TRACE, we see that the estimated value after 10 years is $1,959.

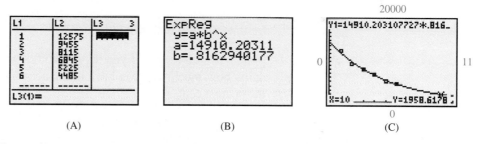

| (A) | (B) | (C) |

Figure 6

Matched Problem 4   Table 3 gives the market value of a midsize sedan (in dollars) x years after its purchase. Find an exponential regression model of the form $y = ab^x$ for this data set. Estimate the purchase price of the sedan. Estimate the value of the sedan 10 years after its purchase. Round answers to the nearest dollar.

Table 3

x	Value ($)
1	23,125
2	19,050
3	15,625
4	11,875
5	9,450
6	7,125

Compound Interest

The fee paid to use another's money is called **interest**. It is usually computed as a percent (called **interest rate**) of the principal over a given period of time. If, at the end of a payment period, the interest due is reinvested at the same rate, then the interest earned as well as the principal will earn interest during the next payment period. Interest paid on interest reinvested is called **compound interest** and may be calculated using the following compound interest formula:

If a **principal P (present value)** is invested at an annual **rate r** (expressed as a decimal) compounded m times a year, then the **amount A (future value)** in the account at the end of t years is given by

$$A = P\left(1 + \frac{r}{m}\right)^{mt} \quad \text{Compound interest formula}$$

For given r and m, the amount A is equal to the principal P multiplied by the exponential function b^t, where $b = (1 + r/m)^m$.

EXAMPLE 5 Compound Growth If $1,000 is invested in an account paying 10% compounded monthly, how much will be in the account at the end of 10 years? Compute the answer to the nearest cent.

SOLUTION We use the compound interest formula as follows:

$$A = P\left(1 + \frac{r}{m}\right)^{mt}$$

$$= 1,000\left(1 + \frac{0.10}{12}\right)^{(12)(10)} \quad \text{Use a calculator.}$$

$$= \$2,707.04$$

The graph of

$$A = 1,000\left(1 + \frac{0.10}{12}\right)^{12t}$$

for $0 \le t \le 20$ is shown in Figure 7.

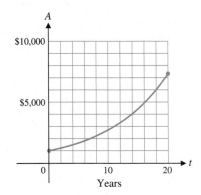

Figure 7

Matched Problem 5 If you deposit $5,000 in an account paying 9% compounded daily, how much will you have in the account in 5 years? Compute the answer to the nearest cent.

Explore and Discuss 2   Suppose that $1,000 is deposited in a savings account at an annual rate of 5%. Guess the amount in the account at the end of 1 year if interest is compounded (1) quarterly, (2) monthly, (3) daily, (4) hourly. Use the compound interest formula to compute the amounts at the end of 1 year to the nearest cent. Discuss the accuracy of your initial guesses.

Explore and Discuss 2 suggests that if $1,000 were deposited in a savings account at an annual interest rate of 5%, then the amount at the end of 1 year would be less than $1,051.28, even if interest were compounded every minute or every second. The limiting value, approximately $1,051.271096, is said to be the amount in the account if interest were compounded continuously.

If a principal, P, is invested at an annual rate, r, and compounded continuously, then the amount in the account at the end of t years is given by

$$A = Pe^{rt} \quad \text{Continuous compound interest formula}$$

where the constant $e \approx 2.71828$ is the base of the exponential function.

EXAMPLE 6 Continuous Compound Interest If $1,000 is invested in an account paying 10% compounded continuously, how much will be in the account at the end of 10 years? Compute the answer to the nearest cent.

SOLUTION We use the continuous compound interest formula:

$$A = Pe^{rt} = 1000e^{0.10(10)} = 1000e = \$2,718.28$$

Compare with the answer to Example 5.

Matched Problem 6 | If you deposit $5,000 in an account paying 9% compounded continuously, how much will you have in the account in 5 years? Compute the answer to the nearest cent.

The formulas for compound interest and continuous compound interest are summarized below for convenient reference.

SUMMARY

Compound Interest: $A = P\left(1 + \dfrac{r}{m}\right)^{mt}$

Continuous Compound Interest: $A = Pe^{rt}$

where A = amount (future value) at the end of t years

P = principal (present value)

r = annual rate (expressed as a decimal)

m = number of compounding periods per year

t = time in years

Exercises 2.5

1. Match each equation with the graph of f, g, h, or k in the figure.

(A) $y = 2^x$ (B) $y = (0.2)^x$

(C) $y = 4^x$ (D) $y = \left(\frac{1}{3}\right)^x$

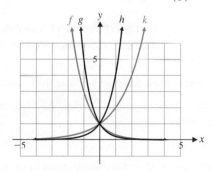

2. Match each equation with the graph of f, g, h, or k in the figure.

(A) $y = \left(\frac{1}{4}\right)^x$ (B) $y = (0.5)^x$

(C) $y = 5^x$ (D) $y = 3^x$

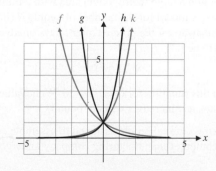

Graph each function in Problems 3–10 over the indicated interval.

3. $y = 5^x$; $[-2, 2]$ **4.** $y = 3^x$; $[-3, 3]$

5. $y = \left(\frac{1}{5}\right)^x = 5^{-x}$; $[-2, 2]$ **6.** $y = \left(\frac{1}{3}\right)^x = 3^{-x}$; $[-3, 3]$

7. $f(x) = -5^x$; $[-2, 2]$ **8.** $g(x) = -3^{-x}$; $[-3, 3]$

9. $y - -e^{-x}$; $[-3, 3]$ **10.** $y = -e^x$; $[-3, 3]$

In Problems 11–18, describe verbally the transformations that can be used to obtain the graph of g from the graph of f (see Section 2.2).

11. $g(x) = -2^x$; $f(x) = 2^x$

12. $g(x) = 2^{x-2}$; $f(x) = 2^x$

13. $g(x) = 3^{x+1}$; $f(x) = 3^x$

14. $g(x) = -3^x$; $f(x) = 3^x$

15. $g(x) = e^x + 1$; $f(x) = e^x$

16. $g(x) = e^x - 2$; $f(x) = e^x$

17. $g(x) = 2e^{-(x+2)}$; $f(x) = e^{-x}$

18. $g(x) = 0.5e^{-(x-1)}$; $f(x) = e^{-x}$

19. Use the graph of f shown in the figure to sketch the graph of each of the following.

(A) $y = f(x) - 1$ (B) $y = f(x + 2)$

(C) $y = 3f(x) - 2$ (D) $y = 2 - f(x - 3)$

20. Use the graph of f shown in the figure to sketch the graph of each of the following.

(A) $y = f(x) + 2$ (B) $y = f(x - 3)$

(C) $y = 2f(x) - 4$ (D) $y = 4 - f(x + 2)$

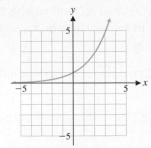

Figure for 19 and 20

In Problems 21–26, graph each function over the indicated interval.

21. $f(t) = 2^{t/10}$; $[-30, 30]$ **22.** $G(t) = 3^{t/100}$; $[-200, 200]$

23. $y = -3 + e^{1+x}$; $[-4, 2]$ **24.** $y = 2 + e^{x-2}$; $[-1, 5]$

25. $y = e^{|x|}$; $[-3, 3]$ **26.** $y = e^{-|x|}$; $[-3, 3]$

27. Find all real numbers a such that $a^2 = a^{-2}$. Explain why this does not violate the second exponential function property in Theorem 2 on page 98.

28. Find real numbers a and b such that $a \neq b$ but $a^4 = b^4$. Explain why this does not violate the third exponential function property in Theorem 2 on page 98.

Solve each equation in Problems 29–34 for x.

29. $10^{2-3x} = 10^{5x-6}$ **30.** $5^{3x} = 5^{4x-2}$

31. $4^{5x-x^2} = 4^{-6}$ **32.** $7^{x^2} = 7^{2x+3}$

33. $5^3 = (x + 2)^3$ **34.** $(1 - x)^5 = (2x - 1)^5$

In Problems 35–42, solve each equation for x. (Remember: $e^x \neq 0$ and $e^{-x} \neq 0$ for all values of x).

35. $xe^{-x} + 7e^{-x} = 0$ **36.** $10xe^x - 5e^x = 0$

37. $2x^2e^x - 8e^x = 0$ **38.** $x^2e^{-x} - 9e^{-x} = 0$

39. $e^{4x} - e = 0$ **40.** $e^{4x} + e = 0$

41. $e^{3x-1} + e = 0$ **42.** $e^{3x-1} - e = 0$

Graph each function in Problems 43–46 over the indicated interval.

43. $h(x) = x(2^x)$; $[-5, 0]$ **44.** $m(x) = x(3^{-x})$; $[0, 3]$

45. $N = \dfrac{100}{1 + e^{-t}}$; $[0, 5]$ **46.** $N = \dfrac{200}{1 + 3e^{-t}}$; $[0, 5]$

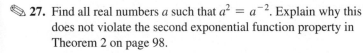

Applications

In all problems involving days, a 365-day year is assumed.

47. Continuous compound interest. Find the value of an investment of $10,000 in 12 years if it earns an annual rate of 3.95% compounded continuously.

48. Continuous compound interest. Find the value of an investment of $24,000 in 7 years if it earns an annual rate of 4.35% compounded continuously.

49. Compound growth. Suppose that $2,500 is invested at 7% compounded quarterly. How much money will be in the account in

(A) $\frac{3}{4}$ year?   (B) 15 years?

Compute answers to the nearest cent.

50. Compound growth. Suppose that $4,000 is invested at 6% compounded weekly. How much money will be in the account in

(A) $\frac{1}{2}$ year?   (B) 10 years?

Compute answers to the nearest cent.

51. Finance. A person wishes to have $15,000 cash for a new car 5 years from now. How much should be placed in an account now, if the account pays 6.75% compounded weekly? Compute the answer to the nearest dollar.

52. Finance. A couple just had a baby. How much should they invest now at 5.5% compounded daily in order to have $40,000 for the child's education 17 years from now? Compute the answer to the nearest dollar.

53. Money growth. BanxQuote operates a network of websites providing real-time market data from leading financial providers. The following rates for 12-month certificates of deposit were taken from the websites:

(A) Stonebridge Bank, 0.95% compounded monthly

(B) DeepGreen Bank, 0.80% compounded daily

(C) Provident Bank, 0.85% compounded quarterly

Compute the value of $10,000 invested in each account at the end of 1 year.

54. Money growth. Refer to Problem 53. The following rates for 60-month certificates of deposit were also taken from BanxQuote websites:

(A) Oriental Bank & Trust, 1.35% compounded quarterly

(B) BMW Bank of North America, 1.30% compounded monthly

(C) BankFirst Corporation, 1.25% compounded daily

Compute the value of $10,000 invested in each account at the end of 5 years.

55. Advertising. A company is trying to introduce a new product to as many people as possible through television advertising in a large metropolitan area with 2 million possible viewers. A model for the number of people N (in millions) who are aware of the product after t days of advertising was found to be

$$N = 2(1 - e^{-0.037t})$$

Graph this function for $0 \leq t \leq 50$. What value does N approach as t increases without bound?

56. Learning curve. People assigned to assemble circuit boards for a computer manufacturing company undergo on-the-job training. From past experience, the learning curve for the average employee is given by

$$N = 40(1 - e^{-0.12t})$$

where N is the number of boards assembled per day after t days of training. Graph this function for $0 \leq t \leq 30$. What is the maximum number of boards an average employee can be expected to produce in 1 day?

57. Sports salaries. Table 4 shows the average salaries for players in Major League Baseball (MLB) and the National Basketball Association (NBA) in selected years since 1990.

(A) Let x represent the number of years since 1990 and find an exponential regression model $(y = ab^x)$ for the average salary in MLB. Use the model to estimate the average salary (to the nearest thousand dollars) in 2022.

(B) The average salary in MLB in 2000 was 1.984 million. How does this compare with the value given by the model of part (A)?

Table 4 **Average Salary (thousand $)**

Year	MLB	NBA
1990	589	750
1993	1,062	1,300
1996	1,101	2,000
1999	1,724	2,400
2002	2,300	4,500
2005	2,633	5,000
2008	3,155	5,585
2011	3,298	4,755

58. Sports salaries. Refer to Table 4.

(A) Let x represent the number of years since 1990 and find an exponential regression model $(y = ab^x)$ for the average salary in the NBA. Use the model to estimate the average salary (to the nearest thousand dollars) in 2022.

(B) The average salary in the NBA in 1997 was $2.2 million. How does this compare with the value given by the model of part (A)?

59. Marine biology. Marine life depends on the microscopic plant life that exists in the photic zone, a zone that goes to a depth where only 1% of surface light remains. In some waters with a great deal of sediment, the photic zone may go down only 15 to 20 feet. In some murky harbors, the intensity of light d feet below the surface is given approximately by

$$I = I_0 e^{-0.23d}$$

What percentage of the surface light will reach a depth of

(A) 10 feet?   (B) 20 feet?

60. Marine biology. Refer to Problem 59. Light intensity I relative to depth d (in feet) for one of the clearest bodies of water in the world, the Sargasso Sea, can be approximated by

$$I = I_0 e^{-0.00942d}$$

where I_0 is the intensity of light at the surface. What percentage of the surface light will reach a depth of

(A) 50 feet?   (B) 100 feet?

61. World population growth. From the dawn of humanity to 1830, world population grew to one billion people. In 100 more years (by 1930) it grew to two billion, and 3 billion more were added in only 60 years (by 1990). In 2013, the estimated world population was 7.1 billion with a relative growth rate of 1.1%.

(A) Write an equation that models the world population growth, letting 2013 be year 0.

(B) Based on the model, what is the expected world population (to the nearest hundred million) in 2025? In 2035?

62. Population growth in Ethiopia. In 2012, the estimated population in Ethiopia was 94 million people with a relative growth rate of 3.2%.

(A) Write an equation that models the population growth in Ethiopia, letting 2012 be year 0.

(B) Based on the model, what is the expected population in Ethiopia (to the nearest million) in 2025? In 2035?

63. Internet growth. The number of Internet hosts grew very rapidly from 1994 to 2012 (Table 5).

(A) Let x represent the number of years since 1994. Find an exponential regression model $(y = ab^x)$ for this data set and estimate the number of hosts in 2022 (to the nearest million).

(B) Discuss the implications of this model if the number of Internet hosts continues to grow at this rate.

Table 5 **Internet Hosts (Millions)**

Year	Hosts
1994	2.4
1997	16.1
2000	72.4
2003	171.6
2006	394.0
2009	625.2
2012	888.2

Source: Internet Software Consortium

64. Life expectancy. Table 6 shows the life expectancy (in years) at birth for residents of the United States from 1970 to 2010. Let x represent years since 1970. Find an exponential regression model for this data and use it to estimate the life expectancy for a person born in 2025.

Table 6

Year of Birth	Life Expectancy
1970	70.8
1975	72.6
1980	73.7
1985	74.7
1990	75.4
1995	75.9
2000	76.9
2005	77.7
2010	78.2

1.

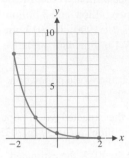

2. (A) 179 bacteria   (B) 1,271,659 bacteria

3. 233 mg

4. Purchase price: $30,363; value after 10 yr: $2,864

5. $7,841.13

6. $7,841.56

2.6   Logarithmic Functions

- Inverse Functions

- Logarithmic Functions

- Properties of Logarithmic Functions

- Calculator Evaluation of Logarithms

- Applications

Find the exponential function keys $\boxed{10^x}$ and $\boxed{e^x}$ on your calculator. Close to these keys you will find the $\boxed{\text{LOG}}$ and $\boxed{\text{LN}}$ keys. The latter two keys represent *logarithmic functions,* and each is closely related to its nearby exponential function. In fact, the exponential function and the corresponding logarithmic function are said to be *inverses* of each other. In this section we will develop the concept of inverse functions and use it to define a logarithmic function as the inverse of an exponential function. We will then investigate basic properties of logarithmic functions, use a calculator to evaluate them for particular values of x, and apply them to real-world problems.

Logarithmic functions are used in modeling and solving many types of problems. For example, the decibel scale is a logarithmic scale used to measure sound intensity, and the Richter scale is a logarithmic scale used to measure the strength of the force of an earthquake. An important business application has to do with finding the time it takes money to double if it is invested at a certain rate compounded a given number of times a year or compounded continuously. This requires the solution of an exponential equation, and logarithms play a central role in the process.

Inverse Functions

Look at the graphs of $f(x) = \dfrac{x}{2}$ and $g(x) = \dfrac{|x|}{2}$ in Figure 1:

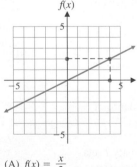

(A) $f(x) = \dfrac{x}{2}$

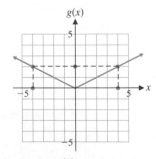

(B) $g(x) = \dfrac{|x|}{2}$

Figure 1

Because both f and g are functions, each domain value corresponds to exactly one range value. For which function does each range value correspond to exactly one domain value? This is the case only for function f. Note that for function f, the range value 2 corresponds to the domain value 4. For function g the range value 2 corresponds to both -4 and 4. Function f is said to be *one-to-one*.

> **DEFINITION** One-to-One Functions
>
> A function f is said to be **one-to-one** if each range value corresponds to exactly one domain value.

It can be shown that any continuous function that is either increasing* or decreasing for all domain values is one-to-one. If a continuous function increases for some domain values and decreases for others, then it cannot be one-to-one. Figure 1 shows an example of each case.

Explore and Discuss 1

Graph $f(x) = 2^x$ and $g(x) = x^2$. For a range value of 4, what are the corresponding domain values for each function? Which of the two functions is one-to-one? Explain why.

Starting with a one-to-one function f, we can obtain a new function called the *inverse* of f.

> **DEFINITION** Inverse of a Function
>
> If f is a one-to-one function, then the **inverse** of f is the function formed by interchanging the independent and dependent variables for f. Thus, if (a, b) is a point on the graph of f, then (b, a) is a point on the graph of the inverse of f.
>
> *Note:* If f is not one-to-one, then f **does not have an inverse.**

In this course, we are interested in the inverses of exponential functions, called *logarithmic functions*.

Logarithmic Functions

If we start with the exponential function f defined by

$$y = 2^x \tag{1}$$

and interchange the variables, we obtain the inverse of f:

$$x = 2^y \tag{2}$$

We call the inverse the **logarithmic function with base 2**, and write

$$y = \log_2 x \quad \text{if and only if} \quad x = 2^y$$

We can graph $y = \log_2 x$ by graphing $x = 2^y$ since they are equivalent. Any ordered pair of numbers on the graph of the exponential function will be on the graph of the logarithmic function if we interchange the order of the components. For example, $(3, 8)$ satisfies equation (1) and $(8, 3)$ satisfies equation (2). The graphs of $y = 2^x$ and $y = \log_2 x$ are shown in Figure 2. Note that if we fold the paper along the dashed line $y = x$ in Figure 2, the two graphs match exactly. The line $y = x$ is a line of symmetry for the two graphs.

---

*Formally, we say that the function f is **increasing** on an interval (a, b) if $f(x_2) > f(x_1)$ whenever $a < x_1 < x_2 < b$; and f is **decreasing** on (a, b) if $f(x_2) < f(x_1)$ whenever $a < x_1 < x_2 < b$.

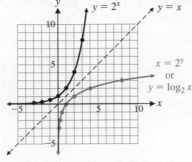

Figure 2

Exponential Function		Logarithmic Function	
x	$y = 2^x$	$x = 2^y$	y
-3	$\frac{1}{8}$	$\frac{1}{8}$	-3
-2	$\frac{1}{4}$	$\frac{1}{4}$	-2
-1	$\frac{1}{2}$	$\frac{1}{2}$	-1
0	1	1	0
1	2	2	1
2	4	4	2
3	8	8	3

$$\left[\begin{array}{c}\text{Ordered}\\ \text{pairs}\\ \text{reversed}\end{array}\right]$$

In general, since the graphs of all exponential functions of the form $f(x) = b^x$, $b \neq 1, b > 0$, are either increasing or decreasing, exponential functions have inverses.

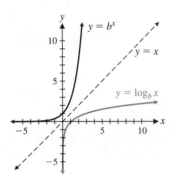

Figure 3

DEFINITION Logarithmic Functions

The inverse of an exponential function is called a **logarithmic function**. For $b > 0$ and $b \neq 1$,

Logarithmic form		Exponential form
$y = \log_b x$	is equivalent to	$x = b^y$

The **log to the base b of x** is the exponent to which b must be raised to obtain x. [*Remember:* A logarithm is an exponent.] The **domain** of the logarithmic function is the set of all positive real numbers, which is also the range of the corresponding exponential function; and the **range** of the logarithmic function is the set of all real numbers, which is also the domain of the corresponding exponential function. Typical graphs of an exponential function and its inverse, a logarithmic function, are shown in Figure 3.

CONCEPTUAL **INSIGHT**

Because the domain of a logarithmic function consists of the positive real numbers, the entire graph of a logarithmic function lies to the right of the y axis. In contrast, the graphs of polynomial and exponential functions intersect every vertical line, and the graphs of rational functions intersect all but a finite number of vertical lines.

The following examples involve converting logarithmic forms to equivalent exponential forms, and vice versa.

EXAMPLE 1 Logarithmic–Exponential Conversions Change each logarithmic form to an equivalent exponential form:

(A) $\log_5 25 = 2$ (B) $\log_9 3 = \frac{1}{2}$ (C) $\log_2\left(\frac{1}{4}\right) = -2$

SOLUTION

(A) $\log_5 25 = 2$ is equivalent to $25 = 5^2$

(B) $\log_9 3 = \frac{1}{2}$ is equivalent to $3 = 9^{1/2}$

(C) $\log_2\left(\frac{1}{4}\right) = -2$ is equivalent to $\frac{1}{4} = 2^{-2}$

Matched Problem 1 Change each logarithmic form to an equivalent exponential form:

(A) $\log_3 9 = 2$ (B) $\log_4 2 = \frac{1}{2}$ (C) $\log_3\left(\frac{1}{9}\right) = -2$

EXAMPLE 2 Exponential–Logarithmic Conversions Change each exponential form to an equivalent logarithmic form:

(A) $64 = 4^3$ (B) $6 = \sqrt{36}$ (C) $\frac{1}{8} = 2^{-3}$

SOLUTION

(A) $64 = 4^3$ is equivalent to $\log_4 64 = 3$
(B) $6 = \sqrt{36}$ is equivalent to $\log_{36} 6 = \frac{1}{2}$
(C) $\frac{1}{8} = 2^{-3}$ is equivalent to $\log_2\left(\frac{1}{8}\right) = -3$

Matched Problem 2 Change each exponential form to an equivalent logarithmic form:

(A) $49 = 7^2$ (B) $3 = \sqrt{9}$ (C) $\frac{1}{3} = 3^{-1}$

To gain a deeper understanding of logarithmic functions and their relationship to exponential functions, we consider a few problems where we want to find x, b, or y in $y = \log_b x$, given the other two values. All values are chosen so that the problems can be solved exactly without a calculator.

EXAMPLE 3 Solutions of the Equation $y = \log_b x$ Find y, b, or x, as indicated.

(A) Find y: $y = \log_4 16$ (B) Find x: $\log_2 x = -3$
(C) Find b: $\log_b 100 = 2$

SOLUTION

(A) $y = \log_4 16$ is equivalent to $16 = 4^y$. So,
$$y = 2$$
(B) $\log_2 x = -3$ is equivalent to $x = 2^{-3}$. So,
$$x = \frac{1}{2^3} = \frac{1}{8}$$
(C) $\log_b 100 = 2$ is equivalent to $100 = b^2$. So,
$$b = 10 \quad \text{Recall that } b \text{ cannot be negative.}$$

Matched Problem 3 Find y, b, or x, as indicated.

(A) Find y: $y = \log_9 27$ (B) Find x: $\log_3 x = -1$
(C) Find b: $\log_b 1,000 = 3$

Properties of Logarithmic Functions

The properties of exponential functions (Section 2.5) lead to properties of logarithmic functions. For example, consider the exponential property $b^x b^y = b^{x+y}$. Let $M = b^x$, $N = b^y$. Then

$$\log_b MN = \log_b(b^x b^y) = \log_b b^{x+y} = x + y = \log_b M + \log_b N$$

So $\log_b MN = \log_b M + \log_b N$, that is, the logarithm of a product is the sum of the logarithms. Similarly, the logarithm of a quotient is the difference of the logarithms. These properties are among the eight useful properties of logarithms that are listed in Theorem 1.

THEOREM 1   Properties of Logarithmic Functions

If b, M, and N are positive real numbers, $b \neq 1$, and p and x are real numbers, then

1. $\log_b 1 = 0$

2. $\log_b b = 1$

3. $\log_b b^x = x$

4. $b^{\log_b x} = x, \quad x > 0$

5. $\log_b MN = \log_b M + \log_b N$

6. $\log_b \dfrac{M}{N} = \log_b M - \log_b N$

7. $\log_b M^p = p \log_b M$

8. $\log_b M = \log_b N$ if and only if $M = N$

EXAMPLE 4   Using Logarithmic Properties

(A) $\log_b \dfrac{wx}{yz}$ $\quad \boxed{\begin{aligned} &= \log_b wx - \log_b yz \\ &= \log_b w + \log_b x - (\log_b y + \log_b z) \end{aligned}}$

$= \log_b w + \log_b x - \log_b y - \log_b z$

(B) $\log_b (wx)^{3/5}$ $\quad \boxed{= \tfrac{3}{5} \log_b wx}$ $= \tfrac{3}{5}(\log_b w + \log_b x)$

(C) $e^{x \log_e b} = e^{\log_e b^x} = b^x$

(D) $\dfrac{\log_e x}{\log_e b} = \dfrac{\log_e (b^{\log_b x})}{\log_e b} = \dfrac{(\log_b x)(\log_e b)}{\log_e b} = \log_b x$

Matched Problem 4   Write in simpler forms, as in Example 4.

(A) $\log_b \dfrac{R}{ST}$ $\qquad$ (B) $\log_b \left(\dfrac{R}{S}\right)^{2/3}$ $\qquad$ (C) $2^{u \log_2 b}$ $\qquad$ (D) $\dfrac{\log_2 x}{\log_2 b}$

The following examples and problems will give you additional practice in using basic logarithmic properties.

EXAMPLE 5   Solving Logarithmic Equations   Find x so that

$$\tfrac{3}{2}\log_b 4 - \tfrac{2}{3}\log_b 8 + \log_b 2 = \log_b x$$

SOLUTION

$$\tfrac{3}{2}\log_b 4 - \tfrac{2}{3}\log_b 8 + \log_b 2 = \log_b x$$

$$\log_b 4^{3/2} - \log_b 8^{2/3} + \log_b 2 = \log_b x \quad \text{Property 7}$$

$$\log_b 8 - \log_b 4 + \log_b 2 = \log_b x$$

$$\log_b \frac{8 \cdot 2}{4} = \log_b x \quad \text{Properties 5 and 6}$$

$$\log_b 4 = \log_b x$$

$$x = 4 \quad \text{Property 8}$$

Matched Problem 5   Find x so that $3 \log_b 2 + \tfrac{1}{2}\log_b 25 - \log_b 20 = \log_b x$.

EXAMPLE 6   Solving Logarithmic Equations

Solve: $\log_{10} x + \log_{10}(x + 1) = \log_{10} 6$.

SOLUTION

$$\log_{10} x + \log_{10}(x + 1) = \log_{10} 6$$

$$\log_{10}[x(x + 1)] = \log_{10} 6 \quad \text{Property 5}$$

$$x(x + 1) = 6 \quad \text{Property 8}$$

$$x^2 + x - 6 = 0 \quad \text{Solve by factoring.}$$

$$(x + 3)(x - 2) = 0$$

$$x = -3, 2$$

We must exclude $x = -3$, since the domain of the function $\log_{10}(x + 1)$ is $x > -1$ or $(-1, \infty)$; so $x = 2$ is the only solution.

Matched Problem 6 Solve: $\log_3 x + \log_3(x - 3) = \log_3 10$.

Calculator Evaluation of Logarithms

Of all possible logarithmic bases, e and 10 are used almost exclusively. Before we can use logarithms in certain practical problems, we need to be able to approximate the logarithm of any positive number either to base 10 or to base e. And conversely, if we are given the logarithm of a number to base 10 or base e, we need to be able to approximate the number. Historically, tables were used for this purpose, but now calculators make computations faster and far more accurate.

Common logarithms are logarithms with base 10. **Natural logarithms** are logarithms with base e. Most calculators have a key labeled "log" (or "LOG") and a key labeled "ln" (or "LN"). The former represents a common (base 10) logarithm and the latter a natural (base e) logarithm. In fact, "log" and "ln" are both used extensively in mathematical literature, and whenever you see either used in this book without a base indicated, they will be interpreted as follows:

> Common logarithm: $\log x$ means $\log_{10} x$
> Natural logarithm: $\ln x$ means $\log_e x$

Finding the common or natural logarithm using a calculator is very easy. On some calculators, you simply enter a number from the domain of the function and press $\boxed{\text{LOG}}$ or $\boxed{\text{LN}}$. On other calculators, you press either $\boxed{\text{LOG}}$ or $\boxed{\text{LN}}$, enter a number from the domain, and then press $\boxed{\text{ENTER}}$. Check the user's manual for your calculator.

EXAMPLE 7 Calculator Evaluation of Logarithms Use a calculator to evaluate each to six decimal places:

(A) log 3,184   (B) ln 0.000 349   (C) log (-3.24)

SOLUTION

(A) $\log 3{,}184 = 3.502\ 973$

(B) $\ln 0.000\ 349 = -7.960\ 439$

(C) $\log(-3.24) = $ Error* -3.24 is not in the domain of the log function.

Matched Problem 7 Use a calculator to evaluate each to six decimal places:

(A) log 0.013 529   (B) ln 28.693 28   (C) ln (-0.438)

Given the logarithm of a number, how do you find the number? We make direct use of the logarithmic-exponential relationships, which follow from the definition of logarithmic function given at the beginning of this section.

> $\log x = y$ **is equivalent to** $x = 10^y$
> $\ln x = y$ **is equivalent to** $x = e^y$

EXAMPLE 8 Solving $\log_b x = y$ for x Find x to four decimal places, given the indicated logarithm:

(A) $\log x = -2.315$ (B) $\ln x = 2.386$

SOLUTION

(A) $\log x = -2.315$ Change to equivalent exponential form.

$\quad\quad x = 10^{-2.315}$ Evaluate with a calculator.

$\quad\quad\ \ = 0.0048$

---

*Some calculators use a more advanced definition of logarithms involving complex numbers and will display an ordered pair of real numbers as the value of log (-3.24). You should interpret such a result as an indication that the number entered is not in the domain of the logarithm function as we have defined it.

(B) $\ln x = 2.386$ Change to equivalent exponential form.

$\quad\quad x = e^{2.386}$ Evaluate with a calculator.

$\quad\quad\quad = 10.8699$

Matched Problem 8 Find x to four decimal places, given the indicated logarithm:

(A) $\ln x = -5.062$ (B) $\log x = 2.0821$

We can use logarithms to solve exponential equations.

EXAMPLE 9 Solving Exponential Equations Solve for x to four decimal places:

(A) $10^x = 2$ (B) $e^x = 3$ (C) $3^x = 4$

SOLUTION

(A) $10^x = 2$ Take common logarithms of both sides.

$\quad \log 10^x = \log 2$ Property 3

$\quad\quad\quad x = \log 2$ Use a calculator.

$\quad\quad\quad\quad = 0.3010$ To four decimal places

(B) $e^x = 3$ Take natural logarithms of both sides.

$\quad \ln e^x = \ln 3$ Property 3

$\quad\quad\quad x = \ln 3$ Use a calculator.

$\quad\quad\quad\quad = 1.0986$ To four decimal places

(C) $3^x = 4$ Take either natural or common logarithms of both sides. (We choose common logarithms.)

$\quad \log 3^x = \log 4$ Property 7

$\quad x \log 3 = \log 4$ Solve for x.

$\quad\quad\quad x = \dfrac{\log 4}{\log 3}$ Use a calculator.

$\quad\quad\quad\quad = 1.2619$ To four decimal places

Matched Problem 9 Solve for x to four decimal places:

(A) $10^x = 7$ (B) $e^x = 6$ (C) $4^x = 5$

Exponential equations can also be solved graphically by graphing both sides of an equation and finding the points of intersection. Figure 4 illustrates this approach for the equations in Example 9.

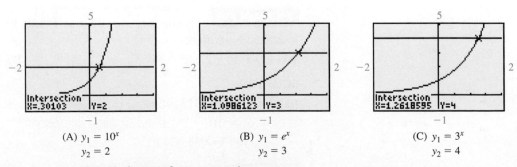

(A) $y_1 = 10^x$ (B) $y_1 = e^x$ (C) $y_1 = 3^x$

$\quad\quad y_2 = 2$ $y_2 = 3$ $y_2 = 4$

Figure 4 **Graphical solution of exponential equations**

Explore and Discuss 2   Discuss how you could find $y = \log_5 38.25$ using either natural or common logarithms on a calculator. [*Hint:* Start by rewriting the equation in exponential form.]

Remark   In the usual notation for natural logarithms, the simplifications of Example 4, parts (C) and (D) on page 110, become

$$e^{x \ln b} = b^x \qquad \text{and} \qquad \frac{\ln x}{\ln b} = \log_b x$$

With these formulas, we can change an exponential function with base b, or a logarithmic function with base b, to expressions involving exponential or logarithmic functions, respectively, to the base e. Such **change-of-base formulas** are useful in calculus.

Applications

A convenient and easily understood way of comparing different investments is to use their **doubling times**—the length of time it takes the value of an investment to double. Logarithm properties, as you will see in Example 10, provide us with just the right tool for solving some doubling-time problems.

EXAMPLE 10   Doubling Time for an Investment   How long (to the next whole year) will it take money to double if it is invested at 20% compounded annually?

SOLUTION   We use the compound interest formula discussed in Section 2.5:

$$A = P\left(1 + \frac{r}{m}\right)^{mt} \qquad \text{Compound interest}$$

The problem is to find t, given $r = 0.20$, $m = 1$, and $A = 2p$; that is,

$$2P = P(1 + 0.2)^t$$

$$2 = 1.2^t \qquad \text{Solve for } t \text{ by taking the natural or}$$
$$1.2^t = 2 \qquad \text{common logarithm of both sides (we choose}$$
$$\ln 1.2^t = \ln 2 \qquad \text{the natural logarithm).}$$
$$t \ln 1.2 = \ln 2 \qquad \text{Property 7}$$
$$t = \frac{\ln 2}{\ln 1.2} \qquad \text{Use a calculator.}$$
$$\approx 3.8 \text{ years} \qquad [\textit{Note: } (\ln 2)/(\ln 1.2) \neq \ln 2 - \ln 1.2]$$
$$\approx 4 \text{ years} \qquad \text{To the next whole year}$$

When interest is paid at the end of 3 years, the money will not be doubled; when paid at the end of 4 years, the money will be slightly more than doubled.

Example 10 can also be solved graphically by graphing both sides of the equation $2 = 1.2^t$, and finding the intersection point (Fig. 5).

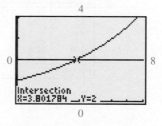

Figure 5 $y_1 = 1.2^x$, $y_2 = 2$

Matched Problem 10)   How long (to the next whole year) will it take money to triple if it is invested at 13% compounded annually?

It is interesting and instructive to graph the doubling times for various rates compounded annually. We proceed as follows:

$$A = P(1 + r)^t$$
$$2P = P(1 + r)^t$$
$$2 = (1 + r)^t$$
$$(1 + r)^t = 2$$
$$\ln(1 + r)^t = \ln 2$$
$$t \ln(1 + r) = \ln 2$$
$$t = \frac{\ln 2}{\ln(1 + r)}$$

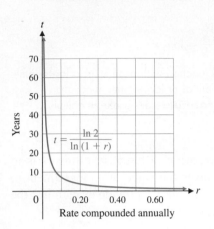

Figure 6

Figure 6 shows the graph of this equation (doubling time in years) for interest rates compounded annually from 1 to 70% (expressed as decimals). Note the dramatic change in doubling time as rates change from 1 to 20% (from 0.01 to 0.20).

Among increasing functions, the logarithmic functions (with bases $b > 1$) increase much more slowly for large values of x than either exponential or polynomial functions. When a visual inspection of the plot of a data set indicates a slowly increasing function, a logarithmic function often provides a good model. We use **logarithmic regression** on a graphing calculator to find the function of the form $y = a + b \ln x$ that best fits the data.

EXAMPLE 11 Home Ownership Rates The U.S. Census Bureau published the data in Table 1 on home ownership rates. Let x represent time in years with $x = 0$ representing 1900. Use logarithmic regression to find the best model of the form $y = a + b \ln x$ for the home ownership rate y as a function of time x. Use the model to predict the home ownership rate in the United States in 2025 (to the nearest tenth of a percent).

Table 1 **Home Ownership Rates**

Year	Rate (%)
1950	55.0
1960	61.9
1970	62.9
1980	64.4
1990	64.2
2000	67.4
2010	66.9

SOLUTION Enter the data in a graphing calculator (Fig. 7A) and find the logarithmic regression equation (Fig. 7B). The data set and the regression equation are graphed in Figure 7C. Using TRACE, we predict that the home ownership rate in 2025 would be 69.8%.

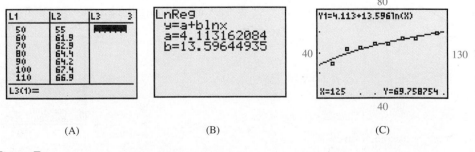

(A)   (B)   (C)

Figure 7

Matched Problem 11) Refer to Example 11. Use the model to predict the home ownership rate in the United States in 2030 (to the nearest tenth of a percent).

⚠ **CAUTION** Note that in Example 11 we let $x = 0$ represent 1900. If we let $x = 0$ represent 1940, for example, we would obtain a different logarithmic regression equation, but the prediction for 2025 would be the same. We would *not* let $x = 0$ represent 1950 (the first year in Table 1) or any later year, because logarithmic functions are undefined at 0. ▲

Exercises 2.6

For Problems 1–6, rewrite in equivalent exponential form.

1. $\log_3 27 = 3$

2. $\log_2 32 = 5$

3. $\log_{10} 1 = 0$

4. $\log_e 1 = 0$

5. $\log_4 8 = \frac{3}{2}$

6. $\log_9 27 = \frac{3}{2}$

For Problems 7–12, rewrite in equivalent logarithmic form.

7. $49 = 7^2$

8. $36 = 6^2$

9. $8 = 4^{3/2}$

10. $9 = 27^{2/3}$

11. $A = b^u$

12. $M = b^x$

In Problems 13–20, evaluate the expression without using a calculator.

13. $\log_{10} 100$

14. $\log_{10} 100{,}000$

15. $\log_2 16$

16. $\log_3 \dfrac{1}{3}$

17. $\log_5 \dfrac{1}{25}$

18. $\log_4 1$

19. $\ln \dfrac{1}{e^4}$

20. $\ln e^{-5}$

For Problems 21–26, write in simpler forms, as in Example 4.

21. $\log_b \dfrac{P}{Q}$

22. $\log_b FG$

23. $\log_b L^5$

24. $\log_b w^{15}$

25. $3^{p \log_3 q}$

26. $\dfrac{\log_3 P}{\log_3 R}$

For Problems 27–34, find x, y, or b without a calculator.

27. $\log_3 x = 2$

28. $\log_2 x = 2$

29. $\log_7 49 = y$

30. $\log_3 27 = y$

31. $\log_b 10^{-4} = -4$

32. $\log_b e^{-2} = -2$

33. $\log_4 x = \frac{1}{2}$

34. $\log_{25} x = \frac{1}{2}$

✎ *In Problems 35–42, discuss the validity of each statement. If the statement is always true, explain why. If not, give a counter example.*

35. Every polynomial function is one-to-one.

36. Every polynomial function of odd degree is one-to-one.

37. If g is the inverse of a function f, then g is one-to-one.

38. The graph of a one-to-one function intersects each vertical line exactly once.

39. The inverse of $f(x) = 2x$ is $g(x) = x/2$.

40. The inverse of $f(x) = x^2$ is $g(x) = \sqrt{x}$.

41. If f is one-to-one, then the domain of f is equal to the range of f.

42. If g is the inverse of a function f, then f is the inverse of g.

Find x in Problems 43–50.

43. $\log_b x = \frac{2}{3}\log_b 8 + \frac{1}{2}\log_b 9 - \log_b 6$

44. $\log_b x = \frac{2}{3}\log_b 27 + 2\log_b 2 - \log_b 3$

45. $\log_b x = \frac{3}{2}\log_b 4 - \frac{2}{3}\log_b 8 + 2\log_b 2$

46. $\log_b x = 3\log_b 2 + \frac{1}{2}\log_b 25 - \log_b 20$

47. $\log_b x + \log_b (x - 4) = \log_b 21$

48. $\log_b (x + 2) + \log_b x = \log_b 24$

49. $\log_{10}(x - 1) - \log_{10}(x + 1) = 1$

50. $\log_{10}(x + 6) - \log_{10}(x - 3) = 1$

Graph Problems 51 and 52 by converting to exponential form first.

51. $y = \log_2 (x - 2)$

52. $y = \log_3 (x + 2)$

✎ **53.** Explain how the graph of the equation in Problem 51 can be obtained from the graph of $y = \log_2 x$ using a simple transformation (see Section 2.2).

✎ **54.** Explain how the graph of the equation in Problem 52 can be obtained from the graph of $y = \log_3 x$ using a simple transformation (see Section 2.2).

55. What are the domain and range of the function defined by $y = 1 + \ln(x + 1)$?

56. What are the domain and range of the function defined by $y = \log (x - 1) - 1$?

For Problems 57 and 58, evaluate to five decimal places using a calculator.

57. (A) $\log 3{,}527.2$ (B) $\log 0.006\,913\,2$

(C) $\ln 277.63$ (D) $\ln 0.040\,883$

58. (A) log 72.604   (B) log 0.033 041

      (C) In 40,257   (D) In 0.005 926 3

For Problems 59 and 60, find x to four decimal places.

59. (A) $\log x = 1.1285$ (B) $\log x = -2.0497$

      (C) $\ln x = 2.7763$ (D) $\ln x = -1.8879$

60. (A) $\log x = 2.0832$ (B) $\log x = -1.1577$

      (C) $\ln x = 3.1336$ (D) $\ln x = -4.3281$

For Problems 61–66, solve each equation to four decimal places.

61. $10^x = 12$ **62.** $10^x = 153$

63. $e^x = 4.304$ **64.** $e^x = 0.3059$

65. $1.005^{12t} = 3$ **66.** $1.02^{4t} = 2$

Graph Problems 67–74 using a calculator and point-by-point plotting. Indicate increasing and decreasing intervals.

67. $y = \ln x$ **68.** $y = -\ln x$

69. $y = |\ln x|$ **70.** $y = \ln|x|$

71. $y = 2\ln(x + 2)$ **72.** $y = 2\ln x + 2$

73. $y = 4\ln x - 3$ **74.** $y = 4\ln(x - 3)$

75. Explain why the logarithm of 1 for any permissible base is 0.

76. Explain why 1 is not a suitable logarithmic base.

77. Let $p(x) = \ln x$, $q(x) = \sqrt{x}$, and $r(x) = x$. Use a graphing calculator to draw graphs of all three functions in the same viewing window for $1 \le x \le 16$. Discuss what it means for one function to be larger than another on an interval, and then order the three functions from largest to smallest for $1 < x \le 16$.

78. Let $p(x) = \log x$, $q(x) = \sqrt[3]{x}$, and $r(x) = x$. Use a graphing calculator to draw graphs of all three functions in the same viewing window for $1 \le x \le 16$. Discuss what it means for one function to be smaller than another on an interval, and then order the three functions from smallest to largest for $1 < x \le 16$.

Applications

79. Doubling time. In its first 10 years the Gabelli Growth Fund produced an average annual return of 21.36%. Assume that money invested in this fund continues to earn 21.36% compounded annually. How long will it take money invested in this fund to double?

80. Doubling time. In its first 10 years the Janus Flexible Income Fund produced an average annual return of 9.58%. Assume that money invested in this fund continues to earn 9.58% compounded annually. How long will it take money invested in this fund to double?

81. Investing. How many years (to two decimal places) will it take $1,000 to grow to $1,800 if it is invested at 6% compounded quarterly? Compounded daily?

82. Investing. How many years (to two decimal places) will it take $5,000 to grow to $7,500 if it is invested at 8% compounded semiannually? Compounded monthly?

83. Continuous compound interest. How many years (to two decimal places) will it take an investment of $35,000 to grow to $50,000 if it is invested at 4.75% compounded continuously?

84. Continuous compound interest. How many years (to two decimal places) will it take an investment of $17,000 to grow to $41,000 if it is invested at 2.95% compounded continuously?

85. Supply and demand. A cordless screwdriver is sold through a national chain of discount stores. A marketing company established price–demand and price–supply tables (Tables 2 and 3), where x is the number of screwdrivers people are willing to buy and the store is willing to sell each month at a price of p dollars per screwdriver.

    (A) Find a logarithmic regression model $(y = a + b \ln x)$ for the data in Table 2. Estimate the demand (to the nearest unit) at a price level of $50.

Table 2 **Price–Demand**

x	$p = D(x)(\$)$
1,000	91
2,000	73
3,000	64
4,000	56
5,000	53

    (B) Find a logarithmic regression model $(y = a + b \ln x)$ for the data in Table 3. Estimate the supply (to the nearest unit) at a price level of $50.

Table 3 **Price–Supply**

x	$p = S(x)(\$)$
1,000	9
2,000	26
3,000	34
4,000	38
5,000	41

    (C) Does a price level of $50 represent a stable condition, or is the price likely to increase or decrease? Explain.

86. Equilibrium point. Use the models constructed in Problem 85 to find the equilibrium point. Write the equilibrium price to the nearest cent and the equilibrium quantity to the nearest unit.

87. Sound intensity: decibels. Because of the extraordinary range of sensitivity of the human ear (a range of over 1,000 million millions to 1), it is helpful to use a logarithmic scale, rather than an absolute scale, to measure sound intensity over this range. The unit of measure is called the *decibel*, after the inventor of the telephone, Alexander Graham Bell. If we let N be the number of decibels, I the power of the sound in question (in watts per square centimeter), and I_0 the power of sound just below the threshold of hearing (approximately 10^{-16} watt per square centimeter), then

$$I = I_0 10^{N/10}$$

Show that this formula can be written in the form

$$N - 10 \log \frac{I}{I_0}$$

88. **Sound intensity: decibels.** Use the formula in Problem 87 (with $I_0 = 10^{-16}$ W/cm^2) to find the decibel ratings of the following sounds:

(A) Whisper: 10^{-13} W/cm^2

(B) Normal conversation: 3.16×10^{-10} W/cm^2

(C) Heavy traffic: 10^{-8} W/cm^2

(D) Jet plane with afterburner: 10^{-1} W/cm^2

89. **Agriculture.** Table 4 shows the yield (in bushels per acre) and the total production (in millions of bushels) for corn in the United States for selected years since 1950. Let x

Table 4 **United States Corn Production**

Year	x	Yield (bushels per acre)	Total Production (million bushels)
1950	50	38	2,782
1960	60	56	3,479
1970	70	81	4,802
1980	80	98	6,867
1990	90	116	7,802
2000	100	140	10,192
2010	110	153	12,447

represent years since 1900. Find a logarithmic regression model ($y = a + b \ln x$) for the yield. Estimate (to the nearest bushel per acre) the yield in 2024.

90. **Agriculture.** Refer to Table 4. Find a logarithmic regression model ($y = a + b \ln x$) for the total production. Estimate (to the nearest million) the production in 2024.

91. **World population.** If the world population is now 7.1 billion people and if it continues to grow at an annual rate of 1.1% compounded continuously, how long (to the nearest year) would it take before there is only 1 square yard of land per person? (The Earth contains approximately 1.68×10^{14} square yards of land.)

92. **Archaeology: carbon-14 dating.** The radioactive carbon-14 $\left({}^{14}C \right)$ in an organism at the time of its death decays according to the equation

$$A = A_0 e^{-0.000124t}$$

where t is time in years and A_0 is the amount of ^{14}C present at time $t = 0$. (See Example 3 in Section 2.5.) Estimate the age of a skull uncovered in an archaeological site if 10% of the original amount of ^{14}C is still present. [*Hint:* Find t such that $A = 0.1A_0$.]

Answers to Matched Problems

1. (A) $9 = 3^2$ (B) $2 = 4^{1/2}$ (C) $\frac{1}{9} = 3^{-2}$

2. (A) $\log_7 49 = 2$ (B) $\log_9 3 = \frac{1}{2}$ (C) $\log_3\left(\frac{1}{3}\right) = -1$

3. (A) $y = \frac{3}{2}$ (B) $x = \frac{1}{3}$ (C) $b = 10$

4. (A) $\log_b R - \log_b S - \log_b T$ (B) $\frac{2}{3}\left(\log_b R \ \ \log_b S\right)$
   (C) b^u (D) $\log_b x$

5. $x = 2$ 6. $x = 5$

7. (A) $-1.868\,734$ (B) $3.356\,663$ (C) Not defined

8. (A) 0.0063 (B) 120.8092

9. (A) 0.8451 (B) 1.7918 (C) 1.1610

10. 9 yr

11. 70.3%

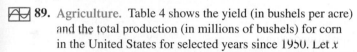

Chapter 2   Summary and Review

Important Terms, Symbols, and Concepts

2.1 Functions

EXAMPLES

- **Point-by-point plotting** may be used to **sketch the graph** of an equation in two variables: Plot enough points from its solution set in a rectangular coordinate system so that the total graph is apparent and then connect these points with a smooth curve.

Ex. 1, p. 43

- A **function** is a correspondence between two sets of elements such that to each element in the first set there corresponds one and only one element in the second set. The first set is called the **domain** and the set of corresponding elements in the second set is called the **range**.

2.1 Functions (*Continued*)

- If x is a placeholder for the elements in the domain of a function, then x is called the **independent variable** or the **input**. If y is a placeholder for the elements in the range, then y is called the **dependent variable** or the **output**.

Ex. 2, p. 46

- If in an equation in two variables we get exactly one output for each input, then the equation specifies a function. The graph of such a function is just the graph of the specifying equation. If we get more than one output for a given input, then the equation does not specify a function.

- The **vertical-line test** can be used to determine whether or not an equation in two variables specifies a function (Theorem 1, p. 47).

- The functions specified by equations of the form $y = mx + b$, where $m \neq 0$, are called **linear functions**. Functions specified by equations of the form $y = b$ are called **constant functions**.

- If a function is specified by an equation and the domain is not indicated, we agree to assume that the domain is the set of all inputs that produce outputs that are real numbers.

Ex. 3, p. 48
Ex. 5, p. 49

- The symbol $f(x)$ represents the element in the range of f that corresponds to the element x of the domain.

Ex. 4, p. 49
Ex. 6, p. 50

- **Break-even** and **profit–loss** analysis use a cost function C and a revenue function R to determine when a company will have a loss $(R < C)$, will break even $(R = C)$, or will have a profit $(R > C)$. Typical **cost, revenue, profit**, and **price–demand functions** are given on pages 50 and 51.

Ex. 7, p. 51

2.2 Elementary Functions: Graphs and Transformations

- The graphs of **six basic elementary functions** (the identity function, the square and cube functions, the square root and cube root functions, and the absolute value function) are shown on page 59.

Ex. 1, p. 58

- Performing an operation on a function produces a **transformation** of the graph of the function. The basic graph transformations, **vertical and horizontal translations** (shifts), **reflection in the x axis**, and **vertical stretches and shrinks**, are summarized on page 63.

- A **piecewise-defined function** is a function whose definition involves more than one rule.

Ex. 2, p. 60
Ex. 3, p. 61
Ex. 4, p. 62
Ex. 5, p. 63
Ex. 6, p. 64

2.3 Quadratic Functions

- If a, b, and c are real numbers with $a \neq 0$, then the function

$$f(x) = ax^2 + bx + c \quad \text{Standard form}$$

is a **quadratic function** in **standard form** and its graph is a **parabola**.

- The quadratic formula

Ex. 1, p. 71

$$x = \frac{-b \pm \sqrt{b^2 - 4ac}}{2a} \quad b^2 - 4ac \geq 0$$

can be used to find the x intercepts of a quadratic function.

- Completing the square in the standard form of a quadratic function produces the **vertex form**

$$f(x) = a(x - h)^2 + k \quad \text{Vertex form}$$

- From the vertex form of a quadratic function, we can read off the vertex, axis of symmetry, maximum or minimum, and range, and can easily sketch the graph (pages 74 and 75).

Ex. 2, p. 75
Ex. 3, p. 76
Ex. 4, p. 77

- If a revenue function $R(x)$ and a cost function $C(x)$ intersect at a point $(x_0\, y_0)$, then both this point and its x coordinate x_0 are referred to as **break-even points**.

- **Quadratic regression** on a graphing calculator produces the function of the form $y = ax^2 + bx + c$ that best fits a data set.

Ex. 5, p. 79

2.4 Polynomial and Rational Functions

- A **polynomial function** is a function that can be written in the form

$$f(x) = a_n x^n + a_{n-1} x^{n-1} + \cdots + a_1 x + a_0$$

for n a nonnegative integer called the **degree** of the polynomial. The coefficients $a_0, a_1, \ldots, a_n$ are real numbers with **leading coefficient** $a_n \neq 0$. The **domain** of a polynomial function is the set of all real numbers. Graphs of representative polynomial functions are shown on page 85 and inside the front cover.

- The graph of a polynomial function of degree n can intersect the x axis at most n times. An x intercept is also called a **zero** or **root**.

- The graph of a polynomial function has no sharp corners and is **continuous**, that is, it has no holes or breaks.

- **Polynomial regression** produces a polynomial of specified degree that best fits a data set.   Ex. 1, p. 86

- A **rational function** is any function that can be written in the form

$$f(x) = \frac{n(x)}{d(x)} \quad d(x) \neq 0$$

where $n(x)$ and $d(x)$ are polynomials. The **domain** is the set of all real numbers such that $d(x) \neq 0$. Graphs of representative rational functions are shown on page 88 and inside the back cover.

- Unlike polynomial functions, a rational function can have vertical asymptotes [but not more than the degree of the denominator $d(x)$] and at most one horizontal asymptote.   Ex. 2, p. 88

- A procedure for finding the vertical and horizontal asymptotes of a rational function is given on page 90.   Ex. 3, p. 90

2.5 Exponential Functions

- An **exponential function** is a function of the form

$$f(x) = b^x$$

where $b \neq 1$ is a positive constant called the **base**. The **domain** of f is the set of all real numbers, and the **range** is the set of positive real numbers.

- The graph of an exponential function is continuous, passes through $(0, 1)$, and has the x axis as a horizontal asymptote. If $b > 1$, then b^x increases as x increases; if $0 < b < 1$, then b^x decreases as x increases (Theorem 1, p. 97).   Ex. 1, p. 97

- Exponential functions obey the familiar laws of exponents and satisfy additional properties (Theorem 2, p. 98).

- The base that is used most frequently in mathematics is the irrational number $e \approx 2.7183$.   Ex. 2, p. 99

- Exponential functions can be used to model population growth and radioactive decay.   Ex. 3, p. 100

- **Exponential regression** on a graphing calculator produces the function of the form $y = ab^x$ that best fits a data set.   Ex. 4, p. 101

- Exponential functions are used in computations of **compound interest** and **continuous compound interest**:

$$A = P\left(1 + \frac{r}{m}\right)^{mt} \quad \text{Compound interest}$$

$$A = Pe^{rt} \quad\quad\quad \text{Continuous compound interest}$$

(see summary on page 103).

2.6 Logarithmic Functions

- A function is said to be **one-to-one** if each range value corresponds to exactly one domain value.

- The **inverse** of a one-to-one function f is the function formed by interchanging the independent and dependent variables of f. That is, (a, b) is a point on the graph of f if and only if (b, a) is a point on the graph of the inverse of f. A function that is not one-to-one does not have an inverse.

- The inverse of the exponential function with base b is called the **logarithmic function with base b**, denoted $y = \log_b x$. The **domain** of $\log_b x$ is the set of all positive real numbers (which is the range of b^x), and the range of $\log_b x$ is the set of all real numbers (which is the domain of b^x).

- Because $\log_b x$ is the inverse of the function b^x,

Ex. 1, p. 108

Logarithmic form		Exponential form
$y = \log_b x$	is equivalent to	$x = b^y$

Ex. 2, p. 109

Ex. 3, p. 109

- Properties of logarithmic functions can be obtained from corresponding properties of exponential functions (Theorem 1, p. 110).

Ex. 4, p. 110

Ex. 5, p. 110

- Logarithms to the base 10 are called **common logarithms**, often denoted simply by $\log x$. Logarithms to the base e are called **natural logarithms**, often denoted by $\ln x$.

Ex. 6, p. 110

Ex. 7, p. 111

Ex. 8, p. 111

Ex. 9, p. 112

- Logarithms can be used to find an investment's **doubling time**—the length of time it takes for the value of an investment to double.

Ex. 10, p. 113

- **Logarithmic regression** on a graphing calculator produces the function of the form $y = a + b \ln x$ that best fits a data set.

Ex. 11, p. 114

Review Exercises

Work through all the problems in this chapter review and check your answers in the back of the book. Answers to all review problems are there along with section numbers in italics to indicate where each type of problem is discussed. Where weaknesses show up, review appropriate sections in the text.

In Problems 1–3, use point-by-point plotting to sketch the graph of each equation.

1. $y = 5 - x^2$

2. $x^2 = y^2$

3. $y^2 = 4x^2$

4. Indicate whether each graph specifies a function:

   (A)   (B)

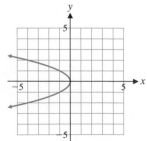

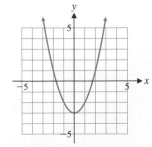

(C)   (D)

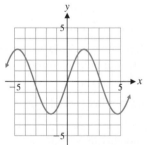

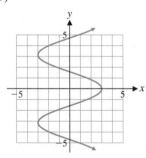

5. For $f(x) = 2x - 1$ and $g(x) = x^2 - 2x$, find:

   (A) $f(-2) + g(-1)$ (B) $f(0) \cdot g(4)$

   (C) $\dfrac{g(2)}{f(3)}$ (D) $\dfrac{f(3)}{g(2)}$

6. Write in logarithmic form using base e: $u = e^v$.

7. Write in logarithmic form using base 10: $x = 10^y$.

8. Write in exponential form using base e: $\ln M = N$.

9. Write in exponential form using base 10: $\log u = v$.

Solve Problems 10–12 for x exactly without using a calculator.

10. $\log_3 x = 2$ 11. $\log_x 36 = 2$

12. $\log_2 16 = x$

Solve Problems 13–16 for x to three decimal places.

13. $10^x = 143.7$ **14.** $e^x = 503,000$

15. $\log x = 3.105$ **16.** $\ln x = -1.147$

17. Use the graph of function f in the figure to determine (to the nearest integer) x or y as indicated.

(A) $y = f(0)$ (B) $4 = f(x)$
(C) $y = f(3)$ (D) $3 = f(x)$
(E) $y = f(-6)$ (F) $-1 = f(x)$

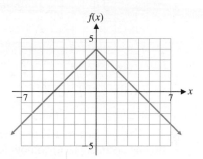

18. Sketch a graph of each of the functions in parts (A)–(D) using the graph of function f in the figure below.

(A) $y = -f(x)$ (B) $y = f(x) + 4$
(C) $y = f(x - 2)$ (D) $y = -f(x + 3) - 3$

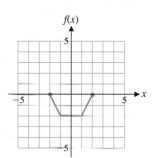

19. Complete the square and find the standard form for the quadratic function

$$f(x) = -x^2 + 4x$$

Then write a brief verbal description of the relationship between the graph of f and the graph of $y = x^2$.

20. Match each equation with a graph of one of the functions f, g, m, or n in the figure.

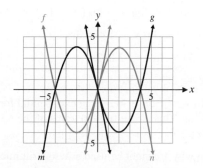

(A) $y = (x - 2)^2 - 4$ (B) $y = -(x + 2)^2 + 4$
(C) $y = -(x - 2)^2 + 4$ (D) $y = (x + 2)^2 - 4$

21. Referring to the graph of function f in the figure for Problem 20 and using known properties of quadratic functions, find each of the following to the nearest integer:

(A) Intercepts   (B) Vertex
(C) Maximum or minimum   (D) Range

In Problems 22–25, each equation specifies a function. Determine whether the function is linear, quadratic, constant, or none of these.

22. $y = 4 - x + 3x^2$ **23.** $y = \dfrac{1 + 5x}{6}$

24. $y = \dfrac{7 - 4x}{2x}$ **25.** $y = 8x + 2(10 - 4x)$

Solve Problems 26–33 for x exactly without using a calculator.

26. $\log(x + 5) = \log(2x - 3)$ **27.** $2\ln(x - 1) = \ln(x^2 - 5)$

28. $9^{x-1} = 3^{1+x}$ **29.** $e^{2x} = e^{x^2-3}$

30. $2x^2 e^x = 3xe^x$ **31.** $\log_{1/3} 9 = x$

32. $\log_x 8 = -3$ **33.** $\log_9 x = \frac{3}{2}$

Solve Problems 34–41 for x to four decimal places.

34. $x = 3(e^{1.49})$ **35.** $x = 230(10^{-0.161})$

36. $\log x = -2.0144$ **37.** $\ln x = 0.3618$

38. $35 = 7(3^x)$ **39.** $0.01 = e^{-0.05x}$

40. $8,000 = 4,000(1.08^x)$ **41.** $5^{2x-3} = 7.08$

42. Find the domain of each function:

(A) $f(x) = \dfrac{2x - 5}{x^2 - x - 6}$ (B) $g(x) = \dfrac{3x}{\sqrt{5 - x}}$

43. Find the vertex form for $f(x) = 4x^2 + 4x - 3$ and then find the intercepts, the vertex, the maximum or minimum, and the range.

44. Let $f(x) = e^x - 1$ and $g(x) = \ln(x + 2)$. Find all points of intersection for the graphs of f and g. Round answers to two decimal places.

In Problems 45 and 46, use point-by-point plotting to sketch the graph of each function.

45. $f(x) = \dfrac{50}{x^2 + 1}$ **46.** $f(x) = \dfrac{-66}{2 + x^2}$

If $f(x) = 5x + 1$, find and simplify each of the following in Problems 47–50.

47. $f(f(0))$ **48.** $f(f(-1))$

49. $f(2x - 1)$ **50.** $f(4 - x)$

51. Let $f(x) = 3 - 2x$. Find

(A) $f(2)$ (B) $f(2 + h)$

(C) $f(2 + h) - f(2)$ (D) $\dfrac{f(2 + h) - f(2)}{h}, h \neq 0$

52. Let $f(x) = x^2 - 3x + 1$. Find

(A) $f(a)$ (B) $f(a + h)$

(C) $f(a + h) - f(a)$ (D) $\dfrac{f(a + h) - f(a)}{h}, h \neq 0$

✎ **53.** Explain how the graph of $m(x) = -|x - 4|$ is related to the graph of $y = |x|$.

✎ **54.** Explain how the graph of $g(x) = 0.3x^3 + 3$ is related to the graph of $y = x^3$.

✎ **55.** The following graph is the result of applying a sequence of transformations to the graph of $y = x^2$. Describe the transformations verbally and write an equation for the graph.

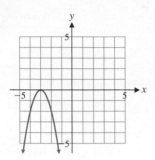

56. The graph of a function f is formed by vertically stretching the graph of $y = \sqrt{x}$ by a factor of 2, and shifting it to the left 3 units and down 1 unit. Find an equation for function f and graph it for $-5 \le x \le 5$ and $-5 \le y \le 5$.

In Problems 57–59, find the equation of any horizontal asymptote.

57. $f(x) = \dfrac{5x + 4}{x^2 - 3x + 1}$ **58.** $f(x) = \dfrac{3x^2 + 2x - 1}{4x^2 - 5x + 3}$

59. $f(x) = \dfrac{x^2 + 4}{100x + 1}$

In Problems 60 and 61, find the equations of any vertical asymptotes.

60. $f(x) = \dfrac{x^2 + 100}{x^2 - 100}$ **61.** $f(x) = \dfrac{x^2 + 3x}{x^2 + 2x}$

In Problems 62–67, discuss the validity of each statement. If the statement is always true, explain why. If not, give a counter example.

✎ **62.** Every polynomial function is a rational function.

✎ **63.** Every rational function is a polynomial function.

✎ **64.** The graph of every rational function has at least one vertical asymptote.

✎ **65.** The graph of every exponential function has a horizontal asymptote.

✎ **66.** The graph of every logarithmic function has a vertical asymptote.

✎ **67.** There exists a rational function that has both a vertical and horizontal asymptote.

68. Sketch the graph of f for $x \ge 0$.

$$f(x) = \begin{cases} 9 + 0.3x & \text{if } 0 \le x \le 20 \\ 5 + 0.2x & \text{if } x > 20 \end{cases}$$

69. Sketch the graph of g for $x \ge 0$.

$$f(x) = \begin{cases} 0.5x + 5 & \text{if } 0 \le x \le 10 \\ 1.2x - 2 & \text{if } 10 < x \le 30 \\ 2x - 26 & \text{if } x > 30 \end{cases}$$

70. Write an equation for the graph shown in the form $y = a(x - h)^2 + k$, where a is either -1 or $+1$ and h and k are integers.

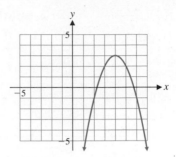

71. Given $f(x) = -0.4x^2 + 3.2x + 1.2$, find the following algebraically (to one decimal place) without referring to a graph:

(A) Intercepts   (B) Vertex

(C) Maximum or minimum   (D) Range

72. Graph $f(x) = -0.4x^2 + 3.2x + 1.2$ in a graphing calculator and find the following (to one decimal place) using TRACE and appropriate commands:

(A) Intercepts   (B) Vertex

(C) Maximum or minimum   (D) Range

73. Noting that $\pi = 3.141\,592\,654\ldots$ and $\sqrt{2} = 1.414\,213\,562\ldots$ explain why the calculator results shown here are obvious. Discuss similar connections between the natural logarithmic function and the exponential function with base e.

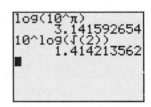

Solve Problems 74–77 exactly without using a calculator.

74. $\log x - \log 3 = \log 4 - \log(x + 4)$

75. $\ln(2x - 2) - \ln(x - 1) = \ln x$

76. $\ln(x + 3) - \ln x = 2 \ln 2$

77. $\log 3x^2 = 2 + \log 9x$

78. Write $\ln y = -5t + \ln c$ in an exponential form free of logarithms. Then solve for y in terms of the remaining variables.

✎ **79.** Explain why 1 cannot be used as a logarithmic base.

80. The following graph is the result of applying a sequence of transformations to the graph of $y = \sqrt[3]{x}$. Describe the transformations verbally, and write an equation for the graph.

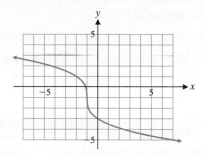

81. Given $G(x) = 0.3x^2 + 1.2\,x - 6.9$, find the following algebraically (to one decimal place) without the use of a graph:

(A) Intercepts   (B) Vertex

(C) Maximum or minimum   (D) Range

82. Graph $G(x) = 0.3x^2 + 1.2x - 6.9$ in a standard viewing window. Then find each of the following (to one decimal place) using appropriate commands.

(A) Intercepts   (B) Vertex

(C) Maximum or minimum   (D) Range

Applications

In all problems involving days, a 365-day year is assumed.

83. Electricity rates. The table shows the electricity rates charged by Easton Utilities in the summer months.

(A) Write a piecewise definition of the monthly charge $S(x)$ (in dollars) for a customer who uses x kWh in a summer month.

(B) Graph $S(x)$.

Energy Charge (June–September)

\$3.00 for the first 20 kWh or less
5.70¢ per kWh for the next 180 kWh
3.46¢ per kWh for the next 800 kWh
2.17¢ per kWh for all over 1,000 kWh

84. Money growth. Provident Bank of Cincinnati, Ohio, offered a certificate of deposit that paid 1.25% compounded quarterly. If a \$5,000 CD earns this rate for 5 years, how much will it be worth?

85. Money growth. Capital One Bank of Glen Allen, Virginia, offered a certificate of deposit that paid 1.05% compounded daily. If a \$5,000 CD earns this rate for 5 years, how much will it be worth?

86. Money growth. How long will it take for money invested at 6.59% compounded monthly to triple?

87. Money growth. How long will it take for money invested at 7.39% compounded continuously to double?

88. Break-even analysis. The research department in a company that manufactures AM/FM clock radios established the following price-demand, cost, and revenue functions:

$$p(x) = 50 - 1.25x \qquad \text{Price–demand function}$$
$$C(x) = 160 + 10x \qquad \text{Cost function}$$
$$R(x) = xp(x)$$
$$\qquad = x(50 - 1.25x) \quad \text{Revenue function}$$

where x is in thousands of units, and $C(x)$ and $R(x)$ are in thousands of dollars. All three functions have domain $1 \le x \le 40$.

(A) Graph the cost function and the revenue function simultaneously in the same coordinate system.

(B) Determine algebraically when $R = C$. Then, with the aid of part (A), determine when $R < C$ and $R > C$ to the nearest unit.

(C) Determine algebraically the maximum revenue (to the nearest thousand dollars) and the output (to the nearest unit) that produces the maximum revenue. What is the wholesale price of the radio (to the nearest dollar) at this output?

89. Profit–loss analysis. Use the cost and revenue functions from Problem 88.

(A) Write a profit function and graph it in a graphing calculator.

(B) Determine graphically when $P = 0$, $P < 0$, and $P > 0$ to the nearest unit.

(C) Determine graphically the maximum profit (to the nearest thousand dollars) and the output (to the nearest unit) that produces the maximum profit. What is the wholesale price of the radio (to the nearest dollar) at this output? [Compare with Problem 88C.]

90. Construction. A construction company has 840 feet of chain-link fence that is used to enclose storage areas for equipment and materials at construction sites. The supervisor wants to set up two identical rectangular storage areas sharing a common fence (see the figure).

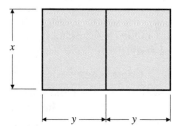

Assuming that all fencing is used,

(A) Express the total area $A(x)$ enclosed by both pens as a function of x.

(B) From physical considerations, what is the domain of the function A?

(C) Graph function A in a rectangular coordinate system.

(D) Use the graph to discuss the number and approximate locations of values of x that would produce storage areas with a combined area of 25,000 square feet.

(E) Approximate graphically (to the nearest foot) the values of x that would produce storage areas with a combined area of 25,000 square feet.

(F) Determine algebraically the dimensions of the storage areas that have the maximum total combined area. What is the maximum area?

91. Equilibrium point. A company is planning to introduce a 10-piece set of nonstick cookware. A marketing company established price–demand and price–supply tables for selected prices (Tables 1 and 2), where x is the number of cookware sets people are willing to buy and the company is willing to sell each month at a price of p dollars per set.

Table 1 **Price–Demand**

x	$p = D(x)\,(\$)$
985	330
2,145	225
2,950	170
4,225	105
5,100	50

Table 2 **Price–Supply**

x	$p = S(x)\,(\$)$
985	30
2,145	75
2,950	110
4,225	155
5,100	190

(A) Find a quadratic regression model for the data in Table 1. Estimate the demand at a price level of $180.

(B) Find a linear regression model for the data in Table 2. Estimate the supply at a price level of $180.

✎(C) Does a price level of $180 represent a stable condition, or is the price likely to increase or decrease? Explain.

(D) Use the models in parts (A) and (B) to find the equilibrium point. Write the equilibrium price to the nearest cent and the equilibrium quantity to the nearest unit.

92. Crime statistics. According to data published by the FBI, the crime index in the United States has shown a downward trend since the early 1990s (Table 3).

Table 3 **Crime Index**

Year	Crimes per 100,000 Inhabitants
1987	5,550
1992	5,660
1997	4,930
2002	4,125
2007	3,749
2009	3,466

(A) Find a cubic regression model for the crime index if $x = 0$ represents 1987.

(B) Use the cubic regression model to predict the crime index in 2022.

93. Medicine. One leukemic cell injected into a healthy mouse will divide into 2 cells in about $\frac{1}{2}$ day. At the end of the day these 2 cells will divide into 4. This doubling continues until

1 billion cells are formed; then the animal dies with leukemic cells in every part of the body.

(A) Write an equation that will give the number N of leukemic cells at the end of t days.

(B) When, to the nearest day, will the mouse die?

94. Marine biology. The intensity of light entering water is reduced according to the exponential equation

$$I = I_0 e^{-kd}$$

where I is the intensity d feet below the surface, I_0 is the intensity at the surface, and k is the coefficient of extinction. Measurements in the Sargasso Sea have indicated that half of the surface light reaches a depth of 73.6 feet. Find k (to five decimal places), and find the depth (to the nearest foot) at which 1% of the surface light remains.

95. Agriculture. The number of dairy cows on farms in the United States is shown in Table 4 for selected years since 1950. Let 1940 be year 0.

Table 4 **Dairy Cows on Farms in the United States**

Year	Dairy Cows (thousands)
1950	23,853
1960	19,527
1970	12,091
1980	10,758
1990	10,015
2000	9,190
2010	9,117

(A) Find a logarithmic regression model ($y = a + b \ln x$) for the data. Estimate (to the nearest thousand) the number of dairy cows in 2023.

✎(B) Explain why it is not a good idea to let 1950 be year 0.

96. Population growth. The population of some countries has a relative growth rate of 3% (or more) per year. At this rate, how many years (to the nearest tenth of a year) will it take a population to double?

97. Medicare. The annual expenditures for Medicare (in billions of dollars) by the U.S. government for selected years since 1980 are shown in Table 5. Let x represent years since 1980.

Table 5 **Medicare Expenditures**

Year	Billion $
1980	37
1985	72
1990	111
1995	181
2000	197
2005	299
2010	452

(A) Find an exponential regression model ($y = ab^x$) for the data. Estimate (to the nearest billion) the annual expenditures in 2022.

(B) When will the annual expenditures reach two trillion dollars?

PART 2

FINITE MATHEMATICS

3 Mathematics of Finance

3.1 Simple Interest

3.2 Compound and Continuous Compound Interest

3.3 Future Value of an Annuity; Sinking Funds

3.4 Present Value of an Annuity; Amortization

Chapter 3 Summary and Review

Review Exercises

Introduction

How do I choose the right loan for college? Would it be better to take the dealer's financing or the rebate for my new car? Should my parents refinance their home mortgage? To make wise decisions in such matters, you need a basic understanding of the mathematics of finance.

In Chapter 3 we study the mathematics of simple and compound interest, ordinary annuities, auto loans, and home mortage loans (see Problems 37–40 in Section 3.4). You will need a calculator with logarithmic and exponential keys. A graphing calculator would be even better: It can help you visualize the rate at which an investment grows or the rate at which principal on a loan is amortized.

You may wish to review arithmetic and geometric sequences, discussed in Appendix B.2, before beginning this chapter.

Finally, to avoid repeating the following reminder many times, we emphasize it here: Throughout the chapter, **interest rates are to be converted to decimal form before they are used in a formula.**

3.1 Simple Interest

- The Simple Interest Formula
- Simple Interest and Investments

The Simple Interest Formula

Simple interest is used on short-term notes—often of duration less than 1 year. The concept of simple interest, however, forms the basis of much of the rest of the material developed in this chapter, for which time periods may be much longer than a year.

If you deposit a sum of money P in a savings account or if you borrow a sum of money P from a lender, then P is referred to as the **principal**. When money is borrowed—whether it is a savings institution borrowing from you when you deposit money in your account, or you borrowing from a lender—a fee is charged for the money borrowed. This fee is rent paid for the use of another's money, just as rent is paid for the use of another's house. The fee is called **interest**. It is usually computed as a percentage (called the **interest rate**)* of the principal over a given period of time. The interest rate, unless otherwise stated, is an annual rate. **Simple interest** is given by the following formula:

> **DEFINITION** Simple Interest
>
> $$I = Prt \tag{1}$$
>
> where $I =$ interest
> $P =$ principal
> $r =$ annual simple interest rate (written as a decimal)
> $t =$ time in years

For example, the interest on a loan of $100 at 12% for 9 months would be

$$I = Prt$$
$$= (100)(0.12)(0.75) \quad \text{Convert 12\% to a decimal (0.12)}$$
$$= \$9 \qquad\qquad\qquad \text{and 9 months to years } \left(\tfrac{9}{12} = 0.75\right).$$

At the end of 9 months, the borrower would repay the principal ($100) plus the interest ($9), or a total of $109.

In general, if a principal P is borrowed at a rate r, then after t years, the borrower will owe the lender an amount A that will include the principal P plus the interest I. Since P is the amount that is borrowed now and A is the amount that must be paid back in the future, P is often referred to as the **present value** and A as the **future value**. The formula relating A and P follows:

> **THEOREM 1** Simple Interest
>
> $$A = P + Prt$$
> $$= P(1 + rt) \tag{2}$$
>
> where $A =$ amount, or future value
> $P =$ principal, or present value
> $r =$ annual simple interest rate (written as a decimal)
> $t =$ time in years

Given any three of the four variables A, P, r, and t in (2), we can solve for the fourth. The following examples illustrate several types of common problems that can be solved by using formula (2).

---

*If r is the interest rate written as a decimal, then $100r\%$ is the rate using %. For example, if $r = 0.12$, then $100r\% = 100(0.12)\% = 12\%$. The expressions 0.12 and 12% are equivalent.

EXAMPLE 1   Total Amount Due on a Loan   Find the total amount due on a loan of $800 at 9% simple interest at the end of 4 months.

SOLUTION   To find the amount A (future value) due in 4 months, we use formula (2) with $P = 800$, $r = 0.09$, and $t = \frac{4}{12} = \frac{1}{3}$ year. Thus,

$$
\begin{aligned}
A &= P(1 + rt) \\
&= 800\left[1 + 0.09\left(\tfrac{1}{3}\right)\right] \\
&= 800(1.03) \\
&= \$824
\end{aligned}
$$

Matched Problem 1   Find the total amount due on a loan of $500 at 12% simple interest at the end of 30 months.

Explore and Discuss 1   (A) Your sister has loaned you $1,000 with the understanding that you will repay the principal plus 4% simple interest when you can. How much would you owe her if you repaid the loan after 1 year? After 2 years? After 5 years? After 10 years?

(B) How is the interest after 10 years related to the interest after 1 year? After 2 years? After 5 years?

(C) Explain why your answers are consistent with the fact that for simple interest, the graph of future value as a function of time is a straight line (Fig. 1).

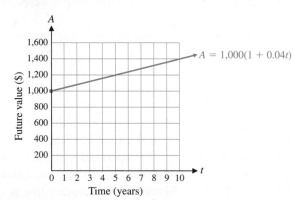

Figure 1

EXAMPLE 2   Present Value of an Investment   If you want to earn an annual rate of 10% on your investments, how much (to the nearest cent) should you pay for a note that will be worth $5,000 in 9 months?

SOLUTION   We again use formula (2), but now we are interested in finding the principal P (present value), given $A = \$5,000$, $r = 0.1$, and $t = \frac{9}{12} = 0.75$ year. Thus,

$$
\begin{aligned}
A &= P(1 + rt) \\
5{,}000 &= P[1 + 0.1(0.75)] \qquad \text{\small Replace } A, r, \text{ and } t \text{ with the} \\
5{,}000 &= (1.075)P \qquad\qquad\quad \text{\small given values, and solve for } P. \\
P &= \$4{,}651.16
\end{aligned}
$$

Matched Problem 2   Repeat Example 2 with a time period of 6 months.

> ## CONCEPTUAL INSIGHT
>
> If we consider future value A as a function of time t with the present value P and the annual rate r being fixed, then $A = P + Prt$ is a linear function of t with y intercept P and slope Pr. For example, if $P = 1,000$ and $r = 0.04$ (Fig. 1), then
>
> $$A = 1,000(1 + 0.04t) = 1,000 + 40t$$
>
> is a linear function with y intercept 1,000 and slope 40.

Simple Interest and Investments

EXAMPLE 3 Interest Rate Earned on a Note T-bills (Treasury bills) are one of the instruments that the U.S. Treasury Department uses to finance the public debt. If you buy a 180-day T-bill with a maturity value of $10,000 for $9,893.78, what annual simple interest rate will you earn? (Express answer as a percentage, correct to three decimal places.)

SOLUTION Again we use formula (2), but this time we are interested in finding r, given $P = \$9,893.78$, $A = \$10,000$, and $t = 180/360 = 0.5$ year.*

$$A = P(1 + rt)$$

$$10,000 = 9,893.78(1 + 0.5r)$$

$$10,000 = 9,893.78 + 4,946.89r$$

$$106.22 = 4,946.89r$$

$$r = \frac{106.22}{4,946.89} \approx 0.02147 \quad \text{or} \quad 2.147\%$$

Replace P, A, and t with the given values, and solve for r.

Matched Problem 3 Repeat Example 3, assuming that you pay $9,828.74 for the T-bill.

EXAMPLE 4 Interest Rate Earned on an Investment Suppose that after buying a new car you decide to sell your old car to a friend. You accept a 270-day note for $3,500 at 10% simple interest as payment. (Both principal and interest are paid at the end of 270 days.) Sixty days later you find that you need the money and sell the note to a third party for $3,550. What annual interest rate will the third party receive for the investment? Express the answer as a percentage, correct to three decimal places.

SOLUTION

Step 1 Find the amount that will be paid at the end of 270 days to the holder of the note.

$$A = P(1 + rt)$$
$$= \$3,500\left[1 + (0.1)\left(\tfrac{270}{360}\right)\right]$$
$$= \$3,762.50$$

Step 2 For the third party, we are to find the annual rate of interest r required to make $3,550 grow to $3,762.50 in 210 days $(270 - 60)$; that is, we are to

---

*In situations that involve days, some institutions use a 360-day year, called a banker's year, to simplify calculations. In this section, we will use a 360-day year. In other sections, we will use a 365-day year. The choice will always be clearly stated.

find r (which is to be converted to $100r\%$), given $A = \$3{,}762.50$, $P = \$3{,}550$, and $t = \frac{210}{360}$.

$$A = P + Prt \qquad \text{Solve for } r.$$

$$r = \frac{A - P}{Pt}$$

$$r = \frac{3{,}762.50 - 3{,}550}{(3{,}550)\left(\frac{210}{360}\right)} \approx 0.102\,62 \quad \text{or} \quad 10.262\%$$

Matched Problem 4 Repeat Example 4 assuming that 90 days after it was initially signed, the note was sold to a third party for $3,500.

Some online discount brokerage firms offer flat rates for trading stock, but many still charge commissions based on the transaction amount (principal). Table 1 shows the commission schedule for one of these firms.

Table 1 **Commission Schedule**

Principal	Commission
$0–$2,499	$29 + 1.6% of principal
$2,500–$9,999	$49 + 0.8% of principal
$10,000+	$99 + 0.3% of principal

EXAMPLE 5   Interest on an Investment   An investor purchases 50 shares of a stock at $47.52 per share. After 200 days, the investor sells the stock for $52.19 per share. Using Table 1, find the annual rate of interest earned by this investment. Express the answer as a percentage, correct to three decimal places.

SOLUTION   The principal referred to in Table 1 is the value of the stock. The total cost for the investor is the cost of the stock plus the commission:

$$47.52(50) = \$2{,}376 \qquad \text{Principal}$$
$$29 + 0.016(2{,}376) = \$67.02 \qquad \text{Commission, using line 1 of Table 1}$$
$$2{,}376 + 67.02 = \$2{,}443.02 \qquad \text{Total investment}$$

When the stock is sold, the commission is subtracted from the proceeds of the sale and the remainder is returned to the investor:

$$52.19(50) = \$2{,}609.50 \qquad \text{Principal}$$
$$49 + 0.008(2{,}609.50) = \$69.88 \qquad \text{Commission, using line 2 of Table 1}$$
$$2{,}609.50 - 69.88 = \$2{,}539.62 \qquad \text{Total return}$$

Now using formula (2) with $A = 2{,}539.62$, $P = 2{,}443.02$, and $t = \frac{200}{360} = \frac{5}{9}$, we have

$$A = P(1 + rt)$$

$$2{,}539.62 = 2{,}443.02\left(1 + \frac{5}{9}r\right)$$

$$= 2{,}443.02 + 1{,}357.23r$$

$$96.60 = 1{,}357.23r$$

$$r = \frac{96.60}{1{,}357.23} \approx 0.07117 \quad \text{or} \quad 7.117\%$$

Matched Problem 5 Repeat Example 5 if 500 shares of stock were purchased for $17.64 per share and sold 270 days later for $22.36 per share.

CONCEPTUAL INSIGHT

The commission schedule in Table 1 specifies a piecewise-defined function C with independent variable p, the principal (see Section 2.2).

$$C = \begin{cases} 29 + 0.016p & \text{if } 0 \le p < 2{,}500 \\ 49 + 0.008p & \text{if } 2{,}500 \le p < 10{,}000 \\ 99 + 0.003p & \text{if } 10{,}000 \le p \end{cases}$$

Two credit card accounts may differ in a number of ways, including annual interest rates, credit limits, minimum payments, annual fees, billing cycles, and even the methods for calculating interest. A common method for calculating the interest owed on a credit card is the **average daily balance method**. In this method, a balance is calculated at the end of each day, incorporating any purchases, credits, or payments that were made that day. Interest is calculated at the end of the billing cycle on the average of those daily balances. The average daily balance method is considered in Example 6.

EXAMPLE 6   Credit Card Accounts   A credit card has an annual interest rate of 21.99%, and interest is calculated by the average daily balance method. In a 30-day billing cycle, purchases of $56.75, $184.36, and $49.19 were made on days 12, 19, and 24, respectively, and a payment of $100.00 was credited to the account on day 10. If the unpaid balance at the start of the billing cycle was $842.67, how much interest will be charged at the end of the billing cycle? What will the unpaid balance be at the start of the next billing cycle?

SOLUTION   First calculate the unpaid balance on each day of the billing cycle:

$$
\begin{array}{ll}
\text{Days } 1\text{--}9\text{:} & \$842.67 \\
\text{Days } 10\text{--}11\text{:} & \$842.67 - \$100.00 = \$742.67 \\
\text{Days } 12\text{--}18\text{:} & \$742.67 + \$56.75 \quad - \$799.42 \\
\text{Days } 19\text{--}23\text{:} & \$799.42 + \$184.36 = \$983.78 \\
\text{Days } 24\text{--}30\text{:} & \$983.78 + \$49.19 \quad = \$1{,}032.97
\end{array}
$$

So the unpaid balance was $842.67 for the first 9 days of the billing cycle, $742.67 for the next 2 days, $799.42 for the next 7 days, and so on. To calculate the average daily balance, we find the sum of the 30 daily balances, and then divide by 30:

Sum: $9(\$842.67) + 2(\$742.67) + 7(\$799.42) + 5(\$983.78)$
$+ 7(\$1032.97) = \$26{,}815.00$

Average daily balance: $\$26{,}815.00/30 = \$893.83$

To calculate the interest, use the formula $I = Prt$ with $P = \$893.83$, $r = .2199$, and $t = 30/360$:

$$I = Prt = \$893.83(0.2199)(30/360) = \$16.38$$

Therefore, the interest charged at the end of the billing cycle is $16.38, and the unpaid balance at the start of the next cycle is $1{,}032.97 + \$16.38 = \$1{,}049.35$.

Matched Problem 6   A credit card has an annual interest rate of 16.99%, and interest is calculated by the average daily balance method. In a 30-day billing cycle, purchases of $345.86 and $246.71 were made on days 9 and 16, respectively, and a payment of $500.00 was credited to the account on day 15. If the unpaid balance at the start of the billing cycle was $1,792.19, how much interest will be charged at the end of the billing cycle? What will the unpaid balance be at the start of the next billing cycle?

Exercises 3.1

Skills Warm-up Exercises

In Problems 1–4, if necessary, review Section A.1.

1. If your state sales tax rate is 5.65%, how much tax will you pay on a bicycle that sells for $449.99?

2. If your state sales tax rate is 8.25%, what is the total cost of a motor scooter that sells for $1,349.95?

3. A basketball team has a 16–5 won-loss record. Find its winning percentage to the nearest percentage point.

4. A football team played 16 games with a winning percentage of 62.5%. How many games did it lose?

In Problems 5–8, give the slope and y intercept of each line. (If necessary, review Section 1.2.)

5. $y = 12,000 + 120x$

6. $y = 15,000 + 300x$

7. $y = 2,000(1 + 0.025x)$

8. $y = 5,000(1 + 0.035x)$

In Problems 9–16, convert the given interest rate to decimal form if it is given as a percentage, and to a percentage if it is given in decimal form.

9. 1.5%   10. 2.75%

11. 0.006   12. 0.0075

13. 0.4%   14. 0.9%

15. 0.2499   16. 0.165

In Problems 17–24, convert the given time period to years, in fraction form, assuming a 360-day year [this assumption does not affect the number of quarters (4), months (12), or weeks (52) in a year].

17. 4 months   18. 39 weeks

19. 240 days   20. 6 quarters

21. 12 weeks   22. 10 months

23. 2 quarters   24. 30 days

In Problems 25–32, use formula (1) for simple interest to find each of the indicated quantities.

25. $P = \$300$; $r = 7\%$; $t = 2$ years; $I = ?$

26. $P = \$950$; $r = 9\%$; $t = 1$ year; $I = ?$

27. $I = \$36$; $r = 4\%$; $t = 6$ months; $P = ?$

28. $I = \$15$; $r = 8\%$; $t = 3$ quarters; $P = ?$

29. $I = \$48$; $P = \$600$; $t = 240$ days; $r = ?$

30. $I = \$28$; $P = \$700$; $t = 13$ weeks; $r = ?$

31. $I = \$60$; $P = \$2,400$; $r = 5\%$; $t = ?$

32. $I = \$96$; $P = \$3,200$; $r = 4\%$; $t = ?$

In Problems 33–40, use formula (2) for the amount to find each of the indicated quantities.

33. $P = \$4,500$; $r = 10\%$; $t = 1$ quarter; $A = ?$

34. $P = \$3,000$; $r = 4.5\%$; $t = 30$ days; $A = ?$

35. $A = \$910$; $r = 16\%$; $t = 13$ weeks; $P = ?$

36. $A = \$6,608$; $r = 24\%$; $t = 3$ quarters; $P = ?$

37. $A = \$14,560$; $P = \$13,000$; $t = 4$ months; $r = ?$

38. $A = \$22,135$; $P = \$19,000$; $t = 39$ weeks; $r = ?$

39. $A = \$736$; $P = \$640$; $r = 15\%$; $t = ?$

40. $A = \$410$; $P = \$400$; $r = 10\%$; $t = ?$

In Problems 41–46, solve each formula for the indicated variable.

41. $I = Prt$; for r 42. $I = Prt$; for P

43. $A = P + Prt$; for P 44. $A = P + Prt$; for r

45. $A = P(1 + rt)$; for t 46. $I = Prt$; for t

47. Discuss the similarities and differences in the graphs of future value A as a function of time t if $1,000 is invested at simple interest at rates of 4%, 8%, and 12%, respectively (see the figure).

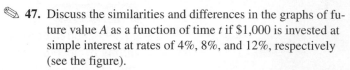
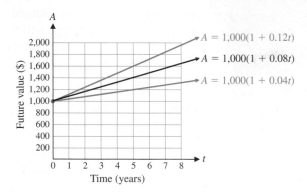

48. Discuss the similarities and differences in the graphs of future value A as a function of time t for loans of $400, $800, and $1,200, respectively, each at 7.5% simple interest (see the figure).

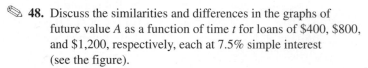
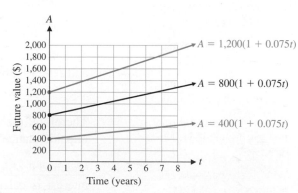

Applications*

In all problems involving days, a 360-day year is assumed. When annual rates are requested as an answer, express the rate as a percentage, correct to three decimal places. Round dollar amounts to the nearest cent.

49. If $3,000 is loaned for 4 months at a 4.5% annual rate, how much interest is earned?

50. If $5,000 is loaned for 9 months at a 6.2% annual rate, how much interest is earned?

51. How much interest will you have to pay for a credit card balance of $554 that is 1 month overdue, if a 20% annual rate is charged?

52. A department store charges an 18% annual rate for overdue accounts. How much interest will be owed on an $835 account that is 2 months overdue?

53. A loan of $7,260 was repaid at the end of 8 months. What size repayment check (principal and interest) was written, if an 8% annual rate of interest was charged?

54. A loan of $10,000 was repaid at the end of 6 months. What amount (principal and interest) was repaid, if a 6.5% annual rate of interest was charged?

55. A loan of $4,000 was repaid at the end of 10 months with a check for $4,270. What annual rate of interest was charged?

56. A check for $3,097.50 was used to retire a 5-month $3,000 loan. What annual rate of interest was charged?

57. If you paid $30 to a loan company for the use of $1,000 for 60 days, what annual rate of interest did they charge?

58. If you paid $120 to a loan company for the use of $2,000 for 90 days, what annual rate of interest did they charge?

59. A radio commercial for a loan company states: "You only pay 29¢ a day for each $500 borrowed." If you borrow $1,500 for 120 days, what amount will you repay, and what annual interest rate is the company charging?

60. George finds a company that charges 59¢ per day for each $1,000 borrowed. If he borrows $3,000 for 60 days, what amount will he repay, and what annual interest rate will he pay the company?

61. What annual interest rate is earned by a 13-week T-bill with a maturity value of $1,000 that sells for $989.37?

62. What annual interest rate is earned by a 33-day T-bill with a maturity value of $1,000 that sells for $996.16?

63. What is the purchase price of a 50-day T-bill with a maturity value of $1,000 that earns an annual interest rate of 5.53%?

64. What is the purchase price of a 26-week T-bill with a maturity value of $1,000 that earns an annual interest rate of 4.903%?

In Problems 65–68, assume that the minimum payment on a credit card is the greater of $20 or 2% of the unpaid balance.

65. Find the minimum payment on an unpaid balance of $1,215.45.

66. Find the minimum payment on an unpaid balance of $936.24.

67. If the annual interest rate is 16.99%, find the difference between the minimum payment and the interest owed on an unpaid balance of $869.89 that is 1 month overdue.

68. If the annual interest rate is 20.99%, find the difference between the minimum payment and the interest owed on an unpaid balance of $1,350.84 that is 1 month overdue.

69. For services rendered, an attorney accepts a 90-day note for $5,500 at 8% simple interest from a client. (Both interest and principal are repaid at the end of 90 days.) Wishing to use her money sooner, the attorney sells the note to a third party for $5,560 after 30 days. What annual interest rate will the third party receive for the investment?

70. To complete the sale of a house, the seller accepts a 180-day note for $10,000 at 7% simple interest. (Both interest and principal are repaid at the end of 180 days.) Wishing to use the money sooner for the purchase of another house, the seller sells the note to a third party for $10,124 after 60 days. What annual interest rate will the third party receive for the investment?

Use the commission schedule from Company A shown in Table 2 to find the annual rate of interest earned by each investment in Problems 71 and 72.

Table 2 **Company A**

Principal	Commission
Under $3,000	$25 + 1.8% of principal
$3,000–$10,000	$37 + 1.4% of principal
Over $10,000	$107 + 0.7% of principal

71. An investor purchases 200 shares at $14.20 a share, holds the stock for 39 weeks, and then sells the stock for $15.75 a share.

72. An investor purchases 450 shares at $21.40 a share, holds the stock for 26 weeks, and then sells the stock for $24.60 a share.

Use the commission schedule from Company B shown in Table 3 to find the annual rate of interest earned by each investment in Problems 73 and 74.

Table 3 **Company B**

Principal	Commission
Under $3,000	$32 + 1.8% of principal
$3,000–$10,000	$56 + 1% of principal
Over $10,000	$106 + 0.5% of principal

73. An investor purchases 215 shares at $45.75 a share, holds the stock for 300 days, and then sells the stock for $51.90 a share.

*The authors wish to thank Professor Roy Luke of Pierce College and Professor Dennis Pence of Western Michigan University for their many useful suggestions of applications for this chapter.

74. An investor purchases 75 shares at $37.90 a share, holds the stock for 150 days, and then sells the stock for $41.20 a share.

Many tax preparation firms offer their clients a refund anticipation loan (RAL). For a fee, the firm will give a client his refund when the return is filed. The loan is repaid when the IRS refund is sent to the firm. The RAL fee is equivalent to the interest charge for a loan. The schedule in Table 4 is from a major RAL lender. Use this schedule to find the annual rate of interest for the RALs in Problems 75–78.

Table 4

RAL Amount	RAL Fee
$0−$500	$29.00
$501−$1,000	$39.00
$1,001−$1,500	$49.00
$1,501−$2,000	$69.00
$2,001−$5,000	$89.00

75. A client receives a $475 RAL, which is paid back in 20 days.

76. A client receives a $1,100 RAL, which is paid back in 30 days.

77. A client receives a $1,900 RAL, which is paid back in 15 days.

78. A client receives a $3,000 RAL, which is paid back in 25 days.

In problems 79–82, assume that the annual interest rate on a credit card is 19.99%, and that interest is calculated by the average daily balance method.

79. The unpaid balance at the start of a 30-day billing cycle was $2,000.00. A $1,000.00 purchase was made on the first day of the billing cycle, and a $1,500.00 payment was credited to the account on the last day of the billing cycle. How much interest will be charged at the end of the billing cycle?

80. The unpaid balance at the start of a 30-day billing cycle was $2,000.00. A $1,500.00 payment was credited to the account on the first day of the billing cycle, and a $1,000.00 purchase was made on the last day of the billing cycle. How much interest will be charged at the end of the billing cycle?

81. The unpaid balance at the start of a 28-day billing cycle was $523.18. Purchases of $147.98 and $36.27 were made on days 12 and 25, respectively, and a payment of $200.00 was credited to the account on day 17. How much interest will be charged at the end of the billing cycle?

82. The unpaid balance at the start of a 28-day billing cycle was $696.21. Purchases of $25.59, $19.95, and $97.26 were made on days 6, 13, and 21, respectively, and a payment of $140.00 was credited to the account on day 8. How much interest will be charged at the end of the billing cycle?

Answers to Matched Problems

1. $650	**2.** $4,761.90	**3.** 3.485%
4. 15.0%	**5.** 31.439%	**6.** $26.94; $1,911.70

3.2 Compound and Continuous Compound Interest

- Compound Interest
- Continuous Compound Interest
- Growth and Time
- Annual Percentage Yield

Compound Interest

If at the end of a payment period the interest due is reinvested at the same rate, then the interest as well as the original principal will earn interest during the next payment period. Interest paid on interest reinvested is called **compound interest**.

For example, suppose you deposit $1,000 in a bank that pays 8% compounded quarterly. How much will the bank owe you at the end of a year? *Compounding quarterly* means that earned interest is paid to your account at the end of each 3-month period and that interest as well as the principal earns interest for the next quarter. Using the simple interest formula (2) from the preceding section, we compute the amount in the account at the end of the first quarter after interest has been paid:

$$A = P(1 + rt)$$
$$= 1{,}000\left[1 + 0.08\left(\tfrac{1}{4}\right)\right]$$
$$= 1{,}000(1.02) = \$1{,}020$$

Now, $1,020 is your new principal for the second quarter. At the end of the second quarter, after interest is paid, the account will have

$$A = \$1{,}020\left[1 + 0.08\left(\tfrac{1}{4}\right)\right]$$
$$= \$1{,}020(1.02) = \$1{,}040.40$$

Similarly, at the end of the third quarter, you will have

$$A = \$1{,}040.40\left[1 + 0.08\left(\tfrac{1}{4}\right)\right]$$
$$= \$1{,}040.40(1.02) = \$1{,}061.21$$

Finally, at the end of the fourth quarter, the account will have

$$A = \$1{,}061.21\left[1 + 0.08\left(\tfrac{1}{4}\right)\right]$$
$$= \$1{,}061.21(1.02) = \$1{,}082.43$$

How does this compounded amount compare with simple interest? The amount with simple interest would be

$$A = P(1 + rt)$$
$$= \$1{,}000[1 + 0.08(1)]$$
$$= \$1{,}000(1.08) = \$1{,}080$$

We see that compounding quarterly yields $2.43 more than simple interest would provide.

Let's look over the calculations for compound interest above to see if we can uncover a pattern that might lead to a general formula for computing compound interest:

$A = 1{,}000(1.02)$	End of first quarter
$A = [1{,}000(1.02)](1.02) = 1{,}000(1.02)^2$	End of second quarter
$A = [1{,}000(1.02)^2](1.02) = 1{,}000(1.02)^3$	End of third quarter
$A = [1{,}000(1.02)^3](1.02) = 1{,}000(1.02)^4$	End of fourth quarter

It appears that at the end of n quarters, we would have

$$A = 1{,}000(1.02)^n \quad \text{End of } n\text{th quarter}$$

or

$$A = 1{,}000\left[1 + 0.08\left(\tfrac{1}{4}\right)\right]^n$$
$$= 1{,}000\left[1 + \tfrac{0.08}{4}\right]^n$$

where $\frac{0.08}{4} = 0.02$ is the interest rate per quarter. Since interest rates are generally quoted as *annual nominal rates*, the **rate per compounding period** is found by dividing the annual nominal rate by the number of compounding periods per year.

In general, if P is the principal earning interest compounded m times a year at an annual rate of r, then (by repeated use of the simple interest formula, using $i = r/m$, the rate per period) the amount A at the end of each period is

$A = P(1 + i)$	End of the first period
$A = [P(1 + i)](1 + i) = P(1 + i)^2$	End of second period
$A = [P(1 + i)^2](1 + i) = P(1 + i)^3$	End of third period
$\vdots$	
$A = [P(1 + i)^{n-1}](1 + i) = P(1 + i)^n$	End of nth period

We summarize this important result in Theorem 1:

THEOREM 1 Compound Interest

$$A = P(1 + i)^n \qquad (1)$$

where $i = r/m$ and $A =$ amount (future value) at the end of n periods

$P =$ principal (present value)

$r =$ annual nominal rate*

$m =$ number of compounding periods per year

$i =$ rate per compounding period

$n =$ total number of compounding periods

---

*This is often shortened to "annual rate" or just "rate."

CONCEPTUAL INSIGHT

Formula (1) of Theorem 1 is equivalent to the formula

$$A = P\left(1 + \frac{r}{m}\right)^{mt} \tag{2}$$

where t is the time, in years, that the principal is invested. For a compound interest calculation, formula (2) may seem more natural to use than (1), if r (the annual interest rate) and t (time in years) are given. On the other hand, if i (the interest rate per period), and n (the number of compounding periods) are given, formula (1) may seem easier to use. It is not necessary to memorize both formulas, but it is important to understand how they are related.

EXAMPLE 1   Comparing Interest for Various Compounding Periods   If $1,000 is invested at 8% compounded

(A) annually,   (B) semiannually,

(C) quarterly,   (D) monthly,

what is the amount after 5 years? Write answers to the nearest cent.

SOLUTION

(A) Compounding annually means that there is one interest payment period per year. So, $n = 5$ and $i = r = 0.08$.

$$\begin{aligned} A &= P(1 + i)^n \\ &= 1,000(1 + 0.08)^5 \quad \text{Use a calculator.} \\ &= 1,000(1.469\ 328) \\ &= \$1,469.33 \quad\quad \text{Interest earned} = A - p = \$469.33. \end{aligned}$$

(B) Compounding semiannually means that there are two interest payment periods per year. The number of payment periods in 5 years is $n = 2(5) = 10$, and the interest rate per period is

$$i = \frac{r}{m} = \frac{0.08}{2} = 0.04$$

$$\begin{aligned} A &= P(1 + i)^n \\ &= 1,000(1 + 0.04)^{10} \quad \text{Use a calculator.} \\ &= 1,000(1.480\ 244) \\ &= \$1,480.24 \quad\quad \text{Interest earned} = A - P = \$480.24. \end{aligned}$$

(C) Compounding quarterly means that there are four interest payments per year. So, $n = 4(5) = 20$ and $i = \frac{0.08}{4} = 0.02$.

$$\begin{aligned} A &= P(1 + i)^n \\ &= 1,000(1 + 0.02)^{20} \quad \text{Use a calculator.} \\ &= 1,000(1.485\ 947) \\ &= \$1,485.95 \quad\quad \text{Interest earned} = A - P = \$485.95. \end{aligned}$$

(D) Compounding monthly means that there are twelve interest payments per year. So, $n = 12(5) = 60$ and $i = \frac{0.08}{12} = 0.006\ 66\overline{6}$.*

---

*Recall that the bar over the 6 indicates a repeating decimal expansion. Rounding i to a small number of decimal places, such as 0.007 or 0.0067, can result in round-off errors. To avoid this, use as many decimal places for i as your calculator is capable of displaying.

$$A = P(1 + i)^n$$

$$= 1{,}000\left(1 + \frac{0.08}{12}\right)^{60} \qquad \text{Use a calculator.}$$

$$= 1{,}000(1.489\ 846)$$

$$= \$1{,}489.85 \qquad \text{Interest earned} = A - P = \$489.85.$$

Matched Problem 1 Repeat Example 1 with an annual interest rate of 6% over an 8-year period.

Continuous Compound Interest

In Example 1, we considered an investment of $1,000 at an annual rate of 8%. We calculated the amount after 5 years for interest compounded annually, semiannually, quarterly, and monthly. What would happen to the amount if interest were compounded daily, or every minute, or every second?

Although the difference in amounts in Example 1 between compounding semiannually and annually is $1,480.24 − $1,469.33 = $10.91, the difference between compounding monthly and quarterly is only $1,489.85 − $1,485.95 = $3.90. This suggests that as the number m of compounding periods per year increases without bound, the amount will approach some limiting value. To see that this is indeed the case, we rewrite the amount A as follows:

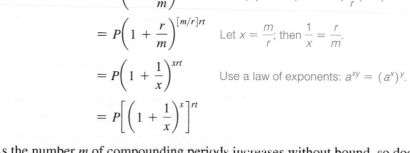

$$A = P(1 + i)^n \qquad \text{Substitute } i = \frac{r}{m}, n = mt.$$

$$= P\left(1 + \frac{r}{m}\right)^{mt} \qquad \text{Multiply the exponent by } \frac{r}{r} (=1).$$

$$= P\left(1 + \frac{r}{m}\right)^{[m/r]rt} \qquad \text{Let } x = \frac{m}{r}; \text{ then } \frac{1}{x} = \frac{r}{m}.$$

$$= P\left(1 + \frac{1}{x}\right)^{xrt} \qquad \text{Use a law of exponents: } a^{xy} = (a^x)^y.$$

$$= P\left[\left(1 + \frac{1}{x}\right)^x\right]^{rt}$$

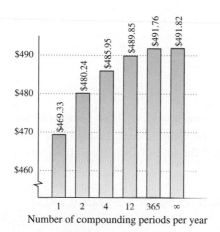

$490 $485.95 $489.85 $491.76 $491.82

$480 $480.24

$470 $469.33

$460

1   2   4   12   365   ∞

Number of compounding periods per year

Figure 1 **Interest on $1,000 for 5 years at 8% with various compounding periods**

As the number m of compounding periods increases without bound, so does x. So the expression in square brackets gets close to the irrational number $e \approx 2.7183$, and the amount approaches the limiting value

$$A = Pe^{rt} = 1{,}000e^{0.08(5)} \approx \$1{,}491.8247$$

In other words, no matter how often interest is compounded, the amount in the account after 5 years will never equal or exceed $1,491.83. Therefore, the interest $I(= A - P)$ will never equal or exceed $491.83 (Fig. 1).

CONCEPTUAL INSIGHT

One column in Figure 1 is labeled with the symbol ∞, read as "infinity." This symbol does not represent a real number. We use ∞ to denote the process of allowing m, the number of compounding periods per year, to get larger and larger with no upper limit on its size.

The formula we have obtained, $A = Pe^{rt}$, is known as the **continuous compound interest formula**. It is used when interest is **compounded continuously**; that is, when the number of compounding periods per year increases without bound.

> **THEOREM 2** Continuous Compound Interest Formula
>
> If a principal P is invested at an annual rate r (expressed as a decimal) compounded continuously, then the amount A in the account at the end of t years is given by
>
> $$A = Pe^{rt} \qquad\qquad (3)$$

EXAMPLE 2 Compounding Daily and Continuously What amount will an account have after 2 years if $5,000 is invested at an annual rate of 8%

(A) compounded daily?   (B) compounded continuously?

Compute answers to the nearest cent.

SOLUTION

(A) Use the compound interest formula

$$A = P\left(1 + \frac{r}{m}\right)^{mt}$$

with $P = 5{,}000$, $r = 0.08$, $m = 365$, and $t = 2$:

$$A = 5{,}000\left(1 + \frac{0.08}{365}\right)^{(365)(2)} \qquad \text{Use a calculator.}$$

$$= \$5{,}867.45$$

(B) Use the continuous compound interest formula

$$A = Pe^{rt}$$

with $P = 5{,}000$, $r = 0.08$, and $t = 2$:

$$A = 5{,}000e^{(0.08)(2)} \qquad \text{Use a calculator.}$$

$$= \$5{,}867.55$$

⚠️ **CAUTION** In Example 2B, do not use the approximation 2.7183 for e; it is not accurate enough to compute the correct amount to the nearest cent. Instead, use your calculator's built-in e. Avoid any rounding off until the end of the calculation, when you round the amount to the nearest cent.   ▲

Matched Problem 2 What amount will an account have after 1.5 years if $8,000 is invested at an annual rate of 9%

(A) compounded weekly?   (B) compounded continuously?

Compute answers to the nearest cent.

> **CONCEPTUAL INSIGHT**
>
> The continuous compound interest formula $A = Pe^{rt}$ is identical, except for the names of the variables, to the equation $y = ce^{kt}$ that we used to model population growth in Section 2.5. Like the growth of an investment that earns continuous compound interest, we usually consider the population growth of a country to be continuous: Births and deaths occur all the time, not just at the end of a month or quarter.

Growth and Time

How much should you invest now to have a given amount at a future date? What annual rate of return have your investments earned? How long will it take your investment to double in value? The formulas for compound interest and continuous compound interest can be used to answer such questions. If the values of all but one of the variables in the formula are known, then we can solve for the remaining variable.

EXAMPLE 3   Finding Present Value   How much should you invest now at 10% to have $8,000 toward the purchase of a car in 5 years if interest is

(A) compounded quarterly?   (B) compounded continuously?

SOLUTION

(A) We are given a future value $A = \$8,000$ for a compound interest investment, and we need to find the present value P given $i = \frac{0.10}{4} = 0.025$ and $n = 4(5) = 20$.

$$A = P(1 + i)^n$$
$$8,000 = P(1 + 0.025)^{20}$$
$$P = \frac{8,000}{(1 + 0.025)^{20}} \qquad \text{Use a calculator.}$$
$$= \frac{8,000}{1.638\ 616}$$
$$= \$4,882.17$$

Your initial investment of $4,882.17 will grow to $8,000 in 5 years.

(B) We are given $A = \$8,000$ for an investment at continuous compound interest, and we need to find the present value P given $r = 0.10$ and $t = 5$.

$$A = Pe^{rt}$$
$$8,000 = Pe^{0.10(5)}$$
$$P = \frac{8,000}{e^{0.10(5)}} \qquad \text{Use a calculator.}$$
$$P = \$4,852.25$$

Your initial investment of $4,852.25 will grow to $8,000 in 5 years.

Matched Problem 3   How much should new parents invest at 8% to have $80,000 toward their child's college education in 17 years if interest is

(A) compounded semiannually?   (B) compounded continuously?

A graphing calculator is a useful tool for studying compound interest. In Figure 2, we use a spreadsheet to illustrate the growth of the investment in Example 3A both numerically and graphically. Similar results can be obtained from most graphing calculators.

	A	B	C
1	Period	Interest	Amount
2	0		$4,882.17
3	1	$122.05	$5,004.22
4	2	$125.11	$5,129.33
5	3	$128.23	$5,257.56
6	4	$131.44	$5,389.00
7	5	$134.73	$5,523.73
8	6	$138.09	$5,661.82
9	7	$141.55	$5,803.37
10	8	$145.08	$5,948.45
11	9	$148.71	$6,097.16
12	10	$152.43	$6,249.59
13	11	$156.24	$6,405.83
14	12	$160.15	$6,565.98
15	13	$164.15	$6,730.13
16	14	$168.25	$6,898.38
17	15	$172.46	$7,070.84
18	16	$176.77	$7,247.61
19	17	$181.19	$7,428.80
20	18	$185.72	$7,614.52
21	19	$190.36	$7,804.88
22	20	$195.12	$8,000.00

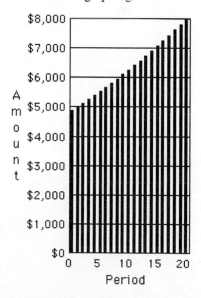

Figure 2 **Growth of $4,882.17 at 10% compounded quarterly for 5 years**

Solving the compound interest formula or the continuous compound interest formula for r enables us to determine the rate of growth of an investment.

EXAMPLE 4   Computing Growth Rate   Figure 3 shows that a $10,000 investment in a growth-oriented mutual fund over a 10-year period would have grown to $126,000. What annual nominal rate would produce the same growth if interest was:

(A) compounded annually?   (B) compounded continuously?

Express answers as percentages, rounded to three decimal places.

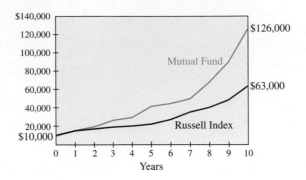

Figure 3   **Growth of a $10,000 investment**

SOLUTION

(A) $126,000 = 10,000(1 + r)^{10}$

$\qquad 12.6 = (1 + r)^{10}$

$\qquad \sqrt[10]{12.6} = 1 + r$

$\qquad r = \sqrt[10]{12.6} - 1 = 0.28836$ or   28.836%

(B) $126,000 = 10,000e^{r(10)}$

$\qquad 12.6 = e^{10r}$ Take ln of both sides.

$\qquad \ln 12.6 = 10r$

$\qquad r = \dfrac{\ln 12.6}{10} = 0.25337$ or   25.337%

Matched Problem 4   The Russell Index tracks the average performance of various groups of stocks. Figure 3 shows that, on average, a $10,000 investment in midcap growth funds over a 10-year period would have grown to $63,000. What annual nominal rate would produce the same growth if interest were

(A) compounded annually?   (B) compounded continuously?

Express answers as percentages, rounded to three decimal places.

CONCEPTUAL INSIGHT

We can solve $A = P(1 + i)^n$ for n using a property of logarithms:

$$\log_b M^p = p \log_b M$$

Theoretically, any base can be used for the logarithm, but most calculators only evaluate logarithms with base 10 (denoted log) or base e (denoted ln).

Finally, if we solve the compound interest formula for n (or the continuous compound interest formula for t), we can determine the **growth time** of an investment—the time it takes a given principal to grow to a particular value (the shorter the time, the greater the return on the investment).

Example 5 illustrates three methods for solving for growth time.

EXAMPLE 5   Computing Growth Time   How long will it take $10,000 to grow to $12,000 if it is invested at 9% compounded monthly?

SOLUTION

Method 1. Use logarithms and a calculator:

$$A = P(1 + i)^n$$

$$12{,}000 = 10{,}000\left(1 + \frac{0.09}{12}\right)^n$$

$$1.2 = 1.0075^n$$

Now, solve for n by taking logarithms of both sides:

$$\ln 1.2 = \ln 1.0075^n \quad \text{We choose the natural logarithm (base } e\text{)}$$

$$\ln 1.2 = n \ln 1.0075 \quad \text{and use the property ln } M^p = p \ln M.$$

$$n = \frac{\ln 1.2}{\ln 1.0075}$$

$$\approx 24.40 \approx 25 \text{ months} \quad \text{or} \quad 2 \text{ years and 1 month}$$

Note: 24.40 is rounded up to 25 to guarantee reaching $12,000 since interest is paid at the end of each month.

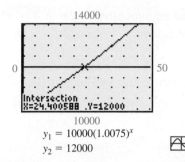

14000

0 ······································· 50

Intersection
X=24.40058B Y=12000

10000

$y_1 = 10000(1.0075)^x$
$y_2 = 12000$

Figure 4

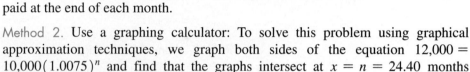 Method 2. Use a graphing calculator: To solve this problem using graphical approximation techniques, we graph both sides of the equation $12{,}000 = 10{,}000(1.0075)^n$ and find that the graphs intersect at $x = n = 24.40$ months (Fig. 4). So, the growth time is 25 months.

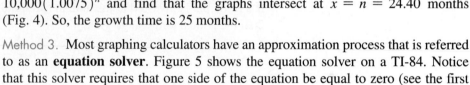 Method 3. Most graphing calculators have an approximation process that is referred to as an **equation solver**. Figure 5 shows the equation solver on a TI-84. Notice that this solver requires that one side of the equation be equal to zero (see the first line in Fig. 5). After entering values for three of the four variables, the solver will approximate the value of the remaining variable. Once again, we see that the growth time is 25 months (Fig. 5).

```
A-P(1+I)^N=0
  A=12000
  P=10000
  I=.0075
• N=24.400588158…
  bound={-1ᴇ99,1…
• left-rt=0
```

Figure 5 **TI-84 equation solver**

Matched Problem 5   How long will it take $10,000 to grow to $25,000 if it is invested at 8% compounded quarterly?

Annual Percentage Yield

Table 1 lists the rate and compounding period for certificates of deposit (CDs) offered by four banks. How can we tell which of these CDs has the best return?

Table 1 **Certificates of Deposit (CDs)**

Bank	Rate	Compounded
Advanta	4.93%	monthly
DeepGreen	4.95%	daily
Charter One	4.97%	quarterly
Liberty	4.94%	continuously

Explore and Discuss 1   Determine the value after 1 year of a $1,000 CD purchased from each of the banks in Table 1. Which CD offers the greatest return? Which offers the least return?

If a principal P is invested at an annual rate r compounded m times a year, then the amount after 1 year is

$$A = P\left(1 + \frac{r}{m}\right)^m$$

The simple interest rate that will produce the same amount A in 1 year is called the **annual percentage yield** (APY). To find the APY, we proceed as follows:

$$\begin{pmatrix} \text{amount at} \\ \text{simple interest} \\ \text{after 1 year} \end{pmatrix} = \begin{pmatrix} \text{amount at} \\ \text{compound interest} \\ \text{after 1 year} \end{pmatrix}$$

$$P(1 + \text{APY}) = P\left(1 + \frac{r}{m}\right)^m \qquad \text{Divide both sides by } P.$$

$$1 + \text{APY} = \left(1 + \frac{r}{m}\right)^m \qquad \text{Isolate APY on the left side.}$$

$$\text{APY} = \left(1 + \frac{r}{m}\right)^m - 1$$

If interest is compounded continuously, then the amount after 1 year is $A = Pe^r$. So to find the annual percentage yield, we solve the equation

$$P(1 + \text{APY}) = Pe^r$$

for APY, obtaining $\text{APY} = e^r - 1$. We summarize our results in Theorem 3.

THEOREM 3   Annual Percentage Yield

If a principal is invested at the annual (nominal) rate r compounded m times a year, then the annual percentage yield is

$$\text{APY} = \left(1 + \frac{r}{m}\right)^m - 1$$

If a principal is invested at the annual (nominal) rate r compounded continuously, then the annual percentage yield is

$$\text{APY} = e^r - 1$$

The annual percentage yield is also referred to as the **effective rate** or **true interest rate**.

Compound rates with different compounding periods cannot be compared directly (see Explore and Discuss 1). But since the annual percentage yield is a simple interest rate, the annual percentage yields for two different compound rates can be compared.

EXAMPLE 6   Using APY to Compare Investments   Find the APYs (expressed as a percentage, correct to three decimal places) for each of the banks in Table 1 and compare these CDs.

SOLUTION   Advanta: $\text{APY} = \left(1 + \dfrac{0.0493}{12}\right)^{12} - 1 = 0.05043$ or 5.043%

DeepGreen: $\text{APY} = \left(1 + \dfrac{0.0495}{365}\right)^{365} - 1 = 0.05074$ or 5.074%

Charter One: $\text{APY} = \left(1 + \dfrac{0.0497}{4}\right)^{4} - 1 = 0.05063$ or 5.063%

Liberty: $\text{APY} = e^{0.0494} - 1 = 0.05064$ or 5.064%

Comparing these APYs, we conclude that the DeepGreen CD will have the largest return and the Advanta CD will have the smallest.

Matched Problem 6) Southern Pacific Bank offered a 1-year CD that paid 4.8% compounded daily and Washington Savings Bank offered one that paid 4.85% compounded quarterly. Find the APY (expressed as a percentage, correct to three decimal places) for each CD. Which has the higher return?

EXAMPLE 7   Computing the Annual Nominal Rate Given the APY A savings and loan wants to offer a CD with a monthly compounding rate that has an APY of 7.5%. What annual nominal rate compounded monthly should they use?

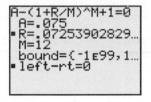 Check with a graphing calculator.

SOLUTION

$$APY = \left(1 + \frac{r}{m}\right)^m - 1$$

$$0.075 = \left(1 + \frac{r}{12}\right)^{12} - 1$$

$$1.075 = \left(1 + \frac{r}{12}\right)^{12}$$

$$\sqrt[12]{1.075} = 1 + \frac{r}{12}$$

$$\sqrt[12]{1.075} - 1 = \frac{r}{12}$$

$$r = 12\left(\sqrt[12]{1.075} - 1\right) \qquad \text{Use a calculator.}$$

$$= 0.072\ 539 \quad \text{or} \quad 7.254\%$$

So, an annual nominal rate of 7.254% compounded monthly is equivalent to an APY of 7.5%.

```
A-(1+R/M)^M+1=0
 A=.075
•R=.07253902829...
 M=12
 bound={-1ᴇ99,1...
•left-rt=0
```

Figure 6 **TI-84 equation solver**

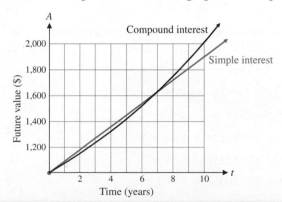

CHECK We use an equation solver on a graphing calculator to check this result (Fig. 6).

Matched Problem 7) What is the annual nominal rate compounded quarterly for a bond that has an APY of 5.8%?

⚠ **CAUTION** Each compound interest problem involves two interest rates. Referring to Example 5, $r = 0.09$ or 9% is the annual nominal compounding rate, and $i = r/12 = 0.0075$ or 0.75% is the interest rate per month. Do not confuse these two rates by using r in place of i in the compound interest formula. If interest is compounded annually, then $i = r/1 = r$. In all other cases, r and i are not the same. ▲

Explore and Discuss 2   (A) Which would be the better way to invest $1,000: at 9% simple interest for 10 years, or at 7% compounded monthly for 10 years?

(B) Explain why the graph of future value as a function of time is a straight line for simple interest, but for compound interest the graph curves upward (see Fig. 7).

Figure 7

CONCEPTUAL INSIGHT

The two curves in Figure 7 intersect at $t = 0$ and again near $t = 7$. The t coordinate of each intersection point is a solution of the equation

$$1{,}000(1 + 0.09t) = 1{,}000(1 + 0.07/12)^{12t}$$

Don't try to use algebra to solve this equation. It can't be done. But the solutions are easily approximated on a graphing calculator (Fig. 8).

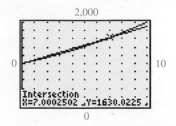

Figure 8

Exercises 3.2

Find all dollar amounts to the nearest cent. When an interest rate is requested as an answer, express the rate as a percentage correct to two decimal places, unless directed otherwise. In all problems involving days, use a 365-day year.

W Skills Warm-up Exercises

In Problems 1–8, solve the equation for the unknown quantity. (If necessary, review section A.7.)

1. $1{,}641.6 = P(1.2)^3$ **2.** $2{,}652.25 = P(1.03)^2$

3. $12x^3 = 58{,}956$ **4.** $100x^4 = 15{,}006.25$

5. $6.75 = 3(1 + i)^2$ **6.** $13.72 = 5(1 + i)^3$

7. $14{,}641 = 10{,}000(1.1)^n$ **8.** $2{,}488.32 = 1{,}000(1.2)^n$

In Problems 9–12, use compound interest formula (1) to find each of the indicated values.

9. $P = \$5{,}000; i = 0.005; n = 36; A = ?$

10. $P = \$2{,}800; i = 0.003; n = 24; A = ?$

11. $A = \$8{,}000; i = 0.02; n = 32; P = ?$

12. $A = \$15{,}000; i = 0.01; n = 28; P = ?$

In Problems 13–20, use the continuous compound interest formula (3) to find each of the indicated values.

13. $P = \$2{,}450; r = 8.12\%; t = 3 \text{ years}; A = ?$

14. $P = \$995; r = 22\%; t = 2 \text{ years}; A = ?$

15. $A = \$6{,}300; r = 9.45\%; t = 8 \text{ years}; P = ?$

16. $A = \$19{,}000; r = 7.69\%; t = 5 \text{ years}; P = ?$

17. $A = \$88{,}000; P = \$71{,}153; r = 8.5\%; t = ?$

18. $A = \$32{,}982; P = \$27{,}200; r = 5.93\%; t = ?$

19. $A = \$15{,}875; P = \$12{,}100; t = 48 \text{ months}; r = ?$

20. $A = \$23{,}600; P = \$19{,}150; t = 60 \text{ months}; r = ?$

In Problems 21–26, use the given annual interest rate r and the compounding period to find i, the interest rate per compounding period.

21. 9% compounded monthly

22. 6% compounded quarterly

23. 14.6% compounded daily

24. 15% compounded monthly

25. 4.8% compounded quarterly

26. 3.2% compounded semiannually

In Problems 27–32, use the given interest rate i per compounding period to find r, the annual rate.

27. 0.395% per month

28. 0.012% per day

29. 0.9% per quarter

30. 0.175% per month

31. 2.1% per half year

32. 1.4% per quarter

33. If \$100 is invested at 6% compounded

   (A) annually   (B) quarterly   (C) monthly

   what is the amount after 4 years? How much interest is earned?

34. If \$2,000 is invested at 7% compounded

   (A) annually   (B) quarterly   (C) monthly

   what is the amount after 5 years? How much interest is earned?

35. If $5,000 is invested at 5% compounded monthly, what is the amount after

(A) 2 years?   (B) 4 years?

36. If $20,000 is invested at 4% compounded monthly, what is the amount after

(A) 5 years?   (B) 8 years?

37. If $8,000 is invested at 7% compounded continuously, what is the amount after 6 years?

38. If $23,000 is invested at 13.5% compounded continuously, what is the amount after 15 years?

39. Discuss the similarities and the differences in the graphs of future value A as a function of time t if $1,000 is invested for 8 years and interest is compounded monthly at annual rates of 4%, 8%, and 12%, respectively (see the figure).

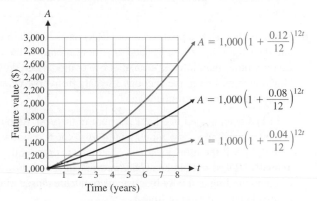

40. Discuss the similarities and differences in the graphs of future value A as a function of time t for loans of $4,000, $8,000, and $12,000, respectively, each at 7.5% compounded monthly for 8 years (see the figure).

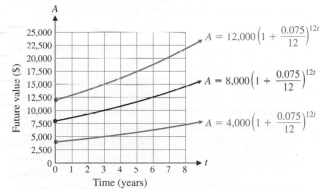

41. If $1,000 is invested in an account that earns 9.75% compounded annually for 6 years, find the interest earned during each year and the amount in the account at the end of each year. Organize your results in a table.

42. If $2,000 is invested in an account that earns 8.25% compounded annually for 5 years, find the interest earned during each year and the amount in the account at the end of each year. Organize your results in a table.

43. If an investment company pays 6% compounded semiannually, how much should you deposit now to have $10,000

(A) 5 years from now?   (B) 10 years from now?

44. If an investment company pays 8% compounded quarterly, how much should you deposit now to have $6,000

(A) 3 years from now?   (B) 6 years from now?

45. If an investment earns 9% compounded continuously, how much should you deposit now to have $25,000

(A) 36 months from now?   (B) 9 years from now?

46. If an investment earns 12% compounded continuously, how much should you deposit now to have $4,800

(A) 48 months from now?   (B) 7 years from now?

47. What is the annual percentage yield (APY) for money invested at an annual rate of

(A) 3.9% compounded monthly?

(B) 2.3% compounded quarterly?

48. What is the annual percentage yield (APY) for money invested at an annual rate of

(A) 4.32% compounded monthly?

(B) 4.31% compounded daily?

49. What is the annual percentage yield (APY) for money invested at an annual rate of

(A) 5.15% compounded continuously?

(B) 5.20% compounded semiannually?

50. What is the annual percentage yield (APY) for money invested at an annual rate of

(A) 3.05% compounded quarterly?

(B) 2.95% compounded continuously?

51. How long will it take $4,000 to grow to $9,000 if it is invested at 7% compounded monthly?

52. How long will it take $5,000 to grow to $7,000 if it is invested at 6% compounded quarterly?

53. How long will it take $6,000 to grow to $8,600 if it is invested at 9.6% compounded continuously?

54. How long will it take $42,000 to grow to $60,276 if it is invested at 4.25% compounded continuously?

In Problems 55 and 56, use compound interest formula (1) to find n to the nearest larger integer value.

55. $A = 2P$; $i = 0.06$; $n = ?$

56. $A = 2P$; $i = 0.05$; $n = ?$

57. How long will it take money to double if it is invested at

(A) 10% compounded quarterly?

(B) 12% compounded quarterly?

58. How long will it take money to double if it is invested at

(A) 8% compounded semiannually?

(B) 7% compounded semiannually?

59. How long will it take money to double if it is invested at

(A) 9% compounded continuously?

(B) 11% compounded continuously?

60. How long will it take money to double if it is invested at

(A) 21% compounded continuously?

(B) 33% compounded continuously?

Applications

61. A newborn child receives a $20,000 gift toward college from her grandparents. How much will the $20,000 be worth in 17 years if it is invested at 7% compounded quarterly?

62. A person with $14,000 is trying to decide whether to purchase a car now, or to invest the money at 6.5% compounded semiannually and then buy a more expensive car. How much will be available for the purchase of a car at the end of 3 years?

63. What will a $210,000 house cost 10 years from now if the inflation rate over that period averages 3% compounded annually?

64. If the inflation rate averages 4% per year compounded annually for the next 5 years, what will a car that costs $17,000 now cost 5 years from now?

65. Rental costs for office space have been going up at 4.8% per year compounded annually for the past 5 years. If office space rent is now $25 per square foot per month, what were the rental rates 5 years ago?

66. In a suburb, housing costs have been increasing at 5.2% per year compounded annually for the past 8 years. A house worth $260,000 now would have had what value 8 years ago?

67. If the population in a particular country is growing at 1.7% compounded continuously, how long will it take the population to double? (Round up to the next-higher year if not exact.)

68. If the world population is now about 7.5 billion people and is growing at 1.1% compounded continuously, how long will it take the population to grow to 10 billion people? (Round up to the next-higher year if not exact.)

✎ **69.** (A) If an investment of $100 were made in 1776, and if it earned 3% compounded quarterly, how much would it be worth in 2026?

(B) Discuss the effect of compounding interest monthly, daily, and continuously (rather than quarterly) on the $100 investment.

▦ (C) Use a graphing calculator to graph the growth of the investment of part (A).

✎ **70.** (A) Starting with formula (1), derive each of the following formulas:

$$P = \frac{A}{(1 + i)^n}, \quad i = \left(\frac{A}{P}\right)^{1/n} - 1, \quad n = \frac{\ln A - \ln P}{\ln(1 + i)}$$

(B) Explain why it is unnecessary to memorize the formulas above for P, i, and n if you know formula (1).

71. A promissory note will pay $50,000 at maturity 6 years from now. If you pay $28,000 for the note now, what rate compounded continuously would you earn?

72. If you deposit $10,000 in a savings account now, what rate compounded continuously would be required for you to withdraw $12,500 at the end of 4 years?

73. You have saved $7,000 toward the purchase of a car costing $9,000. How long will the $7,000 have to be invested at 9% compounded monthly to grow to $9,000? (Round up to the next-higher month if not exact.)

74. A married couple has $15,000 toward the purchase of a house. For the house that the couple wants to buy, a down payment of $20,000 is required. How long will the money have to be invested at 7% compounded quarterly to grow to $20,000? (Round up to the next-higher quarter if not exact.)

75. An Individual Retirement Account (IRA) has $20,000 in it, and the owner decides not to add any more money to the account other than interest earned at 6% compounded daily. How much will be in the account 35 years from now when the owner reaches retirement age?

76. If $1 had been placed in a bank account in the year 1066 and forgotten until now, how much would be in the account at the end of 2026 if the money earned 2% interest compounded annually? 2% simple interest? (Now you can see the power of compounding and why inactive accounts are closed after a relatively short period of time.)

77. How long will it take money to double if it is invested at 7% compounded daily? 8.2% compounded continuously?

78. How long will it take money to triple if it is invested at 5% compounded daily? 6% compounded continuously?

79. In a conversation with a friend, you note that you have two real estate investments, one that has doubled in value in the past 9 years and another that has doubled in value in the past 12 years. Your friend says that the first investment has been growing at approximately 8% compounded annually and the second at 6% compounded annually. How did your friend make these estimates? The **rule of 72** states that the annual compound rate of growth r of an investment that doubles in n years can be approximated by $r = 72/n$. Construct a table comparing the exact rate of growth and the approximate rate provided by the rule of 72 for doubling times of $n = 6, 7, \ldots, 12$ years. Round both rates to one decimal place.

▦ **80.** Refer to Problem 79. Show that the exact annual compound rate of growth of an investment that doubles in n years is given by $r = 100(2^{1/n} - 1)$. Graph this equation and the rule of 72 on a graphing calculator for $5 \leq n \leq 20$.

▦ *Solve Problems 81–84 using graphical approximation techniques on a graphing calculator.*

81. How long does it take for a $2,400 investment at 13% compounded quarterly to be worth more than a $3,000 investment at 6% compounded quarterly?

82. How long does it take for a $4,800 investment at 8% compounded monthly to be worth more than a $5,000 investment at 5% compounded monthly?

83. One investment pays 10% simple interest and another pays 7% compounded annually. Which investment would you choose? Why?

84. One investment pays 9% simple interest and another pays 6% compounded monthly. Which investment would you choose? Why?

85. What is the annual nominal rate compounded daily for a bond that has an annual percentage yield of 6.8%?

86. What is the annual nominal rate compounded monthly for a CD that has an annual percentage yield of 5.9%?

87. What annual nominal rate compounded monthly has the same annual percentage yield as 7% compounded continuously?

88. What annual nominal rate compounded continuously has the same annual percentage yield as 6% compounded monthly?

*Problems 89–92 refer to zero coupon bonds. A **zero coupon bond** is a bond that is sold now at a discount and will pay its **face value** at some time in the future when it matures—no interest payments are made.*

89. A zero coupon bond with a face value of $30,000 matures in 15 years. What should the bond be sold for now if its rate of return is to be 4.348% compounded annually?

90. A zero coupon bond with a face value of $20,000 matures in 10 years. What should the bond be sold for now if its rate of return is to be 4.194% compounded annually?

91. If you pay $4,126 for a 20-year zero coupon bond with a face value of $10,000, what is your annual compound rate of return?

92. If you pay $32,000 for a 5-year zero coupon bond with a face value of $40,000, what is your annual compound rate of return?

93. An online bank listed a money market account that earns 1.02% compounded daily. Find the APY as a percentage, rounded to three decimal places.

94. An online bank listed a 1-year CD that earns 1.25% compounded monthly. Find the APY as a percentage, rounded to three decimal places.

The buying and selling commission schedule shown in the table is from an online discount brokerage firm. Taking into consideration the buying and selling commissions in this schedule, find the annual compound rate of interest earned by each investment in Problems 95–98.

Transaction Size	Commission Rate
$0–$1,500	$29 + 2.5% of principal
$1,501–$6,000	$57 + 0.6% of principal
$6,001–$22,000	$75 + 0.30% of principal
$22,001–$50,000	$97 + 0.20% of principal
$50,001–$500,000	$147 + 0.10% of principal
$500,001+	$247 + 0.08% of principal

95. An investor purchases 100 shares of stock at $65 per share, holds the stock for 5 years, and then sells the stock for $125 a share.

96. An investor purchases 300 shares of stock at $95 per share, holds the stock for 3 years, and then sells the stock for $156 a share.

97. An investor purchases 200 shares of stock at $28 per share, holds the stock for 4 years, and then sells the stock for $55 a share.

98. An investor purchases 400 shares of stock at $48 per share, holds the stock for 6 years, and then sells the stock for $147 a share.

Answers to Matched Problems

1. (A) $1,593.85   (B) $1,604.71
   (C) $1,610.32   (D) $1,614.14
2. (A) $9,155.23   (B) $9,156.29
3. (A) $21,084.17   (B) $20,532.86
4. (A) 20.208%   (B) 18.405%
5. 47 quarters, or 11 years and 3 quarters
6. Southern Pacific Bank: 4.917%
   Washington Savings Bank: 4.939%
   Washington Savings Bank has the higher return.
7. 5.678%

3.3 Future Value of an Annuity; Sinking Funds

- Future Value of an Annuity
- Sinking Funds
- Approximating Interest Rates

Future Value of an Annuity

An **annuity** is any sequence of equal periodic payments. If payments are made at the end of each time interval, then the annuity is called an **ordinary annuity**. We consider only ordinary annuities in this book. The amount, or **future value**, of an annuity is the sum of all payments plus all interest earned.

Suppose you decide to deposit $100 every 6 months into an account that pays 6% compounded semiannually. If you make six deposits, one at the end of each interest payment period, over 3 years, how much money will be in the account after the last deposit is made? To solve this problem, let's look at it in terms of a time line. Using the compound amount formula $A = P(1 + i)^n$, we can find the value of each deposit after it has earned compound interest up through the sixth deposit, as shown in Figure 1.

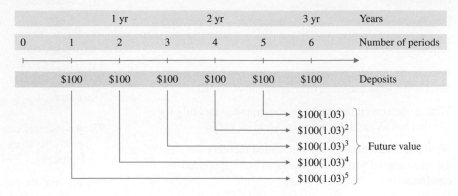

Figure 1

We could, of course, evaluate each of the future values in Figure 1 using a calculator and then add the results to find the amount in the account at the time of the sixth deposit—a tedious project at best. Instead, we take another approach, which leads directly to a formula that will produce the same result in a few steps (even when the number of deposits is very large). We start by writing the total amount in the account after the sixth deposit in the form

$$S = 100 + 100(1.03) + 100(1.03)^2 + 100(1.03)^3 + 100(1.03)^4 + 100(1.03)^5 \quad (1)$$

We would like a simple way to sum these terms. Let us multiply each side of (1) by 1.03 to obtain

$$1.03S = 100(1.03) + 100(1.03)^2 + 100(1.03)^3 + 100(1.03)^4 + 100(1.03)^5 + 100(1.03)^6 \quad (2)$$

Subtracting equation (1) from equation (2), left side from left side and right side from right side, we obtain

$$1.03S - S = 100(1.03)^6 - 100 \qquad \text{Notice how many terms drop out.}$$
$$0.03S = 100[(1.03)^6 - 1]$$
$$S = 100\frac{(1 + 0.03)^6 - 1}{0.03} \qquad \text{We write } S \text{ in this form to observe a general pattern.} \quad (3)$$

In general, if R is the periodic deposit, i the rate per period, and n the number of periods, then the future value is given by

$$S = R + R(1 + i) + R(1 + i)^2 + \cdots + R(1 + i)^{n-1} \qquad \text{Note how this compares to (1).} \quad (4)$$

and proceeding as in the above example, we obtain the general formula for the future value of an ordinary annuity:

$$S = R\frac{(1 + i)^n - 1}{i} \qquad \text{Note how this compares to (3).} \quad (5)$$

Returning to the example above, we use a calculator to complete the problem:

$$S = 100\frac{(1.03)^6 - 1}{0.03} \qquad \text{For improved accuracy, keep all values in the calculator until the end; round to the required number of decimal places.}$$
$$= \$646.84$$

CONCEPTUAL INSIGHT

In general, an expression of the form

$$a + ar + ar^2 + \cdots + ar^{n-1}$$

is called a finite geometric series (each term is obtained from the preceding term by multiplying by r). The sum of the terms of a finite geometric series is (see Section B.2)

$$a + ar + ar^2 + \cdots + ar^{n-1} = a\frac{r^n - 1}{r - 1}$$

If $a = R$ and $r = 1 + i$, then equation (4) is the sum of the terms of a finite geometric series and, using the preceding formula, we have

$$S = R + R(1 + i) + R(1 + i)^2 + \cdots + R(1 + i)^{n-1}$$

$$= R\frac{(1 + i)^n - 1}{1 + i - 1} \qquad a = R, r = 1 + i$$

$$= R\frac{(1 + i)^n - 1}{i} \tag{5}$$

So formula (5) is a direct consequence of the sum formula for a finite geometric series.

It is common to use FV (future value) for S and PMT (payment) for R in formula (5). Making these changes, we have the formula in Theorem 1.

THEOREM 1 Future Value of an Ordinary Annuity

$$FV = PMT\frac{(1 + i)^n - 1}{i} \tag{6}$$

where FV = future value (amount)
   PMT = periodic payment
   i = rate per period
   n = number of payments (periods)

Note: Payments are made at the end of each period.

EXAMPLE 1   Future Value of an Ordinary Annuity   What is the value of an annuity at the end of 20 years if $2,000 is deposited each year into an account earning 8.5% compounded annually? How much of this value is interest?

SOLUTION   To find the value of the annuity, use formula (6) with PMT = $2,000, $i = r = 0.085$, and $n = 20$.

$$FV = PMT\frac{(1 + i)^n - 1}{i}$$

$$= 2,000\frac{(1.085)^{20} - 1}{0.085} = \$96,754.03 \quad \text{Use a calculator.}$$

	A	B	C	D
1	Period	Payment	Interest	Balance
2	1	$2,000.00	$0.00	$2,000.00
3	2	$2,000.00	$170.00	$4,170.00
4	3	$2,000.00	$354.45	$6,524.45
5	4	$2,000.00	$554.58	$9,079.03
6	5	$2,000.00	$771.72	$11,850.75
7	6	$2,000.00	$1,007.31	$14,858.06
8	7	$2,000.00	$1,262.94	$18,120.99
9	8	$2,000.00	$1,540.28	$21,661.28
10	9	$2,000.00	$1,841.21	$25,502.49
11	10	$2,000.00	$2,167.71	$29,670.20
12	11	$2,000.00	$2,521.97	$34,192.17
13	12	$2,000.00	$2,906.33	$39,098.50
14	13	$2,000.00	$3,323.37	$44,421.87
15	14	$2,000.00	$3,775.86	$50,197.73
16	15	$2,000.00	$4,266.81	$56,464.54
17	16	$2,000.00	$4,799.49	$63,264.02
18	17	$2,000.00	$5,377.44	$70,641.47
19	18	$2,000.00	$6,004.52	$78,645.99
20	19	$2,000.00	$6,684.91	$87,330.90
21	20	$2,000.00	$7,423.13	$96,754.03
22	Totals	$40,000.00	$56,754.03	

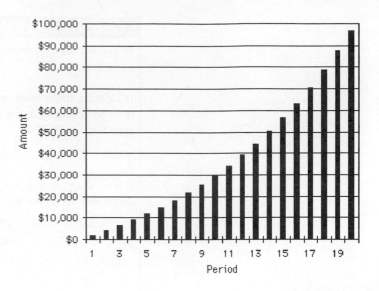

Figure 2   **Ordinary annuity at 8.5% compounded annually for 20 years**

To find the amount of interest earned, subtract the total amount deposited in the annuity (20 payments of $2,000) from the total value of the annuity after the 20th payment.

$$\text{Deposits} = 20(2,000) = \$40,000$$

$$\text{Interest} = \text{value} - \text{deposits} = 96,754.03 - 40,000 = \$56,754.03$$

Figure 2, which was generated using a spreadsheet, illustrates the growth of this account over 20 years.

Matched Problem 1   What is the value of an annuity at the end of 10 years if $1,000 is deposited every 6 months into an account earning 8% compounded semiannually? How much of this value is interest?

The table in Figure 2 is called a **balance sheet**. Let's take a closer look at the construction of this table. The first line is a special case because the payment is made at the end of the period and no interest is earned. Each subsequent line of the table is computed as follows:

$$\text{payment} + \text{interest} \quad + \text{old balance} = \text{new balance}$$
$$2,000 \quad + 0.085(2,000) + 2,000 \quad = 4,170 \qquad \text{Period 2}$$
$$2,000 \quad + 0.085(4,170) + 4,170 \quad = 6,524.45 \qquad \text{Period 3}$$

And so on. The amounts at the bottom of each column in the balance sheet agree with the results we obtained by using formula (6), as you would expect. Although balance sheets are appropriate for certain situations, we will concentrate on applications of formula (6). There are many important problems that can be solved only by using this formula.

Explore and Discuss 1   (A) Discuss the similarities and differences in the graphs of future value FV as a function of time t for ordinary annuities in which $100 is deposited each month for 8 years and interest is compounded monthly at annual rates of 4%, 8%, and 12%, respectively (Fig. 3).

(B) Discuss the connections between the graph of the equation $y = 100t$, where t is time in months, and the graphs of part (A).

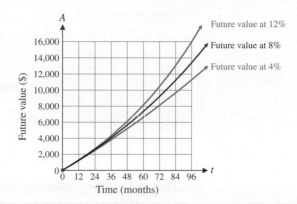

Figure 3

Sinking Funds

The formula for the future value of an ordinary annuity has another important application. Suppose the parents of a newborn child decide that on each of the child's birthdays up to the 17th year, they will deposit $\$PMT$ in an account that pays 6% compounded annually. The money is to be used for college expenses. What should the annual deposit ($\$PMT$) be in order for the amount in the account to be $80,000 after the 17th deposit?

We are given FV, i, and n in formula (6), and we must find PMT:

$$FV = PMT\frac{(1 + i)^n - 1}{i}$$

$$80{,}000 = PMT\frac{(1.06)^{17} - 1}{0.06} \qquad \text{Solve for } PMT.$$

$$PMT = 80{,}000\frac{0.06}{(1.06)^{17} - 1} \qquad \text{Use a calculator.}$$

$$= \$2{,}835.58 \text{ per year}$$

An annuity of 17 annual deposits of $2,835.58 at 6% compounded annually will amount to $80,000 in 17 years.

This is an example of a *sinking fund problem*. In general, any account that is established for accumulating funds to meet future obligations or debts is called a **sinking fund**. If the payments are to be made in the form of an ordinary annuity, then we have only to solve formula (6) for the **sinking fund payment** PMT:

$$PMT = FV\frac{i}{(1 + i)^n - 1} \qquad (7)$$

It is important to understand that formula (7), which is convenient to use, is simply a variation of formula (6). You can always find the sinking fund payment by first substituting the appropriate values into formula (6) and then solving for PMT, as we did in the college fund example discussed above. Or you can substitute directly into formula (7), as we do in the next example. Use whichever method is easier for you.

EXAMPLE 2 Computing the Payment for a Sinking Fund A company estimates that it will have to replace a piece of equipment at a cost of $800,000 in 5 years. To have this money available in 5 years, a sinking fund is established by making equal monthly payments into an account paying 6.6% compounded monthly.

(A) How much should each payment be?

(B) How much interest is earned during the last year?

SOLUTION

(A) To find *PMT*, we can use either formula (6) or (7). We choose formula (7) with $FV = \$800{,}000$, $i = \frac{0.066}{12} = 0.0055$, and $n = 12 \cdot 5 = 60$:

$$PMT = FV \frac{i}{(1+i)^n - 1}$$

$$= 800{,}000 \frac{0.0055}{(1.0055)^{60} - 1}$$

$$= \$11{,}290.42 \text{ per month}$$

(B) To find the interest earned during the fifth year, we first use formula (6) with $PMT = \$11{,}290.42$, $i = 0.0055$, and $n = 12 \cdot 4 = 48$ to find the amount in the account after 4 years:

$$FV = PMT \frac{(1+i)^n - 1}{i}$$

$$= 11{,}290.42 \frac{(1.0055)^{48} - 1}{0.0055}$$

$$= \$618{,}277.04 \qquad \text{Amount after 4 years}$$

During the 5th year, the amount in the account grew from \$618,277.04 to \$800,000. A portion of this growth was due to the 12 monthly payments of \$11,290.42. The remainder of the growth was interest. Thus,

$$800{,}000 - 618{,}277.04 = 181{,}722.96 \quad \text{Growth in the 5th year}$$

$$12 \cdot 11{,}290.42 = 135{,}485.04 \quad \text{Payments during the 5th year}$$

$$181{,}722.96 - 135{,}485.04 = \$46{,}237.92 \quad \text{Interest during the 5th year}$$

Matched Problem 2 A bond issue is approved for building a marina in a city. The city is required to make regular payments every 3 months into a sinking fund paying 5.4% compounded quarterly. At the end of 10 years, the bond obligation will be retired with a cost of \$5,000,000.

(A) What should each payment be?

(B) How much interest is earned during the 10th year?

EXAMPLE 3   Growth in an IRA   Jane deposits \$2,000 annually into a Roth IRA that earns 6.85% compounded annually. (The interest earned by a Roth IRA is tax free.) Due to a change in employment, these deposits stop after 10 years, but the account continues to earn interest until Jane retires 25 years after the last deposit was made. How much is in the account when Jane retires?

SOLUTION   First, we use the future value formula with $PMT = \$2{,}000$, $i = 0.0685$, and $n = 10$ to find the amount in the account after 10 years:

$$FV = PMT \frac{(1+i)^n - 1}{i}$$

$$= 2{,}000 \frac{(1.0685)^{10} - 1}{0.0685}$$

$$= \$27{,}437.89$$

Now we use the compound interest formula from Section 3.2 with $P = \$27{,}437.89$, $i = 0.0685$, and $n = 25$ to find the amount in the account when Jane retires:

$$A = P(1+i)^n$$

$$= 27{,}437.89(1.0685)^{25}$$

$$= \$143{,}785.10$$

Matched Problem 3 ⌋ Refer to Example 3. Mary starts a Roth IRA earning the same rate of interest at the time Jane stops making payments into her IRA. How much must Mary deposit each year for the next 25 years in order to have the same amount at retirement as Jane?

Explore and Discuss 2 Refer to Example 3 and Matched Problem 3. What was the total amount Jane deposited in order to have $143,785.10 at retirement? What was the total amount Mary deposited in order to have the same amount at retirement? Do you think it is advisable to start saving for retirement as early as possible?

 ## Approximating Interest Rates

Algebra can be used to solve the future value formula (6) for *PMT* or *n* but not for *i*. However, graphical techniques or equation solvers can be used to approximate *i* to as many decimal places as desired.

EXAMPLE 4 Approximating an Interest Rate A person makes monthly deposits of $100 into an ordinary annuity. After 30 years, the annuity is worth $160,000. What annual rate compounded monthly has this annuity earned during this 30-year period? Express the answer as a percentage, correct to two decimal places.

SOLUTION Substituting $FV = \$160{,}000$, $PMT = \$100$, and $n = 30(12) = 360$ in (6) produces the following equation:

$$160{,}000 = 100\frac{(1 + i)^{360} - 1}{i}$$

We can approximate the solution to this equation by using graphical techniques (Figs. 4A, 4B) or an equation solver (Fig. 4C). From Figure 4B or 4C, we see that $i = 0.006\ 956\ 7$ and $12(i) = 0.083\ 480\ 4$. So the annual rate (to two decimal places) is $r = 8.35\%$.

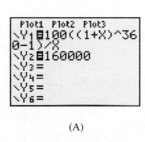
(A)

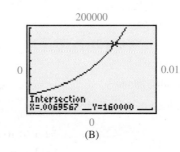
(B)

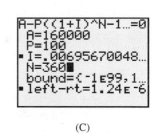
(C)

Figure 4

 Matched Problem 4 ⌋ A person makes annual deposits of $1,000 into an ordinary annuity. After 20 years, the annuity is worth $55,000. What annual compound rate has this annuity earned during this 20-year period? Express the answer as a percentage, correct to two decimal places.

Exercises 3.3

W ⌈ **Skills Warm-up Exercises**

In Problems 1–8, find the sum of the finite geometric series $a + ar + ar^2 + \cdots + ar^{n-1}$. (If necessary, review Section B.2.)

1. $1 + 2 + 4 + 8 + \cdots + 2^9$

2. $1 + 5 + 25 + 125 + \cdots + 5^8$

3. $a = 30, r = 1, n = 100$

4. $a = 25, r = -1, n = 81$

5. $a = 10, r = 3, n = 15$

6. $a = 4, r = 10, n = 6$

In Problems 7–14, find i (the rate per period) and n (the number of periods) for each annuity.

7. Quarterly deposits of $500 are made for 20 years into an annuity that pays 8% compounded quarterly.

8. Monthly deposits of $350 are made for 6 years into an annuity that pays 6% compounded monthly.

9. Semiannual deposits of $900 are made for 12 years into an annuity that pays 7.5% compounded semiannually.

10. Annual deposits of $2,500 are made for 15 years into an annuity that pays 6.25% compounded annually.

11. Monthly deposits of $235 are made for 4 years into an annuity that pays 9% compounded monthly.

12. Semiannual deposits of $1,900 are made for 7 years into an annuity that pays 8.5% compounded semiannually.

13. Annual deposits of $3,100 are made for 12 years into an annuity that pays 5.95% compounded annually.

14. Quarterly deposits of $1,200 are made for 18 years into an annuity that pays 7.6% compounded quarterly.

In Problems 15–22, use the future value formula (6) to find each of the indicated values.

15. $n = 20; i = 0.03; PMT = \$500; FV = ?$

16. $n = 25; i = 0.04; PMT = \$100; FV = ?$

17. $FV = \$5,000; n = 15; i = 0.01; PMT = ?$

18. $FV = \$2,500; n = 10; i = 0.08; PMT = ?$

19. $FV = \$4,000; i = 0.02; PMT = 200; n = ?$

20. $FV = \$8,000; i = 0.04; PMT = 500; n = ?$

21. $FV = \$7,600; PMT = \$500; n = 10; i = ?$
    (Round answer to two decimal places.)

22. $FV = \$4,100; PMT = \$100; n = 20; i = ?$
    (Round answer to two decimal places.)

23. Explain what is meant by an ordinary annuity.

24. Explain why no interest is credited to an ordinary annuity at the end of the first period.

25. Solve the future value formula (6) for *n*.

26. Solve the future value formula (6) for *i* if *n* = 2.

Applications

27. Guaranty Income Life offered an annuity that pays 6.65% compounded monthly. If $500 is deposited into this annuity every month, how much is in the account after 10 years? How much of this is interest?

28. USG Annuity and Life offered an annuity that pays 7.25% compounded monthly. If $1,000 is deposited into this annuity every month, how much is in the account after 15 years? How much of this is interest?

29. In order to accumulate enough money for a down payment on a house, a couple deposits $300 per month into an account

paying 6% compounded monthly. If payments are made at the end of each period, how much money will be in the account in 5 years?

30. A self-employed person has a Keogh retirement plan. (This type of plan is free of taxes until money is withdrawn.) If deposits of $7,500 are made each year into an account paying 8% compounded annually, how much will be in the account after 20 years?

31. Sun America offered an annuity that pays 6.35% compounded monthly. What equal monthly deposit should be made into this annuity in order to have $200,000 in 15 years?

32. The Hartford offered an annuity that pays 5.5% compounded monthly. What equal monthly deposit should be made into this annuity in order to have $100,000 in 10 years?

33. A company estimates that it will need $100,000 in 8 years to replace a computer. If it establishes a sinking fund by making fixed monthly payments into an account paying 7.5% compounded monthly, how much should each payment be?

34. Parents have set up a sinking fund in order to have $120,000 in 15 years for their children's college education. How much should be paid semiannually into an account paying 6.8% compounded semiannually?

35. If $1,000 is deposited at the end of each year for 5 years into an ordinary annuity earning 8.32% compounded annually, construct a balance sheet showing the interest earned during each year and the balance at the end of each year.

36. If $2,000 is deposited at the end of each quarter for 2 years into an ordinary annuity earning 7.9% compounded quarterly, construct a balance sheet showing the interest earned during each quarter and the balance at the end of each quarter.

37. Beginning in January, a person plans to deposit $100 at the end of each month into an account earning 6% compounded monthly. Each year taxes must be paid on the interest earned during that year. Find the interest earned during each year for the first 3 years.

38. If $500 is deposited each quarter into an account paying 8% compounded quarterly for 3 years, find the interest earned during each of the 3 years.

39. Bob makes his first $1,000 deposit into an IRA earning 6.4% compounded annually on his 24th birthday and his last $1,000 deposit on his 35th birthday (12 equal deposits in all). With no additional deposits, the money in the IRA continues to earn 6.4% interest compounded annually until Bob retires on his 65th birthday. How much is in the IRA when Bob retires?

40. Refer to Problem 39. John procrastinates and does not make his first $1,000 deposit into an IRA until he is 36, but then he continues to deposit $1,000 each year until he is 65 (30 deposits in all). If John's IRA also earns 6.4% compounded annually, how much is in his IRA when he makes his last deposit on his 65th birthday?

41. Refer to Problems 39 and 40. How much would John have to deposit each year in order to have the same amount at retirement as Bob has?

42. Refer to Problems 39 and 40. Suppose that Bob decides to continue to make $1,000 deposits into his IRA every year until his 65th birthday. If John still waits until he is 36 to start his IRA, how much must he deposit each year in order to have the same amount at age 65 as Bob has?

43. Compubank, an online banking service, offered a money market account with an APY of 1.551%.

(A) If interest is compounded monthly, what is the equivalent annual nominal rate?

(B) If you wish to have $10,000 in this account after 4 years, what equal deposit should you make each month?

44. American Express's online banking division offered a money market account with an APY of 2.243%.

(A) If interest is compounded monthly, what is the equivalent annual nominal rate?

(B) If a company wishes to have $1,000,000 in this account after 8 years, what equal deposit should be made each month?

45. You can afford monthly deposits of $200 into an account that pays 5.7% compounded monthly. How long will it be until you have $7,000? (Round to the next-higher month if not exact.)

46. A company establishes a sinking fund for upgrading office equipment with monthly payments of $2,000 into an account paying 6.6% compounded monthly. How long will it be before the account has $100,000? (Round up to the next-higher month if not exact.)

In Problems 47–50, use graphical approximation techniques or an equation solver to approximate the desired interest rate. Express each answer as a percentage, correct to two decimal places.

47. A person makes annual payments of $1,000 into an ordinary annuity. At the end of 5 years, the amount in the annuity is $5,840. What annual nominal compounding rate has this annuity earned?

48. A person invests $2,000 annually in an IRA. At the end of 6 years, the amount in the fund is $14,000. What annual nominal compounding rate has this fund earned?

49. At the end of each month, an employee deposits $50 into a Christmas club fund. At the end of the year, the fund contains $620. What annual nominal rate compounded monthly has this fund earned?

50. At the end of each month, an employee deposits $80 into a credit union account. At the end of 2 years, the account contains $2,100. What annual nominal rate compounded monthly has this account earned?

In Problems 51 and 52, use graphical approximation techniques to answer the questions.

51. When would an ordinary annuity consisting of quarterly payments of $500 at 6% compounded quarterly be worth more than a principal of $5,000 invested at 4% simple interest?

52. When would an ordinary annuity consisting of monthly payments of $200 at 5% compounded monthly be worth more than a principal of $10,000 invested at 7.5% compounded monthly?

Answers to Matched Problems

1. Value: $29,778.08; interest: $9,778.08

2. (A) $95,094.67   (B) $248,628.88

3. $2,322.73   4. 9.64%

3.4 Present Value of an Annuity; Amortization

- Present Value of an Annuity
- Amortization
- Amortization Schedules
- General Problem-Solving Strategy

Present Value of an Annuity

How much should you deposit in an account paying 6% compounded semiannually in order to be able to withdraw $1,000 every 6 months for the next 3 years? (After the last payment is made, no money is to be left in the account.)

Actually, we are interested in finding the present value of each $1,000 that is paid out during the 3 years. We can do this by solving for P in the compound interest formula:

$$A = P(1 + i)^n$$

$$P = \frac{A}{(1 + i)^n} = A(1 + i)^{-n}$$

The rate per period is $i = \frac{0.06}{2} = 0.03$. The present value P of the first payment is $1,000(1.03)^{-1}$, the present value of the second payment is $1,000(1.03)^{-2}$, and so on. Figure 1 shows this in terms of a time line.

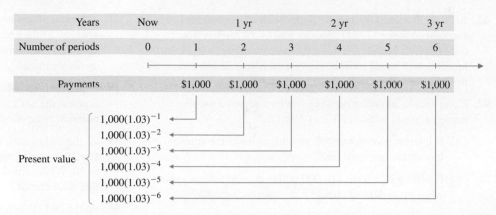

Figure 1

We could evaluate each of the present values in Figure 1 using a calculator and add the results to find the total present values of all the payments (which will be the amount needed now to buy the annuity). Since this is a tedious process, particularly when the number of payments is large, we will use the same device we used in the preceding section to produce a formula that will accomplish the same result in a couple of steps. We start by writing the sum of the present values in the form

$$P = 1{,}000(1.03)^{-1} + 1{,}000(1.03)^{-2} + \cdots + 1{,}000(1.03)^{-6} \qquad (1)$$

Multiplying both sides of equation (1) by 1.03, we obtain

$$1.03P = 1{,}000 + 1{,}000(1.03)^{-1} + \cdots + 1{,}000(1.03)^{-5} \qquad (2)$$

Now subtract equation (1) from equation (2):

$$1.03P - P = 1{,}000 - 1{,}000(1.03)^{-6} \qquad \text{Notice how many terms drop out.}$$

$$0.03P = 1{,}000[1 - (1 + 0.03)^{-6}]$$

$$P = 1{,}000\frac{1 - (1 + 0.03)^{-6}}{0.03} \qquad \begin{array}{l}\text{We write } P \text{ in this form to} \\ \text{observe a general pattern.}\end{array} \qquad (3)$$

In general, if R is the periodic payment, i the rate per period, and n the number of periods, then the present value of all payments is given by

$$P = R(1 + i)^{-1} + R(1 + i)^{-2} + \cdots + R(1 + i)^{-n} \qquad \begin{array}{l}\text{Note how this} \\ \text{compares to (1).}\end{array}$$

Proceeding as in the above example, we obtain the general formula for the present value of an ordinary annuity:

$$P = R\frac{1 - (1 + i)^{-n}}{i} \qquad \text{Note how this compares to (3).} \qquad (4)$$

Returning to the preceding example, we use a calculator to complete the problem:

$$P = 1{,}000\frac{1 - (1.03)^{-6}}{0.03}$$

$$= \$5{,}417.19$$

CONCEPTUAL INSIGHT

Formulas (3) and (4) can also be established by using the sum formula for a finite geometric series (see Section B.2):

$$a + ar + ar^2 + \cdots + ar^{n-1} = a\frac{r^n - 1}{r - 1}$$

It is common to use *PV* (present value) for *P* and *PMT* (payment) for *R* in formula (4). Making these changes, we have the following:

THEOREM 1   Present Value of an Ordinary Annuity

$$PV = PMT\frac{1 - (1 + i)^{-n}}{i}$$ (5)

where PV = present value of all payments
   PMT = periodic payment
   i = rate per period
   n = number of periods

Note: Payments are made at the end of each period.

EXAMPLE 1   Present Value of an Annuity   What is the present value of an annuity that pays $200 per month for 5 years if money is worth 6% compounded monthly?

SOLUTION   To solve this problem, use formula (5) with $PMT = \$200$, $i = \frac{0.06}{12} = 0.005$, and $n = 12(5) = 60$:

$$PV = PMT\frac{1 - (1 + i)^{-n}}{i}$$

$$= 200\frac{1 - (1.005)^{-60}}{0.005} \qquad \text{Use a calculator.}$$

$$= \$10,345.11$$

Matched Problem 1   How much should you deposit in an account paying 8% compounded quarterly in order to receive quarterly payments of $1,000 for the next 4 years?

EXAMPLE 2   Retirement Planning   Lincoln Benefit Life offered an ordinary annuity that earned 6.5% compounded annually. A person plans to make equal annual deposits into this account for 25 years and then make 20 equal annual withdrawals of $25,000, reducing the balance in the account to zero. How much must be deposited annually to accumulate sufficient funds to provide for these payments? How much total interest is earned during this entire 45-year process?

SOLUTION   This problem involves both future and present values. Figure 2 illustrates the flow of money into and out of the annuity.

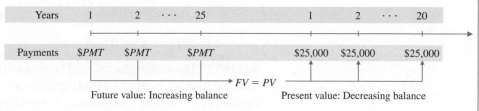

Figure 2

Since we are given the required withdrawals, we begin by finding the present value necessary to provide for these withdrawals. Using formula (5) with $PMT = \$25,000$, $i = 0.065$, and $n = 20$, we have

$$PV = PMT\frac{1 - (1 + i)^{-n}}{i}$$

$$= 25{,}000\frac{1 - (1.065)^{-20}}{0.065} \qquad \text{Use a calculator.}$$

$$= \$275{,}462.68$$

Now we find the deposits that will produce a future value of $\$275{,}462.68$ in 25 years. Using formula (7) from Section 3.3 with $FV = \$275{,}462.68$, $i = 0.065$, and $n = 25$, we have

$$PMT = FV\frac{i}{(1 + i)^n - 1}$$

$$= 275{,}462.68\frac{0.065}{(1.065)^{25} - 1} \qquad \text{Use a calculator.}$$

$$= \$4{,}677.76$$

Thus, depositing $\$4{,}677.76$ annually for 25 years will provide for 20 annual withdrawals of $\$25{,}000$. The interest earned during the entire 45-year process is

$$\text{interest} = (\text{total withdrawals}) - (\text{total deposits})$$
$$= 20(25{,}000) - 25(\$4{,}677.76)$$
$$= \$383{,}056$$

Matched Problem 2⟩   Refer to Example 2. If $\$2,000$ is deposited annually for the first 25 years, how much can be withdrawn annually for the next 20 years?

Amortization

The present value formula for an ordinary annuity, formula (5), has another important use. Suppose that you borrow $\$5,000$ from a bank to buy a car and agree to repay the loan in 36 equal monthly payments, including all interest due. If the bank charges 1% per month on the unpaid balance (12% per year compounded monthly), how much should each payment be to retire the total debt, including interest, in 36 months?

Actually, the bank has bought an annuity from you. The question is: If the bank pays you $\$5,000$ (present value) for an annuity paying them $\$PMT$ per month for 36 months at 12% interest compounded monthly, what are the monthly payments (PMT)? (Note that the value of the annuity at the end of 36 months is zero.) To find PMT, we have only to use formula (5) with $PV = \$5,000$, $i = 0.01$, and $n = 36$:

$$PV = PMT\frac{1 - (1 + i)^{-n}}{i}$$

$$5{,}000 = PMT\frac{1 - (1.01)^{-36}}{0.01} \qquad \text{Solve for } PMT \text{ and use a calculator.}$$

$$PMT = \$166.07 \text{ per month}$$

At $\$166.07$ per month, the car will be yours after 36 months. That is, you have *amortized* the debt in 36 equal monthly payments. (*Mort* means "death" you have "killed" the loan in 36 months.) In general, **amortizing a debt** means that the debt is retired in a given length of time by equal periodic payments that include compound interest. We are interested in computing the equal periodic payments. Solving the present value formula (5) for *PMT* in terms of the other variables, we obtain the following **amortization formula**:

$$PMT = PV\frac{i}{1 - (1 + i)^{-n}} \tag{6}$$

Formula (6) is simply a variation of formula (5), and either formula can be used to find the periodic payment *PMT*.

EXAMPLE 3   Monthly Payment and Total Interest on an Amortized Debt   Assume that you buy a TV for $800 and agree to pay for it in 18 equal monthly payments at $1\frac{1}{2}\%$ interest per month on the unpaid balance.

(A) How much are your payments?

(B) How much interest will you pay?

SOLUTION

(A) Use formula (5) or (6) with $PV = \$800$, $i = 0.015$, $n = 18$, and solve for *PMT*:

$$PMT = PV\frac{i}{1 - (1 + i)^{-n}}$$

$$= 800\frac{0.015}{1 - (1.015)^{-18}} \qquad \text{Use a calculator.}$$

$$\approx \$51.04 \text{ per month}$$

(B) Total interest paid $=$ (amount of all payments) $-$ (initial loan)

$$= 18(\$51.04) - \$800$$

$$= \$118.72$$

Matched Problem 3   If you sell your car to someone for $2,400 and agree to finance it at 1% per month on the unpaid balance, how much should you receive each month to amortize the loan in 24 months? How much interest will you receive?

Explore and Discuss 1   To purchase a home, a family plans to sign a mortgage of $70,000 at 8% on the unpaid balance. Discuss the advantages and disadvantages of a 20-year mortgage as opposed to a 30-year mortgage. Include a comparison of monthly payments and total interest paid.

Amortization Schedules

What happens if you are amortizing a debt with equal periodic payments and later decide to pay off the remainder of the debt in one lump-sum payment? This occurs each time a home with an outstanding mortgage is sold. In order to understand what happens in this situation, we must take a closer look at the amortization process. We begin with an example that allows us to examine the effect each payment has on the debt.

EXAMPLE 4   Constructing an Amortization Schedule   If you borrow $500 that you agree to repay in six equal monthly payments at 1% interest per month on the unpaid balance, how much of each monthly payment is used for interest and how much is used to reduce the unpaid balance?

SOLUTION First, we compute the required monthly payment using formula (5) or (6). We choose formula (6) with $PV = \$500$, $i = 0.01$, and $n = 6$:

$$PMT = PV\frac{i}{1 - (1 + i)^{-n}}$$

$$= 500\frac{0.01}{1 - (1.01)^{-6}} \quad \text{Use a calculator.}$$

$$= \$86.27 \text{ per month}$$

At the end of the first month, the interest due is

$$\$500(0.01) = \$5.00$$

The amortization payment is divided into two parts, payment of the interest due and reduction of the unpaid balance (repayment of principal):

Monthly payment		Interest due		Unpaid balance reduction
$86.27	=	$5.00	+	$81.27

The unpaid balance for the next month is

Previous Unpaid balance		Unpaid balance reduction		New unpaid balance
$500.00	−	$81.27	=	$418.73

At the end of the second month, the interest due on the unpaid balance of $418.73 is

$$\$418.73(0.01) = \$4.19$$

Thus, at the end of the second month, the monthly payment of $86.27 covers interest and unpaid balance reduction as follows:

$$\$86.27 = \$4.19 + \$82.08$$

and the unpaid balance for the third month is

$$\$418.73 - \$82.08 = \$336.65$$

This process continues until all payments have been made and the unpaid balance is reduced to zero. The calculations for each month are listed in Table 1, often referred to as an **amortization schedule**.

CONCEPTUAL INSIGHT

In Table 1, notice that the last payment had to be increased by $0.03 in order to reduce the unpaid balance to zero. This small discrepancy is due to rounding the monthly payment and the entries in the interest column to two decimal places.

Table 1 **Amortization Schedule**

Payment Number	Payment	Interest	Unpaid Balance Reduction	Unpaid Balance
0				$500.00
1	$86.27	$5.00	$81.27	418.73
2	86.27	4.19	82.08	336.65
3	86.27	3.37	82.90	253.75
4	86.27	2.54	83.73	170.02
5	86.27	1.70	84.57	85.45
6	86.30	0.85	85.45	0.00
Totals	$517.65	$17.65	$500.00	

Matched Problem 4 Construct the amortization schedule for a $1,000 debt that is to be amortized in six equal monthly payments at 1.25% interest per month on the unpaid balance.

EXAMPLE 5 Equity in a Home A family purchased a home 10 years ago for $80,000. The home was financed by paying 20% down and signing a 30-year mortgage at 9% on the unpaid balance. The net market value of the house (amount received after subtracting all costs involved in selling the house) is now $120,000, and the family wishes to sell the house. How much equity (to the nearest dollar) does the family have in the house now after making 120 monthly payments?

[**Equity** = (current net market value) − (unpaid loan balance).]*

SOLUTION How can we find the unpaid loan balance after 10 years or 120 monthly payments? One way to proceed would be to construct an amortization schedule, but this would require a table with 120 lines. Fortunately, there is an easier way. The unpaid balance after 120 payments is the amount of the loan that can be paid off with the remaining 240 monthly payments (20 remaining years on the loan). Since the lending institution views a loan as an annuity that they bought from the family, **the unpaid balance of a loan with n remaining payments is the present value of that annuity and can be computed by using formula (5).** Since formula (5) requires knowledge of the monthly payment, we compute *PMT* first using formula (6).

Step 1 Find the monthly payment:

$$PMT = PV\frac{i}{1 - (1 + i)^{-n}}$$

$PV = (0.80)(\$80,000) = \$64,000$

$i - \frac{0.09}{12} = 0.0075$

$$= 64,000\frac{0.0075}{1 - (1.0075)^{-360}}$$

$n = 12(30) = 360$

$$= \$514.96 \text{ per month}$$

Use a calculator.

Step 2 Find the present value of a $514.96 per month, 20-year annuity:

$$PV = PMT\frac{1 - (1 + i)^{-n}}{i}$$

$PMT = \$514.96$

$n = 12(20) = 240$

$$= 514.96\frac{1 - (1.0075)^{-240}}{0.0075}$$

$i = \frac{0.09}{12} - 0.0075$

Use a calculator.

$$- \$57,235$$

Unpaid loan balance

Step 3 Find the equity:

$$\text{equity} = (\text{current net market value}) - (\text{unpaid loan balance})$$
$$= \$120,000 - \$57,235$$
$$= \$62,765$$

If the family sells the house for $120,000 net, the family will have $62,765 after paying off the unpaid loan balance of $57,235.

Matched Problem 5 A couple purchased a home 20 years ago for $65,000. The home was financed by paying 20% down and signing a 30-year mortgage at 8% on the unpaid balance. The net market value of the house is now $130,000, and the couple wishes to sell the house. How much equity (to the nearest dollar) does the couple have in the house now after making 240 monthly payments?

---

*If a family wants to sell a house and buy another, more expensive, house, then the price of a new house that the family can afford to buy will often depend on the family's *equity* in the first house, where equity is defined by the equation given here. In refinancing a house or taking out an "equity loan," the new mortgage (or second mortgage) often will be based on the equity in the house.

The unpaid loan balance in Example 5 may seem a surprisingly large amount to owe after having made payments for 10 years, but long-term amortizations start out with very small reductions in the unpaid balance. For example, the interest due at the end of the very first period of the loan in Example 5 was

$$\$64{,}000(0.0075) = \$480.00$$

The first monthly payment was divided as follows:

$$
\begin{array}{ccc}
\text{Monthly} & \text{Interest} & \text{Unpaid balance} \\
\text{payment} & \text{due} & \text{reduction} \\
\$514.96 & - \ \$480.00 & = \ \$34.96
\end{array}
$$

Only $34.96 was applied to the unpaid balance.

Explore and Discuss 2

(A) A family has an $85,000, 30-year mortgage at 9.6% compounded monthly. Show that the monthly payments are $720.94.

(B) Explain why the equation

$$y = 720.94 \frac{1 - (1.008)^{-12(30-x)}}{0.008}$$

gives the unpaid balance of the loan after x years.

(C) Find the unpaid balance after 5 years, after 10 years, and after 15 years.

(D) When does the unpaid balance drop below half of the original $85,000?

(E) Solve part (D) using graphical approximation techniques on a graphing calculator (see Fig. 3).

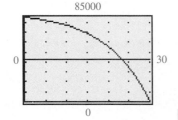

85000

0 ⊢——————————— 30

0

Figure 3

EXAMPLE 6   Automobile Financing   You have negotiated a price of $25,200 for a new Bison pickup truck. Now you must choose between 0% financing for 48 months or a $3,000 rebate. If you choose the rebate, you can obtain a credit union loan for the balance at 4.5% compounded monthly for 48 months. Which option should you choose?

SOLUTION   If you choose 0% financing, your monthly payment will be

$$PMT_1 = \frac{25{,}200}{48} = \$525$$

If you choose the $3,000 rebate, and borrow $22,200 at 4.5% compounded monthly for 48 months, the monthly payment is

$$PMT_2 = PV \frac{i}{1 - (1 + i)^{-n}} \qquad PV = \$22{,}200$$

$$= 22{,}200 \frac{0.00375}{1 - 1.00375^{-48}} \qquad i = \frac{.045}{12} = 0.00375$$

$$= \$506.24 \qquad n = 48$$

ALL 2016 BISONS
$3,000 rebate
or 0% APR for 4 years/ 48 months

You should choose the $3,000 rebate. You will save $525 - 506.24 = \$18.76$ monthly or $48(18.76) = \$900.48$ over the life of the loan.

Matched Problem 6   Which option should you choose if your credit union raises its loan rate to 7.5% compounded monthly and all other data remain the same?

EXAMPLE 7 Credit Cards The annual interest rate on a credit card is 18.99%. How long will it take to pay off an unpaid balance of $847.29 if no new purchases are made and the minimum payment of $20.00 is made each month?

SOLUTION It is necessary to make some simplifying assumptions because the lengths of the billing cycles, the days on which payments are credited, and the method for calculating interest are not specified. So we assume that there are 12 equal billing cycles per year, and that the $20 payments are credited at the end of each cycle. With these assumptions, it is reasonable to use the present value formula with $PV = \$847.29$, $PMT = \$20.00$, and $i = 0.1899/12$ in order to solve for n, the number of payments:

$$PV = PMT\frac{1-(1+i)^{-n}}{i} \qquad \text{Multiply by } i \text{ and divide by } PMT.$$

$$i(PV/PMT) = 1 - (1+i)^{-n} \qquad \text{Solve for } (1+i)^{-n}.$$

$$(1+i)^{-n} = 1 - i(PV/PMT) \qquad \text{Take the ln of both sides.}$$

$$-n\ln(1+i) = \ln(1 - i(PV/PMT)) \qquad \text{Solve for } n.$$

$$n = -\frac{\ln(1 - i(PV/PMT))}{\ln(1+i)} \qquad \text{Substitute } i = 0.1899/12,$$

$$n \approx 70.69 \qquad PV = \$847.29, \text{ and } PMT = \$20.$$

We conclude that the unpaid balance will be paid off in 71 months.

Matched Problem 7 The annual interest rate on a credit card is 24.99%. How long will it take to pay off an unpaid balance of $1,485.73 if no new purchases are made and a $50.00 payment is made each month?

General Problem-Solving Strategy

After working the problems in Exercises 3.4, it is important to work the problems in the Review Exercises. This will give you valuable experience in distinguishing among the various types of problems we have considered in this chapter. It is impossible to completely categorize all the problems you will encounter, but you may find the following guidelines helpful in determining which of the four basic formulas is involved in a particular problem. Be aware that some problems may involve more than one of these formulas and others may not involve any of them.

SUMMARY Strategy for Solving Mathematics of Finance Problems

Step 1 Determine whether the problem involves a single payment or a sequence of equal periodic payments. Simple and compound interest problems involve a single present value and a single future value. Ordinary annuities may be concerned with a present value or a future value but always involve a sequence of equal periodic payments.

Step 2 If a single payment is involved, determine whether simple or compound interest is used. Often simple interest is used for durations of a year or less and compound interest for longer periods.

Step 3 If a sequence of periodic payments is involved, determine whether the payments are being made into an account that is increasing in value—a future value problem—or the payments are being made out of an account that is decreasing in value—a present value problem. Remember that amortization problems always involve the present value of an ordinary annuity.

Steps 1–3 will help you choose the correct formula for a problem, as indicated in Figure 4. Then you must determine the values of the quantities in the formula that are given in the problem and those that must be computed, and solve the problem.

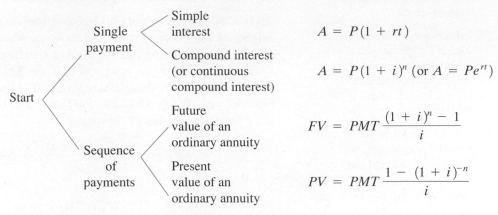

Figure 4 **Selecting the correct formula for a problem**

Exercises 3.4

W **Skills Warm-up Exercises**

In problems 1–6, find the sum of the finite geometric series $a + ar + ar^2 + \cdots + ar^{n-1}$. Write the answer as a quotient of integers. (If necessary, review Section B.2).

1. $1 + \dfrac{1}{2} + \dfrac{1}{4} + \dfrac{1}{8} + \cdots + \dfrac{1}{2^8}$

2. $1 + \dfrac{1}{5} + \dfrac{1}{25} + \dfrac{1}{125} + \cdots + \dfrac{1}{5^7}$

3. $30 + 3 + \dfrac{3}{10} + \dfrac{3}{100} + \cdots + \dfrac{3}{1,000,000}$

4. $10,000 + 1,000 + 100 + 10 + \cdots + \dfrac{1}{10,000}$

5. $1 - \dfrac{1}{2} + \dfrac{1}{4} - \dfrac{1}{8} + \cdots + \dfrac{1}{2^8}$

6. $1 - \dfrac{1}{10} + \dfrac{1}{100} - \dfrac{1}{1,000} + \dfrac{1}{10,000} - \dfrac{1}{100,000}$

In Problems 7–14, find i (the rate per period) and n (the number of periods) for each loan at the given annual rate.

7. Monthly payments of $245.65 are made for 4 years to repay a loan at 7.2% compounded monthly.

8. Semiannual payments of $3,200 are made for 12 years to repay a loan at 9.9% compounded semiannually.

9. Quarterly payments of $975 are made for 10 years to repay a loan at 9.9% compounded quarterly.

10. Annual payments of $1,045 are made for 5 years to repay a loan at 4.75% compounded annually.

11. Semiannual payments of $4,500 are made for 16 years to repay a loan at 5.05% compounded semiannually.

12. Quarterly payments of $610 are made for 6 years to repay loan at 8.24% compounded quarterly.

13. Annual payments of $5,195 are made for 9 years to repay a loan at 5.48% compounded annually.

14. Monthly payments of $433 are made for 3 years to repay a loan at 10.8% compounded monthly.

In Problems 15–26, use formula (5) or (6) to solve each problem.

15. $n = 30$; $i = 0.04$; $PMT = \$200$; $PV = ?$

16. $n = 40$; $i = 0.01$; $PMT = \$400$; $PV = ?$

17. $PV = \$40,000$; $n = 96$; $i = 0.0075$; $PMT = ?$

18. $PV = \$14,000$; $n = 72$; $i = 0.005$; $PMT = ?$

19. $PV = \$5,000$; $i = 0.01$; $PMT = \$200$; $n = ?$

20. $PV = \$20,000$; $i = 0.0175$; $PMT = \$500$; $n = ?$

21. $PV = \$9,000$; $PMT = \$600$; $n = 20$; $i = ?$ (Round answer to three decimal places.)

22. $PV = \$12,000$; $PMT = \$400$; $n = 40$; $i = ?$ (Round answer to three decimal places.)

23. Explain what is meant by the present value of an ordinary annuity.

24. Solve the present value formula (5) for *n*.

25. Explain how an ordinary annuity is involved when you take out an auto loan from a bank.

26. Explain why the last payment in an amortization schedule might differ from the other payments.

Applications

27. American General offers a 10-year ordinary annuity with a guaranteed rate of 6.65% compounded annually. How much should you pay for one of these annuities if you want to receive payments of $5,000 annually over the 10-year period?

28. American General offers a 7-year ordinary annuity with a guaranteed rate of 6.35% compounded annually. How much should you pay for one of these annuities if you want to receive payments of $10,000 annually over the 7-year period?

29. E-Loan, an online lending service, offers a 36-month auto loan at 7.56% compounded monthly to applicants with good credit ratings. If you have a good credit rating and can afford monthly payments of $350, how much can you borrow from E-Loan? What is the total interest you will pay for this loan?

30. E-Loan offers a 36-month auto loan at 9.84% compounded monthly to applicants with fair credit ratings. If you have a fair credit rating and can afford monthly payments of $350, how much can you borrow from E-Loan? What is the total interest you will pay for this loan?

31. If you buy a computer directly from the manufacturer for $2,500 and agree to repay it in 48 equal installments at 1.25% interest per month on the unpaid balance, how much are your monthly payments? How much total interest will be paid?

32. If you buy a computer directly from the manufacturer for $3,500 and agree to repay it in 60 equal installments at 1.75% interest per month on the unpaid balance, how much are your monthly payments? How much total interest will be paid?

In problems 33–36, assume that no new purchases are made with the credit card.

33. The annual interest rate on a credit card is 16.99%. If a payment of $100.00 is made each month, how long will it take to pay off an unpaid balance of $2,487.56?

34. The annual interest rate on a credit card is 24.99%. If a payment of $100.00 is made each month, how long will it take to pay off an unpaid balance of $2,487.56?

35. The annual interest rate on a credit card is 14.99%. If the minimum payment of $20 is made each month, how long will it take to pay off an unpaid balance of $937.14?

36. The annual interest rate on a credit card is 22.99%. If the minimum payment of $25 is made each month, how long will it take to pay off an unpaid balance of $860.22?

Problems 37 and 38 refer to the following ads.

37. Use the information given in the Bison sedan ad to determine if this is really 0% financing. If not, explain why and determine what rate a consumer would be charged for financing one of these sedans.

2016 BISON SEDAN
Zero down - 0% financing
$179 per month*
Buy for $9,330

* Bison sedan, 0% down, 0% for 72 months

2016 BISON WAGON
Zero down - 0% financing
$222 per month*
Buy for $12,690

* Bison wagon, 0% down, 0% for 72 months

38. Use the information given in the Bison wagon ad to determine if this is really 0% financing. If not, explain why and determine what rate a consumer would be charged for financing one of these wagons.

39. You want to purchase an automobile for $27,300. The dealer offers you 0% financing for 60 months or a $5,000 rebate. You can obtain 6.3% financing for 60 months at the local bank. Which option should you choose? Explain.

40. You want to purchase an automobile for $28,500. The dealer offers you 0% financing for 60 months or a $6,000 rebate. You can obtain 6.2% financing for 60 months at the local bank. Which option should you choose? Explain.

41. A sailboat costs $35,000. You pay 20% down and amortize the rest with equal monthly payments over a 12-year period. If you must pay 8.75% compounded monthly, what is your monthly payment? How much interest will you pay?

42. A recreational vehicle costs $80,000. You pay 10% down and amortize the rest with equal monthly payments over a 7-year period. If you pay 9.25% compounded monthly, what is your monthly payment? How much interest will you pay?

43. Construct the amortization schedule for a $5,000 debt that is to be amortized in eight equal quarterly payments at 2.8% interest per quarter on the unpaid balance.

44. Construct the amortization schedule for a $10,000 debt that is to be amortized in six equal quarterly payments at 2.6% interest per quarter on the unpaid balance.

45. A woman borrows $6,000 at 9% compounded monthly, which is to be amortized over 3 years in equal monthly payments. For tax purposes, she needs to know the amount of interest paid during each year of the loan. Find the interest paid during the first year, the second year, and the third year of the loan. [*Hint:* Find the unpaid balance after 12 payments and after 24 payments.]

46. A man establishes an annuity for retirement by depositing $50,000 into an account that pays 7.2% compounded monthly. Equal monthly withdrawals will be made each month for 5 years, at which time the account will have a zero balance. Each year taxes must be paid on the interest earned by the account during that year. How much interest was earned during the first year? [*Hint:* The amount in the account at the end of the first year is the present value of a 4-year annuity.]

47. Some friends tell you that they paid $25,000 down on a new house and are to pay $525 per month for 30 years. If interest is 7.8% compounded monthly, what was the selling price of the house? How much interest will they pay in 30 years?

48. A family is thinking about buying a new house costing $120,000. The family must pay 20% down, and the rest is to be amortized over 30 years in equal monthly payments. If money costs 7.5% compounded monthly, what will the monthly payment be? How much total interest will be paid over 30 years?

49. A student receives a federally backed student loan of $6,000 at 3.5% interest compounded monthly. After finishing college in 2 years, the student must amortize the loan in the next 4 years by making equal monthly payments. What will the payments be and what total interest will the student pay? [*Hint:* This is a two-part problem. First, find the amount of the debt at the end of the first 2 years; then amortize this amount over the next 4 years.]

50. A person establishes a sinking fund for retirement by contributing $7,500 per year at the end of each year for 20 years. For the next 20 years, equal yearly payments are withdrawn, at the end of which time the account will have a zero balance. If money is worth 9% compounded annually, what yearly payments will the person receive for the last 20 years?

51. A family has a $150,000, 30-year mortgage at 6.1% compounded monthly. Find the monthly payment. Also find the unpaid balance after

   (A) 10 years   (B) 20 years

   (C) 25 years

52. A family has a $210,000, 20-year mortgage at 6.75% compounded monthly. Find the monthly payment. Also find the unpaid balance after

   (A) 5 years   (B) 10 years

   (C) 15 years

53. A family has a $129,000, 20-year mortgage at 7.2% compounded monthly.

   (A) Find the monthly payment and the total interest paid.

   (B) Suppose the family decides to add an extra $102.41 to its mortgage payment each month starting with the very first payment. How long will it take the family to pay off the mortgage? How much interest will be saved?

54. At the time they retire, a couple has $200,000 in an account that pays 8.4% compounded monthly.

   (A) If the couple decides to withdraw equal monthly payments for 10 years, at the end of which time the account will have a zero balance, how much should the couple withdraw each month?

   (B) If the couple decides to withdraw $3,000 a month until the balance in the account is zero, how many withdrawals can the couple make?

55. An ordinary annuity that earns 7.5% compounded monthly has a current balance of $500,000. The owner of the account is about to retire and has to decide how much to withdraw from the account each month. Find the number of withdrawals under each of the following options:

   (A) $5,000 monthly   (B) $4,000 monthly

   (C) $3,000 monthly

56. Refer to Problem 55. If the account owner decides to withdraw $3,000 monthly, how much is in the account after 10 years? After 20 years? After 30 years?

57. An ordinary annuity pays 7.44% compounded monthly.

   (A) A person deposits $100 monthly for 30 years and then makes equal monthly withdrawals for the next 15 years, reducing the balance to zero. What are the monthly withdrawals? How much interest is earned during the entire 45-year process?

   (B) If the person wants to make withdrawals of $2,000 per month for the last 15 years, how much must be deposited monthly for the first 30 years?

58. An ordinary annuity pays 6.48% compounded monthly.

   (A) A person wants to make equal monthly deposits into the account for 15 years in order to then make equal monthly withdrawals of $1,500 for the next 20 years, reducing the balance to zero. How much should be deposited each month for the first 15 years? What is the total interest earned during this 35-year process?

   (B) If the person makes monthly deposits of $1,000 for the first 15 years, how much can be withdrawn monthly for the next 20 years?

59. A couple wishes to borrow money using the equity in its home for collateral. A loan company will loan the couple up to 70% of their equity. The couple purchased the home 12 years ago for $179,000. The home was financed by paying 20% down and signing a 30-year mortgage at 8.4% on the unpaid balance. Equal monthly payments were made to amortize the loan over the 30-year period. The net market value of the house is now $215,000. After making the 144th payment, the couple applied to the loan company for the maximum loan. How much (to the nearest dollar) will the couple receive?

60. A person purchased a house 10 years ago for $160,000. The house was financed by paying 20% down and signing a 30-year mortgage at 7.75% on the unpaid balance. Equal monthly payments were made to amortize the loan over a 30-year period. The owner now (after the 120th payment) wishes to refinance the house due to a need for additional cash. If the loan company agrees to a new 30-year mortgage of 80% of the new appraised value of the house, which is $225,000, how much cash (to the nearest dollar) will the owner receive after repaying the balance of the original mortgage?

61. A person purchased a $145,000 home 10 years ago by paying 20% down and signing a 30-year mortgage at 7.9% compounded monthly. Interest rates have dropped and the owner wants to refinance the unpaid balance by signing a new 20-year mortgage at 5.5% compounded monthly. How much interest will refinancing save?

62. A person purchased a $200,000 home 20 years ago by paying 20% down and signing a 30-year mortgage at 13.2% compounded monthly. Interest rates have dropped and the owner wants to refinance the unpaid balance by signing a new 10-year mortgage at 8.2% compounded monthly. How much interest will refinancing save?

✎ **63.** Discuss the similarities and differences in the graphs of unpaid balance as a function of time for 30-year mortgages of $50,000, $75,000, and $100,000, respectively, each at 9% compounded monthly (see the figure). Include computations of the monthly payment and total interest paid in each case.

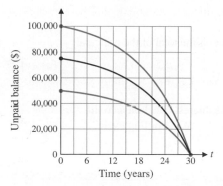

Figure for 63

✎ **64.** Discuss the similarities and differences in the graphs of unpaid balance as a function of time for 30-year mortgages of $60,000 at rates of 7%, 10%, and 13%, respectively (see the figure). Include computations of the monthly payment and total interest paid in each case.

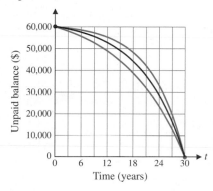

⊞ *In Problems 65–68, use graphical approximation techniques or an equation solver to approximate the desired interest rate. Express each answer as a percentage, correct to two decimal places.*

65. A discount electronics store offers to let you pay for a $1,000 stereo in 12 equal $90 installments. The store claims that since you repay $1,080 in 1 year, the $80 finance charge represents an 8% annual rate. This would be true if you repaid the loan in a single payment at the end of the year. But since you start repayment after 1 month, this is an amortized loan, and 8% is not the correct rate. What is the annual nominal compounding rate for this loan?

66. A $2,000 computer can be financed by paying $100 per month for 2 years. What is the annual nominal compounding rate for this loan?

67. The owner of a small business has received two offers of purchase. The first prospective buyer offers to pay the owner $100,000 in cash now. The second offers to pay the owner $10,000 now and monthly payments of $1,200 for 10 years. In effect, the second buyer is asking the owner for a $90,000 loan. If the owner accepts the second offer, what annual nominal compounding rate will the owner receive for financing this purchase?

68. At the time they retire, a couple has $200,000 invested in an annuity. The couple can take the entire amount in a single payment, or receive monthly payments of $2,000 for 15 years. If the couple elects to receive the monthly payments, what annual nominal compounding rate will the couple earn on the money invested in the annuity?

Answers to Matched Problems

1. $13,577.71   **2.** $10,688.87

3. *PMT* = $112.98/mo; total interest = $311.52

4.

Payment Number	Payment	Interest	Unpaid Balance Reduction	Unpaid Balance
0				$1,000.00
1	$174.03	$12.50	$161.53	838.47
2	174.03	10.48	163.55	674.92
3	174.03	8.44	165.59	509.33
4	174.03	6.37	167.66	341.67
5	174.03	4.27	169.76	171.91
6	174.06	2.15	171.91	0.00
Totals	$1,044.21	$44.21	$1,000.00	

5. $98,551   **6.** Choose the 0% financing.   **7.** 47 months

Chapter 3   Summary and Review

Important Terms, Symbols, and Concepts

3.1 Simple Interest

EXAMPLES

- **Interest** is the fee paid for the use of a sum of money P, called the **principal**. **Simple interest** is given by

$$I = Prt$$

where I = interest

         P = principal

         r = annual simple interest rate (written as a decimal)

         t = time in years

3.1 Simple Interest (*Continued*)

- If a principal P (**present value**) is borrowed, then the **amount** A (**future value**) is the total of the principal and the interest:

$$A = P + Prt$$
$$= P(1 + rt)$$

Ex. 1, p. 128

Ex. 2, p. 128

Ex. 3, p. 129

Ex. 4, p. 129

Ex. 5, p. 130

Ex. 6, p. 131

- The **average daily balance method** is a common method for calculating the interest owed on a credit card. The formula $I = Prt$ is used, but a daily balance is calculated for each day of the billing cycle, and P is the average of those daily balances.

3.2 Compound and Continuous Compound Interest

- **Compound interest** is interest paid on the principal plus reinvested interest. The future and present values are related by

$$A = P(1 + i)^n$$

Ex. 1, p. 136

where $i = r/m$ and

$A =$ amount or future value

$P =$ principal or present value

$r =$ annual nominal rate (or just rate)

$m =$ number of compounding periods per year

$i =$ rate per compounding period

$n =$ total number of compounding periods

- If a principal P is invested at an annual rate r earning **continuous compound interest**, then the amount A after t years is given by

$$A = Pe^n$$

Ex. 2, p. 138

Ex. 3, p. 139

Ex. 4, p. 140

Ex. 5, p. 141

- The **growth time** of an investment is the time it takes for a given principal to grow to a particular amount. Three methods for finding the growth time are as follows:

1. Use logarithms and a calculator.

2. Use graphical approximation on a graphing calculator.

3. Use an **equation solver** on a graphing calculator or a computer.

- The **annual percentage yield** (APY; also called the **effective rate** or **true interest rate**) is the simple interest rate that would earn the same amount as a given annual rate for which interest is compounded.

- If a principal is invested at the annual rate r compounded m times a year, then the annual percentage yield is given by

$$APY = \left(1 + \frac{r}{m}\right)^m - 1$$

Ex. 6, p. 142

- If a principal is invested at the annual rate r compounded continuously, then the annual percentage yield is given by

$$APY = e^r - 1$$

Ex. 7, p. 143

- A **zero coupon bond** is a bond that is sold now at a discount and will pay its **face value** at some time in the future when it matures.

3.3 Future Value of an Annuity; Sinking Funds

- An **annuity** is any sequence of equal periodic payments. If payments are made at the end of each time interval, then the annuity is called an **ordinary annuity**. The amount, or **future value**, of an annuity is the sum of all payments plus all interest earned and is given by

$$FV = PMT \frac{(1 + i)^n - 1}{i}$$

Ex. 1, p. 149

Ex. 3, p. 152

Ex. 4, p. 153

3.3 Future Value of an Annuity; Sinking Funds (*Continued*)

where FV = future value (amount)

PMT = periodic payment

i = rate per period

n = number of payments (periods)

- A **balance sheet** is a table that shows the interest and balance for each payment of an annuity.

- An account that is established to accumulate funds to meet future obligations or debts is called a **sinking fund**. The **sinking fund payment** can be found by solving the future value formula for PMT:

Ex. 2, p. 151

$$PMT = FV\frac{i}{(1+i)^n - 1}$$

3.4 Present Value of an Annuity; Amortization

- If equal payments are made from an account until the amount in the account is 0, the payment and the **present value** are related by the following formula:

Ex. 1, p. 157
Ex. 2, p. 157

$$PV = PMT\frac{1 - (1+i)^{-n}}{i}$$

where PV = present value of all payments

PMT = periodic payment

i = rate per period

n = number of periods

- **Amortizing** a debt means that the debt is retired in a given length of time by equal periodic payments that include compound interest. Solving the present value formula for the payment gives us the **amortization formula**:

Ex. 3, p. 159
Ex. 6, p. 162

$$PMT = PV\frac{i}{1 - (1+i)^{-n}}$$

- An **amortization schedule** is a table that shows the interest due and the balance reduction for each payment of a loan.

Ex. 4, p. 159

- The **equity** in a property is the difference between the current net market value and the unpaid loan balance. The unpaid balance of a loan with n **remaining payments** is given by the present value formula.

Ex. 5, p. 161

- A strategy for solving problems in the mathematics of finance is presented on page 163.

Review Exercises

Work through all the problems in this chapter review and check your answers in the back of the book. Answers to all review problems are there along with section numbers in italics to indicate where each type of problem is discussed. Where weaknesses show up, review appropriate sections in the text.

In Problems 1–4, find the indicated quantity, given $A = P(1 + rt)$.

1. $A = ?; P = \$100; r = 9\%; t = 6$ months

2. $A = \$808; P = ?; r = 12\%; t = 1$ month

3. $A = \$212; P = \$200; r = 8\%; t = ?$

4. $A = \$4,120; P = \$4,000; r = ?; t = 6$ months

In Problems 5 and 6, find the indicated quantity, given $A = P(1 + i)^n$.

5. $A = ?; P = \$1,200; i = 0.005; n = 30$

6. $A = \$5,000; P = ?; i = 0.0075; n = 60$

In Problems 7 and 8, find the indicated quantity, given $A = Pe^{rt}$.

7. $A = ?; P = \$4,750; r = 6.8\%; t = 3$ years

8. $A = \$36,000; P = ?; r = 9.3\%; t = 60$ months

In Problems 9 and 10, find the indicated quantity, given

$$FV = PMT\frac{(1+i)^n - 1}{i}$$

9. $FV = ?; PMT = \$1,000; i = 0.005; n = 60$

10. $FV = \$8,000; PMT = ?; i = 0.015; n = 48$

In Problems 11 and 12, find the indicated quantity, given

$$PV = PMT\frac{1 - (1+i)^{-n}}{i}$$

11. $PV = ?; PMT = \$2,500; i = 0.02; n = 16$

12. $PV = \$8,000; PMT = ?; i = 0.0075; n = 60$

13. Solve the equation $2,500 = 1,000(1.06)^n$ for n to the nearest integer using:

(A) Logarithms

(B) Graphical approximation techniques or an equation solver on a graphing calculator

14. Solve the equation

$$5,000 = 100 \frac{(1.01)^n - 1}{0.01}$$

for n to the nearest integer using:

(A) Logarithms

(B) Graphical approximation techniques or an equation solver on a graphing calculator.

Applications

Find all dollar amounts correct to the nearest cent. When an interest rate is requested as an answer, express the rate as a percentage, correct to two decimal places.

15. If you borrow $3,000 at 14% simple interest for 10 months, how much will you owe in 10 months? How much interest will you pay?

16. Grandparents deposited $6,000 into a grandchild's account toward a college education. How much money (to the nearest dollar) will be in the account 17 years from now if the account earns 7% compounded monthly?

17. How much should you pay for a corporate bond paying 6.6% compounded monthly in order to have $25,000 in 10 years?

18. A savings account pays 5.4% compounded annually. Construct a balance sheet showing the interest earned during each year and the balance at the end of each year for 4 years if

(A) A single deposit of $400 is made at the beginning of the first year.

(B) Four deposits of $100 are made at the end of each year.

19. One investment pays 13% simple interest and another 9% compounded annually. Which investment would you choose? Why?

20. A $10,000 retirement account is left to earn interest at 7% compounded daily. How much money will be in the account 40 years from now when the owner reaches 65? (Use a 365-day year and round answer to the nearest dollar.)

21. A couple wishes to have $40,000 in 6 years for the down payment on a house. At what rate of interest compounded continuously must $25,000 be invested now to accomplish this goal?

22. Which is the better investment and why: 9% compounded quarterly or 9.25% compounded annually?

23. What is the value of an ordinary annuity at the end of 8 years if $200 per month is deposited into an account earning 7.2% compounded monthly? How much of this value is interest?

24. A credit card company charges a 22% annual rate for overdue accounts. How much interest will be owed on a $635 account 1 month overdue?

25. What will a $23,000 car cost (to the nearest dollar) 5 years from now if the inflation rate over that period averages 5% compounded annually?

26. What would the $23,000 car in Problem 25 have cost (to the nearest dollar) 5 years ago if the inflation rate over that period had averaged 5% compounded annually?

27. A loan of $2,500 was repaid at the end of 10 months with a check for $2,812.50. What annual rate of interest was charged?

28. You want to purchase an automobile for $21,600. The dealer offers you 0% financing for 48 months or a $3,000 rebate. You can obtain 4.8% financing for 48 months at the local bank. Which option should you choose? Explain.

29. Find the annual percentage yield on a CD earning 6.25% if interest is compounded

(A) monthly.

(B) continuously.

30. You have $5,000 toward the purchase of a boat that will cost $6,000. How long will it take the $5,000 to grow to $6,000 if it is invested at 9% compounded quarterly? (Round up to the next-higher quarter if not exact.)

31. How long will it take money to double if it is invested at 6% compounded monthly? 9% compounded monthly? (Round up to the next-higher month if not exact.)

32. Starting on his 21st birthday, and continuing on every birthday up to and including his 65th, John deposits $2,000 a year into an IRA. How much (to the nearest dollar) will be in the account on John's 65th birthday, if the account earns:

(A) 7% compounded annually?

(B) 11% compounded annually?

33. If you just sold a stock for $17,388.17 (net) that cost you $12,903.28 (net) 3 years ago, what annual compound rate of return did you make on your investment?

34. The table shows the fees for refund anticipation loans (RALs) offered by an online tax preparation firm. Find the annual rate of interest for each of the following loans. Assume a 360-day year.

(A) A $400 RAL paid back in 15 days

(B) A $1,800 RAL paid back in 21 days

RAL Amount	RAL Fee
$10–$500	$29.00
$501–$1,000	$39.00
$1,001–$1,500	$49.00
$1,501–$2,000	$69.00
$2,001–$5,000	$82.00

35. Lincoln Benefit Life offered an annuity that pays 5.5% compounded monthly. What equal monthly deposit should be made into this annuity in order to have $50,000 in 5 years?

36. A person wants to establish an annuity for retirement purposes. He wants to make quarterly deposits for 20 years so that he can then make quarterly withdrawals of $5,000 for 10 years. The annuity earns 7.32% interest compounded quarterly.

   (A) How much will have to be in the account at the time he retires?

   (B) How much should be deposited each quarter for 20 years in order to accumulate the required amount?

   (C) What is the total amount of interest earned during the 30-year period?

37. If you borrow $4,000 from an online lending firm for the purchase of a computer and agree to repay it in 48 equal installments at 0.9% interest per month on the unpaid balance, how much are your monthly payments? How much total interest will be paid?

38. A company decides to establish a sinking fund to replace a piece of equipment in 6 years at an estimated cost of $50,000. To accomplish this, they decide to make fixed monthly payments into an account that pays 6.12% compounded monthly. How much should each payment be?

39. How long will it take money to double if it is invested at 7.5% compounded daily? 7.5% compounded annually?

40. A student receives a student loan for $8,000 at 5.5% interest compounded monthly to help her finish the last 1.5 years of college. Starting 1 year after finishing college, the student must amortize the loan in the next 5 years by making equal monthly payments. What will the payments be and what total interest will the student pay?

41. If you invest $5,650 in an account paying 8.65% compounded continuously, how much money will be in the account at the end of 10 years?

42. A company makes a payment of $1,200 each month into a sinking fund that earns 6% compounded monthly. Use graphical approximation techniques on a graphing calculator to determine when the fund will be worth $100,000.

43. A couple has a $50,000, 20-year mortgage at 9% compounded monthly. Use graphical approximation techniques on a graphing calculator to determine when the unpaid balance will drop below $10,000.

44. A loan company advertises in the paper that you will pay only 8¢ a day for each $100 borrowed. What annual rate of interest are they charging? (Use a 360-day year.)

45. Construct the amortization schedule for a $1,000 debt that is to be amortized in four equal quarterly payments at 2.5% interest per quarter on the unpaid balance.

46. You can afford monthly deposits of only $300 into an account that pays 7.98% compounded monthly. How long will it be until you will have $9,000 to purchase a used car? (Round to the next-higher month if not exact.)

47. A company establishes a sinking fund for plant retooling in 6 years at an estimated cost of $850,000. How much should be invested semiannually into an account paying 8.76% compounded semiannually? How much interest will the account earn in the 6 years?

48. What is the annual nominal rate compounded monthly for a CD that has an annual percentage yield of 2.50%?

49. If you buy a 13-week T-bill with a maturity value of $5,000 for $4,922.15 from the U.S. Treasury Department, what annual interest rate will you earn?

50. In order to save enough money for the down payment on a condominium, a young couple deposits $200 each month into an account that pays 7.02% interest compounded monthly. If the couple needs $10,000 for a down payment, how many deposits will the couple have to make?

51. A business borrows $80,000 at 9.42% interest compounded monthly for 8 years.

   (A) What is the monthly payment?

   (B) What is the unpaid balance at the end of the first year?

   (C) How much interest was paid during the first year?

52. You unexpectedly inherit $10,000 just after you have made the 72nd monthly payment on a 30-year mortgage of $60,000 at 8.2% compounded monthly. Discuss the relative merits of using the inheritance to reduce the principal of the loan or to buy a certificate of deposit paying 7% compounded monthly.

53. Your parents are considering a $75,000, 30-year mortgage to purchase a new home. The bank at which they have done business for many years offers a rate of 7.54% compounded monthly. A competitor is offering 6.87% compounded monthly. Would it be worthwhile for your parents to switch banks? Explain.

54. How much should a $5,000 face value zero coupon bond, maturing in 5 years, be sold for now, if its rate of return is to be 5.6% compounded annually?

55. If you pay $5,695 for a $10,000 face value zero coupon bond that matures in 10 years, what is your annual compound rate of return?

56. If an investor wants to earn an annual interest rate of 6.4% on a 26-week T-bill with a maturity value of $5,000, how much should the investor pay for the T-bill?

57. Two years ago you borrowed $10,000 at 12% interest compounded monthly, which was to be amortized over 5 years. Now you have acquired some additional funds and decide that you want to pay off this loan. What is the unpaid balance after making equal monthly payments for 2 years?

58. What annual nominal rate compounded monthly has the same annual percentage yield as 7.28% compounded quarterly?

59. (A) A man deposits $2,000 in an IRA on his 21st birthday and on each subsequent birthday up to, and including, his 29th (nine deposits in all). The account earns 8%

compounded annually. If he leaves the money in the account without making any more deposits, how much will he have on his 65th birthday, assuming the account continues to earn the same rate of interest?

(B) How much would be in the account (to the nearest dollar) on his 65th birthday if he had started the deposits on his 30th birthday and continued making deposits on each birthday until (and including) his 65th birthday?

60. A promissory note will pay $27,000 at maturity 10 years from now. How much money should you be willing to pay now if money is worth 5.5% compounded continuously?

61. In a new housing development, the houses are selling for $100,000 and require a 20% down payment. The buyer is given a choice of 30-year or 15-year financing, both at 7.68% compounded monthly.

(A) What is the monthly payment for the 30-year choice? For the 15-year choice?

(B) What is the unpaid balance after 10 years for the 30-year choice? For the 15-year choice?

62. A loan company will loan up to 60% of the equity in a home. A family purchased their home 8 years ago for $83,000. The home was financed by paying 20% down and signing a 30-year mortgage at 8.4% for the balance. Equal monthly payments were made to amortize the loan over the 30-year period. The market value of the house is now $95,000. After making the 96th payment, the family applied to the loan company for the maximum loan. How much (to the nearest dollar) will the family receive?

63. A $600 stereo is financed for 6 months by making monthly payments of $110. What is the annual nominal compounding rate for this loan?

64. A person deposits $2,000 each year for 25 years into an IRA. When she retires immediately after making the 25th deposit, the IRA is worth $220,000.

(A) Find the interest rate earned by the IRA over the 25-year period leading up to retirement.

(B) Assume that the IRA continues to earn the interest rate found in part (A). How long can the retiree withdraw $30,000 per year? How long can she withdraw $24,000 per year?

4 Systems of Linear Equations; Matrices

4.1 Review: Systems of Linear Equations in Two Variables

4.2 Systems of Linear Equations and Augmented Matrices

4.3 Gauss–Jordan Elimination

4.4 Matrices: Basic Operations

4.5 Inverse of a Square Matrix

4.6 Matrix Equations and Systems of Linear Equations

4.7 Leontief Input–Output Analysis

Chapter 4 Summary and Review

Review Exercises

Introduction

Systems of linear equations can be used to solve resource allocation problems in business and economics (see Problems 73 and 76 in Section 4.3 on production schedules for boats and leases for airplanes). Such systems can involve many equations in many variables. So after reviewing methods for solving two linear equations in two variables, we use matrices and matrix operations to develop procedures that are suitable for solving linear systems of any size. We also discuss Wassily Leontief's Nobel prizewinning application of matrices to economic planning for industrialized countries.

4.1   Review: Systems of Linear Equations in Two Variables

- Systems of Linear Equations in Two Variables
- Graphing
- Substitution
- Elimination by Addition
- Applications

Systems of Linear Equations in Two Variables

To establish basic concepts, let's consider the following simple example: If 2 adult tickets and 1 child ticket cost $32, and if 1 adult ticket and 3 child tickets cost $36, what is the price of each?

$$\text{Let:} \quad x = \text{price of adult ticket}$$
$$y = \text{price of child ticket}$$
$$\text{Then:} \quad 2x + y = 32$$
$$x + 3y = 36$$

Now we have a system of two linear equations in two variables. It is easy to find ordered pairs (x, y) that satisfy one or the other of these equations. For example, the ordered pair $(16, 0)$ satisfies the first equation but not the second, and the ordered pair $(24, 4)$ satisfies the second but not the first. To solve this system, we must find all ordered pairs of real numbers that satisfy both equations at the same time. In general, we have the following definition:

> **DEFINITION** Systems of Two Linear Equations in Two Variables
> Given the **linear system**
>
> $$ax + by = h$$
> $$cx + dy = k$$
>
> where a, b, c, d, h, and k are real constants, a pair of numbers $x = x_0$ and $y = y_0$ [also written as an ordered pair (x_0, y_0)] is a **solution** of this system if each equation is satisfied by the pair. The set of all such ordered pairs is called the **solution set** for the system. To **solve** a system is to find its solution set.

We will consider three methods of solving such systems: *graphing, substitution,* and *elimination by addition.* Each method has its advantages, depending on the situation.

Graphing

Recall that the graph of a line is a graph of all the ordered pairs that satisfy the equation of the line. To solve the ticket problem by graphing, we graph both equations in the same coordinate system. The coordinates of any points that the graphs have in common must be solutions to the system since they satisfy both equations.

EXAMPLE 1 Solving a System by Graphing Solve the ticket problem by graphing:

$$2x + y = 32$$
$$x + 3y = 36$$

SOLUTION An easy way to find two distinct points on the first line is to find the x and y intercepts. Substitute $y = 0$ to find the x intercept ($2x = 32$, so $x = 16$), and substitute $x = 0$ to find the y intercept ($y = 32$). Then draw the line through

$(16, 0)$ and $(0, 32)$. After graphing both lines in the same coordinate system (Fig. 1), estimate the coordinates of the intersection point:

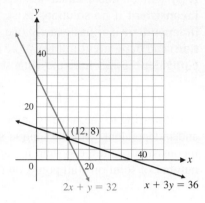

$x = \$12$ Adult ticket
$y = \$8$ Child ticket

$(12, 8)$

Figure 1 $2x + y = 32$ $x + 3y = 36$

CHECK

$$2x + y = 32 \qquad\qquad x + 3y = 36$$
$$2(12) + 8 \overset{?}{=} 32 \qquad 12 + 3(8) \overset{?}{=} 36 \qquad \text{Check that } (12, 8) \text{ satisfies}$$
$$32 \overset{\checkmark}{=} 32 \qquad\qquad 36 \overset{\checkmark}{=} 36 \qquad \text{each of the original equations.}$$

Matched Problem 1   Solve by graphing and check:

$$2x - y = -3$$
$$x + 2y = -4$$

It is clear that Example 1 has exactly one solution since the lines have exactly one point in common. In general, lines in a rectangular coordinate system are related to each other in one of the three ways illustrated in the next example.

EXAMPLE 2   Solving a System by Graphing   Solve each of the following systems by graphing:

(A) $x - 2y = 2$ (B) $x + 2y = -4$ (C) $2x + 4y = 8$
     $x + \ y = 5$ $2x + 4y = \ 8$ $x + 2y = 4$

SOLUTION

(A)   (B)   (C)

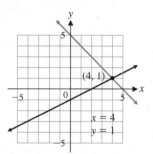   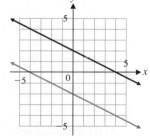   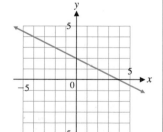

$x = 4$
$y = 1$

Intersection at one point   Lines are parallel (each   Lines coincide—infinite
only—exactly one solution   has slope $-\frac{1}{2}$)—no solutions   number of solutions

Matched Problem 2   Solve each of the following systems by graphing:

(A) $x + y = 4$ (B) $6x - 3y = 9$ (C) $2x - \ y = \ \ 4$
     $2x - y = 2$ $2x - \ y = 3$ $6x - 3y = -18$

We introduce some terms that describe the different types of solutions to systems of equations.

> **DEFINITION**   Systems of Linear Equations: Basic Terms
> A system of linear equations is **consistent** if it has one or more solutions and **inconsistent** if no solutions exist. Furthermore, a consistent system is said to be **independent** if it has exactly one solution (often referred to as the **unique solution**) and **dependent** if it has more than one solution. Two systems of equations are **equivalent** if they have the same solution set.

Referring to the three systems in Example 2, the system in part (A) is consistent and independent with the unique solution $x = 4$, $y = 1$. The system in part (B) is inconsistent. And the system in part (C) is consistent and dependent with an infinite number of solutions (all points on the two coinciding lines).

⚠ **CAUTION**   Given a system of equations, do not confuse the *number of variables* with the *number of solutions*. The systems of Example 2 involve two variables, x and y. A solution to such a system is a *pair* of numbers, one for x and one for y. So the system in Example 2A has two variables, but exactly one solution, namely $x = 4$, $y = 1$. ▲

Explore and Discuss 1   Can a consistent and dependent system have exactly two solutions? Exactly three solutions? Explain.

By graphing a system of two linear equations in two variables, we gain useful information about the solution set of the system. In general, any two lines in a coordinate plane must intersect in exactly one point, be parallel, or coincide (have identical graphs). So the systems in Example 2 illustrate the only three possible types of solutions for systems of two linear equations in two variables. These ideas are summarized in Theorem 1.

> **THEOREM 1**   Possible Solutions to a Linear System
> The linear system
> $$ax + by = h$$
> $$cx + dy = k$$
> must have
> (A) Exactly one solution   Consistent and independent
> or
> (B) No solution   Inconsistent
> or
> (C) Infinitely many solutions   Consistent and dependent
> There are no other possibilities.

   In the past, one drawback to solving systems by graphing was the inaccuracy of hand-drawn graphs. Graphing calculators have changed that. Graphical solutions on a graphing calculator provide an accurate approximation of the solution to a system of linear equations in two variables. Example 3 demonstrates this.

EXAMPLE 3 Solving a System Using a Graphing Calculator Solve to two decimal places using graphical approximation techniques on a graphing calculator:

$$5x + 2y = 15$$
$$2x - 3y = 16$$

SOLUTION First, solve each equation for y.

$$5x + 2y = 15$$
$$2y = -5x + 15$$
$$y = -2.5x + 7.5$$

$$2x - 3y = 16$$
$$-3y = -2x + 16$$
$$y = \frac{2}{3}x - \frac{16}{3}$$

Next, enter each equation in the graphing calculator (Fig. 2A), graph in an appropriate viewing window, and approximate the intersection point (Fig. 2B).

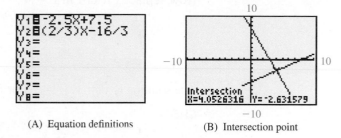

Figure 2   (A) Equation definitions   (B) Intersection point

Rounding the values in Figure 2B to two decimal places, we see that the solution is $x = 4.05$ and $y = -2.63$, or $(4.05, -2.63)$.

CHECK
$$5x + 2y = 15$$
$$5(4.05) + 2(-2.63) \overset{?}{=} 15$$
$$14.99 \overset{\checkmark}{\approx} 15$$

$$2x - 3y = 16$$
$$2(4.05) - 3(-2.63) \overset{?}{=} 16$$
$$15.99 \overset{\checkmark}{\approx} 16$$

The checks are sufficiently close but, due to rounding, not exact.

Matched Problem 3 Solve to two decimal places using graphical approximation techniques on a graphing calculator:

$$2x - 5y = -25$$
$$4x + 3y = 5$$

Graphical methods help us to visualize a system and its solutions, reveal relationships that might otherwise be hidden, and, with the assistance of a graphing calculator, provide accurate approximations to solutions.

Substitution

Now we review an algebraic method that is easy to use and provides exact solutions to a system of two equations in two variables, provided that solutions exist. In this method, first we choose one of two equations in a system and solve for one variable in terms of the other. (We make a choice that avoids fractions, if possible.) Then we **substitute** the result into the other equation and solve the resulting linear equation in one variable. Finally, we substitute this result back into the results of the first step to find the second variable.

EXAMPLE 4 Solving a System by Substitution Solve by substitution:

$$5x + y = 4$$
$$2x - 3y = 5$$

SOLUTION Solve either equation for one variable in terms of the other; then substitute into the remaining equation. In this problem, we avoid fractions by choosing the first equation and solving for y in terms of x:

$$5x + y = 4 \qquad \text{Solve the first equation for } y \text{ in terms of } x.$$
$$y = 4 - 5x \qquad \text{Substitute into the second equation.}$$

$$2x - 3y = \;\; 5 \qquad \text{Second equation}$$
$$2x - 3(4 - 5x) = \;\; 5 \qquad \text{Solve for } x.$$
$$2x - 12 + 15x = \;\; 5$$
$$17x = 17$$
$$x = \;\; 1$$

Now, replace x with 1 in $y = 4 - 5x$ to find y:

$$y = 4 - 5x$$
$$y = 4 - 5(1)$$
$$y = -1$$

The solution is $x = 1, y = -1$ or $(1, -1)$.

CHECK

$$5x + \qquad y = 4 \qquad\qquad 2x - \qquad 3y = 5$$
$$5(1) + \;\; (-1) \overset{?}{=} 4 \qquad\qquad 2(1) - 3(-1) \overset{?}{=} 5$$
$$4 \overset{\checkmark}{=} 4 \qquad\qquad\qquad\qquad 5 \overset{\checkmark}{=} 5$$

Matched Problem 4 **Solve by substitution:**

$$3x + 2y = -2$$
$$2x - \;\; y = -6$$

Explore and Discuss 2   Return to Example 2 and solve each system by substitution. Based on your results, describe how you can recognize a dependent system or an inconsistent system when using substitution.

Elimination by Addition

The methods of graphing and substitution both work well for systems involving two variables. However, neither is easily extended to larger systems. Now we turn to **elimination by addition**. This is probably the most important method of solution. It readily generalizes to larger systems and forms the basis for computer-based solution methods.

To solve an equation such as $2x - 5 = 3$, we perform operations on the equation until we reach an equivalent equation whose solution is obvious (see Appendix A, Section A.7).

$$2x - 5 = 3 \qquad \text{Add 5 to both sides.}$$
$$2x = 8 \qquad \text{Divide both sides by 2.}$$
$$x = 4$$

Theorem 2 indicates that we can solve systems of linear equations in a similar manner.

> **THEOREM 2** Operations That Produce Equivalent Systems
>
> A system of linear equations is transformed into an equivalent system if
> (A) Two equations are interchanged.
> (B) An equation is multiplied by a nonzero constant.
> (C) A constant multiple of one equation is added to another equation.

Any one of the three operations in Theorem 2 can be used to produce an equivalent system, but the operations in parts (B) and (C) will be of most use to us now. Part (A) becomes useful when we apply the theorem to larger systems. The use of Theorem 2 is best illustrated by examples.

EXAMPLE 5 Solving a System Using Elimination by Addition Solve the following system using elimination by addition:

$$3x - 2y = 8$$
$$2x + 5y = -1$$

SOLUTION We use Theorem 2 to eliminate one of the variables, obtaining a system with an obvious solution:

$$3x - 2y = 8$$
$$2x + 5y = -1$$

Multiply the top equation by 5 and the bottom equation by 2 (Theorem 2B).

$$5(3x - 2y) = 5(8)$$
$$2(2x + 5y) = 2(-1)$$

$$\begin{aligned}15x - 10y &= 40 \\ \underline{4x + 10y} &= \underline{-2} \\ 19x &= 38\end{aligned}$$

Add the top equation to the bottom equation (Theorem 2C), eliminating the y terms.
Divide both sides by 19, which is the same as multiplying the equation by $\frac{1}{19}$ (Theorem 2B).

$$x = 2$$

This equation paired with either of the two original equations produces a system equivalent to the original system.

Knowing that $x = 2$, we substitute this number back into either of the two original equations (we choose the second) to solve for y:

$$2(2) + 5y = -1$$
$$5y = -5$$
$$y = -1$$

The solution is $x = 2$, $y = -1$ or $(2, -1)$.

CHECK

$$3x - 2y = 8 \qquad\qquad 2x + 5y = -1$$
$$3(2) - 2(-1) \overset{?}{=} 8 \qquad\qquad 2(2) + 5(-1) \overset{?}{=} -1$$
$$8 \overset{\checkmark}{=} 8 \qquad\qquad -1 \overset{\checkmark}{=} -1$$

Matched Problem 5 Solve the following system using elimination by addition:

$$5x - 2y = 12$$
$$2x + 3y = 1$$

Let's see what happens in the elimination process when a system has either no solution or infinitely many solutions. Consider the following system:

$$2x + 6y = -3$$
$$x + 3y = 2$$

Multiplying the second equation by -2 and adding, we obtain

$$2x + 6y = -3$$
$$\underline{-2x - 6y = -4}$$
$$0 = -7 \quad \text{Not possible}$$

We have obtained a contradiction. The assumption that the original system has solutions must be false. So the system has no solutions, and its solution set is the empty set. The graphs of the equations are parallel lines, and the system is inconsistent.

Now consider the system

$$x - \tfrac{1}{2}y = 4$$
$$-2x + y = -8$$

If we multiply the top equation by 2 and add the result to the bottom equation, we obtain

$$2x - y = 8$$
$$\underline{-2x + y = -8}$$
$$0 = 0$$

Obtaining $0 = 0$ implies that the equations are equivalent; that is, their graphs coincide and the system is dependent. If we let $x = k$, where k is any real number, and solve either equation for y, we obtain $y = 2k - 8$. So $(k, 2k - 8)$ is a solution to this system for any real number k. The variable k is called a **parameter** and replacing k with a real number produces a **particular solution** to the system. For example, some particular solutions to this system are

$k = -1$	$k = 2$	$k = 5$	$k = 9.4$
$(-1, -10)$	$(2, -4)$	$(5, 2)$	$(9.4, 10.8)$

Applications

Many real-world problems are solved readily by constructing a mathematical model consisting of two linear equations in two variables and applying the solution methods that we have discussed. We shall examine two applications in detail.

EXAMPLE 6 Diet Jasmine wants to use milk and orange juice to increase the amount of calcium and vitamin A in her daily diet. An ounce of milk contains 37 milligrams of calcium and 57 micrograms* of vitamin A. An ounce of orange juice contains 5 milligrams of calcium and 65 micrograms of vitamin A. How many ounces of milk and orange juice should Jasmine drink each day to provide exactly 500 milligrams of calcium and 1,200 micrograms of vitamin A?

SOLUTION The first step in solving an application problem is to introduce the proper variables. Often, the question asked in the problem will guide you in this decision. Reading the last sentence in Example 6, we see that we must determine a certain number of ounces of milk and orange juice. So we introduce variables to represent these unknown quantities:

$$x = \text{number of ounces of milk}$$
$$y = \text{number of ounces of orange juice}$$

---

*A microgram (μg) is one millionth (10^{-6}) of a gram.

Next, we summarize the given information using a table. It is convenient to organize the table so that the quantities represented by the variables correspond to columns in the table (rather than to rows) as shown.

	Milk	Orange Juice	Total Needed
Calcium	37 mg/oz	5 mg/oz	500 mg
Vitamin A	57 μg/oz	65 μg/oz	1,200 μg

Now we use the information in the table to form equations involving x and y:

$$\left(\begin{array}{c}\text{calcium in } x \text{ oz} \\ \text{of milk}\end{array}\right) + \left(\begin{array}{c}\text{calcium in } y \text{ oz} \\ \text{of orange juice}\end{array}\right) = \left(\begin{array}{c}\text{total calcium} \\ \text{needed (mg)}\end{array}\right)$$

$$37x \qquad + \qquad 5y \qquad = \qquad 500$$

$$\left(\begin{array}{c}\text{vitamin A in } x \text{ oz} \\ \text{of milk}\end{array}\right) + \left(\begin{array}{c}\text{vitamin A in } y \text{ oz} \\ \text{of orange juice}\end{array}\right) = \left(\begin{array}{c}\text{total vitamin A} \\ \text{needed } (\mu g)\end{array}\right)$$

$$57x \qquad + \qquad 65y \qquad = \qquad 1,200$$

So we have the following model to solve:

$$37x + 5y = 500$$
$$57x + 65y = 1,200$$

We can multiply the first equation by -13 and use elimination by addition:

$$
\begin{aligned}
-481x - 65y &= -6,500 \\
57x + 65y &= 1,200 \\
\hline
-424x &= -5,300 \\
x &= 12.5
\end{aligned}
\qquad\qquad
\begin{aligned}
37(12.5) + 5y &= 500 \\
5y &= 37.5 \\
y &= 7.5
\end{aligned}
$$

Drinking 12.5 ounces of milk and 7.5 ounces of orange juice each day will provide Jasmine with the required amounts of calcium and vitamin A.

CHECK
$$
\begin{aligned}
37x + 5y &= 500 \\
37(12.5) + 5(7.5) &\overset{?}{=} 500 \\
500 &\overset{\checkmark}{=} 500
\end{aligned}
\qquad\qquad
\begin{aligned}
57x + 65y &= 1,200 \\
57(12.5) + 65(7.5) &\overset{?}{=} 1,200 \\
1,200 &\overset{\checkmark}{=} 1,200
\end{aligned}
$$

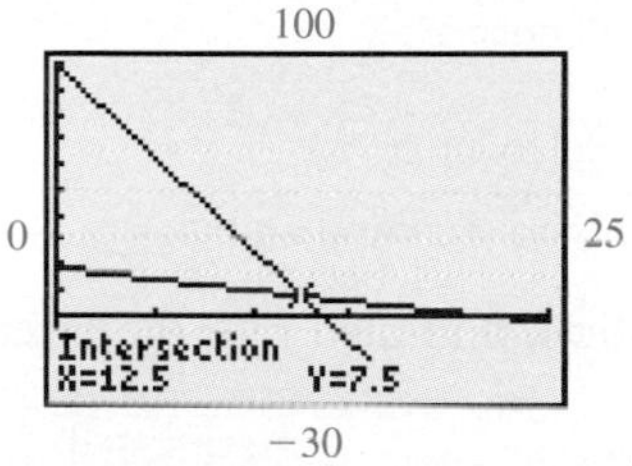

Figure 3 illustrates a solution to Example 6 using graphical approximation techniques.

Figure 3
$y_1 = (500 - 37x)/5$
$y_2 = (1,200 - 57x)/65$

Matched Problem 6 Dennis wants to use cottage cheese and yogurt to increase the amount of protein and calcium in his daily diet. An ounce of cottage cheese contains 3 grams of protein and 15 milligrams of calcium. An ounce of yogurt contains 1 gram of protein and 41 milligrams of calcium. How many ounces of cottage cheese and yogurt should Dennis eat each day to provide exactly 62 grams of protein and 760 milligrams of calcium?

In a free market economy, the price of a product is determined by the relationship between supply and demand. Suppliers are more willing to supply a product at higher prices. So when the price is high, the supply is high. If the relationship between price and supply is linear, then the graph of the price–supply equation is a line with positive slope. On the other hand, consumers of a product are generally less willing to buy a product at higher prices. So when the price is high, demand is low. If the relationship between price and demand is linear, the graph of the price–demand equation is a line with negative slope. In a free competitive market, the price of a product tends to move toward an **equilibrium price**, in which the supply and demand are equal; that common value of the supply and demand is the **equilibrium quantity**. To find the equilibrium price, we solve the system consisting of the price–supply and price–demand equations.

EXAMPLE 7 Supply and Demand At a price of $1.88 per pound, the supply for cherries in a large city is 16,000 pounds, and the demand is 10,600 pounds. When the price drops to $1.46 per pound, the supply decreases to 10,000 pounds, and the demand increases to 12,700 pounds. Assume that the price–supply and price–demand equations are linear.

(A) Find the price–supply equation.

(B) Find the price–demand equation.

(C) Find the supply and demand at a price of $2.09 per pound.

(D) Find the supply and demand at a price of $1.32 per pound.

(E) Use the substitution method to find the equilibrium price and equilibrium demand.

SOLUTION

(A) Let p be the price per pound, and let x be the quantity in thousands of pounds. Then $(16, 1.88)$ and $(10, 1.46)$ are solutions of the price–supply equation. Use the point–slope form for the equation of a line, $y - y_1 = m(x - x_1)$, to obtain the price–supply equation:

$$p - 1.88 = \frac{1.46 - 1.88}{10 - 16}(x - 16) \quad \text{Simplify.}$$

$$p - 1.88 = 0.07(x - 16) \qquad \text{Solve for } p.$$

$$p = 0.07x + 0.76 \qquad \text{Price–supply equation}$$

(B) Again, let p be the price per pound, and let x be the quantity in thousands of pounds. Then $(10.6, 1.88)$ and $(12.7, 1.46)$ are solutions of the price–demand equation.

$$p - 1.88 = \frac{1.46 - 1.88}{12.7 - 10.6}(x - 10.6) \quad \text{Simplify.}$$

$$p - 1.88 = -0.2(x - 10.6) \qquad \text{Solve for } p.$$

$$p = -0.2x + 4 \qquad \text{Price–demand equation}$$

(C) Substitute $p = 2.09$ into the price–supply equation, and also into the price–demand equation, and solve for x:

Price–supply equation	Price–demand equation
$p = 0.07x + 0.76$	$p = -0.2x + 4$
$2.09 = 0.07x + 0.76$	$2.09 = -0.2x + 4$
$x = 19$	$x = 9.55$

At a price of $2.09 per pound, the supply is 19,000 pounds of cherries and the demand is 9,550 pounds. (The supply is greater than the demand, so the price will tend to come down.)

(D) Substitute $p = 1.32$ in each equation and solve for x:

Price–supply equation	Price–demand equation
$p = 0.07x + 0.76$	$p = -0.2x + 4$
$1.32 = 0.07x + 0.76$	$1.32 = -0.2x + 4$
$x = 8$	$x = 13.4$

At a price of $1.32 per pound, the supply is 8,000 pounds of cherries, and the demand is 13,400 pounds. (The demand is greater than the supply, so the price will tend to go up.)

(E) We solve the linear system

$$p = 0.07x + 0.76 \quad \text{Price–supply equation}$$

$$p = -0.2x + 4 \qquad \text{Price–demand equation}$$

using substitution (substitute $p = -0.2x + 4$ in the first equation):

$$-0.2x + 4 = 0.07x + 0.76$$
$$-0.27x = -3.24$$

Equilibrium quantity

$$x = 12 \text{ thousand pounds}$$

Now substitute $x = 12$ into the price–demand equation:

$$p = -0.2(12) + 4$$

Equilibrium price

$$p = \$1.60 \text{ per pound}$$

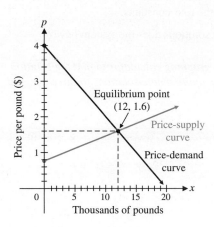

Figure 4

The results are interpreted graphically in Figure 4 (it is customary to refer to the graphs of price–supply and price–demand equations as "curves" even when they are lines). Note that if the price is above the equilibrium price of $1.60 per pound, the supply will exceed the demand and the price will come down. If the price is below the equilibrium price of $1.60 per pound, the demand will exceed the supply and the price will go up. So the price will stabilize at $1.60 per pound. At this equilibrium price, suppliers will supply 12,000 pounds of cherries and consumers will purchase 12,000 pounds.

Matched Problem 7 Find the equilibrium quantity and equilibrium price, and graph the following price–supply and price–demand equations:

$$p = 0.08q + 0.66 \quad \text{Price–supply equation}$$
$$p = -0.1q + 3 \quad \text{Price–demand equation}$$

Exercises 4.1

Skills Warm-up Exercises

W *In Problems 1–6, find the x and y coordinates of the intersection of the given lines. (If necessary, review Section 1.2).*

1. $y = 5x + 7$ and the y axis

2. $y = 5x + 7$ and the x axis.

3. $3x + 4y = 72$ and the x axis

4. $3x + 4y = 72$ and the y axis

5. $6x - 5y = 120$ and $x = 5$

6. $6x - 5y = 120$ and $y = 3$

In Problems 7 and 8, find an equation in point–slope form, $y - y_1 = m(x - x_1)$, of the line through the given points.

7. $(2, 7)$ and $(4, -5)$.   **8.** $(3, 20)$ and $(-5, 4)$.

Match each system in Problems 9–12 with one of the following graphs, and use the graph to solve the system.

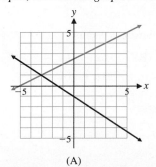

(A)

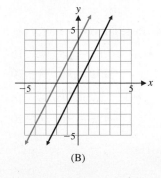

(B)

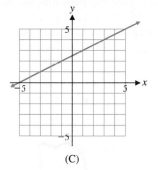

(C)

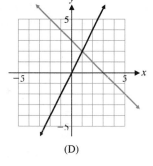

(D)

9. $-4x + 2y = 8$
    $2x - y = 0$

10. $x + y = 3$
     $2x - y = 0$

11. $-x + 2y = 5$
     $2x + 3y = -3$

12. $2x - 4y = -10$
     $-x + 2y = 5$

Solve Problems 13–16 by graphing.

13. $3x - y = 2$
     $x + 2y = 10$

14. $3x - 2y = 12$
     $7x + 2y = 8$

15. $m + 2n = 4$
     $2m + 4n = -8$

16. $3u + 5v = 15$
     $6u + 10v = -30$

Solve Problems 17–20 using substitution.

17. $y = 2x - 3$
     $x + 2y = 14$

18. $y = x - 4$
     $x + 3y = 12$

19. $2x + y = 6$
$x - y = -3$

20. $3x - y = 7$
$2x + 3y = 1$

Solve Problems 21–24 using elimination by addition.

21. $3u - 2v = 12$
$7u + 2v = 8$

22. $2x - 3y = -8$
$5x + 3y = 1$

23. $2m - n = 10$
$m - 2n = -4$

24. $2x + 3y = 1$
$3x - y = 7$

Solve Problems 25–34 using substitution or elimination by addition.

25. $9x - 3y = 24$
$11x + 2y = 1$

26. $4x + 3y = 26$
$3x - 11y = -7$

27. $2x - 3y = -2$
$-4x + 6y = 7$

28. $3x - 6y = -9$
$-2x + 4y = 12$

29. $3x + 8y = 4$
$15x + 10y = -10$

30. $7m + 12n = -1$
$5m - 3n = 7$

31. $-6x + 10y = -30$
$3x - 5y = 15$

32. $2x + 4y = -8$
$x + 2y = 4$

33. $x + y = 1$
$0.3x - 0.4y = 0$

34. $x + y = 1$
$0.5x - 0.4y = 0$

In Problems 35–42, solve the system. Note that each solution can be found mentally, without the use of a calculator or pencil-and-paper calculation; try to visualize the graphs of both lines.

35. $x + 0y = 7$
$0x + y = 3$

36. $x + 0y = -4$
$0x + y = 9$

37. $5x + 0y = 4$
$0x + 3y = -2$

38. $6x + 0y = 7$
$0x + 4y = 9$

39. $x + y = 0$
$x - y = 0$

40. $-2x + y = 0$
$5x - y = 0$

41. $x - 2y = 4$
$0x + y = 5$

42. $x + 3y = 9$
$0x + y = -2$

43. In a free competitive market, if the supply of a good is greater than the demand, will the price tend to go up or come down?

44. In a free competitive market, if the demand for a good is greater than the supply, will the price tend to go up or come down?

✎ *Problems 45–48 are concerned with the linear system*

$$y = mx + b$$
$$y = nx + c$$

where m, b, n, and c are nonzero constants.

45. If the system has a unique solution, discuss the relationships among the four constants.

46. If the system has no solution, discuss the relationships among the four constants.

47. If the system has an infinite number of solutions, discuss the relationships among the four constants.

48. If $m = 0$, how many solutions does the system have?

📊 *In Problems 49–56, use a graphing calculator to find the solution to each system. Round any approximate solutions to three decimal places.*

49. $y = 2x - 9$
$y = 3x + 5$

50. $y = -3x + 3$
$y = 5x + 8$

51. $y = 2x + 1$
$y = 2x + 7$

52. $y = -3x + 6$
$y = -3x + 9$

53. $3x - 2y = 15$
$4x + 3y = 13$

54. $3x - 7y = -20$
$2x + 5y = 8$

55. $-2.4x + 3.5y = 0.1$
$-1.7x + 2.6y = -0.2$

56. $4.2x + 5.4y = -12.9$
$6.4x + 3.7y = -4.5$

In Problems 57–62, graph the equations in the same coordinate system. Find the coordinates of any points where two or more lines intersect and discuss the nature of the solution set.

57. $x - 2y = -6$
$2x + y = 8$
$x + 2y = -2$

58. $x + y = 3$
$x + 3y = 15$
$3x - y = 5$

59. $x + y = 1$
$x - 2y = -8$
$3x + y = -3$

60. $x - y = 6$
$x - 2y = 8$
$x + 4y = -4$

61. $4x - 3y = -24$
$2x + 3y = 12$
$8x - 6y = 24$

62. $2x + 3y = 18$
$2x - 6y = -6$
$4x + 6y = -24$

✎ **63.** The coefficients of the three systems given below are similar. One might guess that the solution sets to the three systems would be nearly identical. Develop evidence for or against this guess by considering graphs of the systems and solutions obtained using substitution or elimination by addition.

(A) $5x + 4y = 4$
$11x + 9y = 4$

(B) $5x + 4y = 4$
$11x + 8y = 4$

(C) $5x + 4y = 4$
$10x + 8y = 4$

✎ **64.** Repeat Problem 63 for the following systems:

(A) $6x - 5y = 10$
$-13x + 11y = -20$

(B) $6x - 5y = 10$
$-13x + 10y = -20$

(C) $6x - 5y = 10$
$-12x + 10y = -20$

Applications

65. Supply and demand for T-shirts. Suppose that the supply and demand equations for printed T-shirts for a particular week are

$$p = 0.7q + 3 \quad \text{Price–supply equation}$$
$$p = -1.7q + 15 \quad \text{Price–demand equation}$$

where p is the price in dollars and q is the quantity in hundreds.

✎ (A) Find the supply and demand (to the nearest unit) if T-shirts are $4 each. Discuss the stability of the T-shirt market at this price level.

✎ (B) Find the supply and demand (to the nearest unit) if T-shirts are $9 each. Discuss the stability of the T-shirt market at this price level.

(C) Find the equilibrium price and quantity.

(D) Graph the two equations in the same coordinate system and identify the equilibrium point, supply curve, and demand curve.

66. Supply and demand for baseball caps. Suppose that the supply and demand for printed baseball caps for a particular week are

$$p = 0.4q + 3.2 \quad \text{Price–supply equation}$$
$$p = -1.9q + 17 \quad \text{Price–demand equation}$$

where p is the price in dollars and q is the quantity in hundreds.

✎ (A) Find the supply and demand (to the nearest unit) if baseball caps are $4 each. Discuss the stability of the baseball cap market at this price level.

✎ (B) Find the supply and demand (to the nearest unit) if baseball caps are $9 each. Discuss the stability of the baseball cap market at this price level.

(C) Find the equilibrium price and quantity.

(D) Graph the two equations in the same coordinate system and identify the equilibrium point, supply curve, and demand curve.

67. Supply and demand for soybeans. At $4.80 per bushel, the annual supply for soybeans in the Midwest is 1.9 billion bushels, and the annual demand is 2.0 billion bushels. When the price increases to $5.10 per bushel, the annual supply increases to 2.1 billion bushels, and the annual demand decreases to 1.8 billion bushels. Assume that the price–supply and price–demand equations are linear. (*Source:* U.S. Census Bureau)

(A) Find the price–supply equation.

(B) Find the price–demand equation.

(C) Find the equilibrium price and quantity.

(D) Graph the two equations in the same coordinate system and identify the equilibrium point, supply curve, and demand curve.

68. Supply and demand for corn. At $2.13 per bushel, the annual supply for corn in the Midwest is 8.9 billion bushels and the annual demand is 6.5 billion bushels. When the price falls to $1.50 per bushel, the annual supply decreases to 8.2 billion bushels and the annual demand increases to 7.4 billion bushels. Assume that the price–supply and price–demand equations are linear. (*Source:* U.S. Census Bureau)

(A) Find the price–supply equation.

(B) Find the price–demand equation.

(C) Find the equilibrium price and quantity.

(D) Graph the two equations in the same coordinate system and identify the equilibrium point, supply curve, and demand curve.

69. Break-even analysis. A small plant manufactures riding lawn mowers. The plant has fixed costs (leases, insurance, etc.) of $48,000 per day and variable costs (labor, materials, etc.) of $1,400 per unit produced. The mowers are sold for $1,800 each. So the cost and revenue equations are

$$y = 48,000 + 1,400x \quad \text{Cost equation}$$
$$y = 1,800x \quad \text{Revenue equation}$$

where x is the total number of mowers produced and sold each day. The daily costs and revenue are in dollars.

(A) How many units must be manufactured and sold each day for the company to break even?

✎ (B) Graph both equations in the same coordinate system and show the break-even point. Interpret the regions between the lines to the left and to the right of the break-even point.

70. Break-even analysis. Repeat Problem 69 with the cost and revenue equations

$$y = 65,000 + 1,100x \quad \text{Cost equation}$$
$$y = 1,600x \quad \text{Revenue equation}$$

71. Break-even analysis. A company markets exercise DVDs that sell for $19.95, including shipping and handling. The monthly fixed costs (advertising, rent, etc.) are $24,000 and the variable costs (materials, shipping, etc) are $7.45 per DVD.

(A) Find the cost equation and the revenue equation.

(B) How many DVDs must be sold each month for the company to break even?

(C) Graph the cost and revenue equations in the same coordinate system and show the break-even point. Interpret the regions between the lines to the left and to the right of the break-even point.

72. Break-even analysis. Repeat Problem 71 if the monthly fixed costs increase to $27,200, the variable costs increase to $9.15, and the company raises the selling price of the DVDs to $21.95.

73. Delivery charges. United Express, a national package delivery service, charges a base price for overnight delivery of packages weighing 1 pound or less and a surcharge for each additional pound (or fraction thereof). A customer is billed $27.75 for shipping a 5-pound package and $64.50 for a 20-pound package. Find the base price and the surcharge for each additional pound.

74. Delivery charges. Refer to Problem 73. Federated Shipping, a competing overnight delivery service, informs the customer in Problem 73 that they would ship the 5-pound package for $29.95 and the 20-pound package for $59.20.

(A) If Federated Shipping computes its cost in the same manner as United Express, find the base price and the surcharge for Federated Shipping.

(B) Devise a simple rule that the customer can use to choose the cheaper of the two services for each package shipped. Justify your answer.

75. Coffee blends. A coffee company uses Colombian and Brazilian coffee beans to produce two blends, robust and mild. A pound of the robust blend requires 12 ounces of Colombian beans and 4 ounces of Brazilian beans. A pound of the mild blend requires 6 ounces of Colombian beans and 10 ounces of Brazilian beans. Coffee is shipped in 132-pound burlap bags. The company has 50 bags of Colombian beans and 40 bags of Brazilian beans on hand. How many pounds of each blend should the company produce in order to use all the available beans?

76. Coffee blends. Refer to Problem 75.

(A) If the company decides to discontinue production of the robust blend and produce only the mild blend, how many pounds of the mild blend can the company produce? How many beans of each type will the company use? Are there any beans that are not used?

(B) Repeat part (A) if the company decides to discontinue production of the mild blend and produce only the robust blend.

77. Animal diet. Animals in an experiment are to be kept under a strict diet. Each animal should receive 20 grams of protein and 6 grams of fat. The laboratory technician is able to purchase two food mixes: Mix A has 10% protein and 6% fat; mix B has 20% protein and 2% fat. How many grams of each mix should be used to obtain the right diet for one animal?

78. Fertilizer. A fruit grower uses two types of fertilizer in an orange grove, brand A and brand B. Each bag of brand A contains 8 pounds of nitrogen and 4 pounds of phosphoric acid. Each bag of brand B contains 7 pounds of nitrogen and 6 pounds of phosphoric acid. Tests indicate that the grove needs 720 pounds of nitrogen and 500 pounds of phosphoric acid. How many bags of each brand should be used to provide the required amounts of nitrogen and phosphoric acid?

79. Electronics. A supplier for the electronics industry manufactures keyboards and screens for graphing calculators at plants in Mexico and Taiwan. The hourly production rates at each plant are given in the table. How many hours should each plant be operated to exactly fill an order for 4,000 keyboards and 4,000 screens?

Plant	Keyboards	Screens
Mexico	40	32
Taiwan	20	32

80. Sausage. A company produces Italian sausages and bratwursts at plants in Green Bay and Sheboygan. The hourly production rates at each plant are given in the table. How many hours should each plant operate to exactly fill an order for 62,250 Italian sausages and 76,500 bratwursts?

Plant	Italian Sausage	Bratwurst
Green Bay	800	800
Sheboygan	500	1,000

81. Physics. An object dropped off the top of a tall building falls vertically with constant acceleration. If s is the distance of the object above the ground (in feet) t seconds after its release, then s and t are related by an equation of the form $s = a + bt^2$, where a and b are constants. Suppose the object is 180 feet above the ground 1 second after its release and 132 feet above the ground 2 seconds after its release.

(A) Find the constants a and b.

(B) How tall is the building?

(C) How long does the object fall?

82. Physics. Repeat Problem 81 if the object is 240 feet above the ground after 1 second and 192 feet above the ground after 2 seconds.

83. Earthquakes. An earthquake emits a primary wave and a secondary wave. Near the surface of the Earth the primary wave travels at 5 miles per second and the secondary wave at 3 miles per second. From the time lag between the two waves arriving at a given receiving station, it is possible to estimate the distance to the quake. Suppose a station measured a time difference of 16 seconds between the arrival of the two waves. How long did each wave travel, and how far was the earthquake from the station?

84. Sound waves. A ship using sound-sensing devices above and below water recorded a surface explosion 6 seconds sooner by its underwater device than its above-water device. Sound travels in air at 1,100 feet per second and in seawater at 5,000 feet per second. How long did it take each sound wave to reach the ship? How far was the explosion from the ship?

85. Psychology. People approach certain situations with "mixed emotions." For example, public speaking often brings forth the positive response of recognition and the negative response of failure. Which dominates? J. S. Brown, in an experiment on approach and avoidance, trained rats by feeding them from a goal box. The rats received mild electric shocks from the same goal box. This established an approach—avoidance conflict relative to the goal box. Using an appropriate apparatus, Brown arrived at the following relationships:

$$p = -\tfrac{1}{5}d + 70 \quad \text{Approach equation}$$
$$p = -\tfrac{4}{3}d + 230 \quad \text{Avoidance equation}$$

where $30 \le d \le 172.5$. The approach equation gives the pull (in grams) toward the food goal box when the rat is placed d centimeters away from it. The avoidance equation gives the pull (in grams) away from the shock goal box when the rat is placed d centimeters from it.

(A) Graph the approach equation and the avoidance equation in the same coordinate system.

(B) Find the value of d for the point of intersection of these two equations.

(C) What do you think the rat would do when placed the distance d from the box found in part (B)?

(*Source: Journal of Comparative and Physiological Psychology,* 41:450–465.)

Answers to Matched Problems

1. $x = -2, y = -1$

$$2x - \quad y = -3$$
$$2(-2) - (-1) \overset{?}{=} -3$$
$$-3 \overset{\checkmark}{=} -3$$
$$x + \quad 2y = -4$$
$$(-2) + 2(-1) \overset{?}{=} -4$$
$$-4 \overset{\checkmark}{=} -4$$

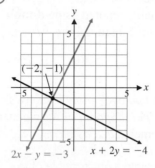

2. (A) $x = 2, y = 2$

   (B) Infinitely many solutions

   (C) No solution

3. $x = -1.92, y = 4.23$

4. $x = -2, y = 2$

5. $x = 2, y = -1$

6. 16.5 oz of cottage cheese, 12.5 oz of yogurt

7. Equilibrium quantity $= 13$ thousand pounds; equilibrium price $= \$1.70$ per pound

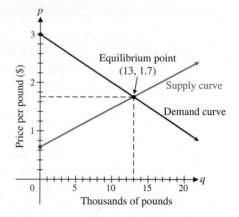

4.2 Systems of Linear Equations and Augmented Matrices

- Matrices
- Solving Linear Systems Using Augmented Matrices
- Summary

Most linear systems of any consequence involve large numbers of equations and variables. It is impractical to try to solve such systems by hand. In the past, these complex systems could be solved only on large computers. Now there are a wide array of approaches to solving linear systems, ranging from graphing calculators to software and spreadsheets. In the rest of this chapter, we develop several *matrix methods* for solving systems with the understanding that these methods are generally used with a graphing calculator. It is important to keep in mind that we are not presenting these techniques as efficient methods for solving linear systems by hand. Instead, we emphasize formulation of mathematical models and interpretation of the results—two activities that graphing calculators cannot perform for you.

Matrices

In solving systems of equations using elimination by addition, the coefficients of the variables and the constant terms played a central role. The process can be made more efficient for generalization and computer work by the introduction of a mathematical form called a *matrix*. A **matrix** is a rectangular array of numbers written within brackets. Two examples are

$$A = \begin{bmatrix} 1 & -4 & 5 \\ 7 & 0 & -2 \end{bmatrix} \qquad B = \begin{bmatrix} -4 & 5 & 12 \\ 0 & 1 & 8 \\ -3 & 10 & 9 \\ -6 & 0 & -1 \end{bmatrix} \qquad (1)$$

Each number in a matrix is called an **element** of the matrix. Matrix A has 6 elements arranged in 2 rows and 3 columns. Matrix B has 12 elements arranged in 4 rows and 3 columns. If a matrix has m rows and n columns, it is called an **$m \times n$ matrix** (read "m by n matrix"). The expression $m \times n$ is called the **size** of the matrix, and the numbers m and n are called the **dimensions** of the matrix. It is important to note that the number of rows is always given first. Referring to equations (1), A is a 2×3 matrix and B is a 4×3 matrix. A matrix with n rows and n columns is called a **square matrix of order n**. A matrix with only 1 column is called a **column matrix**, and a matrix with only 1 row is called a **row matrix**.

$$3 \times 3 \qquad\qquad 4 \times 1 \qquad\qquad 1 \times 4$$

$$\begin{bmatrix} 0.5 & 0.2 & 1.0 \\ 0.0 & 0.3 & 0.5 \\ 0.7 & 0.0 & 0.2 \end{bmatrix} \qquad \begin{bmatrix} 3 \\ -2 \\ 1 \\ 0 \end{bmatrix} \qquad \begin{bmatrix} 2 & \frac{1}{2} & 0 & -\frac{2}{3} \end{bmatrix}$$

Square matrix of order 3   Column matrix   Row matrix

The **position** of an element in a matrix is given by the row and column containing the element. This is usually denoted using **double subscript notation** a_{ij}, where i is the row and j is the column containing the element a_{ij}, as illustrated below:

$$A = \begin{bmatrix} 1 & -4 & 5 \\ 7 & 0 & -2 \end{bmatrix} \qquad \begin{matrix} a_{11} = 1, & a_{12} = -4, & a_{13} = 5 \\ a_{21} = 7, & a_{22} = 0, & a_{23} = -2 \end{matrix}$$

Note that a_{12} is read "a sub one two" (*not* "a sub twelve"). The elements $a_{11} = 1$ and $a_{22} = 0$ make up the *principal diagonal* of A. In general, the **principal diagonal** of a matrix A consists of the elements $a_{11}, a_{22}, a_{33}, \ldots$.

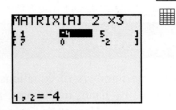

Remark—Most graphing calculators are capable of storing and manipulating matrices. Figure 1 shows matrix A displayed in the editing screen of a graphing calculator. The size of the matrix is given at the top of the screen. The position and value of the currently selected element is given at the bottom. Note that a comma is used in the notation for the position. This is common practice on many graphing calculators but not in mathematical literature. In a spreadsheet, matrices are referred to by their location (upper left corner to lower right corner), using either row and column numbers (Fig. 2A) or row numbers and column letters (Fig. 2B).

Figure 1 **Matrix notation on a graphing calculator**

	1	2	3
1	1	-4	5
2	7	0	-2

	A	B	C	D	E	F
1						
2						
3						
4						
5					1	-4	5
6					7	0	-2

(A) Location of matrix A: (B) Location of matrix A:
R1C1:R2C3   D5:F6

Figure 2 **Matrix notation in a spreadsheet**

Matrices serve as a shorthand for solving systems of linear equations. Associated with the system

$$2x - 3y = 5 \tag{2}$$
$$x + 2y = -3$$

are its **coefficient matrix, constant matrix,** and **augmented matrix**:

Coefficient   Constant   Augmented
matrix   matrix   matrix

$$\begin{bmatrix} 2 & -3 \\ 1 & 2 \end{bmatrix} \qquad \begin{bmatrix} 5 \\ -3 \end{bmatrix} \qquad \left[\begin{array}{cc|c} 2 & -3 & 5 \\ 1 & 2 & -3 \end{array}\right]$$

Note that the augmented matrix is just the coefficient matrix, augmented by the constant matrix. The vertical bar is included only as a visual aid to separate the coefficients from the constant terms. The augmented matrix contains all of the essential information about the linear system—everything but the names of the variables.

For ease of generalization to the larger systems in later sections, we will change the notation for the variables in system (2) to a subscript form. That is, in place of x and y, we use x_1 and x_2, respectively, and system (2) is rewritten as

$$
\begin{aligned}
2x_1 - 3x_2 &= 5 \\
x_1 + 2x_2 &= -3
\end{aligned}
$$

In general, associated with each linear system of the form

$$
\begin{aligned}
a_{11}x_1 + a_{12}x_2 &= k_1 \\
a_{21}x_1 + a_{22}x_2 &= k_2
\end{aligned}
\tag{3}
$$

where x_1 and x_2 are variables, is the *augmented matrix* of the system:

$$
\begin{array}{c}
\text{Column 1 } (C_1) \\
\text{Column 2 } (C_2) \\
\text{Column 3 } (C_3) \\
\begin{bmatrix}
a_{11} & a_{12} & k_1 \\
a_{21} & a_{22} & k_2
\end{bmatrix}
\begin{array}{l}
\leftarrow \text{Row 1 } (R_1) \\
\leftarrow \text{Row 2 } (R_2)
\end{array}
\end{array}
$$

This matrix contains the essential parts of system (3). Our objective is to learn how to manipulate augmented matrices in order to solve system (3), if a solution exists. The manipulative process is closely related to the elimination process discussed in Section 4.1.

Recall that two linear systems are said to be equivalent if they have the same solution set. In Theorem 2, Section 4.1, we used the operations listed below to transform linear systems into equivalent systems:

(A) Two equations are interchanged.

(B) An equation is multiplied by a nonzero constant.

(C) A constant multiple of one equation is added to another equation.

Paralleling the earlier discussion, we say that two augmented matrices are **row equivalent**, denoted by the symbol ~ placed between the two matrices, if they are augmented matrices of equivalent systems of equations. How do we transform augmented matrices into row-equivalent matrices? We use Theorem 1, which is a direct consequence of the operations listed in Section 4.1.

THEOREM 1 Operations That Produce Row-Equivalent Matrices

An augmented matrix is transformed into a row-equivalent matrix by performing any of the following **row operations**:

(A) Two rows are interchanged $(R_i \leftrightarrow R_j)$.

(B) A row is multiplied by a nonzero constant $(kR_i \rightarrow R_i)$.

(C) A constant multiple of one row is added to another row $(kR_j + R_i \rightarrow R_i)$.

Note: The arrow $\rightarrow$ means "replaces."

Solving Linear Systems Using Augmented Matrices

We illustrate the use of Theorem 1 by several examples.

EXAMPLE 1 Solving a System Using Augmented Matrix Methods Solve using augmented matrix methods:

$$
\begin{aligned}
3x_1 + 4x_2 &= 1 \\
x_1 - 2x_2 &= 7
\end{aligned}
\tag{4}
$$

SOLUTION We start by writing the augmented matrix corresponding to system (4):

$$\left[\begin{array}{cc|c} 3 & 4 & 1 \\ 1 & -2 & 7 \end{array}\right] \tag{5}$$

Our objective is to use row operations from Theorem 1 to try to transform matrix (5) into the form

$$\left[\begin{array}{cc|c} 1 & 0 & m \\ 0 & 1 & n \end{array}\right] \tag{6}$$

where m and n are real numbers. Then the solution to system (4) will be obvious, since matrix (6) will be the augmented matrix of the following system (a row in an augmented matrix always corresponds to an equation in a linear system):

$$\begin{aligned} x_1 &= m & x_1 + 0x_2 &= m \\ x_2 &= n & 0x_1 + x_2 &= n \end{aligned}$$

Now we use row operations to transform matrix (5) into form (6).

Step 1 To get a 1 in the upper left corner, we interchange R_1 and R_2 (Theorem 1A):

$$\left[\begin{array}{cc|c} 3 & 4 & 1 \\ 1 & -2 & 7 \end{array}\right] \overset{R_1 \leftrightarrow R_2}{\sim} \left[\begin{array}{cc|c} 1 & -2 & 7 \\ 3 & 4 & 1 \end{array}\right]$$

Step 2 To get a 0 in the lower left corner, we multiply R_1 by (-3) and add to R_2 (Theorem 1C)—this changes R_2 but not R_1. Some people find it useful to write $(-3R_1)$ outside the matrix to help reduce errors in arithmetic, as shown:

$$\left[\begin{array}{cc|c} 1 & -2 & 7 \\ 3 & 4 & 1 \end{array}\right] \overset{(-3)R_1 + R_2 \to R_2}{\sim} \left[\begin{array}{cc|c} 1 & -2 & 7 \\ 0 & 10 & -20 \end{array}\right]$$

$$\begin{array}{ccc} -3 & 6 & -21 \end{array}$$

Step 3 To get a 1 in the second row, second column, we multiply R_2 by $\frac{1}{10}$ (Theorem 1B):

$$\left[\begin{array}{cc|c} 1 & -2 & 7 \\ 0 & 10 & -20 \end{array}\right] \overset{\frac{1}{10}R_2 \to R_2}{\sim} \left[\begin{array}{cc|c} 1 & -2 & 7 \\ 0 & 1 & -2 \end{array}\right]$$

Step 4 To get a 0 in the first row, second column, we multiply R_2 by 2 and add the result to R_1 (Theorem 1C)—this changes R_1 but not R_2:

$$\begin{array}{ccc} 0 & 2 & -4 \end{array}$$

$$\left[\begin{array}{cc|c} 1 & -2 & 7 \\ 0 & 1 & -2 \end{array}\right] \overset{2R_2 + R_1 \to R_1}{\sim} \left[\begin{array}{cc|c} 1 & 0 & 3 \\ 0 & 1 & -2 \end{array}\right]$$

We have accomplished our objective! The last matrix is the augmented matrix for the system

$$\begin{aligned} x_1 &= 3 & x_1 + 0x_2 &= 3 \\ x_2 &= -2 & 0x_1 + x_2 &= -2 \end{aligned} \tag{7}$$

Since system (7) is equivalent to system (4), our starting system, we have solved system (4); that is, $x_1 = 3$ and $x_2 = -2$.

CHECK

$$\begin{array}{ll} 3x_1 + 4x_2 = 1 & x_1 - 2x_2 = 7 \\ 3(3) + 4(-2) \overset{?}{=} 1 & 3 - 2(-2) \overset{?}{=} 7 \\ 1 \overset{\checkmark}{=} 1 & 7 \overset{\checkmark}{=} 7 \end{array}$$

The preceding process may be written more compactly as follows:

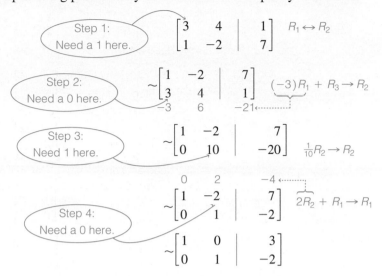

Step 1:
Need a 1 here.

$$\begin{bmatrix} 3 & 4 & | & 1 \\ 1 & -2 & | & 7 \end{bmatrix} \quad R_1 \leftrightarrow R_2$$

Step 2:
Need a 0 here.

$$\sim \begin{bmatrix} 1 & -2 & | & 7 \\ 3 & 4 & | & 1 \\ -3 & 6 & -21 \end{bmatrix} \quad (-3)R_1 + R_3 \rightarrow R_2$$

Step 3:
Need 1 here.

$$\sim \begin{bmatrix} 1 & -2 & | & 7 \\ 0 & 10 & | & -20 \end{bmatrix} \quad \tfrac{1}{10}R_2 \rightarrow R_2$$

$$\begin{matrix} 0 & 2 & -4 \end{matrix}$$

$$\sim \begin{bmatrix} 1 & -2 & | & 7 \\ 0 & 1 & | & -2 \end{bmatrix} \quad 2R_2 + R_1 \rightarrow R_1$$

Step 4:
Need a 0 here.

$$\sim \begin{bmatrix} 1 & 0 & | & 3 \\ 0 & 1 & | & -2 \end{bmatrix}$$

Therefore, $x_1 = 3$ and $x_2 = -2$.

Matched Problem 1) Solve using augmented matrix methods:

$$2x_1 - x_2 = -7$$
$$x_1 + 2x_2 = 4$$

Many graphing calculators can perform row operations. Figure 3 shows the results of performing the row operations used in the solution of Example 1. Consult your manual for the details of performing row operations on your graphing calculator.

[A]
```
      [[3 4  1]]
       [1 -2 7]]
rowSwap([A],1,2)
→[A]
      [[1 -2 7]
       [3 4  1]]
```

(A) $R_1 \leftrightarrow R_2$

[A]
```
      [[1 -2 7]
       [3 4  1]]
*row+(-3,[A],1,2
)→[A]
      [[1 -2 7  ]
       [0 10 -20]]
```

(B) $(-3)R_1 + R_2 \rightarrow R_2$

[A]
```
      [[1 -2 7  ]
       [0 10 -20]]
*row(.1,[A],2)→[
A]
      [[1 -2 7 ]
       [0 1 -2]]
```

(C) $\tfrac{1}{10}R_2 \rightarrow R_2$

[A]
```
      [[1 -2 7 ]
       [0 1 -2]]
*row+(2,[A],2,1)
      [[1 0 3 ]
       [0 1 -2]]
```

(D) $2R_2 + R_1 \rightarrow R_1$

Figure 3 **Row operations on a graphing calculator**

Explore and Discuss 1   The summary following the solution of Example 1 shows five augmented matrices. Write the linear system that each matrix represents, solve each system graphically, and discuss the relationships among these solutions.

EXAMPLE 2   Solving a System Using Augmented Matrix Methods   Solve using augmented matrix methods:

$$2x_1 - 3x_2 = 6$$
$$3x_1 + 4x_2 = \frac{1}{2}$$

SOLUTION

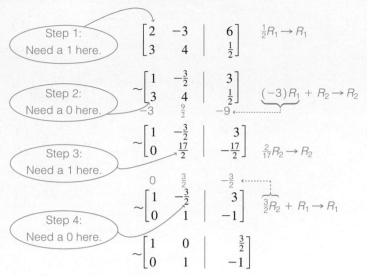

So, $x_1 = \frac{3}{2}$ and $x_2 = -1$. The check is left for you.

Matched Problem 2 ⟩ Solve using augmented matrix methods:

$$5x_1 - 2x_2 = 11$$

$$2x_1 + 3x_2 = \frac{5}{2}$$

EXAMPLE 3   Solving a System Using Augmented Matrix Methods   **Solve using augmented matrix methods:**

$$2x_1 - x_2 = 4 \tag{8}$$
$$-6x_1 + 3x_2 = -12$$

SOLUTION

$$\begin{bmatrix} 2 & -1 \\ -6 & 3 \end{bmatrix} \begin{bmatrix} 4 \\ -12 \end{bmatrix} \quad \begin{array}{l} \frac{1}{2}R_1 \rightarrow R_1 \text{ (to get a 1 in the upper left corner)} \\ \frac{1}{3}R_2 \rightarrow R_2 \text{ (this simplifies } R_2) \end{array}$$

$$\sim \begin{bmatrix} 1 & -\frac{1}{2} \\ -2 & 1 \end{bmatrix} \begin{bmatrix} 2 \\ -4 \end{bmatrix} \quad 2R_1 + R_2 \rightarrow R_2 \text{ (to get a 0 in the lower left corner)}$$

$$\begin{array}{ccc} 2 & -1 & 4 \end{array}$$

$$\sim \begin{bmatrix} 1 & -\frac{1}{2} \\ 0 & 0 \end{bmatrix} \begin{bmatrix} 2 \\ 0 \end{bmatrix}$$

The last matrix corresponds to the system

$$x_1 - \frac{1}{2}x_2 = 2 \qquad x_1 - \frac{1}{2}x_2 = 2 \tag{9}$$

$$0 = 0 \quad 0x_1 + 0x_2 = 0$$

This system is equivalent to the original system. Geometrically, the graphs of the two original equations coincide, and there are infinitely many solutions. In general, if we end up with a row of zeros in an augmented matrix for a two-equation, two-variable system, the system is dependent, and there are infinitely many solutions.

We represent the infinitely many solutions using the same method that was used in Section 4.1; that is, by introducing a parameter. We start by solving

$x_1 - \frac{1}{2}x_2 = 2$, the first equation in system (9), for either variable in terms of the other. We choose to solve for x_1 in terms of x_2 because it is easier:

$$x_1 = \frac{1}{2}x_2 + 2 \qquad (10)$$

Now we introduce a parameter t (we can use other letters, such as k, s, p, q, and so on, to represent a parameter also). If we let $x_2 = t$, then for any real number t,

$$x_1 = \frac{1}{2}t + 2 \qquad (11)$$

$$x_2 = t$$

represents a solution of system (8). Using ordered-pair notation, we write: For any real number t,

$$\left(\frac{1}{2}t + 2, t\right) \qquad (12)$$

is a solution of system (8). More formally, we write

$$\text{solution set} = \left\{\left(\frac{1}{2}t + 2, t\right) \,\middle|\, t \in R\right\} \qquad (13)$$

Typically we use the less formal notations (11) or (12) to represent the solution set for problems of this type.

CHECK The following is a check that system (11) provides a solution to system (8) for any real number t:

$$2x_1 - x_2 = 4 \qquad\qquad -6x_1 + 3x_2 = -12$$

$$2\left(\frac{1}{2}t + 2\right) - t \overset{?}{=} 4 \qquad\qquad -6\left(\frac{1}{2}t + 2\right) + 3t \overset{?}{=} -12$$

$$t + 4 - t \overset{?}{=} 4 \qquad\qquad -3t - 12 + 3t \overset{?}{=} -12$$

$$4 \overset{\checkmark}{=} 4 \qquad\qquad -12 \overset{\checkmark}{=} -12$$

Matched Problem 3) Solve using augmented matrix methods:

$$2x_1 + 6x_2 = 6$$
$$3x_1 - 9x_2 = -9$$

Explore and Discuss 2 The solution of Example 3 involved three augmented matrices. Write the linear system that each matrix represents, solve each system graphically, and discuss the relationships among these solutions.

EXAMPLE 4 Solving a System Using Augmented Matrix Methods Solve using augmented matrix methods:

$$2x_1 + 6x_2 = -3$$
$$x_1 + 3x_2 = 2$$

SOLUTION

$$\begin{bmatrix} 2 & 6 & | & -3 \\ 1 & 3 & | & 2 \end{bmatrix} \quad R_1 \leftrightarrow R_2$$

$$\sim \begin{bmatrix} 1 & 3 & | & 2 \\ 2 & 6 & | & -3 \end{bmatrix} \quad (-2)R_1 + R_2 \rightarrow R_2$$

$$\begin{matrix} -2 & -6 & & -4 \end{matrix}$$

$$\sim \begin{bmatrix} 1 & 3 & | & 2 \\ 0 & 0 & | & -7 \end{bmatrix} \quad R_2 \text{ implies the contradiction } 0 = -7.$$

This is the augmented matrix of the system

$$\begin{aligned} x_1 + 3x_2 &= 2 & x_1 + 3x_2 &= 2 \\ 0 &= -7 & 0x_1 + 0x_2 &= -7 \end{aligned}$$

The second equation is not satisfied by any ordered pair of real numbers. As we saw in Section 4.1, the original system is inconsistent and has no solution. If in a row of an augmented matrix we obtain all zeros to the left of the vertical bar and a nonzero number to the right, the system is inconsistent and there are no solutions.

Matched Problem 4   Solve using augmented matrix methods:

$$\begin{aligned} 2x_1 - x_2 &= 3 \\ 4x_1 - 2x_2 &= -1 \end{aligned}$$

Summary

Examples 2, 3, and 4 illustrate the three possible solution types for a system of two linear equations in two variables, as discussed in Theorem 1, Section 4.1. Examining the final matrix form in each of these solutions leads to the following summary.

SUMMARY
Possible Final Matrix Forms for a System of Two Linear Equations in Two Variables

Form 1: Exactly one solution (consistent and independent)

$$\begin{bmatrix} 1 & 0 & | & m \\ 0 & 1 & | & n \end{bmatrix}$$

Form 2: Infinitely many solutions (consistent and dependent)

$$\begin{bmatrix} 1 & m & | & n \\ 0 & 0 & | & 0 \end{bmatrix}$$

Form 3: No solution (inconsistent)

$$\begin{bmatrix} 1 & m & | & n \\ 0 & 0 & | & p \end{bmatrix}$$

m, n, p are real numbers; $p \neq 0$

The process of solving systems of equations described in this section is referred to as **Gauss–Jordan elimination**. We formalize this method in the next section so that it will apply to systems of any size, including systems where the number of equations and the number of variables are not the same.

Exercises 4.2

Skills Warm-up Exercises

W   *Problems 1–14 refer to the following matrices: (If necessary, review the terminology at the beginning of section 4.2.)*

$$A = \begin{bmatrix} 2 & -4 & 0 \\ 6 & 1 & -5 \end{bmatrix} \quad B = \begin{bmatrix} -1 & 9 & 0 \\ -4 & 8 & 7 \\ 2 & 4 & 0 \end{bmatrix} \quad C = \begin{bmatrix} 2 & -3 & 0 \end{bmatrix} \quad D = \begin{bmatrix} -5 \\ 8 \end{bmatrix}$$

1. How many elements are there in A? In C?

2. How many elements are there in B? In D?

3. What is the size of B? Of D?

4. What is the size of A? Of C?

5. Which of the matrices is a column matrix?

6. Which of the matrices is a row matrix?

7. Which of the matrices is a square matrix?

8. Which of the matrices does not contain the element 0?

9. List the elements on the principal diagonal of A.

10. List the elements on the principal diagonal of B.

11. For matrix B, list the elements b_{31}, b_{22}, b_{13}.

12. For matrix A, list the elements a_{21}, a_{12}.

13. For matrix C, find $c_{11} + c_{12} + c_{13}$.

14. For matrix D, find $d_{11} + d_{21}$.

In Problems 15–18, write the coefficient matrix and the aug-mented matrix of the given system of linear equations.

15. $3x_1 + 5x_2 = 8$
$2x_1 - 4x_2 = -7$

16. $-8x_1 + 3x_2 = 10$
$6x_1 + 5x_2 = 13$

17. $x_1 + 4x_2 = 15$
$6x_1 \qquad = 18$

18. $5x_1 - x_2 = 10$
$3x_2 = 21$

In Problems 19–22, write the system of linear equations that is represented by the given augmented matrix. Assume that the vari-ables are x_1 and x_2.

19. $\begin{bmatrix} 2 & 5 & | & 7 \\ 1 & 4 & | & 9 \end{bmatrix}$

20. $\begin{bmatrix} 0 & 3 & | & 15 \\ -8 & 2 & | & 25 \end{bmatrix}$

21. $\begin{bmatrix} 4 & 0 & | & -10 \\ 0 & 8 & | & 40 \end{bmatrix}$

22. $\begin{bmatrix} 1 & -2 & | & 12 \\ 0 & 1 & | & 6 \end{bmatrix}$

Perform the row operations indicated in Problems 23–34 on the following matrix:

$$\begin{bmatrix} 1 & -3 & | & 2 \\ 4 & -6 & | & -8 \end{bmatrix}$$

23. $R_1 \leftrightarrow R_2$

24. $\frac{1}{2}R_2 \rightarrow R_2$

25. $-4R_1 \rightarrow R_1$

26. $-2R_1 \rightarrow R_1$

27. $2R_2 \rightarrow R_2$

28. $-1R_2 \rightarrow R_2$

29. $(-4)R_1 + R_2 \rightarrow R_2$

30. $\left(-\frac{1}{2}\right)R_2 + R_1 \rightarrow R_1$

31. $(-2)R_1 + R_2 \rightarrow R_2$

32. $(-3)R_1 + R_2 \rightarrow R_2$

33. $(-1)R_1 + R_2 \rightarrow R_2$

34. $R_1 + R_2 \rightarrow R_2$

Each of the matrices in Problems 35–42 is the result of performing a single row operation on the matrix A shown below. Identify the row operation.

$$A = \begin{bmatrix} -1 & 2 & | & -3 \\ 6 & -3 & | & 12 \end{bmatrix}$$

35. $\begin{bmatrix} -1 & 2 & | & -3 \\ 2 & -1 & | & 4 \end{bmatrix}$

36. $\begin{bmatrix} -2 & 4 & | & -6 \\ 6 & -3 & | & 12 \end{bmatrix}$

37. $\begin{bmatrix} -1 & 2 & | & -3 \\ 0 & 9 & | & -6 \end{bmatrix}$

38. $\begin{bmatrix} 3 & 0 & | & 5 \\ 6 & -3 & | & 12 \end{bmatrix}$

39. $\begin{bmatrix} 1 & 1 & | & 1 \\ 6 & -3 & | & 12 \end{bmatrix}$

40. $\begin{bmatrix} -1 & 2 & | & -3 \\ 2 & 5 & | & 0 \end{bmatrix}$

41. $\begin{bmatrix} 6 & -3 & | & 12 \\ -1 & 2 & | & -3 \end{bmatrix}$

42. $\begin{bmatrix} -1 & 2 & | & -3 \\ 0 & 9 & | & -6 \end{bmatrix}$

✎ *Solve Problems 43–46 using augmented matrix methods. Graph each solution set. Discuss the differences between the graph of an equation in the system and the graph of the system's solution set.*

43. $3x_1 - 2x_2 = 6$
$4x_1 - 3x_2 = 6$

44. $x_1 - 2x_2 = 5$
$-2x_1 + 4x_2 = -10$

45. $3x_1 - 2x_2 = -3$
$-6x_1 + 4x_2 = 6$

46. $x_1 - 2x_2 = 1$
$-2x_1 + 5x_2 = 2$

✎ *Solve Problems 47 and 48 using augmented matrix methods. Write the linear system represented by each augmented matrix in your solution, and solve each of these systems graphically. Discuss the relationships among the solutions of these systems.*

47. $x_1 + x_2 = 5$
$x_1 - x_2 = 1$

48. $x_1 - x_2 = 2$
$x_1 + x_2 = 6$

Each of the matrices in Problems 49–54 is the final matrix form for a system of two linear equations in the variables x_1 and x_2. Write the solution of the system.

49. $\begin{bmatrix} 1 & 0 & | & -4 \\ 0 & 1 & | & 6 \end{bmatrix}$

50. $\begin{bmatrix} 1 & 0 & | & 3 \\ 0 & 1 & | & -5 \end{bmatrix}$

51. $\begin{bmatrix} 1 & 3 & | & 2 \\ 0 & 0 & | & 4 \end{bmatrix}$

52. $\begin{bmatrix} 1 & -2 & | & 7 \\ 0 & 0 & | & -9 \end{bmatrix}$

53. $\begin{bmatrix} 1 & -2 & | & 15 \\ 0 & 0 & | & 0 \end{bmatrix}$

54. $\begin{bmatrix} 1 & 5 & | & 10 \\ 0 & 0 & | & 0 \end{bmatrix}$

Solve Problems 55–74 using augmented matrix methods.

55. $x_1 - 2x_2 = 1$
$2x_1 - x_2 = 5$

56. $x_1 + 3x_2 = 1$
$3x_1 - 2x_2 = 14$

57. $x_1 - 4x_2 = -2$
$-2x_1 + x_2 = -3$

58. $x_1 - 3x_2 = -5$
$-3x_1 - x_2 = 5$

59. $3x_1 - x_2 = 2$
$x_1 + 2x_2 = 10$

60. $2x_1 + x_2 = 0$
$x_1 - 2x_2 = -5$

61. $x_1 + 2x_2 = 4$
$2x_1 + 4x_2 = -8$

62. $2x_1 - 3x_2 = -2$
$-4x_1 + 6x_2 = 7$

63. $2x_1 + x_2 = 6$
$x_1 - x_2 = -3$

64. $3x_1 - x_2 = -5$
$x_1 + 3x_2 = 5$

65. $3x_1 - 6x_2 = -9$
$-2x_1 + 4x_2 = 6$

66. $2x_1 - 4x_2 = -2$
$-3x_1 + 6x_2 = 3$

67. $\quad 4x_1 - 2x_2 = 2$
$\quad -6x_1 + 3x_2 = -3$

68. $-6x_1 + 2x_2 = 4$
$\quad 3x_1 - x_2 = -2$

69. $2x_1 + x_2 = 1$
$\quad 4x_1 - x_2 = -7$

70. $2x_1 - x_2 = -8$
$\quad 2x_1 + x_2 = 8$

71. $\quad 4x_1 - 6x_2 = 8$
$\quad -6x_1 + 9x_2 = -10$

72. $\quad 2x_1 - 4x_2 = -4$
$\quad -3x_1 + 6x_2 = 4$

73. $-4x_1 + 6x_2 = -8$
$\quad 6x_1 - 9x_2 = 12$

74. $-2x_1 + 4x_2 = 4$
$\quad 3x_1 - 6x_2 = -6$

Solve Problems 75–80 using augmented matrix methods.

75. $3x_1 - x_2 = 7$
$\quad 2x_1 + 3x_2 = 1$

76. $2x_1 - 3x_2 = -8$
$\quad 5x_1 + 3x_2 = 1$

77. $3x_1 + 2x_2 = 4$
$\quad 2x_1 - x_2 = 5$

78. $4x_1 + 3x_2 = 26$
$\quad 3x_1 - 11x_2 = -7$

79. $0.2x_1 - 0.5x_2 = 0.07$
$\quad 0.8x_1 - 0.3x_2 = 0.79$

80. $0.3x_1 - 0.6x_2 = 0.18$
$\quad 0.5x_1 - 0.2x_2 = 0.54$

Solve Problems 81–84 using augmented matrix methods. Use a graphing calculator to perform the row operations.

81. $\quad 0.8x_1 + 2.88x_2 = 4$
$\quad 1.25x_1 + 4.34x_2 = 5$

82. $\quad 2.7x_1 - 15.12x_2 = 27$
$\quad 3.25x_1 - 18.52x_2 = 33$

83. $\quad 4.8x_1 - 40.32x_2 = 295.2$
$\quad -3.75x_1 + 28.7x_2 = -211.2$

84. $5.7x_1 - 8.55x_2 = -35.91$
$\quad 4.5x_1 + 5.73x_2 = 76.17$

Answers to Matched Problems

1. $x_1 = -2, x_2 = 3$

2. $x_1 = 2, x_2 = -\frac{1}{2}$

3. The system is dependent. For t any real number, a solution is $x_1 = 3t - 3, x_2 = t$.

4. Inconsistent—no solution

4.3 Gauss–Jordan Elimination

- Reduced Matrices
- Solving Systems by Gauss–Jordan Elimination
- Application

Now that you have had some experience with row operations on simple augmented matrices, we consider systems involving more than two variables. We will not require a system to have the same number of equations as variables. Just as for systems of two linear equations in two variables, any linear system, regardless of the number of equations or number of variables, has either

1. Exactly one solution (consistent and independent), or
2. Infinitely many solutions (consistent and dependent), or
3. No solution (inconsistent).

Reduced Matrices

In the preceding section we used row operations to transform the augmented matrix for a system of two equations in two variables,

$$\left[\begin{array}{cc|c} a_{11} & a_{12} & k_1 \\ a_{21} & a_{22} & k_2 \end{array}\right] \qquad \begin{array}{l} a_{11}x_1 + a_{12}x_2 = k_1 \\ a_{21}x_1 + a_{22}x_2 = k_2 \end{array}$$

into one of the following simplified forms:

$$\underset{\text{Form 1}}{\left[\begin{array}{cc|c} 1 & 0 & m \\ 0 & 1 & n \end{array}\right]} \quad \underset{\text{Form 2}}{\left[\begin{array}{cc|c} 1 & m & n \\ 0 & 0 & 0 \end{array}\right]} \quad \underset{\text{Form 3}}{\left[\begin{array}{cc|c} 1 & m & n \\ 0 & 0 & p \end{array}\right]} \qquad (1)$$

where m, n, and p are real numbers, $p \neq 0$. Each of these reduced forms represents a system that has a different type of solution set, and no two of these forms are row equivalent.

For large linear systems, it is not practical to list all such simplified forms; there are too many of them. Instead, we give a general definition of a simplified form called a **reduced matrix**, which can be applied to all matrices and systems, regardless of size.

DEFINITION Reduced Form

A matrix is said to be in **reduced row echelon form**, or, more simply, in **reduced form**, if

1. Each row consisting entirely of zeros is below any row having at least one nonzero element.
2. The leftmost nonzero element in each row is 1.
3. All other elements in the column containing the leftmost 1 of a given row are zeros.
4. The leftmost 1 in any row is to the right of the leftmost 1 in the row above.

The following matrices are in reduced form. Check each one carefully to convince yourself that the conditions in the definition are met.

$$\left[\begin{array}{cc|c} 1 & 0 & 2 \\ 0 & 1 & -3 \end{array}\right] \quad \left[\begin{array}{ccc|c} 1 & 0 & 0 & 2 \\ 0 & 1 & 0 & -1 \\ 0 & 0 & 1 & 3 \end{array}\right] \quad \left[\begin{array}{cc|c} 1 & 0 & 3 \\ 0 & 1 & -1 \\ 0 & 0 & 0 \end{array}\right]$$

$$\left[\begin{array}{cccc|c} 1 & 4 & 0 & 0 & -3 \\ 0 & 0 & 1 & 0 & 2 \\ 0 & 0 & 0 & 1 & 6 \end{array}\right] \quad \left[\begin{array}{ccc|c} 1 & 0 & 4 & 0 \\ 0 & 1 & 3 & 0 \\ 0 & 0 & 0 & 1 \end{array}\right]$$

EXAMPLE 1 Reduced Forms The following matrices are not in reduced form. Indicate which condition in the definition is violated for each matrix. State the row operation(s) required to transform the matrix into reduced form, and find the reduced form.

(A) $\left[\begin{array}{cc|c} 0 & 1 & -2 \\ 1 & 0 & 3 \end{array}\right]$ (B) $\left[\begin{array}{ccc|c} 1 & 2 & -2 & 3 \\ 0 & 0 & 1 & -1 \end{array}\right]$

(C) $\left[\begin{array}{cc|c} 1 & 0 & -3 \\ 0 & 0 & 0 \\ 0 & 1 & -2 \end{array}\right]$ (D) $\left[\begin{array}{ccc|c} 1 & 0 & 0 & -1 \\ 0 & 2 & 0 & 3 \\ 0 & 0 & 1 & -5 \end{array}\right]$

SOLUTION

(A) Condition 4 is violated: The leftmost 1 in row 2 is not to the right of the leftmost 1 in row 1. Perform the row operation $R_1 \leftrightarrow R_2$ to obtain

$$\left[\begin{array}{cc|c} 1 & 0 & 3 \\ 0 & 1 & -2 \end{array}\right]$$

(B) Condition 3 is violated: The column containing the leftmost 1 in row 2 has a nonzero element above the 1. Perform the row operation $2R_2 + R_1 \rightarrow R_1$ to obtain
$$\left[\begin{array}{ccc|c} 1 & 2 & 0 & 1 \\ 0 & 0 & 1 & -1 \end{array}\right]$$

(C) Condition 1 is violated: The second row contains all zeros and is not below any row having at least one nonzero element. Perform the row operation $R_2 \leftrightarrow R_3$ to obtain

$$\begin{bmatrix} 1 & 0 & | & -3 \\ 0 & 1 & | & -2 \\ 0 & 0 & | & 0 \end{bmatrix}$$

(D) Condition 2 is violated: The leftmost nonzero element in row 2 is not a 1. Perform the row operation $\frac{1}{2}R_2 \rightarrow R_2$ to obtain

$$\begin{bmatrix} 1 & 0 & 0 & | & -1 \\ 0 & 1 & 0 & | & \frac{3}{2} \\ 0 & 0 & 1 & | & -5 \end{bmatrix}$$

Matched Problem 1 The matrices below are not in reduced form. Indicate which condition in the definition is violated for each matrix. State the row operation(s) required to transform the matrix into reduced form, and find the reduced form.

(A) $\begin{bmatrix} 1 & 0 & | & 2 \\ 0 & 3 & | & -6 \end{bmatrix}$

(B) $\begin{bmatrix} 1 & 5 & 4 & | & 3 \\ 0 & 1 & 2 & | & -1 \\ 0 & 0 & 0 & | & 0 \end{bmatrix}$

(C) $\begin{bmatrix} 0 & 1 & 0 & | & -3 \\ 1 & 0 & 0 & | & 0 \\ 0 & 0 & 1 & | & 2 \end{bmatrix}$

(D) $\begin{bmatrix} 1 & 2 & 0 & | & 3 \\ 0 & 0 & 0 & | & 0 \\ 0 & 0 & 1 & | & 4 \end{bmatrix}$

Solving Systems by Gauss–Jordan Elimination

We are now ready to outline the Gauss–Jordan method for solving systems of linear equations. The method systematically transforms an augmented matrix into a reduced form. The system corresponding to a reduced augmented matrix is called a **reduced system**. As we shall see, reduced systems are easy to solve.

The Gauss–Jordan elimination method is named after the German mathematician Carl Friedrich Gauss (1777–1885) and the German geodesist Wilhelm Jordan (1842–1899). Gauss, one of the greatest mathematicians of all time, used a method of solving systems of equations in his astronomical work that was later generalized by Jordan to solve problems in large-scale surveying.

EXAMPLE 2 Solving a System Using Gauss–Jordan Elimination   Solve by Gauss–Jordan elimination:

$$\begin{aligned} 2x_1 - 2x_2 + x_3 &= 3 \\ 3x_1 + x_2 - x_3 &= 7 \\ x_1 - 3x_2 + 2x_3 &= 0 \end{aligned}$$

SOLUTION Write the augmented matrix and follow the steps indicated at the right.

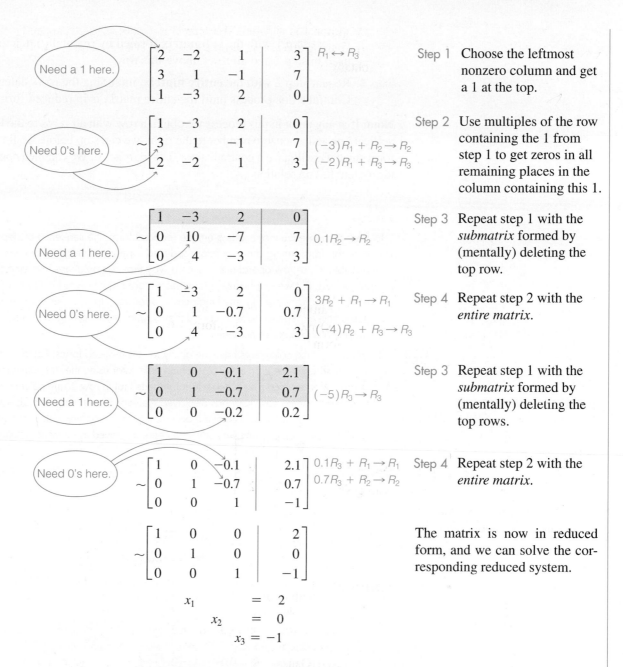

$$\begin{bmatrix} 2 & -2 & 1 & | & 3 \\ 3 & 1 & -1 & | & 7 \\ 1 & -3 & 2 & | & 0 \end{bmatrix} \begin{matrix} R_1 \leftrightarrow R_3 \end{matrix}$$

Need a 1 here.

Step 1 Choose the leftmost nonzero column and get a 1 at the top.

$$\sim \begin{bmatrix} 1 & -3 & 2 & | & 0 \\ 3 & 1 & -1 & | & 7 \\ 2 & -2 & 1 & | & 3 \end{bmatrix} \begin{matrix} \\ (-3)R_1 + R_2 \rightarrow R_2 \\ (-2)R_1 + R_3 \rightarrow R_3 \end{matrix}$$

Need 0's here.

Step 2 Use multiples of the row containing the 1 from step 1 to get zeros in all remaining places in the column containing this 1.

$$\sim \begin{bmatrix} 1 & -3 & 2 & | & 0 \\ 0 & 10 & -7 & | & 7 \\ 0 & 4 & -3 & | & 3 \end{bmatrix} \begin{matrix} \\ 0.1R_2 \rightarrow R_2 \end{matrix}$$

Need a 1 here.

Step 3 Repeat step 1 with the *submatrix* formed by (mentally) deleting the top row.

$$\sim \begin{bmatrix} 1 & -3 & 2 & | & 0 \\ 0 & 1 & -0.7 & | & 0.7 \\ 0 & 4 & -3 & | & 3 \end{bmatrix} \begin{matrix} 3R_2 + R_1 \rightarrow R_1 \\ \\ (-4)R_2 + R_3 \rightarrow R_3 \end{matrix}$$

Need 0's here.

Step 4 Repeat step 2 with the *entire matrix*.

$$\sim \begin{bmatrix} 1 & 0 & -0.1 & | & 2.1 \\ 0 & 1 & -0.7 & | & 0.7 \\ 0 & 0 & -0.2 & | & 0.2 \end{bmatrix} \begin{matrix} \\ \\ (-5)R_3 \rightarrow R_3 \end{matrix}$$

Need a 1 here.

Step 3 Repeat step 1 with the *submatrix* formed by (mentally) deleting the top rows.

$$\sim \begin{bmatrix} 1 & 0 & -0.1 & | & 2.1 \\ 0 & 1 & -0.7 & | & 0.7 \\ 0 & 0 & 1 & | & -1 \end{bmatrix} \begin{matrix} 0.1R_3 + R_1 \rightarrow R_1 \\ 0.7R_3 + R_2 \rightarrow R_2 \end{matrix}$$

Need 0's here.

Step 4 Repeat step 2 with the *entire matrix*.

$$\sim \begin{bmatrix} 1 & 0 & 0 & | & 2 \\ 0 & 1 & 0 & | & 0 \\ 0 & 0 & 1 & | & -1 \end{bmatrix}$$

The matrix is now in reduced form, and we can solve the corresponding reduced system.

$$\begin{aligned} x_1 &= 2 \\ x_2 &= 0 \\ x_3 &= -1 \end{aligned}$$

The solution to this system is $x_1 = 2$, $x_2 = 0$, $x_3 = -1$. You should check this solution in the original system.

Matched Problem 2 Solve by Gauss–Jordan elimination:

$$\begin{aligned} 3x_1 + x_2 - 2x_3 &= 2 \\ x_1 - 2x_2 + x_3 &= 3 \\ 2x_1 - x_2 - 3x_3 &= 3 \end{aligned}$$

PROCEDURE Gauss–Jordan Elimination

Step 1 Choose the leftmost nonzero column and use appropriate row operations to get a 1 at the top.

Step 2 Use multiples of the row containing the 1 from step 1 to get zeros in all remaining places in the column containing this 1.

(Continued)

Step 3   Repeat step 1 with the **submatrix** formed by (mentally) deleting the row used in step 2 and all rows above this row.

Step 4   Repeat step 2 with the **entire matrix,** including the rows deleted mentally. Continue this process until the entire matrix is in reduced form.

Note: If at any point in this process we obtain a row with all zeros to the left of the vertical line and a nonzero number to the right, we can stop before we find the reduced form since we will have a contradiction: $0 = n, n \neq 0$. We can then conclude that the system has no solution.

Remarks

1. Even though each matrix has a unique reduced form, the sequence of steps presented here for transforming a matrix into a reduced form is not unique. For example, it is possible to use row operations in such a way that computations involving fractions are minimized. But we emphasize again that we are not interested in the most efficient hand methods for transforming small matrices into reduced forms. Our main interest is in giving you a little experience with a method that is suitable for solving large-scale systems on a graphing calculator or computer.

2. Most graphing calculators have the ability to find reduced forms. Figure 1 illustrates the solution of Example 2 on a TI-86 graphing calculator using the rref command (rref is an acronym for reduced row echelon form). Notice that in row 2 and column 4 of the reduced form the graphing calculator has displayed the very small number $-3.5\text{E} - 13$, instead of the exact value 0. This is a common occurrence on a graphing calculator and causes no problems. Just replace any very small numbers displayed in scientific notation with 0.

```
A
          [[2  -2   1    3]
           [3   1  -1    7]
           [1  -3   2    0]]
rref A
          [[1   0   0    2       ]
           [0   1   0  -3.5E-13]
           [0   0   1   -1       ]]
```

Figure 1 **Gauss–Jordan elimination on a graphing calculator**

EXAMPLE 3   Solving a System Using Gauss–Jordan Elimination   Solve by Gauss–Jordan elimination:

$$2x_1 - 4x_2 + x_3 = -4$$
$$4x_1 - 8x_2 + 7x_3 = 2$$
$$-2x_1 + 4x_2 - 3x_3 = 5$$

SOLUTION
$$\begin{bmatrix} 2 & -4 & 1 & | & -4 \\ 4 & -8 & 7 & | & 2 \\ -2 & 4 & -3 & | & 5 \end{bmatrix}$$ $0.5R_1 \rightarrow R_1$

$$\sim \begin{bmatrix} 1 & -2 & 0.5 & | & -2 \\ 4 & -8 & 7 & | & 2 \\ -2 & 4 & -3 & | & 5 \end{bmatrix}$$ $(-4)R_1 + R_2 \rightarrow R_2$
                                          $2R_1 + R_3 \rightarrow R_3$

$$\sim \begin{bmatrix} 1 & -2 & 0.5 & | & -2 \\ 0 & 0 & 5 & | & 10 \\ 0 & 0 & -2 & | & 1 \end{bmatrix}$$ $0.2R_2 \rightarrow R_2$ Note that column 3 is the leftmost nonzero column in this submatrix.

$$\sim \begin{bmatrix} 1 & -2 & 0.5 & | & -2 \\ 0 & 0 & 1 & | & 2 \\ 0 & 0 & -2 & | & 1 \end{bmatrix}$$ $(-0.5)R_2 + R_1 \rightarrow R_1$
                                          $2R_2 + R_3 \rightarrow R_3$

$$\sim \begin{bmatrix} 1 & -2 & 0 & | & -3 \\ 0 & 0 & 1 & | & 2 \\ 0 & 0 & 0 & | & 5 \end{bmatrix}$$ We stop the Gauss–Jordan elimination, even though the matrix is not in reduced form, since the last row produces a contradiction.

The system has no solution.

Matched Problem 3 Solve by Gauss–Jordan elimination:

$$2x_1 - 4x_2 - x_3 = -8$$
$$4x_1 - 8x_2 + 3x_3 = 4$$
$$-2x_1 + 4x_2 + x_3 = 11$$

⚠ CAUTION Figure 2 shows the solution to Example 3 on a graphing calculator with a built-in reduced-form routine. Notice that the graphing calculator does not stop when a contradiction first occurs, but continues on to find the reduced form. Nevertheless, the last row in the reduced form still produces a contradiction. Do not confuse this type of reduced form with one that represents a consistent system (see Fig. 1). ▲

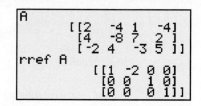

Figure 2 **Recognizing contradictions on a graphing calculator**

EXAMPLE 4 Solving a System Using Gauss–Jordan Elimination Solve by Gauss–Jordan elimination:

$$3x_1 + 6x_2 - 9x_3 = 15$$
$$2x_1 + 4x_2 - 6x_3 = 10$$
$$-2x_1 - 3x_2 + 4x_3 = -6$$

SOLUTION
$$\begin{bmatrix} 3 & 6 & -9 & | & 15 \\ 2 & 4 & -6 & | & 10 \\ -2 & -3 & 4 & | & -6 \end{bmatrix} \quad \tfrac{1}{3}R_1 \to R_1$$

$$\sim \begin{bmatrix} 1 & 2 & -3 & | & 5 \\ 2 & 4 & -6 & | & 10 \\ -2 & -3 & 4 & | & -6 \end{bmatrix} \quad \begin{array}{l}(-2)R_1 + R_2 \to R_2 \\ 2R_1 + R_3 \to R_3\end{array}$$

$$\sim \begin{bmatrix} 1 & 2 & -3 & | & 5 \\ 0 & 0 & 0 & | & 0 \\ 0 & 1 & -2 & | & 4 \end{bmatrix} \quad R_2 \leftrightarrow R_3$$

Note that we must interchange rows 2 and 3 to obtain a nonzero entry at the top of the second column of this submatrix.

$$\sim \begin{bmatrix} 1 & 2 & -3 & | & 5 \\ 0 & 1 & -2 & | & 4 \\ 0 & 0 & 0 & | & 0 \end{bmatrix} \quad (-2)R_2 + R_1 \to R_1$$

$$\sim \begin{bmatrix} 1 & 0 & 1 & | & -3 \\ 0 & 1 & -2 & | & 4 \\ 0 & 0 & 0 & | & 0 \end{bmatrix}$$

The matrix is now in reduced form. Write the corresponding reduced system and solve.

$$x_1 + x_3 = -3$$
$$x_2 - 2x_3 = 4$$

We discard the equation corresponding to the third (all zero) row in the reduced form, since it is satisfied by all values of x_1, x_2, and x_3.

Note that the leftmost variable in each equation appears in one and only one equation. We solve for the leftmost variables x_1 and x_2 in terms of the remaining variable, x_3 :

$$x_1 = -x_3 - 3$$
$$x_2 = 2x_3 + 4$$

If we let $x_3 = t$, then for any real number t,

$$x_1 = -t - 3$$
$$x_2 = 2t + 4$$
$$x_3 = t$$

You should check that $(-t - 3, 2t + 4, t)$ is a solution of the original system for any real number t. Some particular solutions are

$t = 0$	$t = -2$	$t = 3.5$
$(-3, 4, 0)$	$(-1, 0, -2)$	$(-6.5, 11, 3.5)$

In general,

> If the number of leftmost 1's in a reduced augmented coefficient matrix is less than the number of variables in the system and there are no contradictions, then the system is dependent and has infinitely many solutions.

Describing the solution set to this type of system is not difficult. In a reduced system, the **leftmost variables** correspond to the leftmost 1's in the corresponding reduced augmented matrix. The definition of reduced form for an augmented matrix ensures that each leftmost variable in the corresponding reduced system appears in one and only one equation of the system. Solving for each leftmost variable in terms of the remaining variables and writing a general solution to the system is usually easy. Example 5 illustrates a slightly more involved case.

Matched Problem 4) Solve by Gauss–Jordan elimination:

$$2x_1 - 2x_2 - 4x_3 = -2$$
$$3x_1 - 3x_2 - 6x_3 = -3$$
$$-2x_1 + 3x_2 + x_3 = 7$$

Explore and Discuss 1   Explain why the definition of reduced form ensures that each leftmost variable in a reduced system appears in one and only one equation and no equation contains more than one leftmost variable. Discuss methods for determining whether a consistent system is independent or dependent by examining the reduced form.

EXAMPLE 5   Solving a System Using Gauss–Jordan Elimination   Solve by Gauss–Jordan elimination:

$$x_1 + 2x_2 + 4x_3 + x_4 - x_5 = 1$$
$$2x_1 + 4x_2 + 8x_3 + 3x_4 - 4x_5 = 2$$
$$x_1 + 3x_2 + 7x_3 + 3x_5 = -2$$

SOLUTION

$$\begin{bmatrix} 1 & 2 & 4 & 1 & -1 & | & 1 \\ 2 & 4 & 8 & 3 & -4 & | & 2 \\ 1 & 3 & 7 & 0 & 3 & | & -2 \end{bmatrix}$$ $\begin{matrix} (-2)R_1 + R_2 \rightarrow R_2 \\ (-1)R_1 + R_3 \rightarrow R_3 \end{matrix}$

$$\sim \begin{bmatrix} 1 & 2 & 4 & 1 & -1 & | & 1 \\ 0 & 0 & 0 & 1 & -2 & | & 0 \\ 0 & 1 & 3 & -1 & 4 & | & -3 \end{bmatrix}$$ $R_2 \leftrightarrow R_3$

$$\sim \begin{bmatrix} 1 & 2 & 4 & 1 & -1 & | & 1 \\ 0 & 1 & 3 & -1 & 4 & | & -3 \\ 0 & 0 & 0 & 1 & -2 & | & 0 \end{bmatrix}$$ $(-2)R_2 + R_1 \rightarrow R_1$

$$\sim \begin{bmatrix} 1 & 0 & -2 & 3 & -9 & | & 7 \\ 0 & 1 & 3 & -1 & 4 & | & -3 \\ 0 & 0 & 0 & 1 & -2 & | & 0 \end{bmatrix}$$ $\begin{matrix} (-3)R_3 + R_1 \rightarrow R_1 \\ R_3 + R_2 \rightarrow R_2 \end{matrix}$

$$\sim \begin{bmatrix} 1 & 0 & -2 & 0 & -3 & | & 7 \\ 0 & 1 & 3 & 0 & 2 & | & -3 \\ 0 & 0 & 0 & 1 & -2 & | & 0 \end{bmatrix}$$ Matrix is in reduced form.

$$\begin{aligned} x_1 \quad - 2x_3 \quad - 3x_5 &= 7 \\ x_2 + 3x_3 \quad + 2x_5 &= -3 \\ x_4 - 2x_5 &= 0 \end{aligned}$$

Solve for the leftmost variables x_1, x_2, and x_4 in terms of the remaining variables x_3 and x_5:

$$\begin{aligned} x_1 &= \quad 2x_3 + 3x_5 + 7 \\ x_2 &= -3x_3 - 2x_5 - 3 \\ x_4 &= \qquad\qquad 2x_5 \end{aligned}$$

If we let $x_3 = s$ and $x_5 = t$, then for any real numbers s and t,

$$\begin{aligned} x_1 &= 2s + 3t + 7 \\ x_2 &= -3s - 2t - 3 \\ x_3 &= s \\ x_4 &= 2t \\ x_5 &= t \end{aligned}$$

is a solution. The check is left for you.

Matched Problem 5 Solve by Gauss–Jordan elimination:

$$\begin{aligned} x_1 - x_2 + 2x_3 \quad - 2x_5 &= 3 \\ -2x_1 + 2x_2 - 4x_3 - x_4 + x_5 &= -5 \\ 3x_1 - 3x_2 + 7x_3 + x_4 - 4x_5 &= 6 \end{aligned}$$

Application

Dependent systems of linear equations provide an excellent opportunity to discuss mathematical modeling in more detail. The process of using mathematics to solve real-world problems can be broken down into three steps (Fig. 3):

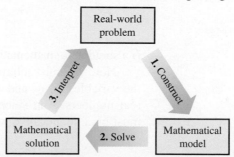

Figure 3

Step 1   *Construct* a mathematical model whose solution will provide information about the real-world problem.

Step 2   *Solve* the mathematical model.

Step 3   *Interpret* the solution to the mathematical model in terms of the original real-world problem.

In more complex problems, this cycle may have to be repeated several times to obtain the required information about the real-world problem.

EXAMPLE 6   Purchasing   A company that rents small moving trucks wants to purchase 25 trucks with a combined capacity of 28,000 cubic feet. Three different types of trucks are available: a 10-foot truck with a capacity of 350 cubic feet, a 14-foot truck with a capacity of 700 cubic feet, and a 24-foot truck with a capacity of 1,400 cubic feet. How many of each type of truck should the company purchase?

SOLUTION   The question in this example indicates that the relevant variables are the number of each type of truck:

$$x_1 = \text{number of 10-foot trucks}$$
$$x_2 = \text{number of 14-foot trucks}$$
$$x_3 = \text{number of 24-foot trucks}$$

Next we form the mathematical model:

$$x_1 + x_2 + x_3 = 25 \quad \text{Total number of trucks} \qquad (2)$$
$$350x_1 + 700x_2 + 1{,}400x_3 = 28{,}000 \quad \text{Total capacity}$$

Now we form the augmented matrix of the system and solve by Gauss–Jordan elimination:

$$\begin{bmatrix} 1 & 1 & 1 & | & 25 \\ 350 & 700 & 1{,}400 & | & 28{,}000 \end{bmatrix} \quad \tfrac{1}{350}R_2 \to R_2$$

$$\sim \begin{bmatrix} 1 & 1 & 1 & | & 25 \\ 1 & 2 & 4 & | & 80 \end{bmatrix} \quad -R_1 + R_2 \to R_2$$

$$\sim \begin{bmatrix} 1 & 1 & 1 & | & 25 \\ 0 & 1 & 3 & | & 55 \end{bmatrix} \quad -R_2 + R_1 \to R_1$$

$$\sim \begin{bmatrix} 1 & 0 & -2 & | & -30 \\ 0 & 1 & 3 & | & 55 \end{bmatrix} \quad \text{Matrix is in reduced form.}$$

$$x_1 - 2x_3 = -30 \quad \text{or} \quad x_1 = 2x_3 - 30$$
$$x_2 + 3x_3 = 55 \quad \text{or} \quad x_2 = -3x_3 + 55$$

Let $x_3 = t$. Then for t any real number,

$$x_1 = 2t - 30$$
$$x_2 = -3t + 55 \qquad (3)$$
$$x_3 = t$$

is a solution to mathematical model (2).

Now we must interpret this solution in terms of the original problem. Since the variables x_1, x_2, and x_3 represent numbers of trucks, they must be nonnegative real numbers. And since we can't purchase a fractional number of trucks, each must be a nonnegative whole number. Since $t = x_3$, it follows that t must also be a nonnegative whole number. The first and second equations in model (3) place additional restrictions on the values that t can assume:

$$x_1 = 2t - 30 \geq 0 \quad \text{implies that} \quad t \geq 15$$

$$x_2 = -3t + 55 \geq 0 \quad \text{implies that} \quad t \leq \frac{55}{3} = 18\tfrac{1}{3}$$

So the only possible values of t that will produce meaningful solutions to the original problem are 15, 16, 17, and 18. That is, the only combinations of 25 trucks that will result in a combined capacity of 28,000 cubic feet are $x_1 = 2t - 30$ 10-foot trucks, $x_2 = -3t + 55$ 14-foot trucks, and $x_3 = t$ 24-foot trucks, where $t = 15$, 16, 17, or 18. A table is a convenient way to display these solutions:

	10-Foot Truck	14-Foot Truck	24-Foot Truck
t	x_1	x_2	x_3
15	0	10	15
16	2	7	16
17	4	4	17
18	6	1	18

Matched Problem 6 A company that rents small moving trucks wants to purchase 16 trucks with a combined capacity of 19,200 cubic feet. Three different types of trucks are available: a cargo van with a capacity of 300 cubic feet, a 15-foot truck with a capacity of 900 cubic feet, and a 24-foot truck with a capacity of 1,500 cubic feet. How many of each type of truck should the company purchase?

Explore and Discuss 2 Refer to Example 6. The rental company charges $19.95 per day for a 10-foot truck, $29.95 per day for a 14-foot truck, and $39.95 per day for a 24-foot truck. Which of the four possible choices in the table would produce the largest daily income from truck rentals?

Exercises 4.3

Skills Warm-up Exercises

W *In Problems 1–4, write the augmented matrix of the system of linear equations. (If necessary, review the terminology of Section 4.2.)*

1. $x_1 + 2x_2 + 3x_3 = 12$
$x_1 + 7x_2 - 5x_3 = 15$

2. $4x_1 + x_2 = 8$
$3x_1 - 5x_2 = 6$
$x_1 + 9x_2 = 4$

3. $x_1 \quad\quad + 6x_3 = 2$
$\quad x_2 - x_3 = 5$
$x_1 + 3x_2 \quad\quad = 7$

4. $3x_1 + 4x_2 \quad\quad = 10$
$x_1 \quad\quad + 5x_3 = 15$
$\quad -x_2 + x_3 = 20$

In Problems 5–8, write the system of linear equations that is represented by the augmented matrix. Assume that the variables are $x_1, x_2, \dots$.

5. $\begin{bmatrix} 1 & -3 & | & 4 \\ 3 & 2 & | & 5 \\ -1 & 6 & | & 3 \end{bmatrix}$

6. $\begin{bmatrix} -1 & 5 & 2 & | & 8 \\ 4 & 0 & -3 & | & 7 \end{bmatrix}$

7. $\begin{bmatrix} 5 & -2 & 0 & 8 & | & 4 \end{bmatrix}$

8. $\begin{bmatrix} 1 & 0 & -1 & | & 1 \\ -1 & 1 & 0 & | & 3 \\ 0 & 2 & 1 & | & -5 \end{bmatrix}$

In Problems 9–18, if a matrix is in reduced form, say so. If not, explain why and indicate the row operation(s) necessary to transform the matrix into reduced form.

9. $\begin{bmatrix} 1 & 0 & | & 2 \\ 0 & 1 & | & -1 \end{bmatrix}$

10. $\begin{bmatrix} 0 & 1 & | & 2 \\ 1 & 0 & | & -1 \end{bmatrix}$ Not Reduced

11. $\begin{bmatrix} 1 & 0 & 2 & | & 3 \\ 0 & 0 & 0 & | & 0 \\ 0 & 1 & -1 & | & 4 \end{bmatrix}$ but not reduced

12. $\begin{bmatrix} 1 & 0 & 0 & | & -2 \\ 0 & 1 & 0 & | & 0 \\ 0 & 0 & 1 & | & 1 \end{bmatrix}$ Reduced

13. $\begin{bmatrix} 0 & 1 & 0 & | & 2 \\ 0 & 0 & 3 & | & -1 \\ 0 & 0 & 0 & | & 0 \end{bmatrix}$

14. $\begin{bmatrix} 1 & 2 & -3 & | & 1 \\ 0 & 0 & 1 & | & 4 \\ 0 & 0 & 0 & | & 0 \end{bmatrix}$ Reduced

15. $\begin{bmatrix} 1 & 1 & 0 & | & 1 \\ 0 & 0 & 1 & | & 1 \\ 0 & 0 & 0 & | & 0 \end{bmatrix}$

16. $\begin{bmatrix} 1 & 0 & -1 & | & 3 \\ 0 & 2 & 1 & | & 1 \\ 0 & 0 & 0 & | & 0 \end{bmatrix}$ Reduced

17. $\begin{bmatrix} 1 & 0 & -2 & 0 & | & 1 \\ 0 & 0 & 1 & 1 & | & 0 \end{bmatrix}$ **18.** $\begin{bmatrix} 1 & -2 & 0 & 0 & | & 1 \\ 0 & 0 & 1 & 1 & | & 0 \end{bmatrix}$

Write the linear system corresponding to each reduced augmented matrix in Problems 19–28 and write the solution of the system.

19. $\begin{bmatrix} 1 & 0 & 0 & | & -2 \\ 0 & 1 & 0 & | & 3 \\ 0 & 0 & 1 & | & 0 \end{bmatrix}$

20. $\begin{bmatrix} 1 & 0 & 0 & 0 & | & -2 \\ 0 & 1 & 0 & 0 & | & 0 \\ 0 & 0 & 1 & 0 & | & 1 \\ 0 & 0 & 0 & 1 & | & 3 \end{bmatrix}$

21. $\begin{bmatrix} 1 & 0 & -2 & | & 3 \\ 0 & 1 & 1 & | & -5 \\ 0 & 0 & 0 & | & 0 \end{bmatrix}$

22. $\begin{bmatrix} 1 & -2 & 0 & | & -3 \\ 0 & 0 & 1 & | & 5 \\ 0 & 0 & 0 & | & 0 \end{bmatrix}$

23. $\begin{bmatrix} 1 & 0 & | & 0 \\ 0 & 1 & | & 0 \\ 0 & 0 & | & 1 \end{bmatrix}$

24. $\begin{bmatrix} 1 & 0 & | & 5 \\ 0 & 1 & | & -3 \\ 0 & 0 & | & 0 \end{bmatrix}$

25. $\begin{bmatrix} 1 & 0 & -3 & | & 5 \\ 0 & 1 & 2 & | & -7 \end{bmatrix}$

26. $\begin{bmatrix} 1 & 0 & 1 & | & -4 \\ 0 & 1 & -1 & | & 6 \end{bmatrix}$

27. $\begin{bmatrix} 1 & -2 & 0 & -3 & | & -5 \\ 0 & 0 & 1 & 3 & | & 2 \end{bmatrix}$

28. $\begin{bmatrix} 1 & 0 & -2 & 3 & | & 4 \\ 0 & 1 & -1 & 2 & | & -1 \end{bmatrix}$

29. In which of Problems 19, 21, 23, 25, and 27 is the number of leftmost ones equal to the number of variables?

30. In which of Problems 20, 22, 24, 26, and 28 is the number of leftmost ones equal to the number of variables?

31. In which of Problems 19, 21, 23, 25, and 27 is the number of leftmost ones less than the number of variables?

32. In which of Problems 20, 22, 24, 26, and 28 is the number of leftmost ones less than the number of variables?

In Problems 33–38, discuss the validity of each statement about linear systems. If the statement is always true, explain why. If not, give a counterexample.

33. If the number of leftmost ones is equal to the number of variables, then the system has exactly one solution.

34. If the number of leftmost ones is less than the number of variables, then the system has infinitely many solutions.

35. If the number of leftmost ones is equal to the number of variables and the system is consistent, then the system has exactly one solution.

36. If the number of leftmost ones is less than the number of variables and the system is consistent, then the system has infinitely many solutions.

37. The number of equations is less than or equal to the number of variables.

38. The number of leftmost ones is less than or equal to the number of equations.

Use row operations to change each matrix in Problems 39–46 to reduced form.

39. $\begin{bmatrix} 1 & 2 & | & -1 \\ 0 & 1 & | & 3 \end{bmatrix}$ **40.** $\begin{bmatrix} 1 & 3 & | & 1 \\ 0 & 2 & | & -4 \end{bmatrix}$

41. $\begin{bmatrix} 1 & 1 & 1 & | & 16 \\ 2 & 3 & 4 & | & 25 \end{bmatrix}$ **42.** $\begin{bmatrix} 1 & 1 & 1 & | & 8 \\ 3 & 5 & 7 & | & 30 \end{bmatrix}$

43. $\begin{bmatrix} 1 & 0 & -3 & | & 1 \\ 0 & 1 & 2 & | & 0 \\ 0 & 0 & 3 & | & -6 \end{bmatrix}$ **44.** $\begin{bmatrix} 1 & 0 & 4 & | & 0 \\ 0 & 1 & -3 & | & -1 \\ 0 & 0 & -2 & | & 2 \end{bmatrix}$

45. $\begin{bmatrix} 1 & 2 & -2 & | & -1 \\ 0 & 3 & -6 & | & 1 \\ 0 & -1 & 2 & | & -\frac{1}{3} \end{bmatrix}$ **46.** $\begin{bmatrix} 0 & -2 & 8 & | & 1 \\ 2 & -2 & 6 & | & -4 \\ 0 & -1 & 4 & | & \frac{1}{2} \end{bmatrix}$

Solve Problems 47–62 using Gauss–Jordan elimination.

47. $2x_1 + 4x_2 - 10x_3 = -2$
$3x_1 + 9x_2 - 21x_3 = 0$
$x_1 + 5x_2 - 12x_3 = 1$

48. $3x_1 + 5x_2 - x_3 = -7$
$x_1 + x_2 + x_3 = -1$
$2x_1 + 11x_3 = 7$

49. $3x_1 + 8x_2 - x_3 = -18$
$2x_1 + x_2 + 5x_3 = 8$
$2x_1 + 4x_2 + 2x_3 = -4$

50. $2x_1 + 6x_2 + 15x_3 = -12$
$4x_1 + 7x_2 + 13x_3 = -10$
$3x_1 + 6x_2 + 12x_3 = -9$

51. $2x_1 - x_2 - 3x_3 = 8$
$x_1 - 2x_2 = 7$

52. $2x_1 + 4x_2 - 6x_3 = 10$
$3x_1 + 3x_2 - 3x_3 = 6$

53. $2x_1 - x_2 = 0$
$3x_1 + 2x_2 = 7$
$x_1 - x_2 = -1$

54. $2x_1 - x_2 = 0$
$3x_1 + 2x_2 = 7$
$x_1 - x_2 = -2$

55. $3x_1 - 4x_2 - x_3 = 1$
$2x_1 - 3x_2 + x_3 = 1$
$x_1 - 2x_2 + 3x_3 = 2$

56. $3x_1 + 7x_2 - x_3 = 11$
$x_1 + 2x_2 - x_3 = 3$
$2x_1 + 4x_2 - 2x_3 = 10$

57. $3x_1 - 2x_2 + x_3 = -7$
$2x_1 + x_2 - 4x_3 = 0$
$x_1 + x_2 - 3x_3 = 1$

58. $2x_1 + 3x_2 + 5x_3 = 21$
$x_1 - x_2 - 5x_3 = -2$
$2x_1 + x_2 - x_3 = 11$

59. $2x_1 + 4x_2 - 2x_3 = 2$
$-3x_1 - 6x_2 + 3x_3 = -3$

60. $3x_1 - 9x_2 + 12x_3 = 6$
$-2x_1 + 6x_2 - 8x_3 = -4$

61. $4x_1 - x_2 + 2x_3 = 3$
$-4x_1 + x_2 - 3x_3 = -10$
$8x_1 - 2x_2 + 9x_3 = -1$

62. $4x_1 - 2x_2 + 2x_3 = 5$
$-6x_1 + 3x_2 - 3x_3 = -2$
$10x_1 - 5x_2 + 9x_3 = 4$

63. Consider a consistent system of three linear equations in three variables. Discuss the nature of the system and its solution set if the reduced form of the augmented coefficient matrix has

(A) One leftmost 1   (B) Two leftmost 1's

(C) Three leftmost 1's   (D) Four leftmost 1's

64. Consider a system of three linear equations in three variables. Give examples of two reduced forms that are not row-equivalent if the system is

(A) Consistent and dependent   (B) Inconsistent

Solve Problems 65–70 using Gauss–Jordan elimination.

65. $x_1 + 2x_2 - 4x_3 - x_4 = 7$
$2x_1 + 5x_2 - 9x_3 - 4x_4 = 16$
$x_1 + 5x_2 - 7x_3 - 7x_4 = 13$

66. $2x_1 + 4x_2 + 5x_3 + 4x_4 = 8$
$x_1 + 2x_2 + 2x_3 + x_4 = 3$

67. $x_1 - x_2 + 3x_3 - 2x_4 = 1$
$-2x_1 + 4x_2 - 3x_3 + x_4 = 0.5$
$3x_1 - x_2 + 10x_3 - 4x_4 = 2.9$
$4x_1 - 3x_2 + 8x_3 - 2x_4 = 0.6$

68. $x_1 + x_2 + 4x_3 + x_4 = 1.3$
$-x_1 + x_2 - x_3 = 1.1$
$2x_1 + x_3 + 3x_4 = -4.4$
$2x_1 + 5x_2 + 11x_3 + 3x_4 = 5.6$

69. $x_1 - 2x_2 + x_3 + x_4 + 2x_5 = 2$
$-2x_1 + 4x_2 + 2x_3 + 2x_4 - 2x_5 = 0$
$3x_1 - 6x_2 + x_3 + x_4 + 5x_5 = 4$
$-x_1 + 2x_2 + 3x_3 + x_4 + x_5 = 3$

70. $x_1 - 3x_2 + x_3 + x_4 + 2x_5 = 2$
$-x_1 + 5x_2 + 2x_3 + 2x_4 - 2x_5 = 0$
$2x_1 - 6x_2 + 2x_3 + 2x_4 + 4x_5 = 4$
$-x_1 + 3x_2 - x_3 + x_5 = -3$

71. Find a, b, and c so that the graph of the quadratic equation $y = ax^2 + bx + c$ passes through the points $(-2, 9)$, $(1, -9)$, and $(4, 9)$.

72. Find a, b, and c so that the graph of the quadratic equation $y = ax^2 + bx + c$ passes through the points $(-1, -5)$, $(2, 7)$, and $(5, 1)$.

Applications

Construct a mathematical model for each of the following problems. (The answers in the back of the book include both the mathematical model and the interpretation of its solution.) Use Gauss–Jordan elimination to solve the model and then interpret the solution.

73. **Boat production.** A small manufacturing plant makes three types of inflatable boats: one-person, two-person, and four-person models. Each boat requires the services of three departments, as listed in the table. The cutting, assembly, and packaging departments have available a maximum of 380, 330, and 120 labor-hours per week, respectively.

Department	One-Person Boat	Two-Person Boat	Four-Person Boat
Cutting	0.5 hr	1.0 hr	1.5 hr
Assembly	0.6 hr	0.9 hr	1.2 hr
Packaging	0.2 hr	0.3 hr	0.5 hr

(A) How many boats of each type must be produced each week for the plant to operate at full capacity?

(B) How is the production schedule in part (A) affected if the packaging department is no longer used?

(C) How is the production schedule in part (A) affected if the four-person boat is no longer produced?

74. **Production scheduling.** Repeat Problem 73 assuming that the cutting, assembly, and packaging departments have available a maximum of 350, 330, and 115 labor-hours per week, respectively.

75. **Tank car leases.** A chemical manufacturer wants to lease a fleet of 24 railroad tank cars with a combined carrying capacity of 520,000 gallons. Tank cars with three different carrying capacities are available: 8,000 gallons, 16,000 gallons, and 24,000 gallons. How many of each type of tank car should be leased?

76. **Airplane leases.** A corporation wants to lease a fleet of 12 airplanes with a combined carrying capacity of 220 passengers. The three available types of planes carry 10, 15, and 20 passengers, respectively. How many of each type of plane should be leased?

77. **Tank car leases.** Refer to Problem 75. The cost of leasing an 8,000-gallon tank car is $450 per month, a 16,000-gallon tank car is $650 per month, and a 24,000-gallon tank car is $1,150 per month. Which of the solutions to Problem 75 would minimize the monthly leasing cost?

78. Airplane leases. Refer to Problem 76. The cost of leasing a 10-passenger airplane is $8,000 per month, a 15-passenger airplane is $14,000 per month, and a 20-passenger airplane is $16,000 per month. Which of the solutions to Problem 76 would minimize the monthly leasing cost?

79. Income tax. A corporation has a taxable income of $7,650,000. At this income level, the federal income tax rate is 50%, the state tax rate is 20%, and the local tax rate is 10%. If each tax rate is applied to the total taxable income, the resulting tax liability for the corporation would be 80% of taxable income. However, it is customary to deduct taxes paid to one agency before computing taxes for the other agencies. Assume that the federal taxes are based on the income that remains after the state and local taxes are deducted, and that state and local taxes are computed in a similar manner. What is the tax liability of the corporation (as a percentage of taxable income) if these deductions are taken into consideration?

80. Income tax. Repeat Problem 79 if local taxes are not allowed as a deduction for federal and state taxes.

81. Taxable income. As a result of several mergers and acquisitions, stock in four companies has been distributed among the companies. Each row of the following table gives the percentage of stock in the four companies that a particular company owns and the annual net income of each company (in millions of dollars):

| | Percentage of Stock Owned in Company | | | | Annual Net Income |
Company	A	B	C	D	Million $
A	71	8	3	7	3.2
B	12	81	11	13	2.6
C	11	9	72	8	3.8
D	6	2	14	72	4.4

So company A holds 71% of its own stock, 8% of the stock in company B, 3% of the stock in company C, etc. For the purpose of assessing a state tax on corporate income, the taxable income of each company is defined to be its share of its own annual net income plus its share of the taxable income of each of the other companies, as determined by the percentages in the table. What is the taxable income of each company (to the nearest thousand dollars)?

82. Taxable income. Repeat Problem 81 if tax law is changed so that the taxable income of a company is defined to be all of its own annual net income plus its share of the taxable income of each of the other companies.

83. Nutrition. A dietitian in a hospital is to arrange a special diet composed of three basic foods. The diet is to include exactly 340 units of calcium, 180 units of iron, and 220 units of vitamin A. The number of units per ounce of each special ingredient for each of the foods is indicated in the table.

| | Units per Ounce | | |
	Food A	Food B	Food C
Calcium	30	10	20
Iron	10	10	20
Vitamin A	10	30	20

(A) How many ounces of each food must be used to meet the diet requirements?

(B) How is the diet in part (A) affected if food C is not used?

(C) How is the diet in part (A) affected if the vitamin A requirement is dropped?

84. Nutrition. Repeat Problem 83 if the diet is to include exactly 400 units of calcium, 160 units of iron, and 240 units of vitamin A.

85. Plant food. A farmer can buy four types of plant food. Each barrel of mix A contains 30 pounds of phosphoric acid, 50 pounds of nitrogen, and 30 pounds of potash; each barrel of mix B contains 30 pounds of phosphoric acid, 75 pounds of nitrogen, and 20 pounds of potash; each barrel of mix C contains 30 pounds of phosphoric acid, 25 pounds of nitrogen, and 20 pounds of potash; and each barrel of mix D contains 60 pounds of phosphoric acid, 25 pounds of nitrogen, and 50 pounds of potash. Soil tests indicate that a particular field needs 900 pounds of phosphoric acid, 750 pounds of nitrogen, and 700 pounds of potash. How many barrels of each type of food should the farmer mix together to supply the necessary nutrients for the field?

86. Animal feed. In a laboratory experiment, rats are to be fed 5 packets of food containing a total of 80 units of vitamin E. There are four different brands of food packets that can be used. A packet of brand A contains 5 units of vitamin E, a packet of brand B contains 10 units of vitamin E, a packet of brand C contains 15 units of vitamin E, and a packet of brand D contains 20 units of vitamin E. How many packets of each brand should be mixed and fed to the rats?

87. Plant food. Refer to Problem 85. The costs of the four mixes are Mix A, $46; Mix B, $72; Mix C, $57; and Mix D, $63. Which of the solutions to Problem 85 would minimize the cost of the plant food?

88. Animal feed. Refer to Problem 86. The costs of the four brands are Brand A, $1.50; Brand B, $3.00; Brand C, $3.75; and Brand D, $2.25. Which of the solutions to Problem 86 would minimize the cost of the rat food?

89. Population growth. The U.S. population was approximately 75 million in 1900, 150 million in 1950, and 275 million in 2000. Construct a model for this data by finding a quadratic equation whose graph passes through the points $(0, 75)$, $(50, 150)$, and $(100, 275)$. Use this model to estimate the population in 2050.

90. Population growth. The population of California was approximately 24 million in 1980, 30 million in 1990, and 34 million in 2000. Construct a model for this data by finding a quadratic equation whose graph passes through the points $(0, 24)$, $(10, 30)$, and $(20, 34)$. Use this model to estimate the population in 2020. Do you think the estimate is plausible? Explain.

91. Female life expectancy. The life expectancy for females born during 1980–1985 was approximately 77.6 years. This grew to 78 years during 1985–1990 and to 78.6 years during 1990–1995. Construct a model for this data by finding a quadratic equation whose graph passes through the points $(0, 77.6)$, $(5, 78)$, and $(10, 78.6)$. Use this model to estimate the life expectancy for females born between 1995 and 2000 and for those born between 2000 and 2005.

92. **Male life expectancy.** The life expectancy for males born during 1980–1985 was approximately 70.7 years. This grew to 71.1 years during 1985–1990 and to 71.8 years during 1990–1995. Construct a model for this data by finding a quadratic equation whose graph passes through the points $(0, 70.7)$, $(5, 71.1)$, and $(10, 71.8)$. Use this model to estimate the life expectancy for males born between 1995 and 2000 and for those born between 2000 and 2005.

93. **Female life expectancy.** Refer to Problem 91. Subsequent data indicated that life expectancy grew to 79.1 years for females born during 1995–2000 and to 79.7 years for females born during 2000–2005. Add the points $(15, 79.1)$ and $(20, 79.7)$ to the data set in Problem 91. Use a graphing calculator to find a quadratic regression model for all five data points. Graph the data and the model in the same viewing window.

94. **Male life expectancy.** Refer to Problem 92. Subsequent data indicated that life expectancy grew to 73.2 years for males born during 1995–2000 and to 74.3 years for males born during 2000–2005. Add the points $(15, 73.2)$ and $(20, 74.3)$ to the data set in Problem 92. Use a graphing calculator to find a quadratic regression model for all five data points. Graph the data and the model in the same viewing window.

95. **Sociology.** Two sociologists have grant money to study school busing in a particular city. They wish to conduct an opinion survey using 600 telephone contacts and 400 house contacts. Survey company A has personnel to do 30 telephone and 10 house contacts per hour; survey company B can handle 20 telephone and 20 house contacts per hour. How many hours should be scheduled for each firm to produce exactly the number of contacts needed?

96. **Sociology.** Repeat Problem 95 if 650 telephone contacts and 350 house contacts are needed.

97. **Traffic flow.** The rush-hour traffic flow for a network of four one-way streets in a city is shown in the figure. The numbers next to each street indicate the number of vehicles per hour that enter and leave the network on that street. The variables x_1, x_2, x_3, and x_4 represent the flow of traffic between the four intersections in the network.

(A) For a smooth traffic flow, the number of vehicles entering each intersection should always equal the number leaving. For example, since 1,500 vehicles enter the intersection of 5th Street and Washington Avenue each hour and $x_1 + x_4$ vehicles leave this intersection, we see that $x_1 + x_4 = 1,500$. Find the equations determined by the traffic flow at each of the other three intersections.

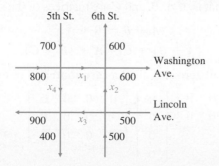

(B) Find the solution to the system in part (A).

(C) What is the maximum number of vehicles that can travel from Washington Avenue to Lincoln Avenue on 5th Street? What is the minimum number?

(D) If traffic lights are adjusted so that 1,000 vehicles per hour travel from Washington Avenue to Lincoln Avenue on 5th Street, determine the flow around the rest of the network.

98. **Traffic flow.** Refer to Problem 97. Closing Washington Avenue east of 6th Street for construction changes the traffic flow for the network as indicated in the figure. Repeat parts (A)–(D) of Problem 97 for this traffic flow.

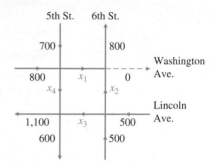

Answers to Matched Problems

1. (A) Condition 2 is violated: The 3 in row 2 and column 2 should be a 1. Perform the operation $\frac{1}{3}R_2 \to R_2$ to obtain
$$\begin{bmatrix} 1 & 0 & | & 2 \\ 0 & 1 & | & -2 \end{bmatrix}$$

(B) Condition 3 is violated: The 5 in row 1 and column 2 should be a 0. Perform the operation $(-5)R_2 + R_1 \to R_1$ to obtain
$$\begin{bmatrix} 1 & 0 & -6 & | & 8 \\ 0 & 1 & 2 & | & -1 \\ 0 & 0 & 0 & | & 0 \end{bmatrix}$$

(C) Condition 4 is violated. The leftmost 1 in the second row is not to the right of the leftmost 1 in the first row. Perform the operation $R_1 \leftrightarrow R_2$ to obtain
$$\begin{bmatrix} 1 & 0 & 0 & | & 0 \\ 0 & 1 & 0 & | & -3 \\ 0 & 0 & 1 & | & 2 \end{bmatrix}$$

(D) Condition 1 is violated: The all-zero second row should be at the bottom. Perform the operation $R_2 \leftrightarrow R_3$ to obtain
$$\begin{bmatrix} 1 & 2 & 0 & | & 3 \\ 0 & 0 & 1 & | & 4 \\ 0 & 0 & 0 & | & 0 \end{bmatrix}$$

2. $x_1 = 1, x_2 = -1, x_3 = 0$

3. Inconsistent; no solution

4. $x_1 = 5t + 4, x_2 = 3t + 5, x_3 = t$, t any real number

5. $x_1 = s + 7, x_2 = s, x_3 = t - 2, x_4 = -3t - 1, x_5 = t$, s and t any real numbers

6. $t - 8$ cargo vans, $-2t + 24$ 15-foot trucks, and t 24-foot trucks, where $t = 8, 9, 10, 11,$ or 12

4.4 Matrices: Basic Operations

- Addition and Subtraction

- Product of a Number k and a Matrix M

- Matrix Product

In the two preceding sections we introduced the important idea of matrices. In this and following sections, we develop this concept further. Matrices are both an ancient and a current mathematical concept. References to matrices and systems of equations can be found in Chinese manuscripts dating back to about 200 B.C. More recently, computers have made matrices a useful tool for a wide variety of applications. Most graphing calculators and computers are capable of performing calculations with matrices.

As we will see, matrix addition and multiplication are similar to real number addition and multiplication in many respects, but there are some important differences. A brief review of Appendix A, Section A.1, where real number operations are discussed, will help you understand the similarities and the differences.

Addition and Subtraction

Before we can discuss arithmetic operations for matrices, we have to define equality for matrices. Two matrices are **equal** if they have the same size and their corresponding elements are equal. For example,

$$\underset{2 \times 3}{\begin{bmatrix} a & b & c \\ d & e & f \end{bmatrix}} = \underset{2 \times 3}{\begin{bmatrix} u & v & w \\ x & y & z \end{bmatrix}} \quad \text{if and only if} \quad \begin{matrix} a = u & b = v & c = w \\ d = x & e = y & f = z \end{matrix}$$

The **sum of two matrices of the same size** is the matrix with elements that are the sum of the corresponding elements of the two given matrices. Addition is not defined for matrices of different sizes.

EXAMPLE 1 Matrix Addition

(A) $\begin{bmatrix} a & b \\ c & d \end{bmatrix} + \begin{bmatrix} w & x \\ y & z \end{bmatrix} = \begin{bmatrix} (a + w) & (b + x) \\ (c + y) & (d + z) \end{bmatrix}$

(B) $\begin{bmatrix} 2 & -3 & 0 \\ 1 & 2 & -5 \end{bmatrix} + \begin{bmatrix} 3 & 1 & 2 \\ -3 & 2 & 5 \end{bmatrix} = \begin{bmatrix} 5 & -2 & 2 \\ -2 & 4 & 0 \end{bmatrix}$

(C) $\begin{bmatrix} 5 & 0 & -2 \\ 1 & -3 & 8 \end{bmatrix} + \begin{bmatrix} -1 & 7 \\ 0 & 6 \\ -2 & 8 \end{bmatrix}$ Not defined

Matched Problem 1 Add: $\begin{bmatrix} 3 & 2 \\ -1 & -1 \\ 0 & 3 \end{bmatrix} + \begin{bmatrix} -2 & 3 \\ 1 & -1 \\ 2 & -2 \end{bmatrix}$

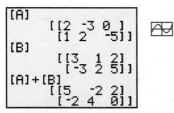

Figure 1 **Addition on a graphing calculator**

Graphing calculators can be used to solve problems involving matrix operations. Figure 1 illustrates the solution to Example 1B on a TI-84.

Because we add two matrices by adding their corresponding elements, it follows from the properties of real numbers that matrices of the same size are commutative and associative relative to addition. That is, if A, B, and C are matrices of the same size, then

Commutative: $A + B = B + A$

Associative: $(A + B) + C = A + (B + C)$

A matrix with elements that are all zeros is called a **zero matrix**. For example,

$$\begin{bmatrix} 0 & 0 & 0 \end{bmatrix} \quad \begin{bmatrix} 0 & 0 \\ 0 & 0 \end{bmatrix} \quad \begin{bmatrix} 0 \\ 0 \\ 0 \\ 0 \end{bmatrix} \quad \begin{bmatrix} 0 & 0 & 0 & 0 \\ 0 & 0 & 0 & 0 \\ 0 & 0 & 0 & 0 \end{bmatrix}$$

are zero matrices of different sizes. [*Note:* The simpler notation "0" is often used to denote the zero matrix of an arbitrary size.] The **negative of a matrix M**, denoted by $-M$, is a matrix with elements that are the negatives of the elements in M. Thus, if

$$M = \begin{bmatrix} a & b \\ c & d \end{bmatrix} \quad \text{then} \quad -M = \begin{bmatrix} -a & -b \\ -c & -d \end{bmatrix}$$

Note that $M + (-M) = 0$ (a zero matrix).

If A and B are matrices of the same size, we define **subtraction** as follows:

$$A - B = A + (-B)$$

So to subtract matrix B from matrix A, we simply add the negative of B to A.

EXAMPLE 2 Matrix Subtraction

$$\begin{bmatrix} 3 & -2 \\ 5 & 0 \end{bmatrix} - \begin{bmatrix} -2 & 2 \\ 3 & 4 \end{bmatrix} = \begin{bmatrix} 3 & -2 \\ 5 & 0 \end{bmatrix} + \begin{bmatrix} 2 & -2 \\ -3 & -4 \end{bmatrix} = \begin{bmatrix} 5 & -4 \\ 2 & -4 \end{bmatrix}$$

Matched Problem 2 Subtract: $[2 \quad -3 \quad 5] - [3 \quad -2 \quad 1]$

EXAMPLE 3 Matrix Equations Find a, b, c, and d so that

$$\begin{bmatrix} a & b \\ c & d \end{bmatrix} - \begin{bmatrix} 2 & -1 \\ -5 & 6 \end{bmatrix} = \begin{bmatrix} 4 & 3 \\ -2 & 4 \end{bmatrix}$$

SOLUTION
$$\begin{bmatrix} a & b \\ c & d \end{bmatrix} - \begin{bmatrix} 2 & -1 \\ -5 & 6 \end{bmatrix} = \begin{bmatrix} 4 & 3 \\ -2 & 4 \end{bmatrix}$$ Subtract the matrices on the left side.

$$\begin{bmatrix} a-2 & b-(-1) \\ c-(-5) & d-6 \end{bmatrix} = \begin{bmatrix} 4 & 3 \\ -2 & 4 \end{bmatrix}$$ Remove parentheses.

$$\begin{bmatrix} a-2 & b+1 \\ c+5 & d-6 \end{bmatrix} = \begin{bmatrix} 4 & 3 \\ -2 & 4 \end{bmatrix}$$ Use the definition of *equality* to change this matrix equation into four real number equations.

$$\begin{array}{cccc} a-2=4 & b+1=3 & c+5=-2 & d-6=4 \\ a=6 & b=2 & c=-7 & d=10 \end{array}$$

Matched Problem 3 Find a, b, c, and d so that

$$\begin{bmatrix} a & b \\ c & d \end{bmatrix} - \begin{bmatrix} -4 & 2 \\ 1 & -3 \end{bmatrix} = \begin{bmatrix} -2 & 5 \\ 8 & 2 \end{bmatrix}$$

Product of a Number *k* and a Matrix *M*

The **product of a number *k* and a matrix *M***, denoted by **kM**, is a matrix formed by multiplying each element of M by k.

EXAMPLE 4 Multiplication of a Matrix by a Number

$$-2 \begin{bmatrix} 3 & -1 & 0 \\ -2 & 1 & 3 \\ 0 & -1 & -2 \end{bmatrix} = \begin{bmatrix} -6 & 2 & 0 \\ 4 & -2 & -6 \\ 0 & 2 & 4 \end{bmatrix}$$

Matched Problem 4 Find: $10 \begin{bmatrix} 1.3 \\ 0.2 \\ 3.5 \end{bmatrix}$

The next example illustrates the use of matrix operations in an applied setting.

EXAMPLE 5 Sales Commissions Ms. Smith and Mr. Jones are salespeople in a new-car agency that sells only two models. August was the last month for this year's models, and next year's models were introduced in September. Gross dollar sales for each month are given in the following matrices:

	August sales			September sales	
	Compact	Luxury		Compact	Luxury
Ms. Smith	$54,000	$88,000	$= A$	$228,000	$368,000	$= B$
Mr. Jones	$126,000	0		$304,000	$322,000	

For example, Ms. Smith had $54,000 in compact sales in August, and Mr. Jones had $322,000 in luxury car sales in September.

(A) What were the combined dollar sales in August and September for each salesperson and each model?

(B) What was the increase in dollar sales from August to September?

(C) If both salespeople receive 5% commissions on gross dollar sales, compute the commission for each person for each model sold in September.

SOLUTION

(A) $A + B =$

	Compact	Luxury	
	$282,000	$456,000	Ms. Smith
	$430,000	$322,000	Mr. Jones

(B) $B - A =$

	Compact	Luxury	
	$174,000	$280,000	Ms. Smith
	$178,000	$322,000	Mr. Jones

(C) $0.05B =$

$$\begin{bmatrix} (0.05)(\$228,000) & (0.05)(\$368,000) \\ (0.05)(\$304,000) & (0.05)(\$322,000) \end{bmatrix}$$

$$= \begin{bmatrix} \$11,400 & \$18,400 \\ \$15,200 & \$16,100 \end{bmatrix} \begin{matrix} \text{Ms. Smith} \\ \text{Mr. Jones} \end{matrix}$$

Matched Problem 5 Repeat Example 5 with

$$A = \begin{bmatrix} \$45,000 & \$77,000 \\ \$106,000 & \$22,000 \end{bmatrix} \quad \text{and} \quad B = \begin{bmatrix} \$190,000 & \$345,000 \\ \$266,000 & \$276,000 \end{bmatrix}$$

 Figure 2 illustrates a solution for Example 5 on a spreadsheet.

	1	2	3	4	5	6	7
1		August Sales		September Sales		September Commissions	
2		Compact	Luxury	Compact	Luxury	Compact	Luxury
3	Smith	$54,000	$88,000	$228,000	$368,000	$11,400	$18,400
4	Jones	$126,000	$0	$304,000	$322,000	$15,200	$16,100
5		Combined Sales		Sales Increase			
6	Smith	$282,000	$456,000	$174,000	$280,000		
7	Jones	$430,000	$322,000	$178,000	$322,000		

Figure 2

Matrix Product

Matrix multiplication was introduced by the English mathematician Arthur Cayley (1821–1895) in studies of systems of linear equations and linear transformations. Although this multiplication may seem strange at first, it is extremely useful in many practical problems.

We start by defining the product of two special matrices, a row matrix and a column matrix.

DEFINITION Product of a Row Matrix and a Column Matrix
The **product** of a $1 \times n$ row matrix and an $n \times 1$ column matrix is a 1×1 matrix given by

$$
\underset{1 \times n}{[a_1 \ a_2 \ \cdots \ a_n]} \overset{n \times 1}{\begin{bmatrix} b_1 \\ b_2 \\ \vdots \\ b_n \end{bmatrix}} = [a_1 b_1 + a_2 b_2 + \cdots + a_n b_n]
$$

Note that the number of elements in the row matrix and in the column matrix must be the same for the product to be defined.

EXAMPLE 6 Product of a Row Matrix and a Column Matrix

$$
[2 \quad -3 \quad 0] \begin{bmatrix} -5 \\ 2 \\ -2 \end{bmatrix} = [(2)(-5) + (-3)(2) + (0)(-2)]
$$

$$
= [-10 - 6 + 0] = [-16]
$$

Matched Problem 6 $\quad [-1 \quad 0 \quad 3 \quad 2] \begin{bmatrix} 2 \\ 3 \\ 4 \\ -1 \end{bmatrix} = ?$

Refer to Example 6. The distinction between the real number -16 and the 1×1 matrix $[-16]$ is a technical one, and it is common to see 1×1 matrices written as real numbers without brackets. In the work that follows, we will frequently refer to 1×1 matrices as real numbers and omit the brackets whenever it is convenient to do so.

EXAMPLE 7 Labor Costs A factory produces a slalom water ski that requires 3 labor-hours in the assembly department and 1 labor-hour in the finishing department. Assembly personnel receive $9 per hour and finishing personnel receive $6 per hour. Total labor cost per ski is given by the product:

$$
[3 \quad 1] \begin{bmatrix} 9 \\ 6 \end{bmatrix} = [(3)(9) + (1)(6)] = [27 + 6] = [33] \quad \text{or} \quad \$33 \text{ per ski}
$$

Matched Problem 7 If the factory in Example 7 also produces a trick water ski that requires 5 labor-hours in the assembly department and 1.5 labor-hours in the finishing department, write a product between appropriate row and column matrices that will give the total labor cost for this ski. Compute the cost.

We now use the product of a $1 \times n$ row matrix and an $n \times 1$ column matrix to extend the definition of matrix product to more general matrices.

DEFINITION Matrix Product
If A is an $m \times p$ matrix and B is a $p \times n$ matrix, then the **matrix product** of A and B, denoted AB, is an $m \times n$ matrix whose element in the ith row and jth column is the real number obtained from the product of the ith row of A and the jth column of B. If the number of columns in A does not equal the number of rows in B, the matrix product AB is **not defined**.

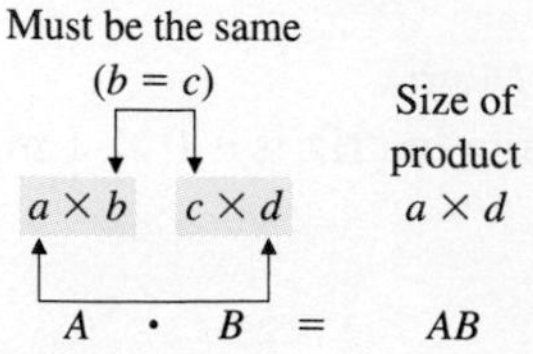

Must be the same
(b = c)

$a \times b$ $c \times d$

Size of
product
$a \times d$

$A \quad \cdot \quad B \quad = \quad AB$

Figure 3

It is important to check sizes before starting the multiplication process. If A is an $a \times b$ matrix and B is a $c \times d$ matrix, then if $b = c$, the product AB will exist and will be an $a \times d$ matrix (see Fig. 3). If $b \neq c$, the product AB does not exist. The definition is not as complicated as it might first seem. An example should help clarify the process.

For

$$A = \begin{bmatrix} 2 & 3 & -1 \\ -2 & 1 & 2 \end{bmatrix} \quad \text{and} \quad B = \begin{bmatrix} 1 & 3 \\ 2 & 0 \\ -1 & 2 \end{bmatrix}$$

A is 2×3 and B is 3×2, so AB is 2×2. To find the first row of AB, we take the product of the first row of A with every column of B and write each result as a real number, not as a 1×1 matrix. The second row of AB is computed in the same manner. The four products of row and column matrices used to produce the four elements in AB are shown in the following dashed box. These products are usually calculated mentally or with the aid of a calculator, and need not be written out. The shaded portions highlight the steps involved in computing the element in the first row and second column of AB.

$$\underset{2 \times 3}{\begin{bmatrix} 2 & 3 & -1 \\ -2 & 1 & 2 \end{bmatrix}} \underset{3 \times 2}{\begin{bmatrix} 1 & 3 \\ 2 & 0 \\ -1 & 2 \end{bmatrix}} = \begin{bmatrix} [2 \quad 3 \quad -1]\begin{bmatrix} 1 \\ 2 \\ -1 \end{bmatrix} & [2 \quad 3 \quad -1]\begin{bmatrix} 3 \\ 0 \\ 2 \end{bmatrix} \\ [-2 \quad 1 \quad 2]\begin{bmatrix} 1 \\ 2 \\ -1 \end{bmatrix} & [-2 \quad 1 \quad 2]\begin{bmatrix} 3 \\ 0 \\ 2 \end{bmatrix} \end{bmatrix}$$

$$= \begin{bmatrix} (2)(1) + (3)(2) + (-1)(-1) & (2)(3) + (3)(0) + (-1)(2) \\ (-2)(1) + (1)(2) + (2)(-1) & (-2)(3) + (1)(0) + (2)(2) \end{bmatrix} = \underset{2 \times 2}{\begin{bmatrix} 9 & 4 \\ -2 & -2 \end{bmatrix}}$$

EXAMPLE 8   Matrix Multiplication   Find the indicated matrix product, if it exists, where:

$$A = \begin{bmatrix} 2 & 1 \\ 1 & 0 \\ -1 & 2 \end{bmatrix} \quad B = \begin{bmatrix} 1 & -1 & 0 & 1 \\ 2 & 1 & 2 & 0 \end{bmatrix} \quad C = \begin{bmatrix} 2 & 6 \\ -1 & -3 \end{bmatrix}$$

$$D = \begin{bmatrix} 1 & 2 \\ 3 & 6 \end{bmatrix} \quad E = \begin{bmatrix} 2 & -3 & 0 \end{bmatrix} \quad F = \begin{bmatrix} -5 \\ 2 \\ -2 \end{bmatrix}$$

(A) $AB = \underset{3 \times 2}{\begin{bmatrix} 2 & 1 \\ 1 & 0 \\ -1 & 2 \end{bmatrix}} \underset{2 \times 4}{\begin{bmatrix} 1 & -1 & 0 & 1 \\ 2 & 1 & 2 & 0 \end{bmatrix}}$

$$= \begin{bmatrix} (2)(1)+(1)(2) & (2)(-1)+(1)(1) & (2)(0)+(1)(2) & (2)(1)+(1)(0) \\ (1)(1)+(0)(2) & (1)(-1)+(0)(1) & (1)(0)+(0)(2) & (1)(1)+(0)(0) \\ (-1)(1)+(2)(2) & (-1)(-1)+(2)(1) & (-1)(0)+(2)(2) & (-1)(1)+(2)(0) \end{bmatrix}$$

$$= \underset{3 \times 4}{\begin{bmatrix} 4 & -1 & 2 & 2 \\ 1 & -1 & 0 & 1 \\ 3 & 3 & 4 & -1 \end{bmatrix}}$$

(B) $BA = \begin{matrix} 2 \times 4 \\ \begin{bmatrix} 1 & -1 & 0 & 1 \\ 2 & 1 & 2 & 0 \end{bmatrix} \end{matrix} \begin{matrix} 3 \times 2 \\ \begin{bmatrix} 2 & 1 \\ 1 & 0 \\ -1 & 2 \end{bmatrix} \end{matrix}$ Not defined

(C) $CD = \begin{bmatrix} 2 & 6 \\ -1 & -3 \end{bmatrix} \begin{bmatrix} 1 & 2 \\ 3 & 6 \end{bmatrix} = \begin{bmatrix} (2)(1)+(6)(3) & (2)(2)+(6)(6) \\ (-1)(1) + (3)(3) & (-1)(2)+(-3)(6) \end{bmatrix}$

$= \begin{bmatrix} 20 & 40 \\ -10 & -20 \end{bmatrix}$

(D) $DC = \begin{bmatrix} 1 & 2 \\ 3 & 6 \end{bmatrix} \begin{bmatrix} 2 & 6 \\ -1 & -3 \end{bmatrix} = \begin{bmatrix} (1)(2)+(2)(-1) & (1)(6)+(2)(-3) \\ (3)(2)+(6)(-1) & (3)(6)+(6)(-3) \end{bmatrix}$

$= \begin{bmatrix} 0 & 0 \\ 0 & 0 \end{bmatrix}$

(E) $EF = \begin{bmatrix} 2 & -3 & 0 \end{bmatrix} \begin{bmatrix} -5 \\ 2 \\ -2 \end{bmatrix} = [(2)(-5)+(-3)(2)+(0)(-2)] = [-16]$

(F) $FE = \begin{bmatrix} -5 \\ 2 \\ -2 \end{bmatrix} \begin{bmatrix} 2 & -3 & 0 \end{bmatrix} = \begin{bmatrix} (-5)(2) & (-5)(-3) & (-5)(0) \\ (2)(2) & (2)(-3) & (2)(0) \\ (-2)(2) & (-2)(-3) & (-2)(0) \end{bmatrix}$

$= \begin{bmatrix} -10 & 15 & 0 \\ 4 & -6 & 0 \\ -4 & 6 & 0 \end{bmatrix}$

(G) $A^{2*} = AA = \begin{matrix} 3 \times 2 \\ \begin{bmatrix} 2 & 1 \\ 1 & 0 \\ -1 & 2 \end{bmatrix} \end{matrix} \begin{matrix} 3 \times 2 \\ \begin{bmatrix} 2 & 1 \\ 1 & 0 \\ -1 & 2 \end{bmatrix} \end{matrix}$ Not defined

(H) $C^2 = CC = \begin{bmatrix} 2 & 6 \\ -1 & -3 \end{bmatrix} \begin{bmatrix} 2 & 6 \\ -1 & -3 \end{bmatrix}$

$= \begin{bmatrix} (2)(2)+(6)(-1) & (2)(6)+(6)(-3) \\ (-1)(2)+(-3)(-1) & (-1)(6)+(-3)(-3) \end{bmatrix}$

$= \begin{bmatrix} -2 & -6 \\ 1 & 3 \end{bmatrix}$

Matched Problem 8 Find each product, if it is defined:

(A) $\begin{bmatrix} -1 & 0 & 3 & -2 \\ 1 & 2 & 2 & 0 \end{bmatrix} \begin{bmatrix} -1 & 1 \\ 2 & 3 \\ 1 & 0 \end{bmatrix}$

(B) $\begin{bmatrix} -1 & 1 \\ 2 & 3 \\ 1 & 0 \end{bmatrix} \begin{bmatrix} -1 & 0 & 3 & -2 \\ 1 & 2 & 2 & 0 \end{bmatrix}$

(C) $\begin{bmatrix} 1 & 2 \\ -1 & -2 \end{bmatrix} \begin{bmatrix} -2 & 4 \\ 1 & -2 \end{bmatrix}$

(D) $\begin{bmatrix} -2 & 4 \\ 1 & -2 \end{bmatrix} \begin{bmatrix} 1 & 2 \\ -1 & -2 \end{bmatrix}$

(E) $\begin{bmatrix} 3 & -2 & 1 \end{bmatrix} \begin{bmatrix} 4 \\ 2 \\ 3 \end{bmatrix}$

(F) $\begin{bmatrix} 4 \\ 2 \\ 3 \end{bmatrix} \begin{bmatrix} 3 & -2 & 1 \end{bmatrix}$

*Following standard algebraic notation, we write $A^2 = AA$, $A^3 = AAA$, and so on.

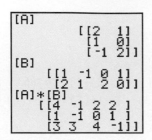

Figure 4 **Multiplication on a graphing calculator**

Figure 4 illustrates a graphing calculator solution to Example 8A. What would you expect to happen if you tried to solve Example 8B on a graphing calculator?

CONCEPTUAL INSIGHT

In the arithmetic of real numbers, it does not matter in which order we multiply. For example, $5 \times 7 = 7 \times 5$. In matrix multiplication, however, it does make a difference. That is, AB does not always equal BA, even if both multiplications are defined and both products are the same size (see Examples 8C and 8D).

Matrix multiplication is not commutative.

The zero property of real numbers states that if the product of two real numbers is 0, then one of the numbers must be 0 (see Appendix A, Section A.1). This property is very important when solving equations. For example,

$$x^2 - 4x + 3 = 0$$
$$(x - 1)(x - 3) = 0$$
$$x - 1 = 0 \quad \text{or} \quad x - 3 = 0$$
$$x = 1 \qquad\qquad x = 3$$

For matrices, it is possible to find nonzero matrices A and B such that AB is a zero matrix (see Example 8D).

The zero property does not hold for matrix multiplication.

Explore and Discuss 1   In addition to the commutative and zero properties, there are other significant differences between real number multiplication and matrix multiplication.

(A) In real number multiplication, the only real number whose square is 0 is the real number 0 ($0^2 = 0$). Find at least one 2×2 matrix A with all elements nonzero such that $A^2 = 0$, where 0 is the 2×2 zero matrix.

(B) In real number multiplication, the only nonzero real number that is equal to its square is the real number 1 ($1^2 = 1$). Find at least one 2×2 matrix B with all elements nonzero such that $B^2 = B$.

EXAMPLE 9   Matrix Multiplication   Find a, b, c, and d so that

$$\begin{bmatrix} 2 & -1 \\ 5 & 3 \end{bmatrix}\begin{bmatrix} a & b \\ c & d \end{bmatrix} = \begin{bmatrix} -6 & 17 \\ 7 & 4 \end{bmatrix}$$

SOLUTION   The product of the matrices on the left side of the equation is

$$\begin{bmatrix} 2 & -1 \\ 5 & 3 \end{bmatrix}\begin{bmatrix} a & b \\ c & d \end{bmatrix} = \begin{bmatrix} 2a - c & 2b - d \\ 5a + 3c & 5b + 3d \end{bmatrix}$$

Therefore,

$$2a - c = -6 \qquad 2b - d = 17$$
$$5a + 3c = 7 \qquad 5b + 3d = 4$$

This gives a system of two equations in the variables a and c, and a second system of two equations in the variables b and d. Each system can be solved by substitution, or elimination by addition, or Gauss–Jordan elimination (the details are omitted). The solution of the first system is $a = -1$, $c = 4$, and the solution of the second system is $b = 5$, $d = -7$.

Matched Problem 9) Find a, b, c, and d so that

$$\begin{bmatrix} 6 & -5 \\ 0 & 3 \end{bmatrix} \begin{bmatrix} a & b \\ c & d \end{bmatrix} = \begin{bmatrix} -16 & 64 \\ 24 & -6 \end{bmatrix}$$

Now we consider an application of matrix multiplication.

EXAMPLE 10 *Labor Costs* We can combine the time requirements for slalom and trick water skis discussed in Example 7 and Matched Problem 7 into one matrix:

Labor-hours per ski

	Assembly department	Finishing department	
Trick ski	5 hr	1.5 hr	$= L$
Slalom ski	3 hr	1 hr	

Now suppose that the company has two manufacturing plants, one in California and the other in Maryland, and that their hourly rates for each department are given in the following matrix:

Hourly wages

	California	Maryland	
Assembly department	\$12	\$13	$= H$
Finishing department	\$7	\$8	

Since H and L are both 2×2 matrices, we can take the product of H and L in either order and the result will be a 2×2 matrix:

$$HL = \begin{bmatrix} 12 & 13 \\ 7 & 8 \end{bmatrix} \begin{bmatrix} 5 & 1.5 \\ 3 & 1 \end{bmatrix} = \begin{bmatrix} 99 & 31 \\ 59 & 18.5 \end{bmatrix}$$

$$LH = \begin{bmatrix} 5 & 1.5 \\ 3 & 1 \end{bmatrix} \begin{bmatrix} 12 & 13 \\ 7 & 8 \end{bmatrix} = \begin{bmatrix} 70.5 & 77 \\ 43 & 47 \end{bmatrix}$$

How can we interpret the elements in these products? Let's begin with the product HL. The element 99 in the first row and first column of HL is the product of the first row matrix of H and the first column matrix of L:

CA MD

$$[12 \quad 13] \begin{bmatrix} 5 \\ 3 \end{bmatrix} \begin{matrix} \text{Trick} \\ \text{Slalom} \end{matrix} = 12(5) + 13(3) = 60 + 39 = 99$$

Notice that \$60 is the labor cost for assembling a trick ski at the California plant and \$39 is the labor cost for assembling a slalom ski at the Maryland plant. Although both numbers represent labor costs, it makes no sense to add them together. They do not pertain to the same type of ski or to the same plant. So even though the product HL happens to be defined mathematically, it has no useful interpretation in this problem.

Now let's consider the product LH. The element 70.5 in the first row and first column of LH is given by the following product:

Assembly Finishing

$$[5 \qquad 1.5] \begin{bmatrix} 12 \\ 7 \end{bmatrix} \begin{matrix} \text{Assembly} \\ \text{Finishing} \end{matrix} = 5(12) + 1.5(7) = 60 + 10.5 = 70.5$$

This time, \$60 is the labor cost for assembling a trick ski at the California plant and \$10.50 is the labor cost for finishing a trick ski at the California plant. So the sum is the total labor cost for producing a trick ski at the California plant. The other elements in *LH* also represent total labor costs, as indicated by the row and column labels shown below:

Labor costs per ski

CA   MD

$$LH = \begin{bmatrix} \$70.50 & \$77 \\ \$43 & \$47 \end{bmatrix} \begin{array}{l} \text{Trick} \\ \text{Slalom} \end{array}$$

Figure 5 shows a solution to Example 9 on a spreadsheet.

	A	B	C	D	E	F
1		Labor-hours per ski			Hourly wages	
2		Assembly	Finishing		California	Maryland
3	Trick ski	5	1.5	Assembly	$12	$13
4	Slalom ski	3	1	Finishing	$7	$8
5		Labor costs per ski				
6		California	Maryland			
7	Trick ski	$70.50	$77.00			
8	Slalom ski	$43.00	$47.00			

Figure 5 **Matrix multiplication in a spreadsheet: The command MMULT(B3:C4, E3:F4) produces the matrix in B7:C8**

Matched Problem 10 Refer to Example 10. The company wants to know how many hours to schedule in each department in order to produce 2,000 trick skis and 1,000 slalom skis. These production requirements can be represented by either of the following matrices:

Trick   Slalom
skis   skis

$$P = [2{,}000 \quad 1{,}000] \qquad Q = \begin{bmatrix} 2{,}000 \\ 1{,}000 \end{bmatrix} \begin{array}{l} \text{Trick skis} \\ \text{Slalom skis} \end{array}$$

Using the labor-hour matrix *L* from Example 10, find *PL* or *LQ*, whichever has a meaningful interpretation for this problem, and label the rows and columns accordingly.

CONCEPTUAL **INSIGHT**

Example 10 and Matched Problem 10 illustrate an important point about matrix multiplication. Even if you are using a graphing calculator to perform the calculations in a matrix product, it is still necessary for you to know the definition of matrix multiplication so that you can interpret the results correctly.

Exercises 4.4

Skills Warm-up Exercises

W   *In Problems 1–14, perform the indicated operation, if possible. (If necessary, review the definitions at the beginning of Section 4.4.)*

1. $[1 \quad 5] + [3 \quad 10]$

2. $\begin{bmatrix} 3 \\ 2 \end{bmatrix} - \begin{bmatrix} 7 \\ -4 \end{bmatrix}$

3. $\begin{bmatrix} 2 & 0 \\ -3 & 6 \end{bmatrix} - \begin{bmatrix} 0 & -4 \\ -1 & 0 \end{bmatrix}$

4. $\begin{bmatrix} -9 & 2 \\ 8 & 0 \end{bmatrix} + \begin{bmatrix} 9 & 0 \\ 0 & 8 \end{bmatrix}$

5. $\begin{bmatrix} 3 \\ 6 \end{bmatrix} + [-1 \quad 9]$

6. $4 \begin{bmatrix} 2 & -6 & 1 \\ 8 & 5 & -3 \end{bmatrix}$

7. $7 [3 \quad -5 \quad 9 \quad 4]$

8. $[10 \quad 12] + \begin{bmatrix} 4 \\ 3 \end{bmatrix}$

9. $\begin{bmatrix} 3 & 4 \\ -1 & -2 \end{bmatrix} \begin{bmatrix} -1 \\ 2 \end{bmatrix}$

10. $\begin{bmatrix} -1 & 1 \\ 2 & -3 \end{bmatrix} \begin{bmatrix} 4 \\ -2 \end{bmatrix}$

11. $\begin{bmatrix} 2 & -3 \\ 1 & 2 \end{bmatrix}\begin{bmatrix} 1 & -1 \\ 0 & -2 \end{bmatrix}$

12. $\begin{bmatrix} -3 & 2 \\ 4 & -1 \end{bmatrix}\begin{bmatrix} -2 & 5 \\ -1 & 3 \end{bmatrix}$

13. $\begin{bmatrix} 1 & -1 \\ 0 & -2 \end{bmatrix}\begin{bmatrix} 2 & -3 \\ 1 & 2 \end{bmatrix}$

14. $\begin{bmatrix} -2 & 5 \\ -1 & 3 \end{bmatrix}\begin{bmatrix} -3 & 2 \\ 4 & -1 \end{bmatrix}$

In Problems 15–22, find the matrix product. Note that each product can be found mentally, without the use of a calculator or pencil-and-paper calculations.

15. $\begin{bmatrix} 5 & 0 \\ 0 & 5 \end{bmatrix}\begin{bmatrix} 1 & 2 \\ 3 & 4 \end{bmatrix}$

16. $\begin{bmatrix} 3 & 0 \\ 0 & 3 \end{bmatrix}\begin{bmatrix} 1 & 3 \\ 5 & 7 \end{bmatrix}$

17. $\begin{bmatrix} 1 & 2 \\ 3 & 4 \end{bmatrix}\begin{bmatrix} 5 & 0 \\ 0 & 5 \end{bmatrix}$

18. $\begin{bmatrix} 1 & 3 \\ 5 & 7 \end{bmatrix}\begin{bmatrix} 3 & 0 \\ 0 & 3 \end{bmatrix}$

19. $\begin{bmatrix} 0 & 1 \\ 0 & 0 \end{bmatrix}\begin{bmatrix} 3 & 5 \\ 7 & 9 \end{bmatrix}$

20. $\begin{bmatrix} 0 & 0 \\ 1 & 0 \end{bmatrix}\begin{bmatrix} 2 & 4 \\ 6 & 8 \end{bmatrix}$

21. $\begin{bmatrix} 3 & 5 \\ 7 & 9 \end{bmatrix}\begin{bmatrix} 0 & 1 \\ 0 & 0 \end{bmatrix}$

22. $\begin{bmatrix} 2 & 4 \\ 6 & 8 \end{bmatrix}\begin{bmatrix} 0 & 0 \\ 1 & 0 \end{bmatrix}$

Find the products in Problems 23–30.

23. $[5 \quad -2]\begin{bmatrix} -3 \\ -4 \end{bmatrix}$

24. $[-4 \quad 3]\begin{bmatrix} -2 \\ 1 \end{bmatrix}$

25. $\begin{bmatrix} -3 \\ -4 \end{bmatrix}[5 \quad -2]$

26. $\begin{bmatrix} -2 \\ 1 \end{bmatrix}[-4 \quad 3]$

27. $[3 \quad -2 \quad -4]\begin{bmatrix} 1 \\ 2 \\ -3 \end{bmatrix}$

28. $[1 \quad -2 \quad 2]\begin{bmatrix} 2 \\ -1 \\ 1 \end{bmatrix}$

29. $\begin{bmatrix} 1 \\ 2 \\ -3 \end{bmatrix}[3 \quad -2 \quad -4]$

30. $\begin{bmatrix} 2 \\ -1 \\ 1 \end{bmatrix}[1 \quad -2 \quad 2]$

Problems 31–48 refer to the following matrices:

$$A = \begin{bmatrix} 2 & -1 & 3 \\ 0 & 4 & -2 \end{bmatrix} \quad B = \begin{bmatrix} -3 & 1 \\ 2 & 5 \end{bmatrix}$$

$$C = \begin{bmatrix} -1 & 0 & 2 \\ 4 & -3 & 1 \\ -2 & 3 & 5 \end{bmatrix} \quad D = \begin{bmatrix} 3 & -2 \\ 0 & -1 \\ 1 & 2 \end{bmatrix}$$

Perform the indicated operations, if possible.

31. AC

32. CA

33. AB

34. BA

35. B^2

36. C^2

37. $B + AD$

38. $C + DA$

39. $(0.1)DB$

40. $(0.2)CD$

41. $(3)BA + (4)AC$

42. $(2)DB + (5)CD$

43. $(-2)BA + (6)CD$

44. $(-1)AC + (3)DB$

45. ACD

46. CDA

47. DBA

48. BAD

49. If a and b are nonzero real numbers,

$$A = \begin{bmatrix} a & a \\ b & b \end{bmatrix}, \quad \text{and} \quad B = \begin{bmatrix} a & a \\ -a & -a \end{bmatrix}$$

find AB and BA.

50. If a and b are nonzero real numbers,

$$A = \begin{bmatrix} a & b \\ -a & -b \end{bmatrix}, \quad \text{and} \quad B = \begin{bmatrix} a & a \\ a & a \end{bmatrix}$$

find AB and BA.

51. If a and b are nonzero real numbers and

$$A = \begin{bmatrix} ab & b^2 \\ -a^2 & -ab \end{bmatrix}$$

find A^2.

52. If a and b are nonzero real numbers and

$$A = \begin{bmatrix} ab & b - ab^2 \\ a & 1 - ab \end{bmatrix}$$

find A^2.

In Problems 53 and 54, use a graphing calculator to calculate B, B^2, B^3, . . . and AB, AB^2, AB^3, ... Describe any patterns you observe in each sequence of matrices.

53. $A = [0.3 \quad 0.7]$ and $B = \begin{bmatrix} 0.4 & 0.6 \\ 0.2 & 0.8 \end{bmatrix}$

54. $A = [0.4 \quad 0.6]$ and $B = \begin{bmatrix} 0.9 & 0.1 \\ 0.3 & 0.7 \end{bmatrix}$

55. Find a, b, c, and d so that

$$\begin{bmatrix} a & b \\ c & d \end{bmatrix} + \begin{bmatrix} 2 & -3 \\ 0 & 1 \end{bmatrix} = \begin{bmatrix} 1 & -2 \\ 3 & -4 \end{bmatrix}$$

56. Find w, x, y, and z so that

$$\begin{bmatrix} 4 & -2 \\ -3 & 0 \end{bmatrix} + \begin{bmatrix} w & x \\ y & z \end{bmatrix} = \begin{bmatrix} 2 & -3 \\ 0 & 5 \end{bmatrix}$$

57. Find a, b, c, and d so that

$$\begin{bmatrix} 1 & -2 \\ 2 & -3 \end{bmatrix}\begin{bmatrix} a & b \\ c & d \end{bmatrix} = \begin{bmatrix} 1 & 0 \\ 3 & 2 \end{bmatrix}$$

58. Find a, b, c, and d so that

$$\begin{bmatrix} 1 & 3 \\ 1 & 4 \end{bmatrix}\begin{bmatrix} a & b \\ c & d \end{bmatrix} = \begin{bmatrix} 6 & -5 \\ 7 & -7 \end{bmatrix}$$

In Problems 59–62, determine whether the statement is true or false.

59. There exist two 1×1 matrices A and B such that $AB \neq BA$.

60. There exist two 2×2 matrices A and B such that $AB \neq BA$.

61. There exist two nonzero 2×2 matrices A and B such that AB is the 2×2 zero matrix.

62. There exist two nonzero 1×1 matrices A and B such that AB is the 1×1 zero matrix.

63. A square matrix is a **diagonal matrix** if all elements not on the principal diagonal are zero. So a 2×2 diagonal matrix has the form

$$A = \begin{bmatrix} a & 0 \\ 0 & d \end{bmatrix}$$

where a and d are real numbers. Discuss the validity of each of the following statements. If the statement is always true, explain why. If not, give examples.

(A) If A and B are 2×2 diagonal matrices, then $A + B$ is a 2×2 diagonal matrix.

(B) If A and B are 2×2 diagonal matrices, then $A + B = B + A$.

(C) If A and B are 2×2 diagonal matrices, then AB is a 2×2 diagonal matrix.

(D) If A and B are 2×2 diagonal matrices, then $AB = BA$.

✎ 64. A square matrix is an **upper triangular matrix** if all elements below the principal diagonal are zero. So a 2×2 upper triangular matrix has the form

$$A = \begin{bmatrix} a & b \\ 0 & d \end{bmatrix}$$

where a, b, and d are real numbers. Discuss the validity of each of the following statements. If the statement is always true, explain why. If not, give examples.

(A) If A and B are 2×2 upper triangular matrices, then $A + B$ is a 2×2 upper triangular matrix.

(B) If A and B are 2×2 upper triangular matrices, then $A + B = B + A$.

(C) If A and B are 2×2 upper triangular matrices, then AB is a 2×2 upper triangular matrix.

(D) If A and B are 2×2 upper triangular matrices, then $AB = BA$.

Applications

65. **Cost analysis.** A company with two different plants manufactures guitars and banjos. Its production costs for each instrument are given in the following matrices:

	Plant X Guitar	Plant X Banjo		Plant Y Guitar	Plant Y Banjo	
Materials	$47	$39	$= A$	$56	$42	$= B$
Labor	$90	$125		$84	$115	

Find $\frac{1}{2}(A + B)$, the average cost of production for the two plants.

66. **Cost analysis.** If both labor and materials at plant X in Problem 65 are increased by 20%, find $\frac{1}{2}(1.2A + B)$, the new average cost of production for the two plants.

67. **Markup.** An import car dealer sells three models of a car. The retail prices and the current dealer invoice prices (costs) for the basic models and options indicated are given in the following two matrices (where "Air" means air-conditioning):

Retail price

	Basic Car	Air	AM/FM radio	Cruise control	
Model A	$35,075	$2,560	$1,070	$640	
Model B	$39,045	$1,840	$770	$460	$= M$
Model C	$45,535	$3,400	$1,415	$850	

Dealer invoice price

	Basic Car	Air	AM/FM radio	Cruise control	
Model A	$30,996	$2,050	$850	$510	
Model B	$34,857	$1,585	$660	$395	$= N$
Model C	$41,667	$2,890	$1,200	$725	

We define the markup matrix to be $M - N$ (**markup** is the difference between the retail price and the dealer invoice price). Suppose that the value of the dollar has had a sharp decline and the dealer invoice price is to have an across-the-board 15% increase next year. To stay competitive with domestic cars, the dealer increases the retail prices 10%. Calculate a markup matrix for next year's models and the options indicated. (Compute results to the nearest dollar.)

68. **Markup.** Referring to Problem 67, what is the markup matrix resulting from a 20% increase in dealer invoice prices and an increase in retail prices of 15%? (Compute results to the nearest dollar.)

69. **Labor costs.** A company with manufacturing plants located in Massachusetts (MA) and Virginia (VA) has labor-hour and wage requirements for the manufacture of three types of inflatable boats as given in the following two matrices:

Labor-hours per boat

	Cutting department	Assembly department	Packaging department	
$M =$	0.6 hr	0.6 hr	0.2 hr	One-person boat
	1.0 hr	0.9 hr	0.3 hr	Two-person boat
	1.5 hr	1.2 hr	0.4 hr	Four-person boat

Hourly wages

	MA	VA	
$N =$	$17.30	$14.65	Cutting department
	$12.22	$10.29	Assembly department
	$10.63	$9.66	Packaging department

(A) Find the labor costs for a one-person boat manufactured at the Massachusetts plant.

(B) Find the labor costs for a four-person boat manufactured at the Virginia plant.

✎ (C) Discuss possible interpretations of the elements in the matrix products MN and NM.

(D) If either of the products MN or NM has a meaningful interpretation, find the product and label its rows and columns.

70. **Inventory value.** A personal computer retail company sells five different computer models through three stores. The inventory of each model on hand in each store is summarized in matrix M. Wholesale (W) and retail (R) values of each model computer are summarized in matrix N.

Model

	A	B	C	D	E	
$M =$	4	2	3	7	1	Store 1
	2	3	5	0	6	Store 2
	10	4	3	4	3	Store 3

	W	R	
$N =$	$700	$840	A
	$1,400	$1,800	B
	$1,800	$2,400	C Model
	$2,700	$3,300	D
	$3,500	$4,900	E

(A) What is the retail value of the inventory at store 2?

(B) What is the wholesale value of the inventory at store 3?

✎ (C) Discuss possible interpretations of the elements in the matrix products MN and NM.

(D) If either product MN or NM has a meaningful interpretation, find the product and label its rows and columns.

✎ (E) Discuss methods of matrix multiplication that can be used to find the total inventory of each model on hand at all three stores. State the matrices that can be used and perform the necessary operations.

✎ (F) Discuss methods of matrix multiplication that can be used to find the total inventory of all five models at each store. State the matrices that can be used and perform the necessary operations.

71. Cereal. A nutritionist for a cereal company blends two cereals in three different mixes. The amounts of protein, carbohydrate, and fat (in grams per ounce) in each cereal are given by matrix M. The amounts of each cereal used in the three mixes are given by matrix N.

$$M = \begin{matrix} & \text{Cereal A} & \text{Cereal B} \\ & \begin{bmatrix} 4\,\text{g/oz} & 2\,\text{g/oz} \\ 20\,\text{g/oz} & 16\,\text{g/oz} \\ 3\,\text{g/oz} & 1\,\text{g/oz} \end{bmatrix} & \begin{matrix} \text{Protein} \\ \text{Carbohydrate} \\ \text{Fat} \end{matrix} \end{matrix}$$

$$N = \begin{matrix} & \text{Mix X} & \text{Mix Y} & \text{Mix Z} \\ & \begin{bmatrix} 15\,\text{oz} & 10\,\text{oz} & 5\,\text{oz} \\ 5\,\text{oz} & 10\,\text{oz} & 15\,\text{oz} \end{bmatrix} & \begin{matrix} \text{Cereal A} \\ \text{Cereal B} \end{matrix} \end{matrix}$$

(A) Find the amount of protein in mix X.

(B) Find the amount of fat in mix Z.

✎ (C) Discuss possible interpretations of the elements in the matrix products MN and NM.

(D) If either of the products MN or NM has a meaningful interpretation, find the product and label its rows and columns.

✎ **72. Heredity.** Gregor Mendel (1822–1884) made discoveries that revolutionized the science of genetics. In one experiment, he crossed dihybrid yellow round peas (yellow and round are dominant characteristics; the peas also contained genes for the recessive characteristics green and wrinkled) and obtained peas of the types indicated in the matrix:

$$\begin{matrix} & \text{Round} & \text{Wrinkled} \\ \text{Yellow} & \begin{bmatrix} 315 & 101 \\ 108 & 32 \end{bmatrix} = M \\ \text{Green} & \end{matrix}$$

Suppose he carried out a second experiment of the same type and obtained peas of the types indicated in this matrix:

$$\begin{matrix} & \text{Round} & \text{Wrinkled} \\ \text{Yellow} & \begin{bmatrix} 370 & 128 \\ 110 & 36 \end{bmatrix} = N \\ \text{Green} & \end{matrix}$$

If the results of the two experiments are combined, discuss matrix multiplication methods that can be used to find the following quantities. State the matrices that can be used and perform the necessary operations.

(A) The total number of peas in each category

(B) The total number of peas in all four categories

(C) The percentage of peas in each category

73. Politics. In a local California election, a public relations firm promoted its candidate in three ways: telephone calls, house calls, and letters. The cost per contact is given in matrix M, and the number of contacts of each type made in two adjacent cities is given in matrix N.

$$\begin{matrix} & \text{Cost per contact} \\ M = & \begin{bmatrix} \$1.20 \\ \$3.00 \\ \$1.45 \end{bmatrix} & \begin{matrix} \text{Telephone call} \\ \text{House call} \\ \text{Letter} \end{matrix} \end{matrix}$$

$$N = \begin{matrix} & \begin{matrix} \text{Telephone} \\ \text{call} \end{matrix} & \begin{matrix} \text{House} \\ \text{call} \end{matrix} & \text{Letter} \\ & \begin{bmatrix} 1{,}000 & 500 & 5{,}000 \\ 2{,}000 & 800 & 8{,}000 \end{bmatrix} & \begin{matrix} \text{Berkeley} \\ \text{Oakland} \end{matrix} \end{matrix}$$

(A) Find the total amount spent in Berkeley.

(B) Find the total amount spent in Oakland.

✎ (C) Discuss possible interpretations of the elements in the matrix products MN and NM.

(D) If either product MN or NM has a meaningful interpretation, find the product and label its rows and columns.

✎ (E) Discuss methods of matrix multiplication that can be used to find the total number of telephone calls, house calls, and letters. State the matrices that can be used and perform the necessary operations.

✎ (F) Discuss methods of matrix multiplication that can be used to find the total number of contacts in Berkeley and in Oakland. State the matrices that can be used and perform the necessary operations.

✎ **74. Test averages.** A teacher has given four tests to a class of five students and stored the results in the following matrix:

$$\begin{matrix} & & & \text{Tests} \\ & & 1 & 2 & 3 & 4 \\ \text{Ann} & & \begin{bmatrix} 78 & 84 & 81 & 86 \\ 91 & 65 & 84 & 92 \\ 95 & 90 & 92 & 91 \\ 75 & 82 & 87 & 91 \\ 83 & 88 & 81 & 76 \end{bmatrix} = M \\ \text{Bob} \\ \text{Carol} \\ \text{Dan} \\ \text{Eric} \end{matrix}$$

Discuss methods of matrix multiplication that the teacher can use to obtain the information indicated below. In each case, state the matrices to be used and then perform the necessary operations.

(A) The average on all four tests for each student, assuming that all four tests are given equal weight

(B) The average on all four tests for each student, assuming that the first three tests are given equal weight and the fourth is given twice this weight

(C) The class average on each of the four tests

Answers to Matched Problems

1. $\begin{bmatrix} 1 & 5 \\ 0 & -2 \\ 2 & 1 \end{bmatrix}$

2. $[-1 \quad -1 \quad 4]$

3. $a = -6$
$b = 7$
$c = 9$
$d = -1$

4. $\begin{bmatrix} 13 \\ 2 \\ 35 \end{bmatrix}$

5. (A) $\begin{bmatrix} \$235,000 & \$422,000 \\ \$372,000 & \$298,000 \end{bmatrix}$

(B) $\begin{bmatrix} \$145,000 & \$268,000 \\ \$160,000 & \$254,000 \end{bmatrix}$

(C) $\begin{bmatrix} \$9,500 & \$17,250 \\ \$13,300 & \$13,800 \end{bmatrix}$

6. [8]

7. $[5 \quad 1.5]\begin{bmatrix} 9 \\ 6 \end{bmatrix} = [54]$, or \$54

8. (A) Not defined

(B) $\begin{bmatrix} 2 & 2 & -1 & 2 \\ 1 & 6 & 12 & -4 \\ -1 & 0 & 3 & -2 \end{bmatrix}$

(C) $\begin{bmatrix} 0 & 0 \\ 0 & 0 \end{bmatrix}$

(D) $\begin{bmatrix} -6 & -12 \\ 3 & 6 \end{bmatrix}$

(E) [11]

(F) $\begin{bmatrix} 12 & -8 & 4 \\ 6 & -4 & 2 \\ 9 & -6 & 3 \end{bmatrix}$

9. $a = 4, c = 8, b = 9, d = -2$

10. $PL = [13{,}000 \quad 4{,}000]$ Labor-hours, with columns labeled Assembly and Finishing.

4.5 Inverse of a Square Matrix

- Identity Matrix for Multiplication
- Inverse of a Square Matrix
- Application: Cryptography

Identity Matrix for Multiplication

Does the set of all matrices of a given size have an identity element for multiplication? That is, if M is an arbitrary $m \times n$ matrix, does there exist an identity element I such that $IM = MI = M$? The answer, in general, is no. However, the set of all **square matrices of order n** (matrices with n rows and n columns) does have an identity element.

> **DEFINITION** Identity Matrix
> The **identity element for multiplication** for the set of all square matrices of order n is the square matrix of order n, denoted by I, with 1's along the principal diagonal (from the upper left corner to the lower right) and 0's elsewhere.

For example,

$$\begin{bmatrix} 1 & 0 \\ 0 & 1 \end{bmatrix} \quad \text{and} \quad \begin{bmatrix} 1 & 0 & 0 \\ 0 & 1 & 0 \\ 0 & 0 & 1 \end{bmatrix}$$

are the identity matrices for all square matrices of order 2 and 3, respectively.

 Most graphing calculators have a built-in command for generating the identity matrix of a given order (see Fig. 1).

```
identity 2
          [[1 0]
           [0 1]]
identity 3
          [[1 0 0]
           [0 1 0]
           [0 0 1]]
```

Figure 1 **Identity matrices**

EXAMPLE 1 Identity Matrix Multiplication

(A) $\begin{bmatrix} 1 & 0 & 0 \\ 0 & 1 & 0 \\ 0 & 0 & 1 \end{bmatrix} \begin{bmatrix} 3 & -2 & 5 \\ 0 & 2 & -3 \\ -1 & 4 & -2 \end{bmatrix} = \begin{bmatrix} 3 & -2 & 5 \\ 0 & 2 & -3 \\ -1 & 4 & -2 \end{bmatrix}$

(B) $\begin{bmatrix} 3 & -2 & 5 \\ 0 & 2 & -3 \\ -1 & 4 & -2 \end{bmatrix} \begin{bmatrix} 1 & 0 & 0 \\ 0 & 1 & 0 \\ 0 & 0 & 1 \end{bmatrix} = \begin{bmatrix} 3 & -2 & 5 \\ 0 & 2 & -3 \\ -1 & 4 & -2 \end{bmatrix}$

(C) $\begin{bmatrix} 1 & 0 \\ 0 & 1 \end{bmatrix} \begin{bmatrix} 2 & -1 & 3 \\ -2 & 0 & 4 \end{bmatrix} = \begin{bmatrix} 2 & -1 & 3 \\ -2 & 0 & 4 \end{bmatrix}$

(D) $\begin{bmatrix} 2 & -1 & 3 \\ -2 & 0 & 4 \end{bmatrix} \begin{bmatrix} 1 & 0 & 0 \\ 0 & 1 & 0 \\ 0 & 0 & 1 \end{bmatrix} = \begin{bmatrix} 2 & -1 & 3 \\ -2 & 0 & 4 \end{bmatrix}$

Matched Problem 1) Multiply:

(A) $\begin{bmatrix} 1 & 0 \\ 0 & 1 \end{bmatrix} \begin{bmatrix} 2 & -3 \\ 5 & 7 \end{bmatrix}$ and $\begin{bmatrix} 2 & -3 \\ 5 & 7 \end{bmatrix} \begin{bmatrix} 1 & 0 \\ 0 & 1 \end{bmatrix}$

(B) $\begin{bmatrix} 1 & 0 & 0 \\ 0 & 1 & 0 \\ 0 & 0 & 1 \end{bmatrix} \begin{bmatrix} 4 & 2 \\ 3 & -5 \\ 6 & 8 \end{bmatrix}$ and $\begin{bmatrix} 4 & 2 \\ 3 & -5 \\ 6 & 8 \end{bmatrix} \begin{bmatrix} 1 & 0 \\ 0 & 1 \end{bmatrix}$

In general, we can show that if M is a square matrix of order n and I is the identity matrix of order n, then

$$IM = MI = M$$

If M is an $m \times n$ matrix that is not square ($m \neq n$), it is still possible to multiply M on the left and on the right by an identity matrix, but not with the same size identity matrix (see Example 1C and D). To avoid the complications involved with associating two different identity matrices with each nonsquare matrix, we restrict our attention in this section to square matrices.

Explore and Discuss 1 The only real number solutions to the equation $x^2 = 1$ are $x = 1$ and $x = -1$.

(A) Show that $A = \begin{bmatrix} 0 & 1 \\ 1 & 0 \end{bmatrix}$ satisfies $A^2 = I$, where I is the 2×2 identity.

(B) Show that $B = \begin{bmatrix} 0 & -1 \\ -1 & 0 \end{bmatrix}$ satisfies $B^2 = I$.

(C) Find a 2×2 matrix with all elements nonzero whose square is the 2×2 identity matrix.

Inverse of a Square Matrix

If r is an arbitrary real number, then its **additive inverse** is the solution x to the equation $r + x = 0$. So the additive inverse of 3 is -3, and the additive inverse of -7 is 7. Similarly, if M is an arbitrary $m \times n$ matrix, then M has an additive inverse $-M$, whose elements are just the additive inverses of the elements of M.

The situation is more complicated for *multiplicative* inverses. The **multiplicative inverse** of an arbitrary real number r is the solution x to the equation $r \cdot x = 1$. So the multiplicative inverse of 3 is $\dfrac{1}{3}$, and the multiplicative inverse of $\dfrac{-15}{4}$ is $\dfrac{-4}{15}$. Every real number has a multiplicative inverse except for 0. Because the equation $0 \cdot x = 1$ has no real solution, 0 does not have a multiplicative inverse.

Can we extend the multiplicative inverse concept to matrices? That is, given a matrix M, can we find another matrix N such that $MN = NM = I$, the matrix identity for multiplication? To begin, we consider the size of these matrices. Let M be an $n \times m$ matrix and N a $p \times q$ matrix. If both MN and NM are defined, then $m = p$ and $q = n$ (Fig. 2). If $MN = NM$, then $n = p$ and $q = m$ (Fig. 3). Thus, we have $m = p = n = q$. In other words, M and N must be square matrices of the same order. Later we will see that not all square matrices have inverses.

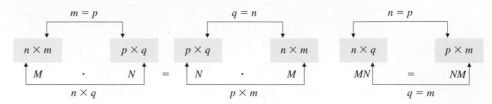

Figure 2   Figure 3

DEFINITION   Inverse of a Square Matrix

Let M be a square matrix of order n and I be the identity matrix of order n. If there exists a matrix M^{-1} (read "M inverse") such that

$$M^{-1}M = MM^{-1} = I$$

then M^{-1} is called the **multiplicative inverse of M** or, more simply, the **inverse of M**. If no such matrix exists, then M is said to be a **singular matrix**.

Let us use the definition above to find M^{-1} for

$$M = \begin{bmatrix} 2 & 3 \\ 1 & 2 \end{bmatrix}$$

We are looking for

$$M^{-1} = \begin{bmatrix} a & c \\ b & d \end{bmatrix}$$

such that

$$MM^{-1} = M^{-1}M = I$$

So we write

$$\overset{M}{\begin{bmatrix} 2 & 3 \\ 1 & 2 \end{bmatrix}} \overset{M^{-1}}{\begin{bmatrix} a & c \\ b & d \end{bmatrix}} = \overset{I}{\begin{bmatrix} 1 & 0 \\ 0 & 1 \end{bmatrix}}$$

and try to find a, b, c, and d so that the product of M and M^{-1} is the identity matrix I. Multiplying M and M^{-1} on the left side, we obtain

$$\begin{bmatrix} (2a + 3b) & (2c + 3d) \\ (a + 2b) & (c + 2d) \end{bmatrix} = \begin{bmatrix} 1 & 0 \\ 0 & 1 \end{bmatrix}$$

which is true only if

$$2a + 3b = 1 \qquad\qquad 2c + 3d = 0$$
$$a + 2b = 0 \qquad\qquad c + 2d = 1$$

Use Gauss–Jordan elimination to solve each system.

$$\begin{bmatrix} 2 & 3 & | & 1 \\ 1 & 2 & | & 0 \end{bmatrix} R_1 \leftrightarrow R_2 \qquad\qquad \begin{bmatrix} 2 & 3 & | & 0 \\ 1 & 2 & | & 1 \end{bmatrix} R_1 \leftrightarrow R_2$$

$$\begin{bmatrix} 1 & 2 & | & 0 \\ 2 & 3 & | & 1 \end{bmatrix} (-2)R_1 + R_2 \rightarrow R_2 \qquad\qquad \begin{bmatrix} 1 & 2 & | & 1 \\ 2 & 3 & | & 0 \end{bmatrix} (-2)R_1 + R_2 \rightarrow R_2$$

$$\begin{bmatrix} 1 & 2 & | & 0 \\ 0 & -1 & | & 1 \end{bmatrix} (-1)R_2 \rightarrow R_2 \qquad\qquad \begin{bmatrix} 1 & 2 & | & 1 \\ 0 & -1 & | & -2 \end{bmatrix} (-1)R_2 \rightarrow R_2$$

$$\begin{bmatrix} 1 & 2 & | & 0 \\ 0 & 1 & | & -1 \end{bmatrix} (-2)R_2 + R_1 \rightarrow R_1 \qquad\qquad \begin{bmatrix} 1 & 2 & | & 1 \\ 0 & 1 & | & 2 \end{bmatrix} (-2)R_2 + R_1 \rightarrow R_1$$

$$\begin{bmatrix} 1 & 0 & | & 2 \\ 0 & 1 & | & -1 \end{bmatrix} \qquad\qquad \begin{bmatrix} 1 & 0 & | & -3 \\ 0 & 1 & | & 2 \end{bmatrix}$$

$$a = 2, b = -1 \qquad\qquad c = -3, d = 2$$

$$M^{-1} = \begin{bmatrix} a & c \\ b & d \end{bmatrix} = \begin{bmatrix} 2 & -3 \\ -1 & 2 \end{bmatrix}$$

CHECK

$$\overset{M}{\begin{bmatrix} 2 & 3 \\ 1 & 2 \end{bmatrix}} \overset{M^{-1}}{\begin{bmatrix} 2 & -3 \\ -1 & 2 \end{bmatrix}} = \overset{I}{\begin{bmatrix} 1 & 0 \\ 0 & 1 \end{bmatrix}} = \overset{M^{-1}}{\begin{bmatrix} 2 & -3 \\ -1 & 2 \end{bmatrix}} \overset{M}{\begin{bmatrix} 2 & 3 \\ 1 & 2 \end{bmatrix}}$$

Unlike nonzero real numbers, inverses do not always exist for square matrices. For example, if

$$N = \begin{bmatrix} 2 & 1 \\ 4 & 2 \end{bmatrix}$$

then, using the previous process, we are led to the systems

$$2a + b = 1 \qquad\qquad 2c + d = 0$$
$$4a + 2b = 0 \qquad\qquad 4c + 2d = 1$$

Use Gauss–Jordan elimination to solve each system.

$$\begin{bmatrix} 2 & 1 & | & 1 \\ 4 & 2 & | & 0 \end{bmatrix} (-2)R_1 + R_2 \rightarrow R_2 \qquad\qquad \begin{bmatrix} 2 & 1 & | & 0 \\ 4 & 2 & | & 1 \end{bmatrix} (-2)R_1 + R_2 \rightarrow R_2$$

$$\begin{bmatrix} 2 & 1 & | & 0 \\ 0 & 0 & | & -2 \end{bmatrix} \qquad\qquad \begin{bmatrix} 2 & 1 & | & 0 \\ 0 & 0 & | & 1 \end{bmatrix}$$

The last row of each augmented matrix contains a contradiction. So each system is inconsistent and has no solution. We conclude that N^{-1} does not exist and N is a singular matrix.

Being able to find inverses, when they exist, leads to direct and simple solutions to many practical problems. In the next section, we show how inverses can be used to solve systems of linear equations.

The method outlined previously for finding M^{-1}, if it exists, gets very involved for matrices of order larger than 2. Now that we know what we are looking for, we can use augmented matrices (see Sections 4.2 and 4.3) to make the process more efficient.

EXAMPLE 2   Finding the Inverse of a Matrix   Find the inverse, if it exists, of the matrix

$$M = \begin{bmatrix} 1 & -1 & 1 \\ 0 & 2 & -1 \\ 2 & 3 & 0 \end{bmatrix}$$

SOLUTION   We start as before and write

$$\overset{M}{\begin{bmatrix} 1 & -1 & 1 \\ 0 & 2 & -1 \\ 2 & 3 & 0 \end{bmatrix}} \overset{M^{-1}}{\begin{bmatrix} a & d & g \\ b & e & h \\ c & f & i \end{bmatrix}} = \overset{I}{\begin{bmatrix} 1 & 0 & 0 \\ 0 & 1 & 0 \\ 0 & 0 & 1 \end{bmatrix}}$$

which is true only if

$$\begin{array}{ccc} a - b + c = 1 & \quad d - e + f = 0 & \quad g - h + i = 0 \\ 2b - c = 0 & \quad 2e - f = 1 & \quad 2h - i = 0 \\ 2a + 3b = 0 & \quad 2d + 3e = 0 & \quad 2g + 3h = 1 \end{array}$$

Now we write augmented matrices for each of the three systems:

$$\overset{\text{First}}{\left[\begin{array}{ccc|c} 1 & -1 & 1 & 1 \\ 0 & 2 & -1 & 0 \\ 2 & 3 & 0 & 0 \end{array}\right]} \quad \overset{\text{Second}}{\left[\begin{array}{ccc|c} 1 & -1 & 1 & 0 \\ 0 & 2 & -1 & 1 \\ 2 & 3 & 0 & 0 \end{array}\right]} \quad \overset{\text{Third}}{\left[\begin{array}{ccc|c} 1 & -1 & 1 & 0 \\ 0 & 2 & -1 & 0 \\ 2 & 3 & 0 & 1 \end{array}\right]}$$

Since each matrix to the left of the vertical bar is the same, exactly the same row operations can be used on each augmented matrix to transform it into a reduced form. We can speed up the process substantially by combining all three augmented matrices into the single augmented matrix form below:

$$\left[\begin{array}{ccc|ccc} 1 & -1 & 1 & 1 & 0 & 0 \\ 0 & 2 & -1 & 0 & 1 & 0 \\ 2 & 3 & 0 & 0 & 0 & 1 \end{array}\right] = [M\,|\,I] \qquad (1)$$

We now try to perform row operations on matrix (1) until we obtain a row-equivalent matrix of the form

$$\left[\begin{array}{ccc|ccc} \overset{I}{1} & 0 & 0 & \overset{B}{a} & d & g \\ 0 & 1 & 0 & b & e & h \\ 0 & 0 & 1 & c & f & i \end{array}\right] = [I\,|\,B] \qquad (2)$$

If this can be done, the new matrix B to the right of the vertical bar will be M^{-1}. Now let's try to transform matrix (1) into a form like matrix (2). We follow the same sequence of steps as we did in the solution of linear systems by Gauss–Jordan elimination (see Section 4.3).

$$
\begin{array}{c}

\left[\begin{array}{ccc|ccc}
1 & -1 & 1 & 1 & 0 & 0 \\
0 & 2 & -1 & 0 & 1 & 0 \\
2 & 3 & 0 & 0 & 0 & 1
\end{array}\right] \quad (-2)R_1 + R_3 \to R_3 \\[24pt]
\sim
\left[\begin{array}{ccc|ccc}
1 & -1 & 1 & 1 & 0 & 0 \\
0 & 2 & -1 & 0 & 1 & 0 \\
0 & 5 & -2 & -2 & 0 & 1
\end{array}\right] \quad \tfrac{1}{2}R_2 \to R_2 \\[24pt]
\sim
\left[\begin{array}{ccc|ccc}
1 & -1 & 1 & 1 & 0 & 0 \\
0 & 1 & -\frac{1}{2} & 0 & \frac{1}{2} & 0 \\
0 & 5 & -2 & -2 & 0 & 1
\end{array}\right]
\begin{array}{l} R_2 + R_1 \to R_1 \\[12pt] (-5)R_2 + R_3 \to R_3 \end{array} \\[24pt]
\sim
\left[\begin{array}{ccc|ccc}
1 & 0 & \frac{1}{2} & 1 & \frac{1}{2} & 0 \\
0 & 1 & -\frac{1}{2} & 0 & \frac{1}{2} & 0 \\
0 & 0 & \frac{1}{2} & -2 & -\frac{5}{2} & 1
\end{array}\right] \quad 2R_3 \to R_3 \\[24pt]
\sim
\left[\begin{array}{ccc|ccc}
1 & 0 & \frac{1}{2} & 1 & \frac{1}{2} & 0 \\
0 & 1 & -\frac{1}{2} & 0 & \frac{1}{2} & 0 \\
0 & 0 & 1 & -4 & -5 & 2
\end{array}\right]
\begin{array}{l} (-\frac{1}{2})R_3 + R_1 \to R_1 \\[12pt] \frac{1}{2}R_3 + R_2 \to R_2 \end{array} \\[24pt]
\sim
\left[\begin{array}{ccc|ccc}
1 & 0 & 0 & 3 & 3 & -1 \\
0 & 1 & 0 & -2 & -2 & 1 \\
0 & 0 & 1 & -4 & -5 & 2
\end{array}\right] = [I \,|\, B]
\end{array}
$$

Converting back to systems of equations equivalent to our three original systems, we have

$$
\begin{array}{ccc}
a = 3 & d = 3 & g = -1 \\
b = -2 & e = -2 & h = 1 \\
c = -4 & f = -5 & i = 2
\end{array}
$$

And these are just the elements of M^{-1} that we are looking for!

$$
M^{-1} = \begin{bmatrix} 3 & 3 & -1 \\ -2 & -2 & 1 \\ -4 & -5 & 2 \end{bmatrix}
$$

Note that this is the matrix to the right of the vertical line in the last augmented matrix. That is, $M^{-1} = B$.

Since the definition of matrix inverse requires that

$$
M^{-1} M = I \quad \text{and} \quad MM^{-1} = I \tag{3}
$$

it appears that we must compute both $M^{-1} M$ and MM^{-1} to check our work. However, it can be shown that if one of the equations in (3) is satisfied, the other is also satisfied. So to check our answer it is sufficient to compute either $M^{-1} M$ or MM^{-1}; we do not need to do both.

CHECK

$$
M^{-1}M = \begin{bmatrix} 3 & 3 & -1 \\ -2 & -2 & 1 \\ -4 & -5 & 2 \end{bmatrix} \begin{bmatrix} 1 & -1 & 1 \\ 0 & 2 & -1 \\ 2 & 3 & 0 \end{bmatrix} = \begin{bmatrix} 1 & 0 & 0 \\ 0 & 1 & 0 \\ 0 & 0 & 1 \end{bmatrix} = I
$$

Matched Problem 2 Let $M = \begin{bmatrix} 3 & -1 & 1 \\ -1 & 1 & 0 \\ 1 & 0 & 1 \end{bmatrix}$.

(A) Form the augmented matrix $[M \,|\, I]$.

(B) Use row operations to transform $[M \,|\, I]$ into $[I \,|\, B]$.

(C) Verify by multiplication that $B = M^{-1}$ (that is, show that $BM = I$).

The procedure shown in Example 2 can be used to find the inverse of any square matrix, if the inverse exists, and will also indicate when the inverse does not exist. These ideas are summarized in Theorem 1.

THEOREM 1 Inverse of a Square Matrix M

If $[M \,|\, I]$ is transformed by row operations into $[I \,|\, B]$, then the resulting matrix B is M^{-1}. However, if we obtain all 0's in one or more rows to the left of the vertical line, then M^{-1} does not exist.

Explore and Discuss 2 (A) Suppose that the square matrix M has a row of all zeros. Explain why M has no inverse.

(B) Suppose that the square matrix M has a column of all zeros. Explain why M has no inverse.

EXAMPLE 3 Finding a Matrix Inverse   Find M^{-1}, given $M = \begin{bmatrix} 4 & -1 \\ -6 & 2 \end{bmatrix}$.

SOLUTION

$$\left[\begin{array}{cc|cc} 4 & -1 & 1 & 0 \\ -6 & 2 & 0 & 1 \end{array}\right] \quad \tfrac{1}{4}R_1 \to R_1$$

$$\sim \left[\begin{array}{cc|cc} 1 & -\tfrac{1}{4} & \tfrac{1}{4} & 0 \\ -6 & 2 & 0 & 1 \end{array}\right] \quad 6R_1 + R_2 \to R_2$$

$$\sim \left[\begin{array}{cc|cc} 1 & -\tfrac{1}{4} & \tfrac{1}{4} & 0 \\ 0 & \tfrac{1}{2} & \tfrac{3}{2} & 1 \end{array}\right] \quad 2R_2 \to R_2$$

$$\sim \left[\begin{array}{cc|cc} 1 & -\tfrac{1}{4} & \tfrac{1}{4} & 0 \\ 0 & 1 & 3 & 2 \end{array}\right] \quad \tfrac{1}{4}R_2 + R_1 \to R_1$$

$$\sim \left[\begin{array}{cc|cc} 1 & 0 & 1 & \tfrac{1}{2} \\ 0 & 1 & 3 & 2 \end{array}\right]$$

Therefore,

$$M^{-1} = \begin{bmatrix} 1 & \tfrac{1}{2} \\ 3 & 2 \end{bmatrix}$$

Check by showing that $M^{-1} M = I$.

Matched Problem 3 Find M^{-1}, given $M = \begin{bmatrix} 2 & -6 \\ 1 & -2 \end{bmatrix}$.

   Most graphing calculators and spreadsheets can compute matrix inverses, as illustrated in Figure 4 for the solution to Example 3.

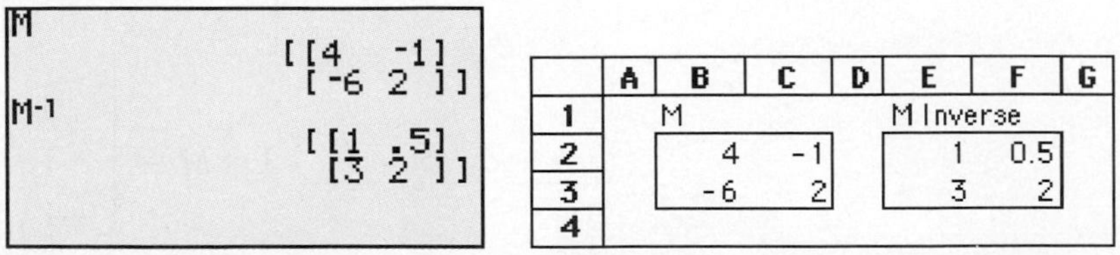

```
M
          [[4  -1]
           [-6  2]]
M-1
          [[1  .5]
           [3  2]]
```

	A	B	C	D	E	F	G
		M				M Inverse	
1							
2		4	-1			1	0.5
3		-6	2			3	2
4							

(A) The command M^{-1} produces the inverse on this graphing calculator

(B) The command MINVERSE (B2:C3) produces the inverse in this spreadsheet

Figure 4 **Finding a matrix inverse**

Explore and Discuss 3 The inverse of

$$A = \begin{bmatrix} a & b \\ c & d \end{bmatrix}$$

is

$$A^{-1} = \begin{bmatrix} \dfrac{d}{ad-bc} & \dfrac{-b}{ad-bc} \\ \dfrac{-c}{ad-bc} & \dfrac{a}{ad-bc} \end{bmatrix} = \dfrac{1}{D}\begin{bmatrix} d & -b \\ -c & a \end{bmatrix} \qquad D = ad - bc$$

provided that $D \neq 0$.

(A) Use matrix multiplication to verify this formula. What can you conclude about A^{-1} if $D = 0$?

(B) Use this formula to find the inverse of matrix M in Example 3.

EXAMPLE 4

Finding a Matrix Inverse Find M^{-1}, given $M = \begin{bmatrix} 2 & -4 \\ -3 & 6 \end{bmatrix}$.

SOLUTION

$$\begin{bmatrix} 2 & -4 & | & 1 & 0 \\ -3 & 6 & | & 0 & 1 \end{bmatrix} \quad \tfrac{1}{2}R_1 \to R_1$$

$$\sim \begin{bmatrix} 1 & -2 & | & \tfrac{1}{2} & 0 \\ -3 & 6 & | & 0 & 1 \end{bmatrix} \quad 3R_1 + R_2 \to R_2$$

$$\sim \begin{bmatrix} 1 & -2 & | & \tfrac{1}{2} & 0 \\ 0 & 0 & | & \tfrac{3}{2} & 1 \end{bmatrix}$$

We have all 0's in the second row to the left of the vertical bar; therefore, the inverse does not exist.

Matched Problem 4 Find N^{-1}, given $N = \begin{bmatrix} 3 & 1 \\ 6 & 2 \end{bmatrix}$.

Square matrices that do not have inverses are called *singular matrices*. Graphing calculators and spreadsheets recognize singular matrices and generally respond with some type of error message, as illustrated in Figure 5 for the solution to Example 4.

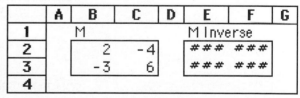

(A) A graphing calculator displays a clear error message

(B) A spreadsheet displays a more cryptic error message

Figure 5

Application: Cryptography

Matrix inverses can provide a simple and effective procedure for encoding and decoding messages. To begin, assign the numbers 1–26 to the letters in the alphabet, as shown below. Also assign the number 0 to a blank to provide for space between words. (A more sophisticated code could include both capital and lowercase letters and punctuation symbols.)

Blank	A	B	C	D	E	F	G	H	I	J	K	L	M	N	O	P	Q	R	S	T	U	V	W	X	Y	Z
0	1	2	3	4	5	6	7	8	9	10	11	12	13	14	15	16	17	18	19	20	21	22	23	24	25	26

The message "SECRET CODE" corresponds to the sequence

$$19 \quad 5 \quad 3 \quad 18 \quad 5 \quad 20 \quad 0 \quad 3 \quad 15 \quad 4 \quad 5$$

Any matrix whose elements are positive integers and whose inverse exists can be used as an **encoding matrix**. For example, to use the 2×2 matrix

$$A = \begin{bmatrix} 4 & 3 \\ 1 & 1 \end{bmatrix}$$

to encode the preceding message, first we divide the numbers in the sequence into groups of 2 and use these groups as the columns of a matrix B with 2 rows:

$$B = \begin{bmatrix} 19 & 3 & 5 & 0 & 15 & 5 \\ 5 & 18 & 20 & 3 & 4 & 0 \end{bmatrix} \quad \text{Proceed down the columns, not across the rows.}$$

Notice that we added an extra blank at the end of the message to make the columns come out even. Then we multiply this matrix on the left by A:

$$AB = \begin{bmatrix} 4 & 3 \\ 1 & 1 \end{bmatrix} \begin{bmatrix} 19 & 3 & 5 & 0 & 15 & 5 \\ 5 & 18 & 20 & 3 & 4 & 0 \end{bmatrix}$$

$$= \begin{bmatrix} 91 & 66 & 80 & 9 & 72 & 20 \\ 24 & 21 & 25 & 3 & 19 & 5 \end{bmatrix}$$

The coded message is

$$91 \quad 24 \quad 66 \quad 21 \quad 80 \quad 25 \quad 9 \quad 3 \quad 72 \quad 19 \quad 20 \quad 5$$

This message can be decoded simply by putting it back into matrix form and multiplying on the left by the **decoding matrix** A^{-1}. Since A^{-1} is easily determined if A is known, the encoding matrix A is the only key needed to decode messages that are encoded in this manner.

EXAMPLE 5   Cryptography   The message

$$46 \quad 84 \quad 85 \quad 28 \quad 47 \quad 46 \quad 4 \quad 5 \quad 10 \quad 30 \quad 48 \quad 72 \quad 29 \quad 57 \quad 38 \quad 38 \quad 57 \quad 95$$

was encoded with the matrix A shown next. Decode this message.

$$A = \begin{bmatrix} 1 & 1 & 1 \\ 2 & 1 & 2 \\ 2 & 3 & 1 \end{bmatrix}$$

SOLUTION   Since the encoding matrix A is 3×3, we begin by entering the coded message in the columns of a matrix C with three rows:

$$C = \begin{bmatrix} 46 & 28 & 4 & 30 & 29 & 38 \\ 84 & 47 & 5 & 48 & 57 & 57 \\ 85 & 46 & 10 & 72 & 38 & 95 \end{bmatrix}$$

If B is the matrix containing the uncoded message, then B and C are related by $C = AB$. To recover B, we find A^{-1} (details omitted) and multiply both sides of the equation $C = AB$ by A^{-1} :

$$B = A^{-1}C$$

$$= \begin{bmatrix} -5 & 2 & 1 \\ 2 & -1 & 0 \\ 4 & -1 & -1 \end{bmatrix} \begin{bmatrix} 46 & 28 & 4 & 30 & 29 & 38 \\ 84 & 47 & 5 & 48 & 57 & 57 \\ 85 & 46 & 10 & 72 & 38 & 95 \end{bmatrix}$$

$$= \begin{bmatrix} 23 & 0 & 0 & 18 & 7 & 19 \\ 8 & 9 & 3 & 12 & 1 & 19 \\ 15 & 19 & 1 & 0 & 21 & 0 \end{bmatrix}$$

Writing the numbers in the columns of this matrix in sequence and using the correspondence between numbers and letters noted earlier produces the decoded message:

23	8	15	0	9	19	0	3	1	18	12	0	7	1	21	19	19	0
W	H	O		I	S		C	A	R	L		G	A	U	S	S	

The answer to this question can be found earlier in this chapter.

Matched Problem 5) The message below was also encoded with the matrix *A* in
Example 5. Decode this message:

46 84 85 28 47 46 32 41 78 25 42 53 25 37 63 43 71 83 19 37 25

Exercises 4.5

Skills Warm-up Exercises

W *In Problems 1–4, find the additive inverse and the multiplicative
inverse, if defined, of each real number. (If necessary, review
Section A.1).*

1. (A) 4   (B) −3   (C) 0

2. (A) −7   (B) 2   (C) −1

3. (A) $\dfrac{2}{3}$ (B) $\dfrac{-1}{7}$ (C) 1.6

4. (A) $\dfrac{4}{5}$ (B) $\dfrac{12}{7}$ (C) −2.5

*In Problems 5–8, does the given matrix have a multiplicative
inverse? Explain your answer.*

5. $\begin{bmatrix} 2 & 5 \end{bmatrix}$

6. $\begin{bmatrix} 4 \\ 8 \end{bmatrix}$

7. $\begin{bmatrix} 0 & 0 \\ 0 & 0 \end{bmatrix}$

8. $\begin{bmatrix} 1 & 0 \\ 0 & 1 \end{bmatrix}$

*In Problems 9–18, find the matrix products. Note that each prod-
uct can be found mentally, without the use of a calculator or
pencil-and-paper calculations.*

9. (A) $\begin{bmatrix} 1 & 0 \\ 0 & 0 \end{bmatrix}\begin{bmatrix} 2 & -3 \\ 4 & 5 \end{bmatrix}$ (B) $\begin{bmatrix} 2 & -3 \\ 4 & 5 \end{bmatrix}\begin{bmatrix} 1 & 0 \\ 0 & 0 \end{bmatrix}$

10. (A) $\begin{bmatrix} 1 & 0 \\ 0 & 0 \end{bmatrix}\begin{bmatrix} -1 & 6 \\ 5 & 2 \end{bmatrix}$ (B) $\begin{bmatrix} -1 & 6 \\ 5 & 2 \end{bmatrix}\begin{bmatrix} 1 & 0 \\ 0 & 0 \end{bmatrix}$

11. (A) $\begin{bmatrix} 0 & 0 \\ 0 & 1 \end{bmatrix}\begin{bmatrix} 2 & -3 \\ 4 & 5 \end{bmatrix}$ (B) $\begin{bmatrix} 2 & -3 \\ 4 & 5 \end{bmatrix}\begin{bmatrix} 0 & 0 \\ 0 & 1 \end{bmatrix}$

12. (A) $\begin{bmatrix} 0 & 0 \\ 0 & 1 \end{bmatrix}\begin{bmatrix} -1 & 6 \\ 5 & 2 \end{bmatrix}$ (B) $\begin{bmatrix} -1 & 6 \\ 5 & 2 \end{bmatrix}\begin{bmatrix} 0 & 0 \\ 0 & 1 \end{bmatrix}$

13. (A) $\begin{bmatrix} 1 & 0 \\ 0 & 1 \end{bmatrix}\begin{bmatrix} 2 & -3 \\ 4 & 5 \end{bmatrix}$ (B) $\begin{bmatrix} 2 & -3 \\ 4 & 5 \end{bmatrix}\begin{bmatrix} 1 & 0 \\ 0 & 1 \end{bmatrix}$

14. (A) $\begin{bmatrix} 1 & 0 \\ 0 & 1 \end{bmatrix}\begin{bmatrix} -1 & 6 \\ 5 & 2 \end{bmatrix}$ (B) $\begin{bmatrix} -1 & 6 \\ 5 & 2 \end{bmatrix}\begin{bmatrix} 1 & 0 \\ 0 & 1 \end{bmatrix}$

15. $\begin{bmatrix} 1 & 0 & 0 \\ 0 & 1 & 0 \\ 0 & 0 & 1 \end{bmatrix}\begin{bmatrix} -2 & 1 & 3 \\ 2 & 4 & -2 \\ 5 & 1 & 0 \end{bmatrix}$

16. $\begin{bmatrix} 1 & 0 & 0 \\ 0 & 1 & 0 \\ 0 & 0 & 1 \end{bmatrix}\begin{bmatrix} 3 & -4 & 0 \\ 1 & 2 & -5 \\ 6 & -3 & -1 \end{bmatrix}$

17. $\begin{bmatrix} -2 & 1 & 3 \\ 2 & 4 & -2 \\ 5 & 1 & 0 \end{bmatrix}\begin{bmatrix} 1 & 0 & 0 \\ 0 & 1 & 0 \\ 0 & 0 & 1 \end{bmatrix}$

18. $\begin{bmatrix} 3 & -4 & 0 \\ 1 & 2 & -5 \\ 6 & -3 & -1 \end{bmatrix}\begin{bmatrix} 1 & 0 & 0 \\ 0 & 1 & 0 \\ 0 & 0 & 1 \end{bmatrix}$

*In Problems 19–28, examine the product of the two matrices to
determine if each is the inverse of the other.*

19. $\begin{bmatrix} 3 & -4 \\ -2 & 3 \end{bmatrix}$; $\begin{bmatrix} 3 & 4 \\ 2 & 3 \end{bmatrix}$

20. $\begin{bmatrix} -2 & -1 \\ -4 & 2 \end{bmatrix}$; $\begin{bmatrix} 1 & -1 \\ 2 & -2 \end{bmatrix}$

21. $\begin{bmatrix} 2 & 2 \\ -1 & -1 \end{bmatrix}$; $\begin{bmatrix} 1 & 1 \\ -1 & -1 \end{bmatrix}$

22. $\begin{bmatrix} 5 & -7 \\ -2 & 3 \end{bmatrix}$; $\begin{bmatrix} 3 & 7 \\ 2 & 5 \end{bmatrix}$

23. $\begin{bmatrix} -5 & 2 \\ -8 & 3 \end{bmatrix}$; $\begin{bmatrix} 3 & -2 \\ 8 & -5 \end{bmatrix}$

24. $\begin{bmatrix} 7 & 4 \\ -5 & -3 \end{bmatrix}$; $\begin{bmatrix} 3 & 4 \\ -5 & -7 \end{bmatrix}$

25. $\begin{bmatrix} 1 & 2 & 0 \\ 0 & 1 & 0 \\ -1 & -1 & 1 \end{bmatrix}$; $\begin{bmatrix} 1 & -2 & 0 \\ 0 & 1 & 0 \\ 1 & -1 & 0 \end{bmatrix}$

26. $\begin{bmatrix} 1 & 0 & 1 \\ -3 & 1 & -2 \\ 0 & 0 & 1 \end{bmatrix}$; $\begin{bmatrix} 1 & 0 & -1 \\ 3 & 1 & -1 \\ 0 & 0 & 1 \end{bmatrix}$

27. $\begin{bmatrix} 1 & -1 & 1 \\ 0 & 2 & -1 \\ 2 & 3 & 0 \end{bmatrix}$; $\begin{bmatrix} 3 & 3 & -1 \\ -2 & -2 & 1 \\ -4 & -5 & 2 \end{bmatrix}$

28. $\begin{bmatrix} 1 & 0 & -1 \\ 3 & 1 & -1 \\ 0 & 0 & 0 \end{bmatrix}$; $\begin{bmatrix} 1 & 0 & -1 \\ -3 & 1 & -2 \\ 0 & 0 & 1 \end{bmatrix}$

✎ *Without performing any row operations, explain why each of the matrices in Problems 29–38 does not have an inverse.*

29. $\begin{bmatrix} 1 & 2 & 0 \\ -3 & 2 & -1 \end{bmatrix}$

30. $\begin{bmatrix} -2 & 3 & -1 \\ 4 & 0 & 1 \end{bmatrix}$

31. $\begin{bmatrix} 1 & -2 \\ 3 & 0 \\ 2 & -1 \end{bmatrix}$

32. $\begin{bmatrix} 0 & -1 \\ 2 & -2 \\ 1 & -3 \end{bmatrix}$

33. $\begin{bmatrix} -1 & 2 \\ 0 & 0 \end{bmatrix}$

34. $\begin{bmatrix} 0 & 0 \\ 1 & 2 \end{bmatrix}$

35. $\begin{bmatrix} 0 & 2 \\ 0 & -1 \end{bmatrix}$

36. $\begin{bmatrix} 1 & 0 \\ 3 & 0 \end{bmatrix}$

37. $\begin{bmatrix} 1 & 2 \\ 3 & 6 \end{bmatrix}$

38. $\begin{bmatrix} -2 & -3 \\ 4 & 6 \end{bmatrix}$

Given M in Problems 39–48, find M^{-1} and show that $M^{-1}M = I$.

39. $\begin{bmatrix} -1 & 0 \\ -3 & 1 \end{bmatrix}$

40. $\begin{bmatrix} 1 & -5 \\ 0 & -1 \end{bmatrix}$

41. $\begin{bmatrix} 1 & 2 \\ 1 & 3 \end{bmatrix}$

42. $\begin{bmatrix} 2 & 1 \\ 5 & 3 \end{bmatrix}$

43. $\begin{bmatrix} 1 & 3 \\ 2 & 7 \end{bmatrix}$

44. $\begin{bmatrix} 2 & 1 \\ 1 & 1 \end{bmatrix}$

45. $\begin{bmatrix} 1 & -3 & 0 \\ 0 & 1 & 1 \\ 2 & -1 & 4 \end{bmatrix}$

46. $\begin{bmatrix} 2 & 3 & 0 \\ 1 & 2 & 3 \\ 0 & -1 & -5 \end{bmatrix}$

47. $\begin{bmatrix} 1 & 1 & 0 \\ 2 & 3 & -1 \\ 1 & 0 & 2 \end{bmatrix}$

48. $\begin{bmatrix} 1 & 0 & -1 \\ 2 & -1 & 0 \\ 1 & 1 & -2 \end{bmatrix}$

Find the inverse of each matrix in Problems 49–54, if it exists.

49. $\begin{bmatrix} 4 & 3 \\ -3 & -2 \end{bmatrix}$

50. $\begin{bmatrix} -4 & 3 \\ -5 & 4 \end{bmatrix}$

51. $\begin{bmatrix} 2 & 6 \\ 3 & 9 \end{bmatrix}$

52. $\begin{bmatrix} 2 & -4 \\ -3 & 6 \end{bmatrix}$

53. $\begin{bmatrix} 2 & 1 \\ 4 & 3 \end{bmatrix}$

54. $\begin{bmatrix} -5 & 3 \\ 2 & -2 \end{bmatrix}$

In Problems 55–60, find the inverse. Note that each inverse can be found mentally, without the use of a calculator or pencil-and-paper calculations.

55. $\begin{bmatrix} 2 & 0 \\ 0 & 2 \end{bmatrix}$

56. $\begin{bmatrix} 4 & 0 \\ 0 & 4 \end{bmatrix}$

57. $\begin{bmatrix} 3 & 0 \\ 0 & -5 \end{bmatrix}$

58. $\begin{bmatrix} -2 & 0 \\ 0 & \frac{1}{2} \end{bmatrix}$

59. $\begin{bmatrix} 4 & 0 & 0 \\ 0 & 2 & 0 \\ 0 & 0 & -8 \end{bmatrix}$

60. $\begin{bmatrix} 3 & 0 & 0 \\ 0 & -6 & 0 \\ 0 & 0 & -5 \end{bmatrix}$

Find the inverse of each matrix in Problems 61–68, if it exists.

61. $\begin{bmatrix} -5 & -2 & -2 \\ 2 & 1 & 0 \\ 1 & 0 & 1 \end{bmatrix}$

62. $\begin{bmatrix} 2 & -2 & 4 \\ 1 & 1 & 1 \\ 1 & 0 & 1 \end{bmatrix}$

63. $\begin{bmatrix} 2 & 1 & 1 \\ 1 & 1 & 0 \\ -1 & -1 & 0 \end{bmatrix}$

64. $\begin{bmatrix} 1 & -1 & 0 \\ 2 & -1 & 1 \\ 0 & 1 & 1 \end{bmatrix}$

65. $\begin{bmatrix} -1 & -2 & 2 \\ 4 & 3 & 0 \\ 4 & 0 & 4 \end{bmatrix}$

66. $\begin{bmatrix} 4 & 2 & 2 \\ 4 & 2 & 0 \\ 5 & 0 & 5 \end{bmatrix}$

67. $\begin{bmatrix} 2 & -1 & -2 \\ -4 & 2 & 8 \\ 6 & -2 & -1 \end{bmatrix}$

68. $\begin{bmatrix} -1 & -1 & 4 \\ 3 & 3 & -22 \\ -2 & -1 & 19 \end{bmatrix}$

69. Show that $(A^{-1})^{-1} = A$ for: $A = \begin{bmatrix} 4 & 3 \\ 3 & 2 \end{bmatrix}$

70. Show that $(AB)^{-1} = B^{-1}A^{-1}$ for

$$A = \begin{bmatrix} 4 & 3 \\ 3 & 2 \end{bmatrix} \quad \text{and} \quad B = \begin{bmatrix} 2 & 5 \\ 3 & 7 \end{bmatrix}$$

✎ **71.** Discuss the existence of M^{-1} for 2×2 diagonal matrices of the form

$$M = \begin{bmatrix} a & 0 \\ 0 & d \end{bmatrix}$$

Generalize your conclusions to $n \times n$ diagonal matrices.

✎ **72.** Discuss the existence of M^{-1} for 2×2 upper triangular matrices of the form

$$M = \begin{bmatrix} a & b \\ 0 & d \end{bmatrix}$$

Generalize your conclusions to $n \times n$ upper triangular matrices.

In Problems 73–75, find A^{-1} and A^2.

73. $A = \begin{bmatrix} 3 & 2 \\ -4 & -3 \end{bmatrix}$ **74.** $A = \begin{bmatrix} -2 & -1 \\ 3 & 2 \end{bmatrix}$

75. $A = \begin{bmatrix} 4 & 3 \\ -5 & -4 \end{bmatrix}$

76. Based on your observations in Problems 73–75, if $A = A^{-1}$ for a square matrix A, what is A^2? Give a mathematical argument to support your conclusion.

Applications

Problems 77–80 refer to the encoding matrix

$$A = \begin{bmatrix} 1 & 2 \\ 1 & 3 \end{bmatrix}$$

77. Cryptography. Encode the message "WINGARDIUM LEVIOSA" using matrix A.

78. Cryptography. Encode the message "FINITE INCANTATEM" using matrix A.

79. Cryptography. The following message was encoded with matrix A. Decode this message:

52 70 17 21 5   5 29 43 4 4 52 70 25
35 29 33 15 18 5 5

80. Cryptography. The following message was encoded with matrix A. Decode this message:

36 44 5 5 38 56 55 75 18 23 56 75
22 33 37 55 27 40 53 79 59 81

Problems 81–84 require the use of a graphing calculator or a computer. Use the 4×4 encoding matrix B given below. Form a matrix with 4 rows and as many columns as necessary to accommodate the message.

$$B = \begin{bmatrix} 2 & 2 & 1 & 3 \\ 1 & 2 & 2 & 1 \\ 1 & 1 & 0 & 1 \\ 2 & 3 & 2 & 3 \end{bmatrix}$$

81. Cryptography. Encode the message "DEPART ISTANBUL ORIENT EXPRESS" using matrix B.

82. Cryptography. Encode the message "SAIL FROM LISBON IN MORNING" using matrix B.

83. Cryptography. The following message was encoded with matrix B. Decode this message:

85 74 27 109 31 27 13 40 139 73 58 154
61 70 18 93 69 59 23 87 18 13 9 22

84. Cryptography. The following message was encoded with matrix B. Decode this message:

75 61 28 94 35 22 13 40 49 21 16 52
42 45 19 64 38 55 10 65 69 75 24 102
67 49 19 82 10 5 5 10

Problems 85–88 require the use of a graphing calculator or a computer. Use the 5×5 encoding matrix C given below. Form a matrix with 5 rows and as many columns as necessary to accommodate the message.

$$C = \begin{bmatrix} 1 & 0 & 1 & 0 & 1 \\ 0 & 1 & 1 & 0 & 3 \\ 2 & 1 & 1 & 1 & 1 \\ 0 & 0 & 1 & 0 & 2 \\ 1 & 1 & 1 & 2 & 1 \end{bmatrix}$$

85. Cryptography. Encode the message "THE EAGLE HAS LANDED" using matrix C.

86. Cryptography. Encode the message "ONE IF BY LAND AND TWO IF BY SEA" using matrix C.

87. Cryptography. The following message was encoded with matrix C. Decode this message:

37 72 58 45 56 30 67 50 46 60 27 77
41 45 39 28 24 52 14 37 32 58 70 36
76 22 38 70 12 67

88. Cryptography. The following message was encoded with matrix C. Decode this message:

25 75 55 35 50 43 83 54 60 53 25 13
59 9 53 15 35 40 15 45 33 60 60 36
51 15 7 37 0 22

Answers to Matched Problems

1. (A) $\begin{bmatrix} 2 & -3 \\ 5 & 7 \end{bmatrix}$ **(B)** $\begin{bmatrix} 4 & 2 \\ 3 & -5 \\ 6 & 8 \end{bmatrix}$

2. (A) $\left[\begin{array}{rrr|rrr} 3 & -1 & 1 & 1 & 0 & 0 \\ -1 & 1 & 0 & 0 & 1 & 0 \\ 1 & 0 & 1 & 0 & 0 & 1 \end{array}\right]$

(B) $\left[\begin{array}{rrr|rrr} 1 & 0 & 0 & 1 & 1 & -1 \\ 0 & 1 & 0 & 1 & 2 & -1 \\ 0 & 0 & 1 & -1 & -1 & 2 \end{array}\right]$

(C) $\begin{bmatrix} 1 & 1 & -1 \\ 1 & 2 & -1 \\ -1 & -1 & 2 \end{bmatrix}\begin{bmatrix} 3 & -1 & 1 \\ -1 & 1 & 0 \\ 1 & 0 & 1 \end{bmatrix} = \begin{bmatrix} 1 & 0 & 0 \\ 0 & 1 & 0 \\ 0 & 0 & 1 \end{bmatrix}$

3. $\begin{bmatrix} -1 & 3 \\ -\frac{1}{2} & 1 \end{bmatrix}$

4. Does not exist

5. WHO IS WILHELM JORDAN

4.6   Matrix Equations and Systems of Linear Equations

- Matrix Equations
- Matrix Equations and Systems of Linear Equations
- Application

The identity matrix and inverse matrix discussed in the preceding section can be put to immediate use in the solution of certain simple matrix equations. Being able to solve a matrix equation gives us another important method of solving systems of equations, provided that the system is independent and has the same number of variables as equations. If the system is dependent or if it has either fewer or more variables than equations, we must return to the Gauss–Jordan method of elimination.

Matrix Equations

Solving simple matrix equations is similar to solving real number equations but with two important differences:

1. there is *no* operation of division for matrices, and

2. matrix multiplication is *not* commutative.

Compare the real number equation $4x = 9$ and the matrix equation $AX = B$. The real number equation can be solved by dividing both sides of the equation by 4. However, that approach cannot be used for $AX = B$, because there is no operation of division for matrices. Instead, we note that $4x = 9$ can be solved by multiplying both sides of the equation by $\frac{1}{4}$, the multiplicative inverse of 4. So we solve $AX = B$ by multiplying both sides of the equation, *on the left*, by A^{-1}, the inverse of A. Because matrix multiplication is not commutative, multiplying both sides of an equation on the left by A^{-1} is different from multiplying both sides of an equation on the right by A^{-1}. In the case of $AX = B$, it is multiplication on the left that is required. The details are presented in Example 1.

In solving matrix equations, we will be guided by the properties of matrices summarized in Theorem 1.

THEOREM 1   Basic Properties of Matrices

Assuming that all products and sums are defined for the indicated matrices A, B, C, I, and 0, then

Addition Properties

Associative:	$(A + B) + C = A + (B + C)$
Commutative:	$A + B = B + A$
Additive identity:	$A + 0 = 0 + A = A$
Additive inverse:	$A + (-A) = (-A) + A = 0$

Multiplication Properties

Associative property:	$A(BC) = (AB)C$
Multiplicative identity:	$AI = IA = A$
Multiplicative inverse:	If A is a square matrix and A^{-1} exists, then $AA^{-1} = A^{-1}A = I$.

Combined Properties

Left distributive:	$A(B + C) = AB + AC$
Right distributive:	$(B + C)A = BA + CA$

Equality

Addition:	If $A = B$, then $A + C = B + C$.
Left multiplication:	If $A = B$, then $CA = CB$.
Right multiplication:	If $A = B$, then $AC = BC$.

EXAMPLE 1 Solving a Matrix Equation Given an $n \times n$ matrix A and $n \times 1$ column matrices B and X, solve $AX = B$ for X. Assume that all necessary inverses exist.

SOLUTION We are interested in finding a column matrix X that satisfies the matrix equation $AX = B$. To solve this equation, we multiply both sides on the left by A^{-1} to isolate X on the left side.

$$AX = B \qquad \text{Use the left multiplication property.}$$
$$A^{-1}(AX) = A^{-1}B \qquad \text{Use the associative property.}$$
$$(A^{-1}A)X = A^{-1}B \qquad A^{-1}A = I$$
$$IX = A^{-1}B \qquad IX = X$$
$$X = A^{-1}B$$

Matched Problem 1 Given an $n \times n$ matrix A and $n \times 1$ column matrices B, C, and X, solve $AX + C = B$ for X. Assume that all necessary inverses exist.

⚠ **CAUTION** Do not mix the left multiplication property and the right multiplication property. If $AX = B$, then

$$A^{-1}(AX) \neq BA^{-1}$$ ▲

Matrix Equations and Systems of Linear Equations

Now we show how independent systems of linear equations with the same number of variables as equations can be solved. First, convert the system into a matrix equation of the form $AX = B$, and then use $X = A^{-1}B$ as obtained in Example 1.

EXAMPLE 2 Using Inverses to Solve Systems of Equations Use matrix inverse methods to solve the system:

$$\begin{aligned} x_1 - x_2 + x_3 &= 1 \\ 2x_2 - x_3 &= 1 \\ 2x_1 + 3x_2 \quad\ &= 1 \end{aligned} \qquad (1)$$

SOLUTION The inverse of the coefficient matrix

$$A = \begin{bmatrix} 1 & -1 & 1 \\ 0 & 2 & -1 \\ 2 & 3 & 0 \end{bmatrix}$$

provides an efficient method for solving this system. To see how, we convert system (1) into a matrix equation:

$$\overset{A}{\begin{bmatrix} 1 & -1 & 1 \\ 0 & 2 & -1 \\ 2 & 3 & 0 \end{bmatrix}} \overset{X}{\begin{bmatrix} x_1 \\ x_2 \\ x_3 \end{bmatrix}} = \overset{B}{\begin{bmatrix} 1 \\ 1 \\ 1 \end{bmatrix}} \qquad (2)$$

Check that matrix equation (2) is equivalent to system (1) by finding the product of the left side and then equating corresponding elements on the left with those on the right.

We are interested in finding a column matrix X that satisfies the matrix equation $AX = B$. In Example 1 we found that if A^{-1} exists, then

$$X = A^{-1}B$$

The inverse of A was found in Example 2, Section 4.5, to be

$$A^{-1} = \begin{bmatrix} 3 & 3 & -1 \\ -2 & -2 & 1 \\ -4 & -5 & 2 \end{bmatrix}$$

Therefore,

$$\underset{X}{\begin{bmatrix} x_1 \\ x_2 \\ x_3 \end{bmatrix}} = \underset{A^{-1}}{\begin{bmatrix} 3 & 3 & -1 \\ -2 & -2 & 1 \\ -4 & -5 & 2 \end{bmatrix}} \underset{B}{\begin{bmatrix} 1 \\ 1 \\ 1 \end{bmatrix}} = \begin{bmatrix} 5 \\ -3 \\ -7 \end{bmatrix}$$

and we can conclude that $x_1 = 5, x_2 = -3$, and $x_3 = -7$. Check this result in system (1).

Matched Problem 2 Use matrix inverse methods to solve the system:

$$\begin{aligned} 3x_1 - x_2 + x_3 &= 1 \\ -x_1 + x_2 \phantom{{}+ x_3} &= 3 \\ x_1 \phantom{{}+ x_2} + x_3 &= 2 \end{aligned}$$

[*Note:* The inverse of the coefficient matrix was found in Matched Problem 2, Section 4.5.]

At first glance, using matrix inverse methods seems to require the same amount of effort as using Gauss–Jordan elimination. In either case, row operations must be applied to an augmented matrix involving the coefficients of the system. The advantage of the inverse matrix method becomes readily apparent when solving a number of systems with a common coefficient matrix and different constant terms.

EXAMPLE 3 Using Inverses to Solve Systems of Equations Use matrix inverse methods to solve each of the following systems:

(A) $\begin{aligned} x_1 - x_2 + x_3 &= 3 \\ 2x_2 - x_3 &= 1 \\ 2x_1 + 3x_2 \phantom{{}- x_3} &= 4 \end{aligned}$ (B) $\begin{aligned} x_1 - x_2 + x_3 &= -5 \\ 2x_2 - x_3 &= 2 \\ 2x_1 + 3x_2 \phantom{{}- x_3} &= -3 \end{aligned}$

SOLUTION Notice that both systems have the same coefficient matrix A as system (1) in Example 2. Only the constant terms have changed. We can use A^{-1} to solve these systems just as we did in Example 2.

(A)

$$\underset{X}{\begin{bmatrix} x_1 \\ x_2 \\ x_3 \end{bmatrix}} = \underset{A^{-1}}{\begin{bmatrix} 3 & 3 & -1 \\ -2 & -2 & 1 \\ -4 & -5 & 2 \end{bmatrix}} \underset{B}{\begin{bmatrix} 3 \\ 1 \\ 4 \end{bmatrix}} = \begin{bmatrix} 8 \\ -4 \\ -9 \end{bmatrix}$$

$x_1 = 8, x_2 = -4$, and $x_3 = -9$.

(B)

$$\underset{X}{\begin{bmatrix} x_1 \\ x_2 \\ x_3 \end{bmatrix}} = \underset{A^{-1}}{\begin{bmatrix} 3 & 3 & -1 \\ -2 & -2 & 1 \\ -4 & -5 & 2 \end{bmatrix}} \underset{B}{\begin{bmatrix} -5 \\ 2 \\ -3 \end{bmatrix}} = \begin{bmatrix} -6 \\ 3 \\ 4 \end{bmatrix}$$

$x_1 = -6, x_2 = 3$, and $x_3 = 4$.

Matched Problem 3 Use matrix inverse methods to solve each of the following systems (see Matched Problem 2):

(A) $\begin{aligned} 3x_1 - x_2 + x_3 &= 3 \\ -x_1 + x_2 \phantom{{}+ x_3} &= -3 \\ x_1 + x_3 &= 2 \end{aligned}$ (B) $\begin{aligned} 3x_1 - x_2 + x_3 &= -5 \\ -x_1 + x_2 \phantom{{}+ x_3} &= 1 \\ x_1 \phantom{{}+ x_2} + x_3 &= -4 \end{aligned}$

As Examples 2 and 3 illustrate, inverse methods are very convenient for hand calculations because once the inverse is found, it can be used to solve any new system formed by changing only the constant terms. Since most graphing calculators and computers can compute the inverse of a matrix, this method also adapts readily to graphing calculator and spreadsheet solutions (Fig. 1). However, if your graphing calculator (or spreadsheet) also has a built-in procedure for finding the reduced form of an augmented matrix, it is just as convenient to use Gauss–Jordan elimination. Furthermore, Gauss–Jordan elimination can be used in all cases and, as noted below, matrix inverse methods cannot always be used.

	A	B	C	D	E	F	G	H	I	J
1		A				B	X		B	X
2		1	-1	1		3	8		-5	-6
3		0	2	-1		1	-4		2	3
4		2	3	0		4	-9		-3	4

Figure 1 **Using inverse methods on a spreadsheet: The values in G2:G4 are produced by the command MMULT (MINVERSE(B2:D4),F2:F4)**

SUMMARY Using Inverse Methods to Solve Systems of Equations
If the number of equations in a system equals the number of variables and the coefficient matrix has an inverse, then the system will always have a unique solution that can be found by using the inverse of the coefficient matrix to solve the corresponding matrix equation.

Matrix equation   Solution
$$AX = B \qquad X = A^{-1}B$$

CONCEPTUAL INSIGHT

There are two cases where inverse methods will not work:

Case 1. The coefficient matrix is singular.
Case 2. The number of variables is not the same as the number of equations.

In either case, use Gauss–Jordan elimination.

Application

The following application illustrates the usefulness of the inverse matrix method for solving systems of equations.

EXAMPLE 4 Investment Analysis An investment advisor currently has two types of investments available for clients: a conservative investment A that pays 10% per year and a higher risk investment B that pays 20% per year. Clients may divide their investments between the two to achieve any total return desired between 10% and 20%. However, the higher the desired return, the higher the risk. How should each client invest to achieve the indicated return?

	Client			
	1	2	3	k
Total investment	$20,000	$50,000	$10,000	k_1
Annual return desired	$ 2,400	$ 7,500	$ 1,300	k_2
	(12%)	(15%)	(13%)	

SOLUTION The answer to this problem involves six quantities, two for each client. Utilizing inverse matrices provides an efficient way to find these quantities. We will solve the problem for an arbitrary client k with unspecified amounts k_1 for the total investment and k_2 for the annual return. (Do not confuse k_1 and k_2 with variables. Their values are known—they just differ for each client.)

$$\text{Let} \quad x_1 = \text{amount invested in } A \text{ by a given client}$$
$$x_2 = \text{amount invested in } B \text{ by a given client}$$

Then we have the following mathematical model:

$$x_1 + \quad x_2 = k_1 \quad \text{Total invested}$$
$$0.1x_1 + 0.2x_2 = k_2 \quad \text{Total annual return desired}$$

Write as a matrix equation:

$$\overset{A}{\begin{bmatrix} 1 & 1 \\ 0.1 & 0.2 \end{bmatrix}} \overset{X}{\begin{bmatrix} x_1 \\ x_2 \end{bmatrix}} = \overset{B}{\begin{bmatrix} k_1 \\ k_2 \end{bmatrix}}$$

If A^{-1} exists, then

$$X = A^{-1}B$$

We now find A^{-1} by starting with the augmented matrix $[A \,|\, I]$ and proceeding as discussed in Section 4.5:

$$\begin{bmatrix} 1 & 1 & | & 1 & 0 \\ 0.1 & 0.2 & | & 0 & 1 \end{bmatrix} \quad 10R_2 \rightarrow R_2$$

$$\sim \begin{bmatrix} 1 & 1 & | & 1 & 0 \\ 1 & 2 & | & 0 & 10 \end{bmatrix} \quad (-1)R_1 + R_2 \rightarrow R_2$$

$$\sim \begin{bmatrix} 1 & 1 & | & 1 & 0 \\ 0 & 1 & | & -1 & 10 \end{bmatrix} \quad (-1)R_2 + R_1 \rightarrow R_1$$

$$\sim \begin{bmatrix} 1 & 0 & | & 2 & -10 \\ 0 & 1 & | & -1 & 10 \end{bmatrix}$$

Therefore,

$$A^{-1} = \begin{bmatrix} 2 & -10 \\ -1 & 10 \end{bmatrix} \qquad \text{Check:} \quad \overset{A^{-1}}{\begin{bmatrix} 2 & -10 \\ -1 & 10 \end{bmatrix}} \overset{A}{\begin{bmatrix} 1 & 1 \\ 0.1 & 0.2 \end{bmatrix}} = \overset{I}{\begin{bmatrix} 1 & 0 \\ 0 & 1 \end{bmatrix}}$$

and

$$\overset{X}{\begin{bmatrix} x_1 \\ x_2 \end{bmatrix}} = \overset{A^{-1}}{\begin{bmatrix} 2 & -10 \\ -1 & 10 \end{bmatrix}} \overset{B}{\begin{bmatrix} k_1 \\ k_2 \end{bmatrix}}$$

To solve each client's investment problem, we replace k_1 and k_2 with appropriate values from the table and multiply by A^{-1}:

Client 1

$$\begin{bmatrix} x_1 \\ x_2 \end{bmatrix} = \begin{bmatrix} 2 & -10 \\ -1 & 10 \end{bmatrix} \begin{bmatrix} 20{,}000 \\ 2{,}400 \end{bmatrix} = \begin{bmatrix} 16{,}000 \\ 4{,}000 \end{bmatrix}$$

Solution: $x_1 = \$16{,}000$ in investment A, $x_2 = \$4{,}000$ in investment B

Client 2

$$\begin{bmatrix} x_1 \\ x_2 \end{bmatrix} = \begin{bmatrix} 2 & -10 \\ -1 & 10 \end{bmatrix} \begin{bmatrix} 50{,}000 \\ 7{,}500 \end{bmatrix} = \begin{bmatrix} 25{,}000 \\ 25{,}000 \end{bmatrix}$$

Solution: $x_1 = \$25{,}000$ in investment A, $x_2 = \$25{,}000$ in investment B

Client 3

$$\begin{bmatrix} x_1 \\ x_2 \end{bmatrix} = \begin{bmatrix} 2 & -10 \\ -1 & 10 \end{bmatrix} \begin{bmatrix} 10{,}000 \\ 1{,}300 \end{bmatrix} = \begin{bmatrix} 7{,}000 \\ 3{,}000 \end{bmatrix}$$

Solution: $x_1 = \$7{,}000$ in investment A, $x_2 = \$3{,}000$ in investment B

Matched Problem 4) Repeat Example 4 with investment A paying 8% and investment B paying 24%.

Figure 2 illustrates a solution to Example 4 on a spreadsheet.

	A	B	C	D	E	F	G
1			Clients				
2		1	2	3		A	
3	Total Investment	$20,000	$50,000	$10,000		1	1
4	Annual Return	$2,400	$7,500	$1,300		0.1	0.2
5	Amount Invested in A	$16,000	$25,000	$7,000			
6	Amount Invested in B	$4,000	$25,000	$3,000			

Figure 2

Explore and Discuss 1   Refer to the mathematical model in Example 4:

$$\overset{A}{\begin{bmatrix} 1 & 1 \\ 0.1 & 0.2 \end{bmatrix}} \overset{X}{\begin{bmatrix} x_1 \\ x_2 \end{bmatrix}} = \overset{B}{\begin{bmatrix} k_1 \\ k_2 \end{bmatrix}} \tag{3}$$

(A) Does matrix equation (3) always have a solution for any constant matrix B?

(B) Do all these solutions make sense for the original problem? If not, give examples.

(C) If the total investment is $k_1 = \$10{,}000$, describe all possible annual returns k_2.

Exercises 4.6

Skills Warm-up Exercises

W   *In Problems 1–8, solve each equation for x, where x represents a real number. (If necessary, review Section 1.1).*

1. $5x = -3$

2. $4x = 9$

3. $4x = 8x + 7$

4. $6x = -3x + 14$

5. $6x + 8 = -2x + 17$

6. $-4x + 3 = 5x + 12$

7. $10 - 3x = 7x + 9$

8. $2x + 7x + 1 = 8x + 3 - x$

Write Problems 9–12 as systems of linear equations without matrices.

9. $\begin{bmatrix} 3 & 1 \\ 2 & -1 \end{bmatrix} \begin{bmatrix} x_1 \\ x_2 \end{bmatrix} = \begin{bmatrix} 5 \\ -4 \end{bmatrix}$

10. $\begin{bmatrix} -2 & 1 \\ -3 & 4 \end{bmatrix} \begin{bmatrix} x_1 \\ x_2 \end{bmatrix} = \begin{bmatrix} -5 \\ 7 \end{bmatrix}$

11. $\begin{bmatrix} -3 & 1 & 0 \\ 2 & 0 & 1 \\ -1 & 3 & -2 \end{bmatrix} \begin{bmatrix} x_1 \\ x_2 \\ x_3 \end{bmatrix} = \begin{bmatrix} 3 \\ -4 \\ 2 \end{bmatrix}$

12. $\begin{bmatrix} 2 & -1 & 0 \\ -2 & 3 & -1 \\ 4 & 0 & 3 \end{bmatrix} \begin{bmatrix} x_1 \\ x_2 \\ x_3 \end{bmatrix} = \begin{bmatrix} 6 \\ -4 \\ 7 \end{bmatrix}$

Write each system in Problems 13–16 as a matrix equation of the form AX = B.

13. $\begin{aligned} 3x_1 - 4x_2 &= 1 \\ 2x_1 + x_2 &= 5 \end{aligned}$

14. $\begin{aligned} 2x_1 + x_2 &= 8 \\ -5x_1 + 3x_2 &= -4 \end{aligned}$

15. $\begin{aligned} x_1 - 3x_2 + 2x_3 &= -3 \\ -2x_1 + 3x_2 &= 1 \\ x_1 + x_2 + 4x_3 &= -2 \end{aligned}$

16. $\begin{aligned} 3x_1 + 2x_3 &= 9 \\ -x_1 + 4x_2 + x_3 &= -7 \\ -2x_1 + 3x_2 &= 6 \end{aligned}$

Find x_1 and x_2 in Problems 17–20.

17. $\begin{bmatrix} x_1 \\ x_2 \end{bmatrix} = \begin{bmatrix} 3 & -2 \\ 1 & 4 \end{bmatrix} \begin{bmatrix} -2 \\ 1 \end{bmatrix}$

18. $\begin{bmatrix} x_1 \\ x_2 \end{bmatrix} = \begin{bmatrix} -2 & 1 \\ -1 & 2 \end{bmatrix} \begin{bmatrix} 3 \\ -2 \end{bmatrix}$

19. $\begin{bmatrix} x_1 \\ x_2 \end{bmatrix} = \begin{bmatrix} -2 & 3 \\ 2 & -1 \end{bmatrix} \begin{bmatrix} 3 \\ 2 \end{bmatrix}$

20. $\begin{bmatrix} x_1 \\ x_2 \end{bmatrix} = \begin{bmatrix} 3 & -1 \\ 0 & 2 \end{bmatrix} \begin{bmatrix} -2 \\ 1 \end{bmatrix}$

In Problems 21–24, find x_1 and x_2.

21. $\begin{bmatrix} 1 & -1 \\ 1 & -2 \end{bmatrix} \begin{bmatrix} x_1 \\ x_2 \end{bmatrix} = \begin{bmatrix} 5 \\ 7 \end{bmatrix}$ **22.** $\begin{bmatrix} 1 & 3 \\ 1 & 4 \end{bmatrix} \begin{bmatrix} x_1 \\ x_2 \end{bmatrix} = \begin{bmatrix} 9 \\ 6 \end{bmatrix}$

23. $\begin{bmatrix} 1 & 1 \\ 2 & -3 \end{bmatrix} \begin{bmatrix} x_1 \\ x_2 \end{bmatrix} = \begin{bmatrix} 15 \\ 10 \end{bmatrix}$ **24.** $\begin{bmatrix} 1 & 1 \\ 3 & -2 \end{bmatrix} \begin{bmatrix} x_1 \\ x_2 \end{bmatrix} = \begin{bmatrix} 10 \\ 20 \end{bmatrix}$

In Problems 25–30, solve for x_1 and x_2.

25. $\begin{bmatrix} 1 & 2 \\ 1 & 1 \end{bmatrix} \begin{bmatrix} x_1 \\ x_2 \end{bmatrix} + \begin{bmatrix} 3 \\ 4 \end{bmatrix} = \begin{bmatrix} 9 \\ 9 \end{bmatrix}$

26. $\begin{bmatrix} 3 & 1 \\ 2 & 1 \end{bmatrix} \begin{bmatrix} x_1 \\ x_2 \end{bmatrix} + \begin{bmatrix} 4 \\ 1 \end{bmatrix} = \begin{bmatrix} 7 \\ 8 \end{bmatrix}$

27. $\begin{bmatrix} 2 & 2 \\ 2 & 3 \end{bmatrix} \begin{bmatrix} x_1 \\ x_2 \end{bmatrix} + \begin{bmatrix} -5 \\ 2 \end{bmatrix} = \begin{bmatrix} 2 \\ 3 \end{bmatrix}$

28. $\begin{bmatrix} 3 & -4 \\ -6 & 8 \end{bmatrix} \begin{bmatrix} x_1 \\ x_2 \end{bmatrix} + \begin{bmatrix} 1 \\ 0 \end{bmatrix} = \begin{bmatrix} 2 \\ 1 \end{bmatrix}$

29. $\begin{bmatrix} 3 & -2 \\ 6 & -4 \end{bmatrix} \begin{bmatrix} x_1 \\ x_2 \end{bmatrix} + \begin{bmatrix} 0 \\ -3 \end{bmatrix} = \begin{bmatrix} 1 \\ -2 \end{bmatrix}$

30. $\begin{bmatrix} 3 & -1 \\ 6 & -4 \end{bmatrix} \begin{bmatrix} x_1 \\ x_2 \end{bmatrix} + \begin{bmatrix} -2 \\ -3 \end{bmatrix} = \begin{bmatrix} -2 \\ -3 \end{bmatrix}$

In Problems 31–38, write each system as a matrix equation and solve using inverses. [Note: The inverses were found in Problems 41–48, Exercises 4.5.]

31. $x_1 + 2x_2 = k_1$
$x_1 + 3x_2 = k_2$
(A) $k_1 = 1, k_2 = 3$
(B) $k_1 = 3, k_2 = 5$
(C) $k_1 = -2, k_2 = 1$

32. $2x_1 + x_2 = k_1$
$5x_1 + 3x_2 = k_2$
(A) $k_1 = 2, k_2 = 13$
(B) $k_1 = 2, k_2 = 4$
(C) $k_1 = 1, k_2 = -3$

33. $x_1 + 3x_2 = k_1$
$2x_1 + 7x_2 = k_2$
(A) $k_1 = 2, k_2 = -1$
(B) $k_1 = 1, k_2 = 0$
(C) $k_1 = 3, k_2 = -1$

34. $2x_1 + x_2 = k_1$
$x_1 + x_2 = k_2$
(A) $k_1 = -1, k_2 = -2$
(B) $k_1 = 2, k_2 = 3$
(C) $k_1 = 2, k_2 = 0$

35. $x_1 - 3x_2 \qquad = k_1$
$x_2 + x_3 = k_2$
$2x_1 - x_2 + 4x_3 = k_3$
(A) $k_1 = 1, k_2 = 0, k_3 = 2$
(B) $k_1 = -1, k_2 = 1, k_3 = 0$
(C) $k_1 = 2, k_2 = -2, k_3 = 1$

36. $2x_1 + 3x_2 \qquad = k_1$
$x_1 + 2x_2 + 3x_3 = k_2$
$-x_2 - 5x_3 = k_3$
(A) $k_1 = 0, k_2 = 2, k_3 = 1$
(B) $k_1 = -2, k_2 = 0, k_3 = 1$
(C) $k_1 = 3, k_2 = 1, k_3 = 0$

37. $x_1 + x_2 \qquad = k_1$
$2x_1 + 3x_2 - x_3 = k_2$
$x_1 \qquad + 2x_3 = k_3$
(A) $k_1 = 2, k_2 = 0, k_3 = 4$
(B) $k_1 = 0, k_2 = 4, k_3 = -2$
(C) $k_1 = 4, k_2 = 2, k_3 = 0$

38. $x_1 \qquad - x_3 = k_1$
$2x_1 - x_2 \qquad = k_2$
$x_1 + x_2 - 2x_3 = k_3$
(A) $k_1 = 4, k_2 = 8, k_3 = 0$
(B) $k_1 = 4, k_2 = 0, k_3 = -4$
(C) $k_1 = 0, k_2 = 8, k_3 = -8$

✎ *In Problems 39–44, the matrix equation is not solved correctly. Explain the mistake and find the correct solution. Assume that the indicated inverses exist.*

39. $AX = B, X = \dfrac{B}{A}$

40. $XA = B, X = \dfrac{B}{A}$

41. $XA = B, X = A^{-1}B$

42. $AX = B, X = BA^{-1}$

43. $AX = BA, X = A^{-1}BA, X = B$

44. $XA = AB, X = AB\,A^{-1}, X = B$

✎ *In Problems 45–50, explain why the system cannot be solved by matrix inverse methods. Discuss methods that could be used and then solve the system.*

45. $-2x_1 + 4x_2 = -5$
$6x_1 - 12x_2 = 15$

46. $-2x_1 + 4x_2 = 5$
$6x_1 - 12x_2 = 15$

47. $x_1 - 3x_2 - 2x_3 = -1$
$-2x_1 + 6x_2 + 4x_3 = 3$

48. $x_1 - 3x_2 - 2x_3 = -1$
$-2x_1 + 7x_2 + 3x_3 = 3$

49. $x_1 - 2x_2 + 3x_3 = 1$
$2x_1 - 3x_2 - 2x_3 = 3$
$x_1 - x_2 - 5x_3 = 2$

50. $x_1 - 2x_2 + 3x_3 = 1$
$2x_1 - 3x_2 - 2x_3 = 3$
$x_1 - x_2 - 5x_3 = 4$

For n × n matrices A and B, and n × 1 column matrices C, D, and X, solve each matrix equation in Problems 51–56 for X. Assume that all necessary inverses exist.

51. $AX - BX = C$

52. $AX + BX = C$

53. $AX + X = C$

54. $AX - X = C$

55. $AX - C = D - BX$

56. $AX + C = BX + D$

57. Use matrix inverse methods to solve the following system for the indicated values of k_1 and k_2.

$$x_1 + 2.001x_2 = k_1$$
$$x_1 + 2x_2 = k_2$$

(A) $k_1 = 1, k_2 = 1$

(B) $k_1 = 1, k_2 = 0$

(C) $k_1 = 0, k_2 = 1$

Discuss the effect of small changes in the constant terms on the solution set of this system.

58. Repeat Problem 57 for the following system:

$$x_1 - 3.001x_2 = k_1$$
$$x_1 - 3x_2 = k_2$$

In Problems 59–62, write each system as a matrix equation and solve using the inverse coefficient matrix. Use a graphing calculator or computer to perform the necessary calculations.

59. $x_1 + 8x_2 + 7x_3 = 135$
$6x_1 + 6x_2 + 8x_3 = 155$
$3x_1 + 4x_2 + 6x_3 = 75$

60. $5x_1 + 3x_2 - 2x_3 = 112$
$7x_1 + 5x_2 \quad = 70$
$3x_1 + x_2 - 9x_3 = 96$

61. $6x_1 + 9x_2 + 7x_3 + 5x_4 = 250$
$6x_1 + 4x_2 + 7x_3 + 3x_4 = 195$
$4x_1 + 5x_2 + 3x_3 + 2x_4 = 145$
$4x_1 + 3x_2 + 8x_3 + 2x_4 = 125$

62. $3x_1 + 3x_2 + 6x_3 + 5x_4 = 10$
$4x_1 + 5x_2 + 8x_3 + 2x_4 = 15$
$3x_1 + 6x_2 + 7x_3 + 4x_4 = 30$
$4x_1 + x_2 + 6x_3 + 3x_4 = 25$

Applications

Construct a mathematical model for each of the following problems. (The answers in the back of the book include both the mathematical model and the interpretation of its solution.) Use matrix inverse methods to solve the model and then interpret the solution.

63. Concert tickets. A concert hall has 10,000 seats and two categories of ticket prices, $25 and $35. Assume that all seats in each category can be sold.

	Concert		
	1	2	3
Tickets sold	10,000	10,000	10,000
Return required	$275,000	$300,000	$325,000

(A) How many tickets of each category should be sold to bring in each of the returns indicated in the table?

(B) Is it possible to bring in a return of $200,000? Of $400,000? Explain.

(C) Describe all the possible returns.

64. Parking receipts. Parking fees at a zoo are $5.00 for local residents and $7.50 for all others. At the end of each day, the total number of vehicles parked that day and the gross receipts for the day are recorded, but the number of vehicles in each category is not. The following table contains the relevant information for a recent 4-day period:

	Day			
	1	2	3	4
Vehicles parked	1,200	1,550	1,740	1,400
Gross receipts	$7,125	$9,825	$11,100	$8,650

(A) How many vehicles in each category used the zoo's parking facilities each day?

(B) If 1,200 vehicles are parked in one day, is it possible to take in gross receipts of $5,000? Of $10,000? Explain.

(C) Describe all possible gross receipts on a day when 1,200 vehicles are parked.

65. Production scheduling. A supplier manufactures car and truck frames at two different plants. The production rates (in frames per hour) for each plant are given in the table:

Plant	Car Frames	Truck Frames
A	10	5
B	8	8

How many hours should each plant be scheduled to operate to exactly fill each of the orders in the following table?

	Orders		
	1	2	3
Car frames	3,000	2,800	2,600
Truck frames	1,600	2,000	2,200

66. Production scheduling. Labor and material costs for manufacturing two guitar models are given in the table:

Guitar Model	Labor Cost	Material Cost
A	$30	$20
B	$40	$30

(A) If a total of $3,000 a week is allowed for labor and material, how many of each model should be produced each week to use exactly each of the allocations of the $3,000 indicated in the following table?

	Weekly Allocation		
	1	2	3
Labor	$1,800	$1,750	$1,720
Material	$1,200	$1,250	$1,280

(B) Is it possible to use an allocation of $1,600 for labor and $1,400 for material? Of $2,000 for labor and $1,000 for material? Explain.

67. Incentive plan. A small company provides an incentive plan for its top executives. Each executive receives as a bonus a percentage of the portion of the annual profit that remains after the bonuses for the other executives have been deducted (see the table). If the company has an annual profit of $2 million, find the bonus for each executive. Round each bonus to the nearest hundred dollars.

Officer	Bonus
President	3%
Executive vice-president	2.5%
Associate vice-president	2%
Assistant vice-president	1.5%

68. Incentive plan. Repeat Problem 67 if the company decides to include a 1% bonus for the sales manager in the incentive plan.

69. Diets. A biologist has available two commercial food mixes containing the percentage of protein and fat given in the table.

Mix	Protein(%)	Fat(%)
A	20	4
B	14	3

(A) How many ounces of each mix should be used to prepare each of the diets listed in the following table?

	Diet		
	1	2	3
Protein	80 oz	90 oz	100 oz
Fat	17 oz	18 oz	21 oz

(B) Is it possible to prepare a diet consisting of 100 ounces of protein and 22 ounces of fat? Of 80 ounces of protein and 15 ounces of fat? Explain.

70. Education. A state university system is planning to hire new faculty at the rank of lecturer or instructor for several of its two-year community colleges. The number of sections taught and the annual salary (in thousands of dollars) for each rank are given in the table.

	Rank	
	Lecturer	Instructor
Sections taught	3	4
Annual salary (thousand $)	20	25

The number of sections taught by new faculty and the amount budgeted for salaries (in thousands of dollars) at each of the colleges are given in the following table. How many faculty of each rank should be hired at each college to exactly meet the demand for sections and completely exhaust the salary budget?

	Community College		
	1	2	3
Demand for sections	30	33	35
Salary budget (thousand $)	200	210	220

Answers to Matched Problems

1.
$$AX + C = B$$
$$(AX + C) - C = B - C$$
$$AX + (C - C) = B - C$$
$$AX + 0 = B - C$$
$$AX = B - C$$
$$A^{-1}(AX) = A^{-1}(B - C)$$
$$(A^{-1}A)X = A^{-1}(B - C)$$
$$IX = A^{-1}(B - C)$$
$$X = A^{-1}(B - C)$$

2. $x_1 = 2, x_2 = 5, x_3 = 0$

3. (A) $x_1 = -2, x_2 = -5, x_3 = 4$
   (B) $x_1 = 0, x_2 = 1, x_3 = -4$

4. $A^{-1} = \begin{bmatrix} 1.5 & -6.25 \\ -0.5 & 6.25 \end{bmatrix}$; client 1: $15,000 in A and $5,000 in B; client 2: $28,125 in A and $21,875 in B; client 3: $6,875 in A and $3,125 in B

4.7   Leontief Input–Output Analysis

- Two-Industry Model
- Three-Industry Model

An important application of matrices and their inverses is **input–output analysis**. Wassily Leontief (1905–1999), the primary force behind this subject, was awarded the Nobel Prize in economics in 1973 because of the significant impact his work had on economic planning for industrialized countries. Among other things, he conducted a comprehensive study of how 500 sectors of the U.S. economy interacted with each other. Of course, large-scale computers played a crucial role in this analysis.

Our investigation will be more modest. In fact, we start with an economy comprised of only two industries. From these humble beginnings, ideas and definitions will evolve that can be readily generalized for more realistic economies. Input–output analysis attempts to establish equilibrium conditions under which industries in an economy have just enough output to satisfy each other's demands in addition to final (outside) demands.

Two-Industry Model

We start with an economy comprised of only two industries, electric company E and water company W. Output for both companies is measured in dollars. The electric company uses both electricity and water (inputs) in the production of electricity (output), and the water company uses both electricity and water (inputs) in the production of water (output). Suppose that the production of each dollar's worth of electricity requires \$0.30 worth of electricity and \$0.10 worth of water, and the production of each dollar's worth of water requires \$0.20 worth of electricity and \$0.40 worth of water. If the final demand (the demand from all other users of electricity and water) is

$$d_1 = \$12 \text{ million for electricity}$$
$$d_2 = \$8 \text{ million for water}$$

how much electricity and water should be produced to meet this final demand?

To begin, suppose that the electric company produces \$12 million worth of electricity and the water company produces \$8 million worth of water. Then the production processes of the companies would require

| Electricity required to produce electricity | Electricity required to produce water |

$$0.3(12) + 0.2(8) = \$5.2 \text{ million of electricity}$$

and

| Water required to produce electricity | Water required to produce water |

$$0.1(12) + 0.4(8) = \$4.4 \text{ million of water}$$

leaving only \$6.8 million of electricity and \$3.6 million of water to satisfy the final demand. To meet the internal demands of both companies and to end up with enough electricity for the final outside demand, both companies must produce more than just the final demand. In fact, they must produce exactly enough to meet their own internal demands plus the final demand. To determine the total output that each company must produce, we set up a system of equations.

If

$$x_1 = \text{total output from electric company}$$
$$x_2 = \text{total output from water company}$$

then, reasoning as before, the internal demands are

$$0.3x_1 + 0.2x_2 \quad \text{Internal demand for electricity}$$
$$0.1x_1 + 0.4x_2 \quad \text{Internal demand for water}$$

Combining the internal demand with the final demand produces the following system of equations:

| Total output | Internal demand | Final demand |

$$x_1 = 0.3x_1 + 0.2x_2 + d_1 \tag{1}$$
$$x_2 = 0.1x_1 + 0.4x_2 + d_2$$

or, in matrix form,

$$\begin{bmatrix} x_1 \\ x_2 \end{bmatrix} = \begin{bmatrix} 0.3 & 0.2 \\ 0.1 & 0.4 \end{bmatrix} \begin{bmatrix} x_1 \\ x_2 \end{bmatrix} + \begin{bmatrix} d_1 \\ d_2 \end{bmatrix}$$

or

$$X = MX + D \tag{2}$$

where

$$D = \begin{bmatrix} d_1 \\ d_2 \end{bmatrix} \quad \text{Final demand matrix}$$

$$X = \begin{bmatrix} x_1 \\ x_2 \end{bmatrix} \quad \text{Output matrix}$$

$$M = \begin{matrix} & E & W \\ E & \\ W & \end{matrix} \begin{bmatrix} 0.3 & 0.2 \\ 0.1 & 0.4 \end{bmatrix} \quad \text{Technology matrix}$$

The **technology matrix** is the heart of input–output analysis. The elements in the technology matrix are determined as follows (read from left to right and then up):

CONCEPTUAL INSIGHT

Labeling the rows and columns of the technology matrix with the first letter of each industry is an important part of the process. The same order must be used for columns as for rows, and that same order must be used for the entries of D (the final demand matrix) and the entries of X (the output matrix). In this book we normally label the rows and columns in alphabetical order.

Now we solve equation (2) for X. We proceed as in Section 4.6:

$$X = MX + D$$
$$X - MX = D$$
$$IX - MX = D \qquad I = \begin{bmatrix} 1 & 0 \\ 0 & 1 \end{bmatrix}$$
$$(I - M)X = D$$
$$X = (I - M)^{-1}D \quad \text{Assuming } I - M \text{ has an inverse} \qquad (3)$$

Omitting the details of the calculations, we find

$$I - M = \begin{bmatrix} 0.7 & -0.2 \\ -0.1 & 0.6 \end{bmatrix} \quad \text{and} \quad (I - M)^{-1} = \begin{bmatrix} 1.5 & 0.5 \\ 0.25 & 1.75 \end{bmatrix}$$

Then we have

$$\begin{bmatrix} x_1 \\ x_2 \end{bmatrix} = \begin{bmatrix} 1.5 & 0.5 \\ 0.25 & 1.75 \end{bmatrix}\begin{bmatrix} d_1 \\ d_2 \end{bmatrix} = \begin{bmatrix} 1.5 & 0.5 \\ 0.25 & 1.75 \end{bmatrix}\begin{bmatrix} 12 \\ 8 \end{bmatrix} = \begin{bmatrix} 22 \\ 17 \end{bmatrix} \qquad (4)$$

Therefore, the electric company must produce an output of $22 million and the water company must produce an output of $17 million so that each company can meet both internal and final demands.

CHECK   We use equation (2) to check our work:

$$X = MX + D$$

$$\begin{bmatrix} 22 \\ 17 \end{bmatrix} \overset{?}{=} \begin{bmatrix} 0.3 & 0.2 \\ 0.1 & 0.4 \end{bmatrix} \begin{bmatrix} 22 \\ 17 \end{bmatrix} + \begin{bmatrix} 12 \\ 8 \end{bmatrix}$$

$$\begin{bmatrix} 22 \\ 17 \end{bmatrix} \overset{?}{=} \begin{bmatrix} 10 \\ 9 \end{bmatrix} + \begin{bmatrix} 12 \\ 8 \end{bmatrix}$$

$$\begin{bmatrix} 22 \\ 17 \end{bmatrix} \overset{\checkmark}{=} \begin{bmatrix} 22 \\ 17 \end{bmatrix}$$

To solve this input–output problem on a graphing calculator, simply store matrices M, D, and I in memory; then use equation (3) to find X and equation (2) to check your results. Figure 1 illustrates this process on a graphing calculator.

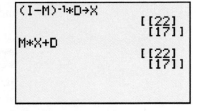

(A) Store M, D, and I in the graphing calculator's memory	(B) Compute X and check in equation (2)

Figure 1

Actually, equation (4) solves the original problem for arbitrary final demands d_1 and d_2. This is very useful, since equation (4) gives a quick solution not only for the final demands stated in the original problem, but also for various other projected final demands. If we had solved system (1) by Gauss–Jordan elimination, then we would have to start over for each new set of final demands.

Suppose that in the original problem the projected final demands 5 years from now are $d_1 = 24$ and $d_2 = 16$. To determine each company's output for this projection, we simply substitute these values into equation (4) and multiply:

$$\begin{bmatrix} x_1 \\ x_2 \end{bmatrix} = \begin{bmatrix} 1.5 & 0.5 \\ 0.25 & 1.75 \end{bmatrix} \begin{bmatrix} 24 \\ 16 \end{bmatrix} = \begin{bmatrix} 44 \\ 34 \end{bmatrix}$$

We summarize these results for convenient reference.

SUMMARY   Solution to a Two-Industry Input–Output Problem

Given two industries, C_1 and C_2, with

$$
\begin{array}{ccc}
\text{Technology matrix} & \text{Output matrix} & \text{Final demand matrix}
\end{array}
$$

$$
M = \begin{array}{c} \\ C_1 \\ C_2 \end{array}\!\!\begin{array}{c} \overset{C_1 \quad\ C_2}{\begin{bmatrix} a_{11} & a_{12} \\ a_{21} & a_{22} \end{bmatrix}} \end{array} \qquad X = \begin{bmatrix} x_1 \\ x_2 \end{bmatrix} \qquad D = \begin{bmatrix} d_1 \\ d_2 \end{bmatrix}
$$

where a_{ij} is the input required from C_i to produce a dollar's worth of output for C_j, the solution to the input–output matrix equation

$$
\begin{array}{cccc}
\text{Total} & \text{Internal} & \text{Final} \\
\text{output} & \text{demand} & \text{demand} \\
X & = \quad MX & + \quad D
\end{array}
$$

is

$$X = (I - M)^{-1}D \tag{3}$$

assuming that $I - M$ has an inverse.

Three-Industry Model

Equations (2) and (3) in the solution to a two-industry input–output problem are the same for a three-industry economy, a four-industry economy, or an economy with n industries (where n is any natural number). The steps we took going from equation (2) to equation (3) hold for arbitrary matrices as long as the matrices have the correct sizes and $(I - M)^{-1}$ exists.

Explore and Discuss 1   If equations (2) and (3) are valid for an economy with n industries, discuss the size of all the matrices in each equation.

The next example illustrates the application of equations (2) and (3) to a three-industry economy.

EXAMPLE 1   Input–Output Analysis An economy is based on three sectors, agriculture (A), energy (E), and manufacturing (M). Production of a dollar's worth of agriculture requires an input of \$0.20 from the agriculture sector and \$0.40 from the energy sector. Production of a dollar's worth of energy requires an input of \$0.20 from the energy sector and \$0.40 from the manufacturing sector. Production of a dollar's worth of manufacturing requires an input of \$0.10 from the agriculture sector, \$0.10 from the energy sector, and \$0.30 from the manufacturing sector. Find the output from each sector that is needed to satisfy a final demand of \$20 billion for agriculture, \$10 billion for energy, and \$30 billion for manufacturing.

SOLUTION Since this is a three-industry problem, the technology matrix will be a 3 × 3 matrix, and the output and final demand matrices will be 3 × 1 column matrices.

To begin, we form a blank 3 × 3 technology matrix and label the rows and columns in alphabetical order.

$$
\begin{array}{c}
\text{Technology matrix} \\
\text{Output} \\
\begin{array}{ccc} A & E & M \end{array} \\
\text{Input} \begin{array}{c} A \\ E \\ M \end{array}
\left[\phantom{\begin{array}{ccc} A & E & M \end{array}} \right] = M
\end{array}
$$

Now we analyze the production information given in the problem, beginning with agriculture.

> **"Production of a dollar's worth of agriculture requires an input of \$0.20 from the agriculture sector and \$0.40 from the energy sector."**

We organize this information in a table and then insert it in the technology matrix. Since manufacturing is not mentioned in the agriculture production information, the input from manufacturing is \$0.

$$
\begin{array}{c}
\text{Agriculture} \\
\begin{array}{cc} \text{Input} & \text{Output} \end{array} \\
\begin{array}{ccc}
A & \xrightarrow{0.2} & A \\
E & \xrightarrow{0.4} & A \\
M & \xrightarrow{0} & A
\end{array}
\qquad
\begin{array}{c}
\begin{array}{ccc} A & E & M \end{array} \\
\begin{array}{c} A \\ E \\ M \end{array}
\left[\begin{array}{c} 0.2 \\ 0.4 \\ 0 \end{array} \phantom{\begin{array}{cc} & \end{array}} \right]
\end{array}
\end{array}
$$

"Production of a dollar's worth of energy requires an input of $0.20 from the energy sector and $0.40 from the manufacturing sector."

Energy

Input	Output			A	E	M
A	$\xrightarrow{0}$	E	A	0.2	0	
E	$\xrightarrow{0.2}$	E	E	0.4	0.2	
M	$\xrightarrow{0.4}$	E	M	0	0.4	

"Production of a dollar's worth of manufacturing requires an input of $0.10 from the agriculture sector, $0.10 from the energy sector and $0.30 from the manufacturing sector."

Manufacturing

Input	Output			A	E	M
A	$\xrightarrow{0.1}$	M	A	0.2	0	0.1
E	$\xrightarrow{0.1}$	M	E	0.4	0.2	0.1
M	$\xrightarrow{0.3}$	M	M	0	0.4	0.3

Therefore,

$$
\begin{array}{ccc}
\text{Technology matrix} & \text{Final demand} & \text{Output} \\
\end{array}
$$

$$
M = \begin{array}{c} A \\ E \\ M \end{array}
\begin{bmatrix} 0.2 & 0 & 0.1 \\ 0.4 & 0.2 & 0.1 \\ 0 & 0.4 & 0.3 \end{bmatrix}
\qquad
D = \begin{bmatrix} 20 \\ 10 \\ 30 \end{bmatrix}
\qquad
X = \begin{bmatrix} x_1 \\ x_2 \\ x_3 \end{bmatrix}
$$

where M, X, and D satisfy the input–output equation $X = MX + D$. Since the solution to this equation is $X = (I - M)^{-1}D$, we must first find $I - M$ and then $(I - M)^{-1}$. Omitting the details of the calculations, we have

$$
I - M = \begin{bmatrix} 0.8 & 0 & -0.1 \\ -0.4 & 0.8 & -0.1 \\ 0 & -0.4 & 0.7 \end{bmatrix}
$$

and

$$
(I - M)^{-1} = \begin{bmatrix} 1.3 & 0.1 & 0.2 \\ 0.7 & 1.4 & 0.3 \\ 0.4 & 0.8 & 1.6 \end{bmatrix}
$$

So the output matrix X is given by

$$
\begin{array}{cccc}
X & (I - M)^{-1} & D \\
\end{array}
$$

$$
\begin{bmatrix} x_1 \\ x_2 \\ x_3 \end{bmatrix} = \begin{bmatrix} 1.3 & 0.1 & 0.2 \\ 0.7 & 1.4 & 0.3 \\ 0.4 & 0.8 & 1.6 \end{bmatrix} \begin{bmatrix} 20 \\ 10 \\ 30 \end{bmatrix} = \begin{bmatrix} 33 \\ 37 \\ 64 \end{bmatrix}
$$

An output of $33 billion for agriculture, $37 billion for energy, and $64 billion for manufacturing will meet the given final demands. You should check this result in equation (2).

Figure 2 illustrates a spreadsheet solution for Example 1.

	A	B	C	D	E	F	G	H	I	J	K	L	M
1	Technology Matrix M									Final		Output	
2		A	E	M			I – M			Demand			
3	A	0.2	0	0.1		0.8	0	-0.1		20		33	
4	E	0.4	0.2	0.1		-0.4	0.8	-0.1		10		37	
5	M	0	0.4	0.3		0	-0.4	0.7		30		64	

Figure 2 **The command MMULT(MINVERSE(F3:H5), J3:J5) produces the output in L3:L5**

Matched Problem 1 An economy is based on three sectors, coal, oil, and transportation. Production of a dollar's worth of coal requires an input of $0.20 from the coal sector and $0.40 from the transportation sector. Production of a dollar's worth of oil requires an input of $0.10 from the oil sector and $0.20 from the transportation sector. Production of a dollar's worth of transportation requires an input of $0.40 from the coal sector, $0.20 from the oil sector, and $0.20 from the transportation sector.

(A) Find the technology matrix M.

(B) Find $(I - M)^{-1}$.

(C) Find the output from each sector that is needed to satisfy a final demand of $30 billion for coal, $10 billion for oil, and $20 billion for transportation.

Exercises 4.7

Skills Warm-up Exercises

W *In Problems 1–8, solve each equation for x, where x represents a real number. (If necessary, review Section 1.1).*

1. $x = 3x + 6$

2. $x = 4x - 5$

3. $x = 0.9x + 10$

4. $x = 0.6x + 84$

5. $x = 0.2x + 3.2$

6. $x = 0.3x + 4.2$

7. $x = 0.68x + 2.56$

8. $x = 0.98x + 8.24$

Problems 9–14 pertain to the following input–output model: Assume that an economy is based on two industrial sectors, agriculture (A) and energy (E). The technology matrix M and final demand matrices (in billions of dollars) are

$$\begin{array}{cc} & A \quad E \end{array}$$
$$\begin{array}{c} A \\ E \end{array} \begin{bmatrix} 0.4 & 0.2 \\ 0.2 & 0.1 \end{bmatrix} = M$$

$$D_1 = \begin{bmatrix} 6 \\ 4 \end{bmatrix} \quad D_2 = \begin{bmatrix} 8 \\ 5 \end{bmatrix} \quad D_3 = \begin{bmatrix} 12 \\ 9 \end{bmatrix}$$

9. How much input from A and E are required to produce a dollar's worth of output for A?

10. How much input from A and E are required to produce a dollar's worth of output for E?

11. Find $I - M$ and $(I - M)^{-1}$.

12. Find the output for each sector that is needed to satisfy the final demand D_1.

13. Repeat Problem 12 for D_2.

14. Repeat Problem 12 for D_3.

Problems 15–20 pertain to the following input–output model: Assume that an economy is based on three industrial sectors: agriculture (A), building (B), and energy (E). The technology matrix M and final demand matrices (in billions of dollars) are

$$\begin{array}{ccc} & A \quad B \quad E \end{array}$$
$$\begin{array}{c} A \\ B \\ E \end{array} \begin{bmatrix} 0.3 & 0.2 & 0.2 \\ 0.1 & 0.1 & 0.1 \\ 0.2 & 0.1 & 0.1 \end{bmatrix} = M$$

$$D_1 = \begin{bmatrix} 5 \\ 10 \\ 15 \end{bmatrix} \quad D_2 = \begin{bmatrix} 20 \\ 15 \\ 10 \end{bmatrix}$$

15. How much input from A, B, and E are required to produce a dollar's worth of output for B?

16. How much of each of B's output dollars is required as input for each of the three sectors?

17. Show that

$$I - M = \begin{bmatrix} 0.7 & -0.2 & -0.2 \\ -0.1 & 0.9 & -0.1 \\ -0.2 & -0.1 & 0.9 \end{bmatrix}$$

18. Given

$$(I - M)^{-1} = \begin{bmatrix} 1.6 & 0.4 & 0.4 \\ 0.22 & 1.18 & 0.18 \\ 0.38 & 0.22 & 1.22 \end{bmatrix}$$

show that $(I - M)^{-1}(I - M) = I$.

19. Use $(I - M)^{-1}$ in Problem 18 to find the output for each sector that is needed to satisfy the final demand D_1.

20. Repeat Problem 19 for D_2.

In Problems 21–26, find $(I - M)^{-1}$ and X.

21. $M = \begin{bmatrix} 0.2 & 0.2 \\ 0.3 & 0.3 \end{bmatrix}; \quad D = \begin{bmatrix} 10 \\ 25 \end{bmatrix}$

22. $M = \begin{bmatrix} 0.4 & 0.2 \\ 0.6 & 0.8 \end{bmatrix}; \quad D = \begin{bmatrix} 30 \\ 50 \end{bmatrix}$

23. $M = \begin{bmatrix} 0.7 & 0.8 \\ 0.3 & 0.2 \end{bmatrix}; \quad D = \begin{bmatrix} 25 \\ 75 \end{bmatrix}$

24. $M = \begin{bmatrix} 0.4 & 0.1 \\ 0.2 & 0.3 \end{bmatrix}; \quad D = \begin{bmatrix} 15 \\ 20 \end{bmatrix}$

25. $M = \begin{bmatrix} 0.3 & 0.1 & 0.3 \\ 0.2 & 0.1 & 0.2 \\ 0.1 & 0.1 & 0.1 \end{bmatrix}; \quad D = \begin{bmatrix} 20 \\ 5 \\ 10 \end{bmatrix}$

26. $M = \begin{bmatrix} 0.3 & 0.2 & 0.3 \\ 0.1 & 0.1 & 0.1 \\ 0.1 & 0.2 & 0.1 \end{bmatrix}; \quad D = \begin{bmatrix} 10 \\ 25 \\ 15 \end{bmatrix}$

27. The technology matrix for an economy based on agriculture (*A*) and manufacturing (*M*) is

$$M = \begin{array}{c} \\ A \\ M \end{array} \begin{array}{cc} A & M \\ \begin{bmatrix} 0.3 & 0.25 \\ 0.1 & 0.25 \end{bmatrix} \end{array}$$

(A) Find the output for each sector that is needed to satisfy a final demand of $40 million for agriculture and $40 million for manufacturing.

(B) Discuss the effect on the final demand if the agriculture output in part (A) is increased by $20 million and manufacturing output remains unchanged.

28. The technology matrix for an economy based on energy (*E*) and transportation (*T*) is

$$M = \begin{array}{c} \\ E \\ T \end{array} \begin{array}{cc} E & T \\ \begin{bmatrix} 0.25 & 0.25 \\ 0.4 & 0.2 \end{bmatrix} \end{array}$$

(A) Find the output for each sector that is needed to satisfy a final demand of $50 million for energy and $50 million for transportation.

(B) Discuss the effect on the final demand if the transportation output in part (A) is increased by $40 million and the energy output remains unchanged.

29. Refer to Problem 27. Fill in the elements in the following technology matrix.

$$T = \begin{array}{c} \\ M \\ A \end{array} \begin{array}{cc} M & A \\ \begin{bmatrix} & \\ & \end{bmatrix} \end{array}$$

Use this matrix to solve Problem 27. Discuss any differences in your calculations and in your answers.

30. Refer to Problem 28. Fill in the elements in the following technology matrix.

$$T = \begin{array}{c} \\ T \\ E \end{array} \begin{array}{cc} T & E \\ \begin{bmatrix} & \\ & \end{bmatrix} \end{array}$$

Use this matrix to solve Problem 28. Discuss any differences in your calculations and in your answers.

31. The technology matrix for an economy based on energy (*E*) and mining (*M*) is

$$M = \begin{array}{c} \\ E \\ M \end{array} \begin{array}{cc} E & M \\ \begin{bmatrix} 0.2 & 0.3 \\ 0.4 & 0.3 \end{bmatrix} \end{array}$$

The management of these two sectors would like to set the total output level so that the final demand is always 40% of the total output. Discuss methods that could be used to accomplish this objective.

32. The technology matrix for an economy based on automobiles (*A*) and construction (*C*) is

$$M = \begin{array}{c} \\ A \\ C \end{array} \begin{array}{cc} A & C \\ \begin{bmatrix} 0.1 & 0.4 \\ 0.1 & 0.1 \end{bmatrix} \end{array}$$

The management of these two sectors would like to set the total output level so that the final demand is always 70% of the total output. Discuss methods that could be used to accomplish this objective.

33. All the technology matrices in the text have elements between 0 and 1. Why is this the case? Would you ever expect to find an element in a technology matrix that is negative? That is equal to 0? That is equal to 1? That is greater than 1?

34. The sum of the elements in a column of any of the technology matrices in the text is less than 1. Why is this the case? Would you ever expect to find a column with a sum equal to 1? Greater than 1? How would you describe an economic system where the sum of the elements in every column of the technology matrix is 1?

Applications

35. Coal, steel. An economy is based on two industrial sectors, coal and steel. Production of a dollar's worth of coal requires an input of $0.10 from the coal sector and $0.20 from the steel sector. Production of a dollar's worth of steel requires an input of $0.20 from the coal sector and $0.40 from the steel sector. Find the output for each sector that is needed to satisfy a final demand of $20 billion for coal and $10 billion for steel.

36. Transportation, manufacturing. An economy is based on two sectors, transportation and manufacturing. Production of a dollar's worth of transportation requires an input of $0.10 from each sector and production of a dollar's worth of manufacturing requires an input of $0.40 from each sector. Find the output for each sector that is needed to satisfy a final demand of $5 billion for transportation and $20 billion for manufacturing.

37. Agriculture, tourism. The economy of a small island nation is based on two sectors, agriculture and tourism. Production of a dollar's worth of agriculture requires an input of $0.20 from agriculture and $0.15 from tourism. Production of a dollar's worth of tourism requires an input of $0.40 from agriculture and $0.30 from tourism. Find the output from each sector that is needed to satisfy a final demand of $60 million for agriculture and $80 million for tourism.

38. Agriculture, oil. The economy of a country is based on two sectors, agriculture and oil. Production of a dollar's worth of agriculture requires an input of $0.40 from agriculture and $0.35 from oil. Production of a dollar's worth of oil requires an input of $0.20 from agriculture and $0.05 from oil. Find the output from each sector that is needed to satisfy a final demand of $40 million for agriculture and $250 million for oil.

39. Agriculture, manufacturing, energy. An economy is based on three sectors, agriculture, manufacturing, and energy. Production of a dollar's worth of agriculture requires inputs of $0.20 from agriculture, $0.20 from manufacturing, and $0.20 from energy. Production of a dollar's worth of manufacturing requires inputs of $0.40 from agriculture, $0.10 from manufacturing, and $0.10 from energy. Production of a dollar's worth of energy requires inputs of $0.30 from agriculture, $0.10 from manufacturing, and $0.10 from energy. Find the output for each sector that is needed to satisfy a final demand of $10 billion for agriculture, $15 billion for manufacturing, and $20 billion for energy.

40. Electricity, natural gas, oil. A large energy company produces electricity, natural gas, and oil. The production of a dollar's worth of electricity requires inputs of $0.30 from electricity, $0.10 from natural gas, and $0.20 from oil. Production of a dollar's worth of natural gas requires inputs of $0.30 from electricity, $0.10 from natural gas, and $0.20 from oil. Production of a dollar's worth of oil requires inputs of $0.10 from each sector. Find the output for each sector that is needed to satisfy a final demand of $25 billion for electricity, $15 billion for natural gas, and $20 billion for oil.

41. Four sectors. An economy is based on four sectors, agriculture (A), energy (E), labor (L), and manufacturing (M). The table gives the input requirements for a dollar's worth of output for each sector, along with the projected final demand (in billions of dollars) for a 3-year period. Find the output for each sector that is needed to satisfy each of these final demands. Round answers to the nearest billion dollars.

		Output				Final Demand		
		A	E	L	M	1	2	3
Input	A	0.05	0.17	0.23	0.09	23	32	55
	E	0.07	0.12	0.15	0.19	41	48	62
	L	0.25	0.08	0.03	0.32	18	21	25
	M	0.11	0.19	0.28	0.16	31	33	35

42. Repeat Problem 41 with the following table:

		Output				Final Demand		
		A	E	L	M	1	2	3
Input	A	0.07	0.09	0.27	0.12	18	22	37
	E	0.14	0.07	0.21	0.24	26	31	42
	L	0.17	0.06	0.02	0.21	12	19	28
	M	0.15	0.13	0.31	0.19	41	45	49

Answers to Matched Problems

1. (A) $\begin{bmatrix} 0.2 & 0 & 0.4 \\ 0 & 0.1 & 0.2 \\ 0.4 & 0.2 & 0.2 \end{bmatrix}$ **(B)** $\begin{bmatrix} 1.7 & 0.2 & 0.9 \\ 0.2 & 1.2 & 0.4 \\ 0.9 & 0.4 & 1.8 \end{bmatrix}$

(C) $71 billion for coal, $26 billion for oil, and $67 billion for transportation

---

Chapter 4   Summary and Review

Important Terms, Symbols, and Concepts

4.1 Review: Systems of Linear Equations in Two Variables

- The **solution** of a system is an ordered pair of real numbers that satisfies each equation in the system. Solution by **graphing** is one method that can be used to find a solution.

- A linear system is **consistent** and **independent** if it has a unique solution, **consistent** and **dependent** if it has more than one solution, and **inconsistent** if it has no solution. A linear system that is consistent and dependent actually has an infinite number of solutions.

- A **graphing calculator** provides accurate solutions to a linear system.

- The **substitution** method can also be used to solve linear systems.

- The **method of elimination by addition** is easily extended to larger systems.

4.2 Systems of Linear Equations and Augmented Matrices

- A **matrix** is a rectangular array of real numbers. **Row operations** performed on an **augmented matrix** produce equivalent systems (Theorem 1, page 189).

- There are only three possible final forms for the augmented matrix for a linear system of two equations in two variables (p. 194).

4.3 Gauss–Jordan Elimination

- There are many possibilities for the final **reduced form** of the augmented matrix of a larger system of linear equations. Reduced form is defined on page 197.

- The **Gauss–Jordan elimination procedure** is described on pages 199 and 200.

EXAMPLES

Ex. 1, p. 174
Ex. 2, p. 175

Ex. 3, p. 177

Ex. 4, p. 177

Ex. 5, p. 179

Ex. 1, p. 189
Ex. 2, p. 191

Ex. 3, p. 192
Ex. 4, p. 193

Ex. 1, p. 197
Ex. 2, p. 198

Ex. 3, p. 200

Ex. 4, p. 201

Ex. 5, p. 202

4.4 Matrices: Basic Operations

- Two matrices are **equal** if they are the same size and their corresponding elements are equal. The **sum** of two matrices of the same size is the matrix with elements that are the sum of the corresponding elements of the two given matrices.
- The **negative of a matrix** is the matrix with elements that are the negatives of the given matrix. If A and B are matrices of the same size, then B can subtracted from A by adding the negative of B to A.
- Matrix equations involving addition and subtraction are solved much like real number equations.
- The product of a real number k and a matrix M is the matrix formed by multiplying each element of M by k.
- The product of a row matrix and a column matrix is defined on page 213.
- The matrix product of an $m \times p$ matrix with a $p \times n$ is defined on page 213.

Ex. 1, p. 210
Ex. 2, p. 211
Ex. 3, p. 211
Ex. 4, p. 211
Ex. 6, p. 213
Ex. 8, p. 214

4.5 Inverse of a Square Matrix

- The **identity matrix** for multiplication is defined on page 222.
- The **inverse** of a square matrix is defined on page 224.

Ex. 1, p. 223
Ex. 2, p. 226
Ex. 3, p. 228
Ex. 4, p. 229

4.6 Matrix Equations and Systems of Linear Equations

- Basic properties of matrices are summarized in Theorem 1 on page 234.
- **Matrix inverse methods** for solving systems of equations are described in the Summary on page 237.

Ex. 1, p. 235
Ex. 2, p. 235
Ex. 3, p. 236

4.7 Leontief Input–Output Analysis

- Leontief **input–output** analysis is summarized on page 245.

Ex. 1, p. 246

Review Exercises

Work through all the problems in this chapter review and check your answers in the back of the book. Answers to all problems are there along with section numbers in italics to indicate where each type of problem is discussed. Where weaknesses show up, review appropriate sections in the text.

1. Solve the following system by graphing:

$$2x - y = 4$$
$$x - 2y = -4$$

2. Solve the system in Problem 1 by substitution.

3. If a matrix is in reduced form, say so. If not, explain why and state the row operation(s) necessary to transform the matrix into reduced form.

(A) $\left[\begin{array}{cc|c} 0 & 1 & 2 \\ 1 & 0 & 3 \end{array}\right]$ (B) $\left[\begin{array}{cc|c} 1 & 0 & 2 \\ 0 & 3 & 3 \end{array}\right]$

(C) $\left[\begin{array}{ccc|c} 1 & 0 & 1 & 2 \\ 0 & 1 & 1 & 3 \end{array}\right]$ (D) $\left[\begin{array}{ccc|c} 1 & 1 & 0 & 2 \\ 0 & 1 & 1 & 3 \end{array}\right]$

4. Given matrices A and B,

$$A = \left[\begin{array}{ccccc} 5 & 3 & -1 & 0 & 2 \\ -4 & 8 & 1 & 3 & 0 \end{array}\right] \quad B = \left[\begin{array}{cc} -3 & 2 \\ 0 & 4 \\ -1 & 7 \end{array}\right]$$

(A) What is the size of A? Of B?

(B) Find a_{24}, a_{15}, b_{31}, and b_{22}.

(C) Is AB defined? Is BA defined?

5. Find x_1 and x_2:

(A) $\left[\begin{array}{cc} 1 & -2 \\ 1 & -3 \end{array}\right]\left[\begin{array}{c} x_1 \\ x_2 \end{array}\right] = \left[\begin{array}{c} 4 \\ 2 \end{array}\right]$

(B) $\left[\begin{array}{cc} 5 & 3 \\ 1 & 1 \end{array}\right]\left[\begin{array}{c} x_1 \\ x_2 \end{array}\right] + \left[\begin{array}{c} 25 \\ 14 \end{array}\right] = \left[\begin{array}{c} 18 \\ 22 \end{array}\right]$

In Problems 6–14, perform the operations that are defined, given the following matrices:

$$A = \left[\begin{array}{cc} 1 & 2 \\ 3 & 1 \end{array}\right] \quad B = \left[\begin{array}{cc} 2 & 1 \\ 1 & 1 \end{array}\right] \quad C = [2 \ 3] \quad D = \left[\begin{array}{c} 1 \\ 2 \end{array}\right]$$

6. $A + B$ **7.** $B + D$

8. $A - 2B$ **9.** AB

10. AC **11.** AD

12. DC **13.** CD

14. $C + D$

15. Find the inverse of the matrix A given below by appropriate row operations on $[A \,|\, I]$. Show that $A^{-1}A = I$.

$$A = \begin{bmatrix} 4 & 3 \\ 3 & 2 \end{bmatrix}$$

16. Solve the following system using elimination by addition:

$$\begin{aligned} 4x_1 + 3x_2 &= 3 \\ 3x_1 + 2x_2 &= 5 \end{aligned}$$

17. Solve the system in Problem 16 by performing appropriate row operations on the augmented matrix of the system.

18. Solve the system in Problem 16 by writing the system as a matrix equation and using the inverse of the coefficient matrix (see Problem 15). Also, solve the system if the constants 3 and 5 are replaced by 7 and 10, respectively. By 4 and 2, respectively.

In Problems 19–24, perform the operations that are defined, given the following matrices:

$$A = \begin{bmatrix} 2 & -2 \\ 1 & 0 \\ 3 & 2 \end{bmatrix} \quad B = \begin{bmatrix} -1 \\ 2 \\ 3 \end{bmatrix} \quad C = [2 \;\; 1 \;\; 3]$$

$$D = \begin{bmatrix} 3 & -2 & 1 \\ -1 & 1 & 2 \end{bmatrix} \quad E = \begin{bmatrix} 3 & -4 \\ -1 & 0 \end{bmatrix}$$

19. $A + D$ **20.** $E + DA$

21. $DA - 3E$ **22.** BC

23. CB **24.** $AD - BC$

25. Find the inverse of the matrix A given below by appropriate row operations on $[A \,|\, I]$. Show that $A^{-1}A = I$.

$$A = \begin{bmatrix} 1 & 2 & 3 \\ 2 & 3 & 4 \\ 1 & 2 & 1 \end{bmatrix}$$

26. Solve by Gauss–Jordan elimination:

(A) $\begin{aligned} x_1 + 2x_2 + 3x_3 &= 1 \\ 2x_1 + 3x_2 + 4x_3 &= 3 \\ x_1 + 2x_2 + x_3 &= 3 \end{aligned}$

(B) $\begin{aligned} x_1 + 2x_2 - x_3 &= 2 \\ 2x_1 + 3x_2 + x_3 &= -3 \\ 3x_1 + 5x_2 \phantom{{}+ x_3} &= -1 \end{aligned}$

(C) $\begin{aligned} x_1 + x_2 + x_3 &= 8 \\ 3x_1 + 2x_2 + 4x_3 &= 21 \end{aligned}$

27. Solve the system in Problem 26A by writing the system as a matrix equation and using the inverse of the coefficient matrix (see Problem 25). Also, solve the system if the constants 1, 3, and 3 are replaced by 0, 0, and −2, respectively. By −3, −4, and 1, respectively.

28. Discuss the relationship between the number of solutions of the following system and the constant k.

$$\begin{aligned} 2x_1 - 6x_2 &= 4 \\ -x_1 + kx_2 &= -2 \end{aligned}$$

29. An economy is based on two sectors, agriculture and energy. Given the technology matrix M and the final demand matrix D (in billions of dollars), find $(I - M)^{-1}$ and the output matrix X:

$$M = \begin{matrix} \\ A \\ E \end{matrix} \begin{matrix} A & E \\ \begin{bmatrix} 0.2 & 0.15 \\ 0.4 & 0.3 \end{bmatrix} \end{matrix} \quad D = \begin{matrix} \\ A \\ E \end{matrix} \begin{bmatrix} 30 \\ 20 \end{bmatrix}$$

30. Use the matrix M in Problem 29 to fill in the elements in the following technology matrix.

$$T = \begin{matrix} \\ E \\ A \end{matrix} \begin{matrix} E & A \\ \begin{bmatrix} & \\ & \end{bmatrix} \end{matrix}$$

Use this matrix to solve Problem 29. Discuss any differences in your calculations and in your answers.

31. An economy is based on two sectors, coal and steel. Given the technology matrix M and the final demand matrix D (in billions of dollars), find $(I - M)^{-1}$ and the output matrix X:

$$M = \begin{matrix} \\ C \\ S \end{matrix} \begin{matrix} C & S \\ \begin{bmatrix} 0.45 & 0.65 \\ 0.55 & 0.35 \end{bmatrix} \end{matrix} \quad D = \begin{matrix} \\ C \\ S \end{matrix} \begin{bmatrix} 40 \\ 10 \end{bmatrix}$$

32. Use graphical approximation techniques on a graphing calculator to find the solution of the following system to two decimal places:

$$\begin{aligned} x - 5y &= -5 \\ 2x + 3y &= 12 \end{aligned}$$

33. Find the inverse of the matrix A given below. Show that $A^{-1}A = I$.

$$A = \begin{bmatrix} 4 & 5 & 6 \\ 4 & 5 & -4 \\ 1 & 1 & 1 \end{bmatrix}$$

34. Solve the system

$$\begin{aligned} 0.04x_1 + 0.05x_2 + 0.06x_3 &= 360 \\ 0.04x_1 + 0.05x_2 - 0.04x_3 &= 120 \\ x_1 + x_2 + x_3 &= 7{,}000 \end{aligned}$$

by writing it as a matrix equation and using the inverse of the coefficient matrix. (Before starting, multiply the first two equations by 100 to eliminate decimals. Also, see Problem 33.)

35. Solve Problem 34 by Gauss–Jordan elimination.

36. Given the technology matrix M and the final demand matrix D (in billions of dollars), find $(I - M)^{-1}$ and the output matrix X:

$$M = \begin{bmatrix} 0.2 & 0 & 0.4 \\ 0.1 & 0.3 & 0.1 \\ 0 & 0.4 & 0.2 \end{bmatrix} \quad D = \begin{bmatrix} 40 \\ 20 \\ 30 \end{bmatrix}$$

37. Discuss the number of solutions for a system of n equations in n variables if the coefficient matrix

(A) Has an inverse.

(B) Does not have an inverse.

38. Discuss the number of solutions for the system corresponding to the reduced form shown below if

(A) $m \neq 0$

(B) $m = 0$ and $n \neq 0$

(C) $m = 0$ and $n = 0$

$$\begin{bmatrix} 1 & 0 & -2 & | & 5 \\ 0 & 1 & 3 & | & 3 \\ 0 & 0 & m & | & n \end{bmatrix}$$

39. One solution to the input–output equation $X = MX + D$ is given by $X = (I - M)^{-1}D$. Discuss the validity of each step in the following solutions of this equation. (Assume that all necessary inverses exist.) Are both solutions correct?

(A)
$$X = MX + D$$
$$X - MX = D$$
$$X(I - M) = D$$
$$X = D(I - M)^{-1}$$

(B)
$$X - MX + D$$
$$-D = MX - X$$
$$-D = (M - I)X$$
$$X = (M - I)^{-1}(-D)$$

Applications

40. Break-even analysis. A cookware manufacturer is preparing to market a new pasta machine. The company's fixed costs for research, development, tooling, etc., are $243,000 and the variable costs are $22.45 per machine. The company sells the pasta machine for $59.95.

(A) Find the cost and revenue equations.

(B) Find the break-even point.

(C) Graph both equations in the same coordinate system and show the break-even point. Use the graph to determine the production levels that will result in a profit and in a loss.

41. Resource allocation. An international mining company has two mines in Voisey's Bay and Hawk Ridge. The composition of the ore from each field is given in the table. How many tons of ore from each mine should be used to obtain exactly 6 tons of nickel and 8 tons of copper?

Mine	Nickel (%)	Copper (%)
Voisey's Bay	2	4
Hawk Ridge	3	2

42. Resource allocation.

(A) Set up Problem 41 as a matrix equation and solve using the inverse of the coefficient matrix.

(B) Solve Problem 41 as in part (A) if 7.5 tons of nickel and 7 tons of copper are needed.

43. Business leases. A grain company wants to lease a fleet of 20 covered hopper railcars with a combined capacity of 108,000 cubic feet. Hoppers with three different carrying capacities are available: 3,000 cubic feet, 4,500 cubic feet, and 6,000 cubic feet.

(A) How many of each type of hopper should they lease?

(B) The monthly rates for leasing these hoppers are $180 for 3,000 cubic feet, $225 for 4,500 cubic feet, and $325 for 6,000 cubic feet. Which of the solutions in part (A) would minimize the monthly leasing costs?

44. Material costs. A manufacturer wishes to make two different bronze alloys in a metal foundry. The quantities of copper, tin, and zinc needed are indicated in matrix M. The costs for these materials (in dollars per pound) from two suppliers are summarized in matrix N. The company must choose one supplier or the other.

$$M = \begin{bmatrix} \text{Copper} & \text{Tin} & \text{Zinc} \\ 4{,}800 \text{ lb} & 600 \text{ lb} & 300 \text{ lb} \\ 6{,}000 \text{ lb} & 1{,}400 \text{ lb} & 700 \text{ lb} \end{bmatrix} \begin{matrix} \text{Alloy 1} \\ \text{Alloy 2} \end{matrix}$$

$$N = \begin{bmatrix} \text{Supplier A} & \text{Supplier B} \\ \$0.75 & \$0.70 \\ \$6.50 & \$6.70 \\ \$0.40 & \$0.50 \end{bmatrix} \begin{matrix} \text{Copper} \\ \text{Tin} \\ \text{Zinc} \end{matrix}$$

(A) Discuss possible interpretations of the elements in the matrix products MN and NM.

(B) If either product MN or NM has a meaningful interpretation, find the product and label its rows and columns.

(C) Discuss methods of matrix multiplication that can be used to determine the supplier that will provide the necessary materials at the lowest cost.

45. Labor costs. A company with manufacturing plants in California and Texas has labor-hour and wage requirements for the manufacture of two inexpensive calculators as given in matrices M and N below:

	Fabricating department	Assembly department	Packaging department
$M =$	0.15 hr	0.10 hr	0.05 hr	Model A
	0.25 hr	0.20 hr	0.05 hr	Model B

Labor-hours per calculator

	California plant	Texas plant	
$N =$	\$12	\$10	Fabricating department
	\$15	\$12	Assembly department
	\$ 7	\$ 6	Packaging department

Hourly wages

(A) Find the labor cost for producing one model B calculator at the California plant.

(B) Discuss possible interpretations of the elements in the matrix products MN and NM.

(C) If either product MN or NM has a meaningful interpretation, find the product and label its rows and columns.

46. Investment analysis. A person has $5,000 to invest, part at 5% and the rest at 10%. How much should be invested at each rate to yield $400 per year? Solve using augmented matrix methods.

47. Investment analysis. Solve Problem 46 by using a matrix equation and the inverse of the coefficient matrix.

48. Investment analysis. In Problem 46, is it possible to have an annual yield of $200? Of $600? Describe all possible annual yields.

49. Ticket prices. An outdoor amphitheater has 25,000 seats. Ticket prices are $8, $12, and $20, and the number of tickets priced at $8 must equal the number priced at $20. How many tickets of each type should be sold (assuming that all seats can be sold) to bring in each of the returns indicated in the table? Solve using the inverse of the coefficient matrix.

	Concert		
	1	**2**	**3**
Tickets sold	25,000	25,000	25,000
Return required	$320,000	$330,000	$340,000

50. Ticket prices. Discuss the effect on the solutions to Problem 49 if it is no longer required to have an equal number of $8 tickets and $20 tickets.

51. Input–output analysis. An economy is based on two industrial sectors, agriculture and fabrication. Production of a dollar's worth of agriculture requires an input of $0.30 from the agriculture sector and $0.20 from the fabrication sector. Production of a dollar's worth of fabrication requires $0.10 from the agriculture sector and $0.40 from the fabrication sector.

(A) Find the output for each sector that is needed to satisfy a final demand of $50 billion for agriculture and $20 billion for fabrication.

(B) Find the output for each sector that is needed to satisfy a final demand of $80 billion for agriculture and $60 billion for fabrication.

52. Cryptography. The following message was encoded with the matrix B shown below. Decode the message.

7 25 30 19 6 24 20 8 28 5 14 14

9 23 28 15 6 21 13 1 14 21 26 29

$$B = \begin{bmatrix} 1 & 1 & 0 \\ 1 & 0 & 1 \\ 1 & 1 & 1 \end{bmatrix}$$

53. Traffic flow. The rush-hour traffic flow (in vehicles per hour) for a network of four one-way streets is shown in the figure.

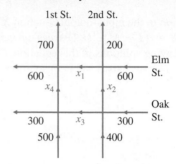

(A) Write the system of equations determined by the flow of traffic through the four intersections.

(B) Find the solution of the system in part (A).

(C) What is the maximum number of vehicles per hour that can travel from Oak Street to Elm Street on 1st Street? What is the minimum number?

(D) If traffic lights are adjusted so that 500 vehicles per hour travel from Oak Street to Elm Street on 1st Street, determine the flow around the rest of the network.

5 Linear Inequalities and Linear Programming

5.1 Linear Inequalities in Two Variables

5.2 Systems of Linear Inequalities in Two Variables

5.3 Linear Programming in Two Dimensions: A Geometric Approach

Chapter 5 Summary and Review

Review Exercises

Introduction

Real-world problems often involve limitations on materials, time, and money. To express such constraints mathematically, we formulate systems of inequalities. In Chapter 5 we discuss systems of inequalities in two variables and introduce a relatively new mathematical tool called linear programming. Linear programming can be used to determine how resources should be allocated in order to maximize profit (see Problem 39 in Section 5.3 on manufacturing water skis).

5.1   Linear Inequalities in Two Variables

- Graphing Linear Inequalities in Two Variables
- Application

Graphing Linear Inequalities in Two Variables

We know how to graph first-degree equations such as

$$y = 2x - 3 \quad \text{and} \quad 2x - 3y = 5$$

but how do we graph first-degree inequalities such as the following?

$$y \le 2x - 3 \quad \text{and} \quad 2x - 3y > 5$$

We will find that graphing these inequalities is similar to graphing the equations, but first we must discuss some important subsets of a plane in a rectangular coordinate system.

A line divides the plane into two regions called **half-planes**. A vertical line divides it into **left** and **right half-planes**; a nonvertical line divides it into **upper** and **lower half-planes**. In either case, the dividing line is called the **boundary line** of each half-plane, as indicated in Figure 1.

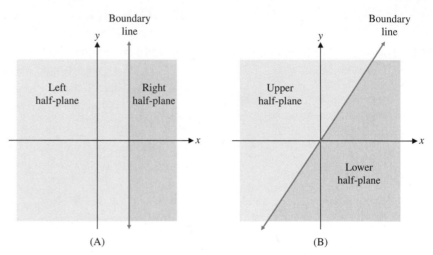

Figure 1   (A)   (B)

To find the half-planes determined by a linear equation such as $y - x = -2$, we rewrite the equation as $y = x - 2$. For any given value of x, there is exactly one value for y such that (x, y) lies on the line. For example, for $x = 4$, we have $y = 4 - 2 = 2$. For the same x and smaller values of y, the point (x, y) will lie below the line since $y < x - 2$. So the lower half-plane corresponds to the solution of the inequality $y < x - 2$. Similarly, the upper half-plane corresponds to $y > x - 2$, as shown in Figure 2.

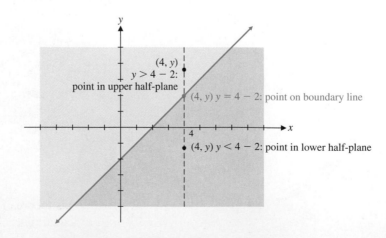

Figure 2

The four inequalities formed from $y = x - 2$, replacing the $=$ sign by $>$, $\geq$, $<$, and $\leq$, respectively, are

$$y > x - 2 \qquad y \geq x - 2 \qquad y < x - 2 \qquad y \leq x - 2$$

The graph of each is a half-plane, excluding the boundary line for $<$ and $>$ and including the boundary line for $\leq$ and $\geq$. In Figure 3, the half-planes are indicated with small arrows on the graph of $y = x - 2$ and then graphed as shaded regions. Excluded boundary lines are shown as dashed lines, and included boundary lines are shown as solid lines.

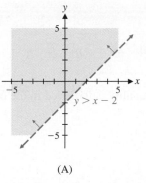
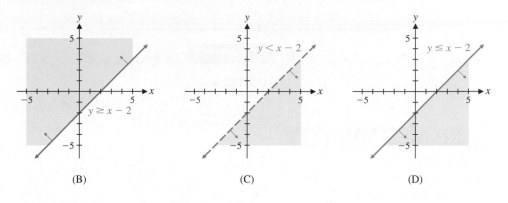

Figure 3

Figure 4 shows the graphs of Figures 3B and 3D on a graphing calculator. Note that it is impossible to show a dotted boundary line when using shading on a calculator.

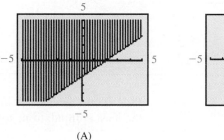

Figure 4   (A)   (B)

The preceding discussion suggests the following theorem, which is stated without proof:

THEOREM 1 Graphs of Linear Inequalities

The graph of the linear inequality

$$Ax + By < C \qquad \text{or} \qquad Ax + By > C$$

with $B \neq 0$, is either the upper half-plane or the lower half-plane (but not both) determined by the line $Ax + By = C$.

   If $B = 0$ and $A \neq 0$, the graph of

$$Ax < C \qquad \text{or} \qquad Ax > C$$

is either the left half-plane or the right half-plane (but not both) determined by the line $Ax = C$.

As a consequence of this theorem, we state a simple and fast mechanical procedure for graphing linear inequalities.

PROCEDURE   Graphing Linear Inequalities

Step 1   First graph $Ax + By = C$ as a dashed line if equality is not included in the original statement, or as a solid line if equality is included.

Step 2   Choose a test point anywhere in the plane not on the line [the origin $(0, 0)$ usually requires the least computation], and substitute the coordinates into the inequality.

Step 3   Does the test point satisfy the original inequality? If so, shade the half-plane that contains the test point. If not, shade the opposite half-plane.

EXAMPLE 1   Graphing a Linear Inequality   Graph $2x - 3y \leq 6$.

SOLUTION

Step 1   Graph $2x - 3y = 6$ as a solid line, since equality is included in the original statement (Fig. 5).

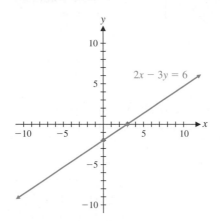

Figure 5

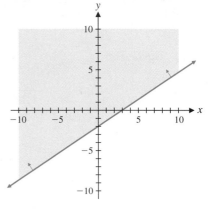

Figure 6

CONCEPTUAL INSIGHT

Recall that the line $2x - 3y = 6$ can be graphed by finding any two points on the line. The x and y intercepts are usually a good choice (see Fig. 5).

x	y
0	−2
3	0

Step 2   Pick a convenient test point above or below the line. The origin $(0, 0)$ requires the least computation, so substituting $(0, 0)$ into the inequality, we get

$$2x - 3y \leq 6$$
$$2(0) - 3(0) = 0 \leq 6$$

This is a true statement; therefore, the point $(0, 0)$ is in the solution set.

Step 3   The line $2x - 3y = 6$ and the half-plane containing the origin form the graph of $2x - 3y \leq 6$, as shown in Figure 6.

Matched Problem 1   Graph $6x - 3y > 18$.

EXAMPLE 2   Graphing Inequalities   Graph

(A) $y > -3$ (B) $2x \leq 5$ (C) $x \leq 3y$

SOLUTION

(A)   Step 1   Graph the horizontal line $y = -3$ as a dashed line, since equality is not included in the original statement (Fig. 7).

Step 2   Substituting $x = 0$ and $y = 0$ in the inequality produces a true statement, so the point $(0, 0)$ is in the solution set.

Step 3   The graph of the solution set is the upper half-plane, excluding the boundary line (Fig. 8).

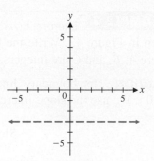

Figure 7

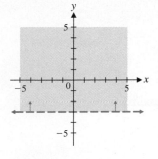

Figure 8

(B) **Step 1** Graph the vertical line $2x = 5$ as a solid line, since equality is included in the original statement (Fig. 9).

Step 2 Substituting $x = 0$ and $y = 0$ in the inequality produces a true statement, so the point $(0, 0)$ is in the solution set.

Step 3 The graph of the solution set is the left half-plane, including the boundary line (Fig. 10).

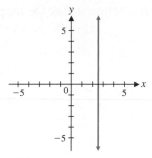

Figure 9

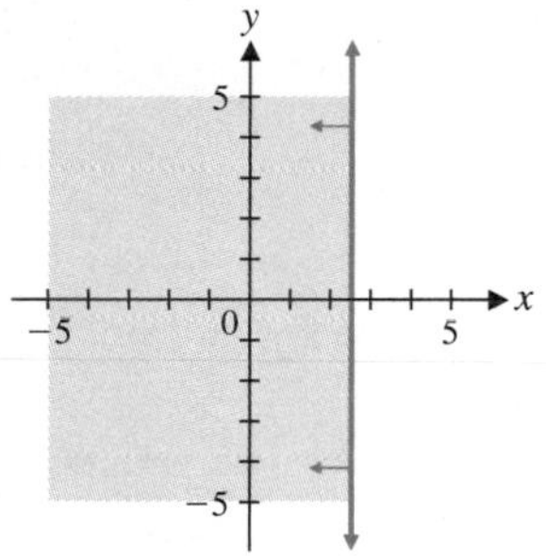

Figure 10

(C) **Step 1** Graph the line $x = 3y$ as a solid line, since equality is included in the original statement (Fig. 11).

Step 2 Since the line passes through the origin, we must use a different test point. We choose $(0, 2)$ for a test point and conclude that this point is in the solution set.

Step 3 The graph of the solution set is the upper half-plane, including the boundary line (Fig. 12).

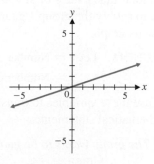

Figure 11

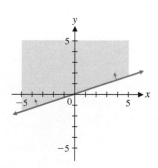

Figure 12

Matched Problem 2 Graph

(A) $y < 4$ (B) $4x \geq -9$ (C) $3x \geq 2y$

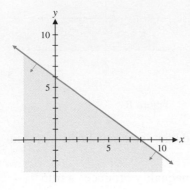

Figure 13

EXAMPLE 3   Interpreting a Graph   Find the linear inequality whose graph is given in Figure 13. Write the boundary line equation in the form $Ax + By = C$, where A, B, and C are integers, before stating the inequality.

SOLUTION   The boundary line (Fig. 13) passes through the points $(0, 6)$ and $(8, 0)$. We use the slope-intercept form to find the equation of this line:

$$\text{Slope: } m = \frac{0 - 6}{8 - 0} = -\frac{6}{8} = -\frac{3}{4}$$

$$y \text{ intercept: } b = 6$$

$$\text{Boundary line equation: } y = -\frac{3}{4}x + 6 \qquad \text{Multiply both sides by 4.}$$

$$4y = -3x + 24 \qquad \text{Add } 3x \text{ to both sides.}$$

$$3x + 4y = 24 \qquad \text{Form: } Ax + By = C$$

Since $(0, 0)$ is in the shaded region in Figure 13 and the boundary line is solid, the graph in Figure 13 is the graph of $3x + 4y \leq 24$.

Matched Problem 3   Find the linear inequality whose graph is given in Figure 14. Write the boundary line equation in the form $Ax + By = C$, where A, B, and C are integers, before stating the inequality.

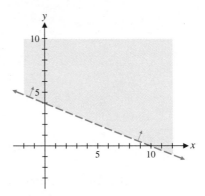

Figure 14

Application

EXAMPLE 4   Sales   A concert promoter wants to book a rock group for a stadium concert. A ticket for admission to the stadium playing field will cost $125, and a ticket for a seat in the stands will cost $175. The group wants to be guaranteed total ticket sales of at least $700,000. How many tickets of each type must be sold to satisfy the group's guarantee? Express the answer as a linear inequality and draw its graph.

SOLUTION   Let $x =$ Number of tickets sold for the playing field

$y =$ Number of tickets sold for seats in the stands

We use these variables to translate the following statement from the problem into a mathematical statement:

The group wants to be guaranteed total ticket sales of at least $700,000.

$$\begin{pmatrix} \text{Sales for the} \\ \text{playing field} \end{pmatrix} + \begin{pmatrix} \text{Sales for seats} \\ \text{in the stands} \end{pmatrix} \quad \begin{pmatrix} \text{At} \\ \text{least} \end{pmatrix} \quad \begin{pmatrix} \text{Total sales} \\ \text{guaranteed} \end{pmatrix}$$

$$125x \quad + \quad 175y \quad \geq \quad 700,000$$

Dividing both sides of this inequality by 25, x and y must satisfy

$$5x + 7y \geq 28,000$$

We use the three-step procedure to graph this inequality.

Step 1 Graph $5x + 7y = 28{,}000$ as a solid line (Fig. 15).

Step 2 Substituting $x = 0$ and $y = 0$ in the inequality produces a false statement, so the point $(0,0)$ is not in the solution set.

Step 3 The graph of the inequality is the upper half-plane including the boundary line (Fig. 16), but does this graph really represent ticket sales?

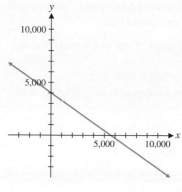

Figure 15

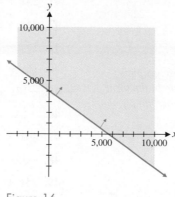

Figure 16   Figure 17

The shaded region in Figure 16 contains points in the second quadrant (where $x < 0$) and the fourth quadrant (where $y < 0$). It is not possible to sell a negative number of tickets, so we must restrict both x and y to the first quadrant. With this restriction, the solution becomes

$$5x + 7y \geq 28{,}000$$
$$x \geq 0, \; y \geq 0$$

and the graph is shown in Figure 17. There is yet another restriction on x and y. It is not possible to sell a fractional number of tickets, so both x and y must be integers. So the solutions of the original problem correspond to those points of the shaded region in Figure 17 that have integer coordinates. This restriction is not indicated in Figure 17, because the points with integer coordinates are too close together (about 9,000 such points per inch) to be visually distinguishable from other points.

Matched Problem 4 A food vendor at a rock concert sells hot dogs for $4 and hamburgers for $5. How many of these sandwiches must be sold to produce sales of at least $1,000? Express the answer as a linear inequality and draw its graph.

Exercises 5.1

Skills Warm-up Exercises

W *In Problems 1–8, if necessary, review Section 1.2.*

1. Is the point $(3,5)$ on the line $y = 2x + 1$?

2. Is the point $(7,9)$ on the line $y = 3x - 11$?

3. Is the point $(3,5)$ in the solution set of $y \leq 2x + 1$?

4. Is the point $(7,9)$ in the solution set of $y \leq 3x - 11$?

5. Is the point $(10,12)$ on the line $13x - 11y = 2$?

6. Is the point $(21,25)$ on the line $30x - 27y = 1$?

7. Is the point $(10,12)$ in the solution set of $13x - 11y \geq 2$?

8. Is the point $(21,25)$ in the solution set of $30x - 27y \leq 1$?

Graph each inequality in Problems 9–18.

9. $y \leq x - 1$

10. $y > x + 1$

11. $3x - 2y > 6$

12. $2x - 5y \leq 10$

13. $x \geq -4$

14. $y < 5$

15. $6x + 4y \geq 24$

16. $4x + 8y \geq 32$

17. $5x \leq -2y$

18. $6x \geq 4y$

In Problems 19–22,

(A) *graph the set of points that satisfy the inequality.*

(B) *graph the set of points that do not satisfy the inequality.*

19. $2x + 3y < 18$

20. $3x + 4y > 24$

21. $5x - 2y \geq 20$

22. $3x - 5y \leq 30$

In Problems 23–28, define the variable and translate the sentence into an inequality.

23. The number of overtime hours is less than 20.

24. Fixed costs are less than $8,000.

25. The annual salary is at least $65,000.

26. Full-time status requires at least 12 credit hours.

27. No more than 1,700 freshmen are admitted.

28. The annual deficit exceeds $600 billion.

In Exercises 29–34, state the linear inequality whose graph is given in the figure. Write the boundary-line equation in the form Ax + By = C, where A, B, and C are integers, before stating the inequality.

29.
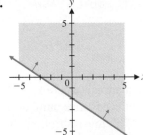

30.
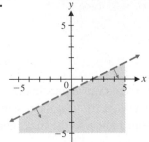

31.
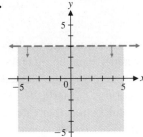

32.
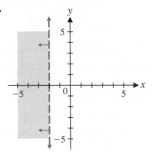

33.
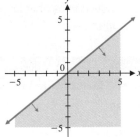

34.
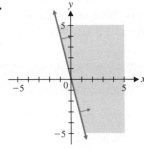

In Problems 35–40, define two variables and translate the sentence into an inequality.

35. Enrollment in finite mathematics plus enrollment in calculus is less than 300.

36. New-car sales and used-car sales combined are at most $500,000.

37. Revenue is at least $20,000 under the cost.

38. The Democratic candidate beat the Republican by at least seven percentage points.

39. The number of grams of saturated fat is more than three times the number of grams of unsaturated fat.

40. The plane is at least 500 miles closer to Chicago than to Denver.

In Problems 41–50, graph each inequality subject to the non-negative restrictions.

41. $25x + 40y \le 3{,}000$, $x \ge 0$, $y \ge 0$

42. $24x + 30y > 7{,}200$, $x \ge 0$, $y \ge 0$

43. $15x - 50y < 1{,}500$, $x \ge 0$, $y \ge 0$

44. $16x - 12y \ge 4{,}800$, $x \ge 0$, $y \ge 0$

45. $-18x + 30y \ge 2{,}700$, $x \ge 0$, $y \ge 0$

46. $-14x + 22y < 1{,}540$, $x \ge 0$, $y \ge 0$

47. $40x - 55y > 0$, $x \ge 0$, $y \ge 0$

48. $-35x + 75y \le 0$, $x \ge 0$, $y \ge 0$

49. $25x + 75y < -600$, $x \ge 0$, $y \ge 0$

50. $75x + 25y > -600$, $x \ge 0$, $y \ge 0$

Applications

In Problems 51–62, express your answer as a linear inequality with appropriate nonnegative restrictions and draw its graph.

51. **Seed costs.** Seed costs for a farmer are $40 per acre for corn and $32 per acre for soybeans. How many acres of each crop should the farmer plant if he wants to spend no more than $5,000 on seed?

52. **Labor costs.** Labor costs for a farmer are $55 per acre for corn and $45 per acre for soybeans. How many acres of each crop should he plant if he wants to spend no more than $6,900 on labor?

53. **Fertilizer.** A farmer wants to use two brands of fertilizer for his corn crop. Brand *A* contains 26% nitrogen, 3% phosphate, and 3% potash. Brand *B* contains 16% nitrogen, 8% phosphate, and 8% potash.
    (*Source:* Spectrum Analytic, Inc.)

    (A) How many pounds of each brand of fertilizer should he add to each acre if he wants to add at least 120 pounds of nitrogen to each acre?

    (B) How many pounds of each brand of fertilizer should he add to each acre if he wants to add at most 28 pounds of phosphate to each acre?

54. **Fertilizer.** A farmer wants to use two brands of fertilizer for his soybean crop. Brand *A* contains 18% nitrogen, 24% phosphate, and 12% potash. Brand *B* contains 5% nitrogen, 10% phosphate, and 15% potash.
    (*Source:* Spectrum Analytic, Inc.)

    (A) How many pounds of each brand of fertilizer should he add to each acre if he wants to add at least 50 pounds of phosphate to each acre?

(B) How many pounds of each brand of fertilizer should he add to each acre if he wants to add at most 60 pounds of potash to each acre?

55. **Textiles.** A textile mill uses two blended yarns—a standard blend that is 30% acrylic, 30% wool, and 40% nylon and a deluxe blend that is 9% acrylic, 39% wool, and 52% nylon —to produce various fabrics. How many pounds of each yarn should the mill use to produce a fabric that is at least 20% acrylic?

56. **Textiles.** Refer to Exercise 55. How many pounds of each yarn should the mill use to produce a fabric that is at least 45% nylon?

57. **Customized vehicles.** A company uses sedans and minivans to produce custom vehicles for transporting hotel guests to and from airports. Plant A can produce 10 sedans and 8 minivans per week, and Plant B can produce 8 sedans and 6 minivans per week. How many weeks should each plant operate in order to produce at least 400 sedans?

58. **Customized vehicles.** Refer to Exercise 57. How many weeks should each plant operate in order to produce at least 480 minivans?

59. **Political advertising.** A candidate has budgeted $10,000 to spend on radio and television advertising. A radio ad costs $200 per 30-second spot, and a television ad costs $800 per 30-second spot. How many radio and television spots can the candidate purchase without exceeding the budget?

60. **Political advertising.** Refer to Problem 59. The candidate decides to replace the television ads with newspaper ads that cost $500 per ad. How many radio spots and newspaper ads can the candidate purchase without exceeding the budget?

61. **Mattresses.** A company produces foam mattresses in two sizes: regular and king. It takes 5 minutes to cut the foam for a regular mattress and 6 minutes for a king mattress. If the cutting department has 50 labor-hours available each day, how many regular and king mattresses can be cut in one day?

62. **Mattresses.** Refer to Problem 61. It takes 15 minutes to cover a regular mattress and 20 minutes to cover a king mattress. If the covering department has 160 labor-hours available each day, how many regular and king mattresses can be covered in one day?

Answers to Matched Problems

1.

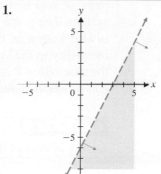

2. (A)

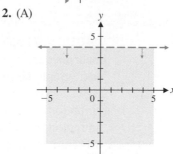

(B)   (C)

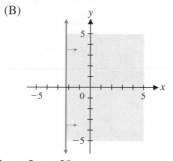   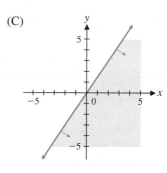

3. $2x + 5y > 20$

4. Let x = Number of hot dogs sold
   y = Number of hamburgers sold
   $4x + 5y \geq 1,000 \qquad x \geq 0, y \geq 0$

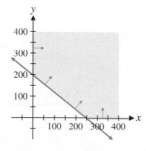

5.2 Systems of Linear Inequalities in Two Variables

- Solving Systems of Linear Inequalities Graphically
- Applications

Solving Systems of Linear Inequalities Graphically

We now consider systems of linear inequalities such as

$$x + y \geq 6 \quad \text{and} \quad 2x + y \leq 22$$
$$2x - y \geq 0 \qquad\qquad\qquad x + y \leq 13$$
$$2x + 5y \leq 50$$
$$x \geq 0$$
$$y \geq 0$$

We wish to **solve** such systems **graphically**—that is, to find the graph of all ordered pairs of real numbers (x, y) that simultaneously satisfy all the inequalities in the system. The graph is called the **solution region** for the system (the solution region is also known as the **feasible region**). To find the solution region, we graph each inequality in the system and then take the intersection of all the graphs. To simplify the discussion that follows, *we consider only systems of linear inequalities where equality is included in each statement in the system.*

EXAMPLE 1   Solving a System of Linear Inequalities Graphically   Solve the following system of linear inequalities graphically:

$$x + y \geq 6$$
$$2x - y \geq 0$$

SOLUTION   Graph the line $x + y = 6$ and shade the region that satisfies the linear inequality $x + y \geq 6$. This region is shaded with gray lines in Figure 1A. Next, graph the line $2x - y = 0$ and shade the region that satisfies the inequality $2x - y \geq 0$. This region is shaded with blue lines in Figure 1A. The solution region for the system of inequalities is the intersection of these two regions. This is the region shaded in both gray and blue (crosshatched) in Figure 1A and redrawn in Figure 1B with only the solution region shaded. The coordinates of any point in the shaded region of Figure 1B specify a solution to the system. For example, the points $(2,4)$, $(6,3)$, and $(7.43, 8.56)$ are three of infinitely many solutions, as can be easily checked. The intersection point $(2,4)$ is obtained by solving the equations $x + y = 6$ and $2x - y = 0$ simultaneously using any of the techniques discussed in Chapter 4.

CONCEPTUAL INSIGHT

To check that you have shaded a solution region correctly, choose a test point in the region and check that it satisfies each inequality in the system. For example, choosing the point $(5, 5)$ in the shaded region in Figure 1B, we have

$$x + y \geq 6 \qquad 2x - y \geq 0$$
$$5 + 5 \overset{?}{\geq} 6 \qquad 10 - 5 \overset{?}{\geq} 0$$
$$10 \overset{\checkmark}{\geq} 6 \qquad 5 \overset{\checkmark}{\geq} 0$$

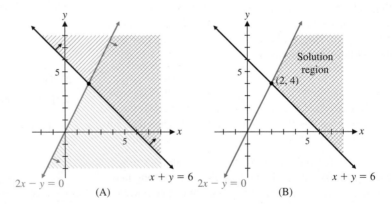

Figure 1   (A)   (B)

Matched Problem 1   Solve the following system of linear inequalities graphically:

$$3x + y \leq 21$$
$$x - 2y \leq 0$$

The points of intersection of the lines that form the boundary of a solution region will play a fundamental role in the solution of linear programming problems, which are discussed in the next section.

DEFINITION   Corner Point
A **corner point** of a solution region is a point in the solution region that is the intersection of two boundary lines.

For example, the point $(2, 4)$ is the only corner point of the solution region in Example 1 (Fig. 1B).

EXAMPLE 2 Solving a System of Linear Inequalities Graphically Solve the following system of linear inequalities graphically and find the corner points:

$$2x + y \leq 22$$
$$x + y \leq 13$$
$$2x + 5y \leq 50$$
$$x \geq 0$$
$$y \geq 0$$

SOLUTION The inequalities $x \geq 0$ and $y \geq 0$ indicate that the solution region will lie in the first quadrant. So we can restrict our attention to that portion of the plane. First, we graph the lines

$$2x + y = 22$$ Find the x and y intercepts of each line; then sketch
$$x + y = 13$$ the line through these points.
$$2x + 5y = 50$$

Next, choosing $(0,0)$ as a test point, we see that the graph of each of the first three inequalities in the system consists of its corresponding line and the half-plane lying below the line, as indicated by the small arrows in Figure 2. The solution region of the system consists of the points in the first quadrant that simultaneously lie on or below all three of these lines (see the shaded region in Fig. 2).

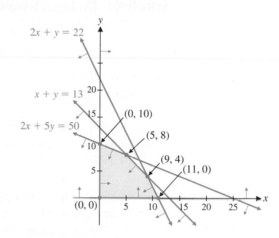

Figure 2

The corner points $(0,0)$, $(0,10)$, and $(11,0)$ can be determined from the graph. The other two corner points are determined as follows:

Solve the system	Solve the system
$2x + 5y = 50$	$2x + y = 22$
$x + y = 13$	$x + y = 13$
to obtain $(5,8)$.	to obtain $(9,4)$.

Note that the lines $2x + 5y = 50$ and $2x + y = 22$ also intersect, but the intersection point is not part of the solution region and so is not a corner point.

Matched Problem 2 Solve the following system of linear inequalities graphically and find the corner points:

$$5x + y \geq 20$$
$$x + y \geq 12$$
$$x + 3y \geq 18$$
$$x \geq 0$$
$$y \geq 0$$

If we compare the solution regions of Examples 1 and 2, we see that there is a fundamental difference between these two regions. We can draw a circle around the solution region in Example 2; however, it is impossible to include all the points in the solution region in Example 1 in any circle, no matter how large we draw it. This leads to the following definition:

> **DEFINITION   Bounded and Unbounded Solution Regions**
> A solution region of a system of linear inequalities is **bounded** if it can be enclosed within a circle. If it cannot be enclosed within a circle, it is **unbounded**.

The solution region for Example 2 is bounded, and the solution region for Example 1 is unbounded. This definition will be important in the next section.

Applications

EXAMPLE 3   Nutrition   A patient on a brown rice and skim milk diet is required to have at least 800 calories and at least 32 grams of protein per day. Each serving of brown rice contains 200 calories and 5 grams of protein. Each serving of skim milk contains 80 calories and 8 grams of protein. How many servings of each food should be eaten per day to meet the minimum daily requirements?

SOLUTION   To answer the question, we need to solve for x and y, where

$$x = \text{number of daily servings of brown rice}$$
$$y = \text{number of daily servings of skim milk}$$

We arrange the information given in the problem in a table, with columns corresponding to x and y.

	Brown Rice	**Skim Milk**	**Minimum Daily Requirement**
Calories	200 cal/svg	80 cal/svg	800 cal
Protein	5 g/svg	8 g/svg	32 g

The number of calories in x servings of brown rice is $200x$, and the number of calories in y servings of skim milk is $80y$. So, to meet the minimum daily requirement for calories, $200x + 80y$ must be greater than or equal to 800. This gives the first of the inequalities below. The second inequality expresses the condition that the minimum daily requirement for protein is met. The last two inequalities express the fact that the number of servings of each food cannot be a negative number.

$$200x + 80y \geq 800 \qquad \text{Requirement for calories}$$
$$5x + 8y \geq 32 \qquad \text{Requirement for protein}$$
$$x \geq 0 \qquad \text{Nonnegative restriction on } x$$
$$y \geq 0 \qquad \text{Nonnegative restriction on } y$$

We graph this system of inequalities, and shade the solution region (Figure 3). Each point in the shaded area, including the straight-line boundaries, will meet the minimum daily requirements for calories and protein; any point outside the shaded area will not. For example, 4 servings of brown rice and 2 servings of skim milk will meet the minimum daily requirements, but 3 servings of brown rice and 2 servings of skim milk will not. Note that the solution region is unbounded.

Matched Problem 3 ⌡ A manufacturing plant makes two types of inflatable boats—a two-person boat and a four-person boat. Each two-person boat requires 0.9 labor-hour in the cutting department and 0.8 labor-hour in the assembly

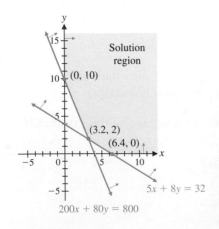

Figure 3

department. Each four-person boat requires 1.8 labor-hours in the cutting department and 1.2 labor-hours in the assembly department. The maximum labor-hours available each month in the cutting and assembly departments are 864 and 672, respectively.

(A) Summarize this information in a table.

(B) If x two-person boats and y four-person boats are manufactured each month, write a system of linear inequalities that reflects the conditions indicated. Graph the feasible region.

Exercises 5.2

Skills Warm-up Exercises

W *For Problems 1–8, if necessary, review Section 1.2. Problems 1–4 refer to the following system of linear inequalities:*

$$4x + y \leq 20$$
$$3x + 5y \leq 37$$
$$x \geq 0$$
$$y \geq 0$$

1. Is the point $(3, 5)$ in the solution region?

2. Is the point $(4, 5)$ in the solution region?

3. Is the point $(3, 6)$ in the solution region?

4. Is the point $(2, 6)$ in the solution region?

Problems 5–8 refer to the following system of linear inequalities:

$$5x + y \leq 32$$
$$7x + 4y \geq 45$$
$$x \geq 0$$
$$y \geq 0$$

5. Is the point $(4, 3)$ in the solution region?

6. Is the point $(5, 3)$ in the solution region?

7. Is the point $(6, 2)$ in the solution region?

8. Is the point $(5, 2)$ in the solution region?

In Problems 9–12, match the solution region of each system of linear inequalities with one of the four regions shown in the figure.

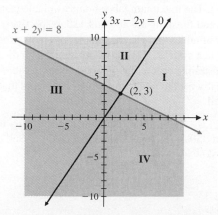

9. $x + 2y \leq 8$
   $3x - 2y \geq 0$

10. $x + 2y \geq 8$
    $3x - 2y \leq 0$

11. $x + 2y \geq 8$
    $3x - 2y \geq 0$

12. $x + 2y \leq 8$
    $3x - 2y \leq 0$

In Problems 13–16, solve each system of linear inequalities graphically.

13. $3x + y \geq 6$
    $x \leq 4$

14. $3x + 4y \leq 12$
    $y \geq -3$

15. $x - 2y \leq 12$
    $2x + y \geq 4$

16. $2x + 5y \leq 20$
    $x - 5y \geq -5$

In Problems 17–20, match the solution region of each system of linear inequalities with one of the four regions shown in the figure. Identify the corner points of each solution region.

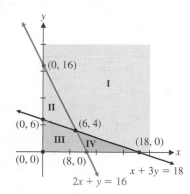

17. $x + 3y \leq 18$
    $2x + y \geq 16$
    $x \geq 0$
    $y \geq 0$

18. $x + 3y \leq 18$
    $2x + y \leq 16$
    $x \geq 0$
    $y \geq 0$

19. $x + 3y \geq 18$
    $2x + y \geq 16$
    $x \geq 0$
    $y \geq 0$

20. $x + 3y \geq 18$
    $2x + y \leq 16$
    $x \geq 0$
    $y \geq 0$

Solve the systems in Problems 21–30 graphically and indicate whether each solution region is bounded or unbounded. Find the coordinates of each corner point.

21. $2x + 3y \leq 12$
$\quad\quad x \geq 0$
$\quad\quad y \geq 0$

22. $3x + 4y \leq 24$
$\quad\quad x \geq 0$
$\quad\quad y \geq 0$

23. $2x + y \leq 10$
$\quad\quad x + 2y \leq 8$
$\quad\quad\quad x \geq 0$
$\quad\quad\quad y \geq 0$

24. $6x + 3y \leq 24$
$\quad\quad 3x + 6y \leq 30$
$\quad\quad\quad x \geq 0$
$\quad\quad\quad y \geq 0$

25. $2x + y \geq 10$
$\quad\quad x + 2y \geq 8$
$\quad\quad\quad x \geq 0$
$\quad\quad\quad y \geq 0$

26. $4x + 3y \geq 24$
$\quad\quad 3x + 4y \geq 8$
$\quad\quad\quad x \geq 0$
$\quad\quad\quad y \geq 0$

27. $2x + y \leq 10$
$\quad\quad x + y \leq 7$
$\quad\quad x + 2y \leq 12$
$\quad\quad\quad x \geq 0$
$\quad\quad\quad y \geq 0$

28. $3x + y \leq 21$
$\quad\quad x + y \leq 9$
$\quad\quad x + 3y \leq 21$
$\quad\quad\quad x \geq 0$
$\quad\quad\quad y \geq 0$

29. $2x + y \geq 16$
$\quad\quad x + y \geq 12$
$\quad\quad x + 2y \geq 14$
$\quad\quad\quad x \geq 0$
$\quad\quad\quad y \geq 0$

30. $3x + y \geq 24$
$\quad\quad x + y \geq 16$
$\quad\quad x + 3y \geq 30$
$\quad\quad\quad x \geq 0$
$\quad\quad\quad y \geq 0$

Solve the systems in Problems 31–40 graphically and indicate whether each solution region is bounded or unbounded. Find the coordinates of each corner point.

31. $x + 4y \leq 32$
$\quad\quad 3x + y \leq 30$
$\quad\quad 4x + 5y \geq 51$

32. $x + y \leq 11$
$\quad\quad x + 5y \geq 15$
$\quad\quad 2x + y \geq 12$

33. $4x + 3y \leq 48$
$\quad\quad 2x + y \geq 24$
$\quad\quad\quad\quad\quad x \leq 9$

34. $2x + 3y \geq 24$
$\quad\quad x + 3y \leq 15$
$\quad\quad\quad\quad\quad y \geq 4$

35. $x - y \leq 0$
$\quad\quad 2x - y \leq 4$
$\quad\quad 0 \leq x \leq 8$

36. $2x + 3y \geq 12$
$\quad\quad -x + 3y \leq 3$
$\quad\quad 0 \leq y \leq 5$

37. $-x + 3y \geq 1$
$\quad\quad 5x - y \geq 9$
$\quad\quad x + y \leq 9$
$\quad\quad\quad\quad\quad x \leq 5$

38. $x + y \leq 10$
$\quad\quad 5x + 3y \geq 15$
$\quad\quad -2x + 3y \leq 15$
$\quad\quad 2x - 5y \leq 6$

39. $16x + 13y \leq 120$
$\quad\quad 3x + 4y \geq 25$
$\quad\quad -4x + 3y \leq 11$

40. $2x + 2y \leq 21$
$\quad\quad -10x + 5y \leq 24$
$\quad\quad 3x + 5y \geq 37$

Problems 41 and 42 introduce an algebraic process for finding the corner points of a solution region without drawing a graph. We will discuss this process later in the chapter.

41. Consider the following system of inequalities and corresponding boundary lines:

$$3x + 4y \leq 36 \quad\quad 3x + 4y = 36$$
$$3x + 2y \leq 30 \quad\quad 3x + 2y = 30$$
$$x \geq 0 \quad\quad\quad\quad\quad x = 0$$
$$y \geq 0 \quad\quad\quad\quad\quad y = 0$$

(A) Use algebraic methods to find the intersection points (if any exist) for each possible pair of boundary lines. (There are six different possible pairs.)

(B) Test each intersection point in all four inequalities to determine which are corner points.

42. Repeat Problem 41 for

$$2x + y \leq 16 \quad\quad 2x + y = 16$$
$$2x + 3y \leq 36 \quad\quad 2x + 3y = 36$$
$$x \geq 0 \quad\quad\quad\quad\quad x = 0$$
$$y \geq 0 \quad\quad\quad\quad\quad y = 0$$

Applications

43. Water skis. A manufacturing company makes two types of water skis, a trick ski and a slalom ski. The trick ski requires 6 labor-hours for fabricating and 1 labor-hour for finishing. The slalom ski requires 4 labor-hours for fabricating and 1 labor-hour for finishing. The maximum labor-hours available per day for fabricating and finishing are 108 and 24, respectively. If x is the number of trick skis and y is the number of slalom skis produced per day, write a system of linear inequalities that indicates appropriate restraints on x and y. Find the set of feasible solutions graphically for the number of each type of ski that can be produced.

44. Furniture. A furniture manufacturing company manufactures dining-room tables and chairs. A table requires 8 labor-hours for assembling and 2 labor-hours for finishing. A chair requires 2 labor-hours for assembling and 1 labor-hour for finishing. The maximum labor-hours available per day for assembly and finishing are 400 and 120, respectively. If x is the number of tables and y is the number of chairs produced per day, write a system of linear inequalities that indicates appropriate restraints on x and y. Find the set of feasible solutions graphically for the number of tables and chairs that can be produced.

45. Water skis. Refer to Problem 43. The company makes a profit of $50 on each trick ski and a profit of $60 on each slalom ski.

(A) If the company makes 10 trick skis and 10 slalom skis per day, the daily profit will be $1,100. Are there other production schedules that will result in a daily profit of $1,100? How are these schedules related to the graph of the line $50x + 60y = 1,100$?

(B) Find a production schedule that will produce a daily profit greater than $1,100 and repeat part (A) for this schedule.

(C) Discuss methods for using lines like those in parts (A) and (B) to find the largest possible daily profit.

46. Furniture. Refer to Problem 44. The company makes a profit of $50 on each table and a profit of $15 on each chair.

(A) If the company makes 20 tables and 20 chairs per day, the daily profit will be $1,300. Are there other production schedules that will result in a daily profit of $1,300? How are these schedules related to the graph of the line $50x + 15y = 1,300$?

(B) Find a production schedule that will produce a daily profit greater than $1,300 and repeat part (A) for this schedule.

(C) Discuss methods for using lines like those in parts (A) and (B) to find the largest possible daily profit.

47. Plant food. A farmer can buy two types of plant food, mix A and mix B. Each cubic yard of mix A contains 20 pounds of phosphoric acid, 30 pounds of nitrogen, and 5 pounds of potash. Each cubic yard of mix B contains 10 pounds of phosphoric acid, 30 pounds of nitrogen, and 10 pounds of potash. The minimum monthly requirements are 460 pounds of phosphoric acid, 960 pounds of nitrogen, and 220 pounds of potash. If x is the number of cubic yards of mix A used and y is the number of cubic yards of mix B used, write a system of linear inequalities that indicates appropriate restraints on x and y. Find the set of feasible solutions graphically for the amounts of mix A and mix B that can be used.

48. Nutrition. A dietitian in a hospital is to arrange a special diet using two foods. Each ounce of food M contains 30 units of calcium, 10 units of iron, and 10 units of vitamin A. Each ounce of food N contains 10 units of calcium, 10 units of iron, and 30 units of vitamin A. The minimum requirements in the diet are 360 units of calcium, 160 units of iron, and 240 units of vitamin A. If x is the number of ounces of food M used and y is the number of ounces of food N used, write a system of linear inequalities that reflects the conditions indicated. Find the set of feasible solutions graphically for the amount of each kind of food that can be used.

49. Psychology. A psychologist uses two types of boxes when studying mice and rats. Each mouse spends 10 minutes per day in box A and 20 minutes per day in box B. Each rat spends 20 minutes per day in box A and 10 minutes per day in box B. The total maximum time available per day

is 800 minutes for box A and 640 minutes for box B. If x is the number of mice used and y the number of rats used, write a system of linear inequalities that indicates appropriate restrictions on x and y. Find the set of feasible solutions graphically.

Answers to Matched Problems

1.

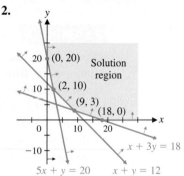

2.

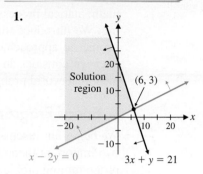

3. (A)

| | Labor-Hours Required | | Maximum Labor-Hours |
	Two-Person Boat	Four-Person Boat	Available per Month
Cutting Department	0.9	1.8	864
Assembly Department	0.8	1.2	672

(B) $0.9x + 1.8y \leq 864$
$0.8x + 1.2y \leq 672$
$x \geq 0$
$y \geq 0$

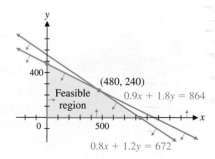

5.3   Linear Programming in Two Dimensions: A Geometric Approach

- A Linear Programming Problem
- General Description of Linear programming
- Geometric Method for Solving Linear Programming Problems
- Applications

Several problems discussed in the preceding section are related to a more general type of problem called a *linear programming problem*. **Linear programming** is a mathematical process that has been developed to help management in decision making. We introduce this topic by considering an example in detail, using an intuitive geometric approach. Insight gained from this approach will prove invaluable when later we consider an algebraic approach that is less intuitive but necessary to solve most real-world problems.

A Linear Programming Problem

We begin our discussion with a concrete example. The solution method will suggest two important theorems and a simple general geometric procedure for solving linear programming problems in two variables.

> **EXAMPLE 1**   Production Scheduling   A manufacturer of lightweight mountain tents makes a standard model and an expedition model. Each standard tent requires 1 labor-hour from the cutting department and 3 labor-hours from the assembly department. Each expedition tent requires 2 labor-hours from the cutting department and 4 labor-hours from the assembly department. The maximum labor-hours available per day in the cutting and assembly departments are 32 and 84, respectively. If the company makes a profit of $50 on each standard tent and $80 on each expedition tent, how many tents of each type should be manufactured each day to maximize the total daily profit (assuming that all tents can be sold)?

SOLUTION   This is an example of a linear programming problem. We begin by analyzing the question posed in this example.

According to the question, the *objective* of management is to maximize profit. Since the profits for standard and expedition tents differ, management must decide how many of each type of tent to manufacture. So it is reasonable to introduce the following **decision variables**:

Let x = number of standard tents produced per day

y = number of expedition tents produced per day

Now we summarize the manufacturing requirements, objectives, and restrictions in Table 1, with the decision variables related to the columns in the table.

Table 1

	Labor-Hours per Tent		Maximum Labor-Hours
	Standard Model	Expedition Model	Available per Day
Cutting department	1	2	32
Assembly department	3	4	84
Profit per tent	$50	$80	

Using the last line of Table 1, we form the **objective function**, in this case the profit P, in terms of the decision variables (we assume that all tents manufactured are sold):

$$P = 50x + 80y \quad \text{Objective function}$$

The **objective** is to find values of the decision variables that produce the **optimal value** (in this case, maximum value) of the objective function.

The form of the objective function indicates that the profit can be made as large as we like, simply by producing enough tents. But any manufacturing company has limits imposed by available resources, plant capacity, demand, and so on. These limits are referred to as **problem constraints**. Using the information in Table 1, we can determine two problem constraints.

$$\begin{pmatrix} \text{daily cutting} \\ \text{time for } x \\ \text{standard tents} \end{pmatrix} + \begin{pmatrix} \text{daily cutting} \\ \text{time for } y \\ \text{expedition tents} \end{pmatrix} \le \begin{pmatrix} \text{maximum labor-} \\ \text{hours available} \\ \text{per day} \end{pmatrix} \quad \begin{array}{l} \text{Cutting} \\ \text{department} \\ \text{constraint} \end{array}$$

$$1x \quad + \quad 2y \quad \le \quad 32$$

$$\begin{pmatrix} \text{daily assembly} \\ \text{time for } x \\ \text{standard tents} \end{pmatrix} + \begin{pmatrix} \text{daily assembly} \\ \text{time for } y \\ \text{expedition tents} \end{pmatrix} \le \begin{pmatrix} \text{maximum labor-} \\ \text{hours available} \\ \text{per day} \end{pmatrix} \quad \begin{array}{l} \text{Assembly} \\ \text{department} \\ \text{constraint} \end{array}$$

$$3x \quad + \quad 4y \quad \le \quad 84$$

It is not possible to manufacture a negative number of tents; thus, we have the **nonnegative constraints**

$$x \ge 0 \text{ and } y \ge 0$$

which we usually write in the form

$$x, y \ge 0 \quad \text{Nonnegative constraints}$$

We now have a **mathematical model** for the problem under consideration:

$$\begin{array}{ll} \text{Maximize} & P = 50x + 80y \quad \text{Objective function} \\ \text{subject to} & \left. \begin{array}{l} x + 2y \le 32 \\ 3x + 4y \le 84 \end{array} \right\} \quad \text{Problem constraints} \\ & x, y \ge 0 \quad \text{Nonnegative constraints} \end{array}$$

Solving the set of linear inequality constraints **graphically**, we obtain the feasible region for production schedules (Fig. 1).

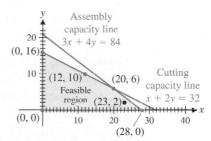

Figure 1

By choosing a production schedule (x, y) from the feasible region, a profit can be determined using the objective function

$$P = 50x + 80y$$

For example, if $x = 12$ and $y = 10$, the profit for the day would be

$$P = 50(12) + 80(10)$$
$$= \$1,400$$

Or if $x = 23$ and $y = 2$, the profit for the day would be

$$P = 50(23) + 80(2)$$
$$= \$1,310$$

Out of all possible production schedules (x, y) from the feasible region, which schedule(s) produces the *maximum* profit? This is a **maximization problem**. Since point-by-point checking is impossible (there are infinitely many points to check), we must find another way.

By assigning P in $P = 50x + 80y$ a particular value and plotting the resulting equation in the coordinate system shown in Figure 1, we obtain a **constant-profit line**. Every point in the feasible region on this line represents a production schedule that will produce the same profit. By doing this for a number of values for P, we obtain a family of constant-profit lines (Fig. 2) that are parallel to each other, since they all have the same slope. To see this, we write $P = 50x + 80y$ in the slope-intercept form

$$y = -\frac{5}{8}x + \frac{P}{80}$$

and note that for any profit P, the constant-profit line has slope $-\frac{5}{8}$. We also observe that as the profit P increases, the y intercept $(P/80)$ increases, and the line moves away from the origin.

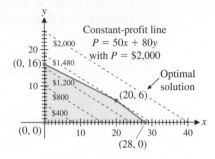

Figure 2 **Constant-profit lines**

Therefore, the maximum profit occurs at a point where a constant-profit line is the farthest from the origin but still in contact with the feasible region; in this example, at $(20, 6)$ (see Fig. 2). So profit is maximized if the manufacturer makes 20 standard tents and 6 expedition tents per day, and the maximum profit is

$$P = 50(20) + 80(6)$$
$$= \$1,480$$

The point $(20, 6)$ is called an **optimal solution** to the problem because it maximizes the objective (profit) function and is in the feasible region. In general, it appears that a maximum profit occurs at one of the corner points. We also note that the minimum profit $(P = 0)$ occurs at the corner point $(0, 0)$.

Matched Problem 1 A manufacturing plant makes two types of inflatable boats—a two-person boat and a four-person boat. Each two-person boat requires 0.9 labor-hour from the cutting department and 0.8 labor-hour from the assembly department. Each four-person boat requires 1.8 labor-hours from the cutting department and 1.2 labor-hours from the assembly department. The maximum labor-hours available per month in the cutting department and the assembly department are 864 and 672, respectively. The company makes a profit of $25 on each two-person boat and $40 on each four-person boat.

(A) Identify the decision variables.

(B) Summarize the relevant material in a table similar to Table 1 in Example 1.

(C) Write the objective function P.

(D) Write the problem constraints and nonnegative constraints.

(E) Graph the feasible region. Include graphs of the objective function for $P = \$5,000$, $P = \$10,000$, $P = \$15,000$, and $P = \$21,600$.

(F) From the graph and constant-profit lines, determine how many boats should be manufactured each month to maximize the profit. What is the maximum profit?

Before proceeding further, let's summarize the steps we used to form the model in Example 1.

> **PROCEDURE** Constructing a Model for an Applied Linear Programming Problem
>
> Step 1 Introduce decision variables.
>
> Step 2 Summarize relevant material in table form, relating columns to the decision variables, if possible (see Table 1).
>
> Step 3 Determine the objective and write a linear objective function.
>
> Step 4 Write problem constraints using linear equations and/or inequalities.
>
> Step 5 Write nonnegative constraints.

Explore and Discuss 1 Refer to the feasible region S shown in Figure 3.

(A) Let $P = x + y$. Graph the constant-profit lines through the points $(5, 5)$ and $(10, 10)$. Place a straightedge along the line with the smaller profit and slide it in the direction of increasing profit, without changing its slope. What is the maximum value of P? Where does this maximum value occur?

(B) Repeat part (A) for $P = x + 10y$.

(C) Repeat part (A) for $P = 10x + y$.

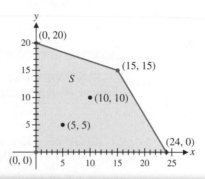

Figure 3

General Description of Linear Programming

In Example 1 and Matched Problem 1, the optimal solution occurs at a corner point of the feasible region. Is this always the case? The answer is a qualified yes, as we will see in Theorem 1. First, we give a few general definitions.

A **linear programming problem** is one that is concerned with finding the **optimal value** (maximum or minimum value) of a linear **objective function** z of the form

$$z = ax + by, \text{ where } a \text{ and } b \text{ do not both} = 0$$

and the **decision variables** x and y are subject to **problem constraints** in the form of $\leq$ or $\geq$ linear inequalities and equations. In addition, the decision variables must satisfy the **nonnegative constraints** $x \geq 0, y \geq 0$. The set of points satisfying both the problem constraints and the nonnegative constraints is called the **feasible region** for the problem. Any point in the feasible region that produces the optimal value of the objective function over the feasible region is called an **optimal solution**.

> **THEOREM 1 Fundamental Theorem of Linear Programming**
>
> If the optimal value of the objective function in a linear programming problem exists, then that value must occur at one or more of the corner points of the feasible region.

Theorem 1 provides a simple procedure for solving a linear programming problem, *provided that the problem has an optimal solution—not all do.* In order to use Theorem 1, we must know that the problem under consideration has an optimal solution. Theorem 2 provides some conditions that will ensure that a linear programming problem has an optimal solution.

> **THEOREM 2 Existence of Optimal Solutions**
>
> (A) If the feasible region for a linear programming problem is bounded, then both the maximum value and the minimum value of the objective function always exist.
>
> (B) If the feasible region is unbounded and the coefficients of the objective function are positive, then the minimum value of the objective function exists but the maximum value does not.
>
> (C) If the feasible region is empty (that is, there are no points that satisfy all the constraints), then both the maximum value and the minimum value of the objective function do not exist.

Geometric Method for Solving Linear Programming Problems

The preceding discussion leads to the following procedure for the geometric solution of linear programming problems with two decision variables:

> **PROCEDURE Geometric Method for Solving a Linear Programming Problem with Two Decision Variables**
>
> Step 1 Graph the feasible region. Then, if an optimal solution exists according to Theorem 2, find the coordinates of each corner point.
>
> Step 2 Construct a **corner point table** listing the value of the objective function at each corner point.
>
> Step 3 Determine the optimal solution(s) from the table in Step 2.
>
> Step 4 For an applied problem, interpret the optimal solution(s) in terms of the original problem.

Before we consider more applications, let's use this procedure to solve some linear programming problems where the model has already been determined.

EXAMPLE 2 Solving a Linear Programming Problem

(A) Minimize and maximize $z = 3x + y$

subject to

$$2x + y \leq 20$$
$$10x + y \geq 36$$
$$2x + 5y \geq 36$$
$$x, y \geq 0$$

(B) Minimize and maximize $z = 10x + 20y$

subject to $6x + 2y \geq 36$

$2x + 4y \geq 32$

$y \leq 20$

$x, y \geq 0$

SOLUTION

(A) Step 1   Graph the feasible region S (Fig. 4). Then, after checking Theorem 2 to determine whether an optimal solution exists, find the coordinates of each corner point. Since S is bounded, z will have both a maximum and a minimum value on S (Theorem 2A) and these will both occur at corner points (Theorem 1).

Step 2   Evaluate the objective function at each corner point, as shown in the table.

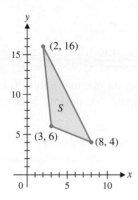
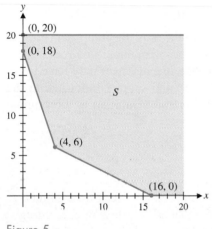

Figure 4

Corner Point	
(x, y)	$z = 3x + y$
$(3, 6)$	15
$(2, 16)$	22
$(8, 4)$	28

Step 3   Determine the optimal solutions from Step 2. Examining the values in the table, we see that the minimum value of z is 15 at $(3, 6)$ and the maximum value of z is 28 at $(8, 4)$.

(B) Step 1   Graph the feasible region S (Fig. 5). Then, after checking Theorem 2 to determine whether an optimal solution exists, find the coordinates of each corner point. Since S is unbounded and the coefficients of the objective function are positive, z has a minimum value on S but no maximum value (Theorem 2B).

Step 2   Evaluate the objective function at each corner point, as shown in the table.

Figure 5

Corner Point	
(x, y)	$z = 10x + 20y$
$(0, 20)$	400
$(0, 18)$	360
$(4, 6)$	160
$(16, 0)$	160

Step 3   Determine the optimal solution from Step 2. The minimum value of z is 160 at $(4, 6)$ and at $(16, 0)$.

The solution to Example 2B is a **multiple optimal solution**. In general, if two corner points are both optimal solutions to a linear programming problem, then any point on the line segment joining them is also an optimal solution. This is the only way that optimal solutions can occur at noncorner points.

Matched Problem 2

(A) Maximize and minimize $z = 4x + 2y$ subject to the constraints given in Example 2A.

(B) Maximize and minimize $z = 20x + 5y$ subject to the constraints given in Example 2B.

CONCEPTUAL INSIGHT

Determining that an optimal solution exists is a critical step in the solution of a linear programming problem. If you skip this step, you may examine a corner point table like the one in the solution of Example 2B and erroneously conclude that the maximum value of the objective function is 400.

Explore and Discuss 2 In Example 2B we saw that there was no optimal solution for the problem of maximizing the objective function z over the feasible region S. We want to add an additional constraint to modify the feasible region so that an optimal solution for the maximization problem does exist. Which of the following constraints will accomplish this objective?

(A) $x \leq 20$ (B) $y \geq 4$ (C) $x \leq y$ (D) $y \leq x$

For an illustration of Theorem 2C, consider the following:

$$\text{Maximize} \quad P = 2x + 3y$$
$$\text{subject to} \quad x + y \geq 8$$
$$x + 2y \leq 8$$
$$2x + y \leq 10$$
$$x, y \geq 0$$

The intersection of the graphs of the constraint inequalities is the empty set (Fig. 6); so the *feasible region is empty.* If this happens, the problem should be reexamined to see if it has been formulated properly. If it has, the management may have to reconsider items such as labor-hours, overtime, budget, and supplies allocated to the project in order to obtain a nonempty feasible region and a solution to the original problem.

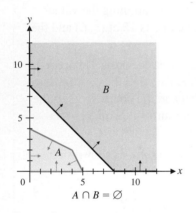

$A \cap B = \varnothing$

Figure 6

Applications

EXAMPLE 3 Medication A hospital patient is required to have at least 84 units of drug A and 120 units of drug B each day (assume that an overdose of either drug is harmless). Each gram of substance M contains 10 units of drug A and 8 units of drug B, and each gram of substance N contains 2 units of drug A and 4 units of drug B. Now, suppose that both M and N contain an undesirable drug D: 3 units per gram in M and 1 unit per gram in N. How many grams of each of substances M and N should be mixed to meet the minimum daily requirements and simultaneously minimize the intake of drug D? How many units of the undesirable drug D will be in this mixture?

SOLUTION First we construct the mathematical model.

Step 1 Introduce decision variables. According to the questions asked, we must decide how many grams of substances M and N should be mixed to form the daily dose of medication. These two quantities are the decision variables:

$$x = \text{number of grams of substance } M \text{ used}$$
$$y = \text{number of grams of substance } N \text{ used}$$

Step 2 Summarize relevant material in a table, relating the columns to substances M and N.

	Amount of Drug per Gram		Minimum Daily Requirement
	Substance M	Substance N	
Drug A	10 units/gram	2 units/gram	84 units
Drug B	8 units/gram	4 units/gram	120 units
Drug D	3 units/gram	1 unit/gram	

Step 3 Determine the objective and the objective function. The objective is to minimize the amount of drug D in the daily dose of medication. Using the decision variables and the information in the table, we form the linear objective function

$$C = 3x + y$$

Step 4 Write the problem constraints. The constraints in this problem involve minimum requirements, so the inequalities will take a different form:

$$10x + 2y \geq 84 \quad \text{Drug } A \text{ constraint}$$
$$8x + 4y \geq 120 \quad \text{Drug } B \text{ constraint}$$

Step 5 Add the nonnegative constraints and summarize the model.

$$\begin{aligned} \text{Minimize} \quad & C = 3x + y & \text{Objective function} \\ \text{subject to} \quad & 10x + 2y \geq 84 & \text{Drug A constraint} \\ & 8x + 4y \geq 120 & \text{Drug B constraint} \\ & x, y \geq 0 & \text{Nonnegative constraints} \end{aligned}$$

Now we use the geometric method to solve the problem.

Step 1 Graph the feasible region (Fig. 7). Then, after checking Theorem 2 to determine whether an optimal solution exists, find the coordinates of each corner point. Since the feasible region is unbounded and the coefficients of the objective function are positive, this minimization problem has a solution.

Step 2 Evaluate the objective function at each corner point, as shown in the table.

Step 3 Determine the optimal solution from Step 2. The optimal solution is $C = 34$ at the corner point $(4, 22)$.

Step 4 Interpret the optimal solution in terms of the original problem. If we use 4 grams of substance M and 22 grams of substance N, we will supply the minimum daily requirements for drugs A and B and minimize the intake of the undesirable drug D at 34 units. (Any other combination of M and N from the feasible region will result in a larger amount of the undesirable drug D.)

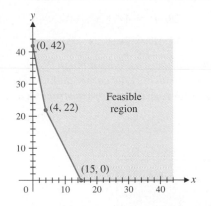

Figure 7

Corner Point

(x, y)	$C = 3x + y$
$(0, 42)$	42
$(4, 22)$	34
$(15, 0)$	45

Matched Problem 3 A chicken farmer can buy a special food mix A at 20¢ per pound and a special food mix B at 40¢ per pound. Each pound of mix A contains 3,000 units of nutrient N_1 and 1,000 units of nutrient N_2; each pound of mix B contains 4,000 units of nutrient N_1 and 4,000 units of nutrient N_2. If the minimum daily requirements for the chickens collectively are 36,000 units of nutrient N_1 and 20,000 units of nutrient N_2, how many pounds of each food mix should be used each day to minimize daily food costs while meeting (or exceeding) the minimum daily nutrient requirements? What is the minimum daily cost? Construct a mathematical model and solve using the geometric method.

> **CONCEPTUAL INSIGHT**
>
> Refer to Example 3. If we change the minimum requirement for drug B from 120 to 125, the optimal solution changes to 3.6 grams of substance M and 24.1 grams of substance N, correct to one decimal place.
>
> Now refer to Example 1. If we change the maximum labor-hours available per day in the assembly department from 84 to 79, the solution changes to 15 standard tents and 8.5 expedition tents.
>
> We can measure 3.6 grams of substance M and 24.1 grams of substance N, but how can we make 8.5 tents? Should we make 8 tents? Or 9 tents? If the solutions to a problem must be integers and the optimal solution found graphically involves decimals, then rounding the decimal value to the nearest integer does not always produce the *optimal integer solution* (see Problem 44, Exercises 5.3). Finding optimal integer solutions to a linear programming problem is called *integer programming* and requires special techniques that are beyond the scope of this book. As mentioned earlier, if we encounter a solution like 8.5 tents per day, we will interpret this as an *average* value over many days of production.

Exercises 5.3

W **Skills Warm-up Exercises**

In problems 1–8, if necessary, review Theorem 1. In Problems 1–4, the feasible region is the set of points on and inside the rectangle with vertices $(0, 0)$, $(12, 0)$, $(0, 5)$, and $(12, 5)$. Find the maximum and minimum values of the objective function Q over the feasible region.

1. $Q = 7x + 14y$ **2.** $Q = 3x + 15y$

3. $Q = 10x - 12y$ **4.** $Q = -9x + 20y$

In Problems 5–8, the feasible region is the set of points on and inside the triangle with vertices $(0, 0)$, $(8, 0)$, and $(0, 10)$. Find the maximum and minimum values of the objective function Q over the feasible region.

5. $Q = -4x - 3y$ **6.** $Q = 3x + 2y$

7. $Q = -6x + 4y$ **8.** $Q = 10x - 8y$

In Problems 9–12, graph the constant-profit lines through $(3, 3)$ and $(6, 6)$. Use a straightedge to identify the corner point where the maximum profit occurs (see Explore and Discuss 1). Confirm your answer by constructing a corner-point table.

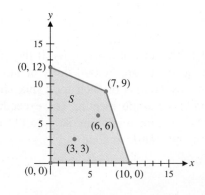

9. $P = x + y$ **10.** $P = 4x + y$

11. $P = 3x + 7y$ **12.** $P = 9x + 3y$

In Problems 13–16, graph the constant-cost lines through $(9, 9)$ and $(12, 12)$. Use a straightedge to identify the corner point where the minimum cost occurs. Confirm your answer by constructing a corner-point table.

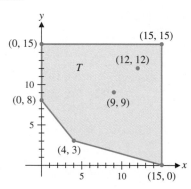

13. $C = 7x + 4y$ **14.** $C = 7x + 9y$

15. $C = 3x + 8y$ **16.** $C = 2x + 11y$

Solve the linear programming problems stated in Problems 17–34.

17. Maximize $P = 5x + 5y$
   subject to $2x + y \le 10$
                 $x + 2y \le 8$
                 $x, y \ge 0$

18. Maximize $P = 3x + 2y$
   subject to $6x + 3y \le 24$
                 $3x + 6y \le 30$
                 $x, y \ge 0$

19. Minimize and maximize
$z = 2x + 3y$
subject to
$$2x + y \geq 10$$
$$x + 2y \geq 8$$
$$x, y \geq 0$$

20. Minimize and maximize
$z = 8x + 7y$
subject to
$$4x + 3y \geq 24$$
$$3x + 4y \geq 8$$
$$x, y \geq 0$$

21. Maximize $P = 30x + 40y$
subject to
$$2x + y \leq 10$$
$$x + y \leq 7$$
$$x + 2y \leq 12$$
$$x, y \geq 0$$

22. Maximize $P = 20x + 10y$
subject to
$$3x + y \leq 21$$
$$x + y \leq 9$$
$$x + 3y \leq 21$$
$$x, y \geq 0$$

23. Minimize and maximize
$z = 10x + 30y$
subject to
$$2x + y \geq 16$$
$$x + y \geq 12$$
$$x + 2y \geq 14$$
$$x, y \geq 0$$

24. Minimize and maximize
$z = 400x + 100y$
subject to
$$3x + y \geq 24$$
$$x + y \geq 16$$
$$x + 3y \geq 30$$
$$x, y \geq 0$$

25. Minimize and maximize
$P = 30x + 10y$
subject to
$$2x + 2y \geq 4$$
$$6x + 4y \leq 36$$
$$2x + y \leq 10$$
$$x, y \geq 0$$

26. Minimize and maximize
$P = 2x + y$
subject to
$$x + y \geq 2$$
$$6x + 4y \leq 36$$
$$4x + 2y \leq 20$$
$$x, y \geq 0$$

27. Minimize and maximize
$P = 3x + 5y$
subject to
$$x + 2y \leq 6$$
$$x + y \leq 4$$
$$2x + 3y \geq 12$$
$$x, y \geq 0$$

28. Minimize and maximize
$P = -x + 3y$
subject to
$$2x - y \geq 4$$
$$-x + 2y \leq 4$$
$$y \leq 6$$
$$x, y \geq 0$$

29. Minimize and maximize
$P = 20x + 10y$
subject to
$$2x + 3y \geq 30$$
$$2x + y \leq 26$$
$$-2x + 5y \leq 34$$
$$x, y \geq 0$$

30. Minimize and maximize
$P = 12x + 14y$
subject to
$$-2x + y \geq 6$$
$$x + y \leq 15$$
$$3x - y \geq 0$$
$$x, y \geq 0$$

31. Maximize $P = 20x + 30y$
subject to
$$0.6x + 1.2y \leq 960$$
$$0.03x + 0.04y \leq 36$$
$$0.3x + 0.2y \leq 270$$
$$x, y \geq 0$$

32. Minimize $C = 30x + 10y$
subject to
$$1.8x + 0.9y \geq 270$$
$$0.3x + 0.2y \geq 54$$
$$0.01x + 0.03y \geq 3.9$$
$$x, y \geq 0$$

33. Maximize $P = 525x + 478y$
subject to
$$275x + 322y \leq 3,381$$
$$350x + 340y \leq 3,762$$
$$425x + 306y \leq 4,114$$
$$x, y \geq 0$$

34. Maximize $P = 300x + 460y$
subject to
$$245x + 452y \leq 4,181$$
$$290x + 379y \leq 3,888$$
$$390x + 299y \leq 4,407$$
$$x, y \geq 0$$

✎ *In Problems 35 and 36, explain why Theorem 2 cannot be used to conclude that a maximum or minimum value exists. Graph the feasible regions and use graphs of the objective function $z = x - y$ for various values of z to discuss the existence of a maximum value and a minimum value.*

35. Minimize and maximize
$z = x - y$
subject to
$$x - 2y \leq 0$$
$$2x - y \leq 6$$
$$x, y \geq 0$$

36. Minimize and maximize

$$z = x - y$$

subject to

$$x - 2y \geq -6$$
$$2x - y \geq 0$$
$$x, y \geq 0$$

37. The corner points for the bounded feasible region determined by the system of linear inequalities

$$x + 2y \leq 10$$
$$3x + y \leq 15$$
$$x, y \geq 0$$

are $O = (0, 0)$, $A = (0, 5)$, $B = (4, 3)$, and $C = (5, 0)$. If $P = ax + by$ and $a, b > 0$, determine conditions on a and b that will ensure that the maximum value of P occurs

(A) only at A

(B) only at B

(C) only at C

(D) at both A and B

(E) at both B and C

38. The corner points for the feasible region determined by the system of linear inequalities

$$x + 4y \geq 30$$
$$3x + y \geq 24$$
$$x, y \geq 0$$

are $A = (0, 24)$, $B = (6, 6)$, and $D = (30, 0)$. If $C = ax + by$ and $a, b > 0$, determine conditions on a and b that will ensure that the minimum value of C occurs

(A) only at A

(B) only at B

(C) only at D

(D) at both A and B

(E) at both B and D

Applications

In Problems 39–54, construct a mathematical model in the form of a linear programming problem. (The answers in the back of the book for these application problems include the model.) Then solve by the geometric method.

39. Water skis. A manufacturing company makes two types of water skis—a trick ski and a slalom ski. The relevant manufacturing data are given in the table below.

Department	Labor-Hours per Ski Trick Ski	Slalom Ski	Maximum Labor-Hours Available per Day
Fabricating	6	4	108
Finishing	1	1	24

(A) If the profit on a trick ski is $40 and the profit on a slalom ski is $30, how many of each type of ski should be manufactured each day to realize a maximum profit? What is the maximum profit?

(B) Discuss the effect on the production schedule and the maximum profit if the profit on a slalom ski decreases to $25.

(C) Discuss the effect on the production schedule and the maximum profit if the profit on a slalom ski increases to $45.

40. Furniture. A furniture manufacturing company manufactures dining-room tables and chairs. The relevant manufacturing data are given in the table below.

Department	Labor-Hours per Unit Table	Chair	Maximum Labor-Hours Available per Day
Assembly	8	2	400
Finishing	2	1	120
Profit per unit	$90	$25	

(A) How many tables and chairs should be manufactured each day to realize a maximum profit? What is the maximum profit?

(B) Discuss the effect on the production schedule and the maximum profit if the marketing department of the company decides that the number of chairs produced should be at least four times the number of tables produced.

41. Production scheduling. A furniture company has two plants that produce the lumber used in manufacturing tables and chairs. In 1 day of operation, plant A can produce the lumber required to manufacture 20 tables and 60 chairs, and plant B can produce the lumber required to manufacture 25 tables and 50 chairs. The company needs enough lumber to manufacture at least 200 tables and 500 chairs.

(A) If it costs $1,000 to operate plant A for 1 day and $900 to operate plant B for 1 day, how many days should each plant be operated to produce a sufficient amount of lumber at a minimum cost? What is the minimum cost?

(B) Discuss the effect on the operating schedule and the minimum cost if the daily cost of operating plant A is reduced to $600 and all other data in part (A) remain the same.

(C) Discuss the effect on the operating schedule and the minimum cost if the daily cost of operating plant B is reduced to $800 and all other data in part (A) remain the same.

42. Computers. An electronics firm manufactures two types of personal computers—a standard model and a portable model. The production of a standard computer requires a capital expenditure of $400 and 40 hours of labor. The production of a portable computer requires a capital expenditure of $250 and 30 hours of labor. The firm has $20,000 capital and 2,160 labor-hours available for production of standard and portable computers.

(A) What is the maximum number of computers the company is capable of producing?

(B) If each standard computer contributes a profit of $320 and each portable model contributes a profit of $220, how much profit will the company make by producing the maximum number of computers determined in part (A)? Is this the maximum profit? If not, what is the maximum profit?

43. Transportation. The officers of a high school senior class are planning to rent buses and vans for a class trip. Each bus can transport 40 students, requires 3 chaperones, and costs $1,200 to rent. Each van can transport 8 students, requires 1 chaperone, and costs $100 to rent. Since there are 400 students in the senior class that may be eligible to go on the trip, the officers must plan to accommodate at least 400 students. Since only 36 parents have volunteered to serve as chaperones, the officers must plan to use at most 36 chaperones. How many vehicles of each type should the officers rent in order to minimize the transportation costs? What are the minimal transportation costs?

44. Transportation. Refer to Problem 43. If each van can transport 7 people and there are 35 available chaperones, show that the optimal solution found graphically involves decimals. Find all feasible solutions with integer coordinates and identify the one that minimizes the transportation costs. Can this optimal integer solution be obtained by rounding the optimal decimal solution? Explain.

45. Investment. An investor has $60,000 to invest in a CD and a mutual fund. The CD yields 5% and the mutual fund yields an average of 9%. The mutual fund requires a minimum investment of $10,000, and the investor requires that at least twice as much should be invested in CDs as in the mutual fund. How much should be invested in CDs and how much in the mutual fund to maximize the return? What is the maximum return?

46. Investment. An investor has $24,000 to invest in bonds of AAA and B qualities. The AAA bonds yield an average of 6%, and the B bonds yield 10%. The investor requires that at least three times as much money should be invested in AAA bonds as in B bonds. How much should be invested in each type of bond to maximize the return? What is the maximum return?

47. Pollution control. Because of new federal regulations on pollution, a chemical plant introduced a new, more expensive process to supplement or replace an older process used in the production of a particular chemical. The older process emitted 20 grams of sulfur dioxide and 40 grams of particulate matter into the atmosphere for each gallon of chemical produced. The new process emits 5 grams of sulfur dioxide and 20 grams of particulate matter for each gallon produced. The company makes a profit of 60¢ per gallon and 20¢ per gallon on the old and new processes, respectively.

(A) If the government allows the plant to emit no more than 16,000 grams of sulfur dioxide and 30,000 grams of particulate matter daily, how many gallons of the chemical should be produced by each process to maximize daily profit? What is the maximum daily profit?

(B) Discuss the effect on the production schedule and the maximum profit if the government decides to restrict emissions of sulfur dioxide to 11,500 grams daily and all other data remain unchanged.

(C) Discuss the effect on the production schedule and the maximum profit if the government decides to restrict emissions of sulfur dioxide to 7,200 grams daily and all other data remain unchanged.

48. Capital expansion. A fast-food chain plans to expand by opening several new restaurants. The chain operates two types of restaurants, drive-through and full-service. A drive-through restaurant costs $100,000 to construct, requires 5 employees, and has an expected annual revenue of $200,000. A full-service restaurant costs $150,000 to construct, requires 15 employees, and has an expected annual revenue of $500,000. The chain has $2,400,000 in capital available for expansion. Labor contracts require that they hire no more than 210 employees, and licensing restrictions require that they open no more than 20 new restaurants. How many restaurants of each type should the chain open in order to maximize the expected revenue? What is the maximum expected revenue? How much of their capital will they use and how many employees will they hire?

49. Fertilizer. A fruit grower can use two types of fertilizer in his orange grove, brand A and brand B. The amounts (in pounds) of nitrogen, phosphoric acid, and chloride in a bag of each brand are given in the table. Tests indicate that the grove needs at least 1,000 pounds of phosphoric acid and at most 400 pounds of chloride.

	Pounds per Bag	
	Brand A	Brand B
Nitrogen	8	3
Phosphoric acid	4	4
Chloride	2	1

(A) If the grower wants to maximize the amount of nitrogen added to the grove, how many bags of each mix should be used? How much nitrogen will be added?

(B) If the grower wants to minimize the amount of nitrogen added to the grove, how many bags of each mix should be used? How much nitrogen will be added?

50. Nutrition. A dietitian is to arrange a special diet composed of two foods, M and N. Each ounce of food M contains 30 units of calcium, 10 units of iron, 10 units of vitamin A, and 8 units of cholesterol. Each ounce of food N contains 10 units of calcium, 10 units of iron, 30 units of vitamin A, and 4 units of cholesterol. If the minimum daily requirements are 360 units of calcium, 160 units of iron, and 240 units of vitamin A, how many ounces of each food should be used to meet the minimum requirements and at the same time minimize the cholesterol intake? What is the minimum cholesterol intake?

51. Plant food. A farmer can buy two types of plant food, mix A and mix B. Each cubic yard of mix A contains 20 pounds of phosphoric acid, 30 pounds of nitrogen, and 5 pounds of potash. Each cubic yard of mix B contains 10 pounds of phosphoric acid, 30 pounds of nitrogen, and 10 pounds of potash. The minimum monthly requirements are 460 pounds of phosphoric acid, 960 pounds of nitrogen, and 220 pounds of potash. If mix A costs $30 per cubic yard and mix B costs $35 per cubic yard, how many cubic yards of each mix should the farmer blend to meet the minimum monthly requirements at a minimal cost? What is this cost?

52. Animal food. A laboratory technician in a medical research center is asked to formulate a diet from two commercially packaged foods, food A and food B, for a group of animals.

Each ounce of food *A* contains 8 units of fat, 16 units of carbohydrate, and 2 units of protein. Each ounce of food *B* contains 4 units of fat, 32 units of carbohydrate, and 8 units of protein. The minimum daily requirements are 176 units of fat, 1,024 units of carbohydrate, and 384 units of protein. If food *A* costs 5¢ per ounce and food *B* costs 5¢ per ounce, how many ounces of each food should be used to meet the minimum daily requirements at the least cost? What is the cost for this amount of food?

53. *Psychology.* A psychologist uses two types of boxes with mice and rats. The amount of time (in minutes) that each mouse and each rat spends in each box per day is given in the table. What is the maximum number of mice and rats that can be used in this experiment? How many mice and how many rats produce this maximum?

	Time		Maximum Time
	Mice	Rats	Available per Day
Box *A*	10 min	20 min	800 min
Box *B*	20 min	10 min	640 min

54. *Sociology.* A city council voted to conduct a study on inner-city community problems using sociologists and research assistants from a nearby university. Allocation of time and costs per week are given in the table. How many sociologists and how many research assistants should be hired to minimize the cost and meet the weekly labor-hour requirements? What is the minimum weekly cost?

	Labor-Hours		Minimum Labor-Hours
	Sociologist	Research Assistant	Needed per Week
Fieldwork	10	30	180
Research center	30	10	140
Costs per week	$500	$300	

Answers to Matched Problems

1. (A) *x* = number of two-person boats produced each month
       y = number of four-person boats produced each month

(B)

	Labor-Hours Required		Maximum Labor-Hours
	Two-Person Boat	Four-Person Boat	Available per Month
Cutting department	0.9	1.8	864
Assembly department	0.8	1.2	672
Profit per boat	$25	$40	

(C) $P = 25x + 40y$

(D) $0.9x + 1.8y \le 864$
     $0.8x + 1.2y \le 672$
     $x, y \ge 0$

(E)

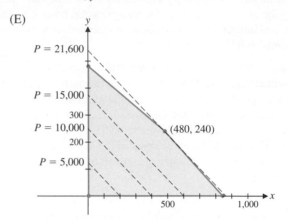

(F) 480 two-person boats, 240 four-person boats;
     Max $P = \$21{,}600$ per month

2. (A) Min $z = 24$ at $(3, 6)$; Max $z = 40$ at $(2, 16)$ and $(8, 4)$ (multiple optimal solution)

(B) Min $z = 90$ at $(0, 18)$; no maximum value

3. Min $C = 0.2x + 0.4y$
    subject to $3{,}000x + 4{,}000y \ge 36{,}000$
    $1{,}000x + 4{,}000y \ge 20{,}000$
    $x, y \ge 0$
    8 lb of mix *A*, 3 lb of mix *B*; Min $C = \$2.80$ per day

---

Chapter 5   Summary and Review

Important Terms, Symbols, and Concepts

5.1 Linear Inequalities in Two Variables

EXAMPLES

- A line divides the plane into two regions called **half-planes**. A vertical line divides the plane into **left** and **right half-planes**; a nonvertical line divides it into **upper** and **lower half-planes**. In either case, the dividing line is called the **boundary line** of each half-plane.

- The **graph of a linear inequality** is the half-plane obtained by following the procedure on page 258.

  Ex. 1, p. 258
  Ex. 2, p. 258
  Ex. 3, p. 260

- The variables in an applied problem are often required to be nonnegative.

  Ex. 4, p. 260

5.2 System of Linear Inequalities in Two Variables

- The **solution region** (also called the **feasible region**) of a system of linear inequalities is the graph of all ordered pairs that simultaneously satisfy all the inequalities in the system.   Ex. 1, p. 264

- A **corner point** of a solution region is a point in the region that is the intersection of two boundary lines.   Ex. 2, p. 265

- A solution region is **bounded** if it can be enclosed in a circle and **unbounded** if it cannot.   Ex. 3, p. 266

5.3 Linear Programming in Two Dimensions: A Geometric Approach

- The problem of finding the optimal (maximum or minimum) value of a linear objective function on a feasible region is called a **linear programming problem**.   Ex. 1, p. 270

- The optimal value (if it exists) of the objective function in a linear programming problem must occur at one (or more) of the corner points of the feasible region (Theorem 1, page 274). Existence criteria are described in Theorem 2, page 274, and a solution procedure is listed on page 274.   Ex. 2, p. 274

    Ex. 3, p. 276

Review Exercises

Work through all the problems in this chapter review and check answers in the back of the book. Answers to all review problems are there, and following each answer is a number in italics indicating the section in which that type of problem is discussed. Where weaknesses show up, review appropriate sections in the text.

Graph each inequality.

1. $x > 2y - 3$

2. $3y - 5x \leq 30$

Graph the systems in Problems 3–6 and indicate whether each solution region is bounded or unbounded. Find the coordinates of each corner point.

3. $5x + 9y \leq 90$
    $x, y \geq 0$

4. $15x + 16y \geq 1{,}200$
    $x, y \geq 0$

5. $2x + y \leq 8$
    $3x + 9y \leq 27$
    $x, y \geq 0$

6. $3x + y \geq 9$
    $2x + 4y \geq 16$
    $x, y \geq 0$

In Exercises 7 and 8, state the linear inequality whose graph is given in the figure. Write the boundary line equation in the form $Ax + By = C$, with A, B, and C integers, before stating the inequality.

7.

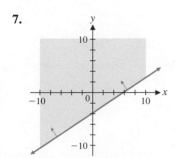

8.

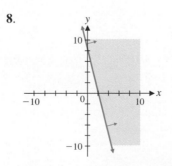

Solve the linear programming problems in Problems 9–13.

9. Maximize $P = 2x + 6y$
    subject to $\quad x + 2y \leq 8$
    $\quad 2x + y \leq 10$
    $\quad x, y \geq 0$

10. Minimize $C = 5x + 2y$
    subject to $\quad x + 3y \geq 15$
    $\quad 2x + y \geq 20$
    $\quad x, y \geq 0$

11. Maximize $P = 3x + 4y$
    subject to $\quad x + 2y \leq 12$
    $\quad x + y \leq 7$
    $\quad 2x + y \leq 10$
    $\quad x, y \geq 0$

12. Minimize $C = 8x + 3y$
    subject to $\quad x + y \geq 10$
    $\quad 2x + y \geq 15$
    $\quad x \geq 3$
    $\quad x, y \geq 0$

13. Maximize $P = 3x + 2y$
    subject to $\quad 2x + y \leq 22$
    $\quad x + 3y \leq 26$
    $\quad x \leq 10$
    $\quad y \leq 10$
    $\quad x, y \geq 0$

Applications

14. Electronics. A company uses two machines to solder circuit boards, an oven and a wave soldering machine. A circuit board for a calculator needs 4 minutes in the oven and 2 minutes on the wave machine, while a circuit board for a toaster requires 3 minutes in the oven and 1 minute on the wave machine. (*Source*: Universal Electronics)

(A) How many circuit boards for calculators and toasters can be produced if the oven is available for 5 hours? Express your answer as a linear inequality with appropriate non-negative restrictions and draw its graph.

(B) How many circuit boards for calculators and toasters can be produced if the wave machine is available for 2 hours? Express your answer as a linear inequality with appropriate nonnegative restrictions and draw its graph.

In Problems 15 and 16, construct a mathematical model in the form of a linear programming problem. (The answers in the back of the book for these application problems include the model.) Then solve the problem by the indicated method.

15. Sail manufacture. South Shore Sail Loft manufactures regular and competition sails. Each regular sail takes 2 hours to cut and 4 hours to sew. Each competition sail takes 3 hours to cut and 10 hours to sew. There are 150 hours available in the cutting department and 380 hours available in the sewing department.

(A) If the Loft makes a profit of $100 on each regular sail and $200 on each competition sail, how many sails of each type should the company manufacture to maximize their profit? What is the maximum profit?

(B) An increase in the demand for competition sails causes the profit on a competition sail to rise to $260. Discuss the effect of this change on the number of sails manufactured and on the maximum profit.

(C) A decrease in the demand for competition sails causes the profit on a competition sail to drop to $140. Discuss the effect of this change on the number of sails manufactured and on the maximum profit.

16. Animal food. A special diet for laboratory animals is to contain at least 850 units of vitamins, 800 units of minerals, and 1,150 calories. There are two feed mixes available, mix *A* and mix *B*. A gram of mix *A* contains 2 units of vitamins, 2 units of minerals, and 4 calories. A gram of mix *B* contains 5 units of vitamins, 4 units of minerals, and 5 calories.

(A) If mix *A* costs $0.04 per gram and mix *B* costs $0.09 per gram, how many grams of each mix should be used to satisfy the requirements of the diet at minimal cost? What is the minimum cost?

(B) If the price of mix *B* decreases to $0.06 per gram, discuss the effect of this change on the solution in part (A).

(C) If the price of mix *B* increases to $0.12 per gram, discuss the effect of this change on the solution in part (A).

6

Linear Programming: The Simplex Method

6.1 The Table Method: An Introduction to the Simplex Method

6.2 The Simplex Method: Maximization with Problem Constraints of the Form ≤

6.3 The Dual Problem: Minimization with Problem Constraints of the Form ≥

6.4 Maximization and Minimization with Mixed Problem Constraints

Chapter 6 Summary and Review

Review Exercises

Introduction

The geometric method of solving linear programming problems (presented in Chapter 5) provides an overview of linear programming. But, practically speaking, the geometric method is useful only for problems involving two decision variables and relatively few problem constraints. What happens when we need more decision variables and more problem constraints? We use an algebraic method called the *simplex method*, developed by George B. Dantzig (1914–2005) in 1947. Ideally suited to computer use, the simplex method is used routinely on applied problems involving thousands of variables and problem constraints. In this chapter, we move from an introduction to the simplex method to the *big M method*, which can be used to solve linear programming problems with a large number of variables and constraints. We also explore many applications. For example, we determine allocations of resources that maximize profit, and we determine production schedules that minimize cost (see Problem 44 in Section 6.2 on bicycle manufacturing and Problem 45 in Section 6.3 on ice-cream production).

6.1   The Table Method: An Introduction to the Simplex Method

- Standard Maximization Problems in Standard Form
- Slack Variables
- The Table Method: Basic Solutions and Basic Feasible Solutions
- Basic and Nonbasic Variables
- Summary

In Chapter 5, we denoted variables by single letters such as x and y. In this chapter, we will use letters with subscripts, for example, x_1, x_2, x_3, to denote variables. With this new notation, we will lay the groundwork for solving linear programming problems algebraically, by means of the *simplex method*. In this section, we introduce the *table method* to provide an introduction to the simplex method. Both methods, the table method and the simplex method, solve linear programming problems without the necessity of drawing a graph of the feasible region.

Standard Maximization Problems in Standard Form

The tent production problem that we considered in Section 5.3 is an example of a *standard maximization problem in standard form*. We restate the tent production problem below using subscript notation for the variables.

$$\begin{aligned} \text{Maximize} \quad & P = 50x_1 + 80x_2 \quad && \text{Objective function} \\ \text{subject to} \quad & x_1 + 2x_2 \le 32 \quad && \text{Cutting department constraint} \\ & 3x_1 + 4x_2 \le 84 \quad && \text{Assembly department constraint} \\ & x_1, x_2 \ge 0 \quad && \text{Nonnegative constraints} \end{aligned} \tag{1}$$

The decision variables x_1 and x_2 are the number of standard and expedition tents, respectively, produced each day.

Notice that the problem constraints involve $\le$ inequalities with positive constants to the right of the inequality. Maximization problems that satisfy this condition are called *standard maximization problems*. In this and the next section, we restrict our attention to standard maximization problems.

DEFINITION   Standard Maximization Problem in Standard Form
A linear programming problem is said to be a **standard maximization problem in standard form** if its mathematical model is of the following form:

Maximize the objective function

$$P = c_1 x_1 + c_2 x_2 + \cdots + c_k x_k$$

subject to problem constraints of the form

$$a_1 x_1 + a_2 x_2 + \cdots + a_k x_k \le b \quad b \ge 0$$

with nonnegative constraints

$$x_1, x_2, \ldots, x_k \ge 0$$

Note: Mathematical model (1) is a standard maximization problem in standard form. The coefficients of the objective function can be any real numbers.

Explore and Discuss 1   Find an example of a standard maximization problem in standard form involving two variables and one problem constraint such that

(A) The feasible region is bounded.

(B) The feasible region is unbounded.

Is it possible for a standard maximization problem to have no solution? Explain.

Slack Variables

To adapt a linear programming problem to the matrix methods used in the simplex process (as discussed in the next section), we convert the problem constraint inequalities into a system of linear equations using *slack variables*. In particular, to convert the system of inequalities from model (1),

$$x_1 + 2x_2 \leq 32 \quad \text{Cutting department constraint}$$
$$3x_1 + 4x_2 \leq 84 \quad \text{Assembly department constraint} \tag{2}$$
$$x_1, x_2 \geq 0 \quad \text{Nonnegative constraints}$$

into a system of equations, we add variables s_1 and s_2 to the left sides of the problem constraint inequalities in (2) to obtain

$$x_1 + 2x_2 + s_1 \qquad = 32$$
$$3x_1 + 4x_2 \qquad + s_2 = 84 \tag{3}$$

The variables s_1 and s_2 are called **slack variables** because each makes up the difference (takes up the slack) between the left and right sides of an inequality in system (2). For example, if we produced 20 standard tents ($x_1 = 20$) and 5 expedition tents ($x_2 = 5$), then the number of labor-hours used in the cutting department would be $20 + 2(5) = 30$, leaving a slack of 2 unused labor-hours out of the 32 available. So s_1 would have the value of 2.

Notice that if the decision variables x_1 and x_2 satisfy the system of constraint inequalities (2), then the slack variables s_1 and s_2 are nonnegative.

The Table Method: Basic Solutions and Basic Feasible Solutions

We call the system of inequalities of a linear programming problem an *i*-**system** ("*i*" for inequality), and we call the associated system of linear equations, obtained via slack variables, an *e*-**system** ("*e*" for equation).

The **solutions** of the *i*-system (2) are the points in the feasible region of Figure 1 (the feasible region was graphed earlier in Section 5.3). For example, $(20, 4)$ is a solution of the *i*-system, but $(20, 8)$ is not (Fig. 1).

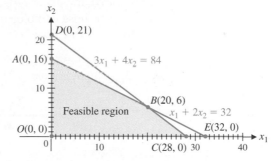

Figure 1

The solutions of the *e*-system (3) are quadruples (x_1, x_2, s_1, s_2). Any point $P = (x_1, x_2)$ in the plane corresponds to a unique solution of the *e*-system (3). For example, $(20, 4)$ corresponds to $(20, 4, 4, 8)$; we say that $(20, 4, 4, 8)$ are the **expanded coordinates** of $(20, 4)$. Similarly, the expanded coordinates of $(20, 8)$ are $(20, 8, -4, -8)$. Both $(20, 4, 4, 8)$ and $(20, 8, -4, -8)$ are solutions of (3).

Systems (2) and (3) do not have the same solutions. However, the solutions of the *i*-system (2) correspond to the solutions of the *e*-system (3) in which the variables x_1, x_2, s_1, and s_2 are all nonnegative. Such a solution of the *e*-system, in which the values of all decision variables and slack variables are nonnegative, is called a **feasible solution**. So $(20, 4, 4, 8)$ is a feasible solution of the *e*-system (3), but the solution $(20, 8, -4, -8)$ is not feasible. Note that $(20, 4)$ is in the feasible region (Fig. 1), but $(20, 8)$ is not. To summarize, the feasible solutions of (3) correspond to the points of the feasible region shown in Figure 1.

Any line segment that forms a boundary for the feasible region of (2) lies on one of the following four lines [replace "≤" by "=" in each inequality of i-system (2)]:

$$x_1 + 2x_2 = 32$$
$$3x_1 + 4x_2 = 84$$
$$x_1 \qquad = 0$$
$$\qquad x_2 = 0$$

So any corner point of the feasible region must lie on at least two of the four lines. If P is the intersection of the first two lines, then $s_1 = 0$ and $s_2 = 0$ in the expanded coordinates for P. If P is the intersection of the first and third lines, then $s_1 = 0$ and $x_1 = 0$ in the expanded coordinates for P. Continuing in this way, we conclude that for any corner point P of the feasible region, at least two of the variables x_1, x_2, s_1, and s_2 must equal 0 in the expanded coordinates of P.

The reasoning above gives a procedure, called the **table method**, for solving a standard maximization problem in standard form.

PROCEDURE   The Table Method (Two Decision Variables)

Assume that a standard maximization problem in standard form has two decision variables x_1 and x_2 and m problem constraints.

Step 1   Use slack variables s_1, s_2, ..., s_m to convert the i-system to an e-system.

Step 2   Form a table with $(m + 2)(m + 1)/2$ rows and $m + 2$ columns labeled x_1, x_2, s_1, s_2, ..., s_m. In the first row, assign 0 to x_1 and x_2. In the second row, assign 0 to x_1 and s_1. Continue until the rows contain all possible combinations of assigning two 0's to the variables.

Step 3   Complete each row to a solution of the e-system, if possible. Because two of the variables have the value 0, this involves solving a system of m linear equations in m variables. Use the Gauss–Jordan method, or another method if you find it easier. If the system has no solution or infinitely many solutions, do not complete the row.

Step 4   Solve the linear programming problem by finding the maximum value of P over those completed rows that have no negative values.

Table 1 shows the table of step 2 for the tent production problem. Note that $m = 2$, so there are $(m + 2)(m + 1)/2 = 6$ rows and $m + 2 = 4$ columns.

Table 1

x_1	x_2	s_1	s_2
0	0		
0		0	
0			0
	0	0	
	0		0
		0	0

Table 2 shows the table of step 3: Each row of Table 1 has been completed to a solution of the e-system (3). This involves solving the e-system six different times. Of course, the solution in the first row is easy: If x_1 and x_2 are both assigned the value 0 in (3), then clearly $s_1 = 32$ and $s_2 = 84$. The remaining rows are completed similarly: After two variables are assigned the value 0, the resulting system of two equations in two variables can be solved by the Gauss–Jordan method, or by substitution, or by elimination by addition. Such solutions of (3), in which two of the variables have been assigned the value 0, are called **basic solutions**. So there are six basic solutions in Table 2. Note that the six basic solutions correspond to the six intersection points O, A, B, C, D, and E of Figure 1.

Table 2 **Basic Solutions**

x_1	x_2	s_1	s_2
0	0	32	84
0	16	0	20
0	21	−10	0
32	0	0	−12
28	0	4	0
20	6	0	0

We use step 4 of the procedure to obtain the solution of the linear programming problem. The value of $P = 50x_1 + 80x_2$ is calculated for each row of Table 2 that has no negative values. See Table 3. We conclude that the solution of the linear programming problem is

$$\text{Max } P = \$1{,}480 \text{ at } x_1 = 20, x_2 = 6$$

Table 3 **The Table Method**

x_1	x_2	s_1	s_2	$P = 50x_1 + 80x_2$
0	0	32	84	0
0	16	0	20	1,280
0	21	−10	0	−
32	0	0	−12	−
28	0	4	0	1,400
20	6	0	0	1,480

We have ignored the third and fourth rows because $s_1 = -10$ in the third row and $s_2 = -12$ in the fourth. Those basic solutions are not feasible. The four remaining basic solutions are feasible. We call them **basic feasible solutions**. Note that the basic feasible solutions correspond to the four corner points O, A, B, C of the feasible region (Fig. 1).

Because basic feasible solutions correspond to the corner points of the feasible region, we can reformulate the fundamental theorem (Theorem 1 in Section 5.3):

THEOREM 1 Fundamental Theorem of Linear Programming: Version 2

If the optimal value of the objective function in a linear programming problem exists, then that value must occur at one or more of the basic feasible solutions.

EXAMPLE 1 Slack Variables and Basic Solutions

(A) Use slack variables s_1 and s_2 to convert the following i-system (system of inequalities) to an e-system (system of equations).

$$3x_1 + 2x_2 \leq 21$$
$$x_1 + 5x_2 \leq 20$$
$$x_1, x_2 \geq 0$$

(B) Find the basic solution for which $x_1 = 0$ and $s_1 = 0$.

(C) Find the basic solution for which $s_1 = 0$ and $s_2 = 0$.

SOLUTION

(A) Introduce slack variables s_1 and s_2 (one slack variable for each of the problem constraint inequalities):

$$3x_1 + 2x_2 + s_1 \qquad = 21 \tag{4}$$
$$x_1 + 5x_2 \qquad + s_2 = 20$$

(B) Substitute $x_1 = 0$ and $s_1 = 0$ into e-system (4):

$$3(0) + 2x_2 + 0 \qquad = 21$$
$$0 + 5x_2 \qquad + s_2 = 20$$

Divide the first equation by 2 to get $x_2 = 10.5$. Substitute $x_2 = 10.5$ in the second equation and solve for s_2:

$$s_2 = 20 - 5(10.5) = -32.5$$

The basic solution is

$$(x_1, x_2, s_1, s_2) = (0, 10.5, 0, -32.5)$$

Note that this basic solution is not feasible because at least one of the variables (s_2 in this case) has a negative value.

(C) Substitute $s_1 = 0$ and $s_2 = 0$ into e-system (4):

$$3x_1 + 2x_2 + 0 \qquad = 21$$
$$x_1 + 5x_2 \qquad + 0 = 20$$

This system of two equations in two variables can be solved by Gauss–Jordan elimination, or by another of our standard methods. Multiplying the second equation by -3, and adding the two equations, gives $-13x_2 = -39$, so $x_2 = 3$. Substituting $x_2 = 3$ in the second equation gives $x_1 = 5$.

The basic solution is

$$(x_1, x_2, s_1, s_2) = (5, 3, 0, 0)$$

Note that this basic solution is feasible because none of the variables has a negative value.

Matched Problem 1 Refer to Example 1. Find the basic solution for which $x_2 = 0$ and $s_1 = 0$.

EXAMPLE 2 The Table Method Construct the table of basic solutions and use it to solve the following linear programming problem:

$$\text{Maximize } P = 10x_1 + 25x_2$$
$$\text{subject to} \quad 3x_1 + 2x_2 \leq 21$$
$$x_1 + 5x_2 \leq 20$$
$$x_1, x_2 \geq 0$$

SOLUTION The system of inequalities is identical to the i-system of Example 1. The number of problem constraints is $m = 2$, so there will be $(m + 2)(m + 1)/2 = 6$ rows in the table of basic solutions. Because there are two decision variables, x_1 and x_2, we assign two zeros to each row of the table in all possible combinations. We also include the basic solution that was found in Example 1B (row 2 of table) and the basic solution that was found in Example 1C (last row).

x_1	x_2	s_1	s_2
0	0		
0	10.5	0	-32.5
0			0
	0	0	
	0		0
5	3	0	0

We complete the table working one row at a time. We substitute 0's for the two variables indicated by the row, in the e-system

$$3x_1 + 2x_2 + s_1 = 21$$
$$x_1 + 5x_2 + s_2 = 20$$

The result is a system of two equations in two variables, which can be solved by Gauss–Jordan elimination or another of our standard methods. Table 4 shows all six basic solutions, and the values of the objective function $P = 10x_1 + 25x_2$ at the four basic feasible solutions.

Table 4 **The Table Method**

x_1	x_2	s_1	s_2	$P = 10x_1 + 25x_2$
0	0	21	20	0
0	10.5	0	−32.5	−
0	4	13	0	100
7	0	0	13	70
20	0	−39	0	−
5	3	0	0	125

We conclude that

$$\text{Max } P = 125 \text{ at } x_1 = 5, x_2 = 3$$

Matched Problem 2) Construct the table of basic solutions and use it to solve the following linear programming problem:

$$\text{Maximize}\quad P = 30x_1 + 40x_2$$
$$\text{subject to}\quad 2x_1 + 3x_2 \leq 24$$
$$4x_1 + 3x_2 \leq 36$$
$$x_1, x_2 \geq 0$$

Explore and Discuss 1   The following linear programming problem has only one problem constraint:

$$\text{Maximize}\quad P = 2x_1 + 3x_2$$
$$\text{subject to}\quad 4x_1 + 5x_2 \leq 20$$
$$x_1, x_2 \geq 0$$

Solve it by the table method, then solve it by graphing, and compare the two solutions.

EXAMPLE 3   The Table Method   Construct the table of basic solutions and use it to solve the following linear programming problem:

$$\text{Maximize}\quad P = 40x_1 + 50x_2$$
$$\text{subject to}\quad x_1 + 6x_2 \leq 72$$
$$x_1 + 3x_2 \leq 45$$
$$2x_1 + 3x_2 \leq 72$$
$$x_1, x_2 \geq 0$$

SOLUTION   There are $m = 3$ problem constraints, so we use slack variables s_1, s_2, and s_3 to convert the i-system to the e-system (5):

$$x_1 + 6x_2 + s_1 \qquad\qquad = 72$$
$$x_1 + 3x_2 \qquad + s_2 \qquad = 45 \qquad\qquad (5)$$
$$2x_1 + 3x_2 \qquad\qquad + s_3 = 72$$

There will be $(m + 2)(m + 1)/2 = 10$ rows in the table of basic solutions. Because there are two decision variables, x_1 and x_2, we assign two zeros to each row of the table in all possible combinations.

x_1	x_2	s_1	s_2	s_3
0	0			
0		0		
0			0	
0				0
	0	0		
	0		0	
	0			0
		0	0	
		0		0
			0	0

We complete the table working one row at a time. We substitute 0's for the two variables indicated by the row, in the e-system (5). The result is a system of three equations in three variables, which can be solved by Gauss–Jordan elimination, or by another of our standard methods. Table 5 shows all ten basic solutions, and the values of the objective function, $P = 40x_1 + 50x_2$, at the five basic feasible solutions.

Table 5 **The Table Method**

x_1	x_2	s_1	s_2	s_3	$P = 40x_1 + 50x_2$
0	0	72	45	72	0
0	12	0	9	36	600
0	15	−18	0	27	–
0	24	−72	−27	0	–
72	0	0	−27	−72	–
45	0	27	0	−18	–
36	0	36	9	0	1,440
18	9	0	0	9	1,170
24	8	0	−3	0	–
27	6	9	0	0	1,380

We conclude that

$$\text{Max } P = 1{,}440 \text{ at } x_1 = 36, x_2 = 0$$

Matched Problem 3 | Construct the table of basic solutions and use it to solve the following linear programming problem:

$$\text{Maximize} \quad P = 36x_1 + 24x_2$$
$$\text{subject to} \quad x_1 + 2x_2 \le 8$$
$$x_1 + x_2 \le 5$$
$$2x_1 + x_2 \le 8$$
$$x_1, x_2 \ge 0$$

Basic and Nonbasic Variables

The basic solutions associated with a linear programming problem are found by assigning the value 0 to certain decision variables (the x_i's) and slack variables (the s_i's). Consider, for example, row 2 of Table 5. That row shows the basic solution $(x_1, x_2, s_1, s_2, s_3) = (0, 12, 0, 9, 36)$. It is customary to refer to the variables that are assigned the value 0 as **nonbasic variables**, and to the others as **basic variables**. So for the basic solution of row 2, the basic variables are x_2, s_2, and s_3; the nonbasic variables are x_1 and s_1.

Note that the classification of variables as basic or nonbasic depends on the basic solution. Row 8 of Table 5 shows the basic solution $(x_1, x_2, s_1, s_2, s_3) = (18, 9, 0, 0, 9)$. For row 8, the basic variables are x_1, x_2, and s_3; the nonbasic variables are s_1 and s_2.

EXAMPLE 4 Basic and Nonbasic Variables Refer to Table 5. For the basic solution $(x_1, x_2, s_1, s_2, s_3) = (36, 0, 36, 9, 0)$ in row 7 of Table 5, classify the variables as basic or nonbasic.

SOLUTION The basic variables are x_1, s_1, and s_2. The other variables, x_2 and s_3, were assigned the value 0, and therefore are nonbasic.

Matched Problem 4) Refer to Table 5. For the basic solution $(x_1, x_2, s_1, s_2, s_3) = (27, 6, 9, 0, 0)$ of row 10, classify the variables as basic or nonbasic.

Explore and Discuss 2 Use the table method to solve the following linear programming problem, and explain why one of the rows in the table cannot be completed to a basic solution:

$$\text{Maximize} \quad P = 10x_1 + 12x_2$$
$$\text{subject to} \quad x_1 + x_2 \le 2$$
$$x_1 + x_2 \le 3$$
$$x_1, x_2 \ge 0$$

Summary

The examples in this section illustrate the table method when there are two decision variables. But the method can be used when there are k decision variables, where k is any positive integer.

The number of ways in which r objects can be chosen from a set of n objects, without regard to order, is denoted by $_nC_r$ and given by the formula

$$_nC_r = \frac{n!}{r!(n - r)!}$$

(The formula, giving the number of combinations of n distinct objects taken r at a time, is explained and derived in Chapter 7). If there are k decision variables and m problem constraints in a linear programming problem, then the number of rows in the table of basic solutions is $_{k+m}C_k$, because this is the number of ways of selecting k of the $k + m$ variables to be assigned the value 0.

PROCEDURE The Table Method (k Decision Variables)

Assume that a standard maximization problem in standard form has k decision variables $x_1, x_2, \ldots, x_k$, and m problem constraints.

Step 1 Use slack variables $s_1, s_2, \ldots, s_m$ to convert the i-system to an e-system.

Step 2 Form a table with $_{k+m}C_k$ rows and $k + m$ columns labeled $x_1, x_2, \ldots, x_k$, $s_1, s_2, \ldots, s_m$. In the first row, assign 0 to $x_1, x_2, \ldots, x_k$. Continue until the rows contain all possible combinations of assigning k 0's to the variables.

(Continued)

Step 3 Complete each row to a solution of the *e*-system, if possible. Because *k* of the variables have the value 0, this involves solving a system of *m* linear equations in *m* variables. Use the Gauss–Jordan method, or another method if you find it easier. If the system has no solutions, or infinitely many solutions, do not complete the row.

Step 4 Solve the linear programming problem by finding the maximum value of *P* over those completed rows that have no negative values (that is, over the basic feasible solutions).

The benefit of the table method is that it gives a procedure for **finding all corner points of the feasible region without drawing a graph**.

Unfortunately, the number of rows in the table becomes too large to be practical, even for computers, when the number of decision variables and problem constraints is large. For example, with $k = 30$ decision variables and $m = 35$ problem constraints, the number of rows is

$$_{65}C_{30} \approx 3 \times 10^{18}$$

We need a procedure for finding the optimal solution of a linear programming problem without having to find every corner point. The *simplex method*, discussed in the next section, is such a procedure. It gives a practical method for solving large linear programming problems.

Exercises 6.1

Skills Warm-up Exercises

W *In Problems 1–4, if necessary, review Section B.3.*

1. In how many ways can two variables be chosen from x_1, x_2, s_1, s_2, s_3 and assigned the value 0?

2. In how many ways can two variables be chosen from x_1, x_2, s_1, s_2 and assigned the value 0?

3. In how many ways can two variables be chosen from $x_1, x_2, x_3, s_1, s_2, s_3$ and assigned the value 0?

4. In how many ways can three variables be chosen from $x_1, x_2, x_3, s_1, s_2, s_3, s_4, s_5$ and assigned the value 0?

Problems 5–8 refer to the system

$$2x_1 + 5x_2 + s_1 \qquad = 10$$
$$x_1 + 3x_2 \qquad + s_2 = 8$$

5. Find the solution of the system for which $x_1 = 0, s_1 = 0$.

6. Find the solution of the system for which $x_1 = 0, s_2 = 0$.

7. Find the solution of the system for which $x_2 = 0, s_2 = 0$.

8. Find the solution of the system for which $x_2 = 0, s_1 = 0$.

In Problems 9–16, write the e-system obtained via slack variables for the given linear programming problem.

9. Maximize $P = 5x_1 + 7x_2$
subject to
$$2x_1 + 3x_2 \leq 9$$
$$6x_1 + 7x_2 \leq 13$$
$$x_1, x_2 \geq 0$$

10. Maximize $P = 35x_1 + 25x_2$
subject to
$$10x_1 + 15x_2 \leq 100$$
$$5x_1 + 20x_2 \leq 120$$
$$x_1, x_2 \geq 0$$

11. Maximize $P = 3x_1 + 5x_2$
subject to
$$12x_1 - 14x_2 \leq 55$$
$$19x_1 + 5x_2 \leq 40$$
$$-8x_1 + 11x_2 \leq 64$$
$$x_1, x_2 \geq 0$$

12. Maximize $P = 13x_1 + 25x_2$
subject to
$$3x_1 + 5x_2 \leq 27$$
$$8x_1 + 3x_2 \leq 19$$
$$4x_1 + 9x_2 \leq 34$$
$$x_1, x_2 \geq 0$$

13. Maximize $P = 4x_1 + 7x_2$

subject to $\qquad 6x_1 + 5x_2 \leq 18$

$\qquad\qquad x_1, x_2 \geq 0$

14. Maximize $P = 13x_1 + 8x_2$

subject to $\qquad x_1 + 2x_2 \leq 20$

$\qquad\qquad x_1, x_2 \geq 0$

15. Maximize $P = x_1 + 2x_2$

subject to $\qquad 4x_1 - 3x_2 \leq 12$

$\qquad\qquad 5x_1 + 2x_2 \leq 25$

$\qquad\qquad -3x_1 + 7x_2 \leq 32$

$\qquad\qquad 2x_1 + x_2 \leq 9$

$\qquad\qquad x_1, x_2 \geq 0$

16. Maximize $P = 8x_1 + 9x_2$

subject to $\qquad 30x_1 - 25x_2 \leq 75$

$\qquad\qquad 10x_1 + 13x_2 \leq 30$

$\qquad\qquad 5x_1 + 18x_2 \leq 40$

$\qquad\qquad 40x_1 + 36x_2 \leq 85$

$\qquad\qquad x_1, x_2 \geq 0$

Problems 17–26 refer to the table below of the six basic solutions to the e-system

$$2x_1 + 3x_2 + s_1 \qquad\quad = 24$$
$$4x_1 + 3x_2 \qquad\;\; + s_2 = 36$$

	x_1	x_2	s_1	s_2
(A)	0	0	24	36
(B)	0	8	0	12
(C)	0	12	-12	0
(D)	12	0	0	-12
(E)	9	0	6	0
(F)	6	4	0	0

17. In basic solution (A), which variables are basic?

18. In basic solution (B), which variables are nonbasic?

19. In basic solution (C), which variables are nonbasic?

20. In basic solution (D), which variables are basic?

21. In basic solution (E), which variables are nonbasic?

22. In basic solution (F), which variables are basic?

23. Which of the six basic solutions are feasible? Explain.

24. Which of the basic solutions are not feasible? Explain.

25. Use the basic feasible solutions to maximize $P = 2x_1 + 5x_2$.

26. Use the basic feasible solutions to maximize $P = 8x_1 + 5x_2$.

Problems 27–36 refer to the partially completed table below of the 10 basic solutions to the e-system

$$x_1 + x_2 + s_1 \qquad\qquad = 24$$
$$2x_1 + x_2 \qquad + s_2 \qquad = 30$$
$$4x_1 + x_2 \qquad\qquad + s_3 = 48$$

	x_1	x_2	s_1	s_2	s_3
(A)	0	0	24	30	48
(B)	0	24	0	6	24
(C)	0	30	-6	0	18
(D)	0	48	-24	-18	0
(E)	24	0	0	-18	-48
(F)	15	0	9	0	-12
(G)		0			0
(H)			0	0	
(I)			0		0
(J)				0	0

27. In basic solution (C), which variables are basic?

28. In basic solution (E), which variables are nonbasic?

29. In basic solution (G), which variables are nonbasic?

30. In basic solution (I), which variables are basic?

31. Which of the basic solutions (A) through (F) are not feasible? Explain.

32. Which of the basic solutions (A) through (F) are feasible? Explain.

33. Find basic solution (G).

34. Find basic solution (H).

35. Find basic solution (I).

36. Find basic solution (J).

In Problems 37–44, convert the given i-system to an e-system using slack variables. Then construct a table of all basic solutions of the e-system. For each basic solution, indicate whether or not it is feasible.

37. $4x_1 + 5x_2 \leq 20$

$\qquad x_1, x_2 \geq 0$

38. $3x_1 + 8x_2 \leq 24$

$\qquad x_1, x_2 \geq 0$

39. $x_1 + x_2 \leq 6$

$\quad x_1 + 4x_2 \leq 12$

$\qquad x_1, x_2 \geq 0$

40. $5x_1 + x_2 \leq 15$

$\quad x_1 + x_2 \leq 7$

$\qquad x_1, x_2 \geq 0$

41. $2x_1 + 5x_2 \leq 20$

$\quad x_1 + 2x_2 \leq 9$

$\qquad x_1, x_2 \geq 0$

42. $x_1 + 3x_2 \leq 18$

$\quad 5x_1 + 4x_2 \leq 35$

$\qquad x_1, x_2 \geq 0$

43. $x_1 + 2x_2 \leq 24$

$\quad x_1 + x_2 \leq 15$

$\quad 2x_1 + x_2 \leq 24$

$\qquad x_1, x_2 \geq 0$

44. $5x_1 + 4x_2 \leq 240$

$\quad 5x_1 + 2x_2 \leq 150$

$\quad 5x_1 + x_2 \leq 120$

$\qquad x_1, x_2 \geq 0$

In Problems 45–50, graph the system of inequalities from the given problem, and list the corner points of the feasible region. Verify that the corner points of the feasible region correspond to the basic feasible solutions of the associated e-system.

45. Problem 37

46. Problem 38

47. Problem 39

48. Problem 40

49. Problem 41

50. Problem 42

In Problems 51–58, solve the given linear programming problem using the table method (the table of basic solutions was constructed in Problems 37–44).

51. Maximize $P = 10x_1 + 9x_2$
   subject to $4x_1 + 5x_2 \leq 20$
                            $x_1, x_2 \geq 0$

52. Maximize $P = 4x_1 + 7x_2$
   subject to $3x_1 + 8x_2 \leq 24$
                            $x_1, x_2 \geq 0$

53. Maximize $P = 15x_1 + 20x_2$
   subject to $x_1 + x_2 \leq 6$
                            $x_1 + 4x_2 \leq 12$
                            $x_1, x_2 \geq 0$

54. Maximize $P = 5x_1 + 20x_2$
   subject to $5x_1 + x_2 \leq 15$
                            $x_1 + x_2 \leq 7$
                            $x_1, x_2 \geq 0$

55. Maximize $P = 25x_1 + 10x_2$
   subject to $2x_1 + 5x_2 \leq 20$
                            $x_1 + 2x_2 \leq 9$
                            $x_1, x_2 \geq 0$

56. Maximize $P = 40x_1 + 50x_2$
   subject to $x_1 + 3x_2 \leq 18$
                            $5x_1 + 4x_2 \leq 35$
                            $x_1, x_2 \geq 0$

57. Maximize $P = 30x_1 + 40x_2$
   subject to $x_1 + 2x_2 \leq 24$
                            $x_1 + x_2 \leq 15$
                            $2x_1 + x_2 \leq 24$
                            $x_1, x_2 \geq 0$

58. Maximize $P = x_1 + x_2$
   subject to $5x_1 + 4x_2 \leq 240$
                            $5x_1 + 2x_2 \leq 150$
                            $5x_1 + x_2 \leq 120$
                            $x_1, x_2 \geq 0$

59. A linear programming problem has four decision variables x_1, x_2, x_3, x_4, and six problem constraints. How many rows are there in the table of basic solutions of the associated *e*-system?

60. A linear programming problem has five decision variables x_1, x_2, x_3, x_4, x_5 and six problem constraints. How many rows are there in the table of basic solutions of the associated *e*-system?

61. A linear programming problem has 30 decision variables $x_1, x_2, \ldots, x_{30}$ and 42 problem constraints. How many rows are there in the table of basic solutions of the associated *e*-system? (Write the answer using scientific notation.)

62. A linear programming problem has 40 decision variables $x_1, x_2, \ldots, x_{40}$ and 85 problem constraints. How many rows are there in the table of basic solutions of the associated *e*-system? (Write the answer using scientific notation.)

Answers to Matched Problems

1. $(x_1, x_2, s_1, s_2) = (7, 0, 0, 13)$

2.

x_1	x_2	s_1	s_2	$P = 30x_1 + 40x_2$
0	0	24	36	0
0	8	0	12	320
0	12	−12	0	–
12	0	0	−12	–
9	0	6	0	270
6	4	0	0	340

Max $P = 340$ at $x_1 = 6, x_2 = 4$

3.

x_1	x_2	s_1	s_2	s_3	$P = 36x_1 + 24x_2$
0	0	8	5	8	0
0	4	0	1	4	96
0	5	−2	0	3	–
0	8	−8	−3	0	–
8	0	0	−3	−8	–
5	0	3	0	−2	–
4	0	4	1	0	144
2	3	0	0	1	144
8/3	8/3	0	−1/3	0	–
3	2	1	0	0	156

Max $P = 156$ at $x_1 = 3, x_2 = 2$

4. $x_1, x_2,$ and s_1 are basic; s_2 and s_3 are nonbasic

6.2 The Simplex Method: Maximization with Problem Constraints of the Form $\leq$

- Initial System
- Simplex Tableau
- Pivot Operation
- Interpreting the Simplex Process Geometrically
- Simplex Method Summarized
- Application

Now we can develop the simplex method for a standard maximization problem. The simplex method is most useful when used with computers. Consequently, it is not intended that you become an expert in manually solving linear programming problems using the simplex method. But it is important that you become proficient in constructing the models for linear programming problems so that they can be solved using a computer, and it is also important that you develop skill in interpreting the results. One way to gain this proficiency and interpretive skill is to set up and manually solve a number of fairly simple linear programming problems using the simplex method. This is the main goal in this section and in Sections 6.3 and 6.4. To assist you in learning to develop the models, the answer sections for Exercises 6.2, 6.3, and 6.4 contain both the model and its solution.

Initial System

We will introduce the concepts and procedures involved in the simplex method through an example—the tent production example discussed earlier. We restate the problem here in standard form for convenient reference:

$$\text{Maximize } P = 50x_1 + 80x_2 \quad \text{Objective function}$$
$$\text{subject to} \quad \left.\begin{array}{r} x_1 + 2x_2 \leq 32 \\ 3x_1 + 4x_2 \leq 84 \end{array}\right\} \quad \text{Problem constraints} \tag{1}$$
$$x_1, x_2 \geq 0 \quad \text{Nonnegative constraints}$$

Introducing slack variables s_1 and s_2, we convert the problem constraint inequalities in problem (1) into the following system of problem constraint equations:

$$\begin{array}{rcl} x_1 + 2x_2 + s_1 & = & 32 \\ 3x_1 + 4x_2 + s_2 & = & 84 \\ x_1, x_2, s_1, s_2 & \geq & 0 \end{array} \tag{2}$$

Since a basic solution of system (2) is not feasible if it contains any negative values, we have also included the nonnegative constraints for both the decision variables x_1 and x_2 and the slack variables s_1 and s_2. From our discussion in Section 6.1, we know that out of the infinitely many solutions to system (2), an optimal solution is one of the basic feasible solutions, which correspond to the corner points of the feasible region.

As part of the simplex method we add the objective function equation $P = 50x_1 + 80x_2$ in the form $-50x_1 - 80x_2 + P = 0$ to system (2) to create what is called the **initial system**:

$$\begin{array}{rcl} x_1 + 2x_2 + s_1 & = & 32 \\ 3x_1 + 4x_2 + s_2 & = & 84 \\ -50x_1 - 80x_2 + P & = & 0 \\ x_1, x_2, s_1, s_2 & \geq & 0 \end{array} \tag{3}$$

When we add the objective function equation to system (2), we must slightly modify the earlier definitions of basic solution and basic feasible solution so that they apply to the initial system (3).

DEFINITION Basic Solutions and Basic Feasible Solutions for Initial Systems

1. The objective function variable P is always selected as a basic variable.
2. Note that a basic solution of system (3) is also a basic solution of system (2) after P is deleted.

(Continued)

3. If a basic solution of system (3) is a basic feasible solution of system (2) after deleting P, then the basic solution of system (3) is called a **basic feasible solution** of system (3).
4. A basic feasible solution of system (3) can contain a negative number, but only if it is the value of P, the objective function variable.

These changes lead to a small change in the second version of the fundamental theorem (see Theorem 1, Section 6.1).

THEOREM 1   Fundamental Theorem of Linear Programming: Version 3

If the optimal value of the objective function in a linear programming problem exists, then that value must occur at one or more of the basic feasible solutions of the initial system.

With these adjustments understood, we start the simplex process with a basic feasible solution of the initial system (3), which we will refer to as an **initial basic feasible solution**. An initial basic feasible solution that is easy to find is the one associated with the origin.

Since system (3) has three equations and five variables, it has three basic variables and two nonbasic variables. Looking at the system, we see that x_1 and x_2 appear in all equations, and s_1, s_2, and P each appear only once and each in a different equation. A basic solution can be found by inspection by selecting s_1, s_2, and P as the basic variables (remember, P is always selected as a basic variable) and x_1 and x_2 as the nonbasic variables to be set equal to 0. Setting x_1 and x_2 equal to 0 and solving for the basic variables, we obtain the basic solution:

$$x_1 = 0, \quad x_2 = 0, \quad s_1 = 32, \quad s_2 = 84, \quad P = 0$$

This basic solution is feasible since none of the variables (excluding P) are negative. This is the initial basic feasible solution that we seek.

Now you can see why we wanted to add the objective function equation to system (2): A basic feasible solution of system (3) not only includes a basic feasible solution of system (2), but, in addition, it includes the value of P for that basic feasible solution of system (2).

The initial basic feasible solution we just found is associated with the origin. Of course, if we do not produce any tents, we do not expect a profit, so $P = \$0$. Starting with this easily obtained initial basic feasible solution, the simplex process moves through each iteration (repetition) to another basic feasible solution, each time improving the profit. The process continues until the maximum profit is reached, then the process stops.

Simplex Tableau

To facilitate the search for the optimal solution, we turn to matrix methods discussed in Chapter 4. Our first step is to write the augmented matrix for the initial system (3). This matrix is called the **initial simplex tableau**,* and it is simply a tabulation of the coefficients in system (3).

$$
\begin{array}{c}
 \\
s_1 \\
s_2 \\
P
\end{array}
\begin{array}{c}
\begin{array}{ccccc}
x_1 & x_2 & s_1 & s_2 & P
\end{array} \\
\left[
\begin{array}{ccccc|c}
1 & 2 & 1 & 0 & 0 & 32 \\
3 & 4 & 0 & 1 & 0 & 84 \\
\hline
-50 & -80 & 0 & 0 & 1 & 0
\end{array}
\right]
\end{array}
\qquad \text{Initial simplex tableau} \qquad (4)
$$

---

*The format of a simplex tableau can vary. Some authors place the objective function coefficients in the first row rather than the last. Others do not include a column for P.

In tableau (4), the row below the dashed line always corresponds to the objective function. Each of the basic variables we selected above, s_1, s_2, and P, is also placed on the left of the tableau so that the intersection element in its row and column is not 0. For example, we place the basic variable s_1 on the left so that the intersection element of the s_1 row and the s_1 column is 1 and not 0. The basic variable s_2 is similarly placed. The objective function variable P is always placed at the bottom. The reason for writing the basic variables on the left in this way is that this placement makes it possible to read certain basic feasible solutions directly from the tableau. If $x_1 = 0$ and $x_2 = 0$, the basic variables on the left of tableau (4) are lined up with their corresponding values, 32, 84, and 0, to the right of the vertical line.

Looking at tableau (4) relative to the choice of s_1, s_2, and P as basic variables, we see that each basic variable is above a column that has all 0 elements except for a single 1 and that no two such columns contain 1's in the same row. These observations lead to a formalization of the process of selecting basic and nonbasic variables that is an important part of the simplex method:

> **PROCEDURE** Selecting Basic and Nonbasic Variables for the Simplex Method
> Given a simplex tableau,
>
> Step 1 *Numbers of variables.* Determine the number of basic variables and the number of nonbasic variables. These numbers do not change during the simplex process.
>
> Step 2 *Selecting basic variables.* A variable can be selected as a basic variable only if it corresponds to a column in the tableau that has exactly one nonzero element (usually 1) and the nonzero element in the column is not in the same row as the nonzero element in the column of another basic variable. This procedure always selects P as a basic variable, since the P column never changes during the simplex process.
>
> Step 3 *Selecting nonbasic variables.* After the basic variables are selected in step 2, the remaining variables are selected as the nonbasic variables. The tableau columns under the nonbasic variables usually contain more than one nonzero element.

The earlier selection of s_1, s_2, and P as basic variables and x_1 and x_2 as nonbasic variables conforms to this prescribed convention of selecting basic and nonbasic variables for the simplex process.

Pivot Operation

The simplex method swaps one of the nonbasic variables, x_1 or x_2, for one of the basic variables, s_1 or s_2 (but not P), as a step toward improving the profit. For a nonbasic variable to be classified as a basic variable, we need to perform appropriate row operations on the tableau so that the newly selected basic variable will end up with exactly one nonzero element in its column. In this process, the old basic variable will usually gain additional nonzero elements in its column as it becomes nonbasic.

Which nonbasic variable should we select to become basic? It makes sense to select the nonbasic variable that will increase the profit the most per unit change in that variable. Looking at the objective function

$$P = 50x_1 + 80x_2$$

we see that if x_1 stays a nonbasic variable (set equal to 0) and if x_2 becomes a new basic variable, then

$$P = 50(0) + 80x_2 = 80x_2$$

and for each unit increase in x_2, P will increase \$80. If x_2 stays a nonbasic variable and x_1 becomes a new basic variable, then (reasoning in the same way) for each unit increase in x_1, P will increase only \$50. So, we select the nonbasic variable x_2 to enter the set of basic variables, and call it the **entering variable**. (The basic variable leaving the set of basic variables to become a nonbasic variable is called the **exiting variable**, which will be discussed shortly.)

The column corresponding to the entering variable is called the **pivot column**. Looking at the bottom row in tableau (4)—the objective function row below the dashed line—we see that the pivot column is associated with the column to the left of the P column that has the most negative bottom element. In general, the most negative element in the bottom row to the left of the P column indicates the variable above it that will produce the greatest increase in P for a unit increase in that variable. For this reason, we call the elements in the bottom row of the tableau, to the left of the P column, **indicators**.

We illustrate the indicators, the pivot column, the entering variable, and the initial basic feasible solution below:

$$
\begin{array}{c}
\text{Entering} \\
\text{variable} \\
\downarrow
\end{array}
$$

$$
\begin{array}{c}
\\
s_1 \\
s_2 \\
P
\end{array}
\begin{array}{c}
x_1 \quad x_2 \quad s_1 \quad s_2 \quad P \\
\left[\begin{array}{ccccc|c}
1 & 2 & 1 & 0 & 0 & 32 \\
3 & 4 & 0 & 1 & 0 & 84 \\
\hline
-50 & -80 & 0 & 0 & 1 & 0
\end{array}\right]
\end{array}
\quad
\begin{array}{l}
\text{Initial simplex tableau} \\
\\
\text{Indicators are shown in color.}
\end{array}
\qquad (5)
$$

$$
\begin{array}{c}
\uparrow \\
\text{Pivot} \\
\text{column}
\end{array}
$$

$$x_1 = 0, \quad x_2 = 0, \quad s_1 = 32, \quad s_2 = 84, \quad P = 0 \quad \text{Initial basic feasible solution}$$

Now that we have chosen the nonbasic variable x_2 as the entering variable (the nonbasic variable to become basic), which of the two basic variables, s_1 or s_2, should we choose as the exiting variable (the basic variable to become nonbasic)? We saw above that for $x_1 = 0$, each unit increase in the entering variable x_2 results in an increase of \$80 for P. Can we increase x_2 without limit? No! A limit is imposed by the nonnegative requirements for s_1 and s_2. (Remember that if any of the basic variables except P become negative, we no longer have a feasible solution.) So we rephrase the question and ask: How much can x_2 be increased when $x_1 = 0$ without causing s_1 or s_2 to become negative? To see how much x_2 can be increased, we refer to tableau (5) or system (3) and write the two problem constraint equations with $x_1 = 0$:

$$2x_2 + s_1 = 32$$
$$4x_2 + s_2 = 84$$

Solving for s_1 and s_2, we have

$$s_1 = 32 - 2x_2$$
$$s_2 = 84 - 4x_2$$

For s_1 and s_2 to be nonnegative, x_2 must be chosen so that both $32 - 2x_2$ and $84 - 4x_2$ are nonnegative. That is, so that

$$
\begin{array}{ccc}
32 - 2x_2 \geq 0 & \quad\text{and}\quad & 84 - 4x_2 \geq 0 \\
-2x_2 \geq -32 & & -4x_2 \geq -84 \\
x_2 \leq \frac{32}{2} = 16 & & x_2 \leq \frac{84}{4} = 21
\end{array}
$$

For both inequalities to be satisfied, x_2 must be less than or equal to the smaller of the values, which is 16. So x_2 can increase to 16 without either s_1 or s_2 becoming negative. Now, observe how each value (16 and 21) can be obtained directly from the following tableau:

$$
\begin{array}{c}
\text{Entering} \\
\text{variable} \\
\downarrow
\end{array}
$$

$$
\begin{array}{c}
 & x_1 & x_2 & s_1 & s_2 & P & \\
s_1 & \begin{bmatrix} 1 & 2 & 1 & 0 & 0 & 32 \\ 3 & 4 & 0 & 1 & 0 & 84 \\ \hdashline -50 & -80 & 0 & 0 & 1 & 0 \end{bmatrix} & & & & & \begin{array}{l} \frac{32}{2} = 16 \quad \text{(smallest)} \\ \\ \frac{84}{4} = 21 \end{array}
\end{array}
$$

$$
\uparrow \\
\text{Pivot} \\
\text{column}
$$

(6)

From tableau (6) we can determine the amount that the entering variable can increase by choosing the smallest of the quotients obtained by dividing each element in the last column above the dashed line by the corresponding *positive* element in the pivot column. The row with the smallest quotient is called the **pivot row**, and the variable to the left of the pivot row is the exiting variable. In this case, s_1 will be the exiting variable, and the roles of x_2 and s_1 will be interchanged. The element at the intersection of the pivot column and the pivot row is called the **pivot element**, and we circle this element for ease of recognition. Since a negative or 0 element in the pivot column places no restriction on the amount that an entering variable can increase, it is not necessary to compute the quotient for negative or 0 values in the pivot column.

A negative or 0 element is never selected for the pivot element.

The following tableau illustrates this process, which is summarized in the next box.

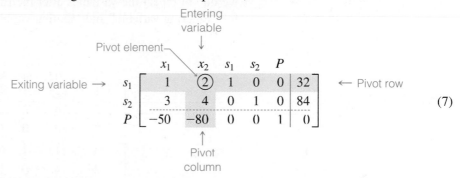

(7)

PROCEDURE Selecting the Pivot Element

Step 1   Locate the most negative indicator in the bottom row of the tableau to the left of the P column (the negative number with the largest absolute value). The column containing this element is the *pivot column*. If there is a tie for the most negative indicator, choose either column.

Step 2   Divide each *positive* element in the pivot column above the dashed line into the corresponding element in the last column. The *pivot row* is the row corresponding to the smallest quotient obtained. If there is a tie for the smallest quotient, choose either row. If the pivot column above the dashed line has no positive elements, there is no solution, and we stop.

Step 3   The *pivot* (or *pivot element*) is the element at the intersection of the pivot column and pivot row.

Note: The pivot element is always positive and never appears in the bottom row.

Remember: The entering variable is at the top of the pivot column, and the exiting variable is at the left of the pivot row.

In order for x_2 to be classified as a basic variable, we perform row operations on tableau (7) so that the pivot element is transformed into 1 and all other elements in the column into 0's. This procedure for transforming a nonbasic variable into a basic variable is called a *pivot operation,* or *pivoting,* and is summarized in the following box.

PROCEDURE Performing a Pivot Operation

A **pivot operation**, or **pivoting**, consists of performing row operations as follows:

Step 1   Multiply the pivot row by the reciprocal of the pivot element to transform the pivot element into a 1. (If the pivot element is already a 1, omit this step.)

Step 2   Add multiples of the pivot row to other rows in the tableau to transform all other nonzero elements in the pivot column into 0's.

CONCEPTUAL INSIGHT

A pivot operation uses some of the same row operations as those used in Gauss–Jordan elimination, but there is one essential difference. **In a pivot operation, you can never interchange two rows.**

Performing a pivot operation has the following effects:

1. The (entering) nonbasic variable becomes a basic variable.
2. The (exiting) basic variable becomes a nonbasic variable.
3. The value of the objective function is increased, or, in some cases, remains the same.

We now carry out the pivot operation on tableau (7). (To facilitate the process, we do not repeat the variables after the first tableau, and we use "Enter" and "Exit" for "Entering variable" and "Exiting variable," respectively.)

$$
\begin{array}{c}
\text{Enter} \\
\downarrow
\end{array}
$$

$$
\begin{array}{c}
\\
\text{Exit} \rightarrow \ s_1 \\
s_2 \\
P
\end{array}
\begin{array}{cccccc}
x_1 & x_2 & s_1 & s_2 & P & \\
\end{array}
\left[\begin{array}{ccccc|c}
1 & ② & 1 & 0 & 0 & 32 \\
3 & 4 & 0 & 1 & 0 & 84 \\
\hline
-50 & -80 & 0 & 0 & 1 & 0
\end{array}\right]
\begin{array}{l}
\frac{1}{2}R_1 \rightarrow R_1
\end{array}
$$

$$
\sim
\left[\begin{array}{ccccc|c}
\frac{1}{2} & ① & \frac{1}{2} & 0 & 0 & 16 \\
3 & 4 & 0 & 1 & 0 & 84 \\
\hline
-50 & -80 & 0 & 0 & 1 & 0
\end{array}\right]
\begin{array}{l}
(-4)R_1 + R_2 \rightarrow R_2 \\
80R_1 + R_3 \rightarrow R_3
\end{array}
$$

$$
\sim
\left[\begin{array}{ccccc|c}
\frac{1}{2} & 1 & \frac{1}{2} & 0 & 0 & 16 \\
1 & 0 & -2 & 1 & 0 & 20 \\
\hline
-10 & 0 & 40 & 0 & 1 & 1{,}280
\end{array}\right]
$$

We have completed the pivot operation, and now we must insert appropriate variables for this new tableau. Since x_2 replaced s_1, the basic variables are now x_2, s_2, and P, as indicated by the labels on the left side of the new tableau. Note that this selection of basic variables agrees with the procedure outlined on page 299 for selecting basic variables. We write the new basic feasible solution by setting the nonbasic variables x_1 and s_1 equal to 0 and solving for the basic variables by inspection. (Remember, the values of the basic variables listed on the left are the corresponding numbers to the right of the vertical line. To see this, substitute $x_1 = 0$ and $s_1 = 0$ in the corresponding system shown next to the simplex tableau.)

$$
\begin{array}{c}
\\
x_2 \\
s_2 \\
P
\end{array}
\begin{array}{ccccc}
x_1 & x_2 & s_1 & s_2 & P \\
\end{array}
\left[\begin{array}{ccccc|c}
\frac{1}{2} & 1 & \frac{1}{2} & 0 & 0 & 16 \\
1 & 0 & -2 & 1 & 0 & 20 \\
\hline
-10 & 0 & 40 & 0 & 1 & 1{,}280
\end{array}\right]
\begin{array}{l}
\frac{1}{2}x_1 + x_2 + \frac{1}{2}s_1 \qquad\quad = 16 \\
x_1 \qquad\quad - 2s_1 + s_2 \quad\;\; = 20 \\
-10x_1 + \qquad 40s_1 \qquad + P = 1{,}280
\end{array}
$$

$$
x_1 = 0, \quad x_2 = 16, \quad s_1 = 0, \quad s_2 = 20, \quad P = \$1{,}280
$$

A profit of \$1,280 is a marked improvement over the \$0 profit produced by the initial basic feasible solution. But we can improve P still further, since a negative indicator still remains in the bottom row. To see why, we write out the objective function:

$$-10x_1 + 40s_1 + P = 1,280$$

or

$$P = 10x_1 - 40s_1 + 1,280$$

If s_1 stays a nonbasic variable (set equal to 0) and x_1 becomes a new basic variable, then

$$P = 10x_1 - 40(0) + 1,280 = 10x_1 + 1,280$$

and for each unit increase in x_1, P will increase \$10.

We now go through another iteration of the simplex process using another pivot element. The pivot element and the entering and exiting variables are shown in the following tableau:

$$
\begin{array}{c}
\text{Enter} \\
\downarrow
\end{array}
$$

	x_1	x_2	s_1	s_2	P		
x_2	$\frac{1}{2}$	1	$\frac{1}{2}$	0	0	16	$\frac{16}{1/2} = 32$
Exit → s_2	①	0	-2	1	0	20	$\frac{20}{1} = 20$
P	-10	0	40	0	1	1,280	

We now pivot on (the circled) 1. That is, we perform a pivot operation using this 1 as the pivot element. Since the pivot element is 1, we do not need to perform the first step in the pivot operation, so we proceed to the second step to get 0's above and below the pivot element 1. As before, to facilitate the process, we omit writing the variables, except for the first tableau.

$$
\begin{array}{c}
\text{Enter} \\
\downarrow
\end{array}
$$

	x_1	x_2	s_1	s_2	P		
x_2	$\frac{1}{2}$	1	$\frac{1}{2}$	0	0	16	$\left(-\frac{1}{2}\right)R_2 + R_1 \rightarrow R_1$
Exit → s_2	①	0	-2	1	0	20	
P	-10	0	40	0	1	1,280	$10R_2 + R_3 \rightarrow R_3$

$$
\sim
\begin{bmatrix}
0 & 1 & \frac{3}{2} & -\frac{1}{2} & 0 & 6 \\
1 & 0 & -2 & 1 & 0 & 20 \\
\hline
0 & 0 & 20 & 10 & 1 & 1,480
\end{bmatrix}
$$

Since there are no more negative indicators in the bottom row, we are done. Let us insert the appropriate variables for this last tableau and write the corresponding basic feasible solution. The basic variables are now x_1, x_2, and P, so to get the corresponding basic feasible solution, we set the nonbasic variables s_1 and s_2 equal to 0 and solve for the basic variables by inspection.

	x_1	x_2	s_1	s_2	P	
x_2	0	1	$\frac{3}{2}$	$-\frac{1}{2}$	0	6
x_1	1	0	-2	1	0	20
P	0	0	20	10	1	1,480

$$x_1 = 20, \quad x_2 = 6, \quad s_1 = 0, \quad s_2 = 0, \quad P = 1,480$$

To see why this is the maximum, we rewrite the objective function from the bottom row:

$$20s_1 + 10s_2 + P = 1,480$$

$$P = 1,480 - 20s_1 - 10s_2$$

Since s_1 and s_2 cannot be negative, any increase of either variable from 0 will make the profit smaller.

Finally, returning to our original problem, we conclude that a production schedule of 20 standard tents and 6 expedition tents will produce a maximum profit of \$1,480 per day, just as was found by the geometric method in Section 5.3. The fact that the slack variables are both 0 means that for this production schedule, the plant will operate at full capacity—there is no slack in either the cutting department or the assembly department.

Interpreting the Simplex Process Geometrically

We can interpret the simplex process geometrically in terms of the feasible region graphed in the preceding section. Table 1 lists the three basic feasible solutions we just found using the simplex method (in the order they were found). Table 1 also includes the corresponding corner points of the feasible region illustrated in Figure 1.

Table 1 **Basic Feasible Solution (obtained above)**

x_1	x_2	s_1	s_2	$P(\$)$	Corner Point
0	0	32	84	0	$O(0, 0)$
0	16	0	20	1,280	$A(0, 16)$
20	6	0	0	1,480	$B(20, 6)$

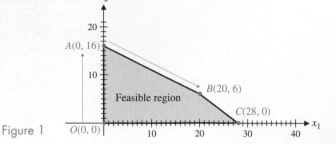

Figure 1

Looking at Table 1 and Figure 1, we see that the simplex process started at the origin, moved to the adjacent corner point $A(0,16)$, and then to the optimal solution $B(20, 6)$ at the next adjacent corner point. This is typical of the simplex process.

Simplex Method Summarized

Before presenting additional examples, we summarize the important parts of the simplex method schematically in Figure 2.

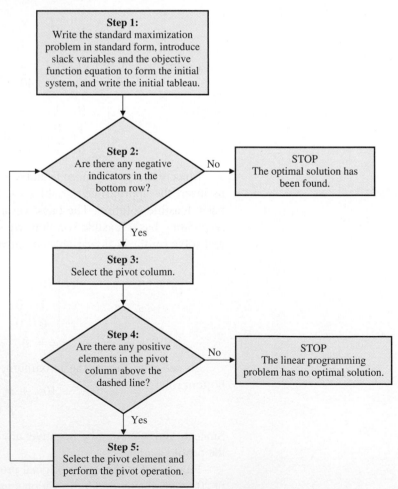

Figure 2 **Simplex algorithm for standard maximization problems (Problem constraints are of the $\leq$ form with nonnegative constants on the right. The coefficients of the objective function can be any real numbers.)**

EXAMPLE 1 Using the Simplex Method Solve the following linear programming problem using the simplex method:

$$\text{Maximize} \quad P = 10x_1 + 5x_2$$
$$\text{subject to} \quad 4x_1 + x_2 \le 28$$
$$2x_1 + 3x_2 \le 24$$
$$x_1, x_2 \ge 0$$

SOLUTION Introduce slack variables s_1 and s_2, and write the initial system:

$$4x_1 + x_2 + s_1 \qquad\qquad = 28$$
$$2x_1 + 3x_2 \qquad + s_2 \qquad = 24$$
$$-10x_1 - 5x_2 \qquad\qquad + P = 0$$
$$x_1, x_2, s_1, s_2 \ge 0$$

Write the simplex tableau, and identify the first pivot element and the entering and exiting variables:

$$
\begin{array}{c}
\quad\quad\quad \overset{\text{Enter}}{\underset{\downarrow}{}} \\
\end{array}
$$

	x_1	x_2	s_1	s_2	P		
Exit → s_1	④	1	1	0	0	28	$\frac{28}{4} = 7$
s_2	2	3	0	1	0	24	$\frac{24}{2} = 12$
P	−10	−5	0	0	1	0	

Perform the pivot operation:

$$
\begin{array}{c}
\quad\quad\quad \overset{\text{Enter}}{\underset{\downarrow}{}} \\
\end{array}
$$

	x_1	x_2	s_1	s_2	P		
Exit → s_1	④	1	1	0	0	28	$\frac{1}{4}R_1 \to R_1$
s_2	2	3	0	1	0	24	
P	−10	−5	0	0	1	0	

~	①	0.25	0.25	0	0	7	
	2	3	0	1	0	24	$(-2)R_1 + R_2 \to R_2$
	−10	−5	0	0	1	0	$10R_1 + R_3 \to R_3$

	x_1	x_2	s_1	s_2	P		
x_1	1	0.25	0.25	0	0	7	
~ s_2	0	2.5	−0.5	1	0	10	
P	0	−2.5	2.5	0	1	70	

Since there is still a negative indicator in the last row, we repeat the process by finding a new pivot element:

$$
\begin{array}{c}
\quad\quad\quad \overset{\text{Enter}}{\underset{\downarrow}{}} \\
\end{array}
$$

	x_1	x_2	s_1	s_2	P		
x_1	1	0.25	0.25	0	0	7	$\frac{7}{0.25} = 28$
Exit → s_2	0	②.5	−0.5	1	0	10	$\frac{10}{2.5} = 4$
P	0	−2.5	2.5	0	1	70	

Performing the pivot operation, we obtain

$$
\begin{array}{c}
 \\
\text{Enter} \\
\downarrow \\
\begin{array}{cccccc}
 & x_1 & x_2 & s_1 & s_2 & P \\
\end{array} \\
\begin{array}{c}
x_1 \\
\text{Exit} \rightarrow \; s_2 \\
P
\end{array}
\left[
\begin{array}{ccccc|c}
1 & 0.25 & 0.25 & 0 & 0 & 7 \\
0 & \enclose{circle}{2.5} & -0.5 & 1 & 0 & 10 \\
\hline
0 & -2.5 & 2.5 & 0 & 1 & 70
\end{array}
\right]
\begin{array}{l}
\\
\frac{1}{2.5}R_2 \rightarrow R_2 \\
\\
\end{array}
\end{array}
$$

$$
\sim
\left[
\begin{array}{ccccc|c}
1 & 0.25 & 0.25 & 0 & 0 & 7 \\
0 & \enclose{circle}{1} & -0.2 & 0.4 & 0 & 4 \\
\hline
0 & -2.5 & 2.5 & 0 & 1 & 70
\end{array}
\right]
\begin{array}{l}
(-0.25)R_2 + R_1 \rightarrow R_1 \\
\\
2.5R_2 + R_3 \rightarrow R_3
\end{array}
$$

$$
\sim
\begin{array}{c}
x_1 \\
x_2 \\
P
\end{array}
\left[
\begin{array}{ccccc|c}
1 & 0 & 0.3 & -0.1 & 0 & 6 \\
0 & 1 & -0.2 & 0.4 & 0 & 4 \\
\hline
0 & 0 & 2 & 1 & 1 & 80
\end{array}
\right]
$$

Since all the indicators in the last row are nonnegative, we stop and read the optimal solution:

$$\text{Max } P = 80 \quad \text{at} \quad x_1 = 6, \quad x_2 = 4, \quad s_1 = 0, \quad s_2 = 0$$

(To see why this makes sense, write the objective function corresponding to the last row to see what happens to P when you try to increase s_1 or s_2.)

Matched Problem 1 Solve the following linear programming problem using the simplex method:

$$
\begin{aligned}
\text{Maximize} \quad & P = 2x_1 + x_2 \\
\text{subject to} \quad & 5x_1 + x_2 \le 9 \\
& x_1 + x_2 \le 5 \\
& x_1, x_2 \le 0
\end{aligned}
$$

Explore and Discuss 1   Graph the feasible region for the linear programming problem in Example 1 and trace the path to the optimal solution determined by the simplex method.

EXAMPLE 2 Using the Simplex Method Solve using the simplex method:

$$
\begin{aligned}
\text{Maximize} \quad & P = 6x_1 + 3x_2 \\
\text{subject to} \quad & -2x_1 + 3x_2 \le 9 \\
& -x_1 + 3x_2 \le 12 \\
& x_1, x_2 \ge 0
\end{aligned}
$$

SOLUTION Write the initial system using the slack variables s_1 and s_2:

$$
\begin{aligned}
-2x_1 + 3x_2 + s_1   &= 9 \\
-x_1 + 3x_2 + s_2 &= 12 \\
-6x_1 - 3x_2   + P &= 0
\end{aligned}
$$

Write the simplex tableau and identify the first pivot element:

$$
\begin{array}{c}
 \\
\begin{array}{ccccc}
x_1 & x_2 & s_1 & s_2 & P
\end{array} \\
\begin{array}{c}
s_1 \\
s_2 \\
P
\end{array}
\left[
\begin{array}{ccccc|c}
-2 & 3 & 1 & 0 & 0 & 9 \\
-1 & 3 & 0 & 1 & 0 & 12 \\
\hline
-6 & -3 & 0 & 0 & 1 & 0
\end{array}
\right] \\
\begin{array}{c}
\uparrow \\
\text{Pivot column}
\end{array}
\end{array}
$$

Since both elements in the pivot column above the dashed line are negative, we are unable to select a pivot row. We stop and conclude that there is no optimal solution.

Matched Problem 2) Solve using the simplex method:

$$\text{Maximize} \quad P = 2x_1 + 3x_2$$
$$\text{subject to} \quad -3x_1 + 4x_2 \le 12$$
$$x_2 \le 2$$
$$x_1, x_2 \ge 0$$

Refer to Examples 1 and 2. In Example 1 we concluded that we had found the optimal solution because we could not select a pivot column. In Example 2 we concluded that the problem had no optimal solution because we selected a pivot column and then could not select a pivot row. Notice that we do not try to continue with the simplex method by selecting a negative pivot element or using a different column for the pivot column. Remember:

> If it is not possible to select a pivot column, then the simplex method stops and we conclude that the optimal solution has been found. If the pivot column has been selected and it is not possible to select a pivot row, then the simplex method stops and we conclude that there is no optimal solution.

Application

EXAMPLE 3 Agriculture A farmer owns a 100-acre farm and plans to plant at most three crops. The seed for crops A, B, and C costs \$40, \$20, and \$30 per acre, respectively. A maximum of \$3,200 can be spent on seed. Crops A, B, and C require one, two, and one work days per acre, respectively, and there are a maximum of 160 work days available. If the farmer can make a profit of \$100 per acre on crop A, \$300 per acre on crop B, and \$200 per acre on crop C, how many acres of each crop should be planted to maximize profit?

SOLUTION The farmer must decide on the number of acres of each crop that should be planted. So the decision variables are

$$x_1 = \text{number of acres of crop } A$$
$$x_2 = \text{number of acres of crop } B$$
$$x_3 = \text{number of acres of crop } C$$

The farmer's objective is to maximize profit:

$$P = 100x_1 + 300x_2 + 200x_3$$

The farmer is limited by the number of acres available for planting, the money available for seed, and the available work days. These limitations lead to the following constraints:

$$x_1 + x_2 + x_3 \le 100 \quad \text{Acreage constraint}$$
$$40x_1 + 20x_2 + 30x_3 \le 3{,}200 \quad \text{Monetary constraint}$$
$$x_1 + 2x_2 + x_3 \le 160 \quad \text{Labor constraint}$$

Adding the nonnegative constraints, we have the following model for a linear programming problem:

$$\text{Maximize} \quad P = 100x_1 + 300x_2 + 200x_3 \quad \text{Objective function}$$
$$\text{subject to} \quad x_1 + x_2 + x_3 \le 100$$
$$40x_1 + 20x_2 + 30x_3 \le 3{,}200 \quad \text{Problem constraints}$$
$$x_1 + 2x_2 + x_3 \le 160$$
$$x_1, x_2, x_3 \ge 0 \quad \text{Nonnegative constraints}$$

Next, we introduce slack variables and form the initial system:

$$
\begin{aligned}
x_1 + \quad x_2 + \quad x_3 + s_1 \qquad\qquad\qquad\qquad &= \quad 100 \\
40x_1 + \; 20x_2 + \; 30x_3 \qquad\quad + s_2 \qquad\qquad &= 3{,}200 \\
x_1 + \quad 2x_2 + \quad x_3 \qquad\qquad\quad + s_3 \qquad &= \quad 160 \\
-100x_1 - 300x_2 - 200x_3 \qquad\qquad\qquad\qquad + P = \quad &\quad 0 \\
x_1, x_2, x_3, s_1, s_2, s_3 \geq 0 &
\end{aligned}
$$

Notice that the initial system has $7 - 4 = 3$ nonbasic variables and 4 basic variables. Now we form the simplex tableau and solve by the simplex method:

Enter
↓

	x_1	x_2	x_3	s_1	s_2	s_3	P		
s_1	1	1	1	1	0	0	0	100	
s_2	40	20	30	0	1	0	0	3,200	
Exit → s_3	1	②	1	0	0	1	0	160	$0.5R_3 \rightarrow R_3$
P	−100	−300	−200	0	0	0	1	0	

	x_1	x_2	x_3	s_1	s_2	s_3	P		
~	1	1	1	1	0	0	0	100	$(-1)R_3 + R_1 \rightarrow R_1$
	40	20	30	0	1	0	0	3,200	$(-20)R_3 + R_2 \rightarrow R_2$
	0.5	①	0.5	0	0	0.5	0	80	
	−100	−300	−200	0	0	0	1	0	$300R_3 + R_4 \rightarrow R_4$

Enter
↓

	x_1	x_2	x_3	s_1	s_2	s_3	P		
Exit → s_1	0.5	0	⓪.5	1	0	−0.5	0	20	$2R_1 \rightarrow R_1$
s_2	30	0	20	0	1	−10	0	1,600	
x_2	0.5	1	0.5	0	0	0.5	0	80	
P	50	0	−50	0	0	150	1	24,000	

	x_1	x_2	x_3	s_1	s_2	s_3	P		
~	1	0	①	2	0	−1	0	40	
	30	0	20	0	1	−10	0	1,600	$(-20)R_1 + R_2 \rightarrow R_2$
	0.5	1	0.5	0	0	0.5	0	80	$(-0.5)R_1 + R_3 \rightarrow R_3$
	50	0	−50	0	0	150	1	24,000	$50R_1 + R_4 \rightarrow R_4$

	x_1	x_2	x_3	s_1	s_2	s_3	P	
x_3	1	0	1	2	0	−1	0	40
~ s_2	10	0	0	−40	1	10	0	800
x_2	0	1	0	−1	0	1	0	60
P	100	0	0	100	0	100	1	26,000

All indicators in the bottom row are nonnegative, and now we can read the optimal solution:

$$x_1 = 0, \quad x_2 = 60, \quad x_3 = 40, \quad s_1 = 0, \quad s_2 = 800, \quad s_3 = 0, \quad P = \$26{,}000$$

So if the farmer plants 60 acres in crop B, 40 acres in crop C, and no crop A, the maximum profit of \$26,000 will be realized. The fact that $s_2 = 800$ tells us (look at the second row in the equations at the start) that this maximum profit is reached by using only \$2,400 of the \$3,200 available for seed; that is, we have a slack of \$800 that can be used for some other purpose.

There are many types of software that can be used to solve linear programming problems by the simplex method. Figure 3 illustrates a solution to Example 3 in Excel, a popular spreadsheet for personal computers.

	A	B	C	D	E	F
1	Resources	Crop A	Crop B	Crop C	Available	Used
2	Acres	1	1	1	100	100
3	Seed($)	40	20	30	3,200	2,400
4	Workdays	1	2	1	160	160
5	Profit Per Acre	100	300	200	26,000	<-Total
6	Acres to plant	0	60	40		Profit

Figure 3

Matched Problem 3) Repeat Example 3 modified as follows:

	Investment per Acre			
	Crop A	Crop B	Crop C	Maximum Available
Seed cost	$24	$40	$30	$3,600
Work days	1	2	2	160
Profit	$140	$200	$160	

CONCEPTUAL INSIGHT

If you are solving a system of linear equations or inequalities, then you can multiply both sides of an equation by any nonzero number and both sides of an inequality by any positive number without changing the solution set. This is still the case for the simplex method, but you must be careful when you interpret the results. For example, consider the second problem constraint in the model for Example 3:

$$40x_1 + 20x_2 + 30x_3 \leq 3,200$$

Multiplying both sides by $\frac{1}{10}$ before introducing slack variables simplifies subsequent calculations. However, performing this operation has a side effect—it changes the units of the slack variable from dollars to tens of dollars. Compare the following two equations:

$$40x_1 + 20x_2 + 30x_3 + s_2 = 3,200 \quad \text{s_2 represents dollars}$$
$$4x_1 + 2x_2 + 3x_3 + s_2' = 320 \quad \text{s_2' represents tens of dollars}$$

In general, if you multiply a problem constraint by a positive number, remember to take this into account when you interpret the value of the slack variable for that constraint.

The feasible region for the linear programming problem in Example 3 has eight corner points, but the simplex method found the solution in only two steps. In larger problems, the difference between the total number of corner points and the number of steps required by the simplex method is even more dramatic. A feasible region may have hundreds or even thousands of corner points, yet the simplex method will often find the optimal solution in 10 or 15 steps.

To simplify this introduction to the simplex method, we have purposely avoided certain degenerate cases that lead to difficulties. Discussion and resolution of these problems is left to a more advanced treatment of the subject.

Exercises 6.2

For the simplex tableaux in Problems 1–4,

(A) *Identify the basic and nonbasic variables.*

(B) *Find the corresponding basic feasible solution.*

(C) *Determine whether the optimal solution has been found, an additional pivot is required, or the problem has no optimal solution.*

1.
$$\begin{array}{ccccc|c}
x_1 & x_2 & s_1 & s_2 & P & \\
2 & 1 & 0 & 3 & 0 & 12 \\
3 & 0 & 1 & -2 & 0 & 15 \\
\hline
-4 & 0 & 0 & 4 & 1 & 50
\end{array}$$

2.
$$\begin{array}{ccccc|c}
x_1 & x_2 & s_1 & s_2 & P & \\
1 & 4 & -2 & 0 & 0 & 10 \\
0 & 2 & 3 & 1 & 0 & 25 \\
\hline
0 & 5 & 6 & 0 & 1 & 35
\end{array}$$

3.
$$\begin{array}{cccccc|c}
x_1 & x_2 & x_3 & s_1 & s_2 & s_3 & P & \\
-2 & 0 & 1 & 3 & 1 & 0 & 0 & 5 \\
0 & 1 & 0 & -2 & 0 & 0 & 0 & 15 \\
-1 & 0 & 0 & 4 & 1 & 1 & 0 & 12 \\
\hline
-4 & 0 & 0 & 2 & 4 & 0 & 1 & 45
\end{array}$$

4.
$$\begin{array}{cccccc|c}
x_1 & x_2 & x_3 & s_1 & s_2 & s_3 & P & \\
0 & 2 & -1 & 1 & 4 & 0 & 0 & 5 \\
0 & 1 & 2 & 0 & -2 & 1 & 0 & 2 \\
1 & 3 & 0 & 0 & 5 & 0 & 0 & 11 \\
\hline
0 & -5 & 4 & 0 & -3 & 0 & 1 & 27
\end{array}$$

In Problems 5–8, find the pivot element, identify the entering and exiting variables, and perform one pivot operation.

5.
$$\begin{array}{ccccc|c}
x_1 & x_2 & s_1 & s_2 & P & \\
1 & 4 & 1 & 0 & 0 & 4 \\
3 & 5 & 0 & 1 & 0 & 24 \\
\hline
-8 & -5 & 0 & 0 & 1 & 0
\end{array}$$

6.
$$\begin{array}{ccccc|c}
x_1 & x_2 & s_1 & s_2 & P & \\
1 & 6 & 1 & 0 & 0 & 36 \\
3 & 1 & 0 & 1 & 0 & 5 \\
\hline
-1 & -2 & 0 & 0 & 1 & 0
\end{array}$$

7.
$$\begin{array}{cccccc|c}
x_1 & x_2 & s_1 & s_2 & s_3 & P & \\
2 & 1 & 1 & 0 & 0 & 0 & 4 \\
3 & 0 & 1 & 1 & 0 & 0 & 8 \\
0 & 0 & 2 & 0 & 1 & 0 & 2 \\
\hline
-4 & 0 & -3 & 0 & 0 & 1 & 5
\end{array}$$

8.
$$\begin{array}{cccccc|c}
x_1 & x_2 & s_1 & s_2 & s_3 & P & \\
0 & 0 & 2 & 1 & 1 & 0 & 2 \\
1 & 0 & -4 & 0 & 1 & 0 & 3 \\
0 & 1 & 5 & 0 & 2 & 0 & 11 \\
\hline
0 & 0 & -6 & 0 & -5 & 1 & 18
\end{array}$$

In Problems 9–12,

(A) *Using slack variables, write the initial system for each linear programming problem.*

(B) *Write the simplex tableau, circle the first pivot, and identify the entering and exiting variables.*

(C) *Use the simplex method to solve the problem.*

9. Maximize $P = 15x_1 + 10x_2$
   subject to
   $$2x_1 + x_2 \leq 10$$
   $$x_1 + 3x_2 \leq 10$$
   $$x_1, x_2 \geq 0$$

10. Maximize $P = 3x_1 + 2x_2$
    subject to
    $$5x_1 + 2x_2 \leq 20$$
    $$3x_1 + 2x_2 \leq 16$$
    $$x_1, x_2 \geq 0$$

11. Repeat Problem 9 with the objective function changed to $P = 30x_1 + x_2$.

12. Repeat Problem 10 with the objective function changed to $P = x_1 + 3x_2$.

Solve the linear programming problems in Problems 13–28 using the simplex method.

13. Maximize $P = 30x_1 + 40x_2$
    subject to
    $$2x_1 + x_2 \leq 10$$
    $$x_1 + x_2 \leq 7$$
    $$x_1 + 2x_2 \leq 12$$
    $$x_1, x_2 \geq 0$$

14. Maximize $P = 15x_1 + 20x_2$
    subject to
    $$2x_1 + x_2 \leq 9$$
    $$x_1 + x_2 \leq 6$$
    $$x_1 + 2x_2 \leq 10$$
    $$x_1, x_2 \geq 0$$

15. Maximize $P = 2x_1 + 3x_2$
    subject to
    $$-2x_1 + x_2 \leq 2$$
    $$-x_1 + x_2 \leq 5$$
    $$x_2 \leq 6$$
    $$x_1, x_2 \geq 0$$

16. Repeat Problem 15 with $P = -x_1 + 3x_2$.

17. Maximize $P = -x_1 + 2x_2$
    subject to
    $$-x_1 + x_2 \leq 2$$
    $$-x_1 + 3x_2 \leq 12$$
    $$x_1 - 4x_2 \leq 4$$
    $$x_1, x_2 \geq 0$$

18. Repeat Problem 17 with $P = x_1 + 2x_2$.

19. Maximize $P = 5x_1 + 2x_2 - x_3$
    subject to
    $$x_1 + x_2 - x_3 \leq 10$$
    $$2x_1 + 4x_2 + 3x_3 \leq 30$$
    $$x_1, x_2, x_3 \geq 0$$

20. Maximize $P = 4x_1 - 3x_2 + 2x_3$
    subject to
    $$x_1 + 2x_2 - x_3 \leq 5$$
    $$3x_1 + 2x_2 + 2x_3 \leq 22$$
    $$x_1, x_2, x_3 \geq 0$$

21. Maximize $P = 2x_1 + 3x_2 + 4x_3$
    subject to
    $$x_1 + x_3 \leq 4$$
    $$x_2 + x_3 \leq 3$$
    $$x_1, x_2, x_3 \geq 0$$

22. Maximize $P = x_1 + x_2 + 2x_3$
    subject to $x_1 - 2x_2 + x_3 \le 9$
             $2x_1 + x_2 + 2x_3 \le 28$
                  $x_1, x_2, x_3 \ge 0$

23. Maximize $P = 4x_1 + 3x_2 + 2x_3$
    subject to $3x_1 + 2x_2 + 5x_3 \le 23$
             $2x_1 + x_2 + x_3 \le 8$
              $x_1 + x_2 + 2x_3 \le 7$
                 $x_1, x_2, x_3 \ge 0$

24. Maximize $P = 4x_1 + 2x_2 + 3x_3$
    subject to $x_1 + x_2 + x_3 \le 11$
             $2x_1 + 3x_2 + x_3 \le 20$
             $x_1 + 3x_2 + 2x_3 \le 20$
                $x_1, x_2, x_3 \ge 0$

25. Maximize $P = 20x_1 + 30x_2$
    subject to $0.6x_1 + 1.2x_2 \le 960$
            $0.03x_1 + 0.04x_2 \le 36$
            $0.3x_1 + 0.2x_2 \le 270$
                $x_1, x_2 \ge 0$

26. Repeat Problem 25 with $P = 20x_1 + 20x_2$.

27. Maximize $P = x_1 + 2x_2 + 3x_3$
    subject to $2x_1 + 2x_2 + 8x_3 \le 600$
           $x_1 + 3x_2 + 2x_3 \le 600$
          $3x_1 + 2x_2 + x_3 \le 400$
               $x_1, x_2, x_3 \ge 0$

28. Maximize $P = 10x_1 + 50x_2 + 10x_3$
    subject to $3x_1 + 3x_2 + 3x_3 \le 66$
             $6x_1 - 2x_2 + 4x_3 \le 48$
             $3x_1 + 6x_2 + 9x_3 \le 108$
                $x_1, x_2, x_3 \ge 0$

✐ *In Problems 29 and 30, first solve the linear programming problem by the simplex method, keeping track of the basic feasible solutions at each step. Then graph the feasible region and illustrate the path to the optimal solution determined by the simplex method.*

29. Maximize $P = 2x_1 + 5x_2$
    subject to $x_1 + 2x_2 \le 40$
             $x_1 + 3x_2 \le 48$
             $x_1 + 4x_2 \le 60$
                  $x_2 \le 14$
               $x_1, x_2 \ge 0$

30. Maximize $P = 5x_1 + 3x_2$
    subject to $5x_1 + 4x_2 \le 100$
           $2x_1 + x_2 \le 28$
           $4x_1 + x_2 \le 42$
                 $x_1 \le 10$
               $x_1, x_2 \ge 0$

✐ *Solve Problems 31 and 32 by the simplex method and also by graphing (the geometric method). Compare and contrast the results.*

31. Maximize $P = 2x_1 + 3x_2$
    subject to $-2x_1 + x_2 \le 4$
                  $x_2 \le 10$
               $x_1, x_2 \ge 0$

32. Maximize $P = 2x_1 + 3x_2$
    subject to $-x_1 + x_2 \le 2$
                  $x_2 \le 4$
               $x_1, x_2 \ge 0$

✐ *In Problems 33–36, there is a tie for the choice of the first pivot column. Use the simplex method to solve each problem two differ-ent ways: first by choosing column 1 as the first pivot column, and then by choosing column 2 as the first pivot column. Discuss the relationship between these two solutions.*

33. Maximize $P = x_1 + x_2$
    subject to $2x_1 + x_2 \le 16$
                  $x_1 \le 6$
                  $x_2 \le 10$
               $x_1, x_2 \ge 0$

34. Maximize $P = x_1 + x_2$
    subject to $x_1 + 2x_2 \le 10$
                  $x_1 \le 6$
                  $x_2 \le 4$
               $x_1, x_2 \ge 0$

35. Maximize $P = 3x_1 + 3x_2 + 2x_3$
    subject to $x_1 + x_2 + 2x_3 \le 20$
           $2x_1 + x_2 + 4x_3 \le 32$
                 $x_1, x_2, x_3 \ge 0$

36. Maximize $P = 2x_1 + 2x_2 + x_3$
    subject to $x_1 + x_2 + 3x_3 \le 10$
          $2x_1 + 4x_2 + 5x_3 \le 24$
                 $x_1, x_2, x_3 \ge 0$

Applications

In Problems 37–52, construct a mathematical model in the form of a linear programming problem. (The answers in the back of the book for these application problems include the model.) Then solve the problem using the simplex method. Include an interpre-tation of any nonzero slack variables in the optimal solution.

37. Manufacturing: resource allocation. A small company man-ufactures three different electronic components for computers. Component A requires 2 hours of fabrication and 1 hour of as-sembly; component B requires 3 hours of fabrication and 1 hour of assembly; and component C requires 2 hours of fab-rication and 2 hours of assembly. The company has up to 1,000 labor-hours of fabrication time and 800 labor-hours of assem-bly time available per week. The profit on each component, A, B, and C, is $7, $8, and $10, respectively. How many compo-nents of each type should the company manufacture each week in order to maximize its profit (assuming that all components manufactured can be sold)? What is the maximum profit?

✐ **38. Manufacturing: resource allocation.** Solve Problem 37 with the additional restriction that the combined total number of components produced each week cannot exceed 420. Discuss the effect of this restriction on the solution to Problem 37.

39. Investment. An investor has at most $100,000 to invest in government bonds, mutual funds, and money market funds. The average yields for government bonds, mutual funds, and money market funds are 8%, 13%, and 15%, respectively. The investor's policy requires that the total amount invested in mutual and money market funds not exceed the amount invested in government bonds. How much should be invested in each type of investment in order to maximize the return? What is the maximum return?

40. Investment. Repeat Problem 39 under the additional assumption that no more than $30,000 can be invested in money market funds.

41. Advertising. A department store has up to $20,000 to spend on television advertising for a sale. All ads will be placed with one television station. A 30-second ad costs $1,000 on daytime TV and is viewed by 14,000 potential customers, $2,000 on prime-time TV and is viewed by 24,000 potential customers, and $1,500 on late-night TV and is viewed by 18,000 potential customers. The television station will not accept a total of more than 15 ads in all three time periods. How many ads should be placed in each time period in order to maximize the number of potential customers who will see the ads? How many potential customers will see the ads? (Ignore repeated viewings of the ad by the same potential customer.)

42. Advertising. Repeat Problem 41 if the department store increases its budget to $24,000 and requires that at least half of the ads be placed during prime-time.

43. Home construction. A contractor is planning a new housing development consisting of colonial, split-level, and ranch-style houses. A colonial house requires $\frac{1}{2}$ acre of land, $60,000 capital, and 4,000 labor-hours to construct, and returns a profit of $20,000. A split-level house requires $\frac{1}{2}$ acre of land, $60,000 capital, and 3,000 labor-hours to construct, and returns a profit of $18,000. A ranch house requires 1 acre of land, $80,000 capital, and 4,000 labor-hours to construct, and returns a profit of $24,000. The contractor has 30 acres of land, $3,200,000 capital, and 180,000 labor-hours available. How many houses of each type should be constructed to maximize the contractor's profit? What is the maximum profit?

44. Bicycle manufacturing. A company manufactures three-speed, five-speed, and ten-speed bicycles. Each bicycle passes through three departments: fabrication, painting & plating, and final assembly. The relevant manufacturing data are given in the table.

	Labor-Hours per Bicycle			Maximum Labor-Hours Available per Day
	Three-speed	Five-Speed	Ten-Speed	
Fabrication	3	4	5	120
Painting & plating	5	3	5	130
Final assembly	4	3	5	120
Profit per bicycle ($)	80	70	100	

How many bicycles of each type should the company manufacture per day in order to maximize its profit? What is the maximum profit?

45. Home building. Repeat Problem 43 if the profit on a colonial house decreases from $20,000 to $17,000 and all other data remain the same. If the slack associated with any problem constraint is nonzero, find it.

46. Bicycle manufacturing. Repeat Problem 44 if the profit on a ten-speed bicycle increases from $100 to $110 and all other data remain the same. If the slack associated with any problem constraint is nonzero, find it.

47. Home building. Repeat Problem 43 if the profit on a colonial house increases from $20,000 to $25,000 and all other data remain the same. If the slack associated with any problem constraint is nonzero, find it.

48. Bicycle manufacturing. Repeat Problem 44 if the profit on a five-speed bicycle increases from $70 to $110 and all other data remain the same. If the slack associated with any problem constraint is nonzero, find it.

49. Animal nutrition. The natural diet of a certain animal consists of three foods: A, B, and C. The number of units of calcium, iron, and protein in 1 gram of each food and the average daily intake are given in the table. A scientist wants to investigate the effect of increasing the protein in the animal's diet while not allowing the units of calcium and iron to exceed their average daily intakes. How many grams of each food should be used to maximize the amount of protein in the diet? What is the maximum amount of protein?

	Units per Gram			Average Daily Intake (units)
	Food A	Food B	Food C	
Calcium	1	3	2	30
Iron	2	1	2	24
Protein	3	4	5	60

50. Animal nutrition. Repeat Problem 49 if the scientist wants to maximize the daily calcium intake while not allowing the intake of iron or protein to exceed the average daily intake.

51. Opinion survey. A political scientist received a grant to fund a research project on voting trends. The budget includes $3,200 for conducting door-to-door interviews on the day before an election. Undergraduate students, graduate students, and faculty members will be hired to conduct the interviews. Each undergraduate student will conduct 18 interviews for $100. Each graduate student will conduct 25 interviews for $150. Each faculty member will conduct 30 interviews for $200. Due to limited transportation facilities, no more than 20 interviewers can be hired. How many undergraduate students, graduate students, and faculty members should be hired in order to maximize the number of interviews? What is the maximum number of interviews?

52. Opinion survey. Repeat Problem 51 if one of the requirements of the grant is that at least 50% of the interviewers be undergraduate students.

Answers to Matched Problems

1. Max $P = 6$ when $x_1 = 1$ and $x_2 = 4$

2. No optimal solution

3. 40 acres of crop A, 60 acres of crop B, no crop C; max $P = $17,600$ (since $s_2 = 240$, $240 out of the $3,600 will not be spent).

6.3 The Dual Problem: Minimization with Problem Constraints of the Form ≥

- Formation of the Dual Problem
- Solution of Minimization Problems
- Application: Transportation Problem
- Summary of Problem Types and Solution Methods

In the preceding section, we restricted attention to standard maximization problems (problem constraints of the form ≤, with nonnegative constants on the right and any real numbers as objective function coefficients). Now we will consider minimization problems with ≥ problem constraints. These two types of problems turn out to be very closely related.

Formation of the Dual Problem

Associated with each minimization problem with ≥ constraints is a maximization problem called the **dual problem**. To illustrate the procedure for forming the dual problem, consider the following minimization problem:

$$\begin{aligned}
\text{Minimize} \quad & C = 16x_1 + 45x_2 \\
\text{subject to} \quad & 2x_1 + 5x_2 \geq 50 \\
& x_1 + 3x_2 \geq 27 \\
& x_1, x_2 \geq 0
\end{aligned} \tag{1}$$

The first step in forming the dual problem is to construct a matrix using the problem constraints and the objective function written in the following form:

$$\begin{aligned}
2x_1 + 5x_2 &\geq 50 \\
x_1 + 3x_2 &\geq 27 \\
16x_1 + 45x_2 &= C
\end{aligned} \qquad A = \begin{bmatrix} 2 & 5 & | & 50 \\ 1 & 3 & | & 27 \\ 16 & 45 & | & 1 \end{bmatrix}$$

⚠ **CAUTION** Do not confuse matrix A with the simplex tableau. We use a solid horizontal line in matrix A to help distinguish the dual matrix from the simplex tableau. No slack variables are involved in matrix A, and the coefficient of C is in the same column as the constants from the problem constraints. ▲

Now we will form a second matrix called the *transpose of A*. In general, the **transpose** of a given matrix A is the matrix A^T formed by interchanging the rows and corresponding columns of A (first row with first column, second row with second column, and so on).

$$A = \begin{bmatrix} 2 & 5 & | & 50 \\ 1 & 3 & | & 27 \\ 16 & 45 & | & 1 \end{bmatrix} \quad \begin{array}{l} R_1 \text{ in } A = C_1 \text{ in } A^T \\ R_2 \text{ in } A = C_2 \text{ in } A^T \\ R_3 \text{ in } A = C_3 \text{ in } A^T \end{array}$$

$$A^T = \begin{bmatrix} 2 & 1 & | & 16 \\ 5 & 3 & | & 45 \\ 50 & 27 & | & 1 \end{bmatrix} \qquad A^T \text{ is the transpose of } A.$$

We can use the rows of A^T to define a new linear programming problem. This new problem will always be a maximization problem with ≤ problem constraints. To avoid confusion, we will use different variables in this new problem:

$$\begin{aligned}
2y_1 + y_2 &\leq 16 \\
5y_1 + 3y_2 &\leq 45 \\
50y_1 + 27y_2 &= P
\end{aligned} \qquad A^T = \begin{array}{cc} \begin{array}{cc} y_1 & y_2 \end{array} & \\ \begin{bmatrix} 2 & 1 & | & 16 \\ 5 & 3 & | & 45 \\ 50 & 27 & | & 1 \end{bmatrix} \end{array}$$

The dual of the minimization problem (1) is the following maximization problem:

$$\begin{aligned}
\text{Maximize} \quad & P = 50y_1 + 27y_2 \\
\text{subject to} \quad & 2y_1 + y_2 \le 16 \\
& 5y_1 + 3y_2 \le 45 \\
& y_1, y_2 \ge 0
\end{aligned} \tag{2}$$

Explore and Discuss 1   Excluding the nonnegative constraints, the components of a linear programming problem can be divided into three categories: the coefficients of the objective function, the coefficients of the problem constraints, and the constants on the right side of the problem constraints. Write a verbal description of the relationship between the components of the original minimization problem (1) and the dual maximization problem (2).

The procedure for forming the dual problem is summarized in the following box:

PROCEDURE Formation of the Dual Problem

Given a minimization problem with $\ge$ problem constraints,

Step 1 Use the coefficients and constants in the problem constraints and the objective function to form a matrix A with the coefficients of the objective function in the last row.

Step 2 Interchange the rows and columns of matrix A to form the matrix A^T, the transpose of A.

Step 3 Use the rows of A^T to form a maximization problem with $\le$ problem constraints.

EXAMPLE 1   Forming the Dual Problem Form the dual problem:

$$\begin{aligned}
\text{Minimize} \quad & C = 40x_1 + 12x_2 + 40x_3 \\
\text{subject to} \quad & 2x_1 + x_2 + 5x_3 \ge 20 \\
& 4x_1 + x_2 + x_3 \ge 30 \\
& x_1, x_2, x_3 \ge 0
\end{aligned}$$

SOLUTION

Step 1 Form the matrix A:

$$A = \left[\begin{array}{ccc|c}
2 & 1 & 5 & 20 \\
4 & 1 & 1 & 30 \\
40 & 12 & 40 & 1
\end{array}\right]$$

Step 2 Form the matrix A^T, the transpose of A:

$$A^T = \left[\begin{array}{cc|c}
2 & 4 & 40 \\
1 & 1 & 12 \\
5 & 1 & 40 \\
20 & 30 & 1
\end{array}\right]$$

Step 3 State the dual problem:

$$\begin{aligned}
\text{Maximize} \quad & P = 20y_1 + 30y_2 \\
\text{subject to} \quad & 2y_1 + 4y_2 \le 40 \\
& y_1 + y_2 \le 12 \\
& 5y_1 + y_2 \le 40 \\
& y_1, y_2 \ge 0
\end{aligned}$$

Matched Problem 1 | Form the dual problem:

$$\text{Minimize} \quad C = 16x_1 + 9x_2 + 21x_3$$
$$\text{subject to} \quad x_1 + x_2 + 3x_3 \geq 12$$
$$2x_1 + x_2 + x_3 \geq 16$$
$$x_1, x_2, x_3 \geq 0$$

Solution of Minimization Problems

The following theorem establishes the relationship between the solution of a minimization problem and the solution of its dual problem:

> **THEOREM 1 Fundamental Principle of Duality**
>
> A minimization problem has a solution if and only if its dual problem has a solution. If a solution exists, then the optimal value of the minimization problem is the same as the optimal value of the dual problem.

The proof of Theorem 1 is beyond the scope of this text. However, we can illustrate Theorem 1 by solving minimization problem (1) and its dual maximization problem (2) geometrically.

ORIGINAL PROBLEM (1)	*DUAL PROBLEM* (2)
Minimize $C = 16x_1 + 45x_2$	Maximize $P = 50y_1 + 27y_2$
subject to $2x_1 + 5x_2 \geq 50$	subject to $2y_1 + y_2 \leq 16$
$x_1 + 3x_2 \geq 27$	$5y_1 + 3y_2 \leq 45$
$x_1, x_2 \geq 0$	$y_1, y_2 \geq 0$

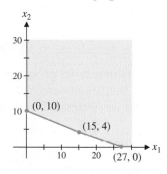
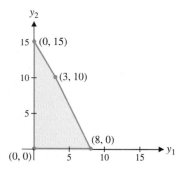

Corner Point

(x_1, x_2)	$C = 16x_1 + 45x_2$
$(0, 10)$	450
$(15, 4)$	420
$(27, 0)$	432

Min $C = 420$ at $(15, 4)$

Corner Point

(y_1, y_2)	$P = 50y_1 + 27y_2$
$(0, 0)$	0
$(0, 15)$	405
$(3, 10)$	420
$(8, 0)$	400

Max $P = 420$ at $(3, 10)$

Note that the minimum value of C in problem (1) is the same as the maximum value of P in problem (2). The optimal solutions producing this optimal value are different: $(15, 4)$ is the optimal solution for problem (1), and $(3, 10)$ is the optimal solution for problem (2). Theorem 1 only guarantees that the optimal values of a minimization problem and its dual are equal, not that the optimal solutions are the same. In general, it is not possible to determine an optimal solution for a minimization problem by examining the feasible set for the dual problem. However, it is possible to apply the simplex method to the dual problem and find both the optimal value and an optimal solution to the original minimization problem. To see how this is done, we will solve problem (2) using the simplex method.

For reasons that will become clear later, we will use the variables x_1 and x_2 from the original problem as the slack variables in the dual problem:

$$\begin{aligned}
2y_1 + y_2 + x_1 \phantom{{}+ x_2 + P} &= 16 \\
5y_1 + 3y_2 \phantom{{}+ x_1} + x_2 \phantom{{}+ P} &= 45 \qquad \text{Initial system for the dual problem} \\
-50y_1 - 27y_2 \phantom{{}+ x_1 + x_2} + P &= 0
\end{aligned}$$

$$
\begin{array}{c}
 \\
x_1 \\
x_2 \\
P
\end{array}
\begin{array}{c}
y_1 \quad\; y_2 \quad\; x_1 \quad\; x_2 \quad P \\
\left[\begin{array}{ccccc|c}
② & 1 & 1 & 0 & 0 & 16 \\
5 & 3 & 0 & 1 & 0 & 45 \\
\hline
-50 & -27 & 0 & 0 & 1 & 0
\end{array}\right]
\end{array}
\begin{array}{l}
0.5R_1 \to R_1 \\
\\
\\
\end{array}
$$

$$
\sim
\left[\begin{array}{ccccc|c}
① & 0.5 & 0.5 & 0 & 0 & 8 \\
5 & 3 & 0 & 1 & 0 & 45 \\
\hline
-50 & -27 & 0 & 0 & 1 & 0
\end{array}\right]
\begin{array}{l}
\\
(-5)R_1 + R_2 \to R_2 \\
50R_1 + R_3 \to R_3
\end{array}
$$

$$
\begin{array}{c}
y_1 \\
\sim\; x_2 \\
P
\end{array}
\left[\begin{array}{ccccc|c}
1 & 0.5 & 0.5 & 0 & 0 & 8 \\
0 & ⓪.⑤ & -2.5 & 1 & 0 & 5 \\
\hline
0 & -2 & 25 & 0 & 1 & 400
\end{array}\right]
\begin{array}{l}
\\
2R_2 \to R_2 \\
\end{array}
$$

$$
\sim
\left[\begin{array}{ccccc|c}
1 & 0.5 & 0.5 & 0 & 0 & 8 \\
0 & ① & -5 & 2 & 0 & 10 \\
\hline
0 & -2 & 25 & 0 & 1 & 400
\end{array}\right]
\begin{array}{l}
(-0.5)R_2 + R_1 \to R_1 \\
\\
2R_2 + R_3 \to R_3
\end{array}
$$

$$
\begin{array}{c}
y_1 \\
\sim\; y_2 \\
P
\end{array}
\left[\begin{array}{ccccc|c}
1 & 0 & 3 & -1 & 0 & 3 \\
0 & 1 & -5 & 2 & 0 & 10 \\
\hline
0 & 0 & 15 & 4 & 1 & 420
\end{array}\right]
$$

Since all indicators in the bottom row are nonnegative, the solution to the dual problem is

$$y_1 = 3, \quad y_2 = 10, \quad x_1 = 0, \quad x_2 = 0, \quad P = 420$$

which agrees with our earlier geometric solution. Furthermore, examining the bottom row of the final simplex tableau, we see the same optimal solution to the minimization problem that we obtained directly by the geometric method:

$$\text{Min } C = 420 \qquad \text{at} \qquad x_1 = 15, \quad x_2 = 4$$

This is no accident.

> An optimal solution to a minimization problem always can be obtained from the bottom row of the final simplex tableau for the dual problem.

We can see that using x_1 and x_2 as slack variables in the dual problem makes it easy to identify the solution of the original problem.

Explore and Discuss 2 The simplex method can be used to solve any standard maximization problem. Which of the following minimization problems have dual problems that are standard maximization problems? (Do not solve the problems.)

(A) Minimize $C = 2x_1 + 3x_2$

       subject to $\qquad 2x_1 - 5x_2 \geq 4$

$$x_1 - 3x_2 \geq -6$$

$$x_1, x_2 \geq 0$$

(B) Minimize $C = 2x_1 - 3x_2$
    subject to $-2x_1 + 5x_2 \geq 4$
                 $-x_1 + 3x_2 \geq 6$
                 $x_1, x_2 \geq 0$

What conditions must a minimization problem satisfy so that its dual problem is a standard maximization problem?

The procedure for solving a minimization problem by applying the simplex method to its dual problem is summarized in the following box:

PROCEDURE Solution of a Minimization Problem

Given a minimization problem with nonnegative coefficients in the objective function,

Step 1 Write all problem constraints as $\geq$ inequalities. (This may introduce negative numbers on the right side of some problem constraints.)

Step 2 Form the dual problem.

Step 3 Write the initial system of the dual problem, using the variables from the minimization problem as slack variables.

Step 4 Use the simplex method to solve the dual problem.

Step 5 Read the solution of the minimization problem from the bottom row of the final simplex tableau in step 4.

Note: If the dual problem has no optimal solution, the minimization problem has no optimal solution.

EXAMPLE 2 Solving a Minimization Problem Solve the following minimization problem by maximizing the dual problem:

$$\text{Minimize}\quad C = 40x_1 + 12x_2 + 40x_3$$
$$\text{subject to}\quad 2x_1 + x_2 + 5x_3 \geq 20$$
$$4x_1 + x_2 + x_3 > 30$$
$$x_1, x_2, x_3 \geq 0$$

SOLUTION From Example 1, the dual problem is

$$\text{Maximize}\quad P = 20y_1 + 30y_2$$
$$\text{subject to}\quad 2y_1 + 4y_2 \leq 40$$
$$y_1 + y_2 \leq 12$$
$$5y_1 + y_2 \leq 40$$
$$y_1, y_2 \geq 0$$

Using x_1, x_2, and x_3 for slack variables, we obtain the initial system for the dual problem:

$$2y_1 + 4y_2 + x_1 \qquad\qquad\qquad = 40$$
$$y_1 + y_2 \quad\; + x_2 \qquad\qquad = 12$$
$$5y_1 + y_2 \qquad\quad + x_3 \qquad = 40$$
$$-20y_1 - 30y_2 \qquad\qquad\quad + P = 0$$

Now we form the simplex tableau and solve the dual problem:

$$
\begin{array}{c}
\begin{array}{ccccccc} y_1 & y_2 & x_1 & x_2 & x_3 & P & \end{array} \\
\begin{array}{c} x_1 \\ x_2 \\ x_3 \\ P \end{array}
\left[
\begin{array}{cccccc|c}
2 & ④ & 1 & 0 & 0 & 0 & 40 \\
1 & 1 & 0 & 1 & 0 & 0 & 12 \\
5 & 1 & 0 & 0 & 1 & 0 & 40 \\
\hdashline
-20 & -30 & 0 & 0 & 0 & 1 & 0
\end{array}
\right]
\begin{array}{l} \frac{1}{4}R_1 \to R_1 \end{array}
\end{array}
$$

$$
\sim
\left[
\begin{array}{cccccc|c}
\frac{1}{2} & ① & \frac{1}{4} & 0 & 0 & 0 & 10 \\
1 & 1 & 0 & 1 & 0 & 0 & 12 \\
5 & 1 & 0 & 0 & 1 & 0 & 40 \\
\hdashline
-20 & -30 & 0 & 0 & 0 & 1 & 0
\end{array}
\right]
\begin{array}{l} \\ (-1)R_1 + R_2 \to R_2 \\ (-1)R_1 + R_3 \to R_3 \\ 30R_1 + R_4 \to R_4 \end{array}
$$

$$
\sim
\begin{array}{c} y_2 \\ x_2 \\ x_3 \\ P \end{array}
\left[
\begin{array}{cccccc|c}
\frac{1}{2} & 1 & \frac{1}{4} & 0 & 0 & 0 & 10 \\
⑦\frac{1}{2} & 0 & -\frac{1}{4} & 1 & 0 & 0 & 2 \\
\frac{9}{2} & 0 & -\frac{1}{4} & 0 & 1 & 0 & 30 \\
\hdashline
-5 & 0 & \frac{15}{2} & 0 & 0 & 1 & 300
\end{array}
\right]
\begin{array}{l} \\ 2R_2 \to R_2 \end{array}
$$

$$
\sim
\left[
\begin{array}{cccccc|c}
\frac{1}{2} & 1 & \frac{1}{4} & 0 & 0 & 0 & 10 \\
① & 0 & -\frac{1}{2} & 2 & 0 & 0 & 4 \\
\frac{9}{2} & 0 & -\frac{1}{4} & 0 & 1 & 0 & 30 \\
\hdashline
-5 & 0 & \frac{15}{2} & 0 & 0 & 1 & 300
\end{array}
\right]
\begin{array}{l} (-\frac{1}{2})R_2 + R_1 \to R_1 \\ \\ (-\frac{9}{2})R_2 + R_3 \to R_3 \\ 5R_2 + R_4 \to R_4 \end{array}
$$

$$
\sim
\begin{array}{c} y_2 \\ y_1 \\ x_3 \\ P \end{array}
\left[
\begin{array}{cccccc|c}
0 & 1 & \frac{1}{2} & -1 & 0 & 0 & 8 \\
1 & 0 & -\frac{1}{2} & 2 & 0 & 0 & 4 \\
0 & 0 & 2 & -9 & 1 & 0 & 12 \\
\hdashline
0 & 0 & 5 & 10 & 0 & 1 & 320
\end{array}
\right]
$$

From the bottom row of this tableau, we see that

$$\text{Min } C = 320 \quad \text{at} \quad x_1 = 5, \quad x_2 = 10, \quad x_3 = 0$$

Matched Problem 2) Solve the following minimization problem by maximizing the dual problem (see Matched Problem 1):

$$
\begin{aligned}
\text{Minimize } \quad & C = 16x_1 + 9x_2 + 21x_3 \\
\text{subject to} \quad & x_1 + x_2 + 3x_3 \ge 12 \\
& 2x_1 + x_2 + x_3 \ge 16 \\
& x_1, x_2, x_3 \ge 0
\end{aligned}
$$

CONCEPTUAL INSIGHT

In Section 6.2, we noted that multiplying a problem constraint by a number changes the units of the slack variable. This requires special interpretation of the value of the slack variable in the optimal solution, but causes no serious problems. However, when using the dual method, multiplying a problem constraint in the dual problem by a number can have serious consequences—the bottom row of the final simplex tableau may no longer give the correct solution to the minimization problem. To see this, refer to the first problem constraint of the dual problem in Example 2:

$$2y_1 + 4y_2 \le 40$$

If we multiply this constraint by $\frac{1}{2}$ and then solve, the final tableau is:

$$
\begin{array}{cccccc} y_1 & y_2 & x_1 & x_2 & x_3 & P \end{array}
\left[
\begin{array}{cccccc|c}
0 & 1 & 1 & -1 & 0 & 0 & 8 \\
1 & 0 & -1 & 2 & 0 & 0 & 4 \\
0 & 0 & 4 & -9 & 1 & 0 & 12 \\
0 & 0 & 10 & 10 & 0 & 1 & 320
\end{array}
\right]
$$

The bottom row of this tableau indicates that the optimal solution to the minimization problem is $C = 320$ at $x_1 = 10$ and $x_2 = 10$. This is not the correct answer ($x_1 = 5$ is the correct answer). Thus, **you should never multiply a problem constraint in a maximization problem by a number if that maximization problem is being used to solve a minimization problem.** You may still simplify problem constraints in a minimization problem before forming the dual problem.

EXAMPLE 3 Solving a Minimization Problem Solve the following minimization problem by maximizing the dual problem:

$$\text{Minimize} \quad C = 5x_1 + 10x_2$$
$$\text{subject to} \quad x_1 - x_2 \geq 1$$
$$-x_1 + x_2 \geq 2$$
$$x_1, x_2 \geq 0$$

SOLUTION $A = \begin{bmatrix} 1 & -1 & | & 1 \\ -1 & 1 & | & 2 \\ \hline 5 & 10 & | & 1 \end{bmatrix}$ $A^T = \begin{bmatrix} 1 & -1 & | & 5 \\ -1 & 1 & | & 10 \\ \hline 1 & 2 & | & 1 \end{bmatrix}$

The dual problem is

$$\text{Maximize} \quad P = y_1 + 2y_2$$
$$\text{subject to} \quad y_1 - y_2 \leq 5$$
$$-y_1 + y_2 \leq 10$$
$$y_1, y_2 \geq 0$$

Introduce slack variables x_1 and x_2, and form the initial system for the dual problem:

$$y_1 - y_2 + x_1 \qquad\qquad = 5$$
$$-y_1 + y_2 \qquad + x_2 \qquad = 10$$
$$-y_1 - 2y_2 \qquad\qquad + P = 0$$

Form the simplex tableau and solve:

$$
\begin{array}{c}
 \\
x_1 \\
x_2 \\
P
\end{array}
\begin{array}{c}
\begin{array}{ccccc}
y_1 & y_2 & x_1 & x_2 & P
\end{array} \\
\left[\begin{array}{ccccc|c}
1 & -1 & 1 & 0 & 0 & 5 \\
-1 & ① & 0 & 1 & 0 & 10 \\
\hline
-1 & -2 & 0 & 0 & 1 & 0
\end{array}\right]
\begin{array}{l}
R_2 + R_1 \rightarrow R_1 \\
\\
2R_2 + R_3 \rightarrow R_3
\end{array}
\end{array}
$$

$$
\sim \left[\begin{array}{ccccc|c}
0 & 0 & 1 & 1 & 0 & 15 \\
-1 & 1 & 0 & 1 & 0 & 10 \\
\hline
-3 & 0 & 0 & 2 & 1 & 20
\end{array}\right]
\begin{array}{l}
\text{No positive elements} \\
\text{above dashed line in} \\
\text{pivot column}
\end{array}
$$

$$\uparrow$$
Pivot column

The -3 in the bottom row indicates that column 1 is the pivot column. Since no positive elements appear in the pivot column above the dashed line, we are unable to select a pivot row. We stop the pivot operation and conclude that this maximization problem has no optimal solution (see Figure 2, Section 6.2). Theorem 1 now implies that the original minimization problem has no solution. The graph of the inequalities in the minimization problem (Fig. 1) shows that the feasible region is empty; so it is not surprising that an optimal solution does not exist.

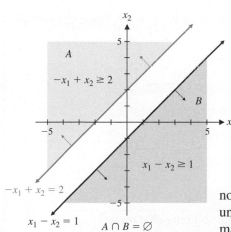

Figure 1

Matched Problem 3 Solve the following minimization problem by maximizing the dual problem:

$$\text{Minimize} \quad C = 2x_1 + 3x_2$$
$$\text{subject to} \quad x_1 - 2x_2 \geq 2$$
$$-x_1 + x_2 \geq 1$$
$$x_1, x_2 \geq 0$$

Application: Transportation Problem

One of the first applications of linear programming was to minimize the cost of transporting materials. Problems of this type are referred to as **transportation problems**.

EXAMPLE 4   Transportation Problem   A computer manufacturing company has two assembly plants, plant A and plant B, and two distribution outlets, outlet I and outlet II. Plant A can assemble at most 700 computers a month, and plant B can assemble at most 900 computers a month. Outlet I must have at least 500 computers a month, and outlet II must have at least 1,000 computers a month. Transportation costs for shipping one computer from each plant to each outlet are as follows: \$6 from plant A to outlet I; \$5 from plant A to outlet II; \$4 from plant B to outlet I; \$8 from plant B to outlet II. Find a shipping schedule that minimizes the total cost of shipping the computers from the assembly plants to the distribution outlets. What is this minimum cost?

SOLUTION   To form a shipping schedule, we must decide how many computers to ship from either plant to either outlet (Fig. 2). This will involve four decision variables:

$$x_1 = \text{number of computers shipped from plant } A \text{ to outlet I}$$
$$x_2 = \text{number of computers shipped from plant } A \text{ to outlet II}$$
$$x_3 = \text{number of computers shipped from plant } B \text{ to outlet I}$$
$$x_4 = \text{number of computers shipped from plant } B \text{ to outlet II}$$

Next, we summarize the relevant data in a table. Note that we do not follow the usual technique of associating each variable with a column of the table. Instead, sources are associated with the rows, and destinations are associated with the columns.

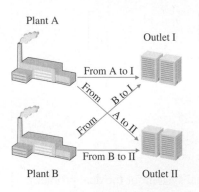

Plant A

Outlet I

From A to I

From B to I

From A to II

From B to II

Plant B

Outlet II

Figure 2

	Distribution Outlet		Assembly Capacity
	I	II	
Plant *A*	\$6	\$5	700
Plant *B*	\$4	\$8	900
Minimum required	500	1,000	

The total number of computers shipped from plant A is $x_1 + x_2$. Since this cannot exceed the assembly capacity at A, we have

$$x_1 + x_2 \leq 700 \quad \text{Number shipped from plant } A$$

Similarly, the total number shipped from plant B must satisfy

$$x_3 + x_4 \leq 900 \quad \text{Number shipped from plant } B$$

The total number shipped to each outlet must satisfy

$$x_1 + x_3 \geq 500 \quad \text{Number shipped to outlet I}$$

and

$$x_2 + x_4 \geq 1,000 \quad \text{Number shipped to outlet II}$$

Using the shipping charges in the table, the total shipping charges are

$$C = 6x_1 + 5x_2 + 4x_3 + 8x_4$$

We must solve the following linear programming problem:

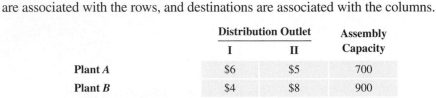

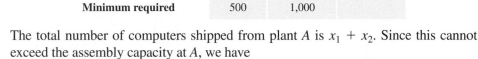

$$\text{Minimize} \quad C = 6x_1 + 5x_2 + 4x_3 + 8x_4$$
$$\text{subject to} \quad x_1 + x_2 \qquad\qquad \leq \quad 700 \qquad \text{Available from } A$$
$$x_3 + x_4 \leq \quad 900 \qquad \text{Available from } B$$
$$x_1 \qquad + x_3 \qquad \geq \quad 500 \qquad \text{Required at I}$$
$$x_2 \qquad\quad + x_4 \geq 1{,}000 \qquad \text{Required at II}$$
$$x_1, x_2, x_3, x_4 \geq \qquad 0$$

Before we can solve this problem, we must multiply the first two constraints by -1 so that all the problem constraints are of the $\geq$ type. This will introduce negative constants into the minimization problem but not into the dual problem. Since the coefficients of C are nonnegative, the constants in the dual problem will be nonnegative and the dual will be a standard maximization problem. The problem can now be stated as

$$\text{Minimize} \quad C = 6x_1 + 5x_2 + 4x_3 + 8x_4$$
$$\text{subject to} \quad -x_1 - x_2 \qquad\qquad \geq -700$$
$$\qquad\qquad\qquad - x_3 - x_4 \geq -900$$
$$\qquad x_1 \qquad + x_3 \qquad \geq \quad 500$$
$$\qquad\qquad x_2 \qquad + x_4 \geq 1{,}000$$
$$\qquad\qquad x_1, x_2, x_3, x_4 \geq \quad 0$$

$$A = \begin{bmatrix} -1 & -1 & 0 & 0 & -700 \\ 0 & 0 & -1 & -1 & -900 \\ 1 & 0 & 1 & 0 & 500 \\ 0 & 1 & 0 & 1 & 1{,}000 \\ \hline 6 & 5 & 4 & 8 & 1 \end{bmatrix}$$

$$A^{\mathrm{T}} = \begin{bmatrix} -1 & 0 & 1 & 0 & 6 \\ -1 & 0 & 0 & 1 & 5 \\ 0 & -1 & 1 & 0 & 4 \\ 0 & -1 & 0 & 1 & 8 \\ \hline -700 & -900 & 500 & 1{,}000 & 1 \end{bmatrix}$$

The dual problem is

$$\text{Maximize} \quad P = -700y_1 - 900y_2 + 500y_3 + 1{,}000y_4$$
$$\text{subject to} \quad -y_1 \quad + y_3 \qquad \leq 6$$
$$\qquad -y_1 \qquad\qquad + y_4 \leq 5$$
$$\qquad\qquad -y_2 + y_3 \qquad \leq 4$$
$$\qquad\qquad -y_2 \qquad + y_4 \leq 8$$
$$\qquad\qquad y_1, y_2, y_3, y_4 \geq 0$$

Introduce slack variables $x_1, x_2, x_3,$ and x_4, and form the initial system for the dual problem:

$$-y_1 \qquad + \quad y_3 \qquad\qquad + x_1 \qquad\qquad\qquad = 6$$
$$-y_1 \qquad\qquad\qquad + \quad y_4 \qquad + x_2 \qquad\qquad = 5$$
$$\qquad -y_2 + \quad y_3 \qquad\qquad\qquad + x_3 \qquad = 4$$
$$\qquad -y_2 \qquad\qquad + \quad y_4 \qquad\qquad + x_4 = 8$$
$$700y_1 + 900y_2 - 500y_3 - 1{,}000y_4 \qquad\qquad\qquad + P = 0$$

Form the simplex tableau and solve:

	y_1	y_2	y_3	y_4	x_1	x_2	x_3	x_4	P	
x_1	-1	0	1	0	1	0	0	0	0	6
x_2	-1	0	0	$①$	0	1	0	0	0	5
x_3	0	-1	1	0	0	0	1	0	0	4
x_4	0	-1	0	1	0	0	0	1	0	8
P	700	900	-500	$-1{,}000$	0	0	0	0	1	0

$(-1)R_2 + R_4 \rightarrow R_4$
$1{,}000R_2 + R_5 \rightarrow R_5$

	y_1	y_2	y_3	y_4	x_1	x_2	x_3	x_4	P	
x_1	-1	0	1	0	1	0	0	0	0	6
y_4	-1	0	0	1	0	1	0	0	0	5
x_3	0	-1	$①$	0	0	0	1	0	0	4
x_4	1	-1	0	0	0	-1	0	1	0	3
P	-300	900	-500	0	0	$1{,}000$	0	0	1	$5{,}000$

$(-1)R_3 + R_1 \rightarrow R_1$

$500R_3 + R_5 \rightarrow R_5$

$$
\begin{array}{c}
\begin{array}{cccccccccc}
\quad y_1 & y_2 & y_3 & \quad y_4 & x_1 & x_2 & x_3 & x_4 & P
\end{array}\\
\begin{array}{c}
x_1\\ y_4\\ \sim y_3\\ x_4\\ P
\end{array}
\left[
\begin{array}{ccccccccc|c}
-1 & 1 & 0 & 0 & 1 & 0 & -1 & 0 & 0 & 2\\
-1 & 0 & 0 & 1 & 0 & 1 & 0 & 0 & 0 & 5\\
0 & -1 & 1 & 0 & 0 & 0 & 1 & 0 & 0 & 4\\
\boxed{1} & -1 & 0 & 0 & 0 & -1 & 0 & 1 & 0 & 3\\
\hline
-300 & 400 & 0 & 0 & 0 & 1{,}000 & 500 & 0 & 1 & 7{,}000
\end{array}
\right]
\begin{array}{l}
R_4 + R_1 \rightarrow R_1\\
R_4 + R_2 \rightarrow R_2\\
\\
\\
300R_4 + R_5 \rightarrow R_5
\end{array}
\end{array}
$$

$$
\begin{array}{c}
\begin{array}{c}
x_1\\ y_4\\ \sim y_3\\ y_1\\ P
\end{array}
\left[
\begin{array}{ccccccccc|c}
0 & 0 & 0 & 0 & 1 & -1 & -1 & 1 & 0 & 5\\
0 & -1 & 0 & 1 & 0 & 0 & 0 & 1 & 0 & 8\\
0 & -1 & 1 & 0 & 0 & 0 & 1 & 0 & 0 & 4\\
1 & -1 & 0 & 0 & 0 & -1 & 0 & 1 & 0 & 3\\
\hline
0 & 100 & 0 & 0 & 0 & 700 & 500 & 300 & 1 & 7{,}900
\end{array}
\right]
\end{array}
$$

From the bottom row of this tableau, we have

$$\text{Min } C = 7{,}900 \qquad \text{at} \qquad x_1 = 0, \quad x_2 = 700, \quad x_3 = 500, \quad x_4 = 300$$

The shipping schedule that minimizes the shipping charges is 700 from plant A to outlet II, 500 from plant B to outlet I, and 300 from plant B to outlet II. The total shipping cost is $7,900.

Figure 3 shows a solution to Example 4 in Excel. Notice that Excel permits the user to organize the original data and the solution in a format that is clear and easy to read. This is one of the main advantages of using spreadsheets to solve linear programming problems.

	A	B	C	D	E	F	G	H	I
1		DATA					SHIPPING SCHEDULE		
2		DISTRIBUTION					DISTRIBUTION		
3		OUTLET		ASSEMBLY			OUTLET		
4		I	II	CAPACITY			I	II	TOTAL
5	PLANT A	$6	$5	700		PLANT A	0	700	700
6	PLANT B	$4	$8	900		PLANT B	500	300	800
7	MINIMUM					TOTAL	500	1,000	
8	REQUIRED	500	1,000					TOTAL COST	$7,900

Figure 3

Matched Problem 4) Repeat Example 4 if the shipping charge from plant A to outlet I is increased to $7 and the shipping charge from plant B to outlet II is decreased to $3.

Summary of Problem Types and Solution Methods

In this and the preceding sections, we have solved both maximization and minimization problems, but with certain restrictions on the problem constraints, constants on the right, and/or objective function coefficients. Table 1 summarizes the types of problems and methods of solution we have considered so far.

Table 1 **Summary of Problem Types and Simplex Solution Methods**

Problem Type	Problem Constraints	Right-Side Constants	Coefficients of Objective Function	Method of Solution
1. Maximization	$\leq$	Nonnegative	Any real numbers	Simplex method with slack variables
2. Minimization	$\geq$	Any real numbers	Nonnegative	Form dual problem and solve by simplex method with slack variables

The next section develops a generalized version of the simplex method that can handle both maximization and minimization problems with any combination of $\leq$, $\geq$, and $=$ problem constraints.

Exercises 6.3

In Problems 1–8, find the transpose of each matrix.

1. $[-5 \quad 0 \quad 3 \quad -1 \quad 8]$

2. $[1 \quad 0 \quad -7 \quad 3 \quad -2]$

3. $\begin{bmatrix} 1 \\ -2 \\ 0 \\ 4 \end{bmatrix}$

4. $\begin{bmatrix} 9 \\ 5 \\ -4 \\ 0 \end{bmatrix}$

5. $\begin{bmatrix} 2 & 1 & -6 & 0 & -1 \\ 5 & 2 & 0 & 1 & 3 \end{bmatrix}$

6. $\begin{bmatrix} 7 & 3 & -1 & 3 \\ -6 & 1 & 0 & -9 \end{bmatrix}$

7. $\begin{bmatrix} 1 & 2 & -1 \\ 0 & 2 & -7 \\ 8 & 0 & 1 \\ 4 & -1 & 3 \end{bmatrix}$

8. $\begin{bmatrix} 1 & -1 & 3 & 2 \\ 1 & -4 & 0 & 2 \\ 4 & -5 & 6 & 1 \\ -3 & 8 & 0 & -1 \\ 2 & 7 & -3 & 1 \end{bmatrix}$

In Problems 9 and 10,

(A) Form the dual problem.

(B) Write the initial system for the dual problem.

(C) Write the initial simplex tableau for the dual problem and label the columns of the tableau.

9. Minimize $C = 8x_1 + 9x_2$
subject to $\quad x_1 + 3x_2 \geq 4$
$\qquad\qquad 2x_1 + x_2 \geq 5$
$\qquad\qquad\quad x_1, x_2 \geq 0$

10. Minimize $C = 12x_1 + 5x_2$
subject to $\quad 2x_1 + x_2 \geq 7$
$\qquad\qquad 3x_1 + x_2 \geq 9$
$\qquad\qquad\quad x_1, x_2 \geq 0$

In Problems 11 and 12, a minimization problem, the corresponding dual problem, and the final simplex tableau in the solution of the dual problem are given.

(A) Find the optimal solution of the dual problem.

(B) Find the optimal solution of the minimization problem.

11. Minimize $C = 21x_1 + 50x_2$
subject to $\quad 2x_1 + 5x_2 \geq 12$
$\qquad\qquad 3x_1 + 7x_2 \geq 17$
$\qquad\qquad\quad x_1, x_2 \geq 0$

Maximize $P = 12y_1 + 17y_2$
subject to $\quad 2y_1 + 3y_2 \leq 21$
$\qquad\qquad 5y_1 + 7y_2 \leq 50$
$\qquad\qquad\quad y_1, y_2 \geq 0$

y_1	y_2	x_1	x_2	P	
0	1	5	-2	0	5
1	0	-7	3	0	3
0	0	1	2	1	121

12. Minimize $C = 16x_1 + 25x_2$
subject to $\quad 3x_1 + 5x_2 \geq 30$
$\qquad\qquad 2x_1 + 3x_2 \geq 19$
$\qquad\qquad\quad x_1, x_2 \geq 0$

Maximize $P = 30y_1 + 19y_2$
subject to $\quad 3y_1 + 2y_2 \leq 16$
$\qquad\qquad 5y_1 + 3y_2 \leq 25$
$\qquad\qquad\quad y_1, y_2 \geq 0$

y_1	y_2	x_1	x_2	P	
0	1	5	-3	0	5
1	0	-3	2	0	2
0	0	5	3	1	155

In Problems 13–20,

(A) Form the dual problem.

(B) Find the solution to the original problem by applying the simplex method to the dual problem.

13. Minimize $C = 9x_1 + 2x_2$
subject to $\quad 4x_1 + x_2 \geq 13$
$\qquad\qquad 3x_1 + x_2 \geq 12$
$\qquad\qquad\quad x_1, x_2 \geq 0$

14. Minimize $C = x_1 + 4x_2$
subject to $\quad x_1 + 2x_2 > 5$
$\qquad\qquad x_1 + 3x_2 \geq 6$
$\qquad\qquad\quad x_1, x_2 \geq 0$

15. Minimize $C = 7x_1 + 12x_2$
subject to $\quad 2x_1 + 3x_2 \geq 15$
$\qquad\qquad\; x_1 + 2x_2 \geq 8$
$\qquad\qquad\quad x_1, x_2 \geq 0$

16. Minimize $C = 3x_1 + 5x_2$
subject to $\quad 2x_1 + 3x_2 \geq 7$
$\qquad\qquad\; x_1 + 2x_2 \geq 4$
$\qquad\qquad\quad x_1, x_2 \geq 0$

17. Minimize $C = 11x_1 + 4x_2$
subject to $\quad\; 2x_1 + x_2 \geq 8$
$\qquad\qquad -2x_1 + 3x_2 \geq 4$
$\qquad\qquad\quad\; x_1, x_2 \geq 0$

18. Minimize $C = 40x_1 + 10x_2$
subject to $\quad 2x_1 + x_2 \geq 12$
$\qquad\qquad 3x_1 - x_2 \geq 3$
$\qquad\qquad\quad x_1, x_2 \geq 0$

19. Minimize $C = 7x_1 + 9x_2$
subject to $\quad -3x_1 + x_2 \geq 6$
$\qquad\qquad\;\; x_1 - 2x_2 \geq 4$
$\qquad\qquad\quad\; x_1, x_2 \geq 0$

20. Minimize $C = 10x_1 + 15x_2$
subject to $\quad -4x_1 + x_2 \geq 12$
$\qquad\qquad\; 12x_1 - 3x_2 \geq 10$
$\qquad\qquad\quad\; x_1, x_2 \geq 0$

Solve the linear programming problems in Problems 21–32 by applying the simplex method to the dual problem.

21. Minimize $C = 3x_1 + 9x_2$
  subject to $2x_1 + x_2 \geq 8$
      $x_1 + 2x_2 \geq 8$
      $x_1, x_2 \geq 0$

22. Minimize $C = 2x_1 + x_2$
  subject to $x_1 + x_2 \geq 8$
      $x_1 + 2x_2 \geq 4$
      $x_1, x_2 \geq 0$

23. Minimize $C = 7x_1 + 5x_2$
  subject to $x_1 + x_2 \geq 4$
      $x_1 - 2x_2 \geq -8$
      $-2x_1 + x_2 \geq -8$
      $x_1, x_2 \geq 0$

24. Minimize $C = 10x_1 + 4x_2$
  subject to $2x_1 + x_2 \geq 6$
      $x_1 - 4x_2 \geq -24$
      $-8x_1 + 5x_2 \geq -24$
      $x_1, x_2 \geq 0$

25. Minimize $C = 10x_1 + 30x_2$
  subject to $2x_1 + x_2 \geq 16$
      $x_1 + x_2 \geq 12$
      $x_1 + 2x_2 \geq 14$
      $x_1, x_2 \geq 0$

26. Minimize $C = 40x_1 + 10x_2$
  subject to $3x_1 + x_2 \geq 24$
      $x_1 + x_2 \geq 16$
      $x_1 + 4x_2 \geq 30$
      $x_1, x_2 \geq 0$

27. Minimize $C = 5x_1 + 7x_2$
  subject to $x_1 \geq 4$
      $x_1 + x_2 \geq 8$
      $x_1 + 2x_2 \geq 10$
      $x_1, x_2 \geq 0$

28. Minimize $C = 4x_1 + 5x_2$
  subject to $2x_1 + x_2 \geq 12$
      $x_1 + x_2 \geq 9$
      $x_2 \geq 4$
      $x_1, x_2 \geq 0$

29. Minimize $C = 10x_1 + 7x_2 + 12x_3$
  subject to $x_1 + x_2 + 2x_3 \geq 7$
      $2x_1 + x_2 + x_3 \geq 4$
      $x_1, x_2, x_3 \geq 0$

30. Minimize $C = 14x_1 + 8x_2 + 20x_3$
  subject to $x_1 + x_2 + 3x_3 \geq 6$
      $2x_1 + x_2 + x_3 \geq 9$
      $x_1, x_2, x_3 \geq 0$

31. Minimize $C = 5x_1 + 2x_2 + 2x_3$
  subject to $x_1 - 4x_2 + x_3 \geq 6$
      $-x_1 + x_2 - 2x_3 \geq 4$
      $x_1, x_2, x_3 \geq 0$

32. Minimize $C = 6x_1 + 8x_2 + 3x_3$
  subject to $-3x_1 - 2x_2 + x_3 \geq 4$
      $x_1 + x_2 - x_3 \geq 2$
      $x_1, x_2, x_3 \geq 0$

33. A minimization problem has 4 variables and 2 problem constraints. How many variables and problem constraints are in the dual problem?

34. A minimization problem has 3 variables and 5 problem constraints. How many variables and problem constraints are in the dual problem?

35. If you want to solve a minimization problem by applying the geometric method to the dual problem, how many variables and problem constraints must be in the original problem?

36. If you want to solve a minimization problem by applying the geometric method to the original problem, how many variables and problem constraints must be in the original problem?

In Problems 37–40, determine whether a minimization problem with the indicated condition can be solved by applying the simplex method to the dual problem. If your answer is yes, describe any necessary modifications that must be made before forming the dual problem. If your answer is no, explain why.

37. A coefficient of the objective function is negative.

38. A coefficient of a problem constraint is negative.

39. A problem constraint is of the $\leq$ form.

40. A problem constraint has a negative constant on the right side.

Solve the linear programming problems in Problems 41–44 by applying the simplex method to the dual problem.

41. Minimize $C = 16x_1 + 8x_2 + 4x_3$
  subject to $3x_1 + 2x_2 + 2x_3 \geq 16$
      $4x_1 + 3x_2 + x_3 \geq 14$
      $5x_1 + 3x_2 + x_3 \geq 12$
      $x_1, x_2, x_3 \geq 0$

42. Minimize $C = 6x_1 + 8x_2 + 12x_3$
  subject to $x_1 + 3x_2 + 3x_3 \geq 6$
      $x_1 + 5x_2 + 5x_3 \geq 4$
      $2x_1 + 2x_2 + 3x_3 \geq 8$
      $x_1, x_2, x_3 \geq 0$

43. Minimize $C = 5x_1 + 4x_2 + 5x_3 + 6x_4$
  subject to $x_1 + x_2 \leq 12$
      $x_3 + x_4 \leq 25$
      $x_1 + x_3 \geq 20$
      $x_2 + x_4 \geq 15$
      $x_1, x_2, x_3, x_4 \geq 0$

44. Repeat Problem 43 with $C = 4x_1 + 7x_2 + 5x_3 + 6x_4$.

Applications

In Problems 45–54, construct a mathematical model in the form of a linear programming problem. (The answers in the back of the book for these application problems include the model.) Then solve the problem by applying the simplex method to the dual problem.

45. Ice cream. A food processing company produces regular and deluxe ice cream at three plants. Per hour of operation, the Cedarburg plant produces 20 gallons of regular ice cream and 10 gallons of deluxe ice cream. The Grafton plant produces 10 gallons of regular and 20 gallons of deluxe, and the West Bend plant produces 20 gallons of regular and 20 gallons of deluxe. It costs $70 per hour to operate the Cedarburg plant, $75 per hour to operate the Grafton plant, and $90 per hour to operate the West Bend plant. The company needs to produce at least 300 gallons of regular ice cream and at least 200 gallons of deluxe ice cream each day. How many hours per day should each plant operate in order to produce the required amounts of ice cream and minimize the cost of production? What is the minimum production cost?

46. Mining. A mining company operates two mines, each producing three grades of ore. The West Summit mine can produce 2 tons of low-grade ore, 3 tons of medium-grade ore, and 1 ton of high-grade ore in one hour of operation. The North Ridge mine can produce 2 tons of low-grade ore, 1 ton of medium-grade ore, and 2 tons of high-grade ore in one hour of operation. To satisfy existing orders, the company needs to produce at least 100 tons of low-grade ore, 60 tons of medium-grade ore, and 80 tons of high-grade ore. The cost of operating each mine varies, depending on conditions while extracting the ore. If it costs $400 per hour to operate the West Summit mine and $600 per hour to operate the North Ridge mine, how many hours should each mine operate to supply the required amounts of ore and minimize the cost of production? What is the minimum production cost?

47. Ice cream. Repeat Problem 45 if the demand for deluxe ice cream increases from 200 gallons to 300 gallons per day and all other data remain the same.

48. Mining. Repeat Problem 46 if it costs $300 per hour to operate the West Summit mine and $700 per hour to operate the North Ridge mine and all other data remain the same.

49. Ice cream. Repeat Problem 45 if the demand for deluxe ice cream increases from 200 gallons to 400 gallons per day and all other data remain the same.

50. Mining. Repeat Problem 46 if it costs $800 per hour to operate the West Summit mine and $200 per hour to operate the North Ridge mine and all other data remain the same.

51. Human nutrition. A dietitian arranges a special diet using three foods: *L*, *M*, and *N*. Each ounce of food *L* contains 20 units of calcium, 10 units of iron, 10 units of vitamin A, and 20 units of cholesterol. Each ounce of food *M* contains 10 units of calcium, 10 units of iron, 15 units of vitamin A, and 24 units of cholesterol. Each ounce of food *N* contains 10 units of calcium, 10 units of iron, 10 units of vitamin A,

and 18 units of cholesterol. If the minimum daily requirements are 300 units of calcium, 200 units of iron, and 240 units of vitamin A, how many ounces of each food should be used to meet the minimum requirements and simultaneously minimize the cholesterol intake? What is the minimum cholesterol intake?

52. Plant food. A farmer can buy three types of plant food: mix *A*, mix *B*, and mix *C*. Each cubic yard of mix *A* contains 20 pounds of phosphoric acid, 10 pounds of nitrogen, and 10 pounds of potash. Each cubic yard of mix *B* contains 10 pounds of phosphoric acid, 10 pounds of nitrogen, and 15 pounds of potash. Each cubic yard of mix *C* contains 20 pounds of phosphoric acid, 20 pounds of nitrogen, and 5 pounds of potash. The minimum monthly requirements are 480 pounds of phosphoric acid, 320 pounds of nitrogen, and 225 pounds of potash. If mix *A* costs $30 per cubic yard, mix *B* costs $36 per cubic yard, and mix *C* costs $39 per cubic yard, how many cubic yards of each mix should the farmer blend to meet the minimum monthly requirements at a minimal cost? What is the minimum cost?

53. Education: resource allocation. A metropolitan school district has two overcrowded high schools and two underenrolled high schools. To balance the enrollment, the school board decided to bus students from the overcrowded schools to the underenrolled schools. North Division High School has 300 more students than normal, and South Division High School has 500 more students than normal. Central High School can accommodate 400 additional students, and Washington High School can accommodate 500 additional students. The weekly cost of busing a student from North Division to Central is $5, from North Division to Washington is $2, from South Division to Central is $3, and from South Division to Washington is $4. Determine the number of students that should be bused from each overcrowded school to each underenrolled school in order to balance the enrollment and minimize the cost of busing the students. What is the minimum cost?

54. Education: resource allocation. Repeat Problem 53 if the weekly cost of busing a student from North Division to Washington is $7 and all other data remain the same.

Answers to Matched Problems

1. Maximize $P = 12y_1 + 16y_2$
subject to $\quad y_1 + 2y_2 \leq 16$
$\quad\quad\quad y_1 + \ y_2 \leq 9$
$\quad\quad\quad 3y_1 + \ y_2 \leq 21$
$\quad\quad\quad\quad y_1, y_2 \geq 0$

2. Min $C = 136$ at $x_1 = 4, x_2 = 8, x_3 = 0$

3. Dual problem:
Maximize $P = 2y_1 + y_2$
subject to $\quad y_1 - y_2 \leq 2$
$\quad\quad\quad -2y_1 + y_2 \leq 3$
$\quad\quad\quad\quad y_1, y_2 \geq 0$
No optimal solution

4. 600 from plant *A* to outlet II, 500 from plant *B* to outlet I, 400 from plant *B* to outlet II; total shipping cost is $6,200.

6.4 Maximization and Minimization with Mixed Problem Constraints

- Introduction to the Big M Method
- Big M Method
- Minimization by the Big M Method
- Summary of Solution Methods
- Larger Problems: Refinery Application

In the preceding two sections, we have seen how to solve both maximization and minimization problems, but with rather severe restrictions on problem constraints, right-side constants, and/or objective function coefficients (see the summary in Table 1 of Section 6.3). In this section we present a generalized version of the simplex method that will solve both maximization and minimization problems with any combination of $\leq$, $\geq$, and $=$ problem constraints. The only requirement is that each problem constraint must have a nonnegative constant on the right side. (This restriction is easily accommodated, as you will see.)

Introduction to the Big M Method

We introduce the *big M method* through a simple maximization problem with mixed problem constraints. The key parts of the method will then be summarized and applied to more complex problems.

Consider the following problem:

$$
\begin{aligned}
\text{Maximize} \quad & P = 2x_1 + x_2 \\
\text{subject to} \quad & x_1 + x_2 \leq 10 \\
& -x_1 + x_2 \geq 2 \\
& x_1, x_2 \geq 0
\end{aligned}
\tag{1}
$$

To form an equation out of the first inequality, we introduce a slack variable s_1, as before, and write

$$x_1 + x_2 + s_1 = 10$$

How can we form an equation out of the second inequality? We introduce a second variable s_2 and subtract it from the left side:

$$-x_1 + x_2 - s_2 = 2$$

The variable s_2 is called a **surplus variable** because it is the amount (surplus) by which the left side of the inequality exceeds the right side.

Next, we express the linear programming problem (1) as a system of equations:

$$
\begin{aligned}
x_1 + x_2 + s_1 \qquad\qquad\quad &= 10 \\
-x_1 + x_2 \qquad - s_2 \qquad\quad &= 2 \\
-2x_1 - x_2 \qquad\qquad + P &= 0 \\
x_1, x_2, s_1, s_2 &\geq 0
\end{aligned}
\tag{2}
$$

It can be shown that a basic solution of system (2) is not feasible if any of the variables (excluding P) are negative. So **a surplus variable is required to satisfy the nonnegative constraint**.

The basic solution found by setting the nonbasic variables x_1 and x_2 equal to 0 is

$$x_1 = 0, \quad x_2 = 0, \quad s_1 = 10, \quad s_2 = -2, \quad P = 0$$

But this basic solution is not feasible, since the surplus variable s_2 is negative (which is a violation of the nonnegative requirements of all variables except P). The simplex method works only when the basic solution for a tableau is feasible, so we cannot solve this problem simply by writing the tableau for (2) and starting pivot operations.

To use the simplex method on problems with mixed constraints, we turn to an ingenious device called an *artificial variable*. This variable has no physical meaning in the original problem. It is introduced solely for the purpose of obtaining a basic feasible solution so that we can apply the simplex method. An **artificial variable** is a variable introduced into each equation that has a surplus variable. As before, to ensure that we consider only basic feasible solutions, **an artificial variable is required to satisfy the nonnegative constraint**. (As we will see later, artificial variables are also used to augment equality problem constraints when they are present.)

Returning to the problem at hand, we introduce an artificial variable a_1 into the equation involving the surplus variable s_2:

$$-x_1 + x_2 - s_2 + a_1 = 2$$

To prevent an artificial variable from becoming part of an optimal solution to the original problem, a very large "penalty" is introduced into the objective function. This penalty is created by choosing a positive constant M so large that the artificial variable is forced to be 0 in any final optimal solution of the original problem. (Since the constant M can be as large as we wish in computer solutions, M is often selected as the largest number the computer can hold!) Then we add the term $-Ma_1$ to the objective function:

$$P = 2x_1 + x_2 - Ma_1$$

We now have a new problem, which we call the **modified problem**:

$$\begin{aligned}
\text{Maximize} \quad & P = 2x_1 + x_2 - Ma_1 \\
\text{subject to} \quad & x_1 + x_2 + s_1 & = 10 \\
& -x_1 + x_2 \quad\quad - s_2 + a_1 = 2 \\
& x_1, x_2, s_1, s_2, a_1 \geq 0
\end{aligned} \tag{3}$$

The initial system for the modified problem (3) is

$$\begin{aligned}
x_1 + x_2 + s_1 & = 10 \\
-x_1 + x_2 \quad\quad - s_2 + a_1 & = 2 \\
-2x_1 - x_2 \quad\quad\quad\quad + Ma_1 + P & = 0 \\
x_1, x_2, s_1, s_2, a_1 & \geq 0
\end{aligned} \tag{4}$$

We next write the augmented matrix for system (4), which we call the **preliminary simplex tableau** for the modified problem. (The reason we call it the "preliminary" simplex tableau instead of the "initial" simplex tableau will be made clear shortly.)

$$\begin{array}{cccccc}
x_1 & x_2 & s_1 & s_2 & a_1 & P \\
\end{array}$$
$$\left[\begin{array}{cccccc|c}
1 & 1 & 1 & 0 & 0 & 0 & 10 \\
-1 & 1 & 0 & -1 & 1 & 0 & 2 \\
\hline
-2 & -1 & 0 & 0 & M & 1 & 0
\end{array}\right] \tag{5}$$

To start the simplex process, including any necessary pivot operations, the preliminary simplex tableau should either meet the two requirements given in the following box or be transformed by row operations into a tableau that meets these two requirements.

DEFINITION Initial Simplex Tableau

For a system tableau to be considered an **initial simplex tableau**, it must satisfy the following two requirements:

1. The requisite number of basic variables must be selectable by the process described in Section 6.2. That is, a variable can be selected as a basic variable only if it corresponds to a column in the tableau that has exactly one nonzero element and the nonzero element in the column is not in the same row as the nonzero element in the column of another basic variable. The remaining variables are then selected as nonbasic variables to be set equal to 0 in determining a basic solution.

2. The basic solution found by setting the nonbasic variables equal to 0 is feasible.

Tableau (5) satisfies the first initial simplex tableau requirement, since s_1, s_2, and P can be selected as basic variables according to the first requirement. (Not all preliminary simplex tableaux satisfy the first requirement; see Example 2.) However, tableau (5) does not satisfy the second initial simplex tableau requirement since the basic solution is not feasible ($s_2 = -2$). To use the simplex method, we must use row operations to transform tableau (5) into an equivalent matrix that satisfies both initial simplex tableau requirements. **Note that this transformation is not a pivot operation.**

To get an idea of how to proceed, notice in tableau (5) that -1 in the s_2 column is in the same row as 1 in the a_1 column. This is not an accident! The artificial variable a_1 was introduced so that this would happen. If we eliminate M from the bottom of the a_1 column, the nonbasic variable a_1 will become a basic variable and the troublesome basic variable s_2 will become a nonbasic variable. We proceed to eliminate M from the a_1 column using row operations:

$$
\begin{array}{cccccc|c}
x_1 & x_2 & s_1 & s_2 & a_1 & P & \\
\end{array}
$$

$$
\left[\begin{array}{cccccc|c}
1 & 1 & 1 & 0 & 0 & 0 & 10 \\
-1 & 1 & 0 & -1 & 1 & 0 & 2 \\
\hline
-2 & -1 & 0 & 0 & M & 1 & 0
\end{array}\right] \quad (-M)R_2 + R_3 \rightarrow R_3
$$

$$
\sim \left[\begin{array}{cccccc|c}
1 & 1 & 1 & 0 & 0 & 0 & 10 \\
-1 & 1 & 0 & -1 & 1 & 0 & 2 \\
\hline
M-2 & -M-1 & 0 & M & 0 & 1 & -2M
\end{array}\right]
$$

From this last matrix we see that the basic variables are s_1, a_1, and P. The basic solution found by setting the nonbasic variables x_1, x_2, and s_2 equal to 0 is

$$x_1 = 0, \quad x_2 = 0, \quad s_1 = 10, \quad s_2 = 0, \quad a_1 = 2, \quad P = -2M$$

The basic solution is feasible (P can be negative), and both requirements for an initial simplex tableau are met. We perform pivot operations to find an optimal solution.

The pivot column is determined by the most negative indicator in the bottom row of the tableau. Since M is a positive number, $-M - 1$ is certainly a negative indicator. What about the indicator $M - 2$? Remember that M is a very large positive number. We will assume that M is so large that any expression of the form $M - k$ is positive. So the only negative indicator in the bottom row is $-M - 1$.

$$
\begin{array}{cccccc|c}
x_1 & x_2 & s_1 & s_2 & a_1 & P & \\
\end{array}
$$

Pivot row $\rightarrow$
$$
\left[\begin{array}{cccccc|c}
1 & 1 & 1 & 0 & 0 & 0 & 10 \\
-1 & ① & 0 & -1 & 1 & 0 & 2 \\
\hline
M-2 & -M-1 & 0 & M & 0 & 1 & -2M
\end{array}\right]
\begin{array}{l}
\frac{10}{1} = 10 \\
\frac{2}{1} = 2
\end{array}
$$

$\uparrow$ Pivot column

Having identified the pivot element, we now begin pivoting:

$$
\begin{array}{cccccc|c}
 & x_1 & x_2 & s_1 & s_2 & a_1 & P & \\
\end{array}
$$

$$
\begin{array}{c}
s_1 \\ a_1 \\ P
\end{array}
\left[\begin{array}{cccccc|c}
1 & 1 & 1 & 0 & 0 & 0 & 10 \\
-1 & ① & 0 & -1 & 1 & 0 & 2 \\
\hline
M-2 & -M-1 & 0 & M & 0 & 1 & -2M
\end{array}\right]
\begin{array}{l}
(-1)R_2 + R_1 \rightarrow R_1 \\
\\
(M+1)R_2 + R_3 \rightarrow R_3
\end{array}
$$

$$
\begin{array}{c}
s_1 \\ \sim x_2 \\ P
\end{array}
\left[\begin{array}{cccccc|c}
② & 0 & 1 & 1 & -1 & 0 & 8 \\
-1 & 1 & 0 & -1 & 1 & 0 & 2 \\
\hline
-3 & 0 & 0 & -1 & M+1 & 1 & 2
\end{array}\right]
\quad \frac{1}{2}R_1 \rightarrow R_1
$$

$$
\sim
\left[\begin{array}{cccccc|c}
① & 0 & \frac{1}{2} & \frac{1}{2} & -\frac{1}{2} & 0 & 4 \\
-1 & 1 & 0 & -1 & 1 & 0 & 2 \\
\hline
-3 & 0 & 0 & -1 & M+1 & 1 & 2
\end{array}\right]
\begin{array}{l}
\\
R_1 + R_2 \rightarrow R_2 \\
3R_1 + R_3 \rightarrow R_3
\end{array}
$$

$$
\begin{array}{c}
x_1 \\ \sim x_2 \\ P
\end{array}
\left[\begin{array}{cccccc|c}
1 & 0 & \frac{1}{2} & \frac{1}{2} & -\frac{1}{2} & 0 & 4 \\
0 & 1 & \frac{1}{2} & -\frac{1}{2} & \frac{1}{2} & 0 & 6 \\
\hline
0 & 0 & \frac{3}{2} & \frac{1}{2} & M-\frac{1}{2} & 1 & 14
\end{array}\right]
$$

Since all the indicators in the last row are nonnegative ($M - \frac{1}{2}$ is nonnegative because M is a very large positive number), we can stop and write the optimal solution:

$$\text{Max } P = 14 \quad \text{at} \quad x_1 = 4, \quad x_2 = 6, \quad s_1 = 0, \quad s_2 = 0, \quad a_1 = 0$$

This is an optimal solution to the modified problem (3). How is it related to the original problem (2)? Since $a_1 = 0$ in this solution,

$$x_1 = 4, \quad x_2 = 6, \quad s_1 = 0, \quad s_2 = 0, \quad P = 14 \tag{6}$$

is certainly a feasible solution for system (2). [You can verify this by direct substitution into system (2).] Surprisingly, it turns out that solution (6) is an optimal solution to the original problem. To see that this is true, suppose we were able to find feasible values of $x_1, x_2, s_1,$ and s_2 that satisfy the original system (2) and produce a value of $P > 14$. Then by using these same values in problem (3) along with $a_1 = 0$, we would have a feasible solution of problem (3) with $P > 14$. This contradicts the fact that $P = 14$ is the maximum value of P for the modified problem. Solution (6) is an optimal solution for the original problem.

As this example illustrates, if $a_1 = 0$ is an optimal solution for the modified problem, then deleting a_1 produces an optimal solution for the original problem. What happens if $a_1 \neq 0$ in the optimal solution for the modified problem? In this case, it can be shown that the original problem has no optimal solution because its feasible set is empty.

In larger problems, each $\geq$ problem constraint will require the introduction of a surplus variable and an artificial variable. If one of the problem constraints is an equation rather than an inequality, then there is no need to introduce a slack or surplus variable. However, each $=$ problem constraint will require the introduction of another artificial variable to prevent the initial basic solution from violating the equality constraint—the decision variables are often 0 in the initial basic solution (see Example 2). Finally, each artificial variable also must be included in the objective function for the modified problem. The same constant M can be used for each artificial variable. Because of the role that the constant M plays in this approach, this method is often called the **big M method**.

Big M Method

We summarize the key steps of the big M method and use them to solve several problems.

PROCEDURE Big M Method: Introducing Slack, Surplus, and Artificial Variables to Form the Modified Problem

Step 1 If any problem constraints have negative constants on the right side, multiply both sides by -1 to obtain a constraint with a nonnegative constant. (If the constraint is an inequality, this will reverse the direction of the inequality.)

Step 2 Introduce a slack variable in each $\leq$ constraint.

Step 3 Introduce a surplus variable and an artificial variable in each $\geq$ constraint.

Step 4 Introduce an artificial variable in each $=$ constraint.

Step 5 For each artificial variable a_i, add $-Ma_i$ to the objective function. Use the same constant M for all artificial variables.

EXAMPLE 1 Finding the Modified Problem Find the modified problem for the following linear programming problem. (Do not attempt to solve the problem.)

$$\begin{aligned}
\text{Maximize} \quad & P = 2x_1 + 5x_2 + 3x_3 \\
\text{subject to} \quad & x_1 + 2x_2 - x_3 \leq 7 \\
& -x_1 + x_2 - 2x_3 \leq -5 \\
& x_1 + 4x_2 + 3x_3 \geq 1 \\
& 2x_1 - x_2 + 4x_3 = 6 \\
& x_1, x_2, x_3 \geq 0
\end{aligned}$$

SOLUTION   First, we multiply the second constraint by -1 to change -5 to 5:

$$(-1)(-x_1 + x_2 - 2x_3) \geq (-1)(-5)$$
$$x_1 - x_2 + 2x_3 \geq 5$$

Next, we introduce the slack, surplus, and artificial variables according to the procedure stated in the box:

$$
\begin{aligned}
x_1 + 2x_2 - x_3 + s_1 \quad &= 7 \\
x_1 - x_2 + 2x_3 \quad - s_2 + a_1 \quad &= 5 \\
x_1 + 4x_2 + 3x_3 \quad - s_3 + a_2 \quad &= 1 \\
2x_1 - x_2 + 4x_3 \quad + a_3 &= 6
\end{aligned}
$$

Finally, we add $-Ma_1$, $-Ma_2$, and $-Ma_3$ to the objective function:

$$P = 2x_1 + 5x_2 + 3x_3 - Ma_1 - Ma_2 - Ma_3$$

The modified problem is

$$
\begin{aligned}
\text{Maximize} \quad & P = 2x_1 + 5x_2 + 3x_3 - Ma_1 - Ma_2 - Ma_3 \\
\text{subject to} \quad & x_1 + 2x_2 - x_3 + s_1 = 7 \\
& x_1 - x_2 + 2x_3 - s_2 + a_1 = 5 \\
& x_1 + 4x_2 + 3x_3 - s_3 + a_2 = 1 \\
& 2x_1 - x_2 + 4x_3 + a_3 = 6 \\
& x_1, x_2, x_3, s_1, s_2, s_3, a_1, a_2, a_3 \geq 0
\end{aligned}
$$

Matched Problem 1   Repeat Example 1 for

$$
\begin{aligned}
\text{Maximize} \quad & P = 3x_1 - 2x_2 + x_3 \\
\text{subject to} \quad & x_1 - 2x_2 + x_3 \geq 5 \\
& -x_1 - 3x_2 + 4x_3 \leq -10 \\
& 2x_1 + 4x_2 + 5x_3 \leq 20 \\
& 3x_1 - x_2 - x_3 = -15 \\
& x_1, x_2, x_3 \geq 0
\end{aligned}
$$

Now we can list the key steps for solving a problem using the big M method. The various steps and remarks are based on a number of important theorems, which we assume without proof. In particular, step 2 is based on the fact that (except for some degenerate cases not considered here) if the modified linear programming problem has an optimal solution, then the preliminary simplex tableau will be transformed into an initial simplex tableau by eliminating the M's from the columns corresponding to the artificial variables in the preliminary simplex tableau. Having obtained an initial simplex tableau, we perform pivot operations.

PROCEDURE   Big M Method: Solving the Problem

Step 1   Form the preliminary simplex tableau for the modified problem.

Step 2   Use row operations to eliminate the M's in the bottom row of the preliminary simplex tableau in the columns corresponding to the artificial variables. The resulting tableau is the initial simplex tableau.

Step 3   Solve the modified problem by applying the simplex method to the initial simplex tableau found in step 2.

Step 4 Relate the optimal solution of the modified problem to the original problem.

(A) If the modified problem has no optimal solution, then the original problem has no optimal solution.

(B) If all artificial variables are 0 in the optimal solution to the modified problem, then delete the artificial variables to find an optimal solution to the original problem.

(C) If any artificial variables are nonzero in the optimal solution to the modified problem, then the original problem has no optimal solution.

EXAMPLE 2 Using the Big M Method Solve the following linear programming problem using the big M method:

$$\text{Maximize} \quad P = x_1 - x_2 + 3x_3$$
$$\text{subject to} \quad x_1 + x_2 \qquad\qquad \le 20$$
$$x_1 \qquad\quad + x_3 = 5$$
$$x_2 + x_3 \ge 10$$
$$x_1, x_2, x_3 \ge 0$$

SOLUTION State the modified problem:

$$\text{Maximize} \quad P = x_1 - x_2 + 3x_3 - Ma_1 - Ma_2$$
$$\text{subject to} \quad x_1 + x_2 \qquad\quad + s_1 \qquad\qquad\qquad = 20$$
$$x_1 \qquad\quad + x_3 \qquad\quad + a_1 \qquad\quad = 5$$
$$x_2 + x_3 \qquad\qquad - s_2 + a_2 = 10$$
$$x_1, x_2, x_3, s_1, s_2, a_1, a_2 \ge 0$$

Write the preliminary simplex tableau for the modified problem, and find the initial simplex tableau by eliminating the M's from the artificial variable columns:

$$
\begin{array}{c}
\begin{array}{ccccccc}
x_1 & x_2 & x_3 & s_1 & a_1 & s_2 & a_2 \quad P
\end{array}\\
\left[\begin{array}{ccccccc|c}
1 & 1 & 0 & 1 & 0 & 0 & 0 \quad 0 & 20\\
1 & 0 & 1 & 0 & 1 & 0 & 0 \quad 0 & 5\\
0 & 1 & 1 & 0 & 0 & -1 & 1 \quad 0 & 10\\
\hline
-1 & 1 & -3 & 0 & M & 0 & M \quad 1 & 0
\end{array}\right]
\end{array}
\quad
\begin{array}{l}
\text{Eliminate } M \text{ from the } a_1\\
\text{column}\\[1.5em]
(-M)R_2 + R_4 \to R_4
\end{array}
$$

$$
\sim
\left[\begin{array}{ccccccc|c}
1 & 1 & 0 & 1 & 0 & 0 & 0 \quad 0 & 20\\
1 & 0 & 1 & 0 & 1 & 0 & 0 \quad 0 & 5\\
0 & 1 & 1 & 0 & 0 & -1 & 1 \quad 0 & 10\\
\hline
-M-1 & 1 & -M-3 & 0 & 0 & 0 & M \quad 1 & -5M
\end{array}\right]
\quad
\begin{array}{l}
\text{Eliminate } M \text{ from the } a_2 \text{ column}\\[1.5em]
(-M)R_3 + R_4 \to R_4
\end{array}
$$

$$
\sim
\left[\begin{array}{ccccccc|c}
1 & 1 & 0 & 1 & 0 & 0 & 0 \quad 0 & 20\\
1 & 0 & 1 & 0 & 1 & 0 & 0 \quad 0 & 5\\
0 & 1 & 1 & 0 & 0 & -1 & 1 \quad 0 & 10\\
\hline
-M-1 & -M+1 & -2M-3 & 0 & 0 & M & 0 \quad 1 & -15M
\end{array}\right]
$$

From this last matrix we see that the basic variables are s_1, a_1, a_2, and P. The basic solution found by setting the nonbasic variables x_1, x_2, x_3, and s_2 equal to 0 is

$$x_1 = 0, \quad x_2 = 0, \quad x_3 = 0, \quad s_1 = 20, \quad a_1 = 5, \quad s_2 = 0, \quad a_2 = 10, \quad P = -15M$$

The basic solution is feasible, and both requirements for an initial simplex tableau are met. We perform pivot operations to find the optimal solution.

	x_1	x_2	x_3	s_1	a_1	s_2	a_2	P	
s_1	1	1	0	1	0	0	0	0	20
a_1	1	0	①	0	1	0	0	0	5
a_2	0	1	1	0	0	-1	1	0	10
P	$-M-1$	$-M+1$	$-2M-3$	0	0	M	0	1	$-15M$

$(-1)R_2 + R_3 \to R_3$
$(2M+3)R_2 + R_4 \to R_4$

	x_1	x_2	x_3	s_1	a_1	s_2	a_2	P	
s_1	1	1	0	1	0	0	0	0	20
x_3	1	0	1	0	1	0	0	0	5
a_2	-1	①	0	0	-1	-1	1	0	5
P	$M+2$	$-M+1$	0	0	$2M+3$	M	0	1	$-5M+15$

$(-1)R_3 + R_1 \to R_1$

$(M-1)R_3 + R_4 \to R_4$

	x_1	x_2	x_3	s_1	a_1	s_2	a_2	P	
s_1	2	0	0	1	1	1	-1	0	15
x_3	1	0	1	0	1	0	0	0	5
x_2	-1	1	0	0	-1	-1	1	0	5
P	3	0	0	0	$M+4$	1	$M-1$	1	10

Since the bottom row has no negative indicators, we can stop and write the optimal solution to the modified problem:

$$x_1 = 0, \quad x_2 = 5, \quad x_3 = 5, \quad s_1 = 15, \quad a_1 = 0, \quad s_2 = 0, \quad a_2 = 0, \quad P = 10$$

Since $a_1 = 0$ and $a_2 = 0$, the solution to the original problem is

$$\text{Max } P = 10 \quad \text{at} \quad x_1 = 0, \quad x_2 = 5, \quad x_3 = 5$$

Matched Problem 2 Solve the following linear programming problem using the big M method:

$$\text{Maximize} \quad P = x_1 + 4x_2 + 2x_3$$
$$\text{subject to} \quad x_2 + x_3 \le 4$$
$$x_1 \quad\quad - x_3 = 6$$
$$x_1 - x_2 - x_3 \ge 1$$
$$x_1, x_2, x_3 \ge 0$$

EXAMPLE 3 Using the Big M Method Solve the following linear programming problem using the big M method:

$$\text{Maximize} \quad P = 3x_1 + 5x_2$$
$$\text{subject to} \quad 2x_1 + \quad x_2 \le 4$$
$$x_1 + 2x_2 \ge 10$$
$$x_1, x_2 \ge 0$$

SOLUTION Introducing slack, surplus, and artificial variables, we obtain the modified problem:

$$2x_1 + \quad x_2 + s_1 \quad\quad\quad\quad = 4$$
$$x_1 + 2x_2 \quad\quad - s_2 + \quad a_1 \quad\quad = 10 \quad \text{Modified problem}$$
$$-3x_1 - 5x_2 \quad\quad\quad\quad\quad + Ma_1 + P = 0$$

Preliminary simplex tableau

	x_1	x_2	s_1	s_2	a_1	P	
	2	1	1	0	0	0	4
	1	2	0	-1	1	0	10
	-3	-5	0	0	M	1	0

Eliminate M in the a_1 column.

$(-M)R_2 + R_3 \to R_3$

Initial simplex tableau

	x_1	x_2	s_1	s_2	a_1	P	
s_1	2	①	1	0	0	0	4
$\sim a_1$	1	2	0	-1	1	0	10
P	$-M-3$	$-2M-5$	0	M	0	1	$-10M$

Begin pivot operations.

$(-2)R_1 + R_2 \to R_2$

$(2M+5)R_1 + R_3 \to R_3$

	x_1	x_2	s_1	s_2	a_1	P	
x_2	2	1	1	0	0	0	4
$\sim a_1$	-3	0	-2	-1	1	0	2
P	$3M+7$	0	$2M+5$	M	0	1	$-2M+20$

The optimal solution of the modified problem is

$$x_1 = 0, \quad x_2 = 4, \quad s_1 = 0, \quad s_2 = 0,$$
$$a_1 = 2, \quad P = -2M + 20$$

Since $a_1 \neq 0$, the original problem has no optimal solution. Figure 1 shows that the feasible region for the original problem is empty.

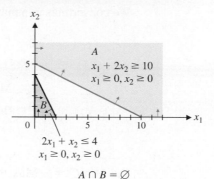

Figure 1 $A \cap B = \varnothing$

Matched Problem 3 Solve the following linear programming problem using the big M method:

$$\text{Maximize} \quad P = 3x_1 + 2x_2$$
$$\text{subject to} \quad x_1 + 5x_2 \leq 5$$
$$\qquad\qquad\qquad 2x_1 + x_2 \geq 12$$
$$\qquad\qquad\qquad x_1, x_2 \geq 0$$

Minimization by the Big M Method

In addition to solving any maximization problem, the big M method can be used to solve minimization problems. To minimize an objective function, we need only to maximize its negative. Figure 2 illustrates the fact that the minimum value of a function f occurs at the same point as the maximum value of the function $-f$. Furthermore, if m is the minimum value of f, then $-m$ is the maximum value of $-f$, and conversely. So we can find the minimum value of a function f by finding the maximum value of $-f$ and then changing the sign of the maximum value.

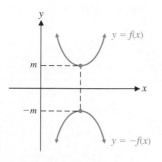

Figure 2

EXAMPLE 4 Production Scheduling: Minimization Problem   A small jewelry manufacturer hires a highly skilled gem cutter to work at least 6 hours per day. On the other hand, the polishing facilities can be used for at most 10 hours per day. The company specializes in three kinds of semiprecious gemstones, J, K, and L. Relevant cutting, polishing, and cost requirements are listed in the table. How many gemstones of each type should be processed each day to minimize the cost of the finished stones? What is the minimum cost?

	J	K	L
Cutting	1 hr	1 hr	1 hr
Polishing	2 hr	1 hr	2 hr
Cost per stone	$30	$30	$10

SOLUTION Since we must decide how many gemstones of each type should be processed each day, the decision variables are

$$x_1 = \text{number of type } J \text{ gemstones processed each day}$$
$$x_2 = \text{number of type } K \text{ gemstones processed each day}$$
$$x_3 = \text{number of type } L \text{ gemstones processed each day}$$

Since the data is already summarized in a table, we can proceed directly to the model:

$$\text{Minimize} \quad C = 30x_1 + 30x_2 + 10x_3 \quad \text{Objective function}$$

$$\text{subject to} \quad \left.\begin{array}{r} x_1 + x_2 + x_3 \geq 6 \\ 2x_1 + x_2 + 2x_3 \leq 10 \end{array}\right\} \quad \text{Problem constraints}$$

$$x_1, x_2, x_3 \geq 0 \quad \text{Nonnegative constraints}$$

We convert this to a maximization problem by letting

$$P = -C = -30x_1 - 30x_2 - 10x_3$$

We get

$$\text{Maximize} \quad P = -30x_1 - 30x_2 - 10x_3$$

$$\text{subject to} \quad x_1 + x_2 + x_3 \geq 6$$

$$2x_1 + x_2 + 2x_3 \leq 10$$

$$x_1, x_2, x_3 \geq 0$$

and Min $C = -\text{Max } P$. To solve, we first state the modified problem:

$$
\begin{array}{rcl}
x_1 + x_2 + x_3 - s_1 + a_1 & = & 6 \\
2x_1 + x_2 + 2x_3 + s_2 & = & 10 \\
30x_1 + 30x_2 + 10x_3 + Ma_1 + P & = & 0 \\
\end{array}
$$

$$x_1, x_2, x_3, s_1, s_2, a_1 \geq 0$$

$$
\begin{array}{c}
\begin{array}{ccccccc}
x_1 & x_2 & x_3 & s_1 & a_1 & s_2 & P \\
\end{array} \\
\left[\begin{array}{ccccccc|c}
1 & 1 & 1 & -1 & 1 & 0 & 0 & 6 \\
2 & 1 & 2 & 0 & 0 & 1 & 0 & 10 \\
\hline
30 & 30 & 10 & 0 & M & 0 & 1 & 0 \\
\end{array}\right]
\end{array}
\quad
\begin{array}{l}
\text{Eliminate } M \text{ in the } a_1 \\
\text{column} \\
(-M)R_1 + R_3 \to R_3
\end{array}
$$

Begin pivot operations. Assume M is so large
that $-M + 30$ and $-M + 10$ are negative

$$
\begin{array}{c}
\begin{array}{c} a_1 \\ \sim s_2 \\ P \end{array}
\left[\begin{array}{ccccccc|c}
1 & 1 & 1 & -1 & 1 & 0 & 0 & 6 \\
2 & 1 & ② & 0 & 0 & 1 & 0 & 10 \\
\hline
-M+30 & -M+30 & -M+10 & M & 0 & 0 & 1 & -6M \\
\end{array}\right]
\begin{array}{l} \\ 0.5R_2 \to R_2 \\ \\ \end{array}
\end{array}
$$

$$
\begin{array}{c}
\sim
\left[\begin{array}{ccccccc|c}
1 & 1 & 1 & -1 & 1 & 0 & 0 & 6 \\
1 & 0.5 & ① & 0 & 0 & 0.5 & 0 & 5 \\
\hline
-M+30 & -M+30 & -M+10 & M & 0 & 0 & 1 & -6M \\
\end{array}\right]
\begin{array}{l} (-1)R_2 + R_1 \to R_1 \\ \\ (M-10)R_2 + R_3 \to R_3 \end{array}
\end{array}
$$

$$
\begin{array}{c}
\begin{array}{c} a_1 \\ \sim x_3 \\ P \end{array}
\left[\begin{array}{ccccccc|c}
0 & ⓪.5 & 0 & -1 & 1 & -0.5 & 0 & 1 \\
1 & 0.5 & 1 & 0 & 0 & 0.5 & 0 & 5 \\
\hline
20 & -0.5M+25 & 0 & M & 0 & 0.5M-5 & 1 & -M-50 \\
\end{array}\right]
\begin{array}{l} 2R_1 \to R_1 \\ \\ \\ \end{array}
\end{array}
$$

$$
\begin{array}{c}
\sim
\left[\begin{array}{ccccccc|c}
0 & ① & 0 & -2 & 2 & -1 & 0 & 2 \\
1 & 0.5 & 1 & 0 & 0 & 0.5 & 0 & 5 \\
\hline
20 & -0.5M+25 & 0 & M & 0 & 0.5M-5 & 1 & -M-50 \\
\end{array}\right]
\begin{array}{l} \\ (-0.5)R_1 + R_2 \to R_2 \\ (0.5M-25)R_1 + R_3 \to R_3 \end{array}
\end{array}
$$

$$
\begin{array}{c}
\begin{array}{c} x_2 \\ \sim x_3 \\ P \end{array}
\left[\begin{array}{ccccccc|c}
0 & 1 & 0 & -2 & 2 & -1 & 0 & 2 \\
1 & 0 & 1 & 1 & -1 & 1 & 0 & 4 \\
\hline
20 & 0 & 0 & 50 & M-50 & 20 & 1 & -100 \\
\end{array}\right]
\end{array}
$$

The bottom row has no negative indicators, so the optimal solution for the modified problem is

$$x_1 = 0, \quad x_2 = 2, \quad x_3 = 4, \quad s_1 = 0, \quad a_1 = 0, \quad s_2 = 0, \quad P = -100$$

Since $a_1 = 0$, deleting a_1 produces the optimal solution to the original maximization problem and also to the minimization problem. Thus,

$$\text{Min } C = -\text{Max } P = -(-100) = 100 \quad \text{at} \quad x_1 = 0, \quad x_2 = 2, \quad x_3 = 4$$

That is, a minimum cost of $100 for gemstones will be realized if no type J, 2 type K, and 4 type L stones are processed each day.

Matched Problem 4) Repeat Example 4 if the gem cutter works at least 8 hours a day and all other data remain the same.

Summary of Solution Methods

The big M method can be used to solve any minimization problems, including those that can be solved by the dual method. (Note that Example 4 could have been solved by the dual method.) Both methods of solving minimization problems are important. You will be instructed to solve most minimization problems in Exercises 6.4 by the big M method in order to gain more experience with this method. However, if the method of solution is not specified, the dual method is usually easier.

Table 1 should help you select the proper method of solution for any linear programming problem.

Table 1 **Summary of Problem Types and Simplex Solution Methods**

Problem Type	Problem Constraints	Right-Side Constants	Coefficients of Objective Function	Method of Solution
1. Maximization	$\leq$	Nonnegative	Any real numbers	Simplex method with slack variables
2. Minimization	$\geq$	Any real numbers	Nonnegative	Form dual problem and solve by the preceding method
3. Maximization	Mixed $(\leq, \geq, =)$	Nonnegative	Any real numbers	Form modified problem with slack, surplus, and artificial variables, and solve by the big M method
4. Minimization	Mixed $(\leq, \geq, =)$	Nonnegative	Any real numbers	Maximize negative of objective function by the preceding method

Larger Problems: Refinery Application

Up to this point, all the problems we have considered could be solved by hand. However, the real value of the simplex method lies in its ability to solve problems with a large number of variables and constraints, where a computer is generally used to perform the actual pivot operations. As a final application, we consider a problem that would require the use of a computer to complete the solution.

EXAMPLE 5 Petroleum Blending   A refinery produces two grades of gasoline, regular and premium, by blending together two components, A and B. Component A has an octane rating of 90 and costs $28 a barrel. Component B has an octane rating of 110 and costs $32 a barrel. The octane rating for regular gasoline must be at least 95, and the octane rating for premium must be at least 105. Regular gasoline sells for $34 a barrel and premium sells for $40 a barrel. Currently, the company has 30,000 barrels of component A and 20,000 barrels of component B. It also has orders for 20,000 barrels of regular and 10,000 barrels of premium that must be filled. Assuming that all the gasoline produced can be sold, determine the maximum possible profit.

SOLUTION This problem is similar to the transportation problem in Section 6.3. That is, to maximize the profit, we must decide how much of each component must be used to produce each grade of gasoline. Thus, the decision variables are

$x_1 =$ number of barrels of component A used in regular gasoline

$x_2 =$ number of barrels of component A used in premium gasoline

$x_3 =$ number of barrels of component B used in regular gasoline

$x_4 =$ number of barrels of component B used in premium gasoline

Table 2

Component	Octane Rating	Cost ($)	Available Supply
A	90	28	30,000 barrels
B	110	32	20,000 barrels
Grade	Minimum Octane Rating	Selling Price($)	Existing Orders
Regular	95	34	20,000 barrels
Premium	105	40	10,000 barrels

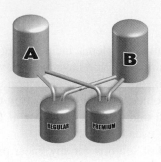

Next, we summarize the data in table form (Table 2). Once again, we have to adjust the form of the table to fit the data.

The total amount of component A used is $x_1 + x_2$. This cannot exceed the available supply. Thus, one constraint is

$$x_1 + x_2 \le 30{,}000$$

The corresponding inequality for component B is

$$x_3 + x_4 \le 20{,}000$$

The amounts of regular and premium gasoline produced must be sufficient to meet the existing orders:

$$x_1 + x_3 \ge 20{,}000 \quad \text{Regular}$$
$$x_2 + x_4 \ge 10{,}000 \quad \text{Premium}$$

Now consider the octane ratings. The octane rating of a blend is simply the proportional average of the octane ratings of the components. So the octane rating for regular gasoline is

$$90\frac{x_1}{x_1 + x_3} + 110\frac{x_3}{x_1 + x_3}$$

where $x_1/(x_1 + x_3)$ is the percentage of component A used in regular gasoline and $x_3/(x_1 + x_3)$ is the percentage of component B. The final octane rating of regular gasoline must be at least 95; so

$$90\frac{x_1}{x_1 + x_3} + 110\frac{x_3}{x_1 + x_3} \ge 95 \qquad \text{Multiply by } x_1 + x_3.$$
$$90x_1 + 110x_3 \ge 95(x_1 + x_3) \qquad \text{Collect like terms on the right side.}$$
$$0 \ge 5x_1 - 15x_3 \qquad \text{Octane rating for regular}$$

The corresponding inequality for premium gasoline is

$$90\frac{x_2}{x_2 + x_4} + 110\frac{x_4}{x_2 + x_4} \ge 105$$
$$90x_2 + 110x_4 \ge 105(x_2 + x_4)$$
$$0 \ge 15x_2 - 5x_4 \qquad \text{Octane rating for premium}$$

The cost of the components used is

$$C = 28(x_1 + x_2) + 32(x_3 + x_4)$$

The revenue from selling all the gasoline is

$$R = 34(x_1 + x_3) + 40(x_2 + x_4)$$

and the profit is

$$\begin{aligned}
P &= R - C \\
&= 34(x_1 + x_3) + 40(x_2 + x_4) - 28(x_1 + x_2) - 32(x_3 + x_4) \\
&= (34 - 28)x_1 + (40 - 28)x_2 + (34 - 32)x_3 + (40 - 32)x_4 \\
&= 6x_1 + 12x_2 + 2x_3 + 8x_4
\end{aligned}$$

To find the maximum profit, we must solve the following linear programming problem:

$$\text{Maximize} \quad P = 6x_1 + 12x_2 + 2x_3 + 8x_4 \qquad \text{Profit}$$

$$
\begin{array}{llll}
\text{subject to} & x_1 + x_2 & \leq 30{,}000 & \text{Available } A \\
& x_3 + x_4 \leq 20{,}000 & & \text{Available } B \\
& x_1 + x_3 \geq 20{,}000 & & \text{Required regular} \\
& x_2 + x_4 \geq 10{,}000 & & \text{Required premium} \\
& 5x_1 - 15x_3 \leq 0 & & \text{Octane for regular} \\
& 15x_2 - 5x_4 \leq 0 & & \text{Octane for premium} \\
& x_1, x_2, x_3, x_4 \geq 0 &
\end{array}
$$

We will use technology to solve this large problem. There are many types of software that use the big M method to solve linear programming problems, including Java applets, graphing calculator programs, and spreadsheets. Because you are likely to use different software than we did, we will simply display the initial and final tableaux. Notice that in the last row of the initial tableau, we entered a large number, 10^6, instead of the symbol M. This is typical of software implementations of the big M method.

x_1	x_2	x_3	x_4	s_1	s_2	s_3	a_1	s_4	a_2	s_5	s_6	P	
1	1	0	0	1	0	0	0	0	0	0	0	0	30,000
0	0	1	1	0	1	0	0	0	0	0	0	0	20,000
1	0	1	0	0	0	−1	1	0	0	0	0	0	20,000
0	1	0	1	0	0	0	0	−1	1	0	0	0	10,000
5	0	−15	0	0	0	0	0	0	0	1	0	0	0
0	15	0	−5	0	0	0	0	0	0	0	1	0	0
−6	−12	−2	−8	0	0	0	10^6	0	10^6	0	0	1	0

The final table produced by the software is

x_1	x_2	x_3	x_4	s_1	s_2	s_3	a_1	s_4	a_2	s_5	s_6	P	
0	0	0	0	1.5	−0.5	1	−1	0	0	−0.1	−0.1	0	15,000
0	0	0	0	−0.5	1.5	0	0	1	−1	0.1	0.1	0	5,000
0	0	1	0	0.375	−0.125	0	0	0	0	−0.075	−0.025	0	8,750
0	0	0	1	−0.375	1.125	0	0	0	0	0.075	0.025	0	11,250
1	0	0	0	1.125	−0.375	0	0	0	0	−0.025	−0.075	0	26,250
0	1	0	0	−0.125	0.375	0	0	0	0	0.025	0.075	0	3,750
0	0	0	0	3	11	0	10^6	0	10^6	0.6	0.6	1	310,000

From the final tableau, we see that the refinery should blend 26,250 barrels of component A and 8,750 barrels of component B to produce 35,000 barrels of regular gasoline. They should blend 3,750 barrels of component A and 11,250 barrels of component B to produce 15,000 barrels of premium gasoline. This will result in a maximum profit of $310,000.

Explore and Discuss 1 Interpret the values of the slack and surplus variables in the computer solution to Example 5.

Matched Problem 5 Suppose that the refinery in Example 5 has 35,000 barrels of component A, which costs $25 a barrel, and 15,000 barrels of component B, which costs $35 a barrel. If all other data remain the same, formulate a linear programming problem to find the maximum profit. Do not attempt to solve the problem (unless you have access to software that can solve linear programming problems).

Exercises 6.4

In Problems 1–8,

(A) *Introduce slack, surplus, and artificial variables and form the modified problem.*

(B) *Write the preliminary simplex tableau for the modified problem and find the initial simplex tableau.*

(C) *Find the optimal solution of the modified problem by applying the simplex method to the initial simplex tableau.*

(D) *Find the optimal solution of the original problem, if it exists.*

1. Maximize $P = 5x_1 + 2x_2$
   subject to $x_1 + 2x_2 \le 12$
   $$x_1 + x_2 \ge 4$$
   $$x_1, x_2 \ge 0$$

2. Maximize $P = 3x_1 + 7x_2$
   subject to $2x_1 + x_2 \le 16$
   $$x_1 + x_2 \ge 6$$
   $$x_1, x_2 \ge 0$$

3. Maximize $P = 3x_1 + 5x_2$
   subject to $2x_1 + x_2 \le 8$
   $$x_1 + x_2 = 6$$
   $$x_1, x_2 \ge 0$$

4. Maximize $P = 4x_1 + 3x_2$
   subject to $x_1 + 3x_2 \le 24$
   $$x_1 + x_2 = 12$$
   $$x_1, x_2 \ge 0$$

5. Maximize $P = 4x_1 + 3x_2$
   subject to $-x_1 + 2x_2 \le 2$
   $$x_1 + x_2 \ge 4$$
   $$x_1, x_2 \ge 0$$

6. Maximize $P = 3x_1 + 4x_2$
   subject to $x_1 - 2x_2 \le 2$
   $$x_1 + x_2 \ge 5$$
   $$x_1, x_2 \ge 0$$

7. Maximize $P = 5x_1 + 10x_2$
   subject to $x_1 + x_2 \le 3$
   $$2x_1 + 3x_2 \ge 12$$
   $$x_1, x_2 \ge 0$$

8. Maximize $P = 4x_1 + 6x_2$
   subject to $x_1 + x_2 \le 2$
   $$3x_1 + 5x_2 \ge 15$$
   $$x_1, x_2 \ge 0$$

Use the big M method to solve Problems 9–22.

9. Minimize and maximize $P = 2x_1 - x_2$
   subject to $x_1 + x_2 \le 8$
   $$5x_1 + 3x_2 \ge 30$$
   $$x_1, x_2 \ge 0$$

10. Minimize and maximize $P = -4x_1 + 16x_2$
    subject to $3x_1 + x_2 \le 28$
    $$x_1 + 2x_2 \ge 16$$
    $$x_1, x_2 \ge 0$$

11. Maximize $P = 2x_1 + 5x_2$
    subject to $x_1 + 2x_2 \le 18$
    $$2x_1 + x_2 \le 21$$
    $$x_1 + x_2 \ge 10$$
    $$x_1, x_2 \ge 0$$

12. Maximize $P = 6x_1 + 2x_2$
    subject to $x_1 + 2x_2 \le 20$
    $$2x_1 + x_2 \le 16$$
    $$x_1 + x_2 \ge 9$$
    $$x_1, x_2 \ge 0$$

13. Maximize $P = 10x_1 + 12x_2 + 20x_3$
    subject to $3x_1 + x_2 + 2x_3 \ge 12$
    $$x_1 - x_2 + 2x_3 = 6$$
    $$x_1, x_2, x_3 \ge 0$$

14. Maximize $P = 5x_1 + 7x_2 + 9x_3$
    subject to $x_1 - x_2 + x_3 \ge 20$
    $$2x_1 + x_2 + 5x_3 = 35$$
    $$x_1, x_2, x_3 \ge 0$$

15. Minimize $C = -5x_1 - 12x_2 + 16x_3$
    subject to $x_1 + 2x_2 + x_3 \le 10$
    $$2x_1 + 3x_2 + x_3 \ge 6$$
    $$2x_1 + x_2 - x_3 = 1$$
    $$x_1, x_2, x_3 \ge 0$$

16. Maximize $C = -3x_1 + 15x_2 - 4x_3$
    subject to $2x_1 + x_2 + 3x_3 \le 24$
    $$x_1 + 2x_2 + x_3 \ge 6$$
    $$x_1 - 3x_2 + x_3 = 2$$
    $$x_1, x_2, x_3 \ge 0$$

17. Maximize $P = 3x_1 + 5x_2 + 6x_3$
    subject to $2x_1 + x_2 + 2x_3 \le 8$
    $$2x_1 + x_2 - 2x_3 = 0$$
    $$x_1, x_2, x_3 \ge 0$$

18. Maximize $P = 3x_1 + 6x_2 + 2x_3$
    subject to $2x_1 + 2x_2 + 3x_3 \le 12$
    $$2x_1 - 2x_2 + x_3 = 0$$
    $$x_1, x_2, x_3 \ge 0$$

19. Maximize $P = 2x_1 + 3x_2 + 4x_3$
    subject to $x_1 + 2x_2 + x_3 \le 25$
    $$2x_1 + x_2 + 2x_3 \le 60$$
    $$x_1 + 2x_2 - x_3 \ge 10$$
    $$x_1, x_2, x_3 \ge 0$$

20. Maximize $P = 5x_1 + 2x_2 + 9x_3$
    subject to $2x_1 + 4x_2 + x_3 \le 150$
    $$3x_1 + 3x_2 + x_3 \le 90$$
    $$-x_1 + 5x_2 + x_3 \ge 120$$
    $$x_1, x_2, x_3 \ge 0$$

21. Maximize $P = x_1 + 2x_2 + 5x_3$
    subject to $x_1 + 3x_2 + 2x_3 \le 60$
    $$2x_1 + 5x_2 + 2x_3 \ge 50$$
    $$x_1 - 2x_2 + x_3 \ge 40$$
    $$x_1, x_2, x_3 \ge 0$$

22. Maximize $P = 2x_1 + 4x_2 + x_3$
    subject to $2x_1 + 3x_2 + 5x_3 \le 280$
    $$2x_1 + 2x_2 + x_3 \ge 140$$
    $$2x_1 + x_2 \ge 150$$
    $$x_1, x_2, x_3 \ge 0$$

23. Solve Problems 5 and 7 by graphing (the geometric method).

24. Solve Problems 6 and 8 by graphing (the geometric method).

Problems 25–32 are mixed. Some can be solved by the methods presented in Sections 6.2 and 6.3 while others must be solved by the big M method.

25. Minimize $C = 10x_1 - 40x_2 - 5x_3$
subject to $x_1 + 3x_2 \quad\quad \le 6$
$\quad\quad\quad 4x_2 + x_3 \le 3$
$\quad\quad\quad x_1, x_2, x_3 \ge 0$

26. Maximize $P = 7x_1 - 5x_2 + 2x_3$
subject to $x_1 - 2x_2 + x_3 \ge -8$
$\quad\quad\quad x_1 - x_2 + x_3 \le 10$
$\quad\quad\quad x_1, x_2, x_3 \ge 0$

27. Maximize $P = -5x_1 + 10x_2 + 15x_3$
subject to $2x_1 + 3x_2 + x_3 \le 24$
$\quad\quad\quad x_1 - 2x_2 - 2x_3 \ge 1$
$\quad\quad\quad x_1, x_2, x_3 \ge 0$

28. Minimize $C = -5x_1 + 10x_2 + 15x_3$
subject to $2x_1 + 3x_2 + x_3 \le 24$
$\quad\quad\quad x_1 - 2x_2 - 2x_3 \ge 1$
$\quad\quad\quad x_1, x_2, x_3 \ge 0$

29. Minimize $C = 10x_1 + 40x_2 + 5x_3$
subject to $x_1 + 3x_2 \quad\quad \ge 6$
$\quad\quad\quad 4x_2 + x_3 \ge 3$
$\quad\quad\quad x_1, x_2, x_3 \ge 0$

30. Maximize $P = 8x_1 + 2x_2 - 10x_3$
subject to $x_1 + x_2 - 3x_3 \le 6$
$\quad\quad\quad 4x_1 - x_2 + 2x_3 \le -7$
$\quad\quad\quad x_1, x_2, x_3 \ge 0$

31. Maximize $P = 12x_1 + 9x_2 + 5x_3$
subject to $x_1 + 3x_2 + x_3 \le 40$
$\quad\quad\quad 2x_1 + x_2 + 3x_3 \le 60$
$\quad\quad\quad x_1, x_2, x_3 \ge 0$

32. Minimize $C = 10x_1 + 12x_2 + 28x_3$
subject to $4x_1 + 2x_2 + 3x_3 \ge 20$
$\quad\quad\quad 3x_1 - x_2 - 4x_3 \le 10$
$\quad\quad\quad x_1, x_2, x_3 \ge 0$

Applications

In Problems 33–38, construct a mathematical model in the form of a linear programming problem. (The answers in the back of the book for these application problems include the model.) Then solve the problem using the big M method.

33. **Advertising.** An advertising company wants to attract new customers by placing a total of at most 10 ads in 3 newspapers. Each ad in the *Sentinel* costs $200 and will be read by 2,000 people. Each ad in the *Journal* costs $200 and will be read by 500 people. Each ad in the *Tribune* costs $100 and will be read by 1,500 people. The company wants at least 16,000 people to read its ads. How many ads should it place in each paper in order to minimize the advertising costs? What is the minimum cost?

34. **Advertising.** Discuss the effect on the solution to Problem 33 if the *Tribune* will not accept more than 4 ads from the company.

35. **Human nutrition.** A person on a high-protein, low-carbohydrate diet requires at least 100 units of protein and at most 24 units of carbohydrates daily. The diet will consist entirely of three special liquid diet foods: A, B, and C. The contents and costs of the diet foods are given in the table. How many bottles of each brand of diet food should be consumed daily in order to meet the protein and carbohydrate requirements at minimal cost? What is the minimum cost?

	Units per Bottle		
	A	**B**	**C**
Protein	10	10	20
Carbohydrates	2	3	4
Cost per bottle ($)	0.60	0.40	0.90

36. **Human nutrition.** Discuss the effect on the solution to Problem 35 if the cost of brand C liquid diet food increases to $1.50 per bottle.

37. **Plant food.** A farmer can use three types of plant food: mix A, mix B, and mix C. The amounts (in pounds) of nitrogen, phosphoric acid, and potash in a cubic yard of each mix are given in the table. Tests performed on the soil indicate that the field needs at least 800 pounds of potash. The tests also indicate that no more than 700 pounds of phosphoric acid should be added to the field. The farmer plans to plant a crop that requires a great deal of nitrogen. How many cubic yards of each mix should be added to the field in order to satisfy the potash and phosphoric acid requirements and maximize the amount of nitrogen? What is the maximum amount of nitrogen?

	Pounds per Cubic Yard		
	A	**B**	**C**
Nitrogen	12	16	8
Phosphoric acid	12	8	16
Potash	16	8	16

38. **Plant food.** Discuss the effect on the solution to Problem 37 if the limit on phosphoric acid is increased to 1,000 pounds.

In Problems 39–47, construct a mathematical model in the form of a linear programming problem. Do not solve.

39. **Manufacturing.** A company manufactures car and truck frames at plants in Milwaukee and Racine. The Milwaukee plant has a daily operating budget of $50,000 and can produce at most 300 frames daily in any combination. It costs $150 to manufacture a car frame and $200 to manufacture a truck frame at the Milwaukee plant. The Racine plant has a daily operating budget of $35,000, and can produce a maximum combined total of 200 frames daily. It costs $135 to manufacture a car frame and $180 to manufacture a truck frame at the Racine plant. Based on past demand, the company wants to limit production to a maximum of 250 car frames and 350 truck frames per day. If the company realizes a profit of $50 on each car frame and $70 on each truck frame, how many frames of each type should be produced at each plant to maximize the daily profit?

40. Loan distributions. A savings and loan company has $3 million to lend. The types of loans and annual returns offered are given in the table. State laws require that at least 50% of the money loaned for mortgages must be for first mortgages and that at least 30% of the total amount loaned must be for either first or second mortgages. Company policy requires that the amount of signature and automobile loans cannot exceed 25% of the total amount loaned and that signature loans cannot exceed 15% of the total amount loaned. How much money should be allocated to each type of loan in order to maximize the company's return?

Type of Loan	Annual Return (%)
Signature	18
First mortgage	12
Second mortgage	14
Automobile	16

41. Oil refining. A refinery produces two grades of gasoline, regular and premium, by blending together three components: A, B, and C. Component A has an octane rating of 90 and costs $28 a barrel, component B has an octane rating of 100 and costs $30 a barrel, and component C has an octane rating of 110 and costs $34 a barrel. The octane rating for regular must be at least 95 and the octane rating for premium must be at least 105. Regular gasoline sells for $38 a barrel and premium sells for $46 a barrel. The company has 40,000 barrels of component A, 25,000 barrels of component B, and 15,000 barrels of component C. It must produce at least 30,000 barrels of regular and 25,000 barrels of premium. How should the components be blended in order to maximize profit?

42. Trail mix. A company makes two brands of trail mix, regular and deluxe, by mixing dried fruits, nuts, and cereal. The recipes for the mixes are given in the table. The company has 1,200 pounds of dried fruits, 750 pounds of nuts, and 1,500 pounds of cereal for the mixes. The company makes a profit of $0.40 on each pound of regular mix and $0.60 on each pound of deluxe mix. How many pounds of each ingredient should be used in each mix in order to maximize the company's profit?

Type of Mix	Ingredients
Regular	At least 20% nuts
	At most 40% cereal
Deluxe	At least 30% nuts
	At most 25% cereal

43. Investment strategy. An investor is planning to divide her investments among high-tech mutual funds, global mutual funds, corporate bonds, municipal bonds, and CDs. Each of these investments has an estimated annual return and a risk factor (see the table). The risk level for each choice is the product of its risk factor and the percentage of the total funds invested in that choice. The total risk level is the sum of the risk levels for all the investments. The investor wants at least 20% of her investments to be in CDs and does not want the risk level to exceed 1.8. What percentage of her total investments should be invested in each choice to maximize the return?

Investment	Annual Return (%)	Risk Factor
High-tech funds	11	2.7
Global funds	10	1.8
Corporate bonds	9	1.2
Muncipal bonds	8	0.5
CDs	5	0

44. Investment strategy. Refer to Problem 43. Suppose the investor decides that she would like to minimize the total risk factor, as long as her return does not fall below 9%. What percentage of her total investments should be invested in each choice to minimize the total risk level?

45. Human nutrition. A dietitian arranges a special diet using foods L, M, and N. The table gives the nutritional contents and cost of 1 ounce of each food. The diet's daily requirements are at least 400 units of calcium, at least 200 units of iron, at least 300 units of vitamin A, at most 150 units of cholesterol, and at most 900 calories. How many ounces of each food should be used in order to meet the diet's requirements at a minimal cost?

	Units per Bottle		
	L	M	N
Calcium	30	10	30
Iron	10	10	10
Vitamin A	10	30	20
Cholesterol	8	4	6
Calories	60	40	50
Cost per ounce ($)	0.40	0.60	0.80

46. Mixing feed. A farmer grows three crops: corn, oats, and soybeans. He mixes them to feed his cows and pigs. At least 40% of the feed mix for the cows must be corn. The feed mix for the pigs must contain at least twice as much soybeans as corn. He has harvested 1,000 bushels of corn, 500 bushels of oats, and 1,000 bushels of soybeans. He needs 1,000 bushels of each feed mix for his livestock. The unused corn, oats, and soybeans can be sold for $4, $3.50, and $3.25 a bushel, respectively (thus, these amounts also represent the cost of the crops used to feed the livestock). How many bushels of each crop should be used in each feed mix in order to produce sufficient food for the livestock at a minimal cost?

47. Transportation. Three towns are forming a consolidated school district with two high schools. Each high school has a maximum capacity of 2,000 students. Town A has 500 high school students, town B has 1,200, and town C has 1,800. The weekly costs of transporting a student from each town to each school are given in the table. In order to balance the enrollment, the school board decided that each high school must enroll at least 40% of the total student population. Furthermore, no more than 60% of the students in any town should be sent to the same high school. How many students from each town should be enrolled in each school in order to meet these requirements and minimize the cost of transporting the students?

	Weekly Transportation Cost per Student ($)	
	School I	School II
Town A	4	8
Town B	6	4
Town C	3	9

1. Maximize $P = 3x_1 - 2x_2 + x_3 - Ma_1 - Ma_2 - Ma_3$

subject to
$$x_1 - 2x_2 + x_3 - s_1 + a_1 \qquad\qquad\qquad = 5$$
$$x_1 + 3x_2 - 4x_3 \qquad\qquad - s_2 + a_2 \qquad\qquad = 10$$
$$2x_1 + 4x_2 + 5x_3 \qquad\qquad\qquad\quad + s_3 \qquad = 20$$
$$-3x_1 + x_2 + x_3 \qquad\qquad\qquad\qquad\qquad + a_3 = 15$$
$$x_1, x_2, x_3, s_1, a_1, s_2, a_2, s_3, a_3 \geq 0$$

2. Max $P = 22$ at $x_1 = 6$, $x_2 = 4$, $x_3 = 0$

3. No optimal solution

4. A minimum cost of \$200 is realized when no type J, 6 type K, and 2 type L stones are processed each day.

5. Maximize $P = 9x_1 + 15x_2 - x_3 + 5x_4$

subject to
$$x_1 + x_2 \qquad\qquad\qquad \leq 35,000$$
$$x_3 + x_4 \leq 15,000$$
$$x_1 \qquad + x_3 \qquad \geq 20,000$$
$$x_2 \qquad + x_4 \geq 10,000$$
$$5x_1 \qquad - 15x_3 \qquad \leq 0$$
$$15x_2 \qquad - 5x_4 \leq 0$$
$$x_1, x_2, x_3, x_4 \geq 0$$

Chapter 6   Summary and Review

Important Terms, Symbols, and Concepts

6.1 The Table Method: An Introduction to the Simplex Method

EXAMPLES

- A linear programming problem is said to be a **standard maximization problem in standard form** if its mathematical model is of the following form: Maximize the objective function

$$P = c_1x_1 + c_2x_2 + \cdots + c_kx_k$$

subject to problem constraints of the form

$$a_1x_1 + a_2x_2 + \cdots + a_kx_k \leq b \qquad b \geq 0$$

with nonnegative constraints

$$x_1, x_2, \ldots, x_k \geq 0$$

- The system of inequalities (*i*-system) of a linear programming problem is converted to a system of equations (*e*-system) by means of **slack variables**. A solution of the *e*-system is a **feasible solution** if the values of all decision variables and slack variables are nonnegative. The feasible solutions of the *e*-system correspond to the points in the feasible region of the *i*-system. A **basic solution** of the *e*-system is found by setting k of the variables equal to 0, where k is the number of decision variables $x_1, x_2, \ldots, x_k$. A solution of the *e*-system that is both basic and feasible is called a **basic feasible solution**. The **table method** for solving a linear programming problem consists of constructing a table of all basic solutions, determining which of the basic solutions are feasible, and then maximizing the objective function over the basic feasible solutions. A procedure for carrying out the table method in the case of $k = 2$ decision variables is given on page 288. For an arbitrary number of decision variables, see the procedure on pages 293 and 294.

Ex. 1, p. 289

Ex. 2, p. 290

Ex. 3, p. 291

- The **fundamental theorem of linear programming** can be formulated in terms of basic feasible solutions. It states that an optimal solution to the linear programming problem, if one exists, must occur at one or more of the basic feasible solutions.

- The k variables that are assigned the value 0, in order to generate a basic solution, are called **nonbasic variables**. The remaining variables are called **basic variables**. So the classification of variables as basic or nonbasic depends on the basic solution under consideration.

Ex. 4, p. 293

- The benefit of the table method is that it gives a procedure for **finding all corner points of the feasible region without drawing a graph**. But the table method has a drawback: If the number of decision variables and problem constraints is large, then the number of rows in the table (that is, the number of basic solutions) becomes too large for practical implementation. The *simplex method,* discussed in Section 6.2, gives a practical method for solving large linear programming problems.

6.2 The Simplex Method: Maximization with Problem Constraints of the Form ≤

- Adding the objective function to the system of constraint equations produces the **initial system**. Negative values of the objective function variable are permitted in a basic feasible solution as long as all other variables are nonnegative. The fundamental theorem of linear programming also applies to initial systems.

- The augmented matrix of the initial system is called the **initial simplex tableau**. The **simplex method** consists of performing **pivot operations**, starting with the initial simplex tableau, until an optimal solution is found (if one exists). The procedure is illustrated in Figure 2 (p. 304).

Ex. 1, p. 305
Ex. 2, p. 306
Ex. 3, p. 307

6.3 The Dual Problem: Minimization with Problem Constraints of the Form ≥

- By the **Fundamental Principle of Duality**, a linear programming problem that asks for the minimum of the objective function over a region described by ≥ problem constraints can be solved by first forming the **dual problem** and then using the simplex method.

Ex. 1, p. 314
Ex. 2, p. 317
Ex. 3, p. 319

6.4 Maximization and Minimization with Mixed Problem Constraints

- The **big M method** can be used to find the maximum of any objective function on any feasible region. The solution process involves the introduction of two new types of variables, **surplus variables** and **artificial variables**, and a modification of the objective function. The result is an initial tableau that can be transformed into the tableau of a **modified problem**.

Ex. 1, p. 329

- Applying the simplex method to the modified problem produces a solution to the original problem, if one exists.

- The dual method can be used to solve *only* certain minimization problems. But *all* minimization problems can be solved by using the big M method to find the maximum of the negative of the objective function. The big M method also lends itself to computer implementation.

Ex. 2, p. 331
Ex. 3, p. 332
Ex. 4, p. 333
Ex. 5, p. 335

Review Exercises

Work through all the problems in this chapter review and check your answers in the back of the book. Answers to all review problems are there along with section numbers in italics to indicate where each type of problem is discussed. Where weaknesses show up, review appropriate sections in the text.

1. Given the linear programming problem

$$\text{Maximize} \quad P = 6x_1 + 2x_2$$
$$\text{subject to} \quad 2x_1 + x_2 \le 8$$
$$x_1 + 2x_2 \le 10$$
$$x_1, x_2 \ge 0$$

Convert the problem constraints into a system of equations using slack variables.

2. How many basic variables and how many nonbasic variables are associated with the system in Problem 1?

3. Find all basic solutions for the system in Problem 1, and determine which basic solutions are feasible.

4. Write the simplex tableau for Problem 1, and circle the pivot element. Indicate the entering and exiting variables.

5. Solve Problem 1 using the simplex method.

6. For the simplex tableau below, identify the basic and nonbasic variables. Find the pivot element, the entering and exiting variables, and perform one pivot operation.

$$\begin{bmatrix} x_1 & x_2 & x_3 & s_1 & s_2 & s_3 & P & \\ 2 & 1 & 3 & -1 & 0 & 0 & 0 & 20 \\ 3 & 0 & 4 & 1 & 1 & 0 & 0 & 30 \\ 2 & 0 & 5 & 2 & 0 & 1 & 0 & 10 \\ -8 & 0 & -5 & 3 & 0 & 0 & 1 & 50 \end{bmatrix}$$

7. Find the basic solution for each tableau. Determine whether the optimal solution has been reached, additional pivoting is required, or the problem has no optimal solution.

(A)
$$\begin{bmatrix} x_1 & x_2 & s_1 & s_2 & P & \\ 4 & 1 & 0 & 0 & 0 & 2 \\ 2 & 0 & 1 & 1 & 0 & 5 \\ -2 & 0 & 3 & 0 & 1 & 12 \end{bmatrix}$$

(B)
$$\begin{bmatrix} x_1 & x_2 & s_1 & s_2 & P & \\ -1 & 3 & 0 & 1 & 0 & 7 \\ 0 & 2 & 1 & 0 & 0 & 0 \\ -2 & 1 & 0 & 0 & 1 & 22 \end{bmatrix}$$

(C)
$$\begin{bmatrix} x_1 & x_2 & s_1 & s_2 & P & \\ 1 & -2 & 0 & 4 & 0 & 6 \\ 0 & 2 & 1 & 6 & 0 & 15 \\ 0 & 3 & 0 & 2 & 1 & 10 \end{bmatrix}$$

8. Form the dual problem of

$$\begin{aligned} \text{Minimize} \quad & C = 5x_1 + 2x_2 \\ \text{subject to} \quad & x_1 + 3x_2 \geq 15 \\ & 2x_1 + x_2 \geq 20 \\ & x_1, x_2 \geq 0 \end{aligned}$$

9. Write the initial system for the dual problem in Problem 8.

10. Write the first simplex tableau for the dual problem in Problem 8 and label the columns.

11. Use the simplex method to find the optimal solution of the dual problem in Problem 8.

12. Use the final simplex tableau from Problem 11 to find the optimal solution of the linear programming problem in Problem 8.

13. Solve the linear programming problem using the simplex method.

$$\begin{aligned} \text{Maximize} \quad & P = 3x_1 + 4x_2 \\ \text{subject to} \quad & 2x_1 + 4x_2 \leq 24 \\ & 3x_1 + 3x_2 \leq 21 \\ & 4x_1 + 2x_2 \leq 20 \\ & x_1, x_2 \geq 0 \end{aligned}$$

14. Form the dual problem of the linear programming problem

$$\begin{aligned} \text{Minimize} \quad & C = 3x_1 + 8x_2 \\ \text{subject to} \quad & x_1 + x_2 \geq 10 \\ & x_1 + 2x_2 \geq 15 \\ & x_2 \geq 3 \\ & x_1, x_2 \geq 0 \end{aligned}$$

15. Solve Problem 14 by applying the simplex method to the dual problem.

Solve the linear programming Problems 16 and 17.

16. Maximize $P = 5x_1 + 3x_2 - 3x_3$
   subject to $x_1 - x_2 - 2x_3 \leq 3$
   $2x_1 + 2x_2 - 5x_3 \leq 10$
   $x_1, x_2, x_3 \geq 0$

17. Maximize $P = 5x_1 + 3x_2 - 3x_3$
   subject to $x_1 - x_2 - 2x_3 \leq 3$
   $x_1 + x_2 \leq 5$
   $x_1, x_2, x_3 \geq 0$

In Problems 18 and 19,

(A) *Introduce slack, surplus, and artificial variables and form the modified problem.*

(B) *Write the preliminary simplex tableau for the modified problem and find the initial simplex tableau.*

(C) *Find the optimal solution of the modified problem by applying the simplex method to the initial simplex tableau.*

(D) *Find the optimal solution of the original problem, if it exists.*

18. Maximize $P = x_1 + 3x_2$
   subject to $x_1 + x_2 \geq 6$
   $x_1 + 2x_2 \leq 8$
   $x_1, x_2 \geq 0$

19. Maximize $P = x_1 + x_2$
   subject to $x_1 + x_2 \geq 5$
   $x_1 + 2x_2 \leq 4$
   $x_1, x_2 \geq 0$

20. Find the modified problem for the following linear programming problem. (Do not solve.)

$$\begin{aligned} \text{Maximize} \quad & P = 2x_1 + 3x_2 + x_3 \\ \text{subject to} \quad & x_1 - 3x_2 + x_3 \leq 7 \\ & -x_1 - x_2 + 2x_3 \leq -2 \\ & 3x_1 + 2x_2 - x_3 = 4 \\ & x_1, x_2, x_3 \geq 0 \end{aligned}$$

✎ *Write a brief verbal description of the type of linear programming problem that can be solved by the method indicated in Problems 21–23. Include the type of optimization, the number of variables, the type of constraints, and any restrictions on the coefficients and constants.*

21. Basic simplex method with slack variables

22. Dual problem method

23. Big M method

24. Solve the following linear programming problem by the simplex method, keeping track of the obvious basic solution at each step. Then graph the feasible region and illustrate the path to the optimal solution determined by the simplex method.

$$\begin{aligned} \text{Maximize} \quad & P = 2x_1 + 3x_2 \\ \text{subject to} \quad & x_1 + 2x_2 \leq 22 \\ & 3x_1 + x_2 \leq 26 \\ & x_1 \leq 8 \\ & x_2 \leq 10 \\ & x_1, x_2 \geq 0 \end{aligned}$$

25. Solve by the dual problem method:

$$\begin{aligned} \text{Minimize} \quad & C = 3x_1 + 2x_2 \\ \text{subject to} \quad & 2x_1 + x_2 \leq 20 \\ & 2x_1 + x_2 \geq 9 \\ & x_1 + x_2 \geq 6 \\ & x_1, x_2 \geq 0 \end{aligned}$$

26. Solve Problem 25 by the big M method.

27. Solve by the dual problem method:

$$\begin{aligned} \text{Minimize} \quad & C = 15x_1 + 12x_2 + 15x_3 + 18x_4 \\ \text{subject to} \quad & x_1 + x_2 \leq 240 \\ & x_3 + x_4 \leq 500 \\ & x_1 + x_3 \geq 400 \\ & x_2 + x_4 \geq 300 \\ & x_1, x_2, x_3, x_4 \geq 0 \end{aligned}$$

Applications

In Problems 28–31, construct a mathematical model in the form of a linear programming problem. (The answers in the back of the book for these application problems include the model.) Then solve the problem by the simplex, dual problem, or big M methods.

28. Investment. An investor has $150,000 to invest in oil stock, steel stock, and government bonds. The bonds are guaranteed to yield 5%, but the yield for each stock can vary. To protect against major losses, the investor decides that the amount invested in oil stock should not exceed $50,000 and that the total amount invested in stock cannot exceed the amount invested in bonds by more than $25,000.

(A) If the oil stock yields 12% and the steel stock yields 9%, how much money should be invested in each alternative in order to maximize the return? What is the maximum return?

(B) Repeat part (A) if the oil stock yields 9% and the steel stock yields 12%.

29. Manufacturing. A company manufactures outdoor furniture consisting of regular chairs, rocking chairs, and chaise lounges. Each piece of furniture passes through three different production departments: fabrication, assembly, and finishing. Each regular chair takes 1 hour to fabricate, 2 hours to assemble, and 3 hours to finish. Each rocking chair takes 2 hours to fabricate, 2 hours to assemble, and 3 hours to finish. Each chaise lounge takes 3 hours to fabricate, 4 hours to assemble, and 2 hours to finish. There are 2,500 labor-hours available in the fabrication department, 3,000 in the assembly department, and 3,500 in the finishing department. The company makes a profit of $17 on each regular chair, $24 on each rocking chair, and $31 on each chaise lounge.

(A) How many chairs of each type should the company produce in order to maximize profit? What is the maximum profit?

(B) Discuss the effect on the optimal solution in part (A) if the profit on a regular chair is increased to $25 and all other data remain the same.

(C) Discuss the effect on the optimal solution in part (A) if the available hours on the finishing department are reduced to 3,000 and all other data remain the same.

30. Shipping schedules. A company produces motors for washing machines at factory A and factory B. The motors are then shipped to either plant X or plant Y, where the washing machines are assembled. The maximum number of motors that can be produced at each factory monthly, the minimum number required monthly for each plant to meet anticipated demand, and the shipping charges for one motor are given in the table. Determine a shipping schedule that will minimize the cost of transporting the motors from the factories to the assembly plants.

	Plant X	Plant Y	Maximum Production
Factory A	$5	$8	1,500
Factory B	$9	$7	1,000
Minimum Requirement	900	1,200	

31. Blending–food processing. A company blends long-grain rice and wild rice to produce two brands of rice mixes: brand A, which is marketed under the company's name; and brand B, which is marketed as a generic brand. Brand A must contain at least 10% wild rice, and brand B must contain at least 5% wild rice. Long-grain rice costs $0.70 per pound, and wild rice costs $3.40 per pound. The company sells brand A for $1.50 a pound and brand B for $1.20 a pound. The company has 8,000 pounds of long-grain rice and 500 pounds of wild rice on hand. How should the company use the available rice to maximize its profit? What is the maximum profit?

7

Logic, Sets, and Counting

7.1 Logic

7.2 Sets

7.3 Basic Counting Principles

7.4 Permutations and Combinations

Chapter 7
Summary and Review

Review Exercises

Introduction

Logic and sets form the foundation of mathematics. That ancient foundation was rebuilt in the 19th and early 20th centuries. George Boole (1815–1864) introduced mathematical methods to logic in 1847, and today those methods are applied to the design of electronic circuits, computer programming, and Internet searches.

We use logic to formulate precise mathematical statements and make correct deductions. We use sets to build mathematical objects such as the functions studied in Chapters 1 and 2 and the solution sets of systems of equations and inequalities studied in Chapters 4, 5, and 6. In those chapters, the foundation of logic and sets remained largely out of view. In this chapter, we examine these topics with an eye toward the study of probability in Chapter 8.

In Section 7.1 we introduce the symbolic logic of propositions. Sets are studied in Section 7.2, and various counting techniques are considered in Sections 7.3 and 7.4. Determining the number of 5-card hands that contain only aces and kings is just one application explored in this chapter (see Example 6 in Section 7.4).

7.1 Logic

- Propositions and Connectives
- Truth Tables
- Logical Implications and Equivalences

Consider the kind of logical reasoning that is used in everyday language. For example, suppose that the following two statements are true:

"If today is Saturday, then Derek plays soccer" and

"Today is Saturday."

From the truth of these two statements, we can conclude that

"Derek plays soccer."

Similarly, suppose that the following two mathematical statements are true:

"If the sum of the digits of 71,325 is divisible by 9, then 71,325 is divisible by 9" and

"The sum of the digits of 71,325 is divisible by 9."

From the truth of these two statements, we can conclude that

"71,325 is divisible by 9."

Logic is the study of the form of arguments. Each of the preceding arguments, the first about Derek and the second about divisibility by 9, has the same form, namely,

$$[(p \rightarrow q) \land p] \Rightarrow q$$

In this section, we introduce the notation that is required to represent an argument in compact form, and we establish precision in the use of logical deduction that forms the foundation for the proof of mathematical theorems.

Propositions and Connectives

A **proposition** is a statement (not a question or command) that is either true or false. So the statement

"There is a prime number between 2,312 and 2,325"

is a proposition. It is a proposition even if we do not know or cannot determine whether it is true. We use lowercase letters such as p, q, and r to denote propositions.

If p and q are propositions, then the compound propositions

$$\neg p, \quad p \lor q, \quad p \land q, \quad \text{and} \quad p \rightarrow q$$

can be formed using the negation symbol $\neg$ and the connectives $\lor$, $\land$, and $\rightarrow$. These propositions are called "not p," "p or q," "p and q," and "if p then q," respectively. We use a *truth table* to specify each of these compound propositions. A truth table gives the proposition's truth value, T (true) or F (false), for all possible values of its variables.

DEFINITION Negation

If p is a proposition, then the proposition $\neg p$, read **not p**, or the **negation** of p, is false if p is true, and true if p is false.

p	$\neg p$
T	F
F	T

DEFINITION   Disjunction

If p and q are propositions, then the proposition $p \vee q$, read **p or q**, or the **disjunction** of p and q, is true if p is true, or if q is true, or if both are true, and is false otherwise.

p	q	$p \vee q$
T	T	T
T	F	T
F	T	T
F	F	F

Note that *or* is used in the inclusive sense; that is, it includes the possibility that both p and q are true. This mathematical usage differs from the way that *or* is sometimes used in everyday language ("I will order chicken or I will order fish") when we intend to exclude the possibility that both are true.

DEFINITION   Conjunction

If p and q are propositions, then the proposition $p \wedge q$, read **p and q**, or the **conjunction** of p and q, is true if both p and q are true, and is false otherwise.

p	q	$p \wedge q$
T	T	T
T	F	F
F	T	F
F	F	F

The truth table for $p \wedge q$ is just what you would expect, based on the use of "and" in everyday language. Note that there is just one T in the third column of the truth table for conjunction; both p and q must be true in order for $p \wedge q$ to be true.

CONCEPTUAL INSIGHT

It is helpful to think of the conditional as a guarantee. For example, an instructor of a mathematics course might give a student the guarantee: "If you score at least 90%, then you will get an A." Suppose the student scores less than 90%, so the hypothesis is false. The guarantee remains in effect even though it is not applicable. We say that the conditional statement is *vacuously true*. In fact, there is only one circumstance in which the conditional statement could be false: The student scores at least 90% (that is, the hypothesis is true), but the grade is not an A (the conclusion is false).

DEFINITION   Conditional

If p and q are propositions, then the proposition $p \rightarrow q$, read **if p then q**, or the **conditional with hypothesis p and conclusion q**, is false if p is true and q is false, but is true otherwise.

p	q	$p \rightarrow q$
T	T	T
T	F	F
F	T	T
F	F	T

Note that the definition of $p \rightarrow q$ differs somewhat from the use of "if p then q" in everyday language. For example, we might question whether the proposition

"If Paris is in Switzerland, then Queens is in New York"

is true on the grounds that there is no apparent connection between p ("Paris is in Switzerland") and q ("Queens is in New York"). We consider it to be true, however, in accordance with the definition of $p \rightarrow q$, because p is false. Whenever the hypothesis p is false, we say that the conditional $p \rightarrow q$ is **vacuously true**.

EXAMPLE 1   Compound Propositions   Consider the propositions p and q:

$$p: \quad \text{"}4 + 3 \text{ is even."}$$
$$q: \quad \text{"}4^2 + 3^2 \text{ is odd."}$$

Express each of the following propositions as an English sentence and determine whether it is true or false.

(A) $\neg p$ (B) $\neg q$ (C) $p \vee q$ (D) $p \wedge q$ (E) $p \rightarrow q$

SOLUTION

(A) $\neg p$: "$4 + 3$ is not even"

(Note that we modified the wording of "Not $4 + 3$ is even" to standard English usage.) Because $4 + 3 = 7$ is odd, p is false, and therefore $\neg p$ is true.

(B) $\neg q$: "$4^2 + 3^2$ is not odd"

Because $4^2 + 3^2 = 25$ is odd, q is true, and therefore $\neg q$ is false.

(C) $p \vee q$: "$4 + 3$ is even or $4^2 + 3^2$ is odd"

Because q is true, $p \vee q$ is true.

(D) $p \wedge q$: "$4 + 3$ is even and $4^2 + 3^2$ is odd"

Because p is false, $p \wedge q$ is false.

(E) $p \rightarrow q$: "if $4 + 3$ is even, then $4^2 + 3^2$ is odd"

Because p is false, $p \rightarrow q$ is (vacuously) true.

Matched Problem 1   Consider the propositions p and q:

$$p: \quad \text{"}14^2 < 200\text{"}$$
$$q: \quad \text{"}23^2 < 500\text{"}$$

Express each of the following propositions as an English sentence and determine whether it is true or false.

(A) $\neg p$ (B) $\neg q$ (C) $p \vee q$ (D) $p \wedge q$ (E) $p \rightarrow q$

With any conditional $p \rightarrow q$, we associate two other propositions: the *converse* of $p \rightarrow q$ and the *contrapositive* of $p \rightarrow q$.

DEFINITION   Converse and Contrapositive

Let $p \rightarrow q$ be a conditional proposition. The proposition $q \rightarrow p$ is called the **converse** of $p \rightarrow q$. The proposition $\neg q \rightarrow \neg p$ is called the **contrapositive** of $p \rightarrow q$.

EXAMPLE 2   Converse and Contrapositive   Consider the propositions p and q:

$$p: \quad \text{"}2 + 2 = 4\text{"}$$
$$q: \quad \text{"}9 \text{ is a prime"}$$

Express each of the following propositions as an English sentence and determine whether it is true or false.

(A) $p \rightarrow q$ (B) The converse of $p \rightarrow q$ (C) The contrapositive of $p \rightarrow q$

SOLUTION

(A) $p \rightarrow q$: "if $2 + 2 = 4$, then 9 is a prime"

Because p is true and q is false, $p \rightarrow q$ is false.

(B) $q \rightarrow p$: "if 9 is a prime, then $2 + 2 = 4$"

Because q is false, $q \rightarrow p$ is vacuously true.

(C) $\neg q \rightarrow \neg p$: "if 9 is not prime, then $2 + 2$ is not equal to 4"

Because q is false and p is true, $\neg q$ is true and $\neg p$ is false, so $\neg q \rightarrow \neg p$ is false.

Matched Problem 2 Consider the propositions p and q:

$$p: \text{"}5^2 + 12^2 = 13^2\text{"}$$
$$q: \text{"}7^2 + 24^2 = 25^2\text{"}$$

Express each of the following propositions as an English sentence and determine whether it is true or false.

(A) $p \rightarrow q$ (B) The converse of $p \rightarrow q$ (C) The contrapositive of $p \rightarrow q$

Truth Tables

A **truth table** for a compound proposition specifies whether it is true or false for any assignment of truth values to its variables. Such a truth table can be constructed for any compound proposition by referring to the truth tables in the definitions of $\neg$, $\vee$, $\wedge$, and $\rightarrow$.

EXAMPLE 3 Constructing Truth Tables Construct the truth table for $\neg p \vee q$.

SOLUTION The proposition contains two variables, p and q, so the truth table will consist of four rows, one for each possible assignment of truth values to two variables (TT, TF, FT, FF). Although the truth table itself consists of three columns, one labeled p, another labeled q, and the third labeled $\neg p \vee q$, it is helpful to insert an additional column labeled $\neg p$. The entries in that additional column are obtained from the first column, changing any T to F, and vice versa, in accordance with the definition of $\neg$. The entries in the last column are obtained from the third and second columns, in accordance with the definition of $\vee$ (in a given row, if either entry in those columns is a T, then the entry in the last column is T; if both are F, the entry in the last column is F).

p	q	$\neg p$	$\neg p \vee q$
T	T	F	T
T	F	F	F
F	T	T	T
F	F	T	T

Note that the truth table for $\neg p \vee q$ is identical to the truth table for $p \rightarrow q$ (see the definition of a conditional).

Matched Problem 3 Construct the truth table for $p \wedge \neg q$.

EXAMPLE 4 Constructing a Truth Table Construct the truth table for $[(p \rightarrow q) \wedge p] \rightarrow q$.

SOLUTION It is helpful to insert a third column labeled $p \rightarrow q$ and a fourth labeled $(p \rightarrow q) \wedge p$. We complete the first three columns. Then we use the third and first columns to complete the fourth, and we use the fourth and second columns to complete the last column.

p	q	$p \rightarrow q$	$(p \rightarrow q) \wedge p$	$[(p \rightarrow q) \wedge p] \rightarrow q$
T	T	T	T	T
T	F	F	F	T
F	T	T	F	T
F	F	T	F	T

Note that $[(p \rightarrow q) \wedge p] \rightarrow q$ is always true, regardless of the truth values of p and q.

Matched Problem 4 Construct the truth table for $[(p \rightarrow q) \wedge \neg q] \rightarrow \neg p$.

Any proposition is either a *tautology*, or a *contradiction*, or a *contingency*. The proposition $[(p \rightarrow q) \wedge p] \rightarrow q$ of Example 4, which is always true, is a tautology. The proposition $\neg p \vee q$ of Example 3, which may be true or false, is a contingency.

DEFINITION Tautology, Contradiction, and Contingency
A proposition is a **tautology** if each entry in its column of the truth table is T, a **contradiction** if each entry is F, and a **contingency** if at least one entry is T and at least one entry is F.

EXAMPLE 5 Constructing a Truth Table Construct the truth table for $p \wedge \neg (p \vee q)$.

SOLUTION It is helpful to insert a third column labeled $p \vee q$ and a fourth labeled $\neg(p \vee q)$. We complete the first three columns. Then we use the third column to complete the fourth, and we use the first and fourth columns to complete the last column.

p	q	$p \vee q$	$\neg(p \vee q)$	$p \wedge \neg(p \vee q)$
T	T	T	F	F
T	F	T	F	F
F	T	T	F	F
F	F	F	T	F

Note that $p \wedge \neg(p \vee q)$ is a contradiction; it is always false, regardless of the truth values of p and q.

Matched Problem 5 Construct the truth table for $(p \rightarrow q) \wedge (p \wedge \neg q)$.

Explore and Discuss 1

The LOGIC menu on the TI-84 Plus contains the operators "and," "or," "xor," and "not." The truth table for $p \wedge q$, for example, can be displayed by entering all combinations of truth values for p and q in lists L_1 and L_2 (using 1 to represent T and 0 to represent F), and then, for L_3, entering "L_1 and L_2" (see Figure 1).

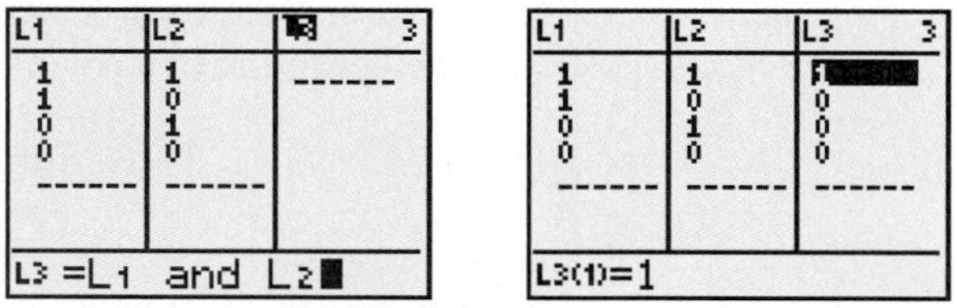

Figure 1

(A) Use a calculator to display the truth table for the exclusive or operator "xor."
(B) Show that the truth table for "xor" is identical to the truth table for the proposition $(p \vee q) \wedge \neg (p \wedge q)$.

Logical Implications and Equivalences

At the beginning of the section, we noted that the notation $[(p \rightarrow q) \wedge p] \Rightarrow q$ could be used to represent the form of a familiar logical deduction. The notion of such a deduction can be defined in terms of truth tables. Consider the truth tables for the propositions $(p \rightarrow q) \wedge p$ and q:

Table 1

p	q	$p \rightarrow q$	$(p \rightarrow q) \wedge p$
T	T	T	T
T	F	F	F
F	T	T	F
F	F	T	F

Whenever the proposition $(p \rightarrow q) \wedge p$ is true (in this case, in the first row only), the proposition q is also true. We say that $(p \rightarrow q) \wedge p$ *implies* q, or that $[(p \rightarrow q) \wedge p] \Rightarrow q$ is a *logical implication*.

DEFINITION Logical Implication
Consider the rows of the truth tables for the compound propositions P and Q. If whenever P is true, Q is also true, we say that P **logically implies** Q and write $P \Rightarrow Q$. We call $P \Rightarrow Q$ a **logical implication**.

CONCEPTUAL INSIGHT

If $P \Rightarrow Q$, then whenever P is true, so is Q. So to say that P implies Q is the same as saying that the proposition $P \rightarrow Q$ is a tautology. This gives us two methods of verifying that P implies Q: We can check the rows of the truth tables for P and Q as we did previously, or we can construct the truth table for $P \rightarrow Q$ to check that it is a tautology. Compare the truth tables in Table 1 with the truth table of Example 4 to decide which of the two methods you prefer.

EXAMPLE 6 Verifying a Logical Implication Show that $[(p \rightarrow q) \wedge \neg q] \Rightarrow \neg p$.

SOLUTION To construct the truth table for $(p \rightarrow q) \wedge \neg q$, it is helpful to insert a third column labeled $p \rightarrow q$ and a fourth labeled $\neg q$. We complete the first three columns. Then we use the second column to complete the fourth, and we use the third and fourth columns to complete the fifth column.

p	q	$p \rightarrow q$	$\neg q$	$(p \rightarrow q) \wedge \neg q$	$\neg p$
T	T	T	F	F	F
T	F	F	T	F	F
F	T	T	F	F	T
F	F	T	T	T	T

Now we compare the fifth column to the sixth: Whenever $(p \rightarrow q) \wedge \neg q$ is true (in the fourth row only), $\neg p$ is also true. We conclude that $[(p \rightarrow q) \wedge \neg q] \Rightarrow \neg p$.

Matched Problem 6 Show that $[(p \rightarrow q) \rightarrow p)] \Rightarrow (q \rightarrow p)$.

If the compound propositions P and Q have identical truth tables, then $P \Rightarrow Q$ and $Q \Rightarrow P$. In this case, we say that P and Q are *logically equivalent*.

DEFINITION Logical Equivalence
If the compound propositions P and Q have identical truth tables, we say that P and Q are **logically equivalent** and write $P \equiv Q$. We call $P \equiv Q$ a **logical equivalence**.

In Example 3, we noted that $p \rightarrow q$ and $\neg p \vee q$ have identical truth tables. Therefore, $p \rightarrow q \equiv \neg p \vee q$. This is formula (4) in Table 2, which lists several logical equivalences. The first three equivalences of Table 2 are obvious, and the last three equivalences are established in Example 7, Matched Problem 7, and Example 8, respectively.

CONCEPTUAL INSIGHT

Formulas (5) and (6) in Table 2 are known as **De Morgan's laws.** They may remind you of the way a negative sign is distributed over a binomial in algebra:

$$-(a + b) = (-a) + (-b)$$

But there is an important difference: Formula (5) has a disjunction ($\lor$) on the left side but a conjunction ($\land$) on the right side. Similarly, formula (6) has a conjunction on the left side but a disjunction on the right side.

Table 2 **Some Logical Equivalences**

$\neg(\neg p) \equiv p$	(1)
$p \lor q \equiv q \lor p$	(2)
$p \land q \equiv q \land p$	(3)
$p \rightarrow q \equiv \neg p \lor q$	(4)
$\neg(p \lor q) \equiv \neg p \land \neg q$	(5)
$\neg(p \land q) \equiv \neg p \lor \neg q$	(6)
$p \rightarrow q \equiv \neg q \rightarrow \neg p$	(7)

EXAMPLE 7 Verifying a Logical Equivalence Show that $\neg(p \lor q) \equiv \neg p \land \neg q$.

SOLUTION We construct truth tables for $\neg(p \lor q)$ and $\neg p \land \neg q$.

p	q	$p \lor q$	$\neg p \lor q$	$\neg p$	$\neg q$	$\neg p \land \neg q$
T	T	T	F	F	F	F
T	F	T	F	F	T	F
F	T	T	F	T	F	F
F	F	F	T	T	T	T

The fourth and seventh columns are identical, so $\neg(p \lor q) \equiv \neg p \land \neg q$.

Matched Problem 7 Show that $\neg(p \land q) \equiv \neg p \lor \neg q$.

One way to show that two propositions are logically equivalent is to check that their truth tables are identical (as in Example 7). Another way is to convert one to the other by a sequence of steps, where a known logical equivalence is used at each step to replace part or all of the proposition by an equivalent proposition. This procedure, analogous to simplifying an algebraic expression, is illustrated in Example 8 to show that **any conditional proposition is logically equivalent to its contrapositive**.

EXAMPLE 8 Any Conditional and Its Contrapositive Are Logically Equivalent
Show that $p \rightarrow q \equiv \neg q \rightarrow \neg p$.

SOLUTION Each step is justified by a reference to one of the formulas in Table 2:

$$\begin{aligned} p \rightarrow q &\equiv \neg p \lor q & \text{By (4)} \\ &\equiv q \lor \neg p & \text{By (2)} \\ &\equiv \neg(\neg q) \lor \neg p & \text{By (1)} \\ &\equiv \neg q \rightarrow \neg p & \text{By (4)} \end{aligned}$$

Therefore, $p \rightarrow q \equiv \neg q \rightarrow \neg p$.

Matched Problem 8 Use equivalences from Table 2 to show that $p \rightarrow q \equiv \neg(\neg q \land p)$.

Explore and Discuss 2 If a compound proposition contains three variables p, q, and r, then its truth table will have eight rows, one for each of the eight ways of assigning truth values to p, q, and r (TTT, TTF, TFT, TFF, FTT, FTF, FFT, FFF). Construct truth tables to verify the following logical implication and equivalences:

(A) $(p \rightarrow q) \land (q \rightarrow r) \Rightarrow (p \rightarrow r)$

(B) $p \land (q \lor r) \equiv (p \land q) \lor (p \land r)$

(C) $p \lor (q \land r) \equiv (p \lor q) \land (p \lor r)$

Exercises 7.1

Skills Warm-up Exercises

In Problems 1–6, refer to the footnote for the definitions of divisor, multiple, prime, even, and odd.*

1. List the positive integers that are divisors of 20.

2. List the positive integers that are divisors of 24.

3. List the positive multiples of 11 that are less than 60.

4. List the positive multiples of 9 that are less than 50.

5. List the primes between 20 and 30.

6. List the primes between 10 and 20.

7. Explain why the sum of any two odd integers is even.

8. Explain why the product of any two odd integers is odd.

In Problems 9–14, express each proposition as an English sentence and determine whether it is true or false, where p and q are the propositions

$$p: \text{"91 is prime"} \qquad q: \text{"91 is odd"}$$

9. $\neg q$

10. $p \lor q$

11. $p \land q$

12. $p \to q$

13. The converse of $p \to q$

14. The contrapositive of $p \to q$

In Problems 15–20, express each proposition as an English sentence and determine whether it is true or false, where r and s are the propositions

$$r: \text{"}2^9 + 2^8 + 2^7 > 987\text{"}$$
$$s: \text{"}9 \cdot 10^2 + 8 \cdot 10 + 7 = 987\text{"}$$

15. $r \to s$

16. $r \land s$

17. $r \lor s$

18. $\neg r$

19. The contrapositive of $r \to s$

20. The converse of $r \to s$

In Problems 21–28, describe each proposition as a negation, disjunction, conjunction, or conditional, and determine whether the proposition is true or false.

21. $-3 < 0$ or $-3 > 0$

22. $-3 < 0$ and $-3 > 0$

23. If $-3 < 0$, then $(-3)^2 < 0$

24. -3 is not greater than 0

25. 11 is not prime

26. 9 is even or 9 is prime

27. 7 is odd and 7 is prime

28. If 4 is even, then 4 is prime

In Problems 29–34, state the converse and the contrapositive of the given proposition.

29. If triangle ABC is equilateral, then triangle ABC is equiangular.

30. If triangle ABC is isosceles, then the base angles of triangle ABC are congruent.

31. If $f(x)$ is a linear function with positive slope, then $f(x)$ is an increasing function.

32. If $g(x)$ is a quadratic function, then $g(x)$ is a function that is neither increasing nor decreasing.

33. If n is an integer that is a multiple of 8, then n is an integer that is a multiple of 2 and a multiple of 4.

34. If n is an integer that is a multiple of 6, then n is an integer that is a multiple of 2 and a multiple of 3.

In Problems 35–52, construct a truth table for the proposition and determine whether the proposition is a contingency, tautology, or contradiction.

35. $\neg p \land q$

36. $p \lor \neg q$

37. $\neg p \to q$

38. $p \to \neg q$

39. $q \land (p \lor q)$

40. $q \lor (p \land q)$

41. $p \lor (p \to q)$

42. $p \land (p \to q)$

43. $p \to (p \land q)$

44. $p \to (p \lor q)$

45. $(p \to q) \to \neg p$

46. $(p \to q) \to \neg q$

47. $\neg p \to (p \lor q)$

48. $\neg p \to (p \land q)$

49. $q \to (\neg p \land q)$

50. $q \to (p \lor \neg q)$

51. $(\neg p \land q) \land (q \to p)$

52. $(p \to \neg q) \land (p \land q)$

In Problems 53–58, construct a truth table to verify each implication.

53. $p \Rightarrow p \lor q$

54. $\neg p \Rightarrow p \to q$

55. $\neg p \land q \Rightarrow p \lor q$

56. $p \land q \Rightarrow p \to q$

57. $\neg p \to (q \land \neg q) \Rightarrow p$

58. $(p \land \neg p) \Rightarrow q$

In Problems 59–64, construct a truth table to verify each equivalence.

59. $\neg p \to (p \lor q) \equiv p \lor q$

60. $q \to (\neg p \land q) \equiv \neg(p \land q)$

61. $q \land (p \lor q) \equiv q \lor (p \land q)$

62. $p \land (p \to q) \equiv p \land q$

63. $p \lor (p \to q) \equiv p \to (p \lor q)$

64. $p \to (p \land q) \equiv p \to q$

In Problems 65–68, verify each equivalence using formulas from Table 2.

65. $p \to \neg q \equiv \neg(p \land q)$

66. $\neg p \to q \equiv p \lor q$

67. $\neg(p \to q) \equiv p \land \neg q$

68. $\neg(\neg p \to \neg q) \equiv q \land \neg p$

✎ **69.** If the conditional proposition p is a contingency, is $\neg p$ also a contingency? Explain.

✎ **70.** If the conditional proposition p is a contradiction, is $\neg p$ also a contradiction? Explain.

---

*An integer d is ***a divisor*** of an integer n (and n is a ***multiple*** of d) if $n = kd$ for some integer k. An integer n is ***even*** if 2 is a divisor of n; otherwise, n is odd. An integer $p > 1$ is ***prime*** if its only positive divisors are 1 and p.

71. Can a conditional proposition be false if its contrapositive is true? Explain.

72. Can a conditional proposition be false if its converse is true? Explain.

Answers to Matched Problems

1. (A) "14^2 is not less than 200"; false

(B) "23^2 is not less than 500"; true

(C) "14^2 is less than 200 or 23^2 is less than 500"; true

(D) "14^2 is less than 200 and 23^2 is less than 500"; false

(E) "If 14^2 is less than 200, then 23^2 is less than 500"; false

2. (A) "If $5^2 + 12^2 = 13^2$, then $7^2 + 24^2 = 25^2$"; true

(B) "If $7^2 + 24^2 = 25^2$, then $5^2 + 12^2 = 13^2$"; true

(C) "If $7^2 + 24^2 \neq 25^2$, then $5^2 + 12^2 \neq 13^2$"; true

3.

p	q	$p \wedge \neg q$
T	T	F
T	F	T
F	T	F
F	F	F

4.

p	q	$[(p \rightarrow q) \wedge \neg q] \rightarrow \neg p$
T	T	T
T	F	T
F	T	T
F	F	T

5.

p	q	$(p \rightarrow q) \wedge (p \wedge \neg q)$
T	T	F
T	F	F
F	T	F
F	F	F

6.

p	q	$(p \rightarrow q) \rightarrow p$	$q \rightarrow p$
T	T	T	T
T	F	T	T
F	T	F	F
F	F	F	T

Whenever $(p \rightarrow q) \rightarrow p$ is true, so is $q \rightarrow p$, so $[(p \rightarrow q) \rightarrow p)] \Rightarrow (q \rightarrow p)$.

7.

p	q	$\neg(p \wedge q)$	$\neg p \vee \neg q$
T	T	F	F
T	F	T	T
F	T	T	T
F	F	T	T

The third and fourth columns are identical, so $\neg(p \wedge q) \equiv \neg p \vee \neg q$.

8. $p \rightarrow q \equiv \neg p \vee q$ By (4)

$\equiv q \vee \neg p$ By (2)

$\equiv \neg(\neg q) \vee \neg p$ By (1)

$\equiv \neg(\neg q \wedge p)$ By (6)

7.2   Sets

- Set Properties and Set Notation
- Venn Diagrams and Set Operations
- Application

In this section, we review a few key ideas about sets. Set concepts and notation help us talk about certain mathematical ideas with greater clarity and precision, and they are indispensable to a clear understanding of probability.

Set Properties and Set Notation

We can think of a **set** as any collection of objects specified in such a way that we can tell whether any given object is or is not in the collection. Capital letters, such as A, B, and C, are often used to designate particular sets. Each object in a set is called a **member**, or **element**, of the set. Symbolically,

$a \in A$ means   "a is an element of set A"

$a \notin A$ means   "a is not an element of set A"

A set without any elements is called the **empty**, or **null**, **set**. For example, the set of all people over 20 feet tall is an empty set. Symbolically,

$\varnothing$ denotes the empty set

A set is described either by listing all its elements between braces $\{\ \}$ (the listing method) or by enclosing a rule within braces that determines the elements of the set (the rule method). So if $P(x)$ is a statement about x, then

$S = \{x \mid P(x)\}$ means   "S is the set of all x such that $P(x)$ is true"

Recall that the vertical bar within the braces is read "such that." The following example illustrates the rule and listing methods of representing sets.

EXAMPLE 1 Representing Sets

| Rule method | Listing method |

$$\{x \mid x \text{ is a weekend day}\} = \{\text{Saturday, Sunday}\}$$
$$\{x \mid x^2 = 4\} = \{-2, 2\}$$
$$\{x \mid x \text{ is an odd positive counting number}\} = \{1, 3, 5, \ldots\}$$

The three dots $(\ldots)$ in the last set of Example 1 indicate that the pattern established by the first three entries continues indefinitely. The first two sets in Example 1 are **finite sets** (the elements can be counted, and there is an end); the last set is an **infinite set** (there is no end when counting the elements). When listing the elements in a set, we do not list an element more than once, and the order in which the elements are listed does not matter.

Matched Problem 1 Let G be the set of all numbers such that $x^2 = 9$.

(A) Denote G by the rule method.

(B) Denote G by the listing method.

(C) Indicate whether the following are true or false: $3 \in G, 9 \notin G$.

CONCEPTUAL INSIGHT

To conclude that $\varnothing$ is a subset of any set A, we must show that the conditional proposition "if $x \in \varnothing$, then $x \in A$" is true. But the set $\varnothing$ has no element, so the hypothesis of the conditional is false, and therefore the conditional itself is vacuously true (see Section 7.1). We correctly conclude that $\varnothing$ is a subset of every set. Note, however, that $\varnothing$ is *not* an *element* of every set.

If each element of a set A is also an element of set B, we say that A is a **subset** of B. For example, the set of all women students in a class is a subset of the whole class. Note that the definition implies that every set is a subset of itself. If set A and set B have exactly the same elements, then the two sets are said to be **equal**. Symbolically,

$A \subset B$	means	"A is a subset of B"
$A = B$	means	"A and B have exactly the same elements"
$A \not\subset B$	means	"A is not a subset of B"
$A \neq B$	means	"A and B do not have exactly the same elements"

From the definition of subset, we conclude that

$\varnothing$ **is a subset of every set, and**

if $A \subset B$ **and** $B \subset A$**, then** $A = B$**.**

EXAMPLE 2 Set Notation If $A = \{-3, -1, 1, 3\}$, $B = \{3, -3, 1, -1\}$, and $C = \{-3, -2, -1, 0, 1, 2, 3\}$, then each of the following statements is true:

$$A = B \quad A \subset C \quad A \subset B$$
$$C \neq A \quad C \not\subset A \quad B \subset A$$
$$\varnothing \subset A \quad \varnothing \subset C \quad \varnothing \notin A$$

Matched Problem 2 Given $A = \{0, 2, 4, 6\}$, $B = \{0, 1, 2, 3, 4, 5, 6\}$, and $C = \{2, 6, 0, 4\}$, indicate whether the following relationships are true (T) or false (F):

(A) $A \subset B$ (B) $A \subset C$ (C) $A = C$

(D) $C \subset B$ (E) $B \not\subset A$ (F) $\varnothing \subset B$

EXAMPLE 3 Subsets List all subsets of the set $\{a, b, c\}$.

SOLUTION

$$\{a, b, c\}, \quad \{a, b\}, \quad \{a, c\}, \quad \{b, c\}, \quad \{a\}, \quad \{b\}, \quad \{c\}, \quad \varnothing$$

Matched Problem 3 List all subsets of the set $\{1, 2\}$.

Which of the following statements are true?

(A) $\varnothing \subset \varnothing$ (B) $\varnothing \in \varnothing$

(C) $\varnothing = \{0\}$ (D) $\varnothing \subset \{0\}$

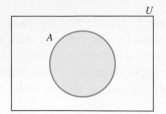

Figure 1 **A is the shaded region**

Venn Diagrams and Set Operations

If the set U is the set of all rental units in a city, and A is the set of all rental units within one mile of the college campus, then it is natural to picture A by Figure 1, called a *Venn diagram.*

The circle is an imaginary boundary that separates the elements of A (inside the circle) from the elements of U that are not in A (outside the circle).

The Venn diagram of Figure 1 can be used to picture any set A and *universal set U.* The **universal set** is the set of all elements under consideration. It is customary to place the label A near the circle itself, but the elements of U are imagined to be arranged so that the elements of A lie inside the circle, and the elements of U that are not in A lie outside the circle.

Let U be the set of all 26 lowercase letters of the English alphabet, let $A = \{a, b, c, d, e\}$, and let $B = \{d, e, f, g, h, i\}$. Sets A and B are pictured in the Venn diagram of Figure 2. Note that the elements d and e belong to both A and B, and that the elements a, b, c belong to A but not to B. Of course sets may have hundreds or thousands of elements, so we seldom write the names of the elements as in Figure 2; instead, we represent the situation by the Venn diagram of Figure 3, or by the Venn diagram of Figure 4.

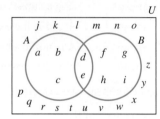

Figure 2

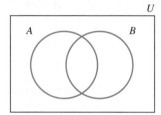

Figure 3

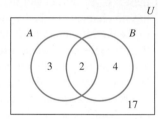

Figure 4

The numbers in Figure 4 are interpreted as follows: There are 2 elements in the overlap of A and B, 3 elements that belong to A but not B, 4 elements that belong to B but not A, and 17 elements that belong to neither A nor B. Using such a diagram, it is easy to calculate the number of elements in A (just add 2 and 3) or the number of elements in B (just add 2 and 4).

The **union** of sets A and B, denoted by $A \cup B$, is the set of elements formed by combining all the elements of A and all the elements of B into one set.

> ### DEFINITION   Union
> $$A \cup B = \{x \,|\, x \in A \text{ **or** } x \in B\}$$
>
> Here we use the word **or** in the way it is always used in mathematics; that is, x may be an element of set A or set B or both.

The union of two sets A and B is pictured in the Venn diagram of Figure 5. Note that
$$A \subset A \cup B \qquad \text{and} \qquad B \subset A \cup B$$

The **intersection** of sets A and B, denoted by $A \cap B$, is the set of elements in set A that are also in set B.

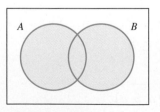

Figure 5 **A ∪ B is the shaded region.**

> ### DEFINITION   Intersection
> $$A \cap B = \{x \,|\, x \in A \text{ **and** } x \in B\}$$

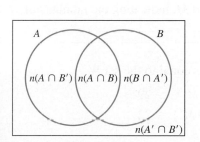

Figure 6 **A ∩ B is the shaded region.**

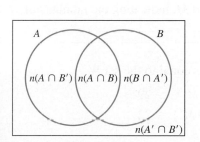

Figure 7 **A ∩ B = ∅; A and B are disjoint.**

The intersection of two sets A and B is pictured in the Venn diagram of Figure 6. Note that

$$A \cap B \subset A \quad \text{and} \quad A \cap B \subset B$$

If $A \cap B = \varnothing$, then the sets A and B are said to be **disjoint**, as shown in Figure 7. We now define an operation on sets called the *complement*. The **complement** of A (relative to a universal set U), denoted by A', is the set of elements in U that are not in A (Fig. 8).

DEFINITION Complement

$$A' = \{x \in U \mid x \notin A\}$$

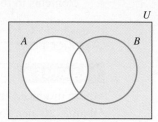

Figure 8 **The shaded region is A', the complement of A.**

EXAMPLE 4 Union, Intersection, and Complement   If $A = \{3, 6, 9\}$, $B = \{3, 4, 5, 6, 7\}$, $C = \{4, 5, 7\}$, and $U = \{1, 2, 3, 4, 5, 6, 7, 8, 9\}$, then

$$A \cup B = \{3, 4, 5, 6, 7, 9\}$$
$$A \cap B = \{3, 6\}$$
$$A \cap C = \varnothing \qquad \textit{A and C are disjoint.}$$
$$B' = \{1, 2, 8, 9\}$$

Matched Problem 4   If $R = \{1, 2, 3, 4\}$, $S = \{1, 3, 5, 7\}$, $T = \{2, 4\}$,   and $U = \{1, 2, 3, 4, 5, 6, 7, 8, 9\}$, find

(A) $R \cup S$ (B) $R \cap S$ (C) $S \cap T$ (D) S'

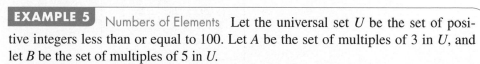

Figure 9

The **number of elements** in a set A is denoted by $n(A)$. So if A and B are sets, then the numbers that are often shown in a Venn diagram, as in Figure 4, are $n(A \cap B')$, $n(A \cap B)$, $n(B \cap A')$, and $n(A' \cap B')$ (see Fig. 9).

EXAMPLE 5 Numbers of Elements   Let the universal set U be the set of positive integers less than or equal to 100. Let A be the set of multiples of 3 in U, and let B be the set of multiples of 5 in U.

(A) Find $n(A \cap B)$, $n(A \cap B')$, $n(B \cap A')$, and $n(A' \cap B')$.

(B) Draw a Venn diagram with circles labeled A and B, indicating the numbers of elements in the subsets of part (A).

SOLUTION

(A) $A = \{3, 6, 9, \ldots, 99\}$, so $n(A) = 33$.

$B = \{5, 10, 15, \ldots, 100\}$, so $n(B) = 20$.

$A \cap B = \{15, 30, 45, \ldots, 90\}$, so $n(A \cap B) = 6$.

$n(A \cap B') = 33 - 6 = 27$

$n(B \cap A') = 20 - 6 = 14$

$n(A' \cap B') = 100 - (6 + 27 + 14) = 53$

(B)

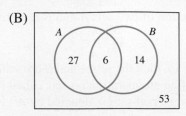

Matched Problem 5⟩ Let the universal set U be the set of positive integers less than or equal to 100. Let A be the set of multiples of 3 in U, and let B be the set of multiples of 7 in U.

(A) Find $n(A \cap B)$, $n(A \cap B')$, $n(B \cap A')$, and $n(A' \cap B')$.

(B) Draw a Venn diagram with circles labeled A and B, indicating the numbers of elements in the subsets of part (A).

Application

EXAMPLE 6   Exit Polling   In the 2012 presidential election, an exit poll of 100 voters produced the results in the table (23 men voted for Obama, 29 women for Obama, 26 men for Romney, and 22 women for Romney).

	Men	Women
Obama	23	29
Romney	26	22

Let the universal set U be the set of 100 voters, O the set of voters for Obama, R the set of voters for Romney, M the set of male voters, and W the set of female voters.

(A) Find $n(O \cap M)$, $n(O \cap M')$, $n(M \cap O')$, and $n(O' \cap M')$.

(B) Draw a Venn diagram with circles labeled O and M, indicating the numbers of elements in the subsets of part (A).

SOLUTION

(A) The set O' is equal to the set R, and the set M' is equal to the set W.

$$n(O \cap M) = 23, n(O \cap M') = 29$$
$$n(M \cap O') = 26, n(O' \cap M') = 22$$

(B)

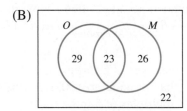

Matched Problem 6⟩ Refer to Example 6.

(A) Find $n(R \cap W)$, $n(R \cap W')$, $n(W \cap R')$, and $n(R' \cap W')$.

(B) Draw a Venn diagram with circles labeled R and W, indicating the numbers of elements in the subsets of part (A).

Explore and Discuss 2   In Example 6, find the number of voters in the set $(O \cup M) \cap M'$. Describe this set verbally and with a Venn diagram.

Exercises 7.2

Skills Warm-up Exercises

In Problems 1–6, answer yes or no. (If necessary, review Section A.1).

1. Is the set of even integers a subset of the set of odd integers?

2. Is the set of rational numbers a subset of the set of integers?

3. Is the set of integers the intersection of the set of even integers and the set of odd integers?

4. Is the set of integers the union of the set of even integers and the set of odd integers?

5. If the universal set is the set of integers, is the set of positive integers the complement of the set of negative integers?

6. If the universal set is the set of integers, is the set of even integers the complement of the set of odd integers?

In Problems 7–14, indicate true (T) or false (F).

7. $\{1, 2\} \subset \{2, 1\}$
8. $\{3, 2, 1\} \subset \{1, 2, 3, 4\}$
9. $\{5, 10\} = \{10, 5\}$
10. $1 \in \{10, 11\}$
11. $\{0\} \in \{0, \{0\}\}$
12. $\{0, 6\} = \{6\}$
13. $8 \in \{1, 2, 4\}$
14. $\varnothing \subset \{1, 2, 3\}$

In Problems 15–28, write the resulting set using the listing method.

15. $\{1, 2, 3\} \cap \{2, 3, 4\}$
16. $\{1, 2, 4\} \cup \{4, 8, 16\}$
17. $\{1, 2, 3\} \cup \{2, 3, 4\}$
18. $\{1, 2, 4\} \cap \{4, 8, 16\}$
19. $\{1, 4, 7\} \cup \{10, 13\}$
20. $\{-3, -1\} \cap \{1, 3\}$
21. $\{1, 4, 7\} \cap \{10, 13\}$
22. $\{-3, -1,\} \cup \{1, 3\}$
23. $\{x \mid x^2 = 25\}$
24. $\{x \mid x^2 = 36\}$
25. $\{x \mid x^3 = -27\}$
26. $\{x \mid x^4 = 16\}$
27. $\{x \mid x \text{ is an odd number between 1 and 9, inclusive}\}$
28. $\{x \mid x \text{ is a month starting with } M\}$

29. For $U = \{1, 2, 3, 4, 5\}$ and $A = \{2, 3, 4\}$, find A'.

30. For $U = \{7, 8, 9, 10, 11\}$ and $A = \{7, 11\}$, find A'.

In Problems 31–44, refer to the Venn diagram below and find the indicated number of elements.

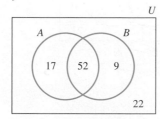

31. $n(U)$
32. $n(A)$
33. $n(B)$
34. $n(A \cap B)$
35. $n(A \cup B)$
36. $n(B')$

37. $n(A')$
38. $n(A \cap B')$
39. $n(B \cap A')$
40. $n((A \cap B)')$
41. $n((A \cup B)')$
42. $n(A' \cap B')$
43. $n(A \cup A')$
44. $n(A \cap A')$

45. If $R = \{1, 2, 3, 4\}$ and $T = \{2, 4, 6\}$, find

   (A) $\{x \mid x \in R \text{ or } x \in T\}$

   (B) $R \cup T$

46. If $R = \{1, 3, 4\}$ and $T = \{2, 4, 6\}$, find

   (A) $\{x \mid x \in R \text{ and } x \in T\}$

   (B) $R \cap T$

47. For $P = \{1, 2, 3, 4\}$, $Q = \{2, 4, 6\}$, and $R = \{3, 4, 5, 6\}$, find $P \cup (Q \cap R)$.

48. For P, Q, and R in Problem 47, find $P \cap (Q \cup R)$.

In Problems 49–58, determine whether the given set is finite or infinite. Consider the set N of positive integers to be the universal set, and let

$$H = \{n \in N \mid \quad n > 100\}$$
$$T = \{n \in N \mid \quad n < 1{,}000\}$$
$$E = \{n \in N \mid \quad n \text{ is even}\}$$
$$P = \{n \in N \mid \quad n \text{ is prime}\}$$

49. H'
50. T'
51. $E \cup P$
52. $H \cap T$
53. $E \cap P$
54. $H \cup T$
55. E'
56. P'
57. $H' \cup T$
58. $(E \cup P)'$

In Problems 59–62, are the given sets disjoint? For the definitions of H, T, E, and P, refer to the instructions for Problems 49–58.

59. H' and T'
60. E and P
61. P' and H
62. E and E'

In Problems 63–72, discuss the validity of each statement. Venn diagrams may be helpful. If the statement is true, explain why. If not, give a counterexample.

63. If $A \subset B$, then $A \cap B = A$.

64. If $A \subset B$, then $A \cup B = A$.

65. If $A \cup B = A$, then $A \subset B$.

66. If $A \cap B = A$, then $A \subset B$.

67. If $A \cap B = \varnothing$, then $A = \varnothing$.

68. If $A = \varnothing$, then $A \cap B = \varnothing$.

69. If $A \subset B$, then $A' \subset B'$.

70. If $A \subset B$, then $B' \subset A'$.

71. The empty set is an element of every set.

72. The empty set is a subset of the empty set.

73. How many subsets does each of the following sets contain?

(A) $\{a\}$

(B) $\{a, b\}$

(C) $\{a, b, c\}$

(D) $\{a, b, c, d\}$

74. Let A be a set that contains exactly n elements. Find a formula in terms of n for the number of subsets of A.

Applications

Enrollments *In Problems 75–88, find the indicated number of elements by referring to the following table of enrollments in a finite mathematics class:*

	Freshmen	Sophomores
Arts & Sciences	19	14
Business	66	21

Let the universal set U be the set of all 120 students in the class, A the set of students from the College of Arts & Sciences, B the set of students from the College of Business, F the set of freshmen, and S the set of sophomores.

75. $n(F)$ **76.** $n(S)$

77. $n(A)$ **78.** $n(B)$

79. $n(A \cap S)$ **80.** $n(A \cap F)$

81. $n(B \cap F)$ **82.** $n(B \cap S)$

83. $n(A \cup S)$ **84.** $n(A \cup F)$

85. $n(B \cup F)$ **86.** $n(B \cup S)$

87. $n(A \cap B)$ **88.** $n(F \cup S)$

89. Committee selection. A company president and three vice-presidents, denoted by the set $\{P, V_1, V_2, V_3\}$, wish to select a committee of 2 people from among themselves. How many ways can this committee be formed? That is, how many 2-person subsets can be formed from a set of 4 people?

90. Voting coalition. The company's leaders in Problem 89 decide for or against certain measures as follows: The president has 2 votes and each vice-president has 1 vote. Three favorable votes are needed to pass a measure. List all minimal winning coalitions; that is, list all subsets of $\{P, V_1, V_2, V_3\}$ that represent exactly 3 votes.

Blood types. *When receiving a blood transfusion, a recipient must have all the antigens of the donor. A person may have one or more of the three antigens A, B, and Rh, or none at all. Eight blood types are possible, as indicated in the following Venn diagram, where U is the set of all people under consideration:*

An $A-$ person has A antigens but no B or Rh antigens; an $O+$ person has Rh but neither A nor B; an $AB-$ person has A and B but no Rh; and so on.

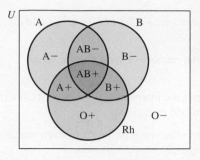

In Problems 91–98, use the Venn diagram to indicate which of the eight blood types are included in each set.

91. $A \cap Rh$ **92.** $A \cap B$

93. $A \cup Rh$ **94.** $A \cup B$

95. $(A \cup B)'$ **96.** $(A \cup B \cup Rh)'$

97. $A' \cap B$ **98.** $Rh' \cap A$

Answers to Matched Problems

1. (A) $\{x \mid x^2 = 9\}$ (B) $\{-3, 3\}$ (C) True; true

2. All are true.

3. $\{1, 2\}, \{1\}, \{2\}, \varnothing$

4. (A) $\{1, 2, 3, 4, 5, 7\}$ (B) $\{1, 3\}$

(C) $\varnothing$ (D) $\{2, 4, 6, 8, 9\}$

5. (A) $n(A \cap B) = 4$; $n(A \cap B') = 29$; $n(B \cap A') = 10$; $n(A' \cap B') = 57$

(B)

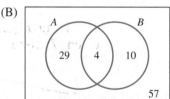

6. (A) $n(R \cap W) = 22$; $n(R \cap W') = 26$; $n(W \cap R') = 29$; $n(R' \cap W') = 23$

(B)

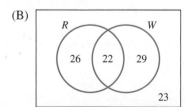

7.3 Basic Counting Principles

- Addition Principle
- Venn Diagrams
- Multiplication Principle

Addition Principle

If the enrollment in a college chemistry class consists of 13 males and 15 females, then there are a total of 28 students enrolled in the class. This is a simple example of a **counting technique**, a method for determining the number of elements in a set without actually enumerating the elements one by one. Set operations play a fundamental role in many counting techniques. For example, if M is the set of male students in the chemistry class and F is the set of female students, then the *union* of sets M and F, denoted $M \cup F$, is the set of all students in the class. Since these sets have no elements in common, the *intersection* of sets M and F, denoted $M \cap F$, is the *empty set* $\varnothing$; we then say that M and F are *disjoint sets*. The total number of students enrolled in the class is the number of elements in $M \cup F$, denoted by $n(M \cup F)$ and given by

$$n(M \cup F) = n(M) + n(F)$$
$$= 13 + 15 = 28$$

In this example, the number of elements in the union of sets M and F is the sum of the number of elements in M and in F. However, this does not work for all pairs of sets. To see why, consider another example. Suppose that the enrollment in a mathematics class consists of 22 math majors and 16 physics majors, and that 7 of these students have majors in both subjects. If M represents the set of math majors and P represents the set of physics majors, then $M \cap P$ represents the set of double majors. It is tempting to proceed as before and conclude that there are $22 + 16 = 38$ students in the class, but this is incorrect. We have counted the double majors twice, once as math majors and again as physics majors. To correct for this double counting, we subtract the number of double majors from this sum. Thus, the total number of students enrolled in this class is given by

$$n(M \cup P) = n(M) + n(P) - n(M \cap P) \tag{1}$$
$$= 22 + 16 - 7 = 31$$

Equation (1) illustrates the *addition principle* for counting the elements in the union of two sets.

THEOREM 1 Addition Principle (for Counting)

For any two sets A and B,

$$n(A \cup B) = n(A) + n(B) - n(A \cap B) \tag{2}$$

Note that if A and B are disjoint, then $n(A \cap B) = 0$, and equation (2) becomes $n(A \cup B) = n(A) + n(B)$.

EXAMPLE 1 Employee Benefits According to a survey of business firms in a certain city, 750 firms offer their employees health insurance, 640 offer dental insurance, and 280 offer health insurance and dental insurance. How many firms offer their employees health insurance or dental insurance?

SOLUTION If H is the set of firms that offer their employees health insurance and D is the set that offer dental insurance, then

$H \cap D =$ set of firms that offer health insurance **and** dental insurance

$H \cup D =$ set of firms that offer health insurance **or** dental insurance

So

$$n(H) = 750 \qquad n(D) = 640 \qquad n(H \cap D) = 280$$

and

$$n(H \cup D) = n(H) + n(D) - n(H \cap D)$$
$$= 750 + 640 - 280 = 1{,}110$$

Therefore, 1,110 firms offer their employees health insurance or dental insurance.

Matched Problem 1 The survey in Example 1 also indicated that 345 firms offer their employees group life insurance, 285 offer long-term disability insurance, and 115 offer group life insurance and long-term disability insurance. How many firms offer their employees group life insurance or long-term disability insurance?

EXAMPLE 2 Market Research A city has two daily newspapers, the *Sentinel* and the *Journal*. The following information was obtained from a survey of 100 city residents: 35 people subscribe to the *Sentinel*, 60 subscribe to the *Journal*, and 20 subscribe to both newspapers.

(A) How many people subscribe to the *Sentinel* but not to the *Journal?*

(B) How many subscribe to the *Journal* but not to the *Sentinel?*

(C) How many do not subscribe to either paper?

(D) Organize this information in a table.

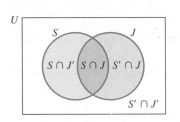

Figure 1 **Venn diagram for the newspaper survey**

SOLUTION Let U be the group of people surveyed. Let S be the set of people who subscribe to the *Sentinel,* and let J be the set of people who subscribe to the *Journal.* Since U contains all the elements under consideration, it is the *universal set* for this problem. The *complement* of S, denoted by S', is the set of people in the survey group U who do not subscribe to the *Sentinel.* Similarly, J' is the set of people in the group who do not subscribe to the *Journal.* Using the sets S and J, their complements, and set intersection, we can divide U into the four disjoint subsets defined below and illustrated as a Venn diagram in Figure 1.

$$S \cap J = \text{set of people who subscribe to both papers}$$
$$S \cap J' = \text{set of people who subscribe to the } \textit{Sentinel} \text{ but not the } \textit{Journal}$$
$$S' \cap J = \text{set of people who subscribe to the } \textit{Journal} \text{ but not the } \textit{Sentinel}$$
$$S' \cap J' = \text{set of people who do not subscribe to either paper}$$

The given survey information can be expressed in terms of set notation as

$$n(U) = 100 \qquad n(S) = 35 \qquad n(J) = 60 \qquad n(S \cap J) = 20$$

We can use this information and a Venn diagram to answer parts (A)–(C). First, we place 20 in $S \cap J$ in the diagram (see Fig. 2). As we proceed with parts (A) through (C), we add each answer to the diagram.

(A) Since 35 people subscribe to the *Sentinel* and 20 subscribe to both papers, the number of people who subscribe to the *Sentinel* but not to the *Journal* is

$$n(S \cap J') = 35 - 20 = 15$$

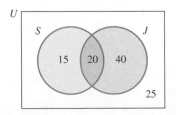

Figure 2 **Newspaper survey results**

(B) In a similar manner, the number of people who subscribe to the *Journal* but not to the *Sentinel* is

$$n(S' \cap J) = 60 - 20 = 40$$

(C) The total number of newspaper subscribers is $20 + 15 + 40 = 75$. So the number of people who do not subscribe to either paper is

$$n(S' \cap J') = 100 - 75 = 25$$

CONCEPTUAL INSIGHT

Carefully compare Table 1 and Figure 2 of Example 2, and note that we do *not* include any of the totals in Table 1 in the Venn diagram. Instead, the numbers in the Venn diagram give the numbers of elements in the four disjoint sets $S \cap J$, $S \cap J'$, $S' \cap J$, and $S' \cap J'$. From Figure 2, it is easy to construct Table 1 (the totals are easily calculated from the Venn diagram). And from Table 1, it is easy to construct the Venn diagram of Figure 2 (simply disregard the totals).

(D) Venn diagrams are useful tools for determining the number of elements in the various sets of a survey, but often the results must be presented in the form of a table, rather than a diagram. Table 1 contains the information of Figure 2 and also includes totals that give the numbers of elements in the sets S, S', J, J', and U.

Table 1

		Journal		
		Subscriber, *J*	Nonsubscriber, *J'*	Totals
Sentinel	Subscriber, *S*	20	15	35
	Nonsubscriber, *S'*	40	25	65
	Totals	60	40	100

Matched Problem 2 A small town has two radio stations: an AM station and an FM station. A survey of 100 town residents produced the following results: In the last 30 days, 65 people have listened to the AM station, 45 have listened to the FM station, and 30 have listened to both stations.

(A) During this 30-day period, how many people in the survey have listened to the AM station but not to the FM station?

(B) How many have listened to the FM station but not to the AM station?

(C) How many have not listened to either station?

(D) Organize this information in a table.

Explore and Discuss 1 Let A, B, and C be three sets. Use a Venn diagram (Fig. 3) to explain the following equation:

$$n(A \cup B \cup C) = n(A) + n(B) + n(C) - n(A \cap B) - n(A \cap C)$$
$$- n(B \cap C) + n(A \cap B \cap C)$$

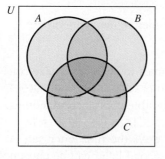

Figure 3

Multiplication Principle

As we have just seen, if the elements of a set are determined by the union operation, addition and subtraction are used to count the number of elements in the set. Now we want to consider sets whose elements are determined by a sequence of operations. We will see that multiplication is used to count the number of elements in sets formed this way.

EXAMPLE 3 Product Mix A retail store stocks windbreaker jackets in small, medium, large, and extra large. All are available in blue or red. What are the combined choices, and how many combined choices are there?

SOLUTION To solve the problem, we use a tree diagram:

SIZE CHOICES (OUTCOMES)	COLOR CHOICES (OUTCOMES)	COMBINED CHOICES (OUTCOMES)
S	B	(S, B)
	R	(S, R)
M	B	(M, B)
	R	(M, R)
L	B	(L, B)
	R	(L, R)
XL	B	(XL, B)
	R	(XL, R)

Start

There are 8 possible combined choices (outcomes). There are 4 ways that a size can be chosen and 2 ways that a color can be chosen. The first element in the ordered pair represents a size choice, and the second element represents a color choice.

Matched Problem 3 ⌉ A company offers its employees health plans from three different companies: *R*, *S*, and *T*. Each company offers two levels of coverage, *A* and *B*, with one level requiring additional employee contributions. What are the combined choices, and how many choices are there? Solve using a tree diagram.

Suppose that you asked, "From the 26 letters in the alphabet, how many ways can 3 letters appear on a license plate if no letter is repeated?" To try to count the possibilities using a tree diagram would be extremely tedious. The following **multiplication principle** will enable us to solve this problem easily. In addition, it forms the basis for developing other counting devices in the next section.

THEOREM 2 Multiplication Principle (for Counting)

1. If two operations O_1 and O_2 are performed in order, with N_1 possible outcomes for the first operation and N_2 possible outcomes for the second operation, then there are

$$N_1 \cdot N_2$$

possible combined outcomes of the first operation followed by the second.

2. In general, if *n* operations $O_1, O_2, \ldots, O_n$ are performed in order, with possible number of outcomes $N_1, N_2, \ldots, N_n$, respectively, then there are

$$N_1 \cdot N_2 \cdot \cdots \cdot N_n$$

possible combined outcomes of the operations performed in the given order.

In Example 3, we see that there are 4 possible outcomes in choosing a size (the first operation) and 2 possible outcomes in choosing a color (the second operation). So by the multiplication principle, there are $4 \cdot 2 = 8$ possible combined outcomes. Use the multiplication principle to solve Matched Problem 3. [*Answer:* $3 \cdot 2 = 6$]

To answer the license plate question: There are 26 ways the first letter can be chosen; after a first letter is chosen, there are 25 ways a second letter can be chosen; and after 2 letters are chosen, there are 24 ways a third letter can be chosen. So, using the multiplication principle, there are $26 \cdot 25 \cdot 24 = 15,600$ possible 3-letter license plates if no letter is repeated.

EXAMPLE 4 Computer-Assisted Testing Many colleges and universities use computer-assisted testing. Suppose that a screening test is to consist of 5 questions, and a computer stores 5 comparable questions for the first test question, 8 for the second, 6 for the third, 5 for the fourth, and 10 for the fifth. How many different 5-question tests can the computer select? (Two tests are considered different if they differ in one or more questions.)

SOLUTION

O_1:	Selecting the first question	N_1:	5 ways
O_2:	Selecting the second question	N_2:	8 ways
O_3:	Selecting the third question	N_3:	6 ways
O_4:	Selecting the fourth question	N_4:	5 ways
O_5:	Selecting the fifth question	N_5:	10 ways

The computer can generate

$$5 \cdot 8 \cdot 6 \cdot 5 \cdot 10 = 12,000 \text{ different tests}$$

Matched Problem 4) Each question on a multiple-choice test has 5 choices. If there are 5 such questions on a test, how many different responses are possible if only 1 choice is marked for each question?

EXAMPLE 5 Code Words How many 3-letter code words are possible using the first 8 letters of the alphabet if

(A) No letter can be repeated?

(B) Letters can be repeated?

(C) Adjacent letters cannot be alike?

SOLUTION To form 3-letter code words from the 8 letters available, we select a letter for the first position, one for the second position, and one for the third position. Altogether, there are three operations.

(A) No letter can be repeated:

O_1:	Selecting the first letter	N_1:	8 ways	
O_2:	Selecting the second letter	N_2:	7 ways	Since 1 letter has been used
O_3:	Selecting the third letter	N_3:	6 ways	Since 2 letters have been used

There are

$$8 \cdot 7 \cdot 6 = 336 \text{ possible code words} \quad \text{Possible combined operations}$$

(B) Letters can be repeated:

O_1:	Selecting the first letter	N_1:	8 ways	
O_2:	Selecting the second letter	N_2:	8 ways	Repeats allowed
O_3:	Selecting the third letter	N_3:	8 ways	Repeats allowed

There are
$$8 \cdot 8 \cdot 8 = 8^3 = 512 \text{ possible code words}$$

(C) Adjacent letters cannot be alike:

O_1:	Selecting the first letter	N_1:	8 ways	
O_2:	Selecting the second letter	N_2:	7 ways	Cannot be the same as the first
O_3:	Selecting the third letter	N_3:	7 ways	Cannot be the same as the second, but can be the same as the first

There are
$$8 \cdot 7 \cdot 7 = 392 \text{ possible code words}$$

Matched Problem 5) How many 4-letter code words are possible using the first 10 letters of the alphabet under the three different conditions stated in Example 5?

Exercises 7.3

Skills Warm-up Exercises

In Problems 1–6, Solve for x. (If necessary, review section 1.1).

1. $50 = 34 + 29 - x$

2. $124 = 73 + 87 - x$

3. $4x = 23 + 42 - x$

4. $7x = 51 + 45 - x$

5. $3(x + 11) = 65 + x - 14$

6. $12(x + 5) = x + 122 - 29$

Solve Problems 7–10 two ways: (A) using a tree diagram, and (B) using the multiplication principle.

7. How many ways can 2 coins turn up—heads, H, or tails, T—if the combined outcome (H, T) is to be distinguished from the outcome (T, H)?

8. How many 2-letter code words can be formed from the first 3 letters of the alphabet if no letter can be used more than once?

9. A coin is tossed with possible outcomes of heads H, or tails T. Then a single die is tossed with possible outcomes 1, 2, 3, 4, 5, or 6. How many combined outcomes are there?

10. In how many ways can 3 coins turn up—heads H, or tails T—if combined outcomes such as (H, T, H), (H, H, T), and (T, H, H) are considered as different?

11. An entertainment guide recommends 6 restaurants and 3 plays that appeal to a couple.

   (A) If the couple goes to dinner or a play, but not both, how many selections are possible?

   (B) If the couple goes to dinner and then to a play, how many combined selections are possible?

12. A college offers 2 introductory courses in history, 3 in science, 2 in mathematics, 2 in philosophy, and 1 in English.

   (A) If a freshman takes one course in each area during her first semester, how many course selections are possible?

   (B) If a part-time student can afford to take only one introductory course, how many selections are possible?

13. How many 3-letter code words can be formed from the letters A, B, C, D, E if no letter is repeated? If letters can be repeated? If adjacent letters must be different?

14. How many 4-letter code words can be formed from the letters A, B, C, D, E, F, G if no letter is repeated? If letters can be repeated? If adjacent letters must be different?

15. A county park system rates its 20 golf courses in increasing order of difficulty as bronze, silver, or gold. There are only two gold courses and twice as many bronze as silver courses.

   (A) If a golfer decides to play a round at a silver or gold course, how many selections are possible?

   (B) If a golfer decides to play one round per week for 3 weeks, first on a bronze course, then silver, then gold, how many combined selections are possible?

16. The 14 colleges of interest to a high school senior include 6 that are expensive (tuition more than $30,000 per year), 7 that are far from home (more than 200 miles away), and 2 that are both expensive and far from home.

   (A) If the student decides to select a college that is not expensive and within 200 miles of home, how many selections are possible?

   (B) If the student decides to attend a college that is not expensive and within 200 miles from home during his first two years of college, and then will transfer to a college that is not expensive but is far from home, how many selections of two colleges are possible?

In Problems 17–24, use the given information to determine the number of elements in each of the four disjoint subsets in the following Venn diagram.

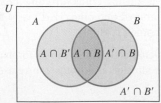

17. $n(A) = 80$, $n(B) = 50$,
    $n(A \cap B) = 20$, $n(U) = 200$

18. $n(A) = 45$, $n(B) = 35$,
    $n(A \cap B) = 15$, $n(U) = 100$

19. $n(A) = 25$, $n(B) = 55$,
    $n(A \cup B) = 60$, $n(U) = 100$

20. $n(A) = 70$, $n(B) = 90$,
    $n(A \cup B) = 120$, $n(U) = 200$

21. $n(A') = 65$, $n(B') = 40$,
    $n(A' \cap B') = 25$, $n(U) = 150$

22. $n(A') = 35$, $n(B') = 75$,
    $n(A' \cup B') = 95$, $n(U) = 120$

23. $n(A) = 48$, $n(B) = 62$,
    $n(A \cap B) = 0$, $n(U) = 180$

24. $n(A) = 57$, $n(B) = 36$,
    $n(A \cap B) = 36$, $n(U) = 180$

In Problems 25–32, use the given information to complete the following table.

	A	A'	Totals
B	?	?	?
B'	?	?	?
Totals	?	?	?

25. $n(A) = 70$, $n(B) = 90$,
    $n(A \cap B) = 30$, $n(U) = 200$

26. $n(A) = 55$, $n(B) = 65$,
    $n(A \cap B) = 35$, $n(U) = 100$

27. $n(A) = 45$, $n(B) = 55$,
    $n(A \cup B) = 80$, $n(U) = 100$

28. $n(A) = 80$, $n(B) = 70$,
    $n(A \cup B) = 110$, $n(U) = 200$

29. $n(A') = 15$, $n(B') = 24$,
    $n(A' \cup B') = 32$, $n(U) = 90$

30. $n(A') = 81$, $n(B') = 90$,
    $n(A' \cap B') = 63$, $n(U) = 180$

31. $n(A) = 110$, $n(B) = 145$,
    $n(A \cup B) = 255$, $n(U) = 300$

32. $n(A) = 175$, $n(B) = 125$,
    $n(A \cup B) = 300$, $n(U) = 300$

In Problems 33 and 34, discuss the validity of each statement. If the statement is always true, explain why. If not, give a counterexample.

33. (A) If A or B is the empty set, then A and B are disjoint.

   (B) If A and B are disjoint, then A or B is the empty set.

34. (A) If A and B are disjoint, then $n(A \cap B) = n(A) + n(B)$.

   (B) If $n(A \cup B) = n(A) + n(B)$, then A and B are disjoint.

35. A particular new car model is available with 5 choices of color, 3 choices of transmission, 4 types of interior, and 2 types of engine. How many different variations of this model are possible?

36. A delicatessen serves meat sandwiches with the following options: 3 kinds of bread, 5 kinds of meat, and lettuce or sprouts. How many different sandwiches are possible, assuming that one item is used out of each category?

37. Using the English alphabet, how many 5-character case-sensitive passwords are possible?

38. Using the English alphabet, how many 5-character case-sensitive passwords are possible if each character is a letter or a digit?

39. A combination lock has 5 wheels, each labeled with the 10 digits from 0 to 9. How many 5-digit opening combinations are possible if no digit is repeated? If digits can be repeated? If successive digits must be different?

40. A small combination lock has 3 wheels, each labeled with the 10 digits from 0 to 9. How many 3-digit combinations are possible if no digit is repeated? If digits can be repeated? If successive digits must be different?

41. How many different license plates are possible if each contains 3 letters (out of the alphabet's 26 letters) followed by 3 digits (from 0 to 9)? How many of these license plates contain no repeated letters and no repeated digits?

42. How many 5-digit ZIP code numbers are possible? How many of these numbers contain no repeated digits?

43. In Example 3, does it make any difference in which order the selection operations are performed? That is, if we select a jacket color first and then select a size, are there as many combined choices available as selecting a size first and then a color? Justify your answer using tree diagrams and the multiplication principle.

44. Explain how three sets, A, B, and C, can be related to each other in order for the following equation to hold true (Venn diagrams may be helpful):

$$n(A \cup B \cup C) = n(A) + n(B) + n(C)$$
$$- n(A \cap C) - n(B \cap C)$$

Problems 45–48 refer to the following Venn diagram.

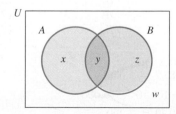

45. Which of the numbers x, y, z, or w must equal 0 if $B \subset A$?

46. Which of the numbers x, y, z, or w must equal 0 if A and B are disjoint?

47. Which of the numbers x, y, z, or w must equal 0 if $A \cup B = A \cap B$?

48. Which of the numbers x, y, z, or w must equal 0 if $A \cup B = U$?

49. A group of 75 people includes 32 who play tennis, 37 who play golf, and 8 who play both tennis and golf. How many people in the group play neither sport?

50. A class of 30 music students includes 13 who play the piano, 16 who play the guitar, and 5 who play both the piano and the guitar. How many students in the class play neither instrument?

51. A group of 100 people touring Europe includes 42 people who speak French, 55 who speak German, and 17 who speak neither language. How many people in the group speak both French and German?

52. A high school football team with 40 players includes 16 players who played offense last year, 17 who played defense, and 12 who were not on last year's team. How many players from last year played both offense and defense?

Applications

53. Management. A management selection service classifies its applicants (using tests and interviews) as high-IQ, middle-IQ, or low-IQ and as aggressive or passive. How many combined classifications are possible?

(A) Solve using a tree diagram.

(B) Solve using the multiplication principle.

54. Management. A corporation plans to fill 2 different positions for vice-president, V_1 and V_2, from administrative officers in 2 of its manufacturing plants. Plant A has 6 officers and plant B has 8. How many ways can these 2 positions be filled if the V_1 position is to be filled from plant A and the V_2 position from plant B? How many ways can the 2 positions be filled if the selection is made without regard to plant?

55. Transportation. A sales representative who lives in city A wishes to start from home and fly to 3 different cities: B, C, and D. If there are 2 choices of local transportation (drive her own car or use a taxi), and if all cities are interconnected by airlines, how many travel plans can be constructed to visit each city exactly once and return home?

56. Transportation. A manufacturing company in city A wishes to truck its product to 4 different cities: B, C, D, and E. If roads interconnect all 4 cities, how many different route plans can be constructed so that a single truck, starting from A, will visit each city exactly once, then return home?

57. Market research. A survey of 1,200 people indicates that 850 own HDTV's, 740 own DVD players, and 580 own HDTV's and DVD players.

(A) How many people in the survey own either an HDTV or a DVD player?

(B) How many own neither an HDTV nor a DVD player?

(C) How many own an HDTV and do not own a DVD player?

58. Market research. A survey of 800 small businesses indicates that 250 own a video conferencing system, 420 own projection equipment, and 180 own a video conferencing system and projection equipment.

(A) How many businesses in the survey own either a video conferencing system or projection equipment?

(B) How many own neither a video conferencing system nor projection equipment?

(C) How many own projection equipment and do not own a video conferencing system?

59. Communications. A cable television company has 8,000 subscribers in a suburban community. The company offers two premium channels: HBO and Showtime. If 2,450 subscribers receive HBO, 1,940 receive Showtime, and 5,180 do not receive any premium channel, how many subscribers receive both HBO and Showtime?

60. Communications. A cable company offers its 10,000 customers two special services: high-speed internet and digital phone. If 3,770 customers use high-speed internet, 3,250 use digital phone, and 4,530 do not use either of these services, how many customers use both high-speed internet and digital phone?

61. Minimum wage. The table below gives the number of male and female workers earning at or below the minimum wage for several age categories.

(A) How many males are of age 20–24 and earn below minimum wage?

Workers per Age Group (thousands)

	16–19	20–24	25+	Totals
Males at Minimum Wage	343	154	237	734
Males below Minimum Wage	118	102	159	379
Females at Minimum Wage	367	186	503	1,056
Females below Minimum Wage	251	202	540	993
Totals	1,079	644	1,439	3,162

(B) How many females are of age 20 or older and earn minimum wage?

(C) How many workers are of age 16–19 or males earning minimum wage?

(D) How many workers earn below minimum wage?

62. Minimum wage. Refer to the table in Problem 61.

(A) How many females are of age 16–19 and earn minimum wage?

(B) How many males are of age 16–24 and earn below minimum wage?

(C) How many workers are of age 20–24 or females earning below minimum wage?

(D) How many workers earn minimum wage?

63. Medicine. A medical researcher classifies subjects according to male or female; smoker or nonsmoker; and underweight, average weight, or overweight. How many combined classifications are possible?

(A) Solve using a tree diagram.

(B) Solve using the multiplication principle.

64. Family planning. A couple is planning to have 3 children. How many boy–girl combinations are possible? Distinguish between combined outcomes such as (B, B, G), (B, G, B), and (G, B, B).

(A) Solve using a tree diagram.

(B) Solve using the multiplication principle.

65. Politics. A politician running for a third term is planning to contact all contributors to her first two campaigns. If 1,475 individuals contributed to the first campaign, 2,350 contributed to the second campaign, and 920 contributed to the first and second campaigns, how many individuals have contributed to the first or second campaign?

66. Politics. If 12,457 people voted for a politician in his first election, 15,322 voted for him in his second election, and 9,345 voted for him in the first and second elections, how many people voted for this politician in the first or second election?

Answers to Matched Problems

1. 515

2. (A) 35   (B) 15   (C) 20

(D)

		FM		
		Listener	Nonlistener	Totals
AM	Listener	30	35	65
	Nonlistener	15	20	35
	Totals	45	55	100

3. There are 6 combined choices:

COMPANY CHOICES (OUTCOMES)	COVERAGE CHOICES (OUTCOMES)	COMBINED CHOICES (OUTCOMES)
R	A / B	(R, A) / (R, B)
Start — S	A / B	(S, A) / (S, B)
T	A / B	(T, A) / (T, B)

4. 5^5, or 3,125

5. (A) $10 \cdot 9 \cdot 8 \cdot 7 = 5,040$

(B) $10 \cdot 10 \cdot 10 \cdot 10 = 10,000$

(C) $10 \cdot 9 \cdot 9 \cdot 9 = 7,290$

7.4 Permutations and Combinations

- Factorials
- Permutations
- Combinations
- Applications

The multiplication principle discussed in the preceding section can be used to develop two additional counting devices that are extremely useful in more complicated counting problems. Both of these devices use *factorials*.

Factorials

When using the multiplication principle, we encountered expressions such as

$$26 \cdot 25 \cdot 24 \quad \text{or} \quad 8 \cdot 7 \cdot 6$$

where each natural number factor is decreased by 1 as we move from left to right. The factors in the following product continue to decrease by 1 until a factor of 1 is reached:

$$5 \cdot 4 \cdot 3 \cdot 2 \cdot 1$$

Products like this are encountered so frequently in counting problems that it is useful to express them in a concise notation. The product of the first n natural numbers is called **n factorial** and is denoted by **$n!$** Also, we define **zero factorial, 0!**, to be 1.

> **DEFINITION** Factorial*
> For a natural number n,
>
> $$n! = n(n-1)(n-2) \cdot \cdots \cdot 2 \cdot 1 \qquad 4! = 4 \cdot 3 \cdot 2 \cdot 1$$
> $$0! = 1$$
> $$n! = n \cdot (n-1)!$$

EXAMPLE 1   Computing Factorials

(A) $5! = 5 \cdot 4 \cdot 3 \cdot 2 \cdot 1 = 120$

(B) $\dfrac{7!}{6!} = \dfrac{7 \cdot 6!}{6!} = 7$

(C) $\dfrac{8!}{5!} = \dfrac{8 \cdot 7 \cdot 6 \cdot 5!}{5!} = 8 \cdot 7 \cdot 6 = 336$

(D) $\dfrac{52!}{5!47!} = \dfrac{52 \cdot 51 \cdot 50 \cdot 49 \cdot 48 \cdot 47!}{5 \cdot 4 \cdot 3 \cdot 2 \cdot 1 \cdot 47!} = 2,598,960$

Matched Problem 1   Find

(A) $6!$ (B) $\dfrac{10!}{9!}$ (C) $\dfrac{10!}{7!}$ (D) $\dfrac{5!}{0!3!}$ (E) $\dfrac{20!}{3!17!}$

It is interesting and useful to note that $n!$ grows very rapidly. Compare the following:

$$5! = 120 \qquad 10! = 3,628,800 \qquad 15! = 1,307,674,368,000$$

Try 69!, 70!, and 71! on your calculator.

Permutations

A particular (horizontal or vertical) arrangement of a set of paintings on a wall is called a *permutation* of the set of paintings.

---

*Many calculators have an n! key or its equivalent.

DEFINITION   Permutation of a Set of Objects
A **permutation** of a set of distinct objects is an arrangement of the objects in a specific order without repetition.

Suppose that 4 pictures are to be arranged from left to right on one wall of an art gallery. How many permutations (ordered arrangements) are possible? Using the multiplication principle, there are 4 ways of selecting the first picture; after the first picture is selected, there are 3 ways of selecting the second picture. After the first 2 pictures are selected, there are 2 ways of selecting the third picture, and after the first 3 pictures are selected, there is only 1 way to select the fourth. So the number of permutations (ordered arrangements) of the set of 4 pictures is

$$4 \cdot 3 \cdot 2 \cdot 1 = 4! = 24$$

In general, how many permutations of a set of n distinct objects are possible? Reasoning as above, there are n ways in which the first object can be chosen; there are $n - 1$ ways in which the second object can be chosen, and so on. Using the multiplication principle, we have the following:

THEOREM 1   Number of Permutations of n Objects
The number of permutations of n distinct objects without repetition, denoted by $_nP_n$, is

$$_nP_n = n(n - 1) \cdot \cdots \cdot 2 \cdot 1 = n! \quad n \text{ factors}$$

Example: The number of permutations of 7 objects is

$$_7P_7 = 7 \cdot 6 \cdot 5 \cdot 4 \cdot 3 \cdot 2 \cdot 1 = 7! \quad 7 \text{ factors}$$

Now suppose that the director of the art gallery decides to use only 2 of the 4 available paintings, and they will be arranged on the wall from left to right. We are now talking about a particular arrangement of 2 paintings out of the 4, which is called a *permutation of 4 objects taken 2 at a time*. In general,

DEFINITION   Permutation of n Objects Taken r at a Time
A permutation of a set of n distinct objects taken r at a time without repetition is an arrangement of r of the n objects in a specific order.

How many ordered arrangements of 2 pictures can be formed from the 4? That is, how many permutations of 4 objects taken 2 at a time are there? There are 4 ways that the first picture can be selected; after selecting the first picture, there are 3 ways that the second picture can be selected. So the number of permutations of a set of 4 objects taken 2 at a time, which is denoted by $_4P_2$, is given by

$$_4P_2 = 4 \cdot 3$$

In terms of factorials, we have

$$_4P_2 = 4 \cdot 3 = \frac{4 \cdot 3 \cdot 2!}{2!} = \frac{4!}{2!} \quad \text{Multiplying } 4 \cdot 3 \text{ by 1 in the form } 2!/2!$$

Reasoning in the same way as in the example, we find that the number of permutations of n distinct objects taken r at a time without repetition $(0 \leq r \leq n)$ is given by

$$_nP_r = n(n - 1)(n - 2) \cdot \cdots \cdot (n - r + 1) \quad r \text{ factors}$$
$$_9P_6 = 9(9 - 1)(9 - 2) \cdot \cdots \cdot (9 - 6 + 1) \quad 6 \text{ factors}$$
$$= 9 \cdot 8 \cdot 7 \cdot 6 \cdot 5 \cdot 4$$

Multiplying the right side of the equation for $_nP_r$ by 1 in the form $(n - r)!/(n - r)!$, we obtain a factorial form for $_nP_r$:

$$_nP_r = n(n - 1)(n - 2) \cdot \cdots \cdot (n - r + 1)\frac{(n - r)!}{(n - r)!}$$

But, since

$$n(n - 1)(n - 2) \cdot \cdots \cdot (n - r + 1)(n - r)! = n!$$

the expression above simplifies to

$$_nP_r = \frac{n!}{(n - r)!}$$

We summarize these results in Theorem 2.

THEOREM 2 Number of Permutations of n Objects Taken r at a Time

The number of permutations of n distinct objects taken r at a time without repetition is given by*

$$_nP_r = n(n - 1)(n - 2) \cdot \cdots \cdot (n - r + 1) \quad \text{r factors}$$

$$_5P_2 = 5 \cdot 4 \quad \text{2 factors}$$

or

$$_nP_r = \frac{n!}{(n - r)!} \quad 0 \le r \le n \quad _5P_2 = \frac{5!}{(5 - 2)!} = \frac{5!}{3!}$$

Note: $_nP_n = \dfrac{n!}{(n - n)!} = \dfrac{n!}{0!} = n!$ permutations of n objects taken n at a time.

Remember, by definition, $0! = 1$.

---

*In place of the symbol $_nP_r$, the symbols P_r^n, $P_{n,r}$, and $P(n, r)$ are often used.

EXAMPLE 2 Permutations Given the set $\{A, B, C\}$, how many permutations are possible for this set of 3 objects taken 2 at a time? Answer the question

(A) Using a tree diagram

(B) Using the multiplication principle

(C) Using the two formulas for $_nP_r$

SOLUTION

(A) Using a tree diagram:

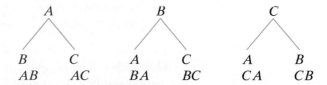

There are 6 permutations of 3 objects taken 2 at a time.

(B) Using the multiplication principle:

$$O_1: \quad \text{Fill the first position} \quad N_1: \quad \text{3 ways}$$
$$O_2: \quad \text{Fill the second position} \quad N_2: \quad \text{2 ways}$$

There are

$$3 \cdot 2 = 6 \text{ permutations of 3 objects taken 2 at a time}$$

(C) Using the two formulas for $_nP_r$:

2 factors

↓

$$_3P_2 = 3 \cdot 2 = 6 \qquad \text{or} \qquad _3P_2 = \frac{3!}{(3-2)!} = \frac{3 \cdot 2 \cdot 1}{1} = 6$$

There are 6 permutations of 3 objects taken 2 at a time. Of course, all three methods produce the same answer.

Matched Problem 2 ⌋ Given the set $\{A, B, C, D\}$, how many permutations are possible for this set of 4 objects taken 2 at a time? Answer the question

(A) Using a tree diagram

(B) Using the multiplication principle

(C) Using the two formulas for $_nP_r$

In Example 2 you probably found the multiplication principle to be the easiest method to use. But for large values of n and r, you will find that the factorial formula is more convenient. In fact, many calculators have functions that compute $n!$ and $_nP_r$ directly.

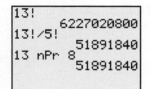

Figure 1

EXAMPLE 3 Permutations Find the number of permutations of 13 objects taken 8 at a time. Compute the answer using a calculator.

SOLUTION We use the factorial formula for $_nP_r$:

$$_{13}P_8 = \frac{13!}{(13-8)!} = \frac{13!}{5!} = 51{,}891{,}840$$

Using a tree diagram to solve this problem would involve a monumental effort. Using the multiplication principle would mean multiplying $13 \cdot 12 \cdot 11 \cdot 10 \cdot 9 \cdot 8 \cdot 7 \cdot 6$ (8 factors), which is not too bad. However, a calculator can provide instant results (see Fig. 1).

Matched Problem 3 ⌋ Find the number of permutations of 30 objects taken 4 at a time. Compute the answer using a calculator.

Combinations

Suppose that an art museum owns 8 paintings by a given artist and another art museum wishes to borrow 3 of these paintings for a special show. In selecting 3 of the 8 paintings for shipment, the order would not matter, and we would simply be selecting a 3-element subset from the set of 8 paintings. That is, we would be selecting what is called *a combination of 8 objects taken 3 at a time*.

DEFINITION Combination of n Objects Taken r at a Time

A **combination** of a set of n distinct objects taken r at a time without repetition is an r-element subset of the set of n objects. The arrangement of the elements in the subset does not matter.

How many ways can the 3 paintings be selected out of the 8 available? That is, what is the number of combinations of 8 objects taken 3 at a time? To answer this question, and to get a better insight into the general problem, we return to Example 2.

In Example 2, we were given the set $\{A, B, C\}$ and found the number of permutations of 3 objects taken 2 at a time using a tree diagram. From this tree diagram, we can determine the number of combinations of 3 objects taken 2 at a time (the number of 2-element subsets from a 3-element set), and compare it with the number of permutations (see Fig. 2).

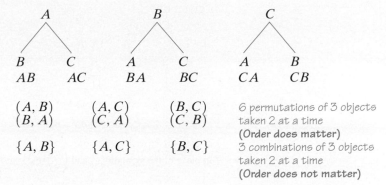

(A, B)	(A, C)	(B, C)	6 permutations of 3 objects
(B, A)	(C, A)	(C, B)	taken 2 at a time
			(Order does matter)
$\{A, B\}$	$\{A, C\}$	$\{B, C\}$	3 combinations of 3 objects
			taken 2 at a time

Figure 2   **(Order does not matter)**

There are fewer combinations than permutations, as we would expect. To each subset (combination), there corresponds two ordered pairs (permutations). We denote the number of combinations in Figure 2 by

$$_3C_2 \quad \text{or} \quad \binom{3}{2}$$

Our final goal is to find a factorial formula for $_nC_r$, the number of combinations of n objects taken r at a time. But first, we will develop a formula for $_3C_2$, and then we will generalize from this experience.

We know the number of permutations of 3 objects taken 2 at a time is given by $_3P_2$, and we have a formula for computing this number. Now, suppose we think of $_3P_2$ in terms of two operations:

O_1: Selecting a subset of 2 elements N_1: $_3C_2$ ways
O_2: Arranging the subset in a given order N_2: 2! ways

The combined operation, O_1 followed by O_2, produces a permutation of 3 objects taken 2 at a time. Thus,

$$_3P_2 = {_3C_2} \cdot 2! \quad \text{or} \quad _3C_2 = \frac{_3P_2}{2!}$$

To find $_3C_2$, the number of combinations of 3 objects taken 2 at a time, we substitute

$$_3P_2 = \frac{3!}{(3-2)!}$$

and solve for $_3C_2$:

$$_3C_2 = \frac{3!}{2!(3-2)!} = \frac{3 \cdot 2 \cdot 1}{(2 \cdot 1)(1)} = 3$$

This result agrees with the result obtained by using a tree diagram. Note that the number of combinations of 3 objects taken 2 at a time is the same as the number of permutations of 3 objects taken 2 at a time divided by the number of permutations of the elements in a 2-element subset. Figure 2 also reveals this observation.

Reasoning the same way as in the example, the number of combinations of n objects taken r at a time $(0 \le r \le n)$ is given by

$$_nC_r = \frac{_nP_r}{r!}$$

$$= \frac{n!}{r!(n-r)!} \quad \text{Since } _nP_r = \frac{n!}{(n-r)!}$$

THEOREM 3 Number of Combinations of n Objects Taken r at a Time

The number of combinations of n distinct objects taken r at a time without repetition is given by*

$$_nC_r = \binom{n}{r}$$

$$= \frac{_nP_r}{r!}$$

$$= \frac{n!}{r!(n-r)!} \quad 0 \le r \le n$$

$$_{52}C_5 = \binom{52}{5}$$

$$= \frac{_{52}P_5}{5!}$$

$$= \frac{52!}{5!(52-5)!}$$

_____
*In place of the symbols $_nC_r$ and $\binom{n}{r}$, the symbols C_r^n, $C_{n,r}$ and $C(n, r)$ are often used.

Now we can answer the question posed earlier in the museum example. There are

$$_8C_3 = \frac{8!}{3!(8-3)!} = \frac{8!}{3!5!} = \frac{8 \cdot 7 \cdot 6 \cdot 5!}{3 \cdot 2 \cdot 1 \cdot 5!} = 56$$

ways that 3 paintings can be selected for shipment. That is, there are 56 combinations of 8 objects taken 3 at a time.

EXAMPLE 4 Permutations and Combinations From a committee of 10 people,

(A) In how many ways can we choose a chairperson, a vice-chairperson, and a secretary, assuming that one person cannot hold more than one position?

(B) In how many ways can we choose a subcommittee of 3 people?

SOLUTION Note how parts (A) and (B) differ. In part (A), order of choice makes a difference in the selection of the officers. In part (B), the ordering does not matter in choosing a 3-person subcommittee. In part (A), we are interested in the number of *permutations* of 10 objects taken 3 at a time; and in part (B), we are interested in the number of *combinations* of 10 objects taken 3 at a time. These quantities are computed as follows (and since the numbers are not large, we do not need to use a calculator):

(A) $_{10}P_3 = \dfrac{10!}{(10-3)!} = \dfrac{10!}{7!} = \dfrac{10 \cdot 9 \cdot 8 \cdot 7!}{7!} = 720$ ways

(B) $_{10}C_3 = \dfrac{10!}{3!(10-3)!} = \dfrac{10!}{3!7!} = \dfrac{10 \cdot 9 \cdot 8 \cdot 7!}{3 \cdot 2 \cdot 1 \cdot 7!} = 120$ ways

Matched Problem 4 From a committee of 12 people,

(A) In how many ways can we choose a chairperson, a vice-chairperson, a secretary, and a treasurer, assuming that one person cannot hold more than one position?

(B) In how many ways can we choose a subcommittee of 4 people?

If n and r are large numbers, a calculator is useful in evaluating expressions involving factorials. Many calculators have a function that computes $_nC_r$ directly (see Fig. 3).

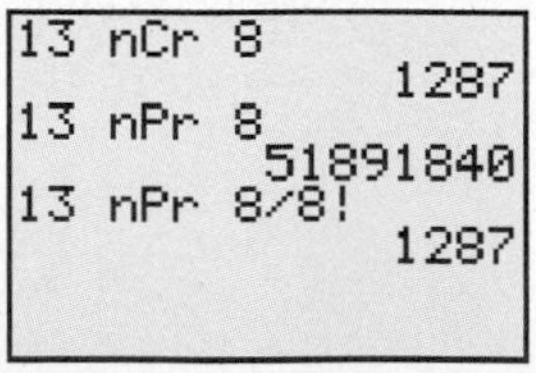

```
13 nCr 8
              1287
13 nPr 8
         51891840
13 nPr 8/8!
              1287
```

Figure 3

EXAMPLE 5 Combinations Find the number of combinations of 13 objects taken 8 at a time. Compute the answer using a calculator.

SOLUTION $_{13}C_8 = \binom{13}{8} = \dfrac{13!}{8!(13-8)!} = \dfrac{13!}{8!5!} = 1{,}287$

Compare the result in Example 5 with that obtained in Example 3, and note that $_{13}C_8$ is substantially smaller than $_{13}P_8$ (see Fig. 3).

Matched Problem 5 Find the number of combinations of 30 objects taken 4 at a time. Compute the answer using a calculator.

CONCEPTUAL INSIGHT

Permutations and combinations are similar in that both are selections in which repetition is *not* allowed. But there is a crucial distinction between the two:

> In a permutation, order is vital.
>
> In a combination, order is irrelevant.

To determine whether a given selection is a permutation or combination, see if rearranging the elements of the selection would produce a different object. If so, the selection is a permutation; if not, the selection is a combination.

Explore and Discuss 1 (A) List alphabetically by the first letter, all 3-letter license plate codes consisting of 3 different letters chosen from M, A, T, H. Discuss how this list relates to $_nP_r$.

(B) Reorganize the list from part (A) so that all codes without M come first, then all codes without A, then all codes without T, and finally all codes without H. Discuss how this list illustrates the formula $_nP_r = r! \,_nC_r$.

Applications

We now consider some applications of permutations and combinations. Several applications in this section involve a standard 52-card deck of playing cards.

Standard 52-Card Deck of Playing Cards

A standard deck of 52 cards (see Fig. 4) has four 13-card suits: diamonds, hearts, clubs, and spades. The diamonds and hearts are red, and the clubs and spades are black. Each 13-card suit contains cards numbered from 2 to 10, a jack, a queen, a king, and an ace. The jack, queen, and king are called *face cards*. (The ace is *not* a face card). Depending on the game, the ace may be counted as the lowest and/or the highest card in the suit.

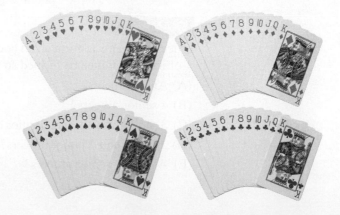

Figure 4

EXAMPLE 6   Counting Techniques   How many 5-card hands have 3 aces and 2 kings?

SOLUTION   The solution involves both the multiplication principle and combinations. Think of selecting the 5-card hand in terms of the following two operations:

O_1: Choosing 3 aces out of 4 possible N_1: $_4C_3$
          (order is not important)

O_2: Choosing 2 kings out of 4 possible N_2: $_4C_2$
          (order is not important)

Using the multiplication principle, we have

$$\text{number of hands} = {_4C_3} \cdot {_4C_2}$$
$$= \frac{4!}{3!(4-3)!} \cdot \frac{4!}{2!(4-2)!}$$
$$= 4 \cdot 6 = 24$$

Matched Problem 6   How many 5-card hands have 3 hearts and 2 spades?

EXAMPLE 7   Counting Techniques   Serial numbers for a product are made using 2 letters followed by 3 numbers. If the letters are taken from the first 8 letters of the alphabet with no repeats and the numbers are taken from the 10 digits (0–9) with no repeats, how many serial numbers are possible?

SOLUTION   The solution involves both the multiplication principle and permutations. Think of selecting a serial number in terms of the following two operations:

O_1: Choosing 2 letters out of 8 available N_1: $_8P_2$
          (order is important)

O_2: Choosing 3 numbers out of 10 available N_2: $_{10}P_3$
          (order is important)

Using the multiplication principle, we have

$$\text{number of serial numbers} = {_8P_2} \cdot {_{10}P_3}$$
$$= \frac{8!}{(8-2)!} \cdot \frac{10!}{(10-3)!}$$
$$= 56 \cdot 720 = 40,320$$

Matched Problem 7   Repeat Example 7 under the same conditions, except that the serial numbers will now have 3 letters followed by 2 digits (no repeats).

EXAMPLE 8   Counting Techniques   A company has 7 senior and 5 junior officers. It wants to form an ad hoc legislative committee. In how many ways can a 4-officer committee be formed so that it is composed of

(A) Any 4 officers?

(B) 4 senior officers?

(C) 3 senior officers and 1 junior officer?

(D) 2 senior and 2 junior officers?

(E) At least 2 senior officers?

SOLUTION

(A) Since there are a total of 12 officers in the company, the number of different 4-member committees is

$$_{12}C_4 = \frac{12!}{4!(12-4)!} = \frac{12!}{4!8!} = 495$$

(B) If only senior officers can be on the committee, the number of different committees is

$$_7C_4 = \frac{7!}{4!(7-4)!} = \frac{7!}{4!3!} = 35$$

(C) The 3 senior officers can be selected in $_7C_3$ ways, and the 1 junior officer can be selected in $_5C_1$ ways. Applying the multiplication principle, the number of ways that 3 senior officers and 1 junior officer can be selected is

$$_7C_3 \cdot {}_5C_1 = \frac{7!}{3!(7-3)!} \cdot \frac{5!}{1!(5-1)!} = \frac{7!5!}{3!4!1!4!} = 175$$

(D) $_7C_2 \cdot {}_5C_2 = \dfrac{7!}{2!(7-2)!} \cdot \dfrac{5!}{2!(5-2)!} = \dfrac{7!5!}{2!5!2!3!} = 210$

(E) The committees with *at least* 2 senior officers can be divided into three disjoint collections:

    1. Committees with 4 senior officers and 0 junior officers

    2. Committees with 3 senior officers and 1 junior officer

    3. Committees with 2 senior officers and 2 junior officers

The number of committees of types 1, 2, and 3 is computed in parts (B), (C), and (D), respectively. The total number of committees of all three types is the sum of these quantities:

    Type 1   Type 2   Type 3

$$_7C_4 + {}_7C_3 \cdot {}_5C_1 + {}_7C_2 \cdot {}_5C_2 = 35 + 175 + 210 = 420$$

Matched Problem 8 Given the information in Example 8, answer the following questions:

(A) How many 4-officer committees with 1 senior officer and 3 junior officers can be formed?

(B) How many 4-officer committees with 4 junior officers can be formed?

(C) How many 4-officer committees with at least 2 junior officers can be formed?

EXAMPLE 9 Counting Techniques From a standard 52-card deck, how many 3-card hands have all cards from the same suit?

SOLUTION There are 13 cards in each suit, so the number of 3-card hands having all hearts, for example, is

$$_{13}C_3 = \frac{13!}{3!(13-3)!} = \frac{13!}{3!10!} = 286$$

Similarly, there are 286 3-card hands having all diamonds, 286 having all clubs, and 286 having all spades. So the total number of 3-card hands having all cards from the same suit is

$$4 \cdot {}_{13}C_3 = 1{,}144$$

Matched Problem 9 From a standard 52-card deck, how many 5-card hands have all cards from the same suit?

Exercises 7.4

Skills Warm-up Exercises

In Problems 1–6, evaluate the given expression without using a calculator. (If necessary, review Section A.4).

1. $\dfrac{12 \cdot 11 \cdot 10}{3 \cdot 2 \cdot 1}$

2. $\dfrac{12 \cdot 10 \cdot 8}{6 \cdot 4 \cdot 2}$

3. $\dfrac{10 \cdot 9 \cdot 8 \cdot 7 \cdot 6}{5 \cdot 4 \cdot 3 \cdot 2 \cdot 1}$

4. $\dfrac{8 \cdot 7 \cdot 6 \cdot 5}{4 \cdot 3 \cdot 2 \cdot 1}$

5. $\dfrac{100 \cdot 99 \cdot 98 \cdot \ldots \cdot 3 \cdot 2 \cdot 1}{98 \cdot 97 \cdot 96 \cdot \ldots \cdot 3 \cdot 2 \cdot 1}$

6. $\dfrac{11 \cdot 10 \cdot 9 \cdot \ldots \cdot 3 \cdot 2 \cdot 1}{8 \cdot 7 \cdot 6 \cdot 5 \cdot 4 \cdot 3 \cdot 2 \cdot 1}$

In Problems 7–26, evaluate the expression. If the answer is not an integer, round to four decimal places.

7. $7!$

8. $10!$

9. $(5 + 6)!$

10. $(7 + 2)!$

11. $5! + 6!$

12. $7! + 2!$

13. $\dfrac{8!}{4!}$

14. $\dfrac{10!}{5!}$

15. $\dfrac{8!}{4!(8 - 4)!}$

16. $\dfrac{10!}{5!(10 - 5)!}$

17. $\dfrac{500!}{498!}$

18. $\dfrac{601!}{599!}$

19. $_{13}C_8$

20. $_{15}C_{10}$

21. $_{18}P_6$

22. $_{10}P_7$

23. $\dfrac{_{12}P_7}{12^7}$

24. $\dfrac{_{365}P_{25}}{365^{25}}$

25. $\dfrac{_{39}C_5}{_{52}C_5}$

26. $\dfrac{_{26}C_4}{_{52}C_4}$

In Problems 27–30, simplify each expression assuming that n is an integer and $n \geq 2$.

27. $\dfrac{n!}{(n - 2)!}$

28. $\dfrac{(n + 1)!}{3!(n - 2)!}$

29. $\dfrac{(n + 1)!}{2!(n - 1)!}$

30. $\dfrac{(n + 3)!}{(n + 1)!}$

In Problems 31–36, would you consider the selection to be a permutation, a combination, or neither? Explain your reasoning.

31. The university president named 3 new officers: a vice-president of finance, a vice-president of academic affairs, and a vice-president of student affairs.

32. The university president selected 2 of her vice-presidents to attend the dedication ceremony of a new branch campus.

33. A student checked out 4 novels from the library.

34. A student bought 4 books: 1 for his father, 1 for his mother, 1 for his younger sister, and 1 for his older brother.

35. A father ordered an ice cream cone (chocolate, vanilla, or strawberry) for each of his 4 children.

36. A book club meets monthly at the home of one of its 10 members. In December, the club selects a host for each meeting of the next year.

37. In a horse race, how many different finishes among the first 3 places are possible if 10 horses are running? (Exclude ties.)

38. In a long-distance foot race, how many different finishes among the first 5 places are possible if 50 people are running? (Exclude ties.)

39. How many ways can a 3-person subcommittee be selected from a committee of 7 people? How many ways can a president, vice-president, and secretary be chosen from a committee of 7 people?

40. Nine cards are numbered with the digits from 1 to 9. A 3-card hand is dealt, 1 card at a time. How many hands are possible in which

(A) Order is taken into consideration?

(B) Order is not taken into consideration?

41. Discuss the relative growth rates of $x!$, 3^x, and x^3.

42. Discuss the relative growth rates of $x!$, 2^x, and x^2.

43. From a standard 52-card deck, how many 6-card hands consist entirely of red cards?

44. From a standard 52-card deck, how many 6-card hands consist entirely of clubs?

45. From a standard 52-card deck, how many 5-card hands consist entirely of face cards?

46. From a standard 52-card deck, how many 5-card hands consist entirely of queens?

47. From a standard 52-card deck, how many 7-card hands contain four kings?

48. From a standard 52-card deck, how many 7-card hands consist of 3 hearts and 4 diamonds?

49. From a standard 52-card deck, how many 4-card hands contain a card from each suit?

50. From a standard 52-card deck, how many 4-card hands consist of cards from the same suit?

51. A catering service offers 8 appetizers, 10 main courses, and 7 desserts. A banquet committee selects 3 appetizers, 4 main courses, and 2 desserts. How many ways can this be done?

52. Three departments have 12, 15, and 18 members, respectively. If each department selects a delegate and an alternate to represent the department at a conference, how many ways can this be done?

In Problems 53 and 54, refer to the table in the graphing calculator display below, which shows $y_1 = {}_nP_r$ and $y_2 = {}_nC_r$ for $n = 6$.

X	Y₁	Y₂
0	1	1
1	6	6
2	30	15
3	120	20
4	360	15
5	720	6
6	720	1

Y₂■6 nCr X

53. Discuss and explain the symmetry of the numbers in the y_2 column of the table.

54. Explain how the table illustrates the formula

$$_nP_r = r! \, _nC_r$$

✏ *In Problems 55–60, discuss the validity of each statement. If the statement is true, explain why. If not, give a counterexample.*

55. If n is a positive integer, then $n! < (n + 1)!$

56. If n is a positive integer greater than 3, then $n! > 2^n$.

57. If n and r are positive integers and $1 < r < n$, then $_nP_r < \, _nP_{r+1}$.

58. If n and r are positive integers and $1 < r < n$, then $_nC_r < \, _nC_{r+1}$.

59. If n and r are positive integers and $1 < r < n$, then $_nC_r = \, _nC_{n-r}$.

60. If n and r are positive integers and $1 < r < n$, then $_nP_r = \, _nP_{n-r}$.

61. Eight distinct points are selected on the circumference of a circle.

(A) How many line segments can be drawn by joining the points in all possible ways?

(B) How many triangles can be drawn using these 8 points as vertices?

(C) How many quadrilaterals can be drawn using these 8 points as vertices?

62. Five distinct points are selected on the circumference of a circle.

(A) How many line segments can be drawn by joining the points in all possible ways?

(B) How many triangles can be drawn using these 5 points as vertices?

63. In how many ways can 4 people sit in a row of 6 chairs?

64. In how many ways can 3 people sit in a row of 7 chairs?

65. A basketball team has 5 distinct positions. Out of 8 players, how many starting teams are possible if

(A) The distinct positions are taken into consideration?

(B) The distinct positions are not taken into consideration?

(C) The distinct positions are not taken into consideration, but either Mike or Ken (but not both) must start?

66. How many 4-person committees are possible from a group of 9 people if

(A) There are no restrictions?

(B) Both Jim and Mary must be on the committee?

(C) Either Jim or Mary (but not both) must be on the committee?

67. Find the largest integer k such that your calculator can compute $k!$ without an overflow error.

68. Find the largest integer k such that your calculator can compute $_{2k}C_k$ without an overflow error.

69. Note from the table in the graphing calculator display below that the largest value of $_nC_r$ when $n = 20$ is $_{20}C_{10} = 184{,}756$. Use a similar table to find the largest value of $_nC_r$ when $n = 24$.

70. Note from the table in the graphing calculator display that the largest value of $_nC_r$ when $n = 21$ is $_{21}C_{10} = \, _{21}C_{11} = 352{,}716$. Use a similar table to find the largest value of $_nC_r$ when $n = 17$.

Applications

71. Quality control. An office supply store receives a shipment of 24 high-speed printers, including 5 that are defective. Three of these printers are selected for a store display.

(A) How many selections can be made?

(B) How many of these selections will contain no defective printers?

72. Quality control. An electronics store receives a shipment of 30 graphing calculators, including 6 that are defective. Four of these calculators are selected for a local high school.

(A) How many selections can be made?

(B) How many of these selections will contain no defective calculators?

73. Business closings. A jewelry store chain with 8 stores in Georgia, 12 in Florida, and 10 in Alabama is planning to close 10 of these stores.

(A) How many ways can this be done?

(B) The company decides to close 2 stores in Georgia, 5 in Florida, and 3 in Alabama. In how many ways can this be done?

74. Employee layoffs. A real estate company with 14 employees in their central office, 8 in their north office, and 6 in their south office is planning to lay off 12 employees.

(A) How many ways can this be done?

(B) The company decides to lay off 5 employees from the central office, 4 from the north office, and 3 from the south office. In how many ways can this be done?

75. Personnel selection. Suppose that 6 female and 5 male applicants have been successfully screened for 5 positions. In how many ways can the following compositions be selected?

(A) 3 females and 2 males

(B) 4 females and 1 male

(C) 5 females

(D) 5 people regardless of sex

(E) At least 4 females

76. Committee selection. A 4-person grievance committee is selected out of 2 departments *A* and *B*, with 15 and 20 people, respectively. In how many ways can the following committees be selected?

(A) 3 from *A* and 1 from *B*

(B) 2 from *A* and 2 from *B*

(C) All from *A*

(D) 4 people regardless of department

(E) At least 3 from department *A*

77. Medicine. There are 8 standard classifications of blood type. An examination for prospective laboratory technicians consists of having each candidate determine the type for 3 blood samples. How many different examinations can be given if no 2 of the samples provided for the candidate have the same type? If 2 or more samples have the same type?

78. Medical research. Because of limited funds, 5 research centers are chosen out of 8 suitable ones for a study on heart disease. How many choices are possible?

79. Politics. A nominating convention will select a president and vice-president from among 4 candidates. Campaign buttons, listing a president and a vice-president, will be designed for each possible outcome before the convention. How many different kinds of buttons should be designed?

80. Politics. In how many different ways can 6 candidates for an office be listed on a ballot?

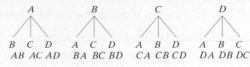

Answers to Matched Problems

1. (A) 720   (B) 10
   (C) 720   (D) 20
   (E) 1,140

2. (A)

```
    A          B          C          D
   /|\        /|\        /|\        /|\
  B C D      A C D      A B D      A B C
 AB AC AD   BA BC BD   CA CB CD   DA DB DC
```

12 *permutations of 4 objects taken 2 at a time*

(B) O_1: Fill first position N_1: 4 ways
   O_2: Fill second position N_2: 3 ways
   $4 \cdot 3 = 12$

(C) $_4P_2 = 4 \cdot 3 = 12$; $_4P_2 = \dfrac{4!}{(4-2)!} = 12$

3. $_{30}P_4 = \dfrac{30!}{(30-4)!} = 657,720$

4. (A) $_{12}P_4 = \dfrac{12!}{(12-4)!} = 11,880$ ways

   (B) $_{12}C_4 = \dfrac{12!}{4!(12-4)!} = 495$ ways

5. $_{30}C_4 = \dfrac{30!}{4!(30-4)!} = 27,405$

6. $_{13}C_3 \cdot _{13}C_2 = 22,308$

7. $_8P_3 \cdot _{10}P_2 = 30,240$

8. (A) $_7C_1 \cdot _5C_3 = 70$
   (B) $_5C_4 = 5$
   (C) $_7C_2 \cdot _5C_2 + _7C_1 \cdot _5C_3 + _5C_4 = 285$

9. $4 \cdot _{13}C_5 = 5,148$

Chapter 7   Summary and Review

Important Terms, Symbols, and Concepts

7.1 Logic

EXAMPLES

- A **proposition** is a statement (not a question or command) that is either true or false.

- If *p* and *q* are propositions, then the compound propositions,

Ex. 1, p. 348

$$\neg p, p \vee q, p \wedge q, \text{ and } p \rightarrow q$$

can be formed using the negation symbol $\neg$ and the connectives $\vee$, $\wedge$, and $\rightarrow$. These propositions are called **not p, p or q, p and q**, and **if p then q**, respectively (or **negation, disjunction, conjunction,** and **conditional**, respectively). Each of these compound propositions is specified by a truth table (see pages 346 and 347).

- Given any conditional proposition $p \rightarrow q$, the proposition $q \rightarrow p$ is called the **converse** of $p \rightarrow q$, and the proposition $\neg q \rightarrow \neg p$, is called the **contrapositive** of $p \rightarrow q$.

Ex. 2, p. 348

- A **truth table** for a compound proposition specifies whether it is true or false for any assignment of truth values to its variables. A proposition is a **tautology** if each entry in its column of the truth table is T; a **contradiction** if each entry is F, and a **contingency** if at least one entry is T and at least one entry is F.

Ex. 3, p. 349
Ex. 4, p. 349
Ex. 5, p. 350

- Consider the rows of the truth tables for the compound propositions P and Q. If whenever P is true, Q is also true, we say that P **logically implies** Q and write $P \Rightarrow Q$. We call $P \Rightarrow Q$ a **logical implication**. If the compound propositions P and Q have identical truth tables, we say that P and Q are **logically equivalent** and write $P \equiv Q$. We call $P \equiv Q$ a **logical equivalence**.

Ex. 6, p. 351
Ex. 7, p. 352

- Several logical equivalences are given in Table 2 on page 352. The last of these implies that **any conditional proposition is logically equivalent to its contrapositive**.

7.2 Sets

- A **set** is a collection of objects specified in such a way that we can tell whether any given object is or is not in the collection.

- Each object in a set is called a **member**, or **element**, of the set. If a is an element of the set A, we write $a \in A$.

- A set without any elements is called the **empty**, or **null**, set, denoted by $\varnothing$.

- A set can be described by listing its elements, or by giving a rule that determines the elements of the set. If $P(x)$ is a statement about x, then $\{x \mid P(x)\}$ denotes the set of all x such that $P(x)$ is true.

Ex. 1, p. 355

- A set is **finite** if its elements can be counted and there is an end; a set such as the positive integers, in which there is no end in counting its elements, is **infinite**.

- We write $A \subset B$, and say that A is a **subset** of B, if each element of A is an element of B. We write $A = B$, and say that sets A and B are equal, if they have exactly the same elements. The empty set $\varnothing$ is a subset of every set.

Ex. 2, p. 355
Ex. 3, p. 355

- If A and B are sets, then

$$A \cup B = \{x \mid x \in A \text{ or } x \in B\}$$

is called the **union** of A and B, and

$$A \cap B = \{x \mid x \in A \text{ and } x \in B\}$$

is called the **intersection** of A and B.

Ex. 4, p. 357

- **Venn diagrams** are useful in visualizing set relationships.

Ex. 5, p. 357

- If $A \cap B = \varnothing$, the sets A and B are said to be **disjoint**.

- The set of all elements under consideration in a given discussion is called the **universal set** U. The set $A' = \{x \in U \mid x \notin A\}$ is called the **complement** of A (relative to U).

- The **number of elements** in set A is denoted by $n(A)$. So if A and B are sets, then the numbers that are often shown in a Venn diagram, as in Figure 4 on page 356, are $n(A \cap B')$, $n(A \cap B)$, $n(B \cap A')$, and $n(A' \cap B')$ (see Fig. 9 on page 357).

7.3 Basic Counting Principles

- If A and B are sets, then the number of elements in the union of A and B is given by the **addition principle** for counting (Theorem 1, page 361).

Ex. 1, p. 361
Ex. 2, p. 362

- If the elements of a set are determined by a sequence of operations, tree diagrams can be used to list all combined outcomes. To count the number of combined outcomes without using a tree diagram, use the **multiplication principle** for counting (Theorem 2, page 364).

Ex. 3, p. 363
Ex. 4, p. 364
Ex. 5, p. 365

7.4 Permutations and Combinations

- The product of the first n natural numbers, denoted $n!$, is called n **factorial**:

Ex. 1, p. 369

$$n! = n(n-1)(n-2) \cdot \cdots \cdot 2 \cdot 1$$
$$0! = 1$$
$$n! = n \cdot (n-1)!$$

- A **permutation** of a set of distinct objects is an arrangement of the objects in a specific order without repetition. The number of permutations of a set of n distinct objects is given by $_nP_n = n!$. A permutation of a set of n distinct objects taken r at a time without repetition is an arrangement of r of the n objects in a specific order. The number of permutations of n distinct objects taken r at a time without repetition is given by

$$_nP_r = \frac{n!}{(n-r)!} \qquad 0 \le r \le n$$

Ex. 2, p. 371
Ex. 3, p. 372

- A **combination** of a set of n distinct objects taken r at a time without repetition is an r-element subset of the set of n objects. The arrangement of the elements in the subset is irrelevant. The number of combinations of n distinct objects taken r at a time without repetition is given by

$$_nC_r = \binom{n}{r} = \frac{_nP_r}{r!} = \frac{n!}{r!(n-r)!} \qquad 0 \le r \le n$$

Ex. 4, p. 374
Ex. 5, p. 375
Ex. 6, p. 376
Ex. 7, p. 376
Ex. 8, p. 376

Review Exercises

Work through all the problems in this chapter review and check your answers in the back of the book. Answers to all problems are there along with section numbers in italics to indicate where each type of problem is discussed. Where weaknesses show up, review appropriate sections in the text.

In Problems 1–6, express each proposition in an English sentence and determine whether it is true or false, where p and q are the propositions

$$p: \text{``}2^3 < 3^2\text{''} \qquad q: \text{``}3^4 < 4^3\text{''}$$

1. $\neg q$

2. $p \vee q$

3. $p \wedge q$

4. $p \rightarrow q$

5. The converse of $p \rightarrow q$

6. The contrapositive of $p \rightarrow q$

In Problems 7–10, indicate true (T) or false (F).

7. $\{5, 6, 7\} = \{6, 7, 5\}$

8. $5 \in \{55, 555\}$

9. $\{9, 27\} \subset \{3, 9, 27, 81\}$

10. $\{1, 2\} \subset \{1, \{1, 2\}\}$

In Problems 11–14, describe each proposition as a negation, disjunction, conjunction, or conditional, and determine whether the proposition is true or false.

11. If 9 is prime, then 10 is odd.

12. 7 is even or 8 is odd.

13. 53 is prime and 57 is prime.

14. 51 is not prime.

In Problems 15–16, state the converse and the contrapositive of the given proposition.

15. If the square matrix A has a row of zeros, then the square matrix A has no inverse.

16. If the square matrix A is an identity matrix, then the square matrix A has an inverse.

In Problems 17–19, write the resulting set using the listing method.

17. $\{1, 2, 3, 4\} \cup \{2, 3, 4, 5\}$

18. $\{1, 2, 3, 4\} \cap \{2, 3, 4, 5\}$

19. $\{1, 2, 3, 4\} \cap \{5, 6\}$

20. A single die is rolled, and a coin is flipped. How many combined outcomes are possible? Solve:

(A) Using a tree diagram

(B) Using the multiplication principle

21. Use the Venn diagram to find the number of elements in each of the following sets:

(A) A (B) B

(C) $A \cap B$ (D) $A \cup B$

(E) U (F) A'

(G) $(A \cap B)'$ (H) $(A \cup B)'$

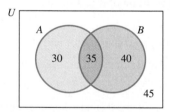

Evaluate the expressions in Problems 22–28.

22. $11!$

23. $(10 - 6)!$

24. $10! - 6!$

25. $\dfrac{15!}{10!}$

26. $\dfrac{15!}{10!5!}$

27. $_8C_5$

28. $_8P_5$

29. How many seating arrangements are possible with 6 people and 6 chairs in a row? Solve using the multiplication principle.

30. Solve Problem 29 using permutations or combinations, whichever is applicable.

In Problems 31–36, construct a truth table for the proposition and determine whether the proposition is a contingency, tautology, or contradiction.

31. $(p \rightarrow q) \wedge (q \rightarrow p)$

32. $p \vee (q \rightarrow p)$

33. $(p \lor \neg p) \rightarrow (q \land \neg q)$ **34.** $\neg q \land (p \rightarrow q)$

35. $\neg p \rightarrow (p \rightarrow q)$ **36.** $\neg(p \lor \neg q)$

In Problems 37–40, determine whether the given set is finite or infinite. Consider the set Z of integers to be the universal set, and let

$$M = \{n \in Z \,|\, n < 10^6\}$$
$$K = \{n \in Z \,|\, n > 10^3\}$$
$$E = \{n \in Z \,|\, n \text{ is even}\}$$

37. $E \cup K$ **38.** $M \cap K$

39. K' **40.** $E \cap M$

In Problems 41–43, determine whether or not the given sets are disjoint. For the definitions of M, K, and E, refer to the instructions for Problems 37–40.

41. M' and K'

42. M and E'

43. K and K'

44. A man has 5 children. Each of those children has 3 children, who in turn each have 2 children. Discuss the number of descendants that the man has.

45. How many 3-letter code words are possible using the first 8 letters of the alphabet if no letter can be repeated? If letters can be repeated? If adjacent letters cannot be alike?

46. Solve the following problems using $_nP_r$ or $_nC_r$:

(A) How many 3-digit opening combinations are possible on a combination lock with 6 digits if the digits cannot be repeated?

(B) Five tennis players have made the finals. If each of the 5 players is to play every other player exactly once, how many games must be scheduled?

47. Use graphical techniques on a graphing calculator to find the largest value of $_nC_r$ when $n = 25$.

48. If 3 operations O_1, O_2, O_3 are performed in order, with possible number of outcomes N_1, N_2, N_3, respectively, determine the number of branches in the corresponding tree diagram.

In Problems 49–51, write the resulting set using the listing method.

49. $\{x \,|\, x^3 - x = 0\}$

50. $\{x \,|\, x \text{ is a positive integer and } x! < 100\}$

51. $\{x \,|\, x \text{ is a positive integer that is a perfect square and } x < 50\}$

52. A software development department consists of 6 women and 4 men.

(A) How many ways can the department select a chief programmer, a backup programmer, and a programming librarian?

(B) How many of the selections in part (A) consist entirely of women?

(C) How many ways can the department select a team of 3 programmers to work on a particular project?

53. A group of 150 people includes 52 who play chess, 93 who play checkers, and 28 who play both chess and checkers. How many people in the group play neither game?

Problems 54 and 55 refer to the following Venn diagram.

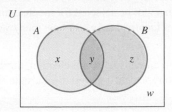

54. Which of the numbers x, y, z, or w must equal 0 if $A \subset B$?

55. Which of the numbers x, y, z, or w must equal 0 if $A \cap B = U$?

In Problems 56–58, discuss the validity of each statement. If the statement is true, explain why. If not, give a counterexample.

56. If n and r are positive integers and $1 < r < n$, then $_nC_r < {_nP_r}$.

57. If n and r are positive integers and $1 < r < n$, then $_nP_r < n!$.

58. If n and r are positive integers and $1 < r < n$, then $_nC_r < n!$.

In Problems 59–64, construct a truth table to verify the implication or equivalence.

59. $p \land q \Rightarrow p$ **60.** $q \Rightarrow p \rightarrow q$

61. $\neg p \rightarrow (q \land \neg q) \equiv p$ **62.** $p \lor q \equiv \neg p \rightarrow q$

63. $p \land (p \rightarrow q) \Rightarrow q$ **64.** $\neg(p \land \neg q) \equiv p \rightarrow q$

65. How many different 5-child families are possible where the gender of the children in the order of their births is taken into consideration [that is, birth sequences such as (B, G, G, B, B) and (G, B, G, B, B) produce different families]? How many families are possible if the order pattern is not taken into account?

66. Can a selection of r objects from a set of n distinct objects, where n is a positive integer, be a combination and a permutation simultaneously? Explain.

Applications

67. Transportation. A distribution center A wishes to send its products to five different retail stores: B, C, D, E, and F. How many different route plans can be constructed so that a single truck, starting from A, will deliver to each store exactly once, and then return to the center?

68. Market research. A survey of 1,000 people indicates that 340 have invested in stocks, 480 have invested in bonds, and 210 have invested in stocks and bonds.

(A) How many people in the survey have invested in stocks or bonds?

(B) How many have invested in neither stocks nor bonds?

(C) How many have invested in bonds and not stocks?

69. Medical research. In a study of twins, a sample of 6 pairs of identical twins will be selected for medical tests from a group of 40 pairs of identical twins. In how many ways can this be done?

70. Elections. In an unusual recall election, there are 67 candidates to replace the governor of a state. To negate the advantage that might accrue to candidates whose names appear near the top of the ballot, it is proposed that equal numbers of ballots be printed for each possible order in which the candidates' names can be listed.

(A) In how many ways can the candidates' names be listed?

(B) Explain why the proposal is not feasible, and discuss possible alternatives.

8 Probability

8.1 Sample Spaces, Events, and Probability

8.2 Union, Intersection, and Complement of Events; Odds

8.3 Conditional Probability, Intersection, and Independence

8.4 Bayes' Formula

8.5 Random Variable, Probability Distribution, and Expected Value

Chapter 8 Summary and Review

Review Exercises

Introduction

Like other branches of mathematics, probability evolved out of practical considerations. Girolamo Cardano (1501–1576), a gambler and physician, produced some of the best mathematics of his time, including a systematic analysis of gambling problems. In 1654, another gambler, Chevalier de Méré, approached the well-known French philosopher and mathematician Blaise Pascal (1623–1662) regarding certain dice problems. Pascal became interested in these problems, studied them, and discussed them with Pierre de Fermat (1601–1665), another French mathematician. So out of the gaming rooms of western Europe, the study of probability was born.

Despite this lowly birth, probability has matured into a highly respected and immensely useful branch of mathematics. It is used in practically every field. In this chapter, we use the counting techniques developed in Chapter 7 to assign probabilities to events. Then we consider many situations in which probability is used to measure and manage risk (see, for instance, Example 2 in Section 8.4 on tuberculosis screening).

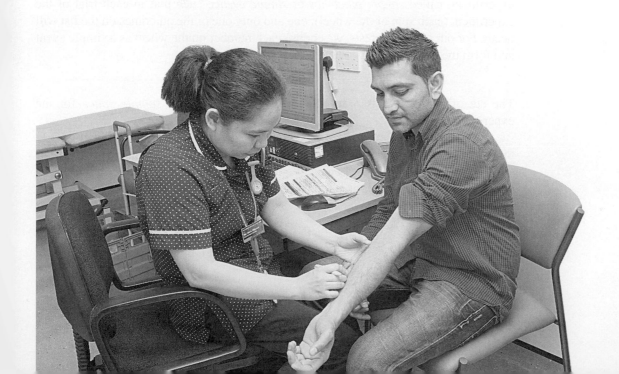

8.1   Sample Spaces, Events, and Probability

- Experiments
- Sample Spaces and Events
- Probability of an Event
- Equally Likely Assumption

This section provides a brief and relatively informal introduction to probability. It assumes a familiarity with the basics of set theory as presented in Chapter 7, including the union, intersection, and complement of sets, as well as various techniques for counting the number of elements in a set. Probability studies involve many subtle ideas, and care must be taken at the beginning to understand the fundamental concepts.

Experiments

Some experiments do not yield the same results each time that they are performed, no matter how carefully they are repeated under the same conditions. These experiments are called **random experiments**. Familiar examples of random experiments are flipping coins, rolling dice, observing the frequency of defective items from an assembly line, or observing the frequency of deaths in a certain age group.

Probability theory is a branch of mathematics that has been developed to deal with outcomes of random experiments, both real and conceptual. In the work that follows, we simply use the word **experiment** to mean a random experiment.

Figure 1

Sample Spaces and Events

Associated with outcomes of experiments are *sample spaces* and *events.* Consider the experiment, "A wheel with 18 numbers on the perimeter (Fig. 1) spins and comes to rest so that a pointer points within a numbered sector."

What outcomes might we observe? When the wheel stops, we might be interested in which number is next to the pointer, or whether that number is an odd number, or whether that number is divisible by 5, or whether that number is prime, or whether the pointer is in a shaded or white sector, and so on. The list of possible outcomes appears endless. In general, there is no unique method of analyzing all possible outcomes of an experiment. Therefore, before conducting an experiment, it is important to decide just what outcomes are of interest.

Suppose we limit our interest to the set of numbers on the wheel and to various subsets of these numbers, such as the set of prime numbers or the set of odd numbers on the wheel. Having decided what to observe, we make a list of outcomes of the experiment, called *simple outcomes* or *simple events,* such that in each trial of the experiment (each spin of the wheel), one and only one of the outcomes on the list will occur. For our stated interests, we choose each number on the wheel as a simple event and form the set

$$S = \{1, 2, 3, \ldots, 17, 18\}$$

The set of simple events S for the experiment is called a *sample space* for the experiment.

Now consider the outcome, "When the wheel comes to rest, the number next to the pointer is divisible by 4." This outcome is not a simple outcome (or simple event) since it is not associated with one and only one element in the sample space S. The outcome will occur whenever any one of the simple events 4, 8, 12, or 16 occurs, that is, whenever an element in the subset

$$E = \{4, 8, 12, 16\}$$

occurs. Subset E is called a *compound event* (and the outcome, a *compound outcome*).

DEFINITION Sample Spaces and Events

If we formulate a set S of outcomes (events) of an experiment in such a way that in each trial of the experiment one and only one of the outcomes (events) in the set will occur, we call the set S a **sample space** for the experiment. Each element in S is called a **simple outcome**, or **simple event**.

An **event E** is defined to be any subset of S (including the empty set $\varnothing$ and the sample space S). Event E is a **simple event** if it contains only one element and a **compound event** if it contains more than one element. We say that **an event E occurs** if any of the simple events in E occurs.

We use the terms *event* and *outcome of an experiment* interchangeably. Technically, an event is the mathematical counterpart of an outcome of an experiment, but we will not insist on strict adherence to this distinction in our development of probability.

Real World	Mathematical Model
Experiment (real or conceptual)	Sample space (set S)
Outcome (simple or compound)	Event (subset of S; simple or compound)

EXAMPLE 1 Simple and Compound Events Relative to the number wheel experiment (Fig. 1) and the sample space

$$S = \{1, 2, 3, \ldots, 17, 18\}$$

what is the event E (subset of the sample space S) that corresponds to each of the following outcomes? Indicate whether the event is a simple event or a compound event.

(A) The outcome is a prime number.   (B) The outcome is the square of 4.

SOLUTION

(A) The outcome is a prime number if any of the simple events 2, 3, 5, 7, 11, 13, or 17 occurs.* To say "A prime number occurs" is the same as saying that the experiment has an outcome in the set

$$E = \{2, 3, 5, 7, 11, 13, 17\}$$

Since event E has more than one element, it is a compound event.

(B) The outcome is the square of 4 if 16 occurs. To say "The square of 4 occurs" is the same as saying that the experiment has an outcome in the set

$$E = \{16\}$$

Since E has only one element, it is a simple event.

Matched Problem 1 Repeat Example 1 for

(A) The outcome is a number divisible by 12.

(B) The outcome is an even number greater than 15.

---

*Technically, we should write $\{2\}$, $\{3\}$, $\{5\}$, $\{7\}$, $\{11\}$, $\{13\}$, and $\{17\}$ for the simple events since there is a logical distinction between an element of a set and a subset consisting of only that element. But we will keep this in mind and drop the braces for simple events to simplify the notation.

EXAMPLE 2   Sample Spaces   A nickel and a dime are tossed. How do we identify a sample space for this experiment? There are a number of possibilities, depending on our interest. We will consider three.

(A) If we are interested in whether each coin falls heads (H) or tails (T), then, using a tree diagram, we can easily determine an appropriate sample space for the experiment:

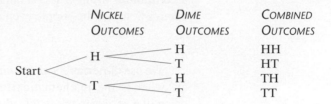

$$S_1 = \{HH, HT, TH, TT\}$$

and there are 4 simple events in the sample space.

(B) If we are interested only in the number of heads that appear on a single toss of the two coins, we can let

$$S_2 = \{0, 1, 2\}$$

and there are 3 simple events in the sample space.

(C) If we are interested in whether the coins match (M) or do not match (D), we can let

$$S_3 = \{M, D\}$$

and there are only 2 simple events in the sample space.

> **CONCEPTUAL INSIGHT**
>
> There is no single correct sample space for a given experiment. When specifying a sample space for an experiment, we include as much detail as necessary to answer all questions of interest regarding the outcomes of the experiment. If in doubt, we choose a sample space that contains more elements rather than fewer.

In Example 2, which sample space would be appropriate for all three interests? Sample space S_1 contains more information than either S_2 or S_3. If we know which outcome has occurred in S_1, then we know which outcome has occurred in S_2 and S_3. However, the reverse is not true. (Note that the simple events in S_2 and S_3 are simple or compound events in S_1.) In this sense, we say that S_1 is a more **fundamental sample space** than either S_2 or S_3. Thus, we would choose S_1 as an appropriate sample space for all three expressed interests.

Matched Problem 2   An experiment consists of recording the boy–girl composition of a two-child family. What would be an appropriate sample space

(A) If we are interested in the genders of the children in the order of their births? Draw a tree diagram.

(B) If we are interested only in the number of girls in a family?

(C) If we are interested only in whether the genders are alike (A) or different (D)?

(D) For all three interests expressed in parts (A) to (C)?

EXAMPLE 3   Sample Spaces and Events   Consider an experiment of rolling two dice. Figure 2 shows a convenient sample space that will enable us to answer many questions about interesting events. Let S be the set of all ordered pairs in the figure. The simple event (3, 2) is distinguished from the simple event (2, 3). The former indicates that a 3 turned up on the first die and a 2 on the second, while the latter indicates that a 2 turned up on the first die and a 3 on the second.

Second Die

		(1, 1)	(1, 2)	(1, 3)	(1, 4)	(1, 5)	(1, 6)
		(2, 1)	(2, 2)	(2, 3)	(2, 4)	(2, 5)	(2, 6)
First Die		(3, 1)	(3, 2)	(3, 3)	(3, 4)	(3, 5)	(3, 6)
		(4, 1)	(4, 2)	(4, 3)	(4, 4)	(4, 5)	(4, 6)
		(5, 1)	(5, 2)	(5, 3)	(5, 4)	(5, 5)	(5, 6)
		(6, 1)	(6, 2)	(6, 3)	(6, 4)	(6, 5)	(6, 6)

Figure 2

What is the event (subset of the sample space S) that corresponds to each of the following outcomes?

(A) A sum of 7 turns up.

(B) A sum of 11 turns up.

(C) A sum less than 4 turns up.

(D) A sum of 12 turns up.

SOLUTION

(A) By "A sum of 7 turns up," we mean that the sum of all dots on both turned-up faces is 7. This outcome corresponds to the event

$$\{(6, 1), (5, 2), (4, 3), (3, 4), (2, 5), (1, 6)\}$$

(B) "A sum of 11 turns up" corresponds to the event

$$\{(6, 5), (5, 6)\}$$

(C) "A sum less than 4 turns up" corresponds to the event

$$\{(1, 1), (2, 1), (1, 2)\}$$

(D) "A sum of 12 turns up" corresponds to the event

$$\{(6, 6)\}$$

Matched Problem 3   Refer to the sample space shown in Figure 2. What is the event that corresponds to each of the following outcomes?

(A) A sum of 5 turns up.

(B) A sum that is a prime number greater than 7 turns up.

As indicated earlier, we often use the terms *event* and *outcome of an experiment* interchangeably. In Example 3, we might say "the event, 'A sum of 11 turns up'" in place of "the outcome, 'A sum of 11 turns up,'" or even write

$$E = \text{a sum of 11 turns up} = \{(6, 5), (5, 6)\}$$

Probability of an Event

The next step in developing our mathematical model for probability studies is the introduction of a *probability function*. This is a function that assigns to an arbitrary event associated with a sample space a real number between 0 and 1, inclusive. We start by discussing ways in which probabilities are assigned to simple events in the sample space S.

DEFINITION   Probabilities for Simple Events
Given a sample space

$$S = \{e_1, e_2, \ldots, e_n\}$$

with n simple events, to each simple event e_i we assign a real number, denoted by $P(e_i)$, called the **probability of the event e_i**. These numbers can be assigned in an arbitrary manner as long as the following two conditions are satisfied:

Condition 1. The probability of a simple event is a number between 0 and 1, inclusive. That is,

$$0 \le P(e_i) \le 1$$

Condition 2. The sum of the probabilities of all simple events in the sample space is 1. That is,

$$P(e_1) + P(e_2) + \cdots + P(e_n) = 1$$

Any probability assignment that satisfies Conditions 1 and 2 is said to be an **acceptable probability assignment**.

Our mathematical theory does not explain how acceptable probabilities are assigned to simple events. These assignments are generally based on the expected or actual percentage of times that a simple event occurs when an experiment is repeated a large number of times. Assignments based on this principle are called **reasonable**.

Let an experiment be the flipping of a single coin, and let us choose a sample space S to be

$$S = \{H, T\}$$

If a coin appears to be fair, we are inclined to assign probabilities to the simple events in S as follows:

$$P(H) = \frac{1}{2} \quad \text{and} \quad P(T) = \frac{1}{2}$$

These assignments are based on reasoning that, since there are 2 ways a coin can land, in the long run, a head will turn up half the time and a tail will turn up half the time. These probability assignments are acceptable since both conditions for acceptable probability assignments stated in the preceding box are satisfied:

1. $0 \le P(H) \le 1, \qquad 0 \le P(T) \le 1$

2. $P(H) + P(T) = \frac{1}{2} + \frac{1}{2} = 1$

If we were to flip a coin 1,000 times, we would expect a head to turn up approximately, but not exactly, 500 times. The random number feature on a graphing calculator can be used to simulate 1,000 flips of a coin. Figure 3 shows the results of 3 such simulations: 497 heads the first time, 495 heads the second, and 504 heads the third.

If, however, we get only 376 heads in 1,000 flips of a coin, we might suspect that the coin is not fair. Then we might assign the simple events in the sample space S the following probabilities, based on our experimental results:

$$P(H) = .376 \quad \text{and} \quad P(T) = .624$$

This is also an acceptable assignment. However, the probability assignment

$$P(H) = 1 \quad \text{and} \quad P(T) = 0$$

although acceptable, is not reasonable (unless the coin has 2 heads). And the assignment

$$P(H) = .6 \quad \text{and} \quad P(T) = .8$$

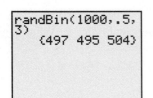

Figure 3

is not acceptable, since $.6 + .8 = 1.4$, which violates Condition 2 in the box on page 390.*

It is important to keep in mind that out of the infinitely many possible acceptable probability assignments to simple events in a sample space, we are generally inclined to choose one assignment over another based on reasoning or experimental results.

Given an acceptable probability assignment for simple events in a sample space S, how do we define the probability of an arbitrary event E associated with S?

DEFINITION Probability of an Event E

Given an acceptable probability assignment for the simple events in a sample space S, we define the **probability of an arbitrary event E**, denoted by $P(E)$, as follows:

(A) If E is the empty set, then $P(E) = 0$.

(B) If E is a simple event, then $P(E)$ has already been assigned.

(C) If E is a compound event, then $P(E)$ is the sum of the probabilities of all the simple events in E.

(D) If E is the sample space S, then $P(E) = P(S) = 1$ [this is a special case of part (C)].

EXAMPLE 4 Probabilities of Events Let us return to Example 2, the tossing of a nickel and a dime, and the sample space

$$S = \{HH, HT, TH, TT\}$$

Since there are 4 simple outcomes and the coins are assumed to be fair, it would appear that each outcome would occur 25% of the time, in the long run. Let us assign the same probability of $\frac{1}{4}$ to each simple event in S:

Simple Event				
e_i	**HH**	**HT**	**TH**	**TT**
$P(e_i)$	$\frac{1}{4}$	$\frac{1}{4}$	$\frac{1}{4}$	$\frac{1}{4}$

This is an acceptable assignment according to Conditions 1 and 2, and it is a reasonable assignment for ideal (perfectly balanced) coins or coins close to ideal.

(A) What is the probability of getting 1 head (and 1 tail)?

(B) What is the probability of getting at least 1 head?

(C) What is the probability of getting at least 1 head or at least 1 tail?

(D) What is the probability of getting 3 heads?

SOLUTION

(A) $E_1 = $ getting 1 head $ = \{HT, TH\}$

Since E_1 is a compound event, we use part (C) in the box and find $P(E_1)$ by adding the probabilities of the simple events in E_1:

$$P(E_1) = P(HT) + P(TH) = \frac{1}{4} + \frac{1}{4} = \frac{1}{2}$$

(B) $E_2 = $ getting at least 1 head $ = \{HH, HT, TH\}$

$$P(E_2) = P(HH) + P(HT) + P(TH) = \frac{1}{4} + \frac{1}{4} + \frac{1}{4} = \frac{3}{4}$$

---

*In probability studies, the 0 to the left of the decimal is usually omitted. So we write .6 and .8 instead of 0.6 and 0.8.

(C) $E_3 = \{HH, HT, TH, TT\} = S$

$$P(E_3) = P(S) = 1 \qquad \tfrac{1}{4} + \tfrac{1}{4} + \tfrac{1}{4} + \tfrac{1}{4} = 1$$

(D) $E_4 = $ getting 3 heads $= \varnothing$ Empty set

$$P(\varnothing) = 0$$

PROCEDURE   Steps for Finding the Probability of an Event E

Step 1   Set up an appropriate sample space S for the experiment.

Step 2   Assign acceptable probabilities to the simple events in S.

Step 3   To obtain the probability of an arbitrary event E, add the probabilities of the simple events in E.

The function P defined in Steps 2 and 3 is a **probability function** whose domain is all possible events (subsets) in the sample space S and whose range is a set of real numbers between 0 and 1, inclusive.

Matched Problem 4)   Suppose in Example 4 that after flipping the nickel and dime 1,000 times, we find that HH turns up 273 times, HT turns up 206 times, TH turns up 312 times, and TT turns up 209 times. On the basis of this evidence, we assign probabilities to the simple events in S as follows:

Simple Event				
e_i	**HH**	**HT**	**TH**	**TT**
$P(e_i)$	.273	.206	.312	.209

This is an acceptable and reasonable probability assignment for the simple events in S. What are the probabilities of the following events?

(A) $E_1 = $ getting at least 1 tail

(B) $E_2 = $ getting 2 tails

(C) $E_3 = $ getting at least 1 head or at least 1 tail

Example 4 and Matched Problem 4 illustrate two important ways in which acceptable and reasonable probability assignments are made for simple events in a sample space S. Each approach has its advantage in certain situations:

1. *Theoretical Approach.* We use assumptions and a deductive reasoning process to assign probabilities to simple events. No experiments are actually conducted. This is what we did in Example 4.

2. *Empirical Approach.* We assign probabilities to simple events based on the results of actual experiments. This is what we did in Matched Problem 4.

Empirical probability concepts are stated more precisely as follows: If we conduct an experiment n times and event E occurs with **frequency $f(E)$**, then the ratio $f(E)/n$ is called the **relative frequency** of the occurrence of event E in n trials. We define the **empirical probability** of E, denoted by $P(E)$, by the number (if it exists) that the relative frequency $f(E)/n$ approaches as n gets larger and larger. Therefore,

$$P(E) \approx \frac{\text{frequency of occurrence of } E}{\text{total number of trials}} = \frac{f(E)}{n}$$

For any particular n, the relative frequency $f(E)/n$ is also called the **approximate empirical probability** of event E.

For most of this section, we emphasize the theoretical approach. In the next section, we return to the empirical approach.

Equally Likely Assumption

In tossing a nickel and a dime (Example 4), we assigned the same probability, $\frac{1}{4}$, to each simple event in the sample space $S = \{HH, HT, TH, TT\}$. By assigning the same probability to each simple event in S, we are actually making the assumption that each simple event is as likely to occur as any other. We refer to this as an **equally likely assumption**. If, in a sample space

$$S = \{e_1, e_2, \ldots, e_n\}$$

with n elements, we assume that each simple event e_i is as likely to occur as any other, then we assign the probability $1/n$ to each. That is,

$$P(e_i) = \frac{1}{n}$$

Under an equally likely assumption, we can develop a very useful formula for finding probabilities of arbitrary events associated with a sample space S. Consider the following example:

If a single die is rolled and we assume that each face is as likely to come up as any other, then for the sample space

$$S = \{1, 2, 3, 4, 5, 6\}$$

we assign a probability of $\frac{1}{6}$ to each simple event since there are 6 simple events. The probability of

$$E = \text{rolling a prime number} = \{2, 3, 5\}$$

is

Number of elements in E
↓

$$P(E) = P(2) + P(3) + P(5) = \frac{1}{6} + \frac{1}{6} + \frac{1}{6} = \frac{3}{6} = \frac{1}{2}$$

↑
Number of elements in S

Under the assumption that each simple event is as likely to occur as any other, the computation of the probability of the occurrence of any event E in a sample space S is the number of elements in E divided by the number of elements in S.

THEOREM 1 Probability of an Arbitrary Event under an Equally Likely Assumption

If we assume that each simple event in sample space S is as likely to occur as any other, then the probability of an arbitrary event E in S is given by

$$P(E) = \frac{\text{number of elements in } E}{\text{number of elements in } S} = \frac{n(E)}{n(S)}$$

EXAMPLE 5 Probabilities and Equally Likely Assumptions Let us again consider rolling two dice, and assume that each simple event in the sample space shown in Figure 2 (page 389) is as likely as any other. Find the probabilities of the following events:

(A) E_1 = a sum of 7 turns up

(B) E_2 = a sum of 11 turns up

(C) E_3 = a sum less than 4 turns up

(D) E_4 = a sum of 12 turns up

SOLUTION Referring to Figure 2 (page 389) and the results found in Example 3, we find

(A) $P(E_1) = \dfrac{n(E_1)}{n(S)} = \dfrac{6}{36} = \dfrac{1}{6}$ $E_1 = \{(6, 1), (5, 2), (4, 3), (3, 4), (2, 5), (1, 6)\}$

(B) $P(E_2) = \dfrac{n(E_2)}{n(S)} = \dfrac{2}{36} = \dfrac{1}{18}$ $E_2 = \{(6, 5), (5, 6)\}$

(C) $P(E_3) = \dfrac{n(E_3)}{n(S)} = \dfrac{3}{36} = \dfrac{1}{12}$ $E_3 = \{(1, 1), (2, 1), (1, 2)\}$

(D) $P(E_4) = \dfrac{n(E_4)}{n(S)} = \dfrac{1}{36}$ $E_4 = \{(6, 6)\}$

Matched Problem 5   Under the conditions in Example 5, find the probabilities of the following events (each event refers to the sum of the dots facing up on both dice):

(A) $E_5 = $ a sum of 5 turns up

(B) $E_6 = $ a sum that is a prime number greater than 7 turns up

---

EXAMPLE 6   Simulation and Empirical Probabilities   Use output from the random number feature of a graphing calculator to simulate 100 rolls of two dice. Determine the empirical probabilities of the following events, and compare with the theoretical probabilities:

(A) $E_1 = $ a sum of 7 turns up

(B) $E_2 = $ a sum of 11 turns up

SOLUTION   A graphing calculator can be used to select a random integer from 1 to 6. Each of the six integers in the given range is equally likely to be selected. Therefore, by selecting a random integer from 1 to 6 and adding it to a second random integer from 1 to 6, we simulate rolling two dice and recording the sum (see the first command in Figure 4A). The second command in Figure 4A simulates 100 rolls of two dice; the sums are stored in list L_1. From the statistical plot of L_1 in Figure 4B we obtain the empirical probabilities.*

(A) The empirical probability of E_1 is $\frac{15}{100} = .15$; the theoretical probability of E_1 (see Example 5A) is $\frac{6}{36} = .167$.

(B) The empirical probability of E_2 is $\frac{6}{100} = .06$; the theoretical probability of E_2 (see Example 5B) is $\frac{2}{36} = .056$.

   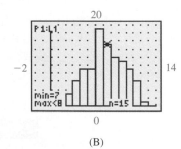

Figure 4   (A)   (B)

Matched Problem 6   Use the graphing calculator output in Figure 4B to determine the empirical probabilities of the following events, and compare with the theoretical probabilities:

(A) $E_3 = $ a sum less than 4 turns up

(B) $E_4 = $ a sum of 12 turns up

---

*If you simulate this experiment on your graphing calculator, you should not expect to get the same empirical probabilities.

Explore and Discuss 1   A shipment box contains 12 graphing calculators, out of which 2 are defective. A calculator is drawn at random from the box and then, without replacement, a second calculator is drawn. Discuss whether the equally likely assumption would be appropriate for the sample space $S = \{GG, GD, DG, DD\}$, where G is a good calculator and D is a defective one.

We now turn to some examples that make use of the counting techniques developed in Chapter 7.

EXAMPLE 7   Probability and Equally Likely Assumption   In drawing 5 cards from a 52-card deck without replacement, what is the probability of getting 5 spades?

SOLUTION   Let the sample space S be the set of all 5-card hands from a 52-card deck. Since the order in a hand does not matter, $n(S) = {}_{52}C_5$. Let event E be the set of all 5-card hands from 13 spades. Again, the order does not matter and $n(E) = {}_{13}C_5$. Assuming that each 5-card hand is as likely as any other,

$$P(E) = \frac{n(E)}{n(S)} = \frac{{}_{13}C_5}{{}_{52}C_5} = \frac{1,287}{2,598,960} \approx .0005$$

(Some calculators display the answer as 4.951980792E–4. This means the same thing as the scientific notation $4.951980792 \times 10^{-4}$, so the answer, rounded to 4 decimal places, is .0005).

Matched Problem 7   In drawing 7 cards from a 52-card deck without replacement, what is the probability of getting 7 hearts?

EXAMPLE 8   Probability and Equally Likely Assumption   The board of regents of a university is made up of 12 men and 16 women. If a committee of 6 is chosen at random, what is the probability that it will contain 3 men and 3 women?

SOLUTION   Let S be the set of all 6-person committees out of 28 people. Then

$$n(S) = {}_{28}C_6$$

Let E be the set of all 6-person committees with 3 men and 3 women. To find $n(E)$, we use the multiplication principle and the following two operations:

O_1:   Select 3 men out of the 12 available N_1: ${}_{12}C_3$

O_2:   Select 3 women out of the 16 available N_2: ${}_{16}C_3$

Therefore,

$$n(E) = N_1 \cdot N_2 = {}_{12}C_3 \cdot {}_{16}C_3$$

and

$$P(E) = \frac{n(E)}{n(S)} = \frac{{}_{12}C_3 \cdot {}_{16}C_3}{{}_{28}C_6} \approx .327$$

Matched Problem 8   What is the probability that the committee in Example 8 will have 4 men and 2 women?

There are many counting problems for which it is not possible to produce a simple formula that will yield the number of possible cases. In situations of this type, we often revert back to tree diagrams and counting branches.

Exercises 8.1

Skills Warm-up Exercises

In Problems 1–6, without using a calculator, determine which event, E or F, is more likely to occur. (If necessary, review Section A.1.)

1. $P(E) = \dfrac{5}{6}; P(F) = \dfrac{4}{5}$

2. $P(E) = \dfrac{2}{7}; P(F) = \dfrac{1}{3}$

3. $P(E) = \dfrac{3}{8}; P(F) = .4$

4. $P(E) = .9; P(F) = \dfrac{7}{8}$

5. $P(E) = .15; P(F) = \dfrac{1}{6}$

6. $P(E) = \dfrac{5}{7}; P(F) = \dfrac{6}{11}$

A circular spinner is divided into 12 sectors of equal area: 5 red sectors, 4 blue, 2 yellow, and 1 green. In Problems 7–14, consider the experiment of spinning the spinner once. Find the probability that the spinner lands on:

7. Blue

8. Yellow

9. Yellow or green

10. Red or blue

11. Orange

12. Yellow, red, or green

13. Blue, red, yellow, or green

14. Purple

Refer to the description of a standard deck of 52 cards and Figure 4 on page 375. An experiment consists of drawing 1 card from a standard 52-card deck. In Problems 15–24, what is the probability of drawing

15. A club

16. A black card

17. A heart or diamond

18. A numbered card

19. The jack of clubs

20. An ace

21. A king or spade

22. A red queen

23. A black diamond

24. A six or club

25. In a family with 2 children, excluding multiple births, what is the probability of having 2 children of the opposite gender? Assume that a girl is as likely as a boy at each birth.

26. In a family with 2 children, excluding multiple births, what is the probability of having 2 girls? Assume that a girl is as likely as a boy at each birth.

27. A store carries four brands of DVD players: *J, G, P,* and *S*. From past records, the manager found that the relative frequency of brand choice among customers varied. Which of the following probability assignments for a particular customer choosing a particular brand of DVD player would have to be rejected? Why?

(A) $P(J) = .15, P(G) = -.35, P(P) = .50, P(S) = .70$

(B) $P(J) = .32, P(G) = .28, P(P) = .24, P(S) = .30$

(C) $P(J) = .26, P(G) = .14, P(P) = .30, P(S) = .30$

28. Using the probability assignments in Problem 27C, what is the probability that a random customer will not choose brand *S?*

29. Using the probability assignments in Problem 27C, what is the probability that a random customer will choose brand *J* or brand *P?*

30. Using the probability assignments in Problem 27C, what is the probability that a random customer will not choose brand *J* or brand *P?*

31. In a family with 3 children, excluding multiple births, what is the probability of having 2 boys and 1 girl, in that order? Assume that a boy is as likely as a girl at each birth.

32. In a family with 3 children, excluding multiple births, what is the probability of having 2 boys and 1 girl, in any order? Assume that a boy is as likely as a girl at each birth.

33. A small combination lock has 3 wheels, each labeled with the 10 digits from 0 to 9. If an opening combination is a particular sequence of 3 digits with no repeats, what is the probability of a person guessing the right combination?

34. A combination lock has 5 wheels, each labeled with the 10 digits from 0 to 9. If an opening combination is a particular sequence of 5 digits with no repeats, what is the probability of a person guessing the right combination?

Refer to the description of a standard deck of 52 cards and Figure 4 on page 375. An experiment consists of dealing 5 cards from a standard 52-card deck. In Problems 35–38, what is the probability of being dealt

35. 5 black cards?

36. 5 hearts?

37. 5 face cards?

38. 5 nonface cards?

39. Twenty thousand students are enrolled at a state university. A student is selected at random, and his or her birthday (month and day, not year) is recorded. Describe an appropriate sample space for this experiment and assign acceptable probabilities to the simple events. What are your assumptions in making this assignment?

40. In a three-way race for the U.S. Senate, polls indicate that the two leading candidates are running neck-and-neck, while the third candidate is receiving half the support of either of the others. Registered voters are chosen at random and asked which of the three will get their vote. Describe an appropriate sample space for this random survey experiment and assign acceptable probabilities to the simple events.

41. Suppose that 5 thank-you notes are written and 5 envelopes are addressed. Accidentally, the notes are randomly inserted into the envelopes and mailed without checking the addresses. What is the probability that all the notes will be inserted into the correct envelopes?

42. Suppose that 6 people check their coats in a checkroom. If all claim checks are lost and the 6 coats are randomly returned, what is the probability that all the people will get their own coats back?

An experiment consists of rolling two fair dice and adding the dots on the two sides facing up. Using the sample space shown in Figure 2 (page 389) and assuming each simple event is as likely as any other, find the probability of the sum of the dots indicated in Problems 43–56.

43. Sum is 6.

44. Sum is 8.

45. Sum is less than 5.

46. Sum is greater than 8.

47. Sum is not 7 or 11.

48. Sum is not 2, 4, or 6.

49. Sum is 1.

50. Sum is 13.

51. Sum is divisible by 3.

52 Sum is divisible by 4.

53. Sum is 7 or 11 (a "natural").

54. Sum is 2, 3, or 12 ("craps").

55. Sum is divisible by 2 or 3.

56. Sum is divisible by 2 and 3.

An experiment consists of tossing three fair (not weighted) coins, except that one of the three coins has a head on both sides. Compute the probability of obtaining the indicated results in Problems 57–62.

57. 1 head

58. 2 heads

59. 3 heads

60. 0 heads

61. More than 1 head

62. More than 1 tail

In Problems 63–68, a sample space S is described. Would it be reasonable to make the equally likely assumption? Explain.

63. A single card is drawn from a standard deck. We are interested in whether or not the card drawn is a heart, so an appropriate sample space is $S = \{H, N\}$.

64. A single fair coin is tossed. We are interested in whether the coin falls heads or tails, so an appropriate sample space is $S = \{H, T\}$.

65. A single fair die is rolled. We are interested in whether or not the number rolled is even or odd, so an appropriate sample space is $S = \{E, O\}$.

66. A nickel and dime are tossed. We are interested in the number of heads that appear, so an appropriate sample space is $S = \{0, 1, 2\}$.

67. A wheel of fortune has seven sectors of equal area colored red, orange, yellow, green, blue, indigo, and violet. We are interested in the color that the pointer indicates when the wheel stops, so an appropriate sample space is $S = \{R, O, Y, G, B, I, V\}$.

68. A wheel of fortune has seven sectors of equal area colored red, orange, yellow, red, orange, yellow, and red. We are interested in the color that the pointer indicates when the wheel stops, so an appropriate sample space is $S = \{R, O, Y\}$.

69. (A) Is it possible to get 19 heads in 20 flips of a fair coin? Explain.

(B) If you flipped a coin 40 times and got 37 heads, would you suspect that the coin was unfair? Why or why not?

If you suspect an unfair coin, what empirical probabilities would you assign to the simple events of the sample space?

70. (A) Is it possible to get 7 double 6's in 10 rolls of a pair of fair dice? Explain.

(B) If you rolled a pair of dice 36 times and got 11 double 6's, would you suspect that the dice were unfair? Why or why not? If you suspect loaded dice, what empirical probability would you assign to the event of rolling a double 6?

An experiment consists of rolling two fair (not weighted) dice and adding the dots on the two sides facing up. Each die has the number 1 on two opposite faces, the number 2 on two opposite faces, and the number 3 on two opposite faces. Compute the probability of obtaining the indicated sums in Problems 71–78.

71. 2

72. 3

73. 4

74. 5

75. 6

76. 7

77. An odd sum

78. An even sum

In Problems 79–86, find the probability of being dealt the given hand from a standard 52-card deck. Refer to the description of a standard 52-card deck on page 375.

79. A 5-card hand that consists entirely of red cards

80. A 5-card hand that consists entirely of face cards

81. A 6-card hand that contains exactly two face cards

82. A 6-card hand that contains exactly two clubs

83. A 4-card hand that contains no aces

84. A 4-card hand that contains no face cards

85. A 7-card hand that contains exactly 2 diamonds and exactly 2 spades

86. A 7-card hand that contains exactly 1 king and exactly 2 jacks

In Problems 87–90, several experiments are simulated using the random number feature on a graphing calculator. For example, the roll of a fair die can be simulated by selecting a random integer from 1 to 6, and 50 rolls of a fair die by selecting 50 random integers from 1 to 6 (see Fig. A for Problem 87 and your user's manual).

87. From the statistical plot of the outcomes of rolling a fair die 50 times (see Fig. B), we see, for example, that the number 4 was rolled exactly 11 times.

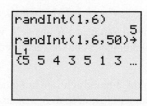

(A)

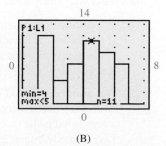

(B)

(A) What is the empirical probability that the number 6 was rolled?

(B) What is the probability that a 6 is rolled under the equally likely assumption?

(C) Use a graphing calculator to simulate 100 rolls of a fair die and determine the empirical probabilities of the six outcomes.

88. Use a graphing calculator to simulate 200 tosses of a nickel and dime, representing the outcomes HH, HT, TH, and TT by 1, 2, 3, and 4, respectively.

(A) Find the empirical probabilities of the four outcomes.

(B) What is the probability of each outcome under the equally likely assumption?

89. (A) Explain how a graphing calculator can be used to simulate 500 tosses of a coin.

(B) Carry out the simulation and find the empirical probabilities of the two outcomes.

(C) What is the probability of each outcome under the equally likely assumption?

90. From a box containing 12 balls numbered 1 through 12, one ball is drawn at random.

(A) Explain how a graphing calculator can be used to simulate 400 repetitions of this experiment.

(B) Carry out the simulation and find the empirical probability of drawing the 8 ball.

(C) What is the probability of drawing the 8 ball under the equally likely assumption?

Applications

91. Consumer testing. Twelve popular brands of beer are used in a blind taste study for consumer recognition.

(A) If 4 distinct brands are chosen at random from the 12 and if a consumer is not allowed to repeat any answers, what is the probability that all 4 brands could be identified by just guessing?

(B) If repeats are allowed in the 4 brands chosen at random from the 12 and if a consumer is allowed to repeat answers, what is the probability that all 4 brands are identified correctly by just guessing?

92. Consumer testing. Six popular brands of cola are to be used in a blind taste study for consumer recognition.

(A) If 3 distinct brands are chosen at random from the 6 and if a consumer is not allowed to repeat any answers, what is the probability that all 3 brands could be identified by just guessing?

(B) If repeats are allowed in the 3 brands chosen at random from the 6 and if a consumer is allowed to repeat answers, what is the probability that all 3 brands are identified correctly by just guessing?

93. Personnel selection. Suppose that 6 female and 5 male applicants have been successfully screened for 5 positions. If the 5 positions are filled at random from the 11 finalists, what is the probability of selecting

(A) 3 females and 2 males?

(B) 4 females and 1 male?

(C) 5 females?

(D) At least 4 females?

94. Committee selection. A 4-person grievance committee is to include employees in 2 departments, A and B, with 15 and 20 employees, respectively. If the 4 people are selected at random from the 35 employees, what is the probability of selecting

(A) 3 from A and 1 from B?

(B) 2 from A and 2 from B?

(C) All 4 from A?

(D) At least 3 from A?

95. Medicine. A laboratory technician is to be tested on identifying blood types from 8 standard classifications.

(A) If 3 distinct samples are chosen at random from the 8 types and if the technician is not allowed to repeat any answers, what is the probability that all 3 could be correctly identified by just guessing?

(B) If repeats are allowed in the 3 blood types chosen at random from the 8 and if the technician is allowed to repeat answers, what is the probability that all 3 are identified correctly by just guessing?

96. Medical research. Because of limited funds, 5 research centers are to be chosen out of 8 suitable ones for a study on heart disease. If the selection is made at random, what is the probability that 5 particular research centers will be chosen?

97. Politics. A town council has 9 members: 5 Democrats and 4 Republicans. A 3-person zoning committee is selected at random.

(A) What is the probability that all zoning committee members are Democrats?

(B) What is the probability that a majority of zoning committee members are Democrats?

98. Politics. There are 10 senators (half Democrats, half Republicans) and 16 representatives (half Democrats, half Republicans) who wish to serve on a joint congressional committee on tax reform. An 8-person committee is chosen at random from those who wish to serve.

(A) What is the probability that the joint committee contains equal numbers of senators and representatives?

(B) What is the probability that the joint committee contains equal numbers of Democrats and Republicans?

Answers to Matched Problems

1. (A) $E = \{12\}$; simple event

   (B) $E = \{16, 18\}$; compound event

2. (A) $S_1 = \{BB, BG, GB, GG\}$;

GENDER OF FIRST CHILD	GENDER OF SECOND CHILD	COMBINED OUTCOMES
B	B	BB
B	G	BG
G	B	GB
G	G	GG

   (B) $S_2 = \{0, 1, 2\}$ (C) $S_3 = \{A, D\}$ (D) S_1

3. (A) $\{(4, 1), (3, 2), (2, 3), (1, 4)\}$

   (B) $\{(6, 5), (5, 6)\}$

4. (A) .727   (B) .209   (C) 1

5. (A) $P(E_5) = \dfrac{1}{9}$ (B) $P(E_6) = \dfrac{1}{18}$

6. (A) $\dfrac{9}{100} = .09$ (empirical); $\dfrac{3}{36} = .083$ (theoretical)

   (B) $\dfrac{1}{100} = .01$ (empirical); $\dfrac{1}{36} = .028$ (theoretical)

7. $_{13}C_7/_{52}C_7 \approx 1.3 \times 10^{-5}$

8. $_{12}C_4 \cdot {}_{16}C_2/_{28}C_6 \approx .158$

8.2 Union, Intersection, and Complement of Events; Odds

- Union and Intersection
- Complement of an Event
- Odds
- Applications to Empirical Probability

Recall from Section 8.1 that given a sample space

$$S = \{e_1, e_2, \ldots, e_n\}$$

any function P defined on S such that

$$0 \le P(e_i) \le 1 \qquad i = 1, 2, \ldots, n$$

and

$$P(e_1) + P(e_2) + \cdots + P(e_n) = 1$$

is called a *probability function*. In addition, any subset of S is called an *event E*, and the probability of E is the sum of the probabilities of the simple events in E.

Union and Intersection

Since events are subsets of a sample space, the union and intersection of events are simply the union and intersection of sets as defined in the following box. In this section, we concentrate on the union of events and consider only simple cases of intersection. More complicated cases of intersection will be investigated in the next section.

DEFINITION Union and Intersection of Events*

If A and B are two events in a sample space S, then the **union** of A and B, denoted by $A \cup B$, and the **intersection** of A and B, denoted by $A \cap B$, are defined as follows:

$$A \cup B = \{e \in S \mid e \in A \text{ or } e \in B\} \qquad A \cap B = \{e \in S \mid e \in A \text{ and } e \in B\}$$

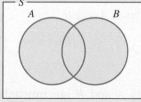

$A \cup B$ is shaded

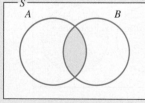

$A \cap B$ is shaded

Furthermore, we define

   The **event A or B** to be $A \cup B$

   The **event A and B** to be $A \cap B$

---

*See Section 7.2 for a discussion of set notation.

EXAMPLE 1   Probability Involving Union and Intersection Consider the sample space of equally likely events for the rolling of a single fair die:

$$S = \{1, 2, 3, 4, 5, 6\}$$

(A) What is the probability of rolling a number that is odd **and** exactly divisible by 3?

(B) What is the probability of rolling a number that is odd **or** exactly divisible by 3?

SOLUTION

(A) Let A be the event of rolling an odd number, B the event of rolling a number divisible by 3, and F the event of rolling a number that is odd **and** divisible by 3. Then

$$A = \{1, 3, 5\} \qquad B = \{3, 6\} \qquad F = A \cap B = \{3\}$$

The probability of rolling a number that is odd **and** exactly divisible by 3 is

$$P(F) = P(A \cap B) = \frac{n(A \cap B)}{n(S)} = \frac{1}{6}$$

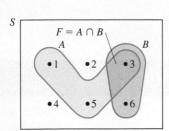

(B) Let A and B be the same events as in part (A), and let E be the event of rolling a number that is odd **or** divisible by 3. Then

$$A = \{1, 3, 5\} \qquad B = \{3, 6\} \qquad E = A \cup B = \{1, 3, 5, 6\}$$

The probability of rolling a number that is odd **or** exactly divisible by 3 is

$$P(E) = P(A \cup B) = \frac{n(A \cup B)}{n(S)} = \frac{4}{6} = \frac{2}{3}$$

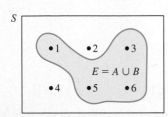

<u>Matched Problem 1</u> Use the sample space in Example 1 to answer the following:

(A) What is the probability of rolling an odd number **and** a prime number?

(B) What is the probability of rolling an odd number **or** a prime number?

Suppose that A and B are events in a sample space S. How is the probability of $A \cup B$ related to the individual probabilities of A and of B? Think of the probability of an element of S as being its weight. To find the total weight of the elements of $A \cup B$, we weigh all of the elements of A, then weigh all of the elements of B, and subtract from the sum the weight of all of the elements that were weighed twice. This gives the formula for $P(A \cup B)$ in Theorem 1.

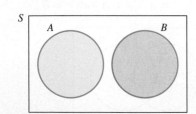

Figure 1 **Mutually exclusive:**
$A \cap B = \varnothing$

THEOREM 1   Probability of the Union of Two Events

For any events A and B,

$$P(A \cup B) = P(A) + P(B) - P(A \cap B) \qquad (1)$$

Events A and B are **mutually exclusive (disjoint)** if their intersection is the empty set (Fig. 1). In that case, equation (1) simplifies, because the probability of the empty set is 0. So, if $A \cap B = \varnothing$, then $P(A \cup B) = P(A) + P(B)$.

CONCEPTUAL INSIGHT

Note the similarity between equation (1) in Theorem 1 and the formula obtained in Section 7–3 for counting the number of elements in $A \cup B$:

$$n(A \cup B) = n(A) + n(B) - n(A \cap B)$$

To find the number of elements in $A \cup B$, we count all of the elements of A, then count all of the elements of B, and subtract from the sum the number of elements that were counted twice.

EXAMPLE 2 Probability Involving Union and Intersection Suppose that two fair dice are rolled.

(A) What is the probability that a sum of 7 or 11 turns up?

(B) What is the probability that both dice turn up the same or that a sum less than 5 turns up?

SOLUTION

(A) If A is the event that a sum of 7 turns up and B is the event that a sum of 11 turns up, then (Fig. 2) the event that a sum of 7 or 11 turns up is $A \cup B$, where

$$A = \{(1,6), (2,5), (3,4), (4,3), (5,2), (6,1)\}$$
$$B = \{(5,6), (6,5)\}$$

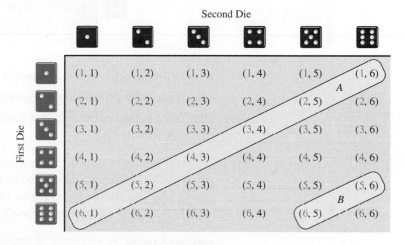

Figure 2

Since events A and B are mutually exclusive, we can use the simplified version of equation (1) to calculate $P(A \cup B)$:

$$P(A \cup B) = P(A) + P(B)$$

$$= \frac{6}{36} + \frac{2}{36}$$

$$= \frac{8}{36} = \frac{2}{9}$$ In this equally likely sample space, $n(A) = 6$, $n(B) = 2$, and $n(S) = 36$.

(B) If A is the event that both dice turn up the same and B is the event that the sum is less than 5, then (Fig. 3) the event that both dice turn up the same or the sum is less than 5 is $A \cup B$, where

$$A = \{(1,1), (2,2), (3,3), (4,4), (5,5), (6,6)\}$$
$$B = \{(1,1), (1,2), (1,3), (2,1), (2,2), (3,1)\}$$

Since $A \cap B = \{(1,1), (2,2)\}$, A and B are not mutually exclusive, and we use equation (1) to calculate $P(A \cup B)$:

$$P(A \cup B) = P(A) + P(B) - P(A \cap B)$$

$$= \frac{6}{36} + \frac{6}{36} - \frac{2}{36}$$

$$= \frac{10}{36} = \frac{5}{18}$$

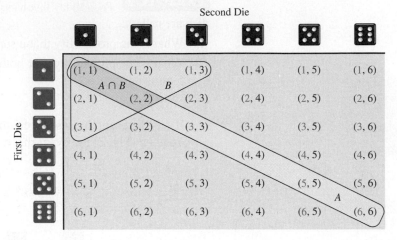

Figure 3

Matched Problem 2 Use the sample space in Example 2 to answer the following:

(A) What is the probability that a sum of 2 or 3 turns up?

(B) What is the probability that both dice turn up the same or that a sum greater than 8 turns up?

You no doubt noticed in Example 2 that we actually did not have to use equation (1). We could have proceeded as in Example 1 and simply counted sample points in $A \cup B$. The following example illustrates the use of equation (1) in a situation where visual representation of sample points is not practical.

EXAMPLE 3   Probability Involving Union and Intersection What is the probability that a number selected at random from the first 500 positive integers is (exactly) divisible by 3 or 4?

SOLUTION Let A be the event that a drawn integer is divisible by 3 and B the event that a drawn integer is divisible by 4. Note that events A and B are not mutually exclusive because multiples of 12 are divisible by both 3 and 4. Since each of the positive integers from 1 to 500 is as likely to be drawn as any other, we can use $n(A), n(B)$, and $n(A \cap B)$ to determine $P(A \cup B)$, where

$$n(A) = \text{the largest integer less than or equal to } \frac{500}{3} = 166$$

$$n(B) = \text{the largest integer less than or equal to } \frac{500}{4} = 125$$

$$n(A \cap B) = \text{the largest integer less than or equal to } \frac{500}{12} = 41$$

Now we can compute $P(A \cup B)$:

$$P(A \cup B) = P(A) + P(B) - P(A \cap B)$$
$$= \frac{n(A)}{n(S)} + \frac{n(B)}{n(S)} - \frac{n(A \cap B)}{n(S)}$$
$$- \frac{166}{500} + \frac{125}{500} - \frac{41}{500} = \frac{250}{500} = .5$$

Matched Problem 3 ⌋ What is the probability that a number selected at random from the first 140 positive integers is (exactly) divisible by 4 or 6?

Complement of an Event

Suppose that we divide a finite sample space

$$S = \{e_1, \ldots, e_n\}$$

into two subsets E and E' such that

$$E \cap E' = \varnothing$$

that is, E and E' are mutually exclusive and

$$E \cup E' = S$$

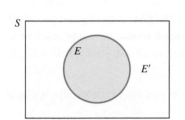

Figure 4

Then E' is called the **complement of E** relative to S. Note that E' contains all the elements of S that are not in E (Fig. 4). Furthermore,

$$P(S) = P(E \cup E')$$
$$= P(E) + P(E') = 1$$

Therefore,

$$P(E) = 1 - P(E') \qquad P(E') = 1 - P(E) \tag{2}$$

If the probability of rain is .67, then the probability of no rain is $1 - .67 = .33$; if the probability of striking oil is .01, then the probability of not striking oil is .99. If the probability of having at least 1 boy in a 2-child family is .75, what is the probability of having 2 girls? [*Answer:*.25.]

Explore and Discuss 1   (A) Suppose that E and F are complementary events. Are E and F necessarily mutually exclusive? Explain why or why not.

(B) Suppose that E and F are mutually exclusive events. Are E and F necessarily complementary? Explain why or why not.

In looking for $P(E)$, there are situations in which it is easier to find $P(E')$ first, and then use equations (2) to find $P(E)$. The next two examples illustrate two such situations.

EXAMPLE 4   Quality Control   A shipment of 45 precision parts, including 9 that are defective, is sent to an assembly plant. The quality control division selects 10 at random for testing and rejects the entire shipment if 1 or more in the sample are found to be defective. What is the probability that the shipment will be rejected?

SOLUTION   If E is the event that 1 or more parts in a random sample of 10 are defective, then E', the complement of E, is the event that no parts in a random sample of 10 are defective. It is easier to compute $P(E')$ than to compute $P(E)$ directly. Once $P(E')$ is found, we will use $P(E) = 1 - P(E')$ to find $P(E)$.

The sample space S for this experiment is the set of all subsets of 10 elements from the set of 45 parts shipped. Thus, since there are $45 - 9 = 36$ nondefective parts,

$$P(E') = \frac{n(E')}{n(S)} = \frac{_{36}C_{10}}{_{45}C_{10}} \approx .08$$

and

$$P(E) = 1 - P(E') \approx 1 - .08 = .92$$

Matched Problem 4 A shipment of 40 precision parts, including 8 that are defective, is sent to an assembly plant. The quality control division selects 10 at random for testing and rejects the entire shipment if 1 or more in the sample are found to be defective. What is the probability that the shipment will be rejected?

EXAMPLE 5 Birthday Problem In a group of n people, what is the probability that at least 2 people have the same birthday (same month and day, excluding February 29)? Make a guess for a class of 40 people, and check your guess with the conclusion of this example.

SOLUTION If we form a list of the birthdays of all the people in the group, we have a simple event in the sample space

$$S = \text{set of all lists of } n \text{ birthdays}$$

For any person in the group, we will assume that any birthday is as likely as any other, so that the simple events in S are equally likely. How many simple events are in the set S? Since any person could have any one of 365 birthdays (excluding February 29), the multiplication principle implies that the number of simple events in S is

$$
\begin{array}{ccccccc}
& \text{1st} & \text{2nd} & \text{3rd} & & \text{nth} \\
& \text{person} & \text{person} & \text{person} & & \text{person} \\
n(S) = & 365 & \cdot\; 365 & \cdot\; 365 & \cdots\cdot & 365
\end{array}
$$
$$= 365^n$$

Now, let E be the event that at least 2 people in the group have the same birthday. Then E' is the event that no 2 people have the same birthday. The multiplication principle can be used to determine the number of simple events in E':

$$
\begin{array}{ccccccc}
& \text{1st} & \text{2nd} & \text{3rd} & & \text{nth} \\
& \text{person} & \text{person} & \text{person} & & \text{person} \\
n(E') = & 365 & \cdot\; 364 & \cdot\; 363 & \cdots\cdot & (366 - n)
\end{array}
$$

$$= \frac{[365 \cdot 364 \cdot 363 \cdot \cdots \cdot (366 - n)](365 - n)!}{(365 - n)!}$$
$\quad$ Multiply numerator and denominator by $(365 - n)!$.

$$= \frac{365!}{(365 - n)!}$$

Since we have assumed that S is an equally likely sample space,

$$P(E') = \frac{n(E')}{n(S)} = \frac{\dfrac{365!}{(365 - n)!}}{365^n} = \frac{365!}{365^n(365 - n)!}$$

Therefore,

$$P(E) = 1 - P(E')$$

$$= 1 - \frac{365!}{365^n(365 - n)!} \tag{3}$$

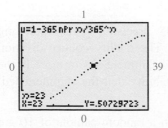

Figure 5

Table 1 Birthday Problem

Number of People in Group n	Probability That 2 or More Have Same Birthday $P(E)$
5	.027
10	.117
15	.253
20	.411
23	.507
30	.706
40	.891
50	.970
60	.994
70	.999

Equation (3) is valid for any n satisfying $1 \leq n \leq 365$. [What is $P(E)$ if $n > 365$?] For example, in a group of 5 people,

$$P(E) = 1 - \frac{365!}{(365)^5 \, 360!}$$

$$= 1 - \frac{365 \cdot 364 \cdot 363 \cdot 362 \cdot 361 \cdot 360!}{365 \cdot 365 \cdot 365 \cdot 365 \cdot 365 \cdot 360!}$$

$$= .027$$

It is interesting to note that as the size of the group increases, $P(E)$ increases more rapidly than you might expect. Figure 5* shows the graph of $P(E)$ for $1 \leq n \leq 39$. Table 1 gives the value of $P(E)$ for selected values of n. If $n = 5$, Table 1 gives $P(E) = .027$, as calculated above. Notice that for a group of only 23 people, the probability that 2 or more have the same birthday is greater than $\frac{1}{2}$.

Matched Problem 5 Use equation (3) to evaluate $P(E)$ for $n = 4$.

Explore and Discuss 2 Determine the smallest number n such that in a group of n people, the probability that 2 or more have a birthday in the same month is greater than .5. Discuss the assumptions underlying your computation.

Odds

When the probability of an event E is known, it is often customary (especially in gaming situations) to speak of odds for or against E rather than the probability of E. For example, if you roll a fair die once, then the *odds for* rolling a 2 are 1 to 5 (also written 1 : 5), and the *odds against* rolling a 2 are 5 to 1 (or 5 : 1). This is consistent with the following instructions for converting probabilities to odds.

DEFINITION From Probabilities to Odds

If $P(E)$ is the probability of the event E, then

(A) **Odds for E** $= \dfrac{P(E)}{1 - P(E)} = \dfrac{P(E)}{P(E')}$ $P(E) \neq 1$

(B) **Odds against E** $= \dfrac{P(E')}{P(E)}$ $P(E) \neq 0$

The ratio $\dfrac{P(E)}{P(E')}$, giving odds for E, is usually expressed as an equivalent ratio $\dfrac{a}{b}$ of whole numbers (by multiplying numerator and denominator by the same number), and written "a to b" or "$a : b$." In this case, the odds against E are written "b to a" or "$b : a$."

Odds have a natural interpretation in terms of **fair games**. Let's return to the experiment of rolling a fair die once. Recall that the odds for rolling a 2 are 1 to 5. Turn the experiment into a fair game as follows: If you bet \$1 on rolling a 2, then the house pays you \$5 (and returns your \$1 bet) if you roll a 2; otherwise, you lose the \$1 bet.

More generally, consider any experiment and an associated event E. If the odds for E are a to b, then the experiment can be turned into a fair game as follows: If you bet \$$a$ on event E, then the house pays you \$$b$ (and returns your \$$a$ bet) if E occurs; otherwise, you lose the \$$a$ bet.

---

*See Problem 71 in Exercises 8.2 for a discussion of the form of equation (3) used to produce the graph in Figure 5.

EXAMPLE 6   Probability and Odds

(A) What are the odds for rolling a sum of 7 in a single roll of two fair dice?

(B) If you bet $1 on rolling a sum of 7, what should the house pay (plus returning your $1 bet) if you roll a sum of 7 in order for the game to be fair?

SOLUTION

(A) Let E denote the event of rolling a sum of 7. Then $P(E) = \frac{6}{36} = \frac{1}{6}$. So the odds for E are

$$\frac{P(E)}{P(E')} = \frac{\frac{1}{6}}{\frac{5}{6}} = \frac{1}{5} \qquad \text{Also written as "1 to 5" or "1 : 5."}$$

(B) The odds for rolling a sum of 7 are 1 to 5. The house should pay $5 (and return your $1 bet) if you roll a sum of 7 for the game to be fair.

Matched Problem 6

(A) What are the odds for rolling a sum of 8 in a single roll of two fair dice?

(B) If you bet $5 that a sum of 8 will turn up, what should the house pay (plus returning your $5 bet) if a sum of 8 does turn up in order for the game to be fair?

Now we will go in the other direction: If we are given the odds for an event, what is the probability of the event? (The verification of the following formula is left to Problem 75 in Exercises 8.2.)

If the odds for event E are a/b, then the probability of E is

$$P(E) = \frac{a}{a + b}$$

EXAMPLE 7   Odds and Probability   If in repeated rolls of two fair dice, the odds for rolling a 5 before rolling a 7 are 2 to 3, then the probability of rolling a 5 before rolling a 7 is

$$P(E) = \frac{a}{a + b} = \frac{2}{2 + 3} = \frac{2}{5}$$

Matched Problem 7   If in repeated rolls of two fair dice, the odds against rolling a 6 before rolling a 7 are 6 to 5, then what is the probability of rolling a 6 before rolling a 7?

Applications to Empirical Probability

In the following discussions, the term *empirical probability* will mean the probability of an event determined by a sample that is used to approximate the probability of the corresponding event in the total population. How does the approximate empirical probability of an event determined from a sample relate to the actual probability of an event relative to the total population? In mathematical statistics an important theorem called the **law of large numbers** (or the **law of averages**) is proved. Informally, it states that the approximate empirical probability can be made as close to the actual probability as we please by making the sample sufficiently large.

EXAMPLE 8   Market Research   From a survey of 1,000 people in Springfield, it was found that 500 people had tried a certain brand of diet cola, 600 had tried a certain brand of regular cola, and 200 had tried both types of cola. If a person from Springfield is selected at random, what is the (empirical) probability that

(A) He or she has tried the diet cola or the regular cola? What are the (empirical) odds for this event?

(B) He or she has tried one of the colas but not both? What are the (empirical) odds against this event?

SOLUTION Let D be the event that a person has tried the diet cola and R the event that a person has tried the regular cola. The events D and R can be used to partition the population of Springfield into four mutually exclusive subsets (a collection of subsets is mutually exclusive if the intersection of any two of them is the empty set):

$D \cap R$ = set of people who have tried both colas

$D \cap R'$ = set of people who have tried the diet cola but not the regular cola

$D' \cap R$ = set of people who have tried the regular cola but not the diet cola

$D' \cap R'$ = set of people who have not tried either cola

These sets are displayed in the Venn diagram in Figure 6.

The sample population of 1,000 persons is also partitioned into four mutually exclusive sets, with $n(D) = 500$, $n(R) = 600$, and $n(D \cap R) = 200$. By using a Venn diagram (Fig. 7), we can determine the number of sample points in the sets $D \cap R'$, $D' \cap R$, and $D' \cap R'$ (see Example 2, Section 7.3).

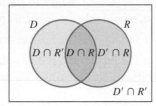

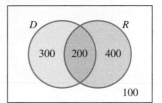

Figure 6 **Total population**   Figure 7 **Sample population**

These numbers are displayed in a table:

		Regular R	No Regular R'	Totals
Diet	D	200	300	500
No Diet	D'	400	100	500
Totals		600	400	1,000

Assuming that each sample point is equally likely, we form a probability table by dividing each entry in this table by 1,000, the total number surveyed. These are empirical probabilities for the sample population, which we can use to approximate the corresponding probabilities for the total population.

		Regular R	No Regular R'	Totals
Diet	D	.2	.3	.5
No Diet	D'	.4	.1	.5
Totals		.6	.4	1.0

Now we are ready to compute the required probabilities.

(A) The event that a person has tried the diet cola or the regular cola is $E = D \cup R$. We compute $P(E)$ two ways:

Method 1. Directly:

$$P(E) = P(D \cup R)$$
$$= P(D) + P(R) - P(D \cap R)$$
$$= .5 + .6 - .2 = .9$$

Method 2. Using the complement of E:

$$P(E) = 1 - P(E')$$
$$= 1 - P(D' \cap R') \quad E' = (D \cup R)' = D' \cap R' \text{ (see Fig. 7)}$$
$$= 1 - .1 = .9$$

In either case,

$$\text{odds for } E = \frac{P(E)}{P(E')} = \frac{.9}{.1} = \frac{9}{1} \quad \text{or} \quad 9:1$$

(B) The event that a person has tried one cola but not both is the event that the person has tried diet cola and not regular cola or has tried regular cola and not diet cola. In terms of sets, this is event $E = (D \cap R') \cup (D' \cap R)$. Since $D \cap R'$ and $D' \cap R$ are mutually exclusive (Fig. 6),

$$P(E) = P[(D \cap R') \cup (D' \cap R)]$$
$$= P(D \cap R') + P(D' \cap R)$$
$$= .3 + .4 = .7$$

$$\text{odds against } E = \frac{P(E')}{P(E)} = \frac{.3}{.7} = \frac{3}{7} \quad \text{or} \quad 3:7$$

Matched Problem 8 Refer to Example 8. If a person from Springfield is selected at random, what is the (empirical) probability that

(A) He or she has not tried either cola? What are the (empirical) odds for this event?

(B) He or she has tried the diet cola or has not tried the regular cola? What are the (empirical) odds against this event?

Exercises 8.2

Skills Warm-up Exercises

W   *In Problems 1–6, write the expression as a quotient of integers, reduced to lowest terms. (If necessary, review Section A.1).*

1. $\dfrac{\frac{3}{10}}{\frac{9}{10}}$

2. $\dfrac{\frac{5}{12}}{\frac{7}{12}}$

3. $\dfrac{\frac{1}{8}}{\frac{3}{7}}$

4. $\dfrac{\frac{4}{5}}{\frac{5}{6}}$

5. $\dfrac{\frac{2}{9}}{1 - \frac{2}{9}}$

6. $\dfrac{\frac{3}{16}}{1 - \frac{3}{16}}$

Problems 7–12 refer to the Venn diagram below for events A and B in an equally likely sample space S. Find each of the indicated probabilities.

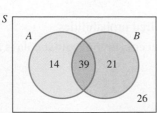

7. $P(A \cap B)$

8. $P(A \cup B)$

9. $P(A' \cup B)$

10. $P(A \cap B')$

11. $P((A \cup B)')$

12. $P((A \cap B)')$

A single card is drawn from a standard 52-card deck. Let D be the event that the card drawn is a diamond, and let F be the event that the card drawn is a face card. In Problems 13–24, find the indicated probabilities.

13. $P(D)$

14. $P(F)$

15. $P(F')$

16. $P(D')$

17. $P(D \cap F)$

18. $P(D' \cap F)$

19. $P(D \cup F)$

20. $P(D' \cup F)$

21. $P(D \cap F')$

22. $P(D' \cap F')$

23. $P(D \cup F')$

24. $P(D' \cup F')$

In a lottery game, a single ball is drawn at random from a container that contains 25 identical balls numbered from 1 through 25. In Problems 25–32, use equation (1) to compute the probability that the number drawn is

25. Odd or a multiple of 4

26. Even or a multiple of 7

27. Prime or greater than 20

28. Less than 10 or greater than 10

29. A multiple of 2 or a multiple of 5

30. A multiple of 3 or a multiple of 4

31. Less than 5 or greater than 20

32. Prime or less than 14

33. If the probability is .51 that a candidate wins the election, what is the probability that he loses?

34. If the probability is .03 that an automobile tire fails in less than 50,000 miles, what is the probability that the tire does not fail in 50,000 miles?

In Problems 35–38, use the equally likely sample space in Example 2 to compute the probability of the following events:

35. A sum that is less than or equal to 5

36. A sum that is greater than 9

37. The number on the first die is a 6 or the number on the second die is a 3.

38. The number on the first die is even or the number on the second die is even.

39. Given the following probabilities for an event E, find the odds for and against E:

(A) $\dfrac{3}{8}$ (B) $\dfrac{1}{4}$ (C) .4   (D) .55

40. Given the following probabilities for an event E, find the odds for and against E:

(A) $\dfrac{3}{5}$ (B) $\dfrac{1}{7}$ (C) .6   (D) .35

41. Compute the probability of event E if the odds in favor of E are

(A) $\dfrac{3}{8}$ (B) $\dfrac{11}{7}$ (C) $\dfrac{4}{1}$ (D) $\dfrac{49}{51}$

42. Compute the probability of event E if the odds in favor of E are

(A) $\dfrac{5}{9}$ (B) $\dfrac{4}{3}$ (C) $\dfrac{3}{7}$ (D) $\dfrac{23}{77}$

✎ *In Problems 43–48, discuss the validity of each statement. If the statement is always true, explain why. If not, give a counterexample.*

43. If the odds for E equal the odds against E', then $P(E) = \dfrac{1}{2}$.

44. If the odds for E are a: b, then the odds against E are b: a.

45. If $P(E) + P(F) = P(E \cup F) + P(E \cap F)$, then E and F are mutually exclusive events.

46. The theoretical probability of an event is less than or equal to its empirical probability.

47. If E and F are complementary events, then E and F are mutually exclusive.

48. If E and F are mutually exclusive events, then E and F are complementary.

In Problems 49–52, compute the odds in favor of obtaining

49. A head in a single toss of a coin

50. A number divisible by 3 in a single roll of a die

51. At least 1 head when a single coin is tossed 3 times

52. 1 head when a single coin is tossed twice

In Problems 53–56, compute the odds against obtaining

53. A number greater than 4 in a single roll of a die

54. 2 heads when a single coin is tossed twice

55. A 3 or an even number in a single roll of a die

56. An odd number or a number divisible by 3 in a single roll of a die

57. (A) What are the odds for rolling a sum of 5 in a single roll of two fair dice?

   (B) If you bet $1 that a sum of 5 will turn up, what should the house pay (plus returning your $1 bet) if a sum of 5 turns up in order for the game to be fair?

58. (A) What are the odds for rolling a sum of 10 in a single roll of two fair dice?

   (B) If you bet $1 that a sum of 10 will turn up, what should the house pay (plus returning your $1 bet) if a sum of 10 turns up in order for the game to be fair?

A pair of dice are rolled 1,000 times with the following frequencies of outcomes:

Sum	2	3	4	5	6	7	8	9	10	11	12
Frequency	10	30	50	70	110	150	170	140	120	80	70

Use these frequencies to calculate the approximate empirical probabilities and odds for the events in Problems 59 and 60.

59. (A) The sum is less than 4 or greater than 9.

   (B) The sum is even or exactly divisible by 5.

60. (A) The sum is a prime number or is exactly divisible by 4.

   (B) The sum is an odd number or exactly divisible by 3.

In Problems 61–64, a single card is drawn from a standard 52-card deck. Calculate the probability of and odds for each event.

61. A face card or a club is drawn.

62. A king or a heart is drawn.

63. A black card or an ace is drawn.

64. A heart or a number less than 7 (count an ace as 1) is drawn.

65. What is the probability of getting at least 1 diamond in a 5-card hand dealt from a standard 52-card deck?

66. What is the probability of getting at least 1 black card in a 7-card hand dealt from a standard 52-card deck?

67. What is the probability that a number selected at random from the first 1,000 positive integers is (exactly) divisible by 6 or 8?

68. What is the probability that a number selected at random from the first 600 positive integers is (exactly) divisible by 6 or 9?

69. Explain how the three events A, B, and C from a sample space S are related to each other in order for the following equation to hold true:

$$P(A \cup B \cup C) = P(A) + P(B) + P(C) - P(A \cap B)$$

70. Explain how the three events A, B, and C from a sample space S are related to each other in order for the following equation to hold true:

$$P(A \cup B \cup C) = P(A) + P(B) + P(C)$$

71. Show that the solution to the birthday problem in Example 5 can be written in the form

$$P(E) = 1 - \frac{_{365}P_n}{365^n}$$

For a calculator that has a $_nP_r$ function, explain why this form may be better for direct evaluation than the other form used in the solution to Example 5. Try direct evaluation of both forms on a calculator for $n = 25$.

72. Many (but not all) calculators experience an overflow error when computing $_{365}P_n$ for $n > 39$ and when computing 365^n. Explain how you would evaluate $P(E)$ for any $n > 39$ on such a calculator.

73. In a group of n people ($n \le 12$), what is the probability that at least 2 of them have the same birth month? (Assume that any birth month is as likely as any other.)

74. In a group of n people ($n \le 100$), each person is asked to select a number between 1 and 100, write the number on a slip of paper and place the slip in a hat. What is the probability that at least 2 of the slips in the hat have the same number written on them?

75. If the odds in favor of an event E occurring are a to b, show that

$$P(E) = \frac{a}{a + b}$$

[*Hint:* Solve the equation $P(E)/P(E') = a/b$ for $P(E)$.]

76. If $P(E) = c/d$, show that odds in favor of E occurring are c to $d - c$.

77. The command in Figure A was used on a graphing calculator to simulate 50 repetitions of rolling a pair of dice and recording their sum. A statistical plot of the results is shown in Figure B.

(A) Use Figure B to find the empirical probability of rolling a 7 or 8.

(B) What is the theoretical probability of rolling a 7 or 8?

(C) Use a graphing calculator to simulate 200 repetitions of rolling a pair of dice and recording their sum, and find the empirical probability of rolling a 7 or 8.

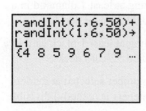

(A)

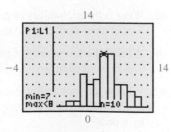

(B)

78. Consider the command in Figure A and the associated statistical plot in Figure B.

(A) Explain why the command does not simulate 50 repetitions of rolling a pair of dice and recording their sum.

(B) Describe an experiment that is simulated by this command.

(C) Simulate 200 repetitions of the experiment you described in part (B), and find the empirical probability of recording a 7 or 8.

(D) What is the theoretical probability of recording a 7 or 8?

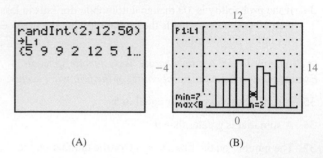

(A)   (B)

Applications

79. Market research. From a survey involving 1,000 university students, a market research company found that 750 students owned laptops, 450 owned cars, and 350 owned cars and laptops. If a university student is selected at random, what is the (empirical) probability that

(A) The student owns either a car or a laptop?

(B) The student owns neither a car nor a laptop?

80. Market research. Refer to Problem 79. If a university student is selected at random, what is the (empirical) probability that

(A) The student does not own a car?

(B) The student owns a car but not a laptop?

81. Insurance. By examining the past driving records of city drivers, an insurance company has determined the following (empirical) probabilities:

	Miles Driven per Year			
	Less than 10,000, M_1	10,000–15,000, inclusive, M_2	More than 15,000, M_3	Totals
Accident A	.05	.1	.15	.3
No Accident A'	.15	.2	.35	.7
Totals	.2	.3	.5	1.0

If a city driver is selected at random, what is the probability that

(A) He or she drives less than 10,000 miles per year or has an accident?

(B) He or she drives 10,000 or more miles per year and has no accidents?

82. Insurance. Use the (empirical) probabilities in Problem 81 to find the probability that a city driver selected at random

(A) Drives more than 15,000 miles per year or has an accident

(B) Drives 15,000 or fewer miles per year and has an accident

83. Quality control. A shipment of 60 game players, including 9 that are defective, is sent to a retail store. The receiving department selects 10 at random for testing and rejects the whole shipment if 1 or more in the sample are found to be defective. What is the probability that the shipment will be rejected?

84. Quality control. An assembly plant produces 40 outboard motors, including 7 that are defective. The quality control department selects 10 at random (from the 40 produced) for testing and will shut down the plant for trouble shooting if 1 or more in the sample are found to be defective. What is the probability that the plant will be shut down?

85. Medicine. In order to test a new drug for adverse reactions, the drug was administered to 1,000 test subjects with the following results: 60 subjects reported that their only adverse reaction was a loss of appetite, 90 subjects reported that their only adverse reaction was a loss of sleep, and 800 subjects reported no adverse reactions at all. If this drug is released for general use, what is the (empirical) probability that a person using the drug will suffer both a loss of appetite and a loss of sleep?

86. Product testing. To test a new car, an automobile manufacturer wants to select 4 employees to test drive the car for 1 year. If 12 management and 8 union employees volunteer to be test drivers and the selection is made at random, what is the probability that at least 1 union employee is selected?

Problems 87 and 88 refer to the data in the following table, obtained from a random survey of 1,000 residents of a state. The participants were asked their political affiliations and

their preferences in an upcoming election. (In the table, D = Democrat, R = Republican, and U = Unaffiliated.)

		D	R	U	Totals
Candidate A	A	200	100	85	385
Candidate B	B	250	230	50	530
No Preference	N	50	20	15	85
Totals		500	350	150	1,000

87. Politics. If a state resident is selected at random, what is the (empirical) probability that the resident is

(A) Not affiliated with a political party or has no preference? What are the odds for this event?

(B) Affiliated with a political party and prefers candidate A? What are the odds against this event?

88. Politics. If a state resident is selected at random, what is the (empirical) probability that the resident is

(A) A Democrat or prefers candidate B? What are the odds for this event?

(B) Not a Democrat and has no preference? What are the odds against this event?

Answers to Matched Problems

1. (A) $\frac{1}{3}$ (B) $\frac{2}{3}$ **2.** (A) $\frac{1}{12}$ (B) $\frac{7}{?}$

3. $\frac{47}{140} \approx .336$ **4.** .92   **5.** .016

6. (A) $5:31$ (B) $\$31$ **7.** $\frac{5}{11} \approx$

8. (A) $P(D' \cap R') = .1$; odds for D'

(B) $P(D \cup R') = .6$; odds aga.

8.3 Conditional Probability, Intersection, and Independence

- Conditional Probability
- Intersection of Events: Product Rule
- Probability Trees
- Independent Events
- Summary

In Section 8.2, we learned that the probability of the union of two events is related to the sum of the probabilities of the individual events (Theorem 1, p. 400):

$$P(A \cup B) = P(A) + P(B) - P(A \cap B)$$

In this section, we will learn how the probability of the intersection of two events is related to the product of the probabilities of the individual events. But first we must investigate the concept of *conditional probability*.

Conditional Probability

The probability of an event may change if we are told of the occurrence of another event. For example, if an adult (21 years or older) is selected at random from all adults in the United States, the probability of that person having lung cancer would not be high. However, if we are told that the person is also a heavy smoker, we would want to revise the probability upward.

In general, the probability of the occurrence of an event A, given the occurrence of another event B, is called a **conditional probability** and is denoted by $P(A|B)$.

In the preceding situation, events A and B would be

$$A = \text{adult has lung cancer}$$
$$B = \text{adult is a heavy smoker}$$

and $P(A|B)$ would represent the probability of an adult having lung cancer, given that he or she is a heavy smoker.

Our objective now is to try to formulate a precise definition of $P(A|B)$. It is helpful to start with a relatively simple problem, solve it intuitively, and then generalize from this experience.

What is the probability of rolling a prime number (2, 3, or 5) in a single roll of a fair die? Let

$$S = \{1, 2, 3, 4, 5, 6\}$$

Then the event of rolling a prime number is (Fig. 1)

$$A = \{2, 3, 5\}$$

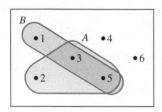

Figure 1

Thus, since we assume that each simple event in the sample space is equally likely,

$$P(A) = \frac{n(A)}{n(S)} = \frac{3}{6} = \frac{1}{2}$$

Now suppose you are asked, "In a single roll of a fair die, what is the probability that a prime number has turned up if we are given the additional information that an odd number has turned up?" The additional knowledge that another event has occurred, namely,

$$B = \text{odd number turns up}$$

puts the problem in a new light. We are now interested only in the part of event A (rolling a prime number) that is in event B (rolling an odd number). Event B, since we know it has occurred, becomes the new sample space. The Venn diagrams in Figure 2 illustrate the various relationships. Thus, the probability of A given B is the number of A elements in B divided by the total number of elements in B. Symbolically,

$$P(A|B) = \frac{n(A \cap B)}{n(B)} = \frac{2}{3}$$

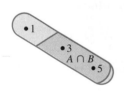

Figure 2   **B is the new sample space.**

Dividing the numerator and denominator of $n(A \cap B)/n(B)$ by $n(S)$, the number of elements in the original sample space, we can express $P(A|B)$ in terms of $P(A \cap B)$ and $P(B)$:*

$$P(A|B) = \frac{n(A \cap B)}{n(B)} = \frac{\dfrac{n(A \cap B)}{n(S)}}{\dfrac{n(B)}{n(S)}} = \frac{P(A \cap B)}{P(B)}$$

Using the right side to compute $P(A|B)$ for the preceding example, we obtain the same result:

$$P(A|B) = \frac{P(A \cap B)}{P(B)} = \frac{\frac{2}{6}}{\frac{3}{6}} = \frac{2}{3}$$

We use the formula above to motivate the following definition of *conditional probability,* which applies to any sample space, including those having simple events that are not equally likely (see Example 1).

DEFINITION   Conditional Probability

For events A and B in an arbitrary sample space S, we define the **conditional probability of A given B** by

$$P(A|B) = \frac{P(A \cap B)}{P(B)} \qquad P(B) \neq 0 \tag{1}$$

---

*Note that $P(A|B)$ is a probability based on the new sample space B, while $P(A \cap B)$ and $P(B)$ are both probabilities based on the original sample space S.

Figure 3

EXAMPLE 1 Conditional Probability A pointer is spun once on a circular spinner (Fig. 3). The probability assigned to the pointer landing on a given integer (from 1 to 6) is the ratio of the area of the corresponding circular sector to the area of the whole circle, as given in the table:

e_i	1	2	3	4	5	6
$P(e_i)$	.1	.2	.1	.1	.3	.2

$S = \{1, 2, 3, 4, 5, 6\}$

(A) What is the probability of the pointer landing on a prime number?

(B) What is the probability of the pointer landing on a prime number, given that it landed on an odd number?

SOLUTION Let the events E and F be defined as follows:

$$E = \text{pointer lands on a prime number} = \{2, 3, 5\}$$
$$F = \text{pointer lands on an odd number} = \{1, 3, 5\}$$

(A) $P(E) = P(2) + P(3) + P(5)$
$$= .2 + .1 + .3 = .6$$

(B) First note that $E \cap F = \{3, 5\}$.

$$P(E|F) = \frac{P(E \cap F)}{P(F)} = \frac{P(3) + P(5)}{P(1) + P(3) + P(5)}$$
$$= \frac{.1 + .3}{.1 + .1 + .3} = \frac{.4}{.5} = .8$$

Matched Problem 1 Refer to Example 1.

(A) What is the probability of the pointer landing on a number greater than 4?

(B) What is the probability of the pointer landing on a number greater than 4, given that it landed on an even number?

EXAMPLE 2 Safety Research Suppose that city records produced the following probability data on a driver being in an accident on the last day of a Memorial Day weekend:

		Accident A	No Accident A'	Totals	
Rain	R	.025	.335	.360	
No Rain	R'	.015	.625	.640	$S = \{RA, RA', R'A, R'A'\}$
Totals		.040	.960	1.000	

(A) Find the probability of an accident, rain or no rain.

(B) Find the probability of rain, accident or no accident.

(C) Find the probability of an accident and rain.

(D) Find the probability of an accident, given rain.

SOLUTION

(A) Let $A = \{RA, R'A\}$ Event: "accident"

$$P(A) = P(RA) + P(R'A) = .025 + .015 = .040$$

(B) Let $R = \{RA, RA'\}$ Event: "rain"

$$P(R) = P(RA) + P(RA') = .025 + .335 = .360$$

(C) $A \cap R = \{RA\}$ Event: "accident and rain"

$$P(A \cap R) = P(RA) = .025$$

(D) $P(A|R) = \dfrac{P(A \cap R)}{P(R)} = \dfrac{.025}{.360} = .069$ Event: "accident, given rain"

Compare the result in part (D) with that in part (A). Note that $P(A|R) \neq P(A)$, and the probability of an accident, given rain, is higher than the probability of an accident without the knowledge of rain.

Matched Problem 2   Referring to the table in Example 2, determine the following:

(A) Probability of no rain

(B) Probability of an accident and no rain

(C) Probability of an accident, given no rain [Use formula (1) and the results of parts (A) and (B).]

Intersection of Events: Product Rule

Let's return to the original problem of this section, that is, representing the probability of an intersection of two events in terms of the probabilities of the individual events. If $P(A) \neq 0$ and $P(B) \neq 0$, then using formula (1), we can write

$$P(A|B) = \dfrac{P(A \cap B)}{P(B)} \quad \text{and} \quad P(B|A) = \dfrac{P(B \cap A)}{P(A)}$$

Solving the first equation for $P(A \cap B)$ and the second equation for $P(B \cap A)$, we have

$$P(A \cap B) = P(B)P(A|B) \quad \text{and} \quad P(B \cap A) = P(A)P(B|A)$$

Since $A \cap B = B \cap A$ for any sets A and B, it follows that

$$P(A \cap B) = P(B)P(A|B) = P(A)P(B|A)$$

and we have the **product rule:**

THEOREM 1   Product Rule

For events A and B with nonzero probabilities in a sample space S,

$$P(A \cap B) = P(A)P(B|A) = P(B)P(A|B) \qquad (2)$$

and we can use either $P(A)P(B|A)$ or $P(B)P(A|B)$ to compute $P(A \cap B)$.

EXAMPLE 3   Consumer Survey   If 60% of a department store's customers are female and 75% of the female customers have credit cards at the store, what is the probability that a customer selected at random is a female and has a store credit card?

SOLUTION   Let

$$S = \text{all store customers}$$
$$F = \text{female customers}$$
$$C = \text{customers with a store credit card}$$

If 60% of the customers are female, then the probability that a customer selected at random is a female is

$$P(F) = .60$$

Since 75% of the female customers have store credit cards, the probability that a customer has a store credit card, given that the customer is a female, is

$$P(C|F) = .75$$

Using equation (2), the probability that a customer is a female and has a store credit card is

$$P(F \cap C) = P(F)P(C|F) = (.60)(.75) = .45$$

Matched Problem 3 If 80% of the male customers of the department store in Example 3 have store credit cards, what is the probability that a customer selected at random is a male and has a store credit card?

Probability Trees

We used tree diagrams in Section 7.3 to help us count the number of combined outcomes in a sequence of experiments. In a similar way, we will use probability trees to help us compute the probabilities of combined outcomes in a sequence of experiments.

EXAMPLE 4 Probability Tree Two balls are drawn in succession, without replacement, from a box containing 3 blue and 2 white balls (Fig. 4). What is the probability of drawing a white ball on the second draw?

SOLUTION We start with a tree diagram (Fig. 5) showing the combined outcomes of the two experiments (first draw and second draw). Then we assign a probability to each branch of the tree (Fig. 6). For example, we assign the probability $\frac{2}{5}$ to the branch Sw_1, since this is the probability of drawing a white ball on the first draw (there are 2 white balls and 3 blue balls in the box). What probability should be assigned to the branch w_1w_2? This is the conditional probability $P(w_2|w_1)$, that is, the probability of drawing a white ball on the second draw given that a white ball was drawn on the first draw and not replaced. Since the box now contains 1 white ball and 3 blue balls, the probability is $\frac{1}{4}$. Continuing in the same way, we assign probabilities to the other branches of the tree and obtain Figure 6.

What is the probability of the combined outcome $w_1 \cap w_2$, that is, the probability of drawing a white ball on the first draw and a white ball on the second draw?* Using the product rule (2), we have

$$P(w_1 \cap w_2) = P(w_1)P(w_2|w_1)$$
$$= \left(\frac{2}{5}\right)\left(\frac{1}{4}\right) = \frac{1}{10}$$

The combined outcome $w_1 \cap w_2$ corresponds to the unique path Sw_1w_2 in the tree diagram, and we see that the probability of reaching w_2 along this path is the product of the probabilities assigned to the branches on the path. Reasoning in this way, we obtain the probability of each remaining combined outcome by multiplying the probabilities assigned to the branches on the path corresponding to the given combined outcomes. These probabilities are often written at the ends of the paths to which they correspond (Fig. 7).

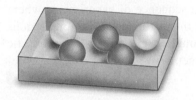
Figure 4

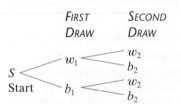
Figure 5

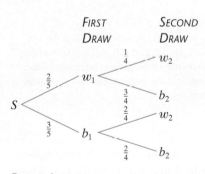
Figure 6

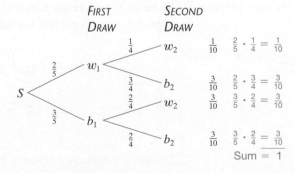
Figure 7

*The sample space for the combined outcomes is $S = \{w_1w_2, w_1b_2, b_1w_2, b_1b_2\}$. If we let $w_1 = \{w_1w_2, w_1b_2\}$ and $w_2 = \{w_1w_2, b_1w_2\}$, then $w_1 \cap w_2 = \{w_1w_2\}$.

Now we can complete the problem. A white ball drawn on the second draw corresponds to either the combined outcome $w_1 \cap w_2$ or $b_1 \cap w_2$ occurring. Thus, since these combined outcomes are mutually exclusive,

$$P(w_2) = P(w_1 \cap w_2) + P(b_1 \cap w_2)$$
$$= \frac{1}{10} + \frac{3}{10} = \frac{4}{10} = \frac{2}{5}$$

which is the sum of the probabilities listed at the ends of the two paths terminating in w_2.

Matched Problem 4) Two balls are drawn in succession without replacement from a box containing 4 red and 2 white balls. What is the probability of drawing a red ball on the second draw?

The sequence of two experiments in Example 4 is an example of a *stochastic process*. In general, a **stochastic process** involves a sequence of experiments where the outcome of each experiment is not certain. Our interest is in making predictions about the process as a whole. The analysis in Example 4 generalizes to stochastic processes involving any finite sequence of experiments. We summarize the procedures used in Example 4 for general application:

PROCEDURE Constructing Probability Trees

Step 1 Draw a tree diagram corresponding to all combined outcomes of the sequence of experiments.

Step 2 Assign a probability to each tree branch. (This is the probability of the occurrence of the event on the right end of the branch subject to the occurrence of all events on the path leading to the event on the right end of the branch. The probability of the occurrence of a combined outcome that corresponds to a path through the tree is the product of all branch probabilities on the path.*)

Step 3 Use the results in Steps 1 and 2 to answer various questions related to the sequence of experiments as a whole.

---

*If we form a sample space S such that each simple event in S corresponds to one path through the tree, and if the probability assigned to each simple event in S is the product of the branch probabilities on the corresponding path, then it can be shown that this is not only an acceptable assignment (all probabilities for the simple events in S are nonnegative and their sum is 1), but it is the only assignment consistent with the method used to assign branch probabilities within the tree.

Explore and Discuss 1   Refer to the table on rain and accidents in Example 2 and use formula (1), where appropriate, to complete the following probability tree:

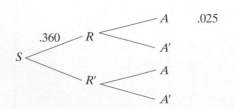

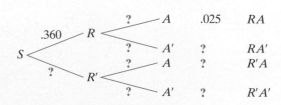

Discuss the difference between $P(R \cap A)$ and $P(A \mid R)$.

EXAMPLE 5 Product Defects An auto company A subcontracts the manufacturing of its onboard computers to two companies: 40% to company B and 60% to company C. Company B in turn subcontracts 70% of the orders it receives from company A to company D and the remaining 30% to company E, both subsidiaries of company B. When the onboard computers are completed by companies D, E, and C, they are shipped to company A to be used in various car models. It has been found that 1.5%, 1%, and .5% of the boards from D, E, and C, respectively, prove defective during the 3-year warranty period after a car is first sold. What is the probability that a given onboard computer will be defective during the 3-year warranty period?

SOLUTION Draw a tree diagram and assign probabilities to each branch (Fig. 8):

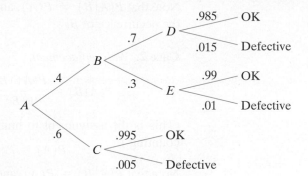

Figure 8

There are three paths leading to defective (the onboard computer will be defective within the 3-year warranty period). We multiply the branch probabilities on each path and add the three products:

$$P(\text{defective}) = (.4)(.7)(.015) + (.4)(.3)(.01) + (.6)(.005)$$
$$= .0084$$

Matched Problem 5 In Example 5, what is the probability that a given onboard computer came from company E or C?

Independent Events

We return to Example 4, which involved drawing two balls in succession without replacement from a box of 3 blue and 2 white balls. What difference does "without replacement" and "with replacement" make? Figure 9 shows probability trees corresponding to each case. Go over the probability assignments for the branches in Figure 9B to convince yourself of their correctness.

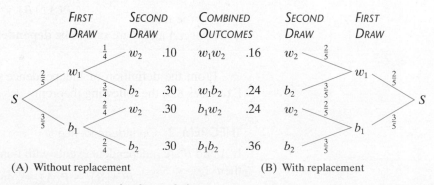

Figure 9 $S = \{w_1w_2, w_1b_2, b_1w_2, b_1b_2\}$

Let
$$A = \text{white ball on second draw} = \{w_1w_2, b_1w_2\}$$
$$B = \text{white ball on first draw} = \{w_1w_2, w_1b_2\}$$

We now compute $P(A|B)$ and $P(A)$ for each case in Figure 9.

Case 1. *Without replacement:*

$$P(A|B) = \frac{P(A \cap B)}{P(B)} = \frac{P\{w_1w_2\}}{P\{w_1w_2, w_1b_2\}} = \frac{.10}{.10 + .30} = .25$$

(This is the assignment to branch w_1w_2 that we made by looking in the box and counting.)
$$P(A) = P\{w_1w_2, b_1w_2\} = .10 + .30 = .40$$

Note that $P(A|B) \neq P(A)$, and we conclude that the probability of A is affected by the occurrence of B.

Case 2. *With replacement:*

$$P(A|B) = \frac{P(A \cap B)}{P(B)} = \frac{P\{w_1w_2\}}{P\{w_1w_2, w_1b_2\}} = \frac{.16}{.16 + .24} = .40$$

(This is the assignment to branch w_1w_2 that we made by looking in the box and counting.)
$$P(A) = P\{w_1w_2, b_1w_2\} = .16 + .24 = .40$$

Note that $P(A|B) = P(A)$, and we conclude that the probability of A is not affected by the occurrence of B.

Intuitively, if $P(A|B) = P(A)$, then it appears that event A is "independent" of B. Let us pursue this further. If events A and B are such that

$$P(A|B) = P(A)$$

then replacing the left side by its equivalent from formula (1), we obtain

$$\frac{P(A \cap B)}{P(B)} = P(A)$$

After multiplying both sides by $P(B)$, the last equation becomes

$$P(A \cap B) = P(A)P(B)$$

This result motivates the following definition of *independence:*

DEFINITION   Independence

If A and B are any events in a sample space S, we say that **A and B are independent** if

$$P(A \cap B) = P(A)P(B) \tag{3}$$

Otherwise, A and B are said to be **dependent**.

From the definition of independence one can prove (see Problems 75 and 76, Exercises 8.3) the following theorem:

THEOREM 2   On Independence

If A and B are independent events with nonzero probabilities in a sample space S, then
$$P(A|B) = P(A) \quad \text{and} \quad P(B|A) = P(B) \tag{4}$$

If either equation in (4) holds, then A and B are independent.

CONCEPTUAL INSIGHT

Sometimes intuitive reasoning can be helpful in deciding whether or not two events are independent. Suppose that a fair coin is tossed five times. What is the probability of a head on the fifth toss, given that the first four tosses are all heads? Our intuition tells us that a coin has no memory, so the probability of a head on the fifth toss given four previous heads should be equal to the probability of a head on the fifth toss, namely, 1/2. In other words, the first equation of Theorem 2 holds intuitively, so "heads on the fifth toss" and "heads on the first four tosses" are independent events.

Often, unfortunately, intuition is *not* a reliable guide to the notion of independence. Independence is a technical concept. So in all cases, an appropriate sample space should be chosen, and either equation (3) or equation (4) should be tested, to confirm that two events are (or are not) independent.

EXAMPLE 6   Testing for Independence   In two tosses of a single fair coin, show that the events "A head on the first toss" and "A head on the second toss" are independent.

SOLUTION   Consider the sample space of equally likely outcomes for the tossing of a fair coin twice,

$$S = \{HH, HT, TH, TT\}$$

and the two events,

$$A = \text{a head on the first toss} = \{HH, HT\}$$
$$B = \text{a head on the second toss} = \{HH, TH\}$$

Then

$$P(A) = \frac{2}{4} = \frac{1}{2} \quad P(B) = \frac{2}{4} = \frac{1}{2} \quad P(A \cap B) = \frac{1}{4}$$

Thus,

$$P(A \cap B) = \frac{1}{4} = \frac{1}{2} \cdot \frac{1}{2} = P(A)P(B)$$

and the two events are independent. (The theory agrees with our intuition—a coin has no memory.)

Matched Problem 6   In Example 6, compute $P(B|A)$ and compare with $P(B)$.

EXAMPLE 7   Testing for Independence   A single card is drawn from a standard 52-card deck. Test the following events for independence (try guessing the answer to each part before looking at the solution):

(A) E = the drawn card is a spade.

   F = the drawn card is a face card.

(B) G = the drawn card is a club.

   H = the drawn card is a heart.

SOLUTION

(A) To test E and F for independence, we compute $P(E \cap F)$ and $P(E)P(F)$. If they are equal, then events E and F are independent; if they are not equal, then events E and F are dependent.

$$P(E \cap F) = \frac{3}{52} \quad P(E)P(F) = \left(\frac{13}{52}\right)\left(\frac{12}{52}\right) = \frac{3}{52}$$

Events E and F are independent. (Did you guess this?)

(B) Proceeding as in part (A), we see that

$$P(G \cap H) = P(\varnothing) = 0 \qquad P(G)P(H) = \left(\frac{13}{52}\right)\left(\frac{13}{52}\right) = \frac{1}{16}$$

Events G and H are dependent. (Did you guess this?)

⚠ **CAUTION**   Students often confuse *mutually exclusive (disjoint) events* with *independent events*. One does not necessarily imply the other. In fact, it is not difficult to show (see Problem 79, Exercises 8.3) that any two mutually exclusive events A and B, with nonzero probabilities, are always dependent.   ▲

Matched Problem 7   A single card is drawn from a standard 52-card deck. Test the following events for independence:

(A) $E =$ the drawn card is a red card
    $F =$ the drawn card's number is divisible by 5 (face cards are not assigned values)

(B) $G =$ the drawn card is a king
    $H =$ the drawn card is a queen

Explore and Discuss 2   In college basketball, would it be reasonable to assume that the following events are independent? Explain why or why not.

$A =$ the Golden Eagles win in the first round of the NCAA tournament.

$B =$ the Golden Eagles win in the second round of the NCAA tournament.

The notion of independence can be extended to more than two events:

DEFINITION   Independent Set of Events
A set of events is said to be **independent** if for each finite subset $\{E_1, E_2, \ldots, E_k\}$

$$P(E_1 \cap E_2 \cap \cdots \cap E_k) = P(E_1)P(E_2) \cdot \cdots \cdot P(E_k) \qquad (5)$$

The next example makes direct use of this definition.

EXAMPLE 8   Computer Control Systems   A space shuttle has four independent computer control systems. If the probability of failure (during flight) of any one system is .001, what is the probability of failure of all four systems?

SOLUTION   Let

$$E_1 = \text{failure of system 1} \qquad E_3 = \text{failure of system 3}$$
$$E_2 = \text{failure of system 2} \qquad E_4 = \text{failure of system 4}$$

Then, since events E_1, E_2, E_3, and E_4 are given to be independent,

$$P(E_1 \cap E_2 \cap E_3 \cap E_4) = P(E_1)P(E_2)P(E_3)P(E_4)$$
$$= (.001)^4$$
$$= .000\ 000\ 000\ 001$$

Matched Problem 8   A single die is rolled 6 times. What is the probability of getting the sequence 1, 2, 3, 4, 5, 6?

Summary

The key results in this section are summarized in the following box:

> **SUMMARY** Key Concepts
> **Conditional Probability**
>
> $$P(A|B) = \frac{P(A \cap B)}{P(B)} \qquad P(B|A) = \frac{P(B \cap A)}{P(A)}$$
>
> **Note:** $P(A|B)$ is a probability based on the new sample space B, while $P(A \cap B)$ and $P(B)$ are probabilities based on the original sample space S.
>
> **Product Rule**
>
> $$P(A \cap B) = P(A)P(B|A) = P(B)P(A|B)$$
>
> **Independent Events**
>
> - A and B are **independent** if
>
> $$P(A \cap B) = P(A)P(B)$$
>
> - If A and B are independent events with nonzero probabilities, then
>
> $$P(A|B) = P(A) \quad \text{and} \quad P(B|A) = P(B)$$
>
> - If A and B are events with nonzero probabilities and either $P(A|B) = P(A)$ or $P(B|A) = P(B)$, then A and B are independent.
> - If $E_1, E_2, \ldots, E_n$ are independent, then
>
> $$P(E_1 \cap E_2 \cap \cdots \cap E_n) = P(E_1)P(E_2) \cdot \cdots \cdot P(E_n)$$

Exercises 8.3

Skills Warm-up Exercises

In Problems 1–6, use a tree diagram to represent a factorization of the given integer into primes, so that there are two branches at each number that is not prime. For example, the factorization $24 = 4 \cdot 6 = (2 \cdot 2) \cdot (2 \cdot 3)$ is represented by:

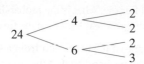

(If necessary, review Section A.3.)

1. 100
2. 120
3. 180
4. 225
5. 315
6. 360

A single card is drawn from a standard 52-card deck. In Problems 7–14, find the conditional probability that

7. The card is an ace, given that it is a heart.

8. The card is red, given that it is a face card.

9. The card is a heart, given that it is an ace.

10. The card is a face card, given that it is red.

11. The card is black, given that it is a club.

12. The card is a jack, given that it is red.

13. The card is a club, given that it is black.

14. The card is red, given that it is a jack.

In Problems 15–22, find the conditional probability, in a single roll of two fair dice, that

15. The sum is less than 6, given that the sum is even.

16. The sum is 10, given that the roll is doubles.

17. The sum is even, given that the sum is less than 6.

18. The roll is doubles, given that the sum is 10.

19. The sum is greater than 7, given that neither die is a six.

20. The sum is odd, given that at least one die is a six.

21. Neither die is a six, given that the sum is greater than 7.

22. At least one die is a six, given that the sum is odd.

In Problems 23–42, use the table below. Events A, B, and C are mutually exclusive; so are D, E, and F.

	A	B	C	Totals
D	.20	.03	.07	.30
E	.28	.05	.07	.40
F	.22	.02	.06	.30
Totals	.70	.10	.20	1.00

In Problems 23–26, find each probability directly from the table.

23. $P(B)$ **24.** $P(E)$

25. $P(B \cap D)$ **26.** $P(C \cap E)$

In Problems 27–34, compute each probability using formula (1) on page 412 and appropriate table values.

27. $P(D|B)$ **28.** $P(C|E)$

29. $P(B|D)$ **30.** $P(E|C)$

31. $P(D|C)$ **32.** $P(E|A)$

33. $P(A|C)$ **34.** $P(B|B)$

In Problems 35–42, test each pair of events for independence.

35. A and D **36.** A and E

37. B and D **38.** B and E

39. B and F **40.** C and F

41. A and B **42.** D and F

43. A fair coin is tossed 8 times.

(A) What is the probability of tossing a head on the 8th toss, given that the preceding 7 tosses were heads?

(B) What is the probability of getting 8 heads or 8 tails?

44. A fair die is rolled 5 times.

(A) What is the probability of getting a 6 on the 5th roll, given that a 6 turned up on the preceding 4 rolls?

(B) What is the probability that the same number turns up every time?

45. A pointer is spun once on the circular spinner shown below. The probability assigned to the pointer landing on a given integer (from 1 to 5) is the ratio of the area of the corresponding circular sector to the area of the whole circle, as given in the table:

e_i	1	2	3	4	5
$P(e_i)$	.3	.1	.2	.3	.1

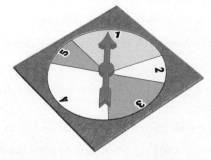

Given the events

E = pointer lands on an even number

F = pointer lands on a number less than 4

(A) Find $P(F|E)$.

(B) Test events E and F for independence.

46. Repeat Problem 45 with the following events:

E = pointer lands on an odd number

F = pointer lands on a prime number

Compute the indicated probabilities in Problems 47 and 48 by referring to the following probability tree:

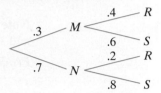

47. (A) $P(M \cap S)$ (B) $P(R)$

48. (A) $P(N \cap R)$ (B) $P(S)$

49. A fair coin is tossed twice. Consider the sample space $S = \{HH, HT, TH, TT\}$ of equally likely simple events. We are interested in the following events:

E_1 = a head on the first toss

E_2 = a tail on the first toss

E_3 = a tail on the second toss

E_4 = a head on the second toss

For each pair of events, discuss whether they are independent and whether they are mutually exclusive.

(A) E_1 and E_4

(B) E_1 and E_2

50. For each pair of events (see Problem 49), discuss whether they are independent and whether they are mutually exclusive.

(A) E_1 and E_3

(B) E_3 and E_4

51. In 2 throws of a fair die, what is the probability that you will get an even number on each throw? An even number on the first or second throw?

52. In 2 throws of a fair die, what is the probability that you will get at least 5 on each throw? At least 5 on the first or second throw?

53. Two cards are drawn in succession from a standard 52-card deck. What is the probability that the first card is a club and the second card is a heart

(A) If the cards are drawn without replacement?

(B) If the cards are drawn with replacement?

54. Two cards are drawn in succession from a standard 52-card deck. What is the probability that both cards are red

(A) If the cards are drawn without replacement?

(B) If the cards are drawn with replacement?

55. A card is drawn at random from a standard 52-card deck. Events G and H are

G = the drawn card is black.

H = the drawn card is divisible by 3

(face cards are not valued).

(A) Find $P(H|G)$.

(B) Test H and G for independence.

56. A card is drawn at random from a standard 52-card deck. Events M and N are

$$M = \text{the drawn card is a diamond.}$$
$$N = \text{the drawn card is even}$$
$$(\text{face cards are not valued}).$$

(A) Find $P(N|M)$.

(B) Test M and N for independence.

57. Let A be the event that all of a family's children are the same gender, and let B be the event that the family has at most 1 boy. Assuming the probability of having a girl is the same as the probability of having a boy (both .5), test events A and B for independence if

(A) The family has 2 children.

(B) The family has 3 children.

58. An experiment consists of tossing n coins. Let A be the event that at least 2 heads turn up, and let B be the event that all the coins turn up the same. Test A and B for independence if

(A) 2 coins are tossed.

(B) 3 coins are tossed.

Problems 59–62 refer to the following experiment: 2 balls are drawn in succession out of a box containing 2 red and 5 white balls. Let R_i be the event that the ith ball is red, and let W_i be the event that the ith ball is white.

59. Construct a probability tree for this experiment and find the probability of each of the events $R_1 \cap R_2$, $R_1 \cap W_2$, $W_1 \cap R_2$, $W_1 \cap W_2$, given that the first ball drawn was

(A) Replaced before the second draw

(B) Not replaced before the second draw

60. Find the probability that the second ball was red, given that the first ball was

(A) Replaced before the second draw

(B) Not replaced before the second draw

61. Find the probability that at least 1 ball was red, given that the first ball was

(A) Replaced before the second draw

(B) Not replaced before the second draw

62. Find the probability that both balls were the same color, given that the first ball was

(A) Replaced before the second draw

(B) Not replaced before the second draw

In Problems 63–70, discuss the validity of each statement. If the statement is always true, explain why. If not, give a counterexample.

63. If $P(A|B) = P(B)$, then A and B are independent.

64. If A and B are independent, then $P(A|B) = P(B|A)$.

65. If A is nonempty and $A \subset B$, then $P(A|B) \geq P(A)$.

66. If A and B are events, then $P(A|B) \leq P(B)$.

67. If A and B are mutually exclusive, then A and B are independent.

68. If A and B are independent, then A and B are mutually exclusive.

69. If two balls are drawn in succession, with replacement, from a box containing m red and n white balls ($m \geq 1$ and $n \geq 1$), then

$$P(W_1 \cap R_2) = P(R_1 \cap W_2)$$

70. If two balls are drawn in succession, without replacement, from a box containing m red and n white balls ($m \geq 1$ and $n \geq 1$), then

$$P(W_1 \cap R_2) = P(R_1 \cap W_2)$$

71. A box contains 2 red, 3 white, and 4 green balls. Two balls are drawn out of the box in succession without replacement. What is the probability that both balls are the same color?

72. For the experiment in Problem 71, what is the probability that no white balls are drawn?

73. An urn contains 2 one-dollar bills, 1 five-dollar bill, and 1 ten-dollar bill. A player draws bills one at a time without replacement from the urn until a ten-dollar bill is drawn. Then the game stops. All bills are kept by the player.

(A) What is the probability of winning $16?

(B) What is the probability of winning all bills in the urn?

(C) What is the probability of the game stopping at the second draw?

74. Ann and Barbara are playing a tennis match. The first player to win 2 sets wins the match. For any given set, the probability that Ann wins that set is $\frac{2}{3}$. Find the probability that

(A) Ann wins the match.

(B) 3 sets are played.

(C) The player who wins the first set goes on to win the match.

75. Show that if A and B are independent events with nonzero probabilities in a sample space S, then

$$P(A|B) = P(A) \quad \text{and} \quad P(B|A) = P(B)$$

76. Show that if A and B are events with nonzero probabilities in a sample space S, and either $P(A|B) = P(A)$ or $P(B|A) = P(B)$, then events A and B are independent.

77. Show that $P(A|A) = 1$ when $P(A) \neq 0$.

78. Show that $P(A|B) + P(A'|B) = 1$.

79. Show that A and B are dependent if A and B are mutually exclusive and $P(A) \neq 0, P(B) \neq 0$.

80. Show that $P(A|B) = 1$ if B is a subset of A and $P(B) \neq 0$.

Applications

81. Labor relations. In a study to determine employee voting patterns in a recent strike election, 1,000 employees were selected at random and the following tabulation was made:

		Salary Classification			
		Hourly (*H*)	**Salary** (*S*)	**Sal. + Bonus** (*B*)	**Totals**
To	**Yes (*Y*)**	400	180	20	600
Strike	**No (*N*)**	150	120	130	400
Totals		550	300	150	1,000

(A) Convert this table to a probability table by dividing each entry by 1,000.

(B) What is the probability of an employee voting to strike, given that the person is paid hourly?

(C) What is the probability of an employee voting to strike, given that the person receives a salary plus bonus?

(D) What is the probability of an employee being on straight salary (*S*)? Of being on straight salary given that he or she voted in favor of striking?

(E) What is the probability of an employee being paid hourly? Of being paid hourly given that he or she voted in favor of striking?

(F) What is the probability of an employee being in a salary plus bonus position and voting against striking?

(G) Are events *S* and *Y* independent?

(H) Are events *H* and *Y* independent?

(I) Are events *B* and *N* independent?

82. Quality control. An automobile manufacturer produces 37% of its cars at plant *A*. If 5% of the cars manufactured at plant *A* have defective emission control devices, what is the probability that one of this manufacturer's cars was manufactured at plant *A* and has a defective emission control device?

83. Bonus incentives. If a salesperson has gross sales of over $600,000 in a year, then he or she is eligible to play the company's bonus game: A black box contains 1 twenty-dollar bill, 2 five-dollar bills, and 1 one-dollar bill. Bills are drawn out of the box one at a time without replacement until a twenty-dollar bill is drawn. Then the game stops. The salesperson's bonus is 1,000 times the value of the bills drawn.

(A) What is the probability of winning a $26,000 bonus?

(B) What is the probability of winning the maximum bonus, $31,000, by drawing out all bills in the box?

(C) What is the probability of the game stopping at the third draw?

84. Personnel selection. To transfer into a particular technical department, a company requires an employee to pass a screening test. A maximum of 3 attempts are allowed at 6-month intervals between trials. From past records it is found that 40% pass on the first trial; of those that fail the first trial and take the test a second time, 60% pass; and of those that fail on the second trial and take the test a third time, 20% pass. For an employee wishing to transfer:

(A) What is the probability of passing the test on the first or second try?

(B) What is the probability of failing on the first 2 trials and passing on the third?

(C) What is the probability of failing on all 3 attempts?

85. U.S. Food and Drug Administration. An ice cream company wishes to use a red dye to enhance the color in its strawberry ice cream. The U.S. Food and Drug Administration (FDA) requires the dye to be tested for cancer-producing potential using laboratory rats. The results of one test on 1,000 rats are summarized in the following table:

		Cancer *C*	**No Cancer** *C'*	**Totals**
Ate Red Dye	*R*	60	440	500
No Red Dye	*R'*	20	480	500
Totals		80	920	1,000

(A) Convert the table into a probability table by dividing each entry by 1,000.

(B) Are "developing cancer" and "eating red dye" independent events?

(C) Should the FDA approve or ban the use of the dye? Explain why or why not using $P(C|R)$ and $P(C)$.

(D) Suppose the number of rats that ate red dye and developed cancer was 20, but the number that developed cancer was still 80 and the number that ate red dye was still 500. What should the FDA do based on these results? Explain why.

86. Genetics. In a study to determine frequency and dependency of color-blindness relative to females and males, 1,000 people were chosen at random, and the following results were recorded:

		Female *F*	**Male** *F'*	**Totals**
Color-Blind	*C*	2	24	26
Normal	*C'*	518	456	974
Totals		520	480	1,000

(A) Convert this table to a probability table by dividing each entry by 1,000.

(B) What is the probability that a person is a woman, given that the person is color-blind?

(C) What is the probability that a person is color-blind, given that the person is a male?

(D) Are the events color-blindness and male independent?

(E) Are the events color-blindness and female independent?

87. Psychology. In a study to determine the frequency and dependency of IQ ranges relative to males and females, 1,000 people were chosen at random and the following results were recorded:

		Below 90 (A)	90–120 (B)	Above 120 (C)	Totals
Female	F	130	286	104	520
Male	F'	120	264	96	480
Totals		250	550	200	1,000

(A) Convert this table to a probability table by dividing each entry by 1,000.

(B) What is the probability of a person having an IQ below 90, given that the person is a female? A male?

(C) What is the probability of a person having an IQ above 120, given that the person is a female? A male?

(D) What is the probability of a person having an IQ below 90?

(E) What is the probability of a person having an IQ between 90 and 120? Of a person having an IQ between 90 and 120, given that the person is a male?

(F) What is the probability of a person being female and having an IQ above 120?

(G) Are events A and F dependent? B and F? C and F?

88. Voting patterns. A survey of a precinct's residents revealed that 55% of the residents were members of the Democratic party and 60% of the Democratic party members voted in the last election. What is the probability that a person selected at random from this precinct is a member of the Democratic party and voted in the last election?

Answers to Matched Problems

1. (A) .5 (B) .4
2. (A) $P(R') = .640$ (B) $P(A \cap R') = .015$
   (C) $P(A|R') = \dfrac{P(A \cap R')}{P(R')} = .023$
3. $P(M \cap C) = P(M)P(C|M) = .32$
4. $\dfrac{2}{3}$ 5. .72
6. $P(B|A) = \dfrac{P(A \cap B)}{P(A)} = \dfrac{\frac{1}{4}}{\frac{1}{2}} = \dfrac{1}{2} = P(B)$
7. (A) E and F are independent.
   (B) G and H are dependent.
8. $\left(\dfrac{1}{6}\right)^6 \approx .000\ 021\ 4$

8.4 Bayes' Formula

In the preceding section, we discussed the conditional probability of the occurrence of an event, given the occurrence of an earlier event. Now we will reverse the problem and try to find the probability of an earlier event conditioned on the occurrence of a later event. As you will see, a number of practical problems have this form. First, let us consider a relatively simple problem that will provide the basis for a generalization.

EXAMPLE 1 Probability of an Earlier Event Given a Later Event One urn has 3 blue and 2 white balls; a second urn has 1 blue and 3 white balls (Fig. 1). A single fair die is rolled and if 1 or 2 comes up, a ball is drawn out of the first urn; otherwise, a ball is drawn out of the second urn. If the drawn ball is blue, what is the probability that it came out of the first urn? Out of the second urn?

Figure 1

SOLUTION We form a probability tree, letting U_1 represent urn 1, U_2 urn 2, B a blue ball, and W a white ball. Then, on the various outcome branches, we assign appropriate probabilities. For example, $P(U_1) = \frac{1}{3}$, $P(B|U_1) = \frac{3}{5}$, and so on:

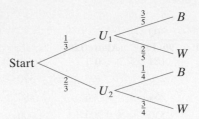

Now we are interested in finding $P(U_1|B)$; that is, the probability that the ball came out of urn 1, given that the drawn ball is blue. Using equation (1) from Section 8.3, we can write

$$P(U_1|B) = \frac{P(U_1 \cap B)}{P(B)} \tag{1}$$

If we look at the tree diagram, we can see that B is at the end of two different branches; thus,

$$P(B) = P(U_1 \cap B) + P(U_2 \cap B) \tag{2}$$

After substituting equation (2) into equation (1), we get

$$P(U_1|B) = \frac{P(U_1 \cap B)}{P(U_1 \cap B) + P(U_2 \cap B)} \qquad\qquad P(A \cap B) = P(A)P(B|A)$$

$$= \frac{P(U_1)P(B|U_1)}{P(U_1)P(B|U_1) + P(U_2)P(B|U_2)}$$

$$= \frac{P(B|U_1)P(U_1)}{P(B|U_1)P(U_1) + P(B|U_2)P(U_2)} \tag{3}$$

Equation (3) is really a lot simpler to use than it looks. You do not need to memorize it; you simply need to understand its form relative to the probability tree above. Referring to the probability tree, we see that

$P(B|U_1)P(U_1) =$ product of branch probabilities leading to B through U_1

$$= \left(\frac{3}{5}\right)\left(\frac{1}{3}\right) \qquad \text{We usually start at } B \text{ and work back through } U_1.$$

$P(B|U_2)P(U_2) =$ product of branch probabilities leading to B through U_2

$$= \left(\frac{1}{4}\right)\left(\frac{2}{3}\right) \qquad \text{We usually start at } B \text{ and work back through } U_2.$$

Equation (3) now can be interpreted in terms of the probability tree as follows:

$$P(U_1|B) = \frac{\text{product of branch probabilities leading to } B \text{ through } U_1}{\text{sum of all branch products leading to } B}$$

$$= \frac{\left(\frac{3}{5}\right)\left(\frac{1}{3}\right)}{\left(\frac{3}{5}\right)\left(\frac{1}{3}\right) + \left(\frac{1}{4}\right)\left(\frac{2}{3}\right)} = \frac{6}{11} \approx .55$$

Similarly,

$$P(U_2|B) = \frac{\text{product of branch probabilities leading to } B \text{ through } U_2}{\text{sum of all branch products leading to } B}$$

$$= \frac{\left(\frac{1}{4}\right)\left(\frac{2}{3}\right)}{\left(\frac{3}{5}\right)\left(\frac{1}{3}\right) + \left(\frac{1}{4}\right)\left(\frac{2}{3}\right)} = \frac{5}{11} \approx .45$$

Note: We could have obtained $P(U_2|B)$ by subtracting $P(U_1|B)$ from 1. Why?

Matched Problem 1) Repeat Example 1, but find $P(U_1|W)$ and $P(U_2|W)$.

Explore and Discuss 1   Study the probability tree below:

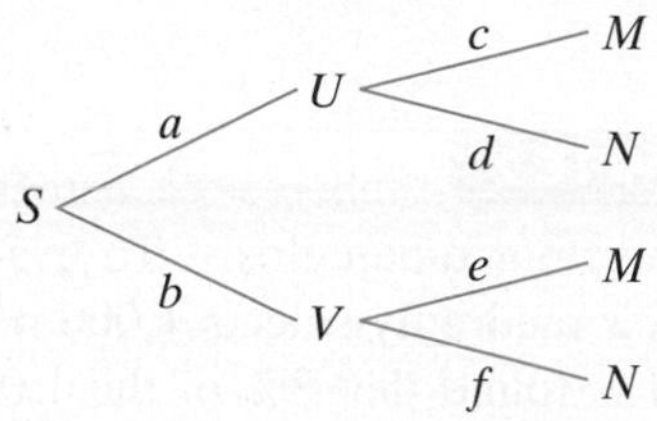

$$S \begin{cases} a & U \begin{cases} c & M \\ d & N \end{cases} \\ b & V \begin{cases} e & M \\ f & N \end{cases} \end{cases}$$

$c + d = 1$

$a + b = 1 \qquad a, b, c, d, e, f \neq 0$

$e + f = 1$

(A) Discuss the difference between $P(M|U)$ and $P(U|M)$, and between $P(N|V)$ and $P(V|N)$, in terms of a, b, c, d, e, and f.

(B) Show that $ac + ad + be + bf = 1$. What is the significance of this result?

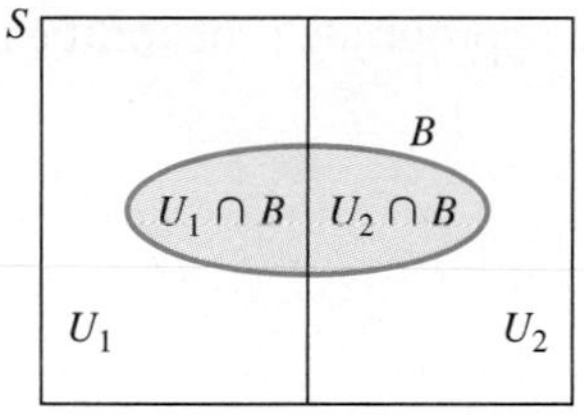

In generalizing the results in Example 1, it is helpful to look at its structure in terms of the Venn diagram shown in Figure 2. We note that U_1 and U_2 are mutually exclusive (disjoint), and their union forms S. The following two equations can now be interpreted in terms of this diagram:

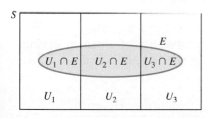

Figure 2

$$P(U_1|B) = \frac{P(U_1 \cap B)}{P(B)} = \frac{P(U_1 \cap B)}{P(U_1 \cap B) + P(U_2 \cap B)}$$

$$P(U_2|B) = \frac{P(U_2 \cap B)}{P(B)} = \frac{P(U_2 \cap B)}{P(U_1 \cap B) + P(U_2 \cap B)}$$

Look over the equations and the diagram carefully.

Of course, there is no reason to stop here. Suppose that U_1, U_2, and U_3 are three mutually exclusive events whose union is the whole sample space S. Then, for an arbitrary event E in S, with $P(E) \neq 0$, the corresponding Venn diagram looks like Figure 3, and

Figure 3

$$P(U_1|E) = \frac{P(U_1 \cap E)}{P(E)} = \frac{P(U_1 \cap E)}{P(U_1 \cap E) + P(U_2 \cap E) + P(U_3 \cap E)}$$

Similar results hold for U_2 and U_3.

Using the same reasoning, we arrive at the following famous theorem, which was first stated by Thomas Bayes (1702–1763):

THEOREM 1 Bayes' Formula

Let $U_1, U_2, \ldots, U_n$ be n mutually exclusive events whose union is the sample space S. Let E be an arbitrary event in S such that $P(E) \neq 0$. Then,

$$P(U_1|E) = \frac{P(U_1 \cap E)}{P(E)} = \frac{P(U_1 \cap E)}{P(U_1 \cap E) + P(U_2 \cap E) + \cdots + P(U_n \cap E)}$$

$$= \frac{P(E|U_1)P(U_1)}{P(E|U_1)P(U_1) + P(E|U_2)P(U_2) + \cdots + P(E|U_n)P(U_n)}$$

Similar results hold for $U_2, U_3, \ldots, U_n$.

You do not need to memorize Bayes' formula. In practice, it is often easier to draw a probability tree and use the following:

$$P(U_1|E) = \frac{\text{product of branch probabilities leading to } E \text{ through } U_1}{\text{sum of all branch products leading to } E}$$

Similar results hold for $U_2, U_3, \ldots, U_n$.

EXAMPLE 2   Tuberculosis Screening   A new, inexpensive skin test is devised for detecting tuberculosis. To evaluate the test before it is used, a medical researcher randomly selects 1,000 people. Using precise but more expensive methods, it is found that 8% of the 1,000 people tested have tuberculosis. Now each of the 1,000 subjects is given the new skin test, and the following results are recorded: The test indicates tuberculosis in 96% of those who have it and in 2% of those who do not. Based on these results, what is the probability of a randomly chosen person having tuberculosis given that the skin test indicates the disease? What is the probability of a person not having tuberculosis given that the skin test indicates the disease? (That is, what is the probability of the skin test giving a *false positive result?*)

SOLUTION   To start, we form a tree diagram and place appropriate probabilities on each branch:

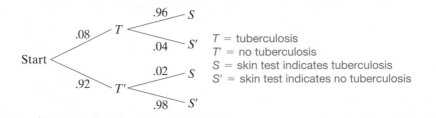

We are interested in finding $P(T|S)$, that is, the probability of a person having tuberculosis given that the skin test indicates the disease. Bayes' formula for this case is

$$P(T|S) = \frac{\text{product of branch probabilities leading to } S \text{ through } T}{\text{sum of all branch products leading to } S}$$

Substituting appropriate values from the probability tree, we obtain

$$P(T|S) = \frac{(.08)(.96)}{(.08)(.96) + (.92)(.02)} = .81$$

The probability of a person not having tuberculosis given that the skin test indicates the disease, denoted by $P(T'|S)$, is

$$P(T'|S) = 1 - P(T|S) = 1 - .81 = .19 \quad P(T|S) + P(T'|S) = 1$$

Matched Problem 2)   What is the probability that a person has tuberculosis given that the test indicates no tuberculosis is present? (That is, what is the probability of the skin test giving a *false negative result?*) What is the probability that a person does not have tuberculosis given that the test indicates no tuberculosis is present?

CONCEPTUAL INSIGHT

From a public health standpoint, which is the more serious error in Example 2 and Matched Problem 2: a false positive result or a false negative result? A false positive result will certainly be unsettling to the subject, who might lose sleep thinking that she has tuberculosis. But she will be sent for further testing and will be relieved to learn it was a false alarm; she does not have the disease. A false negative result, on the other hand, is more serious. A person who has tuberculosis will be unaware that he has a communicable disease and so may pose a considerable risk to public health.

In designing an inexpensive skin test, we would expect a tradeoff between cost and false results. We might be willing to tolerate a moderate number of false positives if we could keep the number of false negatives at a minimum.

EXAMPLE 3   Product Defects   A company produces 1,000 refrigerators a week at three plants. Plant A produces 350 refrigerators a week, plant B produces 250 refrigerators a week, and plant C produces 400 refrigerators a week. Production records indicate that 5% of the refrigerators produced at plant A will be defective, 3% of those produced at plant B will be defective, and 7% of those produced at plant C will be defective. All the refrigerators are shipped to a central warehouse. If a refrigerator at the warehouse is found to be defective, what is the probability that it was produced at plant A?

SOLUTION   We begin by constructing a tree diagram:

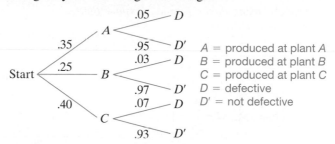

A = produced at plant A
B = produced at plant B
C = produced at plant C
D = defective
D' = not defective

The probability that a defective refrigerator was produced at plant A is $P(A|D)$. Bayes' formula for this case is

$$P(A|D) = \frac{\text{product of branch probabilities leading to } D \text{ through } A}{\text{sum of all branch products leading to } D}$$

Using the values from the probability tree, we have

$$P(A|D) = \frac{(.35)(.05)}{(.35)(.05) + (.25)(.03) + (.40)(.07)}$$
$$\approx .33$$

Matched Problem 3   In Example 3, what is the probability that a defective refrigerator in the warehouse was produced at plant B? At plant C?

Exercises 8.4

Skills Warm-up Exercises

In Problems 1–6, write each expression as a quotient of integers, reduced to lowest terms. (If necessary, review Section A.1.)

1. $\dfrac{\frac{1}{3}}{\frac{1}{3} + \frac{1}{2}}$

2. $\dfrac{\frac{2}{7}}{\frac{1}{4} + \frac{2}{7}}$

3. $\dfrac{1}{3} \div \dfrac{1}{3} + \dfrac{1}{2}$

4. $\dfrac{2}{7} \div \dfrac{1}{4} + \dfrac{2}{7}$

5. $\dfrac{\frac{4}{5} \cdot \frac{3}{4}}{\frac{1}{5} \cdot \frac{1}{3} + \frac{4}{5} \cdot \frac{3}{4}}$

6. $\dfrac{\frac{1}{5} \cdot \frac{2}{3}}{\frac{1}{5} \cdot \frac{2}{3} + \frac{4}{5} \cdot \frac{1}{4}}$

Find the probabilities in Problems 7–12 by referring to the tree diagram below.

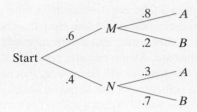

7. $P(M \cap A) = P(M)P(A|M)$

8. $P(N \cap B) = P(N)P(B|N)$

9. $P(A) = P(M \cap A) + P(N \cap A)$

10. $P(B) = P(M \cap B) + P(N \cap B)$

11. $P(M|A) = \dfrac{P(M \cap A)}{P(M \cap A) + P(N \cap A)}$

12. $P(N|B) = \dfrac{P(N \cap B)}{P(N \cap B) + P(M \cap B)}$

Find the probabilities in Problems 13–16 by referring to the following Venn diagram and using Bayes' formula (assume that the simple events in S are equally likely):

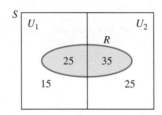

13. $P(U_1|R)$ **14.** $P(U_2|R)$

15. $P(U_1|R')$ **16.** $P(U_2|R')$

Find the probabilities in Problems 17–22 by referring to the following tree diagram and using Bayes' formula. Round answers to three decimal places.

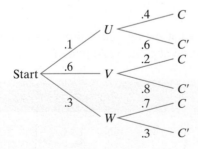

17. $P(U|C)$ **18.** $P(V|C')$

19. $P(W|C)$ **20.** $P(U|C')$

21. $P(V|C)$ **22.** $P(W|C')$

Find the probabilities in Problems 23–28 by referring to the following Venn diagram and using Bayes' formula (assume that the simple events in S are equally likely):

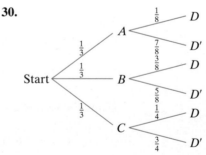

23. $P(U_1|R)$ **24.** $P(U_2|R')$

25. $P(U_3|R)$ **26.** $P(U_1|R')$

27. $P(U_2|R)$ **28.** $P(U_3|R')$

In Problems 29 and 30, use the probabilities in the first tree diagram to find the probability of each branch of the second tree diagram.

29.

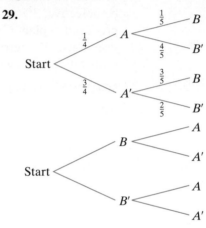

30.

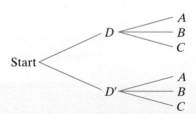

In Problems 31–34, one of two urns is chosen at random with one as likely to be chosen as the other. Then a ball is withdrawn from the chosen urn. Urn 1 contains 1 white and 4 red balls, and urn 2 has 3 white and 2 red balls.

31. If a white ball is drawn, what is the probability that it came from urn 1?

32. If a white ball is drawn, what is the probability that it came from urn 2?

33. If a red ball is drawn, what is the probability that it came from urn 2?

34. If a red ball is drawn, what is the probability that it came from urn 1?

In Problems 35 and 36, an urn contains 4 red and 5 white balls. Two balls are drawn in succession without replacement.

35. If the second ball is white, what is the probability that the first ball was white?

36. If the second ball is red, what is the probability that the first ball was red?

In Problems 37 and 38, urn 1 contains 7 red and 3 white balls. Urn 2 contains 4 red and 5 white balls. A ball is drawn from urn 1 and placed in urn 2. Then a ball is drawn from urn 2.

37. If the ball drawn from urn 2 is red, what is the probability that the ball drawn from urn 1 was red?

38. If the ball drawn from urn 2 is white, what is the probability that the ball drawn from urn 1 was white?

In Problems 39 and 40, refer to the following probability tree:

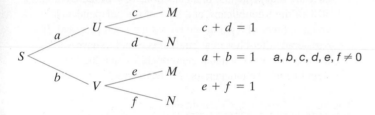

$$c + d = 1$$
$$a + b = 1 \quad a, b, c, d, e, f \neq 0$$
$$e + f = 1$$

39. Suppose that $c = e$. Discuss the dependence or independence of events U and M.

40. Suppose that $c = d = e = f$. Discuss the dependence or independence of events M and N.

In Problems 41 and 42, two balls are drawn in succession from an urn containing m blue balls and n white balls ($m \geq 2$ and $n \geq 2$). Discuss the validity of each statement. If the statement is always true, explain why. If not, give a counterexample.

41. (A) If the two balls are drawn with replacement, then
$$P(B_1|B_2) = P(B_2|B_1).$$

(B) If the two balls are drawn without replacement, then
$$P(B_1|B_2) = P(B_2|B_1).$$

42. (A) If the two balls are drawn with replacement, then
$$P(B_1|W_2) = P(W_2|B_1).$$

(B) If the two balls are drawn without replacement, then
$$P(B_1|W_2) = P(W_2|B_1).$$

43. If 2 cards are drawn in succession from a standard 52-card deck without replacement and the second card is a heart, what is the probability that the first card is a heart?

44. A box contains 10 balls numbered 1 through 10. Two balls are drawn in succession without replacement. If the second ball drawn has the number 4 on it, what is the probability that the first ball had a smaller number on it? An even number on it?

In Problems 45–50, a 3-card hand is dealt from a standard 52-card deck, and then one of the 3 cards is chosen at random.

45. If only one of the cards in the hand is a club, what is the probability that the chosen card is a club?

46. If only two of the cards in the hand are clubs, what is the probability that the chosen card is a club?

47. If the chosen card is a club, what is the probability that it is the only club in the hand?

48. If the chosen card is a club, what is the probability that exactly two of the cards in the hand are clubs?

49. If the chosen card is a club, what is the probability that all of the cards in the hand are clubs?

50. If the chosen card is not a club, what is the probability that none of the cards in the hand is a club?

51. Show that $P(U_1|R) + P(U_1'|R) = 1$.

52. If U_1 and U_2 are two mutually exclusive events whose union is the equally likely sample space S and if E is an arbitrary event in S such that $P(E) \neq 0$, show that

$$P(U_1|E) = \frac{n(U_1 \cap E)}{n(U_1 \cap E) + n(U_2 \cap E)}$$

Applications

In the following applications, the word "probability" is often understood to mean "approximate empirical probability."

53. Employee screening. The management of a company finds that 30% of the administrative assistants hired are unsatisfactory. The personnel director is instructed to devise a test that will improve the situation. One hundred employed administrative assistants are chosen at random and are given the newly constructed test. Out of these, 90% of the satisfactory administrative assistants pass the test and 20% of the unsatisfactory administrative assistants pass. Based on these results, if a person applies for a job, takes the test, and passes it, what is the probability that he or she is a satisfactory administrative assistant? If the applicant fails the test, what is the probability that he or she is a satisfactory administrative assistant?

54. Employee rating. A company has rated 75% of its employees as satisfactory and 25% as unsatisfactory. Personnel records indicate that 80% of the satisfactory workers had previous work experience, while only 40% of the unsatisfactory workers had any previous work experience. If a person with previous work experience is hired, what is the probability that this person will be a satisfactory employee? If a person with no previous work experience is hired, what is the probability that this person will be a satisfactory employee?

55. Product defects. A manufacturer obtains GPS systems from three different subcontractors: 20% from A, 40% from B, and 40% from C. The defective rates for these subcontractors are 1%, 3%, and 2%, respectively. If a defective GPS system is returned by a customer, what is the probability that it came from subcontractor A? From B? From C?

56. Product defects. A store sells three types of flash drives: brand *A*, brand *B*, and brand *C*. Of the flash drives it sells, 60% are brand *A*, 25% are brand *B*, and 15% are brand *C*. The store has found that 20% of the brand *A* flash drives, 15% of the brand *B* flash drives, and 5% of the brand *C* flash drives are returned as defective. If a flash drive is returned as defective, what is the probability that it is a brand *A* flash drive? A brand *B* flash drive? A brand *C* flash drive?

57. Cancer screening. A new, simple test has been developed to detect a particular type of cancer. The test must be evaluated before it is used. A medical researcher selects a random sample of 1,000 adults and finds (by other means) that 2% have this type of cancer. Each of the 1,000 adults is given the test, and it is found that the test indicates cancer in 98% of those who have it and in 1% of those who do not. Based on these results, what is the probability of a randomly chosen person having cancer given that the test indicates cancer? Of a person having cancer given that the test does not indicate cancer?

58. Pregnancy testing. In a random sample of 200 women who suspect that they are pregnant, 100 turn out to be pregnant. A new pregnancy test given to these women indicated pregnancy in 92 of the 100 pregnant women and in 12 of the 100 nonpregnant women. If a woman suspects she is pregnant and this test indicates that she is pregnant, what is the probability that she is pregnant? If the test indicates that she is not pregnant, what is the probability that she is not pregnant?

59. Medical research. In a random sample of 1,000 people, it is found that 7% have a liver ailment. Of those who have a liver ailment, 40% are heavy drinkers, 50% are moderate drinkers, and 10% are nondrinkers. Of those who do not have a liver ailment, 10% are heavy drinkers, 70% are moderate drinkers, and 20% are nondrinkers. If a person is chosen at random and he or she is a heavy drinker, what is the probability of that person having a liver ailment? What is the probability for a nondrinker?

60. Tuberculosis screening. A test for tuberculosis was given to 1,000 subjects, 8% of whom were known to have tuberculosis. For the subjects who had tuberculosis, the test indicated tuberculosis in 90% of the subjects, was inconclusive for 7%, and indicated no tuberculosis in 3%. For the subjects who did not have tuberculosis, the test indicated tuberculosis in 5% of the subjects, was inconclusive for 10%, and indicated no tuberculosis in the remaining 85%. What is the probability of a randomly selected person having tuberculosis given that the test indicates tuberculosis? Of not having tuberculosis given that the test was inconclusive?

61. Police science. A new lie-detector test has been devised and must be tested before it is used. One hundred people are selected at random, and each person draws a card from a box of 100 cards. Half the cards instruct the person to lie, and the others instruct the person to tell the truth. Of those who lied, 80% fail the new lie-detector test (that is, the test indicates lying). Of those who told the truth, 5% failed the test. What is the probability that a randomly chosen subject will have lied given that the subject failed the test? That the subject will not have lied given that the subject failed the test?

62. Politics. In a given county, records show that of the registered voters, 45% are Democrats, 35% are Republicans, and 20% are independents. In an election, 70% of the Democrats, 40% of the Republicans, and 80% of the independents voted in favor of a parks and recreation bond proposal. If a registered voter chosen at random is found to have voted in favor of the bond, what is the probability that the voter is a Republican? An independent? A Democrat?

Answers to Matched Problems

1. $P(U_1 \mid W) = \dfrac{4}{19} \approx .21; P(U_2 \mid W) = \dfrac{15}{19} \approx .79$

2. $P(T \mid S') = .004; P(T' \mid S') = .996$

3. $P(B \mid D) \approx .14; P(C \mid D) \approx .53$

8.5 Random Variable, Probability Distribution, and Expected Value

- Random Variable and Probability Distribution
- Expected Value of a Random Variable
- Decision Making and Expected Value

Random Variable and Probability Distribution

When performing a random experiment, a sample space *S* is selected in such a way that all probability problems of interest relative to the experiment can be solved. In many situations we may not be interested in each simple event in the sample space *S* but in some numerical value associated with the event. For example, if 3 coins are tossed, we may be interested in the number of heads that turn up rather than in the particular pattern that turns up. Or, in selecting a random sample of students, we may be interested in the proportion that are women rather than which particular students are women. In the same way, a "craps" player is usually interested in the sum of the dots on the showing faces of the dice rather than the pattern of dots on each face.

In each of these examples, there is a rule that assigns to each simple event in *S* a single real number. Mathematically speaking, we are dealing with a function (see Section 2.1). Historically, this particular type of function has been called a "random variable."

> **DEFINITION** Random Variable
> A **random variable** is a function that assigns a numerical value to each simple event in a sample space S.

Table 1 Number of Heads in the Toss of 3 Coins

Sample Space S		Number of Heads $X(e_i)$
e_1:	TTT	0
e_2:	TTH	1
e_3:	THT	1
e_4:	HTT	1
e_5:	THH	2
e_6:	HTH	2
e_7:	HHT	2
e_8:	HHH	3

The term *random variable* is an unfortunate choice, since it is neither random nor a variable—it is a function with a numerical value, and it is defined on a sample space. But the terminology has stuck and is now standard. Capital letters, such as X, are used to represent random variables.

Let us return to the experiment of tossing 3 coins. A sample space S of equally likely simple events is indicated in Table 1. Suppose that we are interested in the number of heads $(0, 1, 2,$ or $3)$ appearing on each toss of the 3 coins and the probability of each of these events. We introduce a random variable X (a function) that indicates the number of heads for each simple event in S (see the second column in Table 1). For example, $X(e_1) = 0, X(e_2) = 1$, and so on. The random variable X assigns a numerical value to each simple event in the sample space S.

We are interested in the probability of the occurrence of each image or range value of X; that is, in the probability of the occurrence of 0 heads, 1 head, 2 heads, or 3 heads in the single toss of 3 coins. We indicate this probability by

$$p(x) \quad \text{where} \quad x \in \{0, 1, 2, 3\}$$

The function p is called the **probability distribution* of the random variable X.**

What is $p(2)$, the probability of getting exactly 2 heads on the single toss of 3 coins? "Exactly 2 heads occur" is the event

$$E = \{\text{THH}, \text{HTH}, \text{HHT}\}$$

Thus,

$$p(2) = \frac{n(E)}{n(S)} = \frac{3}{8}$$

Table 2 Probability Distribution

Number of Heads x	0	1	2	3
Probability $p(x)$	$\dfrac{1}{8}$	$\dfrac{3}{8}$	$\dfrac{3}{8}$	$\dfrac{1}{8}$

Proceeding similarly for $p(0)$, $p(1)$, and $p(3)$, we obtain the probability distribution of the random variable X presented in Table 2. Probability distributions are also represented graphically, as shown in Figure 1. The graph of a probability distribution is often called a **histogram**.

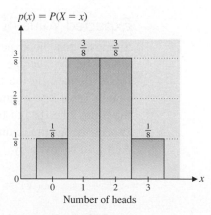

Figure 1 **Histogram for a probability distribution**

---

*The probability distribution p of the random variable X is defined by $p(x) = P(\{e_i \in S \mid X(e_i) = x\})$, which, because of its cumbersome nature, is usually simplified to $p(x) = P(X = x)$ or simply $p(x)$. We will use the simplified notation.

Note from Table 2 or Figure 1 that

1. $0 \le p(x) \le 1, \quad x \in \{0, 1, 2, 3\}$

2. $p(0) + p(1) + p(2) + p(3) = \frac{1}{8} + \frac{3}{8} + \frac{3}{8} + \frac{1}{8} = 1$

These are general properties that any probability distribution of a random variable X associated with a finite sample space must have.

THEOREM 1   Probability Distribution of a Random Variable X

The **probability distribution of a random variable X**, denoted by $P(X = x) = p(x)$, satisfies

1. $0 \le p(x) \le 1, \quad x \in \{x_1, x_2, \ldots, x_n\}$

2. $p(x_1) + p(x_2) + \cdots + p(x_n) = 1$

where $\{x_1, x_2, \ldots, x_n\}$ are the (range) values of X (see Fig. 2).

Figure 2 illustrates the process of forming a probability distribution of a random variable.

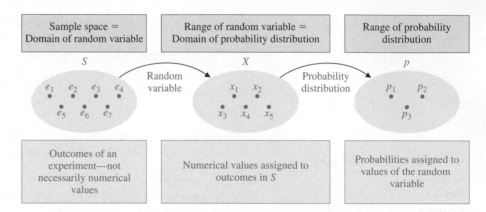

Figure 2 **Probability distribution of a random variable for a finite sample space**

Expected Value of a Random Variable

Suppose that the experiment of tossing 3 coins was repeated many times. What would be the average number of heads per toss (the total number of heads in all tosses divided by the total number of tosses)? Consulting the probability distribution in Table 2 or Figure 1, we would expect to toss 0 heads $\frac{1}{8}$ of the time, 1 head $\frac{3}{8}$ of the time, 2 heads $\frac{3}{8}$ of the time, and 3 heads $\frac{1}{8}$ of the time. In the long run, we would expect the average number of heads per toss of the 3 coins, or the *expected value* $E(X)$, to be given by

$$E(X) = 0\left(\frac{1}{8}\right) + 1\left(\frac{3}{8}\right) + 2\left(\frac{3}{8}\right) + 3\left(\frac{1}{8}\right) = \frac{12}{8} = 1.5$$

It is important to note that the expected value is not a value that will necessarily occur in a single experiment (1.5 heads cannot occur in the toss of 3 coins), but it is an average of what occurs over a large number of experiments. Sometimes we will toss more than 1.5 heads and sometimes less, but if the experiment is repeated many times, the average number of heads per experiment should be close to 1.5.

We now make the preceding discussion more precise through the following definition of expected value:

> **DEFINITION**   Expected Value of a Random Variable X
>
> Given the probability distribution for the random variable X,
>
> | x_i | x_1 | x_2 | $\cdots$ | x_n |
> |-------|-------|-------|----------|-------|
> | p_i | p_1 | p_2 | $\cdots$ | p_n |
>
> where $p_i = p(x_i)$, we define the **expected value of X**, denoted $E(X)$, by the formula
>
> $$E(X) = x_1 p_1 + x_2 p_2 + \cdots + x_n p_n$$

We again emphasize that the expected value is not the outcome of a single experiment, but a long-run average of outcomes of repeated experiments. The expected value is the weighted average of the possible outcomes, each weighted by its probability.

> **PROCEDURE**   Steps for Computing the Expected Value of a Random Variable X
>
> Step 1   Form the probability distribution of the random variable X.
>
> Step 2   Multiply each image value of X, x_i, by its corresponding probability of occurrence p_i; then add the results.

EXAMPLE 1   Expected Value   What is the expected value (long-run average) of the number of dots facing up for the roll of a single die?

SOLUTION   If we choose

$$S = \{1, 2, 3, 4, 5, 6\}$$

as our sample space, then each simple event is a numerical outcome reflecting our interest, and each is equally likely. The random variable X in this case is just the identity function (each number is associated with itself). The probability distribution for X is

x_i	1	2	3	4	5	6
p_i	$\frac{1}{6}$	$\frac{1}{6}$	$\frac{1}{6}$	$\frac{1}{6}$	$\frac{1}{6}$	$\frac{1}{6}$

Therefore,

$$E(X) = 1\left(\frac{1}{6}\right) + 2\left(\frac{1}{6}\right) + 3\left(\frac{1}{6}\right) + 4\left(\frac{1}{6}\right) + 5\left(\frac{1}{6}\right) + 6\left(\frac{1}{6}\right)$$

$$= \frac{21}{6} = 3.5$$

Matched Problem 1)   Suppose that the die in Example 1 is not fair and we obtain (empirically) the following probability distribution for X:

x_i	1	2	3	4	5	6	[Note: Sum = 1.]
p_i	.14	.13	.18	.20	.11	.24	

What is the expected value of X?

Explore and Discuss 1   From Example 1 we can conclude that the probability is 0 that a single roll of a fair die will equal the expected value for a roll of a die (the number of dots facing up is never 3.5). What is the probability that the sum for a single roll of a pair of dice will equal the expected value of the sum for a roll of a pair of dice?

EXAMPLE 2   Expected Value   A carton of 20 laptop batteries contains 2 defective ones. A random sample of 3 is selected from the 20 and tested. Let X be the random variable associated with the number of defective batteries found in a sample.

(A) Find the probability distribution of X.

(B) Find the expected number of defective batteries in a sample.

SOLUTION

(A) The number of ways of selecting a sample of 3 from 20 (order is not important) is $_{20}C_3$. This is the number of simple events in the experiment, each as likely as the other. A sample will have either 0, 1, or 2 defective batteries. These are the values of the random variable in which we are interested. The probability distribution is computed as follows:

$$p(0) = \frac{_{18}C_3}{_{20}C_3} \approx .716 \quad p(1) = \frac{_2C_1 \cdot {_{18}C_2}}{_{20}C_3} \approx .268 \quad p(2) = \frac{_2C_2 \cdot {_{18}C_1}}{_{20}C_3} \approx .016$$

We summarize these results in a table:

x_i	0	1	2
p_i	.716	.268	.016

[Note: .716 + .268 + .016 = 1.]

(B) The expected number of defective batteries in a sample is readily computed as follows:

$$E(X) = (0)(.716) + (1)(.268) + (2)(.016) = .3$$

The expected value is not one of the random variable values; rather, it is a number that the average number of defective batteries in a sample would approach as the experiment is repeated without end.

Matched Problem 2   Repeat Example 2 using a random sample of 4.

EXAMPLE 3   Expected Value of a Game   A spinner device is numbered from 0 to 5, and each of the 6 numbers is as likely to come up as any other. A player who bets $1 on any given number wins $4 (and gets the $1 bet back) if the pointer comes to rest on the chosen number; otherwise, the $1 bet is lost. What is the expected value of the game (long-run average gain or loss per game)?

SOLUTION   The sample space of equally likely events is

$$S = \{0, 1, 2, 3, 4, 5\}$$

Each sample point occurs with a probability of $\frac{1}{6}$. The random variable X assigns $4 to the winning number and $-$1 to each of the remaining numbers. So the probability of winning $4 is $\frac{1}{6}$ and of losing $1 is $\frac{5}{6}$. We form the probability distribution for X, called a **payoff table**, and compute the expected value of the game:

Payoff Table (Probability Distribution for X)

x_i	$4	$-$1
p_i	$\frac{1}{6}$	$\frac{5}{6}$

$$E(X) = \$4\left(\frac{1}{6}\right) + (-\$1)\left(\frac{5}{6}\right) = -\$\frac{1}{6} \approx -\$0.1667 \approx -17¢ \text{ per game}$$

In the long run, the player will lose an average of about 17¢ per game.

Matched Problem 3   Repeat Example 3 with the player winning $5 instead of $4 if the chosen number turns up. The loss is still $1 if any other number turns up. Is this a fair game?

The game in Example 3 is *not* fair: The player tends to lose money in the long run. A game is **fair** if the expected value $E(X)$ is equal to 0; that is, the player neither wins nor loses money in the long run. The fair games discussed in Section 8.2 are fair according to this definition, because their payoff tables have the following form:

Payoff Table

x_i	$\$b$	$\$a$
p_i	$\dfrac{a}{a+b}$	$\dfrac{b}{a+b}$

So $E(X) = b\left(\dfrac{a}{a+b}\right) + (-a)\dfrac{b}{a+b} = 0$

EXAMPLE 4 Expected Value and Insurance Suppose you are interested in insuring a car video system for $2,000 against theft. An insurance company charges a premium of $225 for coverage for 1 year, claiming an empirically determined probability of .1 that the system will be stolen sometime during the year. What is your expected return from the insurance company if you take out this insurance?

SOLUTION This is actually a game of chance in which your stake is $225. You have a .1 chance of receiving $1,775 from the insurance company ($2,000 minus your stake of $225) and a .9 chance of losing your stake of $225. What is the expected value of this "game"? We form a payoff table (the probability distribution for X) and compute the expected value:

Payoff Table

x_i	$\$1,775$	$-\$225$
p_i	.1	.9

$$E(X) = (\$1,775)(.1) + (-\$225)(.9) = -\$25$$

This means that if you insure with this company over many years and circumstances remain the same, you would have an average net loss to the insurance company of $25 per year.

Matched Problem 4 Find the expected value in Example 4 from the insurance company's point of view.

CONCEPTUAL INSIGHT

Suppose that in a class of 10 students, the scores on the first exam are 85, 73, 82, 65, 95, 85, 73, 75, 85, and 75. To compute the class average (mean), we add the scores and divide by the number of scores:

$$\frac{85 + 73 + 82 + 65 + 95 + 85 + 73 + 75 + 85 + 75}{10} = \frac{793}{10} = 79.3$$

Because 1 student scored 95, 3 scored 85, 1 scored 82, 2 scored 75, 2 scored 73, and 1 scored 65, the probability distribution of an exam score, for a student chosen at random from the class, is as follows:

x_i	65	73	75	82	85	95
p_i	.1	.2	.2	.1	.3	.1

The expected value of the probability distribution is

$$65\left(\frac{1}{10}\right) + 73\left(\frac{2}{10}\right) + 75\left(\frac{2}{10}\right) + 82\left(\frac{1}{10}\right) + 85\left(\frac{3}{10}\right) + 95\left(\frac{1}{10}\right) = \frac{793}{10} = 79.3$$

By comparing the two computations, we see that the mean of a data set is just the expected value of the corresponding probability distribution.

Decision Making and Expected Value

We conclude this section with an example in decision making.

EXAMPLE 5   Decision Analysis   An outdoor concert featuring a popular musical group is scheduled for a Sunday afternoon in a large open stadium. The promoter, worrying about being rained out, contacts a long-range weather forecaster who predicts the chance of rain on that Sunday to be .24. If it does not rain, the promoter is certain to net $100,000; if it does rain, the promoter estimates that the net will be only $10,000. An insurance company agrees to insure the concert for $100,000 against rain at a premium of $20,000. Should the promoter buy the insurance?

SOLUTION   The promoter has a choice between two courses of action; A_1: Insure and A_2: Do not insure. As an aid in making a decision, the expected value is computed for each course of action. Probability distributions are indicated in the payoff table (read vertically):

Payoff Table

p_i	A_1: Insure x_i	A_2: Do Not Insure x_i
.24 (rain)	$90,000	$10,000
.76 (no rain)	$80,000	$100,000

Note that the $90,000 entry comes from the insurance company's payoff ($100,000) minus the premium ($20,000) plus gate receipts ($10,000). The reasons for the other entries should be obvious. The expected value for each course of action is computed as follows:

A_1: Insure

$$E(X) = x_1 p_1 + x_2 p_2$$
$$= (\$90,000)(.24) + (\$80,000)(.76)$$
$$= \$82,400$$

A_2: Do Not Insure

$$E(X) = (\$10,000)(.24) + (\$100,000)(.76)$$
$$= \$78,400$$

It appears that the promoter's best course of action is to buy the insurance at $20,000. The promoter is using a long-run average to make a decision about a single event—a common practice in making decisions in areas of uncertainty.

Matched Problem 5   In Example 5, what is the insurance company's expected value if it writes the policy?

Exercises 8.5

Skills Warm-up Exercises

W   *In Problems 1–8, if necessary, review Section B.1.*

1. Find the average (mean) of the exam scores 73, 89, 45, 82, and 66.

2. Find the average (mean) of the exam scores 78, 64, 97, 60, 86, and 83.

3. Find the average (mean) of the exam scores in problem 1, if 4 points are added to each score.

4. Find the average (mean) of the exam scores in problem 2, if 3 points are subtracted from each score.

5. Find the average (mean) of the exam scores in problem 1, if each score is multiplied by 2.

6. Find the average (mean) of the exam scores in problem 2, if each score is divided by 2.

7. If the probability distribution for the random variable X is given in the table, what is the expected value of X?

x_i	−3	0	4
p_i	.3	.5	.2

8. If the probability distribution for the random variable X is given in the table, what is the expected value of X?

x_i	−2	−1	0	1	2
p_i	.1	.2	.4	.2	.1

9. You draw and keep a single bill from a hat that contains a $5, $20, $50, and $100 bill. What is the expected value of the game to you?

10. You draw and keep a single bill from a hat that contains a $1, $10, $20, $50, and $100 bill. What is the expected value of the game to you?

11. You draw and keep a single coin from a bowl that contains 15 pennies, 10 dimes, and 25 quarters. What is the expected value of the game to you?

12. You draw and keep a single coin from a bowl that contains 120 nickels and 80 quarters. What is the expected value of the game to you?

13. You draw a single card from a standard 52-card deck. If it is red, you win $50. Otherwise you get nothing. What is the expected value of the game to you?

14. You draw a single card from a standard 52-card deck. If it is an ace, you win $104. Otherwise you get nothing. What is the expected value of the game to you?

15. In tossing 2 fair coins, what is the expected number of heads?

16. In a family with 2 children, excluding multiple births and assuming that a boy is as likely as a girl at each birth, what is the expected number of boys?

17. A fair coin is flipped. If a head turns up, you win $1. If a tail turns up, you lose $1. What is the expected value of the game? Is the game fair?

18. Repeat Problem 17, assuming an unfair coin with the probability of a head being .55 and a tail being .45.

19. After paying $4 to play, a single fair die is rolled, and you are paid back the number of dollars corresponding to the number of dots facing up. For example, if a 5 turns up, $5 is returned to you for a net gain, or payoff, of $1; if a 1 turns up, $1 is returned for a net gain of −$3; and so on. What is the expected value of the game? Is the game fair?

20. Repeat Problem 19 with the same game costing $3.50 for each play.

21. Two coins are flipped. You win $2 if either 2 heads or 2 tails turn up; you lose $3 if a head and a tail turn up. What is the expected value of the game?

22. In Problem 21, for the game to be fair, how much *should* you lose if a head and a tail turn up?

23. A friend offers the following game: She wins $1 from you if, on four rolls of a single die, a 6 turns up at least once; otherwise, you win $1 from her. What is the expected value of the game to you? To her?

24. On three rolls of a single die, you will lose $10 if a 5 turns up at least once, and you will win $7 otherwise. What is the expected value of the game?

✎ 25. A single die is rolled once. You win $5 if a 1 or 2 turns up and $10 if a 3, 4, or 5 turns up. How much should you lose if a 6 turns up in order for the game to be fair? Describe the steps you took to arrive at your answer.

✎ 26. A single die is rolled once. You lose $12 if a number divisible by 3 turns up. How much should you win if a number not divisible by 3 turns up in order for the game to be fair? Describe the process and reasoning used to arrive at your answer.

27. A pair of dice is rolled once. Suppose you lose $10 if a 7 turns up and win $11 if an 11 or 12 turns up. How much should you win or lose if any other number turns up in order for the game to be fair?

28. A coin is tossed three times. Suppose you lose $3 if 3 heads appear, lose $2 if 2 heads appear, and win $3 if 0 heads appear. How much should you win or lose if 1 head appears in order for the game to be fair?

29. A card is drawn from a standard 52-card deck. If the card is a king, you win $10; otherwise, you lose $1. What is the expected value of the game?

30. A card is drawn from a standard 52-card deck. If the card is a diamond, you win $10; otherwise, you lose $4. What is the expected value of the game?

31. A 5-card hand is dealt from a standard 52-card deck. If the hand contains at least one king, you win $10; otherwise, you lose $1. What is the expected value of the game?

32. A 5-card hand is dealt from a standard 52-card deck. If the hand contains at least one diamond, you win $10; otherwise, you lose $4. What is the expected value of the game?

33. The payoff table for two courses of action, A_1 or A_2, is given below. Which of the two actions will produce the largest expected value? What is it?

p_i	A_1 x_i	A_2 x_i
.1	−$200	−$100
.2	$100	$200
.4	$400	$300
.3	$100	$200

34. The payoff table for three possible courses of action is given below. Which of the three actions will produce the largest expected value? What is it?

P_i	A_1 x_i	A_2 x_i	A_3 x_i
.2	$ 500	$ 400	$ 300
.4	$1,200	$1,100	$1,000
.3	$1,200	$1,800	$1,700
.1	$1,200	$1,800	$2,400

35. Roulette wheels in Nevada generally have 38 equally spaced slots numbered 00, 0, 1, 2, . . . , 36. A player who bets $1 on any given number wins $35 (and gets the bet back) if the ball comes to rest on the chosen number; otherwise, the $1 bet is lost. What is the expected value of this game?

36. In roulette (see Problem 35), the numbers from 1 to 36 are evenly divided between red and black. A player who bets $1 on black wins $1 (and gets the $1 bet back) if the ball comes to rest on black; otherwise (if the ball lands on red, 0, or 00), the $1 bet is lost. What is the expected value of the game?

✎ 37. A game has an expected value to you of $100. It costs $100 to play, but if you win, you receive $100,000 (including your $100 bet) for a net gain of $99,900. What is the probability of winning? Would you play this game? Discuss the factors that would influence your decision.

✎ 38. A game has an expected value to you of −$0.50. It costs $2 to play, but if you win, you receive $20 (including your $2 bet) for a net gain of $18. What is the probability of winning? Would you play this game? Discuss the factors that would influence your decision.

39. Five thousand tickets are sold at $1 each for a charity raffle. Tickets will be drawn at random and monetary prizes awarded as follows: 1 prize of $500; 3 prizes of $100, 5 prizes of $20, and 20 prizes of $5. What is the expected value of this raffle if you buy 1 ticket?

40. Ten thousand raffle tickets are sold at $2 each for a local library benefit. Prizes are awarded as follows: 2 prizes of $1,000, 4 prizes of $500, and 10 prizes of $100. What is the expected value of this raffle if you purchase 1 ticket?

41. A box of 10 flashbulbs contains 3 defective bulbs. A random sample of 2 is selected and tested. Let X be the random variable associated with the number of defective bulbs in the sample.

(A) Find the probability distribution of X.

(B) Find the expected number of defective bulbs in a sample.

42. A box of 8 flashbulbs contains 3 defective bulbs. A random sample of 2 is selected and tested. Let X be the random variable associated with the number of defective bulbs in a sample.

(A) Find the probability distribution of X.

(B) Find the expected number of defective bulbs in a sample.

43. One thousand raffle tickets are sold at $1 each. Three tickets will be drawn at random (without replacement), and each will pay $200. Suppose you buy 5 tickets.

(A) Create a payoff table for 0, 1, 2, and 3 winning tickets among the 5 tickets you purchased. (If you do not have any winning tickets, you lose $5; if you have 1 winning ticket, you net $195 since your initial $5 will not be returned to you; and so on.)

(B) What is the expected value of the raffle to you?

44. Repeat Problem 43 with the purchase of 10 tickets.

45. To simulate roulette on a graphing calculator, a random integer between −1 and 36 is selected (−1 represents 00; see Problem 35). The command in Figure A simulates 200 games.

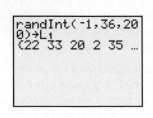

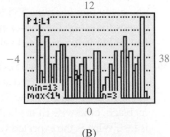

| (A) | (B) |

(A) Use the statistical plot in Figure B to determine the net gain or loss of placing a $1 bet on the number 13 in each of the 200 games.

(B) Compare the results of part (A) with the expected value of the game.

(C) Use a graphing calculator to simulate betting $1 on the number 7 in each of 500 games of roulette and compare the simulated and expected gains or losses.

46. Use a graphing calculator to simulate the results of placing a $1 bet on black in each of 400 games of roulette (see Problems 36 and 45) and compare the simulated and expected gains or losses.

47. A 3-card hand is dealt from a standard deck. You win $20 for each diamond in the hand. If the game is fair, how much should you lose if the hand contains no diamonds?

48. A 3-card hand is dealt from a standard deck. You win $100 for each king in the hand. If the game is fair, how much should you lose if the hand contains no kings?

Applications

49. Insurance. The annual premium for a $5,000 insurance policy against the theft of a painting is $150. If the (empirical) probability that the painting will be stolen during the year is .01, what is your expected return from the insurance company if you take out this insurance?

50. Insurance. Repeat Problem 49 from the point of view of the insurance company.

51. Decision analysis. After careful testing and analysis, an oil company is considering drilling in two different sites. It is estimated that site A will net $30 million if successful (probability .2) and lose $3 million if not (probability .8); site B will net $70 million if successful (probability .1) and lose $4 million if not (probability .9). Which site should the company choose according to the expected return for each site?

52. Decision analysis. Repeat Problem 51, assuming that additional analysis caused the estimated probability of success in field B to be changed from .1 to .11.

53. Genetics. Suppose that at each birth, having a girl is not as likely as having a boy. The probability assignments for the number of boys in a 3-child family are approximated empirically from past records and are given in the table. What is the expected number of boys in a 3-child family?

Number of Boys

x_i	p_i
0	.12
1	.36
2	.38
3	.14

54. Genetics. A pink-flowering plant is of genotype RW. If two such plants are crossed, we obtain a red plant (RR) with probability .25, a pink plant (RW or WR) with probability .50, and a white plant (WW) with probability .25, as shown in the table. What is the expected number of W genes present in a crossing of this type?

Number of W Genes Present

x_i	p_i
0	.25
1	.50
2	.25

55. Politics. A money drive is organized by a campaign committee for a candidate running for public office. Two approaches are considered:

A_1: A general mailing with a follow-up mailing

A_2: Door-to-door solicitation with follow-up telephone calls

From campaign records of previous committees, average donations and their corresponding probabilities are estimated and shown in the table:

A_1		A_2	
x_i (Return per Person)	p_i	x_i (Return per Person)	p_i
$10	.3	$15	.3
5	.2	3	.1
0	.5	0	.6
	1.0		1.0

What are the expected returns? Which course of action should be taken according to the expected returns?

Answers to Matched Problems

1. $E(X) = 3.73$

2. (A)

x_i	0	1	2
p_i	.632	.337	.032*

Note: Due to roundoff error, sum = 1.001 ≈ 1.

(B) .4

3. $E(X) = \$0$; the game is fair

4. $E(X) = (-\$1,775)(.1) + (\$225)(.9) = \$25$ (This amount, of course, is necessary to cover expenses and profit.)

5. $E(X) = (-\$80,000)(.24) + (\$20,000)(.76) = -\$4,000$ (This means that the insurance company had other information regarding the weather than the promoter had; otherwise, the company would not have written this policy.)

Chapter 8   Summary and Review

Important Terms, Symbols, and Concepts

8.1 Sample Spaces, Events, and Probability

EXAMPLES

- Probability theory is concerned with **random experiments** (such as tossing a coin or rolling a pair of dice) for which different outcomes are obtained no matter how carefully the experiment is repeated under the same conditions.

- The set S of all outcomes of a random experiment is called a **sample space**. The subsets of S are called **events**. An event that contains only one outcome is called a **simple event** or **simple outcome**. Events that contain more than one outcome are **compound events**. We say that **an event E occurs** if any of the simple events in E occurs.

  Ex. 1, p. 387
  Ex. 2, p. 388
  Ex. 3, p. 388

- If $S = \{e_1, e_2, \ldots, e_n\}$ is a sample space for an experiment, an **acceptable probability assignment** is an assignment of real numbers $P(e_i)$ to simple events such that

$$0 \le P(e_i) \le 1 \quad \text{and} \quad P(e_1) + P(e_2) + \cdots + P(e_n) = 1$$

- Each number $P(e_i)$ is called the **probability of the event e_i**. The **probability of an arbitrary event E**, denoted by $P(E)$, is the sum of the probabilities of the simple events in E. If E is the empty set, then $P(E) = 0$.

  Ex. 4, p. 391

- Acceptable probability assignments can be made using a theoretical approach or an empirical approach. If an experiment is conducted n times and event E occurs with frequency $f(E)$, then the ratio $f(E)/n$ is called the **relative frequency** of the occurrence of E in n trials, or the **approximate empirical probability of E**. The **empirical probability** of E is the number (if it exists) that $f(E)/n$ approaches as n gets larger and larger.

  Ex. 6, p. 394

- If the **equally likely assumption** is made, each simple event of the sample space $S = \{e_1, e_2, \ldots, e_n\}$ is assigned the same probability, namely $1/n$. Theorem 1 (p. 393) gives the probability of arbitrary events under the equally likely assumption.

  Ex. 5, p. 393
  Ex. 7, p. 395
  Ex. 8, p. 395

8.2 Union, Intersection, and Complement of Events; Odds

- Let A and B be two events in a sample space. Then $A \cup B = \{x \mid x \in A \text{ or } x \in B\}$ is the **union** of A and B; $A \cap B = \{x \mid x \in A \text{ and } x \in B\}$ is the **intersection** of A and B.

Ex. 1, p. 400

Ex. 2, p. 401

Ex. 3, p. 402

- Events whose intersection is the empty set are said to be **mutually exclusive** or **disjoint**.
- The probability of the union of two events is given by

$$P(A \cup B) = P(A) + P(B) - P(A \cap B)$$

- The **complement** of event E, denoted by E', consists of those elements of S that do not belong to E:

$$P(E') = 1 - P(E)$$

- The language of **odds** is sometimes used, as an alternative to the language of probability, to describe the likelihood of an event. If $P(E)$ is the probability of E, then the **odds for E** are $P(E)/P(E')$ [usually expressed as a ratio of whole numbers and read as "$P(E)$ to $P(E')$"], and the **odds against E** are $P(E')/P(E)$.

Ex. 4, p. 403

Ex. 5, p. 404

Ex. 8, p. 406

- If the odds for an event E are a/b, then

Ex. 6, p. 406

Ex. 7, p. 406

$$P(E) = \frac{a}{a + b}$$

8.3 Conditional Probability, Intersection, and Independence

- If A and B are events in a sample space S, and $P(B) \neq 0$, then the **conditional probability of A given B** is defined by

Ex. 1, p. 413

Ex. 2, p. 413

$$P(A \mid B) = \frac{P(A \cap B)}{P(B)}$$

- By solving this equation for $P(A \cap B)$ we obtain the **product rule** (Theorem 1, p. 414):

$$P(A \cap B) = P(B)P(A \mid B) = P(A)P(B \mid A)$$

Ex. 3, p. 414

Ex. 4, p. 415

Ex. 5, p. 417

Ex. 6, p. 419

Ex. 7, p. 419

Ex. 8, p. 420

- Events A and B are **independent** if $P(A \cap B) = P(A)P(B)$.
- Theorem 2 (p. 418) gives a test for independence.

8.4 Bayes' Formula

- Let $U_1, U_2, \ldots, U_n$ be n mutually exclusive events whose union is the sample space S. Let E be an arbitrary event in S such that $P(E) \neq 0$. Then

Ex. 1, p. 425

Ex. 2, p. 428

Ex. 3, p. 429

$$P(U_1 \mid E) = \frac{P(E \mid U_1)P(U_1)}{P(E \mid U_1)P(U_1) + P(E \mid U_2)P(U_2) + \cdots + P(E \mid U_n)P(U_n)}$$

$$= \frac{\left(\begin{array}{c} \text{product of branch probabilities leading} \\ \text{to } E \text{ through } U_1 \end{array} \right)}{\text{sum of all branch products leading to } E}$$

- Similar results hold for $U_2, U_3, \ldots, U_n$. This formula is called **Bayes' formula**.

8.5 Random Variable, Probability Distribution, and Expected Value

- A **random variable** X is a function that assigns a numerical value to each simple event in a sample space S.
- The **probability distribution of X** assigns a probability $p(x)$ to each range element x of X: $p(x)$ is the sum of the probabilities of the simple events in S that are assigned the numerical value x.
- If a random variable X has range values $x_1, x_2, \ldots, x_n$ that have probabilities $p_1, p_2, \ldots, p_n$, respectively, then the **expected value** of X, denoted $E(X)$, is defined by

Ex. 1, p. 435

Ex. 2, p. 436

Ex. 3, p. 436

$$E(X) = x_1 p_1 + x_2 p_2 + \cdots + x_n p_n$$

- Suppose the x_i's are payoffs in a game of chance. If the game is played a large number of times, the expected value approximates the average win per game.

Ex. 4, p. 437

Ex. 5, p. 438

Review Exercises

Work through all the problems in this chapter review and check your answers in the back of the book. Answers to all problems are there along with section numbers in italics to indicate where each type of problem is discussed. Where weaknesses show up, review appropriate sections in the text.

1. In a single deal of 5 cards from a standard 52-card deck, what is the probability of being dealt 5 clubs?

2. Brittani and Ramon are members of a 15-person ski club. If the president and treasurer are selected by lottery, what is the probability that Brittani will be president and Ramon will be treasurer? (A person cannot hold more than one office.)

3. Each of the first 10 letters of the alphabet is printed on a separate card. What is the probability of drawing 3 cards and getting the code word *dig* by drawing *d* on the first draw, *i* on the second draw, and *g* on the third draw? What is the probability of being dealt a 3-card hand containing the letters *d*, *i*, and *g* in any order?

4. A drug has side effects for 50 out of 1,000 people in a test. What is the approximate empirical probability that a person using the drug will have side effects?

5. A spinning device has 5 numbers, 1, 2, 3, 4, and 5, each as likely to turn up as the other. A person pays $3 and then receives back the dollar amount corresponding to the number turning up on a single spin. What is the expected value of the game? Is the game fair?

6. If A and B are events in a sample space S and $P(A) = .3$, $P(B) = .4$, and $P(A \cap B) = .1$, find

   (A) $P(A')$

   (B) $P(A \cup B)$

7. A spinner lands on R with probability .3, on G with probability .5, and on B with probability .2. Find the probability and odds for the spinner landing on either R or G.

8. If in repeated rolls of two fair dice the odds for rolling a sum of 8 before rolling a sum of 7 are 5 to 6, then what is the probability of rolling a sum of 8 before rolling a sum of 7?

Answer Problems 9–17 using the table of probabilities shown below.

	X	Y	Z	Totals
S	.10	.25	.15	.50
T	.05	.20	.02	.27
R	.05	.15	.03	.23
Totals	.20	.60	.20	1.00

9. Find $P(T)$.

10. Find $P(Z)$.

11. Find $P(T \cap Z)$.

12. Find $P(R \cap Z)$.

13. Find $P(R|Z)$.

14. Find $P(Z|R)$.

15. Find $P(T|Z)$.

16. Are T and Z independent?

17. Are S and X independent?

Answer Problems 18–25 using the following probability tree:

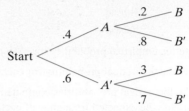

18. $P(A)$

19. $P(B|A)$

20. $P(B|A')$

21. $P(A \cap B)$

22. $P(A' \cap B)$

23. $P(B)$

24. $P(A|B)$

25. $P(A|B')$

26. (A) If 10 out of 32 students in a class were born in June, July, or August, what is the approximate empirical probability of any student being born in June, July, or August?

    (B) If one is as likely to be born in any of the 12 months of a year as any other, what is the theoretical probability of being born in either June, July, or August?

    ✎ (C) Discuss the discrepancy between the answers to parts (A) and (B).

✎ *In Problems 27 and 28, a sample space S is described. Would it be reasonable to make the equally likely assumption? Explain.*

27. A 3-card hand is dealt from a standard deck. We are interested in the number of red cards in the hand, so an appropriate sample space is $S = \{0, 1, 2, 3\}$.

28. A 3-card hand is dealt from a standard deck. We are interested in whether there are more red cards or more black cards in the hand, so an appropriate sample space is $S = \{R, B\}$.

29. A player tosses two coins and receives $5 if 2 heads turn up, loses $4 if 1 head turns up, and wins $2 if 0 heads turn up. Compute the expected value of the game. Is the game fair?

30. A spinning device has 3 numbers, 1, 2, and 3, each as likely to turn up as the other. If the device is spun twice, what is the probability that

    (A) The same number turns up both times?

    (B) The sum of the numbers turning up is 5?

31. In a single draw from a standard 52-card deck, what are the probability and odds for drawing

    (A) A jack or a queen?

    (B) A jack or a spade?

    (C) A card other than an ace?

32. (A) What are the odds for rolling a sum of 5 on the single roll of two fair dice?

    (B) If you bet $1 that a sum of 5 will turn up, what should the house pay (plus return your $1 bet) in order for the game to be fair?

33. Two coins are flipped 1,000 times with the following frequencies:

2 heads	210
1 head	480
0 heads	310

(A) Compute the empirical probability for each outcome.

(B) Compute the theoretical probability for each outcome.

(C) Using the theoretical probabilities computed in part (B), compute the expected frequency of each outcome, assuming fair coins.

34. A fair coin is tossed 10 times. On each of the first 9 tosses the outcome is heads. Discuss the probability of a head on the 10th toss.

35. An experiment consists of rolling a pair of fair dice. Let X be the random variable associated with the sum of the values that turn up.

(A) Find the probability distribution for X.

(B) Find the expected value of X.

36. Two dice are rolled. The sample space is chosen as the set of all ordered pairs of integers taken from $\{1, 2, 3, 4, 5, 6\}$. What is the event A that corresponds to the sum being divisible by 4? What is the event B that corresponds to the sum being divisible by 6? What are $P(A), P(B), P(A \cap B)$, and $P(A \cup B)$?

37. A person tells you that the following approximate empirical probabilities apply to the sample space $\{e_1, e_2, e_3, e_4\}$: $P(e_1) \approx .1, P(e_2) \approx -.2, P(e_3) \approx .6, P(e_4) \approx 2$. There are three reasons why P cannot be a probability function. Name them.

38. Use the following information to complete the frequency table below:

$$n(A) = 50, \quad n(B) = 45,$$
$$n(A \cup B) = 80, \quad n(U) = 100$$

	A	A'	Totals
B			
B'			
Totals			

39. A pointer is spun on a circular spinner. The probabilities of the pointer landing on the integers from 1 to 5 are given in the table below.

e_i	1	2	3	4	5
p_i	.2	.1	.3	.3	.1

(A) What is the probability of the pointer landing on an odd number?

(B) What is the probability of the pointer landing on a number less than 4 given that it landed on an odd number?

40. A card is drawn at random from a standard 52-card deck. If E is the event "The drawn card is red" and F is the event "The drawn card is an ace," then

(A) Find $P(F|E)$.   (B) Test E and F for independence.

In Problems 41–45, urn U_1 contains 2 white balls and 3 red balls; urn U_2 contains 2 white balls and 1 red ball.

41. Two balls are drawn out of urn U_1 in succession. What is the probability of drawing a white ball followed by a red ball if the first ball is

(A) Replaced?   (B) Not replaced?

42. Which of the two parts in Problem 41 involve dependent events?

43. In Problem 41, what is the expected number of red balls if the first ball is

(A) Replaced?   (B) Not replaced?

44. An urn is selected at random by flipping a fair coin; then a ball is drawn from the urn. Compute:

(A) $P(R|U_1)$ (B) $P(R|U_2)$

(C) $P(R)$ (D) $P(U_1|R)$

(E) $P(U_2|W)$ (F) $P(U_1 \cap R)$

45. In Problem 44, are the events "Selecting urn U_1" and "Drawing a red ball" independent?

46. From a standard deck of 52 cards, what is the probability of obtaining a 5-card hand

(A) Of all diamonds?

(B) Of 3 diamonds and 2 spades?

Write answers in terms of $_nC_r$ or $_nP_r$; do not evaluate.

47. A group of 10 people includes one married couple. If 4 people are selected at random, what is the probability that the married couple is selected?

48. A 5-card hand is drawn from a standard deck. Discuss how you can tell that the following two events are dependent without any computation.

$$S = \text{hand consists entirely of spades}$$
$$H = \text{hand consists entirely of hearts}$$

49. The command in Figure A was used on a graphing calculator to simulate 50 repetitions of rolling a pair of dice and recording the minimum of the two numbers. A statistical plot of the results is shown in Figure B.

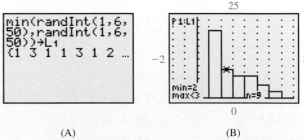

(A)	(B)

(A) Use Figure B to find the empirical probability that the minimum is 2.

(B) What is the theoretical probability that the minimum is 2?

(C) Using a graphing calculator to simulate 200 rolls of a pair of dice, determine the empirical probability that the minimum is 4 and compare with the theoretical probability.

50. A card is drawn at random from a standard 52-card deck. Using a graphing calculator to simulate 800 such draws, determine the empirical probability that the card is a black jack and compare with the theoretical probability.

In Problems 51–56, discuss the validity of each statement. If the statement is always true, explain why. If not, give a counterexample.

51. If $P(E) = 1$, then the odds for E are $1 : 1$.

52. If $E = F'$, then $P(E \cup F) = P(E) + P(F)$.

53. If E and F are complementary events, then E and F are independent.

54. If $P(E \cup F) = 1$, then E and F are complementary events.

55. If E and F are independent events, then $P(E)P(F) = P(E \cap F)$.

56. If E and F are mutually exclusive events, then $P(E)P(F) = P(E \cap F)$.

57. Three fair coins are tossed 1,000 times with the following frequencies of outcomes:

Number of Heads	0	1	2	3
Frequency	120	360	350	170

(A) What is the empirical probability of obtaining 2 heads?

(B) What is the theoretical probability of obtaining 2 heads?

(C) What is the expected frequency of obtaining 2 heads?

58. You bet a friend $1 that you will get 1 or more double 6's on 24 rolls of a pair of fair dice. What is your expected value for this game? What is your friend's expected value? Is the game fair?

59. If 3 people are selected from a group of 7 men and 3 women, what is the probability that at least 1 woman is selected?

Two cards are drawn in succession without replacement from a standard 52-card deck. In Problems 60 and 61, compute the indicated probabilities.

60. The second card is a heart given that the first card is a heart.

61. The first card is a heart given that the second card is a heart.

62. Two fair (not weighted) dice are each numbered with a 3 on one side, a 2 on two sides, and a 1 on three sides. The dice are rolled, and the numbers on the two up faces are added. If X is the random variable associated with the sample space $S = \{2, 3, 4, 5, 6\}$:

(A) Find the probability distribution of X.

(B) Find the expected value of X.

63. If you pay $3.50 to play the game in Problem 62 (the dice are rolled once) and you are returned the dollar amount corresponding to the sum on the faces, what is the expected value of the game? Is the game fair? If it is not fair, how much should you pay in order to make the game fair?

64. Suppose that 3 white balls and 1 black ball are placed in a box. Balls are drawn in succession without replacement until a black ball is drawn, and then the game is over. You win if the black ball is drawn on the fourth draw.

(A) What are the probability and odds for winning?

(B) If you bet $1, what should the house pay you for winning (plus return your $1 bet) if the game is to be fair?

65. If each of 5 people is asked to identify his or her favorite book from a list of 10 best-sellers, what is the probability that at least 2 of them identify the same book?

66. Let A and B be events with nonzero probabilities in a sample space S. Under what conditions is $P(A|B)$ equal to $P(B|A)$?

Applications

67. Market research. From a survey of 100 city residents, it was found that 40 read the daily newspaper, 70 watch the evening news, and 30 do both. What is the (empirical) probability that a resident selected at random

(A) Reads the daily paper or watches the evening news?

(B) Does neither?

(C) Does one but not the other?

68. Market research. A market research firm has determined that 40% of the people in a certain area have seen the advertising for a new product and that 85% of those who have seen the advertising have purchased the product. What is the probability that a person in this area has seen the advertising and purchased the product?

69. Market analysis. A clothing company selected 1,000 persons at random and surveyed them to determine a relationship between the age of the purchaser and the annual purchases of jeans. The results are given in the table.

Age	Jeans Purchased Annually				
	0	1	2	Above 2	Totals
Under 12	60	70	30	10	170
12–18	30	100	100	60	290
19–25	70	110	120	30	330
Over 25	100	50	40	20	210
Totals	260	330	290	120	1,000

Given the events

$A =$ person buys 2 pairs of jeans

$B =$ person is between 12 and 18 years old

$C =$ person does not buy more than 2 pairs of jeans

$D =$ person buys more than 2 pairs of jeans

(A) Find $P(A), P(B), P(A \cap B), P(A|B)$, and $P(B|A)$.

(B) Are events A and B independent? Explain.

(C) Find $P(C), P(D), P(C \cap D), P(C|D)$, and $P(D|C)$.

(D) Are events C and D mutually exclusive? Independent? Explain.

70. Decision analysis. A company sales manager, after careful analysis, presents two sales plans. It is estimated that plan A will net $10 million if successful (probability .8) and lose $2 million if not (probability .2); plan B will net $12 million if successful (probability .7) and lose $2 million if not (probability .3). What is the expected return for each plan? Which plan should be chosen based on the expected return?

71. Insurance. A $2,000 bicycle is insured against theft for an annual premium of $170. If the probability that the bicycle will be stolen during the year is .08 (empirically determined), what is the expected value of the policy?

72. Quality control. Twelve precision parts, including 2 that are substandard, are sent to an assembly plant. The plant will select 4 at random and will return the entire shipment if 1 or more of the sample are found to be substandard. What is the probability that the shipment will be returned?

73. Quality control. A dozen tablet computers, including 2 that are defective, are sent to a computer service center. A random sample of 3 is selected and tested. Let X be the random variable associated with the number of tablet computers in a sample that are defective.

(A) Find the probability distribution of X.

(B) Find the expected number of defective tablet computers in a sample.

74. Medicine: cardiogram test. By testing a large number of individuals, it has been determined that 82% of the population have normal hearts, 11% have some minor heart problems, and 7% have severe heart problems. Ninety-five percent of the persons with normal hearts, 30% of those with minor problems, and 5% of those with severe problems will pass a cardiogram test. What is the probability that a person who passes the cardiogram test has a normal heart?

75. Genetics. Six men in 100 and 1 woman in 100 are color-blind. A person is selected at random and is found to be color-blind. What is the probability that this person is a man? (Assume that the total population contains the same number of women as men.)

76. Voter preference. In a straw poll, 30 students in a mathematics class are asked to indicate their preference for president of student government. Approximate empirical probabilities are assigned on the basis of the poll: candidate A should receive 53% of the vote, candidate B should receive 37%, and candidate C should receive 10%. One week later, candidate B wins the election. Discuss the factors that may account for the discrepancy between the poll and the election results.

9 Markov Chains

9.1 Properties of Markov Chains

9.2 Regular Markov Chains

9.3 Absorbing Markov Chains

    Chapter 9
    Summary and Review

    Review Exercises

Introduction

In this chapter, we consider a mathematical model that combines probability and matrices to analyze certain sequences. The model is called a *Markov chain*, after the Russian mathematician Andrei Markov (1856–1922). Recent applications of Markov chains involve a wide variety of topics, including finance, market research, genetics, medicine, demographics, psychology, and political science. Problem 63 in Section 9.2, for example, studies the changing percentage of commuters that take rapid transit.

In Section 9.1 we introduce the basic properties of Markov chains. In the remaining sections, we discuss the long-term behavior of two different types of Markov chains.

9.1   Properties of Markov Chains

- Introduction
- Transition and State Matrices
- Powers of Transition Matrices
- Application

Introduction

In this section, we explore physical *systems* and their possible *states*. To understand what this means, consider the following examples:

1. A stock listed on the New York Stock Exchange either increases, decreases, or does not change in price each day that the exchange is open. The stock can be thought of as a physical system with three possible states: increase, decrease, or no change.

2. A commuter, relative to a rapid transit system, can be thought of as a physical system with two states, a user or a nonuser.

3. During each congressional election, a voting precinct casts a simple majority vote for a Republican candidate, a Democratic candidate, or a third-party candidate. The precinct, relative to all congressional elections past, present, and future, constitutes a physical system that is in one (and only one) of three states after each election: Republican, Democratic, or other.

If a system evolves from one state to another in such a way that chance elements are involved, then the system's progression through a sequence of states is called a **stochastic process** (*stochos* is the Greek word for "guess"). We will consider a simple example of a stochastic process, and out of it will arise further definitions and methodology.

A toothpaste company markets a product (brand *A*) that currently has 10% of the toothpaste market. The company hires a market research firm to estimate the percentage of the market that it might acquire in the future if it launches an aggressive sales campaign. The research firm uses test marketing and extensive surveys to predict the effect of the campaign. They find that if a person is using brand *A*, the probability is .8 that this person will buy it again when he or she runs out of toothpaste. On the other hand, a person using another brand will switch to brand *A* with a probability of .6 when he or she runs out of toothpaste. So each toothpaste consumer can be considered to be in one of two possible states:

$$A = \text{uses brand } A \qquad \text{or} \qquad A' = \text{uses another brand}$$

The probabilities determined by the market research firm can be represented graphically in a **transition diagram** (Fig. 1).

We can also represent this information numerically in a **transition probability matrix**:

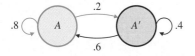

Figure 1   **Transition diagram**

$$\begin{array}{c} & \text{Next state} \\ & \begin{array}{cc} A & A' \end{array} \\ \text{Current state} \quad \begin{array}{c} A \\ A' \end{array} \begin{bmatrix} .8 & .2 \\ .6 & .4 \end{bmatrix} = P \end{array}$$

Explore and Discuss 1

(A) Refer to the transition diagram in Figure 1. What is the probability that a person using brand *A* will switch to another brand when he or she runs out of toothpaste?

(B) Refer to transition probability matrix *P*. What is the probability that a person who is not using brand *A* will not switch to brand *A* when he or she runs out of toothpaste?

(C) In Figure 1, the sum of the probabilities on the arrows leaving each state is 1. Will this be true for any transition diagram? Explain your answer.

(D) In transition probability matrix *P*, the sum of the probabilities in each row is 1. Will this be true for any transition probability matrix? Explain your answer.

The toothpaste company's 10% share of the market at the beginning of the sales campaign can be represented as an **initial-state distribution matrix**:

$$\begin{array}{cc} A & A' \end{array}$$
$$S_0 = [.1 \quad .9]$$

If a person is chosen at random, the probability that this person uses brand A (state A) is .1, and the probability that this person does not use brand A (state A') is .9. Thus, S_0 also can be interpreted as an **initial-state probability matrix**.

What are the probabilities of a person being in state A or A' on the first purchase after the start of the sales campaign? Let us look at the probability tree given below.

Note: A_0 represents state A at the beginning of the campaign, A'_1 represents state A' on the first purchase after the campaign, and so on.

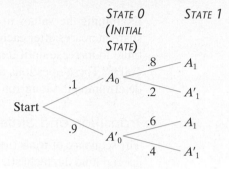

Proceeding as in Chapter 8, we can read the required probabilities directly from the tree:

$$\begin{aligned} P(A_1) &= P(A_0 \cap A_1) + P(A'_0 \cap A_1) \\ &= (.1)(.8) + (.9)(.6) = .62 \\ P(A'_1) &= P(A_0 \cap A'_1) + P(A'_0 \cap A'_1) \\ &= (.1)(.2) + (.9)(.4) = .38 \end{aligned}$$

Note: $P(A_1) + P(A'_1) = 1$, as expected.

The **first-state matrix** is

$$\begin{array}{cc} A & A' \end{array}$$
$$S_1 = [.62 \quad .38]$$

This matrix gives us the probabilities of a randomly chosen person being in state A or A' on the first purchase after the start of the campaign. We see that brand A's market share has increased from 10% to 62%.

Now, if you were asked to find the probabilities of a person being in state A or state A' on the tenth purchase after the start of the campaign, you might start to draw additional branches on the probability tree. However, you would soon become discouraged because the number of branches doubles for each successive purchase. By the tenth purchase, there would be $2^{11} = 2,048$ branches! Fortunately, we can convert the summing of branch products to matrix multiplication. In particular, if we multiply the initial-state matrix S_0 by the transition matrix P, we obtain the first-state matrix S_1:

$$S_0 P = \begin{array}{c} \begin{array}{cc} A & A' \end{array} \\ [.1 \quad .9] \end{array} \begin{bmatrix} .8 & .2 \\ .6 & .4 \end{bmatrix} = [\underbrace{(.1)(.8) + (.9)(.6)} \quad \underbrace{(.1)(.2) + (.9)(.4)}] = \begin{array}{c} \begin{array}{cc} A & A' \end{array} \\ [.62 \quad .38] \end{array} = S_1$$

Initial state   Transition matrix   Compare with the tree computations above   First state

As you might guess, we can get the second-state matrix S_2 (for the second purchase) by multiplying the first-state matrix by the transition matrix:

$$
S_1 P = \begin{matrix} A & A' \\ [.62 & .38] \end{matrix} \begin{bmatrix} .8 & .2 \\ .6 & .4 \end{bmatrix} = \begin{matrix} A & A' \\ [.724 & .276] \end{matrix} = S_2
$$

First state   Second state

The third-state matrix S_3 is computed in a similar manner:

$$
S_2 P = \begin{matrix} A & A' \\ [.724 & .276] \end{matrix} \begin{bmatrix} .8 & .2 \\ .6 & .4 \end{bmatrix} = \begin{matrix} A & A' \\ [.7448 & .2552] \end{matrix} = S_3
$$

Second state   Third state

Examining the values in the first three state matrices, we see that brand A's market share increases after each toothpaste purchase. Will the market share for brand A continue to increase until it approaches 100%, or will it level off at some value less than 100%? These questions are answered in Section 9.2 when we develop techniques for determining the long-run behavior of state matrices.

Transition and State Matrices

The sequence of trials (toothpaste purchases) with the constant transition matrix P is a special kind of stochastic process called a *Markov chain*. In general, a **Markov chain** is a sequence of experiments, trials, or observations such that the transition probability matrix from one state to the next is constant. A Markov chain has no memory. The various matrices associated with a Markov chain are defined in the next box.

> **DEFINITION** Markov Chains
> Given a Markov chain with n states, a **kth-state matrix** is a matrix of the form
>
> $$S_k = \begin{bmatrix} s_{k1} & s_{k2} & \cdots & s_{kn} \end{bmatrix}$$

Each entry s_{ki} is the proportion of the population that is in state i after the kth trial, or, equivalently, the probability of a randomly selected element of the population being in state i after the kth trial. The sum of all the entries in the kth state matrix S_k must be 1.

A **transition matrix** is a constant square matrix P of order n such that the entry in the ith row and jth column indicates the probability of the system moving from the ith state to the jth state on the next observation or trial. The sum of the entries in each row must be 1.

> **CONCEPTUAL INSIGHT**
>
> 1. Since the entries in a kth-state matrix or transition matrix are probabilities, they must be real numbers between 0 and 1, inclusive.
>
> 2. Rearranging the various states and corresponding transition probabilities in a transition matrix will produce a different, but equivalent, transition matrix. For example, both of the following matrices are transition matrices for the toothpaste company discussed earlier:
>
> $$P = \begin{matrix} & A & A' \\ A & \begin{bmatrix} .8 & .2 \\ A' & .6 & .4 \end{bmatrix} \end{matrix} \qquad P' = \begin{matrix} & A' & A \\ A & \begin{bmatrix} .4 & .6 \\ A' & .2 & .8 \end{bmatrix} \end{matrix}$$
>
> Such rearrangements will affect the form of the matrices used in the solution of a problem but will not affect any of the information obtained from these matrices. In Section 9.3, we encounter situations where it will be helpful to select a transition matrix that has a special form. For now, you can choose any order for the states in a transition matrix.

As we indicated in the preceding discussion, matrix multiplication can be used to compute the various state matrices of a Markov chain:

If S_0 is the initial-state matrix and P is the transition matrix for a Markov chain, then the subsequent state matrices are given by

$$S_1 = S_0 P \qquad \text{First-state matrix}$$
$$S_2 = S_1 P \qquad \text{Second-state matrix}$$
$$S_3 = S_2 P \qquad \text{Third-state matrix}$$
$$\vdots$$
$$S_k = S_{k-1} P \qquad \text{kth-state matrix}$$

EXAMPLE 1 **Insurance** An insurance company found that on average, over a period of 10 years, 23% of the drivers in a particular community who were involved in an accident one year were also involved in an accident the following year. They also found that only 11% of the drivers who were not involved in an accident one year were involved in an accident the following year. Use these percentages as approximate empirical probabilities for the following:

(A) Draw a transition diagram.

(B) Find the transition matrix P.

(C) If 5% of the drivers in the community are involved in an accident this year, what is the probability that a driver chosen at random from the community will be involved in an accident next year? Year after next?

SOLUTION

(A)

A = accident
A' = no accident

(B)

$$\begin{array}{cc} & \text{Next year} \\ & \begin{array}{cc} A & A' \end{array} \\ \begin{array}{c} \text{This} \quad A \\ \text{year} \quad A' \end{array} & \begin{bmatrix} .23 & .77 \\ .11 & .89 \end{bmatrix} \end{array} = P \quad \text{Transition matrix}$$

(C) The initial-state matrix S_0 is

$$\begin{array}{cc} & \begin{array}{cc} A & A' \end{array} \\ S_0 = & \begin{bmatrix} .05 & .95 \end{bmatrix} \end{array} \quad \text{Initial-state matrix}$$

Thus,

$$S_0 P = \begin{array}{c} A \quad\quad A' \\ \begin{bmatrix} .05 & .95 \end{bmatrix} \end{array} \begin{bmatrix} .23 & .77 \\ .11 & .89 \end{bmatrix} = \begin{array}{c} A \quad\quad A' \\ \begin{bmatrix} .116 & .884 \end{bmatrix} \end{array} = S_1$$

This year (initial state)   Next year (first state)

$$S_1 P = \begin{array}{c} A \quad\quad A' \\ \begin{bmatrix} .116 & .884 \end{bmatrix} \end{array} \begin{bmatrix} .23 & .77 \\ .11 & .89 \end{bmatrix} = \begin{array}{c} A \quad\quad\quad A' \\ \begin{bmatrix} .12392 & .87608 \end{bmatrix} \end{array} = S_2$$

Next year (first state)   Year after next (second state)

The probability that a driver chosen at random from the community will have an accident next year is .116, and the year after next is .12392. That is, it is expected that 11.6% of the drivers in the community will have an accident next year and 12.392% the year after.

Matched Problem 1 An insurance company classifies drivers as low-risk if they are accident-free for one year. Past records indicate that 98% of the drivers in the low-risk category (L) one year will remain in that category the next year, and 78% of the drivers who are not in the low-risk category (L') one year will be in the low-risk category the next year.

(A) Draw a transition diagram.

(B) Find the transition matrix P.

(C) If 90% of the drivers in the community are in the low-risk category this year, what is the probability that a driver chosen at random from the community will be in the low-risk category next year? Year after next?

Powers of Transition Matrices

Next we investigate the powers of a transition matrix.

The state matrices for a Markov chain are defined **recursively**; that is, each state matrix is defined in terms of the preceding state matrix. For example, to find the fourth-state matrix S_4, it is necessary to compute the preceding three state matrices:

$$S_1 = S_0P \qquad S_2 = S_1P \qquad S_3 = S_2P \qquad S_4 = S_3P$$

Is there any way to compute a given state matrix directly without first computing all the preceding state matrices? If we substitute the equation for S_1 into the equation for S_2, substitute this new equation for S_2 into the equation for S_3, and so on, a definite pattern emerges:

$$S_1 = S_0P$$
$$S_2 = S_1P = (S_0P)P = S_0P^2$$
$$S_3 = S_2P = (S_0P^2)P = S_0P^3$$
$$S_4 = S_3P = (S_0P^3)P = S_0P^4$$
$$\vdots$$

In general, it can be shown that the kth-state matrix is given by $S_k = S_0P^k$. We summarize this important result in Theorem 1.

THEOREM 1   Powers of a Transition Matrix

If P is the transition matrix and S_0 is an initial-state matrix for a Markov chain, then the kth-state matrix is given by

$$S_k = S_0P^k$$

The entry in the ith row and jth column of P^k indicates the probability of the system moving from the ith state to the jth state in k observations or trials. The sum of the entries in each row of P^k is 1.

EXAMPLE 2 Using P^k to Compute S_k Find P^4 and use it to find S_4 for

$$P = \begin{matrix} & A & A' \\ A & \\ A' \end{matrix}\begin{bmatrix} .1 & .9 \\ .6 & .4 \end{bmatrix} \quad \text{and} \quad S_0 = \begin{matrix} A & A' \end{matrix}[.2 \quad .8]$$

SOLUTION $P^2 = PP = \begin{bmatrix} .1 & .9 \\ .6 & .4 \end{bmatrix}\begin{bmatrix} .1 & .9 \\ .6 & .4 \end{bmatrix} = \begin{bmatrix} .55 & .45 \\ .3 & .7 \end{bmatrix}$

$P^4 = P^2P^2 = \begin{bmatrix} .55 & .45 \\ .3 & .7 \end{bmatrix}\begin{bmatrix} .55 & .45 \\ .3 & .7 \end{bmatrix} = \begin{bmatrix} .4375 & .5625 \\ .375 & .625 \end{bmatrix}$

$S_4 = S_0P^4 = [.2 \quad .8]\begin{bmatrix} .4375 & .5625 \\ .375 & .625 \end{bmatrix} = [.3875 \quad .6125]$

Matched Problem 2 Find P^4 and use it to find S_4 for

$$P = \begin{array}{c} \\ A \\ A' \end{array} \begin{array}{cc} A & A' \\ \left[\begin{array}{cc} .8 & .2 \\ .3 & .7 \end{array}\right] \end{array} \quad \text{and} \quad S_0 = \begin{array}{c} A \quad A' \\ [.8 \quad .2] \end{array}$$

If a graphing calculator or a computer is available for computing matrix products and powers of a matrix, finding state matrices for any number of trials becomes a routine calculation.

EXAMPLE 3   Using a Graphing Calculator and P^k to Compute S_k Use P^8 and a graphing calculator to find S_8 for P and S_0 as given in Example 2. Round values in S_8 to six decimal places.

SOLUTION After storing the matrices P and S_0 in the graphing calculator's memory, we use the equation

$$S_8 = S_0 P^8$$

to compute S_8. Figure 2 shows the result on a typical graphing calculator. We see that (to six decimal places)

$$S_8 = [.399219 \quad .600781]$$

```
P
            [[.1 .9],
             [.6 .4]]
S0
            [[.2 .8]]
S0*P^8
   [[.399219 .600781]]
```

Figure 2

Matched Problem 3 Use P^8 and a graphing calculator to find S_8 for P and S_0 as given in Matched Problem 2. Round values in S_8 to six decimal places.

Application

The next example illustrates the use of Theorem 1 in an application.

EXAMPLE 4   Student Retention Part-time students in a university MBA program are considered to be entry-level students until they complete 15 credits successfully. Then they are classified as advanced-level students and can take more advanced courses and work on the thesis required for graduation. Past records indicate that at the end of each year, 10% of the entry-level students (E) drop out of the program (D) and 30% become advanced-level students (A). Also, 10% of the advanced-level students drop out of the program and 40% graduate (G) each year. Students that graduate or drop out never return to the program.

(A) Draw a transition diagram.   (B) Find the transition matrix P.

(C) What is the probability that an entry-level student graduates within 4 years? Drops out within 4 years?

SOLUTION

(A) If 10% of entry-level students drop out and 30% become advanced-level students, then the remaining 60% must continue as entry-level students for another year (see the diagram). Similarly, 50% of advanced-level students must continue as advanced-level students for another year. Since students who drop out never return, all students in state D in one year will continue in that state the next year. We indicate this by placing a 1 on the arrow from D back to D. State G is labeled in the same manner.

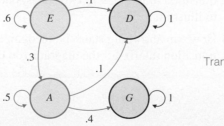

Transition diagram

(B)

$$P = \begin{array}{c} \\ E \\ D \\ A \\ G \end{array} \begin{array}{cccc} E & D & A & G \\ \begin{bmatrix} .6 & .1 & .3 & 0 \\ 0 & 1 & 0 & 0 \\ 0 & .1 & .5 & .4 \\ 0 & 0 & 0 & 1 \end{bmatrix} \end{array}$$ Transition matrix

(C) The probability that an entry-level student moves from state E to state G within 4 years is the entry in row 1 and column 4 of P^4 (Theorem 1). Hand computation of P^4 requires two multiplications:

$$P^2 = \begin{bmatrix} .6 & .1 & .3 & 0 \\ 0 & 1 & 0 & 0 \\ 0 & .1 & .5 & .4 \\ 0 & 0 & 0 & 1 \end{bmatrix} \begin{bmatrix} .6 & .1 & .3 & 0 \\ 0 & 1 & 0 & 0 \\ 0 & .1 & .5 & .4 \\ 0 & 0 & 0 & 1 \end{bmatrix} = \begin{bmatrix} .36 & .19 & .33 & .12 \\ 0 & 1 & 0 & 0 \\ 0 & .15 & .25 & .6 \\ 0 & 0 & 0 & 1 \end{bmatrix}$$

$$P^4 = P^2 P^2 = \begin{bmatrix} .36 & .19 & .33 & .12 \\ 0 & 1 & 0 & 0 \\ 0 & .15 & .25 & .6 \\ 0 & 0 & 0 & 1 \end{bmatrix} \begin{bmatrix} .36 & .19 & .33 & .12 \\ 0 & 1 & 0 & 0 \\ 0 & .15 & .25 & .6 \\ 0 & 0 & 0 & 1 \end{bmatrix}$$

$$= \begin{bmatrix} .1296 & .3079 & .2013 & .3612 \\ 0 & 1 & 0 & 0 \\ 0 & .1875 & .0625 & .75 \\ 0 & 0 & 0 & 1 \end{bmatrix}$$

The probability that an entry-level student has graduated within 4 years is .3612. Similarly, the probability that an entry-level student has dropped out within 4 years is .3079 (the entry in row 1 and column 2 of P^4).

Matched Problem 4 Refer to Example 4. At the end of each year the faculty examines the progress that each advanced-level student has made on the required thesis. Past records indicate that 30% of advanced-level students (A) complete the thesis requirement (C) and 10% are dropped from the program for insufficient progress (D), never to return. The remaining students continue to work on their theses.

(A) Draw a transition diagram.

(B) Find the transition matrix P.

(C) What is the probability that an advanced-level student completes the thesis requirement within 4 years? Is dropped from the program for insufficient progress within 4 years?

Explore and Discuss 2 Refer to Example 4. States D and G are referred to as *absorbing states,* because a student who enters either one of these states never leaves it. Absorbing states are discussed in detail in Section 9.3.

(A) How can absorbing states be recognized from a transition diagram? Draw a transition diagram with two states, one that is absorbing and one that is not, to illustrate.

(B) How can absorbing states be recognized from a transition matrix? Write the transition matrix for the diagram you drew in part (A) to illustrate.

Exercises 9.1

Skills Warm-up Exercises

In Problems 1–8, find the matrix product, if it is defined. (If necessary, review Section 4.4.)

1. $\begin{bmatrix} 2 & 5 \\ 4 & 1 \end{bmatrix}\begin{bmatrix} 3 \\ 2 \end{bmatrix}$

2. $\begin{bmatrix} 4 & 5 \end{bmatrix}\begin{bmatrix} 6 & 9 \\ 3 & 7 \end{bmatrix}$

3. $\begin{bmatrix} 3 \\ 2 \end{bmatrix}\begin{bmatrix} 2 & 5 \\ 4 & 1 \end{bmatrix}$

4. $\begin{bmatrix} 6 & 9 \\ 3 & 7 \end{bmatrix}\begin{bmatrix} 4 & 5 \end{bmatrix}$

5. $\begin{bmatrix} 3 & 2 \end{bmatrix}\begin{bmatrix} 2 & 5 \\ 4 & 1 \end{bmatrix}$

6. $\begin{bmatrix} 4 \\ 5 \end{bmatrix}\begin{bmatrix} 6 & 9 \\ 3 & 7 \end{bmatrix}$

7. $\begin{bmatrix} 2 & 5 \\ 4 & 1 \end{bmatrix}\begin{bmatrix} 3 & 2 \end{bmatrix}$

8. $\begin{bmatrix} 6 & 9 \\ 3 & 7 \end{bmatrix}\begin{bmatrix} 4 \\ 5 \end{bmatrix}$

Problems 9–16 refer to the following transition matrix:

$$P = \begin{array}{c} A \\ B \end{array}\begin{bmatrix} .8 & .2 \\ .4 & .6 \end{bmatrix}\begin{array}{c} \end{array}$$

with column labels $A\quad B$.

In Problems 9–12, find S_1 for the indicated initial-state matrix S_0 and interpret with a tree diagram.

9. $S_0 = \begin{bmatrix} 1 & 0 \end{bmatrix}$

10. $S_0 = \begin{bmatrix} 0 & 1 \end{bmatrix}$

11. $S_0 = \begin{bmatrix} .5 & .5 \end{bmatrix}$

12. $S_0 = \begin{bmatrix} .3 & .7 \end{bmatrix}$

In Problems 13–16, find S_2 for the indicated initial-state matrix S_0, and explain what it represents.

13. $S_0 = \begin{bmatrix} 1 & 0 \end{bmatrix}$

14. $S_0 = \begin{bmatrix} 0 & 1 \end{bmatrix}$

15. $S_0 = \begin{bmatrix} .5 & .5 \end{bmatrix}$

16. $S_0 = \begin{bmatrix} .3 & .7 \end{bmatrix}$

In Problems 17–20, use the transition diagram to find S_1 for the indicated initial-state matrix S_0.

17. $S_0 = \begin{bmatrix} .2 & .8 \end{bmatrix}$

18. $S_0 = \begin{bmatrix} .6 & .4 \end{bmatrix}$

19. $S_0 = \begin{bmatrix} .9 & .1 \end{bmatrix}$

20. $S_0 = \begin{bmatrix} .7 & .3 \end{bmatrix}$

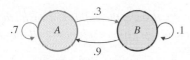

Figure for Problems 17–24

In Problems 21–24, use the transition diagram to find S_2 for the indicated initial-state matrix S_0.

21. $S_0 = \begin{bmatrix} .2 & .8 \end{bmatrix}$

22. $S_0 = \begin{bmatrix} .6 & .4 \end{bmatrix}$

23. $S_0 = \begin{bmatrix} .9 & .1 \end{bmatrix}$

24. $S_0 = \begin{bmatrix} .7 & .3 \end{bmatrix}$

In Problems 25–32, could the given matrix be the transition matrix of a Markov chain?

25. $\begin{bmatrix} .3 & .7 \\ 1 & 0 \end{bmatrix}$

26. $\begin{bmatrix} .9 & .1 \\ .4 & .8 \end{bmatrix}$

27. $\begin{bmatrix} .5 & .5 \\ .7 & -.3 \end{bmatrix}$

28. $\begin{bmatrix} 0 & 1 \\ 1 & 0 \end{bmatrix}$

29. $\begin{bmatrix} .1 & .3 & .6 \\ .2 & .4 & .4 \end{bmatrix}$

30. $\begin{bmatrix} .2 & .8 \\ .5 & .5 \\ .9 & .1 \end{bmatrix}$

31. $\begin{bmatrix} .5 & .1 & .4 \\ 0 & .5 & .5 \\ .2 & .1 & .7 \end{bmatrix}$

32. $\begin{bmatrix} .3 & .3 & .4 \\ .7 & .2 & .2 \\ .1 & .8 & .1 \end{bmatrix}$

In Problems 33–38, is there a unique way of filling in the missing probabilities in the transition diagram? If so, complete the transition diagram and write the corresponding transition matrix. If not, explain why.

33.

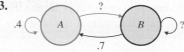

34.

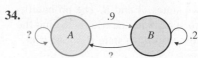

35.

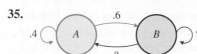

36.

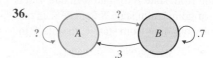

37.

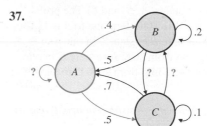

38.

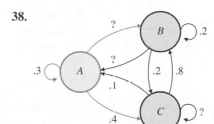

In Problems 39–44, are there unique values of a, b, and c that make P a transition matrix? If so, complete the transition matrix and draw the corresponding transition diagram. If not, explain why.

39. $P = \begin{array}{c} A \\ B \\ C \end{array}\begin{bmatrix} 0 & .5 & a \\ b & 0 & .4 \\ .2 & c & .1 \end{bmatrix}$ with columns $A\ B\ C$

40. $P = \begin{array}{c} A \\ B \\ C \end{array}\begin{bmatrix} a & 0 & .9 \\ .2 & .3 & b \\ .6 & c & 0 \end{bmatrix}$ with columns $A\ B\ C$

41. $P = \begin{array}{c} A \\ B \\ C \end{array}\begin{bmatrix} 0 & a & .3 \\ 0 & b & 0 \\ c & .8 & 0 \end{bmatrix}$ with columns $A\ B\ C$

42. $P = \begin{array}{c} A \\ B \\ C \end{array}\begin{bmatrix} 0 & 1 & a \\ 0 & 0 & b \\ c & .5 & 0 \end{bmatrix}$ with columns $A\ B\ C$

455

$$43.\ P = \begin{array}{c} \\ A \\ B \\ C \end{array} \begin{array}{ccc} A & B & C \\ \left[\begin{array}{ccc} .2 & .1 & .7 \\ a & .4 & c \\ .5 & b & .4 \end{array}\right] \end{array}$$

$$44.\ P = \begin{array}{c} \\ A \\ B \\ C \end{array} \begin{array}{ccc} A & B & C \\ \left[\begin{array}{ccc} a & .8 & .1 \\ .3 & b & .4 \\ .6 & .5 & c \end{array}\right] \end{array}$$

In Problems 45–48, use the given information to draw the transition diagram and find the transition matrix.

45. A Markov chain has two states, *A* and *B*. The probability of going from state *A* to state *B* in one trial is .7, and the probability of going from state *B* to state *A* in one trial is .9.

46. A Markov chain has two states, *A* and *B*. The probability of going from state *A* to state *A* in one trial is .6, and the probability of going from state *B* to state *B* in one trial is .2.

47. A Markov chain has three states, *A*, *B*, and *C*. The probability of going from state *A* to state *B* in one trial is .1, and the probability of going from state *A* to state *C* in one trial is .3. The probability of going from state *B* to state *A* in one trial is .2, and the probability of going from state *B* to state *C* in one trial is .5. The probability of going from state *C* to state *C* in one trial is 1.

48. A Markov chain has three states, *A*, *B*, and *C*. The probability of going from state *A* to state *B* in one trial is 1. The probability of going from state *B* to state *A* in one trial is .5, and the probability of going from state *B* to state *C* in one trial is .5. The probability of going from state *C* to state *A* in one trial is 1.

Problems 49–58 refer to the following transition matrix P and its powers:

$$P = \begin{array}{c} \\ A \\ B \\ C \end{array} \begin{array}{ccc} A & B & C \\ \left[\begin{array}{ccc} .6 & .3 & .1 \\ .2 & .5 & .3 \\ .1 & .2 & .7 \end{array}\right] \end{array} \quad P^2 = \begin{array}{c} \\ A \\ B \\ C \end{array} \begin{array}{ccc} A & B & C \\ \left[\begin{array}{ccc} .43 & .35 & .22 \\ .25 & .37 & .38 \\ .17 & .27 & .56 \end{array}\right] \end{array}$$

$$P^3 = \begin{array}{c} \\ A \\ B \\ C \end{array} \begin{array}{ccc} A & B & C \\ \left[\begin{array}{ccc} .35 & .348 & .302 \\ .262 & .336 & .402 \\ .212 & .298 & .49 \end{array}\right] \end{array}$$

49. Find the probability of going from state *A* to state *B* in two trials.

50. Find the probability of going from state *B* to state *C* in two trials.

51. Find the probability of going from state *C* to state *A* in three trials.

52. Find the probability of going from state *B* to state *B* in three trials.

53. Find S_2 for $S_0 = [1 \quad 0 \quad 0]$ and explain what it represents.

54. Find S_2 for $S_0 = [0 \quad 1 \quad 0]$ and explain what it represents.

55. Find S_3 for $S_0 = [0 \quad 0 \quad 1]$ and explain what it represents.

56. Find S_3 for $S_0 = [1 \quad 0 \quad 0]$ and explain what it represents.

57. Using a graphing calculator to compute powers of *P*, find the smallest positive integer *n* such that the corresponding entries in P^n and P^{n+1} are equal when rounded to two decimal places.

58. Using a graphing calculator to compute powers of *P*, find the smallest positive integer *n* such that the corresponding entries in P^n and P^{n+1} are equal when rounded to three decimal places.

In Problems 59–62, given the transition matrix P and initial-state matrix S_0, find P^4 and use P^4 to find S_4.

$$59.\ P = \begin{array}{c} \\ A \\ B \end{array} \begin{array}{cc} A & B \\ \left[\begin{array}{cc} .1 & .9 \\ .6 & .4 \end{array}\right] \end{array};\quad S_0 = [.8 \quad .2]$$

$$60.\ P = \begin{array}{c} \\ A \\ B \end{array} \begin{array}{cc} A & B \\ \left[\begin{array}{cc} .8 & .2 \\ .3 & .7 \end{array}\right] \end{array};\quad S_0 = [.4 \quad .6]$$

$$61.\ P = \begin{array}{c} \\ A \\ B \\ C \end{array} \begin{array}{ccc} A & B & C \\ \left[\begin{array}{ccc} 0 & .4 & .6 \\ 0 & 0 & 1 \\ 1 & 0 & 0 \end{array}\right] \end{array};\quad S_0 = [.2 \quad .3 \quad .5]$$

$$62.\ P = \begin{array}{c} \\ A \\ B \\ C \end{array} \begin{array}{ccc} A & B & C \\ \left[\begin{array}{ccc} 0 & 1 & 0 \\ .8 & 0 & .2 \\ 1 & 0 & 0 \end{array}\right] \end{array};\quad S_0 = [.4 \quad .2 \quad .4]$$

63. A Markov chain with two states has transition matrix *P*. If the initial-state matrix is $S_0 = [1 \quad 0]$, discuss the relationship between the entries in the *k*th-state matrix and the entries in the *k*th power of *P*.

64. Repeat Problem 63 if the initial-state matrix is $S_0 = [0 \quad 1]$.

In Problems 65–70, discuss the validity of each statement. If the statement is always true, explain why. If not, give a counterexample.

65. If *P* is a transition matrix for a Markov chain, then the sum of the entries in each column of *P* must be 1.

66. If *P* is a transition matrix for a Markov chain, then the sum of the entries in each row of *P* must be 1.

67. If *P* is a transition matrix for a Markov chain, then the product of the entries in each column of *P* must be ≥ 0.

68. If *P* is a transition matrix for a Markov chain, then the product of the entries in each row of *P* must be ≤ 1.

69. If *C* is a state in the transition diagram for a Markov chain, then the sum of the probabilities on all arrows going into *C* is equal to 1.

70. If *C* is a state in the transition diagram for a Markov chain, then the sum of the probabilities on all arrows going out of *C* is equal to 1.

71. Given the transition matrix

$$P = \begin{array}{c} \\ A \\ B \\ C \\ D \end{array} \begin{array}{c} \begin{array}{cccc} A & B & C & D \end{array} \\ \begin{bmatrix} .2 & .2 & .3 & .3 \\ 0 & 1 & 0 & 0 \\ .2 & .2 & .1 & .5 \\ 0 & 0 & 0 & 1 \end{bmatrix} \end{array}$$

(A) Find P^4.

(B) Find the probability of going from state A to state D in four trials.

(C) Find the probability of going from state C to state B in four trials.

(D) Find the probability of going from state B to state A in four trials.

72. Repeat Problem 71 for the transition matrix

$$P = \begin{array}{c} \\ A \\ B \\ C \\ D \end{array} \begin{array}{c} \begin{array}{cccc} A & B & C & D \end{array} \\ \begin{bmatrix} .5 & .3 & .1 & .1 \\ 0 & 1 & 0 & 0 \\ 0 & 0 & 1 & 0 \\ .1 & .2 & .3 & .4 \end{bmatrix} \end{array}$$

A matrix is called a **probability matrix** *if all its entries are real numbers between 0 and 1, inclusive, and the sum of the entries in each row is 1. So transition matrices are square probability matrices and state matrices are probability matrices with one row.*

73. Show that if

$$P = \begin{bmatrix} a & 1-a \\ 1-b & b \end{bmatrix}$$

is a probability matrix, then P^2 is a probability matrix.

74. Show that if

$$P = \begin{bmatrix} a & 1-a \\ 1-b & b \end{bmatrix} \quad \text{and} \quad S = \begin{bmatrix} c & 1-c \end{bmatrix}$$

are probability matrices, then SP is a probability matrix.

Use a graphing calculator and the formula $S_k = S_0 P^k$ (Theorem 1) to compute the required state matrices in Problems 75–78.

75. The transition matrix for a Markov chain is

$$P = \begin{bmatrix} .4 & .6 \\ .2 & .8 \end{bmatrix}$$

(A) If $S_0 = \begin{bmatrix} 0 & 1 \end{bmatrix}$, find $S_2, S_4, S_8, \ldots$. Can you identify a state matrix S that the matrices S_k seem to be approaching?

(B) Repeat part (A) for $S_0 = \begin{bmatrix} 1 & 0 \end{bmatrix}$.

(C) Repeat part (A) for $S_0 = \begin{bmatrix} .5 & .5 \end{bmatrix}$.

(D) Find SP for any matrix S you identified in parts (A)–(C).

(E) Write a brief verbal description of the long-term behavior of the state matrices of this Markov chain based on your observations in parts (A)–(D).

76. Repeat Problem 75 for $P = \begin{bmatrix} .9 & .1 \\ .4 & .6 \end{bmatrix}$.

77. Refer to Problem 75. Find P^k for $k = 2, 4, 8, \ldots$. Can you identify a matrix Q that the matrices P^k are approaching? If so, how is Q related to the results you discovered in Problem 75?

78. Refer to Problem 76. Find P^k for $k = 2, 4, 8, \ldots$. Can you identify a matrix Q that the matrices P^k are approaching? If so, how is Q related to the results you discovered in Problem 76?

Applications

79. **Scheduling.** An outdoor restaurant in a summer resort closes only on rainy days. From past records, it is found that from May through September, when it rains one day, the probability of rain for the next day is .4; when it does not rain one day, the probability of rain for the next day is .06.

(A) Draw a transition diagram.

(B) Write the transition matrix.

(C) If it rains on Thursday, what is the probability that the restaurant will be closed on Saturday? On Sunday?

80. **Scheduling.** Repeat Problem 79 if the probability of rain following a rainy day is .6 and the probability of rain following a nonrainy day is .1.

81. **Advertising.** A television advertising campaign is conducted during the football season to promote a well-known brand X shaving cream. For each of several weeks, a survey is made, and it is found that each week, 80% of those using brand X continue to use it and 20% switch to another brand. It is also found that of those not using brand X, 20% switch to brand X while the other 80% continue using another brand.

(A) Draw a transition diagram.

(B) Write the transition matrix.

(C) If 20% of the people are using brand X at the start of the advertising campaign, what percentage will be using it 1 week later? 2 weeks later?

82. **Car rental.** A car rental agency has facilities at both JFK and LaGuardia airports. Assume that a car rented at either airport must be returned to one or the other airport. If a car is rented at LaGuardia, the probability that it will be returned there is .8; if a car is rented at JFK, the probability that it will be returned there is .7. Assume that the company rents all its 100 cars each day and that each car is rented (and returned) only once a day. If we start with 50 cars at each airport, then

(A) What is the expected distribution on the next day?

(B) What is the expected distribution 2 days later?

83. **Homeowner's insurance.** In a given city, the market for homeowner's insurance is dominated by two companies: National Property and United Family. Currently, National Property insures 50% of homes in the city, United Family insures 30%, and the remainder are insured by a collection of smaller companies. United Family decides to offer rebates to increase its market share. This has the following effects on insurance purchases for the next several years: each year 25% of National Property's customers switch to United Family and 10% switch to other companies; 10% of United Family's customers switch to National Property and 5% switch to other companies; 15% of the customers of other companies switch to National Property and 35% switch to United Family.

(A) Draw a transition diagram.

(B) Write the transition matrix.

(C) What percentage of homes will be insured by National Property next year? The year after next?

(D) What percentage of homes will be insured by United Family next year? The year after next?

84. Service contracts. A small community has two heating services that offer annual service contracts for home heating: Alpine Heating and Badger Furnaces. Currently, 25% of homeowners have service contracts with Alpine, 30% have service contracts with Badger, and the remainder do not have service contracts. Both companies launch aggressive advertising campaigns to attract new customers, with the following effects on service contract purchases for the next several years: each year 35% of homeowners with no current service contract decide to purchase a contract from Alpine and 40% decide to purchase one from Badger. In addition, 10% of the previous customers at each company decide to switch to the other company, and 5% decide they do not want a service contract.

(A) Draw a transition diagram.

(B) Write the transition matrix.

(C) What percentage of homes will have service contracts with Alpine next year? The year after next?

(D) What percentage of homes will have service contracts with Badger next year? The year after next?

85. Travel agent training. A chain of travel agencies maintains a training program for new travel agents. Initially, all new employees are classified as beginning agents requiring extensive supervision. Every 6 months, the performance of each agent is reviewed. Past records indicate that after each semiannual review, 40% of the beginning agents are promoted to intermediate agents requiring only minimal supervision, 10% are terminated for unsatisfactory performance, and the remainder continue as beginning agents. Furthermore, 30% of the intermediate agents are promoted to qualified travel agents requiring no supervision, 10% are terminated for unsatisfactory performance, and the remainder continue as intermediate agents.

(A) Draw a transition diagram.

(B) Write the transition matrix.

(C) What is the probability that a beginning agent is promoted to qualified agent within 1 year? Within 2 years?

86. Welder training. All welders in a factory begin as apprentices. Every year the performance of each apprentice is reviewed. Past records indicate that after each review, 10% of the apprentices are promoted to professional welder, 20% are terminated for unsatisfactory performance, and the remainder continue as apprentices.

(A) Draw a transition diagram.

(B) Write the transition matrix.

(C) What is the probability that an apprentice is promoted to professional welder within 2 years? Within 4 years?

87. Health plans. A midwestern university offers its employees three choices for health care: a clinic-based health maintenance organization (HMO), a preferred provider organization (PPO), and a traditional fee-for-service program (FFS). Each year, the university designates an open enrollment period during which employees may change from one health plan to another. Prior to the last open enrollment period, 20% of employees were enrolled in the HMO, 25% in the PPO, and the remainder in the FFS. During the open enrollment period, 15% of employees in the HMO switched to the PPO and 5% switched to the FFS; 20% of the employees in the PPO switched to the HMO and 10% to the FFS; and 25% of the employees in the FFS switched to the HMO and 30% switched to the PPO.

(A) Write the transition matrix.

(B) What percentage of employees were enrolled in each health plan after the last open enrollment period?

(C) If this trend continues, what percentage of employees will be enrolled in each plan after the next open enrollment period?

88. Dental insurance. Refer to Problem 87. During the open enrollment period, university employees can switch between two available dental care programs: the low-option plan (LOP) and the high-option plan (HOP). Prior to the last open enrollment period, 40% of employees were enrolled in the LOP and 60% in the HOP. During the open enrollment program, 30% of employees in the LOP switched to the HOP and 10% of employees in the HOP switched to the LOP.

(A) Write the transition matrix.

(B) What percentage of employees were enrolled in each dental plan after the last open enrollment period?

(C) If this trend continues, what percentage of employees will be enrolled in each dental plan after the next open enrollment period?

89. Housing trends. The 2000 census reported that 41.9% of the households in the District of Columbia were homeowners and the remainder were renters. During the next decade, 15.3% of homeowners became renters, and the rest continued to be homeowners. Similarly, 17.4% of renters became homeowners, and the rest continued to rent.

(A) Write the appropriate transition matrix.

(B) According to this transition matrix, what percentage of households were homeowners in 2010?

(C) If the transition matrix remains the same, what percentage of households will be homeowners in 2030?

90. Housing trends. The 2000 census reported that 66.4% of the households in Alaska were homeowners, and the remainder were renters. During the next decade, 37.2% of the homeowners became renters, and the rest continued to be homeowners. Similarly, 71.5% of the renters became homeowners, and the rest continued to rent.

(A) Write the appropriate transition matrix.

(B) According to this transition matrix, what percentage of households were homeowners in 2010?

(C) If the transition matrix remains the same, what percentage of households will be homeowners in 2030?

Answers to Matched Problems

1. (A)

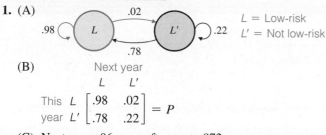

L = Low-risk
L' = Not low-risk

(B)

$$
\begin{array}{c}
 & \text{Next year} \\
 & \begin{array}{cc} L & L' \end{array} \\
\text{This} \\ \text{year}
\begin{array}{c} L \\ L' \end{array}
\begin{bmatrix}
.98 & .02 \\
.78 & .22
\end{bmatrix} = P
\end{array}
$$

(C) Next year: .96; year after next: .972

2. $P^4 = \begin{bmatrix} .625 & .375 \\ .5625 & .4375 \end{bmatrix}; \quad S_4 = [.6125 \quad .3875]$

3. $S_8 = [.600781 \quad .399219]$

4. (A)

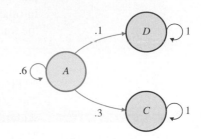

(B) $P = \begin{array}{c} \\ A \\ C \\ D \end{array} \begin{array}{ccc} A & C & D \\ \begin{bmatrix} .6 & .3 & .1 \\ 0 & 1 & 0 \\ 0 & 0 & 1 \end{bmatrix} \end{array}$

(C) .6528; .2176

9.2 Regular Markov Chains

- Stationary Matrices
- Regular Markov Chains
- Applications
- Graphing Calculator Approximations

Given a Markov chain with transition matrix P and initial-state matrix S_0, the entries in the state matrix S_k are the probabilities of being in the corresponding states after k trials. What happens to these probabilities as the number of trials k increases? In this section, we establish conditions on the transition matrix P that enable us to determine the long-run behavior of both the state matrices S_k and the powers of the transition matrix P^k.

Stationary Matrices

We begin by considering a concrete example—the toothpaste company discussed earlier. Recall that the transition matrix was given by

$$
P = \begin{array}{c} \\ A \\ A' \end{array} \begin{array}{cc} A & A' \\ \begin{bmatrix} .8 & .2 \\ .6 & .4 \end{bmatrix} \end{array} \begin{array}{l} A = \text{uses brand } A \text{ toothpaste} \\ A' = \text{uses another brand} \end{array}
$$

Initially, this company had a 10% share of the toothpaste market. If the probabilities in the transition matrix P remain valid over a long period of time, what will happen to the company's market share? Examining the first several state matrices will give us some insight into this situation (matrix multiplication details are omitted):

$$S_0 = [.1 \quad .9]$$
$$S_1 = S_0 P = [.62 \quad .38]$$
$$S_2 = S_1 P = [.724 \quad .276]$$
$$S_3 = S_2 P = [.7448 \quad .2552]$$
$$S_4 = S_3 P = [.74896 \quad .25104]$$
$$S_5 = S_4 P = [.749792 \quad .250208]$$
$$S_6 = S_5 P = [.7499584 \quad .2500416]$$

It appears that the state matrices are getting closer and closer to $S = [.75 \quad .25]$ as we proceed to higher states. Let us multiply the matrix S (the matrix that the other state matrices appear to be approaching) by the transition matrix:

$$SP = [.75 \quad .25] \begin{bmatrix} .8 & .2 \\ .6 & .4 \end{bmatrix} = [.75 \quad .25] = S$$

No change occurs! The matrix $[.75 \quad .25]$ is called a **stationary matrix**. If we reach this state or are very close to it, the system is said to be at a steady state; that is, later states either will not change or will not change very much. In terms of this example, this means that in the long run a person will purchase brand A with a probability of .75. In other words, the company can expect to capture 75% of the market, assuming that the transition matrix does not change.

DEFINITION   Stationary Matrix for a Markov Chain

The state matrix $S = [s_1 \quad s_2 \quad \ldots \quad s_n]$ is a **stationary matrix** for a Markov chain with transition matrix P if

$$SP = S$$

where $s_i \geq 0$, $i = 1, \ldots, n$, and $s_1 + s_2 + \cdots + s_n = 1$.

Explore and Discuss 1
 (A) Suppose that the toothpaste company started with only 5% of the market instead of 10%. Write the initial-state matrix and find the next six state matrices. Discuss the behavior of these state matrices as you proceed to higher states.

 (B) Repeat part (A) if the company started with 90% of the toothpaste market.

Regular Markov Chains

Does every Markov chain have a unique stationary matrix? And if a Markov chain has a unique stationary matrix, will the successive state matrices always approach this stationary matrix? Unfortunately, the answer to both these questions is no (see Problems 43–46, Exercises 9.2). However, there is one important type of Markov chain for which both questions always can be answered in the affirmative. These are called *regular Markov chains*.

DEFINITION   Regular Markov Chains

A transition matrix P is **regular** if some power of P has only positive entries. A Markov chain is a **regular Markov chain** if its transition matrix is regular.

EXAMPLE 1   Recognizing Regular Matrices   Which of the following matrices are regular?

(A) $P = \begin{bmatrix} .8 & .2 \\ .6 & .4 \end{bmatrix}$ (B) $P = \begin{bmatrix} 0 & 1 \\ 1 & 0 \end{bmatrix}$ (C) $P = \begin{bmatrix} .5 & .5 & 0 \\ 0 & .5 & .5 \\ 1 & 0 & 0 \end{bmatrix}$

SOLUTION

(A) This is the transition matrix for the toothpaste company. Since all the entries in P are positive, we can immediately conclude that P is regular.

(B) P has two 0 entries, so we must examine higher powers of P:

$$P^2 = \begin{bmatrix} 1 & 0 \\ 0 & 1 \end{bmatrix} \quad P^3 = \begin{bmatrix} 0 & 1 \\ 1 & 0 \end{bmatrix} \quad P^4 = \begin{bmatrix} 1 & 0 \\ 0 & 1 \end{bmatrix} \quad P^5 = \begin{bmatrix} 0 & 1 \\ 1 & 0 \end{bmatrix}$$

Since the powers of P oscillate between P and I, the 2×2 identity, all powers of P will contain 0 entries. Hence, P is not regular.

(C) Again, we examine higher powers of P:

$$P^2 = \begin{bmatrix} .25 & .5 & .25 \\ .5 & .25 & .25 \\ .5 & .5 & 0 \end{bmatrix} \qquad P^3 = \begin{bmatrix} .375 & .375 & .25 \\ .5 & .375 & .125 \\ .25 & .5 & .25 \end{bmatrix}$$

Since all the entries in P^3 are positive, P is regular.

Matched Problem 1) Which of the following matrices are regular?

(A) $P = \begin{bmatrix} .3 & .7 \\ 1 & 0 \end{bmatrix}$ 　　(B) $P = \begin{bmatrix} 1 & 0 \\ 1 & 0 \end{bmatrix}$ 　　(C) $P = \begin{bmatrix} 0 & 1 & 0 \\ .5 & 0 & .5 \\ .5 & 0 & .5 \end{bmatrix}$

The relationships among successive state matrices, powers of the transition matrix, and the stationary matrix for a regular Markov chain are given in Theorem 1. The proof of this theorem is left to more advanced courses.

THEOREM 1 Properties of Regular Markov Chains

Let P be the transition matrix for a regular Markov chain.

(A) There is a unique stationary matrix S that can be found by solving the equation

$$SP = S$$

(B) Given any initial-state matrix S_0, the state matrices S_k approach the stationary matrix S.

(C) The matrices P^k approach a **limiting matrix** $\overline{P}$, where each row of $\overline{P}$ is equal to the stationary matrix S.

EXAMPLE 2　Finding the Stationary Matrix　The transition matrix for a Markov chain is

$$P = \begin{bmatrix} .7 & .3 \\ .2 & .8 \end{bmatrix}$$

(A) Find the stationary matrix S.

(B) Discuss the long-run behavior of S_k and P^k.

SOLUTION

(A) Since P is regular, the stationary matrix S must exist. To find it, we must solve the equation $SP = S$. Let

$$S = \begin{bmatrix} s_1 & s_2 \end{bmatrix}$$

and write

$$\begin{bmatrix} s_1 & s_2 \end{bmatrix} \begin{bmatrix} .7 & .3 \\ .2 & .8 \end{bmatrix} = \begin{bmatrix} s_1 & s_2 \end{bmatrix}$$

After multiplying the left side, we obtain

$$\begin{bmatrix} (.7s_1 + .2s_2) & (.3s_1 + .8s_2) \end{bmatrix} = \begin{bmatrix} s_1 & s_2 \end{bmatrix}$$

which is equivalent to the system

$$\begin{array}{llll} .7s_1 + .2s_2 = s_1 & \text{or} & -.3s_1 + .2s_2 = 0 \\ .3s_1 + .8s_2 = s_2 & \text{or} & .3s_1 - .2s_2 = 0 \end{array} \qquad (1)$$

System (1) is dependent and has an infinite number of solutions. However, we are looking for a solution that is also a state matrix. This gives us another equation that we can add to system (1) to obtain a system with a unique solution.

$$-.3s_1 + .2s_2 = 0$$
$$.3s_1 - .2s_2 = 0 \qquad (2)$$
$$s_1 + s_2 = 1$$

System (2) can be solved using matrix methods or elimination to obtain

$$s_1 = .4 \qquad \text{and} \qquad s_2 = .6$$

Therefore,

$$S = [.4 \quad .6]$$

is the stationary matrix.
Check:

$$SP = [.4 \quad .6]\begin{bmatrix} .7 & .3 \\ .2 & .8 \end{bmatrix} = [.4 \quad .6] = S$$

(B) Given any initial-state matrix S_0, Theorem 1 guarantees that the state matrices S_k will approach the stationary matrix S. Furthermore,

$$P^k = \begin{bmatrix} .7 & .3 \\ .2 & .8 \end{bmatrix}^k \qquad \text{approaches the limiting matrix} \qquad \overline{P} = \begin{bmatrix} .4 & .6 \\ .4 & .6 \end{bmatrix}$$

Matched Problem 2 The transition matrix for a Markov chain is

$$P = \begin{bmatrix} .6 & .4 \\ .1 & .9 \end{bmatrix}$$

Find the stationary matrix S and the limiting matrix $\overline{P}$.

Applications

EXAMPLE 3 Insurance Refer to Example 1 in Section 9.1, where we found the following transition matrix for an insurance company:

$$P = \begin{matrix} & A & A' \\ A & \begin{bmatrix} .23 & .77 \\ .11 & .89 \end{bmatrix} & \begin{matrix} A = \text{accident} \\ A' = \text{no accident} \end{matrix} \\ A' & \end{matrix}$$

If these probabilities remain valid over a long period of time, what percentage of drivers are expected to have an accident during any given year?

SOLUTION To determine what happens in the long run, we find the stationary matrix by solving the following system:

$$[s_1 \quad s_2]\begin{bmatrix} .23 & .77 \\ .11 & .89 \end{bmatrix} = [s_1 \quad s_2] \qquad \text{and} \qquad s_1 + s_2 = 1$$

which is equivalent to

$$.23s_1 + .11s_2 = s_1 \quad \text{or} \quad -.77s_1 + .11s_2 = 0$$
$$.77s_1 + .89s_2 = s_2 \qquad\qquad .77s_1 - .11s_2 = 0$$
$$s_1 + s_2 = 1 \qquad\qquad s_1 + s_2 = 1$$

Solving this system, we obtain

$$s_1 = .125 \qquad \text{and} \qquad s_2 = .875$$

The stationary matrix is [.125 .875], which means that in the long run, assuming that the transition matrix does not change, about 12.5% of drivers in the community will have an accident during any given year.

Matched Problem 3 Refer to Matched Problem 1 in Section 9.1, where we found the following transition matrix for an insurance company:

$$P = \begin{array}{c} L \\ L' \end{array} \begin{bmatrix} .98 & .02 \\ .78 & .22 \end{bmatrix} \begin{array}{l} L = \text{low-risk} \\ L' = \text{not low-risk} \end{array}$$

If these probabilities remain valid for a long period of time, what percentage of drivers are expected to be in the low-risk category during any given year?

EXAMPLE 4 Employee Evaluation A company rates every employee as below average, average, or above average. Past performance indicates that each year, 10% of the below-average employees will raise their rating to average, and 25% of the average employees will raise their rating to above average. On the other hand, 15% of the average employees will lower their rating to below average, and 15% of the above-average employees will lower their rating to average. Company policy prohibits rating changes from below average to above average, or conversely, in a single year. Over the long run, what percentage of employees will receive below-average ratings? Average ratings? Above-average ratings?

SOLUTION First we find the transition matrix:

$$\begin{array}{c} \text{This} \\ \text{year} \end{array} \begin{array}{c} \\ A^- \\ A \\ A^+ \end{array} \overset{\begin{array}{ccc} A^- & A & A^+ \end{array}}{\begin{bmatrix} .9 & .1 & 0 \\ .15 & .6 & .25 \\ 0 & .15 & .85 \end{bmatrix}} \begin{array}{l} A^- = \text{below average} \\ A = \text{average} \\ A^+ = \text{above average} \end{array}$$

To determine what happens over the long run, we find the stationary matrix by solving the following system:

$$[s_1 \quad s_2 \quad s_3] \begin{bmatrix} .9 & .1 & 0 \\ .15 & .6 & .25 \\ 0 & .15 & .85 \end{bmatrix} = [s_1 \quad s_2 \quad s_3] \quad \text{and} \quad s_1 + s_2 + s_3 = 1$$

which is equivalent to

$$\begin{aligned} .9s_1 + .15s_2 \qquad\quad &= s_1 \\ .1s_1 + .6s_2 + .15s_3 &= s_2 \\ .25s_2 + .85s_3 &= s_3 \\ s_1 + s_2 + s_3 &= 1 \end{aligned} \quad \text{or} \quad \begin{aligned} -.1s_1 + .15s_2 \qquad\quad &= 0 \\ .1s_1 - .4s_2 + .15s_3 &= 0 \\ .25s_2 - .15s_3 &= 0 \\ s_1 + s_2 + s_3 &= 1 \end{aligned}$$

Using Gauss–Jordan elimination to solve this system of four equations with three variables, we obtain

$$s_1 = .36 \qquad s_2 = .24 \qquad s_3 = .4$$

In the long run, 36% of employees will be rated as below average, 24% as average, and 40% as above average.

Matched Problem 4 A mail-order company classifies its customers as preferred, standard, or infrequent, depending on the number of orders placed in a year. Past records indicate that each year, 5% of preferred customers are reclassified as standard and 12% as infrequent; 5% of standard customers are reclassified as preferred and 5% as infrequent; and 9% of infrequent customers are reclassified as preferred and 10% as standard. Assuming that these percentages remain valid, what percentage of customers are expected to be in each category in the long run?

Graphing Calculator Approximations

If P is the transition matrix for a regular Markov chain, then the powers of P approach the limiting matrix $\overline{P}$, where each row of $\overline{P}$ is equal to the stationary matrix S (Theorem 1C). We can use this result to approximate S by computing P^k for sufficiently large values of k. The next example illustrates this approach on a graphing calculator.

 EXAMPLE 5   Approximating the Stationary Matrix   Compute powers of the transition matrix P to approximate $\overline{P}$ and S to four decimal places. Check the approximation in the equation $SP = S$.

$$P = \begin{bmatrix} .5 & .2 & .3 \\ .7 & .1 & .2 \\ .4 & .1 & .5 \end{bmatrix}$$

SOLUTION   To approximate $\overline{P}$ to four decimal places, we store P in a graphing calculator (Fig. 1A), set the decimal display to four places, and compute powers of P until all three rows of P^k are identical. Examining the output in Figure 1B, we conclude that

$$\overline{P} = \begin{bmatrix} .4943 & .1494 & .3563 \\ .4943 & .1494 & .3563 \\ .4943 & .1494 & .3563 \end{bmatrix} \quad \text{and} \quad S = \begin{bmatrix} .4943 & .1494 & .3563 \end{bmatrix}$$

Entering S in the graphing calculator and computing SP shows that these matrices are correct to four decimal places (Fig. 1C).

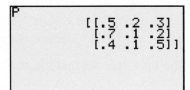   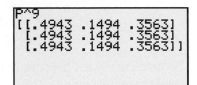   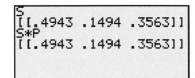

(A) P (B) P^9 (C) Check: $SP = S$

Figure 1

 Matched Problem 5   Repeat Example 5 for

$$P = \begin{bmatrix} .3 & .6 & .1 \\ .2 & .3 & .5 \\ .1 & .2 & .7 \end{bmatrix}$$

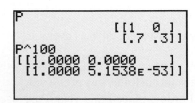

Figure 2

> ## CONCEPTUAL INSIGHT
>
> 1. We used a relatively small value of k to approximate $\overline{P}$ in Example 5. Many graphing calculators will compute P^k for large values of k almost as rapidly as for small values. However, round-off errors can occur in these calculations. A safe procedure is to start with a relatively small value of k, such as $k = 8$, and then keep doubling k until the rows of P^k are identical to the specified number of decimal places.
>
> 2. If any of the entries of P^k are approaching 0, then the graphing calculator may use scientific notation to display these entries as very small numbers. Figure 2 shows the 100th power of a transition matrix P. The entry in row 2 and column 2 of P^{100} is approaching 0, but the graphing calculator displays it as 5.1538×10^{-53}. If this occurs, simply change this value to 0 in the corresponding entry in $\overline{P}$. Thus, from the output in Figure 2 we conclude that
>
> $$P^k = \begin{bmatrix} 1 & 0 \\ .7 & .3 \end{bmatrix}^k \quad \text{approaches} \quad \overline{P} = \begin{bmatrix} 1 & 0 \\ 1 & 0 \end{bmatrix}$$

Exercises 9.2

W

Skills Warm-up Exercises

In Problems 1–8, without using a calculator, find P^{100}. (If necessary, review Section 4.4.) [Hint: First find P^2.]

1. $\begin{bmatrix} 1 & 0 \\ 0 & 0 \end{bmatrix}$ **2.** $\begin{bmatrix} 0 & 0 \\ 1 & 0 \end{bmatrix}$

3. $\begin{bmatrix} 1 & 0 \\ 0 & 1 \end{bmatrix}$ **4.** $\begin{bmatrix} 0 & 1 \\ 1 & 0 \end{bmatrix}$

5. $\begin{bmatrix} 1 & 0 & 0 \\ 0 & 0 & 0 \\ 0 & 0 & 1 \end{bmatrix}$ **6.** $\begin{bmatrix} 1 & 0 & 0 \\ 0 & 1 & 0 \\ 0 & 0 & 1 \end{bmatrix}$

7. $\begin{bmatrix} 0 & 0 & 1 \\ 0 & 0 & 0 \\ 0 & 0 & 0 \end{bmatrix}$ **8.** $\begin{bmatrix} 0 & 1 & 1 \\ 0 & 0 & 1 \\ 0 & 0 & 0 \end{bmatrix}$

In Problems 9–22, could the given matrix be the transition matrix of a regular Markov chain?

9. $\begin{bmatrix} .6 & .4 \\ .4 & .6 \end{bmatrix}$ **10.** $\begin{bmatrix} .3 & .7 \\ .2 & .6 \end{bmatrix}$

11. $\begin{bmatrix} .1 & .9 \\ .5 & .4 \end{bmatrix}$ **12.** $\begin{bmatrix} .5 & .5 \\ .8 & .2 \end{bmatrix}$

13. $\begin{bmatrix} .4 & .6 \\ 0 & 1 \end{bmatrix}$ **14.** $\begin{bmatrix} .4 & .6 \\ 1 & 0 \end{bmatrix}$

15. $\begin{bmatrix} 0 & 1 \\ .8 & .2 \end{bmatrix}$ **16.** $\begin{bmatrix} .3 & .7 \\ .2 & .6 \end{bmatrix}$

17. $\begin{bmatrix} .6 & .4 \\ .1 & .9 \\ .3 & .7 \end{bmatrix}$ **18.** $\begin{bmatrix} .2 & .5 & .3 \\ .6 & .3 & .1 \end{bmatrix}$

19. $\begin{bmatrix} 0 & 1 & 0 \\ 0 & 0 & 1 \\ .5 & .5 & 0 \end{bmatrix}$ **20.** $\begin{bmatrix} .2 & 0 & .8 \\ 0 & 0 & 1 \\ .7 & 0 & .3 \end{bmatrix}$

21. $\begin{bmatrix} .1 & .3 & .6 \\ .8 & .1 & .1 \\ 0 & 0 & 1 \end{bmatrix}$ **22.** $\begin{bmatrix} 0 & 0 & 1 \\ .9 & 0 & .1 \\ 0 & 1 & 0 \end{bmatrix}$

For each transition matrix P in Problems 23–30, solve the equation $SP = S$ to find the stationary matrix S and the limiting matrix $\overline{P}$.

23. $P = \begin{bmatrix} .1 & .9 \\ .6 & .4 \end{bmatrix}$ **24.** $P = \begin{bmatrix} .8 & .2 \\ .3 & .7 \end{bmatrix}$

25. $P = \begin{bmatrix} .5 & .5 \\ .3 & .7 \end{bmatrix}$ **26.** $P = \begin{bmatrix} .9 & .1 \\ .7 & .3 \end{bmatrix}$

27. $P = \begin{bmatrix} .5 & .1 & .4 \\ .3 & .7 & 0 \\ 0 & .6 & .4 \end{bmatrix}$ **28.** $P = \begin{bmatrix} .4 & .1 & .5 \\ .2 & .8 & 0 \\ 0 & .5 & .5 \end{bmatrix}$

29. $P = \begin{bmatrix} .8 & .2 & 0 \\ .5 & .1 & .4 \\ 0 & .6 & .4 \end{bmatrix}$ **30.** $P = \begin{bmatrix} .2 & .8 & 0 \\ .6 & .1 & .3 \\ 0 & .9 & .1 \end{bmatrix}$

In Problems 31–36, discuss the validity of each statement. If the statement is always true, explain why. If not, give a counterexample.

31. The $n \times n$ identity matrix is the transition matrix for a regular Markov chain.

32. The $n \times n$ matrix in which each entry equals $\frac{1}{n}$ is the transition matrix for a regular Markov chain.

33. If the 2×2 matrix P is the transition matrix for a regular Markov chain, then, at most, one of the entries of P is equal to 0.

34. If the 3×3 matrix P is the transition matrix for a regular Markov chain, then, at most, two of the entries of P are equal to 0.

35. If a transition matrix P for a Markov chain has a stationary matrix S, then P is regular.

36. If P is the transition matrix for a Markov chain, then P has a unique stationary matrix.

In Problems 37–40, approximate the stationary matrix S for each transition matrix P by computing powers of the transition matrix P. Round matrix entries to four decimal places.

37. $P = \begin{bmatrix} .51 & .49 \\ .27 & .73 \end{bmatrix}$ **38.** $P = \begin{bmatrix} .68 & .32 \\ .19 & .81 \end{bmatrix}$

39. $P = \begin{bmatrix} .5 & .5 & 0 \\ 0 & .5 & .5 \\ .8 & .1 & .1 \end{bmatrix}$ **40.** $P = \begin{bmatrix} .2 & .2 & .6 \\ .5 & 0 & .5 \\ .5 & 0 & .5 \end{bmatrix}$

41. A red urn contains 2 red marbles and 3 blue marbles, and a blue urn contains 1 red marble and 4 blue marbles. A marble is selected from an urn, the color is noted, and the marble is returned to the urn from which it was drawn. The next marble is drawn from the urn whose color is the same as the marble just drawn. Thus, this is a Markov process with two states: draw from the red urn or draw from the blue urn.

(A) Draw a transition diagram for this process.

(B) Write the transition matrix.

(C) Find the stationary matrix and describe the long-run behavior of this process.

42. Repeat Problem 41 if the red urn contains 5 red and 3 blue marbles, and the blue urn contains 1 red and 3 blue marbles.

43. Given the transition matrix

$$P = \begin{bmatrix} 0 & 1 \\ 1 & 0 \end{bmatrix}$$

(A) Discuss the behavior of the state matrices $S_1, S_2, S_3, \ldots$ for the initial-state matrix $S_0 = [.2 \quad .8]$.

(B) Repeat part (A) for $S_0 = [.5 \quad .5]$.

(C) Discuss the behavior of P^k, $k = 2, 3, 4, \ldots$.

(D) Which of the conclusions of Theorem 1 are not valid for this matrix? Why is this not a contradiction?

44. Given the transition matrix

$$P = \begin{bmatrix} 0 & 1 & 0 \\ 0 & 0 & 1 \\ 1 & 0 & 0 \end{bmatrix}$$

(A) Discuss the behavior of the state matrices $S_1, S_2, S_3, \ldots$ for the initial-state matrix $S_0 = [.2 \quad .3 \quad .5]$.

(B) Repeat part (A) for $S_0 = \begin{bmatrix} \frac{1}{3} & \frac{1}{3} & \frac{1}{3} \end{bmatrix}$.

(C) Discuss the behavior of P^k, $k = 2, 3, 4, \ldots$.

(D) Which of the conclusions of Theorem 1 are not valid for this matrix? Why is this not a contradiction?

45. The transition matrix for a Markov chain is

$$P = \begin{bmatrix} 1 & 0 & 0 \\ .2 & .2 & .6 \\ 0 & 0 & 1 \end{bmatrix}$$

(A) Show that $R = [1 \quad 0 \quad 0]$ and $S = [0 \quad 0 \quad 1]$ are both stationary matrices for P. Explain why this does not contradict Theorem 1A.

(B) Find another stationary matrix for P. [*Hint*: Consider $T = aR + (1-a)S$, where $0 < a < 1$.]

(C) How many different stationary matrices does P have?

46. The transition matrix for a Markov chain is

$$P = \begin{bmatrix} .7 & 0 & .3 \\ 0 & 1 & 0 \\ .2 & 0 & .8 \end{bmatrix}$$

(A) Show that $R = [.4 \quad 0 \quad .6]$ and $S = [0 \quad 1 \quad 0]$ are both stationary matrices for P. Explain why this does not contradict Theorem 1A.

(B) Find another stationary matrix for P. [*Hint*: Consider $T = aR + (1-a)S$, where $0 < a < 1$.]

(C) How many different stationary matrices does P have?

Problems 47 and 48 require the use of a graphing calculator.

47. Refer to the transition matrix P in Problem 45. What matrix $\bar{P}$ do the powers of P appear to be approaching? Are the rows of $\bar{P}$ stationary matrices for P?

48. Refer to the transition matrix P in Problem 46. What matrix $\bar{P}$ do the powers of P appear to be approaching? Are the rows of $\bar{P}$ stationary matrices for P?

49. The transition matrix for a Markov chain is

$$P = \begin{bmatrix} .1 & .5 & .4 \\ .3 & .2 & .5 \\ .7 & .1 & .2 \end{bmatrix}$$

Let M_k denote the maximum entry in the second column of P^k. Note that $M_1 = .5$.

(A) Find M_2, M_3, M_4, and M_5 to three decimal places.

(B) Explain why $M_k \geq M_{k+1}$ for all positive integers k.

50. The transition matrix for a Markov chain is

$$P = \begin{bmatrix} 0 & .2 & .8 \\ .3 & .3 & .4 \\ .6 & .1 & .3 \end{bmatrix}$$

Let m_k denote the minimum entry in the third column of P^k. Note that $m_1 = .3$.

(A) Find m_2, m_3, m_4, and m_5 to three decimal places.

(B) Explain why $m_k \leq m_{k+1}$ for all positive integers k.

Applications

51. Transportation. Most railroad cars are owned by individual railroad companies. When a car leaves its home railroad's tracks, it becomes part of a national pool of cars and can be used by other railroads. The rules governing the use of these pooled cars are designed to eventually return the car to the home tracks. A particular railroad found that each month, 11% of its boxcars on the home tracks left to join the national pool, and 29% of its boxcars in the national pool were returned to the home tracks. If these percentages remain valid for a long period of time, what percentage of its boxcars can this railroad expect to have on its home tracks in the long run?

52. Transportation. The railroad in Problem 51 also has a fleet of tank cars. If 14% of the tank cars on the home tracks enter the national pool each month, and 26% of the tank cars in the national pool are returned to the home tracks each month, what percentage of its tank cars can the railroad expect to have on its home tracks in the long run?

53. Labor force. Table 1 gives the percentage of the U.S. female population who were members of the civilian labor force in the indicated years. The following transition matrix P is proposed as a model for the data, where L represents females who are in the labor force and L' represents females who are not in the labor force:

$$\begin{array}{c} \\ \text{Current} \\ \text{decade} \end{array} \begin{array}{c} \\ L \\ L' \end{array} \overset{\begin{array}{c} \text{Next decade} \\ L \quad\quad L' \end{array}}{\begin{bmatrix} .92 & .08 \\ .2 & .8 \end{bmatrix}} = P$$

(A) Let $S_0 = [.433 \quad .567]$, and find S_1, S_2, S_3, and S_4. Compute the matrices exactly and then round entries to three decimal places.

(B) Construct a new table comparing the results from part (A) with the data in Table 1.

(C) According to this transition matrix, what percentage of the U.S. female population will be in the labor force in the long run?

Table 1

Year	Percent
1970	43.3
1980	51.5
1990	57.5
2000	59.8
2010	58.5

54. Home ownership. The U.S. Census Bureau published the home ownership rates given in Table 2.

Table 2

Year	Percent
1996	65.4
2000	67.4
2004	69.0
2008	67.8

The following transition matrix P is proposed as a model for the data, where H represents the households that own their home.

$$\begin{array}{cc} & \text{Four years later} \\ & \begin{array}{cc} H & H' \end{array} \\ \begin{array}{c} \text{Current } H \\ \text{year } H' \end{array} & \begin{bmatrix} .95 & .05 \\ .15 & .85 \end{bmatrix} = P \end{array}$$

(A) Let $S_0 = [.654 \quad .346]$ and find S_1, S_2, and S_3. Compute both matrices exactly and then round entries to three decimal places.

(B) Construct a new table comparing the results from part (A) with the data in Table 2.

(C) According to this transition matrix, what percentage of households will own their home in the long run?

55. Market share. Consumers can choose between three long-distance telephone services: GTT, NCJ, and Dash. Aggressive marketing by all three companies results in a continual shift of customers among the three services. Each year, GTT loses 5% of its customers to NCJ and 20% to Dash; NCJ loses 15% of its customers to GTT and 10% to Dash; and Dash loses 5% of its customers to GTT and 10% to NCJ. Assuming that these percentages remain valid over a long period of time, what is each company's expected market share in the long run?

56. Market share. Consumers in a certain area can choose between three package delivery services: APS, GX, and WWP. Each week, APS loses 10% of its customers to GX and 20% to WWP; GX loses 15% of its customers to APS and 10% to WWP; and WWP loses 5% of its customers to APS and 5% to GX. Assuming that these percentages remain valid over a long period of time, what is each company's expected market share in the long run?

57. Insurance. An auto insurance company classifies its customers in three categories: poor, satisfactory, and preferred. Each year, 40% of those in the poor category are moved to satisfactory, and 20% of those in the satisfactory category are moved to preferred. Also, 20% in the preferred category are moved to the satisfactory category, and 20% in the satisfactory category are moved to the poor category. Customers are never moved from poor to preferred, or conversely, in a single year. Assuming that these percentages remain valid over a long period of time, how many customers are expected in each category in the long run?

58. Insurance. Repeat Problem 57 if 40% of preferred customers are moved to the satisfactory category each year, and all other information remains the same.

Problems 59 and 60 require the use of a graphing calculator.

59. Market share. Acme Soap Company markets one brand of soap, called Standard Acme (*SA*), and Best Soap Company markets two brands, Standard Best (*SB*) and Deluxe Best (*DB*). Currently, Acme has 40% of the market, and the remainder is divided equally between the two Best brands. Acme is considering the introduction of a second brand to get a larger share of the market. A proposed new brand, called brand *X*, was test-marketed in several large cities, producing the following transition matrix for the consumers' weekly buying habits:

$$P = \begin{array}{c} \\ SB \\ DB \\ SA \\ X \end{array} \begin{array}{c} \begin{array}{cccc} SB & DB & SA & X \end{array} \\ \begin{bmatrix} .4 & .1 & .3 & .2 \\ .3 & .2 & .2 & .3 \\ .1 & .2 & .2 & .5 \\ .3 & .3 & .1 & .3 \end{bmatrix} \end{array}$$

Assuming that P represents the consumers' buying habits over a long period of time, use this transition matrix and the initial-state matrix $S_0 = [.3 \quad .3 \quad .4 \quad 0]$ to compute successive state matrices in order to approximate the elements in the stationary matrix correct to two decimal places. If Acme decides to market this new soap, what is the long-run expected total market share for their two soaps?

60. Market share. Refer to Problem 59. The chemists at Acme Soap Company have developed a second new soap, called brand *Y*. Test-marketing this soap against the three established brands produces the following transition matrix:

$$P = \begin{array}{c} \\ SB \\ DB \\ SA \\ Y \end{array} \begin{array}{c} \begin{array}{cccc} SB & DB & SA & Y \end{array} \\ \begin{bmatrix} .3 & .2 & .2 & .3 \\ .2 & .2 & .2 & .4 \\ .2 & .2 & .4 & .2 \\ .1 & .2 & .3 & .4 \end{bmatrix} \end{array}$$

Proceed as in Problem 59 to approximate the elements in the stationary matrix correct to two decimal places. If Acme decides to market brand *Y*, what is the long-run expected total market share for Standard Acme and brand *Y*? Should Acme market brand *X* or brand *Y*?

61. Genetics. A given plant species has red, pink, or white flowers according to the genotypes RR, RW, and WW, respectively. If each of these genotypes is crossed with a pink-flowering plant (genotype RW), then the transition matrix is

$$\begin{array}{c} \\ \\ \text{This} \\ \text{generation} \end{array} \begin{array}{c} \begin{array}{ccc} & \text{Next generation} & \\ Red & Pink & White \end{array} \\ \begin{array}{c} Red \\ Pink \\ White \end{array} \begin{bmatrix} .5 & .5 & 0 \\ .25 & .5 & .25 \\ 0 & .5 & .5 \end{bmatrix} \end{array}$$

Assuming that the plants of each generation are crossed only with pink plants to produce the next generation, show that regardless of the makeup of the first generation, the genotype composition will eventually stabilize at 25% red, 50% pink, and 25% white. (Find the stationary matrix.)

62. Gene mutation. Suppose that a gene in a chromosome is of type A or type B. Assume that the probability that a gene of type A will mutate to type B in one generation is 10^{-4} and that a gene of type B will mutate to type A is 10^{-6}.

(A) What is the transition matrix?

(B) After many generations, what is the probability that the gene will be of type A? Of type B? (Find the stationary matrix.)

63. Rapid transit. A new rapid transit system has just started operating. In the first month of operation, it is found that 25% of commuters are using the system, while 75% still travel by car. The following transition matrix was determined from records of other rapid transit systems:

$$\begin{array}{cc} & \begin{array}{c} \text{Next month} \\ \text{Rapid} \\ \text{transit\quad Car} \end{array} \\ \begin{array}{c} \text{Current} \\ \text{month} \end{array} \begin{array}{c} \text{Rapid transit} \\ \text{Car} \end{array} & \begin{bmatrix} .8 & .2 \\ .3 & .7 \end{bmatrix} \end{array}$$

(A) What is the initial-state matrix?

(B) What percentage of commuters will be using the new system after 1 month? After 2 months?

(C) Find the percentage of commuters using each type of transportation after the new system has been in service for a long time.

64. Politics: filibuster. The Senate is in the middle of a floor debate, and a filibuster is threatened. Senator Hanks, who is still vacillating, has a probability of .1 of changing his mind during the next 5 minutes. If this pattern continues for each 5 minutes that the debate continues, and if a 24-hour filibuster takes place before a vote is taken, what is the probability that Senator Hanks will cast a yes vote? A no vote?

(A) Complete the following transition matrix:

$$\begin{array}{cc} & \begin{array}{c} \text{Next 5 minutes} \\ \text{Yes\quad No} \end{array} \\ \begin{array}{c} \text{Current} \\ \text{5 minutes} \end{array} \begin{array}{c} \text{Yes} \\ \text{No} \end{array} & \begin{bmatrix} .9 & .1 \\ & \end{bmatrix} \end{array}$$

(B) Find the stationary matrix and answer the two questions.

(C) What is the stationary matrix if the probability of Senator Hanks changing his mind (.1) is replaced with an arbitrary probability p?

The population center of the 48 contiguous states of the United States is the point where a flat, rigid map of the contiguous states would balance if the location of each person was represented on the map by a weight of equal measure. In 1790, the population center was 23 miles east of Baltimore, Maryland. By 1990, the center had shifted about 800 miles west and 100 miles south to a point in southeast Missouri. To study this shifting population, the U.S. Census Bureau divides the states into four regions as shown in the figure. Problems 65 and 66 deal with population shifts among these regions.

Figure for 65 and 66: Regions of the United States and the center of population

65. Population shifts. Table 3 gives the percentage of the U.S. population living in the south region during the indicated years.

Table 3

Year	Percent
1970	30.9
1980	33.3
1990	34.4
2000	35.6
2010	37.1

The following transition matrix P is proposed as a model for the data, where S represents the population that lives in the south region:

$$\begin{array}{cc} & \begin{array}{c} \text{Next decade} \\ S \qquad S' \end{array} \\ \begin{array}{c} \text{Current} \\ \text{decade} \end{array} \begin{array}{c} S \\ S' \end{array} & \begin{bmatrix} .61 & .39 \\ .21 & .79 \end{bmatrix} = P \end{array}$$

(A) Let $S_0 = [.309 \quad .691]$ and find $S_1, S_2, S_3,$ and S_4. Compute the matrices exactly and then round entries to three decimal places.

(B) Construct a new table comparing the results from part (A) with the data in Table 3.

(C) According to this transition matrix, what percentage of the population will live in the south region in the long run?

66. Population shifts. Table 4 gives the percentage of the U.S. population living in the northeast region during the indicated years.

Table 4

Year	Percent
1970	24.1
1980	21.7
1990	20.4
2000	19.0
2010	17.9

The following transition matrix P is proposed as a model for the data, where N represents the population that lives in the northeast region:

$$\begin{array}{cc} & \begin{array}{c} \text{Next decade} \\ N \qquad N' \end{array} \\ \begin{array}{c} \text{Current} \\ \text{decade} \end{array} \begin{array}{c} N \\ N' \end{array} & \begin{bmatrix} .61 & .39 \\ .09 & .91 \end{bmatrix} = P \end{array}$$

(A) Let $S_0 = [.241 \quad .759]$ and find $S_1, S_2, S_3,$ and S_4.
Compute the matrices exactly and then round entries to
three decimal places.

(B) Construct a new table comparing the results from part
(A) with the data in Table 4.

(C) According to this transition matrix, what percentage of
the population will live in the northeast region in the
long run?

Answers to Matched Problems

1. (A) Regular   (B) Not regular   (C) Regular

2. $S = [.2 \quad .8];$ $\bar{P} = \begin{bmatrix} .2 & .8 \\ .2 & .8 \end{bmatrix}$ 3. 97.5%

4. 28% preferred, 43% standard, 29% infrequent

5. $\bar{P} = \begin{bmatrix} .1618 & .2941 & .5441 \\ .1618 & .2941 & .5441 \\ .1618 & .2941 & .5441 \end{bmatrix}$;

$S = [.1618 \quad .2941 \quad .5441]$

9.3 Absorbing Markov Chains

- Absorbing States and Absorbing Chains
- Standard Form
- Limiting Matrix
- Graphing Calculator Approximations

In Section 9.2, we saw that the powers of a regular transition matrix always approach
a limiting matrix. Not all transition matrices have this property. In this section, we
discuss another type of Markov chain, called an *absorbing Markov chain*. Although
regular and absorbing Markov chains have some differences, they have one important
similarity: the powers of the transition matrix for an absorbing Markov chain also
approach a limiting matrix. After introducing basic concepts, we develop methods
for finding the limiting matrix and discuss the relationship between the states in the
Markov chain and the entries in the limiting matrix.

Absorbing States and Absorbing Chains

A state in a Markov chain is called an **absorbing state** if once the state is entered, it
is impossible to leave.

EXAMPLE 1   Recognizing Absorbing States   Identify any absorbing states for
the following transition matrices:

(A)
$$\begin{array}{c} \\ P = \begin{array}{c} A \\ B \\ C \end{array} \end{array} \begin{array}{ccc} A & B & C \\ \begin{bmatrix} 1 & 0 & 0 \\ .5 & .5 & 0 \\ 0 & .5 & .5 \end{bmatrix} \end{array}$$

(B)
$$\begin{array}{c} \\ P = \begin{array}{c} A \\ B \\ C \end{array} \end{array} \begin{array}{ccc} A & B & C \\ \begin{bmatrix} 0 & 0 & 1 \\ 0 & 1 & 0 \\ 1 & 0 & 0 \end{bmatrix} \end{array}$$

SOLUTION

(A) The probability of going from state A to state A is 1, and the probability of
going from state A to either state B or state C is 0. Once state A is entered,
it is impossible to leave; so A is an absorbing state. Since the probability of
going from state B to state A is nonzero, it is possible to leave B, and B is not
an absorbing state. Similarly, the probability of going from state C to state B is
nonzero, so C is not an absorbing state.

(B) Reasoning as before, the 1 in row 2 and column 2 indicates that state B is an
absorbing state. The probability of going from state A to state C and the probability of going from state C to state A are both nonzero. So A and C are not
absorbing states.

Matched Problem 1   Identify any absorbing states for the following transition
matrices:

(A)
$$\begin{array}{c} \\ P = \begin{array}{c} A \\ B \\ C \end{array} \end{array} \begin{array}{ccc} A & B & C \\ \begin{bmatrix} .5 & 0 & .5 \\ 0 & 1 & 0 \\ 0 & .5 & .5 \end{bmatrix} \end{array}$$

(B)
$$\begin{array}{c} \\ P = \begin{array}{c} A \\ B \\ C \end{array} \end{array} \begin{array}{ccc} A & B & C \\ \begin{bmatrix} 0 & 1 & 0 \\ 1 & 0 & 0 \\ 0 & 0 & 1 \end{bmatrix} \end{array}$$

The reasoning used to identify absorbing states in Example 1 is generalized in Theorem 1.

THEOREM 1   Absorbing States and Transition Matrices
A state in a Markov chain is **absorbing** if and only if the row of the transition matrix corresponding to the state has a 1 on the main diagonal and 0's elsewhere.

The presence of an absorbing state in a transition matrix does not guarantee that the powers of the matrix approach a limiting matrix nor that the state matrices in the corresponding Markov chain approach a stationary matrix. For example, if we square the matrix P from Example 1B, we obtain

$$P^2 = \begin{bmatrix} 0 & 0 & 1 \\ 0 & 1 & 0 \\ 1 & 0 & 0 \end{bmatrix} \begin{bmatrix} 0 & 0 & 1 \\ 0 & 1 & 0 \\ 1 & 0 & 0 \end{bmatrix} = \begin{bmatrix} 1 & 0 & 0 \\ 0 & 1 & 0 \\ 0 & 0 & 1 \end{bmatrix} = I$$

Since $P^2 = I$, the 3×3 identity matrix, it follows that

$$P^3 = PP^2 = PI = P \quad \text{Since } P^2 = I$$
$$P^4 = PP^3 = PP = I \quad \text{Since } P^3 = P \text{ and } PP = P^2 = I$$

In general, the powers of this transition matrix P oscillate between P and I and do not approach a limiting matrix.

Explore and Discuss 1   (A) For the initial-state matrix $S_0 = [a \quad b \quad c]$, find the first four state matrices, S_1, S_2, S_3, and S_4, in the Markov chain with transition matrix

$$P = \begin{bmatrix} 0 & 0 & 1 \\ 0 & 1 & 0 \\ 1 & 0 & 0 \end{bmatrix}$$

(B) Do the state matrices appear to be approaching a stationary matrix? Discuss.

To ensure that transition matrices for Markov chains with one or more absorbing states have limiting matrices, it is necessary to require the chain to satisfy one additional condition, as stated in the following definition.

DEFINITION   Absorbing Markov Chains
A Markov chain is an **absorbing chain** if

1. There is at least one absorbing state; and
2. It is possible to go from each nonabsorbing state to at least one absorbing state in a finite number of steps.

As we saw earlier, absorbing states are identified easily by examining the rows of a transition matrix. It is also possible to use a transition matrix to determine whether a Markov chain is an absorbing chain, but this can be a difficult task, especially if the matrix is large. A transition diagram is often a more appropriate tool for determining whether a Markov chain is absorbing. The next example illustrates this approach for the two matrices discussed in Example 1.

EXAMPLE 2 Recognizing Absorbing Markov Chains Use a transition diagram to determine whether P is the transition matrix for an absorbing Markov chain.

(A)
$$P = \begin{array}{c} \\ A \\ B \\ C \end{array} \begin{array}{ccc} A & B & C \\ \left[\begin{array}{ccc} 1 & 0 & 0 \\ .5 & .5 & 0 \\ 0 & .5 & .5 \end{array}\right] \end{array}$$

(B)
$$P = \begin{array}{c} \\ A \\ B \\ C \end{array} \begin{array}{ccc} A & B & C \\ \left[\begin{array}{ccc} 0 & 0 & 1 \\ 0 & 1 & 0 \\ 1 & 0 & 0 \end{array}\right] \end{array}$$

SOLUTION

(A) From Example 1A, we know that A is the only absorbing state. The second condition in the definition of an absorbing Markov chain is satisfied if we can show that it is possible to go from the nonabsorbing states B and C to the absorbing state A in a finite number of steps. This is easily determined by drawing a transition diagram (Fig. 1). Examining the diagram, we see that it is possible to go from state B to the absorbing state A in one step and from state C to the absorbing state A in two steps. So P is the transition matrix for an absorbing Markov chain.

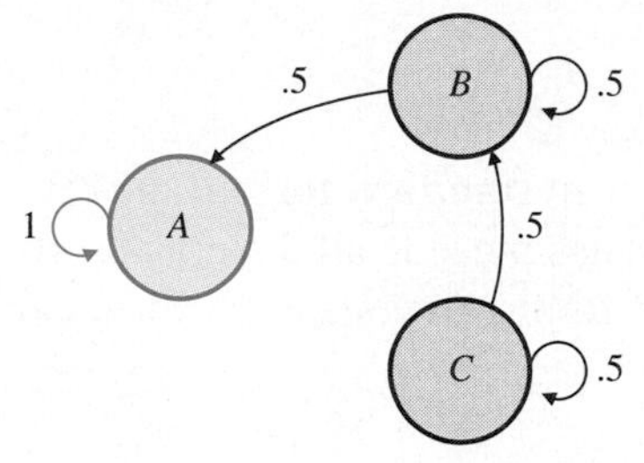

(B) Again, we draw the transition diagram (Fig. 2). From this diagram it is clear that it is impossible to go from either state A or state C to the absorbing state B. So P is not the transition matrix for an absorbing Markov chain.

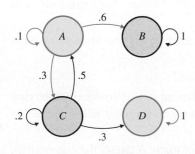

Figure 1

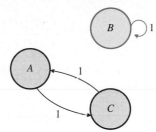

Figure 2

Matched Problem 2 Use a transition diagram to determine whether P is the transition matrix for an absorbing Markov chain.

(A)
$$P = \begin{array}{c} \\ A \\ B \\ C \end{array} \begin{array}{ccc} A & B & C \\ \left[\begin{array}{ccc} .5 & 0 & .5 \\ 0 & 1 & 0 \\ 0 & .5 & .5 \end{array}\right] \end{array}$$

(B)
$$P = \begin{array}{c} \\ A \\ B \\ C \end{array} \begin{array}{ccc} A & B & C \\ \left[\begin{array}{ccc} 0 & 1 & 0 \\ 1 & 0 & 0 \\ 0 & 0 & 1 \end{array}\right] \end{array}$$

Explore and Discuss 2 Determine whether each statement is true or false. Use examples and verbal arguments to support your conclusions.

(A) A Markov chain with two states, one nonabsorbing and one absorbing, is always an absorbing chain.

(B) A Markov chain with two states, both of which are absorbing, is always an absorbing chain.

(C) A Markov chain with three states, one nonabsorbing and two absorbing, is always an absorbing chain.

Standard Form

The transition matrix for a Markov chain is not unique. Consider the transition diagram in Figure 3. Since there are $4! = 24$ different ways to arrange the four states in this diagram, there are 24 different ways to write a transition matrix. (Some of these matrices may have identical entries, but all are different when the row and column

Figure 3

labels are taken into account.) For example, the following matrices M, N, and P are three different transition matrices for this diagram.

$$
M = \begin{array}{c} \\ A \\ B \\ C \\ D \end{array}
\begin{array}{cccc} A & B & C & D \\ \left[\begin{array}{cccc} .1 & .6 & .3 & 0 \\ 0 & 1 & 0 & 0 \\ .5 & 0 & .2 & .3 \\ 0 & 0 & 0 & 1 \end{array}\right] \end{array}
\quad
N = \begin{array}{c} \\ D \\ B \\ C \\ A \end{array}
\begin{array}{cccc} D & B & C & A \\ \left[\begin{array}{cccc} 1 & 0 & 0 & 0 \\ 0 & 1 & 0 & 0 \\ .3 & 0 & .2 & .5 \\ 0 & .6 & .3 & .1 \end{array}\right] \end{array}
\quad
P = \begin{array}{c} \\ B \\ D \\ A \\ C \end{array}
\begin{array}{cccc} B & D & A & C \\ \left[\begin{array}{cccc} 1 & 0 & 0 & 0 \\ 0 & 1 & 0 & 0 \\ .6 & 0 & .1 & .3 \\ 0 & .3 & .5 & .2 \end{array}\right] \end{array}
\quad (1)
$$

In matrices N and P, notice that all the absorbing states precede all the nonabsorbing states. A transition matrix written in this form is said to be in *standard form*. We will find standard forms very useful in determining limiting matrices for absorbing Markov chains.

> **DEFINITION**   Standard Forms for Absorbing Markov Chains
> A transition matrix for an absorbing Markov chain is in **standard form** if the rows and columns are labeled so that all the absorbing states precede all the nonabsorbing states. (There may be more than one standard form.) Any standard form can always be partitioned into four submatrices:
>
> $$
> \begin{array}{c} \\ A \\ N \end{array}
> \begin{array}{cc} A & N \\ \left[\begin{array}{c:c} I & 0 \\ \hdashline R & Q \end{array}\right] \end{array}
> \quad
> \begin{array}{l} A = \text{all absorbing states} \\ N = \text{all nonabsorbing states} \end{array}
> $$
>
> where I is an identity matrix and 0 is a zero matrix.

Referring to the matrix P in (1), we see that the submatrices in this standard form are

$$
I = \begin{bmatrix} 1 & 0 \\ 0 & 1 \end{bmatrix} \qquad 0 = \begin{bmatrix} 0 & 0 \\ 0 & 0 \end{bmatrix}
$$

$$
R = \begin{bmatrix} .6 & 0 \\ 0 & .3 \end{bmatrix} \qquad Q = \begin{bmatrix} .1 & .3 \\ .5 & .2 \end{bmatrix}
\qquad
P = \begin{array}{c} \\ B \\ D \\ A \\ C \end{array}
\begin{array}{cccc} B & D & A & C \\ \left[\begin{array}{cc:cc} 1 & 0 & 0 & 0 \\ 0 & 1 & 0 & 0 \\ \hdashline .6 & 0 & .1 & .3 \\ 0 & .3 & .5 & .2 \end{array}\right] \end{array}
$$

Limiting Matrix

We will now discuss the long-run behavior of absorbing Markov chains.

EXAMPLE 3   Real Estate Development   Two competing real estate companies are trying to buy all the farms in a particular area for future housing development. Each year, 20% of the farmers decide to sell to company A, 30% decide to sell to company B, and the rest continue to farm their land. Neither company ever sells any of the farms they purchase.

(A) Draw a transition diagram and determine whether or not the Markov chain is absorbing.

(B) Write a transition matrix that is in standard form.

(C) If neither company owns any farms at the beginning of this competitive buying process, estimate the percentage of farms that each company will purchase in the long run.

(D) If company A buys 50% of the farms before company B enters the competitive buying process, estimate the percentage of farms that each company will purchase in the long run.

SOLUTION

(A)

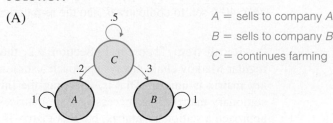

A = sells to company A
B = sells to company B
C = continues farming

The associated Markov chain is absorbing since there are two absorbing states, A and B. It is possible to go from the nonabsorbing state C to either A or B in one step.

(B) We use the transition diagram to write a transition matrix that is in standard form:

$$P = \begin{array}{c} \\ A \\ B \\ C \end{array} \begin{array}{ccc} A & B & C \\ \left[\begin{array}{ccc} 1 & 0 & 0 \\ 0 & 1 & 0 \\ .2 & .3 & .5 \end{array} \right] \end{array} \quad \text{Standard form}$$

(C) At the beginning of the competitive buying process all the farmers are in state C (own a farm). Thus, $S_0 = [0 \quad 0 \quad 1]$. The successive state matrices are (multiplication details omitted):

$$S_1 = S_0 P = [.2 \quad .3 \quad .5]$$
$$S_2 = S_1 P = [.3 \quad .45 \quad .25]$$
$$S_3 = S_2 P = [.35 \quad .525 \quad .125]$$
$$S_4 = S_3 P = [.375 \quad .5625 \quad .0625]$$
$$S_5 = S_4 P = [.3875 \quad .58125 \quad .03125]$$
$$S_6 = S_5 P = [.39375 \quad .590625 \quad .015625]$$
$$S_7 = S_6 P = [.396875 \quad .5953125 \quad .0078125]$$
$$S_8 = S_7 P = [.3984375 \quad .59765625 \quad .00390625]$$
$$S_9 = S_8 P = [.39921875 \quad .598828125 \quad .001953125]$$

It appears that these state matrices are approaching the matrix

$$\begin{array}{ccc} A & B & C \end{array}$$
$$S = [.4 \quad .6 \quad 0]$$

This indicates that in the long run, company A will acquire approximately 40% of the farms and company B will acquire the remaining 60%.

(D) This time, at the beginning of the competitive buying process 50% of farmers are already in state A and the rest are in state C. So $S_0 = [.5 \quad 0 \quad .5]$. The successive state matrices are (multiplication details omitted):

$$S_1 = S_0 P = [.6 \quad .15 \quad .25]$$
$$S_2 = S_1 P = [.65 \quad .225 \quad .125]$$
$$S_3 = S_2 P = [.675 \quad .2625 \quad .0625]$$
$$S_4 = S_3 P = [.6875 \quad .28125 \quad .03125]$$
$$S_5 = S_4 P = [.69375 \quad .290625 \quad .015625]$$
$$S_6 = S_5 P = [.696875 \quad .2953125 \quad .0078125]$$
$$S_7 = S_6 P = [.6984375 \quad .29765625 \quad .00390625]$$
$$S_8 = S_7 P = [.69921875 \quad .298828125 \quad .001953125]$$

These state matrices approach a matrix different from the one in part (C):

$$\begin{array}{ccc} A & B & C \end{array}$$
$$S' = [.7 \quad .3 \quad 0]$$

Because of its head start, company A will now acquire approximately 70% of the farms and company B will acquire the remaining 30%.

Matched Problem 3 Repeat Example 3 if 10% of farmers sell to company *A* each year, 40% sell to company *B*, and the remainder continue farming.

Recall from Theorem 1, Section 9.2, that the successive state matrices of a regular Markov chain always approach a stationary matrix. Furthermore, this stationary matrix is unique. That is, changing the initial-state matrix does not change the stationary matrix. The successive state matrices for an absorbing Markov chain also approach a stationary matrix, but this matrix is not unique. To confirm this, consider the transition matrix *P* and the state matrices *S* and *S'* from Example 3:

$$P = \begin{matrix} & \begin{matrix} A & B & C \end{matrix} \\ \begin{matrix} A \\ B \\ C \end{matrix} & \begin{bmatrix} 1 & 0 & 0 \\ 0 & 1 & 0 \\ .2 & .3 & .5 \end{bmatrix} \end{matrix} \qquad \begin{matrix} \begin{matrix} A & B & C \end{matrix} \\ S = \begin{bmatrix} .4 & .6 & 0 \end{bmatrix} \end{matrix} \qquad \begin{matrix} \begin{matrix} A & B & C \end{matrix} \\ S' = \begin{bmatrix} .7 & .3 & 0 \end{bmatrix} \end{matrix}$$

It turns out that *S* and *S'* are both stationary matrices, as the following multiplications verify:

$$SP = \begin{bmatrix} .4 & .6 & 0 \end{bmatrix} \begin{bmatrix} 1 & 0 & 0 \\ 0 & 1 & 0 \\ .2 & .3 & .5 \end{bmatrix} = \begin{bmatrix} .4 & .6 & 0 \end{bmatrix} = S$$

$$S'P = \begin{bmatrix} .7 & .3 & 0 \end{bmatrix} \begin{bmatrix} 1 & 0 & 0 \\ 0 & 1 & 0 \\ .2 & .3 & .5 \end{bmatrix} = \begin{bmatrix} .7 & .3 & 0 \end{bmatrix} = S'$$

In fact, this absorbing Markov chain has an infinite number of stationary matrices (see Problems 41–48, Exercises 9.3).

Changing the initial-state matrix for an absorbing Markov chain can cause the successive state matrices to approach a different stationary matrix.

In Section 9.2, we used the unique stationary matrix for a regular Markov chain to find the limiting matrix $\overline{P}$. Since an absorbing Markov chain can have many different stationary matrices, we cannot expect this approach to work for absorbing chains. However, it turns out that transition matrices for absorbing chains do have limiting matrices, and they are not very difficult to find. Theorem 2 gives us the necessary tools. The proof of this theorem is left for more advanced courses.

THEOREM 2   Limiting Matrices for Absorbing Markov Chains

If a standard form *P* for an absorbing Markov chain is partitioned as

$$P = \begin{bmatrix} I & 0 \\ \hline R & Q \end{bmatrix}$$

then P^k approaches a limiting matrix $\overline{P}$ as *k* increases, where

$$\overline{P} = \begin{bmatrix} I & 0 \\ \hline FR & 0 \end{bmatrix}$$

The matrix *F* is given by $F = (I - Q)^{-1}$ and is called the **fundamental matrix** for *P*.

The identity matrix used to form the fundamental matrix *F* must be the same size as the matrix *Q*.

EXAMPLE 4   Finding the Limiting Matrix

(A) Find the limiting matrix $\overline{P}$ for the standard form *P* found in Example 3.

(B) Use $\overline{P}$ to find the limit of the successive state matrices for $S_0 = \begin{bmatrix} 0 & 0 & 1 \end{bmatrix}$.

(C) Use $\overline{P}$ to find the limit of the successive state matrices for $S_0 = \begin{bmatrix} .5 & 0 & .5 \end{bmatrix}$.

SOLUTION

(A) From Example 3, we have

$$P = \begin{bmatrix} 1 & 0 & | & 0 \\ 0 & 1 & | & 0 \\ \hline .2 & .3 & | & .5 \end{bmatrix} \quad \begin{bmatrix} I & | & 0 \\ \hline R & | & Q \end{bmatrix}$$

where

$$I = \begin{bmatrix} 1 & 0 \\ 0 & 1 \end{bmatrix} \quad 0 = \begin{bmatrix} 0 \\ 0 \end{bmatrix} \quad R = [.2 \quad .3] \quad Q = [.5]$$

If $I = [1]$ is the 1×1 identity matrix, then $I - Q$ is also a 1×1 matrix; $F = (I - Q)^{-1}$ is simply the multiplicative inverse of the single entry in $I - Q$. So

$$F = ([1] - [.5])^{-1} = [.5]^{-1} = [2]$$
$$FR = [2][.2 \quad .3] = [.4 \quad .6]$$

and the limiting matrix is

$$\overline{P} = \begin{array}{c} \\ A \\ B \\ C \end{array} \begin{array}{c} \begin{array}{ccc} A & B & C \end{array} \\ \begin{bmatrix} 1 & 0 & 0 \\ 0 & 1 & 0 \\ .4 & .6 & 0 \end{bmatrix} \end{array} \quad \begin{bmatrix} I & | & 0 \\ \hline FR & | & 0 \end{bmatrix}$$

(B) Since the successive state matrices are given by $S_k = S_0 P^k$ (Theorem 1, Section 9.1) and P^k approaches $\overline{P}$, it follows that S_k approaches

$$S_0\overline{P} = \begin{bmatrix} 0 & 0 & 1 \end{bmatrix} \begin{bmatrix} 1 & 0 & 0 \\ 0 & 1 & 0 \\ .4 & .6 & 0 \end{bmatrix} = \begin{bmatrix} .4 & .6 & 0 \end{bmatrix}$$

which agrees with the results in part (C) of Example 3.

(C) This time, the successive state matrices approach

$$S_0\overline{P} = \begin{bmatrix} .5 & 0 & .5 \end{bmatrix} \begin{bmatrix} 1 & 0 & 0 \\ 0 & 1 & 0 \\ .4 & .6 & 0 \end{bmatrix} = \begin{bmatrix} .7 & .3 & 0 \end{bmatrix}$$

which agrees with the results in part (D) of Example 3.

Matched Problem 4 Repeat Example 4 for the standard form P found in Matched Problem 3.

Recall that the limiting matrix for a regular Markov chain contains the long-run probabilities of going from any state to any other state. This is also true for the limiting matrix of an absorbing Markov chain. Let's compare the transition matrix P and its limiting matrix $\overline{P}$ from Example 4:

$$P = \begin{array}{c} \\ A \\ B \\ C \end{array} \begin{array}{c} \begin{array}{ccc} A & B & C \end{array} \\ \begin{bmatrix} 1 & 0 & 0 \\ 0 & 1 & 0 \\ .2 & .3 & .5 \end{bmatrix} \end{array} \quad \text{approaches} \quad \overline{P} = \begin{array}{c} \\ A \\ B \\ C \end{array} \begin{array}{c} \begin{array}{ccc} A & B & C \end{array} \\ \begin{bmatrix} 1 & 0 & 0 \\ 0 & 1 & 0 \\ .4 & .6 & 0 \end{bmatrix} \end{array}$$

The rows of P and $\overline{P}$ corresponding to the absorbing states A and B are identical. That is, if the probability of going from state A to state A is 1 at the beginning of the chain, then this probability will remain 1 for all trials in the chain and for the limiting matrix. The entries in the third row of $\overline{P}$ give the long-run probabilities of going from the nonabsorbing state C to states A, B, or C.

The fundamental matrix F provides some additional information about an absorbing chain. Recall from Example 4 that $F = [2]$. It can be shown that the entries in F determine the average number of trials that it takes to go from a given nonabsorbing state to an absorbing state. In the case of Example 4, the single entry 2 in F indicates that it will take an average of 2 years for a farmer to go from state C (owns a farm) to one of the absorbing states (sells the farm). Some will reach an absorbing state in 1 year, and some will take more than 2 years. But the average will be 2 years. These observations are summarized in Theorem 3, which we state without proof.

THEOREM 3   Properties of the Limiting Matrix $\overline{P}$

If P is a transition matrix in standard form for an absorbing Markov chain, F is the fundamental matrix, and $\overline{P}$ is the limiting matrix, then

(A) The entry in row i and column j of $\overline{P}$ is the long-run probability of going from state i to state j. For the nonabsorbing states, these probabilities are also the entries in the matrix FR used to form $\overline{P}$.

(B) The sum of the entries in each row of the fundamental matrix F is the average number of trials it will take to go from each nonabsorbing state to some absorbing state.

(Note that the rows of both F and FR correspond to the nonabsorbing states in the order given in the standard form P.)

CONCEPTUAL INSIGHT

1. The zero matrix in the lower right corner of the limiting matrix $\overline{P}$ in Theorem 2 indicates that the long-run probability of going from any nonabsorbing state to any other nonabsorbing state is always 0. That is, in the long run, all elements in an absorbing Markov chain end up in one of the absorbing states.

2. If the transition matrix for an absorbing Markov chain is not in standard form, it is still possible to find a limiting matrix (see Problems 47 and 48, Exercises 9.3). However, it is customary to use standard form when investigating the limiting behavior of an absorbing chain.

Now that we have developed the necessary tools for analyzing the long-run behavior of an absorbing Markov chain, we apply these tools to an earlier application (see Example 4, Section 9.1).

EXAMPLE 5   Student Retention   The following transition diagram is for part-time students enrolled in a university MBA program:

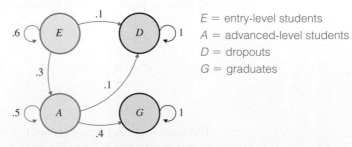

E = entry-level students
A = advanced-level students
D = dropouts
G = graduates

(A) In the long run, what percentage of entry-level students will graduate? What percentage of advanced-level students will not graduate?

(B) What is the average number of years that an entry-level student will remain in this program? An advanced-level student?

SOLUTION

(A) First, notice that this is an absorbing Markov chain with two absorbing states, state D and state G. A standard form for this absorbing chain is

$$P = \begin{array}{c} \\ D \\ G \\ E \\ A \end{array}\begin{array}{cc} \begin{array}{cccc} D & G & E & A \end{array} \\ \left[\begin{array}{cc|cc} 1 & 0 & 0 & 0 \\ 0 & 1 & 0 & 0 \\ \hline .1 & 0 & .6 & .3 \\ .1 & .4 & 0 & .5 \end{array}\right] \end{array} \quad \left[\begin{array}{c|c} I & 0 \\ \hline R & Q \end{array}\right]$$

The submatrices in this partition are

$$I = \begin{bmatrix} 1 & 0 \\ 0 & 1 \end{bmatrix} \quad 0 = \begin{bmatrix} 0 & 0 \\ 0 & 0 \end{bmatrix} \quad R = \begin{bmatrix} .1 & 0 \\ .1 & .4 \end{bmatrix} \quad Q = \begin{bmatrix} .6 & .3 \\ 0 & .5 \end{bmatrix}$$

Therefore,

$$F = (I - Q)^{-1} = \left(\begin{bmatrix} 1 & 0 \\ 0 & 1 \end{bmatrix} - \begin{bmatrix} .6 & .3 \\ 0 & .5 \end{bmatrix}\right)^{-1}$$

$$= \begin{bmatrix} .4 & -.3 \\ 0 & .5 \end{bmatrix}^{-1} \qquad \text{Use row operations to find this matrix inverse.}$$

$$= \begin{bmatrix} 2.5 & 1.5 \\ 0 & 2 \end{bmatrix}$$

and

$$FR = \begin{bmatrix} 2.5 & 1.5 \\ 0 & 2 \end{bmatrix}\begin{bmatrix} .1 & 0 \\ .1 & .4 \end{bmatrix} = \begin{bmatrix} .4 & .6 \\ .2 & .8 \end{bmatrix}$$

The limiting matrix is

$$\overline{P} = \begin{array}{c} \\ D \\ G \\ E \\ A \end{array}\begin{array}{cc} \begin{array}{cccc} D & G & E & A \end{array} \\ \left[\begin{array}{cccc} 1 & 0 & 0 & 0 \\ 0 & 1 & 0 & 0 \\ .4 & .6 & 0 & 0 \\ .2 & .8 & 0 & 0 \end{array}\right] \end{array} \quad \left[\begin{array}{c|c} I & 0 \\ \hline FR & 0 \end{array}\right]$$

From this limiting form, we see that in the long run 60% of the entry-level students will graduate and 20% of the advanced-level students will not graduate.

(B) The sum of the first-row entries of the fundamental matrix F is $2.5 + 1.5 = 4$. According to Theorem 3, this indicates that an entry-level student will spend an average of 4 years in the transient states E and A before reaching one of the absorbing states, D or G. The sum of the second-row entries of F is $0 + 2 = 2$. So an advanced-level student spends an average of 2 years in the program before either graduating or dropping out.

Matched Problem 5) Repeat Example 5 for the following transition diagram:

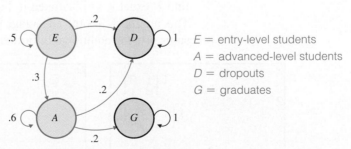

E = entry-level students
A = advanced-level students
D = dropouts
G = graduates

Graphing Calculator Approximations

Just as was the case for regular Markov chains, the limiting matrix $\overline{P}$ for an absorbing Markov chain with transition matrix P can be approximated by computing P^k on a graphing calculator for sufficiently large values of k. For example, computing P^{50} for the standard form P in Example 5 produces the following results:

$$P^{50} = \begin{bmatrix} 1 & 0 & 0 & 0 \\ 0 & 1 & 0 & 0 \\ .1 & 0 & .6 & .3 \\ .1 & .4 & 0 & .5 \end{bmatrix}^{50} = \begin{bmatrix} 1 & 0 & 0 & 0 \\ 0 & 1 & 0 & 0 \\ .4 & .6 & 0 & 0 \\ .2 & .8 & 0 & 0 \end{bmatrix} = \overline{P}$$

where we have replaced very small numbers displayed in scientific notation with 0 (see the Conceptual Insight on page 464, Section 9.2).

⚠ CAUTION Before you use P^k to approximate $\overline{P}$, be certain to determine that $\overline{P}$ exists. If you attempt to approximate a limiting matrix when none exists, the results can be misleading. For example, consider the transition matrix

$$P = \begin{bmatrix} 1 & 0 & 0 & 0 & 0 \\ .2 & .2 & 0 & .3 & .3 \\ 0 & 0 & 0 & .5 & .5 \\ 0 & 0 & 1 & 0 & 0 \\ 0 & 0 & 1 & 0 & 0 \end{bmatrix}$$

Computing P^{50} on a graphing calculator produces the following matrix:

$$P^{50} = \begin{bmatrix} 1 & 0 & 0 & 0 & 0 \\ .25 & 0 & .625 & .0625 & .0625 \\ 0 & 0 & 1 & 0 & 0 \\ 0 & 0 & 0 & .5 & .5 \\ 0 & 0 & 0 & .5 & .5 \end{bmatrix} \qquad (2)$$

It is tempting to stop at this point and conclude that the matrix in (2) must be a good approximation for $\overline{P}$. But to do so would be incorrect! If P^{50} approximates a limiting matrix $\overline{P}$, then P^{51} should also approximate the same matrix. However, computing P^{51} produces quite a different matrix:

$$P^{51} = \begin{bmatrix} 1 & 0 & 0 & 0 & 0 \\ .25 & 0 & .125 & .3125 & .3125 \\ 0 & 0 & 0 & .5 & .5 \\ 0 & 0 & 1 & 0 & 0 \\ 0 & 0 & 1 & 0 & 0 \end{bmatrix} \qquad (3)$$

Computing additional powers of P shows that the even powers of P approach matrix (2) while the odd powers approach matrix (3). The transition matrix P does not have a limiting matrix.   ▲

A graphing calculator can be used to perform the matrix calculations necessary to find $\overline{P}$ exactly, as illustrated in Figure 4 for the transition matrix P from Example 5. This approach has the advantage of producing the fundamental matrix F, whose row sums provide additional information about the long-run behavior of the chain.

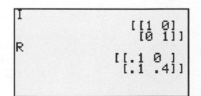

(A) Store I and R in the
graphing calculator memory

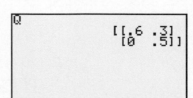

(B) Store Q in the
graphing calculator memory

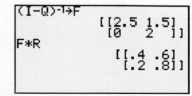

(C) Compute F and FR

Figure 4 **Matrix calculations**

Exercises 9.3

In Problems 1–6, identify the absorbing states in the indicated transition matrix.

$$1. \ P = \begin{array}{c} A \\ B \\ C \end{array} \begin{array}{ccc} A & B & C \\ \left[\begin{array}{ccc} .6 & .3 & .1 \\ 0 & 1 & 0 \\ 0 & 0 & 0 \end{array} \right] \end{array}$$

$$2. \ P = \begin{array}{c} A \\ B \\ C \end{array} \begin{array}{ccc} A & B & C \\ \left[\begin{array}{ccc} 0 & 1 & 0 \\ .3 & .2 & .5 \\ 0 & 0 & 1 \end{array} \right] \end{array}$$

$$3. \ P = \begin{array}{c} A \\ B \\ C \end{array} \begin{array}{ccc} A & B & C \\ \left[\begin{array}{ccc} 0 & 0 & 1 \\ 1 & 0 & 0 \\ 0 & 1 & 0 \end{array} \right] \end{array}$$

$$4. \ P = \begin{array}{c} A \\ B \\ C \end{array} \begin{array}{ccc} A & B & C \\ \left[\begin{array}{ccc} 1 & 0 & 0 \\ .3 & .4 & .3 \\ 0 & 0 & 1 \end{array} \right] \end{array}$$

$$5. \ P = \begin{array}{c} A \\ B \\ C \\ D \end{array} \begin{array}{cccc} A & B & C & D \\ \left[\begin{array}{cccc} 1 & 0 & 0 & 0 \\ 0 & 0 & 1 & 0 \\ .1 & .1 & .5 & .3 \\ 0 & 0 & 0 & 1 \end{array} \right] \end{array}$$

$$6. \ P = \begin{array}{c} A \\ B \\ C \\ D \end{array} \begin{array}{cccc} A & B & C & D \\ \left[\begin{array}{cccc} 0 & 1 & 0 & 0 \\ 1 & 0 & 0 & 0 \\ .1 & .2 & .3 & .4 \\ .7 & .1 & .1 & .1 \end{array} \right] \end{array}$$

In Problems 7–10, identify the absorbing states for each transition diagram, and determine whether or not the diagram represents an absorbing Markov chain.

7.
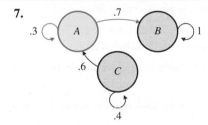

8.
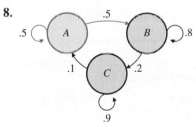

9.
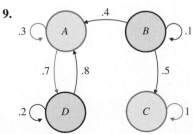

10.
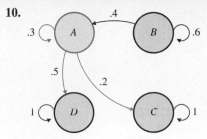

In Problems 11–20, could the given matrix be the transition matrix of an absorbing Markov chain?

$$11. \ \begin{bmatrix} 0 & 1 \\ 1 & 0 \end{bmatrix}$$

$$12. \ \begin{bmatrix} 1 & 0 \\ 0 & 1 \end{bmatrix}$$

$$13. \ \begin{bmatrix} .3 & .7 \\ 0 & 1 \end{bmatrix}$$

$$14. \ \begin{bmatrix} .6 & .4 \\ 1 & 0 \end{bmatrix}$$

$$15. \ \begin{bmatrix} 1 & 0 & 0 \\ 0 & 1 & 0 \\ 0 & 0 & 1 \end{bmatrix}$$

$$16. \ \begin{bmatrix} 0 & 1 & 0 \\ 0 & 0 & 1 \\ 1 & 0 & 0 \end{bmatrix}$$

$$17. \ \begin{bmatrix} .9 & .1 & 0 \\ .1 & .9 & 0 \\ 0 & 0 & 1 \end{bmatrix}$$

$$18. \ \begin{bmatrix} .5 & .5 & 0 \\ .4 & .3 & .3 \\ 0 & 0 & 1 \end{bmatrix}$$

$$19. \ \begin{bmatrix} .9 & 0 & .1 \\ 0 & 1 & 0 \\ 0 & .2 & .8 \end{bmatrix}$$

$$20. \ \begin{bmatrix} 1 & 0 & 0 \\ 0 & 0 & 1 \\ 0 & .7 & .3 \end{bmatrix}$$

In Problems 21–24, find a standard form for the absorbing Markov chain with the indicated transition diagram.

21.
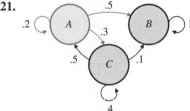

22.
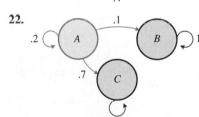

23.
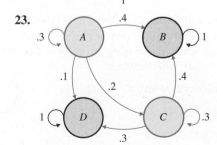

479

24.

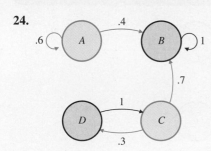

In Problems 25–28, find a standard form for the absorbing Markov chain with the indicated transition matrix.

25. $P = \begin{array}{c} \\ A \\ B \\ C \end{array} \begin{array}{ccc} A & B & C \\ \left[\begin{array}{ccc} .2 & .3 & .5 \\ 1 & 0 & 0 \\ 0 & 0 & 1 \end{array} \right] \end{array}$

26. $P = \begin{array}{c} \\ A \\ B \\ C \end{array} \begin{array}{ccc} A & B & C \\ \left[\begin{array}{ccc} 0 & 0 & 1 \\ 0 & 1 & 0 \\ .7 & .2 & .1 \end{array} \right] \end{array}$

27. $P = \begin{array}{c} \\ A \\ B \\ C \\ D \end{array} \begin{array}{cccc} A & B & C & D \\ \left[\begin{array}{cccc} .1 & .2 & .3 & .4 \\ 0 & 1 & 0 & 0 \\ .5 & .2 & .2 & .1 \\ 0 & 0 & 0 & 1 \end{array} \right] \end{array}$

28. $P = \begin{array}{c} \\ A \\ B \\ C \\ D \end{array} \begin{array}{cccc} A & B & C & D \\ \left[\begin{array}{cccc} 0 & .3 & .3 & .4 \\ 0 & 1 & 0 & 0 \\ 0 & 0 & 1 & 0 \\ .8 & .1 & .1 & 0 \end{array} \right] \end{array}$

In Problems 29–34, find the limiting matrix for the indicated standard form. Find the long-run probability of going from each nonabsorbing state to each absorbing state and the average number of trials needed to go from each nonabsorbing state to an absorbing state.

29. $P = \begin{array}{c} \\ A \\ B \\ C \end{array} \begin{array}{ccc} A & B & C \\ \left[\begin{array}{ccc} 1 & 0 & 0 \\ 0 & 1 & 0 \\ .1 & .4 & .5 \end{array} \right] \end{array}$

30. $P = \begin{array}{c} \\ A \\ B \\ C \end{array} \begin{array}{ccc} A & B & C \\ \left[\begin{array}{ccc} 1 & 0 & 0 \\ 0 & 1 & 0 \\ .3 & .2 & .5 \end{array} \right] \end{array}$

31. $P = \begin{array}{c} \\ A \\ B \\ C \end{array} \begin{array}{ccc} A & B & C \\ \left[\begin{array}{ccc} 1 & 0 & 0 \\ .2 & .6 & .2 \\ .4 & .2 & .4 \end{array} \right] \end{array}$

32. $P = \begin{array}{c} \\ A \\ B \\ C \end{array} \begin{array}{ccc} A & B & C \\ \left[\begin{array}{ccc} 1 & 0 & 0 \\ .1 & .6 & .3 \\ .2 & .2 & .6 \end{array} \right] \end{array}$

33. $P = \begin{array}{c} \\ A \\ B \\ C \\ D \end{array} \begin{array}{cccc} A & B & C & D \\ \left[\begin{array}{cccc} 1 & 0 & 0 & 0 \\ 0 & 1 & 0 & 0 \\ .1 & .2 & .6 & .1 \\ .2 & .2 & .3 & .3 \end{array} \right] \end{array}$

34. $P = \begin{array}{c} \\ A \\ B \\ C \\ D \end{array} \begin{array}{cccc} A & B & C & D \\ \left[\begin{array}{cccc} 1 & 0 & 0 & 0 \\ 0 & 1 & 0 & 0 \\ .1 & .1 & .7 & .1 \\ .3 & .1 & .4 & .2 \end{array} \right] \end{array}$

Problems 35–40 refer to the matrices in Problems 29–34. Use the limiting matrix $\overline{P}$ found for each transition matrix P in Problems 29–34 to determine the long-run behavior of the successive state matrices for the indicated initial-state matrices.

35. For matrix P from Problem 29 with

(A) $S_0 = \begin{bmatrix} 0 & 0 & 1 \end{bmatrix}$ (B) $S_0 = \begin{bmatrix} .2 & .5 & .3 \end{bmatrix}$

36. For matrix P from Problem 30 with

(A) $S_0 = \begin{bmatrix} 0 & 0 & 1 \end{bmatrix}$ (B) $S_0 = \begin{bmatrix} .2 & .5 & .3 \end{bmatrix}$

37. For matrix P from Problem 31 with

(A) $S_0 = \begin{bmatrix} 0 & 0 & 1 \end{bmatrix}$ (B) $S_0 = \begin{bmatrix} .2 & .5 & .3 \end{bmatrix}$

38. For matrix P from Problem 32 with

(A) $S_0 = \begin{bmatrix} 0 & 0 & 1 \end{bmatrix}$

(B) $S_0 = \begin{bmatrix} .2 & .5 & .3 \end{bmatrix}$

39. For matrix P from Problem 33 with

(A) $S_0 = \begin{bmatrix} 0 & 0 & 0 & 1 \end{bmatrix}$

(B) $S_0 = \begin{bmatrix} 0 & 0 & 1 & 0 \end{bmatrix}$

(C) $S_0 = \begin{bmatrix} 0 & 0 & .4 & .6 \end{bmatrix}$

(D) $S_0 = \begin{bmatrix} .1 & .2 & .3 & .4 \end{bmatrix}$

40. For matrix P from Problem 34 with

(A) $S_0 = \begin{bmatrix} 0 & 0 & 0 & 1 \end{bmatrix}$

(B) $S_0 = \begin{bmatrix} 0 & 0 & 1 & 0 \end{bmatrix}$

(C) $S_0 = \begin{bmatrix} 0 & 0 & .4 & .6 \end{bmatrix}$

(D) $S_0 = \begin{bmatrix} .1 & .2 & .3 & .4 \end{bmatrix}$

In Problems 41–48, discuss the validity of each statement. If the statement is always true, explain why. If not, give a counterexample.

41. If a Markov chain has an absorbing state, then it is an absorbing chain.

42. If a Markov chain has exactly two states and at least one absorbing state, then it is an absorbing chain.

43. If a Markov chain has exactly three states, one absorbing and two nonabsorbing, then it is an absorbing chain.

44. If a Markov chain has exactly three states, one nonabsorbing and two absorbing, then it is an absorbing chain.

45. If every state of a Markov chain is an absorbing state, then it is an absorbing chain.

46. If a Markov chain is absorbing, then it has a unique stationary matrix.

47. If a Markov chain is absorbing, then it is regular.

48. If a Markov chain is regular, then it is absorbing.

In Problems 49–52, use a graphing calculator to approximate the limiting matrix for the indicated standard form.

49. $P = \begin{array}{c} \\ A \\ B \\ C \\ D \end{array} \begin{array}{cccc} A & B & C & D \\ \left[\begin{array}{cccc} 1 & 0 & 0 & 0 \\ 0 & 1 & 0 & 0 \\ .5 & .3 & .1 & .1 \\ .6 & .2 & .1 & .1 \end{array} \right] \end{array}$

50. $P = \begin{array}{c} \\ A \\ B \\ C \\ D \end{array} \begin{array}{cccc} A & B & C & D \\ \left[\begin{array}{cccc} 1 & 0 & 0 & 0 \\ 0 & 1 & 0 & 0 \\ .1 & .1 & .5 & .3 \\ 0 & .2 & .3 & .5 \end{array} \right] \end{array}$

51. $P = \begin{array}{c} \\ A \\ B \\ C \\ D \\ E \end{array} \begin{array}{ccccc} A & B & C & D & E \\ \left[\begin{array}{ccccc} 1 & 0 & 0 & 0 & 0 \\ 0 & 1 & 0 & 0 & 0 \\ 0 & .4 & .5 & 0 & .1 \\ 0 & .4 & 0 & .3 & .3 \\ .4 & .4 & 0 & .2 & 0 \end{array} \right] \end{array}$

52. $P = $

$$\begin{array}{c} \\ A \\ B \\ C \\ D \\ E \end{array} \begin{array}{c} \begin{array}{ccccc} A & B & C & D & E \end{array} \\ \left[\begin{array}{ccccc} 1 & 0 & 0 & 0 & 0 \\ 0 & 1 & 0 & 0 & 0 \\ .5 & 0 & 0 & 0 & .5 \\ 0 & .4 & 0 & .2 & .4 \\ 0 & 0 & .1 & .7 & .2 \end{array}\right] \end{array}$$

53. The following matrix P is a nonstandard transition matrix for an absorbing Markov chain:

$$P = \begin{array}{c} \\ A \\ B \\ C \\ D \end{array} \begin{array}{c} \begin{array}{cccc} A & B & C & D \end{array} \\ \left[\begin{array}{cccc} .2 & .2 & .6 & 0 \\ 0 & 1 & 0 & 0 \\ .5 & .1 & 0 & .4 \\ 0 & 0 & 0 & 1 \end{array}\right] \end{array}$$

To find a limiting matrix for P, follow the steps outlined below.

Step 1 Using a transition diagram, rearrange the columns and rows of P to produce a standard form for this chain.

Step 2 Find the limiting matrix for this standard form.

Step 3 Using a transition diagram, reverse the process used in Step 1 to produce a limiting matrix for the original matrix P.

54. Repeat Problem 53 for

$$P = \begin{array}{c} \\ A \\ B \\ C \\ D \end{array} \begin{array}{c} \begin{array}{cccc} A & B & C & D \end{array} \\ \left[\begin{array}{cccc} 1 & 0 & 0 & 0 \\ .3 & .6 & 0 & .1 \\ .2 & .3 & .5 & 0 \\ 0 & 0 & 0 & 1 \end{array}\right] \end{array}$$

55. Verify the results in Problem 53 by computing P^k on a graphing calculator for large values of k.

56. Verify the results in Problem 54 by computing P^k on a graphing calculator for large values of k.

57. Show that $S = \begin{bmatrix} x & 1-x & 0 \end{bmatrix}, 0 \le x \le 1$, is a stationary matrix for the transition matrix

$$P = \begin{array}{c} \\ A \\ B \\ C \end{array} \begin{array}{c} \begin{array}{ccc} A & B & C \end{array} \\ \left[\begin{array}{ccc} 1 & 0 & 0 \\ 0 & 1 & 0 \\ .1 & .5 & .4 \end{array}\right] \end{array}$$

Discuss the generalization of this result to any absorbing Markov chain with two absorbing states and one nonabsorbing state.

58. Show that $S = \begin{bmatrix} x & 1-x & 0 & 0 \end{bmatrix}, 0 \le x \le 1$, is a stationary matrix for the transition matrix

$$P = \begin{array}{c} \\ A \\ B \\ C \\ D \end{array} \begin{array}{c} \begin{array}{cccc} A & B & C & D \end{array} \\ \left[\begin{array}{cccc} 1 & 0 & 0 & 0 \\ 0 & 1 & 0 & 0 \\ .1 & .2 & .3 & .4 \\ .6 & .2 & .1 & .1 \end{array}\right] \end{array}$$

Discuss the generalization of this result to any absorbing Markov chain with two absorbing states and two nonabsorbing states.

59. An absorbing Markov chain has the following matrix P as a standard form:

$$P = \begin{array}{c} \\ A \\ B \\ C \\ D \end{array} \begin{array}{c} \begin{array}{cccc} A & B & C & D \end{array} \\ \left[\begin{array}{cccc} 1 & 0 & 0 & 0 \\ .2 & .3 & .1 & .4 \\ 0 & .5 & .3 & .2 \\ 0 & .1 & .6 & .3 \end{array}\right] \end{array} \qquad \begin{bmatrix} I & 0 \\ \hline R & Q \end{bmatrix}$$

Let w_k denote the maximum entry in Q^k. Note that $w_1 = .6$

(A) Find w_2, w_4, w_8, w_{16}, and w_{32} to three decimal places.

(B) Describe Q^k when k is large.

60. Refer to the matrices P and Q of Problem 59. For k a positive integer, let $T_k = I + Q + Q^2 + \cdots + Q^k$.

(A) Explain why $T_{k+1} = T_k Q + I$.

(B) Using a graphing calculator and part (A) to quickly compute the matrices T_k, discover and describe the connection between $(I - Q)^{-1}$ and T_k when k is large.

Applications

61. Loans. A credit union classifies car loans into one of four categories: the loan has been paid in full (F), the account is in good standing (G) with all payments up to date, the account is in arrears (A) with one or more missing payments, or the account has been classified as a bad debt (B) and sold to a collection agency. Past records indicate that each month 10% of the accounts in good standing pay the loan in full, 80% remain in good standing, and 10% become in arrears. Furthermore, 10% of the accounts in arrears are paid in full, 40% become accounts in good standing, 40% remain in arrears, and 10% are classified as bad debts.

(A) In the long run, what percentage of the accounts in arrears will pay their loan in full?

(B) In the long run, what percentage of the accounts in good standing will become bad debts?

(C) What is the average number of months that an account in arrears will remain in this system before it is either paid in full or classified as a bad debt?

62. Employee training. A chain of car muffler and brake repair shops maintains a training program for its mechanics. All new mechanics begin training in muffler repairs. Every 3 months, the performance of each mechanic is reviewed. Past records indicate that after each quarterly review, 30% of the muffler repair trainees are rated as qualified to repair mufflers and begin training in brake repairs, 20% are terminated for unsatisfactory performance, and the remainder continue as muffler repair trainees. Also, 30% of the brake repair trainees are rated as fully qualified mechanics requiring no further training, 10% are terminated for unsatisfactory performance, and the remainder continue as brake repair trainees.

(A) In the long run, what percentage of muffler repair trainees will become fully qualified mechanics?

(B) In the long run, what percentage of brake repair trainees will be terminated?

(C) What is the average number of quarters that a muffler repair trainee will remain in the training program before being either terminated or promoted to fully qualified mechanic?

63. Marketing. Three electronics firms are aggressively marketing their graphing calculators to high school and college mathematics departments by offering volume discounts, complimentary display equipment, and assistance with curriculum development. Due to the amount of equipment involved and the necessary curriculum changes, once a department decides to use a particular calculator in their courses, they never switch to another brand or stop using calculators. Each year, 6% of the departments decide to use calculators from company A, 3% decide to use calculators from company B, 11% decide to use calculators from company C, and the remainder decide not to use any calculators in their courses.

(A) In the long run, what is the market share of each company?

(B) On average, how many years will it take a department to decide to use calculators from one of these companies in their courses?

64. Pensions. Once a year company employees are given the opportunity to join one of three pension plans: A, B, or C. Once an employee decides to join one of these plans, the employee cannot drop the plan or switch to another plan. Past records indicate that each year 4% of employees elect to join plan A, 14% elect to join plan B, 7% elect to join plan C, and the remainder do not join any plan.

(A) In the long run, what percentage of the employees will elect to join plan A? Plan B? Plan C?

(B) On average, how many years will it take an employee to decide to join a plan?

65. Medicine. After bypass surgery, patients are placed in an intensive care unit (ICU) until their condition stabilizes. Then they are transferred to a cardiac care ward (CCW), where they remain until they are released from the hospital. In a particular metropolitan area, a study of hospital records produced the following data: each day 2% of the patients in the ICU died, 52% were transferred to the CCW, and the remainder stayed in the ICU. Furthermore, each day 4% of the patients in the CCW developed complications and were returned to the ICU, 1% died while in the CCW, 22% were released from the hospital, and the remainder stayed in the CCW.

(A) In the long run, what percentage of the patients in the ICU are released from the hospital?

(B) In the long run, what percentage of the patients in the CCW die without ever being released from the hospital?

(C) What is the average number of days that a patient in the ICU will stay in the hospital?

66. Medicine. The study discussed in Problem 65 also produced the following data for patients who underwent aortic valve replacements: each day 2% of the patients in the ICU died, 60% were transferred to the CCW, and the remainder stayed

in the ICU. Furthermore, each day 5% of the patients in the CCW developed complications and were returned to the ICU, 1% died while in the CCW, 19% were released from the hospital, and the remainder stayed in the CCW.

(A) In the long run, what percentage of the patients in the CCW are released from the hospital?

(B) In the long run, what percentage of the patients in the ICU die without ever being released from the hospital?

(C) What is the average number of days a patient in the CCW will stay in the hospital?

67. Psychology. A rat is placed in room F or room B of the maze shown in the figure. The rat wanders from room to room until it enters one of the rooms containing food, L or R. Assume that the rat chooses an exit from a room at random and that once it enters a room with food it never leaves.

(A) What is the long-run probability that a rat placed in room B ends up in room R?

(B) What is the average number of exits that a rat placed in room B will choose until it finds food?

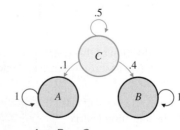

Figure for 67 and 68

68. Psychology. Repeat Problem 67 if the exit from room B to room R is blocked.

Answers to Matched Problems

1. (A) State B is absorbing.

   (B) State C is absorbing.

2. (A) Absorbing Markov chain

   (B) Not an absorbing Markov chain

3. (A)

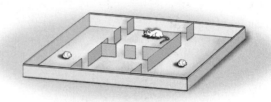

(B) $P = \begin{array}{c} \\ A \\ B \\ C \end{array} \begin{array}{ccc} A & B & C \\ \left[\begin{array}{ccc} 1 & 0 & 0 \\ 0 & 1 & 0 \\ .1 & .4 & .5 \end{array}\right] \end{array}$

(C) Company A will purchase 20% of the farms, and company B will purchase 80%.

(D) Company A will purchase 60% of the farms, and company B will purchase 40%.

4. (A) $\bar{P} = \begin{array}{c} \\ A \\ B \\ C \end{array}\begin{array}{ccc} A & B & C \\ \left[\begin{array}{ccc} 1 & 0 & 0 \\ 0 & 1 & 0 \\ .2 & .8 & 0 \end{array}\right] \end{array}$

   (B) $[.2 \quad .8 \quad 0]$

   (C) $[.6 \quad .4 \quad 0]$

5. (A) 30% of entry-level students will graduate; 50% of advanced-level students will not graduate.

   (B) An entry-level student will spend an average of 3.5 years in the program; an advanced-level student will spend an average of 2.5 years in the program.

Chapter 9 Summary and Review

Important Terms, Symbols, and Concepts

9.1 Properties of Markov Chains

EXAMPLES

- The progression of a system through a sequence of states is called a **stochastic process** if chance elements are involved in the transition from one state to the next.

- A **transition diagram** or **transition probability matrix** can be used to represent the probabilities of moving from one state to another. If those probabilities do not change with time, the stochastic process is called a **Markov chain**.

- If a Markov chain has n states, then the entry s_{ki} of the **kth-state matrix**

$$S_k = [s_{k1} \quad s_{k2} \cdots s_{kn}]$$

gives the probability of being in state i after the kth trial. The sum of the entries in S_k is 1.

- The entry $p_{i,j}$ of the $n \times n$ **transition matrix** P gives the probability of moving from state i to state j on the next trial. The sum of the entries in each row of P is 1.

Ex. 1, p. 451

- If S_0 is an initial-state matrix for a Markov chain, then $S_k = S_0 P^k$ (Theorem 1, page 452).

Ex. 2, p. 452
Ex. 3, p. 453
Ex. 4, p. 453

9.2 Regular Markov Chains

- A transition matrix P is **regular** if some power of P has only positive entries.

Ex. 1, p. 460

- A Markov chain is a **regular Markov chain** if its transition matrix is regular.

- A state matrix S is **stationary** if $SP = S$.

Ex. 2, p. 461

- The state matrices for a regular Markov chain approach a unique stationary matrix S (Theorem 1, page 461).

Ex. 3, p. 462

- If P is the transition matrix for a regular Markov chain, then the matrices P^k approach a **limiting matrix** $\bar{P}$, where each row of $\bar{P}$ is equal to the unique stationary matrix S (Theorem 1, page 461).

Ex. 4, p. 463
Ex. 5, p. 464

9.3 Absorbing Markov Chains

- A state in a Markov chain is an **absorbing state** if once the state is entered it is impossible to leave. A state is absorbing if and only if its row in the transition matrix has a 1 on the main diagonal and 0's elsewhere.

Ex. 1, p. 469

- A Markov chain is an **absorbing Markov chain** if there is at least one absorbing state and it is possible to go from each nonabsorbing state to at least one absorbing state in a finite number of steps.

Ex. 2, p. 471

- A transition matrix for an absorbing Markov chain is in **standard form** if the rows and columns are labeled so that all the absorbing states precede all the nonabsorbing states.

- If a standard form P for an absorbing Markov chain is partitioned as

Ex. 3, p. 472
Ex. 4, p. 474

$$P = \left[\begin{array}{c|c} I & 0 \\ \hline R & Q \end{array}\right]$$

then P^k approaches a limiting matrix $\bar{P}$ as k increases, where

$$\bar{P} = \left[\begin{array}{c|c} I & 0 \\ \hline FR & 0 \end{array}\right]$$

The matrix $F = (I - Q)^{-1}$, where I is the identity matrix of the same size as Q, is called the **fundamental matrix** for P (Theorem 2, page 474).

Ex. 5. p. 476

- The entry in row i and column j of $\overline{P}$ is the long-run probability of going from state i to state j. The sum of the entries in each row of F is the average number of trials that it will take to go from each nonabsorbing state to some absorbing state (Theorem 3, page 476).

Review Exercises

Work through all the problems in this chapter review and check your answers in the back of the book. Answers to all review problems are there along with section numbers in italics to indicate where each type of problem is discussed. Where weaknesses show up, review appropriate sections in the text.

1. Given the transition matrix P and initial-state matrix S_0 shown below, find S_1 and S_2 and explain what each represents:

$$P = \begin{array}{c} A \\ B \end{array}\begin{array}{c} \begin{array}{cc} A & B \end{array} \\ \begin{bmatrix} .6 & .4 \\ .2 & .8 \end{bmatrix} \end{array} \quad S_0 = [.3 \quad .7]$$

In Problems 2–6, P is a transition matrix for a Markov chain. Identify any absorbing states and classify the chain as regular, absorbing, or neither.

2. $P = \begin{array}{c} A \\ B \end{array}\begin{array}{c} \begin{array}{cc} A & B \end{array} \\ \begin{bmatrix} 1 & 0 \\ .7 & .3 \end{bmatrix} \end{array}$
 3. $P = \begin{array}{c} A \\ B \end{array}\begin{array}{c} \begin{array}{cc} A & B \end{array} \\ \begin{bmatrix} 0 & 1 \\ .7 & .3 \end{bmatrix} \end{array}$

4. $P = \begin{array}{c} A \\ B \end{array}\begin{array}{c} \begin{array}{cc} A & B \end{array} \\ \begin{bmatrix} 0 & 1 \\ 1 & 0 \end{bmatrix} \end{array}$
 5. $P = \begin{array}{c} A \\ B \\ C \end{array}\begin{array}{c} \begin{array}{ccc} A & B & C \end{array} \\ \begin{bmatrix} .8 & 0 & .2 \\ 0 & 1 & 0 \\ 0 & 0 & 1 \end{bmatrix} \end{array}$

6. $P = \begin{array}{c} A \\ B \\ C \\ D \end{array}\begin{array}{c} \begin{array}{cccc} A & B & C & D \end{array} \\ \begin{bmatrix} 1 & 0 & 0 & 0 \\ 0 & 1 & 0 & 0 \\ 0 & 0 & .3 & .7 \\ 0 & 0 & .6 & .4 \end{bmatrix} \end{array}$

In Problems 7–10, write a transition matrix for the transition diagram indicated, identify any absorbing states, and classify each Markov chain as regular, absorbing, or neither.

7.

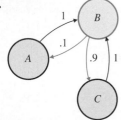

8.

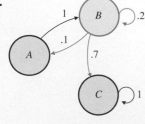

9.

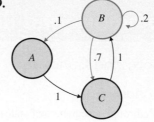

10.

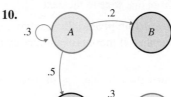

11. A Markov chain has three states, A, B, and C. The probability of going from state A to state B in one trial is .2, the probability of going from state A to state C in one trial is .5, the probability of going from state B to state A in one trial is .8, the probability of going from state B to state C in one trial is .2, the probability of going from state C to state A in one trial is .1, and the probability of going from state C to state B in one trial is .3. Draw a transition diagram and write a transition matrix for this chain.

12. Given the transition matrix

$$P = \begin{array}{c} A \\ B \end{array}\begin{array}{c} \begin{array}{cc} A & B \end{array} \\ \begin{bmatrix} .4 & .6 \\ .9 & .1 \end{bmatrix} \end{array}$$

find the probability of

(A) Going from state A to state B in two trials

(B) Going from state B to state A in three trials

In Problems 13 and 14, solve the equation $SP = S$ to find the stationary matrix S and the limiting matrix $\overline{P}$.

13. $P = \begin{array}{c} A \\ B \end{array}\begin{array}{c} \begin{array}{cc} A & B \end{array} \\ \begin{bmatrix} .4 & .6 \\ .2 & .8 \end{bmatrix} \end{array}$

14. $P = \begin{array}{c} A \\ B \\ C \end{array}\begin{array}{c} \begin{array}{ccc} A & B & C \end{array} \\ \begin{bmatrix} .4 & .6 & 0 \\ .5 & .3 & .2 \\ 0 & .8 & .2 \end{bmatrix} \end{array}$

In Problems 15 and 16, find the limiting matrix for the indicated standard form. Find the long-run probability of going from each nonabsorbing state to each absorbing state and the average number of trials needed to go from each nonabsorbing state to an absorbing state.

15. $P = \begin{array}{c} \\ A \\ B \\ C \end{array} \begin{array}{ccc} A & B & C \\ \left[\begin{array}{ccc} 1 & 0 & 0 \\ 0 & 1 & 0 \\ .3 & .1 & .6 \end{array}\right] \end{array}$ 16. $P = \begin{array}{c} \\ A \\ B \\ C \\ D \end{array} \begin{array}{cccc} A & B & C & D \\ \left[\begin{array}{cccc} 1 & 0 & 0 & 0 \\ 0 & 1 & 0 & 0 \\ .1 & .5 & .2 & .2 \\ .1 & .1 & .4 & .4 \end{array}\right] \end{array}$

In Problems 17–20, use a graphing calculator to approximate the limiting matrix for the indicated transition matrix.

17. Matrix P from Problem 13

18. Matrix P from Problem 14

19. Matrix P from Problem 15

20. Matrix P from Problem 16

21. Find a standard form for the absorbing Markov chain with transition matrix

$$P = \begin{array}{c} \\ A \\ B \\ C \\ D \end{array} \begin{array}{cccc} A & B & C & D \\ \left[\begin{array}{cccc} .6 & .1 & .2 & .1 \\ 0 & 1 & 0 & 0 \\ .3 & .2 & .3 & .2 \\ 0 & 0 & 0 & 1 \end{array}\right] \end{array}$$

In Problems 22 and 23, determine the long-run behavior of the successive state matrices for the indicated transition matrix and initial-state matrices.

22. $P = \begin{array}{c} \\ A \\ B \\ C \end{array} \begin{array}{ccc} A & B & C \\ \left[\begin{array}{ccc} 0 & 1 & 0 \\ 0 & 0 & 1 \\ .2 & .6 & .2 \end{array}\right] \end{array}$

   (A) $S_0 = [0 \quad 0 \quad 1]$

   (B) $S_0 = [.5 \quad .3 \quad .2]$

23. $P = \begin{array}{c} \\ A \\ B \\ C \end{array} \begin{array}{ccc} A & B & C \\ \left[\begin{array}{ccc} 1 & 0 & 0 \\ 0 & 1 & 0 \\ .2 & .6 & .2 \end{array}\right] \end{array}$

   (A) $S_0 = [0 \quad 0 \quad 1]$

   (B) $S_0 = [.5 \quad .3 \quad .2]$

24. Let P be a 2×2 transition matrix for a Markov chain. Can P be regular if two of its entries are 0? Explain.

25. Let P be a 3×3 transition matrix for a Markov chain. Can P be regular if three of its entries are 0? If four of its entries are 0? Explain.

26. A red urn contains 2 red marbles, 1 blue marble, and 1 green marble. A blue urn contains 1 red marble, 3 blue marbles, and 1 green marble. A green urn contains 6 red marbles, 3 blue marbles, and 1 green marble. A marble is selected from an urn, the color is noted, and the marble

is returned to the urn from which it was drawn. The next marble is drawn from the urn whose color is the same as the marble just drawn. Thus, this is a Markov process with three states: draw from the red urn, draw from the blue urn, or draw from the green urn.

   (A) Draw a transition diagram for this process.

   (B) Write the transition matrix P.

   (C) Determine whether this chain is regular, absorbing, or neither.

   (D) Find the limiting matrix $\overline{P}$, if it exists, and describe the long-run behavior of this process.

27. Repeat Problem 26 if the blue and green marbles are removed from the red urn.

28. Show that $S = [x \quad y \quad z \quad 0]$, where $0 \le x \le 1$, $0 \le y \le 1$, $0 \le z \le 1$, and $x + y + z = 1$, is a stationary matrix for the transition matrix

$$P = \begin{array}{c} \\ A \\ B \\ C \\ D \end{array} \begin{array}{cccc} A & B & C & D \\ \left[\begin{array}{cccc} 1 & 0 & 0 & 0 \\ 0 & 1 & 0 & 0 \\ 0 & 0 & 1 & 0 \\ .1 & .3 & .4 & .2 \end{array}\right] \end{array}$$

Discuss the generalization of this result to any absorbing chain with three absorbing states and one nonabsorbing state.

In Problems 29–35, either give an example of a Markov chain with the indicated properties or explain why no such chain can exist.

29. A regular Markov chain with an absorbing state

30. An absorbing Markov chain that is regular

31. A regular Markov chain with two different stationary matrices

32. An absorbing Markov chain with two different stationary matrices

33. A Markov chain with no limiting matrix

34. A regular Markov chain with no limiting matrix

35. An absorbing Markov chain with no limiting matrix

In Problems 36 and 37, use a graphing calculator to approximate the entries (to three decimal places) of the limiting matrix, if it exists, of the indicated transition matrix.

36. $P = \begin{array}{c} \\ A \\ B \\ C \\ D \end{array} \begin{array}{cccc} A & B & C & D \\ \left[\begin{array}{cccc} .2 & .3 & .1 & .4 \\ 0 & 0 & 1 & 0 \\ 0 & .8 & 0 & .2 \\ 0 & 0 & 1 & 0 \end{array}\right] \end{array}$

37. $P = \begin{array}{c} \\ A \\ B \\ C \\ D \end{array} \begin{array}{cccc} A & B & C & D \\ \left[\begin{array}{cccc} .1 & 0 & .3 & .6 \\ .2 & .4 & .1 & .3 \\ .3 & .5 & 0 & .2 \\ .9 & .1 & 0 & 0 \end{array}\right] \end{array}$

Applications

38. **Product switching.** A company's brand (X) has 20% of the market. A market research firm finds that if a person uses brand X, the probability is .7 that he or she will buy it next time. On the other hand, if a person does not use brand X (represented by X′), the probability is .5 that he or she will switch to brand X next time.

(A) Draw a transition diagram.

(B) Write a transition matrix.

(C) Write the initial-state matrix.

(D) Find the first-state matrix and explain what it represents.

(E) Find the stationary matrix.

(F) What percentage of the market will brand X have in the long run if the transition matrix does not change?

39. **Marketing.** Recent technological advances have led to the development of three new milling machines: brand A, brand B, and brand C. Due to the extensive retooling and startup costs, once a company converts its machine shop to one of these new machines, it never switches to another brand. Each year, 6% of the machine shops convert to brand A machines, 8% convert to brand B machines, 11% convert to brand C machines, and the remainder continue to use their old machines.

(A) In the long run, what is the market share of each brand?

(B) What is the average number of years that a company waits before converting to one of the new milling machines?

40. **Internet.** Table 1 gives the percentage of U.S. adults who at least occasionally used the Internet in the given year.

Table 1

Year	Percent
1995	14
2000	49
2005	68
2010	79

Source: Pew Internet & American Life Project Surveys

The following transition matrix P is proposed as a model for the data, where I represents the population of Internet users.

$$\begin{array}{cc} & \text{Five years later} \\ & \begin{array}{cc} I & I' \end{array} \\ \begin{array}{c} \text{Current} \\ \text{Year} \end{array} \begin{array}{c} I \\ I' \end{array} & \begin{bmatrix} .95 & .05 \\ .40 & .60 \end{bmatrix} = P \end{array}$$

(A) Let $S_0 = [.14 \quad .86]$ and find S_1, S_2, and S_3. Compute both matrices exactly and then round entries to two decimal places.

(B) Construct a new table comparing the results from part (A) with the data in Table 1.

(C) According to this transition matrix, what percentage of the adult U.S. population will use the Internet in the long run?

41. **Employee training.** In order to become a fellow of the Society of Actuaries, a person must pass a series of ten examinations. Passage of the first two preliminary exams is a prerequisite for employment as a trainee in the actuarial department of a large insurance company. Each year, 15% of the trainees complete the next three exams and become associates of the Society of Actuaries, 5% leave the company, never to return, and the remainder continue as trainees. Furthermore, each year, 17% of the associates complete the remaining five exams and become fellows of the Society of Actuaries, 3% leave the company, never to return, and the remainder continue as associates.

(A) In the long run, what percentage of the trainees will become fellows?

(B) In the long run, what percentage of the associates will leave the company?

(C) What is the average number of years that a trainee remains in this program before either becoming a fellow or being discharged?

42. **Genetics.** A given plant species has red, pink, or white flowers according to the genotypes RR, RW, and WW, respectively. If each of these genotypes is crossed with a red-flowering plant, the transition matrix is

$$\begin{array}{cc} & \text{Next generation} \\ & \begin{array}{ccc} Red & Pink & White \end{array} \\ \begin{array}{c} \text{This} \\ \text{generation} \end{array} \begin{array}{c} Red \\ Pink \\ White \end{array} & \begin{bmatrix} 1 & 0 & 0 \\ .5 & .5 & 0 \\ 0 & 1 & 0 \end{bmatrix} \end{array}$$

If each generation of the plant is crossed only with red plants to produce the next generation, show that eventually all the flowers produced by the plants will be red. (Find the limiting matrix.)

43. **Smoking.** Table 2 gives the percentage of U.S. adults who were smokers in the given year.

Table 2

Year	Percent
1985	30.1
1995	24.7
2005	20.9
2010	19.3

Source: American Lung Association

The following transition matrix P is proposed as a model for the data, where S represents the population of U.S. adult smokers.

$$\begin{array}{cc} & \text{Five years later} \\ & \begin{array}{cc} S & S' \end{array} \\ \begin{array}{c} \text{Current} \\ \text{Year} \end{array} \begin{array}{c} S \\ S' \end{array} & \begin{bmatrix} .74 & .26 \\ .03 & .97 \end{bmatrix} = P \end{array}$$

(A) Let $S_0 = [.301 \quad .699]$, and find S_1, S_2, and S_3. Compute the matrices exactly and then round entries to three decimal places.

(B) Construct a new table comparing the results from part (A) with the data in Table 2.

(C) According to this transition matrix, what percentage of the adult U.S. population will be smokers in the long run?

PART 3

CALCULUS

10 Limits and the Derivative

10.1 Introduction to Limits

10.2 Infinite Limits and Limits at Infinity

10.3 Continuity

10.4 The Derivative

10.5 Basic Differentiation Properties

10.6 Differentials

10.7 Marginal Analysis in Business and Economics

Chapter 10
Summary and Review

Review Exercises

Introduction

How do algebra and calculus differ? The two words *static* and *dynamic* probably come as close as any to expressing the difference between the two disciplines. In algebra, we solve equations for a particular value of a variable—a static notion. In calculus, we are interested in how a change in one variable affects another variable—a dynamic notion.

Isaac Newton (1642–1727) of England and Gottfried Wilhelm von Leibniz (1646–1716) of Germany developed calculus independently to solve problems concerning motion. Today calculus is used not just in the physical sciences, but also in business, economics, life sciences, and social sciences—any discipline that seeks to understand dynamic phenomena.

In Chapter 10 we introduce the *derivative*, one of the two key concepts of calculus. The second, the *integral*, is the subject of Chapter 13. Both key concepts depend on the notion of *limit*, which is explained in Sections 10.1 and 10.2. We consider many applications of limits and derivatives. See, for example, Problems 89 and 90 in Section 10.2 on the concentration of a drug in the bloodstream.

10.1 Introduction to Limits

- Functions and Graphs: Brief Review
- Limits: A Graphical Approach
- Limits: An Algebraic Approach
- Limits of Difference Quotients

Basic to the study of calculus is the concept of a *limit*. This concept helps us to describe, in a precise way, the behavior of $f(x)$ when x is close, but not equal, to a particular value c. In this section, we develop an intuitive and informal approach to evaluating limits.

Functions and Graphs: Brief Review

The graph of the function $y = f(x) = x + 2$ is the graph of the set of all ordered pairs $(x, f(x))$. For example, if $x = 2$, then $f(2) = 4$ and $(2, f(2)) = (2, 4)$ is a point on the graph of f. Figure 1 shows $(-1, f(-1))$, $(1, f(1))$, and $(2, f(2))$ plotted on the graph of f. Notice that the domain values -1, 1, and 2 are associated with the x axis and the range values $f(-1) = 1, f(1) = 3$, and $f(2) = 4$ are associated with the y axis.

Given x, it is sometimes useful to read $f(x)$ directly from the graph of f. Example 1 reviews this process.

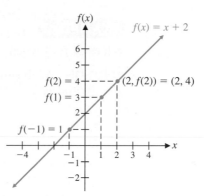

Figure 1

EXAMPLE 1 Finding Values of a Function from Its Graph Complete the following table, using the given graph of the function g.

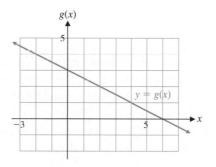

x	$g(x)$
-2	
1	
3	
4	

SOLUTION To determine $g(x)$, proceed vertically from the x value on the x axis to the graph of g and then horizontally to the corresponding y value $g(x)$ on the y axis (as indicated by the dashed lines).

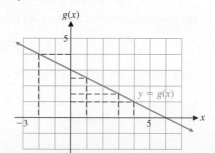

x	$g(x)$
-2	4.0
1	2.5
3	1.5
4	1.0

Matched Problem 1 Complete the following table, using the given graph of the function h.

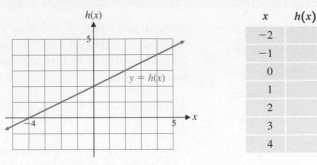

x	h(x)
−2	
−1	
0	
1	
2	
3	
4	

Limits: A Graphical Approach

We introduce the important concept of a *limit* through an example, which leads to an intuitive definition of the concept.

EXAMPLE 2 Analyzing a Limit   Let $f(x) = x + 2$. Discuss the behavior of the values of $f(x)$ when x is close to 2.

SOLUTION We begin by drawing a graph of f that includes the domain value $x = 2$ (Fig. 2).

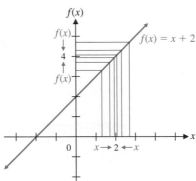

Figure 2

In Figure 2, we are using a static drawing to describe a dynamic process. This requires careful interpretation. The thin vertical lines in Figure 2 represent values of x that are close to 2. The corresponding horizontal lines identify the value of $f(x)$ associated with each value of x. [Example 1 dealt with the relationship between x and $f(x)$ on a graph.] The graph in Figure 2 indicates that as the values of x get closer and closer to 2 on either side of 2, the corresponding values of $f(x)$ get closer and closer to 4. Symbolically, we write

$$\lim_{x \to 2} f(x) = 4$$

This equation is read as "The limit of $f(x)$ as x approaches 2 is 4." Note that $f(2) = 4$. That is, the value of the function at 2 and the limit of the function as x approaches 2 are the same. This relationship can be expressed as

$$\lim_{x \to 2} f(x) = f(2) = 4$$

Graphically, this means that there is no hole or break in the graph of f at $x = 2$.

Matched Problem 2 Let $f(x) = x + 1$. Discuss the behavior of the values of $f(x)$ when x is close to 1.

We now present an informal definition of the important concept of a limit. A precise definition is not needed for our discussion, but one is given in a footnote.*

DEFINITION Limit

We write

$$\lim_{x \to c} f(x) = L \quad \text{or} \quad f(x) \to L \text{ as } x \to c$$

if the functional value $f(x)$ is close to the single real number L whenever x is close, but not equal, to c (on either side of c).

Note: The existence of a limit at c has nothing to do with the value of the function at c. In fact, c may not even be in the domain of f. However, the function must be defined on both sides of c.

The next example involves the **absolute value function**:

$$f(x) = |x| = \begin{cases} -x & \text{if } x < 0 \\ x & \text{if } x \geq 0 \end{cases} \quad \begin{array}{l} f(-2) = |-2| = -(-2) = 2 \\ f(3) = |3| = 3 \end{array}$$

The graph of f is shown in Figure 3.

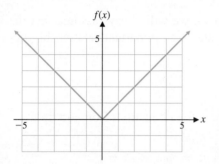

Figure 3 $f(x) = |x|$

EXAMPLE 3 Analyzing a Limit Let $h(x) = |x|/x$. Explore the behavior of $h(x)$ for x near, but not equal, to 0. Find $\lim_{x \to 0} h(x)$ if it exists.

SOLUTION The function h is defined for all real numbers except 0 [$h(0) = |0|/0$ is undefined]. For example,

$$h(-2) = \frac{|-2|}{-2} = \frac{2}{-2} = -1$$

Note that if x is any negative number, then $h(x) = -1$ (if $x < 0$, then the numerator $|x|$ is positive but the denominator x is negative, so $h(x) = |x|/x = -1$). If x is any positive number, then $h(x) = 1$ (if $x > 0$, then the numerator $|x|$ is equal to the denominator x, so $h(x) = |x|/x = 1$). Figure 4 illustrates the behavior of $h(x)$ for x near 0. Note that the absence of a solid dot on the vertical axis indicates that h is not defined when $x = 0$.

When x is near 0 (on either side of 0), is $h(x)$ near one specific number? The answer is "No," because $h(x)$ is -1 for $x < 0$ and 1 for $x > 0$. Consequently, we say that

$$\lim_{x \to 0} \frac{|x|}{x} \text{ does not exist}$$

Neither $h(x)$ nor the limit of $h(x)$ exists at $x = 0$. However, the limit from the left and the limit from the right both exist at 0, but they are not equal.

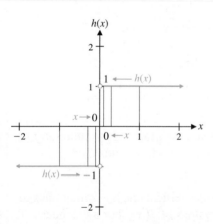

Figure 4

Matched Problem 3 Graph

$$h(x) = \frac{x - 2}{|x - 2|}$$

and find $\lim_{x \to 2} h(x)$ if it exists.

In Example 3, we see that the values of the function $h(x)$ approach two different numbers, depending on the direction of approach, and it is natural to refer to these values as "the limit from the left" and "the limit from the right." These experiences suggest that the notion of **one-sided limits** will be very useful in discussing basic limit concepts.

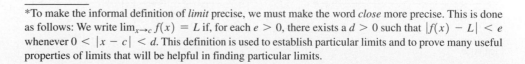

*To make the informal definition of *limit* precise, we must make the word *close* more precise. This is done as follows: We write $\lim_{x \to c} f(x) = L$ if, for each $e > 0$, there exists a $d > 0$ such that $|f(x) - L| < e$ whenever $0 < |x - c| < d$. This definition is used to establish particular limits and to prove many useful properties of limits that will be helpful in finding particular limits.

DEFINITION   One-Sided Limits

We write

$$\lim_{x \to c^-} f(x) = K$$ $x \to c^-$ is read "x approaches c from the left" and means $x \to c$ and $x < c$.

and call K the **limit from the left** or the **left-hand limit** if $f(x)$ is close to K whenever x is close to, but to the left of, c on the real number line. We write

$$\lim_{x \to c^+} f(x) = L$$ $x \to c^+$ is read "x approaches c from the right" and means $x \to c$ and $x > c$.

and call L the **limit from the right** or the **right-hand limit** if $f(x)$ is close to L whenever x is close to, but to the right of, c on the real number line.

If no direction is specified in a limit statement, we will always assume that the limit is **two-sided** or **unrestricted**. Theorem 1 states an important relationship between one-sided limits and unrestricted limits.

THEOREM 1   On the Existence of a Limit

For a (two-sided) limit to exist, the limit from the left and the limit from the right must exist and be equal. That is,

$$\lim_{x \to c} f(x) = L \text{ if and only if } \lim_{x \to c^-} f(x) = \lim_{x \to c^+} f(x) = L$$

In Example 3,

$$\lim_{x \to 0^-} \frac{|x|}{x} = -1 \qquad \text{and} \qquad \lim_{x \to 0^+} \frac{|x|}{x} = 1$$

Since the left- and right-hand limits are *not* the same,

$$\lim_{x \to 0} \frac{|x|}{x} \text{ does not exist}$$

EXAMPLE 4   Analyzing Limits Graphically   Given the graph of the function f in Figure 5, discuss the behavior of $f(x)$ for x near (A) -1, (B) 1, and (C) 2.

SOLUTION

(A) Since we have only a graph to work with, we use vertical and horizontal lines to relate the values of x and the corresponding values of $f(x)$. For any x near -1 on either side of -1, we see that the corresponding value of $f(x)$, determined by a horizontal line, is close to 1.

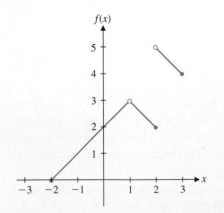

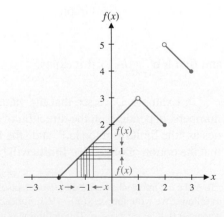

$$\lim_{x \to -1^-} f(x) = 1$$

$$\lim_{x \to -1^+} f(x) = 1$$

$$\lim_{x \to -1} f(x) = 1$$

$$f(-1) = 1$$

Figure 5

(B) Again, for any x near, but not equal to, 1, the vertical and horizontal lines indicate that the corresponding value of $f(x)$ is close to 3. The open dot at $(1, 3)$, together with the absence of a solid dot anywhere on the vertical line through $x = 1$, indicates that $f(1)$ is not defined.

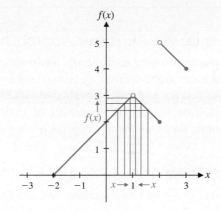

$$\lim_{x \to 1^-} f(x) = 3$$

$$\lim_{x \to 1^+} f(x) = 3$$

$$\lim_{x \to 1} f(x) = 3$$

$f(1)$ not defined

(C) The abrupt break in the graph at $x = 2$ indicates that the behavior of the graph near $x = 2$ is more complicated than in the two preceding cases. If x is close to 2 on the left side of 2, the corresponding horizontal line intersects the y axis at a point close to 2. If x is close to 2 on the right side of 2, the corresponding horizontal line intersects the y axis at a point close to 5. This is a case where the one-sided limits are different.

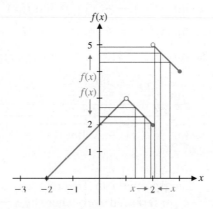

$$\lim_{x \to 2^-} f(x) = 2$$

$$\lim_{x \to 2^+} f(x) = 5$$

$$\lim_{x \to 2} f(x) \text{ does not exist}$$

$f(2) = 2$

Matched Problem 4 | Given the graph of the function f shown in Figure 6, discuss the following, as we did in Example 4:

(A) Behavior of $f(x)$ for x near 0

(B) Behavior of $f(x)$ for x near 1

(C) Behavior of $f(x)$ for x near 3

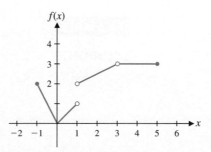

Figure 6

CONCEPTUAL INSIGHT

In Example 4B, note that $\lim\limits_{x \to 1} f(x)$ exists even though f is not defined at $x = 1$ and the graph has a hole at $x = 1$. In general, the value of a function at $x = c$ has no effect on the limit of the function as x approaches c.

Limits: An Algebraic Approach

Graphs are very useful tools for investigating limits, especially if something unusual happens at the point in question. However, many of the limits encountered in calculus are routine and can be evaluated quickly with a little algebraic simplification, some intuition, and basic properties of limits. The following list of properties of limits forms the basis for this approach:

THEOREM 2   Properties of Limits

Let f and g be two functions, and assume that

$$\lim_{x \to c} f(x) = L \qquad \lim_{x \to c} g(x) = M$$

where L and M are real numbers (both limits exist). Then

1. $\lim\limits_{x \to c} k = k$ for any constant k

2. $\lim\limits_{x \to c} x = c$

3. $\lim\limits_{x \to c} [f(x) + g(x)] = \lim\limits_{x \to c} f(x) + \lim\limits_{x \to c} g(x) = L + M$

4. $\lim\limits_{x \to c} [f(x) - g(x)] = \lim\limits_{x \to c} f(x) - \lim\limits_{x \to c} g(x) = L - M$

5. $\lim\limits_{x \to c} kf(x) = k \lim\limits_{x \to c} f(x) = kL$ for any constant k

6. $\lim\limits_{x \to c} [f(x) \cdot g(x)] = [\lim\limits_{x \to c} f(x)][\lim\limits_{x \to c} g(x)] = LM$

7. $\lim\limits_{x \to c} \dfrac{f(x)}{g(x)} = \dfrac{\lim\limits_{x \to c} f(x)}{\lim\limits_{x \to c} g(x)} = \dfrac{L}{M}$ if $M \neq 0$

8. $\lim\limits_{x \to c} \sqrt[n]{f(x)} = \sqrt[n]{\lim\limits_{x \to c} f(x)} = \sqrt[n]{L}$ $L > 0$ for n even

Each property in Theorem 2 is also valid if $x \to c$ is replaced everywhere by $x \to c^-$ or replaced everywhere by $x \to c^+$.

Explore and Discuss 1   The properties listed in Theorem 2 can be paraphrased in brief verbal statements. For example, property 3 simply states that *the limit of a sum is equal to the sum of the limits*. Write brief verbal statements for the remaining properties in Theorem 2.

EXAMPLE 5   Using Limit Properties   Find $\lim\limits_{x \to 3} (x^2 - 4x)$.

SOLUTION

$$\lim_{x \to 3} (x^2 - 4x) = \lim_{x \to 3} x^2 - \lim_{x \to 3} 4x \qquad \text{Property 4}$$

$$= \left(\lim_{x \to 3} x\right) \cdot \left(\lim_{x \to 3} x\right) - 4 \lim_{x \to 3} x \qquad \text{Properties 5 and 6}$$

$$= \left(\lim_{x \to 3} x\right)^2 - 4 \lim_{x \to 3} x \qquad \text{Definition of exponent}$$

$$= 3^2 - 4 \cdot 3 = -3$$

So, omitting the steps in the dashed boxes,

$$\lim_{x \to 3} (x^2 - 4x) = 3^2 - 4 \cdot 3 = -3$$

Matched Problem 5) Find $\lim_{x \to -2} (x^2 + 5x)$.

If $f(x) = x^2 - 4$ and c is any real number, then, just as in Example 5

$$\lim_{x \to c} f(x) = \lim_{x \to c} (x^2 - 4x) = c^2 - 4c = f(c)$$

So the limit can be found easily by evaluating the function f at c.

This simple method for finding limits is very useful, because there are many functions that satisfy the property

$$\lim_{x \to c} f(x) = f(c) \tag{1}$$

Any polynomial function

$$f(x) = a_n x^n + a_{n-1} x^{n-1} + \cdots + a_0$$

satisfies (1) for any real number c. Also, any rational function

$$r(x) = \frac{n(x)}{d(x)}$$

where $n(x)$ and $d(x)$ are polynomials, satisfies (1) provided c is a real number for which $d(c) \neq 0$.

THEOREM 3   Limits of Polynomial and Rational Functions

1. $\lim_{x \to c} f(x) = f(c)$ for f any polynomial function.

2. $\lim_{x \to c} r(x) = r(c)$ for r any rational function with a nonzero denominator at $x = c$.

If Theorem 3 is applicable, the limit is easy to find: *Simply evaluate the function at c.*

EXAMPLE 6   Evaluating Limits   Find each limit.

(A) $\lim_{x \to 2} (x^3 - 5x - 1)$ (B) $\lim_{x \to -1} \sqrt{2x^2 + 3}$ (C) $\lim_{x \to 4} \dfrac{2x}{3x + 1}$

SOLUTION

(A) $\lim_{x \to 2} (x^3 - 5x - 1) = 2^3 - 5 \cdot 2 - 1 = -3$ Theorem 3

(B) $\lim_{x \to -1} \sqrt{2x^2 + 3} = \sqrt{\lim_{x \to -1} (2x^2 + 3)}$ Property 8

$$= \sqrt{2(-1)^2 + 3} \qquad \text{Theorem 3}$$

$$= \sqrt{5}$$

(C) $\lim_{x \to 4} \dfrac{2x}{3x + 1} = \dfrac{2 \cdot 4}{3 \cdot 4 + 1}$ Theorem 3

$$= \frac{8}{13}$$

Matched Problem 6) Find each limit.

(A) $\lim_{x \to -1} (x^4 - 2x + 3)$ (B) $\lim_{x \to 2} \sqrt{3x^2 - 6}$ (C) $\lim_{x \to -2} \dfrac{x^2}{x^2 + 1}$

EXAMPLE 7 Evaluating Limits Let

$$f(x) = \begin{cases} x^2 + 1 & \text{if } x < 2 \\ x - 1 & \text{if } x > 2 \end{cases}$$

Find:

(A) $\lim\limits_{x \to 2^-} f(x)$ (B) $\lim\limits_{x \to 2^+} f(x)$ (C) $\lim\limits_{x \to 2} f(x)$ (D) $f(2)$

SOLUTION

(A) $\lim\limits_{x \to 2^-} f(x) = \lim\limits_{x \to 2^-} (x^2 + 1)$ If $x < 2$, $f(x) = x^2 + 1$.

$= 2^2 + 1 = 5$

(B) $\lim\limits_{x \to 2^+} f(x) = \lim\limits_{x \to 2^+} (x - 1)$ If $x > 2$, $f(x) = x - 1$.

$= 2 - 1 = 1$

(C) Since the one-sided limits are not equal, $\lim\limits_{x \to 2} f(x)$ does not exist.

(D) Because the definition of f does not assign a value to f for $x = 2$, only for $x < 2$ and $x > 2$, $f(2)$ does not exist.

Matched Problem 7 Let

$$f(x) = \begin{cases} 2x + 3 & \text{if } x < 5 \\ -x + 12 & \text{if } x > 5 \end{cases}$$

Find:

(A) $\lim\limits_{x \to 5^-} f(x)$ (B) $\lim\limits_{x \to 5^+} f(x)$ (C) $\lim\limits_{x \to 5} f(x)$ (D) $f(5)$

It is important to note that there are restrictions on some of the limit properties. In particular, if

$$\lim\limits_{x \to c} f(x) = 0 \quad \text{and} \quad \lim\limits_{x \to c} g(x) = 0, \quad \text{then finding} \quad \lim\limits_{x \to c} \frac{f(x)}{g(x)}$$

may present some difficulties, since limit property 7 (the limit of a quotient) does not apply when $\lim\limits_{x \to c} g(x) = 0$. The next example illustrates some techniques that can be useful in this situation.

EXAMPLE 8 Evaluating Limits Find each limit.

(A) $\lim\limits_{x \to 2} \dfrac{x^2 - 4}{x - 2}$ (B) $\lim\limits_{x \to -1} \dfrac{x|x + 1|}{x + 1}$

SOLUTION

(A) Note that $\lim\limits_{x \to 2} x^2 - 4 = 2^2 - 4 = 0$ and $\lim\limits_{x \to 2} x - 2 = 2 - 2 = 0$. Algebraic simplification is often useful in such a case when the numerator and denominator both have limit 0.

$$\lim\limits_{x \to 2} \frac{x^2 - 4}{x - 2} = \lim\limits_{x \to 2} \frac{(x - 2)(x + 2)}{x - 2} = \lim\limits_{x \to 2} (x + 2) = 4$$

(B) One-sided limits are helpful for limits involving the absolute value function.

$$\lim\limits_{x \to -1^+} \frac{x|x + 1|}{x + 1} = \lim\limits_{x \to -1^+} (x) = -1 \quad \text{If } x > -1, \text{ then } \frac{|x + 1|}{x + 1} = 1.$$

$$\lim\limits_{x \to -1^-} \frac{x|x + 1|}{x + 1} = \lim\limits_{x \to -1^-} (-x) = 1 \quad \text{If } x < -1, \text{ then } \frac{|x + 1|}{x + 1} = -1.$$

Since the limit from the left and the limit from the right are not the same, we conclude that

$$\lim\limits_{x \to -1} \frac{x|x + 1|}{x + 1} \quad \text{does not exist}$$

Matched Problem 8 Find each limit.

(A) $\displaystyle\lim_{x \to -3} \frac{x^2 + 4x + 3}{x + 3}$

(B) $\displaystyle\lim_{x \to 4} \frac{x^2 - 16}{|x - 4|}$

CONCEPTUAL INSIGHT

In the solution to Example 8A we used the following algebraic identity:

$$\frac{x^2 - 4}{x - 2} = \frac{(x - 2)(x + 2)}{x - 2} = x + 2, \quad x \neq 2$$

The restriction $x \neq 2$ is necessary here because the first two expressions are not defined at $x = 2$. Why didn't we include this restriction in the solution? When x approaches 2 in a limit problem, it is assumed that x is close, but not equal, to 2. It is important that you understand that both of the following statements are valid:

$$\lim_{x \to 2} \frac{x^2 - 4}{x - 2} = \lim_{x \to 2} (x + 2) \quad \text{and} \quad \frac{x^2 - 4}{x - 2} = x + 2, \quad x \neq 2$$

Limits like those in Example 8 occur so frequently in calculus that they are given a special name.

DEFINITION Indeterminate Form

If $\displaystyle\lim_{x \to c} f(x) = 0$ and $\displaystyle\lim_{x \to c} g(x) = 0$, then $\displaystyle\lim_{x \to c} \frac{f(x)}{g(x)}$ is said to be **indeterminate**, or, more specifically, a **0/0 indeterminate form**.

The term *indeterminate* is used because the limit of an indeterminate form may or may not exist (see Example 8A and 8B).

⚠ **CAUTION** The expression $0/0$ does not represent a real number and should never be used as the value of a limit. If a limit is a $0/0$ indeterminate form, further investigation is always required to determine whether the limit exists and to find its value if it does exist.   ▲

If the denominator of a quotient approaches 0 and the numerator approaches a nonzero number, then the limit of the quotient is not an indeterminate form. In fact, in this case the limit of the quotient does not exist.

THEOREM 4 Limit of a Quotient

If $\displaystyle\lim_{x \to c} f(x) = L, L \neq 0$, and $\displaystyle\lim_{x \to c} g(x) = 0$,

then

$$\lim_{x \to c} \frac{f(x)}{g(x)} \qquad \text{does not exist}$$

EXAMPLE 9 Indeterminate Forms Is the limit expression a $0/0$ indeterminate form? Find the limit or explain why the limit does not exist.

(A) $\displaystyle\lim_{x \to 1} \frac{x - 1}{x^2 + 1}$

(B) $\displaystyle\lim_{x \to 1} \frac{x - 1}{x^2 - 1}$

(C) $\displaystyle\lim_{x \to 1} \frac{x + 1}{x^2 - 1}$

SOLUTION

(A) $\lim\limits_{x \to 1} (x - 1) = 0$ but $\lim\limits_{x \to 1} (x^2 + 1) = 2$. So no, the limit expression is not a $0/0$ indeterminate form. By property 7 of Theorem 2,

$$\cdot \lim_{x \to 1} \frac{x - 1}{x^2 + 1} = \frac{0}{2} = 0$$

(B) $\lim\limits_{x \to 1} (x - 1) = 0$ and $\lim\limits_{x \to 1} (x^2 - 1) = 0$. So yes, the limit expression is a $0/0$ indeterminate form. We factor $x^2 - 1$ to simplify the limit expression and find the limit:

$$\lim_{x \to 1} \frac{x - 1}{x^2 - 1} = \lim_{x \to 1} \frac{x - 1}{(x - 1)(x + 1)} = \lim_{x \to 1} \frac{1}{x + 1} = \frac{1}{2}$$

(C) $\lim\limits_{x \to 1} (x + 1) = 2$ and $\lim\limits_{x \to 1} (x^2 - 1) = 0$. So no, the limit expression is not a $0/0$ indeterminate form. By Theorem 4,

$$\lim_{x \to 1} \frac{x + 1}{x^2 - 1} \quad \text{does not exist}$$

Matched Problem 9 Is the limit expression a $0/0$ indeterminate form? Find the limit or explain why the limit does not exist.

(A) $\lim\limits_{x \to 3} \dfrac{x + 1}{x + 3}$ (B) $\lim\limits_{x \to 3} \dfrac{x - 3}{x^2 + 9}$ (C) $\lim\limits_{x \to 3} \dfrac{x^2 - 9}{x - 3}$

Limits of Difference Quotients

Let the function f be defined in an open interval containing the number a. One of the most important limits in calculus is the limit of the **difference quotient**,

$$\lim_{h \to 0} \frac{f(a + h) - f(a)}{h} \tag{2}$$

If

$$\lim_{h \to 0} [f(a + h) - f(a)] = 0$$

as it often does, then limit (2) is an indeterminate form.

EXAMPLE 10 Limit of a Difference Quotient Find the following limit for $f(x) = 4x - 5$:

$$\lim_{h \to 0} \frac{f(3 + h) - f(3)}{h}$$

SOLUTION

$$\lim_{h \to 0} \frac{f(3 + h) - f(3)}{h} = \lim_{h \to 0} \frac{[4(\mathbf{3 + h}) - 5] - [4(\mathbf{3}) - 5]}{h}$$

Since this is a 0/0 indeterminate form and property 7 in Theorem 2 does not apply, we proceed with algebraic simplification.

$$= \lim_{h \to 0} \frac{12 + 4h - 5 - 12 + 5}{h}$$

$$= \lim_{h \to 0} \frac{4h}{h} = \lim_{h \to 0} 4 = 4$$

Matched Problem 10 Find the following limit for $f(x) = 7 - 2x$:

$$\lim_{h \to 0} \frac{f(4 + h) - f(4)}{h}.$$

Explore and Discuss 2   If $f(x) = \dfrac{1}{x}$, explain why $\displaystyle\lim_{h \to 0} \dfrac{f(3+h) - f(3)}{h} = -\dfrac{1}{9}$.

Exercises 10.1

Skills Warm-up Exercises

W   In Problems 1–8, factor each polynomial into the product of first-degree factors with integer coefficients. (If necessary, review Section A.3).

1. $x^2 - 81$

2. $x^2 - 64$

3. $x^2 - 4x - 21$

4. $x^2 + 5x - 36$

5. $x^3 - 7x^2 + 12x$

6. $x^3 + 15x^2 + 50x$

7. $6x^2 - x - 1$

8. $20x^2 + 11x - 3$

In Problems 9–16, use the graph of the function f shown to estimate the indicated limits and function values.

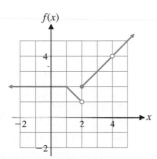

Figure for 9–16

9. $f(-0.5)$

10. $f(-1.5)$

11. $f(1.75)$

12. $f(1.25)$

13. (A) $\displaystyle\lim_{x \to 0^-} f(x)$ 　　(B) $\displaystyle\lim_{x \to 0^+} f(x)$

　　(C) $\displaystyle\lim_{x \to 0} f(x)$ 　　(D) $f(0)$

14. (A) $\displaystyle\lim_{x \to 1^-} f(x)$ 　　(B) $\displaystyle\lim_{x \to 1^+} f(x)$

　　(C) $\displaystyle\lim_{x \to 1} f(x)$ 　　(D) $f(1)$

15. (A) $\displaystyle\lim_{x \to 2^-} f(x)$ 　　(B) $\displaystyle\lim_{x \to 2^+} f(x)$

　　(C) $\displaystyle\lim_{x \to 2} f(x)$ 　　(D) $f(2)$

　　(E) Is it possible to redefine $f(2)$ so that $\displaystyle\lim_{x \to 2} f(x) = f(2)$? Explain.

16. (A) $\displaystyle\lim_{x \to 4^-} f(x)$ 　　(B) $\displaystyle\lim_{x \to 4^+} f(x)$

　　(C) $\displaystyle\lim_{x \to 4} f(x)$ 　　(D) $f(4)$

　　(E) Is it possible to define $f(4)$ so that $\displaystyle\lim_{x \to 4} f(x) = f(4)$? Explain.

In Problems 17–24, use the graph of the function g shown to estimate the indicated limits and function values.

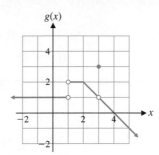

Figure for 17–24

17. $g(1.9)$

18. $g(0.1)$

19. $g(3.5)$

20. $g(2.5)$

21. (A) $\displaystyle\lim_{x \to 1^-} g(x)$ 　　(B) $\displaystyle\lim_{x \to 1^+} g(x)$

　　(C) $\displaystyle\lim_{x \to 1} g(x)$ 　　(D) $g(1)$

　　(E) Is it possible to define $g(1)$ so that $\displaystyle\lim_{x \to 1} g(x) = g(1)$? Explain.

22. (A) $\displaystyle\lim_{x \to 2^-} g(x)$ 　　(B) $\displaystyle\lim_{x \to 2^+} g(x)$

　　(C) $\displaystyle\lim_{x \to 2} g(x)$ 　　(D) $g(2)$

23. (A) $\displaystyle\lim_{x \to 3^-} g(x)$ 　　(B) $\displaystyle\lim_{x \to 3^+} g(x)$

　　(C) $\displaystyle\lim_{x \to 3} g(x)$ 　　(D) $g(3)$

　　(E) Is it possible to redefine $g(3)$ so that $\displaystyle\lim_{x \to 3} g(x) = g(3)$? Explain.

24. (A) $\displaystyle\lim_{x \to 4^-} g(x)$ 　　(B) $\displaystyle\lim_{x \to 4^+} g(x)$

　　(C) $\displaystyle\lim_{x \to 4} g(x)$ 　　(D) $g(4)$

In Problems 25–28, use the graph of the function f shown to estimate the indicated limits and function values.

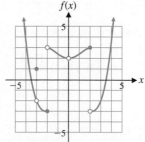

Figure for 25–28

25. (A) $\lim_{x \to -3^+} f(x)$ (B) $\lim_{x \to -3^-} f(x)$

    (C) $\lim_{x \to -3} f(x)$ (D) $f(-3)$

✎ (E) Is it possible to redefine $f(-3)$ so that
      $\lim_{x \to -3} f(x) = f(-3)$? Explain.

26. (A) $\lim_{x \to -2^+} f(x)$ (B) $\lim_{x \to -2^-} f(x)$

    (C) $\lim_{x \to -2} f(x)$ (D) $f(-2)$

✎ (E) Is it possible to redefine $f(-2)$ so that
      $\lim_{x \to -2} f(x) = f(-2)$? Explain.

27. (A) $\lim_{x \to 0^+} f(x)$ (B) $\lim_{x \to 0^-} f(x)$

    (C) $\lim_{x \to 0} f(x)$ (D) $f(0)$

✎ (E) Is it possible to define $f(0)$ so that $\lim_{x \to 0} f(x) = f(0)$?
      Explain.

28. (A) $\lim_{x \to 2^+} f(x)$ (B) $\lim_{x \to 2^-} f(x)$

    (C) $\lim_{x \to 2} f(x)$ (D) $f(2)$

✎ (E) Is it possible to redefine $f(2)$ so that $\lim_{x \to 2} f(x) = f(2)$?
      Explain.

In Problems 29–38, find each limit if it exists.

29. $\lim_{x \to 3} 4x$ **30.** $\lim_{x \to -2} 3x$

31. $\lim_{x \to -4} (x + 5)$ **32.** $\lim_{x \to 5} (x - 3)$

33. $\lim_{x \to 2} x(x - 4)$ **34.** $\lim_{x \to -1} x(x + 3)$

35. $\lim_{x \to -3} \dfrac{x}{x + 5}$ **36.** $\lim_{x \to 4} \dfrac{x - 2}{x}$

37. $\lim_{x \to 1} \sqrt{5x + 4}$ **38.** $\lim_{x \to 0} \sqrt{16 - 7x}$

Given that $\lim_{x \to 1} f(x) = -5$ and $\lim_{x \to 1} g(x) = 4$, find the indicated limits in Problems 39–46.

39. $\lim_{x \to 1} (-3)f(x)$ **40.** $\lim_{x \to 1} 2g(x)$

41. $\lim_{x \to 1} [2f(x) + g(x)]$ **42.** $\lim_{x \to 1} [g(x) - 3f(x)]$

43. $\lim_{x \to 1} \dfrac{2 - f(x)}{x + g(x)}$ **44.** $\lim_{x \to 1} \dfrac{3 - f(x)}{1 - 4g(x)}$

45. $\lim_{x \to 1} \sqrt{g(x) - f(x)}$ **46.** $\lim_{x \to 1} \sqrt[3]{2x + 2f(x)}$

In Problems 47–50, sketch a possible graph of a function that satisfies the given conditions.

47. $f(0) = 1$; $\lim_{x \to 0^-} f(x) = 3$; $\lim_{x \to 0^+} f(x) = 1$

48. $f(1) = -2$; $\lim_{x \to 1^-} f(x) = 2$; $\lim_{x \to 1^+} f(x) = -2$

49. $f(-2) = 2$; $\lim_{x \to -2^-} f(x) = 1$; $\lim_{x \to -2^+} f(x) = 1$

50. $f(0) = -1$; $\lim_{x \to 0^-} f(x) = 2$; $\lim_{x \to 0^+} f(x) = 2$

In Problems 51–66, find each indicated quantity if it exists.

51. Let $f(x) = \begin{cases} 1 - x^2 & \text{if } x \le 0 \\ 1 + x^2 & \text{if } x > 0 \end{cases}$. Find

    (A) $\lim_{x \to 0^+} f(x)$ (B) $\lim_{x \to 0^-} f(x)$

    (C) $\lim_{x \to 0} f(x)$ (D) $f(0)$

52. Let $f(x) = \begin{cases} 2 + x & \text{if } x \le 0 \\ 2 - x & \text{if } x > 0 \end{cases}$. Find

    (A) $\lim_{x \to 0^+} f(x)$ (B) $\lim_{x \to 0^-} f(x)$

    (C) $\lim_{x \to 0} f(x)$ (D) $f(0)$

53. Let $f(x) = \begin{cases} x^2 & \text{if } x < 1 \\ 2x & \text{if } x > 1 \end{cases}$. Find

    (A) $\lim_{x \to 1^+} f(x)$ (B) $\lim_{x \to 1^-} f(x)$

    (C) $\lim_{x \to 1} f(x)$ (D) $f(1)$

54. Let $f(x) = \begin{cases} x + 3 & \text{if } x < -2 \\ \sqrt{x + 2} & \text{if } x > -2 \end{cases}$. Find

    (A) $\lim_{x \to -2^+} f(x)$ (B) $\lim_{x \to -2^-} f(x)$

    (C) $\lim_{x \to -2} f(x)$ (D) $f(-2)$

55. Let $f(x) = \begin{cases} \dfrac{x^2 - 9}{x + 3} & \text{if } x < 0 \\ \dfrac{x^2 - 9}{x - 3} & \text{if } x > 0 \end{cases}$. Find

    (A) $\lim_{x \to -3} f(x)$ (B) $\lim_{x \to 0} f(x)$

    (C) $\lim_{x \to 3} f(x)$

56. Let $f(x) = \begin{cases} \dfrac{x}{x + 3} & \text{if } x < 0 \\ \dfrac{x}{x - 3} & \text{if } x > 0 \end{cases}$. Find

    (A) $\lim_{x \to -3} f(x)$ (B) $\lim_{x \to 0} f(x)$

    (C) $\lim_{x \to 3} f(x)$

57. Let $f(x) = \dfrac{|x - 1|}{x - 1}$. Find

    (A) $\lim_{x \to 1^+} f(x)$ (B) $\lim_{x \to 1^-} f(x)$

    (C) $\lim_{x \to 1} f(x)$ (D) $f(1)$

58. Let $f(x) = \dfrac{x - 3}{|x - 3|}$. Find

    (A) $\lim_{x \to 3^+} f(x)$ (B) $\lim_{x \to 3^-} f(x)$

    (C) $\lim_{x \to 3} f(x)$ (D) $f(3)$

59. Let $f(x) = \dfrac{x - 2}{x^2 - 2x}$. Find

    (A) $\lim_{x \to 0} f(x)$ (B) $\lim_{x \to 2} f(x)$

    (C) $\lim_{x \to 4} f(x)$

60. Let $f(x) = \dfrac{x + 3}{x^2 + 3x}$. Find

    (A) $\lim_{x \to -3} f(x)$ (B) $\lim_{x \to 0} f(x)$

    (C) $\lim_{x \to 3} f(x)$

61. Let $f(x) = \dfrac{x^2 - x - 6}{x + 2}$. Find

    (A) $\lim_{x \to -2} f(x)$ (B) $\lim_{x \to 0} f(x)$

    (C) $\lim_{x \to 3} f(x)$

62. Let $f(x) = \dfrac{x^2 + x - 6}{x + 3}$. Find

    (A) $\lim_{x \to -3} f(x)$ (B) $\lim_{x \to 0} f(x)$

    (C) $\lim_{x \to 2} f(x)$

63. Let $f(x) = \dfrac{(x + 2)^2}{x^2 - 4}$. Find

   (A) $\lim\limits_{x \to -2} f(x)$ (B) $\lim\limits_{x \to 0} f(x)$

   (C) $\lim\limits_{x \to 2} f(x)$

64. Let $f(x) = \dfrac{x^2 - 1}{(x + 1)^2}$. Find

   (A) $\lim\limits_{x \to -1} f(x)$ (B) $\lim\limits_{x \to 0} f(x)$

   (C) $\lim\limits_{x \to 1} f(x)$

65. Let $f(x) = \dfrac{2x^2 - 3x - 2}{x^2 + x - 6}$. Find

   (A) $\lim\limits_{x \to 2} f(x)$ (B) $\lim\limits_{x \to 0} f(x)$

   (C) $\lim\limits_{x \to 1} f(x)$

66. Let $f(x) = \dfrac{3x^2 + 2x - 1}{x^2 + 3x + 2}$. Find

   (A) $\lim\limits_{x \to -3} f(x)$ (B) $\lim\limits_{x \to -1} f(x)$

   (C) $\lim\limits_{x \to 2} f(x)$

✎ *In Problems 67–72, discuss the validity of each statement. If the statement is always true, explain why. If not, give a counterexample.*

67. If $\lim\limits_{x \to 1} f(x) = 0$ and $\lim\limits_{x \to 1} g(x) = 0$, then $\lim\limits_{x \to 1} \dfrac{f(x)}{g(x)} = 0$.

68. If $\lim\limits_{x \to 1} f(x) = 1$ and $\lim\limits_{x \to 1} g(x) = 1$, then $\lim\limits_{x \to 1} \dfrac{f(x)}{g(x)} = 1$.

69. If f is a polynomial, then, as x approaches 0, the right-hand limit exists and is equal to the left-hand limit.

70. If f is a rational function, then, as x approaches 0, the right-hand limit exists and is equal to the left-hand limit.

71. If f is a function such that $\lim\limits_{x \to 0} f(x)$ exists, then $f(0)$ exists.

72. If f is a function such that $f(0)$ exists, then $\lim\limits_{x \to 0} f(x)$ exists.

In Problems 73–80, is the limit expression a 0/0 indeterminate form? Find the limit or explain why the limit does not exist.

73. $\lim\limits_{x \to 7} \dfrac{(x - 7)^2}{x^2 - 4x - 21}$ 74. $\lim\limits_{x \to 2} \dfrac{x - 5}{x + 2}$

75. $\lim\limits_{x \to 4} \dfrac{x^2 + 4}{(x + 4)^2}$ 76. $\lim\limits_{x \to 9} \dfrac{x^2 - 5x - 36}{x - 9}$

77. $\lim\limits_{x \to -6} \dfrac{x^2 + 36}{x + 6}$ 78. $\lim\limits_{x \to 10} \dfrac{x^2 - 15x + 50}{(x - 10)^2}$

79. $\lim\limits_{x \to 8} \dfrac{x - 8}{x^2 - 64}$ 80. $\lim\limits_{x \to -3} \dfrac{x + 3}{x - 3}$

Compute the following limit for each function in Problems 81–88.

$$\lim_{h \to 0} \dfrac{f(2 + h) - f(2)}{h}$$

81. $f(x) = 3x + 1$ 82. $f(x) = 5x - 1$

83. $f(x) = x^2 + 1$ 84. $f(x) = x^2 - 2$

85. $f(x) = -7x + 9$ 86. $f(x) = -4x + 13$

87. $f(x) = |x + 1|$ 88. $f(x) = -3|x|$

89. Let f be defined by

$$f(x) = \begin{cases} 1 + mx & \text{if } x \le 1 \\ 4 - mx & \text{if } x > 1 \end{cases}$$

where m is a constant.

   (A) Graph f for $m = 1$, and find

   $$\lim\limits_{x \to 1^-} f(x) \qquad \text{and} \qquad \lim\limits_{x \to 1^+} f(x)$$

   (B) Graph f for $m = 2$, and find

   $$\lim\limits_{x \to 1^-} f(x) \qquad \text{and} \qquad \lim\limits_{x \to 1^+} f(x)$$

   (C) Find m so that

   $$\lim\limits_{x \to 1^-} f(x) = \lim\limits_{x \to 1^+} f(x)$$

   and graph f for this value of m.

✎ (D) Write a brief verbal description of each graph. How does the graph in part (C) differ from the graphs in parts (A) and (B)?

90. Let f be defined by

$$f(x) = \begin{cases} -3m + 0.5x & \text{if } x \le 2 \\ 3m - x & \text{if } x > 2 \end{cases}$$

where m is a constant.

   (A) Graph f for $m = 0$, and find

   $$\lim\limits_{x \to 2^-} f(x) \qquad \text{and} \qquad \lim\limits_{x \to 2^+} f(x)$$

   (B) Graph f for $m = 1$, and find

   $$\lim\limits_{x \to 2^-} f(x) \qquad \text{and} \qquad \lim\limits_{x \to 2^+} f(x)$$

   (C) Find m so that

   $$\lim\limits_{x \to 2^-} f(x) = \lim\limits_{x \to 2^+} f(x)$$

   and graph f for this value of m.

✎ (D) Write a brief verbal description of each graph. How does the graph in part (C) differ from the graphs in parts (A) and (B)?

Applications

91. **Telephone rates.** A long-distance telephone service charges $0.99 for the first 20 minutes or less of a call and $0.07 per minute thereafter.

   (A) Write a piecewise definition of the charge $F(x)$ for a long-distance call lasting x minutes.

   (B) Graph $F(x)$ for $0 < x \le 40$.

   (C) Find $\lim\limits_{x \to 20^-} F(x)$, $\lim\limits_{x \to 20^+} F(x)$, and $\lim\limits_{x \to 20} F(x)$, whichever exist.

92. **Telephone rates.** A second long-distance telephone service charges $0.09 per minute for calls lasting 10 minutes or more and $0.18 per minute for calls lasting less than 10 minutes.

   (A) Write a piecewise definition of the charge $G(x)$ for a long-distance call lasting x minutes.

   (B) Graph $G(x)$ for $0 < x \le 40$.

   (C) Find $\lim\limits_{x \to 10^-} G(x)$, $\lim\limits_{x \to 10^+} G(x)$, and $\lim\limits_{x \to 10} G(x)$, whichever exist.

93. Telephone rates. Refer to Problems 91 and 92. Write a brief verbal comparison of the two services described for calls lasting 20 minutes or less.

94. Telephone rates. Refer to Problems 91 and 92. Write a brief verbal comparison of the two services described for calls lasting more than 20 minutes.

A company sells custom embroidered apparel and promotional products. Table 1 shows the volume discounts offered by the company, where x is the volume of a purchase in dollars. Problems 95 and 96 deal with two different interpretations of this discount method.

Table 1 **Volume Discount (Excluding Tax)**

Volume (x)	Discount Amount
$300 \le x < \$1,000$	3%
$1,000 \le x < \$3,000$	5%
$3,000 \le x < \$5,000$	7%
$5,000 \le x$	10%

95. Volume discount. Assume that the volume discounts in Table 1 apply to the entire purchase. That is, if the volume x satisfies $300 \le x < \$1,000$, then the entire purchase is discounted 3%. If the volume x satisfies $1,000 \le x < \$3,000$, the entire purchase is discounted 5%, and so on.

(A) If x is the volume of a purchase before the discount is applied, then write a piecewise definition for the discounted price $D(x)$ of this purchase.

(B) Use one-sided limits to investigate the limit of $D(x)$ as x approaches $1,000. As x approaches $3,000.

96. Volume discount. Assume that the volume discounts in Table 1 apply only to that portion of the volume in each interval. That is, the discounted price for a $4,000 purchase would be computed as follows:

$$300 + 0.97(700) + 0.95(2,000) + 0.93(1,000) = 3,809$$

(A) If x is the volume of a purchase before the discount is applied, then write a piecewise definition for the discounted price $P(x)$ of this purchase.

(B) Use one-sided limits to investigate the limit of $P(x)$ as x approaches $1,000. As x approaches $3,000.

(C) Compare this discount method with the one in Problem 95. Does one always produce a lower price than the other? Discuss.

97. Pollution A state charges polluters an annual fee of $20 per ton for each ton of pollutant emitted into the atmosphere, up to a maximum of 4,000 tons. No fees are charged for emissions beyond the 4,000-ton limit. Write a piecewise definition of the fees $F(x)$ charged for the emission of x tons of pollutant in a year. What is the limit of $F(x)$ as x approaches 4,000 tons? As x approaches 8,000 tons?

98. Pollution Refer to Problem 97. The average fee per ton of pollution is given by $A(x) = F(x)/x$. Write a piecewise definition of $A(x)$. What is the limit of $A(x)$ as x approaches 4,000 tons? As x approaches 8,000 tons?

99. Voter turnout. Statisticians often use piecewise-defined functions to predict outcomes of elections. For the following functions f and g, find the limit of each function as x approaches 5 and as x approaches 10.

$$f(x) = \begin{cases} 0 & \text{if } x \le 5 \\ 0.8 - 0.08x & \text{if } 5 < x < 10 \\ 0 & \text{if } 10 \le x \end{cases}$$

$$g(x) = \begin{cases} 0 & \text{if } x \le 5 \\ 0.8x - 0.04x^2 - 3 & \text{if } 5 < x < 10 \\ 1 & \text{if } 10 \le x \end{cases}$$

Answers to Matched Problems

1.

x	-2	-1	0	1	2	3	4
$h(x)$	1.0	1.5	2.0	2.5	3.0	3.5	4.0

2. $\lim\limits_{x \to 1} f(x) = 2$

3.

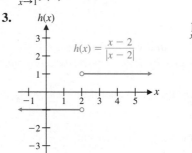

$\lim\limits_{x \to 2} \dfrac{x - 2}{|x - 2|}$ does not exist

4. (A) $\lim\limits_{x \to 0^-} f(x) = 0$

$\lim\limits_{x \to 0^+} f(x) = 0$

$\lim\limits_{x \to 0} f(x) = 0$

$f(0) = 0$

(B) $\lim\limits_{x \to 1^-} f(x) = 1$

$\lim\limits_{x \to 1^+} f(x) = 2$

$\lim\limits_{x \to 1} f(x)$ does not exist

$f(1)$ not defined

(C) $\lim\limits_{x \to 3^-} f(x) = 3$

$\lim\limits_{x \to 3^+} f(x) = 3$

$\lim\limits_{x \to 3} f(x) = 3$ $f(3)$ not defined

5. -6

6. (A) 6

(B) $\sqrt{6}$

(C) $\frac{4}{5}$

7. (A) 13

(B) 7

(C) Does not exist

(D) Not defined

8. (A) -2

(B) Does not exist

9. (A) No; $\dfrac{2}{3}$

(B) No; 0

(C) Yes; 6

10. -2

10.2 Infinite Limits and Limits at Infinity

- Infinite Limits
- Locating Vertical Asymptotes
- Limits at Infinity
- Finding Horizontal Asymptotes

In this section, we consider two new types of limits: infinite limits and limits at infinity. Infinite limits and vertical asymptotes are used to describe the behavior of functions that are unbounded near $x = a$. Limits at infinity and horizontal asymptotes are used to describe the behavior of functions as x assumes arbitrarily large positive values or arbitrarily large negative values. Although we will include graphs to illustrate basic concepts, we postpone a discussion of graphing techniques until Chapter 12.

Infinite Limits

The graph of $f(x) = \dfrac{1}{x - 1}$ (Fig. 1) indicates that

$$\lim_{x \to 1^+} \frac{1}{x - 1}$$

does not exist. There does not exist a real number L that the values of $f(x)$ approach as x approaches 1 from the right. Instead, as x approaches 1 from the right, the values of $f(x)$ are positive and become larger and larger; that is, $f(x)$ increases without bound (Table 1). We express this behavior symbolically as

$$\lim_{x \to 1^+} \frac{1}{x - 1} = \infty \quad \text{or} \quad f(x) = \frac{1}{x - 1} \to \infty \quad \text{as} \quad x \to 1^+ \tag{1}$$

Since ∞ is a not a real number, *the limit in (1) does not exist*. We are using the symbol ∞ to describe the manner in which the limit fails to exist, and we call this situation an **infinite limit**. If x approaches 1 from the left, the values of $f(x)$ are negative and become larger and larger in absolute value; that is, $f(x)$ decreases through negative values without bound (Table 2). We express this behavior symbolically as

$$\lim_{x \to 1^-} \frac{1}{x - 1} = -\infty \quad \text{or} \quad f(x) = \frac{1}{x - 1} \to -\infty \quad \text{as} \quad x \to 1^- \tag{2}$$

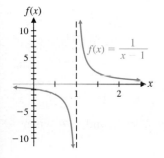

Figure 1

Table 1

x	$f(x) = \dfrac{1}{x - 1}$
1.1	10
1.01	100
1.001	1,000
1.0001	10,000
1.00001	100,000
1.000001	1,000,000

Table 2

x	$f(x) = \dfrac{1}{x - 1}$
0.9	−10
0.99	−100
0.999	−1,000
0.9999	−10,000
0.99999	−100,000
0.999999	−1,000,000

The one-sided limits in (1) and (2) describe the behavior of the graph as $x \to 1$ (Fig. 1). Does the two-sided limit of $f(x)$ as $x \to 1$ exist? No, because neither of the one-sided limits exists. Also, there is no reasonable way to use the symbol ∞ to describe the behavior of $f(x)$ as $x \to 1$ on both sides of 1. We say that

$$\lim_{x \to 1} \frac{1}{x - 1} \text{ does not exist}$$

Explore and Discuss 1

Let $g(x) = \dfrac{1}{(x - 1)^2}$.

Construct tables for $g(x)$ as $x \to 1^+$ and as $x \to 1^-$. Use these tables and infinite limits to discuss the behavior of $g(x)$ near $x = 1$.

We used the dashed vertical line $x = 1$ in Figure 1 to illustrate the infinite limits as x approaches 1 from the right and from the left. We call this line a *vertical asymptote*.

> **DEFINITION**   Infinite Limits and Vertical Asymptotes
> The vertical line $x = a$ is a **vertical asymptote** for the graph of $y = f(x)$ if
>
> $$f(x) \to \infty \quad \text{or} \quad f(x) \to -\infty \quad \text{as} \quad x \to a^+ \quad \text{or} \quad x \to a^-$$
>
> [That is, if $f(x)$ either increases or decreases without bound as x approaches a from the right or from the left].

Locating Vertical Asymptotes

How do we locate vertical asymptotes? If f is a polynomial function, then $\lim_{x \to a} f(x)$ is equal to the real number $f(a)$ [Theorem 3, Section 10.1]. So *a polynomial function has no vertical asymptotes*. Similarly (again by Theorem 3, Section 10.1), *a vertical asymptote of a rational function can occur only at a zero of its denominator*. Theorem 1 provides a simple procedure for locating the vertical asymptotes of a rational function.

> **THEOREM 1**   Locating Vertical Asymptotes of Rational Functions
> If $f(x) = n(x)/d(x)$ is a rational function, $d(c) = 0$ and $n(c) \neq 0$, then the line $x = c$ is a vertical asymptote of the graph of f.

If $f(x) = n(x)/d(x)$ and both $n(c) = 0$ and $d(c) = 0$, then the limit of $f(x)$ as x approaches c involves an indeterminate form and Theorem 1 does not apply:

$$\lim_{x \to c} f(x) = \lim_{x \to c} \frac{n(x)}{d(x)} \quad \frac{0}{0} \text{ indeterminate form}$$

Algebraic simplification is often useful in this situation.

> **EXAMPLE 1**   Locating Vertical Asymptotes   Let $f(x) = \dfrac{x^2 + x - 2}{x^2 - 1}$.

Describe the behavior of f at each zero of the denominator. Use ∞ and $-\infty$ when appropriate. Identify all vertical asymptotes.

SOLUTION   Let $n(x) = x^2 + x - 2$ and $d(x) = x^2 - 1$. Factoring the denominator, we see that

$$d(x) = x^2 - 1 = (x - 1)(x + 1)$$

has two zeros: $x = -1$ and $x = 1$.

First, we consider $x = -1$. Since $d(-1) = 0$ and $n(-1) = -2 \neq 0$, Theorem 1 tells us that the line $x = -1$ is a vertical asymptote. So at least one of the one-sided limits at $x = -1$ must be either ∞ or $-\infty$. Examining tables of values of f for x near -1 or a graph on a graphing calculator will show which is the case. From Tables 3 and 4, we see that

$$\lim_{x \to -1^-} \frac{x^2 + x - 2}{x^2 - 1} = -\infty \quad \text{and} \quad \lim_{x \to -1^+} \frac{x^2 + x - 2}{x^2 - 1} = \infty$$

Table 3

x	$f(x) = \dfrac{x^2 + x - 2}{x^2 - 1}$
-1.1	-9
-1.01	-99
-1.001	-999
-1.0001	$-9,999$
-1.00001	$-99,999$

Table 4

x	$f(x) = \dfrac{x^2 + x - 2}{x^2 - 1}$
-0.9	11
-0.99	101
-0.999	$1,001$
-0.9999	$10,001$
-0.99999	$100,001$

Now we consider the other zero of $d(x)$, $x = 1$. This time $n(1) = 0$ and Theorem 1 does not apply. We use algebraic simplification to investigate the behavior of the function at $x = 1$:

$$\lim_{x \to 1} f(x) = \lim_{x \to 1} \frac{x^2 + x - 2}{x^2 - 1} \qquad \frac{0}{0}\ \text{indeterminate form}$$

$$= \lim_{x \to 1} \frac{(x - 1)(x + 2)}{(x - 1)(x + 1)}$$

$$= \lim_{x \to 1} \frac{x + 2}{x + 1} \qquad \text{Reduced to lowest terms (see Appendix A.4)}$$

$$= \frac{3}{2}$$

Since the limit exists as x approaches 1, f does not have a vertical asymptote at $x = 1$. The graph of f (Fig. 2) shows the behavior at the vertical asymptote $x = -1$ and also at $x = 1$.

Figure 2 $f(x) = \dfrac{x^2 + x - 2}{x^2 - 1}$

Matched Problem 1 Let $f(x) = \dfrac{x - 3}{x^2 - 4x + 3}$.

Describe the behavior of f at each zero of the denominator. Use ∞ and $-\infty$ when appropriate. Identify all vertical asymptotes.

EXAMPLE 2 Locating Vertical Asymptotes   Let $f(x) = \dfrac{x^2 + 20}{5(x - 2)^2}$.

Describe the behavior of f at each zero of the denominator. Use ∞ and $-\infty$ when appropriate. Identify all vertical asymptotes.

SOLUTION Let $n(x) = x^2 + 20$ and $d(x) = 5(x - 2)^2$. The only zero of $d(x)$ is $x = 2$. Since $n(2) = 24 \neq 0$, f has a vertical asymptote at $x = 2$ (Theorem 1). Tables 5 and 6 show that $f(x) \to \infty$ as $x \to 2$ from either side, and we have

$$\lim_{x \to 2^+} \frac{x^2 + 20}{5(x - 2)^2} = \infty \quad \text{and} \quad \lim_{x \to 2^-} \frac{x^2 + 20}{5(x - 2)^2} = \infty$$

Table 5

x	$f(x) = \dfrac{x^2 + 20}{5(x - 2)^2}$
2.1	488.2
2.01	48,080.02
2.001	4,800,800.2

Table 6

x	$f(x) = \dfrac{x^2 + 20}{5(x - 2)^2}$
1.9	472.2
1.99	47,920.02
1.999	4,799,200.2

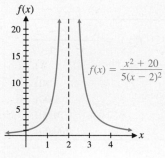

Figure 3

The denominator d has no other zeros, so f does not have any other vertical asymptotes. The graph of f (Fig. 3) shows the behavior at the vertical asymptote $x = 2$. Because the left- and right-hand limits are both infinite, we write

$$\lim_{x \to 2} \frac{x^2 + 20}{5(x - 2)^2} = \infty$$

Matched Problem 2 Let $f(x) = \dfrac{x - 1}{(x + 3)^2}$.

Describe the behavior of f at each zero of the denominator. Use ∞ and $-\infty$ when appropriate. Identify all vertical asymptotes.

CONCEPTUAL INSIGHT

When is it correct to say that a limit does not exist, and when is it correct to use $\pm\infty$? It depends on the situation. Table 7 lists the infinite limits that we discussed in Examples 1 and 2.

Table 7

Right-Hand Limit	Left-Hand Limit	Two-Sided Limit
$\displaystyle\lim_{x \to -1^+} \frac{x^2 + x - 2}{x^2 - 1} = \infty$	$\displaystyle\lim_{x \to -1^-} \frac{x^2 + x - 2}{x^2 - 1} = -\infty$	$\displaystyle\lim_{x \to -1} \frac{x^2 + x - 2}{x^2 - 1}$ does not exist
$\displaystyle\lim_{x \to -2^+} \frac{x^2 + 20}{5(x - 2)^2} = \infty$	$\displaystyle\lim_{x \to 2^-} \frac{x^2 + 20}{5(x - 2)^2} = \infty$	$\displaystyle\lim_{x \to 2} \frac{x^2 + 20}{5(x - 2)^2} = \infty$

The instructions in Examples 1 and 2 said that we should use infinite limits to describe the behavior at vertical asymptotes. If we had been asked to *evaluate* the limits, with no mention of ∞ or asymptotes, then the correct answer would be that **all of these limits do not exist**. Remember, ∞ is a symbol used to describe the behavior of functions at vertical asymptotes.

Limits at Infinity

The symbol ∞ can also be used to indicate that an independent variable is increasing or decreasing without bound. We write $x \to \infty$ to indicate that x is increasing without bound through positive values and $x \to -\infty$ to indicate that x is decreasing without bound through negative values. We begin by considering power functions of the form x^p and $1/x^p$ where p is a positive real number.

If p is a positive real number, then x^p increases as x increases. There is no upper bound on the values of x^p. We indicate this behavior by writing

$$\lim_{x \to \infty} x^p = \infty \quad \text{or} \quad x^p \to \infty \quad \text{as} \quad x \to \infty$$

Since the reciprocals of very large numbers are very small numbers, it follows that $1/x^p$ approaches 0 as x increases without bound. We indicate this behavior by writing

$$\lim_{x \to \infty} \frac{1}{x^p} = 0 \quad \text{or} \quad \frac{1}{x^p} \to 0 \quad \text{as} \quad x \to \infty$$

Figure 4 illustrates the preceding behavior for $f(x) = x^2$ and $g(x) = 1/x^2$, and we write

$$\lim_{x \to \infty} f(x) = \infty \quad \text{and} \quad \lim_{x \to \infty} g(x) = 0$$

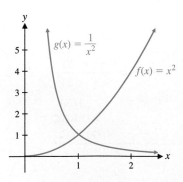

Figure 4

Limits of power forms as x decreases without bound behave in a similar manner, with two important differences. First, if x is negative, then x^p is not defined for all values of p. For example, $x^{1/2} = \sqrt{x}$ is not defined for negative values of x. Second, if x^p is defined, then it may approach ∞ or $-\infty$, depending on the value of p. For example,

$$\lim_{x \to -\infty} x^2 = \infty \quad \text{but} \quad \lim_{x \to -\infty} x^3 = -\infty$$

For the function g in Figure 4, the line $y = 0$ (the x axis) is called a *horizontal asymptote*. In general, a line $y = b$ is a **horizontal asymptote** of the graph of $y = f(x)$ if $f(x)$ approaches b as either x increases without bound or x decreases without bound. Symbolically, $y = b$ is a horizontal asymptote if either

$$\lim_{x \to -\infty} f(x) = b \quad \text{or} \quad \lim_{x \to \infty} f(x) = b$$

In the first case, the graph of f will be close to the horizontal line $y = b$ for large (in absolute value) negative x. In the second case, the graph will be close to the horizontal line $y = b$ for large positive x. Figure 5 shows the graph of a function with two horizontal asymptotes: $y = 1$ and $y = -1$.

Theorem 2 summarizes the various possibilities for limits of power functions as x increases or decreases without bound.

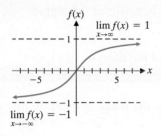

$$\lim_{x \to \infty} f(x) = 1$$
$$\lim_{x \to -\infty} f(x) = -1$$

Figure 5

THEOREM 2 Limits of Power Functions at Infinity

If p is a positive real number and k is any real number except 0, then

1. $\displaystyle\lim_{x \to -\infty} \frac{k}{x^p} = 0$ **2.** $\displaystyle\lim_{x \to \infty} \frac{k}{x^p} = 0$

3. $\displaystyle\lim_{x \to -\infty} kx^p = \pm\infty$ **4.** $\displaystyle\lim_{x \to \infty} kx^p = \pm\infty$

provided that x^p is a real number for negative values of x. The limits in 3 and 4 will be either $-\infty$ or ∞, depending on k and p.

How can we use Theorem 2 to evaluate limits at infinity? It turns out that the limit properties listed in Theorem 2, Section 10.1, are also valid if we replace the statement $x \to c$ with $x \to \infty$ or $x \to -\infty$.

EXAMPLE 3 Limit of a Polynomial Function at Infinity Let $p(x) = 2x^3 - x^2 - 7x + 3$. Find the limit of $p(x)$ as x approaches ∞ and as x approaches $-\infty$.

SOLUTION Since limits of power functions of the form $1/x^p$ approach 0 as x approaches ∞ or $-\infty$, it is convenient to work with these reciprocal forms whenever possible. If we factor out the term involving the highest power of x, then we can write $p(x)$ as

$$p(x) = 2x^3\left(1 - \frac{1}{2x} - \frac{7}{2x^2} + \frac{3}{2x^3}\right)$$

Using Theorem 2 above and Theorem 2 in Section 10.1, we write

$$\lim_{x \to \infty}\left(1 - \frac{1}{2x} - \frac{7}{2x^2} + \frac{3}{2x^3}\right) = 1 - 0 - 0 + 0 = 1$$

For large values of x,

$$\left(1 - \frac{1}{2x} - \frac{7}{2x^2} + \frac{3}{2x^3}\right) \approx 1$$

and

$$p(x) = 2x^3\left(1 - \frac{1}{2x} - \frac{7}{2x^2} + \frac{3}{2x^3}\right) \approx 2x^3$$

Since $2x^3 \to \infty$ as $x \to \infty$, it follows that

$$\lim_{x \to \infty} p(x) = \lim_{x \to \infty} 2x^3 = \infty$$

Similarly, $2x^3 \to -\infty$ as $x \to -\infty$ implies that

$$\lim_{x \to -\infty} p(x) = \lim_{x \to -\infty} 2x^3 = -\infty$$

So the behavior of $p(x)$ for large values is the same as the behavior of the highest-degree term, $2x^3$.

Matched Problem 3) Let $p(x) = -4x^4 + 2x^3 + 3x$. Find the limit of $p(x)$ as x approaches ∞ and as x approaches $-\infty$.

The term with highest degree in a polynomial is called the **leading term**. In the solution to Example 3, the limits at infinity of $p(x) = 2x^3 - x^2 - 7x + 3$ were the same as the limits of the leading term $2x^3$. Theorem 3 states that this is true for any polynomial of degree greater than or equal to 1.

THEOREM 3   Limits of Polynomial Functions at Infinity

If

$$p(x) = a_n x^n + a_{n-1} x^{n-1} + \cdots + a_1 x + a_0, a_n \neq 0, n \geq 1$$

then

$$\lim_{x \to \infty} p(x) = \lim_{x \to \infty} a_n x^n = \pm\infty$$

and

$$\lim_{x \to -\infty} p(x) = \lim_{x \to -\infty} a_n x^n = \pm\infty$$

Each limit will be either $-\infty$ or ∞, depending on a_n and n.

A polynomial of degree 0 is a constant function $p(x) = a_0$, and its limit as x approaches ∞ or $-\infty$ is the number a_0. For any polynomial of degree 1 or greater, Theorem 3 states that the limit as x approaches ∞ or $-\infty$ cannot be equal to a number. This means that **polynomials of degree 1 or greater never have horizontal asymptotes.**

A pair of limit expressions of the form

$$\lim_{x \to \infty} f(x) = A, \quad \lim_{x \to -\infty} f(x) = B$$

where A and B are ∞, $-\infty$, or real numbers, describes the **end behavior** of the function f. The first of the two limit expressions describes the **right end behavior** and the second describes the **left end behavior**. By Theorem 3, the end behavior of any nonconstant polynomial function is described by a pair of infinite limits.

EXAMPLE 4   End Behavior of a Polynomial   Give a pair of limit expressions that describe the end behavior of each polynomial.

(A) $p(x) = 3x^3 - 500x^2$ (B) $p(x) = 3x^3 - 500x^4$

SOLUTION

(A) By Theorem 3,

$$\lim_{x \to \infty} (3x^3 - 500x^2) = \lim_{x \to \infty} 3x^3 = \infty \qquad \text{Right end behavior}$$

and

$$\lim_{x \to -\infty} (3x^3 - 500x^2) = \lim_{x \to -\infty} 3x^3 = -\infty \qquad \text{Left end behavior}$$

(B) By Theorem 3,

$$\lim_{x \to \infty} (3x^3 - 500x^4) = \lim_{x \to \infty} (-500x^4) = -\infty \qquad \text{Right end behavior}$$

and

$$\lim_{x \to -\infty} (3x^3 - 500x^4) = \lim_{x \to -\infty} (-500x^4) = -\infty \qquad \text{Left end behavior}$$

Matched Problem 4) Give a pair of limit expressions that describe the end behavior of each polynomial.

(A) $p(x) = 300x^2 - 4x^5$ \qquad (B) $p(x) = 300x^6 - 4x^5$

Finding Horizontal Asymptotes

Since a rational function is the ratio of two polynomials, it is not surprising that reciprocals of powers of x can be used to analyze limits of rational functions at infinity. For example, consider the rational function

$$f(x) = \frac{3x^2 - 5x + 9}{2x^2 + 7}$$

Factoring the highest-degree term out of the numerator and the highest-degree term out of the denominator, we write

$$f(x) = \frac{3x^2}{2x^2} \cdot \frac{1 - \dfrac{5}{3x} + \dfrac{3}{x^2}}{1 + \dfrac{7}{2x^2}}$$

$$\lim_{x \to \infty} f(x) = \lim_{x \to \infty} \frac{3x^2}{2x^2} \cdot \lim_{x \to \infty} \frac{1 - \dfrac{5}{3x} + \dfrac{3}{x^2}}{1 + \dfrac{7}{2x^2}} = \frac{3}{2} \cdot \frac{1 - 0 + 0}{1 + 0} = \frac{3}{2}$$

The behavior of this rational function as x approaches infinity is determined by the ratio of the highest-degree term in the numerator ($3x^2$) to the highest-degree term in the denominator ($2x^2$). Theorem 2 can be used to generalize this result to any rational function. Theorem 4 lists the three possible outcomes.

THEOREM 4   Limits of Rational Functions at Infinity and Horizontal Asymptotes of Rational Functions

(A) If $f(x) = \dfrac{a_m x^m + a_{m-1}x^{m-1} + \cdots + a_1 x + a_0}{b_n x^n + b_{n-1}x^{n-1} + \cdots + b_1 x + b_0}$, $a_m \neq 0, b_n \neq 0$

then $\displaystyle\lim_{x \to \infty} f(x) = \lim_{x \to \infty} \frac{a_m x^m}{b_n x^n}$ and $\displaystyle\lim_{x \to -\infty} f(x) = \lim_{x \to -\infty} \frac{a_m x^m}{b_n x^n}$

(B) There are three possible cases for these limits:

1. If $m < n$, then $\displaystyle\lim_{x \to \infty} f(x) = \lim_{x \to -\infty} f(x) = 0$, and the line $y = 0$ (the x axis) is a horizontal asymptote of $f(x)$.

2. If $m = n$, then $\displaystyle\lim_{x \to \infty} f(x) = \lim_{x \to -\infty} f(x) = \frac{a_m}{b_n}$, and the line $y = \frac{a_m}{b_n}$ is a horizontal asymptote of $f(x)$.

3. If $m > n$, then each limit will be ∞ or $-\infty$, depending on m, n, a_m, and b_n, and $f(x)$ does not have a horizontal asymptote.

Notice that in cases 1 and 2 of Theorem 4, the limit is the same if x approaches ∞ or $-\infty$. So, **a rational function can have at most one horizontal asymptote** (see Fig. 6).

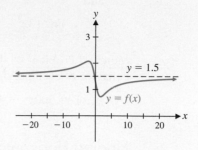

Figure 6 $f(x) = \dfrac{3x^2 - 5x + 9}{2x^2 + 7}$

CONCEPTUAL INSIGHT

The graph of f in Figure 6 dispels the misconception that the graph of a function cannot cross a horizontal asymptote. Horizontal asymptotes give us information about the graph of a function only as $x \to \infty$ and $x \to -\infty$, not at any specific value of x.

EXAMPLE 5 Finding Horizontal Asymptotes Find all horizontal asymptotes, if any, of each function.

(A) $f(x) = \dfrac{5x^3 - 2x^2 + 1}{4x^3 + 2x - 7}$

(B) $f(x) = \dfrac{3x^4 - x^2 + 1}{8x^6 - 10}$

(C) $f(x) = \dfrac{2x^5 - x^3 - 1}{6x^3 + 2x^2 - 7}$

SOLUTION We will make use of part A of Theorem 4.

(A) $\displaystyle \lim_{x \to \infty} f(x) = \lim_{x \to \infty} \frac{5x^3 - 2x^2 + 1}{4x^3 + 2x - 7} = \lim_{x \to \infty} \frac{5x^3}{4x^3} = \frac{5}{4}$

The line $y = 5/4$ is a horizontal asymptote of $f(x)$.

(B) $\displaystyle \lim_{x \to \infty} f(x) = \lim_{x \to \infty} \frac{3x^4 - x^2 + 1}{8x^6 - 10} = \lim_{x \to \infty} \frac{3x^4}{8x^6} = \lim_{x \to \infty} \frac{3}{8x^2} = 0$

The line $y = 0$ (the x axis) is a horizontal asymptote of $f(x)$.

(C) $\displaystyle \lim_{x \to \infty} f(x) = \lim_{x \to \infty} \frac{2x^5 - x^3 - 1}{6x^3 + 2x^2 - 7} = \lim_{x \to \infty} \frac{2x^5}{6x^3} = \lim_{x \to \infty} \frac{x^2}{3} = \infty$

The function $f(x)$ has no horizontal asymptotes.

Matched Problem 5) Find all horizontal asymptotes, if any, of each function.

(A) $f(x) = \dfrac{4x^3 - 5x + 8}{2x^4 - 7}$

(B) $f(x) = \dfrac{5x^6 + 3x}{2x^5 - x - 5}$

(C) $f(x) = \dfrac{2x^3 - x + 7}{4x^3 + 3x^2 - 100}$

An accurate sketch of the graph of a rational function requires knowledge of both vertical and horizontal asymptotes. As we mentioned earlier, we are postponing a detailed discussion of graphing techniques until Section 12.4.

EXAMPLE 6 Find all vertical and horizontal asymptotes of the function
$$f(x) = \frac{2x^2 - 5}{x^2 + 5x + 4}$$

SOLUTION Let $n(x) = 2x^2 - 5$ and $d(x) = x^2 + 5x + 4 = (x + 1)(x + 4)$. The denominator $d(x) = 0$ at $x = -1$ and $x = -4$. Since the numerator $n(x)$ is not zero at these values of x [$n(-1) = -3$ and $n(-4) = 27$], by Theorem 1 there are two vertical asymptotes of f: the line $x = -1$ and the line $x = -4$. Since

$$\lim_{x \to \infty} f(x) = \lim_{x \to \infty} \frac{2x^2 - 5}{x^2 + 5x + 4} = \lim_{x \to \infty} \frac{2x^2}{x^2} = 2$$

the horizontal asymptote is the line $y = 2$ (Theorem 3).

Matched Problem 6 | Find all vertical and horizontal asymptotes of the function
$$f(x) = \frac{x^2 - 9}{x^2 - 4}.$$

Exercises 10.2

Skills Warm-up Exercises

W *In Problems 1–8, find an equation of the form $Ax + By = C$ for the given line. (If necessary, review Section 1.2).*

1. The horizontal line through $(0, 4)$

2. The vertical line through $(5, 0)$

3. The vertical line through $(-6, 3)$

4. The horizontal line through $(7, 1)$

5. The line through $(-2, 9)$ that has slope 2

6. The line through $(8, -4)$ that has slope -3

7. The line through $(9, 0)$ and $(0, 7)$

8. The line through $(-1, 20)$ and $(1, 30)$

Problems 9–16 refer to the following graph of $y = f(x)$.

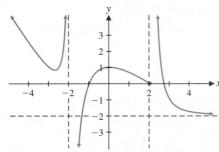

Figure for 9–16

9. $\lim_{x \to \infty} f(x) = ?$

10. $\lim_{x \to -\infty} f(x) = ?$

11. $\lim_{x \to -2^+} f(x) = ?$

12. $\lim_{x \to -2^-} f(x) = ?$

13. $\lim_{x \to -2} f(x) = ?$

14. $\lim_{x \to 2^+} f(x) = ?$

15. $\lim_{x \to 2} f(x) = ?$

16. $\lim_{x \to 2} f(x) = ?$

In Problems 17–24, find each limit. Use $-\infty$ and ∞ when appropriate.

17. $f(x) = \dfrac{x}{x - 5}$

   (A) $\lim_{x \to 5^-} f(x)$ (B) $\lim_{x \to 5^+} f(x)$ (C) $\lim_{x \to 5} f(x)$

18. $f(x) = \dfrac{x^2}{x + 3}$

   (A) $\lim_{x \to -3^-} f(x)$ (B) $\lim_{x \to -3^+} f(x)$ (C) $\lim_{x \to -3} f(x)$

19. $f(x) = \dfrac{2x - 4}{(x - 4)^2}$

   (A) $\lim_{x \to 4^-} f(x)$ (B) $\lim_{x \to 4^+} f(x)$ (C) $\lim_{x \to 4} f(x)$

20. $f(x) = \dfrac{2x + 2}{(x + 2)^2}$

   (A) $\lim_{x \to -2^-} f(x)$ (B) $\lim_{x \to -2^+} f(x)$ (C) $\lim_{x \to -2} f(x)$

21. $f(x) = \dfrac{x^2 + x - 2}{x - 1}$

   (A) $\lim_{x \to 1^-} f(x)$ (B) $\lim_{x \to 1^+} f(x)$ (C) $\lim_{x \to 1} f(x)$

22. $f(x) = \dfrac{x^2 + x + 2}{x - 1}$

   (A) $\lim_{x \to 1^-} f(x)$ (B) $\lim_{x \to 1^+} f(x)$ (C) $\lim_{x \to 1} f(x)$

23. $f(x) = \dfrac{x^2 - 3x + 2}{x + 2}$

   (A) $\lim_{x \to -2^-} f(x)$ (B) $\lim_{x \to -2^+} f(x)$ (C) $\lim_{x \to -2} f(x)$

24. $f(x) = \dfrac{x^2 + x - 2}{x + 2}$

   (A) $\lim_{x \to -2^-} f(x)$ (B) $\lim_{x \to -2^+} f(x)$ (C) $\lim_{x \to -2} f(x)$

In Problems 25–32, find (A) the leading term of the polynomial, (B) the limit as x approaches ∞, and (C) the limit as x approaches −∞.

25. $p(x) = 15 + 3x^2 - 5x^3$

26. $p(x) = 10 - x^6 + 7x^3$

27. $p(x) = 9x^2 - 6x^4 + 7x$

28. $p(x) = -x^5 + 2x^3 + 9x$

29. $p(x) = x^2 + 7x + 12$

30. $p(x) = 5x + x^3 - 8x^2$

31. $p(x) = x^4 + 2x^5 - 11x$

32. $p(x) = 1 + 4x^2 + 4x^4$

In Problems 33–42, use −∞ or ∞ where appropriate to describe the behavior at each zero of the denominator and identify all vertical asymptotes.

33. $f(x) = \dfrac{1}{x + 3}$ **34.** $g(x) = \dfrac{x}{4 - x}$

35. $h(x) = \dfrac{x^2 + 4}{x^2 - 4}$ **36.** $k(x) = \dfrac{x^2 - 9}{x^2 + 9}$

37. $F(x) = \dfrac{x^2 - 4}{x^2 + 4}$ **38.** $G(x) = \dfrac{x^2 + 9}{9 - x^2}$

39. $H(x) = \dfrac{x^2 - 2x - 3}{x^2 - 4x + 3}$ **40.** $K(x) = \dfrac{x^2 + 2x - 3}{x^2 - 4x + 3}$

41. $T(x) = \dfrac{8x - 16}{x^4 - 8x^3 + 16x^2}$ **42.** $S(x) = \dfrac{6x + 9}{x^4 + 6x^3 + 9x^2}$

In Problems 43–50, find each function value and limit. Use −∞ or ∞ where appropriate.

43. $f(x) = \dfrac{4x + 7}{5x - 9}$

   (A) $f(10)$ (B) $f(100)$ (C) $\lim\limits_{x \to \infty} f(x)$

44. $f(x) = \dfrac{2 - 3x^3}{7 + 4x^3}$

   (A) $f(5)$ (B) $f(10)$ (C) $\lim\limits_{x \to \infty} f(x)$

45. $f(x) = \dfrac{5x^2 + 11}{7x - 2}$

   (A) $f(20)$ (B) $f(50)$ (C) $\lim\limits_{x \to \infty} f(x)$

46. $f(x) = \dfrac{5x + 11}{7x^3 - 2}$

   (A) $f(-8)$ (B) $f(-16)$ (C) $\lim\limits_{x \to -\infty} f(x)$

47. $f(x) = \dfrac{7x^4 - 14x^2}{6x^5 + 3}$

   (A) $f(-6)$ (B) $f(-12)$ (C) $\lim\limits_{x \to -\infty} f(x)$

48. $f(x) = \dfrac{4x^7 - 8x}{6x^4 + 9x^2}$

   (A) $f(-3)$ (B) $f(-6)$ (C) $\lim\limits_{x \to -\infty} f(x)$

49. $f(x) = \dfrac{10 - 7x^3}{4 + x^3}$

   (A) $f(-10)$ (B) $f(-20)$ (C) $\lim\limits_{x \to -\infty} f(x)$

50. $f(x) = \dfrac{3 + x}{5 + 4x}$

   (A) $f(-50)$ (B) $f(-100)$ (C) $\lim\limits_{x \to -\infty} f(x)$

In Problems 51–64, find all horizontal and vertical asymptotes.

51. $f(x) = \dfrac{2x}{x + 2}$ **52.** $f(x) = \dfrac{3x + 2}{x - 4}$

53. $f(x) = \dfrac{x^2 + 1}{x^2 - 1}$ **54.** $f(x) = \dfrac{x^2 - 1}{x^2 + 2}$

55. $f(x) = \dfrac{x^3}{x^2 + 6}$ **56.** $f(x) = \dfrac{x}{x^2 - 4}$

57. $f(x) = \dfrac{x}{x^2 + 4}$ **58.** $f(x) = \dfrac{x^2 + 9}{x}$

59. $f(x) = \dfrac{x^2}{x - 3}$ **60.** $f(x) = \dfrac{x + 5}{x^2}$

61. $f(x) = \dfrac{2x^2 + 3x - 2}{x^2 - x - 2}$ **62.** $f(x) = \dfrac{2x^2 + 7x + 12}{2x^2 + 5x - 12}$

63. $f(x) = \dfrac{2x^2 - 5x + 2}{x^2 - x - 2}$ **64.** $f(x) = \dfrac{x^2 - x - 12}{2x^2 + 5x - 12}$

In Problems 65–68, give a limit expression that describes the right end behavior of the function.

65. $f(x) = \dfrac{x + 3}{x^2 - 5}$ **66.** $f(x) = \dfrac{3 + 4x + x^2}{5 - x}$

67. $f(x) = \dfrac{x^2 - 5}{x + 3}$ **68.** $f(x) = \dfrac{4x + 1}{5x - 7}$

In Problems 69–72, give a limit expression that describes the left end behavior of the function.

69. $f(x) = \dfrac{5 - 2x^2}{1 + 8x^2}$ **70.** $f(x) = \dfrac{2x + 3}{x^2 - 1}$

71. $f(x) = \dfrac{x^2 + 4x}{3x + 2}$ **72.** $f(x) = \dfrac{6 - x^4}{1 + 2x}$

In Problems 73–76, give a pair of limit expressions that describe the end behavior of the function.

73. $f(x) = x^3 - 3x + 1$ **74.** $f(x) = 4 - 5x - x^3$

75. $f(x) = \dfrac{2 + 5x}{1 - x}$ **76.** $f(x) = \dfrac{9x^2 + 6x + 1}{4x^2 + 4x + 1}$

In Problems 77–82, discuss the validity of each statement. If the statement is always true, explain why. If not, give a counterexample.

77. A rational function has at least one vertical asymptote.

78. A rational function has at most one vertical asymptote.

79. A rational function has at least one horizontal asymptote.

80. A rational function has at most one horizontal asymptote.

81. A polynomial function of degree ≥ 1 has neither horizontal nor vertical asymptotes.

82. The graph of a rational function cannot cross a horizontal asymptote.

83. Theorem 3 states that

$$\lim_{x \to \infty} (a_n x^n + a_{n-1} x^{n-1} + \cdots + a_0) = \pm \infty.$$

What conditions must n and a_n satisfy for the limit to be ∞? For the limit to be $-\infty$?

84. Theorem 3 also states that

$$\lim_{x \to -\infty} (a_n x^n + a_{n-1} x^{n-1} + \cdots + a_0) = \pm \infty.$$

What conditions must n and a_n satisfy for the limit to be ∞? For the limit to be $-\infty$?

Applications

85. Average cost. A company manufacturing snowboards has fixed costs of $200 per day and total costs of $3,800 per day for a daily output of 20 boards.

(A) Assuming that the total cost per day $C(x)$ is linearly related to the total output per day x, write an equation for the cost function.

(B) The average cost per board for an output of x boards is given by $\overline{C}(x) = C(x)/x$. Find the average cost function.

(C) Sketch a graph of the average cost function, including any asymptotes, for $1 \le x \le 30$.

(D) What does the average cost per board tend to as production increases?

86. Average cost. A company manufacturing surfboards has fixed costs of $300 per day and total costs of $5,100 per day for a daily output of 20 boards.

(A) Assuming that the total cost per day $C(x)$ is linearly related to the total output per day x, write an equation for the cost function.

(B) The average cost per board for an output of x boards is given by $\overline{C}(x) = C(x)/x$. Find the average cost function.

(C) Sketch a graph of the average cost function, including any asymptotes, for $1 \le x \le 30$.

(D) What does the average cost per board tend to as production increases?

87. Energy costs. Most appliance manufacturers produce conventional and energy-efficient models. The energy-efficient models are more expensive to make but cheaper to operate. The costs of purchasing and operating a 23-cubic-foot refrigerator of each type are given in Table 8. These costs do not include maintenance charges or changes in electricity prices.

Table 8 **23-ft³ Refrigerators**

	Energy-Efficient Model	Conventional Model
Initial cost	$950	$900
Total volume	23 ft³	23 ft³
Annual cost of electricity	$56	$66

(A) Express the total cost $C_e(x)$ and the average cost $\overline{C}_e(x) = C_e(x)/x$ of purchasing and operating an energy-efficient model for x years.

(B) Express the total cost $C_c(x)$ and the average cost $\overline{C}_c(x) = C_c(x)/x$ of purchasing and operating a conventional model for x years.

(C) Are the total costs for an energy-efficient model and for a conventional model ever the same? If so, when?

(D) Are the average costs for an energy-efficient model and for a conventional model ever the same? If so, when?

(E) Find the limit of each average cost function as $x \to \infty$ and discuss the implications of the results.

88. Energy costs. Most appliance manufacturers produce conventional and energy-efficient models. The energy-efficient models are more expensive to make but cheaper to operate. The costs of purchasing and operating a 36,000-Btu central air conditioner of each type are given in Table 9. These costs do not include maintenance charges or changes in electricity prices.

Table 9 **36,000 Btu Central Air Conditioner**

	Energy-Efficient Model	Conventional Model
Initial cost	$4,000	$2,700
Total capacity	36,000 Btu	36,000 Btu
Annual cost of electricity	$932	$1,332

(A) Express the total cost $C_e(x)$ and the average cost $\overline{C}_e(x) = C_e(x)/x$ of purchasing and operating an energy-efficient model for x years.

(B) Express the total cost $C_c(x)$ and the average cost $\overline{C}_c(x) = C_c(x)/x$ of purchasing and operating a conventional model for x years.

(C) Are the total costs for an energy-efficient model and for a conventional model ever the same? If so, when?

(D) Are the average costs for an energy-efficient model and for a conventional model ever the same? If so, when?

(E) Find the limit of each average cost function as $x \to \infty$ and discuss the implications of the results.

89. Drug concentration. A drug is administered to a patient through an injection. The drug concentration (in milligrams/milliliter) in the bloodstream t hours after the injection is given by $C(t) = \dfrac{5t^2(t + 50)}{t^3 + 100}$. Find and interpret $\lim_{t \to \infty} C(t)$.

90. Drug concentration. A drug is administered to a patient through an IV drip. The drug concentration (in milligrams/milliliter) in the bloodstream t hours after the drip was started is given by $C(t) = \dfrac{5t(t + 50)}{t^3 + 100}$. Find and interpret $\lim_{t \to \infty} C(t)$.

91. Pollution. In Silicon Valley, a number of computer-related manufacturing firms were contaminating underground water supplies with toxic chemicals stored in leaking underground containers. A water quality control agency ordered the companies to take immediate corrective action and contribute to a monetary pool for the testing and cleanup of the underground contamination. Suppose that the monetary pool (in millions of dollars) for the testing and cleanup is given by

$$P(x) = \frac{2x}{1 - x} \qquad 0 \le x < 1$$

where x is the percentage (expressed as a decimal) of the total contaminant removed.

(A) How much must be in the pool to remove 90% of the contaminant?

(B) How much must be in the pool to remove 95% of the contaminant?

✎ (C) Find $\lim_{x \to 1^-} P(x)$ and discuss the implications of this limit.

92. **Employee training.** A company producing computer components has established that, on average, a new employee can assemble $N(t)$ components per day after t days of on-the-job training, as given by

$$N(t) = \frac{100t}{t + 9} \qquad t \ge 0$$

(A) How many components per day can a new employee assemble after 6 days of on-the-job training?

(B) How many days of on-the-job training will a new employee need to reach the level of 70 components per day?

✎ (C) Find $\lim_{t \to \infty} N(t)$ and discuss the implications of this limit.

93. **Biochemistry.** In 1913, biochemists Leonor Michaelis and Maude Menten proposed the rational function model (see figure)

$$v(s) = \frac{V_{\max}\, s}{K_M + s}$$

for the velocity of the enzymatic reaction v, where s is the substrate concentration. The constants $V_{\max}$ and K_M are determined from experimental data.

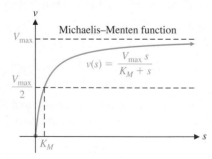

(A) Show that $\lim_{s \to \infty} v(s) = V_{\max}$.

(B) Show that $v(K_M) = \dfrac{V_{\max}}{2}$.

(C) Table 10* lists data for the substrate saccharose treated with an enzyme.
    Plot the points in Table 10 on graph paper and estimate $V_{\max}$ to the nearest integer. To estimate K_M, add the horizontal line $v = \dfrac{V_{\max}}{2}$ to your graph, connect successive points on the graph with straight-line segments, and

Table 10

s	v
5.2	0.866
10.4	1.466
20.8	2.114
41.6	2.666
83.3	3.236
167	3.636
333	3.636

estimate the value of s (to the nearest multiple of 10) that satisfies $v(s) = \dfrac{V_{\max}}{2}$.

(D) Use the constants $V_{\max}$ and K_M from part (C) to form a Michaelis–Menten function for the data in Table 10.

(E) Use the function from part (D) to estimate the velocity of the enzyme reaction when the saccharose is 15 and to estimate the saccharose when the velocity is 3.

94. **Biochemistry.** Table 11* lists data for the substrate sucrose treated with the enzyme invertase. We want to model these data with a Michaelis–Menten function.

Table 11

s	v
2.92	18.2
5.84	26.5
8.76	31.1
11.7	33
14.6	34.9
17.5	37.2
23.4	37.1

(A) Plot the points in Table 11 on graph paper and estimate $V_{\max}$ to the nearest integer. To estimate K_M, add the horizontal line $v = \dfrac{V_{\max}}{2}$ to your graph, connect successive points on the graph with straight-line segments, and estimate the value of s (to the nearest integer) that satisfies $v(s) = \dfrac{V_{\max}}{2}$.

(B) Use the constants $V_{\max}$ and K_M from part (A) to form a Michaelis–Menten function for the data in Table 11.

(C) Use the function from part (B) to estimate the velocity of the enzyme reaction when the sucrose is 9 and to estimate the sucrose when the velocity is 32.

95. **Physics.** The coefficient of thermal expansion (CTE) is a measure of the expansion of an object subjected to extreme temperatures. To model this coefficient, we use a Michaelis–Menten function of the form

$$C(T) = \frac{C_{\max}\, T}{M + T} \qquad \text{(Problem 93)}$$

where $C = $ CTE, T is temperature in K (degrees Kelvin), and $C_{\max}$ and M are constants. Table 12[†] lists the coefficients of thermal expansion for nickel and for copper at various temperatures.

*Michaelis and Menten (1913) *Biochem. Z.* 49, 333–369.

*Institute of Chemistry, Macedonia.
†National Physical Laboratory

Table 12 **Coefficients of Thermal Expansion**

T (K)	Nickel	Copper
100	6.6	10.3
200	11.3	15.2
293	13.4	16.5
500	15.3	18.3
800	16.8	20.3
1,100	17.8	23.7

(A) Plot the points in columns 1 and 2 of Table 12 on graph paper and estimate C_{max} to the nearest integer. To estimate M, add the horizontal line CTE $= \dfrac{C_{max}}{2}$ to your graph, connect successive points on the graph with straight-line segments, and estimate the value of T (to the nearest multiple of fifty) that satisfies $C(T) = \dfrac{C_{max}}{2}$.

(B) Use the constants $\dfrac{C_{max}}{2}$ and M from part (A) to form a Michaelis–Menten function for the CTE of nickel.

(C) Use the function from part (B) to estimate the CTE of nickel at 600 K and to estimate the temperature when the CTE of nickel is 12.

96. **Physics.** Repeat Problem 95 for the CTE of copper (column 3 of Table 12).

Answers to Matched Problems

1. Vertical asymptote: $x = 1$; $\lim\limits_{x \to 1^+} f(x) = \infty$, $\lim\limits_{x \to 1^-} f(x) = -\infty$
   $\lim\limits_{x \to 3} f(x) = 1/2$ so f does not have a vertical asymptote at $x = 3$

2. Vertical asymptote: $x = -3$; $\lim\limits_{x \to -3^+} f(x) = \lim\limits_{x \to -3^-} f(x) = -\infty$

3. $\lim\limits_{x \to \infty} p(x) = \lim\limits_{x \to -\infty} p(x) = -\infty$

4. (A) $\lim\limits_{x \to \infty} p(x) = -\infty$, $\lim\limits_{x \to -\infty} p(x) = \infty$
   (B) $\lim\limits_{x \to \infty} p(x) = \infty$, $\lim\limits_{x \to -\infty} p(x) = \infty$

5. (A) $y = 0$ (B) No horizontal asymptotes
   (C) $y = 1/2$

6. Vertical asymptotes: $x = -2$, $x = 2$;
   horizontal asymptote: $y = 1$

10.3 Continuity

- Continuity
- Continuity Properties
- Solving Inequalities Using Continuity Properties

Theorem 3 in Section 10.1 states that if f is a polynomial function or a rational function with a nonzero denominator at $x = c$, then

$$\lim_{x \to c} f(x) = f(c) \tag{1}$$

Functions that satisfy equation (1) are said to be *continuous* at $x = c$. A firm understanding of continuous functions is essential for sketching and analyzing graphs. We will also see that continuity properties provide a simple and efficient method for solving inequalities—a tool that we will use extensively in later sections.

Continuity

Compare the graphs shown in Figure 1. Notice that two of the graphs are broken; that is, they cannot be drawn without lifting a pen off the paper. Informally, a function is *continuous over an interval* if its graph over the interval can be drawn without removing a pen from the paper. A function whose graph is broken (disconnected) at $x = c$ is said to be *discontinuous* at $x = c$. Function f (Fig. 1A) is continuous for all x. Function g (Fig. 1B) is discontinuous at $x = 2$ but is continuous over any interval that does not include 2. Function h (Fig. 1C) is discontinuous at $x = 0$ but is continuous over any interval that does not include 0.

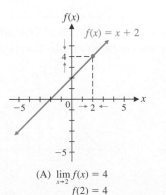

(A) $\lim\limits_{x \to 2} f(x) = 4$
$f(2) = 4$

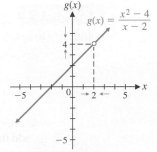

(B) $\lim\limits_{x \to 2} g(x) = 4$
$g(2)$ is not defined

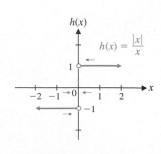

(C) $\lim\limits_{x \to 0} h(x)$ does not exist
$h(0)$ is not defined

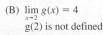

Figure 1

Most graphs of natural phenomena are continuous, whereas many graphs in business and economics applications have discontinuities. Figure 2A illustrates temperature variation over a 24-hour period—a continuous phenomenon. Figure 2B illustrates warehouse inventory over a 1-week period—a discontinuous phenomenon.

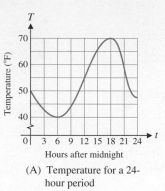

(A) Temperature for a 24-hour period

(B) Inventory in a warehouse during 1 week

Figure 2

Explore and Discuss 1

(A) Write a brief verbal description of the temperature variation illustrated in Figure 2A, including estimates of the high and low temperatures during the period shown and the times at which they occurred.

(B) Write a brief verbal description of the changes in inventory illustrated in Figure 2B, including estimates of the changes in inventory and the times at which those changes occurred.

The preceding discussion leads to the following formal definition of continuity:

DEFINITION Continuity

A function f is **continuous at the point $x = c$ if**

1. $\lim_{x \to c} f(x)$ exists   2. $f(c)$ exists   3. $\lim_{x \to c} f(x) = f(c)$

A function is **continuous on the open interval*** (a, b) if it is continuous at each point on the interval.

_____

*See Table 1 in section 1.1 for a review of interval notation.

If one or more of the three conditions in the definition fails, then the function is **discontinuous** at $x = c$.

EXAMPLE 1   Continuity of a Function Defined by a Graph   Use the definition of continuity to discuss the continuity of the function whose graph is shown in Figure 3.

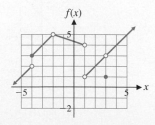

Figure 3

SOLUTION We begin by identifying the points of discontinuity. Examining the graph, we see breaks or holes at $x = -4, -2, 1,$ and 3. Now we must determine which conditions in the definition of continuity are not satisfied at each of these points. In each case, we find the value of the function and the limit of the function at the point in question.

Discontinuity at $x = -4$:

$$\lim_{x \to -4^-} f(x) = 2 \qquad \text{Since the one-sided limits are different,}$$
$$\lim_{x \to -4^+} f(x) = 3 \qquad \text{the limit does not exist (Section 10.1).}$$
$$\lim_{x \to -4} f(x) \text{ does not exist}$$
$$f(-4) = 3$$

So, f is not continuous at $x = -4$ because condition 1 is not satisfied.

Discontinuity at $x = -2$:

$$\lim_{x \to -2^-} f(x) = 5 \qquad \text{The hole at } (-2, 5) \text{ indicates that 5 is not the value of } f$$
$$\lim_{x \to -2^+} f(x) = 5 \qquad \text{at } -2. \text{ Since there is no solid dot elsewhere on the}$$
$$\lim_{x \to -2} f(x) = 5 \qquad \text{vertical line } x = -2, f(-2) \text{ is not defined.}$$
$$f(-2) \text{ does not exist}$$

So even though the limit as x approaches -2 exists, f is not continuous at $x = -2$ because condition 2 is not satisfied.

Discontinuity at $x = 1$:

$$\lim_{x \to 1^-} f(x) = 4$$
$$\lim_{x \to 1^+} f(x) = 1$$
$$\lim_{x \to 1} f(x) \text{ does not exist}$$
$$f(1) \text{ does not exist}$$

This time, f is not continuous at $x = 1$ because neither of conditions 1 and 2 is satisfied.

Discontinuity at $x = 3$:

$$\lim_{x \to 3^-} f(x) = 3 \qquad \text{The solid dot at } (3, 1) \text{ indicates that } f(3) = 1.$$
$$\lim_{x \to 3^+} f(x) = 3$$
$$\lim_{x \to 3} f(x) = 3$$
$$f(3) = 1$$

Conditions 1 and 2 are satisfied, but f is not continuous at $x = 3$ because condition 3 is not satisfied.

Having identified and discussed all points of discontinuity, we can now conclude that f is continuous except at $x = -4, -2, 1,$ and 3.

CONCEPTUAL INSIGHT

Rather than list the points where a function is discontinuous, sometimes it is useful to state the intervals on which the function is continuous. Using the set operation **union**, denoted by $\cup$, we can express the set of points where the function in Example 1 is continuous as follows:

$$(-\infty, -4) \cup (-4, -2) \cup (-2, 1) \cup (1, 3) \cup (3, \infty)$$

Matched Problem 1) Use the definition of continuity to discuss the continuity of the function whose graph is shown in Figure 4.

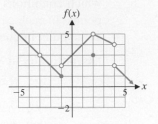

Figure 4

For functions defined by equations, it is important to be able to locate points of discontinuity by examining the equation.

EXAMPLE 2   Continuity of Functions Defined by Equations   Using the definition of continuity, discuss the continuity of each function at the indicated point(s).

(A) $f(x) = x + 2$ at $x = 2$

(B) $g(x) = \dfrac{x^2 - 4}{x - 2}$ at $x = 2$

(C) $h(x) = \dfrac{|x|}{x}$ at $x = 0$ and at $x = 1$

SOLUTION

(A) f is continuous at $x = 2$, since

$$\lim_{x \to 2} f(x) = 4 = f(2) \quad \text{See Figure 1A.}$$

(B) g is not continuous at $x = 2$, since $g(2) = 0/0$ is not defined (see Fig. 1B).

(C) h is not continuous at $x = 0$, since $h(0) = |0|/0$ is not defined; also, $\lim_{x \to 0} h(x)$ does not exist.

h is continuous at $x = 1$, since

$$\lim_{x \to 1} \frac{|x|}{x} = 1 = h(1) \quad \text{See Figure 1C.}$$

Matched Problem 2) Using the definition of continuity, discuss the continuity of each function at the indicated point(s).

(A) $f(x) = x + 1$ at $x = 1$

(B) $g(x) = \dfrac{x^2 - 1}{x - 1}$ at $x = 1$

(C) $h(x) = \dfrac{x - 2}{|x - 2|}$ at $x = 2$ and at $x = 0$

We can also talk about one-sided continuity, just as we talked about one-sided limits. For example, a function is said to be **continuous on the right** at $x = c$ if $\lim_{x \to c^+} f(x) = f(c)$ and **continuous on the left** at $x = c$ if $\lim_{x \to c^-} f(x) = f(c)$. A function is **continuous on the closed interval [a, b]** if it is continuous on the open interval (a, b) and is continuous both on the right at a and on the left at b.

Figure 5A illustrates a function that is continuous on the closed interval $[-1, 1]$. Figure 5B illustrates a function that is continuous on the half-closed interval $[0, \infty)$.

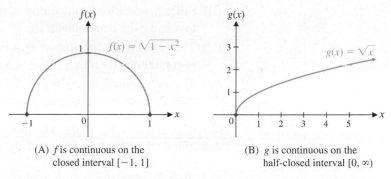

(A) f is continuous on the closed interval $[-1, 1]$

(B) g is continuous on the half-closed interval $[0, \infty)$

Figure 5 **Continuity on closed and half-closed intervals**

Continuity Properties

Functions have some useful **general continuity properties**:

> If two functions are continuous on the same interval, then their sum, difference, product, and quotient are continuous on the same interval except for values of x that make a denominator 0.

These properties, along with Theorem 1, enable us to determine intervals of continuity for some important classes of functions without having to look at their graphs or use the three conditions in the definition.

THEOREM 1 Continuity Properties of Some Specific Functions

(A) A constant function $f(x) = k$, where k is a constant, is continuous for all x.
$f(x) = 7$ is continuous for all x.

(B) For n a positive integer, $f(x) = x^n$ is continuous for all x.
$f(x) = x^5$ is continuous for all x.

(C) A polynomial function is continuous for all x.
$2x^3 - 3x^2 + x - 5$ is continuous for all x.

(D) A rational function is continuous for all x except those values that make a denominator 0.
$\dfrac{x^2 + 1}{x - 1}$ is continuous for all x except $x = 1$, a value that makes the denominator 0.

(E) For n an odd positive integer greater than 1, $\sqrt[n]{f(x)}$ is continuous wherever $f(x)$ is continuous.
$\sqrt[3]{x^2}$ is continuous for all x.

(F) For n an even positive integer, $\sqrt[n]{f(x)}$ is continuous wherever $f(x)$ is continuous and nonnegative.
$\sqrt[4]{x}$ is continuous on the interval $[0, \infty)$.

Parts (C) and (D) of Theorem 1 are the same as Theorem 3 in Section 10.1. They are repeated here to emphasize their importance.

EXAMPLE 3 Using Continuity Properties Using Theorem 1 and the general properties of continuity, determine where each function is continuous.

(A) $f(x) = x^2 - 2x + 1$

(B) $f(x) = \dfrac{x}{(x + 2)(x - 3)}$

(C) $f(x) = \sqrt[3]{x^2 - 4}$

(D) $f(x) = \sqrt{x - 2}$

SOLUTION

(A) Since f is a polynomial function, f is continuous for all x.

(B) Since f is a rational function, f is continuous for all x except -2 and 3 (values that make the denominator 0).

(C) The polynomial function $x^2 - 4$ is continuous for all x. Since $n = 3$ is odd, f is continuous for all x.

(D) The polynomial function $x - 2$ is continuous for all x and nonnegative for $x \geq 2$. Since $n = 2$ is even, f is continuous for $x \geq 2$, or on the interval $[2, \infty)$.

Matched Problem 3 ⌋ Using Theorem 1 and the general properties of continuity, determine where each function is continuous.

(A) $f(x) = x^4 + 2x^2 + 1$ (B) $f(x) = \dfrac{x^2}{(x + 1)(x - 4)}$

(C) $f(x) = \sqrt{x - 4}$ (D) $f(x) = \sqrt[3]{x^3 + 1}$

Solving Inequalities Using Continuity Properties

One of the basic tools for analyzing graphs in calculus is a special line graph called a *sign chart*. We will make extensive use of this type of chart in later sections. In the discussion that follows, we use continuity properties to develop a simple and efficient procedure for constructing sign charts.

Suppose that a function f is continuous over the interval $(1, 8)$ and $f(x) \neq 0$ for any x in $(1, 8)$. Suppose also that $f(2) = 5$, a positive number. Is it possible for $f(x)$ to be negative for any x in the interval $(1, 8)$? The answer is "no." If $f(7)$ were -3, for example, as shown in Figure 6, then how would it be possible to join the points $(2, 5)$ and $(7, -3)$ with the graph of a continuous function without crossing the x axis between 1 and 8 at least once? [Crossing the x axis would violate our assumption that $f(x) \neq 0$ for any x in $(1, 8)$.] We conclude that $f(x)$ must be positive for all x in $(1, 8)$. If $f(2)$ were negative, then, using the same type of reasoning, $f(x)$ would have to be negative over the entire interval $(1, 8)$.

In general, **if f is continuous and $f(x) \neq 0$ on the interval (a, b), then $f(x)$ cannot change sign on (a, b).** This is the essence of Theorem 2.

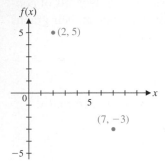

$f(x)$

$\bullet\,(2, 5)$

$(7, -3)$
$\bullet$

Figure 6

THEOREM 2   Sign Properties on an Interval (a, b)

If f is continuous on (a, b) and $f(x) \neq 0$ for all x in (a, b), then either $f(x) > 0$ for all x in (a, b) or $f(x) < 0$ for all x in (a, b).

Theorem 2 provides the basis for an effective method of solving many types of inequalities. Example 4 illustrates the process.

EXAMPLE 4   Solving an Inequality   Solve $\dfrac{x + 1}{x - 2} > 0$.

SOLUTION   We start by using the left side of the inequality to form the function f.

$$f(x) = \frac{x + 1}{x - 2}$$

The denominator is equal to 0 if $x = 2$, and the numerator is equal to 0 if $x = -1$. So the rational function f is discontinuous at $x = 2$, and $f(x) = 0$ for $x = -1$ (a fraction is 0 when the numerator is 0 and the denominator is not 0). We plot

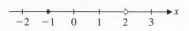

Figure 7

Test Numbers

x	$f(x)$
-2	$\frac{1}{4}$ (+)
0	$-\frac{1}{2}$ (−)
3	4 (+)

$x = 2$ and $x = -1$, which we call *partition numbers,* on a real number line (Fig. 7). (Note that the dot at 2 is open because the function is not defined at $x = 2$.) The partition numbers 2 and -1 determine three open intervals: $(-\infty, -1), (-1, 2),$ and $(2, \infty)$. The function f is continuous and nonzero on each of these intervals. From Theorem 2, we know that $f(x)$ does not change sign on any of these intervals. We can find the sign of $f(x)$ on each of the intervals by selecting a **test number** in each interval and evaluating $f(x)$ at that number. Since any number in each subinterval will do, we choose test numbers that are easy to evaluate: $-2, 0,$ and 3. The table in the margin shows the results.

The sign of $f(x)$ at each test number is the same as the sign of $f(x)$ over the interval containing that test number. Using this information, we construct a **sign chart** for $f(x)$ as shown in Figure 8.

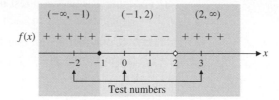

Figure 8

From the sign chart, we can easily write the solution of the given nonlinear inequality:

$$f(x) > 0 \quad \text{for} \quad \begin{array}{ll} x < -1 \quad \text{or} \quad x > 2 & \text{Inequality notation} \\ (-\infty, -1) \cup (2, \infty) & \text{Interval notation} \end{array}$$

Matched Problem 4 Solve $\dfrac{x^2 - 1}{x - 3} < 0$.

Most of the inequalities we encounter will involve strict inequalities ($>$ or $<$). If it is necessary to solve inequalities of the form $\geq$ or $\leq$, we simply include the endpoint x of any interval if f is defined at x and $f(x)$ satisfies the given inequality. For example, from the sign chart in Figure 8, the solution of the inequality

$$\frac{x + 1}{x - 2} \geq 0 \quad \text{is} \quad \begin{array}{ll} x \leq -1 \quad \text{or} \quad x > 2 & \text{Inequality notation} \\ (-\infty, -1] \cup (2, \infty) & \text{Interval notation} \end{array}$$

Example 4 illustrates a general procedure for constructing sign charts.

DEFINITION

A real number x is a **partition number** for a function f if f is discontinuous at x or $f(x) = 0$.

Suppose that p_1 and p_2 are consecutive partition numbers for f, that is, there are no partition numbers in the open interval (p_1, p_2). Then f is continuous on (p_1, p_2) [since there are no points of discontinuity in that interval], so f does not change sign on (p_1, p_2) [since $f(x) \neq 0$ for x in that interval]. In other words, **partition numbers determine open intervals on which f does not change sign.** By using a test number from each interval, we can construct a sign chart for f on the real number line. It is then easy to solve the inequality $f(x) < 0$ or the inequality $f(x) > 0$.

We summarize the procedure for constructing sign charts in the following box.

PROCEDURE   Constructing Sign Charts

Given a function f,

Step 1   Find all partition numbers of f:

(A) Find all numbers x such that f is discontinuous at x. (Rational functions are discontinuous at values of x that make a denominator 0.)

(B) Find all numbers x such that $f(x) = 0$. (For a rational function, this occurs where the numerator is 0 and the denominator is not 0.)

Step 2   Plot the numbers found in step 1 on a real number line, dividing the number line into intervals.

Step 3   Select a test number in each open interval determined in step 2 and evaluate $f(x)$ at each test number to determine whether $f(x)$ is positive $(+)$ or negative $(-)$ in each interval.

Step 4   Construct a sign chart, using the real number line in step 2. This will show the sign of $f(x)$ on each open interval.

There is an alternative to step 3 in the procedure for constructing sign charts that may save time if the function $f(x)$ is written in factored form. The key is to determine the sign of each factor in the numerator and denominator of $f(x)$. We will illustrate with Example 4. The partition numbers -1 and 2 divide the x axis into three open intervals. If $x > 2$, then both the numerator and denominator are positive, so $f(x) > 0$. If $-1 < x < 2$, then the numerator is positive but the denominator is negative, so $f(x) < 0$. If $x < -1$, then both the numerator and denominator are negative, so $f(x) > 0$. Of course both approaches, the test number approach and the sign of factors approach, give the same sign chart.

Exercises 10.3

Skills Warm-up Exercises

W   *In Problems 1–8, use interval notation to specify the given interval. (If necessary, review Table 1 in Section 1.1).*

1. The set of all real numbers from -3 to 5, including -3 and 5

2. The set of all real numbers from -8 to -4, excluding -8 but including -4

3. $\{x \mid -10 < x < 100\}$ 4. $\{x \mid 0.1 \le x \le 0.3\}$

5. $\{x \mid x^2 > 25\}$ 6. $\{x \mid x^2 \ge 16\}$

7. $\{x \mid x \le -1 \text{ or } x > 2\}$

8. $\{x \mid x < 6 \text{ or } x \ge 9\}$

In Problems 9–14, sketch a possible graph of a function that satisfies the given conditions at $x = 1$ and discuss the continuity of f at $x = 1$.

9. $f(1) = 2$ and $\lim_{x \to 1} f(x) = 2$

10. $f(1) = -2$ and $\lim_{x \to 1} f(x) = 2$

11. $f(1) = 2$ and $\lim_{x \to 1} f(x) = -2$

12. $f(1) = -2$ and $\lim_{x \to 1} f(x) = -2$

13. $f(1) = -2$, $\lim_{x \to 1^-} f(x) = 2$, and $\lim_{x \to 1^+} f(x) = -2$

14. $f(1) = 2$, $\lim_{x \to 1^-} f(x) = 2$, and $\lim_{x \to 1^+} f(x) = -2$

Problems 15–22 refer to the function f shown in the figure. Use the graph to estimate the indicated function values and limits.

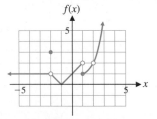

Figure for 15–22

15. $f(0.9)$ 16. $f(0.1)$

17. $f(-1.9)$ 18. $f(-0.9)$

19. (A) $\lim_{x \to 1^-} f(x)$ (B) $\lim_{x \to 1^+} f(x)$
    (C) $\lim_{x \to 1} f(x)$ (D) $f(1)$
    ✐ (E) Is f continuous at $x = 1$? Explain.

20. (A) $\lim_{x \to 2^-} f(x)$ (B) $\lim_{x \to 2^+} f(x)$
    (C) $\lim_{x \to 2} f(x)$ (D) $f(2)$
    ✐ (E) Is f continuous at $x = 2$? Explain.

21. (A) $\lim_{x \to -2^-} f(x)$ (B) $\lim_{x \to -2^+} f(x)$

   (C) $\lim_{x \to -2} f(x)$ (D) $f(-2)$

✎ (E) Is f continuous at $x = -2$? Explain.

22. (A) $\lim_{x \to -1^-} f(x)$ (B) $\lim_{x \to -1^+} f(x)$

   (C) $\lim_{x \to -1} f(x)$ (D) $f(-1)$

✎ (E) Is f continuous at $x = -1$? Explain.

Problems 23–30 refer to the function g shown in the figure. Use the graph to estimate the indicated function values and limits.

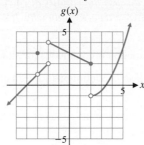

Figure for 23–30

23. $g(-3.1)$ 24. $g(-2.1)$

25. $g(1.9)$ 26. $g(-1.9)$

27. (A) $\lim_{x \to -3^-} g(x)$ (B) $\lim_{x \to -3^+} g(x)$

   (C) $\lim_{x \to -3} g(x)$ (D) $g(-3)$

✎ (E) Is g continuous at $x = -3$? Explain.

28. (A) $\lim_{x \to -2^-} g(x)$ (B) $\lim_{x \to -2^+} g(x)$

   (C) $\lim_{x \to -2} g(x)$ (D) $g(-2)$

✎ (E) Is g continuous at $x = -2$? Explain.

29. (A) $\lim_{x \to 2^-} g(x)$ (B) $\lim_{x \to 2^+} g(x)$

   (C) $\lim_{x \to 2} g(x)$ (D) $g(2)$

✎ (E) Is g continuous at $x = 2$? Explain.

30. (A) $\lim_{x \to 4^-} g(x)$ (B) $\lim_{x \to 4^+} g(x)$

   (C) $\lim_{x \to 4} g(x)$ (D) $g(4)$

✎ (E) Is g continuous at $x = 4$? Explain.

Use Theorem 1 to determine where each function in Problems 31–40 is continuous.

31. $f(x) = 3x - 4$ 32. $h(x) = 4 - 2x$

33. $g(x) = \dfrac{3x}{x + 2}$ 34. $k(x) = \dfrac{2x}{x - 4}$

35. $m(x) = \dfrac{x + 1}{x^2 + 3x - 4}$ 36. $n(x) = \dfrac{x - 2}{x^2 - 2x - 3}$

37. $F(x) = \dfrac{2x}{x^2 + 9}$ 38. $G(x) = \dfrac{1 - x^2}{x^2 + 1}$

39. $M(x) = \dfrac{x - 1}{4x^2 - 9}$ 40. $N(x) = \dfrac{x^2 + 4}{4 - 25x^2}$

In Problems 41–46, find all partition numbers of the function.

41. $f(x) = \dfrac{3x + 8}{x - 4}$ 42. $f(x) = \dfrac{2x + 7}{5x - 1}$

43. $f(x) = \dfrac{1 - x^2}{1 + x^2}$ 44. $f(x) = \dfrac{x^2 + 4}{x^2 - 9}$

45. $f(x) = \dfrac{x^2 + 4x - 45}{x^2 + 6x}$ 46. $f(x) = \dfrac{x^3 + x}{x^2 - x - 42}$

In Problems 47–54, use a sign chart to solve each inequality. Express answers in inequality and interval notation.

47. $x^2 - x - 12 < 0$ 48. $x^2 - 2x - 8 < 0$

49. $x^2 + 21 > 10x$ 50. $x^2 + 7x > -10$

51. $x^3 < 4x$ 52. $x^4 - 9x^2 > 0$

53. $\dfrac{x^2 + 5x}{x - 3} > 0$ 54. $\dfrac{x - 4}{x^2 + 2x} < 0$

55. Use the graph of f to determine where

   (A) $f(x) > 0$ (B) $f(x) < 0$

Express answers in interval notation.

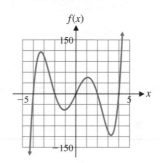

56. Use the graph of g to determine where

   (A) $g(x) > 0$ (B) $g(x) < 0$

Express answers in interval notation.

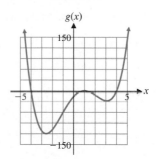

▱ *In Problems 57–60, use a graphing calculator to approximate the partition numbers of each function f(x) to four decimal places. Then solve the following inequalities:*

   (A) $f(x) > 0$ (B) $f(x) < 0$

Express answers in interval notation.

57. $f(x) = x^4 - 6x^2 + 3x + 5$

58. $f(x) = x^4 - 4x^2 - 2x + 2$

59. $f(x) = \dfrac{3 + 6x - x^3}{x^2 - 1}$ 60. $f(x) = \dfrac{x^3 - 5x + 1}{x^2 - 1}$

Use Theorem 1 to determine where each function in Problems 61–68 is continuous. Express the answer in interval notation.

61. $\sqrt{x - 6}$ 62. $\sqrt{7 - x}$

63. $\sqrt[3]{5 - x}$ 64. $\sqrt[3]{x - 8}$

65. $\sqrt{x^2 - 9}$

66. $\sqrt{4 - x^2}$

67. $\sqrt{x^2 + 1}$

68. $\sqrt[3]{x^2 + 2}$

In Problems 69–74, graph f, locate all points of discontinuity, and discuss the behavior of f at these points.

69. $f(x) = \begin{cases} 1 + x & \text{if } x < 1 \\ 5 - x & \text{if } x \ge 1 \end{cases}$

70. $f(x) = \begin{cases} x^2 & \text{if } x \le 1 \\ 2x & \text{if } x > 1 \end{cases}$

71. $f(x) = \begin{cases} 1 + x & \text{if } x \le 2 \\ 5 - x & \text{if } x > 2 \end{cases}$

72. $f(x) = \begin{cases} x^2 & \text{if } x \le 2 \\ 2x & \text{if } x > 2 \end{cases}$

73. $f(x) = \begin{cases} -x & \text{if } x < 0 \\ 1 & \text{if } x = 0 \\ x & \text{if } x > 0 \end{cases}$

74. $f(x) = \begin{cases} 1 & \text{if } x < 0 \\ 0 & \text{if } x = 0 \\ 1 + x & \text{if } x > 0 \end{cases}$

*Problems 75 and 76 refer to the **greatest integer function**, which is denoted by $[\![x]\!]$ and is defined as*

$$[\![x]\!] = greatest\ integer \le x$$

For example,

$$[\![-3.6]\!] = greatest\ integer \le -3.6 = -4$$

$$[\![2]\!] = greatest\ integer \le 2 = 2$$

$$[\![2.5]\!] = greatest\ integer \le 2.5 = 2$$

The graph of $f(x) = [\![x]\!]$ is shown. There, we can see that

$$[\![x]\!] = -2 \quad for \quad -2 \le x < -1$$

$$[\![x]\!] = -1 \quad for \quad -1 \le x < 0$$

$$[\![x]\!] = 0 \quad for \quad 0 \le x < 1$$

$$[\![x]\!] = 1 \quad for \quad 1 \le x < 2$$

$$[\![x]\!] = 2 \quad for \quad 2 \le x < 3$$

and so on.

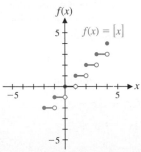

Figure for 75 and 76

75. (A) Is f continuous from the right at $x = 0$?

(B) Is f continuous from the left at $x = 0$?

(C) Is f continuous on the open interval $(0, 1)$?

(D) Is f continuous on the closed interval $[0, 1]$?

(E) Is f continuous on the half-closed interval $[0, 1)$?

76. (A) Is f continuous from the right at $x = 2$?

(B) Is f continuous from the left at $x = 2$?

(C) Is f continuous on the open interval $(1, 2)$?

(D) Is f continuous on the closed interval $[1, 2]$?

(E) Is f continuous on the half-closed interval $[1, 2)$?

In Problems 77–82, discuss the validity of each statement. If the statement is always true, explain why. If not, give a counterexample.

77. A polynomial function is continuous for all real numbers.

78. A rational function is continuous for all but finitely many real numbers.

79. If f is a function that is continuous at $x = 0$ and $x = 2$, then f is continuous at $x = 1$.

80. If f is a function that is continuous on the open interval $(0, 2)$, then f is continuous at $x = 1$.

81. If f is a function that has no partition numbers in the interval (a, b), then f is continuous on (a, b).

82. The greatest integer function (see Problem 75) is a rational function.

In Problems 83–86, sketch a possible graph of a function f that is continuous for all real numbers and satisfies the given conditions. Find the x intercepts of f.

83. $f(x) < 0$ on $(-\infty, -5)$ and $(2, \infty)$; $f(x) > 0$ on $(-5, 2)$

84. $f(x) > 0$ on $(-\infty, -4)$ and $(3, \infty)$; $f(x) < 0$ on $(-4, 3)$

85. $f(x) < 0$ on $(-\infty, -6)$ and $(-1, 4)$; $f(x) > 0$ on $(-6, -1)$ and $(4, \infty)$

86. $f(x) > 0$ on $(-\infty, -3)$ and $(2, 7)$; $f(x) < 0$ on $(-3, 2)$ and $(7, \infty)$

87. The function $f(x) = 2/(1 - x)$ satisfies $f(0) = 2$ and $f(2) = -2$. Is f equal to 0 anywhere on the interval $(-1, 3)$? Does this contradict Theorem 2? Explain.

88. The function $f(x) = 6/(x - 4)$ satisfies $f(2) = -3$ and $f(7) = 2$. Is f equal to 0 anywhere on the interval $(0, 9)$? Does this contradict Theorem 2? Explain.

Applications

89. *Postal rates.* First-class postage in 2009 was \$0.44 for the first ounce (or any fraction thereof) and \$0.17 for each additional ounce (or fraction thereof) up to a maximum weight of 3.5 ounces.

(A) Write a piecewise definition of the first-class postage $P(x)$ for a letter weighing x ounces.

(B) Graph $P(x)$ for $0 < x \le 3.5$.

(C) Is $P(x)$ continuous at $x = 2.5$? At $x = 3$? Explain.

90. **Telephone rates.** A long-distance telephone service charges $0.07 for the first minute (or any fraction thereof) and $0.05 for each additional minute (or fraction thereof).

(A) Write a piecewise definition of the charge $R(x)$ for a long-distance call lasting x minutes.

(B) Graph $R(x)$ for $0 < x \le 6$.

(C) Is $R(x)$ continuous at $x = 3.5$? At $x = 3$? Explain.

91. **Postal rates.** Discuss the differences between the function $Q(x) = 0.44 + 0.17[\![x]\!]$ and the function $P(x)$ defined in Problem 89.

92. **Telephone rates.** Discuss the differences between the function $S(x) = 0.07 + 0.05[\![x]\!]$ and the function $R(x)$ defined in Problem 90.

93. **Natural-gas rates.** Table 1 shows the rates for natural gas charged by the Middle Tennessee Natural Gas Utility District during summer months. The base charge is a fixed monthly charge, independent of the amount of gas used per month.

Table 1 **Summer (May–September)**

Base charge	$5.00
First 50 therms	0.63 per therm
Over 50 therms	0.45 per therm

(A) Write a piecewise definition of the monthly charge $S(x)$ for a customer who uses x therms* in a summer month.

(B) Graph $S(x)$.

(C) Is $S(x)$ continuous at $x = 50$? Explain.

94. **Natural-gas rates.** Table 2 shows the rates for natural gas charged by the Middle Tennessee Natural Gas Utility District during winter months. The base charge is a fixed monthly charge, independent of the amount of gas used per month.

Table 2 **Winter (October– April)**

Base charge	$5.00
First 5 therms	0.69 per therm
Next 45 therms	0.65 per therm
Over 50 therms	0.63 per therm

---

*A British thermal unit (Btu) is the amount of heat required to raise the temperature of 1 pound of water 1 degree Fahrenheit, and a therm is 100,000 Btu.

(A) Write a piecewise definition of the monthly charge $S(x)$ for a customer who uses x therms in a winter month.

(B) Graph $S(x)$.

(C) Is $S(x)$ continuous at $x = 5$? At $x = 50$? Explain.

95. **Income.** A personal-computer salesperson receives a base salary of $1,000 per month and a commission of 5% of all sales over $10,000 during the month. If the monthly sales are $20,000 or more, then the salesperson is given an additional $500 bonus. Let $E(s)$ represent the person's earnings per month as a function of the monthly sales s.

(A) Graph $E(s)$ for $0 \le s \le 30,000$.

(B) Find $\lim_{s \to 10,000} E(s)$ and $E(10,000)$.

(C) Find $\lim_{s \to 20,000} E(s)$ and $E(20,000)$.

(D) Is E continuous at $s = 10,000$? At $s = 20,000$?

96. **Equipment rental.** An office equipment rental and leasing company rents copiers for $10 per day (and any fraction thereof) or for $50 per 7-day week. Let $C(x)$ be the cost of renting a copier for x days.

(A) Graph $C(x)$ for $0 \le x \le 10$.

(B) Find $\lim_{x \to 4.5} C(x)$ and $C(4.5)$.

(C) Find $\lim_{x \to 8} C(x)$ and $C(8)$.

(D) Is C continuous at $x = 4.5$? At $x = 8$?

97. **Animal supply.** A medical laboratory raises its own rabbits. The number of rabbits $N(t)$ available at any time t depends on the number of births and deaths. When a birth or death occurs, the function N generally has a discontinuity, as shown in the figure.

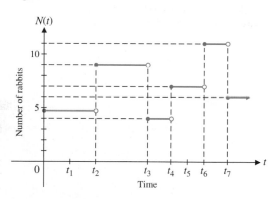

(A) Where is the function N discontinuous?

(B) $\lim_{t \to t_5} N(t) = ?$; $N(t_5) = ?$

(C) $\lim_{t \to t_3} N(t) = ?$; $N(t_3) = ?$

98. **Learning.** The graph shown represents the history of a person learning the material on limits and continuity in this book. At time t_2, the student's mind goes blank during a quiz. At time t_4, the instructor explains a concept particularly well, then suddenly a big jump in understanding takes place.

(A) Where is the function p discontinuous?

(B) $\lim_{t \to t_1} p(t) = ?$; $p(t_1) = ?$

(C) $\lim_{t \to t_2} p(t) = ?$; $p(t_2) = ?$

(D) $\lim_{t \to t_4} p(t) = ?$; $p(t_4) = ?$

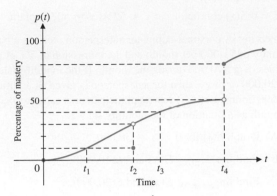

Answers to Matched Problems

1. f is not continuous at $x = -3, -1, 2,$ and 4.

 $x = -3$: $\lim_{x \to -3} f(x) = 3$, but $f(-3)$ does not exist

 $x = -1$: $f(-1) = 1$, but $\lim_{x \to -1} f(x)$ does not exist

$x = 2$: $\lim_{x \to 2} f(x) = 5$, but $f(2) = 3$

$x = 4$: $\lim_{x \to 4} f(x)$ does not exist, and $f(4)$ does not exist

2. (A) f is continuous at $x = 1$, since $\lim_{x \to 1} f(x) = 2 = f(1)$.

 (B) g is not continuous at $x = 1$, since $g(1)$ is not defined.

 (C) h is not continuous at $x = 2$ for two reasons: $h(2)$ does not exist and $\lim_{x \to 2} h(x)$ does not exist.

 h is continuous at $x = 0$, since $\lim_{x \to 0} h(x) = -1 = h(0)$.

3. (A) Since f is a polynomial function, f is continuous for all x.

 (B) Since f is a rational function, f is continuous for all x except -1 and 4 (values that make the denominator 0).

 (C) The polynomial function $x - 4$ is continuous for all x and nonnegative for $x \geq 4$. Since $n = 2$ is even, f is continuous for $x \geq 4$, or on the interval $[4, \infty)$.

 (D) The polynomial function $x^3 + 1$ is continuous for all x. Since $n = 3$ is odd, f is continuous for all x.

4. $-\infty < x < -1$ or $1 < x < 3$; $(-\infty, -1) \cup (1, 3)$

10.4 The Derivative

- Rate of Change
- Slope of the Tangent Line
- The Derivative
- Nonexistence of the Derivative

We will now make use of the limit concepts developed in Sections 10.1, 10.2, and 10.3 to solve the two important problems illustrated in Figure 1. The solution of each of these apparently unrelated problems involves a common concept called the *derivative*.

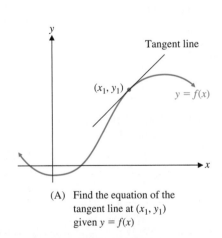

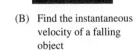

(A) Find the equation of the tangent line at (x_1, y_1) given $y = f(x)$

(B) Find the instantaneous velocity of a falling object

Figure 1 **Two basic problems of calculus**

Rate of Change

If you pass mile marker 120 on the interstate highway at 9 a.m. and mile marker 250 at 11 a.m., then the *average rate of change* of distance with respect to time, also known as *average velocity*, is

$$\frac{250 - 120}{11 - 9} = \frac{130}{2} = 65 \text{ miles per hour}$$

Of course your speedometer reading, that is, the *instantaneous rate of change*, or *instantaneous velocity*, might well have been 75 mph at some moment between 9 a.m. and 11 a.m.

We will define the concepts of average rate of change and instantaneous rate of change more generally, and will apply them in situations that are unrelated to velocity.

> **DEFINITION** Average Rate of Change
> For $y = f(x)$, the **average rate of change from $x = a$ to $x = a + h$** is
> $$\frac{f(a + h) - f(a)}{(a + h) - a} = \frac{f(a + h) - f(a)}{h} \qquad h \neq 0 \qquad (1)$$

Note that the numerator and denominator in (1) are differences, so (1) is a **difference quotient** (see Section 2.1).

EXAMPLE 1 Revenue Analysis The revenue (in dollars) from the sale of x plastic planter boxes is given by

$$R(x) = 20x - 0.02x^2 \qquad 0 \leq x \leq 1{,}000$$

and is graphed in Figure 2.

(A) What is the change in revenue if production is changed from 100 planters to 400 planters?

(B) What is the average rate of change in revenue for this change in production?

SOLUTION

(A) The change in revenue is given by

$$R(400) - R(100) = 20(400) - 0.02(400)^2 - [20(100) - 0.02(100)^2]$$
$$= 4{,}800 - 1{,}800 = \$3{,}000$$

Increasing production from 100 planters to 400 planters will increase revenue by \$3,000.

(B) To find the average rate of change in revenue, we divide the change in revenue by the change in production:

$$\frac{R(400) - R(100)}{400 - 100} = \frac{3{,}000}{300} = \$10$$

The average rate of change in revenue is \$10 per planter when production is increased from 100 to 400 planters.

Matched Problem 1 Refer to the revenue function in Example 1.

(A) What is the change in revenue if production is changed from 600 planters to 800 planters?

(B) What is the average rate of change in revenue for this change in production?

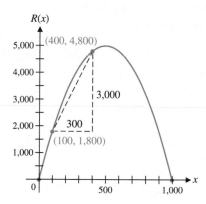

Figure 2 $R(x) = 20x - 0.02x^2$

EXAMPLE 2 Velocity A small steel ball dropped from a tower will fall a distance of y feet in x seconds, as given approximately by the formula

$$y = f(x) = 16x^2$$

Figure 3 shows the position of the ball on a coordinate line (positive direction down) at the end of 0, 1, 2, and 3 seconds.

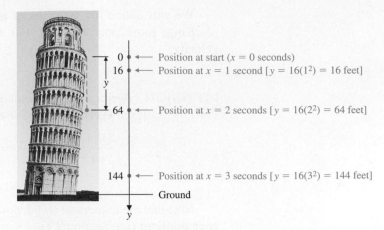

0 ←— Position at start ($x = 0$ seconds)
16 ←— Position at $x = 1$ second [$y = 16(1^2) = 16$ feet]

64 ←— Position at $x = 2$ seconds [$y = 16(2^2) = 64$ feet]

144 ←— Position at $x = 3$ seconds [$y = 16(3^2) = 144$ feet]

Ground

Figure 3 **Note: Positive y direction is down.**

(A) Find the average velocity from $x = 2$ seconds to $x = 3$ seconds.
(B) Find and simplify the average velocity from $x = 2$ seconds to $x = 2 + h$ seconds, $h \neq 0$.
(C) Find the limit of the expression from part (B) as $h \to 0$ if that limit exists.
(D) Discuss possible interpretations of the limit from part (C).

SOLUTION

(A) Recall the formula $d = rt$, which can be written in the form

$$r = \frac{d}{t} = \frac{\text{Distance covered}}{\text{Elapsed time}} = \text{Average velocity}$$

For example, if a person drives from San Francisco to Los Angeles (a distance of about 420 miles) in 7 hours, then the average velocity is

$$r = \frac{d}{t} = \frac{420}{7} = 60 \text{ miles per hour}$$

Sometimes the person will be traveling faster and sometimes slower, but the average velocity is 60 miles per hour. In our present problem, the average velocity of the steel ball from $x = 2$ seconds to $x = 3$ seconds is

$$\text{Average velocity} = \frac{\text{Distance covered}}{\text{Elapsed time}}$$

$$= \frac{f(3) - f(2)}{3 - 2}$$

$$= \frac{16(3)^2 - 16(2)^2}{1} = 80 \text{ feet per second}$$

We see that if $y = f(x)$ is the position of the falling ball, then the average velocity is simply the average rate of change of $f(x)$ with respect to time x.

(B) Proceeding as in part (A), we have

$$\text{Average velocity} = \frac{\text{Distance covered}}{\text{Elapsed time}}$$

$$= \frac{f(2 + h) - f(2)}{h} \qquad \text{Difference quotient}$$

$$= \frac{16(2 + h)^2 - 16(2)^2}{h} \qquad \begin{array}{l}\text{Simplify this O/O}\\\text{indeterminate form.}\end{array}$$

$$= \frac{64 + 64h + 16h^2 - 64}{h}$$

$$= \frac{h(64 + 16h)}{h} = 64 + 16h \qquad h \neq 0$$

Notice that if $h = 1$, the average velocity is 80 feet per second, which is the result in part (A).

(C) The limit of the average velocity expression from part (B) as $h \to 0$ is

$$\lim_{h \to 0} \frac{f(2 + h) - f(2)}{h} = \lim_{h \to 0}(64 + 16h)$$

$$= 64 \text{ feet per second}$$

(D) The average velocity over smaller and smaller time intervals approaches 64 feet per second. This limit can be interpreted as the velocity of the ball at the *instant* that the ball has been falling for exactly 2 seconds. Therefore, 64 feet per second is referred to as the **instantaneous velocity** at $x = 2$ seconds, and we have solved one of the basic problems of calculus (see Fig. 1B).

Matched Problem 2 For the falling steel ball in Example 2, find

(A) The average velocity from $x = 1$ second to $x = 2$ seconds

(B) The average velocity (in simplified form) from $x = 1$ second to $x = 1 + h$ seconds, $h \neq 0$

(C) The instantaneous velocity at $x = 1$ second

The ideas in Example 2 can be applied to the average rate of change of any function.

DEFINITION Instantaneous Rate of Change
For $y = f(x)$, the **instantaneous rate of change at $x = a$** is

$$\lim_{h \to 0} \frac{f(a + h) - f(a)}{h} \tag{2}$$

if the limit exists.

The adjective *instantaneous* is often omitted with the understanding that the phrase **rate of change** always refers to the instantaneous rate of change and not the average rate of change. Similarly, **velocity** always refers to the instantaneous rate of change of distance with respect to time.

Slope of the Tangent Line

So far, our interpretations of the difference quotient have been numerical in nature. Now we want to consider a geometric interpretation.

In geometry, a line that intersects a circle in two points is called a *secant line*, and a line that intersects a circle in exactly one point is called a *tangent line* (Fig. 4, page 530). If the point Q in Figure 4 is moved closer and closer to the point P, then the angle between the secant line PQ and the tangent line at P gets smaller and smaller. We will generalize the geometric concepts of secant line and tangent line of a circle and will use them to study graphs of functions.

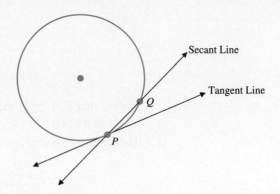

Figure 4 **Secant line and tangent line of a circle**

A line through two points on the graph of a function is called a **secant line**. If $(a, f(a))$ and $(a + h, f(a + h))$ are two points on the graph of $y = f(x)$, then we can use the slope formula from Section 1.2 to find the slope of the secant line through these points (Fig. 5).

$$\textbf{Slope of secant line} = \frac{y_2 - y_1}{x_2 - x_1} = \frac{f(a + h) - f(a)}{(a + h) - a}$$

$$= \frac{f(a + h) - f(a)}{h} \qquad \text{Difference quotient}$$

The difference quotient can be interpreted as both the average rate of change and the slope of the secant line.

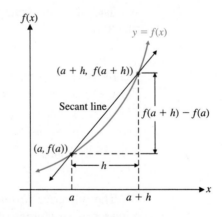

Figure 5 **Secant line**

EXAMPLE 3   Slope of a Secant Line   Given $f(x) = x^2$,

(A) Find the slope of the secant line for $a = 1$ and $h = 2$ and 1, respectively. Graph $y = f(x)$ and the two secant lines.

(B) Find and simplify the slope of the secant line for $a = 1$ and h any nonzero number.

(C) Find the limit of the expression in part (B).

(D) Discuss possible interpretations of the limit in part (C).

SOLUTION

(A) For $a = 1$ and $h = 2$, the secant line goes through $(1, f(1)) = (1, 1)$ and $(3, f(3)) = (3, 9)$, and its slope is

$$\frac{f(1 + 2) - f(1)}{2} = \frac{3^2 - 1^2}{2} = 4$$

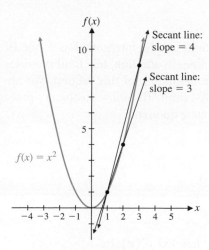

Figure 6 **Secant lines**

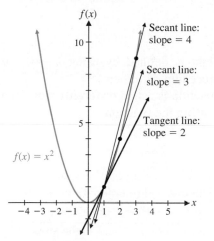

Figure 7 **Tangent line**

For $a = 1$ and $h = 1$, the secant line goes through $(1, f(1)) = (1, 1)$ and $(2, f(2)) = (2, 4)$, and its slope is

$$\frac{f(1 + 1) - f(1)}{1} = \frac{2^2 - 1^2}{1} = 3$$

The graphs of $y = f(x)$ and the two secant lines are shown in Figure 6.

(B) For $a = 1$ and h any nonzero number, the secant line goes through $(1, f(1)) = (1, 1)$ and $(1 + h, f(1 + h)) = (1 + h, (1 + h)^2)$, and its slope is

$$\frac{f(1 + h) - f(1)}{h} = \frac{(1 + h)^2 - 1^2}{h} \qquad \text{Square the binomial.}$$

$$= \frac{1 + 2h + h^2 - 1}{h} \qquad \text{Combine like terms and factor the numerator.}$$

$$= \frac{h(2 + h)}{h} \qquad \text{Cancel.}$$

$$= 2 + h \qquad h \neq 0$$

(C) The limit of the secant line slope from part (B) is

$$\lim_{h \to 0} \frac{f(1 + h) - f(1)}{h} = \lim_{h \to 0} (2 + h)$$

$$= 2$$

(D) In part (C), we saw that the limit of the slopes of the secant lines through the point $(1, f(1))$ is 2. If we graph the line through $(1, f(1))$ with slope 2 (Fig. 7), then this line is the limit of the secant lines. The slope obtained from the limit of slopes of secant lines is called the *slope of the graph* at $x = 1$. The line through the point $(1, f(1))$ with this slope is called the *tangent line*. We have solved another basic problem of calculus (see Fig. 1A on page 526).

Matched Problem 3 Given $f(x) = x^2$,

(A) Find the slope of the secant line for $a = 2$ and $h = 2$ and 1, respectively.

(B) Find and simplify the slope of the secant line for $a = 2$ and h any nonzero number.

(C) Find the limit of the expression in part (B).

(D) Find the slope of the graph and the slope of the tangent line at $a = 2$.

The ideas introduced in the preceding example are summarized next:

DEFINITION Slope of a Graph and Tangent Line
Given $y = f(x)$, the **slope of the graph** at the point $(a, f(a))$ is given by

$$\lim_{h \to 0} \frac{f(a + h) - f(a)}{h} \qquad (3)$$

provided the limit exists. In this case, the **tangent line** to the graph is the line through $(a, f(a))$ with slope given by (3).

CONCEPTUAL INSIGHT

If the function f is continuous at a, then

$$\lim_{h \to 0} f(a + h) = f(a)$$

and limit (3) will be a $0/0$ indeterminate form. As we saw in Examples 2 and 3, evaluating this type of limit typically involves algebraic simplification.

The Derivative

We have seen that the limit of a difference quotient can be interpreted as a rate of change, as a velocity, or as the slope of a tangent line. In addition, this limit provides solutions to the two basic problems stated at the beginning of this section. We are now ready to introduce some terms that refer to that limit. To follow customary practice, we use x in place of a and think of the difference quotient

$$\frac{f(x + h) - f(x)}{h}$$

as a function of h, with x held fixed as h tends to 0.

DEFINITION The Derivative

For $y = f(x)$, we define the **derivative of f at x**, denoted $f'(x)$, by

$$f'(x) = \lim_{h \to 0} \frac{f(x + h) - f(x)}{h} \quad \text{if the limit exists}$$

If $f'(x)$ exists for each x in the open interval (a, b), then f is said to be **differentiable** over (a, b).

(Differentiability from the left or from the right is defined by using $h \to 0^-$ or $h \to 0^+$, respectively, in place of $h \to 0$ in the preceding definition.)

The process of finding the derivative of a function is called **differentiation**. The derivative of a function is obtained by **differentiating** the function.

SUMMARY Interpretations of the Derivative

The derivative of a function f is a new function f'. The domain of f' is a subset of the domain of f. The derivative has various applications and interpretations, including the following:

1. *Slope of the tangent line.* For each x in the domain of f', $f'(x)$ is the slope of the line tangent to the graph of f at the point $(x, f(x))$.
2. *Instantaneous rate of change.* For each x in the domain of f', $f'(x)$ is the instantaneous rate of change of $y = f(x)$ with respect to x.
3. *Velocity.* If $f(x)$ is the position of a moving object at time x, then $v = f'(x)$ is the velocity of the object at that time.

Example 4 illustrates the **four-step process** that we use to find derivatives in this section. In subsequent sections, we develop rules for finding derivatives that do not involve limits. However, it is important that you master the limit process in order to fully comprehend and appreciate the various applications we will consider.

EXAMPLE 4 Finding a Derivative Find $f'(x)$, the derivative of f at x, for $f(x) = 4x - x^2$.

SOLUTION To find $f'(x)$, we use a four-step process.

Step 1 Find $f(x + h)$.
$$f(x + h) = 4(x + h) - (x + h)^2$$
$$= 4x + 4h - x^2 - 2xh - h^2$$

Step 2 Find $f(x + h) - f(x)$.
$$f(x + h) - f(x) = 4x + 4h - x^2 - 2xh - h^2 - (4x - x^2)$$
$$= 4h - 2xh - h^2$$

Step 3 Find $\dfrac{f(x+h)-f(x)}{h}$.

$$\frac{f(x+h)-f(x)}{h}=\frac{4h-2xh-h^2}{h}=\frac{h(4-2x-h)}{h}$$

$$=4-2x-h,\quad h\neq 0$$

Step 4 Find $f'(x)=\lim\limits_{h\to 0}\dfrac{f(x+h)-f(x)}{h}$.

$$f'(x)=\lim_{h\to 0}\frac{f(x+h)-f(x)}{h}=\lim_{h\to 0}(4-2x-h)=4-2x$$

So if $f(x)=4x-x^2$, then $f'(x)=4-2x$. The function f' is a new function derived from the function f.

Matched Problem 4) Find $f'(x)$, the derivative of f at x, for $f(x)=8x-2x^2$.

The four-step process used in Example 4 is summarized as follows for easy reference:

PROCEDURE The four-step process for finding the derivative of a function f:

Step 1 Find $f(x+h)$.

Step 2 Find $f(x+h)-f(x)$.

Step 3 Find $\dfrac{f(x+h)-f(x)}{h}$.

Step 4 Find $\lim\limits_{h\to 0}\dfrac{f(x+h)-f(x)}{h}$.

EXAMPLE 5 Finding Tangent Line Slopes In Example 4, we started with the function $f(x)=4x-x^2$ and found the derivative of f at x to be $f'(x)=4-2x$. So the slope of a line tangent to the graph of f at any point $(x, f(x))$ on the graph is

$$m=f'(x)=4-2x$$

(A) Find the slope of the graph of f at $x=0$, $x=2$, and $x=3$.

(B) Graph $y=f(x)=4x-x^2$ and use the slopes found in part (A) to make a rough sketch of the lines tangent to the graph at $x=0$, $x=2$, and $x=3$.

SOLUTION

(A) Using $f'(x)=4-2x$, we have

$$f'(0)=4-2(0)=4 \qquad \text{Slope at } x=0$$
$$f'(2)=4-2(2)=0 \qquad \text{Slope at } x=2$$
$$f'(3)=4-2(3)=-2 \quad \text{Slope at } x=3$$

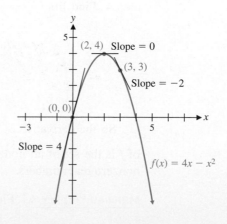

Matched Problem 5) In Matched Problem 4, we started with the function $f(x) = 8x - 2x^2$. Using the derivative found there,

(A) Find the slope of the graph of f at $x = 1$, $x = 2$, and $x = 4$.

(B) Graph $y = f(x) = 8x - 2x^2$, and use the slopes from part (A) to make a rough sketch of the lines tangent to the graph at $x = 1$, $x = 2$, and $x = 4$.

Explore and Discuss 1   In Example 4, we found that the derivative of $f(x) = 4x - x^2$ is $f'(x) = 4 - 2x$. In Example 5, we graphed $f(x)$ and several tangent lines.

(A) Graph f and f' on the same set of axes.

(B) The graph of f' is a straight line. Is it a tangent line for the graph of f? Explain.

(C) Find the x intercept for the graph of f'. What is the slope of the line tangent to the graph of f for this value of x? Write a verbal description of the relationship between the slopes of the tangent lines of a function and the x intercepts of the derivative of the function.

EXAMPLE 6   Finding a Derivative   Find $f'(x)$, the derivative of f at x, for $f(x) = \dfrac{1}{x}$.

SOLUTION

Step 1   Find $f(x + h)$.

$$f(x + h) = \frac{1}{x + h}$$

Step 2   Find $f(x + h) - f(x)$.

$$f(x + h) - f(x) = \frac{1}{x + h} - \frac{1}{x} \qquad \text{Add fractions. (Section A.4)}$$

$$= \frac{x - (x + h)}{x(x + h)} \qquad \text{Simplify.}$$

$$= \frac{-h}{x(x + h)}$$

Step 3   Find $\dfrac{f(x + h) - f(x)}{h}$

$$\frac{f(x + h) - f(x)}{h} = \frac{\dfrac{-h}{x(x + h)}}{h} \qquad \text{Simplify.}$$

$$= \frac{-1}{x(x + h)} \qquad h \neq 0$$

Step 4   Find $\lim\limits_{h \to 0} \dfrac{f(x + h) - f(x)}{h}$.

$$\lim_{h \to 0} \frac{f(x + h) - f(x)}{h} = \lim_{h \to 0} \frac{-1}{x(x + h)}$$

$$= \frac{-1}{x^2} \qquad x \neq 0$$

So the derivative of $f(x) = \dfrac{1}{x}$ is $f'(x) = \dfrac{-1}{x^2}$, a new function. The domain of f is the set of all nonzero real numbers. The domain of f' is also the set of all nonzero real numbers.

Matched Problem 6) Find $f'(x)$ for $f(x) = \dfrac{1}{x + 2}$.

EXAMPLE 7 Finding a Derivative Find $f'(x)$, the derivative of f at x, for $f(x) = \sqrt{x} + 2$.

SOLUTION We use the four-step process to find $f'(x)$.

Step 1 Find $f(x + h)$.
$$f(x + h) = \sqrt{x + h} + 2$$

Step 2 Find $f(x + h) - f(x)$.
$$f(x + h) - f(x) = \sqrt{x + h} + 2 - (\sqrt{x} + 2) \quad \text{Combine like terms.}$$
$$= \sqrt{x + h} - \sqrt{x}$$

Step 3 Find $\dfrac{f(x + h) - f(x)}{h}$.

$$\frac{f(x + h) - f(x)}{h} = \frac{\sqrt{x + h} - \sqrt{x}}{h}$$

We rationalize the numerator (Appendix A, Section A.6) to change the form of this fraction.

$$= \frac{\sqrt{x + h} - \sqrt{x}}{h} \cdot \frac{\sqrt{x + h} + \sqrt{x}}{\sqrt{x + h} + \sqrt{x}}$$

$$= \frac{x + h - x}{h(\sqrt{x + h} + \sqrt{x})}$$

Combine like terms.

$$= \frac{h}{h(\sqrt{x + h} + \sqrt{x})}$$

Cancel.

$$= \frac{1}{\sqrt{x + h} + \sqrt{x}} \quad h \neq 0$$

Step 4 Find $f'(x) = \lim\limits_{h \to 0} \dfrac{f(x + h) - f(x)}{h}$.

$$\lim_{h \to 0} \frac{f(x + h) - f(x)}{h} = \lim_{h \to 0} \frac{1}{\sqrt{x + h} + \sqrt{x}}$$

$$= \frac{1}{\sqrt{x} + \sqrt{x}} = \frac{1}{2\sqrt{x}} \quad x > 0$$

So the derivative of $f(x) = \sqrt{x} + 2$ is $f'(x) = 1/(2\sqrt{x})$, a new function. The domain of f is $[0, \infty)$. Since $f'(0)$ is not defined, the domain of f' is $(0, \infty)$, a subset of the domain of f.

Matched Problem 7 Find $f'(x)$ for $f(x) = \sqrt{x} + 4$.

EXAMPLE 8 Sales Analysis A company's total sales (in millions of dollars) t months from now are given by $S(t) = \sqrt{t} + 2$. Find and interpret $S(25)$ and $S'(25)$. Use these results to estimate the total sales after 26 months and after 27 months.

SOLUTION The total sales function S has the same form as the function f in Example 7. Only the letters used to represent the function and the independent variable have been changed. It follows that S' and f' also have the same form:

$$S(t) = \sqrt{t} + 2 \qquad f(x) = \sqrt{x} + 2$$
$$S'(t) = \frac{1}{2\sqrt{t}} \qquad f'(x) = \frac{1}{2\sqrt{x}}$$

Evaluating S and S' at $t = 25$, we have

$$S(25) = \sqrt{25} + 2 = 7 \qquad S'(25) = \frac{1}{2\sqrt{25}} = 0.1$$

So 25 months from now, the total sales will be $7 million and will be increasing at the rate of $0.1 million ($100,000) per month. If this instantaneous rate of change of sales remained constant, the sales would grow to $7.1 million after 26 months, $7.2 million after 27 months, and so on. Even though $S'(t)$ is not a constant function in this case, these values provide useful estimates of the total sales.

Matched Problem 8⟩ A company's total sales (in millions of dollars) t months from now are given by $S(t) = \sqrt{t + 4}$. Find and interpret $S(12)$ and $S'(12)$. Use these results to estimate the total sales after 13 months and after 14 months. (Use the derivative found in Matched Problem 7.)

In Example 8, we can compare the estimates of total sales by using the derivative with the corresponding exact values of $S(t)$:

Exact values   Estimated values

$$S(26) = \sqrt{26} + 2 = 7.099\ldots \approx 7.1$$
$$S(27) = \sqrt{27} + 2 = 7.196\ldots \approx 7.2$$

For this function, the estimated values provide very good approximations to the exact values of $S(t)$. For other functions, the approximation might not be as accurate.

Using the instantaneous rate of change of a function at a point to estimate values of the function at nearby points is an important application of the derivative.

Nonexistence of the Derivative

The existence of a derivative at $x = a$ depends on the existence of a limit at $x = a$, that is, on the existence of

$$f'(a) = \lim_{h \to 0} \frac{f(a + h) - f(a)}{h} \tag{4}$$

If the limit does not exist at $x = a$, we say that the function f is **nondifferentiable at $x = a$, or $f'(a)$ does not exist.**

Explore and Discuss 2 Let $f(x) = |x - 1|$.

(A) Graph f.

(B) Complete the following table:

h	−0.1	−0.01	−0.001	→0←	0.001	0.01	0.1
$\dfrac{f(1 + h) - f(1)}{h}$	?	?	?	→?←	?	?	?

(C) Find the following limit if it exists:

$$\lim_{h \to 0} \frac{f(1 + h) - f(1)}{h}$$

(D) Use the results of parts (A)–(C) to discuss the existence of $f'(1)$.

(E) Repeat parts (A)–(D) for $\sqrt[3]{x - 1}$.

How can we recognize the points on the graph of f where $f'(a)$ does not exist? It is impossible to describe all the ways that the limit of a difference quotient can fail to exist. However, we can illustrate some common situations where $f'(a)$ fails to exist (see Fig. 8):

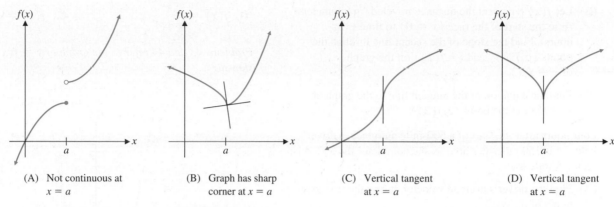

(A) Not continuous at $x = a$

(B) Graph has sharp corner at $x = a$

(C) Vertical tangent at $x = a$

(D) Vertical tangent at $x = a$

Figure 8 **The function f is nondifferentiable at $x = a$.**

1. If the graph of f has a hole or a break at $x = a$, then $f'(a)$ does not exist (Fig. 8A).

2. If the graph of f has a sharp corner at $x = a$, then $f'(a)$ does not exist, and the graph has no tangent line at $x = a$ (Fig. 8B). (In Fig. 8B, the left- and right-hand derivatives exist but are not equal.)

3. If the graph of f has a vertical tangent line at $x = a$, then $f'(a)$ does not exist (Fig. 8C and D).

Exercises 10.4

Skills Warm-up Exercises

W

In Problems 1–4, find the slope of the line through the given points. Write the slope as a reduced fraction, and also give its decimal form. (If necessary, review Section 1.2).

1. $(2, 7)$ and $(6, 16)$

2. $(-1, 11)$ and $(1, 8)$

3. $(10, 14)$ and $(0, 68)$

4. $(-12, -3)$ and $(4, 3)$

In Problems 5–8, write the expression in the form $a + b\sqrt{n}$ where a and b are reduced fractions and n is an integer. (If necessary, review Section A.6).

5. $\dfrac{1}{\sqrt{3}}$

6. $\dfrac{2}{\sqrt{5}}$

7. $\dfrac{5}{3 + \sqrt{7}}$

8. $\dfrac{1 - \sqrt{2}}{5 + \sqrt{2}}$

In Problems 9 and 10, find the indicated quantity for $y = f(x) = 5 - x^2$ and interpret that quantity in terms of the following graph.

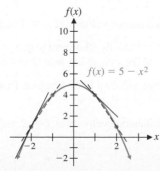

9. (A) $\dfrac{f(2) - f(1)}{2 - 1}$ (B) $\dfrac{f(1 + h) - f(1)}{h}$

(C) $\displaystyle\lim_{h \to 0} \dfrac{f(1 + h) - f(1)}{h}$

10. (A) $\dfrac{f(-1) - f(-2)}{-1 - (-2)}$ (B) $\dfrac{f(-2 + h) - f(-2)}{h}$

(C) $\displaystyle\lim_{h \to 0} \dfrac{f(-2 + h) - f(-2)}{h}$

11. Find the indicated quantities for $f(x) = 3x^2$.

(A) The slope of the secant line through the points $(1, f(1))$ and $(4, f(4))$ on the graph of $y = f(x)$.

(B) The slope of the secant line through the points $(1, f(1))$ and $(1 + h, f(1 + h))$, $h \neq 0$. Simplify your answer.

(C) The slope of the graph at $(1, f(1))$.

12. Find the indicated quantities for $f(x) = 3x^2$.

(A) The slope of the secant line through the points $(2, f(2))$ and $(5, f(5))$ on the graph of $y = f(x)$.

(B) The slope of the secant line through the points $(2, f(2))$ and $(2 + h, f(2 + h))$, $h \neq 0$. Simplify your answer.

(C) The slope of the graph at $(2, f(2))$.

13. Two hours after the start of a 100-kilometer bicycle race, a cyclist passes the 80-kilometer mark while riding at a velocity of 45 kilometers per hour.

(A) Find the cyclist's average velocity during the first two hours of the race.

(B) Let $f(x)$ represent the distance traveled (in kilometers) from the start of the race $(x = 0)$ to time x (in hours). Find the slope of the secant line through the points $(0, f(0))$ and $(2, f(2))$ on the graph of $y = f(x)$.

(C) Find the equation of the tangent line to the graph of $y = f(x)$ at the point $(2, f(2))$.

14. Four hours after the start of a 600-mile auto race, a driver's velocity is 150 miles per hour as she completes the 352nd lap on a 1.5-mile track.

(A) Find the driver's average velocity during the first four hours of the race.

(B) Let $f(x)$ represent the distance traveled (in miles) from the start of the race $(x = 0)$ to time x (in hours). Find the slope of the secant line through the points $(0, f(0))$ and $(4, f(4))$ on the graph of $y = f(x)$.

(C) Find the equation of the tangent line to the graph of $y = f(x)$ at the point $(4, f(4))$.

15. For $f(x) = \frac{1}{1+x^2}$, the slope of the graph of $y = f(x)$ is known to be $-\frac{1}{2}$ at the point with x coordinate 1. Find the equation of the tangent line at that point.

16. For $f(x) = \frac{1}{1+x^2}$, the slope of the graph of $y = f(x)$ is known to be -0.16 at the point with x coordinate 2. Find the equation of the tangent line at that point.

17. For $f(x) = x^4$, the instantaneous rate of change is known to be -32 at $x = -2$. Find the equation of the tangent line to the graph of $y = f(x)$ at the point with x coordinate -2.

18. For $f(x) = x^4$, the instantaneous rate of change is known to be -4 at $x = -1$. Find the equation of the tangent line to the graph of $y = f(x)$ at the point with x coordinate -1.

In Problems 19–42, use the four-step process to find $f'(x)$ and then find $f'(1), f'(2)$, and $f'(3)$.

19. $f(x) = -5$ 20. $f(x) = 9$

21. $f(x) = 3x - 7$ 22. $f(x) = 4 - 6x$

23. $f(x) = 2 - 3x^2$ 24. $f(x) = 2x^2 + 8$

25. $f(x) = x^2 + 6x - 10$ 26. $f(x) = x^2 + 4x + 7$

27. $f(x) = 2x^2 - 7x + 3$ 28. $f(x) = 2x^2 + 5x + 1$

29. $f(x) = -x^2 + 4x - 9$ 30. $f(x) = -x^2 + 9x - 2$

31. $f(x) = 2x^3 + 1$ 32. $f(x) = -2x^3 + 5$

33. $f(x) = 4 + \dfrac{4}{x}$ 34. $f(x) = \dfrac{6}{x} - 2$

35. $f(x) = 5 + 3\sqrt{x}$ 36. $f(x) = 3 - 7\sqrt{x}$

37. $f(x) = 10\sqrt{x + 5}$ 38. $f(x) = 16\sqrt{x + 9}$

39. $f(x) = \dfrac{1}{x - 4}$ 40. $f(x) = \dfrac{1}{x + 4}$

41. $f(x) = \dfrac{x}{x + 1}$ 42. $f(x) = \dfrac{x}{x + 2}$

Problems 43 and 44 refer to the graph of $y = f(x) = x^2 + x$ shown.

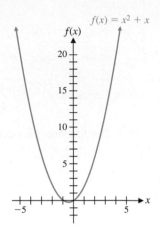

43. (A) Find the slope of the secant line joining $(1, f(1))$ and $(3, f(3))$.

(B) Find the slope of the secant line joining $(1, f(1))$ and $(1 + h, f(1 + h))$.

(C) Find the slope of the tangent line at $(1, f(1))$.

(D) Find the equation of the tangent line at $(1, f(1))$.

44. (A) Find the slope of the secant line joining $(2, f(2))$ and $(4, f(4))$.

(B) Find the slope of the secant line joining $(2, f(2))$ and $(2 + h, f(2 + h))$.

(C) Find the slope of the tangent line at $(2, f(2))$.

(D) Find the equation of the tangent line at $(2, f(2))$.

In Problems 45 and 46, suppose an object moves along the y axis so that its location is $y = f(x) = x^2 + x$ at time x (y is in meters and x is in seconds). Find

45. (A) The average velocity (the average rate of change of y with respect to x) for x changing from 1 to 3 seconds

(B) The average velocity for x changing from 1 to $1 + h$ seconds

(C) The instantaneous velocity at $x = 1$ second

46. (A) The average velocity (the average rate of change of y with respect to x) for x changing from 2 to 4 seconds

(B) The average velocity for x changing from 2 to $2 + h$ seconds

(C) The instantaneous velocity at $x = 2$ seconds

Problems 47–54, refer to the function F in the graph shown. Use the graph to determine whether F'(x) exists at each indicated value of x.

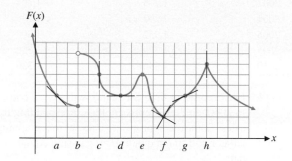

47. $x = a$

48. $x = b$

49. $x = c$

50. $x = d$

51. $x = e$

52. $x = f$

53. $x = g$

54. $x = h$

55. Given $f(x) = x^2 - 4x$,

(A) Find $f'(x)$.

(B) Find the slopes of the lines tangent to the graph of f at $x = 0, 2,$ and 4.

(C) Graph f and sketch in the tangent lines at $x = 0, 2,$ and 4.

56. Given $f(x) = x^2 + 2x$,

(A) Find $f'(x)$.

(B) Find the slopes of the lines tangent to the graph of f at $x = -2, -1,$ and 1.

(C) Graph f and sketch in the tangent lines at $x = -2, -1,$ and 1.

57. If an object moves along a line so that it is at $y = f(x) = 4x^2 - 2x$ at time x (in seconds), find the instantaneous velocity function $v = f'(x)$ and find the velocity at times $x = 1, 3,$ and 5 seconds (y is measured in feet).

58. Repeat Problem 57 with $f(x) = 8x^2 - 4x$.

59. Let $f(x) = x^2$, $g(x) = x^2 - 1$, and $h(x) = x^2 + 2$.

(A) How are the graphs of these functions related? How would you expect the derivatives of these functions to be related?

(B) Use the four-step process to find the derivative of $m(x) = x^2 + C$, where C is any real constant.

60. Let $f(x) = -x^2$, $g(x) = -x^2 - 1$, and $h(x) = -x^2 + 2$.

(A) How are the graphs of these functions related? How would you expect the derivatives of these functions to be related?

(B) Use the four-step process to find the derivative of $m(x) = -x^2 + C$, where C is any real constant.

In Problems 61–66, discuss the validity of each statement. If the statement is always true, explain why. If not, give a counterexample.

61. If $f(x) = C$ is a constant function, then $f'(x) = 0$.

62. If $f(x) = mx + b$ is a linear function, then $f'(x) = m$.

63. If a function f is continuous on the interval (a, b), then f is differentiable on (a, b).

64. If a function f is differentiable on the interval (a, b), then f is continuous on (a, b).

65. The average rate of change of a function f from $x = a$ to $x = a + h$ is less than the instantaneous rate of change at $x = a + \dfrac{h}{2}$.

66. If the graph of f has a sharp corner at $x = a$, then f is not continuous at $x = a$.

In Problems 67–70, sketch the graph of f and determine where f is nondifferentiable.

67. $f(x) = \begin{cases} 2x & \text{if } x < 1 \\ 2 & \text{if } x \geq 1 \end{cases}$
68. $f(x) = \begin{cases} 2x & \text{if } x < 2 \\ 6 - x & \text{if } x \geq 2 \end{cases}$

69. $f(x) = \begin{cases} x^2 + 1 & \text{if } x < 0 \\ 1 & \text{if } x \geq 0 \end{cases}$

70. $f(x) = \begin{cases} 2 - x^2 & \text{if } x \leq 0 \\ 2 & \text{if } x > 0 \end{cases}$

In Problems 71–76, determine whether f is differentiable at x = 0 by considering

$$\lim_{h \to 0} \frac{f(0 + h) - f(0)}{h}$$

71. $f(x) = |x|$

72. $f(x) = 1 - |x|$

73. $f(x) = x^{1/3}$

74. $f(x) = x^{2/3}$

75. $f(x) = \sqrt{1 - x^2}$

76. $f(x) = \sqrt{1 + x^2}$

77. A ball dropped from a balloon falls $y = 16x^2$ feet in x seconds. If the balloon is 576 feet above the ground when the ball is dropped, when does the ball hit the ground? What is the velocity of the ball at the instant it hits the ground?

78. Repeat Problem 77 if the balloon is 1,024 feet above the ground when the ball is dropped.

Applications

79. **Revenue.** The revenue (in dollars) from the sale of x infant car seats is given by

$$R(x) = 60x - 0.025x^2 \qquad 0 \leq x \leq 2,400$$

(A) Find the average change in revenue if production is changed from 1,000 car seats to 1,050 car seats.

(B) Use the four-step process to find $R'(x)$.

(C) Find the revenue and the instantaneous rate of change of revenue at a production level of 1,000 car seats, and write a brief verbal interpretation of these results.

80. Profit. The profit (in dollars) from the sale of x infant car seats is given by

$$P(x) = 45x - 0.025x^2 - 5,000 \qquad 0 \le x \le 2,400$$

(A) Find the average change in profit if production is changed from 800 car seats to 850 car seats.

(B) Use the four-step process to find $P'(x)$.

✎ (C) Find the profit and the instantaneous rate of change of profit at a production level of 800 car seats, and write a brief verbal interpretation of these results.

81. Sales analysis. A company's total sales (in millions of dollars) t months from now are given by

$$S(t) = 2\sqrt{t + 10}$$

(A) Use the four-step process to find $S'(t)$.

✎ (B) Find $S(15)$ and $S'(15)$. Write a brief verbal interpretation of these results.

(C) Use the results in part (B) to estimate the total sales after 16 months and after 17 months.

82. Sales analysis. A company's total sales (in millions of dollars) t months from now are given by

$$S(t) = 2\sqrt{t + 6}$$

(A) Use the four-step process to find $S'(t)$.

✎ (B) Find $S(10)$ and $S'(10)$. Write a brief verbal interpretation of these results.

(C) Use the results in part (B) to estimate the total sales after 11 months and after 12 months.

83. Mineral consumption. The U.S. consumption of tungsten (in metric tons) is given approximately by

$$p(t) = 138t^2 + 1,072t + 14,917$$

where t is time in years and $t = 0$ corresponds to 2010.

(A) Use the four-step process to find $p'(t)$.

✎ (B) Find the annual consumption in 2020 and the instantaneous rate of change of consumption in 2020, and write a brief verbal interpretation of these results.

84. Mineral consumption. The U.S. consumption of refined copper (in thousands of metric tons) is given approximately by

$$p(t) = 48t^2 - 37t + 1,698$$

where t is time in years and $t = 0$ corresponds to 2010.

(A) Use the four-step process to find $p'(t)$.

✎ (B) Find the annual consumption in 2022 and the instantaneous rate of change of consumption in 2022, and write a brief verbal interpretation of these results.

85. Electricity consumption. Table 1 gives the retail sales of electricity (in billions of kilowatt-hours) for the residential and commercial sectors in the United States. (*Source:* Energy Information Administration)

Table 1 **Electricity Sales**

Year	Residential	Commercial
2000	1,192	1,055
2002	1,265	1,104
2004	1,292	1,230
2006	1,352	1,300
2008	1,379	1,336
2010	1,446	1,330

(A) Let x represent time (in years) with $x = 0$ corresponding to 2000, and let y represent the corresponding residential sales. Enter the appropriate data set in a graphing calculator and find a quadratic regression equation for the data.

(B) If $y = R(x)$ denotes the regression equation found in part (A), find $R(20)$ and $R'(20)$, and write a brief verbal interpretation of these results. Round answers to the nearest tenth of a billion.

86. Electricity consumption. Refer to the data in Table 1.

(A) Let x represent time (in years) with $x = 0$ corresponding to 2000, and let y represent the corresponding commercial sales. Enter the appropriate data set in a graphing calculator and find a quadratic regression equation for the data.

✎ (B) If $y = C(x)$ denotes the regression equation found in part (A), find $C(20)$ and $C'(20)$, and write a brief verbal interpretation of these results. Round answers to the nearest tenth of a billion.

87. Air pollution. The ozone level (in parts per billion) on a summer day in a metropolitan area is given by

$$P(t) = 80 + 12t - t^2$$

where t is time in hours and $t = 0$ corresponds to 9 A.M.

(A) Use the four-step process to find $P'(t)$.

✎ (B) Find $P(3)$ and $P'(3)$. Write a brief verbal interpretation of these results.

88. Medicine. The body temperature (in degrees Fahrenheit) of a patient t hours after taking a fever-reducing drug is given by

$$F(t) = 98 + \frac{4}{t + 1}$$

(A) Use the four-step process to find $F'(t)$.

✎ (B) Find $F(3)$ and $F'(3)$. Write a brief verbal interpretation of these results.

Answers to Matched Problems

1. (A) $-\$1,600$ (B) $-\$8$ per planter
2. (A) 48 ft/s   (B) $32 + 16h$
   (C) 32 ft/s
3. (A) 6, 5   (B) $4 + h$
   (C) 4   (D) Both are 4
4. $f'(x) = 8 - 4x$

5. (A) $f'(1) = 4, f'(2) = 0, f'(4) = -8$

(B) $f(x)$

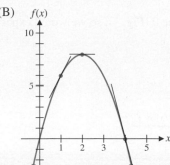

6. $f'(x) = -1/(x + 2)^2, x \neq -2$

7. $f'(x) = 1/(2\sqrt{x + 4}), x \geq -4$

8. $S(12) = 4, S'(12) = 0.125$; 12 months from now, the total sales will be \$4 million and will be increasing at the rate of \$0.125 million (\$125,000) per month. The estimated total sales are \$4.125 million after 13 months and \$4.25 million after 14 months.

10.5 Basic Differentiation Properties

- Constant Function Rule
- Power Rule
- Constant Multiple Property
- Sum and Difference Properties
- Applications

In Section 10.4, we defined the derivative of f at x as

$$f'(x) = \lim_{h \to 0} \frac{f(x + h) - f(x)}{h}$$

if the limit exists, and we used this definition and a four-step process to find the derivatives of several functions. Now we want to develop some rules of differentiation. These rules will enable us to find the derivative of many functions without using the four-step process.

Before exploring these rules, we list some symbols that are often used to represent derivatives.

NOTATION The Derivative

If $y = f(x)$, then

$$f'(x) \qquad y' \qquad \frac{dy}{dx}$$

all represent the derivative of f at x.

Each of these derivative symbols has its particular advantage in certain situations. All of them will become familiar to you after a little experience.

Constant Function Rule

If $f(x) = C$ is a constant function, then the four-step process can be used to show that $f'(x) = 0$. Therefore,

> The derivative of any constant function is 0.

THEOREM 1 Constant Function Rule

If $y = f(x) = C$, then

$$f'(x) = 0$$

Also, $y' = 0$ and $dy/dx = 0$.

Note: When we write $C' = 0$ or $\dfrac{d}{dx}C = 0$, we mean that $y' = \dfrac{dy}{dx} = 0$ when $y = C$.

CONCEPTUAL INSIGHT

The graph of $f(x) = C$ is a horizontal line with slope 0 (Fig. 1), so we would expect that $f'(x) = 0$.

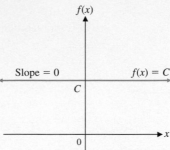

Figure 1

EXAMPLE 1   Differentiating Constant Functions

(A) If $f(x) = 3$, then $f'(x) = 0$.   (B) If $y = -1.4$, then $y' = 0$.

(C) If $y = \pi$, then $\dfrac{dy}{dx} = 0$.   (D) $\dfrac{d}{dx} 23 = 0$

Matched Problem 1   Find

(A) $f'(x)$ for $f(x) = -24$ (B) y' for $y = 12$

(C) $\dfrac{dy}{dx}$ for $y = -\sqrt{7}$ (D) $\dfrac{d}{dx}(-\pi)$

Power Rule

A function of the form $f(x) = x^k$, where k is a real number, is called a **power function**. The following elementary functions are examples of power functions:

$$f(x) = x \qquad h(x) = x^2 \qquad m(x) = x^3 \tag{1}$$
$$n(x) = \sqrt{x} \qquad p(x) = \sqrt[3]{x}$$

Explore and Discuss 1   (A) It is clear that the functions f, h, and m in (1) are power functions. Explain why the functions n and p are also power functions.

(B) The domain of a power function depends on the power. Discuss the domain of each of the following power functions:

$$r(x) = x^4 \qquad s(x) = x^{-4} \qquad t(x) = x^{1/4}$$
$$u(x) = x^{-1/4} \qquad v(x) = x^{1/5} \qquad w(x) = x^{-1/5}$$

The definition of the derivative and the four-step process introduced in Section 10.4 can be used to find the derivatives of many power functions. For example, it can be shown that

$$\begin{aligned} \text{If} \quad & f(x) = x^2, \quad \text{then} \quad f'(x) = 2x. \\ \text{If} \quad & f(x) = x^3, \quad \text{then} \quad f'(x) = 3x^2. \\ \text{If} \quad & f(x) = x^4, \quad \text{then} \quad f'(x) = 4x^3. \\ \text{If} \quad & f(x) = x^5, \quad \text{then} \quad f'(x) = 5x^4. \end{aligned}$$

Notice the pattern in these derivatives. In each case, the power in f becomes the coefficient in f' and the power in f' is 1 less than the power in f. In general, for any positive integer n,

$$\text{If} \quad f(x) = x^n, \quad \text{then} \quad f'(x) = nx^{n-1}. \tag{2}$$

In fact, more advanced techniques can be used to show that (2) holds for *any* real number n. We will assume this general result for the remainder of the book.

THEOREM 2 Power Rule

If $y = f(x) = x^n$, where n is a real number, then

$$f'(x) = nx^{n-1}$$

Also, $y' = nx^{n-1}$ and $dy/dx = nx^{n-1}$.

EXAMPLE 2   Differentiating Power Functions

(A) If $f(x) = x^5$, then $f'(x) = 5x^{5-1} = 5x^4$.

(B) If $y = x^{25}$, then $y' = 25x^{25-1} = 25x^{24}$.

(C) If $y = t^{-3}$, then $\dfrac{dy}{dt} = -3t^{-3-1} = -3t^{-4} = -\dfrac{3}{t^4}$.

(D) $\dfrac{d}{dx}x^{5/3} = \dfrac{5}{3}x^{(5/3)-1} = \dfrac{5}{3}x^{2/3}$.

Matched Problem 2   Find

(A) $f'(x)$ for $f(x) = x^6$

(B) y' for $y = x^{30}$

(C) $\dfrac{dy}{dt}$ for $y = t^{-2}$

(D) $\dfrac{d}{dx}x^{3/2}$

In some cases, properties of exponents must be used to rewrite an expression before the power rule is applied.

EXAMPLE 3   Differentiating Power Functions

(A) If $f(x) = 1/x^4$, we can write $f(x) = x^{-4}$ and

$$f'(x) = -4x^{-4-1} = -4x^{-5}, \quad \text{or} \quad \frac{-4}{x^5}$$

(B) If $y = \sqrt{u}$, we can write $y = u^{1/2}$ and

$$y' = \frac{1}{2}u^{(1/2)-1} = \frac{1}{2}u^{-1/2}, \quad \text{or} \quad \frac{1}{2\sqrt{u}}$$

(C) $\dfrac{d}{dx}\dfrac{1}{\sqrt[3]{x}} = \dfrac{d}{dx}x^{-1/3} = -\dfrac{1}{3}x^{(-1/3)-1} = -\dfrac{1}{3}x^{-4/3}$, or $\dfrac{-1}{3\sqrt[3]{x^4}}$

Matched Problem 3   Find

(A) $f'(x)$ for $f(x) = \dfrac{1}{x}$

(B) y' for $y = \sqrt[3]{u^2}$

(C) $\dfrac{d}{dx}\dfrac{1}{\sqrt{x}}$

Constant Multiple Property

Let $f(x) = ku(x)$, where k is a constant and u is differentiable at x. Using the four-step process, we have the following:

Step 1 $f(x + h) = ku(x + h)$

Step 2 $f(x + h) - f(x) = ku(x + h) - ku(x) = k[u(x + h) - u(x)]$

Step 3 $\dfrac{f(x + h) - f(x)}{h} = \dfrac{k[u(x + h) - u(x)]}{h} = k\left[\dfrac{u(x + h) - u(x)}{h}\right]$

Step 4 $f'(x) = \lim\limits_{h \to 0} \dfrac{f(x + h) - f(x)}{h}$

$\qquad = \lim\limits_{h \to 0} k\left[\dfrac{u(x + h) - u(x)}{h}\right]$ $\lim\limits_{x \to c} kg(x) = k \lim\limits_{x \to c} g(x)$

$\qquad = k \lim\limits_{h \to 0} \left[\dfrac{u(x + h) - u(x)}{h}\right]$ Definition of $u'(x)$

$\qquad = ku'(x)$

Therefore,

> The derivative of a constant times a differentiable function is the constant times the derivative of the function.

THEOREM 3 Constant Multiple Property

If $y = f(x) = ku(x)$, then

$$f'(x) = ku'(x)$$

Also,

$$y' = ku' \qquad \frac{dy}{dx} = k\frac{du}{dx}$$

EXAMPLE 4   Differentiating a Constant Times a Function

(A) If $f(x) = 3x^2$, then $f'(x) = 3 \cdot 2x^{2-1} = 6x$.

(B) If $y = \dfrac{t^3}{6} = \dfrac{1}{6}t^3$, then $\dfrac{dy}{dt} = \dfrac{1}{6} \cdot 3t^{3-1} = \dfrac{1}{2}t^2$.

(C) If $y = \dfrac{1}{2x^4} = \dfrac{1}{2}x^{-4}$, then $y' = \dfrac{1}{2}(-4x^{-4-1}) = -2x^{-5}$, or $\dfrac{-2}{x^5}$.

(D) $\dfrac{d}{dx} \dfrac{0.4}{\sqrt{x^3}} = \dfrac{d}{dx} \dfrac{0.4}{x^{3/2}} = \dfrac{d}{dx} 0.4x^{-3/2} = 0.4\left[-\dfrac{3}{2}x^{(-3/2)-1}\right]$

$\qquad\qquad\qquad\qquad\qquad\qquad = -0.6x^{-5/2}$, or $-\dfrac{0.6}{\sqrt{x^5}}$

Matched Problem 4   Find

(A) $f'(x)$ for $f(x) = 4x^5$

(B) $\dfrac{dy}{dt}$ for $y = \dfrac{t^4}{12}$

(C) y' for $y = \dfrac{1}{3x^3}$

(D) $\dfrac{d}{dx} \dfrac{0.9}{\sqrt[3]{x}}$

Sum and Difference Properties

Let $f(x) = u(x) + v(x)$, where $u'(x)$ and $v'(x)$ exist. Using the four-step process (see Problems 87 and 88 in Exercises 10.5):

$$f'(x) = u'(x) + v'(x)$$

Therefore,

> The derivative of the sum of two differentiable functions is the sum of the derivatives of the functions.

Similarly, we can show that

> The derivative of the difference of two differentiable functions is the difference of the derivatives of the functions.

Together, we have the **sum and difference property** for differentiation:

THEOREM 4 Sum and Difference Property

If $y = f(x) = u(x) \pm v(x)$, then

$$f'(x) = u'(x) \pm v'(x)$$

Also,

$$y' = u' \pm v' \qquad \frac{dy}{dx} = \frac{du}{dx} \pm \frac{dv}{dx}$$

Note: This rule generalizes to the sum and difference of any given number of functions.

With Theorems 1 through 4, we can compute the derivatives of all polynomials and a variety of other functions.

EXAMPLE 5 Differentiating Sums and Differences

(A) If $f(x) = 3x^2 + 2x$,   then

$$f'(x) = (3x^2)' + (2x)' = 3(2x) + 2(1) = 6x + 2$$

(B) If $y = 4 + 2x^3 - 3x^{-1}$,   then

$$y' = (4)' + (2x^3)' - (3x^{-1})' = 0 + 2(3x^2) - 3(-1)x^{-2} = 6x^2 + 3x^{-2}$$

(C) If $y = \sqrt[3]{w} - 3w$,   then

$$\frac{dy}{dw} = \frac{d}{dw}w^{1/3} - \frac{d}{dw}3w = \frac{1}{3}w^{-2/3} - 3 = \frac{1}{3w^{2/3}} - 3$$

(D) $$\frac{d}{dx}\left(\frac{5}{3x^2} - \frac{2}{x^4} + \frac{x^3}{9}\right) = \frac{d}{dx}\frac{5}{3}x^{-2} - \frac{d}{dx}2x^{-4} + \frac{d}{dx}\frac{1}{9}x^3$$

$$= \frac{5}{3}(-2)x^{-3} - 2(-4)x^{-5} + \frac{1}{9} \cdot 3x^2$$

$$= -\frac{10}{3x^3} + \frac{8}{x^5} + \frac{1}{3}x^2$$

Matched Problem 5 Find

(A) $f'(x)$ for $f(x) = 3x^4 - 2x^3 + x^2 - 5x + 7$

(B) y' for $y = 3 - 7x^{-2}$

(C) $\dfrac{dy}{dv}$ for $y = 5v^3 - \sqrt[4]{v}$

(D) $\dfrac{d}{dx}\left(-\dfrac{3}{4x} + \dfrac{4}{x^3} - \dfrac{x^4}{8}\right)$

Some algebraic rewriting of a function is sometimes required before we can apply the rules for differentiation.

EXAMPLE 6   Rewrite before Differentiating   Find the derivative of
$$f(x) = \frac{1 + x^2}{x^4}.$$

SOLUTION   It is helpful to rewrite $f(x) = \dfrac{1 + x^2}{x^4}$, expressing $f(x)$ as the sum of terms, each of which can be differentiated by applying the power rule.

$$f(x) = \frac{1 + x^2}{x^4} \qquad \text{Write as a sum of two terms}$$

$$= \frac{1}{x^4} + \frac{x^2}{x^4} \qquad \text{Write each term as a power of } x$$

$$= x^{-4} + x^{-2}$$

Note that we have rewritten $f(x)$, but we have not used any rules of differentiation. Now, however, we can apply those rules to find the derivative:

$$f'(x) = -4x^{-5} - 2x^{-3}$$

Matched Problem 6   Find the derivative of $f(x) = \dfrac{5 - 3x + 4x^2}{x}$.

Applications

EXAMPLE 7   Instantaneous Velocity   An object moves along the y axis (marked in feet) so that its position at time x (in seconds) is
$$f(x) = x^3 - 6x^2 + 9x$$

(A) Find the instantaneous velocity function v.

(B) Find the velocity at $x = 2$ and $x = 5$ seconds.

(C) Find the time(s) when the velocity is 0.

SOLUTION

(A) $v = f'(x) = (x^3)' - (6x^2)' + (9x)' = 3x^2 - 12x + 9$

(B) $f'(2) = 3(2)^2 - 12(2) + 9 = -3$ feet per second

    $f'(5) = 3(5)^2 - 12(5) + 9 = 24$ feet per second

(C) $v = f'(x) = 3x^2 - 12x + 9 = 0$ Factor 3 out of each term.

            $3(x^2 - 4x + 3) = 0$ Factor the quadratic term.

          $3(x - 1)(x - 3) = 0$ Use the zero property.

                  $x = 1, 3$

So, $v = 0$ at $x = 1$ and $x = 3$ seconds.

Matched Problem 7   Repeat Example 7 for $f(x) = x^3 - 15x^2 + 72x$.

EXAMPLE 8   Tangents   Let $f(x) = x^4 - 6x^2 + 10$.

(A) Find $f'(x)$.

(B) Find the equation of the tangent line at $x = 1$.

(C) Find the values of x where the tangent line is horizontal.

SOLUTION

(A) $f'(x)$ $= (x^4)' - (6x^2)' + (10)'$

$\qquad = 4x^3 - 12x$

(B) $y - y_1 = m(x - x_1)$ $y_1 = f(x_1) = f(1) = (1)^4 - 6(1)^2 + 10 = 5$

$\qquad y - 5 = -8(x - 1)$ $m = f'(x_1) = f'(1) = 4(1)^3 - 12(1) = -8$

$\qquad y = -8x + 13$ Tangent line at $x = 1$

(C) Since a horizontal line has 0 slope, we must solve $f'(x) = 0$ for x:

$\qquad f'(x) = 4x^3 - 12x = 0$ Factor $4x$ out of each term.

$\qquad 4x(x^2 - 3) = 0$ Factor the difference of two squares.

$\qquad 4x(x + \sqrt{3})(x - \sqrt{3}) = 0$ Use the zero property.

$\qquad x = 0, -\sqrt{3}, \sqrt{3}$

Matched Problem 8   Repeat Example 8 for $f(x) = x^4 - 8x^3 + 7$.

Exercises 10.5

Skills Warm-up Exercises

W In Problems 1–8, write the expression in the form x^n. (If necessary, review Section A.6).

1. $\sqrt{x}$ **2.** $\sqrt[3]{x}$ **3.** $\dfrac{1}{x^5}$ **4.** $\dfrac{1}{x}$

5. $(x^4)^3$ **6.** $\dfrac{1}{(x^5)^2}$ **7.** $\dfrac{1}{\sqrt[4]{x}}$ **8.** $\dfrac{1}{\sqrt[5]{x}}$

Find the indicated derivatives in Problems 9–26.

9. $f'(x)$ for $f(x) = 7$ **10.** $\dfrac{d}{dx}3$

11. $\dfrac{dy}{dx}$ for $y = x^9$ **12.** y' for $y = x^6$

13. $\dfrac{d}{dx}x^3$ **14.** $g'(x)$ for $g(x) = x^5$

15. y' for $y = x^{-4}$ **16.** $\dfrac{dy}{dx}$ for $y = x^{-8}$

17. $g'(x)$ for $g(x) = x^{8/3}$ **18.** $f'(x)$ for $f(x) = x^{9/2}$

19. $\dfrac{dy}{dx}$ for $y = \dfrac{1}{x^{10}}$ **20.** y' for $y = \dfrac{1}{x^{12}}$

21. $f'(x)$ for $f(x) = 5x^2$ **22.** $\dfrac{d}{dx}(-2x^3)$

23. y' for $y = 0.4x^7$

24. $f'(x)$ for $f(x) = 0.8x^4$

25. $\dfrac{d}{dx}\left(\dfrac{x^3}{18}\right)$ **26.** $\dfrac{dy}{dx}$ for $y = \dfrac{x^5}{25}$

Problems 27–32 refer to functions f and g that satisfy $f'(2) = 3$ and $g'(2) = -1$. In each problem, find $h'(2)$ for the indicated function h.

27. $h(x) = 4f(x)$ **28.** $h(x) = 5g(x)$

29. $h(x) = f(x) + g(x)$ **30.** $h(x) = g(x) - f(x)$

31. $h(x) = 2f(x) - 3g(x) + 7$

32. $h(x) = -4f(x) + 5g(x) - 9$

Find the indicated derivatives in Problems 33–56.

33. $\dfrac{d}{dx}(2x - 5)$ **34.** $\dfrac{d}{dx}(-4x + 9)$

35. $f'(t)$ if $f(t) = 2t^2 - 3t + 1$

36. $\dfrac{dy}{dt}$ if $y = 2 + 5t - 8t^3$

37. y' for $y = 5x^{-2} + 9x^{-1}$

38. $g'(x)$ if $g(x) = 5x^{-7} - 2x^{-4}$

39. $\dfrac{d}{du}(5u^{0.3} - 4u^{2.2})$

40. $\dfrac{d}{du}(2u^{4.5} - 3.1u + 13.2)$

41. $h'(t)$ if $h(t) = 2.1 + 0.5t - 1.1t^3$

42. $F'(t)$ if $F(t) = 0.2t^3 - 3.1t + 13.2$

43. y' if $y = \dfrac{2}{5x^4}$

44. w' if $w = \dfrac{7}{5u^2}$

45. $\dfrac{d}{dx}\left(\dfrac{3x^2}{2} - \dfrac{7}{5x^2}\right)$

46. $\dfrac{d}{dx}\left(\dfrac{5x^3}{4} - \dfrac{2}{5x^3}\right)$

47. $G'(w)$ if $G(w) = \dfrac{5}{9w^4} + 5\sqrt[3]{w}$

48. $H'(w)$ if $H(w) = \dfrac{5}{w^6} - 2\sqrt{w}$

49. $\dfrac{d}{du}(3u^{2/3} - 5u^{1/3})$

50. $\dfrac{d}{du}(8u^{3/4} + 4u^{-1/4})$

51. $h'(t)$ if $h(t) = \dfrac{3}{t^{3/5}} - \dfrac{6}{t^{1/2}}$

52. $F'(t)$ if $F(t) = \dfrac{5}{t^{1/5}} - \dfrac{8}{t^{3/2}}$

53. y' if $y = \dfrac{1}{\sqrt[3]{x}}$

54. w' if $w = \dfrac{10}{\sqrt[5]{u}}$

55. $\dfrac{d}{dx}\left(\dfrac{1.2}{\sqrt{x}} - 3.2x^{-2} + x\right)$

56. $\dfrac{d}{dx}\left(2.8x^{-3} - \dfrac{0.6}{\sqrt[3]{x^2}} + 7\right)$

For Problems 57–60, find

(A) $f'(x)$

(B) *The slope of the graph of f at $x = 2$ and $x = 4$*

(C) *The equations of the tangent lines at $x = 2$ and $x = 4$*

(D) *The value(s) of x where the tangent line is horizontal*

57. $f(x) = 6x - x^2$ **58.** $f(x) = 2x^2 + 8x$

59. $f(x) = 3x^4 - 6x^2 - 7$ **60.** $f(x) = x^4 - 32x^2 + 10$

If an object moves along the y axis (marked in feet) so that its position at time x (in seconds) is given by the indicated functions in Problems 61–64, find

(A) *The instantaneous velocity function $v = f'(x)$*

(B) *The velocity when $x = 0$ and $x = 3$ seconds*

(C) *The time(s) when $v = 0$*

61. $f(x) = 176x - 16x^2$ **62.** $f(x) = 80x - 10x^2$

63. $f(x) = x^3 - 9x^2 + 15x$ **64.** $f(x) = x^3 - 9x^2 + 24x$

Problems 65–72 require the use of a graphing calculator. For each problem, find $f'(x)$ and approximate (to four decimal places) the value(s) of x where the graph of f has a horizontal tangent line.

65. $f(x) = x^2 - 3x - 4\sqrt{x}$

66. $f(x) = x^2 + x - 10\sqrt{x}$

67. $f(x) = 3\sqrt[3]{x^4} - 1.5x^2 - 3x$

68. $f(x) = 3\sqrt[3]{x^4} - 2x^2 + 4x$

69. $f(x) = 0.05x^4 + 0.1x^3 - 1.5x^2 - 1.6x + 3$

70. $f(x) = 0.02x^4 - 0.06x^3 - 0.78x^2 + 0.94x + 2.2$

71. $f(x) = 0.2x^4 - 3.12x^3 + 16.25x^2 - 28.25x + 7.5$

72. $f(x) = 0.25x^4 - 2.6x^3 + 8.1x^2 - 10x + 9$

73. Let $f(x) = ax^2 + bx + c, a \neq 0$. Recall that the graph of $y = f(x)$ is a parabola. Use the derivative $f'(x)$ to derive a formula for the x coordinate of the vertex of this parabola.

74. Now that you know how to find derivatives, explain why it is no longer necessary for you to memorize the formula for the x coordinate of the vertex of a parabola.

75. Give an example of a cubic polynomial function that has

(A) No horizontal tangents

(B) One horizontal tangent

(C) Two horizontal tangents

76. Can a cubic polynomial function have more than two horizontal tangents? Explain.

Find the indicated derivatives in Problems 77–82.

77. $f'(x)$ if $f(x) = (2x - 1)^2$

78. y' if $y = (2x - 5)^2$

79. $\dfrac{d}{dx}\dfrac{10x + 20}{x}$

80. $\dfrac{dy}{dx}$ if $y = \dfrac{x^2 + 25}{x^2}$

81. $\dfrac{dy}{dx}$ if $y = \dfrac{3x - 4}{12x^2}$

82. $f'(x)$ if $f(x) = \dfrac{2x^5 - 4x^3 + 2x}{x^3}$

In Problems 83–86, discuss the validity of each statement. If the statement is always true, explain why. If not, give a counterexample.

83. The derivative of a product is the product of the derivatives.

84. The derivative of a quotient is the quotient of the derivatives.

85. The derivative of a constant is 0.

86. The derivative of a constant times a function is 0.

87. Let $f(x) = u(x) + v(x)$, where $u'(x)$ and $v'(x)$ exist. Use the four-step process to show that $f'(x) = u'(x) + v'(x)$.

88. Let $f(x) = u(x) - v(x)$, where $u'(x)$ and $v'(x)$ exist. Use the four-step process to show that $f'(x) = u'(x) - v'(x)$.

Applications

89. **Sales analysis.** A company's total sales (in millions of dollars) t months from now are given by

$$S(t) = 0.03t^3 + 0.5t^2 + 2t + 3$$

(A) Find $S'(t)$.

(B) Find $S(5)$ and $S'(5)$ (to two decimal places). Write a brief verbal interpretation of these results.

(C) Find $S(10)$ and $S'(10)$ (to two decimal places). Write a brief verbal interpretation of these results.

90. Sales analysis. A company's total sales (in millions of dollars) t months from now are given by

$$S(t) = 0.015t^4 + 0.4t^3 + 3.4t^2 + 10t - 3$$

(A) Find $S'(t)$.

(B) Find $S(4)$ and $S'(4)$ (to two decimal places). Write a brief verbal interpretation of these results.

(C) Find $S(8)$ and $S'(8)$ (to two decimal places). Write a brief verbal interpretation of these results.

91. Advertising. A marine manufacturer will sell $N(x)$ power boats after spending $\$x$ thousand on advertising, as given by

$$N(x) = 1,000 - \frac{3,780}{x} \qquad 5 \le x \le 30$$

(see figure).

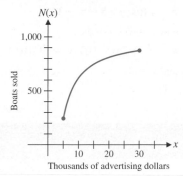

Thousands of advertising dollars

(A) Find $N'(x)$.

(B) Find $N'(10)$ and $N'(20)$. Write a brief verbal interpretation of these results.

92. Price–demand equation. Suppose that, in a given gourmet food store, people are willing to buy x pounds of chocolate candy per day at $\$p$ per quarter pound, as given by the price–demand equation

$$x = 10 + \frac{180}{p} \qquad 2 \le p \le 10$$

This function is graphed in the figure. Find the demand and the instantaneous rate of change of demand with respect to price when the price is $5. Write a brief verbal interpretation of these results.

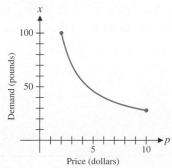

Price (dollars)

93. College enrollment. The percentages of male high-school graduates who enrolled in college are given in the second column of Table 1.

Table 1 **College enrollment percentages**

Year	Male	Female
1970	41.0	25.5
1980	33.5	30.3
1990	40.0	38.3
2000	40.8	45.6
2010	45.9	50.5

(A) Let x represent time (in years) since 1970, and let y represent the corresponding percentage of male high-school graduates who enrolled in college. Enter the data in a graphing calculator and find a cubic regression equation for the data.

(B) If $y = M(x)$ denotes the regression equation found in part (A), find $M(50)$ and $M'(50)$ (to the nearest tenth), and write a brief verbal interpretation of these results.

94. College enrollment. The percentages of female high-school graduates who enrolled in college are given in the third column of Table 1.

(A) Let x represent time (in years) since 1970, and let y represent the corresponding percentage of female high-school graduates who enrolled in college. Enter the data in a graphing calculator and find a cubic regression equation for the data.

(B) If $y = F(x)$ denotes the regression equation found in part (A), find $F(50)$ and $F'(50)$ (to the nearest tenth), and write a brief verbal interpretation of these results.

95. Medicine. A person x inches tall has a pulse rate of y beats per minute, as given approximately by

$$y = 590x^{-1/2} \qquad 30 \le x \le 75$$

What is the instantaneous rate of change of pulse rate at the

(A) 36-inch level?

(B) 64-inch level?

96. Ecology. A coal-burning electrical generating plant emits sulfur dioxide into the surrounding air. The concentration $C(x)$, in parts per million, is given approximately by

$$C(x) = \frac{0.1}{x^2}$$

where x is the distance from the plant in miles. Find the instantaneous rate of change of concentration at

(A) $x = 1$ mile

(B) $x = 2$ miles

97. Learning. Suppose that a person learns y items in x hours, as given by

$$y = 50\sqrt{x} \qquad 0 \le x \le 9$$

(see figure). Find the rate of learning at the end of

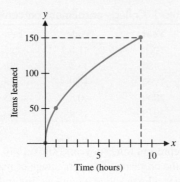

(A) 1 hour

(B) 9 hours

98. Learning. If a person learns y items in x hours, as given by

$$y = 21\sqrt[3]{x^2} \qquad 0 \le x \le 8$$

find the rate of learning at the end of

(A) 1 hour

(B) 8 hours

Answers to Matched Problems

1. All are 0.

2. (A) $6x^5$ (B) $30x^{29}$
   (C) $-2t^{-3} = -2/t^3$ (D) $\frac{3}{2}x^{1/2}$

3. (A) $-x^{-2}$, or $-1/x^2$ (B) $\frac{2}{3}u^{-1/3}$, or $2/(3\sqrt[3]{u})$
   (C) $-\frac{1}{2}x^{-3/2}$, or $-1/(2\sqrt{x^3})$

4. (A) $20x^4$ (B) $t^3/3$
   (C) $-x^{-4}$, or $-1/x^4$ (D) $-0.3x^{-4/3}$, or $-0.3/\sqrt[3]{x^4}$

5. (A) $12x^3 - 6x^2 + 2x - 5$ (B) $14x^{-3}$, or $14/x^3$
   (C) $15v^2 - \frac{1}{4}v^{-3/4}$, or $15v^2 - 1/(4v^{3/4})$
   (D) $3/(4x^2) - (12/x^4) - (x^3/2)$

6. $f'(x) = -5x^{-2} + 4$

7. (A) $v = 3x^2 - 30x + 72$
   (B) $f'(2) = 24$ ft/s; $f'(5) = -3$ ft/s
   (C) $x = 4$ and $x = 6$ seconds

8. (A) $f'(x) = 4x^3 - 24x^2$ (B) $y = -20x + 20$
   (C) $x = 0$ and $x = 6$

10.6 Differentials

- Increments
- Differentials
- Approximations Using Differentials

In this section, we introduce increments and differentials. Increments are useful and they provide an alternative notation for defining the derivative. Differentials are often easier to compute than increments and can be used to approximate increments.

Increments

In Section 10.4, we defined the derivative of f at x as the limit of the difference quotient

$$f'(x) = \lim_{h \to 0} \frac{f(x + h) - f(x)}{h}$$

We considered various interpretations of this limit, including slope, velocity, and instantaneous rate of change. Increment notation enables us to interpret the numerator and denominator of the difference quotient separately.

Given $y = f(x) = x^3$, if x changes from 2 to 2.1, then y will change from $y = f(2) = 2^3 = 8$ to $y = f(2.1) = 2.1^3 = 9.261$. The change in x is called the *increment in x* and is denoted by Δx (read as "delta x").* Similarly, the change in y is called the *increment in y* and is denoted by Δy. In terms of the given example, we write

$$\Delta x = 2.1 - 2 = 0.1 \qquad \text{Change in } x$$
$$\Delta y = f(2.1) - f(2) \qquad f(x) = x^3$$
$$= 2.1^3 - 2^3 \qquad \text{Use a calculator.}$$
$$= 9.261 - 8$$
$$= 1.261 \qquad \text{Corresponding change in } y$$

> **CONCEPTUAL INSIGHT**
>
> The symbol Δx does not represent the product of Δ and x but is the symbol for a single quantity: the *change in x*. Likewise, the symbol Δy represents a single quantity: the *change in y*.

*Δ is the uppercase Greek letter delta.

DEFINITION Increments

For $y = f(x)$, $\Delta x = x_2 - x_1$,   so $x_2 = x_1 + \Delta x$, and

$$\begin{aligned}\Delta y &= y_2 - y_1 \\ &= f(x_2) - f(x_1) \\ &= f(x_1 + \Delta x) - f(x_1)\end{aligned}$$

Δy represents the change in y corresponding to a change Δx in x. Δx can be either positive or negative.

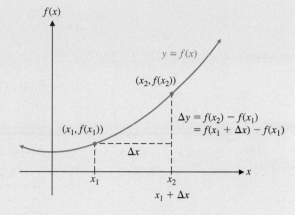

$y = f(x)$

$(x_2, f(x_2))$

$(x_1, f(x_1))$

$\Delta y = f(x_2) - f(x_1)$
$= f(x_1 + \Delta x) - f(x_1)$

Δx

x_1 x_2
$x_1 + \Delta x$

[**Note:** Δy depends on the function f, the input x_1, and the increment Δx.]

EXAMPLE 1   Increments Given the function
$$y = f(x) = \frac{x^2}{2},$$

(A) Find Δx, Δy, and $\Delta y / \Delta x$ for $x_1 = 1$ and $x_2 = 2$.

(B) Find $\dfrac{f(x_1 + \Delta x) - f(x_1)}{\Delta x}$ for $x_1 = 1$ and $\Delta x = 2$.

SOLUTION

(A) $\Delta x = x_2 - x_1 = 2 - 1 = 1$

$\Delta y = f(x_2) - f(x_1)$

$\quad = f(2) - f(1) = \dfrac{4}{2} - \dfrac{1}{2} = \dfrac{3}{2}$

$\dfrac{\Delta y}{\Delta x} = \dfrac{f(x_2) - f(x_1)}{x_2 - x_1} = \dfrac{\frac{3}{2}}{1} = \dfrac{3}{2}$

(B) $\dfrac{f(x_1 + \Delta x) - f(x_1)}{\Delta x} = \dfrac{f(1 + 2) - f(1)}{2}$

$\qquad = \dfrac{f(3) - f(1)}{2} = \dfrac{\frac{9}{2} - \frac{1}{2}}{2} = \dfrac{4}{2} = 2$

Matched Problem 1 Given the function $y = f(x) = x^2 + 1$,

(A) Find Δx, Δy, and $\Delta y / \Delta x$ for $x_1 = 2$ and $x_2 = 3$.

(B) Find $\dfrac{f(x_1 + \Delta x) - f(x_1)}{\Delta x}$ for $x_1 = 1$ and $\Delta x = 2$.

In Example 1, we observe another notation for the difference quotient

$$\frac{f(x + h) - f(x)}{h} \qquad (1)$$

It is common to refer to h, the change in x, as Δx. Then the difference quotient (1) takes on the form

$$\frac{f(x + \Delta x) - f(x)}{\Delta x} \qquad \text{or} \qquad \frac{\Delta y}{\Delta x} \quad \Delta y = f(x + \Delta x) - f(x)$$

and the derivative is defined by

$$f'(x) = \lim_{\Delta x \to 0} \frac{f(x + \Delta x) - f(x)}{\Delta x}$$

or

$$f'(x) = \lim_{\Delta x \to 0} \frac{\Delta y}{\Delta x} \qquad (2)$$

if the limit exists.

Explore and Discuss 1   Suppose that $y = f(x)$ defines a function whose domain is the set of all real numbers. If every increment Δy is equal to 0, then what is the range of f?

Differentials

Assume that the limit in equation (2) exists. Then, for small Δx, the difference quotient $\Delta y / \Delta x$ provides a good approximation for $f'(x)$. Also, $f'(x)$ provides a good approximation for $\Delta y / \Delta x$. We write

$$\frac{\Delta y}{\Delta x} \approx f'(x) \qquad \Delta x \text{ is small, but} \neq 0 \qquad (3)$$

Multiplying both sides of (3) by Δx gives us

$$\Delta y \approx f'(x)\,\Delta x \qquad \Delta x \text{ is small, but} \neq 0 \qquad (4)$$

From equation (4), we see that $f'(x)\,\Delta x$ provides a good approximation for Δy when Δx is small.

Because of the practical and theoretical importance of $f'(x)\,\Delta x$, we give it the special name **differential** and represent it with the special symbol dy or df:

$$dy = f'(x)\,\Delta x \qquad \text{or} \qquad df = f'(x)\,\Delta x$$

For example,

$$d(2x^3) = (2x^3)'\,\Delta x = 6x^2\,\Delta x$$
$$d(x) = (x)'\,\Delta x = 1\,\Delta x = \Delta x$$

In the second example, we usually drop the parentheses in $d(x)$ and simply write

$$dx = \Delta x$$

In summary, we have the following:

DEFINITION   Differentials

If $y = f(x)$ defines a differentiable function, then the **differential dy, or df**, is defined as the product of $f'(x)$ and dx, where $dx = \Delta x$. Symbolically,

$$dy = f'(x)\,dx, \qquad \text{or} \qquad df = f'(x)\,dx$$

where

$$dx = \Delta x$$

Note: The differential dy (or df) is actually a function involving two independent variables, x and dx. A change in either one or both will affect dy (or df).

EXAMPLE 2 Differentials Find dy for $f(x) = x^2 + 3x$. Evaluate dy for

(A) $x = 2$ and $dx = 0.1$

(B) $x = 3$ and $dx = 0.1$

(C) $x = 1$ and $dx = 0.02$

SOLUTION

$dy = f'(x)\,dx$

$\quad = (2x + 3)\,dx$

(A) When $x = 2$ and $dx = 0.1$,　　　　(B) When $x = 3$ and $dx = 0.1$,

$\quad\quad dy = \big[2(2) + 3\big]0.1 = 0.7$　　　　　$dy = \big[2(3) + 3\big]0.1 = 0.9$

(C) When $x = 1$ and $dx = 0.02$,

$\quad\quad dy = \big[2(1) + 3\big]0.02 = 0.1$

<u>Matched Problem 2</u> Find dy for $f(x) = \sqrt{x} + 3$. Evaluate dy for

(A) $x = 4$ and $dx = 0.1$

(B) $x = 9$ and $dx = 0.12$

(C) $x = 1$ and $dx = 0.01$

We now have two interpretations of the symbol dy/dx. Referring to the function $y = f(x) = x^2 + 3x$ in Example 2 with $x = 2$ and $dx = 0.1$, we have

$$\frac{dy}{dx} = f'(2) = 7 \quad \text{Derivative}$$

and

$$\frac{dy}{dx} = \frac{0.7}{0.1} = 7 \quad \text{Ratio of differentials}$$

Approximations Using Differentials

Earlier, we noted that for small Δx,

$$\frac{\Delta y}{\Delta x} \approx f'(x) \quad \text{and} \quad \Delta y \approx f'(x)\Delta x$$

Also, since

$$dy = f'(x)\,dx$$

it follows that

$$\Delta y \approx dy$$

and dy can be used to approximate Δy.

To interpret this result geometrically, we need to recall a basic property of the slope. The vertical change in a line is equal to the product of the slope and the horizontal change, as shown in Figure 1 on page 554.

Now consider the line tangent to the graph of $y = f(x)$, as shown in Figure 2 on page 554. Since $f'(x)$ is the slope of the tangent line and dx is the horizontal change in the tangent line, it follows that the vertical change in the tangent line is given by $dy = f'(x)\,dx$, as indicated in Figure 2.

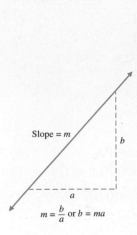

$$m = \frac{b}{a} \text{ or } b = ma$$

Figure 1

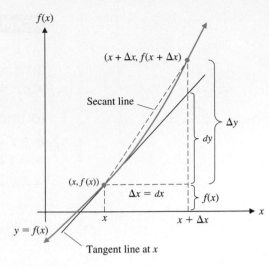

Figure 2

EXAMPLE 3   Comparing Increments and Differentials   Let $y = f(x) = 6x - x^2$.

(A) Find Δy and dy when $x = 2$.

(B) Compare Δy and dy from part (A) for $\Delta x = 0.1, 0.2,$ and 0.3.

SOLUTION

(A) $\Delta y = f(2 + \Delta x) - f(2)$

$\quad\quad = 6(2 + \Delta x) - (2 + \Delta x)^2 - (6 \cdot 2 - 2^2)$ Remove parentheses.

$\quad\quad = 12 + 6\Delta x - 4 - 4\Delta x - \Delta x^2 - 12 + 4$ Collect like terms.

$\quad\quad = 2\Delta x - \Delta x^2$

Since $f'(x) = 6 - 2x, f'(2) = 2$, and $dx = \Delta x, dy = f'(2) \, dx = 2\Delta x$

(B) Table 1 compares the values of Δx and dy for the indicated values of Δx.

Table 1

Δx	Δy	dy
0.1	0.19	0.2
0.2	0.36	0.4
0.3	0.51	0.6

Matched Problem 3   Repeat Example 3 for $x = 4$ and $\Delta x = dx = -0.1, -0.2,$ and -0.3.

EXAMPLE 4   Cost–Revenue   A company manufactures and sells x transistor radios per week. If the weekly cost and revenue equations are

$$C(x) = 5,000 + 2x \quad\quad R(x) = 10x - \frac{x^2}{1,000} \quad\quad 0 \le x \le 8,000$$

then use differentials to approximate the changes in revenue and profit if production is increased from 2,000 to 2,010 units per week.

SOLUTION   We will approximate ΔR and ΔP with dR and dP, respectively, using $x = 2,000$ and $dx = 2,010 - 2,000 = 10$.

$$R(x) = 10x - \frac{x^2}{1,000} \qquad P(x) = R(x) - C(x) = 10x - \frac{x^2}{1,000} - 5,000 - 2x$$

$$dR = R'(x)\, dx \qquad\qquad = 8x - \frac{x^2}{1,000} - 5,000$$

$$= \left(10 - \frac{x}{500}\right) dx \qquad dP = P'(x)\, dx$$

$$= \left(10 - \frac{2,000}{500}\right) 10 \qquad = \left(8 - \frac{x}{500}\right) dx$$

$$= \$60 \text{ per week} \qquad = \left(8 - \frac{2,000}{500}\right) 10$$

$$= \$40 \text{ per week}$$

Matched Problem 4 Repeat Example 4 with production increasing from 6,000 to 6,010.

Comparing the results in Example 4 and Matched Problem 4, we see that an increase in production results in a revenue and profit increase at the 2,000 production level but a revenue and profit loss at the 6,000 production level.

Exercises 10.6

Skills Warm-up Exercises

W In Problems 1–4, let $f(x) = 0.1x + 3$ and find the given values without using a calculator. (If necessary, review Section 2.1.)

1. $f(0); f(0.1)$

2. $f(7); f(7.1)$

3. $f(-2); f(-2.01)$

4. $f(-10); f(-10.01)$

In Problems 5–8, let $g(x) = x^2$ and find the given values without using a calculator.

5. $g(0); g(0.1)$

6. $g(1); g(1.1)$

7. $g(10); g(10.1)$

8. $g(5); g(4.9)$

In Problems 9–14, find the indicated quantities for $y = f(x) = 3x^2$.

9. $\Delta x, \Delta y,$ and $\Delta y/\Delta x$; given $x_1 = 1$ and $x_2 = 4$

10. $\Delta x, \Delta y,$ and $\Delta y/\Delta x$; given $x_1 = 2$ and $x_2 = 5$

11. $\dfrac{f(x_1 + \Delta x) - f(x_1)}{\Delta x}$; given $x_1 = 1$ and $\Delta x = 2$

12. $\dfrac{f(x_1 + \Delta x) - f(x_1)}{\Delta x}$; given $x_1 = 2$ and $\Delta x = 1$

13. $\Delta y/\Delta x$; given $x_1 = 1$ and $x_2 = 3$

14. $\Delta y/\Delta x$; given $x_1 = 2$ and $x_2 = 3$

In Problems 15–20, find dy for each function.

15. $y = 30 + 12x^2 - x^3$

16. $y = 200x - \dfrac{x^2}{30}$

17. $y = x^2\left(1 - \dfrac{x}{9}\right)$

18. $y = x^3(60 - x)$

19. $y = \dfrac{590}{\sqrt{x}}$

20. $y = 52\sqrt{x}$

In Problems 21 and 22, find the indicated quantities for $y = f(x) = 3x^2$.

21. (A) $\dfrac{f(2 + \Delta x) - f(2)}{\Delta x}$ (simplify)

(B) What does the quantity in part (A) approach as Δx approaches 0?

22. (A) $\dfrac{f(3 + \Delta x) - f(3)}{\Delta x}$ (simplify)

(B) What does the quantity in part (A) approach as Δx approaches 0?

In Problems 23–26, find dy for each function.

23. $y = (2x + 1)^2$

24. $y = (3x + 5)^2$

25. $y = \dfrac{x^2 + 9}{x}$

26. $y = \dfrac{(x - 1)^2}{x^2}$

In Problems 27–30, evaluate dy and Δy for each function for the indicated values.

27. $y = f(x) = x^2 - 3x + 2; x = 5, dx = \Delta x = 0.2$

28. $y = f(x) = 30 + 12x^2 - x^3; x = 2, dx = \Delta x = 0.1$

29. $y = f(x) = 75\left(1 - \dfrac{2}{x}\right); x = 5, dx = \Delta x = -0.5$

30. $y = f(x) = 100\left(x - \dfrac{4}{x^2}\right); x = 2, dx = \Delta x = -0.1$

31. A cube with 10-inch sides is covered with a coat of fiberglass 0.2 inch thick. Use differentials to estimate the volume of the fiberglass shell.

32. A sphere with a radius of 5 centimeters is coated with ice 0.1 centimeter thick. Use differentials to estimate the volume of the ice. $\left[\text{Recall that } V = \frac{4}{3}\pi r^3.\right]$

In Problems 33–36,

(A) Find Δy and dy for the function f at the indicated value of x.

(B) Graph Δy and dy from part (A) as functions of Δx.

(C) Compare the values of Δy and dy from part (A) at the indicated values of Δx.

33. $f(x) = x^2 + 2x + 3; x = -0.5, \Delta x = dx = 0.1, 0.2, 0.3$

34. $f(x) = x^2 + 2x + 3; x = -2, \Delta x = dx = -0.1, -0.2, -0.3$

35. $f(x) = x^3 - 2x^2; x = 1, \Delta x = dx = 0.05, 0.10, 0.15$

36. $f(x) = x^3 - 2x^2; x = 2, \Delta x = dx = -0.05, -0.10, -0.15$

In Problems 37–40, discuss the validity of each statement. If the statement is always true, explain why. If not, give a counterexample.

37. If the graph of the function $y = f(x)$ is a line, then the functions Δy and dy (of the independent variable $\Delta x = dx$) for $f(x)$ at $x = 3$ are identical.

38. If the graph of the function $y = f(x)$ is a parabola, then the functions Δy and dy (of the independent variable $\Delta x = dx$) for $f(x)$ at $x = 0$ are identical.

39. Suppose that $y = f(x)$ defines a differentiable function whose domain is the set of all real numbers. If every differential dy at $x = 2$ is equal to 0, then $f(x)$ is a constant function.

40. Suppose that $y = f(x)$ defines a function whose domain is the set of all real numbers. If every increment at $x = 2$ is equal to 0, then $f(x)$ is a constant function.

41. Find dy if $y = (1 - 2x)\sqrt[3]{x^2}$.

42. Find dy if $y = (2x^2 - 4)\sqrt{x}$.

43. Find dy and Δy for $y = 52\sqrt{x}, x = 4,$ and $\Delta x = dx = 0.3$.

44. Find dy and Δy for $y = 590/\sqrt{x}, x = 64,$ and $\Delta x = dx = 1$.

Applications

Use differential approximations in the following problems.

45. Advertising. A company will sell N units of a product after spending $\$x$ thousand in advertising, as given by

$$N = 60x - x^2 \qquad 5 \le x \le 30$$

Approximately what increase in sales will result by increasing the advertising budget from \$10,000 to \$11,000? From \$20,000 to \$21,000?

46. Price–demand. Suppose that the daily demand (in pounds) for chocolate candy at $\$x$ per pound is given by

$$D = 1,000 - 40x^2 \qquad 1 \le x \le 5$$

If the price is increased from \$3.00 per pound to \$3.20 per pound, what is the approximate change in demand?

47. Average cost. For a company that manufactures tennis rackets, the average cost per racket $\overline{C}$ is

$$\overline{C} = \frac{400}{x} + 5 + \frac{1}{2}x \qquad x \ge 1$$

where x is the number of rackets produced per hour. What will the approximate change in average cost per racket be if production is increased from 20 per hour to 25 per hour? From 40 per hour to 45 per hour?

48. Revenue and profit. A company manufactures and sells x televisions per month. If the cost and revenue equations are

$$C(x) = 72,000 + 60x$$
$$R(x) = 200x - \frac{x^2}{30} \qquad 0 \le x \le 6,000$$

what will the approximate changes in revenue and profit be if production is increased from 1,500 to 1,510? From 4,500 to 4,510?

49. Pulse rate. The average pulse rate y (in beats per minute) of a healthy person x inches tall is given approximately by

$$y = \frac{590}{\sqrt{x}} \qquad 30 \le x \le 75$$

Approximately how will the pulse rate change for a change in height from 36 to 37 inches? From 64 to 65 inches?

50. Measurement. An egg of a particular bird is nearly spherical. If the radius to the inside of the shell is 5 millimeters and the radius to the outside of the shell is 5.3 millimeters, approximately what is the volume of the shell? [Remember that $V = \frac{4}{3}\pi r^3$.]

51. Medicine. A drug is given to a patient to dilate her arteries. If the radius of an artery is increased from 2 to 2.1 millimeters, approximately how much is the cross-sectional area increased? [Assume that the cross section of the artery is circular; that is, $A = \pi r^2$.]

52. Drug sensitivity. One hour after x milligrams of a particular drug are given to a person, the change in body temperature T (in degrees Fahrenheit) is given by

$$T = x^2\left(1 - \frac{x}{9}\right) \qquad 0 \le x \le 6$$

Approximate the changes in body temperature produced by the following changes in drug dosages:

(A) From 2 to 2.1 milligrams

(B) From 3 to 3.1 milligrams

(C) From 4 to 4.1 milligrams

53. Learning. A particular person learning to type has an achievement record given approximately by

$$N = 75\left(1 - \frac{2}{t}\right) \qquad 3 \le t \le 20$$

where N is the number of words per minute typed after t weeks of practice. What is the approximate improvement from 5 to 5.5 weeks of practice?

54. Learning. If a person learns y items in x hours, as given approximately by

$$y = 52\sqrt{x} \qquad 0 \le x \le 9$$

what is the approximate increase in the number of items learned when x changes from 1 to 1.1 hours? From 4 to 4.1 hours?

55. Politics. In a new city, the voting population (in thousands) is given by

$$N(t) = 30 + 12t^2 - t^3 \qquad 0 \le t \le 8$$

where t is time in years. Find the approximate change in votes for the following changes in time:

(A) From 1 to 1.1 years

(B) From 4 to 4.1 years

(C) From 7 to 7.1 years

Answers to Matched Problems

1. (A) $\Delta x = 1, \Delta y = 5, \Delta y / \Delta x = 5$ (B) 4

2. $dy = \dfrac{1}{2\sqrt{x}}\,dx$

   (A) 0.025   (B) 0.02   (C) 0.005

3. (A) $\Delta y = -2\Delta x - \Delta x^2; dy = -2\Delta x$

   (B)

Δx	Δy	dy
-0.1	0.19	0.2
-0.2	0.36	0.4
-0.3	0.51	0.6

4. $dR = -\$20/\text{wk}; dP = -\$40/\text{wk}$

10.7 Marginal Analysis in Business and Economics

- Marginal Cost, Revenue, and Profit
- Application
- Marginal Average Cost, Revenue, and Profit

Marginal Cost, Revenue, and Profit

One important application of calculus to business and economics involves *marginal analysis*. In economics, the word *marginal* refers to a rate of change—that is, to a derivative. Thus, if $C(x)$ is the total cost of producing x items, then $C'(x)$ is called the *marginal cost* and represents the instantaneous rate of change of total cost with respect to the number of items produced. Similarly, the *marginal revenue* is the derivative of the total revenue function, and the *marginal profit* is the derivative of the total profit function.

> **DEFINITION** Marginal Cost, Revenue, and Profit
>
> If x is the number of units of a product produced in some time interval, then
>
> $$\text{total cost} = C(x)$$
> $$\textbf{marginal cost} = C'(x)$$
> $$\text{total revenue} = R(x)$$
> $$\textbf{marginal revenue} = R'(x)$$
> $$\text{total profit} = P(x) = R(x) - C(x)$$
> $$\textbf{marginal profit} = P'(x) = R'(x) - C'(x)$$
> $$= (\text{marginal revenue}) - (\text{marginal cost})$$
>
> Marginal cost (or revenue or profit) is the instantaneous rate of change of cost (or revenue or profit) relative to production at a given production level.

To begin our discussion, we consider a cost function $C(x)$. It is important to remember that $C(x)$ represents the *total* cost of producing x items, not the cost of producing a *single* item. To find the cost of producing a single item, we use the difference of two successive values of $C(x)$:

$$\text{Total cost of producing } x + 1 \text{ items} = C(x + 1)$$
$$\text{Total cost of producing } x \text{ items} = C(x)$$
$$\text{Exact cost of producing the } (x + 1)\text{st item} = C(x + 1) - C(x)$$

EXAMPLE 1 Cost Analysis A company manufactures fuel tanks for cars. The total weekly cost (in dollars) of producing x tanks is given by

$$C(x) = 10,000 + 90x - 0.05x^2$$

(A) Find the marginal cost function.

(B) Find the marginal cost at a production level of 500 tanks per week.

(C) Interpret the results of part (B).

(D) Find the exact cost of producing the 501st item.

SOLUTION

(A) $C'(x) = 90 - 0.1x$

(B) $C'(500) = 90 - 0.1(500) = \$40$ Marginal cost

(C) At a production level of 500 tanks per week, the total production costs are increasing at the rate of \$40 per tank.

(D) $C(501) = 10,000 + 90(501) - 0.05(501)^2$

                $= \$42,539.95$ Total cost of producing 501 tanks per week

        $C(500) = 10,000 + 90(500) - 0.05(500)^2$

                $= \$42,500.00$ Total cost of producing 500 tanks per week

    $C(501) - C(500) = 42,539.95 - 42,500.00$

                $= \$39.95$ Exact cost of producing the 501st tank

Matched Problem 1 A company manufactures automatic transmissions for cars. The total weekly cost (in dollars) of producing x transmissions is given by

$$C(x) = 50,000 + 600x - 0.75x^2$$

(A) Find the marginal cost function.

(B) Find the marginal cost at a production level of 200 transmissions per week.

(C) Interpret the results of part (B).

(D) Find the exact cost of producing the 201st transmission.

In Example 1, we found that the cost of the 501st tank and the marginal cost at a production level of 500 tanks differ by only a nickel. Increments and differentials will help us understand the relationship between marginal cost and the cost of a single item. If $C(x)$ is any total cost function, then

$$C'(x) \approx \frac{C(x + \Delta x) - C(x)}{\Delta x}$$ See Section 10.6

$$C'(x) \approx C(x + 1) - C(x)$$ $\Delta x = 1$

We see that the marginal cost $C'(x)$ approximates $C(x + 1) - C(x)$, the exact cost of producing the $(x + 1)$st item. These observations are summarized next and are illustrated in Figure 1.

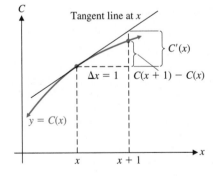

Figure 1 $C'(x) \approx C(x + 1) - C(x)$

THEOREM 1 Marginal Cost and Exact Cost

If $C(x)$ is the total cost of producing x items, then the marginal cost function approximates the exact cost of producing the $(x + 1)$st item:

            Marginal cost   Exact cost

$$C'(x) \approx C(x + 1) - C(x)$$

Similar statements can be made for total revenue functions and total profit functions.

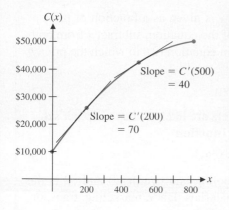

Figure 2
$C(x) = 10{,}000 + 90x - 0.05x^2$

Theorem 1 states that the marginal cost at a given production level x approximates the cost of producing the $(x + 1)$st, or *next,* item. In practice, the marginal cost is used more frequently than the exact cost. One reason for this is that the marginal cost is easily visualized when one is examining the graph of the total cost function. Figure 2 shows the graph of the cost function discussed in Example 1, with tangent lines added at $x = 200$ and $x = 500$. The graph clearly shows that as production increases, the slope of the tangent line decreases. Thus, the cost of producing the next tank also decreases, a desirable characteristic of a total cost function. We will have much more to say about graphical analysis in Chapter 12.

EXAMPLE 2 Exact Cost and Marginal Cost The total cost of producing x bicycles is given by the cost function

$$C(x) = 10{,}000 + 150x - 0.2x^2$$

(A) Find the exact cost of producing the 121st bicycle.

(B) Use marginal cost to approximate the cost of producing the 121st bicycle.

SOLUTION

(A) The cost of producing 121 bicycles is

$$C(121) = 10{,}000 + 150(121) - 0.2(121)^2 = \$25{,}221.80$$

and the cost of producing 120 bicycles is

$$C(120) = 10{,}000 + 150(120) - 0.2(120)^2 = \$25{,}120.00$$

So the exact cost of producing the 121st bicycle is

$$C(121) - C(120) = \$25{,}221.80 - 25{,}120.00 = \$101.80$$

(B) By Theorem 1, the marginal cost function $C'(x)$, evaluated at $x = 120$, approximates the cost of producing the 121st bicycle:

$$C'(x) = 150 - 0.4x$$
$$C'(120) = 150 - 0.4(120) = \$102.00$$

Note that the marginal cost, $\$102.00$, at a production level of 120 bicycles, is a good approximation to the exact cost, $\$101.80$, of producing the 121st bicycle.

Matched Problem 2 For the cost function $C(x)$ in Example 2

(A) Find the exact cost of producing the 141st bicycle.

(B) Use marginal cost to approximate the cost of producing the 141st bicycle.

Application

Now we discuss how price, demand, revenue, cost, and profit are tied together in typical applications. Although either price or demand can be used as the independent variable in a price–demand equation, it is common to use demand as the independent variable when marginal revenue, cost, and profit are also involved.

EXAMPLE 3 Production Strategy A company's market research department recommends the manufacture and marketing of a new headphone set for MP3 players. After suitable test marketing, the research department presents the following **price–demand equation**:

$$x = 10{,}000 - 1{,}000p \quad \text{x is demand at price p.} \tag{1}$$

In the price–demand equation (1), the demand x is given as a function of price p. By solving (1) for p (add 1,000p to both sides of the equation, subtract x from both sides, and divide both sides by 1,000), we obtain equation (2), in which the price p is given as a function of demand x:

$$p = 10 - 0.001x \tag{2}$$

where x is the number of headphones that retailers are likely to buy at \$$p$ per set.

The financial department provides the **cost function**

$$C(x) = 7,000 + 2x \tag{3}$$

where \$7,000 is the estimate of fixed costs (tooling and overhead) and \$2 is the estimate of variable costs per headphone set (materials, labor, marketing, transportation, storage, etc.).

(A) Find the domain of the function defined by the price–demand equation (2).

(B) Find and interpret the marginal cost function $C'(x)$.

(C) Find the revenue function as a function of x and find its domain.

(D) Find the marginal revenue at $x = 2{,}000$, 5,000, and 7,000. Interpret these results.

(E) Graph the cost function and the revenue function in the same coordinate system. Find the intersection points of these two graphs and interpret the results.

(F) Find the profit function and its domain and sketch the graph of the function.

(G) Find the marginal profit at $x = 1{,}000$, 4,000, and 6,000. Interpret these results.

SOLUTION

(A) Since price p and demand x must be nonnegative, we have $x \geq 0$ and

$$p = 10 - 0.001x \geq 0$$
$$10 \geq 0.001x$$
$$10{,}000 \geq x$$

Thus, the permissible values of x are $0 \leq x \leq 10{,}000$.

(B) The marginal cost is $C'(x) = 2$. Since this is a constant, it costs an additional \$2 to produce one more headphone set at any production level.

(C) The **revenue** is the amount of money R received by the company for manufacturing and selling x headphone sets at \$$p$ per set and is given by

$$R = (\text{number of headphone sets sold})(\text{price per headphone set}) = xp$$

In general, the revenue R can be expressed as a function of p using equation (1) or as a function of x using equation (2). As we mentioned earlier, when using marginal functions, we will always use the number of items x as the independent variable. Thus, the **revenue function** is

$$R(x) = xp = x(10 - 0.001x) \quad \text{Using equation (2)} \tag{4}$$
$$= 10x - 0.001x^2$$

Since equation (2) is defined only for $0 \leq x \leq 10{,}000$, it follows that the domain of the revenue function is $0 \leq x \leq 10{,}000$.

(D) The **marginal revenue** is

$$R'(x) = 10 - 0.002x$$

For production levels of $x = 2{,}000$, 5,000, and 7,000, we have

$$R'(2{,}000) = 6 \qquad R'(5{,}000) = 0 \qquad R'(7{,}000) = -4$$

This means that at production levels of 2,000, 5,000, and 7,000, the respective approximate changes in revenue per unit change in production are \$6, \$0, and $-\$4$. That is, at the 2,000 output level, revenue increases as production increases; at the 5,000 output level, revenue does not change with a "small" change in production; and at the 7,000 output level, revenue decreases with an increase in production.

(E) Graphing $R(x)$ and $C(x)$ in the same coordinate system results in Figure 3 on page 562. The intersection points are called the **break-even points**, because revenue equals cost at these production levels. The company neither makes nor loses money, but just breaks even. The break-even points are obtained as follows:

$$C(x) = R(x)$$
$$7{,}000 + 2x = 10x - 0.001x^2$$
$$0.001x^2 - 8x + 7{,}000 = 0 \quad \text{Solve by the quadratic formula}$$
$$x^2 - 8{,}000x + 7{,}000{,}000 = 0 \quad \text{(see Appendix A.7).}$$
$$x = \frac{8{,}000 \pm \sqrt{8{,}000^2 - 4(7{,}000{,}000)}}{2}$$
$$= \frac{8{,}000 \pm \sqrt{36{,}000{,}000}}{2}$$
$$= \frac{8{,}000 \pm 6{,}000}{2}$$
$$= 1{,}000, \quad 7{,}000$$
$$R(1{,}000) = 10(1{,}000) - 0.001(1{,}000)^2 = 9{,}000$$
$$C(1{,}000) = 7{,}000 + 2(1{,}000) = 9{,}000$$
$$R(7{,}000) = 10(7{,}000) - 0.001(7{,}000)^2 = 21{,}000$$
$$C(7{,}000) = 7{,}000 + 2(7{,}000) = 21{,}000$$

The break-even points are $(1{,}000, 9{,}000)$ and $(7{,}000, 21{,}000)$, as shown in Figure 3. Further examination of the figure shows that cost is greater than revenue for production levels between 0 and 1,000 and also between 7,000 and 10,000. Consequently, the company incurs a loss at these levels. By contrast, for production levels between 1,000 and 7,000, revenue is greater than cost, and the company makes a profit.

(F) The **profit function** is

$$\begin{aligned} P(x) &= R(x) - C(x) \\ &= (10x - 0.001x^2) - (7{,}000 + 2x) \\ &= -0.001x^2 + 8x - 7{,}000 \end{aligned}$$

The domain of the cost function is $x \geq 0$, and the domain of the revenue function is $0 \leq x \leq 10{,}000$. The domain of the profit function is the set of x values for which both functions are defined—that is, $0 \leq x \leq 10{,}000$. The graph of the profit function is shown in Figure 4 on page 562. Notice that the x coordinates of the break-even points in Figure 3 are the x intercepts of the profit function. Furthermore, the intervals on which cost is greater than revenue and on which revenue is greater than cost correspond, respectively, to the intervals on which profit is negative and on which profit is positive.

(G) The **marginal profit** is

$$P'(x) = -0.002x + 8$$

For production levels of 1,000, 4,000, and 6,000, we have

$$P'(1{,}000) = 6 \qquad P'(4{,}000) = 0 \qquad P'(6{,}000) = -4$$

This means that at production levels of 1,000, 4,000, and 6,000, the respective approximate changes in profit per unit change in production are $6, $0, and $-$4. That is, at the 1,000 output level, profit will be increased if production is increased; at the 4,000 output level, profit does not change for "small" changes in production;

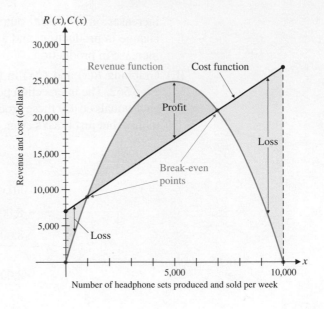

Figure 3

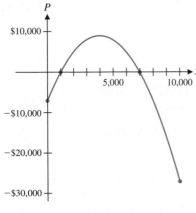

Figure 4

and at the 6,000 output level, profits will decrease if production is increased. It seems that the best production level to produce a maximum profit is 4,000.

Example 3 requires careful study since a number of important ideas in economics and calculus are involved. In the next chapter, we will develop a systematic procedure for finding the production level (and, using the demand equation, the selling price) that will maximize profit.

Matched Problem 3 Refer to the revenue and profit functions in Example 3.

(A) Find $R'(3,000)$ and $R'(6,000)$. Interpret the results.

(B) Find $P'(2,000)$ and $P'(7,000)$. Interpret the results.

Marginal Average Cost, Revenue, and Profit

Sometimes it is desirable to carry out marginal analysis relative to **average cost (cost per unit), average revenue (revenue per unit),** and **average profit (profit per unit).**

> **DEFINITION** Marginal Average Cost, Revenue, and Profit
>
> If x is the number of units of a product produced in some time interval, then
>
> **Cost per unit:**   average cost $= \overline{C}(x) = \dfrac{C(x)}{x}$
>
>   **marginal average cost** $= \overline{C}'(x) = \dfrac{d}{dx}\overline{C}(x)$
>
> **Revenue per unit:**   average revenue $= \overline{R}(x) = \dfrac{R(x)}{x}$
>
>   **marginal average revenue** $= \overline{R}'(x) = \dfrac{d}{dx}\overline{R}(x)$
>
> **Profit per unit:**   average profit $= \overline{P}(x) = \dfrac{P(x)}{x}$
>
>   **marginal average profit** $= \overline{P}'(x) = \dfrac{d}{dx}\overline{P}(x)$

EXAMPLE 4 Cost Analysis   A small machine shop manufactures drill bits used in the petroleum industry. The manager estimates that the total daily cost (in dollars) of producing x bits is

$$C(x) = 1,000 + 25x - 0.1x^2$$

(A) Find $\overline{C}(x)$ and $\overline{C}'(x)$.

(B) Find $\overline{C}(10)$ and $\overline{C}'(10)$. Interpret these quantities.

(C) Use the results in part (B) to estimate the average cost per bit at a production level of 11 bits per day.

SOLUTION

(A) $\overline{C}(x) = \dfrac{C(x)}{x} = \dfrac{1,000 + 25x - 0.1x^2}{x}$

$\qquad = \dfrac{1,000}{x} + 25 - 0.1x$ $\qquad\qquad$ Average cost function

$\qquad \overline{C}'(x) = \dfrac{d}{dx}\overline{C}(x) = -\dfrac{1,000}{x^2} - 0.1$ $\qquad$ Marginal average cost function

(B) $\overline{C}(10) = \dfrac{1,000}{10} + 25 - 0.1(10) = \$124$

$\qquad \overline{C}'(10) = -\dfrac{1,000}{10^2} - 0.1 = -\$10.10$

At a production level of 10 bits per day, the average cost of producing a bit is \$124. This cost is decreasing at the rate of \$10.10 per bit.

(C) If production is increased by 1 bit, then the average cost per bit will decrease by approximately \$10.10. So, the average cost per bit at a production level of 11 bits per day is approximately $\$124 - \$10.10 = \$113.90$.

Matched Problem 4 Consider the cost function for the production of headphone sets from Example 3: $\qquad C(x) = 7,000 + 2x$

(A) Find $\overline{C}(x)$ and $\overline{C}'(x)$.

(B) Find $\overline{C}(100)$ and $\overline{C}'(100)$. Interpret these quantities.

(C) Use the results in part (B) to estimate the average cost per headphone set at a production level of 101 headphone sets.

Explore and Discuss 1   A student produced the following solution to Matched Problem 4:

$$C(x) = 7,000 + 2x \quad \text{Cost}$$

$$C'(x) = 2 \qquad\qquad\qquad \text{Marginal cost}$$

$$\dfrac{C'(x)}{x} = \dfrac{2}{x} \qquad\qquad \text{"Average" of the marginal cost}$$

Explain why the last function is not the same as the marginal average cost function.

⚠ CAUTION

1. The marginal average cost function is computed by first finding the average cost function and then finding its derivative. As Explore and Discuss 1 illustrates, reversing the order of these two steps produces a different function that does not have any useful economic interpretations.

2. Recall that the marginal cost function has two interpretations: the usual interpretation of any derivative as an instantaneous rate of change and the special interpretation as an approximation to the exact cost of the $(x + 1)$st item. This special interpretation does not apply to the marginal average cost function. Referring to Example 4, we would be incorrect to interpret $\overline{C}'(10) = -\$10.10$ to mean that the average cost of the next bit is approximately $-\$10.10$. In fact, the phrase "average cost of the next bit" does not even make sense. Averaging is a concept applied to a collection of items, not to a single item.

These remarks also apply to revenue and profit functions. ▲

Exercises 10.7

Skills Warm-up Exercises

W *In Problems 1–8, let $C(x) = 10,000 + 150x - 0.2x^2$ be the total cost in dollars of producing x bicycles. (If necessary, review Section 2.1).*

1. Find the total cost of producing 99 bicycles.

2. Find the total cost of producing 100 bicycles.

3. Find the cost of producing the 100th bicycle.

4. Find the total cost of producing 199 bicycles.

5. Find the total cost of producing 200 bicycles.

6. Find the cost of producing the 200th bicycle.

7. Find the average cost per bicycle of producing 100 bicycles.

8. Find the average cost per bicycle of producing 200 bicycles.

In Problems 9–12, find the marginal cost function.

9. $C(x) = 175 + 0.8x$ 10. $C(x) = 4,500 + 9.5x$

11. $C(x) = 210 + 4.6x - 0.01x^2$

12. $C(x) = 790 + 13x - 0.2x^2$

In Problems 13–16, find the marginal revenue function.

13. $R(x) = 4x - 0.01x^2$ 14. $R(x) = 36x - 0.03x^2$

15. $R(x) = x(12 - 0.04x)$ 16. $R(x) = x(25 - 0.05x)$

In Problems 17–20, find the marginal profit function if the cost and revenue, respectively, are those in the indicated problems.

17. Problem 9 and Problem 13

18. Problem 10 and Problem 14

19. Problem 11 and Problem 15

20. Problem 12 and Problem 16

In Problems 21–28, find the indicated function if cost and revenue are given by $C(x) = 145 + 1.1x$ and $R(x) = 5x - 0.02x^2$, respectively.

21. Average cost function

22. Average revenue function

23. Marginal average cost function

24. Marginal average revenue function

25. Profit function

26. Marginal profit function

27. Average profit function

28. Marginal average profit function

In Problems 29–32, discuss the validity of each statement. If the statement is always true, explain why. If not, give a counterexample.

29. If a cost function is linear, then the marginal cost is a constant.

30. If a price–demand equation is linear, then the marginal revenue function is linear.

31. Marginal profit is equal to marginal cost minus marginal revenue.

32. Marginal average cost is equal to average marginal cost.

Applications

33. **Cost analysis.** The total cost (in dollars) of producing x food processors is

$$C(x) = 2,000 + 50x - 0.5x^2$$

(A) Find the exact cost of producing the 21st food processor.

(B) Use marginal cost to approximate the cost of producing the 21st food processor.

34. **Cost analysis.** The total cost (in dollars) of producing x electric guitars is

$$C(x) = 1,000 + 100x - 0.25x^2$$

(A) Find the exact cost of producing the 51st guitar.

(B) Use marginal cost to approximate the cost of producing the 51st guitar.

35. Cost analysis. The total cost (in dollars) of manufacturing x auto body frames is

$$C(x) = 60,000 + 300x$$

(A) Find the average cost per unit if 500 frames are produced.

(B) Find the marginal average cost at a production level of 500 units and interpret the results.

(C) Use the results from parts (A) and (B) to estimate the average cost per frame if 501 frames are produced.

36. Cost analysis. The total cost (in dollars) of printing x dictionaries is

$$C(x) = 20,000 + 10x$$

(A) Find the average cost per unit if 1,000 dictionaries are produced.

(B) Find the marginal average cost at a production level of 1,000 units and interpret the results.

(C) Use the results from parts (A) and (B) to estimate the average cost per dictionary if 1,001 dictionaries are produced.

37. Profit analysis. The total profit (in dollars) from the sale of x skateboards is

$$P(x) = 30x - 0.3x^2 - 250 \qquad 0 \le x \le 100$$

(A) Find the exact profit from the sale of the 26th skateboard.

(B) Use marginal profit to approximate the profit from the sale of the 26th skateboard.

38. Profit analysis. The total profit (in dollars) from the sale of x calendars is

$$P(x) - 22x - 0.2x^2 - 400 \qquad 0 \le x \le 100$$

(A) Find the exact profit from the sale of the 41st calendar.

(B) Use the marginal profit to approximate the profit from the sale of the 41st calendar.

39. Profit analysis. The total profit (in dollars) from the sale of x DVDs is

$$P(x) = 5x - 0.005x^2 - 450 \qquad 0 \le x \le 1,000$$

Evaluate the marginal profit at the given values of x, and interpret the results.

(A) $x = 450$ 　　　　　　(B) $x = 750$

40. Profit analysis. The total profit (in dollars) from the sale of x cameras is

$$P(x) = 12x - 0.02x^2 - 1,000 \qquad 0 \le x \le 600$$

Evaluate the marginal profit at the given values of x, and interpret the results.

(A) $x = 200$ 　　　　　　(B) $x = 350$

41. Profit analysis. The total profit (in dollars) from the sale of x lawn mowers is

$$P(x) = 30x - 0.03x^2 - 750 \qquad 0 \le x \le 1,000$$

(A) Find the average profit per mower if 50 mowers are produced.

(B) Find the marginal average profit at a production level of 50 mowers and interpret the results.

(C) Use the results from parts (A) and (B) to estimate the average profit per mower if 51 mowers are produced.

42. Profit analysis. The total profit (in dollars) from the sale of x gas grills is

$$P(x) = 20x - 0.02x^2 - 320 \qquad 0 \le x \le 1,000$$

(A) Find the average profit per grill if 40 grills are produced.

(B) Find the marginal average profit at a production level of 40 grills and interpret the results.

(C) Use the results from parts (A) and (B) to estimate the average profit per grill if 41 grills are produced.

43. Revenue analysis. The price p (in dollars) and the demand x for a brand of running shoes are related by the equation

$$x = 4,000 - 40p$$

(A) Express the price p in terms of the demand x, and find the domain of this function.

(B) Find the revenue $R(x)$ from the sale of x pairs of running shoes. What is the domain of R?

(C) Find the marginal revenue at a production level of 1,600 pairs and interpret the results.

(D) Find the marginal revenue at a production level of 2,500 pairs, and interpret the results.

44. Revenue analysis. The price p (in dollars) and the demand x for a particular steam iron are related by the equation

$$x = 1,000 - 20p$$

(A) Express the price p in terms of the demand x, and find the domain of this function.

(B) Find the revenue $R(x)$ from the sale of x steam irons. What is the domain of R?

(C) Find the marginal revenue at a production level of 400 steam irons and interpret the results.

(D) Find the marginal revenue at a production level of 650 steam irons and interpret the results.

45. Revenue, cost, and profit. The price–demand equation and the cost function for the production of table saws are given, respectively, by

$$x = 6,000 - 30p \qquad \text{and} \qquad C(x) = 72,000 + 60x$$

where x is the number of saws that can be sold at a price of $\$p$ per saw and $C(x)$ is the total cost (in dollars) of producing x saws.

(A) Express the price p as a function of the demand x, and find the domain of this function.

(B) Find the marginal cost.

(C) Find the revenue function and state its domain.

(D) Find the marginal revenue.

✎ (E) Find $R'(1,500)$ and $R'(4,500)$ and interpret these quantities.

(F) Graph the cost function and the revenue function on the same coordinate system for $0 \leq x \leq 6,000$. Find the break-even points, and indicate regions of loss and profit.

(G) Find the profit function in terms of x.

(H) Find the marginal profit.

✎ (I) Find $P'(1,500)$ and $P'(3,000)$ and interpret these quantities.

46. Revenue, cost, and profit. The price–demand equation and the cost function for the production of HDTVs are given, respectively, by

$$x = 9,000 - 30p \quad \text{and} \quad C(x) = 150,000 + 30x$$

where x is the number of HDTVs that can be sold at a price of $\$p$ per TV and $C(x)$ is the total cost (in dollars) of producing x TVs.

(A) Express the price p as a function of the demand x, and find the domain of this function.

(B) Find the marginal cost.

(C) Find the revenue function and state its domain.

(D) Find the marginal revenue.

✎ (E) Find $R'(3,000)$ and $R'(6,000)$ and interpret these quantities.

(F) Graph the cost function and the revenue function on the same coordinate system for $0 \leq x \leq 9,000$. Find the break-even points and indicate regions of loss and profit.

(G) Find the profit function in terms of x.

(H) Find the marginal profit.

✎ (I) Find $P'(1,500)$ and $P'(4,500)$ and interpret these quantities.

47. Revenue, cost, and profit. A company is planning to manufacture and market a new two-slice electric toaster. After conducting extensive market surveys, the research department provides the following estimates: a weekly demand of 200 toasters at a price of $16 per toaster and a weekly demand of 300 toasters at a price of $14 per toaster. The financial department estimates that weekly fixed costs will be $1,400 and variable costs (cost per unit) will be $4.

(A) Assume that the relationship between price p and demand x is linear. Use the research department's estimates to express p as a function of x and find the domain of this function.

(B) Find the revenue function in terms of x and state its domain.

(C) Assume that the cost function is linear. Use the financial department's estimates to express the cost function in terms of x.

(D) Graph the cost function and revenue function on the same coordinate system for $0 \leq x \leq 1,000$. Find the break-even points and indicate regions of loss and profit.

(E) Find the profit function in terms of x.

✎ (F) Evaluate the marginal profit at $x = 250$ and $x = 475$ and interpret the results.

48. Revenue, cost, and profit. The company in Problem 47 is also planning to manufacture and market a four-slice toaster. For this toaster, the research department's estimates are a weekly demand of 300 toasters at a price of $25 per toaster and a weekly demand of 400 toasters at a price of $20. The financial department's estimates are fixed weekly costs of $5,000 and variable costs of $5 per toaster.

(A) Assume that the relationship between price p and demand x is linear. Use the research department's estimates to express p as a function of x, and find the domain of this function.

(B) Find the revenue function in terms of x and state its domain.

(C) Assume that the cost function is linear. Use the financial department's estimates to express the cost function in terms of x.

(D) Graph the cost function and revenue function on the same coordinate system for $0 \leq x \leq 800$. Find the break-even points and indicate regions of loss and profit.

(E) Find the profit function in terms of x.

✎ (F) Evaluate the marginal profit at $x = 325$ and $x = 425$ and interpret the results.

49. Revenue, cost, and profit. The total cost and the total revenue (in dollars) for the production and sale of x ski jackets are given, respectively, by

$$C(x) = 24x + 21,900 \quad \text{and} \quad R(x) = 200x - 0.2x^2$$
$$0 \leq x \leq 1,000$$

(A) Find the value of x where the graph of $R(x)$ has a horizontal tangent line.

(B) Find the profit function $P(x)$.

(C) Find the value of x where the graph of $P(x)$ has a horizontal tangent line.

(D) Graph $C(x)$, $R(x)$, and $P(x)$ on the same coordinate system for $0 \leq x \leq 1,000$. Find the break-even points. Find the x intercepts of the graph of $P(x)$.

50. Revenue, cost, and profit. The total cost and the total revenue (in dollars) for the production and sale of x hair dryers are given, respectively, by

$$C(x) = 5x + 2,340 \quad \text{and} \quad R(x) = 40x - 0.1x^2$$
$$0 \leq x \leq 400$$

(A) Find the value of x where the graph of $R(x)$ has a horizontal tangent line.

(B) Find the profit function $P(x)$.

(C) Find the value of x where the graph of $P(x)$ has a horizontal tangent line.

(D) Graph $C(x)$, $R(x)$, and $P(x)$ on the same coordinate system for $0 \le x \le 400$. Find the break-even points. Find the x intercepts of the graph of $P(x)$.

51. Break-even analysis. The price–demand equation and the cost function for the production of garden hoses are given, respectively, by

$$p = 20 - \sqrt{x} \quad \text{and} \quad C(x) = 500 + 2x$$

where x is the number of garden hoses that can be sold at a price of $\$p$ per unit and $C(x)$ is the total cost (in dollars) of producing x garden hoses.

(A) Express the revenue function in terms of x.

(B) Graph the cost function and revenue function in the same viewing window for $0 \le x \le 400$. Use approximation techniques to find the break-even points correct to the nearest unit.

52. Break-even analysis. The price–demand equation and the cost function for the production of handwoven silk scarves are given, respectively, by

$$p = 60 - 2\sqrt{x} \quad \text{and} \quad C(x) = 3{,}000 + 5x$$

where x is the number of scarves that can be sold at a price of $\$p$ per unit and $C(x)$ is the total cost (in dollars) of producing x scarves.

(A) Express the revenue function in terms of x.

(B) Graph the cost function and the revenue function in the same viewing window for $0 \le x \le 900$. Use approximation techniques to find the break-even points correct to the nearest unit.

53. Break-even analysis. Table 1 contains price–demand and total cost data for the production of projectors, where p is the wholesale price (in dollars) of a projector for an annual demand of x projectors and C is the total cost (in dollars) of producing x projectors.

Table 1

x	$p(\$)$	$C(\$)$
3,190	581	1,130,000
4,570	405	1,241,000
5,740	181	1,410,000
7,330	85	1,620,000

(A) Find a quadratic regression equation for the price–demand data, using x as the independent variable.

(B) Find a linear regression equation for the cost data, using x as the independent variable. Use this equation to estimate the fixed costs and variable costs per projector. Round answers to the nearest dollar.

(C) Find the break-even points. Round answers to the nearest integer.

(D) Find the price range for which the company will make a profit. Round answers to the nearest dollar.

54. Break-even analysis. Table 2 contains price–demand and total cost data for the production of treadmills, where p is the wholesale price (in dollars) of a treadmill for an annual demand of x treadmills and C is the total cost (in dollars) of producing x treadmills.

Table 2

x	$p(\$)$	$C(\$)$
2,910	1,435	3,650,000
3,415	1,280	3,870,000
4,645	1,125	4,190,000
5,330	910	4,380,000

(A) Find a linear regression equation for the price–demand data, using x as the independent variable.

(B) Find a linear regression equation for the cost data, using x as the independent variable. Use this equation to estimate the fixed costs and variable costs per treadmill. Round answers to the nearest dollar.

(C) Find the break-even points. Round answers to the nearest integer.

(D) Find the price range for which the company will make a profit. Round answers to the nearest dollar.

Answers to Matched Problems

1. (A) $C'(x) = 600 - 1.5x$

   (B) $C'(200) = 300$.

   (C) At a production level of 200 transmissions, total costs are increasing at the rate of $\$300$ per transmission.

   (D) $C(201) - C(200) = \$299.25$

2. (A) $\$93.80$ (B) $\$94.00$

3. (A) $R'(3{,}000) = 4$. At a production level of 3,000, a unit increase in production will increase revenue by approximately $\$4$.
   $R'(6{,}000) = -2$. At a production level of 6,000, a unit increase in production will decrease revenue by approximately $\$2$.

   (B) $P'(2{,}000) = 4$. At a production level of 2,000, a unit increase in production will increase profit by approximately $\$4$.
   $P'(7{,}000) = -6$. At a production level of 7,000, a unit increase in production will decrease profit by approximately $\$6$.

4. (A) $\overline{C}(x) = \dfrac{7{,}000}{x} + 2$; $\overline{C}'(x) = -\dfrac{7{,}000}{x^2}$

   (B) $\overline{C}(100) = \$72$; $\overline{C}'(100) = -\$0.70$. At a production level of 100 headphone sets, the average cost per headphone set is $\$72$. This average cost is decreasing at a rate of $\$0.70$ per headphone set.

   (C) Approx. $\$71.30$.

Chapter 10   Summary and Review

Important Terms, Symbols, and Concepts

10.1 Introduction to Limits

- The graph of the function $y = f(x)$ is the graph of the set of all ordered pairs $(x, f(x))$.
- The limit of the function $y = f(x)$ as x approaches c is L, written as $\lim_{x \to c} f(x) = L$, if the functional value $f(x)$ is close to the single real number L whenever x is close, but not equal, to c (on either side of c).
- The limit of the function $y = f(x)$ as x approaches c from the left is K, written as $\lim_{x \to c^-} f(x) = K$, if $f(x)$ is close to K whenever x is close to, but to the left of, c on the real-number line.
- The limit of the function $y = f(x)$ as x approaches c from the right is L, written as $\lim_{x \to c^+} f(x) = L$, if $f(x)$ is close to L whenever x is close to, but to the right of, c on the real-number line.
- The limit of the difference quotient $[f(a + h) - f(a)]/h$ is often a $0/0$ indeterminate form. Algebraic simplification is often required to evaluate this type of limit.

Ex. 1, p. 489
Ex. 2, p. 490
Ex. 3, p. 491
Ex. 4, p. 492
Ex. 5, p. 494
Ex. 6, p. 495
Ex. 7, p. 496
Ex. 8, p. 496
Ex. 9, p. 497
Ex. 10, p. 498

10.2 Infinite Limits and Limits at Infinity

- If $f(x)$ increases or decreases without bound as x approaches a from either side of a, then the line $x = a$ is a **vertical asymptote** of the graph of $y = f(x)$.
- If $f(x)$ gets close to L as x increases without bound or decreases without bound, then L is called the limit of f at ∞ or $-\infty$.
- The **end behavior** of a function is described by its limits at infinity.
- If $f(x)$ approaches L as $x \to \infty$ or as $x \to -\infty$, then the line $y = L$ is a **horizontal asymptote** of the graph of $y = f(x)$. Polynomial functions never have horizontal asymptotes. A rational function can have at most one.

Ex. 1, p. 504
Ex. 2, p. 505
Ex. 3, p. 507
Ex. 4, p. 508
Ex. 5, p. 510
Ex. 6, p. 511

10.3 Continuity

- Intuitively, the graph of a continuous function can be drawn without lifting a pen off the paper. By definition, a function f is **continuous at c** if

  1. $\lim_{x \to c} f(x)$ exists,   **2.** $f(c)$ exists, and   **3.** $\lim_{x \to c} f(x) = f(c)$

- Continuity properties are useful for determining where a function is continuous and where it is discontinuous.
- Continuity properties are also useful for solving inequalities.

Ex. 1, p. 516
Ex. 2, p. 518
Ex. 3, p. 519
Ex. 4, p. 520

10.4 The Derivative

- Given a function $y = f(x)$, the **average rate of change** is the ratio of the change in y to the change in x.
- The **instantaneous rate of change** is the limit of the average rate of change as the change in x approaches 0.
- The slope of the secant line through two points on the graph of a function $y = f(x)$ is the ratio of the change in y to the change in x. The **slope of the graph** at the point $(a, f(a))$ is the limit of the slope of the secant line through the points $(a, f(a))$ and $(a + h, f(a + h))$ as h approaches 0, provided the limit exists. In this case, the **tangent line** to the graph is the line through $(a, f(a))$ with slope equal to the limit.
- The **derivative of $y = f(x)$ at x**, denoted $f'(x)$, is the limit of the difference quotient $[f(x + h) - f(x)]/h$ as $h \to 0$ (if the limit exists).
- The four-step process is used to find derivatives.
- If the limit of the difference quotient does not exist at $x = a$, then f is **nondifferentiable at a** and $f'(a)$ does not exist.

Ex. 1, p. 527
Ex. 2, p. 527
Ex. 3, p. 530
Ex. 4, p. 532
Ex. 5, p. 533
Ex. 6, p. 534

10.5 Basic Differentiation Properties

- The derivative of a constant function is 0.
- For any real number n, the derivative of $f(x) = x^n$ is nx^{n-1}.
- If f is a differentiable function, then the derivative of $kf(x)$ is $kf'(x)$.
- The derivative of the sum or difference of two differentiable functions is the sum or difference of the derivatives of the functions.

Ex. 1, p. 542
Ex. 2, p. 543
Ex. 3, p. 543
Ex. 4, p. 544
Ex. 5, p. 545

10.6 Differentials

- Given the function $y = f(x)$, the change in x is also called the **increment of x** and is denoted as Δx. The corresponding change in y is called the **increment of y** and is given by $\Delta y = f(x + \Delta x) - f(x)$.
- If $y = f(x)$ is differentiable at x, then the **differential of x** is $dx = \Delta x$ and the **differential of $y = f(x)$** is $dy = f'(x)dx$, or $df = f'(x)dx$. In this context, x and dx are both independent variables.

Ex. 1, p. 551
Ex. 2, p. 553
Ex. 3, p. 554

10.7 Marginal Analysis in Business and Economics

- If $y = C(x)$ is the total cost of producing x items, then $y = C'(x)$ is the **marginal cost** and $C(x + 1) - C(x)$ is the exact cost of producing item $x + 1$. Furthermore, $C'(x) \approx C(x + 1) - C(x)$. Similar statements can be made regarding total revenue and total profit functions.
- If $y = C(x)$ is the total cost of producing x items, then the **average cost**, or cost per unit, is $\overline{C}(x) = \dfrac{C(x)}{x}$ and the **marginal average cost** is $\overline{C}'(x) = \dfrac{d}{dx}\overline{C}(x)$. Similar statements can be made regarding total revenue and total profit functions.

Ex. 1, p. 558
Ex. 2, p. 559
Ex. 3, p. 559

Review Exercises

Work through all the problems in this chapter review, and check your answers in the back of the book. Answers to all review problems are there, along with section numbers in italics to indicate where each type of problem is discussed. Where weaknesses show up, review appropriate sections of the text.

Many of the problems in this exercise set ask you to find a derivative. Most of the answers to these problems contain both an unsimplified form and a simplified form of the derivative. When checking your work, first check that you applied the rules correctly, and then check that you performed the algebraic simplification correctly.

1. Find the indicated quantities for $y = f(x) = 2x^2 + 5$:

(A) The change in y if x changes from 1 to 3

(B) The average rate of change of y with respect to x if x changes from 1 to 3

(C) The slope of the secant line through the points $(1, f(1))$ and $(3, f(3))$ on the graph of $y = f(x)$

(D) The instantaneous rate of change of y with respect to x at $x = 1$

(E) The slope of the line tangent to the graph of $y = f(x)$ at $x = 1$

(F) $f'(1)$

2. Use the four-step process to find $f'(x)$ for $f(x) = -3x + 2$.

3. If $\lim\limits_{x \to 1} f(x) = 2$ and $\lim\limits_{x \to 1} g(x) = 4$, find

(A) $\lim\limits_{x \to 1} (5f(x) + 3g(x))$

(B) $\lim\limits_{x \to 1} [f(x)g(x)]$

(C) $\lim\limits_{x \to 1} \dfrac{g(x)}{f(x)}$

(D) $\lim\limits_{x \to 1} [5 + 2x - 3g(x)]$

In Problems 4–10, use the graph of f to estimate the indicated limits and function values.

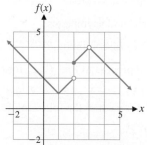

Figure for 4–10

4. $f(1.5)$ **5.** $f(2.5)$ **6.** $f(2.75)$ **7.** $f(3.25)$

8. (A) $\lim\limits_{x \to 1^-} f(x)$

(B) $\lim\limits_{x \to 1^+} f(x)$

(C) $\lim\limits_{x \to 1} f(x)$

(D) $f(1)$

9. (A) $\lim\limits_{x \to 2^-} f(x)$

(B) $\lim\limits_{x \to 2^+} f(x)$

(C) $\lim\limits_{x \to 2} f(x)$

(D) $f(2)$

10. (A) $\lim\limits_{x \to 3^-} f(x)$

(B) $\lim\limits_{x \to 3^+} f(x)$

(C) $\lim\limits_{x \to 3} f(x)$

(D) $f(3)$

In Problems 11–13, use the graph of the function f shown in the figure to answer each question.

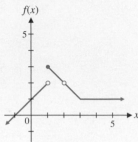

Figure for 11–13

11. (A) $\lim_{x \to 1} f(x) = ?$　　　　(B) $f(1) = ?$

(C) Is f continuous at $x = 1$?

12. (A) $\lim_{x \to 2} f(x) = ?$　　　　(B) $f(2) = ?$

(C) Is f continuous at $x = 2$?

13. (A) $\lim_{x \to 3} f(x) = ?$　　　　(B) $f(3) = ?$

(C) Is f continuous at $x = 3$?

In Problems 14–23, refer to the following graph of $y = f(x)$:

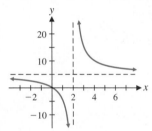

Figure for 14–23

14. $\lim_{x \to \infty} f(x) = ?$　　　　**15.** $\lim_{x \to -\infty} f(x) = ?$

16. $\lim_{x \to 2^+} f(x) = ?$　　　　**17.** $\lim_{x \to 2^-} f(x) = ?$

18. $\lim_{x \to 0^-} f(x) = ?$　　　　**19.** $\lim_{x \to 0^+} f(x) = ?$

20. $\lim_{x \to 0} f(x) = ?$

21. Identify any vertical asymptotes.

22. Identify any horizontal asymptotes.

23. Where is $y = f(x)$ discontinuous?

24. Use the four-step process to find $f'(x)$ for $f(x) = 5x^2$.

25. If $f(5) = 4, f'(5) = -1, g(5) = 2$, and $g'(5) = -3$, then find $h'(5)$ for each of the following functions:

(A) $h(x) = 3f(x)$

(B) $h(x) = -2g(x)$

(C) $h(x) = 2f(x) + 5$

(D) $h(x) = -g(x) - 1$

(E) $h(x) = 2f(x) + 3g(x)$

In Problems 26–31, find $f'(x)$ and simplify.

26. $f(x) = \frac{1}{3}x^3 - 5x^2 + 1$　　**27.** $f(x) = 2x^{1/2} - 3x$

28. $f(x) = 5$　　　　　　　　　**29.** $f(x) = \frac{3}{2x} + \frac{5x^3}{4}$

30. $f(x) = \frac{0.5}{x^4} + 0.25x^4$

31. $f(x) = (3x^3 - 2)(x + 1)$ (*Hint:* Multiply and then differentiate.)

In Problems 32–35, find the indicated quantities for $y = f(x) = x^2 + x$.

32. $\Delta x, \Delta y,$ and $\Delta y / \Delta x$ for $x_1 = 1$ and $x_2 = 3$.

33. $[f(x_1 + \Delta x) - f(x_1)] / \Delta x$ for $x_1 = 1$ and $\Delta x = 2$.

34. dy for $x_1 = 1$ and $x_2 = 3$.

35. Δy and dy for $x = 1, \Delta x = dx = 0.2$.

Problems 36–38 refer to the function.

$$f(x) = \begin{cases} x^2 & \text{if } 0 \le x < 2 \\ 8 - x & \text{if } x \ge 2 \end{cases}$$

which is graphed in the figure.

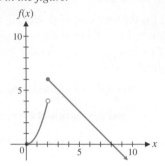

Figure for 36–38

36. (A) $\lim_{x \to 2^-} f(x) = ?$　　　(B) $\lim_{x \to 2^+} f(x) = ?$

(C) $\lim_{x \to 2} f(x) = ?$　　　　(D) $f(2) = ?$

(E) Is f continuous at $x = 2$?

37. (A) $\lim_{x \to 5^-} f(x) = ?$　　　(B) $\lim_{x \to 5^+} f(x) = ?$

(C) $\lim_{x \to 5} f(x) = ?$　　　　(D) $f(5) = ?$

(E) Is f continuous at $x = 5$?

38. Solve each inequality. Express answers in interval notation.

(A) $f(x) < 0$　　　　　　(B) $f(x) \ge 0$

In Problems 39–41, solve each inequality. Express the answer in interval notation. Use a graphing calculator in Problem 41 to approximate partition numbers to four decimal places.

39. $x^2 - x < 12$　　　　　**40.** $\frac{x - 5}{x^2 + 3x} > 0$

41. $x^3 + x^2 - 4x - 2 > 0$

42. Let $f(x) = 0.5x^2 - 5$.

(A) Find the slope of the secant line through $(2, f(2))$ and $(4, f(4))$.

(B) Find the slope of the secant line through $(2, f(2))$ and $(2 + h, f(2 + h)), h \ne 0$.

(C) Find the slope of the tangent line at $x = 2$.

In Problems 43–46, find the indicated derivative and simplify.

43. $\frac{dy}{dx}$ for $y = \frac{1}{3}x^{-3} - 5x^{-2} + 1$

44. y' for $y = \dfrac{3\sqrt{x}}{2} + \dfrac{5}{3\sqrt{x}}$

45. $g'(x)$ for $g(x) = 1.8\sqrt[3]{x} + \dfrac{0.9}{\sqrt[3]{x}}$

46. $\dfrac{dy}{dx}$ for $y = \dfrac{2x^3 - 3}{5x^3}$

47. For $y = f(x) = x^2 + 4$, find

    (A) The slope of the graph at $x = 1$

    (B) The equation of the tangent line at $x = 1$ in the form $y = mx + b$

In Problems 48 and 49, find the value(s) of x where the tangent line is horizontal.

48. $f(x) = 10x - x^2$

49. $f(x) = x^3 + 3x^2 - 45x - 135$

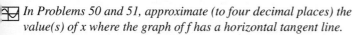 *In Problems 50 and 51, approximate (to four decimal places) the value(s) of x where the graph of f has a horizontal tangent line.*

50. $f(x) = x^4 - 2x^3 - 5x^2 + 7x$

51. $f(x) = x^5 - 10x^3 - 5x + 10$

52. If an object moves along the y axis (scale in feet) so that it is at $y = f(x) = 8x^2 - 4x + 1$ at time x (in seconds), find

    (A) The instantaneous velocity function

    (B) The velocity at time $x = 3$ seconds

53. An object moves along the y axis (scale in feet) so that at time x (in seconds) it is at $y = f(x) = -5x^2 + 16x + 3$. Find

    (A) The instantaneous velocity function

    (B) The time(s) when the velocity is 0

54. Let $f(x) = x^3$, $g(x) = (x - 4)^3$, and $h(x) = (x + 3)^3$.

    (A) How are the graphs of f, g, and h related? Illustrate your conclusion by graphing f, g, and h on the same coordinate axes.

    (B) How would you expect the graphs of the derivatives of these functions to be related? Illustrate your conclusion by graphing f', g', and h' on the same coordinate axes.

In Problems 55–59, determine where f is continuous. Express the answer in interval notation.

55. $f(x) = x^2 - 4$

56. $f(x) = \dfrac{x + 1}{x - 2}$

57. $f(x) = \dfrac{x + 4}{x^2 + 3x - 4}$

58. $f(x) = \sqrt[3]{4 - x^2}$

59. $f(x) = \sqrt{4 - x^2}$

In Problems 60–69, evaluate the indicated limits if they exist.

60. Let $f(x) = \dfrac{2x}{x^2 - 3x}$. Find

    (A) $\lim\limits_{x \to 1} f(x)$ (B) $\lim\limits_{x \to 3} f(x)$ (C) $\lim\limits_{x \to 0} f(x)$

61. Let $f(x) = \dfrac{x + 1}{(3 - x)^2}$. Find

    (A) $\lim\limits_{x \to 1} f(x)$ (B) $\lim\limits_{x \to -1} f(x)$ (C) $\lim\limits_{x \to 3} f(x)$

62. Let $f(x) = \dfrac{|x - 4|}{x - 4}$. Find

    (A) $\lim\limits_{x \to 4^-} f(x)$ (B) $\lim\limits_{x \to 4^+} f(x)$ (C) $\lim\limits_{x \to 4} f(x)$

63. Let $f(x) = \dfrac{x - 3}{9 - x^2}$. Find

    (A) $\lim\limits_{x \to 3} f(x)$ (B) $\lim\limits_{x \to -3} f(x)$ (C) $\lim\limits_{x \to 0} f(x)$

64. Let $f(x) = \dfrac{x^2 - x - 2}{x^2 - 7x + 10}$. Find

    (A) $\lim\limits_{x \to -1} f(x)$ (B) $\lim\limits_{x \to 2} f(x)$ (C) $\lim\limits_{x \to 5} f(x)$

65. Let $f(x) = \dfrac{2x}{3x - 6}$. Find

    (A) $\lim\limits_{x \to \infty} f(x)$ (B) $\lim\limits_{x \to -\infty} f(x)$ (C) $\lim\limits_{x \to 2} f(x)$

66. Let $f(x) = \dfrac{2x^3}{3(x - 2)^2}$. Find

    (A) $\lim\limits_{x \to \infty} f(x)$ (B) $\lim\limits_{x \to -\infty} f(x)$ (C) $\lim\limits_{x \to 2} f(x)$

67. Let $f(x) = \dfrac{2x}{3(x - 2)^3}$. Find

    (A) $\lim\limits_{x \to \infty} f(x)$ (B) $\lim\limits_{x \to -\infty} f(x)$ (C) $\lim\limits_{x \to 2} f(x)$

68. $\lim\limits_{h \to 0} \dfrac{f(2 + h) - f(2)}{h}$ for $f(x) = x^2 + 4$

69. $\lim\limits_{h \to 0} \dfrac{f(x + h) - f(x)}{h}$ for $f(x) = \dfrac{1}{x + 2}$

In Problems 70 and 71, use the definition of the derivative and the four-step process to find $f'(x)$.

70. $f(x) = x^2 - x$ **71.** $f(x) = \sqrt{x} - 3$

Problems 72–77 refer to the function f in the figure. Determine whether f is differentiable at the indicated value of x.

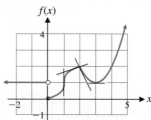

72. $x = -1$ **73.** $x = 0$ **74.** $x = 1$

75. $x = 2$ **76.** $x = 3$ **77.** $x = 4$

In Problems 78–82, find all horizontal and vertical asymptotes.

78. $f(x) = \dfrac{5x}{x - 7}$

79. $f(x) = \dfrac{-2x + 5}{(x - 4)^2}$

80. $f(x) = \dfrac{x^2 + 9}{x - 3}$

81. $f(x) = \dfrac{x^2 - 9}{x^2 + x - 2}$

82. $f(x) = \dfrac{x^3 - 1}{x^3 - x^2 - x + 1}$

83. The domain of the power function $f(x) = x^{1/5}$ is the set of all real numbers. Find the domain of the derivative $f'(x)$. Discuss the nature of the graph of $y = f(x)$ for any x values excluded from the domain of $f'(x)$.

84. Let f be defined by

$$f(x) = \begin{cases} x^2 - m & \text{if } x \le 1 \\ -x^2 + m & \text{if } x > 1 \end{cases}$$

where m is a constant.

(A) Graph f for $m = 0$, and find

$$\lim_{x \to 1^-} f(x) \quad \text{and} \quad \lim_{x \to 1^+} f(x)$$

(B) Graph f for $m = 2$, and find

$$\lim_{x \to 1^-} f(x) \quad \text{and} \quad \lim_{x \to 1^+} f(x)$$

(C) Find m so that

$$\lim_{x \to 1^-} f(x) = \lim_{x \to 1^+} f(x)$$

and graph f for this value of m.

(D) Write a brief verbal description of each graph. How does the graph in part (C) differ from the graphs in parts (A) and (B)?

85. Let $f(x) = 1 - |x - 1|, 0 \le x \le 2$ (see the figure).

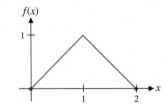

(A) $\displaystyle\lim_{h \to 0^-} \dfrac{f(1 + h) - f(1)}{h} = ?$

(B) $\displaystyle\lim_{h \to 0^+} \dfrac{f(1 + h) - f(1)}{h} = ?$

(C) $\displaystyle\lim_{h \to 0} \dfrac{f(1 + h) - f(1)}{h} = ?$

(D) Does $f'(1)$ exist?

Applications

86. Natural-gas rates. Table 1 shows the winter rates for natural gas charged by the Bay State Gas Company. The base charge is a fixed monthly charge, independent of the amount of gas used per month.

Table 1 **Natural Gas Rates**

Base charge	$7.47
First 90 therms	$0.4000 per therm
All usage over 90 therms	$0.2076 per therm

(A) Write a piecewise definition of the monthly charge $S(x)$ for a customer who uses x therms in a winter month.

(B) Graph $S(x)$.

(C) Is $S(x)$ continuous at $x = 90$? Explain.

87. Cost analysis. The total cost (in dollars) of producing x HDTVs is

$$C(x) = 10,000 + 200x - 0.1x^2$$

(A) Find the exact cost of producing the 101st TV.

(B) Use the marginal cost to approximate the cost of producing the 101st TV.

88. Cost analysis. The total cost (in dollars) of producing x bicycles is

$$C(x) = 5,000 + 40x + 0.05x^2$$

(A) Find the total cost and the marginal cost at a production level of 100 bicycles and interpret the results.

(B) Find the average cost and the marginal average cost at a production level of 100 bicycles and interpret the results.

89. Cost analysis. The total cost (in dollars) of producing x laser printers per week is shown in the figure. Which is greater, the approximate cost of producing the 201st printer or the approximate cost of producing the 601st printer? Does this graph represent a manufacturing process that is becoming more efficient or less efficient as production levels increase? Explain.

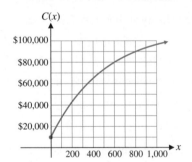

90. Cost analysis. Let

$$p = 25 - 0.01x \quad \text{and} \quad C(x) = 2x + 9,000$$
$$0 \le x \le 2,500$$

be the price–demand equation and cost function, respectively, for the manufacture of umbrellas.

(A) Find the marginal cost, average cost, and marginal average cost functions.

(B) Express the revenue in terms of x, and find the marginal revenue, average revenue, and marginal average revenue functions.

(C) Find the profit, marginal profit, average profit, and marginal average profit functions.

(D) Find the break-even point(s).

(E) Evaluate the marginal profit at $x = 1,000, 1,150,$ and $1,400$, and interpret the results.

(F) Graph $R = R(x)$ and $C = C(x)$ on the same coordinate system, and locate regions of profit and loss.

91. Employee training. A company producing computer components has established that, on average, a new employee can assemble $N(t)$ components per day after t days of on-the-job training, as given by

$$N(t) = \frac{40t - 80}{t}, t \geq 2$$

(A) Find the average rate of change of $N(t)$ from 2 days to 5 days.

(B) Find the instantaneous rate of change of $N(t)$ at 2 days.

92. Sales analysis. The total number of swimming pools, N (in thousands), sold during a year is given by

$$N(t) = 2t + \frac{1}{3} t^{3/2}$$

where t is the number of months since the beginning of the year. Find $N(9)$ and $N'(9)$, and interpret these quantities.

93. Natural-gas consumption. The data in Table 2 give the U.S. consumption of natural gas in trillions of cubic feet.

Table 2

Year	Natural-Gas Consumption
1960	12.0
1970	21.1
1980	19.9
1990	18.7
2000	21.9
2010	24.1

(A) Let x represent time (in years), with $x = 0$ corresponding to 1960, and let y represent the corresponding U.S. consumption of natural gas. Enter the data set in a graphing calculator and find a cubic regression equation for the data.

(B) If $y = N(x)$ denotes the regression equation found in part (A), find $N(60)$ and $N'(60)$, and write a brief verbal interpretation of these results.

94. Break-even analysis. Table 3 contains price–demand and total cost data from a bakery for the production of kringles (a Danish pastry), where p is the price (in dollars) of a kringle for a daily demand of x kringles and C is the total cost (in dollars) of producing x kringles.

Table 3

x	$p(\$)$	$C(\$)$
125	9	740
140	8	785
170	7	850
200	6	900

(A) Find a linear regression equation for the price–demand data, using x as the independent variable.

(B) Find a linear regression equation for the cost data, using x as the independent variable. Use this equation to estimate the fixed costs and variable costs per kringle.

(C) Find the break-even points.

(D) Find the price range for which the bakery will make a profit.

95. Pollution. A sewage treatment plant uses a pipeline that extends 1 mile toward the center of a large lake. The concentration of effluent $C(x)$ in parts per million, x meters from the end of the pipe is given approximately by

$$C(x) = \frac{500}{x^2}, x \geq 1$$

What is the instantaneous rate of change of concentration at 10 meters? At 100 meters?

96. Medicine. The body temperature (in degrees Fahrenheit) of a patient t hours after taking a fever-reducing drug is given by

$$F(t) = 0.16t^2 - 1.6t + 102$$

Find $F(4)$ and $F'(4)$. Write a brief verbal interpretation of these quantities.

97. Learning. If a person learns N items in t hours, as given by

$$N(t) = 20\sqrt{t}$$

find the rate of learning after

(A) 1 hour   (B) 4 hours

98. Physics: The coefficient of thermal expansion (CTE) is a measure of the expansion of an object subjected to extreme temperatures. We want to use a Michaelis–Menten function of the form

$$C(T) = \frac{C_{max}T}{M + T}$$

where $C = $ CTE, T is temperature in K (degrees Kelvin), and C_{max} and M are constants. Table 4 lists the coefficients of thermal expansion for titanium at various temperatures.

Table 4 **Coefficients of Thermal Expansion**

$T(K)$	Titanium
100	4.5
200	7.4
293	8.6
500	9.9
800	11.1
1100	11.7

(A) Plot the points in columns 1 and 2 of Table 4 on graph paper and estimate C_{max} to the nearest integer. To estimate M, add the horizontal line CTE $= \frac{C_{max}}{2}$ to your graph, connect successive points on the graph with straight-line segments, and estimate the value of T (to the nearest multiple of fifty) that satisfies

$$C(T) = \frac{C_{max}}{2}.$$

(B) Use the constants $\frac{C_{max}}{2}$ and M from part (A) to form a Michaelis–Menten function for the CTE of titanium.

(C) Use the function from part (B) to estimate the CTE of titanium at 600 K and to estimate the temperature when the CTE of titanium is 10.

11 Additional Derivative Topics

11.1 The Constant *e* and Continuous Compound Interest

11.2 Derivatives of Exponential and Logarithmic Functions

11.3 Derivatives of Products and Quotients

11.4 The Chain Rule

11.5 Implicit Differentiation

11.6 Related Rates

11.7 Elasticity of Demand

Chapter 11 Summary and Review

Review Exercises

Introduction

In this chapter, we develop techniques for finding derivatives of a wide variety of functions, including exponential and logarithmic functions. There are straightforward procedures—the product rule, quotient rule, and chain rule—for writing down the derivative of any function that is the product, quotient, or composite of functions whose derivatives are known. With the ability to calculate derivatives easily, we consider a wealth of applications involving rates of change. For example, we apply the derivative to study population growth, radioactive decay, elasticity of demand, and environmental crises (see Problem 39 in Section 11.6 or Problem 87 in Section 11.7). Before starting this chapter, you may find it helpful to review the basic properties of exponential and logarithmic functions in Sections 2.5 and 2.6.

11.1 The Constant *e* and Continuous Compound Interest

- The Constant *e*
- Continuous Compound Interest

In Chapter 2, both the exponential function with base *e* and continuous compound interest were introduced informally. Now, with an understanding of limit concepts, we can give precise definitions of *e* and continuous compound interest.

The Constant *e*

The irrational number *e* is a particularly suitable base for both exponential and logarithmic functions. The reasons for choosing this number as a base will become clear as we develop differentiation formulas for the exponential function e^x and the natural logarithmic function ln *x*.

In precalculus treatments (Chapter 2), the number *e* is defined informally as the irrational number that can be approximated by the expression $[1 + (1/n)]^n$ for *n* sufficiently large. Now we will use the limit concept to formally define *e* as either of the following two limits. [*Note:* If $s = 1/n$, then as $n \to \infty$, $s \to 0$.]

> **DEFINITION** The Number *e*
>
> $$e = \lim_{n \to \infty} \left(1 + \frac{1}{n}\right)^n \qquad \text{or, alternatively,} \qquad e = \lim_{s \to 0} (1 + s)^{1/s}$$

Both limits are equal to $e = 2.718\ 281\ 828\ 459\ldots$

Proof that the indicated limits exist and represent an irrational number between 2 and 3 is not easy and is omitted.

> ## CONCEPTUAL INSIGHT
>
> The two limits used to define *e* are unlike any we have encountered so far. Some people reason (incorrectly) that both limits are 1, since $1 + s \to 1$ as $s \to 0$ and 1 to any power is 1. An ordinary scientific calculator with a y^x key can convince you otherwise. Consider the following table of values for *s* and $f(s) = (1 + s)^{1/s}$ and Figure 1 for *s* close to 0. Compute the table values with a calculator yourself, and try several values of *s* even closer to 0. Note that the function is discontinuous at $s = 0$.
>
>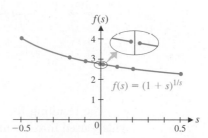
>
> Figure 1
>
> s approaches 0 from the left $\to 0 \leftarrow$ s approaches 0 from the right
>
> | *s* | −0.5 | −0.2 | −0.1 | −0.01 $\to 0 \leftarrow$ 0.01 | 0.1 | 0.2 | 0.5 |
> |---|---|---|---|---|---|---|---|
> | $(1+s)^{1/s}$ | 4.0000 | 3.0518 | 2.8680 | 2.7320 $\to e \leftarrow$ 2.7048 | 2.5937 | 2.4883 | 2.2500 |

Continuous Compound Interest

Now we can see how *e* appears quite naturally in the important application of compound interest. Let us start with simple interest, move on to compound interest, and then proceed on to continuous compound interest.

On one hand, if a principal P is borrowed at an annual rate r,* then after t years at simple interest, the borrower will owe the lender an amount A given by

$$A = P + Prt = P(1 + rt) \qquad \text{Simple interest} \qquad (1)$$

On the other hand, if interest is compounded m times a year, then the borrower will owe the lender an amount A given by

$$A = P\left(1 + \frac{r}{m}\right)^{mt} \qquad \text{Compound interest} \qquad (2)$$

where r/m is the interest rate per compounding period and mt is the number of compounding periods. Suppose that P, r, and t in equation (2) are held fixed and m is increased. Will the amount A increase without bound, or will it tend to approach some limiting value?

Let us perform a calculator experiment before we attack the general limit problem. If $P = \$100$, $r = 0.06$, and $t = 2$ years, then

$$A = 100\left(1 + \frac{0.06}{m}\right)^{2m}$$

We compute A for several values of m in Table 1. The biggest gain appears in the first step, then the gains slow down as m increases. The amount A appears to approach $\$112.75$ as m gets larger and larger.

Table 1

Compounding Frequency	m	$A = 100\left(1 + \dfrac{0.06}{m}\right)^{2m}$
Annually	1	\$112.3600
Semiannually	2	112.5509
Quarterly	4	112.6493
Monthly	12	112.7160
Weekly	52	112.7419
Daily	365	112.7486
Hourly	8,760	112.7496

Keeping P, r, and t fixed in equation (2), we compute the following limit and observe an interesting and useful result:

$$\lim_{m \to \infty} P\left(1 + \frac{r}{m}\right)^{mt} = P \lim_{m \to \infty} \left(1 + \frac{r}{m}\right)^{(m/r)rt} \quad \begin{array}{l}\text{Insert } r/r \text{ in the exponent and}\\ \text{let } s = r/m. \text{ Note that}\\ m \to \infty \text{ implies } s \to 0.\end{array}$$

$$= P \lim_{s \to 0}[(1 + s)^{1/s}]^{rt} \qquad \text{Use a limit property.}^{\dagger}$$

$$= P[\lim_{s \to 0}(1 + s)^{1/s}]^{rt} \qquad \lim_{s \to 0}(1 + s)^{1/s} = e$$

$$= Pe^{rt}$$

The resulting formula is called the **continuous compound interest formula**, a widely used formula in business and economics.

> ### THEOREM 1   Continuous Compound Interest Formula
>
> If a principal P is invested at an annual rate r (expressed as a decimal) compounded continuously, then the amount A in the account at the end of t years is given by
>
> $$A = Pe^{rt}$$

*If r is the interest rate written as a decimal, then $100r\%$ is the rate in percent. For example, if $r = 0.12$, then $100r\% = 100(0.12)\% = 12\%$. The expressions 0.12 and 12% are equivalent. Unless stated otherwise, all formulas in this book use r in decimal form.
†The following new limit property is used: If $\lim_{x \to c} f(x)$ exists, then $\lim_{x \to c}[f(x)]^p = [\lim_{x \to c} f(x)]^p$, provided that the last expression names a real number.

EXAMPLE 1   Computing Continuously Compounded Interest   If $100 is invested at 6% compounded continuously,* what amount will be in the account after 2 years? How much interest will be earned?

SOLUTION

$$A = Pe^{rt}$$
$$= 100e^{(0.06)(2)} \quad \text{6\% is equivalent to } r = 0.06.$$
$$\approx \$112.7497$$

Compare this result with the values calculated in Table 1. The interest earned is $112.7497 − $100 = $12.7497.

Matched Problem 1)   What amount (to the nearest cent) will be in an account after 5 years if $100 is invested at an annual nominal rate of 8% compounded annually? Semiannually? Continuously?

EXAMPLE 2   Graphing the Growth of an Investment   Union Savings Bank offers a 5-year certificate of deposit (CD) that earns 5.75% compounded continuously. If $1,000 is invested in one of these CDs, graph the amount in the account as a function of time for a period of 5 years.

SOLUTION   We want to graph

$$A = 1,000e^{0.0575t} \quad 0 \le t \le 5$$

Using a calculator, we construct a table of values (Table 2). Then we graph the points from the table and join the points with a smooth curve (Fig. 2).

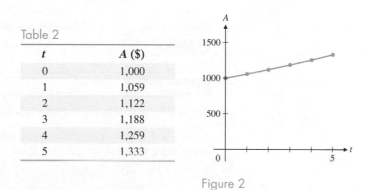

Table 2

t	A ($)
0	1,000
1	1,059
2	1,122
3	1,188
4	1,259
5	1,333

Figure 2

CONCEPTUAL INSIGHT

Depending on the domain, the graph of an exponential function can appear to be linear. Table 2 shows that the graph in Figure 2 is *not* linear. The slope determined by the first two points (for $t = 0$ and $t = 1$) is 59 but the slope determined by the first and third points (for $t = 0$ and $t = 2$) is 61. For a linear graph, the slope determined by any two points is constant.

Matched Problem 2)   If $5,000 is invested in a Union Savings Bank 4-year CD that earns 5.61% compounded continuously, graph the amount in the account as a function of time for a period of 4 years.

*Following common usage, we will often write "at 6% compounded continuously," understanding that this means "at an annual rate of 6% compounded continuously."

EXAMPLE 3   Computing Growth Time   How long will it take an investment of $5,000 to grow to $8,000 if it is invested at 5% compounded continuously?

SOLUTION   Starting with the continous compound interest formula $A = Pe^{rt}$, we must solve for t:

$$A = Pe^{rt}$$

$$8,000 = 5,000e^{0.05t} \qquad \text{Divide both sides by 5,000 and}$$

$$e^{0.05t} = 1.6 \qquad \text{reverse the equation.}$$

$$\ln e^{0.05t} = \ln 1.6 \qquad \text{Take the natural logarithm of both}$$

$$0.05t = \ln 1.6 \qquad \text{sides—recall that } \log_b b^x = x.$$

$$t = \frac{\ln 1.6}{0.05}$$

$$t \approx 9.4 \text{ years}$$

Figure 3 shows an alternative method for solving Example 3 on a graphing calculator.

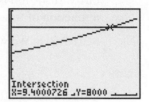

Figure 3

$$y_1 = 5,000e^{0.05x}$$
$$y_2 = 8,000$$

Matched Problem 3   How long will it take an investment of $10,000 to grow to $15,000 if it is invested at 9% compounded continuously?

EXAMPLE 4   Computing Doubling Time   How long will it take money to double if it is invested at 6.5% compounded continuously?

SOLUTION   Starting with the continuous compound interest formula $A = Pe^{rt}$, we solve for t, given $A = 2P$ and $R = 0.065$:

$$2P = Pe^{0.065t} \qquad \text{Divide both sides by } P \text{ and reverse the equation.}$$

$$e^{0.065t} = 2 \qquad \text{Take the natural logarithm of both sides.}$$

$$\ln e^{0.065t} = \ln 2$$

$$0.065t = \ln 2$$

$$t = \frac{\ln 2}{0.065}$$

$$t \approx 10.66 \text{ years}$$

Matched Problem 4   How long will it take money to triple if it is invested at 5.5% compounded continuously?

Explore and Discuss 1   You are considering three options for investing $10,000: at 7% compounded annually, at 6% compounded monthly, and at 5% compounded continuously.

(A) Which option would be the best for investing $10,000 for 8 years?

(B) How long would you need to invest your money for the third option to be the best?

Skills Warm-up Exercise

In Problems 1–8, solve for the variable to two decimal places. (If necessary, review Section 2.5).

1. $A = 1,200e^{0.04(5)}$

2. $A = 3,000e^{0.07(10)}$

3. $9827.30 = Pe^{0.025(3)}$

4. $50,000 = Pe^{0.054(7)}$

5. $6,000 = 5,000e^{0.0325t}$

6. $10,0000 = 7,500e^{0.085t}$

7. $956 = 900e^{1.5r}$

8. $4,840 = 3,750e^{4.25r}$

Use a calculator to evaluate A to the nearest cent in Problems 9 and 10.

9. $A = \$1,000e^{0.1t}$ for $t = 2, 5,$ and 8

10. $A = \$5,000e^{0.08t}$ for $t = 1, 4,$ and 10

11. If \$6,000 is invested at 10% compounded continuously, graph the amount in the account as a function of time for a period of 8 years.

12. If \$4,000 is invested at 8% compounded continuously, graph the amount in the account as a function of time for a period of 6 years.

In Problems 13–18, solve for t or r to two decimal places.

13. $2 = e^{0.06t}$

14. $2 = e^{0.03t}$

15. $3 = e^{0.1r}$

16. $3 = e^{0.25t}$

17. $2 = e^{5r}$

18. $3 = e^{10r}$

In Problems 19 and 20, use a calculator to complete each table to five decimal places.

19.

n	$[1 + (1/n)]^n$
10	2.593 74
100	
1,000	
10,000	
100,000	
1,000,000	
10,000,000	
↓	↓
∞	$e = 2.718281828459\ldots$

20.

s	$(1 + s)^{1/s}$
0.01	2.704 81
−0.01	
0.001	
−0.001	
0.000 1	
−0.000 1	
0.000 01	
−0.000 01	
↓	↓
0	$e = 2.718281828459\ldots$

21. Use a calculator and a table of values to investigate

$$\lim_{n \to \infty} (1 + n)^{1/n}$$

Do you think this limit exists? If so, what do you think it is?

22. Use a calculator and a table of values to investigate

$$\lim_{s \to 0^+} \left(1 + \frac{1}{s}\right)^s$$

Do you think this limit exists? If so, what do you think it is?

23. It can be shown that the number e satisfies the inequality

$$\left(1 + \frac{1}{n}\right)^n < e < \left(1 + \frac{1}{n}\right)^{n+1} \qquad n \geq 1$$

Illustrate this condition by graphing

$$y_1 = (1 + 1/n)^n$$
$$y_2 = 2.718\,281\,828 \approx e$$
$$y_3 = (1 + 1/n)^{n+1}$$

in the same viewing window, for $1 \leq n \leq 20$.

24. It can be shown that

$$e^s = \lim_{n \to \infty} \left(1 + \frac{s}{n}\right)^n$$

for any real number s. Illustrate this equation graphically for $s = 2$ by graphing

$$y_1 = (1 + 2/n)^n$$
$$y_2 = 7.389\,056\,099 \approx e^2$$

in the same viewing window, for $1 \leq n \leq 50$.

Applications

25. **Continuous compound interest.** Provident Bank offers a 10-year CD that earns 2.15% compounded continuously.

   (A) If \$10,000 is invested in this CD, how much will it be worth in 10 years?

   (B) How long will it take for the account to be worth \$18,000?

26. **Continuous compound interest.** Provident Bank also offers a 3-year CD that earns 1.64% compounded continuously.

   (A) If \$10,000 is invested in this CD, how much will it be worth in 3 years?

   (B) How long will it take for the account to be worth \$11,000?

27. **Present value.** A note will pay \$20,000 at maturity 10 years from now. How much should you be willing to pay for the note now if money is worth 5.2% compounded continuously?

28. Present value. A note will pay $50,000 at maturity 5 years from now. How much should you be willing to pay for the note now if money is worth 6.4% compounded continuously?

29. Continuous compound interest. An investor bought stock for $20,000. Five years later, the stock was sold for $30,000. If interest is compounded continuously, what annual nominal rate of interest did the original $20,000 investment earn?

30. Continuous compound interest. A family paid $99,000 cash for a house. Fifteen years later, the house was sold for $195,000. If interest is compounded continuously, what annual nominal rate of interest did the original $99,000 investment earn?

31. Present value. Solving $A = Pe^{rt}$ for P, we obtain

$$P = Ae^{-rt}$$

which is the present value of the amount A due in t years if money earns interest at an annual nominal rate r compounded continuously.

(A) Graph $P = 10,000e^{-0.08t}$, $0 \le t \le 50$.

(B) $\lim\limits_{t \to \infty} 10,000e^{-0.08t} = ?$ [Guess, using part (A).]

[*Conclusion:* The longer the time until the amount A is due, the smaller is its present value, as we would expect.]

32. Present value. Referring to Problem 31, in how many years will the $10,000 be due in order for its present value to be $5,000?

33. Doubling time. How long will it take money to double if it is invested at 4% compounded continuously?

34. Doubling time. How long will it take money to double if it is invested at 5% compounded continuously?

35. Doubling rate. At what nominal rate compounded continuously must money be invested to double in 8 years?

36. Doubling rate. At what nominal rate compounded continuously must money be invested to double in 10 years?

37. Growth time. A man with $20,000 to invest decides to diversify his investments by placing $10,000 in an account that earns 7.2% compounded continuously and $10,000 in an account that earns 8.4% compounded annually. Use graphical approximation methods to determine how long it will take for his total investment in the two accounts to grow to $35,000.

38. Growth time. A woman invests $5,000 in an account that earns 8.8% compounded continuously and $7,000 in an account that earns 9.6% compounded annually. Use graphical approximation methods to determine how long it will take for her total investment in the two accounts to grow to $20,000.

39. Doubling times

(A) Show that the doubling time t (in years) at an annual rate r compounded continuously is given by

$$t = \frac{\ln 2}{r}$$

(B) Graph the doubling-time equation from part (A) for $0.02 \le r \le 0.30$. Is this restriction on r reasonable? Explain.

(C) Determine the doubling times (in years, to two decimal places) for r = 5%, 10%, 15%, 20%, 25%, and 30%.

40. Doubling rates

(A) Show that the rate r that doubles an investment at continuously compounded interest in t years is given by

$$r = \frac{\ln 2}{t}$$

(B) Graph the doubling-rate equation from part (A) for $1 \le t \le 20$. Is this restriction on t reasonable? Explain.

(C) Determine the doubling rates for t = 2, 4, 6, 8, 10, and 12 years.

41. Radioactive decay. A mathematical model for the decay of radioactive substances is given by

$$Q = Q_0 e^{rt}$$

where

Q_0 = amount of the substance at time $t = 0$

r = continuous compound rate of decay

t = time in years

Q = amount of the substance at time t

If the continuous compound rate of decay of radium per year is $r = -0.000\ 433\ 2$, how long will it take a certain amount of radium to decay to half the original amount? (This period is the *half-life* of the substance.)

42. Radioactive decay. The continuous compound rate of decay of carbon-14 per year is $r = -0.000\ 123\ 8$. How long will it take a certain amount of carbon-14 to decay to half the original amount? (Use the radioactive decay model in Problem 41.)

43. Radioactive decay. A cesium isotope has a half-life of 30 years. What is the continuous compound rate of decay? (Use the radioactive decay model in Problem 41.)

44. Radioactive decay. A strontium isotope has a half-life of 90 years. What is the continuous compound rate of decay? (Use the radioactive decay model in Problem 41.)

45. World population. A mathematical model for world population growth over short intervals is given by

$$P = P_0 e^{rt}$$

where

P_0 = population at time $t = 0$

r = continuous compound rate of growth

t = time in years

P = population at time t

How long will it take world population to double if it continues to grow at its current continuous compound rate of 1.3% per year?

46. **U.S. population.** How long will it take for the U.S. population to double if it continues to grow at a rate of 0.975% per year?

47. **Population growth.** Some underdeveloped nations have population doubling times of 50 years. At what continuous compound rate is the population growing? (Use the population growth model in Problem 45.)

48. **Population growth.** Some developed nations have population doubling times of 200 years. At what continuous compound rate is the population growing? (Use the population growth model in Problem 45.)

1. $146.93; $148.02; $149.18
2. $A = 5,000e^{0.0561t}$

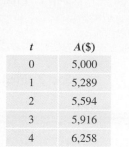

t	$A(\$)$
0	5,000
1	5,289
2	5,594
3	5,916
4	6,258

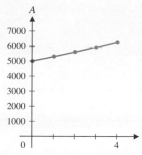

3. 4.51 yr
4. 19.97 yr

11.2 Derivatives of Exponential and Logarithmic Functions

- The Derivative of e^x
- The Derivative of ln x
- Other Logarithmic and Exponential Functions
- Exponential and Logarithmic Models

In this section, we find formulas for the derivatives of logarithmic and exponential functions. A review of Sections 2.5 and 2.6 may prove helpful. In particular, recall that $f(x) = e^x$ is the exponential function with base $e \approx 2.718$, and the inverse of the function e^x is the natural logarithm function ln x. More generally, if b is a positive real number, $b \neq 1$, then the exponential function b^x with base b, and the logarithmic function $\log_b x$ with base b, are inverses of each other.

The Derivative of e^x

In the process of finding the derivative of e^x, we use (without proof) the fact that

$$\lim_{h \to 0} \frac{e^h - 1}{h} = 1 \tag{1}$$

Explore and Discuss 1 Complete Table 1.

Table 1

h	−0.1	−0.01	−0.001	→ 0 ←	0.001	0.01	0.1
$\dfrac{e^h - 1}{h}$							

Do your calculations make it reasonable to conclude that

$$\lim_{h \to 0} \frac{e^h - 1}{h} = 1?$$

Discuss.

We now apply the four-step process (Section 10.4) to the exponential function $f(x) = e^x$.

Step 1 Find $f(x + h)$.

$$f(x + h) = e^{x+h} = e^x e^h \qquad \text{See Section 2.5.}$$

Step 2 Find $f(x + h) - f(x)$.

$$f(x + h) - f(x) = e^x e^h - e^x \quad \text{Factor out } e^x.$$
$$= e^x(e^h - 1)$$

Step 3 Find $\dfrac{f(x + h) - f(x)}{h}$.

$$\frac{f(x + h) - f(x)}{h} = \frac{e^x(e^h - 1)}{h} = e^x\left(\frac{e^h - 1}{h}\right)$$

Step 4 Find $f'(x) = \lim\limits_{h \to 0} \dfrac{f(x + h) - f(x)}{h}$.

$$f'(x) = \lim_{h \to 0} \frac{f(x + h) - f(x)}{h}$$
$$= \lim_{h \to 0} e^x\left(\frac{e^h - 1}{h}\right)$$
$$= e^x \lim_{h \to 0} \left(\frac{e^h - 1}{h}\right) \qquad \text{Use the limit in (1).}$$
$$= e^x \cdot 1 = e^x$$

Therefore,

$$\frac{d}{dx} e^x = e^x \qquad \begin{array}{l}\text{The derivative of the exponential} \\ \text{function is the exponential function.}\end{array}$$

EXAMPLE 1   Finding Derivatives Find $f'(x)$ for

(A) $f(x) = 5e^x - 3x^4 + 9x + 16$ (B) $f(x) = -7x^e + 2e^x + e^2$

SOLUTIONS

(A) $f'(x) = 5e^x - 12x^3 + 9$ (B) $f'(x) = -7ex^{e-1} + 2e^x$

Remember that e is a real number, so the power rule (Section 2.5) is used to find the derivative of x^e. The derivative of the exponential function e^x, however, is e^x. Note that $e^2 \approx 7.389$ is a constant, so its derivative is 0.

Matched Problem 1 Find $f'(x)$ for

(A) $f(x) = 4e^x + 8x^2 + 7x - 14$ (B) $f(x) = x^7 - x^5 + e^3 - x + e^x$

⚠ **CAUTION**

$$\frac{d}{dx} e^x \neq xe^{x-1} \qquad \frac{d}{dx} e^x = e^x$$

The power rule cannot be used to differentiate the exponential function. The power rule applies to exponential forms x^n, where the exponent is a constant and the base is a variable. In the exponential form e^x, the base is a constant and the exponent is a variable. ▲

The Derivative of ln x

We summarize some important facts about logarithmic functions from Section 2.6:

SUMMARY

Recall that the inverse of an exponential function is called a **logarithmic function**. For $b > 0$ and $b \neq 1$,

Logarithmic form		Exponential form
$y = \log_b x$	is equivalent to	$x = b^y$
Domain: $(0, \infty)$		Domain: $(-\infty, \infty)$
Range: $(-\infty, \infty)$		Range: $(0, \infty)$

The graphs of $y = \log_b x$ and $y = b^x$ are symmetric with respect to the line $y = x$. (See Figure 1.)

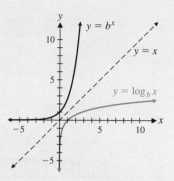

Figure 1

Of all the possible bases for logarithmic functions, the two most widely used are

$\log x = \log_{10} x$ Common logarithm (base 10)

$\ln x = \log_e x$ Natural logarithm (base e)

We are now ready to use the definition of the derivative and the four-step process discussed in Section 10.4 to find a formula for the derivative of $\ln x$. Later we will extend this formula to include $\log_b x$ for any base b.

Let $f(x) = \ln x, x > 0$.

Step 1 Find $f(x + h)$.

$$f(x + h) = \ln(x + h)$$ $\ln(x + h)$ cannot be simplified.

Step 2 Find $f(x + h) - f(x)$.

$$f(x + h) - f(x) = \ln(x + h) - \ln x$$ Use $\ln A - \ln B = \ln \dfrac{A}{B}$.

$$= \ln \frac{x + h}{x}$$

Step 3 Find $\dfrac{f(x + h) - f(x)}{h}$.

$$\frac{f(x + h) - f(x)}{h} = \frac{\ln(x + h) - \ln x}{h}$$

$$= \frac{1}{h} \ln \frac{x + h}{x}$$ Multiply by $1 = x/x$ to change form.

$$= \frac{x}{x} \cdot \frac{1}{h} \ln \frac{x + h}{x}$$

$$= \frac{1}{x} \left[\frac{x}{h} \ln \left(1 + \frac{h}{x} \right) \right]$$ Use $p \ln A = \ln A^p$.

$$= \frac{1}{x} \ln \left(1 + \frac{h}{x} \right)^{x/h}$$

Step 4   Find $f'(x) = \lim\limits_{h \to 0} \dfrac{f(x + h) - f(x)}{h}$.

$$
\begin{aligned}
f'(x) &= \lim_{h \to 0} \frac{f(x + h) - f(x)}{h} \\[2mm]
&= \lim_{h \to 0} \left[\frac{1}{x} \ln\left(1 + \frac{h}{x}\right)^{x/h} \right] && \text{Let } s = h/x. \text{ Note that } h \to 0 \text{ implies } s \to 0. \\[2mm]
&= \frac{1}{x} \lim_{s \to 0} \left[\ln(1 + s)^{1/s} \right] && \text{Use a new limit property.}^* \\[2mm]
&= \frac{1}{x} \ln\left[\lim_{s \to 0} (1 + s)^{1/s} \right] && \text{Use the definition of } e. \\[2mm]
&= \frac{1}{x} \ln e && \ln e = \log_e e = 1 \\[2mm]
&= \frac{1}{x}
\end{aligned}
$$

Therefore,

$$
\frac{d}{dx} \ln x = \frac{1}{x} \quad (x > 0)
$$

CONCEPTUAL INSIGHT

In finding the derivative of $\ln x$, we used the following properties of logarithms:

$$
\ln \frac{A}{B} = \ln A - \ln B \qquad \ln A^p = p \ln A
$$

We also noted that there is no property that simplifies $\ln(A + B)$. (See Theorem 1 in Section 2.6 for a list of properties of logarithms.)

EXAMPLE 2   Finding Derivatives   Find y' for

(A) $y = 3e^x + 5 \ln x$ (B) $y = x^4 - \ln x^4$

SOLUTIONS

(A) $y' = 3e^x + \dfrac{5}{x}$

(B) Before taking the derivative, we use a property of logarithms (see Theorem 1, Section 2.6) to rewrite y.

$$
\begin{aligned}
y &= x^4 - \ln x^4 && \text{Use } \ln M^p = p \ln M. \\[2mm]
y &= x^4 - 4 \ln x && \text{Now take the derivative of both sides.} \\[2mm]
y' &= 4x^3 - \frac{4}{x}
\end{aligned}
$$

Matched Problem 2   Find y' for

(A) $y = 10x^3 - 100 \ln x$ (B) $y = \ln x^5 + e^x - \ln e^2$

Other Logarithmic and Exponential Functions

In most applications involving logarithmic or exponential functions, the number e is the preferred base. However, in some situations it is convenient to use a base other than e. Derivatives of $y = \log_b x$ and $y = b^x$ can be obtained by expressing these functions in terms of the natural logarithmic and exponential functions.

---

*The following new limit property is used: If $\lim_{x \to c} f(x)$ exists and is positive, then $\lim_{x \to c} [\ln f(x)] = \ln[\lim_{x \to c} f(x)]$.

We begin by finding a relationship between $\log_b x$ and $\ln x$ for any base b such that $b > 0$ and $b \neq 1$.

$$y = \log_b x \qquad \text{Change to exponential form.}$$
$$b^y = x \qquad \text{Take the natural logarithm of both sides.}$$
$$\ln b^y = \ln x \qquad \text{Recall that } \ln b^y = y \ln b.$$
$$y \ln b = \ln x \qquad \text{Solve for } y.$$
$$y = \frac{1}{\ln b} \ln x$$

Therefore,

$$\log_b x = \frac{1}{\ln b} \ln x \qquad \text{Change-of-base formula for logarithms*} \qquad (2)$$

Similarly, we can find a relationship between b^x and e^x for any base b such that $b > 0, b \neq 1$.

$$y = b^x \qquad \text{Take the natural logarithm of both sides.}$$
$$\ln y = \ln b^x \qquad \text{Recall that } \ln b^x = x \ln b.$$
$$\ln y = x \ln b \qquad \text{Take the exponential function of both sides.}$$
$$y = e^{x \ln b}$$

Therefore,

$$b^x = e^{x \ln b} \qquad \text{Change-of-base formula for exponential functions} \qquad (3)$$

Differentiating both sides of equation (2) gives

$$\frac{d}{dx} \log_b x = \frac{1}{\ln b} \frac{d}{dx} \ln x = \frac{1}{\ln b} \left(\frac{1}{x} \right) \qquad (x > 0)$$

It can be shown that the derivative of the function e^{cx}, where c is a constant, is the function ce^{cx} (see Problems 61–62 in Exercise 11.2 or the more general results of Section 11.4). Therefore, differentiating both sides of equation (3), we have

$$\frac{d}{dx} b^x = e^{x \ln b} \ln b = b^x \ln b$$

For convenience, we list the derivative formulas for exponential and logarithmic functions:

Derivatives of Exponential and Logarithmic Functions

For $b > 0, b \neq 1$,

$$\frac{d}{dx} e^x = e^x \qquad \frac{d}{dx} b^x = b^x \ln b$$

For $b > 0, b \neq 1$, and $x > 0$,

$$\frac{d}{dx} \ln x = \frac{1}{x} \qquad \frac{d}{dx} \log_b x = \frac{1}{\ln b} \left(\frac{1}{x} \right)$$

EXAMPLE 3 Finding Derivatives   Find $g'(x)$ for

(A) $g(x) = 2^x - 3^x$

(B) $g(x) = \log_4 x^5$

---

*Equation (2) is a special case of the **general change-of-base formula** for logarithms (which can be derived in the same way): $\log_b x = (\log_a x)/(\log_a b)$.

SOLUTIONS

(A) $g'(x) = 2^x \ln 2 - 3^x \ln 3$

(B) First, use a property of logarithms to rewrite $g(x)$.

$$g(x) = \log_4 x^5 \qquad \text{Use } \log_b M^p = p \log_b M.$$
$$g(x) = 5 \log_4 x \qquad \text{Take the derivative of both sides.}$$
$$g'(x) = \frac{5}{\ln 4}\left(\frac{1}{x}\right)$$

Matched Problem 3 | Find $g'(x)$ for

(A) $g(x) = x^{10} + 10^x$ (B) $g(x) = \log_2 x - 6 \log_5 x$

Explore and Discuss 2 (A) The graphs of $f(x) = \log_2 x$ and $g(x) = \log_4 x$ are shown in Figure 2. Which graph belongs to which function?

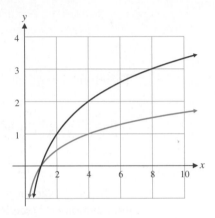

Figure 2

(B) Sketch graphs of $f'(x)$ and $g'(x)$.

(C) The function $f(x)$ is related to $g(x)$ in the same way that $f'(x)$ is related to $g'(x)$. What is that relationship?

Exponential and Logarithmic Models

EXAMPLE 4 Price–Demand Model An Internet store sells Australian wool blankets. If the store sells x blankets at a price of $\$p$ per blanket, then the price–demand equation is $p = 350(0.999)^x$. Find the rate of change of price with respect to demand when the demand is 800 blankets and interpret the result.

SOLUTION

$$\frac{dp}{dx} = 350(0.999)^x \ln 0.999$$

If $x = 800$, then

$$\frac{dp}{dx} = 350(0.999)^{800} \ln 0.999 \approx -0.157, \text{ or } -\$0.16$$

When the demand is 800 blankets, the price is decreasing by $0.16 per blanket.

Matched Problem 4 | The store in Example 4 also sells a reversible fleece blanket. If the price–demand equation for reversible fleece blankets is $p = 200(0.998)^x$, find the rate of change of price with respect to demand when the demand is 400 blankets and interpret the result.

EXAMPLE 5 Continuous Compound Interest An investment of $1,000 earns interest at an annual rate of 4% compounded continuously.

(A) Find the instantaneous rate of change of the amount in the account after 2 years.

(B) Find the instantaneous rate of change of the amount in the account at the time the amount is equal to $2,000.

SOLUTION

(A) The amount $A(t)$ at time t (in years) is given by $A(t) = 1,000e^{0.04t}$. Note that $A(t) = 1,000b^t$, where $b = e^{0.04}$. The instantaneous rate of change is the derivative $A'(t)$, which we find by using the formula for the derivative of the exponential function with base b:

$$A'(t) = 1,000b^t \ln \quad b = 1,000e^{0.04t}(0.04) = 40e^{0.04t}$$

After 2 years, $A'(2) = 40e^{0.04(2)} = \$43.33$ per year.

(B) From the calculation of the derivative in part (A), we note that $A'(t) = (0.04)1,000e^{0.04t} = 0.04A(t)$. In other words, the instantaneous rate of change of the amount is always equal to 4% of the amount. So if the amount is $2,000, then the instantaneous rate of change is $(0.04)\$2,000 = \$80$ per year.

Matched Problem 5 An investment of $5,000 earns interest at an annual rate of 6% compounded continuously.

(A) Find the instantaneous rate of change of the amount in the account after 3 years.

(B) Find the instantaneous rate of change of the amount in the account at the time the amount is equal to $8,000.

EXAMPLE 6 Cable TV Subscribers A statistician used data from the U.S. Census Bureau to construct the model

$$S(t) = 11 \ln t + 29$$

where $S(t)$ is the number of cable TV subscribers (in millions) in year ($t = 0$) corresponds to 1980). Use this model to estimate the number of cable TV subscribers in 2020 and the rate of change of the number of subscribers in 2020. Round both to the nearest tenth of a million. Interpret these results.

SOLUTION Since 2020 corresponds to $t = 40$, we must find $S(40)$ and $S'(40)$.

$$S(40) = 11 \ln 40 + 29 = 69.6 \text{ million}$$

$$S'(t) = 11\frac{1}{t} = \frac{11}{t}$$

$$S'(40) = \frac{11}{40} = 0.3 \text{ million}$$

In 2020 there will be approximately 69.6 million subscribers, and this number is growing at the rate of 0.3 million subscribers per year.

Matched Problem 6 A model for a newspaper's circulation is

$$C(t) = 83 - 9 \ln t$$

where $C(t)$ is the circulation (in thousands) in year t ($t = 0$ corresponds to 1980). Use this model to estimate the circulation and the rate of change of circulation in 2020. Round both to the nearest hundred. Interpret these results.

CONCEPTUAL INSIGHT

On most graphing calculators, exponential regression produces a function of the form $y = a \cdot b^x$. Formula (3) on page 585 allows you to change the base b (chosen by the graphing calculator) to the more familiar base e:

$$y = a \cdot b^x = a \cdot e^{x \ln b}$$

On most graphing calculators, logarithmic regression produces a function of the form $y = a + b \ln x$. Formula (2) on page 585 allows you to write the function in terms of logarithms to any base d that you may prefer:

$$y = a + b \ln x = a + b(\ln d) \log_d x$$

Exercises 11.2

Skills Warm-up Exercises

In Problems 1–8, solve for the variable without using a calculator. (If necessary, review Section 2.6).

1. $y = \log_2 128$

2. $y = \ln e^5$

3. $\log_3 x = 4$

4. $\log_5 x = -2$

5. $\log_b 64 = 2$

6. $\log_b 5 = \dfrac{1}{3}$

7. $y = \ln \sqrt{e}$

8. $y = \ln (\ln e)$

In Problems 9–26, find $f'(x)$.

9. $f(x) = 5e^x + 3x + 1$

10. $f(x) = -7e^x - 2x + 5$

11. $f(x) = -2 \ln x + x^2 - 4$

12. $f(x) = 6 \ln x - x^3 + 2$

13. $f(x) = x^3 - 6e^x$

14. $f(x) = 9e^x + 2x^2$

15. $f(x) = e^x + x - \ln x$

16. $f(x) = \ln x + 2e^x - 3x^2$

17. $f(x) = \ln x^3$

18. $f(x) = \ln x^8$

19. $f(x) = 5x - \ln x^5$

20. $f(x) = 4 + \ln x^9$

21. $f(x) = \ln x^2 + 4e^x$

22. $f(x) = \ln x^{10} + 2 \ln x$

23. $f(x) = e^x + x^e$

24. $f(x) = 3x^e - 2e^x$

25. $f(x) = xx^e$

26. $f(x) = ee^x$

In Problems 27–34, find the equation of the line tangent to the graph of f at the indicated value of x.

27. $f(x) = 3 + \ln x; x = 1$

28. $f(x) = 2 \ln x; x = 1$

29. $f(x) = 3e^x; x = 0$

30. $f(x) = e^x + 1; x = 0$

31. $f(x) = \ln x^3; x = e$

32. $f(x) = 1 + \ln x^4; x = e$

33. $f(x) = 2 + e^x; x = 1$

34. $f(x) = 5e^x; x = 1$

✎ **35.** A student claims that the line tangent to the graph of $f(x) = e^x$ at $x = 3$ passes through the point $(2, 0)$ (see the figure). Is she correct? Will the line tangent at $x = 4$ pass through $(3, 0)$? Explain.

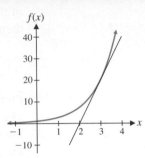

$f(x)$

✎ **36.** Refer to Problem 35. Does the line tangent to the graph of $f(x) = e^x$ at $x = 1$ pass through the origin? Are there any other lines tangent to the graph of f that pass through the origin? Explain.

✎ **37.** A student claims that the line tangent to the graph of $g(x) = \ln x$ at $x = 3$ passes through the origin (see the figure). Is he correct? Will the line tangent at $x = 4$ pass through the origin? Explain.

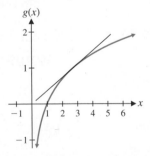

$g(x)$

✎ **38.** Refer to Problem 37. Does the line tangent to the graph of $f(x) = \ln x$ at $x = e$ pass through the origin? Are there any other lines tangent to the graph of f that pass through the origin? Explain.

In Problems 39–42, first use appropriate properties of logarithms to rewrite $f(x)$, and then find $f'(x)$.

39. $f(x) = 10x + \ln 10x$

40. $f(x) = 2 + 3 \ln \dfrac{1}{x}$

41. $f(x) = \ln \dfrac{4}{x^3}$

42. $f(x) = x + 5 \ln 6x$

In Problems 43–54, find $\dfrac{dy}{dx}$ for the indicated function y.

43. $y = \log_2 x$

44. $y = 3 \log_5 x$

45. $y = 3^x$

46. $y = 4^x$

47. $y = 2x - \log x$

48. $y = \log x + 4x^2 + 1$

49. $y = 10 + x + 10^x$

50. $y = x^5 - 5^x$

51. $y = 3 \ln x + 2 \log_3 x$

52. $y = -\log_2 x + 10 \ln x$

53. $y = 2^x + e^2$

54. $y = e^3 - 3^x$

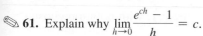 *In Problems 55–60, use graphical approximation methods to find the points of intersection of $f(x)$ and $g(x)$ (to two decimal places).*

55. $f(x) = e^x$; $g(x) = x^4$

[Note that there are three points of intersection and that e^x is greater than x^4 for large values of x.]

56. $f(x) = e^x$; $g(x) = x^5$

[Note that there are two points of intersection and that e^x is greater than x^5 for large values of x.]

57. $f(x) = (\ln x)^2$; $g(x) = x$

58. $f(x) = (\ln x)^3$; $g(x) = x$

59. $f(x) = \ln x$; $g(x) = x^{1/5}$

60. $f(x) = \ln x$; $g(x) = x^{1/4}$

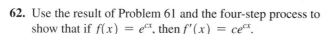 **61.** Explain why $\displaystyle\lim_{h \to 0} \frac{e^{ch} - 1}{h} = c$.

62. Use the result of Problem 61 and the four-step process to show that if $f(x) = e^{cx}$, then $f'(x) = ce^{cx}$.

Applications

63. Salvage value. The estimated salvage value S (in dollars) of a company airplane after t years is given by

$$S(t) = 300,000(0.9)^t$$

What is the rate of depreciation (in dollars per year) after 1 year? 5 years? 10 years?

64. Resale value. The estimated resale value R (in dollars) of a company car after t years is given by

$$R(t) = 20,000(0.86)^t$$

What is the rate of depreciation (in dollars per year) after 1 year? 2 years? 3 years?

65. Bacterial growth. A single cholera bacterium divides every 0.5 hour to produce two complete cholera bacteria. If we start with a colony of 5,000 bacteria, then after t hours, there will be

$$A(t) = 5,000 \cdot 2^{2t} = 5,000 \cdot 4^t$$

bacteria. Find $A'(t), A'(1)$, and $A'(5)$, and interpret the results.

66. Bacterial growth. Repeat Problem 65 for a starting colony of 1,000 bacteria such that a single bacterium divides every 0.25 hour.

67. Blood pressure. An experiment was set up to find a relationship between weight and systolic blood pressure in children. Using hospital records for 5,000 children, the experimenters found that the systolic blood pressure was given approximately by

$$P(x) = 17.5(1 + \ln x) \qquad 10 \le x \le 100$$

where $P(x)$ is measured in millimeters of mercury and x is measured in pounds. What is the rate of change of blood pressure with respect to weight at the 40-pound weight level? At the 90-pound weight level?

68. Blood pressure. Refer to Problem 67. Find the weight (to the nearest pound) at which the rate of change of blood pressure with respect to weight is 0.3 millimeter of mercury per pound.

69. Psychology: stimulus/response. In psychology, the Weber–Fechner law for the response to a stimulus is

$$R = k \ln \frac{S}{S_0}$$

where R is the response, S is the stimulus, and S_0 is the lowest level of stimulus that can be detected. Find dR/dS.

70. Psychology: learning. A mathematical model for the average of a group of people learning to type is given by

$$N(t) = 10 + 6 \ln t \qquad t \ge 1$$

where $N(t)$ is the number of words per minute typed after t hours of instruction and practice (2 hours per day, 5 days per week). What is the rate of learning after 10 hours of instruction and practice? After 100 hours?

71. Continuous Compound Interest. An investment of $10,000 earns interest at an annual rate of 7.5% compounded continuously.

(A) Find the instantaneous rate of change of the amount in the account after 1 year.

(B) Find the instantaneous rate of change of the amount in the account at the time the amount is equal to $12,500.

72. Continuous Compound Interest. An investment of $25,000 earns interest at an annual rate of 8.4% compounded continuously.

(A) Find the instantaneous rate of change of the amount in the account after 2 years.

(B) Find the instantaneous rate of change of the amount in the account at the time the amount is equal to $30,000.

Answers to Matched Problems

1. (A) $4e^x + 16x + 7$ (B) $7x^6 - 5x^4 - 1 + e^x$

2. (A) $30x^2 - \dfrac{100}{x}$ (B) $\dfrac{5}{x} + e^x$

3. (A) $10x^9 + 10^x \ln 10$ (B) $\left(\dfrac{1}{\ln 2} - \dfrac{6}{\ln 5}\right)\dfrac{1}{x}$

4. The price is decreasing at the rate of $0.18 per blanket.

5. (A) $359.17 per year   (B) $480 per year

6. The circulation in 2020 is approximately 49,800 and is decreasing at the rate of 200 per year.

11.3 Derivatives of Products and Quotients

- Derivatives of Products
- Derivatives of Quotients

The derivative properties discussed in Section 10.5 add substantially to our ability to compute and apply derivatives to many practical problems. In this and the next two sections, we add a few more properties that will increase this ability even further.

Derivatives of Products

In Section 10.5, we found that the derivative of a sum is the sum of the derivatives. Is the derivative of a product the product of the derivatives?

Explore and Discuss 1   Let $F(x) = x^2$, $S(x) = x^3$, and $f(x) = F(x)S(x) = x^5$. Which of the following is $f'(x)$?

(A) $F'(x)S'(x)$ (B) $F(x)S'(x)$

(C) $F'(x)S(x)$ (D) $F(x)S'(x) + F'(x)S(x)$

Comparing the various expressions computed in Explore and Discuss 1, we see that the derivative of a product is not the product of the derivatives.

Using the definition of the derivative and the four-step process, we can show that

The derivative of the product of two functions is the first times the derivative of the second, plus the second times the derivative of the first.

This **product rule** is expressed more compactly in Theorem 1, with notation chosen to aid memorization (F for "first", S for "second").

THEOREM 1   Product Rule

If

$$y = f(x) = F(x)S(x)$$

and if $F'(x)$ and $S'(x)$ exist, then

$$f'(x) = F(x)S'(x) + S(x)F'(x)$$

Using simplified notation,

$$y' = FS' + SF' \qquad \text{or} \qquad \frac{dy}{dx} = F\frac{dS}{dx} + S\frac{dF}{dx}$$

EXAMPLE 1   Differentiating a Product Use two different methods to find $f'(x)$ for

$$f(x) = 2x^2(3x^4 - 2).$$

SOLUTION

Method 1. Use the product rule with $F(x) = 2x^2$ and $S(x) = 3x^4 - 2$:

$$f'(x) = 2x^2(3x^4 - 2)' + (3x^4 - 2)(2x^2)' \qquad \text{First times derivative of}$$
$$= 2x^2(12x^3) + (3x^4 - 2)(4x) \qquad \text{second, plus second times}$$
$$= 24x^5 + 12x^5 - 8x \qquad \text{derivative of first}$$
$$= 36x^5 - 8x$$

Method 2. Multiply first; then find the derivative:

$$f(x) = 2x^2(3x^4 - 2) = 6x^6 - 4x^2$$
$$f'(x) = 36x^5 - 8x$$

Matched Problem 1 Use two different methods to find $f'(x)$ for
$$f(x) = 3x^3(2x^2 - 3x + 1).$$

Some products we encounter can be differentiated by either method illustrated in Example 1. In other situations, the product rule *must* be used. Unless instructed otherwise, you should use the product rule to differentiate all products in this section in order to gain experience with this important differentiation rule.

EXAMPLE 2 Tangent Lines Let $f(x) = (2x - 9)(x^2 + 6)$.

(A) Find the equation of the line tangent to the graph of $f(x)$ at $x = 3$.

(B) Find the value(s) of x where the tangent line is horizontal.

SOLUTION

(A) First, find $f'(x)$:

$$f'(x) = (2x - 9)(x^2 + 6)' + (x^2 + 6)(2x - 9)'$$
$$= (2x - 9)(2x) + (x^2 + 6)(2)$$

Then, find $f(3)$ and $f'(3)$:

$$f(3) = [2(3) - 9](3^2 + 6) = (-3)(15) = -45$$
$$f'(3) = [2(3) - 9]2(3) + (3^2 + 6)(2) = -18 + 30 = 12$$

Now, find the equation of the tangent line at $x = 3$:

$$y - y_1 = m(x - x_1) \qquad y_1 = f(x_1) = f(3) = -45$$
$$y - (-45) = 12(x - 3) \qquad m = f'(x_1) = f'(3) = 12$$
$$y = 12x - 81 \qquad \text{Tangent line at } x = 3$$

(B) The tangent line is horizontal at any value of x such that $f'(x) = 0$, so

$$f'(x) = (2x - 9)2x + (x^2 + 6)2 = 0$$
$$6x^2 - 18x + 12 = 0$$
$$x^2 - 3x + 2 = 0$$
$$(x - 1)(x - 2) = 0$$
$$x = 1, 2$$

The tangent line is horizontal at $x = 1$ and at $x = 2$.

Matched Problem 2 Repeat Example 2 for $f(x) = (2x + 9)(x^2 - 12)$.

CONCEPTUAL INSIGHT

As Example 2 illustrates, the way we write $f'(x)$ depends on what we want to do. If we are interested only in evaluating $f'(x)$ at specified values of x, then the form in part (A) is sufficient. However, if we want to solve $f'(x) = 0$, we must multiply and collect like terms, as we did in part (B).

EXAMPLE 3 Finding Derivatives Find $f'(x)$ for

(A) $f(x) = 2x^3 e^x$ (B) $f(x) = 6x^4 \ln x$

SOLUTIONS

(A) $f'(x) = 2x^3(e^x)' + e^x(2x^3)'$

$= 2x^3 e^x + e^x(6x^2)$

$= 2x^2 e^x(x + 3)$

(B) $f'(x) = 6x^4(\ln x)' + (\ln x)(6x^4)'$

$= 6x^4 \dfrac{1}{x} + (\ln x)(24x^3)$

$= 6x^3 + 24x^3 \ln x$

$= 6x^3(1 + 4 \ln x)$

Matched Problem 3 Find $f'(x)$ for

(A) $f(x) = 5x^8 e^x$

(B) $f(x) = x^7 \ln x$

Derivatives of Quotients

The derivative of a quotient of two functions is not the quotient of the derivatives of the two functions.

Explore and Discuss 2 Let $T(x) = x^5$, $B(x) = x^2$, and

$$f(x) = \frac{T(x)}{B(x)} = \frac{x^5}{x^2} = x^3$$

Which of the following is $f'(x)$?

(A) $\dfrac{T'(x)}{B'(x)}$

(B) $\dfrac{T'(x)B(x)}{[B(x)]^2}$

(C) $\dfrac{T(x)B'(x)}{[B(x)]^2}$

(D) $\dfrac{T'(x)B(x)}{[B(x)]^2} - \dfrac{T(x)B'(x)}{[B(x)]^2} = \dfrac{B(x)T'(x) - T(x)B'(x)}{[B(x)]^2}$

The expressions in Explore and Discuss 2 suggest that the derivative of a quotient leads to a more complicated quotient than expected.

If $T(x)$ and $B(x)$ are any two differentiable functions and

$$f(x) = \frac{T(x)}{B(x)}$$

then

$$f'(x) = \frac{B(x)T'(x) - T(x)B'(x)}{[B(x)]^2}$$

Therefore,

The derivative of the quotient of two functions is the denominator times the derivative of the numerator, minus the numerator times the derivative of the denominator, divided by the denominator squared.

This **quotient rule** is expressed more compactly in Theorem 2, with notation chosen to aid memorization (T for "top", B for "bottom").

THEOREM 2 Quotient Rule

If

$$y = f(x) = \frac{T(x)}{B(x)}$$

and if $T'(x)$ and $B'(x)$ exist, then

$$f'(x) = \frac{B(x)T'(x) - T(x)B'(x)}{[B(x)]^2}$$

Using simplified notation,

$$y' = \frac{BT' - TB'}{B^2} \qquad \text{or} \qquad \frac{dy}{dx} = \frac{B\dfrac{dT}{dx} - T\dfrac{dB}{dx}}{B^2}$$

EXAMPLE 4 Differentiating Quotients

(A) If $f(x) = \dfrac{x^2}{2x - 1}$, find $f'(x)$.   (B) If $y = \dfrac{t^2 - t}{t^3 + 1}$, find y'.

(C) Find $\dfrac{d}{dx}\dfrac{x^2 - 3}{x^2}$ by using the quotient rule and also by splitting the fraction into two fractions.

SOLUTION

(A) Use the quotient rule with $T(x) = x^2$ and $B(x) = 2x - 1$;

$$f'(x) = \frac{(2x - 1)(x^2)' - x^2(2x - 1)'}{(2x - 1)^2}$$

The denominator times the derivative of the numerator, minus the numerator times the derivative of the denominator, divided by the square of the denominator

$$= \frac{(2x - 1)(2x) - x^2(2)}{(2x - 1)^2}$$

$$= \frac{4x^2 - 2x - 2x^2}{(2x - 1)^2}$$

$$= \frac{2x^2 - 2x}{(2x - 1)^2}$$

(B) $y' = \dfrac{(t^3 + 1)(t^2 - t)' - (t^2 - t)(t^3 + 1)'}{(t^3 + 1)^2}$

$$= \frac{(t^3 + 1)(2t - 1) - (t^2 - t)(3t^2)}{(t^3 + 1)^2}$$

$$= \frac{2t^4 - t^3 + 2t - 1 - 3t^4 + 3t^3}{(t^3 + 1)^2}$$

$$= \frac{-t^4 + 2t^3 + 2t - 1}{(t^3 + 1)^2}$$

(C) Method 1. Use the quotient rule:

$$\frac{d}{dx}\frac{x^2 - 3}{x^2} = \frac{x^2\dfrac{d}{dx}(x^2 - 3) - (x^2 - 3)\dfrac{d}{dx}x^2}{(x^2)^2}$$

$$= \frac{x^2(2x) - (x^2 - 3)2x}{x^4}$$

$$= \frac{2x^3 - 2x^3 + 6x}{x^4} = \frac{6x}{x^4} = \frac{6}{x^3}$$

Method 2. Split into two fractions:

$$\frac{x^2 - 3}{x^2} = \frac{x^2}{x^2} - \frac{3}{x^2} = 1 - 3x^{-2}$$

$$\frac{d}{dx}(1 - 3x^{-2}) = 0 - 3(-2)x^{-3} = \frac{6}{x^3}$$

Comparing methods 1 and 2, we see that it often pays to change an expression algebraically before choosing a differentiation formula.

Matched Problem 4 ⌋ Find

(A) $f'(x)$ for $f(x) = \dfrac{2x}{x^2 + 3}$

(B) y' for $y = \dfrac{t^3 - 3t}{t^2 - 4}$

(C) $\dfrac{d}{dx} \dfrac{2 + x^3}{x^3}$ in two ways

EXAMPLE 5   Finding Derivatives   Find $f'(x)$ for

(A) $f(x) = \dfrac{3e^x}{1 + e^x}$

(B) $f(x) = \dfrac{\ln x}{2x + 5}$

SOLUTIONS

(A) $f'(x) = \dfrac{(1 + e^x)(3e^x)' - 3e^x(1 + e^x)'}{(1 + e^x)^2}$

$= \dfrac{(1 + e^x)3e^x - 3e^x e^x}{(1 + e^x)^2}$

$= \dfrac{3e^x}{(1 + e^x)^2}$

(B) $f'(x) = \dfrac{(2x + 5)(\ln x)' - (\ln x)(2x + 5)'}{(2x + 5)^2}$

$= \dfrac{(2x + 5) \cdot \dfrac{1}{x} - (\ln x)(2)}{(2x + 5)^2}$ Multiply by $\dfrac{x}{x}$

$= \dfrac{2x + 5 - 2x \ln x}{x(2x + 5)^2}$

Matched Problem 5 ⌋ Find $f'(x)$ for

(A) $f(x) = \dfrac{x^3}{e^x + 2}$

(B) $f(x) = \dfrac{4x}{1 + \ln x}$

EXAMPLE 6   Sales Analysis   The total sales S (in thousands of games) of a video game t months after the game is introduced are given by

$$S(t) = \dfrac{125t^2}{t^2 + 100}$$

(A) Find $S'(t)$.

(B) Find $S(10)$ and $S'(10)$. Write a brief interpretation of these results.

(C) Use the results from part (B) to estimate the total sales after 11 months.

SOLUTION

(A) $S'(t) = \dfrac{(t^2 + 100)(125t^2)' - 125t^2(t^2 + 100)'}{(t^2 + 100)^2}$

$= \dfrac{(t^2 + 100)(250t) - 125t^2(2t)}{(t^2 + 100)^2}$

$= \dfrac{250t^3 + 25{,}000t - 250t^3}{(t^2 + 100)^2}$

$= \dfrac{25{,}000t}{(t^2 + 100)^2}$

(B) $S(10) = \dfrac{125(10)^2}{10^2 + 100} = 62.5$ and $S'(10) = \dfrac{25{,}000(10)}{(10^2 + 100)^2} = 6.25.$

Total sales after 10 months are 62,500 games, and sales are increasing at the rate of 6,250 games per month.

(C) Total sales will increase by approximately 6,250 games during the next month, so the estimated total sales after 11 months are $62{,}500 + 6{,}250 = 68{,}750$ games.

Matched Problem 6 Refer to Example 6. Suppose that the total sales S (in thousands of games) t months after the game is introduced are given by

$$S(t) = \frac{150t}{t + 3}$$

(A) Find $S'(t)$.

(B) Find $S(12)$ and $S'(12)$. Write a brief interpretation of these results.

(C) Use the results from part (B) to estimate the total sales after 13 months.

Exercises 11.3

Skills Warm-up Exercises

W In Problems 1–4, find functions $F(x)$ and $S(x)$, neither a constant function, such that the given function $f(x)$ is equal to the product $F(x)S(x)$. (If necessary, review Section A.2).

1. $f(x) = 5x^3 - 4x^3 \ln x$

2. $f(x) = 6e^x + 2x^2e^x + 3x^4e^x$

3. $f(x) = x^3e^x + 2x^3 + 3e^x + 6$

4. $f(x) = 20 + 5 \ln x + 4x^2 + x^2 \ln x$

In Problems 5–8, find functions $T(x)$ and $B(x)$, neither a constant function, such that the given function $f(x)$ is equal to the quotient $T(x)/B(x)$. (If necessary, review Section A.2).

5. $f(x) = 9x^2e^{-5x}$ 6. $f(x) = 8x^{-2} \ln x$

7. $f(x) = \dfrac{3}{x^2} + \dfrac{e^x}{x^4}$ 8. $f(x) = \dfrac{1}{x} + \dfrac{2}{x^2} + \dfrac{4}{x^3}$

Answers to most of the following problems in this exercise set contain both an unsimplified form and a simplified form of the derivative. When checking your work, first check that you applied the rules correctly and then check that you performed the algebraic simplification correctly. Unless instructed otherwise, when differentiating a product, use the product rule rather than performing the multiplication first.

In Problems 9–34, find $f'(x)$ and simplify.

9. $f(x) = 2x^3(x^2 - 2)$ 10. $f(x) = 5x^2(x^3 + 2)$

11. $f(x) = (x - 3)(2x - 1)$

12. $f(x) = (3x + 2)(4x - 5)$

13. $f(x) = \dfrac{x}{x - 3}$ 14. $f(x) = \dfrac{3x}{2x + 1}$

15. $f(x) = \dfrac{2x + 3}{x - 2}$ 16. $f(x) = \dfrac{3x - 4}{2x + 3}$

17. $f(x) = 3xe^x$ 18. $f(x) = x^2e^x$

19. $f(x) = x^3 \ln x$ 20. $f(x) = 5x \ln x$

21. $f(x) = (x^2 + 1)(2x - 3)$

22. $f(x) = (3x + 5)(x^2 - 3)$

23. $f(x) = (0.4x + 2)(0.5x - 5)$

24. $f(x) = (0.5x - 4)(0.2x + 1)$

25. $f(x) = \dfrac{x^2 + 1}{2x - 3}$ 26. $f(x) = \dfrac{3x + 5}{x^2 - 3}$

27. $f(x) = (x^2 + 2)(x^2 - 3)$

28. $f(x) = (x^2 - 4)(x^2 + 5)$

29. $f(x) = \dfrac{x^2 + 2}{x^2 - 3}$ 30. $f(x) = \dfrac{x^2 - 4}{x^2 + 5}$

31. $f(x) = \dfrac{e^x}{x^2 + 1}$ 32. $f(x) = \dfrac{1 - e^x}{1 + e^x}$

33. $f(x) = \dfrac{\ln x}{1 + x}$ 34. $f(x) = \dfrac{2x}{1 + \ln x}$

In Problems 35–46, find $h'(x)$, where $f(x)$ is an unspecified differentiable function.

35. $h(x) = xf(x)$ 36. $h(x) = x^2f(x)$

37. $h(x) = x^3f(x)$ 38. $h(x) = \dfrac{f(x)}{x}$

39. $h(x) = \dfrac{f(x)}{x^2}$ 40. $h(x) = \dfrac{f(x)}{x^3}$

41. $h(x) = \dfrac{x}{f(x)}$ 42. $h(x) = \dfrac{x^2}{f(x)}$

43. $h(x) = e^x f(x)$

44. $h(x) = \dfrac{e^x}{f(x)}$

45. $h(x) = \dfrac{\ln x}{f(x)}$

46. $h(x) = \dfrac{f(x)}{\ln x}$

In Problems 47–56, find the indicated derivatives and simplify.

47. $f'(x)$ for $f(x) = (2x + 1)(x^2 - 3x)$

48. y' for $y = (x^3 + 2x^2)(3x - 1)$

49. $\dfrac{dy}{dt}$ for $y = (2.5t - t^2)(4t + 1.4)$

50. $\dfrac{d}{dt}[(3 - 0.4t^3)(0.5t^2 - 2t)]$

51. y' for $y = \dfrac{5x - 3}{x^2 + 2x}$

52. $f'(x)$ for $f(x) = \dfrac{3x^2}{2x - 1}$

53. $\dfrac{d}{dw}\dfrac{w^2 - 3w + 1}{w^2 - 1}$

54. $\dfrac{dy}{dw}$ for $y = \dfrac{w^4 - w^3}{3w - 1}$

55. y' for $y = (1 + x - x^2)e^x$

56. $\dfrac{dy}{dt}$ for $y = (1 + e^t)\ln t$

In Problems 57–60:

(A) *Find $f'(x)$ using the quotient rule, and*

(B) *Explain how $f'(x)$ can be found easily without using the quotient rule.*

57. $f(x) = \dfrac{1}{x}$

58. $f(x) = \dfrac{-1}{x^2}$

59. $f(x) = \dfrac{-3}{x^4}$

60. $f(x) = \dfrac{2}{x^3}$

In Problems 61–66, find $f'(x)$ and find the equation of the line tangent to the graph of f at $x = 2$.

61. $f(x) = (1 + 3x)(5 - 2x)$

62. $f(x) = (7 - 3x)(1 + 2x)$

63. $f(x) = \dfrac{x - 8}{3x - 4}$

64. $f(x) = \dfrac{2x - 5}{2x - 3}$

65. $f(x) = \dfrac{x}{2^x}$

66. $f(x) = (x - 2)\ln x$

In Problems 67–70, find $f'(x)$ and find the value(s) of x where $f'(x) = 0$.

67. $f(x) = (2x - 15)(x^2 + 18)$

68. $f(x) = (2x - 3)(x^2 - 6)$

69. $f(x) = \dfrac{x}{x^2 + 1}$

70. $f(x) = \dfrac{x}{x^2 + 9}$

In Problems 71–74, find $f'(x)$ in two ways: (1) using the product or quotient rule and (2) simplifying first.

71. $f(x) = x^3(x^4 - 1)$

72. $f(x) = x^4(x^3 - 1)$

73. $f(x) = \dfrac{x^3 + 9}{x^3}$

74. $f(x) = \dfrac{x^4 + 4}{x^4}$

In Problems 75–92, find each indicated derivative and simplify.

75. $f(w) = (w + 1)2^w$

76. $g(w) = (w - 5)\log_3 w$

77. $\dfrac{dy}{dx}$ for $y = 9x^{1/3}(x^3 + 5)$

78. $\dfrac{d}{dx}[(4x^{1/2} - 1)(3x^{1/3} + 2)]$

79. y' for $y = \dfrac{\log_2 x}{1 + x^2}$

80. $\dfrac{dy}{dx}$ for $y = \dfrac{10^x}{1 + x^4}$

81. $f'(x)$ for $f(x) = \dfrac{6^3\sqrt{x}}{x^2 - 3}$

82. y' for $y = \dfrac{2\sqrt{x}}{x^2 - 3x + 1}$

83. $g'(t)$ if $g(t) = \dfrac{0.2t}{3t^2 - 1}$

84. $h'(t)$ if $h(t) = \dfrac{-0.05t^2}{2t + 1}$

85. $\dfrac{d}{dx}[4x \log x^5]$

86. $\dfrac{d}{dt}[10^t \log t]$

87. $\dfrac{d}{dx}\dfrac{x^3 - 2x^2}{\sqrt[3]{x^2}}$

88. $\dfrac{dy}{dx}$ for $y = \dfrac{x^2 - 3x + 1}{\sqrt[4]{x}}$

89. $f'(x)$ for $f(x) = \dfrac{(2x^2 - 1)(x^2 + 3)}{x^2 + 1}$

90. y' for $y = \dfrac{2x - 1}{(x^3 + 2)(x^2 - 3)}$

91. $\dfrac{dy}{dt}$ for $y = \dfrac{t \ln t}{e^t}$

92. $\dfrac{dy}{du}$ for $y = \dfrac{u^2 e^u}{1 + \ln u}$

Applications

93. Sales analysis. The total sales S (in thousands of DVDs) of a DVD are given by

$$S(t) = \dfrac{90t^2}{t^2 + 50}$$

where t is the number of months since the release of the DVD.

(A) Find $S'(t)$.

(B) Find $S(10)$ and $S'(10)$. Write a brief interpretation of these results.

(C) Use the results from part (B) to estimate the total sales after 11 months.

94. Sales analysis. A communications company has installed a new cable television system in a city. The total number N (in thousands) of subscribers t months after the installation of the system is given by

$$N(t) = \frac{180t}{t + 4}$$

(A) Find $N'(t)$.

(B) Find $N(16)$ and $N'(16)$. Write a brief interpretation of these results.

(C) Use the results from part (B) to estimate the total number of subscribers after 17 months.

95. Price–demand equation. According to economic theory, the demand x for a quantity in a free market decreases as the price p increases (see the figure). Suppose that the number x of DVD players people are willing to buy per week from a retail chain at a price of $\$p$ is given by

$$x = \frac{4,000}{0.1p + 1} \qquad 10 \le p \le 70$$

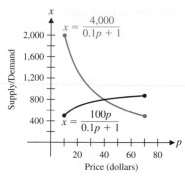

Figure for 95 and 96

(A) Find dx/dp.

(B) Find the demand and the instantaneous rate of change of demand with respect to price when the price is $40. Write a brief interpretation of these results.

(C) Use the results from part (B) to estimate the demand if the price is increased to $41.

96. Price–supply equation. According to economic theory, the supply x of a quantity in a free market increases as the price p increases (see the figure). Suppose that the number x of DVD players a retail chain is willing to sell per week at a price of $\$p$ is given by

$$x = \frac{100p}{0.1p + 1} \qquad 10 \le p \le 70$$

(A) Find dx/dp.

(B) Find the supply and the instantaneous rate of change of supply with respect to price when the price is $40. Write a brief verbal interpretation of these results.

(C) Use the results from part (B) to estimate the supply if the price is increased to $41.

97. Medicine. A drug is injected into a patient's bloodstream through her right arm. The drug concentration (in milligrams per cubic centimeter) in the bloodstream of the left arm t hours after the injection is given by

$$C(t) = \frac{0.14t}{t^2 + 1}$$

(A) Find $C'(t)$.

(B) Find $C'(0.5)$ and $C'(3)$, and interpret the results.

98. Drug sensitivity. One hour after a dose of x milligrams of a particular drug is administered to a person, the change in body temperature $T(x)$, in degrees Fahrenheit, is given approximately by

$$T(x) = x^2\left(1 - \frac{x}{9}\right) \qquad 0 \le x \le 7$$

The rate $T'(x)$ at which T changes with respect to the size of the dosage x is called the *sensitivity* of the body to the dosage.

(A) Use the product rule to find $T'(x)$.

(B) Find $T'(1), T'(3)$, and $T'(6)$.

Answers to Matched Problems

1. $30x^4 - 36x^3 + 9x^2$

2. (A) $y = 84x - 297$ (B) $x = -4, x = 1$

3. (A) $5x^8 e^x + e^x(40x^7) = 5x^7(x + 8)e^x$

   (B) $x^7 \cdot \dfrac{1}{x} + \ln x \,(7x^6) = x^6\,(1 + 7 \ln x)$

4. (A) $\dfrac{(x^2 + 3)2 - (2x)(2x)}{(x^2 + 3)^2} = \dfrac{6 - 2x^2}{(x^2 + 3)^2}$

   (B) $\dfrac{(t^2 - 4)(3t^2 - 3) - (t^3 - 3t)(2t)}{(t^2 - 4)^2} = \dfrac{t^4 - 9t^2 + 12}{(t^2 - 4)^2}$

   (C) $-\dfrac{6}{x^4}$

5. (A) $\dfrac{(e^x + 2)\,3x^2 - x^3 e^x}{(e^x + 2)^2}$

   (B) $\dfrac{(1 + \ln x)\,4 - 4x\,\dfrac{1}{x}}{(1 + \ln x)^2} = \dfrac{4 \ln x}{(1 + \ln x)^2}$

6. (A) $S'(t) = \dfrac{450}{(t + 3)^2}$

   (B) $S(12) = 120; S'(12) = 2$. After 12 months, the total sales are 120,000 games, and sales are increasing at the rate of 2,000 games per month.

   (C) 122,000 games

11.4 The Chain Rule

- Composite Functions
- General Power Rule
- The Chain Rule

The word *chain* in the name "chain rule" comes from the fact that a function formed by composition involves a chain of functions—that is, a function of a function. The *chain rule* enables us to compute the derivative of a composite function in terms of the derivatives of the functions making up the composite. In this section, we review composite functions, introduce the chain rule by means of a special case known as the *general power rule,* and then discuss the chain rule itself.

Composite Functions

The function $m(x) = (x^2 + 4)^3$ is a combination of a quadratic function and a cubic function. To see this more clearly, let

$$y = f(u) = u^3 \quad \text{and} \quad u = g(x) = x^2 + 4$$

We can express y as a function of x:

$$y = f(u) = f[g(x)] = [x^2 + 4]^3 = m(x)$$

The function m is the *composite* of the two functions f and g.

> **DEFINITION**　Composite Functions
> A function m is a **composite** of functions f and g if
> $$m(x) = f[g(x)]$$
> The domain of m is the set of all numbers x such that x is in the domain of g, and $g(x)$ is in the domain of f.

The composite m of functions f and g is pictured in Figure 1. The domain of m is the shaded subset of the domain of g (Fig. 1); it consists of all numbers x such that x is in the domain of g and $g(x)$ is in the domain of f. Note that the functions f and g play different roles. The function g, which is on the *inside* or *interior* of the square brackets in $f[g(x)]$, is applied first to x. Then function f, which appears on the *outside* or *exterior* of the square brackets, is applied to $g(x)$, provided $g(x)$ is in the domain of f. Because f and g play different roles, the composite of f and g is usually a different function than the composite of g and f, as illustrated by Example 1.

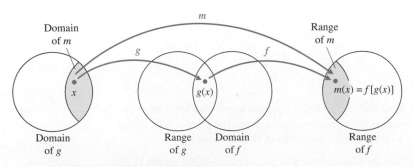

Figure 1　**The composite *m* of *f* and *g***

EXAMPLE 1 Composite Functions Let $f(u) = e^u$ and $g(x) = -3x$. Find $f[g(x)]$ and $g[f(u)]$.

SOLUTION

$$f[g(x)] = f(-3x) = e^{-3x}$$
$$g[f(u)] = g(e^u) = -3e^u$$

Matched Problem 1 Let $f(u) = 2u$ and $g(x) = e^x$. Find $f[g(x)]$ and $g[f(u)]$.

EXAMPLE 2 Composite Functions Write each function as a composite of two simpler functions.

(A) $y = 100e^{0.04x}$ (B) $y = \sqrt{4 - x^2}$

SOLUTION

(A) Let

$$y = f(u) = 100e^u$$
$$u = g(x) = 0.04x$$

Check: $y = f[g(x)] = f(0.04x) = 100e^{0.04x}$

(B) Let

$$y = f(u) = \sqrt{u}$$
$$u = g(x) = 4 - x^2$$

Check: $y = f[g(x)] = f(4 - x^2) = \sqrt{4 - x^2}$

Matched Problem 2 Write each function as a composite of two simpler functions.

(A) $y = 50e^{-2x}$ (B) $y = \sqrt[3]{1 + x^3}$

CONCEPTUAL INSIGHT

There can be more than one way to express a function as a composite of simpler functions. Choosing $y = f(u) = 100u$ and $u = g(x) = e^{0.04x}$ in Example 2A produces the same result:

$$y = f[g(x)] = 100g(x) = 100e^{0.04x}$$

Since we will be using composition as a means to an end (finding a derivative), usually it will not matter which functions you choose for the composition.

General Power Rule

We have already made extensive use of the power rule,

$$\frac{d}{dx}x^n = nx^{n-1} \qquad (1)$$

Can we apply rule (1) to find the derivative of the composite function $m(x) = p[u(x)] = [u(x)]^n$, where p is the power function $p(u) = u^n$ and $u(x)$ is a differentiable function? In other words, is rule (1) valid if x is replaced by $u(x)$?

Explore and Discuss 1 Let $u(x) = 2x^2$ and $m(x) = [u(x)]^3 = 8x^6$. Which of the following is $m'(x)$?

(A) $3[u(x)]^2$ (B) $3[u'(x)]^2$ (C) $3[u(x)]^2u'(x)$

The calculations in Explore and Discuss 1 show that we cannot find the derivative of $[u(x)]^n$ simply by replacing x with $u(x)$ in equation (1).

How can we find a formula for the derivative of $[u(x)]^n$, where $u(x)$ is an arbitrary differentiable function? Let's begin by considering the derivatives of $[u(x)]^2$ and $[u(x)]^3$ to see if a general pattern emerges. Since $[u(x)]^2 = u(x)u(x)$, we use the product rule to write

$$\frac{d}{dx}[u(x)]^2 = \frac{d}{dx}[u(x)u(x)]$$

$$= u(x)u'(x) + u(x)u'(x)$$

$$= 2u(x)u'(x) \qquad\qquad (2)$$

Because $[u(x)]^3 = [u(x)]^2 u(x)$, we use the product rule and the result in equation (2) to write

$$\frac{d}{dx}[u(x)]^3 = \frac{d}{dx}\{[u(x)]^2 u(x)\} \qquad \text{Use equation (2) to substitute for}$$

$$= [u(x)]^2 \frac{d}{dx}u(x) + u(x)\frac{d}{dx}[u(x)]^2 \qquad \frac{d}{dx}[u(x)]^2.$$

$$= [u(x)]^2 u'(x) + u(x)[2u(x)u'(x)]$$

$$= 3[u(x)]^2 u'(x)$$

Continuing in this fashion, we can show that

$$\frac{d}{dx}[u(x)]^n = n[u(x)]^{n-1}u'(x) \qquad n \text{ a positive integer} \qquad (3)$$

Using more advanced techniques, we can establish formula (3) for all real numbers n, obtaining the **general power rule**.

THEOREM 1   General Power Rule

If $u(x)$ is a differentiable function, n is any real number, and

$$y = f(x) = [u(x)]^n$$

then

$$f'(x) = n[u(x)]^{n-1}u'(x)$$

Using simplified notation,

$$y' = nu^{n-1}u' \qquad \text{or} \qquad \frac{d}{dx}u^n = nu^{n-1}\frac{du}{dx} \qquad \text{where } u = u(x)$$

EXAMPLE 3   Using the General Power Rule   Find the indicated derivatives:

(A) $f'(x)$ if $f(x) = (3x + 1)^4$

(B) y' if $y = (x^3 + 4)^7$

(C) $\dfrac{d}{dt}\dfrac{1}{(t^2 + t + 4)^3}$

(D) $\dfrac{dh}{dw}$ if $h(w) = \sqrt{3 - w}$

SOLUTION

(A) $f(x) = (3x + 1)^4$ Let $u = 3x + 1$, $n = 4$.

$\boxed{f'(x) = 4(3x + 1)^3(3x + 1)'}$ $nu^{n-1}\dfrac{du}{dx}$

$\qquad = 4(3x + 1)^3\, 3$ $\dfrac{du}{dx} = 3$

$\qquad = 12(3x + 1)^3$

(B) $y = (x^3 + 4)^7$ Let $u = (x^3 + 4)$, $n = 7$.

$\boxed{y' = 7(x^3 + 4)^6(x^3 + 4)'}$ $nu^{n-1}\dfrac{du}{dx}$

$\qquad = 7(x^3 + 4)^6\, 3x^2$ $\dfrac{du}{dx} = 3x^2$

$\qquad = 21x^2(x^3 + 4)^6$

(C) $\dfrac{d}{dt}\dfrac{1}{(t^2 + t + 4)^3}$

$\qquad = \dfrac{d}{dt}(t^2 + t + 4)^{-3}$ Let $u = t^2 + t + 4$, $n = -3$.

$\qquad \boxed{= -3(t^2 + t + 4)^{-4}(t^2 + t + 4)'}$ $nu^{n-1}\dfrac{du}{dt}$

$\qquad = -3(t^2 + t + 4)^{-4}(2t + 1)$ $\dfrac{du}{dt} = 2t + 1$

$\qquad = \dfrac{-3(2t + 1)}{(t^2 + t + 4)^4}$

(D) $h(w) = \sqrt{3 - w} = (3 - w)^{1/2}$ Let $u = 3 - w$, $n = \dfrac{1}{2}$.

$\boxed{\dfrac{dh}{dw} = \dfrac{1}{2}(3 - w)^{-1/2}(3 - w)'}$ $nu^{n-1}\dfrac{du}{dw}$

$\qquad = \dfrac{1}{2}(3 - w)^{-1/2}(-1)$ $\dfrac{du}{dw} = -1$

$\qquad = -\dfrac{1}{2(3 - w)^{1/2}}$ or $-\dfrac{1}{2\sqrt{3 - w}}$

Matched Problem 3   Find the indicated derivatives:

(A) $h'(x)$ if $h(x) = (5x + 2)^3$

(B) y' if $y = (x^4 - 5)^5$

(C) $\dfrac{d}{dt}\dfrac{1}{(t^2 + 4)^2}$

(D) $\dfrac{dg}{dw}$ if $g(w) = \sqrt{4 - w}$

Notice that we used two steps to differentiate each function in Example 3. First, we applied the general power rule, and then we found du/dx. As you gain experience with the general power rule, you may want to combine these two steps. If you do this, be certain to multiply by du/dx. For example,

$$\dfrac{d}{dx}(x^5 + 1)^4 = 4(x^5 + 1)^3 5x^4 \quad \text{Correct}$$

$$\dfrac{d}{dx}(x^5 + 1)^4 \neq 4(x^5 + 1)^3 \quad du/dx = 5x^4 \text{ is missing}$$

CONCEPTUAL INSIGHT

If we let $u(x) = x$, then $du/dx = 1$, and the general power rule reduces to the (ordinary) power rule discussed in Section 10.5. Compare the following:

$$\frac{d}{dx}x^n = nx^{n-1} \qquad \text{Yes—power rule}$$

$$\frac{d}{dx}u^n = nu^{n-1}\frac{du}{dx} \qquad \text{Yes—general power rule}$$

$$\frac{d}{dx}u^n \neq nu^{n-1} \qquad \text{Unless } u(x) = x + k, \text{ so that } du/dx = 1$$

The Chain Rule

We have used the general power rule to find derivatives of composite functions of the form $f[g(x)]$, where $f(u) = u^n$ is a power function. But what if f is not a power function? Then a more general rule, the *chain rule*, enables us to compute the derivatives of many composite functions of the form $f[g(x)]$.

Suppose that

$$y = m(x) = f[g(x)]$$

is a composite of f and g, where

$$y = f(u) \qquad \text{and} \qquad u = g(x)$$

To express the derivative dy/dx in terms of the derivatives of f and g, we use the definition of a derivative (see Section 10.4).

$$m'(x) = \lim_{h \to 0} \frac{m(x+h) - m(x)}{h} \qquad \begin{array}{l} \text{Substitute } m(x+h) = f[g(x+h)] \\ \text{and } m(x) = f[g(x)]. \end{array}$$

$$= \lim_{h \to 0} \frac{f[g(x+h)] - f[g(x)]}{h} \qquad \text{Multiply by } 1 = \frac{g(x+h) - g(x)}{g(x+h) - g(x)}.$$

$$= \lim_{h \to 0} \left[\frac{f[g(x+h)] - f[g(x)]}{h} \cdot \frac{g(x+h) - g(x)}{g(x+h) - g(x)} \right]$$

$$= \lim_{h \to 0} \left[\frac{f[g(x+h)] - f[g(x)]}{g(x+h) - g(x)} \cdot \frac{g(x+h) - g(x)}{h} \right] \qquad (4)$$

We recognize the second factor in equation (4) as the difference quotient for $g(x)$. To interpret the first factor as the difference quotient for $f(u)$, we let $k = g(x+h) - g(x)$. Since $u = g(x)$, we write

$$u + k = g(x) + g(x+h) - g(x) = g(x+h)$$

Substituting in equation (4), we have

$$m'(x) = \lim_{h \to 0} \left[\frac{f(u+k) - f(u)}{k} \cdot \frac{g(x+h) - g(x)}{h} \right] \qquad (5)$$

If we assume that $k = [g(x+h) - g(x)] \to 0$ as $h \to 0$, we can find the limit of each difference quotient in equation (5):

$$m'(x) = \left[\lim_{k \to 0} \frac{f(u+k) - f(u)}{k} \right]\left[\lim_{h \to 0} \frac{g(x+h) - g(x)}{h} \right]$$

$$= f'(u)g'(x)$$

$$= f'[g(x)]g'(x)$$

Therefore, referring to f and g in the composite function $f[g(x)]$ as the exterior function and interior function, respectively,

The derivative of the composite of two functions is the derivative of the exterior, evaluated at the interior, times the derivative of the interior.

This **chain rule** is expressed more compactly in Theorem 2, with notation chosen to aid memorization (*E* for "exterior", *I* for "interior").

THEOREM 2 Chain Rule

If $m(x) = E[I(x)]$ is a composite function, then

$$m'(x) = E'[I(x)]I'(x)$$

provided that $E'[I(x)]$ and $I'(x)$ exist.
Equivalently, if $y = E(u)$ and $u = I(x)$, then

$$\frac{dy}{dx} = \frac{dy}{du}\frac{du}{dx}$$

provided that $\dfrac{dy}{du}$ and $\dfrac{du}{dx}$ exist.

EXAMPLE 4 Using the Chain Rule Find the derivative $m'(x)$ of the composite function $m(x)$.

(A) $m(x) = (3x^2 + 1)^{3/2}$ (B) $m(x) = e^{2x^3 + 5}$ (C) $m(x) = \ln(x^2 - 4x + 2)$

SOLUTION

(A) The function m is the composite of $E(u) = u^{3/2}$ and $I(x) = 3x^2 + 1$. Then
$E'(u) = \frac{3}{2}u^{1/2}$ and $I'(x) = 6x$, so by the chain rule,
$m'(x) = \frac{3}{2}(3x^2 + 1)^{1/2}(6x) = 9x(3x^2 + 1)^{1/2}$.

(B) The function m is the composite of $E(u) = e^u$ and $I(x) = 2x^3 + 5$. Then
$E'(u) = e^u$ and $I'(x) = 6x^2$, so by the chain rule,
$m'(x) = e^{2x^3 + 5}(6x^2) = 6x^2 e^{2x^3 + 5}$.

(C) The function m is the composite of $E(u) = \ln u$ and $I(x) = x^2 - 4x + 2$.
Then $E'(u) = \frac{1}{u}$ and $I'(x) = 2x - 4$, so by the chain rule,
$m'(x) = \frac{1}{x^2 - 4x + 2}(2x - 4) = \frac{2x - 4}{x^2 - 4x + 2}$.

Matched Problem 4 Find the derivative $m'(x)$ of the composite function $m(x)$.
(A) $m(x) = (2x^3 + 4)^{-5}$ (B) $m(x) = e^{3x^4 + 6}$ (C) $m(x) = \ln(x^2 + 9x + 4)$

Explore and Discuss 2 Let $m(x) = f[g(x)]$. Use the chain rule and Figures 2 and 3 to find

(A) $f(4)$ (B) $g(6)$ (C) $m(6)$
(D) $f'(4)$ (E) $g'(6)$ (F) $m'(6)$

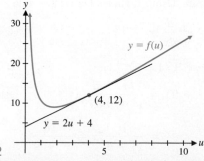

Figure 2

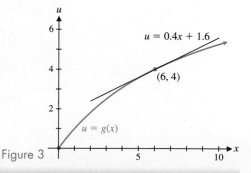

Figure 3

The chain rule can be extended to compositions of three or more functions. For example, if $y = f(w)$, $w = g(u)$, and $u = h(x)$, then

$$\frac{dy}{dx} = \frac{dy}{dw}\frac{dw}{du}\frac{du}{dx}$$

EXAMPLE 5 Using the Chain Rule For $y = h(x) = e^{1 + (\ln x)^2}$, find dy/dx.

SOLUTION Note that h is of the form $y = e^w$, where $w = 1 + u^2$ and $u = \ln x$.

$$\frac{dy}{dx} = \frac{dy}{dw}\frac{dw}{du}\frac{du}{dx}$$

$$= e^w(2u)\left(\frac{1}{x}\right)$$

$$= e^{1 + u^2}(2u)\left(\frac{1}{x}\right) \qquad \text{Since } w = 1 + u^2$$

$$= e^{1 + (\ln x)^2}(2 \ln x)\left(\frac{1}{x}\right) \qquad \text{Since } u = \ln x$$

$$= \frac{2}{x}(\ln x)e^{1 + (\ln x)^2}$$

Matched Problem 5 For $y = h(x) = [\ln(1 + e^x)]^3$, find dy/dx.

The chain rule generalizes basic derivative rules. We list three general derivative rules here for convenient reference [the first, equation (6), is the general power rule of Theorem 1].

General Derivative Rules

$$\frac{d}{dx}[f(x)]^n = n[f(x)]^{n-1}f'(x) \qquad (6)$$

$$\frac{d}{dx}\ln[f(x)] = \frac{1}{f(x)}f'(x) \qquad (7)$$

$$\frac{d}{dx}e^{f(x)} = e^{f(x)}f'(x) \qquad (8)$$

Unless directed otherwise, you now have a choice between the chain rule and the general derivative rules. However, practicing with the chain rule will help prepare you for concepts that appear later in the text. Examples 4 and 5 illustrate the chain rule method, and the next example illustrates the general derivative rules method.

EXAMPLE 6 Using General Derivative Rules

(A) $\dfrac{d}{dx}e^{2x} = e^{2x}\dfrac{d}{dx}2x$ \qquad Using equation (8)

$\qquad = e^{2x}(2) = 2e^{2x}$

(B) $\dfrac{d}{dx}\ln(x^2 + 9) = \dfrac{1}{x^2 + 9}\dfrac{d}{dx}(x^2 + 9)$ \qquad Using equation (7)

$\qquad = \dfrac{1}{x^2 + 9}2x = \dfrac{2x}{x^2 + 9}$

(C) $\dfrac{d}{dx}(1 + e^{x^2})^3 = 3(1 + e^{x^2})^2 \dfrac{d}{dx}(1 + e^{x^2})$ Using equation (6)

$= 3(1 + e^{x^2})^2 e^{x^2} \dfrac{d}{dx} x^2$ Using equation (8)

$= 3(1 + e^{x^2})^2 e^{x^2}(2x)$

$= 6xe^{x^2}(1 + e^{x^2})^2$

Matched Problem 6 Find

(A) $\dfrac{d}{dx} \ln(x^3 + 2x)$ (B) $\dfrac{d}{dx} e^{3x^2 + 2}$ (C) $\dfrac{d}{dx}(2 + e^{-x^2})^4$

Exercises 11.4

For many of the problems in this exercise set, the answers in the back of the book include both an unsimplified form and a simplified form. When checking your work, first check that you applied the rules correctly, and then check that you performed the algebraic simplification correctly.

Skills Warm-up Exercises

W In Problems 1–4, find $f[g(x)]$. (If necessary, review Section 2.1).

1. $f(u) = 3u + 5; g(x) = x^3$

2. $f(u) = u^2 + u; g(x) = 2x - 7$

3. $f(u) = 2u + \ln u; g(x) = x^2 e^x$

4. $f(u) = 4ue^u; g(x) = x^3 + 1$

A In Problems 5–8, find functions $E(u)$ and $I(x)$ so that $y = E[I(x)]$.

5. $y = \ln(x^3 - 6x + 10)$ **6.** $y = (2x - 9)^8$

7. $y = \sqrt{x^2 + 4}$ **8.** $y = e^{2x} + 3e^x - 10$

In Problems 9–16, replace? with an expression that will make the indicated equation valid.

9. $\dfrac{d}{dx}(3x + 4)^4 = 4(3x + 4)^3 \underline{\ ?\ }$

10. $\dfrac{d}{dx}(5 - 2x)^6 = 6(5 - 2x)^5 \underline{\ ?\ }$

11. $\dfrac{d}{dx}(4 - 2x^2)^3 = 3(4 - 2x^2)^2 \underline{\ ?\ }$

12. $\dfrac{d}{dx}(3x^2 + 7)^5 = 5(3x^2 + 7)^4 \underline{\ ?\ }$

13. $\dfrac{d}{dx} e^{x^2 + 1} = e^{x^2 + 1} \underline{\ ?\ }$

14. $\dfrac{d}{dx} e^{4x - 2} = e^{4x - 2} \underline{\ ?\ }$

15. $\dfrac{d}{dx} \ln(x^4 + 1) = \dfrac{1}{x^4 + 1} \underline{\ ?\ }$

16. $\dfrac{d}{dx} \ln(x - x^3) = \dfrac{1}{x - x^3} \underline{\ ?\ }$

In Problems 17–38, find $f'(x)$ and simplify.

17. $f(x) = (5 - 2x)^4$ **18.** $f(x) = (9 - 5x)^2$

19. $f(x) = (4 + 0.2x)^5$ **20.** $f(x) = (6 - 0.5x)^4$

21. $f(x) = (3x^2 + 5)^5$ **22.** $f(x) = (5x^2 - 3)^6$

23. $f(x) = 5e^x$ **24.** $f(x) = 10 - 4e^x$

25. $f(x) = e^{5x}$ **26.** $f(x) = 6e^{-2x}$

27. $f(x) = 3e^{-6x}$ **28.** $f(x) = e^{x^2 + 3x + 1}$

29. $f(x) = (2x - 5)^{1/2}$ **30.** $f(x) = (4x + 3)^{1/2}$

31. $f(x) = (x^4 + 1)^{-2}$ **32.** $f(x) = (x^5 + 2)^{-3}$

33. $f(x) = 4 - 2 \ln x$ **34.** $f(x) = 8 \ln x$

35. $f(x) = 3 \ln(1 + x^2)$

36. $f(x) = 2 \ln(x^2 - 3x + 4)$

37. $f(x) = (1 + \ln x)^3$

38. $f(x) = (x - 2 \ln x)^4$

In Problems 39–44, find $f'(x)$ and the equation of the line tangent to the graph of f at the indicated value of x. Find the value(s) of x where the tangent line is horizontal.

39. $f(x) = (2x - 1)^3; \quad x = 1$

40. $f(x) = (3x - 1)^4; \quad x = 1$

41. $f(x) = (4x - 3)^{1/2}; \quad x = 3$

42. $f(x) = (2x + 8)^{1/2}; \quad x = 4$

43. $f(x) = 5e^{x^2 - 4x + 1}; \quad x = 0$

44. $f(x) = \ln(1 - x^2 + 2x^4); \quad x = 1$

In Problems 45–60, find the indicated derivative and simplify.

45. y' if $y = 3(x^2 - 2)^4$

46. y' if $y = 2(x^3 + 6)^5$

47. $\dfrac{d}{dt} 2(t^2 + 3t)^{-3}$

48. $\dfrac{d}{dt} 3(t^3 + t^2)^{-2}$

49. $\dfrac{dh}{dw}$ if $h(w) = \sqrt{w^2 + 8}$

50. $\dfrac{dg}{dw}$ if $g(w) = \sqrt[3]{3w - 7}$

51. $g'(x)$ if $g(x) = 4xe^{3x}$

52. $h'(x)$ if $h(x) = \dfrac{e^{2x}}{x^2 + 9}$

53. $\dfrac{d}{dx} \dfrac{\ln(1 + x)}{x^3}$

54. $\dfrac{d}{dx}[x^4 \ln(1 + x^4)]$

55. $F'(t)$ if $F(t) = (e^{t^2+1})^3$

56. $G'(t)$ if $G(t) = (1 - e^{2t})^2$

57. y' if $y = \ln(x^2 + 3)^{3/2}$

58. y' if $y = [\ln(x^2 + 3)]^{3/2}$

59. $\dfrac{d}{dw} \dfrac{1}{(w^3 + 4)^5}$

60. $\dfrac{d}{dw} \dfrac{1}{(w^2 - 2)^6}$

In Problems 61–66, find $f'(x)$ and find the equation of the line tangent to the graph of f at the indicated value of x.

61. $f(x) = x(4 - x)^3$; $x = 2$

62. $f(x) = x^2(1 - x)^4$; $x = 2$

63. $f(x) = \dfrac{x}{(2x - 5)^3}$; $x = 3$

64. $f(x) = \dfrac{x^4}{(3x - 8)^2}$; $x = 4$

65. $f(x) = \sqrt{\ln x}$; $x = e$

66. $f(x) = e^{\sqrt{x}}$; $x = 1$

In Problems 67–72, find $f'(x)$ and find the value(s) of x where the tangent line is horizontal.

67. $f(x) = x^2(x - 5)^3$

68. $f(x) = x^3(x - 7)^4$

69. $f(x) = \dfrac{x}{(2x + 5)^2}$

70. $f(x) = \dfrac{x - 1}{(x - 3)^3}$

71. $f(x) = \sqrt{x^2 - 8x + 20}$

72. $f(x) = \sqrt{x^2 + 4x + 5}$

73. A student reasons that the functions $f(x) = \ln[5(x^2 + 3)^4]$ and $g(x) = 4 \ln(x^2 + 3)$ must have the same derivative since he has entered $f(x)$, $g(x)$, $f'(x)$, and $g'(x)$ into a graphing calculator, but only three graphs appear (see the figure). Is his reasoning correct? Are $f'(x)$ and $g'(x)$ the same function? Explain.

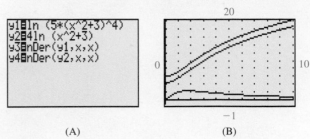

(A)   (B)

Figure for 73

74. A student reasons that the functions $f(x) = (x + 1) \ln (x + 1) - x$ and $g(x) = (x + 1)^{1/3}$ must have the same derivative since she has entered $f(x)$, $g(x)$, $f'(x)$, and $g'(x)$ into a graphing calculator, but only three graphs appear (see the figure). Is her reasoning correct? Are $f'(x)$ and $g'(x)$ the same function? Explain.

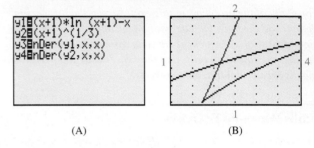

(A)   (B)

In Problems 75–78, give the domain of f, the domain of g, and the domain of m, where $m(x) = f[g(x)]$.

75. $f(u) = \ln u$; $g(x) = 4 - x^2$

76. $f(u) = \ln u$; $g(x) = 2x + 10$

77. $f(u) = \dfrac{1}{u^2 - 1}$; $g(x) = \ln x$

78. $f(u) = \dfrac{1}{u}$; $g(x) = x^2 - 9$

In Problems 79–90, find each derivative and simplify.

79. $\dfrac{d}{dx}[3x(x^2 + 1)^3]$

80. $\dfrac{d}{dx}[2x^2(x^3 - 3)^4]$

81. $\dfrac{d}{dx} \dfrac{(x^3 - 7)^4}{2x^3}$

82. $\dfrac{d}{dx} \dfrac{3x^2}{(x^2 + 5)^3}$

83. $\dfrac{d}{dx} \log_2(3x^2 - 1)$

84. $\dfrac{d}{dx} \log(x^3 - 1)$

85. $\dfrac{d}{dx} 10^{x^2+x}$

86. $\dfrac{d}{dx} 8^{1-2x^2}$

87. $\dfrac{d}{dx} \log_3(4x^3 + 5x + 7)$

88. $\dfrac{d}{dx} \log_5(5^{x^2-1})$

89. $\dfrac{d}{dx} 2^{x^3-x^2+4x+1}$

90. $\dfrac{d}{dx} 10^{\ln x}$

Applications

91. Cost function. The total cost (in hundreds of dollars) of producing x cell phones per day is

$$C(x) = 10 + \sqrt{2x + 16} \qquad 0 \le x \le 50$$

(see the figure).

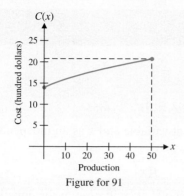

Figure for 91

(A) Find $C'(x)$.

✎ (B) Find $C'(24)$ and $C'(42)$. Interpret the results.

92. Cost function. The total cost (in hundreds of dollars) of producing x cameras per week is

$$C(x) = 6 + \sqrt{4x + 4} \qquad 0 \le x \le 30$$

(A) Find $C'(x)$.

✎ (B) Find $C'(15)$ and $C'(24)$. Interpret the results.

93. Price–supply equation. The number x of bicycle helmets a retail chain is willing to sell per week at a price of $\$p$ is given by

$$x = 80\sqrt{p + 25} - 400 \qquad 20 \le p \le 100$$

(see the figure).

(A) Find dx/dp.

✎ (B) Find the supply and the instantaneous rate of change of supply with respect to price when the price is $75. Write a brief interpretation of these results.

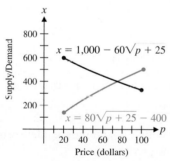

Figure for 93 and 94

94. Price–demand equation. The number x of bicycle helmets people are willing to buy per week from a retail chain at a price of $\$p$ is given by

$$x = 1,000 - 60\sqrt{p + 25} \qquad 20 \le p \le 100$$

(see the figure).

(A) Find dx/dp.

✎ (B) Find the demand and the instantaneous rate of change of demand with respect to price when the price is $75. Write a brief interpretation of these results.

95. Drug concentration. The drug concentration in the bloodstream t hours after injection is given approximately by

$$C(t) = 4.35e^{-t} \qquad 0 \le t \le 5$$

where $C(t)$ is concentration in milligrams per milliliter.

(A) What is the rate of change of concentration after 1 hour? After 4 hours?

(B) Graph C.

96. Water pollution. The use of iodine crystals is a popular way of making small quantities of water safe to drink. Crystals placed in a 1-ounce bottle of water will dissolve until the solution is saturated. After saturation, half of the solution is poured into a quart container of water, and after about an hour, the water is usually safe to drink. The half-empty 1-ounce bottle is then refilled, to be used again in the same way. Suppose that the concentration of iodine in the 1-ounce bottle t minutes after the crystals are introduced can be approximated by

$$C(t) = 250(1 - e^{-t}) \qquad t \ge 0$$

where $C(t)$ is the concentration of iodine in micrograms per milliliter.

(A) What is the rate of change of the concentration after 1 minute? After 4 minutes?

(B) Graph C for $0 \le t \le 5$.

97. Blood pressure and age. A research group using hospital records developed the following mathematical model relating systolic blood pressure and age:

$$P(x) = 40 + 25 \ln(x + 1) \qquad 0 \le x \le 65$$

$P(x)$ is pressure, measured in millimeters of mercury, and x is age in years. What is the rate of change of pressure at the end of 10 years? At the end of 30 years? At the end of 60 years?

98. Biology. A yeast culture at room temperature (68°F) is placed in a refrigerator set at a constant temperature of 38°F. After t hours, the temperature T of the culture is given approximately by

$$T = 30e^{-0.58t} + 38 \qquad t \ge 0$$

What is the rate of change of temperature of the culture at the end of 1 hour? At the end of 4 hours?

Answers to Matched Problems

1. $f[g(x)] = 2e^x, \quad g[f(u)] = e^{2u}$

2. (A) $f(u) = 50e^u, \quad u = -2x$
   (B) $f(u) = \sqrt[3]{u}, \quad u = 1 + x^3$
   [*Note:* There are other correct answers.]

3. (A) $15(5x + 2)^2$ (B) $20x^3(x^4 - 5)^4$
   (C) $-4t/(t^2 + 4)^3$ (D) $-1/(2\sqrt{4 - w})$

4. (A) $m'(x) = -30x^2(2x^3 + 4)^{-6}$
   (B) $m'(x) = 12x^3e^{3x^4 + 6}$ (C) $m'(x) = \dfrac{2x + 9}{x^2 + 9x + 4}$

5. $\dfrac{3e^x[\ln(1 + e^x)]^2}{1 + e^x}$

6. (A) $\dfrac{3x^2 + 2}{x^3 + 2x}$ (B) $6xe^{3x^2 + 2}$
   (C) $-8xe^{-x^2}(2 + e^{-x^2})^3$

11.5 Implicit Differentiation

- Special Function Notation
- Implicit Differentiation

Special Function Notation

The equation

$$y = 2 - 3x^2 \tag{1}$$

defines a function f with y as a dependent variable and x as an independent variable. Using function notation, we would write

$$y = f(x) \quad \text{or} \quad f(x) = 2 - 3x^2$$

In order to minimize the number of symbols, we will often write equation (1) in the form

$$y = 2 - 3x^2 = y(x)$$

where y is *both* a dependent variable and a function symbol. This is a convenient notation, and no harm is done as long as one is aware of the double role of y. Other examples are

$$x = 2t^2 - 3t + 1 = x(t)$$
$$z = \sqrt{u^2 - 3u} = z(u)$$
$$r = \frac{1}{(s^2 - 3s)^{2/3}} = r(s)$$

Until now, we have considered functions involving only one independent variable. There is no reason to stop there: The concept can be generalized to functions involving two or more independent variables, and this will be done in detail in Chapter 15. For now, we will "borrow" the notation for a function involving two independent variables. For example,

$$F(x, y) = x^2 - 2xy + 3y^2 - 5$$

specifies a function F involving two independent variables.

Implicit Differentiation

Consider the equation

$$3x^2 + y - 2 = 0 \tag{2}$$

and the equation obtained by solving equation (2) for y in terms of x,

$$y = 2 - 3x^2 \tag{3}$$

Both equations define the same function with x as the independent variable and y as the dependent variable. For equation (3), we write

$$y = f(x)$$

where

$$f(x) = 2 - 3x^2 \tag{4}$$

and we have an **explicit** (directly stated) rule that enables us to determine y for each value of x. On the other hand, the y in equation (2) is the same y as in equation (3), and equation (2) **implicitly** gives (implies, though does not directly express) y as a function of x. We say that equations (3) and (4) define the function f explicitly and equation (2) defines f implicitly.

Using an equation that defines a function implicitly to find the derivative of the function is called **implicit differentiation**. Let's differentiate equation (2) implicitly and equation (3) directly, and compare results.

Starting with

$$3x^2 + y - 2 = 0$$

we think of y as a function of x and write

$$3x^2 + y(x) - 2 = 0$$

Then we differentiate both sides with respect to x:

$$\frac{d}{dx}[(3x^2 + y(x) - 2)] = \frac{d}{dx}0 \qquad \text{Since } y \text{ is a function of } x, \text{ but is not}$$
$$\text{explicitly given, we simply write}$$

$$\frac{d}{dx}3x^2 + \frac{d}{dx}y(x) - \frac{d}{dx}2 = 0 \qquad \frac{d}{dx}y(x) = y' \text{ to indicate its derivative.}$$

$$6x + y' - 0 = 0$$

Now we solve for y':

$$y' = -6x$$

Note that we get the same result if we start with equation (3) and differentiate directly:

$$y = 2 - 3x^2$$
$$y' = -6x$$

Why are we interested in implicit differentiation? Why not solve for y in terms of x and differentiate directly? The answer is that there are many equations of the form

$$F(x, y) = 0 \qquad\qquad (5)$$

that are either difficult or impossible to solve for y explicitly in terms of x (try it for $x^2y^5 - 3xy + 5 = 0$ or for $e^y - y = 3x$, for example). But it can be shown that, under fairly general conditions on F, equation (5) will define one or more functions in which y is a dependent variable and x is an independent variable. To find y' under these conditions, we differentiate equation (5) implicitly.

Explore and Discuss 1 (A) How many tangent lines are there to the graph in Figure 1 when $x = 0$? When $x = 1$? When $x = 2$? When $x = 4$? When $x = 6$?

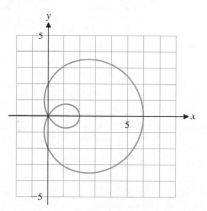

Figure 1

(B) Sketch the tangent lines referred to in part (A), and estimate each of their slopes.

(C) Explain why the graph in Figure 1 is not the graph of a function.

EXAMPLE 1 Differentiating Implicitly Given

$$F(x, y) = x^2 + y^2 - 25 = 0 \qquad\qquad (6)$$

find y' and the slope of the graph at $x = 3$.

SOLUTION We start with the graph of $x^2 + y^2 - 25 = 0$ (a circle, as shown in Fig. 2) so that we can interpret our results geometrically. From the graph, it is clear that equation (6) does not define a function. But with a suitable restriction on the variables, equation (6) can define two or more functions. For example, the upper half and the lower half of the circle each define a function. On each half-circle, a point that corresponds to $x = 3$ is found by substituting $x = 3$ into equation (6) and solving for y:

$$x^2 + y^2 - 25 = 0$$
$$(3)^2 + y^2 = 25$$
$$y^2 = 16$$
$$y = \pm 4$$

The point $(3, 4)$ is on the upper half-circle, and the point $(3, -4)$ is on the lower half-circle. We will use these results in a moment. We now differentiate equation (6) implicitly, treating y as a function of x [i.e., $y = y(x)$]:

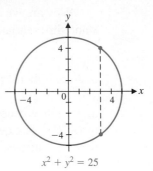

Figure 2 $x^2 + y^2 = 25$

$$x^2 + y^2 - 25 = 0$$
$$x^2 + [y(x)]^2 - 25 = 0$$
$$\frac{d}{dx}\{x^2 + [y(x)]^2 - 25\} = \frac{d}{dx}0$$
$$\frac{d}{dx}x^2 + \frac{d}{dx}[y(x)]^2 - \frac{d}{dx}25 = 0 \qquad \text{Use the chain rule.}$$
$$2x + 2[y(x)]^{2-1}y'(x) - 0 = 0$$
$$2x + 2yy' = 0 \qquad \text{Solve for } y' \text{ in terms of } x \text{ and } y.$$
$$y' = -\frac{2x}{2y}$$
$$y' = -\frac{x}{y} \qquad \text{Leave the answer in terms of } x \text{ and } y.$$

We have found y' without first solving $x^2 + y^2 - 25 = 0$ for y in terms of x. And by leaving y' in terms of x and y, we can use $y' = -x/y$ to find y' for *any* point on the graph of $x^2 + y^2 - 25 = 0$ (except where $y = 0$). In particular, for $x = 3$, we found that $(3, 4)$ and $(3, -4)$ are on the graph. The slope of the graph at $(3, 4)$ is

$$y'|_{(3,4)} = -\tfrac{3}{4} \qquad \text{The slope of the graph at } (3, 4)$$

and the slope at $(3, -4)$ is

$$y'|_{(3,-4)} = -\tfrac{3}{-4} = \tfrac{3}{4} \qquad \text{The slope of the graph at } (3, -4)$$

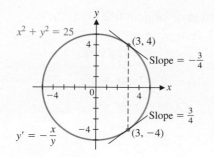

$x^2 + y^2 = 25$

(3, 4)

Slope $= -\frac{3}{4}$

Slope $= \frac{3}{4}$

$y' = -\dfrac{x}{y}$

(3, −4)

Figure 3

The symbol

$$y'\big|_{(a,\, b)}$$

is used to indicate that we are evaluating y' at $x = a$ and $y = b$.

The results are interpreted geometrically in Figure 3 on the original graph.

Matched Problem 1 ⟩ Graph $x^2 + y^2 - 169 = 0$, find y' by implicit differentiation, and find the slope of the graph when $x = 5$.

CONCEPTUAL INSIGHT

When differentiating implicitly, the derivative of y^2 is $2yy'$, not just $2y$. This is because y represents a function of x, so the chain rule applies. Suppose, for example, that y represents the function $y = 5x + 4$. Then

$$(y^2)' = [(5x + 4)^2]' = 2(5x + 4) \cdot 5 = 2yy'$$

So, when differenting implicitly, the derivative of y is y', the derivative of y^2 is $2yy'$, the derivative of y^3 is $3y^2y'$, and so on.

EXAMPLE 2   Differentiating Implicitly   Find the equation(s) of the tangent line(s) to the graph of

$$y - xy^2 + x^2 + 1 = 0 \qquad (7)$$

at the point(s) where $x = 1$.

SOLUTION   We first find y when $x = 1$:

$$y - xy^2 + x^2 + 1 = 0$$
$$y - (1)y^2 + (1)^2 + 1 = 0$$
$$y - y^2 + 2 = 0$$
$$y^2 - y - 2 = 0$$
$$(y - 2)(y + 1) = 0$$
$$y = -1 \quad \text{or} \quad 2$$

So there are two points on the graph of (7) where $x = 1$, namely, $(1, -1)$ and $(1, 2)$. We next find the slope of the graph at these two points by differentiating equation (7) implicitly:

$$y - xy^2 + x^2 + 1 = 0 \qquad \text{Use the product rule and the}$$

$$\frac{d}{dx}y - \frac{d}{dx}xy^2 + \frac{d}{dx}x^2 + \frac{d}{dx}1 = \frac{d}{dx}0 \qquad \text{chain rule for } \frac{d}{dx}xy^2.$$

$$y' - (x \cdot 2yy' + y^2) + 2x = 0$$

$$y' - 2xyy' - y^2 + 2x = 0 \qquad \text{Solve for } y' \text{ by getting all}$$

$$y' - 2xyy' = y^2 - 2x \qquad \text{terms involving } y' \text{ on one side.}$$

$$(1 - 2xy)y' = y^2 - 2x \qquad \text{Factor out } y'.$$

$$y' = \frac{y^2 - 2x}{1 - 2xy}$$

Now find the slope at each point:

$$y'\big|_{(1,\,-1)} = \frac{(-1)^2 - 2(1)}{1 - 2(1)(-1)} = \frac{1 - 2}{1 + 2} = \frac{-1}{3} = -\frac{1}{3}$$

$$y'\big|_{(1,\,2)} = \frac{(2)^2 - 2(1)}{1 - 2(1)(2)} = \frac{4 - 2}{1 - 4} = \frac{2}{-3} = -\frac{2}{3}$$

Equation of tangent line at $(1, -1)$:

$$y - y_1 = m(x - x_1)$$
$$y + 1 = -\tfrac{1}{3}(x - 1)$$
$$y + 1 = -\tfrac{1}{3}x + \tfrac{1}{3}$$
$$y = -\tfrac{1}{3}x - \tfrac{2}{3}$$

Equation of tangent line at $(1, 2)$:

$$y - y_1 = m(x - x_1)$$
$$y - 2 = -\tfrac{2}{3}(x - 1)$$
$$y - 2 = -\tfrac{2}{3}x + \tfrac{2}{3}$$
$$y = -\tfrac{2}{3}x + \tfrac{8}{3}$$

Matched Problem 2 | Repeat Example 2 for $x^2 + y^2 - xy - 7 = 0$ at $x = 1$.

EXAMPLE 3 Differentiating Implicitly Find x' for $x = x(t)$ defined implicitly by

$$t \ln x = xe^t - 1$$

and evaluate x' at $(t, x) = (0, 1)$.

SOLUTION It is important to remember that x is the dependent variable and t is the independent variable. Therefore, we differentiate both sides of the equation with respect to t (using product and chain rules where appropriate) and then solve for x':

$$t \ln x = xe^t - 1 \qquad \text{Differentiate implicitly with respect to } t.$$

$$\frac{d}{dt}(t \ln x) = \frac{d}{dt}(xe^t) - \frac{d}{dt}1 \qquad \text{Use the product rule twice.}$$

$$t\frac{x'}{x} + \ln x = xe^t + e^t x' \qquad \text{Clear fractions.}$$

$$x \cdot t\frac{x'}{x} + x \cdot \ln x = x \cdot xe^t + x \cdot e^t x' \qquad x \neq 0$$

$$tx' + x \ln x = x^2 e^t + xe^t x' \qquad \text{Solve for } x'.$$

$$tx' - xe^t x' = x^2 e^t - x \ln x \qquad \text{Factor out } x'.$$

$$(t - xe^t)x' = x^2 e^t - x \ln x$$

$$x' = \frac{x^2 e^t - x \ln x}{t - xe^t}$$

Now we evaluate x' at $(t, x) = (0, 1)$, as requested:

$$x'\big|_{(0,1)} = \frac{(1)^2 e^0 - 1 \ln 1}{0 - 1e^0}$$

$$= \frac{1}{-1} = -1$$

Matched Problem 3 | Find x' for $x = x(t)$ defined implicitly by

$$1 + x \ln t = te^x$$

and evaluate x' at $(t, x) = (1, 0)$.

Exercises 11.5

Skills Warm-up Exercises

W   *In Problems 1–8, if it is possible to solve for y in terms of x, do so. If not, write "Impossible". (If necessary, review Section 1.1).*

1. $3x + 2y - 20 = 0$

2. $-4x^2 + 3y + 12 = 0$

3. $\dfrac{x^2}{9} + \dfrac{y^2}{16} = 1$

4. $4y^2 - x^2 = 36$

5. $x^2 + xy + y^2 = 1$

6. $2 \ln y + y \ln x = 3x$

7. $5x + 3y = e^y$

8. $y^2 + e^x y + x^3 = 0$

In Problems 9–12, find y' in two ways:

(A) *Differentiate the given equation implicitly and then solve for y'.*

(B) *Solve the given equation for y and then differentiate directly.*

9. $3x + 5y + 9 = 0$

10. $-2x + 6y - 4 = 0$

11. $3x^2 - 4y - 18 = 0$ **12.** $2x^3 + 5y - 2 = 0$

In Problems 13–30, use implicit differentiation to find y' and evaluate y' at the indicated point.

13. $y - 5x^2 + 3 = 0$; $(1, 2)$

14. $5x^3 - y - 1 = 0$; $(1, 4)$

15. $x^2 - y^3 - 3 = 0$; $(2, 1)$

16. $y^2 + x^3 + 4 = 0$; $(-2, 2)$

17. $y^2 + 2y + 3x = 0$; $(-1, 1)$

18. $y^2 - y - 4x = 0$; $(0, 1)$

19. $xy - 6 = 0$; $(2, 3)$

20. $3xy - 2x - 2 = 0$; $(2, 1)$

21. $2xy + y + 2 = 0$; $(-1, 2)$

22. $2y + xy - 1 = 0$; $(-1, 1)$

23. $x^2y - 3x^2 - 4 = 0$; $(2, 4)$

24. $2x^3y - x^3 + 5 = 0$; $(-1, 3)$

25. $e^y = x^2 + y^2$; $(1, 0)$ **26.** $x^2 - y = 4e^y$; $(2, 0)$

27. $x^3 - y = \ln y$; $(1, 1)$ **28.** $\ln y = 2y^2 - x$; $(2, 1)$

29. $x \ln y + 2y = 2x^3$; $(1, 1)$ **30.** $xe^y - y = x^2 - 2$; $(2, 0)$

In Problems 31 and 32, find x' for $x = x(t)$ defined implicitly by the given equation. Evaluate x' at the indicated point.

31. $x^2 - t^2x + t^3 + 11 = 0$; $(-2, 1)$

32. $x^3 - tx^2 - 4 = 0$; $(-3, -2)$

Problems 33 and 34 refer to the equation and graph shown in the figure.

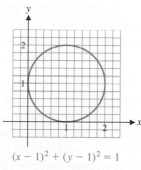

$(x - 1)^2 + (y - 1)^2 = 1$

Figure for 33 and 34

33. Use implicit differentiation to find the slopes of the tangent lines at the points on the graph where $x = 1.6$. Check your answers by visually estimating the slopes on the graph in the figure.

34. Find the slopes of the tangent lines at the points on the graph where $x = 0.2$. Check your answers by visually estimating the slopes on the graph in the figure.

In Problems 35–38, find the equation(s) of the tangent line(s) to the graphs of the indicated equations at the point(s) with the given value of x.

35. $xy - x - 4 = 0$; $x = 2$ **36.** $3x + xy + 1 = 0$; $x = -1$

37. $y^2 - xy - 6 = 0$; $x = 1$ **38.** $xy^2 - y - 2 = 0$; $x = 1$

39. If $xe^y = 1$, find y' in two ways, first by differentiating implicitly and then by solving for y explicitly in terms of x. Which method do you prefer? Explain.

40. Explain the difficulty that arises in solving $x^3 + y + xe^y = 1$ for y as an explicit function of x. Find the slope of the tangent line to the graph of the equation at the point $(0, 1)$.

In Problems 41–48, find y' and the slope of the tangent line to the graph of each equation at the indicated point.

41. $(1 + y)^3 + y = x + 7$; $(2, 1)$

42. $(y - 3)^4 - x = y$; $(-3, 4)$

43. $(x - 2y)^3 = 2y^2 - 3$; $(1, 1)$

44. $(2x - y)^4 - y^3 = 8$; $(-1, -2)$

45. $\sqrt{7 + y^2} - x^3 + 4 = 0$; $(2, 3)$

46. $6\sqrt{y^3 + 1} - 2x^{3/2} - 2 = 0$; $(4, 2)$

47. $\ln(xy) = y^2 - 1$; $(1, 1)$

48. $e^{xy} - 2x = y + 1$; $(0, 0)$

49. Find the equation(s) of the tangent line(s) at the point(s) on the graph of the equation

$$y^3 - xy - x^3 = 2$$

where $x = 1$. Round all approximate values to two decimal places.

50. Refer to the equation in Problem 49. Find the equation(s) of the tangent line(s) at the point(s) on the graph where $y = -1$. Round all approximate values to two decimal places.

Applications

For the demand equations in Problems 51–54, find the rate of change of p with respect to x by differentiating implicitly (x is the number of items that can be sold at a price of $p).

51. $x = p^2 - 2p + 1{,}000$ **52.** $x = p^3 - 3p^2 + 200$

53. $x = \sqrt{10{,}000 - p^2}$ **54.** $x = \sqrt[3]{1{,}500 - p^3}$

55. Biophysics. In biophysics, the equation

$$(L + m)(V + n) = k$$

is called the *fundamental equation of muscle contraction*, where m, n, and k are constants and V is the velocity of the shortening of muscle fibers for a muscle subjected to a load L. Find dL/dV by implicit differentiation.

56. Biophysics. In Problem 55, find dV/dL by implicit differentiation.

57. Speed of sound. The speed of sound in air is given by the formula

$$v = k\sqrt{T}$$

where v is the velocity of sound, T is the temperature of the air, and k is a constant. Use implicit differentiation to find $\dfrac{dT}{dv}$.

58. Gravity. The equation

$$F = G\frac{m_1 m_2}{r^2}$$

is Newton's law of universal gravitation. G is a constant and F is the gravitational force between two objects having masses m_1 and m_2 that are a distance r from each other. Use implicit differentiation to find $\dfrac{dr}{dF}$. Assume that m_1 and m_2 are constant.

59. Speed of sound. Refer to Problem 57. Find $\dfrac{dv}{dT}$ and discuss the connection between $\dfrac{dv}{dT}$ and $\dfrac{dT}{dv}$.

60. Gravity. Refer to Problem 58. Find $\dfrac{dF}{dr}$ and discuss the connection between $\dfrac{dF}{dr}$ and $\dfrac{dr}{dF}$.

Answers to Matched Problems

1. $y' = -x/y$. When $x = 5$, $y = \pm 12$; thus, $y'|_{(5,12)} = -\frac{5}{12}$ and $y'|_{(5,-12)} = \frac{5}{12}$

2. $y' = \dfrac{y - 2x}{2y - x}$; $y = \frac{4}{5}x - \frac{14}{5}$, $y = \frac{1}{5}x + \frac{14}{5}$

3. $x' = \dfrac{te^x - x}{t \ln t - t^2 e^x}$; $x'|_{(1,0)} = -1$

11.6 Related Rates

Union workers are concerned that the rate at which wages are increasing is lagging behind the rate of increase in the company's profits. An automobile dealer wants to predict how much an anticipated increase in interest rates will decrease his rate of sales. An investor is studying the connection between the rate of increase in the Dow Jones average and the rate of increase in the gross domestic product over the past 50 years.

In each of these situations, there are two quantities—wages and profits, for example—that are changing with respect to time. We would like to discover the precise relationship between the rates of increase (or decrease) of the two quantities. We begin our discussion of such *related rates* by considering familiar situations in which the two quantities are distances and the two rates are velocities.

EXAMPLE 1   Related Rates and Motion A 26-foot ladder is placed against a wall (Fig. 1). If the top of the ladder is sliding down the wall at 2 feet per second, at what rate is the bottom of the ladder moving away from the wall when the bottom of the ladder is 10 feet away from the wall?

SOLUTION Many people think that since the ladder is a constant length, the bottom of the ladder will move away from the wall at the rate that the top of the ladder is moving down the wall. This is not the case, however.

At any moment in time, let x be the distance of the bottom of the ladder from the wall and let y be the distance of the top of the ladder from the ground (see Fig. 1). Both x and y are changing with respect to time and can be thought of as functions of time; that is, $x = x(t)$ and $y = y(t)$. Furthermore, x and y are related by the Pythagorean relationship:

$$x^2 + y^2 = 26^2 \tag{1}$$

Differentiating equation (1) implicitly with respect to time t and using the chain rule where appropriate, we obtain

$$2x\frac{dx}{dt} + 2y\frac{dy}{dt} = 0 \tag{2}$$

The rates dx/dt and dy/dt are related by equation (2). This is a **related-rates problem**.

Our problem is to find dx/dt when $x = 10$ feet, given that $dy/dt = -2$ (y is decreasing at a constant rate of 2 feet per second). We have all the quantities we

26 ft

Figure 1

need in equation (2) to solve for dx/dt, except y. When $x = 10$, y can be found from equation (1):

$$10^2 + y^2 = 26^2$$
$$y = \sqrt{26^2 - 10^2} = 24 \text{ feet}$$

Substitute $dy/dt = -2$, $x = 10$, and $y = 24$ into (2). Then solve for dx/dt:

$$2(10)\frac{dx}{dt} + 2(24)(-2) = 0$$

$$\frac{dx}{dt} = \frac{-2(24)(-2)}{2(10)} = 4.8 \text{ feet per second}$$

The bottom of the ladder is moving away from the wall at a rate of 4.8 feet per second.

CONCEPTUAL INSIGHT

In the solution to Example 1, we used equation (1) in two ways: first, to find an equation relating dy/dt and dx/dt, and second, to find the value of y when $x = 10$. These steps must be done in this order. Substituting $x = 10$ and then differentiating does not produce any useful results:

$$x^2 + y^2 = 26^2 \qquad \text{Substituting 10 for } x \text{ has the effect of}$$
$$100 + y^2 = 26^2 \qquad \text{stopping the ladder.}$$
$$0 + 2yy' = 0 \qquad \text{The rate of change of a stationary object}$$
$$y' = 0 \qquad \text{is always 0, but that is not the rate of}$$
$$\text{change of the moving ladder.}$$

Matched Problem 1 Again, a 26-foot ladder is placed against a wall (Fig. 1). If the bottom of the ladder is moving away from the wall at 3 feet per second, at what rate is the top moving down when the top of the ladder is 24 feet above ground?

Explore and Discuss 1

(A) For which values of x and y in Example 1 is dx/dt equal to 2 (i.e., the same rate that the ladder is sliding down the wall)?

(B) When is dx/dt greater than 2? Less than 2?

DEFINITION Suggestions for Solving Related-Rates Problems

Step 1 Sketch a figure if helpful.

Step 2 Identify all relevant variables, including those whose rates are given and those whose rates are to be found.

Step 3 Express all given rates and rates to be found as derivatives.

Step 4 Find an equation connecting the variables identified in step 2.

Step 5 Implicitly differentiate the equation found in step 4, using the chain rule where appropriate, and substitute in all given values.

Step 6 Solve for the derivative that will give the unknown rate.

EXAMPLE 2 Related Rates and Motion Suppose that two motorboats leave from the same point at the same time. If one travels north at 15 miles per hour and the other travels east at 20 miles per hour, how fast will the distance between them be changing after 2 hours?

Figure 2

SOLUTION First, draw a picture, as shown in Figure 2.

All variables, x, y, and z, are changing with time. They can be considered as functions of time: $x = x(t), y = y(t)$, and $z = z(t)$, given implicitly. It now makes sense to find derivatives of each variable with respect to time. From the Pythagorean theorem,

$$z^2 = x^2 + y^2 \qquad (3)$$

We also know that

$$\frac{dx}{dt} = 20 \text{ miles per hour} \qquad \text{and} \qquad \frac{dy}{dt} = 15 \text{ miles per hour}$$

We want to find dz/dt at the end of 2 hours—that is, when $x = 40$ miles and $y = 30$ miles. To do this, we differentiate both sides of equation (3) with respect to t and solve for dz/dt:

$$2z\frac{dz}{dt} = 2x\frac{dx}{dt} + 2y\frac{dy}{dt} \qquad (4)$$

We have everything we need except z. From equation (3), when $x = 40$ and $y = 30$, we find z to be 50. Substituting the known quantities into equation (4), we obtain

$$2(50)\frac{dz}{dt} = 2(40)(20) + 2(30)(15)$$

$$\frac{dz}{dt} = 25 \text{ miles per hour}$$

The boats will be separating at a rate of 25 miles per hour.

Matched Problem 2 │ Repeat Example 2 for the same situation at the end of 3 hours.

EXAMPLE 3 Related Rates and Motion Suppose that a point is moving along the graph of $x^2 + y^2 = 25$ (Fig. 3). When the point is at $(-3, 4)$, its x coordinate is increasing at the rate of 0.4 unit per second. How fast is the y coordinate changing at that moment?

SOLUTION Since both x and y are changing with respect to time, we can consider each as a function of time, namely,

$$x = x(t) \qquad \text{and} \qquad y = y(t)$$

but restricted so that

$$x^2 + y^2 = 25 \qquad (5)$$

We want to find dy/dt, given $x = -3$, $y = 4$, and $dx/dt = 0.4$. Implicitly differentiating both sides of equation (5) with respect to t, we have

$$x^2 + y^2 = 25$$

$$2x\frac{dx}{dt} + 2y\frac{dy}{dt} = 0 \qquad \text{Divide both sides by 2.}$$

$$x\frac{dx}{dt} + y\frac{dy}{dt} = 0 \qquad \begin{array}{l}\text{Substitute } x = -3, y = 4, \text{ and} \\ dx/dt = 0.4, \text{ and solve for } dy/dt.\end{array}$$

$$(-3)(0.4) + 4\frac{dy}{dt} = 0$$

$$\frac{dy}{dt} = 0.3 \text{ unit per second}$$

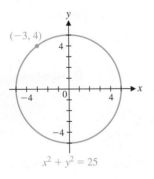

Figure 3

Matched Problem 3 │ A point is moving on the graph of $y^3 = x^2$. When the point is at $(-8, 4)$, its y coordinate is decreasing by 2 units per second. How fast is the x coordinate changing at that moment?

EXAMPLE 4 Related Rates and Business Suppose that for a company manufacturing flash drives, the cost, revenue, and profit equations are given by

$$C = 5,000 + 2x \qquad \text{Cost equation}$$
$$R = 10x - 0.001x^2 \qquad \text{Revenue equation}$$
$$P = R - C \qquad \text{Profit equation}$$

where the production output in 1 week is x flash drives. If production is increasing at the rate of 500 flash drives per week when production is 2,000 flash drives, find the rate of increase in

(A) Cost (B) Revenue (C) Profit

SOLUTION If production x is a function of time (it must be, since it is changing with respect to time), then C, R, and P must also be functions of time. These functions are given implicitly (rather than explicitly). Letting t represent time in weeks, we differentiate both sides of each of the preceding three equations with respect to t and then substitute $x = 2,000$ and $dx/dt = 500$ to find the desired rates.

(A) $C = 5,000 + 2x$ Think: $C = C(t)$ and $x = x(t)$.

$$\frac{dC}{dt} = \frac{d}{dt}(5,000) + \frac{d}{dt}(2x) \qquad \text{Differentiate both sides with respect to } t.$$

$$\frac{dC}{dt} = 0 + 2\frac{dx}{dt} = 2\frac{dx}{dt}$$

Since $dx/dt = 500$ when $x = 2,000$,

$$\frac{dC}{dt} = 2(500) = \$1,000 \text{ per week}$$

Cost is increasing at a rate of $1,000 per week.

(B) $R = 10x - 0.001x^2$

$$\frac{dR}{dt} = \frac{d}{dt}(10x) - \frac{d}{dt}0.001x^2$$

$$\frac{dR}{dt} = 10\frac{dx}{dt} - 0.002x\frac{dx}{dt}$$

$$\frac{dR}{dt} = (10 - 0.002x)\frac{dx}{dt}$$

Since $dx/dt = 500$ when $x = 2,000$,

$$\frac{dR}{dt} = [10 - 0.002(2,000)](500) = \$3,000 \text{ per week}$$

Revenue is increasing at a rate of $3,000 per week.

(C) $P = R - C$

$$\frac{dP}{dt} = \frac{dR}{dt} - \frac{dC}{dt} \qquad \text{Results from parts (A) and (B)}$$

$$= \$3,000 - \$1,000$$

$$= \$2,000 \text{ per week}$$

Profit is increasing at a rate of $2,000 per week.

Matched Problem 4 Repeat Example 4 for a production level of 6,000 flash drives per week.

Exercises 11.6

Skills Warm-up Exercises

W *For Problems 1–8, review the geometric formulas in Appendix C, if necessary.*

1. A circular flower bed has an area of 300 square feet. Find its diameter to the nearest tenth of a foot.

2. A central pivot irrigation system covers a circle of radius 400 meters. Find the area of the circle to the nearest square meter.

3. The hypotenuse of a right triangle has length 50 meters, and another side has length 20 meters. Find the length of the third side to the nearest meter.

4. The legs of a right triangle have lengths 54 feet and 69 feet. Find the length of the hypotenuse to the nearest foot.

5. A person 69 inches tall stands 40 feet from the base of a streetlight. The streetlight casts a shadow of length 96 inches. How far above the ground is the streetlight?

6. The radius of a spherical balloon is 3 meters. Find its volume to the nearest tenth of a cubic meter.

7. A right circular cylinder and a sphere both have radius 12 feet. If the volume of the cylinder is twice the volume of the sphere, find the height of the cylinder.

8. The height of a right circular cylinder is twice its radius. If the volume is 1,000 cubic meters, find the radius and height to the nearest hundredth of a meter.

In Problems 9–14, assume that $x = x(t)$ and $y = y(t)$. Find the indicated rate, given the other information.

9. $y = x^2 + 2$; $dx/dt = 3$ when $x = 5$; find dy/dt

10. $y = x^3 - 3$; $dx/dt = -2$ when $x = 2$; find dy/dt

11. $x^2 + y^2 = 1$; $dy/dt = -4$ when $x = -0.6$ and $y = 0.8$; find dx/dt

12. $x^2 + y^2 = 4$; $dy/dt = 5$ when $x = 1.2$ and $y = -1.6$; find dx/dt

13. $x^2 + 3xy + y^2 = 11$; $dx/dt = 2$ when $x = 1$ and $y = 2$; find dy/dt

14. $x^2 - 2xy - y^2 = 7$; $dy/dt = -1$ when $x = 2$ and $y = -1$; find dx/dt

15. A point is moving on the graph of $xy = 36$. When the point is at $(4, 9)$, its x coordinate is increasing by 4 units per second. How fast is the y coordinate changing at that moment?

16. A point is moving on the graph of $4x^2 + 9y^2 = 36$. When the point is at $(3, 0)$, its y coordinate is decreasing by 2 units per second. How fast is its x coordinate changing at that moment?

17. A boat is being pulled toward a dock as shown in the figure. If the rope is being pulled in at 3 feet per second, how fast is the distance between the dock and the boat decreasing when it is 30 feet from the dock?

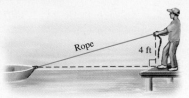

Figure for 17 and 18

18. Refer to Problem 17. Suppose that the distance between the boat and the dock is decreasing by 3.05 feet per second. How fast is the rope being pulled in when the boat is 10 feet from the dock?

19. A rock thrown into a still pond causes a circular ripple. If the radius of the ripple is increasing by 2 feet per second, how fast is the area changing when the radius is 10 feet?

20. Refer to Problem 19. How fast is the circumference of a circular ripple changing when the radius is 10 feet?

21. The radius of a spherical balloon is increasing at the rate of 3 centimeters per minute. How fast is the volume changing when the radius is 10 centimeters?

22. Refer to Problem 21. How fast is the surface area of the sphere increasing when the radius is 10 centimeters?

23. Boyle's law for enclosed gases states that if the volume is kept constant, the pressure P and temperature T are related by the equation

$$\frac{P}{T} = k$$

where k is a constant. If the temperature is increasing at 3 kelvins per hour, what is the rate of change of pressure when the temperature is 250 kelvins and the pressure is 500 pounds per square inch?

24. Boyle's law for enclosed gases states that if the temperature is kept constant, the pressure P and volume V of a gas are related by the equation

$$VP = k$$

where k is a constant. If the volume is decreasing by 5 cubic inches per second, what is the rate of change of pressure when the volume is 1,000 cubic inches and the pressure is 40 pounds per square inch?

25. A 10-foot ladder is placed against a vertical wall. Suppose that the bottom of the ladder slides away from the wall at a constant rate of 3 feet per second. How fast is the top of the ladder sliding down the wall when the bottom is 6 feet from the wall?

26. A weather balloon is rising vertically at the rate of 5 meters per second. An observer is standing on the ground 300 meters from where the balloon was released. At what rate is the distance between the observer and the balloon changing when the balloon is 400 meters high?

27. A streetlight is on top of a 20-foot pole. A person who is 5 feet tall walks away from the pole at the rate of 5 feet per second. At what rate is the tip of the person's shadow moving away from the pole when he is 20 feet from the pole?

28. Refer to Problem 27. At what rate is the person's shadow growing when he is 20 feet from the pole?

29. Helium is pumped into a spherical balloon at a constant rate of 4 cubic feet per second. How fast is the radius increasing after 1 minute? After 2 minutes? Is there any time at which the radius is increasing at a rate of 100 feet per second? Explain.

30. A point is moving along the x axis at a constant rate of 5 units per second. At which point is its distance from $(0, 1)$ increasing at a rate of 2 units per second? At 4 units per second? At 5 units per second? At 10 units per second? Explain.

31. A point is moving on the graph of $y = e^x + x + 1$ in such a way that its x coordinate is always increasing at a rate of 3 units per second. How fast is the y coordinate changing when the point crosses the x axis?

32. A point is moving on the graph of $x^3 + y^2 = 1$ in such a way that its y coordinate is always increasing at a rate of 2 units per second. At which point(s) is the x coordinate increasing at a rate of 1 unit per second?

Applications

33. **Cost, revenue, and profit rates.** Suppose that for a company manufacturing calculators, the cost, revenue, and profit equations are given by

$$C = 90,000 + 30x \qquad R = 300x - \frac{x^2}{30}$$

$$P = R - C$$

where the production output in 1 week is x calculators. If production is increasing at a rate of 500 calculators per week when production output is 6,000 calculators, find the rate of increase (decrease) in

(A) Cost   (B) Revenue   (C) Profit

34. **Cost, revenue, and profit rates.** Repeat Problem 33 for

$$C = 72,000 + 60x \qquad R = 200x - \frac{x^2}{30}$$

$$P = R - C$$

where production is increasing at a rate of 500 calculators per week at a production level of 1,500 calculators.

35. **Advertising.** A retail store estimates that weekly sales s and weekly advertising costs x (both in dollars) are related by

$$s = 60,000 - 40,000e^{-0.0005x}$$

The current weekly advertising costs are $2,000, and these costs are increasing at the rate of $300 per week. Find the current rate of change of sales.

36. **Advertising.** Repeat Problem 35 for

$$s = 50,000 - 20,000e^{-0.0004x}$$

37. **Price–demand.** The price p (in dollars) and demand x for a product are related by

$$2x^2 + 5xp + 50p^2 = 80,000$$

(A) If the price is increasing at a rate of $2 per month when the price is $30, find the rate of change of the demand.

(B) If the demand is decreasing at a rate of 6 units per month when the demand is 150 units, find the rate of change of the price.

38. **Price–demand.** Repeat Problem 37 for

$$x^2 + 2xp + 25p^2 = 74,500$$

39. **Pollution.** An oil tanker aground on a reef is forming a circular oil slick about 0.1 foot thick (see the figure). To estimate the rate dV/dt (in cubic feet per minute) at which the oil is leaking from the tanker, it was found that the radius of the slick was increasing at 0.32 foot per minute $(dR/dt = 0.32)$ when the radius R was 500 feet. Find dV/dt.

40. **Learning.** A person who is new on an assembly line performs an operation in T minutes after x performances of the operation, as given by

$$T = 6\left(1 + \frac{1}{\sqrt{x}}\right)$$

If $dx/dt = 6$ operations per hours, where t is time in hours, find dT/dt after 36 performances of the operation.

Answers to Matched Problems

1. $dy/dt = -1.25$ ft/sec   2. $dz/dt = 25$ mi/hr
3. $dx/dt = 6$ units/sec
4. (A) $dC/dt = \$1,000/\text{wk}$ (B) $dR/dt = -\$1,000/\text{wk}$
   (C) $dP/dt = -\$2,000/\text{wk}$

11.7 Elasticity of Demand

- Relative Rate of Change
- Elasticity of Demand

When will a price increase lead to an increase in revenue? To answer this question and study relationships among price, demand, and revenue, economists use the notion of *elasticity of demand*. In this section, we define the concepts of *relative rate of change*, *percentage rate of change*, and *elasticity of demand*.

Relative Rate of Change

Explore and Discuss 1

A broker is trying to sell you two stocks: Biotech and Comstat. The broker estimates that Biotech's price per share will increase $2 per year over the next several years, while Comstat's price per share will increase only $1 per year. Is this sufficient information for you to choose between the two stocks? What other information might you request from the broker to help you decide?

Interpreting rates of change is a fundamental application of calculus. In Explore and Discuss 1, Biotech's price per share is increasing at twice the rate of Comstat's, but that does not automatically make Biotech the better buy. The obvious information that is missing is the current price of each stock. If Biotech costs $100 a share and Comstat costs $25 a share, then which stock is the better buy? To answer this question, we introduce two new concepts: *relative rate of change* and *percentage rate of change*.

> **DEFINITION** Relative and Percentage Rates of Change
>
> The **relative rate of change** of a function $f(x)$ is $\dfrac{f'(x)}{f(x)}$, or equivalently, $\dfrac{d}{dx} \ln f(x)$.
>
> The **percentage rate of change** is $100 \times \dfrac{f'(x)}{f(x)}$, or equivalently, $100 \times \dfrac{d}{dx} \ln f(x)$.

The alternative form for the relative rate of change, $\dfrac{d}{dx} \ln f(x)$, is called the **logarithmic derivative** of $f(x)$.

Note that

$$\frac{d}{dx} \ln f(x) = \frac{f'(x)}{f(x)}$$

by the chain rule. So the relative rate of change of a function $f(x)$ is its logarithmic derivative, and the percentage rate of change is 100 times the logarithmic derivative.

Returning to Explore and Discuss 1, the table shows the relative rate of change and percentage rate of change for Biotech and Comstat. We conclude that Comstat is the better buy.

	Relative rate of change	Percentage rate of change
Biotech	$\dfrac{2}{100} = 0.02$	2%
Comstat	$\dfrac{1}{25} = 0.04$	4%

EXAMPLE 1 Percentage Rate of Change Table 1 lists the GDP (gross domestic product expressed in billions of 2005 dollars) and U.S. population from 2000 to 2012. A model for the GDP is

$$f(t) = 209.5t + 11{,}361$$

where t is years since 2000. Find and graph the percentage rate of change of $f(t)$ for $0 \le t \le 12$.

Table 1

Year	Real GDP (billions of 2005 dollars)	Population (in millions)
2000	$11,226	282.2
2004	$12,264	292.9
2008	$13,312	304.1
2012	$13,670	313.9

SOLUTION If $p(t)$ is the percentage rate of change of $f(t)$, then

$$p(t) = 100 \times \frac{d}{dx} \ln\left(209.5t + 11{,}361\right)$$

$$= \frac{20{,}950}{209.5t + 11{,}361}$$

The graph of $p(t)$ is shown in Figure 1 (graphing details omitted). Notice that $p(t)$ is decreasing, even though the GDP is increasing.

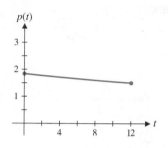

Figure 1

Matched Problem 1 A model for the population data in Table 1 is

$$f(t) = 2.7t + 282$$

where t is years since 2000. Find and graph $p(t)$, the percentage rate of change of $f(t)$ for $0 \le t \le 12$.

CONCEPTUAL INSIGHT

If $10,000 is invested at an annual rate of 4.5% compounded continuously, what is the relative rate of change of the amount in the account? The answer is the logarithmic derivative of $A(t) = 10{,}000e^{0.045t}$, namely

$$\frac{d}{dx} \ln\left(10{,}000e^{0.045t}\right) = \frac{10{,}000e^{0.045t}(0.045)}{10{,}000e^{0.045t}} = 0.045$$

So the relative rate of change of $A(t)$ is 0.045, and the percentage rate of change is just the annual interest rate, 4.5%.

Elasticity of Demand

Explore and Discuss 2 In both parts below, assume that increasing the price per unit by $1 will decrease the demand by 500 units. If your objective is to increase revenue, should you increase the price by $1 per unit?

(A) At the current price of $8.00 per baseball cap, there is a demand for 6,000 caps.

(B) At the current price of $12.00 per baseball cap, there is a demand for 4,000 caps.

In Explore and Discuss 2, the rate of change of demand with respect to price was assumed to be −500 units per dollar. But in one case, part (A), you should increase the price, and in the other, part (B), you should not. Economists use the concept of *elasticity of demand* to answer the question "When does an increase in price lead to an increase in revenue?"

DEFINITION   Elasticity of Demand

Let the price p and demand x for a product be related by a price–demand equation of the form $x = f(p)$. Then the **elasticity of demand at price p**, denoted by $E(p)$, is

$$E(p) = -\frac{\text{relative rate of change of demand}}{\text{relative rate of change of price}}$$

Using the definition of relative rate of change, we can find a formula for $E(p)$:

$$E(p) = -\frac{\text{relative rate of change of demand}}{\text{relative rate of change of price}} = -\frac{\dfrac{d}{dp}\ln f(p)}{\dfrac{d}{dp}\ln p}$$

$$= -\frac{\dfrac{f'(p)}{f(p)}}{\dfrac{1}{p}}$$

$$= -\frac{pf'(p)}{f(p)}$$

THEOREM 1   Elasticity of Demand

If price and demand are related by $x = f(p)$, then the elasticity of demand is given by

$$E(p) = -\frac{pf'(p)}{f(p)}$$

CONCEPTUAL **INSIGHT**

Since p and $f(p)$ are nonnegative and $f'(p)$ is negative (demand is usually a decreasing function of price), $E(p)$ is nonnegative. This is why elasticity of demand is defined as the negative of a ratio.

EXAMPLE 2   Elasticity of Demand   The price p and the demand x for a product are related by the price–demand equation

$$x + 500p = 10,000 \tag{1}$$

Find the elasticity of demand, $E(p)$, and interpret each of the following:
(A) $E(4)$ (B) $E(16)$ (C) $E(10)$

SOLUTION   To find $E(p)$, we first express the demand x as a function of the price p by solving (1) for x:

$$x = 10,000 - 500p$$
$$= 500(20 - p) \quad \text{Demand as a function of price}$$

or

$$x = f(p) = 500(20 - p) \qquad 0 \leq p \leq 20 \tag{2}$$

Since x and p both represent nonnegative quantities, we must restrict p so that $0 \le p \le 20$. Note that the demand is a decreasing function of price. That is, a price increase results in lower demand, and a price decrease results in higher demand (see Figure 2).

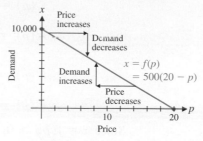

Figure 2

$$E(p) = -\frac{pf'(p)}{f(p)} = -\frac{p(-500)}{500(20-p)} = \frac{p}{20-p}$$

In order to interpret values of $E(p)$, we must recall the definition of elasticity:

$$E(p) = -\frac{\text{relative rate of change of demand}}{\text{relative rate of change of price}}$$

or

$$-\left(\begin{array}{c} \text{relative rate of} \\ \text{change of demand} \end{array} \right) \approx E(p) \left(\begin{array}{c} \text{relative rate of} \\ \text{change of price} \end{array} \right)$$

(A) $E(4) = \frac{4}{16} = 0.25 < 1$. If the \$4 price changes by 10%, then the demand will change by approximately $0.25(10\%) = 2.5\%$.

(B) $E(16) = \frac{16}{4} = 4 > 1$. If the \$16 price changes by 10%, then the demand will change by approximately $4(10\%) = 40\%$.

(C) $E(10) = \frac{10}{10} = 1$. If the \$10 price changes by 10%, then the demand will also change by approximately 10%.

Matched Problem 2 ⌋ Find $E(p)$ for the price–demand equation

$$x = f(p) = 1{,}000(40 - p)$$

Find and interpret each of the following:

(A) $E(8)$ (B) $E(30)$ (C) $E(20)$

The three cases illustrated in the solution to Example 2 are referred to as **inelastic demand**, **elastic demand**, and **unit elasticity**, as indicated in Table 2.

Table 2

$E(p)$	Demand	Interpretation	Revenue
$0 < E(p) < 1$	Inelastic	Demand is not sensitive to changes in price, that is, percentage change in price produces a smaller percentage change in demand.	A price increase will increase revenue.
$E(p) > 1$	Elastic	Demand is sensitive to changes in price, that is, a percentage change in price produces a larger percentage change in demand.	A price increase will decrease revenue.
$E(p) = 1$	Unit	A percentage change in price produces the same percentage change in demand.	

To justify the connection between elasticity of demand and revenue as given in the fourth column of Table 2, we recall that revenue R is the demand x (number of items sold) multiplied by p (price per item). Assume that the price–demand equation is written in the form $x = f(p)$. Then

$$R(p) = xp = f(p)p \qquad \text{Use the product rule.}$$

$$R'(p) = f(p) \cdot 1 + pf'(p) \qquad \text{Multiply and divide by } f(p).$$

$$R'(p) = f(p) + pf'(p)\frac{f(p)}{f(p)} \qquad \text{Factor out } f(p).$$

$$R'(p) = f(p)\left[1 + \frac{pf'(p)}{f(p)}\right] \qquad \text{Use Theorem 1.}$$

$$R'(p) = f(p)[1 - E(p)]$$

Since $x = f(p) > 0$, it follows that $R'(p)$ and $1 - E(p)$ have the same sign. So if $E(p) < 1$, then $R'(p)$ is positive and revenue is increasing (Fig. 3). Similarly, if $E(p) > 1$, then $R'(p)$ is negative, and revenue is decreasing (Fig. 3).

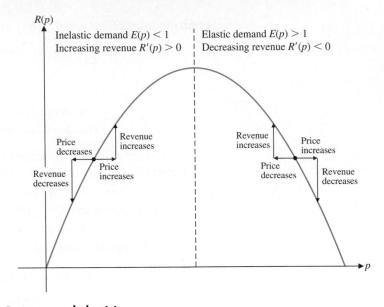

Figure 3 **Revenue and elasticity**

EXAMPLE 3   Elasticity and Revenue   A manufacturer of sunglasses currently sells one type for \$15 a pair. The price p and the demand x for these glasses are related by

$$x = f(p) = 9,500 - 250p$$

If the current price is increased, will revenue increase or decrease?

SOLUTION
$$E(p) = -\frac{pf'(p)}{f(p)}$$

$$= -\frac{p(-250)}{9,500 - 250p}$$

$$= \frac{p}{38 - p}$$

$$E(15) = \frac{15}{23} \approx 0.65$$

At the \$15 price level, demand is inelastic and a price increase will increase revenue.

Matched Problem 3   Repeat Example 3 if the current price for sunglasses is \$21 a pair.

In summary, if demand is inelastic, then a price increase will increase revenue. But if demand is elastic, then a price increase will decrease revenue.

Exercises 11.7

Skills Warm-up Exercises

W   *In Problems 1–8, use the given equation, which expresses price p as a function of demand x, to find a function f(p) that expresses demand x as a function of price p. Give the domain of f(p). (If necessary, review Section 2.1).*

1. $p = 42 - 0.4x, 0 \le x \le 105$

2. $p = 125 - 0.02x, 0 \le x \le 6{,}250$

3. $p = 50 - 0.5x^2, 0 \le x \le 10$

4. $p = 180 - 0.8x^2, 0 \le x \le 15$

5. $p = 25e^{-x/20}, 0 \le x \le 20$

6. $p = 45 - e^{x/4}, 0 \le x \le 12$

7. $p = 80 - 10 \ln x, 1 \le x \le 30$

8. $p = \ln(500 - 5x), 0 \le x \le 90$

In Problems 9–14, find the relative rate of change of f(x).

9. $f(x) = 35x - 0.4x^2$

10. $f(x) = 60x - 1.2x^2$

11. $f(x) = 7 + 4e^{-x}$

12. $f(x) = 15 - 3e^{-0.5x}$

13. $f(x) = 12 + 5 \ln x$

14. $f(x) = 25 - 2 \ln x$

In Problems 15–24, find the relative rate of change of f(x) at the indicated value of x. Round to three decimal places.

15. $f(x) = 45; x = 100$

16. $f(x) = 580; x = 300$

17. $f(x) = 420 - 5x; x = 25$

18. $f(x) = 500 - 6x; x = 40$

19. $f(x) = 420 - 5x; x = 55$

20. $f(x) = 500 - 6x; x = 75$

21. $f(x) = 4x^2 - \ln x; x = 2$

22. $f(x) = 9x - 5 \ln x; x = 3$

23. $f(x) = 4x^2 - \ln x; x = 5$

24. $f(x) = 9x - 5 \ln x; x = 7$

In Problems 25–32, find the percentage rate of change of f(x) at the indicated value of x. Round to the nearest tenth of a percent.

25. $f(x) = 225 + 65x; x = 5$

26. $f(x) = 75 + 110x; x = 4$

27. $f(x) = 225 + 65x; x = 15$

28. $f(x) = 75 + 110x; x = 16$

29. $f(x) = 5{,}100 - 3x^2; x = 35$

30. $f(x) = 3{,}000 - 8x^2; x = 12$

31. $f(x) = 5{,}100 - 3x^2; x = 41$

32. $f(x) = 3{,}000 - 8x^2; x = 18$

In Problems 33–38, use the price–demand equation to find E(p), the elasticity of demand.

33. $x = f(p) = 25{,}000 - 450p$

34. $x = f(p) = 10{,}000 - 190p$

35. $x = f(p) = 4{,}800 - 4p^2$

36. $x = f(p) = 8{,}400 - 7p^2$

37. $x = f(p) = 98 - 0.6e^p$

38. $x = f(p) = 160 - 35 \ln p$

In Problems 39–46, find the logarithmic derivative.

39. $A(t) = 500e^{0.07t}$ **40.** $A(t) = 2{,}000e^{0.052t}$

41. $A(t) = 3{,}500e^{0.15t}$ **42.** $A(t) = 900e^{0.24t}$

43. $f(x) = xe^x$ **44.** $f(x) = x^2e^x$

45. $f(x) = \ln x$ **46.** $f(x) = x \ln x$

In Problems 47–50, use the price–demand equation to determine whether demand is elastic, is inelastic, or has unit elasticity at the indicated values of p.

47. $x = f(p) = 12{,}000 - 10p^2$

    (A) $p = 10$ (B) $p = 20$

    (C) $p = 30$

48. $x = f(p) = 1{,}875 - p^2$

    (A) $p = 15$ (B) $p = 25$

    (C) $p = 40$

49. $x = f(p) = 950 - 2p - 0.1p^2$

    (A) $p = 30$ (B) $p = 50$

    (C) $p = 70$

50. $x = f(p) = 875 - p - 0.05p^2$

    (A) $p = 50$ (B) $p = 70$

    (C) $p = 100$

51. Given the price–demand equation

$$p + 0.005x = 30$$

    (A) Express the demand x as a function of the price p.

    (B) Find the elasticity of demand, $E(p)$.

    (C) What is the elasticity of demand when $p = \$10$? If this price is increased by 10%, what is the approximate percentage change in demand?

    (D) What is the elasticity of demand when $p = \$25$? If this price is increased by 10%, what is the approximate percentage change in demand?

    (E) What is the elasticity of demand when $p = \$15$? If this price is increased by 10%, what is the approximate percentage change in demand?

52. Given the price–demand equation

$$p + 0.01x = 50$$

(A) Express the demand x as a function of the price p.

(B) Find the elasticity of demand, $E(p)$.

(C) What is the elasticity of demand when $p = \$10$? If this price is decreased by 5%, what is the approximate change in demand?

(D) What is the elasticity of demand when $p = \$45$? If this price is decreased by 5%, what is the approximate change in demand?

(E) What is the elasticity of demand when $p = \$25$? If this price is decreased by 5%, what is the approximate change in demand?

53. Given the price–demand equation

$$0.02x + p = 60$$

(A) Express the demand x as a function of the price p.

(B) Express the revenue R as a function of the price p.

(C) Find the elasticity of demand, $E(p)$.

(D) For which values of p is demand elastic? Inelastic?

(E) For which values of p is revenue increasing? Decreasing?

(F) If $p = \$10$ and the price is decreased, will revenue increase or decrease?

(G) If $p = \$40$ and the price is decreased, will revenue increase or decrease?

54. Repeat Problem 53 for the price–demand equation

$$0.025x + p = 50$$

In Problems 55–62, use the price–demand equation to find the values of p for which demand is elastic and the values for which demand is inelastic. Assume that price and demand are both positive.

55. $x = f(p) = 210 - 30p$ **56.** $x = f(p) = 480 - 8p$

57. $x = f(p) = 3{,}125 - 5p^2$ **58.** $x = f(p) = 2{,}400 - 6p^2$

59. $x = f(p) = \sqrt{144 - 2p}$ **60.** $x = f(p) = \sqrt{324 - 2p}$

61. $x = f(p) = \sqrt{2{,}500 - 2p^2}$

62. $x = f(p) = \sqrt{3{,}600 - 2p^2}$

In Problems 63–68, use the demand equation to find the revenue function. Sketch the graph of the revenue function, and indicate the regions of inelastic and elastic demand on the graph.

63. $x = f(p) = 20(10 - p)$ **64.** $x = f(p) = 10(16 - p)$

65. $x = f(p) = 40(p - 15)^2$ **66.** $x = f(p) = 10(p - 9)^2$

67. $x = f(p) = 30 - 10\sqrt{p}$ **68.** $x = f(p) = 30 - 5\sqrt{p}$

If a price–demand equation is solved for p, then price is expressed as $p = g(x)$ and x becomes the independent variable. In this case, it can be shown that the elasticity of demand is given by

$$E(x) = -\frac{g(x)}{xg'(x)}$$

In Problems 69–72, use the price–demand equation to find $E(x)$ at the indicated value of x.

69. $p = g(x) = 50 - 0.1x, x = 200$

70. $p = g(x) = 30 - 0.05x, x = 400$

71. $p = g(x) = 50 - 2\sqrt{x}, x = 400$

72. $p = g(x) = 20 - \sqrt{x}, x = 100$

In Problems 73–76, use the price–demand equation to find the values of x for which demand is elastic and for which demand is inelastic.

73. $p = g(x) = 180 - 0.3x$ **74.** $p = g(x) = 640 - 0.4x$

75. $p = g(x) = 90 - 0.1x^2$ **76.** $p = g(x) = 540 - 0.2x^2$

77. Find $E(p)$ for $x = f(p) = Ap^{-k}$, where A and k are positive constants.

78. Find $E(p)$ for $x = f(p) = Ae^{-kp}$, where A and k are positive constants.

Applications

79. Rate of change of cost. A fast-food restaurant can produce a hamburger for \$2.50. If the restaurant's daily sales are increasing at the rate of 30 hamburgers per day, how fast is its daily cost for hamburgers increasing?

80. Rate of change of cost. The fast-food restaurant in Problem 79 can produce an order of fries for \$0.80. If the restaurant's daily sales are increasing at the rate of 45 orders of fries per day, how fast is its daily cost for fries increasing?

81. Revenue and elasticity. The price–demand equation for hamburgers at a fast-food restaurant is

$$x + 400p = 3{,}000$$

Currently, the price of a hamburger is \$3.00. If the price is increased by 10%, will revenue increase or decrease?

82. Revenue and elasticity. Refer to Problem 81. If the current price of a hamburger is \$4.00, will a 10% price increase cause revenue to increase or decrease?

83. Revenue and elasticity. The price–demand equation for an order of fries at a fast-food restaurant is

$$x + 1{,}000p = 2{,}500$$

Currently, the price of an order of fries is \$0.99. If the price is decreased by 10%, will revenue increase or decrease?

84. Revenue and elasticity. Refer to Problem 83. If the current price of an order of fries is \$1.49; will a 10% price decrease cause revenue to increase or decrease?

85. Maximum revenue. Refer to Problem 81. What price will maximize the revenue from selling hamburgers?

86. Maximum revenue. Refer to Problem 83. What price will maximize the revenue from selling fries?

87. Population growth. A model for Canada's population (Table 3) is

$$f(t) = 0.31t + 18.5$$

where t is years since 1960. Find and graph the percentage rate of change of $f(t)$ for $0 \leq t \leq 50$.

Table 3 **Population**

Year	Canada (millions)	Mexico (millions)
1960	18	39
1970	22	53
1980	25	68
1990	28	85
2000	31	100
2010	34	112

88. Population growth. A model for Mexico's population (Table 3) is

$$f(t) = 1.49t + 38.8$$

where t is years since 1960. Find and graph the percentage rate of change of $f(t)$ for $0 \leq t \leq 50$.

89. Crime. A model for the number of robberies in the United States (Table 4) is

$$r(t) = 3.3 - 0.7 \ln t$$

where t is years since 1990. Find the relative rate of change for robberies in 2020.

Table 4 **Number of Victimizations per 1,000 Population**

	Robbery	Aggravated Assault
1995	2.21	4.18
2000	1.45	3.24
2005	1.41	2.91
2010	1.19	2.52

90. Crime. A model for the number of assaults in the United States (Table 4) is

$$a(t) = 6.0 - 1.2 \ln t$$

where t is years since 1990. Find the relative rate of change for assaults in 2020.

Answers to Matched Problems

1. $p(t) = \dfrac{270}{2.7t + 282}$

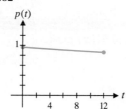

2. $E(p) = \dfrac{p}{40 - p}$

(A) $E(8) = 0.25$; demand is inelastic.

(B) $E(30) = 3$; demand is elastic.

(C) $E(20) = 1$; demand has unit elasticity.

3. $E(21) = \dfrac{21}{17} \approx 1.2$; demand is elastic. Increasing price will decrease revenue.

Chapter 11   Summary and Review

Important Terms, Symbols, and Concepts

11.1 The Constant e and Continuous Compound Interest

EXAMPLES

- The number e is defined as

$$\lim_{x \to \infty} \left(1 + \frac{1}{n}\right)^{n} = \lim_{x \to 0} (1 + s)^{1/s} = 2.718\ 281\ 828\ 459 \ldots$$

- If a principal P is invested at an annual rate r (expressed as a decimal) compounded continuously, then the amount A in the account at the end of t years is given by the **compound interest formula**

$$A = Pe^{rt}$$

Ex. 1, p. 577
Ex. 2, p. 577
Ex. 3, p. 578
Ex. 4, p. 578

11.2 Derivatives of Exponential and Logarithmic Functions

- For $b > 0, b \neq 1$,

$$\frac{d}{dx} e^x = e^x \qquad \frac{d}{dx} b^x = b^x \ln b$$

For $b > 0, b \neq 1$, and $x > 0$,

$$\frac{d}{dx} \ln x = \frac{1}{x} \qquad \frac{d}{dx} \log_b x = \frac{1}{\ln b} \frac{1}{x}$$

Ex. 1, p. 582
Ex. 2, p. 584
Ex. 3, p. 585
Ex. 4, p. 586
Ex. 5, p. 587

11.2 Derivatives of Exponential and Logarithmic Functions (*Continued*)

- The **change-of-base formulas** allow conversion from base e to any base $b, b > 0, b \neq 1$:

$$b^x = e^{x \ln b} \qquad \log_b x = \frac{\ln x}{\ln b}$$

11.3 Derivatives of Products and Quotients

- Product rule. If $y = f(x) = F(x) \, S(x)$, then $f'(x) = F(x)S'(x) + S(x)F'(x)$, provided that both $F'(x)$ and $S'(x)$ exist.

- Quotient rule. If $y = f(x) = \dfrac{T(x)}{B(x)}$, then $f'(x) = \dfrac{B(x) \, T'(x) - T(x) \, B'(x)}{[B(x)]^2}$ provided that both $T'(x)$ and $B'(x)$ exist.

Ex. 1, p. 590
Ex. 2, p. 591
Ex. 3, p. 591
Ex. 4, p. 593
Ex. 5, p. 594
Ex. 6, p. 594

11.4 The Chain Rule

- A function m is a **composite** of functions f and g if $m(x) = f[g(x)]$.

- The **chain rule** gives a formula for the derivative of the composite function $m(x) = E[I(x)]$:

$$m'(x) = E'[I(x)]I'(x)$$

- A special case of the chain rule is called the **general power rule**:

$$\frac{d}{dx}[f(x)]^n = n[f(x)]^{n-1}f'(x)$$

- Other special cases of the chain rule are the following **general derivative rules**:

$$\frac{d}{dx} \ln [f(x)] = \frac{1}{f(x)}f'(x)$$

$$\frac{d}{dx} e^{f(x)} = e^{f(x)}f'(x)$$

Ex. 1, p. 599
Ex. 2, p. 599
Ex. 4, p. 603
Ex. 5, p. 604
Ex. 3, p. 600

Ex. 6, p. 604

11.5 Implicit Differentiation

- If $y = y(x)$ is a function defined implicitly by the equation $F(x, y) = 0$, then we use **implicit differentiation** to find an equation in x, y, and y'.

Ex. 1, p. 609
Ex. 2, p. 611
Ex. 3, p. 612

11.6 Related Rates

- If x and y represent quantities that are changing with respect to time and are related by the equation $F(x, y) = 0$, then implicit differentiation produces an equation that relates x, y, dy/dt, and dx/dt. Problems of this type are called **related-rates problems**.

- Suggestions for solving related-rates problems are given on page 221.

Ex. 1, p. 614
Ex. 2, p. 615
Ex. 3, p. 616

11.7 Elasticity of Demand

- The **relative rate of change**, or the **logarithmic derivative**, of a function $f(x)$ is $f'(x)/f(x)$, and the **percentage rate of change** is $100 \times [f'(x)/f(x)]$.

- If price and demand are related by $x = f(p)$, then the **elasticity of demand** is given by

$$E(p) = -\frac{pf'(p)}{f(p)} = -\frac{\text{relative rate of change of demand}}{\text{relative rate of change of price}}$$

- **Demand is inelastic** if $0 < E(p) < 1$. (Demand is not sensitive to changes in price; a percentage change in price produces a smaller percentage change in demand.) **Demand is elastic** if $E(p) > 1$. (Demand is sensitive to changes in price; a percentage change in price produces a larger percentage change in demand.) **Demand has unit elasticity** if $E(p) = 1$. (A percentage change in price produces the same percentage change in demand.)

- If $R(p) = pf(p)$ is the revenue function, then $R'(p)$ and $[1 - E(p)]$ always have the same sign. If demand is inelastic, then a price increase will increase revenue. If demand is elastic, then a price increase will decrease revenue.

Ex. 1, p. 621

Ex. 2, p. 622

Ex. 3, p. 624

Review Exercises

Work through all the problems in this chapter review, and check your answers in the back of the book. Answers to all review problems are there, along with section numbers in italics to indicate where each type of problem is discussed. Where weaknesses show up, review appropriate sections of the text.

1. Use a calculator to evaluate $A = 2,000e^{0.09t}$ to the nearest cent for $t = 5$, 10, and 20.

In Problems 2–4, find functions $E(u)$ and $I(x)$ so that $f(x) = E[I(x)]$.

2. $f(x) = (6x + 5)^{3/2}$ **3.** $f(x) = \ln(x^2 + 4)$

4. $f(x) = e^{0.02x}$

In Problems 5–8, find the indicated derivative.

5. $\dfrac{d}{dx}(2\ln x + 3e^x)$ **6.** $\dfrac{d}{dx}e^{2x-3}$

7. y' for $y = \ln(2x + 7)$

8. $f'(x)$ for $f(x) = \ln(3 + e^x)$

9. Find y' for $y = y(x)$ defined implicity by the equation $2y^2 - 3x^3 - 5 = 0$, and evaluate at $(x, y) = (1, 2)$.

10. For $y = 3x^2 - 5$, where $x = x(t)$ and $y = y(t)$, find dy/dt if $dx/dt = 3$ when $x = 12$.

11. Given the demand equation $25p + x = 1,000$,

  (A) Express the demand x as a function of the price p.

  (B) Find the elasticity of demand, $E(p)$.

  (C) Find $E(15)$ and interpret.

  (D) Express the revenue function as a function of price p.

  (E) If $p = \$25$, what is the effect of a small price cut on revenue?

12. Find the slope of the line tangent to $y = 100e^{-0.1x}$ when $x = 0$.

13. Use a calculator and a table of values to investigate

$$\lim_{n \to \infty}\left(1 + \frac{2}{n}\right)^n$$

  Do you think the limit exists? If so, what do you think it is?

Find the indicated derivatives in Problems 14–19.

14. $\dfrac{d}{dz}[(\ln z)^7 + \ln z^7]$ **15.** $\dfrac{d}{dx}(x^6 \ln x)$

16. $\dfrac{d}{dx}\dfrac{e^x}{x^6}$ **17.** y' for $y = \ln(2x^3 - 3x)$

18. $f'(x)$ for $f(x) = e^{x^3 - x^2}$ **19.** dy/dx for $y = e^{-2x}\ln 5x$

20. Find the equation of the line tangent to the graph of $y = f(x) = 1 + e^{-x}$ at $x = 0$. At $x = -1$.

21. Find y' for $y = y(x)$ defined implicitly by the equation $x^2 - 3xy + 4y^2 = 23$, and find the slope of the graph at $(-1, 2)$.

22. Find x' for $x = x(t)$ defined implicitly by $x^3 - 2t^2x + 8 = 0$, and evaluate at $(t, x) = (-2, 2)$.

23. Find y' for $y = y(x)$ defined implicitly by $x - y^2 = e^y$, and evaluate at $(1, 0)$.

24. Find y' for $y = y(x)$ defined implicitly by $\ln y = x^2 - y^2$, and evaluate at $(1, 1)$.

In Problems 25–27, find the logarithmic derivatives.

25. $A(t) = 400e^{0.049t}$ **26.** $f(p) = 100 - 3p$

27. $f(x) = 1 + x^2$

28. A point is moving on the graph of $y^2 - 4x^2 = 12$ so that its x coordinate is decreasing by 2 units per second when $(x, y) = (1, 4)$. Find the rate of change of the y coordinate.

29. A 17-foot ladder is placed against a wall. If the foot of the ladder is pushed toward the wall at 0.5 foot per second, how fast is the top of the ladder rising when the foot is 8 feet from the wall?

30. Water is leaking onto a floor. The resulting circular pool has an area that is increasing at the rate of 24 square inches per minute. How fast is the radius R of the pool increasing when the radius is 12 inches?

31. Find the values of p for which demand is elastic and the values for which demand is inelastic if the price–demand equation is

$$x = f(p) = 20(p - 15)^2 \qquad 0 < p \le 15$$

32. Graph the revenue function as a function of price p, and indicate the regions of inelastic and elastic demand if the price–demand equation is

$$x = f(p) = 5(20 - p) \qquad 0 \le p \le 20$$

33. Let $y = w^3$, $w = \ln u$, and $u = 4 - e^x$.

  (A) Express y in terms of x.

  (B) Use the chain rule to find dy/dx.

Find the indicated derivatives in Problems 34–36.

34. y' for $y = 5^{x^2-1}$ **35.** $\dfrac{d}{dx}\log_5(x^2 - x)$

36. $\dfrac{d}{dx}\sqrt{\ln(x^2 + x)}$

37. Find y' for $y = y(x)$ defined implicitly by the equation $e^{xy} = x^2 + y + 1$, and evaluate at $(0, 0)$.

38. A rock thrown into a still pond causes a circular ripple. The radius is increasing at a constant rate of 3 feet per second. Show that the area does not increase at a constant rate. When is the rate of increase of the area the smallest? The largest? Explain.

39. A point moves along the graph of $y = x^3$ in such a way that its y coordinate is increasing at a constant rate of 5 units per second. Does the x coordinate ever increase at a faster rate than the y coordinate? Explain.

Applications

40. Doubling time. How long will it take money to double if it is invested at 5% interest compounded

(A) Annually?   (B) Continuously?

41. Continuous compound interest. If $100 is invested at 10% interest compounded continuously, then the amount (in dollars) at the end of t years is given by

$$A = 100e^{0.1t}$$

Find $A'(t)$, $A'(1)$, and $A'(10)$.

42. Continuous compound interest. If $12,000 is invested in an account that earns 3.95% compounded continuously, find the instantaneous rate of change of the amount when the account is worth $25,000.

43. Marginal analysis. The price–demand equation for 14-cubic-foot refrigerators at an appliance store is

$$p(x) = 1,000e^{-0.02x}$$

where x is the monthly demand and p is the price in dollars. Find the marginal revenue equation.

44. Demand equation. Given the demand equation

$$x = \sqrt{5,000 - 2p^3}$$

find the rate of change of p with respect to x by implicit differentiation (x is the number of items that can be sold at a price of $\$p$ per item).

45. Rate of change of revenue. A company is manufacturing kayaks and can sell all that it manufactures. The revenue (in dollars) is given by

$$R = 750x - \frac{x^2}{30}$$

where the production output in 1 day is x kayaks. If production is increasing at 3 kayaks per day when production is 40 kayaks per day, find the rate of increase in revenue.

46. Revenue and elasticity. The price–demand equation for home-delivered large pizzas is

$$p = 38.2 - 0.002x$$

where x is the number of pizzas delivered weekly. The current price of one pizza is $21. In order to generate additional revenue from the sale of large pizzas, would you recommend a price increase or a price decrease? Explain.

47. Average income. A model for the average income per household before taxes are paid is

$$f(t) = 1,700t + 20,500$$

where t is years since 1980. Find the relative rate of change of household income in 2015.

48. Drug concentration. The drug concentration in the bloodstream t hours after injection is given approximately by

$$C(t) = 5e^{-0.3t}$$

where $C(t)$ is concentration in milligrams per milliliter. What is the rate of change of concentration after 1 hour? After 5 hours?

49. Wound healing. A circular wound on an arm is healing at the rate of 45 square millimeters per day (the area of the wound is decreasing at this rate). How fast is the radius R of the wound decreasing when $R = 15$ millimeters?

50. Psychology: learning. In a computer assembly plant, a new employee, on the average, is able to assemble

$$N(t) = 10(1 - e^{-0.4t})$$

units after t days of on-the-job training.

(A) What is the rate of learning after 1 day? After 5 days?

(B) Find the number of days (to the nearest day) after which the rate of learning is less than 0.25 unit per day.

51. Learning. A new worker on the production line performs an operation in T minutes after x performances of the operation, as given by

$$T = 2\left(1 + \frac{1}{x^{3/2}}\right)$$

If, after performing the operation 9 times, the rate of improvement is $dx/dt = 3$ operations per hour, find the rate of improvement in time dT/dt in performing each operation.

12

Graphing and Optimization

12.1 First Derivative and Graphs

12.2 Second Derivative and Graphs

12.3 L' Hôpital's Rule

12.4 Curve-Sketching Techniques

12.5 Absolute Maxima and Minima

12.6 Optimization

Chapter 12
Summary and Review

Review Exercises

Introduction

Since the derivative is associated with the slope of the graph of a function at a point, we might expect that it is also related to other properties of a graph. As we will see in this chapter, the derivative can tell us a great deal about the shape of the graph of a function. In particular, we will study methods for finding absolute maximum and minimum values. Manufacturing companies can use these methods to find production levels that will minimize cost or maximize profit, pharmacologists can use them to find levels of drug dosages that will produce maximum sensitivity, and advertisers can use them to determine the number of ads that will maximize the rate of change of sales (see, for example, Problem 93 in Section 12.2).

12.1 First Derivative and Graphs

- Increasing and Decreasing Functions
- Local Extrema
- First-Derivative Test
- Economics Applications

Increasing and Decreasing Functions

Sign charts will be used throughout this chapter. You may find it helpful to review the terminology and techniques for constructing sign charts in Section 10.3.

Explore and Discuss 1 Figure 1 shows the graph of $y = f(x)$ and a sign chart for $f'(x)$, where

$$f(x) = x^3 - 3x$$

and

$$f'(x) = 3x^2 - 3 = 3(x + 1)(x - 1)$$

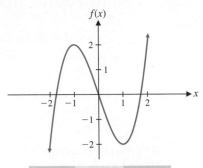

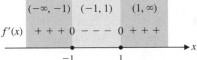

Figure 1

Discuss the relationship between the graph of f and the sign of $f'(x)$ over each interval on which $f'(x)$ has a constant sign. Also, describe the behavior of the graph of f at each partition number for f'.

As they are scanned from left to right, graphs of functions generally have rising and falling sections. If you scan the graph of $f(x) = x^3 - 3x$ in Figure 1 from left to right, you will observe the following:

- On the interval $(-\infty, -1)$, the graph of f is rising, $f(x)$ is increasing,* and tangent lines have positive slope $[f'(x) > 0]$.
- On the interval $(-1, 1)$, the graph of f is falling, $f(x)$ is decreasing, and tangent lines have negative slope $[f'(x) < 0]$.
- On the interval $(1, \infty)$, the graph of f is rising, $f(x)$ is increasing, and tangent lines have positive slope $[f'(x) > 0]$.
- At $x = -1$ and $x = 1$, the slope of the graph is 0 $[f'(x) = 0]$.

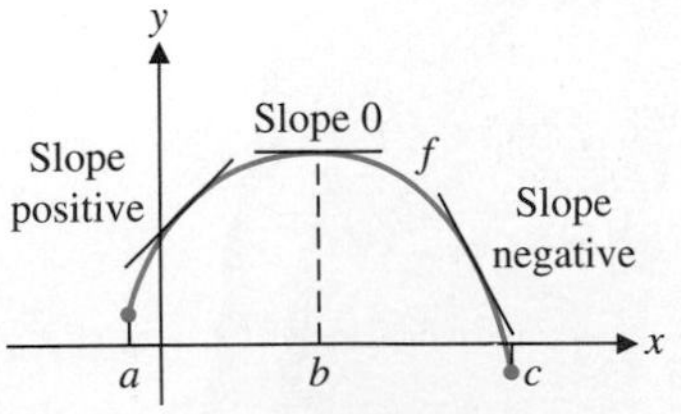

If $f'(x) > 0$ (is positive) on the interval (a, b) (Fig. 2), then $f(x)$ increases ($\nearrow$) and the graph of f rises as we move from left to right over the interval. If $f'(x) < 0$ (is negative) on an interval (a, b), then $f(x)$ decreases ($\searrow$) and the graph of f falls as we move from left to right over the interval. We summarize these important results in Theorem 1.

Figure 2

*Formally, we say that the function f is **increasing** on an interval (a, b) if $f(x_2) > f(x_1)$ whenever $a < x_1 < x_2 < b$, and f is **decreasing** on (a, b) if $f(x_2) < f(x_1)$ whenever $a < x_1 < x_2 < b$.

THEOREM 1   Increasing and Decreasing Functions

For the interval (a, b), if $f' > 0$, then f is increasing, and if $f' < 0$, then f is decreasing.

$f'(x)$	$f(x)$	Graph of f	Examples
$+$	Increases ↗	Rises ↗	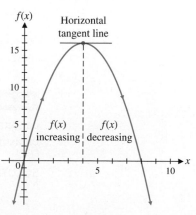
$-$	Decreases ↘	Falls ↘	

EXAMPLE 1   Finding Intervals on Which a Function Is Increasing or Decreasing

Given the function $f(x) = 8x - x^2$,

(A) Which values of x correspond to horizontal tangent lines?

(B) For which values of x is $f(x)$ increasing? Decreasing?

(C) Sketch a graph of f. Add any horizontal tangent lines.

SOLUTION

(A) $f'(x) = 8 - 2x = 0$

$$x = 4$$

So, a horizontal tangent line exists at $x = 4$ only.

(B) We will construct a sign chart for $f'(x)$ to determine which values of x make $f'(x) > 0$ and which values make $f'(x) < 0$. Recall from Section 2.3 that the partition numbers for a function are the numbers at which the function is 0 or discontinuous. When constructing a sign chart for $f'(x)$, we must locate all points where $f'(x) = 0$ or $f'(x)$ is discontinuous. From part (A), we know that $f'(x) = 8 - 2x = 0$ at $x = 4$. Since $f'(x) = 8 - 2x$ is a polynomial, it is continuous for all x. So, 4 is the only partition number for f'. We construct a sign chart for the intervals $(-\infty, 4)$ and $(4, \infty)$, using test numbers 3 and 5:

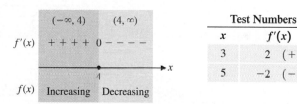

	Test Numbers	
x	$f'(x)$	
3	2	$(+)$
5	-2	$(-)$

Therefore, $f(x)$ is increasing on $(-\infty, 4)$ and decreasing on $(4, \infty)$.

(C)

x	$f(x)$
0	0
2	12
4	16
6	12
8	0

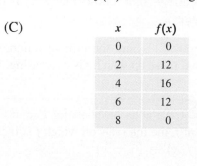

Matched Problem 1   Repeat Example 1 for $f(x) = x^2 - 6x + 10$.

As Example 1 illustrates, the construction of a sign chart will play an important role in using the derivative to analyze and sketch the graph of a function f. The partition numbers for f' are central to the construction of these sign charts and also to the analysis of the graph of $y = f(x)$. The partition numbers for f' that belong to the domain of f are called **critical numbers** of f.*

> **DEFINITION** Critical Numbers
> A real number x in the domain of f such that $f'(x) = 0$ or $f'(x)$ does not exist is called a **critical number** of f.

> **CONCEPTUAL INSIGHT**
>
> The critical numbers of f belong to the domain of f and are partition numbers for f'. But f' may have partition numbers that do not belong to the domain of f, so are not critical numbers of f.
>
> If f is a polynomial, then both the partition numbers for f' and the critical numbers of f are the solutions of $f'(x) = 0$.

EXAMPLE 2 Partition Numbers for f' and Critical Numbers of f Find the critical numbers of f, the intervals on which f is increasing, and those on which f is decreasing, for $f(x) = 1 + x^3$.

SOLUTION Begin by finding the partition numbers for $f'(x)$ [since $f'(x) = 3x^2$ is continuous we just need to solve $f'(x) = 0$]

$$f'(x) = 3x^2 = 0 \quad \text{only if } x = 0$$

The partition number 0 for f' is in the domain of f, so 0 is the only critical number of f.

The sign chart for $f'(x) = 3x^2$ (partition number is 0) is

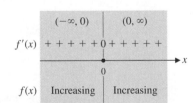

Test Numbers	
x	$f'(x)$
-1	3 $(+)$
1	3 $(+)$

The sign chart indicates that $f(x)$ is increasing on $(-\infty, 0)$ and $(0, \infty)$. Since f is continuous at $x = 0$, it follows that $f(x)$ is increasing for all x. The graph of f is shown in Figure 3.

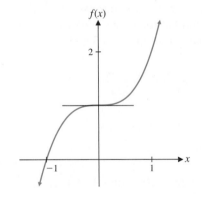

Figure 3

Matched Problem 2 Find the critical numbers of f, the intervals on which f is increasing, and those on which f is decreasing, for $f(x) = 1 - x^3$.

EXAMPLE 3 Partition Numbers for f' and Critical Numbers of f Find the critical numbers of f, the intervals on which f is increasing, and those on which f is decreasing, for $f(x) = (1 - x)^{1/3}$.

SOLUTION
$$f'(x) = -\frac{1}{3}(1 - x)^{-2/3} = \frac{-1}{3(1 - x)^{2/3}}$$

---

*We are assuming that $f'(c)$ does not exist at any point of discontinuity of f'. There do exist functions f such that f' is discontinuous at $x = c$, yet $f'(c)$ exists. However, we do not consider such functions in this book.

To find the partition numbers for f', we note that f' is continuous for all x, except for values of x for which the denominator is 0; that is, $f'(1)$ does not exist and f' is discontinuous at $x = 1$. Since the numerator of f' is the constant -1, $f'(x) \neq 0$ for any value of x. Thus, $x = 1$ is the only partition number for f'. Since 1 is in the domain of f, $x = 1$ is also the only critical number of f. When constructing the sign chart for f' we use the abbreviation ND to note the fact that $f'(x)$ is *not defined* at $x = 1$.

The sign chart for $f'(x) = -1/[3(1 - x)^{2/3}]$ (partition number for f' is 1) is as follows:

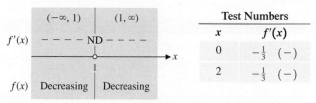

The sign chart indicates that f is decreasing on $(-\infty, 1)$ and $(1, \infty)$. Since f is continuous at $x = 1$, it follows that $f(x)$ is decreasing for all x. **A continuous function can be decreasing (or increasing) on an interval containing values of x where $f'(x)$ does not exist.** The graph of f is shown in Figure 4. Notice that the undefined derivative at $x = 1$ results in a vertical tangent line at $x = 1$. **A vertical tangent will occur at $x = c$ if f is continuous at $x = c$ and if $|f'(x)|$ becomes larger and larger as x approaches c.**

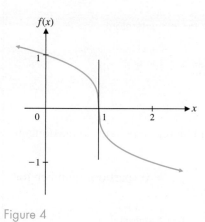

Figure 4

Matched Problem 3 Find the critical numbers of f, the intervals on which f is increasing, and those on which f is decreasing, for $f(x) = (1 + x)^{1/3}$.

EXAMPLE 4 Partition Numbers for f' and Critical Numbers of f Find the critical numbers of f, the intervals on which f is increasing, and those on which f is decreasing, for $f(x) = \dfrac{1}{x - 2}$.

SOLUTION
$$f(x) = \frac{1}{x - 2} = (x - 2)^{-1}$$

$$f'(x) = -(x - 2)^{-2} = \frac{-1}{(x - 2)^2}$$

To find the partition numbers for f', note that $f'(x) \neq 0$ for any x and f' is not defined at $x = 2$. Thus, $x = 2$ is the only partition number for f'. However, $x = 2$ is *not* in the domain of f. Consequently, $x = 2$ is *not* a critical number of f. The function f has no critical numbers.

The sign chart for $f'(x) = -1/(x - 2)^2$ (partition number for f' is 2) is as follows:

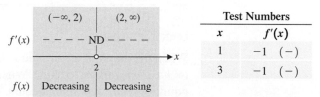

Therefore, f is decreasing on $(-\infty, 2)$ and $(2, \infty)$. The graph of f is shown in Figure 5.

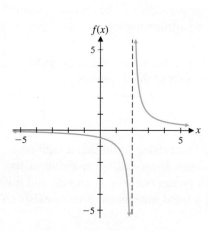

Figure 5

Matched Problem 4 Find the critical numbers of f, the intervals on which f is increasing, and those on which f is decreasing, for $f(x) = \dfrac{1}{x}$.

EXAMPLE 5 Partition Numbers for f' and Critical Numbers of f Find the critical numbers of f, the intervals on which f is increasing, and those on which f is decreasing, for $f(x) = 8 \ln x - x^2$.

SOLUTION The natural logarithm function $\ln x$ is defined on $(0, \infty)$, or $x > 0$, so $f(x)$ is defined only for $x > 0$.

$$f(x) = 8 \ln x - x^2, x > 0$$

$$f'(x) = \frac{8}{x} - 2x \qquad \text{Find a common denominator.}$$

$$= \frac{8}{x} - \frac{2x^2}{x} \qquad \text{Subtract numerators.}$$

$$= \frac{8 - 2x^2}{x} \qquad \text{Factor numerator.}$$

$$= \frac{2(2 - x)(2 + x)}{x}, \quad x > 0$$

The only partition number for f' that is positive, and therefore belongs to the domain of f, is 2. So 2 is the only critical number of f.

The sign chart for $f'(x) = \dfrac{2(2 - x)(2 + x)}{x}, x > 0$ (partition number for f' is 2), is as follows:

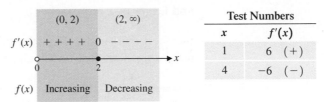

	Test Numbers
x	$f'(x)$
1	6 (+)
4	−6 (−)

Therefore, f is increasing on $(0, 2)$ and decreasing on $(2, \infty)$. The graph of f is shown in Figure 6.

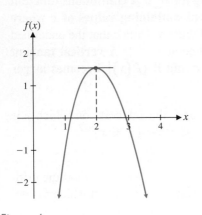

Figure 6

Matched Problem 5 Find the critical numbers of f, the intervals on which f is increasing, and those on which f is decreasing, for $f(x) = 5 \ln x - x$.

CONCEPTUAL INSIGHT

Examples 4 and 5 illustrate two important ideas:

1. Do not assume that all partition numbers for the derivative f' are critical numbers of the function f. To be a critical number of f, a partition number for f' must also be in the domain of f.

2. The intervals on which a function f is increasing or decreasing must always be expressed in terms of open intervals that are subsets of the domain of f.

Local Extrema

When the graph of a continuous function changes from rising to falling, a high point, or *local maximum,* occurs. When the graph changes from falling to rising, a low point, or *local minimum,* occurs. In Figure 7, high points occur at c_3 and c_6, and low points occur at c_2 and c_4. In general, we call $f(c)$ a **local maximum** if there exists an interval (m, n) containing c such that

$$f(x) \leq f(c) \qquad \text{for all } x \text{ in } (m, n)$$

Note that this inequality need hold only for numbers x near c, which is why we use the term *local.* So the y coordinate of the high point $(c_3, f(c_3))$ in Figure 7 is a local maximum, as is the y coordinate of $(c_6, f(c_6))$.

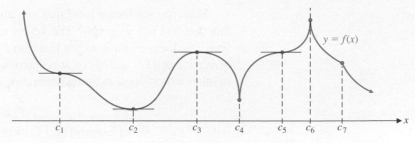

Figure 7

The value $f(c)$ is called a **local minimum** if there exists an interval (m, n) containing c such that

$$f(x) \geq f(c) \qquad \text{for all } x \text{ in } (m, n)$$

The value $f(c)$ is called a **local extremum** if it is either a local maximum or a local minimum. A point on a graph where a local extremum occurs is also called a **turning point**. In Figure 7 we see that local maxima occur at c_3 and c_6, local minima occur at c_2 and c_4, and all four values produce local extrema. Also, the local maximum $f(c_3)$ is not the largest y coordinate of points on the graph in Figure 7. Later in this chapter, we consider the problem of finding *absolute extrema,* the y coordinates of the highest and lowest points on a graph. For now, we are concerned only with locating *local* extrema.

EXAMPLE 6 Analyzing a Graph Use the graph of f in Figure 8 to find the intervals on which f is increasing, those on which f is decreasing, any local maxima, and any local minima.

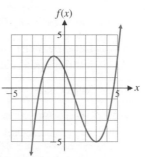

Figure 8

SOLUTION The function f is increasing (the graph is rising) on $(-\infty, -1)$ and on $(3, \infty)$ and is decreasing (the graph is falling) on $(-1, 3)$. Because the graph changes from rising to falling at $x = -1$, $f(-1) = 3$ is a local maximum. Because the graph changes from falling to rising at $x = 3$, $f(3) = -5$ is a local minimum.

Matched Problem 6 Use the graph of g in Figure 9 to find the intervals on which g is increasing, those on which g is decreasing, any local maxima, and any local minima.

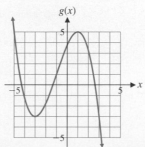

Figure 9

How can we locate local maxima and minima if we are given the equation of a function and not its graph? The key is to examine the critical numbers of the function. The local extrema of the function f in Figure 7 occur either at points where the derivative is 0 (c_2 and c_3) or at points where the derivative does not exist (c_4 and c_6). In other words, local extrema occur only at critical numbers of f.

THEOREM 2 Local Extrema and Critical Numbers

If $f(c)$ is a local extremum of the function f, then c is a critical number of f.

Theorem 2 states that a local extremum can occur only at a critical number, but it does not imply that every critical number produces a local extremum. In Figure 7, c_1 and c_5 are critical numbers (the slope is 0), but the function does not have a local maximum or local minimum at either of these numbers.

Our strategy for finding local extrema is now clear: We find all critical numbers of f and test each one to see if it produces a local maximum, a local minimum, or neither.

First-Derivative Test

If $f'(x)$ exists on both sides of a critical number c, the sign of $f'(x)$ can be used to determine whether the point $(c, f(c))$ is a local maximum, a local minimum, or neither. The various possibilities are summarized in the following box and are illustrated in Figure 10:

PROCEDURE First-Derivative Test for Local Extrema

Let c be a critical number of f [$f(c)$ is defined and either $f'(c) = 0$ or $f'(c)$ is not defined]. Construct a sign chart for $f'(x)$ close to and on either side of c.

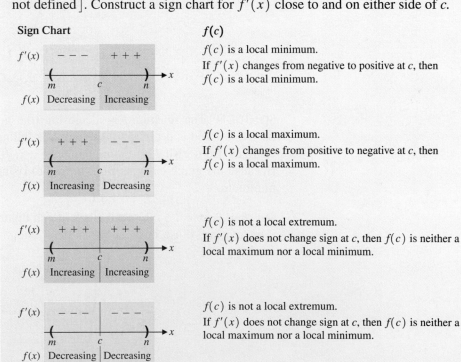

Sign Chart

$f'(x)$ $- - -$ $+ + +$
m c n $\to x$
$f(x)$ Decreasing Increasing

$f(c)$
$f(c)$ is a local minimum.
If $f'(x)$ changes from negative to positive at c, then $f(c)$ is a local minimum.

$f'(x)$ $+ + +$ $- - -$
m c n $\to x$
$f(x)$ Increasing Decreasing

$f(c)$ is a local maximum.
If $f'(x)$ changes from positive to negative at c, then $f(c)$ is a local maximum.

$f'(x)$ $+ + +$ $+ + +$
m c n $\to x$
$f(x)$ Increasing Increasing

$f(c)$ is not a local extremum.
If $f'(x)$ does not change sign at c, then $f(c)$ is neither a local maximum nor a local minimum.

$f'(x)$ $- - -$ $- - -$
m c n $\to x$
$f(x)$ Decreasing Decreasing

$f(c)$ is not a local extremum.
If $f'(x)$ does not change sign at c, then $f(c)$ is neither a local maximum nor a local minimum.

$f'(c) = 0$: Horizontal tangent

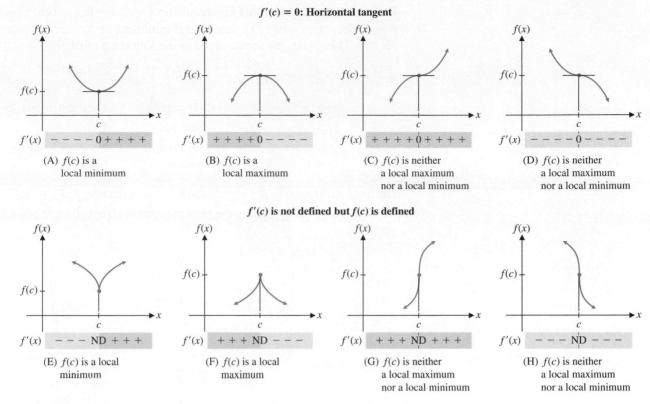

(A) $f(c)$ is a local minimum

(B) $f(c)$ is a local maximum

(C) $f(c)$ is neither a local maximum nor a local minimum

(D) $f(c)$ is neither a local maximum nor a local minimum

$f'(c)$ is not defined but $f(c)$ is defined

(E) $f(c)$ is a local minimum

(F) $f(c)$ is a local maximum

(G) $f(c)$ is neither a local maximum nor a local minimum

(H) $f(c)$ is neither a local maximum nor a local minimum

Figure 10 **Local extrema**

EXAMPLE 7 Locating Local Extrema Given $f(x) = x^3 - 6x^2 + 9x + 1$,

(A) Find the critical numbers of f.

(B) Find the local maxima and local minima of f.

(C) Sketch the graph of f.

SOLUTION

(A) Find all numbers x in the domain of f where $f'(x) = 0$ or $f'(x)$ does not exist.

$$f'(x) = 3x^2 - 12x + 9 = 0$$
$$3(x^2 - 4x + 3) = 0$$
$$3(x - 1)(x - 3) = 0$$
$$x = 1 \quad \text{or} \quad x = 3$$

$f'(x)$ exists for all x; the critical numbers of f are $x = 1$ and $x = 3$.

(B) The easiest way to apply the first-derivative test for local maxima and minima is to construct a sign chart for $f'(x)$ for all x. Partition numbers for $f'(x)$ are $x = 1$ and $x = 3$ (which also happen to be critical numbers of f).

Sign chart for $f'(x) = 3(x - 1)(x - 3)$:

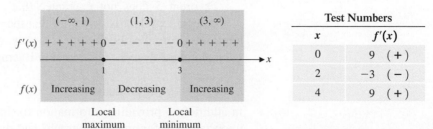

The sign chart indicates that f increases on $(-\infty, 1)$, has a local maximum at $x = 1$, decreases on $(1, 3)$, has a local minimum at $x = 3$, and increases on $(3, \infty)$. These facts are summarized in the following table:

x	$f'(x)$	$f(x)$	Graph of f
$(-\infty, 1)$	$+$	Increasing	Rising
$x = 1$	0	Local maximum	Horizontal tangent
$(1, 3)$	$-$	Decreasing	Falling
$x = 3$	0	Local minimum	Horizontal tangent
$(3, \infty)$	$+$	Increasing	Rising

The local maximum is $f(1) = 5$; the local minimum is $f(3) = 1$.

(C) We sketch a graph of f, using the information from part (B) and point-by-point plotting.

x	$f(x)$
0	1
1	5
2	3
3	1
4	5

Matched Problem 7 Given $f(x) = x^3 - 9x^2 + 24x - 10$,

(A) Find the critical numbers of f.

(B) Find the local maxima and local minima of f.

(C) Sketch a graph of f.

How can you tell if you have found all the local extrema of a function? In general, this can be a difficult question to answer. However, in the case of a polynomial function, there is an easily determined upper limit on the number of local extrema. Since the local extrema are the x intercepts of the derivative, this limit is a consequence of the number of x intercepts of a polynomial. The relevant information is summarized in the following theorem, which is stated without proof:

THEOREM 3   Intercepts and Local Extrema of Polynomial Functions

If $f(x) = a_n x^n + a_{n-1} x^{n-1} + \cdots + a_1 x + a_0, a_n \neq 0$, is a polynomial function of degree $n \geq 1$, then f has at most n x intercepts and at most $n - 1$ local extrema.

Theorem 3 does not guarantee that every nth-degree polynomial has exactly $n - 1$ local extrema; it says only that there can never be more than $n - 1$ local extrema. For example, the third-degree polynomial in Example 7 has two local extrema, while the third-degree polynomial in Example 2 does not have any.

Economics Applications

In addition to providing information for hand-sketching graphs, the derivative is an important tool for analyzing graphs and discussing the interplay between a function and its rate of change. The next two examples illustrate this process in the context of economics.

EXAMPLE 8   Agricultural Exports and Imports   Over the past few decades, the United States has exported more agricultural products than it has imported, maintaining a positive balance of trade in this area. However, the trade balance fluctuated considerably during that period. The graph in Figure 11 approximates the rate of change of the balance of trade over a 15-year period, where $B(t)$ is the balance of trade (in billions of dollars) and t is time (in years).

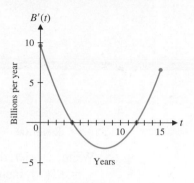

Figure 11 **Rate of change of the balance of trade**

✎ (A) Write a brief description of the graph of $y = B(t)$, including a discussion of any local extrema.

(B) Sketch a possible graph of $y = B(t)$.

SOLUTION

(A) The graph of the derivative $y = B'(t)$ contains the same essential information as a sign chart. That is, we see that $B'(t)$ is positive on $(0, 4)$, 0 at $t = 4$, negative on $(4, 12)$, 0 at $t = 12$, and positive on $(12, 15)$. The trade balance increases for the first 4 years to a local maximum, decreases for the next 8 years to a local minimum, and then increases for the final 3 years.

(B) Without additional information concerning the actual values of $y = B(t)$, we cannot produce an accurate graph. However, we can sketch a possible graph that illustrates the important features, as shown in Figure 12. The absence of a scale on the vertical axis is a consequence of the lack of information about the values of $B(t)$.

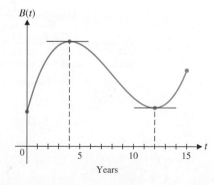

Figure 12 **Balance of trade**

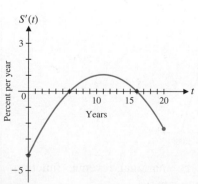

Figure 13

Matched Problem 8 The graph in Figure 13 approximates the rate of change of the U.S. share of the total world production of motor vehicles over a 20-year period, where $S(t)$ is the U.S. share (as a percentage) and t is time (in years).

✎ (A) Write a brief description of the graph of $y = S(t)$, including a discussion of any local extrema.

(B) Sketch a possible graph of $y = S(t)$.

EXAMPLE 9   Revenue Analysis   The graph of the total revenue $R(x)$ (in dollars) from the sale of x bookcases is shown in Figure 14.

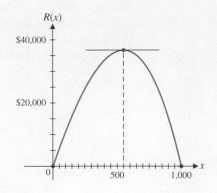

Figure 14 **Revenue**

✎(A) Write a brief description of the graph of the marginal revenue function $y = R'(x)$, including a discussion of any x intercepts.

(B) Sketch a possible graph of $y = R'(x)$.

SOLUTION

(A) The graph of $y = R(x)$ indicates that $R(x)$ increases on $(0, 550)$, has a local maximum at $x = 550$, and decreases on $(550, 1,000)$. Consequently, the marginal revenue function $R'(x)$ must be positive on $(0, 550)$, 0 at $x = 550$, and negative on $(550, 1,000)$.

(B) A possible graph of $y = R'(x)$ illustrating the information summarized in part (A) is shown in Figure 15.

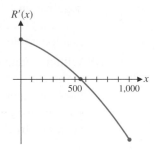

Figure 15 **Marginal revenue**

Matched Problem 9 ⌡ The graph of the total revenue $R(x)$ (in dollars) from the sale of x desks is shown in Figure 16.

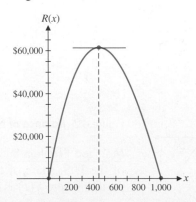

Figure 16

✎(A) Write a brief description of the graph of the marginal revenue function $y = R'(x)$, including a discussion of any x intercepts.

(B) Sketch a possible graph of $y = R'(x)$.

Comparing Examples 8 and 9, we see that we were able to obtain more information about the function from the graph of its derivative (Example 8) than we were when the process was reversed (Example 9). In the next section, we introduce some ideas that will help us obtain additional information about the derivative from the graph of the function.

Exercises 12.1

Skills Warm-up Exercises

W   In Problems 1–8, inspect the graph of the function to determine whether it is increasing or decreasing on the given interval. (If necessary, review Section 2.2).

1. $g(x) = |x|$ on $(-\infty, 0)$ 2. $m(x) = x^3$ on $(0, \infty)$

3. $f(x) = x$ on $(-\infty, \infty)$ 4. $k(x) = -x^2$ on $(0, \infty)$

5. $p(x) = \sqrt[3]{x}$ on $(-\infty, 0)$

6. $h(x) = x^2$ on $(-\infty, 0)$

7. $r(x) = 4 - \sqrt{x}$ on $(0, \infty)$

8. $g(x) = |x|$ on $(0, \infty)$

Problems 9–16, refer to the following graph of $y = f(x)$:

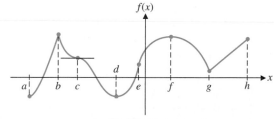

Figure for 9–16

9. Identify the intervals on which $f(x)$ is increasing.

10. Identify the intervals on which $f(x)$ is decreasing.

11. Identify the intervals on which $f'(x) < 0$.

12. Identify the intervals on which $f'(x) > 0$.

13. Identify the x coordinates of the points where $f'(x) = 0$.

14. Identify the x coordinates of the points where $f'(x)$ does not exist.

15. Identify the x coordinates of the points where $f(x)$ has a local maximum.

16. Identify the x coordinates of the points where $f(x)$ has a local minimum.

In Problems 17 and 18, $f(x)$ is continuous on $(-\infty, \infty)$ and has critical numbers at $x = a, b, c,$ and d. Use the sign chart for $f'(x)$ to determine whether f has a local maximum, a local minimum, or neither at each critical number.

17.

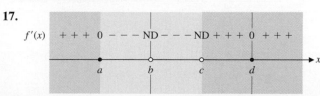

18.

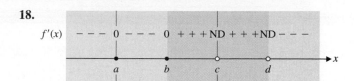

In Problems 19–26, give the local extrema of f and match the graph of f with one of the sign charts a–h in the figure on page 644.

19.
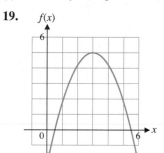

20.
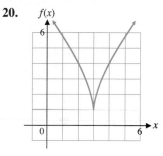

21.
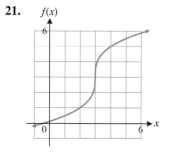

22.
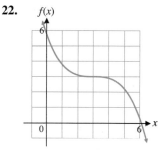

23.
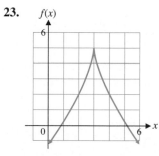

24.
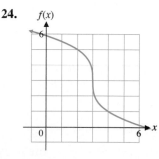

25.
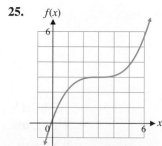

26.
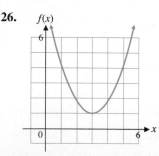

(a)
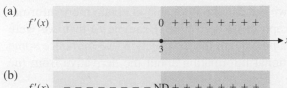

(b)

(c)
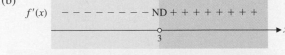

(d)

(e)
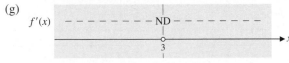

(f)

(g)

(h)

Figure for 19–26

In Problems 27–32, find (A) $f'(x)$, (B) the partition numbers for f', and (C) the critical numbers of f.

27. $f(x) = x^3 - 12x + 8$

28. $f(x) = x^3 - 27x + 30$

29. $f(x) = \dfrac{6}{x + 2}$ **30.** $f(x) = \dfrac{5}{x - 4}$

31. $f(x) = |x|$ **32.** $f(x) = |x + 3|$

In Problems 33–46, find the intervals on which $f(x)$ is increasing, the intervals on which $f(x)$ is decreasing, and the local extrema.

33. $f(x) = 2x^2 - 4x$

34. $f(x) = -3x^2 - 12x$

35. $f(x) = -2x^2 - 16x - 25$

36. $f(x) = -3x^2 + 12x - 5$

37. $f(x) = x^3 + 4x - 5$

38. $f(x) = -x^3 - 4x + 8$

39. $f(x) = 2x^3 - 3x^2 - 36x$

40. $f(x) = -2x^3 + 3x^2 + 120x$

41. $f(x) = 3x^4 - 4x^3 + 5$

42. $f(x) = x^4 + 2x^3 + 5$

43. $f(x) = (x - 1)e^{-x}$ **44.** $f(x) = x \ln x - x$

45. $f(x) = 4x^{1/3} - x^{2/3}$ **46.** $f(x) = (x^2 - 9)^{2/3}$

In Problems 47–52, use a graphing calculator to approximate the critical numbers of $f(x)$ to two decimal places. Find the intervals on which $f(x)$ is increasing, the intervals on which $f(x)$ is decreasing, and the local extrema.

47. $f(x) = x^4 - 4x^3 + 9x$ **48.** $f(x) = x^4 + 5x^3 - 15x$

49. $f(x) = x \ln x - (x - 2)^3$ **50.** $f(x) = e^{-x} - 3x^2$

51. $f(x) = e^x - 2x^2$ **52.** $f(x) = \dfrac{\ln x}{x} - 5x + x^2$

In Problems 53–60, find the intervals on which $f(x)$ is increasing and the intervals on which $f(x)$ is decreasing. Then sketch the graph. Add horizontal tangent lines.

53. $f(x) = 4 + 8x - x^2$

54. $f(x) = 2x^2 - 8x + 9$

55. $f(x) = x^3 - 3x + 1$

56. $f(x) = x^3 - 12x + 2$

57. $f(x) = 10 - 12x + 6x^2 - x^3$

58. $f(x) = x^3 + 3x^2 + 3x$

59. $f(x) = x^4 - 18x^2$

60. $f(x) = -x^4 + 50x^2$

In Problems 61–68, $f(x)$ is continuous on $(-\infty, \infty)$. Use the given information to sketch the graph of f.

61.

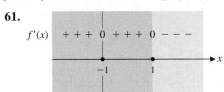

x	-2	-1	0	1	2
$f(x)$	-1	1	2	3	1

62.

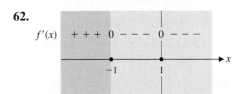

x	-2	-1	0	1	2
$f(x)$	1	3	2	1	-1

63.

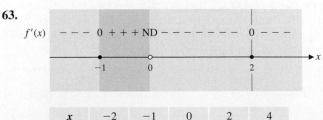

x	-2	-1	0	2	4
$f(x)$	2	1	2	1	0

64.

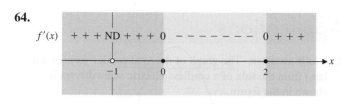

x	-2	-1	0	2	3
$f(x)$	-3	0	2	-1	0

65. $f(-2) = 4, f(0) = 0, f(2) = -4;$

$f'(-2) = 0, f'(0) = 0, f'(2) = 0;$

$f'(x) > 0$ on $(-\infty, -2)$ and $(2, \infty);$

$f'(x) < 0$ on $(-2, 0)$ and $(0, 2)$

66. $f(-2) = -1, f(0) = 0, f(2) = 1;$

$f'(-2) = 0, f'(2) = 0;$

$f'(x) > 0$ on $(-\infty, -2), (-2, 2),$ and $(2, \infty)$

67. $f(-1) = 2, f(0) = 0, f(1) = -2;$

$f'(-1) = 0, f'(1) = 0, f'(0)$ is not defined;

$f'(x) > 0$ on $(-\infty, -1)$ and $(1, \infty);$

$f'(x) < 0$ on $(-1, 0)$ and $(0, 1)$

68. $f(-1) = 2, f(0) = 0, f(1) = 2;$

$f'(-1) = 0, f'(1) = 0, f'(0)$ is not defined;

$f'(x) > 0$ on $(-\infty, -1)$ and $(0, 1);$

$f'(x) < 0$ on $(-1, 0)$ and $(1, \infty)$

Problems 69–74 involve functions f_1–f_6 and their derivatives, g_1–g_6. Use the graphs shown in figures (A) and (B) to match each function f_i with its derivative g_j.

69. f_1 **70.** f_2 **71.** f_3

72. f_4 **73.** f_5 **74.** f_6

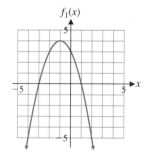
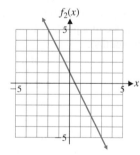

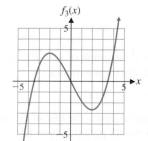
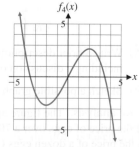

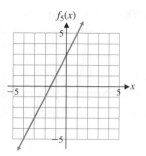
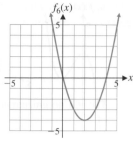

Figure (A) for 69–74

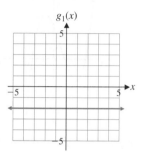
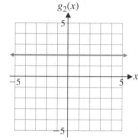

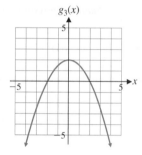
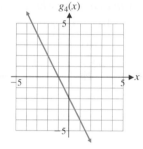

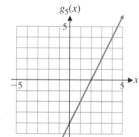
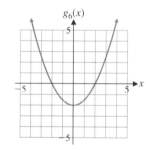

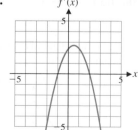
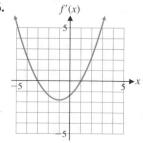

Figure (B) for 69–74

In Problems 75–80, use the given graph of $y = f'(x)$ to find the intervals on which f is increasing, the intervals on which f is decreasing, and the x coordinates of the local extrema of f. Sketch a possible graph of $y = f(x)$.

75. $f'(x)$ **76.** $f'(x)$

12.2 Second Derivative and Graphs

- Using Concavity as a Graphing Tool
- Finding Inflection Points
- Analyzing Graphs
- Curve Sketching
- Point of Diminishing Returns

In Section 12.1, we saw that the derivative can be used to determine when a graph is rising or falling. Now we want to see what the *second derivative* (the derivative of the derivative) can tell us about the shape of a graph.

Using Concavity as a Graphing Tool

Consider the functions

$$f(x) = x^2 \quad \text{and} \quad g(x) = \sqrt{x}$$

for x in the interval $(0, \infty)$. Since

$$f'(x) = 2x > 0 \quad \text{for } 0 < x < \infty$$

and

$$g'(x) = \frac{1}{2\sqrt{x}} > 0 \quad \text{for } 0 < x < \infty$$

both functions are increasing on $(0, \infty)$.

Explore and Discuss 1   (A) Discuss the difference in the shapes of the graphs of f and g shown in Figure 1.

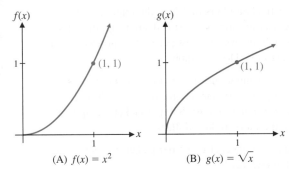

Figure 1

(A) $f(x) = x^2$ (B) $g(x) = \sqrt{x}$

(B) Complete the following table, and discuss the relationship between the values of the derivatives of f and g and the shapes of their graphs:

x	0.25	0.5	0.75	1
$f'(x)$				
$g'(x)$				

We use the term *concave upward* to describe a graph that opens upward and *concave downward* to describe a graph that opens downward. Thus, the graph of f in Figure 1A is concave upward, and the graph of g in Figure 1B is concave downward. Finding a mathematical formulation of concavity will help us sketch and analyze graphs.

We examine the slopes of f and g at various points on their graphs (see Fig. 2) and make two observations about each graph:

1. Looking at the graph of f in Figure 2A, we see that $f'(x)$ (the slope of the tangent line) is *increasing* and that the graph lies *above* each tangent line;

2. Looking at Figure 2B, we see that $g'(x)$ is *decreasing* and that the graph lies *below* each tangent line.

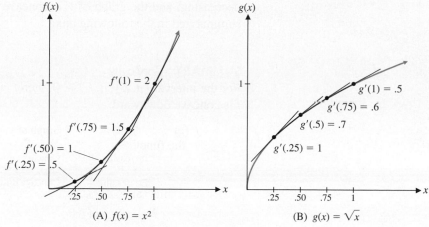

(A) $f(x) = x^2$ (B) $g(x) = \sqrt{x}$

Figure 2

DEFINITION Concavity

The graph of a function f is **concave upward** on the interval (a, b) if $f'(x)$ is *increasing* on (a, b) and is **concave downward** on the interval (a, b) if $f'(x)$ is *decreasing* on (a, b).

Geometrically, the graph is concave upward on (a, b) if it lies above its tangent lines in (a, b) and is concave downward on (a, b) if it lies below its tangent lines in (a, b).

How can we determine when $f'(x)$ is increasing or decreasing? In Section 12.1, we used the derivative to determine when a function is increasing or decreasing. To determine when the function $f'(x)$ is increasing or decreasing, we use the derivative of $f'(x)$. The derivative of the derivative of a function is called the *second derivative* of the function. Various notations for the second derivative are given in the following box:

NOTATION Second Derivative

For $y = f(x)$, the **second derivative** of f, provided that it exists, is

$$f''(x) = \frac{d}{dx} f'(x)$$

Other notations for $f''(x)$ are

$$\frac{d^2y}{dx^2} \quad \text{and} \quad y''$$

Returning to the functions f and g discussed at the beginning of this section, we have

$$f(x) = x^2 \qquad\qquad g(x) = \sqrt{x} = x^{1/2}$$

$$f'(x) = 2x \qquad\qquad g'(x) = \frac{1}{2}x^{-1/2} = \frac{1}{2\sqrt{x}}$$

$$f''(x) = \frac{d}{dx}2x = 2 \qquad g''(x) = \frac{d}{dx}\frac{1}{2}x^{-1/2} = -\frac{1}{4}x^{-3/2} = -\frac{1}{4\sqrt{x^3}}$$

For $x > 0$, we see that $f''(x) > 0$; so, $f'(x)$ is increasing, and the graph of f is concave upward (see Fig. 2A). For $x > 0$, we also see that $g''(x) < 0$; so, $g'(x)$ is decreasing, and the graph of g is concave downward (see Fig. 2B). These ideas are summarized in the following box:

SUMMARY   Concavity

For the interval (a, b), if $f'' > 0$, then f is concave upward, and if $f'' < 0$, then f is concave downward.

$f''(x)$	$f'(x)$	Graph of $y = f(x)$	Examples
$+$	Increasing	Concave upward	$\smile$ $\searprofile$ $\profile$
$-$	Decreasing	Concave downward	$\frown$ $\profile$ $\profile$

CONCEPTUAL INSIGHT

Be careful not to confuse concavity with falling and rising. A graph that is concave upward on an interval may be falling, rising, or both falling and rising on that interval. A similar statement holds for a graph that is concave downward. See Figure 3.

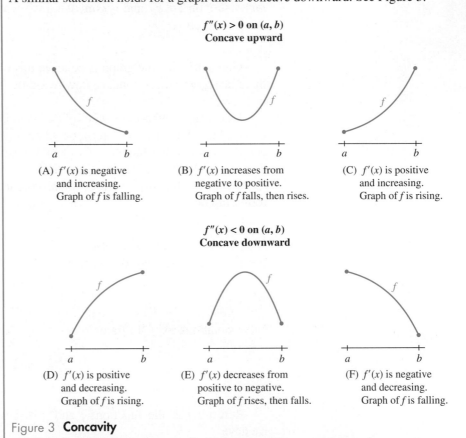

Figure 3 Concavity

EXAMPLE 1   Concavity of Graphs   Determine the intervals on which the graph of each function is concave upward and the intervals on which it is concave downward. Sketch a graph of each function.

(A) $f(x) = e^x$

(B) $g(x) = \ln x$

(C) $h(x) = x^3$

SOLUTION

(A) $f(x) = e^x$

$f'(x) = e^x$

$f''(x) = e^x$

Since $f''(x) > 0$ on $(-\infty, \infty)$, the graph of $f(x) = e^x$ [Fig. 4(A)] is concave upward on $(-\infty, \infty)$.

(B) $g(x) = \ln x$

$g'(x) = \dfrac{1}{x}$

$g''(x) = -\dfrac{1}{x^2}$

The domain of $g(x) = \ln x$ is $(0, \infty)$ and $g''(x) < 0$ on this interval, so the graph of $g(x) = \ln x$ [Fig. 4(B)] is concave downward on $(0, \infty)$.

(C) $h(x) = x^3$

$h'(x) = 3x^2$

$h''(x) = 6x$

Since $h''(x) < 0$ when $x < 0$ and $h''(x) > 0$ when $x > 0$, the graph of $h(x) = x^3$ [Fig. 4(C)] is concave downward on $(-\infty, 0)$ and concave upward on $(0, \infty)$.

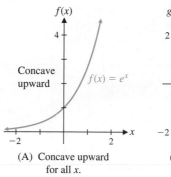

(A) Concave upward for all x.

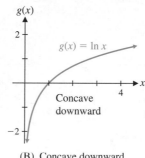

(B) Concave downward for $x > 0$.

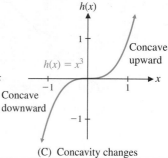

(C) Concavity changes at the origin.

Figure 4

Matched Problem 1 Determine the intervals on which the graph of each function is concave upward and the intervals on which it is concave downward. Sketch a graph of each function.

(A) $f(x) = -e^{-x}$ (B) $g(x) = \ln \dfrac{1}{x}$ (C) $h(x) = x^{1/3}$

Refer to Example 1. The graphs of $f(x) = e^x$ and $g(x) = \ln x$ never change concavity. But the graph of $h(x) = x^3$ changes concavity at $(0, 0)$. This point is called an *inflection point*.

Finding Inflection Points

An **inflection point** is a point on the graph of a function where the concavity changes (from upward to downward or from downward to upward). For the concavity to change at a point, $f''(x)$ must change sign at that point. But in Section 2.2, we saw that the partition numbers identify the points where a function can change sign.

THEOREM 1 Inflection Points

If $(c, f(c))$ is an inflection point of f, then c is a partition number for f''.

Our strategy for finding inflection points is clear: We find all partition numbers c for f'' and ask

1. Does $f''(x)$ change sign at c?

2. Is c is in the domain of f?

If both answers are "yes", then $(c, f(c))$ is an inflection point of f. Figure 5 illustrates several typical cases.

If $f'(c)$ exists and $f''(x)$ changes sign at $x = c$, then the tangent line at an inflection point $(c, f(c))$ will always lie below the graph on the side that is concave upward and above the graph on the side that is concave downward (see Fig. 5A, B, and C).

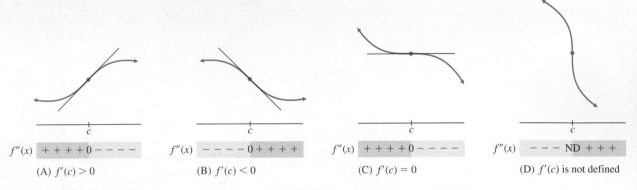

$f''(x)$ $+\ +\ +\ +\ 0\ -\ -\ -\ -$ $f''(x)$ $-\ -\ -\ -\ 0\ +\ +\ +\ +$ $f''(x)$ $+\ +\ +\ +\ 0\ -\ -\ -\ -$ $f''(x)$ $-\ -\ -\ \text{ND}\ +\ +\ +$

(A) $f'(c) > 0$ (B) $f'(c) < 0$ (C) $f'(c) = 0$ (D) $f'(c)$ is not defined

Figure 5 **Inflection points**

EXAMPLE 2   Locating Inflection Points   Find the inflection point(s) of

$$f(x) = x^3 - 6x^2 + 9x + 1$$

SOLUTION Since inflection points occur at values of x where $f''(x)$ changes sign, we construct a sign chart for $f''(x)$.

$$f(x) = x^3 - 6x^2 + 9x + 1$$
$$f'(x) = 3x^2 - 12x + 9$$
$$f''(x) = 6x - 12 = 6(x - 2)$$

The sign chart for $f''(x) = 6(x - 2)$ (partition number is 2) is as follows:

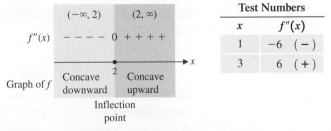

Test Numbers	
x	$f''(x)$
1	-6 $(-)$
3	6 $(+)$

From the sign chart, we see that the graph of f has an inflection point at $x = 2$. That is, the point

$$(2, f(2)) = (2, 3)   \quad f(2) = 2^3 - 6 \cdot 2^2 + 9 \cdot 2 + 1 = 3$$

is an inflection point on the graph of f.

Matched Problem 2   Find the inflection point(s) of

$$f(x) = x^3 - 9x^2 + 24x - 10$$

EXAMPLE 3   Locating Inflection Points   Find the inflection point(s) of

$$f(x) = \ln(x^2 - 4x + 5)$$

SOLUTION   First we find the domain of f. Since $\ln x$ is defined only for $x > 0$, f is defined only for

$$x^2 - 4x + 5 > 0 \quad \text{Complete the square (Section A.7).}$$
$$(x - 2)^2 + 1 > 0 \quad \text{True for all } x.$$

So the domain of f is $(-\infty, \infty)$. Now we find $f''(x)$ and construct a sign chart for it.

$$f(x) = \ln(x^2 - 4x + 5)$$

$$f'(x) = \frac{2x - 4}{x^2 - 4x + 5}$$

$$f''(x) = \frac{(x^2 - 4x + 5)(2x - 4)' - (2x - 4)(x^2 - 4x + 5)'}{(x^2 - 4x + 5)^2}$$

$$= \frac{(x^2 - 4x + 5)2 - (2x - 4)(2x - 4)}{(x^2 - 4x + 5)^2}$$

$$= \frac{2x^2 - 8x + 10 - 4x^2 + 16x - 16}{(x^2 - 4x + 5)^2}$$

$$= \frac{-2x^2 + 8x - 6}{(x^2 - 4x + 5)^2}$$

$$= \frac{-2(x - 1)(x - 3)}{(x^2 - 4x + 5)^2}$$

The partition numbers for $f''(x)$ are $x = 1$ and $x = 3$.
Sign chart for $f''(x)$:

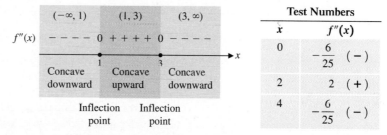

The sign chart shows that the graph of f has inflection points at $x = 1$ and $x = 3$.
Since $f(1) = \ln 2$ and $f(3) = \ln 2$, the inflection points are $(1, \ln 2)$ and $(3, \ln 2)$.

Matched Problem 3   Find the inflection point(s) of

$$f(x) = \ln(x^2 - 2x + 5)$$

CONCEPTUAL INSIGHT

It is important to remember that the partition numbers for f'' are only *candidates* for inflection points. The function f must be defined at $x = c$, and the second derivative must change sign at $x = c$ in order for the graph to have an inflection point at $x = c$. For example, consider

$$f(x) = x^4 \qquad g(x) = \frac{1}{x}$$

$$f'(x) = 4x^3 \qquad g'(x) = -\frac{1}{x^2}$$

$$f''(x) = 12x^2 \qquad g''(x) = \frac{2}{x^3}$$

In each case, $x = 0$ is a partition number for the second derivative, but neither the graph of $f(x)$ nor the graph of $g(x)$ has an inflection point at $x = 0$. Function f does not have an inflection point at $x = 0$ because $f''(x)$ does not change sign at $x = 0$ (see Fig. 6A). Function g does not have inflection point at $x = 0$ because $g(0)$ is not defined (see Fig. 6B).

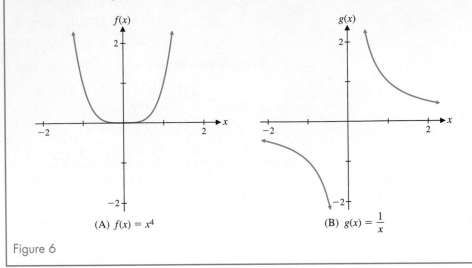

(A) $f(x) = x^4$ (B) $g(x) = \dfrac{1}{x}$

Figure 6

Analyzing Graphs

In the next example, we combine increasing/decreasing properties with concavity properties to analyze the graph of a function.

EXAMPLE 4   Analyzing a Graph   Figure 7 shows the graph of the derivative of a function f. Use this graph to discuss the graph of f. Include a sketch of a possible graph of f.

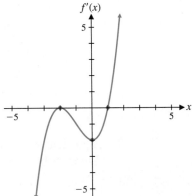

Figure 7

SOLUTION The sign of the derivative determines where the original function is increasing and decreasing, and the increasing/decreasing properties of the derivative determine the concavity of the original function. The relevant information obtained from the graph of f' is summarized in Table 1, and a possible graph of f is shown in Figure 8.

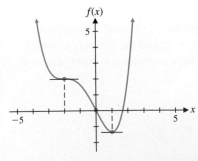

Figure 8

Table 1

x	$f'(x)$ (Fig. 7)	$f(x)$ (Fig. 8)
$-\infty < x < -2$	Negative and increasing	Decreasing and concave upward
$x = -2$	Local maximum	Inflection point
$-2 < x < 0$	Negative and decreasing	Decreasing and concave downward
$x = 0$	Local minimum	Inflection point
$0 < x < 1$	Negative and increasing	Decreasing and concave upward
$x = 1$	x intercept	Local minimum
$1 < x < \infty$	Positive and increasing	Increasing and concave upward

Matched Problem 4) Figure 9 shows the graph of the derivative of a function f. Use this graph to discuss the graph of f. Include a sketch of a possible graph of f.

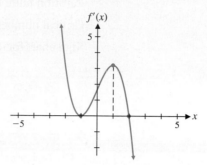

Figure 9

Curve Sketching

Graphing calculators and computers produce the graph of a function by plotting many points. However, key points on a plot many be difficult to identify. Using information gained from the function $f(x)$ and its derivatives, and plotting the key points—intercepts, local extrema, and inflection points—we can sketch by hand a very good representation of the graph of $f(x)$. This graphing process is called **curve sketching**.

PROCEDURE Graphing Strategy (First Version)*

Step 1 *Analyze $f(x)$.* Find the domain and the intercepts. The x intercepts are the solutions of $f(x) = 0$, and the y intercept is $f(0)$.

Step 2 *Analyze $f'(x)$.* Find the partition numbers for f' and the critical numbers of f. Construct a sign chart for $f'(x)$, determine the intervals on which f is increasing and decreasing, and find the local maxima and minima of f.

Step 3 *Analyze $f''(x)$.* Find the partition numbers for $f''(x)$. Construct a sign chart for $f''(x)$, determine the intervals on which the graph of f is concave upward and concave downward, and find the inflection points of f.

Step 4 *Sketch the graph of f.* Locate intercepts, local maxima and minima, and inflection points. Sketch in what you know from steps 1–3. Plot additional points as needed and complete the sketch.

EXAMPLE 5 Using the Graphing Strategy Follow the graphing strategy and analyze the function

$$f(x) = x^4 - 2x^3$$

State all the pertinent information and sketch the graph of f.

SOLUTION

Step 1 *Analyze $f(x)$.* Since f is a polynomial, its domain is $(-\infty, \infty)$.

x intercept: $f(x) = 0$
$$x^4 - 2x^3 = 0$$
$$x^3(x - 2) = 0$$
$$x = 0, 2$$

y intercept: $f(0) = 0$

---

*We will modify this summary in Section 12.4 to include additional information about the graph of f.

Step 2 *Analyze $f'(x)$.* $f'(x) = 4x^3 - 6x^2 = 4x^2(x - \frac{3}{2})$

Partition numbers for $f'(x)$: 0 and $\frac{3}{2}$

Critical numbers of $f(x)$: 0 and $\frac{3}{2}$

Sign chart for $f'(x)$:

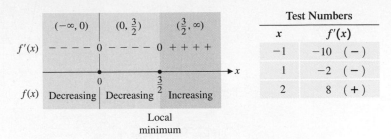

	x	$f'(x)$
Test Numbers		
	-1	-10 $(-)$
	1	-2 $(-)$
	2	8 $(+)$

So $f(x)$ is decreasing on $(-\infty, \frac{3}{2})$, is increasing on $(\frac{3}{2}, \infty)$, and has a local minimum at $x = \frac{3}{2}$. The local minimum is $f(\frac{3}{2}) = -\frac{27}{16}$.

Step 3 *Analyze $f''(x)$.* $f''(x) = 12x^2 - 12x = 12x(x - 1)$

Partition numbers for $f''(x)$: 0 and 1

Sign chart for $f''(x)$:

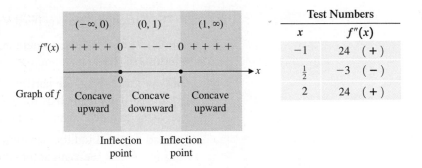

	x	$f''(x)$
Test Numbers		
	-1	24 $(+)$
	$\frac{1}{2}$	-3 $(-)$
	2	24 $(+)$

So the graph of f is concave upward on $(-\infty, 0)$ and $(1, \infty)$, is concave downward on $(0, 1)$, and has inflection points at $x = 0$ and $x = 1$. Since $f(0) = 0$ and $f(1) = -1$, the inflection points are $(0, 0)$ and $(1, -1)$.

Step 4 *Sketch the graph of f.*

Key Points	
x	$f(x)$
0	0
1	-1
$\frac{3}{2}$	$-\frac{27}{16}$
2	0

Matched Problem 5) Follow the graphing strategy and analyze the function $f(x) = x^4 + 4x^3$. State all the pertinent information and sketch the graph of f.

CONCEPTUAL INSIGHT

Refer to the solution of Example 5. Combining the sign charts for $f'(x)$ and $f''(x)$ (Fig. 10) partitions the real-number line into intervals on which neither $f'(x)$ nor $f''(x)$ changes sign. On each of these intervals, the graph of $f(x)$ must have one of four basic shapes (see also Fig. 3, parts A, C, D, and F on page 650). This reduces sketching the graph of a function to plotting the points identified in the graphing strategy and connecting them with one of the basic shapes.

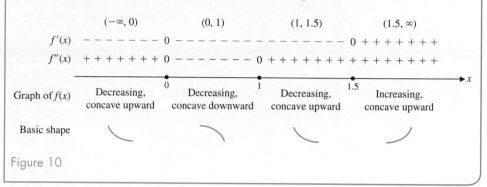

Figure 10

EXAMPLE 6 Using the Graphing Strategy Follow the graphing strategy and analyze the function

$$f(x) = 3x^{5/3} - 20x$$

State all the pertinent information and sketch the graph of f. Round any decimal values to two decimal places.

SOLUTION

Step 1 *Analyze $f(x)$.* $f(x) = 3x^{5/3} - 20x$

Since x^p is defined for any x and any positive p, the domain of f is $(-\infty, \infty)$.

$$\text{x intercepts: Solve } f(x) = 0$$
$$3x^{5/3} - 20x = 0$$
$$3x\left(x^{2/3} - \frac{20}{3}\right) = 0 \quad (a^2 - b^2) = (a - b)(a + b)$$
$$3x\left(x^{1/3} - \sqrt{\frac{20}{3}}\right)\left(x^{1/3} + \sqrt{\frac{20}{3}}\right) = 0$$

The x intercepts of f are

$$x = 0, \quad x = \left(\sqrt{\frac{20}{3}}\right)^3 \approx 17.21, \quad x = \left(-\sqrt{\frac{20}{3}}\right)^3 \approx -17.21$$

y intercept: $f(0) = 0$.

Step 2 Analyze $f'(x)$.

$$f'(x) = 5x^{2/3} - 20$$
$$= 5(x^{2/3} - 4) \qquad \text{Again, } a^2 - b^2 = (a - b)(a + b)$$
$$= 5(x^{1/3} - 2)(x^{1/3} + 2)$$

Partition numbers for f': $x = 2^3 = 8$ and $x = (-2)^3 = -8$.

Critical numbers of f: $-8, 8$

Sign chart for $f'(x)$:

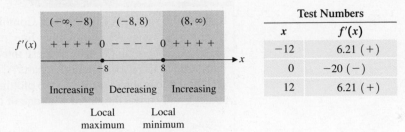

So f is increasing on $(-\infty, -8)$ and $(8, \infty)$ and decreasing on $(-8, 8)$. Therefore, $f(-8) = 64$ is a local maximum, and $f(8) = -64$ is a local minimum.

Step 3   *Analyze* $f''(x)$.

$$f'(x) = 5x^{2/3} - 20$$
$$f''(x) = \frac{10}{3}x^{-1/3} = \frac{10}{3x^{1/3}}$$

Partition number for f'': 0

Sign chart for $f''(x)$:

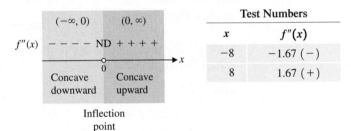

So f is concave downward on $(-\infty, 0)$, is concave upward on $(0, \infty)$, and has an inflection point at $x = 0$. Since $f(0) = 0$, the inflection point is $(0, 0)$.

Step 4   *Sketch the graph of f.*

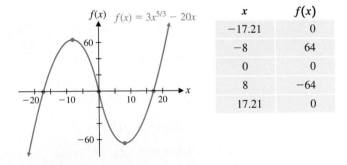

x	$f(x)$
-17.21	0
-8	64
0	0
8	-64
17.21	0

Matched Problem 6   Follow the graphing strategy and analyze the function $f(x) = 3x^{2/3} - x$. State all the pertinent information and sketch the graph of f. Round any decimal values to two decimal places.

Point of Diminishing Returns

If a company decides to increase spending on advertising, it would expect sales to increase. At first, sales will increase at an increasing rate and then increase at a decreasing rate. The dollar amount x at which the rate of change of sales goes from increasing to decreasing is called the **point of diminishing returns**. This is also the amount at which the rate of change has a maximum value. Money spent beyond this amount may increase sales but at a lower rate.

EXAMPLE 7 Maximum Rate of Change Currently, a discount appliance store is selling 200 large-screen TVs monthly. If the store invests $x thousand in an advertising campaign, the ad company estimates that monthly sales will be given by

$$N(x) = 3x^3 - 0.25x^4 + 200 \qquad 0 \le x \le 9$$

When is the rate of change of sales increasing and when is it decreasing? What is the point of diminishing returns and the maximum rate of change of sales? Graph N and N' on the same coordinate system.

SOLUTION The rate of change of sales with respect to advertising expenditures is

$$N'(x) = 9x^2 - x^3 = x^2(9 - x)$$

To determine when $N'(x)$ is increasing and decreasing, we find $N''(x)$, the derivative of $N'(x)$:

$$N''(x) = 18x - 3x^2 = 3x(6 - x)$$

The information obtained by analyzing the signs of $N'(x)$ and $N''(x)$ is summarized in Table 2 (sign charts are omitted).

Table 2

x	$N''(x)$	$N'(x)$	$N'(x)$	$N(x)$
$0 < x < 6$	+	+	Increasing	Increasing, concave upward
$x = 6$	0	+	Local maximum	Inflection point
$6 < x < 9$	−	+	Decreasing	Increasing, concave downward

Examining Table 2, we see that $N'(x)$ is increasing on $(0, 6)$ and decreasing on $(6, 9)$. The point of diminishing returns is $x = 6$ and the maximum rate of change is $N'(6) = 108$. Note that $N'(x)$ has a local maximum and $N(x)$ has an inflection point at $x = 6$ [the inflection point of $N(x)$ is $(6, 524)$].

So if the store spends $6,000 on advertising, monthly sales are expected to be 524 TVs, and sales are expected to increase at a rate of 108 TVs per thousand dollars spent on advertising. Money spent beyond the $6,000 would increase sales, but at a lower rate.

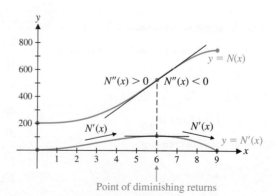

Matched Problem 7 Repeat Example 7 for

$$N(x) = 4x^3 - 0.25x^4 + 500 \qquad 0 \le x \le 12$$

Exercises 12.2

Skills Warm-up Exercises

W *In Problems 1–8, inspect the graph of the function to determine whether it is concave up, concave down, or neither, on the given interval. (If necessary, review Section 2.2).*

1. The square function, $h(x) = x^2$, on $(-\infty, \infty)$

2. The identity function, $f(x) = x$, on $(-\infty, \infty)$

3. The cube function, $m(x) = x^3$, on $(-\infty, 0)$

4. The cube function, $m(x) = x^3$, on $(0, \infty)$

5. The square root function, $n(x) = \sqrt{x}$, on $(0, \infty)$

6. The cube root function, $p(x) = \sqrt[3]{x}$, on $(-\infty, 0)$

7. The absolute value function, $g(x) = |x|$, on $(-\infty, 0)$

8. The cube root function, $p(x) = \sqrt[3]{x}$, on $(0, \infty)$

9. Use the graph of $y = f(x)$, assuming $f''(x) > 0$ if $x = b$ or f, to identify

(A) Intervals on which the graph of f is concave upward

(B) Intervals on which the graph of f is concave downward

(C) Intervals on which $f''(x) < 0$

(D) Intervals on which $f''(x) > 0$

(E) Intervals on which $f'(x)$ is increasing

(F) Intervals on which $f'(x)$ is decreasing

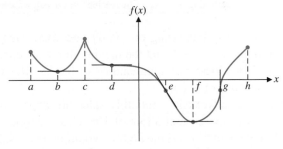

10. Use the graph of $y = g(x)$, assuming $g''(x) > 0$ if $x = c$ or g, to identify

(A) Intervals on which the graph of g is concave upward

(B) Intervals on which the graph of g is concave downward

(C) Intervals on which $g''(x) < 0$

(D) Intervals on which $g''(x) > 0$

(E) Intervals on which $g'(x)$ is increasing

(F) Intervals on which $g'(x)$ is decreasing

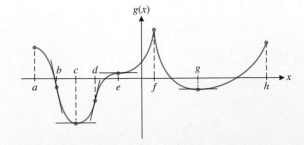

11. Use the graph of $y = f(x)$ to identify

(A) The local extrema of $f(x)$.

(B) The inflection points of $f(x)$.

(C) The numbers u for which $f'(u)$ is a local extremum of $f'(x)$.

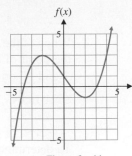 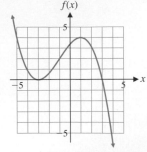

Figure for 11   Figure for 12

12. Use the graph of $y = f(x)$ to identify

(A) The local extrema of $f(x)$.

(B) The inflection points of $f(x)$.

(C) The numbers u for which $f'(u)$ is a local extremum of $f'(x)$.

In Problems 13–16, match the indicated conditions with one of the graphs (A)–(D) shown in the figure.

13. $f'(x) > 0$ and $f''(x) > 0$ on (a, b)

14. $f'(x) > 0$ and $f''(x) < 0$ on (a, b)

15. $f'(x) < 0$ and $f''(x) > 0$ on (a, b)

16. $f'(x) < 0$ and $f''(x) < 0$ on (a, b)

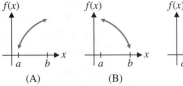 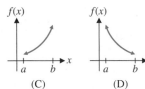

In Problems 17–24, find the indicated derivative for each function.

17. $f''(x)$ for $f(x) = 2x^3 - 4x^2 + 5x - 6$

18. $g''(x)$ for $g(x) = -x^3 + 2x^2 - 3x + 9$

19. $h''(x)$ for $h(x) = 2x^{-1} - 3x^{-2}$

20. $k''(x)$ for $k(x) = -6x^{-2} + 12x^{-3}$

21. d^2y/dx^2 for $y = x^2 - 18x^{1/2}$

22. d^2y/dx^2 for $y = x^3 - 24x^{1/3}$

23. y'' for $y = (x^2 + 9)^4$

24. y'' for $y = (x^2 - 16)^5$

In Problems 25–30, find the x and y coordinates of all inflection points.

25. $f(x) = x^3 + 30x^2$ **26.** $f(x) = x^3 - 24x^2$

27. $f(x) = x^{5/3} + 2$

28. $f(x) = 5 - x^{4/3}$

29. $f(x) = 1 + x + x^{2/5}$

30. $f(x) = x^{3/5} - 6x + 7$

In Problems 31–40, find the intervals on which the graph of f is concave upward, the intervals on which the graph of f is concave downward, and the x coordinates of the inflection points.

31. $f(x) = x^4 + 6x^2$

32. $f(x) = x^4 + 6x$

33. $f(x) = x^3 - 4x^2 + 5x - 2$

34. $f(x) = -x^3 - 5x^2 + 4x - 3$

35. $f(x) = -x^4 + 12x^3 - 12x + 24$

36. $f(x) = x^4 - 2x^3 - 36x + 12$

37. $f(x) = \ln(x^2 - 2x + 10)$

38. $f(x) = \ln(x^2 + 6x + 13)$

39. $f(x) = 8e^x - e^{2x}$

40. $f(x) = e^{3x} - 9e^x$

In Problems 41–48, f(x) is continuous on $(-\infty, \infty)$. Use the given information to sketch the graph of f.

41.

x	−4	−2	−1	0	2	4
f(x)	0	3	1.5	0	−1	−3

42.

x	−4	−2	−1	0	2	4
f(x)	0	−2	−1	0	1	3

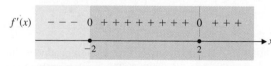

43.

x	−3	0	1	2	4	5
f(x)	−4	0	2	1	−1	0

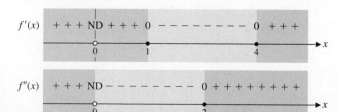

44.

x	−4	−2	0	2	4	6
f(x)	0	3	0	−2	0	3

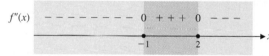

45. $f(0) = 2, f(1) = 0, f(2) = -2$;

$f'(0) = 0, f'(2) = 0$;

$f'(x) > 0$ on $(-\infty, 0)$ and $(2, \infty)$;

$f'(x) < 0$ on $(0, 2)$;

$f''(1) = 0$;

$f''(x) > 0$ on $(1, \infty)$;

$f''(x) < 0$ on $(-\infty, 1)$

46. $f(-2) = -2, f(0) = 1, f(2) = 4$;

$f'(-2) = 0, f'(2) = 0$;

$f'(x) > 0$ on $(-2, 2)$;

$f'(x) < 0$ on $(-\infty, -2)$ and $(2, \infty)$;

$f''(0) = 0$;

$f''(x) > 0$ on $(-\infty, 0)$;

$f''(x) < 0$ on $(0, \infty)$

47. $f(-1) = 0, f(0) = -2, f(1) = 0$;

$f'(0) = 0, f'(-1)$ and $f'(1)$ are not defined;

$f'(x) > 0$ on $(0, 1)$ and $(1, \infty)$;

$f'(x) < 0$ on $(-\infty, -1)$ and $(-1, 0)$;

$f''(-1)$ and $f''(1)$ are not defined;

$f''(x) > 0$ on $(-1, 1)$;

$f''(x) < 0$ on $(-\infty, -1)$ and $(1, \infty)$

48. $f(0) = -2, f(1) = 0, f(2) = 4$;

$f'(0) = 0, f'(2) = 0, f'(1)$ is not defined;

$f'(x) > 0$ on $(0, 1)$ and $(1, 2)$;

$f'(x) < 0$ on $(-\infty, 0)$ and $(2, \infty)$;

$f''(1)$ is not defined;

$f''(x) > 0$ on $(-\infty, 1)$;

$f''(x) < 0$ on $(1, \infty)$

In Problems 49–70, summarize the pertinent information obtained by applying the graphing strategy and sketch the graph of $y = f(x)$.

49. $f(x) = (x - 2)(x^2 - 4x - 8)$

50. $f(x) = (x - 3)(x^2 - 6x - 3)$

51. $f(x) = (x + 1)(x^2 - x + 2)$

52. $f(x) = (1 - x)(x^2 + x + 4)$

53. $f(x) = -0.25x^4 + x^3$

54. $f(x) = 0.25x^4 - 2x^3$

55. $f(x) = 16x(x - 1)^3$

56. $f(x) = -4x(x + 2)^3$

57. $f(x) = (x^2 + 3)(9 - x^2)$

58. $f(x) = (x^2 + 3)(x^2 - 1)$

59. $f(x) = (x^2 - 4)^2$

60. $f(x) = (x^2 - 1)(x^2 - 5)$

61. $f(x) = 2x^6 - 3x^5$

62. $f(x) = 3x^5 - 5x^4$

63. $f(x) = 1 - e^{-x}$

64. $f(x) = 2 - 3e^{-2x}$

65. $f(x) = e^{0.5x} + 4e^{-0.5x}$

66. $f(x) = 2e^{0.5x} + e^{-0.5x}$

67. $f(x) = -4 + 2\ln x$

68. $f(x) = 5 - 3\ln x$

69. $f(x) = \ln(x + 4) - 2$

70. $f(x) = 1 - \ln(x - 3)$

In Problems 71–74, use the graph of $y = f'(x)$ to discuss the graph of $y = f(x)$. Organize your conclusions in a table (see Example 4), and sketch a possible graph of $y = f(x)$.

71.

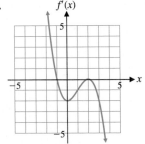

72.

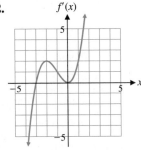

73.

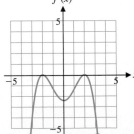

74.

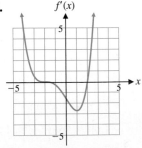

In Problems 75–82, apply steps 1–3 of the graphing strategy to $f(x)$. Use a graphing calculator to approximate (to two decimal places) x intercepts, critical numbers, and inflection points. Summarize all the pertinent information.

75. $f(x) = x^4 - 5x^3 + 3x^2 + 8x - 5$

76. $f(x) = x^4 + 2x^3 - 5x^2 - 4x + 4$

77. $f(x) = x^4 - 21x^3 + 100x^2 + 20x + 100$

78. $f(x) = x^4 - 12x^3 + 28x^2 + 76x - 50$

79. $f(x) = -x^4 - x^3 + 2x^2 - 2x + 3$

80. $f(x) = -x^4 + x^3 + x^2 + 6$

81. $f(x) = 0.1x^5 + 0.3x^4 - 4x^3 - 5x^2 + 40x + 30$

82. $f(x) = x^5 + 4x^4 - 7x^3 - 20x^2 + 20x - 20$

Applications

83. Inflation. One commonly used measure of inflation is the annual rate of change of the Consumer Price Index (CPI). A TV news story says that the annual rate of change of the CPI is increasing. What does this say about the shape of the graph of the CPI?

84. Inflation. Another commonly used measure of inflation is the annual rate of change of the Producer Price Index (PPI). A government report states that the annual rate of change of the PPI is decreasing. What does this say about the shape of the graph of the PPI?

85. Cost analysis. A company manufactures a variety of camp stoves at different locations. The total cost $C(x)$ (in dollars) of producing x camp stoves per week at plant A is shown in the figure. Discuss the graph of the marginal cost function $C'(x)$ and interpret the graph of $C'(x)$ in terms of the efficiency of the production process at this plant.

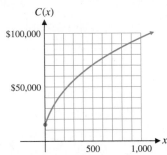

Production costs at plant A

86. Cost analysis. The company in Problem 85 produces the same camp stove at another plant. The total cost $C(x)$ (in dollars) of producing x camp stoves per week at plant B is shown in the figure. Discuss the graph of the marginal cost function $C'(x)$ and interpret the graph of $C'(x)$ in terms of the efficiency of the production process at plant B. Compare the production processes at the two plants.

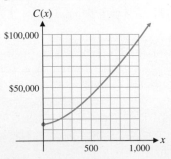

Production costs at plant B

87. Revenue. The marketing research department of a computer company used a large city to test market the firm's new laptop. The department found that the relationship between price p (dollars per unit) and the demand x (units per week) was given approximately by

$$p = 1,296 - 0.12x^2 \qquad 0 < x < 80$$

So, weekly revenue can be approximated by

$$R(x) = xp = 1,296x - 0.12x^3 \qquad 0 < x < 80$$

(A) Find the local extrema for the revenue function.

(B) On which intervals is the graph of the revenue function concave upward? Concave downward?

88. Profit. Suppose that the cost equation for the company in Problem 87 is

$$C(x) = 830 + 396x$$

(A) Find the local extrema for the profit function.

(B) On which intervals is the graph of the profit function concave upward? Concave downward?

89. Revenue. A dairy is planning to introduce and promote a new line of organic ice cream. After test marketing the new line in a large city, the marketing research department found that the demand in that city is given approximately by

$$p = 10e^{-x} \qquad 0 \le x \le 5$$

where x thousand quarts were sold per week at a price of $p each.

(A) Find the local extrema for the revenue function.

(B) On which intervals is the graph of the revenue function concave upward? Concave downward?

90. Revenue. A national food service runs food concessions for sporting events throughout the country. The company's marketing research department chose a particular football stadium to test market a new jumbo hot dog. It was found that the demand for the new hot dog is given approximately by

$$p = 8 - 2 \ln x \qquad 5 \le x \le 50$$

where x is the number of hot dogs (in thousands) that can be sold during one game at a price of $p.

(A) Find the local extrema for the revenue function.

(B) On which intervals is the graph of the revenue function concave upward? Concave downward?

91. Production: point of diminishing returns. A T-shirt manufacturer is planning to expand its workforce. It estimates that the number of T-shirts produced by hiring x new workers is given by

$$T(x) = -0.25x^4 + 5x^3 \qquad 0 \le x \le 15$$

When is the rate of change of T-shirt production increasing and when is it decreasing? What is the point of diminishing returns and the maximum rate of change of T-shirt production? Graph T and T' on the same coordinate system.

92. Production: point of diminishing returns. A baseball cap manufacturer is planning to expand its workforce. It estimates that the number of baseball caps produced by hiring x new workers is given by

$$T(x) = -0.25x^4 + 6x^3 \qquad 0 \le x \le 18$$

When is the rate of change of baseball cap production increasing and when is it decreasing? What is the point of diminishing returns and the maximum rate of change of baseball cap production? Graph T and T' on the same coordinate system.

93. Advertising: point of diminishing returns. A company estimates that it will sell $N(x)$ units of a product after spending $x thousand on advertising, as given by

$$N(x) = -0.25x^4 + 23x^3 - 540x^2 + 80,000 \qquad 24 \le x \le 45$$

When is the rate of change of sales increasing and when is it decreasing? What is the point of diminishing returns and the maximum rate of change of sales? Graph N and N' on the same coordinate system.

94. Advertising: point of diminishing returns. A company estimates that it will sell $N(x)$ units of a product after spending $x thousand on advertising, as given by

$$N(x) = -0.25x^4 + 13x^3 - 180x^2 + 10,000 \qquad 15 \le x \le 24$$

When is the rate of change of sales increasing and when is it decreasing? What is the point of diminishing returns and the maximum rate of change of sales? Graph N and N' on the same coordinate system.

95. Advertising. An automobile dealer uses TV advertising to promote car sales. On the basis of past records, the dealer arrived at the following data, where x is the number of ads placed monthly and y is the number of cars sold that month:

Number of Ads x	Number of Cars y
10	325
12	339
20	417
30	546
35	615
40	682
50	795

(A) Enter the data in a graphing calculator and find a cubic regression equation for the number of cars sold monthly as a function of the number of ads.

(B) How many ads should the dealer place each month to maximize the rate of change of sales with respect to the number of ads, and how many cars can the dealer expect to sell with this number of ads? Round answers to the nearest integer.

96. Advertising. A sporting goods chain places TV ads to promote golf club sales. The marketing director used past records to determine the following data, where x is the number of ads placed monthly and y is the number of golf clubs sold that month.

Number of Ads x	Number of Golf Clubs y
10	345
14	488
20	746
30	1,228
40	1,671
50	1,955

(A) Enter the data in a graphing calculator and find a cubic regression equation for the number of golf clubs sold monthly as a function of the number of ads.

(B) How many ads should the store manager place each month to maximize the rate of change of sales with respect to the number of ads, and how many golf clubs can the manager expect to sell with this number of ads? Round answers to the nearest integer.

97. Population growth: bacteria. A drug that stimulates reproduction is introduced into a colony of bacteria. After t minutes, the number of bacteria is given approximately by

$$N(t) = 1,000 + 30t^2 - t^3 \qquad 0 \le t \le 20$$

(A) When is the rate of growth, $N'(t)$, increasing? Decreasing?

(B) Find the inflection points for the graph of N.

(C) Sketch the graphs of N and N' on the same coordinate system.

(D) What is the maximum rate of growth?

98. Drug sensitivity. One hour after x milligrams of a particular drug are given to a person, the change in body temperature $T(x)$, in degrees Fahrenheit, is given by

$$T(x) = x^2\left(1 - \frac{x}{9}\right) \qquad 0 \le x \le 6$$

The rate $T'(x)$ at which $T(x)$ changes with respect to the size of the dosage x is called the *sensitivity* of the body to the dosage.

(A) When is $T'(x)$ increasing? Decreasing?

(B) Where does the graph of T have inflection points?

(C) Sketch the graphs of T and T' on the same coordinate system.

(D) What is the maximum value of $T'(x)$?

99. Learning. The time T (in minutes) it takes a person to learn a list of length n is

$$T(n) = 0.08n^3 - 1.2n^2 + 6n \qquad n \ge 0$$

(A) When is the rate of change of T with respect to the length of the list increasing? Decreasing?

(B) Where does the graph of T have inflection points?

(C) Graph T and T' on the same coordinate system.

(D) What is the minimum value of $T'(n)$?

Answers to Matched Problems

1. (A) Concave downward on $(-\infty, \infty)$

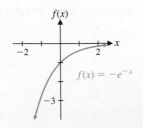

(B) Concave upward on $(0, \infty)$

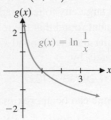

(C) Concave upward on $(-\infty, 0)$ and concave downward on $(0, \infty)$

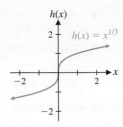

2. The only inflection point is $(3, f(3)) = (3, 8)$.

3. The inflection points are $(-1, f(-1)) = (-1, \ln 8)$ and $(3, f(3)) = (3, \ln 8)$.

4.

x	$f'(x)$	$f(x)$
$-\infty < x < -1$	Positive and decreasing	Increasing and concave downward
$x = -1$	Local minimum	Inflection point
$-1 < x < 1$	Positive and increasing	Increasing and concave upward
$x = 1$	Local maximum	Inflection point
$1 < x < 2$	Positive and decreasing	Increasing and concave downward
$x = 2$	x intercept	Local maximum
$2 < x < \infty$	Negative and decreasing	Decreasing and concave downward

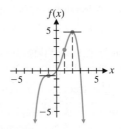

5. Domain: All real numbers
x intercepts: $-4, 0$; y intercept: $f(0) = 0$

Decreasing on $(-\infty, -3)$; increasing on $(-3, \infty)$; local minimum: $f(-3) = -27$

Concave upward on $(-\infty, -2)$ and $(0, \infty)$; concave downward on $(-2, 0)$

Inflection points: $(-2, -16)$, $(0, 0)$

x	$f(x)$
-4	0
-3	-27
-2	-16
0	0

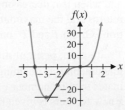

6. Domain: All real numbers

x intercepts: 0, 27; y intercept: $f(0) = 0$

Decreasing on $(-\infty, 0)$ and $(8, \infty)$; increasing on $(0, 8)$; local minimum: $f(0) = 0$; local maximum: $f(8) = 4$

Concave downward on $(-\infty, 0)$ and $(0, \infty)$; no inflection points

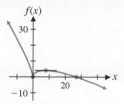

x	f(x)
0	0
8	4
27	0

7. $N'(x)$ is increasing on $(0, 8)$ and decreasing on $(8, 12)$.

The point of diminishing returns is $x = 8$ and the maximum rate of change is $N'(8) = 256$.

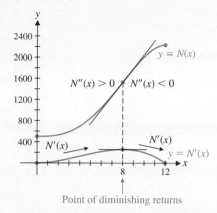

12.3 L'Hôpital's Rule

- Introduction
- L'Hôpital's Rule and the Indeterminate Form 0/0
- One-Sided Limits and Limits at ∞
- L'Hôpital's Rule and the Indeterminate Form ∞/∞

Introduction

The ability to evaluate a wide variety of different types of limits is one of the skills that are necessary to apply the techniques of calculus successfully. Limits play a fundamental role in the development of the derivative and are an important graphing tool. In order to deal effectively with graphs, we need to develop some more methods for evaluating limits.

In this section, we discuss a powerful technique for evaluating limits of quotients called *L'Hôpital's rule*. The rule is named after the French mathematician Marquis de L'Hôpital (1661–1704). To use L'Hôpital's rule, it is necessary to be familiar with the limit properties of some basic functions. Figure 1 reviews some limits involving powers of x that were discussed earlier.

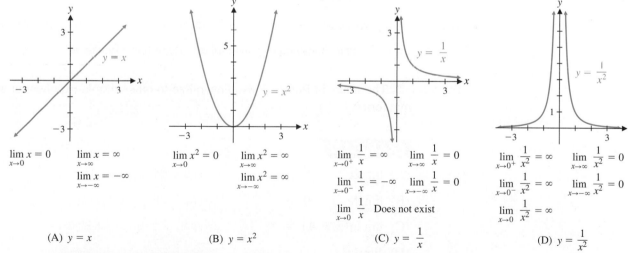

Figure 1 **Limits involving powers of x**

The limits in Figure 1 are easily extended to functions of the form $f(x) = (x - c)^n$ and $g(x) = 1/(x - c)^n$. In general, if n is an odd integer, then limits involving $(x - c)^n$ or $1/(x - c)^n$ as x approaches c (or $\pm\infty$) behave, respectively, like the limits of x and $1/x$ as x approaches 0 (or $\pm\infty$). If n is an even integer, then limits involving these expressions behave, respectively, like the limits of x^2 and $1/x^2$ as x approaches 0 (or $\pm\infty$).

EXAMPLE 1   Limits Involving Powers of $x - c$

(A) $\lim\limits_{x \to 2} \dfrac{5}{(x - 2)^4} = \infty$ Compare with $\lim\limits_{x \to 0} \dfrac{1}{x^2}$ in Figure 1.

(B) $\lim\limits_{x \to -1^-} \dfrac{4}{(x + 1)^3} = -\infty$ Compare with $\lim\limits_{x \to 0} \dfrac{1}{x}$ in Figure 1.

(C) $\lim\limits_{x \to \infty} \dfrac{4}{(x - 9)^6} = 0$ Compare with $\lim\limits_{x \to \infty} \dfrac{1}{x^2}$ in Figure 1.

(D) $\lim\limits_{x \to -\infty} 3x^3 = -\infty$ Compare with $\lim\limits_{x \to -\infty} x$ in Figure 1.

Matched Problem 1   Evaluate each limit.

(A) $\lim\limits_{x \to 3^+} \dfrac{7}{(x - 3)^5}$

(B) $\lim\limits_{x \to -4} \dfrac{6}{(x + 4)^6}$

(C) $\lim\limits_{x \to -\infty} \dfrac{3}{(x + 2)^3}$

(D) $\lim\limits_{x \to \infty} 5x^4$

Figure 2 reviews limits of exponential and logarithmic functions.

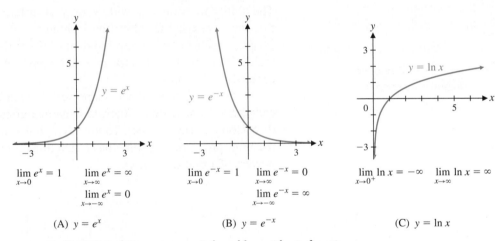

$\lim\limits_{x \to 0} e^x = 1$ $\lim\limits_{x \to \infty} e^x = \infty$

$\lim\limits_{x \to -\infty} e^x = 0$

$\lim\limits_{x \to 0} e^{-x} = 1$ $\lim\limits_{x \to \infty} e^{-x} = 0$

$\lim\limits_{x \to -\infty} e^{-x} = \infty$

$\lim\limits_{x \to 0^+} \ln x = -\infty$ $\lim\limits_{x \to \infty} \ln x = \infty$

(A) $y = e^x$ (B) $y = e^{-x}$ (C) $y = \ln x$

Figure 2 **Limits involving exponential and logarithmic functions**

The limits in Figure 2 also generalize to other simple exponential and logarithmic forms.

EXAMPLE 2   Limits Involving Exponential and Logarithmic Forms

(A) $\lim\limits_{x \to \infty} 2e^{3x} = \infty$ Compare with $\lim\limits_{x \to \infty} e^x$ in Figure 2.

(B) $\lim\limits_{x \to \infty} 4e^{-5x} = 0$ Compare with $\lim\limits_{x \to \infty} e^{-x}$ in Figure 2.

(C) $\lim\limits_{x \to \infty} \ln(x + 4) = \infty$ Compare with $\lim\limits_{x \to \infty} \ln x$ in Figure 2.

(D) $\lim\limits_{x \to 2^+} \ln(x - 2) = -\infty$ Compare with $\lim\limits_{x \to 0^+} \ln x$ in Figure 2.

Matched Problem 2   Evaluate each limit.

(A) $\lim\limits_{x \to -\infty} 2e^{-6x}$

(B) $\lim\limits_{x \to -\infty} 3e^{2x}$

(C) $\lim\limits_{x \to -4^+} \ln(x + 4)$

(D) $\lim\limits_{x \to \infty} \ln(x - 10)$

Now that we have reviewed the limit properties of some basic functions, we are ready to consider the main topic of this section: L'Hôpital's rule.

L'Hôpital's Rule and the Indeterminate Form 0/0

Recall that the limit

$$\lim_{x \to c} \frac{f(x)}{g(x)}$$

is a 0/0 indeterminate form if

$$\lim_{x \to c} f(x) = 0 \quad \text{and} \quad \lim_{x \to c} g(x) = 0$$

The quotient property for limits in Section 2.1 does not apply since $\lim_{x \to c} g(x) = 0$.

If we are dealing with a 0/0 indeterminate form, the limit may or may not exist, and we cannot tell which is true without further investigation.

Each of the following is a 0/0 indeterminate form:

$$\lim_{x \to 2} \frac{x^2 - 4}{x - 2} \quad \text{and} \quad \lim_{x \to 1} \frac{e^x - e}{x - 1} \tag{1}$$

The first limit can be evaluated by performing an algebraic simplification:

$$\lim_{x \to 2} \frac{x^2 - 4}{x - 2} = \lim_{x \to 2} \frac{(x - 2)(x + 2)}{x - 2} = \lim_{x \to 2}(x + 2) = 4$$

The second cannot. Instead, we turn to the powerful **L'Hôpital's rule**, which we state without proof. This rule can be used whenever a limit is a 0/0 indeterminate form, so can be used to evaluate both of the limits in (1).

THEOREM 1 L'Hôpital's Rule for 0/0 Indeterminate Forms: Version 1

For c a real number,
if $\lim_{x \to c} f(x) = 0$ and $\lim_{x \to c} g(x) = 0$, then

$$\lim_{x \to c} \frac{f(x)}{g(x)} = \lim_{x \to c} \frac{f'(x)}{g'(x)}$$

provided that the second limit exists or is ∞ or $-\infty$.

By L'Hôpital's rule,

$$\lim_{x \to 2} \frac{x^2 - 4}{x - 2} = \lim_{x \to 2} \frac{2x}{1} = 4$$

which agrees with the result obtained by algebraic simplification.

EXAMPLE 3 L'Hôpital's Rule Evaluate $\lim_{x \to 1} \frac{e^x - e}{x - 1}$.

SOLUTION

Step 1 *Check to see if L'Hôpital's rule applies:*

$$\lim_{x \to 1}(e^x - e) = e^1 - e = 0 \quad \text{and} \quad \lim_{x \to 1}(x - 1) = 1 - 1 = 0$$

L'Hôpital's rule does apply.

Step 2 *Apply L'Hôpital's rule:*

0/0 form

$$\lim_{x \to 1} \frac{e^x - e}{x - 1} = \lim_{x \to 1} \frac{\dfrac{d}{dx}(e^x - e)}{\dfrac{d}{dx}(x - 1)}$$ Apply L'Hôpital's rule.

$$= \lim_{x \to 1} \frac{e^x}{1}$$ e^x is continuous at $x = 1$.

$$= \frac{e^1}{1} = e$$

Matched Problem 3 Evaluate $\displaystyle\lim_{x \to 4} \frac{e^x - e^4}{x - 4}$.

⚠ **CAUTION** In L'Hôpital's rule, the symbol $f'(x)/g'(x)$ represents the derivative of $f(x)$ divided by the derivative of $g(x)$, not the derivative of the quotient $f(x)/g(x)$.

> When applying L'Hôpital's rule to a **0/0** indeterminate form, do not use the quotient rule. Instead, evaluate the limit of the derivative of the numerator divided by the derivative of the denominator.   ▲

The functions

$$y_1 = \frac{e^x - e}{x - 1} \quad \text{and} \quad y_2 = \frac{e^x}{1}$$

of Example 3 are different functions (see Fig. 3), but both functions have the same limit e as x approaches 1. Although y_1 is undefined at $x = 1$, the graph of y_1 provides a check of the answer to Example 3.

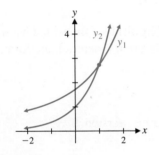

Figure 3

EXAMPLE 4 L'Hôpital's Rule Evaluate $\displaystyle\lim_{x \to 0} \frac{\ln(1 + x^2)}{x^4}$.

SOLUTION

Step 1 *Check to see if L'Hôpital's rule applies:*

$$\lim_{x \to 0} \ln(1 + x^2) = \ln 1 = 0 \quad \text{and} \quad \lim_{x \to 0} x^4 = 0$$

L'Hôpital's rule does apply.

Step 2 *Apply L'Hôpital's rule:*

0/0 form

$$\lim_{x \to 0} \frac{\ln(1 + x^2)}{x^4} = \lim_{x \to 0} \frac{\dfrac{d}{dx} \ln(1 + x^2)}{\dfrac{d}{dx} x^4}$$ Apply L'Hôpital's rule.

$$\lim_{x \to 0} \frac{\ln(1 + x^2)}{x^4} = \lim_{x \to 0} \frac{\dfrac{2x}{1 + x^2}}{4x^3}$$ Multiply numerator and denominator by $1/4x^3$.

$$= \lim_{x \to 0} \frac{\dfrac{2x}{1 + x^2} \dfrac{1}{4x^3}}{4x^3 \dfrac{1}{4x^3}}$$ Simplify.

$$= \lim_{x \to 0} \frac{1}{2x^2(1 + x^2)}$$

Apply Theorem 1 in Section 2.2 and compare with Fig. 1(D).

$$= \infty$$

Matched Problem 4 Evaluate $\lim_{x \to 1} \dfrac{\ln x}{(x - 1)^3}$.

EXAMPLE 5 L'Hôpital's Rule May Not Be Applicable Evaluate $\lim_{x \to 1} \dfrac{\ln x}{x}$.

SOLUTION

Step 1 *Check to see if L'Hôpital's rule applies:*

$$\lim_{x \to 1} \ln x = \ln 1 = 0, \quad \text{but} \quad \lim_{x \to 1} x = 1 \neq 0$$

L'Hôpital's rule does not apply.

Step 2 *Evaluate by another method.* The quotient property for limits from Section 2.1 does apply, and we have

$$\lim_{x \to 1} \frac{\ln x}{x} = \frac{\lim_{x \to 1} \ln x}{\lim_{x \to 1} x} = \frac{\ln 1}{1} = \frac{0}{1} = 0$$

Note that applying L'Hôpital's rule would give us an incorrect result:

$$\lim_{x \to 1} \frac{\ln x}{x} \neq \lim_{x \to 1} \frac{\frac{d}{dx} \ln x}{\frac{d}{dx} x} = \lim_{x \to 1} \frac{1/x}{1} = 1$$

Matched Problem 5 Evaluate $\lim_{x \to 0} \dfrac{x}{e^x}$.

⚠ **CAUTION** As Example 5 illustrates, some limits involving quotients are not $0/0$ indeterminate forms.

You must always check to see if L'Hôpital's rule applies before you use it. ▲

EXAMPLE 6 Repeated Application of L'Hôpital's Rule Evaluate

$$\lim_{x \to 0} \frac{x^2}{e^x - 1 - x}$$

SOLUTION

Step 1 *Check to see if L'Hôpital's rule applies:*

$$\lim_{x \to 0} x^2 = 0 \quad \text{and} \quad \lim_{x \to 0} (e^x - 1 - x) = 0$$

L'Hôpital's rule does apply.

Step 2 Apply L'Hôpital's rule:

0/0 form

$$\lim_{x \to 0} \frac{x^2}{e^x - 1 - x} = \lim_{x \to 0} \frac{\frac{d}{dx} x^2}{\frac{d}{dx}(e^x - 1 - x)} = \lim_{x \to 0} \frac{2x}{e^x - 1}$$

Since $\lim_{x \to 0} 2x = 0$ and $\lim_{x \to 0}(e^x - 1) = 0$, the new limit obtained is also a $0/0$ indeterminate form, and L'Hôpital's rule can be applied again.

Step 3 *Apply L'Hôpital's rule again:*

0/0 form

$$\lim_{x \to 0} \frac{2x}{e^x - 1} = \lim_{x \to 0} \frac{\frac{d}{dx} 2x}{\frac{d}{dx}(e^x - 1)} = \lim_{x \to 0} \frac{2}{e^x} = \frac{2}{e^0} = 2$$

Therefore,

$$\lim_{x \to 0} \frac{x^2}{e^x - 1 - x} = \lim_{x \to 0} \frac{2x}{e^x - 1} = \lim_{x \to 0} \frac{2}{e^x} = 2$$

Matched Problem 6   Evaluate $\lim_{x \to 0} \dfrac{e^{2x} - 1 - 2x}{x^2}$

One-Sided Limits and Limits at ∞

In addition to examining the limit as x approaches c, we have discussed one-sided limits and limits at ∞ in Chapter 10. L'Hôpital's rule is valid in these cases also.

THEOREM 2   L'Hôpital's Rule for 0/0 Indeterminate Forms: Version 2 (for one-sided limits and limits at infinity)

The first version of L'Hôpital's rule (Theorem 1) remains valid if the symbol $x \to c$ is replaced everywhere it occurs with one of the following symbols:

$$x \to c^+ \qquad x \to c^- \qquad x \to \infty \qquad x \to -\infty$$

For example, if $\lim\limits_{x \to \infty} f(x) = 0$ and $\lim\limits_{x \to \infty} g(x) = 0$, then

$$\lim_{x \to \infty} \frac{f(x)}{g(x)} = \lim_{x \to \infty} \frac{f'(x)}{g'(x)}$$

provided that the second limit exists or is $+\infty$ or $-\infty$. Similar rules can be written for $x \to c^+$, $x \to c^-$, and $x \to -\infty$.

EXAMPLE 7   L'Hôpital's Rule for One-Sided Limits   Evaluate $\lim\limits_{x \to 1^+} \dfrac{\ln x}{(x - 1)^2}$.

SOLUTION

Step 1 *Check to see if L'Hôpital's rule applies:*

$$\lim_{x \to 1^+} \ln x = 0 \qquad \text{and} \qquad \lim_{x \to 1^+} (x - 1)^2 = 0$$

L'Hôpital's rule does apply.

Step 2 *Apply L'Hôpital's rule:*

0/0 form

$$\lim_{x \to 1^+} \frac{\ln x}{(x - 1)^2} = \lim_{x \to 1^+} \frac{\frac{d}{dx}(\ln x)}{\frac{d}{dx}(x - 1)^2} \qquad \text{Apply L'Hôpital's rule.}$$

$$= \lim_{x \to 1^+} \frac{1/x}{2(x - 1)} \qquad \text{Simplify.}$$

$$= \lim_{x \to 1^+} \frac{1}{2x(x - 1)}$$

$$= \infty$$

The limit as $x \to 1^+$ is ∞ because $1/2x(x - 1)$ has a vertical asymptote at $x = 1$ (Theorem 1, Section 10.2) and $x(x - 1) > 0$ for $x > 1$.

Matched Problem 7 ⟩ Evaluate $\displaystyle\lim_{x \to 1^-} \frac{\ln x}{(x - 1)^2}$.

EXAMPLE 8 L'Hôpital's Rule for Limits at Infinity Evaluate $\displaystyle\lim_{x \to \infty} \frac{\ln\left(1 + e^{-x}\right)}{e^{-x}}$.

SOLUTION

Step 1 *Check to see if L'Hôpital's rule applies:*

$$\lim_{x \to \infty} \ln(1 + e^{-x}) = \ln(1 + 0) = \ln 1 = 0 \text{ and } \lim_{x \to \infty} e^{-x} = 0$$

L'Hôpital's rule does apply.

Step 2 *Apply L'Hôpital's rule:*

$$\underset{\text{0/0 form}}{\lim_{x \to \infty} \frac{\ln\left(1 + e^{-x}\right)}{e^{-x}}} = \lim_{x \to \infty} \frac{\dfrac{d}{dx}\left[\ln\left(1 + e^{-x}\right)\right]}{\dfrac{d}{dx} e^{-x}} \qquad \text{Apply L'Hôpital's rule.}$$

$$= \lim_{x \to \infty} \frac{-e^{-x}/\left(1 + e^{-x}\right)}{-e^{-x}} \qquad \begin{array}{l}\text{Multiply numerator and} \\ \text{denominator by } -e^x.\end{array}$$

$$= \lim_{x \to \infty} \frac{1}{1 + e^{-x}} \qquad \lim_{x \to \infty} e^{-x} = 0$$

$$= \frac{1}{1 + 0} = 1$$

Matched Problem 8 ⟩ Evaluate $\displaystyle\lim_{x \to -\infty} \frac{\ln(1 + 2e^x)}{e^x}$.

L'Hôpital's Rule and the Indeterminate Form ∞/∞

In Section 10.2, we discussed techniques for evaluating limits of rational functions such as

$$\lim_{x \to \infty} \frac{2x^2}{x^3 + 3} \qquad \lim_{x \to \infty} \frac{4x^3}{2x^2 + 5} \qquad \lim_{x \to \infty} \frac{3x^3}{5x^3 + 6} \qquad (2)$$

Each of these limits is an ∞/∞ *indeterminate form*. In general, if $\lim_{x \to c} f(x) = \pm\infty$ and $\lim_{x \to c} g(x) = \pm\infty$, then

$$\lim_{x \to c} \frac{f(x)}{g(x)}$$

is called an **∞/∞ indeterminate form**. Furthermore, $x \to c$ can be replaced in all three limits above with $x \to c^+, x \to c^-, x \to \infty$, or $x \to -\infty$. It can be shown that L'Hôpital's rule also applies to these ∞/∞ indeterminate forms.

THEOREM 3 L'Hôpital's Rule for the Indeterminate Form ∞/∞: Version 3

Versions 1 and 2 of L'Hôpital's rule for the indeterminate form $0/0$ are also valid if the limit of f and the limit of g are both infinite; that is, both $+\infty$ and $-\infty$ are permissible for either limit.

For example, if $\lim_{x \to c^+} f(x) = \infty$ and $\lim_{x \to c^+} g(x) = -\infty$, then L'Hôpital's rule can be applied to $\lim_{x \to c^+} [f(x)/g(x)]$.

Explore and Discuss 1

Evaluate each of the limits in (2) on page 671 in two ways:

1. Use Theorem 4 in Section 10.2.
2. Use L'Hôpital's rule.

Given a choice, which method would you choose? Why?

EXAMPLE 9   L'Hôpital's Rule for the Indeterminate Form ∞/∞ Evaluate

$$\lim_{x \to \infty} \frac{\ln x}{x^2}.$$

SOLUTION

Step 1   *Check to see if L'Hôpital's rule applies:*

$$\lim_{x \to \infty} \ln x = \infty \qquad \text{and} \qquad \lim_{x \to \infty} x^2 = \infty$$

L'Hôpital's rule does apply.

Step 2   *Apply L'Hôpital's rule:*

∞/∞ form

$$\lim_{x \to \infty} \frac{\ln x}{x^2} = \lim_{x \to \infty} \frac{\dfrac{d}{dx}(\ln x)}{\dfrac{d}{dx} x^2} \qquad \text{Apply L'Hôpital's rule.}$$

$$= \lim_{x \to \infty} \frac{1/x}{2x} \qquad \text{Simplify.}$$

$$\lim_{x \to \infty} \frac{\ln x}{x^2} = \lim_{x \to \infty} \frac{1}{2x^2} \qquad \text{See Figure 1(D).}$$

$$= 0$$

Matched Problem 9   Evaluate $\lim_{x \to \infty} \dfrac{\ln x}{x}$.

EXAMPLE 10   L'Hôpital's Rule for the Indeterminate Form ∞/∞

Evaluate $\lim_{x \to \infty} \dfrac{e^x}{x^2}$.

SOLUTION

Step 1   *Check to see if L'Hôpital's rule applies:*

$$\lim_{x \to \infty} e^x = \infty \qquad \text{and} \qquad \lim_{x \to \infty} x^2 = \infty$$

L'Hôpital's rule does apply.

Step 2   *Apply L'Hôpital's rule:*

∞/∞ form

$$\lim_{x \to \infty} \frac{e^x}{x^2} = \lim_{x \to \infty} \frac{\dfrac{d}{dx} e^x}{\dfrac{d}{dx} x^2} = \lim_{x \to \infty} \frac{e^x}{2x}$$

Since $\lim_{x \to \infty} e^x = \infty$ and $\lim_{x \to \infty} 2x = \infty$, this limit is an ∞/∞ indeterminate form and L'Hôpital's rule can be applied again.

Step 3 *Apply L'Hôpital's rule again:*

$$\underbrace{\lim_{x\to\infty} \frac{e^x}{2x}}_{\infty/\infty \text{ form}} = \lim_{x\to\infty} \frac{\frac{d}{dx} e^x}{\frac{d}{dx} 2x} = \lim_{x\to\infty} \frac{e^x}{2} = \infty$$

Therefore,

$$\lim_{x\to\infty} \frac{e^x}{x^2} = \lim_{x\to\infty} \frac{e^x}{2x} = \lim_{x\to\infty} \frac{e^x}{2} = \infty$$

Matched Problem 10) Evaluate $\lim\limits_{x\to\infty} \dfrac{e^{2x}}{x^2}$.

CONCEPTUAL INSIGHT

The three versions of L'Hôpital's rule cover a multitude of limits—far too many to remember case by case. Instead, we suggest you use the following pattern, common to all versions, as a memory aid:

1. All versions involve three limits: $\lim [f(x)/g(x)]$, $\lim f(x)$, and $\lim g(x)$.

2. The independent variable x must behave the same way in all three limits. The acceptable behaviors are $x \to c$, $x \to c^+$, $x \to c^-$, $x \to \infty$, or $x \to -\infty$.

3. The form of $\lim [f(x)/g(x)]$ must be $\frac{0}{0}$ or $\frac{\pm\infty}{\pm\infty}$ and both $\lim f(x)$ and $\lim g(x)$ must approach 0 or both must approach $\pm\infty$.

Exercises 12.3

Skills Warm-up Exercises

W   *In Problems 1–8, round each expression to the nearest integer without using a calculator. (If necessary, review Section A.1).*

1. $\dfrac{5}{0.01}$

2. $\dfrac{8}{0.002}$

3. $\dfrac{3}{1,000}$

4. $\dfrac{2^8}{8}$

5. $\dfrac{1}{2(1.01 - 1)}$

6. $\dfrac{47}{106}$

7. $\dfrac{\ln 100}{100}$

8. $\dfrac{e^5 + 5^2}{e^5}$

In Problems 9–16, even though the limit can be found using algebraic simplification as in Section 10.1, use L'Hôpital's rule to find the limit.

9. $\lim\limits_{x\to 3} \dfrac{x^2 - 9}{x - 3}$

10. $\lim\limits_{x\to -3} \dfrac{x^2 - 9}{x + 3}$

11. $\lim\limits_{x\to -5} \dfrac{x + 5}{x^2 - 25}$

12. $\lim\limits_{x\to 4} \dfrac{x - 4}{x^2 - 16}$

13. $\lim\limits_{x\to 1} \dfrac{x^2 + 5x - 6}{x - 1}$

14. $\lim\limits_{x\to 10} \dfrac{x^2 - 5x - 50}{x - 10}$

15. $\lim\limits_{x\to -9} \dfrac{x + 9}{x^2 + 13x + 36}$

16. $\lim\limits_{x\to -1} \dfrac{x + 1}{x^2 - 7x - 8}$

In Problems 17–24, even though the limit can be found using Theorem 4 of Section 10.2, use L'Hôpital's rule to find the limit.

17. $\lim\limits_{x\to\infty} \dfrac{2x + 3}{5x - 1}$

18. $\lim\limits_{x\to\infty} \dfrac{6x - 7}{7x - 6}$

19. $\lim\limits_{x\to\infty} \dfrac{3x^2 - 1}{x^3 + 4}$

20. $\lim\limits_{x\to\infty} \dfrac{5x^2 + 10x + 1}{x^4 + x^2 + 1}$

21. $\lim\limits_{x\to -\infty} \dfrac{x^2 - 9}{x - 3}$

22. $\lim\limits_{x\to -\infty} \dfrac{x^4 - 16}{x^2 + 4}$

23. $\lim\limits_{x\to\infty} \dfrac{2x^2 + 3x + 1}{3x^2 - 2x + 1}$

24. $\lim\limits_{x\to\infty} \dfrac{5 - 4x^3}{1 + 7x^3}$

In Problems 25–32, use L'Hôpital's rule to find the limit. Note that in these problems, neither algebraic simplification nor Theorem 4 of Section 10.2 provides an alternative to L'Hôpital's rule.

25. $\lim\limits_{x\to 0} \dfrac{e^x - 1}{2x}$

26. $\lim\limits_{x\to 0} \dfrac{5x}{e^x - 1}$

27. $\lim\limits_{x\to 1} \dfrac{x - 1}{\ln x}$

28. $\lim\limits_{x\to 1} \dfrac{x - 1}{\ln x^4}$

29. $\lim\limits_{x\to\infty} \dfrac{x^2}{e^x}$

30. $\lim\limits_{x\to\infty} \dfrac{x^3}{\ln x}$

31. $\lim\limits_{x\to 0} \dfrac{e^{4x} - 1}{x}$

32. $\lim\limits_{x\to 0} \dfrac{\ln(1 - 3x)}{x}$

✎ *In Problems 33–36, explain why L'Hôpital's rule does not apply. If the limit exists, find it by other means.*

33. $\lim\limits_{x\to 1} \dfrac{x^2 + 5x + 4}{x^3 + 1}$

34. $\lim\limits_{x\to\infty} \dfrac{e^{-x}}{\ln x}$

35. $\lim\limits_{x\to 2} \dfrac{x + 2}{(x - 2)^4}$

36. $\lim\limits_{x\to -3} \dfrac{x^2}{(x + 3)^5}$

Find each limit in Problems 37–60. Note that L'Hôpital's rule does not apply to every problem, and some problems will require more than one application of L'Hôpital's rule.

37. $\lim\limits_{x \to 0} \dfrac{e^{4x} - 1 - 4x}{x^2}$

38. $\lim\limits_{x \to 0} \dfrac{3x + 1 - e^{3x}}{x^2}$

39. $\lim\limits_{x \to 2} \dfrac{\ln(x - 1)}{x - 1}$

40. $\lim\limits_{x \to -1} \dfrac{\ln(x + 2)}{x + 2}$

41. $\lim\limits_{x \to 0^+} \dfrac{\ln(1 + x^2)}{x^3}$

42. $\lim\limits_{x \to 0^-} \dfrac{\ln(1 + 2x)}{x^2}$

43. $\lim\limits_{x \to 0^+} \dfrac{\ln(1 + \sqrt{x})}{x}$

44. $\lim\limits_{x \to 0^+} \dfrac{\ln(1 + x)}{\sqrt{x}}$

45. $\lim\limits_{x \to -2} \dfrac{x^2 + 2x + 1}{x^2 + x + 1}$

46. $\lim\limits_{x \to 1} \dfrac{2x^3 - 3x^2 + 1}{x^3 - 3x + 2}$

47. $\lim\limits_{x \to -1} \dfrac{x^3 + x^2 - x - 1}{x^3 + 4x^2 + 5x + 2}$

48. $\lim\limits_{x \to 3} \dfrac{x^3 + 3x^2 - x - 3}{x^2 + 6x + 9}$

49. $\lim\limits_{x \to 2} \dfrac{x^3 - 12x + 16}{x^3 - 6x^2 + 12x - 8}$

50. $\lim\limits_{x \to 1^+} \dfrac{x^3 + x^2 - x + 1}{x^3 + 3x^2 + 3x - 1}$

51. $\lim\limits_{x \to \infty} \dfrac{3x^2 + 5x}{4x^3 + 7}$

52. $\lim\limits_{x \to \infty} \dfrac{4x^2 + 9x}{5x^2 + 8}$

53. $\lim\limits_{x \to \infty} \dfrac{x^2}{e^{2x}}$

54. $\lim\limits_{x \to \infty} \dfrac{e^{3x}}{x^3}$

55. $\lim\limits_{x \to \infty} \dfrac{1 + e^{-x}}{1 + x^2}$

56. $\lim\limits_{x \to -\infty} \dfrac{1 + e^{-x}}{1 + x^2}$

57. $\lim\limits_{x \to \infty} \dfrac{e^{-x}}{\ln(1 + 4e^{-x})}$

58. $\lim\limits_{x \to \infty} \dfrac{\ln(1 + 2e^{-x})}{\ln(1 + e^{-x})}$

59. $\lim\limits_{x \to 0} \dfrac{e^x - e^{-x} - 2x}{x^3}$

60. $\lim\limits_{x \to 0} \dfrac{e^{2x} - 1 - 2x - 2x^2}{x^3}$

61. Find $\lim\limits_{x \to 0^+} (x \ln x)$.

   [*Hint*: Write $x \ln x = (\ln x)/x^{-1}$.]

62. Find $\lim\limits_{x \to 0^+} (\sqrt{x} \ln x)$.

   [*Hint*: Write $\sqrt{x} \ln x = (\ln x)/x^{-1/2}$.]

In Problems 63–66, n is a positive integer. Find each limit.

63. $\lim\limits_{x \to \infty} \dfrac{\ln x}{x^n}$

64. $\lim\limits_{x \to \infty} \dfrac{x^n}{\ln x}$

65. $\lim\limits_{x \to \infty} \dfrac{e^x}{x^n}$

66. $\lim\limits_{x \to \infty} \dfrac{x^n}{e^x}$

In Problems 67–70, show that the repeated application of L'Hôpital's rule does not lead to a solution. Then use algebraic manipulation to evaluate each limit. [Hint: If $x > 0$ and $n > 0$, then $\sqrt[n]{x^n} = x$.]

67. $\lim\limits_{x \to \infty} \dfrac{\sqrt{1 + x^2}}{x}$

68. $\lim\limits_{x \to -\infty} \dfrac{x}{\sqrt{4 + x^2}}$

69. $\lim\limits_{x \to -\infty} \dfrac{\sqrt[3]{x^3 + 1}}{x}$

70. $\lim\limits_{x \to \infty} \dfrac{x^2}{\sqrt[3]{(x^3 + 1)^2}}$

Answers to Matched Problems

1. (A) ∞ (B) ∞ (C) 0 (D) ∞

2. (A) ∞ (B) 0 (C) $-\infty$ (D) ∞

3. e^4 **4.** ∞ **5.** 0 **6.** 2

7. $-\infty$ **8.** 2 **9.** 0 **10.** ∞

12.4 Curve-Sketching Techniques

- Modifying the Graphing Strategy
- Using the Graphing Strategy
- Modeling Average Cost

When we summarized the graphing strategy in Section 12.2, we omitted one important topic: asymptotes. Polynomial functions do not have any asymptotes. Asymptotes of rational functions were discussed in Section 10.2, but what about all the other functions, such as logarithmic and exponential functions? Since investigating asymptotes always involves limits, we can now use L'Hôpital's rule (Section 12.3) as a tool for finding asymptotes of many different types of functions.

Modifying the Graphing Strategy

The first version of the graphing strategy in Section 12.2 made no mention of asymptotes. Including information about asymptotes produces the following (and final) version of the graphing strategy.

> **PROCEDURE** Graphing Strategy (Final Version)
>
> Step 1   *Analyze $f(x)$.*
>
>   (A) Find the domain of f.
>
>   (B) Find the intercepts.
>
>   (C) Find asymptotes.

Step 2 *Analyze $f'(x)$.* Find the partition numbers for f' and the critical numbers of f. Construct a sign chart for $f'(x)$, determine the intervals on which f is increasing and decreasing, and find local maxima and minima of f.

Step 3 *Analyze $f''(x)$.* Find the partition numbers for $f''(x)$. Construct a sign chart for $f''(x)$, determine the intervals on which the graph of f is concave upward and concave downward, and find the inflection points of f.

Step 4 *Sketch the graph of f.* Draw asymptotes and locate intercepts, local maxima and minima, and inflection points. Sketch in what you know from steps 1–3. Plot additional points as needed and complete the sketch.

Using the Graphing Strategy

We will illustrate the graphing strategy with several examples. From now on, you should always use the final version of the graphing strategy. If a function does not have any asymptotes, simply state this fact.

EXAMPLE 1 Using the Graphing Strategy Use the graphing strategy to analyze the function $f(x) = (x - 1)/(x - 2)$. State all the pertinent information and sketch the graph of f.

SOLUTION

Step 1 *Analyze $f(x)$.* $\quad f(x) = \dfrac{x - 1}{x - 2}$

(A) Domain: All real x, except $x = 2$

(B) y intercept: $\quad f(0) = \dfrac{0 - 1}{0 - 2} = \dfrac{1}{2}$

x intercepts: Since a fraction is 0 when its numerator is 0 and its denominator is not 0, the x intercept is $x = 1$.

(C) Horizontal asymptote: $\quad \dfrac{a_m x^m}{b_n x^n} = \dfrac{x}{x} = 1$

So the line $y = 1$ is a horizontal asymptote.
Vertical asymptote: The denominator is 0 for $x = 2$, and the numerator is not 0 for this value. Therefore, the line $x = 2$ is a vertical asymptote.

Step 2 *Analyze $f'(x)$.* $\quad f'(x) = \dfrac{(x - 2)(1) - (x - 1)(1)}{(x - 2)^2} = \dfrac{-1}{(x - 2)^2}$

Partition number for $f'(x)$: $\quad x = 2$
Critical numbers of $f(x)$: None
Sign chart for $f'(x)$:

	Test Numbers	
	x	$f'(x)$
	1	$-1 \ (-)$
	3	$-1 \ (-)$

So $f(x)$ is decreasing on $(-\infty, 2)$ and $(2, \infty)$. There are no local extrema.

Step 3 *Analyze $f''(x)$.* $\quad f''(x) = \dfrac{2}{(x - 2)^3}$

Partition number for $f''(x)$: $\quad x = 2$
Sign chart for $f''(x)$:

	$(-\infty, 2)$	$(2, \infty)$
$f''(x)$	$- - - -$ ND	$+ + + +$

Test Numbers	
x	$f''(x)$
1	-2 $(-)$
3	2 $(+)$

Graph of f — Concave downward | Concave upward

The graph of f is concave downward on $(-\infty, 2)$ and concave upward on $(2, \infty)$. Since $f(2)$ is not defined, there is no inflection point at $x = 2$, even though $f''(x)$ changes sign at $x = 2$.

Step 4 *Sketch the graph of f.* Insert intercepts and asymptotes, and plot a few additional points (for functions with asymptotes, plotting additional points is often helpful). Then sketch the graph.

x	$f(x)$
-2	$\frac{3}{4}$
0	$\frac{1}{2}$
1	0
$\frac{3}{2}$	-1
$\frac{5}{2}$	3
3	2
4	$\frac{3}{2}$

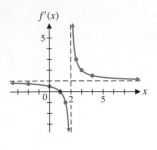

Matched Problem 1 Follow the graphing strategy and analyze the function $f(x) = 2x/(1 - x)$. State all the pertinent information and sketch the graph of f.

EXAMPLE 2 Using the Graphing Strategy   Use the graphing strategy to analyze the function

$$g(x) = \frac{2x - 1}{x^2}$$

State all pertinent information and sketch the graph of g.

SOLUTION

Step 1 *Analyze $g(x)$.*

(A) Domain: All real x, except $x = 0$

(B) x intercept: $x = \dfrac{1}{2} = 0.5$

y intercept: Since 0 is not in the domain of g, there is no y intercept.

(C) Horizontal asymptote: $y = 0$ (the x axis)

Vertical asymptote: The denominator of $g(x)$ is 0 at $x = 0$ and the numerator is not. So the line $x = 0$ (the y axis) is a vertical asymptote.

Step 2 *Analyze $g'(x)$.*

$$g(x) = \frac{2x - 1}{x^2} = \frac{2}{x} - \frac{1}{x^2} = 2x^{-1} - x^{-2}$$

$$g'(x) = -2x^{-2} + 2x^{-3} = -\frac{2}{x^2} + \frac{2}{x^3} = \frac{-2x + 2}{x^3}$$

$$= \frac{2(1 - x)}{x^3}$$

Partition numbers for $g'(x)$: $x = 0, x = 1$
Critical number of $g(x)$: $x = 1$
Sign chart for $g'(x)$:

	$(-\infty, 0)$	$(0, 1)$	$(1, \infty)$
$g'(x)$	$- - - -$	ND $+ + + +$	$0 \; - - - -$

Function $f(x)$ is decreasing on $(-\infty, 0)$ and $(1, \infty)$, is increasing on $(0, 1)$, and has a local maximum at $x = 1$. The local maximum is $g(1) = 1$.

Step 3 *Analyze $g''(x)$.*

$$g'(x) = -2x^{-2} + 2x^{-3}$$

$$g''(x) = 4x^{-3} - 6x^{-4} = \frac{4}{x^3} - \frac{6}{x^4} = \frac{4x - 6}{x^4} = \frac{2(2x - 3)}{x^4}$$

Partition numbers for $g''(x)$: $x = 0, x = \dfrac{3}{2} = 1.5$

Sign chart for $g''(x)$:

	$(-\infty, 0)$	$(0, 1.5)$	$(1.5, \infty)$
$g''(x)$	$- - - -$	ND $- - - - - -$	$0 \; + + + +$

Function $g(x)$ is concave downward on $(-\infty, 0)$ and $(0, 1.5)$, is concave upward on $(1.5, \infty)$, and has an inflection point at $x = 1.5$. Since $g(1.5) = 0.89$, the inflection point is $(1.5, 0.89)$.

Step 4 *Sketch the graph of g.* Plot key points, note that the coordinate axes are asymptotes, and sketch the graph.

x	$g(x)$
-10	-0.21
-1	-3
0.5	0
1	1
1.5	0.89
10	0.19

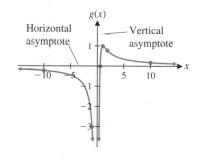

Matched Problem 2 Use the graphing strategy to analyze the function

$$h(x) = \frac{4x + 3}{x^2}$$

State all pertinent information and sketch the graph of h.

EXAMPLE 3 Graphing Strategy Follow the steps of the graphing strategy and analyze the function $f(x) = xe^x$. State all the pertinent information and sketch the graph of f.

SOLUTION

Step 1   *Analyze* $f(x)$: $f(x) = xe^x$.

(A)   Domain: All real numbers

(B) y intercept: $f(0) = 0$

x intercept: $xe^x = 0$ for $x = 0$ only, since $e^x > 0$ for all x.

(C)   Vertical asymptotes: None

Horizontal asymptotes: We use tables to determine the nature of the graph of f as $x \to \infty$ and $x \to -\infty$:

x	1	5	10	$\to \infty$
$f(x)$	2.72	742.07	220,264.66	$\to \infty$

x	-1	-5	-10	$\to -\infty$
$f(x)$	-0.37	-0.03	$-0.000\,45$	$\to 0$

Step 2   *Analyze* $f'(x)$:

$$f'(x) = x\frac{d}{dx}e^x + e^x\frac{d}{dx}x$$
$$= xe^x + e^x = e^x(x+1)$$

Partition number for $f'(x)$: -1
Critical number of $f(x)$: -1
Sign chart for $f'(x)$:

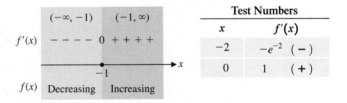

Test Numbers	
x	$f'(x)$
-2	$-e^{-2}$ $(-)$
0	1 $(+)$

So $f(x)$ decreases on $(-\infty, -1)$, has a local minimum at $x = -1$, and increases on $(-1, \infty)$. The local minimum is $f(-1) = -0.37$.

Step 3   *Analyze* $f''(x)$:

$$f''(x) = e^x\frac{d}{dx}(x+1) + (x+1)\frac{d}{dx}e^x$$
$$= e^x + (x+1)e^x = e^x(x+2)$$

Sign chart for $f''(x)$ (partition number is -2):

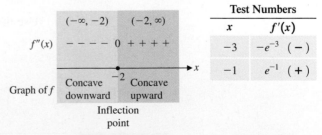

Test Numbers	
x	$f'(x)$
-3	$-e^{-3}$ $(-)$
-1	e^{-1} $(+)$

The graph of f is concave downward on $(-\infty, -2)$, has an inflection point at $x = -2$, and is concave upward on $(-2, \infty)$. Since $f(-2) = -0.27$, the inflection point is $(-2, -0.27)$.

Step 4   *Sketch the graph of f, using the information from steps 1 to 3:*

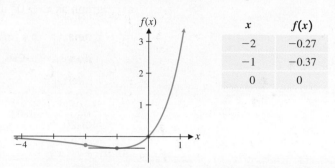

x	f(x)
−2	−0.27
−1	−0.37
0	0

<u>Matched Problem 3</u>   Analyze the function $f(x) = xe^{-0.5x}$. State all the pertinent information and sketch the graph of f.

Explore and Discuss 1   Refer to the discussion of asymptotes in the solution of Example 3. We used tables of values to estimate limits at infinity and determine horizontal asymptotes. In some cases, the functions involved in these limits can be written in a form that allows us to use L'Hôpital's rule.

$$\lim_{x \to -\infty} f(x) = \overset{-\infty \,\cdot\, 0 \text{ form}}{\lim_{x \to -\infty} xe^x} \qquad \text{Rewrite as a fraction.}$$

$$= \overset{-\infty/\infty \text{ form}}{\lim_{x \to -\infty} \frac{x}{e^{-x}}} \qquad \text{Apply l 'Hôpital's rule.}$$

$$= \lim_{x \to -\infty} \frac{1}{-e^{-x}} \qquad \text{Simplify.}$$

$$= \lim_{x \to -\infty} (-e^x) \qquad \text{Property of } e^x$$

$$= 0$$

Use algebraic manipulation and L'Hôpital's rule to verify the value of each of the following limits:

(A) $\lim_{x \to \infty} xe^{-0.5x} = 0$

(B) $\lim_{x \to 0^+} x^2(\ln x - 0.5) = 0$

(C) $\lim_{x \to 0^+} x \ln x = 0$

EXAMPLE 4   Graphing Strategy   Let $f(x) = x^2 \ln x - 0.5x^2$. Follow the steps in the graphing strategy and analyze this function. State all the pertinent information and sketch the graph of f.

SOLUTION

Step 1   *Analyze $f(x)$:* $f(x) = x^2 \ln x - 0.5x^2 = x^2(\ln x - 0.5)$.

(A) Domain: $(0, \infty)$

(B) y intercept: None [$f(0)$ is not defined.]

   x intercept: Solve $x^2(\ln x - 0.5) = 0$

   $\ln x - 0.5 = 0$ or $x^2 = 0$ Discard, since 0 is not in the domain of f.

   $\ln x = 0.5$ $\ln x = a$ if and only if $x = e^a$.

   $x = e^{0.5}$ x intercept

(C) Asymptotes: None. The following tables suggest the nature of the graph as $x \rightarrow 0^+$ and as $x \rightarrow \infty$:

x	0.1	0.01	0.001	$\rightarrow 0^+$
$f(x)$	-0.0280	-0.00051	-0.000007	$\rightarrow 0$

See Explore & Discuss 1(B).

x	10	100	1,000	$\rightarrow \infty$
$f(x)$	180	41,000	6,400,000	$\rightarrow \infty$

Step 2 *Analyze $f'(x)$:*

$$f'(x) = x^2 \frac{d}{dx} \ln x + (\ln x) \frac{d}{dx} x^2 - 0.5 \frac{d}{dx} x^2$$

$$= x^2 \frac{1}{x} + (\ln x) \, 2x - 0.5(2x)$$

$$= x + 2x \ln x - x$$

$$= 2x \ln x$$

Partition number for $f'(x)$: 1
Critical number of $f(x)$: 1
Sign chart for $f'(x)$:

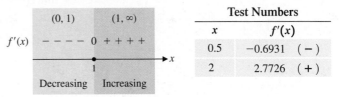

	Test Numbers	
x		$f'(x)$
0.5	-0.6931	$(-)$
2	2.7726	$(+)$

The function $f(x)$ decreases on $(0, 1)$, has a local minimum at $x = 1$, and increases on $(1, \infty)$. The local minimum is $f(1) = -0.5$.

Step 3 *Analyze $f''(x)$:*

$$f''(x) = 2x \frac{d}{dx}(\ln x) + (\ln x) \frac{d}{dx}(2x)$$

$$= 2x \frac{1}{x} + (\ln x) \, 2$$

$$= 2 + 2 \ln x = 0$$

$$2 \ln x = -2$$

$$\ln x = -1$$

$$x = e^{-1} \approx 0.3679$$

Sign chart for $f''(x)$ (partition number is e^{-1}):

	$(0, e^{-1})$	(e^{-1}, ∞)
$f''(x)$	$----$	$0 ++++$

	Test Numbers	
x		$f''(x)$
0.2	-1.2189	$(-)$
1	2	$(+)$

Concave downward · e^{-1} · Concave upward

The graph of $f(x)$ is concave downward on $(0, e^{-1})$, has an inflection point at $x = e^{-1}$, and is concave upward on (e^{-1}, ∞). Since $f(e^{-1}) = -1.5e^{-2} \approx -0.20$, the inflection point is $(0.37, -0.20)$.

Step 4 *Sketch the graph of f, using the information from steps 1 to 3:*

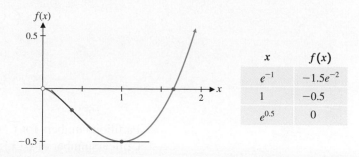

x	$f(x)$
e^{-1}	$-1.5e^{-2}$
1	-0.5
$e^{0.5}$	0

Matched Problem 4   Analyze the function $f(x) = x \ln x$. State all pertinent information and sketch the graph of f.

Modeling Average Cost

EXAMPLE 5   Average Cost   Given the cost function $C(x) = 5,000 + 0.5x^2$, where x is the number of items produced, use the graphing strategy to analyze the graph of the average cost function. State all the pertinent information and sketch the graph of the average cost function. Find the marginal cost function and graph it on the same set of coordinate axes.

SOLUTION   The average cost function is

$$\overline{C}(x) = \frac{5,000 + 0.5x^2}{x} = \frac{5,000}{x} + 0.5x$$

Step 1   *Analyze* $\overline{C}(x)$.

(A) Domain: Since negative values of x do not make sense and $\overline{C}(0)$ is not defined, the domain is the set of positive real numbers.

(B) Intercepts: None

(C) Horizontal asymptote: $\dfrac{a_m x^m}{b_n x^n} = \dfrac{0.5x^2}{x} = 0.5x$

So there is no horizontal asymptote.

Vertical asymptote: The line $x = 0$ is a vertical asymptote since the denominator is 0 and the numerator is not 0 for $x = 0$.

Oblique asymptotes: If a graph approaches a line that is neither horizontal nor vertical as x approaches ∞ or $-\infty$, then that line is called an **oblique asymptote**. If x is a large positive number, then $5,000/x$ is very small and

$$\overline{C}(x) = \frac{5,000}{x} + 0.5x \approx 0.5x$$

That is,

$$\lim_{x \to \infty}\left[\overline{C}(x) - 0.5x\right] = \lim_{x \to \infty} \frac{5,000}{x} = 0$$

This implies that the graph of $y = \overline{C}(x)$ approaches the line $y = 0.5x$ as x approaches ∞. That line is an oblique asymptote for the graph of $y = \overline{C}(x)$.*

----

*If $f(x) = n(x)/d(x)$ is a rational function for which the degree of $n(x)$ is 1 more than the degree of $d(x)$, then we can use polynomial long division to write $f(x) = mx + b + r(x)/d(x)$, where the degree of $r(x)$ is less than the degree of $d(x)$. The line $y = mx + b$ is then an oblique asymptote for the graph of $y = f(x)$.

Step 2   *Analyze* $\overline{C}'(x)$.

$$\overline{C}'(x) = -\frac{5,000}{x^2} + 0.5$$

$$= \frac{0.5x^2 - 5,000}{x^2}$$

$$= \frac{0.5(x - 100)(x + 100)}{x^2}$$

Partition numbers for $\overline{C}'(x)$:   0 and 100
Critical number of $\overline{C}(x)$:   100
Sign chart for $\overline{C}'(x)$:

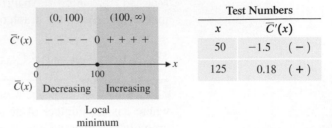

Test Numbers	
x	$\overline{C}'(x)$
50	-1.5 $(-)$
125	0.18 $(+)$

So $\overline{C}(x)$ is decreasing on (0, 100), is increasing on $(100, \infty)$, and has a local minimum at $x = 100$. The local minimum is $\overline{C}(100) = 100$.

Step 3   *Analyze* $\overline{C}''(x)$: $\overline{C}''(x) = \dfrac{10,000}{x^3}$.

$\overline{C}''(x)$ is positive for all positive x, so the graph of $y = \overline{C}(x)$ is concave upward on $(0, \infty)$.

Step 4   *Sketch the graph of* $\overline{C}$. The graph of $\overline{C}$ is shown in Figure 1.

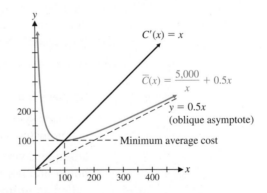

Figure 1

The marginal cost function is $C'(x) = x$. The graph of this linear function is also shown in Figure 1.

Figure 1 illustrates an important principle in economics:

> The minimum average cost occurs when the average cost is equal to the marginal cost.

Matched Problem 5   Given the cost function $C(x) = 1,600 + 0.25x^2$, where x is the number of items produced,

(A) Use the graphing strategy to analyze the graph of the average cost function. State all the pertinent information and sketch the graph of the average cost function. Find the marginal cost function and graph it on the same set of coordinate axes. Include any oblique asymptotes.

(B) Find the minimum average cost.

Exercises 12.4

Skills Warm-up Exercises

W   *In Problems 1–8, find the domain of the function and all x or y
     intercepts. (If necessary, review Section 2.1).*

1. $f(x) = 3x + 36$

2. $f(x) = -4x - 28$

3. $f(x) = \sqrt{25 - x}$

4. $f(x) = \sqrt{9 - x^2}$

5. $f(x) = \dfrac{x + 1}{x - 2}$

6. $f(x) = \dfrac{x^2 - 4}{x + 3}$

7. $f(x) = \dfrac{3}{x^2 - 1}$

8. $f(x) = \dfrac{x}{x^2 + 5x + 4}$

9. Use the graph of f in the figure to identify the following
(assume that $f''(0) < 0, f''(b) > 0$, and $f''(g) > 0$):

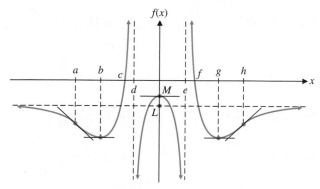

(A) the intervals on which $f'(x) < 0$

(B) the intervals on which $f'(x) > 0$

(C) the intervals on which $f(x)$ is increasing

(D) the intervals on which $f(x)$ is decreasing

(E) the x coordinate(s) of the point(s) where $f(x)$ has a local
maximum

(F) the x coordinate(s) of the point(s) where $f(x)$ has a local
minimum

(G) the intervals on which $f''(x) < 0$

(H) the intervals on which $f''(x) > 0$

(I) the intervals on which the graph of f is concave upward

(J) the intervals on which the graph of f is concave downward

(K) the x coordinate(s) of the inflection point(s)

(L) the horizontal asymptote(s)

(M) the vertical asymptote(s)

10. Repeat Problem 9 for the following graph of f (assume that
$f''(d) < 0$):

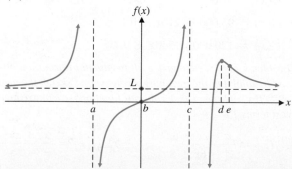

*In Problems 11–18, use the given information to sketch the graph
of f. Assume that f is continuous on its domain and that all intercepts are included in the table of values.*

11. Domain: All real x; $\lim\limits_{x \to \pm\infty} f(x) = 2$

x	-4	-2	0	2	4
$f(x)$	0	-2	0	-2	0

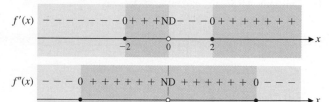

12. Domain: All real x; $\lim\limits_{x \to -\infty} f(x) = -3$; $\lim\limits_{x \to \infty} f(x) = 3$

x	-2	-1	0	1	2
$f(x)$	0	2	0	-2	0

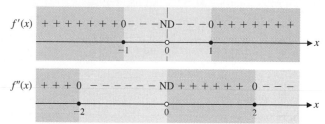

13. Domain: All real x, except $x = -2$;
$\lim\limits_{x \to -2^-} f(x) = \infty$; $\lim\limits_{x \to -2^+} f(x) = -\infty$; $\lim\limits_{x \to \infty} f(x) = 1$

x	4	0	4	6
$f(x)$	0	0	3	2

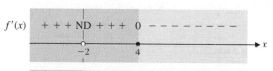

14. Domain: All real x, except $x = 1$;
$\lim\limits_{x \to 1^-} f(x) = \infty$; $\lim\limits_{x \to 1^+} f(x) = \infty$; $\lim\limits_{x \to \infty} f(x) = -2$

x	-4	-2	0	2
$f(x)$	0	-2	0	0

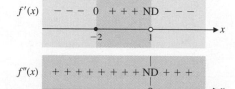

15. Domain: All real x, except $x = -1$;

$f(-3) = 2, f(-2) = 3, f(0) = -1, f(1) = 0$;

$f'(x) > 0$ on $(-\infty, -1)$ and $(-1, \infty)$;

$f''(x) > 0$ on $(-\infty, -1); f''(x) < 0$ on $(-1, \infty)$;

vertical asymptote: $x = -1$;

horizontal asymptote: $y = 1$

16. Domain: All real x, except $x = 1$;

$f(0) = -2, f(2) = 0$;

$f'(x) < 0$ on $(-\infty, 1)$ and $(1, \infty)$;

$f''(x) < 0$ on $(-\infty, 1)$;

$f''(x) > 0$ on $(1, \infty)$;

vertical asymptote: $x = 1$;

horizontal asymptote: $y = -1$

17. Domain: All real x, except $x = -2$ and $x = 2$;

$f(-3) = -1, f(0) = 0, f(3) = 1$;

$f'(x) < 0$ on $(-\infty, -2)$ and $(2, \infty)$;

$f'(x) > 0$ on $(-2, 2)$;

$f''(x) < 0$ on $(-\infty, -2)$ and $(-2, 0)$;

$f''(x) > 0$ on $(0, 2)$ and $(2, \infty)$;

vertical asymptotes: $x = -2$ and $x = 2$;

horizontal asymptote: $y = 0$

18. Domain: All real x, except $x = -1$ and $x = 1$;

$f(-2) = 1, f(0) = 0, f(2) = 1$;

$f'(x) > 0$ on $(-\infty, -1)$ and $(0, 1)$;

$f'(x) < 0$ on $(-1, 0)$ and $(1, \infty)$;

$f''(x) > 0$ on $(-\infty, -1), (-1, 1)$, and $(1, \infty)$;

vertical asymptotes: $x = -1$ and $x = 1$;

horizontal asymptote: $y = 0$

In Problems 19–58, summarize the pertinent information obtained by applying the graphing strategy and sketch the graph of $y = f(x)$.

19. $f(x) = \dfrac{x + 3}{x - 3}$

20. $f(x) = \dfrac{2x - 4}{x + 2}$

21. $f(x) = \dfrac{x}{x - 2}$

22. $f(x) = \dfrac{2 + x}{3 - x}$

23. $f(x) = 5 + 5e^{-0.1x}$

24. $f(x) = 3 + 7e^{-0.2x}$

25. $f(x) = 5xe^{-0.2x}$

26. $f(x) = 10xe^{-0.1x}$

27. $f(x) = \ln(1 - x)$

28. $f(x) = \ln(2x + 4)$

29. $f(x) = x - \ln x$

30. $f(x) = \ln(x^2 + 4)$

31. $f(x) = \dfrac{x}{x^2 - 4}$

32. $f(x) = \dfrac{1}{x^2 - 4}$

33. $f(x) = \dfrac{1}{1 + x^2}$

34. $f(x) = \dfrac{x^2}{1 + x^2}$

35. $f(x) = \dfrac{2x}{1 - x^2}$

36. $f(x) = \dfrac{2x}{x^2 - 9}$

37. $f(x) = \dfrac{-5x}{(x - 1)^2}$

38. $f(x) = \dfrac{x}{(x - 2)^2}$

39. $f(x) = \dfrac{x^2 + x - 2}{x^2}$

40. $f(x) = \dfrac{x^2 - 5x - 6}{x^2}$

41. $f(x) = \dfrac{x^2}{x - 1}$

42. $f(x) = \dfrac{x^2}{2 + x}$

43. $f(x) = \dfrac{3x^2 + 2}{x^2 - 9}$

44. $f(x) = \dfrac{2x^2 + 5}{4 - x^2}$

45. $f(x) = \dfrac{x^3}{x - 2}$

46. $f(x) = \dfrac{x^3}{4 - x}$

47. $f(x) = (3 - x)e^x$

48. $f(x) = (x - 2)e^x$

49. $f(x) = e^{-0.5x^2}$

50. $f(x) = e^{-2x^2}$

51. $f(x) = x^2 \ln x$

52. $f(x) = \dfrac{\ln x}{x}$

53. $f(x) = (\ln x)^2$

54. $f(x) = \dfrac{x}{\ln x}$

55. $f(x) = \dfrac{1}{x^2 + 2x - 8}$

56. $f(x) = \dfrac{1}{3 - 2x - x^2}$

57. $f(x) = \dfrac{x^3}{3 - x^2}$

58. $f(x) = \dfrac{x^3}{x^2 - 12}$

In Problems 59–66, show that the line $y = x$ is an oblique asymptote for the graph of $y = f(x)$, summarize all pertinent information obtained by applying the graphing strategy, and sketch the graph of $y = f(x)$.

59. $f(x) = x + \dfrac{4}{x}$

60. $f(x) = x - \dfrac{9}{x}$

61. $f(x) = x - \dfrac{4}{x^2}$

62. $f(x) = x + \dfrac{32}{x^2}$

63. $f(x) = x - \dfrac{9}{x^3}$

64. $f(x) = x + \dfrac{27}{x^3}$

65. $f(x) = x + \dfrac{1}{x} + \dfrac{4}{x^3}$

66. $f(x) = x - \dfrac{16}{x^3}$

In Problems 67–70, for the given cost function $C(x)$, find the oblique asymptote of the average cost function $\overline{C}(x)$.

67. $C(x) = 10{,}000 + 90x + 0.02x^2$

68. $C(x) = 7{,}500 + 65x + 0.01x^2$

69. $C(x) = 95{,}000 + 210x + 0.1x^2$

70. $C(x) = 120{,}000 + 340x + 0.4x^2$

In Problems 71–78, summarize all pertinent information obtained by applying the graphing strategy, and sketch the graph of $y = f(x)$. [Note: These rational functions are not reduced to lowest terms.]

71. $f(x) = \dfrac{x^2 + x - 6}{x^2 - 6x + 8}$

72. $f(x) = \dfrac{x^2 + x - 6}{x^2 - x - 12}$

73. $f(x) = \dfrac{2x^2 + x - 15}{x^2 - 9}$

74. $f(x) = \dfrac{2x^2 + 11x + 14}{x^2 - 4}$

75. $f(x) = \dfrac{x^3 - 5x^2 + 6x}{x^2 - x - 2}$

76. $f(x) = \dfrac{x^3 - 5x^2 - 6x}{x^2 + 3x + 2}$

77. $f(x) = \dfrac{x^2 + x - 2}{x^2 - 2x + 1}$

78. $f(x) - \dfrac{x^2 + x - 2}{x^2 + 4x + 4}$

Applications

79. Revenue. The marketing research department for a computer company used a large city to test market the firm's new laptop. The department found that the relationship between price p (dollars per unit) and demand x (units sold per week) was given approximately by

$$p = 1,296 - 0.12x^2 \qquad 0 \le x \le 80$$

So, weekly revenue can be approximated by

$$R(x) = xp = 1,296x - 0.12x^3 \qquad 0 \le x \le 80$$

Graph the revenue function R.

80. Profit. Suppose that the cost function $C(x)$ (in dollars) for the company in Problem 79 is

$$C(x) = 830 + 396x$$

(A) Write an equation for the profit $P(x)$.

(B) Graph the profit function P.

81. Pollution. In Silicon Valley, a number of computer firms were found to be contaminating underground water supplies with toxic chemicals stored in leaking underground containers. A water quality control agency ordered the companies to take immediate corrective action and contribute to a monetary pool for the testing and cleanup of the underground contamination. Suppose that the required monetary pool (in millions of dollars) is given by

$$P(x) = \dfrac{2x}{1 - x} \qquad 0 \le x < 1$$

where x is the percentage (expressed as a decimal fraction) of the total contaminant removed.

(A) Where is $P(x)$ increasing? Decreasing?

(B) Where is the graph of P concave upward? Downward?

(C) Find any horizontal and vertical asymptotes.

(D) Find the x and y intercepts.

(E) Sketch a graph of P.

82. Employee training. A company producing dive watches has established that, on average, a new employee can assemble $N(t)$ dive watches per day after t days of on-the-job training, as given by

$$N(t) = \dfrac{100t}{t + 9} \qquad t \ge 0$$

(A) Where is $N(t)$ increasing? Decreasing?

(B) Where is the graph of N concave upward? Downward?

(C) Find any horizontal and vertical asymptotes.

(D) Find the intercepts.

(E) Sketch a graph of N.

83. Replacement time. An outboard motor has an initial price of $3,200. A service contract costs $300 for the first year and increases $100 per year thereafter. The total cost of the outboard motor (in dollars) after n years is given by

$$C(n) = 3,200 + 250n + 50n^2 \qquad n \ge 1$$

(A) Write an expression for the average cost per year, $\overline{C}(n)$, for n years.

(B) Graph the average cost function found in part (A).

(C) When is the average cost per year at its minimum? (This time is frequently referred to as the **replacement time** for this piece of equipment.)

84. Construction costs. The management of a manufacturing plant wishes to add a fenced-in rectangular storage yard of 20,000 square feet, using a building as one side of the yard (see the figure). If x is the distance (in feet) from the building to the fence, show that the length of the fence required for the yard is given by

$$L(x) = 2x + \dfrac{20,000}{x} \qquad x > 0$$

Storage yard

(A) Graph L.

(B) What are the dimensions of the rectangle requiring the least amount of fencing?

85. Average and marginal costs. The total daily cost (in dollars) of producing x mountain bikes is given by

$$C(x) = 1,000 + 5x + 0.1x^2$$

(A) Sketch the graphs of the average cost function and the marginal cost function on the same set of coordinate axes. Include any oblique asymptotes.

(B) Find the minimum average cost.

86. Average and marginal costs. The total daily cost (in dollars) of producing x city bikes is given by

$$C(x) = 500 + 2x + 0.2x^2$$

(A) Sketch the graphs of the average cost function and the marginal cost function on the same set of coordinate axes. Include any oblique asymptotes.

(B) Find the minimum average cost.

87. Minimizing average costs. The table gives the total daily costs y (in dollars) of producing x pepperoni pizzas at various production levels.

Number of Pizzas	Total Costs
x	y
50	395
100	475
150	640
200	910
250	1,140
300	1,450

(A) Enter the data into a graphing calculator and find a quadratic regression equation for the total costs.

(B) Use the regression equation from part (A) to find the minimum average cost (to the nearest cent) and the corresponding production level (to the nearest integer).

88. Minimizing average costs. The table gives the total daily costs y (in dollars) of producing x deluxe pizzas at various production levels.

Number of Pizzas	Total Costs
x	y
50	595
100	755
150	1,110
200	1,380
250	1,875
300	2,410

(A) Enter the data into a graphing calculator and find a quadratic regression equation for the total costs.

(B) Use the regression equation from part (A) to find the minimum average cost (to the nearest cent) and the corresponding production level (to the nearest integer).

89. Medicine. A drug is injected into the bloodstream of a patient through her right arm. The drug concentration in the bloodstream of the left arm t hours after the injection is given by

$$C(t) = \frac{0.14t}{t^2 + 1}$$

Graph C.

90. Physiology. In a study on the speed of muscle contraction in frogs under various loads, researchers found that the speed of contraction decreases with increasing loads. More precisely, they found that the relationship between speed of contraction, S (in centimeters per second), and load w (in grams) is given approximately by

$$S(w) = \frac{26 + 0.06w}{w} \qquad w \geq 5$$

Graph S.

91. Psychology: retention. Each student in a psychology class is given one day to memorize the same list of 30 special characters. The lists are turned in at the end of the day, and for each succeeding day for 30 days, each student is asked to turn in a list of as many of the symbols as can be recalled. Averages are taken, and it is found that

$$N(t) = \frac{5t + 20}{t} \qquad t \geq 1$$

provides a good approximation of the average number $N(t)$ of symbols retained after t days. Graph N.

Answers to Matched Problems

1. Domain: All real x, except $x = 1$

   y intercept: $f(0) = 0$; x intercept: 0

   Horizontal asymptote: $y = -2$;

   Vertical asymptote: $x = 1$

   Increasing on $(-\infty, 1)$ and $(1, \infty)$

   Concave upward on $(-\infty, 1)$;

   Concave downward on $(1, \infty)$

x	$f(x)$
-1	-1
0	0
$\frac{1}{2}$	2
$\frac{3}{2}$	-6
2	-4
5	$-\frac{5}{2}$

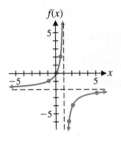

2. Domain: All real x, except $x = 0$

   x intercept: $= -\frac{3}{4} = -0.75$

   $h(0)$ is not defined

   Vertical asymptote: $x = 0$ (the y axis)

   Horizontal asymptote: $y = 0$ (the x axis)

   Increasing on $(-1.5, 0)$

   Decreasing on $(-\infty, -1.5)$ and $(0, \infty)$

   Local minimum: $f(-1.5) = -1.33$

   Concave upward on $(-2.25, 0)$ and $(0, \infty)$

   Concave downward on $(-\infty, -2.25)$

   Inflection point: $(-2.25, -1.19)$

x	$h(x)$
-10	-0.37
-2.25	-1.19
-1.5	-1.33
-0.75	0
2	2.75
10	0.43

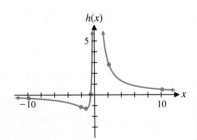

3. Domain: $(-\infty, \infty)$

   y intercept: $f(0) = 0$

   x intercept: $x = 0$

   Horizontal asymptote: $y = 0$ (the x axis)

   Increasing on $(-\infty, 2)$

   Decreasing on $(2, \infty)$

Local maximum: $f(2) = 2e^{-1} \approx 0.736$

Concave downward on $(-\infty, 4)$

Concave upward on $(4, \infty)$

Inflection point: $(4, 0.541)$

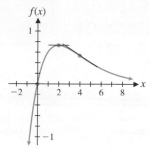

4. Domain: $(0, \infty)$

y intercept: None $[f(0)$ is not defined$]$

x intercept: $x = 1$

Increasing on (e^{-1}, ∞)

Decreasing on $(0, e^{-1})$

Local minimum: $f(e^{-1}) = -e^{-1} \approx -0.368$

Concave upward on $(0, \infty)$

x	5	10	100	$\rightarrow \infty$
$f(x)$	8.05	23.03	460.52	$\rightarrow \infty$

x	0.1	0.01	0.001	0.000 1	$\rightarrow 0$
$f(x)$	-0.23	-0.046	$-0.006\ 9$	$-0.000\ 92$	$\rightarrow 0$

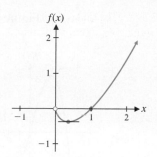

5. (A) Domain: $(0, \infty)$

Intercepts: None

Vertical asymptote: $x = 0$; oblique asymptote: $y = 0.25x$

Decreasing on $(0, 80)$; increasing on $(80, \infty)$;
local minimum at $x = 80$

Concave upward on $(0, \infty)$

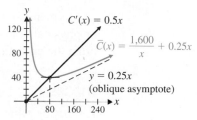

(B) Minimum average cost is 40 at $x = 80$.

12.5 Absolute Maxima and Minima

- Absolute Maxima and Minima
- Second Derivative and Extrema

One of the most important applications of the derivative is to find the absolute maximum or minimum value of a function. An economist may be interested in the price or production level of a commodity that will bring a maximum profit; a doctor may be interested in the time it takes for a drug to reach its maximum concentration in the bloodstream after an injection; and a city planner might be interested in the location of heavy industry in a city in order to produce minimum pollution in residential and business areas. In this section, we develop the procedures needed to find the absolute maximum and absolute minimum values of a function.

Absolute Maxima and Minima

Recall that $f(c)$ is a local maximum if $f(x) \leq f(c)$ for x near c and a local minimum if $f(x) \geq f(c)$ for x near c. Now we are interested in finding the largest and the smallest values of $f(x)$ throughout the domain of f.

> **DEFINITION** Absolute Maxima and Minima
>
> If $f(c) \geq f(x)$ for all x in the domain of f, then $f(c)$ is called the **absolute maximum** of f. If $f(c) \leq f(x)$ for all x in the domain of f, then $f(c)$ is called the **absolute minimum** of f. An absolute maximum or absolute minimum is called an **absolute extremum**.

Figure 1 illustrates some typical examples.

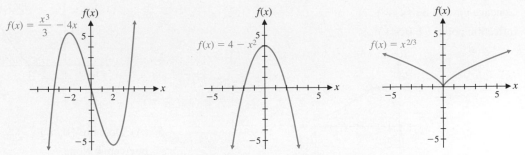

(A) No absolute maximum or minimum
$f(-2) = \frac{16}{3}$ is a local maximum
$f(2) = -\frac{16}{3}$ is a local minimum

(B) $f(0) = 4$ is the absolute maximum
No absolute minimum

(C) $f(0) = 0$ is the absolute minimum
No absolute maximum

Figure 1

In many applications, the domain of a function is restricted because of practical or physical considerations. If the domain is restricted to some closed interval, as is often the case, then Theorem 1 applies.

THEOREM 1 Extreme Value Theorem

A function f that is continuous on a closed interval $[a, b]$ has both an absolute maximum and an absolute minimum on that interval.

It is important to understand that the absolute maximum and absolute minimum depend on both the function f and the interval $[a, b]$. Figure 2 illustrates four cases.

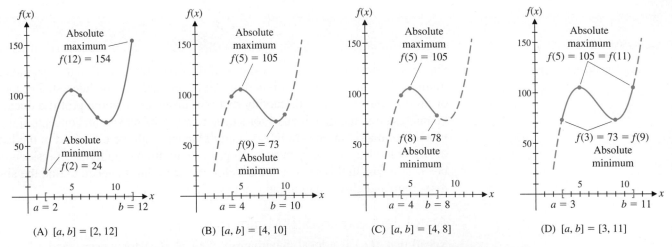

(A) $[a, b] = [2, 12]$

(B) $[a, b] = [4, 10]$

(C) $[a, b] = [4, 8]$

(D) $[a, b] = [3, 11]$

Figure 2 **Absolute extrema for $f(x) = x^3 - 21x^2 + 135x - 170$ on various closed intervals**

In all four cases illustrated in Figure 2, the absolute maximum and absolute minimum occur at a critical number or an endpoint. This property is generalized in Theorem 2. Note that both the absolute maximum and the absolute minimum are unique, but each can occur at more than one point in the interval (Fig. 2D).

THEOREM 2 Locating Absolute Extrema

Absolute extrema (if they exist) must occur at critical numbers or at endpoints.

To find the absolute maximum and minimum of a continuous function on a closed interval, we simply identify the endpoints and critical numbers in the interval, evaluate the function at each, and choose the largest and smallest values.

PROCEDURE Finding Absolute Extrema on a Closed Interval

Step 1 Check to make certain that f is continuous over $[a, b]$.

Step 2 Find the critical numbers in the interval (a, b).

Step 3 Evaluate f at the endpoints a and b and at the critical numbers found in step 2.

Step 4 The absolute maximum of f on $[a, b]$ is the largest value found in step 3.

Step 5 The absolute minimum of f on $[a, b]$ is the smallest value found in step 3.

EXAMPLE 1 Finding Absolute Extrema Find the absolute maximum and absolute minimum of

$$f(x) = x^3 + 3x^2 - 9x - 7$$

on each of the following intervals:

(A) $[-6, 4]$ (B) $[-4, 2]$ (C) $[-2, 2]$

SOLUTION

(A) The function is continuous for all values of x.

$$f'(x) = 3x^2 + 6x - 9 = 3(x - 1)(x + 3)$$

So, $x = -3$ and $x = 1$ are the critical numbers in the interval $(-6, 4)$. Evaluate f at the endpoints and critical numbers $(-6, -3, 1, \text{and } 4)$, and choose the largest and smallest values.

$$f(-6) = -61 \quad \text{Absolute minimum}$$
$$f(-3) = 20$$
$$f(1) = -12$$
$$f(4) = 69 \quad \text{Absolute maximum}$$

The absolute maximum of f on $[-6, 4]$ is 69, and the absolute minimum is -61.

(B) Interval: $[-4, 2]$

x	$f(x)$	
-4	13	
-3	20	Absolute maximum
1	-12	Absolute minimum
2	-5	

The absolute maximum of f on $[-4, 2]$ is 20, and the absolute minimum is -12.

(C) Interval: $[-2, 2]$

x	$f(x)$	
-2	15	Absolute maximum
1	-12	Absolute minimum
2	-5	

Note that the critical number $x = -3$ is not included in the table, because it is not in the interval $[-2, 2]$. The absolute maximum of f on $[-2, 2]$ is 15, and the absolute minimum is -12.

Matched Problem 1 | Find the absolute maximum and absolute minimum of

$$f(x) = x^3 - 12x$$

on each of the following intervals:

(A) $[-5, 5]$ (B) $[-3, 3]$ (C) $[-3, 1]$

Now, suppose that we want to find the absolute maximum or minimum of a function that is continuous on an interval that is not closed. Since Theorem 1 no longer applies, we cannot be certain that the absolute maximum or minimum value exists. Figure 3 illustrates several ways that functions can fail to have absolute extrema.

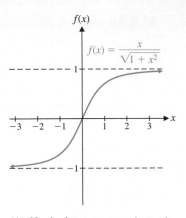

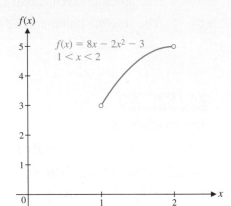

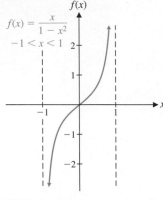

(A) No absolute extrema on $(-\infty, \infty)$:
$-1 < f(x) < 1$ for all x
$[f(x) \neq 1$ or -1 for any $x]$

(B) No absolute extrema on $(1, 2)$:
$3 < f(x) < 5$ for $x \in (1, 2)$
$[f(x) \neq 3$ or 5 for any $x \in (1, 2)]$

(C) No absolute extrema on $(-1, 1)$:
Graph has vertical asymptotes
at $x = -1$ and $x = 1$

Figure 3 **Functions with no absolute extrema**

In general, the best procedure to follow in searching for absolute extrema on an interval that is not of the form $[a, b]$ is to sketch a graph of the function. However, many applications can be solved with a new tool that does not require any graphing.

Second Derivative and Extrema

The second derivative can be used to classify the local extrema of a function. Suppose that f is a function satisfying $f'(c) = 0$ and $f''(c) > 0$. First, note that if $f''(c) > 0$, then it follows from the properties of limits* that $f''(x) > 0$ in some interval (m, n) containing c. Thus, the graph of f must be concave upward in this interval. But this implies that $f'(x)$ is increasing in the interval. Since $f'(c) = 0$, $f'(x)$ must change from negative to positive at $x = c$, and $f(c)$ is a local minimum (see Fig. 4). Reasoning in the same fashion, we conclude that if $f'(c) = 0$ and $f''(c) < 0$, then $f(c)$ is a local maximum. Of course, it is possible that both $f'(c) = 0$ and $f''(c) = 0$. In this case, the second derivative cannot be used to determine the shape of the graph around $x = c$; $f(c)$ may be a local minimum, a local maximum, or neither.

The sign of the second derivative provides a simple test for identifying local maxima and minima. This test is most useful when we do not want to draw the graph of the function. If we are interested in drawing the graph and have already constructed the sign chart for $f'(x)$, then the first-derivative test can be used to identify the local extrema.

---

*Actually, we are assuming that $f''(x)$ is continuous in an interval containing c. It is unlikely that we will encounter a function for which $f''(c)$ exists but $f''(x)$ is not continuous in an interval containing c.

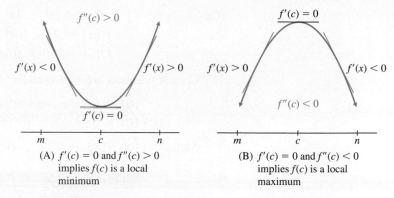

Figure 4 **Second derivative and local extrema**

RESULT Second-Derivative Test for Local Extrema

Let c be a critical number of $f(x)$ such that $f'(c) = 0$. If the second derivative $f''(c) > 0$, then $f(c)$ is a local minimum. If $f''(c) < 0$, then $f(c)$ is a local maximum.

$f'(c)$	$f''(c)$	Graph of f is:	$f(c)$	Example
0	+	Concave upward	Local minimum	⌣
0	−	Concave downward	Local maximum	⌢
0	0	?	Test does not apply	

EXAMPLE 2 Testing Local Extrema Find the local maxima and minima for each function. Use the second-derivative test for local extrema when it applies.

(A) $f(x) = x^3 - 6x^2 + 9x + 1$

(B) $f(x) = xe^{-0.2x}$

(C) $f(x) = \frac{1}{6}x^6 - 4x^5 + 25x^4$

SOLUTION

(A) Find first and second derivatives and determine critical numbers:

$$f(x) = x^3 - 6x^2 + 9x + 1$$
$$f'(x) = 3x^2 - 12x + 9 = 3(x - 1)(x - 3)$$
$$f''(x) = 6x - 12 = 6(x - 2)$$

Critical numbers are $x = 1$ and $x = 3$.

$$f''(1) = -6 < 0 \quad \text{\small f has a local maximum at $x = 1$.}$$
$$f''(3) = 6 > 0 \quad \text{\small f has a local minimum at $x = 3$.}$$

Substituting $x = 1$ in the expression for $f(x)$, we find that $f(1) = 5$ is a local maximum. Similarly, $f(3) = 1$ is a local minimum.

(B)
$$f(x) = xe^{-0.2x}$$
$$f'(x) = e^{-0.2x} + xe^{-0.2x}(-0.2)$$
$$= e^{-0.2x}(1 - 0.2x)$$
$$f''(x) = e^{-0.2x}(-0.2)(1 - 0.2x) + e^{-0.2x}(-0.2)$$
$$= e^{-0.2x}(0.04x - 0.4)$$

Critical number: $x = 1/0.2 = 5$

$$f''(5) = e^{-1}(-0.2) < 0 \quad \text{\small f has a local maximum at $x = 5$.}$$

So $f(5) = 5e^{-0.2(5)} \approx 1.84$ is a local maximum.

(C)
$$f(x) = \tfrac{1}{6}x^6 - 4x^5 + 25x^4$$
$$f'(x) = x^5 - 20x^4 + 100x^3 = x^3(x - 10)^2$$
$$f''(x) = 5x^4 - 80x^3 + 300x^2$$

Critical numbers are $x = 0$ and $x = 10$.

$$f''(0) = 0$$ The second-derivative test fails at both critical numbers, so
$$f''(10) = 0$$ the first-derivative test must be used.

Sign chart for $f'(x) = x^3(x - 10)^2$ (partition numbers for f' are 0 and 10):

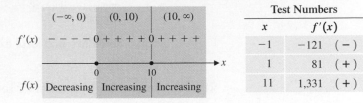

Test Numbers	
x	$f'(x)$
-1	-121 $(-)$
1	81 $(+)$
11	$1{,}331$ $(+)$

From the chart, we see that $f(x)$ has a local minimum at $x = 0$ and does not have a local extremum at $x = 10$. So $f(0) = 0$ is a local minimum.

Matched Problem 2 Find the local maxima and minima for each function. Use the second-derivative test when it applies.

(A) $f(x) = x^3 - 9x^2 + 24x - 10$

(B) $f(x) = e^x - 5x$

(C) $f(x) = 10x^6 - 24x^5 + 15x^4$

CONCEPTUAL INSIGHT

The second-derivative test for local extrema does not apply if $f''(c) = 0$ or if $f''(c)$ is not defined. As Example 2C illustrates, if $f''(c) = 0$, then $f(c)$ may or may not be a local extremum. Some other method, such as the first-derivative test, must be used when $f''(c) = 0$ or $f''(c)$ does not exist.

The solution of many optimization problems involves searching for an absolute extremum. If the function in question has only one critical number, then the second-derivative test for local extrema not only classifies the local extremum but also guarantees that the local extremum is, in fact, the absolute extremum.

THEOREM 3 Second-Derivative Test for Absolute Extrema on an Open Interval

Let f be continuous on an open interval I with only one critical number c in I.

If $f'(c) = 0$ and $f''(c) > 0$, then $f(c)$ is the absolute minimum of f on I.

If $f'(c) = 0$ and $f''(c) < 0$, then $f(c)$ is the absolute maximum of f on I.

Since the second-derivative test for local extrema cannot be applied when $f''(c) = 0$ or $f''(c)$ does not exist, Theorem 3 makes no mention of these cases.

EXAMPLE 3 Finding Absolute Extrema on an Open Interval Find the absolute extrema of each function on $(0, \infty)$.

(A) $f(x) = x + \dfrac{4}{x}$ (B) $f(x) = (\ln x)^2 - 3 \ln x$

SOLUTION

(A) $f(x) = x + \dfrac{4}{x}$

$f'(x) = 1 - \dfrac{4}{x^2} = \dfrac{x^2 - 4}{x^2} = \dfrac{(x-2)(x+2)}{x^2}$ Critical numbers are $x = -2$ and $x = 2$.

$f''(x) = \dfrac{8}{x^3}$

The only critical number in the interval $(0, \infty)$ is $x = 2$. Since $f''(2) = 1 > 0$, $f(2) = 4$ is the absolute minimum of f on $(0, \infty)$.

(B) $f(x) = (\ln x)^2 - 3 \ln x$

$f'(x) = (2 \ln x)\dfrac{1}{x} - \dfrac{3}{x} = \dfrac{2 \ln x - 3}{x}$ Critical number is $x = e^{3/2}$.

$f''(x) = \dfrac{x\dfrac{2}{x} - (2 \ln x - 3)}{x^2} = \dfrac{5 - 2 \ln x}{x^2}$

The only critical number in the interval $(0, \infty)$ is $x = e^{3/2}$. Since $f''(e^{3/2}) = 2/e^3 > 0, f(e^{3/2}) = -2.25$ is the absolute minimum of f on $(0, \infty)$.

Matched Problem 3 Find the absolute extrema of each function on $(0, \infty)$.

(A) $f(x) = 12 - x - \dfrac{5}{x}$

(B) $f(x) = 5 \ln x - x$

Exercises 12.5

Skills Warm-up Exercises

W

In Problems 1–8, by inspecting the graph of the function, find the absolute maximum and absolute minimum on the given interval. (If necessary, review Section 2.2).

1. $f(x) = x$ on $[-2, 3]$

2. $g(x) = |x|$ on $[-1, 4]$

3. $h(x) = x^2$ on $[-5, 3]$

4. $m(x) = x^3$ on $[-3, 1]$

5. $n(x) = \sqrt{x}$ on $[3, 4]$

6. $p(x) = \sqrt[3]{x}$ on $[-125, 216]$

7. $q(x) = -\sqrt[3]{x}$ on $[27, 64]$

8. $r(x) = -x^2$ on $[-10, 11]$

Problems 9–18 refer to the graph of $y = f(x)$ shown here. Find the absolute minimum and the absolute maximum over the indicated interval.

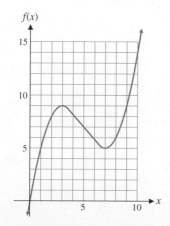

9. $[0, 10]$ **10.** $[2, 8]$ **11.** $[0, 8]$ **12.** $[2, 10]$

13. $[1, 10]$ **14.** $[0, 9]$ **15.** $[1, 9]$ **16.** $[0, 2]$

17. $[2, 5]$ **18.** $[5, 8]$

In Problems 19–22, find the absolute maximum and absolute minimum of each function on the indicated intervals.

19. $f(x) = 2x - 5$

(A) $[0, 4]$ (B) $[0, 10]$ (C) $[-5, 10]$

20. $f(x) = 8 - x$

(A) $[0, 1]$ (B) $[-1, 1]$ (C) $[-1, 6]$

21. $f(x) = x^2$

(A) $[-1, 1]$ (B) $[1, 5]$ (C) $[-5, 5]$

22. $f(x) = 100 - x^2$

(A) $[-10, 10]$ (B) $[0, 10]$ (C) $[10, 11]$

In Problems 23–26, find the absolute maximum and absolute minimum of each function on the given interval.

23. $f(x) = e^{-x}$ on $[-1, 1]$

24. $f(x) = \ln x$ on $[1, 2]$

25. $f(x) = 9 - x^2$ on $[-4, 4]$

26. $f(x) = x^2 - 6x + 7$ on $[0, 10]$

In Problems 27–42, find the absolute maximum and minimum, if either exists, for each function.

27. $f(x) = x^2 - 2x + 3$

28. $f(x) = x^2 + 4x - 3$

29. $f(x) = -x^2 - 6x + 9$

30. $f(x) = -x^2 + 2x + 4$

31. $f(x) = x^3 + x$

32. $f(x) = -x^3 - 2x$

33. $f(x) = 8x^3 - 2x^4$

34. $f(x) = x^4 - 4x^3$

35. $f(x) = x + \dfrac{16}{x}$

36. $f(x) = x + \dfrac{25}{x}$

37. $f(x) = \dfrac{x^2}{x^2 + 1}$

38. $f(x) = \dfrac{1}{x^2 + 1}$

39. $f(x) = \dfrac{2x}{x^2 + 1}$

40. $f(x) = \dfrac{-8x}{x^2 + 4}$

41. $f(x) = \dfrac{x^2 - 1}{x^2 + 1}$

42. $f(x) = \dfrac{9 - x^2}{x^2 + 4}$

In Problems 43–66, find the indicated extremum of each function on the given interval.

43. Absolute minimum value on $[0, \infty)$ for
$$f(x) = 2x^2 - 8x + 6$$

44. Absolute maximum value on $[0, \infty)$ for
$$f(x) = 6x - x^2 + 4$$

45. Absolute maximum value on $[0, \infty)$ for
$$f(x) = 3x^2 - x^3$$

46. Absolute minimum value on $[0, \infty)$ for
$$f(x) = x^3 - 6x^2$$

47. Absolute minimum value on $[0, \infty)$ for
$$f(x) = (x + 4)(x - 2)^2$$

48. Absolute minimum value on $[0, \infty)$ for
$$f(x) = (2 - x)(x + 1)^2$$

49. Absolute maximum value on $(0, \infty)$ for
$$f(x) = 2x^4 - 8x^3$$

50. Absolute maximum value on $(0, \infty)$ for
$$f(x) = 4x^3 - 8x^4$$

51. Absolute maximum value on $(0, \infty)$ for
$$f(x) = 20 - 3x - \dfrac{12}{x}$$

52. Absolute minimum value on $(0, \infty)$ for
$$f(x) = 4 + x + \dfrac{9}{x}$$

53. Absolute minimum value on $(0, \infty)$ for
$$f(x) = 10 + 2x + \dfrac{64}{x^2}$$

54. Absolute maximum value on $(0, \infty)$ for
$$f(x) = 20 - 4x - \dfrac{250}{x^2}$$

55. Absolute minimum value on $(0, \infty)$ for
$$f(x) = x + \dfrac{1}{x} + \dfrac{30}{x^3}$$

56. Absolute minimum value on $(0, \infty)$ for
$$f(x) = 2x + \dfrac{5}{x} + \dfrac{4}{x^3}$$

57. Absolute minimum value on $(0, \infty)$ for
$$f(x) = \dfrac{e^x}{x^2}$$

58. Absolute maximum value on $(0, \infty)$ for
$$f(x) = \dfrac{x^4}{e^x}$$

59. Absolute maximum value on $(0, \infty)$ for
$$f(x) = \dfrac{x^3}{e^x}$$

60. Absolute minimum value on $(0, \infty)$ for
$$f(x) = \dfrac{e^x}{x}$$

61. Absolute maximum value on $(0, \infty)$ for
$$f(x) = 5x - 2x \ln x$$

62. Absolute minimum value on $(0, \infty)$ for
$$f(x) = 4x \ln x - 7x$$

63. Absolute maximum value on $(0, \infty)$ for
$$f(x) = x^2(3 - \ln x)$$

64. Absolute minimum value on $(0, \infty)$ for
$$f(x) = x^3(\ln x - 2)$$

65. Absolute maximum value on $(0, \infty)$ for
$$f(x) = \ln(xe^{-x})$$

66. Absolute maximum value on $(0, \infty)$ for
$$f(x) = \ln(x^2 e^{-x})$$

In Problems 67–72, find the absolute maximum and minimum, if either exists, for each function on the indicated intervals.

67. $f(x) = x^3 - 6x^2 + 9x - 6$

(A) $[-1, 5]$ (B) $[-1, 3]$ (C) $[2, 5]$

68. $f(x) = 2x^3 - 3x^2 - 12x + 24$

(A) $[-3, 4]$ (B) $[-2, 3]$ (C) $[-2, 1]$

69. $f(x) = (x - 1)(x - 5)^3 + 1$

(A) $[0, 3]$ (B) $[1, 7]$ (C) $[3, 6]$

70. $f(x) = x^4 - 8x^2 + 16$

(A) $[-1, 3]$ (B) $[0, 2]$ (C) $[-3, 4]$

71. $f(x) = x^4 - 4x^3 + 5$

(A) $[-1, 2]$ (B) $[0, 4]$ (C) $[-1, 1]$

72. $f(x) = x^4 - 18x^2 + 32$

(A) $[-4, 4]$ (B) $[-1, 1]$ (C) $[1, 3]$

In Problems 73–80, describe the graph of f at the given point relative to the existence of a local maximum or minimum with one of the following phrases: "Local maximum," "Local minimum," "Neither," or "Unable to determine from the given information." Assume that f(x) is continuous on $(-\infty, \infty)$.

73. $(2, f(2))$ if $f'(2) = 0$ and $f''(2) > 0$

74. $(4, f(4))$ if $f'(4) = 1$ and $f''(4) < 0$

75. $(-3, f(-3))$ if $f'(-3) = 0$ and $f''(-3) = 0$

76. $(-1, f(-1))$ if $f'(-1) = 0$ and $f''(-1) < 0$

77. $(6, f(6))$ if $f'(6) = 1$ and $f''(6)$ does not exist

78. $(5, f(5))$ if $f'(5) = 0$ and $f''(5)$ does not exist

79. $(-2, f(-2))$ if $f'(-2) = 0$ and $f''(-2) < 0$

80. $(1, f(1))$ if $f'(1) = 0$ and $f''(1) > 0$

Answers to Matched Problems

1. (A) Absolute maximum: $f(5) = 65$; absolute minimum: $f(-5) = -65$

(B) Absolute maximum: $f(-2) = 16$; absolute minimum: $f(2) = -16$

(C) Absolute maximum: $f(-2) = 16$; absolute minimum: $f(1) = -11$

2. (A) $f(2) = 10$ is a local maximum; $f(4) = 6$ is a local minimum.

(B) $f(\ln 5) = 5 - 5 \ln 5$ is a local minimum.

(C) $f(0) = 0$ is a local minimum; there is no local extremum at $x = 1$.

3. (A) $f(\sqrt{5}) = 12 - 2\sqrt{5}$ (B) $f(5) = 5 \ln 5 - 5$

12.6 Optimization

- Area and Perimeter
- Maximizing Revenue and Profit
- Inventory Control

Now we can use calculus to solve **optimization problems**—problems that involve finding the absolute maximum or the absolute minimum of a function. As you work through this section, note that the statement of the problem does not usually include the function to be optimized. Often, it is your responsibility to find the function and then to find the relevant absolute extremum.

Area and Perimeter

The techniques used to solve optimization problems are best illustrated through examples.

EXAMPLE 1   Maximizing Area   A homeowner has $320 to spend on building a fence around a rectangular garden. Three sides of the fence will be constructed with wire fencing at a cost of $2 per linear foot. The fourth side will be constructed with wood fencing at a cost of $6 per linear foot. Find the dimensions and the area of the largest garden that can be enclosed with $320 worth of fencing.

SOLUTION   To begin, we draw a figure (Fig. 1), introduce variables, and look for relationships among the variables.

Since we don't know the dimensions of the garden, the lengths of fencing are represented by the variables x and y. The costs of the fencing materials are fixed and are represented by constants.

Now we look for relationships among the variables. The area of the garden is

$$A = xy$$

while the cost of the fencing is

$$C = 2y + 2x + 2y + 6x$$
$$= 8x + 4y$$

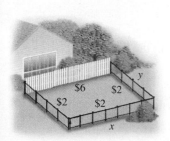

Figure 1

The problem states that the homeowner has $320 to spend on fencing. We assume that enclosing the largest area will use all the money available for fencing. The problem has now been reduced to

Maximize $A = xy$ subject to $8x + 4y = 320$

Before we can use calculus to find the maximum area A, we must express A as a function of a single variable. We use the cost equation to eliminate one of the variables in the area expression (we choose to eliminate y—either will work).

$$8x + 4y = 320$$
$$4y = 320 - 8x$$
$$y = 80 - 2x$$
$$A = xy = x(80 - 2x) = 80x - 2x^2$$

Now we consider the permissible values of x. Because x is one of the dimensions of a rectangle, x must satisfy

$$x \geq 0 \quad \text{Length is always nonnegative.}$$

And because $y = 80 - 2x$ is also a dimension of a rectangle, y must satisfy

$$y = 80 - 2x \geq 0 \quad \text{Width is always nonnegative.}$$
$$80 \geq 2x$$
$$40 \geq x \quad \text{or} \quad x \leq 40$$

We summarize the preceding discussion by stating the following model for this optimization problem:

$$\text{Maximize} \quad A(x) = 80x - 2x^2 \quad \text{for } 0 \leq x \leq 40$$

Next, we find any critical numbers of A:

$$A'(x) = 80 - 4x = 0$$
$$80 = 4x$$
$$x = \frac{80}{4} = 20 \quad \text{Critical number}$$

Since $A(x)$ is continuous on $[0, 40]$, the absolute maximum of A, if it exists, must occur at a critical number or an endpoint. Evaluating A at these numbers (Table 1), we see that the maximum area is 800 when

$$x = 20 \quad \text{and} \quad y = 80 - 2(20) = 40$$

Finally, we must answer the questions posed in the problem. The dimensions of the garden with the maximum area of 800 square feet are 20 feet by 40 feet, with one 20-foot side of wood fencing.

Matched Problem 1 Repeat Example 1 if the wood fencing costs $8 per linear foot and all other information remains the same.

Table 1

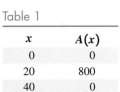

x	$A(x)$
0	0
20	800
40	0

We summarize the steps in the solution of Example 1 in the following box:

PROCEDURE Strategy for Solving Optimization Problems

Step 1 Introduce variables, look for relationships among the variables, and construct a mathematical model of the form

Maximize (or minimize) $f(x)$ on the interval I

Step 2 Find the critical numbers of $f(x)$.

Step 3 Use the procedures developed in Section 12.5 to find the absolute maximum (or minimum) of $f(x)$ on the interval I and the numbers x where this occurs.

Step 4 Use the solution to the mathematical model to answer all the questions asked in the problem.

EXAMPLE 2 Minimizing Perimeter Refer to Example 1. The homeowner judges that an area of 800 square feet for the garden is too small and decides to increase the area to 1,250 square feet. What is the minimum cost of building a fence that will enclose a garden with an area of 1,250 square feet? What are the dimensions of this garden? Assume that the cost of fencing remains unchanged.

SOLUTION Refer to Figure 1 and the solution of Example 1. This time we want to minimize the cost of the fencing that will enclose 1,250 square feet. The problem can be expressed as

$$\text{Minimize} \quad C = 8x + 4y \quad \text{subject to} \quad xy = 1{,}250$$

Since x and y represent distances, we know that $x \geq 0$ and $y \geq 0$. But neither variable can equal 0 because their product must be 1,250.

$$xy = 1{,}250 \qquad\qquad \text{Solve the area equation for } y.$$

$$y = \frac{1{,}250}{x}$$

$$C(x) = 8x + 4\frac{1{,}250}{x} \qquad\qquad \text{Substitute for } y \text{ in the cost equation.}$$

$$= 8x + \frac{5{,}000}{x} \qquad x > 0$$

The model for this problem is

$$\text{Minimize} \quad C(x) = 8x + \frac{5{,}000}{x} \qquad \text{for } x > 0$$

$$= 8x + 5{,}000x^{-1}$$

$$C'(x) = 8 - 5{,}000x^{-2}$$

$$= 8 - \frac{5{,}000}{x^2} = 0$$

$$8 = \frac{5{,}000}{x^2}$$

$$x^2 = \frac{5{,}000}{8} = 625$$

$$x = \sqrt{625} = 25 \quad \begin{array}{l}\text{The negative square} \\ \text{root is discarded,} \\ \text{since } x > 0.\end{array}$$

We use the second derivative to determine the behavior at $x = 25$.

$$C'(x) = 8 - 5{,}000x^{-2}$$

$$C''(x) = 0 + 10{,}000x^{-3} = \frac{10{,}000}{x^3}$$

$$C''(25) = \frac{10{,}000}{25^3} = 0.64 > 0$$

The second-derivative test for local extrema shows that $C(x)$ has a local minimum at $x = 25$, and since $x = 25$ is the only critical number of $C(x)$ for $x > 0$, then $C(25)$ must be the absolute minimum for $x > 0$. When $x = 25$, the cost is

$$C(25) = 8(25) + \frac{5{,}000}{25} = 200 + 200 = \$400$$

and

$$y = \frac{1{,}250}{25} = 50$$

The minimum cost for enclosing a 1,250-square-foot garden is $400, and the dimensions are 25 feet by 50 feet, with one 25-foot side of wood fencing.

Matched Problem 2 Repeat Example 2 if the homeowner wants to enclose an 1,800-square-foot garden and all other data remain unchanged.

CONCEPTUAL INSIGHT

The restrictions on the variables in the solutions of Examples 1 and 2 are typical of problems involving areas or perimeters (or the cost of the perimeter):

$$8x + 4y = 320 \quad \text{Cost of fencing (Example 1)}$$
$$xy = 1,250 \quad \text{Area of garden (Example 2)}$$

The equation in Example 1 restricts the values of x to

$$0 \leq x \leq 40 \quad \text{or} \quad [0, 40]$$

The endpoints are included in the interval for our convenience (a closed interval is easier to work with than an open one). The area function is defined at each endpoint, so it does no harm to include them.

The equation in Example 2 restricts the values of x to

$$x > 0 \quad \text{or} \quad (0, \infty)$$

Neither endpoint can be included in this interval. We cannot include 0 because the area is not defined when $x = 0$, and we can never include ∞ as an endpoint. Remember, ∞ is not a number; it is a symbol that indicates the interval is unbounded.

Maximizing Revenue and Profit

EXAMPLE 3   Maximizing Revenue   An office supply company sells x permanent markers per year at $\$p$ per marker. The price–demand equation for these markers is $p = 10 - 0.001x$. What price should the company charge for the markers to maximize revenue? What is the maximum revenue?

SOLUTION

$$\text{Revenue} = \text{price} \times \text{demand}$$
$$R(x) = (10 - 0.001x)x$$
$$= 10x - 0.001x^2$$

Both price and demand must be nonnegative, so

$$x \geq 0 \quad \text{and} \quad p = 10 - 0.001x \geq 0$$
$$10 \geq 0.001x$$
$$10,000 \geq x$$

The mathematical model for this problem is

$$\text{Maximize} \quad R(x) = 10x - 0.001x^2 \quad 0 \leq x \leq 10,000$$
$$R'(x) = 10 - 0.002x$$
$$10 - 0.002x = 0$$
$$10 = 0.002x$$
$$x = \frac{10}{0.002} = 5,000 \qquad \text{Critical number}$$

Use the second-derivative test for absolute extrema:

$$R''(x) = -0.002 < 0 \quad \text{for all } x$$
$$\text{Max } R(x) = R(5,000) = \$25,000$$

When the demand is $x = 5,000$, the price is

$$10 - 0.001(5,000) = \$5 \quad p = 10 - 0.001x$$

The company will realize a maximum revenue of $25,000 when the price of a marker is $5.

Matched Problem 3) An office supply company sells x heavy-duty paper shredders per year at $\$p$ per shredder. The price–demand equation for these shredders is

$$p = 300 - \frac{x}{30}$$

What price should the company charge for the shredders to maximize revenue? What is the maximum revenue?

EXAMPLE 4 Maximizing Profit The total annual cost of manufacturing x permanent markers for the office supply company in Example 3 is

$$C(x) = 5,000 + 2x$$

What is the company's maximum profit? What should the company charge for each marker, and how many markers should be produced?

SOLUTION Using the revenue model in Example 3, we have

$$\text{Profit} = \text{Revenue} - \text{Cost}$$
$$P(x) = R(x) - C(x)$$
$$= 10x - 0.001x^2 - 5,000 - 2x$$
$$= 8x - 0.001x^2 - 5,000$$

The mathematical model for profit is

$$\text{Maximize} \quad P(x) = 8x - 0.001x^2 - 5,000 \qquad 0 \leq x \leq 10,000$$

The restrictions on x come from the revenue model in Example 3.

$$P'(x) = 8 - 0.002x = 0$$
$$8 = 0.002x$$
$$x = \frac{8}{0.002} = 4,000 \quad \text{Critical number}$$
$$P''(x) = -0.002 < 0 \quad \text{for all } x$$

Since $x = 4,000$ is the only critical number and $P''(x) < 0$,

$$\text{Max } P(x) = P(4,000) = \$11,000$$

Using the price–demand equation from Example 3 with $x = 4,000$, we find that

$$p = 10 - 0.001(4,000) = \$6 \quad p = 10 - 0.001x$$

A maximum profit of $11,000 is realized when 4,000 markers are manufactured annually and sold for $6 each.

The results in Examples 3 and 4 are illustrated in Figure 2.

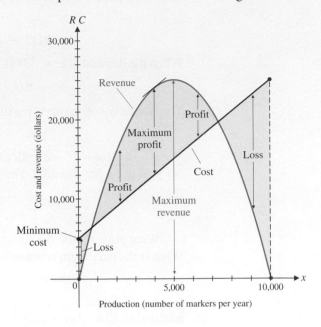

Figure 2

CONCEPTUAL INSIGHT

In Figure 2, notice that the maximum revenue and the maximum profit occur at different production levels. The maximum profit occurs when

$$P'(x) = R'(x) - C'(x) = 0$$

that is, when the marginal revenue is equal to the marginal cost. Notice that the slopes of the revenue function and the cost function are the same at this production level.

Matched Problem 4 The annual cost of manufacturing x paper shredders for the office supply company in Matched Problem 3 is $C(x) = 90{,}000 + 30x$. What is the company's maximum profit? What should it charge for each shredder, and how many shredders should it produce?

EXAMPLE 5 Maximizing Profit The government decides to tax the company in Example 4 $2 for each marker produced. Taking into account this additional cost, how many markers should the company manufacture annually to maximize its profit? What is the maximum profit? How much should the company charge for the markers to realize the maximum profit?

SOLUTION The tax of $2 per unit changes the company's cost equation:

$$C(x) = \text{original cost} + \text{tax}$$
$$= 5{,}000 + 2x + 2x$$
$$= 5{,}000 + 4x$$

The new profit function is

$$P(x) = R(x) - C(x)$$
$$= 10x - 0.001x^2 - 5{,}000 - 4x$$
$$= 6x - 0.001x^2 - 5{,}000$$

So, we must solve the following equation:

$$\text{Maximize} \quad P(x) = 6x - 0.001x^2 - 5{,}000 \qquad 0 \le x \le 10{,}000$$
$$P'(x) = 6 - 0.002x$$
$$6 - 0.002x = 0$$
$$x = 3{,}000 \quad \text{Critical number}$$
$$P''(x) = -0.002 < 0 \quad \text{for all } x$$
$$\text{Max } P(x) = P(3{,}000) = \$4{,}000$$

Using the price–demand equation (Example 3) with $x = 3{,}000$, we find that

$$p = 10 - 0.001(3{,}000) = \$7 \quad p = 10 - 0.001x$$

The company's maximum profit is $4,000 when 3,000 markers are produced and sold annually at a price of $7.

Even though the tax caused the company's cost to increase by $2 per marker, the price that the company should charge to maximize its profit increases by only $1. The company must absorb the other $1, with a resulting decrease of $7,000 in maximum profit.

Matched Problem 5) The government decides to tax the office supply company in Matched Problem 4 $20 for each shredder produced. Taking into account this additional cost, how many shredders should the company manufacture annually to maximize its profit? What is the maximum profit? How much should the company charge for the shredders to realize the maximum profit?

EXAMPLE 6 Maximizing Revenue When a management training company prices its seminar on management techniques at $400 per person, 1,000 people will attend the seminar. The company estimates that for each $5 reduction in price, an additional 20 people will attend the seminar. How much should the company charge for the seminar in order to maximize its revenue? What is the maximum revenue?

SOLUTION Let x represent the number of $5 price reductions.

$$400 - 5x = \text{price per customer}$$
$$1{,}000 + 20x = \text{number of customers}$$
$$\text{Revenue} = (\text{price per customer})(\text{number of customers})$$
$$R(x) = (400 - 5x) \times (1{,}000 + 20x)$$

Since price cannot be negative, we have

$$400 - 5x \ge 0$$
$$400 \ge 5x$$
$$80 \ge x \quad \text{or} \quad x \le 80$$

A negative value of x would result in a price increase. Since the problem is stated in terms of price reductions, we must restrict x so that $x \ge 0$. Putting all this together, we have the following model:

$$\text{Maximize} \quad R(x) = (400 - 5x)(1{,}000 + 20x) \quad \text{for } 0 \le x \le 80$$
$$R(x) = 400{,}000 + 3{,}000x - 100x^2$$
$$R'(x) = 3{,}000 - 200x = 0$$
$$3{,}000 = 200x$$
$$x = 15 \quad \text{Critical number}$$

Table 2

x	R(x)
0	400,000
15	422,500
80	0

Since $R(x)$ is continuous on the interval [0, 80], we can determine the behavior of the graph by constructing a table. Table 2 shows that $R(15) = \$422,500$ is the absolute maximum revenue. The price of attending the seminar at $x = 15$ is $400 - 5(15) = \$325$. The company should charge $325 for the seminar in order to receive a maximum revenue of $422,500.

Matched Problem 6 A walnut grower estimates from past records that if 20 trees are planted per acre, then each tree will average 60 pounds of nuts per year. If, for each additional tree planted per acre, the average yield per tree drops 2 pounds, then how many trees should be planted to maximize the yield per acre? What is the maximum yield?

EXAMPLE 7 Maximizing Revenue After additional analysis, the management training company in Example 6 decides that its estimate of attendance was too high. Its new estimate is that only 10 additional people will attend the seminar for each $5 decrease in price. All other information remains the same. How much should the company charge for the seminar now in order to maximize revenue? What is the new maximum revenue?

SOLUTION Under the new assumption, the model becomes

$$\text{Maximize} \quad R(x) = (400 - 5x)(1,000 + 10x) \quad 0 \le x \le 80$$
$$= 400,000 - 1,000x - 50x^2$$
$$R'(x) = -1,000 - 100x = 0$$
$$-1,000 = 100x$$
$$x = -10 \quad \text{Critical number}$$

Note that $x = -10$ is not in the interval [0, 80]. Since $R(x)$ is continuous on [0, 80], we can use a table to find the absolute maximum revenue. Table 3 shows that the maximum revenue is $R(0) = \$400,000$. The company should leave the price at $400. Any $5 decreases in price will lower the revenue.

Table 3

x	R(x)
0	400,000
80	0

Matched Problem 7 After further analysis, the walnut grower in Matched Problem 6 determines that each additional tree planted will reduce the average yield by 4 pounds. All other information remains the same. How many additional trees per acre should the grower plant now in order to maximize the yield? What is the new maximum yield?

CONCEPTUAL INSIGHT

The solution in Example 7 is called an **endpoint solution** because the optimal value occurs at the endpoint of an interval rather than at a critical number in the interior of the interval.

Inventory Control

EXAMPLE 8 Inventory Control A multimedia company anticipates that there will be a demand for 20,000 copies of a certain DVD during the next year. It costs the company $0.50 to store a DVD for one year. Each time it must make additional DVDs, it costs $200 to set up the equipment. How many DVDs should the company make during each production run to minimize its total storage and setup costs?

SOLUTION This type of problem is called an **inventory control problem**. One of the basic assumptions made in such problems is that the demand is uniform. For example, if there are 250 working days in a year, then the daily demand would be $20{,}000 \div 250 = 80$ DVDs. The company could decide to produce all 20,000 DVDs at the beginning of the year. This would certainly minimize the setup costs but would result in very large storage costs. At the other extreme, the company could produce 80 DVDs each day. This would minimize the storage costs but would result in very large setup costs. Somewhere between these two extremes is the optimal solution that will minimize the total storage and setup costs. Let

x = number of DVDs manufactured during each production run

y = number of production runs

It is easy to see that the total setup cost for the year is $200y$, but what is the total storage cost? If the demand is uniform, then the number of DVDs in storage between production runs will decrease from x to 0, and the average number in storage each day is $x/2$. This result is illustrated in Figure 3.

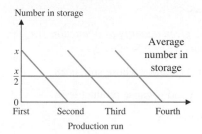

Figure 3

Since it costs \$0.50 to store a DVD for one year, the total storage cost is $0.5(x/2) = 0.25x$ and the total cost is

$$\text{total cost} = \text{setup cost} + \text{storage cost}$$
$$C = 200y + 0.25x$$

In order to write the total cost C as a function of one variable, we must find a relationship between x and y. If the company produces x DVDs in each of y production runs, then the total number of DVDs produced is xy.

$$xy = 20{,}000$$
$$y = \frac{20{,}000}{x}$$

Certainly, x must be at least 1 and cannot exceed 20,000. We must solve the following equation:

$$\text{Minimize} \quad C(x) = 200\left(\frac{20{,}000}{x}\right) + 0.25x \qquad 1 \le x \le 20{,}000$$

$$C(x) = \frac{4{,}000{,}000}{x} + 0.25x$$

$$C'(x) = -\frac{4{,}000{,}000}{x^2} + 0.25$$

$$-\frac{4{,}000{,}000}{x^2} + 0.25 = 0$$

$$x^2 = \frac{4{,}000{,}000}{0.25}$$

$$x^2 = 16{,}000{,}000 \qquad \text{$-4{,}000$ is not a critical number, since}$$

$$x = 4{,}000 \qquad \text{$1 \le x \le 20{,}000$.}$$

$$C''(x) = \frac{8{,}000{,}000}{x^3} > 0 \qquad \text{for } x \in (1, 20{,}000)$$

Therefore,

$$\text{Min } C(x) = C(4,000) = 2,000$$

$$y = \frac{20,000}{4,000} = 5$$

The company will minimize its total cost by making 4,000 DVDs five times during the year.

Matched Problem 8   Repeat Example 8 if it costs $250 to set up a production run and $0.40 to store a DVD for one year.

Exercises 12.6

Skills Warm-up Exercises

W   *In Problems 1–8, express the given quantity as a function $f(x)$ of one variable x. (If necessary, review Section 2.1).*

1. The product of two numbers x and y whose sum is 28

2. The sum of two numbers x and y whose product is 36

3. The area of a circle of diameter x

4. The volume of a sphere of diameter x

5. The volume of a right circular cylinder of radius x and height equal to the radius

6. The volume of a right circular cylinder of diameter x and height equal to twice the diameter

7. The area of a rectangle of length x and width y that has a perimeter of 120 feet

8. The perimeter of a rectangle of length x and width y that has an area of 200 square meters

9. Find two numbers whose sum is 15 and whose product is a maximum.

10. Find two numbers whose sum is 21 and whose product is a maximum.

11. Find two numbers whose difference is 15 and whose product is a minimum.

12. Find two numbers whose difference is 21 and whose product is a minimum.

13. Find two positive numbers whose product is 15 and whose sum is a minimum.

14. Find two positive numbers whose product is 21 and whose sum is a minimum.

15. Find the dimensions of a rectangle with an area of 200 square feet that has the minimum perimeter.

16. Find the dimensions of a rectangle with an area of 108 square feet that has the minimum perimeter.

17. Find the dimensions of a rectangle with a perimeter of 148 feet that has the maximum area.

18. Find the dimensions of a rectangle with a perimeter of 76 feet that has the maximum area.

19. Maximum revenue and profit. A company manufactures and sells x smartphones per week. The weekly price–demand and cost equations are, respectively,

$$p = 500 - 0.5x \quad \text{and} \quad C(x) = 20,000 + 135x$$

(A) What price should the company charge for the phones, and how many phones should be produced to maximize the weekly revenue? What is the maximum weekly revenue?

(B) What is the maximum weekly profit? How much should the company charge for the phones, and how many phones should be produced to realize the maximum weekly profit?

20. Maximum revenue and profit. A company manufactures and sells x digital cameras per week. The weekly price–demand and cost equations are, respectively,

$$p = 400 - 0.4x \quad \text{and} \quad C(x) = 2,000 + 160x$$

(A) What price should the company charge for the cameras, and how many cameras should be produced to maximize the weekly revenue? What is the maximum revenue?

(B) What is the maximum weekly profit? How much should the company charge for the cameras, and how many cameras should be produced to realize the maximum weekly profit?

21. Maximum revenue and profit. A company manufactures and sells x television sets per month. The monthly cost and price–demand equations are

$$C(x) = 72,000 + 60x$$

$$p = 200 - \frac{x}{30} \quad 0 \le x \le 6,000$$

(A) Find the maximum revenue.

(B) Find the maximum profit, the production level that will realize the maximum profit, and the price the company should charge for each television set.

(C) If the government decides to tax the company $5 for each set it produces, how many sets should the company manufacture each month to maximize its profit? What is the maximum profit? What should the company charge for each set?

22. Maximum revenue and profit. Repeat Problem 21 for

$$C(x) = 60,000 + 60x$$

$$p = 200 - \frac{x}{50} \quad 0 \le x \le 10,000$$

23. Maximum profit. The following table contains price–demand and total cost data for the production of extreme-cold sleeping bags, where p is the wholesale price (in dollars) of a sleeping bag for an annual demand of x sleeping bags and C is the total cost (in dollars) of producing x sleeping bags:

(A) Find a quadratic regression equation for the price–demand data, using x as the independent variable.

x	p	C
950	240	130,000
1,200	210	150,000
1,800	160	180,000
2,050	120	190,000

(B) Find a linear regression equation for the cost data, using x as the independent variable.

(C) What is the maximum profit? What is the wholesale price per extreme-cold sleeping bag that should be charged to realize the maximum profit? Round answers to the nearest dollar.

24. Maximum profit. The following table contains price–demand and total cost data for the production of regular sleeping bags, where p is the wholesale price (in dollars) of a sleeping bag for an annual demand of x sleeping bags and C is the total cost (in dollars) of producing x sleeping bags:

x	p	C
2,300	98	145,000
3,300	84	170,000
4,500	67	190,000
5,200	51	210,000

(A) Find a quadratic regression equation for the price–demand data, using x as the independent variable.

(B) Find a linear regression equation for the cost data, using x as the independent variable.

(C) What is the maximum profit? What is the wholesale price per regular sleeping bag that should be charged to realize the maximum profit? Round answers to the nearest dollar.

25. Maximum revenue. A deli sells 640 sandwiches per day at a price of $8 each.

(A) A market survey shows that for every $0.10 reduction in price, 40 more sandwiches will be sold. How much should the deli charge for a sandwich in order to maximize revenue?

(B) A different market survey shows that for every $0.20 reduction in the original $8 price, 15 more sandwiches

will be sold. Now how much should the deli charge for a sandwich in order to maximize revenue?

26. Maximum revenue. A university student center sells 1,600 cups of coffee per day at a price of $2.40.

(A) A market survey shows that for every $0.05 reduction in price, 50 more cups of coffee will be sold. How much should the student center charge for a cup of coffee in order to maximize revenue?

(B) A different market survey shows that for every $0.10 reduction in the original $2.40 price, 60 more cups of coffee will be sold. Now how much should the student center charge for a cup of coffee in order to maximize revenue?

27. Car rental. A car rental agency rents 200 cars per day at a rate of $30 per day. For each $1 increase in rate, 5 fewer cars are rented. At what rate should the cars be rented to produce the maximum income? What is the maximum income?

28. Rental income. A 300-room hotel in Las Vegas is filled to capacity every night at $80 a room. For each $1 increase in rent, 3 fewer rooms are rented. If each rented room costs $10 to service per day, how much should the management charge for each room to maximize gross profit? What is the maximum gross profit?

29. Agriculture. A commercial cherry grower estimates from past records that if 30 trees are planted per acre, then each tree will yield an average of 50 pounds of cherries per season. If, for each additional tree planted per acre (up to 20), the average yield per tree is reduced by 1 pound, how many trees should be planted per acre to obtain the maximum yield per acre? What is the maximum yield?

30. Agriculture. A commercial pear grower must decide on the optimum time to have fruit picked and sold. If the pears are picked now, they will bring 30¢ per pound, with each tree yielding an average of 60 pounds of salable pears. If the average yield per tree increases 6 pounds per tree per week for the next 4 weeks, but the price drops 2¢ per pound per week, when should the pears be picked to realize the maximum return per tree? What is the maximum return?

31. Manufacturing. A candy box is to be made out of a piece of cardboard that measures 8 by 12 inches. Squares of equal size will be cut out of each corner, and then the ends and sides will be folded up to form a rectangular box. What size square should be cut from each corner to obtain a maximum volume?

32. Packaging. A parcel delivery service will deliver a package only if the length plus girth (distance around) does not exceed 108 inches.

(A) Find the dimensions of a rectangular box with square ends that satisfies the delivery service's restriction and has maximum volume. What is the maximum volume?

(B) Find the dimensions (radius and height) of a cylindrical container that meets the delivery service's requirement and has maximum volume. What is the maximum volume?

Figure for 32

33. Construction costs. A fence is to be built to enclose a rectangular area of 800 square feet. The fence along three sides is to be made of material that costs $6 per foot. The material for the fourth side costs $18 per foot. Find the dimensions of the rectangle that will allow for the most economical fence to be built.

34. Construction costs. If a builder has only $840 to spend on a fence, but wants to use both $6 and $18 per foot fencing as in Problem 33, what is the maximum area that can be enclosed? What are its dimensions?

35. Construction costs. The owner of a retail lumber store wants to construct a fence to enclose an outdoor storage area adjacent to the store, using all of the store as part of one side of the area (see the figure). Find the dimensions that will enclose the largest area if

(A) 240 feet of fencing material are used.

(B) 400 feet of fencing material are used.

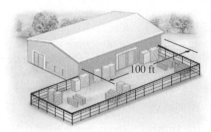

36. Construction costs. If the owner wants to enclose a rectangular area of 12,100 square feet as in Problem 35, what are the dimensions of the area that requires the least fencing? How many feet of fencing are required?

37. Inventory control. A paint manufacturer has a uniform annual demand for 16,000 cans of automobile primer. It costs $4 to store one can of paint for one year and $500 to set up the plant for production of the primer. How many times a year should the company produce this primer in order to minimize the total storage and setup costs?

38. Inventory control. A pharmacy has a uniform annual demand for 200 bottles of a certain antibiotic. It costs $10 to store one bottle for one year and $40 to place an order. How many times during the year should the pharmacy order the antibiotic in order to minimize the total storage and reorder costs?

39. Inventory control. A publishing company sells 50,000 copies of a certain book each year. It costs the company $1 to store a book for one year. Each time that it prints additional copies, it costs the company $1,000 to set up the presses. How many books should the company produce during each printing in order to minimize its total storage and setup costs?

40. Inventory control. A tool company has a uniform annual demand for 9,000 premium chainsaws. It costs $5 to store a chainsaw for a year and $2,500 to set up the plant for manufacture of the premium model. How many chainsaws should be manufactured in each production run in order to minimize the total storage and setup costs?

41. Operational costs. The cost per hour for fuel to run a train is $v^2/4$ dollars, where v is the speed of the train in miles per hour. (Note that the cost goes up as the square of the speed.) Other costs, including labor, are $300 per hour. How fast should the train travel on a 360-mile trip to minimize the total cost for the trip?

42. Operational costs. The cost per hour for fuel to drive a rental truck from Chicago to New York, a distance of 800 miles, is given by

$$f(v) = 0.03v^2 - 2.2v + 72$$

where v is the speed of the truck in miles per hour. Other costs are $40 per hour. How fast should you drive to minimize the total cost?

43. Construction costs. A freshwater pipeline is to be run from a source on the edge of a lake to a small resort community on an island 5 miles offshore, as indicated in the figure.

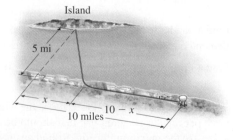

(A) If it costs 1.4 times as much to lay the pipe in the lake as it does on land, what should x be (in miles) to minimize the total cost of the project?

(B) If it costs only 1.1 times as much to lay the pipe in the lake as it does on land, what should x be to minimize the total cost of the project? [*Note:* Compare with Problem 46.]

44. Drug concentration. The concentration $C(t)$, in milligrams per cubic centimeter, of a particular drug in a patient's bloodstream is given by

$$C(t) = \frac{0.16t}{t^2 + 4t + 4}$$

where t is the number of hours after the drug is taken. How many hours after the drug is taken will the concentration be maximum? What is the maximum concentration?

45. Bacteria control. A lake used for recreational swimming is treated periodically to control harmful bacteria growth. Suppose that t days after a treatment, the concentration of bacteria per cubic centimeter is given by

$$C(t) = 30t^2 - 240t + 500 \qquad 0 \le t \le 8$$

How many days after a treatment will the concentration be minimal? What is the minimum concentration?

46. Bird flights. Some birds tend to avoid flights over large bodies of water during daylight hours. Suppose that an adult bird with this tendency is taken from its nesting area on the edge of a large lake to an island 5 miles offshore and is then released (see the figure).

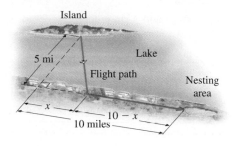

(A) If it takes 1.4 times as much energy to fly over water as land, how far up the shore $(x$, in miles$)$ should the bird head to minimize the total energy expended in returning to the nesting area?

(B) If it takes only 1.1 times as much energy to fly over water as land, how far up the shore should the bird head to minimize the total energy expended in returning to the nesting area? [*Note:* Compare with Problem 43.]

47. Botany. If it is known from past experiments that the height (in feet) of a certain plant after t months is given approximately by

$$H(t) = 4t^{1/2} - 2t \qquad 0 \le t \le 2$$

then how long, on average, will it take a plant to reach its maximum height? What is the maximum height?

48. Pollution. Two heavily industrial areas are located 10 miles apart, as shown in the figure. If the concentration of particulate matter (in parts per million) decreases as the reciprocal of the square of the distance from the source, and if area A_1 emits eight times the particulate matter as A_2, then the concentration of particulate matter at any point between the two areas is given by

$$C(x) = \frac{8k}{x^2} + \frac{k}{(10 - x)^2} \qquad 0.5 < x \le 9.5, \quad k > 0$$

How far from A_1 will the concentration of particulate matter between the two areas be at a minimum?

49. Politics. In a newly incorporated city, the voting population (in thousands) is estimated to be

$$N(t) = 30 + 12t^2 - t^3 \qquad 0 \le t \le 8$$

where t is time in years. When will the rate of increase of $N(t)$ be most rapid?

50. Learning. A large grocery chain found that, on average, a checker can recall $P\%$ of a given price list x hours after starting work, as given approximately by

$$P(x) = 96x - 24x^2 \qquad 0 \le x \le 3$$

At what time x does the checker recall a maximum percentage? What is the maximum?

Answers to Matched Problems

1. The dimensions of the garden with the maximum area of 640 square feet are 16 feet by 40 feet, with one 16-foot side with wood fencing.

2. The minimum cost for enclosing a 1,800-square-foot garden is $480, and the dimensions are 30 feet by 60 feet, with one 30-foot side with wood fencing.

3. The company will realize a maximum revenue of $675,000 when the price of a shredder is $150.

4. A maximum profit of $456,750 is realized when 4,050 shredders are manufactured annually and sold for $165 each.

5. A maximum profit of $378,750 is realized when 3,750 shredders are manufactured annually and sold for $175 each.

6. The maximum yield is 1,250 pounds per acre when 5 additional trees are planted on each acre.

7. The maximum yield is 1,200 pounds when no additional trees are planted.

8. The company should produce 5,000 DVDs four times a year.

Important Terms, Symbols, and Concepts

12.1 First Derivative and Graphs

EXAMPLES

- A function f is **increasing** on an interval (a, b) if $f(x_2) > f(x_1)$ whenever $a < x_1 < x_2 < b$, and f is **decreasing** on (a, b) if $f(x_2) < f(x_1)$ whenever $a < x_1 < x_2 < b$.

  Ex. 1, p. 633

- For the interval (a, b), if $f' > 0$, then f is increasing, and if $f' < 0$, then f is decreasing. So a sign chart for f' can be used to tell where f is increasing or decreasing.

  Ex. 2, p. 634

  Ex. 3, p. 634

- A real number x in the domain of f such that $f'(x) = 0$ or $f'(x)$ does not exist is called a **critical number** of f. So a critical number of f is a partition number for f' that also belongs to the domain of f.

  Ex. 4, p. 635

  Ex. 5, p. 636

- A value $f(c)$ is a **local maximum** if there is an interval (m, n) containing c such that $f(x) \leq f(c)$ for all x in (m, n). A value $f(c)$ is a **local minimum** if there is an interval (m, n) containing c such that $f(x) \geq f(c)$ for all x in (m, n). A local maximum or local minimum is called a **local extremum**.

  Ex. 6, p. 637

  Ex. 7, p. 639

- If $f(c)$ is a local extremum, then c is a critical number of f.

  Ex. 8, p. 641

- The **first-derivative test for local extrema** identifies local maxima and minima of f by means of a sign chart for f'.

  Ex. 9, p. 642

12.2 Second Derivative and Graphs

- The graph of f is **concave upward** on (a, b) if f' is increasing on (a, b), and is **concave downward** on (a, b) if f' is decreasing on (a, b).

- For the interval (a, b), if $f'' > 0$, then f is concave upward, and if $f'' < 0$, then f is concave downward. So a sign chart for f'' can be used to tell where f is concave upward or concave downward.

  Ex. 1, p. 650

  Ex. 2, p. 652

- An **inflection point** of f is a point $(c, f(c))$ on the graph of f where the concavity changes.

  Ex. 3, p. 653

- The graphing strategy on page 655 is used to organize the information obtained from f' and f'' in order to sketch the graph of f.

  Ex. 4, p. 654

  Ex. 5, p. 655

- If sales $N(x)$ are expressed as a function of the amount x spent on advertising, then the dollar amount at which $N'(x)$, the rate of change of sales, goes from increasing to decreasing is called the **point of diminishing returns**. If d is the point of diminishing returns, then $(d, N(d))$ is an inflection point of $N(x)$.

  Ex. 6, p. 657

  Ex. 7, p. 659

12.3 L'Hôpital's Rule

- L'Hôpital's rule for $0/0$ indeterminate forms: If $\lim_{x \to c} f(x) = 0$ and $\lim_{x \to c} g(x) = 0$, then

  Ex. 1, p. 666

  Ex. 2, p. 666

  $$\lim_{x \to c} \frac{f(x)}{g(x)} = \lim_{x \to c} \frac{f'(x)}{g'(x)}$$

  Ex. 3, p. 667

  Ex. 4, p. 668

  provided the second limit exists or is ∞ or $-\infty$.

  Ex. 5, p. 669

- Always check to make sure that L'Hôpital's rule is applicable before using it.

  Ex. 6, p. 669

- L'Hôpital's rule remains valid if the symbol $x \to c$ is replaced everywhere it occurs by one of

  Ex. 7, p. 670

  $$x \to c^+ \quad x \to c^- \quad x \to \infty \quad x \to -\infty$$

  Ex. 8, p. 671

  Ex. 9, p. 672

- L'Hôpital's rule is also valid for indeterminate forms $\dfrac{\pm \infty}{\pm \infty}$.

  Ex. 10, p. 672

12.4 Curve-Sketching Techniques

- The graphing strategy on pages 674 and 675 incorporates analyses of f, f', and f'' in order to sketch a graph of f, including intercepts and asymptotes.

  Ex. 1, p. 675

  Ex. 2, p. 676

- If $f(x) = n(x)/d(x)$ is a rational function and the degree of $n(x)$ is 1 more than the degree of $d(x)$, then the graph of $f(x)$ has an **oblique asymptote** of the form $y = mx + b$.

  Ex. 3, p. 677

  Ex. 4, p. 679

  Ex. 5, p. 681

12.5 Absolute Maxima and Minima

- If $f(c) \geq f(x)$ for all x in the domain of f, then $f(c)$ is called the **absolute maximum** of f. If $f(c) \leq f(x)$ for all x in the domain of f, then $f(c)$ is called the **absolute minimum** of f. An absolute maximum or absolute minimum is called an **absolute extremum**.

- A function that is continuous on a closed interval $[a, b]$ has both an absolute maximum and an absolute minimum on that interval.

  Ex. 1, p. 689

- Absolute extrema, if they exist, must occur at critical numbers or endpoints.
- To find the absolute maximum and absolute minimum of a continuous function f on a closed interval, identify the endpoints and critical numbers in the interval, evaluate the function f at each of them, and choose the largest and smallest values of f.
- **Second-derivative test for local extrema:** If $f'(c) = 0$ and $f''(c) > 0$, then $f(c)$ is a local minimum. If $f'(c) = 0$ and $f''(c) < 0$, then $f(c)$ is a local maximum. No conclusion can be drawn if $f''(c) = 0$. Ex. 2, p. 691
- The **second-derivative test for absolute extrema on an open interval** is applicable when there is only one critical number c in an open interval I and $f'(c) = 0$ and $f''(c) \neq 0$. Ex. 3, p. 692

12.6 Optimization

- The procedure on page 696 for solving optimization problems involves finding the absolute maximum or absolute minimum of a function $f(x)$ on an interval I. If the absolute maximum or absolute minimum occurs at an endpoint, not at a critical number in the interior of I, the extremum is called an **endpoint solution**. The procedure is effective in solving problems in business, including **inventory control problems**, manufacturing, construction, engineering, and many other fields.

Ex. 1, p. 695
Ex. 2, p. 697
Ex. 3, p. 698
Ex. 4, p. 699
Ex. 5, p. 700
Ex. 6, p. 701
Ex. 7, p. 702
Ex. 8, p. 702

Review Exercises

Work through all the problems in this chapter review, and check your answers in the back of the book. Answers to all review problems are there, along with section numbers in italics to indicate where each type of problem is discussed. Where weaknesses show up, review appropriate sections in the text.

Problems 1–8 refer to the following graph of $y = f(x)$. Identify the points or intervals on the x axis that produce the indicated behavior.

1. $f(x)$ is increasing.   **2.** $f'(x) < 0$

3. The graph of f is concave downward.

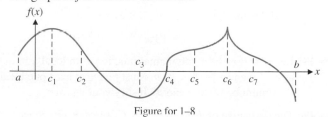

Figure for 1–8

4. Local minima   **5.** Absolute maxima

6. $f'(x)$ appears to be 0.

7. $f'(x)$ does not exist.

8. Inflection points

In Problems 9 and 10, use the given information to sketch the graph of f. Assume that f is continuous on its domain and that all intercepts are included in the information given.

9. Domain: All real x

x	-3	-2	-1	0	2	3
$f(x)$	0	3	2	0	-3	0

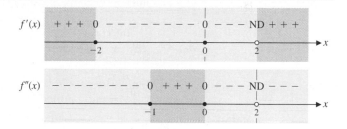

10. Domain: All real x

$$f(-2) = 1, f(0) = 0, f(2) = 1;$$
$$f'(0) = 0; f'(x) < 0 \text{ on } (-\infty, 0);$$
$$f'(x) > 0 \text{ on } (0, \infty);$$
$$f''(-2) = 0, f''(2) = 0;$$
$$f''(x) < 0 \text{ on } (-\infty, -2) \text{ and } (2, \infty);$$
$$f''(x) > 0 \text{ on } (-2, 2);$$
$$\lim_{x \to -\infty} f(x) = 2; \lim_{x \to \infty} f(x) = 2$$

11. Find $f''(x)$ for $f(x) = x^4 + 5x^3$.

12. Find y'' for $y = 3x + \dfrac{4}{x}$.

In Problems 13 and 14, find the domain and intercepts.

13. $f(x) = \dfrac{5 + x}{4 - x}$ **14.** $f(x) = \ln(x + 2)$

In Problems 15 and 16, find the horizontal and vertical asymptotes.

15. $f(x) = \dfrac{x + 3}{x^2 - 4}$ **16.** $f(x) = \dfrac{2x - 7}{3x + 10}$

In Problems 17 and 18, find the x and y coordinates of all inflection points.

17. $f(x) = x^4 - 12x^2$ **18.** $f(x) = (2x + 1)^{1/3} - 6$

In Problems 19 and 20, find (A) $f'(x)$, (B) the partition numbers for f', and (C) the critical numbers of f.

19. $f(x) = x^{1/5}$ **20.** $f(x) = x^{-1/5}$

In Problems 21–30, summarize all the pertinent information obtained by applying the final version of the graphing strategy (Section 4–4) to f, and sketch the graph of f.

21. $f(x) = x^3 - 18x^2 + 81x$

22. $f(x) = (x + 4)(x - 2)^2$

23. $f(x) = 8x^3 - 2x^4$ **24.** $f(x) = (x - 1)^3(x + 3)$

25. $f(x) = \dfrac{3x}{x + 2}$ **26.** $f(x) = \dfrac{x^2}{x^2 + 27}$

27. $f(x) = \dfrac{x}{(x + 2)^2}$ **28.** $f(x) = \dfrac{x^3}{x^2 + 3}$

29. $f(x) = 5 - 5e^{-x}$ **30.** $f(x) = x^3 \ln x$

Find each limit in Problems 31–40.

31. $\displaystyle\lim_{x \to 0} \frac{e^{3x} - 1}{x}$ **32.** $\displaystyle\lim_{x \to 2} \frac{x^2 - 5x + 6}{x^2 + x - 6}$

33. $\displaystyle\lim_{x \to 0^-} \frac{\ln(1 + x)}{x^2}$ **34.** $\displaystyle\lim_{x \to 0} \frac{\ln(1 + x)}{1 + x}$

35. $\displaystyle\lim_{x \to \infty} \frac{e^{4x}}{x^2}$ **36.** $\displaystyle\lim_{x \to 0} \frac{e^x + e^{-x} - 2}{x^2}$

37. $\displaystyle\lim_{x \to 0^+} \frac{\sqrt{1 + x} - 1}{\sqrt{x}}$ **38.** $\displaystyle\lim_{x \to \infty} \frac{\ln x}{x^5}$

39. $\displaystyle\lim_{x \to \infty} \frac{\ln(1 + 6x)}{\ln(1 + 3x)}$ **40.** $\displaystyle\lim_{x \to 0} \frac{\ln(1 + 6x)}{\ln(1 + 3x)}$

41. Use the graph of $y = f'(x)$ shown here to discuss the graph of $y = f(x)$. Organize your conclusions in a table (see Example 4, Section 12.2). Sketch a possible graph of $y = f(x)$.

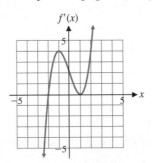

f'(x)

Figure for 41 and 42

42. Refer to the above graph of $y = f'(x)$. Which of the following could be the graph of $y = f''(x)$?

(A)

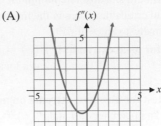

f''(x)

(B)

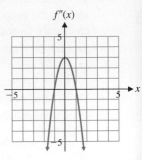

f''(x)

(C)

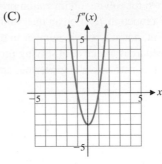

f''(x)

43. Use the second-derivative test to find any local extrema for
$$f(x) = x^3 - 6x^2 - 15x + 12$$

44. Find the absolute maximum and absolute minimum, if either exists, for
$$y = f(x) = x^3 - 12x + 12 \qquad -3 \leq x \leq 5$$

45. Find the absolute minimum, if it exists, for
$$y = f(x) = x^2 + \frac{16}{x^2} \qquad x > 0$$

46. Find the absolute maximum, if it exists, for
$$f(x) = 11x - 2x \ln x \qquad x > 0$$

47. Find the absolute maximum, if it exists, for
$$f(x) = 10xe^{-2x} \qquad x > 0$$

48. Let $y = f(x)$ be a polynomial function with local minima at $x = a$ and $x = b$, $a < b$. Must f have at least one local maximum between a and b? Justify your answer.

49. The derivative of $f(x) = x^{-1}$ is $f'(x) = -x^{-2}$. Since $f'(x) < 0$ for $x \neq 0$, is it correct to say that $f(x)$ is decreasing for all x except $x = 0$? Explain.

50. Discuss the difference between a partition number for $f'(x)$ and a critical number of $f(x)$, and illustrate with examples.

51. Find the absolute maximum for $f'(x)$ if
$$f(x) = 6x^2 - x^3 + 8$$
Graph f and f' on the same coordinate system for $0 \leq x \leq 4$.

52. Find two positive numbers whose product is 400 and whose sum is a minimum. What is the minimum sum?

In Problems 53 and 54, apply the graphing strategy and summarize the pertinent information. Round any approximate values to two decimal places.

53. $f(x) = x^4 + x^3 - 4x^2 - 3x + 4$

54. $f(x) = 0.25x^4 - 5x^3 + 31x^2 - 70x$

55. Find the absolute maximum, if it exists, for
$$f(x) = 3x - x^2 + e^{-x} \quad x > 0$$

56. Find the absolute maximum, if it exists, for
$$f(x) = \frac{\ln x}{e^x} \quad x > 0$$

Applications

57. Price analysis. The graph in the figure approximates the rate of change of the price of tomatoes over a 60-month period, where $p(t)$ is the price of a pound of tomatoes and t is time (in months).

(A) Write a brief description of the graph of $y = p(t)$, including a discussion of local extrema and inflection points.

(B) Sketch a possible graph of $y = p(t)$.

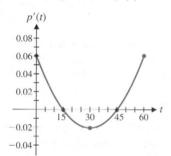

58. Maximum revenue and profit. A company manufactures and sells x e-book readers per month. The monthly cost and price–demand equations are, respectively,

$$C(x) = 350x + 50,000$$
$$p = 500 - 0.025x \quad 0 \le x \le 20,000$$

(A) Find the maximum revenue.

(B) How many readers should the company manufacture each month to maximize its profit? What is the maximum monthly profit? How much should the company charge for each reader?

(C) If the government decides to tax the company $20 for each reader it produces, how many readers should the company manufacture each month to maximize its profit? What is the maximum monthly profit? How much should the company charge for each reader?

59. Construction. A fence is to be built to enclose a rectangular area. The fence along three sides is to be made of material that costs $5 per foot. The material for the fourth side costs $15 per foot.

(A) If the area is 5,000 square feet, find the dimensions of the rectangle that will allow for the most economical fence.

(B) If $3,000 is available for the fencing, find the dimensions of the rectangle that will enclose the most area.

60. Rental income. A 200-room hotel in Reno is filled to capacity every night at a rate of $40 per room. For each $1 increase in the nightly rate, 4 fewer rooms are rented. If each rented room costs $8 a day to service, how much should the management charge per room in order to maximize gross profit? What is the maximum gross profit?

61. Inventory control. A computer store sells 7,200 boxes of storage disks annually. It costs the store $0.20 to store a box of disks for one year. Each time it reorders disks, the store must pay a $5.00 service charge for processing the order. How many times during the year should the store order disks to minimize the total storage and reorder costs?

62. Average cost. The total cost of producing x dorm refrigerators per day is given by

$$C(x) = 4,000 + 10x + 0.1x^2$$

Find the minimum average cost. Graph the average cost and the marginal cost functions on the same coordinate system. Include any oblique asymptotes.

63. Average cost. The cost of producing x wheeled picnic coolers is given by

$$C(x) = 200 + 50x - 50 \ln x \quad x \ge 1$$

Find the minimum average cost.

64. Marginal analysis. The price–demand equation for a GPS device is

$$p(x) = 1,000e^{-0.02x}$$

where x is the monthly demand and p is the price in dollars. Find the production level and price per unit that produce the maximum revenue. What is the maximum revenue?

65. Maximum revenue. Graph the revenue function from Problem 64 for $0 \le x \le 100$.

66. Maximum profit. Refer to Problem 64. If the GPS devices cost the store $220 each, find the price (to the nearest cent) that maximizes the profit. What is the maximum profit (to the nearest dollar)?

67. Maximum profit. The data in the table show the daily demand x for cream puffs at a state fair at various price levels p. If it costs \$1 to make a cream puff, use logarithmic regression $(p = a + b \ln x)$ to find the price (to the nearest cent) that maximizes profit.

Demand x	Price per Cream Puff(\$) p
3,125	1.99
3,879	1.89
5,263	1.79
5,792	1.69
6,748	1.59
8,120	1.49

68. Construction costs. The ceiling supports in a new discount department store are 12 feet apart. Lights are to be hung from these supports by chains in the shape of a "Y." If the lights are 10 feet below the ceiling, what is the shortest length of chain that can be used to support these lights?

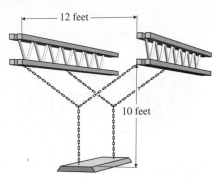

— 12 feet —

10 feet

69. Average cost. The table gives the total daily cost y (in dollars) of producing x dozen chocolate chip cookies at various production levels.

Dozens of Cookies x	Total Cost y
50	119
100	187
150	248
200	382
250	505
300	695

(A) Enter the data into a graphing calculator and find a quadratic regression equation for the total cost.

(B) Use the regression equation from part (A) to find the minimum average cost (to the nearest cent) and the corresponding production level (to the nearest integer).

70. Advertising—point of diminishing returns. A company estimates that it will sell $N(x)$ units of a product after spending \$x thousand on advertising, as given by

$$N(x) = -0.25x^4 + 11x^3 - 108x^2 + 3{,}000$$
$$9 \le x \le 24$$

When is the rate of change of sales increasing and when is it decreasing? What is the point of diminishing returns and the maximum rate of change of sales? Graph N and N' on the same coordinate system.

71. Advertising. A chain of appliance stores uses TV ads to promote its HDTV sales. Analyzing past records produced the data in the following table, where x is the number of ads placed monthly and y is the number of HDTVs sold that month:

Number of Ads x	Number of HDTVs y
10	271
20	427
25	526
30	629
45	887
48	917

(A) Enter the data into a graphing calculator, set the calculator to display two decimal places, and find a cubic regression equation for the number of HDTVs sold monthly as a function of the number of ads.

(B) How many ads should be placed each month to maximize the rate of change of sales with respect to the number of ads, and how many HDTVs can be expected to be sold with that number of ads? Round answers to the nearest integer.

72. Bacteria control. If t days after a treatment the bacteria count per cubic centimeter in a body of water is given by

$$C(t) = 20t^2 - 120t + 800 \qquad 0 \le t \le 9$$

then in how many days will the count be a minimum?

73. Politics. In a new suburb, the number of registered voters is estimated to be

$$N(t) = 10 + 6t^2 - t^3 \qquad 0 \le t \le 5$$

where t is time in years and N is in thousands. When will the rate of increase of $N(t)$ be at its maximum?

13 Integration

13.1 Antiderivatives and Indefinite Integrals

13.2 Integration by Substitution

13.3 Differential Equations; Growth and Decay

13.4 The Definite Integral

13.5 The Fundamental Theorem of Calculus

Chapter 13 Summary and Review

Review Exercises

Introduction

In the preceding three chapters, we studied the *derivative* and its applications. In Chapter 13, we introduce the *integral*, the second key concept of calculus. The integral can be used to calculate areas, volumes, the index of income concentration, and consumers' surplus. At first glance, the integral may appear to be unrelated to the derivative. There is, however, a close connection between these two concepts, which is made precise by the *fundamental theorem of calculus* (Section 13.5). We consider many applications of integrals and differential equations in Chapter 13. See, for example, Problems 77 and 78 in Section 13.2 on price–demand and price–supply equations.

13.1 Antiderivatives and Indefinite Integrals

- Antiderivatives
- Indefinite Integrals: Formulas and Properties
- Applications

Many operations in mathematics have reverses—addition and subtraction, multiplication and division, powers and roots. We now know how to find the derivatives of many functions. The reverse operation, *antidifferentiation* (the reconstruction of a function from its derivative), will receive our attention in this and the next two sections.

Antiderivatives

A function F is an **antiderivative** of a function f if $F'(x) = f(x)$.

The function $F(x) = \dfrac{x^3}{3}$ is an antiderivative of the function $f(x) = x^2$ because

$$\frac{d}{dx}\left(\frac{x^3}{3}\right) = x^2$$

However, $F(x)$ is not the only antiderivative of x^2. Note also that

$$\frac{d}{dx}\left(\frac{x^3}{3} + 2\right) = x^2 \qquad \frac{d}{dx}\left(\frac{x^3}{3} - \pi\right) = x^2 \qquad \frac{d}{dx}\left(\frac{x^3}{3} + \sqrt{5}\right) = x^2$$

Therefore,

$$\frac{x^3}{3} + 2 \qquad \frac{x^3}{3} - \pi \qquad \frac{x^3}{3} + \sqrt{5}$$

are also antiderivatives of x^2 because each has x^2 as a derivative. In fact, it appears that

$$\frac{x^3}{3} + C \qquad \text{for any real number } C$$

is an antiderivative of x^2 because

$$\frac{d}{dx}\left(\frac{x^3}{3} + C\right) = x^2$$

Antidifferentiation of a given function does not give a unique function, but an entire family of functions.

Does the expression

$$\frac{x^3}{3} + C \qquad \text{with } C \text{ any real number}$$

include all antiderivatives of x^2? Theorem 1 (stated without proof) indicates that the answer is yes.

THEOREM 1   Antiderivatives

If the derivatives of two functions are equal on an open interval (a, b), then the functions differ by at most a constant. Symbolically, if F and G are differentiable functions on the interval (a, b) and $F'(x) = G'(x)$ for all x in (a, b), then $F(x) = G(x) + k$ for some constant k.

CONCEPTUAL INSIGHT

Suppose that $F(x)$ is an antiderivative of $f(x)$. If $G(x)$ is any other antiderivative of $f(x)$, then by Theorem 1, the graph of $G(x)$ is a vertical translation of the graph of $F(x)$ (see Section 2.2).

EXAMPLE 1 A Family of Antiderivatives Note that

$$\frac{d}{dx}\left(\frac{x^2}{2}\right) = x$$

(A) Find all antiderivatives of $f(x) = x$.

(B) Graph the antiderivative of $f(x) = x$ that passes through the point $(0,0)$; through the point $(0,1)$; through the point $(0,2)$.

(C) How are the graphs of the three antiderivatives in part (B) related?

SOLUTION

(A) By Theorem 1, any antiderivative of $f(x)$ has the form

$$F(x) = \frac{x^2}{2} + k$$

where k is a real number.

(B) Because $F(0) = (0^2/2) + k = k$, the functions

$$F_0(x) = \frac{x^2}{2}, \quad F_1(x) = \frac{x^2}{2} + 1, \quad \text{and} \quad F_2(x) = \frac{x^2}{2} + 2$$

pass through the points $(0,0)$, $(0,1)$, and $(0,2)$, respectively (see Fig. 1).

(C) The graphs of the three antiderivatives are vertical translations of each other.

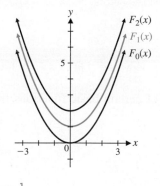

Figure 1

Matched Problem 1 Note that

$$\frac{d}{dx}(x^3) = 3x^2$$

(A) Find all antiderivatives of $f(x) = 3x^2$.

(B) Graph the antiderivative of $f(x) = 3x^2$ that passes through the point $(0,0)$; through the point $(0,1)$; through the point $(0,2)$.

(C) How are the graphs of the three antiderivatives in part (B) related?

Indefinite Integrals: Formulas and Properties

Theorem 1 states that if the derivatives of two functions are equal, then the functions differ by at most a constant. We use the symbol

$$\int f(x)\,dx$$

called the **indefinite integral**, to represent the family of all antiderivatives of $f(x)$, and we write

$$\int f(x)\,dx = F(x) + C \quad \text{if} \quad F'(x) = f(x)$$

The symbol $\int$ is called an **integral sign**, and the function $f(x)$ is called the **integrand**. The symbol dx indicates that the antidifferentiation is performed with respect to the variable x. (We will have more to say about the symbols $\int$ and dx later in the chapter.) The arbitrary constant C is called the **constant of integration**. Referring to the preceding discussion, we can write

$$\int x^2\,dx = \frac{x^3}{3} + C \quad \text{since} \quad \frac{d}{dx}\left(\frac{x^3}{3} + C\right) = x^2$$

Of course, variables other than x can be used in indefinite integrals. For example,

$$\int t^2\,dt = \frac{t^3}{3} + C \quad \text{since} \quad \frac{d}{dt}\left(\frac{t^3}{3} + C\right) = t^2$$

or

$$\int u^2 \, du = \frac{u^3}{3} + C \qquad \text{since} \qquad \frac{d}{du}\left(\frac{u^3}{3} + C\right) = u^2$$

The fact that indefinite integration and differentiation are reverse operations, except for the addition of the constant of integration, can be expressed symbolically as

$$\frac{d}{dx}\left[\int f(x) \, dx\right] = f(x) \qquad \text{The derivative of the indefinite integral of } f(x) \text{ is } f(x).$$

and

$$\int F'(x) \, dx = F(x) + C \qquad \text{The indefinite integral of the derivative of } F(x) \text{ is } F(x) + C.$$

We can develop formulas for the indefinite integrals of certain basic functions from the formulas for derivatives in Chapters 2 and 3.

> **FORMULAS**   Indefinite Integrals of Basic Functions
> For C a constant,
>
> **1.** $\displaystyle\int x^n \, dx = \frac{x^{n+1}}{n+1} + C, \qquad n \neq -1$
>
> **2.** $\displaystyle\int e^x \, dx = e^x + C$
>
> **3.** $\displaystyle\int \frac{1}{x} \, dx = \ln|x| + C, \qquad x \neq 0$

Formula 3 involves the natural logarithm of the absolute value of x. Although the natural logarithm function is only defined for $x > 0$, $f(x) = \ln|x|$ is defined for all $x \neq 0$. Its graph is shown in Figure 2A. Note that $f(x)$ is decreasing for $x < 0$ but is increasing for $x > 0$. Therefore the derivative of f, which by formula 3 is $f'(x) = \dfrac{1}{x}$, is negative for $x < 0$ and positive for $x > 0$ (see Fig. 2B).

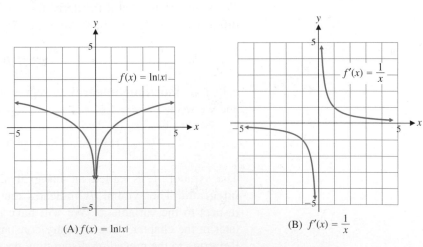

(A) $f(x) = \ln|x|$ (B) $f'(x) = \dfrac{1}{x}$

Figure 2

To justify the three formulas, show that the derivative of the right-hand side is the integrand of the left-hand side (see Problems 75–78 in Exercise 13.1). Note that formula 1 does not give the antiderivative of x^{-1} (because $x^{n+1}/(n+1)$ is undefined when $n = -1$), but formula 3 does.

Explore and Discuss 1   Formulas 1, 2, and 3 do *not* provide a formula for the indefinite integral of the function $\ln x$. Show that if $x > 0$, then

$$\int \ln x \, dx = x \ln x - x + C$$

by differentiating the right-hand side.

We can obtain properties of the indefinite integral from derivative properties that were established in Chapter 2.

> **PROPERTIES** Indefinite Integrals
> For k a constant,
>
> **4.** $\displaystyle\int k f(x) \, dx = k \int f(x) \, dx$
>
> **5.** $\displaystyle\int [f(x) \pm g(x)] \, dx = \int f(x) \, dx \pm \int g(x) \, dx$

Property 4 states that

> **The indefinite integral of a constant times a function is the constant times the indefinite integral of the function.**

Property 5 states that

> **The indefinite integral of the sum of two functions is the sum of the indefinite integrals, and the indefinite integral of the difference of two functions is the difference of the indefinite integrals.**

To establish property 4, let F be a function such that $F'(x) = f(x)$. Then

$$k \int f(x) \, dx = k \int F'(x) \, dx = k[F(x) + C_1] = kF(x) + kC_1$$

and since $[kF(x)]' = kF'(x) = kf(x)$, we have

$$\int kf(x) \, dx = \int kF'(x) \, dx = kF(x) + C_2$$

But $kF(x) + kC_1$ and $kF(x) + C_2$ describe the same set of functions, because C_1 and C_2 are arbitrary real numbers. Property 4 is established. Property 5 can be established in a similar manner (see Problems 79 and 80 in Exercise 13.1).

⚠ **CAUTION** Property 4 states that **a constant factor can be moved across an integral sign. A variable factor cannot be moved across an integral sign:**

CONSTANT FACTOR

$$\int 5x^{1/2} \, dx = 5 \int x^{1/2} \, dx$$

VARIABLE FACTOR

$$\int xx^{1/2} \, dx \neq x \int x^{1/2} \, dx$$ ▲

Indefinite integral formulas and properties can be used together to find indefinite integrals for many frequently encountered functions. If $n = 0$, then formula 1 gives

$$\int dx = x + C$$

Therefore, by property 4,

$$\int k\,dx = k(x + C) = kx + kC$$

Because kC is a constant, we replace it with a single symbol that denotes an arbitrary constant (usually C), and write

$$\int k\,dx = kx + C$$

In words,

The indefinite integral of a constant function with value k is $kx + C$.

Similarly, using property 5 and then formulas 2 and 3, we obtain

$$\int \left(e^x + \frac{1}{x}\right) dx = \int e^x\,dx + \int \frac{1}{x}\,dx$$

$$= e^x + C_1 + \ln|x| + C_2$$

Because $C_1 + C_2$ is a constant, we replace it with the symbol C and write

$$\int \left(e^x + \frac{1}{x}\right) dx = e^x + \ln|x| + C$$

EXAMPLE 2 Using Indefinite Integral Properties and Formulas

(A) $\displaystyle\int 5\,dx = 5x + C$

(B) $\displaystyle\int 9e^x\,dx = 9\int e^x\,dx = 9e^x + C$

(C) $\displaystyle\int 5t^7\,dt = 5\int t^7\,dt = 5\frac{t^8}{8} + C = \frac{5}{8}t^8 + C$

(D) $\displaystyle\int (4x^3 + 2x - 1)\,dx = \int 4x^3\,dx + \int 2x\,dx - \int dx$

$$= 4\int x^3\,dx + 2\int x\,dx - \int dx$$

$$= \frac{4x^4}{4} + \frac{2x^2}{2} - x + C$$

$$= x^4 + x^2 - x + C$$

Property 4 can be extended to the sum and difference of an arbitrary number of functions.

(E) $\displaystyle\int \left(2e^x + \frac{3}{x}\right) dx = 2\int e^x\,dx + 3\int \frac{1}{x}\,dx$

$$= 2e^x + 3\ln|x| + C$$

To check any of the results in Example 2, we differentiate the final result to obtain the integrand in the original indefinite integral. When you evaluate an indefinite integral, do not forget to include the arbitrary constant C.

Matched Problem 2 Find each indefinite integral:

(A) $\displaystyle\int 2\,dx$ (B) $\displaystyle\int 16e^t\,dt$ (C) $\displaystyle\int 3x^4\,dx$

(D) $\displaystyle\int (2x^5 - 3x^2 + 1)\,dx$ (E) $\displaystyle\int \left(\frac{5}{x} - 4e^x\right) dx$

EXAMPLE 3 Using Indefinite Integral Properties and Formulas

(A) $\displaystyle\int \frac{4}{x^3}\,dx = \int 4x^{-3}\,dx = \frac{4x^{-3+1}}{-3+1} + C = -2x^{-2} + C$

(B) $\displaystyle\int 5\sqrt[3]{u^2}\,du = 5\int u^{2/3}\,du = 5\frac{u^{(2/3)+1}}{\frac{2}{3}+1} + C$

$\displaystyle\qquad\qquad = 5\frac{u^{5/3}}{\frac{5}{3}} + C = 3u^{5/3} + C$

(C) $\displaystyle\int \frac{x^3 - 3}{x^2}\,dx = \int \left(\frac{x^3}{x^2} - \frac{3}{x^2}\right)dx$

$\displaystyle\qquad\qquad = \int (x - 3x^{-2})\,dx$

$\displaystyle\qquad\qquad = \int x\,dx - 3\int x^{-2}\,dx$

$\displaystyle\qquad\qquad = \frac{x^{1+1}}{1+1} - 3\frac{x^{-2+1}}{-2+1} + C$

$\displaystyle\qquad\qquad = \tfrac{1}{2}x^2 + 3x^{-1} + C$

(D) $\displaystyle\int \left(\frac{2}{\sqrt[3]{x}} - 6\sqrt{x}\right)dx = \int (2x^{-1/3} - 6x^{1/2})\,dx$

$\displaystyle\qquad\qquad = 2\int x^{-1/3}\,dx - 6\int x^{1/2}\,dx$

$\displaystyle\qquad\qquad = 2\frac{x^{(-1/3)+1}}{-\frac{1}{3}+1} - 6\frac{x^{(1/2)+1}}{\frac{1}{2}+1} + C$

$\displaystyle\qquad\qquad = 2\frac{x^{2/3}}{\frac{2}{3}} - 6\frac{x^{3/2}}{\frac{3}{2}} + C$

$\displaystyle\qquad\qquad = 3x^{2/3} - 4x^{3/2} + C$

(E) $\displaystyle\int x(x^2 + 2)\,dx = \int (x^3 + 2x)\,dx = \frac{x^4}{4} + x^2 + C$

Matched Problem 3 Find each indefinite integral:

(A) $\displaystyle\int \left(2x^{2/3} - \frac{3}{x^4}\right)dx$

(B) $\displaystyle\int 4\sqrt[5]{w^3}\,dw$

(C) $\displaystyle\int \frac{x^4 - 8x^3}{x^2}\,dx$

(D) $\displaystyle\int \left(8\sqrt[3]{x} - \frac{6}{\sqrt{x}}\right)dx$

(E) $\displaystyle\int (x^2 - 2)(x + 3)\,dx$

⚠ CAUTION

1. Note from Example 3E that

$$\int x(x^2 + 2)\,dx \neq \frac{x^2}{2}\left(\frac{x^3}{3} + 2x\right) + C$$

In general, the **indefinite integral of a product is not the product of the indefinite integrals.** (This is expected because the derivative of a product is not the product of the derivatives.)

2.
$$\int e^x \, dx \neq \frac{e^{x+1}}{x+1} + C$$

The power rule applies only to power functions of the form x^n, where the exponent n is a real constant not equal to -1 and the base x is the variable. The function e^x is an exponential function with variable exponent x and constant base e. The correct form is

$$\int e^x \, dx = e^x + C$$

3. Not all elementary functions have elementary antiderivatives. It is impossible, for example, to give a formula for the antiderivative of $f(x) = e^{x^2}$ in terms of elementary functions. Nevertheless, finding such a formula, when it exists, can markedly simplify the solution of certain problems.   ▲

Applications

Let's consider some applications of the indefinite integral.

EXAMPLE 4   Curves   Find the equation of the curve that passes through $(2, 5)$ if the slope of the curve is given by $dy/dx = 2x$ at any point x.

SOLUTION   We want to find a function $y = f(x)$ such that

$$\frac{dy}{dx} = 2x \tag{1}$$

and

$$y = 5 \quad \text{when} \quad x = 2 \tag{2}$$

If $dy/dx = 2x$, then

$$y = \int 2x \, dx \tag{3}$$

$$= x^2 + C$$

Since $y = 5$ when $x = 2$, we determine the *particular value of C* so that

$$5 = 2^2 + C$$

So $C = 1$, and

$$y = x^2 + 1$$

is the *particular antiderivative* out of all those possible from equation (3) that satisfies both equations (1) and (2) (see Fig. 3).

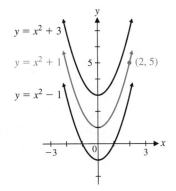

$y = x^2 + 3$

$y = x^2 + 1$

$y = x^2 - 1$

$(2, 5)$

Figure 3 $y = x^2 + C$

Matched Problem 4   Find the equation of the curve that passes through $(2, 6)$ if the slope of the curve is given by $dy/dx = 3x^2$ at any point x.

In certain situations, it is easier to determine the rate at which something happens than to determine how much of it has happened in a given length of time (for example, population growth rates, business growth rates, the rate of healing of a wound, rates of learning or forgetting). If a rate function (derivative) is given and we know the value of the dependent variable for a given value of the independent variable, then we can often find the original function by integration.

EXAMPLE 5   Cost Function   If the marginal cost of producing x units of a commodity is given by

$$C'(x) = 0.3x^2 + 2x$$

and the fixed cost is \$2,000, find the cost function $C(x)$ and the cost of producing 20 units.

SOLUTION   Recall that marginal cost is the derivative of the cost function and that fixed cost is cost at a zero production level. So we want to find $C(x)$, given

$$C'(x) = 0.3x^2 + 2x \qquad C(0) = 2,000$$

We find the indefinite integral of $0.3x^2 + 2x$ and determine the arbitrary integration constant using $C(0) = 2,000$:

$$C'(x) = 0.3x^2 + 2x$$

$$C(x) = \int (0.3x^2 + 2x)\, dx$$

$$= 0.1x^3 + x^2 + K \qquad \text{Since } C \text{ represents the cost, we use } K \text{ for the constant of integration.}$$

But

$$C(0) = (0.1)0^3 + 0^2 + K = 2,000$$

So $K = 2,000$, and the cost function is

$$C(x) = 0.1x^3 + x^2 + 2,000$$

We now find $C(20)$, the cost of producing 20 units:

$$C(20) = (0.1)20^3 + 20^2 + 2,000$$

$$= \$3,200$$

See Figure 4 for a geometric representation.

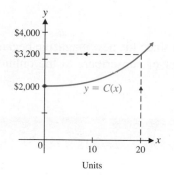

Figure 4

Matched Problem 5   Find the revenue function $R(x)$ when the marginal revenue is

$$R'(x) = 400 - 0.4x$$

and no revenue results at a zero production level. What is the revenue at a production level of 1,000 units?

EXAMPLE 6   Advertising   A satellite radio station is launching an aggressive advertising campaign in order to increase the number of daily listeners. The station currently has 27,000 daily listeners, and management expects the number of daily listeners, $S(t)$, to grow at the rate of

$$S'(t) = 60t^{1/2}$$

listeners per day, where t is the number of days since the campaign began. How long should the campaign last if the station wants the number of daily listeners to grow to 41,000?

SOLUTION   We must solve the equation $S(t) = 41,000$ for t, given that

$$S'(t) = 60t^{1/2} \qquad \text{and} \qquad S(0) = 27,000$$

First, we use integration to find $S(t)$:

$$S(t) = \int 60t^{1/2}\, dt$$

$$= 60\frac{t^{3/2}}{\frac{3}{2}} + C$$

$$= 40t^{3/2} + C$$

Since

$$S(0) = 40(0)^{3/2} + C = 27{,}000$$

we have $C = 27{,}000$ and

$$S(t) = 40t^{3/2} + 27{,}000$$

Now we solve the equation $S(t) = 41{,}000$ for t:

$$40t^{3/2} + 27{,}000 = 41{,}000$$
$$40t^{3/2} = 14{,}000$$
$$t^{3/2} = 350$$
$$t = 350^{2/3} \qquad \text{Use a calculator.}$$
$$= 49.664\,419\ldots$$

The advertising campaign should last approximately 50 days.

Matched Problem 6) There are 64,000 subscribers to an online fashion magazine. Due to competition from a new magazine, the number $C(t)$ of subscribers is expected to decrease at the rate of

$$C'(t) = -600t^{1/3}$$

subscribers per month, where t is the time in months since the new magazine began publication. How long will it take until the number of subscribers to the online fashion magazine drops to 46,000?

Exercises 13.1

Skills Warm-up Exercises

W *In Problems 1–8, write each function as a sum of terms of the form ax^n, where a is a constant. (If necessary, review Section A.6).*

1. $f(x) = \dfrac{5}{x^4}$

2. $f(x) = -\dfrac{6}{x^9}$

3. $f(x) = \dfrac{3x - 2}{x^5}$

4. $f(x) = \dfrac{x^2 + 5x - 1}{x^3}$

5. $f(x) = \sqrt{x} + \dfrac{5}{\sqrt{x}}$

6. $f(x) = \sqrt[3]{x} - \dfrac{4}{\sqrt[3]{x}}$

7. $f(x) = \sqrt[3]{x}(4 + x - 3x^2)$

8. $f(x) = \sqrt{x}(1 - 5x + x^3)$

In Problems 9–24, find each indefinite integral. Check by differentiating.

9. $\displaystyle\int 7\,dx$

10. $\displaystyle\int 10\,dx$

11. $\displaystyle\int 8x\,dx$

12. $\displaystyle\int 14x\,dx$

13. $\displaystyle\int 9x^2\,dx$

14. $\displaystyle\int 15x^2\,dx$

15. $\displaystyle\int x^5\,dx$

16. $\displaystyle\int x^8\,dx$

17. $\displaystyle\int x^{-3}\,dx$

18. $\displaystyle\int x^{-4}\,dx$

19. $\displaystyle\int 10x^{3/2}\,dx$

20. $\displaystyle\int 8x^{1/3}\,dx$

21. $\displaystyle\int \dfrac{3}{z}\,dz$

22. $\displaystyle\int \dfrac{7}{z}\,dz$

23. $\displaystyle\int 16e^u\,du$

24. $\displaystyle\int 5e^u\,du$

25. Is $F(x) = (x + 1)(x + 2)$ an antiderivative of $f(x) = 2x + 3$? Explain.

26. Is $F(x) = (2x + 5)(x - 6)$ an antiderivative of $f(x) = 4x - 7$? Explain.

27. Is $F(x) = 1 + x \ln x$ an antiderivative of $f(x) = 1 + \ln x$? Explain.

28. Is $F(x) = x \ln x - x + e$ an antiderivative of $f(x) = \ln x$? Explain.

29. Is $F(x) = \dfrac{(2x + 1)^3}{3}$ an antiderivative of $f(x) = (2x + 1)^2$? Explain.

30. Is $F(x) = \dfrac{(3x - 2)^4}{4}$ an antiderivative of $f(x) = (3x - 2)^3$? Explain.

31. Is $F(x) = e^{x^3/3}$ an antiderivative of $f(x) = e^{x^2}$? Explain.

32. Is $F(x) = (e^x - 10)(e^x + 10)$ an antiderivative of $f(x) = 2e^{2x}$? Explain.

In Problems 33–38, discuss the validity of each statement. If the statement is always true, explain why. If not, give a counterexample.

33. The constant function $f(x) = \pi$ is an antiderivative of the constant function $k(x) = 0$.

34. The constant function $k(x) = 0$ is an antiderivative of the constant function $f(x) = \pi$.

35. If n is an integer, then $x^{n+1}/(n+1)$ is an antiderivative of x^n.

36. The constant function $k(x) = 0$ is an antiderivative of itself.

37. The function $h(x) = 5e^x$ is an antiderivative of itself.

38. The constant function $g(x) = 5e^\pi$ is an antiderivative of itself.

In Problems 39–42, could the three graphs in each figure be antiderivatives of the same function? Explain.

39.

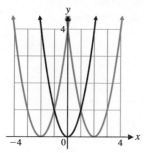

40.

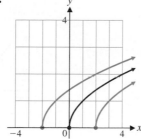

41.

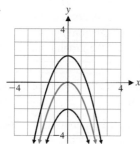

42.

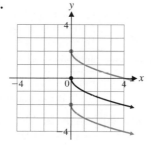

In Problems 43–54, find each indefinite integral. (Check by differentiation.)

43. $\displaystyle \int 5x(1-x)\,dx$

44. $\displaystyle \int x^2(1+x^3)\,dx$

45. $\displaystyle \int \frac{du}{\sqrt{u}}$

46. $\displaystyle \int \frac{dt}{\sqrt[3]{t}}$

47. $\displaystyle \int \frac{dx}{4x^3}$

48. $\displaystyle \int \frac{6\,dm}{m^2}$

49. $\displaystyle \int \frac{4+u}{u}\,du$

50. $\displaystyle \int \frac{1-y^2}{3y}\,dy$

51. $\displaystyle \int (5e^z + 4)\,dz$

52. $\displaystyle \int \frac{e^t - t}{2}\,dt$

53. $\displaystyle \int \left(3x^2 - \frac{2}{x^2}\right)dx$

54. $\displaystyle \int \left(4x^3 + \frac{2}{x^3}\right)dx$

In Problems 55–62, find the particular antiderivative of each derivative that satisfies the given condition.

55. $C'(x) = 6x^2 - 4x;\ C(0) = 3{,}000$

56. $R'(x) = 600 - 0.6x;\ R(0) = 0$

57. $\dfrac{dx}{dt} = \dfrac{20}{\sqrt{t}};\ x(1) = 40$

58. $\dfrac{dR}{dt} = \dfrac{100}{t^2};\ R(1) = 400$

59. $\dfrac{dy}{dx} = 2x^{-2} + 3x^{-1} - 1;\ y(1) = 0$

60. $\dfrac{dy}{dx} = 3x^{-1} + x^{-2};\ y(1) = 1$

61. $\dfrac{dx}{dt} = 4e^t - 2;\ x(0) = 1$

62. $\dfrac{dy}{dt} = 5e^t - 4;\ y(0) = -1$

63. Find the equation of the curve that passes through $(2,3)$ if its slope is given by
$$\frac{dy}{dx} = 4x - 3$$
for each x.

64. Find the equation of the curve that passes through $(1,3)$ if its slope is given by
$$\frac{dy}{dx} = 12x^2 - 12x$$
for each x.

In Problems 65–70, find each indefinite integral.

65. $\displaystyle \int \frac{2x^4 - x}{x^3}\,dx$

66. $\displaystyle \int \frac{x^{-1} - x^4}{x^2}\,dx$

67. $\displaystyle \int \frac{x^5 - 2x}{x^4}\,dx$

68. $\displaystyle \int \frac{1 - 3x^4}{x^2}\,dx$

69. $\displaystyle \int \frac{x^2 e^x - 2x}{x^2}\,dx$

70. $\displaystyle \int \frac{1 - xe^x}{x}\,dx$

In Problems 71–74, find the derivative or indefinite integral as indicated.

71. $\dfrac{d}{dx}\left(\displaystyle\int x^3\,dx\right)$

72. $\dfrac{d}{dt}\left(\displaystyle\int \frac{\ln t}{t}\,dt\right)$

73. $\displaystyle \int \frac{d}{dx}(x^4 + 3x^2 + 1)\,dx$

74. $\displaystyle \int \frac{d}{du}(e^{u^2})\,du$

75. Use differentiation to justify the formula
$$\int x^n\,dx = \frac{x^{n+1}}{n+1} + C$$
provided that $n \neq -1$.

76. Use differentiation to justify the formula
$$\int e^x\,dx = e^x + C$$

77. Assuming that $x > 0$, use differentiation to justify the formula

$$\int \frac{1}{x}\, dx = \ln|x| + C$$

78. Assuming that $x < 0$, use differentiation to justify the formula

$$\int \frac{1}{x}\, dx = \ln|x| + C$$

[*Hint:* Use the chain rule after noting that $\ln|x| = \ln(-x)$ for $x < 0$.]

79. Show that the indefinite integral of the sum of two functions is the sum of the indefinite integrals.

[*Hint:* Assume that $\int f(x)\, dx = F(x) + C_1$ and $\int g(x)\, dx = G(x) + C_2$. Using differentiation, show that $F(x) + C_1 + G(x) + C_2$ is the indefinite integral of the function $s(x) = f(x) + g(x)$.]

80. Show that the indefinite integral of the difference of two functions is the difference of the indefinite integrals.

Applications

81. **Cost function.** The marginal average cost of producing x sports watches is given by

$$\overline{C}'(x) = -\frac{1,000}{x^2} \qquad \overline{C}(100) = 25$$

where $\overline{C}(x)$ is the average cost in dollars. Find the average cost function and the cost function. What are the fixed costs?

82. **Renewable energy.** In 2012, U.S. consumption of renewable energy was 8.45 quadrillion Btu (or 8.45×10^{15} Btu). Since the 1960s, consumption has been growing at a rate (in quadrillion Btu per year) given by

$$f'(t) = 0.004t + 0.062$$

where t is years after 1960. Find $f(t)$ and estimate U.S. consumption of renewable energy in 2024.

83. **Production costs.** The graph of the marginal cost function from the production of x thousand bottles of sunscreen per month [where cost $C(x)$ is in thousands of dollars per month] is given in the figure.

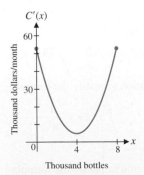

Thousand dollars/month

Thousand bottles

(A) Using the graph shown, describe the shape of the graph of the cost function $C(x)$ as x increases from 0 to 8,000 bottles per month.

(B) Given the equation of the marginal cost function,

$$C'(x) = 3x^2 - 24x + 53$$

find the cost function if monthly fixed costs at 0 output are $30,000. What is the cost of manufacturing 4,000 bottles per month? 8,000 bottles per month?

(C) Graph the cost function for $0 \le x \le 8$. [Check the shape of the graph relative to the analysis in part (A).]

(D) Why do you think that the graph of the cost function is steeper at both ends than in the middle?

84. **Revenue.** The graph of the marginal revenue function from the sale of x sports watches is given in the figure.

(A) Using the graph shown, describe the shape of the graph of the revenue function $R(x)$ as x increases from 0 to 1,000.

(B) Find the equation of the marginal revenue function (the linear function shown in the figure).

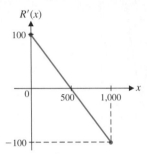

(C) Find the equation of the revenue function that satisfies $R(0) = 0$. Graph the revenue function over the interval $[0, 1,000]$. [Check the shape of the graph relative to the analysis in part (A).]

(D) Find the price–demand equation and determine the price when the demand is 700 units.

85. **Sales analysis.** Monthly sales of an SUV model are expected to increase at the rate of

$$S'(t) = -24t^{1/3}$$

SUVs per month, where t is time in months and $S(t)$ is the number of SUVs sold each month. The company plans to stop manufacturing this model when monthly sales reach 300 SUVs. If monthly sales now $(t = 0)$ are 1,200 SUVs, find $S(t)$. How long will the company continue to manufacture this model?

86. **Sales analysis.** The rate of change of the monthly sales of a newly released football game is given by

$$S'(t) = 500t^{1/4} \qquad S(0) = 0$$

where t is the number of months since the game was released and $S(t)$ is the number of games sold each month. Find $S(t)$. When will monthly sales reach 20,000 games?

87. **Sales analysis.** Repeat Problem 85 if $S'(t) = -24t^{1/3} - 70$ and all other information remains the same. Use a graphing calculator to approximate the solution of the equation $S(t) = 300$ to two decimal places.

88. Sales analysis. Repeat Problem 86 if $S'(t) = 500t^{1/4} + 300$ and all other information remains the same. Use a graphing calculator to approximate the solution of the equation $S(t) = 20,000$ to two decimal places.

89. Labor costs. A defense contractor is starting production on a new missile control system. On the basis of data collected during the assembly of the first 16 control systems, the production manager obtained the following function describing the rate of labor use:

$$L'(x) = 2,400x^{-1/2}$$

For example, after assembly of 16 units, the rate of assembly is 600 labor-hours per unit, and after assembly of 25 units, the rate of assembly is 480 labor-hours per unit. The more units assembled, the more efficient the process. If 19,200 labor-hours are required to assemble of the first 16 units, how many labor-hours $L(x)$ will be required to assemble the first x units? The first 25 units?

90. Labor costs. If the rate of labor use in Problem 89 is

$$L'(x) = 2,000x^{-1/3}$$

and if the first 8 control units require 12,000 labor-hours, how many labor-hours, $L(x)$, will be required for the first x control units? The first 27 control units?

91. Weight–height. For an average person, the rate of change of weight W (in pounds) with respect to height h (in inches) is given approximately by

$$\frac{dW}{dh} = 0.0015h^2$$

Find $W(h)$ if $W(60) = 108$ pounds. Find the weight of an average person who is 5 feet, 10 inches, tall.

92. Wound healing. The area A of a healing wound changes at a rate given approximately by

$$\frac{dA}{dt} = -4t^{-3} \quad 1 \le t \le 10$$

where t is time in days and $A(1) = 2$ square centimeters. What will the area of the wound be in 10 days?

93. Urban growth. The rate of growth of the population $N(t)$ of a new city t years after its incorporation is estimated to be

$$\frac{dN}{dt} = 400 + 600\sqrt{t} \quad 0 \le t \le 9$$

If the population was 5,000 at the time of incorporation, find the population 9 years later.

94. Learning. A college language class was chosen for an experiment in learning. Using a list of 50 words, the experiment involved measuring the rate of vocabulary memorization at different times during a continuous 5-hour study session. It was found that the average rate of learning for the entire class was inversely proportional to the time spent studying and was given approximately by

$$V'(t) = \frac{15}{t} \quad 1 \le t \le 5$$

If the average number of words memorized after 1 hour of study was 15 words, what was the average number of words memorized after t hours of study for $1 \le t \le 5$? After 4 hours of study? Round answer to the nearest whole number.

Answers to Matched Problems

1. (A) $x^3 + C$

(B)

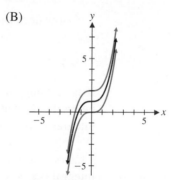

(C) The graphs are vertical translations of each other.

2. (A) $2x + C$ (B) $16e^t + C$ (C) $\frac{3}{5}x^5 + C$
(D) $\frac{1}{3}x^6 - x^3 + x + C$ (E) $5\ln|x| - 4e^x + C$

3. (A) $\frac{6}{5}x^{5/3} + x^{-3} + C$ (B) $\frac{5}{2}w^{8/5} + C$
(C) $\frac{1}{3}x^3 - 4x^2 + C$ (D) $6x^{4/3} - 12x^{1/2} + C$
(E) $\frac{1}{4}x^4 + x^3 - x^2 - 6x + C$

4. $y = x^3 - 2$

5. $R(x) = 400x - 0.2x^2; R(1,000) = \$200,000$

6. $t = (40)^{3/4} \approx 16$ mo

13.2 Integration by Substitution

- Reversing the Chain Rule
- Integration by Substitution
- Additional Substitution Techniques
- Application

Many of the indefinite integral formulas introduced in the preceding section are based on corresponding derivative formulas studied earlier. We now consider indefinite integral formulas and procedures based on the chain rule for differentiation.

Reversing the Chain Rule

Recall the chain rule:

$$\frac{d}{dx} f[g(x)] = f'[g(x)]g'(x)$$

The expression on the right is formed from the expression on the left by taking the derivative of the outside function f and multiplying it by the derivative of the inside function g. If we recognize an integrand as a chain-rule form $E'[I(x)]I'(x)$, we can easily find an antiderivative and its indefinite integral:

$$\int E'[I(x)]I'(x)\,dx = E[I(x)] + C \qquad (1)$$

We are interested in finding the indefinite integral

$$\int 3x^2 e^{x^3-1}\,dx \qquad (2)$$

The integrand appears to be the chain-rule form $e^{g(x)}g'(x)$, which is the derivative of $e^{g(x)}$. Since

$$\frac{d}{dx}e^{x^3-1} = 3x^2 e^{x^3-1}$$

it follows that

$$\int 3x^2 e^{x^3-1}\,dx = e^{x^3-1} + C \qquad (3)$$

How does the following indefinite integral differ from integral (2)?

$$\int x^2 e^{x^3-1}\,dx \qquad (4)$$

It is missing the constant factor 3. That is, $x^2 e^{x^3-1}$ is within a constant factor of being the derivative of e^{x^3-1}. But because a constant factor can be moved across the integral sign, this causes us little trouble in finding the indefinite integral of $x^2 e^{x^3-1}$. We introduce the constant factor 3 and at the same time multiply by $\frac{1}{3}$ and move the $\frac{1}{3}$ factor outside the integral sign. This is equivalent to multiplying the integrand in integral (4) by 1:

$$\int x^2 e^{x^3-1}\,dx = \int \frac{3}{3}x^2 e^{x^3-1}\,dx \qquad (5)$$

$$= \frac{1}{3}\int 3x^2 e^{x^3-1}\,dx = \frac{1}{3}e^{x^3-1} + C$$

The derivative of the rightmost side of equation (5) is the integrand of the indefinite integral (4). Check this.

How does the following indefinite integral differ from integral (2)?

$$\int 3x e^{x^3-1}\,dx \qquad (6)$$

It is missing a variable factor x. This is more serious. As tempting as it might be, we *cannot* adjust integral (6) by introducing the variable factor x and moving $1/x$ outside the integral sign, as we did with the constant 3 in equation (5).

⚠ CAUTION A constant factor can be moved across an integral sign, but a variable factor cannot.   ▲

There is nothing wrong with educated guessing when you are looking for an antiderivative of a given function. You have only to check the result by differentiation. If you are right, you go on your way; if you are wrong, you simply try another approach.

In Section 3.4, we saw that the chain rule extends the derivative formulas for x^n, e^x, and $\ln x$ to derivative formulas for $[f(x)]^n$, $e^{f(x)}$, and $\ln[f(x)]$. The chain rule can also be used to extend the indefinite integral formulas discussed in Section 13.1. Some general formulas are summarized in the following box:

FORMULAS General Indefinite Integral Formulas

1. $\displaystyle\int [f(x)]^n f'(x)\, dx = \frac{[f(x)]^{n+1}}{n+1} + C,\, n \neq -1$

2. $\displaystyle\int e^{f(x)} f'(x)\, dx = e^{f(x)} + C$

3. $\displaystyle\int \frac{1}{f(x)} f'(x)\, dx = \ln|f(x)| + C$

We can verify each formula by using the chain rule to show that the derivative of the function on the right is the integrand on the left. For example,

$$\frac{d}{dx}\left[e^{f(x)} + C\right] = e^{f(x)} f'(x)$$

verifies formula 2.

EXAMPLE 1 Reversing the Chain Rule

(A) $\displaystyle\int (3x+4)^{10}(3)\, dx = \frac{(3x+4)^{11}}{11} + C$
  Formula 1 with $f(x) = 3x + 4$ and $f'(x) = 3$

Check:

$$\frac{d}{dx}\frac{(3x+4)^{11}}{11} = 11\frac{(3x+4)^{10}}{11}\frac{d}{dx}(3x+4) = (3x+4)^{10}(3)$$

(B) $\displaystyle\int e^{x^2}(2x)\, dx = e^{x^2} + C$
  Formula 2 with $f(x) = x^2$ and $f'(x) = 2x$

Check:

$$\frac{d}{dx}e^{x^2} = e^{x^2}\frac{d}{dx}x^2 = e^{x^2}(2x)$$

(C) $\displaystyle\int \frac{1}{1+x^3}3x^2\, dx = \ln|1 + x^3| + C$
  Formula 3 with $f(x) = 1 + x^3$ and $f'(x) = 3x^2$

Check:

$$\frac{d}{dx}\ln|1 + x^3| = \frac{1}{1+x^3}\frac{d}{dx}(1+x^3) = \frac{1}{1+x^3}3x^2$$

Matched Problem 1 Find each indefinite integral.

(A) $\displaystyle\int (2x^3 - 3)^{20}(6x^2)\, dx$ (B) $\displaystyle\int e^{5x}(5)\, dx$ (C) $\displaystyle\int \frac{1}{4+x^2}2x\, dx$

Integration by Substitution

The key step in using formulas 1, 2, and 3 is recognizing the form of the integrand. Some people find it difficult to identify $f(x)$ and $f'(x)$ in these formulas and prefer to use a *substitution* to simplify the integrand. The *method of substitution,* which we now discuss, becomes increasingly useful as one progresses in studies of integration.

We start by recalling the definition of the *differential* (see Section 10.6, p. 552). We represent the derivative by the symbol dy/dx taken as a whole and now define dy and dx as two separate quantities with the property that their ratio is still equal to $f'(x)$:

> **DEFINITION** Differentials
>
> If $y = f(x)$ defines a differentiable function, then
>
> 1. The **differential dx** of the independent variable x is an arbitrary real number.
> 2. The **differential dy** of the dependent variable y is defined as the product of $f'(x)$ and dx:
>
> $$dy = f'(x)\,dx$$

Differentials involve mathematical subtleties that are treated carefully in advanced mathematics courses. Here, we are interested in them mainly as a book-keeping device to aid in the process of finding indefinite integrals. We can always check an indefinite integral by differentiating.

EXAMPLE 2 Differentials

(A) If $y = f(x) = x^2$, then

$$dy = f'(x)\,dx = 2x\,dx$$

(B) If $u = g(x) = e^{3x}$, then

$$du = g'(x)\,dx = 3e^{3x}\,dx$$

(C) If $w = h(t) = \ln(4 + 5t)$, then

$$dw = h'(t)\,dt = \frac{5}{4 + 5t}\,dt$$

Matched Problem 2

(A) Find dy for $y = f(x) = x^3$.

(B) Find du for $u = h(x) = \ln(2 + x^2)$.

(C) Find dv for $v = g(t) = e^{-5t}$.

The **method of substitution** is developed through Examples 3–6.

EXAMPLE 3 Using Substitution   Find $\int (x^2 + 2x + 5)^5(2x + 2)\,dx$.

SOLUTION If

$$u = x^2 + 2x + 5$$

then the differential of u is

$$du = (2x + 2)\,dx$$

Notice that du is one of the factors in the integrand. Substitute u for $x^2 + 2x + 5$ and du for $(2x + 2)\,dx$ to obtain

$$\int (x^2 + 2x + 5)^5(2x + 2)\,dx = \int u^5\,du$$

$$= \frac{u^6}{6} + C$$

$$= \frac{1}{6}(x^2 + 2x + 5)^6 + C \quad \text{Since } u = x^2 + 2x + 5$$

Check:

$$\frac{d}{dx}\frac{1}{6}(x^2 + 2x + 5)^6 = \frac{1}{6}(6)(x^2 + 2x + 5)^5\frac{d}{dx}(x^2 + 2x + 5)$$

$$= (x^2 + 2x + 5)^5(2x + 2)$$

Matched Problem 3 Find $\int (x^2 - 3x + 7)^4(2x - 3)\, dx$ by substitution.

The substitution method is also called the **change-of-variable method** since u replaces the variable x in the process. Substituting $u = f(x)$ and $du = f'(x)\, dx$ in formulas 1, 2, and 3 produces the general indefinite integral formulas 4, 5, and 6:

FORMULAS General Indefinite Integral Formulas

4. $\int u^n\, du = \dfrac{u^{n+1}}{n + 1} + C, \quad n \neq -1$

5. $\int e^u\, du = e^u + C$

6. $\int \dfrac{1}{u}\, du = \ln|u| + C$

These formulas are valid if u is an independent variable, or if u is a function of another variable and du is the differential of u with respect to that variable.

The substitution method for evaluating certain indefinite integrals is outlined as follows:

PROCEDURE Integration by Substitution

Step 1 Select a substitution that appears to simplify the integrand. In particular, try to select u so that du is a factor in the integrand.

Step 2 Express the integrand entirely in terms of u and du, completely eliminating the original variable and its differential.

Step 3 Evaluate the new integral if possible.

Step 4 Express the antiderivative found in step 3 in terms of the original variable.

EXAMPLE 4 Using Substitution Use a substitution to find each indefinite integral.

(A) $\int (3x + 4)^6(3)\, dx$ 　　　　　　(B) $\int e^{t^2}(2t)\, dt$

SOLUTION

(A) If we let $u = 3x + 4$, then $du = 3\, dx$, and

$$\int (3x + 4)^6(3)\, dx = \int u^6\, du \qquad \text{Use formula 4.}$$

$$= \frac{u^7}{7} + C$$

$$= \frac{(3x + 4)^7}{7} + C \qquad \text{Since } u = 3x + 4$$

Check:

$$\frac{d}{dx}\frac{(3x+4)^7}{7} = \frac{7(3x+4)^6}{7}\frac{d}{dx}(3x+4) = (3x+4)^6(3)$$

(B) If we let $u = t^2$, then $du = 2t\,dt$, and

$$\int e^{t^2}(2t)\,dt = \int e^u\,du \quad \text{Use formula 5.}$$

$$= e^u + C$$

$$= e^{t^2} + C \quad \text{Since } u = t^2$$

Check:

$$\frac{d}{dt}e^{t^2} = e^{t^2}\frac{d}{dt}t^2 = e^{t^2}(2t)$$

Matched Problem 4   Use a substitution to find each indefinite integral.

(A) $\displaystyle\int (2x^3 - 3)^4(6x^2)\,dx$ (B) $\displaystyle\int e^{5w}(5)\,dw$

Additional Substitution Techniques

In order to use the substitution method, **the integrand must be expressed entirely in terms of u and du.** In some cases, the integrand must be modified before making a substitution and using one of the integration formulas. Example 5 illustrates this process.

EXAMPLE 5   Substitution Techniques   Integrate.

(A) $\displaystyle\int \frac{1}{4x+7}\,dx$ (B) $\displaystyle\int te^{-t^2}\,dt$

(C) $\displaystyle\int 4x^2\sqrt{x^3+5}\,dx$

SOLUTION

(A) If $u = 4x + 7$, then $du = 4\,dx$ and, dividing both sides of the equation $du = 4\,dx$ by 4, we have $dx = \frac{1}{4}\,du$. In the integrand, replace $4x + 7$ by u and replace dx by $\frac{1}{4}\,du$:

$$\int \frac{1}{4x+7}\,dx = \int \frac{1}{u}\left(\frac{1}{4}\,du\right) \quad \begin{array}{l}\text{Move constant factor across}\\ \text{the integral sign.}\end{array}$$

$$= \frac{1}{4}\int \frac{1}{u}\,du \quad \text{Use formula 6.}$$

$$= \frac{1}{4}\ln|u| + C$$

$$= \frac{1}{4}\ln|4x+7| + C \quad \text{Since } u = 4x + 7$$

Check:

$$\frac{d}{dx}\frac{1}{4}\ln|4x+7| = \frac{1}{4}\frac{1}{4x+7}\frac{d}{dx}(4x+7) = \frac{1}{4}\frac{1}{4x+7}4 = \frac{1}{4x+7}$$

(B) If $u = -t^2$, then $du = -2t\,dt$ and, dividing both sides by -2, $-\frac{1}{2}du = t\,dt$. In the integrand, replace $-t^2$ by u and replace $t\,dt$ by $-\frac{1}{2}du$:

$$\int te^{-t^2}dt = \int e^u\left(-\frac{1}{2}du\right) \qquad \text{Move constant factor across the integral sign.}$$

$$= -\frac{1}{2}\int e^u\,du \qquad \text{Use formula 5.}$$

$$= -\frac{1}{2}e^u + C$$

$$= -\frac{1}{2}e^{-t^2} + C \qquad \text{Since } u = -t^2$$

Check:

$$\frac{d}{dt}\left(-\frac{1}{2}e^{-t^2}\right) = -\frac{1}{2}e^{-t^2}\frac{d}{dt}(-t^2) = -\frac{1}{2}e^{-t^2}(-2t) = te^{-t^2}$$

(C) If $u = x^3 + 5$, then $du = 3x^2\,dx$ and, dividing both sides by 3, $\frac{1}{3}du = x^2\,dx$. In the integrand, replace $x^3 + 5$ by u and replace $x^2\,dx$ by $\frac{1}{3}du$:

$$\int 4x^2\sqrt{x^3 + 5}\,dx = \int 4\sqrt{u}\left(\frac{1}{3}du\right) \qquad \text{Move constant factors across the integral sign.}$$

$$= \frac{4}{3}\int \sqrt{u}\,du$$

$$= \frac{4}{3}\int u^{1/2}\,du \qquad \text{Use formula 4.}$$

$$= \frac{4}{3}\cdot\frac{u^{3/2}}{\frac{3}{2}} + C$$

$$= \frac{8}{9}u^{3/2} + C$$

$$= \frac{8}{9}(x^3 + 5)^{3/2} + C \qquad \text{Since } u = x^3 + 5$$

Check:

$$\frac{d}{dx}\left[\frac{8}{9}(x^3 + 5)^{3/2}\right] = \frac{4}{3}(x^3 + 5)^{1/2}\frac{d}{dx}(x^3 + 5)$$

$$= \frac{4}{3}(x^3 + 5)^{1/2}(3x^2) = 4x^2\sqrt{x^3 + 5}$$

Matched Problem 5) Integrate.

(A) $\int e^{-3x}\,dx$

(B) $\int \frac{x}{x^2 - 9}\,dx$

(C) $\int 5t^2(t^3 + 4)^{-2}\,dt$

Even if it is not possible to find a substitution that makes an integrand match one of the integration formulas exactly, a substitution may simplify the integrand sufficiently so that other techniques can be used.

EXAMPLE 6   Substitution Techniques   Find $\displaystyle\int \frac{x}{\sqrt{x+2}}\,dx$.

SOLUTION   Proceeding as before, if we let $u = x + 2$, then $du = dx$ and

$$\int \frac{x}{\sqrt{x+2}}\,dx = \int \frac{x}{\sqrt{u}}\,du$$

Notice that this substitution is not complete because we have not expressed the integrand entirely in terms of u and du. As we noted earlier, only a constant factor can be moved across an integral sign, so we cannot move x outside the integral sign. Instead, we must return to the original substitution, solve for x in terms of u, and use the resulting equation to complete the substitution:

$$u = x + 2 \quad \text{Solve for } x \text{ in terms of } u.$$
$$u - 2 = x \quad \text{Substitute this expression for } x.$$

Thus,

$$\int \frac{x}{\sqrt{x+2}}\,dx = \int \frac{u-2}{\sqrt{u}}\,du \qquad\qquad \text{Simplify the integrand.}$$

$$= \int \frac{u-2}{u^{1/2}}\,du$$

$$= \int (u^{1/2} - 2u^{-1/2})\,du$$

$$\boxed{= \int u^{1/2}\,du - 2\int u^{-1/2}\,du}$$

$$= \frac{u^{3/2}}{\frac{3}{2}} - 2\frac{u^{1/2}}{\frac{1}{2}} + C$$

$$= \tfrac{2}{3}(x+2)^{3/2} - 4(x+2)^{1/2} + C \quad \text{Since } u = x + 2$$

Check:

$$\frac{d}{dx}\left[\tfrac{2}{3}(x+2)^{3/2} - 4(x+2)^{1/2}\right] = (x+2)^{1/2} - 2(x+2)^{-1/2}$$

$$= \frac{x+2}{(x+2)^{1/2}} - \frac{2}{(x+2)^{1/2}}$$

$$= \frac{x}{(x+2)^{1/2}}$$

Matched Problem 6   Find $\int x\sqrt{x+1}\,dx$.

We can find the indefinite integral of some functions in more than one way. For example, we can use substitution to find

$$\int x(1+x^2)^2\,dx$$

by letting $u = 1 + x^2$. As a second approach, we can expand the integrand, obtaining

$$\int (x + 2x^3 + x^5)\,dx$$

for which we can easily calculate an antiderivative. In such a case, choose the approach that you prefer.

There are also some functions for which substitution is not an effective approach to finding the indefinite integral. For example, substitution is not helpful in finding

$$\int e^{x^2}\, dx \quad \text{or} \quad \int \ln x\, dx$$

Application

EXAMPLE 7 Price–Demand The market research department of a supermarket chain has determined that, for one store, the marginal price $p'(x)$ at x tubes per week for a certain brand of toothpaste is given by

$$p'(x) = -0.015e^{-0.01x}$$

Find the price–demand equation if the weekly demand is 50 tubes when the price of a tube is \$4.35. Find the weekly demand when the price of a tube is \$3.89.

SOLUTION

$$p(x) = \int -0.015e^{-0.01x}\, dx$$

$$= -0.015 \int e^{-0.01x}\, dx$$

$$= -0.015 \int e^{-0.01x}\frac{-0.01}{-0.01}\, dx$$

$$= \frac{-0.015}{-0.01} \int e^{-0.01x}(-0.01)\, dx \quad \begin{array}{l}\text{Substitute } u = -0.01x \\ \text{and } du = -0.01\ dx.\end{array}$$

$$= 1.5 \int e^{u}\, du$$

$$= 1.5e^{u} + C$$

$$= 1.5e^{-0.01x} + C \qquad\qquad \text{Since } u = -0.01x$$

We find C by noting that

$$p(50) = 1.5e^{-0.01(50)} + C = \$4.35$$

$$C = \$4.35 - 1.5e^{-0.5} \quad \text{Use a calculator.}$$

$$= \$4.35 - 0.91$$

$$= \$3.44$$

So,

$$p(x) = 1.5e^{-0.01x} + 3.44$$

To find the demand when the price is \$3.89, we solve $p(x) = \$3.89$ for x:

$$1.5e^{-0.01x} + 3.44 = 3.89$$

$$1.5e^{-0.01x} = 0.45$$

$$e^{-0.01x} = 0.3$$

$$-0.01x = \ln 0.3$$

$$x = -100 \ln 0.3 \approx 120 \text{ tubes}$$

Matched Problem 7) The marginal price $p'(x)$ at a supply level of x tubes per week for a certain brand of toothpaste is given by

$$p'(x) = 0.001e^{0.01x}$$

Find the price–supply equation if the supplier is willing to supply 100 tubes per week at a price of $3.65 each. How many tubes would the supplier be willing to supply at a price of $3.98 each?

We conclude with two final cautions. The first was stated earlier, but it is worth repeating.

⚠ CAUTION

1. A variable cannot be moved across an integral sign.
2. An integral must be expressed entirely in terms of u and du before applying integration formulas 4, 5, and 6.   ▲

Exercises 13.2

Skills Warm-up Exercises

W   In Problems 1–8, use the chain rule to find the derivative of each function. (If necessary, review Section 11.4).

1. $f(x) = (5x + 1)^{10}$ 2. $f(x) = (4x - 3)^6$

3. $f(x) = (x^2 + 1)^7$ 4. $f(x) = (x^3 - 4)^5$

5. $f(x) = e^{x^2}$ 6. $f(x) = 6e^{x^3}$

7. $f(x) = \ln(x^4 - 10)$ 8. $f(x) = \ln(x^2 + 5x + 4)$

In Problems 9–44, find each indefinite integral and check the result by differentiating.

9. $\int (3x + 5)^2(3)\ dx$ 10. $\int (6x - 1)^3(6)\ dx$

11. $\int (x^2 - 1)^5(2x)\ dx$ 12. $\int (x^6 + 1)^4(6x^5)\ dx$

13. $\int (5x^3 + 1)^{-3}(15x^2)\ dx$ 14. $\int (4x^2 - 3)^{-6}(8x)\ dx$

15. $\int e^{5x}(5)\ dx$ 16. $\int e^{x^3}(3x^2)\ dx$

17. $\int \dfrac{1}{1 + x^2}(2x)\ dx$ 18. $\int \dfrac{1}{5x - 7}(5)\ dx$

19. $\int \sqrt{1 + x^4}\,(4x^3)dx$ 20. $\int (x^2 + 9)^{-1/2}(2x)dx$

21. $\int (x + 3)^{10}\ dx$ 22. $\int (x - 3)^{-4}\ dx$

23. $\int (6t - 7)^{-2}\ dt$ 24. $\int (5t + 1)^3\ dt$

25. $\int (t^2 + 1)^5\,t\ dt$ 26. $\int (t^3 + 4)^{-2}\,t^2\ dt$

27. $\int xe^{x^2}\ dx$ 28. $\int e^{-0.01x}\ dx$

29. $\int \dfrac{1}{5x + 4}\,dx$ 30. $\int \dfrac{x}{1 + x^2}\,dx$

31. $\int e^{1-t}\ dt$ 32. $\int \dfrac{3}{2 - t}\,dt$

33. $\int \dfrac{t}{(3t^2 + 1)^4}\ dt$ 34. $\int \dfrac{t^2}{(t^3 - 2)^5}\ dt$

35. $\int x\sqrt{x + 4}\ dx$ 36. $\int x\sqrt{x - 9}\ dx$

37. $\int \dfrac{x}{\sqrt{x - 3}}\,dx$ 38. $\int \dfrac{x}{\sqrt{x + 5}}\,dx$

39. $\int x(x - 4)^9\ dx$ 40. $\int x(x + 6)^8\ dx$

41. $\int e^{2x}(1 + e^{2x})^3\ dx$ 42. $\int e^{-x}(1 - e^{-x})^4\ dx$

43. $\int \dfrac{1 + x}{4 + 2x + x^2}\,dx$ 44. $\int \dfrac{x^2 - 1}{x^3 - 3x + 7}\,dx$

In Problems 45–50, the indefinite integral can be found in more than one way. First use the substitution method to find the indefinite integral. Then find it without using substitution. Check that your answers are equivalent.

45. $\int 5(5x + 3)\ dx$ 46. $\int -7(4 - 7x)\ dx$

47. $\int 2x(x^2 - 1)\ dx$ 48. $\int 3x^2(x^3 + 1)\ dx$

49. $\int 5x^4(x^5)^4\ dx$ 50. $\int 8x^7(x^8)^3\ dx$

51. Is $F(x) = x^2 e^x$ an antiderivative of $f(x) = 2xe^x$? Explain.

52. Is $F(x) = \dfrac{1}{x}$ an antiderivative of $f(x) = \ln x$? Explain.

53. Is $F(x) = (x^2 + 4)^6$ an antiderivative of $f(x) = 12x(x^2 + 4)^5$? Explain.

54. Is $F(x) = (x^2 - 1)^{100}$ an antiderivative of $f(x) = 200x(x^2 - 1)^{99}$? Explain.

55. Is $F(x) = e^{2x} + 4$ an antiderivative of $f(x) = e^{2x}$? Explain.

56. Is $F(x) = 1 - 0.2e^{-5x}$ an antiderivative of $f(x) = e^{-5x}$? Explain.

57. Is $F(x) = 0.5(\ln x)^2 + 10$ an antiderivative of $f(x) = \dfrac{\ln x}{x}$? Explain.

58. Is $F(x) = \ln(\ln x)$ an antiderivative of $f(x) = \dfrac{l}{x \ln x}$? Explain.

In Problems 59–70, find each indefinite integral and check the result by differentiating.

59. $\displaystyle \int x\sqrt{3x^2 + 7}\, dx$

60. $\displaystyle \int x^2 \sqrt{2x^3 + 1}\, dx$

61. $\displaystyle \int x(x^3 + 2)^2\, dx$

62. $\displaystyle \int x(x^2 + 2)^2\, dx$

63. $\displaystyle \int x^2(x^3 + 2)^2\, dx$

64. $\displaystyle \int (x^2 + 2)^2\, dx$

65. $\displaystyle \int \frac{x^3}{\sqrt{2x^4 + 3}}\, dx$

66. $\displaystyle \int \frac{x^2}{\sqrt{4x^3 - 1}}\, dx$

67. $\displaystyle \int \frac{(\ln x)^3}{x}\, dx$

68. $\displaystyle \int \frac{e^x}{1 + e^x}\, dx$

69. $\displaystyle \int \frac{1}{x^2} e^{-1/x}\, dx$

70. $\displaystyle \int \frac{1}{x \ln x}\, dx$

In Problems 71–76, find the family of all antiderivatives of each derivative.

71. $\dfrac{dx}{dt} = 7t^2(t^3 + 5)^6$

72. $\dfrac{dm}{dn} = 10n(n^2 - 8)^7$

73. $\dfrac{dy}{dt} = \dfrac{3t}{\sqrt{t^2 - 4}}$

74. $\dfrac{dy}{dx} = \dfrac{5x^2}{(x^3 - 7)^4}$

75. $\dfrac{dp}{dx} = \dfrac{e^x + e^{-x}}{(e^x - e^{-x})^2}$

76. $\dfrac{dm}{dt} = \dfrac{\ln(t - 5)}{t - 5}$

Applications

77. **Price–demand equation.** The marginal price for a weekly demand of x bottles of shampoo in a drugstore is given by

$$p'(x) = \frac{-6,000}{(3x + 50)^2}$$

Find the price–demand equation if the weekly demand is 150 when the price of a bottle of shampoo is $8. What is the weekly demand when the price is $6.50?

78. **Price–supply equation.** The marginal price at a supply level of x bottles of laundry detergent per week is given by

$$p'(x) = \frac{300}{(3x + 25)^2}$$

Find the price–supply equation if the distributor of the detergent is willing to supply 75 bottles a week at a price of $5.00 per bottle. How many bottles would the supplier be willing to supply at a price of $5.15 per bottle?

79. **Cost function.** The weekly marginal cost of producing x pairs of tennis shoes is given by

$$C'(x) = 12 + \frac{500}{x + 1}$$

where $C(x)$ is cost in dollars. If the fixed costs are $2,000 per week, find the cost function. What is the average cost per pair of shoes if 1,000 pairs of shoes are produced each week?

80. **Revenue function.** The weekly marginal revenue from the sale of x pairs of tennis shoes is given by

$$R'(x) = 40 - 0.02x + \frac{200}{x + 1} \qquad R(0) = 0$$

where $R(x)$ is revenue in dollars. Find the revenue function. Find the revenue from the sale of 1,000 pairs of shoes.

81. **Marketing.** An automobile company is ready to introduce a new line of hybrid cars through a national sales campaign. After test marketing the line in a carefully selected city, the marketing research department estimates that sales (in millions of dollars) will increase at the monthly rate of

$$S'(t) = 10 - 10e^{-0.1t} \qquad 0 \le t \le 24$$

t months after the campaign has started.

(A) What will be the total sales $S(t)$ t months after the beginning of the national campaign if we assume no sales at the beginning of the campaign?

(B) What are the estimated total sales for the first 12 months of the campaign?

(C) When will the estimated total sales reach $100 million? Use a graphing calculator to approximate the answer to two decimal places.

82. **Marketing.** Repeat Problem 81 if the monthly rate of increase in sales is found to be approximated by

$$S'(t) = 20 - 20e^{-0.05t} \qquad 0 \le t \le 24$$

83. **Oil production.** Using production and geological data, the management of an oil company estimates that oil will be pumped from a field producing at a rate given by

$$R(t) = \frac{100}{t + 1} + 5 \qquad 0 \le t \le 20$$

where $R(t)$ is the rate of production (in thousands of barrels per year) t years after pumping begins. How many barrels of oil $Q(t)$ will the field produce in the first t years if $Q(0) = 0$? How many barrels will be produced in the first 9 years?

84. Oil production. Assume that the rate in Problem 83 is found to be

$$R(t) = \frac{120t}{t^2 + 1} + 3 \qquad 0 \le t \le 20$$

(A) When is the rate of production greatest?

(B) How many barrels of oil $Q(t)$ will the field produce in the first t years if $Q(0) = 0$? How many barrels will be produced in the first 5 years?

(C) How long (to the nearest tenth of a year) will it take to produce a total of a quarter of a million barrels of oil?

85. Biology. A yeast culture is growing at the rate of $w'(t) = 0.2e^{0.1t}$ grams per hour. If the starting culture weighs 2 grams, what will be the weight of the culture $W(t)$ after t hours? After 8 hours?

86. Medicine. The rate of healing for a skin wound (in square centimeters per day) is approximated by $A'(t) = -0.9e^{-0.1t}$. If the initial wound has an area of 9 square centimeters, what will its area $A(t)$ be after t days? After 5 days?

87. Pollution. A contaminated lake is treated with a bactericide. The rate of increase in harmful bacteria t days after the treatment is given by

$$\frac{dN}{dt} = -\frac{2,000t}{1 + t^2} \qquad 0 \le t \le 10$$

where $N(t)$ is the number of bacteria per milliliter of water. Since dN/dt is negative, the count of harmful bacteria is decreasing.

(A) Find the minimum value of dN/dt.

(B) If the initial count was 5,000 bacteria per milliliter, find $N(t)$ and then find the bacteria count after 10 days.

(C) When (to two decimal places) is the bacteria count 1,000 bacteria per milliliter?

88. Pollution. An oil tanker aground on a reef is losing oil and producing an oil slick that is radiating outward at a rate given approximately by

$$\frac{dR}{dt} = \frac{60}{\sqrt{t + 9}} \qquad t \ge 0$$

where R is the radius (in feet) of the circular slick after t minutes. Find the radius of the slick after 16 minutes if the radius is 0 when $t = 0$.

89. Learning. An average student enrolled in an advanced typing class progressed at a rate of $N'(t) = 6e^{-0.1t}$ words per minute per week t weeks after enrolling in a 15-week course. If, at the beginning of the course, a student could type 40 words per minute, how many words per minute $N(t)$ would the student be expected to type t weeks into the course? After completing the course?

90. Learning. An average student enrolled in a stenotyping class progressed at a rate of $N'(t) = 12e^{-0.06t}$ words per minute per week t weeks after enrolling in a 15-week course. If, at the beginning of the course, a student could stenotype at zero words per minute, how many words per minute $N(t)$ would the student be expected to handle t weeks into the course? After completing the course?

91. College enrollment. The projected rate of increase in enrollment at a new college is estimated by

$$\frac{dE}{dt} = 5,000(t + 1)^{-3/2} \qquad t \ge 0$$

where $E(t)$ is the projected enrollment in t years. If enrollment is 2,000 now $(t = 0)$, find the projected enrollment 15 years from now.

Answers to Matched Problems

1. (A) $\frac{1}{21}(2x^3 - 3)^{21} + C$ (B) $e^{5x} + C$

    (C) $\ln|4 + x^2| + C$ or $\ln(4 + x^2) + C$, since $4 + x^2 > 0$

2. (A) $dy = 3x^2\, dx$

    (B) $du = \frac{2x}{2 + x^2}dx$

    (C) $dv = -5e^{-5t}\, dt$

3. $\frac{1}{5}(x^2 - 3x + 7)^5 + C$

4. (A) $\frac{1}{5}(2x^3 - 3)^5 + C$ (B) $e^{5w} + C$

5. (A) $-\frac{1}{3}e^{-3x} + C$ (B) $\frac{1}{2}\ln|x^2 - 9| + C$

    (C) $-\frac{5}{3}(t^3 + 4)^{-1} + C$

6. $\frac{2}{5}(x + 1)^{5/2} - \frac{2}{3}(x + 1)^{3/2} + C$

7. $p(x) = 0.1e^{0.01x} + 3.38$; 179 tubes

13.3 Differential Equations; Growth and Decay

- Differential Equations and Slope Fields
- Continuous Compound Interest Revisited
- Exponential Growth Law
- Population Growth, Radioactive Decay, and Learning
- Comparison of Exponential Growth Phenomena

In the preceding section, we considered equations of the form

$$\frac{dy}{dx} = 6x^2 - 4x \qquad y' = -400e^{-0.04x}$$

These are examples of *differential equations*. In general, an equation is a **differential equation** if it involves an unknown function and one or more of its derivatives. Other examples of differential equations are

$$\frac{dy}{dx} = ky \qquad y'' - xy' + x^2 = 5 \qquad \frac{dy}{dx} = 2xy$$

The first and third equations are called **first-order** equations because each involves a first derivative but no higher derivative. The second equation is called a **second-order** equation because it involves a second derivative but no higher derivative.

A **solution** of a differential equation is a function $f(x)$ which, when substituted for y, satisfies the equation; that is, the left side and right side of the equation are the same function. Finding a solution of a given differential equation may be very difficult. However, it is easy to determine whether or not a given function is a solution of a given differential equation. Just substitute and check whether both sides of the differential equation are equal as functions. For example, even if you have trouble finding a function y that satisfies the differential equation

$$(x - 3)\frac{dy}{dx} = y + 4 \tag{1}$$

it is easy to determine whether or not the function $y = 5x - 19$ is a solution: Since $dy/dx = 5$, the left side of (1) is $(x - 3)5$ and the right side is $(5x - 19) + 4$, so the left and right sides are equal and $y = 5x - 19$ is a solution.

In this section, we emphasize a few special first-order differential equations that have immediate and significant applications. We start by looking at some first-order equations geometrically, in terms of *slope fields*. We then consider continuous compound interest as modeled by a first-order differential equation. From this treatment, we can generalize our approach to a wide variety of other types of growth phenomena.

Differential Equations and Slope Fields

We introduce the concept of *slope field* through an example. Consider the first-order differential equation

$$\frac{dy}{dx} = 0.2y \tag{2}$$

A function f is a solution of equation (2) if $y = f(x)$ satisfies equation (2) for all values of x in the domain of f. Geometrically interpreted, equation (2) gives us the slope of a solution curve that passes through the point (x, y). For example, if $y = f(x)$ is a solution of equation (2) that passes through the point $(0, 2)$, then the slope of f at $(0, 2)$ is given by

$$\frac{dy}{dx} = 0.2(2) = 0.4$$

We indicate this relationship by drawing a short segment of the tangent line at the point $(0, 2)$, as shown in Figure 1A. The procedure is repeated for points $(-3, 1)$ and $(2, 3)$. Assuming that the graph of f passes through all three points, we sketch an approximate graph of f in Figure 1B.

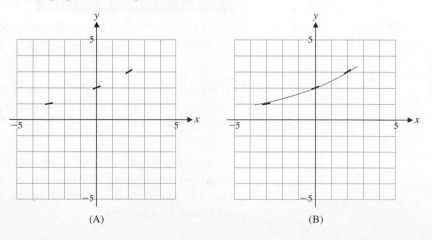

(A)   (B)

Figure 1

If we continue the process of drawing tangent line segments at each point grid in Figure 1—a task easily handled by computers, but not by hand—we obtain a *slope field*. A slope field for differential equation (2), drawn by a computer, is shown in Figure 2. In general, a **slope field** for a first-order differential equation is obtained by drawing tangent line segments determined by the equation at each point in a grid.

Explore and Discuss 1

(A) In Figure 1A (or a copy), draw tangent line segments for a solution curve of differential equation (2) that passes through $(-3, -1), (0, -2)$, and $(2, -3)$.

(B) In Figure 1B (or a copy), sketch an approximate graph of the solution curve that passes through the three points given in part (A). Repeat the tangent line segments first.

(C) What type of function, of all the elementary functions discussed in the first two chapters, appears to be a solution of differential equation (2)?

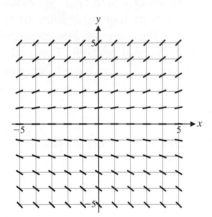

Figure 2

In Explore & Discuss 1, if you guessed that all solutions of equation (2) are exponential functions, you are to be congratulated. We now show that

$$y = Ce^{0.2x} \tag{3}$$

is a solution of equation (2) for any real number C. We substitute $y = Ce^{0.2x}$ into equation (2) to see if the left side is equal to the right side for all real x:

$$\frac{dy}{dx} = 0.2y$$

$$\textit{Left side:} \quad \frac{dy}{dx} = \frac{d}{dx}(Ce^{0.2x}) = 0.2Ce^{0.2x}$$

$$\textit{Right side:} \quad 0.2y = 0.2Ce^{0.2x}$$

So equation (3) is a solution of equation (2) for C any real number. Which values of C will produce solution curves that pass through $(0, 2)$ and $(0, -2)$, respectively? Substituting the coordinates of each point into equation (3) and solving for C, we obtain

$$y = 2e^{0.2x} \quad \text{and} \quad y = -2e^{0.2x} \tag{4}$$

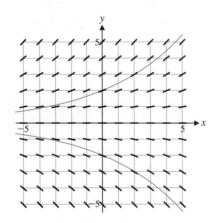

Figure 3

The graphs of equations (4) are shown in Figure 3, and they confirm the results shown in Figure 1B. We say that (3) is the **general solution** of the differential equation (2), and the functions in (4) are the **particular solutions** that satisfy $y(0) = 2$ and $y(0) = -2$, respectively.

CONCEPTUAL INSIGHT

For a complicated first-order differential equation, say,

$$\frac{dy}{dx} = \frac{3 + \sqrt{xy}}{x^2 - 5y^4}$$

it may be impossible to find a formula analogous to (3) for its solutions. Nevertheless, it is routine to evaluate the right-hand side at each point in a grid. The resulting slope field provides a graphical representation of the solutions of the differential equation.

Drawing slope fields by hand is not a task for human beings: A 20-by-20 grid would require drawing 400 tangent line segments! Repetitive tasks of this type are what computers are for. A few problems in Exercises 13.3 involve interpreting slope fields, not drawing them.

Continuous Compound Interest Revisited

Let P be the initial amount of money deposited in an account, and let A be the amount in the account at any time t. Instead of assuming that the money in the account earns a particular rate of interest, suppose we say that the rate of growth of the amount of money in the account at any time t is proportional to the amount present at that time. Since dA/dt is the rate of growth of A with respect to t, we have

$$\frac{dA}{dt} = rA \qquad A(0) = P \qquad A, P > 0 \tag{5}$$

where r is an appropriate constant. We would like to find a function $A = A(t)$ that satisfies these conditions. Multiplying both sides of equation (5) by $1/A$, we obtain

$$\frac{1}{A}\frac{dA}{dt} = r$$

Now we integrate each side with respect to t:

$$\int \frac{1}{A}\frac{dA}{dt}\,dt = \int r\,dt \qquad \frac{dA}{dt}\,dt = A'(t)\,dt = dA$$

$$\int \frac{1}{A}\,dA = \int r\,dt$$

$$\ln|A| = rt + C \quad |A| = A,\ \text{since } A > 0$$

$$\ln A = rt + C$$

We convert this last equation into the equivalent exponential form

$$A = e^{rt+C} \quad \begin{array}{l}\text{Definition of logarithmic function:}\\ y = \ln x \text{ if and only if } x = e^{y}\end{array}$$

$$= e^{C}e^{rt} \quad \text{Property of exponents: } b^{m}b^{n} = b^{m+n}$$

Since $A(0) = P$, we evaluate $A(t) = e^{C}e^{rt}$ at $t = 0$ and set the result equal to P:

$$A(0) = e^{C}e^{0} = e^{C} = P$$

Hence, $e^{C} = P$, and we can rewrite $A = e^{C}e^{rt}$ in the form

$$A = Pe^{rt}$$

This is the same continuous compound interest formula obtained in Section 4.1, where the principal P is invested at an annual nominal rate r compounded continuously for t years.

Exponential Growth Law

In general, if the rate of change of a quantity Q with respect to time is proportional to the amount of Q present and $Q(0) = Q_0$, then, proceeding in exactly the same way as we just did, we obtain the following theorem:

THEOREM 1 Exponential Growth Law

If $\dfrac{dQ}{dt} = rQ$ and $Q(0) = Q_0$, then $Q = Q_0 e^{rt}$,

where

$\qquad Q_0 =$ amount of Q at $t = 0$
$\qquad r =$ relative growth rate (expressed as a decimal)
$\qquad t =$ time
$\qquad Q =$ quantity at time t

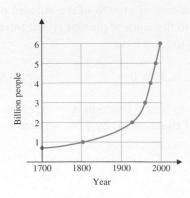

Figure 4 **World population growth**

The constant r in the exponential growth law is called the **relative growth rate**. If the relative growth rate is $r = 0.02$, then the quantity Q is growing at a rate $dQ/dt = 0.02Q$ (that is, 2% of the quantity Q per unit of time t). Note the distinction between the relative growth rate r and the rate of growth dQ/dt of the quantity Q. If $r < 0$, then $dQ/dt < 0$ and Q is decreasing. This type of growth is called **exponential decay**.

Once we know that the rate of growth is proportional to the amount present, we recognize exponential growth and can use Theorem 1 without solving the differential equation each time. The exponential growth law applies not only to money invested at interest compounded continuously, but also to many other types of problems—population growth, radioactive decay, the depletion of a natural resource, and so on.

Population Growth, Radioactive Decay, and Learning

The world population passed 1 billion in 1804, 2 billion in 1927, 3 billion in 1960, 4 billion in 1974, 5 billion in 1987, and 6 billion in 1999, as illustrated in Figure 4. **Population growth** over certain periods often can be approximated by the exponential growth law of Theorem 1.

EXAMPLE 1   Population Growth India had a population of about 1.2 billion in 2010 $(t = 0)$. Let P represent the population (in billions) t years after 2010, and assume a growth rate of 1.5% compounded continuously.

(A) Find an equation that represents India's population growth after 2010, assuming that the 1.5% growth rate continues.

(B) What is the estimated population (to the nearest tenth of a billion) of India in the year 2030?

(C) Graph the equation found in part (A) from 2010 to 2030.

SOLUTION

(A) The exponential growth law applies, and we have

$$\frac{dP}{dt} = 0.015P \qquad P(0) = 1.2$$

Therefore,

$$P = 1.2e^{0.015t} \qquad\qquad (6)$$

(B) Using equation (6), we can estimate the population in India in 2030 $(t = 20)$:

$$P = 1.2e^{0.015(20)} = 1.6 \text{ billion people}$$

(C) The graph is shown in Figure 5.

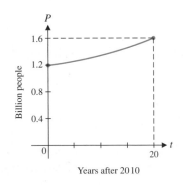

Figure 5 **Population of India**

Matched Problem 1 Assuming the same continuous compound growth rate as in Example 1, what will India's population be (to the nearest tenth of a billion) in the year 2020?

EXAMPLE 2   Population Growth If the exponential growth law applies to Canada's population growth, at what continuous compound growth rate will the population double over the next 100 years?

SOLUTION We must find r, given that $P = 2P_0$ and $t = 100$:

$$P = P_0e^{rt}$$
$$2P_0 = P_0e^{100r}$$
$$2 = e^{100r} \qquad \text{Take the natural logarithm}$$
$$100r = \ln 2 \qquad \text{of both sides and reverse}$$
$$\qquad\qquad \text{the equation.}$$
$$r = \frac{\ln 2}{100}$$
$$\approx 0.0069 \quad \text{or} \quad 0.69\%$$

Matched Problem 2⌋ If the exponential growth law applies to population growth in Nigeria, find the doubling time (to the nearest year) of the population if it grows at 2.1% per year compounded continuously.

We now turn to another type of exponential growth: **radioactive decay**. In 1946, Willard Libby (who later received a Nobel Prize in chemistry) found that as long as a plant or animal is alive, radioactive carbon-14 is maintained at a constant level in its tissues. Once the plant or animal is dead, however, the radioactive carbon-14 diminishes by radioactive decay at a rate proportional to the amount present.

$$\frac{dQ}{dt} = rQ \qquad Q(0) = Q_0$$

This is another example of the exponential growth law. The continuous compound rate of decay for radioactive carbon-14 is 0.000 123 8, so $r = -0.000\ 123\ 8$, since decay implies a negative continuous compound growth rate.

EXAMPLE 3 Archaeology A human bone fragment was found at an archaeological site in Africa. If 10% of the original amount of radioactive carbon-14 was present, estimate the age of the bone (to the nearest 100 years).

SOLUTION By the exponential growth law for

$$\frac{dQ}{dt} = -0.000\ 123\ 8Q \qquad Q(0) = Q_0$$

we have

$$Q = Q_0 e^{-0.0001238t}$$

We must find t so that $Q = 0.1Q_0$ (since the amount of carbon-14 present now is 10% of the amount Q_0 present at the death of the person).

$$0.1Q_0 = Q_0 e^{-0.0001238t}$$
$$0.1 = e^{-0.0001238t}$$
$$\ln 0.1 = \ln e^{-0.0001238t}$$
$$t = \frac{\ln 0.1}{-0.000\ 123\ 8} \approx 18{,}600 \text{ years}$$

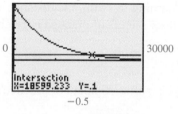

Figure 6 $y_1 = e^{-0.0001238x}$; $y_2 = 0.1$

See Figure 6 for a graphical solution to Example 3.

Matched Problem 3⌋ Estimate the age of the bone in Example 3 (to the nearest 100 years) if 50% of the original amount of carbon-14 is present.

In learning certain skills, such as typing and swimming, one often assumes that there is a maximum skill attainable—say, M—and the rate of improvement is proportional to the difference between what has been achieved y and the maximum attainable M. Mathematically,

$$\frac{dy}{dt} = k(M - y) \qquad y(0) = 0$$

We solve this type of problem with the same technique used to obtain the exponential growth law. First, multiply both sides of the first equation by $1/(M - y)$ to get

$$\frac{1}{M - y}\frac{dy}{dt} = k$$

and then integrate each side with respect to t:

$$\int \frac{1}{M-y}\frac{dy}{dt}dt = \int k\,dt$$

$$-\int \frac{1}{M-y}\left(-\frac{dy}{dt}\right)dt = \int k\,dt \qquad \text{Substitute } u = M - y \text{ and}$$

$$-\int \frac{1}{u}du = \int k\,dt \qquad du = -dy = -\frac{dy}{dt}dt.$$

$$-\ln|u| = kt + C \qquad \text{Substitute } M - y, \text{ which is } > 0, \text{ for } u.$$

$$-\ln(M-y) = kt + C \qquad \text{Multiply both sides by } -1.$$

$$\ln(M-y) = -kt - C$$

Change this last equation to an equivalent exponential form:

$$M - y = e^{-kt-C}$$

$$M - y = e^{-C}e^{-kt}$$

$$y = M - e^{-C}e^{-kt}$$

Now, $y(0) = 0$; hence,

$$y(0) = M - e^{-C}e^{0} = 0$$

Solving for e^{-C}, we obtain

$$e^{-C} = M$$

and our final solution is

$$y = M - Me^{-kt} = M(1 - e^{-kt})$$

EXAMPLE 4   Learning   For a particular person learning to swim, the distance y (in feet) that the person is able to swim in 1 minute after t hours of practice is given approximately by

$$y = 50(1 - e^{-0.04t})$$

What is the rate of improvement (to two decimal places) after 10 hours of practice?

SOLUTION

$$y = 50 - 50e^{-0.04t}$$

$$y'(t) = 2e^{-0.04t}$$

$$y'(10) = 2e^{-0.04(10)} \approx 1.34 \text{ feet per hour of practice}$$

Matched Problem 4⎦ In Example 4, what is the rate of improvement (to two decimal places) after 50 hours of practice?

Comparison of Exponential Growth Phenomena

Table 1 compares four widely used growth models. Each model (column 2) consists of a first-order differential equation and an **initial condition** that specifies $y(0)$, the value of a solution y when $x = 0$. The differential equation has a family of solutions, but there is only one solution (the particular solution in column 3) that also satisfies the initial condition [just as there is a family, $y = x^2 + k$, of antiderivatives of $g(x) = 2x$, but only one antiderivative (the particular antiderivative $y = x^2 + 5$) that also satisfies the condition $y(0) = 5$]. A graph of the model's solution is shown in column 4 of Table 1, followed by a short (and necessarily incomplete) list of areas in which the model is used.

Table 1 **Exponential Growth**

Description	Model	Solution	Graph	Uses
Unlimited growth: Rate of growth is proportional to the amount present	$\dfrac{dy}{dt} = ky$ $k, t > 0$ $y(0) = c$	$y = ce^{kt}$		• Short-term population growth (people, bacteria, etc.) • Growth of money at continuous compound interest • Price–supply curves
Exponential decay: Rate of growth is proportional to the amount present	$\dfrac{dy}{dt} = -ky$ $k, t > 0$ $y(0) = c$	$y = ce^{-kt}$		• Depletion of natural resources • Radioactive decay • Absorption of light in water • Price–demand curves • Atmospheric pressure (t is altitude)
Limited growth: Rate of growth is proportional to the difference between the amount present and a fixed limit	$\dfrac{dy}{dt} = k(M - y)$ $k, t > 0$ $y(0) = 0$	$y = M(1 - e^{-kt})$		• Sales fads (for example, skateboards) • Depreciation of equipment • Company growth • Learning
Logistic growth: Rate of growth is proportional to the amount present and to the difference between the amount present and a fixed limit	$\dfrac{dy}{dt} = ky(M - y)$ $k, t > 0$ $y(0) = \dfrac{M}{1 + c}$	$y = \dfrac{M}{1 + ce^{-kMt}}$		• Long-term population growth • Epidemics • Sales of new products • Spread of a rumor • Company growth

Exercises 13.3

Skills Warm-up Exercises

In Problems 1–8, express the relationship between $f'(x)$ and $f(x)$ in words, and write a differential equation that $f(x)$ satisfies. For example, the derivative of $f(x) = e^{3x}$ is 3 times $f(x)$; $y' = 3y$. (If necessary, review Section 11.4).

1. $f(x) = e^{5x}$

2. $f(x) - e^{-2x}$

3. $f(x) = 10e^{-x}$

4. $f(x) = 25e^{0.04x}$

5. $f(x) = 3.2e^{x^2}$

6. $f(x) = e^{-x^2}$

7. $f(x) = 1 - e^{-x}$

8. $f(x) = 1 - e^{-3x}$

In Problems 9–20, find the general or particular solution, as indicated, for each first-order differential equation.

9. $\dfrac{dy}{dx} = 6x$

10. $\dfrac{dy}{dx} = 3x^{-2}$

11. $\dfrac{dy}{dx} = \dfrac{7}{x}$

12. $\dfrac{dy}{dx} = e^{0.1x}$

13. $\dfrac{dy}{dx} = e^{0.02x}$

14. $\dfrac{dy}{dx} = 8x^{-1}$

15. $\dfrac{dy}{dx} = x^2 - x; y(0) = 0$

16. $\dfrac{dy}{dx} = \sqrt{x}; y(0) = 0$

17. $\dfrac{dy}{dx} = -2xe^{-x^2}; y(0) = 3$

18. $\dfrac{dy}{dx} = e^{x-3}; y(3) = -5$

19. $\dfrac{dy}{dx} = \dfrac{2}{1 + x}; y(0) = 5$

20. $\dfrac{dy}{dx} = \dfrac{1}{4(3 - x)}; y(0) = 1$

In Problems 21–24, give the order (first, second, third, etc.) of each differential equation, where y represents a function of the variable x.

21. $y - 2y' + x^3y'' = 0$

22. $xy' + y^4 = e^x$

23. $y''' - 3y'' + 3y' - y = 0$

24. $y^3 + x^4y'' = \dfrac{5y}{1 + x^2}$

25. Is $y = 5x$ a solution of the differential equation $\dfrac{dy}{dx} = \dfrac{y}{x}$? Explain.

26. Is $y = 8x + 8$ a solution of the differential equation $\dfrac{dy}{dx} = \dfrac{y}{x + 1}$? Explain.

27. Is $y = \sqrt{9 + x^2}$ a solution of the differential equation $y' = \dfrac{x}{y}$? Explain.

28. Is $y = 5e^{x^2/2}$ a solution of the differential equation $y' = xy$? Explain.

29. Is $y = e^{3x}$ a solution of the differential equation $y'' - 4y' + 3y = 0$? Explain.

30. Is $y = -2e^x$ a solution of the differential equation
$y'' - 4y' + 3y = 0$? Explain.

31. Is $y = 100e^{3x}$ a solution of the differential equation
$y'' - 4y' + 3y = 0$? Explain.

32. Is $y = e^{-3x}$ a solution of the differential equation
$y'' - 4y' + 3y = 0$? Explain.

Problems 33–38 refer to the following slope fields:

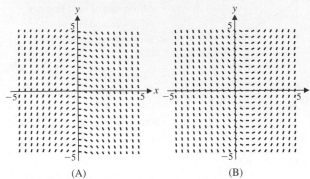

(A)   (B)

Figure for 33–38

33. Which slope field is associated with the differential equation
$dy/dx = x - 1$? Briefly justify your answer.

34. Which slope field is associated with the differential equation
$dy/dx = -x$? Briefly justify your answer.

35. Solve the differential equation $dy/dx = x - 1$ and find the
particular solution that passes through $(0, -2)$.

36. Solve the differential equation $dy/dx = -x$ and find the
particular solution that passes through $(0, 3)$.

37. Graph the particular solution found in Problem 35 in the
appropriate Figure A or B (or a copy).

38. Graph the particular solution found in Problem 36 in the
appropriate Figure A or B (or a copy).

*In Problems 39–46, find the general or particular solution, as in-
dicated, for each differential equation.*

39. $\dfrac{dy}{dt} = 2y$

40. $\dfrac{dy}{dt} = -3y$

41. $\dfrac{dy}{dx} = -0.5y; \; y(0) = 100$

42. $\dfrac{dy}{dx} = 0.1y; \; y(0) = -2.5$

43. $\dfrac{dx}{dt} = -5x$

44. $\dfrac{dx}{dt} = 4t$

45. $\dfrac{dx}{dt} = -5t$

46. $\dfrac{dx}{dt} = 4x$

*In Problems 47–50, does the given differential equation model
unlimited growth, exponential decay, limited growth, or logistic
growth?*

47. $y' = 2.5y(300 - y)$

48. $y' = -0.0152y$

49. $y' = 0.43y$

50. $y' = 10,000 - y$

Problems 51–58 refer to the following slope fields:

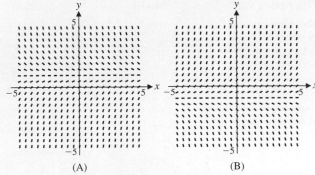

(A)   (B)

Figure for 51–58

51. Which slope field is associated with the differential equation
$dy/dx = 1 - y$? Briefly justify your answer.

52. Which slope field is associated with the differential equation
$dy/dx = y + 1$? Briefly justify your answer.

53. Show that $y = 1 - Ce^{-x}$ is a solution of the differential
equation $dy/dx = 1 - y$ for any real number C. Find the
particular solution that passes through $(0, 0)$.

54. Show that $y = Ce^x - 1$ is a solution of the differential equation $dy/dx = y + 1$ for any real number C. Find the particular solution that passes through $(0, 0)$.

55. Graph the particular solution found in Problem 53 in the
appropriate Figure A or B (or a copy).

56. Graph the particular solution found in Problem 54 in the
appropriate Figure A or B (or a copy).

57. Use a graphing calculator to graph $y = 1 - Ce^{-x}$ for
$C = -2, -1, 1,$ and 2, for $-5 \le x \le 5, -5 \le y \le 5$, all in
the same viewing window. Observe how the solution curves
go with the flow of the tangent line segments in the corresponding slope field shown in Figure A or Figure B.

58. Use a graphing calculator to graph $y = Ce^x - 1$ for
$C = -2, -1, 1,$ and 2, for $-5 \le x \le 5, -5 \le y \le 5$, all in
the same viewing window. Observe how the solution curves
go with the flow of the tangent line segments in the corresponding slope field shown in Figure A or Figure B.

59. Show that $y = \sqrt{C - x^2}$ is a solution of the differential
equation $dy/dx = -x/y$ for any positive real number C. Find
the particular solution that passes through $(3, 4)$.

60. Show that $y = \sqrt{x^2 + C}$ is a solution of the differential
equation $dy/dx = x/y$ for any real number C. Find the particular solution that passes through $(-6, 7)$.

61. Show that $y = Cx$ is a solution of the differential equation
$dy/dx = y/x$ for any real number C. Find the particular
solution that passes through $(-8, 24)$.

62. Show that $y = C/x$ is a solution of the differential equation
$dy/dx = -y/x$ for any real number C. Find the particular
solution that passes through $(2, 5)$.

63. Show that $y = 1/(1 + ce^{-t})$ is a solution of the differential
equation $dy/dt = y(1 - y)$ for any real number c. Find the particular solution that passes through $(0, -1)$.

64. Show that $y = 2/(1 + ce^{-6t})$ is a solution of the differential
equation $dy/dt = 3y(2 - y)$ for any real number c. Find the
particular solution that passes through $(0, 1)$.

In Problems 65–72, use a graphing calculator to graph the given examples of the various cases in Table 1 on page 743.

65. Unlimited growth:

$y = 1{,}000e^{0.08t}$
$0 \le t \le 15$
$0 \le y \le 3{,}500$

66. Unlimited growth:

$y = 5{,}250e^{0.12t}$
$0 \le t \le 10$
$0 \le y \le 20{,}000$

67. Exponential decay:

$p = 100e^{-0.05x}$
$0 \le x \le 30$
$0 \le p \le 100$

68. Exponential decay:

$p = 1{,}000e^{-0.08x}$
$0 \le x \le 40$
$0 \le p \le 1{,}000$

69. Limited growth:

$N = 100(1 - e^{-0.05t})$
$0 \le t \le 100$
$0 \le N \le 100$

70. Limited growth:

$N = 1{,}000(1 - e^{-0.07t})$
$0 \le t \le 70$
$0 \le N \le 1{,}000$

71. Logistic growth:

$N = \dfrac{1{,}000}{1 + 999e^{-0.4t}}$
$0 \le t \le 40$
$0 \le N \le 1{,}000$

72. Logistic growth:

$N = \dfrac{400}{1 + 99e^{-0.4t}}$
$0 \le t \le 30$
$0 \le N \le 400$

73. Show that the rate of logistic growth, $dy/dt = ky(M - y)$, has its maximum value when $y = M/2$.

74. Find the value of t for which the logistic function

$$y = \frac{M}{1 + ce^{-kMt}}$$

is equal to $M/2$.

75. Let $Q(t)$ denote the population of the world at time t. In 1999, the world population was 6.0 billion and increasing at 1.3% per year; in 2009, it was 6.8 billion and increasing at 1.2% per year. In which year, 1999 or 2009, was dQ/dt (the rate of growth of Q with respect to t) greater? Explain.

76. Refer to Problem 75. Explain why the world population function $Q(t)$ does not satisfy an exponential growth law.

Applications

77. Continuous compound interest. Find the amount A in an account after t years if

$$\frac{dA}{dt} = 0.03A \quad \text{and} \quad A(0) = 1{,}000$$

78. Continuous compound interest. Find the amount A in an account after t years if

$$\frac{dA}{dt} = 0.02A \quad \text{and} \quad A(0) = 5{,}250$$

79. Continuous compound interest. Find the amount A in an account after t years if

$$\frac{dA}{dt} = rA \quad A(0) = 8{,}000 \quad A(2) = 8{,}260.14$$

80. Continuous compound interest. Find the amount A in an account after t years if

$$\frac{dA}{dt} = rA \quad A(0) = 5{,}000 \quad A(5) = 5{,}581.39$$

81. Price–demand. The marginal price dp/dx at x units of demand per week is proportional to the price p. There is no weekly demand at a price of \$100 per unit $[p(0) = 100]$, and there is a weekly demand of 5 units at a price of \$77.88 per unit $[p(5) = 77.88]$.

(A) Find the price–demand equation.

(B) At a demand of 10 units per week, what is the price?

(C) Graph the price–demand equation for $0 \le x \le 25$.

82. Price–supply. The marginal price dp/dx at x units of supply per day is proportional to the price p. There is no supply at a price of \$10 per unit $[p(0) = 10]$, and there is a daily supply of 50 units at a price of \$12.84 per unit $[p(50) = 12.84]$.

(A) Find the price–supply equation.

(B) At a supply of 100 units per day, what is the price?

(C) Graph the price–supply equation for $0 \le x \le 250$.

83. Advertising. A company is trying to expose a new product to as many people as possible through TV ads. Suppose that the rate of exposure to new people is proportional to the number of those who have not seen the product out of L possible viewers (limited growth). No one is aware of the product at the start of the campaign, and after 10 days, 40% of L are aware of the product. Mathematically,

$$\frac{dN}{dt} = k(L - N) \quad N(0) = 0 \quad N(10) = 0.4L$$

(A) Solve the differential equation.

(B) What percent of L will have been exposed after 5 days of the campaign?

(C) How many days will it take to expose 80% of L?

(D) Graph the solution found in part (A) for $0 \le t \le 90$.

84. Advertising. Suppose that the differential equation for Problem 83 is

$$\frac{dN}{dt} = k(L - N) \quad N(0) = 0 \quad N(10) = 0.1L$$

(A) Explain what the equation $N(10) = 0.1L$ means.

(B) Solve the differential equation.

(C) How many days will it take to expose 50% of L?

(D) Graph the solution found in part (B) for $0 \le t \le 300$.

85. Biology. For relatively clear bodies of water, the intensity of light is reduced according to

$$\frac{dI}{dx} = -kI \quad I(0) = I_0$$

where I is the intensity of light at x feet below the surface. For the Sargasso Sea off the West Indies, $k = 0.00942$. Find I in terms of x, and find the depth at which the light is reduced to half of that at the surface.

86. Blood pressure. Under certain assumptions, the blood pressure P in the largest artery in the human body (the aorta) changes between beats with respect to time t according to

$$\frac{dP}{dt} = -aP \quad P(0) = P_0$$

where a is a constant. Find $P = P(t)$ that satisfies both conditions.

87. Drug concentration. A single injection of a drug is administered to a patient. The amount Q in the body then decreases at a rate proportional to the amount present. For a particular drug, the rate is 4% per hour. Thus,

$$\frac{dQ}{dt} = -0.04Q \qquad Q(0) = Q_0$$

where t is time in hours.

(A) If the initial injection is 3 milliliters $[Q(0) = 3]$, find $Q = Q(t)$ satisfying both conditions.

(B) How many milliliters (to two decimal places) are in the body after 10 hours?

(C) How many hours (to two decimal places) will it take for only 1 milliliter of the drug to be left in the body?

(D) Graph the solution found in part (A).

88. Simple epidemic. A community of 1,000 people is homogeneously mixed. One person who has just returned from another community has influenza. Assume that the home community has not had influenza shots and all are susceptible. One mathematical model assumes that influenza tends to spread at a rate in direct proportion to the number N who have the disease and to the number $1,000 - N$ who have not yet contracted the disease (logistic growth). Mathematically,

$$\frac{dN}{dt} = kN(1,000 - N) \qquad N(0) = 1$$

where N is the number of people who have contracted influenza after t days. For $k = 0.0004$, $N(t)$ is the logistic growth function

$$N(t) = \frac{1,000}{1 + 999e^{-0.4t}}$$

(A) How many people have contracted influenza after 10 days? After 20 days?

(B) How many days will it take until half the community has contracted influenza?

(C) Find $\lim_{t \to \infty} N(t)$.

(D) Graph $N = N(t)$ for $0 \le t \le 30$.

89. Nuclear accident. One of the dangerous radioactive isotopes detected after the Chernobyl nuclear disaster in 1986 was cesium-137. If 93.3% of the cesium-137 emitted during the disaster was still present 3 years later, find the continuous compound rate of decay of this isotope.

90. Insecticides. Many countries have banned the use of the insecticide DDT because of its long-term adverse effects. Five years after a particular country stopped using DDT, the amount of DDT in the ecosystem had declined to 75% of the amount present at the time of the ban. Find the continuous compound rate of decay of DDT.

91. Archaeology. A skull found in an ancient tomb has 5% of the original amount of radioactive carbon-14 present. Estimate the age of the skull. (See Example 3.)

92. Learning. For a person learning to type, the number N of words per minute that the person could type after t hours of practice was given by the limited growth function

$$N = 100(1 - e^{-0.02t})$$

What is the rate of improvement after 10 hours of practice? After 40 hours of practice?

93. Small-group analysis. In a study on small-group dynamics, sociologists found that when 10 members of a discussion group were ranked according to the number of times each participated, the number $N(k)$ of times that the kth-ranked person participated was given by

$$N(k) = N_1 e^{-0.11(k-1)} \qquad 1 \le k \le 10$$

where N_1 is the number of times that the first-ranked person participated in the discussion. If $N_1 = 180$, in a discussion group of 10 people, estimate how many times the sixth-ranked person participated. How about the 10th-ranked person?

94. Perception. The Weber–Fechner law concerns a person's sensed perception of various strengths of stimulation involving weights, sound, light, shock, taste, and so on. One form of the law states that the rate of change of sensed sensation S with respect to stimulus R is inversely proportional to the strength of the stimulus R. So

$$\frac{dS}{dR} = \frac{k}{R}$$

where k is a constant. If we let R_0 be the threshold level at which the stimulus R can be detected (the least amount of sound, light, weight, and so on, that can be detected), then

$$S(R_0) = 0$$

Find a function S in terms of R that satisfies these conditions.

95. Rumor propagation. Sociologists have found that a rumor tends to spread at a rate in direct proportion to the number x who have heard it and to the number $P - x$ who have not, where P is the total population (logistic growth). If a resident of a 400-student dormitory hears a rumor that there is a case of TB on campus, then $P = 400$ and

$$\frac{dx}{dt} = 0.001x(400 - x) \qquad x(0) = 1$$

where t is time (in minutes). From these conditions, it can be shown that $x(t)$ is the logistic growth function

$$x(t) = \frac{400}{1 + 399e^{-0.4t}}$$

(A) How many people have heard the rumor after 5 minutes? after 20 minutes?

(B) Find $\lim_{t \to \infty} x(t)$.

(C) Graph $x = x(t)$ for $0 \le t \le 30$.

96. Rumor propagation. In Problem 95, how long (to the nearest minute) will it take for half of the group of 400 to have heard the rumor?

Answers to Matched Problems

1. 1.4 billion people
2. 33 yr
3. 5,600 yr
4. 0.27 ft/hr

13.4 The Definite Integral

- Approximating Areas by Left and Right Sums

- The Definite Integral as a Limit of Sums

- Properties of the Definite Integral

The first three sections of this chapter focused on the *indefinite integral*. In this section, we introduce the *definite integral*. The definite integral is used to compute areas, probabilities, average values of functions, future values of continuous income streams, and many other quantities. Initially, the concept of the definite integral may seem unrelated to the notion of the indefinite integral. There is, however, a close connection between the two integrals. The fundamental theorem of calculus, discussed in Section 13.5, makes that connection precise.

Approximating Areas by Left and Right Sums

How do we find the shaded area in Figure 1? That is, how do we find the area bounded by the graph of $f(x) = 0.25x^2 + 1$, the x axis, and the vertical lines $x = 1$ and $x = 5$? [This cumbersome description is usually shortened to "the area under the graph of $f(x) = 0.25x^2 + 1$ from $x = 1$ to $x = 5$."] Our standard geometric area formulas do not apply directly, but the formula for the area of a rectangle can be used indirectly. To see how, we look at a method of approximating the area under the graph by using rectangles. This method will give us any accuracy desired, which is quite different from finding the area exactly. Our first area approximation is made by dividing the interval $[1, 5]$ on the x axis into four equal parts, each of length

$$\Delta x = \frac{5 - 1}{4} = 1*$$

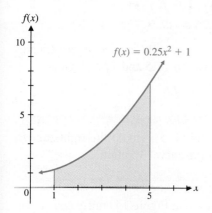

Figure 1 **What is the shaded area?**

We then place a **left rectangle** on each subinterval, that is, a rectangle whose base is the subinterval and whose height is the value of the function at the *left* endpoint of the subinterval (see Fig. 2).

Summing the areas of the left rectangles in Figure 2 results in a **left sum** of four rectangles, denoted by L_4, as follows:

$$L_4 = f(1) \cdot 1 + f(2) \cdot 1 + f(3) \cdot 1 + f(4) \cdot 1$$
$$= 1.25 + 2.00 + 3.25 + 5 = 11.5$$

From Figure 3, since $f(x)$ is increasing, we see that the left sum L_4 underestimates the area, and we can write

$$11.5 = L_4 < \text{Area}$$

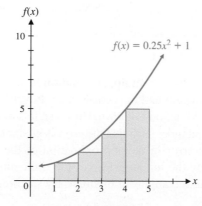

Figure 2 **Left rectangles**

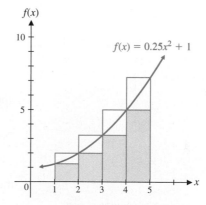

Figure 3 **Left and right rectangles**

---

*It is customary to denote the length of the subintervals by Δx, which is read "delta x," since Δ is the Greek capital letter delta.

Explore and Discuss 1   If $f(x)$ were decreasing over the interval $[1, 5]$, would the left sum L_4 over- or underestimate the actual area under the curve? Explain.

Similarly, we use the *right* endpoint of each subinterval to find the height of the **right rectangle** placed on the subinterval. Superimposing right rectangles on Figure 2, we get Figure 3 on page 747.

Summing the areas of the right rectangles in Figure 3 results in a **right sum** of four rectangles, denoted by R_4, as follows (compare R_4 with L_4 and note that R_4 can be obtained from L_4 by deleting one rectangular area and adding one more):

$$R_4 = f(2) \cdot 1 + f(3) \cdot 1 + f(4) \cdot 1 + f(5) \cdot 1$$
$$= 2.00 + 3.25 + 5.00 + 7.25 = 17.5$$

From Figure 3, since $f(x)$ is increasing, we see that the right sum R_4 overestimates the area, and we conclude that the actual area is between 11.5 and 17.5. That is,

$$11.5 = L_4 < \text{Area} < R_4 = 17.5$$

Explore and Discuss 2   If $f(x)$ in Figure 3 were decreasing over the interval $[1, 5]$, would the right sum R_4 overestimate or underestimate the actual area under the curve? Explain.

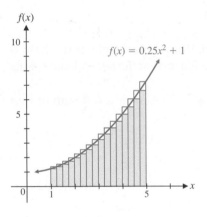

$f(x) = 0.25x^2 + 1$

Figure 4

The first approximation of the area under the curve in Figure 1 is fairly coarse, but the method outlined can be continued with increasingly accurate results by dividing the interval $[1, 5]$ into more and more subintervals of equal horizontal length. Of course, this is not a job for hand calculation, but a job that computers are designed to do.* Figure 4 shows left- and right-rectangle approximations for 16 equal subdivisions.

For this case,

$$\Delta x = \frac{5 - 1}{16} = 0.25$$

$$L_{16} = f(1) \cdot \Delta x + f(1.25) \cdot \Delta x + \cdots + f(4.75) \cdot \Delta x$$
$$= 13.59$$

$$R_{16} = f(1.25) \cdot \Delta x + f(1.50) \cdot \Delta x + \cdots + f(5) \cdot \Delta x$$
$$= 15.09$$

Now we know that the area under the curve is between 13.59 and 15.09. That is,

$$13.59 = L_{16} < \text{Area} < R_{16} = 15.09$$

For 100 equal subdivisions, computer calculations give us

$$14.214 = L_{100} < \text{Area} < R_{100} = 14.454$$

The **error in an approximation** is the absolute value of the difference between the approximation and the actual value. In general, neither the actual value nor the error in an approximation is known. However, it is often possible to calculate an **error bound**—a positive number such that the error is guaranteed to be less than or equal to that number.

The error in the approximation of the area under the graph of f from $x = 1$ to $x = 5$ by the left sum L_{16} (or the right sum R_{16}) is less than the sum of the areas of the small rectangles in Figure 4. By stacking those rectangles (see Fig. 5), we see that

$$\text{Error} = |\text{Area} - L_{16}| < |f(5) - f(1)| \cdot \Delta x = 1.5$$

Therefore, 1.5 is an error bound for the approximation of the area under f by L_{16}. We can apply the same stacking argument to any positive function that is increasing on $[a, b]$ or decreasing on $[a, b]$, to obtain the error bound in Theorem 1.

*The computer software that accompanies this book will perform these calculations (see the preface).

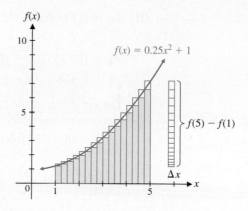

Figure 5

THEOREM 1 Error Bounds for Approximations of Area by Left or Right Sums

If $f(x) > 0$ and is either increasing on $[a, b]$ or decreasing on $[a, b]$, then

$$|f(b) - f(a)| \cdot \frac{b - a}{n}$$

is an error bound for the approximation of the area between the graph of f and the x axis, from $x = a$ to $x = b$, by L_n or R_n.

Because the error bound of Theorem 1 approaches 0 as $n \to \infty$, it can be shown that left and right sums, for certain functions, approach the same limit as $n \to \infty$.

THEOREM 2 Limits of Left and Right Sums

If $f(x) > 0$ and is either increasing on $[a, b]$ or decreasing on $[a, b]$, then its left and right sums approach the same real number as $n \to \infty$.

The number approached as $n \to \infty$ by the left and right sums in Theorem 2 is the area between the graph of f and the x axis from $x = a$ to $x = b$.

EXAMPLE 1 Approximating Areas Given the function $f(x) = 9 - 0.25x^2$, we want to approximate the area under $y = f(x)$ from $x = 2$ to $x = 5$.

(A) Graph the function over the interval $[0, 6]$. Then draw left and right rectangles for the interval $[2, 5]$ with $n = 6$.

(B) Calculate L_6, R_6, and error bounds for each.

(C) How large should n be in order for the approximation of the area by L_n or R_n to be within 0.05 of the true value?

SOLUTION

(A) $\Delta x = 0.5$:

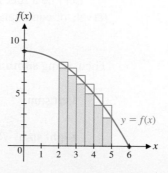

(B) $L_6 = f(2) \cdot \Delta x + f(2.5) \cdot \Delta x + f(3) \cdot \Delta x + f(3.5) \cdot \Delta x + f(4) \cdot \Delta x$
$+ f(4.5) \cdot \Delta x = 18.53$
$R_6 = f(2.5) \cdot \Delta x + f(3) \cdot \Delta x + f(3.5) \cdot \Delta x + f(4) \cdot \Delta x$
$+ f(4.5) \cdot \Delta x + f(5) \cdot \Delta x = 15.91$

The error bound for L_6 and R_6 is

$$\text{error} \le |f(5) - f(2)|\frac{5 - 2}{6} = |2.75 - 8|(0.5) = 2.625$$

(C) For L_n and R_n, find n such that error ≤ 0.05:

$$|f(b) - f(a)|\frac{b - a}{n} \le 0.05$$

$$|2.75 - 8|\frac{3}{n} \le 0.05$$

$$|-5.25|\frac{3}{n} \le 0.05$$

$$15.75 \le 0.05n$$

$$n \ge \frac{15.75}{0.05} = 315$$

Matched Problem 1 Given the function $f(x) = 8 - 0.5x^2$, we want to approximate the area under $y = f(x)$ from $x = 1$ to $x = 3$.

(A) Graph the function over the interval $[0, 4]$. Then draw left and right rectangles for the interval $[1, 3]$ with $n = 4$.

(B) Calculate L_4, R_4, and error bounds for each.

(C) How large should n be in order for the approximation of the area by L_n or R_n to be within 0.5 of the true value?

CONCEPTUAL INSIGHT

Note from Example 1C that a relatively large value of n ($n = 315$) is required to approximate the area by L_n or R_n to within 0.05. In other words, 315 rectangles must be used, and 315 terms must be summed, to guarantee that the error does not exceed 0.05. We can obtain a more efficient approximation of the area (fewer terms are summed to achieve a given accuracy) by replacing rectangles with trapezoids. The resulting **trapezoidal rule** and other methods for approximating areas are discussed in Section 14.4.

The Definite Integral as a Limit of Sums

Left and right sums are special cases of more general sums, called *Riemann sums* [named after the German mathematician Georg Riemann (1826–1866)], that are used to approximate areas by means of rectangles.

Let f be a function defined on the interval $[a, b]$. We partition $[a, b]$ into n subintervals of equal length $\Delta x = (b - a)/n$ with endpoints

$$a = x_0 < x_1 < x_2 < \cdots < x_n = b$$

Then, using **summation notation** (see Appendix B.1), we have

Left sum: $L_n = f(x_0)\Delta x + f(x_1)\Delta x + \cdots + f(x_{n-1})\Delta x = \sum_{k=1}^{n} f(x_{k-1})\Delta x$

Right sum: $R_n = f(x_1)\Delta x + f(x_2)\Delta x + \cdots + f(x_n)\Delta x = \sum_{k=1}^{n} f(x_k)\Delta x$

Riemann sum: $S_n = f(c_1)\Delta x + f(c_2)\Delta x + \cdots + f(c_n)\Delta x = \displaystyle\sum_{k=1}^{n} f(c_k)\Delta x$

In a **Riemann sum,*** each c_k is required to belong to the subinterval $[x_{k-1}, x_k]$. Left and right sums are the special cases of Riemann sums in which c_k is the left endpoint or right endpoint, respectively, of the subinterval. If $f(x) > 0$, then each term of a Riemann sum S_n represents the area of a rectangle having height $f(c_k)$ and width Δx (see Fig. 6). If $f(x)$ has both positive and negative values, then some terms of S_n represent areas of rectangles, and others represent the negatives of areas of rectangles, depending on the sign of $f(c_k)$ (see Fig. 7).

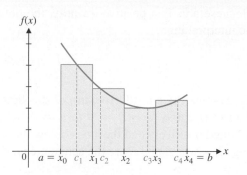

Figure 6

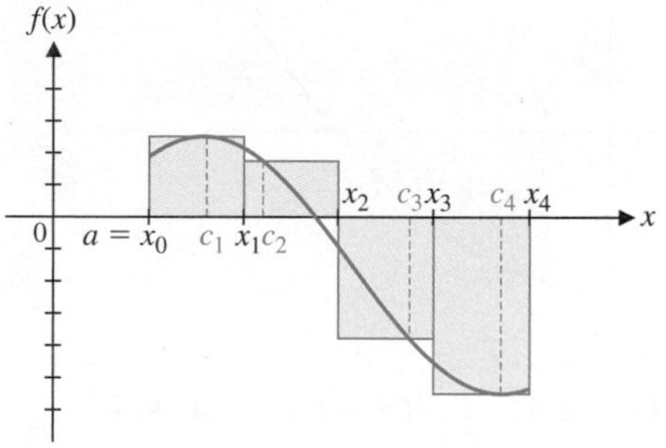

Figure 7

EXAMPLE 2 Riemann Sums Consider the function $f(x) = 15 - x^2$ on $[1, 5]$. Partition the interval $[1, 5]$ into four subintervals of equal length. For each subinterval $[x_{k-1}, x_k]$, let c_k be the midpoint. Calculate the corresponding Riemann sum S_4. (Riemann sums for which the c_k are the midpoints of the subintervals are called **midpoint sums.**)

SOLUTION $\Delta x = \dfrac{5 - 1}{4} = 1$

$$S_4 = f(c_1) \cdot \Delta x + f(c_2) \cdot \Delta x + f(c_3) \cdot \Delta x + f(c_4) \cdot \Delta x$$
$$= f(1.5) \cdot 1 + f(2.5) \cdot 1 + f(3.5) \cdot 1 + f(4.5) \cdot 1$$
$$= 12.75 + 8.75 + 2.75 - 5.25 = 19$$

Matched Problem 2 Consider the function $f(x) = x^2 - 2x - 10$ on $[2, 8]$. Partition the interval $[2, 8]$ into three subintervals of equal length. For each subinterval $[x_{k-1}, x_k]$, let c_k be the midpoint. Calculate the corresponding Riemann sum S_3.

By analyzing properties of a continuous function on a closed interval, it can be shown that the conclusion of Theorem 2 is valid if f is continuous. In that case, not just left and right sums, but Riemann sums, have the same limit as $n \to \infty$.

THEOREM 3 Limit of Riemann Sums

If f is a continuous function on $[a, b]$, then the Riemann sums for f on $[a, b]$ approach a real number limit I as $n \to \infty$.†

---

*The term *Riemann sum* is often applied to more general sums in which the subintervals $[x_{k-1}, x_k]$ are not required to have the same length. Such sums are not considered in this book.
†The precise meaning of this limit statement is as follows: For each $e > 0$, there exists some $d > 0$ such that $|S_n - I| < e$ whenever S_n is a Riemann sum for f on $[a, b]$ for which $\Delta x < d$.

DEFINITION   Definite Integral

Let f be a continuous function on $[a, b]$. The limit I of Riemann sums for f on $[a, b]$, guaranteed to exist by Theorem 3, is called the **definite integral** of f from a to b and is denoted as

$$\int_a^b f(x)\,dx$$

The **integrand** is $f(x)$, the **lower limit of integration** is a, and the **upper limit of integration** is b.

Because area is a positive quantity, the definite integral has the following geometric interpretation:

$$\int_a^b f(x)\,dx$$

represents the cumulative sum of the signed areas between the graph of f and the x axis from $x = a$ to $x = b$, where the areas above the x axis are counted positively and the areas below the x axis are counted negatively (see Fig. 8, where A and B are the actual areas of the indicated regions).

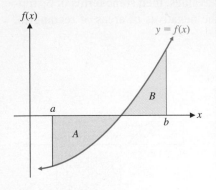

Figure 8 $\displaystyle\int_a^b f(x)\,dx = -A + B$

EXAMPLE 3   Definite Integrals   Calculate the definite integrals by referring to Figure 9.

(A) $\displaystyle\int_a^b f(x)\,dx$

(B) $\displaystyle\int_a^c f(x)\,dx$

(C) $\displaystyle\int_b^c f(x)\,dx$

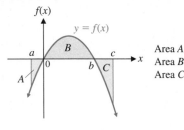

Area $A = 2.33$
Area $B = 10.67$
Area $C = 5.63$

Figure 9

SOLUTION

(A) $\displaystyle\int_a^b f(x)\,dx = -2.33 + 10.67 = 8.34$

(B) $\displaystyle\int_a^c f(x)\,dx = -2.33 + 10.67 - 5.63 = 2.71$

(C) $\displaystyle\int_b^c f(x)\,dx = -5.63$

Matched Problem 3   Referring to the figure for Example 3, calculate the definite integrals.

(A) $\displaystyle\int_a^0 f(x)\,dx$ (B) $\displaystyle\int_0^c f(x)\,dx$ (C) $\displaystyle\int_0^b f(x)\,dx$

Properties of the Definite Integral

Because the definite integral is defined as the limit of Riemann sums, many properties of sums are also properties of the definite integral. Note that Properties 3 and 4 are similar to the indefinite integral properties given in Section 13.1.

Property 5 is illustrated by Figure 9 in Example 3: $2.71 = 8.34 + (-5.63)$. Property 1 follows from the special case of Property 5 in which b and c are both replaced by a. Property 2 follows from the special case of Property 5 in which c is replaced by a.

PROPERTIES Properties of Definite Integrals

1. $\displaystyle\int_a^a f(x)\,dx = 0$

2. $\displaystyle\int_a^b f(x)\,dx = -\int_b^a f(x)\,dx$

3. $\displaystyle\int_a^b kf(x)\,dx = k\int_a^b f(x)\,dx,\ k\text{ a constant}$

4. $\displaystyle\int_a^b [f(x) \pm g(x)]\,dx = \int_a^b f(x)\,dx \pm \int_a^b g(x)\,dx$

5. $\displaystyle\int_a^c f(x)\,dx = \int_a^b f(x)\,dx + \int_b^c f(x)\,dx$

EXAMPLE 4 Using Properties of the Definite Integral If

$$\int_0^2 x\,dx = 2 \qquad \int_0^2 x^2\,dx = \frac{8}{3} \qquad \int_2^3 x^2\,dx = \frac{19}{3}$$

then

(A) $\displaystyle\int_0^2 12x^2\,dx = 12\int_0^2 x^2\,dx = 12\left(\frac{8}{3}\right) = 32$

(B) $\displaystyle\int_0^2 (2x - 6x^2)\,dx = 2\int_0^2 x\,dx - 6\int_0^2 x^2\,dx = 2(2) - 6\left(\frac{8}{3}\right) = -12$

(C) $\displaystyle\int_3^2 x^2\,dx = -\int_2^3 x^2\,dx = -\frac{19}{3}$

(D) $\displaystyle\int_5^5 3x^2\,dx = 0$

(E) $\displaystyle\int_0^3 3x^2\,dx = 3\int_0^2 x^2\,dx + 3\int_2^3 x^2\,dx = 3\left(\frac{8}{3}\right) + 3\left(\frac{19}{3}\right) = 27$

Matched Problem 4 Using the same integral values given in Example 4, find

(A) $\displaystyle\int_2^3 6x^2\,dx$

(B) $\displaystyle\int_0^2 (9x^2 - 4x)\,dx$

(C) $\displaystyle\int_2^0 3x\,dx$

(D) $\displaystyle\int_{-2}^{-2} 3x\,dx$

(E) $\displaystyle\int_0^3 12x^2\,dx$

Skills Warm-up Exercises

In Problems 1–6, perform a mental calculation to find the answer and include the correct units. (If necessary, review Appendix C).

1. Find the total area enclosed by 5 non-overlapping rectangles, if each rectangle is 8 inches high and 2 inches wide.

2. Find the total area enclosed by 6 non-overlapping rectangles, if each rectangle is 10 centimeters high and 3 centimeters wide.

3. Find the total area enclosed by 4 non-overlapping rectangles, if each rectangle has width 2 meters and the heights of the rectangles are 3, 4, 5, and 6 meters, respectively.

4. Find the total area enclosed by 5 non-overlapping rectangles, if each rectangle has width 3 feet and the heights of the rectangles are 2, 4, 6, 8, and 10 feet, respectively.

5. A square is inscribed in a circle of radius 1 meter. Is the area inside the circle but outside the square less than 1 square meter?

6. A square is circumscribed around a circle of radius 1 foot. Is the area inside the square but outside the circle less than 1 square foot?

Problems 7–10 refer to the rectangles A, B, C, D, and E in the following figure.

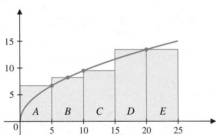

7. Which rectangles are left rectangles?

8. Which rectangles are right rectangles?

9. Which rectangles are neither left nor right rectangles?

10. Which rectangles are both left and right rectangles?

Problems 11–14 refer to the rectangles F, G, H, I, and J in the following figure.

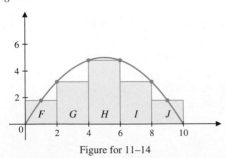

Figure for 11–14

11. Which rectangles are right rectangles?

12. Which rectangles are left rectangles?

13. Which rectangles are both left and right rectangles?

14. Which rectangles are neither left nor right rectangles?

Problems 15–22 involve estimating the area under the curves in Figures A–D from $x = 1$ to $x = 4$. For each figure, divide the interval [1, 4] into three equal subintervals.

15. Draw in left and right rectangles for Figures A and B.

16. Draw in left and right rectangles for Figures C and D.

17. Using the results of Problem 15, compute L_3 and R_3 for Figure A and for Figure B.

18. Using the results of Problem 16, compute L_3 and R_3 for Figure C and for Figure D.

19. Replace the question marks with L_3 and R_3 as appropriate. Explain your choice.

$$? \leq \int_1^4 f(x)\, dx \leq ? \qquad ? \leq \int_1^4 g(x)\, dx \leq ?$$

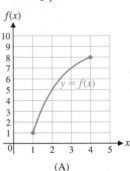

(A)

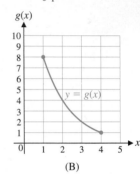

(B)

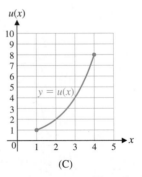

(C)

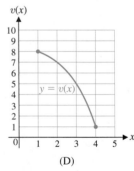

(D)

Figure for 15–22

20. Replace the question marks with L_3 and R_3 as appropriate. Explain your choice.

$$? \leq \int_1^4 u(x)\, dx \leq ? \qquad ? \leq \int_1^4 v(x)\, dx \leq ?$$

21. Compute error bounds for L_3 and R_3 found in Problem 17 for both figures.

22. Compute error bounds for L_3 and R_3 found in Problem 18 for both figures.

In Problems 23–26, calculate the indicated Riemann sum S_n for the function $f(x) = 25 - 3x^2$.

23. Partition $[-2, 8]$ into five subintervals of equal length, and for each subinterval $[x_{k-1}, x_k]$, let $c_k = (x_{k-1} + x_k)/2$.

24. Partition $[0, 12]$ into four subintervals of equal length, and for each subinterval $[x_{k-1}, x_k]$, let $c_k = (x_{k-1} + 2x_k)/3$.

25. Partition $[0, 12]$ into four subintervals of equal length, and for each subinterval $[x_{k-1}, x_k]$, let $c_k = (2x_{k-1} + x_k)/3$.

26. Partition $[-5, 5]$ into five subintervals of equal length, and for each subinterval $[x_{k-1}, x_k]$, let $c_k = (x_{k-1} + x_k)/2$.

In Problems 27–30, calculate the indicated Riemann sum S_n for the function $f(x) = x^2 - 5x - 6$.

27. Partition $[0, 3]$ into three subintervals of equal length, and let $c_1 = 0.7$, $c_2 = 1.8$, and $c_3 = 2.4$.

28. Partition $[0, 3]$ into three subintervals of equal length, and let $c_1 = 0.2$, $c_2 = 1.5$, and $c_3 = 2.8$.

29. Partition $[1, 7]$ into six subintervals of equal length, and let $c_1 = 1$, $c_2 = 3$, $c_3 = 3$, $c_4 = 5$, $c_5 = 5$, and $c_6 = 7$.

30. Partition $[1, 7]$ into six subintervals of equal length, and let $c_1 = 2$, $c_2 = 2$, $c_3 = 4$, $c_4 = 4$, $c_5 = 6$, and $c_6 = 6$.

In Problems 31–42, calculate the definite integral by referring to the figure with the indicated areas.

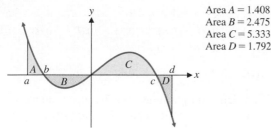

Area $A = 1.408$
Area $B = 2.475$
Area $C = 5.333$
Area $D = 1.792$

Figure for 31–42

31. $\int_b^0 f(x)\, dx$

32. $\int_0^c f(x)\, dx$

33. $\int_a^c f(x)\, dx$

34. $\int_b^d f(x)\, dx$

35. $\int_a^d f(x)\, dx$

36. $\int_0^d f(x)\, dx$

37. $\int_c^0 f(x)\, dx$

38. $\int_d^a f(x)\, dx$

39. $\int_0^a f(x)\, dx$

40. $\int_c^a f(x)\, dx$

41. $\int_d^b f(x)\, dx$

42. $\int_c^b f(x)\, dx$

In Problems 43–54, calculate the definite integral, given that

$$\int_1^4 x\, dx = 7.5 \qquad \int_1^4 x^2\, dx = 21 \qquad \int_4^5 x^2\, dx = \frac{61}{3}$$

43. $\int_1^4 2x\, dx$

44. $\int_1^4 3x^2\, dx$

45. $\int_1^4 (5x + x^2)\, dx$

46. $\int_1^4 (7x - 2x^2)\, dx$

47. $\int_1^4 (x^2 - 10x)\, dx$

48. $\int_1^4 (4x^2 - 9x)\, dx$

49. $\int_1^5 6x^2\, dx$

50. $\int_1^5 -4x^2\, dx$

51. $\int_4^4 (7x - 2)^2\, dx$

52. $\int_5^5 (10 - 7x + x^2)\, dx$

53. $\int_5^4 9x^2\, dx$

54. $\int_4^1 x(1 - x)\, dx$

In Problems 55–60, discuss the validity of each statement. If the statement is always true, explain why. If it is not always true, give a counterexample.

55. If $\int_a^b f(x)\, dx = 0$, then $f(x) = 0$ for all x in $[a, b]$.

56. If $f(x) = 0$ for all x in $[a, b]$, then $\int_a^b f(x)\, dx = 0$.

57. If $f(x) = 2x$ on $[0, 10]$, then there is a positive integer n for which the left sum L_n equals the exact area under the graph of f from $x = 0$ to $x = 10$.

58. If $f(x) = 2x$ on $[0, 10]$ and n is a positive integer, then there is some Riemann sum S_n that equals the exact area under the graph of f from $x = 0$ to $x = 10$.

59. If the area under the graph of f on $[a, b]$ is equal to both the left sum L_n and the right sum R_n for some positive integer n, then f is constant on $[a, b]$.

60. If f is a decreasing function on $[a, b]$, then the area under the graph of f is greater than the left sum L_n and less than the right sum R_n, for any positive integer n.

Problems 61 and 62 refer to the following figure showing two parcels of land along a river:

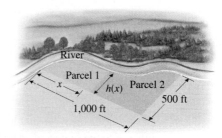

Figure for 61 and 62

61. You want to purchase both parcels of land shown in the figure and make a quick check on their combined area. There is no equation for the river frontage, so you use the average of the left and right sums of rectangles covering the area. The 1,000-foot baseline is divided into 10 equal parts. At the end of each subinterval, a measurement is made from the baseline to the river, and the results are tabulated. Let x be the distance from the left end of the baseline and let $h(x)$ be the distance from the baseline to the river at x. Use L_{10} to estimate the combined area of both parcels, and calculate an error bound for this estimate. How many subdivisions of the baseline would be required so that the error incurred in using L_n would not exceed 2,500 square feet?

x	0	100	200	300	400	500
$h(x)$	0	183	235	245	260	286

x	600	700	800	900	1,000
$h(x)$	322	388	453	489	500

62. Refer to Problem 61. Use R_{10} to estimate the combined area of both parcels, and calculate an error bound for this estimate. How many subdivisions of the baseline would be required so that the error incurred in using R_n would not exceed 1,000 square feet?

Problems 63 and 64 refer to the following figure:

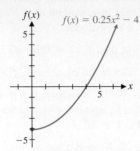

$f(x)$

$f(x) = 0.25x^2 - 4$

Figure for 63 and 64

✎ 63. Use L_6 and R_6 to approximate $\int_2^5 (0.25x^2 - 4)\, dx$. Compute error bounds for each. (Round answers to two decimal places.) Describe in geometric terms what the definite integral over the interval $[2, 5]$ represents.

✎ 64. Use L_5 and R_5 to approximate $\int_1^6 (0.25x^2 - 4)\, dx$. Compute error bounds for each. (Round answers to two decimal places.) Describe in geometric terms what the definite integral over the interval $[1, 6]$ represents.

For Problems 65–68, use a graphing calculator to determine the intervals on which each function is increasing or decreasing.

65. $f(x) = e^{-x^2}$

66. $f(x) = \dfrac{3}{1 + 2e^{-x}}$

67. $f(x) = x^4 - 2x^2 + 3$

68. $f(x) = e^{x^2}$

In Problems 69–72, the left sum L_n or the right sum R_n is used to approximate the definite integral to the indicated accuracy. How large must n be chosen in each case? (Each function is increasing over the indicated interval.)

69. $\displaystyle\int_1^3 \ln x\, dx = R_n \pm 0.1$

70. $\displaystyle\int_0^{10} \ln(x^2 + 1)\, dx = L_n \pm 0.5$

71. $\displaystyle\int_1^3 x^x\, dx = L_n \pm 0.5$

72. $\displaystyle\int_1^4 x^x\, dx = R_n \pm 0.5$

Applications

73. Employee training. A company producing electric motors has established that, on the average, a new employee can assemble $N(t)$ components per day after t days of on-the-job training, as shown in the following table (a new

employee's productivity increases continuously with time on the job):

t	0	20	40	60	80	100	120
$N(t)$	10	51	68	76	81	84	86

Use left and right sums to estimate the area under the graph of $N(t)$ from $t = 0$ to $t = 60$. Use three subintervals of equal length for each. Calculate an error bound for each estimate.

74. Employee training. For a new employee in Problem 73, use left and right sums to estimate the area under the graph of $N(t)$ from $t = 20$ to $t = 100$. Use four equal subintervals for each. Replace the question marks with the values of L_4 or R_4 as appropriate:

$$? \le \int_{20}^{100} N(t)\, dt \le ?$$

75. Medicine. The rate of healing, $A'(t)$ (in square centimeters per day), for a certain type of skin wound is given approximately by the following table:

t	0	1	2	3	4	5
$A'(t)$	0.90	0.81	0.74	0.67	0.60	0.55
t	6	7	8	9	10	
$A'(t)$	0.49	0.45	0.40	0.36	0.33	

(A) Use left and right sums over five equal subintervals to approximate the area under the graph of $A'(t)$ from $t = 0$ to $t = 5$.

(B) Replace the question marks with values of L_5 and R_5 as appropriate:

$$? \le \int_0^5 A'(t)\, dt \le ?$$

76. Medicine. Refer to Problem 75. Use left and right sums over five equal subintervals to approximate the area under the graph of $A'(t)$ from $t = 5$ to $t = 10$. Calculate an error bound for this estimate.

77. Learning. A psychologist found that, on average, the rate of learning a list of special symbols in a code $N'(x)$ after x days of practice was given approximately by the following table values:

x	0	2	4	6	8	10	12
$N'(x)$	29	26	23	21	19	17	15

Use left and right sums over three equal subintervals to approximate the area under the graph of $N'(x)$ from $x = 6$ to $x = 12$. Calculate an error bound for this estimate.

78. Learning. For the data in Problem 77, use left and right sums over three equal subintervals to approximate the area under the graph of $N'(x)$ from $x = 0$ to $x = 6$. Replace the question marks with values of L_3 and R_3 as appropriate:

$$? \le \int_0^6 N'(x)\, dx \le ?$$

Answers to Matched Problems

1. (A) $\Delta x = 0.5$:

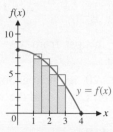

(B) $L_4 = 12.625$, $R_4 = 10.625$; error for L_4 and $R_4 = 2$

(C) $n > 16$ for L_n and R_n

2. $S_3 = 46$

3. (A) -2.33 (B) 5.04 (C) 10.67

4. (A) 38 (B) 16 (C) -6

    (D) 0 (E) 108

13.5 The Fundamental Theorem of Calculus

- Introduction to the Fundamental Theorem
- Evaluating Definite Integrals
- Recognizing a Definite Integral: Average Value

The definite integral of a function f on an interval $[a, b]$ is a number, the area (if $f(x) > 0$) between the graph of f and the x axis from $x = a$ to $x = b$. The indefinite integral of a function is a family of antiderivatives. In this section, we explain the connection between these two integrals, a connection made precise by the fundamental theorem of calculus.

Introduction to the Fundamental Theorem

Suppose that the daily cost function for a small manufacturing firm is given (in dollars) by

$$C(x) = 180x + 200 \qquad 0 \le x \le 20$$

Then the marginal cost function is given (in dollars per unit) by

$$C'(x) = 180$$

What is the change in cost as production is increased from $x = 5$ units to $x = 10$ units? That change is equal to

$$
\begin{aligned}
C(10) - C(5) &= (180 \cdot 10 + 200) - (180 \cdot 5 + 200) \\
&= 180(10 - 5) \\
&= \$900
\end{aligned}
$$

Notice that $180(10 - 5)$ is equal to the area between the graph of $C'(x)$ and the x axis from $x = 5$ to $x = 10$. Therefore,

$$C(10) - C(5) = \int_5^{10} 180 \, dx$$

In other words, the change in cost from $x = 5$ to $x = 10$ is equal to the area between the marginal cost function and the x axis from $x = 5$ to $x = 10$ (see Fig. 1).

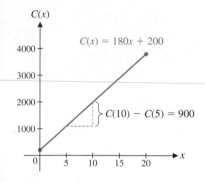

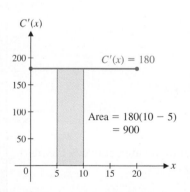

Figure 1

> ### CONCEPTUAL INSIGHT
>
> Consider the formula for the slope of a line:
>
> $$m = \frac{y_2 - y_1}{x_2 - x_1}$$
>
> Multiplying both sides of this equation by $x_2 - x_1$ gives
>
> $$y_2 - y_1 = m(x_2 - x_1)$$
>
> The right-hand side, $m(x_2 - x_1)$, is equal to the area of a rectangle of height m and width $x_2 - x_1$. So the change in y coordinates is equal to the area under the constant function with value m from $x = x_1$ to $x = x_2$.

EXAMPLE 1   Change in Cost vs Area under Marginal Cost   The daily cost function for a company (in dollars) is given by

$$C(x) = -5x^2 + 210x + 400 \qquad 0 \le x \le 20$$

(A) Graph $C(x)$ for $0 \le x \le 20$, calculate the change in cost from $x = 5$ to $x = 10$, and indicate that change in cost on the graph.

(B) Graph the marginal cost function $C'(x)$ for $0 \le x \le 20$, and use geometric formulas (see Appendix C) to calculate the area between $C'(x)$ and the x axis from $x = 5$ to $x = 10$.

(C) Compare the results of the calculations in parts (A) and (B).

SOLUTION

(A) $C(10) - C(5) = 2,000 - 1,325 = 675$, and this change in cost is indicated in Figure 2A.

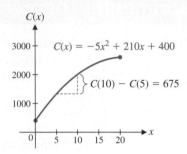

Figure 2A

(B) $C'(x) = -10x + 210$, so the area between $C'(x)$ and the x axis from $x = 5$ to $x = 10$ (see Fig. 2B) is the area of a trapezoid (geometric formulas are given in Appendix C):

$$\text{Area} = \frac{C'(5) + C'(10)}{2}(10 - 5) = \frac{160 + 110}{2}(5) = 675$$

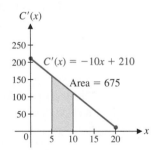

Figure 2B

(C) The change in cost from $x = 5$ to $x = 10$ is equal to the area between the marginal cost function and the x axis from $x = 5$ to $x = 10$.

Matched Problem 1   Repeat Example 1 for the daily cost function

$$C(x) = -7.5x^2 + 305x + 625$$

The connection illustrated in Example 1, between the change in a function from $x = a$ to $x = b$ and the area under the derivative of the function, provides the link between antiderivatives (or indefinite integrals) and the definite integral. This link is known as the fundamental theorem of calculus. (See Problems 67 and 68 in Exercise 13.5 for an outline of its proof.)

THEOREM 1 Fundamental Theorem of Calculus

If f is a continuous function on $[a, b]$, and F is any antiderivative of f, then

$$\int_a^b f(x)\, dx = F(b) - F(a)$$

CONCEPTUAL INSIGHT

Because a definite integral is the limit of Riemann sums, we expect that it would be difficult to calculate definite integrals exactly. The fundamental theorem, however, gives us an easy method for evaluating definite integrals, *provided that we can find an antiderivative $F(x)$ of $f(x)$*: Simply calculate the difference $F(b) - F(a)$. But what if we are unable to find an antiderivative of $f(x)$? In that case, we must resort to left sums, right sums, or other approximation methods to approximate the definite integral. However, it is often useful to remember that such an approximation is also an estimate of the change $F(b) - F(a)$.

Evaluating Definite Integrals

By the fundamental theorem, we can evaluate $\int_a^b f(x)\, dx$ easily and exactly whenever we can find an antiderivative $F(x)$ of $f(x)$. We simply calculate the difference $F(b) - F(a)$. If $G(x)$ is another antiderivative of $f(x)$, then $G(x) = F(x) + C$ for some constant C. So

$$G(b) - G(a) = F(b) + C - [F(a) + C]$$
$$= F(b) - F(a)$$

In other words:

> Any antiderivative of $f(x)$ can be used in the fundamental theorem. One generally chooses the simplest antiderivative by letting $C = 0$, since any other value of C will drop out in computing the difference $F(b) - F(a)$.

Now you know why we studied techniques of indefinite integration before this section—so that we would have methods of finding antiderivatives of large classes of elementary functions for use with the fundamental theorem.

In evaluating definite integrals by the fundamental theorem, it is convenient to use the notation $F(x)\big|_a^b$, which represents the change in $F(x)$ from $x = a$ to $x = b$, as an intermediate step in the calculation. This technique is illustrated in the following examples.

EXAMPLE 2 Evaluating Definite Integrals Evaluate $\displaystyle\int_1^2 \left(2x + 3e^x - \frac{4}{x}\right) dx.$

SOLUTION $\displaystyle\int_1^2 \left(2x + 3e^x - \frac{4}{x}\right) dx = 2\int_1^2 x\, dx + 3\int_1^2 e^x\, dx - 4\int_1^2 \frac{1}{x}\, dx$

$$= 2\frac{x^2}{2}\bigg|_1^2 + 3e^x\bigg|_1^2 - 4\ln|x|\,\bigg|_1^2$$

$$= (2^2 - 1^2) + (3e^2 - 3e^1) - (4\ln 2 - 4\ln 1)$$

$$= 3 + 3e^2 - 3e - 4\ln 2 \approx 14.24$$

Matched Problem 2 Evaluate $\displaystyle\int_1^3 \left(4x - 2e^x + \frac{5}{x}\right) dx.$

The evaluation of a definite integral is a two-step process: First, find an antiderivative. Then find the change in that antiderivative. If *substitution techniques* are required to find the antiderivative, there are two different ways to proceed. The next example illustrates both methods.

EXAMPLE 3   Definite Integrals and Substitution Techniques   Evaluate

$$\int_0^5 \frac{x}{x^2 + 10}\, dx$$

SOLUTION   We solve this problem using substitution in two different ways.

Method 1.   Use substitution in an indefinite integral to find an antiderivative as a function of x. Then evaluate the definite integral.

$$\int \frac{x}{x^2 + 10}\, dx = \frac{1}{2}\int \frac{1}{x^2 + 10} 2x\, dx \qquad \text{Substitute } u = x^2 + 10$$
$$\text{and } du = 2x\, dx.$$
$$= \frac{1}{2}\int \frac{1}{u}\, du$$
$$= \tfrac{1}{2}\ln|u| + C$$
$$= \tfrac{1}{2}\ln(x^2 + 10) + C \qquad \text{Since } u = x^2 + 10 > 0$$

We choose $C = 0$ and use the antiderivative $\frac{1}{2}\ln(x^2 + 10)$ to evaluate the definite integral.

$$\int_0^5 \frac{x}{x^2 + 10}\, dx = \frac{1}{2}\ln(x^2 + 10)\Big|_0^5$$
$$= \tfrac{1}{2}\ln 35 - \tfrac{1}{2}\ln 10 \approx 0.626$$

Method 2.   Substitute directly into the definite integral, changing both the variable of integration and the limits of integration. In the definite integral

$$\int_0^5 \frac{x}{x^2 + 10}\, dx$$

the upper limit is $x = 5$ and the lower limit is $x = 0$. When we make the substitution $u = x^2 + 10$ in this definite integral, we must change the limits of integration to the corresponding values of u:

$$x = 5 \quad \text{implies} \quad u = 5^2 + 10 = 35 \quad \text{New upper limit}$$
$$x = 0 \quad \text{implies} \quad u = 0^2 + 10 = 10 \quad \text{New lower limit}$$

We have

$$\int_0^5 \frac{x}{x^2 + 10}\, dx = \frac{1}{2}\int_0^5 \frac{1}{x^2 + 10} 2x\, dx$$
$$= \frac{1}{2}\int_{10}^{35} \frac{1}{u}\, du$$
$$= \frac{1}{2}\left(\ln|u|\,\Big|_{10}^{35}\right)$$
$$= \tfrac{1}{2}(\ln 35 - \ln 10) \approx 0.626$$

Matched Problem 3   Use both methods described in Example 3 to evaluate

$$\int_0^1 \frac{1}{2x + 4}\, dx.$$

EXAMPLE 4 Definite Integrals and Substitution Use method 2 described in Example 3 to evaluate

$$\int_{-4}^{1} \sqrt{5 - t}\, dt$$

SOLUTION If $u = 5 - t$, then $du = -dt$, and

$t = 1$	implies	$u = 5 - 1 = 4$	New upper limit
$t = -4$	implies	$u = 5 - (-4) = 9$	New lower limit

Notice that the lower limit for u is larger than the upper limit. Be careful not to reverse these two values when substituting into the definite integral:

$$\int_{-4}^{1} \sqrt{5 - t}\, dt = -\int_{-4}^{1} \sqrt{5 - t}\,(-dt)$$

$$= -\int_{9}^{4} \sqrt{u}\, du$$

$$= -\int_{9}^{4} u^{1/2}\, du$$

$$= -\left(\frac{u^{3/2}}{\frac{3}{2}} \Big|_{9}^{4} \right)$$

$$= -\left[\tfrac{2}{3}(4)^{3/2} - \tfrac{2}{3}(9)^{3/2} \right]$$

$$= -\left[\tfrac{16}{3} - \tfrac{54}{3} \right] = \tfrac{38}{3} \approx 12.667$$

Matched Problem 4 Use method 2 described in Example 3 to evaluate

$$\int_{2}^{5} \frac{1}{\sqrt{6 - t}}\, dt.$$

EXAMPLE 5 Change in Profit A company manufactures x HDTVs per month. The monthly marginal profit (in dollars) is given by

$$P'(x) = 165 - 0.1x \qquad 0 \le x \le 4{,}000$$

The company is currently manufacturing 1,500 HDTVs per month, but is planning to increase production. Find the change in the monthly profit if monthly production is increased to 1,600 HDTVs.

SOLUTION

$$P(1{,}600) - P(1{,}500) = \int_{1{,}500}^{1{,}600} (165 - 0.1x)\, dx$$

$$= \left(165x - 0.05x^2 \right)\Big|_{1{,}500}^{1{,}600}$$

$$= \left[165(1{,}600) - 0.05(1{,}600)^2 \right]$$
$$\qquad - \left[165(1{,}500) - 0.05(1{,}500)^2 \right]$$

$$= 136{,}000 - 135{,}000$$

$$= 1{,}000$$

Increasing monthly production from 1,500 units to 1,600 units will increase the monthly profit by $1,000.

Matched Problem 5) Repeat Example 5 if
$$P'(x) = 300 - 0.2x \qquad 0 \le x \le 3,000$$
and monthly production is increased from 1,400 to 1,500 HDTVs.

EXAMPLE 6   Useful Life An amusement company maintains records for each video game installed in an arcade. Suppose that $C(t)$ and $R(t)$ represent the total accumulated costs and revenues (in thousands of dollars), respectively, t years after a particular game has been installed. Suppose also that
$$C'(t) = 2 \qquad R'(t) = 9e^{-0.5t}$$
The value of t for which $C'(t) = R'(t)$ is called the **useful life** of the game.
(A) Find the useful life of the game, to the nearest year.
(B) Find the total profit accumulated during the useful life of the game.

SOLUTION

(A) $R'(t) = C'(t)$

$9e^{-0.5t} = 2$

$e^{-0.5t} = \frac{2}{9}$ Convert to equivalent logarithmic form.

$-0.5t = \ln\frac{2}{9}$

$t = -2\ln\frac{2}{9} \approx 3$ years

Thus, the game has a useful life of 3 years. This is illustrated graphically in Figure 3.

(B) The total profit accumulated during the useful life of the game is

$$P(3) - P(0) = \int_0^3 P'(t)\, dt$$

$$= \int_0^3 [R'(t) - C'(t)]\, dt$$

$$= \int_0^3 (9e^{-0.5t} - 2)\, dt$$

$$= \left(\frac{9}{-0.5}e^{-0.5t} - 2t\right)\Big|_0^3 \qquad \text{Recall: } \int e^{ax}\, dx = \frac{1}{a}e^{ax} + C$$

$$= (-18e^{-0.5t} - 2t)\big|_0^3$$

$$= (-18e^{-1.5} - 6) - (-18e^0 - 0)$$

$$= 12 - 18e^{-1.5} \approx 7.984 \quad \text{or} \quad \$7,984$$

Matched Problem 6) Repeat Example 6 if $C'(t) = 1$ and $R'(t) = 7.5e^{-0.5t}$.

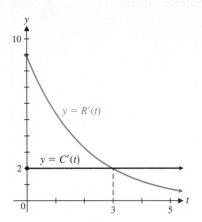

Figure 3 **Useful life**

EXAMPLE 7   Numerical Integration on a Graphing Calculator   Evaluate $\int_{-1}^{2} e^{-x^2}\, dx$ to three decimal places.

SOLUTION The integrand e^{-x^2} does not have an elementary antiderivative, so we are unable to use the fundamental theorem to evaluate the definite integral.

Figure 4

Instead, we use a numerical integration routine that has been preprogrammed into a graphing calculator. (Consult your user's manual for specific details.) Such a routine is an approximation algorithm, more powerful than the left-sum and right-sum methods discussed in Section 13.4. From Figure 4,

$$\int_{-1}^{2} e^{-x^2}\, dx = 1.629$$

Matched Problem 7) Evaluate $\int_{1.5}^{4.3} \frac{x}{\ln x}\, dx$ to three decimal places.

Recognizing a Definite Integral: Average Value

Recall that the derivative of a function f was defined in Section 10.4 by

$$f'(x) = \lim_{h \to 0} \frac{f(x+h) - f(x)}{h}$$

This form is generally not easy to compute directly but is easy to recognize in certain practical problems (slope, instantaneous velocity, rates of change, and so on). Once we know that we are dealing with a derivative, we proceed to try to compute the derivative with the use of derivative formulas and rules.

Similarly, evaluating a definite integral with the use of the definition

$$\int_{a}^{b} f(x)\, dx = \lim_{n \to \infty} [f(c_1)\Delta x_1 + f(c_2)\Delta x_2 + \cdots + f(c_n)\Delta x_n] \tag{1}$$

is generally not easy, but the form on the right occurs naturally in many practical problems. We can use the fundamental theorem to evaluate the definite integral (once it is recognized) if an antiderivative can be found; otherwise, we will approximate it with a rectangle sum. We will now illustrate these points by finding the *average value* of a continuous function.

Suppose that the temperature F (in degrees Fahrenheit) in the middle of a small shallow lake from 8 AM $(t = 0)$ to 6 PM $(t = 10)$ during the month of May is given approximately by $F(t) = -t^2 + 10t + 50$ as shown in Figure 5.

How can we compute the average temperature from 8 AM to 6 PM? We know that the average of a finite number of values $a_1, a_2, \ldots, a_n$ is given by

$$\text{average} = \frac{a_1 + a_2 + \cdots + a_n}{n}$$

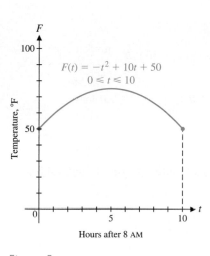

Figure 5

But how can we handle a continuous function with infinitely many values? It would seem reasonable to divide the time interval $[0, 10]$ into n equal subintervals, compute the temperature at a point in each subinterval, and then use the average of the temperatures as an approximation of the average value of the continuous function $F = F(t)$ over $[0, 10]$. We would expect the approximations to improve as n increases. In fact, we would define the limit of the average of n values as $n \to \infty$ as the *average value of F over* $[0, 10]$ if the limit exists. This is exactly what we will do:

$$\left(\begin{array}{c}\text{average temperature} \\ \text{for } n \text{ values}\end{array}\right) = \frac{1}{n}[F(t_1) + F(t_2) + \cdots + F(t_n)] \tag{2}$$

Here t_k is a point in the kth subinterval. We will call the limit of equation (2) as $n \to \infty$ the *average temperature over the time interval* $[0, 10]$.

Form (2) resembles form (1), but we are missing the Δt_k. We take care of this by multiplying equation (2) by $(b - a)/(b - a)$, which will change the form of equation (2) without changing its value:

$$\frac{b - a}{b - a} \cdot \frac{1}{n}\left[F(t_1) + F(t_2) + \cdots + F(t_n)\right] = \frac{1}{b - a} \cdot \frac{b - a}{n}\left[F(t_1) + F(t_2) + \cdots + F(t_n)\right]$$

$$= \frac{1}{b - a}\left[F(t_1)\frac{b - a}{n} + F(t_2)\frac{b - a}{n} + \cdots + F(t_n)\frac{b - a}{n}\right]$$

$$= \frac{1}{b - a}\left[F(t_1)\Delta t + F(t_2)\Delta t + \cdots + F(t_n)\Delta t\right]$$

Therefore,

$$\left(\begin{array}{c}\text{average temperature}\\ \text{over } [a, b] = [0, 10]\end{array}\right) = \lim_{n \to \infty}\left\{\frac{1}{b - a}\left[F(t_1)\Delta t + F(t_2)\Delta t + \cdots + F(t_n)\Delta t\right]\right\}$$

$$= \frac{1}{b - a}\left\{\lim_{n \to \infty}\left[F(t_1)\,\Delta t + F(t_2)\,\Delta t + \cdots + F(t_n)\,\Delta t\right]\right\}$$

The limit inside the braces is of form (1)—that is, a definite integral. So

$$\left(\begin{array}{c}\text{average temperature}\\ \text{over } [a, b] = [0, 10]\end{array}\right) = \frac{1}{b - a}\int_a^b F(t)\,dt$$

We now use the fundamental theorem to evaluate the definite integral.

$$= \frac{1}{10 - 0}\int_0^{10}(-t^2 + 10t + 50)\,dt$$

$$= \frac{1}{10}\left(-\frac{t^3}{3} + 5t^2 + 50t\right)\Big|_0^{10}$$

$$= \frac{200}{3} \approx 67°F$$

Proceeding as before for an arbitrary continuous function f over an interval $[a, b]$, we obtain the following general formula:

DEFINITION   Average Value of a Continuous Function f over $[a, b]$

$$\frac{1}{b - a}\int_a^b f(x)\,dx$$

Explore and Discuss 1   In Figure 6, the rectangle shown has the same area as the area under the graph of $y = f(x)$ from $x = a$ to $x = b$. Explain how the average value of $f(x)$ over the interval $[a, b]$ is related to the height of the rectangle.

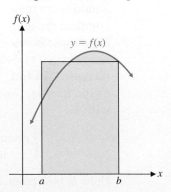

Figure 6

EXAMPLE 8 Average Value of a Function Find the average value of $f(x) = x - 3x^2$ over the interval $[-1, 2]$.

SOLUTION

$$\frac{1}{b-a} \int_a^b f(x)\, dx = \frac{1}{2 - (-1)} \int_{-1}^2 (x - 3x^2)\, dx$$

$$= \frac{1}{3}\left(\frac{x^2}{2} - x^3\right)\Big|_{-1}^2 = -\frac{5}{2}$$

Matched Problem 8 Find the average value of $g(t) = 6t^2 - 2t$ over the interval $[-2, 3]$.

EXAMPLE 9 Average Price Given the demand function

$$p = D(x) = 100e^{-0.05x}$$

find the average price (in dollars) over the demand interval $[40, 60]$.

SOLUTION

$$\text{Average price} = \frac{1}{b-a} \int_a^b D(x)\, dx$$

$$= \frac{1}{60 - 40} \int_{40}^{60} 100e^{-0.05x}\, dx$$

$$= \frac{100}{20} \int_{40}^{60} e^{-0.05x}\, dx \qquad \text{Use } \int e^{ax}\, dx = \frac{1}{a}e^{ax}, a \neq 0.$$

$$= -\frac{5}{0.05} e^{-0.05x}\Big|_{40}^{60}$$

$$= 100(e^{-2} - e^{-3}) \approx \$8.55$$

Matched Problem 9 Given the supply equation

$$p = S(x) = 10e^{0.05x}$$

find the average price (in dollars) over the supply interval $[20, 30]$.

Exercises 13.5

Skills Warm-up Exercises

W In Problems 1–8, use geometric formulas to find the unsigned area between the graph of $y = f(x)$ and the x axis over the indicated interval. (If necessary, review Appendix C.)

1. $f(x) = 100; [1, 6]$

2. $f(x) = -50; [8, 12]$

3. $f(x) = x + 5; [0, 4]$

4. $f(x) = x - 2; [-3, -1]$

5. $f(x) = 3x; [-4, 4]$

6. $f(x) = -10x; [-100, 50]$

7. $f(x) = \sqrt{9 - x^2}; [-3, 3]$

8. $f(x) = -\sqrt{25 - x^2}; [-5, 5]$

In Problems 9–12,

(A) Calculate the change in $F(x)$ from $x = 10$ to $x = 15$.

(B) Graph $F'(x)$ and use geometric formulas (see Appendix C) to calculate the area between the graph of $F'(x)$ and the x axis from $x = 10$ to $x = 15$.

(C) Verify that your answers to (A) and (B) are equal, as is guaranteed by the fundamental theorem of calculus.

9. $F(x) = 3x^2 + 160$

10. $F(x) = 9x + 120$

11. $F(x) = -x^2 + 42x + 240$

12. $F(x) = x^2 + 30x + 210$

Evaluate the integrals in Problems 13–32.

13. $\int_0^{10} 4 \, dx$

14. $\int_0^8 9x \, dx$

15. $\int_0^6 x^2 \, dx$

16. $\int_0^4 x^3 \, dx$

17. $\int_1^4 (5x + 3) \, dx$

18. $\int_2^5 (2x - 1) \, dx$

19. $\int_0^1 e^x \, dx$

20. $\int_0^2 4e^x \, dx$

21. $\int_1^2 \frac{1}{x} \, dx$

22. $\int_1^5 \frac{2}{x} \, dx$

23. $\int_{-2}^2 (x^3 + 7x) \, dx$

24. $\int_0^8 (0.25x - 1) \, dx$

25. $\int_2^5 (2x + 9) \, dx$

26. $\int_1^4 (6x - 5) \, dx$

27. $\int_5^2 (2x + 9) \, dx$

28. $\int_4^1 (6x - 5) \, dx$

29. $\int_2^3 (6 - x^3) \, dx$

30. $\int_6^9 (5 - x^2) \, dx$

31. $\int_6^6 (x^2 - 5x + 1)^{10} \, dx$

32. $\int_{-3}^{-3} (x^2 + 4x + 2)^8 \, dx$

Evaluate the integrals in Problems 33–48.

33. $\int_1^2 (2x^{-2} - 3) \, dx$

34. $\int_1^2 (5 - 16x^{-3}) \, dx$

35. $\int_1^4 3\sqrt{x} \, dx$

36. $\int_4^{25} \frac{2}{\sqrt{x}} \, dx$

37. $\int_2^3 12(x^2 - 4)^5 x \, dx$

38. $\int_0^1 32(x^2 + 1)^7 x \, dx$

39. $\int_3^9 \frac{1}{x - 1} \, dx$

40. $\int_2^8 \frac{1}{x + 1} \, dx$

41. $\int_{-5}^{10} e^{-0.05x} \, dx$

42. $\int_{-10}^{25} e^{-0.01x} \, dx$

43. $\int_1^e \frac{\ln t}{t} \, dt$

44. $\int_e^{e^2} \frac{(\ln t)^2}{t} \, dt$

45. $\int_0^1 xe^{-x^2} \, dx$

46. $\int_0^1 xe^{x^2} \, dx$

47. $\int_1^1 e^{x^2} \, dx$

48. $\int_{-1}^{-1} e^{-x^2} \, dx$

In Problems 49–56,

(A) Find the average value of each function over the indicated interval.

(B) Use a graphing calculator to graph the function and its average value over the indicated interval in the same viewing window.

49. $f(x) = 500 - 50x$; $[0, 10]$

50. $g(x) = 2x + 7$; $[0, 5]$

51. $f(t) = 3t^2 - 2t$; $[-1, 2]$

52. $g(t) = 4t - 3t^2$; $[-2, 2]$

53. $f(x) = \sqrt[3]{x}$; $[1, 8]$

54. $g(x) = \sqrt{x + 1}$; $[3, 8]$

55. $f(x) = 4e^{-0.2x}$; $[0, 10]$

56. $f(x) = 64e^{0.08x}$; $[0, 10]$

Evaluate the integrals in Problems 57–62.

57. $\int_2^3 x\sqrt{2x^2 - 3} \, dx$

58. $\int_0^1 x\sqrt{3x^2 + 2} \, dx$

59. $\int_0^1 \frac{x - 1}{x^2 - 2x + 3} \, dx$

60. $\int_1^2 \frac{x + 1}{2x^2 + 4x + 4} \, dx$

61. $\int_{-1}^1 \frac{e^{-x} - e^x}{(e^{-x} + e^x)^2} \, dx$

62. $\int_6^7 \frac{\ln(t - 5)}{t - 5} \, dt$

Use a numerical integration routine to evaluate each definite integral in Problems 63–66 (to three decimal places).

63. $\int_{1.7}^{3.5} x \ln x \, dx$

64. $\int_{-1}^1 e^{x^2} \, dx$

65. $\int_{-2}^2 \frac{1}{1 + x^2} \, dx$

66. $\int_0^3 \sqrt{9 - x^2} \, dx$

67. The **mean value theorem** states that if $F(x)$ is a differentiable function on the interval $[a, b]$, then there exists some number c between a and b such that

$$F'(c) = \frac{F(b) - F(a)}{b - a}$$

Explain why the mean value theorem implies that if a car averages 60 miles per hour in some 10-minute interval, then the car's instantaneous velocity is 60 miles per hour at least once in that interval.

68. The fundamental theorem of calculus can be proved by showing that, for every positive integer n, there is a Riemann sum for f on $[a, b]$ that is equal to $F(b) - F(a)$. By the mean value theorem (see Problem 67), within each subinterval $[x_{k-1}, x_k]$ that belongs to a partition of $[a, b]$, there is some c_k such that

$$f(c_k) = F'(c_k) = \frac{F(x_k) - F(x_{k-1})}{x_k - x_{k-1}}$$

Multiplying by the denominator $x_k - x_{k-1}$, we get

$$f(c_k)(x_k - x_{k-1}) = F(x_k) - F(x_{k-1})$$

Show that the Riemann sum

$$S_n = \sum_{k=1}^n f(c_k)(x_k - x_{k-1})$$

is equal to $F(b) - F(a)$.

Applications

69. Cost. A company manufactures mountain bikes. The research department produced the marginal cost function

$$C'(x) = 500 - \frac{x}{3} \qquad 0 \le x \le 900$$

where $C'(x)$ is in dollars and x is the number of bikes produced per month. Compute the increase in cost going from a production level of 300 bikes per month to 900 bikes per month. Set up a definite integral and evaluate it.

70. Cost. Referring to Problem 69, compute the increase in cost going from a production level of 0 bikes per month to 600 bikes per month. Set up a definite integral and evaluate it.

71. Salvage value. A new piece of industrial equipment will depreciate in value, rapidly at first and then less rapidly as time goes on. Suppose that the rate (in dollars per year) at which the book value of a new milling machine changes is given approximately by

$$V'(t) = f(t) = 500(t - 12) \qquad 0 \le t \le 10$$

where $V(t)$ is the value of the machine after t years. What is the total loss in value of the machine in the first 5 years? In the second 5 years? Set up appropriate integrals and solve.

72. Maintenance costs. Maintenance costs for an apartment house generally increase as the building gets older. From past records, the rate of increase in maintenance costs (in dollars per year) for a particular apartment complex is given approximately by

$$M'(x) = f(x) = 90x^2 + 5,000$$

where x is the age of the apartment complex in years and $M(x)$ is the total (accumulated) cost of maintenance for x years. Write a definite integral that will give the total maintenance costs from the end of the second year to the end of the seventh year, and evaluate the integral.

73. Employee training. A company producing computer components has established that, on the average, a new employee can assemble $N(t)$ components per day after t days of on-the-job training, as indicated in the following table (a new employee's productivity usually increases with time on the job, up to a leveling-off point):

t	0	20	40	60	80	100	120
$N(t)$	10	51	68	76	81	84	85

(A) Find a quadratic regression equation for the data, and graph it and the data set in the same viewing window.

(B) Use the regression equation and a numerical integration routine on a graphing calculator to approximate the number of units assembled by a new employee during the first 100 days on the job.

74. Employee training. Refer to Problem 73.

(A) Find a cubic regression equation for the data, and graph it and the data set in the same viewing window.

(B) Use the regression equation and a numerical integration routine on a graphing calculator to approximate the number of units assembled by a new employee during the second 60 days on the job.

75. Useful life. The total accumulated costs $C(t)$ and revenues $R(t)$ (in thousands of dollars), respectively, for a photocopying machine satisfy

$$C'(t) = \tfrac{1}{11}t \qquad \text{and} \qquad R'(t) = 5te^{-t^2}$$

where t is time in years. Find the useful life of the machine, to the nearest year. What is the total profit accumulated during the useful life of the machine?

76. Useful life. The total accumulated costs $C(t)$ and revenues $R(t)$ (in thousands of dollars), respectively, for a coal mine satisfy

$$C'(t) = 3 \qquad \text{and} \qquad R'(t) = 15e^{-0.1t}$$

where t is the number of years that the mine has been in operation. Find the useful life of the mine, to the nearest year. What is the total profit accumulated during the useful life of the mine?

77. Average cost. The total cost (in dollars) of manufacturing x auto body frames is $C(x) = 60,000 + 300x$.

(A) Find the average cost per unit if 500 frames are produced. [*Hint*: Recall that $\overline{C}(x)$ is the average cost per unit.]

(B) Find the average value of the cost function over the interval [0, 500].

(C) Discuss the difference between parts (A) and (B).

78. Average cost. The total cost (in dollars) of printing x dictionaries is $C(x) = 20,000 + 10x$.

(A) Find the average cost per unit if 1,000 dictionaries are produced.

(B) Find the average value of the cost function over the interval [0, 1,000].

(C) Discuss the difference between parts (A) and (B).

79. Cost. The marginal cost at various levels of output per month for a company that manufactures sunglasses is shown in the following table, with the output x given in thousands of units per month and the total cost $C(x)$ given in thousands of dollars per month:

x	0	1	2	3	4	5	6	7	8
$C'(x)$	58	30	18	9	5	7	17	33	51

(A) Find a quadratic regression equation for the data, and graph it and the data set in the same viewing window.

(B) Use the regression equation and a numerical integration routine on a graphing calculator to approximate (to the nearest dollar) the increased cost in going from a production level of 2 thousand sunglasses per month to 8 thousand sunglasses per month.

80. Cost. Refer to Problem 79.

(A) Find a cubic regression equation for the data, and graph it and the data set in the same viewing window.

(B) Use the regression equation and a numerical integration routine on a graphing calculator to approximate (to the nearest dollar) the increased cost in going from a production level of 1 thousand sunglasses per month to 7 thousand sunglasses per month.

81. Supply function. Given the supply function

$$p = S(x) = 10(e^{0.02x} - 1)$$

find the average price (in dollars) over the supply interval [20, 30].

82. Demand function. Given the demand function

$$p = D(x) = \frac{1,000}{x}$$

find the average price (in dollars) over the demand interval [400, 600].

83. Labor costs and learning. A defense contractor is starting production on a new missile control system. On the basis of data collected during assembly of the first 16 control systems, the production manager obtained the following function for the rate of labor use:

$$L'(x) = 2,400x^{-1/2}$$

Approximately how many labor-hours will be required to assemble the 17th through the 25th control units? [*Hint:* Let $a = 16$ and $b = 25$.]

84. Labor costs and learning. If the rate of labor use in Problem 83 is

$$L'(x) = 2,000x^{-1/3}$$

then approximately how many labor-hours will be required to assemble the 9th through the 27th control units? [*Hint:* Let $a = 8$ and $b = 27$.]

85. Inventory. A store orders 600 units of a product every 3 months. If the product is steadily depleted to 0 by the end of each 3 months, the inventory on hand I at any time t during the year is shown in the following figure:

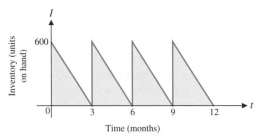

Time (months)

(A) Write an inventory function (assume that it is continuous) for the first 3 months. [The graph is a straight line joining (0, 600) and (3, 0).]

(B) What is the average number of units on hand for a 3-month period?

86. Repeat Problem 85 with an order of 1,200 units every 4 months.

87. Oil production. Using production and geological data, the management of an oil company estimates that oil will be pumped from a producing field at a rate given by

$$R(t) = \frac{100}{t + 1} + 5 \qquad 0 \le t \le 20$$

where $R(t)$ is the rate of production (in thousands of barrels per year) t years after pumping begins. Approximately how many barrels of oil will the field produce during the first

10 years of production? From the end of the 10th year to the end of the 20th year of production?

88. Oil production. In Problem 87, if the rate is found to be

$$R(t) = \frac{120t}{t^2 + 1} + 3 \qquad 0 \le t \le 20$$

then approximately how many barrels of oil will the field produce during the first 5 years of production? The second 5 years of production?

89. Biology. A yeast culture weighing 2 grams is expected to grow at the rate of $W'(t) = 0.2e^{0.1t}$ grams per hour at a higher controlled temperature. How much will the weight of the culture increase during the first 8 hours of growth? How much will the weight of the culture increase from the end of the 8th hour to the end of the 16th hour of growth?

90. Medicine. The rate at which the area of a skin wound is increasing is given (in square centimeters per day) by $A'(t) = -0.9e^{-0.1t}$. The initial wound has an area of 9 square centimeters. How much will the area change during the first 5 days? The second 5 days?

91. Temperature. If the temperature in an aquarium (in degrees Celsius) is given by

$$C(t) = t^3 - 2t + 10 \qquad 0 \le t \le 2$$

over a 2-hour period, what is the average temperature over this period?

92. Medicine. A drug is injected into the bloodstream of a patient through her right arm. The drug concentration in the bloodstream of the left arm t hours after the injection is given by

$$C(t) = \frac{0.14t}{t^2 + 1}$$

What is the average drug concentration in the bloodstream of the left arm during the first hour after the injection? During the first 2 hours after the injection?

93. Politics. Public awareness of a congressional candidate before and after a successful campaign was approximated by

$$P(t) = \frac{8.4t}{t^2 + 49} + 0.1 \qquad 0 \le t \le 24$$

where t is time in months after the campaign started and $P(t)$ is the fraction of the number of people in the congressional district who could recall the candidate's name. What is the average fraction of the number of people who could recall the candidate's name during the first 7 months of the campaign? During the first 2 years of the campaign?

94. Population composition. The number of children in a large city was found to increase and then decrease rather drastically. If the number of children (in millions) over a 6-year period was given by

$$N(t) = -\tfrac{1}{4}t^2 + t + 4 \qquad 0 \le t \le 6$$

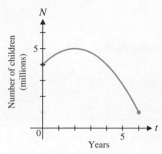

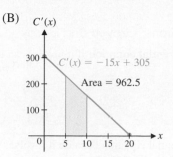

what was the average number of children in the city over the 6-year period? [Assume that $N = N(t)$ is continuous.]

(C) The change in cost from $x = 5$ to $x = 10$ is equal to the area between the marginal cost function and the x axis from $x = 5$ to $x = 10$.

Answers to Matched Problems

1. (A)

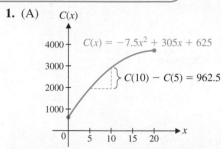

2. $16 + 2e - 2e^3 + 5 \ln 3 \approx -13.241$

3. $\frac{1}{2}(\ln 6 - \ln 4) \approx 0.203$

4. 2

5. \$1,000

6. (A) $-2 \ln \frac{2}{15} \approx 4$ yr   (B) $11 - 15e^{-2} \approx 8.970$ or \$8,970

7. 8.017

8. 13

9. \$35.27

Chapter 13   Summary and Review

Important Terms, Symbols, and Concepts

13.1 Antiderivatives and Indefinite Integrals

EXAMPLES

• A function F is an **antiderivative** of a function f if $F'(x) = f(x)$.

Ex. 1, p. 715

• If F and G are both antiderivatives of f, then F and G differ by a constant; that is, $F(x) = G(x) + k$ for some constant k.

• We use the symbol $\int f(x)\, dx$, called an **indefinite integral,** to represent the family of all antiderivatives of f, and we write

$$\int f(x)\, dx = F(x) + C$$

The symbol $\int$ is called an **integral sign,** $f(x)$ is the **integrand,** and C is the **constant of integration.**

Ex. 2, p. 718

• Indefinite integrals of basic functions are given by the formulas on page 716.

Ex. 3, p. 719

• Properties of indefinite integrals are given on page 717; in particular, a constant factor can be moved across an integral sign. However, a variable factor *cannot* be moved across an integral sign.

Ex. 4, p. 720
Ex. 5, p. 721
Ex. 6, p. 721

13.2 Integration by Substitution

• The **method of substitution** (also called the **change-of-variable method**) is a technique for finding indefinite integrals. It is based on the following formula, which is obtained by reversing the chain rule:

Ex. 1, p. 727

$$\int E'[I(x)]I'(x)\, dx = E[I(x)] + C$$

• This formula implies the general indefinite integral formulas on page 727.

Ex. 2, p. 728
Ex. 3, p. 728

• When using the method of substitution, it is helpful to use differentials as a bookkeeping device:

Ex. 4, p. 729

1. The **differential dx** of the independent variable x is an arbitrary real number.

Ex. 5, p. 730

2. The **differential dy** of the dependent variable y is defined by $dy = f'(x)\, dx$.

Ex. 6, p. 732

• Guidelines for using the substitution method are given by the procedure on page 729.

Ex. 7, p. 733

13.3 Differential Equations; Growth and Decay

- An equation is a **differential equation** if it involves an unknown function and one or more of the function's derivatives.

- The equation

$$\frac{dy}{dx} = 3x(1 + xy^2)$$

  is a **first-order** differential equation because it involves the first derivative of the unknown function y but no second or higher-order derivative.

- A **slope field** can be constructed for the preceding differential equation by drawing a tangent line segment with slope $3x(1 + xy^2)$ at each point (x, y) of a grid. The slope field gives a graphical representation of the functions that are solutions of the differential equation.

- The differential equation

$$\frac{dQ}{dt} = rQ$$

Ex. 1, p. 740
Ex. 2, p. 740
Ex. 3, p. 741

  (in words, the rate at which the unknown function Q increases is proportional to Q) is called the **exponential growth law**. The constant r is called the **relative growth rate**. The solutions of the exponential growth law are the functions

$$Q(t) = Q_0 e^{rt}$$

  where Q_0 denotes $Q(0)$, the amount present at time $t = 0$. These functions can be used to solve problems in population growth, continuous compound interest, radioactive decay, blood pressure, and light absorption.

- Table 1 on page 743 gives the solutions of other first-order differential equations that can be used to model the limited or logistic growth of epidemics, sales, and corporations.

Ex. 4, p. 742

13.4 The Definite Integral

- If the function f is positive on $[a, b]$, then the area between the graph of f and the x axis from $x = a$ to $x = b$ can be approximated by partitioning $[a, b]$ into n subintervals $[x_{k-1}, x_k]$ of equal length $\Delta x = (b - a)/n$ and summing the areas of n rectangles. This can be done using **left sums, right sums,** or, more generally, **Riemann sums:**

Ex. 1, p. 749
Ex. 2, p. 751

Left sum: $\quad L_n = \sum_{k=1}^{n} f(x_{k-1}) \Delta x$

Right sum: $\quad R_n = \sum_{k=1}^{n} f(x_k) \Delta x$

Riemann sum: $\quad S_n = \sum_{k=1}^{n} f(c_k) \Delta x$

  In a Riemann sum, each c_k is required to belong to the subinterval $[x_{k-1}, x_k]$. Left sums and right sums are the special cases of Riemann sums in which c_k is the left endpoint and right endpoint, respectively, of the subinterval.

- The **error in an approximation** is the absolute value of the difference between the approximation and the actual value. An **error bound** is a positive number such that the error is guaranteed to be less than or equal to that number.

- Theorem 1 on page 749 gives error bounds for the approximation of the area between the graph of a positive function f and the x axis from $x = a$ to $x = b$, by left sums or right sums, if f is either increasing or decreasing.

- If $f(x) > 0$ and is either increasing on $[a, b]$ or decreasing on $[a, b]$, then the left and right sums of $f(x)$ approach the same real number as $n \to \infty$ (Theorem 2, page 749).

- If f is a continuous function on $[a, b]$, then the Riemann sums for f on $[a, b]$ approach a real-number limit I as $n \to \infty$ (Theorem 3, page 751).

• Let f be a continuous function on $[a, b]$. Then the limit I of Riemann sums for f on $[a, b]$, guaranteed to exist by Theorem 3, is called the **definite integral** of f from a to b and is denoted

$$\int_a^b f(x)\, dx$$

The **integrand** is $f(x)$, the **lower limit of integration** is a, and the **upper limit of integration** is b.   Ex. 3, p. 752

• Geometrically, the definite integral

$$\int_a^b f(x)\, dx$$

represents the cumulative sum of the signed areas between the graph of f and the x axis from $x = a$ Ex. 4, p. 753
to $x = b$.

• Properties of the definite integral are given on page 753.

13.5 The Fundamental Theorem of Calculus

• If f is a continuous function on $[a, b\}$ and F is any antiderivative of f, then   Ex. 1, p. 758
Ex. 2, p. 759

$$\int_a^b f(x)\, dx = F(b) - F(a)$$

This is the fundamental theorem of calculus (see page 759).   Ex. 3, p. 760

• The fundamental theorem gives an easy and exact method for evaluating definite integrals, provided   Ex. 4, p. 761
that we can find an antiderivative $F(x)$ of $f(x)$. In practice, we first find an antiderivative $F(x)$ Ex. 5, p. 761
(when possible), using techniques for computing indefinite integrals. Then we calculate the difference   Ex. 6, p. 762
$F(b) - F(a)$. If it is impossible to find an antiderivative, we must resort to left or right sums, or other   Ex. 7, p. 762
approximation methods, to evaluate the definite integral. Graphing calculators have a built-in numerical
approximation routine, more powerful than left- or right-sum methods, for this purpose.

• If f is a continuous function on $[a, b]$, then the **average value** of f over $[a, b]$ is defined to be   Ex. 8, p. 765
Ex. 9, p. 765

$$\frac{1}{b - a} \int_a^b f(x)\, dx$$

Review Exercises

*Work through all the problems in this chapter review and check
your answers in the back of the book. Answers to all review problems are there, along with section numbers in italics to indicate
where each type of problem is discussed. Where weaknesses show
up, review appropriate sections of the text.*

Find each integral in Problems 1–6.

1. $\displaystyle\int (6x + 3)\, dx$ **2.** $\displaystyle\int_{10}^{20} 5\, dx$

3. $\displaystyle\int_0^9 (4 - t^2)\, dt$ **4.** $\displaystyle\int (1 - t^2)^3 t\, dt$

5. $\displaystyle\int \frac{1 + u^4}{u}\, du$ **6.** $\displaystyle\int_0^1 xe^{-2x^2}\, dx$

7. Is $F(x) = \ln x^2$ an antiderivative of $f(x) = \ln (2x)$?
Explain.

8. Is $F(x) = \ln x^2$ an antiderivative of $f(x) = \dfrac{2}{x}$? Explain.

9. Is $F(x) = (\ln x)^2$ an antiderivative of $f(x) = 2 \ln x$?
Explain.

10. Is $F(x) = (\ln x)^2$ an antiderivative of $f(x) = \dfrac{2 \ln x}{x}$?
Explain.

11. Is $y = 3x + 17$ a solution of the differential equation
$(x + 5)y' = y - 2$? Explain.

12. Is $y = 4x^3 + 7x^2 - 5x + 2$ a solution of the differential
equation $(x + 2)y''' - 24x = 48$? Explain.

*In Problems 13 and 14, find the derivative or indefinite integral as
indicated.*

13. $\dfrac{d}{dx}\left(\displaystyle\int e^{-x^2}\, dx \right)$ **14.** $\displaystyle\int \dfrac{d}{dx}\left(\sqrt{4 + 5x} \right) dx$

15. Find a function $y = f(x)$ that satisfies both conditions:

$$\frac{dy}{dx} = 3x^2 - 2 \qquad f(0) = 4$$

16. Find all antiderivatives of

(A) $\dfrac{dy}{dx} = 8x^3 - 4x - 1$ (B) $\dfrac{dx}{dt} = e^t - 4t^{-1}$

17. Approximate $\int_1^5 (x^2 + 1)\, dx$, using a right sum with $n = 2$. Calculate an error bound for this approximation.

18. Evaluate the integral in Problem 17, using the fundamental theorem of calculus, and calculate the actual error $|I - R_2|$ produced by using R_2.

19. Use the following table of values and a left sum with $n = 4$ to approximate $\int_1^{17} f(x)\, dx$:

x	1	5	9	13	17
$f(x)$	1.2	3.4	2.6	0.5	0.1

20. Find the average value of $f(x) = 6x^2 + 2x$ over the interval $[-1, 2]$.

21. Describe a rectangle that has the same area as the area under the graph of $f(x) = 6x^2 + 2x$ from $x = -1$ to $x = 2$ (see Problem 20).

In Problems 22 and 23, calculate the indicated Riemann sum S_n for the function $f(x) = 100 - x^2$.

22. Partition $[3, 11]$ into four subintervals of equal length, and for each subinterval $[x_{i-1}, x_i]$, let $c_i = (x_{i-1} + x_i)/2$.

23. Partition $[-5, 5]$ into five subintervals of equal length and let $c_1 = -4$, $c_2 = -1$, $c_3 = 1$, $c_4 = 2$, and $c_5 = 5$.

Use the graph and actual areas of the indicated regions in the figure to evaluate the integrals in Problems 24–31:

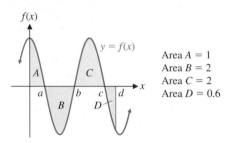

Area $A = 1$
Area $B = 2$
Area $C = 2$
Area $D = 0.6$

Figure for 24–31

24. $\int_a^b 5f(x)\, dx$

25. $\int_b^c \dfrac{f(x)}{5}\, dx$

26. $\int_b^d f(x)\, dx$

27. $\int_a^c f(x)\, dx$

28. $\int_0^d f(x)\, dx$

29. $\int_b^a f(x)\, dx$

30. $\int_c^b f(x)\, dx$

31. $\int_d^0 f(x)\, dx$

Problems 32–37 refer to the slope field shown in the figure:

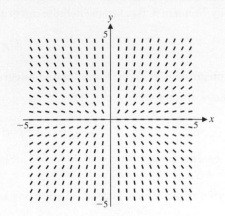

Figure for 32–37

32. (A) For $dy/dx = (2y)/x$, what is the slope of a solution curve at $(2, 1)$? At $(-2, -1)$?

   (B) For $dy/dx = (2x)/y$, what is the slope of a solution curve at $(2, 1)$? At $(-2, -1)$?

33. Is the slope field shown in the figure for $dy/dx = (2x)/y$ or for $dy/dx = (2y)/x$? Explain.

34. Show that $y = Cx^2$ is a solution of $dy/dx = (2y)/x$ for any real number C.

35. Referring to Problem 34, find the particular solution of $dy/dx = (2y)/x$ that passes through $(2, 1)$. Through $(-2, -1)$.

36. Graph the two particular solutions found in Problem 35 in the slope field shown (or a copy).

37. Use a graphing calculator to graph, in the same viewing window, graphs of $y = Cx^2$ for $C = -2, -1, 1$, and 2 for $-5 \le x \le 5$ and $-5 \le y \le 5$.

Find each integral in Problems 38–48.

38. $\int_{-1}^1 \sqrt{1 + x}\, dx$

39. $\int_{-1}^0 x^2 (x^3 + 2)^{-2}\, dx$

40. $\int 5e^{-t}\, dt$

41. $\int_1^e \dfrac{1 + t^2}{t}\, dt$

42. $\int xe^{3x^2}\, dx$

43. $\int_{-3}^1 \dfrac{1}{\sqrt{2 - x}}\, dx$

44. $\int_0^3 \dfrac{x}{1 + x^2}\, dx$

45. $\int_0^3 \dfrac{x}{(1 + x^2)^2}\, dx$

46. $\int x^3 (2x^4 + 5)^5\, dx$

47. $\int \dfrac{e^{-x}}{e^{-x} + 3}\, dx$

48. $\int \dfrac{e^x}{(e^x + 2)^2}\, dx$

49. Find a function $y = f(x)$ that satisfies both conditions:

$$\frac{dy}{dx} = 3x^{-1} - x^{-2} \qquad f(1) = 5$$

50. Find the equation of the curve that passes through (2, 10) if its slope is given by

$$\frac{dy}{dx} = 6x + 1$$

for each x.

51. (A) Find the average value of $f(x) = 3\sqrt{x}$ over the interval [1, 9].

(B) Graph $f(x) = 3\sqrt{x}$ and its average over the interval [1, 9] in the same coordinate system.

Find each integral in Problems 52–56.

52. $\int \frac{(\ln x)^2}{x} dx$

53. $\int x(x^3 - 1)^2 dx$

54. $\int \frac{x}{\sqrt{6 - x}} dx$

55. $\int_0^7 x\sqrt{16 - x}\, dx$

56. $\int_1^1 (x + 1)^9 dx$

57. Find a function $y = f(x)$ that satisfies both conditions:

$$\frac{dy}{dx} = 9x^2 e^{x^3} \qquad f(0) = 2$$

58. Solve the differential equation

$$\frac{dN}{dt} = 0.06N \qquad N(0) = 800 \qquad N > 0$$

☑ *Graph Problems 59–62 on a graphing calculator, and identify each curve as unlimited growth, exponential decay, limited growth, or logistic growth:*

59. $N = 50(1 - e^{-0.07t}); 0 \le t \le 80, 0 \le N \le 60$

60. $p = 500e^{-0.03x}; 0 \le x \le 100, 0 \le p \le 500$

61. $A = 200e^{0.08t}; 0 \le t \le 20, 0 \le A \le 1,000$

62. $N = \dfrac{100}{1 + 9e^{-0.3t}}; 0 \le t \le 25, 0 \le N \le 100$

☑ *Use a numerical integration routine to evaluate each definite integral in Problems 63–65 (to three decimal places).*

63. $\int_{-0.5}^{0.6} \frac{1}{\sqrt{1 - x^2}} dx$

64. $\int_{-2}^3 x^2 e^x dx$

65. $\int_{0.5}^{2.5} \frac{\ln x}{x^2} dx$

Applications

66. Cost. A company manufactures downhill skis. The research department produced the marginal cost graph shown in the accompanying figure, where $C'(x)$ is in dollars and x is the number of pairs of skis produced per week. Estimate the increase in cost going from a production level of 200 to 600 pairs of skis per week. Use left and right sums over two equal subintervals. Replace the question marks with the values of L_2 and R_2 as appropriate:

$$? \le \int_{200}^{600} C'(x)\, dx \le ?$$

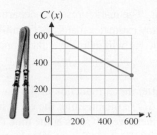

Figure for 66

67. Cost. Assuming that the marginal cost function in Problem 66 is linear, find its equation and write a definite integral that represents the increase in costs going from a production level of 200 to 600 pairs of skis per week. Evaluate the definite integral.

68. Profit and production. The weekly marginal profit for an output of x units is given approximately by

$$P'(x) = 150 - \frac{x}{10} \qquad 0 \le x \le 40$$

What is the total change in profit for a change in production from 10 units per week to 40 units? Set up a definite integral and evaluate it.

69. Profit function. If the marginal profit for producing x units per day is given by

$$P'(x) = 100 - 0.02x \qquad P(0) = 0$$

where $P(x)$ is the profit in dollars, find the profit function P and the profit on 10 units of production per day.

70. Resource depletion. An oil well starts out producing oil at the rate of 60,000 barrels of oil per year, but the production rate is expected to decrease by 4,000 barrels per year. Thus, if $P(t)$ is the total production (in thousands of barrels) in t years, then

$$P'(t) = f(t) = 60 - 4t \qquad 0 \le t \le 15$$

Write a definite integral that will give the total production after 15 years of operation, and evaluate the integral.

71. Inventory. Suppose that the inventory of a certain item t months after the first of the year is given approximately by

$$I(t) = 10 + 36t - 3t^2 \qquad 0 \le t \le 12$$

What is the average inventory for the second quarter of the year?

72. Price–supply. Given the price–supply function

$$p = S(x) = 8(e^{0.05x} - 1)$$

find the average price (in dollars) over the supply interval [40, 50].

73. Useful life. The total accumulated costs $C(t)$ and revenues $R(t)$ (in thousands of dollars), respectively, for a coal mine satisfy

$$C'(t) = 3 \qquad \text{and} \qquad R'(t) = 20e^{-0.1t}$$

where t is the number of years that the mine has been in operation. Find the useful life of the mine, to the nearest year. What is the total profit accumulated during the useful life of the mine?

74. Marketing. The market research department for an automobile company estimates that sales (in millions of dollars) of a new electric car will increase at the monthly rate of

$$S'(t) = 4e^{-0.08t} \qquad 0 \le t \le 24$$

t months after the introduction of the car. What will be the total sales $S(t)$ t months after the car is introduced if we assume that there were 0 sales at the time the car entered the marketplace? What are the estimated total sales during the first 12 months after the introduction of the car? How long will it take for the total sales to reach $40 million?

75. Wound healing. The area of a healing skin wound changes at a rate given approximately by

$$\frac{dA}{dt} = -5t^{-2} \qquad 1 \le t \le 5$$

where t is time in days and $A(1) = 5$ square centimeters. What will be the area of the wound in 5 days?

76. Pollution. An environmental protection agency estimates that the rate of seepage of toxic chemicals from a waste dump (in gallons per year) is given by

$$R(t) = \frac{1,000}{(1 + t)^2}$$

where t is the time in years since the discovery of the seepage. Find the total amount of toxic chemicals that seep from the dump during the first 4 years of its discovery.

77. Population. The population of Mexico was 116 million in 2013 and was growing at a rate of 1.07% per year, compounded continuously.

(A) Assuming that the population continues to grow at this rate, estimate the population of Mexico in the year 2025.

(B) At the growth rate indicated, how long will it take the population of Mexico to double?

78. Archaeology. The continuous compound rate of decay for carbon-14 is $r = -0.000\ 123\ 8$. A piece of animal bone found at an archaeological site contains 4% of the original amount of carbon-14. Estimate the age of the bone.

79. Learning. An average student enrolled in a typing class progressed at a rate of $N'(t) = 7e^{-0.1t}$ words per minute t weeks after enrolling in a 15-week course. If a student could type 25 words per minute at the beginning of the course, how many words per minute $N(t)$ would the student be expected to type t weeks into the course? After completing the course?

14

Additional Integration Topics

14.1 Area Between Curves

14.2 Applications in Business and Economics

14.3 Integration by Parts

14.4 Other Integration Methods

Chapter 14
Summary and Review

Review Exercises

Introduction

In Chapter 14 we explore additional applications and techniques of integration. We use the integral to find probabilities and to calculate several quantities that are important in business and economics: the total income and future value produced by a continuous income stream, consumers' and producers' surplus, and the Gini index of income concentration. The Gini index is a single number that measures the equality of a country's income distribution (see Problems 93 and 94, for example, in Section 14.1).

14.1 Area Between Curves

- Area Between Two Curves
- Application: Income Distribution

In Chapter 13, we found that the definite integral $\int_a^b f(x)\,dx$ represents the sum of the signed areas between the graph of $y = f(x)$ and the x axis from $x = a$ to $x = b$, where the areas above the x axis are counted positively and the areas below the x axis are counted negatively (see Fig. 1). In this section, we are interested in using the definite integral to find the actual area between a curve and the x axis or the actual area between two curves. These areas are always nonnegative quantities—**area measure is never negative**.

Area Between Two Curves

Consider the area bounded by $y = f(x)$ and $y = g(x)$, where $f(x) \geq g(x) \geq 0$, for $a \leq x \leq b$, as shown in Figure 2.

$$\begin{pmatrix} \text{Area } A \text{ between} \\ f(x) \text{ and } g(x) \end{pmatrix} = \begin{pmatrix} \text{area} \\ \text{under } f(x) \end{pmatrix} - \begin{pmatrix} \text{area} \\ \text{under } g(x) \end{pmatrix} \quad \begin{array}{l}\text{Areas are from } x = a \text{ to} \\ x = b \text{ above the } x \text{ axis.}\end{array}$$

$$= \int_a^b f(x)\,dx - \int_a^b g(x)\,dx \quad \begin{array}{l}\text{Use definite integral} \\ \text{property 4 (Section 13.4).}\end{array}$$

$$= \int_a^b [f(x) - g(x)]\,dx$$

It can be shown that the preceding result does not require $f(x)$ or $g(x)$ to remain positive over the interval $[a, b]$. A more general result is stated in the following box:

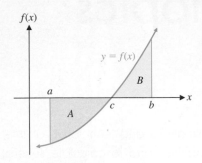

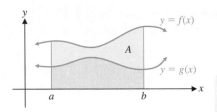

Figure 1 $\int_a^b f(x)\,dx = -A + B$

Figure 2

THEOREM 1   Area Between Two Curves

If f and g are continuous and $f(x) \geq g(x)$ over the interval $[a, b]$, then the area bounded by $y = f(x)$ and $y = g(x)$ for $a \leq x \leq b$ is given exactly by

$$A = \int_a^b [f(x) - g(x)]\,dx$$

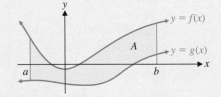

CONCEPTUAL   INSIGHT

Theorem 1 requires the graph of f to be *above* (or equal to) the graph of g throughout $[a, b]$, but f and g can be either positive, negative, or 0. In Section 13.4, we considered the special cases of Theorem 1 in which (1) f is positive and g is the zero function on $[a, b]$; and (2) f is the zero function and g is negative on $[a, b]$:

Special case 1. If f is continuous and positive over $[a, b]$, then the area bounded by the graph of f and the x axis for $a \leq x \leq b$ is given exactly by

$$\int_a^b f(x)\,dx$$

Special case 2. If g is continuous and negative over $[a, b]$, then the area bounded by the graph of g and the x axis for $a \leq x \leq b$ is given exactly by

$$\int_a^b [-g(x)]\,dx$$

EXAMPLE 1 Area Between a Curve and the x Axis Find the area bounded by $f(x) = 6x - x^2$ and $y = 0$ for $1 \le x \le 4$.

SOLUTION We sketch a graph of the region first (Fig. 3). The solution of every area problem should begin with a sketch. Since $f(x) \ge 0$ on $[1, 4]$,

$$A = \int_1^4 (6x - x^2)\, dx = \left(3x^2 - \frac{x^3}{3}\right)\Big|_1^4$$

$$= \left[3(4)^2 - \frac{(4)^3}{3}\right] - \left[3(1)^2 - \frac{(1)^3}{3}\right]$$

$$= 48 - \frac{64}{3} - 3 + \frac{1}{3}$$

$$= 48 - 21 - 3$$

$$= 24$$

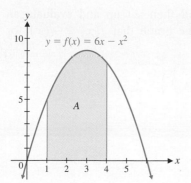

Figure 3

Matched Problem 1 Find the area bounded by $f(x) = x^2 + 1$ and $y = 0$ for $-1 \le x \le 3$.

EXAMPLE 2 Area Between a Curve and the x Axis Find the area between the graph of $f(x) = x^2 - 2x$ and the x axis over the indicated intervals:

(A) $[1, 2]$ (B) $[-1, 1]$

SOLUTION We begin by sketching the graph of f, as shown in Figure 4.

(A) From the graph, we see that $f(x) \le 0$ for $1 \le x \le 2$, so we integrate $-f(x)$:

$$A_1 = \int_1^2 [-f(x)]\, dx$$

$$= \int_1^2 (2x - x^2)\, dx$$

$$= \left(x^2 - \frac{x^3}{3}\right)\Big|_1^2$$

$$= \left[(2)^2 - \frac{(2)^3}{3}\right] - \left[(1)^2 - \frac{(1)^3}{3}\right]$$

$$= 4 - \frac{8}{3} - 1 + \frac{1}{3} = \frac{2}{3} \approx 0.667$$

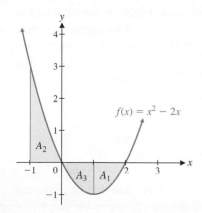

Figure 4

(B) Since the graph shows that $f(x) \ge 0$ on $[-1, 0]$ and $f(x) \le 0$ on $[0, 1]$, the computation of this area will require two integrals:

$$A = A_2 + A_3$$

$$= \int_{-1}^0 f(x)\, dx + \int_0^1 [-f(x)]\, dx$$

$$= \int_{-1}^0 (x^2 - 2x)\, dx + \int_0^1 (2x - x^2)\, dx$$

$$= \left(\frac{x^3}{3} - x^2\right)\Big|_{-1}^0 + \left(x^2 - \frac{x^3}{3}\right)\Big|_0^1$$

$$= \frac{4}{3} + \frac{2}{3} = 2$$

Matched Problem 2 Find the area between the graph of $f(x) = x^2 - 9$ and the x axis over the indicated intervals:

(A) $[0, 2]$ (B) $[2, 4]$

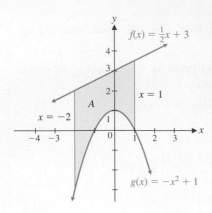

Figure 5

EXAMPLE 3   Area Between Two Curves   Find the area bounded by the graphs of $f(x) = \frac{1}{2}x + 3$, $g(x) = -x^2 + 1$, $x = -2$, and $x = 1$.

SOLUTION   We first sketch the area (Fig. 5) and then set up and evaluate an appropriate definite integral. We observe from the graph that $f(x) \geq g(x)$ for $-2 \leq x \leq 1$, so

$$A = \int_{-2}^{1} [f(x) - g(x)]\, dx = \int_{-2}^{1} \left[\left(\frac{x}{2} + 3 \right) - (-x^2 + 1) \right] dx$$

$$= \int_{-2}^{1} \left(x^2 + \frac{x}{2} + 2 \right) dx$$

$$= \left(\frac{x^3}{3} + \frac{x^2}{4} + 2x \right) \Big|_{-2}^{1}$$

$$= \left(\frac{1}{3} + \frac{1}{4} + 2 \right) - \left(\frac{-8}{3} + \frac{4}{4} - 4 \right) = \frac{33}{4} = 8.25$$

Matched Problem 3   Find the area bounded by $f(x) = x^2 - 1$, $g(x) = -\frac{1}{2}x - 3$, $x = -1$, and $x = 2$.

EXAMPLE 4   Area Between Two Curves   Find the area bounded by $f(x) = 5 - x^2$ and $g(x) = 2 - 2x$.

SOLUTION   First, graph f and g on the same coordinate system, as shown in Figure 6. Since the statement of the problem does not include any limits on the values of x, we must determine the appropriate values from the graph. The graph of f is a parabola and the graph of g is a line. The area bounded by these two graphs extends from the intersection point on the left to the intersection point on the right. To find these intersection points, we solve the equation $f(x) = g(x)$ for x:

$$f(x) = g(x)$$
$$5 - x^2 = 2 - 2x$$
$$x^2 - 2x - 3 = 0$$
$$x = -1, 3$$

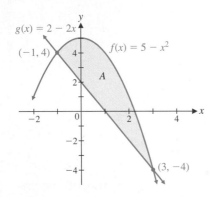

Figure 6

You should check these values in the original equations. (Note that the area between the graphs for $x < -1$ is unbounded on the left, and the area between the graphs for $x > 3$ is unbounded on the right.) Figure 6 shows that $f(x) \geq g(x)$ over the interval $[-1, 3]$, so we have

$$A = \int_{-1}^{3} [f(x) - g(x)]\, dx = \int_{-1}^{3} [5 - x^2 - (2 - 2x)]\, dx$$

$$= \int_{-1}^{3} (3 + 2x - x^2)\, dx$$

$$= \left(3x + x^2 - \frac{x^3}{3} \right) \Big|_{-1}^{3}$$

$$= \left[3(3) + (3)^2 - \frac{(3)^3}{3} \right] - \left[3(-1) + (-1)^2 - \frac{(-1)^3}{3} \right] = \frac{32}{3} \approx 10.667$$

Matched Problem 4   Find the area bounded by $f(x) = 6 - x^2$ and $g(x) = x$.

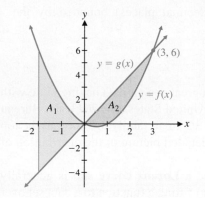

Figure 7

EXAMPLE 5 Area Between Two Curves Find the area bounded by $f(x) = x^2 - x$ and $g(x) = 2x$ for $-2 \leq x \leq 3$.

SOLUTION The graphs of f and g are shown in Figure 7. Examining the graph, we see that $f(x) \geq g(x)$ on the interval $[-2, 0]$, but $g(x) \geq f(x)$ on the interval $[0, 3]$. Thus, two integrals are required to compute this area:

$$A_1 = \int_{-2}^{0} [f(x) - g(x)]\, dx \qquad f(x) \geq g(x) \text{ on } [-2, 0]$$

$$= \int_{-2}^{0} [x^2 - x - 2x]\, dx$$

$$= \int_{-2}^{0} (x^2 - 3x)\, dx$$

$$= \left(\frac{x^3}{3} - \frac{3}{2} x^2 \right) \Big|_{-2}^{0}$$

$$= (0) - \left[\frac{(-2)^3}{3} - \frac{3}{2}(-2)^2 \right] = \frac{26}{3} \approx 8.667$$

$$A_2 = \int_{0}^{3} [g(x) - f(x)]\, dx \qquad g(x) \geq f(x) \text{ on } [0, 3]$$

$$= \int_{0}^{3} [2x - (x^2 - x)]\, dx$$

$$= \int_{0}^{3} (3x - x^2)\, dx$$

$$= \left(\frac{3}{2} x^2 - \frac{x^3}{3} \right) \Big|_{0}^{3}$$

$$= \left[\frac{3}{2}(3)^2 - \frac{(3)^3}{3} \right] - (0) = \frac{9}{2} = 4.5$$

The total area between the two graphs is

$$A = A_1 + A_2 = \tfrac{26}{3} + \tfrac{9}{2} = \tfrac{79}{6} \approx 13.167$$

Matched Problem 5 Find the area bounded by $f(x) = 2x^2$ and $g(x) = 4 - 2x$ for $-2 \leq x \leq 2$.

EXAMPLE 6 Computing Areas with a Numerical Integration Routine Find the area (to three decimal places) bounded by $f(x) = e^{-x^2}$ and $g(x) = x^2 - 1$.

SOLUTION First, we use a graphing calculator to graph the functions f and g and find their intersection points (see Fig. 8A). We see that the graph of f is bell shaped and the graph of g is a parabola. We note that $f(x) \geq g(x)$ on the interval $[-1.131, 1.131]$ and compute the area A by a numerical integration routine (see Fig. 8B):

$$A = \int_{-1.131}^{1.131} \left[e^{-x^2} - (x^2 - 1) \right] dx = 2.876$$

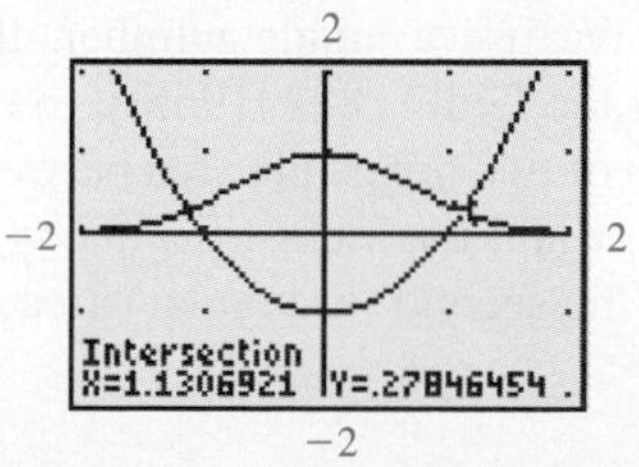

Figure 8   (A)   (B)

Matched Problem 6) Find the area (to three decimal places) bounded by the graphs of $f(x) = x^2 \ln x$ and $g(x) = 3x - 3$.

Application: Income Distribution

The U.S. Census Bureau compiles and analyzes a great deal of data having to do with the distribution of income among families in the United States. For 2011, the Bureau reported that the lowest 20% of families received 3% of all family income and the top 20% received 51%. Table 1 and Figure 9 give a detailed picture of the distribution of family income in 2011.

The graph of $y = f(x)$ in Figure 9 is called a **Lorenz curve** and is generally found by using *regression analysis,* a technique for fitting a function to a data set over a given interval. The variable **x represents the cumulative percentage of families at or below a given income level,** and **y represents the cumulative percentage of total family income received.** For example, data point (0.40, 0.11) in Table 1 indicates that the bottom 40% of families (those with incomes under $39,000) received 11% of the total income for all families in 2011, data point (0.60, 0.26) indicates that the bottom 60% of families received 26% of the total income for all families that year, and so on.

Table 1 **Family Income Distribution in the United States, 2011**

Income Level	x	y
Under $20,000	0.20	0.03
Under $39,000	0.40	0.11
Under $62,000	0.60	0.26
Under $102,000	0.80	0.49

Source: U.S. Census Bureau

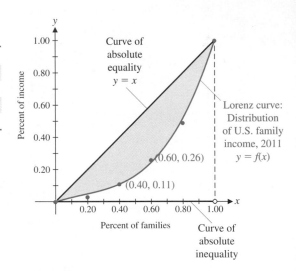

Figure 9 **Lorenz curve**

Absolute equality of income would occur if the area between the Lorenz curve and $y = x$ were 0. In this case, the Lorenz curve would be $y = x$ and all families would receive equal shares of the total income. That is, 5% of the families would receive 5% of the income, 20% of the families would receive 20% of the income, 65% of the families would receive 65% of the income, and so on. The maximum possible area between a Lorenz curve and $y = x$ is $\frac{1}{2}$, the area of the triangle below $y = x$. In this case, we would have **absolute inequality**: All the income would be in the hands of one family and the rest would have none. In actuality, Lorenz curves lie between these two extremes. But as the shaded area increases, the greater is the inequality of income distribution.

We use a single number, the **Gini index** [named after the Italian sociologist Corrado Gini (1884–1965)], to measure income concentration. The Gini index is the ratio of two areas: the area between $y = x$ and the Lorenz curve, and the area between $y = x$ and the x axis, from $x = 0$ to $x = 1$. The first area equals $\int_0^1 [x - f(x)]\, dx$ and the second (triangular) area equals $\frac{1}{2}$, giving the following definition:

DEFINITION Gini Index of Income Concentration

If $y = f(x)$ is the equation of a Lorenz curve, then

$$\textbf{Gini index} = 2 \int_0^1 [x - f(x)]\, dx$$

The Gini index is always a number between 0 and 1:

A Gini index of 0 indicates absolute equality—all people share equally in the income. A Gini index of 1 indicates absolute inequality—one person has all the income and the rest have none.

The closer the index is to 0, the closer the income is to being equally distributed. The closer the index is to 1, the closer the income is to being concentrated in a few hands. The Gini index of income concentration is used to compare income distributions at various points in time, between different groups of people, before and after taxes are paid, between different countries, and so on.

EXAMPLE 7 Distribution of Income The Lorenz curve for the distribution of income in a certain country in 2013 is given by $f(x) = x^{2.6}$. Economists predict that the Lorenz curve for the country in the year 2025 will be given by $g(x) = x^{1.8}$. Find the Gini index of income concentration for each curve, and interpret the results.

SOLUTION The Lorenz curves are shown in Figure 10.

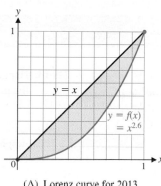

(A) Lorenz curve for 2013

(B) Projected Lorenz curve for 2025

Figure 10

The Gini index in 2013 is (see Fig. 10A)

$$2 \int_0^1 [x - f(x)]\, dx = 2 \int_0^1 [x - x^{2.6}]\, dx = 2\left(\frac{1}{2}x^2 - \frac{1}{3.6}x^{3.6} \right)\Big|_0^1$$

$$= 2\left(\frac{1}{2} - \frac{1}{3.6} \right) \approx 0.444$$

The projected Gini index in 2025 is (see Fig. 10B)

$$2 \int_0^1 [x - g(x)]\, dx = 2 \int_0^1 [x - x^{1.8}]\, dx = 2\left(\frac{1}{2}x^2 - \frac{1}{2.8}x^{2.8} \right)\Big|_0^1$$

$$= 2\left(\frac{1}{2} - \frac{1}{2.8} \right) \approx 0.286$$

If this projection is correct, the Gini index will decrease, and income will be more equally distributed in the year 2025 than in 2013.

Matched Problem 7 Repeat Example 7 if the projected Lorenz curve in the year 2025 is given by $g(x) = x^{3.8}$.

Explore and Discuss 1   Do you agree or disagree with each of the following statements (explain your answers by referring to the data in Table 2):

(A) In countries with a low Gini index, there is little incentive for individuals to strive for success, and therefore productivity is low.

(B) In countries with a high Gini index, it is almost impossible to rise out of poverty, and therefore productivity is low.

Table 2

Country	Gini Index	Per Capita Gross Domestic Product
Brazil	0.52	$12,100
Canada	0.32	43,400
China	0.47	9,300
France	0.33	36,100
Germany	0.27	39,700
India	0.37	3,900
Japan	0.38	36,900
Jordan	0.40	6,100
Mexico	0.48	15,600
Russia	0.42	18,000
Sweden	0.23	41,900
United States	0.45	50,700

Source: The World Factbook, CIA

Exercises 14.1

Skills Warm-up Exercises

W

In Problems 1–8, use geometric formulas to find the area between the graphs of $y = f(x)$ and $y = g(x)$ over the indicated interval. (If necessary, review Appendix C).

1. $f(x) = 60, g(x) = 45; [2, 12]$

2. $f(x) = -30, g(x) = 20; [-3, 6]$

3. $f(x) = 6 + 2x, g(x) = 6 - x; [0, 5]$

4. $f(x) = 0.5x, g(x) = 0.5x - 4; [0, 8]$

5. $f(x) = -3 - x, g(x) = 4 + 2x; [-1, 2]$

6. $f(x) = 100 - 2x, g(x) = 10 + 3x; [5, 10]$

7. $f(x) = x, g(x) = \sqrt{4 - x^2}; [0, \sqrt{2}]$

8. $f(x) = \sqrt{16 - x^2}, g(x) = |x|; [-2\sqrt{2}, 2\sqrt{2}]$

Problems 9–14 refer to Figures A–D. Set up definite integrals in Problems 9–12 that represent the indicated shaded area.

9. Shaded area in Figure B

10. Shaded area in Figure A

11. Shaded area in Figure C

12. Shaded area in Figure D

13. Explain why $\int_a^b h(x)\, dx$ does not represent the area between the graph of $y = h(x)$ and the x axis from $x = a$ to $x = b$ in Figure C.

14. Explain why $\int_a^b [-h(x)]\, dx$ represents the area between the graph of $y = h(x)$ and the x axis from $x = a$ to $x = b$ in Figure C.

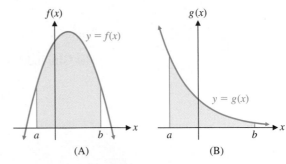
(A)   (B)

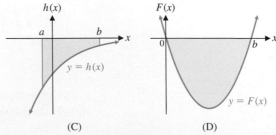
(C)   (D)

Figures for 9–14

In Problems 15–28, find the area bounded by the graphs of the indicated equations over the given interval. Compute answers to three decimal places.

15. $y = x + 4; y = 0; 0 \le x \le 4$

16. $y = -x + 10; y = 0; -2 \le x \le 2$

17. $y = x^2 - 20; y = 0; -3 \le x \le 0$

18. $y = x^2 + 2; y = 0; 0 \le x \le 3$

19. $y = -x^2 + 10; y = 0; -3 \le x \le 3$

20. $y = -2x^2; y = 0; -6 \le x \le 0$

21. $y = x^3 + 1; y = 0; 0 \le x \le 2$

22. $y = -x^3 + 3; y = 0; -2 \le x \le 1$

23. $y = x(1 - x); y = 0; -1 \le x \le 0$

24. $y = -x(3 - x); y = 0; 1 \le x \le 2$

25. $y = -e^x; y = 0; -1 \le x \le 1$

26. $y = e^x; y = 0; 0 \le x \le 1$

27. $y = \dfrac{1}{x}; y = 0; 1 \le x \le e$

28. $y = -\dfrac{1}{x}; y = 0; -1 \le x \le -\dfrac{1}{e}$

In Problems 29–32, base your answers on the Gini index of income concentration (see Table 2, page 782).

29. In which of Canada, Mexico, or the United States is income most equally distributed? Most unequally distributed?

30. In which of France, Germany, or Sweden is income most equally distributed? Most unequally distributed?

31. In which of Brazil, India, or Jordan is income most equally distributed? Most unequally distributed?

32. In which of China, Japan, or Russia is income most equally distributed? Most unequally distributed?

Problems 33–42 refer to Figures A and B. Set up definite integrals in Problems 33–40 that represent the indicated shaded areas over the given intervals.

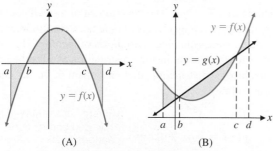

(A)   (B)

Figures for 33–42

33. Over interval $[a, b]$ in Figure A

34. Over interval $[c, d]$ in Figure A

35. Over interval $[b, d]$ in Figure A

36. Over interval $[a, c]$ in Figure A

37. Over interval $[c, d]$ in Figure B

38. Over interval $[a, b]$ in Figure B

39. Over interval $[a, c]$ in Figure B

40. Over interval $[b, d]$ in Figure B

41. Referring to Figure B, explain how you would use definite integrals and the functions f and g to find the area bounded by the two functions from $x = a$ to $x = d$.

42. Referring to Figure A, explain how you would use definite integrals to find the area between the graph of $y = f(x)$ and the x axis from $x = a$ to $x = d$.

In Problems 43–58, find the area bounded by the graphs of the indicated equations over the given intervals (when stated). Compute answers to three decimal places.

43. $y = -x; y = 0; -2 \le x \le 1$

44. $y = -x + 1; y = 0; -1 \le x \le 2$

45. $y = x^2 - 4; y = 0; 0 \le x \le 3$

46. $y = 4 - x^2; y = 0; 0 \le x \le 4$

47. $y = x^2 - 3x; y = 0; -2 \le x \le 2$

48. $y = -x^2 - 2x; y = 0; -2 \le x \le 1$

49. $y = -2x + 8; y = 12; -1 \le x \le 2$

50. $y = 2x + 6; y = 3; -1 \le x \le 2$

51. $y = 3x^2; y = 12$

52. $y = x^2; y = 9$

53. $y = 4 - x^2; y = -5$

54. $y = x^2 - 1; y = 3$

55. $y = x^2 + 1; y = 2x - 2; -1 \le x \le 2$

56. $y = x^2 - 1; y = x - 2; -2 \le x \le 1$

57. $y = e^{0.5x}; y = -\dfrac{1}{x}; 1 \le x \le 2$

58. $y = \dfrac{1}{x}; y = -e^x; 0.5 \le x \le 1$

In Problems 59–64, set up a definite integral that represents the area bounded by the graphs of the indicated equations over the given interval. Find the areas to three decimal places. [Hint: A circle of radius r, with center at the origin, has equation $x^2 + y^2 = r^2$ and area πr^2].

59. $y = \sqrt{9 - x^2}; y = 0; -3 \le x \le 3$

60. $y = \sqrt{25 - x^2}; y = 0; -5 \le x \le 5$

61. $y = -\sqrt{16 - x^2}; y = 0; 0 \le x \le 4$

62. $y = -\sqrt{36 - x^2}; y = 0; -6 \le x \le 0$

63. $y = -\sqrt{4 - x^2}; y = \sqrt{4 - x^2}; -2 \le x \le 2$

64. $y = -\sqrt{100 - x^2}; y = \sqrt{100 - x^2}; -10 \le x \le 10$

In Problems 65–70, find the area bounded by the graphs of the indicated equations over the given interval (when stated). Compute answers to three decimal places.

65. $y = e^x; y = e^{-x}; 0 \le x \le 4$

66. $y = e^x; y = -e^{-x}; 1 \le x \le 2$

67. $y = x^3; y = 4x$

68. $y = x^3 + 1; y = x + 1$

69. $y = x^3 - 3x^2 - 9x + 12; y = x + 12$

70. $y = x^3 - 6x^2 + 9x; y = x$

In Problems 71–76, use a graphing calculator to graph the equations and find relevant intersection points. Then find the area bounded by the curves. Compute answers to three decimal places.

71. $y = x^3 - x^2 + 2; y = -x^3 + 8x - 2$

72. $y = 2x^3 + 2x^2 - x; y = -2x^3 - 2x^2 + 2x$

73. $y = e^{-x}; y = 3 - 2x$

74. $y = 2 - (x + 1)^2; y = e^{x+1}$

75. $y = e^x; y = 5x - x^3$

76. $y = 2 - e^x; y = x^3 + 3x^2$

In Problems 77–80, use a numerical integration routine on a graphing calculator to find the area bounded by the graphs of the indicated equations over the given interval (when stated). Compute answers to three decimal places.

77. $y = e^{-x}; y = \sqrt{\ln x}; 2 \le x \le 5$

78. $y = x^2 + 3x + 1; y = e^{e^x}; -3 \le x \le 0$

79. $y = e^{x^2}; y = x + 2$

80. $y = \ln(\ln x); y = 0.01x$

In Problems 81–84, find the constant c (to 2 decimal places) such that the Lorenz curve $f(x) = x^c$ has the given Gini index of income concentration.

81. 0.52

82. 0.23

83. 0.29

84. 0.65

Applications

In the applications that follow, it is helpful to sketch graphs to get a clearer understanding of each problem and to interpret results. A graphing calculator will prove useful if you have one, but it is not necessary.

85. Oil production. Using production and geological data, the management of an oil company estimates that oil will be pumped from a producing field at a rate given by

$$R(t) = \frac{100}{t + 10} + 10 \qquad 0 \le t \le 15$$

where $R(t)$ is the rate of production (in thousands of barrels per year) t years after pumping begins. Find the area between the graph of R and the t axis over the interval [5, 10] and interpret the results.

86. Oil production. In Problem 85, if the rate is found to be

$$R(t) = \frac{100t}{t^2 + 25} + 4 \qquad 0 \le t \le 25$$

then find the area between the graph of R and the t axis over the interval [5, 15] and interpret the results.

87. Useful life. An amusement company maintains records for each video game it installs in an arcade. Suppose that $C(t)$ and $R(t)$ represent the total accumulated costs and revenues (in thousands of dollars), respectively, t years after a particular game has been installed. If

$$C'(t) = 2 \qquad \text{and} \qquad R'(t) = 9e^{-0.3t}$$

then find the area between the graphs of C' and R' over the interval on the t axis from 0 to the useful life of the game and interpret the results.

88. Useful life. Repeat Problem 87 if

$$C'(t) = 2t \qquad \text{and} \qquad R'(t) = 5te^{-0.1t^2}$$

89. Income distribution. In a study on the effects of World War II on the U.S. economy, an economist used data from the U.S. Census Bureau to produce the following Lorenz curves for the distribution of U.S. income in 1935 and in 1947:

$$f(x) = x^{2.4} \quad \text{Lorenz curve for 1935}$$
$$g(x) = x^{1.6} \quad \text{Lorenz curve for 1947}$$

Find the Gini index of income concentration for each Lorenz curve and interpret the results.

90. Income distribution. Using data from the U.S. Census Bureau, an economist produced the following Lorenz curves for the distribution of U.S. income in 1962 and in 1972:

$$f(x) = \tfrac{3}{10}x + \tfrac{7}{10}x^2 \quad \text{Lorenz curve for 1962}$$
$$g(x) = \tfrac{1}{2}x + \tfrac{1}{2}x^2 \quad \text{Lorenz curve for 1972}$$

Find the Gini index of income concentration for each Lorenz curve and interpret the results.

91. Distribution of wealth. Lorenz curves also can provide a relative measure of the distribution of a country's total assets. Using data in a report by the U.S. Congressional Joint Economic Committee, an economist produced the following Lorenz curves for the distribution of total U.S. assets in 1963 and in 1983:

$$f(x) = x^{10} \quad \text{Lorenz curve for 1963}$$
$$g(x) = x^{12} \quad \text{Lorenz curve for 1983}$$

Find the Gini index of income concentration for each Lorenz curve and interpret the results.

92. Income distribution. The government of a small country is planning sweeping changes in the tax structure in order to provide a more equitable distribution of income. The Lorenz curves for the current income distribution and for the projected income distribution after enactment of the tax changes are as follows:

$$f(x) = x^{2.3} \quad \text{Current Lorenz curve}$$
$$g(x) = 0.4x + 0.6x^2 \quad \text{Projected Lorenz curve after changes in tax laws}$$

Find the Gini index of income concentration for each Lorenz curve. Will the proposed changes provide a more equitable income distribution? Explain.

93. Distribution of wealth. The data in the following table describe the distribution of wealth in a country:

x	0	0.20	0.40	0.60	0.80	1
y	0	0.12	0.31	0.54	0.78	1

(A) Use quadratic regression to find the equation of a Lorenz curve for the data.

(B) Use the regression equation and a numerical integration routine to approximate the Gini index of income concentration.

 94. Distribution of wealth. Refer to Problem 93.

(A) Use cubic regression to find the equation of a Lorenz curve for the data.

(B) Use the cubic regression equation you found in Part (A) and a numerical integration routine to approximate the Gini index of income concentration.

95. Biology. A yeast culture is growing at a rate of $W'(t) = 0.3e^{0.1t}$ grams per hour. Find the area between the graph of W' and the t axis over the interval $[0, 10]$ and interpret the results.

96. Natural resource depletion. The instantaneous rate of change in demand for U.S. lumber since 1970 $(t = 0)$, in billions of cubic feet per year, is given by

$$Q'(t) = 12 + 0.006t^2 \qquad 0 \le t \le 50$$

Find the area between the graph of Q' and the t axis over the interval $[35, 40]$, and interpret the results.

97. Learning. A college language class was chosen for a learning experiment. Using a list of 50 words, the experiment

measured the rate of vocabulary memorization at different times during a continuous 5-hour study session. The average rate of learning for the entire class was inversely proportional to the time spent studying and was given approximately by

$$V'(t) = \frac{15}{t} \qquad 1 \le t \le 5$$

Find the area between the graph of V' and the t axis over the interval $[2, 4]$, and interpret the results.

98. Learning. Repeat Problem 97 if $V'(t) = 13/t^{1/2}$ and the interval is changed to $[1, 4]$.

Answers to Matched Problems

1. $A = \int_{-1}^{3} (x^2 + 1)\, dx = \frac{40}{3} \approx 13.333$
2. (A) $A = \int_{0}^{2} (9 - x^2)\, dx = \frac{46}{3} \approx 15.333$
   (B) $A = \int_{2}^{3} (9 - x^2)\, dx + \int_{3}^{4} (x^2 - 9)\, dx = 6$
3. $A = \int_{-1}^{2} \left[(x^2 - 1) - \left(-\frac{x}{2} - 3 \right) \right] dx = \frac{39}{4} = 9.75$
4. $A = \int_{-3}^{2} [(6 - x^2) - x]\, dx = \frac{125}{6} \approx 20.833$
5. $A = \int_{-2}^{1} [(4 - 2x) - 2x^2]\, dx + \int_{1}^{2} [2x^2 - (4 - 2x)]\, dx = \frac{38}{3} \approx 12.667$
6. 0.443
7. Gini index of income concentration ≈ 0.583; income will be less equally distributed in 2025.

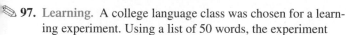

14.2 Applications in Business and Economics

- Probability Density Functions
- Continuous Income Stream
- Future Value of a Continuous Income Stream
- Consumers' and Producers' Surplus

This section contains important applications of the definite integral to business and economics. Included are three independent topics: probability density functions, continuous income streams, and consumers' and producers' surplus. Any of the three may be covered in any order as time and interests dictate.

Probability Density Functions

We now take a brief, informal look at the use of the definite integral to determine probabilities. A more formal treatment of the subject requires the use of the special "improper" integral form $\int_{-\infty}^{\infty} f(x)\, dx$, which we will not discuss.

Suppose that an experiment is designed in such a way that any real number x on the interval $[c, d]$ is a possible outcome. For example, x may represent an IQ score, the height of a person in inches, or the life of a lightbulb in hours. Technically, we refer to x as a *continuous random variable*.

In certain situations, we can find a function f with x as an independent variable such that the function f can be used to determine Probability $(c \le x \le d)$, that is, the probability that the outcome x of an experiment will be in the interval $[c, d]$. Such a function, called a **probability density function**, must satisfy the following three conditions (see Fig. 1):

1. $f(x) \ge 0$ for all real x.
2. The area under the graph of $f(x)$ over the interval $(-\infty, \infty)$ is exactly 1.
3. If $[c, d]$ is a subinterval of $(-\infty, \infty)$, then

$$\text{Probability } (c \le x \le d) = \int_{c}^{d} f(x)\, dx$$

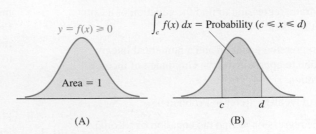

$$\int_{c}^{d} f(x)\, dx = \text{Probability}\ (c \le x \le d)$$

$y = f(x) \ge 0$

Area = 1

(A)   (B)

Figure 1 **Probability density function**

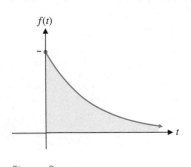

$f(t)$

t

Figure 2

EXAMPLE 1   Duration of Telephone Calls   Suppose that the length of telephone calls (in minutes) is a continuous random variable with the probability density function shown in Figure 2:

$$f(t) = \begin{cases} \frac{1}{4}e^{-t/4} & \text{if } t \ge 0 \\ 0 & \text{otherwise} \end{cases}$$

(A) Determine the probability that a call selected at random will last between 2 and 3 minutes.

(B) Find b (to two decimal places) so that the probability of a call selected at random lasting between 2 and b minutes is .5.

SOLUTION

(A) Probability $(2 \le t \le 3) = \displaystyle\int_{2}^{3} \frac{1}{4}e^{-t/4}\, dt$

$$= \left(-e^{-t/4}\right)\big|_{2}^{3}$$

$$= -e^{-3/4} + e^{-1/2} \approx .13$$

(B) We want to find b such that Probability $(2 \le t \le b) = .5$.

$$\int_{2}^{b} \frac{1}{4}e^{-t/4}\, dt = .5$$

$$-e^{-b/4} + e^{-1/2} = .5 \qquad\qquad \text{Solve for } b.$$

$$e^{-b/4} = e^{-.5} - .5$$

$$-\frac{b}{4} = \ln(e^{-.5} - .5)$$

$$b = 8.96 \text{ minutes}$$

So the probability of a call selected at random lasting from 2 to 8.96 minutes is .5.

Matched Problem 1

(A) In Example 1, find the probability that a call selected at random will last 4 minutes or less.

(B) Find b (to two decimal places) so that the probability of a call selected at random lasting b minutes or less is .9

CONCEPTUAL INSIGHT

The probability that a phone call in Example 1 lasts exactly 2 minutes (not 1.999 minutes, not 1.999 999 minutes) is given by

$$\text{Probability } (2 \le t \le 2) = \int_{2}^{2} \frac{1}{4}e^{-t/4}\, dt \qquad \text{Use Property 1, Section 13.4}$$

$$= 0$$

In fact, for any *continuous* random variable x with probability density function $f(x)$, the probability that x is exactly equal to a constant c is equal to 0:

$$\text{Probability } (c \le x \le c) = \int_{c}^{c} f(x)\, dx \quad \text{Use Property 1, Section 13.4}$$

$$= 0$$

In this respect, a *continuous* random variable differs from a *discrete* random variable. If x, for example, is the discrete random variable that represents the number of dots that appear on the top face when a fair die is rolled, then

$$\text{Probability } (2 \le x \le 2) = \tfrac{1}{6}$$

One of the most important probability density functions, the **normal probability density function**, is defined as follows and graphed in Figure 3:

$$f(x) = \frac{1}{\sigma \sqrt{2\pi}} e^{-(x-\mu)^2 / 2\sigma^2} \qquad \begin{array}{l} \mu \text{ is the mean.} \\ \sigma \text{ is the standard deviation.} \end{array}$$

It can be shown (but not easily) that the area under the normal curve in Figure 3 over the interval $(-\infty, \infty)$ is exactly 1. Since $\int e^{-x^2}\, dx$ is nonintegrable in terms of elementary functions (that is, the antiderivative cannot be expressed as a finite combination of simple functions), probabilities such as

$$\text{Probability } (c \le x \le d) = \frac{1}{\sigma \sqrt{2\pi}} \int_{c}^{d} e^{-(x-\mu)^2 / 2\sigma^2}\, dx$$

can be determined by making an appropriate substitution in the integrand and then using a table of areas under the standard normal curve (that is, the normal curve with $\mu = 0$ and $\sigma = 1$). As an alternative to a table, calculators and computers can be used to compute areas under normal curves.

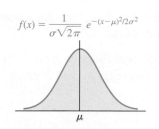

$$f(x) = \frac{1}{\sigma \sqrt{2\pi}} e^{-(x-\mu)^2 / 2\sigma^2}$$

μ

Figure 3 **Normal curve**

Continuous Income Stream

We start with a simple example having an obvious solution and generalize the concept to examples having less obvious solutions.

Suppose that an aunt has established a trust that pays you $2,000 a year for 10 years. What is the total amount you will receive from the trust by the end of the 10th year? Since there are 10 payments of $2,000 each, you will receive

$$10 \times \$2{,}000 = \$20{,}000$$

We now look at the same problem from a different point of view. Let's assume that the income stream is continuous at a rate of $2,000 per year. In Figure 4, the area under the graph of $f(t) = 2{,}000$ from 0 to t represents the income accumulated t years after the start. For example, for $t = \tfrac{1}{4}$ year, the income would be $\tfrac{1}{4}(2{,}000) = \$500$; for $t = \tfrac{1}{2}$ year, the income would be $\tfrac{1}{2}(2{,}000) = \$1{,}000$; for $t = 1$ year, the income would be $1(2{,}000) = \$2{,}000$; for $t = 5.3$ years, the income would be $5.3(2{,}000) = \$10{,}600$; and for $t = 10$ years, the income would be $10(2{,}000) = \$20{,}000$. The total income over a 10-year period—that is, the area under the graph of $f(t) = 2{,}000$ from 0 to 10—is also given by the definite integral

$$\int_{0}^{10} 2{,}000\, dt = 2{,}000t \big|_{0}^{10} = 2{,}000(10) - 2{,}000(0) = \$20{,}000$$

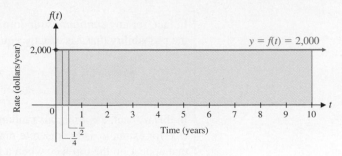

Figure 4 **Continuous income stream**

EXAMPLE 2   Continuous Income Stream   The rate of change of the income produced by a vending machine is given by

$$f(t) = 5,000e^{0.04t}$$

where t is time in years since the installation of the machine. Find the total income produced by the machine during the first 5 years of operation.

SOLUTION   The area under the graph of the rate-of-change function from 0 to 5 represents the total change in income over the first 5 years (Fig. 5), and is given by a definite integral:

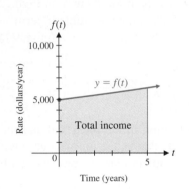

Figure 5 **Continuous income stream**

$$\text{Total income} = \int_0^5 5,000e^{0.04t}\, dt$$

$$= 125,000e^{0.04t}\Big|_0^5$$

$$= 125,000e^{0.04(5)} - 125,000e^{0.04(0)}$$

$$= 152,675 - 125,000$$

$$= \$27,675 \qquad \text{Rounded to the nearest dollar}$$

The vending machine produces a total income of \$27,675 during the first 5 years of operation.

Matched Problem 2   Referring to Example 2, find the total income produced (to the nearest dollar) during the second 5 years of operation.

In Example 2, we assumed that the rate of change of income was given by the continuous function f. The assumption is reasonable because income from a vending machine is often collected daily. In such situations, we assume that income is received in a **continuous stream**; that is, we assume that the rate at which income is received is a continuous function of time. The rate of change is called the **rate of flow** of the continuous income stream.

DEFINITION   Total Income for a Continuous Income Stream
If $f(t)$ is the rate of flow of a continuous income stream, then the **total income** produced during the period from $t = a$ to $t = b$ is

$$\text{Total income} = \int_a^b f(t)\, dt$$

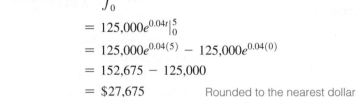

Future Value of a Continuous Income Stream

In Section 11.1, we discussed the continuous compound interest formula

$$A = Pe^{rt}$$

where P is the principal (or present value), A is the amount (or future value), r is the annual rate of continuous compounding (expressed as a decimal), and t is time in years. For example, if money is worth 12% compounded continuously, then the future value of a $10,000 investment in 5 years is (to the nearest dollar)

$$A = 10{,}000e^{0.12(5)} = \$18{,}221$$

We want to apply the future value concept to the income produced by a continuous income stream. Suppose that $f(t)$ is the rate of flow of a continuous income stream, and the income produced by this continuous income stream is invested as soon as it is received at a rate r, compounded continuously. We already know how to find the total income produced after T years, but how can we find the total of the income produced and the interest earned by this income? Since the income is received in a continuous flow, we cannot just use the formula $A = Pe^{rt}$. This formula is valid only for a single deposit P, not for a continuous flow of income. Instead, we use a Riemann sum approach that will allow us to apply the formula $A = Pe^{rt}$ repeatedly. To begin, we divide the time interval $[0, T]$ into n equal subintervals of length Δt and choose an arbitrary point c_k in each subinterval, as shown in Figure 6.

The total income produced during the period from $t = t_{k-1}$ to $t = t_k$ is equal to the area under the graph of $f(t)$ over this subinterval and is approximately equal to $f(c_k)\,\Delta t$, the area of the shaded rectangle in Figure 6. The income received during this period will earn interest for approximately $T - c_k$ years. So, from the future-value formula $A = Pe^{rt}$ with $P = f(c_k)\,\Delta t$ and $t = T - c_k$, the future value of the income produced during the period from $t = t_{k-1}$ to $t = t_k$ is approximately equal to

$$f(c_k)\,\Delta t\,e^{(T-c_k)r}$$

The total of these approximate future values over n subintervals is then

$$f(c_1)\,\Delta t\,e^{(T-c_1)r} + f(c_2)\,\Delta t\,e^{(T-c_2)r} + \cdots + f(c_n)\,\Delta t\,e^{(T-c_n)r} = \sum_{k=1}^{n} f(c_k)e^{r(T-c_k)}\,\Delta t$$

This equation has the form of a Riemann sum, the limit of which is a definite integral. (See the definition of the definite integral in Section 13.4.) Therefore, the *future value FV* of the income produced by the continuous income stream is given by

$$FV = \int_0^T f(t)e^{r(T-t)}\,dt$$

Since r and T are constants, we also can write

$$FV = \int_0^T f(t)e^{rT}e^{-rt}\,dt = e^{rT}\int_0^T f(t)e^{-rt}\,dt \tag{1}$$

This last form is preferable, since the integral is usually easier to evaluate than the first form.

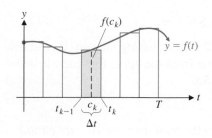

Figure 6

> **DEFINITION** Future Value of a Continuous Income Stream
>
> If $f(t)$ is the rate of flow of a continuous income stream, $0 \le t \le T$, and if the income is continuously invested at a rate r, compounded continuously, then the **future value FV** at the end of T years is given by
>
> $$FV = \int_0^T f(t)e^{r(T-t)}\,dt = e^{rT}\int_0^T f(t)e^{-rt}\,dt$$

The future value of a continuous income stream is the total value of all money produced by the continuous income stream (income and interest) at the end of T years.

We return to the trust that your aunt set up for you. Suppose that the $2,000 per year you receive from the trust is invested as soon as it is received at 8%, compounded continuously. We consider the trust income to be a continuous income stream with a flow rate of $2,000 per year. What is its future value (to the nearest dollar) by the end of the 10th year? Using the definite integral for future value from the preceding box, we have

$$FV = e^{rT} \int_0^T f(t)e^{-rt}\,dt$$

$$FV = e^{0.08(10)} \int_0^{10} 2,000e^{-0.08t}\,dt \qquad r = 0.08,\ T = 10,\ f(t) = 2,000$$

$$= 2,000e^{0.8} \int_0^{10} e^{-0.08t}\,dt$$

$$= 2,000e^{0.8} \left[\frac{e^{-0.08t}}{-0.08} \right]\Big|_0^{10}$$

$$= 2,000e^{0.8}[-12.5e^{-0.8} + 12.5] = \$30,639$$

At the end of 10 years, you will have received $30,639, including interest. How much is interest? Since you received $20,000 in income from the trust, the interest is the difference between the future value and income. So,

$$\$30,639 - \$20,000 = \$10,639$$

is the interest earned by the income received from the trust over the 10-year period.

EXAMPLE 3 Future Value of a Continuous Income Stream   Using the continuous income rate of flow for the vending machine in Example 2, namely,

$$f(t) = 5,000e^{0.04t}$$

find the future value of this income stream at 12%, compounded continuously for 5 years, and find the total interest earned. Compute answers to the nearest dollar.

SOLUTION   Using the formula

$$FV = e^{rT} \int_0^T f(t)e^{-rt}\,dt$$

with $r = 0.12$, $T = 5$, and $f(t) = 5,000e^{0.04t}$, we have

$$FV = e^{0.12(5)} \int_0^5 5,000e^{0.04t}e^{-0.12t}\,dt$$

$$= 5,000e^{0.6} \int_0^5 e^{-0.08t}\,dt$$

$$= 5,000e^{0.6} \left(\frac{e^{-0.08t}}{-0.08} \right)\Big|_0^5$$

$$= 5,000e^{0.6}(-12.5e^{-0.4} + 12.5)$$

$$= \$37,545 \qquad \text{Rounded to the nearest dollar}$$

The future value of the income stream at 12% compounded continuously at the end of 5 years is $37,545.

In Example 2, we saw that the total income produced by this vending machine over a 5-year period was $27,675. The difference between future value and income is interest. So,

$$\$37,545 - \$27,675 = \$9,870$$

is the interest earned by the income produced by the vending machine during the 5-year period.

Matched Problem 3   Repeat Example 3 if the interest rate is 9%, compounded continuously.

Consumers' and Producers' Surplus

Let $p = D(x)$ be the price–demand equation for a product, where x is the number of units of the product that consumers will purchase at a price of $\$p$ per unit. Suppose that $\bar{p}$ is the current price and $\bar{x}$ is the number of units that can be sold at that price. Then the price–demand curve in Figure 7 shows that if the price is higher than $\bar{p}$, the demand x is less than $\bar{x}$, but some consumers are still willing to pay the higher price. Consumers who are willing to pay more than $\bar{p}$, but who are still able to buy the product at $\bar{p}$, have saved money. We want to determine the total amount saved by all the consumers who are willing to pay a price higher than $\bar{p}$ for the product.

To do this, consider the interval $[c_k, c_k + \Delta x]$, where $c_k + \Delta x < \bar{x}$. If the price remained constant over that interval, the savings on each unit would be the difference between $D(c_k)$, the price consumers are willing to pay, and $\bar{p}$, the price they actually pay. Since Δx represents the number of units purchased by consumers over the interval, the total savings to consumers over this interval is approximately equal to

$$[D(c_k) - \bar{p}]\, \Delta x \quad \text{(savings per unit)} \times \text{(number of units)}$$

which is the area of the shaded rectangle shown in Figure 7. If we divide the interval $[0, \bar{x}]$ into n equal subintervals, then the total savings to consumers is approximately equal to

$$[D(c_1) - \bar{p}]\, \Delta x + [D(c_2) - \bar{p}]\, \Delta x + \cdots + [D(c_n) - \bar{p}]\, \Delta x = \sum_{k=1}^{n} [D(c_k) - \bar{p}]\, \Delta x$$

which we recognize as a Riemann sum for the integral

$$\int_{0}^{\bar{x}} [D(x) - \bar{p}]\, dx$$

We define the *consumers' surplus* to be this integral.

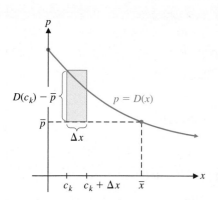

Figure 7

DEFINITION   Consumers' Surplus
If $(\bar{x}, \bar{p})$ is a point on the graph of the price–demand equation $p = D(x)$ for a particular product, then the **consumers' surplus CS** at a price level of $\bar{p}$ is

$$CS = \int_{0}^{\bar{x}} [D(x) - \bar{p}]\, dx$$

which is the area between $p = \bar{p}$ and $p = D(x)$ from $x = 0$ to $x = \bar{x}$, as shown in Figure 8.

The consumers' surplus represents the total savings to consumers who are willing to pay more than $\bar{p}$ for the product but are still able to buy the product for $\bar{p}$.

Figure 8

EXAMPLE 4   Consumers' Surplus   Find the consumers' surplus at a price level of $\$8$ for the price–demand equation

$$p = D(x) = 20 - 0.05x$$

SOLUTION

Step 1   Find $\bar{x}$, the demand when the price is $\bar{p} = 8$:

$$\bar{p} = 20 - 0.05\bar{x}$$
$$8 = 20 - 0.05\bar{x}$$
$$0.05\bar{x} = 12$$
$$\bar{x} = 240$$

Step 2   Sketch a graph, as shown in Figure 9.

Step 3   Find the consumers' surplus (the shaded area in the graph):

$$CS = \int_0^{\bar{x}} [D(x) - \bar{p}]\, dx$$

$$= \int_0^{240} (20 - 0.05x - 8)\, dx$$

$$= \int_0^{240} (12 - 0.05x)\, dx$$

$$= (12x - 0.025x^2)\Big|_0^{240}$$

$$= 2{,}880 - 1{,}440 = \$1{,}440$$

The total savings to consumers who are willing to pay a higher price for the product is $1,440.

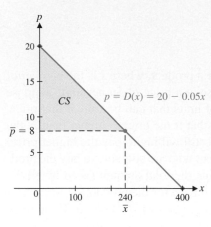

Figure 9

Matched Problem 4   Repeat Example 4 for a price level of $4.

If $p = S(x)$ is the price–supply equation for a product, $\bar{p}$ is the current price, and $\bar{x}$ is the current supply, then some suppliers are still willing to supply some units at a lower price than $\bar{p}$. The additional money that these suppliers gain from the higher price is called the *producers' surplus* and can be expressed in terms of a definite integral (proceeding as we did for the consumers' surplus).

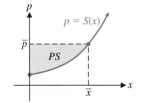

Figure 10

> **DEFINITION   Producers' Surplus**
>
> If $(\bar{x}, \bar{p})$ is a point on the graph of the price–supply equation $p = S(x)$, then the **producers' surplus PS** at a price level of $\bar{p}$ is
>
> $$PS = \int_0^{\bar{x}} [\bar{p} - S(x)]\, dx$$
>
> which is the area between $p = \bar{p}$ and $p = S(x)$ from $x = 0$ to $x = \bar{x}$, as shown in Figure 10.
>
> The producers' surplus represents the total gain to producers who are willing to supply units at a lower price than $\bar{p}$ but are still able to supply units at $\bar{p}$.

EXAMPLE 5   Producers' Surplus   Find the producers' surplus at a price level of $20 for the price–supply equation

$$p = S(x) = 2 + 0.0002x^2$$

SOLUTION

Step 1   Find $\bar{x}$, the supply when the price is $\bar{p} = 20$:

$$\bar{p} = 2 + 0.0002\bar{x}^2$$
$$20 = 2 + 0.0002\bar{x}^2$$
$$0.0002\bar{x}^2 = 18$$
$$\bar{x}^2 = 90{,}000$$
$$\bar{x} = 300 \qquad \text{There is only one solution, since } \bar{x} \ge 0.$$

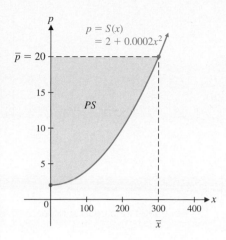

Figure 11

Step 2   Sketch a graph, as shown in Figure 11.

Step 3   Find the producers' surplus (the shaded area in the graph):

$$PS = \int_0^{\bar{x}} [\bar{p} - S(x)]\, dx = \int_0^{300} [20 - (2 + 0.0002x^2)]\, dx$$

$$= \int_0^{300} (18 - 0.0002x^2)\, dx = \left(18x - 0.0002\frac{x^3}{3} \right)\Big|_0^{300}$$

$$= 5{,}400 - 1{,}800 = \$3{,}600$$

The total gain to producers who are willing to supply units at a lower price is $3,600.

Matched Problem 5   Repeat Example 5 for a price level of $4.

In a free competitive market, the price of a product is determined by the relationship between supply and demand. If $p = D(x)$ and $p = S(x)$ are the price–demand and price–supply equations, respectively, for a product and if $(\bar{x}, \bar{p})$ is the point of intersection of these equations, then $\bar{p}$ is called the **equilibrium price** and $\bar{x}$ is called the **equilibrium quantity**. If the price stabilizes at the equilibrium price $\bar{p}$, then this is the price level that will determine both the consumers' surplus and the producers' surplus.

EXAMPLE 6   Equilibrium Price and Consumers' and Producers' Surplus   Find the equilibrium price and then find the consumers' surplus and producers' surplus at the equilibrium price level, if

$$p = D(x) = 20 - 0.05x \qquad \text{and} \qquad p = S(x) = 2 + 0.0002x^2$$

SOLUTION

Step 1   Find the equilibrium quantity. Set $D(x)$ equal to $S(x)$ and solve:

$$D(x) = S(x)$$
$$20 - 0.05x = 2 + 0.0002x^2$$
$$0.0002x^2 + 0.05x - 18 = 0$$
$$x^2 + 250x - 90{,}000 = 0$$
$$x = 200, -450$$

Since x cannot be negative, the only solution is $\bar{x} = 200$. The equilibrium price can be determined by using $D(x)$ or $S(x)$. We will use both to check our work:

$$\bar{p} = D(200) \qquad\qquad \bar{p} = S(200)$$
$$= 20 - 0.05(200) = 10 \qquad = 2 + 0.0002(200)^2 = 10$$

The equilibrium price is $\bar{p} = 10$, and the equilibrium quantity is $\bar{x} = 200$.

Step 2   Sketch a graph, as shown in Figure 12.

Step 3   Find the consumers' surplus:

$$CS = \int_0^{\bar{x}} [D(x) - \bar{p}]\, dx = \int_0^{200} (20 - 0.05x - 10)\, dx$$

$$= \int_0^{200} (10 - 0.05x)\, dx$$

$$= (10x - 0.025x^2)\big|_0^{200}$$

$$= 2{,}000 - 1{,}000 = \$1{,}000$$

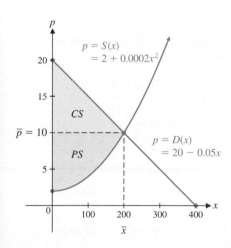

Figure 12

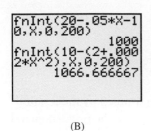

(A)

(B)

Figure 13

Step 4 Find the producers' surplus:

$$PS = \int_0^{\bar{x}} [\bar{p} - S(x)]\, dx$$

$$= \int_0^{200} [10 - (2 + 0.0002x^2)]\, dx$$

$$= \int_0^{200} (8 - 0.0002x^2)\, dx$$

$$= \left(8x - 0.0002\frac{x^3}{3}\right)\Big|_0^{200}$$

$$= 1{,}600 - \frac{1{,}600}{3} \approx \$1{,}067 \qquad \text{Rounded to the nearest dollar}$$

A graphing calculator offers an alternative approach to finding the equilibrium point for Example 6 (Fig. 13A). A numerical integration command can then be used to find the consumers' and producers' surplus (Fig. 13B).

Matched Problem 6 Repeat Example 6 for

$$p = D(x) = 25 - 0.001x^2 \qquad \text{and} \qquad p = S(x) = 5 + 0.1x$$

Exercises 14.2

Skills Warm-up Exercises

In Problems 1–8, find real numbers b and c such that $f(t) = e^b e^{ct}$. (If necessary, review Section 2.5).

1. $f(t) = e^{5(4-t)}$

2. $f(t) = e^{3(15-t)}$

3. $f(t) = e^{0.04(8-t)}$

4. $f(t) = e^{0.02(12-t)}$

5. $f(t) = e^{0.05t}e^{0.08(20-t)}$

6. $f(t) = e^{0.03t}e^{0.09(30-t)}$

7. $f(t) = e^{0.09t}e^{0.07(25-t)}$

8. $f(t) = e^{0.14t}e^{0.11(15-t)}$

In Problems 9–14, evaluate each definite integral to two decimal places.

9. $\int_0^8 e^{0.06(8-t)}\, dt$

10. $\int_1^{10} e^{0.07(10-t)}\, dt$

11. $\int_0^{20} e^{0.08t}e^{0.12(20-t)}\, dt$

12. $\int_0^{15} e^{0.05t}e^{0.06(15-t)}\, dt$

13. $\int_0^{30} 500\, e^{0.02t}e^{0.09(30-t)}\, dt$

14. $\int_0^{25} 900\, e^{0.03t}e^{0.04(25-t)}\, dt$

In Problems 15 and 16, explain which of (A), (B), and (C) are equal before evaluating the expressions. Then evaluate each expression to two decimal places.

15. (A) $\int_0^8 e^{0.07(8-t)}\, dt$ (B) $\int_0^8 (e^{0.56} - e^{0.07t})\, dt$

(C) $e^{0.56} \int_0^8 e^{-0.07t}\, dt$

16. (A) $\int_0^{10} 2{,}000 e^{0.05t}e^{0.12(10-t)}\, dt$

(B) $2{,}000 e^{1.2} \int_0^{10} e^{-0.07t}\, dt$

(C) $2{,}000 e^{0.05} \int_0^{10} e^{0.12(10-t)}\, dt$

In Problems 17–20, use a graphing calculator to graph the normal probability density function

$$f(x) = \frac{1}{\sigma\sqrt{2\pi}}\, e^{-(x-\mu)^2/2\sigma^2}$$

that has the given mean μ and standard deviation σ.

17. $\mu = 0,\ \sigma = 1$

18. $\mu = 20,\ \sigma = 5$

19. $\mu = 500,\ \sigma = 100$

20. $\mu = 300,\ \sigma = 25$

In Problems 21–24, use a numerical integration routine on a graphing calculator.

21. For the normal probability density function of Problem 17, find:

(A) Probability $(-1 \leq x \leq 1)$

(B) Probability $(-2 \leq x \leq 2)$

(C) Probability $(-3 \leq x \leq 3)$

22. For the normal probability density function of Problem 18, find:

(A) Probability $(15 \leq x \leq 25)$

(B) Probability $(10 \leq x \leq 30)$

(C) Probability $(5 \leq x \leq 35)$

23. For the normal probability density function of Problem 19, find:

(A) Probability $(400 \le x \le 600)$

(B) Probability $(300 \le x \le 700)$

(C) Probability $(200 \le x \le 800)$

24. For the normal probability density function of Problem 20, find:

(A) Probability $(275 \le x \le 325)$

(B) Probability $(250 \le x \le 350)$

(C) Probability $(225 \le x \le 375)$

Applications

Unless stated to the contrary, compute all monetary answers to the nearest dollar.

25. The life expectancy (in years) of a microwave oven is a continuous random variable with probability density function

$$f(x) = \begin{cases} 2/(x+2)^2 & \text{if } x \ge 0 \\ 0 & \text{otherwise} \end{cases}$$

(A) Find the probability that a randomly selected microwave oven lasts at most 6 years.

(B) Find the probability that a randomly selected microwave oven lasts from 6 to 12 years.

(C) Graph $y = f(x)$ for [0, 12] and show the shaded region for part (A).

26. The shelf life (in years) of a laser pointer battery is a continuous random variable with probability density function

$$f(x) = \begin{cases} 1/(x+1)^2 & \text{if } x \ge 0 \\ 0 & \text{otherwise} \end{cases}$$

(A) Find the probability that a randomly selected laser pointer battery has a shelf life of 3 years or less.

(B) Find the probability that a randomly selected laser pointer battery has a shelf life of from 3 to 9 years.

(C) Graph $y = f(x)$ for [0, 10] and show the shaded region for part (A).

27. In Problem 25, find d so that the probability of a randomly selected microwave oven lasting d years or less is .8.

28. In Problem 26, find d so that the probability of a randomly selected laser pointer battery lasting d years or less is .5.

29. A manufacturer guarantees a product for 1 year. The time to failure of the product after it is sold is given by the probability density function

$$f(t) = \begin{cases} .01e^{-.01t} & \text{if } t \ge 0 \\ 0 & \text{otherwise} \end{cases}$$

where t is time in months. What is the probability that a buyer chosen at random will have a product failure

(A) During the warranty period?

(B) During the second year after purchase?

30. In a certain city, the daily use of water (in hundreds of gallons) per household is a continuous random variable with probability density function

$$f(x) = \begin{cases} .15e^{-.15x} & \text{if } x \ge 0 \\ 0 & \text{otherwise} \end{cases}$$

Find the probability that a household chosen at random will use

(A) At most 400 gallons of water per day

(B) Between 300 and 600 gallons of water per day

31. In Problem 29, what is the probability that the product will last at least 1 year? [*Hint:* Recall that the total area under the probability density function curve is 1.]

32. In Problem 30, what is the probability that a household will use more than 400 gallons of water per day? [See the hint in Problem 31.]

33. Find the total income produced by a continuous income stream in the first 5 years if the rate of flow is $f(t) = 2,500$.

34. Find the total income produced by a continuous income stream in the first 10 years if the rate of flow is $f(t) = 3,000$.

35. Interpret the results of Problem 33 with both a graph and a description of the graph.

36. Interpret the results of Problem 34 with both a graph and a description of the graph.

37. Find the total income produced by a continuous income stream in the first 3 years if the rate of flow is $f(t) = 400e^{0.05t}$.

38. Find the total income produced by a continuous income stream in the first 2 years if the rate of flow is $f(t) = 600e^{0.06t}$.

39. Interpret the results of Problem 37 with both a graph and a description of the graph.

40. Interpret the results of Problem 38 with both a graph and a description of the graph.

41. Starting at age 25, you deposit $2,000 a year into an IRA account. Treat the yearly deposits into the account as a continuous income stream. If money in the account earns 5%, compounded continuously, how much will be in the account 40 years later, when you retire at age 65? How much of the final amount is interest?

42. Suppose in Problem 41 that you start the IRA deposits at age 30, but the account earns 6%, compounded continuously. Treat the yearly deposits into the account as a continuous income stream. How much will be in the account 35 years later when you retire at age 65? How much of the final amount is interest?

43. Find the future value at 3.25% interest, compounded continuously for 4 years, of the continuous income stream with rate of flow $f(t) = 1,650e^{-0.02t}$.

44. Find the future value, at 2.95% interest, compounded continuously for 6 years, of the continuous income stream with rate of flow $f(t) = 2,000e^{0.06t}$.

45. Compute the interest earned in Problem 43.

46. Compute the interest earned in Problem 44.

47. An investor is presented with a choice of two investments: an established clothing store and a new computer store. Each choice requires the same initial investment and each produces a continuous income stream of 4%, compounded continuously. The rate of flow of income from the clothing store is $f(t) = 12,000$, and the rate of flow of income from the computer store is expected to be $g(t) = 10,000e^{0.05t}$. Compare the future values of these investments to determine which is the better choice over the next 5 years.

48. Refer to Problem 47. Which investment is the better choice over the next 10 years?

49. An investor has $10,000 to invest in either a bond that matures in 5 years or a business that will produce a continuous stream of income over the next 5 years with rate of flow $f(t) = 2,150$. If both the bond and the continuous income stream earn 3.75%, compounded continuously, which is the better investment?

50. Refer to Problem 49. Which is the better investment if the rate of the income from the business is $f(t) = 2,250$?

51. A business is planning to purchase a piece of equipment that will produce a continuous stream of income for 8 years with rate of flow $f(t) = 9,000$. If the continuous income stream earns 6.95%, compounded continuously, what single deposit into an account earning the same interest rate will produce the same future value as the continuous income stream? (This deposit is called the **present value** of the continuous income stream.)

52. Refer to Problem 51. Find the present value of a continuous income stream at 7.65%, compounded continuously for 12 years, if the rate of flow is $f(t) = 1,000e^{0.03t}$.

53. Find the future value at a rate r, compounded continuously for T years, of a continuous income stream with rate of flow $f(t) = k$, where k is a constant.

54. Find the future value at a rate r, compounded continuously for T years, of a continuous income stream with rate of flow $f(t) = ke^{ct}$, where c and k are constants, $c \neq r$.

55. Find the consumers' surplus at a price level of $\bar{p} = \$150$ for the price–demand equation

$$p = D(x) = 400 - 0.05x$$

56. Find the consumers' surplus at a price level of $\bar{p} = \$120$ for the price–demand equation

$$p = D(x) = 200 - 0.02x$$

57. Interpret the results of Problem 55 with both a graph and a description of the graph.

58. Interpret the results of Problem 56 with both a graph and a description of the graph.

59. Find the producers' surplus at a price level of $\bar{p} = \$67$ for the price–supply equation

$$p = S(x) = 10 + 0.1x + 0.0003x^2$$

60. Find the producers' surplus at a price level of $\bar{p} = \$55$ for the price–supply equation

$$p = S(x) = 15 + 0.1x + 0.003x^2$$

61. Interpret the results of Problem 59 with both a graph and a description of the graph.

62. Interpret the results of Problem 60 with both a graph and a description of the graph.

In Problems 63–70, find the consumers' surplus and the producers' surplus at the equilibrium price level for the given price–demand and price–supply equations. Include a graph that identifies the consumers' surplus and the producers' surplus. Round all values to the nearest integer.

63. $p = D(x) = 50 - 0.1x; p = S(x) = 11 + 0.05x$

64. $p = D(x) = 25 - 0.004x^2; p = S(x) = 5 + 0.004x^2$

65. $p = D(x) = 80e^{-0.001x}; p = S(x) = 30e^{0.001x}$

66. $p = D(x) = 185e^{-0.005x}; p = S(x) = 25e^{0.005x}$

67. $p = D(x) = 80 - 0.04x; p = S(x) = 30e^{0.001x}$

68. $p = D(x) = 190 - 0.2x; p = S(x) = 25e^{0.005x}$

69. $p = D(x) = 80e^{-0.001x}; p = S(x) = 15 + 0.0001x^2$

70. $p = D(x) = 185e^{-0.005x}; p = S(x) = 20 + 0.002x^2$

71. The following tables give price–demand and price–supply data for the sale of soybeans at a grain market, where x is the number of bushels of soybeans (in thousands of bushels) and p is the price per bushel (in dollars):

Tables for 71–72

	Price–Demand		Price–Supply
x	$p = D(x)$	x	$p = S(x)$
0	6.70	0	6.43
10	6.59	10	6.45
20	6.52	20	6.48
30	6.47	30	6.53
40	6.45	40	6.62

Use quadratic regression to model the price–demand data and linear regression to model the price–supply data.

(A) Find the equilibrium quantity (to three decimal places) and equilibrium price (to the nearest cent).

(B) Use a numerical integration routine to find the consumers' surplus and producers' surplus at the equilibrium price level.

72. Repeat Problem 71, using quadratic regression to model both sets of data.

Answers to Matched Problems

1. (A) .63   (B) 9.21 min

2. $33,803   **3.** $FV = \$34,691$; interest $= \$7,016$

4. $2,560   **5.** $133

6. $\bar{p} = 15$; $CS = \$667$; $PS = \$500$

14.3 Integration by Parts

In Section 13.1, we promised to return later to the indefinite integral

$$\int \ln x \, dx$$

since none of the integration techniques considered up to that time could be used to find an antiderivative for ln x. We now develop a very useful technique, called *integration by parts,* that will enable us to find not only the preceding integral, but also many others, including integrals such as

$$\int x \ln x \, dx \quad \text{and} \quad \int x e^x \, dx$$

The method of integration by parts is based on the product formula for derivatives. If f and g are differentiable functions, then

$$\frac{d}{dx}[f(x)g(x)] = f(x)g'(x) + g(x)f'(x)$$

which can be written in the equivalent form

$$f(x)g'(x) = \frac{d}{dx}[f(x)g(x)] - g(x)f'(x)$$

Integrating both sides, we obtain

$$\int f(x)g'(x) \, dx = \int \frac{d}{dx}[f(x)g(x)] \, dx - \int g(x)f'(x) \, dx$$

The first integral to the right of the equal sign is $f(x)g(x) + C$. Why? We will leave out the constant of integration for now, since we can add it after integrating the second integral to the right of the equal sign. So,

$$\int f(x)g'(x) \, dx = f(x)g(x) - \int g(x)f'(x) \, dx$$

This equation can be transformed into a more convenient form by letting $u = f(x)$ and $v = g(x)$; then $du = f'(x) \, dx$ and $dv = g'(x) \, dx$. Making these substitutions, we obtain the **integration-by-parts formula**:

Integration-by-Parts Formula

$$\int u \, dv = uv - \int v \, du$$

This formula can be very useful when the integral on the left is difficult or impossible to integrate with standard formulas. If u and dv are chosen with care—this is the crucial part of the process—then the integral on the right side may be easier to integrate than the one on the left. The formula provides us with another tool that is helpful in many, but not all, cases. We are able to easily check the results by differentiating to get the original integrand, a good habit to develop.

EXAMPLE 1 Integration by Parts Find $\int x e^x \, dx$, using integration by parts, and check the result.

SOLUTION First, write the integration-by-parts formula:

$$\int u \, dv = uv - \int v \, du \tag{1}$$

Now try to identify u and dv in $\int xe^x \, dx$ so that $\int v \, du$ on the right side of (1) is easier to integrate than $\int u \, dv = \int xe^x \, dx$ on the left side. There are essentially two reasonable choices in selecting u and dv in $\int xe^x \, dx$:

$$\text{Choice 1} \qquad \text{Choice 2}$$
$$\int \underset{u}{x} \, \underset{dv}{e^x \, dx} \qquad \int \underset{u}{e^x} \, \underset{dv}{x \, dx}$$

We pursue choice 1 and leave choice 2 for you to explore (see Explore and Discuss 1 following this example).

From choice 1, $u = x$ and $dv = e^x \, dx$. Looking at formula (1), we need du and v to complete the right side. Let

$$u = x \qquad dv = e^x \, dx$$

Then,

$$du = dx \qquad \int dv = \int e^x \, dx$$

$$v = e^x$$

Any constant may be added to v, but we will always choose 0 for simplicity. The general arbitrary constant of integration will be added at the end of the process.

Substituting these results into formula (1), we obtain

$$\int u \, dv = uv - \int v \, du$$

$$\int xe^x \, dx = xe^x - \int e^x \, dx \qquad \text{The right integral is easy to integrate.}$$

$$= xe^x - e^x + C \qquad \text{Now add the arbitrary constant } C.$$

Check:

$$\frac{d}{dx}(xe^x - e^x + C) = xe^x + e^x - e^x = xe^x$$

Explore and Discuss 1 Pursue choice 2 in Example 1, using the integration-by-parts formula, and explain why this choice does not work out.

Matched Problem 1 Find $\int xe^{2x} \, dx$.

EXAMPLE 2 Integration by Parts Find $\int x \ln x \, dx$.

SOLUTION As before, we have essentially two choices in choosing u and dv:

$$\text{Choice 1} \qquad \text{Choice 2}$$
$$\int \underset{u}{x} \, \underset{dv}{\ln x \, dx} \qquad \int \underset{u}{\ln x} \, \underset{dv}{x \, dx}$$

Choice 1 is rejected since we do not yet know how to find an antiderivative of $\ln x$. So we move to choice 2 and choose $u = \ln x$ and $dv = x\,dx$. Then we proceed as in Example 1. Let

$$u = \ln x \qquad dv = x\,dx$$

Then,

$$du = \frac{1}{x}dx \qquad \int dv = \int x\,dx$$

$$v = \frac{x^2}{2}$$

Substitute these results into the integration-by-parts formula:

$$\int u\,dv = uv - \int v\,du$$

$$\int x\ln x\,dx = (\ln x)\left(\frac{x^2}{2}\right) - \int \left(\frac{x^2}{2}\right)\left(\frac{1}{x}\right)dx$$

$$= \frac{x^2}{2}\ln x - \int \frac{x}{2}dx \qquad \text{An easy integral to evaluate}$$

$$= \frac{x^2}{2}\ln x - \frac{x^2}{4} + C$$

Check:

$$\frac{d}{dx}\left(\frac{x^2}{2}\ln x - \frac{x^2}{4} + C\right) = x\ln x + \left(\frac{x^2}{2}\cdot\frac{1}{x}\right) - \frac{x}{2} = x\ln x$$

Matched Problem 2) Find $\int x\ln 2x\,dx$.

CONCEPTUAL INSIGHT

As you may have discovered in Explore and Discuss 1, some choices for u and dv will lead to integrals that are more complicated than the original integral. This does not mean that there is an error in either the calculations or the integration-by-parts formula. It simply means that the particular choice of u and dv does not change the problem into one we can solve. When this happens, we must look for a different choice of u and dv. In some problems, it is possible that no choice will work.

   Guidelines for selecting u and dv for integration by parts are summarized in the following box:

SUMMARY Integration by Parts: Selection of u and dv

For $\int u\,dv = uv - \int v\,du,$

1. The product $u\,dv$ must equal the original integrand.
2. It must be possible to integrate dv (preferably by using standard formulas or simple substitutions).
3. The new integral $\int v\,du$ should not be more complicated than the original integral $\int u\,dv$.
4. For integrals involving $x^p e^{ax}$, try

$$u = x^p \qquad \text{and} \qquad dv = e^{ax}\,dx$$

5. For integrals involving $x^p(\ln x)^q$, try

$$u = (\ln x)^q \qquad \text{and} \qquad dv = x^p\,dx$$

In some cases, repeated use of the integration-by-parts formula will lead to the evaluation of the original integral. The next example provides an illustration of such a case.

EXAMPLE 3   Repeated Use of Integration by Parts   Find $\int x^2 e^{-x}\,dx$.

SOLUTION   Following suggestion 4 in the box, we choose

$$u = x^2 \qquad dv = e^{-x}\,dx$$

Then,

$$du = 2x\,dx \qquad v = -e^{-x}$$

and

$$\int x^2 e^{-x}\,dx = x^2(-e^{-x}) - \int (-e^{-x})2x\,dx$$

$$= -x^2 e^{-x} + 2\int x e^{-x}\,dx \qquad (2)$$

The new integral is not one we can evaluate by standard formulas, but it is simpler than the original integral. Applying the integration-by-parts formula to it will produce an even simpler integral. For the integral $\int x e^{-x}\,dx$, we choose

$$u = x \qquad dv = e^{-x}\,dx$$

Then,

$$du = dx \qquad v = -e^{-x}$$

and

$$\int x e^{-x}\,dx = x(-e^{-x}) - \int (-e^{-x})\,dx$$

$$= -x e^{-x} + \int e^{-x}\,dx$$

$$= -x e^{-x} - e^{-x} \qquad \text{Choose 0 for the constant.} \qquad (3)$$

Substituting equation (3) into equation (2), we have

$$\int x^2 e^{-x}\,dx = -x^2 e^{-x} + 2(-x e^{-x} - e^{-x}) + C \quad \text{Add an arbitrary constant here.}$$

$$= -x^2 e^{-x} - 2x e^{-x} - 2e^{-x} + C$$

Check:

$$\frac{d}{dx}(-x^2 e^{-x} - 2x e^{-x} - 2e^{-x} + C) = x^2 e^{-x} - 2x e^{-x} + 2x e^{-x} - 2e^{-x} + 2e^{-x}$$

$$= x^2 e^{-x}$$

Matched Problem 3)   Find $\int x^2 e^{2x}\,dx$.

EXAMPLE 4   Using Integration by Parts   Find $\int_1^e \ln x\,dx$ and interpret the result geometrically.

SOLUTION   First, we find $\int \ln x\,dx$. Then we return to the definite integral. Following suggestion 5 in the box (with $p = 0$), we choose

$$u = \ln x \qquad dv = dx$$

Then,

$$du = \frac{1}{x}dx \qquad v = x$$

$$\int \ln x \, dx = (\ln x)(x) - \int (x)\frac{1}{x}dx$$

$$= x \ln x - x + C$$

This is the important result we mentioned at the beginning of this section. Now we have

$$\int_1^e \ln x \, dx = (x \ln x - x)\Big|_1^e$$

$$= (e \ln e - e) - (1 \ln 1 - 1)$$

$$= (e - e) - (0 - 1)$$

$$= 1$$

The integral represents the area under the curve $y = \ln x$ from $x = 1$ to $x = e$, as shown in Figure 1.

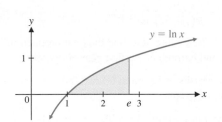

Figure 1

Matched Problem 4 Find $\int_1^2 \ln 3x \, dx$.

Explore and Discuss 2 Try using the integration-by-parts formula on $\int e^{x^2} dx$, and explain why it does not work.

Exercises 14.3

Skills Warm-up Exercises

W

In Problems 1–8, find the derivative of $f(x)$ and the indefinite integral of $g(x)$. (If necessary, review Sections 11.2 and 13.1).

1. $f(x) = 5x; g(x) = x^3$ **2.** $f(x) = x^2; g(x) = e^x$

3. $f(x) = x^3; g(x) = 5x$ **4.** $f(x) = e^x; g(x) = x^2$

5. $f(x) = e^{4x}; g(x) = \dfrac{1}{x}$

6. $f(x) = \sqrt{x}; g(x) = e^{-2x}$

7. $f(x) = \dfrac{1}{x}; g(x) = e^{4x}$

8. $f(x) = e^{-2x}; g(x) = \sqrt{x}$

In Problems 9–12, integrate by parts. Assume that $x > 0$ whenever the natural logarithm function is involved.

9. $\displaystyle\int xe^{3x} \, dx$ **10.** $\displaystyle\int xe^{4x} \, dx$

11. $\displaystyle\int x^2 \ln x \, dx$ **12.** $\displaystyle\int x^3 \ln x \, dx$

13. If you want to use integration by parts to find $\int (x + 1)^5 (x + 2) \, dx$, which is the better choice for u: $u = (x + 1)^5$ or $u = x + 2$? Explain your choice and then integrate.

14. If you want to use integration by parts to find $\int (5x - 7)(x - 1)^4 \, dx$, which is the better choice for u: $u = 5x - 7$ or $u = (x - 1)^4$? Explain your choice and then integrate.

Problems 15–28 are mixed—some require integration by parts, and others can be solved with techniques considered earlier. Integrate as indicated, assuming $x > 0$ whenever the natural logarithm function is involved.

15. $\displaystyle\int xe^{-x} \, dx$ **16.** $\displaystyle\int (x - 1)e^{-x} \, dx$

17. $\displaystyle\int xe^{x^2} \, dx$ **18.** $\displaystyle\int xe^{-x^2} \, dx$

19. $\displaystyle\int_0^1 (x - 3)e^x \, dx$ **20.** $\displaystyle\int_0^1 (x + 1)e^x \, dx$

21. $\displaystyle\int_1^3 \ln 2x \, dx$ **22.** $\displaystyle\int_1^2 \ln\left(\frac{x}{2}\right) dx$

23. $\displaystyle\int \frac{2x}{x^2 + 1} \, dx$ **24.** $\displaystyle\int \frac{x^2}{x^3 + 5} \, dx$

25. $\displaystyle\int \frac{\ln x}{x} \, dx$ **26.** $\displaystyle\int \frac{e^x}{e^x + 1} \, dx$

27. $\displaystyle\int \sqrt{x} \ln x \, dx$ **28.** $\displaystyle\int \frac{\ln x}{\sqrt{x}} \, dx$

In Problems 29–32, the integral can be found in more than one way. First use integration by parts, then use a method that does not involve integration by parts. Which method do you prefer?

29. $\int (x-3)(x+1)^2 \, dx$ **30.** $\int (x+2)(x-1)^2 \, dx$

31. $\int (2x+1)(x-2)^2 \, dx$ **32.** $\int (5x-1)(x+2)^2 \, dx$

In Problems 33–36, illustrate each integral graphically and describe what the integral represents in terms of areas.

33. Problem 19 **34.** Problem 20

35. Problem 21 **36.** Problem 22

Problems 37–58 are mixed—some may require use of the integration-by-parts formula along with techniques we have considered earlier; others may require repeated use of the integration-by-parts formula. Assume that $g(x) > 0$ whenever $\ln g(x)$ is involved.

37. $\int x^2 e^x \, dx$ **38.** $\int x^3 e^x \, dx$

39. $\int xe^{ax} \, dx, a \neq 0$ **40.** $\int \ln(ax) \, dx, a > 0$

41. $\int_1^e \frac{\ln x}{x^2} \, dx$ **42.** $\int_1^2 x^3 e^{x^2} \, dx$

43. $\int_0^2 \ln(x+4) \, dx$ **44.** $\int_0^2 \ln(4-x) \, dx$

45. $\int xe^{x-2} \, dx$ **46.** $\int xe^{x+1} \, dx$

47. $\int x \ln(1+x^2) \, dx$ **48.** $\int x \ln(1+x) \, dx$

49. $\int e^x \ln(1+e^x) \, dx$ **50.** $\int \frac{\ln(1+\sqrt{x})}{\sqrt{x}} \, dx$

51. $\int (\ln x)^2 \, dx$ **52.** $\int x(\ln x)^2 \, dx$

53. $\int (\ln x)^3 \, dx$ **54.** $\int x(\ln x)^3 \, dx$

55. $\int_1^e \ln(x^2) \, dx$ **56.** $\int_1^e \ln(x^4) \, dx$

57. $\int_0^1 \ln(e^{x^2}) \, dx$ **58.** $\int_1^2 \ln(xe^x) \, dx$

In Problems 59–62, use a graphing calculator to graph each equation over the indicated interval and find the area between the curve and the x axis over that interval. Find answers to two decimal places.

59. $y = x - 2 - \ln x; 1 \leq x \leq 4$

60. $y = 6 - x^2 - \ln x; 1 \leq x \leq 4$

61. $y = 5 - xe^x; 0 \leq x \leq 3$

62. $y = xe^x + x - 6; 0 \leq x \leq 3$

Applications

63. Profit. If the marginal profit (in millions of dollars per year) is given by

$$P'(t) = 2t - te^{-t}$$

use an appropriate definite integral to find the total profit (to the nearest million dollars) earned over the first 5 years of operation.

64. Production. An oil field is estimated to produce oil at a rate of $R(t)$ thousand barrels per month t months from now, as given by

$$R(t) = 10te^{-0.1t}$$

Use an appropriate definite integral to find the total production (to the nearest thousand barrels) in the first year of operation.

65. Profit. Interpret the results of Problem 63 with both a graph and a description of the graph.

66. Production. Interpret the results of Problem 64 with both a graph and a description of the graph.

67. Continuous income stream. Find the future value at 3.95%, compounded continuously, for 5 years of a continuous income stream with a rate of flow of

$$f(t) = 1,000 - 200t$$

68. Continuous income stream. Find the interest earned at 4.15%, compounded continuously, for 4 years for a continuous income stream with a rate of flow of

$$f(t) = 1,000 - 250t$$

69. Income distribution. Find the Gini index of income concentration for the Lorenz curve with equation

$$y = xe^{x-1}$$

70. Income distribution. Find the Gini index of income concentration for the Lorenz curve with equation

$$y = x^2 e^{x-1}$$

71. Income distribution. Interpret the results of Problem 69 with both a graph and a description of the graph.

72. Income distribution. Interpret the results of Problem 70 with both a graph and a description of the graph.

73. Sales analysis. Monthly sales of a particular personal computer are expected to increase at the rate of

$$S'(t) = -4te^{0.1t}$$

computers per month, where t is time in months and $S(t)$ is the number of computers sold each month. The company plans to stop manufacturing this computer when monthly sales reach 800 computers. If monthly sales now ($t = 0$) are 2,000 computers, find $S(t)$. How long, to the nearest month, will the company continue to manufacture the computer?

74. Sales analysis. The rate of change of the monthly sales of a new basketball game is given by

$$S'(t) = 350 \ln(t+1) \qquad S(0) = 0$$

where t is the number of months since the game was released and $S(t)$ is the number of games sold each month. Find $S(t)$.

When, to the nearest month, will monthly sales reach 15,000 games?

75. Consumers' surplus. Find the consumers' surplus (to the nearest dollar) at a price level of $\bar{p} = \$2.089$ for the price–demand equation

$$p = D(x) = 9 - \ln(x + 4)$$

Use $\bar{x}$ computed to the nearest higher unit.

76. Producers' surplus. Find the producers' surplus (to the nearest dollar) at a price level of $\bar{p} = \$26$ for the price–supply equation

$$p = S(x) = 5 \ln(x + 1)$$

Use $\bar{x}$ computed to the nearest higher unit.

77. Consumers' surplus. Interpret the results of Problem 75 with both a graph and a description of the graph.

78. Producers' surplus. Interpret the results of Problem 76 with both a graph and a description of the graph.

79. Pollution. The concentration of particulate matter (in parts per million) t hours after a factory ceases operation for the day is given by

$$C(t) = \frac{20 \ln(t + 1)}{(t + 1)^2}$$

Find the average concentration for the period from $t = 0$ to $t = 5$.

80. Medicine. After a person takes a pill, the drug contained in the pill is assimilated into the bloodstream. The rate of assimilation t minutes after taking the pill is

$$R(t) = te^{-0.2t}$$

Find the total amount of the drug that is assimilated into the bloodstream during the first 10 minutes after the pill is taken.

81. Learning. A student enrolled in an advanced typing class progressed at a rate of

$$N'(t) = (t + 6)e^{-0.25t}$$

words per minute per week t weeks after enrolling in a 15-week course. If a student could type 40 words per minute at the beginning of the course, then how many words per minute $N(t)$ would the student be expected to type t weeks into the course? How long, to the nearest week, should it take the student to achieve the 70-word-per-minute level? How many words per minute should the student be able to type by the end of the course?

82. Learning. A student enrolled in a stenotyping class progressed at a rate of

$$N'(t) = (t + 10)e^{-0.1t}$$

words per minute per week t weeks after enrolling in a 15-week course. If a student had no knowledge of stenotyping (that is, if the student could stenotype at 0 words per minute) at the beginning of the course, then how many words per minute $N(t)$ would the student be expected to handle t weeks into the course? How long, to the nearest week, should it take the student to achieve 90 words per minute? How many words per minute should the student be able to handle by the end of the course?

83. Politics. The number of voters (in thousands) in a certain city is given by

$$N(t) = 20 + 4t - 5te^{-0.1t}$$

where t is time in years. Find the average number of voters during the period from $t = 0$ to $t = 5$.

Answers to Matched Problems

1. $\dfrac{x}{2}e^{2x} - \dfrac{1}{4}e^{2x} + C$

2. $\dfrac{x^2}{2} \ln 2x - \dfrac{x^2}{4} + C$

3. $\dfrac{x^2}{2}e^{2x} - \dfrac{x}{2}e^{2x} + \dfrac{1}{4}e^{2x} + C$

4. $2 \ln 6 - \ln 3 - 1 \approx 1.4849$

14.4 Other Integration Methods

- The Trapezoidal Rule
- Simpson's Rule
- Using a Table of Integrals
- Substitution and Integral Tables
- Reduction Formulas
- Application

In Chapter 5 we used left and right sums to approximate the definite integral of a function, and, if an antiderivative could be found, calculated the exact value using the fundamental theorem of calculus. Now we discuss other methods for approximating definite integrals, and a procedure for finding exact values of definite integrals of many standard functions.

Approximation of definite integrals by left sums and right sums is instructive and important, but not efficient. A large number of rectangles must be used, and many terms must be summed, to get good approximations. The *trapezoidal rule* and *Simpson's rule* provide more efficient approximations of definite integrals in the sense that fewer terms must be summed to achieve a given accuracy.

A **table of integrals** can be used to find antiderivatives of many standard functions (see Table II of Appendix C on pages 948–950). Definite integrals of such functions can therefore be found exactly by means of the fundamental theorem of calculus.

The Trapezoidal Rule

The trapezoid in Figure 1 is a more accurate approximation of the area under the graph of f and above the x axis than the left rectangle or the right rectangle. Using Δx to denote $x_1 - x_0$,

$$\text{Area of left rectangle: } f(x_0)\Delta x$$

$$\text{Area of right rectangle: } f(x_1)\Delta x$$

$$\text{Area of trapezoid: } \frac{f(x_0) + f(x_1)}{2}\Delta x \qquad (1)$$

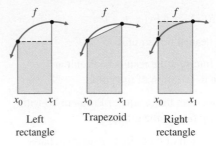

Left rectangle   Trapezoid   Right rectangle

Figure 1

Note that the area of the trapezoid in Figure 1 [also see formula (1)] is the average of the areas of the left and right rectangles. So the average T_4 of the left sum L_4 and the right sum R_4 for a function f on an interval $[a, b]$ is equal to the sum of the areas of four trapezoids (Fig. 2).

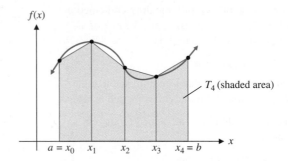

T_4 (shaded area)

Figure 2

Adding L_4 and R_4 and dividing by 2 gives a formula for T_4:

$$L_4 = [f(x_0) + f(x_1) + f(x_2) + f(x_3)]\Delta x$$

$$R_4 = [f(x_1) + f(x_2) + f(x_3) + f(x_4)]\Delta x$$

$$T_4 = [f(x_0) + 2f(x_1) + 2f(x_2) + 2f(x_3) + f(x_4)]\Delta x/2$$

The **trapezoidal sum** T_4 is the case $n = 4$ of the **trapezoidal rule**.

TRAPEZOIDAL RULE

Let f be a function defined on an interval $[a, b]$. Partition $[a, b]$ into n subintervals of equal length $\Delta x = (b - a)/n$ with endpoints

$$a = x_0 < x_1 < x_2 < \cdots < x_n = b.$$

Then

$$T_n = [f(x_0) + 2f(x_1) + 2f(x_2) + \cdots + 2f(x_{n-1}) + f(x_n)]\Delta x/2$$

is an approximation of $\int_a^b f(x)\, dx$.

EXAMPLE 1 Trapezoidal rule Use the trapezoidal rule with $n = 5$ to approximate $\int_2^4 \sqrt{100 + x^2}\, dx$. Round function values to 4 decimal places and the final answer to 2 decimal places.

SOLUTION Partition $[2, 4]$ into 5 equal subintervals of width $(4 - 2)/5 = 0.4$. The endpoints are $x_0 = 2$, $x_1 = 2.4$, $x_2 = 2.8$, $x_3 = 3.2$, $x_4 = 3.6$, and $x_5 = 4$. We calculate the value of the function $f(x) = \sqrt{100 + x^2}$ at each endpoint.

x	$f(x)$
2.0	10.1980
2.4	10.2840
2.8	10.3846
3.2	10.4995
3.6	10.6283
4.0	10.7703

By the trapezoidal rule,

$$T_5 = [f(2) + 2f(2.4) + 2f(2.8) + 2f(3.2) + 2f(3.6) + f(4)](0.4/2)$$
$$= [10.1980 + 2(10.2840) + 2(10.3846) + 2(10.4995) + 2(10.6283) + 10.7703](0.2)$$
$$= 20.91$$

Matched Problem 1 Use the trapezoidal rule with $n = 5$ to approximate $\int_2^4 \sqrt{81 + x^5}\, dx$ (round function values to 4 decimal places and the final answer to 2 decimal places).

Simpson's Rule

The trapezoidal sum provides a better approximation of the definite integral of a function that is increasing (or decreasing) than either the left or right sum. Similarly, the **midpoint sum**,

$$M_n = \left[f\left(\frac{x_0 + x_1}{2}\right) + f\left(\frac{x_1 + x_2}{2}\right) + \cdots + f\left(\frac{x_{n-1} + x_n}{2}\right) \right] \Delta x$$

(see Example 2, Section 5.4) is a better approximation of the definite integral of a function that is increasing (or decreasing) than either the left or right sum. How do T_n and M_n compare? A midpoint sum rectangle has the same area as the corresponding tangent line trapezoid (the larger trapezoid in Fig. 3). It appears from Figure 3, and can be proved in general, that the trapezoidal sum error is about double the midpoint sum error when the graph of the function is concave up or concave down.

$$T_n \leq \int_a^b f(x)\, dx \leq M_n$$

T_n underestimates

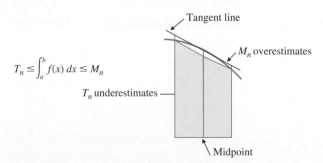

Tangent line

M_n overestimates

Midpoint

Figure 3

This suggests that a weighted average of the two estimates, with the midpoint sum being counted double the trapezoidal sum, might be an even better estimate than either separately. This weighted average,

$$S_{2n} = \frac{2M_n + T_n}{3} \tag{2}$$

leads to a formula called *Simpson's rule*. To simplify notation, we agree to divide the interval $[a, b]$ into $2n$ equal subintervals when Simpson's rule is applied. So, if $n = 2$, for example, then $[a, b]$ is divided into $2n = 4$ equal subintervals, of length Δx, with endpoints

$$a = x_0 < x_1 < x_2 < x_3 < x_4 = b.$$

There are two equal subintervals for M_2 and T_2, each of length $2\Delta x$, with endpoints

$$a = x_0 < x_2 < x_4 = b.$$

Therefore,

$$M_2 = [f(x_1) + f(x_3)](2\Delta x)$$
$$T_2 = [f(x_0) + 2f(x_2) + f(x_4)](2\Delta x)/2$$

We use equation (2) to get a formula for S_4:

$$S_4 = [f(x_0) + 4f(x_1) + 2f(x_2) + 4f(x_3) + f(x_4)]\Delta x/3$$

The formula for S_4 is the case $n = 2$ of **Simpson's rule**.

Simpson's Rule

Let f be a function defined on an interval $[a, b]$. Partition $[a, b]$ into $2n$ subintervals of equal length $\Delta x = (b - a)/n$ with endpoints

$$a = x_0 < x_1 < x_2 < \cdots < x_{2n} = b.$$

Then

$$S_{2n} = [f(x_0) + 4f(x_1) + 2f(x_2) + 4f(x_3) + 2f(x_4) + \cdots + 4f(x_{2n-1}) + f(x_{2n})]\Delta x/3$$

is an approximation of $\int_a^b f(x)\,dx$.

⚠ **CAUTION** Simpson's rule always requires an *even* number of subintervals of $[a, b]$.   ▲

EXAMPLE 2   **Simpson's rule**   Use Simpson's rule with $n = 2$ to approximate $\int_2^{10} \frac{x^4}{\ln x}\,dx$. Round function values to 4 decimal places and the final answer to 2 decimal places.

SOLUTION   Partition the interval $[2, 10]$ into $2n = 4$ equal subintervals of width $(10 - 2)/4 = 2$. The endpoints are $x_0 = 2, x_1 = 4, \; x_2 = 6, \; x_3 = 8,$ and $x_4 = 10$. We calculate the value of the function $f(x) = \frac{x^4}{\ln x}$ at each endpoint:

x	$f(x)$
2	23.0831
4	184.6650
6	723.3114
8	1,969.7596
10	4,342.9448

By Simpson's rule,

$$S_4 = [f(2) + 4f(4) + 2f(6) + 4f(8) + f(10)](2/3)$$
$$= [23.0831 + 4(184.6650) + 2(723.3114) + 4(1,969.7596) + 4,342.9448](2/3)$$
$$= 9,620.23$$

Matched Problem 2   Use Simpson's rule with $n = 2$ to approximate $\int_2^{10} \frac{1}{\ln x}\,dx$ (round function values to 4 decimal places and the final answer to 2 decimal places).

CONCEPTUAL INSIGHT

The trapezoidal rule and Simpson's rule require the values of a function at the end-points of the subintervals of a partition, but neither requires a formula for the function. So either rule can be used on data that give the values of a function at the required points. It is not necessary to use regression techniques to find a formula for the function.

Using a Table of Integrals

The formulas in Table II on pages 948–950 are organized by categories, such as "Integrals Involving $a + bu$," "Integrals Involving $\sqrt{u^2 - a^2}$," and so on. The variable u is the variable of integration. All other symbols represent constants. To use a table to evaluate an integral, you must first find the category that most closely agrees with the form of the integrand and then find a formula in that category that you can make to match the integrand exactly by assigning values to the constants in the formula.

EXAMPLE 3 Integration Using Tables   Use Table II to find

$$\int \frac{x}{(5 + 2x)(4 - 3x)}\,dx$$

SOLUTION Since the integrand

$$f(x) = \frac{x}{(5 + 2x)(4 - 3x)}$$

is a rational function involving terms of the form $a + bu$ and $c + du$, we examine formulas 15 to 20 in Table II on page 949 to see if any of the integrands in these formulas can be made to match $f(x)$ exactly. Comparing the integrand in formula 16 with $f(x)$, we see that this integrand will match $f(x)$ if we let $a = 5$, $b = 2$, $c = 4$, and $d = -3$. Letting $u = x$ and substituting for a, b, c, and d in formula 16, we have

$$\int \frac{u}{(a + bu)(c + du)}\,du = \frac{1}{ad - bc}\left(\frac{a}{b}\ln|a + bu| - \frac{c}{d}\ln|c + du|\right) \quad \text{Formula 16}$$

$$\int \underbrace{\frac{x}{(5 + 2x)(4 - 3x)}}_{a \quad b \quad c \quad d}\,dx = \frac{1}{5 \cdot (-3) - 2 \cdot 4}\left(\frac{5}{2}\ln|5 + 2x| - \frac{4}{-3}\ln|4 - 3x|\right) + C$$
$$a \cdot d - b \cdot c = 5 \cdot (-3) - 2 \cdot 4 = -23$$
$$= -\tfrac{5}{46}\ln|5 + 2x| - \tfrac{4}{69}\ln|4 - 3x| + C$$

Notice that the constant of integration, C, is not included in any of the formulas in Table II. However, you must still include C in all antiderivatives.

Matched Problem 3 Use Table II to find $\displaystyle\int \frac{1}{(5 + 3x)^2(1 + x)}\,dx$.

EXAMPLE 4 Integration Using Tables   Evaluate $\displaystyle\int_3^4 \frac{1}{x\sqrt{25 - x^2}}\,dx$.

SOLUTION First, we use Table II to find

$$\int \frac{1}{x\sqrt{25 - x^2}}\,dx$$

Since the integrand involves the expression $\sqrt{25 - x^2}$, we examine formulas 29 to 31 in Table II and select formula 29 with $a^2 = 25$ and $a = 5$:

$$\int \frac{1}{u\sqrt{a^2 - u^2}}\,du = -\frac{1}{a}\ln\left|\frac{a + \sqrt{a^2 - u^2}}{u}\right| \qquad \text{Formula 29}$$

$$\int \frac{1}{x\sqrt{25 - x^2}}\,dx = -\frac{1}{5}\ln\left|\frac{5 + \sqrt{25 - x^2}}{x}\right| + C$$

So

$$\int_3^4 \frac{1}{x\sqrt{25 - x^2}}\,dx = -\frac{1}{5}\ln\left|\frac{5 + \sqrt{25 - x^2}}{x}\right|\,\Bigg|_3^4$$

$$= -\frac{1}{5}\ln\left|\frac{5 + 3}{4}\right| + \frac{1}{5}\ln\left|\frac{5 + 4}{3}\right|$$

$$= -\tfrac{1}{5}\ln 2 + \tfrac{1}{5}\ln 3 = \tfrac{1}{5}\ln 1.5 \approx 0.0811$$

Matched Problem 4 Evaluate $\displaystyle\int_6^8 \frac{1}{x^2\sqrt{100 - x^2}}\,dx$.

Substitution and Integral Tables

As Examples 3 and 4 illustrate, if the integral we want to evaluate can be made to match one in the table exactly, then evaluating the indefinite integral consists simply of substituting the correct values of the constants into the formula. But what happens if we cannot match an integral with one of the formulas in the table? In many cases, a substitution will change the given integral into one that corresponds to a table entry.

EXAMPLE 5 Integration Using Substitution and Tables Find $\displaystyle\int \frac{x^2}{\sqrt{16x^2 - 25}}\,dx$.

SOLUTION In order to relate this integral to one of the formulas involving $\sqrt{u^2 - a^2}$ (formulas 40 to 45 in Table II), we observe that if $u = 4x$, then

$$u^2 = 16x^2 \qquad \text{and} \qquad \sqrt{16x^2 - 25} = \sqrt{u^2 - 25}$$

So, we will use the substitution $u = 4x$ to change this integral into one that appears in the table:

$$\int \frac{x^2}{\sqrt{16x^2 - 25}}\,dx = \frac{1}{4}\int \frac{\frac{1}{16}u^2}{\sqrt{u^2 - 25}}\,du \qquad \begin{array}{l}\text{Substitution:}\\ u = 4x,\ du = 4\ dx,\ x = \tfrac{1}{4}u\end{array}$$

$$= \frac{1}{64}\int \frac{u^2}{\sqrt{u^2 - 25}}\,du$$

This last integral can be evaluated with the aid of formula 44 in Table II with $a = 5$:

$$\int \frac{u^2}{\sqrt{u^2 - a^2}}\,du = \frac{1}{2}\left(u\sqrt{u^2 - a^2} + a^2 \ln\left|u + \sqrt{u^2 - a^2}\right|\right) \qquad \text{Formula 44}$$

$$\int \frac{x^2}{\sqrt{16x^2 - 25}}\,dx = \frac{1}{64}\int \frac{u^2}{\sqrt{u^2 - 25}}\,du \qquad \text{Use formula 44 with } a = 5.$$

$$= \tfrac{1}{128}\left(u\sqrt{u^2 - 25} + 25 \ln\left|u + \sqrt{u^2 - 25}\right|\right) + C \qquad \text{Substitute } u = 4x.$$

$$= \tfrac{1}{128}\left(4x\sqrt{16x^2 - 25} + 25 \ln\left|4x + \sqrt{16x^2 - 25}\right|\right) + C$$

Matched Problem 5 Find $\displaystyle\int \sqrt{9x^2 - 16}\,dx$.

EXAMPLE 6 Integration Using Substitution and Tables Find $\displaystyle\int \frac{x}{\sqrt{x^4 + 1}}\,dx.$

SOLUTION None of the formulas in Table II involve fourth powers; however, if we let $u = x^2$, then

$$\sqrt{x^4 + 1} = \sqrt{u^2 + 1}$$

and this form does appear in formulas 32 to 39. Thus, we substitute $u = x^2$:

$$\int \frac{1}{\sqrt{x^4 + 1}} x\,dx = \frac{1}{2}\int \frac{1}{\sqrt{u^2 + 1}}\,du \qquad \begin{array}{l}\text{Substitution:}\\ u = x^2,\ du = 2x\,dx\end{array}$$

We recognize the last integral as formula 36 with $a = 1$:

$$\int \frac{1}{\sqrt{u^2 + a^2}}\,du = \ln|u + \sqrt{u^2 + a^2}| \qquad \text{Formula 36}$$

$$\int \frac{x}{\sqrt{x^4 + 1}}\,dx = \frac{1}{2}\int \frac{1}{\sqrt{u^2 + 1}}\,du \qquad \text{Use formula 36 with } a = 1.$$

$$= \tfrac{1}{2}\ln|u + \sqrt{u^2 + 1}| + C \qquad \text{Substitute } u = x^2.$$

$$= \tfrac{1}{2}\ln|x^2 + \sqrt{x^4 + 1}| + C$$

Matched Problem 6) Find $\displaystyle\int x\sqrt{x^4 + 1}\,dx.$

Reduction Formulas

EXAMPLE 7 Using Reduction Formulas Use Table II to find $\displaystyle\int x^2 e^{3x}\,dx.$

SOLUTION Since the integrand involves the function e^{3x}, we examine formulas 46–48 and conclude that formula 47 can be used for this problem. Letting $u = x$, $n = 2$, and $a = 3$ in formula 47, we have

$$\int u^n e^{au}\,du = \frac{u^n e^{au}}{a} - \frac{n}{a}\int u^{n-1} e^{au}\,du \qquad \text{Formula 47}$$

$$\int x^2 e^{3x}\,dx = \frac{x^2 e^{3x}}{3} - \frac{2}{3}\int x e^{3x}\,dx$$

Notice that the expression on the right still contains an integral, but the exponent of x has been reduced by 1. Formulas of this type are called **reduction formulas** and are designed to be applied repeatedly until an integral that can be evaluated is obtained. Applying formula 47 to $\int x e^{3x}\,dx$ with $n = 1$, we have

$$\int x^2 e^{3x}\,dx = \frac{x^2 e^{3x}}{3} - \frac{2}{3}\left(\frac{x e^{3x}}{3} - \frac{1}{3}\int e^{3x}\,dx\right)$$

$$= \frac{x^2 e^{3x}}{3} - \frac{2x e^{3x}}{9} + \frac{2}{9}\int e^{3x}\,dx$$

This last expression contains an integral that is easy to evaluate:

$$\int e^{3x}\,dx = \tfrac{1}{3}e^{3x}$$

After making a final substitution and adding a constant of integration, we have

$$\int x^2 e^{3x}\,dx = \frac{x^2 e^{3x}}{3} - \frac{2x e^{3x}}{9} + \frac{2}{27}e^{3x} + C$$

Matched Problem 7) Use Table II to find $\displaystyle\int (\ln x)^2\,dx.$

Application

EXAMPLE 8   Producers' Surplus   Find the producers' surplus at a price level of $20 for the price–supply equation

$$p = S(x) = \frac{5x}{500 - x}$$

SOLUTION

Step 1   Find $\bar{x}$, the supply when the price is $\bar{p} = 20$:

$$\bar{p} = \frac{5\bar{x}}{500 - \bar{x}}$$

$$20 = \frac{5\bar{x}}{500 - \bar{x}}$$

$$10,000 - 20\bar{x} = 5\bar{x}$$

$$10,000 = 25\bar{x}$$

$$\bar{x} = 400$$

Step 2   Sketch a graph, as shown in Figure 4.

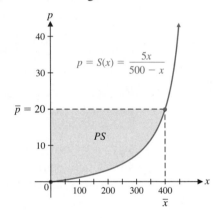

Figure 4

Step 3   Find the producers' surplus (the shaded area of the graph):

$$PS = \int_0^{\bar{x}} [\bar{p} - S(x)] \, dx$$

$$= \int_0^{400} \left(20 - \frac{5x}{500 - x}\right) dx$$

$$= \int_0^{400} \frac{10,000 - 25x}{500 - x} \, dx$$

Use formula 20 with $a = 10,000$, $b = -25$, $c = 500$, and $d = -1$:

$$\int \frac{a + bu}{c + du} \, du = \frac{bu}{d} + \frac{ad - bc}{d^2} \ln|c + du| \qquad \text{Formula 20}$$

$$PS = (25x + 2,500 \ln|500 - x|)\big|_0^{400}$$

$$= 10,000 + 2,500 \ln|100| - 2,500 \ln|500|$$

$$\approx \$5,976$$

Matched Problem 8   Find the consumers' surplus at a price level of $10 for the price–demand equation

$$p = D(x) = \frac{20x - 8,000}{x - 500}$$

Exercises 14.4

In Problems 1–8, round function values to 4 decimal places and the final answer to 2 decimal places.

1. Use the trapezoidal rule with $n = 3$ to approximate
$\int_0^6 \sqrt{1 + x^4}\, dx$.

2. Use the trapezoidal rule with $n = 2$ to approximate
$\int_0^8 \sqrt{1 + x^2}\, dx$.

3. Use the trapezoidal rule with $n = 6$ to approximate
$\int_0^6 \sqrt{1 + x^4}\, dx$.

4. Use the trapezoidal rule with $n = 4$ to approximate
$\int_0^8 \sqrt{1 + x^2}\, dx$.

5. Use Simpson's rule with $n = 1$ (so there are $2n = 2$
subintervals) to approximate $\int_1^3 \dfrac{1}{1 + x^2}\, dx$.

6. Use Simpson's rule with $n = 1$ (so there are $2n = 2$
subintervals) to approximate $\int_2^{10} \dfrac{x^2}{\ln x}\, dx$.

7. Use Simpson's rule with $n = 2$ (so there are $2n = 4$
subintervals) to approximate $\int_1^3 \dfrac{1}{1 + x^2}\, dx$.

8. Use Simpson's rule with $n = 2$ (so there are $2n = 4$
subintervals) to approximate $\int_2^{10} \dfrac{x^2}{\ln x}\, dx$.

Use Table II on pages 948–950 to find each indefinite integral in Problems 9–22.

9. $\displaystyle\int \dfrac{1}{x(1 + x)}\, dx$

10. $\displaystyle\int \dfrac{1}{x^2(1 + x)}\, dx$

11. $\displaystyle\int \dfrac{1}{(3 + x)^2(5 + 2x)}\, dx$

12. $\displaystyle\int \dfrac{x}{(5 + 2x)^2(2 + x)}\, dx$

13. $\displaystyle\int \dfrac{x}{\sqrt{16 + x}}\, dx$

14. $\displaystyle\int \dfrac{1}{x\sqrt{16 + x}}\, dx$

15. $\displaystyle\int \dfrac{1}{x\sqrt{1 - x^2}}\, dx$

16. $\displaystyle\int \dfrac{\sqrt{9 - x^2}}{x}\, dx$

17. $\displaystyle\int \dfrac{1}{x\sqrt{x^2 + 4}}\, dx$

18. $\displaystyle\int \dfrac{1}{x^2\sqrt{x^2 - 16}}\, dx$

19. $\displaystyle\int x^2 \ln x\, dx$

20. $\displaystyle\int x^3 \ln x\, dx$

21. $\displaystyle\int \dfrac{1}{1 + e^x}\, dx$

22. $\displaystyle\int \dfrac{1}{5 + 2e^{3x}}\, dx$

Evaluate each definite integral in Problems 23–28. Use Table II on pages 948–950 to find the antiderivative.

23. $\displaystyle\int_1^3 \dfrac{x^2}{3 + x}\, dx$

24. $\displaystyle\int_2^6 \dfrac{x}{(6 + x)^2}\, dx$

25. $\displaystyle\int_0^7 \dfrac{1}{(3 + x)(1 + x)}\, dx$

26. $\displaystyle\int_0^7 \dfrac{x}{(3 + x)(1 + x)}\, dx$

27. $\displaystyle\int_0^4 \dfrac{1}{\sqrt{x^2 + 9}}\, dx$

28. $\displaystyle\int_4^5 \sqrt{x^2 - 16}\, dx$

29. Use the trapezoidal rule with $n = 5$ to approximate $\int_3^{13} x^2 dx$ and use the fundamental theorem of calculus to find the exact value of the definite integral.

30. Use the trapezoidal rule with $n = 5$ to approximate $\int_1^{11} x^3 dx$ and use the fundamental theorem of calculus to find the exact value of the definite integral.

31. Use Simpson's rule with $n = 4$ (so there are $2n = 8$ subintervals) to approximate $\int_1^5 \dfrac{1}{x}\, dx$ and use the fundamental theorem of calculus to find the exact value of the definite integral.

32. Use Simpson's rule with $n = 4$ (so there are $2n = 8$ subintervals) to approximate $\int_1^5 x^4 dx$ and use the fundamental theorem of calculus to find the exact value of the definite integral.

33. Use the trapezoidal rule with $n = 3$ to approximate $\int_5^8 (4x - 3)\, dx$ and use the fundamental theorem of calculus to find the exact value of the definite integral.

34. Use the trapezoidal rule with $n = 3$ to approximate $\int_7^{10} (2 - 9x)\, dx$ and use the fundamental theorem of calculus to find the exact value of the definite integral.

35. Use Simpson's rule with $n = 2$ (so there are $2n = 4$ subintervals) to approximate $\int_5^9 (3x^2 + 5x + 3)\, dx$ and use the fundamental theorem of calculus to find the exact value of the definite integral.

36. Use Simpson's rule with $n = 2$ (so there are $2n = 4$ subintervals) to approximate $\int_{-2}^2 (x^3 - 3x^2 + 2x + 8)\, dx$ and use the fundamental theorem of calculus to find the exact value of the definite integral.

In Problems 37–48, use substitution techniques and Table II to find each indefinite integral.

37. $\displaystyle\int \dfrac{\sqrt{4x^2 + 1}}{x^2}\, dx$

38. $\displaystyle\int x^2\sqrt{9x^2 - 1}\, dx$

39. $\displaystyle\int \dfrac{x}{\sqrt{x^4 - 16}}\, dx$

40. $\displaystyle\int x\sqrt{x^4 - 16}\, dx$

41. $\displaystyle\int x^2\sqrt{x^6 + 4}\, dx$

42. $\displaystyle\int \dfrac{x^2}{\sqrt{x^6 + 4}}\, dx$

43. $\displaystyle\int \dfrac{1}{x^3\sqrt{4 - x^4}}\, dx$

44. $\displaystyle\int \dfrac{\sqrt{x^4 + 4}}{x}\, dx$

45. $\displaystyle\int \dfrac{e^x}{(2 + e^x)(3 + 4e^x)}\, dx$

46. $\displaystyle\int \dfrac{e^x}{(4 + e^x)^2(2 + e^x)}\, dx$

47. $\displaystyle\int \dfrac{\ln x}{x\sqrt{4 + \ln x}}\, dx$

48. $\displaystyle\int \dfrac{1}{x \ln x\sqrt{4 + \ln x}}\, dx$

In Problems 49–54, use Table II to find each indefinite integral.

49. $\displaystyle\int x^2 e^{5x}\, dx$

50. $\displaystyle\int x^2 e^{-4x}\, dx$

51. $\displaystyle\int x^3 e^{-x}\, dx$

52. $\displaystyle\int x^3 e^{2x}\, dx$

53. $\displaystyle\int (\ln x)^3\, dx$

54. $\displaystyle\int (\ln x)^4\, dx$

Problems 55–62 are mixed—some require the use of Table II, and others can be solved with techniques considered earlier.

55. $\displaystyle\int_3^5 x\sqrt{x^2 - 9}\, dx$

56. $\displaystyle\int_3^5 x^2\sqrt{x^2 - 9}\, dx$

57. $\displaystyle\int_2^4 \frac{1}{x^2 - 1}\, dx$

58. $\displaystyle\int_2^4 \frac{x}{(x^2 - 1)^2}\, dx$

59. $\displaystyle\int \frac{\ln x}{x^2}\, dx$

60. $\displaystyle\int \frac{(\ln x)^2}{x}\, dx$

61. $\displaystyle\int \frac{x}{\sqrt{x^2 - 1}}\, dx$

62. $\displaystyle\int \frac{x^2}{\sqrt{x^2 - 1}}\, dx$

63. If $f(x) = ax^2 + bx + c$, where a, b, and c are any real numbers, use Simpson's rule with $n = 1$ (so there are $2n = 2$ subintervals) to show that

$$S_2 = \int_{-1}^1 f(x)\, dx.$$

64. If $f(x) = ax^3 + bx^2 + cx + d$, where a, b, c, and d are any real numbers, use Simpson's rule with $n = 1$ (so there are $2n = 2$ subintervals) to show that

$$S_2 = \int_{-1}^1 f(x)\, dx.$$

In Problems 65–68, find the area bounded by the graphs of $y = f(x)$ and $y = g(x)$ to two decimal places. Use a graphing calculator to approximate intersection points to two decimal places.

65. $f(x) = \dfrac{10}{\sqrt{x^2 + 1}}$; $g(x) = x^2 + 3x$

66. $f(x) = \sqrt{1 + x^2}$; $g(x) = 5x - x^2$

67. $f(x) = x\sqrt{4 + x}$; $g(x) = 1 + x$

68. $f(x) = \dfrac{x}{\sqrt{x + 4}}$; $g(x) = x - 2$

Applications

Use Table II to evaluate all integrals involved in any solutions of Problems 69–92.

69. Consumers' surplus. Find the consumers' surplus at a price level of $\bar{p} = \$15$ for the price–demand equation

$$p = D(x) = \frac{7{,}500 - 30x}{300 - x}$$

70. Producers' surplus. Find the producers' surplus at a price level of $\bar{p} = \$20$ for the price–supply equation

$$p = S(x) = \frac{10x}{300 - x}$$

71. Consumers' surplus. Graph the price–demand equation and the price-level equation $\bar{p} = 15$ of Problem 69 in the same coordinate system. What region represents the consumers' surplus?

72. Producers' surplus. Graph the price–supply equation and the price-level equation $\bar{p} = 20$ of Problem 70 in the same coordinate system. What region represents the producers' surplus?

73. Cost. A company manufactures downhill skis. It has fixed costs of $25,000 and a marginal cost given by

$$C'(x) = \frac{250 + 10x}{1 + 0.05x}$$

where $C(x)$ is the total cost at an output of x pairs of skis. Find the cost function $C(x)$ and determine the production level (to the nearest unit) that produces a cost of $150,000. What is the cost (to the nearest dollar) for a production level of 850 pairs of skis?

74. Cost. A company manufactures a portable DVD player. It has fixed costs of $11,000 per week and a marginal cost given by

$$C'(x) = \frac{65 + 20x}{1 + 0.4x}$$

where $C(x)$ is the total cost per week at an output of x players per week. Find the cost function $C(x)$ and determine the production level (to the nearest unit) that produces a cost of $52,000 per week. What is the cost (to the nearest dollar) for a production level of 700 players per week?

75. Continuous income stream. Find the future value at 4.4%, compounded continuously, for 10 years for the continuous income stream with rate of flow $f(t) = 50t^2$.

76. Continuous income stream. Find the interest earned at 3.7%, compounded continuously, for 5 years for the continuous income stream with rate of flow $f(t) = 200t$.

77. Income distribution. Find the Gini index of income concentration for the Lorenz curve with equation

$$y = \tfrac{1}{2}x\sqrt{1 + 3x}$$

78. Income distribution. Find the Gini index of income concentration for the Lorenz curve with equation

$$y = \tfrac{1}{2}x^2\sqrt{1 + 3x}$$

79. Income distribution. Graph $y = x$ and the Lorenz curve of Problem 77 over the interval $[0, 1]$. Discuss the effect of the area bounded by $y = x$ and the Lorenz curve getting smaller relative to the equitable distribution of income.

80. Income distribution. Graph $y = x$ and the Lorenz curve of Problem 78 over the interval $[0, 1]$. Discuss the effect of the area bounded by $y = x$ and the Lorenz curve getting larger relative to the equitable distribution of income.

81. Marketing. After test marketing a new high-fiber cereal, the market research department of a major food producer estimates that monthly sales (in millions of dollars) will grow at the monthly rate of

$$S'(t) = \frac{t^2}{(1 + t)^2}$$

t months after the cereal is introduced. If we assume 0 sales at the time the cereal is introduced, find $S(t)$, the total sales *t* months after the cereal is introduced. Find the total sales during the first 2 years that the cereal is on the market.

82. Average price. At a discount department store, the price–demand equation for premium motor oil is given by

$$p = D(x) = \frac{50}{\sqrt{100 + 6x}}$$

where *x* is the number of cans of oil that can be sold at a price of $*p*. Find the average price over the demand interval [50, 250].

83. Marketing. For the cereal of Problem 81, show the sales over the first 2 years geometrically, and describe the geometric representation.

84. Price–demand. For the motor oil of Problem 82, graph the price–demand equation and the line representing the average price in the same coordinate system over the interval [50, 250]. Describe how the areas under the two curves over the interval [50, 250] are related.

85. Profit. The marginal profit for a small car agency that sells *x* cars per week is given by

$$P'(x) = x\sqrt{2 + 3x}$$

where $P(x)$ is the profit in dollars. The agency's profit on the sale of only 1 car per week is $-\$2,000$. Find the profit function and the number of cars that must be sold (to the nearest unit) to produce a profit of $13,000 per week. How much weekly profit (to the nearest dollar) will the agency have if 80 cars are sold per week?

86. Revenue. The marginal revenue for a company that manufactures and sells *x* graphing calculators per week is given by

$$R'(x) = \frac{x}{\sqrt{1 + 2x}} \qquad R(0) = 0$$

where $R(x)$ is the revenue in dollars. Find the revenue function and the number of calculators that must be sold (to the nearest unit) to produce $10,000 in revenue per week. How much weekly revenue (to the nearest dollar) will the company have if 1,000 calculators are sold per week?

87. Pollution. An oil tanker is producing an oil slick that is radiating outward at a rate given approximately by

$$\frac{dR}{dt} = \frac{100}{\sqrt{t^2 + 9}} \qquad t \geq 0$$

where *R* is the radius (in feet) of the circular slick after *t* minutes. Find the radius of the slick after 4 minutes if the radius is 0 when $t = 0$.

88. Pollution. The concentration of particulate matter (in parts per million) during a 24-hour period is given approximately by

$$C(t) = t\sqrt{24 - t} \qquad 0 \leq t \leq 24$$

where *t* is time in hours. Find the average concentration during the period from $t = 0$ to $t = 24$.

89. Learning. A person learns *N* items at a rate given approximately by

$$N'(t) = \frac{60}{\sqrt{t^2 + 25}} \qquad t \geq 0$$

where *t* is the number of hours of continuous study. Determine the total number of items learned in the first 12 hours of continuous study.

90. Politics. The number of voters (in thousands) in a metropolitan area is given approximately by

$$f(t) = \frac{500}{2 + 3e^{-t}} \qquad t \geq 0$$

where *t* is time in years. Find the average number of voters during the period from $t = 0$ to $t = 10$.

91. Learning. Interpret Problem 89 geometrically. Describe the geometric interpretation.

92. Politics. For the voters of Problem 90, graph $y = f(t)$ and the line representing the average number of voters over the interval [0, 10] in the same coordinate system. Describe how the areas under the two curves over the interval [0, 10] are related.

Answers to Matched Problems

1. 38.85   2. 5.20

3. $\frac{1}{2}\left(\frac{1}{5 + 3x}\right) + \frac{1}{4} \ln\left|\frac{1 + x}{5 + 3x}\right| + C$

4. $\frac{7}{1,200} \approx 0.0058$

5. $\frac{1}{6}\left(3x\sqrt{9x^2 - 16} - 16 \ln\left|3x + \sqrt{9x^2 - 16}\right|\right) + C$

6. $\frac{1}{4}\left(x^2\sqrt{x^4 + 1} + \ln\left|x^2 + \sqrt{x^4 + 1}\right|\right) + C$

7. $x(\ln x)^2 - 2x \ln x + 2x + C$

8. $3,000 + 2,000 \ln 200 - 2,000 \ln 500 \approx \$1,167$

Important Terms, Symbols, and Concepts

14.1 Area Between Curves

- If f and g are continuous and $f(x) \geq g(x)$ over the interval $[a, b]$, then the area bounded by $y = f(x)$ and $y = g(x)$ for $a \leq x \leq b$ is given exactly by

$$A = \int_a^b [f(x) - g(x)] \, dx$$

- A graphical representation of the distribution of income among a population can be found by plotting data points (x, y), where **x represents the cumulative percentage of families at or below a given income level** and **y represents the cumulative percentage of total family income received.** Regression analysis can be used to find a particular function $y = f(x)$, called a **Lorenz curve,** that best fits the data.

- A single number, the **Gini index,** measures income concentration:

$$\text{Gini index} = 2 \int_0^1 [x - f(x)] \, dx$$

A Gini index of 0 indicates **absolute equality:** All families share equally in the income. A Gini index of 1 indicates **absolute inequality:** One family has all of the income and the rest have none.

Ex. 1, p. 777
Ex. 2, p. 777
Ex. 3, p. 778
Ex. 4, p. 778
Ex. 5, p. 779
Ex. 6, p. 779

Ex. 7, p. 781

14.2 Applications in Business and Economics

- **Probability Density Functions** If any real number x in an interval is a possible outcome of an experiment, then x is said to be a **continuous random variable.** The probability distribution of a continuous random variable is described by a **probability density function** f that satisfies the following conditions:

1. $f(x) \geq 0$ for all real x.

2. The area under the graph of $f(x)$ over the interval $(-\infty, \infty)$ is exactly 1.

3. If $[c, d]$ is a subinterval of $(-\infty, \infty)$, then

$$\text{Probability} \quad (c \leq x \leq d) = \int_c^d f(x) \, dx$$

- **Continuous Income Stream** If the rate at which income is received—its **rate of flow**—is a continuous function $f(t)$ of time, then the income is said to be a **continuous income stream.** The **total income** produced by a continuous income stream from $t = a$ to $t = b$ is

$$\text{Total income} = \int_a^b f(t) \, dt$$

The **future value** of a continuous income stream that is invested at rate r, compounded continuously, for $0 \leq t \leq T$, is

$$FV = \int_0^T f(t) e^{r(T-t)} \, dt$$

- **Consumers' and Producers' Surplus** If $(\bar{x}, \bar{p})$ is a point on the graph of a price–demand equation $p = D(x)$, then the **consumers' surplus** at a price level of $\bar{p}$ is

$$CS = \int_0^{\bar{x}} [D(x) - \bar{p}] \, dx$$

The consumers' surplus represents the total savings to consumers who are willing to pay more than $\bar{p}$ but are still able to buy the product for $\bar{p}$.

Similarly, for a point $(\bar{x}, \bar{p})$ on the graph of a price–supply equation $p = S(x)$, the **producers' surplus** at a price level of $\bar{p}$ is

Ex. 5, p. 792

$$PS = \int_0^{\bar{x}} [\bar{p} - S(x)]\, dx$$

The producers' surplus represents the total gain to producers who are willing to supply units at a lower price $\bar{p}$, but are still able to supply units at $\bar{p}$.

If $(\bar{x}, \bar{p})$ is the intersection point of a price–demand equation $p = D(x)$ and a price–supply equation $p = S(x)$, then $\bar{p}$ is called the **equilibrium price** and $\bar{x}$ is called the **equilibrium quantity**.

Ex. 6, p. 793

14.3 Integration by Parts

- Some indefinite integrals, but not all, can be found by means of the **integration-by-parts formula**

Ex. 1, p. 797
Ex. 2, p. 798
Ex. 3, p. 800
Ex. 4, p. 800

$$\int u\, dv = uv - \int v\, du$$

- Select u and dv with the help of the guidelines in the summary on page 799.

14.4 Other Integration Methods

- The **trapezoidal rule** and **Simpson's rule** provide approximations of the definite integral that are more efficient than approximations by left or right sums: Fewer terms must be summed to achieve a given accuracy.

- **Trapezoidal Rule** Let f be a function defined on an interval $[a, b]$. Partition $[a, b]$ into n subintervals of equal length $\Delta x = (b - a)/n$ with endpoints

Ex. 1, p. 805

$$a = x_0 < x_1 < x_2 < \cdots < x_n = b.$$

Then

$$T_n = [f(x_0) + 2f(x_1) + 2f(x_2) + \cdots + 2f(x_{n-1}) + f(x_n)]\Delta x/2$$

is an approximation of $\int_a^b f(x)\, dx$.

- **Simpson's Rule** Let f be a function defined on an interval $[a, b]$. Partition $[a, b]$ into $2n$ subintervals of equal length $\Delta x = (b - a)/n$ with endpoints

Ex. 2, p. 806

$$a = x_0 < x_1 < x_2 < \cdots < x_{2n} = b.$$

Then

$$S_{2n} = [f(x_0) + 4f(x_1) + 2f(x_2) + 4f(x_3) + 2f(x_4) + \cdots + 4f(x_{2n-1}) + f(x_{2n})]\Delta x/3$$

is an approximation of $\int_a^b f(x)\, dx$.

- A **table of integrals** is a list of integration formulas that can be used to find indefinite or definite integrals of frequently encountered functions. Such a list appears in Table II of Appendix C on pages 948–950.

Ex. 3, p. 807
Ex. 4, p. 807
Ex. 5, p. 808
Ex. 6, p. 809
Ex. 7, p. 809
Ex. 8, p. 810

Review Exercises

Work through all the problems in this chapter review and check your answers in the back of the book. Answers to all review problems are there, along with section numbers in italics to indicate where each type of problem is discussed. Where weaknesses show up, review appropriate sections of the text.

Compute all numerical answers to three decimal places unless directed otherwise.

In Problems 1–3, set up definite integrals that represent the shaded areas in the figure over the indicated intervals.

1. Interval $[a, b]$ **2.** Interval $[b, c]$

3. Interval $[a, c]$

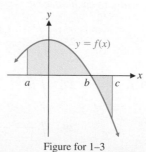

Figure for 1–3

4. Sketch a graph of the area between the graphs of $y = \ln x$ and $y = 0$ over the interval $[0.5, e]$ and find the area.

In Problems 5–10, evaluate each integral.

5. $\displaystyle\int xe^{4x}\, dx$

6. $\displaystyle\int x \ln x\, dx$

7. $\displaystyle\int \frac{\ln x}{x}\, dx$

8. $\displaystyle\int \frac{x}{1 + x^2}\, dx$

9. $\displaystyle\int \frac{1}{x(1 + x)^2}\, dx$

10. $\displaystyle\int \frac{1}{x^2\sqrt{1 + x}}\, dx$

In Problems 11–16, find the area bounded by the graphs of the indicated equations over the given interval.

11. $y = 5 - 2x - 6x^2; y = 0, 1 \le x \le 2$

12. $y = 5x + 7; y = 12, -3 \le x \le 1$

13. $y = -x + 2; y = x^2 + 3, -1 \le x \le 4$

14. $y = \dfrac{1}{x}; y = -e^{-x}, 1 \le x \le 2$

15. $y = x; y = -x^3, -2 \le x \le 2$

16. $y = x^2; y = -x^4; -2 \le x \le 2$

17. The Gini indices of Indonesia and Malaysia are 0.37 and 0.46, respectively. In which country is income more equally distributed?

18. The Gini indices of Thailand and Vietnam are 0.54 and 0.38, respectively. In which country is income more equally distributed?

In Problems 19–22, set up definite integrals that represent the shaded areas in the figure over the indicated intervals.

19. Interval $[a, b]$

20. Interval $[b, c]$

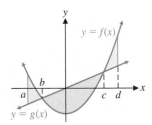

Figure for 19–22

21. Interval $[b, d]$

22. Interval $[a, d]$

23. Sketch a graph of the area bounded by the graphs of $y = x^2 - 6x + 9$ and $y = 9 - x$ and find the area.

In Problems 24–29, evaluate each integral.

24. $\displaystyle\int_0^1 xe^x\, dx$

25. $\displaystyle\int_0^3 \frac{x^2}{\sqrt{x^2 + 16}}\, dx$

26. $\displaystyle\int \sqrt{9x^2 - 49}\, dx$

27. $\displaystyle\int te^{-0.5t}\, dt$

28. $\displaystyle\int x^2 \ln x\, dx$

29. $\displaystyle\int \frac{1}{1 + 2e^x}\, dx$

30. Sketch a graph of the area bounded by the indicated graphs, and find the area. In part (B), approximate intersection points and area to two decimal places.

(A) $y = x^3 - 6x^2 + 9x; y = x$

(B) $y = x^3 - 6x^2 + 9x; y = x + 1$

In Problems 31–34, round function values to 4 decimal places and the final answer to 2 decimal places.

31. Use the trapezoidal rule with $n = 3$ to approximate $\int_0^3 e^{x^2}dx$.

32. Use the trapezoidal rule with $n = 5$ to approximate $\int_0^3 e^{x^2}dx$.

33. Use Simpson's rule with $n = 2$ (so there are $2n = 4$ subintervals) to approximate $\int_1^5 (\ln x)^2 dx$.

34. Use Simpson's rule with $n = 4$ (so there are $2n = 8$ subintervals) to approximate $\int_1^5 (\ln x)^2 dx$.

In Problems 35–42, evaluate each integral.

35. $\displaystyle\int \frac{(\ln x)^2}{x}\, dx$

36. $\displaystyle\int x(\ln x)^2\, dx$

37. $\displaystyle\int \frac{x}{\sqrt{x^2 - 36}}\, dx$

38. $\displaystyle\int \frac{x}{\sqrt{x^4 - 36}}\, dx$

39. $\displaystyle\int_0^4 x \ln(10 - x)\, dx$

40. $\displaystyle\int (\ln x)^2\, dx$

41. $\displaystyle\int xe^{-2x^2}\, dx$

42. $\displaystyle\int x^2 e^{-2x}\, dx$

43. Use a numerical integration routine on a graphing calculator to find the area in the first quadrant that is below the graph of

$$y = \frac{6}{2 + 5e^{-x}}$$

and above the graph of $y = 0.2x + 1.6$.

Applications

44. Product warranty. A manufacturer warrants a product for parts and labor for 1 year and for parts only for a second year. The time to a failure of the product after it is sold is given by the probability density function

$$f(t) = \begin{cases} 0.21e^{-0.21t} & \text{if } t \ge 0 \\ 0 & \text{otherwise} \end{cases}$$

What is the probability that a buyer chosen at random will have a product failure

(A) During the first year of warranty?

(B) During the second year of warranty?

45. Product warranty. Graph the probability density function for Problem 44 over the interval $[0, 3]$, interpret part (B) of Problem 44 geometrically and describe the geometric representation.

46. Revenue function. The weekly marginal revenue from the sale of x hair dryers is given by

$$R'(x) = 65 - 6\ln(x + 1) \qquad R(0) = 0$$

where $R(x)$ is the revenue in dollars. Find the revenue function and the production level (to the nearest unit) for a revenue of $20,000 per week. What is the weekly revenue (to the nearest dollar) at a production level of 1,000 hair dryers per week?

47. Continuous income stream. The rate of flow (in dollars per year) of a continuous income stream for a 5-year period is given by

$$f(t) = 2,500e^{0.05t} \qquad 0 \le t \le 5$$

(A) Graph $y = f(t)$ over $[0, 5]$ and shade the area that represents the total income received from the end of the first year to the end of the fourth year.

(B) Find the total income received, to the nearest dollar, from the end of the first year to the end of the fourth year.

48. Future value of a continuous income stream. The continuous income stream in Problem 47 is invested at 4%, compounded continuously.

(A) Find the future value (to the nearest dollar) at the end of the 5-year period.

(B) Find the interest earned (to the nearest dollar) during the 5-year period.

49. Income distribution. An economist produced the following Lorenz curves for the current income distribution and the projected income distribution 10 years from now in a certain country:

$$f(x) = 0.1x + 0.9x^2 \quad \text{Current Lorenz curve}$$

$$g(x) = x^{1.5} \qquad\qquad \text{Projected Lorenz curve}$$

(A) Graph $y = x$ and the current Lorenz curve on one set of coordinate axes for $[0, 1]$ and graph $y = x$ and the projected Lorenz curve on another set of coordinate axes over the same interval.

(B) Looking at the areas bounded by the Lorenz curves and $y = x$, can you say that the income will be more or less equitably distributed 10 years from now?

(C) Compute the Gini index of income concentration (to one decimal place) for the current and projected curves. What can you say about the distribution of income 10 years from now? Is it more equitable or less?

50. Consumers' and producers' surplus. Find the consumers' surplus and the producers' surplus at the equilibrium price level for each pair of price–demand and price–supply equations. Include a graph that identifies the consumers' surplus and the producers' surplus. Round all values to the nearest integer.

(A) $p = D(x) = 70 - 0.2x$;
$p = S(x) = 13 + 0.0012x^2$

(B) $p = D(x) = 70 - 0.2x$;
$p = S(x) = 13e^{0.006x}$

51. Producers' surplus. The accompanying table gives price–supply data for the sale of hogs at a livestock market, where x is the number of pounds (in thousands) and p is the price per pound (in cents):

Price–Supply	
x	$p = S(x)$
0	43.50
10	46.74
20	50.05
30	54.72
40	59.18

(A) Using quadratic regression to model the data, find the demand at a price of 52.50 cents per pound.

(B) Use a numerical integration routine to find the producers' surplus (to the nearest dollar) at a price level of 52.50 cents per pound.

52. Drug assimilation. The rate at which the body eliminates a certain drug (in milliliters per hour) is given by

$$R(t) = \frac{60t}{(t + 1)^2(t + 2)}$$

where t is the number of hours since the drug was administered. How much of the drug is eliminated during the first hour after it was administered? During the fourth hour?

53. With the aid of a graphing calculator, illustrate Problem 52 geometrically.

54. Medicine. For a particular doctor, the length of time (in hours) spent with a patient per office visit has the probability density function

$$f(t) = \begin{cases} \dfrac{\frac{4}{3}}{(t + 1)^2} & \text{if } 0 \le t \le 3 \\ 0 & \text{otherwise} \end{cases}$$

(A) What is the probability that this doctor will spend less than 1 hour with a randomly selected patient?

(B) What is the probability that this doctor will spend more than 1 hour with a randomly selected patient?

55. Medicine. Illustrate part (B) in Problem 54 geometrically. Describe the geometric interpretation.

56. Politics. The rate of change of the voting population of a city with respect to time t (in years) is estimated to be

$$N'(t) = \frac{100t}{(1 + t^2)^2}$$

where $N(t)$ is in thousands. If $N(0)$ is the current voting population, how much will this population increase during the next 3 years?

57. Psychology. Rats were trained to go through a maze by rewarding them with a food pellet upon successful completion of the run. After the seventh successful run, the probability density function for length of time (in minutes) until success on the eighth trial was given by

$$f(t) = \begin{cases} .5e^{-.5t} & \text{if } t \ge 0 \\ 0 & \text{otherwise} \end{cases}$$

What is the probability that a rat selected at random after seven successful runs will take 2 or more minutes to complete the eighth run successfully? [Recall that the area under a probability density function curve from $-\infty$ to ∞ is 1.]

15

Multivariable Calculus

15.1 Functions of Several Variables

15.2 Partial Derivatives

15.3 Maxima and Minima

15.4 Maxima and Minima Using Lagrange Multipliers

15.5 Method of Least Squares

15.6 Double Integrals over Rectangular Regions

15.7 Double Integrals over More General Regions

Chapter 15
Summary and Review

Review Exercises

Introduction

In previous chapters, we have applied the key concepts of calculus, the derivative and the integral, to functions with one independent variable. The graph of such a function is a curve in the plane. In Chapter 15, we extend the key concepts of calculus to functions with two independent variables. Graphs of such functions are surfaces in a three-dimensional coordinate system. We use functions with two independent variables to study how production depends on both labor and capital; how braking distance depends on both the weight and speed of a car (see Problem 66 in Section 15.1); how resistance in a blood vessel depends on both its length and radius. In Section 15.5, we justify the method of least squares and use the method to construct linear models.

15.1 Functions of Several Variables

- Functions of Two or More Independent Variables
- Examples of Functions of Several Variables
- Three-Dimensional Coordinate Systems

Functions of Two or More Independent Variables

In Section 2.1, we introduced the concept of a function with one independent variable. Now we broaden the concept to include functions with more than one independent variable.

A small manufacturing company produces a standard type of surfboard. If fixed costs are $500 per week and variable costs are $70 per board produced, the weekly cost function is given by

$$C(x) = 500 + 70x \qquad (1)$$

where x is the number of boards produced per week. The cost function is a function of a single independent variable x. For each value of x from the domain of C, there exists exactly one value of $C(x)$ in the range of C.

Now, suppose that the company decides to add a high-performance competition board to its line. If the fixed costs for the competition board are $200 per week and the variable costs are $100 per board, then the cost function (1) must be modified to

$$C(x, y) = 700 + 70x + 100y \qquad (2)$$

where $C(x, y)$ is the cost for a weekly output of x standard boards and y competition boards. Equation (2) is an example of a function with two independent variables x and y. Of course, as the company expands its product line even further, its weekly cost function must be modified to include more and more independent variables, one for each new product produced.

In general, an equation of the form

$$z = f(x, y)$$

describes a **function of two independent variables** if, for each permissible ordered pair (x, y), there is one and only one value of z determined by $f(x, y)$. The variables x and y are **independent variables**, and the variable z is a **dependent variable**. The set of all ordered pairs of permissible values of x and y is the **domain** of the function, and the set of all corresponding values $f(x, y)$ is the **range** of the function. Unless otherwise stated, we will assume that the domain of a function specified by an equation of the form $z = f(x, y)$ is the set of all ordered pairs of real numbers (x, y) such that $f(x, y)$ is also a real number. It should be noted, however, that certain conditions in practical problems often lead to further restrictions on the domain of a function.

We can similarly define functions of three independent variables, $w = f(x, y, z)$; of four independent variables, $u = f(w, x, y, z)$; and so on. In this chapter, we concern ourselves primarily with functions of two independent variables.

EXAMPLE 1   Evaluating a Function of Two Independent Variables   For the cost function $C(x, y) = 700 + 70x + 100y$ described earlier, find $C(10, 5)$.

SOLUTION

$$C(10, 5) = 700 + 70(10) + 100(5)$$
$$= \$1,900$$

Matched Problem 1 ⌋ Find $C(20, 10)$ for the cost function in Example 1.

EXAMPLE 2 Evaluating a Function of Three Independent Variables For the function $f(x, y, z) = 2x^2 - 3xy + 3z + 1$, find $f(3, 0, -1)$.

SOLUTION

$$f(3, 0, -1) = 2(3)^2 - 3(3)(0) + 3(-1) + 1$$
$$= 18 - 0 - 3 + 1 = 16$$

Matched Problem 2 Find $f(-2, 2, 3)$ for f in Example 2.

EXAMPLE 3 Revenue, Cost, and Profit Functions Suppose the surfboard company discussed earlier has determined that the demand equations for its two types of boards are given by

$$p = 210 - 4x + y$$
$$q = 300 + x - 12y$$

where p is the price of the standard board, q is the price of the competition board, x is the weekly demand for standard boards, and y is the weekly demand for competition boards.

(A) Find the weekly revenue function $R(x, y)$, and evaluate $R(20, 10)$.

(B) If the weekly cost function is

$$C(x, y) = 700 + 70x + 100y$$

find the weekly profit function $P(x, y)$ and evaluate $P(20, 10)$.

SOLUTION

(A)

$$\text{Revenue} = \begin{pmatrix} \text{demand for} \\ \text{standard} \\ \text{boards} \end{pmatrix} \times \begin{pmatrix} \text{price of a} \\ \text{standard} \\ \text{board} \end{pmatrix} + \begin{pmatrix} \text{demand for} \\ \text{competition} \\ \text{boards} \end{pmatrix} \times \begin{pmatrix} \text{price of a} \\ \text{competition} \\ \text{board} \end{pmatrix}$$

$$R(x, y) = xp + yq$$
$$= x(210 - 4x + y) + y(300 + x - 12y)$$
$$= 210x + 300y - 4x^2 + 2xy - 12y^2$$
$$R(20, 10) = 210(20) + 300(10) - 4(20)^2 + 2(20)(10) - 12(10)^2$$
$$= \$4,800$$

(B)   Profit $=$ revenue $-$ cost

$$P(x, y) = R(x, y) - C(x, y)$$
$$= 210x + 300y - 4x^2 + 2xy - 12y^2 - 700 - 70x - 100y$$
$$= 140x + 200y - 4x^2 + 2xy - 12y^2 - 700$$
$$P(20, 10) = 140(20) + 200(10) - 4(20)^2 + 2(20)(10) - 12(10)^2 - 700$$
$$= \$1,700$$

Matched Problem 3 Repeat Example 3 if the demand and cost equations are given by

$$p = 220 - 6x + y$$
$$q = 300 + 3x - 10y$$
$$C(x, y) = 40x + 80y + 1,000$$

Examples of Functions of Several Variables

A number of concepts can be considered as functions of two or more variables.

Area of a rectangle	$A(x, y) = xy$	A = area, with sides x and y
Volume of a box	$V(x, y, z) = xyz$	V = volume
Volume of a right circular cylinder	$V(r, h) = \pi r^2 h$	cylinder with radius r and height h
Simple interest	$A(P, r, t) = P(1 + rt)$	A = amount P = principal r = annual rate t = time in years
Compound interest	$A(P, r, t, n) = P\left(1 + \dfrac{r}{n}\right)^{nt}$	A = amount P = principal r = annual rate t = time in years n = number of compound periods per year
IQ	$Q(M, C) = \dfrac{M}{C}(100)$	Q = IQ = intelligence quotient M = MA = mental age C = CA = chronological age
Resistance for blood flow in a vessel (Poiseuille's law)	$R(L, r) = k\dfrac{L}{r^4}$	R = resistance L = length of vessel r = radius of vessel k = constant

EXAMPLE 4 Package Design A company uses a box with a square base and an open top for a bath assortment (see figure). If x is the length (in inches) of each side of the base and y is the height (in inches), find the total amount of material $M(x, y)$ required to construct one of these boxes, and evaluate $M(5, 10)$.

SOLUTION

$$\text{Area of base} = x^2$$
$$\text{Area of one side} = xy$$
$$\text{Total material} = (\text{area of base}) + 4(\text{area of one side})$$
$$M(x, y) = x^2 + 4xy$$
$$M(5, 10) = (5)^2 + 4(5)(10)$$
$$= 225 \text{ square inches}$$

Matched Problem 4 For the box in Example 4, find the volume $V(x, y)$ and evaluate $V(5, 10)$.

The next example concerns the **Cobb–Douglas production function**

$$f(x, y) = kx^m y^n$$

where k, m, and n are positive constants with $m + n = 1$. Economists use this function to describe the number of units $f(x, y)$ produced from the utilization of x units of labor and y units of capital (for equipment such as tools, machinery, buildings, and so on). Cobb–Douglas production functions are also used to describe the productivity of a single industry, of a group of industries producing the same product, or even of an entire country.

EXAMPLE 5   Productivity   The productivity of a steel-manufacturing company is given approximately by the function

$$f(x, y) = 10x^{0.2}y^{0.8}$$

with the utilization of x units of labor and y units of capital. If the company uses 3,000 units of labor and 1,000 units of capital, how many units of steel will be produced?

SOLUTION   The number of units of steel produced is given by

$$f(3,000, 1,000) = 10(3,000)^{0.2}(1,000)^{0.8} \quad \text{Use a calculator.}$$
$$\approx 12,457 \text{ units}$$

Matched Problem 5   Refer to Example 5. Find the steel production if the company uses 1,000 units of labor and 2,000 units of capital.

Three-Dimensional Coordinate Systems

We now take a brief look at graphs of functions of two independent variables. Since functions of the form $z = f(x, y)$ involve two independent variables x and y, and one dependent variable z, we need a *three-dimensional coordinate system* for their graphs. A **three-dimensional coordinate system** is formed by three mutually perpendicular number lines intersecting at their origins (see Fig. 1). In such a system, every ordered **triplet of numbers** (x, y, z) can be associated with a unique point, and conversely.

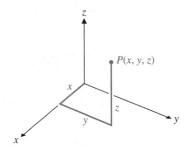

Figure 1   **Rectangular coordinate system**

EXAMPLE 6   Three-Dimensional Coordinates   Locate $(-3, 5, 2)$ in a rectangular coordinate system.

SOLUTION

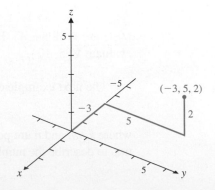

Matched Problem 6 Find the coordinates of the corners A, C, G, and D of the rectangular box shown in the following figure.

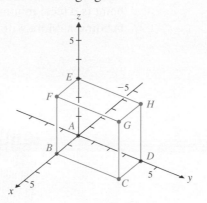

Explore and Discuss 1   Imagine that you are facing the front of a classroom whose rectangular walls meet at right angles. Suppose that the point of intersection of the floor, front wall, and left-side wall is the origin of a three-dimensional coordinate system in which every point in the room has nonnegative coordinates. Then the plane $z = 0$ (or, equivalently, the xy plane) can be described as "the floor," and the plane $z = 2$ can be described as "the plane parallel to, but 2 units above, the floor." Give similar descriptions of the following planes:

(A) $x = 0$ (B) $x = 3$ (C) $y = 0$ (D) $y = 4$ (E) $x = -1$

What does the graph of $z = x^2 + y^2$ look like? If we let $x = 0$ and graph $z = 0^2 + y^2 = y^2$ in the yz plane, we obtain a parabola; if we let $y = 0$ and graph $z = x^2 + 0^2 = x^2$ in the xz plane, we obtain another parabola. The graph of $z = x^2 + y^2$ is either one of these parabolas rotated around the z axis (see Fig. 2). This cup-shaped figure is a *surface* and is called a **paraboloid**.

In general, the graph of any function of the form $z = f(x, y)$ is called a **surface**. The graph of such a function is the graph of all ordered triplets of numbers (x, y, z) that satisfy the equation. Graphing functions of two independent variables is a difficult task, and the general process will not be dealt with in this book. We present only a few simple graphs to suggest extensions of earlier geometric interpretations of the derivative and local maxima and minima to functions of two variables. Note that $z = f(x, y) = x^2 + y^2$ appears (see Fig. 2) to have a local minimum at $(x, y) = (0, 0)$. Figure 3 shows a local maximum at $(x, y) = (0, 0)$.

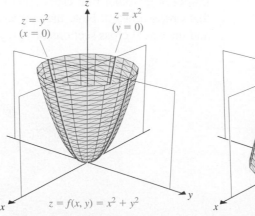

Figure 2 **Paraboloid**

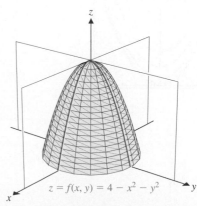

Figure 3 **Local maximum:** $f(0, 0) = 4$

Figure 4 shows a point at $(x, y) = (0, 0)$, called a **saddle point**, that is neither a local minimum nor a local maximum. Note that in the cross section $x = 0$, the saddle point is a local minimum, and in the cross section $y = 0$, the saddle point is a local maximum. More will be said about local maxima and minima in Section 15.3.

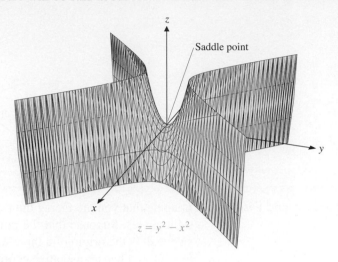

$$z = y^2 - x^2$$

Figure 4 **Saddle point at (0, 0, 0)**

Some graphing calculators are designed to draw graphs (like those of Figs. 2, 3, and 4) of functions of two independent variables. Others, such as the graphing calculator used for the displays in this book, are designed to draw graphs of functions of one independent variable. When using the latter type of calculator, we can graph cross sections produced by cutting surfaces with planes parallel to the xz plane or yz plane to gain insight into the graph of a function of two independent variables.

EXAMPLE 7   Graphing Cross Sections

(A) Describe the cross sections of $f(x, y) = 2x^2 + y^2$ in the planes $y = 0$, $y = 1, y = 2, y = 3$, and $y = 4$.

(B) Describe the cross sections of $f(x, y) = 2x^2 + y^2$ in the planes $x = 0$, $x = 1, x = 2, x = 3$, and $x = 4$.

SOLUTION

(A) The cross section of $f(x, y) = 2x^2 + y^2$ produced by cutting it with the plane $y = 0$ is the graph of the function $f(x, 0) = 2x^2$ in this plane. We can examine the shape of this cross section by graphing $y_1 = 2x^2$ on a graphing calculator (Fig. 5). Similarly, the graphs of $y_2 = f(x, 1) = 2x^2 + 1$, $y_3 = f(x, 2) = 2x^2 + 4$, $y_4 = f(x, 3) = 2x^2 + 9$, and $y_5 = f(x, 4) = 2x^2 + 16$ show the shapes of the other four cross sections (see Fig. 5). Each of these is a parabola that opens upward. Note the correspondence between the graphs in Figure 5 and the actual cross sections of $f(x, y) = 2x^2 + y^2$ shown in Figure 6.

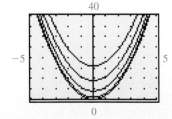

Figure 5

$y_1 = 2x^2$ $y_4 = 2x^2 + 9$

$y_2 = 2x^2 + 1$ $y_5 = 2x^2 + 16$

$y_3 = 2x^2 + 4$

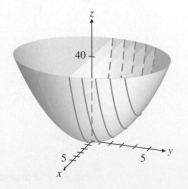

Figure 6

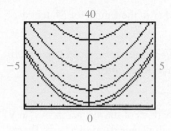

Figure 7

$y_1 = x^2$ $y_4 = 18 + x^2$

$y_2 = 2 + x^2$ $y_5 = 32 + x^2$

$y_3 = 8 + x^2$

(B) The five cross sections are represented by the graphs of the functions $f(0, y) = y^2, f(1, y) = 2 + y^2, f(2, y) = 8 + y^2, f(3, y) = 18 + y^2$, and $f(4, y) = 32 + y^2$. These five functions are graphed in Figure 7. (Note that changing the name of the independent variable from y to x for graphing purposes does not affect the graph displayed.) Each of the five cross sections is a parabola that opens upward.

Matched Problem 7

(A) Describe the cross sections of $g(x, y) = y^2 - x^2$ in the planes $y = 0$, $y = 1, y = 2, y = 3$, and $y = 4$.

(B) Describe the cross sections of $g(x, y) = y^2 - x^2$ in the planes $x = 0$, $x = 1, x = 2, x = 3$, and $x = 4$.

CONCEPTUAL INSIGHT

The graph of the *equation*

$$x^2 + y^2 + z^2 = 4 \tag{3}$$

is the graph of all ordered triplets of numbers (x, y, z) that satisfy the equation. The Pythagorean theorem can be used to show that the distance from the point (x, y, z) to the origin $(0, 0, 0)$ is equal to

$$\sqrt{x^2 + y^2 + z^2}$$

Therefore, the graph of (3) consists of all points that are at a distance 2 from the origin—that is, all points on the sphere of radius 2 and with center at the origin. Recall that a circle in the plane is *not* the graph of a function $y = f(x)$, because it fails the vertical-line test (Section 2.1). Similarly, a sphere is *not* the graph of a *function* $z = f(x, y)$ of two variables.

Exercises 15.1

Skills Warm-up Exercises

In Problems 1–8, find the indicated value of the function of two or three variables. (If necessary, review Appendix C).

1. The height of a trapezoid is 3 feet and the lengths of its parallel sides are 5 feet and 8 feet. Find the area.

2. The height of a trapezoid is 4 meters and the lengths of its parallel sides are 25 meters and 32 meters. Find the area.

3. The length, width, and height of a rectangular box are 12 inches, 5 inches, and 4 inches, respectively. Find the volume.

4. The length, width, and height of a rectangular box are 30 centimeters, 15 centimeters, and 10 centimeters, respectively. Find the volume.

5. The height of a right circular cylinder is 8 meters and the radius is 2 meters. Find the volume.

6. The height of a right circular cylinder is 6 feet and the diameter is also 6 feet. Find the total surface area.

7. The height of a right circular cone is 48 centimeters and the radius is 20 centimeters. Find the total surface area.

8. The height of a right circular cone is 42 inches and the radius is 7 inches. Find the volume.

In Problems 9–16, find the indicated values of the functions

$$f(x, y) = 2x + 7y - 5 \quad \text{and} \quad g(x, y) = \frac{88}{x^2 + 3y}$$

9. $f(4, -1)$

10. $f(0, 10)$

11. $f(8, 0)$

12. $f(5, 6)$

13. $g(1, 7)$

14. $g(-2, 0)$

15. $g(3, -3)$

16. $g(0, 0)$

In Problems 17–20, find the indicated values of

$$f(x, y, z) = 2x - 3y^2 + 5z^3 - 1$$

17. $f(0, 0, 0)$

18. $f(0, 0, 2)$

19. $f(6, -5, 0)$

20. $f(-10, 4, -3)$

In Problems 21–30, find the indicated value of the given function.

21. $P(13, 5)$ for $P(n, r) = \dfrac{n!}{(n - r)!}$

22. $C(13, 5)$ for $C(n, r) = \dfrac{n!}{r!(n - r)!}$

23. $V(4, 12)$ for $V(R, h) = \pi R^2 h$

24. $T(4, 12)$ for $T(R, h) = 2\pi R(R + h)$

25. $S(3, 10)$ for $S(R, h) = \pi R\sqrt{R^2 + h^2}$

26. $W(3, 10)$ for $W(R, h) = \dfrac{1}{3}\pi R^2 h$

27. $A(100, 0.06, 3)$ for $A(P, r, t) = P + Prt$

28. $A(10, 0.04, 3, 2)$ for $A(P, r, t, n) = P\left(1 + \dfrac{r}{n}\right)^{tn}$

29. $P(0.05, 12)$ for $P(r, T) = \displaystyle\int_0^T 4{,}000e^{-rt}\, dt$

30. $F(0.07, 10)$ for $F(r, T) = \displaystyle\int_0^T 4{,}000e^{r(T-t)}\, dt$

In Problems 31–36, find the indicated function f of a single variable.

31. $f(x) = G(x, 0)$ for $G(x, y) = x^2 + 3xy + y^2 - 7$

32. $f(y) = H(0, y)$ for $H(x, y) = x^2 - 5xy - y^2 + 2$

33. $f(y) = K(4, y)$ for $K(x, y) = 10xy + 3x - 2y + 8$

34. $f(x) = L(x, -2)$ for $L(x, y) = 25 - x + 5y - 6xy$

35. $f(y) = M(y, y)$ for $M(x, y) = x^2y - 3xy^2 + 5$

36. $f(x) = N(x, 2x)$ for $N(x, y) = 3xy + x^2 - y^2 + 1$

37. Let $F(x, y) = 2x + 3y - 6$. Find all values of y such that $F(0, y) = 0$.

38. Let $F(x, y) = 5x - 4y + 12$. Find all values of x such that $F(x, 0) = 0$.

39. Let $F(x, y) = 2xy + 3x - 4y - 1$. Find all values of x such that $F(x, x) = 0$.

40. Let $F(x, y) = xy + 2x^2 + y^2 - 25$. Find all values of y such that $F(y, y) = 0$.

41. Let $F(x, y) = x^2 + e^xy - y^2$. Find all values of x such that $F(x, 2) = 0$.

42. Let $G(a, b, c) = a^3 + b^3 + c^3 - (ab + ac + bc) - 6$. Find all values of b such that $G(2, b, 1) = 0$.

43. For the function $f(x, y) = x^2 + 2y^2$, find

$$\frac{f(x + h, y) - f(x, y)}{h}$$

44. For the function $f(x, y) = x^2 + 2y^2$, find

$$\frac{f(x, y + k) - f(x, y)}{k}$$

45. For the function $f(x, y) = 2xy^2$, find

$$\frac{f(x + h, y) - f(x, y)}{h}$$

46. For the function $f(x, y) = 2xy^2$, find

$$\frac{f(x, y + k) - f(x, y)}{k}$$

47. Find the coordinates of E and F in the figure for Matched Problem 6 on page 823.

48. Find the coordinates of B and H in the figure for Matched Problem 6 on page 823.

In Problems 49–54, use a graphing calculator as necessary to explore the graphs of the indicated cross sections.

49. Let $f(x, y) = x^2$.

(A) Explain why the cross sections of the surface $z = f(x, y)$ produced by cutting it with planes parallel to $y = 0$ are parabolas.

(B) Describe the cross sections of the surface in the planes $x = 0$, $x = 1$, and $x = 2$.

(C) Describe the surface $z = f(x, y)$.

50. Let $f(x, y) = \sqrt{4 - y^2}$.

(A) Explain why the cross sections of the surface $z = f(x, y)$ produced by cutting it with planes parallel to $x = 0$ are semicircles of radius 2.

(B) Describe the cross sections of the surface in the planes $y = 0$, $y = 2$, and $y = 3$.

(C) Describe the surface $z = f(x, y)$.

51. Let $f(x, y) = \sqrt{36 - x^2 - y^2}$.

(A) Describe the cross sections of the surface $z = f(x, y)$ produced by cutting it with the planes $y = 1$, $y = 2$, $y = 3$, $y = 4$, and $y = 5$.

(B) Describe the cross sections of the surface in the planes $x = 0$, $x = 1$, $x = 2$, $x = 3$, $x = 4$, and $x = 5$.

(C) Describe the surface $z = f(x, y)$.

52. Let $f(x, y) = 100 + 10x + 25y - x^2 - 5y^2$.

(A) Describe the cross sections of the surface $z = f(x, y)$ produced by cutting it with the planes $y = 0$, $y = 1$, $y = 2$, and $y = 3$.

(B) Describe the cross sections of the surface in the planes $x = 0$, $x = 1$, $x = 2$, and $x = 3$.

(C) Describe the surface $z = f(x, y)$.

53. Let $f(x, y) = e^{-(x^2+y^2)}$.

(A) Explain why $f(a, b) = f(c, d)$ whenever (a, b) and (c, d) are points on the same circle centered at the origin in the xy plane.

(B) Describe the cross sections of the surface $z = f(x, y)$ produced by cutting it with the planes $x = 0$, $y = 0$, and $x = y$.

(C) Describe the surface $z = f(x, y)$.

54. Let $f(x, y) = 4 - \sqrt{x^2 + y^2}$.

(A) Explain why $f(a, b) = f(c, d)$ whenever (a, b) and (c, d) are points on the same circle with center at the origin in the xy plane.

(B) Describe the cross sections of the surface $z = f(x, y)$ produced by cutting it with the planes $x = 0$, $y = 0$, and $x = y$.

(C) Describe the surface $z = f(x, y)$.

Applications

55. Cost function. A small manufacturing company produces two models of a surfboard: a standard model and a competition model. If the standard model is produced at a variable cost of $210 each and the competition model at a variable cost of $300 each, and if the total fixed costs per month are $6,000, then the monthly cost function is given by

$$C(x, y) = 6,000 + 210x + 300y$$

where x and y are the numbers of standard and competition models produced per month, respectively. Find $C(20, 10)$, $C(50, 5)$, and $C(30, 30)$.

56. Advertising and sales. A company spends $\$x$ thousand per week on online advertising and $\$y$ thousand per week on TV advertising. Its weekly sales are found to be given by

$$S(x, y) = 5x^2y^3$$

Find $S(3, 2)$ and $S(2, 3)$.

57. Revenue function. A supermarket sells two brands of coffee: brand A at $\$p$ per pound and brand B at $\$q$ per pound. The daily demand equations for brands A and B are, respectively,

$$x = 200 - 5p + 4q$$
$$y = 300 + 2p - 4q$$

(both in pounds). Find the daily revenue function $R(p, q)$. Evaluate $R(2, 3)$ and $R(3, 2)$.

58. Revenue, cost, and profit functions. A company manufactures 10- and 3-speed bicycles. The weekly demand and cost equations are

$$p = 230 - 9x + y$$
$$q = 130 + x - 4y$$
$$C(x, y) = 200 + 80x + 30y$$

where $\$p$ is the price of a 10-speed bicycle, $\$q$ is the price of a 3-speed bicycle, x is the weekly demand for 10-speed bicycles, y is the weekly demand for 3-speed bicycles, and $C(x, y)$ is the cost function. Find the weekly revenue function $R(x, y)$ and the weekly profit function $P(x, y)$. Evaluate $R(10, 15)$ and $P(10, 15)$.

59. Productivity. The Cobb–Douglas production function for a petroleum company is given by

$$f(x, y) = 20x^{0.4}y^{0.6}$$

where x is the utilization of labor and y is the utilization of capital. If the company uses 1,250 units of labor and 1,700 units of capital, how many units of petroleum will be produced?

60. Productivity. The petroleum company in Problem 59 is taken over by another company that decides to double both the units of labor and the units of capital utilized in the production of petroleum. Use the Cobb–Douglas production function given in Problem 59 to find the amount of petroleum that will be produced by this increased utilization of labor and capital. What is the effect on productivity of doubling both the units of labor and the units of capital?

61. Future value. At the end of each year, $5,000 is invested into an IRA earning 3% compounded annually.

(A) How much will be in the account at the end of 30 years? Use the annuity formula

$$F(P, i, n) = P\frac{(1 + i)^n - 1}{i}$$

where

P = periodic payment
i = rate per period
n = number of payments (periods)
F = FV = future value

(B) Use graphical approximation methods to determine the rate of interest that would produce $300,000 in the account at the end of 30 years.

62. Package design. The packaging department in a company has been asked to design a rectangular box with no top and a partition down the middle (see the figure). Let x, y, and z be the dimensions of the box (in inches). Ignore the thickness of the material from which the box will be made.

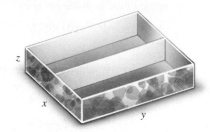

(A) Find the total area of material $M(x, y, z)$ used in constructing one of these boxes, and evaluate $M(10, 12, 6)$.

(B) Suppose that the box will have a square base and a volume of 720 cubic inches. Use graphical approximation methods to determine the dimensions that require the least material.

63. Marine biology. For a diver using scuba-diving gear, a marine biologist estimates the time (duration) of a dive according to the equation

$$T(V, x) = \frac{33V}{x + 33}$$

where

T = time of dive in minutes
V = volume of air, at sea level pressure, compressed into tanks
x = depth of dive in feet

Find $T(70, 47)$ and $T(60, 27)$.

64. Blood flow. Poiseuille's law states that the resistance R for blood flowing in a blood vessel varies directly as the length L of the vessel and inversely as the fourth power of its radius r. Stated as an equation,

$$R(L, r) = k\frac{L}{r^4} \qquad k \text{ a constant}$$

Find $R(8, 1)$ and $R(4, 0.2)$.

65. Physical anthropology. Anthropologists use an index called the *cephalic index*. The cephalic index C varies directly as the width W of the head and inversely as the length L of the head (both viewed from the top). In terms of an equation,

$$C(W, L) = 100\frac{W}{L}$$

where

$W = $ width in inches

$L = $ length in inches

Find $C(6, 8)$ and $C(8.1, 9)$.

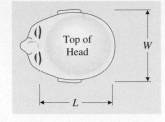

66. Safety research. Under ideal conditions, if a person driving a car slams on the brakes and skids to a stop, the length of the skid marks (in feet) is given by the formula

$$L(w, v) = kwv^2$$

where

$k = $ constant

$w = $ weight of car in pounds

$v = $ speed of car in miles per hour

For $k = 0.000\ 013\ 3$, find $L(2{,}000, 40)$ and $L(3{,}000, 60)$.

67. Psychology. The intelligence quotient (IQ) is defined to be the ratio of mental age (MA), as determined by certain tests, to chronological age (CA), multiplied by 100. Stated as an equation,

$$Q(M, C) = \frac{M}{C} \cdot 100$$

where

$$Q = \text{IQ} \qquad M = \text{MA} \qquad C = \text{CA}$$

Find $Q(12, 10)$ and $Q(10, 12)$.

Answers to Matched Problems

1. $3,100   2. 30

3. (A) $R(x, y) = 220x + 300y - 6x^2 + 4xy - 10y^2$; $R(20, 10) = \$4{,}800$

   (B) $P(x, y) = 180x + 220y - 6x^2 + 4xy - 10y^2 - 1{,}000$; $P(20, 10) = \$2{,}200$

4. $V(x, y) = x^2y$; $V(5, 10) = 250$ in.^{3}

5. 17,411 units

6. $A(0, 0, 0)$; $C(2, 4, 0)$; $G(2, 4, 3)$; $D(0, 4, 0)$

7. (A) Each cross section is a parabola that opens downward.

   (B) Each cross section is a parabola that opens upward.

15.2 Partial Derivatives

- Partial Derivatives
- Second-Order Partial Derivatives

Partial Derivatives

We know how to differentiate many kinds of functions of one independent variable and how to interpret the derivatives that result. What about functions with two or more independent variables? Let's return to the surfboard example considered on page 820.

For the company producing only the standard board, the cost function was

$$C(x) = 500 + 70x$$

Differentiating with respect to x, we obtain the marginal cost function

$$C'(x) = 70$$

Since the marginal cost is constant, $70 is the change in cost for a 1-unit increase in production at any output level.

For the company producing two types of boards—a standard model and a competition model—the cost function was

$$C(x, y) = 700 + 70x + 100y$$

Now suppose that we differentiate with respect to x, holding y fixed, and denote the resulting function by $C_x(x, y)$; or suppose we differentiate with respect to y, holding x fixed, and denote the resulting function by $C_y(x, y)$. Differentiating in this way, we obtain

$$C_x(x, y) = 70 \qquad C_y(x, y) = 100$$

Each of these functions is called a **partial derivative**, and, in this example, each represents marginal cost. The first is the change in cost due to a 1-unit increase in production of the standard board with the production of the competition model held fixed. The second is the change in cost due to a 1-unit increase in production of the competition board with the production of the standard board held fixed.

In general, if $z = f(x, y)$, then the **partial derivative of f with respect to x**, denoted $\partial z/\partial x$, f_x, or $f_x(x, y)$, is defined by

$$\frac{\partial z}{\partial x} = \lim_{h \to 0} \frac{f(x + h, y) - f(x, y)}{h}$$

provided that the limit exists. We recognize this formula as the ordinary derivative of f with respect to x, holding y constant. We can continue to use all the derivative rules and properties discussed in Chapters 10 to 12 and apply them to partial derivatives.

Similarly, the **partial derivative of f with respect to y**, denoted $\partial z/\partial y$, f_y, or $f_y(x, y)$, is defined by

$$\frac{\partial z}{\partial y} = \lim_{k \to 0} \frac{f(x, y + k) - f(x, y)}{k}$$

which is the ordinary derivative with respect to y, holding x constant.

Parallel definitions and interpretations hold for functions with three or more independent variables.

EXAMPLE 1   Partial Derivatives   For $z = f(x, y) = 2x^2 - 3x^2y + 5y + 1$, find

(A) $\partial z/\partial x$ (B) $f_x(2, 3)$

SOLUTION

(A) $z = 2x^2 - 3x^2y + 5y + 1$

Differentiating with respect to x, holding y constant (that is, treating y as a constant), we obtain

$$\frac{\partial z}{\partial x} = 4x - 6xy$$

(B) $f(x, y) = 2x^2 - 3x^2y + 5y + 1$

First, differentiate with respect to x. From part (A), we have

$$f_x(x, y) = 4x - 6xy$$

Then evaluate this equation at $(2, 3)$:

$$f_x(2, 3) = 4(2) - 6(2)(3) = -28$$

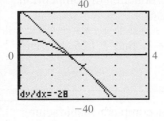

Figure 1 $y_1 = -7x^2 + 16$

In part 1B, an alternative approach would be to substitute $y = 3$ into $f(x, y)$ and graph the function $f(x, 3) = -7x^2 + 16$, which represents the cross section of the surface $z = f(x, y)$ produced by cutting it with the plane $y = 3$. Then determine the slope of the tangent line when $x = 2$. Again, we conclude that $f_x(2, 3) = -28$ (see Fig. 1).

Matched Problem 1   For f in Example 1, find

(A) $\partial z/\partial y$ (B) $f_y(2, 3)$

EXAMPLE 2   Partial Derivatives Using the Chain Rule   For $z = f(x, y) = e^{x^2+y^2}$, find

(A) $\partial z/\partial x$ (B) $f_y(2, 1)$

SOLUTION

(A) Using the chain rule [thinking of $z = e^u$, $u = u(x)$; y is held constant], we obtain

$$\frac{\partial z}{\partial x} = e^{x^2+y^2} \frac{\partial(x^2 + y^2)}{\partial x}$$

$$= 2xe^{x^2+y^2}$$

(B) $f_y(x, y) = e^{x^2+y^2}\dfrac{\partial(x^2 + y^2)}{\partial y} = 2ye^{x^2+y^2}$

$f_y(2, 1) = 2(1)e^{(2)^2 + (1)^2}$

$\qquad\quad = 2e^5$

Matched Problem 2) For $z = f(x, y) = (x^2 + 2xy)^5$, find
(A) $\partial z/\partial y$ (B) $f_x(1, 0)$

EXAMPLE 3 Profit The profit function for the surfboard company in Example 3 of Section 15.1 was

$$P(x, y) = 140x + 200y - 4x^2 + 2xy - 12y^2 - 700$$

Find $P_x(15, 10)$ and $P_x(30, 10)$, and interpret the results.

SOLUTION

$$P_x(x, y) = 140 - 8x + 2y$$
$$P_x(15, 10) = 140 - 8(15) + 2(10) = 40$$
$$P_x(30, 10) = 140 - 8(30) + 2(10) = -80$$

At a production level of 15 standard and 10 competition boards per week, increasing the production of standard boards by 1 unit and holding the production of competition boards fixed at 10 will increase profit by approximately \$40. At a production level of 30 standard and 10 competition boards per week, increasing the production of standard boards by 1 unit and holding the production of competition boards fixed at 10 will decrease profit by approximately \$80.

Matched Problem 3) For the profit function in Example 3, find $P_y(25, 10)$ and $P_y(25, 15)$, and interpret the results.

EXAMPLE 4 Productivity The productivity of a major computer manufacturer is given approximately by the Cobb–Douglas production function

$$f(x, y) = 15x^{0.4}y^{0.6}$$

with the utilization of x units of labor and y units of capital. The partial derivative $f_x(x, y)$ represents the rate of change of productivity with respect to labor and is called the **marginal productivity of labor**. The partial derivative $f_y(x, y)$ represents the rate of change of productivity with respect to capital and is called the **marginal productivity of capital**. If the company is currently utilizing 4,000 units of labor and 2,500 units of capital, find the marginal productivity of labor and the marginal productivity of capital. For the greatest increase in productivity, should the management of the company encourage increased use of labor or increased use of capital?

SOLUTION

$$f_x(x, y) = 6x^{-0.6}y^{0.6}$$
$$f_x(4,000, 2,500) = 6(4,000)^{-0.6}(2,500)^{0.6}$$
$$\approx 4.53 \qquad \text{Marginal productivity of labor}$$
$$f_y(x, y) = 9x^{0.4}y^{-0.4}$$
$$f_y(4,000, 2,500) = 9(4,000)^{0.4}(2,500)^{-0.4}$$
$$\approx 10.86 \qquad \text{Marginal productivity of capital}$$

At the current level of utilization of 4,000 units of labor and 2,500 units of capital, each 1-unit increase in labor utilization (keeping capital utilization fixed at 2,500 units) will increase production by approximately 4.53 units, and each 1-unit increase in capital utilization (keeping labor utilization fixed at 4,000 units) will increase production by approximately 10.86 units. The management of the company should encourage increased use of capital.

Matched Problem 4) The productivity of an airplane-manufacturing company is given approximately by the Cobb–Douglas production function

$$f(x, y) = 40x^{0.3}y^{0.7}$$

(A) Find $f_x(x, y)$ and $f_y(x, y)$.

(B) If the company is currently using 1,500 units of labor and 4,500 units of capital, find the marginal productivity of labor and the marginal productivity of capital.

(C) For the greatest increase in productivity, should the management of the company encourage increased use of labor or increased use of capital?

Partial derivatives have simple geometric interpretations, as shown in Figure 2. If we hold x fixed at $x = a$, then $f_y(a, y)$ is the slope of the curve obtained by intersecting the surface $z = f(x, y)$ with the plane $x = a$. A similar interpretation is given to $f_x(x, b)$.

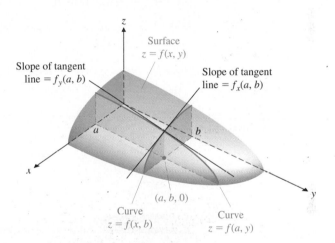

Figure 2

Second-Order Partial Derivatives

The function

$$z = f(x, y) = x^4y^7$$

has two **first-order partial derivatives**:

$$\frac{\partial z}{\partial x} = f_x = f_x(x, y) = 4x^3y^7 \quad \text{and} \quad \frac{\partial z}{\partial y} = f_y = f_y(x, y) = 7x^4y^6$$

Each of these partial derivatives, in turn, has two partial derivatives called **second-order partial derivatives** of $z = f(x, y)$. Generalizing the various notations we have for first-order partial derivatives, we write the four second-order partial derivatives of $z = f(x, y) = x^4y^7$ as

Equivalent notations

$$f_{xx} = f_{xx}(x, y) = \frac{\partial^2 z}{\partial x^2} = \frac{\partial}{\partial x}\left(\frac{\partial z}{\partial x}\right) = \frac{\partial}{\partial x}(4x^3y^7) = 12x^2y^7$$

$$f_{xy} = f_{xy}(x, y) = \frac{\partial^2 z}{\partial y\, \partial x} = \frac{\partial}{\partial y}\left(\frac{\partial z}{\partial x}\right) = \frac{\partial}{\partial y}(4x^3y^7) = 28x^3y^6$$

$$f_{yx} = f_{yx}(x, y) = \frac{\partial^2 z}{\partial x\, \partial y} = \frac{\partial}{\partial x}\left(\frac{\partial z}{\partial y}\right) = \frac{\partial}{\partial x}(7x^4y^6) = 28x^3y^6$$

$$f_{yy} = f_{yy}(x, y) = \frac{\partial^2 z}{\partial y^2} = \frac{\partial}{\partial y}\left(\frac{\partial z}{\partial y}\right) = \frac{\partial}{\partial y}(7x^4y^6) = 42x^4y^5$$

In the mixed partial derivative $\partial^2 z/\partial y\, \partial x = f_{xy}$, we started with $z = f(x, y)$ and first differentiated with respect to x (holding y constant). Then we differentiated with respect to y (holding x constant). In the other mixed partial derivative, $\partial^2 z/\partial x\, \partial y = f_{yx}$, the order of differentiation was reversed; however, the final result was the same—that is, $f_{xy} = f_{yx}$. Although it is possible to find functions for which $f_{xy} \neq f_{yx}$, such functions rarely occur in applications involving partial derivatives. For all the functions in this book, we will assume that $f_{xy} = f_{yx}$.

In general, we have the following definitions:

DEFINITION   Second-Order Partial Derivatives

If $z = f(x, y)$, then

$$f_{xx} = f_{xx}(x, y) = \frac{\partial^2 z}{\partial x^2} = \frac{\partial}{\partial x}\left(\frac{\partial z}{\partial x}\right)$$

$$f_{xy} = f_{xy}(x, y) = \frac{\partial^2 z}{\partial y\, \partial x} = \frac{\partial}{\partial y}\left(\frac{\partial z}{\partial x}\right)$$

$$f_{yx} = f_{yx}(x, y) = \frac{\partial^2 z}{\partial x\, \partial y} = \frac{\partial}{\partial x}\left(\frac{\partial z}{\partial y}\right)$$

$$f_{yy} = f_{yy}(x, y) = \frac{\partial^2 z}{\partial y^2} = \frac{\partial}{\partial y}\left(\frac{\partial z}{\partial y}\right)$$

EXAMPLE 5   Second-Order Partial Derivatives   For $z = f(x, y) = 3x^2 - 2xy^3 + 1$, find

(A) $\dfrac{\partial^2 z}{\partial x\, \partial y},\ \dfrac{\partial^2 z}{\partial y\, \partial x}$ (B) $\dfrac{\partial^2 z}{\partial x^2}$ (C) $f_{yx}(2, 1)$

SOLUTION

(A) First differentiate with respect to y and then with respect to x:

$$\frac{\partial z}{\partial y} = -6xy^2 \qquad \frac{\partial^2 z}{\partial x\, \partial y} = \frac{\partial}{\partial x}\left(\frac{\partial z}{\partial y}\right) = \frac{\partial}{\partial x}(-6xy^2) = -6y^2$$

Now differentiate with respect to x and then with respect to y:

$$\frac{\partial z}{\partial x} = 6x - 2y^3 \qquad \frac{\partial^2 z}{\partial y\, \partial x} = \frac{\partial}{\partial y}\left(\frac{\partial z}{\partial x}\right) = \frac{\partial}{\partial y}(6x - 2y^3) = -6y^2$$

(B) Differentiate with respect to x twice:

$$\frac{\partial z}{\partial x} = 6x - 2y^3 \qquad \frac{\partial^2 z}{\partial x^2} = \frac{\partial}{\partial x}\left(\frac{\partial z}{\partial x}\right) = 6$$

(C) First find $f_{yx}(x, y)$; then evaluate the resulting equation at $(2, 1)$. Again, remember that f_{yx} signifies differentiation first with respect to y and then with respect to x.

$$f_y(x, y) = -6xy^2 \qquad f_{yx}(x, y) = -6y^2$$

and

$$f_{yx}(2, 1) = -6(1)^2 = -6$$

Matched Problem 5 For $z = f(x, y) = x^3y - 2y^4 + 3$, find

(A) $\dfrac{\partial^2 z}{\partial y\, \partial x}$

(B) $\dfrac{\partial^2 z}{\partial y^2}$

(C) $f_{xy}(2, 3)$

(D) $f_{yx}(2, 3)$

CONCEPTUAL INSIGHT

Although the mixed second-order partial derivatives f_{xy} and f_{yx} are equal for all functions considered in this book, it is a good idea to compute both of them, as in Example 5A, as a check on your work. By contrast, the other two second-order partial derivatives, f_{xx} and f_{yy}, are generally not equal to each other. For example, for the function

$$f(x, y) = 3x^2 - 2xy^3 + 1$$

of Example 5,

$$f_{xx} = 6 \qquad \text{and} \qquad f_{yy} = -12xy$$

Exercises 15.2

W Skills Warm-up Exercises

In Problems 1–8, find the indicated derivative. (If necessary, review Sections 11.3 and 11.4).

1. $f'(x)$ if $f(x) = \pi x^3 + x\pi^3$

2. $f'(x)$ if $f(x) = (\pi x + 3)^5 - x^4$

3. $f'(x)$ if $f(x) = x^e + e^x$

4. $f'(x)$ if $f(x) = x \ln \pi + \pi \ln x$

5. $\dfrac{dz}{dx}$ if $z = \dfrac{x}{e} + \dfrac{e}{x}$

6. $\dfrac{dz}{dx}$ if $z = x^3 \ln \pi + 4\pi^2 e^x$

7. $\dfrac{dz}{dx}$ if $z = \ln(x^2 + e^2)$

8. $\dfrac{dz}{dx}$ if $z = e^7 - 2ex^7$

In Problems 9–16, find the indicated first-order partial derivative for each function $z = f(x, y)$.

9. $f_x(x, y)$ if $f(x, y) = 4x - 3y + 6$

10. $f_x(x, y)$ if $f(x, y) = 7x + 8y - 2$

11. $f_y(x, y)$ if $f(x, y) = x^2 - 3xy + 2y^2$

12. $f_y(x, y)$ if $f(x, y) = 3x^2 + 2xy - 7y^2$

13. $\dfrac{\partial z}{\partial x}$ if $z = x^3 + 4x^2y + 2y^3$

14. $\dfrac{\partial z}{\partial y}$ if $z = 4x^2y - 5xy^2$

15. $\dfrac{\partial z}{\partial y}$ if $z = (5x + 2y)^{10}$

16. $\dfrac{\partial z}{\partial x}$ if $z = (2x - 3y)^8$

In Problems 17–24, find the indicated value.

17. $f_x(1, 3)$ if $f(x, y) = 5x^3y - 4xy^2$

18. $f_x(4, 1)$ if $f(x, y) = x^2y^2 - 5xy^3$

19. $f_y(1, 0)$ if $f(x, y) = 3xe^y$

20. $f_y(2, 4)$ if $f(x, y) = x^4 \ln y$

21. $f_y(2, 1)$ if $f(x, y) = e^{x^2} - 4y$

22. $f_y(3, 3)$ if $f(x, y) = e^{3x} - y^2$

23. $f_x(1, -1)$ if $f(x, y) = \dfrac{2xy}{1 + x^2y^2}$

24. $f_x(-1, 2)$ if $f(x, y) = \dfrac{x^2 - y^2}{1 + x^2}$

In Problems 25–30, $M(x, y) = 68 + 0.3x - 0.8y$ gives the mileage (in mpg) of a new car as a function of tire pressure x (in psi) and speed (in mph). Find the indicated quantity (include the appropriate units) and explain what it means.

25. $M(32, 40)$ **26.** $M(22, 40)$

27. $M(32, 50)$ **28.** $M(22, 50)$

29. $M_x(32, 50)$ **30.** $M_y(32, 50)$

In Problems 31–42, find the indicated second-order partial derivative for each function $f(x, y)$.

31. $f_{xx}(x, y)$ if $f(x, y) = 6x - 5y + 3$

32. $f_{yx}(x, y)$ if $f(x, y) = -2x + y + 8$

33. $f_{xy}(x, y)$ if $f(x, y) = 4x^2 + 6y^2 - 10$

34. $f_{yy}(x, y)$ if $f(x, y) = x^2 + 9y^2 - 4$

35. $f_{xy}(x, y)$ if $f(x, y) = e^{xy^2}$

36. $f_{yx}(x, y)$ if $f(x, y) = e^{3x+2y}$

37. $f_{yy}(x, y)$ if $f(x, y) = \dfrac{\ln x}{y}$

38. $f_{xx}(x, y)$ if $f(x, y) = \dfrac{3 \ln x}{y^2}$

39. $f_{xx}(x, y)$ if $f(x, y) = (2x + y)^5$

40. $f_{yx}(x, y)$ if $f(x, y) = (3x - 8y)^6$

41. $f_{xy}(x, y)$ if $f(x, y) = (x^2 + y^4)^{10}$

42. $f_{yy}(x, y)$ if $f(x, y) = (1 + 2xy^2)^8$

In Problems 43–52, find the indicated function or value if $C(x, y) = 3x^2 + 10xy - 8y^2 + 4x - 15y - 120$.

43. $C_x(x, y)$ **44.** $C_y(x, y)$

45. $C_x(3, -2)$ **46.** $C_y(3, -2)$

47. $C_{xx}(x, y)$ **48.** $C_{yy}(x, y)$

49. $C_{xy}(x, y)$ **50.** $C_{yx}(x, y)$

51. $C_{xx}(3, -2)$ **52.** $C_{yy}(3, -2)$

In Problems 53–58, $S(T, r) = 50(T - 40)(5 - r)$ gives an ice cream shop's daily sales as a function of temperature T (in °F) and rain r (in inches). Find the indicated quantity (include the appropriate units) and explain what it means.

53. $S(60, 2)$ **54.** $S(80, 0)$

55. $S_r(90, 1)$ **56.** $S_T(90, 1)$

57. $S_{Tr}(90, 1)$ **58.** $S_{rT}(90, 1)$

59. (A) Let $f(x, y) = y^3 + 4y^2 - 5y + 3$. Show that $\partial f / \partial x = 0$.

(B) Explain why there are an infinite number of functions $g(x, y)$ such that $\partial g / \partial x = 0$.

60. (A) Find an example of a function $f(x, y)$ such that $\partial f / \partial x = 3$ and $\partial f / \partial y = 2$.

(B) How many such functions are there? Explain.

In Problems 61–66, find $f_{xx}(x, y)$, $f_{xy}(x, y)$, $f_{yx}(x, y)$, and $f_{yy}(x, y)$ for each function f.

61. $f(x, y) = x^2y^2 + x^3 + y$

62. $f(x, y) = x^3y^3 + x + y^2$

63. $f(x, y) = \dfrac{x}{y} - \dfrac{y}{x}$ **64.** $f(x, y) = \dfrac{x^2}{y} - \dfrac{y^2}{x}$

65. $f(x, y) = xe^{xy}$ **66.** $f(x, y) = x \ln(xy)$

67. For
$$P(x, y) = -x^2 + 2xy - 2y^2 - 4x + 12y - 5$$
find all values of x and y such that
$$P_x(x, y) = 0 \quad \text{and} \quad P_y(x, y) = 0$$
simultaneously.

68. For
$$C(x, y) = 2x^2 + 2xy + 3y^2 - 16x - 18y + 54$$
find all values of x and y such that
$$C_x(x, y) = 0 \quad \text{and} \quad C_y(x, y) = 0$$
simultaneously.

69. For
$$F(x, y) = x^3 - 2x^2y^2 - 2x - 4y + 10$$
find all values of x and y such that
$$F_x(x, y) = 0 \quad \text{and} \quad F_y(x, y) = 0$$
simultaneously.

70. For
$$G(x, y) = x^2 \ln y - 3x - 2y + 1$$
find all values of x and y such that
$$G_x(x, y) = 0 \quad \text{and} \quad G_y(x, y) = 0$$
simultaneously.

71. Let $f(x, y) = 3x^2 + y^2 - 4x - 6y + 2$.

(A) Find the minimum value of $f(x, y)$ when $y = 1$.

(B) Explain why the answer to part (A) is not the minimum value of the function $f(x, y)$.

72. Let $f(x, y) = 5 - 2x + 4y - 3x^2 - y^2$.

(A) Find the maximum value of $f(x, y)$ when $x = 2$.

(B) Explain why the answer to part (A) is not the maximum value of the function $f(x, y)$.

73. Let $f(x, y) = 4 - x^4y + 3xy^2 + y^5$.

(A) Use graphical approximation methods to find c (to three decimal places) such that $f(c, 2)$ is the maximum value of $f(x, y)$ when $y = 2$.

(B) Find $f_x(c, 2)$ and $f_y(c, 2)$.

74. Let $f(x, y) = e^x + 2e^y + 3xy^2 + 1$.

(A) Use graphical approximation methods to find d (to three decimal places) such that $f(1, d)$ is the minimum value of $f(x, y)$ when $x = 1$.

(B) Find $f_x(1, d)$ and $f_y(1, d)$.

75. For $f(x, y) = x^2 + 2y^2$, find

(A) $\displaystyle\lim_{h \to 0} \frac{f(x + h, y) - f(x, y)}{h}$

(B) $\displaystyle\lim_{k \to 0} \frac{f(x, y + k) - f(x, y)}{k}$

76. For $f(x, y) = 2xy^2$, find

(A) $\displaystyle\lim_{h \to 0} \frac{f(x + h, y) - f(x, y)}{h}$

(B) $\displaystyle\lim_{k \to 0} \frac{f(x, y + k) - f(x, y)}{k}$

Applications

77. **Profit function.** A firm produces two types of calculators each week, x of type A and y of type B. The weekly revenue and cost functions (in dollars) are

$$R(x, y) = 80x + 90y + 0.04xy - 0.05x^2 - 0.05y^2$$
$$C(x, y) = 8x + 6y + 20{,}000$$

Find $P_x(1{,}200, 1{,}800)$ and $P_y(1{,}200, 1{,}800)$, and interpret the results.

78. **Advertising and sales.** A company spends $\$x$ per week on online advertising and $\$y$ per week on TV advertising. Its weekly sales were found to be given by

$$S(x, y) = 10x^{0.4}y^{0.8}$$

Find $S_x(3{,}000, 2{,}000)$ and $S_y(3{,}000, 2{,}000)$, and interpret the results.

79. **Demand equations.** A supermarket sells two brands of coffee: brand A at $\$p$ per pound and brand B at $\$q$ per pound. The daily demands x and y (in pounds) for brands A and B, respectively, are given by

$$x = 200 - 5p + 4q$$
$$y = 300 + 2p - 4q$$

Find $\partial x/\partial p$ and $\partial y/\partial p$, and interpret the results.

80. **Revenue and profit functions.** A company manufactures 10- and 3-speed bicycles. The weekly demand and cost functions are

$$p = 230 - 9x + y$$
$$q = 130 + x - 4y$$
$$C(x, y) = 200 + 80x + 30y$$

where $\$p$ is the price of a 10-speed bicycle, $\$q$ is the price of a 3-speed bicycle, x is the weekly demand for 10-speed

bicycles, y is the weekly demand for 3-speed bicycles, and $C(x, y)$ is the cost function. Find $R_x(10, 5)$ and $P_x(10, 5)$, and interpret the results.

81. **Productivity.** The productivity of a certain third-world country is given approximately by the function

$$f(x, y) = 10x^{0.75}y^{0.25}$$

with the utilization of x units of labor and y units of capital.

(A) Find $f_x(x, y)$ and $f_y(x, y)$.

(B) If the country is now using 600 units of labor and 100 units of capital, find the marginal productivity of labor and the marginal productivity of capital.

(C) For the greatest increase in the country's productivity, should the government encourage increased use of labor or increased use of capital?

82. **Productivity.** The productivity of an automobile-manufacturing company is given approximately by the function

$$f(x, y) = 50\sqrt{xy} = 50x^{0.5}y^{0.5}$$

with the utilization of x units of labor and y units of capital.

(A) Find $f_x(x, y)$ and $f_y(x, y)$.

(B) If the company is now using 250 units of labor and 125 units of capital, find the marginal productivity of labor and the marginal productivity of capital.

(C) For the greatest increase in the company's productivity, should the management encourage increased use of labor or increased use of capital?

*Problems 83–86 refer to the following: If a decrease in demand for one product results in an increase in demand for another product, the two products are said to be **competitive**, or **substitute, products**. (Real whipping cream and imitation whipping cream are examples of competitive, or substitute, products.) If a decrease in demand for one product results in a decrease in demand for another product, the two products are said to be **complementary products**. (Fishing boats and outboard motors are examples of complementary products.) Partial derivatives can be used to test whether two products are competitive, complementary, or neither. We start with demand functions for two products such that the demand for either depends on the prices for both:*

$$x = f(p, q) \qquad \text{Demand function for product } A$$
$$y = g(p, q) \qquad \text{Demand function for product } B$$

The variables x and y represent the number of units demanded of products A and B, respectively, at a price p for 1 unit of product A and a price q for 1 unit of product B. Normally, if the price of A increases while the price of B is held constant, then the demand for A will decrease; that is, $f_p(p, q) < 0$. Then, if A and B are competitive products, the demand for B will increase; that is, $g_p(p, q) > 0$. Similarly, if the price of B increases while the price of A is held constant, the demand for B will decrease; that is, $g_q(p, q) < 0$. Then, if A and B are competitive products, the demand for A will increase; that is,

$f_q(p, q) > 0$. *Reasoning similarly for complementary products, we arrive at the following test:*

Test for Competitive and Complementary Products

Partial Derivatives	Products A and B
$f_q(p, q) > 0$ and $g_p(p, q) > 0$	Competitive (substitute)
$f_q(p, q) < 0$ and $g_p(p, q) < 0$	Complementary
$f_q(p, q) \geq 0$ and $g_p(p, q) \leq 0$	Neither
$f_q(p, q) \leq 0$ and $g_p(p, q) \geq 0$	Neither

Use this test in Problems 83–86 to determine whether the indicated products are competitive, complementary, or neither.

83. Product demand. The weekly demand equations for the sale of butter and margarine in a supermarket are

$$x = f(p, q) = 8{,}000 - 0.09p^2 + 0.08q^2 \quad \text{Butter}$$
$$y = g(p, q) = 15{,}000 + 0.04p^2 - 0.3q^2 \quad \text{Margarine}$$

84. Product demand. The daily demand equations for the sale of brand A coffee and brand B coffee in a supermarket are

$$x = f(p, q) = 200 - 5p + 4q \quad \text{Brand } A \text{ coffee}$$
$$y = g(p, q) = 300 + 2p - 4q \quad \text{Brand } B \text{ coffee}$$

85. Product demand. The monthly demand equations for the sale of skis and ski boots in a sporting goods store are

$$x = f(p, q) = 800 - 0.004p^2 - 0.003q^2 \quad \text{Skis}$$
$$y = g(p, q) = 600 - 0.003p^2 - 0.002q^2 \quad \text{Ski boots}$$

86. Product demand. The monthly demand equations for the sale of tennis rackets and tennis balls in a sporting goods store are

$$x = f(p, q) = 500 - 0.5p - q^2 \quad \text{Tennis rackets}$$
$$y = g(p, q) = 10{,}000 - 8p - 100q^2 \quad \text{Tennis balls (cans)}$$

87. Medicine. The following empirical formula relates the surface area A (in square inches) of an average human body to its weight w (in pounds) and its height h (in inches):

$$A = f(w, h) = 15.64w^{0.425}h^{0.725}$$

(A) Find $f_w(w, h)$ and $f_h(w, h)$.

(B) For a 65-pound child who is 57 inches tall, find $f_w(65, 57)$ and $f_h(65, 57)$, and interpret the results.

88. Blood flow. Poiseuille's law states that the resistance R for blood flowing in a blood vessel varies directly as the length L of the vessel and inversely as the fourth power of its radius r. Stated as an equation,

$$R(L, r) = k\frac{L}{r^4} \quad k \text{ a constant}$$

Find $R_L(4, 0.2)$ and $R_r(4, 0.2)$, and interpret the results.

89. Physical anthropology. Anthropologists use the cephalic index C, which varies directly as the width W of the head and inversely as the length L of the head (both viewed from the top). In terms of an equation,

$$C(W, L) = 100\frac{W}{L}$$

where

$$W = \text{width in inches}$$
$$L = \text{length in inches}$$

Find $C_W(6, 8)$ and $C_L(6, 8)$, and interpret the results.

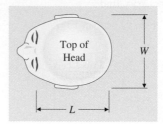

90. Safety research. Under ideal conditions, if a person driving a car slams on the brakes and skids to a stop, the length of the skid marks (in feet) is given by the formula

$$L(w, v) = kwv^2$$

where

$$k = \text{constant}$$
$$w = \text{weight of car in pounds}$$
$$v = \text{speed of car in miles per hour}$$

For $k = 0.000\ 013\ 3$, find $L_w(2{,}500, 60)$ and $L_v(2{,}500, 60)$, and interpret the results.

Answers to Matched Problems

1. (A) $\partial z/\partial y = -3x^2 + 5$ (B) $f_y(2, 3) = -7$
2. (A) $10x(x^2 + 2xy)^4$ (B) 10
3. $P_y(25, 10) = 10$: At a production level of $x = 25$ and $y = 10$, increasing y by 1 unit and holding x fixed at 25 will increase profit by approximately \$10; $P_y(25, 15) = -110$: At a production level of $x = 25$ and $y = 15$, increasing y by 1 unit and holding x fixed at 25 will decrease profit by approximately \$110
4. (A) $f_x(x, y) = 12x^{-0.7}y^{0.7}; f_y(x, y) = 28x^{0.3}y^{-0.3}$
   (B) Marginal productivity of labor ≈ 25.89; marginal productivity of capital ≈ 20.14
   (C) Labor
5. (A) $3x^2$ (B) $-24y^2$ (C) 12 (D) 12

15.3 Maxima and Minima

We are now ready to undertake a brief, but useful, analysis of local maxima and minima for functions of the type $z = f(x, y)$. We will extend the second-derivative test developed for functions of a single independent variable. We assume that all second-order partial derivatives exist for the function f in some circular region in the xy plane. This guarantees that the surface $z = f(x, y)$ has no sharp points, breaks, or ruptures. In other words, we are dealing only with smooth surfaces with no edges (like the edge of a box), breaks (like an earthquake fault), or sharp points (like the bottom point of a golf tee). (See Figure 1.)

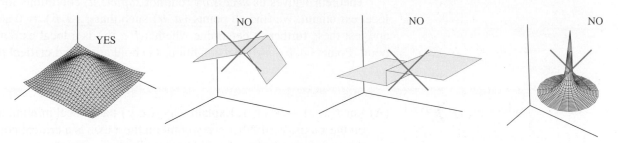

Figure 1

In addition, we will not concern ourselves with boundary points or absolute maxima–minima theory. Despite these restrictions, the procedure we will describe will help us solve a large number of useful problems.

What does it mean for $f(a, b)$ to be a local maximum or a local minimum? We say that $f(a, b)$ **is a local maximum** if there exists a circular region in the domain of f with (a, b) as the center, such that

$$f(a, b) \geq f(x, y)$$

for all (x, y) in the region. Similarly, we say that $f(a, b)$ **is a local minimum** if there exists a circular region in the domain of f with (a, b) as the center, such that

$$f(a, b) \leq f(x, y)$$

for all (x, y) in the region. Figure 2A illustrates a local maximum, Figure 2B a local minimum, and Figure 2C a **saddle point**, which is neither a local maximum nor a local minimum.

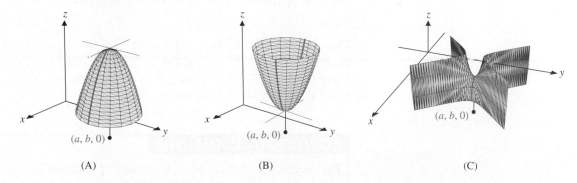

Figure 2

What happens to $f_x(a, b)$ and $f_y(a, b)$ if $f(a, b)$ is a local minimum or a local maximum and the partial derivatives of f exist in a circular region containing (a, b)? Figure 2 suggests that $f_x(a, b) = 0$ and $f_y(a, b) = 0$, since the tangent lines to the given curves are horizontal. Theorem 1 indicates that our intuitive reasoning is correct.

THEOREM 1   Local Extrema and Partial Derivatives

Let $f(a, b)$ be a local extremum (a local maximum or a local minimum) for the function f. If both f_x and f_y exist at (a, b), then

$$f_x(a, b) = 0 \quad \text{and} \quad f_y(a, b) = 0 \qquad (1)$$

The converse of this theorem is false. If $f_x(a, b) = 0$ and $f_y(a, b) = 0$, then $f(a, b)$ may or may not be a local extremum; for example, the point $(a, b, f(a, b))$ may be a saddle point (see Fig. 2C).

Theorem 1 gives us *necessary* (but not *sufficient*) conditions for $f(a, b)$ to be a local extremum. We find all points (a, b) such that $f_x(a, b) = 0$ and $f_y(a, b) = 0$ and test these further to determine whether $f(a, b)$ is a local extremum or a saddle point. Points (a, b) such that conditions (1) hold are called **critical points**.

Explore and Discuss 1

(A) Let $f(x, y) = y^2 + 1$. Explain why $f(x, y)$ has a local minimum at every point on the x axis. Verify that every point on the x axis is a critical point. Explain why the graph of $z = f(x, y)$ could be described as a trough.

(B) Let $g(x, y) = x^3$. Show that every point on the y axis is a critical point. Explain why no point on the y axis is a local extremum. Explain why the graph of $z = g(x, y)$ could be described as a slide.

The next theorem, using second-derivative tests, gives us *sufficient* conditions for a critical point to produce a local extremum or a saddle point.

THEOREM 2   Second-Derivative Test for Local Extrema

If

1. $z = f(x, y)$
2. $f_x(a, b) = 0$ and $f_y(a, b) = 0$ [(a, b) is a critical point]
3. All second-order partial derivatives of f exist in some circular region containing (a, b) as center.
4. $A = f_{xx}(a, b), \quad B = f_{xy}(a, b), \quad C = f_{yy}(a, b)$

Then

Case 1. If $AC - B^2 > 0$ and $A < 0$, then $f(a, b)$ is a local maximum.
Case 2. If $AC - B^2 > 0$ and $A > 0$, then $f(a, b)$ is a local minimum.
Case 3. If $AC - B^2 < 0$, then f has a saddle point at (a, b).
Case 4. If $AC - B^2 = 0$, the test fails.

CONCEPTUAL INSIGHT

The condition $A = f_{xx}(a, b) < 0$ in case 1 of Theorem 2 is analogous to the condition $f''(c) < 0$ in the second-derivative test for local extrema for a function of one variable (Section 12.5), which implies that the function is concave downward and therefore has a local maximum. Similarly, the condition $A = f_{xx}(a, b) > 0$ in case 2 is analogous to the condition $f''(c) > 0$ in the earlier second-derivative test, which implies that the function is concave upward and therefore has a local minimum.

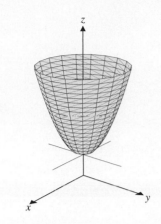

Figure 3

To illustrate the use of Theorem 2, we find the local extremum for a very simple function whose solution is almost obvious: $z = f(x, y) = x^2 + y^2 + 2$. From the function f itself and its graph (Fig. 3), it is clear that a local minimum is found at $(0, 0)$. Let us see how Theorem 2 confirms this observation.

Step 1 Find critical points: Find (x, y) such that $f_x(x, y) = 0$ and $f_y(x, y) = 0$ simultaneously:

$$f_x(x, y) = 2x = 0 \qquad f_y(x, y) = 2y = 0$$
$$x = 0 \qquad\qquad y = 0$$

The only critical point is $(a, b) = (0, 0)$.

Step 2 Compute $A = f_{xx}(0, 0)$, $B = f_{xy}(0, 0)$, and $C = f_{yy}(0, 0)$:

$$f_{xx}(x, y) = 2; \quad \text{so,} \quad A = f_{xx}(0, 0) = 2$$
$$f_{xy}(x, y) = 0; \quad \text{so,} \quad B = f_{xy}(0, 0) = 0$$
$$f_{yy}(x, y) = 2; \quad \text{so,} \quad C = f_{yy}(0, 0) = 2$$

Step 3 Evaluate $AC - B^2$ and try to classify the critical point $(0, 0)$ by using Theorem 2:

$$AC - B^2 = (2)(2) - (0)^2 = 4 > 0 \qquad \text{and} \qquad A = 2 > 0$$

Therefore, case 2 in Theorem 2 holds. That is, $f(0, 0) = 2$ is a local minimum. We will now use Theorem 2 to analyze extrema without the aid of graphs.

EXAMPLE 1 Finding Local Extrema Use Theorem 2 to find local extrema of
$$f(x, y) = -x^2 - y^2 + 6x + 8y - 21$$

SOLUTION

Step 1 Find critical points: Find (x, y) such that $f_x(x, y) = 0$ and $f_y(x, y) = 0$ simultaneously:

$$f_x(x, y) = -2x + 6 = 0 \qquad f_y(x, y) = -2y + 8 = 0$$
$$x = 3 \qquad\qquad\qquad y = 4$$

The only critical point is $(a, b) = (3, 4)$.

Step 2 Compute $A = f_{xx}(3, 4)$, $B = f_{xy}(3, 4)$, and $C = f_{yy}(3, 4)$:

$$f_{xx}(x, y) = -2; \quad \text{so,} \quad A = f_{xx}(3, 4) = -2$$
$$f_{xy}(x, y) = 0; \quad \text{so,} \quad B = f_{xy}(3, 4) = 0$$
$$f_{yy}(x, y) = -2; \quad \text{so,} \quad C = f_{yy}(3, 4) = -2$$

Step 3 Evaluate $AC - B^2$ and try to classify the critical point $(3, 4)$ by using Theorem 2:

$$AC - B^2 = (-2)(-2) - (0)^2 = 4 > 0 \qquad \text{and} \qquad A = -2 < 0$$

Therefore, case 1 in Theorem 2 holds, and $f(3, 4) = 4$ is a local maximum.

Matched Problem 1 Use Theorem 2 to find local extrema of
$$f(x, y) = x^2 + y^2 - 10x - 2y + 36$$

EXAMPLE 2 Finding Local Extrema: Multiple Critical Points Use Theorem 2 to find local extrema of

$$f(x, y) = x^3 + y^3 - 6xy$$

SOLUTION

Step 1 Find critical points of $f(x, y) = x^3 + y^3 - 6xy$:

$$f_x(x, y) = 3x^2 - 6y = 0 \qquad \text{Solve for } y.$$

$$6y = 3x^2$$

$$y = \tfrac{1}{2}x^2 \qquad\qquad\qquad (2)$$

$$f_y(x, y) = 3y^2 - 6x = 0$$

$$3y^2 = 6x \qquad \text{Use equation (2) to eliminate } y.$$

$$3\left(\tfrac{1}{2}x^2\right)^2 = 6x$$

$$\tfrac{3}{4}x^4 = 6x \qquad \text{Solve for } x.$$

$$3x^4 - 24x = 0$$

$$3x(x^3 - 8) = 0$$

$$x = 0 \qquad \text{or} \qquad x = 2$$

$$y = 0 \qquad\qquad\qquad y = \tfrac{1}{2}(2)^2 = 2$$

The critical points are $(0, 0)$ and $(2, 2)$.

Since there are two critical points, steps 2 and 3 must be performed twice.

Test (0, 0)

Step 2 Compute $A = f_{xx}(0, 0)$, $B = f_{xy}(0, 0)$, and $C = f_{yy}(0, 0)$:

$$f_{xx}(x, y) = 6x; \qquad \text{so,} \qquad A = f_{xx}(0, 0) = 0$$

$$f_{xy}(x, y) = -6; \qquad \text{so,} \qquad B = f_{xy}(0, 0) = -6$$

$$f_{yy}(x, y) = 6y; \qquad \text{so,} \qquad C = f_{yy}(0, 0) = 0$$

Step 3 Evaluate $AC - B^2$ and try to classify the critical point $(0, 0)$ by using Theorem 2:

$$AC - B^2 = (0)(0) - (-6)^2 = -36 < 0$$

Therefore, case 3 in Theorem 2 applies. That is, f has a saddle point at $(0, 0)$.

Now we will consider the second critical point, $(2, 2)$:

Test (2, 2)

Step 2 Compute $A = f_{xx}(2, 2)$, $B = f_{xy}(2, 2)$, and $C = f_{yy}(2, 2)$:

$$f_{xx}(x, y) = 6x; \qquad \text{so,} \qquad A = f_{xx}(2, 2) = 12$$

$$f_{xy}(x, y) = -6; \qquad \text{so,} \qquad B = f_{xy}(2, 2) = -6$$

$$f_{yy}(x, y) = 6y; \qquad \text{so,} \qquad C = f_{yy}(2, 2) = 12$$

Step 3 Evaluate $AC - B^2$ and try to classify the critical point $(2, 2)$ by using Theorem 2:

$$AC - B^2 = (12)(12) - (-6)^2 = 108 > 0 \qquad \text{and} \qquad A = 12 > 0$$

So, case 2 in Theorem 2 applies, and $f(2, 2) = -8$ is a local minimum.

Our conclusions in Example 2 may be confirmed geometrically by graphing cross sections of the function f. The cross sections of f in the planes $y = 0$, $x = 0$, $y = x$, and $y = -x$ [each of these planes contains $(0, 0)$] are represented by the graphs of the functions $f(x, 0) = x^3$, $f(0, y) = y^3$, $f(x, x) = 2x^3 - 6x^2$, and $f(x, -x) = 6x^2$, respectively, as shown in Figure 4A (note that the first two functions have the same graph). The cross sections of f in the planes $y = 2$, $x = 2$, $y = x$, and $y = 4 - x$ [each of these planes contains $(2, 2)$] are represented by the graphs of $f(x, 2) = x^3 - 12x + 8$, $f(2, y) = y^3 - 12y + 8$, $f(x, x) = 2x^3 - 6x^2$, and $f(x, 4 - x) = x^3 + (4 - x)^3 + 6x^2 - 24x$, respectively, as shown in Figure 4B (the first two functions have the same graph). Figure 4B illustrates the fact that since f has a local minimum at $(2, 2)$, each of the cross sections of f through $(2, 2)$ has a local minimum of -8 at $(2, 2)$. Figure 4A, by contrast, indicates that some cross sections of f through $(0, 0)$ have a local minimum, some a local maximum, and some neither one, at $(0, 0)$.

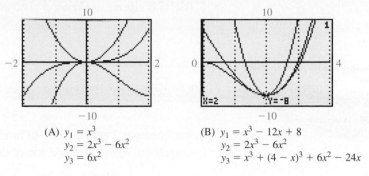

(A) $y_1 = x^3$
    $y_2 = 2x^3 - 6x^2$
    $y_3 = 6x^2$

(B) $y_1 = x^3 - 12x + 8$
    $y_2 = 2x^3 - 6x^2$
    $y_3 = x^3 + (4 - x)^3 + 6x^2 - 24x$

Figure 4

Matched Problem 2 Use Theorem 2 to find local extrema for $f(x, y) = x^3 + y^2 - 6xy$.

EXAMPLE 3 Profit Suppose that the surfboard company discussed earlier has developed the yearly profit equation

$$P(x, y) = -22x^2 + 22xy - 11y^2 + 110x - 44y - 23$$

where x is the number (in thousands) of standard surfboards produced per year, y is the number (in thousands) of competition surfboards produced per year, and P is profit (in thousands of dollars). How many of each type of board should be produced per year to realize a maximum profit? What is the maximum profit?

SOLUTION

Step 1 Find critical points:

$$P_x(x, y) = -44x + 22y + 110 = 0$$
$$P_y(x, y) = 22x - 22y - 44 = 0$$

Solving this system, we obtain $(3, 1)$ as the only critical point.

Step 2 Compute $A = P_{xx}(3, 1), B = P_{xy}(3, 1)$, and $C = P_{yy}(3, 1)$:

$$P_{xx}(x, y) = -44; \quad \text{so,} \quad A = P_{xx}(3, 1) = -44$$
$$P_{xy}(x, y) = 22; \quad \text{so,} \quad B = P_{xy}(3, 1) = 22$$
$$P_{yy}(x, y) = -22; \quad \text{so,} \quad C = P_{yy}(3, 1) = -22$$

Step 3 Evaluate $AC - B^2$ and try to classify the critical point $(3, 1)$ by using Theorem 2:

$$AC - B^2 = (-44)(-22) - 22^2 = 484 > 0 \quad \text{and} \quad A = -44 < 0$$

Therefore, case 1 in Theorem 2 applies. That is, $P(3, 1) = 120$ is a local maximum. A maximum profit of \$120,000 is obtained by producing and selling 3,000 standard boards and 1,000 competition boards per year.

Matched Problem 3 Repeat Example 3 with

$$P(x, y) = -66x^2 + 132xy - 99y^2 + 132x - 66y - 19$$

EXAMPLE 4   Package Design   The packaging department in a company is to design a rectangular box with no top and a partition down the middle. The box must have a volume of 48 cubic inches. Find the dimensions that will minimize the area of material used to construct the box.

SOLUTION   Refer to Figure 5. The area of material used in constructing this box is

$$
\begin{array}{cccc}
 & \text{Front,} & \text{Sides,} \\
\text{Base} & \text{back} & \text{partition} \\
M = xy & + \ 2xz & + \ 3yz
\end{array}
\tag{3}
$$

The volume of the box is

$$V = xyz = 48 \tag{4}$$

Since Theorem 2 applies only to functions with two independent variables, we must use equation (4) to eliminate one of the variables in equation (3):

$$M = xy + 2xz + 3yz \qquad \text{Substitute } z = 48/xy.$$

$$= xy + 2x\left(\frac{48}{xy}\right) + 3y\left(\frac{48}{xy}\right)$$

$$= xy + \frac{96}{y} + \frac{144}{x}$$

So, we must find the minimum value of

$$M(x, y) = xy + \frac{96}{y} + \frac{144}{x} \qquad x > 0 \qquad \text{and} \qquad y > 0$$

Step 1   Find critical points:

$$M_x(x, y) = y - \frac{144}{x^2} = 0$$

$$y = \frac{144}{x^2} \tag{5}$$

$$M_y(x, y) = x - \frac{96}{y^2} = 0$$

$$x = \frac{96}{y^2} \qquad\qquad \text{Solve for } y^2.$$

$$y^2 = \frac{96}{x} \qquad\qquad \text{Use equation (5) to eliminate } y \text{ and solve for } x.$$

$$\left(\frac{144}{x^2}\right)^2 = \frac{96}{x}$$

$$\frac{20{,}736}{x^4} = \frac{96}{x} \qquad\qquad \text{Multiply both sides by } x^4/96 \\ \text{(recall that } x > 0\text{).}$$

$$x^3 = \frac{20{,}736}{96} = 216$$

$$x = 6 \qquad\qquad \text{Use equation (5) to find } y.$$

$$y = \frac{144}{36} = 4$$

Therefore, $(6, 4)$ is the only critical point.

Figure 5

Step 2 Compute $A = M_{xx}(6, 4)$, $B = M_{xy}(6, 4)$, and $C = M_{yy}(6, 4)$:

$$M_{xx}(x, y) = \frac{288}{x^3}; \qquad \text{so,} \qquad A = M_{xx}(6, 4) = \frac{288}{216} = \frac{4}{3}$$

$$M_{xy}(x, y) = 1; \qquad \text{so,} \qquad B = M_{xy}(6, 4) = 1$$

$$M_{yy}(x, y) - \frac{192}{y^3}; \qquad \text{so,} \qquad C = M_{yy}(6, 4) = \frac{192}{64} = 3$$

Step 3 Evaluate $AC - B^2$ and try to classify the critical point $(6, 4)$ by using Theorem 2:

$$AC - B^2 = \left(\tfrac{4}{3}\right)(3) - (1)^2 = 3 > 0 \qquad \text{and} \qquad A = \tfrac{4}{3} > 0$$

Case 2 in Theorem 2 applies, and $M(x, y)$ has a local minimum at $(6, 4)$. If $x = 6$ and $y = 4$, then

$$z = \frac{48}{xy} = \frac{48}{(6)(4)} = 2$$

The dimensions that will require the least material are 6 inches by 4 inches by 2 inches (see Fig. 6).

2 inches

4 inches

6 inches

Figure 6

Matched Problem 4 If the box in Example 4 must have a volume of 384 cubic inches, find the dimensions that will require the least material.

Exercises 15.3

Skills Warm-up Exercises

In Problems 1–8, find $f'(0)$, $f''(0)$, and determine whether f has a local minimum, local maximum, or neither at $x = 0$. (If necessary, review the second derivative test for local extrema in Section 12.5).

1. $f(x) = 2x^3 - 9x^2 + 4$ 2. $f(x) = 4x^3 + 6x^2 + 100$

3. $f(x) = \dfrac{1}{1 - x^2}$ 4. $f(x) = \dfrac{1}{1 + x^2}$

5. $f(x) = e^{x^2}$ 6. $f(x) = e^{x^2}$

7. $f(x) = x^3 - x^2 + x - 1$ 8. $f(x) = (3x + 1)^2$

In Problems 9–12, find $f_x(x, y)$ and $f_y(x, y)$, and explain, using Theorem 1, why $f(x, y)$ has no local extrema.

9. $f(x, y) = 4x + 5y - 6$

10. $f(x, y) = 10 - 2x - 3y + x^2$

11. $f(x, y) = 3.7 - 1.2x + 6.8y + 0.2y^3 + x^4$

12. $f(x, y) = x^3 - y^2 + 7x + 3y + 1$

Use Theorem 2 to find local extrema in Problems 13–32.

13. $f(x, y) = 6 - x^2 - 4x - y^2$

14. $f(x, y) = 3 - x^2 - y^2 + 6y$

15. $f(x, y) = x^2 + y^2 + 2x - 6y + 14$

16. $f(x, y) = x^2 + y^2 - 4x + 6y + 23$

17. $f(x, y) = xy + 2x - 3y - 2$

18. $f(x, y) = x^2 - y^2 + 2x + 6y - 4$

19. $f(x, y) = -3x^2 + 2xy - 2y^2 + 14x + 2y + 10$

20. $f(x, y) = -x^2 + xy - 2y^2 + x + 10y - 5$

21. $f(x, y) = 2x^2 - 2xy + 3y^2 - 4x - 8y + 20$

22. $f(x, y) = 2x^2 - xy + y^2 - x - 5y + 8$

23. $f(x, y) = e^{xy}$

24. $f(x, y) = x^2y - xy^2$

25. $f(x, y) = x^3 + y^3 - 3xy$

26. $f(x, y) = 2y^3 - 6xy - x^2$

27. $f(x, y) = 2x^4 + y^2 - 12xy$

28. $f(x, y) = 16xy - x^4 - 2y^2$

29. $f(x, y) = x^3 - 3xy^2 + 6y^2$

30. $f(x, y) = 2x^2 - 2x^2y + 6y^3$

31. $f(x, y) = y^3 + 2x^2y^2 - 3x - 2y + 8$

32. $f(x, y) = x \ln y + x^2 - 4x - 5y + 3$

33. Explain why $f(x, y) = x^2$ has a local extremum at infinitely many points.

34. (A) Find the local extrema of the functions $f(x, y) = x + y$, $g(x, y) = x^2 + y^2$, and $h(x, y) = x^3 + y^3$.

(B) Discuss the local extrema of the function $k(x, y) = x^n + y^n$, where n is a positive integer.

35. (A) Show that $(0, 0)$ is a critical point of the function
$f(x, y) = x^4e^y + x^2y^4 + 1$, but that the second-derivative test for local extrema fails.

(B) Use cross sections, as in Example 2, to decide whether f has a local maximum, a local minimum, or a saddle point at $(0, 0)$.

36. (A) Show that $(0, 0)$ is a critical point of the function
$g(x, y) = e^{xy^2} + x^2y^3 + 2$, but that the second-derivative test for local extrema fails.

(B) Use cross sections, as in Example 2, to decide whether g has a local maximum, a local minimum, or a saddle point at $(0, 0)$.

Applications

37. Product mix for maximum profit. A firm produces two types of earphones per year: x thousand of type A and y thousand of type B. If the revenue and cost equations for the year are (in millions of dollars)

$$R(x, y) = 2x + 3y$$
$$C(x, y) = x^2 - 2xy + 2y^2 + 6x - 9y + 5$$

determine how many of each type of earphone should be produced per year to maximize profit. What is the maximum profit?

38. Automation–labor mix for minimum cost. The annual labor and automated equipment cost (in millions of dollars) for a company's production of HDTVs is given by

$$C(x, y) = 2x^2 + 2xy + 3y^2 - 16x - 18y + 54$$

where x is the amount spent per year on labor and y is the amount spent per year on automated equipment (both in millions of dollars). Determine how much should be spent on each per year to minimize this cost. What is the minimum cost?

39. Maximizing profit. A store sells two brands of camping chairs. The store pays $60 for each brand A chair and $80 for each brand B chair. The research department has estimated the weekly demand equations for these two competitive products to be

$$x = 260 - 3p + q \quad \text{Demand equation for brand } A$$
$$y = 180 + p - 2q \quad \text{Demand equation for brand } B$$

where p is the selling price for brand A and q is the selling price for brand B.

(A) Determine the demands x and y when $p = \$100$ and $q = \$120$; when $p = \$110$ and $q = \$110$.

(B) How should the store price each chair to maximize weekly profits? What is the maximum weekly profit? [*Hint:* $C = 60x + 80y$, $R = px + qy$, and $P = R - C$.]

40. Maximizing profit. A store sells two brands of laptop sleeves. The store pays $25 for each brand A sleeve and $30 for each brand B sleeve. A consulting firm has estimated the daily demand equations for these two competitive products to be

$$x = 130 - 4p + q \quad \text{Demand equation for brand } A$$
$$y = 115 + 2p - 3q \quad \text{Demand equation for brand } B$$

where p is the selling price for brand A and q is the selling price for brand B.

(A) Determine the demands x and y when $p = \$40$ and $q = \$50$; when $p = \$45$ and $q = \$55$.

(B) How should the store price each brand of sleeve to maximize daily profits? What is the maximum daily profit? [*Hint:* $C = 25x + 30y$, $R = px + qy$, and $P = R - C$.]

41. Minimizing cost. A satellite TV station is to be located at $P(x, y)$ so that the sum of the squares of the distances from P to the three towns A, B, and C is a minimum (see the figure). Find the coordinates of P, the location that will minimize the cost of providing satellite TV for all three towns.

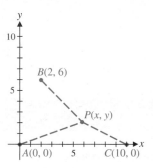

42. Minimizing cost. Repeat Problem 41, replacing the coordinates of B with $B(6, 9)$ and the coordinates of C with $C(9, 0)$.

43. Minimum material. A rectangular box with no top and two parallel partitions (see the figure) must hold a volume of 64 cubic inches. Find the dimensions that will require the least material.

44. Minimum material. A rectangular box with no top and two intersecting partitions (see the figure) must hold a volume of 72 cubic inches. Find the dimensions that will require the least material.

45. Maximum volume. A mailing service states that a rectangular package cannot have the sum of its length and girth exceed 120 inches (see the figure). What are the dimensions of the largest (in volume) mailing carton that can be constructed to meet these restrictions?

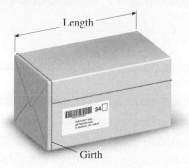

Length

34☐

Girth

46. Maximum shipping volume. A shipping box is to be reinforced with steel bands in all three directions, as shown in the figure. A total of 150 inches of steel tape is to be used, with 6 inches of waste because of a 2-inch overlap in each direction.

Find the dimensions of the box with maximum volume that can be taped as described.

FRAGILE

Answers to Matched Problems

1. $f(5, 1) = 10$ is a local minimum
2. f has a saddle point at $(0, 0)$; $f(6, 18) = -108$ is a local minimum
3. Local maximum for $x = 2$ and $y = 1$; $P(2, 1) = 80$; a maximum profit of \$80,000 is obtained by producing and selling 2,000 standard boards and 1,000 competition boards
4. 12 in. by 8 in. by 4 in.

15.4 Maxima and Minima Using Lagrange Multipliers

- Functions of Two Independent Variables
- Functions of Three Independent Variables

Existing fence

x

y

x

Figure 1

Functions of Two Independent Variables

We now consider a powerful method of solving a certain class of maxima–minima problems. Joseph Louis Lagrange (1736–1813), an eighteenth-century French mathematician, discovered this method, called the **method of Lagrange multipliers**. We introduce the method through an example.

A rancher wants to construct two feeding pens of the same size along an existing fence (see Fig. 1). If the rancher has 720 feet of fencing materials available, how long should x and y be in order to obtain the maximum total area? What is the maximum area?

The total area is given by

$$f(x, y) = xy$$

which can be made as large as we like, provided that there are no restrictions on x and y. But there are restrictions on x and y, since we have only 720 feet of fencing. The variables x and y must be chosen so that

$$3x + y = 720$$

This restriction on x and y, called a **constraint**, leads to the following maxima–minima problem:

Maximize $f(x, y) = xy$ (1)

subject to $3x + y = 720$, or $3x + y - 720 = 0$ (2)

This problem is one of a general class of problems of the form

Maximize (or minimize) $z = f(x, y)$ (3)

subject to $g(x, y) = 0$ (4)

Of course, we could try to solve equation (4) for y in terms of x, or for x in terms of y, then substitute the result into equation (3), and use methods developed in Section 12.5 for functions of a single variable. But what if equation (4) is more complicated than equation (2), and solving for one variable in terms of the other is either very difficult or impossible? In the method of Lagrange multipliers, we will work with $g(x, y)$ directly and avoid solving equation (4) for one variable in terms of the other. In addition, the method generalizes to functions of arbitrarily many variables subject to one or more constraints.

Now to the method: We form a new function F, using functions f and g in equations (3) and (4), as follows:

$$F(x, y, \lambda) = f(x, y) + \lambda g(x, y) \tag{5}$$

Here, λ (the Greek lowercase letter lambda) is called a **Lagrange multiplier**. Theorem 1 gives the basis for the method.

THEOREM 1   Method of Lagrange Multipliers for Functions of Two Variables

Any local maxima or minima of the function $z = f(x, y)$ subject to the constraint $g(x, y) = 0$ will be among those points (x_0, y_0) for which (x_0, y_0, λ_0) is a solution of the system

$$F_x(x, y, \lambda) = 0$$
$$F_y(x, y, \lambda) = 0$$
$$F_\lambda(x, y, \lambda) = 0$$

where $F(x, y, \lambda) = f(x, y) + \lambda g(x, y)$, provided that all the partial derivatives exist.

We now use the method of Lagrange multipliers to solve the fence problem.

Step 1 Formulate the problem in the form of equations (3) and (4):

$$\text{Maximize}\quad f(x, y) = xy$$
$$\text{subject to}\quad g(x, y) = 3x + y - 720 = 0$$

Step 2 Form the function F, introducing the Lagrange multiplier λ:

$$F(x, y, \lambda) = f(x, y) + \lambda g(x, y)$$
$$= xy + \lambda(3x + y - 720)$$

Step 3 Solve the system $F_x = 0, F_y = 0, F_\lambda = 0$ (the solutions are **critical points** of F):

$$F_x = y + 3\lambda = 0$$
$$F_y = x + \lambda = 0$$
$$F_\lambda = 3x + y - 720 = 0$$

From the first two equations, we see that

$$y = -3\lambda$$
$$x = -\lambda$$

Substitute these values for x and y into the third equation and solve for λ:

$$-3\lambda - 3\lambda = 720$$
$$-6\lambda = 720$$
$$\lambda = -120$$

So,

$$y = -3(-120) = 360 \text{ feet}$$
$$x = -(-120) = 120 \text{ feet}$$

and $(x_0, y_0, \lambda_0) = (120, 360, -120)$ is the only critical point of F.

Step 4 According to Theorem 1, if the function $f(x, y)$, subject to the constraint $g(x, y) = 0$, has a local maximum or minimum, that maximum or minimum must occur at $x = 120$, $y = 360$. Although it is possible to develop a test similar to Theorem 2 in Section 15.3 to determine the nature of this local extremum, we will not do so. [Note that Theorem 2 cannot be applied to $f(x, y)$ at (120, 360), since this point is not a critical point of the unconstrained function $f(x, y)$.] We simply assume that the maximum value of $f(x, y)$ must occur for $x = 120$, $y = 360$.

$$\text{Max } f(x, y) = f(120, 360)$$
$$= (120)(360) = 43{,}200 \text{ square feet}$$

The key steps in applying the method of Lagrange multipliers are as follows:

PROCEDURE Method of Lagrange Multipliers: Key Steps

Step 1 Write the problem in the form

$$\text{Maximize (or minimize)} \quad z = f(x, y)$$
$$\text{subject to} \quad g(x, y) = 0$$

Step 2 Form the function F:

$$F(x, y, \lambda) = f(x, y) + \lambda g(x, y)$$

Step 3 Find the critical points of F; that is, solve the system

$$F_x(x, y, \lambda) = 0$$
$$F_y(x, y, \lambda) = 0$$
$$F_\lambda(x, y, \lambda) = 0$$

Step 4 If (x_0, y_0, λ_0) is the only critical point of F, we assume that (x_0, y_0) will always produce the solution to the problems we consider. If F has more than one critical point, we evaluate $z = f(x, y)$ at (x_0, y_0) for each critical point (x_0, y_0, λ_0) of F. For the problems we consider, we assume that the largest of these values is the maximum value of $f(x, y)$, subject to the constraint $g(x, y) = 0$, and the smallest is the minimum value of $f(x, y)$, subject to the constraint $g(x, y) = 0$.

EXAMPLE 1 Minimization Subject to a Constraint   Minimize $f(x, y) = x^2 + y^2$ subject to $x + y = 10$.

SOLUTION

Step 1
$$\text{Minimize} \quad f(x, y) = x^2 + y^2$$
$$\text{subject to} \quad g(x, y) = x + y - 10 = 0$$

Step 2
$$F(x, y, \lambda) = x^2 + y^2 + \lambda(x + y - 10)$$

Step 3
$$F_x = 2x + \lambda = 0$$
$$F_y = 2y + \lambda = 0$$
$$F_\lambda = x + y - 10 = 0$$

From the first two equations, $x = -\lambda/2$ and $y = -\lambda/2$. Substituting these values into the third equation, we obtain

$$-\frac{\lambda}{2} - \frac{\lambda}{2} = 10$$
$$-\lambda = 10$$
$$\lambda = -10$$

The only critical point is $(x_0, y_0, \lambda_0) = (5, 5, -10)$.

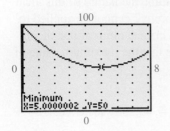

Figure 2 $h(x) = x^2 + (10 - x)^2$

Step 4 Since $(5, 5, -10)$ is the only critical point of F, we conclude that (see step 4 in the box)

$$\text{Min } f(x, y) = f(5, 5) = (5)^2 + (5)^2 = 50$$

Since $g(x, y)$ in Example 1 has a relatively simple form, an alternative to the method of Lagrange multipliers is to solve $g(x, y) = 0$ for y and then substitute into $f(x, y)$ to obtain the function $h(x) = f(x, 10 - x) = x^2 + (10 - x)^2$ in the single variable x. Then we minimize h (see Fig. 2). From Figure 2, we conclude that min $f(x, y) = f(5, 5) = 50$. This technique depends on being able to solve the constraint for one of the two variables and so is not always available as an alternative to the method of Lagrange multipliers.

Matched Problem 1 Maximize $f(x, y) = 25 - x^2 - y^2$ subject to $x + y = 4$.

Figures 3 and 4 illustrate the results obtained in Example 1 and Matched Problem 1, respectively.

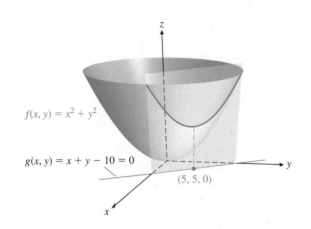

Figure 3

Figure 4

Explore and Discuss 1   Consider the problem of minimizing $f(x, y) = 3x^2 + 5y^2$ subject to the constraint $g(x, y) = 2x + 3y - 6 = 0$.

(A) Compute the value of $f(x, y)$ when x and y are integers, $0 \le x \le 3, 0 \le y \le 2$. Record your answers in the empty boxes next to the points (x, y) in Figure 5.

(B) Graph the constraint $g(x, y) = 0$.

(C) Estimate the minimum value of f on the basis of your graph and the computations from part (A).

(D) Use the method of Lagrange multipliers to solve the minimization problem.

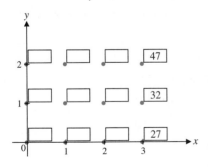

Figure 5

EXAMPLE 2 Productivity The Cobb–Douglas production function for a new product is given by

$$N(x, y) = 16x^{0.25}y^{0.75}$$

where x is the number of units of labor and y is the number of units of capital required to produce $N(x, y)$ units of the product. Each unit of labor costs \$50 and each unit of capital costs \$100. If \$500,000 has been budgeted for the production of this product, how should that amount be allocated between labor and capital in order to maximize production? What is the maximum number of units that can be produced?

SOLUTION The total cost of using x units of labor and y units of capital is $50x + 100y$. Thus, the constraint imposed by the $\$500,000$ budget is

$$50x + 100y = 500,000$$

Step 1   Maximize $N(x, y) = 16x^{0.25}y^{0.75}$

subject to $g(x, y) = 50x + 100y - 500,000 = 0$

Step 2 $F(x, y, \lambda) = 16x^{0.25}y^{0.75} + \lambda(50x + 100y - 500,000)$

Step 3 $F_x = 4x^{-0.75}y^{0.75} + 50\lambda = 0$

$F_y = 12x^{0.25}y^{-0.25} + 100\lambda = 0$

$F_\lambda = 50x + 100y - 500,000 = 0$

From the first two equations,

$$\lambda = -\tfrac{2}{25}x^{-0.75}y^{0.75} \quad \text{and} \quad \lambda = -\tfrac{3}{25}x^{0.25}y^{-0.25}$$

Therefore,

$$-\tfrac{2}{25}x^{-0.75}y^{0.75} = -\tfrac{3}{25}x^{0.25}y^{-0.25} \quad \text{Multiply both sides by } x^{0.75}\,y^{0.25}.$$

$$-\tfrac{2}{25}y = -\tfrac{3}{25}x \qquad \text{(We can assume that } x \neq 0 \text{ and } y \neq 0\text{.)}$$

$$y = \tfrac{3}{2}x$$

Now substitute for y in the third equation and solve for x:

$$50x + 100\left(\tfrac{3}{2}x\right) - 500,000 = 0$$

$$200x = 500,000$$

$$x = 2,500$$

So,

$$y = \tfrac{3}{2}(2,500) = 3,750$$

and

$$\lambda = -\tfrac{2}{25}(2,500)^{-0.75}(3,750)^{0.75} \approx -0.1084$$

The only critical point of F is $(2,500, 3,750, -0.1084)$.

Step 4 Since F has only one critical point, we conclude that maximum productivity occurs when 2,500 units of labor and 3,750 units of capital are used (see step 4 in the method of Lagrange multipliers).

$$\text{Max } N(x, y) = N(2,500, 3,750)$$

$$= 16(2,500)^{0.25}(3,750)^{0.75}$$

$$\approx 54,216 \text{ units}$$

The negative of the value of the Lagrange multiplier found in step 3 is called the **marginal productivity of money** and gives the approximate increase in production for each additional dollar spent on production. In Example 2, increasing the production budget from $\$500,000$ to $\$600,000$ would result in an approximate increase in production of

$$0.1084(100,000) = 10,840 \text{ units}$$

Note that simplifying the constraint equation

$$50x + 100y - 500,000 = 0$$

to

$$x + 2y - 10,000 = 0$$

before forming the function $F(x, y, \lambda)$ would make it difficult to interpret $-\lambda$ correctly. **In marginal productivity problems, the constraint equation should not be simplified.**

Matched Problem 2 The Cobb–Douglas production function for a new product is given by

$$N(x, y) = 20x^{0.5}y^{0.5}$$

where x is the number of units of labor and y is the number of units of capital required to produce $N(x, y)$ units of the product. Each unit of labor costs $40 and each unit of capital costs $120.

(A) If $300,000 has been budgeted for the production of this product, how should that amount be allocated in order to maximize production? What is the maximum production?

(B) Find the marginal productivity of money in this case, and estimate the increase in production if an additional $40,000 is budgeted for production.

Explore and Discuss 2 Consider the problem of maximizing $f(x, y) = 4 - x^2 - y^2$ subject to the constraint $g(x, y) = y - x^2 + 1 = 0$.

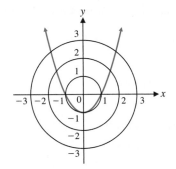

(A) Explain why $f(x, y) = 3$ whenever (x, y) is a point on the circle of radius 1 centered at the origin. What is the value of $f(x, y)$ when (x, y) is a point on the circle of radius 2 centered at the origin? On the circle of radius 3 centered at the origin? (See Fig. 6.)

(B) Explain why some points on the parabola $y - x^2 + 1 = 0$ lie inside the circle $x^2 + y^2 = 1$.

(C) In light of part (B), would you guess that the maximum value of $f(x, y)$ subject to the constraint is greater than 3? Explain.

(D) Use Lagrange multipliers to solve the maximization problem.

Figure 6

Functions of Three Independent Variables

The method of Lagrange multipliers can be extended to functions with arbitrarily many independent variables with one or more constraints. We now state a theorem for functions with three independent variables and one constraint, and we consider an example that demonstrates the advantage of the method of Lagrange multipliers over the method used in Section 15.3.

> **THEOREM 2   Method of Lagrange Multipliers for Functions of Three Variables**
>
> Any local maxima or minima of the function $w = f(x, y, z)$, subject to the constraint $g(x, y, z) = 0$, will be among the set of points (x_0, y_0, z_0) for which $(x_0, y_0, z_0, \lambda_0)$ is a solution of the system
>
> $$F_x(x, y, z, \lambda) = 0$$
> $$F_y(x, y, z, \lambda) = 0$$
> $$F_z(x, y, z, \lambda) = 0$$
> $$F_\lambda(x, y, z, \lambda) = 0$$
>
> where $F(x, y, z, \lambda) = f(x, y, z) + \lambda g(x, y, z)$, provided that all the partial derivatives exist.

EXAMPLE 3 **Package Design** A rectangular box with an open top and one partition is to be constructed from 162 square inches of cardboard (Fig. 7). Find the dimensions that result in a box with the largest possible volume.

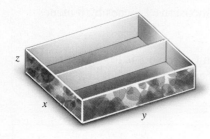

Figure 7

SOLUTION We must maximize

$$V(x, y, z) = xyz$$

subject to the constraint that the amount of material used is 162 square inches. So x, y, and z must satisfy

$$xy + 2xz + 3yz = 162$$

Step 1   Maximize $V(x, y, z) = xyz$

subject to $g(x, y, z) = xy + 2xz + 3yz - 162 = 0$

Step 2 $F(x, y, z, \lambda) = xyz + \lambda(xy + 2xz + 3yz - 162)$

Step 3
$$F_x = yz + \lambda(y + 2z) = 0$$
$$F_y = xz + \lambda(x + 3z) = 0$$
$$F_z = xy + \lambda(2x + 3y) = 0$$
$$F_\lambda = xy + 2xz + 3yz - 162 = 0$$

From the first two equations, we can write

$$\lambda = \frac{-yz}{y + 2z} \qquad \lambda = \frac{-xz}{x + 3z}$$

Eliminating λ, we have

$$\frac{-yz}{y + 2z} = \frac{-xz}{x + 3z}$$
$$-xyz - 3yz^2 = -xyz - 2xz^2$$
$$3yz^2 = 2xz^2 \qquad \text{We can assume that } z \neq 0.$$
$$3y = 2x$$
$$x = \tfrac{3}{2}y$$

From the second and third equations,

$$\lambda = \frac{-xz}{x + 3z} \qquad \lambda = \frac{-xy}{2x + 3y}$$

Eliminating λ, we have

$$\frac{-xz}{x + 3z} = \frac{-xy}{2x + 3y}$$
$$-2x^2z - 3xyz = -x^2y - 3xyz$$
$$2x^2z = x^2y \qquad \text{We can assume that } x \neq 0.$$
$$2z = y$$
$$z = \tfrac{1}{2}y$$

Substituting $x = \tfrac{3}{2}y$ and $z = \tfrac{1}{2}y$ into the fourth equation, we have

$$\left(\tfrac{3}{2}y\right)y + 2\left(\tfrac{3}{2}y\right)\left(\tfrac{1}{2}y\right) + 3y\left(\tfrac{1}{2}y\right) - 162 = 0$$
$$\tfrac{3}{2}y^2 + \tfrac{3}{2}y^2 + \tfrac{3}{2}y^2 = 162$$
$$y^2 = 36 \quad \text{We can assume that } y > 0.$$
$$y = 6$$
$$x = \tfrac{3}{2}(6) = 9 \quad \text{Using } x = \tfrac{3}{2}y$$
$$z = \tfrac{1}{2}(6) = 3 \quad \text{Using } z = \tfrac{1}{2}y$$

and finally,

$$\lambda = \frac{-(6)(3)}{6 + 2(3)} = -\frac{3}{2} \quad \text{Using } \lambda = \frac{-yz}{y + 2z}$$

The only critical point of F with x, y, and z all positive is $\left(9, 6, 3, -\tfrac{3}{2}\right)$.

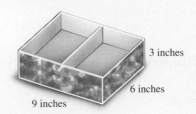

Figure 8

Step 4   The box with the maximum volume has dimensions 9 inches by 6 inches by 3 inches (see Fig. 8).

3 inches

6 inches

9 inches

Matched Problem 3   A box of the same type as in Example 3 is to be constructed from 288 square inches of cardboard. Find the dimensions that result in a box with the largest possible volume.

CONCEPTUAL INSIGHT

An alternative to the method of Lagrange multipliers would be to solve Example 3 by means of Theorem 2 (the second-derivative test for local extrema) in Section 15.3. That approach involves solving the material constraint for one of the variables, say, z:

$$z = \frac{162 - xy}{2x + 3y}$$

Then we would eliminate z in the volume function to obtain a function of two variables:

$$V(x, y) = xy\,\frac{162 - xy}{2x + 3y}$$

The method of Lagrange multipliers allows us to avoid the formidable tasks of calculating the partial derivatives of V and finding the critical points of V in order to apply Theorem 2.

Exercises 15.4

Skills Warm-up Exercises

W   *In Problems 1–6, maximize or minimize subject to the constraint without using the method of Lagrange multipliers; instead, solve the constraint for x or y and substitute into f(x, y). (If necessary, review Section 2.3).*

1. Minimize $f(x, y) = x^2 + xy + y^2$
  subject to $y = 4$

2. Maximize $f(x, y) = 64 + x^2 + 3xy - y^2$
  subject to $x = 6$

3. Minimize $f(x, y) = 4xy$
  subject to $x - y = 2$

4. Maximize $f(x, y) = 3xy$
  subject to $x + y = 1$

5. Maximize $f(x, y) = 2x + y$
  subject to $x^2 + y = 1$

6. Minimize $f(x, y) = 10x - y^2$
  subject to $x^2 + y^2 = 25$

Use the method of Lagrange multipliers in Problems 7–10.

7. Maximize $f(x, y) = 2xy$
  subject to $x + y = 6$

8. Minimize $f(x, y) = 6xy$
  subject to $y - x = 6$

9. Minimize $f(x, y) = x^2 + y^2$
  subject to $3x + 4y = 25$

10. Maximize $f(x, y) = 25 - x^2 - y^2$
  subject to $2x + y = 10$

In Problems 11 and 12, use Theorem 1 to explain why no maxima or minima exist.

11. Minimize $f(x, y) = 4y - 3x$
  subject to $2x + 5y = 3$

12. Maximize $f(x, y) = 6x + 5y + 24$
  subject to $3x + 2y = 4$

Use the method of Lagrange multipliers in Problems 13–22.

13. Find the maximum and minimum of $f(x, y) = 2xy$ subject to $x^2 + y^2 = 18$.

14. Find the maximum and minimum of $f(x, y) = x^2 - y^2$ subject to $x^2 + y^2 = 25$.

15. Maximize the product of two numbers if their sum must be 10.

16. Minimize the product of two numbers if their difference must be 10.

17. Minimize $f(x, y, z) = x^2 + y^2 + z^2$
  subject to $2x - y + 3z = -28$

18. Maximize $f(x, y, z) = xyz$
  subject to $2x + y + 2z = 120$

19. Maximize and Minimize $f(x, y, z) = x + y + z$
  subject to $x^2 + y^2 + z^2 = 12$

20. Maximize and Minimize $f(x, y, z) = 2x + 4y + 4z$
  subject to $x^2 + y^2 + z^2 = 9$

21. Maximize $f(x, y) = y + xy^2$

subject to $x + y^2 = 1$

22. Maximize and Minimize $f(x, y) = x + e^y$

subject to $x^2 + y^2 = 1$

In Problems 23 and 24, use Theorem 1 to explain why no maxima or minima exist.

23. Maximize $f(x, y) = e^x + 3e^y$

subject to $x - 2y = 6$

24. Minimize $f(x, y) = x^3 + 2y^3$

subject to $6x - 2y = 1$

25. Consider the problem of maximizing $f(x, y)$ subject to $g(x, y) = 0$, where $g(x, y) = y - 5$. Explain how the maximization problem can be solved without using the method of Lagrange multipliers.

26. Consider the problem of minimizing $f(x, y)$ subject to $g(x, y) = 0$, where $g(x, y) = 4x - y + 3$. Explain how the minimization problem can be solved without using the method of Lagrange multipliers.

27. Consider the problem of maximizing $f(x, y) = e^{-(x^2 + y^2)}$ subject to the constraint $g(x, y) = x^2 + y - 1 = 0$.

(A) Solve the constraint equation for y, and then substitute into $f(x, y)$ to obtain a function $h(x)$ of the single variable x. Solve the original maximization problem by maximizing h (round answers to three decimal places).

(B) Confirm your answer by the method of Lagrange multipliers.

28. Consider the problem of minimizing

$$f(x, y) = x^2 + 2y^2$$

subject to the constraint $g(x, y) = ye^{x^2} - 1 = 0$.

(A) Solve the constraint equation for y, and then substitute into $f(x, y)$ to obtain a function $h(x)$ of the single variable x. Solve the original minimization problem by minimizing h (round answers to three decimal places).

(B) Confirm your answer by the method of Lagrange multipliers.

Applications

29. Budgeting for least cost. A manufacturing company produces two models of an HDTV per week, x units of model A and y units of model B at a cost (in dollars) of

$$C(x, y) = 6x^2 + 12y^2$$

If it is necessary (because of shipping considerations) that

$$x + y = 90$$

how many of each type of set should be manufactured per week to minimize cost? What is the minimum cost?

30. Budgeting for maximum production. A manufacturing firm has budgeted $60,000 per month for labor and materials.

If $x thousand is spent on labor and $y thousand is spent on materials, and if the monthly output (in units) is given by

$$N(x, y) = 4xy - 8x$$

then how should the $60,000 be allocated to labor and materials in order to maximize N? What is the maximum N?

31. Productivity. A consulting firm for a manufacturing company arrived at the following Cobb–Douglas production function for a particular product:

$$N(x, y) = 50x^{0.8}y^{0.2}$$

In this equation, x is the number of units of labor and y is the number of units of capital required to produce $N(x, y)$ units of the product. Each unit of labor costs $40 and each unit of capital costs $80.

(A) If $400,000 is budgeted for production of the product, determine how that amount should be allocated to maximize production, and find the maximum production.

(B) Find the marginal productivity of money in this case, and estimate the increase in production if an additional $50,000 is budgeted for the production of the product.

32. Productivity. The research department of a manufacturing company arrived at the following Cobb–Douglas production function for a particular product:

$$N(x, y) = 10x^{0.6}y^{0.4}$$

In this equation, x is the number of units of labor and y is the number of units of capital required to produce $N(x, y)$ units of the product. Each unit of labor costs $30 and each unit of capital costs $60.

(A) If $300,000 is budgeted for production of the product, determine how that amount should be allocated to maximize production, and find the maximum production.

(B) Find the marginal productivity of money in this case, and estimate the increase in production if an additional $80,000 is budgeted for the production of the product.

33. Maximum volume. A rectangular box with no top and two intersecting partitions is to be constructed from 192 square inches of cardboard (see the figure). Find the dimensions that will maximize the volume.

34. Maximum volume. A mailing service states that a rectangular package shall have the sum of its length and girth not to exceed 120 inches (see the figure). What are the dimensions of the largest (in volume) mailing carton that can be constructed to meet these restrictions?

Figure for 34

35. Agriculture. Three pens of the same size are to be built along an existing fence (see the figure). If 400 feet of fencing is available, what length should x and y be to produce the maximum total area? What is the maximum area?

36. Diet and minimum cost. A group of guinea pigs is to receive 25,600 calories per week. Two available foods produce $200xy$ calories for a mixture of x kilograms of type M food and y kilograms of type N food. If type M costs \$1 per kilogram and type N costs \$2 per kilogram, how much of each type of food should be used to minimize weekly food costs? What is the minimum cost?

Note: $x \geq 0,\ y \geq 0$

Answers to Matched Problems

1. Max $f(x, y) = f(2, 2) = 17$ (see Fig. 4)
2. (A) 3,750 units of labor and 1,250 units of capital;
      Max $N(x, y) = N(3,750, 1,250) \approx 43,301$ units
   (B) Marginal productivity of money ≈ 0.1443; increase in production $\approx 5,774$ units
3. 12 in. by 8 in. by 4 in.

15.5 Method of Least Squares

- Least Squares Approximation
- Applications

Least Squares Approximation

Regression analysis is the process of fitting an elementary function to a set of data points by the **method of least squares**. The mechanics of using regression techniques were introduced in Chapter 1. Now, using the optimization techniques of Section 15.3, we can develop and explain the mathematical foundation of the method of least squares. We begin with **linear regression**, the process of finding the equation of the line that is the "best" approximation to a set of data points.

Suppose that a manufacturer wants to approximate the cost function for a product. The value of the cost function has been determined for certain levels of production, as listed in Table 1. Although these points do not all lie on a line (see Fig. 1), they are very close to being linear. The manufacturer would like to approximate the cost function by a linear function—that is, determine values a and b so that the line

$$y = ax + b$$

is, in some sense, the "best" approximation to the cost function.

Table 1

Number of Units x (hundreds)	Cost y (thousand \$)
2	4
5	6
6	7
9	8

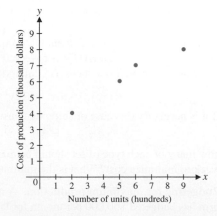

Figure 1

What do we mean by "best"? Since the line $y = ax + b$ will not go through all four points, it is reasonable to examine the differences between the y coordinates of the points listed in the table and the y coordinates of the corresponding points on the line. Each of these differences is called the **residual** at that point (see Fig. 2). For example, at $x = 2$, the point from Table 1 is $(2, 4)$ and the point on the line is $(2, 2a + b)$, so the residual is

$$4 - (2a + b) = 4 - 2a - b$$

All the residuals are listed in Table 2.

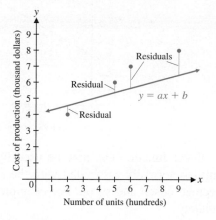

Table 2

x	y	$ax + b$	Residual
2	4	$2a + b$	$4 - 2a - b$
5	6	$5a + b$	$6 - 5a - b$
6	7	$6a + b$	$7 - 6a - b$
9	8	$9a + b$	$8 - 9a - b$

Figure 2

Our criterion for the "best" approximation is the following: Determine the values of a and b that *minimize the sum of the squares* of the residuals. The resulting line is called the **least squares line**, or the **regression line**. To this end, we minimize

$$F(a, b) = (4 - 2a - b)^2 + (6 - 5a - b)^2 + (7 - 6a - b)^2 + (8 - 9a - b)^2$$

Step 1 Find critical points:

$$\begin{aligned} F_a(a, b) &= 2(4 - 2a - b)(-2) + 2(6 - 5a - b)(-5) \\ &\quad + 2(7 - 6a - b)(-6) + 2(8 - 9a - b)(-9) \\ &= -304 + 292a + 44b = 0 \\ F_b(a, b) &= 2(4 - 2a - b)(-1) + 2(6 - 5a - b)(-1) \\ &\quad + 2(7 - 6a - b)(-1) + 2(8 - 9a - b)(-1) \\ &= -50 + 44a + 8b = 0 \end{aligned}$$

After dividing each equation by 2, we solve the system

$$146a + 22b = 152$$
$$22a + 4b = 25$$

obtaining $(a, b) = (0.58, 3.06)$ as the only critical point.

Step 2 Compute $A = F_{aa}(a, b)$, $B = F_{ab}(a, b)$, and $C = F_{bb}(a, b)$:

$$\begin{aligned} F_{aa}(a, b) &= 292; &\text{so,}\quad& A = F_{aa}(0.58, 3.06) = 292 \\ F_{ab}(a, b) &= 44; &\text{so,}\quad& B = F_{ab}(0.58, 3.06) = 44 \\ F_{bb}(a, b) &= 8; &\text{so,}\quad& C = F_{bb}(0.58, 3.06) = 8 \end{aligned}$$

Step 3 Evaluate $AC - B^2$ and try to classify the critical point (a, b) by using Theorem 2 in Section 15.3:

$$AC - B^2 = (292)(8) - (44)^2 = 400 > 0 \quad \text{and} \quad A = 292 > 0$$

Therefore, case 2 in Theorem 2 applies, and $F(a, b)$ has a local minimum at the critical point $(0.58, 3.06)$.

So, the least squares line for the given data is

$$y = 0.58x + 3.06 \qquad \text{Least squares line}$$

The sum of the squares of the residuals is minimized for this choice of a and b (see Fig. 3).

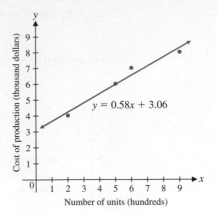

Figure 3

This linear function can now be used by the manufacturer to estimate any of the quantities normally associated with the cost function—such as costs, marginal costs, average costs, and so on. For example, the cost of producing 2,000 units is approximately

$$y = (0.58)(20) + 3.06 = 14.66, \qquad \text{or} \qquad \$14{,}660$$

The marginal cost function is

$$\frac{dy}{dx} = 0.58$$

The average cost function is

$$\bar{y} = \frac{0.58x + 3.06}{x}$$

In general, if we are given a set of n points $(x_1, y_1), (x_2, y_2), \ldots, (x_n, y_n)$, we want to determine the line $y = ax + b$ for which the sum of the squares of the residuals is minimized. Using summation notation, we find that the sum of the squares of the residuals is given by

$$F(a, b) = \sum_{k=1}^{n} (y_k - ax_k - b)^2$$

Note that in this expression the variables are a and b, and the x_k and y_k are all known values. To minimize $F(a, b)$, we thus compute the partial derivatives with respect to a and b and set them equal to 0:

$$F_a(a, b) = \sum_{k=1}^{n} 2(y_k - ax_k - b)(-x_k) = 0$$

$$F_b(a, b) = \sum_{k=1}^{n} 2(y_k - ax_k - b)(-1) = 0$$

Dividing each equation by 2 and simplifying, we see that the coefficients a and b of the least squares line $y = ax + b$ must satisfy the following system of *normal equations*:

$$\left(\sum_{k=1}^{n} x_k^2 \right) a + \left(\sum_{k=1}^{n} x_k \right) b = \sum_{k=1}^{n} x_k y_k$$

$$\left(\sum_{k=1}^{n} x_k \right) a + nb = \sum_{k=1}^{n} y_k$$

Solving this system for a and b produces the formulas given in Theorem 1.

THEOREM 1 Least Squares Approximation

For a set of n points $(x_1, y_1), (x_2, y_2), \dots, (x_n, y_n)$, the coefficients of the least squares line $y = ax + b$ are the solutions of the system of **normal equations**

$$\left(\sum_{k=1}^{n} x_k^2\right)a + \left(\sum_{k=1}^{n} x_k\right)b = \sum_{k=1}^{n} x_k y_k \tag{1}$$

$$\left(\sum_{k=1}^{n} x_k\right)a + nb = \sum_{k=1}^{n} y_k$$

and are given by the formulas

$$a = \frac{n\left(\sum_{k=1}^{n} x_k y_k\right) - \left(\sum_{k=1}^{n} x_k\right)\left(\sum_{k=1}^{n} y_k\right)}{n\left(\sum_{k=1}^{n} x_k^2\right) - \left(\sum_{k=1}^{n} x_k\right)^2} \tag{2}$$

$$b = \frac{\sum_{k=1}^{n} y_k - a\left(\sum_{k=1}^{n} x_k\right)}{n} \tag{3}$$

Now we return to the data in Table 1 and tabulate the sums required for the normal equations and their solution in Table 3.

Table 3

	x_k	y_k	$x_k y_k$	x_k^2
	2	4	8	4
	5	6	30	25
	6	7	42	36
	9	8	72	81
Totals	22	25	152	146

The normal equations (1) are then

$$146a + 22b = 152$$
$$22a + 4b = 25$$

The solution of the normal equations given by equations (2) and (3) is

$$a = \frac{4(152) - (22)(25)}{4(146) - (22)^2} = 0.58$$

$$b = \frac{25 - 0.58(22)}{4} = 3.06$$

Compare these results with step 1 on page 855. Note that Table 3 provides a convenient format for the computation of step 1.

Many graphing calculators have a linear regression feature that solves the system of normal equations obtained by setting the partial derivatives of the sum of squares of the residuals equal to 0. Therefore, in practice, we simply enter the given data points and use the linear regression feature to determine the line $y = ax + b$ that best fits the data (see Fig. 4). There is no need to compute partial derivatives or even to tabulate sums (as in Table 3).

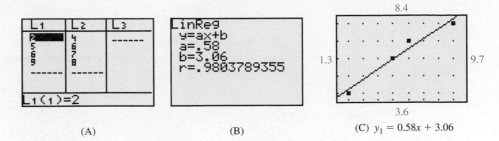

(A)   (B)   (C) $y_1 = 0.58x + 3.06$

Figure 4

Explore and Discuss 1

(A) Plot the four points $(0, 0)$, $(0, 1)$, $(10, 0)$, and $(10, 1)$. Which line would you guess "best" fits these four points? Use formulas (2) and (3) to test your conjecture.

(B) Plot the four points $(0, 0)$, $(0, 10)$, $(1, 0)$ and $(1, 10)$. Which line would you guess "best" fits these four points? Use formulas (2) and (3) to test your conjecture.

(C) If either of your conjectures was wrong, explain how your reasoning was mistaken.

CONCEPTUAL INSIGHT

Formula (2) for a is undefined if the denominator equals 0. When can this happen? Suppose $n = 3$. Then

$$n\left(\sum_{k=1}^{n} x_k^2\right) - \left(\sum_{k=1}^{n} x_k\right)^2$$

$$= 3(x_1^2 + x_2^2 + x_3^2) - (x_1 + x_2 + x_3)^2$$
$$= 3(x_1^2 + x_2^2 + x_3^2) - (x_1^2 + x_2^2 + x_3^2 + 2x_1x_2 + 2x_1x_3 + 2x_2x_3)$$
$$= 2(x_1^2 + x_2^2 + x_3^2) - (2x_1x_2 + 2x_1x_3 + 2x_2x_3)$$
$$= (x_1^2 + x_2^2) + (x_1^2 + x_3^2) + (x_2^2 + x_3^2) - (2x_1x_2 + 2x_1x_3 + 2x_2x_3)$$
$$= (x_1^2 - 2x_1x_2 + x_2^2) + (x_1^2 - 2x_1x_3 + x_3^2) + (x_2^2 - 2x_2x_3 + x_3^2)$$
$$= (x_1 - x_2)^2 + (x_1 - x_3)^2 + (x_2 - x_3)^2$$

and the last expression is equal to 0 if and only if $x_1 = x_2 = x_3$ (i.e., if and only if the three points all lie on the same vertical line). A similar algebraic manipulation works for any integer $n > 1$, showing that, in formula (2) for a, the denominator equals 0 if and only if all n points lie on the same vertical line.

The method of least squares can also be applied to find the quadratic equation $y = ax^2 + bx + c$ that best fits a set of data points. In this case, the sum of the squares of the residuals is a function of three variables:

$$F(a, b, c) = \sum_{k=1}^{n} (y_k - ax_k^2 - bx_k - c)^2$$

There are now three partial derivatives to compute and set equal to 0:

$$F_a(a, b, c) = \sum_{k=1}^{n} 2(y_k - ax_k^2 - bx_k - c)(-x_k^2) = 0$$

$$F_b(a, b, c) = \sum_{k=1}^{n} 2(y_k - ax_k^2 - bx_k - c)(-x_k) = 0$$

$$F_c(a, b, c) = \sum_{k=1}^{n} 2(y_k - ax_k^2 - bx_k - c)(-1) = 0$$

The resulting set of three linear equations in the three variables a, b, and c is called the *set of normal equations for quadratic regression.*

A quadratic regression feature on a calculator is designed to solve such normal equations after the given set of points has been entered. Figure 5 illustrates the computation for the data of Table 1.

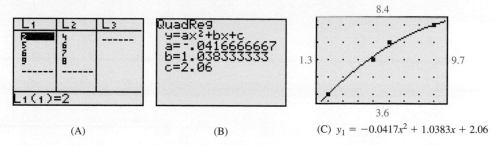

  (A)   (B)   (C) $y_1 = -0.0417x^2 + 1.0383x + 2.06$

Figure 5

Explore and Discuss 2 (A) Use the graphs in Figures 4 and 5 to predict which technique, linear regression or quadratic regression, yields the smaller sum of squares of the residuals for the data of Table 1. Explain.

(B) Confirm your prediction by computing the sum of squares of the residuals in each case.

The method of least squares can also be applied to other regression equations—for example, cubic, quartic, logarithmic, exponential, and power regression models. Details are explored in some of the exercises at the end of this section.

Applications

EXAMPLE 1 Exam Scores Table 4 lists the midterm and final examination scores of 10 students in a calculus course.

Table 4

Midterm	Final	Midterm	Final
49	61	78	77
53	47	83	81
67	72	85	79
71	76	91	93
74	68	99	99

(A) Use formulas (1), (2), and (3) to find the normal equations and the least squares line for the data given in Table 4.

(B) Use the linear regression feature on a graphing calculator to find and graph the least squares line.

(C) Use the least squares line to predict the final examination score of a student who scored 95 on the midterm examination.

SOLUTION

(A) Table 5 shows a convenient way to compute all the sums in the formulas for a and b.

Table 5

x_k	y_k	$x_k y_k$	x_k^2
49	61	2,989	2,401
53	47	2,491	2,809
67	72	4,824	4,489
71	76	5,396	5,041
74	68	5,032	5,476
78	77	6,006	6,084
83	81	6,723	6,889
85	79	6,715	7,225
91	93	8,463	8,281
99	99	9,801	9,801
Totals 750	753	58,440	58,496

From the last line in Table 5, we have

$$\sum_{k=1}^{10} x_k = 750 \qquad \sum_{k=1}^{10} y_k = 753 \qquad \sum_{k=1}^{10} x_k y_k = 58,440 \qquad \sum_{k=1}^{10} x_k^2 = 58,496$$

and the normal equations are

$$58,496a + 750b = 58,440$$
$$750a + 10b = 753$$

Using formulas (2) and (3), we obtain

$$a = \frac{10(58,440) - (750)(753)}{10(58,496) - (750)^2} = \frac{19,650}{22,460} \approx 0.875$$

$$b = \frac{753 - 0.875(750)}{10} = 9.675$$

The least squares line is given (approximately) by

$$y = 0.875x + 9.675$$

(B) We enter the data and use the linear regression feature, as shown in Figure 6. [The discrepancy between values of a and b in the preceding calculations and those in Figure 6B is due to rounding in part (A).]

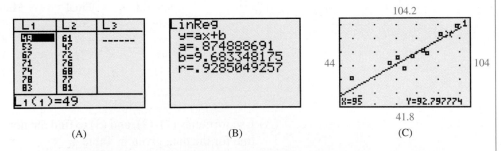

(A)　　　　　　　　(B)　　　　　　　　(C)

Figure 6

(C) If $x = 95$, then $y = 0.875(95) + 9.675 \approx 92.8$ is the predicted score on the final exam. This is also indicated in Figure 6C. If we assume that the exam score must be an integer, then we would predict a score of 93.

Matched Problem 1 Repeat Example 1 for the scores listed in Table 6.

Table 6

Midterm	Final	Midterm	Final
54	50	84	80
60	66	88	95
75	80	89	85
76	68	97	94
78	71	99	86

EXAMPLE 2 Energy Consumption The use of fuel oil for home heating in the United States has declined steadily for several decades. Table 7 lists the percentage of occupied housing units in the United States that were heated by fuel oil for various years between 1960 and 2009. Use the data in the table and linear regression to estimate the percentage of occupied housing units in the United States that were heated by fuel oil in the year 1995.

Table 7 **Occupied Housing Units Heated by Fuel Oil**

Year	Percent	Year	Percent
1960	32.4	1989	13.3
1970	26.0	1999	9.8
1979	19.5	2009	7.3

SOLUTION We enter the data, with $x = 0$ representing 1960, $x = 10$ representing 1970, and so on, and use linear regression as shown in Figure 7.

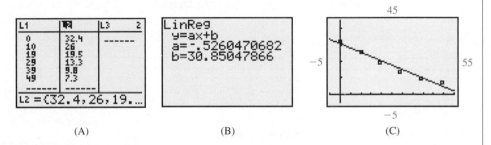

(A)   (B)   (C)

Figure 7

Figure 7 indicates that the least squares line is $y = -0.526x + 30.85$. To estimate the percentage of occupied housing units heated by fuel oil in the year 1995 (corresponding to $x = 35$), we substitute $x = 35$ in the equation of the least squares line: $-0.526(35) + 30.85 = 12.44$. The estimated percentage for 1995 is 12.44%.

Matched Problem 2 In 1950, coal was still a major source of fuel for home energy consumption, and the percentage of occupied housing units heated by fuel oil was only 22.1%. Add the data for 1950 to the data for Example 2, and compute the new least squares line and the new estimate for the percentage of occupied housing units heated by fuel oil in the year 1995. Discuss the discrepancy between the two estimates. (As in Example 2, let $x = 0$ represent 1960.)

Exercises 15.5

Skills Warm-up Exercises

Problems 1–6 refer to the $n = 5$ data points $(x_1, y_1) = (0, 4)$, $(x_2, y_2) = (1, 5)$, $(x_3, y_3) = (2, 7)$, $(x_4, y_4) = (3, 9)$, and $(x_5, y_5) = (4, 13)$. Calculate the indicated sum or product of sums. (If necessary, review Section B.1).

1. $\sum_{k=1}^{5} x_k$ **2.** $\sum_{k=1}^{5} y_k$ **3.** $\sum_{k=1}^{5} x_k y_k$

4. $\sum_{k=1}^{5} x_k^2$ **5.** $\sum_{k=1}^{5} x_k \sum_{k=1}^{5} y_k$ **6.** $\left(\sum_{k=1}^{5} x_k \right)^2$

In Problems 7–12, find the least squares line. Graph the data and the least squares line.

7.

x	y
1	1
2	3
3	4
4	3

8.

x	y
1	-2
2	-1
3	3
4	5

9.

x	y
1	8
2	5
3	4
4	0

10.

x	y
1	20
2	14
3	11
4	3

11.

x	y
1	3
2	4
3	5
4	6

12.

x	y
1	2
2	3
3	3
4	2

In Problems 13–20, find the least squares line and use it to estimate y for the indicated value of x. Round answers to two decimal places.

13.

x	y
1	3
2	1
2	2
3	0

Estimate y when $x = 2.5$.

14.

x	y
1	0
3	1
3	6
3	4

Estimate y when $x = 3$.

15.

x	y
0	10
5	22
10	31
15	46
20	51

Estimate y when $x = 25$.

16.

x	y
-5	60
0	50
5	30
10	20
15	15

Estimate y when $x = 20$.

17.

x	y
-1	14
1	12
3	8
5	6
7	5

Estimate y when $x = 2$.

18.

x	y
2	-4
6	0
10	8
14	12
18	14

Estimate y when $x = 15$.

19.

x	y	x	y
0.5	25	9.5	12
2	22	11	11
3.5	21	12.5	8
5	21	14	5
6.5	18	15.5	1

Estimate y when $x = 8$.

20.

x	y	x	y
0	-15	12	11
2	-9	14	13
4	-7	16	19
6	-7	18	25
8	-1	20	33

Estimate y when $x = 10$.

21. To find the coefficients of the parabola

$$y = ax^2 + bx + c$$

that is the "best" fit to the points $(1, 2)$, $(2, 1)$, $(3, 1)$, and $(4, 3)$, minimize the sum of the squares of the residuals

$$\begin{aligned} F(a, b, c) &= (a + b + c - 2)^2 \\ &+ (4a + 2b + c - 1)^2 \\ &+ (9a + 3b + c - 1)^2 \\ &+ (16a + 4b + c - 3)^2 \end{aligned}$$

by solving the system of normal equations

$$F_a(a, b, c) = 0 \qquad F_b(a, b, c) = 0 \qquad F_c(a, b, c) = 0$$

for a, b, and c. Graph the points and the parabola.

22. Repeat Problem 21 for the points $(-1, -2)$, $(0, 1)$, $(1, 2)$, and $(2, 0)$.

Problems 23 and 24 refer to the system of normal equations and the formulas for a and b given on page 857.

23. Verify formulas (2) and (3) by solving the system of normal equations (1) for a and b.

24. If

$$\bar{x} = \frac{1}{n} \sum_{k=1}^{n} x_k \qquad \text{and} \qquad \bar{y} = \frac{1}{n} \sum_{k=1}^{n} y_k$$

are the averages of the x and y coordinates, respectively, show that the point $(\bar{x}, \bar{y})$ satisfies the equation of the least squares line, $y = ax + b$.

25. (A) Suppose that $n = 5$ and the x coordinates of the data points (x_1, y_1), (x_2, y_2), ..., (x_n, y_n) are $-2, -1, 0, 1, 2$. Show that system (1) in the text implies that

$$a = \frac{\sum x_k y_k}{\sum x_k^2}$$

and that b is equal to the average of the values of y_k.

(B) Show that the conclusion of part (A) holds whenever the average of the x coordinates of the data points is 0.

26. (A) Give an example of a set of six data points such that half of the points lie above the least squares line and half lie below.

(B) Give an example of a set of six data points such that just one of the points lies above the least squares line and five lie below.

27. (A) Find the linear and quadratic functions that best fit the data points $(0, 1.3)$, $(1, 0.6)$, $(2, 1.5)$, $(3, 3.6)$, and $(4, 7.4)$. Round coefficients to two decimal places.

(B) Which of the two functions best fits the data? Explain.

28. (A) Find the linear, quadratic, and logarithmic functions that best fit the data points $(1, 3.2)$, $(2, 4.2)$, $(3, 4.7)$, $(4, 5.0)$, and $(5, 5.3)$. (Round coefficients to two decimal places.)

(B) Which of the three functions best fits the data? Explain.

29. Describe the normal equations for cubic regression. How many equations are there? What are the variables? What techniques could be used to solve the equations?

30. Describe the normal equations for quartic regression. How many equations are there? What are the variables? What techniques could be used to solve the equations?

Applications

31. Crime rate. Data on U.S. property crimes (in number of crimes per 100,000 population) are given in the table for the years 2001 through 2011.

U.S. Property Crime Rates

Year	Rate
2001	3,658
2003	3,591
2005	3,431
2007	3,276
2009	3,041
2011	2,908

(A) Find the least squares line for the data, using $x = 0$ for 2000.

(B) Use the least squares line to predict the property crime rate in 2024.

32. Cable TV revenue. Data for cable TV revenue are given in the table for the years 2002 through 2010.

Cable TV Revenue

Year	Millions of dollars
2002	47,989
2004	58,586
2006	71,887
2008	85,232
2010	93,368

(A) Find the least squares line for the data, using $x = 0$ for 2000.

(B) Use the least squares line to predict cable TV revenue in 2025.

33. Maximizing profit. The market research department for a drugstore chain chose two summer resort areas to test market a new sunscreen lotion packaged in 4-ounce plastic bottles. After a summer of varying the selling price and recording the monthly demand, the research department arrived at the following demand table, where y is the number of bottles purchased per month (in thousands) at x dollars per bottle:

x	y
5.0	2.0
5.5	1.8
6.0	1.4
6.5	1.2
7.0	1.1

(A) Use the method of least squares to find a demand equation.

(B) If each bottle of sunscreen costs the drugstore chain $4, how should the sunscreen be priced to achieve a maximum monthly profit? [*Hint:* Use the result of part (A), with $C = 4y$, $R = xy$, and $P = R - C$.]

34. Maximizing profit. A market research consultant for a supermarket chain chose a large city to test market a new brand of mixed nuts packaged in 8-ounce cans. After a year of varying the selling price and recording the monthly demand, the consultant arrived at the following demand table, where y is the number of cans purchased per month (in thousands) at x dollars per can:

x	y
4.0	4.2
4.5	3.5
5.0	2.7
5.5	1.5
6.0	0.7

(A) Use the method of least squares to find a demand equation.

(B) If each can of nuts costs the supermarket chain $3, how should the nuts be priced to achieve a maximum monthly profit?

35. Olympic Games. The table gives the winning heights in the pole vault in the Olympic Games from 1980 to 2012.

Olympic Pole Vault Winning Height

Year	Height (ft)
1980	18.96
1984	18.85
1988	19.35
1992	19.02
1996	19.42
2000	19.35
2004	19.52
2008	19.56
2012	19.59

(A) Use a graphing calculator to find the least squares line for the data, letting $x = 0$ for 1980.

(B) Estimate the winning height in the pole vault in the Olympic Games of 2024.

36. Biology. In biology, there is an approximate rule, called the *bioclimatic rule for temperate climates*. This rule states that in spring and early summer, periodic phenomena such as the blossoming of flowers, the appearance of insects, and the ripening of fruit usually come about 4 days later for each 500 feet of altitude. Stated as a formula, the rule becomes

$$d = 8h \qquad 0 \le h \le 4$$

where d is the change in days and h is the altitude (in thousands of feet). To test this rule, an experiment was set up to record the difference in blossoming times of the same type of apple tree at different altitudes. A summary of the results is given in the following table:

h	d
0	0
1	7
2	18
3	28
4	33

(A) Use the method of least squares to find a linear equation relating h and d. Does the bioclimatic rule $d = 8h$ appear to be approximately correct?

(B) How much longer will it take this type of apple tree to blossom at 3.5 thousand feet than at sea level? [Use the linear equation found in part (A).]

37. **Global warming.** Average global temperatures from 1885 to 2005 are given in the table.

Average Global Temperatures

Year	°F	Year	°F
1885	56.65	1955	57.06
1895	56.64	1965	57.05
1905	56.52	1975	57.04
1915	56.57	1985	57.36
1925	56.74	1995	57.64
1935	57.00	2005	58.59
1945	57.13		

(A) Find the least squares line for the data, using $x = 0$ for 1885.

(B) Use the least squares line to estimate the average global temperature in 2085.

38. **Organic Farming.** The table gives the number of acres of certified organic farmland in the United States from 2000 to 2008.

Certified Organic Farmland in the United States

Year	Acres
2000	1,776,000
2002	1,926,000
2004	3,045,000
2006	2,936,000
2008	4,816,000

(A) Find the least squares line for the data, using $x = 0$ for the year 2000.

(B) Use the least squares line to estimate the number of acres of certified organic farmland in the United States in 2023.

Answers to Matched Problems

1. (A) $y = 0.85x + 9.47$

   (B)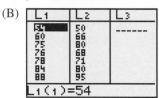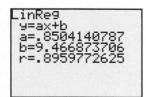

   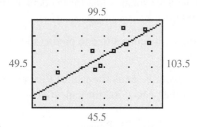

   (C) 90.3

2. $y = -0.37x + 25.86$; 12.83%

15.6 Double Integrals over Rectangular Regions

- Introduction
- Definition of the Double Integral
- Average Value over Rectangular Regions
- Volume and Double Integrals

Introduction

We have generalized the concept of differentiation to functions with two or more independent variables. How can we do the same with integration, and how can we interpret the results? Let's look first at the operation of antidifferentiation. We can antidifferentiate a function of two or more variables with respect to one of the variables by treating all the other variables as though they were constants. Thus, this operation is the reverse operation of partial differentiation, just as ordinary antidifferentiation is the reverse operation of ordinary differentiation. We write $\int f(x, y) \, dx$ to indicate that we are to antidifferentiate $f(x, y)$ with respect to x, holding y fixed; we write $\int f(x, y) \, dy$ to indicate that we are to antidifferentiate $f(x, y)$ with respect to y, holding x fixed.

EXAMPLE 1 Partial Antidifferentiation Evaluate

(A) $\displaystyle\int (6xy^2 + 3x^2)\, dy$ (B) $\displaystyle\int (6xy^2 + 3x^2)\, dx$

SOLUTION

(A) Treating x as a constant and using the properties of antidifferentiation from Section 5.1, we have

$$\int (6xy^2 + 3x^2)\, dy = \int 6xy^2\, dy + \int 3x^2\, dy$$

The dy tells us that we are looking for the antiderivative of $6xy^2 + 3x^2$ with respect to y only, holding x constant.

$$= 6x\int y^2\, dy + 3x^2\int dy$$

$$= 6x\left(\frac{y^3}{3}\right) + 3x^2(y) + C(x)$$

$$= 2xy^3 + 3x^2y + C(x)$$

Note that the constant of integration can be *any function of x alone* since for any such function,

$$\frac{\partial}{\partial y}C(x) = 0$$

Check:

We can verify that our answer is correct by using partial differentiation:

$$\frac{\partial}{\partial y}\left[2xy^3 + 3x^2y + C(x)\right] = 6xy^2 + 3x^2 + 0$$

$$= 6xy^2 + 3x^2$$

(B) We treat y as a constant:

$$\int (6xy^2 + 3x^2)\, dx = \int 6xy^2\, dx + \int 3x^2\, dx$$

$$= 6y^2\int x\, dx + 3\int x^2\, dx$$

$$= 6y^2\left(\frac{x^2}{2}\right) + 3\left(\frac{x^3}{3}\right) + E(y)$$

$$= 3x^2y^2 + x^3 + E(y)$$

The antiderivative contains an arbitrary function $E(y)$ of y alone.

Check:

$$\frac{\partial}{\partial x}\left[3x^2y^2 + x^3 + E(y)\right] = 6xy^2 + 3x^2 + 0$$

$$= 6xy^2 + 3x^2$$

Matched Problem 1 Evaluate

(A) $\displaystyle\int (4xy + 12x^2y^3)\, dy$ (B) $\displaystyle\int (4xy + 12x^2y^3)\, dx$

Now that we have extended the concept of antidifferentiation to functions with two variables, we also can evaluate definite integrals of the form

$$\int_a^b f(x, y)\, dx \quad\text{or}\quad \int_c^d f(x, y)\, dy$$

EXAMPLE 2   Evaluating a Partial Antiderivative   Evaluate, substituting the limits of integration in y if dy is used and in x if dx is used:

(A) $\displaystyle\int_0^2 (6xy^2 + 3x^2)\, dy$ (B) $\displaystyle\int_0^1 (6xy^2 + 3x^2)\, dx$

SOLUTION

(A) From Example 1A, we know that

$$\int (6xy^2 + 3x^2)\, dy = 2xy^3 + 3x^2y + C(x)$$

According to properties of the definite integral for a function of one variable, we can use any antiderivative to evaluate the definite integral. Thus, choosing $C(x) = 0$, we have

$$\int_0^2 (6xy^2 + 3x^2)\, dy = (2xy^3 + 3x^2y)\Big|_{y=0}^{y=2}$$
$$= [2x(2)^3 + 3x^2(2)] - [2x(0)^3 + 3x^2(0)]$$
$$= 16x + 6x^2$$

(B) From Example 1B, we know that

$$\int (6xy^2 + 3x^2)\, dx = 3x^2y^2 + x^3 + E(y)$$

Choosing $E(y) = 0$, we have

$$\int_0^1 (6xy^2 + 3x^2)\, dx = (3x^2y^2 + x^3)\Big|_{x=0}^{x=1}$$
$$= [3y^2(1)^2 + (1)^3] - [3y^2(0)^2 + (0)^3]$$
$$= 3y^2 + 1$$

Matched Problem 2   Evaluate

(A) $\displaystyle\int_0^1 (4xy + 12x^2y^3)\, dy$ (B) $\displaystyle\int_0^3 (4xy + 12x^2y^3)\, dx$

Integrating and evaluating a definite integral with integrand $f(x, y)$ with respect to y produces a function of x alone (or a constant). Likewise, integrating and evaluating a definite integral with integrand $f(x, y)$ with respect to x produces a function of y alone (or a constant). Each of these results, involving at most one variable, can now be used as an integrand in a second definite integral.

EXAMPLE 3   Evaluating Integrals   Evaluate

(A) $\displaystyle\int_0^1 \left[\int_0^2 (6xy^2 + 3x^2)\, dy \right] dx$

(B) $\displaystyle\int_0^2 \left[\int_0^1 (6xy^2 + 3x^2)\, dx \right] dy$

SOLUTION

(A) Example 2A showed that

$$\int_0^2 (6xy^2 + 3x^2)\, dy = 16x + 6x^2$$

Therefore,

$$\int_0^1 \left[\int_0^2 (6xy^2 + 3x^2)\, dy \right] dx = \int_0^1 (16x + 6x^2)\, dx$$

$$= (8x^2 + 2x^3)\big|_{x=0}^{x=1}$$

$$= [8(1)^2 + 2(1)^3] - [8(0)^2 + 2(0)^3] = 10$$

(B) Example 2B showed that

$$\int_0^1 (6xy^2 + 3x^2)\, dx = 3y^2 + 1$$

Therefore,

$$\int_0^2 \left[\int_0^1 (6xy^2 + 3x^2)\, dx \right] dy = \int_0^2 (3y^2 + 1)\, dy$$

$$= (y^3 + y)\big|_{y=0}^{y=2}$$

$$= [(2)^3 + 2] - [(0)^3 + 0] = 10$$

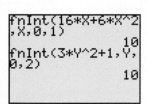

Figure 1

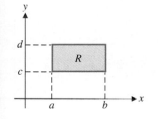

A numerical integration command can be used as an alternative to the fundamental theorem of calculus to evaluate the last integrals in Examples 3A and 3B, $\int_0^1 (16x + 6x^2)\, dx$ and $\int_0^2 (3y^2 + 1)\, dy$, since the integrand in each case is a function of a single variable (see Fig. 1).

Matched Problem 3 Evaluate

(A) $\displaystyle\int_0^3 \left[\int_0^1 (4xy + 12x^2y^3)\, dy \right] dx$ (B) $\displaystyle\int_0^1 \left[\int_0^3 (4xy + 12x^2y^3)\, dx \right] dy$

Definition of the Double Integral

Notice that the answers in Examples 3A and 3B are identical. This is not an accident. In fact, it is this property that enables us to define the *double integral,* as follows:

> **DEFINITION Double Integral**
> The **double integral** of a function $f(x, y)$ over a rectangle
> $$R = \{ (x, y)\,|\,a \le x \le b, c \le y \le d \}$$
> is
> $$\iint_R f(x, y)\, dA = \int_a^b \left[\int_c^d f(x, y)\, dy \right] dx$$
> $$= \int_c^d \left[\int_a^b f(x, y)\, dx \right] dy$$

In the double integral $\iint_R f(x, y)\, dA, f(x, y)$ is called the **integrand**, and R is called the **region of integration**. The expression dA indicates that this is an integral over a two-dimensional region. The integrals

$$\int_a^b \left[\int_c^d f(x, y)\, dy \right] dx \qquad \text{and} \qquad \int_c^d \left[\int_a^b f(x, y)\, dx \right] dy$$

are referred to as **iterated integrals** (the brackets are often omitted), and the order in which dx and dy are written indicates the order of integration. This is not the most general definition of the double integral over a rectangular region; however, it is equivalent to the general definition for all the functions we will consider.

EXAMPLE 4   Evaluating a Double Integral   Evaluate

$$\iint\limits_R (x + y)\, dA \qquad \text{over} \qquad R = \{(x, y)\,|\,1 \le x \le 3, \quad -1 \le y \le 2\}$$

SOLUTION   Region R is illustrated in Figure 2. We can choose either order of iteration. As a check, we will evaluate the integral both ways:

$$\iint\limits_R (x + y)\, dA = \int_1^3 \int_{-1}^2 (x + y)\, dy\, dx$$

$$= \int_1^3 \left[\left(xy + \frac{y^2}{2} \right) \Big|_{y=-1}^{y=2} \right] dx$$

$$= \int_1^3 \left[(2x + 2) - \left(-x + \tfrac{1}{2} \right) \right] dx$$

$$= \int_1^3 \left(3x + \tfrac{3}{2} \right) dx$$

$$= \left(\tfrac{3}{2} x^2 + \tfrac{3}{2} x \right) \Big|_{x=1}^{x=3}$$

$$= \left(\tfrac{27}{2} + \tfrac{9}{2} \right) - \left(\tfrac{3}{2} + \tfrac{3}{2} \right) = 18 - 3 = 15$$

$$\iint\limits_R (x + y)\, dA = \int_{-1}^2 \int_1^3 (x + y)\, dx\, dy$$

$$= \int_{-1}^2 \left[\left(\frac{x^2}{2} + xy \right) \Big|_{x=1}^{x=3} \right] dy$$

$$= \int_{-1}^2 \left[\left(\tfrac{9}{2} + 3y \right) - \left(\tfrac{1}{2} + y \right) \right] dy$$

$$= \int_{-1}^2 (4 + 2y)\, dy$$

$$= (4y + y^2) \Big|_{y=-1}^{y=2}$$

$$= (8 + 4) - (-4 + 1) = 12 - (-3) = 15$$

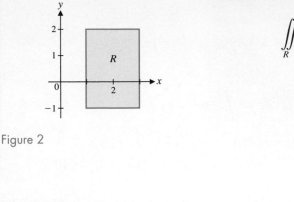

Figure 2

Matched Problem 4   Evaluate

$$\iint\limits_R (2x - y)\, dA \qquad \text{over} \qquad R = \{(x, y)\,|\,-1 \le x \le 5, \quad 2 \le y \le 4\}$$

both ways.

EXAMPLE 5   Double Integral of an Exponential Function   Evaluate

$$\iint\limits_R 2x e^{x^2 + y}\, dA \qquad \text{over} \qquad R = \{(x, y)\,|\,0 \le x \le 1, \quad -1 \le y \le 1\}$$

SOLUTION   Region R is illustrated in Figure 3.

$$\iint\limits_R 2x e^{x^2 + y}\, dA = \int_{-1}^1 \int_0^1 2x e^{x^2 + y}\, dx\, dy$$

$$= \int_{-1}^1 \left[(e^{x^2 + y}) \Big|_{x=0}^{x=1} \right] dy$$

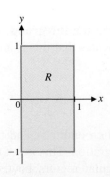

Figure 3

$$= \int_{-1}^{1} (e^{1+y} - e^y)\, dy$$

$$= (e^{1+y} - e^y)|_{y=-1}^{y=1}$$

$$= (e^2 - e) - (e^0 - e^{-1})$$

$$= e^2 - e - 1 + e^{-1}$$

Matched Problem 5 | Evaluate

$$\iint_R \frac{x}{y^2} e^{x/y}\, dA \quad \text{over} \quad R = \{(x, y)\,|\,0 \leq x \leq 1, \;\; 1 \leq y \leq 2\}.$$

Average Value over Rectangular Regions

In Section 5.5, the average value of a function $f(x)$ over an interval $[a, b]$ was defined as

$$\frac{1}{b - a} \int_a^b f(x)\, dx$$

This definition is easily extended to functions of two variables over rectangular regions as follows (notice that the denominator $(b - a)(d - c)$ is simply the area of the rectangle R):

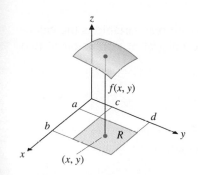

> **DEFINITION** Average Value over Rectangular Regions
> The **average value** of the function $f(x, y)$ over the rectangle
>
> $$R = \{(x, y)\,|\,a \leq x \leq b, \;\; c \leq y \leq d\}$$
>
> is
>
> $$\frac{1}{(b - a)(d - c)} \iint_R f(x, y)\, dA$$

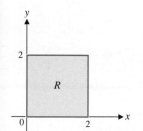

Figure 4

EXAMPLE 6 Average Value Find the average value of $f(x, y) = 4 - \frac{1}{2}x - \frac{1}{2}y$ over the rectangle $R = \{(x, y)\,|\,0 \leq x \leq 2, \;\; 0 \leq y \leq 2\}$.

SOLUTION Region R is illustrated in Figure 4. We have

$$\frac{1}{(b - a)(d - c)} \iint_R f(x, y)\, dA = \frac{1}{(2 - 0)(2 - 0)} \iint_R \left(4 - \frac{1}{2}x - \frac{1}{2}y\right) dA$$

$$= \tfrac{1}{4} \int_0^2 \int_0^2 \left(4 - \tfrac{1}{2}x - \tfrac{1}{2}y\right) dy\, dx$$

$$= \tfrac{1}{4} \int_0^2 \left[\left(4y - \tfrac{1}{2}xy - \tfrac{1}{4}y^2\right)\Big|_{y=0}^{y=2}\right] dx$$

$$= \tfrac{1}{4} \int_0^2 (7 - x)\, dx$$

$$= \tfrac{1}{4}\left(7x - \tfrac{1}{2}x^2\right)|_{x=0}^{x=2}$$

$$= \tfrac{1}{4}(12) = 3$$

Figure 5 illustrates the surface $z = f(x, y)$, and our calculations show that 3 is the average of the z values over the region R.

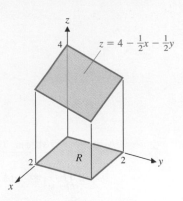

Figure 5

Matched Problem 6) Find the average value of $f(x, y) = x + 2y$ over the rectangle
$R = \{(x, y) \,|\, 0 \leq x \leq 2, \quad 0 \leq y \leq 1\}$

Explore and Discuss 1   (A) Which of the functions $f(x, y) = 4 - x^2 - y^2$ and $g(x, y) = 4 - x - y$
would you guess has the greater average value over the rectangle
$R = \{(x, y) \,|\, 0 \leq x \leq 1, \quad 0 \leq y \leq 1\}$? Explain.

(B) Use double integrals to check the correctness of your guess in part (A).

Volume and Double Integrals

One application of the definite integral of a function with one variable is the calculation of areas, so it is not surprising that the definite integral of a function of two variables can be used to calculate volumes of solids.

THEOREM 1   Volume under a Surface

If $f(x, y) \geq 0$ over a rectangle $R = \{(x, y) \,|\, a \leq x \leq b, \quad c \leq y \leq d\}$, then the
volume of the solid formed by graphing f over the rectangle R is given by

$$V = \iint_R f(x, y)\, dA$$

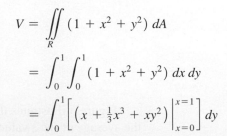

EXAMPLE 7   Volume   Find the volume of the solid under the graph of $f(x, y) = 1 + x^2 + y^2$ over the rectangle $R = \{(x, y) \,|\, 0 \leq x \leq 1, \quad 0 \leq y \leq 1\}$.

SOLUTION   Figure 6 shows the region R, and Figure 7 illustrates the volume under
consideration.

$$V = \iint_R (1 + x^2 + y^2)\, dA$$

$$= \int_0^1 \int_0^1 (1 + x^2 + y^2)\, dx\, dy$$

$$= \int_0^1 \left[\left(x + \tfrac{1}{3}x^3 + xy^2 \right) \Big|_{x=0}^{x=1} \right] dy$$

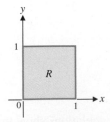

Figure 6

$$= \int_0^1 \left(\tfrac{4}{3} + y^2 \right) dy$$

$$= \left(\tfrac{4}{3} y + \tfrac{1}{3} y^3 \right) \big|_{y=0}^{y=1} = \tfrac{5}{3} \text{ cubic units}$$

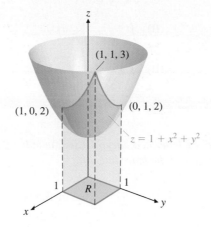

Figure 7

Matched Problem 7 Find the volume of the solid under the graph of $f(x, y) = 1 + x + y$ over the rectangle $R = \{(x, y) \,|\, 0 \le x \le 1, \quad 0 \le y \le 2\}$.

CONCEPTUAL INSIGHT

Double integrals can be defined over regions that are more general than rectangles. For example, let $R > 0$. Then the function $f(x, y) = \sqrt{R^2 - (x^2 + y^2)}$ can be integrated over the circular region $C = \{(x, y) \,|\, x^2 + y^2 \le R^2\}$. In fact, it can be shown that

$$\iint\limits_C \sqrt{R^2 - (x^2 + y^2)} \, dx \, dy = \frac{2\pi R^3}{3}$$

Because $x^2 + y^2 + z^2 = R^2$ is the equation of a sphere of radius R centered at the origin, the double integral over C represents the volume of the upper hemisphere. Therefore, the volume of a sphere of radius R is given by

$$V = \frac{4\pi R^3}{3} \qquad \text{Volume of sphere of radius } R$$

Double integrals can also be used to obtain volume formulas for other geometric figures (see Table 1, Appendix C).

Exercises 15.6

Skills Warm-up Exercises

In Problems 1–6, find each antiderivative. (If necessary, review Sections 13.1 and 13.2).

1. $\displaystyle \int (\pi + x) \, dx$

2. $\displaystyle \int (x\pi^2 + \pi x^2) \, dx$

3. $\displaystyle \int \left(1 + \frac{\pi}{x} \right) dx$

4. $\displaystyle \int \left(1 + \frac{x}{\pi} \right) dx$

5. $\displaystyle \int e^{\pi x} \, dx$

6. $\displaystyle \int \frac{\ln x}{\pi x} \, dx$

In Problems 7–14, find each antiderivative. Then use the antiderivative to evaluate the definite integral.

7. (A) $\displaystyle \int 12x^2 y^3 \, dy$ (B) $\displaystyle \int_0^1 12x^2 y^3 \, dy$

8. (A) $\displaystyle \int 12x^2 y^3 \, dx$ (B) $\displaystyle \int_{-1}^2 12x^2 y^3 \, dx$

9. (A) $\displaystyle \int (4x + 6y + 5) \, dx$ (B) $\displaystyle \int_{-2}^3 (4x + 6y + 5) \, dx$

10. (A) $\displaystyle\int (4x + 6y + 5)\, dy$ **(B)** $\displaystyle\int_1^4 (4x + 6y + 5)\, dy$

11. (A) $\displaystyle\int \frac{x}{\sqrt{y + x^2}}\, dx$ **(B)** $\displaystyle\int_0^2 \frac{x}{\sqrt{y + x^2}}\, dx$

12. (A) $\displaystyle\int \frac{x}{\sqrt{y + x^2}}\, dy$ **(B)** $\displaystyle\int_1^5 \frac{x}{\sqrt{y + x^2}}\, dy$

13. (A) $\displaystyle\int \frac{\ln x}{xy}\, dy$ **(B)** $\displaystyle\int_1^{e^2} \frac{\ln x}{xy}\, dy$

14. (A) $\displaystyle\int \frac{\ln x}{xy}\, dx$ **(B)** $\displaystyle\int_1^e \frac{\ln x}{xy}\, dx$

In Problems 15–22, evaluate each iterated integral. (See the indicated problem for the evaluation of the inner integral.)

15. $\displaystyle\int_{-1}^2 \int_0^1 12x^2 y^3 \, dy\, dx$

(See Problem 7.)

16. $\displaystyle\int_0^1 \int_{-1}^2 12x^2 y^3 \, dx\, dy$

(See Problem 8.)

17. $\displaystyle\int_1^4 \int_{-2}^3 (4x + 6y + 5)\, dx\, dy$

(See Problem 9.)

18. $\displaystyle\int_{-2}^3 \int_1^4 (4x + 6y + 5)\, dy\, dx$

(See Problem 10.)

19. $\displaystyle\int_1^5 \int_0^2 \frac{x}{\sqrt{y + x^2}}\, dx\, dy$

(See Problem 11.)

20. $\displaystyle\int_0^2 \int_1^5 \frac{x}{\sqrt{y + x^2}}\, dy\, dx$

(See Problem 12.)

21. $\displaystyle\int_1^e \int_1^{e^2} \frac{\ln x}{xy}\, dy\, dx$

(See Problem 13.)

22. $\displaystyle\int_1^{e^2} \int_1^e \frac{\ln x}{xy}\, dx\, dy$

(See Problem 14.)

Use both orders of iteration to evaluate each double integral in Problems 23–26.

23. $\displaystyle\iint_R xy\, dA; R = \{(x, y)\,|\,0 \le x \le 2, \ 0 \le y \le 4\}$

24. $\displaystyle\iint_R \sqrt{xy}\, dA; R = \{(x, y)\,|\,1 \le x \le 4, \ 1 \le y \le 9\}$

25. $\displaystyle\iint_R (x + y)^5\, dA; R = \{(x, y)\,|\,-1 \le x \le 1, \ 1 \le y \le 2\}$

26. $\displaystyle\iint_R xe^y\, dA; R = \{(x, y)\,|\,-2 \le x \le 3, \ 0 \le y \le 2\}$

In Problems 27–30, find the average value of each function over the given rectangle.

27. $f(x, y) = (x + y)^2$;

$R = \{(x, y)\,|\,1 \le x \le 5, \ -1 \le y \le 1\}$

28. $f(x, y) = x^2 + y^2$;

$R = \{(x, y)\,|\,-1 \le x \le 2, \ 1 \le y \le 4\}$

29. $f(x, y) = x/y; R = \{(x, y)\,|\,1 \le x \le 4, \ 2 \le y \le 7\}$

30. $f(x, y) = x^2 y^3; R = \{(x, y)\,|\,-1 \le x \le 1, \ 0 \le y \le 2\}$

In Problems 31–34, find the volume of the solid under the graph of each function over the given rectangle.

31. $f(x, y) = 2 - x^2 - y^2$;

$R = \{(x, y)\,|\,0 \le x \le 1, \ 0 \le y \le 1\}$

32. $f(x, y) = 5 - x$;

$R = \{(x, y)\,|\,0 \le x \le 5, \ 0 \le y \le 5\}$

33. $f(x, y) = 4 - y^2; R = \{(x, y)\,|\,0 \le x \le 2, \ 0 \le y \le 2\}$

34. $f(x, y) = e^{-x-y}; R = \{(x, y)\,|\,0 \le x \le 1, \ 0 \le y \le 1\}$

Evaluate each double integral in Problems 35–38. Select the order of integration carefully; each problem is easy to do one way and difficult the other.

35. $\displaystyle\iint_R xe^{xy}\, dA; R = \{(x, y)\,|\,0 \le x \le 1, \ 1 \le y \le 2\}$

36. $\displaystyle\iint_R xye^{x^2 y}\, dA; R = \{(x, y)\,|\,0 \le x \le 1, \ 1 \le y \le 2\}$

37. $\displaystyle\iint_R \frac{2y + 3xy^2}{1 + x^2}\, dA$;

$R = \{(x, y)\,|\,0 \le x \le 1, \ -1 \le y \le 1\}$

38. $\displaystyle\iint_R \frac{2x + 2y}{1 + 4y + y^2}\, dA$;

$R = \{(x, y)\,|\,1 \le x \le 3, \ 0 \le y \le 1\}$

39. Show that $\int_0^2 \int_0^2 (1 - y)\, dx\, dy = 0$. Does the double integral represent the volume of a solid? Explain.

40. (A) Find the average values of the functions $f(x, y) = x + y$, $g(x, y) = x^2 + y^2$, and $h(x, y) = x^3 + y^3$ over the rectangle

$R = \{(x, y)\,|\,0 \le x \le 1, \ 0 \le y \le 1\}$

(B) Does the average value of $k(x, y) = x^n + y^n$ over the rectangle

$R_1 = \{(x, y)\,|\,0 \le x \le 1, \ 0 \le y \le 1\}$

increase or decrease as n increases? Explain.

(C) Does the average value of $k(x, y) = x^n + y^n$ over the rectangle

$R_2 = \{(x, y)\,|\,0 \le x \le 2, \ 0 \le y \le 2\}$

increase or decrease as n increases? Explain.

41. Let $f(x, y) = x^3 + y^2 - e^{-x} - 1$.

   (A) Find the average value of $f(x, y)$ over the rectangle

   $$R = \{(x, y) \mid -2 \le x \le 2, \quad -2 \le y \le 2\}.$$

   (B) Graph the set of all points (x, y) in R for which $f(x, y) = 0$.

   (C) For which points (x, y) in R is $f(x, y)$ greater than 0? Less than 0? Explain.

42. Find the dimensions of the square S centered at the origin for which the average value of $f(x, y) = x^2 e^y$ over S is equal to 100.

Applications

43. Multiplier principle. Suppose that Congress enacts a one-time-only 10% tax rebate that is expected to infuse \y billion, $5 \le y \le 7$, into the economy. If every person and every corporation is expected to spend a proportion x, $0.6 \le x \le 0.8$, of each dollar received, then, by the **multiplier principle** in economics, the total amount of spending S (in billions of dollars) generated by this tax rebate is given by

$$S(x, y) = \frac{y}{1 - x}$$

What is the average total amount of spending for the indicated ranges of the values of x and y? Set up a double integral and evaluate it.

44. Multiplier principle. Repeat Problem 43 if $6 \le y \le 10$ and $0.7 \le x \le 0.9$.

45. Cobb–Douglas production function. If an industry invests x thousand labor-hours, $10 \le x \le 20$, and \y million, $1 \le y \le 2$, in the production of N thousand units of a certain item, then N is given by

$$N(x, y) = x^{0.75}y^{0.25}$$

What is the average number of units produced for the indicated ranges of x and y? Set up a double integral and evaluate it.

46. Cobb–Douglas production function. Repeat Problem 45 for

$$N(x, y) = x^{0.5}y^{0.5}$$

where $10 \le x \le 30$ and $1 \le y \le 3$.

47. Population distribution. In order to study the population distribution of a certain species of insect, a biologist has constructed an artificial habitat in the shape of a rectangle 16 feet long and 12 feet wide. The only food available to the insects in this habitat is located at its center. The biologist has determined that the concentration C of insects per square foot at a point d units from the food supply (see the figure) is given approximately by

$$C = 10 - \tfrac{1}{10}d^2$$

What is the average concentration of insects throughout the habitat? Express C as a function of x and y, set up a double integral, and evaluate it.

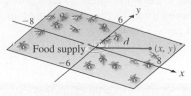

Figure for 47

48. Population distribution. Repeat Problem 47 for a square habitat that measures 12 feet on each side, where the insect concentration is given by

$$C = 8 - \tfrac{1}{10}d^2$$

49. Pollution. A heavy industrial plant located in the center of a small town emits particulate matter into the atmosphere. Suppose that the concentration of particulate matter (in parts per million) at a point d miles from the plant (see the figure) is given by

$$C = 100 - 15d^2$$

If the boundaries of the town form a rectangle 4 miles long and 2 miles wide, what is the average concentration of particulate matter throughout the town? Express C as a function of x and y, set up a double integral, and evaluate it.

50. Pollution. Repeat Problem 49 if the boundaries of the town form a rectangle 8 miles long and 4 miles wide and the concentration of particulate matter is given by

$$C = 100 - 3d^2$$

51. Safety research. Under ideal conditions, if a person driving a car slams on the brakes and skids to a stop, the length of the skid marks (in feet) is given by the formula

$$L = 0.000\ 013\ 3xy^2$$

where x is the weight of the car (in pounds) and y is the speed of the car (in miles per hour). What is the average length of the skid marks for cars weighing between 2,000 and 3,000 pounds and traveling at speeds between 50 and 60 miles per hour? Set up a double integral and evaluate it.

52. Safety research. Repeat Problem 51 for cars weighing between 2,000 and 2,500 pounds and traveling at speeds between 40 and 50 miles per hour.

53. Psychology. The intelligence quotient Q for a person with mental age x and chronological age y is given by

$$Q(x, y) = 100\frac{x}{y}$$

In a group of sixth-graders, the mental age varies between 8 and 16 years and the chronological age varies between

10 and 12 years. What is the average intelligence quotient for this group? Set up a double integral and evaluate it.

54. Psychology. Repeat Problem 53 for a group with mental ages between 6 and 14 years and chronological ages between 8 and 10 years.

15.7 Double Integrals over More General Regions

- Regular Regions
- Double Integrals over Regular Regions
- Reversing the Order of Integration
- Volume and Double Integrals

In this section, we extend the concept of double integration discussed in Section 15.6 to nonrectangular regions. We begin with an example and some new terminology.

Regular Regions

Let R be the region graphed in Figure 1. We can describe R with the following inequalities:

$$R = \{(x, y) \mid x \le y \le 6x - x^2, \quad 0 \le x \le 5\}$$

The region R can be viewed as a union of vertical line segments. For each x in the interval $[0, 5]$, the line segment from the point $(x, g(x))$ to the point $(x, f(x))$ lies in the region R. Any region that can be covered by vertical line segments in this manner is called a *regular x region*.

Now consider the region S in Figure 2. It can be described with the following inequalities:

$$S = \{(x, y) \mid y^2 \le x \le y + 2, \quad -1 \le y \le 2\}$$

The region S can be viewed as a union of horizontal line segments going from the graph of $h(y) = y^2$ to the graph of $k(y) = y + 2$ on the interval $[-1, 2]$. Regions that can be described in this manner are called *regular y regions*.

In general, *regular regions* are defined as follows:

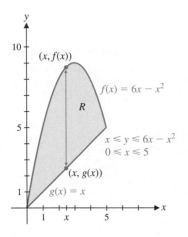

Figure 1

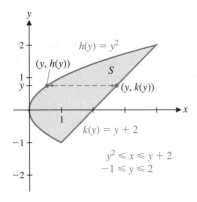

Figure 2

> **DEFINITION   Regular Regions**
>
> A region R in the xy plane is a **regular x region** if there exist functions $f(x)$ and $g(x)$ and numbers a and b such that
>
> $$R = \{(x, y) \mid g(x) \le y \le f(x), \quad a \le x \le b\}$$
>
> A region R in the xy plane is a **regular y region** if there exist functions $h(y)$ and $k(y)$ and numbers c and d such that
>
> $$R = \{(x, y) \mid h(y) \le x \le k(y), \quad c \le y \le d\}$$
>
> See Figure 3 for a geometric interpretation.

CONCEPTUAL **INSIGHT**

If, for some region R, there is a horizontal line that has a nonempty intersection I with R, and if I is neither a closed interval nor a point, then R is *not* a regular y region. Similarly, if, for some region R, there is a vertical line that has a nonempty intersection I with R, and if I is neither a closed interval nor a point, then R is *not* a regular x region (see Fig. 3).

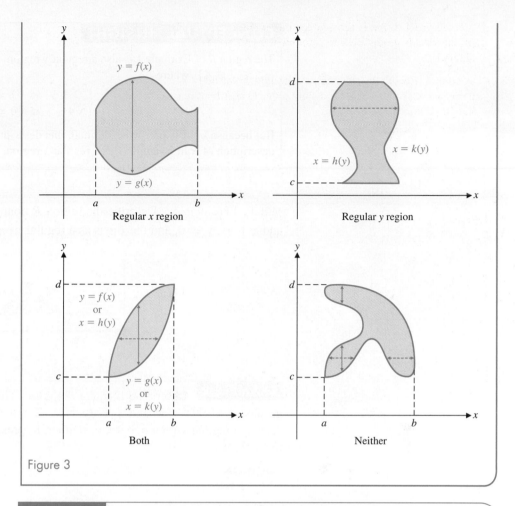

Figure 3

EXAMPLE 1 Describing a Regular x Region The region R is bounded by the graphs of $y = 4 - x^2$ and $y = x - 2$, $x \geq 0$, and the y axis. Graph R and use set notation with double inequalities to describe R as a regular x region.

SOLUTION As the solid line in the following figure indicates, R can be covered by vertical line segments that go from the graph of $y = x - 2$ to the graph of $y = 4 - x^2$. So, R is a regular x region. In terms of set notation with double inequalities, we can write

$$R = \{(x, y) \mid x - 2 \leq y \leq 4 - x^2, \ 0 \leq x \leq 2\}$$

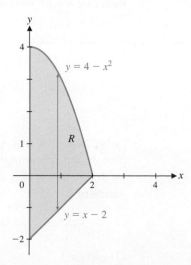

CONCEPTUAL INSIGHT

The region R of Example 1 is also a regular y region, since $R = \{(x, y) \,|\, 0 \le x \le k(y),\ -2 \le y \le 4\}$, where

$$k(y) = \begin{cases} 2 + y & \text{if } -2 \le y \le 0 \\ \sqrt{4 - y} & \text{if } 0 \le y \le 4 \end{cases}$$

But because $k(y)$ is piecewise defined, this description is more complicated than the description of R in Example 1 as a regular x region.

Matched Problem 1 Describe the region R bounded by the graphs of $x = 6 - y$ and $x = y^2$, $y \ge 0$, and the x axis as a regular y region.

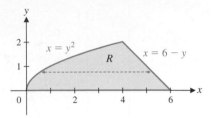

EXAMPLE 2 Describing Regular Regions The region R is bounded by the graphs of $x + y^2 = 9$ and $x + 3y = 9$. Graph R and describe R as a regular x region, a regular y region, both, or neither. Represent R in set notation with double inequalities.

SOLUTION

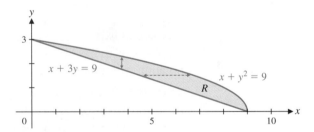

Region R can be covered by vertical line segments that go from the graph of $x + 3y = 9$ to the graph of $x + y^2 = 9$. Thus, R is a regular x region. In order to describe R with inequalities, we must solve each equation for y in terms of x:

$$x + 3y = 9 \qquad\qquad x + y^2 = 9$$
$$3y = 9 - x \qquad\qquad y^2 = 9 - x$$
$$y = 3 - \tfrac{1}{3}x \qquad\qquad y = \sqrt{9 - x}$$

We use the positive square root, since the graph is in the first quadrant.

So,

$$R = \{(x, y) \,|\, 3 - \tfrac{1}{3}x \le y \le \sqrt{9 - x},\ \ 0 \le x \le 9\}$$

Since region R also can be covered by horizontal line segments (see the dashed line in the preceding figure) that go from the graph of $x + 3y = 9$ to the graph of $x + y^2 = 9$, it is a regular y region. Now we must solve each equation for x in terms of y:

$$x + 3y = 9 \qquad\qquad x + y^2 = 9$$
$$x = 9 - 3y \qquad\qquad x = 9 - y^2$$

Therefore,

$$R = \{(x, y) \,|\, 9 - 3y \le x \le 9 - y^2,\ \ 0 \le y \le 3\}$$

Matched Problem 2 ⎤ Repeat Example 2 for the region bounded by the graphs of $2y - x = 4$ and $y^2 - x = 4$, as shown in the following figure:

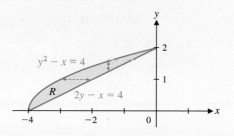

Explore and Discuss 1

A E I O U

Consider the vowels A, E, I, O, U, written in block letters as shown in the margin, to be regions of the plane. One of the vowels is a regular x region, but not a regular y region; one is a regular y region, but not a regular x region; one is both; two are neither. Explain.

Double Integrals over Regular Regions

Now we want to extend the definition of double integration to include regular x regions and regular y regions. The order of integration now depends on the nature of the region R. If R is a regular x region, we integrate with respect to y first, while if R is a regular y region, we integrate with respect to x first.

> Note that the variable limits of integration (when present) are always on the inner integral, and the constant limits of integration are always on the outer integral.

DEFINITION Double Integration over Regular Regions

Regular x Region

If $R = \{(x, y) \,|\, g(x) \le y \le f(x), \quad a \le x \le b\}$, then

$$\iint_R F(x, y)\, dA = \int_a^b \left[\int_{g(x)}^{f(x)} F(x, y)\, dy \right] dx$$

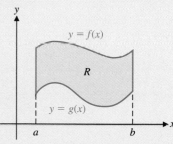

Regular y Region

If $R = \{(x, y) \,|\, h(y) \le x \le k(y), \quad c \le y \le d\}$, then

$$\iint_R F(x, y)\, dA = \int_c^d \left[\int_{h(y)}^{k(y)} F(x, y)\, dx \right] dy$$

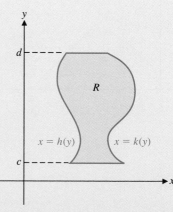

EXAMPLE 3 Evaluating a Double Integral Evaluate $\iint\limits_{R} 2xy\, dA$, where R is the region bounded by the graphs of $y = -x$ and $y = x^2$, $x \geq 0$, and the graph of $x = 1$.

SOLUTION From the graph, we can see that R is a regular x region described by

$$R = \{(x, y)\,|\, -x \leq y \leq x^2,\ \ 0 \leq x \leq 1\}$$

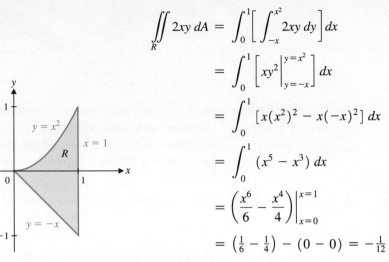

$$\iint\limits_{R} 2xy\, dA = \int_0^1 \left[\int_{-x}^{x^2} 2xy\, dy \right] dx$$

$$= \int_0^1 \left[xy^2 \Big|_{y=-x}^{y=x^2} \right] dx$$

$$= \int_0^1 \left[x(x^2)^2 - x(-x)^2 \right] dx$$

$$= \int_0^1 (x^5 - x^3)\, dx$$

$$= \left(\frac{x^6}{6} - \frac{x^4}{4} \right) \Big|_{x=0}^{x=1}$$

$$= \left(\tfrac{1}{6} - \tfrac{1}{4} \right) - (0 - 0) = -\tfrac{1}{12}$$

Matched Problem 3 Evaluate $\iint\limits_{R} 3xy^2\, dA$, where R is the region in Example 3.

EXAMPLE 4 Evaluating a Double Integral Evaluate $\iint\limits_{R} (2x + y)\, dA$, where R is the region bounded by the graphs of $y = \sqrt{x}$, $x + y = 2$, and $y = 0$.

SOLUTION From the graph, we can see that R is a regular y region. After solving each equation for x, we can write

$$R = \{(x, y)\,|\, y^2 \leq x \leq 2 - y,\ \ 0 \leq y \leq 1\}$$

$$\iint\limits_{R} (2x + y)\, dA = \int_0^1 \left[\int_{y^2}^{2-y} (2x + y)\, dx \right] dy$$

$$= \int_0^1 \left[(x^2 + yx) \Big|_{x=y^2}^{x=2-y} \right] dy$$

$$= \int_0^1 \{ [(2 - y)^2 + y(2 - y)] - [(y^2)^2 + y(y^2)] \}\, dy$$

$$= \int_0^1 (4 - 2y - y^3 - y^4)\, dy$$

$$= \left(4y - y^2 - \tfrac{1}{4}y^4 - \tfrac{1}{5}y^5 \right) \Big|_{y=0}^{y=1}$$

$$= \left(4 - 1 - \tfrac{1}{4} - \tfrac{1}{5} \right) - 0 = \tfrac{51}{20}$$

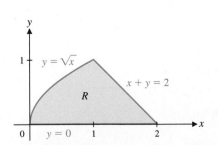

Matched Problem 4 Evaluate $\iint\limits_{R} (y - 4x)\, dA$, where R is the region in Example 4.

EXAMPLE 5 Evaluating a Double Integral The region R is bounded by the graphs of $y = \sqrt{x}$ and $y = \frac{1}{2}x$. Evaluate $\iint\limits_R 4xy^3 dA$ two different ways.

SOLUTION Region R is both a regular x region and a regular y region:

$$R = \{(x, y) \mid \tfrac{1}{2}x \le y \le \sqrt{x}, \ 0 \le x \le 4\} \quad \text{Regular } x \text{ region}$$
$$R = \{(x, y) \mid y^2 \le x \le 2y, \ 0 \le y \le 2\} \quad \text{Regular } y \text{ region}$$

Using the first representation (a regular x region), we obtain

$$\iint\limits_R 4xy^3 \, dA = \int_0^4 \left[\int_{x/2}^{\sqrt{x}} 4xy^3 \, dy \right] dx$$

$$= \int_0^4 \left[xy^4 \Big|_{y=x/2}^{y=\sqrt{x}} \right] dx$$

$$= \int_0^4 \left[x(\sqrt{x})^4 - x\left(\tfrac{1}{2}x\right)^4 \right] dx$$

$$= \int_0^4 \left(x^3 - \tfrac{1}{16}x^5 \right) dx$$

$$= \left(\tfrac{1}{4}x^4 - \tfrac{1}{96}x^6 \right) \Big|_{x=0}^{x=4}$$

$$= \left(64 - \tfrac{128}{3} \right) - 0 = \tfrac{64}{3}$$

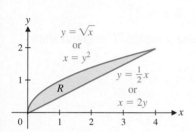

Using the second representation (a regular y region), we obtain

$$\iint\limits_R 4xy^3 \, dA = \int_0^2 \left[\int_{y^2}^{2y} 4xy^3 \, dx \right] dy$$

$$= \int_0^2 \left[2x^2y^3 \Big|_{x=y^2}^{x=2y} \right] dy$$

$$= \int_0^2 \left[2(2y)^2y^3 - 2(y^2)^2y^3 \right] dy$$

$$= \int_0^2 \left(8y^5 - 2y^7 \right) dy$$

$$= \left(\tfrac{4}{3}y^6 - \tfrac{1}{4}y^8 \right) \Big|_{y=0}^{y=2}$$

$$= \left(\tfrac{256}{3} - 64 \right) - 0 = \tfrac{64}{3}$$

Matched Problem 5 The region R is bounded by the graphs of $y = x$ and $y = \frac{1}{2}x^2$. Evaluate $\iint\limits_R 4xy^3 dA$ two different ways.

Reversing the Order of Integration

Example 5 shows that

$$\iint\limits_R 4xy^3 \, dA = \int_0^4 \left[\int_{x/2}^{\sqrt{x}} 4xy^3 \, dy \right] dx = \int_0^2 \left[\int_{y^2}^{2y} 4xy^3 \, dx \right] dy$$

In general, if R is both a regular x region and a regular y region, then the two iterated integrals are equal. In rectangular regions, reversing the order of integration in an iterated integral was a simple matter. As Example 5 illustrates, the process is more complicated in nonrectangular regions. The next example illustrates how to start with an iterated integral and reverse the order of integration. Since we are interested in the reversal process and not in the value of either integral, the integrand will not be specified.

EXAMPLE 6 Reversing the Order of Integration Reverse the order of integration in

$$\int_{1}^{3}\left[\int_{0}^{x-1} f(x, y)\, dy\right] dx$$

SOLUTION The order of integration indicates that the region of integration is a regular x region:

$$R = \{(x, y)\,|\, 0 \le y \le x - 1,\ \ 1 \le x \le 3\}$$

Graph region R to determine whether it is also a regular y region. The graph shows that R is also a regular y region, and we can write

$$R = \{(x, y)\,|\, y + 1 \le x \le 3,\ \ 0 \le y \le 2\}$$

$$\int_{1}^{3}\left[\int_{0}^{x-1} f(x, y)\, dy\right] dx = \int_{0}^{2}\left[\int_{y+1}^{3} f(x, y)\, dx\right] dy$$

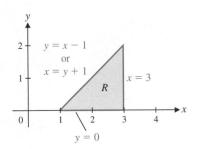

Matched Problem 6 Reverse the order of integration in $\int_{2}^{4}\left[\int_{0}^{4-x} f(x, y)\, dy\right] dx$.

Explore and Discuss 2 Explain the difficulty in evaluating $\int_{0}^{2}\int_{x^2}^{4} x e^{y^2}\, dy\, dx$ and how it can be overcome by reversing the order of integration.

Volume and Double Integrals

In Section 15.6, we used the double integral to calculate the volume of a solid with a rectangular base. In general, if a solid can be described by the graph of a positive function $f(x, y)$ over a regular region R (not necessarily a rectangle), then the double integral of the function f over the region R still represents the volume of the corresponding solid.

EXAMPLE 7 Volume The region R is bounded by the graphs of $x + y = 1$, $y = 0$, and $x = 0$. Find the volume of the solid under the graph of $z = 1 - x - y$ over the region R.

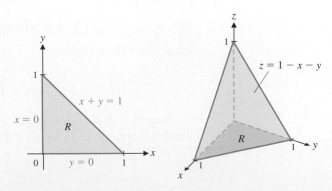

SOLUTION The graph of R shows that R is both a regular x region and a regular y region. We choose to use the regular x region:

$$R = \{(x, y)|0 \le y \le 1 - x, \; 0 \le x \le 1\}$$

The volume of the solid is

$$V = \iint\limits_{R} (1 - x - y) \, dA = \int_0^1 \left[\int_0^{1-x} (1 - x - y) \, dy \right] dx$$

$$= \int_0^1 \left[\left(y - xy - \tfrac{1}{2}y^2 \right) \Big|_{y=0}^{y=1-x} \right] dx$$

$$= \int_0^1 \left[(1 - x) - x(1 - x) - \tfrac{1}{2}(1 - x)^2 \right] dx$$

$$= \int_0^1 \left(\tfrac{1}{2} - x + \tfrac{1}{2}x^2 \right) dx$$

$$= \left(\tfrac{1}{2}x - \tfrac{1}{2}x^2 + \tfrac{1}{6}x^3 \right) \Big|_{x=0}^{x=1}$$

$$= \left(\tfrac{1}{2} - \tfrac{1}{2} + \tfrac{1}{6} \right) - 0 = \tfrac{1}{6}$$

Matched Problem 7 The region R is bounded by the graphs of $y + 2x = 2$, $y = 0$, and $x = 0$. Find the volume of the solid under the graph of $z = 2 - 2x - y$ over the region R. [*Hint:* Sketch the region first; the solid does not have to be sketched.]

Exercises 15.7

In Problems 1–6, graph the region R bounded by the graphs of the equations. Use set notation and double inequalities to describe R as a regular x region and a regular y region in Problems 1 and 2, and as a regular x region or a regular y region, whichever is simpler, in Problems 3–6.

1. $y = 4 - x^2, y = 0, 0 \le x \le 2$

2. $y = x^2, y = 9, 0 \le x \le 3$

3. $y = x^3, y = 12 - 2x, x = 0$

4. $y = 5 - x, y = 1 + x, y = 0$

5. $y^2 = 2x, y = x - 4$

6. $y = 4 + 3x - x^2, x + y = 4$

Evaluate each integral in Problems 7–10.

7. $\int_0^1 \int_0^x (x + y) \, dy \, dx$

8. $\int_0^2 \int_0^y xy \, dx \, dy$

9. $\int_0^1 \int_{y^3}^{\sqrt{y}} (2x + y) \, dx \, dy$

10. $\int_1^4 \int_x^{x^2} (x^2 + 2y) \, dy \, dx$

In Problems 11–14, give a verbal description of the region R and determine whether R is a regular x region, a regular y region, both, or neither.

11. $R = \{(x, y)\,|\,|x| \le 2, \; |y| \le 3\}$

12. $R = \{(x, y)\,|\,1 \le x^2 + y^2 \le 4\}$

13. $R = \{(x, y)\,|\,x^2 + y^2 \ge 1, \; |x| \le 2, \; 0 \le y \le 2\}$

14. $R = \{(x, y)\,|\,|x| + |y| \le 1\}$

In Problems 15–20, use the description of the region R to evaluate the indicated integral.

15. $\iint\limits_{R} (x^2 + y^2) \, dA$;

$R = \{(x, y)|0 \le y \le 2x, \; 0 \le x \le 2\}$

16. $\iint\limits_{R} 2x^2 y \, dA$;

$R = \{(x, y)|0 \le y \le 9 - x^2, \; -3 \le x \le 3\}$

17. $\iint\limits_{R} (x + y - 2)^3 \, dA$;

$R = \{(x, y)|0 \le x \le y + 2, \; 0 \le y \le 1\}$

18. $\iint\limits_{R} (2x + 3y) \, dA$;

$R = \{(x, y)|y^2 - 4 \le x \le 4 - 2y, \; 0 \le y \le 2\}$

19. $\iint\limits_{R} e^{x+y} \, dA$;

$R = \{(x, y)|-x \le y \le x, \; 0 \le x \le 2\}$

20. $\iint\limits_{R} \dfrac{x}{\sqrt{x^2 + y^2}} \, dA$;

$R = \{(x, y)|0 \le x \le \sqrt{4y - y^2}, \; 0 \le y \le 2\}$

In Problems 21–26, graph the region R bounded by the graphs of the indicated equations. Describe R in set notation with double inequalities, and evaluate the indicated integral.

21. $y = x + 1, y = 0, x = 0, x = 1;$ $\iint\limits_R \sqrt{1 + x + y}\, dA$

22. $y = x^2,\ y = \sqrt{x};$ $\iint\limits_R 12xy\, dA$

23. $y = 4x - x^2,\ y = 0;$ $\iint\limits_R \sqrt{y + x^2}\, dA$

24. $x = 1 + 3y,\ x = 1 - y,\ y = 1;$ $\iint\limits_R (x + y + 1)^3\, dA$

25. $y = 1 - \sqrt{x}, y = 1 + \sqrt{x}, x = 4;$ $\iint\limits_R x(y - 1)^2\, dA$

26. $y = \tfrac{1}{2}x, y = 6 - x,\ y = 1;$ $\iint\limits_R \dfrac{1}{x + y}\, dA$

In Problems 27–32, evaluate each integral. Graph the region of integration, reverse the order of integration, and then evaluate the integral with the order reversed.

27. $\int_0^3 \int_0^{3-x} (x + 2y)\, dy\, dx$

28. $\int_0^2 \int_0^y (y - x)^4\, dx\, dy$

29. $\int_0^1 \int_0^{1-x^2} x\sqrt{y}\, dy\, dx$

30. $\int_0^2 \int_{x^3}^{4x} (1 + 2y)\, dy\, dx$

31. $\int_0^4 \int_{x/4}^{\sqrt{x}/2} x\, dy\, dx$

32. $\int_0^4 \int_{y^2/4}^{2\sqrt{y}} (1 + 2xy)\, dx\, dy$

In Problems 33–36, find the volume of the solid under the graph of $f(x, y)$ over the region R bounded by the graphs of the indicated equations. Sketch the region R; the solid does not have to be sketched.

33. $f(x, y) = 4 - x - y$; R is the region bounded by the graphs of $x + y = 4, y = 0, x = 0$

34. $f(x, y) = (x - y)^2$; R is the region bounded by the graphs of $y = x, y = 2, x = 0$

35. $f(x, y) = 4$; R is the region bounded by the graphs of $y = 1 - x^2$ and $y = 0$ for $0 \le x \le 1$

36. $f(x, y) = 4xy$; R is the region bounded by the graphs of $y = \sqrt{1 - x^2}$ and $y = 0$ for $0 \le x \le 1$

In Problems 37–40, reverse the order of integration for each integral. Evaluate the integral with the order reversed. Do not attempt to evaluate the integral in the original form.

37. $\int_0^2 \int_{x^2}^4 \dfrac{4x}{1 + y^2}\, dy\, dx$

38. $\int_0^1 \int_y^1 \sqrt{1 - x^2}\, dx\, dy$

39. $\int_0^1 \int_{y^2}^1 4ye^{x^2}\, dx\, dy$

40. $\int_0^4 \int_{\sqrt{x}}^2 \sqrt{3x + y^2}\, dy\, dx$

In Problems 41–46, use a graphing calculator to graph the region R bounded by the graphs of the indicated equations. Use approximation techniques to find intersection points correct to two decimal places. Describe R in set notation with double inequalities, and evaluate the indicated integral correct to two decimal places.

41. $y = 1 + \sqrt{x},\quad y = x^2,\quad x = 0;$ $\iint\limits_R x\, dA$

42. $y = 1 + \sqrt[3]{x}, y = x, x = 0;$ $\iint\limits_R x\, dA$

43. $y = \sqrt[3]{x}, y = 1 - x, y = 0;$ $\iint\limits_R 24xy\, dA$

44. $y = x^3, y = 1 - x, y = 0;$ $\iint\limits_R 48xy\, dA$

45. $y = e^{-x},\ y = 3 - x;$ $\iint\limits_R 4y\, dA$

46. $y = e^x,\ y = 2 + x;$ $\iint\limits_R 8y\, dA$

Answers to Matched Problems

1. $R = \{(x, y)\,|\,y^2 \le x \le 6 - y,\ 0 \le y \le 2\}$

2. R is both a regular x region and a regular y region:
$R = \{(x, y)\,|\tfrac{1}{2}x + 2 \le y \le \sqrt{x + 4},\ -4 \le x \le 0\}$
$R = \{(x, y)\,|\,y^2 - 4 \le x \le 2y - 4,\ 0 \le y \le 2\}$

3. $\tfrac{13}{40}$ **4.** $-\tfrac{77}{20}$ **5.** $\tfrac{64}{15}$

6. $\int_0^2 \int_2^{4-y} f(x, y)\, dx\, dy$

7. $\tfrac{2}{3}$

Chapter 15   Summary and Review

Important Terms, Symbols, and Concepts

15.1 Functions of Several Variables

- An equation of the form $z = f(x, y)$ describes a **function of two independent variables** if, for each permissible ordered pair (x, y), there is one and only one value of z determined by $f(x, y)$. The variables x and y are **independent variables**, and z is a **dependent variable**. The set of all ordered pairs of permissible values of x and y is the **domain** of the function, and the set of all corresponding values $f(x, y)$ is the **range**. Functions of more than two independent variables are defined similarly.

- The graph of $z = f(x, y)$ consists of all triples (x, y, z) in a **three-dimensional coordinate system** that satisfy the equation. The graphs of the functions $z = f(x, y) = x^2 + y^2$ and $z = g(x, y) = x^2 - y^2$, for example, are **surfaces**; the first has a local minimum, and the second has a **saddle point**, at $(0, 0)$.

EXAMPLES

Ex. 1, p. 819
Ex. 2, p. 820
Ex. 3, p. 820
Ex. 4, p. 821
Ex. 5, p. 822
Ex. 6, p. 822
Ex. 7, p. 824

15.2 Partial Derivatives

- If $z = f(x, y)$, then the **partial derivative of f with respect to x**, denoted as $\partial z/\partial x$, f_x, or $f_x(x, y)$, is

$$\frac{\partial z}{\partial x} = \lim_{h \to 0} \frac{f(x + h, y) - f(x, y)}{h}$$

Ex. 1, p. 829
Ex. 2, p. 829
Ex. 3, p. 830
Ex. 4, p. 830

Similarly, the **partial derivative of f with respect to y**, denoted as $\partial z/\partial y$, f_y, or $f_y(x, y)$, is

$$\frac{\partial z}{\partial y} = \lim_{k \to 0} \frac{f(x, y + k) - f(x, y)}{k}$$

The partial derivatives $\partial z/\partial x$ and $\partial z/\partial y$ are said to be **first-order partial derivatives**.

- There are four **second-order partial derivatives** of $z = f(x, y)$:

Ex. 5, p. 832

$$f_{xx} = f_{xx}(x, y) = \frac{\partial^2 z}{\partial x^2} = \frac{\partial}{\partial x}\left(\frac{\partial z}{\partial x}\right)$$

$$f_{xy} = f_{xy}(x, y) = \frac{\partial^2 z}{\partial y\, \partial x} = \frac{\partial}{\partial y}\left(\frac{\partial z}{\partial x}\right)$$

$$f_{yx} = f_{yx}(x, y) = \frac{\partial^2 z}{\partial x\, \partial y} = \frac{\partial}{\partial x}\left(\frac{\partial z}{\partial y}\right)$$

$$f_{yy} = f_{yy}(x, y) = \frac{\partial^2 z}{\partial y^2} = \frac{\partial}{\partial y}\left(\frac{\partial z}{\partial y}\right)$$

15.3 Maxima and Minima

- If $f(a, b) \geq f(x, y)$ for all (x, y) in a circular region in the domain of f with (a, b) as center, then $f(a, b)$ is a **local maximum**. If $f(a, b) \leq f(x, y)$ for all (x, y) in such a region, then $f(a, b)$ is a **local minimum**.

Ex. 1, p. 839
Ex. 2, p. 839
Ex. 3, p. 841
Ex. 4, p. 842

- If a function $f(x, y)$ has a local maximum or minimum at the point (a, b), and f_x and f_y exist at (a, b), then both first-order partial derivatives equal 0 at (a, b) [Theorem 1, p. 838].

- The second-derivative test for local extrema (Theorem 2, p. 838) gives conditions on the first- and second-order partial derivatives of $f(x, y)$, which guarantee that $f(a, b)$ is a local maximum, local minimum, or saddle point.

15.4 Maxima and Minima Using Lagrange Multipliers

- The **method of Lagrange multipliers** can be used to find local extrema of a function $z = f(x, y)$ subject to the constraint $g(x, y) = 0$. A procedure that lists the key steps in the method is given on page 847.

Ex. 1, p. 847
Ex. 2, p. 848
Ex. 3, p. 850

- The method of Lagrange multipliers can be extended to functions with arbitrarily many independent variables with one or more constraints (see Theorem 1, p. 846, and Theorem 2, p. 850, for the method when there are two and three independent variables, respectively).

15.5 Method of Least Squares

- **Linear regression** is the process of fitting a line $y = ax + b$ to a set of data points $(x_1, y_1), (x_2, y_2), \ldots, (x_n, y_n)$ by using the **method of least squares**.

Ex. 1, p. 859

- We minimize $F(a, b) = \sum_{k=1}^{n} (y_k - ax_k - b)^2$, the **sum of the squares of the residuals**, by computing the first-order partial derivatives of F and setting them equal to 0. Solving for a and b gives the formulas

$$a = \frac{n\left(\sum_{k=1}^{n} x_k y_k\right) - \left(\sum_{k=1}^{n} x_k\right)\left(\sum_{k=1}^{n} y_k\right)}{n\left(\sum_{k=1}^{n} x_k^2\right) - \left(\sum_{k=1}^{n} x_k\right)^2}$$

$$b = \frac{\sum_{k=1}^{n} y_k - a\left(\sum_{k=1}^{n} x_k\right)}{n}$$

- Graphing calculators have built-in routines to calculate linear—as well as quadratic, cubic, quartic, logarithmic, exponential, power, and trigonometric—regression equations.

Ex. 2, p. 861

15.6 Double Integrals over Rectangular Regions

Ex. 1, p. 865
Ex. 2, p. 866
Ex. 3, p. 866
Ex. 4, p. 868
Ex. 5, p. 868

- The **double integral** of a function $f(x, y)$ over a rectangle

$$R = \{(x, y) \,|\, a \leq x \leq b, \quad c \leq y \leq d\}$$

is

$$\iint\limits_{R} f(x, y)\, dA = \int_{a}^{b}\left[\int_{c}^{d} f(x, y)\, dy\right] dx$$

$$= \int_{c}^{d}\left[\int_{a}^{b} f(x, y)\, dx\right] dy$$

- In the double integral $\iint_{R} f(x, y)\, dA, f(x, y)$ is called the **integrand** and R is called the **region of integration.** The expression dA indicates that this is an integral over a two-dimensional region. The integrals

$$\int_{a}^{b}\left[\int_{c}^{d} f(x, y)\, dy\right] dx \quad \text{and} \quad \int_{c}^{d}\left[\int_{a}^{b} f(x, y)\, dx\right] dy$$

are referred to as **iterated integrals** (the brackets are often omitted), and the order in which dx and dy are written indicates the order of integration.

- The **average value** of the function $f(x, y)$ over the rectangle

Ex. 6, p. 869

$$R = \{(x, y) \,|\, a \leq x \leq b, \quad c \leq y \leq d\}$$

is

$$\frac{1}{(b - a)(d - c)} \iint\limits_{R} f(x, y)\, dA$$

- If $f(x, y) \geq 0$ over a rectangle $R = \{(x, y) \,|\, a \leq x \leq b, c \leq y \leq d\}$, then the volume of the solid formed by graphing f over the rectangle R is given by

Ex. 7, p. 870

$$V = \iint\limits_{R} f(x, y)\, dA$$

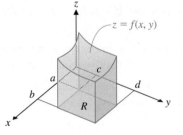

15.7 Double Integrals over More General Regions

- A region R in the xy plane is a **regular x region** if there exist functions $f(x)$ and $g(x)$ and numbers a and b such that

Ex. 1, p. 875

$$R = \{(x, y) \,|\, g(x) \leq y \leq f(x), \quad a \leq x \leq b\}$$

- A region R in the xy plane is a **regular y region** if there exist functions $h(y)$ and $k(y)$ and numbers c and d such that

Ex. 2, p. 876

$$R = \{(x, y) \,|\, h(y) \leq x \leq k(y), \quad c \leq y \leq d\}$$

- The double integral of a function $F(x, y)$ over a regular x region $R = \{(x, y) \,|\, g(x) \leq y \leq f(x), a \leq x \leq b\}$ is

Ex. 3, p. 878
Ex. 4, p. 878
Ex. 5, p. 879
Ex. 6, p. 880
Ex. 7, p. 880

$$\iint\limits_{R} F(x, y)\, dA = \int_{a}^{b}\left[\int_{g(x)}^{f(x)} F(x, y)\, dy\right] dx$$

- The double integral of a function $F(x, y)$ over a regular y region $R = \{(x, y) \,|\, h(y) \leq x \leq k(y), c \leq y \leq d\}$ is

$$\iint\limits_{R} F(x, y)\, dA = \int_{c}^{d}\left[\int_{h(y)}^{k(y)} F(x, y)\, dx\right] dy$$

Review Exercises

Work through all the problems in this chapter review and check your answers in the back of the book. Answers to all review problems are there, along with section numbers in italics to indicate where each type of problem is discussed. Where weaknesses show up, review appropriate sections of the text.

1. For $f(x, y) = 2{,}000 + 40x + 70y$, find $f(5, 10), f_x(x, y)$, and $f_y(x, y)$.

2. For $z = x^3 y^2$, find $\partial^2 z / \partial x^2$ and $\partial^2 z / \partial x\, \partial y$.

3. Evaluate $\int (6xy^2 + 4y)\, dy$.

4. Evaluate $\int (6xy^2 + 4y)\, dx$.

5. Evaluate $\int_0^1 \int_0^1 4xy\, dy\, dx$.

6. For $f(x, y) = 6 + 5x - 2y + 3x^2 + x^3$, find $f_x(x, y)$, and $f_y(x, y)$, and explain why $f(x, y)$ has no local extrema.

7. For $f(x, y) = 3x^2 - 2xy + y^2 - 2x + 3y - 7$, find $f(2, 3)$, $f_y(x, y)$, and $f_y(2, 3)$.

8. For $f(x, y) = -4x^2 + 4xy - 3y^2 + 4x + 10y + 81$, find $[f_{xx}(2, 3)][f_{yy}(2, 3)] - [f_{xy}(2, 3)]^2$.

9. If $f(x, y) = x + 3y$ and $g(x, y) = x^2 + y^2 - 10$, find the critical points of $F(x, y, \lambda) = f(x, y) + \lambda g(x, y)$.

10. Use the least squares line for the data in the following table to estimate y when $x = 10$.

x	y
2	12
4	10
6	7
8	3

11. For $R = \{(x, y)\,|\,-1 \le x \le 1, \ 1 \le y \le 2\}$, evaluate the following in two ways:

$$\iint_R (4x + 6y)\, dA$$

12. For $R = \{(x, y)\,|\,\sqrt{y} \le x \le 1, \ 0 \le y \le 1\}$, evaluate

$$\iint_R (6x + y)\, dA$$

13. For $f(x, y) = e^{x^2 + 2y}$, find f_x, f_y, and f_{xy}.

14. For $f(x, y) = (x^2 + y^2)^5$, find f_x and f_{xy}.

15. Find all critical points and test for extrema for

$$f(x, y) = x^3 - 12x + y^2 - 6y$$

16. Use Lagrange multipliers to maximize $f(x, y) = xy$ subject to $2x + 3y = 24$.

17. Use Lagrange multipliers to minimize $f(x, y, z) = x^2 + y^2 + z^2$ subject to $2x + y + 2z = 9$.

18. Find the least squares line for the data in the following table.

x	y	x	y
10	50	60	80
20	45	70	85
30	50	80	90
40	55	90	90
50	65	100	110

19. Find the average value of $f(x, y) = x^{2/3} y^{1/3}$ over the rectangle

$$R = \{(x, y)\,|\,-8 \le x \le 8, \ 0 \le y \le 27\}$$

20. Find the volume of the solid under the graph of $z = 3x^2 + 3y^2$ over the rectangle

$$R = \{(x, y)\,|\,0 \le x \le 1, \ -1 \le y \le 1\}$$

21. Without doing any computation, predict the average value of $f(x, y) = x + y$ over the rectangle $R = \{(x, y)\,|\,-10 \le x \le 10, \ -10 \le y \le 10\}$. Then check the correctness of your prediction by evaluating a double integral.

22. (A) Find the dimensions of the square S centered at the origin such that the average value of

$$f(x, y) = \frac{e^x}{y + 10}$$

over S is equal to 5.

(B) Is there a square centered at the origin over which

$$f(x, y) = \frac{e^x}{y + 10}$$

has average value 0.05? Explain.

23. Explain why the function $f(x, y) = 4x^3 - 5y^3$, subject to the constraint $3x + 2y = 7$, has no maxima or minima.

24. Find the volume of the solid under the graph of $F(x, y) = 60x^2 y$ over the region R bounded by the graph of $x + y = 1$ and the coordinate axes.

Applications

25. **Maximizing profit.** A company produces x units of product A and y units of product B (both in hundreds per month). The monthly profit equation (in thousands of dollars) is given by

$$P(x, y) = -4x^2 + 4xy - 3y^2 + 4x + 10y + 81$$

(A) Find $P_x(1, 3)$ and interpret the results.

(B) How many of each product should be produced each month to maximize profit? What is the maximum profit?

26. Minimizing material. A rectangular box with no top and six compartments (see the figure) is to have a volume of 96 cubic inches. Find the dimensions that will require the least amount of material.

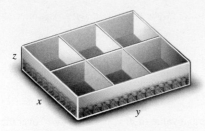

27. Profit. A company's annual profits (in millions of dollars) over a 5-year period are given in the following table. Use the least squares line to estimate the profit for the sixth year.

Year	Profit
1	2
2	2.5
3	3.1
4	4.2
5	4.3

28. Productivity. The Cobb–Douglas production function for a product is

$$N(x, y) = 10x^{0.8}y^{0.2}$$

where x is the number of units of labor and y is the number of units of capital required to produce N units of the product.

(A) Find the marginal productivity of labor and the marginal productivity of capital at $x = 40$ and $y = 50$. For the greatest increase in productivity, should management encourage increased use of labor or increased use of capital?

(B) If each unit of labor costs $100, each unit of capital costs $50, and $10,000 is budgeted for production of this product, use the method of Lagrange multipliers to determine the allocations of labor and capital that will maximize the number of units produced and find the maximum production. Find the marginal productivity of money and approximate the increase in production that would result from an increase of $2,000 in the amount budgeted for production.

(C) If $50 \leq x \leq 100$ and $20 \leq y \leq 40$, find the average number of units produced. Set up a double integral, and evaluate it.

29. Marine biology. When diving using scuba gear, the function used for timing the duration of the dive is

$$T(V, x) = \frac{33V}{x + 33}$$

where T is the time of the dive in minutes, V is the volume of air (in cubic feet, at sea-level pressure) compressed into tanks, and x is the depth of the dive in feet. Find $T_x(70, 17)$ and interpret the results.

30. Pollution. A heavy industrial plant located in the center of a small town emits particulate matter into the atmosphere. Suppose that the concentration of particulate matter (in parts per million) at a point d miles from the plant is given by

$$C = 100 - 24d^2$$

If the boundaries of the town form a square 4 miles long and 4 miles wide, what is the average concentration of particulate matter throughout the town? Express C as a function of x and y, and set up a double integral and evaluate it.

31. Sociology. A sociologist found that the number n of long-distance telephone calls between two cities during a given period varied (approximately) jointly as the populations P_1 and P_2 of the two cities and varied inversely as the distance d between the cities. An equation for a period of 1 week is

$$n(P_1, P_2, d) = 0.001\frac{P_1P_2}{d}$$

Find $n(100,000, 50,000, 100)$.

32. Education. At the beginning of the semester, students in a foreign language course take a proficiency exam. The same exam is given at the end of the semester. The results for 5 students are shown in the following table. Use the least squares line to estimate the second exam score of a student who scored 40 on the first exam.

First Exam	Second Exam
30	60
50	75
60	80
70	85
90	90

33. Population density. The following table gives the U.S. population per square mile for the years 1960–2010:

U.S. Population Density

Year	Population (per square mile)
1960	50.6
1970	57.5
1980	64.1
1990	70.4
2000	79.7
2010	87.4

(A) Find the least squares line for the data, using $x = 0$ for 1960.

(B) Use the least squares line to estimate the population density in the United States in the year 2025.

(C) Now use quadratic regression and exponential regression to obtain the estimate of part (B).

34. Life expectancy. The following table gives life expectancies for males and females in a sample of Central and South American countries:

Life Expectancies for Central and South American Countries

Males	Females	Males	Females
62.30	67.50	70.15	74.10
68.05	75.05	62.93	66.58
72.40	77.04	68.43	74.88
63.39	67.59	66.68	72.80
55.11	59.43		

(A) Find the least squares line for the data.

(B) Use the least squares line to estimate the life expectancy of a female in a Central or South American country in which the life expectancy for males is 60 years.

(C) Now use quadratic regression and logarithmic regression to obtain the estimate of part (B).

Basic Algebra Review

A.1 Real Numbers

A.2 Operations on Polynomials

A.3 Factoring Polynomials

A.4 Operations on Rational Expressions

A.5 Integer Exponents and Scientific Notation

A.6 Rational Exponents and Radicals

A.7 Quadratic Equations

Appendix A reviews some important basic algebra concepts usually studied in earlier courses. The material may be studied systematically before beginning the rest of the book or reviewed as needed.

A.1 Real Numbers

- Set of Real Numbers
- Real Number Line
- Basic Real Number Properties
- Further Properties
- Fraction Properties

The rules for manipulating and reasoning with symbols in algebra depend, in large measure, on properties of the real numbers. In this section we look at some of the important properties of this number system. To make our discussions here and elsewhere in the book clearer and more precise, we occasionally make use of simple *set* concepts and notation.

Set of Real Numbers

Informally, a **real number** is any number that has a decimal representation. Table 1 describes the set of real numbers and some of its important subsets. Figure 1 illustrates how these sets of numbers are related.

The set of integers contains all the natural numbers and something else—their negatives and 0. The set of rational numbers contains all the integers and something else—noninteger ratios of integers. And the set of real numbers contains all the rational numbers and something else—irrational numbers.

Table 1 **Set of Real Numbers**

Symbol	Name	Description	Examples
N	Natural numbers	Counting numbers (also called positive integers)	$1, 2, 3, \ldots$
Z	Integers	Natural numbers, their negatives, and 0	$\ldots, -2, -1, 0, 1, 2, \ldots$
Q	Rational numbers	Numbers that can be represented as a/b, where a and b are integers and $b \neq 0$; decimal representations are repeating or terminating	$-4, 0, 1, 25, \frac{-3}{5}, \frac{2}{3}, 3.67, -0.33\overline{3}, 5.272\,7\overline{27}$*
I	Irrational numbers	Numbers that can be represented as nonrepeating and nonterminating decimal numbers	$\sqrt{2}, \pi, \sqrt[3]{7}, 1.414\,213\ldots, 2.718\,281\,82\ldots$
R	Real numbers	Rational and irrational numbers	

*The overbar indicates that the number (or block of numbers) repeats indefinitely. The space after every third digit is used to help keep track of the number of decimal places.

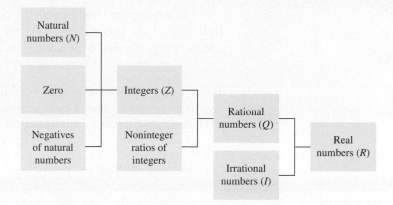

Figure 1 **Real numbers and important subsets**

Real Number Line

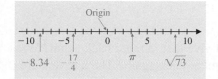

Figure 2 **Real number line**

A one-to-one correspondence exists between the set of real numbers and the set of points on a line. That is, each real number corresponds to exactly one point, and each point corresponds to exactly one real number. A line with a real number associated with each point, and vice versa, as shown in Figure 2, is called a **real number line**, or simply a **real line**. Each number associated with a point is called the coordinate of the point.

The point with coordinate 0 is called the **origin**. The arrow on the right end of the line indicates a positive direction. The coordinates of all points to the right of the origin are called **positive real numbers**, and those to the left of the origin are called **negative real numbers**. The real number 0 is neither positive nor negative.

Basic Real Number Properties

We now take a look at some of the basic properties of the real number system that enable us to convert algebraic expressions into *equivalent forms*.

> **SUMMARY**   Basic Properties of the Set of Real Numbers
> Let a, b, and c be arbitrary elements in the set of real numbers R.
>
> *Addition Properties*
> **Associative:** $(a + b) + c = a + (b + c)$
>
> **Commutative:** $a + b = b + a$
>
> **Identity:**   0 is the additive identity; that is, $0 + a = a + 0 = a$ for all a in R, and 0 is the only element in R with this property.
>
> **Inverse:**   For each a in R, $-a$, is its unique additive inverse; that is, $a + (-a) = (-a) + a = 0$ and $-a$ is the only element in R relative to a with this property.
>
> *Multiplication Properties*
> **Associative:** $(ab)c = a(bc)$
>
> **Commutative:** $ab = ba$
>
> **Identity:**   1 is the multiplicative identity; that is, $(1)a = a(1) = a$ for all a in R, and 1 is the only element in R with this property.
>
> **Inverse:**   For each a in R, $a \neq 0$, $1/a$ is its unique multiplicative inverse; that is, $a(1/a) = (1/a)a = 1$, and $1/a$ is the only element in R relative to a with this property.
>
> *Distributive Properties*
> $$a(b + c) = ab + ac \quad (a + b)c = ac + bc$$

You are already familiar with the **commutative properties** for addition and multiplication. They indicate that the order in which the addition or multiplication of two numbers is performed does not matter. For example,

$$7 + 2 = 2 + 7 \quad \text{and} \quad 3 \cdot 5 = 5 \cdot 3$$

Is there a commutative property relative to subtraction or division? That is, does $a - b = b - a$ or does $a \div b = b \div a$ for all real numbers a and b (division by 0 excluded)? The answer is no, since, for example,

$$8 - 6 \neq 6 - 8 \quad \text{and} \quad 10 \div 5 \neq 5 \div 10$$

When computing

$$3 + 2 + 6 \quad \text{or} \quad 3 \cdot 2 \cdot 6$$

why don't we need parentheses to indicate which two numbers are to be added or multiplied first? The answer is to be found in the **associative properties**. These properties allow us to write

$$(3 + 2) + 6 = 3 + (2 + 6) \quad \text{and} \quad (3 \cdot 2) \cdot 6 = 3 \cdot (2 \cdot 6)$$

so it does not matter how we group numbers relative to either operation. Is there an associative property for subtraction or division? The answer is no, since, for example,

$$(12 - 6) - 2 \neq 12 - (6 - 2) \quad \text{and} \quad (12 \div 6) \div 2 \neq 12 \div (6 \div 2)$$

Evaluate each side of each equation to see why.

What number added to a given number will give that number back again? What number times a given number will give that number back again? The answers are 0 and 1, respectively. Because of this, 0 and 1 are called the **identity elements** for the real numbers. Hence, for any real numbers a and b,

$$0 + 5 = 5 \quad \text{and} \quad (a + b) + 0 = a + b$$
$$1 \cdot 4 = 4 \quad \text{and} \quad (a + b) \cdot 1 = a + b$$

We now consider **inverses**. For each real number a, there is a unique real number $-a$ such that $a + (-a) = 0$. The number $-a$ is called the **additive inverse** of a, or the **negative** of a. For example, the additive inverse (or negative) of 7 is -7, since $7 + (-7) = 0$. The additive inverse (or negative) of -7 is $-(-7) = 7$, since $-7 + [-(-7)] = 0$.

CONCEPTUAL INSIGHT

Do not confuse negation with the sign of a number. If a is a real number, $-a$ is the negative of a and may be positive or negative. Specifically, if a is negative, then $-a$ is positive and if a is positive, then $-a$ is negative.

For each nonzero real number a, there is a unique real number $1/a$ such that $a(1/a) = 1$. The number $1/a$ is called the **multiplicative inverse** of a, or the **reciprocal** of a. For example, the multiplicative inverse (or reciprocal) of 4 is $\frac{1}{4}$, since $4\left(\frac{1}{4}\right) = 1$. $\left(\text{Also note that 4 is the multiplicative inverse of } \frac{1}{4}.\right)$ The number 0 has no multiplicative inverse.

We now turn to the **distributive properties**, which involve both multiplication and addition. Consider the following two computations:

$$5(3 + 4) = 5 \cdot 7 = 35 \qquad 5 \cdot 3 + 5 \cdot 4 = 15 + 20 = 35$$

Thus,

$$5(3 + 4) = 5 \cdot 3 + 5 \cdot 4$$

and we say that multiplication by 5 *distributes* over the sum $(3 + 4)$. In general, **multiplication distributes over addition** in the real number system. Two more illustrations are

$$9(m + n) = 9m + 9n \qquad (7 + 2)u = 7u + 2u$$

EXAMPLE 1   Real Number Properties   State the real number property that justifies the indicated statement.

Statement	Property Illustrated
(A) $x(y + z) = (y + z)x$	Commutative $(\cdot)$
(B) $5(2y) = (5 \cdot 2)y$	Associative $(\cdot)$
(C) $2 + (y + 7) = 2 + (7 + y)$	Commutative $(+)$
(D) $4z + 6z = (4 + 6)z$	Distributive
(E) If $m + n = 0$, then $n = -m$.	Inverse $(+)$

Matched Problem 1   State the real number property that justifies the indicated statement.

(A) $8 + (3 + y) = (8 + 3) + y$

(B) $(x + y) + z = z + (x + y)$

(C) $(a + b)(x + y) = a(x + y) + b(x + y)$

(D) $5xy + 0 = 5xy$

(E) If $xy = 1, x \neq 0$, then $y = 1/x$.

Further Properties

Subtraction and *division* can be defined in terms of addition and multiplication, respectively:

DEFINITION   Subtraction and Division

For all real numbers a and b,

Subtraction: $\qquad a - b = a + (-b) \qquad \begin{aligned} 7 - (-5) &= 7 + [-(-5)] \\ &= 7 + 5 = 12 \end{aligned}$

Division: $\qquad a \div b = a\left(\dfrac{1}{b}\right), b \neq 0 \qquad 9 \div 4 = 9\left(\dfrac{1}{4}\right) = \dfrac{9}{4}$

To subtract b from a, add the negative (the additive inverse) of b to a. To divide a by b, multiply a by the reciprocal (the multiplicative inverse) of b. Note that division by 0 is not defined, since 0 does not have a reciprocal. **0 can never be used as a divisor!**

The following properties of negatives can be proved using the preceding assumed properties and definitions.

THEOREM 1   Negative Properties

For all real numbers a and b,

1. $-(-a) = a$

2. $(-a)b = -(ab)$
   $\qquad = a(-b) = -ab$

3. $(-a)(-b) = ab$

4. $(-1)a = -a$

5. $\dfrac{-a}{b} = -\dfrac{a}{b} = \dfrac{a}{-b}, b \neq 0$

6. $\dfrac{-a}{-b} = -\dfrac{-a}{b} = -\dfrac{a}{-b} = \dfrac{a}{b}, b \neq 0$

We now state two important properties involving 0.

THEOREM 2   Zero Properties

For all real numbers a and b,

1. $a \cdot 0 = 0 \qquad 0 \cdot 0 = 0 \quad (-35)(0) = 0$

2. $ab = 0 \qquad$ if and only if $\quad a = 0 \quad$ or $\quad b = 0$
   $\qquad$ If $(3x + 2)(x - 7) = 0$, then either $3x + 2 = 0$ or $x - 7 = 0$.

Fraction Properties

Recall that the quotient $a \div b (b \neq 0)$ written in the form a/b is called a **fraction**. The quantity a is called the **numerator**, and the quantity b is called the **denominator**.

> **THEOREM 3   Fraction Properties**
>
> For all real numbers a, b, c, d, and k (division by 0 excluded):
>
> 1. $\dfrac{a}{b} = \dfrac{c}{d}$ if and only if $ad = bc$ $\dfrac{4}{6} = \dfrac{6}{9}$ since $4 \cdot 9 = 6 \cdot 6$
>
> 2. $\dfrac{ka}{kb} = \dfrac{a}{b}$ 3. $\dfrac{a}{b} \cdot \dfrac{c}{d} = \dfrac{ac}{bd}$ 4. $\dfrac{a}{b} \div \dfrac{c}{d} = \dfrac{a}{b} \cdot \dfrac{d}{c}$
>
> $\dfrac{7 \cdot 3}{7 \cdot 5} = \dfrac{3}{5}$ $\dfrac{3}{5} \cdot \dfrac{7}{8} = \dfrac{3 \cdot 7}{5 \cdot 8}$ $\dfrac{2}{3} \div \dfrac{5}{7} = \dfrac{2}{3} \cdot \dfrac{7}{5}$
>
> 5. $\dfrac{a}{b} + \dfrac{c}{b} = \dfrac{a + c}{b}$ 6. $\dfrac{a}{b} - \dfrac{c}{b} = \dfrac{a - c}{b}$ 7. $\dfrac{a}{b} + \dfrac{c}{d} = \dfrac{ad + bc}{bd}$
>
> $\dfrac{3}{6} + \dfrac{5}{6} = \dfrac{3 + 5}{6}$ $\dfrac{7}{8} - \dfrac{3}{8} = \dfrac{7 - 3}{8}$ $\dfrac{2}{3} + \dfrac{3}{5} = \dfrac{2 \cdot 5 + 3 \cdot 3}{3 \cdot 5}$

A fraction is a quotient, not just a pair of numbers. So if a and b are real numbers with $b \neq 0$, then $\frac{a}{b}$ corresponds to a point on the real number line. For example, $\frac{17}{2}$ corresponds to the point halfway between $\frac{16}{2} = 8$ and $\frac{18}{2} = 9$. Similarly, $-\frac{21}{5}$ corresponds to the point that is $\frac{1}{5}$ unit to the left of -4.

> **EXAMPLE 2**   Estimation   Round $\frac{22}{7} + \frac{18}{19}$ to the nearest integer.
>
> **SOLUTION**   Note that a calculator is not required: $\frac{22}{7}$ is a little greater than 3, and $\frac{18}{19}$ is a little less than 1. Therefore the sum, rounded to the nearest integer, is 4.
>
> Matched Problem 2   Round $\frac{6}{93}$ to the nearest integer.

Fractions with denominator 100 are called **percentages**. They are used so often that they have their own notation:

$$\frac{3}{100} = 3\% \qquad \frac{7.5}{100} = 7.5\% \qquad \frac{110}{100} = 110\%$$

So 3% is equivalent to 0.03, 7.5% is equivalent to 0.075, and so on.

> **EXAMPLE 3**   State Sales Tax   Find the sales tax that is owed on a purchase of $947.69 if the tax rate is 6.5%.
>
> **SOLUTION** $6.5\%(\$947.69) = 0.065(947.69) = \$61.60$
>
> Matched Problem 3   You intend to give a 20% tip, rounded to the nearest dollar, on a restaurant bill of $78.47. How much is the tip?

Exercises A.1

All variables represent real numbers.

In Problems 1–6, replace each question mark with an appropriate expression that will illustrate the use of the indicated real number property.

1. Commutative property $(\cdot)$: $uv = ?$

2. Commutative property $(+)$: $x + 7 = ?$

3. Associative property $(+)$: $3 + (7 + y) = ?$

4. Associative property $(\cdot)$: $x(yz) = ?$

5. Identity property $(\cdot)$: $1(u + v) = ?$

6. Identity property $(+)$: $0 + 9m = ?$

In Problems 7–26, indicate true (T) or false (F).

7. $5(8m) = (5 \cdot 8)m$

8. $a + cb = a + bc$

9. $5x + 7x = (5 + 7)x$

10. $uv(w + x) = uvw + uvx$

11. $-2(-a)(2x - y) = 2a(-4x + y)$

12. $8 \div (-5) = 8\left(\dfrac{1}{-5}\right)$

13. $(x + 3) + 2x = 2x + (x + 3)$

14. $\dfrac{x}{3y} \div \dfrac{5y}{x} = \dfrac{15y^2}{x^2}$

15. $\dfrac{2x}{-(x + 3)} = -\dfrac{2x}{x + 3}$

16. $-\dfrac{2x}{-(x - 3)} = \dfrac{2x}{x - 3}$

17. $(-3)\left(\dfrac{1}{-3}\right) = 1$

18. $(-0.5) + (0.5) = 0$

19. $-x^2y^2 = (-1)x^2y^2$

20. $[-(x + 2)](-x) = (x + 2)x$

21. $\dfrac{a}{b} + \dfrac{c}{d} = \dfrac{a + c}{b + d}$

22. $\dfrac{k}{k + b} = \dfrac{1}{1 + b}$

23. $(x + 8)(x + 6) = (x + 8)x + (x + 8)6$

24. $u(u - 2v) + v(u - 2v) = (u + v)(u - 2v)$

25. If $(x - 2)(2x + 3) = 0$, then either $x - 2 = 0$ or $2x + 3 = 0$.

26. If either $x - 2 = 0$ or $2x + 3 = 0$, then $(x - 2)(2x + 3) = 0$.

27. If $uv = 1$, does either u or v have to be 1? Explain.

28. If $uv = 0$, does either u or v have to be 0? Explain.

29. Indicate whether the following are true (T) or false (F):

(A) All integers are natural numbers.

(B) All rational numbers are real numbers.

(C) All natural numbers are rational numbers.

30. Indicate whether the following are true (T) or false (F):

(A) All natural numbers are integers.

(B) All real numbers are irrational.

(C) All rational numbers are real numbers.

31. Give an example of a real number that is not a rational number.

32. Give an example of a rational number that is not an integer.

33. Given the sets of numbers N (natural numbers), Z (integers), Q (rational numbers), and R (real numbers), indicate to which set(s) each of the following numbers belongs:

(A) 8 (B) $\sqrt{2}$ (C) -1.414 (D) $\dfrac{-5}{2}$

34. Given the sets of numbers N, Z, Q, and R (see Problem 33), indicate to which set(s) each of the following numbers belongs:

(A) -3 (B) 3.14 (C) π (D) $\dfrac{2}{3}$

35. Indicate true (T) or false (F), and for each false statement find real number replacements for a, b, and c that will provide a counterexample. For all real numbers a, b, and c,

(A) $a(b - c) = ab - c$

(B) $(a - b) - c = a - (b - c)$

(C) $a(bc) = (ab)c$

(D) $(a \div b) \div c = a \div (b \div c)$

36. Indicate true (T) or false (F), and for each false statement find real number replacements for a and b that will provide a counterexample. For all real numbers a and b,

(A) $a + b = b + a$

(B) $a - b = b - a$

(C) $ab = ba$

(D) $a \div b = b \div a$

37. If $c = 0.151515\ldots$, then $100c = 15.1515\ldots$ and

$$100c - c = 15.1515\ldots - 0.151515\ldots$$
$$99c = 15$$
$$c = \frac{15}{99} = \frac{5}{33}$$

Proceeding similarly, convert the repeating decimal $0.090909\ldots$ into a fraction. (All repeating decimals are rational numbers, and all rational numbers have repeating decimal representations.)

38. Repeat Problem 37 for $0.181818\ldots$.

Use a calculator to express each number in Problems 39 and 40 as a decimal to the capacity of your calculator. Observe the repeating decimal representation of the rational numbers and the nonrepeating decimal representation of the irrational numbers.

39. (A) $\dfrac{13}{6}$ (B) $\sqrt{21}$ (C) $\dfrac{7}{16}$ (D) $\dfrac{29}{111}$

40. (A) $\dfrac{8}{9}$ (B) $\dfrac{3}{11}$ (C) $\sqrt{5}$ (D) $\dfrac{11}{8}$

In Problems 41–44, without using a calculator, round to the nearest integer.

41. (A) $\dfrac{43}{13}$ (B) $\dfrac{37}{19}$

42. (A) $\dfrac{9}{17}$ (B) $-\dfrac{12}{25}$

43. (A) $\dfrac{7}{8} + \dfrac{11}{12}$ (B) $\dfrac{55}{9} - \dfrac{7}{55}$

44. (A) $\dfrac{5}{6} - \dfrac{18}{19}$ (B) $\dfrac{13}{5} + \dfrac{44}{21}$

Applications

45. Sales tax. Find the tax owed on a purchase of $182.39 if the state sales tax rate is 9%. (Round to the nearest cent).

46. Sales tax. If you paid $29.86 in tax on a purchase of $533.19, what was the sales tax rate? (Write as a percentage, rounded to one decimal place).

47. Gasoline prices. If the price per gallon of gas jumped from $4.25 to $4.37, what was the percentage increase? (Round to one decimal place).

48. Gasoline prices. The price of gas increased 4% in one week. If the price last week was $4.30 per gallon, what is the price now? (Round to the nearest cent).

Answers to Matched Problems

1. (A) Associative $(+)$ (B) Commutative $(+)$
     (C) Distributive   (D) Identity $(+)$
     (E) Inverse $(\cdot)$

2. 0   **3.** $16

A.2 Operations on Polynomials

- Natural Number Exponents
- Polynomials
- Combining Like Terms
- Addition and Subtraction
- Multiplication
- Combined Operations

This section covers basic operations on *polynomials*. Our discussion starts with a brief review of natural number exponents. Integer and rational exponents and their properties will be discussed in detail in subsequent sections. (Natural numbers, integers, and rational numbers are important parts of the real number system; see Table 1 and Figure 1 in Appendix A.1.)

Natural Number Exponents

We define a **natural number exponent** as follows:

> **DEFINITION** Natural Number Exponent
> For n a natural number and b any real number,
> $$b^n = b \cdot b \cdot \; \cdots \; \cdot b \qquad n \text{ factors of } b$$
> $$3^5 = 3 \cdot 3 \cdot 3 \cdot 3 \cdot 3 \qquad 5 \text{ factors of } 3$$
> where n is called the exponent and b is called the **base**.

Along with this definition, we state the **first property of exponents**:

> **THEOREM 1** First Property of Exponents
> For any natural numbers m and n, and any real number b:
> $$b^m b^n = b^{m+n} \qquad (2t^4)(5t^3) = 2 \cdot 5t^{4+3} = 10t^7$$

Polynomials

Algebraic expressions are formed by using constants and variables and the algebraic operations of addition, subtraction, multiplication, division, raising to powers, and taking roots. Special types of algebraic expressions are called *polynomials*. A **polynomial in one variable** x is constructed by adding or subtracting constants and terms of the form ax^n, where a is a real number and n is a natural number. A **polynomial in two variables** x and y is constructed by adding and subtracting constants and terms of the form $ax^m y^n$, where a is a real number and m and n are natural numbers. Polynomials in three and more variables are defined in a similar manner.

Polynomials		Not Polynomials	
8	0	$\dfrac{1}{x}$	$\dfrac{x-y}{x^2+y^2}$
$3x^3 - 6x + 7$	$6x + 3$		
$2x^2 - 7xy - 8y^2$	$9y^3 + 4y^2 - y + 4$	$\sqrt{x^3 - 2x}$	$2x^{-2} - 3x^{-1}$
$2x - 3y + 2$	$u^5 - 3u^3v^2 + 2uv^4 - v^4$		

Polynomial forms are encountered frequently in mathematics. For the efficient study of polynomials, it is useful to classify them according to their *degree*. If a term in a polynomial has only one variable as a factor, then the **degree of the term** is the power of the variable. If two or more variables are present in a term as factors, then the **degree of the term** is the sum of the powers of the variables. The **degree of a polynomial** is the degree of the nonzero term with the highest degree in the polynomial. Any nonzero constant is defined to be a **polynomial of degree 0**. The number 0 is also a polynomial but is not assigned a degree.

EXAMPLE 1 Degree

(A) The degree of the first term in $5x^3 + \sqrt{3}x - \frac{1}{2}$ is 3, the degree of the second term is 1, the degree of the third term is 0, and the degree of the whole polynomial is 3 (the same as the degree of the term with the highest degree).

(B) The degree of the first term in $8u^3v^2 - \sqrt{7}uv^2$ is 5, the degree of the second term is 3, and the degree of the whole polynomial is 5.

Matched Problem 1

(A) Given the polynomial $6x^5 + 7x^3 - 2$, what is the degree of the first term? The second term? The third term? The whole polynomial?

(B) Given the polynomial $2u^4v^2 - 5uv^3$, what is the degree of the first term? The second term? The whole polynomial?

In addition to classifying polynomials by degree, we also call a single-term polynomial a **monomial**, a two-term polynomial a **binomial**, and a three-term polynomial a **trinomial**.

Combining Like Terms

The concept of *coefficient* plays a central role in the process of combining *like terms*. A constant in a term of a polynomial, including the sign that precedes it, is called the **numerical coefficient**, or simply, the **coefficient**, of the term. If a constant does not appear, or only a $+$ sign appears, the coefficient is understood to be 1. If only a $-$ sign appears, the coefficient is understood to be -1. Given the polynomial

$$5x^4 - x^3 - 3x^2 + x - 7 \quad = 5x^4 + (-1)x^3 + (-3)x^2 + 1x + (-7)$$

the coefficient of the first term is 5, the coefficient of the second term is -1, the coefficient of the third term is -3, the coefficient of the fourth term is 1, and the coefficient of the fifth term is -7.

The following distributive properties are fundamental to the process of combining *like terms*.

THEOREM 2 Distributive Properties of Real Numbers

1. $a(b + c) = (b + c)a = ab + ac$

2. $a(b - c) = (b - c)a = ab - ac$

3. $a(b + c + \cdots + f) = ab + ac + \cdots + af$

Two terms in a polynomial are called **like terms** if they have exactly the same variable factors to the same powers. The numerical coefficients may or may not be the same. Since constant terms involve no variables, all constant terms are like terms. If a polynomial contains two or more like terms, these terms can be combined into a single term by making use of distributive properties. The following example illustrates the reasoning behind the process:

$$
\begin{aligned}
3x^2y - 5xy^2 + x^2y - 2x^2y &= 3x^2y + x^2y - 2x^2y - 5xy^2 \\
&= (3x^2y + 1x^2y - 2x^2y) - 5xy^2 \\
&= (3 + 1 - 2)x^2y - 5xy^2 \\
&= 2x^2y - 5xy^2
\end{aligned}
$$

Note the use of distributive properties.

Free use is made of the real number properties discussed in Appendix A.1.

How can we simplify expressions such as $4(x - 2y) - 3(2x - 7y)$? We clear the expression of parentheses using distributive properties, and combine like terms:

$$
\begin{aligned}
4(x - 2y) - 3(2x - 7y) &= 4x - 8y - 6x + 21y \\
&= -2x + 13y
\end{aligned}
$$

EXAMPLE 2   Removing Parentheses   Remove parentheses and simplify:

(A) $2(3x^2 - 2x + 5) + (x^2 + 3x - 7)$
$$
\begin{aligned}
&= 2(3x^2 - 2x + 5) + 1(x^2 + 3x - 7) \\
&= 6x^2 - 4x + 10 + x^2 + 3x - 7 \\
&= 7x^2 - x + 3
\end{aligned}
$$

(B) $(x^3 - 2x - 6) - (2x^3 - x^2 + 2x - 3)$
$$
\begin{aligned}
&= 1(x^3 - 2x - 6) + (-1)(2x^3 - x^2 + 2x - 3) \\
&= x^3 - 2x - 6 - 2x^3 + x^2 - 2x + 3 \\
&= -x^3 + x^2 - 4x - 3
\end{aligned}
$$

Be careful with the sign here

(C) $[3x^2 - (2x + 1)] - (x^2 - 1) = [3x^2 - 2x - 1] - (x^2 - 1)$
$$
\begin{aligned}
&= 3x^2 - 2x - 1 - x^2 + 1 \\
&= 2x^2 - 2x
\end{aligned}
$$

Remove inner parentheses first.

Matched Problem 2   Remove parentheses and simplify:

(A) $3(u^2 - 2v^2) + (u^2 + 5v^2)$

(B) $(m^3 - 3m^2 + m - 1) - (2m^3 - m + 3)$

(C) $(x^3 - 2) - [2x^3 - (3x + 4)]$

Addition and Subtraction

Addition and subtraction of polynomials can be thought of in terms of removing parentheses and combining like terms, as illustrated in Example 2. Horizontal and vertical arrangements are illustrated in the next two examples. You should be able to work either way, letting the situation dictate your choice.

EXAMPLE 3   Adding Polynomials   Add horizontally and vertically:
$$
x^4 - 3x^3 + x^2, \quad -x^3 - 2x^2 + 3x, \quad \text{and} \quad 3x^2 - 4x - 5
$$

SOLUTION   Add horizontally:
$$
\begin{aligned}
(x^4 - 3x^3 + x^2) &+ (-x^3 - 2x^2 + 3x) + (3x^2 - 4x - 5) \\
&= x^4 - 3x^3 + x^2 - x^3 - 2x^2 + 3x + 3x^2 - 4x - 5 \\
&= x^4 - 4x^3 + 2x^2 - x - 5
\end{aligned}
$$

Or vertically, by lining up like terms and adding their coefficients:
$$
\begin{array}{r}
x^4 - 3x^3 + x^2 \\
- x^3 - 2x^2 + 3x \\
3x^2 - 4x - 5 \\
\hline
x^4 - 4x^3 + 2x^2 - x - 5
\end{array}
$$

Matched Problem 3 ⎤ Add horizontally and vertically:

$$3x^4 - 2x^3 - 4x^2, \quad x^3 - 2x^2 - 5x, \quad \text{and} \quad x^2 + 7x - 2$$

EXAMPLE 4 Subtracting Polynomials Subtract $4x^2 - 3x + 5$ from $x^2 - 8$, both horizontally and vertically.

SOLUTION
$$(x^2 - 8) - (4x^2 - 3x + 5) \quad \text{or} \quad \begin{array}{r} x^2 \qquad\;\; - \;\; 8 \\ -4x^2 + 3x - 5 \\ \hline -3x^2 + 3x - 13 \end{array}$$
$$= x^2 - 8 - 4x^2 + 3x - 5 \qquad \leftarrow \text{Change}$$
$$= -3x^2 + 3x - 13 \qquad\qquad\quad\;\; \text{signs and add.}$$

Matched Problem 4 ⎤ Subtract $2x^2 - 5x + 4$ from $5x^2 - 6$, both horizontally and vertically.

Multiplication

Multiplication of algebraic expressions involves the extensive use of distributive properties for real numbers, as well as other real number properties.

EXAMPLE 5 Multiplying Polynomials Multiply: $(2x - 3)(3x^2 - 2x + 3)$

SOLUTION
$$(2x - 3)(3x^2 - 2x + 3) \quad \boxed{= 2x(3x^2 - 2x + 3) - 3(3x^2 - 2x + 3)}$$
$$= 6x^3 - 4x^2 + 6x - 9x^2 + 6x - 9$$
$$= 6x^3 - 13x^2 + 12x - 9$$

Or, using a vertical arrangement,

$$\begin{array}{r} 3x^2 - \;\; 2x \; + 3 \\ 2x \; - \;\; 3 \\ \hline 6x^3 - \;\; 4x^2 + \;\; 6x \\ - \;\; 9x^2 + \;\; 6x - 9 \\ \hline 6x^3 - 13x^2 + 12x - 9 \end{array}$$

Matched Problem 5 ⎤ Multiply: $(2x - 3)(2x^2 + 3x - 2)$

Thus, to multiply two polynomials, multiply each term of one by each term of the other, and combine like terms.

Products of binomial factors occur frequently, so it is useful to develop procedures that will enable us to write down their products by inspection. To find the product $(2x - 1)(3x + 2)$ we proceed as follows:

$$(2x - 1)(3x + 2) \quad \boxed{= 6x^2 + 4x - 3x - 2} \quad \text{The inner and outer products}$$
$$= 6x^2 + x - 2 \qquad\qquad\qquad \text{are like terms, so combine into a single term.}$$

To speed the process, we do the step in the dashed box mentally.

Products of certain binomial factors occur so frequently that it is useful to learn formulas for their products. The following formulas are easily verified by multiplying the factors on the left.

THEOREM 3 Special Products

1. $(a - b)(a + b) = a^2 - b^2$
2. $(a + b)^2 = a^2 + 2ab + b^2$
3. $(a - b)^2 = a^2 - 2ab + b^2$

EXAMPLE 6   Special Products   Multiply mentally, where possible.

(A) $(2x - 3y)(5x + 2y)$ (B) $(3a - 2b)(3a + 2b)$

(C) $(5x - 3)^2$ (D) $(m + 2n)^3$

SOLUTION

(A) $(2x - 3y)(5x + 2y)$ $\boxed{= 10x^2 + 4xy - 15xy - 6y^2}$

$\qquad\qquad\qquad\qquad = 10x^2 - 11xy - 6y^2$

(B) $(3a - 2b)(3a + 2b)$ $\boxed{= (3a)^2 - (2b)^2}$

$\qquad\qquad\qquad\qquad = 9a^2 - 4b^2$

(C) $(5x - 3)^2$ $\boxed{= (5x)^2 - 2(5x)(3) + 3^2}$

$\qquad\qquad\quad = 25x^2 - 30x + 9$

(D) $(m + 2n)^3 = (m + 2n)^2(m + 2n)$

$\qquad\qquad\quad = (m^2 + 4mn + 4n^2)(m + 2n)$

$\qquad\qquad\quad = m^2(m + 2n) + 4mn(m + 2n) + 4n^2(m + 2n)$

$\qquad\qquad\quad = m^3 + 2m^2n + 4m^2n + 8mn^2 + 4mn^2 + 8n^3$

$\qquad\qquad\quad = m^3 + 6m^2n + 12mn^2 + 8n^3$

Matched Problem 6   Multiply mentally, where possible.

(A) $(4u - 3v)(2u + v)$ (B) $(2xy + 3)(2xy - 3)$

(C) $(m + 4n)(m - 4n)$ (D) $(2u - 3v)^2$

(E) $(2x - y)^3$

Combined Operations

We complete this section by considering several examples that use all the operations just discussed. Note that in simplifying, we usually remove grouping symbols starting from the inside. That is, we remove parentheses ( ) first, then brackets [ ], and finally braces { }, if present. Also,

DEFINITION   Order of Operations

Multiplication and division precede addition and subtraction, and taking powers precedes multiplication and division.

$$2 \cdot 3 + 4 = 6 + 4 = 10, \quad \text{not} \quad 2 \cdot 7 = 14$$

$$\frac{10^2}{2} = \frac{100}{2} = 50, \quad \text{not} \quad 5^2 = 25$$

EXAMPLE 7   Combined Operations   Perform the indicated operations and simplify:

(A) $3x - \{5 - 3[x - x(3 - x)]\} = 3x - \{5 - 3[x - 3x + x^2]\}$

$\qquad\qquad\qquad\qquad\qquad\qquad = 3x - \{5 - 3x + 9x - 3x^2\}$

$\qquad\qquad\qquad\qquad\qquad\qquad = 3x - 5 + 3x - 9x + 3x^2$

$\qquad\qquad\qquad\qquad\qquad\qquad = 3x^2 - 3x - 5$

(B) $(x - 2y)(2x + 3y) - (2x + y)^2 = 2x^2 - xy - 6y^2 - (4x^2 + 4xy + y^2)$

$\qquad\qquad\qquad\qquad\qquad\qquad\qquad = 2x^2 - xy - 6y^2 - 4x^2 - 4xy - y^2$

$\qquad\qquad\qquad\qquad\qquad\qquad\qquad = -2x^2 - 5xy - 7y^2$

Matched Problem 7   Perform the indicated operations and simplify:

(A) $2t - \{7 - 2[t - t(4 + t)]\}$

(B) $(u - 3v)^2 - (2u - v)(2u + v)$

Exercises A.2

Problems 1–8 refer to the following polynomials:

  (A) $2x - 3$ (B) $2x^2 - x + 2$ (C) $x^3 + 2x^2 - x + 3$

1. What is the degree of (C)?

2. What is the degree of (A)?

3. Add (B) and (C).

4. Add (A) and (B).

5. Subtract (B) from (C).

6. Subtract (A) from (B).

7. Multiply (B) and (C).

8. Multiply (A) and (C).

In Problems 9–30, perform the indicated operations and simplify.

9. $2(u - 1) - (3u + 2) - 2(2u - 3)$

10. $2(x - 1) + 3(2x - 3) - (4x - 5)$

11. $4a - 2a[5 - 3(a + 2)]$

12. $2y - 3y[4 - 2(y - 1)]$

13. $(a + b)(a - b)$

14. $(m - n)(m + n)$

15. $(3x - 5)(2x + 1)$

16. $(4t - 3)(t - 2)$

17. $(2x - 3y)(x + 2y)$

18. $(3x + 2y)(x - 3y)$

19. $(3y + 2)(3y - 2)$

20. $(2m - 7)(2m + 7)$

21. $-(2x - 3)^2$

22. $-(5 - 3x)^2$

23. $(4m + 3n)(4m - 3n)$

24. $(3x - 2y)(3x + 2y)$

25. $(3u + 4v)^2$

26. $(4x - y)^2$

27. $(a - b)(a^2 + ab + b^2)$

28. $(a + b)(a^2 - ab + b^2)$

29. $[(x - y) + 3z][(x - y) - 3z]$

30. $[a - (2b - c)][a + (2b - c)]$

In Problems 31–44, perform the indicated operations and simplify.

31. $m - \{m - [m - (m - 1)]\}$

32. $2x - 3\{x + 2[x - (x + 5)] + 1\}$

33. $(x^2 - 2xy + y^2)(x^2 + 2xy + y^2)$

34. $(3x - 2y)^2(2x + 5y)$

35. $(5a - 2b)^2 - (2b + 5a)^2$

36. $(2x - 1)^2 - (3x + 2)(3x - 2)$

37. $(m - 2)^2 - (m - 2)(m + 2)$

38. $(x - 3)(x + 3) - (x - 3)^2$

39. $(x - 2y)(2x + y) - (x + 2y)(2x - y)$

40. $(3m + n)(m - 3n) - (m + 3n)(3m - n)$

41. $(u + v)^3$

42. $(x - y)^3$

43. $(x - 2y)^3$

44. $(2m - n)^3$

45. Subtract the sum of the last two polynomials from the sum of the first two: $2x^2 - 4xy + y^2$, $3xy - y^2$, $x^2 - 2xy - y^2$, $-x^2 + 3xy - 2y^2$

46. Subtract the sum of the first two polynomials from the sum of the last two: $3m^2 - 2m + 5$, $4m^2 - m$, $3m^2 - 3m - 2$, $m^3 + m^2 + 2$

In Problems 47–50, perform the indicated operations and simplify.

47. $[(2x - 1)^2 - x(3x + 1)]^2$

48. $[5x(3x + 1) - 5(2x - 1)^2]^2$

49. $2\{(x - 3)(x^2 - 2x + 1) - x[3 - x(x - 2)]\}$

50. $-3x\{x[x - x(2 - x)] - (x + 2)(x^2 - 3)\}$

51. If you are given two polynomials, one of degree m and the other of degree n, where m is greater than n, what is the degree of their product?

52. What is the degree of the sum of the two polynomials in Problem 51?

53. How does the answer to Problem 51 change if the two polynomials can have the same degree?

54. How does the answer to Problem 52 change if the two polynomials can have the same degree?

55. Show by example that, in general, $(a + b)^2 \neq a^2 + b^2$. Discuss possible conditions on a and b that would make this a valid equation.

56. Show by example that, in general, $(a - b)^2 \neq a^2 - b^2$. Discuss possible conditions on a and b that would make this a valid equation.

Applications

57. *Investment.* You have $10,000 to invest, part at 9% and the rest at 12%. If x is the amount invested at 9%, write an algebraic expression that represents the total annual income from both investments. Simplify the expression.

58. *Investment.* A person has $100,000 to invest. If $x are invested in a money market account yielding 7% and twice that amount in certificates of deposit yielding 9%, and if the rest is invested in high-grade bonds yielding 11%, write an algebraic expression that represents the total annual income from all three investments. Simplify the expression.

59. Gross receipts. Four thousand tickets are to be sold for a musical show. If x tickets are to be sold for $20 each and three times that number for $30 each, and if the rest are sold for $50 each, write an algebraic expression that represents the gross receipts from ticket sales, assuming all tickets are sold. Simplify the expression.

60. Gross receipts. Six thousand tickets are to be sold for a concert, some for $20 each and the rest for $35 each. If x is the number of $20 tickets sold, write an algebraic expression that represents the gross receipts from ticket sales, assuming all tickets are sold. Simplify the expression.

61. Nutrition. Food mix A contains 2% fat, and food mix B contains 6% fat. A 10-kilogram diet mix of foods A and B is formed. If x kilograms of food A are used, write an algebraic expression that represents the total number of kilograms of fat in the final food mix. Simplify the expression.

62. Nutrition. Each ounce of food M contains 8 units of calcium, and each ounce of food N contains 5 units of calcium. A 160-ounce diet mix is formed using foods M and N. If x is the number of ounces of food M used, write an algebraic expression that represents the total number of units of calcium in the diet mix. Simplify the expression.

Answers to Matched Problems

1. (A) $5, 3, 0, 5$ (B) $6, 4, 6$
2. (A) $4u^2 - v^2$ (B) $-m^3 - 3m^2 + 2m - 4$
   (C) $-x^3 + 3x + 2$
3. $3x^4 - x^3 - 5x^2 + 2x - 2$
4. $3x^2 + 5x - 10$ 5. $4x^3 - 13x + 6$
6. (A) $8u^2 - 2uv - 3v^2$ (B) $4x^2y^2 - 9$ (C) $m^2 - 16n^2$
   (D) $4u^2 - 12uv + 9v^2$ (E) $8x^3 - 12x^2y + 6xy^2 - y^3$
7. (A) $-2t^2 - 4t - 7$ (B) $-3u^2 - 6uv + 10v^2$

A.3 Factoring Polynomials

- Common Factors
- Factoring by Grouping
- Factoring Second-Degree Polynomials
- Special Factoring Formulas
- Combined Factoring Techniques

A positive integer is **written in factored form** if it is written as the product of two or more positive integers; for example, $120 = 10 \cdot 12$. A positive integer is **factored completely** if each factor is prime; for example, $120 = 2 \cdot 2 \cdot 2 \cdot 3 \cdot 5$. (Recall that an integer $p > 1$ is **prime** if p cannot be factored as the product of two smaller positive integers. So the first ten primes are 2, 3, 5, 7, 11, 13, 17, 19, 23, and 29). A **tree diagram** is a helpful way to visualize a factorization (Fig 1).

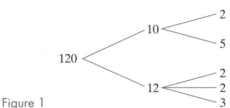

Figure 1

A polynomial is **written in factored form** if it is written as the product of two or more polynomials. The following polynomials are written in factored form:

$$4x^2y - 6xy^2 = 2xy(2x - 3y) \qquad 2x^3 - 8x = 2x(x - 2)(x + 2)$$
$$x^2 - x - 6 = (x - 3)(x + 2) \qquad 5m^2 + 20 = 5(m^2 + 4)$$

Unless stated to the contrary, we will limit our discussion of factoring polynomials to polynomials with integer coefficients.

A polynomial with integer coefficients is said to be **factored completely** if each factor cannot be expressed as the product of two or more polynomials with integer coefficients, other than itself or 1. All the polynomials above, as we will see by the conclusion of this section, are factored completely.

Writing polynomials in completely factored form is often a difficult task. But accomplishing it can lead to the simplification of certain algebraic expressions and to the solution of certain types of equations and inequalities. The distributive properties for real numbers are central to the factoring process.

Common Factors

Generally, a first step in any factoring procedure is to factor out all factors common to all terms.

EXAMPLE 1   Common Factors   Factor out all factors common to all terms.

(A) $3x^3y - 6x^2y^2 - 3xy^3$

(B) $3y(2y + 5) + 2(2y + 5)$

SOLUTION

(A) $3x^3y - 6x^2y^2 - 3xy^3 = (\mathbf{3xy})x^2 - (\mathbf{3xy})2xy - (\mathbf{3xy})y^2$

$\qquad\qquad\qquad\qquad = 3xy(x^2 - 2xy - y^2)$

(B) $3y(2y + 5) + 2(2y + 5) = 3y(\mathbf{2y + 5}) + 2(\mathbf{2y + 5})$

$\qquad\qquad\qquad\qquad = (3y + 2)(2y + 5)$

Matched Problem 1 Factor out all factors common to all terms.

(A) $2x^3y - 8x^2y^2 - 6xy^3$ (B) $2x(3x - 2) - 7(3x - 2)$

Factoring by Grouping

Occasionally, polynomials can be factored by grouping terms in such a way that we obtain results that look like Example 1B. We can then complete the factoring following the steps used in that example. This process will prove useful in the next subsection, where an efficient method is developed for factoring a second-degree polynomial as the product of two first-degree polynomials, if such factors exist.

EXAMPLE 2   Factoring by Grouping   Factor by grouping.

(A) $3x^2 - 3x - x + 1$

(B) $4x^2 - 2xy - 6xy + 3y^2$

(C) $y^2 + xz + xy + yz$

SOLUTION

(A) $3x^2 - 3x - x + 1$

$\qquad = (3x^2 - 3x) - (x - 1)$ Group the first two and the last two terms.

$\qquad = 3x(x - 1) - (x - 1)$ Factor out any common factors from each group. The common factor $(x - 1)$ can be

$\qquad = (x - 1)(3x - 1)$ taken out, and the factoring is complete.

(B) $4x^2 - 2xy - 6xy + 3y^2 = (4x^2 - 2xy) - (6xy - 3y^2)$

$\qquad\qquad\qquad\qquad\qquad\quad = 2x(2x - y) - 3y(2x - y)$

$\qquad\qquad\qquad\qquad\qquad\quad = (2x - y)(2x - 3y)$

(C) If, as in parts (A) and (B), we group the first two terms and the last two terms of $y^2 + xz + xy + yz$, no common factor can be taken out of each group to complete the factoring. However, if the two middle terms are reversed, we can proceed as before:

$$y^2 + xz + xy + yz = y^2 + xy + xz + yz$$
$$= (y^2 + xy) + (xz + yz)$$
$$= y(y + x) + z(x + y)$$
$$= y(x + y) + z(x + y)$$
$$= (x + y)(y + z)$$

Matched Problem 2 Factor by grouping.

(A) $6x^2 + 2x + 9x + 3$

(B) $2u^2 + 6uv - 3uv - 9v^2$

(C) $ac + bd + bc + ad$

Factoring Second-Degree Polynomials

We now turn our attention to factoring second-degree polynomials of the form

$$2x^2 - 5x - 3 \quad \text{and} \quad 2x^2 + 3xy - 2y^2$$

into the product of two first-degree polynomials with integer coefficients. Since many second-degree polynomials with integer coefficients cannot be factored in this way, it would be useful to know ahead of time that the factors we are seeking actually exist. The factoring approach we use, involving the *ac test*, determines at the beginning whether first-degree factors with integer coefficients do exist. Then, if they exist, the test provides a simple method for finding them.

THEOREM 1   *ac* Test for Factorability

If in polynomials of the form

$$ax^2 + bx + c \quad \text{or} \quad ax^2 + bxy + cy^2 \tag{1}$$

the product ac has two integer factors p and q whose sum is the coefficient b of the middle term; that is, if integers p and q exist so that

$$pq = ac \quad \text{and} \quad p + q = b \tag{2}$$

then the polynomials have first-degree factors with integer coefficients. If no integers p and q exist that satisfy equations (2), then the polynomials in equations (1) will not have first-degree factors with integer coefficients.

If integers p and q exist that satisfy equations (2) in the *ac* test, the factoring always can be completed as follows: Using $b = p + q$, split the middle terms in equations (1) to obtain

$$ax^2 + bx + c = ax^2 + px + qx + c$$
$$ax^2 + bxy + cy^2 = ax^2 + pxy + qxy + cy^2$$

Complete the factoring by grouping the first two terms and the last two terms as in Example 2. This process always works, and it does not matter if the two middle terms on the right are interchanged.

Several examples should make the process clear. After a little practice, you will perform many of the steps mentally and will find the process fast and efficient.

EXAMPLE 3   Factoring Second-Degree Polynomials   Factor, if possible, using integer coefficients.

(A) $4x^2 - 4x - 3$ (B) $2x^2 - 3x - 4$ (C) $6x^2 - 25xy + 4y^2$

SOLUTION

(A) $4x^2 - 4x - 3$

Step 1   Use the *ac* test to test for factorability. Comparing $4x^2 - 4x - 3$ with $ax^2 + bx + c$, we see that $a = 4$, $b = -4$, and $c = -3$. Multiply a and c to obtain

$$ac = (4)(-3) = -12$$

List all pairs of integers whose product is -12, as shown in the margin. These are called **factor pairs** of -12. Then try to find a factor pair that sums to $b = -4$, the coefficient of the middle term in $4x^2 - 4x - 3$. (In practice, this part of Step 1 is often done mentally and can be done rather quickly.) Notice that the factor pair 2 and -6 sums to -4. By the *ac* test, $4x^2 - 4x - 3$ has first-degree factors with integer coefficients.

pq	
$(1)(-12)$	All factor pairs of
$(-1)(12)$	$-12 = ac$
$(2)(-6)$	
$(-2)(6)$	
$(3)(-4)$	
$(-3)(4)$	

Step 2 Split the middle term, using $b = p + q$, and complete the factoring by grouping. Using $-4 = 2 + (-6)$, we split the middle term in $4x^2 - 4x - 3$ and complete the factoring by grouping:

$$4x^2 - 4x - 3 = 4x^2 + 2x - 6x - 3$$
$$= (4x^2 + 2x) - (6x + 3)$$
$$= 2x(2x + 1) - 3(2x + 1)$$
$$= (2x + 1)(2x - 3)$$

The result can be checked by multiplying the two factors to obtain the original polynomial.

(B) $2x^2 - 3x - 4$

Step 1 Use the ac test to test for factorability:

$$ac = (2)(-4) = -8$$

pq

(−1)(8)	All factor pairs of
(1)(−8)	$-8 = ac$
(−2)(4)	
(2)(−4)	

Does -8 have a factor pair whose sum is -3? None of the factor pairs listed in the margin sums to $-3 = b$, the coefficient of the middle term in $2x^2 - 3x - 4$. According to the ac test, we can conclude that $2x^2 - 3x - 4$ does not have first-degree factors with integer coefficients, and we say that the polynomial is **not factorable**.

(C) $6x^2 - 25xy + 4y^2$

Step 1 Use the ac test to test for factorability:

$$ac = (6)(4) = 24$$

Mentally checking through the factor pairs of 24, keeping in mind that their sum must be $-25 = b$, we see that if $p = -1$ and $q = -24$, then

$$pq = (-1)(-24) = 24 = ac$$

and

$$p + q = (-1) + (-24) = -25 = b$$

So the polynomial is factorable.

Step 2 Split the middle term, using $b = p + q$, and complete the factoring by grouping. Using $-25 = (-1) + (-24)$, we split the middle term in $6x^2 - 25xy + 4y^2$ and complete the factoring by grouping:

$$6x^2 - 25xy + 4y^2 = 6x^2 - xy - 24xy + 4y^2$$
$$= (6x^2 - xy) - (24xy - 4y^2)$$
$$= x(6x - y) - 4y(6x - y)$$
$$= (6x - y)(x - 4y)$$

The check is left to the reader.

Matched Problem 3 Factor, if possible, using integer coefficients.

(A) $2x^2 + 11x - 6$

(B) $4x^2 + 11x - 6$

(C) $6x^2 + 5xy - 4y^2$

Special Factoring Formulas

The factoring formulas listed in the following box will enable us to factor certain polynomial forms that occur frequently. These formulas can be established by multiplying the factors on the right.

THEOREM 2 Special Factoring Formulas

Perfect square:	1. $u^2 + 2uv + v^2 = (u + v)^2$
Perfect square:	2. $u^2 - 2uv + v^2 = (u - v)^2$
Difference of squares:	3. $u^2 - v^2 = (u - v)(u + v)$
Difference of cubes:	4. $u^3 - v^3 = (u - v)(u^2 + uv + v^2)$
Sum of cubes:	5. $u^3 + v^3 = (u + v)(u^2 - uv + v^2)$

⚠ **CAUTION** Notice that $u^2 + v^2$ is not included in the list of special factoring formulas. In fact,

$$u^2 + v^2 \neq (au + bv)(cu + dv)$$

for any choice of real number coefficients a, b, c, and d. ▲

EXAMPLE 4 Factoring Factor completely.

(A) $4m^2 - 12mn + 9n^2$ (B) $x^2 - 16y^2$ (C) $z^3 - 1$

(D) $m^3 + n^3$ (E) $a^2 - 4(b + 2)^2$

SOLUTION

(A) $4m^2 - 12mn + 9n^2 = (2m - 3n)^2$

(B) $x^2 - 16y^2 \;\boxed{= x^2 - (4y)^2}\; = (x - 4y)(x + 4y)$

(C) $z^3 - 1 = (z - 1)(z^2 + z + 1)$ Use the ac test to verify that $z^2 + z + 1$ cannot be factored.

(D) $m^3 + n^3 = (m + n)(m^2 - mn + n^2)$ Use the ac test to verify that $m^2 - mn + n^2$ cannot be factored.

(E) $a^2 - 4(b + 2)^2 = [a - 2(b + 2)][a + 2(b + 2)]$

Matched Problem 4) Factor completely:

(A) $x^2 + 6xy + 9y^2$ (B) $9x^2 - 4y^2$ (C) $8m^3 - 1$

(D) $x^3 + y^3z^3$ (E) $9(m - 3)^2 - 4n^2$

Combined Factoring Techniques

We complete this section by considering several factoring problems that involve combinations of the preceding techniques.

PROCEDURE Factoring Polynomials

Step 1 Take out any factors common to all terms.

Step 2 Use any of the special formulas listed in Theorem 2 that are applicable.

Step 3 Apply the ac test to any remaining second-degree polynomial factors.

Note: It may be necessary to perform some of these steps more than once. Furthermore, the order of applying these steps can vary.

EXAMPLE 5 Combined Factoring Techniques Factor completely.

(A) $3x^3 - 48x$ (B) $3u^4 - 3u^3v - 9u^2v^2$

(C) $3m^2 - 24mn^3$ (D) $3x^4 - 5x^2 + 2$

SOLUTION

(A) $3x^3 - 48x = 3x(x^2 - 16) = 3x(x - 4)(x + 4)$

(B) $3u^4 - 3u^3v - 9u^2v^2 = 3u^2(u^2 - uv - 3v^2)$

(C) $3m^4 - 24mn^3 = 3m(m^3 - 8n^3) = 3m(m - 2n)(m^2 + 2mn + 4n^2)$

(D) $3x^4 - 5x^2 + 2 = (3x^2 - 2)(x^2 - 1) = (3x^2 - 2)(x - 1)(x + 1)$

Matched Problem 5 | Factor completely.

(A) $18x^3 - 8x$

(B) $4m^3n - 2m^2n^2 + 2mn^3$

(C) $2t^4 - 16t$

(D) $2y^4 - 5y^2 - 12$

Exercises A.3

In Problems 1–8, factor out all factors common to all terms.

1. $6m^4 - 9m^3 - 3m^2$ **2.** $6x^4 - 8m^3 - 2x^2$

3. $8u^3v - 6u^2v^2 + 4uv^3$ **4.** $10x^3y + 20x^2y^2 - 15xy^3$

5. $7m(2m - 3) + 5(2m - 3)$

6. $5x(x + 1) - 3(x + 1)$

7. $4ab(2c + d) - (2c + d)$

8. $12a(b - 2c) - 15b(b - 2c)$

In Problems 9–18, factor by grouping.

9. $2x^2 - x + 4x - 2$ **10.** $x^2 - 3x + 2x - 6$

11. $3y^2 - 3y + 2y - 2$ **12.** $2x^2 - x + 6x - 3$

13. $2x^2 + 8x - x - 4$ **14.** $6x^2 + 9x - 2x - 3$

15. $wy - wz + xy - xz$ **16.** $ac + ad + bc + bd$

17. $am - 3bm + 2na - 6bn$ **18.** $ab + 6 + 2a + 3b$

In Problems 19–56, factor completely. If a polynomial cannot be factored, say so.

19. $3y^2 - y - 2$ **20.** $2x^2 + 5x - 3$

21. $u^2 - 2uv - 15v^2$ **22.** $x^2 - 4xy - 12y^2$

23. $m^2 - 6m - 3$ **24.** $x^2 + x - 4$

25. $w^2x^2 - y^2$ **26.** $25m^2 - 16n^2$

27. $9m^2 - 6mn + n^2$ **28.** $x^2 + 10xy + 25y^2$

29. $y^2 + 16$ **30.** $u^2 + 81$

31. $4z^2 - 28z + 48$ **32.** $6x^2 + 48x + 72$

33. $2x^4 - 24x^3 + 40x^2$ **34.** $2y^3 - 22y^2 + 48y$

35. $4xy^2 - 12xy + 9x$ **36.** $16x^2y - 8xy + y$

37. $6m^2 - mn - 12n^2$ **38.** $6s^2 + 7st - 3t^2$

39. $4u^3v - uv^3$ **40.** $x^3y - 9xy^3$

41. $2x^3 - 2x^2 + 8x$ **42.** $3m^3 - 6m^2 + 15m$

43. $8x^3 - 27y^3$ **44.** $5x^3 + 40y^3$

45. $x^4y + 8xy$ **46.** $8a^3 - 1$

47. $(x + 2)^2 - 9y^2$ **48.** $(a - b)^2 - 4(c - d)^2$

49. $5u^2 + 4uv - 2v^2$ **50.** $3x^2 - 2xy - 4y^2$

51. $6(x - y)^2 + 23(x - y) - 4$

52. $4(A + B)^2 - 5(A + B) - 6$

53. $y^4 - 3y^2 - 4$

54. $m^4 - n^4$

55. $15y(x - y)^3 + 12x(x - y)^2$

56. $15x^2(3x - 1)^4 + 60x^3(3x - 1)^3$

✐ *In Problems 57–60, discuss the validity of each statement. If the statement is true, explain why. If not, give a counterexample.*

57. If n is a positive integer greater than 1, then $u^n - v^n$ can be factored.

58. If m and n are positive integers and $m \neq n$, then $u^m - v^n$ is not factorable.

59. If n is a positive integer greater than 1, then $u^n + v^n$ can be factored.

60. If k is a positive integer, then $u^{2k+1} + v^{2k+1}$ can be factored.

Answers to Matched Problems

1. (A) $2xy(x^2 - 4xy - 3y^2)$ (B) $(2x - 7)(3x - 2)$

2. (A) $(3x + 1)(2x + 3)$ (B) $(u + 3v)(2u - 3v)$
   (C) $(a + b)(c + d)$

3. (A) $(2x - 1)(x + 6)$ (B) Not factorable
   (C) $(3x + 4y)(2x - y)$

4. (A) $(x + 3y)^2$ (B) $(3x - 2y)(3x + 2y)$
   (C) $(2m - 1)(4m^2 + 2m + 1)$
   (D) $(x + yz)(x^2 - xyz + y^2z^2)$
   (E) $[3(m - 3) - 2n][3(m - 3) + 2n]$

5. (A) $2x(3x - 2)(3x + 2)$ (B) $2mn(2m^2 - mn + n^2)$
   (C) $2t(t - 2)(t^2 + 2t + 4)$
   (D) $(2y^2 + 3)(y - 2)(y + 2)$

A.4 Operations on Rational Expressions

- Reducing to Lowest Terms
- Multiplication and Division
- Addition and Subtraction
- Compound Fractions

We now turn our attention to fractional forms. A quotient of two algebraic expressions (division by 0 excluded) is called a **fractional expression**. If both the numerator and the denominator are polynomials, the fractional expression is called a **rational expression**. Some examples of rational expressions are

$$\frac{1}{x^3 + 2x} \qquad \frac{5}{x} \qquad \frac{x + 7}{3x^2 - 5x + 1} \qquad \frac{x^2 - 2x + 4}{1}$$

In this section, we discuss basic operations on rational expressions. Since variables represent real numbers in the rational expressions we will consider, the properties of real number fractions summarized in Appendix A.1 will play a central role.

> **AGREEMENT**   Variable Restriction
> Even though not always explicitly stated, we always assume that variables are restricted so that division by 0 is excluded.

For example, given the rational expression

$$\frac{2x + 5}{x(x + 2)(x - 3)}$$

the variable x is understood to be restricted from being 0, -2, or 3, since these values would cause the denominator to be 0.

Reducing to Lowest Terms

Central to the process of reducing rational expressions to *lowest terms* is the *fundamental property of fractions*, which we restate here for convenient reference:

> **THEOREM 1**   Fundamental Property of Fractions
> If a, b, and k are real numbers with $b, k \neq 0$, then
>
> $$\frac{ka}{kb} = \frac{a}{b} \qquad \frac{5 \cdot 2}{5 \cdot 7} = \frac{2}{7} \qquad \frac{x(x + 4)}{2(x + 4)} = \frac{x}{2}, \quad x \neq -4$$

Using this property from left to right to eliminate all common factors from the numerator and the denominator of a given fraction is referred to as **reducing a fraction to lowest terms**. We are actually dividing the numerator and denominator by the same nonzero common factor.

Using the property from right to left—that is, multiplying the numerator and denominator by the same nonzero factor—is referred to as **raising a fraction to higher terms**. We will use the property in both directions in the material that follows.

EXAMPLE 1   Reducing to Lowest Terms   Reduce each fraction to lowest terms.

(A) $\dfrac{1 \cdot 2 \cdot 3 \cdot 4}{1 \cdot 2 \cdot 3 \cdot 4 \cdot 5 \cdot 6} = \dfrac{\cancel{1} \cdot \cancel{2} \cdot \cancel{3} \cdot \cancel{4}}{\cancel{1} \cdot \cancel{2} \cdot \cancel{3} \cdot \cancel{4} \cdot 5 \cdot 6} = \dfrac{1}{5 \cdot 6} = \dfrac{1}{30}$

(B) $\dfrac{2 \cdot 4 \cdot 6 \cdot 8}{1 \cdot 2 \cdot 3 \cdot 4} = \dfrac{\overset{2}{\cancel{2}} \cdot \overset{2}{\cancel{4}} \cdot \overset{2}{\cancel{6}} \cdot \overset{2}{\cancel{8}}}{\cancel{1} \cdot \cancel{2} \cdot \cancel{3} \cdot \cancel{4}} = 2 \cdot 2 \cdot 2 \cdot 2 = 16$

Matched Problem 1   Reduce each fraction to lowest terms.

(A) $\dfrac{1 \cdot 2 \cdot 3 \cdot 4 \cdot 5}{1 \cdot 2 \cdot 1 \cdot 2 \cdot 3}$ (B) $\dfrac{1 \cdot 4 \cdot 9 \cdot 16}{1 \cdot 2 \cdot 3 \cdot 4}$

CONCEPTUAL INSIGHT

Using Theorem 1 to divide the numerator and denominator of a fraction by a common factor is often referred to as **canceling**. This operation can be denoted by drawing a slanted line through each common factor and writing any remaining factors above or below the common factor. Canceling is often incorrectly applied to individual terms in the numerator or denominator, instead of to common factors. For example,

$$\frac{14 - 5}{2} = \frac{9}{2}$$ Theorem 1 does not apply. There are no common factors in the numerator.

$$\frac{14 - 5}{2} \neq \frac{\overset{7}{\cancel{14}} - 5}{\underset{1}{\cancel{2}}} = 2$$ Incorrect use of Theorem 1. To cancel 2 in the denominator, 2 must be a factor of each term in the numerator.

EXAMPLE 2   Reducing to Lowest Terms   Reduce each rational expression to lowest terms.

(A) $\dfrac{6x^2 + x - 1}{2x^2 - x - 1} = \dfrac{(2x + 1)(3x - 1)}{(2x + 1)(x - 1)}$ Factor numerator and denominator completely.

$$= \frac{3x - 1}{x - 1}$$ Divide numerator and denominator by the common factor $(2x + 1)$.

(B) $\dfrac{x^4 - 8x}{3x^3 - 2x^2 - 8x} = \dfrac{x(x - 2)(x^2 + 2x + 4)}{x(x - 2)(3x + 4)}$

$$= \frac{x^2 + 2x + 4}{3x + 4}$$

Matched Problem 2   Reduce each rational expression to lowest terms.

(A) $\dfrac{x^2 - 6x + 9}{x^2 - 9}$ (B) $\dfrac{x^3 - 1}{x^2 - 1}$

Multiplication and Division

Since we are restricting variable replacements to real numbers, multiplication and division of rational expressions follow the rules for multiplying and dividing real number fractions summarized in Appendix A.1.

THEOREM 2   Multiplication and Division

If a, b, c, and d are real numbers, then

1. $\dfrac{a}{b} \cdot \dfrac{c}{d} = \dfrac{ac}{bd}$, $b, d \neq 0$ $\dfrac{3}{5} \cdot \dfrac{x}{x + 5} = \dfrac{3x}{5(x + 5)}$

2. $\dfrac{a}{b} \div \dfrac{c}{d} = \dfrac{a}{b} \cdot \dfrac{d}{c}$, $b, c, d \neq 0$ $\dfrac{3}{5} \div \dfrac{x}{x + 5} = \dfrac{3}{5} \cdot \dfrac{x + 5}{x}$

EXAMPLE 3   Multiplication and Division   Perform the indicated operations and reduce to lowest terms.

(A) $\dfrac{10x^3y}{3xy + 9y} \cdot \dfrac{x^2 - 9}{4x^2 - 12x}$ Factor numerators and denominators. Then divide any numerator and any denominator with a like common factor.

$$= \frac{\overset{5x^2}{\cancel{10x^3y}}}{\underset{3 \cdot 1}{3\cancel{y}(\cancel{x + 3})}} \cdot \frac{\overset{1 \cdot 1}{(\cancel{x - 3})(\cancel{x + 3})}}{\underset{2 \cdot 1}{4\cancel{x}(\cancel{x - 3})}}$$

$$= \frac{5x^2}{6}$$

(B) $\dfrac{4 - 2x}{4} \div (x - 2) = \dfrac{\overset{1}{2}(2 - x)}{\underset{2}{4}} \cdot \dfrac{1}{x - 2}$ $x - 2 = \dfrac{x - 2}{1}$

$= \dfrac{2 - x}{2(x - 2)} = \dfrac{-\overset{-1}{(x - 2)}}{2\underset{1}{(x - 2)}}$ $b - a = -(a - b)$, a useful change in some problems

$= -\dfrac{1}{2}$

Matched Problem 3) Perform the indicated operations and reduce to lowest terms.

(A) $\dfrac{12x^2y^3}{2xy^2 + 6xy} \cdot \dfrac{y^2 + 6y + 9}{3y^3 + 9y^2}$ (B) $(4 - x) \div \dfrac{x^2 - 16}{5}$

Addition and Subtraction

Again, because we are restricting variable replacements to real numbers, addition and subtraction of rational expressions follow the rules for adding and subtracting real number fractions.

THEOREM 3   Addition and Subtraction

For a, b, and c real numbers,

1. $\dfrac{a}{b} + \dfrac{c}{b} = \dfrac{a + c}{b}$, $b \neq 0$ $\dfrac{x}{x + 5} + \dfrac{8}{x + 5} = \dfrac{x + 8}{x + 5}$

2. $\dfrac{a}{b} - \dfrac{c}{b} = \dfrac{a - c}{b}$, $b \neq 0$ $\dfrac{x}{3x^2y^2} - \dfrac{x + 7}{3x^2y^2} = \dfrac{x - (x + 7)}{3x^2y^2}$

We add rational expressions with the same denominators by adding or subtracting their numerators and placing the result over the common denominator. If the denominators are not the same, we raise the fractions to higher terms, using the fundamental property of fractions to obtain common denominators, and then proceed as described.

Even though any common denominator will do, our work will be simplified if the *least common denominator (LCD)* is used. Often, the LCD is obvious, but if it is not, the steps in the next box describe how to find it.

PROCEDURE   Least Common Denominator

The least common denominator (LCD) of two or more rational expressions is found as follows:

1. Factor each denominator completely, including integer factors.
2. Identify each different factor from all the denominators.
3. Form a product using each different factor to the highest power that occurs in any one denominator. This product is the LCD.

EXAMPLE 4   Addition and Subtraction   Combine into a single fraction and reduce to lowest terms.

(A) $\dfrac{3}{10} + \dfrac{5}{6} - \dfrac{11}{45}$ (B) $\dfrac{4}{9x} - \dfrac{5x}{6y^2} + 1$ (C) $\dfrac{1}{x - 1} - \dfrac{1}{x} - \dfrac{2}{x^2 - 1}$

SOLUTION

(A) To find the LCD, factor each denominator completely:

$$\left.\begin{array}{l} 10 = 2 \cdot 5 \\ 6 = 2 \cdot 3 \\ 45 = 3^2 \cdot 5 \end{array}\right\} \quad LCD = 2 \cdot 3^2 \cdot 5 = 90$$

Now use the fundamental property of fractions to make each denominator 90:

$$\frac{3}{10} + \frac{5}{6} - \frac{11}{45} = \frac{9 \cdot 3}{9 \cdot 10} + \frac{15 \cdot 5}{15 \cdot 6} - \frac{2 \cdot 11}{2 \cdot 45}$$

$$= \frac{27}{90} + \frac{75}{90} - \frac{22}{90}$$

$$= \frac{27 + 75 - 22}{90} = \frac{80}{90} = \frac{8}{9}$$

(B) $$\left.\begin{array}{l} 9x = 3^2 x \\ 6y^2 = 2 \cdot 3y^2 \end{array}\right\} \quad LCD = 2 \cdot 3^2 xy^2 = 18xy^2$$

$$\frac{4}{9x} - \frac{5x}{6y^2} + 1 = \frac{2y^2 \cdot 4}{2y^2 \cdot 9x} - \frac{3x \cdot 5x}{3x \cdot 6y^2} + \frac{18xy^2}{18xy^2}$$

$$= \frac{8y^2 - 15x^2 + 18xy^2}{18xy^2}$$

(C) $$\frac{1}{x - 1} - \frac{1}{x} - \frac{2}{x^2 - 1}$$

$$= \frac{1}{x - 1} - \frac{1}{x} - \frac{2}{(x - 1)(x + 1)} \qquad LCD = x(x - 1)(x + 1)$$

$$= \frac{x(x + 1) - (x - 1)(x + 1) - 2x}{x(x - 1)(x + 1)}$$

$$= \frac{x^2 + x - x^2 + 1 - 2x}{x(x - 1)(x + 1)}$$

$$= \frac{1 - x}{x(x - 1)(x + 1)}$$

$$= \frac{-\overset{-1}{(x - 1)}}{x\underset{1}{(x - 1)}(x + 1)} = \frac{-1}{x(x + 1)}$$

Matched Problem 4 Combine into a single fraction and reduce to lowest terms.

(A) $$\frac{5}{28} - \frac{1}{10} + \frac{6}{35}$$

(B) $$\frac{1}{4x^2} - \frac{2x + 1}{3x^3} + \frac{3}{12x}$$

(C) $$\frac{2}{x^2 - 4x + 4} + \frac{1}{x} - \frac{1}{x - 2}$$

Compound Fractions

A fractional expression with fractions in its numerator, denominator, or both is called a **compound fraction**. It is often necessary to represent a compound fraction as a **simple fraction**—that is (in all cases we will consider), as the quotient of two polynomials. The process does not involve any new concepts. It is a matter of applying old concepts and processes in the correct sequence.

EXAMPLE 5   Simplifying Compound Fractions   Express as a simple fraction reduced to lowest terms:

(A) $\dfrac{\dfrac{1}{5+h} - \dfrac{1}{5}}{h}$

(B) $\dfrac{\dfrac{y}{x^2} - \dfrac{x}{y^2}}{\dfrac{y}{x} - \dfrac{x}{y}}$

SOLUTION   We will simplify the expressions in parts (A) and (B) using two different methods—each is suited to the particular type of problem.

(A) We simplify this expression by combining the numerator into a single fraction and using division of rational forms.

$$\dfrac{\dfrac{1}{5+h} - \dfrac{1}{5}}{h} = \left[\dfrac{1}{5+h} - \dfrac{1}{5} \right] \div \dfrac{h}{1}$$

$$= \dfrac{5-5-h}{5(5+h)} \cdot \dfrac{1}{h}$$

$$= \dfrac{-h}{5(5+h)h} = \dfrac{-1}{5(5+h)}$$

(B) The method used here makes effective use of the fundamental property of fractions in the form

$$\dfrac{a}{b} = \dfrac{ka}{kb} \qquad b, k \neq 0$$

Multiply the numerator and denominator by the LCD of all fractions in the numerator and denominator—in this case, $x^2 y^2$:

$$\dfrac{x^2 y^2 \left(\dfrac{y}{x^2} - \dfrac{x}{y^2} \right)}{x^2 y^2 \left(\dfrac{y}{x} - \dfrac{x}{y} \right)} = \dfrac{x^2 y^2 \dfrac{y}{x^2} - x^2 y^2 \dfrac{x}{y^2}}{x^2 y^2 \dfrac{y}{x} - x^2 y^2 \dfrac{x}{y}} = \dfrac{y^3 - x^3}{xy^3 - x^3 y}$$

$$= \dfrac{\cancel{(y-x)}(y^2 + xy + x^2)}{xy\cancel{(y-x)}(y+x)}$$

$$= \dfrac{y^2 + xy + x^2}{xy(y+x)} \quad \text{or} \quad \dfrac{x^2 + xy + y^2}{xy(x+y)}$$

Matched Problem 5   Express as a simple fraction reduced to lowest terms:

(A) $\dfrac{\dfrac{1}{2+h} - \dfrac{1}{2}}{h}$

(B) $\dfrac{\dfrac{a}{b} - \dfrac{b}{a}}{\dfrac{a}{b} + 2 + \dfrac{b}{a}}$

Exercises A.4

In Problems 1–22, perform the indicated operations and reduce answers to lowest terms.

1. $\dfrac{5 \cdot 9 \cdot 13}{3 \cdot 5 \cdot 7}$

2. $\dfrac{10 \cdot 9 \cdot 8}{3 \cdot 2 \cdot 1}$

3. $\dfrac{12 \cdot 11 \cdot 10 \cdot 9}{4 \cdot 3 \cdot 2 \cdot 1}$

4. $\dfrac{15 \cdot 10 \cdot 5}{20 \cdot 15 \cdot 10}$

5. $\dfrac{d^5}{3a} \div \left(\dfrac{d^2}{6a^2} \cdot \dfrac{a}{4d^3} \right)$

6. $\left(\dfrac{d^5}{3a} \div \dfrac{d^2}{6a^2} \right) \cdot \dfrac{a}{4d^3}$

7. $\dfrac{x^2}{12} + \dfrac{x}{18} - \dfrac{1}{30}$

8. $\dfrac{2y}{18} - \dfrac{-1}{28} - \dfrac{y}{42}$

9. $\dfrac{4m - 3}{18m^3} + \dfrac{3}{4m} - \dfrac{2m - 1}{6m^2}$

10. $\dfrac{3x + 8}{4x^2} - \dfrac{2x - 1}{x^3} - \dfrac{5}{8x}$

11. $\dfrac{x^2 - 9}{x^2 - 3x} \div (x^2 - x - 12)$

12. $\dfrac{2x^2 + 7x + 3}{4x^2 - 1} \div (x + 3)$

13. $\dfrac{2}{x} - \dfrac{1}{x - 3}$

14. $\dfrac{5}{m - 2} - \dfrac{3}{2m + 1}$

15. $\dfrac{2}{(x + 1)^2} - \dfrac{5}{x^2 - x - 2}$

16. $\dfrac{3}{x^2 - 5x + 6} - \dfrac{5}{(x - 2)^2}$

17. $\dfrac{x + 1}{x - 1} - 1$

18. $m - 3 - \dfrac{m - 1}{m - 2}$

19. $\dfrac{3}{a - 1} - \dfrac{2}{1 - a}$

20. $\dfrac{5}{x - 3} - \dfrac{2}{3 - x}$

21. $\dfrac{2x}{x^2 - 16} - \dfrac{x - 4}{x^2 + 4x}$

22. $\dfrac{m + 2}{m^2 - 2m} - \dfrac{m}{m^2 - 4}$

In Problems 23–34, perform the indicated operations and reduce answers to lowest terms. Represent any compound fractions as simple fractions reduced to lowest terms.

23. $\dfrac{x^2}{x^2 + 2x + 1} + \dfrac{x - 1}{3x + 3} - \dfrac{1}{6}$

24. $\dfrac{y}{y^2 - y - 2} - \dfrac{1}{y^2 + 5y - 14} - \dfrac{2}{y^2 + 8y + 7}$

25. $\dfrac{1 - \dfrac{x}{y}}{2 - \dfrac{y}{x}}$

26. $\dfrac{2}{5 - \dfrac{3}{4x + 1}}$

27. $\dfrac{c + 2}{5c - 5} - \dfrac{c - 2}{3c - 3} + \dfrac{c}{1 - c}$

28. $\dfrac{x + 7}{ax - bx} + \dfrac{y + 9}{by - ay}$

29. $\dfrac{1 + \dfrac{3}{x}}{x - \dfrac{9}{x}}$

30. $\dfrac{1 - \dfrac{y^2}{x^2}}{1 - \dfrac{y}{x}}$

31. $\dfrac{\dfrac{1}{2(x + h)} - \dfrac{1}{2x}}{h}$

32. $\dfrac{\dfrac{1}{x + h} - \dfrac{1}{x}}{h}$

33. $\dfrac{\dfrac{x}{y} - 2 + \dfrac{y}{x}}{\dfrac{x}{y} - \dfrac{y}{x}}$

34. $\dfrac{1 + \dfrac{2}{x} - \dfrac{15}{x^2}}{1 + \dfrac{4}{x} - \dfrac{5}{x^2}}$

In Problems 35–42, imagine that the indicated "solutions" were given to you by a student whom you were tutoring in this class.

(A) Is the solution correct? If the solution is incorrect, explain what is wrong and how it can be corrected.

(B) Show a correct solution for each incorrect solution.

35. $\dfrac{x^2 + 4x + 3}{x + 3} = \dfrac{x^2 + 4x}{x} = x + 4$

36. $\dfrac{x^2 - 3x - 4}{x - 4} = \dfrac{x^2 - 3x}{x} = x - 3$

37. $\dfrac{(x + h)^2 - x^2}{h} = (x + 1)^2 - x^2 = 2x + 1$

38. $\dfrac{(x + h)^3 - x^3}{h} = (x + 1)^3 - x^3 = 3x^2 + 3x + 1$

39. $\dfrac{x^2 - 3x}{x^2 - 2x - 3} + x - 3 = \dfrac{x^2 - 3x + x - 3}{x^2 - 2x - 3} = 1$

40. $\dfrac{2}{x - 1} - \dfrac{x + 3}{x^2 - 1} = \dfrac{2x + 2 - x - 3}{x^2 - 1} = \dfrac{1}{x + 1}$

41. $\dfrac{2x^2}{x^2 - 4} - \dfrac{x}{x - 2} = \dfrac{2x^2 - x^2 - 2x}{x^2 - 4} = \dfrac{x}{x + 2}$

42. $x + \dfrac{x - 2}{x^2 - 3x + 2} = \dfrac{x + x - 2}{x^2 - 3x + 2} = \dfrac{2}{x - 2}$

Represent the compound fractions in Problems 43–46 as simple fractions reduced to lowest terms.

43. $\dfrac{\dfrac{1}{3(x + h)^2} - \dfrac{1}{3x^2}}{h}$

44. $\dfrac{\dfrac{1}{(x + h)^2} - \dfrac{1}{x^2}}{h}$

45. $x - \dfrac{2}{1 - \dfrac{1}{x}}$

46. $2 - \dfrac{1}{1 - \dfrac{2}{a + 2}}$

Answers to Matched Problems

1. (A) 10   (B) 24

2. (A) $\dfrac{x - 3}{x + 3}$ (B) $\dfrac{x^2 + x + 1}{x + 1}$

3. (A) $2x$ (B) $\dfrac{-5}{x + 4}$

4. (A) $\dfrac{1}{4}$ (B) $\dfrac{3x^2 - 5x - 4}{12x^3}$ (C) $\dfrac{4}{x(x - 2)^2}$

5. (A) $\dfrac{-1}{2(2 + h)}$ (B) $\dfrac{a - b}{a + b}$

A.5   Integer Exponents and Scientific Notation

- Integer Exponents
- Scientific Notation

We now review basic operations on integer exponents and scientific notation.

Integer Exponents

DEFINITION   Integer Exponents

For n an integer and a a real number:

1. For n a positive integer,

$$a^n = a \cdot a \cdot \,\cdots\, \cdot a \quad n \text{ factors of } a \qquad 5^4 = 5 \cdot 5 \cdot 5 \cdot 5$$

2. For $n = 0$,

$$a^0 = 1 \quad a \neq 0 \qquad 12^0 = 1$$
$$0^0 \text{ is not defined.}$$

3. For n a negative integer,

$$a^n = \frac{1}{a^{-n}} \quad a \neq 0 \qquad a^{-3} = \frac{1}{a^{-(-3)}} = \frac{1}{a^3}$$

$\big[$If n is negative, then $(-n)$ is positive.$\big]$

Note: It can be shown that for *all* integers n,

$$a^{-n} = \frac{1}{a^n} \quad \text{and} \quad a^n = \frac{1}{a^{-n}} \quad a \neq 0 \qquad a^5 = \frac{1}{a^{-5}}, \quad a^{-5} = \frac{1}{a^5}$$

The following properties are very useful in working with integer exponents.

THEOREM 1   Exponent Properties

For n and m integers and a and b real numbers,

1. $a^m a^n = a^{m+n}$ $\qquad\qquad\qquad a^8 a^{-3} = a^{8+(-3)} = a^5$

2. $(a^n)^m = a^{mn}$ $\qquad\qquad\qquad (a^{-2})^3 = a^{3(-2)} = a^{-6}$

3. $(ab)^m = a^m b^m$ $\qquad\qquad\quad (ab)^{-2} = a^{-2}b^{-2}$

4. $\left(\dfrac{a}{b}\right)^m = \dfrac{a^m}{b^m} \quad b \neq 0 \qquad\quad \left(\dfrac{a}{b}\right)^5 = \dfrac{a^5}{b^5}$

5. $\dfrac{a^m}{a^n} = a^{m-n} = \dfrac{1}{a^{n-m}} \quad a \neq 0 \qquad \dfrac{a^{-3}}{a^7} = \dfrac{1}{a^{7-(-3)}} = \dfrac{1}{a^{10}}$

Exponents are frequently encountered in algebraic applications. You should sharpen your skills in using exponents by reviewing the preceding basic definitions and properties and the examples that follow.

EXAMPLE 1   Simplifying Exponent Forms   Simplify, and express the answers using positive exponents only.

(A) $(2x^3)(3x^5) \; \boxed{= 2 \cdot 3x^{3+5}} \; = 6x^8$

(B) $x^5 x^{-9} = x^{-4} = \dfrac{1}{x^4}$

(C) $\dfrac{x^5}{x^7} \; \boxed{= x^{5-7}} \; = x^{-2} = \dfrac{1}{x^2} \quad \text{or} \quad \dfrac{x^5}{x^7} \; \boxed{= \dfrac{1}{x^{7-5}}} \; = \dfrac{1}{x^2}$

(D) $\dfrac{x^{-3}}{y^{-4}} = \dfrac{y^4}{x^3}$

(E) $(u^{-3}v^2)^{-2} = (u^{-3})^{-2}(v^2)^{-2} = u^6v^{-4} = \dfrac{u^6}{v^4}$

(F) $\left(\dfrac{y^{-5}}{y^{-2}}\right)^{-2} = \dfrac{(y^{-5})^{-2}}{(y^{-2})^{-2}} = \dfrac{y^{10}}{y^4} = y^6$

(G) $\dfrac{4m^{-3}n^{-5}}{6m^{-4}n^3} = \dfrac{2m^{-3-(-4)}}{3n^{3-(-5)}} = \dfrac{2m}{3n^8}$

Matched Problem 1 Simplify, and express the answers using positive exponents only.

(A) $(3y^4)(2y^3)$ (B) m^2m^{-6} (C) $(u^3v^{-2})^{-2}$

(D) $\left(\dfrac{y^{-6}}{y^{-2}}\right)^{-1}$ (E) $\dfrac{8x^{-2}y^{-4}}{6x^{-5}y^2}$

EXAMPLE 2 Converting to a Simple Fraction Write $\dfrac{1-x}{x^{-1}-1}$ as a simple fraction with positive exponents.

SOLUTION First note that

$$\dfrac{1-x}{x^{-1}-1} \neq \dfrac{x(1-x)}{-1} \qquad \text{A common error}$$

The original expression is a compound fraction, and we proceed to simplify it as follows:

$$\dfrac{1-x}{x^{-1}-1} = \dfrac{1-x}{\dfrac{1}{x}-1} \qquad \begin{array}{l}\text{Multiply numerator and denominator}\\ \text{by } x \text{ to clear internal fractions.}\end{array}$$

$$= \dfrac{x(1-x)}{x\left(\dfrac{1}{x}-1\right)}$$

$$= \dfrac{x(1-x)}{1-x} = x$$

Matched Problem 2 Write $\dfrac{1+x^{-1}}{1-x^{-2}}$ as a simple fraction with positive exponents.

Scientific Notation

In the real world, one often encounters very large and very small numbers. For example,

- The public debt in the United States in 2013, to the nearest billion dollars, was

  $16,739,000,000,000

- The world population in the year 2025, to the nearest million, is projected to be

  7,947,000,000

- The sound intensity of a normal conversation is

  0.000 000 000 316 watt per square centimeter*

---

*We write 0.000 000 000 316 in place of 0.000000000316, because it is then easier to keep track of the number of decimal places.

It is generally troublesome to write and work with numbers of this type in standard decimal form. The first and last example cannot even be entered into many calculators as they are written. But with exponents defined for all integers, we can now express any finite decimal form as the product of a number between 1 and 10 and an integer power of 10, that is, in the form

$$a \times 10^n \qquad 1 \le a < 10, \quad a \text{ in decimal form}, \quad n \text{ an integer}$$

A number expressed in this form is said to be in **scientific notation**. The following are some examples of numbers in standard decimal notation and in scientific notation:

Decimal and Scientific Notation

$7 = 7 \times 10^0$	$0.5 = 5 \times 10^{-1}$
$67 = 6.7 \times 10$	$0.45 = 4.5 \times 10^{-1}$
$580 = 5.8 \times 10^2$	$0.0032 = 3.2 \times 10^{-3}$
$43,000 = 4.3 \times 10^4$	$0.000\ 045 = 4.5 \times 10^{-5}$
$73,400,000 = 7.34 \times 10^7$	$0.000\ 000\ 391 = 3.91 \times 10^{-7}$

Note that the power of 10 used corresponds to the number of places we move the decimal to form a number between 1 and 10. The power is positive if the decimal is moved to the left and negative if it is moved to the right. Positive exponents are associated with numbers greater than or equal to 10; negative exponents are associated with positive numbers less than 1; and a zero exponent is associated with a number that is 1 or greater, but less than 10.

EXAMPLE 3   Scientific Notation

(A) Write each number in scientific notation:

$$7,320,000 \quad \text{and} \quad 0.000\ 000\ 54$$

(B) Write each number in standard decimal form:

$$4.32 \times 10^6 \quad \text{and} \quad 4.32 \times 10^{-5}$$

SOLUTION

(A) $7,320,000 = 7.320\ 000. \times 10^6 = 7.32 \times 10^6$

    6 places left
    Positive exponent

$0.000\ 000\ 54 = 0.000\ 000\ 5.4 \times 10^{-7} = 5.4 \times 10^{-7}$

    7 places right
    Negative exponent

(B) $4.32 \times 10^6 = 4,320,000$ $4.32 \times 10^{-5} = \dfrac{4.32}{10^5} = 0.000\ 043\ 2$

    6 places right   5 places left
    Positive exponent 6   Negative exponent −5

Matched Problem 3

(A) Write each number in scientific notation: 47,100; 2,443,000,000; 1.45

(B) Write each number in standard decimal form: 3.07×10^8; 5.98×10^{-6}

In Problems 1–14, simplify and express answers using positive exponents only. Variables are restricted to avoid division by 0.

1. $2x^{-9}$

2. $3y^{-5}$

3. $\dfrac{3}{2w^{-7}}$

4. $\dfrac{5}{4x^{-9}}$

5. $2x^{-8}x^5$

6. $3c^{-9}c^4$

7. $\dfrac{w^{-8}}{w^{-3}}$

8. $\dfrac{m^{-11}}{m^{-5}}$

9. $(2a^{-3})^2$

10. $7d^{-4}d^4$

11. $(a^{-3})^2$

12. $(5b^{-2})^2$

13. $(2x^4)^{-3}$

14. $(a^{-3}b^4)^{-3}$

In Problems 15–20, write each number in scientific notation.

15. 82,300,000,000

16. 5,380,000

17. 0.783

18. 0.019

19. 0.000 034

20. 0.000 000 007 832

In Problems 21–28, write each number in standard decimal notation.

21. 4×10^4

22. 9×10^6

23. 7×10^{-3}

24. 2×10^{-5}

25. 6.171×10^7

26. 3.044×10^3

27. 8.08×10^{-4}

28. 1.13×10^{-2}

In Problems 29–38, simplify and express answers using positive exponents only. Assume that variables are nonzero.

29. $(22 + 31)^0$

30. $(2x^3y^4)^0$

31. $\dfrac{10^{-3} \cdot 10^4}{10^{-11} \cdot 10^{-2}}$

32. $\dfrac{10^{-17} \cdot 10^{-5}}{10^{-3} \cdot 10^{-14}}$

33. $(5x^2y^{-3})^2$

34. $(2m^{-3}n^2)^{-3}$

35. $\left(\dfrac{-5}{2x^3}\right)^{-2}$

36. $\left(\dfrac{2a}{3b^2}\right)^{-3}$

37. $\dfrac{8x^{-3}y^{-1}}{6x^2y^{-4}}$

38. $\dfrac{9m^{-4}n^3}{12m^{-1}n^{-1}}$

In Problems 39–42, write each expression in the form $ax^p + bx^q$ or $ax^p + bx^q + cx^r$, where a, b, and c are real numbers and p, q, and r are integers. For example,

$$\frac{2x^4 - 3x^2 + 1}{2x^3} \;\left[= \frac{2x^4}{2x^3} - \frac{3x^2}{2x^3} + \frac{1}{2x^3}\right] = x - \frac{3}{2}x^{-1} + \frac{1}{2}x^{-3}$$

39. $\dfrac{7x^5 - x^2}{4x^5}$

40. $\dfrac{5x^3 - 2}{3x^2}$

41. $\dfrac{5x^4 - 3x^2 + 8}{2x^2}$

42. $\dfrac{2x^3 - 3x^2 + x}{2x^2}$

Write each expression in Problems 43–46 with positive exponents only, and as a single fraction reduced to lowest terms.

43. $\dfrac{3x^2(x - 1)^2 - 2x^3(x - 1)}{(x - 1)^4}$

44. $\dfrac{5x^4(x + 3)^2 - 2x^5(x + 3)}{(x + 3)^4}$

45. $2x^{-2}(x - 1) - 2x^{-3}(x - 1)^2$

46. $2x(x + 3)^{-1} - x^2(x + 3)^{-2}$

In Problems 47–50, convert each number to scientific notation and simplify. Express the answer in both scientific notation and in standard decimal form.

47. $\dfrac{9,600,000,000}{(1,600,000)(0.000\,000\,25)}$

48. $\dfrac{(60,000)(0.000\,003)}{(0.0004)(1,500,000)}$

49. $\dfrac{(1,250,000)(0.000\,38)}{0.0152}$

50. $\dfrac{(0.000\,000\,82)(230,000)}{(625,000)(0.0082)}$

51. What is the result of entering 2^{3^2} on a calculator?

52. Refer to Problem 51. What is the difference between $2^{(3^2)}$ and $(2^3)^2$? Which agrees with the value of 2^{3^2} obtained with a calculator?

53. If $n = 0$, then property 1 in Theorem 1 implies that $a^m a^0 = a^{m+0} = a^m$. Explain how this helps motivate the definition of a^0.

54. If $m = -n$, then property 1 in Theorem 1 implies that $a^{-n}a^n = a^0 = 1$. Explain how this helps motivate the definition of a^{-n}.

Write the fractions in Problems 55–58 as simple fractions reduced to lowest terms.

55. $\dfrac{u + v}{u^{-1} + v^{-1}}$

56. $\dfrac{x^{-2} - y^{-2}}{x^{-1} + y^{-1}}$

57. $\dfrac{b^{-2} - c^{-2}}{b^{-3} - c^{-3}}$

58. $\dfrac{xy^{-2} - yx^{-2}}{y^{-1} - x^{-1}}$

Applications

Problems 59 and 60 refer to Table 1.

Table 1 **U.S. Public Debt, Interest on Debt, and Population**

Year	Public Debt ($)	Interest on Debt ($)	Population
2000	5,674,000,000,000	362,000,000,000	281,000,000
2012	16,066,000,000,000	360,000,000,000	313,000,000

59. **Public debt.** Carry out the following computations using scientific notation, and write final answers in standard decimal form.

(A) What was the per capita debt in 2012 (to the nearest dollar)?

(B) What was the per capita interest paid on the debt in 2012 (to the nearest dollar)?

(C) What was the percentage interest paid on the debt in 2012 (to two decimal places)?

60. Public debt. Carry out the following computations using scientific notation, and write final answers in standard decimal form.

(A) What was the per capita debt in 2000 (to the nearest dollar)?

(B) What was the per capita interest paid on the debt in 2000 (to the nearest dollar)?

(C) What was the percentage interest paid on the debt in 2000 (to two decimal places)?

Air pollution. *Air quality standards establish maximum amounts of pollutants considered acceptable in the air. The amounts are frequently given in parts per million (ppm). A standard of 30 ppm also can be expressed as follows:*

$$30 \text{ ppm} = \frac{30}{1{,}000{,}000} = \frac{3 \times 10}{10^6}$$
$$= 3 \times 10^{-5} = 0.00003 = 0.003\%$$

In Problems 61 and 62, express the given standard:

(A) In scientific notation

(B) In standard decimal notation

(C) As a percent

61. 9 ppm, the standard for carbon monoxide, when averaged over a period of 8 hours

62. 0.03 ppm, the standard for sulfur oxides, when averaged over a year

63. Crime. In 2010, the United States had a violent crime rate of 404 per 100,000 people and a population of 309 million people. How many violent crimes occurred that year? Compute the answer using scientific notation and convert the answer to standard decimal form (to the nearest thousand).

64. Population density. The United States had a 2012 population of 313 million people and a land area of 3,539,000 square miles. What was the population density? Compute the answer using scientific notation and convert the answer to standard decimal form (to one decimal place).

Answers to Matched Problems

1. (A) $6y^7$ (B) $\dfrac{1}{m^4}$ (C) $\dfrac{v^4}{u^6}$ (D) y^4 (E) $\dfrac{4x^3}{3y^6}$

2. $\dfrac{x}{x-1}$

3. (A) 4.7×10^4; 2.443×10^9; 1.45×10^0

(B) 307,000,000; 0.000 005 98

A.6 Rational Exponents and Radicals

- *n*th Roots of Real Numbers
- Rational Exponents and Radicals
- Properties of Radicals

Square roots may now be generalized to *n*th roots, and the meaning of exponent may be generalized to include all rational numbers.

*n*th Roots of Real Numbers

Consider a square of side r with area 36 square inches. We can write

$$r^2 = 36$$

and conclude that side r is a number whose square is 36. We say that r is a **square root** of b if $r^2 = b$. Similarly, we say that r is a **cube root** of b if $r^3 = b$. And, in general,

---

DEFINITION *n*th Root

For any natural number n,

$$r \text{ is an } \textbf{\textit{n}th root} \text{ of } b \text{ if } r^n = b$$

So 4 is a square root of 16, since $4^2 = 16$; -2 is a cube root of -8, since $(-2)^3 = -8$. Since $(-4)^2 = 16$, we see that -4 is also a square root of 16. It can be shown that any positive number has two real square roots, two real 4th roots, and, in general, two real *n*th roots if n is even. Negative numbers have no real square roots, no real 4th roots, and, in general, no real *n*th roots if n is even. The reason is that no real number raised to an even power can be negative. For odd roots, the situation is simpler. Every real number has exactly one real cube root, one real 5th root, and, in general, one real *n*th root if n is odd.

Additional roots can be considered in the *complex number system*. In this book, we restrict our interest to *real roots of real numbers*, and *root* will always be interpreted to mean "real root."

Rational Exponents and Radicals

We now turn to the question of what symbols to use to represent *n*th roots. For *n* a natural number greater than 1, we use

$$b^{1/n} \quad \text{or} \quad \sqrt[n]{b}$$

to represent a **real *n*th root of *b***. The exponent form is motivated by the fact that $(b^{1/n})^n = b$ if exponent laws are to continue to hold for rational exponents. The other form is called an ***n*th root radical**. In the expression below, the symbol $\sqrt{}$ is called a **radical**, *n* is the **index** of the radical, and *b* is the **radicand**:

$$\text{Index} \longrightarrow \quad \underset{\longleftarrow \text{ Radicand}}{\overset{\longleftarrow \text{ Radical}}{\sqrt[n]{b}}}$$

When the index is 2, it is usually omitted. That is, when dealing with square roots, we simply use $\sqrt{b}$ rather than $\sqrt[2]{b}$. If there are two real *n*th roots, both $b^{1/n}$ and $\sqrt[n]{b}$ denote the positive root, called the **principal *n*th root**.

EXAMPLE 1 Finding *n*th Roots Evaluate each of the following:

(A) $4^{1/2}$ and $\sqrt{4}$ (B) $-4^{1/2}$ and $-\sqrt{4}$ (C) $(-4)^{1/2}$ and $\sqrt{-4}$

(D) $8^{1/3}$ and $\sqrt[3]{8}$ (E) $(-8)^{1/3}$ and $\sqrt[3]{-8}$ (F) $-8^{1/3}$ and $-\sqrt[3]{8}$

SOLUTION

(A) $4^{1/2} = \sqrt{4} = 2 \quad (\sqrt{4} \neq \pm 2)$ (B) $-4^{1/2} = -\sqrt{4} = -2$

(C) $(-4)^{1/2}$ and $\sqrt{-4}$ are not real numbers

(D) $8^{1/3} = \sqrt[3]{8} = 2$ (E) $(-8)^{1/3} = \sqrt[3]{-8} = -2$

(F) $-8^{1/3} = -\sqrt[3]{8} = -2$

Matched Problem 1 Evaluate each of the following:

(A) $16^{1/2}$ (B) $-\sqrt{16}$ (C) $\sqrt[3]{-27}$ (D) $(-9)^{1/2}$ (E) $\left(\sqrt[4]{81}\right)^3$

⚠ **CAUTION** The symbol $\sqrt{4}$ represents the single number 2, not ± 2. Do not confuse $\sqrt{4}$ with the solutions of the equation $x^2 = 4$, which are usually written in the form $x = \pm\sqrt{4} = \pm 2$. ▲

We now define b^r for any rational number $r = m/n$.

DEFINITION Rational Exponents

If *m* and *n* are natural numbers without common prime factors, *b* is a real number, and *b* is nonnegative when *n* is even, then

$$b^{m/n} = \begin{cases} (b^{1/n})^m = (\sqrt[n]{b})^m \\ (b^m)^{1/n} = \sqrt[n]{b^m} \end{cases} \qquad \begin{array}{l} 8^{2/3} = (8^{1/3})^2 = (\sqrt[3]{8})^2 = 2^2 = 4 \\ 8^{2/3} = (8^2)^{1/3} = \sqrt[3]{8^2} = \sqrt[3]{64} = 4 \end{array}$$

and

$$b^{-m/n} = \frac{1}{b^{m/n}} \quad b \neq 0 \qquad 8^{-2/3} = \frac{1}{8^{2/3}} = \frac{1}{4}$$

Note that the two definitions of $b^{m/n}$ are equivalent under the indicated restrictions on *m*, *n*, and *b*.

CONCEPTUAL INSIGHT

All the properties for integer exponents listed in Theorem 1 in Section A.5 also hold for rational exponents, provided that b is nonnegative when n is even. This restriction on b is necessary to avoid nonreal results. For example,

$$(-4)^{3/2} = \sqrt{(-4)^3} = \sqrt{-64} \quad \text{Not a real number}$$

To avoid nonreal results, all variables in the remainder of this discussion represent positive real numbers.

EXAMPLE 2   From Rational Exponent Form to Radical Form and Vice Versa   Change rational exponent form to radical form.

(A) $x^{1/7} = \sqrt[7]{x}$

(B) $(3u^2v^3)^{3/5} = \sqrt[5]{(3u^2v^3)^3}$ or $(\sqrt[5]{3u^2v^3})^3$ The first is usually preferred.

(C) $y^{-2/3} = \dfrac{1}{y^{2/3}} = \dfrac{1}{\sqrt[3]{y^2}}$ or $\sqrt[3]{y^{-2}}$ or $\sqrt[3]{\dfrac{1}{y^2}}$

Change radical form to rational exponent form.

(D) $\sqrt[5]{6} = 6^{1/5}$ (E) $-\sqrt[3]{x^2} = -x^{2/3}$

(F) $\sqrt{x^2 + y^2} = (x^2 + y^2)^{1/2}$ Note that $(x^2 + y^2)^{1/2} \neq x + y$. Why?

Matched Problem 2   Convert to radical form.

(A) $u^{1/5}$ (B) $(6x^2y^5)^{2/9}$ (C) $(3xy)^{-3/5}$

Convert to rational exponent form.

(D) $\sqrt[4]{9u}$ (E) $-\sqrt[7]{(2x)^4}$ (F) $\sqrt[3]{x^3 + y^3}$

EXAMPLE 3   Working with Rational Exponents   Simplify each and express answers using positive exponents only. If rational exponents appear in final answers, convert to radical form.

(A) $(3x^{1/3})(2x^{1/2}) = 6x^{1/3+1/2} = 6x^{5/6} = 6\sqrt[6]{x^5}$

(B) $(-8)^{5/3} = [(-8)^{1/3}]^5 = (-2)^5 = -32$

(C) $(2x^{1/3}y^{-2/3})^3 = 8xy^{-2} = \dfrac{8x}{y^2}$

(D) $\left(\dfrac{4x^{1/3}}{x^{1/2}}\right)^{1/2} = \dfrac{4^{1/2}x^{1/6}}{x^{1/4}} = \dfrac{2}{x^{1/4-1/6}} = \dfrac{2}{x^{1/12}} = \dfrac{2}{\sqrt[12]{x}}$

Matched Problem 3   Simplify each and express answers using positive exponents only. If rational exponents appear in final answers, convert to radical form.

(A) $9^{3/2}$ (B) $(-27)^{4/3}$ (C) $(5y^{1/4})(2y^{1/3})$ (D) $(2x^{-3/4}y^{1/4})^4$

(E) $\left(\dfrac{8x^{1/2}}{x^{2/3}}\right)^{1/3}$

EXAMPLE 4   Working with Rational Exponents   Multiply, and express answers using positive exponents only.

(A) $3y^{2/3}(2y^{1/3} - y^2)$ (B) $(2u^{1/2} + v^{1/2})(u^{1/2} - 3v^{1/2})$

SOLUTION

(A) $3y^{2/3}(2y^{1/3} - y^2) = 6y^{2/3+1/3} - 3y^{2/3+2}$

$= 6y - 3y^{8/3}$

(B) $(2u^{1/2} + v^{1/2})(u^{1/2} - 3v^{1/2}) = 2u - 5u^{1/2}v^{1/2} - 3v$

Matched Problem 4) Multiply, and express answers using positive exponents only.

(A) $2c^{1/4}(5c^3 - c^{3/4})$ (B) $(7x^{1/2} - y^{1/2})(2x^{1/2} + 3y^{1/2})$

EXAMPLE 5 Working with Rational Exponents Write the following expression in the form $ax^p + bx^q$, where a and b are real numbers and p and q are rational numbers:

$$\frac{2\sqrt{x} - 3\sqrt[3]{x^2}}{2\sqrt[3]{x}}$$

SOLUTION $\dfrac{2\sqrt{x} - 3\sqrt[3]{x^2}}{2\sqrt[3]{x}} = \dfrac{2x^{1/2} - 3x^{2/3}}{2x^{1/3}}$ Change to rational exponent form.

$$= \frac{2x^{1/2}}{2x^{1/3}} - \frac{3x^{2/3}}{2x^{1/3}}$$ Separate into two fractions.

$$= x^{1/6} - 1.5x^{1/3}$$

Matched Problem 5) Write the following expression in the form $ax^p + bx^q$, where a and b are real numbers and p and q are rational numbers:

$$\frac{5\sqrt[3]{x} - 4\sqrt{x}}{2\sqrt{x^3}}$$

Properties of Radicals

Changing or simplifying radical expressions is aided by several properties of radicals that follow directly from the properties of exponents considered earlier.

THEOREM 1 Properties of Radicals

If n is a natural number greater than or equal to 2, and if x and y are positive real numbers, then

1. $\sqrt[n]{x^n} = x$ $\sqrt[3]{x^3} = x$
2. $\sqrt[n]{xy} = \sqrt[n]{x}\sqrt[n]{y}$ $\sqrt[5]{xy} = \sqrt[5]{x}\,\sqrt[5]{y}$
3. $\sqrt[n]{\dfrac{x}{y}} = \dfrac{\sqrt[n]{x}}{\sqrt[n]{y}}$ $\sqrt[4]{\dfrac{x}{y}} = \dfrac{\sqrt[4]{x}}{\sqrt[4]{y}}$

EXAMPLE 6 Applying Properties of Radicals Simplify using properties of radicals.

(A) $\sqrt[4]{(3x^4y^3)^4}$ (B) $\sqrt[4]{8}\,\sqrt[4]{2}$ (C) $\sqrt[3]{\dfrac{xy}{27}}$

SOLUTION

(A) $\sqrt[4]{(3x^4y^3)^4} = 3x^4y^3$ Property 1

(B) $\sqrt[4]{8}\sqrt[4]{2} = \sqrt[4]{16} = \sqrt[4]{2^4} = 2$ Properties 2 and 1

(C) $\sqrt[3]{\dfrac{xy}{27}} = \dfrac{\sqrt[3]{xy}}{\sqrt[3]{27}} = \dfrac{\sqrt[3]{xy}}{3}$ or $\dfrac{1}{3}\sqrt[3]{xy}$ Properties 3 and 1

Matched Problem 6) Simplify using properties of radicals.

(A) $\sqrt[7]{(x^3 + y^3)^7}$ (B) $\sqrt[3]{8y^3}$ (C) $\dfrac{\sqrt[3]{16x^4y}}{\sqrt[3]{2xy}}$

What is the best form for a radical expression? There are many answers, depending on what use we wish to make of the expression. In deriving certain formulas, it is sometimes useful to clear either a denominator or a numerator of radicals.

The process is referred to as **rationalizing** the denominator or numerator. Examples 7 and 8 illustrate the rationalizing process.

EXAMPLE 7   Rationalizing Denominators   Rationalize each denominator.

(A) $\dfrac{6x}{\sqrt{2x}}$ (B) $\dfrac{6}{\sqrt{7} - \sqrt{5}}$ (C) $\dfrac{x - 4}{\sqrt{x} + 2}$

SOLUTION

(A) $\dfrac{6x}{\sqrt{2x}} = \dfrac{6x}{\sqrt{2x}} \cdot \dfrac{\sqrt{2x}}{\sqrt{2x}} = \dfrac{6x\sqrt{2x}}{2x} = 3\sqrt{2x}$

(B) $\dfrac{6}{\sqrt{7} - \sqrt{5}} = \dfrac{6}{\sqrt{7} - \sqrt{5}} \cdot \dfrac{\sqrt{7} + \sqrt{5}}{\sqrt{7} + \sqrt{5}}$

$\qquad = \dfrac{6(\sqrt{7} + \sqrt{5})}{2} = 3(\sqrt{7} + \sqrt{5})$

(C) $\dfrac{x - 4}{\sqrt{x} + 2} = \dfrac{x - 4}{\sqrt{x} + 2} \cdot \dfrac{\sqrt{x} - 2}{\sqrt{x} - 2}$

$\qquad = \dfrac{(x - 4)(\sqrt{x} - 2)}{x - 4} = \sqrt{x} - 2$

Matched Problem 7   Rationalize each denominator.

(A) $\dfrac{12ab^2}{\sqrt{3ab}}$ (B) $\dfrac{9}{\sqrt{6} + \sqrt{3}}$ (C) $\dfrac{x^2 - y^2}{\sqrt{x} - \sqrt{y}}$

EXAMPLE 8   Rationalizing Numerators   Rationalize each numerator.

(A) $\dfrac{\sqrt{2}}{2\sqrt{3}}$ (B) $\dfrac{3 + \sqrt{m}}{9 - m}$ (C) $\dfrac{\sqrt{2 + h} - \sqrt{2}}{h}$

SOLUTION

(A) $\dfrac{\sqrt{2}}{2\sqrt{3}} = \dfrac{\sqrt{2}}{2\sqrt{3}} \cdot \dfrac{\sqrt{2}}{\sqrt{2}} = \dfrac{2}{2\sqrt{6}} = \dfrac{1}{\sqrt{6}}$

(B) $\dfrac{3 + \sqrt{m}}{9 - m} = \dfrac{3 + \sqrt{m}}{9 - m} \cdot \dfrac{3 - \sqrt{m}}{3 - \sqrt{m}} = \dfrac{9 - m}{(9 - m)(3 - \sqrt{m})} = \dfrac{1}{3 - \sqrt{m}}$

(C) $\dfrac{\sqrt{2 + h} - \sqrt{2}}{h} = \dfrac{\sqrt{2 + h} - \sqrt{2}}{h} \cdot \dfrac{\sqrt{2 + h} + \sqrt{2}}{\sqrt{2 + h} + \sqrt{2}}$

$\qquad = \dfrac{h}{h(\sqrt{2 + h} + \sqrt{2})} = \dfrac{1}{\sqrt{2 + h} + \sqrt{2}}$

Matched Problem 8   Rationalize each numerator.

(A) $\dfrac{\sqrt{3}}{3\sqrt{2}}$ (B) $\dfrac{2 - \sqrt{n}}{4 - n}$ (C) $\dfrac{\sqrt{3 + h} - \sqrt{3}}{h}$

Exercises A.6

Change each expression in Problems 1–6 to radical form. Do not simplify.

1. $6x^{3/5}$ **2.** $7y^{2/5}$ **3.** $(32x^2y^3)^{3/5}$

4. $(7x^2y)^{5/7}$ **5.** $(x^2 + y^2)^{1/2}$ **6.** $x^{1/2} + y^{1/2}$

Change each expression in Problems 7–12 to rational exponent form. Do not simplify.

7. $5\sqrt[4]{x^3}$ **8.** $7m\sqrt[5]{n^2}$ **9.** $\sqrt[5]{(2x^2y)^3}$

10. $\sqrt[7]{(8x^4y)^3}$ **11.** $\sqrt[3]{x} + \sqrt[3]{y}$ **12.** $\sqrt[3]{x^2 + y^3}$

In Problems 13–24, find rational number representations for each, if they exist.

13. $25^{1/2}$ **14.** $64^{1/3}$ **15.** $16^{3/2}$

16. $16^{3/4}$ **17.** $-49^{1/2}$ **18.** $(-49)^{1/2}$

19. $-64^{2/3}$ **20.** $(-64)^{2/3}$ **21.** $\left(\dfrac{4}{25}\right)^{3/2}$

22. $\left(\dfrac{8}{27}\right)^{2/3}$ **23.** $9^{-3/2}$ **24.** $8^{-2/3}$

In Problems 25–34, simplify each expression and write answers using positive exponents only. All variables represent positive real numbers.

25. $x^{4/5}x^{-2/5}$ **26.** $y^{-3/7}y^{4/7}$ **27.** $\dfrac{m^{2/3}}{m^{-1/3}}$

28. $\dfrac{x^{1/4}}{x^{3/4}}$ **29.** $(8x^3y^{-6})^{1/3}$ **30.** $(4u^{-2}v^4)^{1/2}$

31. $\left(\dfrac{4x^{-2}}{y^4}\right)^{-1/2}$ **32.** $\left(\dfrac{w^4}{9x^{-2}}\right)^{-1/2}$

33. $\dfrac{(8x)^{-1/3}}{12x^{1/4}}$ **34.** $\dfrac{6a^{3/4}}{15a^{-1/3}}$

Simplify each expression in Problems 35–40 using properties of radicals. All variables represent positive real numbers.

35. $\sqrt[5]{(2x+3)^5}$ **36.** $\sqrt[3]{(7+2y)^3}$

37. $\sqrt{6x}\sqrt{15x^3}\sqrt{30x^7}$ **38.** $\sqrt[5]{16a^4}\sqrt[5]{4a^2}\sqrt[5]{8a^3}$

39. $\dfrac{\sqrt{6x}\sqrt{10}}{\sqrt{15x}}$ **40.** $\dfrac{\sqrt{8}\sqrt{12y}}{\sqrt{6y}}$

In Problems 41–48, multiply, and express answers using positive exponents only.

41. $3x^{3/4}(4x^{1/4} - 2x^8)$

42. $2m^{1/3}(3m^{2/3} - m^6)$

43. $(3u^{1/2} - v^{1/2})(u^{1/2} - 4v^{1/2})$

44. $(a^{1/2} + 2b^{1/2})(a^{1/2} - 3b^{1/2})$

45. $(6m^{1/2} + n^{-1/2})(6m - n^{-1/2})$

46. $(2x - 3y^{1/3})(2x^{1/3} + 1)$

47. $(3x^{1/2} - y^{1/2})^2$

48. $(x^{1/2} + 2y^{1/2})^2$

Write each expression in Problems 49–54 in the form $ax^p + bx^q$, where a and b are real numbers and p and q are rational numbers.

49. $\dfrac{\sqrt[3]{x^2} + 2}{2\sqrt[3]{x}}$ **50.** $\dfrac{12\sqrt{x} - 3}{4\sqrt{x}}$ **51.** $\dfrac{2\sqrt[4]{x^3} + \sqrt[3]{x}}{3x}$

52. $\dfrac{3\sqrt[3]{x^2} + \sqrt{x}}{5x}$ **53.** $\dfrac{2\sqrt[3]{x} - \sqrt{x}}{4\sqrt{x}}$ **54.** $\dfrac{x^2 - 4\sqrt{x}}{2\sqrt[3]{x}}$

Rationalize the denominators in Problems 55–60.

55. $\dfrac{12mn^2}{\sqrt{3mn}}$ **56.** $\dfrac{14x^2}{\sqrt{7x}}$ **57.** $\dfrac{2(x+3)}{\sqrt{x-2}}$

58. $\dfrac{3(x+1)}{\sqrt{x+4}}$ **59.** $\dfrac{7(x-y)^2}{\sqrt{x} - \sqrt{y}}$ **60.** $\dfrac{3a - 3b}{\sqrt{a} + \sqrt{b}}$

Rationalize the numerators in Problems 61–66.

61. $\dfrac{\sqrt{5xy}}{5x^2y^2}$ **62.** $\dfrac{\sqrt{3mn}}{3mn}$

63. $\dfrac{\sqrt{x+h} - \sqrt{x}}{h}$ **64.** $\dfrac{\sqrt{2(a+h)} - \sqrt{2a}}{h}$

65. $\dfrac{\sqrt{t} - \sqrt{x}}{t^2 - x^2}$ **66.** $\dfrac{\sqrt{x} - \sqrt{y}}{\sqrt{x} + \sqrt{y}}$

Problems 67–70 illustrate common errors involving rational exponents. In each case, find numerical examples that show that the left side is not always equal to the right side.

67. $(x+y)^{1/2} \neq x^{1/2} + y^{1/2}$ **68.** $(x^3+y^3)^{1/3} \neq x+y$

69. $(x+y)^{1/3} \neq \dfrac{1}{(x+y)^3}$ **70.** $(x+y)^{-1/2} \neq \dfrac{1}{(x+y)^2}$

In Problems 71–82, discuss the validity of each statement. If the statement is true, explain why. If not, give a counterexample.

71. $\sqrt{x^2} = x$ for all real numbers x

72. $\sqrt{x^2} = |x|$ for all real numbers x

73. $\sqrt[3]{x^3} = |x|$ for all real numbers x

74. $\sqrt[3]{x^3} = x$ for all real numbers x

75. If $r < 0$, then r has no cube roots.

76. If $r < 0$, then r has no square roots.

77. If $r > 0$, then r has two square roots.

78. If $r > 0$, then r has three cube roots.

79. The fourth roots of 100 are $\sqrt{10}$ and $-\sqrt{10}$.

80. The square roots of $2\sqrt{6} - 5$ are $\sqrt{3} - \sqrt{2}$ and $\sqrt{2} - \sqrt{3}$.

81. $\sqrt{355 - 60\sqrt{35}} = 5\sqrt{7} - 6\sqrt{5}$

82. $\sqrt[3]{7 - 5\sqrt{2}} = 1 - \sqrt{2}$

In Problems 83–88, simplify by writing each expression as a simple or single fraction reduced to lowest terms and without negative exponents.

83. $-\dfrac{1}{2}(x-2)(x+3)^{-3/2} + (x+3)^{-1/2}$

84. $2(x-2)^{-1/2} - \dfrac{1}{2}(2x+3)(x-2)^{-3/2}$

85. $\dfrac{(x-1)^{1/2} - x(\frac{1}{2})(x-1)^{-1/2}}{x-1}$

86. $\dfrac{(2x-1)^{1/2} - (x+2)(\frac{1}{2})(2x-1)^{-1/2}(2)}{2x-1}$

87. $\dfrac{(x+2)^{2/3} - x(\frac{2}{3})(x+2)^{-1/3}}{(x+2)^{4/3}}$

88. $\dfrac{2(3x-1)^{1/3} - (2x+1)(\frac{1}{3})(3x-1)^{-2/3}(3)}{(3x-1)^{2/3}}$

In Problems 89–94, evaluate using a calculator. (Refer to the instruction book for your calculator to see how exponential forms are evaluated.)

89. $22^{3/2}$

90. $15^{5/4}$

91. $827^{-3/8}$

92. $103^{-3/4}$

93. $37.09^{7/3}$

94. $2.876^{8/5}$

In Problems 95 and 96, evaluate each expression on a calculator and determine which pairs have the same value. Verify these results algebraically.

95. (A) $\sqrt{3} + \sqrt{5}$ (B) $\sqrt{2 + \sqrt{3}} + \sqrt{2 - \sqrt{3}}$

(C) $1 + \sqrt{3}$ (D) $\sqrt[3]{10 + 6\sqrt{3}}$

(E) $\sqrt{8 + \sqrt{60}}$ (F) $\sqrt{6}$

96. (A) $2\sqrt[3]{2} + \sqrt{5}$ (B) $\sqrt{8}$

(C) $\sqrt{3} + \sqrt{7}$ (D) $\sqrt{3 + \sqrt{8}} + \sqrt{3 - \sqrt{8}}$

(E) $\sqrt{10 + \sqrt{84}}$ (F) $1 + \sqrt{5}$

Answers to Matched Problems

1. (A) 4   (B) -4

(C) -3 (D) Not a real number   (E) 27

2. (A) $\sqrt[5]{u}$ (B) $\sqrt[9]{(6x^2y^5)^2}$ or $\left(\sqrt[9]{(6x^2y^5)}\right)^2$

(C) $1/\sqrt[5]{(3xy)^3}$ (D) $(9u)^{1/4}$

(E) $-(2x)^{4/7}$ (F) $(x^3 + y^3)^{1/3}$ (not $x + y$)

3. (A) 27   (B) 81

(C) $10y^{7/12} = 10\sqrt[12]{y^7}$ (D) $16y/x^3$

(E) $2/x^{1/18} = 2/\sqrt[18]{x}$

4. (A) $10c^{13/4} - 2c$ (B) $14x + 19x^{1/2}y^{1/2} - 3y$

5. $2.5x^{-7/6} - 2x^{-1}$

6. (A) $x^3 + y^3$ (B) $2y$ (C) $2x$

7. (A) $4b\sqrt{3ab}$ (B) $3(\sqrt{6} - \sqrt{3})$

(C) $(x + y)(\sqrt{x} + \sqrt{y})$

8. (A) $\dfrac{1}{\sqrt{6}}$ (B) $\dfrac{1}{2 + \sqrt{n}}$ (C) $\dfrac{1}{\sqrt{3 + h} + \sqrt{3}}$

A.7   Quadratic Equations

- Solution by Square Root
- Solution by Factoring
- Quadratic Formula
- Quadratic Formula and Factoring
- Other Polynomial Equations
- Application: Supply and Demand

In this section we consider equations involving second-degree polynomials.

> **DEFINITION** Quadratic Equation
> A **quadratic equation** in one variable is any equation that can be written in the form
> $$ax^2 + bx + c = 0 \qquad a \neq 0 \quad \text{Standard form}$$
> where x is a variable and a, b, and c are constants.

The equations

$$5x^2 - 3x + 7 = 0 \qquad \text{and} \qquad 18 = 32t^2 - 12t$$

are both quadratic equations, since they are either in the standard form or can be transformed into this form.

We restrict our review to finding real solutions to quadratic equations.

Solution by Square Root

The easiest type of quadratic equation to solve is the special form where the first-degree term is missing:

$$ax^2 + c = 0 \qquad a \neq 0$$

The method of solution of this special form makes direct use of the square-root property:

> **THEOREM 1** Square-Root Property
> If $a^2 = b$, then $a = \pm\sqrt{b}$.

EXAMPLE 1 Square-Root Method Use the square-root property to solve each equation.

(A) $x^2 - 7 = 0$ (B) $2x^2 - 10 = 0$

(C) $3x^2 + 27 = 0$ (D) $(x - 8)^2 = 9$

SOLUTION

(A) $x^2 - 7 = 0$

$\qquad x^2 = 7$ What real number squared is 7?

$\qquad x = \pm\sqrt{7}$ Short for $\sqrt{7}$ and $-\sqrt{7}$

(B) $2x^2 - 10 = 0$

$\qquad 2x^2 = 10$

$\qquad x^2 = 5$ What real number squared is 5?

$\qquad x = \pm\sqrt{5}$

(C) $3x^2 + 27 = 0$

$\qquad 3x^2 = -27$

$\qquad x^2 = -9$ What real number squared is -9?

No real solution, since no real number squared is negative.

(D) $(x - 8)^2 = 9$

$\qquad x - 8 = \pm\sqrt{9}$

$\qquad x - 8 = \pm 3$

$\qquad x = 8 \pm 3 = 5$ or 11

Matched Problem 1 Use the square-root property to solve each equation.

(A) $x^2 - 6 = 0$ (B) $3x^2 - 12 = 0$

(C) $x^2 + 4 = 0$ (D) $(x + 5)^2 = 1$

Solution by Factoring

If the left side of a quadratic equation when written in standard form can be factored, the equation can be solved very quickly. The method of solution by factoring rests on a basic property of real numbers, first mentioned in Section A.1.

CONCEPTUAL INSIGHT

Theorem 2 in Section A.1 states that if a and b are real numbers, then $ab = 0$ if and only if $a = 0$ or $b = 0$. To see that this property is useful for solving quadratic equations, consider the following:

$$x^2 - 4x + 3 = 0 \tag{1}$$

$$(x - 1)(x - 3) = 0$$

$$x - 1 = 0 \quad \text{or} \quad x - 3 = 0$$

$$x = 1 \quad \text{or} \quad x = 3$$

You should check these solutions in equation (1).

If one side of the equation is not 0, then this method cannot be used. For example, consider

$$x^2 - 4x + 3 = 8 \tag{2}$$

$$(x - 1)(x - 3) = 8$$

$$x - 1 \neq 8 \quad \text{or} \quad x - 3 \neq 8 \qquad \begin{array}{l} ab = 8 \text{ does not imply} \\ \text{that } a = 8 \text{ or } b = 8. \end{array}$$

$$x = 9 \quad \text{or} \quad x = 11$$

Verify that neither $x = 9$ nor $x = 11$ is a solution for equation (2).

EXAMPLE 2   Factoring Method   Solve by factoring using integer coefficients, if possible.

(A) $3x^2 - 6x - 24 = 0$ (B) $3y^2 = 2y$ (C) $x^2 - 2x - 1 = 0$

SOLUTION

(A) $3x^2 - 6x - 24 = 0$ Divide both sides by 3, since 3 is a factor
                                      of each coefficient.

    $x^2 - 2x - 8 = 0$ Factor the left side, if possible.

    $(x - 4)(x + 2) = 0$

    $x - 4 = 0$ or $x + 2 = 0$

    $x = 4$ or $x = -2$

(B) $3y^2 = 2y$

        $3y^2 - 2y = 0$ We lose the solution $y = 0$ if both sides are divided by y
                               ($3y^2 = 2y$ and $3y = 2$ are not equivalent).
        $y(3y - 2) = 0$

    $y = 0$ or $3y - 2 = 0$

                    $3y = 2$

                    $y = \dfrac{2}{3}$

(C) $x^2 - 2x - 1 = 0$

This equation cannot be factored using integer coefficients. We will solve this type of equation by another method, considered below.

Matched Problem 2   Solve by factoring using integer coefficients, if possible.

(A) $2x^2 + 4x - 30 = 0$ (B) $2x^2 = 3x$ (C) $2x^2 - 8x + 3 = 0$

Note that an equation such as $x^2 = 25$ can be solved by either the square-root or the factoring method, and the results are the same (as they should be). Solve this equation both ways and compare.

Also, note that the factoring method can be extended to higher-degree polynomial equations. Consider the following:

$$x^3 - x = 0$$
$$x(x^2 - 1) = 0$$
$$x(x - 1)(x + 1) = 0$$
$$x = 0 \quad \text{or} \quad x - 1 = 0 \quad \text{or} \quad x + 1 = 0$$
$$\text{Solution: } x = 0, 1, -1$$

Check these solutions in the original equation.

The factoring and square-root methods are fast and easy to use when they apply. However, there are quadratic equations that look simple but cannot be solved by either method. For example, as was noted in Example 2C, the polynomial in

$$x^2 - 2x - 1 = 0$$

cannot be factored using integer coefficients. This brings us to the well-known and widely used *quadratic formula*.

Quadratic Formula

There is a method called *completing the square* that will work for all quadratic equations. After briefly reviewing this method, we will use it to develop the quadratic formula, which can be used to solve any quadratic equation.

The method of **completing the square** is based on the process of transforming a quadratic equation in standard form,

$$ax^2 + bx + c = 0$$

into the form

$$(x + A)^2 = B$$

where A and B are constants. Then, this last equation can be solved easily (if it has a real solution) by the square-root method discussed above.

Consider the equation from Example 2C:

$$x^2 - 2x - 1 = 0 \qquad (3)$$

Since the left side does not factor using integer coefficients, we add 1 to each side to remove the constant term from the left side:

$$x^2 - 2x = 1 \qquad (4)$$

Now we try to find a number that we can add to each side to make the left side a square of a first-degree polynomial. Note the following square of a binomial:

$$(x + m)^2 = x^2 + 2mx + m^2$$

We see that the third term on the right is the square of one-half the coefficient of x in the second term on the right. To complete the square in equation (4), we add the square of one-half the coefficient of x, $(-\frac{2}{2})^2 = 1$, to each side. (This rule works only when the coefficient of x^2 is 1, that is, $a = 1$.) Thus,

$$x^2 - 2x + 1 = 1 + 1$$

The left side is the square of $x - 1$, and we write

$$(x - 1)^2 = 2$$

What number squared is 2?

$$x - 1 = \pm\sqrt{2}$$
$$x = 1 \pm \sqrt{2}$$

And equation (3) is solved!

Let us try the method on the general quadratic equation

$$ax^2 + bx + c = 0 \qquad a \neq 0 \qquad (5)$$

and solve it once and for all for x in terms of the coefficients a, b, and c. We start by multiplying both sides of equation (5) by $1/a$ to obtain

$$x^2 + \frac{b}{a}x + \frac{c}{a} = 0$$

Add $-c/a$ to both sides:

$$x^2 + \frac{b}{a}x = -\frac{c}{a}$$

Now we complete the square on the left side by adding the square of one-half the coefficient of x, that is, $(b/2a)^2 = b^2/4a^2$ to each side:

$$x^2 + \frac{b}{a}x + \frac{b^2}{4a^2} = \frac{b^2}{4a^2} - \frac{c}{a}$$

Writing the left side as a square and combining the right side into a single fraction, we obtain

$$\left(x + \frac{b}{2a}\right)^2 = \frac{b^2 - 4ac}{4a^2}$$

Now we solve by the square-root method:

$$x + \frac{b}{2a} = \pm\sqrt{\frac{b^2 - 4ac}{4a^2}}$$

$$x = -\frac{b}{2a} \pm \frac{\sqrt{b^3 - 4ac}}{2a} \qquad \text{Since } \pm\sqrt{4a^2} = \pm 2a \text{ for any real number } a$$

When this is written as a single fraction, it becomes the **quadratic formula**:

Quadratic Formula

If $ax^2 + bx + c = 0, a \neq 0$, then

$$x = \frac{-b \pm \sqrt{b^2 - 4ac}}{2a}$$

This formula is generally used to solve quadratic equations when the square-root or factoring methods do not work. The quantity $b^2 - 4ac$ under the radical is called the **discriminant**, and it gives us the useful information about solutions listed in Table 1.

Table 1

$b^2 - 4ac$	$ax^2 + bx + c = 0$
Positive	Two real solutions
Zero	One real solution
Negative	No real solutions

EXAMPLE 3 Quadratic Formula Method Solve $x^2 - 2x - 1 = 0$ using the quadratic formula.

SOLUTION

$$x^2 - 2x - 1 = 0$$

$$x = \frac{-b \pm \sqrt{b^2 - 4ac}}{2a} \qquad a = 1, b = -2, c = -1$$

$$= \frac{-(-2) \pm \sqrt{(-2)^2 - 4(1)(-1)}}{2(1)}$$

$$= \frac{2 \pm \sqrt{8}}{2} = \frac{2 \pm 2\sqrt{2}}{2} = 1 \pm \sqrt{2} \approx -0.414 \quad \text{or} \quad 2.414$$

CHECK

$$x^2 - 2x - 1 = 0$$
When $x = 1 + \sqrt{2}$,

$$(1 + \sqrt{2})^2 - 2(1 + \sqrt{2}) - 1 = 1 + 2\sqrt{2} + 2 - 2 - 2\sqrt{2} - 1 = 0$$

When $x = 1 - \sqrt{2}$,

$$(1 - \sqrt{2})^2 - 2(1 - \sqrt{2}) - 1 = 1 - 2\sqrt{2} + 2 - 2 + 2\sqrt{2} - 1 = 0$$

Matched Problem 3 Solve $2x^2 - 4x - 3 = 0$ using the quadratic formula.

If we try to solve $x^2 - 6x + 11 = 0$ using the quadratic formula, we obtain

$$x = \frac{6 \pm \sqrt{-8}}{2}$$

which is not a real number. (Why?)

Quadratic Formula and Factoring

As in Section A.3, we restrict our interest in factoring to polynomials with integer coefficients. If a polynomial cannot be factored as a product of lower-degree polynomials with integer coefficients, we say that the polynomial is **not factorable in the integers**.

How can you factor the quadratic polynomial $x^2 - 13x - 2,310$? We start by solving the corresponding quadratic equation using the quadratic formula:

$$x^2 - 13x - 2,310 = 0$$

$$x = \frac{-(-13) \pm \sqrt{(-13)^3 - 4(1)(-2,310)}}{2}$$

$$x = \frac{-(-13) \pm \sqrt{9,409}}{2}$$

$$= \frac{13 \pm 97}{2} = 55 \quad \text{or} \quad -42$$

Now we write

$$x^2 - 13x - 2,310 = [x - 55][x - (-42)] = (x - 55)(x + 42)$$

Multiplying the two factors on the right produces the second-degree polynomial on the left.

What is behind this procedure? The following two theorems justify and generalize the process:

THEOREM 2 Factorability Theorem

A second-degree polynomial, $ax^2 + bx + c$, with integer coefficients can be expressed as the product of two first-degree polynomials with integer coefficients if and only if $\sqrt{b^2 - 4ac}$ is an integer.

THEOREM 3 Factor Theorem

If r_1 and r_2 are solutions to the second-degree equation $ax^2 + bx + c = 0$, then

$$ax^2 + bx + c = a(x - r_1)(x - r_2)$$

EXAMPLE 4 Factoring with the Aid of the Discriminant Factor, if possible, using integer coefficients.

(A) $4x^2 - 65x + 264$ (B) $2x^2 - 33x - 306$

SOLUTION (A) $4x^2 - 65x + 264$

Step 1 Test for factorability:

$$\sqrt{b^2 - 4ac} = \sqrt{(-65)^2 - 4(4)(264)} = 1$$

Since the result is an integer, the polynomial has first-degree factors with integer coefficients.

Step 2 Factor, using the factor theorem. Find the solutions to the corresponding quadratic equation using the quadratic formula:

$$4x^2 - 65x + 264 = 0 \quad \text{From step 1}$$

$$x = \frac{-(-65) \pm 1}{2 \cdot 4} = \frac{33}{4} \quad \text{or} \quad 8$$

Thus,

$$4x^2 - 65x + 264 = 4\left(x - \frac{33}{4}\right)(x - 8)$$

$$= (4x - 33)(x - 8)$$

(B) $2x^2 - 33x - 306$

Step 1 Test for factorability:

$$\sqrt{b^2 - 4ac} = \sqrt{(-33)^2 - 4(2)(-306)} = \sqrt{3{,}537}$$

Since $\sqrt{3{,}537}$ is not an integer, the polynomial is not factorable in the integers.

Matched Problem 4 ⌋ Factor, if possible, using integer coefficients.

(A) $3x^2 - 28x - 464$ (B) $9x^2 + 320x - 144$

Other Polynomial Equations

There are formulas that are analogous to the quadratic formula, but considerably more complicated, that can be used to solve any cubic (degree 3) or quartic (degree 4) polynomial equation. It can be shown that no such general formula exists for solving quintic (degree 5) or polynomial equations of degree greater than five. Certain polynomial equations, however, can be solved easily by taking roots.

EXAMPLE 5   Solving a Quartic Equation   Find all real solutions to $6x^4 - 486 = 0$.

SOLUTION

$$\begin{aligned}
6x^4 - 486 &= 0 &&\text{Add 486 to both sides}\\
6x^4 &= 486 &&\text{Divide both sides by 6}\\
x^4 &= 81 &&\text{Take the 4th root of both sides}\\
x &= \pm 3
\end{aligned}$$

Matched Problem 5 ⌋ Find all real solutions to $6x^5 + 192 = 0$.

Application: Supply and Demand

Supply-and-demand analysis is a very important part of business and economics. In general, producers are willing to supply more of an item as the price of an item increases and less of an item as the price decreases. Similarly, buyers are willing to buy less of an item as the price increases, and more of an item as the price decreases. We have a dynamic situation where the price, supply, and demand fluctuate until a price is reached at which the supply is equal to the demand. In economic theory, this point is called the **equilibrium point**. If the price increases from this point, the supply will increase and the demand will decrease; if the price decreases from this point, the supply will decrease and the demand will increase.

EXAMPLE 6   Supply and Demand   At a large summer beach resort, the weekly supply-and-demand equations for folding beach chairs are

$$p = \frac{x}{140} + \frac{3}{4} \qquad \text{Supply equation}$$

$$p = \frac{5{,}670}{x} \qquad\qquad \text{Demand equation}$$

The supply equation indicates that the supplier is willing to sell x units at a price of p dollars per unit. The demand equation indicates that consumers are willing to buy x units at a price of p dollars per unit. How many units are required for supply to equal demand? At what price will supply equal demand?

SOLUTION Set the right side of the supply equation equal to the right side of the demand equation and solve for x.

$$\frac{x}{140} + \frac{3}{4} = \frac{5{,}670}{x} \qquad \text{Multiply by } 140x, \text{ the LCD.}$$

$$x^2 + 105x = 793{,}800 \qquad \text{Write in standard form.}$$

$$x^2 + 105x - 793{,}800 = 0 \qquad \text{Use the quadratic formula.}$$

$$x = \frac{-105 \pm \sqrt{105^2 - 4(1)(-793{,}800)}}{2}$$

$$x = 840 \text{ units}$$

The negative root is discarded since a negative number of units cannot be produced or sold. Substitute $x = 840$ back into either the supply equation or the demand equation to find the equilibrium price (we use the demand equation).

$$p = \frac{5{,}670}{x} = \frac{5{,}670}{840} = \$6.75$$

At a price of \$6.75 the supplier is willing to supply 840 chairs and consumers are willing to buy 840 chairs during a week.

Matched Problem 6 Repeat Example 6 if near the end of summer, the supply-and-demand equations are

$$p = \frac{x}{80} - \frac{1}{20} \qquad \text{Supply equation}$$

$$p = \frac{1{,}264}{x} \qquad \text{Demand equation}$$

Exercises A.7

Find only real solutions in the problems below. If there are no real solutions, say so.

Solve Problems 1–4 by the square-root method.

1. $2x^2 - 22 = 0$

2. $3m^2 - 21 = 0$

3. $(3x - 1)^2 = 25$

4. $(2x + 1)^2 = 16$

Solve Problems 5–8 by factoring.

5. $2u^2 - 8u - 24 = 0$

6. $3x^2 - 18x + 15 = 0$

7. $x^2 = 2x$

8. $n^2 = 3n$

Solve Problems 9–12 by using the quadratic formula.

9. $x^2 - 6x - 3 = 0$

10. $m^2 + 8m + 3 = 0$

11. $3u^2 + 12u + 6 = 0$

12. $2x^2 - 20x - 6 = 0$

Solve Problems 13–30 by using any method.

13. $\dfrac{2x^2}{3} = 5x$

14. $x^2 = -\dfrac{3}{4}x$

15. $4u^2 - 9 = 0$

16. $9y^2 - 25 = 0$

17. $8x^2 + 20x = 12$

18. $9x^2 - 6 = 15x$

19. $x^2 = 1 - x$

20. $m^2 = 1 - 3m$

21. $2x^2 = 6x - 3$

22. $2x^2 = 4x - 1$

23. $y^2 - 4y = -8$

24. $x^2 - 2x = -3$

25. $(2x + 3)^2 = 11$

26. $(5x - 2)^2 = 7$

27. $\dfrac{3}{p} = p$

28. $x - \dfrac{7}{x} = 0$

29. $2 - \dfrac{2}{m^2} = \dfrac{3}{m}$

30. $2 + \dfrac{5}{u} = \dfrac{3}{u^2}$

In Problems 31–38, factor, if possible, as the product of two first-degree polynomials with integer coefficients. Use the quadratic formula and the factor theorem.

31. $x^2 + 40x - 84$

32. $x^2 - 28x - 128$

33. $x^2 - 32x + 144$

34. $x^2 + 52x + 208$

35. $2x^2 + 15x - 108$

36. $3x^2 - 32x - 140$

37. $4x^2 + 241x - 434$

38. $6x^2 - 427x - 360$

39. Solve $A = P(1 + r)^2$ for r in terms of A and P; that is, isolate r on the left side of the equation (with coefficient 1) and end up with an algebraic expression on the right side involving A and P but not r. Write the answer using positive square roots only.

40. Solve $x^2 + 3mx - 3n = 0$ for x in terms of m and n.

41. Consider the quadratic equation

$$x^2 + 4x + c = 0$$

where c is a real number. Discuss the relationship between the values of c and the three types of roots listed in Table 1 on page 926.

42. Consider the quadratic equation

$$x^2 - 2x + c = 0$$

where c is a real number. Discuss the relationship between the values of c and the three types of roots listed in Table 1 on page 926.

In Problems 43–48, find all real solutions.

43. $x^3 + 8 = 0$

44. $x^3 - 8 = 0$

45. $5x^4 - 500 = 0$

46. $2x^3 + 250 = 0$

47. $x^4 - 8x^2 + 15 = 0$

48. $x^4 - 12x^2 + 32 = 0$

Applications

49. **Supply and demand.** A company wholesales shampoo in a particular city. Their marketing research department established the following weekly supply-and-demand equations:

$$p = \frac{x}{450} + \frac{1}{2} \quad \text{Supply equation}$$

$$p = \frac{6,300}{x} \quad \text{Demand equation}$$

How many units are required for supply to equal demand? At what price per bottle will supply equal demand?

50. **Supply and demand.** An importer sells an automatic camera to outlets in a large city. During the summer, the weekly supply-and-demand equations are

$$p = \frac{x}{6} + 9 \quad \text{Supply equation}$$

$$p = \frac{24,840}{x} \quad \text{Demand equation}$$

How many units are required for supply to equal demand? At what price will supply equal demand?

51. **Interest rate.** If P dollars are invested at $100r$ percent compounded annually, at the end of 2 years it will grow to $A = P(1 + r)^2$. At what interest rate will \$484 grow to \$625 in 2 years? (*Note:* If $A = 625$ and $P = 484$ find r.)

52. **Interest rate.** Using the formula in Problem 51, determine the interest rate that will make \$1,000 grow to \$1,210 in 2 years.

53. **Ecology.** To measure the velocity v (in feet per second) of a stream, we position a hollow L-shaped tube with one end under the water pointing upstream and the other end pointing straight up a couple of feet out of the water. The water will then be pushed up the tube a certain distance h (in feet) above the surface of the stream. Physicists have shown that $v^2 = 64h$. Approximately how fast is a stream flowing if $h = 1$ foot? If $h = 0.5$ foot?

54. **Safety research.** It is of considerable importance to know the least number of feet d in which a car can be stopped, including reaction time of the driver, at various speeds v (in miles per hour). Safety research has produced the formula $d = 0.044v^2 + 1.1v$. If it took a car 550 feet to stop, estimate the car's speed at the moment the stopping process was started.

Answers to Matched Problems

1. (A) $\pm\sqrt{6}$ (B) ± 2

(C) No real solution   (D) $-6, -4$

2. (A) $-5, 3$ (B) $0, \frac{3}{2}$

(C) Cannot be factored using integer coefficients

3. $(2 \pm \sqrt{10})/2$

4. (A) Cannot be factored using integer coefficients

(B) $(9x - 4)(x + 36)$

5. -2

6. 320 chairs at \$3.95 each

B Special Topics

B.1   Sequences, Series, and Summation Notation

B.2   Arithmetic and Geometric Sequences

B.3   Binomial Theorem

B.1   Sequences, Series, and Summation Notation

- Sequences
- Series and Summation Notation

If someone asked you to list all natural numbers that are perfect squares, you might begin by writing

$$1, 4, 9, 16, 25, 36$$

But you would soon realize that it is impossible to actually list all the perfect squares, since there are an infinite number of them. However, you could represent this collection of numbers in several different ways. One common method is to write

$$1, 4, 9, \ldots, n^2, \ldots \quad n \in N$$

where N is the set of natural numbers. A list of numbers such as this is generally called a *sequence*.

Sequences

Consider the function f given by

$$f(n) = 2n + 1 \tag{1}$$

where the domain of f is the set of natural numbers N. Note that

$$f(1) = 3, \quad f(2) = 5, \quad f(3) = 7, \quad \ldots$$

The function f is an example of a sequence. In general, a **sequence** is a function with domain a set of successive integers. Instead of the standard function notation used in equation (1), sequences are usually defined in terms of a special notation.

The range value $f(n)$ is usually symbolized more compactly with a symbol such as a_n. Thus, in place of equation (1), we write

$$a_n = 2n + 1$$

and the domain is understood to be the set of natural numbers unless something is said to the contrary or the context indicates otherwise. The elements in the range are

931

called **terms of the sequence**; a_1 is the first term, a_2 is the second term, and a_n is the **nth term**, or **general term**.

$$a_1 = 2(1) + 1 = 3 \quad \text{First term}$$
$$a_2 = 2(2) + 1 = 5 \quad \text{Second term}$$
$$a_3 = 2(3) + 1 = 7 \quad \text{Third term}$$
$$\vdots$$
$$a_n = 2n + 1 \qquad \text{General term}$$

The ordered list of elements

$$3, 5, 7, \ldots, 2n + 1, \ldots$$

obtained by writing the terms of the sequence in their natural order with respect to the domain values is often informally referred to as a sequence. A sequence also may be represented in the abbreviated form $\{a_n\}$, where a symbol for the nth term is written within braces. For example, we could refer to the sequence $3, 5, 7, \ldots, 2n + 1, \ldots$ as the sequence $\{2n + 1\}$.

If the domain of a sequence is a finite set of successive integers, then the sequence is called a **finite sequence**. If the domain is an infinite set of successive integers, then the sequence is called an **infinite sequence**. The sequence $\{2n + 1\}$ discussed above is an infinite sequence.

EXAMPLE 1 Writing the Terms of a Sequence Write the first four terms of each sequence:

(A) $a_n = 3n - 2$

(B) $\left\{ \dfrac{(-1)^n}{n} \right\}$

SOLUTION

(A) $1, 4, 7, 10$

(B) $-1, \dfrac{1}{2}, \dfrac{-1}{3}, \dfrac{1}{4}$

Matched Problem 1 Write the first four terms of each sequence:

(A) $a_n = -n + 3$

(B) $\left\{ \dfrac{(-1)^n}{2^n} \right\}$

Now that we have seen how to use the general term to find the first few terms in a sequence, we consider the reverse problem. That is, can a sequence be defined just by listing the first three or four terms of the sequence? And can we then use these initial terms to find a formula for the nth term? In general, without other information, the answer to the first question is no. Many different sequences may start off with the same terms. Simply listing the first three terms (or any other finite number of terms) does not specify a particular sequence.

What about the second question? That is, given a few terms, can we find the general formula for at least one sequence whose first few terms agree with the given terms? The answer to this question is a qualified yes. If we can observe a simple pattern in the given terms, we usually can construct a general term that will produce that pattern. The next example illustrates this approach.

EXAMPLE 2 Finding the General Term of a Sequence Find the general term of a sequence whose first four terms are

(A) $3, 4, 5, 6, \ldots$

(B) $5, -25, 125, -625, \ldots$

SOLUTION

(A) Since these terms are consecutive integers, one solution is $a_n = n, n \geq 3$. If we want the domain of the sequence to be all natural numbers, another solution is $b_n = n + 2$.

(B) Each of these terms can be written as the product of a power of 5 and a power of -1:

$$5 = (-1)^0 5^1 = a_1$$
$$-25 = (-1)^1 5^2 = a_2$$
$$125 = (-1)^2 5^3 = a_3$$
$$-625 = (-1)^3 5^4 = a_4$$

If we choose the domain to be all natural numbers, a solution is

$$a_n = (-1)^{n-1} 5^n$$

Matched Problem 2 Find the general term of a sequence whose first four terms are

(A) $3, 6, 9, 12, \ldots$ (B) $1, -2, 4, -8, \ldots$

In general, there is usually more than one way of representing the nth term of a given sequence (see the solution of Example 2A). However, unless something is stated to the contrary, we assume that the domain of the sequence is the set of natural numbers N.

Series and Summation Notation

If $a_1, a_2, a_3, \ldots, a_n, \ldots$ is a sequence, the expression

$$a_1 + a_2 + a_3 + \cdots + a_n + \cdots$$

is called a **series**. If the sequence is finite, the corresponding series is a **finite series**. If the sequence is infinite, the corresponding series is an **infinite series**. We consider only finite series in this section. For example,

$$1, 3, 5, 7, 9 \qquad \text{Finite sequence}$$
$$1 + 3 + 5 + 7 + 9 \qquad \text{Finite series}$$

Notice that we can easily evaluate this series by adding the five terms:

$$1 + 3 + 5 + 7 + 9 = 25$$

Series are often represented in a compact form called **summation notation**. Consider the following examples:

$$\sum_{k=3}^{6} k^2 = 3^2 + 4^2 + 5^2 + 6^2$$

$$= 9 + 16 + 25 + 36 = 86$$

$$\sum_{k=0}^{2} (4k + 1) = (4 \cdot 0 + 1) + (4 \cdot 1 + 1) + (4 \cdot 2 + 1)$$

$$= 1 + 5 + 9 = 15$$

In each case, the terms of the series on the right are obtained from the expression on the left by successively replacing the **summing index k** with integers, starting with the number indicated below the **summation sign** Σ and ending with the number that appears above Σ. The summing index may be represented by letters other than k and may start at any integer and end at any integer greater than or equal to the starting integer. If we are given the finite sequence

$$\frac{1}{2}, \frac{1}{4}, \frac{1}{8}, \ldots, \frac{1}{2^n}$$

the corresponding series is

$$\frac{1}{2} + \frac{1}{4} + \frac{1}{8} + \cdots + \frac{1}{2^n} = \sum_{j=1}^{n} \frac{1}{2^j}$$

where we have used j for the summing index.

EXAMPLE 3   Summation Notation   Write

$$\sum_{k=1}^{5} \frac{k}{k^2 + 1}$$

without summation notation. Do not evaluate the sum.

SOLUTION

$$\sum_{k=1}^{5} \frac{k}{k^2 + 1} = \frac{1}{1^2 + 1} + \frac{2}{2^2 + 1} + \frac{3}{3^2 + 1} + \frac{4}{4^2 + 1} + \frac{5}{5^2 + 1}$$

$$= \frac{1}{2} + \frac{2}{5} + \frac{3}{10} + \frac{4}{17} + \frac{5}{26}$$

Matched Problem 3   Write

$$\sum_{k=1}^{5} \frac{k + 1}{k}$$

without summation notation. Do not evaluate the sum.

If the terms of a series are alternately positive and negative, we call the series an **alternating series**. The next example deals with the representation of such a series.

EXAMPLE 4   Summation Notation   Write the alternating series

$$\frac{1}{2} - \frac{1}{4} + \frac{1}{6} - \frac{1}{8} + \frac{1}{10} - \frac{1}{12}$$

using summation notation with

(A) The summing index k starting at 1

(B) The summing index j starting at 0

SOLUTION

(A) $(-1)^{k+1}$ provides the alternation of sign, and $1/(2k)$ provides the other part of each term. So, we can write

$$\frac{1}{2} - \frac{1}{4} + \frac{1}{6} - \frac{1}{8} + \frac{1}{10} - \frac{1}{12} = \sum_{k=1}^{6} \frac{(-1)^{k+1}}{2k}$$

(B) $(-1)^j$ provides the alternation of sign, and $1/[2(j + 1)]$ provides the other part of each term. So, we can write

$$\frac{1}{2} - \frac{1}{4} + \frac{1}{6} - \frac{1}{8} + \frac{1}{10} - \frac{1}{12} = \sum_{j=0}^{5} \frac{(-1)^j}{2(j + 1)}$$

Matched Problem 4   Write the alternating series

$$1 - \frac{1}{3} + \frac{1}{9} - \frac{1}{27} + \frac{1}{81}$$

using summation notation with

(A) The summing index k starting at 1

(B) The summing index j starting at 0

Summation notation provides a compact notation for the sum of any list of numbers, even if the numbers are not generated by a formula. For example, suppose that the results of an examination taken by a class of 10 students are given in the following list:

$$87, 77, 95, 83, 86, 73, 95, 68, 75, 86$$

If we let $a_1, a_2, a_3, \ldots, a_{10}$ represent these 10 scores, then the average test score is given by

$$\frac{1}{10}\sum_{k=1}^{10} a_k = \frac{1}{10}(87 + 77 + 95 + 83 + 86 + 73 + 95 + 68 + 75 + 86)$$

$$= \frac{1}{10}(825) = 82.5$$

More generally, in statistics, the **arithmetic mean** $\bar{a}$ of a list of n numbers $a_1, a_2, \ldots, a_n$ is defined as

$$\bar{a} = \frac{1}{n}\sum_{k=1}^{n} a_k$$

EXAMPLE 5 Arithmetic Mean Find the arithmetic mean of 3, 5, 4, 7, 4, 2, 3, and 6.

SOLUTION

$$\bar{a} = \frac{1}{8}\sum_{k=1}^{8} a_k = \frac{1}{8}(3 + 5 + 4 + 7 + 4 + 2 + 3 + 6) = \frac{1}{8}(34) = 4.25$$

Matched Problem 5 Find the arithmetic mean of 9, 3, 8, 4, 3, and 6.

Exercises B.1

Write the first four terms for each sequence in Problems 1–6.

1. $a_n = 2n + 3$

2. $a_n = 4n - 3$

3. $a_n = \dfrac{n + 2}{n + 1}$

4. $a_n = \dfrac{2n + 1}{2n}$

5. $a_n = (-3)^{n+1}$

6. $a_n = \left(-\frac{1}{4}\right)^{n-1}$

7. Write the 10th term of the sequence in Problem 1.

8. Write the 15th term of the sequence in Problem 2.

9. Write the 99th term of the sequence in Problem 3.

10. Write the 200th term of the sequence in Problem 4.

In Problems 11–16, write each series in expanded form without summation notation, and evaluate.

11. $\sum_{k=1}^{6} k$

12. $\sum_{k=1}^{5} k^2$

13. $\sum_{k=4}^{7} (2k - 3)$

14. $\sum_{k=0}^{4} (-2)^k$

15. $\sum_{k=0}^{3} \frac{1}{10^k}$

16. $\sum_{k=1}^{4} \frac{1}{2^k}$

Find the arithmetic mean of each list of numbers in Problems 17–20.

17. 5, 4, 2, 1, and 6

18. 7, 9, 9, 2, and 4

19. 96, 65, 82, 74, 91, 88, 87, 91, 77, and 74

20. 100, 62, 95, 91, 82, 87, 70, 75, 87, and 82

Write the first five terms of each sequence in Problems 21–26.

21. $a_n = \dfrac{(-1)^{n+1}}{2^n}$

22. $a_n = (-1)^n(n - 1)^2$

23. $a_n = n[1 + (-1)^n]$

24. $a_n = \dfrac{1 - (-1)^n}{n}$

25. $a_n = \left(-\dfrac{3}{2}\right)^{n-1}$

26. $a_n = \left(-\dfrac{1}{2}\right)^{n+1}$

In Problems 27–42, find the general term of a sequence whose first four terms agree with the given terms.

27. $-2, -1, 0, 1, \ldots$

28. $4, 5, 6, 7, \ldots$

29. $4, 8, 12, 16, \ldots$

30. $-3, -6, -9, -12, \ldots$

31. $\frac{1}{2}, \frac{3}{4}, \frac{5}{6}, \frac{7}{8}, \ldots$

32. $\frac{1}{2}, \frac{2}{3}, \frac{3}{4}, \frac{4}{5}, \ldots$

33. $1, -2, 3, -4, \ldots$

34. $-2, 4, -8, 16, \ldots$

35. $1, -3, 5, -7, \ldots$

36. $3, -6, 9, -12, \ldots$

37. $1, \frac{2}{5}, \frac{4}{25}, \frac{8}{125}, \ldots$

38. $\frac{4}{3}, \frac{16}{9}, \frac{64}{27}, \frac{256}{81}, \ldots$

39. $x, x^2, x^3, x^4, \ldots$

40. $1, 2x, 3x^2, 4x^3, \ldots$

41. $x, -x^3, x^5, -x^7, \ldots$

42. $x, \dfrac{x^2}{2}, \dfrac{x^3}{3}, \dfrac{x^4}{4}, \ldots$

Write each series in Problems 43–50 in expanded form without summation notation. Do not evaluate.

43. $\displaystyle\sum_{k=1}^{5} (-1)^{k+1}(2k-1)^2$

44. $\displaystyle\sum_{k=1}^{4} \dfrac{(-2)^{k+1}}{2k+1}$

45. $\displaystyle\sum_{k=2}^{5} \dfrac{2^k}{2k+3}$

46. $\displaystyle\sum_{k=3}^{7} \dfrac{(-1)^k}{k^2-k}$

47. $\displaystyle\sum_{k=1}^{5} x^{k-1}$

48. $\displaystyle\sum_{k=1}^{3} \dfrac{1}{k} x^{k+1}$

49. $\displaystyle\sum_{k=0}^{4} \dfrac{(-1)^k x^{2k+1}}{2k+1}$

50. $\displaystyle\sum_{k=0}^{4} \dfrac{(-1)^k x^{2k}}{2k+2}$

Write each series in Problems 51–54 using summation notation with

(A) The summing index k starting at $k = 1$

(B) The summing index j starting at $j = 0$

51. $2 + 3 + 4 + 5 + 6$

52. $1^2 + 2^2 + 3^2 + 4^2$

53. $1 - \frac{1}{2} + \frac{1}{3} - \frac{1}{4}$

54. $1 - \frac{1}{3} + \frac{1}{5} - \frac{1}{7} + \frac{1}{9}$

Write each series in Problems 55–58 using summation notation with the summing index k starting at $k = 1$.

55. $2 + \dfrac{3}{2} + \dfrac{4}{3} + \cdots + \dfrac{n+1}{n}$

56. $1 + \dfrac{1}{2^2} + \dfrac{1}{3^2} + \cdots + \dfrac{1}{n^2}$

57. $\dfrac{1}{2} - \dfrac{1}{4} + \dfrac{1}{8} - \cdots + \dfrac{(-1)^{n+1}}{2^n}$

58. $1 - 4 + 9 - \cdots + (-1)^{n+1} n^2$

In Problems 59–62, discuss the validity of each statement. If the statement is true, explain why. If not, give a counterexample.

59. For each positive integer n, the sum of the series

$1 + \dfrac{1}{2} + \dfrac{1}{3} + \cdots + \dfrac{1}{n}$ is less than 4.

60. For each positive integer n, the sum of the series

$\dfrac{1}{2} + \dfrac{1}{4} + \dfrac{1}{8} + \cdots + \dfrac{1}{2^n}$ is less than 1.

61. For each positive integer n, the sum of the series

$\dfrac{1}{2} - \dfrac{1}{4} + \dfrac{1}{8} - \cdots + \dfrac{(-1)^{n+1}}{2^n}$ is greater than or

equal to $\dfrac{1}{4}$.

62. For each positive integer n, the sum of the series

$1 - \dfrac{1}{2} + \dfrac{1}{3} - \dfrac{1}{4} + \cdots + \dfrac{(-1)^{n+1}}{n}$ is greater than or

equal to $\dfrac{1}{2}$.

*Some sequences are defined by a **recursion formula** —that is, a formula that defines each term of the sequence in terms of one or more of the preceding terms. For example, if $\{a_n\}$ is defined by*

$$a_1 = 1 \quad and \quad a_n = 2a_{n-1} + 1 \quad for \quad n \geq 2$$

then

$$a_2 = 2a_1 + 1 = 2 \cdot 1 + 1 = 3$$
$$a_3 = 2a_2 + 1 = 2 \cdot 3 + 1 = 7$$
$$a_4 = 2a_3 + 1 = 2 \cdot 7 + 1 = 15$$

and so on. In Problems 63–66, write the first five terms of each sequence.

63. $a_1 = 2$ and $a_n = 3a_{n-1} + 2$ for $n \geq 2$

64. $a_1 = 3$ and $a_n = 2a_{n-1} - 2$ for $n \geq 2$

65. $a_1 = 1$ and $a_n = 2a_{n-1}$ for $n \geq 2$

66. $a_1 = 1$ and $a_n = -\frac{1}{3} a_{n-1}$ for $n \geq 2$

If A is a positive real number, the terms of the sequence defined by

$$a_1 = \dfrac{A}{2} \quad and \quad a_n = \dfrac{1}{2}\left(a_{n-1} + \dfrac{A}{a_{n-1}}\right) \quad for\ n \geq 2$$

can be used to approximate $\sqrt{A}$ to any decimal place accuracy desired. In Problems 67 and 68, compute the first four terms of this sequence for the indicated value of A, and compare the fourth term with the value of $\sqrt{A}$ obtained from a calculator.

67. $A = 2$

68. $A = 6$

69. The sequence defined recursively by $a_1 = 1, a_2 = 1$, $a_n = a_{n-1} + a_{n-2}$ for $n \geq 3$ is called the *Fibonacci sequence*. Find the first ten terms of the Fibonacci sequence.

70. The sequence defined by $b_n = \dfrac{\sqrt{5}}{5}\left(\dfrac{1+\sqrt{5}}{2}\right)^n$ is related to the Fibonacci sequence. Find the first ten terms (to three decimal places) of the sequence $\{b_n\}$ and describe the relationship.

Answers to Matched Problems

1. (A) $2, 1, 0, -1$ (B) $\frac{-1}{2}, \frac{1}{4}, \frac{-1}{8}, \frac{1}{16}$

2. (A) $a_n = 3n$ (B) $a_n = (-2)^{n-1}$

3. $2 + \frac{3}{2} + \frac{4}{3} + \frac{5}{4} + \frac{6}{5}$

4. (A) $\displaystyle\sum_{k=1}^{5} \dfrac{(-1)^{k-1}}{3^{k-1}}$ (B) $\displaystyle\sum_{j=0}^{4} \dfrac{(-1)^j}{3^j}$

5. 5.5

B.2 Arithmetic and Geometric Sequences

- Arithmetic and Geometric Sequences
- *n*th-Term Formulas
- Sum Formulas for Finite Arithmetic Series
- Sum Formulas for Finite Geometric Series
- Sum Formula for Infinite Geometric Series
- Applications

For most sequences, it is difficult to sum an arbitrary number of terms of the sequence without adding term by term. But particular types of sequences—*arithmetic sequences* and *geometric sequences*—have certain properties that lead to convenient and useful formulas for the sums of the corresponding *arithmetic series* and *geometric series*.

Arithmetic and Geometric Sequences

The sequence $5, 7, 9, 11, 13, \ldots, 5 + 2(n - 1), \ldots$, where each term after the first is obtained by adding 2 to the preceding term, is an example of an arithmetic sequence. The sequence $5, 10, 20, 40, 80, \ldots, 5(2)^{n-1}, \ldots$, where each term after the first is obtained by multiplying the preceding term by 2, is an example of a geometric sequence.

DEFINITION Arithmetic Sequence

A sequence of numbers

$$a_1, a_2, a_3, \ldots, a_n, \ldots$$

is called an **arithmetic sequence** if there is a constant d, called the **common difference**, such that

$$a_n - a_{n-1} = d$$

That is,

$$a_n = a_{n-1} + d \quad \text{for every } n > 1$$

DEFINITION Geometric Sequence

A sequence of numbers

$$a_1, a_2, a_3, \ldots, a_n, \ldots$$

is called a **geometric sequence** if there exists a nonzero constant r, called a **common ratio**, such that

$$\frac{a_n}{a_{n-1}} = r$$

That is,

$$a_n = ra_{n-1} \quad \text{for every } n > 1$$

EXAMPLE 1 Recognizing Arithmetic and Geometric Sequences Which of the following can be the first four terms of an arithmetic sequence? Of a geometric sequence?

(A) $1, 2, 3, 5, \ldots$ (B) $-1, 3, -9, 27, \ldots$

(C) $3, 3, 3, 3, \ldots$ (D) $10, 8.5, 7, 5.5, \ldots$

SOLUTION

(A) Since $2 - 1 \neq 5 - 3$, there is no common difference, so the sequence is not an arithmetic sequence. Since $2/1 \neq 3/2$, there is no common ratio, so the sequence is not geometric either.

(B) The sequence is geometric with common ratio -3. It is not arithmetic.

(C) The sequence is arithmetic with common difference 0, and is also geometric with common ratio 1.

(D) The sequence is arithmetic with common difference -1.5. It is not geometric.

Matched Problem 1) Which of the following can be the first four terms of an arithmetic sequence? Of a geometric sequence?

(A) $8, 2, 0.5, 0.125, \ldots$ (B) $-7, -2, 3, 8, \ldots$ (C) $1, 5, 25, 100, \ldots$

nth-Term Formulas

If $\{a_n\}$ is an arithmetic sequence with common difference d, then

$$a_2 = a_1 + d$$
$$a_3 = a_2 + d = a_1 + 2d$$
$$a_4 = a_3 + d = a_1 + 3d$$

This suggests that

THEOREM 1   nth Term of an Arithmetic Sequence

$$a_n = a_1 + (n - 1)d \quad \text{for all } n > 1 \tag{1}$$

Similarly, if $\{a_n\}$ is a geometric sequence with common ratio r, then

$$a_2 = a_1 r$$
$$a_3 = a_2 r = a_1 r^2$$
$$a_4 = a_3 r = a_1 r^3$$

This suggests that

THEOREM 2   nth Term of a Geometric Sequence

$$a_n = a_1 r^{n-1} \quad \text{for all } n > 1 \tag{2}$$

EXAMPLE 2   Finding Terms in Arithmetic and Geometric Sequences

(A) If the 1st and 10th terms of an arithmetic sequence are 3 and 30, respectively, find the 40th term of the sequence.

(B) If the 1st and 10th terms of a geometric sequence are 3 and 30, find the 40th term to three decimal places.

SOLUTION

(A) First use formula (1) with $a_1 = 3$ and $a_{10} = 30$ to find d:

$$a_n = a_1 + (n - 1)d$$
$$a_{10} = a_1 + (10 - 1)d$$
$$30 = 3 + 9d$$
$$d = 3$$

Now find a_{40}:

$$a_{40} = 3 + 39 \cdot 3 = 120$$

(B) First use formula (2) with $a_1 = 3$ and $a_{10} = 30$ to find r:

$$a_n = a_1 r^{n-1}$$
$$a_{10} = a_1 r^{10-1}$$
$$30 = 3r^9$$
$$r^9 = 10$$
$$r = 10^{1/9}$$

Now find a_{40}:

$$a_{40} = 3(10^{1/9})^{39} = 3(10^{39/9}) = 64{,}633.041$$

Matched Problem 2

(A) If the 1st and 15th terms of an arithmetic sequence are -5 and 23, respectively, find the 73rd term of the sequence.

(B) Find the 8th term of the geometric sequence

$$\frac{1}{64}, \frac{-1}{32}, \frac{1}{16}, \ldots$$

Sum Formulas for Finite Arithmetic Series

If $a_1, a_2, a_3, \ldots, a_n$ is a finite arithmetic sequence, then the corresponding series $a_1 + a_2 + a_3 + \cdots + a_n$ is called a *finite arithmetic series*. We will derive two simple and very useful formulas for the sum of a finite arithmetic series. Let d be the common difference of the arithmetic sequence $a_1, a_2, a_3, \ldots, a_n$ and let S_n denote the sum of the series $a_1 + a_2 + a_3 + \cdots + a_n$. Then

$$S_n = a_1 + (a_1 + d) + \cdots + [a_1 + (n-2)d] + [a_1 + (n-1)d]$$

Reversing the order of the sum, we obtain

$$S_n = [a_1 + (n-1)d] + [a_1 + (n-2)d] + \cdots + (a_1 + d) + a_1$$

Something interesting happens if we combine these last two equations by addition (adding corresponding terms on the right sides):

$$2S_n = [2a_1 + (n-1)d] + [2a_1 + (n-1)d] + \cdots + [2a_1 + (n-1)d] + [2a_1 + (n-1)d]$$

All the terms on the right side are the same, and there are n of them. Thus,

$$2S_n = n[2a_1 + (n-1)d]$$

and we have the following general formula:

> **THEOREM 3 Sum of a Finite Arithmetic Series: First Form**
>
> $$S_n = \frac{n}{2}[2a_1 + (n-1)d] \tag{3}$$

Replacing

$$[a_1 + (n-1)d] \quad \text{in} \quad \frac{n}{2}[a_1 + a_1 + (n-1)d]$$

by a_n from equation (1), we obtain a second useful formula for the sum:

> **THEOREM 4 Sum of a Finite Arithmetic Series: Second Form**
>
> $$S_n = \frac{n}{2}(a_1 + a_n) \tag{4}$$

EXAMPLE 3 Finding a Sum   Find the sum of the first 30 terms in the arithmetic sequence:

$$3, 8, 13, 18, \ldots$$

SOLUTION Use formula (3) with $n = 30$, $a_1 = 3$, and $d = 5$:

$$S_{30} = \frac{30}{2}[2 \cdot 3 + (30-1)5] = 2,265$$

Matched Problem 3   Find the sum of the first 40 terms in the arithmetic sequence:

$$15, 13, 11, 9, \ldots$$

EXAMPLE 4   Finding a Sum   Find the sum of all the even numbers between 31 and 87.

SOLUTION   First, find n using equation (1):

$$a_n = a_1 + (n - 1)d$$
$$86 = 32 + (n - 1)2$$
$$n = 28$$

Now find S_{28} using formula (4):

$$S_n = \frac{n}{2}(a_1 + a_n)$$

$$S_{28} = \frac{28}{2}(32 + 86) = 1,652$$

Matched Problem 4   Find the sum of all the odd numbers between 24 and 208.

Sum Formulas for Finite Geometric Series

If $a_1, a_2, a_3, \ldots, a_n$ is a finite geometric sequence, then the corresponding series $a_1 + a_2 + a_3 + \cdots + a_n$ is called a *finite geometric series*. As with arithmetic series, we can derive two simple and very useful formulas for the sum of a finite geometric series. Let r be the common ratio of the geometric sequence $a_1, a_2, a_3, \ldots, a_n$ and let S_n denote the sum of the series $a_1 + a_2 + a_3 + \cdots + a_n$. Then

$$S_n = a_1 + a_1 r + a_1 r^2 + \cdots + a_1 r^{n-2} + a_1 r^{n-1}$$

If we multiply both sides by r, we obtain

$$r S_n = a_1 r + a_1 r^2 + a_1 r^3 + \cdots + a_1 r^{n-1} + a_1 r^n$$

Now combine these last two equations by subtraction to obtain

$$r S_n - S_n = (a_1 r + a_1 r^2 + a_1 r^3 + \cdots + a_1 r^{n-1} + a_1 r^n) - (a_1 + a_1 r + a_1 r^2 + \cdots + a_1 r^{n-2} + a_1 r^{n-1})$$
$$(r - 1)S_n = a_1 r^n - a_1$$

Notice how many terms drop out on the right side. Solving for S_n, we have

THEOREM 5   Sum of a Finite Geometric Series: First Form

$$S_n = \frac{a_1(r^n - 1)}{r - 1} \quad r \neq 1 \tag{5}$$

Since $a_n = a_1 r^{n-1}$, or $r a_n = a_1 r^n$, formula (5) also can be written in the form

THEOREM 6   Sum of a Finite Geometric Series: Second Form

$$S_n = \frac{r a_n - a_1}{r - 1} \quad r \neq 1 \tag{6}$$

EXAMPLE 5   Finding a Sum   Find the sum (to 2 decimal places) of the first ten terms of the geometric sequence:

$$1, 1.05, 1.05^2, \ldots$$

SOLUTION Use formula (5) with $a_1 = 1$, $r = 1.05$, and $n = 10$:

$$S_n = \frac{a_1(r^n - 1)}{r - 1}$$

$$S_{10} = \frac{1(1.05^{10} - 1)}{1.05 - 1}$$

$$\approx \frac{0.6289}{0.05} \approx 12.58$$

Matched Problem 5 ⌋ Find the sum of the first eight terms of the geometric sequence:

$$100, 100(1.08), 100(1.08)^2, \ldots$$

Sum Formula for Infinite Geometric Series

Given a geometric series, what happens to the sum S_n of the first n terms as n increases without stopping? To answer this question, let us write formula (5) in the form

$$S_n = \frac{a_1 r^n}{r - 1} - \frac{a_1}{r - 1}$$

It is possible to show that if $-1 < r < 1$, then r^n will approach 0 as n increases. The first term above will approach 0 and S_n can be made as close as we please to the second term, $-a_1/(r - 1)$ [which can be written as $a_1/(1 - r)$], by taking n sufficiently large. So, if the common ratio r is between -1 and 1, we conclude that the sum of an infinite geometric series is

THEOREM 7   Sum of an Infinite Geometric Series

$$S_\infty = \frac{a_1}{1 - r} \qquad -1 < r < 1 \tag{7}$$

If $r \leq -1$ or $r \geq 1$, then an infinite geometric series has no sum.

Applications

EXAMPLE 6 Loan Repayment A person borrows \$3,600 and agrees to repay the loan in monthly installments over 3 years. The agreement is to pay 1% of the unpaid balance each month for using the money and \$100 each month to reduce the loan. What is the total cost of the loan over the 3 years?

SOLUTION Let us look at the problem relative to a time line:

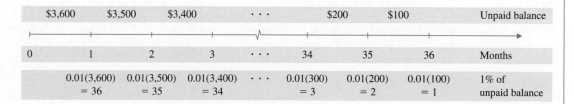

The total cost of the loan is

$$1 + 2 + \cdots + 34 + 35 + 36$$

The terms form a finite arithmetic series with $n = 36$, $a_1 = 1$, and $a_{36} = 36$, so we can use formula (4):

$$S_n = \frac{n}{2}(a_1 + a_n)$$

$$S_{36} = \frac{36}{2}(1 + 36) = \$666$$

We conclude that the total cost of the loan over 3 years is $666.

Matched Problem 6 Repeat Example 6 with a loan of $6,000 over 5 years.

EXAMPLE 7   Economy Stimulation   The government has decided on a tax rebate program to stimulate the economy. Suppose that you receive $1,200 and you spend 80% of this, and each of the people who receive what you spend also spend 80% of what they receive, and this process continues without end. According to the **multiplier principle** in economics, the effect of your $1,200 tax rebate on the economy is multiplied many times. What is the total amount spent if the process continues as indicated?

SOLUTION   We need to find the sum of an infinite geometric series with the first amount spent being $a_1 = (0.8)(\$1,200) = \$960$ and $r = 0.8$. Using formula (7), we obtain

$$S_\infty = \frac{a_1}{1 - r}$$

$$= \frac{\$960}{1 - 0.8} = \$4,800$$

Assuming the process continues as indicated, we would expect the $1,200 tax rebate to result in about $4,800 of spending.

Matched Problem 7 Repeat Example 7 with a tax rebate of $2,000.

Exercises B.2

In Problems 1 and 2, determine whether the indicated sequence can be the first three terms of an arithmetic or geometric sequence, and, if so, find the common difference or common ratio and the next two terms of the sequence.

1. (A) $-11, -16, -21, \ldots$ (B) $2, -4, 8, \ldots$

   (C) $1, 4, 9, \ldots$ (D) $\frac{1}{2}, \frac{1}{6}, \frac{1}{18}, \ldots$

2. (A) $5, 20, 100, \ldots$ (B) $-5, -5, -5, \ldots$

   (C) $7, 6.5, 6, \ldots$ (D) $512, 256, 128, \ldots$

In Problems 3–8, determine whether the finite series is arithmetic, geometric, both, or neither. If the series is arithmetic or geometric, find its sum.

3. $\sum_{k=1}^{101} (-1)^{k+1}$

4. $\sum_{k=1}^{200} 3$

5. $1 + \frac{1}{2} + \frac{1}{3} + \cdots + \frac{1}{50}$

6. $3 - 9 + 27 - \cdots - 3^{20}$

7. $5 + 4.9 + 4.8 + \cdots + 0.1$

8. $1 - \frac{1}{4} + \frac{1}{9} - \cdots - \frac{1}{100^2}$

Let $a_1, a_2, a_3, \ldots, a_n, \ldots$ be an arithmetic sequence. In Problems 9–14, find the indicated quantities.

9. $a_1 = 7; d = 4; a_2 = \; ?; a_3 = \; ?$

10. $a_1 = -2; d = -3; a_2 = \; ?; a_3 = \; ?$

11. $a_1 = 2; d = 4; a_{21} = \; ?; S_{31} = \; ?$

12. $a_1 = 8; d = -10; a_{15} = \; ?; S_{23} = \; ?$

13. $a_1 = 18; a_{20} = 75; S_{20} = \; ?$

14. $a_1 = 203; a_{30} = 261; S_{30} = \; ?$

Let $a_1, a_2, a_3, \ldots, a_n, \ldots$ be a geometric sequence. In Problems 15–24, find the indicated quantities.

15. $a_1 = 3; r = -2; a_2 = \; ?; a_3 = \; ?; a_4 = \; ?$

16. $a_1 = 32; r = -\frac{1}{2}; a_2 = \; ?; a_3 = \; ?; a_4 = \; ?$

17. $a_1 = 1; a_7 = 729; r = -3; S_7 = ?$

18. $a_1 = 3; a_7 = 2,187; r = 3; S_7 = ?$

19. $a_1 = 100; r = 1.08; a_{10} = ?$

20. $a_1 = 240; r = 1.06; a_{12} = ?$

21. $a_1 = 100; a_9 = 200; r = ?$

22. $a_1 = 100; a_{10} = 300; r = ?$

23. $a_1 = 500; r = 0.6; S_{10} = ?; S_\infty = ?$

24. $a_1 = 8,000; r = 0.4; S_{10} = ?; S_\infty = ?$

25. $S_{41} = \sum_{k=1}^{41} (3k + 3) = ?$ **26.** $S_{50} = \sum_{k=1}^{50} (2k - 3) = ?$

27. $S_8 = \sum_{k=1}^{8} (-2)^{k-1} = ?$ **28.** $S_8 = \sum_{k=1}^{8} 2^k = ?$

29. Find the sum of all the odd integers between 12 and 68.

30. Find the sum of all the even integers between 23 and 97.

31. Find the sum of each infinite geometric sequence (if it exists).

  (A) $2, 4, 8, \ldots$ (B) $2, -\frac{1}{2}, \frac{1}{8}, \ldots$

32. Repeat Problem 31 for:

  (A) $16, 4, 1, \ldots$ (B) $1, -3, 9, \ldots$

33. Find $f(1) + f(2) + f(3) + \cdots + f(50)$ if $f(x) = 2x - 3$.

34. Find $g(1) + g(2) + g(3) + \cdots + g(100)$ if $g(t) = 18 - 3t$.

35. Find $f(1) + f(2) + \cdots + f(10)$ if $f(x) = \left(\frac{1}{2}\right)^x$.

36. Find $g(1) + g(2) + \cdots + g(10)$ if $g(x) = 2^x$.

37. Show that the sum of the first n odd positive integers is n^2, using appropriate formulas from this section.

38. Show that the sum of the first n even positive integers is $n + n^2$, using formulas in this section.

39. If $r = 1$, neither the first form nor the second form for the sum of a finite geometric series is valid. Find a formula for the sum of a finite geometric series if $r = 1$.

40. If all of the terms of an infinite geometric series are less than 1, could the sum be greater than 1,000? Explain.

41. Does there exist a finite arithmetic series with $a_1 = 1$ and $a_n = 1.1$ that has sum equal to 100? Explain.

42. Does there exist a finite arithmetic series with $a_1 = 1$ and $a_n = 1.1$ that has sum equal to 105? Explain.

43. Does there exist an infinite geometric series with $a_1 = 10$ that has sum equal to 6? Explain.

44. Does there exist an infinite geometric series with $a_1 = 10$ that has sum equal to 5? Explain.

Applications

45. Loan repayment. If you borrow $4,800 and repay the loan by paying $200 per month to reduce the loan and 1% of the unpaid balance each month for the use of the money, what is the total cost of the loan over 24 months?

46. Loan repayment. If you borrow $5,400 and repay the loan by paying $300 per month to reduce the loan and 1.5% of the unpaid balance each month for the use of the money, what is the total cost of the loan over 18 months?

47. Economy stimulation. The government, through a subsidy program, distributes $5,000,000. If we assume that each person or agency spends 70% of what is received, and 70% of this is spent, and so on, how much total increase in spending results from this government action? (Let $a_1 = \$3,500,000$.)

48. Economy stimulation. Due to reduced taxes, a person has an extra $1,200 in spendable income. If we assume that the person spends 65% of this on consumer goods, and the producers of these goods in turn spend 65% on consumer goods, and that this process continues indefinitely, what is the total amount spent (to the nearest dollar) on consumer goods?

49. Compound interest. If $1,000 is invested at 5% compounded annually, the amount A present after n years forms a geometric sequence with common ratio $1 + 0.05 = 1.05$. Use a geometric sequence formula to find the amount A in the account (to the nearest cent) after 10 years. After 20 years. (*Hint*: Use a time line.)

50. Compound interest. If P is invested at $100r\%$ compounded annually, the amount A present after n years forms a geometric sequence with common ratio $1 + r$. Write a formula for the amount present after n years. (*Hint*: Use a time line.)

Answers to Matched Problems

1. (A) The sequence is geometric with $r = \frac{1}{4}$. It is not arithmetic.

  (B) The sequence is arithmetic with $d = 5$. It is not geometric.

  (C) The sequence is neither arithmetic nor geometric.

2. (A) 139   (B) -2

3. -960 **4.** 10,672   **5.** 1,063.66   **6.** $1,830   **7.** $8,000

B.3 Binomial Theorem

- Factorial
- Development of the Binomial Theorem

The binomial form

$$(a + b)^n$$

where n is a natural number, appears more frequently than you might expect. The coefficients in the expansion play an important role in probability studies. The *binomial formula*, which we will derive informally, enables us to expand $(a + b)^n$ directly for *n* any natural number. Since the formula involves *factorials*, we digress for a moment here to introduce this important concept.

Factorial

For *n* a natural number, **n factorial**, denoted by $n!$, is the product of the first *n* natural numbers. **Zero factorial** is defined to be 1. That is,

---

DEFINITION   *n* Factorial

$$n! = n \cdot (n - 1) \cdot \cdots \cdot 2 \cdot 1$$
$$1! = 1$$
$$0! = 1$$

It is also useful to note that $n!$ can be defined recursively.

DEFINITION   *n* Factorial—Recursive Definition

$$n! = n \cdot (n - 1)! \quad n \geq 1$$

EXAMPLE 1   Factorial Forms   Evaluate.

(A) $5! = 5 \cdot 4 \cdot 3 \cdot 2 \cdot 1 = 120$

(B) $\dfrac{8!}{7!} = \dfrac{8 \cdot \cancel{7!}}{\cancel{7!}} = 8$

(C) $\dfrac{10!}{7!} = \dfrac{10 \cdot 9 \cdot 8 \cdot \cancel{7!}}{\cancel{7!}} = 720$

Matched Problem 1   Evaluate.

(A) $4!$ (B) $\dfrac{7!}{6!}$ (C) $\dfrac{8!}{5!}$

The following formula involving factorials has applications in many areas of mathematics and statistics. We will use this formula to provide a more concise form for the expressions encountered later in this discussion.

THEOREM 1   For *n* and *r* integers satisfying $0 \leq r \leq n$,

$$_nC_r = \frac{n!}{r!(n - r)!}$$

EXAMPLE 2   Evaluating $_nC_r$

(A) $_9C_2 = \dfrac{9!}{2!(9 - 2)!} = \dfrac{9!}{2!7!} = \dfrac{9 \cdot 8 \cdot \cancel{7!}}{2 \cdot \cancel{7!}} = 36$

(B) $_5C_5 = \dfrac{5!}{5!(5 - 5)!} = \dfrac{5!}{5!0!} = \dfrac{5!}{5!} = 1$

Matched Problem 2   Find

(A) $_5C_2$ (B) $_6C_0$

Development of the Binomial Theorem

Let us expand $(a + b)^n$ for several values of n to see if we can observe a pattern that leads to a general formula for the expansion for any natural number n:

$$(a + b)^1 = a + b$$
$$(a + b)^2 = a^2 + 2ab + b^2$$
$$(a + b)^3 = a^3 + 3a^2b + 3ab^2 + b^3$$
$$(a + b)^4 = a^4 + 4a^3b + 6a^2b^2 + 4ab^3 + b^4$$
$$(a + b)^5 = a^5 + 5a^4b + 10a^3b^2 + 10a^2b^3 + 5ab^4 + b^5$$

CONCEPTUAL INSIGHT

1. The expansion of $(a + b)^n$ has $(n + 1)$ terms.

2. The power of a decreases by 1 for each term as we move from left to right.

3. The power of b increases by 1 for each term as we move from left to right.

4. In each term, the sum of the powers of a and b always equals n.

5. Starting with a given term, we can get the coefficient of the next term by multiplying the coefficient of the given term by the exponent of a and dividing by the number that represents the position of the term in the series of terms. For example, in the expansion of $(a + b)^4$ above, the coefficient of the third term is found from the second term by multiplying 4 and 3, and then dividing by 2 [that is, the coefficient of the third term $= (4 \cdot 3)/2 = 6$].

We now postulate these same properties for the general case:

$$(a + b)^n = a^n + \frac{n}{1}a^{n-1}b + \frac{n(n-1)}{1 \cdot 2}a^{n-2}b^2 + \frac{n(n-1)(n-2)}{1 \cdot 2 \cdot 3}a^{n-3}b^3 + \cdots + b^n$$

$$= \frac{n!}{0!(n-0)!}a^n + \frac{n!}{1!(n-1)!}a^{n-1}b + \frac{n!}{2!(n-2)!}a^{n-2}b^2 + \frac{n!}{3!(n-3)!}a^{n-3}b^3 + \cdots + \frac{n!}{n!(n \quad n)!}b^n$$

$$= {}_nC_0a^n + {}_nC_1a^{n-1}b + {}_nC_2a^{n-2}b^2 + {}_nC_3a^{n-3}b^3 + \cdots + {}_nC_nb^n$$

And we are led to the formula in the binomial theorem:

THEOREM 2   Binomial Theorem

For all natural numbers n,

$$(a + b)^n = {}_nC_0a^n + {}_nC_1a^{n-1}b + {}_nC_2a^{n-2}b^2 + {}_nC_3a^{n-3}b^3 + \cdots + {}_nC_nb^n$$

EXAMPLE 3   Using the Binomial Theorem   Use the binomial theorem to expand $(u + v)^6$.

SOLUTION

$$(u + v)^6 = {}_6C_0u^6 + {}_6C_1u^5v + {}_6C_2u^4v^2 + {}_6C_3u^3v^3 + {}_6C_4u^2v^4 + {}_6C_5uv^5 + {}_6C_6v^6$$
$$= u^6 + 6u^5v + 15u^4v^2 + 20u^3v^3 + 15u^2v^4 + 6uv^5 + v^6$$

Matched Problem 3 Use the binomial theorem to expand $(x + 2)^5$.

EXAMPLE 4   Using the Binomial Theorem   Use the binomial theorem to find the sixth term in the expansion of $(x - 1)^{18}$.

SOLUTION   Sixth term $= {}_{18}C_5 x^{13}(-1)^5 = \dfrac{18!}{5!(18 - 5)!} x^{13}(-1)$

$$= -8{,}568 x^{13}$$

Matched Problem 4   Use the binomial theorem to find the fourth term in the expansion of $(x - 2)^{20}$.

Exercises B.3

In Problems 1–20, evaluate each expression.

1. $6!$

2. $7!$

3. $\dfrac{10!}{9!}$

4. $\dfrac{20!}{19!}$

5. $\dfrac{12!}{9!}$

6. $\dfrac{10!}{6!}$

7. $\dfrac{5!}{2!3!}$

8. $\dfrac{7!}{3!4!}$

9. $\dfrac{6!}{5!(6 - 5)!}$

10. $\dfrac{7!}{4!(7 - 4)!}$

11. $\dfrac{20!}{3!17!}$

12. $\dfrac{52!}{50!2!}$

13. ${}_5C_3$

14. ${}_7C_3$

15. ${}_6C_5$

16. ${}_7C_4$

17. ${}_5C_0$

18. ${}_5C_5$

19. ${}_{18}C_{15}$

20. ${}_{18}C_3$

Expand each expression in Problems 21–26 using the binomial theorem.

21. $(a + b)^4$

22. $(m + n)^5$

23. $(x - 1)^6$

24. $(u - 2)^5$

25. $(2a - b)^5$

26. $(x - 2y)^5$

Find the indicated term in each expansion in Problems 27–32.

27. $(x - 1)^{18}$; 5th term

28. $(x - 3)^{20}$; 3rd term

29. $(p + q)^{15}$; 7th term

30. $(p + q)^{15}$; 13th term

31. $(2x + y)^{12}$; 11th term

32. $(2x + y)^{12}$; 3rd term

33. Show that ${}_nC_0 = {}_nC_n$ for $n \geq 0$.

34. Show that ${}_nC_r = {}_nC_{n-r}$ for $n \geq r \geq 0$.

35. The triangle next is called **Pascal's triangle**. Can you guess what the next two rows at the bottom are? Compare these numbers with the coefficients of binomial expansions.

$$
\begin{array}{ccccccccc}
 & & & & 1 & & & & \\
 & & & 1 & & 1 & & & \\
 & & 1 & & 2 & & 1 & & \\
 & 1 & & 3 & & 3 & & 1 & \\
1 & & 4 & & 6 & & 4 & & 1
\end{array}
$$

36. Explain why the sum of the entries in each row of Pascal's triangle is a power of 2. (*Hint:* Let $a = b = 1$ in the binomial theorem.)

37. Explain why the alternating sum of the entries in each row of Pascal's triangle (e.g., $1 - 4 + 6 - 4 + 1$) is equal to 0.

38. Show that ${}_nC_r = \dfrac{n - r + 1}{r} {}_nC_{r-1}$ for $n \geq r \geq 1$.

39. Show that ${}_nC_{r-1} + {}_nC_r = {}_{n+1}C_r$ for $n \geq r \geq 1$.

Answers to Matched Problems

1. (A) 24   (B) 7   (C) 336

2. (A) 10   (B) 1

3. $x^5 + 10x^4 + 40x^3 + 80x^2 + 80x + 32$

4. $-9{,}120x^{17}$

Tables

Table I **Basic Geometric Formulas**

1. Similar Triangles

(A) Two triangles are similar if two angles of one triangle have the same measure as two angles of the other.

(B) If two triangles are similar, their corresponding sides are proportional:

$$\frac{a}{a'} = \frac{b}{b'} = \frac{c}{c'}$$

2. Pythagorean Theorem

$$c^2 = a^2 + b^2$$

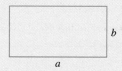

3. Rectangle

$A = ab$ Area

$P = 2a + 2b$ Perimeter

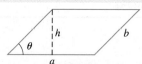

4. Parallelogram

$h =$ height

$A = ah = ab \sin \theta$ Area

$P = 2a + 2b$ Perimeter

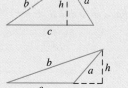

5. Triangle

$h =$ height

$A = \frac{1}{2}hc$ Area

$P = a + b + c$ Perimeter

$s = \frac{1}{2}(a + b + c)$ Semiperimeter

$A = \sqrt{s(s - a)(s - b)(s - c)}$ Area: Heron's formula

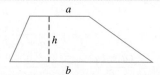

6. Trapezoid

Base a is parallel to base b.

$h =$ height

$A = \frac{1}{2}(a + b)h$ Area

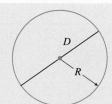

7. Circle

$R =$ radius

$D =$ diameter

$D = 2R$

$A = \pi R^2 = \frac{1}{4}\pi D^2$ Area

$C = 2\pi R = \pi D$ Circumference

$\dfrac{C}{D} = \pi$ For all circles

$\pi \approx 3.14159$

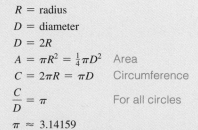

Table I **Continued**

8. **Rectangular Solid**

$V = abc$ Volume

$T = 2ab + 2ac + 2bc$ Total surface area

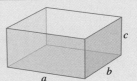

9. **Right Circular Cylinder**

$R =$ radius of base

$h =$ height

$V = \pi R^2 h$ Volume

$S = 2\pi R h$ Lateral surface area

$T = 2\pi R(R + h)$ Total surface area

10. **Right Circular Cone**

$R =$ radius of base

$h =$ height

$s =$ slant height

$V = \frac{1}{3}\pi R^2 h$ Volume

$S = \pi R s = \pi R\sqrt{R^2 + h^2}$ Lateral surface area

$T = \pi R(R + s) = \pi R(R + \sqrt{R^2 + h^2})$ Total surface area

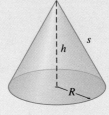

11. **Sphere**

$R =$ radius

$D =$ diameter

$D = 2R$

$V = \frac{4}{3}\pi R^3 = \frac{1}{6}\pi D^3$ Volume

$S = 4\pi R^2 = \pi D^2$ Surface area

Table II **Integration Formulas**

Integrals Involving u^n

1. $\int u^n \, du = \dfrac{u^{n+1}}{n + 1}, \quad n \neq -1$

2. $\int u^{-1} \, du = \int \dfrac{1}{u} \, du = \ln|u|$

Integrals Involving $a + bu, a \neq 0$ and $b \neq 0$

3. $\int \dfrac{1}{a + bu} \, du = \dfrac{1}{b}\ln|a + bu|$

4. $\int \dfrac{u}{a + bu} \, du = \dfrac{u}{b} - \dfrac{a}{b^2}\ln|a + bu|$

5. $\int \dfrac{u^2}{a + bu} \, du = \dfrac{(a + bu)^2}{2b^3} - \dfrac{2a(a + bu)}{b^3} + \dfrac{a^2}{b^3}\ln|a + bu|$

6. $\int \dfrac{u}{(a + bu)^2} \, du = \dfrac{1}{b^2}\left(\ln|a + bu| + \dfrac{a}{a + bu}\right)$

7. $\int \dfrac{u^2}{(a + bu)^2} \, du = \dfrac{(a + bu)}{b^3} - \dfrac{a^2}{b^3(a + bu)} - \dfrac{2a}{b^3}\ln|a + bu|$

8. $\int u(a + bu)^n \, du = \dfrac{(a + bu)^{n+2}}{(n + 2)b^2} - \dfrac{a(a + bu)^{n+1}}{(n + 1)b^2}, \quad n \neq -1, -2$

9. $\int \dfrac{1}{u(a + bu)} \, du = \dfrac{1}{a}\ln\left|\dfrac{u}{a + bu}\right|$

10. $\int \dfrac{1}{u^2(a + bu)} \, du = -\dfrac{1}{au} + \dfrac{b}{a^2}\ln\left|\dfrac{a + bu}{u}\right|$

11. $\int \dfrac{1}{u(a + bu)^2} \, du = \dfrac{1}{a(a + bu)} + \dfrac{1}{a^2}\ln\left|\dfrac{u}{a + bu}\right|$

12. $\int \dfrac{1}{u^2(a + bu)^2} \, du = -\dfrac{a + 2bu}{a^2 u(a + bu)} + \dfrac{2b}{a^3}\ln\left|\dfrac{a + bu}{u}\right|$

Table II **Continued**

Integrals Involving $a^2 - u^2, a > 0$

13. $\displaystyle\int \frac{1}{u^2 - a^2}\, du = \frac{1}{2a} \ln\left|\frac{u - a}{u + a}\right|$

14. $\displaystyle\int \frac{1}{a^2 - u^2}\, du = \frac{1}{2a} \ln\left|\frac{u + a}{u - a}\right|$

Integrals Involving $(a + bu)$ and $(c + du), b \neq 0, d \neq 0,$ and $ad - bc \neq 0$

15. $\displaystyle\int \frac{1}{(a + bu)(c + du)}\, du = \frac{1}{ad - bc} \ln\left|\frac{c + du}{a + bu}\right|$

16. $\displaystyle\int \frac{u}{(a + bu)(c + du)}\, du = \frac{1}{ad - bc}\left(\frac{a}{b} \ln|a + bu| - \frac{c}{d} \ln|c + du|\right)$

17. $\displaystyle\int \frac{u^2}{(a + bu)(c + du)}\, du = \frac{1}{bd} u - \frac{1}{ad - bc}\left(\frac{a^2}{b^2} \ln|a + bu| - \frac{c^2}{d^2} \ln|c + du|\right)$

18. $\displaystyle\int \frac{1}{(a + bu)^2(c + du)}\, du = \frac{1}{ad - bc}\frac{1}{a + bu} + \frac{d}{(ad - bc)^2} \ln\left|\frac{c + du}{a + bu}\right|$

19. $\displaystyle\int \frac{u}{(a + bu)^2(c + du)}\, du = -\frac{a}{b(ad - bc)}\frac{1}{a + bu} - \frac{c}{(ad - bc)^2} \ln\left|\frac{c + du}{a + bu}\right|$

20. $\displaystyle\int \frac{a + bu}{c + du}\, du = \frac{bu}{d} + \frac{ad - bc}{d^2} \ln|c + du|$

Integrals Involving $\sqrt{a + bu}, a \neq 0$ and $b \neq 0$

21. $\displaystyle\int \sqrt{a + bu}\, du = \frac{2\sqrt{(a + bu)^3}}{3b}$

22. $\displaystyle\int u\sqrt{a + bu}\, du = \frac{2(3bu - 2a)}{15b^2}\sqrt{(a + bu)^3}$

23. $\displaystyle\int u^2\sqrt{a + bu}\, du = \frac{2(15b^2u^2 - 12abu + 8a^2)}{105b^3}\sqrt{(a + bu)^3}$

24. $\displaystyle\int \frac{1}{\sqrt{a + bu}}\, du = \frac{2\sqrt{a + bu}}{b}$

25. $\displaystyle\int \frac{u}{\sqrt{a + bu}}\, du = \frac{2(bu - 2a)}{3b^2}\sqrt{a + bu}$

26. $\displaystyle\int \frac{u^2}{\sqrt{a + bu}}\, du = \frac{2(3b^2u^2 - 4abu + 8a^2)}{15b^3}\sqrt{a + bu}$

27. $\displaystyle\int \frac{1}{u\sqrt{a + bu}}\, du = \frac{1}{\sqrt{a}} \ln\left|\frac{\sqrt{a + bu} - \sqrt{a}}{\sqrt{a + bu} + \sqrt{a}}\right|, \quad a > 0$

28. $\displaystyle\int \frac{1}{u^2\sqrt{a + bu}}\, du = -\frac{\sqrt{a + bu}}{au} - \frac{b}{2a\sqrt{a}} \ln\left|\frac{\sqrt{a + bu} - \sqrt{a}}{\sqrt{a + bu} + \sqrt{a}}\right|, \quad a > 0$

Integrals Involving $\sqrt{a^2 - u^2}, a > 0$

29. $\displaystyle\int \frac{1}{u\sqrt{a^2 - u^2}}\, du = -\frac{1}{a} \ln\left|\frac{a + \sqrt{a^2 - u^2}}{u}\right|$

30. $\displaystyle\int \frac{1}{u^2\sqrt{a^2 - u^2}}\, du = -\frac{\sqrt{a^2 - u^2}}{a^2u}$

31. $\displaystyle\int \frac{\sqrt{a^2 - u^2}}{u}\, du = \sqrt{a^2 - u^2} - a \ln\left|\frac{a + \sqrt{a^2 - u^2}}{u}\right|$

Integrals Involving $\sqrt{u^2 + a^2}, a > 0$

32. $\displaystyle\int \sqrt{u^2 + a^2}\, du = \frac{1}{2}\left(u\sqrt{u^2 + a^2} + a^2 \ln|u + \sqrt{u^2 + a^2}|\right)$

33. $\displaystyle\int u^2\sqrt{u^2 + a^2}\, du = \frac{1}{8}\left[u(2u^2 + a^2)\sqrt{u^2 + a^2} - a^4 \ln|u + \sqrt{u^2 + a^2}|\right]$

34. $\displaystyle\int \frac{\sqrt{u^2 + a^2}}{u}\, du = \sqrt{u^2 + a^2} - a \ln\left|\frac{a + \sqrt{u^2 + a^2}}{u}\right|$

35. $\displaystyle\int \frac{\sqrt{u^2 + a^2}}{u^2}\, du = -\frac{\sqrt{u^2 + a^2}}{u} + \ln|u + \sqrt{u^2 + a^2}|$

36. $\displaystyle\int \frac{1}{\sqrt{u^2 + a^2}}\, du = \ln|u + \sqrt{u^2 + a^2}|$

(continued)

Table II Continued

37. $\int \dfrac{1}{u\sqrt{u^2 + a^2}}\, du = \dfrac{1}{a}\ln\left|\dfrac{u}{a + \sqrt{u^2 + a^2}}\right|$

38. $\int \dfrac{u^2}{\sqrt{u^2 + a^2}}\, du = \dfrac{1}{2}\left(u\sqrt{u^2 + a^2} - a^2\ln|u + \sqrt{u^2 + a^2}|\right)$

39. $\int \dfrac{1}{u^2\sqrt{u^2 + a^2}}\, du = -\dfrac{\sqrt{u^2 + a^2}}{a^2 u}$

Integrals Involving $\sqrt{u^2 - a^2}, a > 0$

40. $\int \sqrt{u^2 - a^2}\, du = \dfrac{1}{2}\left(u\sqrt{u^2 - a^2} - a^2\ln|u + \sqrt{u^2 - a^2}|\right)$

41. $\int u^2\sqrt{u^2 - a^2}\, du = \dfrac{1}{8}\left[u(2u^2 - a^2)\sqrt{u^2 - a^2} - a^4\ln|u + \sqrt{u^2 - a^2}|\right]$

42. $\int \dfrac{\sqrt{u^2 - a^2}}{u^2}\, du = -\dfrac{\sqrt{u^2 - a^2}}{u} + \ln|u + \sqrt{u^2 - a^2}|$

43. $\int \dfrac{1}{\sqrt{u^2 - a^2}}\, du = \ln|u + \sqrt{u^2 - a^2}|$

44. $\int \dfrac{u^2}{\sqrt{u^2 - a^2}}\, du = \dfrac{1}{2}\left(u\sqrt{u^2 - a^2} + a^2\ln|u + \sqrt{u^2 - a^2}|\right)$

45. $\int \dfrac{1}{u^2\sqrt{u^2 - a^2}}\, du = \dfrac{\sqrt{u^2 - a^2}}{a^2 u}$

Integrals Involving $e^{au}, a \neq 0$

46. $\int e^{au}\, du = \dfrac{e^{au}}{a}$

47. $\int u^n e^{au}\, du = \dfrac{u^n e^{au}}{a} - \dfrac{n}{a}\int u^{n-1} e^{au}\, du$

48. $\int \dfrac{1}{c + de^{au}}\, du = \dfrac{u}{c} - \dfrac{1}{ac}\ln|c + de^{au}|, \quad c \neq 0$

Integrals Involving $\ln u$

49. $\int \ln u\, du = u\ln u - u$

50. $\int \dfrac{\ln u}{u}\, du = \dfrac{1}{2}(\ln u)^2$

51. $\int u^n \ln u\, du = \dfrac{u^{n+1}}{n + 1}\ln u - \dfrac{u^{n+1}}{(n + 1)^2}, \quad n \neq -1$

52. $\int (\ln u)^n\, du = u(\ln u)^n - n\int (\ln u)^{n-1}\, du$

Integrals Involving Trigonometric Functions of $au, a \neq 0$

53. $\int \sin au\, du = -\dfrac{1}{a}\cos au$

54. $\int \cos au\, du = \dfrac{1}{a}\sin au$

55. $\int \tan au\, du = -\dfrac{1}{a}\ln|\cos au|$

56. $\int \cot au\, du = \dfrac{1}{a}\ln|\sin au|$

57. $\int \sec au\, du = \dfrac{1}{a}\ln|\sec au + \tan au|$

58. $\int \csc au\, du = \dfrac{1}{a}\ln|\csc au - \cot au|$

59. $\int (\sin au)^2\, du = \dfrac{u}{2} - \dfrac{1}{4a}\sin 2au$

60. $\int (\cos au)^2\, du = \dfrac{u}{2} + \dfrac{1}{4a}\sin 2au$

61. $\int (\sin au)^n\, du = -\dfrac{1}{an}(\sin au)^{n-1}\cos au + \dfrac{n-1}{n}\int (\sin au)^{n-2}\, du, \quad n \neq 0$

62. $\int (\cos au)^n\, du = \dfrac{1}{an}\sin au(\cos au)^{n-1} + \dfrac{n-1}{n}\int (\cos au)^{n-2}\, du, \quad n \neq 0$

[*Note:* **The constant of integration is omitted for each integral, but must be included in any particular application of a formula.** The variable u is the variable of integration; all other symbols represent constants.]

ANSWERS

Diagnostic Prerequisite Test

Section references are provided in parentheses following each answer to guide students to the specific content in the book where they can find help or remediation.

1. (A) $(y + z)x$ (B) $(2 + x) + y$ (C) $2x + 3x$ *(A.1)*   **2.** $x^3 - 3x^2 + 4x + 8$ *(A.2)*   **3.** $x^3 + 3x^2 - 2x + 12$ *(A.2)*   **4.** $-3x^5 + 2x^3 - 24x^2 + 16$ *(A.2)*

5. (A) 1   (B) 1   (C) 2   (D) 3 *(A.2)*   **6.** (A) 3   (B) 1   (C) −3   (D) 1 *(A.2)*   **7.** $14x^2 - 30x$ *(A.2)*   **8.** $6x^2 - 5xy - 4y^2$ *(A.2)*   **9.** $(x + 2)(x + 5)$ *(A.3)*

10. $x(x + 3)(x - 5)$ *(A.3)*   **11.** $7/20$ *(A.1)*   **12.** 0.875 *(A.1)*   **13.** (A) 4.065×10^{12} (B) 7.3×10^{-3} *(A.5)*   **14.** (A) 255,000,000   (B) 0,000 406 *(A.5)*

15. (A) T   (B) F *(A.1)*   **16.** 0 and −3 are two examples of infinitely many. *(A.1)*   **17.** $6x^5 y^{15}$ *(A.5)*   **18.** $3u^4/v^2$ *(A.5)*   **19.** 6×10^2 *(A.5)*   **20.** x^6/y^4 *(A.5)*

21. $u^{7/3}$ *(A.6)*   **22.** $3a^2/b$ *(A.6)*   **23.** $\frac{5}{9}$ *(A.5)*   **24.** $x + 2x^{1/2}y^{1/2} + y$ *(A.6)*   **25.** $\frac{a^2 + b^2}{ab}$ *(A.4)*   **26.** $\frac{a^2 - c^2}{abc}$ *(A.4)*   **27.** $\frac{y^5}{x}$ *(A.4)*   **28.** $\frac{1}{xy^2}$ *(A.4)*

29. $\frac{-1}{7(7 + h)}$ *(A.4)*   **30.** $\frac{xy}{y - x}$ *(A.6)*   **31.** (A) Subtraction   (B) Commutative (+)   (C) Distributive   (D) Associative ($\cdot$)   (E) Negatives   (F) Identity (+) *(A.1)*

32. (A) 6   (B) 0 *(A.1)*   **33.** $4x = x - 4; x = -4/3$ *(1.1)*   **34.** $-15/7$ *(1.2)*   **35.** $(4/7, 0)$ *(1.2)*   **36.** $(0, -4)$ *(1.2)*   **37.** $(x - 5y)(x + 2y)$ *(A.3)*

38. $(3x - y)(2x - 5y)$ *(A.3)*   **39.** $3x^{-1} + 4y^{1/2}$ *(A.6)*   **40.** $8x^{-2} - 5y^{-4}$ *(A.5)*   **41.** $\frac{2}{5}x^{-3/4} - \frac{7}{6}y^{-2/3}$ *(A.6)*   **42.** $\frac{1}{3}x^{-1/2} + 9y^{-1/3}$ *(A.6)*

43. $\frac{2}{7} + \frac{1}{14}\sqrt{2}$ *(A.6)*   **44.** $\frac{14}{11} - \frac{5}{11}\sqrt{3}$ *(A.6)*   **45.** $x = 0, 5$ *(A.7)*   **46.** $x = \pm\sqrt{7}$ *(A.7)*   **47.** $x = -4, 5$ *(A.7)*   **48.** $x = 1, \frac{1}{6}$ *(A.7)*

49. $x = -1 \pm \sqrt{2}$ *(A.7)*   **50.** $x = \pm 1, \pm\sqrt{5}$ *(A.7)*

Chapter 1

Exercises 1.1   **1.** $m = 5$ **3.** $x < -\frac{7}{2}$ **5.** $x \le 4$ **7.** $x < -3$ or $(-\infty, -3)$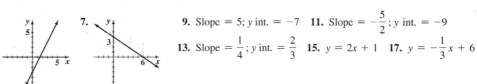

9. $-1 \le x \le 2$ or $[-1, 2]$ **11.** $x = -\frac{3}{2}$ **13.** $y < -\frac{15}{2}$ **15.** $u = -\frac{3}{4}$ **17.** $x = 10$ **19.** $y \ge 3$ **21.** $x = 36$

23. $m < \frac{36}{7}$ **25.** $3 \le x < 7$ or $[3, 7)$ **27.** $-20 \le C \le 20$ or $[-20, 20]$ **29.** $y = \frac{3}{4}x - 3$

31. $y = -(A/B)x + (C/B) = (-Ax + C)/B$ **33.** $C = \frac{5}{9}(F - 32)$ **35.** $-2 < x \le 1$ or $(-2, 1]$

37. (A) and (C): $a > 0$ and $b > 0$, or $a < 0$ and $b < 0$ (B) and (D): $a > 0$ and $b < 0$, or $a < 0$ and $b > 0$ **39.** Negative   **41.** True   **43.** False   **45.** True

47. 4,500 \$35 tickets and 5,000 \$55 tickets   **49.** Fund A: \$180,000; Fund B: \$320,000   **51.** \$14,634   **53.** (A) \$420   (B) \$55   **55.** 34 rounds   **57.** \$32,000

59. 5,851 books   **61.** (B) 6,180 books   (C) At least \$11.50   **63.** 5,000   **65.** 12.6 yr

Exercises 1.2   **1.** (D)   **3.** (C)   **5.**   **7.**   **9.** Slope = 5; y int. = −7   **11.** Slope = $-\frac{5}{2}$; y int. = −9

13. Slope = $\frac{1}{4}$; y int. = $\frac{2}{3}$ **15.** $y = 2x + 1$ **17.** $y = -\frac{1}{3}x + 6$

19. x int.: −1; y int.: −2; $y = -2x - 2$ **23.**   **25.**   **27.**   **29.** −4   **31.** $-\frac{3}{5}$ **33.** 2

21. x int.: −3; y int.: 1; $y = \frac{x}{3} + 1$

35.   **37.** 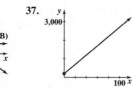   **39.** (A) 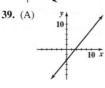   **39.** (B) x int.: 3.5; y int.: −4.2   **39.** (C)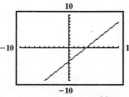

39. (D) x int.: 3.5; y int.: −4.2   **41.** $x = 4, y = -3$ **43.** $x = -1.5, y = -3.5$ **45.** $y = 5x - 15$ **47.** $y = -2x + 7$ **49.** $y = \frac{1}{3}x - \frac{20}{3}$

51. $y = -3.2x + 30.86$ **53.** (A) $m = \frac{2}{3}$ (B) $-2x + 3y = 11$ (C) $y = \frac{2}{3}x + \frac{11}{3}$ **55.** (A) $m = -\frac{5}{4}$ (B) $5x + 4y = -14$ (C) $y = -\frac{5}{4}x - \frac{7}{2}$

57. (A) Not defined   (B) $x = 5$ (C) None   **59.** (A) $m = 0$ (B) $y = 5$ (C) $y = 5$ **61.** The graphs have the same y int., (0, 2).

63. $C = 124 + 0.12x$; 1,050 donuts   **65.** (A) $C = 75x + 1,647$ (B)   (C) The y int., \$1,647, is the fixed cost and the slope, \$75, is the cost per club.   **67.** (A) $R = 1.4C - 7$ (B) \$137

69. (A) $V = -7{,}500t + 157{,}000$ (B) \$112,000 (C) During the 12th year (D)

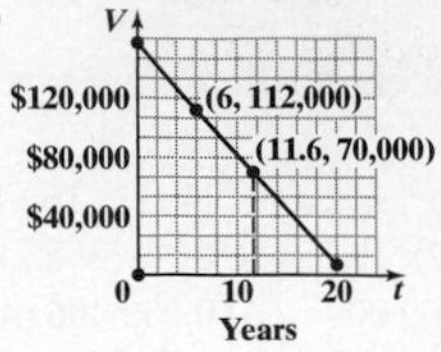

71. (A) $T = -1.84x + 212$ (B) 205.56°F (C) 6,522 ft (D)

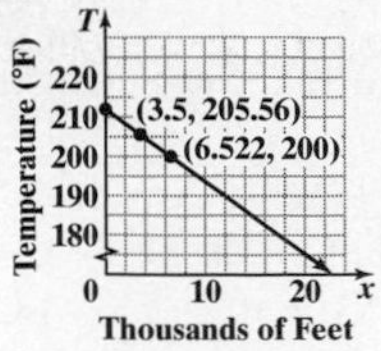

73. (A) $T = 70 - 3.6A$ (B) 10,000 ft **75.** (A) $N = -0.0066t + 2.76$ (B) 2.43 persons **77.** (A) $f = -0.37t + 21$ (B) 2024

79. (A) $p = 0.000225x + 0.5925$ (B) $p = -0.0009x + 9.39$ (C) (7,820, 2.352) (D)

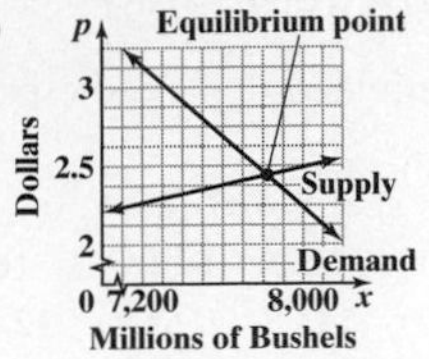

81. (A) $s = \frac{2}{5}w$ (B) 8 in. (C) 9 lb

Exercises 1.3

1. (A) $w = 49 + 1.7h$ (B) The rate of change of weight with respect to height is 1.7 kg/in. (C) 55.8 kg (D) 5′6.5″
3. (A) $P = 0.4\overline{45}d + 14.7$ (B) The rate of change of pressure with respect to depth is $0.4\overline{45}$ lb/in.^{2} per ft. (C) 37 lb/in.^{2} (D) 99 ft
5. (A) $a = 2{,}880 - 24t$ (B) −24 ft/sec (C) 24 ft/sec **7.** $s = 0.6t + 331$; the rate of change of the speed of sound with respect to temperature is 0.6 m/s per °C.

9. (A)

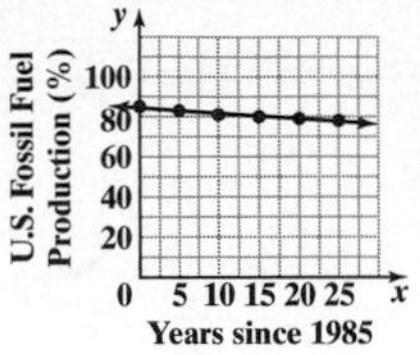

(B) The rate of change of fossil fuel production is −0.3% per year. (C) 72.4% of total production (D) 2034

11. (A)

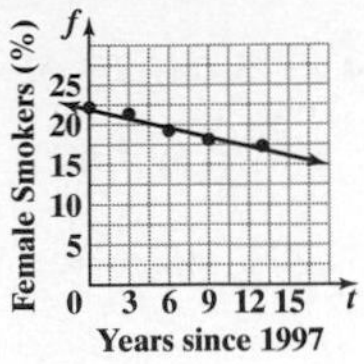

(B) 2028

13. (A)

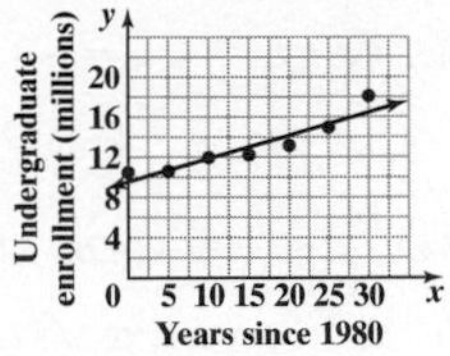

(B) 19,900,000 (C) Undergraduate enrollment is increasing at a rate of 230,000 students per year.

15. (A)

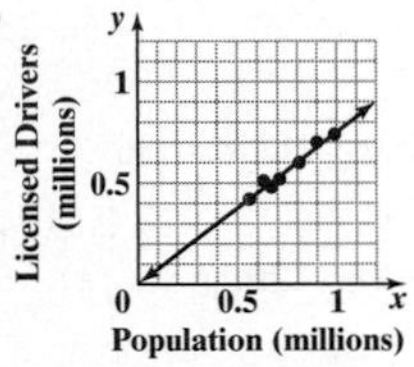

(B) 1,200,000 (C) 1,000,000

17. (A)

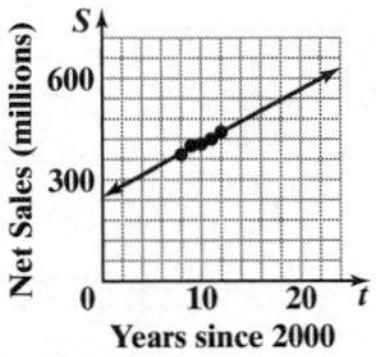

(B) \$599 billion

19. (A)

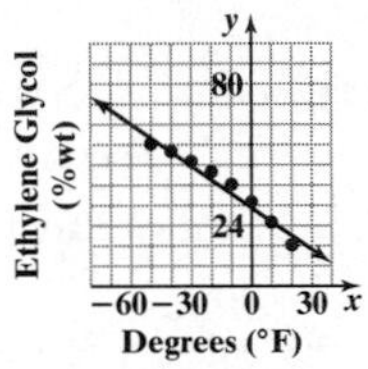

(B) 2°F (C) 22.75%

21. (A) The rate of change of height with respect to Dbh is 1.37 ft/in. (B) Height increases by approximately 1.37 ft. (C) 18 ft (D) 20 in.
23. (A) The monthly price is increasing at a rate of \$1.70 per year. (B) \$71.70
25. (A) Male enrollment is increasing at a rate of 70,000 students per year; female enrollment is increasing at a rate of 180,000 students per year. (B) Male: 8.9 million; female: 13.7 million (C) 2027
27. Men: $y = -0.087x + 49.207$; women: $y = -0.088x + 54.884$; yes **29.** Supply: $y = 0.2x + 0.87$; demand: $y = -0.15x + 3.5$; equilibrium price = \$2.37

Chapter 1 Review Exercises

1. $x = 2.8$ *(1.1)* **2.** $x = 2$ *(1.1)* **3.** $y = 1.8 - 0.4x$ *(1.1)* **4.** $x = \frac{4}{3}y + \frac{7}{3}$ *(1.1)*

5. $y < \frac{13}{4}$ or $\left(-\infty, \frac{13}{4}\right)$ 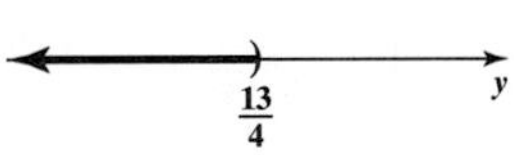 *(1.1)*

6. $1 \le x < 3$ or $[1, 3)$ 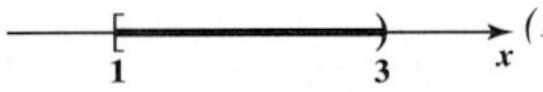 *(1.1)*

7. $x \ge \frac{9}{2}$ or $\left[\frac{9}{2}, \infty\right)$ 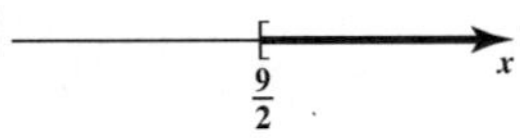 *(1.1)*

8. 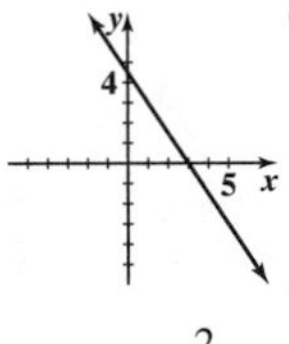 *(1.2)*

9. $2x + 3y = 12$ *(1.2)*

10. x int. = 9; y int. = −6; slope = $\frac{2}{3}$ *(1.2)* 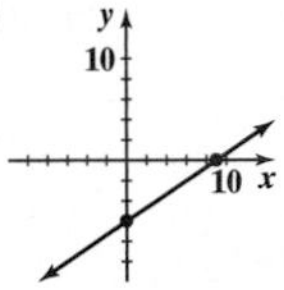

11. $y = -\frac{2}{3}x + 6$ *(1.2)* **12.** Vert. line: $x = -6$; hor. line: $y = 5$ *(1.2)*

13. (A) $y = -\frac{2}{3}x$ (B) $y = 3$ *(1.2)* **14.** (A) $3x + 2y = 1$ (B) $y = 5$ (C) $x = -2$ *(1.2)* **15.** $x = \frac{25}{2}$ *(1.1)* **16.** $u = 36$ *(1.1)*

17. $x = \frac{30}{11}$ *(1.1)* **18.** $x = 21$ *(1.1)* **19.** $x = 4$ *(1.1)* **20.** $x < 4$ or $(-\infty, 4)$ *(1.1)* 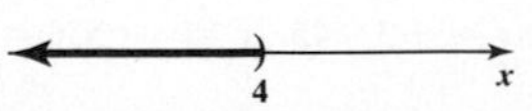

21. $x \ge 1$ or $[1, \infty)$ *(1.1)* 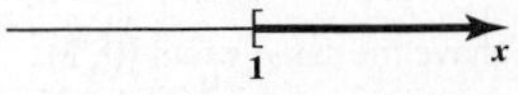

22. $x < -\frac{143}{17}$ or $\left(-\infty, -\frac{143}{17}\right)$ 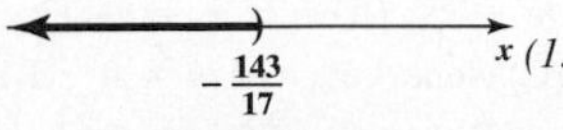 *(1.1)*

23. $1 < x \le 4$ or $(1, 4]$ 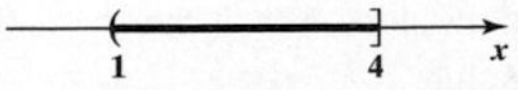 *(1.1)*

24. $\frac{3}{8} \le x \le \frac{7}{8}$ or $\left[\frac{3}{8}, \frac{7}{8}\right]$ 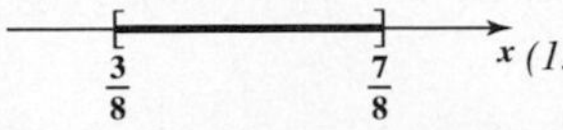 *(1.1)*

25. 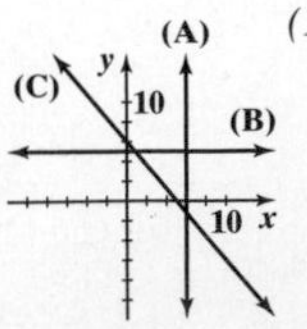 *(1.2)*

26. The graph of $x = -3$ is a vert. line with x int. −3, and the graph of $y = 2$ is a hor. line with y int. 2. *(1.2)*

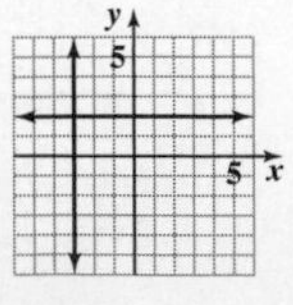

27. (A) An oblique line through the origin with slope −3/4 (B) A vert. line with x int. −4/3 (C) The x axis (D) An oblique line with x int. 12 and y int. 9 *(1.2)* **28.** $\frac{2A - bh}{h}$ *(1.1)* **29.** $\frac{S - P}{St}$ *(1.1)*
30. $a < 0$ and b any real number *(1.1)* **31.** Less than *(1.1)*

32. The graphs appear to be perpendicular to each other. (It can be shown that if the slopes of two slant lines are the negative reciprocals of each other, then the two lines are perpendicular.) *(1.2)*   **33.** $75,000 *(1.1)*   **34.** 9,375 DVDs *(1.1)*   **35.** (A) $m = 132 - 0.6x$ (B) $M = 187 - 0.85x$ (C) Between 120 and 170 beats per minute   (D) Between 102 and 144.5 beats per minute *(1.3)*   **36.** (A) $V = 224,000 - 15,500t$ (B) $38,000 *(1.2)*   **37.** (A) $R = 1.6C$ (B) $192   (C) $110   (D) The slope is 1.6. This is the rate of change of retail price with respect to cost. *(1.2)*   **38.** $400; $800 *(1.1)*   **39.** Demand: $p = 5.24 - 0.00125x$; 1,560 bottles *(1.2)*

40. (A)   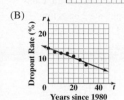   (B) $-30°$F   (C) 45% *(1.3)*   **41.** (A) The dropout rate is decreasing at a rate of 0.198 percentage points per year.

(B)   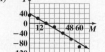   (C) 2027 *(1.3)*   **42.** (A) The CPI is increasing at a rate of 4.75 units per year.   (B) 285 *(1.3)*   **43.** (A) The rate of change of tree height with respect to Dbh is 0.74.   (B) Tree height increases by about 0.74 ft.   (C) 21 ft   (D) 16 in. *(1.3)*

Chapter 2

Exercises 2.1

1.   **3.**   **5.**   **7.**   **9.** A function   **11.** Not a function   **13.** A function   **15.** A function   **17.** Not a function   **19.** A function   **21.** Linear   **23.** Neither   **25.** Constant   **27.** Linear

29.   **31.**   **33.**   **35.** 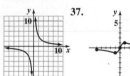   **37.** 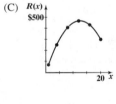   **39.** $y = 0$ **41.** $y = 4$ **43.** $x = -5, 0, 4$ **45.** $x = -6$ **47.** All real numbers   **49.** All real numbers except -4 **51.** $x \le 7$ **53.** Yes; all real numbers   **55.** No; for example, when $x = 0$, $y = \pm 2$

57. Yes; all real numbers except 0   **59.** No; when $x = 1$, $y = \pm 1$ **61.** 12   **63.** $x^2 + 2x - 3$ **65.** $36x^2 - 4$ **67.** $x^6 - 4$ **69.** $h^2 - 4$ **71.** $4h + h^2$ **73.** $4h + h^2$ **75.** (A) $4x + 4h - 3$ (B) $4h$ (C) 4   **77.** (A) $4x^2 + 8xh + 4h^2 - 7x - 7h + 6$ (B) $8xh + 4h^2 - 7h$ (C) $8x + 4h - 7$

79. (A) $20x + 20h - x^2 - 2xh - h^2$ (B) $20h - 2xh - h^2$ (C) $20 - 2x - h$ **81.** $P(w) = 2w + \dfrac{50}{w}, w > 0$ **83.** $A(l) = l(50 - l), 0 < l < 50$

85. (A) 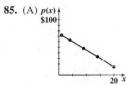   (B) $54, $42   **87.** (A) $R(x) = (75 - 3x)x, 1 \le x \le 20$ (B)

x	$R(x)$
1	72
4	252
8	408
12	468
16	432
20	300

(C)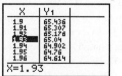

89. (A) $P(x) = 59x - 3x^2 - 125, 1 \le x \le 20$

(B)

x	$P(x)$
1	-69
4	63
8	155
12	151
16	51
20	-145

(C)

91. (A) $V(x) = x(8 - 2x)(12 - 2x)$ (B) $0 < x < 4$

(C)

x	$V(x)$
1	60
2	64
3	36

(D)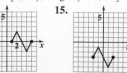

93. (A) The graph indicates that there is a value of x near 2 that will produce a volume of 65.   (B) The table shows $x = 1.9$ to one decimal place:

x	1.8	1.9	2
$V(x)$	66.5	65.4	64

(C) $x = 1.93$ to two decimal places

95. $v = \dfrac{75 - w}{15 + w}$; 1.9032 cm/sec

Exercises 2.2

1. Domain: all real numbers; range: all real numbers   **3.** Domain: $[0, \infty)$; range: $(-\infty, 15]$ **5.** Domain: all real numbers; range: $[7, \infty)$ **7.** Domain: all real numbers; range: all real numbers   **9.**   **11.**   **13.** (graph)   **15.** (graph)   **17.** (graph)   **19.** (graph)

21.

23.

25. The graph of $g(x) = -|x + 3|$ is the graph of $y = |x|$ reflected in the x axis and shifted 3 units to the left.

27. The graph of $f(x) = (x - 4)^2 - 3$ is the graph of $y = x^2$ shifted 4 units to the right and 3 units down.

29. The graph of $f(x) = 7 - \sqrt{x}$ is the graph of $y = \sqrt{x}$ reflected in the x axis and shifted 7 units up.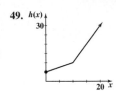

31. The graph of $h(x) = -3|x|$ is the graph of $y = |x|$ reflected in the x axis and vertically stretched by a factor of 3.

33. The graph of the basic function $y = x^2$ is shifted 2 units to the left and 3 units down. Equation: $y = (x + 2)^2 - 3$ **35.** The graph of the basic function $y = x^2$ is reflected in the x axis and shifted 3 units to the right and 2 units up. Equation: $y = 2 - (x - 3)^2$

37. The graph of the basic function $y = \sqrt{x}$ is reflected in the x axis and shifted 4 units up. Equation: $y = 4 - \sqrt{x}$

39. The graph of the basic function $y = x^3$ is shifted 2 units to the left and 1 unit down. Equation: $y = (x + 2)^3 - 1$

41. $g(x) = \sqrt{x - 2} - 3$ **43.** $g(x) = -|x + 3|$ **45.** $g(x) = -(x - 2)^3 - 1$

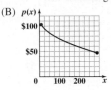   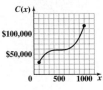

47. 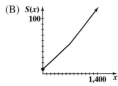 **49.**   **51.**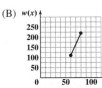

53. The graph of the basic function $y = |x|$ is reflected in the x axis and vertically shrunk by a factor of 0.5. Equation: $y = -0.5|x|$ **55.** The graph of the basic function $y = x^2$ is reflected in the x axis and vertically stretched by a factor of 2. Equation: $y = -2x^2$ **57.** The graph of the basic function $y = \sqrt[3]{x}$ is reflected in the x axis and vertically stretched by a factor of 3. Equation: $y = -3\sqrt[3]{x}$

59. Reversing the order does not change the result.

61. Reversing the order can change the result.

63. Reversing the order can change the result.

65. (A) The graph of the basic function $y = \sqrt{x}$ is reflected in the x axis, vertically expanded by a factor of 4, and shifted up 115 units. (B)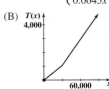

67. (A) The graph of the basic function $y = x^3$ is vertically shrunk by a factor of 0.00048 and shifted right 500 units and up 60,000 units. (B)

69. (A) $S(x) = \begin{cases} 8.5 + 0.065x & \text{if } 0 \leq x \leq 700 \\ -9 + 0.09x & \text{if } x > 700 \end{cases}$

(B)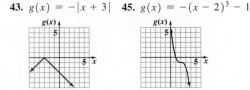

71. (A) $T(x) = \begin{cases} 0.035x & \text{if } 0 \leq x \leq 30{,}000 \\ 0.0625x - 825 & \text{if } 30{,}000 < x \leq 60{,}000 \\ 0.0645x - 945 & \text{if } x > 60{,}000 \end{cases}$

(B) 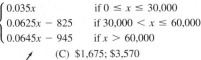 (C) \$1,675; \$3,570

73. (A) The graph of the basic function $y = x$ is vertically stretched by a factor of 5.5 and shifted down 220 units. (B) $w(x)$

75. (A) The graph of the basic function $y = \sqrt{x}$ is vertically stretched by a factor of 7.08. (B) $v(x)$

Exercises 2.3   1. $f(x) = (x - 5)^2 - 25$ **3.** $f(x) = (x + 10)^2 - 50$ **5.** $f(x) = -2(x - 1)^2 - 3$ **7.** $f(x) = 2\left(x + \frac{1}{2}\right)^2 + \frac{1}{2}$

9. The graph of $f(x)$ is the graph of $y = x^2$ shifted right 2 units and down 1 unit.   **11.** The graph of $m(x)$ is the graph of $y = x^2$ reflected in the x axis, then shifted right 3 units and up 5 units.   **13.** (A) m (B) g (C) f (D) n **15.** (A) x int.: 1, 3; y int.: -3 (B) Vertex: $(2, 1)$ (C) Max.: 1 (D) Range: $y \leq 1$ or $(-\infty, 1]$ **17.** (A) x int.: $-3, -1$; y int.: 3 (B) Vertex: $(-2, -1)$ (C) Min.: -1 (D) Range: $y \geq -1$ or $[-1, \infty)$ **19.** (A) x int.: $3 \pm \sqrt{2}$; y int.: -7 (B) Vertex: $(3, 2)$ (C) Max.: 2 (D) Range: $y \leq 2$ or $(-\infty, 2]$ **21.** (A) x int.: $-1 \pm \sqrt{2}$; y int.: -1 (B) Vertex: $(-1, -2)$ (C) Min.: -2 (D) Range: $y \geq -2$ or $[-2, \infty)$ **23.** $y = -[x - (-2)]^2 + 5$ or $y = -(x + 2)^2 + 5$ **25.** $y = (x - 1)^2 - 3$ **27.** Vertex form: $(x - 4)^2 - 4$ (A) x int.: 2, 6; y int.: 12 (B) Vertex: $(4, -4)$ (C) Min.: -4 (D) Range: $y \geq -4$ or $[-4, \infty)$ **29.** Vertex form: $-4(x - 2)^2 + 1$ (A) x int.: 1.5, 2.5; y int.: -15 (B) Vertex: $(2, 1)$ (C) Max.: 1 (D) Range: $y \leq 1$ or $(-\infty, 1]$ **31.** Vertex form: $0.5(x - 2)^2 + 3$ (A) x int.: none; y int.: 5 (B) Vertex: $(2, 3)$ (C) Min.: 3 (D) Range: $y \geq 3$ or $[3, \infty)$ **33.** (A) $-4.87, 8.21$ (B) $-3.44, 6.78$ (C) No solution   **35.** 651.0417   **37.** $g(x) = 0.25(x - 3)^2 - 9.25$ (A) x int.: $-3.08, 9.08$; y int.: -7 (B) Vertex: $(3, -9.25)$ (C) Min.: -9.25 (D) Range: $y \geq -9.25$ or $[-9.25, \infty)$ **39.** $f(x) = -0.12(x - 4)^2 + 3.12$ (A) x int.: $-1.1, 9.1$; y int.: 1.2 (B) Vertex: $(4, 3.12)$ (C) Max.: 3.12 (D) Range: $y \leq 3.12$ or $(-\infty, 3.12]$ **41.** $x = -5.37, 0.37$ **43.** $-1.37 < x < 2.16$ **45.** $x \leq -0.74$ or $x \geq 4.19$ **47.** Axis: $x = 2$; vertex: $(2, 4)$; range: $y \geq 4$ or $[4, \infty)$; no x int.

49. (A) 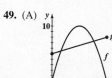   (B) 1.64, 7.61
(C) 1.64 < x < 7.61
(D) 0 ≤ x < 1.64 or 7.61 < x ≤ 10

51. (A) 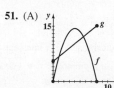   (B) 1.10, 5.57
(C) 1.10 < x < 5.57
(D) 0 ≤ x < 1.10 or 5.57 < x ≤ 8

61. (A)

x	28	30	32	34	36
Mileage	45	52	55	51	47
$f(x)$	45.3	51.8	54.2	52.4	46.5

(B)

(C) $f(31) = 53.50$ thousand miles;
(D) $f(35) = 49.95$ thousand miles

65. (A)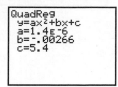

(B) 12.5 (12,500,000 chips);
$468,750,000
(C) $37.50

67. (A)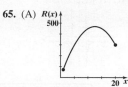

(B) 2,415,000 chips and 17,251,000 chips
(C) Loss: $1 ≤ x < 2.415$ or
$17.251 < x ≤ 20$;
profit: $2.415 < x < 17.251$

69. (A) $P(x) = 59x - 3x^2 - 125$ (C) Intercepts and break-even points: 2,415,000 chips and 17,251,000 chips

(E) Maximum profit is $165,083,000 at a production level of 9,833,000 chips. This is much smaller than the maximum revenue of $468,750,000.

71. $x = 0.14$ cm   **73.** 10.6 mph

Exercises 2.4   1. (A) 1   (B) 10   (C) 50   **3.** (A) 5   (B) 0, 1   (C) 0   **5.** (A) 2   (B) −1, −2   (C) 2   **7.** (A) 4   (B) −1, 1, −3, 3   (C) 9   **9.** (A) 9
(B) −3/2, 5   (C) −253,125   **11.** (A) 4   (B) Negative   **13.** (A) 5   (B) Negative   **15.** (A) 1   (B) Negative   **17.** (A) 6   (B) Positive   **19.** 10   **21.** 1

23. (A) x intercept: −2; y intercept: −1   (E)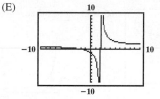
(B) Domain: all real numbers except 2
(C) Vertical asymptote: $x = 2$;
horizontal asymptote: $y = 1$
(D)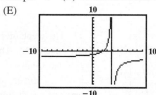

25. (A) x intercept: 0; y intercept: 0   (B) Domain: all real numbers except −2
(C) Vertical asymptote: $x = -2$; horizontal asymptote: $y = 3$
(D)   (E)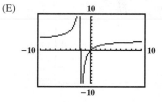

27. (A) x intercept: 2; y intercept: −1   (B) Domain: all real numbers except 4   (C) Vertical asymptote: $x = 4$; horizontal asymptote: $y = -2$
(D) 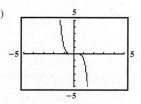   (E)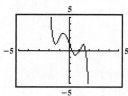

29. (A) 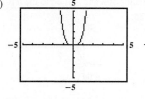   (B)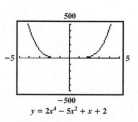

$y = 2x^4$ $y = 2x^4 - 5x^2 + x + 2$

31. (A) 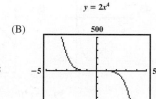   (B)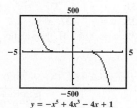

$y = -x^5$ $y = -x^5 + 4x^3 - 4x + 1$

33. $y = \dfrac{5}{6}$ **35.** $y = \dfrac{1}{4}$ **37.** $y = 0$ **39.** None   **41.** $x = -1, x = 1, x = -3, x = 3$ **43.** $x = 5$ **45.** $x = -6, x = 6$

47. (A) x intercept: 0; y intercept: 0
(B) Vertical asymptotes: $x = -2, x = 3$;
horizontal asymptote: $y = 2$
(C)

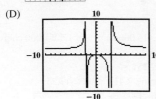

(D)

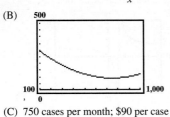

49. (A) x intercept: $\pm\sqrt{3}$; y intercept: $-\dfrac{2}{3}$
(B) Vertical asymptotes: $x = -3, x = 3$;
horizontal asymptote: $y = -2$
(C)

(D)

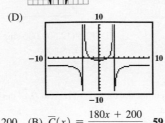

51. (A) x intercept: 6; y intercept: -4
(B) Vertical asymptotes: $x = -3, x = 2$;
horizontal asymptote: $y = 0$
(C)

(D)

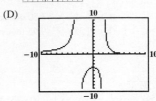

53. $f(x) = x^2 - x - 2$ **57.** (A) $C(x) = 180x + 200$ (B) $\bar{C}(x) = \dfrac{180x + 200}{x}$
55. $f(x) = 4x - x^3$
(C)

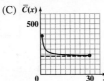

(D) \$180 per board

59. (A) $\bar{C}(n) = \dfrac{2{,}500 + 175n + 25n^2}{n}$ (C) 10 yr; \$675.00 per year
(D) 10 yr; \$675.00 per year
(B)

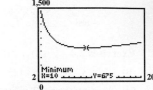

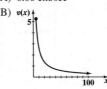

61. (A) $\bar{C}(x) = \dfrac{0.00048(x - 500)^3 + 60{,}000}{x}$
(B)

63. (A)

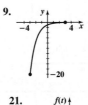

(B) 4.1 lb

65. (A) 0.06 cm/sec
(B) $v(x)$

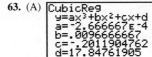

67. (A)

(B) 5.5

(C) 750 cases per month; \$90 per case

Exercises 2.5   **1.** (A) k **3.**   **5.**   **7.**   **9.**   **11.** The graph of g is the graph of f reflected in
(B) g the x axis.   **13.** The graph of g is the graph of
(C) h

f shifted 1 unit to the left.   **15.** The graph of
(D) f g is the graph of f shifted 1 unit up.   **17.** The
graph of g is the graph of f vertically expanded
by a factor of 2 and shifted to the left 2 units.

19. (A)   (B)   (C)   (D)   **21.**   **23.**

      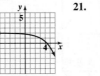

25.   **27.** $a = 1, -1$ **29.** $x = 1$ **43.**   **45.**   **47.** \$16,064.07   **49.** (A) \$2,633.56   (B) \$7,079.54
31. $x = -1, 6$ **33.** $x = 3$ **51.** \$10,706
35. $x = -7$ **37.** $x = -2, 2$ **53.** (A) \$10,095.41   (B) \$10,080.32   (C) \$10,085.27
39. $x = 1/4$ **41.** No solution

55. N approaches 2 as t increases   **57.** (A) \$9,781,000   **59.** (A) 10%   (B) 1%   **63.** (A) 42,772,000,000
without bound.   **61.** (A) $P = 7.1e^{0.011t}$

(B) 2025: 8,100,000,000;
2035: 9,000,000,000

(B) The model gives an annual
salary of \$1,647,000 in 2000.

Exercises 2.6   **1.** $27 = 3^3$ **3.** $10^0 = 1$ **5.** $8 = 4^{3/2}$ **7.** $\log_7 49 = 2$ **9.** $\log_4 8 = \dfrac{3}{2}$ **11.** $\log_b A = u$ **13.** 2   **15.** 4   **17.** -2 **19.** -4

21. $\log_b P - \log_b Q$ **23.** $5 \log_b L$ **25.** q^p **27.** $x = 9$ **29.** $y = 2$ **31.** $b = 10$ **33.** $x = 2$ **35.** False   **37.** True   **39.** True   **41.** False   **43.** $x = 2$

45. $x = 8$ **47.** $x = 7$ **49.** No solution   **51.**   **53.** The graph of $y = \log_2 (x - 2)$ is the graph of $y = \log_2 x$ shifted to the right 2 units.

55. Domain: $(-1, \infty)$; range: all real numbers   **57.** (A) 3.547 43   (B) $-2.160\ 32$
(C) 5.626 29   (D) $-3.197\ 04$ **59.** (A) 13.4431   (B) 0.0089   (C) 16.0595   (D) 0.1514

61. 1.0792   **63.** 1.4595   **65.** 18.3559

67. Increasing: $(0, \infty)$ **69.** Decreasing: $(0, 1]$ **71.** Increasing: $(-2, \infty)$ **73.** Increasing: $(0, \infty)$ **75.** Because $b^0 = 1$ for any permissible base
Increasing: $[1, \infty)$
$b (b > 0, b \neq 1)$.   **77.** $x > \sqrt{x} > \ln x$ for $1 < x \leq 16$

79. 4 yr   **81.** 9.87 yr; 9.80 yr

83. 7.51 yr

85. (A) 5,373 (B) 7,220   **89.** 168 bushels/acre   **91.** 916 yr

Chapter 2 Review Exercises

1. *(2.1)*   **2.** *(2.1)*   **3.** *(2.1)*   **4.** (A) Not a function   (B) A function   (C) A function   (D) Not a function *(2.1)*

5. (A) -2 (B) -8 (C) 0   (D) Not defined *(2.1)*   **6.** $v = \ln u$ *(2.6)*

7. $y = \log x$ *(2.6)*   **8.** $M = e^N$ *(2.6)*   **9.** $u = 10^v$ *(2.6)*   **10.** $x = 9$ *(2.6)*

11. $x = 6$ *(2.6)*   **12.** $x = 4$ *(2.6)*   **13.** $x = 2.157$ *(2.6)*   **14.** $x = 13.128$ *(2.6)*

15. $x = 1,273.503$ *(2.6)*   **16.** $x = 0.318$ *(2.6)*

17. (A) $y = 4$ (B) $x = 0$ (C) $y = 1$ (D) $x = -1$ or 1   (E) $y = -2$ (F) $x = -5$ or 5 *(2.1)*

18. (A)   (B)   (C)   (D)   *(2.2)*

19. $f(x) = -(x - 2)^2 + 4$. The graph of $f(x)$ is the graph of $y = x^2$ reflected in the x axis, then shifted right 2 units and up 4 units. *(2.2)*

20. (A) g (B) m (C) n (D) f *(2.2, 2.3)*   **21.** (A) x intercepts: $-4, 0$; y intercept: 0   (B) Vertex: $(-2, -4)$ (C) Minimum: -4

(D) Range: $y \geq -4$ or $[-4, \infty)$ *(2.3)*   **22.** Quadratic *(2.3)*   **23.** Linear *(2.1)*   **24.** None *(2.1, 2.3)*   **25.** Constant *(2.1)*   **26.** $x = 8$ *(2.6)*

27. $x = 3$ *(2.6)*   **28.** $x = 3$ *(2.5)*   **29.** $x = -1, 3$ *(2.5)*   **30.** $x = 0, \frac{3}{2}$ *(2.5)*   **31.** $x = -2$ *(2.6)*   **32.** $x = \frac{1}{2}$ *(2.6)*   **33.** $x = 27$ *(2.6)*

34. $x = 13.3113$ *(2.6)*   **35.** $x = 158.7552$ *(2.6)*   **36.** $x = 0.0097$ *(2.6)*   **37.** $x = 1.4359$ *(2.6)*   **38.** $x = 1.4650$ *(2.6)*   **39.** $x = 92.1034$ *(2.6)*

40. $x = 9.0065$ *(2.6)*   **41.** $x = 2.1081$ *(2.6)*   **42.** (A) All real numbers except $x = -2$ and 3   (B) $x < 5$ *(2.1)*   **43.** Vertex form: $4\left(x + \dfrac{1}{2}\right)^2 - 4$;

x intercepts: $-\frac{3}{2}$ and $\frac{1}{2}$; y intercept: -3; vertex: $(-\frac{1}{2}, -4)$; minimum: -4; range: $y \geq -4$ or $[-4, \infty)$ *(2.3)*   **44.** $(-1.54, -0.79)$; $(0.69, 0.99)$ *(2.5, 2.6)*

45. *(2.1)*   **47.** 6 *(2.1)*   **48.** -19 *(2.1)*   **49.** $10x - 4$ *(2.1)*   **50.** $21 - 5x$ *(2.1)*   **51.** (A) -1 (B) $-1 - 2h$ (C) $-2h$ (D) -2 *(2.1)*

52. (A) $a^2 - 3a + 1$ (B) $a^2 + 2ah + h^2 - 3a - 3h + 1$ (C) $2ah + h^2 - 3h$ (D) $2a + h - 3$ *(2.1)*

53. The graph of function m is the graph of $y = |x|$ reflected in the x axis and shifted to the right 4 units. *(2.2)*

54. The graph of function g is the graph of $y = x^3$ vertically shrunk by a factor of 0.3 and shifted up 3 units. *(2.2)*

55. The graph of $y = x^2$ is vertically stretched by a factor of 2, reflected in the x axis, and shifted to the left 3 units.
Equation: $y = -2(x + 3)^2$ *(2.2)*

46. *(2.1)*   **56.** $f(x) = 2\sqrt{x + 3} - 1$ *(2.2)*   **57.** $y = 0$ *(2.4)*   **58.** $y = \dfrac{3}{4}$ *(2.4)*   **59.** None *(2.4)*   **60.** $x = -10, x = 10$ *(2.4)*

61. $x = -2$ *(2.4)*   **62.** True *(2.3)*   **63.** False *(2.3)*   **64.** False *(2.3)*   **65.** True *(2.4)*

66. True *(2.5)*   **67.** True *(2.3)*

68.  *(2.2)* **69.** *(2.2)* **70.** $y = -(x - 4)^2 + 3$ *(2.2, 2.3)*

71. $f(x) = -0.4(x - 4)^2 + 7.6$ (A) x intercepts: $-0.4, 8.4$; y intercept: 1.2
   (B) Vertex: $(4.0, 7.6)$ (C) Maximum: 7.6   (D) Range: $y \le 7.6$ or $(-\infty, 7.6]$ *(2.3)*

72.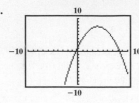

73. $\log 10^\pi = \pi$ and $10^{\log \sqrt{2}} = \sqrt{2}$; $\ln e^\pi = \pi$ and $e^{\ln \sqrt{2}} = \sqrt{2}$ *(2.6)*   **74.** $x = 2$ *(2.6)*
75. $x = 2$ *(2.6)*   **76.** $x = 1$ *(2.6)*   **77.** $x = 300$ *(2.6)*   **78.** $y = ce^{-5t}$ *(2.6)*
79. If $\log_1 x = y$, then $1^y = x$; that is, $1 = x$ for all positive real numbers x, which is not possible. *(2.6)*
80. The graph of $y = \sqrt[3]{x}$ is vertically stretched by a factor of 2, reflected in the x axis, and shifted 1 unit left and 1 unit down. Equation: $y = -2\sqrt[3]{x + 1} - 1$. *(2.2)*
81. $G(x) = 0.3(x + 2)^2 - 8.1$ (A) x intercepts: $-7.2, 3.2$; y intercept: -6.9 (B) Vertex:
$(-2, -8.1)$ (C) Minimum: -8.1 (D) Range: $y \ge -8.1$ or $[-8.1, \infty)$ *(2.3)*

(A) x intercepts: $-.4, 8.4$; y intercept: 1.2
(B) Vertex: $(4.0, 7.6)$ (C) Maximum: 7.6
(D) Range: $y \le 7.6$ or $(-\infty, 7.6]$ *(2.3)*

82.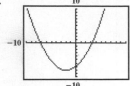

(A) x intercepts: $-7.2, 3.2$; y intercept: -6.9 (B) Vertex: $(-2, -8.1)$ (C) Minimum: -8.1
(D) Range: $y \ge -8.1$ or $[-8.1, \infty)$ *(2.3)*

83. (A) $S(x) = \begin{cases} 3 & \text{if } 0 \le x \le 20 \\ 0.057x + 1.86 & \text{if } 20 < x \le 200 \\ 0.0346x + 6.34 & \text{if } 200 < x \le 1{,}000 \\ 0.0217x + 19.24 & \text{if } x > 1{,}000 \end{cases}$

(B) 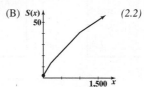 *(2.2)*

84. $5,321.95 *(2.5)*   **85.** $5,269.51 *(2.5)*   **86.** 201 months (≈ 16.7 years) *(2.5)*   **87.** 9.38 yr *(2.5)*

88. (A)   (B) $R = C$ for $x = 4.686$ thousand units (4,686 units) and for $x = 27.314$ thousand units (27,314 units);
   $R < C$ for $1 \le x < 4.686$ or $27.314 < x \le 40$; $R > C$ for $4.686 < x < 27.314$.
   (C) Maximum revenue is 500 thousand dollars ($500,000). This occurs at an output of 20 thousand units (20,000 units).
   At this output, the wholesale price is $p(20) = 25. *(2.3)*

89. (A) $P(x) = R(x) - C(x) = x(50 - 1.25x) - (160 + 10x)$
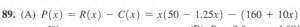
   (B) $P = 0$ for $x = 4.686$ thousand units (4,686 units) and for $x = 27.314$ thousand units (27,314 units);
   $P < 0$ for $1 \le x < 4.686$ or $27.314 < x \le 40$; $p > 0$ for $4.686 < x < 27.314$.
   (C) Maximum profit is 160 thousand dollars ($160,000). This occurs at an output of 16 thousand units (16,000 units).
   At this output, the wholesale price is $p(16) = 30. *(2.3)*

90. (A) $A(x) = -\frac{3}{2}x^2 + 420x$
(B) Domain: $0 \le x \le 280$
(C)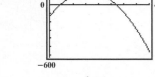

(D) There are two solutions to the equation
   $A(x) = 25{,}000$, one near 90 and another near 190.
(E) 86 ft; 194 ft
(F) Maximum combined area is 29,400 ft^2. This occurs
   for $x = 140$ ft and $y = 105$ ft. *(2.3)*

91. (A) 2,833 sets   (B) 4,836

(C) Equilibrium price: $131.59; equilibrium quantity: 3,587 cookware sets *(2.3)*

92. (A) 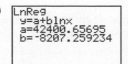   (B) 7,725 *(2.4)*   **93.** (A) $N = 2^{2t}$ or $N = 4^t$
   (B) 15 days *(2.5)*
94. $k = 0.009\,42$; 489 ft *(2.6)*

95. (A) 6,134,000 *(2.6)*    **96.** 23.1 yr *(2.5)*   **97.** (A) $1,203 billion    (B) 2028 *(2.5)*

Chapter 3

Exercises 3.1   **1.** $25.42   **3.** 76%   **5.** Slope = 120; y int. = 12,000   **7.** Slope = 50; y int. = 2,000   **9.** 0.015   **11.** 0.6%   **13.** 0.004   **15.** 24.99%
17. $\frac{1}{3}$ yr   **19.** $\frac{2}{3}$ yr   **21.** $\frac{3}{13}$ yr   **23.** $\frac{1}{2}$ yr   **25.** $42   **27.** $1,800   **29.** 0.12 or 12%   **31.** $\frac{1}{2}$ yr   **33.** $4,612.50   **35.** $875   **37.** 0.36 or 36%   **39.** 1 yr
41. $r = I/Pt$ **43.** $P = A/(1 + rt)$ **45.** $t = \dfrac{A - P}{Pr}$ **47.** The graphs are linear, all with y intercept $1,000; their slopes are 40, 80, and 120, respectively.
49. $45   **51.** $9.23   **53.** $7,647.20   **55.** 8.1%   **57.** 18%   **59.** $1,604.40; 20.88%   **61.** 4.298%   **63.** $992.38   **65.** $24.31   **67.** $7.68   **69.** 5.396%
71. 6.986%   **73.** 12.085%   **75.** 109.895%   **77.** 87.158%   **79.** $49.14   **81.** $8.28

Exercises 3.2   **1.** $P = 950$ **3.** $x = 17$ **5.** $i = 0.5$ **7.** $n = 4$ **9.** $5,983.40   **11.** $4,245.07   **13.** $3,125.79   **15.** $2,958.11   **17.** 2.5 yr
19. 6.79%   **21.** 0.75% per month   **23.** 0.04% per day   **25.** 1.2% per quarter   **27.** 4.74%   **29.** 3.6%   **31.** 4.2%   **33.** (A) $126.25; $26.25
(B) $126.90; $26.90   (C) $127.05; $27.05   **35.** (A) $5,524.71   (B) $6,104.48   **37.** $12,175.69   **39.** All three graphs are increasing, curve upward, and have
the same y intercept; the greater the interest rate, the greater the increase. The amounts at the end of 8 years are $1,376.40, $1,892.46, and $2,599.27, respectively.

41.

Period	Interest	Amount
0		$1,000.00
1	$97.50	$1,097.50
2	$107.01	$1,204.51
3	$117.44	$1,321.95
4	$128.89	$1,450.84
5	$141.46	$1,592.29
6	$155.25	$1,747.54

43. (A) $7,440.94   (B) $5,536.76   **45.** (A) $19,084.49   (B) $11,121.45
47. (A) 3.97%   (B) 2.32%   **49.** (A) 5.28%   (B) 5.27%   **51.** $11\frac{2}{3}$ yr   **53.** 3.75 yr
55. $n \approx 12$ **57.** (A) $7\frac{1}{4}$ yr   (B) 6 yr   **59.** (A) 7.7 yr   (B) 6.3 yr   **61.** $65,068.44   **63.** $282,222.44
65. $19.78 per ft^2 per mo   **67.** 41 yr
69. (A) In 2026, 250 years after the signing, it would be worth $175,814.55.   (B) If interest were
compounded monthly, daily, or continuously, it would be worth $179,119.92, $180,748.53, or
$180,804.24, respectively.

69. (C)

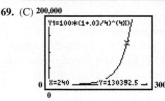

71. 9.66%
73. 2 yr, 10 mo
75. $163,295.21
77. 3,615 days; 8.453 yr

79.

Years	Exact Rate	Rule of 72
6	12.2	12.0
7	10.4	10.3
8	9.1	9.0
9	8.0	8.0
10	7.2	7.2
11	6.5	6.5
12	5.9	6.0

81. 14 quarters   **83.** To maximize
earnings, choose 10% simple interest for
investments lasting fewer than 11 years
and 7% compound interest otherwise.
85. 6.58%   **87.** 7.02%   **89.** $15,843.80
91. 4.53%   **93.** 1.025%   **95.** 13.44%
97. 17.62%

Exercises 3.3   **1.** 1,023   **3.** 3,000   **5.** 71,744,530   **7.** $i = 0.02; n = 80$ **9.** $i = 0.0375; n = 24$ **11.** $i = 0.0075; n = 48$ **13.** $i = 0.0595; n = 12$

15. $FV = $13,435.19$ **17.** $PMT = 310.62 **19.** $n = 17$ **21.** $i = 0.09$ **25.** $n = \dfrac{\ln\left(1 + i\dfrac{FV}{PMT}\right)}{\ln(1 + i)}$ **27.** Value: $84,895.40; interest: $24,895.40

29. $20,931.01   **31.** $667.43   **33.** $763.39

35.

Period	Amount	Interest	Balance
1	$1,000.00	$0.00	$1,000.00
2	$1,000.00	$83.20	$2,083.20
3	$1,000.00	$173.32	$3,256.52
4	$1,000.00	$270.94	$4,527.46
5	$1,000.00	$376.69	$5,904.15

37. First year: $33.56; second year: $109.64; third year: $190.41   **39.** $111,050.77   **41.** $1,308.75
43. (A) 1.540%   (B) $202.12   **45.** 33 months   **47.** 7.77%   **49.** 7.13%   **51.** After 11 quarterly
payments

Exercises 3.4   **1.** 511/256   **3.** 33,333,333/1,000,000   **5.** 171/256   **7.** $i = 0.006; n = 48$ **9.** $i = 0.02475; n = 40$ **11.** $i = 0.02525; n = 32$
13. $i = 0.0548; n = 9$ **15.** $PV = $3,458.41$ **17.** $PMT = 586.01 **19.** $n = 29$ **21.** $i = 0.029$ **27.** $35,693.18   **29.** $11,241.81; $1,358.19
31. $69.58; $839.84   **33.** 31 months   **35.** 71 months   **37.** For 0% financing, the monthly payments should be $129.58, not $179. If a loan of $9,330 is
amortized in 72 payments of $179, the rate is 11.29% compounded monthly.   **39.** The monthly payments with 0% financing are $455. If you take the rebate, the
monthly payments are $434.24. You should choose the rebate.   **41.** $314.72; $17,319.68

43.

Payment Number	Payment	Interest	Unpaid Balance Reduction	Unpaid Balance
0				$5,000.00
1	$706.29	$140.00	$566.29	4,433.71
2	706.29	124.14	582.15	3,851.56
3	706.29	107.84	598.45	3,253.11
4	706.29	91.09	615.20	2,637.91
5	706.29	73.86	632.43	2,005.48
6	706.29	56.15	650.14	1,355.34
7	706.29	37.95	668.34	687.00
8	706.24	19.24	687.00	0.00
Totals	$5,650.27	$650.27	$5,000.00	

45. First year: $466.05; second year: $294.93; third year: $107.82
47. $97,929.78; $116,070.22   **49.** $143.85/mo; $904.80   **51.** Monthly
payment: $908.99   (A) $125,862   (B) $81,507   (C) $46,905
53. (A) Monthly payment: $1,015.68; interest: $114,763   (B) 197 months;
interest saved: $23,499   **55.** (A) 157   (B) 243   (C) The withdrawals
continue forever.   **57.** (A) Monthly withdrawals: $1,229.66; total interest:
$185,338.80   (B) Monthly deposits: $162.65   **59.** $65,584
61. $34,692   **63.** All three graphs are decreasing, curve downward, and
have the same x intercept; the unpaid balances are always in the ratio 2:3:4.
The monthly payments are $402.31, $603.47, and $804.62, with total interest
amounting to $94,831.60, $142,249.20, and $189,663.20, respectively.
65. 14.45%   **67.** 10.21%

Chapter 3 Review Exercises   **1.** $A = \$104.50$ *(3.1)*   **2.** $P = \$800$ *(3.1)*   **3.** $t = 0.75$ yr, or 9 mo *(3.1)*   **4.** $r = 6\%$ *(3.1)*   **5.** $A = \$1,393.68$ *(3.2)*
6. $P = \$3,193.50$ *(3.2)*   **7.** $A = \$5,824.92$ *(3.2)*   **8.** $P = \$22,612.86$ *(3.2)*   **9.** $FV = \$69,770.03$ *(3.3)*   **10.** $PMT = \$115.00$ *(3.3)*
11. $PV = \$33,944.27$ *(3.4)*   **12.** $PMT = \$166.07$ *(3.4)*   **13.** $n \approx 16$ *(3.2)*   **14.** $n \approx 41$ *(3.3)*   **15.** \$3,350.00; \$350.00 *(3.1)*   **16.** \$19,654 *(3.2)*
17. \$12,944.67 *(3.2)*

18. (A)

Period	Interest	Amount
0		\$400.00
1	\$21.60	\$421.60
2	\$22.77	\$444.37
3	\$24.00	\$468.36
4	\$25.29	\$493.65

(3.2)

(B)

Period	Interest	Payment	Balance
1		\$100.00	\$100.00
2	\$5.40	\$100.00	\$205.40
3	\$11.09	\$100.00	\$316.49
4	\$17.09	\$100.00	\$433.58

(3.3)

19. To maximize earnings, choose 13% simple interest for investments lasting less than 9 years and 9% compound interest for investments lasting 9 years
or more. *(3.2)*   **20.** \$164,402 *(3.2)*   **21.** 7.83% *(3.2)*   **22.** 9% compounded quarterly, since its effective rate is 9.31%, while the effective rate of 9.25%
compounded annually is 9.25% *(3.2)*   **23.** \$25,861.65; \$6,661.65 *(3.3)*   **24.** \$11.64 *(3.1)*   **25.** \$29,354 *(3.2)*   **26.** \$18,021 *(3.2)*   **27.** 15% *(3.1)*
28. The monthly payments with 0% financing are \$450. If you take the rebate, the monthly payments are \$426.66. You should choose the rebate. *(3.4)*
29. (A) 6.43% (B) 6.45% *(3.2)*   **30.** 9 quarters or 2 yr, 3 mo *(3.2)*   **31.** 139 mo; 93 mo *(3.2)*   **32.** (A) \$571,499 (B) \$1,973,277 *(3.3)*   **33.** 10.45% *(3.2)*
34. (A) 174% (B) 65.71% *(3.1)*   **35.** \$725.89 *(3.3)*   **36.** (A) \$140,945.57 (B) \$789.65 (C) \$136,828 *(3.3, 3.4)*   **37.** \$102.99; \$943.52 *(3.4)*
38. \$576.48 *(3.3)*   **39.** 3,374 days; 10 yr *(3.2)*   **40.** \$175.28; \$2,516.80 *(3.4)*   **41.** \$13,418.78 *(3.2)*   **42.** 5 yr, 10 mo *(3.3)*   **43.** 18 yr *(3.4)*   **44.** 28.8% *(3.1)*

45.

Payment Number	Payment	Interest	Unpaid Balance Reduction	Unpaid Balance
0				\$1,000.00
1	\$265.82	\$25.00	\$240.82	759.18
2	265.82	18.98	246.84	512.34
3	265.82	12.81	253.01	259.33
4	265.81	6.48	259.33	0.00
Totals	\$1,063.27	\$63.27	\$1,000.00	

(3.4)

46. 28 months *(3.3)*   **47.** \$55,347.48; \$185,830.24 *(3.3)*
48. 2.47% *(3.2)*   **49.** 6.33% *(3.1)*   **50.** 44 deposits *(3.3)*
51. (A) \$1,189.52 (B) \$72,963.07 (C) \$7,237.31 *(3.4)*
52. The certificate would be worth \$53,394.30 when the 360th payment is made. By reducing the principal the loan would be paid off in 252 months. If the monthly payment were then invested at 7% compounded monthly, it would be worth \$67,234.20 at the time of the 360th payment. *(3.2, 3.3, 3.4)*   **53.** The lower rate would save \$12,247.20 in interest payments. *(3.4)*   **54.** \$3,807.59 *(3.2)*
55. 5.79% *(3.2)*   **56.** \$4,844.96 *(3.1)*   **57.** \$6,697.11 *(3.4)*
58. 7.24% *(3.2)*   **59.** (A) \$398,807 (B) \$374,204 *(3.3)*
60. \$15,577.64 *(3.2)*   **61.** (A) 30 yr: \$569.26; 15 yr: \$749.82 (B) 30 yr: \$69,707.99; 15 yr: \$37,260.74 *(3.4)*   **62.** \$20,516 *(3.4)*
63. 33.52% *(3.4)*   **64.** (A) 10.74% (B) 15 yr; 40 yr *(3.3)*

Chapter 4

Exercises 4.1   1. $(0, 7)$ **3.** $(24, 0)$ **5.** $(5, -18)$ **7.** $y - 7 = -6(x - 2)$ **9.** (B); no solution   **11.** (A); $x = -3, y = 1$ **13.** $x = 2, y = 4$
15. No solution (parallel lines)   **17.** $x = 4, y = 5$ **19.** $x = 1, y = 4$ **21.** $u = 2, v = -3$ **23.** $m = 8, n = 6$ **25.** $x = 1, y = -5$ **27.** No solution
(inconsistent)   **29.** $x = -\frac{4}{3}, y = 1$ **31.** Infinitely many solutions (dependent)   **33.** $x = \frac{4}{7}, y = \frac{3}{7}$ **35.** $x = 7, y = 3$ **37.** $x = \frac{4}{5}, y = -\frac{2}{3}$ **39.** $x = 0,$
$y = 0$ **41.** $x = 14, y = 5$ **43.** Price tends to come down.   **49.** $(-14, -37)$ **51.** No solution   **53.** $(4.176, -1.235)$ **55.** $(-3.310, -2.241)$

57.
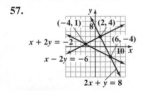

59.
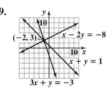

61.
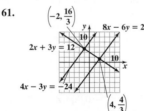

63. (A) $(20, -24)$
(B) $(-4, 6)$
(C) No solution

65. (A) Supply: 143 T-shirts; demand: 647 T-shirts
(B) Supply: 857 T-shirts; demand: 353 T-shirts
(C) Equilibrium price = \$6.50; equilibrium quantity = 500 T-shirts

65. (D)
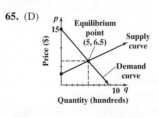

67. (A) $p = 1.5x + 1.95$ (B) $p = -1.5x + 7.8$
(C) Equilibrium price: \$4.875; equilibrium quantity: 1.95 billion bushels

(D)
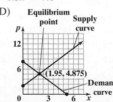

69. (A) 120 mowers
(B)
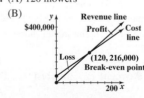

71. (A) $C = 24,000 + 7.45x$; $R = 19.95x$
(B) 1,920 (C)
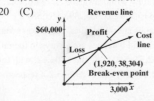

73. Base price = \$17.95; surcharge = \$2.45/lb   **75.** 5,720 lb robust blend; 6,160 lb mild blend
77. Mix *A:* 80 g; mix *B:* 60 g   **79.** Operate the Mexico plant for 75 hours and the Taiwan plant for
50 hours.   **81.** (A) $a = 196, b = -16$ (B) 196 ft (C) 3.5 sec   **83.** 40 sec, 24 sec, 120 mi
85. (A)
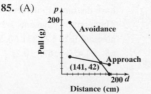
(B) $d = 141$ cm (approx.)
(C) Vacillate

Exercises 4.2   **1.** 6; 3   **3.** 3 × 3; 2 × 1   **5.** *D*   **7.** *B*   **9.** 2, 1   **11.** 2, 8, 0   **13.** −1   **15.** $\begin{bmatrix} 3 & 5 \\ 2 & -4 \end{bmatrix}$; $\begin{bmatrix} 3 & 5 & 8 \\ 2 & -4 & -7 \end{bmatrix}$ **17.** $\begin{bmatrix} 1 & 4 \\ 6 & 0 \end{bmatrix}$; $\begin{bmatrix} 1 & 4 & 15 \\ 6 & 0 & 18 \end{bmatrix}$

19. $\begin{aligned} 2x_1 + 5x_2 &= 7 \\ x_1 + 4x_2 &= 9 \end{aligned}$ **21.** $\begin{aligned} 4x_1 &= -10 \\ 8x_2 &= 40 \end{aligned}$ **23.** $\begin{bmatrix} 4 & -6 & -8 \\ 1 & -3 & 2 \end{bmatrix}$ **25.** $\begin{bmatrix} -4 & 12 & -8 \\ 4 & -6 & -8 \end{bmatrix}$ **27.** $\begin{bmatrix} 1 & -3 & 2 \\ 8 & -12 & -16 \end{bmatrix}$ **29.** $\begin{bmatrix} 1 & -3 & 2 \\ 0 & 6 & -16 \end{bmatrix}$ **31.** $\begin{bmatrix} 1 & -3 & 2 \\ 2 & 0 & -12 \end{bmatrix}$

33. $\begin{bmatrix} 1 & -3 & 2 \\ 3 & -3 & -10 \end{bmatrix}$ **35.** $\frac{1}{3}R_2 \to R_2$ **37.** $6R_1 + R_2 \to R_2$ **39.** $\frac{1}{3}R_2 + R_1 \to R_1$ **41.** $R_1 \leftrightarrow R_2$

43. $\{(6, 6)\}$ **45.** $\left\{ \left(\frac{2}{3}t - 1, t \right) \mid t \text{ is any real number} \right\}$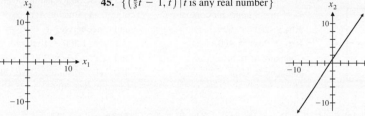

47. $x_1 = 3, x_2 = 2$; each pair of lines has the same intersection point.

  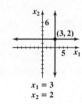

$\begin{aligned} x_1 + x_2 &= 5 \\ x_1 - x_2 &= 1 \end{aligned}$ $\begin{aligned} x_1 + x_2 &= 5 \\ -2x_2 &= -4 \end{aligned}$ $\begin{aligned} x_1 + x_2 &= 5 \\ x_2 &= 2 \end{aligned}$ $\begin{aligned} x_1 &= 3 \\ x_2 &= 2 \end{aligned}$

49. $x_1 = -4, x_2 = 6$ **51.** No solution   **53.** $x_1 = 2t + 15, x_2 = t$ for any real number t **55.** $x_1 = 3, x_2 = 1$ **57.** $x_1 = 2, x_2 = 1$ **59.** $x_1 = 2, x_2 = 4$ **61.** No solution   **63.** $x_1 = 1, x_2 = 4$ **65.** Infinitely many solutions: $x_2 = s$, $x_1 = 2s - 3$ for any real number s **67.** Infinitely many solutions; $x_2 = s$, $x_1 = \frac{1}{2}s + \frac{1}{2}$ for any real number s **69.** $x_1 = -1, x_2 = 3$ **71.** No solution   **73.** Infinitely many solutions: $x_2 = t, x_1 = \frac{3}{2}t + 2$ for any real number t **75.** $x_1 = 2, x_2 = -1$ **77.** $x_1 = 2, x_2 = -1$ **79.** $x_1 = 1.1, x_2 = 0.3$ **81.** $x_1 = -23.125, x_2 = 7.8125$ **83.** $x_1 = 3.225, x_2 = -6.9375$

Exercises 4.3   **1.** $\begin{bmatrix} 1 & 2 & 3 & 12 \\ 1 & 7 & 5 & 15 \end{bmatrix}$ **3.** $\begin{bmatrix} 1 & 0 & 6 & 2 \\ 0 & 1 & -1 & 5 \\ 1 & 3 & 0 & 7 \end{bmatrix}$ **5.** $\begin{aligned} x_1 - 3x_2 &= 4 \\ 3x_1 + 2x_2 &= 5 \\ -x_1 + 6x_2 &= 3 \end{aligned}$ **7.** $\begin{aligned} 5x_1 - 2x_2 \quad + 8x_4 = 4 \end{aligned}$ **9.** Reduced form   **11.** Not reduced form; $R_2 \leftrightarrow R_3$ **13.** Not reduced form; $\frac{1}{3}R_2 \to R_2$ **15.** Reduced form   **17.** Not reduced form; $2R_2 + R_1 \to R_1$ **19.** $x_1 = -2, x_2 = 3, x_3 = 0$

21. $x_1 = 2t + 3, x_2 = -t - 5, x_3 = t$ for t any real number   **23.** No solution   **25.** $x_1 = 3t + 5, x_2 = -2t - 7, x_3 = t$ for t any real number   **27.** $x_1 = 2s + 3t - 5, x_2 = s, x_3 = -3t + 2, x_4 = t$ for s and t any real numbers   **29.** 19   **31.** 21, 25, 27   **33.** False   **35.** True   **37.** False

39. $\begin{bmatrix} 1 & 0 & -7 \\ 0 & 1 & 3 \end{bmatrix}$ **41.** $\begin{bmatrix} 1 & 0 & -1 & 23 \\ 0 & 1 & 2 & -7 \end{bmatrix}$ **43.** $\begin{bmatrix} 1 & 0 & 0 & -5 \\ 0 & 1 & 0 & 4 \\ 0 & 0 & 1 & -2 \end{bmatrix}$ **45.** $\begin{bmatrix} 1 & 0 & 2 & -\frac{5}{3} \\ 0 & 1 & -2 & \frac{1}{3} \\ 0 & 0 & 0 & 0 \end{bmatrix}$ **47.** $x_1 = -2, x_2 = 3, x_3 = 1$ **49.** $x_1 = 0, x_2 = -2, x_3 = 2$

51. $x_1 = 2t + 3, x_2 = t - 2, x_3 = t$ for t any real number   **53.** $x_1 = 1, x_2 = 2$ **55.** No solution   **57.** $x_1 = t - 1, x_2 = 2t + 2, x_3 = t$ for t any real number   **59.** $x_1 = -2s + t + 1, x_2 = s, x_3 = t$ for s and t any real numbers   **61.** No solution   **63.** (A) Dependent system with two parameters and an infinite number of solutions   (B) Dependent system with one parameter and an infinite number of solutions   (C) Independent system with a unique solution   (D) Impossible   **65.** $x_1 = 2s - 3t + 3, x_2 = s + 2t + 2, x_3 = s, x_4 = t$ for s and t any real numbers   **67.** $x_1 = -0.5, x_2 = 0.2, x_3 = 0.3, x_4 = -0.4$ **69.** $x_1 = 2s - 1.5t + 1, x_2 = s, x_3 = -t + 1.5, x_4 = 0.5t - 0.5, x_5 = t$ for s and t any real numbers   **71.** $a = 2, b = -4, c = -7$

73. (A) $x_1 = $ no. of one-person boats
$x_2 = $ no. of two-person boats
$x_3 = $ no. of four-person boats
$0.5x_1 + \quad x_2 + 1.5x_3 = 380$
$0.6x_1 + 0.9x_2 + 1.2x_3 = 330$
$0.2x_1 + 0.3x_2 + 0.5x_3 = 120$
20 one-person boats, 220 two-person boats, and 100 four-person boats

(B) $0.5x_1 + \quad x_2 + 1.5x_3 = 380$
$0.6x_1 + 0.9x_2 + 1.2x_3 = 330$
$(t - 80)$ one-person boats, $(420 - 2t)$ two-person boats, and t four-person boats, where t is an integer satisfying $80 \le t \le 210$

(C) $0.5x_1 + \quad x_2 = 380$
$0.6x_1 + 0.9x_2 = 330$
$0.2x_1 + 0.3x_2 = 120$
There is no production schedule that will use all the labor-hours in all departments.

75. $x_1 = $ no. of 8,000-gal tank cars
$x_2 = $ no. of 16,000-gal tank cars
$x_3 = $ no. of 24,000-gal tank cars
$x_1 + \quad x_2 + \quad x_3 = 24$
$8,000x_1 + 16,000x_2 + 24,000x_3 = 520,000$
$(t - 17)$ 8,000-gal tank cars, $(41 - 2t)$ 16,000-gal tank cars, and t 24,000-gal tank cars, where $t = 17, 18, 19,$ or 20

77. The minimum monthly cost is $24,100 when 716,000-gallon and 17 24,000-gallon tank cars are leased.

79. $x_1 = $ federal income tax
$x_2 = $ state income tax
$x_3 = $ local income tax
$x_1 + 0.5x_2 + 0.5x_3 = 3,825,000$
$0.2x_1 + \quad x_2 + 0.2x_3 = 1,530,000$
$0.1x_1 + 0.1x_2 + \quad x_3 = 765,000$
Tax liability is 57.65%.

81. $x_1 = $ taxable income of company A
$x_2 = $ taxable income of company B
$x_3 = $ taxable income of company C
$x_4 = $ taxable income of company D
$x_1 - 0.08x_2 - 0.03x_3 - 0.07x_4 = 2.272$
$-0.12x_1 + \quad x_2 - 0.11x_3 - 0.13x_4 = 2.106$
$-0.11x_1 - 0.09x_2 + \quad x_3 - 0.08x_4 = 2.736$
$-0.06x_1 - 0.02x_2 - 0.14x_3 + \quad x_4 = 3.168$
Taxable incomes are $2,927,000 for company A, $3,372,000 for company B, $3,675,000 for company C, and $3,926,000 for company D.

83. (A) $x_1 = $ no. of ounces of food A
$x_2 = $ no. of ounces of food B
$x_3 = $ no. of ounces of food C
$30x_1 + 10x_2 + 20x_3 = 340$
$10x_1 + 10x_2 + 20x_3 = 180$
$10x_1 + 30x_2 + 20x_3 = 220$
8 oz of food A, 2 oz of food B, and 4 oz of food C

(B) $30x_1 + 10x_2 = 340$
$10x_1 + 10x_2 = 180$
$10x_1 + 30x_2 = 220$
There is no combination that will meet all the requirements.

(C) $30x_1 + 10x_2 + 20x_3 = 340$
$10x_1 + 10x_2 + 20x_3 = 180$
8 oz of food A, $(10 - 2t)$ oz of food B, and t oz of food C where $0 \le t \le 5$

85. x_1 = no. of barrels of mix A $30x_1 + 30x_2 + 30x_3 + 60x_4 = 900$

x_2 = no. of barrels of mix B $50x_1 + 75x_2 + 25x_3 + 25x_4 = 750$

x_3 = no. of barrels of mix C $30x_1 + 20x_2 + 20x_3 + 50x_4 = 700$

x_4 = no. of barrels of mix D

$(10 - t)$ barrels of mix A, $(t - 5)$ barrels of mix B, $(25 - 2t)$ barrels of

mix C, and t barrels of mix D, where t is an integer satisfying $5 \le t \le 10$

87. 0 barrels of mix A, 5 barrels of mix B, 5 barrels of mix C, and 10 barrels of mix D

89. $y = 0.01x^2 + x + 75,\ 450$ million

91. $y = 0.004x^2 + 0.06x + 77.6$; 1995–2000: 79.4 years; 2000–2005: 80.4 years

93.

```
QuadReg
y=ax²+bx+c
a=8.5714286E-4
b=.0888571429
c=77.58285714
R²=.998378926
```

95. x_1 = no. of hours for company A

x_2 = no. of hours for company B

$30x_1 + 20x_2 = 600$

$10x_1 + 20x_2 = 400$

Company A: 10 hr;

company B: 15 hr

97. (A) 6th St. and Washington Ave.: $x_1 + x_2 = 1,200$; 6th St. and Lincoln Ave.: $x_2 + x_3 = 1,000$; 5th St. and Lincoln Ave.: $x_3 + x_4 = 1,300$

(B) $x_1 = 1,500 - t$, $x_2 = t - 300$, $x_3 = 1,300 - t$, and $x_4 = t$, where $300 \le t \le 1,300$ (C) 1,300; 300   (D) Washington Ave.: 500; 6th St.: 700; Lincoln Ave.: 300

Exercises 4.4

1. $[4 \quad 15]$ **3.** $\begin{bmatrix} 2 & 4 \\ -2 & 6 \end{bmatrix}$ **5.** Not defined   **7.** $[21 \quad -35 \quad 63 \quad 28]$ **9.** $\begin{bmatrix} 5 \\ -3 \end{bmatrix}$ **11.** $\begin{bmatrix} 2 & 4 \\ 1 & -5 \end{bmatrix}$ **13.** $\begin{bmatrix} 1 & -5 \\ -2 & -4 \end{bmatrix}$ **15.** $\begin{bmatrix} 5 & 10 \\ 15 & 20 \end{bmatrix}$

17. $\begin{bmatrix} 5 & 10 \\ 15 & 20 \end{bmatrix}$ **19.** $\begin{bmatrix} 7 & 9 \\ 0 & 0 \end{bmatrix}$ **21.** $\begin{bmatrix} 0 & 3 \\ 0 & 7 \end{bmatrix}$ **23.** $[-7]$ **25.** $\begin{bmatrix} -15 & 6 \\ -20 & 8 \end{bmatrix}$ **27.** $[11]$ **29.** $\begin{bmatrix} 3 & -2 & -4 \\ 6 & -4 & -8 \\ -9 & 6 & 12 \end{bmatrix}$ **31.** $\begin{bmatrix} -12 & 12 & 18 \\ 20 & -18 & -6 \end{bmatrix}$ **33.** Not defined

35. $\begin{bmatrix} 11 & 2 \\ 4 & 27 \end{bmatrix}$ **37.** $\begin{bmatrix} 6 & 4 \\ 0 & -3 \end{bmatrix}$ **39.** $\begin{bmatrix} -1.3 & -0.7 \\ -0.2 & -0.5 \\ 0.1 & 1.1 \end{bmatrix}$ **41.** $\begin{bmatrix} -66 & 69 & 39 \\ 92 & -18 & -36 \end{bmatrix}$ **43.** Not defined   **45.** $\begin{bmatrix} -18 & 48 \\ 54 & -34 \end{bmatrix}$ **47.** $\begin{bmatrix} -26 & -15 & -25 \\ -4 & -18 & 4 \\ 2 & 43 & -19 \end{bmatrix}$

49. $AB = \begin{bmatrix} 0 & 0 \\ 0 & 0 \end{bmatrix}$, $BA = \begin{bmatrix} a^2 + ab & a^2 + ab \\ -a^2 - ab & -a^2 - ab \end{bmatrix}$ **51.** $\begin{bmatrix} 0 & 0 \\ 0 & 0 \end{bmatrix}$ **53.** B^n approaches $\begin{bmatrix} 0.25 & 0.75 \\ 0.25 & 0.75 \end{bmatrix}$; AB^n approaches $[0.25 \ 0.75]$

55. $a = -1, b = 1, c = 3, d = -5$ **57.** $a = 3, b = 4, c = 1, d = 2$ **59.** False   **61.** True   **63.** (A) True   (B) True   (C) True   (D) True

65.

	Guitar	Banjo	
	$51.50	$40.50	Materials
	$87.00	$120.00	Labor

67.

	Basic car	Air	AM/FM radio	Cruise control
Model A	$2,937	$459	$200	$118
Model B	$2,864	$201	$88	$52
Model C	$2,171	$417	$177	$101

69. (A) $19.84   (B) $38.19   (C) MN gives the labor costs at each plant.

(D)

	MA	VA	
$MN =$	$19.84	$16.90	One-person boat
	$31.49	$26.81	Two-person boat
	$44.87	$38.19	Four-person boat

71. (A) 70 g

(B) 30 g

(C) MN gives the amount (in grams) of protein, carbohydrate, and fat in 20 oz of each mix.

(D)

	Mix X	Mix Y	Mix Z	
	70	60	50	Protein
$MN =$	380	360	340	Carbohydrate
	50	40	30	Fat

73. (A) $9,950   (B) $16,400   (C) NM gives the total cost per town.

(D)

	Cost per town	
$NM =$	$9,950	Berkeley
	$16,400	Oakland

(E)

		Telephone call	House call	Letter
$[1 \quad 1]N =$		$[3,000$	$1,300$	$13,000]$

(F)

		Total contacts	
$N\begin{bmatrix}1\\1\\1\end{bmatrix} =$		$6,500$	Berkeley
		$10,800$	Oakland

Exercises 4.5

1. (A) -4; $1/4$ (B) 3; $-1/3$ (C) 0; not defined   **3.** (A) $-2/3$; $3/2$ (B) $1/7$; -7 (C) -1.6; 0.625 **5.** No   **7.** No

9. (A) $\begin{bmatrix} 2 & -3 \\ 0 & 0 \end{bmatrix}$ (B) $\begin{bmatrix} 2 & 0 \\ 4 & 0 \end{bmatrix}$ **11.** (A) $\begin{bmatrix} 0 & 0 \\ 4 & 5 \end{bmatrix}$ (B) $\begin{bmatrix} 0 & -3 \\ 0 & 5 \end{bmatrix}$ **13.** (A) $\begin{bmatrix} 2 & -3 \\ 4 & 5 \end{bmatrix}$ (B) $\begin{bmatrix} 2 & -3 \\ 4 & 5 \end{bmatrix}$ **15.** $\begin{bmatrix} -2 & 1 & 3 \\ 2 & 4 & -2 \\ 5 & 1 & 0 \end{bmatrix}$ **17.** $\begin{bmatrix} -2 & 1 & 3 \\ 2 & 4 & -2 \\ 5 & 1 & 0 \end{bmatrix}$

19. Yes   **21.** No   **23.** Yes   **25.** No   **27.** Yes   **39.** $\begin{bmatrix} -1 & 0 \\ -3 & 1 \end{bmatrix}$ **41.** $\begin{bmatrix} 3 & -2 \\ -1 & 1 \end{bmatrix}$ **43.** $\begin{bmatrix} 7 & -3 \\ -2 & 1 \end{bmatrix}$ **45.** $\begin{bmatrix} -5 & -12 & 3 \\ -2 & -4 & 1 \\ 2 & 5 & -1 \end{bmatrix}$

47. $\begin{bmatrix} 6 & -2 & -1 \\ -5 & 2 & 1 \\ -3 & 1 & 1 \end{bmatrix}$ **49.** $\begin{bmatrix} -2 & -3 \\ 3 & 4 \end{bmatrix}$ **51.** Does not exist   **53.** $\begin{bmatrix} 1.5 & -0.5 \\ -2 & 1 \end{bmatrix}$ **55.** $\begin{bmatrix} \frac{1}{2} & 0 \\ 0 & \frac{1}{2} \end{bmatrix}$ **57.** $\begin{bmatrix} \frac{1}{3} & 0 \\ 0 & -\frac{1}{5} \end{bmatrix}$ **59.** $\begin{bmatrix} \frac{1}{4} & 0 & 0 \\ 0 & \frac{1}{2} & 0 \\ 0 & 0 & -\frac{1}{8} \end{bmatrix}$

61. $\begin{bmatrix} 1 & 2 & 2 \\ -2 & -3 & -4 \\ -1 & -2 & -1 \end{bmatrix}$ **63.** Does not exist   **65.** $\begin{bmatrix} -3 & -2 & 1.5 \\ 4 & 3 & -2 \\ 3 & 2 & -1.25 \end{bmatrix}$ **67.** $\begin{bmatrix} -1.75 & -0.375 & 0.5 \\ -5.5 & -1.25 & 1 \\ 0.5 & 0.25 & 0 \end{bmatrix}$ **71.** M^{-1} exists if and only if all the elements on the main diagonal are nonzero.

73. $A^{-1} = A$; $A^2 = I$ **75.** $A^{-1} = A$; $A^2 = I$

77. 41 50 28 35 37 55 22 31 47 60 24 36 49 71 39 54 21 22   **79.** PRIDE AND PREJUDICE   **81.** 37 47 10 58 103 67 47 123 121 75 53 142 58 68 23 91 90 74 38 117 83 59 39 103 113 97 45 147 76 57 38 95   **83.** RAWHIDE TO WALTER REED   **85.** 30 28 58 15 38 13 19 26 12 30 39 56 48 43 40 9 30 29 12 33   **87.** DOUBLE DOUBLE TOIL AND TROUBLE

Exercises 4.6 **1.** $-3/5$ **3.** $-7/4$ **5.** $9/8$ **7.** $1/10$ **9.** $\begin{aligned}3x_1 + x_2 &= 5 \\ 2x_1 - x_2 &= -4\end{aligned}$ **11.** $\begin{aligned}-3x_1 + x_2 &= 3 \\ 2x_1 + x_3 &= -4 \\ -x_1 + 3x_2 - 2x_3 &= 2\end{aligned}$ **13.** $\begin{bmatrix} 3 & -4 \\ 2 & 1 \end{bmatrix}\begin{bmatrix} x_1 \\ x_2 \end{bmatrix} = \begin{bmatrix} 1 \\ 5 \end{bmatrix}$

15. $\begin{bmatrix} 1 & -3 & 2 \\ -2 & 3 & 0 \\ 1 & 1 & 4 \end{bmatrix}\begin{bmatrix} x_1 \\ x_2 \\ x_3 \end{bmatrix} = \begin{bmatrix} -3 \\ 1 \\ -2 \end{bmatrix}$ **17.** $x_1 = -8, x_2 = 2$ **19.** $x_1 = 0, x_2 = 4$ **21.** $x_1 = 3, x_2 = -2$ **23.** $x_1 = 11, x_2 = 4$ **25.** $x_1 = 4, x_2 = 1$ **27.** $x_1 = 9.5, x_2 = -6$ **29.** No solution **31.** (A) $x_1 = -3, x_2 = 2$ (B) $x_1 = -1, x_2 = 2$ (C) $x_1 = -8, x_2 = 3$

33. (A) $x_1 = 17, x_2 = 5$ **35.** (A) $x_1 - 1, x_2 - 0, x_3 = 0$ **37.** (A) $x_1 = 8, x_2 = -6, x_3 = -2$ **39.** $X = A^{-1}B$ **41.** $X = BA^{-1}$ **43.** $X = A^{-1}BA$
(B) $x_1 = 7, x_2 = -2$ (B) $x_1 = -7, x_2 = -2, x_3 = 3$ (B) $x_1 = -6, x_2 = 6, x_3 = 2$ **45.** $x_1 = 2t + 2.5, x_2 = t, t$ any real number
(C) $x_1 = 24, x_2 = -7$ (C) $x_1 = 17, x_2 = 5, x_3 = -7$ (C) $x_1 = 20, x_2 = -16, x_3 = -10$ **47.** No solution

49. $x_1 = 13t + 3, x_2 = 8t + 1, x_3 = t, t$ any real number **51.** $X = (A - B)^{-1}C$ **53.** $X = (A + I)^{-1}C$ **55.** $X = (A + B)^{-1}(C + D)$
57. (A) $x_1 = 1, x_2 = 0$ (B) $x_1 = -2{,}000, x_2 = 1{,}000$ (C) $x_1 = 2{,}001, x_2 = -1{,}000$ **59.** $x_1 = 18.2, x_2 = 27.9, x_3 = -15.2$
61. $x_1 = 24, x_2 = 5, x_3 = -2, x_4 = 15$

63. $x_1 = $ no. of \$25 tickets
$\quad x_2 = $ no. of \$35 tickets

$\quad \begin{aligned} x_1 + \quad x_2 &= 10{,}000 \quad \text{Seats} \\ 25x_1 + 35x_2 &= k \qquad\quad \text{Return}\end{aligned}$

(A) Concert 1: 7,500 \$25 tickets,
2,500 \$35 tickets
Concert 2: 5,000 \$25 tickets,
5,000 \$35 tickets
Concert 3: 2,500 \$25 tickets,
7,500 \$35 tickets
(B) No
(C) $250{,}000 + 10t, 0 \le t \le 10{,}000$

65. $x_1 = $ no. of hours plant A operates
$\quad x_2 = $ no. of hours plant B operates

$\quad \begin{aligned} 10x_1 + 8x_2 &= k_1 \quad \text{No. of car frames produced} \\ 5x_1 + 8x_2 &= k_2 \quad \text{No. of truck frames produced}\end{aligned}$

Order 1: 280 hr at plant A and 25 hr at plant B
Order 2: 160 hr at plant A and 150 hr at plant B
Order 3: 80 hr at plant A and 225 hr at plant B

67. $x_1 = $ president's bonus
$\quad x_2 = $ executive vice-president's bonus
$\quad x_3 = $ associate vice-president's bonus
$\quad x_4 = $ assistant vice-president's bonus

$\quad \begin{aligned} x_1 + \quad 0.03x_2 + \quad 0.03x_3 + \quad 0.03x_4 &= 60{,}000 \\ 0.025x_1 + \quad\quad x_2 + 0.025x_3 + 0.025x_4 &= 50{,}000 \\ 0.02x_1 + \quad 0.02x_2 + \quad\quad x_3 + \quad 0.02x_4 &= 40{,}000 \\ 0.015x_1 + 0.015x_2 + 0.015x_3 + \quad\quad x_4 &= 30{,}000\end{aligned}$

President: \$56,600; executive vice-president:
\$47,000; associate
vice-president: \$37,400; assistant vice-president:
\$27,900

69. (A) $x_1 = $ no. of ounces of mix A
$\quad x_2 = $ no. of ounces of mix B

$\quad \begin{aligned} 0.20x_1 + 0.14x_2 &= k_1 \quad \text{Protein} \\ 0.04x_1 + 0.03x_2 &= k_2 \quad \text{Fat}\end{aligned}$

Diet 1: 50 oz mix A and 500 oz mix B
Diet 2: 450 oz mix A and 0 oz mix B
Diet 3: 150 oz mix A and 500 oz mix B

(B) No

Exercises 4.7 **1.** -3 **3.** 100 **5.** 4 **7.** 8 **9.** 40¢ from A; 20¢ from E **11.** $\begin{bmatrix} 0.6 & -0.2 \\ -0.2 & 0.9 \end{bmatrix}; \begin{bmatrix} 1.8 & 0.4 \\ 0.4 & 1.2 \end{bmatrix}$ **13.** $X = \begin{bmatrix} x_1 \\ x_2 \end{bmatrix} = \begin{bmatrix} 16.4 \\ 9.2 \end{bmatrix}$

15. 20¢ from A; 10¢ from B; 10¢ from E **19.** Agriculture: \$18 billion; building: \$15.6 billion; energy: \$22.4 billion **21.** $\begin{bmatrix} 1.4 & 0.4 \\ 0.6 & 1.6 \end{bmatrix}; \begin{bmatrix} 24 \\ 46 \end{bmatrix}$ **23.** $I - M$ is

singular; X does not exist. **25.** $\begin{bmatrix} 1.58 & 0.24 & 0.58 \\ 0.4 & 1.2 & 0.4 \\ 0.22 & 0.16 & 1.22 \end{bmatrix}; \begin{bmatrix} 38.6 \\ 18 \\ 17.4 \end{bmatrix}$ **27.** (A) Agriculture: \$80 million; manufacturing: \$64 million.
(B) The final demand for agriculture increases to \$54 million and
the final demand for manufacturing decreases to \$38 million. **29.** $\begin{bmatrix} 0.25 & 0.1 \\ 0.25 & 0.3 \end{bmatrix}$

31. The total output of the energy sector should be 75% of the total output of the mining sector. **33.** Each element should be between 0 and 1, inclusive.
35. Coal: \$28 billion; steel: \$26 billion **37.** Agriculture: \$148 million; tourism: \$146 million **39.** Agriculture: \$40.1 billion; manufacturing: \$29.4 billion;
energy: \$34.4 billion
41. Year 1: agriculture: \$65 billion; energy: \$83 billion; labor: \$71 billion; manufacturing: \$88 billion
Year 2: agriculture: \$81 billion; energy: \$97 billion; labor: \$83 billion; manufacturing: \$99 billion
Year 3: agriculture: \$117 billion; energy: \$124 billion; labor: \$106 billion; manufacturing: \$120 billion

Chapter 4 Review Exercises **1.** $x = 4, y = 4$ *(4.1)* **2.** $x = 4, y = 4$ *(4.1)*
3. (A) Not in reduced form; $R_1 \leftrightarrow R_2$ (B) Not in reduced form; $\frac{1}{3}R_2 \leftrightarrow R_2$ (C) Reduced form (D) Not in reduced form; $(-1)R_2 + R_1 \to R_1$ *(4.3)*
4. (A) $2 \times 5, 3 \times 2$ (B) $a_{24} = 3, a_{15} = 2, b_{31} = -1, b_{22} = 4$ (C) AB is not defined; BA is defined *(4.2, 4.4)*

5. (A) $x_1 = 8, x_2 = 2$ (B) $x_1 = -15.5, x_2 = 23.5$ *(4.6)* **6.** $\begin{bmatrix} 3 & 3 \\ 4 & 2 \end{bmatrix}$ *(4.4)* **7.** Not defined *(4.4)* **8.** $\begin{bmatrix} -3 & 0 \\ 1 & -1 \end{bmatrix}$ *(4.4)*

9. $\begin{bmatrix} 4 & 3 \\ 7 & 4 \end{bmatrix}$ *(4.4)* **10.** Not defined *(4.4)* **11.** $\begin{bmatrix} 5 \\ 5 \end{bmatrix}$ *(4.4)* **12.** $\begin{bmatrix} 2 & 3 \\ 4 & 6 \end{bmatrix}$ *(4.4)* **13.** $\begin{bmatrix} 8 \end{bmatrix}$ *(4.4)* **14.** Not defined *(4.4)* **15.** $\begin{bmatrix} -2 & 3 \\ 3 & -4 \end{bmatrix}$ *(4.5)*

16. $x_1 = 9, x_2 = -11$ *(4.1)* **17.** $x_1 = 9, x_2 = -11$ *(4.2)* **18.** $x_1 = 9, x_2 = -11; x_1 = 16, x_2 = -19; x_1 = -2, x_2 = 4$ *(4.6)* **19.** Not defined *(4.4)*
20. $\begin{bmatrix} 10 & -8 \\ 4 & 6 \end{bmatrix}$ *(4.4)* **21.** $\begin{bmatrix} -2 & 8 \\ 8 & 6 \end{bmatrix}$ *(4.4)* **22.** $\begin{bmatrix} -2 & -1 & -3 \\ 4 & 2 & 6 \\ 6 & 3 & 9 \end{bmatrix}$ *(4.4)* **23.** $\begin{bmatrix} 9 \end{bmatrix}$ *(4.4)* **24.** $\begin{bmatrix} 10 & -5 & 1 \\ -1 & -4 & -5 \\ 1 & -7 & -2 \end{bmatrix}$ *(4.4)* **25.** $\begin{bmatrix} -\frac{5}{2} & 2 & -\frac{1}{2} \\ 1 & -1 & 1 \\ \frac{1}{2} & 0 & -\frac{1}{2} \end{bmatrix}$ *(4.5)*

26. (A) $x_1 = 2, x_2 = 1, x_3 = -1$

(B) $x_1 = -5t - 12, x_2 = 3t + 7, x_3 = t$
for t any real number

(C) $x_1 = -2t + 5, x_2 = t + 3, x_3 = t$ for t any real number *(4.3)*

27. $x_1 = 2, x_2 = 1, x_3 = -1; x_1 = 1, x_2 = -2, x_3 = 1; x_1 = -1, x_2 = 2, x_3 = -2$ *(4.6)*

28. The system has an infinite no. of solutions for $k = 3$ and a unique solution for any other value of k. *(4.3)*

29. $(I - M)^{-1} = \begin{bmatrix} 1.4 & 0.3 \\ 0.8 & 1.6 \end{bmatrix}; X = \begin{bmatrix} 48 \\ 56 \end{bmatrix}$ *(4.7)*

30. $\begin{bmatrix} 0.3 & 0.4 \\ 0.15 & 0.2 \end{bmatrix}$ *(4.7)*

31. $I - M$ is singular; X does not exist. *(4.5)*

32. $x = 3.46, y = 1.69$ *(4.1)*

33. $\begin{bmatrix} -0.9 & -0.1 & 5 \\ 0.8 & 0.2 & -4 \\ 0.1 & -0.1 & 0 \end{bmatrix}$ *(4.5)*

34. $x_1 = 1,400, x_2 = 3,200, x_3 = 2,400$ *(4.6)*

35. $x_1 = 1,400, x_2 = 3,200, x_3 = 2,400$ *(4.3)*

36. $(I - M)^{-1} = \begin{bmatrix} 1.3 & 0.4 & 0.7 \\ 0.2 & 1.6 & 0.3 \\ 0.1 & 0.8 & 1.4 \end{bmatrix}; X = \begin{bmatrix} 81 \\ 49 \\ 62 \end{bmatrix}$ *(4.7)*

37. (A) Unique solution

(B) Either no solution or an infinite no. of solutions *(4.6)*

38. (A) Unique solution

(B) No solution

(C) Infinite no. of solutions *(4.3)*

39. (B) is the only correct solution. *(4.6)*

40. (A) $C = 243,000 + 22.45x; R = 59.95x$

(B) $x = 6,480$ machines; $R = C = \$388,476$

(C) Profit occurs if $x > 6,480$; loss occurs if $x < 6,480$. *(4.1)*

41. $x_1 = $ no. of tons Voisey's Bay ore

$x_2 = $ no. of tons of Hawk Ridge ore

$0.02x_1 + 0.03x_2 = 6$

$0.04x_1 + 0.02x_2 = 8$

$x_1 = 150$ tons of Voisey's Bay ore

$x_2 = 100$ tons of Hawk Ridge ore *(4.3)*

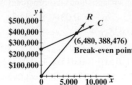

42. (A) $\begin{bmatrix} x_1 \\ x_2 \end{bmatrix} = \begin{bmatrix} -25 & 37.5 \\ 50 & -25 \end{bmatrix}\begin{bmatrix} 6 \\ 8 \end{bmatrix} = \begin{bmatrix} 150 \\ 100 \end{bmatrix}$

$x_1 = 150$ tons of Voisey's Bay ore

$x_2 = 100$ tons of Hawk Ridge ore

(B) $\begin{bmatrix} x_1 \\ x_2 \end{bmatrix} = \begin{bmatrix} -25 & 37.5 \\ 50 & -25 \end{bmatrix}\begin{bmatrix} 7.5 \\ 7 \end{bmatrix} = \begin{bmatrix} 75 \\ 200 \end{bmatrix}$

$x_1 = 75$ tons of Voisey's Bay ore

$x_2 = 200$ tons of Hawk Ridge ore *(4.6)*

43. (A) $x_1 = $ no. of 3,000-ft^3 hoppers

$x_2 = $ no. of 4,500-ft^3 hoppers

$x_3 = $ no. of 6,000-ft^3 hoppers

$x_1 + x_2 + x_3 = 20$

$3,000x_1 + 4,500x_2 + 6,000x_3 = 108,000$

$x_1 = (t - 12)$ 3,000-ft^3 hoppers

$x_2 = (32 - 2t)$ 4,500-ft^3 hoppers

$x_3 = t$ 6,000-ft^3 hoppers

where $t = 12, 13, 14, 15,$ or 16

(B) The minimum monthly cost is $\$5,700$ when 8 4,500-ft^3 and 12 6,000-ft^3 hoppers are leased. *(4.3)*

44. (A) Elements in MN give the cost of materials for each alloy from each supplier.

(B) $\qquad$ Supplier A $\quad$ Supplier B

$MN = \begin{bmatrix} \$7,620 & \$7,530 \\ \$13,880 & \$13,930 \end{bmatrix}$ Alloy 1
Alloy 2

(C) $\qquad$ Supplier A $\quad$ Supplier B

$[11]MN = [\$21,500 \quad \$21,460]$ Total material costs *(4.4)*

45. (A) $\$6.35$

(B) Elements in MN give the total labor costs for each calculator at each plant.

(C) $\qquad$ CA $\qquad$ TX

$MN = \begin{bmatrix} \$3.65 & \$3.00 \\ \$6.35 & \$5.20 \end{bmatrix}$ Model A
Model B *(4.4)*

46. $x_1 = $ amount invested at 5%

$x_2 = $ amount invested at 10%

$x_1 + x_2 = 5,000$

$0.05x_1 + 0.1x_2 = 400$

$\$2,000$ at 5%, $\$3,000$ at 10% *(4.3)*

47. $\$2,000$ at 5% and $\$3,000$ at 10% *(4.6)*

48. No to both. The annual yield must be between $\$250$ and $\$500$ inclusive. *(4.6)*

49. $x_1 = $ no. of $8 tickets

$x_2 = $ no. of $12 tickets

$x_3 = $ no. of $20 tickets

$x_1 + x_2 + x_3 = 25,000$

$8x_1 + 12x_2 + 20x_3 = k_1 \qquad$ Return requested

$x_1 \qquad - x_3 = 0$

Concert 1: 5000 $8 tickets, 15,000 $12 tickets, and 5,000 $20 tickets

Concert 2: 7,500 $8 tickets, 10,000 $12 tickets, and 7,500 $20 tickets

Concert 3: 10,000 $8 tickets, 5,000 $12 tickets, and 10,000 $20 tickets *(4.6)*

50. $x_1 + x_2 + x_3 = 25,000$

$8x_1 + 12x_2 + 20x_3 = k_1 \quad$ Return requested

Concert 1: $(2t - 5,000)$ $8 tickets, $(30,000 - 3t)$ $12 tickets, and t $20 tickets, where t is an integer satisfying $2,500 \le t \le 10,000$

Concert 2: $(2t - 7,500)$ $8 tickets, $(32,500 - 3t)$ $12 tickets, and t $20 tickets, where t is an integer satisfying $3,750 \le t \le 10,833$

Concert 3: $(2t - 10,000)$ $8 tickets, $(35,000 - 3t)$ $12 tickets, and t $20 tickets, where t is an integer satisfying $5,000 \le t \le 11,666$ *(4.3)*

51. (A) Agriculture: $\$80$ billion; fabrication: $\$60$ billion

(B) Agriculture: $\$135$ billion; fabrication: $\$145$ billion *(4.7)*

52. BEWARE THE IDES OF MARCH *(4.5)*

53. (A) 1st & Elm: $x_1 + x_4 = 1,300$

2nd & Elm: $x_1 - x_2 = 400$

2nd & Oak: $x_2 + x_3 = 700$

1st & Oak: $x_3 - x_4 = -200$

(B) $x_1 = 1,300 - t, x_2 = 900 - t, x_3 = t - 200, x_4 = t$, where $200 \le t \le 900$

(C) 900; 200 (D) Elm St.: 800; 2nd St.: 400; Oak St.: 300 *(4.3)*

Chapter 5

Exercises 5.1   **1.** No **3.** Yes **5.** No **7.** No **9.**   **11.**   **13.**   **15.**

17.   **19.** (A) (B)   **21.** (A) (B)   **23.** Let h = no. of overtime hours; $h < 20$

25. Let s = annual salary; $s \geq \$65{,}000$

27. Let a = no. of freshmen admitted; $a \leq 1{,}700$ **29.** $2x + 3y = -6$; $2x + 3y \geq -6$ **31.** $y = 3$; $y < 3$ **33.** $4x - 5y = 0$; $4x - 5y \geq 0$

35. Let x = enrollment in finite mathematics; let y = enrollment in calculus; $x + y < 300$ **37.** Let x = revenue; y = cost; $x \leq y - 20{,}000$

39. Let x = no. of grams of saturated fat; let y = no. of grams of unsaturated fat; $x > 3y$

41. 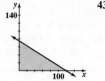 **43.**   **45.**   **47.**   **49.** The solution set is the empty set and has no graph.

51. Let x = no. of acres planted with corn.
Let y = no. of acres planted with soybeans.
$40x + 32y \leq 5{,}000$, $x \geq 0$, $y \geq 0$

53. Let x = no. of lbs of brand A.
Let y = no. of lbs of brand B.
(A) $0.26x + 0.16y \geq 120$, $x \geq 0$, $y \geq 0$ (B) $0.03x + 0.08y \leq 28$, $x \geq 0$, $y \geq 0$

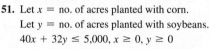

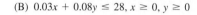

55. Let x = no. of lbs of the standard blend.
Let y = no. of lbs of the deluxe blend.
$0.3x + 0.09y \geq 0.20(x + y)$, $x \geq 0$, $y \geq 0$

57. Let x = no. of weeks Plant A is operated.
Let y = no. of weeks Plant B is operated.
$10x + 8y \geq 400$, $x \geq 0$, $y \geq 0$

59. Let x = no. of radio spots.
Let y = no. of television spots.
$200x + 800y \leq 10{,}000$, $x \geq 0$, $y \geq 0$

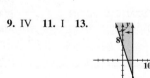

61. Let x = no. of regular mattresses cut per day.
Let y = no. of king mattresses cut per day
$5x + 6y \leq 3{,}000$, $x \geq 0$, $y \geq 0$

Exercises 5.2   **1.** Yes **3.** No **5.** No **7.** Yes **9.** IV **11.** I **13.**   **15.**

17. IV; $(8, 0)$, $(18, 0)$, $(6, 4)$ **21.** Bounded **23.** Bounded **25.** Unbounded **27.** Bounded

19. I; $(0, 16)$, $(6, 4)$, $(18, 0)$

    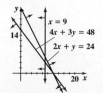

29. Unbounded **31.** Bounded **33.** Empty **35.** Unbounded **37.** Bounded

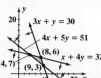   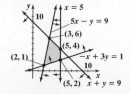

39. Bounded

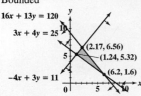

41. (A) $3x + 4y = 36$ and $3x + 2y = 30$ intersect at $(8, 3)$;
$3x + 4y = 36$ and $x = 0$ intersect at $(0, 9)$;
$3x + 4y = 36$ and $y = 0$ intersect at $(12, 0)$;
$3x + 2y = 30$ and $x = 0$ intersect at $(0, 15)$;
$3x + 2y = 30$ and $y = 0$ intersect at $(10, 0)$;
$x = 0$ and $y = 0$ intersect at $(0, 0)$
(B) $(8, 3), (0, 9), (10, 0), (0, 0)$

43. $6x + 4y \le 108$
$x + y \le 24$
$x \ge 0$
$y \ge 0$

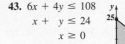

45. (A) All production schedules in the feasible region that are on the graph of $50x + 60y = 1,100$ will result in a profit of $1,100.
(B) There are many possible choices. For example, producing 5 trick skis and 15 slalom skis will produce a profit of $1,150. All the production schedules in the feasible region that are on the graph of $50x + 60y = 1,150$ will result in a profit of $1,150.

47. $20x + 10y \ge 460$
$30x + 30y \ge 960$
$5x + 10y \ge 220$
$x \ge 0$
$y \ge 0$

49. $10x + 20y \le 800$
$20x + 10y \le 640$
$x \ge 0$
$y \ge 0$

Exercises 5.3   1. Max $Q = 154$; Min $Q = 0$ **3.** Max $Q = 120$; Min $Q = -60$ **5.** Max $Q = 0$; Min $Q = -32$ **7.** Max $Q = 40$; Min $Q = -48$
9. Max $P = 16$ at $x = 7$ and $y = 9$ **11.** Max $P = 84$ at $x = 7$ and $y = 9$, at $x = 0$ and $y = 12$, and at every point on the line segment joining the preceding two points. **13.** Min $C = 32$ at $x = 0$ and $y = 8$ **15.** Min $C = 36$ at $x = 4$ and $y = 3$ **17.** Max $P = 30$ at $x = 4$ and $y = 2$ **19.** Min $z = 14$ at $x = 4$ and $y = 2$; no max **21.** Max $P = 260$ at $x = 2$ and $y = 5$ **23.** Min $z = 140$ at $x = 14$ and $y = 0$; no max **25.** Min $P = 20$ at $x = 0$ and $y = 2$; Max $P = 150$ at $x = 5$ and $y = 0$ **27.** Feasible region empty; no optimal solutions **29.** Min $P = 140$ at $x = 3$ and $y = 8$; Max $P = 260$ at $x = 8$ and $y = 10$, at $x = 12$ and $y = 2$, or at any point on the line segment from $(8, 10)$ to $(12, 2)$ **31.** Max $P = 26,000$ at $x = 400$ and $y = 600$
33. Max $P = 5,507$ at $x = 6.62$ and $y = 4.25$ **35.** Max $z = 2$ at $x = 4$ and $y = 2$; min z does not exist

37. (A) $2a < b$ (B) $\frac{1}{3}a < b < 2a$ (C) $b < \frac{1}{3}a$ (D) $b = 2a$ (E) $b = \frac{1}{3}a$

39. (A) Let: x = no. of trick skis
y = no. of slalom skis produced per day.
Maximize $P = 40x + 30y$
subject to $6x + 4y \le 108$
$x + y \le 24$
$x \ge 0, y \ge 0$

Max profit = $780 when 6 trick skis and 18 slalom skis are produced.
(B) Max profit decreases to $720 when 18 trick skis and no slalom skis are produced.
(C) Max profit increases to $1,080 when no trick skis and 24 slalom skis are produced.

41. (A) Let x = no. of days to operate plant A
y = no. of days to operate plant B
Maximize $C = 1000x + 900y$
subject to $20x + 25y \ge 200$
$60x + 50y \ge 500$
$x \ge 0, y \ge 0$
Plant A: 5 days; Plant B: 4 days; min cost $8,600
(B) Plant A: 10 days; Plant B: 0 days; min cost $6,000
(C) Plant A: 0 days; Plant B: 10 days; min cost $8,000

43. Let x = no. of buses
y = no. of vans
Maximize $C = 1,200x + 100y$
subject to $40x + 8y \ge 400$
$3x + y \le 36$
$x \ge 0, y \ge 0$
7 buses, 15 vans; min cost $9,900

45. Let x = amount invested in the CD
y = amount invested in the mutual fund
Maximize $P = 0.05x + 0.09y$
subject to $x + y \le 60,000$
$y \ge 10,000$
$x \ge 2y$
$x, y \ge 0$
$40,000 in the CD and $20,000 in the mutual fund; max return is $3,800

47. (A) Let x = no. of gallons produced by the old process
y = no. of gallons produced by the new process
Maximize $P = 60x + 20y$
subject to $20x + 5y \le 16,000$
$40x + 20y \le 30,000$
$x \ge 0, y \ge 0$
Max $P = $450 when 750 gal are produced using the old process exclusively.
(B) Max $P = $380 when 400 gal are produced using the old process and 700 gal are produced using the new process.
(C) Max $P = $288 when 1,440 gal are produced using the new process exclusively.

49. (A) Let x = no. of bags of brand A
y = no. of bags of brand B
Maximize $N = 8x + 3y$
subject to $4x + 4y \ge 1,000$
$2x + y \le 400$
$x \ge 0, y \ge 0$
150 bags brand A, 100 bags brand B; min nitrogen 1,500 lb
(B) 0 bags brand A, 250 bags brand B; min nitrogen 750 lb

51. Let x = no. of cubic yards of mix A
y = no. of cubic yards of mix B
Minimize $C = 30x + 35y$
subject to $20x + 10y \ge 460$
$30x + 30y \ge 960$
$5x + 10y \ge 220$
$x \ge 0, y \ge 0$
20 yd³ A, 12 yd³ B; $1,020

53. Let x = no. of mice used
y = no. of rats used
Maximize $P = x + y$
subject to $10x + 20y \le 800$
$20x + 10y \le 640$
$x \ge 0, y \ge 0$
48; 16 mice, 32 rats

Chapter 5 Review Exercises   1. *(5.1)*   **2.** *(5.1)*   **3.** Bounded *(5.2)*   **4.** Unbounded *(5.2)*

5. Bounded *(5.2)*

6. Unbounded *(5.2)*

7. $2x - 3y = 12; 2x - 3y \leq 12$ *(5.1)*   **8.** $4x + y = 8; 4x + y \geq 8$ *(5.1)*   **9.** Max $P = 24$ at $x = 0$ and $y = 4$ *(5.3)*   **10.** Min $C = 40$ at $x = 0$ and $y = 20$ *(5.3)*   **11.** Max $P = 26$ at $x = 2$ and $y = 5$ *(5.3)*

12. Min $C = 51$ at $x = 3$ and $y = 9$ *(5.3)*   **13.** Max $P = 36$ at $x = 8$ and $y = 6$ *(5.3)*

14. Let x = the no. of calculator boards. *(5.1)*

Let y = the no. of toaster boards

(A) $4x + 3y \leq 300, x \geq 0, y \geq 0$ (B) $2x + y \leq 120, x \geq 0, y \geq 0$

15. (A) Let x = no. of regular sails

y = no. of competition sails

Maximize $P = 100x + 200y$

subject to $2x + 3y \leq 150$

$4x + 10y \leq 380$

$x, y \geq 0$

Max $P = \$8,500$ when 45 regular and 20 competition sails are produced.

(B) Max profit increases to $9,880 when 38 competition and no regular sails are produced.

(C) Max profit decreases to $7,500 when no competition and 75 regular sails are produced. *(5.3)*

16. (A) Let x = no. of grams of mix A

y = no. of grams of mix B

Minimize $C = 0.04x + 0.09y$

subject to $2x + 5y \geq 850$

$2x + 4y \geq 800$

$4x + 5y \geq 1,150$

$x, y \geq 0$

Min $C = \$16.50$ when 300 g mix A and 50 g mix B are used.

(B) The minimum cost decreases to $13.00 when 100 g mix A and 150 g mix B are used.

(C) The minimum cost increases to $17.00 when 425 g mix A and no mix B are used. *(5.3)*

Chapter 6

Exercises 6.1   **1.** 10   **3.** 15   **5.** $(x_1, x_2, s_1, s_2) = (0, 2, 0, 2)$ **7.** $(x_1, x_2, s_1, s_2) = (8, 0, -6, 0)$

9. $2x_1 + 3x_2 + s_1 \quad\quad = 9$ **11.** $12x_1 - 14x_2 + s_1 \quad\quad\quad = 55$ **13.** $6x_1 + 5x_2 + s_1 = 18$ **15.** $4x_1 - 3x_2 + s_1 \quad\quad\quad\quad = 12$
$6x_1 + 7x_2 \quad\quad + s_2 = 13$ $\quad 19x_1 + 5x_2 \quad\quad + s_2 \quad = 40$ $\quad\quad\quad 5x_1 + 2x_2 \quad\quad + s_2 \quad\quad\quad = -25$
$\quad\quad\quad\quad\quad -8x_1 + 11x_2 \quad\quad\quad + s_3 = 64$ $\quad\quad -3x_1 + 7x_2 \quad\quad\quad + s_3 \quad = 32$
$\quad\quad\quad\quad\quad\quad 2x_1 + x_2 \quad\quad\quad\quad + s_4 = 9$

17. s_1, s_2 **19.** x_1, s_2 **21.** x_2, s_2 **23.** (A), (B), (E), (F)   **25.** Max $P = 40$ at $x_1 = 0, x_2 = 8$ **27.** x_2, s_1, s_3 **29.** x_2, s_3 **31.** (C), (D), (E), (F)

33. $(x_1, x_2, s_1, s_2, s_3) = (12, 0, 12, 6, 0)$ **35.** $(x_1, x_2, s_1, s_2, s_3) = (8, 16, 0, -2, 0)$

37. $4x_1 + 5x_2 + s_1 = 20$

x_1	x_2	s_1	
0	0	20	Feasible
0	4	0	Feasible
5	0	0	Feasible

39. $x_1 + x_2 + s_1 \quad\quad = 6$
$\quad x_1 + 4x_2 \quad\quad + s_2 = 12$

x_1	x_2	s_1	s_2	
0	0	6	12	Feasible
0	6	0	-12	Not feasible
0	3	3	0	Feasible
6	0	0	6	Feasible
12	0	-6	0	Not feasible
4	2	0	0	Feasible

41. $2x_1 + 5x_2 + s_1 \quad\quad = 20$
$\quad x_1 + 2x_2 \quad\quad + s_2 = 9$

x_1	x_2	s_1	s_2	
0	0	20	9	Feasible
0	4	0	1	Feasible
0	9/2	-5/2	0	Not feasible
10	0	0	-1	Not feasible
9	0	2	0	Feasible
5	2	0	0	Feasible

43. $x_1 + 2x_2 + s_1 \quad\quad\quad = 24$
$\quad x_1 + x_2 \quad\quad + s_2 \quad\quad = 15$
$\quad 2x_1 + x_2 \quad\quad\quad + s_3 = 24$

x_1	x_2	s_1	s_2	s_3	
0	0	24	15	24	Feasible
0	12	0	3	12	Feasible
0	15	-6	0	9	Not feasible
0	24	-24	-9	0	Not feasible
24	0	0	-9	-24	Not feasible
15	0	9	0	-6	Not feasible
12	0	12	3	0	Feasible
6	9	0	0	3	Feasible
8	8	0	-1	0	Not feasible
9	6	3	0	0	Feasible

45. Corner points: $(0, 0), (0, 4),$ $(5, 0)$

47. Corner points: $(0, 0), (0, 3),$ $(6, 0), (4, 2)$

49. Corner points: $(0, 0), (0, 4),$ $(9, 0), (5, 2)$

51. Max $P = 50$ at $x_1 = 5, x_2 = 0$ **53.** Max $P = 100$ at $x_1 = 4, x_2 = 2$

55. Max $P = 225$ at $x_1 = 9, x_2 = 0$ **57.** Max $P = 540$ at $x_1 = 6, x_2 = 9$

59. $_{10}C_4 = 210$ **61.** $_{72}C_{30} \approx 1.64 \times 10^{20}$

Exercises 6.2   **1.** (A) Basic: x_2, s_1, P; nonbasic: x_1, s_2 (B) $x_1 = 0, x_2 = 12, s_1 = 15, s_2 = 0, P = 50$ (C) Additional pivot required

3. (A) Basic: x_2, x_3, s_3, P; nonbasic: x_1, s_1, s_2 (B) $x_1 = 0, x_2 = 15, x_3 = 5, s_1 = 0, s_2 = 0, s_3 = 12, P = 45$ (C) No optimal solution

5.
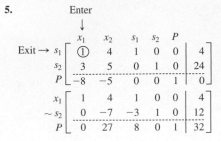

7.
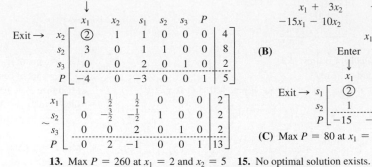

9. (A)
$$2x_1 + x_2 + s_1 \qquad\qquad = 10$$
$$x_1 + 3x_2 \qquad + s_2 \qquad = 10$$
$$-15x_1 - 10x_2 \qquad\qquad + P = 0$$
$$x_1, x_2, s_1, s_2 \geq 0$$

(B)
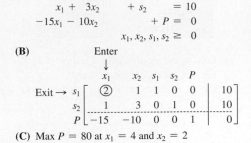

(C) Max $P = 80$ at $x_1 = 4$ and $x_2 = 2$

11. (A)
$$2x_1 + x_2 + s_1 \qquad\qquad = 10$$
$$x_1 + 3x_2 \qquad + s_2 \qquad = 10$$
$$-30x_1 - x_2 \qquad\qquad + P = 0$$
$$x_1, x_2, s_1, s_2 \geq 0$$

(B)   Enter
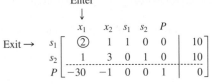

(C) Max $P = 150$ at $x_1 = 5$ and $x_2 = 0$

13. Max $P = 260$ at $x_1 = 2$ and $x_2 = 5$ **15.** No optimal solution exists.

17. Max $P = 7$ at $x_1 = 3$ and $x_2 = 5$ **19.** Max $P = 58$ at $x_1 = 12, x_2 = 0$, and $x_3 = 2$

21. Max $P = 17$ at $x_1 = 4, x_2 = 3$ and $x_3 = 0$ **23.** Max $P = 22$ at $x_1 = 1, x_2 = 6$, and $x_3 = 0$

25. Max $P = 26{,}000$ at $x_1 = 400$ and $x_2 = 600$ **27.** Max $P = 450$ at $x_1 = 0, x_2 = 180$, and $x_2 = 30$

29. Max $P = 88$ at $x_1 = 24$ and $x_2 = 8$

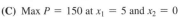

31. No solution   **33.** Choosing either col. produces the same optimal solution: max $P = 13$ at $x_1 = 13$ and $x_2 = 10$.   **35.** Choosing col. 1: max $P = 60$ at $x_1 = 12, x_2 = 8$, and $x_3 = 0$.   Choosing col. 2: max $P = 60$ at $x_1 = 0, x_2 = 20$, and $x_3 = 0$.

37. Let x_1 = no. of A components
x_2 = no. of B components
x_3 = no. of C components
Maximize $P = 7x_1 + 8x_2 + 10x_3$
subject to $2x_1 + 3x_2 + 2x_3 \leq 1{,}000$
$x_1 + x_2 + 2x_3 \leq 800$
$x_1, x_2, x_3 \geq 0$
200 A components, 0 B components, and 300 C components; max profit is \$4,400

39. Let x_1 = amount invested in government bonds
x_2 = amount invested in mutual funds
x_3 = amount invested in money markey funds
Maximize $P = 0.08x_1 + 0.13x_2 + 0.15x_3$
subject to $x_1 + x_2 + x_3 \leq 100{,}000$
$-x_1 + x_2 + x_3 \leq 0$
$x_1, x_2, x_2 \geq 0$
\$50,000 in government bonds, \$0 in mutual funds, and \$50,000 in money market funds; max return is \$11,500

41. Let x_1 = no. of ads placed in daytime shows
x_2 = no. of ads placed in prime-time shows
x_3 = no. of ads placed in late-night shows
Maximize $P = 14{,}000x_1 + 24{,}000x_2 + 18{,}000x_3$
subject to $x_1 + x_2 + x_3 \leq 15$
$1{,}000x_1 + 2{,}000x_2 + 1{,}500x_3 \leq 20{,}000$
$x_1, x_2, x_3 \geq 0$
10 daytime ads, 5 prime-time ads, and 0 late-night ads; max no. of potential customers is 260,000

43. Let x_1 = no. of colonial houses
x_2 = no. of split-level houses
x_3 = no. of ranch houses
Maximize $P = 20{,}000x_1 + 18{,}000x_2 + 24{,}000x_3$
subject to $\frac{1}{2}x_1 + \frac{1}{2}x_2 + x_3 \leq 30$
$60{,}000x_1 + 60{,}000x_2 + 80{,}000x_3 \leq 3{,}200{,}000$
$4{,}000x_1 + 3{,}000x_3 + 4{,}000x_3 \leq 180{,}000$
$x_1, x_2, x_3 \geq 0$
20 colonial, 20 split-level, and 10 ranch houses; max profit is \$1,000,000

45. The model is the same as the model for Problem 43 except that $P = 17{,}000x_1 + 18{,}000x_2 + 24{,}000x_3$. 0 colonial, 40 split-level, and 10 ranch houses; max profit is \$960,000; 20,000 labor-hours are not used

47. The model is the same as the model for Problem 43 except that $P = 25{,}000x_1 + 18{,}000x_2 + 24{,}000x_3$. 45 colonial, 0 split-level, and 0 ranch houses; max profit is \$1,125,000; 7.5 acres of land and \$500,000 of capital are not used

49. Let x_1 = no. of grams of food A
x_2 = no. of grams of food B
x_3 = no. of grams of food C
Maximize $P = 3x_1 + 4x_2 + 5x_3$
subject to $x_1 + 3x_2 + 2x_3 \leq 30$
$2x_1 + x_2 + 2x_3 \leq 24$
$x_1, x_2, x_3 \geq 0$
0 g food A, 3 g food B, and 10.5 g food C; max protein is 64.5 units

51. Let x_1 = no. of undergraduate students
x_2 = no. of graduate students
x_3 = no. of faculty members
Maximize $P = 18x_1 + 25x_2 + 30x_3$
subject to $x_1 + x_2 + x_3 \leq 20$
$100x_1 + 150x_2 + 200x_3 \leq 3{,}200$
$x_1, x_2, x_3 \geq 0$
0 undergraduate students, 16 graduate students, and 4 faculty members; max no. of interviews is 520

Exercises 6.3

1. $\begin{bmatrix} -5 \\ 0 \\ 3 \\ -1 \\ 8 \end{bmatrix}$ **3.** $\begin{bmatrix} 1 & -2 & 0 & 4 \end{bmatrix}$ **5.** $\begin{bmatrix} 2 & 5 \\ 1 & 2 \\ -6 & 0 \\ 0 & 1 \\ -1 & 3 \end{bmatrix}$ **7.** $\begin{bmatrix} 1 & 0 & 8 & 4 \\ 2 & 2 & 0 & -1 \\ -1 & -7 & 1 & 3 \end{bmatrix}$

9. (A) Maximize $P = 4y_1 + 5y_2$
subject to $y_1 + 2y_2 \leq 8$
$3y_1 + y_2 \leq 9$
$y_1, y_2 \geq 0$

(B) $y_1 + 2y_2 + x_1 = 8$
$3y_1 + y_2 + x_2 = 9$
$-4y_1 - 5y_2 + P = 0$

(C) $\begin{bmatrix} y_1 & y_2 & x_1 & x_2 & P & \\ 1 & 2 & 1 & 0 & 0 & 8 \\ 3 & 1 & 0 & 1 & 0 & 9 \\ -4 & -5 & 0 & 0 & 1 & 0 \end{bmatrix}$

11. (A) Max $P = 121$ at $y_1 = 3$ and $y_2 = 5$
(B) Min $C = 121$ at $x_1 = 1$ and $x_2 = 2$

13. (A) Maximize $P = 13y_1 + 12y_2$
subject to $4y_1 + 3y_2 \le 9$
$y_1 + y_2 \le 2$
$y_1, y_2 \ge 0$
(B) Min $C = 26$ at $x_1 = 0$
and $x_2 = 13$

15. (A) Maximize $P = 15y_1 + 8y_2$
subject to $2y_1 + y_2 \le 7$
$3y_1 + 2y_2 \le 12$
$y_1, y_2 \ge 0$
(B) Min $C = 54$ at $x = 6$
and $x_2 = 1$

17. (A) Maximize $P = 8y_1 + 4y_2$
subject to $2y_1 - 2y_2 \le 11$
$y_1 + 3y_2 \le 4$
$y_1, y_2 \ge 0$
(B) Min $C = 32$ at $x_1 = 0$
and $x_2 = 8$

19. (A) Maximize $P = 6y_1 + 4y_2$
subject to $-3y_1 + y_2 \le 7$
$y_1 - 2y_2 \le 9$
$y_1, y_2 \ge 0$
(B) No optimal solution exists.

21. Min $C = 24$ at $x_1 = 8$ and $x_2 = 0$ **23.** Min $C = 20$ at $x_1 = 0$ and $x_2 = 4$ **25.** Min $C = 140$ at $x_1 = 14$ and $x_2 = 0$ **27.** Min $C = 44$ at $x_1 = 6$ and $x_2 = 2$ **29.** Min $C = 43$ at $x_1 = 0, x_2 = 1$, and $x_3 = 3$ **31.** No optimal solution exists.   **33.** 2 variables and 4 problem constraints   **35.** 2 constraints and any no. of variables   **37.** No; the dual problem is not a standard maximization problem.   **39.** Yes; multiply both sides of the inequality by -1.   **41.** Min $C = 44$ at $x_1 = 0, x_2 = 3$, and $x_3 = 5$ **43.** Min $C = 166$ at $x_1 = 0, x_2 = 12, x_3 = 20$, and $x_4 = 3$

45. Let x_1 = no. of hours the Cedarburg plant is operated
x_2 = no. of hours the Grafton plant is operated
x_3 = no. of hours the West Bend plant is operated
Minimize $C = 70x_1 + 75x_2 + 90x_3$
subject to $20x_1 + 10x_2 + 20x_3 \ge 300$
$10x_1 + 20x_2 + 20x_3 \ge 200$
$x_1, x_2, x_3 \ge 0$
Cedarburg plant 10 hr per day, West Bend plant 5 hr per day, Grafton plant not used; min cost is $1,150

47. The model is the same as the model for Problem 45 except that the second constraint (deluxe ice cream) is $10x_1 + 20x_2 + 20x_3 \ge 300$. West Bend Plant 15 hr per day, Cedarburg and Grafton plants not used; min cost is $1,350

49. The model is the same as the model for Problem 45 except that the second constraint (deluxe ice cream) is $10x_1 + 20x_2 + 20x_3 \ge 400$. Grafton plant 10 hr per day, West Bend plant 10 hr per day, Cedarburg plant not used; min cost is $1,650

51. Let x_1 = no. of ounces of food L
x_2 = no. of ounces of food M
x_3 = no. of ounces of food N
Minimize $C = 20x_1 + 24x_2 + 18x_3$
subject to $20x_1 + 10x_2 + 10x_3 \ge 300$
$10x_1 + 10x_2 + 10x_3 \ge 200$
$10x_1 + 15x_2 + 10x_3 \ge 240$
$x_1, x_2, x_3 \ge 0$
10 oz L, 8 oz M, 2 oz N; min cholesterol intake is 428 units

53. Let x_1 = no. of students bused from North Division to Central
x_2 = no. of students bused from North Division to Washington
x_3 = no. of students bused from South Division to Central
x_4 = no. of students bused from South Division to Washington
Minimize $C = 5x_1 + 2x_2 + 3x_3 + 4x_4$
subject to $x_1 + x_2 \ge 300$
$x_3 + x_4 \ge 500$
$x_1 + x_3 \le 400$
$x_2 + x_4 \le 500$
$x_1, x_2, x_3, x_4 \ge 0$
300 students bused from North Division to Washington, 400 from South Division to Central, and 100 from South Division to Washington; min cost is $2,200

Exercises 6.4

1. (A) Maximize $P = 5x_1 + 2x_2 - Ma_1$
subject to $x_1 + 2x_2 + s_1 \qquad\qquad = 12$
$x_1 + x_2 \qquad - s_2 + u_1 = 4$
$x_1, x_2, s_1, s_2, a_1 \ge 0$

(B)

x_1	x_2	s_1	s_2	a_1	P	
1	2	1	0	0	0	12
1	1	0	-1	1	0	4
$-M-5$	$-M-2$	0	M	0	1	$-4M$

(C) $x_1 = 12, x_2 = 0, s_1 = 0, s_2 = 8, a_1 = 0, P = 60$
(D) Max $P = 60$ at $x_1 = 12$ and $x_2 = 0$

3. (A) Maximize $P = 3x_1 + 5x_2 - Ma_1$
subject to $2x_1 + x_2 + s_1 \qquad = 8$
$x_1 + x_2 \qquad + a_1 - 6$
$x_1, x_2, s_1, a_1 \ge 0$

(B)

x_1	x_2	s_1	a_1	P	
2	1	1	0	0	8
1	1	0	1	0	6
$-M-3$	$-M-5$	0	0	1	$-6M$

(C) $x_1 = 0, x_2 = 6, s_1 = 2, a_1 = 0, P = 30$
(D) Max $P - 30$ at $x_1 = 0$ and $x_2 = 6$

5. (A) Maximize $P = 4x_1 + 3x_2 - Ma_1$
subject to $-x_1 + 2x_2 + s_1 \qquad\qquad = 2$
$x_1 + x_2 \qquad - s_2 + a_1 = 4$
$x_1, x_2, s_1, s_2, a_1 \ge 0$

(B)

x_1	x_2	s_1	s_2	a_1	P	
-1	2	1	0	0	0	2
1	1	0	-1	1	0	4
$-M-4$	$-M-3$	0	M	0	1	$-4M$

(C) No optimal solution exists.
(D) No optimal solution exists.

7. (A) Maximize $P = 5x_1 + 10x_2 - Ma_1$
subject to $x_1 + x_2 + s_1 \qquad\qquad = 3$
$2x_1 + 3x_2 \qquad - s_2 + a_1 = 12$
$x_1, x_2, s_1, s_2, a_1 \ge 0$

(B)

x_1	x_2	s_1	s_2	a_1	P	
1	1	1	0	0	0	3
2	3	0	-1	1	0	12
$-2M-5$	$-3M-10$	0	M	0	1	$-12M$

(C) $x_1 = 0, x_2 = 3, s_1 = 0, s_2 = 0, a_1 = 3, P = -3M + 30$
(D) No optimal solution exists.

9. Min $P = 1$ at $x_1 = 3$ and $x_2 = 5$; max $P = 16$ at $x_1 = 8$ and $x_2 = 0$
11. Max $P = 44$ at $x_1 = 2$ and $x_2 = 8$
13. No optimal solution exists.
15. Min $C = -9$ at $x_1 = 0$, $x_2 = \frac{7}{4}$, and $x_3 = \frac{3}{4}$
17. Max $P = 32$ at $x_1 = 0$, $x_2 = 4$, and $x_3 = 2$

19. Max $P = 65$ at $x_1 = \frac{35}{2}, x_2 = 0$, and $x_3 = \frac{15}{2}$ **21.** Max $P = 120$ at $x_1 = 20, x_2 = 0$, and $x_3 = 20$

23. Problem 5: unbounded feasible region:
Problem 7: empty feasible region:

25. Min $C = -30$ at $x_1 = 0, x_2 = \frac{3}{4}$, and $x_3 = 0$
27. Max $P = 17$ at $x_1 = \frac{49}{5}, x_2 = 0$, and $x_3 = \frac{22}{5}$
29. Min $C = \frac{135}{2}$ at $x_1 = \frac{15}{4}, x_2 = \frac{3}{4}$, and $x_3 = 0$
31. Max $P = 372$ at $x_1 = 28, x_2 = 4$, and $x_3 = 0$

33. Let x_1 = no. of ads placed in the *Sentinel*
x_2 = no. of ads placed in the *Journal*
x_3 = no. of ads placed in the *Tribune*
Minimize $C = 200x_1 + 200x_2 + 100x_3$
subject to $x_1 + x_2 + x_3 \le 10$
$2,000x_1 + 500x_2 + 1,500x_3 \ge 16,000$
$x_1, x_2, x_3 \ge 0$
2 ads in the *Sentinel*, 0 ads in the *Journal*, 8 ads in the *Tribune*; min cost is $1,200

35. Let x_1 = no. of bottles of brand A
x_2 = no. of bottles of brand B
x_3 = no. of bottles of brand C
Minimize $C = 0.6x_1 + 0.4x_2 + 0.9x_3$
subject to $10x_1 + 10x_2 + 20x_3 \geq 100$
$2x_1 + 3x_2 + 4x_3 \leq 24$
$x_1, x_2, x_3 \geq 0$
0 bottles of A, 4 bottles of B, 3 bottles
of C; min cost is \$4.30

37. Let x_1 = no. of cubic yards of mix A
x_2 = no. of cubic yards of mix B
x_3 = no. of cubic yards of mix C
Maximize $P = 12x_1 + 16x_2 + 8x_3$
subject to $12x_1 + 8x_2 + 16x_3 \leq 700$
$16x_1 + 8x_2 + 16x_3 \geq 800$
$x_1, x_2, x_3 \geq 0$
25 yd^3 A, 50 yd^3 B, 0 yd^3 C;
max is 1,100 lb

39. Let x_1 = no. of car frames produced at the Milwaukee plant
x_2 = no. of truck frames produced at the Milwaukee plant
x_3 = no. of car frames produced at the Racine plant
x_4 = no. of truck frames produced at the Racine plant
Maximize $P = 50x_1 + 70x_2 + 50x_3 + 70x_4$
subject to
$$x_1 + x_3 \leq 250$$
$$x_2 + x_4 \leq 350$$
$$x_1 + x_2 \leq 300$$
$$x_3 + x_4 \leq 200$$
$$150x_1 + 200x_2 \leq 50,000$$
$$135x_3 + 180x_4 \leq 35,000$$
$$x_1, x_2, x_3, x_4 \geq 0$$

41. Let x_1 = no. of barrels of A used in regular gasoline
x_2 = no. of barrels of A used in premium gasoline
x_3 = no. of barrels of B used in regular gasoline
x_4 = no. of barrels of B used in premium gasoline
x_5 = no. of barrels of C used in regular gasoline
x_6 = no. of barrels of C used in premium gasoline
Maximize $P = 10x_1 + 18x_2 + 8x_3 + 16x_4 + 4x_5 + 12x_6$
subject to
$$x_1 + x_2 \leq 40,000$$
$$x_3 + x_4 \leq 25,000$$
$$x_5 + x_6 \leq 15,000$$
$$x_1 + x_3 + x_5 \geq 30,000$$
$$x_2 + x_4 + x_6 \geq 25,000$$
$$-5x_1 + 5x_3 + 15x_5 \geq 0$$
$$- 15x_2 - 5x_4 + 5x_6 \geq 0$$
$$x_1, x_2, x_3, x_4, x_5, x_6 \geq 0$$

43. Let x_1 = percentage invested in high-tech funds
x_2 = percentage invested in global funds
x_3 = percentage invested in corporate bonds
x_4 = percentage invested in municipal bonds
x_5 = percentage invested in CDs
Maximize $P = 0.11x_1 + 0.1x_2 + 0.09x_3 + 0.08x_4 + 0.05x_5$
subject to $x_1 + x_2 + x_3 + x_4 + x_5 = 1$
$2.7x_1 + 1.8x_2 + 1.2x_3 + 0.5x_4 \leq 1.8$
$x_5 \geq 0.2$
$x_1, x_2, x_3, x_4, x_5 \geq 0$

45. Let x_1 = no. of ounces of food L
x_2 = no. of ounces of food M
x_3 = no. of ounces of food N
Minimize $C = 0.4x_1 + 0.6x_2 + 0.8x_3$
subject to $30x_1 + 10x_2 + 30x_3 \geq 400$
$10x_1 + 10x_2 + 10x_3 \geq 200$
$10x_1 + 30x_2 + 20x_3 \geq 300$
$8x_1 + 4x_2 + 6x_3 \leq 150$
$60x_1 + 40x_2 + 50x_3 \leq 900$
$x_1, x_2, x_3 \geq 0$

47. Let x_1 = no. of students from town A enrolled in school I
x_2 = no. of students from town A enrolled in school II
x_3 = no. of students from town B enrolled in school I
x_4 = no. of students from town B enrolled in school II
x_5 = no. of students from town C enrolled in school I
x_6 = no. of students from town C enrolled in school II

Minimize $C = 4x_1 + 8x_2 + 6x_3 + 4x_4 + 3x_5 + 9x_6$
subject to $x_1 + x_2 = 500$
$$x_3 + x_4 = 1,200$$
$$x_5 + x_6 = 1,800$$
$$x_1 + x_3 + x_5 \leq 2,000$$
$$x_2 + x_4 + x_6 \leq 2,000$$
$$x_1 + x_3 + x_5 \geq 1,400$$
$$x_2 + x_4 + x_6 \geq 1,400$$
$$x_1 \leq 300$$
$$x_2 \leq 300$$
$$x_3 \leq 720$$
$$x_4 \leq 720$$
$$x_5 \leq 1,080$$
$$x_6 \leq 1,080$$
$$x_1, x_2, x_3, x_4, x_5, x_6 \geq 0$$

Chapter 6 Review Exercises

1. $2x_1 + x_2 + s_1 = 8$
$x_1 + 2x_2 + s_2 = 10$ (6.1)

2. 2 basic and 2 nonbasic variables (6.1)

3.

x_1	x_2	s_1	s_2	Feasible?
0	0	8	10	Yes
0	8	0	−6	No
0	5	3	0	Yes
4	0	0	6	Yes
10	0	−12	0	No
2	4	0	0	Yes	(6.1)

4.

Enter
↓

$$\text{Exit} \rightarrow \begin{array}{c} s_1 \\ s_2 \\ P \end{array} \left[\begin{array}{ccccc|c} x_1 & x_2 & s_1 & s_2 & P & \\ ② & 1 & 1 & 0 & 0 & 8 \\ 1 & 2 & 0 & 1 & 0 & 10 \\ \hline -6 & -2 & 0 & 0 & 1 & 0 \end{array} \right] \ (6.2)$$

5. Max $P = 24$ at $x_1 = 4$ and $x_2 = 0$ (6.2)

6. Basic variables: x_2, s_2, s_3, P; nonbasic variables: x_1, x_3, s_1

Enter
↓

$$\begin{array}{c} x_2 \\ s_2 \\ \text{Exit} \rightarrow s_3 \\ P \end{array} \left[\begin{array}{ccccccc|c} x_1 & x_2 & x_3 & s_1 & s_2 & s_3 & P & \\ 2 & 1 & 3 & -1 & 0 & 0 & 0 & 20 \\ 3 & 0 & 4 & 1 & 1 & 0 & 0 & 30 \\ ② & 0 & 5 & 2 & 0 & 1 & 0 & 10 \\ \hline -8 & 0 & -5 & 3 & 0 & 0 & 1 & 50 \end{array} \right] \sim \begin{array}{c} x_2 \\ s_2 \\ x_1 \\ P \end{array} \left[\begin{array}{ccccccc|c} x_1 & x_2 & x_3 & s_1 & s_2 & s_3 & P & \\ 0 & 1 & -2 & -3 & 0 & -1 & 0 & 10 \\ 0 & 0 & -\frac{7}{2} & -2 & 1 & -\frac{3}{2} & 0 & 15 \\ 1 & 0 & \frac{5}{2} & 1 & 0 & \frac{1}{2} & 0 & 5 \\ \hline 0 & 0 & 15 & 11 & 0 & 4 & 1 & 90 \end{array} \right] \ (6.2)$$

7. (A) $x_1 = 0$, $x_2 = 2$, $s_1 = 0$, $s_2 = 5$, $P = 12$; additional pivoting required
(B) $x_1 = 0$, $x_2 = 0$, $s_1 = 0$, $s_2 = 7$, $P = 22$; no optimal solution exists
(C) $x_1 = 6$, $x_2 = 0$, $s_1 = 15$, $s_2 = 0$, $P = 10$; optimal solution *(6.2)*

8. Maximize $P = 15y_1 + 20y_2$
subject to $y_1 + 2y_2 \le 5$
  $3y_1 + y_2 \le 2$
  $y_1, y_2 \ge 0$ *(6.3)*

9.
$$\begin{aligned} y_1 + 2y_2 + x_1 &= 5 \\ 3y_1 + y_2 + x_2 &= 2 \\ -15y_1 - 20y_2 + P &= 0 \end{aligned} \text{ (6.3)}$$

10.

y_1	y_2	x_1	x_2	P	
1	2	1	0	0	5
3	1	0	1	0	2
-15	-20	0	0	1	0

(6.3)

11. Max $P = 40$ at $y_1 = 0$ and $y_2 = 2$ *(6.2)*
12. Min $C = 40$ at $x_1 = 0$ and $x_2 = 20$ *(6.3)*
13. Max $P = 26$ at $x_1 = 2$ and $x_2 = 5$ *(6.2)*

14. Maximize $P = 10y_1 + 15y_2 + 3y_3$
subject to $y_1 + y_2 \le 3$
  $y_1 + 2y_2 + y_3 \le 8$
  $y_1, y_2, y_3 \ge 0$ *(6.3)*

15. Min $C = 51$ at $x_1 = 9$ and $x_2 = 3$ *(6.3)*
16. No optimal solution exists. *(6.2)*
17. Max $P = 23$ at $x_1 = 4$, $x_2 = 1$, and $x_3 = 0$ *(6.2)*

18. (A) Modified problem:
Maximize $P = x_1 + 3x_2 - Ma_1$
subject to $x_1 + x_2 - s_1 + a_1 = 6$
  $x_1 + 2x_2 + s_2 = 8$
  $x_1, x_2, s_1, s_2, a_1 \ge 0$

(B) Preliminary simplex tableau:

	x_1	x_2	s_1	a_1	s_2	P	
	1	1	-1	1	0	0	6
	1	2	0	0	1	0	8
	-1	-3	0	M	0	1	0

Initial simplex tableau:

x_1	x_2	s_1	a_1	s_2	P	
1	1	-1	1	0	0	6
1	2	0	0	1	0	8
$-M-1$	$-M-3$	M	0	0	1	$-6M$

(C) $x_1 = 4$, $x_2 = 2$, $s_1 = 0$, $a_1 = 0$, $s_2 = 0$, $P = 10$ (D) Since $a_1 = 0$, the optimal solution to the original problem is
Max $P = 10$ at $x_1 = 4$ and $x_2 = 2$ *(6.4)*

19. (A) Modified problem:
Maximize $P = x_1 + x_2 - Ma_1$
subject to $x_1 + x_2 - s_1 + a_1 = 5$
  $x_1 + 2x_2 + s_2 = 4$
  $x_1, x_2, s_1, s_2, a_1 \ge 0$

(B) Preliminary simplex tableau:

	x_1	x_2	s_1	a_1	s_2	P	
	1	1	-1	1	0	0	5
	1	2	0	0	1	0	4
	-1	-1	0	M	0	1	0

Initial simplex tableau:

x_1	x_2	s_1	a_1	s_2	P	
1	1	-1	1	0	0	5
1	2	0	0	1	0	4
$-M-1$	$-M-1$	M	0	0	1	$-5M$

(C) $x_1 = 4$, $x_2 = 0$, $s_1 = 0$, $s_2 = 0$, $a_1 = 1$, $P = -M + 4$ (D) Since $a_1 \ne 0$, the original problem has no optimal solution. *(6.4)*

20. Maximize $P = 2x_1 + 3x_2 + x_3 - Ma_1 - Ma_2$
subject to $x_1 - 3x_2 + x_3 + s_1 = 7$
  $x_1 + x_2 - 2x_3 - s_2 + a_1 = 2$
  $3x_1 + 2x_2 - x_3 + a_2 = 4$
  $x_1, x_2, x_3, s_1, s_2, a_1, a_2 \ge 0$ *(6.4)*

21. The basic simplex method with slack variables solves standard maximization problems involving $\le$ constraints with nonnegative constants on the right side. *(6.2)*
22. The dual problem method solves minimization problems with positive coefficients in the objective function. *(6.3)*
23. The big M method solves any linear programming problem. *(6.4)*

24. Max $P = 36$ at $x_1 = 6$, $x_2 = 8$ *(6.2)*

25. Min $C = 15$ at $x_1 = 3$ and $x_2 = 3$ *(6.3)*
26. Min $C = 15$ at $x_1 = 3$ and $x_2 = 3$ *(6.4)*
27. Min $C = 9{,}960$ at $x_1 = 0$, $x_2 = 240$, $x_3 = 400$, and $x_4 = 60$ *(6.3)*

28. (A) Let $x_1 =$ amount invested in oil stock
  $x_2 =$ amount invested in steel stock
  $x_3 =$ amount invested in government bonds
Maximize $P = 0.12x_1 + 0.09x_2 + 0.05x_3$
subject to $x_1 + x_2 + x_3 \le 150{,}000$
  $x_1 \le 50{,}000$
  $x_1 + x_2 - x_3 \le 25{,}000$
  $x_1, x_2, x_3 \ge 0$
Max return is \$12,500 when \$50,000 is invested in oil stock, \$37,500 is invested in steel stock, and \$62,500 in government bonds.
(B) Max return is \$13,625 when \$87,500 is invested in steel stock and \$62,500 in government bonds. *(6.2)*

29. (A) Let $x_1 =$ no. of regular chairs
  $x_2 =$ no. of rocking chairs
  $x_3 =$ no. of chaise lounges
Maximize $P = 17x_1 + 24x_2 + 31x_3$
subject to $x_1 + 2x_2 + 3x_3 \le 2{,}500$
  $2x_1 + 2x_2 + 4x_3 \le 3{,}000$
  $3x_1 + 3x_2 + 2x_3 \le 3{,}500$
  $x_1, x_2, x_3 \ge 0$
Max $P = \$30{,}000$ when 250 regular chairs, 750 rocking chairs, and 250 chaise lounges are produced.
(B) Maximum profit increases to \$32,750 when 1,000 regular chairs, 0 rocking chairs, and 250 chaise lounges are produced.
(C) Maximum profit decreases to \$28,750 when 125 regular chairs, 625 rocking chairs, and 375 chaise lounges are produced. *(6.2)*

30. Let $x_1 = $ no. of motors shipped from factory A to plant X
$x_2 = $ no. of motors shipped from factory A to plant Y
$x_3 = $ no. of motors shipped from factory B to plant X
$x_4 = $ no. of motors shipped from factory B to plant Y

Minimize $C = 5x_1 + 8x_2 + 9x_3 + 7x_4$
subject to $x_1 + x_2 \quad\quad\quad \leq 1,500$
$\quad\quad\quad\quad x_3 + x_4 \leq 1,000$
$\quad x_1 \quad + x_3 \quad\quad \geq 900$
$\quad\quad x_2 + \quad\quad x_4 \geq 1,200$
$\quad\quad x_1, x_2, x_3, x_4 \geq 0$

Min $C = \$13,100$ when 900 motors are shipped from factory A to plant X, 200 motors are shipped from factory A to plant Y, and 1,000 motors are shipped from factory B to plant Y *(6.3)*

31. Let $x_1 = $ no. of lbs of long-grain rice used in brand A
$x_2 = $ no. of lbs of long-grain rice used in brand B
$x_3 = $ no. of lbs of wild rice used in brand A
$x_4 = $ no. of lbs of wild rice used in brand B

Maximize $P = 0.8x_1 + 0.5x_2 - 1.9x_3 - 2.2x_4$
subject to $0.1x_1 \quad\quad - 0.9x_3 \quad\quad \leq 0$
$\quad\quad\quad 0.05x_2 \quad\quad - 0.95x_4 \leq 0$
$\quad x_1 + \quad x_2 \quad\quad\quad \leq 8,000$
$\quad\quad\quad\quad x_3 + \quad\quad x_4 \leq 500$
$\quad\quad x_1, x_2, x_3, x_4 \geq 0$

Max profit is $\$3,350$ when 1,350 lb long-grain rice and 150 lb wild rice are used to produce 1,500 lb brand A, and 6,650 lb long-grain rice and 350 lb wild rice are used to produce 7,000 lb brand B. *(6.2)*

Chapter 7

Exercises 7.1   **1.** 1, 2, 4, 5, 10, 20   **3.** 11, 22, 33, 44, 55   **5.** 23, 29   **9.** 91 is not odd; false   **11.** 91 is prime and 91 is odd; false   **13.** If 91 is odd, then 91 is prime; false   **15.** If $2^9 + 2^8 + 2^7 > 987$, then $9 \cdot 10^2 + 8 \cdot 10 + 7 = 987$; true   **17.** $2^9 + 2^8 + 2^7 > 987$ or $9 \cdot 10^2 + 8 \cdot 10 + 7 = 987$; true   **19.** If $9 \cdot 10^2 + 8 \cdot 10 + 7$ is not equal to 987, then $2^9 + 2^8 + 2^7$ is not greater than 987; true   **21.** Disjunction; true   **23.** Conditional; false   **25.** Negation; false   **27.** Conjunction; true   **29.** Converse: If triangle ABC is equiangular, then triangle ABC is equilateral. Contrapositive: If triangle ABC is not equiangular, then triangle ABC is not equilateral.   **31.** Converse: If $f(x)$ is an increasing function, then $f(x)$ is a linear function with positive slope. Contrapositive: If $f(x)$ is not an increasing function, then $f(x)$ is not a linear function with positive slope.   **33.** Converse: If n is an integer that is a multiple of 2 and a multiple of 4, then n is an integer that is a multiple of 8. Contrapositive: If n is an integer that is not a multiple of 2 or not a multiple of 4, then n is an integer that is not a multiple of 8.

35.

p	q	$\neg p \wedge q$	**Contingency**
T	T	F	
T	F	F	
F	T	T	
F	F	F	

37.

p	q	$\neg p \to q$	**Contingency**
T	T	T	
T	F	T	
F	T	T	
F	F	F	

39.

p	q	$q \wedge (p \vee q)$	**Contingency**
T	T	T	
T	F	F	
F	T	T	
F	F	F	

41.

p	q	$p \vee (p \to q)$	**Tautology**
T	T	T	
T	F	T	
F	T	T	
F	F	T	

43.

p	q	$p \to (p \wedge q)$	**Contingency**
T	T	T	
T	F	F	
F	T	T	
F	F	T	

45.

p	q	$(p \to q) \to \neg p$	**Contingency**
T	T	F	
T	F	T	
F	T	T	
F	F	T	

47.

p	q	$\neg p \to (p \vee q)$	**Contingency**
T	T	T	
T	F	T	
F	T	T	
F	F	F	

49.

p	q	$q \to (\neg p \wedge q)$	**Contingency**
T	T	F	
T	F	T	
F	T	T	
F	F	T	

51.

p	q	$(\neg p \wedge q) \wedge (q \to p)$	**Contradiction**
T	T	F	
T	F	F	
F	T	F	
F	F	F	

53.

p	q	$p \vee q$
T	T	T
T	F	T
F	T	T
F	F	F

55.

p	q	$\neg p \wedge q$	$p \vee q$
T	T	F	T
T	F	F	T
F	T	T	T
F	F	F	F

57.

p	q	$\neg p \to (q \wedge \neg q)$
T	T	T
T	F	T
F	T	F
F	F	F

59.

p	q	$\neg p \to (p \vee q)$	$p \vee q$
T	T	T	T
T	F	T	T
F	T	T	T
F	F	F	F

61.

p	q	$q \wedge (p \vee q)$	$q \vee (p \wedge q)$
T	T	T	T
T	F	F	F
F	T	T	T
F	F	F	F

63.

p	q	$p \vee (p \to q)$	$p \to (p \vee q)$
T	T	T	T
T	F	T	T
F	T	T	T
F	F	T	T

65. $p \to \neg q \equiv \neg p \vee \neg q$ By (4)
$\equiv \neg(p \wedge q)$ By (6)

67. $\neg(p \to q) \equiv \neg(\neg p \vee q)$ By (4)
$\equiv \neg(\neg p) \wedge \neg q$ By (5)
$\equiv p \wedge \neg q$ By (1)

69. Yes   **71.** No

Exercises 7.2   **1.** No   **3.** No   **5.** No   **7.** T   **9.** T   **11.** T   **13.** F   **15.** $\{2,3\}$ **17.** $\{1,2,3,4\}$ **19.** $\{1,4,7,10,13\}$ **21.** $\varnothing$ **23.** $\{5,-5\}$
25. $\{-3\}$ **27.** $\{1,3,5,7,9\}$ **29.** $A' = \{1,5\}$ **31.** 100   **33.** 61   **35.** 78   **37.** 31   **39.** 9   **41.** 22   **43.** 100   **45.** (A) $\{1,2,3,4,6\}$ (B) $\{1,2,3,4,6\}$
47. $\{1,2,3,4,6\}$ **49.** Finite   **51.** Infinite   **53.** Finite   **55.** Infinite   **57.** Finite   **59.** Disjoint   **61.** Not disjoint   **63.** True   **65.** False   **67.** False
69. False   **71.** False   **73.** (A) 2   (B) 4   (C) 8   (D) 16   **75.** 85   **77.** 33   **79.** 14   **81.** 66   **83.** 54   **85.** 106   **87.** 0   **89.** 6   **91.** A+, AB+
93. A−, A+, B+, AB−, AB+, O+   **95.** O+, O−   **97.** B−, B+

Exercises 7.3   **1.** 13   **3.** 13   **5.** 9   **7.** (A) 4 ways:

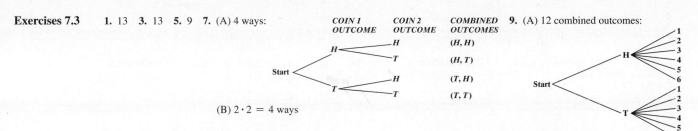

(B) $2 \cdot 2 = 4$ ways

9. (A) 12 combined outcomes:

(B) $2 \cdot 6 = 12$ combined outcomes

11. (A) 9   (B) 18   **13.** 60; 125; 80   **15.** (A) 8   (B) 144   **17.** $n(A \cap B') = 60, n(A \cap B) = 20, n(A' \cap B) = 30, n(A' \cap B') = 90$
19. $n(A \cap B') = 5, n(A \cap B) = 20, n(A' \cap B) = 35, n(A' \cap B') = 40$ **21.** $n(A \cap B') = 15, n(A \cap B) = 70, n(A' \cap B) = 40, n(A' \cap B') = 25$
23. $n(A \cap B') = 48, n(A \cap B) = 0, n(A' \cap B) = 62, n(A' \cap B') = 70$

25.

	A	A'	Totals
B	30	60	90
B'	40	70	110
Totals	70	130	200

27.

	A	A'	Totals
B	20	35	55
B'	25	20	45
Totals	45	55	100

29.

	A	A'	Totals
B	58	8	66
B'	17	7	24
Totals	75	15	90

31.

	A	A'	Totals
B	0	145	145
B'	110	45	155
Totals	110	190	300

33. (A) True   (B) False   **35.** $5 \cdot 3 \cdot 4 \cdot 2 = 120$ **37.** $52^5 = 380,204,032$
39. $10 \cdot 9 \cdot 8 \cdot 7 \cdot 6 = 30,240$; $10 \cdot 10 \cdot 10 \cdot 10 \cdot 10 = 100,000$; $10 \cdot 9 \cdot 9 \cdot 9 \cdot 9 = 65,610$
41. $26 \cdot 26 \cdot 26 \cdot 10 \cdot 10 \cdot 10 = 17,576,000$; $26 \cdot 25 \cdot 24 \cdot 10 \cdot 9 \cdot 8 = 11,232,000$
43. No, the same 8 combined choices are available either way.   **45.** z **47.** x and z **49.** 14   **51.** 14

53. (A) 6 combined outcomes: (B) $3 \cdot 2 = 6$

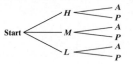

55. 12   **57.** (A) 1,010   (B) 190   (C) 270   **59.** 1,570
61. (A) 102,000   (B) 689,000   (C) 1,470,000   (D) 1,372,000
63. (A) 12 classifications:

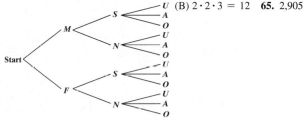

(B) $2 \cdot 2 \cdot 3 = 12$ **65.** 2,905

Exercises 7.4   **1.** 220   **3.** 252   **5.** 9,900   **7.** 5,040   **9.** 39,916,800   **11.** 840   **13.** 1,680   **15.** 70   **17.** 249,500   **19.** 1,287   **21.** 13,366,080
23. 0.1114   **25.** 0.2215   **27.** $n(n-1)$ **29.** $n(n+1)/2$ **31.** Permutation   **33.** Combination   **35.** Neither   **37.** $_{10}P_3 = 10 \cdot 9 \cdot 8 = 720$
39. $_7C_3 = 35$; $_7P_3 = 210$ **41.** The factorial function $x!$ grows much faster than the exponential function 3^x, which in turn grows much faster than the cubic
function x^3.   **43.** $_{26}C_6 = 230,230$ **45.** $_{12}C_5 = 792$ **47.** $_{48}C_3 = 17,296$ **49.** $13^4 = 28,561$ **51.** $_8C_3 \cdot {_{10}C_4} \cdot {_7C_2} = 246,960$ **53.** The numbers are the
same read up or down, since $_nC_r = {_nC_{n-r}}$.   **55.** True   **57.** False   **59.** True   **61.** (A) $_8C_2 = 28$ (B) $_8C_3 = 56$ (C) $_8C_4 = 70$ **63.** $_6P_4 = 360$
65. (A) $_8P_5 = 6,720$ (B) $_8C_5 = 56$ (C) $2 \cdot {_6C_4} = 30$ **67.** For many calculators $k = 69$, but your calculator may be different.   **69.** $_{24}C_{12} = 2,704,156$
71. (A) $_{24}C_3 = 2,024$ (B) $_{19}C_3 = 969$ **73.** (A) $_{30}C_{10} = 30,045,015$ (B) $_8C_2 \cdot {_{12}C_5} \cdot {_{10}C_3} = 2,661,120$ **75.** (A) $_6C_3 \cdot {_5C_2} = 200$ (B) $_6C_4 \cdot {_5C_1} = 75$
(C) $_6C_5 = 6$ (D) $_{11}C_5 = 462$ (E) $_6C_4 \cdot {_5C_1} + {_6C_5} = 81$ **77.** 336; 512   **79.** $_4P_2 = 12$

Chapter 7 Review Exercises   **1.** 3^4 is not less than 4^3; true *(7.1)*   **2.** 2^3 is less than 3^2 or 3^4 is less than 4^3; true *(7.1)*   **3.** 2^3 is less than 3^2 and 3^4 is less
than 4^3; false *(7.1)*   **4.** If 2^3 is less than 3^2, then 3^4 is less than 4^3; false *(7.1)*   **5.** If 3^4 is less than 4^3, then 2^3 is less than 3^2; true *(7.1)*   **6.** If 3^4 is not less than 4^3,
then 2^3 is not less than 3^2; false *(7.1)*   **7.** T *(7.2)*   **8.** F *(7.2)*   **9.** T *(7.2)*   **10.** F *(7.2)*   **11.** Conditional; true *(7.1)*   **12.** Disjunction; false *(7.1)*
13. Conjunction; false *(7.1)*   **14.** Negation; true *(7.1)*   **15.** Converse: If the square matrix A does not have an inverse, then the square matrix A has a row of
zeros. Contrapositive: If the square matrix A has an inverse, then the square matrix A does not have a row of zeros. *(7.1)*   **16.** Converse: If the square matrix A
has an inverse, then the square matrix A is an identity matrix. Contrapositive: If the square matrix A does not have an inverse, then the square matrix A is not an
identity matrix. *(7.1)*   **17.** $\{1, 2, 3, 4, 5\}$ *(7.2)*   **18.** $\{2, 3, 4\}$ *(7.2)*   **19.** $\varnothing$ *(7.2)*   **20.** (A) 12 combined outcomes:
(B) $6 \cdot 2 = 12$ *(7.3)*

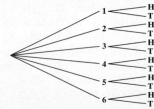

21. (A) 65 (B) 75 (C) 35 (D) 105 (E) 150 (F) 85 (G) 115 (H) 45 *(7.3)* **22.** 39,916,800 *(7.4)* **23.** 24 *(7.4)* **24.** 3,628,080 *(7.4)*
25. 360,360 *(7.4)* **26.** 3,003 *(7.4)* **27.** 56 *(7.4)* **28.** 6720 *(7.4)* **29.** $6 \cdot 5 \cdot 4 \cdot 3 \cdot 2 \cdot 1 = 720$ *(7.3)* **30.** $_6P_6 = 6! = 720$ *(7.4)*

31.

p	q	$(p \to q) \wedge (q \to p)$	Contingency
T	T	T	
T	F	F	
F	T	F	
F	F	T	*(7.1)*

32.

p	q	$p \vee (q \to p)$	Contingency
T	T	T	
T	F	T	
F	T	F	
F	F	T	*(7.1)*

33.

p	q	$(p \vee \neg p) \to (q \wedge \neg q)$	Contradiction
T	T	F	
T	F	F	
F	T	F	
F	F	F	*(7.1)*

34.

p	q	$\neg q \wedge (p \to q)$	Contingency
T	T	F	
T	F	F	
F	T	F	
F	F	T	*(7.1)*

35.

p	q	$\neg p \to (p \to q)$	Tautology
T	T	T	
T	F	T	
F	T	T	
F	F	T	*(7.1)*

36.

p	q	$\neg (p \vee \neg q)$	Contingency
T	T	F	
T	F	F	
F	T	T	
F	F	F	*(7.1)*

37. Infinite *(7.2)* **38.** Finite *(7.2)* **39.** Infinite *(7.2)* **40.** Infinite *(7.2)* **41.** Disjoint *(7.2)* **42.** Not disjoint *(7.2)* **43.** Disjoint *(7.2)* **44.** 5 children, 15 grandchildren, and 30 great grandchildren, for a total of 50 descendants *(7.3, 7.4)* **45.** 336; 512; 392 *(7.3)* **46.** (A) $_6P_3 = 120$ (B) $_5C_2 = 10$ *(7.4)*
47. $_{25}C_{12} = {}_{25}C_{13} = 5{,}200{,}300$ *(7.4)* **48.** $N_1 \cdot N_2 \cdot N_3$ *(7.3)* **49.** $\{-1, 0, 1\}$ *(7.2)* **50.** $\{1, 2, 3, 4\}$ *(7.2, 7.4)* **51.** $\{1, 4, 9, 16, 25, 36, 49\}$ *(7.2)*
52. (A) $_{10}P_3 = 720$ (B) $_6P_3 = 120$ (C) $_{10}C_3 = 120$ *(7.3, 7.4)* **53.** 33 *(7.3)* **54.** x *(7.3)* **55.** $x, z,$ and w *(7.3)* **56.** True *(7.4)* **57.** False *(7.4)*
58. True *(7.4)*

59.

p	q	$p \wedge q$
T	T	T
T	F	F
F	T	F
F	F	F *(7.1)*

60.

p	q	$p \to q$
T	T	T
T	F	F
F	T	T
F	F	T *(7.1)*

61.

p	q	$\neg p \to (q \wedge \neg q)$
T	T	T
T	F	T
F	T	F
F	F	F *(7.1)*

62.

p	q	$p \vee q$	$\neg p \to q$
T	T	T	T
T	F	T	T
F	T	T	T
F	F	F	F *(7.1)*

63.

p	q	$p \wedge (p \to q)$
T	T	T
T	F	F
F	T	F
F	F	F *(7.1)*

64.

p	q	$\neg (p \wedge \neg q)$	$p \to q$
T	T	T	T
T	F	F	F
F	T	T	T
F	F	T	T *(7.1)*

65. $2^5 = 32; 6$ *(7.3)* **66.** Yes, it is both if $r = 0$ or $r = 1$. *(7.4)* **67.** 120 *(7.3)* **68.** (A) 610 (B) 390
(C) 270 *(7.3)* **69.** $_{40}C_6 = 3{,}838{,}380$ *(7.4)* **70.** (A) $67! \approx 3.647 \times 10^{94}$ *(7.4)*

Chapter 8

Exercises 8.1 **1.** E **3.** F **5.** F **7.** $\frac{1}{3}$ **9.** $\frac{1}{4}$ **11.** 0 **13.** 1 **15.** $\frac{1}{4}$ **17.** $\frac{1}{2}$ **19.** $\frac{1}{52}$ **21.** $\frac{4}{13}$ **23.** 0 **25.** $\frac{1}{2}$ **27.** (A) Reject; no probability can be negative (B) Reject; $P(J) + P(G) + P(P) + P(S) \neq 1$ (C) Acceptable **29.** $P(J) + P(P) = .56$ **31.** $\frac{1}{8}$ **33.** $1/{}_{10}P_3 \approx .0014$
35. $_{26}C_5/_{52}C_5 \approx .025$ **37.** $_{12}C_5/_{52}C_5 \approx .000\ 305$ **39.** $S = \{$ All days in a year, 365, excluding leap year $\}$; $\frac{1}{365}$, assuming each day is as likely as any other day for a person to be born **41.** $1/_5P_5 = 1/5! = .008\ 33$ **43.** $\frac{5}{36}$ **45.** $\frac{1}{6}$ **47.** $\frac{7}{9}$ **49.** 0 **51.** $\frac{1}{3}$ **53.** $\frac{2}{9}$ **55.** $\frac{2}{3}$ **57.** $\frac{1}{4}$ **59.** $\frac{1}{4}$ **61.** $\frac{3}{4}$ **63.** No **65.** Yes
67. Yes **69.** (A) Yes (B) Yes, because we would expect, on average, 20 heads in 40 flips; $P(\text{H}) = \frac{37}{40} = .925$; $P(\text{T}) = \frac{3}{40} = .075$ **71.** $\frac{1}{9}$ **73.** $\frac{1}{3}$ **75.** $\frac{1}{9}$
77. $\frac{4}{9}$ **79.** $_{26}C_5/_{52}C_5 \approx .0253$ **81.** $_{12}C_2 \cdot {}_{40}C_4/_{52}C_6 \approx .2963$ **83.** $_{48}C_4/_{52}C_4 \approx .7187$ **85.** $_{13}C_2 \cdot {}_{13}C_2 \cdot {}_{26}C_3/_{52}C_7 \approx .1182$ **87.** (A) $\frac{7}{50} = .14$
(B) $\frac{1}{6} \approx .167$ (C) Answer depends on results of simulation. **89.** (A) Represent the outcomes H and T by 1 and 2, respectively, and select 500 random integers from the integers 1 and 2. (B) Answer depends on results of simulation. (C) Each is $\frac{1}{2} = .5$ **91.** (A) $_{12}P_4 \approx .000\ 084$ (B) $1/12^4 \approx .000\ 048$
93. (A) $_6C_3 \cdot {}_5C_2/_{11}C_5 \approx .433$ (B) $_6C_4 \cdot {}_5C_1/_{11}C_5 \approx .162$ (C) $_6C_5/_{11}C_5 \approx .013$ (D) $(_6C_4 \cdot {}_5C_1 + {}_6C_5)/_{11}C_5 \approx .175$ **95.** (A) $1/_8P_3 \approx .0030$
(B) $1/8^3 \approx .0020$ **97.** (A) $_5C_3/_9C_3 \approx .1190$ (B) $(_5C_3 + {}_5C_2 \cdot {}_4C_1)/_9C_3 \approx .5952$

Exercises 8.2   **1.** $\frac{1}{3}$ **3.** $\frac{7}{24}$ **5.** $\frac{2}{7}$ **7.** .39 **9.** .86 **11.** .26 **13.** $\frac{1}{4}$ **15.** $\frac{10}{13}$ **17.** $\frac{3}{52}$ **19.** $\frac{11}{26}$ **21.** $\frac{5}{26}$ **23.** $\frac{43}{52}$ **25.** .76 **27.** .52 **29.** .6 **31.** .36
33. .49 **35.** $\frac{5}{18}$ **37.** $\frac{11}{36}$ **39.** (A) $\frac{3}{5}; \frac{3}{2}$ (B) $\frac{1}{3}; \frac{3}{1}$ (C) $\frac{2}{3}; \frac{3}{2}$ (D) $\frac{11}{9}; \frac{9}{11}$ **41.** (A) $\frac{3}{11}$ (B) $\frac{11}{18}$ (C) $\frac{4}{5}$ or .8 (D) .49 **43.** False **45.** False **47.** True **49.** 1:1
51. 7:1 **53.** 2:1 **55.** 1:2 **57.** (A) $\frac{1}{8}$ (B) $8 **59.** (A) $.31; \frac{31}{69}$ (B) $.6; \frac{3}{2}$ **61.** $\frac{11}{26}; \frac{11}{15}$ **63.** $\frac{7}{13}; \frac{7}{6}$ **65.** .78 **67.** $\frac{250}{1,000}$ = .25 **69.** Either events A, B, and C
are mutually exclusive, or events A and B are not mutually exclusive and the other pairs of events are mutually exclusive.   **71.** There are fewer calculator steps,
and, in addition, 365! produces an overflow error on many calculators, while $_{365}P_n$ does not produce an overflow error for many values of n.
73. $P(E) = 1 - \frac{12!}{(12 - n)!12^n}$ **77.** (A) $\frac{10}{50} + \frac{10}{50} = \frac{20}{50} = .4$ (B) $\frac{6}{36} + \frac{5}{36} = \frac{11}{36} \approx .306$ (C) Answer depends on results of simulation.
79. (A) $P(C \cup L) = P(C) + P(L) - P(C \cap L) = .45 + .75 - .35 = .85$ (B) $P(C' \cap L') = .15$
81. (A) $P(M_1 \cup A) = P(M_1) + P(A) - P(M_1 \cap A) = .2 + .3 - .05 = .45$ (B) $P[(M_2 \cap A') \cup (M_3 \cap A')] = P(M_2 \cap A') + P(M_3 \cap A') =$
$.2 + .35 = .55$ **83.** .83   **85.** $P(A \cap S) = \frac{50}{1,000} = .05$ **87.** (A) $P(U \cup N) = .22; \frac{11}{39}$ (B) $P[(D \cap A) \cup (R \cap A)] = .3; \frac{7}{3}$

Exercises 8.3

1. One answer is: **3.** One answer is: **5.** One answer is: **7.** 1/13 **9.** 1/4 **11.** 1 **13.** 1/2 **15.** 2/9
17. 2/5 **19.** 6/25 **21.** 2/5 **23.** .10 **25.** .03

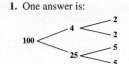 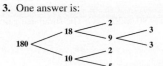 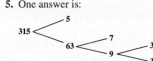

27. .3 **29.** .1 **31.** .35 **33.** 0 **35.** Dependent
37. Independent **39.** Dependent **41.** Dependent
43. (A) $\frac{1}{2}$ (B) $2(\frac{1}{2})^8 \approx .00781$

45. (A) $\frac{1}{4}$ (B) Dependent **47.** (A) .18 (B) .26 **49.** (A) Independent and not mutually exclusive (B) Dependent and mutually exclusive.
51. $(\frac{1}{2})(\frac{1}{2}) = \frac{1}{4}; \frac{1}{2} + \frac{1}{2} - \frac{1}{4} = \frac{3}{4}$ **53.** (A) $(\frac{1}{4})(\frac{13}{51}) \approx .0637$ (B) $(\frac{1}{4})(\frac{1}{4}) = .0625$ **55.** (A) $\frac{3}{13}$ (B) Independent **57.** (A) Dependent (B) Independent

59.

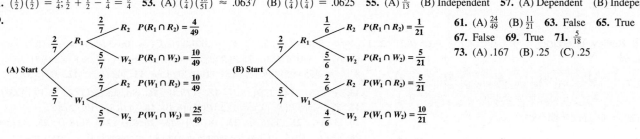

61. (A) $\frac{24}{49}$ (B) $\frac{11}{21}$ **63.** False **65.** True
67. False **69.** True **71.** $\frac{5}{18}$
73. (A) .167 (B) .25 (C) .25

81. (A)

	H	S	B	Totals
Y	.400	.180	.020	.600
N	.150	.120	.130	.400
Totals	.550	.300	.150	1.000

(B) $P(Y|H) = \frac{.400}{.550} \approx .727$ (C) $P(Y|B) = \frac{.020}{.150} \approx .133$
(D) $P(S) = .300; P(S|Y) = .300$ (E) $P(H) = .550; P(H|Y) \approx .667$
(F) $P(B \cap N) = .130$ (G) Yes (H) No (I) No
83. (A) .167 (B) .25 (C) .25

85. (A)

	C	C'	Totals
R	.06	.44	.50
R'	.02	.48	.50
Totals	.08	.92	1.00

(B) Dependent
(C) $P(C|R) = .12$ and $P(C) = .08$; since $P(C|R) > P(C)$, the red dye should be banned.
(D) $P(C|R) = .04$ and $P(C) = .08$; since $P(C|R) < P(C)$, the red dye should not be
banned, since it appears to prevent cancer.

87. (A)

	A	B	C	Totals
F	.130	.286	.104	.520
F'	.120	.264	.096	.480
Totals	.250	.550	.200	1.000

(B) $P(A|F) = \frac{.130}{.520} = .250; P(A|F') = \frac{.120}{.480} = .250$
(C) $P(C|F) = \frac{.104}{.520} = .200; P(C|F') = \frac{.096}{.480} = .200$
(D) $P(A) = .250$ (E) $P(B) = .550; P(B|F') = .550$
(F) $P(F \cap C) = .104$ (G) No; no; no

Exercises 8.4   **1.** 2/5 **3.** 3/2 **5.** 9/10 **7.** $(.6)(.8) = .48$ **9.** $(.6)(.8) + (.4)(.3) = .60$ **11.** .80 **13.** .417 **15.** .375 **17.** .108 **19.** .568
21. .324 **23.** .125 **25.** .50 **27.** .375

29.

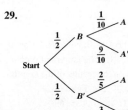

31. .25 **33.** .333 **35.** .50 **37.** .745 **41.** (A) True (B) True **43.** .235 **45.** $\frac{1}{3}$ **47.** .581
49. .052 **53.** .913; .226 **55.** .091; .545; .364 **57.** .667; .000 412 **59.** .231; .036 **61.** .941; .0588

Exercises 8.5   **1.** 71 **3.** 75 **5.** 142 **7.** $E(X) = -.1$ **9.** $43.75 **11.** $0.148 **13.** $25
15. Probability distribution: **17.** Payoff table: **19.** Payoff table:

x_i	0	1	2
p_i	$\frac{1}{4}$	$\frac{1}{2}$	$\frac{1}{4}$

$E(X) = 1$

x_i	$1	-$1
p_i	$\frac{1}{2}$	$\frac{1}{2}$

$E(X) = 0$; game is fair

x_i	-$3	-$2	-$1	$0	$1	$2
p_i	$\frac{1}{6}$	$\frac{1}{6}$	$\frac{1}{6}$	$\frac{1}{6}$	$\frac{1}{6}$	$\frac{1}{6}$

$E(X) = -50$¢; game is not fair

21. $-\$0.50$ **23.** $-\$0.035; \$0.035$ **25.** $\$40$. Let x = amount you should lose if a 6 turns up. Set up a payoff table; then set the expected value of the game equal to zero and solve for x.   **27.** Win $\$1$ **29.** $-\$0.154$ **31.** $\$2.75$ **33.** $A_2; \$210$

35. Payoff table:

x_i	$\$35$	$-\$1$
p_i	$\frac{1}{38}$	$\frac{37}{38}$

$E(X) = -5.26¢$

37. .002   **39.** Payoff table:

x_i	$\$499$	$\$99$	$\$19$	$\$4$	$-\$1$
p_i	.0002	.0006	.001	.004	.9942

$E(X) = -80¢$

41. (A)

x_i	0	1	2
p_i	$\frac{7}{15}$	$\frac{7}{15}$	$\frac{1}{15}$

(B) .60

43. (A)

x_i	$-\$5$	$\$195$	$\$395$	$\$595$
p_i	.985	.0149	.000 059 9	.000 000 06

(B) $E(X) \approx -\$2$

45. (A) $-\$92$ (B) The value per game is $\frac{-\$92}{200} = -\$0.46$, compared with an expected value of $-\$0.0526$.   (C) The simulated gain or loss depends on the results of the simulation; the expected loss is $\$26.32$.   **47.** $\$36.27$

49. Payoff table:

x_i	$\$4,850$	$-\$150$
p_i	.01	.99

$E(X) = -\$100$

51. Site A, with $E(X) = \$3.6$ million   **53.** 1.54
55. For A_1, $E(X) = \$4$, and for A_2, $E(X) = \$4.80$; A_2 is better

Chapter 8 Review Exercises

1. $_{13}C_5/_{52}C_5 \approx .0005$ *(8.1)*   **2.** $1/_{15}P_2 \approx .0048$ *(8.1)*   **3.** $1/_{10}P_3 \approx .0014; 1/_{10}C_3 \approx .0083$ *(8.1)*   **4.** .05 *(8.1)*

5. Payoff table:

x_i	$-\$2$	$-\$1$	$\$0$	$\$1$	$\$2$
p_i	$\frac{1}{5}$	$\frac{1}{5}$	$\frac{1}{5}$	$\frac{1}{5}$	$\frac{1}{5}$

$E(X) = 0$; game is fair *(8.5)*

6. (A) .7   (B) .6 *(8.2)*   **7.** $P(R \cup G) = .8$; odds for $R \cup G$ are 8 to 2 *(8.2)*
8. $\frac{5}{11} \approx .455$ *(8.2)*   **9.** .27 *(8.3)*   **10.** .20 *(8.3)*   **11.** .02 *(8.3)*   **12.** .03 *(8.3)*
13. .15 *(8.3)*   **14.** .1304 *(8.3)*   **15.** .1 *(8.3)*   **16.** No, since $P(T|Z) \neq P(T)$ *(8.3)*
17. Yes, since $P(S \cap X) = P(S)P(X)$ *(8.3)*   **18.** .4 *(8.3)*   **19.** .2 *(8.3)*
20. .3 *(8.3)*   **21.** .08 *(8.3)*   **22.** .18 *(8.3)*   **23.** .26 *(8.3)*   **24.** .31 *(8.4)*   **25.** .43 *(8.4)*
26. (A) $\frac{5}{16}$ (B) $\frac{1}{4}$ (C) As the sample in part (A) increases in size, approximate empirical probabilities should approach the theoretical probabilities. *(8.1)*

27. No *(8.1)*   **28.** Yes *(8.1)*   **29.** Payoff table:

x_i	$\$5$	$-\$4$	$\$2$
p_i	.25	.5	.25

$E(X) = -25¢$; game is not fair *(8.5)*

30. (A) $\frac{1}{3}$ (B) $\frac{2}{9}$ *(8.3)*   **31.** (A) $\frac{2}{13}$; 2 to 11;   (B) $\frac{4}{13}$; 4 to 9;   (C) $\frac{12}{13}$; 12 to 1 *(8.2)*
32. (A) 1 to 8   (B) $\$8$ *(8.2)*
33. (A) $P(2\,\text{heads}) = .21, P(1\,\text{head}) = .48, P(0\,\text{heads}) = .31$
(B) $P(2\,\text{heads}) = .25, P(1\,\text{head}) = .50, P(0\,\text{heads}) = .25$
(C) 2 heads, 250; 1 head, 500; 0 heads, 250 *(8.1, 8.5)*

34. $\frac{1}{2}$; since the coin has no memory, the 10th toss is independent of the preceding 9 tosses. *(8.3)*

35. (A)

x_i	2	3	4	5	6	7	8	9	10	11	12
p_i	$\frac{1}{36}$	$\frac{2}{36}$	$\frac{3}{36}$	$\frac{4}{36}$	$\frac{5}{36}$	$\frac{6}{36}$	$\frac{5}{36}$	$\frac{4}{36}$	$\frac{3}{36}$	$\frac{2}{36}$	$\frac{1}{36}$

(B) $E(X) = 7$ *(8.5)*

36. $A = \{(1,3), (2,2), (3,1), (2,6), (3,5), (4,4), (5,3), (6,2), (6,6)\}$;
$B = \{(1,5), (2,4), (3,3), (4,2), (5,1), (6,6)\}$; $P(A) = \frac{1}{4}$; $P(B) = \frac{1}{6}$; $P(A \cap B) = \frac{1}{36}$; $P(A \cup B) = \frac{7}{18}$ *(8.2)*
37. (1) Probability of an event cannot be negative;   (2) sum of probabilities of simple events must be 1;   (3) probability of an event cannot be greater than 1. *(8.1)*

38.

	A	A'	Totals
B	15	30	45
B'	35	20	55
Totals	50	50	100

(8.2)

39. (A) .6   (B) $\frac{5}{6}$ *(8.3)*   **40.** (A) $\frac{1}{13}$ (B) Independent *(8.3)*   **41.** (A) $\frac{6}{25}$ (B) $\frac{3}{10}$ *(8.3)*
42. Part (B) *(8.3)*   **43.** (A) 1.2   (B) 1.2 *(8.5)*   **44.** (A) $\frac{3}{5}$ (B) $\frac{1}{3}$ (C) $\frac{7}{15}$ (D) $\frac{9}{14}$
(E) $\frac{5}{8}$ (F) $\frac{3}{10}$ *(8.3, 8.4)*   **45.** No *(8.3)*

46. (A) $_{13}C_5/_{52}C_5$ (B) $_{13}C_3 \cdot _{13}C_2/_{52}C_5$ *(8.1)*   **47.** $_8C_2/_{10}C_4 = \frac{2}{15}$ *(8.1)*   **48.** Events S and H are mutually exclusive. Hence, $P(S \cap H) = 0$, while $P(S) \neq 0$ and $P(H) \neq 0$. Therefore, $P(S \cap H) \neq P(S)P(H)$, which implies S and H are dependent. *(8.3)*   **49.** (A) $\frac{9}{50} = .18$ (B) $\frac{9}{36} = .25$ (C) The empirical probability depends on the results of the simulation; the theoretical probability is $\frac{5}{36} \approx .139$. *(8.1)*   **50.** The empirical probability depends on the results of the simulation; the theoretical probability is $\frac{2}{52} \approx .038$. *(8.3)*   **51.** False *(8.2)*   **52.** True *(8.2)*   **53.** False *(8.3)*   **54.** False *(8.2)*   **55.** True *(8.3)*   **56.** False *(8.2)*
57. (A) .350   (B) $\frac{3}{8} = .375$ (C) 375 *(8.1)*   **58.** $-.0172; .0172$; no *(8.5)*   **59.** $1 - _7C_3/_{10}C_3 = \frac{17}{24}$ *(8.1)*   **60.** $\frac{12}{51} \approx .235$ *(8.3)*   **61.** $\frac{12}{51} \approx .235$ *(8.3)*

62. (A)

x_i	2	3	4	5	6
p_i	$\frac{9}{36}$	$\frac{12}{36}$	$\frac{10}{36}$	$\frac{4}{36}$	$\frac{1}{36}$

(B) $E(X) = \frac{10}{3}$ *(8.5)*   **63.** $E(X) \approx -\$0.167$; no; $\$(10/3) \approx \$3.33$ *(8.5)*
64. (A) $\frac{1}{4}$; 1 to 3   (B) $\$3$ *(8.2, 8.4)*   **65.** $1 - 10!/(5!10^5) \approx .70$ *(8.2)*
66. $P(A|B) = P(B|A)$ if and only if $P(A) = P(B)$ or $P(A \cap B) = 0$. *(8.3)*

67. (A) .8   (B) .2   (C) .5 *(8.2)*   **68.** $P(A \cap P) = P(A)P(P|A) = .34$ *(8.3)*
69. (A) $P(A) = .290$; $P(B) = .290$; $P(A \cap B) = .100$; $P(A|B) = .345$; $P(B|A) = .345$
(B) No, since $P(A \cap B) \neq P(A)P(B)$.
(C) $P(C) = .880$; $P(D) = .120$; $P(C \cap D) = 0$; $P(C|D) = 0$; $P(D|C) = 0$
(D) Yes, since $C \cap D = \varnothing$; dependent, since $P(C \cap D) = 0$ and $P(C)P(D) \neq 0$ *(8.3)*

70. Plan A: $E(X) = \$7.6$ million; plan B: $E(X) = \$7.8$ million; plan B *(8.5)*

71. Payoff table:

x_i	\$1,830	−\$170
p_i	.08	.92

$E(X) = -\$10$ *(8.5)*

72. $1 - ({}_{10}C_4/{}_{12}C_4) \approx .576$ *(8.2)*

73. (A)

x_i	0	1	2
p_i	$\frac{12}{22}$	$\frac{9}{22}$	$\frac{1}{22}$

(B) $E(X) = \frac{1}{2}$ *(8.5)*

74. .955 *(8.4)*

75. $\frac{6}{7} \approx .857$ *(8.4)*

Chapter 9

Exercises 9.1   **1.** $\begin{bmatrix} 16 \\ 14 \end{bmatrix}$ **3.** Not defined   **5.** $[14 \quad 17]$ **7.** Not defined

9. $S_1 = \begin{matrix} A & B \\ [.8 & .2] \end{matrix}$; the probability of being in state A after one trial is .8, and the probability of being in state B after one trial is .2.

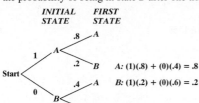

A: (1)(.8) + (0)(.4) = .8
B: (1)(.2) + (0)(.6) = .2

11. $S_1 = \begin{matrix} A & B \\ [.6 & .4] \end{matrix}$; the probability of being in state A after one trial is .6, and the probability of being in state B after one trial is .4.

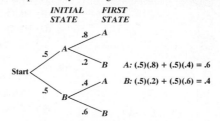

A: (.5)(.8) + (.5)(.4) = .6
B: (.5)(.2) + (.5)(.6) = .4

13. $S_2 = \begin{matrix} A & B \\ [.72 & .28] \end{matrix}$; the probability of being in state A after two trials is .72, and the probability of being in state B after two trials is .28.

15. $S_2 = \begin{matrix} A & B \\ [.64 & .36] \end{matrix}$; the probability of being in state A after two trials is .64, and the probability of being in state B after two trials is .36.

17. $S_1 = [.86 \quad .14]$ **19.** $S_1 = [.72 \quad .28]$ **21.** $S_2 = [.728 \quad .272]$ **23.** $S_2 = [.756 \quad .244]$ **25.** Yes   **27.** No   **29.** No   **31.** Yes

33. 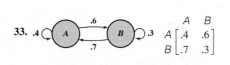 $\begin{matrix} & A & B \\ A & [.4 & .6] \\ B & [.7 & .3] \end{matrix}$

35. No

37. 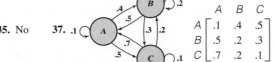 $\begin{matrix} & A & B & C \\ A & [.1 & .4 & .5] \\ B & [.5 & .2 & .3] \\ C & [.7 & .2 & .1] \end{matrix}$

39. $a = .5, b = .6, c = .7$

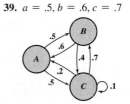

41. $a = .7, b = 1, c = .2$

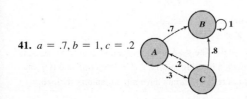

43. No

45. 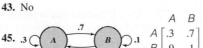 $\begin{matrix} & A & B \\ A & [.3 & .7] \\ B & [.9 & .1] \end{matrix}$

47. 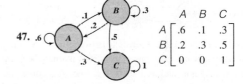 $\begin{matrix} & A & B & C \\ A & [.6 & .1 & .3] \\ B & [.2 & .3 & .5] \\ C & [0 & 0 & 1] \end{matrix}$

49. .35   **51.** .212   **53.** $S_2 = \begin{matrix} A & B & C \\ [.43 & .35 & .22] \end{matrix}$; the probabilities of going from state A to states A, B, and C in two trials   **55.** $S_3 = [.212 \quad .298 \quad .49]$; the probabilities of going from state C to states A, B, and C in three trials   **57.** $n = 9$

59. $P^4 = \begin{matrix} & A & B \\ A & [.4375 & .5625] \\ B & [.375 & .625] \end{matrix}$; $S_4 = [.425 \quad .575]$

61. $P^4 = \begin{matrix} & A & B & C \\ A & [.36 & .16 & .48] \\ B & [.6 & 0 & .4] \\ C & [.4 & .24 & .36] \end{matrix}$; $S_4 = [.452 \quad .152 \quad .396]$ **65.** False   **67.** True   **69.** False

71. (A) $\begin{matrix} & A & B & C & D \\ A & [.0154 & .3534 & .0153 & .6159] \\ B & [0 & 1 & 0 & 0] \\ C & [.0102 & .2962 & .0103 & .6833] \\ D & [0 & 0 & 0 & 1] \end{matrix}$

(B) .6159   (C) .2962   (D) 0

75. (A) $[.25 \quad .75]$ (B) $[.25 \quad .75]$
(C) $[.25 \quad .75]$ (D) $[.25 \quad .75]$
(E) The state matrices appear to approach the same matrix, $S = [.25 \quad .75]$, regardless of the values in the initial-state matrix.

77. $Q = \begin{bmatrix} .25 & .75 \\ .25 & .75 \end{bmatrix}$; the rows of Q are the same as the matrix S from Problem 75

79. (A) R = rain, R' = no rain

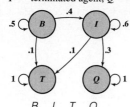

(B) $\begin{array}{c} \\ R \\ R' \end{array}\begin{array}{cc} R & R' \\ \begin{bmatrix} .4 & .6 \\ .06 & .94 \end{bmatrix} \end{array}$

(C) Saturday: .196;
Sunday: .12664

81. (A)

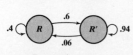

(B) $\begin{array}{c} \\ X \\ X' \end{array}\begin{array}{cc} X & X' \\ \begin{bmatrix} .8 & .2 \\ .2 & .8 \end{bmatrix} \end{array}$

(C) 32%; 39.2%

83. (A) N = National Property, U = United Family, O = other companies

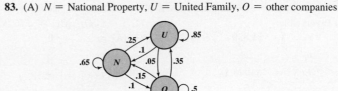

(B) $\begin{array}{c} \\ N \\ U \\ C \end{array}\begin{array}{ccc} N & U & O \\ \begin{bmatrix} .65 & .25 & .1 \\ .1 & .85 & .05 \\ .15 & .35 & .5 \end{bmatrix} \end{array}$ (C) 38.5%; 32%   (D) 45%; 53.65%

85. (A) B = beginning agent, I = intermediate agent, T = terminated agent, Q = qualified agent

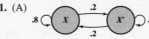

(B) $\begin{array}{c} \\ B \\ I \\ T \\ Q \end{array}\begin{array}{cccc} B & I & T & Q \\ \begin{bmatrix} .5 & .4 & .1 & 0 \\ 0 & .6 & .1 & .3 \\ 0 & 0 & 1 & 0 \\ 0 & 0 & 0 & 1 \end{bmatrix} \end{array}$ (C) .12; .3612

87. (A) $\begin{array}{c} \\ HMO \\ PPO \\ FFS \end{array}\begin{array}{ccc} HMO & PPO & FFS \\ \begin{bmatrix} .8 & .15 & .05 \\ .2 & .7 & .1 \\ .25 & .3 & .45 \end{bmatrix} \end{array}$

(B) HMO: 34.75%; PPO: 37%; FFS: 28.25%
(C) HMO: 42.2625%; PPO: 39.5875%; FFS: 18.15%

89. (A) $\begin{array}{c} \\ H \\ R \end{array}\begin{array}{cc} H & R \\ \begin{bmatrix} .847 & .153 \\ .174 & .826 \end{bmatrix} \end{array}$

(B) 45.6%
(C) 49.8%

Exercises 9.2   **1.** $\begin{bmatrix} 1 & 0 \\ 0 & 0 \end{bmatrix}$ **3.** $\begin{bmatrix} 1 & 0 \\ 0 & 1 \end{bmatrix}$ **5.** $\begin{bmatrix} 1 & 0 & 0 \\ 0 & 0 & 0 \\ 0 & 0 & 1 \end{bmatrix}$ **7.** $\begin{bmatrix} 0 & 0 & 0 \\ 0 & 0 & 0 \\ 0 & 0 & 0 \end{bmatrix}$ **9.** Yes   **11.** No   **13.** No   **15.** Yes   **17.** No   **19.** Yes   **21.** No

23. $S = [.4 \quad .6]$; $\overline{P} = \begin{bmatrix} .4 & .6 \\ .4 & .6 \end{bmatrix}$ **25.** $S = [.375 \quad .625]$; $\overline{P} = \begin{bmatrix} .375 & .625 \\ .375 & .625 \end{bmatrix}$ **27.** $S = [.3 \quad .5 \quad .2]$; $\overline{P} = \begin{bmatrix} .3 & .5 & .2 \\ .3 & .5 & .2 \\ .3 & .5 & .2 \end{bmatrix}$

29. $S = [.6 \quad .24 \quad .16]$; $\overline{P} = \begin{bmatrix} .6 & .24 & .16 \\ .6 & .24 & .16 \\ .6 & .24 & .16 \end{bmatrix}$ **31.** False   **33.** True   **35.** False   **37.** $S = [.3553 \quad .6447]$ **39.** $S = [.3636 \quad .4091 \quad .2273]$

41. (A)

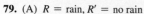

(B) $\begin{array}{c} \\ Red \\ Blue \end{array}\begin{array}{cc} Red & Blue \\ \begin{bmatrix} .4 & .6 \\ .2 & .8 \end{bmatrix} \end{array}$ (C) $[.25 \quad .75]$; in the long run, the red urn will be selected 25% of the time and the blue urn 75% of the time.

43. (A) The state matrices alternate between $[.2 \quad .8]$ and $[.8 \quad .2]$; so they do not approach any one matrix.   (B) The state matrices are all equal to S_0, so S_0 is a stationary matrix.   (C) The powers of P alternate between P and I (the 2×2 identity); so they do not approach a limiting matrix.   (D) Parts (B) and (C) of Theorem 1 are not valid for this matrix. Since P is not regular, this is not a contradiction.   **45.** (A) Since P is not regular, it may have more than one stationary matrix.   (B) $[.5 \quad 0 \quad .5]$ is another stationary matrix.   (C) P has an infinite number of stationary matrices.

47. $\overline{P} = \begin{bmatrix} 1 & 0 & 0 \\ .25 & 0 & .75 \\ 0 & 0 & 1 \end{bmatrix}$; each row of $\overline{P}$ is a stationary matrix for P.   **49.** (A) .39; .3; .284; .277   (B) Each entry of the second column of P^{k+1} is the product of a row of P and the second column of P^k. Each entry of the latter is $\leq M_k$, so the product is $\leq M_k$.   **51.** 72.5%

53. (A) $S_1 = [.512 \quad .488]$; $S_2 = [.568 \quad .432]$; $S_3 = [.609 \quad .391]$; $S_4 = [.639 \quad .361]$

(B)

Year	Data (%)	Model (%)
1970	43.3	43.3
1980	51.5	51.2
1990	57.5	56.8
2000	59.8	60.9
2010	58.5	63.9

(C) 71.4%

55. GTT: 25%; NCJ: 25%; Dash: 50%   **57.** Poor: 20%; satisfactory: 40%; preferred: 40%   **59.** 51%   **61.** Stationary matrix = $[.25 \quad .50 \quad .25]$

63. (A) $[.25 \quad .75]$ (B) 42.5%; 51.25%
(C) 60% rapid transit; 40% automobile

65. (A) $S_1 = [.334 \quad .666]$; $S_2 = [.343 \quad .657]$; $S_3 = [.347 \quad .653]$; $S_4 = [.349 \quad .651]$

(B)

Year	Data (%)	Model (%)
1970	30.9	30.9
1980	33.3	33.4
1990	34.4	34.3
2000	35.6	34.7
2010	37.1	34.1

(C) 35%

Exercises 9.3   **1.** B, C **3.** No absorbing states   **5.** A, D **7.** B is an absorbing state; absorbing chain

9. C is an absorbing state; not an absorbing chain   **11.** No   **13.** Yes   **15.** Yes   **17.** No   **19.** Yes

21. $\begin{array}{c} \\ B \\ A \\ C \end{array}\begin{array}{ccc} B & A & C \\ \begin{bmatrix} 1 & 0 & 0 \\ .5 & .2 & .3 \\ .1 & .5 & .4 \end{bmatrix} \end{array}$

23. $\begin{array}{c} \\ B \\ D \\ A \\ C \end{array}\begin{array}{cccc} B & A & C & D \\ \begin{bmatrix} 1 & 0 & 0 & 0 \\ 0 & 1 & 0 & 0 \\ .4 & .1 & .3 & .2 \\ .4 & .3 & 0 & .3 \end{bmatrix} \end{array}$

25. $\begin{array}{c} \\ C \\ A \\ B \end{array}\begin{array}{ccc} C & A & B \\ \begin{bmatrix} 1 & 0 & 0 \\ .5 & .2 & .3 \\ 0 & 1 & 0 \end{bmatrix} \end{array}$

27. $\begin{array}{c} \\ B \\ D \\ A \\ C \end{array}\begin{array}{cccc} B & C & A & D \\ \begin{bmatrix} 1 & 0 & 0 & 0 \\ 0 & 1 & 0 & 0 \\ .2 & .4 & .1 & .3 \\ .2 & .1 & .5 & .2 \end{bmatrix} \end{array}$

29. $\overline{P} = \begin{array}{c} \\ A \\ B \\ C \end{array}\begin{array}{ccc} A & B & C \\ \begin{bmatrix} 1 & 0 & 0 \\ 0 & 1 & 0 \\ .2 & .8 & 0 \end{bmatrix} \end{array}$; $P(C \text{ to } A) = .2$; $P(C \text{ to } B) = .8$.
It will take an avg. of 2 trials to go from C to either A or B.

31. $\overline{P} = \begin{array}{c} \\ A \\ B \\ C \end{array}\begin{array}{ccc} A & B & C \\ \begin{bmatrix} 1 & 0 & 0 \\ 1 & 0 & 0 \\ 1 & 0 & 0 \end{bmatrix} \end{array}$; $P(B \text{ to } A) = 1$; $P(C \text{ to } A) = 1$. It will take an avg. of 4 trials to go from B to A, and an avg. of 3 trials to go from C to A.

33. $\overline{P} = \begin{array}{c} \\ A \\ B \\ C \\ D \end{array}\begin{array}{cccc} A & B & C & D \\ \begin{bmatrix} 1 & 0 & 0 & 0 \\ 0 & 1 & 0 & 0 \\ .36 & .64 & 0 & 0 \\ .44 & .56 & 0 & 0 \end{bmatrix} \end{array}$; $P(C \text{ to } A) = .36$; $P(C \text{ to } B) = .64$; $P(D \text{ to } A) = .44$; $P(D \text{ to } B) = .56$. It will take an avg. of 3.2 trials to go from C to either A or B, and an avg. of 2.8 trials to go from D to either A or B.

35. (A) $[.2 \quad .8 \quad 0]$ (B) $[.26 \quad .74 \quad 0]$ **37.** (A) $[1 \quad 0 \quad 0]$ (B) $[1 \quad 0 \quad 0]$ **39.** (A) $[.44 \quad .56 \quad 0 \quad 0]$ (B) $[.36 \quad .64 \quad 0 \quad 0]$
(C) $[.408 \quad .592 \quad 0 \quad 0]$ (D) $[.384 \quad .616 \quad 0 \quad 0]$ **41.** False   **43.** False   **45.** True   **47.** False

49. $\begin{array}{c} \\ A \\ B \\ C \\ D \end{array}\begin{array}{cccc} A & B & C & D \\ \begin{bmatrix} 1 & 0 & 0 & 0 \\ 0 & 1 & 0 & 0 \\ .6375 & .3625 & 0 & 0 \\ .7375 & .2625 & 0 & 0 \end{bmatrix} \end{array}$

51. $\begin{array}{c} \\ A \\ B \\ C \\ D \\ E \end{array}\begin{array}{ccccc} A & B & C & D & E \\ \begin{bmatrix} 1 & 0 & 0 & 0 & 0 \\ 0 & 1 & 0 & 0 & 0 \\ .0875 & .9125 & 0 & 0 & 0 \\ .1875 & .8125 & 0 & 0 & 0 \\ .4375 & .5625 & 0 & 0 & 0 \end{bmatrix} \end{array}$

53. $\begin{array}{c} \\ A \\ B \\ C \\ D \end{array}\begin{array}{cccc} A & B & C & D \\ \begin{bmatrix} 0 & .52 & 0 & .48 \\ 0 & 1 & 0 & 0 \\ 0 & .36 & 0 & .64 \\ 0 & 0 & 0 & 1 \end{bmatrix} \end{array}$

59. (A) .370; .297; .227; .132; .045
(B) For large k, all entries of Q^k are close to 0.

61. (A) 75%   (B) 12.5%   (C) 7.5 months   **63.** (A) Company A: 30%; company B: 15%; company C: 55%   (B) 5 yr
65. (A) 91.52%   (B) 4.96%   (C) 6.32 days   **67.** (A) .375   (B) 1.75 exits

Chapter 9 Review Exercises   **1.** $S_1 = \begin{array}{cc} A & B \end{array}[.32 \quad .68]$; $S_2 = \begin{array}{cc} A & B \end{array}[.328 \quad .672]$. The probability of being in state A after one trial is .32 and after two trials is .328; the probability of being in state B after one trial is .68 and after two trials is .672. *(9.1)*   **2.** State A is absorbing; chain is absorbing. *(9.2, 9.3)*
3. No absorbing states; chain is regular. *(9.2, 9.3)*   **4.** No absorbing states; chain is neither. *(9.2, 9.3)*
5. States B and C are absorbing; chain is absorbing. *(9.2, 9.3)*   **6.** States A and B are absorbing; chain is neither. *(9.2, 9.3)*

7. $\begin{array}{c} \\ A \\ B \\ C \end{array}\begin{array}{ccc} A & B & C \\ \begin{bmatrix} 0 & 1 & 0 \\ .1 & 0 & .9 \\ 0 & 1 & 0 \end{bmatrix} \end{array}$; No absorbing states; chain is neither. *(9.1, 9.2, 9.3)*

8. $\begin{array}{c} \\ A \\ B \\ C \end{array}\begin{array}{ccc} A & B & C \\ \begin{bmatrix} 0 & 1 & 0 \\ .1 & .2 & .7 \\ 0 & 0 & 1 \end{bmatrix} \end{array}$; C is absorbing; chain is absorbing. *(9.1, 9.2, 9.3)*

9. $\begin{array}{c} \\ A \\ B \\ C \end{array}\begin{array}{ccc} A & B & C \\ \begin{bmatrix} 0 & 0 & 1 \\ .1 & .2 & .7 \\ 0 & 1 & 0 \end{bmatrix} \end{array}$; No absorbing states; chain is regular. *(9.1, 9.2, 9.3)*

10. $\begin{array}{c} \\ A \\ B \\ C \\ D \end{array}\begin{array}{cccc} A & B & C & D \\ \begin{bmatrix} .3 & .2 & 0 & .5 \\ 0 & 1 & 0 & 0 \\ 0 & 0 & .2 & .8 \\ 0 & 0 & .3 & .7 \end{bmatrix} \end{array}$; B is absorbing; chain is neither. *(9.1, 9.2, 9.3)*

11.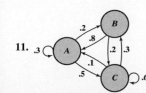

$\begin{array}{c} \\ A \\ B \\ C \end{array}\begin{array}{ccc} A & B & C \\ \begin{bmatrix} .3 & .2 & .5 \\ .8 & 0 & .2 \\ .1 & .3 & .6 \end{bmatrix} \end{array}$ *(9.1)*

12. (A) .3   (B) .675 *(9.1)*   **13.** $S = [.25 \quad .75]$; $\overline{P} = \begin{array}{c} \\ A \\ B \end{array}\begin{array}{cc} A & B \\ \begin{bmatrix} .25 & .75 \\ .25 & .75 \end{bmatrix} \end{array}$ *(9.2)*

14. $S = \begin{bmatrix} A & B & C \\ .4 & .48 & .12 \end{bmatrix}$; $\overline{P} = \begin{matrix} A \\ B \\ C \end{matrix}\begin{bmatrix} A & B & C \\ .4 & .48 & .12 \\ .4 & .48 & .12 \\ .4 & .48 & .12 \end{bmatrix}$ *(9.2)*

15. $\begin{matrix} A \\ B \\ C \end{matrix}\begin{bmatrix} A & B & C \\ 1 & 0 & 0 \\ 0 & 1 & 0 \\ .75 & .25 & 0 \end{bmatrix}$; $P(C \text{ to } A) = .75$; $P(C \text{ to } B) = .25$. It takes an average of 2.5 trials to go from C to an absorbing state. *(9.3)*

16. $\begin{matrix} A \\ B \\ C \\ D \end{matrix}\begin{bmatrix} A & B & C & D \\ 1 & 0 & 0 & 0 \\ 0 & 1 & 0 & 0 \\ .2 & .8 & 0 & 0 \\ .3 & .7 & 0 & 0 \end{bmatrix}$; $P(C \text{ to } A) = .2$; $P(C \text{ to } B) = .8$; $P(D \text{ to } A) = .3$; $P(D \text{ to } B) = .7$. It takes an avg. of 2 trials to go from C to an absorbing state and an avg. of 3 trials to go from D to an absorbing state. *(9.3)*.

21. $\begin{matrix} B \\ D \\ A \\ C \end{matrix}\begin{bmatrix} B & D & A & C \\ 1 & 0 & 0 & 0 \\ 0 & 1 & 0 & 0 \\ .1 & .1 & .6 & .2 \\ .2 & .2 & .3 & .3 \end{bmatrix}$ *(9.3)*

22. (A) $\begin{bmatrix} A & B & C \\ .1 & .4 & .5 \end{bmatrix}$ (B) $\begin{bmatrix} A & B & C \\ .1 & .4 & .5 \end{bmatrix}$ *(9.3)*   **23.** (A) $\begin{bmatrix} A & B & C \\ .25 & .75 & 0 \end{bmatrix}$ (B) $\begin{bmatrix} A & B & C \\ .55 & .45 & 0 \end{bmatrix}$ *(9.3)*

24. No. Each row of P would contain a 0 and a 1, but none of the four matrices with this property is regular. *(9.2)*

25. Yes; for example, $P = \begin{bmatrix} 0 & 0 & 1 \\ 0 & 0 & 1 \\ .2 & .3 & .5 \end{bmatrix}$ is regular. *(9.2)*

26. (A)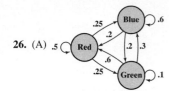

(B) $\begin{matrix} R \\ B \\ G \end{matrix}\begin{bmatrix} R & B & G \\ .5 & .25 & .25 \\ .2 & .6 & .2 \\ .6 & .3 & .1 \end{bmatrix}$

(C) Regular

(D) $\begin{matrix} R \\ B \\ G \end{matrix}\begin{bmatrix} R & B & G \\ .4 & .4 & .2 \\ .4 & .4 & .2 \\ .4 & .4 & .2 \end{bmatrix}$ In the long run, the red urn will be selected 40% of the time, the blue urn 40% of the time, and the green urn 20% of the time. *(9.2)*

27. (A)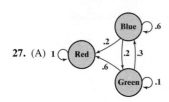

(B) $\begin{matrix} R \\ B \\ G \end{matrix}\begin{bmatrix} R & B & G \\ 1 & 0 & 0 \\ .2 & .6 & .2 \\ .6 & .3 & .1 \end{bmatrix}$

(C) Absorbing

(D) $\begin{matrix} R \\ B \\ G \end{matrix}\begin{bmatrix} R & B & G \\ 1 & 0 & 0 \\ 1 & 0 & 0 \\ 1 & 0 & 0 \end{bmatrix}$ Once the red urn is selected, the blue and green urns will never be selected again. It will take an avg. of 3.67 trials to reach the red urn from the blue urn and an avg. of 2.33 trials to reach the red urn from the green urn. *(9.3)*

29. No such chain exists. *(9.2, 9.3)*   **30.** No such chain exists. *(9.2, 9.3)*   **31.** No such chain exists. *(9.2)*

32. $S = \begin{bmatrix} 1 & 0 & 0 \end{bmatrix}$ and $S' = \begin{bmatrix} 0 & 1 & 0 \end{bmatrix}$ are both stationary matrices for $P = \begin{matrix} A \\ B \\ C \end{matrix}\begin{bmatrix} A & B & C \\ 1 & 0 & 0 \\ 0 & 1 & 0 \\ .6 & .3 & .1 \end{bmatrix}$ *(9.3)*   **33.** $P = \begin{matrix} A \\ B \end{matrix}\begin{bmatrix} A & B \\ 0 & 1 \\ 1 & 0 \end{bmatrix}$ *(9.2, 9.3)*

34. No such chain exists. *(9.2)*   **35.** No such chain exists. *(9.3)*   **36.** No limiting matrix *(9.2, 9.3)*

37. $P = \begin{matrix} A \\ B \\ C \\ D \end{matrix}\begin{bmatrix} A & B & C & D \\ .392 & .163 & .134 & .311 \\ .392 & .163 & .134 & .311 \\ .392 & .163 & .134 & .311 \\ .392 & .163 & .134 & .311 \end{bmatrix}$ *(9.2)*

38. (A)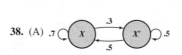

(B) $\begin{matrix} X \\ X' \end{matrix}\begin{bmatrix} X & X' \\ .7 & .3 \\ .5 & .5 \end{bmatrix}$ (C) $\begin{bmatrix} X & X' \\ .2 & .8 \end{bmatrix}$

(D) $\begin{bmatrix} X & X' \\ .54 & .46 \end{bmatrix}$; 54% of the consumers will purchase brand X on the next purchase.

(E) $\begin{bmatrix} X & X' \\ .625 & .375 \end{bmatrix}$ (F) 62.5% *(9.2)*

39. (A) Brand A: 24%; brand B: 32%; brand C: 44%   (B) 4 yr *(9.3)*

40. (A) $S_1 = \begin{bmatrix} .48 & .52 \end{bmatrix}$; $S_2 = \begin{bmatrix} .66 & .34 \end{bmatrix}$; $S_3 = \begin{bmatrix} .76 & .24 \end{bmatrix}$

(B)

Year	Data (%)	Model (%)
1995	14	14
2000	49	48
2005	68	66
2010	79	76

(C) 89% *(9.2)*

41. (A) 63.75%
(B) 15%
(C) 8.75 yr *(9.3)*

42. $\overline{P} = \begin{matrix} Red \\ Pink \\ White \end{matrix}\begin{bmatrix} Red & Pink & White \\ 1 & 0 & 0 \\ 1 & 0 & 0 \\ 1 & 0 & 0 \end{bmatrix}$ *(9.3)*

43. (A) $S_1 = [.244 \quad .756]$; $S_2 = [.203 \quad .797]$; $S_3 = [.174 \quad .826]$

(B)

Year	Data (%)	Model (%)
1985	30.1	30.1
1995	24.7	24.4
2005	20.9	20.3
2010	19.3	17.4

(C) 10.3% *(9.2)*

Chapter 10

Exercises 10.1   **1.** $(x-9)(x+9)$ **3.** $(x-7)(x+3)$ **5.** $x(x-3)(x-4)$ **7.** $(2x-1)(3x+1)$ **9.** 2   **11.** 1.25   **13.** (A) 2   (B) 2
(C) 2   (D) 2   **15.** (A) 1   (B) 2   (C) Does not exist   (D) 2   (E) No   **17.** 2   **19.** 0.5   **21.** (A) 1   (B) 2   (C) Does not exist   (D) Does not exist   (E) No
23. (A) 1   (B) 1   (C) 1   (D) 3   (E) Yes, define $g(3) = 1$ **25.** (A) −2   (B) −2   (C) −2   (D) 1   (E) Yes, define $f(-3) = -2$ **27.** (A) 2   (B) 2   (C) 2
(D) Does not exist   (E) Yes, define $f(0) = 2$ **29.** 12   **31.** 1   **33.** −4   **35.** −1.5   **37.** 3   **39.** 15   **41.** −6   **43.** $\dfrac{7}{5}$ **45.** 3

47. 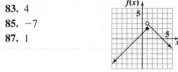   **49.** 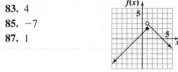   **51.** (A) 1   (B) 1   (C) 1   (D) 1   **53.** (A) 2   (B) 1   (C) Does not exist   (D) Does not exist
55. (A) −6   (B) Does not exist   (C) 6   **57.** (A) 1   (B) −1   (C) Does not exist   (D) Does not exist
59. (A) Does not exist   (B) $\dfrac{1}{2}$ (C) $\dfrac{1}{4}$ **61.** (A) −5   (B) −3   (C) 0   **63.** (A) 0   (B) −1   (C) Does not exist
65. (A) 1   (B) $\dfrac{1}{3}$ (C) $\dfrac{3}{4}$ **67.** False   **69.** True   **71.** False   **73.** Yes; 0   **75.** No; 5/16   **77.** No; does not exist

79. Yes; 1/16   **89.** (A) $\lim_{x\to 1^-} f(x) = 2$ (B) $\lim_{x\to 1^-} f(x) = 3$ (C) $m = 1.5$ (D) The graph in (A) is broken when it jumps from $(1, 2)$ up to
81. 3 $\lim_{x\to 1^+} f(x) = 3$ $\lim_{x\to 1^+} f(x) = 2$ $(1, 3)$. The graph in (B) is also broken when it jumps down
83. 4   from $(1, 3)$ to $(1, 2)$. The graph in (C) is one continuous
85. −7   piece, with no breaks or jumps.
87. 1

91. (A) $F(x) = \begin{cases} 0.99 & \text{if } 0 < x \le 20 \\ 0.07x - 0.41 & \text{if } x \ge 20 \end{cases}$

(B)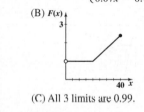

(C) All 3 limits are 0.99.

95. (A) $D(x) = \begin{cases} x & \text{if } 0 \le x < 300 \\ 0.97x & \text{if } 300 \le x < 1{,}000 \\ 0.95x & \text{if } 1{,}000 \le x < 3{,}000 \\ 0.93x & \text{if } 3{,}000 \le x < 5{,}000 \\ 0.9x & \text{if } x \ge 5{,}000 \end{cases}$

(B) $\lim_{x\to 1{,}000} D(x)$ does not exist because
$\lim_{x\to 1{,}000^-} D(x) = 970$ and $\lim_{x\to 1{,}000^+} D(x) = 950$;
$\lim_{x\to 3{,}000} D(x)$ does not exist because
$\lim_{x\to 3{,}000^-} D(x) = 2{,}850$ and $\lim_{x\to 3{,}000^+} D(x) = 2{,}790$

97. $F(x) = \begin{cases} 20x & \text{if } 0 < x \le 4{,}000 \\ 80{,}000 & \text{if } x \ge 4{,}000 \end{cases}$
$\lim_{x\to 4{,}000} F(x) = 80{,}000;$ $\lim_{x\to 8{,}000} F(x) = 80{,}000$

99. $\lim_{x\to 5} f(x)$ does not exit; $\lim_{x\to 10} f(x) = 0;$
$\lim_{x\to 5} g(x) = 0;$ $\lim_{x\to 10} g(x) = 1$

Exercises 10.2   **1.** $y = 4$ **3.** $x = -6$ **5.** $2x - y = -13$ **7.** $7x + 9y = 63$ **9.** −2   **11.** −∞   **13.** Does not exist   **15.** 0   **17.** (A) −∞
(B) ∞   (C) Does not exist   **19.** (A) ∞   (B) ∞   (C) ∞   **21.** (A) 3   (B) 3   (C) 3   **23.** (A) −∞   (B) ∞   (C) Does not exist   **25.** (A) $-5x^3$
(B) −∞   (C) ∞   **27.** (A) $-6x^4$ (B) −∞   (C) −∞   **29.** (A) x^2 (B) ∞   (C) ∞   **31.** (A) $2x^5$ (B) ∞   (C) −∞
33. $\lim_{x\to -3^-} f(x) = -\infty$; $\lim_{x\to -3^+} f(x) = \infty$; $x = -3$ is a vertical asymptote
35. $\lim_{x\to -2^-} h(x) = \infty$; $\lim_{x\to -2^+} h(x) = -\infty$; $\lim_{x\to 2^-} h(x) = -\infty$; $\lim_{x\to 2^+} h(x) = \infty$; $x = -2$ and $x = 2$ are a vertical asymptote
37. No zeros of denominator; no vertical asymptotes   **39.** $\lim_{x\to 1^-} H(x) = -\infty$; $\lim_{x\to 1^+} H(x) = \infty$; $\lim_{x\to 3} H(x) = 2$; $x = 1$ is a vertical asymptote
41. $\lim_{x\to 0} T(x) = -\infty$; $\lim_{x\to 4} T(x) = \infty$; $x = 0$ and $x = 4$ are vertical asymptotes   **43.** (A) $\dfrac{47}{41} \approx 1.146$ (B) $\dfrac{407}{491} \approx 0.829$ (C) $\dfrac{4}{5} = 0.8$
45. (A) $\dfrac{2{,}011}{138} \approx 14.572$ (B) $\dfrac{12{,}511}{348} \approx 35.951$ (C) ∞   **47.** (A) $-\dfrac{8{,}568}{46{,}653} \approx -0.184$ (B) $-\dfrac{143{,}136}{1{,}492{,}989} \approx -0.096$ (C) 0   **49.** (A) $-\dfrac{7{,}010}{996} \approx -7.038$
(B) $-\dfrac{56{,}010}{7{,}996} \approx -7.005$ (C) −7   **51.** Horizontal asymptote: $y = 2$; vertical asymptote: $x = -2$ **53.** Horizontal asymptote: $y = 1$; vertical asymptotes:
$x = -1$ and $x = 1$ **55.** No horizontal asymptotes; no vertical asymptotes   **57.** Horizontal asymptote: $y = 0$; no vertical asymptotes   **59.** No horizontal
asymptotes; vertical asymptote: $x = 3$ **61.** Horizontal asymptote: $y = 2$; vertical asymptotes: $x = -1$ and $x = 2$ **63.** Horizontal asymptote: $y = 2$; vertical
asymptote: $x = -1$ **65.** $\lim_{x\to\infty} f(x) = 0$ **67.** $\lim_{x\to\infty} f(x) = \infty$ **69.** $\lim_{x\to\infty} f(x) = -\dfrac{1}{4}$ **71.** $\lim_{x\to\infty} f(x) = -\infty$ **73.** $\lim_{x\to\infty} f(x) = \infty$, $\lim_{x\to-\infty} f(x) = -\infty$
75. $\lim_{x\to\infty} f(x) = -5$, $\lim_{x\to-\infty} f(x) = -5$ **77.** False   **79.** False   **81.** True   **83.** If $n \ge 1$ and $a_n > 0$, then the limit is ∞. If $n \ge 1$ and $a_n < 0$, then the limit i

85. (A) $C(x) = 180x + 200$ (C) $\overline{C}(x)$ (D) \$180 per board   **87.** (A) $C_e(x) = 950 + 56x$; $\overline{C}_e(x) = \dfrac{950}{x} + 56$ (B) $C_e(x) = 900 + 66x$;

(B) $\overline{C}(x) = \dfrac{180x + 200}{x}$

$\overline{C}_e(x) = \dfrac{900}{x} + 66$ (C) At $x = 5$ years   (D) At $x = 5$ years

(E) $\lim\limits_{x\to\infty} \overline{C}_e(x) = 56$; $\lim\limits_{x\to\infty} \overline{C}_e(x) = 66$ **89.** The long-term drug concentration

is 5 mg/ml.   **91.** (A) \$18 million   (B) \$38 million   (C) $\lim\limits_{x\to 1^-} P(x) = \infty$

93. (A) $V_{\max} = 4$, $K_M = 20$ (B) $v(s) = \dfrac{4s}{20+s}$

(C) $v = \dfrac{12}{7}$ when $s = 15$; $s = 60$ when $v = 3$

95. (A) $C_{\max} = 18$, $M = 150$ (B) $C(T) = \dfrac{18T}{150 + T}$

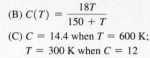

(C) $C = 14.4$ when $T = 600$ K; $T = 300$ K when $C = 12$

Exercises 10.3   1. $[-3, 5]$ **3.** $(-10, 100)$ **5.** $(-\infty, -5) \cup (5, \infty)$ **7.** $(-\infty, -1] \cup (2, \infty)$

9. f is continuous at $x = 1$, since $\lim\limits_{x\to 1} f(x) = f(1)$.

11. f is discontinuous at $x = 1$, since $\lim\limits_{x\to 1} f(x) \neq f(1)$.

13. f is discontinuous at $x = 1$, since $\lim\limits_{x\to 1} f(x)$ does not exist.

15. 1.9   **17.** 0.9   **19.** (A) 2   (B) 1   (C) Does not exist   (D) 1   (E) No

21. (A) 1   (B) 1   (C) 1   (D) 3   (E) No

23. 0.9   **25.** 2.05   **27.** (A) 1   (B) 1   (C) 1   (D) 3   (E) No   **29.** (A) 2   (B) -1 (C) Does not exist   (D) 2   (E) No   **31.** All x **33.** All x, except $x = -2$ **35.** All x, except $x = -4$ and $x = 1$ **37.** All x

39. All x, except $x = \pm\dfrac{3}{2}$ **41.** $-\dfrac{8}{3}, 4$ **43.** $-1, 1$ **45.** $-9, -6, 0, 5$ **47.** $-3 < x < 4$; $(-3, 4)$ **49.** $x < 3$ or $x > 7$; $(-\infty, 3) \cup (7, \infty)$

51. $x < -2$ or $0 < x < 2$; $(-\infty, -2) \cup (0, 2)$ **53.** $-5 < x < 0$ or $x > 3$; $(-5, 0) \cup (3, \infty)$ **55.** (A) $(-4, -2) \cup (0, 2) \cup (4, \infty)$

(B) $(-\infty, -4) \cup (-2, 0) \cup (2, 4)$ **57.** (A) $(-\infty, -2.5308) \cup (-0.7198, \infty)$ (B) $(-2.5308, -0.7198)$

59. (A) $(-\infty, -2.1451) \cup (-1, -0.5240) \cup (1, 2.6691)$ (B) $(-2.1451, -1) \cup (-0.5240, 1) \cup (2.6691, \infty)$ **61.** $[6, \infty)$ **63.** $(-\infty, \infty)$

65. $(-\infty, -3] \cup [3, \infty)$ **67.** $(-\infty, \infty)$

69. Since $\lim\limits_{x\to 1^-} f(x) = 2$ and $\lim\limits_{x\to 1^+} f(x) = 4$, $\lim\limits_{x\to 1} f(x)$ does not exist and f is not continuous at $x = 1$.

71. This function is continuous for all x.

73. Since $\lim\limits_{x\to 0} f(x) = 0$ and $f(0) = 1$, $\lim\limits_{x\to 0} f(x) \neq f(0)$ and f is not continuous at $x = 0$.

75. (A) Yes   (B) No   (C) Yes   (D) No   (E) Yes   **77.** True

79. False   **81.** True

83. x int.: $-5, 2$

85. x int.: $x = -6, -1, 4$

87. No, but this does not contradict Theorem 2, since f is discontinuous at $x = 1$.

89. (A) $P(x) = \begin{cases} 0.44 & \text{if } 0 < x \leq 1 \\ 0.61 & \text{if } 1 < x \leq 2 \\ 0.78 & \text{if } 2 < x \leq 3 \\ 0.95 & \text{if } 3 < x \leq 3.5 \end{cases}$ (B) $P(x)$ (C) Yes; no

93. (A) $S(x) = \begin{cases} 5 + 0.63x & \text{if } 0 \leq x \leq 50 \\ 14 + 0.45x & \text{if } 50 < x \end{cases}$ (B) $S(x)$ (C) Yes

95. (A) $E(s)$ (B) $\lim\limits_{x\to 10,000} E(s) = \$1,000$; $E(10,000) = \$1,000$ (C) $\lim\limits_{x\to 20,000} E(s)$ does not exist; $E(20,000) = \$2,000$ (D) Yes; no

97. (A) t_2, t_3, t_4, t_6, t_7 (B) $\lim\limits_{t\to t_5} N(t) = 7$; $N(t_5) = 7$ (C) $\lim\limits_{t\to t_3} N(t)$ does not exist; $N(t_3) = 4$

  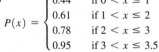 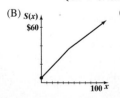

$= 2.25$ **3.** $-\dfrac{27}{5} = -5.4$ **5.** $\dfrac{1}{3}\sqrt{3}$ **7.** $\dfrac{15}{2} - \dfrac{5}{2}\sqrt{7}$ **9.** (A) -3; slope of the secant line through $(1, f(1))$ and $(2, f(2))$

ant line through $(1, f(1))$ and $(1 + h, f(1 + h))$ (C) -2; slope of the tangent line at $(1, f(1))$ **11.** (A) 15   (B) $6 + 3h$ (C) 6

(C) $y - 80 = 45(x - 2)$ or $y = 45x - 10$ **15.** $y - \dfrac{1}{2} = -\dfrac{1}{2}(x - 1)$ or $y = -\dfrac{x}{2} + 1$ **17.** $y - 16 = -32(x + 2)$ or

$(x) = 0$; $f'(1) = 0$, $f'(2) = 0$, $f'(3) = 0$ **21.** $f'(x) = 3$; $f'(1) = 3$, $f'(2) = 3$, $f'(3) = 3$ **23.** $f'(x) = -6x$; $f'(1) = -6$,

-18 **25.** $f'(x) = 2x + 6$; $f'(1) = 8$, $f'(2) = 10$, $f'(3) = 12$ **27.** $f'(x) = 4x - 7$; $f'(1) = -3$, $f'(2) = 1$, $f'(3) = 5$

29. $f'(x) = -2x + 4; f'(1) = 2, f'(2) = 0, f'(3) = -2$ **31.** $f'(x) = 6x^2; f'(1) = 6, f'(2) = 24, f'(3) = 54$ **33.** $f'(x) = -\dfrac{4}{x^2}; f'(1) = -4,$

$f'(2) = -1, f'(3) = -\dfrac{4}{9}$ **35.** $f'(x) = \dfrac{3}{2\sqrt{x}}; f'(1) = \dfrac{3}{2}, f'(2) = \dfrac{3}{2\sqrt{2}}$ or $\dfrac{3\sqrt{2}}{4}, f'(3) = \dfrac{3}{2\sqrt{3}}$ or $\dfrac{\sqrt{3}}{2}$ **37.** $f'(x) = \dfrac{5}{\sqrt{x+5}}; f'(1) = \dfrac{5}{\sqrt{6}}$ or $\dfrac{5\sqrt{6}}{6},$

$f'(2) = \dfrac{5}{\sqrt{7}}$ or $\dfrac{5\sqrt{7}}{7}, f'(3) = \dfrac{5}{2\sqrt{2}}$ or $\dfrac{5\sqrt{2}}{4}$ **39.** $f'(x) = -\dfrac{1}{(x-4)^2}; f'(1) = -\dfrac{1}{9}; f'(2) = -\dfrac{1}{4}; f'(3) = -1$ **41.** $f'(x) = \dfrac{1}{(x+1)^2}; f'(1) = \dfrac{1}{4};$

$f'(2) = \dfrac{1}{9}; f'(3) = \dfrac{1}{16}$ **43.** (A) 5   (B) $3 + h$ (C) 3   (D) $y = 3x - 1$ **45.** (A) 5 m/s   (B) $3 + h$ m/s   (C) 3 m/s   **47.** Yes   **49.** No

51. Yes   **55.** (A) $f'(x) = 2x - 4$ **57.** $v = f'(x) = 8x - 2$; 6 ft/s, 22 ft/s, 38 ft/s   **67.** f is nondifferentiable at $x = 1$ **69.** f is differentiable for all
53. Yes   (B) $-4, 0, 4$ real numbers
        (C)   **59.** (A) The graphs of g and h are vertical
        translations of the graph of f. All three functions
        should have the same derivative.   (B) $2x$
        61. True   **63.** False   **65.** False

71. No   **73.** No   **75.** $f'(0) = 0$ **77.** 6 s; 192 ft/s   **79.** (A) $8.75   (B) $R'(x) = 60 - 0.05x$ (C) $R(1,000) = 35,000; R'(1,000) = 10$; At a production
level of 1,000 car seats, the revenue is $35,000 and is increasing at the rate of $10 per seat.   **81.** (A) $S'(t) = 1/\sqrt{t + 10}$ (B) $S(15) = 10; S'(15) = 0.2;$
After 15 months, the total sales are $10 million and are increasing at the rate of $0.2 million, or $200,000, per month.   (C) The estimated total sales are $10.2
million after 16 months and $10.4 million after 17 months.   **83.** (A) $p'(t) = 276t + 1,072$ (B) $p(10) = 39,437, p'(10) = 3,832$; In 2020, 39,437 metric
tons of tungsten are consumed and this quantity is increasing at the rate of 3,832 metric tons per year.
85. (A)
```
QuadReg
y=ax²+bx+c
a=-.1339285714
b=25.225
c=1199.785714
```
(B) $R(20) = 1650.7$ billion kilowatts, $R'(20) = 19.9$ billion kilowatts per year. In 2020, 1650.7 billion kilowatts will be
sold and the amount sold is increasing at the rate of 19.9 billion kilowatts per year.   **87.** (A) $P'(t) = 12 - 2t$
(B) $P(3) = 107; P'(3) = 6$. After 3 hours, the ozone level is 107 ppb and is increasing at the rate of 6 ppb per hour.

Exercises 10.5   **1.** $x^{1/2}$ **3.** x^{-5} **5.** x^{12} **7.** $x^{-1/4}$ **9.** 0   **11.** $9x^8$ **13.** $3x^2$ **15.** $-4x^{-5}$ **17.** $\dfrac{8}{3}x^{5/3}$ **19.** $-\dfrac{10}{x^{11}}$ **21.** $10x$ **23.** $2.8x^6$ **25.** $\dfrac{x^2}{6}$

27. 12   **29.** 2   **31.** 9   **33.** 2   **35.** $4t - 3$ **37.** $-10x^{-3} - 9x^{-2}$ **39.** $1.5u^{-0.7} - 8.8u^{1.2}$ **41.** $0.5 - 3.3t^2$ **43.** $-\dfrac{8}{5}x^{-5}$ **45.** $3x + \dfrac{14}{5}x^{-3}$

47. $-\dfrac{20}{9}w^{-5} + \dfrac{5}{3}w^{-2/3}$ **49.** $2u^{-1/3} - \dfrac{5}{3}u^{-2/3}$ **51.** $-\dfrac{9}{5}t^{-8/5} + 3t^{-3/2}$ **53.** $-\dfrac{1}{3}x^{-4/3}$ **55.** $-0.6x^{-3/2} + 6.4x^{-3} + 1$ **57.** (A) $f'(x) = 6 - 2x$
(B) $f'(2) = 2; f'(4) = -2$ (C) $y = 2x + 4; y = -2x + 16$ (D) $x = 3$ **59.** (A) $f'(x) = 12x^3 - 12x$ (B) $f'(2) = 72; f'(4) = 720$
(C) $y = 72x - 127; y = 720x - 2,215$ (D) $x = -1, 0, 1$ **61.** (A) $v = f'(x) = 176 - 32x$ (B) $f'(0) = 176$ ft/s; $f'(3) = 80$ ft/s   (C) 5.5 s
63. (A) $v = f'(x) = 3x^2 - 18x + 15$ (B) $f'(0) = 15$ ft/s; $f'(3) = -12$ ft/s   (C) $x = 1$ s, $x = 5$ s

65. $f'(x) = 2x - 3 - 2x^{-1/2} = 2x - 3 - \dfrac{2}{x^{1/2}}; x = 2.1777$ **67.** $f'(x) = 4\sqrt[3]{x} - 3x - 3; x = -2.9018$ **69.** $f'(x) = 0.2x^3 + 0.3x^2 - 3x - 1.6;$

$x = -4.4607, -0.5159, 3.4765$ **71.** $f'(x) = 0.8x^3 - 9.36x^2 + 32.5x - 28.25; x = 1.3050$ **77.** $8x - 4$ **79.** $-20x^{-2}$ **81.** $-\dfrac{1}{4}x^{-2} + \dfrac{2}{3}x^{-3}$

83. False   **85.** True   **89.** (A) $S'(t) = 0.09t^2 + t + 2$ (B) $S(5) = 29.25, S'(5) = 9.25$. After 5 months, sales are $29.25 million and are increasing at the
rate of $9.25 million per month.   (C) $S(10) = 103, S'(10) = 21$. After 10 months, sales are $103 million and are increasing at the rate of $21 million
per month.   **91.** (A) $N'(x) = 3,780/x^2$ (B) $N'(10) = 37.8$. At the $10,000 level of advertising, sales are increasing at the rate of 37.8 boats per $1,000 spent
on advertising. $N'(20) = 9.45$. At the $20,000 level of advertising, sales are increasing at the rate of 9.45 boats per $1,000 spent on advertising.
93. (A)
```
CubicReg
y=ax³+bx²+cx+d
a=-8.083333E-4
b=.0624285714
c=-1.081309524
d=40.57571429
```
(B) In 2020, 41.5% of male high-school graduates enroll in   **95.** (A) -1.37 beats/min   (B) -0.58 beats/min
college and the percentage is decreasing at the rate of   **97.** (A) 25 items/hr   (B) 8.33 items/hr
0.9% per year.

Exercises 10.6   **1.** 3; 3.01   **3.** 2.8; 2.799   **5.** 0; 0.01   **7.** 100; 102.01   **9.** $\Delta x = 3; \Delta y = 45; \Delta y/\Delta x = 15$ **11.** 12   **13.** 12   **15.** $dy = (24x - 3x^2)dx$

17. $dy = \left(2x - \dfrac{x^2}{3}\right)dx$ **19.** $dy = -\dfrac{295}{x^{3/2}}dx$ **21.** (A) $12 + 3\Delta x$ (B) 12   **23.** $dy = (8x + 4)dx$ **25.** $dy = (1 - 9x^{-2})dx$ **27.** $dy = 1.4; \Delta y = 1.44$

29. $dy = -3; \Delta y = -\dfrac{10}{3}$ **31.** 120 in.^{3} **33.** (A) $\Delta y = \Delta x + (\Delta x)^2; dy = \Delta x$ (B) 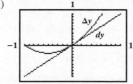   (C)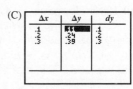

35. (A) $\Delta y = -\Delta x + (\Delta x)^2 + (\Delta x)^3$; $dy = -\Delta x$

(B)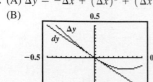

(C)

Δx	Δy	dy
.05	-.0474	-.05
.1	-.089	-.1
.15	-.1241	-.15

37. True   **39.** False   **41.** $dy = \left(\dfrac{2}{3}x^{-1/3} - \dfrac{10}{3}x^{2/3}\right)dx$ **43.** $dy = 3.9$; $\Delta y = 3.83$

45. 40-unit increase; 20-unit increase   **47.** $-\$2.50$; $\$1.25$ **49.** -1.37 beats/min; -0.58 beats/min   **51.** 1.26 mm^2 **53.** 3 wpm   **55.** (A) 2,100 increase   (B) 4,800 increase   (C) 2,100 increase

Exercises 10.7   **1.** $\$22,889.80$ **3.** $\$110.20$ **5.** $\$32,000.00$ **7.** $\$230.00$ **9.** $C'(x) = 0.8$ **11.** $C'(x) = 4.6 - 0.02x$ **13.** $R'(x) = 4 - 0.02x$

15. $R'(x) = 12 - 0.08x$ **17.** $P'(x) = 3.2 - 0.02x$ **19.** $P'(x) = 7.4 - 0.06x$ **21.** $\overline{C}(x) = 1.1 + \dfrac{145}{x}$ **23.** $\overline{C}'(x) = -\dfrac{145}{x^2}$

25. $P(x) = 3.9x - 0.02x^2 - 145$ **27.** $\overline{P}(x) = 3.9 - 0.02x - \dfrac{145}{x}$ **29.** True   **31.** False   **33.** (A) $\$29.50$ (B) $\$30$ **35.** (A) $\$420$ (B) $\overline{C}'(500) = -0.24$. At a production level of 500 frames, average cost is decreasing at the rate of 24¢ per frame.   (C) Approximately $\$419.76$ **37.** (A) $\$14.70$ (B) $\$15$

39. (A) $P'(450) = 0.5$. At a production level of 450 DVDs, profit is increasing at the rate of 50¢ per DVD.   (B) $P'(750) = -2.5$. At a production level of 750 DVDs, profit is decreasing at the rate of $\$2.50$ per DVD.   **41.** (A) $\$13.50$ (B) $\overline{P}'(50) = \$0.27$. At a production level of 50 mowers, the average profit per mower is increasing at the rate of $\$0.27$ per mower.   (C) Approximately $\$13.77$ **43.** (A) $p = 100 - 0.025x$, domain: $0 \le x \le 4{,}000$

(B) $R(x) = 100x - 0.025x^2$, domain: $0 \le x \le 4{,}000$ (C) $R'(1{,}600) = 20$. At a production level of 1,600 pairs of running shoes, revenue is increasing at the rate of $\$20$ per pair.   (D) $R'(2{,}500) = -25$. At a production level of 2,500 pairs of running shoes, revenue is decreasing at the rate of $\$25$ per pair.

45. (A) $p = 200 - \dfrac{1}{30}x$, domain: $0 \le x \le 6{,}000$ (B) $C'(x) = 60$ (C) $R(x) = 200x - (x^2/30)$, domain: $0 \le x \le 6{,}000$

(D) $R'(x) = 200 - (x/15)$ (E) $R'(1{,}500) = 100$. At a production level of 1,500 saws, revenue is increasing at the rate of $\$100$ per saw. $R'(4{,}500) = -100$. At a production level of 4,500 saws, revenue is decreasing at the rate of $\$100$ per saw.

(F) Break-even points: $(600, 108{,}000)$ and $(3{,}600, 288{,}000)$ (G) $P(x) = -(x^2/30) + 140x - 72{,}000$

(H) $P'(x) = -(x/15) + 140$

(I) $P'(1{,}500) = 40$. At a production level of 1,500 saws, profit is increasing at the rate of $\$40$ per saw. $P'(3{,}000) = -60$. At a production level of 3,000 saws, profit is decreasing at the rate of $\$60$ per saw.

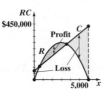

47. (A) $p = 20 - 0.02x$, domain: $0 \le x \le 1{,}000$

(B) $R(x) = 20x - 0.02x^2$, domain: $0 \le x \le 1{,}000$ (C) $C(x) = 4x + 1{,}400$ (D) Break-even points: $(100, 1{,}800)$ and $(700, 4{,}200)$

(E) $P(x) = 16x - 0.02x^2 - 1{,}400$

(F) $P'(250) = 6$. At a production level of 250 toasters, profit is increasing at the rate of $\$6$ per toaster. $P'(475) = -3$. At a production level of 475 toasters, profit is decreasing at the rate of $\$3$ per toaster.

49. (A) $x = 500$

(B) $P(x) = 176x - 0.2x^2 - 21{,}900$

(C) $x = 440$

(D) Break-even points: $(150, 25{,}500)$ and $(730, 39{,}420)$; x intercepts for $P(x)$: $x = 150$ and $x = 730$

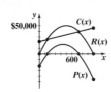

51. (A) $R(x) = 20x - x^{3/2}$

(B) Break-even points: $(44, 588)$, $(258, 1{,}016)$

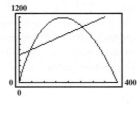

53. (A)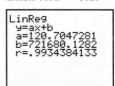

(B) Fixed costs $\approx \$721{,}680$ Variable costs $\approx \$121$

(C) $(713, 807{,}703)$, $(5{,}423, 1{,}376{,}227)$ (D) $\$254 \le p \le \$1{,}133$

Chapter 10 Review Exercises   **1.** (A) 16   (B) 8   (C) 8   (D) 4   (E) 4   (F) 4 *(10.2)*   **2.** $f'(x) = -3$ *(10.2)*   **3.** (A) 22   (B) 8   (C) 2   (D) -5 *(10.1)*

4. 1.5 *(10.1)*   **5.** 3.5 *(10.1)*   **6.** 3.75 *(10.1)*   **7.** 3.75 *(10.1)*   **8.** (A) 1   (B) 1   (C) 1   (D) 1 *(10.1)*   **9.** (A) 2   (B) 3   (C) Does not exist   (D) 3 *(10.1)*   **10.** (A) 4

(B) 4   (C) 4   (D) Does not exist *(10.1)*   **11.** (A) Does not exist   (B) 3   (C) No *(10.3)*   **12.** (A) 2   (B) Not defined   (C) No *(10.3)*   **13.** (A) 1   (B) 1

(C) Yes *(10.3)*   **14.** 5 *(10.2)*   **15.** 5 *(10.2)*   **16.** ∞ *(10.2)*   **17.** $-\infty$ *(10.2)*   **18.** 0 *(10.1)*   **19.** 0 *(10.1)*   **20.** 0 *(10.1)*   **21.** Vertical asymptote: $x = 2$ *(10.3)*

22. Horizontal asymptote: $y = 5$ *(10.2)*   **23.** $x = 2$ *(10.3)*   **24.** $f'(x) = 10x$ *(10.4)*   **25.** (A) -3 (B) 6   (C) -2 (D) 3   (E) -11 **26.** $x^2 - 10x$ *(10.5)*

27. $x^{-1/2} - 3 = \dfrac{1}{x^{1/2}} - 3$ *(10.5)*   **28.** 0 *(10.5)*   **29.** $-\dfrac{3}{2}x^{-2} + \dfrac{15}{4}x^2 = \dfrac{-3}{2x^2} + \dfrac{15x^2}{4}$ *(10.5)*   **30.** $-2x^{-5} + x^3 = \dfrac{-2}{x^5} + x^3$ *(10.5)*

31. $f'(x) = 12x^3 + 9x^2 - 2$ *(10.5)*   **32.** $\Delta x = 2$, $\Delta y = 10$, $\Delta y/\Delta x = 5$ *(10.6)*   **33.** 5 *(10.6)*   **34.** 6 *(10.6)*   **35.** $\Delta y = 0.64$; $dy = 0.6$ *(10.6)*

36. (A) 4   (B) 6   (C) Does not exist   (D) 6   (E) No *(10.3)*   **37.** (A) 3   (B) 3   (C) 3   (D) 3   (E) Yes *(10.3)*   **38.** (A) $(8, \infty)$ (B) $[0, 8]$ *(10.3)*   **39.** $(-3, 4)$ *(10.3)*

40. $(-3, 0) \cup (5, \infty)$ *(10.3)*   **41.** $(-2.3429, -0.4707) \cup (1.8136, \infty)$ *(10.3)*   **42.** (A) 3   (B) $2 + 0.5h$ (C) 2 *(10.4)*   **43.** $-x^{-4} + 10x^{-3}$ *(10.4)*

44. $\dfrac{3}{4}x^{-1/2} - \dfrac{5}{6}x^{-3/2} = \dfrac{3}{4\sqrt{x}} - \dfrac{5}{6\sqrt{x^3}}$ *(10.5)*   **45.** $0.6x^{-2/3} - 0.3x^{-4/3} = \dfrac{0.6}{x^{2/3}} - \dfrac{0.3}{x^{4/3}}$ *(10.4)*   **46.** $-\dfrac{3}{5}(-3)x^{-4} = \dfrac{9}{5x^4}$ *(10.5)*   **47.** (A) $m = f'(1) = 2$

(B) $y = 2x + 3$ *(10.4, 10.5)*   **48.** $x = 5$ *(10.4)*   **49.** $x = -5, x = 3$ *(10.5)*   **50.** $x = -1.3401, 0.5771, 2.2630$ *(10.4)*   **51.** ± 2.4824 *(10.5)*

52. (A) $v = f'(x) = 16x - 4$ (B) 44 ft/sec *(10.5)*   **53.** (A) $v = f'(x) = -10x + 16$ (B) $x = 1.6$ sec *(10.5)*

54. (A) The graph of g is the graph of f shifted 4 units to the right, and the graph of h is the graph of f shifted 3 units to the left:

(B) The graph of g' is the graph of f' shifted 4 units to the right, and the graph of h' is the graph of f' shifted 3 units to the left:

55. $(-\infty, \infty)$ *(10.3)*

56. $(-\infty, 2) \cup (2, \infty)$ *(10.3)*

57. $(-\infty, -4) \cup (-4, 1) \cup (1, \infty)$ *(10.3)*

58. $(-\infty, \infty)$ *(10.3)*   **59.** $[-2, 2]$ *(10.3)*

60. (A) -1 (B) Does not exist   (C) $-\dfrac{2}{3}$ *(10.1)*

61. (A) $\dfrac{1}{2}$ (B) 0   (C) Does not exist *(10.1)*   **62.** (A) -1 (B) 1   (C) Does not exist *(10.1)*   **63.** (A) $-\dfrac{1}{6}$ (B) Does not exist   (C) $-\dfrac{1}{3}$ *(10.1)*   **64.** (A) 0   (B) -1

(C) Does not exist *(10.1)*   **65.** (A) $\dfrac{2}{3}$ (B) $\dfrac{2}{3}$ (C) Does not exist *(10.2)*   **66.** (A) ∞ (B) $-\infty$ (C) ∞ *(10.3)*   **67.** (A) 0   (B) 0   (C) Does not exist *(10.2)*

68. 4 *(10.1)*   **69.** $\dfrac{-1}{(x+2)^2}$ *(10.1)*   **70.** $2x - 1$ *(10.4)*   **71.** $1/(2\sqrt{x})$ *(10.4)*   **72.** Yes *(10.4)*   **73.** No *(10.4)*   **74.** No *(10.4)*   **75.** No *(10.4)*   **76.** Yes *(10.4)*

77. Yes *(10.4)*   **78.** Horizontal asymptote: $y = 5$; vertical asymptote: $x = 7$ *(10.2)*   **79.** Horizontal asymptote: $y = 0$; vertical asymptote: $x = 4$ *(10.2)*

80. No horizontal asymptote; vertical asymptote: $x = 3$ *(10.2)*   **81.** Horizontal asymptote: $y = 1$; vertical asymptotes: $x = -2, x = 1$ *(10.2)*   **82.** Horizontal

asymptote: $y = 1$; vertical asymptotes: $x = -1, x = 1$ *(10.2)*   **83.** The domain of $f'(x)$ is all real numbers except $x = 0$. At $x = 0$, the graph of $y = f(x)$ is

smooth, but it has a vertical tangent. *(10.4)*

84. (A) $\lim_{x \to 1^-} f(x) = 1$;   (B) $\lim_{x \to 1^-} f(x) = -1$;   (C) $m = 1$ (D) The graphs in (A) and (B) have   **85.** (A) 1   (B) -1 (C) Does not exist

$\lim_{x \to 1^+} f(x) = -1$ $\lim_{x \to 1^+} f(x) = 1$ discontinuities at $x = 1$; the graph in   (D) No *(10.4)*

(C) does not. *(10.2)*

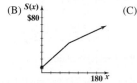

86. (A) $S(x) = \begin{cases} 7.47 & + 0.4x & \text{if } 0 \le x \le 90 \\ 24.786 + 0.2076x & \text{if } 90 < x \end{cases}$

(B)   (C) Yes *(10.2)*

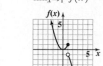

87. (A) \$179.90   (B) \$180 *(10.7)*   **88.** (A) $C(100) = 9,500$; $C'(100) = 50$. At a production level of 100 bicycles, the total cost is \$9,500, and cost is increasing at the rate of \$50 per bicycle.

(B) $\overline{C}(100) = 95$; $\overline{C}'(100) = -0.45$. At a production level of 100 bicycles, the average cost is \$95, and average cost is decreasing at a rate of \$0.45 per bicycle. *(10.7)*   **89.** The approximate cost of

producing the 201st printer is greater than that of the 601st printer. Since these marginal costs are

decreasing, the manufacturing process is becoming more efficient. *(10.7)*

90. (A) $C'(x) = 2$; $\overline{C}(x) = 2 + \dfrac{9,000}{x}$; $\overline{C}'(x) = \dfrac{-9,000}{x^2}$ (B) $R(x) = xp = 25x - 0.01x^2$; $R'(x) = 25 - 0.02x$; $\overline{R}(x) = 25 - 0.01x$; $\overline{R}'(x) = -0.01$

(C) $P(x) = R(x) - C(x) = 23x - 0.01x^2 - 9,000$; $P'(x) = 23 - 0.02x$; $\overline{P}(x) = 23 - 0.01x - \dfrac{9,000}{x}$; $\overline{P}'(x) = -0.01 + \dfrac{9,000}{x^2}$

(D) (500, 10,000) and (1,800, 12,600)   (E) $P'(1,000) = 3$. Profit is increasing at   (F)   **91.** (A) 8   (B) 20 *(10.5)*

the rate of \$3 per umbrella.   **92.** $N(9) = 27$; $N'(t) = 3.5$; After

$P'(1,150) = 0$. Profit is flat.   9 months, 27,000 pools have been sold

$P'(1,400) = -5$. Profit is decreasing   and the total sales are increasing at the

at the rate of \$5 per umbrella.   rate of 3,500 pools per month. *(10.5)*

93. (A)
```
CubicReg
y=ax³+bx²+cx+d
a=5.5277778E-4
b=-.044761905
c=1.084484127
d=12.5452381
```

94. (A)
```
LinReg
y=ax+b
a=-.0384180791
b=13.59887006
r=-.9897782666
```

(B) Fixed costs: \$484.21; variable   (C) (51, 591.15), (248, 1,007.62)

costs per kringle: \$2.11   (D) \$4.07 $< p <$ \$11.64 *(10.7)*

```
LinReg
y=ax+b
a=2.107344633
b=484.2090395
r=.9939318704
```

(B) $N(60) = 36.9$; $N'(60) = 1.7$.

In 2020, natural-gas consump-

tion is 36.9 trillion cubic feet and

is increasing at the rate of 1.7

trillion cubic feet per year *(10.4)*

95. $C'(10) = -1$; $C'(100) = -0.001$ *(10.5)*

96. $F(4) = 98.16$; $F'(4) = -0.32$; After 4 hours the

patient's temperature is 98.16°F and is decreasing at the rate

of 0.32°F per hour. *(10.5)*

97. (A) 10 items/h   (B) 5 items/h *(10.5)*

98. (A)

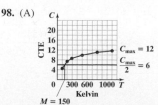

(B) $C(T) = \dfrac{12T}{150 + T}$

(C) $C = 9.6$ at $T = 600$ K,

$T = 750$ K when $C = 10$ *(10.3)*

Chapter 11

Exercises 11.1   **1.** $A = 1,465.68$ **3.** $P = 9,117.21$ **5.** $t = 5.61$ **7.** $r = 0.04$ **9.** $\$1,221.40$; $\$1,648.72$; $\$2,225.54$

11. 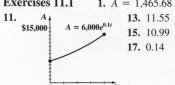   **13.** 11.55   **19.**   **15.** 10.99   **17.** 0.14

21. $\lim_{n \to \infty} (1 + n)^{1/n} = 1$

n	$[1 + (1/n)]^n$
10	2.593 74
100	2.704 81
1,000	2.716 92
10,000	2.718 15
100,000	2.718 27
1,000,000	2.718 28
10,000,000	2.718 28
↓	↓
∞	$e = 2.718\,281\,828\,459 \ldots$

23.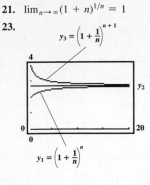

$y_3 = \left(1 + \dfrac{1}{n}\right)^{n+1}$

$y_1 = \left(1 + \dfrac{1}{n}\right)^n$

25. (A) $\$12,398.62$ (B) 27.34 yr   **31.** (A)   (B) $\lim_{t \to \infty} 10,000e^{-0.08t} = 0$ **33.** 17.33 yr   **35.** 8.66%   **37.** 7.3 yr
27. $\$11,890.41$ **29.** 8.11%

39. (A) $A = Pe^{rt}$ (B) 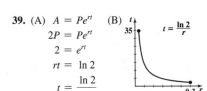   Although r could be any positive number, the restrictions on   **41.** $t = -(\ln 0.5)/0.000\,433\,2 \approx 1,600$ yr
$2P = Pe^{rt}$ r are reasonable in the sense that most investments would be   **43.** $r = (\ln 0.5)/30 \approx -0.0231$
$2 = e^{rt}$ expected to earn a return of between 2% and 30%.   **45.** 53.3 yr
$rt = \ln 2$ (C) The doubling times (in years) are 13.86, 6.93, 4.62,   **47.** 1.39%
$t = \dfrac{\ln 2}{r}$ 3.47, 2.77, and 2.31, respectively.

Exercises 11.2   **1.** $y = 7$ **3.** $x = 81$ **5.** $b = 8$ **7.** $y = \dfrac{1}{2}$ **9.** $5e^x + 3$ **11.** $-\dfrac{2}{x} + 2x$ **13.** $3x^2 - 6e^x$ **15.** $e^x + 1 - \dfrac{1}{x}$ **17.** $\dfrac{3}{x}$ **19.** $5 - \dfrac{5}{x}$

21. $\dfrac{2}{x} + 4e^x$ **23.** $f'(x) = e^x + exe^{-1}$ **25.** $f'(x) = (e + 1)x^e$ **27.** $f'(x) = \dfrac{1}{x}; y = x + 2$ **29.** $f'(x) = 3e^x; y = 3x + 3$ **31.** $f'(x) = \dfrac{3}{x}; y = \dfrac{3x}{e}$

33. $f'(x) = e^x; y = ex + 2$ **35.** Yes; yes   **37.** No; no   **39.** $f(x) = 10x + \ln 10 + \ln x$; $f'(x) = 10 + \dfrac{1}{x}$ **41.** $f(x) = \ln 4 - 3 \ln x; f'(x) = -\dfrac{3}{x}$

43. $\dfrac{1}{x \ln 2}$ **45.** $3^x \ln 3$ **47.** $2 - \dfrac{1}{x \ln 10}$ **49.** $1 + 10^x \ln 10$ **51.** $\dfrac{3}{x} + \dfrac{2}{x \ln 3}$ **53.** $2^x \ln 2$ **55.** $(-0.82, 0.44), (1.43, 4.18), (8.61, 5503.66)$

57. $(0.49, 0.49)$ **59.** $(3.65, 1.30), (332,105.11, 12.71)$ **63.** $\$28,447/\text{yr}$; $\$18,664/\text{yr}$; $\$11,021/\text{yr}$ **65.** $A'(t) = 5,000(\ln 4)4^t$; $A'(1) = 27,726$ bacteria/hr (rate of change at the end of the first hour); $A'(5) = 7,097,827$ bacteria/hr (rate of change at the end of the fifth hour)   **67.** At the 40-lb weight level, blood pressure would increase at the rate of 0.44 mm of mercury per pound of weight gain. At the 90-lb weight level, blood pressure would increase at the rate of 0.19 mm of mercury per pound of weight gain.   **69.** $dR/dS = k/S$ **71.** (A) $\$808.41$ per year   (B) $\$937.50$ per year

Exercises 11.3   **1.** One answer is: $F(x) = x^3, S(x) = 5 - 4 \ln x$ **3.** One answer is: $F(x) = x^3 + 3, S(x) = e^x + 2$ **5.** One answer is:
$T(x) = 9x^2, B(x) = e^{5x}$ **7.** One answer is: $T(x) = 3x^2 + e^x, B(x) = x^4$ **9.** $2x^3(2x) + (x^2 - 2)(6x^2) = 10x^4 - 12x^2$

11. $(x - 3)(2) + (2x - 1)(1) = 4x - 7$ **13.** $\dfrac{(x - 3)(1) - x(1)}{(x - 3)^2} = \dfrac{-3}{(x - 3)^2}$ **15.** $\dfrac{(x - 2)(2) - (2x + 3)(1)}{(x - 2)^2} = \dfrac{-7}{(x - 2)^2}$

17. $3xe^x + 3e^x = 3(x + 1)e^x$ **19.** $x^3\left(\dfrac{1}{x}\right) + 3x^2 \ln x = x^2(1 + 3 \ln x)$ **21.** $(x^2 + 1)(2) + (2x - 3)(2x) = 6x^2 - 6x + 2$

23. $(0.4x + 2)(0.5) + (0.5x - 5)(0.4) = 0.4x - 1$ **25.** $\dfrac{(2x - 3)(2x) - (x^2 + 1)(2)}{(2x - 3)^2} = \dfrac{2x^2 - 6x - 2}{(2x - 3)^2}$ **27.** $(x^2 + 2)2x + (x^2 - 3)2x = 4x^3 - 2x$

29. $\dfrac{(x^2 - 3)2x - (x^2 + 2)2x}{(x^2 - 3)^2} = \dfrac{-10x}{(x^2 - 3)^2}$ **31.** $\dfrac{(x^2 + 1)e^x - e^x(2x)}{(x^2 + 1)^2} = \dfrac{(x - 1)^2 e^x}{(x^2 + 1)^2}$ **33.** $\dfrac{(1 + x)\left(\dfrac{1}{x}\right) - \ln x}{(1 + x)^2} = \dfrac{1 + x - x \ln x}{x(1 + x)^2}$ **35.** $xf'(x) + f(x)$

37. $x^3 f'(x) + 3x^2 f(x)$ **39.** $\dfrac{x^2 f'(x) - 2xf(x)}{x^4}$ **41.** $\dfrac{f(x) - xf'(x)}{[f(x)]^2}$ **43.** $e^x f'(x) + f(x)e^x = e^x[f'(x) + f(x)]$

45. $\dfrac{f(x)\left(\dfrac{1}{x}\right) - (\ln x)f'(x)}{f(x)^2} = \dfrac{f(x) - (x \ln x)f'(x)}{xf(x)^2}$ **47.** $(2x + 1)(2x - 3) + (x^2 - 3x)(2) = 6x^2 - 10x - 3$

49. $(2.5t - t^2)(4) + (4t + 1.4)(2.5 - 2t) = -12t^2 + 17.2t + 3.5$ **51.** $\dfrac{(x^2 + 2x)(5) - (5x - 3)(2x + 2)}{(x^2 + 2x)^2} = \dfrac{-5x^2 + 6x + 6}{(x^2 + 2x)^2}$

53. $\dfrac{(w^2 - 1)(2w - 3) - (w^2 - 3w + 1)(2w)}{(w^2 - 1)^2} = \dfrac{3w^2 - 4w + 3}{(w^2 - 1)^2}$ **55.** $(1 + x - x^2)e^x + e^x(1 - 2x) = (2 - x - x^2)e^x$

57. (A) $f'(x) = \dfrac{x \cdot 0 - 1 \cdot 1}{x^2} = -\dfrac{1}{x^2}$ (B) Note that $f(x) = x^{-1}$ and use the power rule: $f'(x) = -x^{-2} = -\dfrac{1}{x^2}$ **59.** (A) $f'(x) = \dfrac{x^4 \cdot 0 - (-3) \cdot 4x^3}{x^8} = \dfrac{12}{x^5}$

(B) Note that $f(x) = -3x^{-4}$ and use the power rule: $f'(x) = 12x^{-5} = \dfrac{12}{x^5}$ **61.** $f'(x) = (1 + 3x)(-2) + (5 - 2x)(3); y = -11x + 29$

63. $f'(x) = \dfrac{(3x - 4)(1) - (x - 8)(3)}{(3x - 4)^2}; y = 5x - 13$ **65.** $f'(x) = \dfrac{2^x - x(2^x \ln 2)}{2^{2x}}; y = \left(\dfrac{1 - 2 \ln 2}{4}\right)x + \ln 2$

67. $f'(x) = (2x - 15)(2x) + (x^2 + 18)(2) = 6(x - 2)(x - 3); x = 2, x = 3$ **69.** $f'(x) = \dfrac{(x^2 + 1)(1) - x(2x)}{(x^2 + 1)^2} = \dfrac{1 - x^2}{(x^2 + 1)^2}; x = -1, x = 1$

71. $7x^6 - 3x^2$ **73.** $-27x^{-4} = -\dfrac{27}{x^4}$ **75.** $(w + 1)2^w \ln 2 + 2^w = [(w + 1) \ln 2 + 1]2^w$ **77.** $9x^{1/3}(3x^2) + (x^3 + 5)(3x^{-2/3}) = \dfrac{30x^3 + 15}{x^{2/3}}$

79. $\dfrac{(1 + x^2)\dfrac{1}{x \ln 2} - 2x \log_2 x}{(1 + x^2)^2} = \dfrac{1 + x^2 - 2x^2 \ln x}{x(1 + x^2)^2 \ln 2}$ **81.** $\dfrac{(x^2 - 3)(2x^{-2/3}) - 6x^{1/3}(2x)}{(x^2 - 3)^2} = \dfrac{-10x^2 - 6}{(x^2 - 3)^2 x^{2/3}}$

83. $g'(t) = \dfrac{(3t^2 - 1)(0.2) - (0.2t)(6t)}{(3t^2 - 1)^2} = \dfrac{-0.6t^2 - 0.2}{(3t^2 - 1)^2}$ **85.** $(20x)\dfrac{1}{x \ln 10} + 20 \log x = \dfrac{20(1 + \ln x)}{\ln 10}$

87. $x^{-2/3}(3x^2 - 4x) + (x^3 - 2x^2)(-\tfrac{2}{3}x^{-5/3}) = -\tfrac{8}{3}x^{1/3} + \tfrac{7}{3}x^{4/3}$

89. $\dfrac{(x^2 + 1)[(2x^2 - 1)(2x) + (x^2 + 3)(4x)] - (2x^2 - 1)(x^2 + 3)(2x)}{(x^2 + 1)^2} = \dfrac{4x^5 + 8x^3 + 16x}{(x^2 + 1)^2}$ **91.** $\dfrac{e^t(1 + \ln t) - (t \ln t)e^t}{e^{2t}} = \dfrac{1 + \ln t - t \ln t}{e^t}$

93. (A) $S'(t) = \dfrac{(t^2 + 50)(180t) - 90t^2(2t)}{(t^2 + 50)^2} = \dfrac{9{,}000t}{(t^2 + 50)^2}$ (B) $S(10) = 60; S'(10) = 4$. After 10 months, the total sales are 60,000 DVDs and sales are

increasing at the rate of 4,000 DVDs per month.   (C) Approximately 64,000 DVDs

95. (A) $\dfrac{dx}{dp} = \dfrac{(0.1p + 1)(0) - 4{,}000(0.1)}{(0.1p + 1)^2} = \dfrac{-400}{(0.1p + 1)^2}$ (B) $x = 800; dx/dp = -16$. At a price level of $40, the demand is 800 DVD players per week

and demand is decreasing at the rate of 16 players per dollar.   (C) Approximately 784 DVD players

97. (A) $C'(t) = \dfrac{(t^2 + 1)(0.14) - 0.14t(2t)}{(t^2 + 1)^2} = \dfrac{0.14 - 0.14t^2}{(t^2 + 1)^2}$ (B) $C'(0.5) = 0.0672$. After 0.5 hr, concentration is increasing at the rate of

0.0672 mg/cm^3 per hour. $C'(3) = -0.0112$. After 3 hr, concentration is decreasing at the rate of 0.0112 mg/cm^3 per hour.

Exercises 11.4   **1.** $3x^3 + 5$ **3.** $2x^2 e^x + \ln x^2 e^x$ **5.** $E(u) = \ln u, I(x) = x^3 - 6x + 10$ is one answer.   **7.** $E(u) = \sqrt{u}, I(x) = x^2 + 4$ is one answer.
9. 3   **11.** $(-4x)$ **13.** $2x$ **15.** $4x^3$ **17.** $-8(5 - 2x)^3$ **19.** $5(4 + 0.2x)^4(0.2) = (4 + 0.2x)^4$ **21.** $30x(3x^2 + 5)^4$ **23.** $5e^x$ **25.** $5e^{5x}$ **27.** $-18e^{-6x}$

29. $(2x - 5)^{-1/2} = \dfrac{1}{(2x - 5)^{1/2}}$ **31.** $-8x^3(x^4 + 1)^{-3} = \dfrac{-8x^3}{(x^4 + 1)^3}$ **33.** $-\dfrac{2}{x}$ **35.** $\dfrac{6x}{1 + x^2}$ **37.** $\dfrac{3(1 + \ln x)^2}{x}$ **39.** $f'(x) = 6(2x - 1)^2; y = 6x - 5; x = \tfrac{1}{2}$

41. $f'(x) = 2(4x - 3)^{-1/2} = \dfrac{2}{(4x - 3)^{1/2}}; y = \dfrac{2}{3}x + 1;$ none   **43.** $f'(x) = 10(x - 2)e^{x^2 - 4x + 1}; y = -20ex + 5e; x = 2$

45. $12(x^2 - 2)^3(2x) = 24x(x^2 - 2)^3$ **47.** $-6(t^2 + 3t)^{-4}(2t + 3) = \dfrac{-6(2t + 3)}{(t^2 + 3t)^4}$ **49.** $\dfrac{1}{2}(w^2 + 8)^{-1/2}(2w) = \dfrac{w}{\sqrt{w^2 + 8}}$

51. $12xe^{3x} + 4e^{3x} = 4(3x + 1)e^{3x}$ **53.** $\dfrac{x^3\left(\dfrac{1}{1 + x}\right) - 3x^2 \ln (1 + x)}{x^6} = \dfrac{x - 3(1 + x) \ln (1 + x)}{x^4(1 + x)}$ **55.** $6te^{3(t^2 + 1)}$ **57.** $\dfrac{3x}{x^2 + 3}$

59. $-5(w^3 + 4)^{-6}(3w^2) = \dfrac{-15w^2}{(w^3 + 4)^6}$ **61.** $f'(x) = (4 - x)^3 - 3x(4 - x)^2 = 4(4 - x)^2(1 - x); y = -16x + 48$

63. $f'(x) = \dfrac{(2x - 5)^3 - 6x(2x - 5)^2}{(2x - 5)^6} = \dfrac{-4x - 5}{(2x - 5)^4}; y = -17x + 54$ **65.** $f'(x) = \dfrac{1}{2x\sqrt{\ln x}}; y = \dfrac{x}{2e} + \dfrac{1}{2}$

67. $f'(x) = 2x(x - 5)^3 + 3x^2(x - 5)^2 = 5x(x - 5)^2(x - 2); x = 0, 2, 5$ **69.** $f'(x) = \dfrac{(2x + 5)^2 - 4x(2x + 5)}{(2x + 5)^4} = \dfrac{5 - 2x}{(2x + 5)^3}; x = \tfrac{5}{2}$

71. $f'(x) = (x^2 - 8x + 20)^{-1/2}(x - 4) = \dfrac{x - 4}{(x^2 - 8x + 20)^{1/2}}; x = 4$ **73.** No; yes   **75.** Domain of f: $(0, \infty)$; domain of g: $(-\infty, \infty)$; domain of m: $(-2, 2)$

77. Domain of f: all real numbers except ± 1; domain of g: $(0, \infty)$; domain of m: all positive real numbers except e and $\tfrac{1}{e}$

79. $18x^2(x^2 + 1)^2 + 3(x^2 + 1)^3 = 3(x^2 + 1)^2(7x^2 + 1)$ **81.** $\dfrac{24x^5(x^3 - 7)^3 - (x^3 - 7)^4 6x^2}{4x^6} = \dfrac{3(x^3 - 7)^3(3x^3 + 7)}{2x^4}$ **83.** $\dfrac{1}{\ln 2}\left(\dfrac{6x}{3x^2 - 1}\right)$

85. $(2x + 1)(10^{x^2 + x})(\ln 10)$ **87.** $\dfrac{12x^2 + 5}{(4x^3 + 5x + 7) \ln 3}$ **89.** $2^{x^3 - x^2 + 4x + 1}(3x^2 - 2x + 4) \ln 2$ **91.** (A) $C'(x) = (2x + 16)^{-1/2} = \dfrac{1}{(2x + 16)^{1/2}}$

(B) $C'(24) = \tfrac{1}{8}$, or $12.50. At a production level of 24 cell phones, total cost is increasing at the rate of $12.50 per cell phone and the cost of producing the 25th

cell phone is approximately $12.50. $C'(42) = \tfrac{1}{10}$, or $10.00. At a production level of 42 cell phones, total cost is increasing at the rate of $10.00 per cell phone

and the cost of producing the 43rd cell phone is approximately $10.00.   **93.** (A) $\dfrac{dx}{dp} = 40(p + 25)^{-1/2} = \dfrac{40}{(p + 25)^{1/2}}$ (B) $x = 400$ and $dx/dp = 4$.

At a price of $75, the supply is 400 bicycle helmets per week and supply is increasing at the rate of 4 bicycle helmets per dollar.

95. (A) After 1 hr, the concentration is decreasing at the rate of 1.60 mg/mL per hour; after 4 hr, the concentration is decreasing at the rate of 0.08 mg/mL per hour.   **(B)** $C(t)$

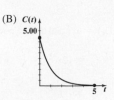

97. 2.27 mm of mercury/yr; 0.81 mm of mercury/yr; 0.41 mm of mercury/yr

Exercises 11.5   **1.** $y = -\dfrac{3}{2}x + 10$ **3.** $y = \pm\dfrac{4}{3}\sqrt{9 - x^2}$ **5.** $y = \dfrac{-x \pm \sqrt{4 - 3x^2}}{2}$ **7.** Impossible   **9.** $y' = -\dfrac{3}{5}$ **11.** $y' = \dfrac{3x}{2}$ **13.** $y' = 10x; 10$

15. $y' = \dfrac{2x}{3y^2}; \dfrac{4}{3}$ **17.** $y' = -\dfrac{3}{2y + 2}; -\dfrac{3}{4}$ **19.** $y' = -\dfrac{y}{x}; -\dfrac{3}{2}$ **21.** $y' = -\dfrac{2y}{2x + 1}; 4$ **23.** $y' = \dfrac{6 - 2y}{x}; -1$ **25.** $y' = \dfrac{2x}{e^y - 2y}; 2$ **27.** $y' = \dfrac{3x^2 y}{y + 1}; \dfrac{3}{2}$

29. $y' = \dfrac{6x^2 y - y \ln y}{x + 2y}; 2$ **31.** $x' = \dfrac{2tx - 3t^2}{2x - t^2}; 8$ **33.** $y'|_{(1.6,1.8)} = -\dfrac{3}{4}; y'|_{(1.6,0.2)} = \dfrac{3}{4}$ **35.** $y = -x + 5$ **37.** $y = \frac{2}{5}x - \frac{12}{5}; y = \frac{3}{5}x + \frac{12}{5}$

39. $y' = -\dfrac{1}{x}$ **41.** $y' = \dfrac{1}{3(1 + y)^2 + 1}; \dfrac{1}{13}$ **43.** $y' = \dfrac{3(x - 2y)^2}{6(x - 2y)^2 + 4y}; \dfrac{3}{10}$ **45.** $y' = \dfrac{3x^2(7 + y^2)^{1/2}}{y}; 16$ **47.** $y' = \dfrac{y}{2xy^2 - x}; 1$

49. $y = 0.63x + 1.04$ **51.** $p' = \dfrac{1}{2p - 2}$ **53.** $p' = -\dfrac{x}{p} = -\dfrac{\sqrt{10,000 - p^2}}{p}$ **55.** $\dfrac{dL}{dV} = \dfrac{-(L + m)}{V + n}$ **57.** $\dfrac{dT}{dv} = \dfrac{2}{k}\sqrt{T}$ **59.** $\dfrac{dv}{dT} = \dfrac{k}{2\sqrt{T}}$

Exercises 11.6   **1.** 19.5 ft   **3.** 46 m   **5.** 34.5 ft   **7.** 32 ft   **9.** 30   **11.** $-\frac{16}{3}$ **13.** $-\frac{16}{7}$ **15.** Decreasing at 9 units/sec   **17.** Approx. -3.03 ft/sec
19. $dA/dt \approx 126$ ft²/sec   **21.** 3,770 cm³/min   **23.** 6 lb/in.²/hr   **25.** $\frac{9}{4}$ ft/sec   **27.** $\frac{20}{3}$ ft/sec   **29.** 0.0214 ft/sec; 0.0135 ft/sec; yes, at $t = 0.000\ 19$ sec
31. 3.835 units/sec   **33. (A)** $dC/dt = \$15,000$/wk   **(B)** $dR/dt = -\$50,000$/wk   **(C)** $dP/dt = -\$65,000$/wk   **35.** $ds/dt = \$2,207$/wk
37. (A) $dx/dt = -12.73$ units/month   **(B)** $dp/dt = \$1.53$/month   **39.** Approximately 100 ft³/min

Exercises 11.7   **1.** $x = f(p) = 105 - 2.5p, 0 \le p \le 42$ **3.** $x = f(p) = \sqrt{100 - 2p}, 0 \le p \le 50$

5. $x = f(p) = 20(\ln 25 - \ln p), 25/e \approx 9.2 \le p \le 25$ **7.** $x = f(p) = e^{8 - 0.1p}, 80 - 10 \ln 30 \approx 46.0 \le p \le 80$ **9.** $\dfrac{35 - 0.8x}{35x - 0.4x^2}$ **11.** $-\dfrac{4e^{-x}}{7 + 4e^{-x}}$
13. $\dfrac{5}{x(12 + 5 \ln x)}$ **15.** 0   **17.** -0.017 **19.** -0.034 **21.** 1.013   **23.** 0.405   **25.** 11.8%   **27.** 5.4%   **29.** -14.7%   **31.** -431.6%

33. $E(p) = \dfrac{450p}{25,000 - 450p}$ **35.** $E(p) = \dfrac{8p^2}{4,800 - 4p^2}$ **37.** $E(p) = \dfrac{0.6pe^p}{98 - 0.6e^p}$ **39.** 0.07   **41.** 0.15   **43.** $\dfrac{x + 1}{x}$ **45.** $\dfrac{1}{x \ln x}$ **47. (A)** Inelastic
(B) Unit elasticity   **(C)** Elastic   **49. (A)** Inelastic   **(B)** Unit elasticity   **(C)** Elastic   **51. (A)** $x = 6,000 - 200p\ \ 0 \le p \le 30$ **(B)** $E(p) = \dfrac{p}{30 - p}$
(C) $E(10) = 0.5; 5$% decrease   **(D)** $E(25) = 5; 50$% decrease   **(E)** $E(15) = 1; 10$% decrease   **53. (A)** $x = 3,000 - 50p\ \ 0 \le p \le 60$
(B) $R(p) = 3,000p - 50p^2$ **(C)** $E(p) = \dfrac{p}{60 - p}$ **(D)** Elastic on $(30, 60)$; inelastic on $(0, 30)$ **(E)** Increasing on $(0, 30)$; decreasing on $(30, 60)$
(F) Decrease   **(G)** Increase   **63.** $R(p) = 20p(10 - p)$ **65.** $R(p) = 40p(p - 15)^2$ **67.** $R(p) = 30p - 10p\sqrt{p}$
55. Elastic on $(3.5, 7)$; inelastic on $(0, 3.5)$
57. Elastic on $(25/\sqrt{3}, 25)$; inelastic on $(0, 25/\sqrt{3})$
59. Elastic on $(48, 72)$; inelastic on $(0, 48)$
61. Elastic on $(25, 25\sqrt{2})$; inelastic on $(0, 25)$

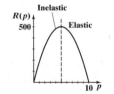
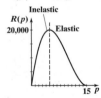
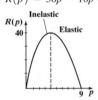

69. $\dfrac{3}{2}$ **71.** $\dfrac{1}{2}$ **73.** Elastic on $(0, 300)$; inelastic on $(300, 600)$ **75.** Elastic on $(0, 10\sqrt{3})$; inelastic on $(10\sqrt{3}, 30)$ **77.** k **79.** \$75 per day

81. Increase   **83.** Decrease   **85.** \$3.75   **87.** $p(t) = \dfrac{31}{0.31t + 18.5}$ **89.** -0.025

Chapter 11 Review Exercises   **1.** \$3,136.62; \$4,919.21; \$12,099.29 *(11.1)*   **2.** $E(u) = u^{3/2}, I(x) = 6x + 5$ is one answer. *(11.4)*

3. $E(u) = \ln u, I(x) = x^2 + 4$ is one answer. *(11.4)*   **4.** $E(u) = e^u, I(x) = 0.02x$ is one answer. *(11.4)*   **5.** $\dfrac{2}{x} + 3e^x$ *(11.2)*   **6.** $2e^{2x - 3}$ *(11.4)*   **7.** $\dfrac{2}{2x + 7}$ *(11.4)*

8. $\dfrac{e^x}{3 + e^x}$ *(11.4)*   **9.** $y' = \dfrac{9x^2}{4y}; \dfrac{9}{8}$ *(11.5)*   **10.** $dy/dt = 216$ *(11.6)*   **11. (A)** $x = 1,000 - 25p$ **(B)** $\dfrac{p}{40 - p}$ **(C)** 0.6; demand is inelastic and insensitive to small

changes in price.   **(D)** $1,000p - 25p^2$ **(E)** Revenue increases *(11.7)*   **12.** -10 *(11.2)*   **13.** $\lim_{n \to \infty}\left(1 + \dfrac{2}{n}\right)^n = e^2 \approx 7.389\ 06$ *(11.1)*

14. $\dfrac{7[(\ln z)^6 + 1]}{z}$ *(11.4)*   **15.** $x^5(1 + 6 \ln x)$ *(11.3)*   **16.** $\dfrac{e^x(x - 6)}{x^7}$ *(11.3)*   **17.** $\dfrac{6x^2 - 3}{2x^3 - 3x}$ *(11.4)*   **18.** $(3x^2 - 2x)e^{x^3 - x^2}$ *(11.4)*   **19.** $\dfrac{1 - 2x \ln 5x}{xe^{2x}}$ *(11.4)*

20. $y = -x + 2; y = -ex + 1$ *(11.4)*   **21.** $y' = \dfrac{3y - 2x}{8y - 3x}; \dfrac{8}{19}$ *(11.5)*   **22.** $x' = \dfrac{4tx}{3x^2 - 2t^2}; -4$ *(11.5)*   **23.** $y' = \dfrac{1}{e^y + 2y}; 1$ *(11.5)*

24. $y' = \dfrac{2xy}{1 + 2y^2}; \dfrac{2}{3}$ *(11.5)*   **25.** 0.049 *(11.7)*   **26.** $-\dfrac{3}{100 - 3p}$ *(11.7)*   **27.** $\dfrac{2x}{1 + x^2}$ *(11.7)*   **28.** $dy/dt = -2$ units/sec *(11.6)*   **29.** 0.27 ft/sec *(11.6)*

30. $dR/dt = 1/\pi \approx 0.318$ in./min *(11.6)*   **31.** Elastic for $5 < p < 15$; inelastic for $0 < p < 5$ *(11.7)*

32.   **33.** (A) $y = \left[\ln(4 - e^x)\right]^3$ (B) $\dfrac{dy}{dx} = \dfrac{-3e^x\left[\ln(4 - e^x)\right]^2}{4 - e^x}$ *(11.4)*   **34.** $2x(5^{x^2-1})(\ln 5)$ *(11.4)*   **35.** $\left(\dfrac{1}{\ln 5}\right)\dfrac{2x - 1}{x^2 - x}$ *(11.4)*

36. $\dfrac{2x + 1}{2(x^2 + x)\sqrt{\ln(x^2 + x)}}$ *(11.4)*   **37.** $y' = \dfrac{2x - e^{xy}y}{xe^{xy} - 1}; 0$ *(11.5)*   **38.** The rate of increase of area is proportional to the radius R,

so the rate is smallest when $R = 0$, and has no largest value. *(11.6)*   **39.** Yes, for $-\sqrt{3}/3 < x < \sqrt{3}/3$ *(11.6)*

40. (A) 15 yr   (B) 13.9 yr *(11.1)*   **41.** $A'(t) = 10e^{0.1t}; A'(1) = \$11.05/\text{yr}; A'(10) = \$27.18/\text{yr}$ *(11.1)*   **42.** $\$987.50/\text{yr}$ *(11.2)*

43. $R'(x) = (1{,}000 - 20x)e^{-0.02x}$ *(11.4)*   **44.** $p' = -\dfrac{x}{3p^2} = \dfrac{-(5{,}000 - 2p^3)^{1/2}}{3p^2}$ *(11.5)*   **45.** $dR/dt = \$2{,}242/\text{day}$ *(11.6)*   **46.** Decrease price *(11.7)*   **47.** 0.02125 *(11.7)*

48. -1.111 mg/mL per hour; -0.335 mg/mL per hour *(11.4)*   **49.** $dR/dt = -3/(2\pi)$; approx. 0.477 mm/day *(11.6)*   **50.** (A) Increasing at the rate of 2.68 units/ day at the end of 1 day of training; increasing at the rate of 0.54 unit/day after 5 days of training   (B) 7 days *(11.4)*   **51.** $dT/dt = -1/27 \approx -0.037$ min/hr *(11.6)*

Chapter 12

Exercises 12.1   **1.** Decreasing   **3.** Increasing   **5.** Increasing   **7.** Decreasing   **9.** $(a, b); (d, f); (g, h)$ **11.** $(b, c); (c, d); (f, g)$ **13.** c, d, f
15. b, f **17.** Local maximum at $x = a$; local minimum at $x = c$; no local extrema at $x = b$ and $x = d$ **19.** $f(3) = 5$ is a local maximum; e
21. No local extremum; d   **23.** $f(3) = 5$ is a local maximum; f   **25.** No local extremum; c   **27.** (A) $f'(x) = 3x^2 - 12$ (B) $-2, 2$ (C) $-2, 2$
29. (A) $f'(x) = -\dfrac{6}{(x + 2)^2}$ (B) -2 (C) None   **31.** (A) $f'(x) = \begin{cases} -1 \text{ if } x < 0 \\ 1 \text{ if } x > 0 \end{cases}$ (B) 0   (C) 0   **33.** Decreasing on $(-\infty, 1)$; increasing on
$(1, \infty); f(1) = -2$ is a local minimum   **35.** Increasing on $(-\infty, -4)$; decreasing on $(-4, \infty); f(-4) = 7$ is a local maximum   **37.** Increasing for all x;
no local extrema   **39.** Increasing on $(-\infty, -2)$ and $(3, \infty)$; decreasing on $(-2, 3); f(-2) = 44$ is a local maximum, $f(3) = -81$ is a local minimum
41. Decreasing on $(-\infty, 1)$; increasing on $(1, \infty); f(1) = 4$ is a local minimum   **43.** Increasing on $(-\infty, 2)$; decreasing on $(2, \infty); f(2) = e^{-2} \approx 0.135$ is a
local maximum   **45.** Increasing on $(-\infty, 8)$; decreasing on $(8, \infty); f(8) = 4$ is a local maximum   **47.** Critical numbers: $x = -0.77, 1.08, 2.69$;
decreasing on $(-\infty, -0.77)$ and $(1.08, 2.69)$; increasing on $(-0.77, 1.08)$ and $(2.69, \infty); f(-0.77) = -4.75$ and $f(2.69) = -1.29$ are local minima; $f(1.08) = 6.04$ is a local maximum   **49.** Critical numbers: $x = 1.34, 2.82$; decreasing on $(0, 1.34)$ and $(2.82, \infty)$; increasing on $(1.34, 2.82)$;
$f(1.34) = 0.68$ is a local minimum; $f(2.82) = 2.37$ is a local maximum   **51.** Critical numbers: 0.36, 2.15; increasing on $(-\infty, 0.36)$ and $(2.15, \infty)$;
decreasing on $(0.36, 2.15); f(0.36) = 1.17$ is a local maximum; $f(2.15) = -0.66$ is a local minimum

53. Increasing on $(-\infty, 4)$ **55.** Increasing on $(-\infty, -1), (1, \infty)$ **57.** Decreasing for all x **59.** Decreasing on $(-\infty, -3)$ and $(0, 3)$;
Decreasing on $(4, \infty)$ Decreasing on $(-1, 1)$ Horizontal tangent at $x = 2$ increasing on $(-3, 0)$ and $(3, \infty)$
Horizontal tangent at $x = 4$ Horizontal tangents at $x = -1, 1$ Horizontal tangents at $x = -3, 0, 3$

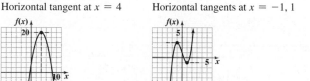

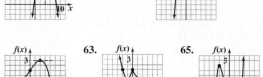

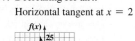

61.   **63.**   **65.**   **67.**   **69.** g_4 **71.** g_6 **73.** g_2

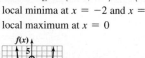

75. Increasing on $(-1, 2)$; decreasing on $(-\infty, -1)$ **77.** Increasing on $(-1, 2)$ and $(2, \infty)$ **79.** Increasing on $(-2, 0)$ and $(3, \infty)$;
and $(2, \infty)$; local minimum at $x = -1$;   decreasing on $(-\infty, -1)$; local   decreasing on $(-\infty, -2)$ and $(0, 3)$;
local maximum at $x = 2$ minimum at $x = -1$ local minima at $x = -2$ and $x = 3$;
local maximum at $x = 0$

81. $f'(x) > 0$ on $(-\infty, -1)$ and   **83.** $f'(x) > 0$ on $(-2, 1)$ and   **85.** Critical numbers: $x = -2, x = 2$; increasing on $(-\infty, -2)$
$(3, \infty); f'(x) < 0$ on $(-1, 3); f'(x) = 0$ $(3, \infty); f'(x) < 0$ on $(-\infty, -2)$ and $(2, \infty)$; decreasing on $(-2, 0)$ and $(0, 2); f(-2) = -4$ is a
at $x = -1$ and $x = 3$ and $(1, 3); f'(x) = 0$ at   local maximum; $f(2) = 4$ is a local minimum
    $x = -2, x = 1$, and $x = 3$ **87.** Critical numbers: $x = -2$; increasing on $(-2, 0)$; decreasing

    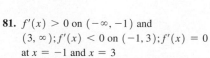   on $(-\infty, -2)$ and $(0, \infty); f(-2) = 0.75$ is a local minimum
    89. Critical numbers: $x = 0, x = 4$; increasing on $(-\infty, 0)$ and
    $(4, \infty)$; decreasing on $(0, 2)$ and $(2, 4); f(0) = 0$ is a local
    maximum; $f(4) = 8$ is a local minimum

91. (A) The marginal profit is positive on $(0, 600)$, 0   (B) $P'(x)$
at $x = 600$, and negative on $(600, 1,000)$.

93. (A) The price decreases for the first 15   (B) $B(t)$
months to a local minimum, increases
for the next 40 months to a local
maximum, and then decreases for the
remaining 15 months.

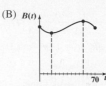

95. (A) $\overline{C}(x) = 0.05x + 20 + \dfrac{320}{x}$ (B) Critical number: $x = 80$; decreasing for $0 < x < 80$; increasing for $80 < x < 150$; $\overline{C}(80) = 28$ is a local minimum

97. Critical number: $t = 2$; increasing on $(0, 2)$; decreasing on $(2, 24)$; $C(2) = 0.07$ is a local maximum.

Exercises 12.2   **1.** Concave up   **3.** Concave down   **5.** Concave down   **7.** Neither   **9.** (A) $(a, c), (c, d), (e, g)$ (B) $(d, e), (g, h)$ (C) $(d, e), (g, h)$
(D) $(a, c), (c, d), (e, g)$ (E) $(a, c), (c, d), (e, g)$ (F) $(d, e), (g, h)$ **11.** (A) $f(-2) = 3$ is a local maximum of f; $f(2) = -1$ is a local minimum of f.

(B) $(0, 1)$ (C) 0   **13.** (C)   **15.** (D)   **17.** $12x - 8$ **19.** $4x^{-3} - 18x^{-4}$ **21.** $2 + \dfrac{9}{2}x^{-3/2}$ **23.** $8(x^2 + 9)^3 + 48x^2(x^2 + 9)^2 = 8(x^2 + 9)^2(7x^2 + 9)$

25. $(-10, 2,000)$ **27.** $(0, 2)$ **29.** None   **31.** Concave upward for all x; no inflection points   **33.** Concave downward on $\left(-\infty, \frac{4}{3}\right)$; concave upward
on $\left(\frac{4}{3}, \infty\right)$; inflection point at $x = \frac{4}{3}$ **35.** Concave downward on $(-\infty, 0)$ and $(6, \infty)$; concave upward on $(0, 6)$; inflection points at $x = 0$ and $x = 6$
37. Concave upward on $(-2, 4)$; concave downward on $(-\infty, -2)$ and $(4, \infty)$; inflection points at $x = -2$ and $x = 4$ **39.** Concave upward on $(-\infty, \ln 2)$;
concave downward on $(\ln 2, \infty)$; inflection point at $x = \ln 2$

41.
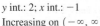

43.

45.

47.

49. Domain: All real numbers
y int.: 16; x int.: $2 - 2\sqrt{3}, 2, 2 + 2\sqrt{3}$
Increasing on $(-\infty, 0)$ and $(4, \infty)$
Decreasing on $(0, 4)$
Local maximum: $f(0) = 16$;
local minimum: $f(4) = -16$
Concave downward on $(-\infty, 2)$
Concave upward on $(2, \infty)$
Inflection point: $(2, 0)$

51. Domain: All real numbers
y int.: 2; x int.: -1
Increasing on $(-\infty, \infty)$
Concave downward on $(-\infty, 0)$
Concave upward on $(0, \infty)$
Inflection point: $(0, 2)$

53. Domain: All real numbers
y int.: 0; x int.: 0, 4
Increasing on $(-\infty, 3)$
Decreasing on $(3, \infty)$
Local maximum: $f(3) = 6.75$
Concave upward on $(0, 2)$
Concave downward on $(-\infty, 0)$ and $(2, \infty)$
Inflection points: $(0, 0), (2, 4)$

55. Domain: All real numbers
y int.: 0; x int.: 0, 1
Increasing on $(0.25, \infty)$
Decreasing on $(-\infty, 0.25)$
Local minimum: $f(0.25) = -1.6875$
Concave upward on $(-\infty, 0.5)$ and $(1, \infty)$
Concave downward on $(0.5, 1)$
Inflection points: $(0.5, -1), (1, 0)$

57. Domain: All real numbers
y int.: 27; x int.: $-3, 3$
Increasing on $(-\infty, -\sqrt{3})$ and $(0, \sqrt{3})$
Decreasing on $(-\sqrt{3}, 0)$ and $(\sqrt{3}, \infty)$
Local maxima: $f(-\sqrt{3}) = 36, f(\sqrt{3}) = 36$
Local minimum: $f(0) = 27$
Concave upward on $(-1, 1)$
Concave downward on $(-\infty, -1)$ and $(1, \infty)$
Inflection points: $(-1, 32), (1, 32)$

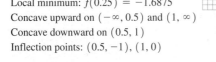

59. Domain: All real numbers
y int.: 16; x int.: $-2, 2$
Decreasing on $(-\infty, -2)$ and $(0, 2)$
Increasing on $(-2, 0)$ and $(2, \infty)$
Local minima: $f(-2) = 0, f(2) = 0$
Local maximum: $f(0) = 16$
Concave upward on $(-\infty, -2\sqrt{3}/3)$ and $(2\sqrt{3}/3, \infty)$
Concave downward on $(-2\sqrt{3}/3, 2\sqrt{3}/3)$
Inflection points: $(-1.15, 7.11), (1.15, 7.11)$

61. Domain: All real numbers
y int.: 0; int.: 0, 1.5
Decreasing on $(-\infty, 0)$ and $(0, 1.25)$
Increasing on $(1.25, \infty)$
Local minimum: $f(1.25) = -1.53$
Concave upward on $(-\infty, 0)$ and $(1, \infty)$
Concave downward on $(0, 1)$
Inflection points: $(0, 0), (1, -1)$

63. Domain: All real numbers
y int.: 0; x int.: 0
Increasing on $(-\infty, \infty)$
Concave downward on $(-\infty, \infty)$

65. Domain: All real numbers
y int.: 5
Decreasing on $(-\infty, \ln 4)$
Increasing on $(\ln 4, \infty)$
Local minimum: $f(\ln 4) = 4$
Concave upward on $(-\infty, \infty)$

67. Domain: $(0, \infty)$
x int.: e^2
Increasing on $(-\infty, \infty)$
Concave downward on $(-\infty, \infty)$

69. Domain: $(-4, \infty)$
y int.: $-2 + \ln 4$; x int.: $e^2 - 4$
Increasing on $(-4, \infty)$
Concave downward on $(-4, \infty)$

71.

x	$f'(x)$	$f(x)$
$-\infty < x < -1$	Positive and decreasing	Increasing and concave downward
$x = -1$	x intercept	Local maximum
$-1 < x < 0$	Negative and decreasing	Decreasing and concave downward
$x = 0$	Local minimum	Inflection point
$0 < x < 2$	Negative and increasing	Decreasing and concave upward
$x = 2$	Local max., x intercept	Inflection point, horiz. tangent
$2 < x < \infty$	Negative and decreasing	Decreasing and concave downward

73.

x	$f'(x)$	$f(x)$
$-\infty < x < -2$	Negative and increasing	Decreasing and concave upward
$x = -2$	Local max., x intercept	Inflection point, horiz. tangent
$-2 < x < 0$	Negative and decreasing	Decreasing and concave downward
$x = 0$	Local minimum	Inflection point
$0 < x < 2$	Negative and increasing	Decreasing and concave upward
$x = 2$	Local max., x intercept	Inflection point, horiz. tangent
$2 < x < \infty$	Negative and decreasing	Decreasing and concave downward

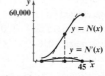

75. Domain: All real numbers
x int.: $-1.18, 0.61, 1.87, 3.71$
y int.: -5
Decreasing on $(-\infty, -0.53)$ and $(1.24, 3.04)$
Increasing on $(-0.53, 1.24)$ and $(3.04, \infty)$
Local minima: $f(-0.53) = -7.57$, $f(3.04) = -8.02$
Local maximum: $f(1.24) = 2.36$
Concave upward on $(-\infty, 0.22)$ and $(2.28, \infty)$
Concave downward on $(0.22, 2.28)$
Inflection points: $(0.22, -3.15)$, $(2.28, -3.41)$

77. Domain: All real numbers
y int.: 100; x int.: $8.01, 13.36$
Increasing on $(-0.10, 4.57)$ and $(11.28, \infty)$
Decreasing on $(-\infty, -0.10)$ and $(4.57, 11.28)$
Local minima: $f(-0.10) = 99.02$, $f(11.28) = -901.18$
Local maximum: $f(4.57) = 711.75$
Concave upward on $(-\infty, 1.95)$ and $(8.55, \infty)$
Concave downward on $(1.95, 8.55)$
Inflection points: $(1.95, 377.82)$, $(8.55, -200.66)$

79. Domain: All real numbers
x int.: $-2.40, 1.16$; y int.: 3
Increasing on $(-\infty, -1.58)$
Decreasing on $(-1.58, \infty)$
Local maximum: $f(-1.58) = 8.87$
Concave downward on $(-\infty, -0.88)$ and $(0.38, \infty)$
Concave upward on $(-0.88, 0.38)$
Inflection points: $(-0.88, 6.39)$, $(0.38, 2.45)$

81. Domain: All real numbers
x int.: $-6.68, -3.64, -0.72$; y int.: 30
Decreasing on $(-5.59, -2.27)$ and $(1.65, 3.82)$
Increasing on $(-\infty, -5.59)$, $(-2.27, 1.65)$, and $(3.82, \infty)$
Local minima: $f(-2.27) = -37.84$, $f(3.82) = 32.09$
Local maxima: $f(-5.59) = 95.97$, $f(1.65) = 67.87$
Concave upward on $(-4.31, -0.40)$ and $(2.91, \infty)$
Concave downward on $(-\infty, -4.31)$ and $(-0.40, 2.91)$
Inflection points: $(-4.31, 39.98)$, $(-0.40, 13.54)$, $(2.91, 47.83)$

83. The graph of the CPI is concave upward.   **85.** The graph of $y = C'(x)$ is positive and decreasing. Since marginal costs are decreasing, the production process is becoming more efficient as production increases.   **87.** (A) Local maximum at $x = 60$ (B) Concave downward on the whole interval $(0, 80)$ **89.** (A) Local maximum at $x = 1$ (B) Concave downward on $(-\infty, 2)$; concave upward on $(2, \infty)$

91. Increasing on $(0, 10)$; decreasing on $(10, 15)$; point of diminishing returns is $x = 10$, max $T'(x) = T'(10) = 500$

93. Increasing on $(24, 36)$; decreasing on $(36, 45)$; point of diminishing returns is $x = 36$, max $N'(x) = N'(36) = 3,888$

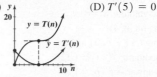

95. (A)

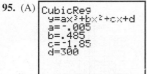

(B) 32 ads to sell 574 cars per month

97. (A) Increasing on $(0, 10)$; decreasing on $(10, 20)$
(B) Inflection point: $(10, 3000)$
(C)

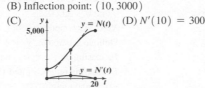

(D) $N'(10) = 300$

99. (A) Increasing on $(5, \infty)$; decreasing on $(0, 5)$
(B) Inflection point: $(5, 10)$
(C)

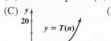

(D) $T'(5) = 0$

Exercises 12.3   **1.** 500 **3.** 0 **5.** 50 **7.** 0 **9.** 6 **11.** $-\frac{1}{10}$ **13.** 7 **15.** $-\frac{1}{5}$ **17.** $\frac{2}{5}$ **19.** 0 **21.** $-\infty$ **23.** $\frac{2}{3}$ **25.** $\frac{1}{2}$ **27.** 1 **29.** 0 **31.** 4
33. 5 **35.** ∞ **37.** 8 **39.** 0 **41.** ∞ **43.** ∞ **45.** $\frac{1}{3}$ **47.** -2 **49.** $-\infty$ **51.** 0 **53.** 0 **55.** 0 **57.** $\frac{1}{4}$ **59.** $\frac{1}{3}$ **61.** 0 **63.** 0 **65.** ∞ **67.** 1 **69.** 1

Exercises 12.4   **1.** Domain: All real numbers; x int.: -12; y int.: 36 **3.** Domain: $(-\infty, 25]$; x int.: 25; y int.: 5 **5.** Domain: All real
numbers except 2; x int.: -1; y int.: $-\frac{1}{2}$ **7.** Domain: All real numbers except -1 and 1; No x intercept; y int.: -3 **9.** (A) $(-\infty, b)$, $(0, e)$, (e, g)
(B) (b, d), $(d, 0)$, (g, ∞) (C) (b, d), $(d, 0)$, (g, ∞) (D) $(-\infty, b)$, $(0, e)$, (e, g) (E) $x = 0$ (F) $x = b, x = g$ (G) $(-\infty, a)$, (d, e), (h, ∞)
(H) (a, d), (e, h) (I) (a, d), (e, h) (J) $(-\infty, a)$, (d, e), (h, ∞) (K) $x = a, x = h$ (L) $y = L$ (M) $x = d, x = e$

11.   **13.**   **15.**   **17.**

19. Domain: All real numbers, except 3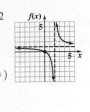
y int.: -1; x int.: -3
Horizontal asymptote: $y = 1$
Vertical asymptote: $x = 3$
Decreasing on $(-\infty, 3)$ and $(3, \infty)$
Concave upward on $(3, \infty)$
Concave downward on $(-\infty, 3)$

21. Domain: All real numbers, except 2
y int.: 0; x int.: 0
Horizontal asymptote: $y = 1$
Vertical asymptote: $x = 2$
Decreasing on $(-\infty, 2)$ and $(2, \infty)$
Concave downward on $(-\infty, 2)$
Concave upward on $(2, \infty)$

23. Domain: $(-\infty, \infty)$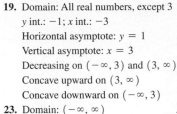
y int.: 10
Horizontal asymptote: $y = 5$
Decreasing on $(-\infty, \infty)$
Concave upward on $(-\infty, \infty)$

25. Domain: $(-\infty, \infty)$
y int.: 0; x int.: 0
Horizontal asymptote: $y = 0$
Increasing on $(-\infty, 5)$
Decreasing on $(5, \infty)$
Local maximum: $f(5) = 9.20$
Concave upward on $(10, \infty)$
Concave downward on $(-\infty, 10)$
Inflection point: $(10, 6.77)$

27. Domain: $(-\infty, 1)$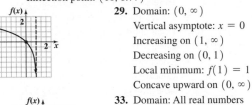
y int.: 0; x int.: 0
Vertical asymptote: $x = 1$
Decreasing on $(-\infty, 1)$
Concave downward on $(-\infty, 1)$

29. Domain: $(0, \infty)$
Vertical asymptote: $x = 0$
Increasing on $(1, \infty)$
Decreasing on $(0, 1)$
Local minimum: $f(1) = 1$
Concave upward on $(0, \infty)$

31. Domain: All real numbers, except ± 2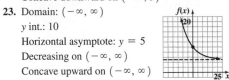
y int.: 0; x int.: 0
Horizontal asymptote: $y = 0$
Vertical asymptotes: $x = -2, x = 2$
Decreasing on $(-\infty, -2)$, $(-2, 2)$, and $(2, \infty)$
Concave upward on $(-2, 0)$ and $(2, \infty)$
Concave downward on $(-\infty, -2)$ and $(0, 2)$
Inflection point: $(0, 0)$

33. Domain: All real numbers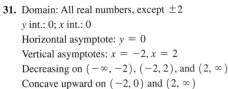
y int.: 1
Horizontal asymptote: $y = 0$
Increasing on $(-\infty, 0)$
Decreasing on $(0, \infty)$
Local maximum: $f(0) = 1$
Concave upward on $(-\infty, -\sqrt{3}/3)$ and $(\sqrt{3}/3, \infty)$
Concave downward on $(-\sqrt{3}/3, \sqrt{3}/3)$
Inflection points: $(-\sqrt{3}/3, 0.75)$, $(\sqrt{3}/3, 0.75)$

35. Domain: All real numbers except -1 and 1
y int.: 0; x int.: 0
Horizontal asymptote: $y = 0$
Vertical asymptote: $x = -1$ and $x = 1$
Increasing on $(-\infty, -1)$, $(-1, 1)$, and $(1, \infty)$
Concave upward on $(-\infty, -1)$ and $(0, 1)$
Concave downward on $(-1, 0)$ and $(1, \infty)$
Inflection point: $(0, 0)$

37. Domain: All real numbers except 1
y int.: 0; x int.: 0
Horizontal asymptote: $y = 0$
Vertical asymptote: $x = 1$
Increasing on $(-\infty, -1)$ and $(1, \infty)$
Decreasing on $(-1, 1)$
Local maximum: $f(-1) = 1.25$
Concave upward on $(-\infty, -2)$
Concave downward on $(-2, 1)$ and $(1, \infty)$
Inflection point: $(-2, 1.11)$

39. Domain: All real numbers except 0
Horizontal asymptote: $y = 1$
Vertical asymptote: $x = 0$
Increasing on $(0, 4)$
Decreasing on $(-\infty, 0)$ and $(4, \infty)$
Local maximum: $f(4) = 1.125$
Concave upward on $(6, \infty)$
Concave downward on $(-\infty, 0)$ and $(0, 6)$
Inflection point: $(6, 1.11)$

41. Domain: All real numbers except 1
y int.: 0; x int.: 0
Vertical asymptote: $x = 1$
Oblique asymptote: $y = x + 1$
Increasing on $(-\infty, 0)$ and $(2, \infty)$
Decreasing on $(0, 1)$ and $(1, 2)$
Local maximum: $f(0) = 0$
Local minimum: $f(2) = 4$
Concave upward on $(1, \infty)$
Concave downward on $(-\infty, 1)$

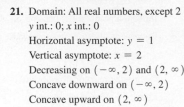
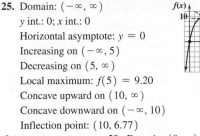
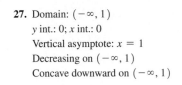
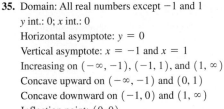
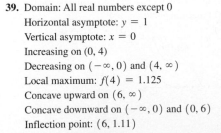
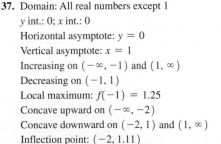
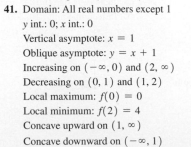

43. Domain: All real numbers except $-3, 3$

y int.: $-\frac{2}{9}$

Horizontal asymptote: $y = 3$

Vertical asymptotes: $x = -3, x = 3$

Increasing on $(-\infty, -3)$ and $(-3, 0)$

Decreasing on $(0, 3)$ and $(3, \infty)$

Local maximum: $f(0) = -0.22$

Concave upward on $(-\infty, -3)$ and $(3, \infty)$

Concave downward on $(-3, 3)$

45. Domain: All real numbers except 2

y int.: 0; x int.: 0

Vertical asymptote: $x = 2$

Increasing on $(3, \infty)$

Decreasing on $(-\infty, 2)$ and $(2, 3)$

Local minimum: $f(3) = 27$

Concave upward on $(-\infty, 0)$ and $(2, \infty)$

Concave downward on $(0, 2)$

Inflection point: $(0, 0)$

47. Domain: All real numbers

y int.: 3; x int.: 3

Horizontal asymptote: $y = 0$

Increasing on $(-\infty, 2)$

Decreasing on $(2, \infty)$

Local maximum: $f(2) = 7.39$

Concave upward on $(-\infty, 1)$

Concave downward on $(1, \infty)$

Inflection point: $(1, 5.44)$

49. Domain: $(-\infty, \infty)$

y int.: 1

Horizontal asymptote: $y = 0$

Increasing on $(-\infty, 0)$

Decreasing on $(0, \infty)$

Local maximum: $f(0) = 1$

Concave upward on $(-\infty, -1)$ and $(1, \infty)$

Concave downward on $(-1, 1)$

Inflection points: $(-1, 0.61), (1, 0.61)$

51. Domain: $(0, \infty)$

x int.: 1

Increasing on $(e^{-1/2}, \infty)$

Decreasing on $(0, e^{-1/2})$

Local minimum: $f(e^{-1/2}) = -0.18$

Concave upward on $(e^{-3/2}, \infty)$

Concave downward on $(0, e^{-3/2})$

Inflection point: $(e^{-3/2}, -0.07)$

53. Domain: $(0, \infty)$

x int.: 1

Vertical asymptote: $x = 0$

Increasing on $(1, \infty)$

Decreasing on $(0, 1)$

Local minimum: $f(1) = 0$

Concave upward on $(0, e)$

Concave downward on (e, ∞)

Inflection point: $(e, 1)$

55. Domain: All real numbers except $-4, 2$

y int.: $-\frac{1}{8}$

Horizontal asymptote: $y = 0$

Vertical asymptote: $x = -4, x = 2$

Increasing on $(-\infty, -4)$ and $(-4, -1)$

Decreasing on $(-1, 2)$ and $(2, \infty)$

Local maximum: $f(-1) = -0.11$

Concave upward on $(-\infty, -4)$ and $(2, \infty)$

Concave downward on $(-4, 2)$

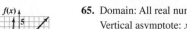

57. Domain: All real numbers except $-\sqrt{3}, \sqrt{3}$

y int.: 0; x int.: 0

Vertical asymptote: $x = -\sqrt{3}, x = \sqrt{3}$

Oblique asymptote: $y = -x$

Increasing on $(-3, -\sqrt{3}), (-\sqrt{3}, \sqrt{3})$, and $(\sqrt{3}, 3)$

Decreasing on $(-\infty, -3)$ and $(3, \infty)$

Local maximum: $f(3) = -4.5$

Local minimum: $f(-3) = 4.5$

Concave upward on $(-\infty, -\sqrt{3})$ and $(0, \sqrt{3})$

Concave downward on $(-\sqrt{3}, 0)$ and $(\sqrt{3}, \infty)$

Inflection point: $(0, 0)$

59. Domain: All real numbers except 0

Vertical asymptote: $x = 0$

Oblique asymptote: $y = x$

Increasing on $(-\infty, -2)$ and $(2, \infty)$

Decreasing on $(-2, 0)$ and $(0, 2)$

Local maximum: $f(-2) = -4$

Local minimum: $f(2) = 4$

Concave upward on $(0, \infty)$

Concave downward on $(-\infty, 0)$

61. Domain: All real numbers except 0

x int.: $\sqrt[3]{4}$

Vertical asymptote: $x = 0$

Oblique asymptote: $y = x$

Increasing on $(-\infty, -2)$ and $(0, \infty)$

Local maximum: $f(-2) = -3$

Decreasing on $(-2, 0)$

Concave downward on $(-\infty, 0)$ and $(0, \infty)$

63. Domain: All real numbers except 0

x int.: $-\sqrt{3}, \sqrt{3}$

Vertical asymptote: $x = 0$

Oblique asymptote: $y = x$

Increasing on $(-\infty, 0)$ and $(0, \infty)$

Concave upward on $(-\infty, 0)$

Concave downward on $(0, \infty)$

65. Domain: All real numbers except 0

Vertical asymptote: $x = 0$

Oblique asymptote: $y = x$

Increasing on $(-\infty, -2)$ and $(2, \infty)$

Decreasing on $(-2, 0)$ and $(0, 2)$

Local maximum: $f(-2) = -3$

Local minimum: $f(2) = 3$

Concave upward on $(0, \infty)$

Concave downward on $(-\infty, 0)$

67. $y = 90 + 0.02x$ **69.** $y = 210 + 0.1x$

71. Domain: All real numbers except 2, 4

   y int.: $-3/4$; x int.: -3

   Vertical asymptote: $x = 4$

   Horizontal asymptote: $y = 1$

   Decreasing on $(-\infty, 2), (2, 4)$, and $(4, \infty)$

   Concave upward on $(4, \infty)$

   Concave downward on $(-\infty, 2)$ and $(2, 4)$

73. Domain: All real numbers except $-3, 3$

   y int.: $5/3$; x int.: 2.5

   Vertical asymptote: $x = 3$

   Horizontal asymptote: $y = 2$

   Decreasing on $(-\infty, -3), (-3, 3)$, and $(3, \infty)$

   Concave upward on $(3, \infty)$

   Concave downward on $(-\infty, -3)$ and $(-3, 3)$

75. Domain: All real numbers except $-1, 2$

   y int.: 0; x int.: 0, 3

   Vertical asymptote: $x = -1$

   Oblique asymptote: $y = x - 4$

   Increasing on $(-\infty, -3), (1, 2)$, and $(2, \infty)$

   Decreasing on $(-3, -1)$ and $(-1, 1)$

   Local maximum: $f(-3) = -9$

   Local minimum: $f(1) = -1$

   Concave upward on $(-1, 2)$ and $(2, \infty)$

   Concave downward on $(-\infty, -1)$

77. Domain: All real numbers except 1

   y int.: -2; x int.: -2

   Vertical asymptote: $x = 1$

   Horizontal asymptote: $y = 1$

   Decreasing on $(-\infty, 1)$ and $(1, \infty)$

   Concave upward on $(1, \infty)$

   Concave downward on $(-\infty, 1)$

79.

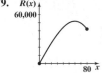

81. (A) Increasing on $(0, 1)$

   (B) Concave upward on $(0, 1)$

   (C) $x = 1$ is a vertical asymptote

   (D) The origin is both an x and a y intercept

   (E) $P(x)$

83. (A) $\overline{C}(n) = \dfrac{3{,}200}{n} + 250 + 50n$

   (B) $\overline{C}(n)$

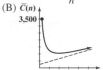

   (C) 8 yr

85. (A)

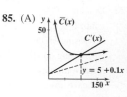

   (B) \$25 at $x = 100$

87. (A)

```
QuadReg
y=ax²+bx+c
a=.0100714286
b=.7835714286
c=316
```

   (B) Minimum average cost is \$4.35 when 177 pizzas are produced daily.

89. $C(t)$

91. $N(t)$

Exercises 12.5   **1.** Max $f(x) = f(3) = 3$; Min $f(x) = f(-2) = -2$ **3.** Max $h(x) = h(-5) = 25$; Min $h(x) = h(0) = 0$

5. Max $n(x) = n(4) = 2$; Min $n(x) = n(3) = \sqrt{3}$ **7.** Max $q(x) = q(27) = -3$; Min $q(x) = q(64) = -4$

9. Min $f(x) = f(0) = 0$; Max $f(x) = f(10) = 14$ **11.** Min $f(x) = f(0) = 0$; Max $f(x) = f(3) = 9$

13. Min $f(x) = f(1) = f(7) = 5$; Max $f(x) = f(10) = 14$ **15.** Min $f(x) = f(1) = f(7) = 5$; Max $f(x) = f(3) = f(9) = 9$

17. Min $f(x) = f(5) = 7$; Max $f(x) = f(3) = 9$ **19.** (A) Max $f(x) = f(4) = 3$; Min $f(x) = f(0) = -5$

(B) Max $f(x) = f(10) = 15$; Min $f(x) = f(0) = -5$ (C) Max $f(x) = f(10) = 15$; Min $f(x) = f(-5) = -15$

21. (A) Max $f(x) = f(-1) = f(1) = 1$; Min $f(x) = f(0) = 0$ (B) Max $f(x) = f(5) = 25$; Min $f(x) = f(1) = 1$

(C) Max $f(x) = f(-5) = f(5) = 25$; Min $f(x) = f(0) = 0$ **23.** Max $f(x) = f(-1) = e \approx 2.718$; Min $f(x) = e^{-1} \approx 0.368$

25. Max $f(x) = f(0) = 9$; Min $f(x) = f(\pm 4) = -7$ **27.** Min $f(x) = f(1) = 2$; no maximum   **29.** Max $f(x) = f(-3) = 18$; no minimum

31. No absolute extrema   **33.** Max $f(x) = f(3) = 54$; no minimum   **35.** No absolute extrema   **37.** Min $f(x) = f(0) = 0$; no maximum

39. Max $f(x) = f(1) = 1$; Min $f(x) = f(-1) = -1$ **41.** Min $f(x) = f(0) = -1$; no maximum   **43.** Min $f(x) = f(2) = -2$

45. Max $f(x) = f(2) = 4$ **47.** Min $f(x) = f(2) = 0$ **49.** No maximum   **51.** Max $f(x) = f(2) = 8$ **53.** Min $f(x) = f(4) = 22$

55. Min $f(x) = f(\sqrt{10}) = 14/\sqrt{10}$ **57.** Min $f(x) = f(2) = \dfrac{e^2}{4} \approx 1.847$ **59.** Max $f(x) = f(3) = \dfrac{27}{e^3} \approx 1.344$

61. Max $f(x) = f(e^{1.5}) = 2e^{1.5} \approx 8.963$ **63.** Max $f(x) = f(e^{2.5}) = \dfrac{e^5}{2} \approx 74.207$ **65.** Max $f(x) = f(1) = -1$

67. (A) Max $f(x) = f(5) = 14$; Min $f(x) = f(-1) = -22$ (B) Max $f(x) = f(1) = -2$; Min $f(x) = f(-1) = -22$

(C) Max $f(x) = f(5) = 14$; Min $f(x) = f(3) = -6$ **69.** (A) Max $f(x) = f(0) = 126$; Min $f(x) = f(2) = -26$

(B) Max $f(x) = f(7) = 49$; Min $f(x) = f(2) = -26$ (C) Max $f(x) = f(6) = 6$; Min $f(x) = f(3) = -15$

71. (A) Max $f(x) = f(-1) = 10$; Min $f(x) = f(2) = -11$ (B) Max $f(x) = f(0) = f(4) = 5$; Min $f(x) = f(3) = -22$

(C) Max $f(x) = f(-1) = 10$; Min $f(x) = f(1) = 2$ **73.** Local minimum   **75.** Unable to determine   **77.** Neither   **79.** Local maximum

---

Exercises 12.6   **1.** $f(x) = x(28 - x)$ **3.** $f(x) = \pi x^2/4$ **5.** $f(x) = \pi x^3$ **7.** $f(x) = x(60 - x)$ **9.** 7.5 and 7.5   **11.** 7.5 and -7.5

13. $\sqrt{15}$ and $\sqrt{15}$ **15.** $10\sqrt{2}$ ft by $10\sqrt{2}$ ft   **17.** 37 ft by 37 ft   **19.** (A) Maximum revenue is \$125,000 when 500 phones are produced and sold for \$250 each.   (B) Maximum profit is \$46,612.50 when 365 phones are produced and sold for \$317.50 each.   **21.** (A) Max $R(x) = R(3{,}000) = \$300{,}000$

(B) Maximum profit is \$75,000 when 2,100 sets are manufactured and sold for \$130 each.   (C) Maximum profit is \$64,687.50 when 2,025 sets are manufactured and sold for \$132.50 each.

23. (A)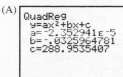
(B)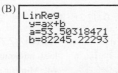
(C) The maximum profit is \$118,996 when the price per sleeping bag is \$195.

25. (A) \$4.80 (B) \$8 **27.** \$35; \$6,125 **29.** 40 trees; 1,600 lb

31. $(10 - 2\sqrt{7})/3 = 1.57$ in. squares **33.** 20 ft by 40 ft (with the expensive side being one of the short sides) **35.** (A) 70 ft by 100 ft (B) 125 ft by 125 ft **37.** 8 production runs per year **39.** 10,000 books in 5 printings **41.** 34.64 mph

43. (A) $x = 5.1$ mi (B) $x = 10$ mi **45.** 4 days; 20 bacteria/cm^3 **47.** 1 month; 2 ft **49.** 4 yr from now

Chapter 12 Review Exercises

1. (a, c_1), (c_3, c_6) *(12.1, 12.2)* **2.** (c_1, c_3), (c_6, b) *(12.1, 12.2)* **3.** (a, c_2), (c_4, c_5), (c_7, b) *(12.1, 12.2)* **4.** c_3 *(12.1)*
5. c_1, c_6 *(12.1)* **6.** c_1, c_3, c_5 *(12.1)* **7.** c_4, c_6 *(12.1)* **8.** c_2, c_4, c_5, c_7 *(12.2)*
9. *(12.2)*

10. *(12.2)*

11. $f''(x) = 12x^2 + 30x$ *(12.2)* **12.** $y'' = 8/x^3$ *(12.2)* **13.** Domain: All real numbers, except 4
y int.: $\frac{5}{4}$; x int.: -5 *(12.2)* **14.** Domain: $(-2, \infty)$ y int.: ln 2; x int.: -1 *(12.2)*
15. Horizontal asymptote: $y = 0$; Vertical asymptotes: $x = -2, x = 2$ *(12.4)* **16.** Horizontal asymptote:
$y = \frac{2}{3}$; Vertical asymptote: $x = -\frac{10}{3}$ *(12.4)* **17.** $(-\sqrt{2}, -20)$, $(\sqrt{2}, -20)$ *(12.2)*

18. $\left(-\frac{1}{2}, -6\right)$ *(12.2)* **19.** (A) $f'(x) = \frac{1}{5}x^{-4/5}$ (B) 0 (C) 0 *(12.1)* **20.** (A) $f'(x) = -\frac{1}{5}x^{-6/5}$ (B) 0 (C) None *(12.1)*

21. Domain: All real numbers
  y int.: 0; x int.: 0, 9
  Increasing on $(-\infty, 3)$ and $(9, \infty)$
  Decreasing on $(3, 9)$
  Local maximum: $f(3) = 108$
  Local minimum: $f(9) = 0$
  Concave upward on $(6, \infty)$
  Concave downward on $(-\infty, 6)$
  Inflection point: $(6, 54)$ *(12.4)*

22. Domain: All real numbers
  y int.: 16; x int.: -4, 2
  Increasing on $(-\infty, -2)$ and $(2, \infty)$
  Decreasing on $(-2, 2)$
  Local maximum: $f(-2) = 32$
  Local minimum: $f(2) = 0$
  Concave upward on $(0, \infty)$
  Concave downward on $(-\infty, 0)$
  Inflection point: $(0, 16)$ *(12.4)*

23. Domain: All real numbers
  y int.: 0; x int.: 0, 4
  Increasing on $(-\infty, 3)$
  Decreasing on $(3, \infty)$
  Local maximum: $f(3) = 54$
  Concave upward on $(0, 2)$
  Concave downward on $(-\infty, 0)$ and $(2, \infty)$
  Inflection points: $(0, 0)$, $(2, 32)$ *(12.4)*

24. Domain: all real numbers
  y int.: -3; x int.: -3, 1
  No vertical or horizontal asymptotes
  Increasing on $(-2, \infty)$
  Decreasing on $(-\infty, -2)$
  Local minimum: $f(-2) = -27$
  Concave upward on $(-\infty, -1)$ and $(1, \infty)$
  Concave downward on $(-1, 1)$
  Inflection points: $(-1, -16)$, $(1, 0)$ *(12.4)*

25. Domain: All real numbers, except -2
  y int.: 0; x int.: 0
  Horizontal asymptote: $y = 3$
  Vertical asymptote: $x = -2$
  Increasing on $(-\infty, -2)$ and $(-2, \infty)$
  Concave upward on $(-\infty, -2)$
  Concave downward on $(-2, \infty)$ *(12.4)*

26. Domain: All real numbers
  y int.: 0; x int.: 0
  Horizontal asymptote: $y = 1$
  Increasing on $(0, \infty)$
  Decreasing on $(-\infty, 0)$
  Local minimum: $f(0) = 0$
  Concave upward on $(-3, 3)$
  Concave downward on $(-\infty, -3)$ and $(3, \infty)$
  Inflection points: $(-3, 0.25)$, $(3, 0.25)$ *(12.4)*

27. Domain: All real numbers except $x = -2$
  y int.: 0; x int.: 0
  Horizontal asymptote: $y = 0$
  Vertical asymptote: $x = -2$
  Increasing on $(-2, 2)$
  Decreasing on $(-\infty, -2)$ and $(2, \infty)$
  Local maximum: $f(2) = 0.125$
  Concave upward on $(4, \infty)$
  Concave downward on $(-\infty, -2)$ and $(-2, 4)$
  Inflection point: $(4, 0.111)$ *(12.4)*

28. Domain: All real numbers
  y int.: 0; x int.: 0
  Oblique asymptote: $y = x$
  Increasing on $(-\infty, \infty)$
  Concave upward on $(-\infty, -3)$ and $(0, 3)$
  Concave downward on $(-3, 0)$ and $(3, \infty)$
  Inflection points: $(-3, -2.25)$, $(0, 0)$, $(3, 2.25)$ *(12.4)*

29. Domain: All real numbers
  y int.: 0; x int.: 0
  Horizontal asymptote: $y = 5$
  Increasing on $(-\infty, \infty)$
  Concave downward on $(-\infty, \infty)$ *(12.4)*

30. Domain: $(0, \infty)$
  x int.: 1
  Increasing on $(e^{-1/3}, \infty)$
  Decreasing on $(0, e^{-1/3})$
  Local minimum: $f(e^{-1/3}) = -0.123$
  Concave upward on $(e^{-5/6}, \infty)$
  Concave downward on $(0, e^{-5/6})$
  Inflection point: $(e^{-5/6}, -0.068)$ *(12.4)*

31. 3 *(12.3)* **32.** $-\dfrac{1}{5}$ *(12.3)* **33.** $-\infty$ *(12.3)* **34.** 0 *(12.3)* **35.** ∞ *(12.3)* **36.** 1 *(12.3)* **37.** 0 *(12.3)* **38.** 0 *(12.3)* **39.** 1 *(12.3)* **40.** 2 *(12.3)*

41.

x	$f'(x)$	$f(x)$
$-\infty < x < -2$	Negative and increasing	Decreasing and concave upward
$x = -2$	x intercept	Local minimum
$-2 < x < -1$	Positive and increasing	Increasing and concave upward
$x = -1$	Local maximum	Inflection point
$-1 < x < 1$	Positive and decreasing	Increasing and concave downward
$x = 1$	Local min., x intercept	Inflection point, horiz. tangent
$1 < x < \infty$	Positive and increasing	Increasing and concave upward

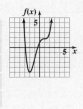
(12.2)

42. (C) *(12.2)* **43.** Local maximum: $f(-1) = 20$; local minimum $f(5) = -88$ *(12.5)*
44. Min $f(x) = f(2) = -4$; Max $f(x) = f(5) = 77$ *(12.5)*
45. Min $f(x) = f(2) = 8$ *(12.5)*
46. Max $f(x) = f(e^{4.5}) = 2e^{4.5} \approx 180.03$ *(12.5)*
47. Max $f(x) = f(0.5) = 5e^{-1} \approx 1.84$ *(12.5)*
48. Yes. Since f is continuous on $[a, b]$, f has an absolute maximum on $[a, b]$. But neither $f(a)$ nor $f(b)$ is an absolute maximum, so the absolute maximum must occur between a and b. *(12.5)*

49. No, increasing/decreasing properties apply to intervals in the domain of f. It is correct to say that $f(x)$ is decreasing on $(-\infty, 0)$ and $(0, \infty)$. *(12.1)*
50. A critical number of $f(x)$ is a partition number for $f'(x)$ that is also in the domain of f. For example, if $f(x) = x^{-1}$, then 0 is a partition number for $f'(x) = -x^{-2}$, but 0 is not a critical number of $f(x)$ since 0 is not in the domain of f. *(12.1)*
51. Max $f'(x) = f'(2) = 12$ *(12.2, 12.5)*

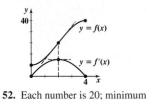

52. Each number is 20; minimum sum is 40 *(12.6)*

53. Domain: All real numbers
x int.: 0.79, 1.64; y int.: 4
Increasing on $(-1.68, -0.35)$ and $(1.28, \infty)$
Decreasing on $(-\infty, -1.68)$ and $(-0.35, 1.28)$
Local minima: $f(-1.68) = 0.97$, $f(1.28) = -1.61$
Local maximum: $f(-0.35) = 4.53$
Concave downward on $(-1.10, 0.60)$
Concave upward on $(-\infty, -1.10)$ and $(0.60, \infty)$
Inflection points: $(-1.10, 2.58)$, $(0.60, 1.08)$

54. Domain: All real numbers
x intercepts: 0, 11.10; y int.: 0
Increasing on $(1.87, 4.19)$ and $(8.94, \infty)$
Decreasing on $(-\infty, 1.87)$ and $(4.19, 8.94)$
Local maximum: $f(4.19) = -39.81$
Local minima: $f(1.87) = -52.14$, $f(8.94) = -123.81$
Concave upward on $(-\infty, 2.92)$ and $(7.08, \infty)$
Concave downward on $(2.92, 7.08)$
Inflection points: $(2.92, -46.41)$, $(7.08, -88.04)$ *(12.4)*

55. Max $f(x) = f(1.373) = 2.487$ *(12.5)* **56.** Max $f(x) = f(1.763) = 0.097$ *(12.5)*

57. (A) For the first 15 months, the graph of the price is increasing and concave downward, with a local maximum at $t = 15$.
For the next 15 months, the graph of the price is decreasing and concave downward, with an inflection point at $t = 30$.
For the next 15 months, the graph of the price is decreasing and concave upward, with a local minimum at $t = 45$.
For the remaining 15 months, the graph of the price is increasing and concave upward.

(B)
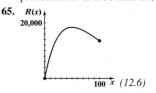
(12.2)

58. (A) Max $R(x) = R(10,000) = \$2,500,000$
(B) Maximum profit is $175,000 when 3,000 readers are manufactured and sold for $425 each.
(C) Maximum profit is $119,000 when 2,600 readers are manufactured and sold for $435 each. *(12.6)*
59. (A) The expensive side is 50 ft; the other side is 100 ft. (B) The expensive side is 75 ft; the other side is 150 ft. *(12.6)* **60.** $49; $6,724 *(12.6)*
61. 12 orders/yr *(12.6)*
66. $549.15; $9,864 *(12.6)*
67. $1.52 *(12.6)*
68. 20.39 feet *(12.6)*

62. Min $\overline{C}(x) = \overline{C}(200) = \$50$ *(12.4)*

63. Min $\overline{C}(x) = \overline{C}(e^5) \approx \$49.66$ *(12.4)*
64. A maximum revenue of $18,394 is realized at a production level of 50 units at $367.88 each. *(12.6)*

65.
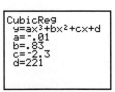
(12.6)

69. (A)
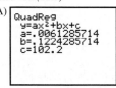

```
QuadReg
y=ax²+bx+c
a=.0061285714
b=.1224285714
c=102.2
```

(B) Min $\overline{C}(x) = \overline{C}(129) = \$1.71$ *(12.4)*

70. Increasing on $(0, 18)$; decreasing on $(18, 24)$; point of diminishing returns is $x = 18$, max $N'(x) = N'(18) = 972$ *(12.2)*

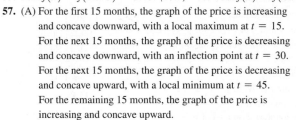

71. (A)
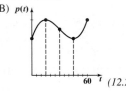

```
CubicReg
y=ax³+bx²+cx+d
a=-.01
b=.83
c=-2.3
d=221
```

(B) 28 ads to sell 588 refrigerators per month *(12.2)*
72. 3 days *(12.1)*
73. 2 yr from now *(12.1)*

Chapter 13

Exercises 13.1   **1.** $f(x) = 5x^{-4}$ **3.** $f(x) = 3x^{-4} - 2x^{-5}$ **5.** $f(x) = x^{1/2} + 5x^{-1/2}$ **7.** $f(x) = 4x^{1/3} + x^{4/3} - 3x^{7/3}$ **9.** $7x + C$ **11.** $4x^2 + C$
13. $3x^3 + C$ **15.** $(x^6/6) + C$ **17.** $(-x^{-2}/2) + C$ **19.** $4x^{5/2} + C$ **21.** $3\ln|x| + C$ **23.** $16e^u + C$ **25.** Yes **27.** Yes **29.** No **31.** No
33. True **35.** False **37.** True **39.** No, since one graph cannot be obtained from another by a vertical translation. **41.** Yes, since one graph can be obtained from another by a vertical translation. **43.** $(5x^2/2) - (5x^3/3) + C$ **45.** $2\sqrt{u} + C$ **47.** $-(x^{-2}/8) + C$ **49.** $4\ln|u| + u + C$ **51.** $5e^z + 4z + C$
53. $x^3 + 2x^{-1} + C$ **55.** $C(x) = 2x^3 - 2x^2 + 3,000$ **57.** $x = 40\sqrt{t}$ **59.** $y = -2x^{-1} + 3\ln|x| - x + 3$ **61.** $x = 4e^t - 2t - 3$
63. $y = 2x^2 - 3x + 1$ **65.** $x^2 + x^{-1} + C$ **67.** $\frac{1}{2}x^2 + x^{-2} + C$ **69.** $e^x - 2\ln|x| + C$ **71.** x^3 **73.** $x^4 + 3x^2 + C$

81. $\overline{C}(x) = 15 + \dfrac{1,000}{x}$; $C(x) = 15x + 1,000$; $C(0) = \$1,000$ **83.** (A) The cost function increases from 0 to 8, is concave downward from 0 to 4, and is

concave upward from 4 to 8. There is an inflection point at $x = 4$.   (B) $C(x) = x^3 - 12x^2 + 53x + 30$; $C(4) = \$114,000$; $C(8) = \$198,000$

(C) 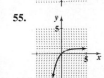   (D) Manufacturing plants are often inefficient at low and high levels of production.

85. $S(t) = 1,200 - 18t^{4/3}$; $50^{3/4} \approx 19$ mo   **87.** $S(t) = 1,200 - 18t^{4/3} - 70t$; $t \approx 8.44$ mo

89. $L(x) = 4,800x^{1/2}$; 24,000 labor-hours   **91.** $W(h) = 0.0005h^3$; 171.5 lb   **93.** 19,400

Exercises 13.2   **1.** $f'(x) = 50(5x + 1)^9$ **3.** $f'(x) = 14x(x^2 + 1)^6$ **5.** $f'(x) = 2xe^{x^2}$ **7.** $f'(x) = \dfrac{4x^3}{x^4 - 10}$ **9.** $\frac{1}{3}(3x + 5)^3 + C$

11. $\frac{1}{6}(x^2 - 1)^6 + C$ **13.** $-\frac{1}{2}(5x^3 + 1)^{-2} + C$ **15.** $e^{5x} + C$ **17.** $\ln|1 + x^2| + C$ **19.** $\frac{2}{3}(1 + x^4)^{3/2} + C$ **21.** $\frac{1}{11}(x + 3)^{11} + C$

23. $-\frac{1}{6}(6t - 7)^{-1} + C$ **25.** $\frac{1}{12}(t^2 + 1)^6 + C$ **27.** $\frac{1}{2}e^{x^2} + C$ **29.** $\frac{1}{5}\ln|5x + 4| + C$ **31.** $-e^{1-t} + C$ **33.** $-\frac{1}{18}(3t^2 + 1)^{-3} + C$

35. $\frac{2}{5}(x + 4)^{5/2} - \frac{8}{3}(x + 4)^{3/2} + C$ **37.** $\frac{2}{3}(x - 3)^{3/2} + 6(x - 3)^{1/2} + C$ **39.** $\frac{1}{11}(x - 4)^{11} + \frac{2}{5}(x - 4)^{10} + C$ **41.** $\frac{1}{8}(1 + e^{2x})^4 + C$

43. $\frac{1}{2}\ln|4 + 2x + x^2| + C$ **45.** $\frac{1}{2}(5x + 3)^2 + C$ **47.** $\frac{1}{2}(x^2 - 1)^2 + C$ **49.** $\frac{1}{5}(x^5)^5 + C$ **51.** No   **53.** Yes   **55.** No   **57.** Yes

59. $\frac{1}{9}(3x^2 + 7)^{3/2} + C$ **61.** $\frac{1}{8}x^8 + \frac{4}{5}x^5 + 2x^2 + C$ **63.** $\frac{1}{9}(x^3 + 2)^3 + C$ **65.** $\frac{1}{4}(2x^4 + 3)^{1/2} + C$ **67.** $\frac{1}{4}(\ln x)^4 + C$ **69.** $e^{-1/x} + C$

71. $x = \frac{1}{3}(t^3 + 5)^7 + C$ **73.** $y = 3(t^2 - 4)^{1/2} + C$ **75.** $p = -(e^x - e^{-x})^{-1} + C$ **77.** $p(x) = \dfrac{2,000}{3x + 50} + 4$; 250 bottles

79. $C(x) = 12x + 500\ln(x + 1) + 2,000$; $\overline{C}(1,000) = \$17.45$ **81.** (A) $S(t) = 10t + 100e^{-0.1t} - 100$, $0 \le t \le 24$ (B) \$50 million   (C) 18.41 mo

83. $Q(t) = 100\ln(t + 1) + 5t$, $0 \le t \le 20$; 275 thousand barrels   **85.** $W(t) = 2e^{0.1t}$; 4.45 g   **87.** (A) $-1,000$ bacteria/mL per day

(B) $N(t) = 5,000 - 1,000\ln(1 + t^2)$; 385 bacteria/mL   (C) 7.32 days   **89.** $N(t) = 100 - 60e^{-0.1t}$, $0 \le t \le 15$; 87 words/min

91. $E(t) = 12,000 - 10,000(t + 1)^{-1/2}$; 9,500 students

Exercises 13.3   **1.** The derivative of $f(x)$ is 5 times $f(x)$; $y' = 5y$ **3.** The derivative of $f(x)$ is -1 times $f(x)$; $y' = -y$

5. The derivative of $f(x)$ is $2x$ times $f(x)$; $y' = 2xy$ **7.** The derivative of $f(x)$ is 1 minus $f(x)$; $y' = 1 - y$ **9.** $y = 3x^2 + C$ **11.** $y = 7\ln|x| + C$

13. $y = 50e^{0.02x} + C$ **15.** $y = \dfrac{x^3}{3} - \dfrac{x^2}{2}$ **17.** $y = e^{-x^2} + 2$ **19.** $y = 2\ln|1 + x| + 5$ **21.** Second-order   **23.** Third-order   **25.** Yes

27. Yes   **29.** Yes   **31.** Yes   **33.** Figure B. When $x = 1$, the slope $dy/dx = 1 - 1 = 0$ for any y. When $x = 0$, the slope $dy/dx = 0 - 1 = -1$

for any y. Both are consistent with the slope field shown in Figure B.   **35.** $y = \dfrac{x^2}{2} - x + C$; $y = \dfrac{x^2}{2} - x - 2$

37. 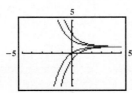   **39.** $y = Ce^{2t}$ **41.** $y = 100e^{-0.5x}$ **43.** $x = Ce^{-5t}$ **45.** $x = -(5t^2/2) + C$ **47.** Logistic growth   **49.** Unlimited growth

51. Figure A. When $y = 1$, the slope $dy/dx = 1 - 1 = 0$ for any x. When $y = 2$, the slope $dy/dx = 1 - 2 = -1$ for any x. Both are

consistent with the slope field shown in Figure A.   **53.** $y = 1 - e^{-x}$

55. 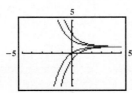   **57.** 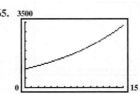   **59.** $y = \sqrt{25 - x^2}$ **65.** 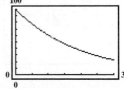   **67.**

61. $y = -3x$

63. $y = 1/(1 - 2e^{-t})$

69. 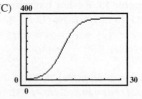   **71.**   **73.** Apply the second-derivative test to $f(y) = ky(M - y)$.   **75.** 2009

77. $A = 1,000e^{0.03t}$ **79.** $A = 8,000e^{0.016t}$ **81.** (A) $p(x) = 100e^{-0.05x}$

(B) \$60.65 per unit   (C)

83. (A) $N = L(1 - e^{-0.051t})$ **85.** $I = I_0 e^{-0.00942x}$; $x \approx 74$ ft   **89.** 0.023 117   **91.** Approx. 24,200 yr   **93.** 104 times; 67 times

(B) 22.5%   **87.** (A) $Q = 3e^{-0.04t}$ **95.** (A) 7 people; 353 people   (B) 400   (C)

(C) 32 days   (B) $Q(10) = 2.01$ mL

(D) 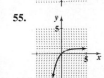   (C) 27.47 hr

(D)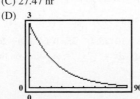

Exercises 13.4   **1.** 80 in.^{2} **3.** 36 m^2 **5.** No, $\pi - 2 > 1$ m^2 **7.** C, E   **9.** B   **11.** H, I   **13.** H

15. 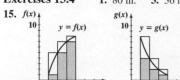   **17.** Figure A: $L_3 = 13$, $R_3 = 20$; Figure B: $L_3 = 14$, $R_3 = 7$

19. $L_3 \leq \int_1^4 f(x)\, dx \leq R_3$; $R_3 \leq \int_1^4 g(x)\, dx \leq L_3$; since $f(x)$ is increasing, L_3 underestimates the area and R_3 overestimates the area; since $g(x)$ is decreasing, the reverse is true.   **21.** In both figures, the error bound for L_3 and R_3 is 7.   **23.** $S_5 = -260$ **25.** $S_4 = -1{,}194$ **27.** $S_3 = -33.01$ **29.** $S_6 = -38$ **31.** -2.475

33. 4.266   **35.** 2.474   **37.** -5.333 **39.** 1.067   **41.** -1.066 **43.** 15   **45.** 58.5   **47.** -54 **49.** 248   **51.** 0   **53.** -183 **55.** False   **57.** False
59. False   **61.** $L_{10} = 286{,}100$ ft^2; error bound is 50,000 ft^2; $n \geq 200$ **63.** $L_6 = -3.53$, $R_6 = -0.91$; error bound for L_6 and R_6 is 2.63. Geometrically, the definite integral over the interval [2, 5] is the sum of the areas between the curve and the x axis from $x = 2$ to $x = 5$, with the areas below the x axis counted negatively and those above the x axis counted positively.   **65.** Increasing on $(-\infty, 0]$; decreasing on $[0, \infty)$ **67.** Increasing on $[-1, 0]$ and $[1, \infty)$; decreasing on $(-\infty, -1]$ and $[0, 1]$ **69.** $n \geq 22$ **71.** $n \geq 104$ **73.** $L_3 = 2{,}580$, $R_3 = 3{,}900$; error bound for L_3 and R_3 is 1,320

75. (A) $L_5 = 3.72$; $R_5 = 3.37$ (B) $R_5 = 3.37 \leq \int_0^5 A'(t)\, dt \leq 3.72 = L_5$ **77.** $L_3 = 114$, $R_3 = 102$; error bound for L_3 and R_3 is 12

Exercises 13.5   **1.** 500   **3.** 28   **5.** 48   **7.** $4.5\pi \approx 14.14$ **9.** (A) $F(15) - F(10) = 375$ **11.** (A) $F(15) - F(10) = 85$ **13.** 40   **15.** 72   **17.** 46.5

(B)   (B)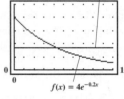

19. $e - 1 \approx 1.718$
21. $\ln 2 \approx 0.693$
23. 0   **25.** 48
27. -48 **29.** -10.25
31. 0   **33.** -2 **35.** 14

37. $5^6 = 15{,}625$ **39.** $\ln 4 \approx 1.386$ **41.** $20(e^{0.25} - e^{-0.5}) \approx 13.550$ **43.** $\frac{1}{2}$ **45.** $\frac{1}{2}(1 - e^{-1}) \approx 0.316$ **47.** 0
49. (A) Average $f(x) = 250$ **51.** (A) Average $f(t) = 2$ **53.** (A) Average $f(x) = \frac{45}{28} \approx 1.61$ **55.** (A) Average $f(x) = 2(1 - e^{-2}) \approx 1.73$

(B) 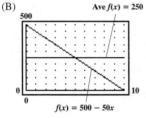   (B) 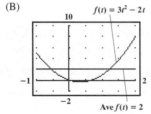   (B) 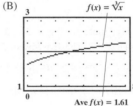   (B)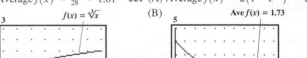

57. $\frac{1}{6}(15^{3/2} - 5^{3/2}) \approx 7.819$ **59.** $\frac{1}{2}(\ln 2 - \ln 3) \approx -0.203$ **61.** 0   **63.** 4.566   **65.** 2.214   **69.** $\int_{300}^{900}\left(500 - \frac{x}{3}\right) dx = \$180{,}000$

71. $\int_0^5 500(t - 12)\, dt = -\$23{,}750$; $\int_5^{10} 500(t - 12)\, dt = -\$11{,}250$

73. (A) 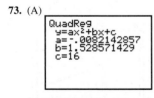   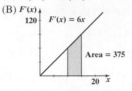   **75.** Useful life $= \sqrt{\ln 55} \approx 2$ yr; total profit $= \frac{51}{22} - \frac{5}{2} e^{-4} \approx 2.272$ or \$2,272
77. (A) \$420   (B) \$135,000

(B) 6,505

79. (A) 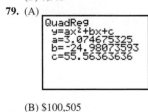   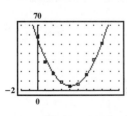   **81.** $50e^{0.6} - 50e^{0.4} - 10 \approx \$6.51$ **83.** 4,800 labor-hours   **85.** (A) $I = -200t + 600$
(B) $\frac{1}{3}\int_0^3 (-200t + 600)\, dt = 300$ **87.** $100 \ln 11 + 50 \approx 290$ thousand barrels;
$100 \ln 21 - 100 \ln 11 + 50 \approx 115$ thousand barrels
89. $2e^{0.8} - 2 \approx 2.45$ g; $2e^{1.6} - 2e^{0.8} \approx 5.45$ g
91. 10°C   **93.** $0.6 \ln 2 + 0.1 \approx 0.516$; $(4.2 \ln 625 + 2.4 - 4.2 \ln 49)/24 \approx 0.546$

(B) \$100,505

Chapter 13 Review Exercises   **1.** $3x^2 + 3x + C$ *(13.1)*   **2.** 50 *(13.5)*   **3.** -207 *(13.5)*   **4.** $-\frac{1}{8}(1 - t^2)^4 + C$ *(13.2)*   **5.** $\ln|u| + \frac{1}{4}u^4 + C$ *(13.1)*
6. 0.216 *(13.5)*   **7.** No *(13.1)*   **8.** Yes *(13.1)*   **9.** No *(13.1)*   **10.** Yes *(13.1)*   **11.** Yes *(13.3)*   **12.** Yes *(13.3)*   **13.** e^{-x^2} *(13.1)*   **14.** $\sqrt{4 + 5x} + C$ *(13.1)*
15. $y = f(x) = x^3 - 2x + 4$ *(13.3)*   **16.** (A) $2x^4 - 2x^2 - x + C$ (B) $e^t - 4\ln|t| + C$ *(13.1)*   **17.** $R_2 = 72$; error bound for R_2 is 48 *(13.4)*
18. $\int_1^5 (x^2 + 1)\, dx = \frac{136}{3} \approx 45.33$; actual error is $\frac{80}{3} \approx 26.67$ *(13.5)*   **19.** $L_4 = 30.8$ *(13.4)*   **20.** 7 *(13.5)*   **21.** Width $= 2 - (-1) = 3$;
height $=$ average $f(x) = 7$ *(13.5)*   **22.** $S_4 = 368$ *(13.4)*   **23.** $S_5 = 906$ *(13.4)*   **24.** -10 *(13.4)*   **25.** 0.4 *(13.4)*   **26.** 1.4 *(13.4)*   **27.** 0 *(13.4)*   **28.** 0.4 *(13.4)*
29. 2 *(13.4)*   **30.** -2 *(13.4)*   **31.** -0.4 *(13.4)*   **32.** (A) 1; 1   (B) 4; 4 *(13.3)*   **33.** $dy/dx = (2y)/x$; at points on the x axis $(y = 0)$ the slopes are 0. *(13.3)*
35. $y = \frac{1}{4}x^2$; $y = -\frac{1}{4}x^2$ *(13.3)*

36. *(13.3)*   **37.** *(13.3)*

          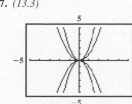

38. $\frac{2}{3}(2)^{3/2} \approx 1.886$ *(13.5)*   **39.** $\frac{1}{6} \approx 0.167$ *(13.5)*   **40.** $-5e^{-t} + C$ *(13.1)*   **41.** $\frac{1}{2}(1 + e^2)$ *(13.1)*

42. $\frac{1}{6}e^{3x^2} + C$ *(13.2)*   **43.** $2(\sqrt{5} - 1) \approx 2.472$ *(13.5)*   **44.** $\frac{1}{2}\ln 10 \approx 1.151$ *(13.5)*

45. 0.45 *(13.5)*   **46.** $\frac{1}{48}(2x^4 + 5)^6 + C$ *(13.2)*   **47.** $-\ln(e^{-x} + 3) + C$ *(13.2)*

48. $-(e^x + 2)^{-1} + C$ *(13.2)*   **49.** $y = f(x) = 3\ln|x| + x^{-1} + 4$ *(13.2, 13.3)*

50. $y = 3x^2 + x - 4$ *(13.3)*

51. (A) Average $f(x) = 6.5$

(B)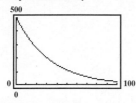

52. $\frac{1}{3}(\ln x)^3 + C$ *(13.2)*   **53.** $\frac{1}{8}x^8 - \frac{2}{5}x^5 + \frac{1}{2}x^2 + C$ *(13.2)*

54. $\frac{2}{3}(6 - x)^{3/2} - 12(6 - x)^{1/2} + C$ *(13.2)*   **55.** $\frac{1,234}{15} \approx 82.267$ *(13.5)*

56. 0 *(13.5)*   **57.** $y = 3e^{x^3} - 1$ *(13.3)*   **58.** $N = 800e^{0.06t}$ *(13.3)*

59. Limited growth

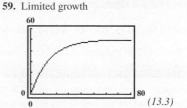

(13.3)

60. Exponential decay   **61.** Unlimited growth   **62.** Logistic growth

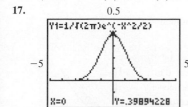 (13.3)   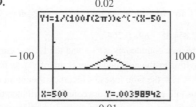 (13.3)   (13.3)

63. 1.167 *(13.5)*   **64.** 99.074 *(13.5)*   **65.** -0.153 *(13.5)*   **66.** $L_2 = \$180,000$; $R_2 = \$140,000$; $\$140,000 \le \int_{200}^{600} C'(x)\,dx \le \$180,000$ *(13.4)*

67. $\int_{200}^{600}\left(600 - \frac{x}{2}\right)dx = \$160,000$ *(13.5)*   **68.** $\int_{10}^{40}\left(150 - \frac{x}{10}\right)dx = \$4,425$ *(13.5)*   **69.** $P(x) = 100x - 0.01x^2$; $P(10) = \$999$ *(13.3)*

70. $\int_0^{15}(60 - 4t)\,dt = 450$ thousand barrels *(13.5)*   **71.** 109 items *(13.5)*   **72.** $16e^{2.5} - 16e^2 - 8 \approx \$68.70$ *(13.5)*   **73.** Useful life $= 10\ln\frac{20}{3} \approx 19$ yr;

total profit $= 143 - 200e^{-1.9} \approx 113.086$ or $\$113,086$ *(13.5)*   **74.** $S(t) = 50 - 50e^{-0.08t}$; $50 - 50e^{-0.96} \approx \$31$ million; $-(\ln 0.2)/0.08 \approx 20$ mo *(13.3)*

75. 1 cm^2 *(13.3)*   **76.** 800 gal *(13.5)*   **77.** (A) 132 million   (B) About 65 years *(13.3)*   **78.** $\frac{-\ln 0.04}{0.000\,123\,8} \approx 26,000$ yr *(13.3)*

79. $N(t) = 95 - 70e^{-0.1t}$; $N(15) \approx 79$ words/min *(13.3)*

Chapter 14

Exercises 14.1   **1.** 150 **3.** 37.5 **5.** 25.5 **7.** $\pi/2$ **9.** $\int_a^b g(x)\,dx$ **11.** $\int_a^b[-h(x)]\,dx$ **13.** Since the shaded region in Figure C is below the x axis, $h(x) \le 0$; so, $\int_a^b h(x)\,dx$ represents the negative of the area.   **15.** 24 **17.** 51 **19.** 42 **21.** 6 **23.** 0.833 **25.** 2.350 **27.** 1 **29.** Canada; Mexico

31. India; Brazil   **33.** $\int_a^b[-f(x)]\,dx$ **35.** $\int_b^c f(x)\,dx + \int_a^b[-f(x)]\,dx$ **37.** $\int_c^d[f(x) - g(x)]\,dx$ **39.** $\int_a^b[f(x) - g(x)]\,dx + \int_b^c[g(x) - f(x)]\,dx$

41. Find the intersection points by solving $f(x) = g(x)$ on the interval $[a, d]$ to determine b and c.

Then observe that $f(x) > g(x)$ over $[a, b]$, $g(x) \ge f(x)$ over $[b, c]$, and $f(x) \ge g(x)$ over $[c, d]$.

Area $= \int_a^b[f(x) - g(x)]\,dx + \int_b^c[g(x) - f(x)]\,dx + \int_c^d[f(x) - g(x)]\,dx$.

43. 2.5 **45.** 7.667 **47.** 12 **49.** 15 **51.** 32 **53.** 36 **55.** 9 **57.** 2.832 **59.** $\int_{-3}^3\sqrt{9 - x^2}\,dx$; 14.137 **61.** $\int_0^4\sqrt{16 - x^2}\,dx$; 12.566

63. $\int_{-2}^2 2\sqrt{4 - x^2}\,dx$; 12.566 **65.** 52.616 **67.** 8 **69.** 101.75 **71.** 17.979 **73.** 5.113 **75.** 8.290 **77.** 3.166 **79.** 1.385 **81.** 3.17 **83.** 1.82

85. Total production from the end of the fifth year to the end of the 10th year is $50 + 100\ln 20 - 100\ln 15 \approx 79$ thousand barrels.   **87.** Total profit over the 5-yr useful life of the game is $20 - 30e^{-1.5} \approx 13.306$, or $\$13,306$.   **89.** 1935: 0.412; 1947: 0.231; income was more equality distributed in 1947.

91. 1963: 0.818; 1983: 0.846; total assets were less equally distributed in 1983.   **93.** (A) $f(x) = 0.3125x^2 + 0.7175x - 0.015$ (B) 0.104

95. Total weight gain during the first 10 hr is $3e - 3 \approx 5.15$ g.   **97.** Average number of words learned from $t = 2$ hr to $t = 4$ hr is $15\ln 4 - 15\ln 2 \approx 10$.

Exercises 14.2   **1.** $b = 20$; $c = -5$ **3.** $b = 0.32$; $c = -0.04$ **5.** $b = 1.6$; $c = -0.03$ **7.** $b = 1.75$; $c = 0.02$ **9.** 10.27 **11.** 151.75

13. $93,268.66$ **15.** (A) 10.72 (B) 3.28 (C) 10.72

17.

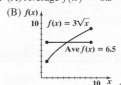

19.

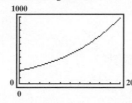

21. (A) 0.6827 (B) 0.9545 (C) 0.9973

23. (A) 0.6827 (B) 0.9545 (C) 0.9973

25. (A) $.75$ (B) $.11$ (C)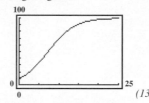

27. 8 yr   **29.** (A) $.11$ (B) $.10$ **31.** $P(t \ge 12) = 1 - P(0 \le t \le 12) = .89$ **33.** $\$12,500$

35. If $f(t)$ is the rate of flow of a continuous income stream, then the total income produced from 0 to 5 yr is the area under the graph of $y = f(t)$ from $t = 0$ to $t = 5$.

37. $8,000(e^{0.15} - 1) \approx \$1,295$

39. If $f(t)$ is the rate of flow of a continuous income stream, then the total income produced from 0 to 3 yr is the area under the graph of $y = f(t)$ from $t = 0$ to $t = 3$.

41. $255,562; $175,562   **43.** $6,780   **45.** $437   **47.** Clothing store: $66,421; computer store: $62,623; the clothing store is the better investment.

49. Bond: $12,062 business: $11,824; the bond is the better investment.   **51.** $55,230   **53.** $\dfrac{k}{r}(e^{rT} - 1)$ **55.** $625,000

57. The shaded area is the consumers' surplus and represents the total savings to consumers who are willing to pay more than $150 for a product but are still able to buy the product for $150.

59. $9,900

61. The area of the region PS is the producers' surplus and represents the total gain to producers who are willing to supply units at a lower price than $67 but are still able to supply the product at $67.

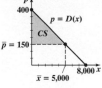

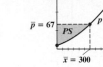

63. $CS = \$3,380; PS = \$1,690$ **65.** $CS = \$6,980; PS = \$5,041$ **67.** $CS = \$7,810; PS = \$8,336$ **69.** $CS = \$8,544; PS = \$11,507$

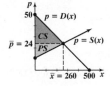

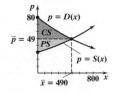

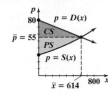

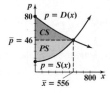

71. (A) $\bar{x} = 21.457; \bar{p} = \$6.51$ (B) $CS = 1.774$ or $1,774; PS = 1.087$ or $1,087$

Exercises 14.3   **1.** $f'(x) = 5; \displaystyle\int g(x)\,dx = \dfrac{x^4}{4} + C$ **3.** $f'(x) = 3x^2; \displaystyle\int g(x)\,dx = \dfrac{5x^2}{2} + C$ **5.** $f'(x) = 4e^{4x}; \displaystyle\int g(x)\,dx = \ln|x| + C$

7. $f'(x) = -x^{-2}; \displaystyle\int g(x)\,dx = \dfrac{1}{4}e^{4x} + C$ **9.** $\frac{1}{3}xe^{3x} - \frac{1}{9}e^{3x} + C$ **11.** $\dfrac{x^3}{3}\ln x - \dfrac{x^3}{9} + C$ **13.** $u = x + 2; \dfrac{(x+2)(x+1)^6}{6} - \dfrac{(x+1)^7}{42} + C$

15. $-xe^{-x} - e^{-x} + C$ **17.** $\frac{1}{2}e^{x^2} + C$ **19.** $(xe^x - 4e^x)|_0^1 = -3e + 4 \approx -4.1548$ **21.** $(x\ln 2x - x)|_1^3 = (3\ln 6 - 3) - (\ln 2 - 1) \approx 2.6821$

23. $\ln(x^2 + 1) + C$ **25.** $(\ln x)^2/2 + C$ **27.** $\frac{2}{3}x^{3/2}\ln x - \frac{4}{9}x^{3/2} + C$ **29.** $\dfrac{(x-3)(x+1)^3}{3} - \dfrac{(x+1)^4}{12} + C$ or $\dfrac{x^4}{4} - \dfrac{x^3}{3} - \dfrac{5x^2}{2} - 3x + C$

31. $\dfrac{(2x+1)(x-2)^3}{3} - \dfrac{(x-2)^4}{6} + C$ or $\dfrac{x^4}{2} - \dfrac{7x^3}{3} + 2x^2 + 4x + C$

33. The integral represents the negative of the area between the graph of $y = (x - 3)e^x$ and the x axis from $x = 0$ to $x = 1$.

35. The integral represents the area between the graph of $y = \ln 2x$ and the x axis from $x = 1$ to $x = 3$.

37. $(x^2 - 2x + 2)e^x + C$

39. $\dfrac{xe^{ax}}{a} - \dfrac{e^{ax}}{a^2} + C$

41. $\left(-\dfrac{\ln x}{x} - \dfrac{1}{x}\right)\Big|_1^e = -\dfrac{2}{e} + 1 \approx 0.2642$

43. $6\ln 6 - 4\ln 4 - 2 \approx 3.205$

45. $xe^{x-2} - e^{x-2} + C$

47. $\frac{1}{2}(1 + x^2)\ln(1 + x^2) - \frac{1}{2}(1 + x^2) + C$ **49.** $(1 + e^x)\ln(1 + e^x) - (1 + e^x) + C$ **51.** $x(\ln x)^2 - 2x\ln x + 2x + C$

53. $x(\ln x)^3 - 3x(\ln x)^2 + 6x\ln x - 6x + C$ **55.** 2   **57.** $\dfrac{1}{3}$ **59.** 1.56   **61.** 34.98   **63.** $\int_0^5 (2t - te^{-t})\,dt = \$24$ million

65. The total profit for the first 5 yr (in millions of dollars) is the same as the area under the marginal profit function, $P'(t) = 2t - te^{-t}$, from $t = 0$ to $t = 5$.

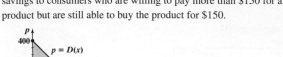

67. $2,854.88   **69.** 0.264

71. The area bounded by $y = x$ and the Lorenz curve $y = xe^{x-1}$, divided by the area under the graph of $y = x$ from $x = 0$ to $x = 1$, is the Gini index of income concentration. The closer this index is to 0, the more equally distributed the income; the closer the index is to 1, the more concentrated the income in a few hands.

73. $S(t) = 1,600 + 400e^{0.1t} - 40te^{0.1t}$; 15 mo   **75.** \$977

77. The area bounded by the price–demand equation,
$p = 9 - \ln(x + 4)$, and the price equation, $y = \bar{p} = 2.089$, from
$x = 0$ to $x = \bar{x} = 1,000$, represents the consumers' surplus. This is
the amount consumers who are willing to pay more than \$2.089 save.

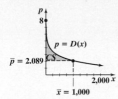

79. 2.1388 ppm   **81.** $N(t) = -4te^{-0.25t} - 40e^{-0.25t} + 80$; 8 wk; 78 words/min

83. 20,980

Exercises 14.4   **1.** 77.32   **3.** 74.15   **5.** 0.47   **7.** 0.46   **9.** $\ln\left|\dfrac{x}{1+x}\right| + C$ **11.** $\dfrac{1}{3+x} + 2\ln\left|\dfrac{5+2x}{3+x}\right| + C$ **13.** $\dfrac{2(x-32)}{3}\sqrt{16+x} + C$

15. $-\ln\left|\dfrac{1+\sqrt{1-x^2}}{x}\right| + C$ **17.** $\dfrac{1}{2}\ln\left|\dfrac{x}{2+\sqrt{x^2+4}}\right| + C$ **19.** $\frac{1}{3}x^3\ln x - \frac{1}{9}x^3 + C$ **21.** $x - \ln|1 + e^x| + C$ **23.** $9\ln\frac{3}{2} - 2 \approx 1.6492$

25. $\frac{1}{2}\ln\frac{12}{5} \approx 0.4377$ **27.** $\ln 3 \approx 1.0986$ **29.** 730; $723\frac{1}{3}$ **31.** 1.61; $\ln 5 \approx 1.61$ **33.** 69; 69   **35.** 756; 756

37. $-\dfrac{\sqrt{4x^2+1}}{x} + 2\ln|2x + \sqrt{4x^2+1}| + C$ **39.** $\frac{1}{2}\ln|x^2 + \sqrt{x^4-16}| + C$ **41.** $\frac{1}{6}(x^3\sqrt{x^6+4} + 4\ln|x^3 + \sqrt{x^6+4}|) + C$

43. $-\dfrac{\sqrt{4-x^4}}{8x^2} + C$ **45.** $\dfrac{1}{5}\ln\left|\dfrac{3+4e^x}{2+e^x}\right| + C$ **47.** $\frac{2}{3}(\ln x - 8)\sqrt{4 + \ln x} + C$ **49.** $\frac{1}{5}x^2e^{5x} - \frac{2}{25}xe^{5x} + \frac{2}{125}e^{5x} + C$

51. $-x^3e^{-x} - 3x^2e^{-x} - 6xe^{-x} - 6e^{-x} + C$ **53.** $x(\ln x)^3 - 3x(\ln x)^2 + 6x\ln x - 6x + C$ **55.** $\frac{64}{3}$ **57.** $\frac{1}{2}\ln\frac{9}{5} \approx 0.2939$ **59.** $\dfrac{-1-\ln x}{x} + C$

61. $\sqrt{x^2 - 1} + C$ **65.** 31.38   **67.** 5.48   **69.** $3,000 + 1,500\ln\dfrac{1}{3} \approx \$1,352$

71.

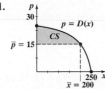

73. $C(x) = 200x + 1,000\ln(1 + 0.05x) + 25,000$; 608; \$198,773   **75.** \$18,673.95   **77.** 0.1407

79. As the area bounded by the two curves gets smaller, the Lorenz
curve approaches $y = x$ and the distribution of income approaches
perfect equality—all persons share equally in the income available.

81. $S(t) = 1 + t - \dfrac{1}{1+t} - 2\ln|1 + t|$; $24.96 - 2\ln 25 \approx \$18.5$ million

83. The total sales (in millions of dollars) over the first 2 yr (24 mo) is the area
under the graph of $y = S'(t)$ from $t = 0$ to $t = 24$.

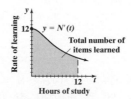

85. $p(x) = \dfrac{2(9x-4)}{135}(2 + 3x)^{3/2} - 2,000.83$; 54; \$37,932   **87.** $100\ln 3 \approx 110$ ft   **89.** $60\ln 5 \approx 97$ items

91. The area under the graph of $y = N'(t)$ from
$t = 0$ to $t = 12$ represents the total number of
items learned in that time interval.

Chapter 14 Review Exercises

1. $\int_a^b f(x)\,dx\ (14.1)$ **2.** $\int_b^c [-f(x)]\,dx\ (14.1)$ **3.** $\int_a^b f(x)\,dx + \int_b^c [-f(x)]\,dx\ (14.1)$

4. Area $= 1.153\ (14.1)$ **5.** $\frac{1}{4}xe^{4x} - \frac{1}{16}e^{4x} + C\ (14.3, 14.4)$ **6.** $\frac{1}{2}x^2\ln x - \frac{1}{4}x^2 + C\ (14.3, 14.4)$ **7.** $\dfrac{(\ln x)^2}{2} + C\ (13.2)$ **8.** $\dfrac{\ln(1+x^2)}{2} + C\ (14.2)$

9. $\dfrac{1}{1+x} + \ln\left|\dfrac{x}{1+x}\right| + C\ (14.4)$ **10.** $-\dfrac{\sqrt{1+x}}{x} - \dfrac{1}{2}\ln\left|\dfrac{\sqrt{1+x}-1}{\sqrt{1+x}+1}\right| + C\ (14.4)$ **11.** $12\ (14.1)$ **12.** $40\ (14.1)$ **13.** $34.167\ (14.1)$

14. $0.926\ (14.1)$ **15.** $12\ (14.1)$ **16.** $18.133\ (14.1)$ **17.** Indonesia (14.1) **18.** Vietnam (14.1) **19.** $\int_a^b [f(x) - g(x)]\,dx\ (14.1)$

20. $\int_b^c [g(x) - f(x)]\,dx\ (14.1)$ **21.** $\int_b^c [g(x) - f(x)]\,dx + \int_c^d [f(x) - g(x)]\,dx\ (14.1)$

22. $\int_a^b[f(x) - g(x)]dx + \int_b^c[g(x) - f(x)]dx + \int_c^d[f(x) - g(x)]dx$ *(14.1)*

23. Area $= 20.833$ *(14.1)*   **24.** 1 *(14.3, 14.4)*   **25.** $\frac{15}{2} - 8\ln 8 + 8\ln 4 \approx 1.955$ *(14.4)*   **26.** $\frac{1}{6}(3x\sqrt{9x^2 - 49} - 49\ln|3x + \sqrt{9x^2 - 49}|) + C$ *(14.4)*

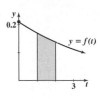

27. $-2te^{-0.5t} - 4e^{-0.5t} + C$ *(14.3, 14.4)*   **28.** $\frac{1}{3}x^3\ln x - \frac{1}{9}x^3 + C$ *(14.3, 14.4)*   **29.** $x - \ln|1 + 2e^x| + C$ *(14.4)*

30. (A) Area $= 8$ (B) Area $= 8.38$ *(14.1)*   **31.** $4,109.36$ *(14.4)*   **32.** $2,640.35$ *(14.4)*   **33.** 4.87 *(14.4)*   **34.** 4.86 *(14.4)*   **35.** $\frac{1}{3}(\ln x)^3 + C$ *(13.2)*

36. $\frac{1}{2}x^2(\ln x)^2 - \frac{1}{2}x^2\ln x + \frac{1}{4}x^2 + C$ *(14.3, 14.4)*   **37.** $\sqrt{x^2 - 36} + C$ *(13.2)*

38. $\frac{1}{2}\ln|x^2 + \sqrt{x^4 - 36}| + C$ *(14.4)*   **39.** $50\ln 10 - 42\ln 6 - 24 \approx 15.875$ *(14.3, 14.4)*

40. $x(\ln x)^2 - 2x\ln x + 2x + C$ *(14.3, 14.4)*   **41.** $-\frac{1}{4}e^{-2x^2} + C$ *(13.2)*

42. $-\frac{1}{2}x^2e^{-2x} - \frac{1}{2}xe^{-2x} - \frac{1}{4}e^{-2x} + C$ *(14.3, 14.4)*   **43.** 1.703 *(14.1)*   **44.** (A) $.189$ (B) $.154$ *(14.2)*

45. The probability that the product will fail during the second year of warranty is the area under the probability density function $y = f(t)$ from $t = 1$ to $t = 2$. *(14.2)*

46. $R(x) = 65x - 6[(x + 1)\ln(x + 1) - x]$; 618/wk; $29,506 *(14.3)*
47. (A)   (B) $8,507 *(14.2)*

48. (A) $15,656   (B) $1,454 *(14.2)*   **49.** (A)   (B) More equitably distributed, since the area bounded by the two curves will have decreased.

(C) Current $= 0.3$; Projected $= 0.2$; income will be more equitably distributed 10 years from now *(14.1)*

50. (A) $CS = $2,250$;   (B) $CS = $2,890$; *(14.2)*   **51.** (A) 25.403 or 25,403 lb   (B) $PS = 121.6$ or $1,216 *(14.2)*   **52.** 4.522 mL; 1.899 mL *(13.5, 14.4)*
    $PS = $2,700 $PS = $2,278   **53.**   **54.** $.667$; $.333$ *(14.2)*

55. The probability that the doctor will spend more than an hour with a randomly selected patient is the area under the probability density function $y = f(t)$ from $t = 1$ to $t = 3$. *(14.2)*

56. 45 thousand *(13.5, 14.1)*   **57.** $.368$ *(14.2)*

Chapter 15

Exercises 15.1   **1.** 19.5 ft^2 **3.** 240 in.^{3} **5.** $32\pi \approx 100.5$ m^3 **7.** $1,440\pi \approx 4,523.9$ cm^2 **9.** -4 **11.** 11 **13.** 4 **15.** Not defined   **17.** -1
19. -64 **21.** $154,440$ **23.** $192\pi \approx 603.2$ **25.** $3\pi\sqrt{109} \approx 98.4$ **27.** 118 **29.** $36,095.07$ **31.** $f(x) = x^2 - 7$ **33.** $f(y) = 38y + 20$

35. $f(y) = -2y^3 + 5$ **37.** $y = 2$ **39.** $x = -\frac{1}{2}, 1$ **41.** $-1.926, 0.599$ **43.** $2x + h$ **45.** $2y^2$ **47.** $E(0, 0, 3); F(2, 0, 3)$ **49.** (A) In the plane $y = c$,

c any constant, $z = x^2$.   (B) The y axis; the line parallel to the y axis and passing through the point $(1, 0, 1)$; the line parallel to the y axis and passing through the point $(2, 0, 4)$ (C) A parabolic "trough" lying on top of the y axis   **51.** (A) Upper semicircles whose centers lie on the y axis   (B) Upper semicircles whose centers lie on the x axis   (C) The upper hemisphere of radius 6 with center at the origin   **53.** (A) $a^2 + b^2$ and $c^2 + d^2$ both equal the square of the radius of the circle.   (B) Bell-shaped curves with maximum values of 1 at the origin   (C) A bell, with maximum value 1 at the origin, extending infinitely far in all directions.
55. $13,200; $18,000; $21,300   **57.** $R(p, q) = -5p^2 + 6pq - 4q^2 + 200p + 300q; R(2, 3) = $1,280; R(3, 2) = $1,175   **59.** 30,065 units
61. (A) $237,877.08   (B) 4.4%   **63.** $T(70, 47) \approx 29$ min; $T(60, 27) = 33$ min   **65.** $C(6, 8) = 75; C(8.1, 9) = 90$ **67.** $Q(12, 10) = 120; Q(10, 12) \approx 83$

Exercises 15.2   **1.** $f'(x) = 3\pi x^2 + \pi^3$ **3.** $f'(x) = exe^{-1} + e^x$ **5.** $\dfrac{dz}{dx} = \dfrac{1}{e} - \dfrac{e}{x^2}$ **7.** $\dfrac{dz}{dx} = \dfrac{2x}{x^2 + e^2}$ **9.** $f_x(x, y) = 4$ **11.** $f_y(x, y) = -3x + 4y$

13. $\dfrac{\partial z}{\partial x} = 3x^2 + 8xy$ **15.** $\dfrac{\partial z}{\partial y} = 20(5x + 2y)^9$ **17.** 9   **19.** 3   **21.** -4 **23.** 0   **25.** 45.6 mpg; mileage is 45.6 mpg at a tire pressure of 32 psi and a speed of

40 mph   **27.** 37.6 mpg; mileage is 37.6 mpg at a tire pressure of 32 psi and a speed of 50 mph   **29.** 0.3 mpg per psi; mileage increases at a rate of

0.3 mpg per psi of tire pressure   **31.** $f_{xx}(x, y) = 0$ **33.** $f_{xy}(x, y) = 0$ **35.** $f_{xy}(x, y) = y^2 e^{xy^2}(2xy) + e^{xy^2}(2y) = 2y(1 + xy^2)e^{xy^2}$ **37.** $f_{yy}(x, y) = \dfrac{2\ln x}{y^3}$

39. $f_{xx}(x, y) = 80(2x + y)^3$ **41.** $f_{xy}(x, y) = 720xy^3(x^2 + y^4)^8$ **43.** $C_x(x, y) = 6x + 10y + 4$ **45.** 2   **47.** $C_{xx}(x, y) = 6$ **49.** $C_{xy}(x, y) = 10$
51. 6   **53.** $3,000; daily sales are $3,000 when the temperature is $60°$ and the rainfall is 2 in.   **55.** $-2,500$ $/in.; daily sales decrease at a rate of $2,500 per inch
of rain when the temperature is $90°$ and rainfall is 1 in.   **57.** -50 $/in. per °F; S_r decreases at a rate of 50 $/in. per degree of temperature
61. $f_{xx}(x, y) = 2y^2 + 6x$; $f_{xy}(x, y) = 4xy = f_{yx}(x, y)$; $f_{yy}(x, y) = 2x^2$ **63.** $f_{xx}(x, y) = -2y/x^3$; $f_{xy}(x, y) = (-1/y^2) + (1/x^2) = f_{yx}(x, y)$; $f_{yy}(x, y) = 2x/y^3$
65. $f_{xx}(x, y) = (2y + xy^2)e^{xy}$; $f_{xy}(x, y) = (2x + x^2y)e^{xy} = f_{yx}(x, y)$; $f_{yy}(x, y) = x^3 e^{xy}$ **67.** $x = 2$ and $y = 4$ **69.** $x = 1.200$ and $y = -0.695$
71. (A) $-\dfrac{13}{3}$ (B) The function $f(0, y)$, for example, has values less than $-\dfrac{3}{13}$.   **73.** (A) $c = 1.145$ (B) $f_x(c, 2) = 0$; $f_y(c, 2) = 92.021$ **75.** (A) $2x$ (B) $4y$
77. $P_x(1,200, 1,800) = 24$; profit will increase approx. $24 per unit increase in production of type A calculators at the $(1,200, 1,800)$ output level;
$P_y(1,200, 1,800) = -48$; profit will decrease approx. $48 per unit increase in production of type B calculators at the $(1,200, 1,800)$ output level
79. $\partial x/\partial p = -5$: a $1 increase in the price of brand A will decrease the demand for brand A by 5 lb at any price level (p, q); $\partial y/\partial p = 2$: a $1 increase in the price
of brand A will increase the demand for brand B by 2 lb at any price level (p, q) **81.** (A) $f_x(x, y) = 7.5x^{-0.25}y^{0.25}$; $f_y(x, y) = 2.5x^{0.75}y^{-0.75}$ (B) Marginal
productivity of labor $= f_x(600, 100) \approx 4.79$; marginal productivity of capital $= f_y(600, 100) \approx 9.58$ (C) Capital   **83.** Competitive   **85.** Complementary
87. (A) $f_w(w, h) = 6.65w^{-0.575}h^{0.725}$; $f_h(w, h) = 11.34w^{0.425}h^{-0.275}$ (B) $f_w(65, 57) = 11.31$: for a 65-lb child 57 in. tall, the rate of change in surface area
is 11.31 in.^{2} for each pound gained in weight (height is held fixed); $f_h(65, 57) = 21.99$: for a child 57 in. tall, the rate of change in surface area is 21.99 in.^{2}
for each inch gained in height (weight is held fixed)   **89.** $C_W(6, 8) = 12.5$: index increases approx. 12.5 units for a 1-in. increase in width of head (length held
fixed) when $W = 6$ and $L = 8$; $C_L(6, 8) = -9.38$: index decreases approx. 9.38 units for a 1-in. increase in length (width held fixed) when $W = 6$ and $L = 8$.

Exercises 15.3   **1.** $f'(0) = 0$; $f''(0) = -18$; local maximum   **3.** $f'(0) = 0$; $f''(0) = 2$; local minimum   **5.** $f'(0) = 0$; $f''(0) = -2$; local maximum
7. $f'(0) = 1$; $f''(0) = -2$; neither   **9.** $f_x(x, y) = 4$; $f_y(x, y) = 5$; the functions $f_x(x, y)$ and $f_y(x, y)$ never have the value 0.   **11.** $f_x(x, y) = -1.2 + 4x^3$;
$f_y(x, y) = 6.8 + 0.6y^2$; the function $f_y(x, y)$ never has the value 0.   **13.** $f(-2, 0) = 10$ is a local maximum.   **15.** $f(-1, 3) = 4$ is a local minimum.
17. f has a saddle point at $(3, -2)$.   **19.** $f(3, 2) = 33$ is a local maximum.   **21.** $f(2, 2) = 8$ is a local minimum.   **23.** f has a saddle point at $(0, 0)$.
25. f has a saddle point at $(0, 0)$; $f(1, 1) = -1$ is a local minimum.   **27.** f has a saddle point at $(0, 0)$; $f(3, 18) = -162$ and $f(-3, -18) = -162$ are
local minima.   **29.** The test fails at $(0, 0)$; f has saddle points at $(2, 2)$ and $(2, -2)$.   **31.** f has a saddle point at $(0.614, -1.105)$.   **33.** $f(x, y)$ is nonnegative
and equals 0 when $x = 0$, so f has the local minimum 0 at each point of the y axis.   **35.** (B) Local minimum   **37.** 2,000 type A and 4,000 type B; max $P = $
$P(2, 4) = 15 million   **39.** (A) When $p = 100 and $q = 120, $x = 80$ and $y = 40$; when $p = 110 and $q = 110, $x = 40$ and $y = 70$ (B) A maximum
weekly profit of $4,800 is realized for $p = 100 and $q = 120.   **41.** $P(x, y) = P(4, 2)$ **43.** 8 in. by 4 in. by 2 in.   **45.** 20 in. by 20 in. by 40 in.

Exercises 15.4   **1.** Min $f(x, y) = f(-2, 4) = 12$ **3.** Min $f(x, y) = f(1, -1) = -4$ **5.** Max $f(x, y) = f(1, 0) = 2$
7. Max $f(x, y) = f(3, 3) = 18$ **9.** Min $f(x, y) = f(3, 4) = 25$ **11.** $F_x = -3 + 2\lambda = 0$ and $F_y = 4 + 5\lambda = 0$ have no simultaneous solution.
13. Max $f(x, y) = f(3, 3) = f(-3, -3) = 18$; min $f(x, y) = f(3, -3) = f(-3, 3) = -18$ **15.** Maximum product is 25 when each number is 5.
17. Min $f(x, y, z) = f(-4, 2, -6) = 56$ **19.** Max $f(x, y, z) = f(2, 2, 2) = 6$; min $f(x, y, z) = f(-2, -2, -2) = -6$
21. Max $f(x, y) = f(0.217, 0.885) = 1.055$ **23.** $F_x = e^x + \lambda = 0$ and $F_y = 3e^y - 2\lambda = 0$ have no simultaneous solution.   **25.** Maximize $f(x, 5)$, a
function of just one independent variable.   **27.** (A) Max $f(x, y) = f(0.707, 0.5) = f(-0.707, 0.5) = 0.47$ **29.** 60 of model A and 30 of model B will yield a
minimum cost of $32,400 per week.   **31.** (A) 8,000 units of labor and 1,000 units of capital; max $N(x, y) = N(8,000, 1,000) \approx 263,902$ units   (B) Marginal
productivity of money ≈ 0.6598; increase in production $\approx 32,990$ units   **33.** 8 in. by 8 in. by $\frac{8}{3}$ in.   **35.** $x = 50$ ft and $y = 200$ ft; maximum area is 10,000 ft^2

Exercises 15.5   **1.** 10   **3.** 98   **5.** 380   **7.** $y = 0.7x + 1$ **9.** $y = -2.5x + 10.5$ **11.** $y = x + 2$ **13.** $y = -1.5x + 4.5$; $y = 0.75$ when $x = 2.5$
15. $y = 2.12x + 10.8$; $y = 63.8$ when $x = 25$
17. $y = -1.2x + 12.6$; $y = 10.2$ when $x = 2$
19. $y = -1.53x + 26.67$; $y = 14.4$ when $x = 8$

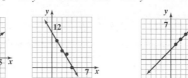

21. $y = 0.75x^2 - 3.45x + 4.75$ **27.** (A) $y = 1.52x - 0.16$; $y = 0.73x^2 - 1.39x + 1.30$ (B) The quadratic function   **29.** The normal equations form a
system of 4 linear equations in the 4 variables a, b, c, and d, which can be solved by Gauss–Jordan elimination.
31. (A) $y = -79.36x + 3793.6$ (B) 1,889 crimes per 100,000 population   **33.** (A) $y = -0.48x + 4.38$
(B) $6.56 per bottle   **35.** (A) $y = 0.0222x + 18.94$ (B) 19.92 ft   **37.** (A) $y = 0.0121x + 56.35$ (B) 58.77°F

Exercises 15.6   **1.** $\pi x + \dfrac{x^2}{2} + C$ **3.** $x + \pi \ln |x| + C$ **5.** $\dfrac{e^{\pi x}}{\pi} + C$ **7.** (A) $3x^2y^4 + C(x)$ (B) $3x^2$ **9.** (A) $2x^2 + 6xy + 5x + E(y)$ (B) $35 + 30y$

11. (A) $\sqrt{y + x^2} + E(y)$ (B) $\sqrt{y + 4} - \sqrt{y}$ **13.** (A) $\dfrac{\ln x \ln y}{x} + C(x)$ (B) $\dfrac{2\ln x}{x}$ **15.** 9   **17.** 330   **19.** $(56 - 20\sqrt{5})/3$ **21.** 1   **23.** 16

25. 49   **27.** $\frac{1}{8}\int_1^5 \int_{-1}^1 (x + y)^2 dy\, dx = \frac{32}{3}$ **29.** $\frac{1}{15}\int_1^4 \int_2^7 (x/y)\, dy\, dx = \frac{1}{2}\ln\frac{7}{2} \approx 0.6264$ **31.** $\frac{4}{3}$ cubic units   **33.** $\frac{32}{3}$ cubic units

35. $\int_0^1 \int_1^2 xe^{xy} \, dy \, dx = \frac{1}{2} + \frac{1}{2}e^2 - e$ **37.** $\int_0^1 \int_{-1}^1 \frac{2y + 3xy^2}{1 + x^2} \, dy \, dx = \ln 2$ **41. (A)** $\frac{1}{3} + \frac{1}{4}e^{-2} - \frac{1}{4}e^2$ **(B)**

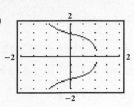

(C) Points to the right of the graph in part (B) are greater than 0; points to the left of the graph are less than 0.

43. $\frac{1}{0.4} \int_{0.6}^{0.8} \int_5^7 \frac{y}{1-x} \, dy \, dx = 30 \ln 2 \approx \$20.8$ billion   **45.** $\frac{1}{10} \int_{10}^{20} \int_1^2 x^{0.75} y^{0.25} dy \, dx = \frac{8}{175}(2^{1.25} - 1)(20^{1.75} - 10^{1.75}) \approx 8.375$ or 8,375 units

47. $\frac{1}{192} \int_{-8}^8 \int_{-6}^6 [10 - \frac{1}{10}(x^2 + y^2)] dy \, dx = \frac{20}{3}$ insects/ft^2 **49.** $\frac{1}{8} \int_{-2}^2 \int_{-1}^1 [100 - 15(x^2 + y^2)] dy \, dx = 75$ ppm

51. $\frac{1}{10,000} \int_{2,000}^{3,000} \int_{50}^{60} 0.000\,013\,3xy^2 dy \, dx \approx 100.86$ ft   **53.** $\frac{1}{16} \int_8^{16} \int_{10}^{12} 100\frac{x}{y} \, dy \, dx = 600 \ln 1.2 \approx 109.4$

Exercises 15.7

1. $R = \{(x,y) \mid 0 \le y \le 4 - x^2, 0 \le x \le 2\}$
$R = \{(x,y) \mid 0 \le x \le \sqrt{4-y}, 0 \le y \le 4\}$

3. R is a regular x region:
$R = \{(x,y) \mid x^3 \le y \le 12 - 2x, 0 \le x \le 2\}$

5. R is a regular y region:
$R = \{(x,y) \mid \frac{1}{2}y^2 \le x \le y + 4, -2 \le y \le 4\}$

7. $\frac{1}{2}$ **9.** $\frac{39}{70}$ **11.** R consists of the points on or inside the rectangle with corners $(\pm 2, \pm 3)$; both   **13.** R is the arch-shaped region consisting of the points on or inside the rectangle with corners $(\pm 2, 0)$ and $(\pm 2, 2)$ that are not inside the circle of radius 1 centered at the origin; regular x region   **15.** $\frac{56}{3}$ **17.** $-\frac{3}{4}$
19. $\frac{1}{2}e^4 - \frac{5}{2}$ **21.** $R = \{(x,y) \mid 0 \le y \le x + 1, 0 \le x \le 1\}$ **23.** $R = \{(x,y) \mid 0 \le y \le 4x - x^2, 0 \le x \le 4\}$

$\int_0^1 \int_0^{x+1} \sqrt{1 + x + y} \, dy \, dx = (68 - 24\sqrt{2})/15$ $\int_0^4 \int_0^{4x-x^2} \sqrt{y + x^2} \, dy \, dx = \frac{128}{5}$

25. $R = \{(x,y) \mid 1 - \sqrt{x} \le y \le 1 + \sqrt{x}, 0 \le x \le 4\}$ **27.** $\int_0^3 \int_0^{3-y}(x + 2y) dx \, dy = \frac{27}{2}$ **29.** $\int_0^1 \int_0^{\sqrt{1-y}} x\sqrt{y} \, dx \, dy = \frac{2}{15}$ **31.** $\int_0^1 \int_{4y^2}^{4y} x \, dx \, dy = \frac{16}{15}$

$\int_0^4 \int_{1-\sqrt{x}}^{1+\sqrt{x}} x(y-1)^2 dy \, dx = \frac{512}{21}$

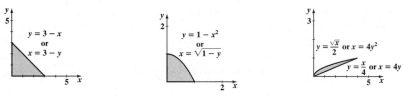

33. $\int_0^4 \int_0^{4-x}(4 - x - y) \, dy \, dx = \frac{32}{3}$ **35.** $\int_0^1 \int_0^{1-x^2} 4 \, dy \, dx = \frac{8}{3}$ **37.** $\int_0^4 \int_0^{\sqrt{y}} \frac{4x}{1 + y^2} \, dx \, dy = \ln 17$ **39.** $\int_0^1 \int_0^{\sqrt{x}} 4ye^{x^2} \, dy \, dx = e - 1$

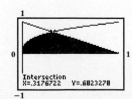

41. $R = \{(x,y) \mid x^2 \le y \le 1 + \sqrt{x}, 0 \le x \le 1.49\}$ **43.** $R = \{(x,y) \mid y^3 \le x \le 1 - y, 0 \le y \le 0.68\}$

$\int_0^{1.49} \int_{x^2}^{1+\sqrt{x}} x \, dy \, dx \approx 0.96$ $\int_0^{0.68} \int_{y^3}^{1-y} 24xy \, dx \, dy \approx 0.83$

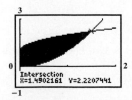

45. $R = \{(x, y)\,|\,e^{-x} \leq y \leq 3 - x, -1.51 \leq x \leq 2.95\}$ Regular x region

$R = \{(x, y)\,|\,-\ln y \leq x \leq 3 - y, 0.05 \leq y \leq 4.51\}$ Regular y region

$\int_{-1.51}^{2.95} \int_{e^{-x}}^{3-x} 4y\,dy\,dx = \int_{0.05}^{4.51} \int_{-\ln y}^{3-y} 4y\,dx\,dy \approx 40.67$

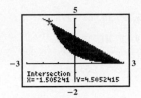

Chapter 15 Review Exercises

1. $f(5, 10) = 2{,}900; f_x(x, y) = 40; f_y(x, y) = 70$ *(15.1, 15.2)*   **2.** $\partial^2 z/\partial x^2 = 6xy^2; \partial^2 z/\partial x\,\partial y = 6x^2 y$ *(15.2)*
3. $2xy^3 + 2y^2 + C(x)$ *(15.6)*   **4.** $3x^2 y^2 + 4xy + E(y)$ *(15.6)*   **5.** 1 *(15.6)*   **6.** $f_x(x, y) = 5 + 6x + 3x^2; f_y(x, y) = -2$; the function $f_y(x, y)$ never has the
value 0. *(15.3)*   **7.** $f(2, 3) = 7; f_y(x, y) = -2x + 2y + 3; f_y(2, 3) = 5$ *(15.1, 15.2)*   **8.** $(-8)(-6) - (4)^2 = 32$ *(15.2)*   **9.** $(1, 3, -\tfrac{1}{2}), (-1, -3, \tfrac{1}{2})$ *(15.4)*
10. $y = -1.5x + 15.5; y = 0.5$ when $x = 10$ *(15.5)*   **11.** 18 *(15.6)*   **12.** $\tfrac{8}{5}$ *(15.7)*   **13.** $f_x(x, y) = 2xe^{x^2 + 2y}; f_y(x, y) = 2e^{x^2 + 2y}; f_{xy}(x, y) = 4xe^{x^2 + 2y}$ *(15.2)*
14. $f_x(x, y) = 10x(x^2 + y^2)^4; f_{xy}(x, y) = 80xy(x^2 + y^2)^3$ *(15.2)*   **15.** $f(2, 3) = -25$ is a local minimum; f has a saddle point at $(-2, 3)$. *(15.3)*
16. Max $f(x, y) = f(6, 4) = 24$ *(15.4)*   **17.** Min $f(x, y, z) = f(2, 1, 2) = 9$ *(15.4)*   **18.** $y = \tfrac{116}{165}x + \tfrac{100}{3}$ *(15.5)*   **19.** $\tfrac{27}{5}$ *(15.6)*   **20.** 4 cubic units *(15.6)*
21. 0 *(15.6)*   **22.** (A) 12.56   (B) No *(15.6)*   **23.** $F_x = 12x^2 + 3\lambda = 0, F_y = -15y^2 + 2\lambda = 0$, and $F_\lambda = 3x + 2y - 7 = 0$ have no simultaneous solution. *(15.4)*
24. 1 *(15.7)*   **25.** (A) $P_x(1, 3) = 8$; profit will increase \$8,000 for a 100-unit increase in product A if the production of product B is held fixed at an output level
of $(1, 3)$.   (B) For 200 units of A and 300 units of B, $P(2, 3) = \$100$ thousand is a local maximum. *(15.2, 15.3)*   **26.** $x = 6$ in., $y = 8$ in., $z = 2$ in. *(15.3)*
27. $y = 0.63x + 1.33$; profit in sixth year is \$5.11 million *(15.4)*   **28.** (A) Marginal productivity of labor ≈ 8.37; marginal productivity of capital ≈ 1.67;
management should encourage increased use of labor.   (B) 80 units of labor and 40 units of capital; max $N(x, y) = N(80, 40) \approx 696$ units; marginal productiv-
ity of money ≈ 0.0696; increase in production ≈ 139 units   (C) $\dfrac{1}{1{,}000} \int_{50}^{100} \int_{20}^{40} 10x^{0.8}y^{0.2}\,dy\,dx = \dfrac{(40^{1.2} - 20^{1.2})(100^{1.8} - 50^{1.8})}{216} = 621$ items *(15.4)*
29. $T_x(70, 17) = -0.924$ min/ft increase in depth when $V = 70$ ft^3 and $x = 17$ ft *(15.2)*   **30.** $\tfrac{1}{16} \int_{-2}^{2} \int_{-2}^{2} [100 - 24(x^2 + y^2)]\,dy\,dx = 36$ ppm *(15.6)*
31. $50{,}000$ *(15.1)*   **32.** $y = \tfrac{1}{2}x + 48; y = 68$ when $x = 40$ *(15.5)*   **33.** (A) $y = 0.734x + 49.93$ (B) 97.64 people/mi^2 (C) 101.10 people/mi^2;
103.70 people/mi^2 *(15.5)*   **34.** (A) $y = 1.069x + 0.522$ (B) 64.68 yr   (C) 64.78 yr; 64.80 yr *(15.5)*

Appendix A

Exercises A.1

1. vu **3.** $(3 + 7) + y$ **5.** $u + v$ **7.** T   **9.** T   **11.** F   **13.** T   **15.** T   **17.** T   **19.** T   **21.** F   **23.** T   **25.** T   **27.** No
29. (A) F   (B) T   (C) T   **31.** $\sqrt{2}$ and π are two examples of infinitely many.   **33.** (A) N, Z, Q, R (B) R (C) Q, R (D) Q, R
35. (A) F, since, for example, $2(3 - 1) \neq 2 \cdot 3 - 1$ (B) F, since, for example, $(8 - 4) - 2 \neq 8 - (4 - 2)$ (C) T   (D) F, since, for example,
$(8 \div 4) \div 2 \neq 8 \div (4 \div 2)$.   **37.** $\dfrac{1}{11}$ **39.** (A) $2.166\,666\,666\ldots$ (B) $4.582\,575\,69\ldots$ (C) $0.437\,500\,000\ldots$ (D) $0.261\,261\,261\ldots$ **41.** (A) 3   (B) 2
43. (A) 2   (B) 6   **45.** \$16.42   **47.** 2.8%

Exercises A.2

1. 3   **3.** $x^3 + 4x^2 - 2x + 5$ **5.** $x^3 + 1$ **7.** $2x^5 + 3x^4 - 2x^3 + 11x^2 - 5x + 6$ **9.** $-5u + 2$ **11.** $6a^2 + 6a$ **13.** $a^2 - b^2$
15. $6x^2 - 7x - 5$ **17.** $2x^2 + xy - 6y^2$ **19.** $9y^2 - 4$ **21.** $-4x^2 + 12x - 9$ **23.** $16m^2 - 9n^2$ **25.** $9u^2 + 24uv + 16v^2$ **27.** $a^3 - b^3$
29. $x^2 - 2xy + y^2 - 9z^2$ **31.** 1   **33.** $x^4 - 2x^2 y^2 + y^4$ **35.** $40ab$ **37.** $4m + 8$ **39.** $6xy$ **41.** $u^3 + 3u^2 v + 3uv^2 + v^3$
43. $x^3 - 6x^2 y + 12xy^2 - 8y^3$ **45.** $2x^2 - 2xy + 3y^2$ **47.** $x^4 - 10x^3 + 27x^2 - 10x + 1$ **49.** $4x^3 - 14x^2 + 8x - 6$ **51.** $m + n$ **53.** No change
55. $(1 + 1)^2 \neq 1^2 + 1^2$; either a or b must be 0 **57.** $0.09x + 0.12(10{,}000 - x) = 1{,}200 - 0.03x$
59. $20x + 30(3x) + 50(4{,}000 - x - 3x) = 200{,}000 - 90x$ **61.** $0.02x + 0.06(10 - x) = 0.6 - 0.04x$

Exercises A.3

1. $3m^2(2m^2 - 3m - 1)$ **3.** $2uv(4u^2 - 3uv + 2v^2)$ **5.** $(7m + 5)(2m - 3)$ **7.** $(4ab - 1)(2c + d)$ **9.** $(2x - 1)(x + 2)$
11. $(y - 1)(3y + 2)$ **13.** $(x + 4)(2x - 1)$ **15.** $(w + x)(y - z)$ **17.** $(a - 3b)(m + 2n)$ **19.** $(3y + 2)(y - 1)$ **21.** $(u - 5v)(u + 3v)$
23. Not factorable   **25.** $(wx - y)(wx + y)$ **27.** $(3m - n)^2$ **29.** Not factorable   **31.** $4(z - 3)(z - 4)$ **33.** $2x^2(x - 2)(x - 10)$ **35.** $x(2y - 3)^2$
37. $(2m - 3n)(3m + 4n)$ **39.** $uv(2u - v)(2u + v)$ **41.** $2x(x^2 - x + 4)$ **43.** $(2x - 3y)(4x^2 + 6xy + 9y^2)$ **45.** $xy(x + 2)(x^2 - 2x + 4)$
47. $[(x + 2) - 3y][(x + 2) + 3y]$ **49.** Not factorable   **51.** $(6x - 6y - 1)(x - y + 4)$ **53.** $(y - 2)(y + 2)(y^2 + 1)$
55. $3(x - y)^2(5xy - 5y^2 + 4x)$ **57.** True   **59.** False

Exercises A.4

1. $39/7$ **3.** 495   **5.** $8d^6$ **7.** $\dfrac{15x^2 + 10x - 6}{180}$ **9.** $\dfrac{15m^2 + 14m - 6}{36m^3}$ **11.** $\dfrac{1}{x(x - 4)}$ **13.** $\dfrac{x - 6}{x(x - 3)}$ **15.** $\dfrac{-3x - 9}{(x - 2)(x + 1)^2}$
17. $\dfrac{2}{x - 1}$ **19.** $\dfrac{5}{a - 1}$ **21.** $\dfrac{x^2 + 8x - 16}{x(x - 4)(x + 4)}$ **23.** $\dfrac{7x^2 - 2x - 3}{6(x + 1)^2}$ **25.** $\dfrac{x(y - x)}{y(2x - y)}$ **27.** $\dfrac{-17c + 16}{15(c - 1)}$ **29.** $\dfrac{1}{x - 3}$ **31.** $\dfrac{-1}{2x(x + h)}$ **33.** $\dfrac{x - y}{x + y}$
35. (A) Incorrect   (B) $x + 1$ **37.** (A) Incorrect   (B) $2x + h$ **39.** (A) Incorrect   (B) $\dfrac{x^2 - x - 3}{x + 1}$ **41.** (A) Correct   **43.** $\dfrac{-2x - h}{3(x + h)^2 x^2}$ **45.** $\dfrac{x(x - 3)}{x - 1}$

Exercises A.5

1. $2/x^9$ **3.** $3w^7/2$ **5.** $2/x^3$ **7.** $1/w^5$ **9.** $4/a^6$ **11.** $1/a^6$ **13.** $1/8x^{12}$ **15.** 8.23×10^{10} **17.** 7.83×10^{-1} **19.** 3.4×10^{-5}
21. 40,000   **23.** 0.007   **25.** 61,710,000   **27.** 0.000 808   **29.** 1   **31.** 10^{14} **33.** $y^6/25x^4$ **35.** $4x^6/25$ **37.** $4y^3/3x^5$ **39.** $\dfrac{7}{4} - \dfrac{1}{4}x^{-3}$

41. $\dfrac{5}{2}x^2 - \dfrac{3}{2} + 4x^{-2}$ **43.** $\dfrac{x^2(x-3)}{(x-1)^3}$ **45.** $\dfrac{2(x-1)}{x^3}$ **47.** 2.4×10^{10}; 24,000,000,000   **49.** 3.125×10^4; 31,250   **51.** 64   **55.** uv **57.** $\dfrac{bc(c+b)}{c^2 + bc + b^2}$

59. (A) \$51,329   (B) \$1,150   (C) 2.24%   **61.** (A) 9×10^{-6} (B) 0.000 009   (C) 0.0009%   **63.** 1,248,000

Exercises A.6   **1.** $6\sqrt[5]{x^3}$ **3.** $\sqrt[5]{(32x^2y^3)^3}$ **5.** $\sqrt{x^2+y^2}$ (not $x+y$)   **7.** $5x^{3/4}$ **9.** $(2x^2y)^{3/5}$ **11.** $x^{1/3} + y^{1/3}$ **13.** 5   **15.** 64   **17.** -7

19. -16 **21.** $\dfrac{8}{125}$ **23.** $\dfrac{1}{27}$ **25.** $x^{2/5}$ **27.** m **29.** $2x/y^2$ **31.** $xy^2/2$ **33.** $1/(24x^{7/12})$ **35.** $2x + 3$ **37.** $30x^5\sqrt{3x}$ **39.** 2   **41.** $12x - 6x^{35/4}$

43. $3u - 13u^{1/2}v^{1/2} + 4v$ **45.** $36m^{3/2} - \dfrac{6m^{1/2}}{n^{1/2}} + \dfrac{6m}{n^{1/2}} - \dfrac{1}{n}$ **47.** $9x - 6x^{1/2}y^{1/2} + y$ **49.** $\dfrac{1}{2}x^{1/3} + x^{-1/3}$ **51.** $\dfrac{2}{3}x^{-1/4} + \dfrac{1}{3}x^{-2/3}$ **53.** $\dfrac{1}{2}x^{-1/6} - \dfrac{1}{4}$

55. $4n\sqrt{3mn}$ **57.** $\dfrac{2(x+3)\sqrt{x-2}}{x-2}$ **59.** $7(x-y)(\sqrt{x} + \sqrt{y})$ **61.** $\dfrac{1}{xy\sqrt{5xy}}$ **63.** $\dfrac{1}{\sqrt{x+h} + \sqrt{x}}$ **65.** $\dfrac{1}{(t+x)(\sqrt{t} + \sqrt{x})}$

67. $x = y = 1$ is one of many choices.   **69.** $x = y = 1$ is one of many choices.   **71.** False   **73.** False   **75.** False   **77.** True   **79.** True   **81.** False

83. $\dfrac{x+8}{2(x+3)^{3/2}}$ **85.** $\dfrac{x-2}{2(x-1)^{3/2}}$ **87.** $\dfrac{x+6}{3(x+2)^{5/3}}$ **89.** 103.2   **91.** 0.0805   **93.** 4,588   **95.** (A) and (E); (B) and (F); (C) and (D)

Exercises A.7   **1.** $\pm\sqrt{11}$ **3.** $-\dfrac{4}{3}, 2$ **5.** $-2, 6$ **7.** 0, 2   **9.** $3 \pm 2\sqrt{3}$ **11.** $-2 \pm \sqrt{2}$ **13.** $0, \dfrac{15}{2}$ **15.** $\pm\dfrac{3}{2}$ **17.** $\dfrac{1}{2}, -3$ **19.** $(-1 \pm \sqrt{5})/2$

21. $(3 \pm \sqrt{3})/2$ **23.** No real solution   **25.** $(-3 \pm \sqrt{11})/2$ **27.** $\pm\sqrt{3}$ **29.** $-\dfrac{1}{2}, 2$ **31.** $(x-2)(x+42)$ **33.** Not factorable in the integers
35. $(2x-9)(x+12)$ **37.** $(4x-7)(x+62)$ **39.** $r = \sqrt{A/P} - 1$ **41.** If $c < 4$, there are two distinct real roots; if $c = 4$, there is one real double root; and if $c > 4$, there are no real roots.   **43.** -2 **45.** $\pm\sqrt{10}$ **47.** $\pm\sqrt{3}, \pm\sqrt{5}$ **49.** 1,575 bottles at \$4 each   **51.** 13.64%   **53.** 8 ft/sec; $4\sqrt{2}$ or 5.66 ft/sec

Appendix B

Exercises B.1   **1.** 5, 7, 9, 11   **3.** $\dfrac{3}{2}, \dfrac{4}{3}, \dfrac{5}{4}, \dfrac{6}{5}$ **5.** $9, -27, 81, -243$ **7.** 23   **9.** $\dfrac{101}{100}$ **11.** $1 + 2 + 3 + 4 + 5 + 6 = 21$ **13.** $5 + 7 + 9 + 11 = 32$

15. $1 + \dfrac{1}{10} + \dfrac{1}{100} + \dfrac{1}{1,000} = \dfrac{1,111}{1,000}$ **17.** 3.6   **19.** 82.5   **21.** $\dfrac{1}{2}, -\dfrac{1}{4}, \dfrac{1}{8}, -\dfrac{1}{16}, \dfrac{1}{32}$ **23.** 0, 4, 0, 8, 0   **25.** $1, -\dfrac{3}{2}, \dfrac{9}{4}, -\dfrac{27}{8}, \dfrac{81}{16}$ **27.** $a_n = n - 3$

29. $a_n = 4n$ **31.** $a_n = (2n-1)/2n$ **33.** $a_n = (-1)^{n+1}n$ **35.** $a_n = (-1)^{n+1}(2n-1)$ **37.** $a_n = \left(\dfrac{2}{5}\right)^{n-1}$ **39.** $a_n = x^n$ **41.** $a_n = (-1)^{n+1}x^{2n-1}$

43. $1 - 9 + 25 - 49 + 81$ **45.** $\dfrac{4}{7} + \dfrac{8}{9} + \dfrac{16}{11} + \dfrac{32}{13}$ **47.** $1 + x + x^2 + x^3 + x^4$ **49.** $x - \dfrac{x^3}{3} + \dfrac{x^5}{5} - \dfrac{x^7}{7} + \dfrac{x^9}{9}$ **51.** (A) $\displaystyle\sum_{k=1}^{5}(k+1)$ (B) $\displaystyle\sum_{j=0}^{4}(j+2)$

53. (A) $\displaystyle\sum_{k=1}^{4}\dfrac{(-1)^{k+1}}{k}$ (B) $\displaystyle\sum_{j=0}^{3}\dfrac{(-1)^j}{j+1}$ **55.** $\displaystyle\sum_{k=1}^{n}\dfrac{k+1}{k}$ **57.** $\displaystyle\sum_{k=1}^{n}\dfrac{(-1)^{k+1}}{2^k}$ **59.** False   **61.** True   **63.** 2, 8, 26, 80, 242   **65.** 1, 2, 4, 8, 16

67. $1, \dfrac{3}{2}, \dfrac{17}{12}, \dfrac{577}{408}; a_4 = \dfrac{577}{408} \approx 1.414\,216, \sqrt{2} \approx 1.414\,214$ **69.** 1, 1, 2, 3, 5, 8, 13, 21, 34, 55

Exercises B.2   **1.** (A) Arithmetic, with $d = -5$; $-26, -31$ (B) Geometric, with $r = -2$; $-16, 32$ (C) Neither   (D) Geometric, with $r = \dfrac{1}{3}$; $\dfrac{1}{54}, \dfrac{1}{162}$

3. Geometric; 1   **5.** Neither   **7.** Arithmetic; 127.5   **9.** $a_2 = 11, a_3 = 15$ **11.** $a_{21} = 82, S_{31} = 1,922$ **13.** $S_{20} = 930$ **15.** $a_2 = -6, a_3 = 12, a_4 = -24$

17. $S_7 = 547$ **19.** $a_{10} = 199.90$ **21.** $r = 1.09$ or -1.09 **23.** $S_{10} = 1,242, S_\infty = 1,250$ **25.** 2,706   **27.** -85 **29.** 1,120   **31.** (A) Does not exist

(B) $S_\infty = \dfrac{8}{5} = 1.6$ **33.** 2,400   **35.** 0.999   **37.** Use $a_1 = 1$ and $d = 2$ in $S_n = (n/2)[2a_1 + (n-1)d]$.   **39.** $S_n = na_1$ **41.** No   **43.** Yes

45. $\$48 + \$46 + \cdots + \$4 + \$2 = \$600$ **47.** About \$11,670,000   **49.** \$1,628.89; \$2,653.30

Exercises B.3   **1.** 720   **3.** 10   **5.** 1,320   **7.** 10   **9.** 6   **11.** 1,140   **13.** 10   **15.** 6   **17.** 1   **19.** 816
21. $_4C_0a^4 + {_4C_1}a^3b + {_4C_2}a^2b^2 + {_4C_3}ab^3 + {_4C_4}b^4 = a^4 + 4a^3b + 6a^2b^2 + 4ab^3 + b^4$ **23.** $x^6 - 6x^5 + 15x^4 - 20x^3 + 15x^2 - 6x + 1$

25. $32a^5 - 80a^4b + 80a^3b^2 - 40a^2b^3 + 10ab^4 - b^5$ **27.** $3,060x^{14}$ **29.** $5,005p^9q^6$ **31.** $264x^2y^{10}$ **33.** $_nC_0 = \dfrac{n!}{0!\,n!} = 1; {_nC_n} = \dfrac{n!}{n!\,0!} = 1$
35. 1 5 10 10 5 1; 1 6 15 20 15 6 1

INDEX

A Abscissa, 13
Absolute equality of income, 780
Absolute extrema, 637, 687. *See also*
    Absolute maxima and minima
  functions with no, 690
  locating, 688–690
  on open interval, finding, 692–693
  second-derivative test for, 692
Absolute inequality of income, 780
Absolute maxima and minima. *See also* Extrema
  on closed intervals, finding, 689
  definition of, 687
  extreme value theorem and, 688
  graphing of, 687–695
  second derivative and, 690–693
Absolute minima. *See* Absolute maxima
  and minima
Absolute values, 27–28, 59, 491
Absorbing Markov chains, 469–483
  absorbing states, 469–471
  definition of, 470
  graphing calculator approximations
    for, 477–478
  limiting matrix and, 472–477
  recognizing, 471
  standard forms for, 471–472
  transition matrix for, 476
Absorbing states, 454, 469–471
Acceptable probability assignment, 390, 392
ac test, 902
Addition
  elimination by, 178–180
  of matrix, 210–211, 234
  of polynomials, 896–897
  principle of, 361–363
  of rational expressions, 908–909
  of real numbers, 889–890
Additive inverse, 910
Algebra, 888–930
  integer exponents, 912–913
  polynomials, 894–905
  quadratic equations, 922–930
  radicals, 919–920
  rational exponents, 916–922
  rational expressions, 906–911
  real numbers, 888–894
  scientific notation, 913–914
Algebraic expressions, 894
Algebra *vs.* calculus, 488
Amortization, 158–159
  of debt, 158–159
  formula for, 159, 161–162, 164
  interest on, 159
  schedules of, 159–163
Amount, 102, 127–129. *See also*
    Future value (FV)
Analysis
  break-even, 9–10, 50

input–output, 242, 246–247
  of investments, 237–239
  of limits, 490–491
  marginal, 557–567
  profit–loss, 50
  of regression, 28, 854
  of revenue, 527
  of second derivative, graphing of, 654–655
Analytic geometry, fundamental theorem of, 13
Annual nominal rates, 135, 143. *See also*
    Interest rate
Annual percentage yield (APY), 141–143
Annuities
  defined, 147
  future value of, 147–155 (*See also*
    Sinking funds)
  ordinary, 147, 149–150, 157
  present value of, 155–167 (*See also*
    Amortization)
Antiderivatives, 714–715, 866
Antidifferentiation, 714, 865
Applications. *See* Index of Applications
Approximation
  of area, 747–750
  error in, 748
  least square, 854–859
  using differentials, 553–555
Arbitrary event, 393–394
Archaeology, 741–742
Area
  approximating, 749–750
  integration for computing, 779–780
  marginal cost *vs,* 758–759
  maximizing, 695–696
  maximum, finding, 845
  perimeter and, 695–698
  between two curves, 776–785
Arithmetic mean, 935
Arithmetic sequences, 937–943
  common differences in, 937
  *n*th-term formulas and, 938–939
  sum formulas and, 939–942
Arithmetic series, 939
Artificial variables, 326–327, 329
Associative matrix, 210
Associative properties, 890
Asymptotes
  graphing strategy and, 674–675
  horizontal, 507–511
  vertical, 504–506, 511
Asymptotes of rational functions
  finding, 90–91
  horizontal, 88–91, 97
  vertical, 88–91
Augmented matrix, 187–189
  row equivalent, 189–191
  solving, 189–194
  summary of, 194

Average cost, curve-sketching techniques
    and, 681–682
Average daily balance method, 131
Average price, 765
Average rate of change, 526–527
Averages, 437. *See also* Mean
  law of, 406
Average value
  of continuous functions, 764
  definite integral and, 763–765
  of functions, 765
  over rectangular regions, 869–870
Average velocity, 526–529
Axis
  coordinate, 13
  horizontal, 12–13
  of parabolas, 73
  vertical, 12–13
  x, 777
  x, reflection in, 62

B Balance sheet, 150
Base
  of exponential functions, 96
Base e, 98–99
Base 2 logarithmic functions, 107–108
Basic algebra. *See* Algebra
Bayes' formula, 425–432
Best fit, 29
Big *M* method. *See also* Modified problem
  definition of, 285, 326, 329
  introduction to, 326–329
  for minimization problem, 333–335
  modified problem, for solving,
    330–331
  procedure for, 329
  slack variables and, 329
  solutions to, 335
  summary of, 329–333
Binomials, 895
  formulas for, 943–944
  theorem for, 943–946
Body surface area (BSA), 27
Boundary line of half-planes, 256–257
Bounded functions, 91
Bounded solution region, 266
Break-even analysis, 9–10, 50
Break-even points, 561
Business. related rates and, 617
b^x with base b, 581

C Calculus. *See also* Multivariable calculus
  algebra *vs,* 488
  fundamental theorem of, 757–769
Canceling, in fractions, 907
Cartesian (rectangular) coordinate
    system, 12–13
Certificates of deposit (CDs), 141

Chain rule, 598–605
  composite functions and, 598–599
  definition of, 603
  general derivative rules and, 604–605
  general power rule and, 599–602
  partial derivatives using, 829–830
  reversing, 725–727
  using, 603–604
Chains. *See* Markov chains
Change-of-base formulas, 113
Change-of-variable method, 729
Circle, 947
Closed intervals, 6
  absolute maxima and minima on, finding, 689
  continuous on, 518–519
Cobb–Douglas production function,
    821–822, 848
Coefficient matrix, 188–189
Coefficients
  concept of, 895
  leading, 85
  numerical, 895
  of objective functions, 314
  of problem constraints, 314
Column matrix, 188
Combinations, 372–375
  of *n* distinct objects taken *r* at a
    time, 372–374
Combined factoring polynomials, techniques
    for, 904–905
Combined matrix, 234
Combined polynomials, 898
Combining like terms, 895–896
Commission schedule, 130–131
Common differences, in arithmetic sequence, 937
Common logarithms, 111
Common ratio, in geometric sequence, 937
Commutative properties, 890
Complement
  of event, 403–405
  of sets, 357
Completely factored numbers, 900
Completing the square, 72–73, 925
Compliment of events, 403
Composite functions, 598–599
Compound events, 386–387
Compound fractions, 909–910
Compound growth rate, 102
Compounding quarterly interest, 134–135
Compound interest, 101–103, 134–137
  annual percentage yield, 141–144
  continuous, 137–138, 575–578, 739
  daily, 138
  definition of, 102, 134
  graphing, 139
  growth of, time and, 138–141
Compound propositions, 348
  truth table for, 346–347, 349, 351–352
Concavity
  definition of, 649
  downward, 649–650
  graphing, 650–651
  for second derivative, graphing of, 648–651
  upward, 649–650

Conceptual Insight, 7, 17, 50, 59, 70, 73, 89, 97,
    108, 129, 131, 136, 137, 138, 140, 144, 149,
    156, 160, 216, 218, 237, 244, 258, 264, 276,
    278, 302, 309, 318–319, 347, 355, 363, 375,
    388, 401, 419, 429, 437, 450, 464, 476, 494,
    497, 506, 510, 517, 531, 542, 550, 559, 575,
    577, 584, 588, 591, 599, 602, 611, 615, 621,
    622, 634, 636, 650, 653–654, 657, 673, 692,
    698, 700, 702, 714, 738, 750, 757, 759, 776,
    786–787, 799, 807, 825, 833, 838, 852, 858,
    871, 874–875, 876
Conditional probability, 411–414, 421
  concept of, 411
  definition of, 411–412
  events for, 414–415, 417–420
  probability trees for, 415–417
  product rule for, 414–415
  summary of, 421
Conditional propositions, 347, 352
Cone, 948
Conjunction, 347
Connectives, 346–349
Consistent systems of linear equations, 176
Constant *e*, 575. *See also* Continuous
    compound interest
Constant functions, 47
  continuity properties of, 519
  rules of, 541–542
Constant of integration, 715
Constant matrix, 188–189
Constant multiple property, 544
Constant-profit line, 270
Constant rate of change, 27
Consumer Price Index (CPI), 10
Consumers' surplus, 791–794
Contingency, 350
Continuity, 515–526
  definition of, 516
  of functions, 516–518
  inequalities and, 520–522
  properties of, 519–522
Continuous on closed intervals, 518–519
Continuous compound interest, 575–578
  computing, 577
  differential equations and, 739
  double time and, 578
  formula for, 576
  graphing, 577
  growth time and, 578
  models for, 587
Continuous functions, 516–518, 635
  average value of, 764
Continuous on half-closed intervals, 519
Continuous income stream, 787–789
  future value of, 789–791
Continuous on the left functions, 518
Continuous on the right functions, 518
Continuous stream, 788
Contradiction, 350
Contrapositive propositions,
    348–349, 352
Converse propositions, 348–349
Coordinate axis, 13
Coordinates, 13, 287

Coordinate systems
  rectangular, 822
  three-dimensional, 822–825
Corner points, 264, 274, 294
Cost functions, 560, 721, 820
Costs, 20–21, 50–51, 77
  average, 681–682
  marginal, 557–559, 758–759
  marginal average, 562
Counting principles, 361–368
  addition principle, 361–363
  multiplication principle, 363–365
Counting technique, 361
Critical numbers, 634–636
  in domain, 634
  local extrema and, 638
Critical points, 838, 846
  local extrema and, 839–840
Cross sections, graphing of, 824–825
Cube root, 916
Cumulative matrix, 210, 216
Curve fitting, 28
Curves, area between two, 776–785
Curve-sketching techniques, 674–687. *See also*
    Graphing strategy
  average cost and, 681–682
  definition of, 655
  for second derivative, graphing of, 655–658
Cylinder, 948

D Daily compound interest, 138
Debt, amortizing, 158–159
Decision variables, 270, 273
Decoding matrix, 230
Decreasing functions, 632–636
Definite integral, 747–757
  approximation of areas by left and right sums
    and, 747–750
  average value and, 763–765
  definition of, 752
  evaluating, 759–763
  as limit of sums, 750–752
  properties of, 752–753
  recognizing, 763–765
  substitution and, 760–761
Degree
  of polynomials, 895
Demand, elasticity of, 620–624
Denominators, 892
  least common, 4
  rationalizing, 920
Dependent events, 418–419
Dependent systems of linear equations, 176
Dependent variables, 46, 819
Derivatives, 526–541
  chain rule for, 598–605
  concept of, 526
  constant *e* and, 575
  continuous compound interest and, 575–578
  definition of, 532
  differentials and, 550–557
  differentiation properties and, 541–550
  elasticity of demand and, 620–624
  of e^x, 581–582

of exponential functions, 585
  first, 632–647
  four-step process for finding, 532–535
  general rules for, 604–605
  implicit differentiation and, 608–612
  interpretations of, 532
  of logarithmic functions, 585, 620
  marginal analysis and, 557–567
  nonexistence of, 536–537
  notation for, 541
  partial, 828–836
  of products, 590–592
  of quotients, 592–595
  rate of change and, 526–529
  related rates and, 614–617
  second, 648–665
  slope of tangent line and, 529–532
Diameter of a tree at breast height (Dbh), 31–32
Difference quotients, 50, 498, 527
Differential equations, 736–746
  archaeology and, 741–742
  continuous compound interest and, 739
  definition of, 736
  exponential growth law and, 739–740
  exponential growth phenomena, comparison
    of, 742–743
  general solution of, 738
  learning rate of improvement and, 742
  particular solution of, 738
  population growth and, 740–741
  slope fields and, 737–738
Differentials, 550–557
  approximations using, 553–555
  definition of, 552, 728
  increments and, 550–552
Differentiation. *See also* Differentiation properties
  of functions, 532, 544
  implicit, 608–612
  of products, 590–591
  of quotients, 593–594
  residual, 855
Differentiation properties, 541–550
  constant function rule and, 541–542
  constant multiple property and, 544
  power rule and, 542–543
  sum and difference properties and,
    545–547
Discontinuous functions, 86, 516–517
Discriminate, 926–928
Disjoint events, 400
Disjoint sets, 361
Disjunction, 347
Distributive properties, 889–890, 895
Division of rational expressions, 907–908
Domains, 45–46
  of composite functions, 598
  critical numbers in, 634
  of elementary functions, 59
  of exponential functions, 96–97
  finding, 48–49
  of functions, 819
  of logarithmic functions, 106–108
  of polynomial functions, 85
  of rational functions, 87

Double inequalities, 6, 8
Double integrals
  definition of, 867–869
  evaluating, 878–879
  of exponential function, 868–869
  over rectangular regions, 864–874
  over regular regions, 874–882
  volume and, 870–871, 880–881
Double subscript notation, 188
Double time, 578
Downward concavity, 649–650
Dual problems, 313–325
  definition of, 313
  formation of, 313–315
  fundamental principle of, 315
  problem constraints of, 318–319

E

Effective rate, 142. *See also* Annual percentage
    yield (APY)
Elastic demand, 623. *See also* Elasticity of demand
Elasticity of demand, 620–624
  definition of, 622
  at price, 622–623
  relative rate of change and, 620–621
  revenue and, 624
Elementary functions, 57–69
  beginning library of, 58–59
  domain of, 59
  evaluating, 58
  graph of, 59
  horizontal shifts, 59–61
  range of, 59
  reflections, 61–65
  shrinks, 61–65
  stretches, 61–65
  vertical shifts, 59–61
Elements
  in events, 401
  identity, 890
  of matrix, 187–188
  pivot, 301
  of sets, 354, 357–358, 361, 363–364
Elimination by addition, 178–180
Empirical probability, 392, 394
  applications to, 406–408
  approximate, 392
  on graphing calculator, 394
  simulation and, 394
Empty sets, 354, 361
Encoding matrix, 230
End behavior, 508–509
Endpoints of intervals, 6–7
Endpoint solution, 702
Entering variables, 300
Equality, absolute, 780
Equality matrix, 234
Equally likely assumption, 393–395
Equal matrix, 210
Equal sets, 355
Equations. *See also* specific types of
  continuity of functions by, 518
  cost, 20–21
  differential, 736–746
  equivalent, 3, 4–5

exponential, 112
  first-order, 737
  functions specified by, 43–48
  graph of, 13
  linear, 3–4, 309
  of lines, 18–20
  logarithmic, 110–111
  matrix, 211, 234–235
  price–demand, 22, 559–560
  price–supply, 21–22
  quadratic, 922–930
  second-order, 737
  solution of, 3
Equilibrium point, 22, 928
Equilibrium price, 793
Equilibrium quantity, 22, 181, 793
Equity, 161
Equivalent equations, 3–5
Equivalent formulas, 4–5
Equivalent inequalities, 6
Equivalent systems of linear
    equations, 179
Error in approximation, 748
Error bound, 748–749
e-system, 287
Events, 387
  arbitrary, 393–394
  compliment of, 403
  compound, 386–387
  dependent, 418–419
  independent, 418–421
  intersection of, 399, 400–403
  mutually exclusive, 400
  odds for, 405–406
  probability of, 389–392, 400–403,
    425–427
  sample spaces and, 386–389
  simple, 386–387, 390
  union of, 399–403
e^x, derivative of, 581–582
Exiting variables, 300
Expanded coordinates, 287
Expected value, 435–436
  decision making and, 438
  of games, 436
  insurance and, 437
  of probability distributions, 437
  of random variables, 434–437
Experiments, 386–387
Explicit rule for evaluating functions, 608
Explore and Discuss, 4, 5, 7, 13, 18, 27, 30, 44,
    45, 60, 62, 63, 70, 72, 91, 99, 102, 107, 113,
    128, 141–142, 143, 150–151, 153, 159, 162,
    176, 178, 191, 193, 202, 205, 216, 223, 228,
    229, 239, 246, 273, 276, 286, 291, 293, 306,
    314, 316–317, 337, 350, 352, 356, 358, 363,
    375, 395, 403, 405, 416, 420, 427, 435, 448,
    454, 460, 470, 471, 494, 499, 503, 516, 534,
    536, 542, 552, 563, 578, 581, 586, 590, 592,
    599, 603, 609, 615, 620, 621, 632, 648, 672,
    679, 717, 738, 748, 764, 782, 798, 801, 823,
    838, 850, 858, 859, 870, 877, 880
Exponential decay, 740
Exponential equations, 112

Exponential functions, 95–106. *See also*
    Logarithmic functions
  base of, 96
  base *e*, 98–99
  b^x with base *b* as, 581
  compound interest, 101–103
  defined, 95–96
  derivatives of, 585
  double integrals of, 868–869
  domain of, 96–97
  e^x as, 581–582
  formula for, 581, 585
  four-step process for, 581–582
  graphs of, 96–97, 577
  inverse of, 583 (*See also* Logarithmic functions)
  limits involving, 669
  logarithmic functions, conversion to, 108–109
  models of, 586–588
  natural, 584
  other, 584–586
  power rule for, 582
  properties of, 98
  range of, 96
  slope of, 577
Exponential growth, 743
  law of, 739–740
  phenomena of, comparison of, 742–743
Exponential growth rate, 99–100
Exponential regression, 101, 588
Exponents
  first property of, 892
  integer, 912–913
  natural number, 892
  properties of, 912
  radicals, properties of, 919–920
  rational, 916–922
  scientific notation and, 913–914
  simplifying, 912–913
Extrapolation, 30
Extrema. *See also* Absolute extrema; Local
    extrema
  second derivative and, graphing of, 690–693
Extreme value theorem, 688

F Factorability, theorem of, 927
Factored polynomials, 900
Factored form of numbers, 900
Factorials, 368–369, 374, 944
Factoring, 927–928
  quadratic equations, solution by, 923–924
  quadratic formula and, 927–928
Factoring polynomials, 900–905
  combined, techniques for, 904–905
  common factors, 900–901
  by grouping, 901
  second-degree polynomials, factoring, 902–903
  special formulas for, 903–904
Fair games, 405, 437
False negative results, 429
False positive results, 429
Feasible region, 264, 273, 276, 288
  corner points of, 294
  for linear programming, 309
Feasible solution, 287, 297–298

Finance, 126–167
  annuities, 147–167
  compound interest, 134–147
  mathematics of, 126–167
  simple interest, 127–134
Finite arithmetic series, 939–940
Finite geometric series, 149, 156
Finite sample space, 434
Finite sequence, 932
Finite series, 933
Finite sets, 355
First derivative, graphing of, 632–647
  increasing and decreasing functions
      and, 632–636
  local extrema and, 636–640
First-derivative test, 638–640
First-order equations, 737
First-order partial derivatives, 831
First-state matrix, 449
Fixed costs, 21, 50
Formulas
  for binomials, 943–944
  for continuous compound interest, 576
  for exponential functions, 581, 585
  geometric, 947–948
  for indefinite integrals, 716–717, 727, 729
  for integration by parts, 797–799
  for logarithmic functions, 581, 585
  quadratic, 926
  reduction, 809
  for slope of line, 757
Formulas. *See also* specific types of
  for amortization, 159, 161–162, 164
  Bayes', 425–432
  change-of-base, 113
  equivalent, 4–5
  quadratic, 70
  of simple interest, 127–129
Fractional expressions, 906
Fractions
  canceling in, 907
  compound, 909–910
  definition of, 892
  fundamental property of, 906
  raising to highest terms, 906
  reducing to lowest terms, 906–907
  with real numbers, 892
  simple, 909, 913
Frequency, 392
  relative, 392
Functions, 42–117. *See also* specific types of
  absolute value, 491
  applications, 50–53
  average value of, 765
  bounded, 91
  Cobb–Douglas production, 821–822, 848
  composite, 598–599
  constant, 47, 519, 541–542
  continuous, 516–518, 635, 764
  cost, 77, 560, 721, 820
  decreasing, 632–636
  definition of, 44–45
  differentiation of, 532, 544
  discontinuous, 86, 516–517

  domain of, 819
  elementary, 57–69
  end behavior of, 508
  equations and, 43–48
  evaluation of, 49–50
  exponential, 95–106
  general notion of, 43
  graph/graphing of, 45
  increasing, 632–636
  of independent variables, 819–820, 845–852
  inverse of, 107
  limits and, 489–490
  linear, 47, 64–65
  logarithmic, 106–117
  of multiple variables, 819–828
  with no absolute extrema, 690
  nondifferentiable, 536–537
  notation for, 48–50
  objective, 270, 273, 297, 314
  one-to-one, 107
  polynomial, 84–86, 495, 507–508, 519, 640
  price–demand, 50, 76
  probability, 389, 399
  probability density, 785–787
  profit, 79, 561, 820
  quadratic, 69–84
  range of, 819
  rational, 84, 87–91, 495, 504–506, 509–510, 519
  revenue, 76–77, 560, 820
  root of, 70, 86
  second-degree, 69
  sharp-corner, 86
  special notation for, 608
  values of, 489–490
  vertical-line test for, 47
  zero of, 70, 86
Fundamental theorem of calculus, 757–769
  definite integrals and, 759–765
Future value (FV), 102, 127–129, 149, 789–791.
    See also Amount
  of annuities, 147–155
$f(x)$ notation
  definition of, 48–49
  graph of, 73–74, 97
  maximum value of, 73

G Games
  fair, 405, 437
Gauss-Jordan elimination, 196–209, 237, 245
  definition of, 194
  linear systems of equations, solving by,
      198–203
  for reduced matrix, 196–198
  using graphing calculator, 201
General derivative rules, 604–605
General power rule, 598–602
General problem-solving strategy, 163–164
General regions, double integrals over
    more, 874–882
General terms of sequence, 931–933
Geometric formulas, 947–948
Geometric sequence, 937–943
  definition of, 937
  *n*th-term formulas, 938–939

Geometric series
  finite, 149, 156, 940
  infinite, 941
  sum formulas for, 940–941
Gini index, 780–781
Graphing. *See also* Graphs
  of absolute maxima and minima, 687–695
  of concavity, 650–651
  of continuity of functions, 516–518
  of continuous compound interest, 577
  of cross sections, 824–825
  curve-sketching techniques for, 674–687
  of exponential functions, 577
  of first derivative, 632–647
  of inflection points, 654–655
  of investment growth, 577
  L'Hôpital's rule for, 665–674
  of limits, 489–494
  of local extrema, 637
  of optimization problems, 695–707
  of second derivative, 648–665
  of triplet of numbers, 825
Graphing calculator, 15
  for displaying truth table, 350
  empirical probability on, 394
  exponential regression on, 588
  factorials on, 374
  Gauss-Jordan elimination using, 201
  half-planes on, 257
  identity element on, 222
  integration on, 762–763
  linear regression on, 30
  logarithmic regression on, 588
  for Markov chains, 464, 477–478
  matrix on, 216, 229, 453, 478
  matrix inverses on, 228
  row operations on, 191
  systems of linear equations, for solving, 177
Graphing strategy
  asymptotes and, 674–675
  modifying, 674–675
  procedure for, 655–658
  using, 675–681
Graphs. *See also* Graphing
  compound interest, 139
  of elementary functions, 59
  of equations, 13
  of exponential equations, 112
  of exponential functions, 96–97
  of functions, 45
  of $f(x)$, 73–74, 97
  horizontal translation of, 61, 63
  line, 7, 14
  linear, 577
  of linear equalities, 256–260
  of linear equations, 14
  of linear programming problem, 270
  of piecewise linear functions, 64–65
  of polynomial functions, 85–86
  of price–demand equations, 22
  of price–supply equations, 22
  of quadratic functions, 70, 72–76
  of rational functions, 88–90, 97
  reflections of, 62

sketching on, 43, 91
slope of, 531
of systems of linear equalities, 257–258, 264–265
of systems of linear equations, 174–177
of systems of linear inequalities, 258–260
transformation of, 59–63
vertical shrink of, 62–63
vertical stretch of, 62–63
vertical translation of, 60–61, 63
Grouping, factoring polynomials by, 901
Growth rate
  compound, 102
  computing, 140
  exponential, 99–100
  relative, 99
Growth time, 138–141, 578

H Half-closed intervals, continuous on, 519
Half-life, 100
Half-planes, 256–257
Histograms
  for probability distributions, 433
Horizontal asymptotes
  limits at infinity and, 507, 509–511
  of rational functions, 509–511
Horizontal asymptotes of rational functions, 88–91, 97
Horizontal axis, 12–13
Hypothesis p and conclusion q, conditional propositions with, 347

I Identity elements, 222, 890
Implicit differentiation, 608–612
  definition of, 608–609
  special function notation and, 608
Implicit rule for evaluating functions, 608
Increasing functions, 632–636
Increments, 550–552
Indefinite integrals, 715–720
  cost function and, 721
  curves in, 720
  definition of, 715, 717
  formula for, 716–717, 727, 729
  of products, 720
  properties of, 717–720
Independent events, 417–420
Independent systems of linear equations, 176
Independent variables, 46
  definition of, 819
  functions of, 819–820, 845–852
Indeterminate form
  0/0, 667–670
  infinity/infinity, 671–673
  limits of, 497–498
Index of radicals, 917
Indicators, 300
Individual retirement account (IRA), 152–153
Inelastic demand, 623
Inequalities
  absolute, 780
  continuity and, 520–522
  double, 6, 8
  equivalent, 6

linear, 3, 5–8, 309
  properties of, 6
  sense of, 6
Infinite geometric series, 941–942
Infinite limits, 503–506
  definition of, 503
  vertical asymptotes and, 504–506
Infinite sequence, 932
Infinite series, 933
Infinite sets, 355
Infinity, limits at, 506–511, 670–671
Inflection points, 651–654
  definition of, 651
  graphing, 654–655
  locating, 652–653
Initial condition, 742
Initial simplex tableau, 298–299, 327
Initial-state distribution matrix, 449
Initial-state probability matrix, 449
Initial system, 297–298
Input, 46
Input–output analysis, 242, 246–247
Instantaneous rate of change, 526–527
Instantaneous velocity, 526, 529
Integer exponents, 912–913
Integers. *See* numbers
Integrals
  table of, 803, 807–809
  definite, 747–757, 759–765
  double, 864–882
  evaluating, 866–867
  formulas for, 948–950
  indefinite, 715–720
  iterated, 867
Integral sign, 715
Integrand, 715, 752, 867
Integration
  antiderivatives and, 714–715
  area, for computing, 779–780
  area between curves and, 776–785
  in business and economic applications, 785–796
  constant of, 715
  in consumers' and producers' surplus, 791–794
  in continuous income stream, 787–791
  definite integral and, 747–757
  differential equations and, 736–746
  formulas, 948–950
  fundamental theorem of calculus and, 757–769
  on graphing calculator, 762–763
  indefinite integrals and, 715–720
  table of integrals and, 807–809
  lower limit of, 752
  other methods of, 803–813
  by parts, 797–803
  in probability density functions, 785–787
  reduction formulas and, 809
  region of, 867
  reversing order of, 879–880
  Simpson's rule and, 805–807
  by substitution, 725–736
  trapezoidal rule and, 804–805
  upper limit of, 752
  using table of integrals, 807–808

Intercepts
    of polynomial functions, 640
Interchanging rows, 302
Interest. *See also* Compound interest
    on amortization, 159
    compound, 575–578, 739
    compounding quarterly, 134–135
    definition of, 101, 127
    on investments, 129–131
    simple, 127–134, 576
Interest rate, 101
    annuities, approximating future value of, 153
    definition of, 127
    on investments, 129–130
    on a note, 129
    true, 142
Interpolation, 30
Intersection
    of events, 399, 400–403, 414–415
    of sets, 356–357, 361
    union and, 399–403
Intervals
    closed, 6, 518–519, 689
    endpoints of, 6–7
    half-closed, 519
    open, 6, 692–693
    sign properties of, 520
Inventory control problem, 702–704
Inverse of exponential functions, 583. *See also*
    Logarithmic functions
Inverses
    additive, 223, 890
    of functions, 106–107
    of *M*, 224, 228
    matrix, 228–229
    multiplicative, 223–224, 890
    of square matrix, 222–233
Investment growth, graphing, 577
Investments
    analysis of, 237–239
    annual percentage yield of, 142–143
    growth time of, 140–141
    interest on, 129–131
    interest rate earned on, 129–130
    present value of, 128
i-system, 287
Iterated integrals, 867

L Lagrange multipliers, 845–854
    definition of, 846
    for functions of three independent
        variables, 850–852
    for functions of two independent
        variables, 845–850
    method of, 845
Large numbers, law of, 406
Larger problems, 335–337
Law of averages, 406
Law of large numbers, 406
Leading coefficients, 85
Leading term, 508
Learning rate of improvement, 742
Least common denominator (LCD), 4, 908
Least square approximation, 854–859

Least squares line, 855
Left end behavior, 508
Left half-planes, 256
Left-hand limits, 492, 506
Leftmost variables, 202
Left rectangle, 747–748
Left and right sums
    approximation of areas by, 747–750
    limits of, 759
Left sum, 747. *See also* Left and right sums
Leontief input–output analysis, 242–250
    three-industry model for, 246–248
    two-industry model for, 243–245
L'Hôpital's rule, 665–674. *See also* Limits
    definition of, 665
    indeterminate form 0/0 and, 667–670
    indeterminate form infinity/infinity and, 671–673
    limits at infinity and, 670–671
    one-sided limits and, 670–671
Like terms, 895–896
Limited growth, 743
Limiting matrix, 461, 472–477
Limits
    algebraic approach to, 494–498
    analysis of, 490–491
    basics of, 489–503
    concept of, 489–490
    constant *e* and, 575
    continuity and, 515–526
    definition of, 491
    of difference quotients, 498
    evaluating, 495–497
    existence of, 492, 506
    functions and, 489–490
    graphing, 490–494
    of indeterminate form, 497–498
    infinite, 503–506
    at infinity, 506–511, 670–671
    involving exponential functions, 669
    involving logarithmic functions, 669
    involving powers of *x*, 665
    involving powers of *x* – *c*, 666
    from the left, 492
    left-hand, 492, 506
    of left and right sums, 759
    one-sided, 491–492, 670–671
    of polynomial functions, 495
    of powers, 665
    properties of, 494–495
    of quotients, 497
    of rational functions, 495
    from the right, 492
    right-hand, 492, 506
    of sums, 750–752
    two-sided, 492, 506
    unrestricted, 492
Limits at infinity, 506–511
    horizontal asymptotes and, 507, 509–511
    of polynomial functions, 507–508
    of power functions, 507
    of rational functions, 509–510
Linear equalities
    systems of, 257–258, 263–269
    in variables, 256–263

Linear equations, 3–5, 13–16, 309
Linear functions, 47, 64
Linear graphs, 577
Linear inequalities, 3, 5–8, 309
Linearly related variables, 27
Linear programming, 270–282, 285–341.
        See also Problems
    components of, 314
    definition of, 270
    feasible region for, 309
    fundamental theorem of, 274, 289, 298
    general description for, 273–274
    geometrically interpreting, 304
    introduction to, 286–296
    for maximization problem, 297–312
    table method for, 286–296
    for minimization problem, 313–325
    simplex method for, 285–341
    slack variables and, 287
    summary of, 304–307
    variables, basic and nonbasic, 292–294
Linear programming problem
    definition of, 270, 273
    graphically solving, 270
    mathematical model for, 270, 273
Linear regression, 26–38, 854
    on graphing calculator, 30
    slope as rate of change and, 27–28
    with spreadsheet, 32
Linear system. *See* Systems of linear equations
Line graph, 7
Lines
    constant-profit, 270
    equation of, 18–20
    graphing intercepts for, 14
    horizontal, 14–16, 20
    least squares, 855
    real number, 889
    regression, 29–30, 855
    secant, 530
    slope of, 16
    tangent, 529–532, 591
    vertical, 14–16, 20
ln *x*, 582–584. *See also* Logarithmic functions
    four-step process for finding, 583–584
    $\log_b x$ and, relationship between, 585
Loans, 128, 161–162
Local extrema, 636–638, 838–840
    critical numbers and, 638
    critical points and, 839–840
    first-derivative test for, 638–640
    graphing, 637
    partial derivatives and, 838
    of polynomial functions, 640
    second derivative and, 691
    second-derivative test for, 691–692, 838
Local maximum, 636–637, 823, 837
Local minimum, 837
Logarithmic equations, 110–111
Logarithmic functions, 106–117
    with base, 2, 107–108
    definition of, 583, 106, 108
    derivatives of, 585, 620
    domains of, 106–108

exponential function, conversion to, 108–109
  formula for, 581, 585
  inverse functions and, 106–107
  limits involving, 669
  ln x as, 582–584
  $\log_b x$ with base b as, 581
  models of, 586–588
  natural, 584
  other, 584–586
  properties of, 109–111
  range of, 106–108
Logarithmic regression, 114, 588
Logarithms, 111–113, 584
Log to the base b of x, 108
$\log_b x$ with base b, 581
Logic, 346–354
  connectives and, 346–349
  logical equivalences/implications,
    350–352
  propositions, 346–349
  truth tables, 349–350
Logical equivalence, 351–352
Logical implication, 351
Logical reasoning, 346
Logistic growth, 743
Lorenz curve, 780–781
Lower half-planes, 256
Lower limit of integration, 752
Lowest terms, reducing to, 906–907

M M, inverses of, 224, 228. *See also*
    Singular matrix
Marginal analysis, 557–567. *See also*
    Rate of change
  of marginal cost, 557–559
  of marginal profit, 557–559
  of marginal revenue, 557–559
Marginal cost, 557–559
  area *vs*, 758–759
  average, 562
  definition of, 562
  exact cost *vs*, 558–559
Marginal productivity, 830, 849
Marginal profit, 557–559, 561–562
Marginal revenue, 557–562
Markov chains, 447–483
  absorbing, 469–483
  definition of, 447, 450
  introduction to, 448–450
  matrices of, 451–452, 460–464
  properties of, 448–459
  regular, 459–469
  state matrices and, 450–452
  transitions matrices and, 450–453
Mathematical modeling, 51, 270, 273
Mathematics of finance, 126–167. *See also*
    Finance
Matrix (matrices)
  addition of, 210–211, 234
  associative, 210
  augmented, 188–191
  coefficient, 188–189
  column, 188
  combined, 234

commutative, 210, 216
constant, 188–189
decoding, 230
definition of, 187
dimensions of, 188
elements of, 187–188
encoding, 230
equal, 210
equality, 234
first-state, 449
on graphing calculator/calculating, 478
initial-state distribution, 449
initial-state probability, 449
Leontief input–output analysis, 242–250
limiting, 461
of Markov chains, 451–452, 460–461, 464
multiplication of, 211–218, 222, 234
$m \times n$, 187–188
negative of, 210
notation for, 188
operations of, 210–222
principal diagonal, 188
product, 211–218
properties of, 234
reduced, 196–198
row, 188
singular, 224, 229
solving systems of linear equations, methods
  for, 187
square, 188, 222–223
state, 450–452, 454, 469–470
subtraction of, 210–211
technology, 244
transitioning, 450–454, 470
transposition of, 313
zero, 210–211, 216, 476
Matrix equations, 211, 234–235
  systems of linear equations and, 234–242
Matrix games. *See* Games
Matrix product, 213
Maxima and minima
  absolute, 687–695
  local extrema and, 838–840
  multivariable calculus and, 837–845
  using Lagrange multipliers, 845–854
Maximization problems, 270, 297–312
  initial system for, 297–298
  with mixed problem constraints, 326–341
  pivot operation for, 299–304
  problem constraints in, 319
  simplex method for, 316–317
  simplex tableau for, 298–299
  in standard form, 286
Maximum area, 845
Maximum rate of change, 659
Maximum value of $f(x)$, 73
Mean
  arithmetic, 935
Member of sets, 354
Method of Lagrange multipliers, 845
Method of least squares, 854–864
  least square approximation and, 854–859
Midpoint sums, 751, 805
Minima. *See* Maxima and minima

Minimization problems, 313–325
  big M method for, 333–335
  with mixed problem constraints, 326–341
  solution of, 315–319
Mixed problem constraints, 326–341. *See also*
    Big M method
Model/modeling, mathematical, 26, 51,
    270, 273
Modified problem, 330–331
Monomials, 895
Motion, related rates and, 614–616
Multiple optimal solutions, 275
Multiplication
  for linear equations, 309
  for linear inequalities, 309
  of matrix, 211–218, 222, 234
  of polynomials, 897–898
  principles of, 363–365
  of problem constraints, 318–319
  of rational expressions, 907–908
  of real numbers, 216, 889–890
  square matrix, identity of, 222–223
Multiplicative inverse, 890
Multivariable calculus
  double integrals over rectangular regions and,
    864–874
  double integrals over regular regions and,
    874–882
  functions of multiple variables, 819–828
  maxima and minima and, 837–854
  method of least squares and, 854–864
  partial derivatives and, 828–836
Mutually exclusive events, 400
$m \times n$ matrix, 187–188

N Natural exponential functions, 584
Natural logarithmic functions, 584
Natural logarithms, 111
Natural number exponents, 892
n distinct objects taken r at a time, permutations
    of, 370–371
Negation, definition of, 346
Negative of matrix, 210
Negative real numbers, 889–890
n factorials, 369, 944
Nondifferentiable functions, 536–537
Nonnegative constraints, 271, 273, 297, 314,
    326–327
Normal curves, 787
Notation
  for derivative, 541
  double subscript, 188
  for functions, 48–50
  $f(x)$, 48–49, 73–74, 97
  inequality, 7
  interval, 6–7
  for matrix, 188
  scientific, 913–914
  for second derivative, graphing of, 649
  for sets, 354–356
  special function, 608
  summation, 750–751, 933–935
Not defined matrix product, 213
Not defined product matrix, 213

Note, interest rate earned on, 129
Not factorable polynomials, 903, 927
nth root, 916–917
nth-term formulas, 938–939
nth terms of sequence, 931–932, 938
Null sets, 354
Numbers. *See also* Real numbers
    completely factored, 900
    critical, 634–636, 638
    of elements of sets, 357–358
    factored form of, 920
    natural, 892
    partition, 521, 634–636, 653–654
    prime, 900
    test, 521
    triplet of, 822, 825
Numerator, 892, 920
Numerical coefficient, 895

O Objective functions, 270, 273, 297, 314
Odds, 405–406
One-to-one functions, 107
One-sided limits, 491–492, 670–671
Open intervals, 6, 692–693
Operations
    canceling, 907
    of matrices, 210–222
    order of, 898
    pivot, 299–304, 328
    on polynomials, 894–900
    row, 191
Optimal solution, 272–276
Optimal values, 270, 273
Optimization problems
    for area and perimeter, 695–698
    graphing of, 695–707
    inventory control and, 702–704
    for maximizing revenue and profit,
        698–702
    strategies for solving, 696
Ordered pair, 13
Ordinary annuities, 147, 149–150
    present value of, 157
Ordinate, 13
Origin, 13, 889
Outcomes
    compound, 386
    of experiments, 387 (*See also* Events)
    simple, 386–387
Output, 46

P Parabolas, 70, 73
Paraboloid, 823
Parallelogram, 947
Parameter, 180
Parentheses, 50, 896
Partial antiderivatives, 866
Partial antidifferentiation, 865
Partial derivatives, 828–836
    definition of, 828
    first-order, 831
    local extrema and, 838
    of f with respect to x, 829
    of f with respect to y, 829

second-order, 831–833
    using chain rule, 829–830
Partition numbers, 521, 634–636, 653–654
Parts, integration by, 797–803
    formula for, 797–799
Payment, 151–152, 159, 161–162
Percentage rate of change, 620–621
Percentages, 892
Perfect squares, 931
Perimeter, 695–698
Permutations, 369–372
    definition of, 369–370
    of n distinct objects taken r at a time, 370–371
    of sets, 370
Piecewise linear functions, 64–65
Pivot element, 301
Pivoting, 302. *See also* Pivot operation
Pivot operation, 299–304
Pivot row, 301
Plot/plotting, 29, 43–44
Point-by-point plotting, 43–44
Points
    break-even, 561
    critical, 838, 839–840, 846
    of diminishing returns, 658–659
    inflection, 651–654
    turning, 637
Point-slope form, 19–20
Polynomial functions
    continuity properties of, 519
    intercepts of, 640
    limits of, 495
    limits at infinity of, 507–508
    local extrema of, 640
Polynomials
    adding, 896–897
    classifying, 895
    combined, 898
    combining like terms in, 895–896
    definition of, 894
    degree of, 895
    end behavior of, 508–509
    equations for, 898
    factoring, 900–905, 927
    functions for, 84–86
    multiplying, 897–898
    natural number exponents and, 894
    operations on, 894–900
    regression of, 86–87
    subtracting, 896–897
    in variables, 894
Population growth, 740–741
Positive real numbers, 889, 890
Power functions, 507, 542–543. *See also* Functions
Power rule
    for exponential functions, 582
    general, 598–602
Powers, 665–666
Powers of transition matrix, 452–453
Predictions, 30
Preliminary simplex tableau, 327
Present value. *See also* Amortization; Principle
    amortization and, 158–163
    of annuities, 155–167, 161, 163–164

of investments, 128
    of ordinary annuities, 157
Price
    average, 765
    elasticity of demand at, 622–623
    equilibrium, 793
Price–demand equation, 22, 559–560
Price–demand functions, 50, 76
Price–demand model, 586–587
Prices
    diamond, 28–29
    equilibrium, 22, 181
    purchase, 8–9
Price–supply equations, 21–22
Prime numbers, 900
Principal diagonal matrix, 188
Principle, 102, 127–129. *See also* Present value
    of addition, 361–363
    counting, 361–368
    finding, 139
Probabilities, 385–441
    Bayes' formula, 425–432
    conditional, 411–414, 421
    empirical, 392, 394
    equally likely assumption and, 393–395
    of events, 389–392, 400–403, 425–427
    of simple events, 390
Probability density functions, 785–787
Probability distributions
    expected value of, 437
    histogram for, 433
    random variables and, 432–434
    of random variables, 433–434
Probability functions, 389, 399
Probability trees, 415–417, 427
Problem constraints, 271, 273, 297
    coefficients of, 314
    of dual problems, 318–319
    in maximization problems, 319
    multiplication of, 318–319
Problems
    dual, 313, 315, 318–319
    linear programming, 270, 273
    maximization, 270, 286, 316–317, 319
    modified, 327, 330–331
    word, 8
Producers' surplus, 791–794
    definition of, 792
Product matrix, 212–218
    definition of, 213
    not defined, 213
    of a number k and a matrix M, 211–212
Product rule, 414–415, 421
Products, 590–592
    differentiating, 590–591
    indefinite integrals of, 720
    rules for, 590
    tangent lines and, 590–591
Profit
    changes in, 761–762
    marginal, 557–559, 561–562
    marginal average, 562
    maximizing, 699–701
Profit, 50–51

Profit functions, 79, 561, 820
Profit–loss analysis, 50
  equality, 3
  of exponential functions, 98
  of inequalities, 6
  of logarithmic functions, 109–111
  of Markov chains, 448–459, 461
  of matrix, 234
  of quadratic functions, 72–76
  of real numbers, 17
  of sets, 354–356
Properties
  associative, 890
  commutative, 890
  constant multiple, 544
  of continuity, 519–522
  of definite integral, 752–753
  differentiation, 541–550
  distributive, 889–890, 895
  of exponents, 912
  of fractions, 906
  of indefinite integrals, 717–720
  of limits, 494–495
  of logarithms, 584
  of radicals, 919–920
  sign, 520
  sum and difference, 545–547
  zero, 891
Propositions, 346–349
  compound, 348
  conditional, 347, 352
  contrapositive, 348–349, 352
  converse, 348–349
  definition of, 346
  truth table for, 346–352, 350–352
  types of, 350
Pythagorean Theorem, 947

Q Quadrants, 13
Quadratic equations, 922–930
  definition of, 922
  factoring, solution by, 923–924
  polynomial and, 928
  quadratic formula for, 924–928
  solving, 928
  square root, solution by, 922–923
Quadratic formula, 70, 924–928
Quadratic functions, 69–84
  analyzing, 75–76
  definition of, 69
  graph of, 70, 74–76
  parabolas of, 70
  properties of, 72–76
  vertex form of, 72–76
Quotients, 592–595
  difference, 498, 527
  differentiating, 593–594
  limits of, 497
  rules for, 592–593

R Radicals, 917
  properties of, 919–920
  rational exponents and, 916–922
Radicand, 917
Radioactive decay, 741

Raising fractions to higher terms, 906
Random experiments, 386
Random variables
  definition of, 432–433
  expective value of, 434–437
  probability distribution of, 432–434
Ranges, 45
  of elementary functions, 59
  of exponential functions, 96
  of functions, 819
  of logarithmic functions, 106–108
Rate of change, 526–529. *See also*
    Marginal analysis
  average, 526–527
  instantaneous, 526–527
  maximum, 659
  percentage, 620–621
  relative, 620–621
Rate of flow, 788
Rates, 102, 614–617. *See also* Rate of change
  of change, 21, 27–28
  of descent, 27–28
  effective, 142
  per compounding period, 135
Rational exponents
  radicals and, 916–922
  working with, 918–919
Rational expressions
  definition of, 906
  operations on, 906–911
Rational functions
  continuity properties of, 519
  horizontal asymptotes of, 509–511
  limits of, 495
  limits at infinity of, 509–510
  vertical asymptotes of, 504–506, 511
Rational functions, 87–91
  asymptotes of, 88–91, 97
  definition of, 84, 87
  domain of, 87
  graphs of, 88–90, 97
Rationalizing denominator, 920
Rationalizing numerator, 920
Real nth root, 917
Real number line, 889
Real numbers, 888–894
  addition of, 899–890
  associative properties of, 890
  commutative properties of, 890
  definition of, 946
  distributive properties of, 889–890, 895
  division of, 891
  fractions with, 892
  identity elements of, 890
  multiplication of, 216, 889–890
  negative, 889–891
  nth root of, 916–917
  positive, 889, 890
  properties of, 17, 889–891
  real number line for, 889
  real roots of, 917
  sets of, 888–890
  subtraction of, 891
  zero, 891

Real roots, 917
Reasonable probability assignment, 390, 392
Rectangle, 747–748, 947
Rectangular coordinate system, 12–13, 822.
    See also Cartesian (rectangular)
    coordinate system
Rectangular regions
  average value over, 869–870
  double integrals over, 864–874
Recursive Markov chains, 452
Reduced matrix, 197
Reduced row echelon form (reduced
    form), 197. *See also* Reduced matrix
Reduced system, 198, 202. *See also*
    Gauss-Jordan elimination
Reducing fractions to lowest terms, 906–907
Reduction formulas, 809
Reflections, of graphs, 62
Region of integration, 867
Regression
  analysis of, 28, 854
  exponential, 101, 588
  linear, 28–32, 854
  logarithmic, 114, 588
  of polynomials, 86–87
Regression line, 855
Regular Markov chains, 459–469
  definition of, 460–461
  properties of, 461
  stationary matrix and, 459–460
Regular x region, 874–877
Regular y region, 874, 877
Related rates, 614–617
  business and, 617
  motion and, 614–616
  suggestions for solving, 615
Related-rates problem, 614
Relative frequency, 392
Relative growth rate, 99
Relative rate of change, 620–621
Removing parentheses, 896
Representing sets, 355
Residual differences, 855
Restriction of variables, 906
Revenue, 50–51, 76–77
  analysis of, 527
  definition of, 560
  elasticity of demand and, 624
  marginal, 557–561
  marginal average, 562
  maximizing, 698–699, 701–702
Revenue function, 560, 820
Reversing chain rule, 725–727
Reversing order of integration, 879–880
Riemann sum, 751
Right end behavior, 508
Right half-planes, 256
Right-hand limits, 492, 506
Right rectangle, 748
Right sum, 748. *See also* Left and right sums
Rise, 16
Roots
  cube, 916
  of functions, 70, 86

Roots (*continued*)
nth, 916–917
real, 917
Row equivalent augmented matrix, 189–191
Row matrix, 188
Rows
interchanging, 302
operations, 191
pivot, 301
Rules
chain, 598–605
general derivative, 604–605
general power, 598–602
L'Hôpital's, 665–674
power, 582, 599–602
for products, 590
for quotients, 592–593
Simpson's, 803, 805–807
trapezoidal, 750, 803–805
Run, 16

S Saddle point, 824, 837
Sample space
definition of, 387
events and, 386–389
finite, 434
Schedules
of amortization, 159–163
commission, 130–131
Scientific notation, 913–914
Secant line, 530–531
Second-degree functions, 69. *See also*
Quadratic functions
Second-degree polynomials, 902–903
Second derivative, graphing of, 648–665
absolute maxima and minima and,
690–693
analysis of, 654–655
concavity for, 648–651
curve sketching technique for, 655–658
extrema and, 690–693
inflection points and, 651–654
local extrema and, 691
notation for, 649
point of diminishing returns and, 658–659
Second-derivative test, 691–692, 838
Second-order equations, 737
Second-order partial derivatives, 831–833
Sequence
arithmetic, 937–943
definition of, 931
finite, 932
geometric, 937–938
infinite, 932
terms of, 931–933
Series
arithmetic, 939
definition of, 933
finite, 933
geometric, 940–941
infinite, 933
Sets, 354–360
complement of, 357
definition of, 354

disjoint, 361
element of, 354, 357–358, 361–364
empty, 354, 361
equal, 355
finite, 355
infinite, 355
intersection of, 356–357, 361
member of, 354
notations for, 354–356
null, 354
permutation of, 370
properties of, 354–356
of real numbers, 888–890
representing, 355
solutions to, 3, 174
subset of, 355
union of, 356, 361
universal, 356
Venn diagrams and, 356–358
Sharp-corner functions, 86
Sign chart, 521–522
Sign properties of intervals, 520
Simple events, 386–387, 390
Simple fractions, 909, 913
Simple interest, 127–134, 576
definition of, 127
formula of, 127–129
investments and, 129–131
Simplex method. *See also* Linear programming
algorithm for, 304
defined, 285–286
for maximization problems, 316–317
variables for, 299
Simplex tableau
initial, 298–299, 327
for maximization problems, 298–299
preliminary, 327
procedure for, 299
Simpson's rule, 803, 805–807
Simulation, empirical probability and, 394
Singular matrix, 224, 229
Sinking funds, 151–153
Slack variables, 287, 318, 329
Slope, 17–18
of exponential functions, 577
formula for, 757
geometric interpretation of, 17
graphing, 531
of line, 16–18
as rate of change, 27–28
of secant line, 530–531
of tangent line, 529–534
Slope fields, 737–738
Slope-intercept form, 18–20
Solution region for system of linear
inequalities, 264, 266
Solutions
basic, 288–289
big *M* method, summary of, 335
of equations, 3
feasible, 287, 289, 298
to initial system, 297–298
to *i*-system, 287
of linear equations, 13

for linear programming, 287–292
of minimization problem, 315–319
optimal, 272–276
to problems, summary of, 322
to sets, 3, 174
by square root, 922–923
to systems of linear equations, 176, 180
unique, 176
Solving
of augmented matrix, 189–194
of double inequalities, 8
geometric method for, 274–276
larger problems, 335–337
logarithmic equations, 110–111
quadratic equations, 928
systems of linear equations, 174–187
(*See also* Substitution in solving system of
linear equations)
Special function notation, 608
Speed, 27–28
Sphere, 948
Spreadsheets
input–output analysis in, 247
inverses matrix in, 228
linear regression with, 32
multiplication of matrix in, 218
notation for matrix in, 188
singular matrix in, 229
Square matrix, 188, 222–223
inverse of, 222–233
multiplication, identity of, 222–223
Square root, solution by, 922–923
Squares, perfect, 931
Standard maximization problem in standard
form, 286
State matrix, 450–452, 454, 469–471
Stationary Markov chains, 460–461, 464
Stationary matrix, 459–460
Stochastic process, 416, 448
Subset of set, 355
Substitution
additional techniques for, 730–733
definite integral and, 760–761
table of integrals and, 808–809
integration by, 727–730
method of, 728–732
reversing chain rule and, 725–727
Substitution in solving system of linear
equations, 174–187
addition, by elimination of, 178–180
graphing calculator for, 177
methods for, 174, 187
solution set for, 174
Subtraction
of matrices, 210–211
of polynomials, 896–897
of rational expressions, 908–909
of real numbers, 891
Sum. *See also* Addition
formulas for, 156, 939–942
of two matrices of the same size, 210
Sum and difference properties, 545–547
Summation, 933–935
Summation notation, 750–751

Summing index, 933
Sums
  left and right, 747–750, 759
  limit of, 750–752
  midpoint, 751, 805
  Riemann, 751
  trapezoidal, 804
Supply and demand, 21–23, 928–929
Surface, 823, 870–871
Surplus variables, 326, 329
Systems of linear equalities, 257–258, 264–265
Systems of linear equations
  augmented matrices and, 187–194
  consistent, 176
  defined, 174
  dependent, 176
  equivalent, 179
  independent, 176
  matrix equations and, 234–242
  solutions to, 176, 180
  substitution, 174–187
  in variables, 174–187
Systems of linear inequalities, 258–260

T Table method for linear programming,
      286–296, 290–292
  definition of, 286, 288
  procedure for, 288, 293–294
  solutions, 287–292
Table of integrals, 803, 807–808
  integration using, 807–808
  substitution and, 808–809
Tables, 947–950
  corner point, 274
  truth, 346–352
Tangent, 635
Tangent lines, 591. *See also* Secant line
  definition of, 531
  slope of, 529–534
Tautology, 350, 351
Technology matrix, 244
Terms of sequence, 931–933
Test/testing
  ac, 902
  for independent events, 419–420
  vertical-line, 47
Test number, 521
Three-dimensional coordinate systems, 822–825
Three-industry model, input–output analysis,
      246–248
Time
  doubling, 113–114
  growth and, 138–141
Total income, 788

Transition diagram, 448
Transition matrix, 450–454, 470
  graphing calculator for, 453
  of Markov chains, 450–453, 476
  powers of, 452–453
  probability of, 448
Transposition of matrix, 313
Trapezoid, 947
Trapezoidal rule, 750, 803–805
Trapezoidal sum, 804
Tree diagram, 900
Triangle, 947
Triplet of numbers, 822, 825
True interest rate, 142. *See also* Annual
      percentage yield (APY)
Truth table, 349–350
  for compound propositions, 346–347, 349,
      351–352
  constructing, 349–350
  definition of, 349
  graphing calculator for displaying, 350
  for propositions, 346–352
Turning points, 637
Two-industry model, input–output analysis,
      243–245
Two-sided limits, 492, 506

U Unbounded solution regions, 266
Union, 517
  of events, 399–403
  intersection and, 399–403
  of sets, 356, 361
Unique solution to system of linear equations, 176
Unit elasticity, 623
Universal set, 356
Unlimited growth, 743
Unrestricted limits, 492
Upper half-planes, 256
Upper limit of integration, 752
Upward concavity, 649–650

V Vacuously true conditional propositions, 347
Values
  absolute, 59, 491
  average, 763–765, 869–870
  expected, 435–437
  of functions, 489–490
  future, 102, 127–129, 147–155, 789–791
  maximum, of *f*(*x*), 73
  optimal, 270, 273
Variable costs, 51
Variables
  artificial, 326–327, 329
  decision, 270, 273

  dependent, 46, 819
  entering, 300
  exiting, 300
  independent, 46, 819–820, 845–852
  leftmost, 202
  linear equalities in, 256–263
  linearly related, 27
  multiple, 819–828
  polynomials in, 894
  random, 432–435
  restriction of, 906
  for simplex method, 299
  slack, 287, 318, 329
  surplus, 326, 329
  in systems of linear equations, 174–187
  three-dimensional coordinate systems
      and, 822–825
  two decision, 290–292
Velocity
  average, 526–529
  instantaneous, 526, 529
Venn diagram
  definition of, 356
  elements of sets, for determining, 363
  set operations and, 356–358
Vertex,
  of parabolas, 73
  of quadratic functions, 72–76
Vertical asymptotes
  infinite limits and, 504–506
  of rational functions, 504–506, 511
Vertical asymptotes of rational
      functions, 88–91
Vertical axis, 12–13
Vertical-line test, 47
Vertical shrink of graphs, 62–63
Vertical stretch of graphs, 62–63
Vertical tangent, 635
Vertical translation of graphs, 60–61, 63
Volume
  double integrals and, 870–871,
      880–881
  under surface, 870–871

X *x* axis, 777. *See also* Horizontal axis
x axis, reflection in, 62
x coordinates, 13

Y *y* coordinates, 13

Z Zero factorials, 369, 944
Zero of a function, 70, 86
Zero matrix, 210–211, 216, 476
Zero properties of real numbers, 891

INDEX OF APPLICATIONS

Business & Economics

Advertising, 104, 312, 339, 457, 549, 556, 619, 663–664, 712, 721–722, 745
  Point of diminishing returns and, 712
  And sales, 827, 835
Agriculture, 117, 124, 249–250, 307–309, 705, 854
  Exports and imports of, 641
Amortization, 159
Annuities, 154
  Future values of, 149–150
  Guaranteed rate of, 164–165
  Interest rate of, 166
  Present value of, 157
Automation-labor mix for minimum cost, 844
Average cost, 513, 556, 647, 681–682, 711, 712, 767 (See also Minimizing average cost)
Average income, 630
Average and marginal costs, 685
Average price, 813

Blending—food processing, 344
Bonus incentives, 424
Break-even analysis, 9–10, 12, 40, 77–79, 82–83, 123, 185, 253, 567, 573
Budgeting
  For least cost, 853
  For maximum production, 853
Business closings, 379
Buying and selling commissions schedule, 147

Cable television, 36
  Revenue from, 863
  Subscribers to, 587
Capital expansion, 281
Car rental, 705
Coal, 249
Cobb-Douglas production function, 873
Coffee blends, 186
Committee selection, 360, 380, 398
Communications, 368
Computers, 280
  Control systems for, 420
  Testing on, 364–365
Construction, 123–124, 711
  Costs of, 685, 706, 712
Consumer financing rate, 165–166
Consumer Price Index (CPI), 10, 41
Consumer's surplus, 791–792, 793–794, 803, 812, 817
Consumer survey/testing, 398, 414–415
Continuous compound interest, 579, 580, 587, 589, 630, 745
Continuous income stream, 788, 790–791, 802, 812, 817
Cost, revenue, and profit rates, 619
Costs, 767, 773, 812
  Analysis of, 24, 220, 563, 564–565, 572, 662

Average, 93–94
  Equation for, 20–21
  Function of, 606–607, 724, 735, 827
  Hospital, 67
  Material, 253
  Minimum average, 94
Credit cards, 131, 163
  Annual interest rate on, 134
  Minimum payment due on, 133

Decision analysis, 440, 446
Delivery charges, 185–186
Demand equation, 630, 835
Demand function, 768
Depreciation, 24–25, 40, 101
Distribution of wealth, 784–785
Doubling rate, 580
Doubling time, 113, 116, 580, 630

Economy, 250, 942, 943
Electricity
  Consumption of, 540
  Natural gas, 250
  Oil, 250
  Rates for, 67, 123
Electronics, 186, 284, 384
Employees
  Benefits for, 361–362
  Evaluation of, 463
  Layoffs for, 379
  Rating of, 431
  Screening of, 431
  Training of, 91, 481–482, 486
Employee training, 514, 573, 685, 756, 767
Energy, 33–34
  Consumption of, 861
  Costs of, 513
Equilibrium point, 116, 124
Equilibrium price, 793–794
Equipment rental, 11, 525
Exit polling, 358

Finance/financing, 82, 104, 162
Furniture, 268–269, 280
Future value, 827
  Of continuous income stream, 817

Gross receipts, 900
Growth
  Compound, 104, 154, 172
  Exponential, 99–100
  in individual retirement account, 152–153
  Internet, 105
  Money, 104, 123
Growth time, 580

Home
  Building/construction of, 312
  Equity in, 161, 166–167, 172

Insurance for, 457–458
  Values of, 11
Housing trends, 458–459

Ice cream, 325
Incentive plan, 242
Income, 41, 525
  Distribution of, 781, 784, 802, 812, 817
  Operating, 35
  State tax on, 68
  Taxable, 208
Individual retirement account (IRA), 11, 152–153, 155, 172
Inflation, 662
Input–output analysis, 254
Insurance, 410–411, 437, 440, 446, 451–452, 462–463, 467
Interest
  Compound, 104, 116, 146–147, 943
  on loans, 165
Interest rate, 154, 170–171, 930
  Annual, 133–134, 165, 171–172
  of annuities, 166
  Approximating, 153
  Earned on Note, 129, 133
  Graphical approximation techniques for, 155, 167
Internet, 105, 486
Inventory, 768, 773
  Control over, 706, 711
Inventory value, 220–221
Investment, 40, 116, 281, 311–312, 344, 899
  Analysis on, 237–239, 254
  Annual percentage yield (APY) on, 142–143, 155
  Interest rate earned on, 129–130
  Present value of, 128
  Strategy for, 340

Labor costs, 725
  Learning and, 768
Labor force, 466
  Costs, 213, 217–218, 220, 253, 262
  Relations with, 424
Leases
  Airplane, 207–208
  Business, 253
  Tank car, 207
Loans, 128, 481
  Distribution of, 340
  Refund anticipation, 134
  Repayment of, 943

Maintenance costs, 767
Management, 367
Manufacturing, 339, 344, 705
  Bicycle, 312
  Resource allocation, 311
  Sail, 284